FRONTIERS OF ENVIRONMENTAL CHEMISTRY

环境化学前沿

江桂斌　刘维屏　主编

科 学 出 版 社
北 京

内 容 简 介

环境化学是一门研究有害化学物质在环境介质中的存在、特性、行为、效应及其控制技术原理和方法的学科。经过40多年的快速发展，我国环境化学在学科建设、人才培养、队伍规模、国家目标和国际影响等方面均取得了长足进步，已成为化学的一个重要分支、环境科学的主流与核心组成部分。

本书邀请我国环境化学领域著名的专家学者撰稿。作者队伍中包括了30多位“国家杰出青年科学基金”获得者、教育部“长江学者”和国家“千人计划”入选者，且所有作者均是在环境化学一线从事相关研究工作并具有深刻见解的专家，他们的论述能够反映和代表我国目前环境化学领域的工作特色和主流发展趋势。

本书内容丰富、前瞻性强，可供环境化学、环境工程以及地学、材料、公共卫生、农业科学等交叉学科领域研究的科研人员、高年级本科生、研究生和政府管理人员阅读和参考。

图书在版编目（CIP）数据

环境化学前沿/江桂斌，刘维屏主编. —北京：科学出版社，2017.9
ISBN 978-7-03-054637-1

Ⅰ. ①环… Ⅱ. ①江… ②刘… Ⅲ. ①环境化学–研究 Ⅳ. ①X13

中国版本图书馆 CIP 数据核字(2017)第 237702 号

责任编辑：朱 丽 杨新改 / 责任校对：韩 杨
责任印制：徐晓晨 / 封面设计：耕者设计工作室

科 学 出 版 社 出版
北京东黄城根北街 16 号
邮政编码：100717
http://www.sciencep.com

北京虎彩文化传播有限公司 印刷
科学出版社发行 各地新华书店经销
*
2017 年 10 月第 一 版 开本：889×1194 1/16
2021 年 3 月第三次印刷 印张：50 3/4
字数：1 590 000

定价：288.00 元

(如有印装质量问题，我社负责调换)

《环境化学前沿》编辑委员会

序

鉴于全国环境化学大会的规模与影响力，根据专家建议并借鉴国内外学术会议成功的经验，中国化学会环境化学专业委员会决定自本届起，编辑出版《环境化学前沿》，将此作为全国环境化学大会的一项成果，奉献给广大读者。《环境化学前沿》将总结自上届会议以来我国环境化学领域所取得的部分重要成果，展望未来发展趋势。由于环境化学涉猎范围非常广泛，研究内容极为丰富，所以每次会议出版的前沿成果或展望将不追求领域的全面覆盖和完整的系统性，而是通过一些重要方面进展与前沿的总结，展示我国环境化学工作者的最新成果，以期进一步围绕国家环境保护的重大需求，提高我国环境化学领域的创新能力与国际影响力。

本书有幸邀请到我国环境化学领域著名的专家学者撰稿，所选内容均由第九届全国环境化学大会各分会负责人牵头，本领域 180 多位中青年专家分工协作而成。作者队伍中包括了 30 多位“国家杰出青年科学基金”获得者、教育部“长江学者”和国家“千人计划”入选者。所有作者均是在环境化学一线从事相关研究工作并具有深刻见解的专家，他们的论述能够反映和代表我国目前环境化学领域的工作特色和主流发展趋势。专家们的倾心支持和高度的责任感，使得书稿能够在短时间内完成，这充分展现了我国环境化学学界空前的凝聚力、向上的精神风貌和充满生机的朝气。在此，谨向所有作者的辛勤工作表示诚挚的谢意。本着学术自由、学术民主和学术平等的原则，本书允许不同风格，鼓励各抒己见，强调文责自负。

环境化学是一门研究有害化学物质在环境介质中的存在、特性、行为、效应及其控制技术原理和方法的学科。经过 40 多年的快速发展历程，我国环境化学在学科建设、人才培养、队伍规模、国家目标和国际影响等方面均取得了长足进步，已经成为化学的一个重要分支，环境科学的主流与核心组成部分。

环境化学学科的形成和发展与环境污染密切相关。第二次世界大战后至 20 世纪 60 年代初，发达国家经济从恢复逐步走向高速发展。当时人们高度关注经济发展，普遍缺乏环境保护意识，各种环境污染事件接连发生，造成了一些严重的生态问题和健康危害，曾经出现了日本水俣病等八大环境污染公害事件。1962 年，美国海洋生物学家蕾切尔·卡逊出版了《寂静的春天》(*Silent Spring*)，书中描述了滴滴涕（DDT）使用对野生动物生殖发育的影响，引发了公众对环境污染的关注，将环境保护问题提到了各国政府面前。

20 世纪 70 年代，各国的环境保护实践为环境化学学科的形成奠定了重要基础。为推动国际重大环境前沿性问题的研究，国际科学联盟理事会 1969 年成立了环境问题专门委员会，1971 年出版了第一部专著《全球环境监测》。1972 年 6 月，联合国在瑞典斯德哥尔摩召开了 113 个国家和地区参加的“人类环境会议”，讨论了保护全球环境的行动计划，通过了《人类环境宣言》，成立了联合国环境规划署（UNEP），成为人类保护环境的重要里程碑。

1973 年 8 月，我国第一次环境保护会议召开。会议确立了环境保护工作的基本方针，通过了《关于保护和改善环境的若干规定》，揭开了我国当代环境保护的序幕，成为新中国环境科学事业标志性起点。我国的环境化学研究也起步于这一时期，并在重金属和有机污染物分析方法、典型地区环境质量评价、环境容量和环境背景值调查、工业“三废”污染普查与治理、酸雨和酸沉降、重金属和有机污染物的迁移转化规律以及污染控制技术等方面开展了大量的基础性研究工作。

进入 20 世纪 80 年代，环境化学开始进入全面发展的阶段。这期间开展了元素生物地球化学循环及其相互作用与研究；重视有机污染物等有毒化学品的安全性评价；加强了污染控制化学的研究。科学家们在判定氯氟烃损耗平流层臭氧的作用研究方面取得重大突破，对加深臭氧层破坏、温室效应等全球性环境问题的认识做出了重大贡献，这一研究成果在随后的 1995 年被授予诺贝尔化学奖。这些理论研究成果因 1985 年南极“臭氧洞”的发现而引起全世界的“震动”，从而促成了 1987 年《蒙特利尔议定书》的签订。

20 世纪 90 年代至今，我国经历了经济飞速发展，在诸多老的环境问题还没有得到解决的情况下，又出现了许多新的环境问题，如复合污染问题、水体富营养化、室内外空气污染、海洋污染、土壤和地下水污染、食品污染等。而环境内分泌干扰物（EEDs），持久性有机污染物（POPs），药物与个人护理用品（PPCPs）、塑料和微生物等污染也引起了全球特别是国际组织的高度重视，逐步成为环境化学研究的前沿。

1992 年 6 月，在巴西里约热内卢召开的联合国环境与发展首脑会议上，明确提出了可持续发展概念。2002 年 9 月，联合国在南非约翰内斯堡召开了地球峰会，通过的“可持续发展首脑执行计划”，为世界环境保护和发展指明了方向。此前的 2001 年，为了在全球范围内控制和消减持久性有机污染物（POPs），联合国环境规划署通过《斯德哥尔摩公约》，并开放各国签署，该公约于 2004 年付诸实施。公约认定的 POPs 是一个开放体系，首批名单中包含了二噁英、多氯联苯、部分有机氯农药等共 12 类化合物。随着科学研究的深入和经济社会的发展，不断有新化合物加入，目前，包括：全氟辛基磺酸及其盐类、全氟辛基磺酰氟、商用五溴二苯醚、商用八溴二苯醚、开蓬、林丹、五氯苯、α-六六六、β-六六六、六溴联苯、硫丹及其硫丹硫酸盐、六溴环十二烷、多氯萘、五氯苯酚、十溴二苯醚、短链氯化石蜡以及六氯丁二烯等在内的十六种新增化学物质被列入公约进行受控。除此之外，正在进行审查的化学品还包括三氯杀螨醇、全氟己基磺酸及其盐类相关化合物和全氟辛酸及其盐类相关化合物等。《斯德哥尔摩公约》的正式生效表明 POPs 污染成为当今世界各国共同面临的全球性重大环境问题。

2013 年 1 月 19 日，联合国环境规划署通过了旨在全球范围内控制和减少汞排放的《关于汞的水俣公约》，就汞的具体限排范围作出详细规定，以减少汞对环境和人类健康造成的损害。2013 年 10 月 10 日，《关于汞的水俣公约》在日本熊本通过。包括中国在内的 87 个国家和地区的代表共同签署该公约，标志着全球减少汞污染迈出了第一步。

2016 年 4 月 25 日，全国人大常委会第二十次会议通过我国加入《关于汞的水俣公约》。2017 年 5 月 18 日，欧盟及其七个成员国批准了《关于汞的水俣公约》。2017 年 8 月 16 日，《关于汞的水俣公约》正式生效。这是近十年来环境与健康领域新增的一项全球性公约。履行《关于汞的水俣公约》，我国面临着汞污染减排的巨大挑战，急需科学技术的支撑。

上述这些环境问题的解决既是环境化学学科面临的极大挑战，也赋予环境化学发展以空前机遇。随着各传统和新兴学科向环境化学渗透融合趋势的进一步加强，世界环境化学进入到飞速发展的新阶段。研究内容、方法、对象和学科框架基本成熟。并在大气对流层自由基化学、气溶胶和矿物质表面等的多相化学反应、重金属的形态及转化过程、水的环境质量标准、有机污染物的降解与分子转化、湖泊富营养化机理、污染物的界面过程、多介质复合污染的过程机理、分子毒理学、内分泌干扰物、化学污染物的结构-活性关系、污染物的原位修复等研究领域取得重要突破。形成了包括环境分析化学、（大气、土壤和水体）环境污染化学、污染生态化学、化学毒理学、环境理论化学、环境质量标准、污染控制化学以及环境污染与健康效应等在内的环境化学学科体系。

与此同时，我国的环境化学研究的水平、深度和广度有了空前的提升，一些研究成果开始在国际学术界产生重要影响。当然，科学的发展永无止境，且与国际环境化学学科发展的先进水平和国内相关学科的快速发展相比，环境化学在学科积累、人才队伍和研究基础等方面差距仍然较大，尤其是在研究的原创性、系统综合性、应用性和产业化等方面差距更加明显。

中国化学会环境化学专业委员会的成立是我国环境化学快速发展的必然产物。专业委员会及环境化学学科赶上了国家改革开放、经济快速增长的大好机遇，直接见证和参与了我国环境化学学科发展的历程，为学科发展提供了交流和研讨的平台，对推动我国环境化学研究的进步发挥了重要作用。

2001 年，徐晓白院士主持成立了中国化学会环境化学专业委员会，挂靠单位为中国科学院生态环境研究中心。第一届专业委员会 24 名委员中既包括了戴树桂、王连生、陈静生、汤鸿霄、魏复盛等著名专家，也吸纳了一批年富力强的中青年学者，如陶澍、江桂斌、赵进才、王春霞、刘维屏、袁东星、余刚、郑明辉、贾金平等。

随后的 2002 年 10 月 24~27 日，第一届全国环境化学大会拉开了环境化学领域学术交流的大幕，此次会议由浙江大学刘维屏教授承办。会议 176 人注册，共收到论文 162 篇，实际参会 270 余人。为期三天会议，开幕式和大会报告各半天，徐晓白院士、戴树桂教授、王晓蓉教授、陶澍教授等作了大会报告，随后两天 4 个分会近 100 人进行了口头报告。

第二届全国环境化学大会于 2004 年 10 月 10~13 日在上海召开。由上海交通大学贾金平教授承办。来自海内外高校、科研院所及相关企业部门的 280 余名代表围绕“可持续发展中的化学问题——环境化学”的主题，进行了高水平的学术交流，会议收录科技论文 245 篇。

第三届全国环境化学大会于 2005 年 11 月 4~7 日在厦门召开，由厦门大学袁东星教授承办。此次会议正式报到的代表 330 多人，绝大多数代表来自高校和科研院所。除了大会报告和主题报告外，会议论文集共收录了 319 篇短论文。内容涵盖环境分析化学/监测新技术方法、土壤污染及修复、大气污染及控制、水体污染及控制、环境化学与毒理学研究、固体废物的处置与利用等。投稿论文质量总体较高，基本反映了我国环境化学界当时研究的最新成果。

第四届全国环境化学大会于 2007 年 10 月 26~29 日在南京钟山宾馆举办，由南京大学环境学院和中国科学院南京土壤研究所联合承办，南京大学孙成教授为承办人。此次会议代表近 500 人。会上，7 篇大会报告和 30 篇主题报告，展示了环境化学研究领域的前沿进展并探讨了一些热点问题。另外，还有 190 多篇口头报告，45 位研究生首次走上了大会安排的研究生论坛，成为大会的传统。会议共收到论文 400 余篇，出版了两本论文摘要集。

第五届全国环境化学大会于 2009 年 5 月 9~12 日在大连世界博览广场举行，由大连理工大学全燮教授和陈景文教授承办。来自中国科学院、中国工程院的魏复盛、陈君石、傅家谟、郝吉明、王文兴、张玉奎等多位院士以及国内外 180 多个高校及科研院所的 1000 多位专家学者（20 位“杰青”和 10 位“长江学者”）参加大会。为期 3 天的大会，除开幕式和 7 个大会报告以外，还安排了五个分会场的 170 余场报告，本次大会共收到论文 760 篇。

第六届全国环境化学大会于 2011 年 9 月 21~24 日在上海举行，由复旦大学陈建民教授、上海交通大学贾金平教授和上海电力大学李和兴教授承办。来自世界各地近 1700 名环境科学及相关领域的专家（包括了十多位中外院士、20 多位国外华裔学者、多位“973”项目首席科学家、50 余位“杰青”和“长江学者”）参与了此次盛会。诺贝尔化学奖得主、美国科学院院士、美国医学科学院院士、墨西哥科学院院士 Molina 教授，美国工程院院士、著名杂志 *ES&T* 主编 Jerry Schnoor 教授，著名杂志 *EHP* 主编 Hugh Tilson 教授，北京大学陶澍院士，中国疾病预防控制中心（CDC）陈君石院士，福州大学付贤智院士，加拿大阿尔伯塔大学 Chris Le 教授等就大气污染、食品安全、污染物控制技术、毒理与健康效应等方面做了精彩的大会报告。本次会议共收到论文摘要 1160 篇。会议共设有 24 个专场，还特别设立了海外专家、青年科学基金获得者、研究生论坛等专场报告。大会邀请著名学术刊物主编与参会人员面对面讨论，这一形式也成为大会的传统。

第七届全国环境化学大会于 2013 年 9 月 22~25 日在贵阳召开，由中国科学院地球化学研究所冯新斌研究员承办。来自国内外近 1500 名环境科学及相关领域的专家参加了此次盛会。王文兴院士、傅家谟院士、郝吉明院士、张玉奎院士和赵进才院士出席会议。美国科学院院士、美国国家海洋和大气管理局地球系统研究实验室化学科学部主任 A. Ravishankara 教授，*ES&T* 副主编、赖斯大学 Pedro Alvarez 教授，*Sustainable Chemistry & Engineering* 主编、得克萨斯大学奥斯汀分校 Dave Allen 教授，*ES&T Letters* 副主编、印第安纳大学 Staci Simonich 教授，*ETC* 主编、密歇根大学 G. Allen Burton 教授，美国赖斯大学副校长 Vicki Colvin 教授等专程参加会议并做了精彩的报告。此次大会共收到论文摘要 1050 份，会议共安排 220 余个口头报告、近 600 份展板报告。会议期间举办了第二届海峡两岸环境化学交流会、美国化学会（ACS）学术研讨会专场、由美国 Rice 大学和南开大学联合组织的页岩气开发过程水问题研讨会专场、环境化学自然科学基金讲座专场及研究生报告专场等专题研讨会。

第八届全国环境化学大会于 2015 年 11 月 5~8 日在广州举行，由华南理工大学党志教授和叶代启教授承办。来自国内外本领域专家和研究生共 4142 人参加了此次盛会，收到论文摘要 1948 篇。会议邀请了国内 13 位院士、128 名“千人”、“杰青”、“长江学者”等国家级人才以及美国科学院院士、加拿大科学院院士、*ES&T* 主编等 20 余位海外学者参会。大会期间，7 位海内外院士或期刊主编做了精彩的大会报告，设分会场 32 个，安排口头报告 992 个，展板 1001 份。会议闭幕式上，7 位不同领域的著名专家就本次大会的若干方面进行了系统的总结与展望，成为这次大会的亮点。

时隔 15 年，第九届全国环境化学大会又回到了其启航之地——杭州。本次会议由浙江大学承办，刘维屏教授再次担任大会总负责人。大会将于 2017 年 10 月 19~22 日在浙江大学紫金港校区隆重召开。本次大会主题是“环境化学的创新与可持续发展”，将充分体现“创新、参与、合作、前瞻”宗旨，加强国内外学术研究的合作，推动我国环境化学学科走向国际，加快环境化学的学科建设与人才培养。会议内容涉及五大研究领域（分析、监测相关领域；毒理、健康相关领域；水、土、气、固废相关领域；环境化学专题；人才及交流专场），共设包括“海外华裔学者论坛”、“ACS Publications Forum”、“大气细颗粒物的毒理与健康效应”等在内的 41 个分会。截至 2017 年 8 月底，大会注册人数为 5760 人，提交会议摘要 3560 份。可以肯定，本次会议参会规模将再创新的纪录，为我国环境化学学科和环境保护事业的发展做出新的贡献！

《环境化学前沿》一书是集体努力的结晶。然而，受编辑时间之匆忙和学识水平之局限，作者在撰稿和修改过程中难免失之偏颇，出现不同学术观点甚至缺点错误。幸运的是环境化学学科一直是在不断学习中提高的，学科的发展是永无止境的，人们对环境问题的认识也总是随着时间的推移而不断提高。我们希望本书能够对广大环境化学工作者、研究生及环境管理专家有所裨益。若能对读者了解并把握环境化学研究的热点和前沿领域起到抛砖引玉作用，引起广大读者的广泛兴趣、讨论、争论和批评指正，都是编者期待和深感欣慰之处。

本书成稿过程中，中国科学院生态环境研究中心的刘稷燕、周群芳、史建波、王亚韡、何滨、胡立刚、阮挺、刘倩，浙江大学的陈宝梁、赵和平、刘璟等在审稿、改稿过程中做出了突出贡献；科学出版社为本书的顺利出版提供了极为方便的条件；朱丽及其团队为本书的策划与出版发挥了不可替代的作用。在此，一并表示诚挚的感谢！

“江城如画里，山晓望晴空。雨水夹明镜，双桥落彩虹”；“空山新雨后，天气晚来秋。明月松间照，清泉石上流”。展望未来，再过若干年，通过大家的共同奋斗，古人描绘的这种如诗如画的美好景象必将重现并遍布于我中华大地！让我们携手努力，开拓进取，再创明日环境化学之辉煌。

江桂斌

2017年初秋於北京

参考文献

王春霞, 朱利中, 江桂斌. 环境化学学科前沿与展望. 北京: 科学出版社, 2011

江桂斌.《环境化学》杂志创刊 30 周年专刊前言. 环境化学, 2011, 30(1)

全书所涉彩图及内容信息请扫描右侧二维码扩展阅读。

目　　录

第1章　环境污染物形态分析研究进展

本章导读

污染物的存在形态决定了其环境行为和生物效应，对污染物进行准确形态分析是深入研究环境过程和代谢机制的关键。采用色谱分离和原子光谱/质谱联用方法可以实现对多种有毒元素形态的灵敏分析，分子质谱可以为不同形态化合物的准确鉴定提供分子结构信息。为了提高分析方法的灵敏度、满足实际样品分析的具体要求，需要发展形态分析样品前处理技术，创建新的联用系统，发展新的分析方法。随着纳米材料的广泛应用，被释放进入环境的纳米材料以不同的物理、化学形态存在于各种环境介质并参与生物地球化学循环。不同形态纳米材料表现出明显不同的毒性和生物活性，仅仅测定样品中纳米材料的总量已不足以充分反映其环境过程和生物效应。本章将针对污染物包括传统污染物（有毒元素）和新型污染物（环境纳米材料）形态分析的方法、技术和应用进行较全面的综述。

关键词

形态分析，有毒元素，纳米材料，联用系统，样品前处理

1 引 言

污染物在实际环境中的存在形态决定了其生物地球化学行为、迁移转化规律和毒理学效应。对有毒元素而言，污染物在环境中的迁移、转化、控制和消除过程，其实质是元素在不同环境介质发生的时空改变和形态转化；在生物体内的吸收、分布、代谢和排泄等毒理学过程，实质上也是有毒物质不断穿透生物膜并发生形态改变的过程。因此，从大尺度环境归趋、中尺度环境行为到微观界面作用等不同层面的环境过程研究而言，对污染物存在形态进行准确鉴定和分析均是揭示其本质的关键。因此，污染物总量分析已远远不能够满足环境科学及生命科学、临床医学等相关学科高速发展的需求，污染物存在形态的准确甄别和灵敏测定是将环境科学研究不断推向深入的钥匙。污染物形态分析是指对特定样品中某一元素或特定污染物不同的物理、化学存在状态的表征和描述。污染物形态分析作为环境化学领域举足轻重的研究方向，经过几代科学家的努力探索和开拓积累，已经逐渐形成了较为成熟的方法体系和明确的研究主题。作为现代环境化学学科的重要组成部分，污染物形态分析不仅为环境科学领域各分支学科提供了分析手段，而且为其他相关学科如生命科学、临床医学、食品科学、农业科学等提供了重要的支撑。随着环境科学深入发展、多学科交叉融合、新理论提出和新物质的出现，污染物形态分析在概念、理论、方法和手段方面也不断发展和完善。材料、生命、食品和医学等相关科学的迅猛发展，一方面为形态分析研究提出了更高的要求，另一方面也为形态分析理论、方法和技术的发展创造了条件。新材料和新技术的不断出现，推动了原有方法和技术的不断革新和完善，使形态分析更加准确、灵敏和快速，同时也拓展了形态分析的研究对象。一些新型污染物如人工纳米材料也被纳入形态分析研究的范畴中，这极大地丰富了形态分析的研究内容，也对形态分析工作者提出了严峻的挑战。

2 有毒元素形态分析

金属元素在化合物中存在的化学价态和键合状态直接影响其环境过程和生物效应，在生物地球化学

循环中表现出不同的环境行为特征和循环代谢途径[1]。自然界普遍存在的砷、汞、硒等元素是最为典型的代表。砷可以无机形态和有机形态存在，也可以不同化学价态存在。无论是价态还是结合状态不同，不同砷化合物的环境行为和毒理学特征均表现出明显差异。

无机砷化合物毒性强烈，且三价砷[As(III)]的毒性明显高于五价砷[As(V)]；五价砷甲基化合物[一甲基胂 MMA(V)和二甲基胂 DMA(V)的毒性远低于无机砷形态；而广泛存在于水生生物体内的砷甜菜碱(AsB)、胂胆碱(AsC)、胂糖(AsS)和胂脂(AsL)等有机胂形态的毒性很小，甚至被认为无毒。

在有机胂形态中，由于砷元素的价态不同，毒性差异巨大。MMA(III)和 DMA(III)等三价砷有机化合物通常是无机砷在生命代谢过程中的中间产物，虽然存在时间短、浓度低，但毒性通常很大，在致毒机理方面起关键作用。汞化合物通常都具有很高的毒性，溶解态的汞离子（Hg^{2+}）和元素汞（Hg^0）的摄入途径、生物有效性和毒性作用机理不同，毒性效应和毒作用的靶器官也不同。无机汞在生物体内可以被甲基化为有机汞化合物，其毒性比无机汞化合物高数千倍，且表现出明显的生物富集效应、神经毒性和遗传毒性。

硒是人体必需元素，摄入缺失或过量均可导致健康损伤。元素硒在抗癌、抗氧化和毒物拮抗方面具有多种功能。与砷和汞一样，硒也可以存在不同的形态。自然界存在的硒（Se）以–2（硒化合物）、0（元素硒）、+4（亚硒酸盐，SeO_3^{2-}），+6（硒酸盐，SeO_4^{2-}）价等不同形态存在；在生物体中，无机硒可以被生物甲基化，以有机化合物形式存在，部分可以和体内的氨基酸结合，生成硒代胱氨酸（SeCys）、硒代蛋氨酸（SeMet）等有机形态。硒的生物有效性和毒性效应与其存在的化学形态密切相关。与无机硒化合物（如 SeO_3^{2-}和 SeO_4^{2-}）相比，有机硒化合物（如硒氨酸类、硒肽类、硒蛋白类化合物等）表现出了更高的生物有效性和较低的生物毒性。不同形态的硒化合物，不仅化学毒理学性质差异明显，物理性质也有明显不同。烷基硒化合物通常具有一定的挥发性。生物体对硒的烷基化过程通常被认为是解毒过程。由此可见，同一元素由于其存在形态差异，在体内的吸收、分布和代谢途径也不同，进而在生物体内表现出明显不同的 I 相和 II 相反应过程和机制。因此，在分子水平的毒理学研究当中，必须对污染物的形态进行准确定性、定量分析才能为阐明毒理学过程和代谢机制提供更加科学的依据。

元素形态分析是污染物形态分析的主要研究内容，经过长期的理论研究和实践应用，其分析方法和样品处理技术已较为成熟。然而，随着环境科学、生命科学、食品安全以及临床医学等快速发展，它们对形态分析在灵敏度、准确性和有效性等方面提出了更高的要求，亟须对现有方法不断完善和创新；另一方面，新材料和新理论的涌现为形态分析方法学的发展提供了契机，创造了前所未有的发展条件。近年来，国内外学者在形态分析领域研究活跃，在分析方法、前处理技术和污染物迁移转化过程研究等方面均获得了重要的研究进展。

2.1　样品前处理技术

相比于污染物总量，样品中不同形态化合物的含量将更低，分析过程基体干扰效应更加明显。在形态分析之前对样品中不同形态化合物进行有效富集可提高方法的灵敏度，大大降低基体干扰。针对痕量元素总量分离富集的前处理技术已有很多报道，但针对不同形态污染物进行分离富集的研究还远未成熟。不同形态化合物通常具有明显不同的物理化学性质，同时还要考虑前处理技术与后继分析手段的兼容性，因此，实现不同形态化合物的同时富集从方法和技术上均具有一定的难度，需要发展一系列新的预富集方法[2]。近年来，固相萃取、液液微萃取、固相微萃取、搅拌棒吸附萃取、浊点萃取和固相悬浮萃取等多种前处理技术已被成功应用于污染物形态分析。

2.1.1　固相萃取

固相萃取是污染物形态分析中应用最广泛的样品前处理技术，通过在线和离线方式，均可以获得很高的富集倍数。Cheng 等[3]采用强阴离子交换柱（SAX）在线富集 HPLC-ICP-MS 分析测定了水样中的 Hg^{+}，$MeHg^{+}$，$EtHg^{+}$和 Hg^{2+}，富集倍数可达 1000 倍以上。随后，该课题组又采用巯基功能化的二氧化硅微球（HS-SiO_2）在线富集了 Hg^{2+}、$MeHg^{+}$和 $EtHg^{+}$ [4]。近几年出现的磁固相萃取因其便于分离受到了更多关注。该技术多通过制备功能化磁性纳米颗粒来实现对不同形态化合物的高效富集和有效分离[5,6]。在实现磁分离之后，还可以结合其他技术实现对不同形态化合物的二次富集，以获得很高富集因子和很好的抗干扰能力。二次富集预处理技术可以用于基体较为复杂的样品分析，如食品和酵母细胞当中痕量硒氨酸化合物的形态分析[7]。

细胞中的元素形态分析在金属组学和生命科学研究方面具有非常重要的意义，但同时面临样品量少、形态复杂、含量低的难题。Chen 等[8]创造性地建立了一套基于微流控芯片的磁固相萃取装置，并用于酵母细胞当中硒形态分析。这项研究将微流控芯片、磁固相萃取技术和 HPLC-ICP-MS 形态分析联用技术很好地结合起来，优势互补。在分析过程中，磺化聚苯乙烯包覆的磁性纳米颗粒（Fe_3O_4@PSS MNPs）作为硒氨酸类和硒肽类化合物的吸附材料。基于芯片的磁性固相萃取较常规的磁性固相萃取而言，样品消耗量更少，更适合于稀有样品分析。Wang 等[9]设计和构建了基于微流控芯片的阵列磁固相萃取装置（图 1-1）并与毛细管 HPLC-ICP-MS 在线联用，实现了细胞中汞形态的高通量分析。在后续的研究中，该技术成功地用于细胞中硒和汞的拮抗作用研究[10]。

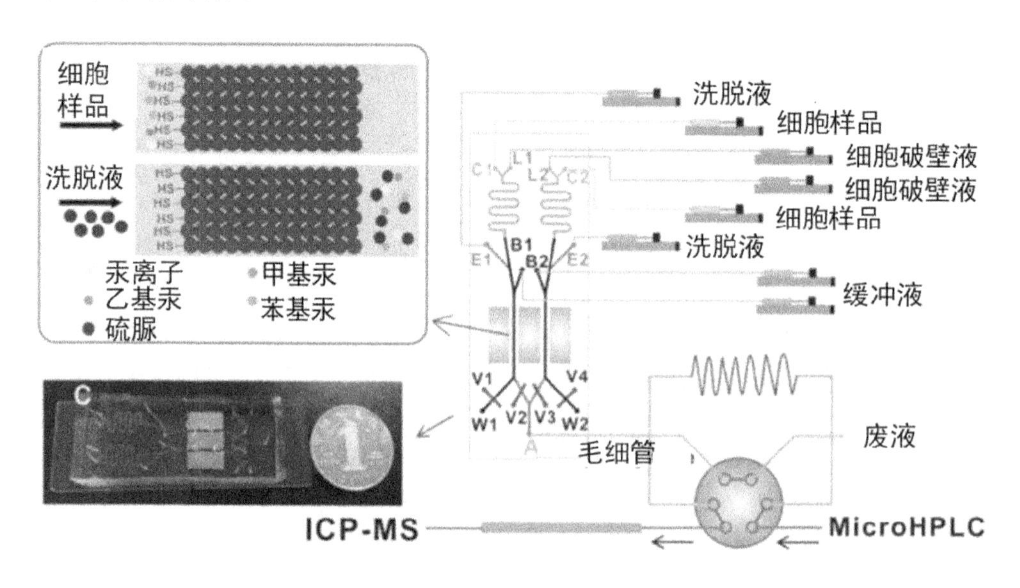

图 1-1 基于微流控芯片的阵列磁固相萃取-MicroHPLC-ICP-MS 联用装置示意图（引自参考文献[9]）

固相萃取用于毛细管电泳形态分析可以使方法获得较高灵敏度。Chen 等[11]以巯基棉为吸附剂，将分散固相萃取(DSPE)和场放大样品堆积注射(field-amplified sample stacking injection，FASI)相结合，通过 CE-ICP-MS 分析测定，建立了高灵敏汞形态分析方法，其对 $MeHg^{+}$、$EtHg^{+}$和 Hg^{2+}的定量限为 0.26~0.45 pg/mL，拓展了毛细管电泳在形态分析领域的应用范围。以纳米氧化铝为吸附剂，制备 SPE 填充微柱，通过大体积样品过滤富集，结合毛细管电泳（CE-UV）可以实现 Se(IV)和 Se(VI)的分离分析，大大提高 CE 分析测定 Se 形态化合物的灵敏度[12]。借助毛细管电泳的强大分离能力，结合有效的预富集技术，还可以对部分元素形态化合物进行手性分离分析。

2.1.2 液液微萃取技术

液液微萃取技术使用溶剂少，成本低，富集倍数高，易于操作，近年来，被成功用于了形态分析样品前处理。液液微萃取一般较适合对非极性化合物分离富集，而无机元素各形态化合物多为极性物质，因此需要采用衍生化技术和载体调控技术（carrier mediated），以期提高对水溶性化合物的萃取效率。Chen 等[13]以 1-(2-吡啶偶氮)-2-萘酚(PAN)为络合剂将无机汞和有机汞化合物络合为疏水化合物以后被同时从水相萃取进入有机相，随后被转移至中空纤维内腔的接收相，以此实现对样品中痕量汞形态化合物（Hg^{2+}、$MeHg^{+}$、$EtHg^{+}$和 $PhHg^{+}$）的分离和富集。分析测定硒化合物形态，特别是食品中硒化合物的形态是研究硒抗癌机制和代谢机理的关键。Duan 等[14]以氯甲酸乙酯作为衍生试剂，建立了中空纤维膜液液微萃取-GC-ICP-MS 分析 4 种硒代氨基酸形态的新方法，并成功地应用于富硒紫花苜蓿、大蒜和洋葱中硒代氨基酸形态分析。除了常规试剂，离子液体也被成功用于中空纤维液液微萃取砷形态化合物和汞形态化合物[15,16]。液液微萃取不仅可以获得理想的分离富集效果，而且有助于实现前处理过程自动化。Liu 等[17]组建了自动化分散液液微萃取装置（DLLME），并与 HPLC-CV-AFS 系统联用分离富集了自然水体中痕量 $MeHg^{+}$、$EtHg^{+}$和 Hg^{2+}。

毛细管电泳的强大分离能力一直被形态分析领域所关注，但因其受进样量和检测器的限制，测定灵敏度低，抗干扰能力差，很难真正满足实际样品的分析要求，应用范围长期以来受到很大限制。如果采用有效的前处理技术，则可以实现毛细管电泳在元素形态分析领域的深入应用，充分发挥其样品需求量少的优点。近年来，国内外学者不断尝试采用微萃取技术结合毛细管电泳分离来发展新的形态分析方法。在微萃取过程中需要采用和毛细管电泳缓冲液相兼容的溶液作为接收相，因此，在进行液相微萃取过程中，一般采用液-液-液三相微萃取模式。获得高富集倍数、高选择性、快的动力学过程以及实现自动化是现代液相微萃取技术的发展目标。相比于静态液相微萃取，动态液相微萃取可以在较短的时间内获得更高的富集倍数和更好的灵敏度。Li 等[18]成功组装了一套简单有效的基于中空纤维液-液-液动态微萃取自动化装置。借助程序化流动注射分析仪，该装置可以同时萃取水体中的有机汞和无机汞形态化合物。以冠醚（18-冠-6）为络合剂，不同汞形态化合物被萃取到有机相（氯苯）当中，随后，被反萃取到 0.1%（*m/v*）3-巯基-1-丙磺酸水溶液接收相。采用大体积样品动态萃取方法，可获得较高的富集倍数，成功实现了毛细管电泳-紫外检测分析测定生物和环境样品中汞形态化合物的目标。所设计的动态富集装置如图 1-2 所示。Chen 等[19]发展了中空纤维液膜微萃取（HF-LLLMME）方法同时高效萃取溶液中无机汞离子和有机汞化合物，对 Hg^{2+}、$MeHg^{+}$、$EtHg^{+}$和 $PhHg^{+}$的富集倍数分别达到了 103 倍、265 倍、511 倍和 683 倍。借助该富集方法可以采用毛细管电泳实现样品中痕量汞的形态分离分析。Yang 等[20]采用分散液液微萃取技术对不同汞化合物（$MeHg^{+}$、$EtHg^{+}$、$PhHg^{+}$和 Hg^{2+}）进行预富集，再用毛细管电泳进行汞形态分析。所建立的方法简单、快速、费用低廉并且对环境友好，可以作为水体中痕量汞形态的有效分析方法。Duan 等[21]成功制备了苯丙氨酸衍生物[l-*N*-(2-hydroxy-propyl)-phenylalanine, 1-HP-Phe]及其 Cu^{2+}螯合物[Cu(Ⅱ)-(l-HP-Phe)$_2$]。后者被用作手性选择剂，采用配体交换手性分离方法和胶束电动毛细管色谱分析了富硒酵母样品中的硒代蛋氨酸（D, L-selenomethionine；SeMet）。

2.1.3　毛细管微萃取技术

毛细管微萃取（也称管内固相微萃取）用于元素形态分析也有报道。Chen 等[22]发展了基于管内中空纤维液相微萃取技术，与 HPLC-ICP-MS 在线联用测定了 As(III)、As(V)、MMA(V)、DMA(V)、AsB 和 AsC 六种砷形态化合物。将磺化聚苯乙烯(PSP)、3-巯丙基三甲氧基硅烷(3-mercapto propyltrimethoxysilane, MPTS)和 *N*-(2-氨乙基)-3-氨丙基三甲氧基硅烷[*N*-(2-aminoethyl)-3-aminopropyltrimethoxysilane, AAPTS]的混合凝胶共同固载到中空纤维孔隙和内壁，实现了对尿液中不同砷形态化合物的在线富集。Liu 等[23]将锐钛矿型纳米 TiO_2 包埋在聚甲基丙烯酸-乙二醇二甲基丙烯酸酯[poly(MAA-EDMA)]整体柱中，合成了纳米 TiO_2 功能化整体柱，建立了纳米 TiO_2 功能化整体柱毛细管微萃取与 ICP-MS 在线联用顺序分析人

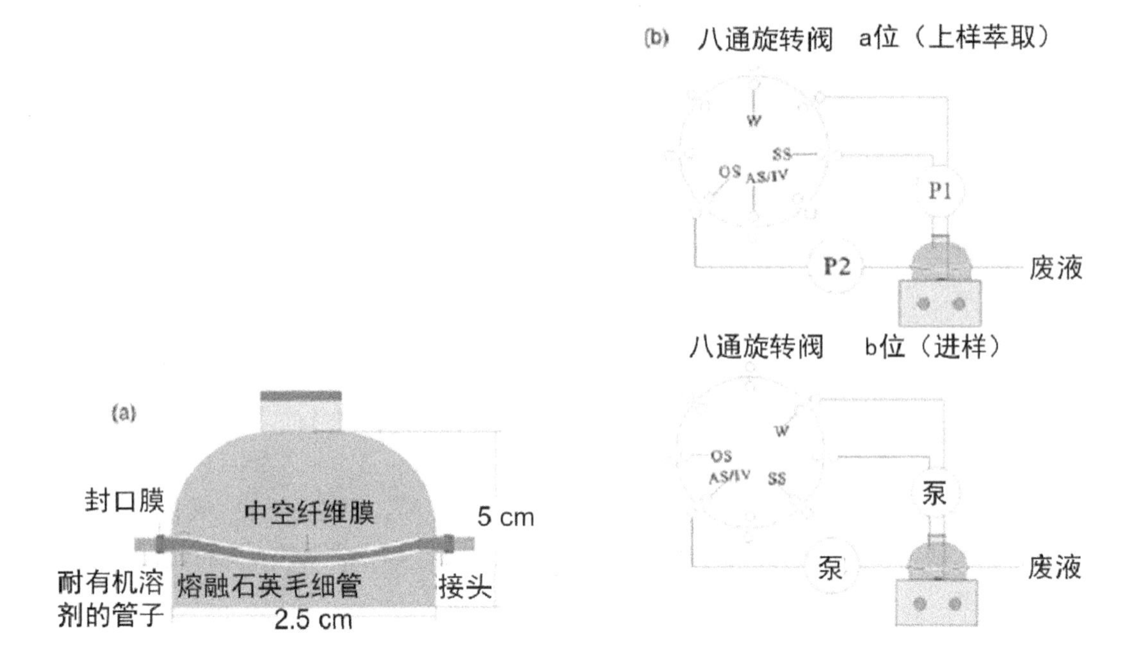

图 1-2 AD-HF-LLLME 自动微萃取装置示意图（引自参考文献[18]）

尿样中 Gd^{3+}和钆造影剂的新方法。磷酸盐缓冲体系中，Gd^{3+}/钆造影剂(如 Gd-DTPA 和 Gd-DTPA-BMA)在 TiO_2 NPs 功能化 poly(MAA-EDMA)整体柱上会表现出不同的吸附行为。Gd^{3+}在 pH 2~9 范围内能被整体柱定量吸附，而钆造影剂只能在 pH 2~3 内被整体柱定量吸附。通过选择 pH 5 和 pH 2.5 作为 Gd^{3+}和钆造影剂的吸附 pH，可以实现 Gd^{3+}和钆造影剂的分离。该方法测定 Gd^{3+}、Gd-DTPA 和 Gd-DTPA-BMA 的检出限分别为 3.6 ng/L、3.2 ng/L 和 4.5 ng/L，是目前文献报道的分析 Gd^{3+}和 Gd 造影剂的最低浓度水平。

2.1.4 搅拌棒吸附萃取技术

搅拌棒吸附萃取技术也是近几年发展起来的有效分离富集技术，在污染物总量富集方面已得到广泛应用。采用 C-18 修饰的搅拌棒微萃取和 40%(*v*/*v*)甲醇解吸的方法，可以实现对环境样品中丁基锡形态化合物(一丁基锡、二丁基锡和三丁基锡)的分离富集[24]。Mao 等[25]采用溶胶-凝胶法制备了磺化聚苯乙烯-二氧化钛（PSP-TiO_2）有机-无机杂化材料包覆的搅拌棒，建立了搅拌棒富集(SBSE)-HPLC-ICP-MS 分析生物样品中甲基硒代半胱氨酸（MeSeCys）、硒代蛋氨酸（SeMet）、硒乙硫基氨基酪酸（SeEt）、硒寡肽类化合物（γ-GluMeSeCys）和硒谷胱甘肽类化合物（GS-Se-SG）等硒氨酸类化合物的新方法。与常规的搅拌棒萃取机理不同，通过修饰高极性有机-无机杂化复合材料，该搅拌棒通过阳离子交换作用来实现高极性硒化合物的高效富集，而不需要采用其他任何衍生化技术。

由此可见，虽然目前较为成熟的形态分析技术为 HPLC-ICP-MS，且该技术为形态分析领域的主流分析手段，灵敏度高、准确性好。但是，由于部分样品基体复杂、形态分布浓度差异较大，为降低干扰获得准确的形态定量定性信息，采用样品预富集技术是实现准确分析的关键环节。通过样品预富集技术，可以拓展毛细管电泳的实际应用范围和分析效果，使得该项有效分离技术在形态分析领域中可以发挥独特的优势。

2.1.5　样品提取技术

与元素总量分析不同，形态分析中所涉及的固体样品前处理技术，主要是采用较为温和的形式（如溶剂提取、吸附富集等）将存在于固体及其他样品当中的不同形态化合物转移至与所选用分析手段相兼容的溶剂当中，以满足随后分离和检测的要求。在形态分析样品前处理过程中，需要尽量保持目标物在原始样品中的化学形态的完整性，尽量避免在前处理过程中由于操作的原因导致目标化合物发生形态变化和明显损失。有关形态分析前处理研究多集中在对生物样品包括动物和植物体内目标化合物的有效提取，利用的提取手段主要包括振荡、超声、微波等。Wu 等[26]采用 25%（*m*/*v*）四甲基氢氧化铵溶液提取生物样品中无机汞，随后采用微波消解测定总汞的方法来实现无机汞和有机汞的形态分析。Zhao 等[27]对比研究了四种不同提取方法对植物样品中的砷形态分析进行样品前处理。结果表明，采用乙醇/水体-超声提取方法效果最佳。发展一些快速、有效而简便的前处理方法，特别是针对少量样品的有效处理方法非常必要。国内学者也积极发展了新的样品处理技术。Deng 等[28]建立了以多壁碳纳米管（MWCNTs）为支载平台的介质固相分散（MSPD）样品前处理方法。由于碳纳米管大的比表面积和良好的机械强度，使得它可以成为良好的介质分散材料。与传统介质固相分散前处理技术相比，采用碳纳米管为支载固体，获得了更好的提取效果和提取效率。

在形态分析样品前处理方面，另一个重要的研究方向是针对一些特殊样品中污染物形态分析进行前处理，使处理后的样品介质和状态满足分析仪器的进样要求。报道显示，石油和天然气中含有一定量的汞。考虑到石油来自于古生物，因此，对石油中汞进行形态分析具有重要意义。虽然目前用于汞形态分析的测定方法较为成熟，但是针对石油当中汞形态分析的方法和样品前处理技术还很稀缺。采用 GC-ICP-MS、GC-AFS 联用技术通过衍生化可以实现石油中汞形态分析，但这种方法较为繁琐，且仪器分析成本较高、耗时长。液相色谱分离不需要对样品进行衍生化处理，但对前处理过程要求较高。如果建立合适的样品前处理方法，即可采用 HPLC-AFS 分析测定石油中的有机汞化合物，降低分析成本，简化操作过程。Yun 等[29]详细研究了石油中有机汞形态包括甲基汞和乙基汞的提取方法，考察了不同提取手段（振荡、超声和微波）和不同提取剂（TMAH、KOH/CH_3OH、HCl 和酸性 $CuSO_4$/KBr）对不同形态汞的提取效果，成功建立了有效的 HPLC-AFS 分析石油中烷基汞形态化合物的方法。

2.2　联用系统研制

2.2.1　液相色谱联用系统

除 HPLC-ICP-MS 联用系统以外，高效液相色谱-氢化物发生-原子荧光光谱联用系统（HPLC-HG-AFS）也是常用形态分析技术。近年来，该联用系统的原子化组件和接口技术研究取得了一系列进展。通常在测定汞形态的时候采用冷原子荧光光谱法，无机汞可以和 KBH_4 反应生成气态元素汞被 AFS 检测。但甲基汞和乙基汞等有机汞化合物会生成对应的挥发性有机汞气态氢化物（MeHgH 和 EtHgH），而非气态元素汞形态（Hg^0）。因此，需要在 CV-AFS 检测之前将气态有机汞氢化物进行原子化以获得对应的灵敏荧光信号。Huang 等[30]建立了一套简单、高效的低温紫外辐照 UV 原子化器，它由缠绕石英管和 UV 灯构成，经液相色谱分离的 Hg^{2+}、$MeHg^+$和 $EtHg^+$离子与 0.5%（*m*/*v*）KBH_4 溶液反应，生成的 MeHgH 和 EtHgH 在 UV 原子化器内可完全实现原子化。Hg^{2+}、$MeHg^+$和 $EtHg^+$的检出限分别可以达到 0.38 mg/L、0.41 mg/L 和 0.56 mg/L。CV-AFS 检测有机汞化合物的另一途径是在柱后将有机汞化合物进行有效降解，使所有经分离后的有机汞化合物形态全部转化为无机汞形式，然后再与硼氢化物反应生成气态汞。一般这一过程都要借助微波或紫外光照的手段来实现。近年来，具有优良催化性能的纳米材料被成功用于在线降解，促进了联用系统的发展，并有利于降低整体分析成本。Ai 等[31]发展了基于 Fe_3O_4 磁性纳米材料（MNPs）的绿色、高效柱后氧化法实现在线转换氢化物非活性形态为活性形态。对无机汞离子（Hg^{2+}）、甲基汞（$MeHg^+$）、乙基汞（$EtHg^+$）和苯基汞（$PhHg^+$）等模型化合物，在 pH 2.0，0.6%（*v/v*）H_2O_2 存在的条件下，该装置可以将经分离后的有机汞化合物在线转化为无机汞离子，避免了微波或紫外装置的使用。

Li 等[32]早在 2012 年就报道了基于 Ag-TiO_2 和 ZrO_2 的半导体纳米材料光催化蒸气发生系统（PCVG），并成功用于原子荧光光谱的进样装置以提高测定 Se 灵敏度和液相色谱与光谱/质谱联用系统的接口装置。在这个过程中，半导体纳米材料的导带电子作为“还原剂”将不同 Se 形态包括 Se（VI）直接转化为挥发性的 H_2Se。与传统的硼氢化物氢化物发生体系相比，该光催化系统可以不经过预还原过程即可将 Se（VI）生成气态氢化物。而在硼氢化物体系中 Se（VI）很难反应生成气态氢化物，这也是通常化学氢化物发生在测定 Se 时所遇到的难题。采用纳米二氧化锆（Nano-ZrO_2）为催化剂，以 UV/nano-ZrO_2/HCOOH 为光化学原子化系统，可以用作 HPLC-AFS 联用系统的接口[33]。该系统可以将不同汞形态化合物还原为汞蒸气（Hg^0），达到原子化的效果。

介质阻挡放电（DBD）是一种低温等离子体技术，其产生的自由基可有效促进化合物分子的分解。DBD 的这一显著特征赋予了该技术在作为新型原子化器方面的条件。Liu 等[34]建立了基于介质阻挡放电等离子体诱导化学蒸气发生（DBD plasma-CVG）的 HPLC-AFS 汞形态分析系统。该系统不用其他化学试剂即可一步完成有机汞形态降解和汞离子还原，其在线联用接口技术简单、环境友好，因此具有很好的应用前景。

除了联用系统接口技术研究以外，形态化合物分离手段研究也有一定的进展。近年来，为满足大量样品筛查，研究人员发展了高效液相色谱快速分离形态分析方法。Chen 等[35]采用阳离子交换色谱为分离手段建立快速分离 HPLC-ICP-MS 联用系统，在 2.5 min 之内即可实现 Hg^{2+}、$MeHg^+$、$EtHg^+$和苯基汞（$PhHg^+$）的有效分离。采用短柱分离，阳离子交换柱在线富集的方法，Jia 等[36]采用 HPLC-ICP-MS 联用系统分离分析了海水当中的 Hg^{2+}、$MeHg^+$和 $EtHg^+$。

2.2.2 气相色谱联用系统

传统形态分析联用方法通常需要较昂贵的分离技术和检测手段。同时，由于现场分析的需求，在形态分析方面的仪器微型化受到了广大研究者的关注。为降低汞形态分析成本，便于仪器小型化和实现现场分析，一些新的检测器也被应用于形态分析仪器的研制。Lin 等[37]采用多孔碳为固相微萃取材料顶空吸附，以气相色谱(GC)-介质阻挡放电发射光谱（DBD-OES）为检测手段，建立了汞形态分析新装置和新方法。在采用 GC-DBD-OES 进行测定之前，不同形态的汞化合物与四苯硼钠（$NaBPh_4$）溶液发生反应，衍生为对应的挥发性气态氢化物并被固相微萃取顶空富集。由于仪器装置体积小、耗能低，相比于传统联用系统而言，其费用更加低廉。以 DBD-OES 为检测器与小型气相色谱联用有望实现汞形态分析系统的小型化。于永亮等[38]以 DBD 为低温原子化器并引入长光程吸收检测池，建立了微型化原子吸收光谱系统。当 DBD 原子化器关闭时，通过冷原子吸收测得无机汞的吸光度，而当 DBD 原子化器开启时，得到无机汞和甲基汞的总吸光度，从而实现无机汞和甲基汞的分别测定。Zeng 等[39]组建了气相色谱-原子荧光光谱联用系统（GC-AFS）。该系统并没有采用复杂的结构组件和技术，而是将气相色谱通过氩氢（Ar-H_2）火焰直接与原子荧光光谱仪相连。气相色谱柱的出口直接接入 Ar-H_2 火焰原子化器的底部。该联用系统可分离分析二甲基硒[$(Met)_2Se$]、二甲基二硒[$(Met)_2Se_2$]、四甲基锡[$Sn(Met)_4$]和四乙基铅[$Pb(Et)_4$]等不同形态烷基金属/类金属化合物。

2.2.3 毛细管电泳联用系统

在生命科学研究过程中，样品量一般较少。由于毛细管电泳分离效果好、样品需求量少，通过将毛细管电泳和电感耦合等离子体质谱进行联用，可以实现部分有机金属化合物的高效分离和灵敏检测，满足生物化学和毒理学研究的需求，特别是在研究不同形态化合物与生物大分子作用过程中充当重要角色。Sun 等[40]利用 CE-ICP-MS 联用系统样品需求量少、灵敏度高的特点，分析测定了三甲基锡（TMT）、三丙基锡（TPrT）、三丁基锡（TBT）和三苯基锡（TPhT）等 4 种有机金属化合物与蛋白分子之间的络合稳定常数。Liu 等[41]采用 CE- ICP-MS 联用系统分离分析了包括无机砷和有机胂在内的 10 种砷形态化合物。随后，通过建立的 CE-ICP-MS 联用系统和有效的接口技术，在 9 min 之内同时基线分离分析了 6 种砷形态化合物和 5 种硒形态化合物[42]。采用微波辅助提取样品前处理技术，CE-ICP-MS 也被成功用于鱼体中无机和有机汞形态分析[43]。

通过设计合适的接口装置，CE 还可以和除 ICP-MS 以外的其他检测器联用。Deng 等[44]采用新的接口装置，以原子吸收光谱为 CE 检测器，成功建立了 CE-HG-ETAAS 形态分析联用系统。Matusiewicz 和 Slachcinski[45]将微流控芯片用作新的接口技术建立了毛细管电泳-氢化物发生-微波诱导等离子发射光谱联用系统（CE-HG-MIP-OES），分析测定了无机砷化合物 As(III)和 As(V)。Niegel 等[46]将快速毛细管电泳（CE）和飞行时间质谱（TOF-MS）联用分离分析了包括胂糖在内的 4 种有机胂化合物。

2.3　色谱分离-原子/分子质谱形态分析联用技术

不同形态有毒元素在人体代谢过程中可以生成各种毒性差异明显的代谢产物。由于分析原理和过程的限制，目前常用的分析方法一次进样仅能对所测生物样品中有限数量的代谢产物进行测定，对一些色谱保留时间相同的组分很难进行准确鉴定，不利于对代谢产物分布特征和毒理学过程进行准确评价。为了满足对代谢产物的分析要求，需要对现有分析系统的色谱分离条件进行优化，尽可能改善色谱分离效果，提高分辨率。Stice 等[47]通过对分离条件的优化，改进了 HPLC-ICP-MS 联用系统的分离效果，实现了人体内可能存在的 13 种砷代谢产物的分离分析，其中包括部分含硫砷化合物形态。采用离子色谱-氢化物发生-原子荧光光谱联用系统（IC-HG-AFS），同样可以实现不同含硫砷形态化合物的分离分析[48]。

尽管通过条件优化可以提高色谱分离效果，但受色谱分离和检测器局限性限制，色谱-原子光（质）谱联用系统不可能实现对所有形态化合物及其代谢产物进行分离分析。串联质谱（ESI/MS/MS）作为液相色谱的检测器可以实现在对元素形态化合物灵敏、准确定量的同时，获得形态化合物的分子结构信息，为发现和鉴定新的化合物形态、代谢产物提供条件[49]。在家禽工业，为了降低家禽疾病风险和促进生长，一些饲料添加剂被长期广泛使用。洛克沙胂（3-硝基-4-羟基苯胂酸，3-nitro-4-hydroxyphenylarsonic acid，Roxarsone，ROX）是一种具有抗生素作用和促进家禽生长的有机胂制剂，在国际上使用时间长达 60 年。目前一些国家已经明确禁用此药，但在部分国家还被广泛使用。被家禽摄入的 ROX 在体内经过代谢可以部分随粪便排出体外；同时，一部分也会残留、富集在体内器官。因此，在还未禁用 ROX 的国家，食用鸡肉中砷的含量和存在形态是政府和各级卫生组织重点关注的问题。有关 ROX 在动物体内的代谢过程和机制还不清楚，主要是因为以前缺乏有效分析手段。代谢产物含量通常会很低，且在体内生物酶作用下可以代谢生成多种结构复杂的中间产物。对砷化合物代谢产物进行分析鉴定的主要挑战是需要有高灵敏度的定量分析方法和准确、可靠的定性手段。实验表明，通过将 HPLC 与 ICP-MS 和 ESI/MS/MS 进行平行联用，可以实现高灵敏定量和准确定性的双重功能[49]，能够满足 ROX 在生物体内代谢产物的研究需求。Yang 等[50]采用 HPLC-ICP-MS/ESI-MS/MS 联用技术以 1600 只鸡为研究对象，对 ROX 在家禽体内的代谢过程和形态分布开展了深入研究。借助双质谱检测手段，研究发现了 ROX 在鸡体内新的代谢产物 *N*-乙酰基-4-羟基-*m*-对氨基苯胂酸（*N*-acetyl-4-hydroxy-*m*-arsanilic acid，*N*-AHAA）。该代谢产物占总砷含量的 3%~12%。除此之外，对砷可能代谢产物进行分析，发现排泄物中还包括 3-氨基-4 羟基苯胂酸（3-amino-4-hydroxyphenylarsonic acid，3-AHPAA）、As(III)、As(V)、MMA(V)、DMA(V)和 ROX。对排泄物中砷形态的鉴定和分析，进一步验证了 ROX 可在家禽体内器官或其肠道微生物作用下进行代谢。Peng 等[51]采用液相色谱和原子/分子双质谱联用的方法，成功分析和鉴定了 ROX 饲养的鸡肝中砷的形态。采用阴离子交换柱为分离柱，ROX 和其他目标砷形态化合物在 12 min 之内实现基线分离。鸡肝中的 8 种砷形态化合物得到了分析测定。随后，该课题组[52]在鸡肝中发现了三种 ROX 甲基化产物：甲基化-3-硝基-4-羟基苯胂酸（methyl-3-nitro-4-hydroxyphenylarsonic acid, methyl-ROX），甲基化-3-氨基-4-羟基苯胂酸（methyl-3-amino-4-hydroxyphenylarsonic acid, methyl-3-AHPAA）和甲基化-3-乙酰氨基-4-羟基苯胂酸（3-methyl-3-acetamido-4-hydroxyphenylarsonic acid 或 methyl-*N*-acetyl-ROX, methyl-*N*-AHPAA）。在甲基化过程中，三价苯胂酸中间产物、*S*-腺苷甲硫氨酸和砷甲基化转移酶（As3MT）参与了反应。通过酶反应过程，将甲基转移到了三价苯胂酸甲基化中间产物上面而得到了最终的甲基化产物。毒理学试验研究结果表明三价苯胂酸的 IC_{50} 值较对应的五价苯胂酸低 300~30000 倍，表明其具有很高的生物毒性。甲基化代谢产物如图 1-3 所示。采用双质谱检测技术，可以避免保留时间一致的共馏组分化合物漏检。ESI/MS/MS 可以在 ICP-MS 测定元素砷含量的同时，提供分子结构信息，为准确鉴定 ROX 在鸡肝中的最终代谢产物提供直接依据。

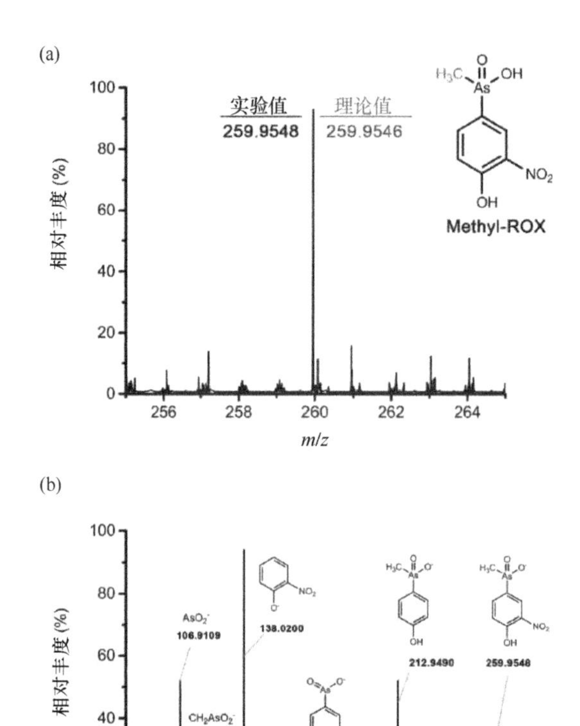

图 1-3　鸡肝中砷形态化合物的 EST-TOF-MS 质谱图（引自参考文献[52]）

海藻中砷形态通常较为复杂，且有可能存在未知形态。Hsieh 和 Jiang[53]采用 HPLC-ICP-MS 和 HPLC-ESI/MS 两套联用系统对海藻中不同砷形态化合物进行了准确的定量分析，并将所检测的未知砷形态鉴定为两种胂糖和四甲基胂化合物。Raab 等[54]采用 HPLC 分离 ICP-MS-ESI/MS 平行双质谱检测，准确鉴定了褐藻中存在的三类脂溶性砷形态化合物。Arroyo-Abad 等[55]采用 RP-HPLC-ICP-MS 联用系统结合 ESI-Q-TOF-MS 分析鉴定了鱼肝甲醇提取物中砷形态。除了含有极性无机砷和有机砷以外，疏水性更强的含砷脂肪酸化合物和含砷碳氢类化合物也被检出。研究总共鉴定了提取物中的 20 种有机胂化合物，其中 10 种为含砷脂肪酸类化合物(AsFA)。采用此联用系统，首次鉴定了鱼肝提取物中的含砷碳氢类化合物(AsHC)。Garcia-Sartal 等[56]采用二维液相色谱(体积排阻和阴离子交换)双质谱联用系统(ICP-MS-ESI/MS/MS)测定了食用藻类中有机胂形态化合物，并采用模拟肠胃液方法对生物可利用的有机胂形态进行了分析鉴定。

液相色谱(包括体积排阻色谱)/质谱(包括原子或分子质谱)联用系统也被用来研究不同金属、不同形态化合物与蛋白分子或其他生命分子的相互作用，为金属组学研究提供了有效手段。Xu 等[57]采用 SEC-ICP-MS 结合 RP-HPLC/ESI-IT-MS 和 MALDI-TOF-MS 系统研究了“Mn，Co，Ni，Cu，Cd”、“Pd，Pt，Au”和“Hg，CH_3Hg，C_2H_5Hg-THI”等三组金属与金属硫蛋白(Zn_7MT-2)的化学亲和作用。通常，有关铬（Cr）的形态分析研究中，主要是针对 Cr(III)和 Cr(VI)进行分析测定。已有研究报道也显示，Cr 可能以多聚体形式存在于环境中，但由于缺乏有效的分析手段，长期以来并未对假设进行准确验证。Hu 等[58]采用 HPLC-ICP-MS 联用技术，在研究 CCA(chromated-copper-arsenate)处理的木材浸出液中 Cr 形态时发现存在未知形态。借助质谱分析对 m/z 进行研判和 HPLC-ICP-MS 分析结果，最终确定了 Cr 可以多聚体形式存在于自然环境中。Cr(III)多聚体形态可以占到 CCA 浸出液未知 Cr 形态的 39.1%~67.4%。随后，实验对表层水体、自来水和废水当中的未知 Cr 形态进行分析。结果发现，Cr(III)多聚体形态可以占到总 Cr 含量的 60%。由此可见，在进行 Cr 形态分析过程中，如果不能够对 Cr(III)多聚体进行准确测定，最终会造成 Cr 含量低估和形态分析结果的偏差。砷(As)可以和溶解性有机质形成络合物，进而影响 As 在自然环境介质中的归趋行为。对 As 及其他元素在自然水体中与溶解性有机质的结合形态进行准确鉴定是深入研究元素生物地球化学循环的关键。但是，因常规液相色谱-光谱/质谱联用技术很难实现对结合物有效分离和灵敏检测，As 与溶解性有机质结合物的准确鉴定一直是形态分析领域所关注的问题。Liu 和 Cai[59]建立了体积排阻色谱-紫外光谱-电感耦合等离子体质谱（SEC-UV-ICP-MS）串联检测的联用分析方法，成功用于研究 As 和溶解性有机质络合形态的分离分析及定量检测。Peng 等[60]采用 SEC-ICP-MS 联用技术和 RP-HPLC-ICP-MS/ESI-QTOF-MS 平行检测技术研究了 4 种 CdSe/ZnS 量子点在不同培育条件(培育时间、培育浓度和清除时间)下在 HepG2 细胞中的存在形态。SEC-ICP-MS 分析 Cd 的形态结果表明，CdSe/ZnS 量子点在不同培育条件的细胞中均以两种形态（QD-1 和 QD-2）存在。通过保留时间比较、透射电镜表征和荧光检测等方法，证明 QD-1 是一种类似量子点的纳米颗粒。通过 RP-HPLC-ICP-MS/ESI-QTOF-MS 平行检测，证明 QD-2 是一类含有 Cd 的金属硫蛋白（Cd-MTs）。而且，QD-1 和 QD-2 的含量随着培育浓度和培育时间的增加而增加，这为从分子水平上揭示量子点的细胞毒性机理提供了新思路。

2.4　非色谱形态分析方法

2.4.1　络合分离

不同形态化合物可以与特定的络合剂发生络合反应，生成不同的亲疏水络合物。采用萃取分离的方式可以将不同形态离子进行区分。虽然该分离机制早已被用于污染物形态分析，但随着萃取分离新手段的出现，近年来一些基于络合分离的形态分析新方法也被不断提出。以 5-硫基-3-苯基-1, 3, 4-噻二唑-2-硫酮钾盐(Bismuthiol II，铋试剂)同时作为络合剂和化学改进剂，以甲醇为分散剂，氯仿为萃取剂进行分散液液微萃取时，在 pH 2.0 的条件下，铋试剂与 Se(IV)的络合物很快即被萃取进入萃取剂，而 Se(VI)被保留在溶液当中，实现了不同价态无机硒化合物的分离。Zhang 等[61]采用该分散液液微萃取技术结合电热蒸发-电感耦合等离子体质谱（ETV-ICP-MS）实现了环境水样中硒形态化合物的分离分析和高灵敏检测。采用同样的研究思路，As(III)、Se(IV)和Te(IV)分别与二乙基二硫代氨基甲酸钠(DDTC)络合，以溴苯为萃取剂，甲醇为分散剂形成微液滴分散体系，对络合物进行了富集分离，Liu 等[62]成功将分散液液微萃取技术(DLLME)和 ETV-ICP-MS 相结合，同时实现了无机砷、硒和碲的形态分析。以 DDTC、吡咯烷二硫代甲酸铵(APDC)和钼酸盐等为络合剂，采用离子液体/微柱分离、浊点萃取、固相悬浮微萃取(SFODME)和固相萃取(SPE)技术可以实现无机砷形态分析[63-66]。

2.4.2 选择性吸附

在进行非色谱分离形态分析方法研究中，通过制备高效吸附剂，根据不同形态化合物在吸附剂表面的吸附行为和特性差异，控制实验吸附条件可实现不同形态化合物的分别检测。Zhang 等[67]采用磷酸铁($FePO_4$)为特效吸附剂，详细研究了不同价态 Cr 在材料表面的吸附行为和特征。结果发现，在 pH 5.9 时，几乎 100%Cr(III)可以被吸附剂吸附。然而，在相同条件下，Cr(VI)几乎不能够被吸附。被吸附的 Cr(III)可以很容易被 0.1%H_2O_2和 0.05 mol L^{-1} NH_3 溶液洗脱，得到令人满意的回收率，借此可以建立 ETAAS 测定 Cr(III)和 Cr(VI)的方法。通过制备修饰多壁碳纳米管微柱，采用 SPE 方法可以同时实现环境水样中无机 As、Cr 和 Se 的形态分析 [68]。以 3-(2-乙氨基)-氨丙基三甲氧基硅烷(AAPTS)修饰制备功能化多壁碳纳米管材料作为吸附剂填充制备固相萃取柱，在优化的 pH 条件下，As(V)、Cr(VI)和 Se(VI)可以有选择地被固相萃取柱吸附，而 As(III)、Cr(III)和 Se(IV)不被保留。采用 ICP-MS 测定，通过预氧化结合差减法可以实现对以上元素进行形态分析。Chen 等[69]成功制备了高枝化聚乙烯亚胺(branch-polyethyleneimine，BPEI)修饰的碳纳米管(MWNTs)复合材料。该吸附材料对 As(V)表现出了高选择性吸附性能。利用顺序流动注射系统，通过填充固相萃取微柱进行在线固相萃取富集，采用 NH_4HCO_3(0.6%，*m*/*v*)洗脱的方法实现了无机砷形态分析。Ma 等[70]成功制备了 γ-巯丙基三甲氧基硅烷(*γ*-mercaptopropyl trimethoxysilane, *γ*-MPTS)修饰的 Fe_3O_4@SiO_2 磁性纳米颗粒，建立了基于磁性纳米材料分离富集 ICP-MS 测定的汞形态分析方法。通过对 3-巯丙基三甲氧基硅烷(3-mercaptopropyl trimethoxysilane, MPTS)进行水解缩合制备了富巯基多面体低聚倍半硅氧烷(thiol-rich polyhedral oligomeric silsesquioxane，POSS-SH)，可以实现无机汞和甲基汞化合物的形态分析[71]。亚氨基二乙酸和壳聚糖修饰的磁性纳米颗粒均被成功用于 Cr(III)和 Cr(VI)形态分析[72,73]。氧化石墨烯二氧化钛复合材料(GO)-TiO_2 被用于 Se(IV)和 Se(VI)形态分析[74]。Chen 等[75]对蛋壳膜进行酯化获得了甲基酯化蛋白膜，酯化蛋白膜对砷的吸附容量较原始蛋壳膜提高了 200 倍，且在 pH 6.0 时，As(V)可以被 100%吸附，而 As(III)不被吸附。Li 等[76]采用溶胶凝胶法，通过 *N*-(*β*-氨乙基)-*γ*-氨丙基三乙氧基硅烷(*N*-(beta-aminoethyl)-gamma-aminopropyltriethoxysilane)构建富含氨基的有机无机杂化毛细管整体柱，建立了顺序固相微萃取系统用于不同形态无机砷的分离和富集新方法。As(III)可以和吡咯烷二硫代氨基甲酸盐生成稳定络合物以后被选择性吸附在聚甲基丙烯酸羟乙酯(PHEMA)微球上[77]，通过将微球填充在移液器枪头中制备了简单易行的砷形态分析固相萃取装置。羧基化石墨烯纳米材料(G-COOH)可以有效吸附和定量回收 As(V)，而对 As(III)吸附效果不明显(<5%)[78]。巯基和氨基功能化多孔硅、分子印迹聚合物材料等也被成功用于了无机砷和汞形态分析[79-81]。

2.4.3 选择性氢化物(气态)发生

考虑到无机砷的毒性高，在食品安全领域需要对大量样品进行筛查，因此，需要建立快速、高效的无机胂和有机胂形态分析方法。苑春刚等[82]建立了脱脂棉填充柱分离 HG-CT-AFS 高灵敏分析测定无机砷、MMA(V)和 DMA(V)的方法，并成功用于了食品调料中痕量砷形态分析。该方法无需对样品进行任何前处理，简单、快速，适合于大量样品的快速筛查。Musil 等[83]根据无机砷和有机胂生成气态氢化物所需实验条件和生成效率的差异，通过 HG-ICP-MS 方法实现了无机砷和有机胂化合物的形态分析。Matousek

等[84]根据 As(III)、As(V)以及甲基砷化合物氢化合物发生实验条件和反应效率的差异，结合冷阱捕集技术，建立了超灵敏的 HG-CT-ICP-MS 砷形态分析方法，对不同形态砷化合物的检出限可以达到 pg 级，该方法可满足人体细胞中砷的形态分析要求。通过设计合适火焰原子化器并用作氢化物发生-冷阱捕集（HG-CT）和原子荧光光谱仪的接口(AFS)，采用选择氢化物发生技术，可以实现有毒砷形态的高灵敏检测[85]。在 HG-AFS 分析测定不同汞形态化合物的过程中，无机汞离子可以采用无火焰冷原子荧光光谱法测定，而甲基汞则只有在火焰模式下才能够被原子化。根据二者原子化条件的差异，可以通过火焰/非火焰原子化模式的切换，分别测定无机汞和甲基汞的含量[86]。

研究发现，电解蒸气发生(electrolytic vapor generation，EVG)和 AFS 联用过程中，*L*-半胱氨酸修饰的石墨电极可以有效地将 Hg^{2+}和 CH_3Hg^+转化为汞蒸气。在不同电流强度下，电极可以选择性转化 Hg^{2+}和 CH_3Hg^+为汞蒸气，以满足随后的 AFS 测定，进而可以通过控制电极电流大小来实现 Hg^{2+}和 CH_3Hg^+的定量分析[87]。Zhang 等[88]发展了一套基于流动注射 UV 辐照光化学/超声蒸气发生原子荧光测定的非色谱、绿色汞形态分析方法。仅利用甲酸借助 UV 辐照或超声既可以产生汞冷蒸汽（Hg^0），随后即可被 AFS 检测。借助 UV 辐照，Hg^{2+}和 MeHg 均可生成汞蒸气（Hg^0），借此可以用来测定样品当中的总汞含量；而超声辐射仅能将 Hg^{2+}转变为 Hg^0，借此仅能测定 Hg^{2+}。基于以上原理，可以通过差减法实现无机汞离子和有机汞化合物的形态分析。

2.4.4　纳米探针技术

近年来，纳米材料探针也被成功用于了污染物的形态分析，逐渐发展成为了形态分析和荧光分析交叉领域的一个重要研究方向，在实现痕量污染物形态可视化检测方面具有很好的研究价值和应用前景。在砷污染较为严重的国家和地区，需要发展适合现场监测的快速形态分析方法。实验室较为成熟的联用系统显然无法满足现场分析的需求。Tan 等[89]将季膦盐离子液体修饰在纳米金表面，发展了可视化检测 As(III)和 As(V)的形态分析新法。如图 1-4 所示，在无砷离子存在时，离子液体功能化纳米金探针保持分散状态，溶液呈红色；在 As(III)存在时，该金探针迅速发生团聚，溶液颜色由红色转为蓝色或无色，通过观察溶液颜色的变化可确定 As(III)的含量。As(V)不会引起本纳米金探针团聚，但可预先向溶液中加入还原剂抗坏血酸将其还原为 As(III)，再用探针测定溶液中 As(V)的量。当 As(V)和 As(III)共存时，通过测定加入还原剂前后 As(III)的量，利用差减法可实现这两种无机砷离子形态分析。利用本探针可肉眼检测 7.5 μg L^{-1} As(III)[或还原后的 As(V)]，方法至少可抵抗水体中 5 倍浓度的其他常见离子的干扰，已成功用于

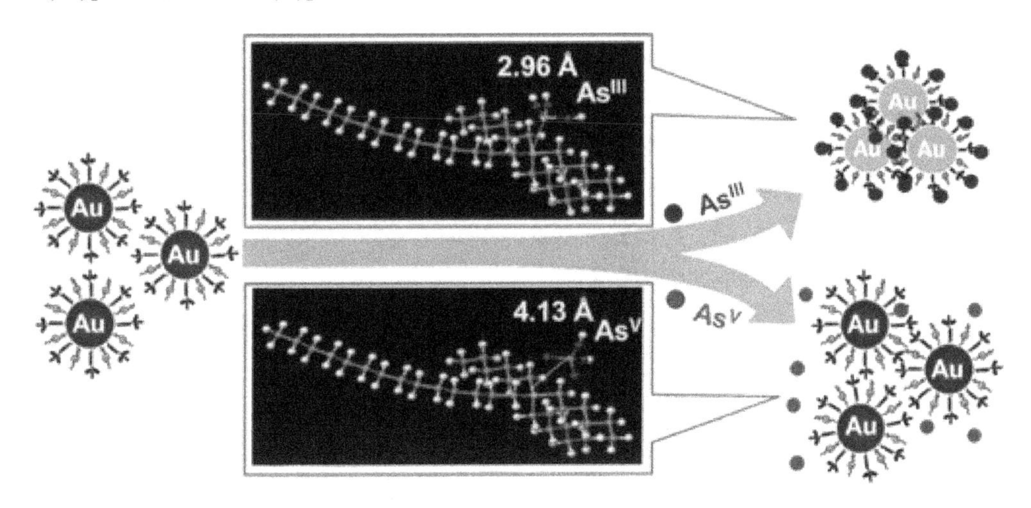

图 1-4　Au-离子液体测定不同价态无机砷原理图（引自参考文献[89]）

湖水和地下水等水体中砷离子的测定。在二乙氨基二硫代甲酸钠（DDTC）存在条件下，目标化合物（汞形态）可以诱导纳米金颗粒的团聚。由于 Hg-DDTC 比 Cu-DDTC 更加稳定，Hg^{2+}和 Cu^{2+}会发生位置交换，导致功能化纳米颗粒团聚，发生颜色改变[90]，可以作为汞形态分析的比色纳米传感器。

通过荧光猝灭或荧光强度差异也可以实现元素形态分析。以谷胱甘肽(GSH)为还原剂和保护剂可以制备具有荧光效果的 GSH-AuNCs 纳米团簇。Cr(III)和 Cr(VI)在不同 pH 表现出了不同的 GSH-AuNCs 的荧光猝灭特性，根据二者猝灭行为的差异可以通过控制 pH 实现不同价态 Cr 的形态分析[91]。采用巯基乙酸包覆的 CdS 量子点(CdS-MAA QDs)荧光猝灭特性，可以实现 As(III)和 As(V)的定量分析[92]。Deng 等[93]借助汞齐原理，采用 DNA 修饰的纳米银颗粒超灵敏识别 CH_3Hg^+。如不经过分离，在光谱分析中 CH_3Hg^+和 Hg^{2+}的识别通常很难实现。在 CH_3Hg^+- DNA 模板上，银原子和汞原子可以形成 Ag/Hg 银汞齐，而银原子与 CH_3Hg^+不能形成汞齐，据此可以实现对不同形态汞化合物的有效识别。金属有机框架材料(MOFs)也被用于形态分析。Xu 等[94]利用金属有机框架材料 ZIF-7 和 ZIF60 纳米材料作为荧光探针，建立了无机汞和甲基汞荧光探针分析新方法。这主要是基于 Hg^{2+}和 CH_3Hg^+与 ZIF-7 和 ZIF-60 相互作用产生的荧光强度变化差异而实现的。比色/荧光探针的方法灵敏、快速，已被成功用于现场和活体细胞中元素形态分析[95,96]。

2.4.5 电化学方法

电化学方法因其费用低廉，便于实现现场分析，也被用于形态分析。大部分地下水表现为弱碱环境，在弱碱性环境条件下，As(III)主要表现为中性亚砷酸分子形式(H_3AsO_3)和离子化的亚砷酸二氢根离子($H_2AsO_3^-$)形态。Yang 等[97]利用 AuNPs/α-MnO_2 纳米复合材料的高效吸附和电催化性能，建立了在弱碱性环境下电化学测定 As(III)的灵敏方法。Saha 和 Sarkar[98]通过壳聚糖-$Fe(OH)_3$ 复合材料和 L-半胱氨酸对电极进行修饰，采用阳极溶出伏安法测定了 As(III)。

2.4.6 原位形态分析技术

以上形态分析方法均是对溶解状态的污染物进行形态分析，而针对固体样品则需要进行前处理，将目标物转移到液相才能够被仪器系统分析。液相色谱-原子光谱/质谱联用技术是目前元素形态分析最为有效和常用的分析手段，但却存在两方面的局限性。

（1）所有样品需要进行样品前处理，通常采用一定的试剂提取固体样品中的目标化合物形态，然而在提取过程中很难保证形态不被改变。

（2）无论采用原子光谱还是电感耦合等离子体质谱为检测器，完全依靠标准化合物的保留时间实现对形态化合物的定性鉴定。

因此，在进行固体样品元素原位形态分析或目标化合物在色谱分离过程中容易发生降解（如含 As—S 键的化合物）时，需要借助其他非色谱方法来作为辅助手段[99]。基于同步辐射的 X 射线光谱技术长期以来被用于固体样品中金属或类金属元素的原位形态分析，但由于受分析方法灵敏度的限制，仅适用于含量较高的样品分析。近年来的有关研究主要是将其作为形态分析的有效补充手段，结合 HPLC-AFS、HPLC-ICP-MS等仪器联用形态分析技术或顺序提取形态分析技术来进一步论证元素在实际固体样品中的矿物学信息和化学结构信息[100-102]，与化学形态分析结果互为补充。X 射线荧光（XFS）和 X 射线吸收光谱技术（XAS）技术在微界面过程研究中显示出很大的优势[103-105]，可以更好诠释目标化合物在植物体内的分布情况，有助于深入研究元素的生物吸收和富集过程。Meng 等[106]结合同步辐射 XRF 和 XANES 深入剖析了水稻中汞的富集部位、存在形态以及分子结合方式，为深入认识汞在稻米中的富集、转化过程研究提供了直接依据。Zeng 等[107]结合 HPLC-HG-AFS 分析和 XANES 分析，研究了 As（III）在输入和输出菌类细胞膜过程中的形态变化和分布特征。对一些砷含量浓度高的地质、矿物样品中砷矿物组分形态分析也可以采用 XFS 和 XAS 来实现[108-110]。Deonarine 等[111]采用 XAS 详细研究了粉煤灰中砷在水体中不同氧化还原条件下的迁移转化过程。X 射线光谱进行形态分析的突出优势是可以提供完整的原位化合物信息，而非提取物种的溶解态形式。相比于其他分析方法更加直观、更接近于目标化合物的真实存在状态。

近年来，表面增强拉曼光谱（SERS）等新的原位分析技术也被用于了元素形态分析[112,113]。Wang 等[114]基于"咖啡环效应"设计了一个可以同时实现高效 SERS 基底和分析物的新型预富集分析平台。如图 1-5 所示，通过将 1 μL 银纳米颗粒（$AgNP_S$）和聚乙烯醇-124 混合物滴在硅晶片上后，随着液滴中水的蒸发，$AgNP_S$ 和分析物逐渐由液滴中心迁移到液滴边缘/硅表面/空气的三相接触线，从而形成富集有 $AgNP_S$ 自组装和目标分析物的环结构，该结构可实现基于 AgNPs 自组装体的 SERS 热点的构建与分析物的富集，对 As(V)可以实现 350 倍的预富集，而 AgNPs 自组装可以进一步获得 10^6 以上的拉曼增强因子，因此目标分析物的拉曼信号可以增强 10^8 以上。此外，咖啡环效应可以有效地消除样品基体效应，实现环境水样的直接分析，对 As(V)的检出限为 0.03 μg/L。当 As(III)和 As(V)共存时，通过测定加入氧化剂前后 As(V)的量，利用差减法可实现这两种无机砷离子的形态分析。

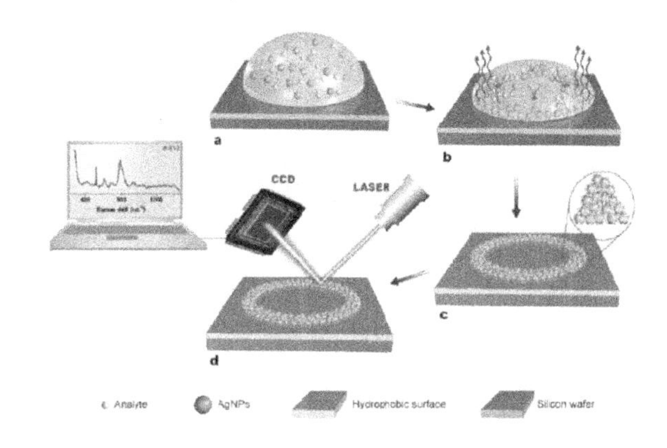

图 1-5　基于咖啡环的新型 SERS 基底构建及其在 As 形态分析中的应用（引自参考文献[114]）

2.4.7　其他形态分析方法

除化学结构意义上的形态分析以外，有关操作定义元素形态分析研究也一直比较活跃。如对大气中的汞不仅可以区分为 Hg^{2+}、$MeHg^+$和 $EtHg^+$，也可以按照存在状态和物理性质，将其区分为气态元素汞（Hg^0）、氧化态汞或活性气态汞（Hg^{2+}）和颗粒态汞（Hg^P）。Zhang 等[115]连续监测了北京郊区大气中的气态元素汞（GEM），活性气态汞（RGM）和颗粒态汞（PBM）。采用同样的定义方法，Xu 等[116]和 Duan 等[117]则分别测定了厦门和上海大气中的汞形态。固体样品中汞的热解释放温度依真实存在的汞化合物种类不同而存在明显差异，据此可发展热解释放-汞形态分析方法，用于分析测定固体样品中汞的化合物存在形式[118]。Liu 等[119]采用程序升温热解方法，详细研究了不同汞化合物（Hg_2SO_4、Hg_2Cl_2、$HgCl_2$、black HgS、HgO、red HgS 和 $HgSO_4$）的差异，成功测定燃煤烟气脱硫石膏中汞的化合物存在形态，并考察了脱硫石膏资源化利用过程中汞的形态转化和潜在释放风险。顺序化学提取也是主要的有毒元素形态分析方法之一，虽然该提取程序不能够反映元素的真实化学存在状态，但可以提供比总量更有价值的元素环境活性、生物有效性等信息[120,121]。近年来，有关操作定义形态分析逐渐趋向于化学提取程序的标准化，以有利于不同实验室之间的比对，因此在顺序提取方法学研究方面报道较少，但顺序提取方案却在不同领域得到了广泛认可和应用。在土壤-植物微界面过程迁移、工业过程污染、环境健康暴露评估研究过程中，结合操作定义元素形态分析可以为研究提供更有参考价值的科学信息[122-125]。

2.5 污染物形态转化与迁移机理研究

近年来，形态迁移转化机理研究也有了新的进展。Yin 等[126]借助 GC-ICP-MS 对不同形态汞在水体中的甲基化产物进行分析，通过实验室模拟，证明了碘甲烷（CH_3I）在自然水体中可以甲基化无机汞为甲基汞化合物。在此项研究中，研究人员采用了同位素 Hg（$^{199}HgCl_2$ 和 $CH_3{}^{201}Hgt$）和氢稳定同位素（CD_3I）添加技术，深入探讨了甲基化机理。研究发现，自然水体中的 Hg^0、$Hg_2{}^{2+}$和 Hg^{2+}在太阳光照的条件下可以被甲基化，但在去离子水中仅 Hg^0 和 $Hg_2{}^{2+}$可以被甲基化。通常认为水体中存在的厌氧菌对无机汞化合物的甲基化是水体中甲基汞的主要来源途径。同时，最近研究结果表明水体中存在的腐殖质物质和低分子量有机化合物的非生物学过程或化学甲基化过程也可能生成甲基汞。Yin 等[127]采用 LC-AFS 联用系统、分子质谱和稳定同位素示踪技术，对酮类、醛类和小分子有机酸存在 UV 辐照下水溶液中 Hg^{2+}的光化学转化过程进行了较深入研究。

众所周知，砷的甲基化过程可以影响其环境行为和生物富集特征，但却很少有关砷在实际环境介质中甲基化过程的研究报道。Zhao 等[128]借助形态分析手段研究了不同土壤类型对无机砷的甲基化过程，并就微生物甲基转移酶基因（arsM）与甲基砷的生成关系进行了探索，指出水稻中甲基砷主要来自于土壤中砷的甲基化产物，而土壤条件对砷的甲基化过程起关键作用。砷在水稻中的吸收、分布和富集过程研究一直是环境科学和食品科学领域共同关注的重要课题[129]。砷在根际环境和植物体内的形态转化过程是该项研究的关键。Wang 等[130]研究了在硫代硫酸盐存在的情况下芥菜对汞的吸收情况，借助形态分析技术详细研究了汞被植物吸收、形态转化和在土壤中再分布过程。研究也发现，自然界中的无机砷可以在生物地球化学循环过程中被甲基化生成气态砷化合物[131]，并可采用 GC-CT-ICP-MS 联用技术和 GC-CT-HG-AFS 联用系统实现对气态砷化合物的高灵敏检测[132,133]。Huang 等[134]研究了无机砷在土壤有机质和微生物作用下的形态转化和生成气态砷化合物的过程，发现有机质的加入明显促进了土壤中无机砷的甲基化和气态砷生成过程。Meng 等[135]利用 MERX 全自动烷基汞/总汞形态分析系统对来自于湖南、贵州和广东三个省份的 10 个矿区的 155 个水稻全植株样品中的根、茎秆、叶、稻壳和米粒中的汞形态分别进行分析，同时对水稻种植的根际土壤也进行了汞形态分析，发现稻米对甲基汞的富集作用最强。水稻中总汞和甲基汞均与土壤中的汞呈显著正相关关系，表明水稻中的总汞和甲基汞的一个重要来源是土壤。Shao 等[136]详细调查了来自青藏高原的鱼体样品中汞和甲基汞的形态，对高原地区淡水水体中汞在鱼体内的形态分布和富集机制进行了研究。

近些年发展起来的梯度扩散薄膜技术（DGT 和 DET）与形态分析技术相结合，可以获得较为准确的生物有效性信息和生物活性信息，为深入研究元素的跨膜运输等环境归趋和毒理学行为提供更加准确可靠的依据[137-139]。另外，同位素稀释也被更为广泛地应用到汞形态分析当中[140]。Laffont 等[141]采用同位素稀释技术结合 GC-ICP-MS 详细研究了人发中汞形态的提取方法。Rahman 等[142]将顶空固相萃取（HS-SPME）与 GC-ICP-MS 相结合，建立了同位素稀释质谱汞形态分析方法。同位素分馏和形态分析相结合，可以更好地研究汞在生物体内和不同环境介质之间的迁移转化[143]。Mao 等[144]采用稳定同位素示踪方法研究了不同来源汞（包括大气汞和土壤中的汞）在植物体吸收、甲基化和迁移转化过程。

2.6 形态分析技术在相关领域的应用研究

在形态分析技术应用方面，已逐渐形成了与其他研究领域交叉融合、互相促进的发展趋势。在相关交叉学科和方向，越来越多的过程和机理研究需要借助形态分析手段来获得更为准确、可靠的分析数据。形态分析研究从单纯的方法学创新逐渐转变成为揭示环境过程机制、毒理学过程必不可少的关键手段。

由于高灵敏检测器的广泛应用和方法灵敏度的提高，形态分析应用领域和所检测的样品种类不断扩展。除水、土、气和生物等传统样品以外，近年来，形态分析在能源利用、司法鉴定、临床医学等领域均有较多的应用报道。由于我国近几年来大气雾霾污染频发，有关大气颗粒物中污染物的形态分析研究也逐渐增多[145-148]。形态分析技术在煤炭、石油、天然气等能源领域逐渐得到了应用，并为推动相关学科发展起到了非常重要的作用[149-154]。Liu 等[155]详细研究了粉煤灰中砷的顺序提取形态和化学形态，探讨了砷在燃煤固废产物中的潜在环境特性。电厂燃煤过程中砷、汞等元素的形态分析和释放也逐渐受到了交叉研究领域的重视[123,125,156,157]。在考古和司法领域，借助 X 射线光谱技术（包括 X 射线荧光、X 射线吸收和 X 射线衍射）可以分析生物检材中的元素存在形态，通过多种分析技术联用可以区分元素的准确来源[158]。形态分析技术在临床医学方面重要作用也逐渐突显。Chen 等[159]借助 HPLC-ICP-MS 联用系统研究了砷制剂（As_2O_3）治疗急性早幼粒细胞白血病患者（APL）的体内代谢动力学过程和形态分布特征，为深入研究临床机理提供了很有价值的参考数据。

3　环境纳米材料形态分析

纳米材料在环境介质和生物体内存在着丰富的转化行为，如物理转化、化学转化、生物转化等，这些转化导致其分散/团聚状态、表面电荷、形貌、粒径分布和化学组成等物理化学形态发生一系列变化。鉴于不同形态的纳米材料表现出明显不同的毒性和生物活性，仅仅测定样品中纳米材料的总量已不足以充分反映其环境过程和生物效应，还需区分纳米材料的诸种物理化学形态，因此纳米材料的形态分析在环境科学中具有重要的意义。本节将对纳米材料组成、结构与分散状态的识别与表征方法进行简要介绍，重点介绍环境基质中纳米材料的分离富集，不同粒径、不同电性纳米材料的分离测定以及纳米材料形态分离装置的研制。

3.1　纳米材料组成、结构与分散状态的识别与表征

纳米材料是指在三维空间中至少有一维处于纳米尺度范围（1~100 nm），或者由它们作为基本单元构成的具有特殊性质的材料[160,161]。纳米材料的识别与表征主要包括其组成、结构、分散状态和表面性质等方面的定性定量测定，如表 1-1 所示。

表 1-1　纳米材料组成、结构与分散状态的识别与表征

特性	表征参数	测试技术
组成	主体化学组成、表面化学组成、原子种类、价态、官能团等	原子吸收（AAS）、原子发射（AES）、质谱（MS）、X 射线荧光光谱（XRF）、紫外可见吸收光谱（UV-vis）、电子能谱（EDS）等
结构	尺寸、形貌、晶体结构、分子/原子空间排列、缺陷、位错、孪晶界等	透射电镜（TEM）、扫描电镜（SEM）、原子力显微镜（AFM）、X 射线衍射（XRD）、选区电子衍射（SAED）、X 射线吸收光谱等（XAS）
分散状态	分散性、团聚度等	TEM、SEM、AFM、动态光散射（DLS）、静态光散射（SLS）等

纳米材料的组成是影响其生物毒性最本质的因素。因此，确定纳米材料的元素组成、测定纳米材料的杂质种类和含量，是纳米材料分析的重要内容之一。纳米材料组成可分为主体化学组成、表面化学组

成和微区化学组成。根据主体化学组成的不同，大致可将纳米材料分为五大类：碳纳米材料、金属氧化物纳米材料、零价金属纳米材料、量子点和枝状大分子材料。纳米材料主体化学组成的分析方法包括 AAS、AES、MS 和 XRF 等。其中，前三种方法需要将样品溶解后再测定，属于破坏性样品分析方法，而 XRF 可以对固体样品直接测定，是一种非破坏性元素分析方法，因此在纳米材料成分分析中具有较大的优势。纳米材料表面化学组成分析方法常用的有电子能谱和电子衍射等。这些方法可以对纳米材料表面化学成分、分布状态与价态、表面或界面的吸附和扩散反应的状况进行分析测定。在实际研究中，将能谱与 SEM、TEM 相结合，还可用于纳米材料微区成分的分析。

纳米材料的结构主要包括两部分内容，一方面是指纳米粒子的尺寸、形貌等，另一方面是指纳米粒子的晶体结构、晶面结构、晶界及各种缺陷，如点缺陷、位错、孪晶界等。根据其结构的不同，大体上可将纳米材料分为四类：零维结构如原子簇或簇组装；一维结构如纳米管、纳米线、纳米棒等；二维结构如纳米片、纳米薄膜等；三维纳米结构如等轴微晶等。纳米材料的结构直接影响着其物理化学特性，因此纳米材料的结构分析也极其重要。目前，纳米材料结构表征方法很多，如高分辨电子显微镜能够以原子级的分辨率显示材料的原子排列，AFM 可以测定纳米材料表面和近表面原子排列和电子结构，XRD 可用于纳米材料晶体结构分析和物相鉴定。

纳米材料的分散状态是指纳米颗粒的粒径、形貌分布均匀程度。因此，纳米材料分散状态的表征主要是基于粒径与形貌的分析，常用的分析方法有 TEM、SEM、AFM 和光散射技术等。

3.2 环境基质中纳米材料的分离富集

由于环境样品中纳米材料含量极低，且复杂基体的存在会干扰纳米材料的测定，直接检测纳米材料存在很大的困难。因此，测定前对环境样品中纳米材料进行分离富集是非常必要的，常用的分离富集方法有膜过滤、浊点萃取和磁固相萃取等。

3.2.1 膜过滤

膜过滤是环境样品中纳米材料分离富集的一种常用方法，具有设备简单、操作方便、容易控制等优点，其主要缺点在于纳米材料往往会吸附在膜孔中导致洗脱困难。Cornelis 等[162]报道了膜过滤用于分离土壤悬浮液中纳米 Ag、纳米 CeO_2 和相应的金属离子的研究。将土壤悬浮液分别经 0.45 μm 微滤膜和 1 kDa 的超滤膜过滤后，金属离子存在于小于 1 kDa 的滤液中，纳米颗粒介于 1 kDa 和 0.45 μm 之间。但是，该方法存在一定的缺陷，对于粒径大于 0.45 μm 的纳米材料团聚物会被排除在纳米颗粒外，造成纳米颗粒测定值偏低。Yu 等[163]在研究水环境中纳米 Ag 的团聚与化学转化行为时，采用滤膜孔径为 30 kDa 的超滤管来分离纳米 Ag 和 Ag^+。经超滤后，收集滤液，用 ICP-MS 测定透过液中 Ag^+含量，用 TEM 表征纳米 Ag 浓缩液。为了考察该方法的可靠性，用 UV-vis 和 TEM 技术分别表征了纳米 Ag 原液和经超滤后的滤液。结果显示，与原液相比，透过液无色澄清，在 400 nm 左右没有纳米 Ag 特征吸收，同时用 TEM 观察也没有发现纳米 Ag 的存在，表明了超滤能很好地分离纳米 Ag 和 Ag^+。

3.2.2 浊点萃取

浊点萃取是一种新兴的液液萃取技术，最早由 Watanabe 等[164,165]提出。该技术基于胶束水溶液的溶解性和浊点现象，通过改变离子强度、pH、温度等实验参数引发相分离，从而将待测物从基质中分离出来，并得到一定程度的富集，目前已应用于金属螯合物、生物大分子的分离纯化及环境样品中污染物分离富集。

浊点萃取作为痕量物质分离富集的有效手段，已被用于环境基质中纳米材料的分离富集。基于非离子型表面活性剂 Triton X-114（TX-114）与纳米材料可通过非共价键作用形成复合体，Liu 等[166]最早提出了浊点萃取可逆分离富集和再分散多种纳米材料，包括纳米 Ag、纳米 Au、纳米 TiO_2、纳米 Fe_3O_4、量子点、富勒烯和单壁碳纳米管等。通过 UV-vis 表征萃取前后的表面活性剂层和水溶液层中纳米 Au，得到经浊点萃取分离后，纳米 Au 进入富 TX-114 相，在水溶液中几乎无残留，表明了浊点萃取纳米材料的分离效率很高。同时，TEM 结果显示，浊点萃取前后，纳米材料的粒径和形貌基本维持不变。

浊点萃取能与多种检测技术联用，通过在萃取前的体系中加入合适的金属离子络合剂，可以实现金属纳米材料和金属离子的形态分析。Liu 等[167,168]通过改进实验条件，以 $Na_2S_2O_3$ 作为 Ag^+络合剂，运用浊点萃取技术选择性地分离富集环境水样中痕量纳米 Ag。经萃取进入富 TX-114 相的纳米 Ag 可以直接采用 TEM、SEM、EDS 和 UV-vis 等技术进行定性表征，经微波消解后采用 ICP-MS 进行定量测定。该方法为分析和追溯环境中纳米 Ag 提供了有效途径。随后，Hartmann 等[169]将浊点萃取与电热蒸发原子吸收（ET-AAS）联用，实现了环境水样中纳米 Au、纳米 Ag 和相应金属离子的富集分离。由于实验中无需对萃取后的纳米材料进行微波消解和 ICP-MS 测样前样品的稀释，因此该方法更为简便、经济。此外，Tsogas 等[170]结合浊点萃取和化学发光法，分离测定了环境水体中痕量混合金属纳米材料（纳米 Au、纳米 Ag 和纳米 Fe_3O_4）和相应的金属离子。

3.2.3　磁固相萃取

磁固相萃取是以磁性纳米材料作为吸附剂的固相萃取技术。近年来，磁固相萃取技术也用于纳米材料的富集分离。Su 等[171]合成了 Al^{3+}固定化亚氨基二乙酸（IDA）包裹的磁性纳米材料 Fe_3O_4@SiO_2@IDA-Al^{3+}，用于纳米 Au 和 $AuCl_4^-$的分离富集。由于该材料可同时吸附纳米 Au 和 $AuCl_4^-$，实验时需要通过分步洗脱法区分纳米 Au 和 $AuCl_4^-$：先采用 $Na_2S_2O_3$ 洗脱 $AuCl_4^-$，再用氨水洗脱纳米 Au。该方法选择性高，可有效消除高浓度 $AuCl_4^-$对纳米 Au 分离测定的干扰，但仅适用于基质简单的环境样品，对于基质复杂的污水和海水中纳米 Au 和 $AuCl_4^-$的回收率仅为 23.0%~72.8%和 0~18.1%。Liu 等[172]比较了不同修饰的 Fe_3O_4 磁性纳米颗粒对纳米 Ag 的选择性萃取，发现裸露的 Fe_3O_4 磁性纳米颗粒选择性最好，并将其用于环境水样中纳米 Ag 的富集分离。Tolessa 等[173]采用共沉淀法和悬浮交联技术合成了壳聚糖包裹的 Fe_3O_4 磁性微球，以其作为吸附剂建立了磁固相萃取富集分离环境水样中纳米 Ag 和 Ag^+的方法。研究中，作者优化了吸附剂用量、溶液 pH、富集因子和重复使用次数等萃取条件，并考察了 Ag^+和环境中普遍存在的腐殖酸对纳米 Ag 萃取率的影响。结果发现，当 Ag^+浓度在纳米 Ag 含量的约 5 倍范围内，共存的 Ag^+对纳米 Ag 的萃取无显著影响。同时也发现，尽管样品中的腐殖酸会影响纳米 Ag 的萃取，但环境水样中无机阳离子如 Ca^{2+}、Mg^{2+}等可以有效消除腐殖酸的干扰。

此外，磁固相萃取还可用于不同形态含 Ag 纳米颗粒和 Ag^+的富集分离。Zhou 等[174]采用水热法合成了 Fe_3O_4 纳米颗粒，将其老化后用作磁固相萃取剂，用于环境水样中纳米 Ag、纳米 AgCl、纳米 Ag_2S 和 Ag^+的分析。老化后的 Fe_3O_4 纳米颗粒可以在 Ag^+存在条件下，选择性吸附含 Ag 纳米颗粒。洗脱时，用醋酸溶液选择性洗脱纳米 Ag 和纳米 AgCl，再用含硫脲的醋酸溶液洗脱纳米 Ag_2S，从而实现不同含 Ag 纳米颗粒的形态分析。该方法为研究纳米 Ag 的环境归趋和毒性提供了有效的分析测试技术。

Zhang 等[175]将亲水性聚（丙烯酰胺-4-乙烯基吡啶-*N*, *N*-亚甲基双丙烯酰胺）[poly(AA-VP-Bis)]整体柱用于环境水样中含羧基的 Au NPs 的富集，该整体柱对 Au NPs 表现出较高吸附容量和选择性。据此，他们建立了在线整体柱毛细管微萃取-ICP-MS 分析环境水样含羧基 Au NPs 的新方法。

对于几种粒径、电性相似的纳米材料混合体系，常用的离心、过滤和浊点萃取均难以将其分离。为此，研究者针对目标纳米粒子的表面化学特性，设计和合成了高选择性捕获特定纳米粒子的磁性纳米材料来实现其分离。Essinger-Hileman 等[176]针对目标纳米材料的表面化学特性，利用一种一端连接生物素相似物的缩氨酸对纳米 Au 进行修饰，通过修饰后的纳米 Au 可与链球菌亲和素包裹的 Fe_2O_3 磁性纳米材料特异性结合，成功分离了纳米 Au 和 CdS 量子点。此外，Shen 等[177]合成了双磷酸基团功能化的磁性纳米材料，用于复杂基质中纳米 TiO_2 和纳米 SiO_2 的高效分离。

3.3 不同粒径纳米材料的分离测定

纳米材料的粒径显著影响纳米材料进入细胞的方式及在体内的分布，进而表现出不同的生物效应。通常认为，小粒径纳米材料的毒性大于大粒径纳米材料[178]。因此，不同粒径纳米材料的分离测定对于科学评价纳米材料的毒性具有重要意义。目前，关于不同粒径纳米材料分离的常用方法有色谱分离、流场流分级和梯度密度离心等。

3.3.1 色谱分离

色谱分离作为痕量物质分离的常用手段，在不同粒径纳米材料分离测定中有广泛的应用。常用的色谱分离方法包括尺寸排阻色谱、水动力学色谱和薄层色谱法等。

尺寸排阻色谱采用表面惰性、具有不同尺寸孔穴或网状结构的硅胶或凝胶填充色谱柱。当纳米材料在流动相的带动下通过色谱柱时，对于粒径不同的纳米材料，可以渗入孔穴的不同深度。大粒径的纳米颗粒可以渗入大孔穴内，但进不了小孔，甚至完全被排斥，而小粒径的纳米颗粒对于大孔小孔均能渗进去，甚至很深，一时不易洗脱出来。因此，大粒径纳米颗粒在色谱柱中的停留时间短，粒径小的纳米材料在色谱柱中停留时间长，从而可以分离不同粒径纳米材料。1994 年，Fischer 等[179]首次利用尺寸排阻色谱技术，以多孔硅胶为固定相，柠檬酸钠为流动相，实现了 3~20 nm 纳米 Au 的分离。但是，分离过程中，纳米颗粒与固定相之间存在很强的吸附作用，容易造成纳米颗粒的损失，影响分离效果。Wei 等[180,181]通过在流动相中加入表面活性剂（十二硫酸烷基钠），很好地改善了纳米材料的稳定性，避免了纳米金在固定相上的吸附，从而大幅度地提高了尺寸排阻色谱的分离效率。近年来，尺寸排阻色谱与电感耦合等子体质谱（ICP-MS）在线联用技术，被用于不同粒径金属纳米颗粒和相应金属离子的分离。Soto-Alvaredo 等[182]将尺寸排阻色谱和 ICP-MS 在线联用，实现了不同粒径纳米 Ag 和 Ag^+的分离。但是，Ag^+与小粒径纳米颗粒（< 10 nm）的分离效果较差。同时，由于纳米颗粒的 ICP-MS 响应与粒径有关，因此无法实现不同粒径纳米颗粒的定量分析。随后，Zhou 等[183]以孔径 500 Å 的氨基柱为分离柱，以含 2 mmol/L $Na_2S_2O_3$ 的 0.1% FL-70（*v/v*）为流动相，首次实现了 1~100 nm 含 Ag 纳米颗粒和 Ag^+的基线分离。该方法 Ag^+检出限低（0.019 μg/L），操作简单，耗时短（5 min），适用粒径范围宽，重现性好，但是不能用于不同粒径纳米颗粒的分离。在此基础上，Zhou 等[184]进一步通过优化色谱柱孔径、流动相组成和流速等，成功实现了 10~40 nm 纳米颗粒的分离。由于保留时间和粒径线性相关，可以将保留时间换算成粒径，得到纳米粒子的粒径分布。同时，该研究发现，粒径相关的 ICP-MS 响应源于不同粒径纳米颗粒经过尺寸排阻色谱柱后的回收率不同，通过在流动相中加入无机盐，可以显著提高大粒径纳米颗粒的回收率，使 SEC 分离时粒径不同的纳米颗粒 ICP-MS 响应值接近，从而实现不同粒径纳米颗粒的分离测定（图 1-6）。

水动力色谱作为粒径分离方法在 20 世纪 70 年代由 Small 等提出[185,186]。该技术采用无孔刚性颗粒填充色谱柱，当纳米粒子在流动相带动下流经色谱柱时，由于不同粒径的纳米颗粒受到的水动力效应不同，大粒径纳米颗粒倾向于远离管壁附近的低流速区而进入高流速的中心区，能更快地洗脱出来，从而可实现不同粒径纳米颗粒的分离。通过将该技术与不同检测器联用，可以进一步实现不同粒径纳米颗粒的准确定量。相比于尺寸排阻色谱，水动力色谱适用粒径范围更宽，可用于粒径范围在几纳米的纳米颗粒至几微米的纳米颗粒团聚物的分离。然而，水动力色谱的粒径分辨能力较差，一定程度上限制了其在纳米颗粒分离领域中的应用。Tiede 等[187,188]将水动力色谱与 ICP-MS 联用，成功分离了复杂污泥悬浮液中 5~300 nm

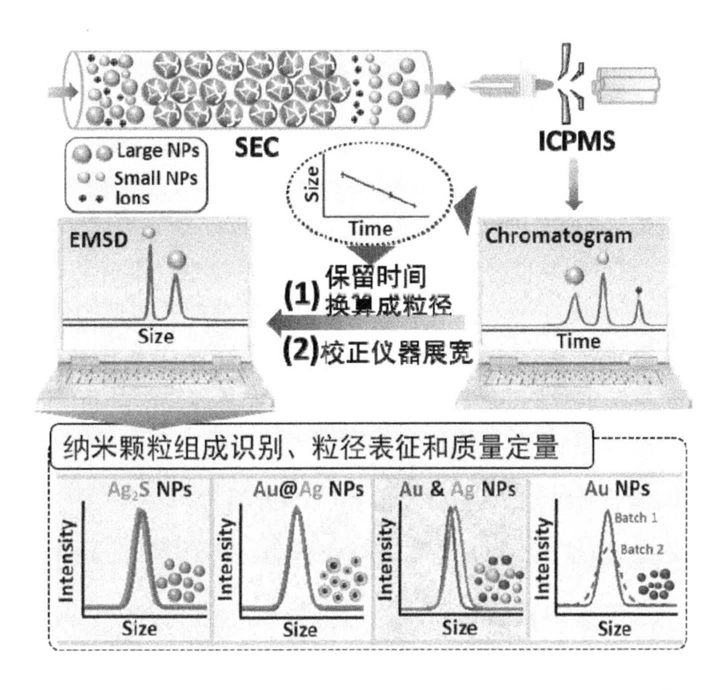

图 1-6　SEC-ICP-MS 测定纳米材料元素质量粒径分布（EMSD）示意图（引自参考文献[184]）

纳米 Ag。该方法操作简单，分析时间短（< 10 min），不需任何的样品前处理过程，显示了水动力色谱在复杂基质中纳米颗粒分离方面的应用前景。但实验结果表明 5~250 nm 纳米 Au 的保留时间比较接近，因此很难得到纳米颗粒的粒径分布信息。近年来，研究者将水动力色谱与 ICP-MS 联用，在单颗粒模式下，获得了不同粒径纳米材料的颗粒浓度和粒径分布信息，并将该技术成功应用于环境中金属纳米颗粒的迁移转化以及环境行为的研究[189-191]。

薄层色谱也是一种分离不同粒径纳米材料的有效手段。由于不同粒径纳米材料与展开剂及薄层色谱板之间的作用力存在差异，导致不同粒径纳米材料在薄层色谱分离过程中的迁移距离不同，从而可以实现不同粒径纳米材料的分离。Yan 等[192]将薄层色谱与激光剥蚀电感耦合等离子体（LA-ICP-MS）联用，实现了纳米 Au 和 $AuCl_4^-$的分离和测定，以及不同粒径纳米 Au 的分离和粒径表征。用乙酰丙酮/丁醇/三乙胺（6∶3∶1，*v*/*v*）为展开剂，可有效分离纳米 Au 和 $AuCl_4^-$；用含 0.2 mol/L 的磷酸缓冲盐（pH 6.8）、0.4%TX-114（*m*/*v*）和 10 mmol/L EDTA 为展开剂时，则可分离 13 nm、34 nm 和 47 nm 的纳米 Au。研究发现，在薄层色谱分离过程中，不同粒径纳米 Au 比移值（R_f）与其粒径有着很好的线性相关性，因此可通过该线性关系来估算纳米颗粒的粒径范围。在此基础上，Yan 等[193]还将薄层色谱与化学发光技术联用，通过纳米 Au 的催化化学发光强度与其浓度之间的线性关系，实现了不同粒径（13 nm、40 nm 和 100 nm）纳米 Au 的分离测定。

3.3.2　流场流分级

场流分级是一种类似于色谱分离的分析方法，最早由 Giddings 在 1966 年提出[194]。该技术将流体和外加场联合作用于样品，利用待分离物质的大小、密度和电荷等差异实现分离。根据外加场的不同，场流分级分为流场流分级、磁场流分级、电场流分级、沉降场流分级和重力场流分级等[195]。

在纳米材料粒径分离领域，使用最多的是流场流分级技术。当待测物通过分离通道时，同时受到水平和垂直方向的流场作用，粒径小的颗粒受垂直方向作用小，向分离通道中心平移扩散；粒径相对较大的颗粒受垂直方向作用力大而靠近积聚壁，从而在垂直方向形成尺寸梯度。而主流体在分离通道内，越靠近中心，流速越快，越靠近边缘，流速越慢，从而可以实现不同粒径纳米颗粒的分离。流场流分级分辨率高，无需固定相，避免了待测物与固定相间的相互作用，从而保持了纳米颗粒的原有形貌和尺寸。目前，该技术已被广泛用于纳米颗粒粒径分离和纯化，如纳米 Ag[196-198]、纳米 Au[199,200]、纳米 TiO_2[201]和碳纳米管[202]等。Schmidt 等[199]通过流场流分级与多角度光散射（MALS）、动态光散射（DLS）和 ICP-MS 联用，实现了 10 nm，20 nm 和 60 nm 纳米 Au 分离后的粒径表征和质量浓度定量。但是在分离过程中，由于纳米颗粒和膜之间存在很强的吸附，10 nm 纳米 Au 的回收率仅为 50%。Mudalige 等[203]通过对纳米 Au 颗粒和流场流分级的分离膜进行功能化修饰，减少了纳米颗粒在分离膜上的残留和积聚，成功地分离了 3 种不同粒径的纳米 Au，同时提高了分离效率、样品回收率以及分离膜的使用寿命。Tan 等[198]以中空纤维膜为分离通道，将中空纤维膜流场流分级-微柱富集与 UV-vis、DLS 和 ICP-MS 等多重检测器联用，实现了环境水体中 Ag^+和不同粒径纳米 Ag（1.4~100 nm）的分离测定（图 1-7），并发现由该方法得到的纳米颗粒粒径与 TEM 结果接近，表明流场流分级在纳米颗粒分离表征上可以作为比较可靠的分析手段。

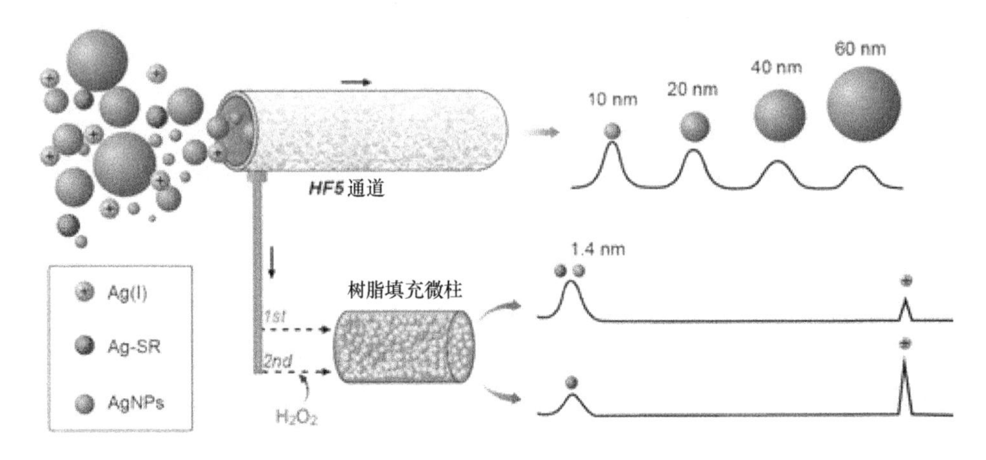

图 1-7　中空纤维膜流场流分级-微柱富集分离不同粒径纳米 Ag 和不同形态 Ag^+示意图
（引自参考文献[198]）

3.3.3　密度梯度离心

密度梯度离心最初用于细胞及生物大分子分离，是一种纯液相分离方法。该技术解决了液-固相分离过程中粘连问题，避免了胶体粒子在分离过程中的损失和结构的破坏。近年来，密度梯度离心被引入纳米材料分离，在一定离心力作用下，依据纳米材料沉降系数差异进行分离。分离过程很简单，将介质在离心管中配置成一个阶梯的或连续的密度梯度，然后把纳米颗粒置于介质顶部，在重力或离心力作用下，不同粒径和形貌的颗粒朝着平衡密度点移动，从而实现分离。在实际研究中，密度梯度离心技术常用的介质有氯化铯，蔗糖和多聚蔗糖等。

Lee 等[204]以蔗糖为分离介质，利用密度梯度离心法成功分离了日本辛夷（*M. kobus*）植物叶子合成的不同形貌、不同粒径的纳米金和纳米银。分离后，收集样品并用 TEM 表征，发现小粒径纳米颗粒主要出现在低密度介质中，而大粒径纳米颗粒集中分布在高密度介质中，且同一区带的纳米颗粒分布较均匀。因此该方法可为单分散纳米颗粒的合成提供了重要的技术手段。除了适用于水相体系中纳米颗粒的分离，密度梯度离心还可拓展到有机相体系中，用以得到高度分散的纳米材料。Sun 等[205,206]利用四氯化碳-环己烷的梯度液，成功分离了油相中不同尺寸大小的 Au、Ag 纳米颗粒，CdSe 和 CdS 量子点。TEM 表征结果显示，相邻区带间纳米颗粒的粒径仅相差 2 nm，表明了密度梯度离心具有非常好的分离效率。

在以往研究中，密度梯度离心主要用于纳米颗粒合成后的分离纯化，经分离后，通过 UV-vis 或 TEM 表征，仅实现了纳米颗粒的定性表征，而不能对分离后的纳米颗粒进行定量测定。为此，Johnson 等[207]用密度梯度离心分离不同粒径纳米 Au 后，收集分离后的组分，采用 ICP-MS，在单颗粒模式下，测定了分离后纳米 Au 的颗粒浓度和粒径分布，首次将梯度密度离心成功应用于环境样品中痕量纳米颗粒的分离测定。

3.3.4　其他分离方法

除了上述分离方法外，还有一些其他的分离方法可用于不同粒径纳米颗粒的分离，如离心，膜过滤、超滤等。有文献报道，通过超高速离心机逐步分梯度离心上层液，可以分离饮用水中~ 50 nm 的纤维多糖和 5~10 nm 的黑色颗粒物[208]。然而，在离心过程中，由于纳米颗粒易与大颗粒结合，容易造成纳米颗粒损失。因此，在纳米颗粒痕量分析时，需要在分离体系中加入一定浓度的表面活性剂以避免纳米颗粒与大颗粒物的结合，从而使得分离结果更为准确。Akbulut 等[209]通过在样品中加入一定浓度的十二烷基聚乙二醇醚，聚（2-乙基-2-噁唑啉）和聚蔗糖混合物，利用差速离心技术成功实现了不同粒径纳米颗粒的分离。

同时，通过微孔膜、超滤膜和纳孔膜逐级过滤，也可以分离不同粒径的纳米颗粒。Liang 等[210]制备了孔径均一的碳素纤维膜，已成功用于两种不同尺寸的纳米 Au 和三种不同尺寸的纳米 Au、纳米 Ag 自组装体的分离。但是，由于极化效应，在膜过滤分离过程中，纳米颗粒容易在膜表面沉积和团聚，堵塞膜孔，降低分离效率[211]。因此在实际应用中需要不断更换滤膜，无形中增加了纳米材料分离的成本。

此外，整体柱固相萃取在不同粒径纳米颗粒分离方面也有相关报道。整体柱是一种用有机或无机聚合法在色谱柱或毛细管柱内进行原位聚合的连续床固定相。与常规的色谱柱相比，该柱具有原位制备、高渗透性、良好机械强度和多孔性等优点。Liu 等[212]报道了聚合物毛细管整体柱与 ICP-MS 在线联用分离测定环境水样中 3~40 nm 含羧基基团的纳米 Au。通过静电或氢键作用，含羧基基团的纳米 Au 吸附在聚合物毛细管柱内。洗脱时，纳米颗粒可以渗透至流通孔内，类似于 SEC，大粒径纳米颗粒洗脱快而小粒径纳米颗粒洗脱慢，从而可以实现不同粒径纳米颗粒的分离。因此，该方法兼具了在线富集和粒径分离测定的优点。

3.4　不同表面电性纳米材料的分离测定

纳米材料自身处于热力学不稳定的状态，在合成过程中通常需要加入稳定剂，通过空间位阻或静电排斥来增加纳米材料的分散性和稳定性，因此纳米材料表面往往会带有一定电荷。研究表明，纳米材料的毒性除了与其自身组成、粒径有关外，还与其表面电性密切相关。因此，分离粒径近似但表面电性不同的纳米材料对于深入研究其毒理作用是十分重要的。但是，目前关于不同电性纳米材料分离测定的研究较少，主要原因有两个，一方面是目前关于纳米材料的分离测定研究尚处于起步阶段，另一方面是纳米材料自身性质的特殊性，纳米材料粒径、溶液 pH、无机盐和大分子物质均能影响其表面电性。

3.4.1　电泳分离

电泳是分离不同电性纳米材料的有效手段。在电场作用下，不同电性的纳米材料以不同速度向着与其电荷相反的方向迁移，从而实现纳米材料的分离。目前，在纳米材料分离中，应用较多的电泳技术有凝胶电泳和毛细管电泳。

凝胶电泳是在电场作用下，利用物质在具有分子筛效应的凝胶介质中的迁移行为的不同而进行分离的技术。根据凝胶介质的不同，可将凝胶电泳分为琼脂糖凝胶电泳和聚丙酰胺凝胶电泳。与聚丙酰胺凝胶相比，琼脂糖凝胶的孔径更大，可以在几十至几百纳米间变化，且孔径分布均匀，制备简单，因此在纳米颗粒分离领域中应用更为广泛。Xu 等[213]用孔径约 100 nm 的琼脂糖凝胶电泳柱，分离了具有不同表面电荷的 5 nm、15 nm 和 20 nm 的纳米 Au 球颗粒。通过在纳米 Au 表面引入羧酸根，不仅使颗粒带负电荷，在电场中向正极迁移，同时增加了颗粒间的斥力，使得纳米 Au 更为稳定。应用该方法，还可实现不同电性纳米 Au 球、纳米 Au 片和纳米 Au 棒的有效分离。纳米颗粒表面化学组成直接影响凝胶电泳分离效果。Helfrich 等[214]将凝胶电泳与 ICP-MS 在线联用，分离和测定了不同表面修饰的纳米 Au。结果显示，表面经巯基琥珀酸修饰的纳米 Au 表现出与柠檬酸修饰纳米 Au 完全不同的电泳行为。

毛细管电泳是 20 世纪 80 年代发展起来的一项分离技术，它是以空心毛细管或经过处理的毛细管为微分离通道，通过在两端施加上高压直流电场，依据样品中不同电性物质在电场作用下的迁移速度和分配系数不同而实现分离的技术。毛细管电泳具有仪器简单、操作方便、高效快速等优点，是一种极其有效的纳米材料分离方法。Liu 等[215-217]先后报道了用毛细管电泳技术分离不同粒径、不同形貌的纳米 Au、纳米 Ag 和 Au@Ag 核壳结构纳米球等。分离时，为了抑制纳米颗粒在管壁上的吸附及纳米颗粒间的团聚，通常需要在分离前对毛细管内壁进行预处理，使管壁带电，以增加管壁和纳米颗粒间的排斥力。同时，还需在电解液中加入适量的表面活性剂，以增加纳米颗粒的分散性和稳定性。Franze 等[218]在缓冲液中加入 60 mM SDS，显著地提高了纳米 Au 和纳米 Ag 在毛细管电泳分离时的回收率。毛细管电泳可以与多种检测器联用，如 UV-vis、ICP-MS、拉曼光谱和二极管阵列等，实现纳米颗粒的定量测定和定性表征。Liu 等[219]将毛细管电泳和 ICP-MS 联用技术，通过 ICP-MS 的多元素同时监测可用于纳米颗粒的组成识别。同时，通过相同包裹剂、不同表面电荷的纳米颗粒的迁移时间与粒径之间的线性关系，得到纳米颗粒的粒径分布信息。需要指出的是，在实际样品中，纳米材料表面电荷容易受所处环境因素的影响。Qu 等[220]研究发现，纳米材料在毛细管电泳分离中的迁移时间高度依赖于纳米材料自身包裹剂和样品基质的组成。因此，用毛细管电泳表征不同电性纳米材料的粒径时，需要消除纳米材料包裹剂和样品基质的影响。

3.4.2 电场流分级

电场流分级作为微粒子分离技术，最早出现于 1972 年，用于多种蛋白质的分离[221]。电场流分级是场流分级的一分支，其纵向场是电场。分离时，不同的粒子由于其电泳淌度和离子表面电荷不同，所受的电场作用力不同。当样品所受的电场作用力与扩散力达到平衡时，不同的粒子距积聚壁不同的距离，因此在流道中有了不同速度，使得不同的微粒在不同的时间出现在分离通道的出口，从而实现不同物质的分离。

电场流分级的分离精度可达 20 nm，主要用于高分子和生物材料的分离。近年来，电场流分级在纳米颗粒的分离中也有一定的应用。Somchue 等[222]用电场流分级，成功分离了电泳淌度不同的 10~40 nm 柠檬酸包裹的纳米 Au。但是，研究发现，对于粒径相同、包裹剂分别为柠檬酸和单宁酸的纳米 Au，尽管它们的电泳淌度相差较大，依旧难以实现混合后两种颗粒的分离。这是由于包裹剂和纳米颗粒作用力较弱，混合时，不同颗粒上包裹剂分子容易发生交换，最终使体系中的纳米颗粒的包裹剂相似，两纳米 Au 电泳淌度接近。在传统的电场流分级中，通常在电极上施加静态电压。在电场流分级装置中，电极和流路中的液体接触，短时间内（~1 min）产生双电层，使得分离电场强度远小于施加的电场强度，显著降低了纳米颗粒的分离效率。为此，Tasci 等[223]改进传统的循环电场流分级，通过修改电路和抵消电压，使得双电层没有充分的时间形成，成功实现了 15~40 nm 不同电性纳米 Au 的高效分离，且该技术能够使用的电压振幅、频率和波形较为宽泛。

3.5 纳米材料形态分离测定装置研制

纳米材料的生物效应很大程度上取决于其尺寸和表面电性等属性。因此，分离不同粒径、不同电性纳米材料是研究其环境安全问题的基础和关键。然而，目前并没有一种通用的技术能同时分离不同粒径、不同电性的纳米材料，这在一定程度上制约了纳米材料环境过程和生物效应的研究。为实现纳米材料的多维分离，中国科学院生态环境研究中心刘景富课题组结合膜分离技术、流场流分离技术、体积排阻色谱分离技术和离子交换色谱分离技术等，经过多年的努力，成功地研发了一套可用于复杂基质中纳米材料分离测定的仪器装置。该装置的关键部件包括中空纤维膜流场流分离管、体积排阻色谱柱、离子交换色谱柱和自动控制模块、泵、阀、管路等。

整个装置由膜过滤分离模块、中空纤维膜流场流分离模块、体积排阻色谱分离模块、离子交换色谱分离模块和自动控制模块等几部分组成（如图 1-8 所示）。经过这些分离模块的多维分离，可以实现复杂基质中不同粒径和表面电性纳米材料的分离纯化。膜过滤分离模块可将纳米材料粗分为大粒径（>10 nm）和小粒径（<10 nm）纳米颗粒。中空纤维膜流场流分离模块兼具膜过滤和流场流分离两种工作模式，主要用于分离纯化大粒径的纳米颗粒。样品经膜过滤模式去除分子、离子和小粒径纳米材料，再由流场流分离模式将粒径大于 10 nm 的纳米材料按粒径大小分离，可在 22 min 内基线分离 46 nm 和 100 nm 的纳米颗粒。体积排阻色谱分离模块主要用于小粒径纳米颗粒的分离，经膜过滤分离得到的小粒径纳米颗粒，在此模块按体积排阻色谱分离原理进一步分离，可在 8 min 内基线分离粒径分别为 2 nm、4 nm 和 10 nm 的纳米材料。离子交换色谱分离模块用于分离粒径相同，但表面电性不同的纳米材料，可在 10 min 分离粒径都是 10 nm、但包裹剂分别为聚吡咯烷酮（PVP）和柠檬酸的两种纳米颗粒。自动控制模块由泵、阀、管路和计算机控制软件组成，可以实现对分离模块的自动控制和数据采集。

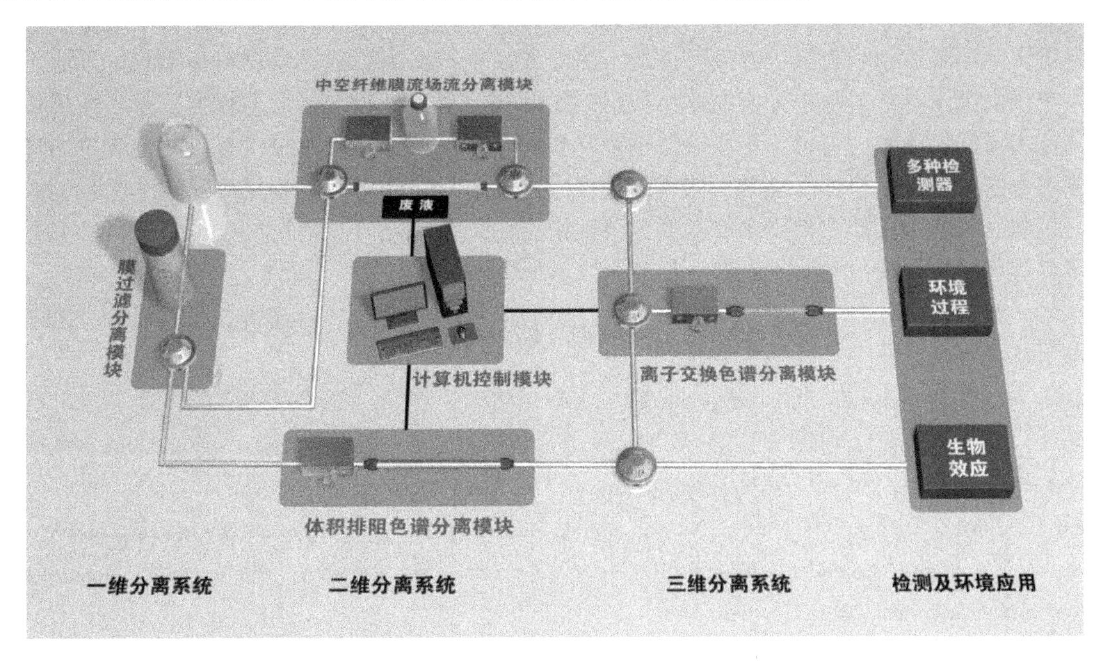

图 1-8　纳米颗粒多维分离系统装置流程图

样品分析测定时，先采用膜过滤和高速离心技术，实现大粒径和小粒径纳米颗粒的粗分。收集膜分离后的组分，再通过中空纤维膜流场流分离模块或体积排阻色谱分离模块，按粒径大小细分纳米颗粒。最后，通过离子交换色谱分离模块分离粒径接近但包裹剂不同的纳米颗粒。该仪器装置可与 DLS、UV-vis 和 ICP-MS 等检测仪器在线联用，通过自主研发的计算机软件控制各模块的切换和数据采集，能够实现纳米颗粒的高效分离和高灵敏检测。该装置分辨率高，性能稳定，操作简单，已被用于纳米材料环境过程和生物效应的研究。

4 展　望

污染物形态分析方法和技术虽然具有很长的发展历史，相关理论和方法较为成熟，但随着新材料、新理论的不断出现，分析方法将会得到持续创新，应用领域也将逐步推向深入。同时，材料科学、生命科学等相关学科的发展也为形态分析研究提出新的挑战、创造了新的机遇。

有毒元素形态分析是揭示元素生物地球化学循环的重要途径。经过国内外学者多年的研究，虽然针对有毒元素的形态分析方法和技术相对成熟，但也存在诸多需要完善之处。首先，发展稳定、灵敏、快速的形态分析方法特别是适合大通量筛查的可视化现场分析技术和小型化仪器设备，无疑是目前研究的主要方向之一。其次，生命科学等相关领域研究对分析方法的灵敏度提出了更高的要求，所面对的样品基体也更加复杂。需要发展新的、更加有效的形态分析样品前处理技术，进一步满足微量样品中痕量元素形态分析。再者，虽然目前已经具备多种仪器联用系统可以用于形态分析，但针对污染物原位形态分析还需要发展新的技术手段以满足微量甚至痕量元素的原位分析。

纳米材料的形态分析是研究其环境过程和生物效应的关键。近年来，人们发展了一些纳米材料的形态分析方法，但仍存在许多问题亟待解决。首先，目前大多数分离测定方法只能针对简单基质，且检出限往往高于实际环境中纳米材料的浓度。因此，需要发展一些高灵敏的纳米材料分离测定方法，以满足复杂基质中痕量纳米材料的形态分析需求。其次，实际样品中纳米材料的形态较为复杂，性质各异，单一的分离技术无法实现不同粒径、不同电性纳米材料的分离。将各种分离技术有机结合构建纳米材料多维分离测定系统，是纳米材料形态分析的重要发展方向。最后，由于定量测定痕量碳纳米材料的方法十分缺乏，目前有关复杂基质中碳纳米材料的分离测定方法的研究还较少，发展碳纳米材料的形态分析方法将为研究其环境过程和毒性效应提供强有力的技术手段。

参 考 文 献

[1] 苑春刚, Le X C. 砷形态分析. 化学进展, 2009, 21(2-3): 467-473.

[2] Wang H, Liu X L, Nan K, Chen B B, He M, Hu B. Sample pre-treatment techniques for use with ICP-MS hyphenated techniques for elemental speciation in biological samples. J Anal Atom Spectrom, 2017, 32(1): 58-77.

[3] Cheng H Y, Wu C L, Shen L H, Liu J H, Xu Z G. Online anion exchange column preconcentration and high performance liquid chromatographic separation with inductively coupled plasma mass spectrometry detection for mercury speciation analysis. Anal Chim Acta, 2014, 828: 9-16.

[4] Cheng H Y, Wu C L, Liu J H, Xu Z G. Thiol-functionalized silica microspheres for online preconcentration and determination of mercury species in seawater by high performance liquid chromatography and inductively coupled plasma mass spectrometry. RSC Adv, 2015, 5(25): 19082-19090.

[5] Zhu S, Chen B, He M, Huang T, Hu B. Speciation of mercury in water and fish samples by HPLC-ICP-MS after magnetic solid phase extraction. Talanta, 2017, 171: 213-219.

[6] Zhang S X, Luo H, Zhang Y Y, Li X Y, Liu J S, Xu Q, Wang Z H. *In situ* rapid magnetic solid-phase extraction coupled with HPLC-ICP-MS for mercury speciation in environmental water. Microchem J, 2016, 126: 25-31.

[7] Guo X, He M, Nan K, Yan H, Chen B, Hu B. A dual extraction technique combined with HPLCICP-MS for speciation of seleno-amino acids in rice and yeast samples. J Anal Atom Spectrom, 2016, 31(2): 406-414.

[8] Chen B, Hu B, He M, Huang Q, Zhang Y, Zhang X. Speciation of selenium in cells by HPLC-ICP-MS after

(on-chip)magnetic solid phase extraction. J Anal Atom Spectrom, 2013, 28(3): 334-343.

[9] Wang H, Chen B B, Zhu S Q, Yu X X, He M, Hu B. Chip-based magnetic solid-phase microextraction online coupled with microHPLC-ICPMS for the determination of mercury species in cells. Anal Chem, 2016, 88(1): 796-802.

[10] Wang H, Chen B B, He M, Yu X, Hu B. Selenocystine against methyl mercury cytotoxicity in HepG2 cells. Sci Rep, 2017, 7. DOI: 10.1038/S41598-017-00231-7.

[11] Chen Y Q, Cheng X, Mo F, Huang L M, Wu Z J, Wu Y N, Xu L J, Fu F F. Ultra-sensitive speciation analysis of mercury by CE-ICP-MS together with field-amplified sample stacking injection and dispersive solid-phase extraction. Electrophoresis, 2016, 37(7-8): 1055-1062.

[12] Duan J, Hu B, He M. Nanometer-sized alumina packed microcolumn solid-phase extraction combined with field-amplified sample stacking-capillary electrophoresis for the speciation analysis of inorganic selenium in environmental water samples. Electrophoresis, 2012, 33: 2953-2960.

[13] Chen B, Wu Y, Guo X, He M, Hu B. Speciation of mercury in various samples from the micro-ecosystem of East Lake by hollow fiber-liquid-liquid-liquid microextraction-HPLC-ICP-MS. J Anal Atom Spectrom, 2015, 30(4): 875-881.

[14] Duan J K, Hu B. Separation and determination of seleno amino acids using gas chromatography hyphenated with inductively coupled plasma mass spectrometry after hollow fiber liquid phase microextraction. J Mass Spectrom, 2009, 44(5): 605-612.

[15] Guo X, Chen B, He M, Hu B, Zhou X. Ionic liquid based carrier mediated hollow fiber liquid liquid liquid microextraction combined with HPLC-ICP-MS for the speciation of phenylarsenic compounds in chicken and feed samples. J Anal Atom Spectrom, 2013, 28(10): 1638-1647.

[16] Wang Z H, Xu Q Z, Li S Y, Luan L Y, Li J, Zhang S X, Dong H H. Hollow fiber supported ionic liquid membrane microextraction for speciation of mercury by high-performance liquid chromatography- inductively coupled plasma mass spectrometry. Anal Methods, 2015, 7(3): 1140-1146.

[17] Liu Y M, Zhang F P, Jiao B Y, Rao J Y, Leng G. Automated dispersive liquid-liquid microextraction coupled to high performance liquid chromatography-cold vapour atomic fluorescence spectroscopy for the determination of mercury species in natural water samples. J Chromatogr A, 2017, 1493: 1-9.

[18] Li P, He M, Chen B, Hu B. Automated dynamic hollow fiber liquid-liquid-liquid microextraction combined with capillary electrophoresis for speciation of mercury in biological and environmental samples. J Chromatogr A, 2015, 1415: 48-56.

[19] Chen C, Peng M T, Hou X D, Zheng C B, Long Z. Improved hollow fiber supported liquid-liquid-liquid membrane microextraction for speciation of inorganic and organic mercury by capillary electrophoresis. Anal Methods, 2013, 5(5): 1185-1191.

[20] Yang F F, Li J H, Lu W H, Wen Y Y, Cai X Q, You J M, Ma J P, Ding Y J, Chen L X. Speciation analysis of mercury in water samples by dispersive liquid-liquid microextraction coupled to capillary electrophoresis. Electrophoresis, 2014, 35(4): 474-481.

[21] Duan J, He M, Hu B. Chiral speciation and determination of selenomethionine enantiomers in selenized yeast by ligand-exchange micellar electrokinetic capillary chromatography after solid phase extraction. J Chromatogr A, 2012, 1268: 173-179.

[22] Chen B, Hu B, He M, Mao X, Zu W. Synthesis of mixed coating with multi-functional groups for in-tube hollow fiber solid phase microextraction-high performance liquid chromatography-inductively coupled plasma mass spectrometry speciation of arsenic in human urine. J Chromatogr A, 2012, 1227: 19-28.

[23] Liu X L, Chen B B, Zhang L, Song S Y, Cai Y B, He M, Hu B. TiO_2 nanoparticles functionalized monolithic capillary microextraction online coupled with inductively coupled plasma mass spectrometry for the analysis of Gd ion and Gd-based contrast agents in human urine. Anal Chem, 2015, 87(17): 8949-8956.

[24] Mao X, Fan W, He M, Chen B, Hu B. C18-coated stir bar sorptive extraction combined with HPLC-ICP-MS for the speciation of butyltins in environmental samples. J Anal Atom Spectrom, 2015, 30(1): 162-171.

[25] Mao X, Hu B, He M, Chen B. High polar organic-inorganic hybrid coating stir bar sorptive extraction combined with high performance liquid chromatography-inductively coupled plasma mass spectrometry for the speciation of seleno-amino acids and seleno-oligopeptides in biological samples. J Chromatogr A, 2012, 1256: 32-39.

[26] Wu Y, Lee Y I, Wu L, Hou X D. Simple mercury speciation analysis by CVG-ICP-MS following TMAH pre-treatment and microwave-assisted digestion. Microchem J, 2012, 103: 105-109.

[27] Zhao D, Li H B, Xu J Y, Luo J, Ma L N Q Y. Arsenic extraction and speciation in plants: Method comparison and development. Sci Total Environ, 2015, 523: 138-145.

[28] Deng D Y, Zhang S, Chen H, Yang L, Yin H, Hou X D, Zheng C B. Online solid sampling platform using multi-wall carbon nanotube assisted matrix solid phase dispersion for mercury speciation in fish by HPLC-ICP-MS. J Anal Atom Spectrom, 2015, 30(4): 882-887.

[29] Yun Z J, He B, Wang Z H, Wang T, Jiang G B. Evaluation of different extraction procedures for determination of organic mercury species in petroleum by high performance liquid chromatography coupled with cold vapor atomic fluorescence spectrometry. Talanta, 2013, 106: 60-65.

[30] Huang K, Xu K L, Hou X D, Jia Y, Zheng C B, Yang L. UV-induced atomization of gaseous mercury hydrides for atomic fluorescence spectrometric detection of inorganic and organic mercury after high performance liquid chromatographic separation. J Anal Atom Spectrom, 2013, 28(4): 510-515

[31] Ai X, Wang Y, Hou X D, Yang L, Zheng C B, Wu L. Advanced oxidation using Fe_3O_4 magnetic nanoparticles and its application in mercury speciation analysis by high performance liquid chromatography-cold vapor generation atomic fluorescence spectrometry. Analyst, 2013, 138: 3494-3501.

[32] Li H M, Luo Y C, Li Z X, Yang L M, Wang Q Q. Nanosemiconductor-based photocatalytic vapor generation systems for subsequent selenium determination and speciation with atomic fluorescence spectrometry and inductively coupled plasma mass spectrometry. Anal Chem, 2012, 84(6): 2974-2981.

[33] Li H M, Xu Z G, Yang L M, Wang Q Q. Determination and speciation of Hg using HPLC-AFS by atomization of this metal on a UV/nano-ZrO_2/HCOOH photocatalytic reduction unit. J Anal Atom Spectrom, 2015, 30(4): 916-921.

[34] Liu Z F, Xing Z, Li Z Y, Zhu Z L, Ke Y Q, Jin L L, Hu S H. The online coupling of high performance liquid chromatography with atomic fluorescence spectrometry based on dielectric barrier discharge induced chemical vapor generation for the speciation of mercury. J Anal Atom Spectrom, 2017, 32(3): 678-685.

[35] Chen X P, Han C, Cheng HY, Wang Y C, Liu J H, Xu Z G, Hu L. Rapid speciation analysis of mercury in seawater and marine fish by cation exchange chromatography hyphenated with inductively coupled plasma mass spectrometry. J Chromatogr A, 2013, 1314: 86-93.

[36] Jia X Y, Gong D R, Han Y, Wei C, Duan T C, Chen H T. Fast speciation of mercury in seawater by short-column high-performance liquid chromatography hyphenated to inductively coupled plasma spectrometry after online cation exchange column preconcentration. Talanta, 2012, 88: 724-729.

[37] Lin Y, Yang Y, Li Y X, Yang L, Hou X D, Feng X B, Zheng C B. Ultrasensitive speciation analysis of mercury in rice by headspace solid phase microextraction using porous carbons and gas chromatography-Dielectric barrier discharge optical emission spectrometry. Environ Sci Technol, 2016, 50(5): 2468-2476.

[38] 于永亮, 高飞, 陈明丽, 王建华. 介质阻挡放电微型化-长光程原子吸收光谱测定汞及甲基汞. 化学学报, 2013, 71(8): 1121-1124.

[39] Zeng Y, Xu K L, Hou X D, Jiang X M. Compact integration of gas chromatographer and atomic fluorescence spectrometer

for speciation analysis of trace alkyl metals/semimetals. Microchem J, 2014, 114: 16-21.

[40] Sun J, He B, Liu Q, Ruan T, Jiang G B. Characterization of interactions between organotin compounds and human serum albumin by capillary electrophoresis coupled with inductively coupled plasma mass spectrometry. Talanta, 2012, 93: 239-244.

[41] Liu L H, He B, Yun Z J, Sun J, Jiang G B. Speciation analysis of arsenic compounds by capillary electrophoresison-line coupled with inductively coupled plasma mass spectrometry using a novel interface. J Chromatogr A, 2013, 1304: 227-233.

[42] Liu L H, Yun Z J, He B, Jiang G B. Efficient interface for online coupling of capillary electrophoresis with inductively coupled plasma-mass spectrometry and its application in simultaneous speciation analysis of arsenic and selenium. Anal Chem, 2014, 86(16): 8167-8175.

[43] Zhao YQ, Zheng J P, Fang L, Lin Q, Wu Y N, Xue Z M, Fu F F. Speciation analysis of mercury in natural water and fish samples by using capillary electrophoresis-inductively coupled plasma mass spectrometry. Talanta, 2012, 89: 280-285.

[44] Deng B Y, Qin X D, Xiao Y, Wang Y Z, Yin H H, Xu X S, Shen C Y. Interface of on line coupling capillary electrophoresis with hydride generation electrothermal atomic absorption spectrometry and its application to arsenic speciation in sediment. Talanta, 2013, 109: 128-132.

[45] Matusiewicz H, Slachcinski M. Development of a new hybrid technique for inorganic arsenic speciation analysis by microchip capillary electrophoresis coupled with hydride generation microwave induced plasma spectrometry. Microchem J, 2012, 102: 61-67.

[46] Niegel C, Pfeiffer S A, Grundmann M, Arroyo-Abad U, Mattusch J, Matysik F M. Fast separations by capillary electrophoresis hyphenated to electrospray ionization time-of-flight mass spectrometry as a tool for arsenic speciation analysis. Analyst, 2012, 137(8): 1956-1962.

[47] Stice S, Liu G L, Matulis S, Boise L H, Cai Y. Determination of multiple human arsenic metabolites employing high performance liquid chromatography inductively coupled plasma mass spectrometry. J Chromatogr B, 2016, 1009: 55-65.

[48] Keller N S, Stefasson A, Sigfuson B. Determination of arsenic speciation in sulfidic waters by ion chromatography hydride-generation atomic fluorescence spectrometry (IC-HG-AFS). Talanta, 2014, 128: 466-472.

[49] Yuan C G, Lu X F, Oro N, Wang Z W, Xia Y J, Wade T J, Mumford J, Le X C. Arsenic speciation analysis in human saliva. Clin Chem, 2008, 54(1): 163-171.

[50] Yang Z, Peng H, Lu X, Liu Q, Huang R, Hu B, Kachanoski G, Zuidhof M J, Le X C. Arsenic metabolites, including *N*-Acetyl-4-hydroxy-m-arsanilic acid, in chicken litter from a Roxarsone-feeding study involving 1600 chickens. Environ Sci Technol, 2016, 50(13): 6737-6743.

[51] Peng H, Hu B, Liu Q, Yang Z, Lu X, Huang R, Li X F, Zuidhof M J, Le X C. Liquid chromatography combined with atomic and molecular mass spectrometry for speciation of arsenic in chicken liver. J Chromatogr A, 2014, 1370: 40-49.

[52] Peng H, Hu B, Liu Q, Li J, Li X F, Zhang H, Le X C. Methylated phenylarsenical metabolites discovered in chicken liver. Angew Chem Int Ed, 2017, 56(24): 6773-6777.

[53] Hsieh Y J, Jiang S J. Application of HPLC-ICP-MS and HPLC-ESI-MS procedures for arsenic speciation in seaweeds. J Agr Food Chem, 2012, 60(9): 2083-2089.

[54] Raab A, Newcombe C, Pitton D, Ebel R, Feldmann J. Comprehensive analysis of lipophilic arsenic species in a brown alga (*Saccharina latissima*). Anal Chem, 2013, 85(5): 2817-2824.

[55] Arroyo-Abad U, Lischka S, Piechotta C, Mattusch J, Reemtsma T. Determination and identification of hydrophilic and hydrophobic arsenic species in methanol extract of fresh cod liver by RP-HPLC with simultaneous ICP-MS and ESI-Q-TOF-MS detection. Food Chem, 2013, 141(3): 3093-3102.

[56] Garcia-Sartal C, Taebunpakul S, Stokes E, Barciela-Alonso M D, Bermejo-Barrera P, Goenaga-Infante H. Two-dimensional

HPLC coupled to ICP-MS and electrospray ionisation (ESI)-MS/MS for investigating the bioavailability *in vitro* of arsenic species from edible seaweed. Anal Bioanal Chem, 2012, 402(10): 3359-3369.

[57] Xu M, Yang L M, Wang Q Q. Chemical interactions of mercury species and some transition and noble metals towards metallothionein (Zn7MT-2) evaluated using SEC/ICP-MS, RP-HPLC/ESI-MS and MALDI-TOF-MS. Metallomics, 2013, 5: 855-860.

[58] Hu L G, Cai Y, Jiang G B. Occurrence and speciation of polymeric chromium(III), monomeric chromium(III)and chromium (VI) in environmental samples. Chemosphere, 2016, 156: 14-20.

[59] Liu G L, Cai Y. Studying arsenite-humic acid complexation using size exclusion chromatography- inductively coupled plasma mass spectrometry. J Hazard Mater, 2013, 262: 1223-1229.

[60] Peng L, He M, Chen B B, QiaoY, Hu B. Metallomics study of CdSe/ZnS quantum dots in HepG2 cells. ACS Nano, 2015, 9(10): 10324-10334.

[61] Zhang Y, Duan J, He M, Chen B, Hu B. Dispersive liquid liquid microextraction combined with electrothermal vaporization inductively coupled plasma mass spectrometry for the speciation of inorganic selenium in environmental water samples. Talanta, 2013, 115: 730-736.

[62] Liu Y, He M, Chen B, Hu B. Simultaneous speciation of inorganic arsenic, selenium and tellurium in environmental water samples by dispersive liquid liquid microextraction combined with electrothermal vaporization inductively coupled plasma mass spectrometry. Talanta, 2015, 142: 213-220.

[63] Escudero L B, Martinis E M, Olsina R A, Wuilloud R G. Arsenic speciation analysis in mono-varietal wines by on-line ionic liquid-based dispersive liquid-liquid microextraction. Food Chem, 2013, 138(1): 484-490.

[64] Li S, Wang M, Zhong Y Z, Zhang Z H, Yang B Y. Cloud point extraction for trace inorganic arsenic speciation analysis in water samples by hydride generation atomic fluorescence spectrometry. Spectrochim Acta Part B, 2015, 111: 74-79.

[65] Asadollahzadeh M, Niksirat N, Tavakoli H, Hemmati A, Rahdari P, Mohammadi M, Fazaeli R. Application of multi-factorial experimental design to successfully model and optimize inorganic arsenic speciation in environmental water samples by ultrasound assisted emulsification of solidified floating organic drop microextraction. Anal Methods, 2014, 6(9): 2973-2981.

[66] Chen S Z, Li J F, Lu D B, Zhang Y. Dual extraction based on solid phase extraction and solidified floating organic drop microextraction for speciation of arsenic and its distribution in tea leaves and tea infusion by electrothermal vaporization ICP-MS. Food Chem, 2016, 211: 741-747.

[67] Zhang X X, Tang S S, Chen M Li, Wang J H. Iron phosphate as a novel sorbent for selective adsorption of chromium(III)and chromium speciation with detection by ETAAS. J Anal Atom Spectrom, 2012, 27: 466-472.

[68] Peng H, Zhang N, He M, Chen B, Hu B. Simultaneous speciation analysis of inorganic arsenic, chromium and selenium in environmental waters by 3-(2-aminoethylamino)propyl trimethoxysilane modified multi-wall carbon nanotubes packed microcolumn solid phase extraction and ICP-MS. Talanta, 2015, 131: 266-272.

[69] Chen M L, Lin Y M, Gu C B, Wang J H. Arsenic sorption and speciation with branch-polyethyleneimine modified carbon nanotubes with detection by atomic fluorescence spectrometry. Talanta, 2013, 104: 53-57.

[70] Ma S, He M, Chen B, Deng W, Zheng Q, Hu B. Magnetic solid phase extraction coupled with inductively coupled plasma mass spectrometry for the speciation of mercury in environmental water and human hair samples. Talanta, 2016, 146: 93-99.

[71] Wang W J, Chen M L, Chen X W, Wang J H. Thiol-rich polyhedral oligomeric silsesquioxane as a novel adsorbent for mercury adsorption and speciation. Chem Eng J, 2014, 242: 62-68.

[72] Wei W, Zhao B, He M, Chen B, Hu B. Iminodiacetic acid functionalized magnetic nanoparticles for speciation of Cr(III) and Cr(VI)followed by graphite furnace atomic absorption spectrometry detection. RSC Adv, 2017, 7: 8504-8511.

[73] Cui C, He M, Chen B, Hu B. Chitosan modified magnetic nanoparticles based solid phase extraction combined with

ICP-OES for the speciation of Cr(III) and Cr(VI). Anal Methods, 2014, 6(21): 8577-8583.

[74] Zhang Y, Chen B, Wu S, He M, Hu B. Grapheneoxide-TiO_2 composite solid phase extraction combined with graphite furnace atomic absorption spectrometry for the speciation of inorganic selenium in water samples. Talanta, 2016, 154: 474-480.

[75] Chen M L, Gu C B, Yang T, Sun Y, Wang J H. A green sorbent of esterified egg-shell membrane for highly selective uptake of arsenate and speciation of inorganic arsenic. Talanta, 2013, 116: 688-694.

[76] Li P, Zhang X Q, Chen Y J, Lian H Z, Hu X. A sequential solid phase microextraction system coupled with inductively coupled plasma mass spectrometry for speciation of inorganic arsenic. Anal Methods, 2014, 6(12): 4205-4211.

[77] Doker S, Uzun L, Denizli A. Arsenic speciation in water and snow samples by adsorption onto PHEMA in a micro-pipette-tip and GFAAS detection applying large-volume injection. Talanta, 2013, 103: 123-129.

[78] Khaligh A, Mousavi H Z, Shirkhanloo H, Rashidi A. Speciation and determination of inorganic arsenic species in water and biological samples by ultrasound assisted-dispersive-micro-solid phase extraction on carboxylated nanoporous graphene coupled with flow injection-hydride generation atomic absorption spectrometry. RSC Adv, 2015, 5(113): 93347-93359.

[79] Li P, Zhang X Q, Chen Y J, Bai T Y, Lian H Z, Hu X. One-pot synthesis of thiol- and amine-bifunctionalized mesoporous silica and applications in uptake and speciation of arsenic. RSC Adv, 2014, 4(90): 49421-49428.

[80] Zhang Z, Li J H, Song X L, Ma J P, Chen L X. Hg^{2+} ion-imprinted polymers sorbents based on dithizone-Hg^{2+} chelation for mercury speciation analysis in environmental and biological samples. RSC Adv, 2014, 4(87): 46444-46453.

[81] Dakova I, Yordanova T, Karadjova I. Non-chromatographic mercury speciation and determination in wine by new core-shell ion-imprinted sorbents. J Hazard Mater, 2012, 231: 49-56.

[82] 苑春刚, 江万平, 祝涛, 袁博, 宋小卫. 基于在线捕集分离联用技术的食品调料中砷形态分析方法研究. 光谱学与光谱分析, 2014, 34(8): 2259-2263.

[83] Musil S, Petursdottir A H, Raab A, Gunnlaugsdottir H, Krupp E, Feldmann J. Speciation without chromatography using selective hydride generation: Inorganic arsenic in rice and samples of marine origin. Anal Chem, 2014, 862(2): 993-999.

[84] Matousek T, Currier J M, Trojankova N, Saunders R J, Ishida M C, Gonzalez-Horta C, Musil S, Mester Z, Styblo M, Dedina J. Selective hydride generation-cryotrapping-ICP-MS for arsenic speciation analysis at picogram levels: Analysis of river and sea water reference materials and human bladder epithelial cells. J Anal Atom Spectrom, 2013, 28(9): 1456-1465.

[85] Musil S, Matousek T, Currier J M, Styblo M, Dedina J. Speciation analysis of arsenic by selective hydride generation-cryotrapping-atomic fluorescence spectrometry with flame-in-gas-shield atomizer: Achieving extremely low detection limits with inexpensive instrumentation. Anal Chem, 2014, 86(20): 10422-10428.

[86] Yin Y G, Liu Y, Liu J F, He B, Jiang G B. Determination of methylmercury and inorganic mercury by volatile species generation-flameless/flame atomization-atomic fluorescence spectrometry without chromatographic separation. Anal Methods, 2012, 4(4): 1122-1125.

[87] Zhang W B, Yang X A, Dong Y P, Xue J J. Speciation of inorganic- and methyl-mercury in biological matrixes by electrochemical vapor generation from an L-cysteine modified graphite electrode with atomic fluorescence spectrometry detection. Anal Chem, 2012, 84(21): 9199-9207.

[88] Zhang R X, Peng M T, Zheng C B, Xu K L, Hou X D. Application of flow injection-green chemical vapor generation-atomic fluorescence spectrometry to ultrasensitive mercury speciation analysis of water and biological samples. Microchem J, 2016, 127: 62-67.

[89] Tan Z Q, Liu J F, Yin Y G, Shi Q T, Jing C Y, Jiang G B. Colorimetric Au nanoparticle probe for speciation test of arsenite and arsenate inspired by selective interaction between phosphonium ionic liquid and arsenite. ACS Appl Mater Inter, 2014, 6: 19833-19839.

[90] Chen L, Li J H, Chen L X. Colorimetric detection of mercury species based on functionalized gold nanoparticles. ACS Appl Mater Inter, 2014, 6(18): 15897-15904.

[91] Zhang H Y, Liu Q, Wang T, Yun Z J, Li G L, Liu J Y, Jiang G B. Facile preparation of glutathione-stabilized gold nanoclusters for selective determination of chromium (III) and chromium (VI) in environmental water samples. Anal Chim Acta, 2013, 770: 140-146.

[92] Butwong N, Srijaranai S, Ngeontae W, Burakham R. Speciation of arsenic (III) and arsenic (V) based on quenching of CdS quantum dots fluorescence using hybrid sequential injection-stopped flow injection gas diffusion system. Spectrochim Acta Part A: Molecular and Biomolecular Spectroscopy, 2012, 97: 17-23.

[93] Deng L, Li Y, Yan X, Xiao J, Ma C, Zheng J, Liu S J, Yang R H. Ultrasensitive and highly selective detection of bioaccumulation of methyl-mercury in fish samples via Ag^0/Hg^0 amalgamation. Anal Chem, 2015, 87(4): 2452-2458.

[94] Xu F J, Kou L, Jia J, Hou X D, Long Z, Wang S L. Metal-organic frameworks of zeolitic imidazolate framework-7 and zeolitic imidazolate framework-60 for fast mercuryand methylmercury speciation analysis. Anal Chim Acta, 2013, 804: 240-245.

[95] Hu S, Lu J S, Jing C Y. A novel colorimetric method for field arsenic speciation analysis. J Environ Sci, 2012, 24(7): 1341-1346.

[96] Jiang J, Liu W, Cheng J, Yang L Z, Jiang H, Bai D C, Liu W S. A sensitive colorimetric and ratiometric fluorescent probe for mercury species in aqueous solution and living cells. Chem Commun, 2012, 48(67): 8371-8373.

[97] Yang M, Chen X, Jiang T J, Guo Z, Liu J H, Huang X J. Electrochemical detection of trace arsenic(III) by nanocomposite of nanorod-like alpha-MnO_2 decorated with similar to 5 nm Au nanoparticles: Considering the change of arsenic speciation. Anal Chem, 2016, 88(19): 9720-9728.

[98] Saha S, Sarkar P. Differential pulse anodic stripping voltammetry for detection of As (III) by Chitosan- $Fe(OH)_3$ modified glassy carbon electrode: A new approach towards speciation of arsenic. Talanta, 2016, 158: 235-245.

[99] Nearing M M, Koch I, Reimer K J. Complementary arsenic speciation methods: A review. Spectrochim Acta Part B, 2014, 99: 150-162.

[100] Kim E J, Yoo J C, Baek K. Arsenic speciation and bioaccessibility in arsenic-contaminated soils: Sequential extraction and mineralogical investigation. Environ Pollut, 2014, 186: 29-35.

[101] Nearing M M, Koch I, Reimer K J. Arsenic speciation in edible mushrooms. Environ Sci Technol, 2014, 48(24): 14203-14210.

[102] Maher W, Foster S, Krikowa F, Donner E, Lombi E. Measurement of inorganic arsenic species in iice after nitric acid extraction by HPLC-ICPMS: Verification using XANES. Environ Sci Technol, 2013, 47(11): 5821-5827.

[103] Yamaguchi N, Ohkura T, Takahashi Y, Maejima Y, Arao T. Arsenic distribution and speciation near rice roots influenced by iron plaques and redox conditions of the soil matrix. Environ Sci Technol, 2014, 48(3): 1549-1556.

[104] Kopittke P M, de Jonge M D, Wang P, McKenna B A, Lombi E, Paterson D J, Howard D L, James S A, Spiers K M, Ryan C G, Johnson A A T, Menzies N W. Laterally resolved speciation of arsenic in roots of wheat and rice using fluorescence-XANES imaging. New Phytol, 2014, 201(4): 1251-1262.

[105] Lin J R, Chen N, Pan Y M. Arsenic speciation in newberyite ($MgHPO_4 \cdot 3H_2O$)determined by synchrotron X-ray absorption and electron paramagnetic resonance spectroscopies: Implications for the fate of arsenic in green fertilizers. Environ Sci Technol, 2014, 48(12): 6938-6946.

[106] Meng B, Feng X B, Qiu G L, Anderson C W N, Wang J X, Zhao L. Localization and speciation of mercury in brown rice with implications for Pan-Asian public health. Environ Sci Technol, 2014, 48(14): 7974-7981.

[107] Zeng X B, Su S M, Feng Q F, Wang X R, Zhang Y Z, Zharig L L, Jiang S, Li A G, Li L F, Wang Y N, Wu C X, Bai L Y,

Duan R. Arsenic speciation transformation and arsenite influx and efflux across the cell membrane of fungi investigated using HPLC-HG-AFS and, in-situ XANES. Chemosphere, 2015, 119: 1163-1168.

[108] Langner P, Mikutta C, Suess E, Marcus M A, Kretzschmar R. Spatial distribution and speciation of arsenic in peat studied with microfocused X-ray fluorescence spectrometry and X-ray absorption spectroscopy. Environ Sci Technol, 2013, 47(17): 9706-9714.

[109] Winkel L H E, Casentini B, Bardelli F, Voegelin A, Nikolaidis N P, Charlet L. Speciation of arsenic in Greek travertines: Co-precipitation of arsenate with calcite. Geochim Cosmochim Ac, 2013, 106: 99-110.

[110] Essilfie-Dughan J, Hendry M J, Warner J, Kotzer T. Arsenic and iron speciation in uranium mine tailings using X-ray absorption spectroscopy. Appl Geochem, 2013, 28: 11-18.

[111] Deonarine A, Kolker A, Foster A L, Doughten M W, Holland J T, Bailoo J D. Arsenic speciation in bituminous coal fly ash and transformations in response to redox conditions. Environ Sci Technol, 2016, 50(11): 6099-6106.

[112] Guerrini L, Rodriguez-Loureiro I, Correa-Duarte M A, Lee Y H, Ling X Y, de Abajo F J G, Alvarez-Puebla R A. Chemical speciation of heavy metals by surface-enhanced Raman scattering spectroscopy: Identification and quantification of inorganic- and methyl-mercury in water. Nanoscale, 2014, 6(140): 8368-8375.

[113] Du J J, Cui J L, Jing CY. Rapid *in situ* identification of arsenic species using a portable Fe_3O_4@Ag SERS sensor. Chem Commun, 2014, 50(3): 347-349.

[114] Wang W D, Yin Y G, Tan Z Q, Liu J F. Coffee-ring effect-based simultaneous SERS substrate fabrication and analyte enrichment for trace analysis. Nanoscale, 2014, 6: 9588-9593.

[115] Zhang L, Wang S X, Wang L, Hao J M. Atmospheric mercury concentration and chemical speciation at a rural site in Beijing, China: Implications of mercury emission sources. Atmos Chem Phys, 2013, 13(20): 10505-10516.

[116] Xu L L, Chen J S, Yang L M, Niu Z C, Tong L, Yin L Q, ChenY T. Characteristics and sources of atmospheric mercury speciation in a coastal city, Xiamen, China. Chemosphere, 2015, 119: 530-539.

[117] Duan L, Wang X H, Wang D F, Duan Y S, Cheng N, Xiu G L. Atmospheric mercury speciation in Shanghai, China. Sci Total Environ, 2017, 578: 460-468.

[118] Reis A T, Coelho J P, Rodrigues S M, Rocha R, Davidson C M, Duarte A C, Pereira E. Development and validation of a simple thermo-desorption technique for mercury speciation in soils and sediments. Talanta, 2012, 99: 363-368.

[119] Liu X L, Wang S X, Zhang L, Wu Y, Duan L, Hao J M. Speciation of mercury in FGD gypsum and mercury emission during the wallboard production in China. Fuel, 2013, 111: 621-627.

[120] Li H B, Li J, Juhasz A L, Ma LQ. Correlation of *in vivo* relative bioavailability to *in vitro* bioaccessibility for arsenic in household dust from China and its implication for human exposure assessment. Environ Sci Technol, 2014, 48(23): 13652-13659.

[121] Li H B, Cui X Y, Li K, Li J, Juhasz A L, Ma L Q. Assessment of *in vitro* lead bioaccessibility in house dust and its relationship to *in vivo* lead relative bioavailability. Environ Sci Technol, 2014, 48(15): 8548-8555.

[122] Jin Y, Yuan C G, Jiang W P, Qi L Q. Evaluation of bioaccessible arsenic in fly ash by an in vitro method and influence of particle-size fraction on arsenic distribution. J Mater Cycles Waste, 2013, 15(4): 516–521.

[123] Yuan C G, Jin Y, Wang S Y. Fractionation and distribution of arsenic in desulfurization gypsum, slag and fly ash from a coal-fired power plant, Fresen Environ Bull, 2013, 22(3A): 884-889.

[124] Yuan C G, Li Q P, Feng Y N, Chang A L. Fractions and leaching characteristics of mercury in coal. Environ Monit Assess, 2010, 167(1-4): 581-586.

[125] Yuan C G. Leaching characteristics of metals in fly ash from coal-fired power plant by sequential extraction procedure. Microchim Acta, 2009, 165(1-2): 91-96.

[126] Yin Y G, Li Y B, Tai C, Cai Y, Jiang G B. Fumigant methyl iodide can methylate inorganic mercury species in natural waters. Nat Commun, 2014, 5.

[127] Yin Y G, Chen B W, Mao Y X, Wang T, Liu J F, Cai Y, Jiang G B. Possible alkylation of inorganic Hg(II) by photochemical processes in the environment. Chemosphere, 2012, 88: 8-16.

[128] Zhao F J, Harris E, Yan J, Ma J C, Wu L Y, Liu W J, McGrath S P, Zhou J Z, Zhu Y G. Arsenic methylation in soils and its relationship with microbial arsM abundance and diversity, and As speciation in rice. Environ Sci Technol, 2013, 47(13): 7147-7154.

[129] Wang X, Peng B, Tan C Y, Ma L N, Rathinasabapathi B. Recent advances in arsenic bioavailability, transport, and speciation in rice. Environ Sci Pollut R, 2015, 22(8): 5742-5750.

[130] Wang J X, Feng X B, Anderson C W N, Wang H, Zheng L R, Hu T D. Implications of mercury speciation in thiosulfate treated plants. Environ Sci Technol, 2012, 46(10): 5361-5368.

[131] Qin J, Lehr C R, Yuan C G, Le X C, McDermott T R, Rosen B P. Biotransformation of arsenic by a Yellowstone thermoacidophilic eukaryotic alga. P Natl Acad Sci USA, 2009, 106 (13): 5213-5217.

[132] Yuan C G, Lu X F, Qin J, Rosen B P, Le X C. Volatile arsenic species released from *Escherichia coli* expressing the AsIIIS-adenosylmethionine methyltransferase gene. Environ Sci Technol, 2008, 42(9): 3201-3206.

[133] Yuan C G, Zhang K G, Wang Z H, Jiang G B. Rapid analysis of volatile arsenic species released from lake sediment by a packed cotton column coupled with atomic fluorescence spectrometry. J Anal Atom Spectrom, 2010, 25(10): 1605-1611.

[134] Huang H, Jia Y, Sun G X, Zhu Y G. Arsenic speciation and volatilization from flooded paddy soils amended with different organic matters. Environ Sci Technol, 2012, 46(4): 2163-2168.

[135] Meng M, Li B, Shao J J, Wang T, He B, Shi J B, Ye Z H, Jiang G B. Accumulation of total mercury and methylmercury in rice plants collected from different mining areas in China. Environ Pollut, 2014, 184: 179-186.

[136] Shao J J, Shi J B, Duo B, Liu C B, Gao Y, Fu J J, Yang R Q, Jiang G B. Mercury in alpine fish from four rivers in the Tibetan Plateau. J Environ Sci, 2016, 39: 22-28.

[137] Bennett W W, Teasdale P R, Panther J G, Welsh D T, Zhao H J, Jolley D F. Investigating arsenic speciation and mobilization in sediments with DGT and DET: A mesocosm evaluation of oxic-anoxic transitions. Environ Sci Technol, 2012, 46(7): 3981-3989.

[138] Rolisola A M C M, Suarez C A, Menegario A A, Gastmans D, Kiang C H, Colaco C D, Garcez D L, Santelli R E. Speciation analysis of inorganic arsenic in river water by Amberlite IRA 910 resin immobilized in a polyacrylamide gel as a selective binding agent for As(V) in diffusive gradient thin film technique. Analyst, 2014, 139(17): 4373-4380.

[139] Guan D X, Williams P N, Luo J, Zheng J L, Xu H C, Cai C, Ma L Q. Novel precipitated zirconia-based DGT technique for high-resolution imaging of oxyanions in waters and sediments. Environ Sci Technol, 2015, 49(6): 3653-3661.

[140] Clemens S, Monperrus M, Donard O F X, Amouroux D, Guerin T. Mercury speciation in seafood using isotope dilution analysis: A review. Talanta, 2012, 89: 12-20.

[141] Laffont L, Maurice L, Amouroux D, Navarro P, Monperrus M, Sonke J E, Behra P. Mercury speciation analysis in human hair by species-specific isotope-dilution using GC-ICP-MS. Anal Bioanal Chem, 2013, 405(9): 3001-3010.

[142] Rahman G M M, Wolle M M, Fahrenholz T, Kingston H M, Pamuku M. Measurement of mercury species in whole blood using speciated isotope dilution methodology integrated with microwave-enhanced solubilization and spike equilibration, headspace-solid-phase microextraction, and GC-ICP-MS analysis. Anal Chem, 2014, 86(12): 6130-6137.

[143] Feng C Y, Pedrero Z, Gentes S, Barre J, Renedo M, Tessier E, Beraitt S, Maury-Brachet R, Mesmer-Dudons N, Baudrimont M, Legeay A, Maurice L, Gonzalez P, Amouroux D. Specific pathways of dietary methylmercury and inorganic mercury determined by mercury speciation and isotopic composition in Zebrafish (*Danio rerio*). Environ Sci

Technol, 2015, 49(21): 12984-12993.

[144] Mao Y X, Li Y B, Richards J, Cai Y. Investigating uptake and translocation of mercury species by sawgrass (*Cladium jamaicense*) using a stable isotope tracer technique. Environ Sci Technol, 2013, 47(17): 9678-9684.

[145] Yang G S, Ma L L, Xu D D, Li J, He T T, Liu L Y, Jia H L, Zhang Y B, Chen Y, Chai Z F. Levels and speciation of arsenic in the atmosphere in Beijing, China. Chemosphere, 2012, 87(8): 845-850.

[146] Huang M J, Chen X W, Zhao Y G, Chan C Y, Wang W, Wang X M, Wong M H. Arsenic speciation in total contents and bioaccessible fractions in atmospheric particles related to human intakes. Environ Pollut, 2014, 188: 37-44.

[147] Tirez K, Vanhoof C, Peters J, Geerts L, Bleux N, Adriaenssens E, Roekens E, Smolek S, Maderitsch A, Steininger R, Gottlicher J, Meirer F, Streli C, Berghmans P. Speciation of inorganic arsenic in particulate matter by combining HPLC/ICP-MS and XANES analyses. J Anal Atom Spectrom, 2015, 30(10): 2074-2088.

[148] Tziaras T, Pergantis S A, Stephanou E G. Investigating the occurrence and environmental significance of methylated arsenic species in atmospheric particles by overcoming analytical method limitations. Environ Sci Technol, 2015, 49(19): 11640-11648.

[149] Bolanz R M, Majzlan J, Jurkovic L, Gottlicher J. Mineralogy, geochemistry, and arsenic speciation in coal combustion waste from Novaky, Slovakia. Fuel, 2012, 94(1): 125-136.

[150] Sun M, Liu G J, Wu Q H, Liu W Q. Speciation analysis of inorganic arsenic in coal samples by microwave-assisted extraction and high performance liquid chromatography coupled to hydride generation atomic fluorescence spectrometry. Talanta, 2013, 106: 8-13.

[151] Gao X B, Wang Y X, Hu Q H. Fractionation and speciation of arsenic in fresh and combusted coal wastes from Yangquan, northern China. Environ Geochem Hlth, 2012, 43(1): 113-122.

[152] Keller N S, Stefansson A, Sigfusson B. Arsenic speciation in natural sulfidic geothermal waters. Geochim Cosmochimi Ac, 2014, 15-26.

[153] Gaulier F, Gibert A, Walls D, Langford M, Baker S, Baudot A, Porcheron F, Lienemann C P. Mercury speciation in liquid petroleum products: Comparison between on-site approach and lab measurement using size exclusion chromatography with high resolution inductively coupled plasma mass spectrometric detection (SEC-ICP-HR MS). Fuel Process Technol, 2015, 131: 254-261.

[154] Ezzeldin M F, Gajdosechova Z, Masod M B, Zald T, Feldmann J, Kruppt E M. Mercury speciation and distribution in an egyptian natural gas processing plant. Energ Fuel, 2016, 30(12): 10236-10243.

[155] Liu G L, Cai Y, Hernandez D, Schrlau J, Allen M. Mobility and speciation of arsenic in the coal fly ashes collected from the Savannah River Site (SRS). Chemosphere, 2016, 151: 138-144.

[156] Tang S L, Wang L N, Feng X B, Feng Z H, Li R Y, Fan H P, Li K. Actual mercury speciation and mercury discharges from coal-fired power plants in Inner Mongolia, Northern China. Fuel, 2016, 180: 194-204.

[157] Yang J P, Ma S M, Zhao Y C, Zhang J Y, Liu Z H, Zhang S H, Zhang Y, Liu Y M, Feng Y X, Xu K, Xiang J, Zheng C G. Mercury emission and speciation in fly ash from a 35 MW_{th} large pilot boiler of oxyfuel combustion with different flue gas recycle. Fuel, 2017, 195: 174-181.

[158] Kakoulli I, Prikhodko S V, Fischer C, Cilluffo M, Uribe M, Bechtel H A, Fakra S C, Marcus M A. Distribution and chemical speciation of arsenic in ancient human hair using synchrotron radiation. Anal Chem, 2014, 86(1): 521-526.

[159] Chen B W, Cao F L, Yuan C G, Lu X F, Shen S W, Zhou J, Le X C. Arsenic speciation in saliva of acute promyelocytic leukemia patients undergoing arsenic trioxide treatment. Anal Bioanal Chem, 2013, 405(6): 1903-1911.

[160] Masciangioli T, Zhang W X. Environmental technologies at the nanoscale. Environ Sci Technol, 2003, 37(5): 102-108.

[161] Klaine S J, Alvarez P J J, Batley G E, Fernandes T F, Handy R D, Lyon D Y, Mahendra S, McLaughlin M J, Lead J R.

Nanomaterials in the environment: Behavior, fate, bioavailability, and effects. Environ Toxicol Chem, 2008, 27(9): 1825-1851.

[162] Cornelis G, Kirby J K, Beak D, Chittleborough D, Mclaughlin M J. A method for determination of retention of silver and cerium oxide manufactured nanoparticles in soils. Environ Chem, 2010, 7(3): 298-308.

[163] Yu S J, Yin Y G, Chao J B, Shen M H, Liu J F. Highly dynamic PVP-coated silver nanoparticles in aquatic environments: Chemical and morphology change induced by oxidation of Ag(0) and reduction of Ag^{+}. Environ Sci Technol, 2014, 48(1): 403-411.

[164] Goto K, Taguchi S, Fukue Y, Ohta K, Watanabe H. Spectrophotometric determination of manganese with 1-(2-pyridylazo)-2-naphthol and a nanionic surfaciant. Talanta, 1977, 24(12): 752-753.

[165] Watanabe H, Tanaka H. Nonionic surfactant as a new solvent for liquid-liquid-extraction of zinc(II) with 1-(2-pyridylazo)-2-naphthol. Talanta, 1978, 25(10): 585-589.

[166] Liu J F, Liu R, Yin Y G, Jiang G B. Triton X-114 based cloud point extraction: A thermoreversible approach for separation/concentration and dispersion of nanomaterials in the aqueous phase. Chem Commun, 2009, 1514-1516.

[167] Liu J F, Chao J B, Liu R, Tan Z Q, Yin Y G, Wu Y, Jiang G B. Cloud point extraction as an advantageous preconcentration approach for analysis of trace silver nanoparticles in environmental waters. Anal Chem, 2009, 81(15): 6496-6502.

[168] Chao J B, Liu J F, Yu S J, Feng Y D, Tan Z Q, Liu R, Yin Y G. Speciation analysis of silver nanoparticles and silver ions in antibacterial products and environmental waters *via* cloud point extraction-based separation. Anal Chem, 2011, 83(17): 6875-6882.

[169] Hartmann G, Hutterer C, Schuster M. Ultra-trace determination of silver nanoparticles in water samples using cloud point extraction and ETAAS. J Anal Atom Spectrom, 2013, 28(4): 567-572.

[170] Tsogas G Z, Giokas D L, Vlessidis A G. Ultratrace determination of silver, gold, and iron oxide nanoparticles by micelle mediated preconcentration/selective back-extraction coupled with flow injection chemiluminescence detection. Anal Chem, 2014, 86(7): 3484-3492.

[171] Su S, Chen B, He M, Xiao Z W, Hu B. A novel strategy for sequential analysis of gold nanoparticles and gold ions in water samples by combining magnetic solid phase extraction with inductively coupled plasma mass spectrometry. J Anal Atom Spectrom, 2014, 29(3): 444-453.

[172] Mwilu S K, Siska E, Baig R B N, Varma R S, Heithmar E, Rogers K R. Separation and measurement of silver nanoparticles and silver ions using magnetic particles. Sci Total Environ, 2014, 472: 316-323.

[173] Tolessa T, Zhou X X, Amde M, Liu J F. Development of reusable magnetic chitosan microspheres adsorbent for selective extraction of trace level silver nanoparticles in environmental waters prior to ICP-MS analysis. Talanta, 2017, 169: 91-97.

[174] Zhou X X, Liu J F, Yuan C G, Chen Y S. Speciation analysis of silver sulfide nanoparticles in environmental waters by magnetic solid-phase extraction coupled with ICP-MS. J Anal Atom Spectrom, 2016, 31(11): 2285-2292.

[175] Zhang L, Chen B B, He M, Liu X L, Hu B. Hydrophilic polymer monolithic capillary microextraction online coupled to ICPMS for the determination of carboxyl group-containing gold nanoparticles in environmental waters. Anal Chem, 2015, 87(3): 1789-1796.

[176] Essinger-Hileman E R, Popczun E J, Schaak R E. Magnetic separation of colloidal nanoparticle mixtures using a material specific peptide. Chem Commun, 2013, 49(48): 5471-5473.

[177] Shen L F, Zhu Y Z, Zhang P, F, Wang H F. Capturing of nano-TiO_2 from complex mixtures by bisphosphonate-functionalized Fe_3O_4 nanoparticles. ACS Sustain Chem Eng, 2017, 5(2): 1704-1710.

[178] Park E J, Bae E, Yi J, Kim Y, Choi K, Lee S H, Yoon J, Lee B C, Park K. Repeated-dose toxicity and inflammatory responses in mice by oral administration of silver nanoparticles. Environ Toxicol Pharmacol, 2010, 30(2): 162-168.

[179] Fischer C H, Giersig M, Siebrands T. Analysis of colloids . 5. Size-exclusion chromatography of colloidal semiconductor particles. J Chromatogr A, 1994, 670(1-2): 89-97.

[180] Wei G T, Liu F K. Separation of nanometer gold particles by size exclusion chromatography. J Chromatogr A, 1999, 836(2): 253-260.

[181] Wei G T, Liu F K, Wang C R C. Shape separation of nanometer cold particles by size-exclusion chromotography. Anal Chem, 1999, 71(11): 2085-2091.

[182] Soto-Alvaredo J, Montes-Bayon M, Bettmer J. Speciation of silver nanoparticles and silver(I) by reversed-phase liquid chromatography coupled to ICPMS. Anal Chem, 2013, 85(3): 1316-1321.

[183] Zhou X X, Liu R, Liu J F. Rapid chromatographic separation of dissoluble Ag(I) and silver-containing nanoparticles of 1-100 nanometer in antibacterial products and environmental waters. Environ Sci Technol, 2014, 48(24): 14516-14524.

[184] Zhou X X, Liu J F, Jiang G B. Elemental mass size distribution for characterization, quantification and identification of trace nanoparticles in serum and environmental waters. Environ Sci Technol, 2017, 51(7): 3892-3901.

[185] Small H. Hydrodynamic chromatography technique for size analysis of collidal particles. J Colloid Interf Sci, 1974, 48(1): 147-161.

[186] Small H, Saunders F L, Solc J. Hydrodynamic chromatography: New apprpach to particle size analysis. Adv Colloid Interf Sci, 1976, 6(4): 237-266.

[187] Tiede K, Boxall A B A, Tiede D, Tear S P, David H, Lewis J. A robust size-characterisation methodology for studying nanoparticle behaviour in 'real' environmental samples, using hydrodynamic chromatography coupled to ICP-MS. J Anal Atom Spectrom, 2009, 24(7): 964-972.

[188] Tiede K, Boxall A B A, Wang X, Gore D, Tiede D, Baxter M, David H, Tear S P, Lewis J. Application of hydrodynamic chromatography-ICP-MS to investigate the fate of silver nanoparticles in activated sludge. J Anal Atom Spectrom, 2010, 25(7): 1149-1154.

[189] Pergantis S A, Jones-Lepp T L, Heithmar E M. Hydrodynamic chromatography online with single particle-inductively coupled plasma mass spectrometry for ultratrace detection of metal-containing nanoparticles. Anal Chem, 2012, 84(15): 6454-6462.

[190] Rakcheev D, Philippe A, Schaumann G E. Hydrodynamic chromatography coupled with single particle-inductively coupled plasma mass spectrometry for investigating nanoparticles agglomerates. Anal Chem, 2013, 85(22): 10643-10647.

[191] Proulx K, Wilkinson K J. Separation, detection and characterisation of engineered nanoparticles in natural waters using hydrodynamic chromatography and multi-method detection (light scattering, analytical ultracentrifugation and single particle ICP-MS). Environ Chem, 2014, 11(4): 392-401.

[192] Yan N, Zhu Z, Jin L, Guo W, Gan Y, Hu S. Quantitative characterization of gold nanoparticles by coupling thin layer chromatography with laser ablation inductively coupled plasma mass spectrometry. Anal Chem, 2015, 87(12): 6079-6087.

[193] Yan N, Zhu Z, He D, Jin L, Zheng H, Hu S. Simultaneous determination of size and quantification of gold nanoparticles by direct coupling thin layer chromatography with catalyzed luminol chemiluminescence. Sci Rep, 2016, 6.

[194] Giddings J C. A new separation concept based on a coupling of concentration and flow nonuniformities. J Sep Sci, 1966, 1: 123-125.

[195] Williams S K R, Runyon J R, Ashames A A. Field-flow fractionation: Addressing the nano challenge. Anal Chem, 2011, 83(3): 634-642.

[196] Hoque M E, Khosravi K, Newman K, Metcalfe C D. Detection and characterization of silver nanoparticles in aqueous matrices using asymmetric-flow field flow fractionation with inductively coupled plasma mass spectrometry. J Chromatogr A, 2012, 1233: 109-115.

[197] Poda A R, Bednar A J, Kennedy A J, Harmon A, Hull M, Mitrano D M, Ranville J F, Steevens J. Characterization of silver nanoparticles using flow-field flow fractionation interfaced to inductively coupled plasma mass spectrometry. J Chromatogr A, 2011, 1218(27): 4219-4225.

[198] Tan Z Q, Liu J F, Guo X R, Yin Y G, Byeon S K, Moon M H, Jiang G B. Toward full spectrum speciation of silver nanoparticles and ionic silver by on-line coupling of hollow fiber flow field-flow fractionation and minicolumn concentration with multiple detectors. Anal Chem, 2015, 87(16): 8441-8447.

[199] Schmidt B, Loeschner K, Hadrup N, Mortensen A, Sloth J J, Koch C B, Larsen E H, Quantitative characterization of gold nanoparticles by field-flow fractionation coupled online with light scattering detection and inductively coupled plasma mass spectrometry. Anal Chem, 2011, 83(7): 2461-2468.

[200] Jochem A R, Ankah G N, Meyer L A, Elsenberg S, Johann C, Kraus T. Colloidal mechanisms of gold Nanoparticle loss in asymmetric flow field-flow fractionation. Anal Chem, 2016, 88(20): 10065-10073.

[201] Contado C, Pagnoni A. TiO_2 in commercial sunscreen lotion: Flow field-flow fractionation and ICP-AES together for size analysis. Anal Chem, 2008, 80(19): 7594-7608.

[202] Chun J, Fagan J A, Hobbie E K, Bauer B J. Size separation of single-wall carbon nanotubes by flow-field flow fractionation. Anal Chem, 2008, 80(7): 2514-2523.

[203] Mudalige T K, Qu H, Sanchez-Pomales G, Sisco P N, Linder S W. Simple functionalization strategies for enhancing nanoparticle separation and recovery with asymmetric flow field flow fractionation. Anal Chem, 2015, 87(3): 1764-1772.

[204] Lee S H, Salunke B K, Kim B S. Sucrose density gradient centrifugation separation of gold and silver nanoparticles synthesized using Magnolia kobus plant leaf extracts. Biotechnol Bioproc E, 2014, 19(1): 169-174.

[205] Bai L, Ma X, Liu J, Sun X, Zhao D, Evans D G. Rapid separation and purification of nanoparticles in organic density gradients. J Am Chem Soc, 2010, 132(7): 2333-2337.

[206] Ma X, Kuang Y, Bai L, Chang Z, Wang F, Sun X, Evans D G. Experimental and mathematical modeling studies of the separation of zinc blende and wurtzite phases of CdS nanorods by density gradient ultracentrifugation. ACS Nano, 2011, 5(4): 3242-3249.

[207] Johnson M E, Montoro Bustos A R, Winchester M R. Practical utilization of spICP-MS to study sucrose density gradient centrifugation for the separation of nanoparticles. Anal Bioanal Chem, 2016, 408(27): 7629-7640.

[208] Paleologos E K, Giokas D L, Karayannis M I. Micelle-mediated separation and cloud-point extraction. Trac-Trend Anal Chem, 2005, 24(5): 426-436.

[209] Akbulut O, Mace C R, Martinez R V, Kumar A A, Nie Z, Patton M R, Whitesides G M. Separation of nanoparticles in aqueous multiphase systems through centrifugation. Nano Lett, 2012, 12(8): 4060-4064.

[210] Liang H W, Liu S, Gong J Y, Wang S B, Wang L, Yu S H. Ultrathin Te nanowires: An excellent platform for controlled synthesis of ultrathin platinum and palladium nanowires/nanotubes with very high aspect ratio. Adv Mater, 2009, 21(18), 1850-1854.

[211] Franklin N M, Rogers N J, Apte S C, Batley G E, Gadd G E, Casey P S. Comparative toxicity of nanoparticulate ZnO, bulk ZnO, and $ZnCl_2$ to a freshwater microalga (*Pseudokirchneriella subcapitata*): The importance of particle solubility. Environ Sci Technol, 2007, 41(24): 8484-8490.

[212] Liu X, Chen B, Cai Y, He M, Hu B. Size-based analysis of AuNPs by online monolithic capillary microextraction-ICPMS. Anal Chem, 2017, 89(1): 560-564

[213] Xu X, Caswell K K, Tucker E, Kabisatpathy S, Brodhacker K L, Scrivens W A. Size and shape separation of gold nanoparticles with preparative gel electrophoresis. J Chromatogr A, 2007, 1167(1): 35-41.

[214] Helfrich A, Bruchert W, Bettmer J. Size characterisation of Au nanoparticles by ICP-MS coupling techniques. J Anal Atom

Spectrom, 2006, 21(4): 431-434.

[215] Liu F K, Ko F H, Huang P W, Wu C H, Chu T C. Studying the size/shape separation and optical properties of silver nanoparticles by capillary electrophoresis. J Chromatogr A, 2005, 1062(1): 139-145.

[216] Liu F K, Tsai M H, Hsu Y C, Chu T C. Analytical separation of Au/Ag core/shell nanoparticles by capillary electrophoresis. J Chromatogr A, 2006, 1133(1-2): 340-346.

[217] Liu K H, Chu T C, Liu F K. On-line enhancement and separation of nanoparticles using capillary electrophoresis. J Chromatogr A, 2007, 1161(1-2): 314-321.

[218] Franze B, Engelhard C. Fast separation, characterization, and speciation of gold and silver nanoparticles and their ionic counterparts with micellar electrokinetic chromatography coupled to ICP-MS. Anal Chem, 2014, 86(12): 5713-5720.

[219] Liu L, He B, Liu Q, Yun Z, Yan X, Long Y, Jiang G. Identification and accurate size characterization of nanoparticles in complex media. Angew Chem Int Ed, 2014, 53(52): 14476-14479.

[220] Qu H, Linder S W, Mudalige T K. Surface coating and matrix effect on the electrophoretic mobility of gold nanoparticles: A capillary electrophoresis-inductively coupled plasma mass spectrometry study. Anal Bioanal Chem, 2017, 409(4): 979-988.

[221] Caldwell K D, Giddings J C, Myers M N, Kesner L F. Electrical field-flow fractionation of proteins. Science, 1972, 176(4032): 296-298.

[222] Somchue W, Siripinyanond A, Gale B K. Electrical field-flow fractionation for metal nanoparticle characterization. Anal Chem, 2012, 84(11): 4993-4998.

[223] Tasci T O, Johnson W P, Fernandez D P, Manangon E, Gale B K. Circuit modification in electrical field flow fractionation systems generating higher resolution separation of nanoparticles. J chromatogr A, 2014, 1365: 164-172.

作者：苑春刚[1]，周小霞[2]，刘景富[2]，胡　斌[3]

[1]华北电力大学（保定）环境科学与工程系，[2]中国科学院生态环境研究中心环境化学与生态毒理学国家重点实验室，[3]武汉大学化学与分子科学学院

第 2 章　短链氯化石蜡的检测、污染特征与暴露评估研究进展

本章导读

短链氯化石蜡（short-chain chlorinated paraffins，SCCPs）因具有环境持久性、远距离迁移性及生物累积性和毒性，被认为是一种全球性的持久性有机污染物。2017 年 SCCPs 被列入《斯德哥尔摩公约》（附件 A）的名单。我国是氯化石蜡的生产大国，摸清我国 SCCPs 生产和污染状况，建立可靠的分析方法，对于我国制定针对 SCCPs 的减排措施及开展 SCCPs 的生态及人类健康风险评估具有十分重要的意义。本章综述我国氯化石蜡的生产及其对周边环境的影响，介绍最新发展的 SCCPs 分析方法，归纳 SCCPs 在大气、土壤和沉积物中的赋存水平，阐述 SCCPs 在食物链中的富集规律，并对 SCCPs 人体暴露水平做了初步评估。

关键词

短链氯化石蜡，分析方法，斯德哥尔摩公约，人体暴露，生物蓄积

1 引　　言

持久性有机污染物（persistent organic pollutants，POPs）在环境中具有持久性和长距离迁移性、生物累积性，并随食物链浓缩和放大，对生物及人体健康构成威胁。POPs 导致的环境污染已成为环境科学领域最受关注和最为重要问题之一。

为了推动 POPs 的削减和控制，减少其对环境所造成的危害，2001 年 5 月 22 日包括我国在内的 152 个国家和地区共同签署了《关于持久性有机污染物的斯德哥尔摩公约》（以下简称《斯德哥尔摩公约》），目前签约的国家和地区已达 181 个。根据《斯德哥尔摩公约》相关条款，公约管控的 POPs 名单是开放的，自公约公布首批优先控制的 12 种 POPs 之后，新增列 POPs 已达 16 种。2007 年欧盟及其成员国提议将短链氯化石蜡（SCCPs）列入 POPs 公约新增候选 POPs 名单（UNEP/POPS/COP.3/12），2017 年 5 月《斯德哥尔摩公约》第八次缔约方大会通过将 SCCPs 列入公约附件 A 受控名单。SCCPs 的污染防控成为了全球性环境保护的重要议题。

我国是氯化石蜡（CPs）生产大国，年产量 100 万吨左右。我国的 CPs 产品中 SCCPs 的浓度水平差异很大。鉴于 CPs 在我国的巨大累积生产量，我国由于生产和广泛使用 CPs 所引起的 SCCPs 环境污染和人体健康风险问题要比其他国家更为严峻。因此开展 SCCPs 的污染特征及其导致的健康风险评估研究十分迫切。近几年，我国针对 SCCPs 的相关研究明显增多，在 SCCPs 分析方法和多种环境介质的污染水平方面取得了一定进展。然而，SCCPs 是一类组成和结构十分复杂的混合物，复杂环境介质中 SCCPs 的分离与定量分析十分困难，来自 CPs 同类物和相似 POPs 的定量干扰仍是目前 SCCPs 定量分析方法面临的巨大挑战。在我国开展 SCCPs 的污染防控不仅是保护我国生态环境和人体健康，同时也是作为一个负责任大国所应当履行的国际义务。摸清我国 SCCPs 生产、污染状况，建立可靠的分析方法，评估我国人群 SCCPs 暴露水平，对我国制定针对 SCCPs 的减排措施具有十分重要的意义。

2 氯化石蜡的生产及其对周边环境的影响

2.1 氯化石蜡的生产

氯化石蜡自 20 世纪 30 年代开始生产，1964 全世界的 CPs 的总产量不足 5 万吨。在 20 世纪 70 年代，

由于环境保护规定的逐步实施，世界上多数国家禁止多氯联苯（PCBs）的生产和使用，CPs 作为替代产品被广泛的生产和使用。1977 年 CPs 产量约 23 万吨，到 20 世纪 90 年代末，CPs 的全球产量达到 50 万吨。1998 年，美国生产 SCCPs、中链氯化石蜡（MCCPs）和长链氯化石蜡（LCCPs）的产量分别为 0.79 万吨、1.78 万吨和 1.27 万吨。近年来，由于美国、加拿大和欧盟等国家陆续出台法令限制 SCCPs 的生产和使用，SCCPs 的产量随之降低，MCCPs 的产量相对提高[1,2]。

CPs 是一种优良的化工产品，广泛用于生产电缆料、地板料、软管、人造革、橡胶等制品以及作为添加剂应用于涂料、润滑油等产品，在我国有较大的需求及生产量。不同碳链长度的 CPs 具体用途又可以细分如下：SCCPs 主要用作金属加工液中的高压添加剂，MCCPs 主要用作 PVC 塑料的增塑剂，LCCPs 主要用作橡胶和纺织品的阻燃剂。我国从 20 世纪 50 年代末开始生产 CPs，1963 年的总产量仅 1859 吨，1980 年达到 1.8 万吨。近年来，随着国内塑料制品工业的迅速发展，邻苯二甲酸二辛酯、邻苯二甲酸二丁酯等主增塑剂的供应满足不了工业需要，而作为替代品的 CPs 原料易得，主要以石油分馏后产生的直链蜡油以及氯碱工业的副产品氯气为原料生产，生产投资成本低，因此我国 CPs 生产发展迅速。到 2015 年，我国 CPs 产量已达 100 万吨，是目前世界上最大的 CPs 生产国和主要出口国[3,4]。CPs 产品广泛存在于我们的日常生活中，市售 CPs 产品就多达 200 种以上。目前我国 CPs 生产厂超过 150 家，主要分布在河北、河南、山东、江苏、浙江、上海、辽宁、黑龙江和广东、福建等省。生产的 CPs 的工业产品主要有 CP-42（氯含量 42%±2%）、CP-52（氯含量 52%±2%）和 CP-70（氯含量 70%±2%），其中 CP-42 和 CP-52 占 CPs 总产量的 80%以上[5,6]。CP-42 和 CP-52 是无色或淡黄色液体，可以部分代替增塑剂，降低产品成本，并使制品具有较好的阻燃性以及更好的相溶性，被广泛用于电缆、塑料、地板和薄膜等制造领域。CP-70 是白色或浅黄色固体粉末，主要用作阻燃剂，常添加于聚氯乙烯及聚苯乙烯等高分子聚合材料中。

由于生产 CPs 的技术门槛低，投资成本少，目前我国 CPs 的生产以中小企业为主，这些中小企业技术力量薄弱，大多沿用 20 世纪五六十年代的生产工艺，氯化时间长、能耗大。这就形成了目前国内 CPs 生产厂家多而杂，产品质量参差不齐的局面。环保部通过对河南、江浙、福建、东北和山东等国内 CPs 主要产区的 19 家 CPs 生产企业调研及样品采集，对现阶段国内 CPs 的生产状况及产品中 SCCPs 的分布情况有了初步认识和了解。我国 CPs 生产相对集中地在河南省荥阳市，该市有 CPs 生产厂近 30 家，生产能力从 4000t/a 到 20000t/a 不等，保守估计荥阳市 CPs 年产能为 25~30 万吨。荥阳市最早生产 CPs 的厂家是华夏化工厂，使用工艺是热氯化与光催化技术相结合。江浙、山东及福建地区是国内经济较发达地区，制造加工企业较多，CPs 作为阻燃剂、润滑剂在当地需求较大，一些 CPs 生产企业在当地应运而生。这些地区生产 CPs 的厂家也是以中小企业为主，分布相对比较分散，原材料价格和人力成本都比荥阳地区高，因此这些发达地区氯化石蜡行业利润低，部分原来生产 CPs 的大型企业已经转产，如杭州电化集团现已停止 CPs 的生产。因为 CPs 产品本身的利润较低（<500 元/吨），作为副产品的盐酸（30%浓度）是利润组成中的重要部分，所以相关的氯化氢吸收技术各厂都做得比较好。CPs 生产中虽然使用氯气，但只要将氯气输送管道管理好，不会造成氯气泄漏，也不会对环境造成直接污染。从实地考察的结果看，CPs 生产厂区周围并没有很严重的异味。

我国 CPs 生产存在以下主要问题：

（1）作坊式生产。大部分 CPs 生产企业没有原材料与产品分析等相关技术支持，产品的检测主要依靠测比重、目视颜色这两个指标判断是否合格。如果客户对颜色要求较高（要求无色），工厂在产品氯化后一般会用荧光剂处理。CP-52 正常颜色是淡黄色，用荧光剂处理后为无色透明液体。

（2）企业产品单一。很多企业都主要生产 CP-52 一种产品。一方面是由于现在市场上对 CP-52 需求最高，另一方面是企业的技术力量不足，尚未掌握生产 CP-70 需要深度氯化的技术工艺。

（3）技术工艺路线落后。目前生产 CPs 的企业大多使用热氯化技术工艺生产 CP-52。有少部分企业使用光催化结合的技术工艺，但由于反应釜中的内置光源容易损坏，更换光源不及时，多数时间还是用热氯化法生产。热氯化对生产设备要求低，并且工艺比较成熟，能较稳定地进行生产，但单纯的热氯化反应时间较长，能耗高，严重影响了生产效率与效益。

（4）对 SCCPs 可能造成的环境污染风险认识不足。由于国内需求 CP-52 的厂家对产品中 SCCPs 含量没有要求，相关管理部门也没有做出相应规定，导致目前生产企业并不关注 CPs 产品中 SCCPs 的含量。多数企业使用含有短碳链烷烃的石蜡作为原料进行氯化，所以产品中普遍含有大量(30%~80%)的 SCCPs。

总之，目前国内绝大多数 CPs 生产厂家并没有控制产品中 SCCPs 的含量，CPs 产品的使用对环境造成污染的风险很大。另外，多数企业现使用的氯化工艺落后，反应时间长、能耗高，不能适应当前国家节能减排的需要。针对现生产技术工艺存在的不足，相关研究首先要建立一套可靠的 CPs 产品中 SCCPs 的定量分析方法，再结合现有工艺开展低 SCCPs 含量（SCCPs 的含量低于 1%）的 CPs 生产新技术工艺研究与新技术的推广应用。

2.2　氯化石蜡生产厂周边 SCCPs 和 MCCPs 的释放及分布特征

国家环境分析测试中心研究了 CPs 生产企业和周边空气中 SCCPs 赋存状况。春、夏、冬三季 CPs 生产厂上、下风向 1km 以外点位空气中 SCCPs 水平差异不明显，均比 CPs 生产厂内低一个数量级。厂区 1km 以外点位冬夏季监测结果差异不大，表明 CPs 生产在周边空气迁移范围在 4 km^2 以内，大气扩散并不明显。该研究者还研究了 SCCPs 在空气中气-粒组成和迁移特征，发现温度是影响空气中 SCCPs 迁移的重要因素之一。冬季颗粒物中以 $C_{11}H_{17}Cl_7$ 和 $C_{10}H_{15}Cl_7$ 丰度最高，其次是 $C_{11}H_{16}Cl_8$、$C_{11}H_{18}Cl_6$、$C_{10}H_{16}C_{16}$；而夏季颗粒物中 $C_{11}H_{16}Cl_8$ 丰度最高，$C_{10}H_{16}Cl_6$ 比例下降，$C_{10}H_{16}C_{16}$ 比例有所上升。也就是说夏季温度较高，导致高分子量的 SCCPs 同系物挥发性增强。气相中冬夏两季 SCCPs 分布差异不明显，都以 $C_{10}H_{16}C_{16}$ 丰度最高，其次是 $C_{10}H_{15}Cl_7$、$C_{11}H_{18}Cl_6$、$C_{11}H_{17}Cl_7$ 以及部分 $C_{10}H_{17}Cl_5$、$C_{10}H_{14}Cl_8$、$C_{11}H_{16}Cl_8$。换言之，气相中 C_{10} 和 C_{11} 组分 SCCPs 较多，占总 SCCPs 的比例高达 75%以上，其中又以 Cl 数为 6~7 的同类物居多。对厂区周边土壤检测发现，SCCPs 在土壤与空气中的分布特征一致，表明区域土壤中 SCCPs 是主要来源于大气沉降，CPs 生产未对厂区 1km 以外上、下风向采样点造成显著影响。随土壤深度增加，SCCPs 浓度逐渐降低，20~25 cm SCCPs 浓度比第一层（0~5 cm）SCCPs 减少了 92%，表明 SCCPs 土壤迁移能力不强。

Xu 等[7]为了考查 CPs 生产厂区周边 SCCPs 和 MCCPs 的污染情况，在 CPs 生产厂周边 7km 范围内采集松针和土壤样品，并采用 HRGC-ECNI-LRMS 方法分析样品中 SCCPs 和 MCCPs 含量。结果表明，土壤中∑SCCPs 浓度为 24.8~1842 ng/g，氯含量为 60.5%~66.6%，∑MCCPs 浓度为 19.3~2074.8 ng/g，氯含量为 54.5~58.4%；松针中∑SCCPs 的浓度为 218.7~2197 ng/g，氯含量为 60.0%~65.2%，∑MCCPs 的浓度为 337.8~4388 ng/g，氯含量为 50.5%~57.1%。土壤和松针中∑SCCPs 与∑MCCPs 整体上均随着离厂区距离的增加，浓度呈减小趋势。在土壤中，SCCPs 和 MCCPs 浓度水平具有很好的相关性。而在松针中，远离厂区而近居民区的采样点∑MCCPs 的浓度升高，推测可能是由于当地居民生活中使用的商品中含有以 MCCPs 为主的 CPs 添加剂，其在长期使用过程中作为污染源经挥发排放到大气中而被松针捕集。土壤和松针中 SCCPs 和 MCCPs 的同类物分布与采样点离厂区的距离有关。就碳链长度而言，所有样品均以 C_{10}、C_{11} 和 C_{14} 为主，但是对于 MCCPs，厂区内样品的 C_{15}，C_{16} 和 C_{17} 的相对丰度明显高于厂区外的样品。就氯取代数而言，厂区内的土壤样品以 Cl_8 和 Cl_9 为主，松针以 Cl_7 和 Cl_8 同类物为主，而在厂区外，所有的样品均以 Cl_6、Cl_7 同类物为主。这一现象表明，氯取代数越少，链长越短的 CPs 更容易从厂区向周围扩散（分馏效应），主要受 SCCPs 和 MCCPs 的 K_{OA} 特性影响。土壤中 CPs 的浓度随着离厂区距离的增加呈现 Boltzmann 分布，这一现象表明厂区中挥发出来的 CPs 附着在空气中的颗粒相中，随着重力的作用大部分迅速沉降，只有少部分随着气-粒分配作用向更远的距离传播。而松针中 CPs 的浓度随厂区距离的变化则呈 Gaussian 分布。

3 SCCPs 的分析方法

CPs 在生产过程中氯化位点选择特异性低，氯化比例不固定，正构烷烃氯化后产生大量氯化同系物、对映及非对映异构体，这些极性差异很小的同类物无法在分析色谱柱上实现完全分离，这给 SCCPs 或 MCCPs 的准确定量带来极大困难。近年来研究人员尝试多种方法对 SCCPs 进行分析检测，并取得了很大进展。目前普遍采用气相色谱（GC）方法对 SCCPs 进行分离，但由于 SCCPs 成分复杂，无法达到各组分完全分离，大量的共流出物常使色谱峰呈现驼峰状，对其准确定量分析产生很大影响。

为了解决大量共流出物对定量分析的困扰，研究人员尝试对常规 GC 方法进行改进。Coelhan[8]采用短柱气相色谱方法，使 SCCPs 作为一个峰流出色谱柱。此方法分析速度快，灵敏度较高，但无法避免 MCCPs 的干扰，且对环境样品的净化处理要求苛刻。Korytár 等[9-12]采用全二维气相色谱（GC×GC）方法分离 CPs，通过改变色谱柱的极性优化分离效果。该方法分析组分单一的 CPs 样品时分离效果较好，而对于复杂样品，相邻色谱峰边界出现明显重叠，难以得到有效分离。尽管目前多数研究者用气相色谱法对 CPs 进行分离，也有文献报道采用高效液相色谱（HPLC）分析 CPs[13]，但是由于液相色谱柱分离能力有限，难以将 SCCPs 与 MCCPs 和其他干扰物质分离，因此应用较少。

SCCPs 最常用检测手段是配备电子捕获负化学源（ECNI）的高/低分辨质谱（HRMS/LRMS）。经过近 15 年的研究，这一方法已日趋完善。在 ECNI 电离模式下，CPs 在质谱中主要产生$[M\text{-}Cl]^-$、$[M\text{-}HCl]^-$、$[M+Cl]^-$以及$[Cl_2]^-$和$[HCl_2]^-$离子[14]，该方法可以获得 SCCPs 同系物组分分布及氯含量信息。其中 HRMS 检测灵敏度高，但无法避免其他有机氯化合物如氯丹、毒杀芬和 PCBs 等对 SCCPs 的干扰，且设备昂贵。LRMS 检测成本相对较低，但其缺点在于分辨率较低，分子量相近的同系物组分彼此存在干扰，如 SCCPs 中的 $C_{10}H_{14}Cl_8$ 和 MCCPs 中的 $C_{15}H_{26}Cl_6$ 的相互干扰难以克服，因此需要选择干扰尽可能小的碎片离子进行定量分析，并且严格控制保留时间和同位素丰度比，避免定量结果偏高。另一方面，采用 LMRS 分析环境样品时，其他有机氯化合物的干扰较 HRMS 更为严重，对前处理要求严格。但无论是 HRMS 还是 LRMS 都无法避免的问题就是在 ECNI 检测模式下，SCCPs 的响应因子依赖氯原子的数量和其在碳链上的位置，有些低氯代组分因响应弱时常检测不到。当采用不同组成的 SCCPs 混合物做定量标样时会对分析结果造成相当大的偏离。为了减小氯原子个数及取代位置的影响，Zencak 等[15]采用甲烷/二氯甲烷作为反应气，该方法在 ECNI 模式下不同氯含量的 SCCPs 组分具有相似的响应因子，可检测到低氯代（3~5 个氯原子）组分。但在电离过程中容易形成炭黑残留物，导致离子源信号快速衰减，因此不适于作为常规分析方法。此外，还有离子阱质谱、三重四级杆串联质谱等方法，可以获得 CPs 的总浓度，但不能消除 SCCPs 和 MCCPs 的互相干扰。

3.1 SCCPs 的样品前处理方法

CPs 的样品前处理过程主要包括提取和净化。总体来说，CPs 的提取方法和其他半挥发性有机氯化物相似。对于固体样品，最常用的是索氏提取法，常用的提取溶剂有正己烷、二氯甲烷、甲苯、二氯甲烷/正己烷或正己烷/丙酮的混合溶液，索氏提取的缺点是费时且需耗费大量的溶剂。因此正在逐渐被一些新的提取技术所取代，如加速溶剂提取（accelerated solvent extraction，ASE）、超声辅助提取（ultrasonic assisted extraction，UAE）和微波辅助提取（microwave assisted extraction，MAE）等。对于液体样品如水样等，提取方法有液液萃取法（liquid-liquid extraction，LLE）和固相萃取法（solid phase extraction，SPE），液液萃取适用于水中有机污染物的富集，而固相萃取法则被广泛应用于萃取环境和生物样品中的有机污染物。

对于生物样品，去除脂肪和大分子物质是净化过程的关键。主要方法包括凝胶渗透色谱法（gel permeation chromatography，GPC）和浓硫酸酸化法等。GPC 主要基于体积排阻的分离机理，通过具有分子筛性质的固定相，去除对 SCCPs 测定有干扰的大分子和小分子物质。相比其他 POPs，由于 CPs 的 $\log K_{ow}$ 范围很宽，不同同类物的物化性质差别较大，使得 CPs 的净化过程较为复杂。许多卤代有机污染物（比如氯代芳烃，毒杀芬、部分 PCBs 同类物等）与 CPs 同类物具有相同的色谱保留时间范围和质量数相近的质谱特征离子，因此会对 CPs 的定量分析造成较大干扰。因此如何最大限度地去除 CPs 分析中存在的干扰化合物是 CPs 分析中的关键步骤。CPs 的净化通常通过柱吸附色谱法进行，常用的净化柱包括硅胶层析柱、氧化铝柱和弗罗里硅土柱等[16,3,17-19]。不同环境介质和不同实验室的净化方法有所不同，在诸多净化方法中，硅胶层析柱法是 CPs 分析最常用的净化分离方法。

3.2　SCCPs 的检测方法

1）气相色谱-电子捕获检测器

电子捕获检测器（ECD），以其价格低廉、普遍性以及对氯化物的高灵敏度等特点成为检测工业品如金属加工削切油中 CPs 总量的常规检测器。史慧娟等[20]采用 GC-ECD 对油漆中的 SCCPs 进行检测，该方法检出限为 0.4mg/kg，定量限是 1.0mg/kg。马贺伟[21]对皮革中的 SCCPs 进行测定，结果表明，SCCPs 的共流出色谱峰保留时间跨度大，仪器响应值表现出对氯含量的依赖。MCCPs 的存在对 SCCPs 的分析过程干扰严重。由于实际皮革基质非常复杂，导致样品中 SCCPs 的定性和定量困难。袁丽凤等[22]对橡胶制品中的 SCCPs 进行检测，测定下限为 20mg/kg，与其峰面积呈线性关系。ECD 的灵敏度往往是质谱（MS）的 1~2 倍，但其选择性远低于 MS，因此不能用于分析复杂环境介质或生物样品中痕量浓度水平的 SCCPs。

2）气相色谱-电子轰击串联质谱

电子轰击（electron impact，EI）离子源作为质谱最常用的电离方式，广泛应用于环境中半挥发性 POPs 的检测。但由于 EI 轰击能量过高，CPs 在电离过程中发生多种断链反应，产生大量离子碎片，缺乏特征性，难以分辨，故通常不用于 SCCPs 的检测[23]。但是 EI-MS 单独或与 ECNI-MS 结合使用，也可用于检测环境样品中的 CPs。Iozza 等[24 将 EI-MS 与 ECNI-LRMS 的检测结果相结合，通过对瑞士境内的 Thun 湖底柱状沉积物的检测，研究了近十年来 CPs 的污染变化趋势。和 ECNI-MS 相比，CPs 在 EI-MS 中的响应不受氯取代质量分数的影响。

3）气相色谱-电子捕获负化学源低分辨/高分辨质谱质谱

气相色谱-电子捕获负化学源是目前测定 SCCPs 和 MCCPs 最常用的方法，被广泛应用于环境介质中 CPs 的检测（表 2-1）。Gao 等[19]通过实验室合成的 SCCPs 标准物质和 SCCPs 标准溶液，筛选 SCCPs 在 GC-ECNI-LRMS 上的定量、定性离子，优化离子源温度等参数，建立了 SCCPs 的定量分析方法。该方法离子源温度为 150℃，基于 SIM 模式检测 SCCPs 各同系物最高丰度离子碎片[M-HCl]$^-$作为定量离子。通过控制保留时间、比较样品与标准溶液中 SCCPs 的色谱峰形状及同位素丰度比等多个方面保证结果的有效性。采用 Reth 等人提出的方法定量计算样品中 SCCPs 的浓度。首先对 5 个氯含量不同的 SCCPs 标准贮备液进行分析，计算各贮备液中 SCCPs 总响应因子和氯含量，并对二者进行线性回归分析，获得回归方程，然后用此方程定量计算待测样品中 SCCPs 的含量。SCCPs 总响应因子与氯含量呈显著线性相关。

表 2-1 环境基质中 CPs 的不同定量分析方法

样品类型	CPs	提取方法	净化方法	仪器分析方法	色谱柱	定量标准品
大气[34]	SCCPs 和 MCCPs	索氏提取	I. Al_2O_3 & SiO_2；II. Al_2O_3，SiO_2：H_2SO_4，& Florisil	GC-ECNI-LRMS	DB-5MS（30×0.25×0.25）	SCCPs（51.5%，55.5% 和 63%），MCCPs（42%，52% 和 57%）
大气[35]	SCCPs，MCCPs 和 LCCPs	索氏提取	I. H_2SO_4；II. SiO_2：H_2SO_4；III. Florisil	APCI-qTOF-HRMS & GC-ECNI-HRMS	RTX-5MS（15×0.25×0.25）	SCCPs（51.5%，55.5% 和 63%），MCCPs（42%，52% 和 57%）LCCPs（40%，49% 和 70%）
灰尘[36]	SCCPs 和 MCCPs	UAE	SiO_2：H_2SO_4（40% *w/w*），SiO_2：NaOH（10%）& SiO_2	GC-ECNI-LRMS	ZB-5MS（30×0.25×0.25）	SCCPs（51.5%，55.5% 和 63%），MCCPs（42%，52% 和 57%）
污泥[5]	SCCPs	ASE	多层：弗罗里硅土，SiO_2 & SiO_2：H_2SO_4（30% *w/w*）	GC-ECNI-LRMS	DB-5MS（30×0.25×0.25）	SCCPs（51.5%，55.5% 和 63%）
土壤[3]	SCCPs	索氏提取	多层：弗罗里硅土，SiO_2 & SiO_2：H_2SO_4（44% *w/w*）	GC-ECNI-LRMS	DB-5MS（30×0.25×0.25）	SCCPs（51.5%，55.5% 和 63%），MCCPs（42%，52% 和 57%）
沉积物[37,38]	SCCPs	ASE	多层：弗罗里硅土，SiO_2 & SiO_2：H_2SO_4（44% *w/w*）	GC-ECNI-QQQMS	DB-5MS（30×0.25×0.25）	SCCPs（51.5%，55.5% 和 63%）
沉积物[4,39]	SCCPs 和 MCCPs	索氏提取	多层：弗罗里硅土，SiO_2 & SiO_2：H_2SO_4（44% *w/w*）	GC-ECNI-QQQMS	DB-5HT（15×0.25×0.25）	SCCPs（63%）和 MCCPs（57%）
贝类[40]	SCCPs	ASE	Al_2O_3，SiO_2：NaOH，SiO_2 & SiO_2：H_2SO_4（30% *w/w*）	GC-ECNI-QQQMS	DB-5MS（15×0.25×0.25）	SCCPs（51.5%，55.5% 和 63%）
鲨鱼肝脏[41]	SCCPs	CSE	I. H_2SO_4；II. SiO_2：H_2SO_4（30% *w/w*）	GC-ECNI-HRMS	DB-5MS（30×0.25×0.25）	SCCPs（60%）
鱼体[42]	SCCPs	索氏提取	活性弗罗里硅土	GC-ECNI-LRMS	DB-5MS（15×0.25×0.25）	SCCPs（51.5%，55.5% 和 63%）
鱼体[43]	SCCPs 和 MCCPs	索氏提取	I. GPC II. SiO_2：H_2SO_4（22% *w/w*）III. SiO_2	GC-ECNI-HRMS	DB-5MS（30×0.25×0.25）	SCCPs（51.5%，55.5% 和 63%），MCCPs（42%，52% 和 57%）
食物[44]	SCCPs	索氏提取	活性弗罗里硅土	GC-ECNI-HRMS &GC-HRTOF-MS	DB-5MS（15×0.25×0.25）	SCCPs 单标
食物[45]	SCCPs	CSE	活性弗罗里硅土	GC-ECNI-HRMS	DB-5MS（15×0.25×0.25）	SCCPs 单标
母乳[46]	SCCPs 和 MCCPs	CSE	I. H_2SO_4；II. SiO_2：H_2SO_4（40%），SiO_2：NaOH & SiO_2	GC-ECNI-LRMS	DB-5MS（15×0.25×0.25）	SCCPs（63%），MCCPs（42%，52%和 57%）
树皮&松针[47]	SCCPs	ASE	I. SiO_2：H_2SO_4（40%）II. 活性弗罗里硅土，SiO_2 & SiO_2：H_2SO_4（30% *w/w*）	GC-ECNI-QQQMS	DB-5MS（30×0.25×0.25）	SCCPs（51.5%，55.5% 和 63%）

计算样品中 SCCPs 同系物的色谱峰面积，并折算成氯含量，代入此线性回归曲线计算出样品中 SCCPs 的总响应因子，然后计算出 SCCPs 的浓度。采用 ECNI–LRMS 分析环境样品中的 SCCPs 和 MCCPs，由于个别同系物分子量相近，且在气相色谱上保留时间部分重叠，因此会彼此干扰从而影响测定结果。Zeng 等提出二元一次方程组方法，通过数学计算减少 SCCPs 和 MCCPs 同系物之间的干扰，该方法通过计算

CPs 干扰组分检测离子碎片的同位素丰度，将干扰组分的真实响应信号以未知量的形式代入方程组中，通过求解获得其真实响应的信号值。该方法有效减少 MCCPs 对 SCCPs 的干扰，灵敏度高，重现性好，仪器检出限（LOD）为 50μg/L[25]。

4）脱氯加氢或氘气相色谱-质谱法

Gao[26]研发了在线催化加氢装置，通过催化剂的作用，将 SCCPs 加氢还原为相应碳链长度的烷烃，并与气相色谱检测技术联用，以氢火焰离子化检测器（FID）分析测定生成的烷烃含量，结合氯含量信息，计算出 SCCPs 的质量浓度。该装置具有如下特点：①高转化率。在线催化加氢装置通过将 SCCPs 转化为相应碳链长度的烷烃，并通过对烷烃的检测进而计算样品中 SCCPs 的含量。因此在转化过程中必须保证所有的 SCCPs 尽可能完全转化成相应的烷烃，要求方法具有高的转化率。通过对催化剂种类、装填量、反应温度、压力、样品中 SCCPs 的浓度范围等条件的考察，对装置的参数进行优化，确定催化剂为 Pd 催化剂，装填量为 3g，反应温度为 280℃，氢气压力为 10psi。在此条件下，SCCPs 的转化率均达到 90% 以上。②高灵敏度。仪器对碳链长度在 10~13 的烷烃的检出限为 16~19μg/L，对 SCCPs 的绝对检出限可达 pg 级，对于 CPs 样品中 SCCPs 的最低检测浓度为 80mg/kg。③高稳定性。该方法相对标准偏差为 3.8%。

基于在线催化加氢和气相色谱的联用，进行在线分析和检测，建立了 SCCPs 的分析方法。该方法以结构与 SCCPs 相似的氯代 2-甲基-十一烷作为进样内标，采用内标法进行定量分析，定量结果准确度高且前处理步骤和计算方法简便，适合常规实验室分析，尤其适合作为检测商品 CPs 中 SCCPs 含量的实验室。该方法避免了质谱法等直接分析 SCCPs 同系物和同分异构体所带来的定量误差和商品 CPs 样品中大量存在的 MCCP 和 LCCPs 对 SCCPs 分析的干扰。

Gao[26]还提出采用氘代还原剂 $LiAlD_4$ 对 SCCPs 进行脱氯加氘，在氯取代的位置将其还原成相应取代位置的氘代烷烃，建立离线加氘还原法。通过气相色谱-质谱联用分析生成的氘代烷烃，进而达到对 SCCPs 质量浓度和同系物分布同时检测的目的，并且避免了直接分析 SCCPs 所面临的分离定量方面的困扰，实现精确定量。针对环境和生物样品，采用具有支链结构的氯代 2–甲基–壬烷和氯代 2–甲基–十一烷作为提取和反应内标，建立 SCCPs 的内标定量方法。该方法分析 SCCPs 标准溶液总浓度比值的最大误差<1.26，相对标准偏差≤10%。采用该方法分析环境和生物样品，可以检测到低氯代组分（$Cl_{1\sim4}$），实测样品中低氯代组分的质量分数占总浓度的 32.4%~62.4%。

5）全二维气相色谱-飞行时间质谱

目前，CPs 分析最广泛使用的分析方法是一维气相色谱-电子捕获负化学源-低分辨质谱法[27,28,29]，典型色谱图如图 2-1 所示。由于 SCCPs 和 MCCPs 的同类物具有相同的色谱保留时间，并且不同同类物之间由于存在质谱 *m/z* 重叠而相互干扰，使得环境样品中 CPs 的总浓度及同类物浓度可能被高估。使用全二维气相色谱（GC×GC）可以有效地实现不同 CPs 的色谱分离。GC×GC 通过线性程序升温，以及通过两根不同色谱柱间固定相极性的改变，可实现待测物质的正交分离，从而显著提高结构相似化合物的分离度。

Xia 等[30]等通过对色谱和质谱各项参数条件的选择与优化，选择的色谱柱组合方式为 DB-5MS×BPX-50，获得工业混合 SCCPs（氯含量 51.5%）和 MCCPs（氯含量 52%）标准品的 GC×GC-HRTOF-MS 总离子流出色谱图，如图 2-2 所示。与一维气相色谱相比，GC×GC 显著提高了 CPs 同类物之间色谱分离，CPs 同类物在全二维色谱图上呈瓦片状分布，GC×GC 实现了链长不同、氯原子数不同的 SCCPs 同类物的组间分离。对于 SCCPs 和 MCCPs 的定量方法，基于 Reth 等[47,66]提出的定量曲线校正法，在优化色谱和质谱条件下，针对 ECNI 模式下 SCCPs 和 MCCPs 产生的主要碎片离子$[M-Cl]^-$进行检测，同位素丰度最高的碎片离子作为该同类物的定量离子，次高为定性离子，检测同类物包括碳链长度 10~17 个碳原子（C_{10}~C_{17}）、氯取代数 5~10 个（Cl_5~Cl_{10}）在内总计 48 种同类物，仪器对 SCCPs 的 LOD 为 20pg/μL（$C_{11}H_{18}Cl_6$），对 MCCPs 的 LOD 为 100pg/μL（$C_{14}H_{23}Cl_7$）。

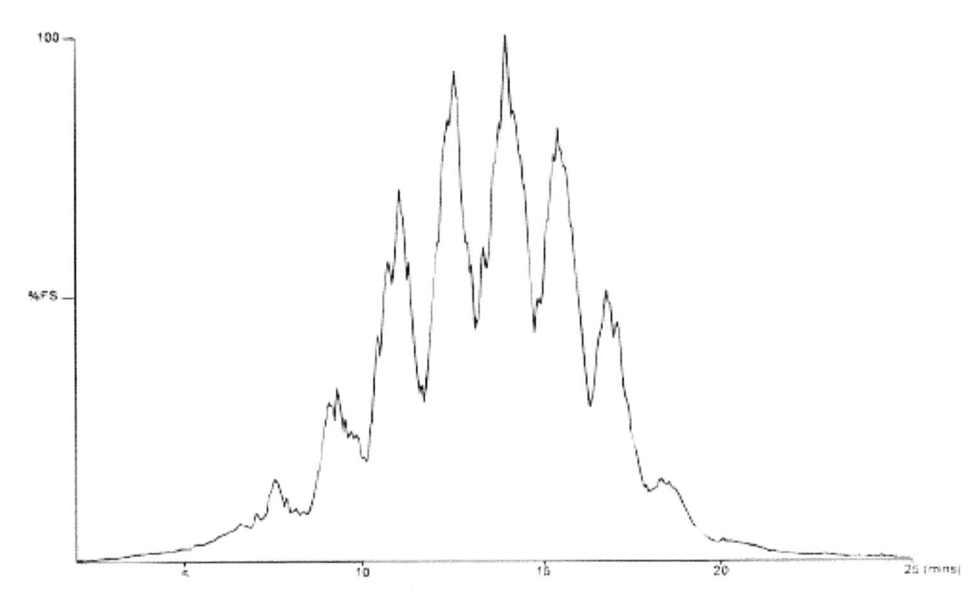

图 2-1 工业 CPs 产品 PCA-60 的 GC-ECNI-MS 总离子流图[29]

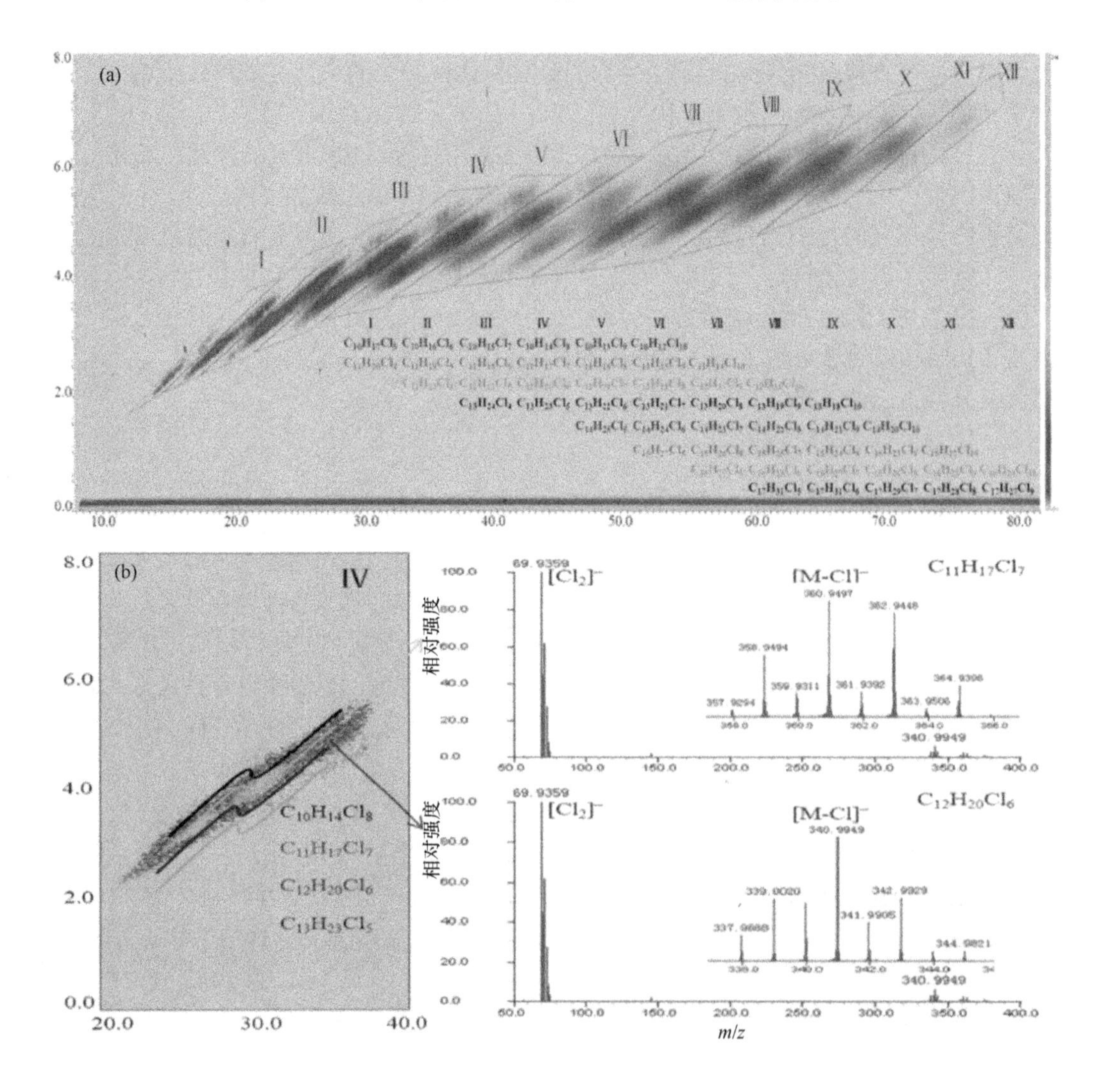

图 2-2 SCCPs 和 MCCPs 在 GC×GC-HRTOF-MS 色谱图及同类物分布模式

由于 SCCPs 组分复杂、不同实验室采用检测方法各异，实验室之间的分析检测结果存在一定差异。到目前为止，一共有七次国际实验室间 SCCPs 测试的分析比对，用以评估不同 SCCPs 分析方法的可靠性[31-33]。1999 年 Tomy 等组织 7 家实验室分别采用气相色谱分离[34]，ECNI-LRMS、ECNI-HRMS 和 ECD 三种检测方法分析了 2 个 SCCPs 标准溶液和 2 个净化处理后的鱼类提取物。7 家实验室的分析结果普遍高于实际值 2~4 倍。2009 年由 Pellizzato 等组织了 6 家实验室对 SCCPs 的不同分析方法进行比对试验[32]，待分析土壤样品经加速溶剂提取、弗罗里硅土净化处理后交由不同实验室分析样品中 SCCPs 的总量。各实验室分别采用气相色谱分离，ECNI-LRMS、EI-MS/MS 以及原子发射检测器（AED）检测方法以及碳骨架反应气相色谱分析方法检测 SCCPs。定量方法主要采用多元线性回归、选择与样品氯含量最接近的 SCCPs 标准样品和氯含量校正的响应因子曲线对 SCCPs 进行定量，碳骨架反应气相色谱法采用内标法标准曲线定量，AED 方法针对不同氯含量标样采用 4 个浓度水平的标准曲线进行定量。其中，4 家实验室结果较为接近，采用 AED 检测方法的 4 号实验室的结果略高，但和其他 4 家的分析结果处于相同数量级。可能是由于 AED 无法分辨 SCCPs 和其他含氯化合物，使得 SCCPs 含量被高估。1 号实验室的结果与其他实验室的分析结果差异较大，最大可达 370 倍。2010 年，欧洲海洋环境监测信息的质量保证项目（QUASMEME）组织比利时的一个工作小组开展了 CPs 的实验室间分析比对研究[33]，参加单位被要求分析异辛烷溶剂中 CPs 的总浓度和 3 种 CPs 同类物标样的浓度。结果表明，绝大多数参加单位对单一组成的 CPs 标样的分析结果是令人满意的，3 种 CPs 同类物浓度的分析结果变异系数为 22%~46%，变异系数对氯化石蜡总浓度分析结果是 56%。随着净化方法的完善，定量标准的统一，采用相同分析方法的实验室间比对结果差异性正在逐渐降低（22%~34%）。

4　SCCPs 的环境污染特征

4.1　SCCPs 在空气中的赋存水平及污染特征

4.1.1　SCCPs 在室外大气中的赋存水平及污染特征

CPs 尤其是 SCCPs 具有长距离迁移能力，在全球各地大气环境中均检测到 SCCPs 的存在，如表 2-2 所示。Li 等[48]采用国际经合组织的特征迁移距离预测软件[40]计算得出，SCCPs 在大气中的迁移距离大约在 500~900km 范围之内，但是在“蒸馏效应”的作用下，SCCPs 仍然可以到达偏远的极地地区[49,50,51]。1992 年在北极高纬度地区埃尔斯米尔岛北端的 Alert 观测站采集到的空气样品中检测出 SCCPs，其浓度范围为 1~8.5 pg/m^3[49]。Borgen 等[50]于 1999 年在挪威斯瓦尔巴德群岛齐伯林山采集的北极空气样本中也检测到 SCCPs，浓度在 9.0~57 pg/m^3，而斯瓦尔巴德群岛与挪威本土之间的熊岛采集到的空气样本中测得 SCCPs 浓度范围为 1.8~10.6ng/m^3，比北极地区要高三个数量级[51]。虽然极地地区 SCCPs 很大程度上来源于远距离大气传输，但这一结果表明当地一次排放源对大气中 SCCPs 具有重要贡献[16]。Ma 等[52]于 2012 年在南极菲尔德斯半岛的格鲁吉亚王岛采集了大气样品并对气相和颗粒相中的 CPs 进行分析，大气中 SCCPs 和 MCCPs 水平分别介于 9.6~20.8 pg/m^3 和 3.7~5.2 pg/m^3 之间，平均浓度分别为 14.9 pg/m^3 和 4.5 pg/m^3，目前还处于较低水平。

随着工业的发展，添加 CPs 的用品逐渐增多，导致大气中 SCCPs 浓度增大。Peters 等[53 54]报道 1997 年在英国兰开斯特乡村地区采集的空气样本中 SCCPs 的浓度为 5.4~1085pg/m^3，在 1997~1998 年的平均值为 99pg/m^3。Barber 等[55]2003 年测得英国大气 SCCPs 和 MCCPs 的浓度分别为 185~3430pg/m^3（平均

表 2-2 不同国家和地区大气中 CPs 浓度水平

国家/地区	CPs	采样时间	采样方法	分析方法	浓度
北极/Ellesmere Island[49]	SCCPs	秋冬季，1992	大体积采样	GC-ECNI-HRMS	1~8.5 pg/m^3
加拿大/Egbert[49]	SCCPs	夏季，1990	大体积采样	GC-ECNI-HRMS	65~924 pg/m^3
挪威/Svalbard [50]	SCCPs	3–6，1999	大体积采样	GC-ECNI-HRMS	9.0~57 pg/m^3
挪威/Bear Island[51]	SCCPs	5–11，2000	大体积采样	GC-ECNI-HRMS	1.8~10.6 ng/m^3
英国/Lancaster[53]	SCCPs	4，1997–5，1998	大体积采样	GC-ECNI-HRMS	5.4~085 pg/m^3
英国/Lancaster[54]	SCCPs	5，1997–1，1998	大体积采样	GC-ECNI-HRMS	平均值：99 pg/m^3
英国[55]	SCCPs MCCPs	4–5，2003	被动采样	GC-ECNI-HRMS	185~3430 pg/m^3 811~14500 pg/m^3
中国/北京[56]	SCCPs	1–2，2011 6–7，2011	大体积采样	GC-ENCI-LRMS	冬季：1.85~33.0 ng/m3 夏季：112~332 ng/m^3
印度[57]	SCCPs MCCPs	冬季，2006	被动采样	GC-ECNI-LRMS	SCCPs：n.d.*~47ng/m^3 MCCPs：n.d.*~38 ng/m^3
巴基斯坦 [57]	SCCPs MCCPs	冬季，2011	被动采样	GC-ECNI-LRMS	SCCPs：0.37~14.2 ng/m^3 MCCPs：0.3~9.4 ng/m^3
中国 [58]	SCCPs	8–10，2008	被动采样	GC-ECNI-LRMS	13.5~517 ng/m^3
韩国 [58]	SCCPs	8–10，2008	被动采样	GC-ECNI-LRMS	0.60~8.96 ng/m^3
日本 [58]	SCCPs	8–10，2008	被动采样	GC-ECNI-LRMS	0.28~14.2 ng/m^3
中国/珠江三角洲 [59]	SCCPs MCCPs	12，2009–9，2010	被动采样	GC-ECNI-LRMS	0.28~31.2 μg 0.03~67.9 μg

* n.d. 未检出

值 1130 pg/m^3）、811~14500pg/m^3（平均值 3040 pg/m^3），较 1997 年在同一地区的浓度高出许多。研究表明，SCCPs 在空气中的浓度还与季节相关，如 Wang 等[56]2012 年报道北京地区大气中 SCCPs 的浓度在冬季达到 1.85~33.0ng/m^3，而夏季则高达 112~332ng/m^3。SCCPs 在气固两相的分配系数与其蒸汽压和辛醇水分配系数线性相关。Chaemfa 等[57]研究印度和巴基斯坦冬季大气中 SCCPs 的平均浓度分别为 10.2 ng/m^3 和 5.13ng/m^3，印度的污染水平略高于巴基斯坦，SCCPs 同系物丰度分布为 C_{10}>C_{11} >C_{12}≈C_{13}，与中国类似。我国大气环境中也存在 SCCPs 污染，对中国、韩国和日本大气中 SCCPs 的浓度[58]范围分别为 13.5~517ng/m^3、0.60~8.96 ng/m^3 和 0.28~14.2ng/m^3，我国东部地区大气中 SCCPs 浓度显著高于日本和韩国。大气中 SCCPs 的含量水平呈现出城市>城郊/乡村地区>偏远地区的趋势。

另外，研究表明 CPs 在大气中更易于存在于气相中。Peters 等[53]研究了 CPs 的气固相分布规律，发现大气中 95%的 CPs 存在于气相中。Barber 等[55]发现了同样的规律，91%的 SCCPs 存在于气相中，9%存在于颗粒相；92%的 MCCPs 存于气相中，8%存在于颗粒相。Wang 等[59]研究了珠江三角洲地区 SCCPs 的污染特征，发现 SCCPs 在气相和颗粒相的分布特征与季节有一定的关系，夏季气相 SCCPs 污染水平较高，而冬季颗粒相 SCCPs 污染水平较高。复杂的环境过程如蒸发和沉降使 SCCPs 在不同的环境介质和采样地点有不同的同系物分布。在珠江三角洲地区，大气中 SCCPs 以 $C_{10\sim11}Cl_{6\sim7}$ 为主要组分，大气沉降形成的颗粒相中 $C_{11\sim12}Cl_{6\sim8}$ 丰度较高。

4.1.2 室内空气和灰尘中 CPs 的污染现状

室内环境是人体 CPs 暴露的主要介质之一，且室内环境中 CPs 的浓度水平普遍高于室外大气，表明室内环境中存在 CPs 的潜在释放源[60-63]。Fridén 等[60]采集了冬季瑞典斯德哥尔摩城市公寓 44 个室内空气样品和 6 个灰尘样品，并分析其中 SCCPs 和 MCCPs 的总含量和同系物分布。结果表明室内空气中 SCCPs 和 MCCPs 的总浓度范围在 5~210ng/m^3，SCCPs 的浓度高于 MCCPs，这可能是由于 SCCPs 具有更高的挥发性。灰尘样品中 SCCPs 和 MCCPs 的总浓度为 3.2~18μg/g，长链组分在灰尘中的比例明显高于空气样品。法国家庭环境灰尘中 SCCPs 的调查研究显示[64]，SCCPs 的浓度相当高，平均浓度达到 45μg/g，仅低于邻苯二甲酸盐，远高于多环芳烃、多溴二苯醚（PBDEs）、PCBs 等。Barber 等[43]采用被动采样方法分析机械加工厂、实验室、办公室和家庭环境中 SCCPs 和 MCCPs 的浓度水平，采样周期约 12 周。结果显示，四个采样点的浓度差异较大，其中工厂空气中 SCCPs 的含量最高，达到 17000 ng/m^3，这可能是由于该工厂曾经使用 CPs 作为切削液。SCCPs 含量最低的是家庭环境，其 SCCPs 含量为 220 ng/m^3。办公室和实验室则分别为 1600 和 5500 ng/m^3，比室外大气中 SCCPs 高 2 倍左右，进一步说明室内环境中可能存在 SCCPs 释放源。MCCPs 与 SCCPs 在不同环境中的浓度相似，如表 2-3 所示，工厂中浓度最高，家庭环境中浓度最低。但实验室与办公室中 MCCPs 的浓度水平比附近室外大气浓度低 1 个数量级，推测是由于附近建筑施工引起 MCCPs 含量较高，室外大气中 CPs 对人体暴露风险也是不容忽视的问题。Hilger 等[62]比较了家庭和公共场所灰尘中 SCCPs 和 MCCPs 的浓度水平差异。家庭环境中 MCCPs 的浓度为 9~892μg/g，高于 SCCPs 的浓度水平（4~27μg/g）。而某公共场所中 SCCPs 的浓度水平非常高，达到 2050μg/g，MCCPs 则未检出。作者同时分析了办公室内混凝土及地板与墙间密封材料，发现 SCCPs 含量非常高，超过 50000μg/g，而 MCCPs 未检出，这也解释了 SCCPs 含量较高的可能原因。与同为阻燃剂的 PBDEs 相比，在美国室内灰尘中 PBDEs 浓度远高于其他国家，其中位浓度水平约为 21μg/g[65,66]，而 SCCPs 和 MCCPs 的浓度水平远高于 PBDEs。我国有关室内环境中 CPs 的污染水平研究较少，对北京建筑物玻璃内外有机膜中 SCCPs 浓度水平的研究表明，人们日常生活中 SCCPs 暴露水平较高，SCCPs 浓度为 337~114360ng/m^2，其中家庭环境玻璃外膜高于玻璃内膜，而办公室环境相反，推测可能是由于家庭装修日益要求健康环保，而办公环境的装修材料以经济型为主[22]。另外低分子量 SCCPs 组分易于累积在玻璃内膜，而玻璃外膜大分子量 SCCPs 组分的相对丰度更高。

表 2-3　不同国家和地区室内空气与灰尘中 CPs 浓度水平

国家/地区	类型	采样时间	采样方法	分析方法	浓度
德国/Bavaria[62]	灰尘	n.a.*	真空吸尘器	GC-ECNI-LRMS	家庭：SCCPs 4~27 μg/g，MCCPs：9~892 μg/g；办公室：SCCPs 2050 μg/g，MCCPs n.d.**
英国/Lancaster[55]	室内空气	4–7，2003	PAS/PUF	HRGC-ECNI-HRMS	SCCPs：220~17000 ng MCCPs：450~19000 ng
中国/北京[61]	玻璃膜	11，2013–2，2014	–	HRGC-ECNI-LRMS	SCCPs：337~114360 ng/m^2
瑞典/Stockholm[60]	室内空气	10，2006–2，2007	小体积采样器	GC-EI-LRMS^2；GC-ECNI-LRMS	SCCPs+MCCPs：5~210 ng/m^3
瑞典/Stockholm[60]	灰尘	10，2006–2，2007	真空吸尘器	GC-EI-LRMS^2；GC-ECNI-LRMS	3.2~18 μg/g

* n.a. 未注明；** n.d. 未检出

4.2　土壤和沉积物中 SCCPs 的污染水平及空间分布

我国是 CPs 生产大国，国内一些研究团队较早的开展了土壤和沉积物中 SCCPs 的研究。Yuan 等[67]在中国浙江台州电子拆解地及其周边和远离此拆解地的土壤样品中均检测到了 SCCPs，结果分别为 2689ng/g dw、990ng/g dw、192ng/g dw，电子拆解为当地 SCCPs 污染源之一。Chen 等[68]发现中国珠三角地区沉积物中 SCCPs 的浓度为 320~6600ng/g dw。该研究还对珠江流域的柱状沉积物中 SCCPs 进行了分析，发现浅层的沉积物中 SCCPs 浓度高，而深层沉积物中 SCCPs 浓度低，说明最近几年内珠三角地区 SCCPs 的使用量大增，这也和中国的氯化石蜡产量相符。Zeng 等对北京市通州区的城市污水灌溉农田进行了 SCCPs 的时空分布研究[69]。在所有表层土壤中都检测到了 SCCPs 的存在，浓度范围 159. 9~1450 ng/g dw，且随土壤层深度增加，SCCPs 浓度呈指数下降趋势，这可能和深层土壤中有机碳含量少有关。

Iion 等[70]对日本的 Arakawa 河、Tamagawa 河和 Yodogawa 河中的沉积物样品进行测定，结果分别为 211.1 ng/g dw（干重）和 484.4 ng/g dw、384.7ng/g dw 和 196.6 ng/g dw 以及 4.9 ng/g dw 和 424.0 ng/g dw。Nicholls 等[71]在英国英格兰和威尔士地区工业区所采集的样品中检测到的 SCCPs 的浓度为：沉积物 0.2-65.1μg/g dw，土壤＜0.1μg/g dw。这两个地区土壤、沉积物样品中 SCCPs 的检出说明 SCCPs 在英国环境中已广泛存在。Pribylová 等[72]对采集于捷克共和国 11 条河流中的 36 个沉积物样品进行检测，SCCPs 浓度范围为从未检出至 347ng/g dw，呈现最高浓度的沉积物样品是来自于靠近化工和电镀工厂的河流中。Stejnarová 等[73]对捷克共和国工业活动程度不同的 3 个地区的沉积物样品进行 SCCPs 的检测分析，Zlín 地区（以橡胶、纺织品和制革业为主的工业区）工业区沉积物中浓度水平为 16.30~180.8 ng/g dw，在 Beroun 地区（以黏结剂和制造业为主的工业区）浓度为 4.58~21.57 ng/g dw，在 Koetice 地区（背景区域）浓度为 24.00~45.78 ng/g dw。这 3 个地区分别位于捷克共和国的中部、东部和西部，此测定结果也表明 SCCPs 在捷克共和国是一种普遍存在的污染物质。

Iozza 等[74]为研究不同年份沉积物中 SCCPs 的浓度与其当年产量和使用量的相关性，对瑞士境内 Thun 湖的泥芯样品进行测定，结果表明这两者之间存在显著性相关，SCCPs 浓度在 1986 年达到最高值(33ng/g dw)。Tomy 等[75]利用沉积柱样品，研究了加拿大随纬度变化（中纬度到高纬度）的 6 个湖泊沉积物中的 SCCPs 浓度（4.52~135.5ng/g dw），表明底泥中 SCCPs 浓度随纬度升高而呈现减少的趋势。同时计算了 Winnipeg 湖（中纬度）和 Hazen 湖（高纬度）的 SCCPs 的沉积通量，分别为 147 和 0.9μg/(m^2·a)。同时发现不同深度泥芯中 SCCPs 的浓度变化趋势与经济发展状况等吻合，其中 20 世纪 80 年代初到 90 年代的沉积物中 SCCPs 浓度最高。Marvin 等[76]在安大略湖设计了 55 个采样点，SCCPs 平均浓度为 49ng/gdw，浓度最大值出现在 20 世纪 70 年代的沉积物层。

Gao 等[77]采用 HRGC–ECNI–LRMS 方法分析 2009 年和 2010 年采集的中国大辽河流域及其入海口沉积物样品中 SCCPs 浓度和各同系物组分的分布特征。辽河沉积物中 SCCPs 的浓度为 39.8~480.3ng/g，入海口沉积物中 SCCPs 的浓度为 64.9~407.0ng/g。全流域 SCCPs 污染水平较高，污染较高的点多集中于城市和工业排污口之下，在大辽河下游和入海口处 SCCPs 的浓度也较高。随着向海洋方向延伸，SCCPs 含量逐渐降低。SCCPs 在城市周边的污染水平最高，采样点分布于污水处理厂下游，钢铁企业排污口附近。通过对数据进行栅格化处理，评估全长约 560km 的大辽河流域表层 20cm 沉积物中 SCCPs 的环境存量约为 30.82t。大辽河沉积物样品中 SCCPs 各同系物组分的分布模式相似，以 C_{10} 和 C_{11} 的 SCCPs 相对丰度最高，分别是 40.4%和 39.4%。氯取代分布模式以 $Cl_{5\sim6}$ 组分为主，高氯代 SCCPs 的相对丰度相对较低。辽东湾海域水中 ΣSCCPs 的浓度范围在 4.1~13.1 ng/L 之间，平均值为 7.7ng/L。表层沉积物中∑SCCPs 的浓度范围在 65~541 ng/g dw 之间，平均值为 299 ng/g dw。相应的有机质归一化浓度范围为 8.9~46.1μg/g，平均值为 23.6μg/g。海水和沉积物样品中 SCCPs 空间分布特征相似，靠近河口位置最高，随着离岸距离的增加，ΣSCCPs 呈现明显的降低趋势，说明河流输入时海水中 SCCPs 的一个主要来源，并且影响 SCCPs 在海洋环境中的空间分布。

4.3 SCCPs 生物累积和生物放大

有文献报道，鱼体会蓄积环境中疏水性含氯有机污染物[78]。鱼作为一种食物，是人体摄入 CPs 的一种重要途径[28]。目前鱼体中 SCCPs 和 MCCPs 浓度的文献数据相对较少[79-83]。Zhou 等最近调查了长江三角洲一种鱼体内 CPs 的总浓度，但没有研究其同系物分布特征[84]。Basconcillo 等讨论了加拿大掠食性鱼体内 SCCPs 和 MCCPs 的来源可能是大气源或城市工业来源[85]。姜国等研究了我国长三角地区 8 种食用鱼，鱼体中 SCCPs 的浓度范围为 36~801ng/g dw，与文献中报道的世界范围的水平相比较。上海食用鱼体内 SCCPs 含量处于中等水平，同系物分布主要以低氯代的 SCCPs 为主，SCCPs 在这几种鱼体内的生物蓄积现象不明显[86]。此外，Houde 等对安大略湖和密歇根湖食物链中 SCCPs 和 MCCPs 的研究发现它们在生物体内发生生物累积和生物放大[87]。Zeng 和 Ma 等人研究了食物链动物体内中 SCCPs 的生物累积和生物放大[84,88]。

Ma 等[88]考察了环渤海地区表层沉积物及底栖双壳类动体内 SCCPs 的含量水平。沉积物中∑SCCPs 的浓度范围在 97.4~1757 ng/g dw 之间，平均值为 650.7ng/g dw。沉积物中 SCCPs 分布在空间上表现出明显的区域特征，工业化程度较高的地区含量水平明显高于其他地区。双壳类软体动物中∑SCCPs 的浓度范围在 476~3270ng/g dw 之间，平均值为 1710ng/g dw。与沉积物空间分布特征相似，SCCPs 同样表现出明显的区域特征，表明底栖软体动物作为环境指示物种可以有效地反映 SCCPs 在沉积物中的污染状况。表层沉积物和生体内 SCCPs 同类物主要以低碳链和低氯代为主，河流输入是近岸海域 SCCPs 的主要来源之一；渤海沉积环境中 SCCPs 的污染主要来源于上游地区 SCCPs 的生产及使用。双壳类底栖动物对 SCCPs 表现出明显的生物富集性，三种生物中∑SCCPs 的生物-沉积物富集因子（BSAF）平均值从高到低依次为四角蛤蜊（1.61）>菲律宾蛤仔（1.25）>青蛤（1.08），造成物种间富集能力差异的主要原因可能是由于其滤食习性及代谢能力所致。不同 SCCPs 同类物的 BSAF 值与碳、氯原子数均表现出显著的正相关系，并且随着 $\log K_{OW}$ 的增加呈显著的增加趋势。SCCPs 的生物富集行为可能受生物降解和生物转化等因素的影响，而该方面的结论有待进一步研究证实。

Ma 等[89]在渤海辽东湾采集浮游动物、无脊椎动物和鱼类样品，采用 HRGC-ECNI/LRMS 分析方法测定样品中的 SCCPs 含量，通过氮同位素分析确定不同生物所处的营养级。结果表明，辽东湾不同生物体中∑SCCPs 的浓度范围为 86~4400ng/g ww（湿重），平均值为 940ng/g ww，相应的脂质归一化浓度范围为 2.3~76.5 μg/g lw（脂肪重量），平均值为 20.1μg/g lw。不同物种间∑SCCPs 含量差异显著，脂质归一化浓度从高到低依次为：双壳类（平均值 37.1μg/g lw）>蟹类（19.8μg/g lw）>鱼类（19.4μg/g lw）>螺类（16.6μg/g lw）>虾类（9.0μg/g lw）>浮游动物（4.8μg/g lw）。SCCPs 的生物累积因子对数值范围在 4.5–5.6 之间，表明海洋生物对 SCCPs 具有明显的生物富集性。8 种底栖类生物中 SCCPs 同类物的沉积物-生物富集因子的范围为 0.1~7.3，并且物种间的差异比较显著。浮游动物-虾-鱼食物网中 SCCPs 的营养级放大因子为 2.38，表明 SCCPs 具有明显的生物放大潜力；同时发现 SCCPs 不同同类物的营养级放大因子与 $\log K_{OW}$ 之间表现出显著的正相关关系。

5 SCCPs 的人体暴露评估

5.1 室内环境 SCCPs 通过的人体暴露研究

已有研究表明，室内空气和灰尘中广泛检出 SCCPs 和 MCCPs。Barber 等[90]对室内空气 CPs 进行研究，发现在一个机械工厂测得的 CPs 浓度最高，约 6324pg/m^3，这可能是由于机械厂使用含有 CPs 的产品。进一步研究表明，SCCPs 在气相和颗粒相的平均分配比例为 91∶9。Friden 等[91]针对瑞典斯德哥尔摩办公室和居民家庭中的室内空气和灰尘样品 SCCPs 和 MCCPs 进行了检测，结果发现，95%的成人和儿童通过呼吸和灰尘摄入的 CPs 总量分别为 56.7ng/(kg BW(体重)·d)和 490ng/(kg BW·d)。该研究表明，呼吸和灰尘摄入可能是室内环境中成人和幼儿的主要 CPs 暴露途径。与人类通过环境间接暴露于 SCCPs 观察到的不良反应水平参考值 100 mg/(kg·d)相比，瑞典居民呼吸和灰尘摄入对 CPs 的暴露处于较低的水平，不会引起显著的健康风险。Coelhan 等[62]对德国巴伐利亚地区 11 个房屋灰尘样品的研究发现，室内灰尘中的 SCCPs 和 MCCPs 浓度较高，平均浓度分别高达 2.41μg/g 和 70.7μg/g。值得注意的是，SCCPs 的生产和使用在欧洲受到限制，然而我国仍然在大量的生产和消费含 CPs 的产品，使得我国室内环境中 SCCPs 和 MCCPs 的暴露水平可能较高。室内大气和灰尘摄入可能也是 CPs 人体暴露的重要污染来源。Gao 等[92]测得北京普通居民室内空气和玻璃膜中 SCCPs 分别达到了 0.06~1.35μg/m^3 和 0.34~54.0μg/m^2。因此，室内空气和灰尘是人体暴露 CPs 的主要途径之一。

5.2 膳食暴露 SCCPs 的研究

膳食摄入被认为是脂溶性 POPs 的主要人体摄入途径。CPs 具有脂溶性，可以通过食物链富集。膳食中 CPs 的污染水平引起了广泛关注。Thomas 和 Jones[93]在来自英国兰卡斯特地区的牛奶样本和来自欧洲多个地区的黄油样本中测出了 SCCPs。在来自丹麦的黄油样本中测出的 SCCPs 浓度为 1.2ng/g，爱尔兰样本的浓度为 2.7ng/g。而对代表美国人食谱上约 5000 个食物品种的 234 种开袋即食食品 SCCPs 浓度水平进行检测发现，在一种营养面包中检测出了 SCCPs，浓度达到 0.13μg/g[89]。Lahaniatis 等[94]发现，在各种生物鱼油中，SCCPs 的浓度平均值为 7.0~206 ng/g。

Iino 等[85]检测了日本 11 种食物中 SCCPs 的含量水平。结果发现，多种食物如谷类（2.5ng/g）、种子和马铃薯（1.4ng/g）、调料和饮料（2.4ng/g）、脂肪（人造黄油、油类等，140ng/g）、蘑菇和海藻（1.7ng/g）、水果（1.5ng/g）、鱼类（16ng/g）、贝类（18ng/g）、肉类（7ng/g）、蛋类（2ng/g）和牛奶（0.75ng/g）中均检测出了 SCCPs。根据不同年龄段人群的食物消费量和体重等调查数据计算 SCCPs 的日摄入总量。结果发现，居民年龄越小，日摄入 SCCPs 总量越高。95%的 1 岁女孩每天的 SCCPs 摄入量为 0.68μg/(kg·d)。膳食是日本人群接触 SCCPs 的主要暴露途径，但尚不会引起健康风险。Harada 等[45]研究了日本、韩国和中国居民日常饮食混合样本中 SCCPs 的含量及变化，发现北京居民饮食摄入 SCCPs 的量为 620ng/(kg BW·d)，比日本和韩国高出一个数量级，并且从 1993 年至 2009 年期间，北京居民的 SCCPs 摄入量增加了两个数量级，而日本居民膳食中 SCCPs 的含量较为稳定。Cao 等[88]发现食用油是我国食品中 SCCPs 的主要来源，食用油中 SCCPs 的饮食摄入量为 0.78~38μg/d。考虑到 SCCPs 和 MCCPs（TMF>1）的食物链放大效应，海产品也可能是人体暴露的重要来源之一。Krogseth[93]利用人体暴露模型预测，通过鱼类消费 SCCPs 摄入量比例高达 80%~100%。

5.3 人体 SCCPs 的内暴露水平

由于 CPs 的组成复杂，目前对于人体血液中的 CPs 仅有一篇关于分析方法学的研究。CPs 容易在人体脂肪组织中富集，母乳中 CPs 浓度可反映 CPs 的人体负荷水平，同时也可评估母婴 CPs 的潜在暴露风险。德国环境机构的一项研究指出，1995 年德国妇女的母乳中 CPs（$C_{10\sim24}$总量）的含量平均值为 45ng/g lw[95]。Tomy[96]对加拿大魁北克地区妇女母乳中的 SCCPs 进行了检测，SCCPs 的浓度为 11~17ng/g lw（平均浓度为 13ng/g lw）。Thomas 等[83]对英国哺乳妇女母乳样本的分析发现，在兰卡斯特地区 8 个样本中，有 5 个样本中发现了 SCCPs，其浓度为 4.6~110ng/g lw；来自伦敦地区的 14 个样本中有 7 个样本中发现了 SCCPs，其浓度为 4.5~43ng/g lw。在后续的研究中，Thomas 等[84]对英国城市和乡村哺乳妇女的母乳中 SCCPs 和 MCCPs 的含量进行调查，发现这两个城市的 SCCPs 的浓度介于 49~820ng/g lw（平均值为 180ng/g lw），MCCPs 的含量分布在 6.2~320ng/g lw 之间（平均值为 21ng/g lw）。城市和乡村人口母乳中 CPs 的浓度没有显著差异。根据 Thomas 的研究，哺乳妇女母乳中 SCCPs 和 MCCPs 的总量约为 55.2~1140ng/g lw，母乳喂养的婴儿每天从母乳中摄取 CPs 的平均浓度约为 900 ng/(kg·d)，处于较低的人体健康风险水平。

我国卫生部门组织开展的有关中国总膳食研究也关注了我国人群的有害化学物质膳食暴露问题。Xia

等[97,98]对我国母乳中的 SCCPs 和 MCCPs 开展了研究。所分析的样品针对我国 2007 年和 2011 年第四次和第五次总膳食研究涉及省市地区的母乳。2007 年共涉及 12 个省份，2011 年共涉及 16 个省份。采样省份包括黑龙江、吉林、辽宁、内蒙古、河北、宁夏、青海、河南、陕西、上海、四川、湖北、江西、福建、广西和广东。母乳样本的采集以世界卫生组织（WHO）和联合国规划署（UNEP）共同制定的第四轮《全球母乳中持久性有机污染物监测导则》为基础[129]，每省市自治区下设 1 个城市采样点，采集至少 50 份母乳样品，设 2 个农村采样点，各采集份母乳样品 30 份。此项研究共采集 1370 份城市母乳样品，1412 份农村母乳样品。对每个采样点样品进行混合，测定 SCCPs 和 MCCPs。28 份城市母乳混合样品中 SCCPs 和 MCCPs 均广泛检出，表明我国母乳中 CPs 的污染具有普遍性。其中，2007 年 12 个省份母乳样品中 SCCPs 的浓度范围为 170~6150ng/g lw，平均值为 681ng/g lw。MCCPs 的浓度范围在 18.7~350ng/g lw 之间，平均值为 60.4ng/g lw。2011 年 SCCPs 的浓度范围和 2007 年在同一水平，浓度范围为 131~16100ng/g lw，平均值为 733ng/g lw。MCCPs 的浓度范围在 22.3~1510ng/g lw 之间，平均值为 137ng/g lw。相比已有文献对母乳中多种 POPs 浓度水平的报道，CPs 在我国城市居民母乳样品中处于较高的暴露水平。我国母乳样品中 SCCPs 和 MCCPs 浓度水平具有显著的相关性（R^2= 0.82，2007；R^2=0.85，2011，p<0.001），这表明 SCCPs 和 MCCPs 可能存在相似的暴露来源和人体吸收途径。河北和河南省母乳中 SCCPs 的浓度最高，其原因可能是因为河北和河南两省是我国氯化石蜡的主要产地，也是国内氯化石蜡生产厂家最为集中地区[134]。SCCPs 和 MCCPs 在农村母乳样品中广泛检出。其中 2007 年 8 个省份母乳样品中 SCCPs 的浓度范围为 68.0~1580ng/g lw，平均值为 304ng/g lw。MCCPs 的浓度范围在 9.05~139ng/g lw 之间，平均值 35.7ng/g lw。2011 年 SCCPs 的浓度范围和 2007 年在同一水平，浓度范围为 65.6~2310ng/g lw，平均值 360ng/g lw。MCCPs 的浓度范围在 9.51~146ng/g lw 之间，平均值 45.7ng/g lw。与城市样品类似，河北省母乳中 SCCPs 和 MCCPs 浓度最高，浓度最低的省份是黑龙江省和广西省。我国母乳中 SCCPs 和 MCCPs 污染水平，城市高于农村城乡居民在暴露水平和异构体组成上没有差异，膳食摄入是普通人群 SCCPs 和 MCCPs 暴露的主要途径。城市地区 CPs 的生产使用是当地人群主要的暴露来源，而大气和室内灰尘可能最主要的暴露途径。

6　SCCPs 的研究展望

氯化石蜡的同系物、同分异构体和对映异构体非常多，组分非常复杂，给 SCCPs 的环境监测和污染控制带来巨大的困难。我国关于 CPs 在不同环境介质中的含量水平及其来源与归趋的报道还相对较少。

由于 SCCPs 分析方法的不同以及缺乏不同实验室间的数据比对分析，不同地域间环境数据的比对缺乏可比性，因此需要确立一套标准的检测方法，同时建议参加国际实验室分析比对研究，确保分析数据的可比性。

国内生产企业 CPs 产品没有严格区分碳链的长度，大多数产品中混有 SCCPs 且比例不确定，应该尽快掌握 CPs 工业品中 SCCPs 的比例。同时未来尽快掌握 CPs 的环境存量、分布情况、环境行为、来源和排放等基础数据，为从源头削减 SCCPs 提供技术依据。

有关 SCCPs 的降解与消除技术鲜有报道。现有废水处理工艺尚未考察对 SCCPs 的去除效果，以致污水处理厂的污泥成为环境中 SCCPs 的二次污染源。今后相当长的时间内，含有 SCCPs 的产品逐步进入报废期，含有 SCCPs 的废弃物无害化处理又将成为新的环保难题。

我国是全世界最大的 CPs 生产国和出口国，CPs 生产和使用可能引起的生态和人体健康风险须给予足够关注。应加强 SCCPs 和 MCCPs 的毒性效应和致毒机理研究，评估人类健康风险。在未来的研究中，关于 SCCPs 和 MCCPs 的人体暴露行为和代谢蓄积规律的研究有待进一步深入开展，人体内暴露和外暴露水平及暴露途径的研究有待系统性深入探讨，以期为 CPs 人体暴露源解析提供科学依据。SCCPs 的长距离迁移能力和生物富集能力等还有待在更大尺度范围和不同环境条件下进行系统性研究。

我国控制 SCCPs 环境污染的工作任重而道远，科学认知我国 SCCPs 的来源、环境过程与人体暴露途径，发展 CPs 替代技术与消除 SCCPs 的实用技术，是有效控制 SCCPs 环境污染的必由之路。

参考文献

[1] van Mourik L M, Gaus C, Leonards P E G, de Boer J. Chlorinated paraffins in the environment: A review on their production, fate, levels and trends between 2010 and 2015. Chemosphere, 2016, 155: 415-428.

[2] Wei G, Liang X , Li D, Zhuo M, Zhang S, Huang Q, Liao Y, Xie Z, Guo T, Yuan Z. Occurrence, fate and ecological risk of chlorinated paraffins in Asia: A review. Environment International, 2016, 92-93: 373-387.

[3] Chen L, Huang Y, Han S, Feng Y, Jiang G, Tang C, Ye Z, Zhan W, Liu M, Zhang S. Sample pretreatment optimization for the analysis of short chain chlorinated paraffins in soil with gas chromatography-electron capture negative ion-mass spectrometry. Journal of Chromatography A, 2013, 1274: 36-43.

[4] Zeng L, Wang T, Ruan T, Liu Q, Wang Y, Jiang G. Levels and distribution patterns of short chain chlorinated paraffins in sewage sludge of wastewater treatment plants in China. Environmental Pollution, 2012, 160: 88-94.

[5] Zeng L, Li H, Wang T, Gao Y, Xiao K, Du Y, Wang Y, Jiang G. Behavior, fate, and mass loading of short chain chlorinated paraffins in an advanced municipal sewage treatment plant. Environmental Science & Technology, 2013, 47(2): 732-740.

[6] Gao Y, Zhang H, Su F, Tian Y, Chen J. Environmental occurrence and distribution of short chain chlorinated paraffins in sediments and soils from the Liaohe River Basin, P. R. China. Environmental Science &Technology, 2012, 46: 3771-3778.

[7] Xu J, Gao Y, Zhang H, Zhan F, Chen J. Dispersion of short- and medium-chain chlorinated paraffins (CPs) from a CP production plant to the surrounding surface soils and coniferous leaves. Environmental Science & Technology2016, 50 (23), 12759-12766.

[8] Coelhan M. Determination of short-chain polychlorinated paraffins in fish samples by short-column GC/ECNI-MS. Analytical Chemistry, 1999, 71: 4498-4505.

[9] Korytár P, Leonards P E G, de Boer J, Brinkman UAT. Group separation of organohalogenated compounds by means of comprehensive two-dimensional gas chromatography. J Chromatogr aphy A, 2005, 1086: 29-44.

[10] Korytár P, Leonards P E G, de Boer J, Brinkman UAT. Quadrupole mass spectrometer operating in the electron-capture negative ion mode as detector for comprehensive two-dimensional gas chromatography. J Chromatogr A, 2005, 1067: 255-264.

[11] Korytár P, Parera J, Leonards P E G, de Boer J, Brinkman. Characterization of polychlorinated n-alkanes using comprehensive two-dimensional gas chromatography-electron-capture negative ionization time-of-flight mass spectrometry. J Chromatography A, 2005, 1086: 71-82.

[12] Xia D, Lirong G, Zhu S, Zheng M. Separation and screening of short chain chlorinated paraffins in environmental samples using comprehensive two-dimensional gas chromatography with micro-electron capture detection.Analytical and Bioanalytical Chemistry, 2014, 406: 7561-7570.

[13] Zencak Z, Oehme M. Chloride-enhanced atmospheric pressure chemical ionization mass spectrometry of polychlorinated n-alkanes. Rapid Communnation Mass Spectrom, 2004, 18: 2235-2240.

[14] Pellizzato F, Ricci M, Held A, Emons H, Böhmer W, Geiss, Iozza S, Mais S, Petersen M, Lepom P. Laboratory intercomparison study on the analysis of short-chain chlorinated paraffins in an extract of industrial soil. TrAC Trends Analalytical Chemistry, 2009, 28: 1029-1035.

[15] Zencak Z, Borgen A, Reth M, Oehme M. Evaluation of four mass spectrometric methods for the gas chromatographic analysis of polychlorinated n-alkanes. J Chromatography A, 2005, 1067: 295-301.

[16] Feo M L, Eljarrat E, Barceló D, Barceló D. Occurrence, fate and analysis of polychlorinated n-alkanes in the environment. Trends in Analytical Chemistry, 2009, 28(6): 778-791.

[17] Gandolfi F, Malleret L, Sergent M, Doumenq P. Parameters optimization using experimental design for headspace solid phase micro-extraction analysis of short-chain chlorinated paraffins in waters under the European water framework directive. Journal of Chromatography A, 2015, 1406: 59-67.

[18] Pribylova P, Klanova J, Holoubek I. Screening of short- and medium-chain chlorinated paraffins in selected riverine sediments and sludge from the Czech Republic. Environmental Pollution, 2006, 144(1): 248-254.

[19] Gao Y, Zhang H, Chen J, Zhang Q, Tian Y, Qi P, Yu Z. Optimized cleanup method for the determination of short chain polychlorinated n-alkanes in sediments by high resolution gas chromatography/electron capture negative ion-low resolution mass spectrometry. Analytica Chimica Acta, 2011, 703(2): 187-193.

[20] 史慧娟, 薛平, 林勤保, 杜利君. 快速溶剂萃取-气相色谱法测定油漆中的短链氯化石蜡.山西大学学报, 2012, 35(1): 108-112.

[21] 马贺伟. 皮革中短链氯化石蜡的 GC-ECD 测定方法探讨.中国皮革, 2012, 41(21): 1-3.

[22] 袁丽凤, 邬蓓蕾, 罗川, 徐善浩, 华正江, 王豪. 气相色谱法测定橡胶制品中短链氯化石蜡含量.理化检验-化学分册, 2012, 48(8): 920-923.

[23] Pellizzato F, Ricci M, Held A, Emons H. Analysis of short-chain chlorinated paraffins : A discussion paper. Journal of Environmental Monitoring, 2007, 9(9): 924-930.

[24] Iozza S, Mueller C E, Schmid P, Bogdal C, Oehme M. Historical profiles of chlorinated paraffins and polychlorinated biphenyls in a dated sediment core from Lake Thun (Switzerland). Environmental Science & Technology, 2008, 42(4): 1045-1050.

[25] Zeng L, Wang T, Han W, Yuan B, Liu Q, Wang Y, Jiang G. Spatial and vertical distribution of short chain chlorinated paraffins in soils from wastewater irrigated farmlands. Environmental Science & Technology, 2011, 45(6): 2100-2106.

[26] Gao Y, Zhang H, Zou L, Wu P, Yu Z, Lu X, Chen J. Quantification of short-chain chlorinated paraffins by deuterodechlorination combined with gas chromatography–mass spectrometry. Environmental Science & Technology, 2016, 50(7) : 3746-3753.

[27] Reth M, Zencak Z, Oehme M. New quantification procedure for the analysis of chlorinated paraffins using electron capture negative ionization mass spectrometry. Journal of Chromatography A, 2005, 1081(2): 225-231.

[28] Reth M, Zencak Z, Oehme M. First study of congener group patterns and concentrations of short- and medium-chain chlorinated paraffins in fish from the North and Baltic Sea. Chemosphere, 2005, 58(7): 847-854.

[29] Tomy G T, Stern G A, Muir D C G, Fisk A T, Cymbalisty C D, Westmore J B. Quantifying C_{10}–C_{13} polychloroalkanes in environmental samples by high-resolution gas chromatography electron capture negative ion high resolution mass spectrometry. Analytical Chemistry, 1997, 69(14): 2762-2771.

[30] Xia D, Gao L, Zheng M, Tian Q, Huang H. A novel method for profiling and quantifying short- and medium-chain chlorinated paraffins in environmental samples using comprehensive two-dimensional gas chromatography−electron apture negative ionization high-resolution time-of-flight mass spectrometry. Environmental Science & Technology, 2016, 50: 7601-7609.

[31]Tomy G T, Westmore J B, Stern G A, Muir D C G, Fisk A T. Interlaboratory study on quantitative methods of analysis of C_{10}–C_{13} polychloro-*n*-alkanes. Analalytical Chemistry, 1999, 71: 446-451.

[32] Pellizzato F, Ricci M, Held A, Emons H, Böhmer W, Geiss S, Iozza S, Mais S, Petersen M, Lepom P.Laboratory intercomparison study on the analysis of short-chain chlorinated paraffins in an extract of industrial soil. TrAC Trends Analytical Chemistry, 2009, 28: 1029-1035.

[33] van Mourik L M, Leonards P E G, Gaus C, de Boer J. Recent developments in capabilities for analyzing chlorinated paraffins in environmental matrices: A review. Chemosphere, 2015, 136: 259-272.

[34] Chaemfa C, Xu Y, Li J, Chakraborty P, Hussain Syed J, Naseem Malik R, Wang Y, Tian C, Zhang G, Jones K C. Screening of atmospheric short- and medium-chain chlorinated paraffins in India and Pakistan using polyurethane foam based passive air sampler. Environmental Science & Technology, 2014, 48(9): 4799-4808.

[35] Bogdal C, Alsberg T, Diefenbacher P S, MacLeod M, Berger U. Fast quantification of chlorinated paraffins in environmental samples by direct injection high-resolution mass spectrometry with pattern deconvolution. Analytical Chemistry, 2015, 87(5): 2852-2860.

[36] Friden U E, McLachlan M S, Berger U. Chlorinated paraffins in indoor air and dust: concentrations, congener patterns, and human exposure. Environment International, 2011, 37(7): 1169-1174.

[37] Zeng L, Chen R, Zhao Z, Wang T, Gao Y, Li A, Wang Y, Jiang G, Sun L. Spatial distributions and deposition chronology of short chain chlorinated paraffins in marine sediments across the Chinese Bohai and Yellow Seas. Environmental Science & Technology, 2013, 47(20): 11449-11456.

[38] Zeng L, Zhao Z, Li H, Thanh W, Liu Q, Xiao K, Du Y, Wang Y, Jiang G. Distribution of short chain chlorinated paraffins in marine sediments of the East China Sea: influencing factors, transport and implications. Environmental Science & Technology, 2012, 46(18): 9898-9906.

[39] Zeng L, Wang T, Han W, Yuan B, Liu Q, Wang Y, Jiang G. Spatial and vertical distribution of short chain chlorinated paraffins in soils from wastewater irrigated farmlands. Environmental Science & Technology, 2011, 45(6): 2100-2106.

[40] Ma X, Chen C, Zhang H, Gao Y, Wang Z, Yao Z, Chen J, Chen J. Congener-specific distribution and bioaccumulation of short-chain chlorinated paraffins in sediments and bivalves of the Bohai Sea, China. Marine Pollution Bulletin, 2014, 79(1-2): 299-304.

[41] Wang Y, Li J, Cheng Z, Li Q, Pan X, Zhang R, Liu D, Luo C, Liu X, Katsoyiannis A, Zhang G. Short- and medium-chain chlorinated paraffins in air and soil of subtropical terrestrial environment in the pearl river delta, South China: Distribution, composition, atmospheric deposition fluxes, and environmental fate. Environmental Science & Technology, 2013, 47(6): 2679-2687.

[42] Halse A K, Schlabach M, Schuster J K, Jones K C, Steinnes E, Breivik K. Endosulfan, pentachlorobenzene and short-chain chlorinated paraffins in background soils from Western Europe. Environmental Pollution, 2015, 196: 21-28.

[43] Luo X, Sun Y, Wu J, Chen S, Mai B. Short-chain chlorinated paraffins in terrestrial bird species inhabiting an e-waste recycling site in South China. Environmental Pollution, 2015, 198: 41-46.

[44] Liu X, Li D, Li J, Rose G, Marriott P J. Organophosphorus pesticide and ester analysis by using comprehensive two-dimensional gas chromatography with flame photometric detection. Journal of Hazardous Materials, 2013, 263(2): 761-767.

[45] Harada K H, Takasuga T, Hitomi T, Wang P, Matsukami H, Koizumi A. Dietary exposure to short-chain chlorinated paraffins has increased in Beijing, China. Environmental Science & Technology, 2011, 45(16): 7019-7027.

[46] Thomas G O, Farrar D, Braekevelt E, Stern G, Kalantzi O I, Martin F L, Jones K C. Short and medium chain length chlorinated paraffins in UK human milk fat [J]. Environment International, 2006, 32(1): 34-40.

[47] Wang T, Yu J, Han S, Wang Y, Jiang G. Levels of short chain chlorinated paraffins in pine needles and bark and their vegetation-air partitioning in urban areas. Environmental Pollution, 2015, 196: 309-312.

[48] Li Q, Li J, Wang Y, et al. Atmospheric short-chain chlorinated paraffins in China, Japan, and South Korea.Environment Science & Technolgy. 2012, 46(21): 11948-11954.

[49] Stern G A, Tomy G T. An overview of the environmental levels and distribution of polychlorinated paraffins. Organohalogen Compounds, 2000, 47: 135-138.

[50] Borgen A R, Schlabach M, Gundersen H. Polychlorinated alkanes in arctic air. Organohalogen Compounds, 2000, 47: 272-274.

[51] Borgen A R, Schlabach M, Kallenborn R, Christensen G, Skotvold T. Polychlorinated alkanes in ambient air from bear Island. Organohalogen Compounds, 2002, 59: 303-306.

[52] Ma X D, Zhang H J, Zhou H Q, Na G, Wang Z, Chen C, Chen J, Chen J. Occurrence and gas/particle partitioning of short- and medium-chain chlorinated paraffins in the atmosphere of Fildes Peninsula of Antarctica. Atmospheric Environment, 2014, 90: 10-15.

[53] Peters A J, Tomy G T, Jones K C, Coleman P, Stern GA. Occurrence of C_{10}–C_{13} polychlorinated *n*–alkanes in the atmosphere of the United Kingdom. Atmospheric Environment, 2000, 34: 3085-3090.

[54] Peters A J, Tomy G T, Stem G A Jones K C. Polychlorinated alkanes in the atmosphere of the United Kingdom and Canada-analytical methodology and evidence of the potential for long range transport. Organohalogen Compounds, 1998, 35: 439-442.

[55] Barber J L, Sweetman A, Thomas G O, Braekevelt E, Stern G A, Jones K C. Spatial and temporal variability in air concentrations of short-chain (C_{10}–C_{13}) and mediu m-chain (C_{14}–C_{17}) chlorinated *n*-alkanes measured in the U.K. atmosphere. Environmental Science & Technology, 2005, 39: 4407-4415.

[56] Wang T, Han S, Yuan B, Zeng L, Li Y, Wang Y, Jiang G. Summer-winter concentrations and gas-particle partitioning of short chain chlorinated paraffins in the atmosphere of an urban setting. Environmental Pollution, 2012, 171: 38-45

[57] Chaemfa C, Xu Y, Li J, Chakraborty P , Hussain Syed J , Naseem Malik R , Wang Y , Tian C, Zhang G , Jones KC. Screening of atmospheric short- and medium-chain chlorinated paraffins in India and Pakistan using polyurethane foam based passive air sampler. Environmental Science & Technology, 2014, 48(9): 4799-4808.

[58] Li Q, Li J, Wang Y, Xu Y, Pan X, Zhang G, Luo C, Kobara Y, Nam J J, Jones K C. Atmospheric short-chain chlorinated paraffins in China, Japan, and South Korea. Environment Science &Technolgy. 2012, 46(21): 11948-11954.

[59] Wang Y, Li J, Cheng Z N, Li Q, Pan X, Zhang R, Liu D, Luo C, Liu X, Katsoyiannis A, and Zhang G. Short- and

medium-chain chlorinated paraffins in air and soil of subtropical terrestrial environment in the Pearl River Delta, South China: distribution, composition, atmospheric deposition fluxes, and environmental fate. Environmental Science & Technology, 2013, 47(6): 26792-687.

[60] Fridén U E, Mclachlan M S, Berger U. Chlorinated paraffins in indoor air and dust: concentrations, congener patterns, and human exposure. Environment Intentional, 2011, 37(7): 1169-1174.

[61] Gao W, Wu J, Wang Y W, Jiang G B. Distribution and congener profiles of short-chain chlorinated paraffins in indoor/outdoor glass window surface films and their fim-air partitioning in Beijing, China. Chemosphere, 2016, 144: 1327-1333.

[62] Hilger B, Fromme H, W Völkel, Coelhan M. Occurrence of chlorinated paraffins in house dust samples from Bavaria Germany. Environmental Pollution, 2013, 175: 16-21.

[63] 赵洋洋, 姚义鸣, 常帅, 等. 室内空气和灰尘中全(多)氟烷基化合物的研究进展. 环境化学, 2015, 34(4): 656-662.

[64] Bonvallot N, Mandin C, Mercier F. Health ranking of ingested semi-volatile organic compounds in house dust: an application to France. Indoor Air, 2010, 20: 458-472.

[65] Batterman S, Chernyak S, Jia C, Godwin C, Charles S. Concentrations and emissions of polybrominated diphenyl ethers from U.S. houses and garages. Environmental Science & Technology, 2009, 43: 2693-2700.

[66] Besis A, Samara C. Polybrominated diphenyl ethers (PBDEs)in the indoor and outdoor environments–a review on occurrence and human exposure. Environmental Pollution, 2012, 169, 217-229.

[67] Yuan B, Wang Y, Fu J, Zhang Q, Jiang G. An analytical method for chlorinated paraffins and their determination in soil samples. Chinese Science Bulletin, 2010, 55(22): 2396-2402.

[68] Chen M, Luo X, Zhang X, He M, Chen S. Chlorinated paraffins in sediments from the Pearl River Delta, South China: Spatial and temporal distributions and implication for processes. Environmental Science & Technology, 2011, 45(23): 9936-9943.

[69] Zeng L, Wang T, Han W, Yuan B, Liu Q. Spatial and vertical distribution of short chain chlorinated paraffins in soils from wastewater irrigated farmlands. Environmental Science & Technology, 2011, 45 (6): 2100-2106.

[70] Iino F, Takasuga T, Senthilkumar K, Nakamura A N, Nakanishi J. Risk assessment of short-chain chlorinated paraffins in Japan based on the first market basket study and species sensitivity distributions. Environmental Science & Technology, 2005, 39(3): 859-866.

[71] Nicholls C R, Allchin C R, Law R J. Levels of short and medium chain length polychlorinated n-alkanes in environmental samples from selected industrial areas in England and Wales. Environmental Pollution, 2001, 114(3): 415-430.

[72] Pribylova P, Klάnovά J, Holoubek I. Screening of short-and medium-chain chlorinated paraffins in selected riverine sediments and sludge from the Czech Republic.Environmental Pollution, 2006, 144 (1): 248-254.

[73] Stejnarova P, Coelhan M, Kostrhounovά R, Parlar H, Holoubek I. Analysis of short chain chlorinated paraffins in sediment samples from the Czech Republic by short-column GC/ECNI-MS. Chemosphere, 2005, 58(3): 253-262.

[74] Iozza S, Müller CE, Schmid P, Bogdal C, Oehme M. Historical profiles of chlorinated paraffins and polychlorinated biphenyls in a dated sediment core from Lake Thun (Switzerland). Environmental Science & Technology, 2008, 42(4): 1045-1050.

[75] Tomy G, Stern G, Lockhart W, Muir D C. Occurrence of C_{10}-C_{13} polychlorinated *n*-alkanes in Canadian midlatitude and Arctic lake sediments. Environmental Science & Technology, 1999, 33(17): 2858-2863.

[76] Marvin C H, Painter S, Tomy G, Stern G, Braekevelt E , Muir D C. Spatial and temporal trends in short-chain chlorinated paraffins in lake ontario sediments. Environmental Science & Technology, 2003, 37(20): 4561-4568.

[77] Gao Y, Zhang H, Su F, Tian Y, Chen J. Environmental occurrence and distribution of short chain chlorinated paraffins in sediments and soils from the Liaohe River Basin, P.R. China. Environmental Science & Technology, 2011, 46: 3771-3778.

[78] Parera J, Abalos M, Santos F J, Galceran M T, Abad E. Polychlorinated dibenzo-*p*-dioxins, dibenzofurans, biphenyls, paraffins and polybrominated diphenyl ethers in marine fish species from Ebro River Delta (Spain). Chemosphere, 2013, 93(3): 499-505.

[79] Nicholls C R, Allchin C R, Law R J. Levels of short and medium chain length polychlorinated *n*-alkanes in environmental samples from selected industrial areas in England and Wales. Environmental Pollution, 2001, 114(3): 415-430.

[80] Reth M, Ciric A, Christensen G N, Heimstad E S, Oehme M. Short- and medium-chain chlorinated paraffins in biota from the European Arctic — differences in homologue group patterns. Science of the Total Environment, 2006, 367(1): 252-260.

[81] Borgen A R, Schlabach M, Kallenborn R, Fjeld E. Polychlorinated alkanes in fish from Norwegian freshwater. The Scientific

World Journal, 2002, 2: 136-140.

[82] Zhou Y, Asplund L, Yin G, Athanassiadis I, Wideqvist U, Bignert A, Qiu Y, Zhu Z, Zhao J, Bergman Å. Extensive organohalogen contamination in wildlife from a site in the Yangtze River Delta. Science of the Total Environment, 2016, 554-555: 320-328.

[83] 姜国, 陈来国, 何秋生, 孟祥周, 封永斌, 黄玉妹, 唐才明. 上海食用鱼中短链氯化石蜡的污染特征. 环境科学, 2013, 34(9): 3374-3380.

[84] Zeng L, Wang T, Wang P, Liu Q, Han S, Yuan B, Zhu N, Wang Y, Jiang G. Distribution and trophic transfer of short-chain chlorinated paraffins in an aquatic ecosystem receiving effluents from a sewage treatment plant. Environmental Science & Technology, 2011, 45(13): 5529-5535.

[85] Iino F, Takasuga T, Senthilkumar K, Nakamura N, Nakanishi J. Risk assessment of short-chain chlorinated paraffins in Japan based on the first market basket study and species sensitivity distributions. Environmental Science & Technology, 2005, 39(3): 859-866.

[86] Hassanvand M S, Naddafi K, Faridi S, Nabizadeh R, Sowlat M H, Momeniha F, Gholampour A, Arhami M, Kashani H, Zare A, Niazi S, Rastkari N, Nazmara S, Ghani M, Yunesian M. Characterization of PAHs and metals in indoor/outdoor PM_{10}/$PM_{2.5}$/PM_1 in a retirement home and a school dormitory. Science of the Total Environment, 2015, 527-528: 100-110.

[87] Harrad S, Hazrati S, Ibarra C. Concentrations of polychlorinated biphenyls in indoor air and polybrominated diphenyl ethers in indoor air and dust in Birmingham, United Kingdom: Implications for human exposure. Environmental Science & Technology, 2006, 40(15): 4633-4638.

[88] Ma X, Chen C, Zhang H, Gao Y, Wang Z, Yao Z, Chen J, Chen J. Congener-specific distribution and bioaccumulation of short-chain chlorinated paraffins in sediments and bivalves of the Bohai Sea, China. Marine Pollution Bulletin, 2014, 79(1-2): 299-304.

[89] Ma X, Zhang H, Wang Z, Yao Z, Chen J, Chen J. Bioaccumulation and trophic transfer of short chain chlorinated paraffins in a marine food web from Liaodong Bay, North China. Environmental Science & Technology, 2014, 48(10): 5964-5971.

[90] Barber J L, Sweetman A J, Thomas G O, Braekevelt E, Stern G A, Jones K C. Spatial and temporal variability in air concentrations of short-chain (C_{10}–C_{13}) and medium-chain (C_{14}–C_{17}) chlorinated *n*-alkanes measured in the U.K. atmosphere. Environmental Science & Technology, 2005, 39(12): 4407-4415.

[91] Fridén U E. Sources, emissions, and occurrence of chlorinated paraffins in Stockholm, Sweden.http://www.diva-portal.org/smash/record.jsf?pid=diva2%3A357952&dswid=-5526.

[92] Gao W, Wu J, Wang Y, Jiang G. Distribution and congener profiles of short-chain chlorinated paraffins in indoor/outdoor glass window surface films and their film-air partitioning in Beijing, China. Chemosphere, 2016, 144: 1327-1333.

[93] Thomas G O, Jones K C. 2002. Chlorinated paraffins in human and bovine milk-fat. Department of Environmental Sciences, Lancaster University, Lancaster, U.K.

[94] Lahaniatis M R, Coelhan M, Parlar H. 2000. Clean-up and quantification of short and medium chain polychlorinated n-alkanes in fish, fish oil, and fish feed. Organohalogen Compounds, 47: 276-279.

[95] Persistent Organic Pollutants Review Committe (PORC). Revised draft risk profile: short-chain chlorinated paraffins; United Nations Environment Programme (UNEP), Geneva, Switzerland, 2011.

[96] Tomy G T. The mass spectrometric characterization of polychlorinated n-alkanes and the methodology for their analysis in the environment. University of Manitoba, winnipeg. 1997.

[97] Xia D, Gao L, Zheng M, Li J, Zhang L, Wu Y, Qiao L, Tian Q, Huang H, Liu W, Su G, Liu G. Health risks posed to infants in rural China by exposure to short- and medium-chain chlorinated paraffins in breast milk. Environment International, 2017, 103: 1-7.

[98] Xia D, Gao L, Zheng M, Li J, Zhang L, Wu Y, Tian Q, Huang H, Qiao L. Human exposure to short- and medium-chain chlorinated paraffins via mothers' milk in Chinese urban population. Environmental Science & Technology, 2017, 51: 608-615.

作者：高丽荣[1]，夏 丹[1]，高 媛[2]，郑明辉[1]，陈吉平[2]，黄业茹[3]

[1]中国科学院生态环境研究中心，[2]中国科学院大连化学物理研究所，[3]国家环境分析测试中心

第3章 环境中微塑料的生物效应及载体作用

本章导读

微米到毫米级别的塑料（微塑料，MP）以及纳米级别的塑料（纳米塑料，NP）所造成的环境污染问题近年来引起了学术界的广泛关注，特别是微塑料对海洋和淡水环境的污染、与环境中其他污染物的复合污染效应及生态风险已经成为环境科学研究的热点。本章将就微塑料的定义及分布、环境检测方法、毒性与危害、载体作用、与污染物共存时的生物富集和降解等研究进行综述，并展望微塑料未来的研究方向。

关键词

微塑料，环境行为，毒性效应，分布和检测，复合污染

1 微塑料的定义、来源及分布

微塑料是指粒径<5 mm 的塑料颗粒。微塑料在全球水生态系统中广泛分布，已经成为新的威胁水生生物甚至人类健康的隐患，其主要分布在水体、沉积物（沙滩和底泥）以及生物体中，并可随水力、风力作用进行长距离运输[1,2]。由于微塑料密度小且具有漂浮性，水体中的微塑料主要集中在表层水中，各大海洋、河流及湖泊中均检测到一定浓度的微塑料（表 3-1）。某些密度比水大的微塑料（如聚氯乙烯）和黏附有其他杂质的微塑料颗粒，会慢慢沉入水底，沉积在底泥中，Vianello 等[3]在威尼斯河底泥中就检测到了微塑料的存在。随着浪潮涌动，表层水的微塑料颗粒会被推送到沙滩或河岸并沉积下来，因此在很多沙滩或河岸均能检测到微塑料（表 3-1）。生活在水中的各类生物由于对食物缺乏选择性，很有可能会误食微塑料颗粒，因此在很多水生生物体内也可以检测到微塑料（表 3-1）。另外，Setälä 等[4]探究了微塑料在浮游生物食物网中的传输，Seltenrich [5]认为食物链中的微塑料污染与海鲜食品安全密切相关，而 Yang 等[6]则直接在食盐中检测出微塑料，这些研究均表明微塑料污染与人体健康息息相关。

过去五十年中，塑料产品的生产和使用量急剧增长。据统计，2014 年全球塑料制品的产量已经达到 3 亿吨[7]，并且仍在持续增加。塑料的大量使用以及随意抛弃导致其在环境中大量积累，由于塑料制品极难降解，其在环境中的存留时间可以达到 100 年甚至更长[8]。残留在环境中的塑料制品通过水流、风力和其他机械作用力以及紫外线，环境微生物等物理、化学和生物的共同作用，可以形成粒径更小的塑料颗粒——微塑料[9,10]。已经有很多研究表明微塑料污染正成为世界性的问题[11]，尤其是人类居住地附近的海岸，甚至在孤岛深海也有发现[12]。

此外，微米塑料以及纳米塑料还是个人护理品中的添加成分，因为对个人护理品的生产和使用，使得微米塑料及纳米塑料可随生活废水直接进入环境[12-14]。另外，研究还发现化纤类衣物在洗涤过程中会有大量衣物纤维通过洗衣机废水排入城市污水，从而进入环境[12]。在人类活动频繁的区域，海口河流中的微塑料含量较高[15]，表明微塑料的环境分布与人类活动密切相关。而微塑料一旦进入海洋或者大气层，将随着洋流以及大气流动而移动，从而分布到全球各地，严重威胁海洋环境和人类健康。总的来说，微塑料进入环境的途径可以归纳为以下几点：

（1）农用土壤中的废弃地膜等农业垃圾与水体中被丢弃的塑料（包装）垃圾在物理化学降解和碎化作用下逐渐形成粒径在 5mm 以下的微塑料颗粒；

（2）微塑料颗粒直接排放进入下水道或地表径流，以及通过堆放等方式进入土壤环境（例如，塑料的原料颗粒、通过珠粒喷漆的船体表面颗粒脱落、工业清洗产品、磨削研磨工艺产物等）；

（3）个人护理用品（例如洗面奶等）中含有的微塑料颗粒、纤维多聚物等，衣物、地毯，以及其他聚酰胺（尼龙）、聚酯和聚丙烯酸纺织品上脱落的塑料颗粒和纤维等，可以进入城市排水系统。尽管在污水处理厂中，一部分微塑料可以通过初级沉降和浮渣清除等处理工艺被清除，并蓄积在污泥中[16]，但仍有相当一部分微塑料颗粒残留在处理后的出水中，被排入地表水体。McCormick 等[17]的最新研究显示：有大量微塑料残留在污水处理厂处理后的出水中，并最终进入河流，是淡水生态系统中微塑料的重要来源。

表 3-1　不同环境介质中微塑料的分布情况

环境介质	区域或物种	主要类型*	浓度	颗粒粒径	参考文献
水体	大西洋	ABS、PE	13~501 颗/m^3	10~1000 μm	[18]
	南韩镇海湾	PP、PE、PS	6~23 颗/m^2	0.2~1 mm	[19]
	太平洋东北部	塑料碎片、纤维、小球	4~190 颗/10^3 m^3	>0.505 mm	[20]
	美国芝加哥城市河流	塑料碎片、薄膜、泡沫、小球、线条等	2~18 颗/m^3	>0.333 mm	[21]
	中国东海	塑料纤维、薄膜、颗粒、小球等	167 颗/10^3 m^3	>0.5 mm	[22]
	中国长江河口	塑料纤维、薄膜、颗粒、小球等	4.14×10^4 颗/m^3	>0.5 mm	[22]
	波罗的海	塑料薄膜、碎片、油漆碎片、颗粒等	0~1 颗/m^3	0.333~5 mm	[23]
	地中海西北部	细丝、塑料颗粒、塑料薄膜等	116 颗/10^3 m^3	0.5~5 mm	[24]
	大西洋西部	塑料碎片、薄膜、油漆碎片	30 颗/10^3 m^3	0.3~5 mm	[25]
	黑海东南部	塑料薄膜、碎片等	0.6~1.2×10^3 颗/m^3	0.2~5 mm	[26]
	葡萄牙沿海	PE、PP、PA	2~36 颗/10^3 m^3	0.18~5 mm	[27]
	南非西南海岸线	聚苯乙烯小球、塑料碎片、纤维等	258~1，215 颗/m^3	0.08~5 mm	[28]
	加拿大温尼伯湖	塑料纤维、薄膜、泡沫等	1.93±1.15×10^4 颗/m^2	0.333~5 mm	[29]
	韩国南部海岸	EPS、PE、PP、PES、PA 等	1.63±1.35×10^4 颗/m^3	≤5 mm	[30]
沉积物	地中海海底	颜料碎片等	0~1 颗/m^2	<1 mm	[31]
	中国西藏高原湖岸	PP、PE、PS、PVC、PET	8~563 颗/m^2	≤5 mm	[32]
	葡萄牙里斯本沙滩	PE、PS、PES	29~393 颗/m^2	50 μm~20 mm	[33]
	意大利威尼斯河河底	PP、PE、PS、PVC 等	2，175~672 颗/kg 干重	<1 mm	[3]
	北海沙滩	PE、PP、PET、PVC、PS、PA	1~3 颗/kg 干重	<1 mm	[34]
	德国波罗的海沙滩	纤维类、塑料颗粒	0~7 颗/kg 干重	55 μm~5 mm	[35]
	巴西东南海湾沙滩	纤维类、塑料碎片、塑料泡沫、塑料颗粒	12~1300 颗/m^2	≤5 mm	[36]
	中国渤海沙滩	PEVA、PE、PS、PP、PET	103~163 颗/kg 干重	≤5 mm	[37]
	中国三峡水库湘西湾河岸	PS、PP、PE	80~864 颗/m^2	0.3~5 mm	[38]
	中国长江入海口沙滩	RY、PES、AC、PET、PS	20~340 颗/kg 干重	< 5 mm	[39]
	英国河流河底	塑料薄膜、纤维、碎片、泡沫等	25~30 颗/100 g 干重	< 5 mm	[40]
生物体	海洋幼虫	聚苯乙烯小球	1~2 颗/只	< 1 mm	[41]
	虾	塑料纤维、小球等	0~2 颗/只	200~1000 μm	[42]
	贻贝	塑料纤维、碎片、小球、薄片等	1~8 颗/只	< 5 mm	[43]
	双壳类	塑料纤维、碎片、颗粒等	4~57 颗/只	< 5 mm	[44]
	鱼类	主要是塑料纤维类	4 颗/条鱼	< 5 mm	[45]
	鱼类	主要是塑料小球	2~6 颗/条鱼	1~5 mm	[46]
	海鸟	塑料颗粒、纤维、薄膜、碎片等	0.385 克/67 只	0.5~5 mm	[47]
	海龟	塑料颗粒、纤维、薄膜、碎片等	3~134 颗/只	< 5 mm	[48]
	鲸	RY、PP、PE、PES	88 颗/条	0.3~7 mm	[49]

* ABS 指丙烯腈—丁二烯—苯乙烯聚合物；PE 指聚乙烯；PP 指聚丙烯；PS 指聚苯乙烯；PET 指聚乙二醇对苯二甲酸酯聚合物；PEVA 指乙烯-醋酸乙烯共聚物；PY 指乙烯-乙酸乙烯共聚物；AC 指丙烯酸聚合物；PES 指聚酯；PA 指聚丙烯酸酯；PVC 指聚氯乙烯

2 环境中微塑料的检测方法

2.1 样品采集

由于微塑料具有漂浮性，环境水体中的微塑料主要漂浮在水体表层[30]。最常见的表层水体采样方式为拖网式，广泛应用于海洋表层水和大型河流表层水的采集[22,37,50]。除此之外，抽滤泵常用来采集深度为0.2~4.5m 处的水体样品[51,52]，抽滤泵上所采用的网筛孔径在不堵塞前提下，用得较多的有 50 μm，100 μm 和 300 μm 的网筛[23]。Reisser 等[53]检测了北大西洋环流海水里微塑料在距水-气界面 0~5m 的垂直分布，发现海水中的微塑料主要集中在海水表层 0~0.5m 处。因此，推荐的水体采样方法是拖网式采样。

随着潮汐涌动，部分水体表层微塑料被推向沙滩，沉积在沙滩或河岸上。沙滩或河岸沉积物的采集主要是用不锈钢工具挖取深度（1~20 cm 不等）和面积（边长为 5 cm、15 cm、20 cm 或 1 m 的正方形）一定的泥沙，采样点一般选择在高位潮汐线[36,54]。

河流底泥样品的采集一般采用不锈钢抓斗[3]，海洋底部样品可以采集一个面积为 25cm^2 的沉积核，再将该核表层的 1 cm 沉积物进行分离分析[31]。

由于微塑料易被生物体吞食，因此生物样品也是监测水生环境中微塑料污染的一项重要指标。主要采集的生物有鱼类[55]、海龟[56]、海鸟[57]、贝壳类[44]、海虫[58]和浮游生物[4]等。野生或者农场饲养的生物样品采集到之后，通常将其内脏各组织器官分离，并进一步提取微塑料。

2.2 分离提取及纯化

密度浮选法是应用最广泛的从泥沙沉积物或其他环境样品中分离微塑料的方法。一般来说，沉积物中的微塑料颗粒（一般约为 0.9~1.4 g/cm^3）比泥沙（一般约为 2.0 g/cm^3）密度低，实验中常选用密度高于微塑料颗粒并低于泥沙的重液使微塑料浮在水面，达到分离微塑料的目的。常用的重液有 NaCl（1.2 g/cm^3）、$ZnCl_2$（1.6~1.7 g/cm^3）、NaI（1.6 g/cm^3）等[59-61]。基于密度浮选法，Nuelle 等[62]结合使用 NaCl 和 NaI 用实验室常用玻璃器皿搭建一组简易装置对沉积物中的微塑料进行分离提取，该方法主要是利用高密度重液和气流推动微塑料上浮，并多次循环提取，达到充分分离微塑料的目的。同样是基于密度浮选法，Claessens 等[61]和 Imhof 等[63]开发新装置，在装置底部放置磁力搅拌器，不停搅拌沉积物，使积压在沉积物底部的微塑料能充分浮出，微塑料的提取率高达 98%~100%。密度浮选初步分离出来的微塑料随后将进行筛分和过滤，肉眼可识别的较大塑料颗粒可用孔径为 500 μm 的筛网筛分出来，余下细小颗粒需借助显微镜进一步挑选出来或用孔径更小的滤纸过滤出[64]。

分离出来的微塑料在仪器定性检测之前需进行纯化，最常用的纯化试剂是 30%H_2O_2 溶液[22,35]。Nuelle 等[62]对比分析了几种纯化试剂（如 30%H_2O_2、35%H_2O_2、20%HCl 和 20%~50%NaOH）去除样品中有机杂质的效率，其中 35%H_2O_2 的效果最佳，但是该试剂会使微塑料样品褪色。蛋白酶水解法也适用于微塑料环境样品中的有机杂质[65,66]。生物样品前处理通常是将生物体的各组织器官解剖分离，用纯化试剂（如 KOH、NaClO、NaOH 或者 HNO_3 溶液）将有机杂质消解，去除效率可高达 95%[67,68]。除此之外，在污水处理中常用的芬顿（Fenton）消解法也可应用于微塑料样品的纯化[29]。有机杂质的去除方法有多种，在确保高效去除有机杂质的同时需考虑是否会对目标物（如微塑料及其所携带的有机污染物）产生影响。

2.3 定性定量分析

分离纯化后的微塑料样品需进行定性分析来确定其组分，常用的几种定性检测仪器主要有热解气相色谱质谱联用仪（Pyr-GC/MS 或 TDS-GC/MS）、傅里叶红外光谱仪（FT-IR）和激光拉曼光谱仪（Raman spectroscopy）等。热解 GC/MS 可以同时分析微塑料材质及塑料添加剂，然而该方法将破坏微塑料样品本身[59]。Fischer 和 Scholz-Böttcher[69]用不同塑料的特征离子作为标志物来对微塑料进行定性分析，无需对微塑料样品进行预挑选，然而该方法只能得到微塑料的质量值，无法测得其颗粒数。傅里叶红外光谱仪（FT-IR）对微塑料进行定性分析的检测限一般为 10~20 μm，广泛应用于粒径大于 10 μm 的微塑料检测[43,70,71]，要用 FT-IR 测定更小的微塑料颗粒仍是一大挑战。全反射红外光谱（ATR-FT-IR）可检测形状不规则的微塑料颗粒，但该方法仅适用于粒径大于 500 μm 的大颗粒。显微红外光谱（micro-FT-IR）可直接测定 60℃烘干后滤膜上的微塑料颗粒[72]，Löder 和 Gerdts[73]发现分辨率调至 8 cm^{-1} 时可得到高质量的分析数据并节省测定时间。激光拉曼光谱仪可检测到粒径低至 1 μm 的微塑料颗粒[18]，然而在仪器购买价格上，拉曼光谱仪大概是 FT-IR 的五倍，检测成本远远大于 FT-IR。

微塑料的定性方法多样，实际操作中可将这几种检测方法结合起来，优势互补，最大限度的准确检测出粒径极小的微塑料颗粒[69]。在微塑料的定性上，未来尚需在检测技术上进行优化，对粒径更小的纳米级塑料颗粒实现准确定性。

微塑料的定量分析主要是基于人工计数或天平称量来确定其颗粒数或质量。微塑料在不同环境介质中的浓度表示单位不同，例如水面微塑料浓度单位一般用“颗/m^2”或“颗/m^3”表示，沉积物中的微塑料浓度单位一般用“颗/m^2”、“颗/kg dw”或“g/m^2 或 mg/m^2”表示，生物体中微塑料的浓度一般用“颗/单位个体”或者“g/单位个体”表示。在微塑料的定量分析中，未来可采用自动计数方法代替人工计数，浓度单位尽可能地用可以相互转换的单位来表示，这样便于对比不同地点的微塑料污染程度。

3 微塑料的毒性与危害

微塑料粒径微小，很容易被水生生物摄食。已经在端足类甲壳动物、沙蚕、藤壶、贻贝、十足甲壳类、海鸟以及鱼类等多种海洋生物体内发现了微塑料[74-78]，这些微塑料颗粒会对水生生物个体造成影响，甚至产生毒性，从而影响生态系统安全[79,80]。微塑料被浮游或底栖动物摄食后，可以妨碍浮游甲壳类生物的附肢运动，堵塞水生生物消化道并对其消化系统形成机械损伤，或者进入循环系统对动物组织造成伤害并且通过食物链影响上一层捕食者[75,77,81]。此外，微塑料能显著影响鲟鱼的摄食率[82]，影响大型蚤的运动能力和生殖健康，导致发育畸形[83,84]。微塑料还有可能影响种群的活性和种群结构进而影响生态系统，研究表明，微塑料的存在可以影响绿藻的光合作用[85]和呼吸速率，从而影响绿藻生长，并导致绿藻活性氧含量的增加[85]，使得生态系统中的建群种生物滤食效率下降，蟹类的摄食和活性下降，最终使得生态系统中的种群分布发生改变[86]。而海盐[6]和水生生物体内的微塑料[13,15]则会通过人类的摄氏或者食物链的传递进入人体，从而危害人类健康[87]。

与微塑料相较，粒径更小的纳米塑料还有可能进入生物细胞内部[88,89]，研究表明，纳米塑料可能积累在细胞溶酶体内，导致溶酶体破裂组织蛋白释放[90]，产生比微塑料更大的危害[79,91]。

4 微塑料的载体作用

除了微塑料自身可能对水生生物造成的物理性伤害以外，水体中的塑料还被检测出携带大量的持久性有机污染物（POPs）。有毒化合物如多氯联苯（PCBs）、多溴联苯醚（PBDEs）、多环芳烃类（PAHs）、双酚 A（BPA）均在海洋塑料废弃物表面被发现[92-95]。这些持久性有机污染物一方面是在塑料生产时为了特定的目的而添加的，如增加塑料弹性的邻苯二甲酸盐以及作为阻燃剂加入的 PCBs、PBDEs 等，这些添加剂在塑料分解过程中析出；另一方面，巨大的比表面积以及强疏水性使得微塑料成为环境中绝佳的疏水有机污染物吸附剂[92,96,97]。Rios 等[93]对北太平洋涡流、美国加州和夏威夷海岸、墨西哥瓜达鲁普岛水域中微塑料富集的 POPs 进行调查发现，微塑料中 PCBs 含量为 27~980 ppm，DDTs 含量为 22~7100 ppm，PAHs 含量为 39~1200 ppm，长链烃类含量为 1.1~8600 ppm。微塑料还兼具类似被动采样器的功能，它们可以从周围水体中富集有机污染物[1,98]。当污染物负载至微塑料后，微塑料可增加这些污染物的持久性[97]，通过“吸附”和“穿梭”作用增强它们的流动性，甚至可以将污染物迁移至北极地区[99,100]。这些吸附在微塑料表面的高浓度有毒物质随微塑料一起被各类动物摄取，并可能进一步在其消化道内解吸而释放到动物体内。这些有毒物质绝大部分还可能通过生物富集与放大作用随食物链积累，最终威胁到人类健康。

微塑料除了可以负载有机污染物，还可以携带多种微生物，其对微生物和化学物质的“载体”作用可以相互影响，并产生潜在的生态风险（图 3-1）。微塑料及其附着生物群落组成了“塑料圈”，由于塑料独特的物理化学结构，其与天然颗粒物如木屑、矿物等不同，研究结果显示微塑料上的生物群落与其周围海水中的群落明显不相同[101,102]，某些附着微生物是条件致病菌，它们漂流扩散，从而改变其自然分布范围，加速了生物的传播和入侵。McCormick 等人在最近的美国地球物理联合会（AGU 会议，2017 年 2 月，美国新奥尔良）上指出：微塑料可以逃脱污水处理系统的拦截进入河流，成为运载细菌（包括病原菌）的良好载体。同时，微塑料上的生物群落还会随着地理位置、温度、塑料类型而发生改变[103]。目前关于微塑料表面的生物群落结构尤其是致病菌的研究还处于起步阶段，研究手段尚不成熟，尤其是微塑料迁移过程中进入不同环境时表面微生物群落结构的变化，以及致病菌可能在生物肠道中释放的环境风险尚无研究。

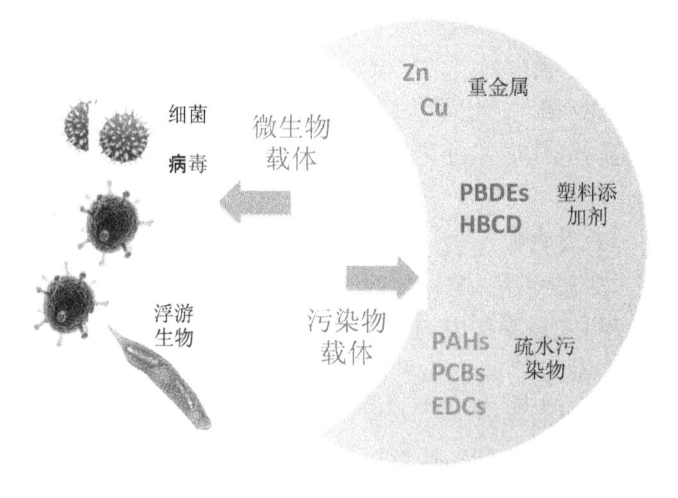

图 3-1 微塑料对污染物和微生物的“载体”作用示意图

表 3-2　微塑料与有机污染物及重金属的联合作用：环境样品检测、吸附一解吸、生物富集和生物毒性等

样品种类	塑料种类	污染物种类	主要发现	参考文献
沙滩样品	聚乙烯	多环芳烃 多环芳烃	PAH 浓度范围为 39~1200 ng/g，65%样品中检测到芘；62%样品中检测到荧蒽；54%样品中检测到萘和菲	[93]
	聚丙烯	多氯联苯	PCBs 浓度范围为 27~980 ng/g，其中最丰富的是 CB-52、101、118 和 170	
		滴滴涕	DDT 代谢物浓度范围为 22 7100 ng/g，样品中只检测到了 4,4-DDT 以及其代谢产物 4,4-DDE 和 4,4-DDD	
	聚乙烯	重金属	沙滩微塑料样品中检测到 Cd，Pb，Se，Cr 等重金属，大部分由塑料生产过程中加入的颜料带入，不同重金属含量与塑料颜色有关。	[104]
水体	聚丙烯	重金属	重金属 Cu 和 Zn 均能显著吸附在聚苯乙烯和聚氯乙烯塑料上，且在老化的聚氯乙烯塑料上吸附更强。	[105]
贻贝	聚乙烯	多环芳烃	微塑料存在时贻贝体内芘的显著积累，并引起了细胞效应和基因表达的改变	[106]
	聚苯乙烯	多环芳烃	聚苯乙烯存在时贻贝体内芘发生显著积累，细胞效应和基因表达发生了改变	
斑马鱼	聚乙烯	多环芳烃	未吸附污染物的微塑料没有引起肠道物理损伤，吸附在微塑料上的污染物可能在鱼肠道中解吸，并转移到肠上皮和肝脏中	[107]
大型溞	聚苯乙烯	多环芳烃	纳米塑料（50 nm）显著增强了大型溞体内菲及其衍生残留物的生物积累，而微塑料（10μm）对菲的生物转化和生物积累无显著影响	[84]
模拟肠道	聚乙烯 聚氯乙烯	多环芳烃、滴滴涕、全氟辛酸、邻苯二甲酸酯、多氯联苯	在肠道条件下污染物从塑料表面解吸速率是海水中的 30 多倍	[108]

5　污染物与微塑料共存时的生物富集和降解研究

有机污染物在水生生态系统中主要通过生物降解、光降解、化学降解和挥发等途径去除。如前所述，疏水性有机污染物（HOCs）极易与具有表面疏水性的塑料碎片，特别是小粒径的微塑料颗粒相结合[109,110]。另外，塑料经过物理化学及生物降解等过程产生不规则形状以及丰富的表面官能团[111]，可以影响有机污染物与微塑料之间吸附和结合的能力，从而改变他们在环境中的迁移能力。污染物在微塑料上的吸附/脱附行为还会严重影响其自由溶解态浓度，减缓污染物的转化[84]。尽管负载有机污染物的微塑料具有较大的潜在生态风险，然而微塑料对水环境中共存污染物的降解、环境归趋，以及生物有效性的影响机制仍不清晰。

目前已经证实：污染物可以经由微塑料的生物摄入过程进入生物体内。这些吸附的有毒物质可以在生物体内脱附，并在生物体内累积[83,112-114]。一些证据还显示塑料摄入和 PCBs 在海鸟体内的浓度呈正相关关系，这提示我们海鸟体内的部分 PCBs 可能是随塑料的摄取而带入的[97,115,116]。但也有许多模型研究指出，尽管微塑料有助于共存污染物在生物体内的生物富集和迁移，但也是随塑料和污染物特性而异的[117]，甚至有研究者认为与直接从水体或者食物中吸收相比，微塑料携带的污染物迁移量几乎可以忽略[118]。因此，微塑料如何影响其携带的有机污染物的环境行为和降解的机理，尤其是微塑料的种类、风化程度和粒径大小对不同种类的污染物的不同影响，以及环境条件如盐度、pH、生物表面活性剂等变化情况下的微塑料对污染物降解的不同影响将是未来研究的重要方向。

6 展　望

目前关于微塑料及其携带污染物的调查研究大部分还局限于水体环境中，尤其是粒径相对较大的毫米和微米级塑料颗粒。由于分析手段的局限，微米以下的塑料颗粒尚无很好的检测和分离方法。同时，对有机溶剂不耐受的聚苯乙烯、聚氯乙烯等材质的塑料颗粒，其表面的有机污染物无法很好的提取和分离，也无法检测。陆地尤其是土壤环境中，随着农业生产中塑料产品的大面积使用，微米或纳米塑料的环境暴露风险逐年增加，但是目前还没有很好的分离土壤或其他陆地环境中微塑料的方法，更难以对土壤中微塑料颗粒携带污染物的环境风险进行研究。在微塑料的环境监测研究中，绝大部分工作都聚焦在某个区域内微塑料的分布上，对微塑料在环境中的整体归趋及其在生物链中的富集与传播等方面的研究还很缺乏。因此，微塑料及其携带污染物的环境风险研究亟需解决的技术瓶颈主要是微塑料的分离、提取和纯化方法，以及提取分离和定量方法的标准化。同时，大尺度环境中微塑料的分布和迁移规律，其携带污染物在食物链和食物网中的传播规律，生物分子水平上的毒性机理研究也是未来需要解决的关键科学问题。

致谢：国家自然科学青年基金项目（编号：21407075）、“十三五”国家重点研发计划“海洋环境安全保障”重点专项（编号：2016YFC1402200）和国家科技部“水体污染控制与治理科技”重大专项（编号：2017ZX07301005）。

参考文献

[1] Thompson R C, Moore C J, von Saal F S, Swan S H. Plastics, the environment and human health: current consensus and future trends. Philosophical Transactions of the Royal Society B, 2009, 363: 2153-2166.

[2] Auta H S, Emenike C U, Fauziah S H. Distribution and importance of microplastics in the marine environment: A review of the sources, fate, effects, and potential solutions. Environment International, 2017, 102: 165-176.

[3] Vianello A, Boldrin A, Guerriero P, Moschino V, Rella R, Sturaro A, Da Ros L. Microplastic particles in sediments of Lagoon of Venice, Italy: First observations on occurrence, spatial patterns and identification. Estuarine, Coastal and Shelf Science, 2013, 130: 54-61.

[4] Setala O, Fleming-Lehtinen V, Lehtiniemi M. Ingestion and transfer of microplastics in the planktonic food web. Environmental Pollution, 2014, 185: 77-83.

[5] Seltenrich N. New link in the food chain? Marine plastic pollution and seafood safety. Environmental Health Perspectives, 2015, 123 (2): A34-A41.

[6] Yang D, Shi H, Li L, Li J, Jabeen K, Kolandhasamy P. Microplastic pollution in table salts from China. Environmental Science and Technology, 2015, 49 (22): 13622-13627.

[7] Anderson J C, Park B J, Palace V P. Microplastics in aquatic environments: Implications for Canadian ecosystems. Environmental Pollution, 2016, 218: 269-280.

[8] Hamer J, Gutow L, Kohler A, Saborowski R. Fate of microplastics in the marine isopod Idotea emarginata. Environmental Science and Technology, 2014, 48 (22): 13451-13458.

[9] Lambert S, Wagner M. Characterisation of nanoplastics during the degradation of polystyrene. Chemosphere, 2016, 145:

265-268.

[10] Stolte A, Forster S, Gerdts G, Schubert H. Microplastic concentrations in beach sediments along the German Baltic coast. Marine Pollution Bulletin, 2015, 99: 216-229.

[11] Mathalon A, Hill P. Microplastic fibers in the intertidal ecosystem surrounding Halifax Harbor, Nova Scotia. Marine Pollution Bulletin, 2014, 81 (1): 69-79.

[12] Browne M A, Crump P, Niven S J, Teuten E, Tonkin A, Galloway T, Thompson R. Accumulation of microplastic on shorelines woldwide: sources and sinks. Environmental Science and Technology, 2011, 45 (21), 9175-9179.

[13] Wardrop P, Shimeta J, Nugegoda D, Morrison P D, Miranda A, Tang M, Clarke B O. Chemical Pollutants Sorbed to Ingested Microbeads from Personal Care Products Accumulate in Fish. Environmental Science and Technology, 2016, 50 (7): 4037-4044.

[14] Hernandez L M, Yousefi N, Tufenkji N. Are there nanoplastics in your personal care products? Environmental Science and Technology Letters, 2017, 4 (7): 280-285.

[15] Su L, Xue Y, Li L, Yang D, Kolandhasamy P, Li D, Shi H. Microplastics in Taihu Lake, China. Environmental Pollution, 2016, 216: 711-719.

[16] Carr S A, Liu J, Tesoro A G. Transport and fate of microplastic particles in wastewater treatment plants. Water Reseash, 2016, 91: 174-182.

[17] McCormick A R, Hoellein T J, London M G, Hittie J, Scott J W, Kelly J J. Microplastic in surface waters of urban rivers: concentration, sources, and associated bacterial assemblages. Ecosphere, 2016, 7 (11): 1-22.

[18] Lenz R, Enders K, Stedmon C A, Mackenzie D M A, Nielsen T G. A critical assessment of visual identification of marine microplastic using Raman spectroscopy for analysis improvement. Marine Pollution Bulletin, 2015, 100 (1): 82-91.

[19] Song Y K, Hong S H, Mi J, Han G M, Shim W J. Occurrence and distribution of microplastics in the sea surface microlayer in Jinhae Bay, South Korea. Archives of Environmental Contamination and Toxicology 2015, 69 (3): 279-287.

[20] Doyle M J, Watson W, Bowlin N M, Sheavly S B. Plastic particles in coastal pelagic ecosystems of the Northeast Pacific ocean. Marine Environmental Research, 2011, 71 (1): 41-52.

[21] Eriksen M, Mason S, Wilson S, Box C, Zellers A, Edwards W, Farley H, Amato S. Microplastic pollution in the surface waters of the Laurentian Great Lakes. Marine Pollution Bulletin, 2013, 77 (1–2): 177-182.

[22] Zhao S Y, Zhu L, Wang T, Li D. Suspended microplastics in the surface water of the Yangtze Estuary System, China: First observations on occurrence, distribution. Marine Pollution Bulletin, 2014, 86 (1–2): 562-568.

[23] Setälä O, Magnusson K, Lehtiniemi M, Noren F. Distribution and abundance of surface water microlitter in the Baltic Sea: A comparison of two sampling methods. Marine Pollution Bulletin, 2016, 110 (1): 177-183.

[24] Collignon A, Hecq J H, Glagani F, Voisin P, Collard F, Goffart A. Neustonic microplastic and zooplankton in the North Western Mediterranean Sea. Marine Pollution Bulletin 2012, 64 (4): 861-864.

[25] Ivar Do Sul J A, Costa M F, Fillmann G. Microplastics in the pelagic environment around oceanic islands of the Western Tropical Atlantic Ocean. Water Air and Soil Pollution, 2014, 225 (7): 1-13.

[26] Aytan U, Valente A, Senturk Y, Usta R, Esensoy Sahin F B, Mazlum R E, Agirbas E. First evaluation of neustonic microplastics in Black Sea waters. Marine Environmental Research, 2016, 119: 22-30.

[27] Frias J P G L, Otero V, Sobral P. Evidence of microplastics in samples of zooplankton from Portuguese coastal waters. Marine Environmental Research, 2014, 95: 89-95.

[28] Nel H A, Froneman P W. A quantitative analysis of microplastic pollution along the south-eastern coastline of South Africa. Marine Pollution Bulletin, 2015, 101 (1): 274-279.

[29] Anderson P J, Warrack S, Langen V, Challis J K, Hanson M L, Rennie M D. Microplastic contamination in Lake Winnipeg,

Canada. Environmental Pollution, 2017, 225: 223-231.

[30] Song Y K, Hong S H, Mi J, Kang J H, Kwon O Y, Han G M, Shim W J. Large accumulation of micro-sized synthetic polymer particles in the sea surface microlayer. Environmental Science and Technology, 2014, 48, (16): 9014-9021.

[31] van Cauwenberghe L, Vanreusel A, Mees J, Janssen C R. Microplastic pollution in deep-sea sediments. Environmental Pollution, 2013, 182: 495-499.

[32] Zhang K, Su J, Xiong X, Wu X, Wu C, Liu J. Microplastic pollution of lakeshore sediments from remote lakes in Tibet plateau, China. Environmental Pollution, 2016, 219: 450-455.

[33] Martins J, Sobral P. Plastic marine debris on the Portuguese coastline: A matter of size? Marine Pollution Bulletin, 2011, 62 (12): 2649-2653.

[34] Dekiff J H, Remy D, Klasmeier J, Fries E. Occurrence and spatial distribution of microplastics in sediments from Norderney. Environmental Pollution, 2014, 186: 248-256.

[35] Stolte A, Forster S, Gerdts G, Schubert H. Microplastic concentrations in beach sediments along the German Baltic coast. Marine Pollution Bulletin, 2015, 99 (1–2): 216-229.

[36] de Carvalho D G, Baptista Neto J A. Microplastic pollution of the beaches of Guanabara Bay, Southeast Brazil. Ocean and Coastal Management, 2016, 128: 10-17.

[37] Yu X, Peng J, Wang J, Wang K, Bao S. Occurrence of microplastics in the beach sand of the Chinese inner sea: The Bohai Sea. Environmental Pollution, 2016, 214: 722-730.

[38] Zhang K, Xiong X, Hu H, Wu C, Bi Y, Wu Y, Zhou B, Lam P K S, Liu J. Occurrence and characteristics of microplastic pollution in Xiangxi Bay of Three Gorges Reservoir, China. Environmental Science and Technology, 2017, 51 (7): 3794-3801.

[39] Peng G, Zhu B, Yang D, Su L, Shi H, Li D. Microplastics in sediments of the Changjiang Estuary, China. Environmental Pollution, 2017, 225: 283-290.

[40] Vaughan R, Turner S D, Rose N L. Microplastics in the sediments of a UK urban lake. Environmental Pollution, 2017, 229: 10-18.

[41] Kaposi K L, Mos B, Kelaher B P, Dworjanyn S A. Ingestion of microplastic has limited impact on a marine larva. Environmental Science and Technology, 2014, 48 (3): 1638-1645.

[42] Devriese L I, van der Meulen M D, Maes T, Bekaert K, Paul-Pont I, Frère L, Robbens J, Vethaak A D. Microplastic contamination in brown shrimp (*Crangon crangon*, Linnaeus 1758) from coastal waters of the Southern North Sea and Channel area. Marine Pollution Bulletin, 2015, 98 (1–2): 179-187.

[43] Cheung P K, Cheung L T O, Fok L. Seasonal variation in the abundance of marine plastic debris in the estuary of a subtropical macro-scale drainage basin in South China. Science of the Total Environment, 2016, 562: 658-665.

[44] Li J, Yang D, Li L, Jabeen K, Shi H. Microplastics in commercial bivalves from China. Environmental Pollution, 2015, 207: 190-195.

[45] Nadal M A, Alomar C, Deudero S. High levels of microplastic ingestion by the semipelagic fish bogue Boops boops (L.) around the Balearic Islands. Environmental Pollution, 2016, 214: 517-523.

[46] Miranda D D A, de Carvalho-Souza G F. Are we eating plastic-ingesting fish? Marine Pollution Bulletin, 2016, 103 (1–2): 109-114.

[47] Avery-Gomm S, O'Hara P D, Kleine L, Bowes V, Wilson L K, Barry K L. Northern fulmars as biological monitors of trends of plastic pollution in the eastern North Pacific. Marine Pollution Bulletin, 2012, 64 (9): 1776-1781.

[48] Tourinho P S, Ivar do Sul J A, Fillmann G. Is marine debris ingestion still a problem for the coastal marine biota of southern Brazil? Marine Pollution Bulletin, 2010, 60 (3): 396-401.

[49] Lusher A L, Hernandez-Milian G, O'Brien J, Berrow S, O'Connor I, Officer R. Microplastic and macroplastic ingestion by a deep diving, oceanic cetacean: The True's beaked whale Mesoplodon mirus. Environmental Pollution, 2015, 199: 185-191.

[50] Eriksen M, Lebreton L C M, Carson H S, Thiel M, Moore C J, Borerro J C, Galgani F, Ryan P G, Reisser J. Plastic pollution in the world's oceans: more than 5 trillion plastic pieces weighing over 250, 000 tons afloat at sea. PLoS ONE, 2014, 9 (12): 1-15.

[51] Desforges J-P W, Galbraith M, Dangerfield N, Ross P S. Widespread distribution of microplastics in subsurface seawater in the NE Pacific Ocean. Marine Pollution Bulletin, 2014, 79 (1-2): 94-99.

[52] Enders K, Lenz R, Stedmon C A, Nielsen T G. Abundance, size and polymer composition of marine microplastics ≥ 10 μm in the Atlantic Ocean and their modeled vertical distribution. Marine Pollution Bulletin, 2015, 100 (1): 70-81.

[53] Reisser J, Slat B, Noble K, Plessis K D, Epp M, Proietti M, De Sonneville J, Becker T, Pattiaratchi C. The vertical distribution of buoyant plastics at sea: an observational study in the North Atlantic Gyre. Biogeosciences, 2015, 12 (4): 1249-1256.

[54] Zhang K, Su J, Xiong X, Wu X, Wu C, Liu J. Microplastic pollution of lakeshore sediments from remote lakes in Tibet plateau, China. Environmental Pollution, 2016, 219: 450-455.

[55] Peters C A, Bratton S P. Urbanization is a major influence on microplastic ingestion by sunfish in the Brazos River Basin, Central Texas, USA. Environmental Pollution, 2016, 210: 380–387.

[56] Wedemeyer-Strombel K R, Balazs G H, Johnson J B, Peterson T D, Wicksten M K, Plotkin P T. High frequency of occurrence of anthropogenic debris ingestion by sea turtles in the North Pacific Ocean. Marine Biology, 2015, 162 (10): 2079-2091.

[57] van Franeker J A, Law K L. Seabirds, gyres and global trends in plastic pollution. Environmental Pollution, 2015, 203: 89-96.

[58] Wright S L, Rowe D, Thompson R C, Galloway T S. Microplastic ingestion decreases energy reserves in marine worms. Current Biology, 2013, 23: R1031-R1033.

[59] Fries E, Dekiff J H, Willmeyer J, Nuelle M-T, Ebert M, Remy D. Identification of polymer types and additives in marine microplastic particles using pyrolysis-GC/MS and scanning electron microscopy. Environmental Science: Processes and Impacts, 2013, 15 (10): 1949-1956.

[60] Liebezeit G, Dubaish F. Microplastics in beaches of the East Frisian Islands Spiekeroog and Kachelotplate. Bulletin of Environmental Contamination and Toxicology, 2012, 89 (1): 213-217.

[61] Claessens M, Van Cauwenberghe L, Vandegehuchte M B, Janssen C R. New techniques for the detection of microplastics in sediments and field collected organisms. Marine Pollution Bulletin, 2013, 70 (1-2): 227-233.

[62] Nuelle M T, Dekiff J H, Remy D, Fries E. A new analytical approach for monitoring microplastics in marine sediments. Environmental Pollution, 2014, 184 (1): 161-169.

[63] Imhof H K, Schmid J, Niessner R, Ivleva N P, Laforsch C. A novel, highly efficient method for the separation and quantification of plastic particles in sediments of aquatic environments. Limnology and Oceanography Methods, 2012, 10 (7): 524-537.

[64] Hidalgo-Ruz V, Gutow L, Thompson R C, Thiel M. Microplastics in the marine environment: A review of the methods used for identification and quantification. Environmental Science and Technology, 2012, 46 (6): 3060-3075.

[65] Cole M, Webb H, Lindeque P K, Fileman E S, Halsband C, Galloway T S. Isolation of microplastics in biota-rich seawater samples and marine organisms. Scientific Reports, 2014, 4(3): 2231-2236.

[66] Catarino A I, Thompson R, Sanderson W, Henry T B. Development and optimization of a standard method for extraction of microplastics in mussels by enzyme digestion of soft tissues. Environmental toxicology and chemistry, 2017, 36 (4):

947-951.

[67] Roch S, Brinker A. A rapid and efficient method for the detection of microplastic in the gastrointestinal tract of fishes. Environmental Science and Technology, 2017, 51 (8): 4522-4530.

[68] Collard F, Gilbert B, Eppe G, Parmentier E, Das K. Detection of anthropogenic particles in fish stomachs: An isolation method adapted to identification by Raman spectroscopy. Archives of Environmental Contamination and Toxicology, 2015, 69 (3): 331-339.

[69] Fischer M, Scholz-Böttcher B M. Simultaneous trace identification and quantification of common types of microplastics in environmental samples by pyrolysis-gas chromatography-mass spectrometry. Environmental Science and Technology, 2017, 51 (9): 5052-5060.

[70] Löder M G J, Kuczera M, Mintenig S, Lorenz C, Gerdts G. Focal plane array detector-based micro-Fourier-transform infrared imaging for the analysis of microplastics in environmental samples. Environmental Chemistry, 2015, 12 (5): 563-581.

[71] Harrison J P, Ojeda J J, Romero-González M E. The applicability of reflectance micro-Fourier-transform infrared spectroscopy for the detection of synthetic microplastics in marine sediments. Science of the Total Environment, 2012, 416 (2): 455-463.

[72] Song Y K, Hong S H, Jang M, Han G M, Rani M, Lee J, Shim W J. A comparison of microscopic and spectroscopic identification methods for analysis of microplastics in environmental samples. Marine Pollution Bulletin, 2015, 93 (1–2): 202-209.

[73] Löder M G J, Gerdts G. Methodology used for the detection and identification of microplastics—a critical appraisal. Springer International Publishing, 2015, 201-227.

[74] Boerger C M, Lattin G L, Moore S L, Moore C J. Plastic ingestion by planktivorous fishes in the North Pacific Central Gyre. Marine Pollution Bulletin, 2010, 60 (12): 2275-2278.

[75] Browne M A, Dissanayake A, Galloway T S, Lowe D M, Thompson R C. Ingested microscopic plastic translocates to the circulatory system of the mussel, *Mytilus edulis* (L.). Environmental Science and Technology, 2008, 42 (13): 5026-5031.

[76] Davidson T M. Boring crustaceans damage polystyrene floats under docks polluting marine waters with microplastic. Marine Pollution Bulletin, 2012, 64 (9): 1821-1828.

[77] Murray F, Cowie P R. Plastic contamination in the decapod crustacean Nephrops norvegicus (Linnaeus, 1758). Marine Pollution Bulletin, 2011, 62 (6): 1207-1217.

[78] Thompson R C, Olsen Y, Mitchell R P, Davis A, Rowland S J, John A W G, McGonigle D, Russell A E. Lost at sea: Where is all the plastic? Science, 2004, 304 (5672): 838-838.

[79] Lu Y, Zhang Y, Deng Y, Jiang W, Zhao Y, Geng J, Ding L, Ren H. Uptake and accumulation of polystyrene microplastics in Zebrafish (*Danio rerio*)and toxic effects in liver. Environmental Science and Technology 2016, 50 (7): 4054-4060.

[80] Browne M A, Niven S J, Galloway T S, Rowland S J, Thompson R C. Microplastic moves pollutants and additives to worms, reducing functions linked to health and biodiversity. Current biology: CB, 2013, 23 (23): 2388-2392.

[81] Barnes D K A, Galgani F, Thompson R C, Barlaz M. Accumulation and fragmentation of plastic debris in global environments. Philosophical Transactions of the Royal Society B-Biological Sciences, 2009, 364 (1526): 1985-1998.

[82] Liboiron M, Liboiron F, Wells E, Richard N, Zahara A, Mather C, Bradshaw H, Murichi J. Low plastic ingestion rate in Atlantic cod (*Gadus morhua*) from Newfoundland destined for human consumption collected through citizen science methods. Marine Pollution Bulletin, 2016, 113: 428-437.

[83] Besseling E, Wang B, Lurling M, Koelmans A A. Nanoplastic affects growth of S. obliquus and reproduction of D. magna. Environmental Science and Technology, 2014, 48 (20): 12336-12343.

[84] Ma Y, Huang A, Cao S, Sun F, Wang L, Guo H, Ji R. Effects of nanoplastics and microplastics on toxicity, bioaccumulation, and environmental fate of phenanthrene in fresh water. Environmental Pollution, 2016, 219: 166-173.

[85] Bhattacharya P, Lin S J, Turner J P, Ke P C. Physical adsorption of charged plastic nanoparticles affects algal photosynthesis. Journal of Physical Chemistry C, 2010, 114 (39): 16556-16561.

[86] Green D S, Boots B, O'Connor N E, Thompson R. Microplastics affect the ecological functioning of an important biogenic habitat. Environmental Science and Technology, 2017, 51 (1): 68-77.

[87] Bouwmeester H, Hollman P C, Peters R J. Potential health impact of environmentally released micro- and nanoplastics in the human food production chain: experiences from nanotoxicology. Environmental Science and Technology, 2015, 49 (15): 8932-8947.

[88] Wegner A, Besseling E, Foekema E M, Kamermans P, Koelmans A A. Effects of nanopolystyrene on the feeding behavior of the blue mussel (Mytilus edulis L.). Environmental toxicology and chemistry/SETAC 2012, 31 (11): 2490-2497.

[89] Cole M, Galloway T S. Ingestion of nanoplastics and microplastics by pacific oyster larvae. Environmental Science and Technology, 2015, 49 (24): 14625-14632.

[90] Wang F, Bexiga M G, Anguissola S, Boya P, Simpson J C, Salvati A, Dawson K A. Time resolved study of cell death mechanisms induced by amine-modified polystyrene nanoparticles. Nanoscale, 2013, 5 (22): 10868-10876.

[91] Ma Y, Huang A, Cao S, Sun F, Wang L, Guo H, Ji R. Effects of nanoplastics and microplastics on toxicity, bioaccumulation, and environmental fate of phenanthrene in fresh water. Environmental Pollution, 2016, 219: 166-173.

[92] Mato Y, Isobe T, Takada H, Kanehiro H, Ohtake C, Kaminuma T. Plastic resin pellets as a transport medium for toxic chemicals in the marine environment. Environmental Science and Technology, 2001, 35 (2): 318-324.

[93] Rios L M, Moore C, Jones P R. Persistent organic pollutants carried by Synthetic polymers in the ocean environment. Marine Pollution Bulletin, 2007, 54 (8): 1230-1237.

[94] Hirai H, Takada H, Ogata Y, Yamashita R, Mizukawa K, Saha M, Kwan C, Moore C, Gray H, Laursen D, Zettler E R, Farrington J W, Reddy C M, Peacock E E, Ward M W. Organic micropollutants in marine plastics debris from the open ocean and remote and urban beaches. Marine Pollution Bulletin, 2011, 62 (8): 1683-1692.

[95] Zhang W, Ma X, Zhang Z, Wang Y, Wang J, Wang J, Ma D. Persistent organic pollutants carried on plastic resin pellets from two beaches in China. Marine Pollution Bulletin, 2015, 99 (1–2): 28-34.

[96] Talsness C E, Andrade A J M, Kuriyama S N, Taylor J A, vom Saal F S. Components of plastic: experimental studies in animals and relevance for human health. Philosophical Transactions of the Royal Society B-Biological Sciences 2009, 364 (1526): 2079-2096.

[97] Teuten E L, Saquing J M, Knappe D R U, Barlaz M A, Jonsson S, Bjorn A, Rowland S J, Thompson R C, Galloway T S, Yamashita R, Ochi D, Watanuki Y, Moore C, Pham H V, Tana T S, Prudente M, Boonyatumanond R, Zakaria M P, Akkhavong K, Ogata Y, Hirai H, Iwasa S, Mizukawa K, Hagino Y, Imamura A, Saha M, Takada H. Transport and release of chemicals from plastics to the environment and to wildlife. Philosophical Transactions of the Royal Society B, 2009, 364 (1526): 2027-2045.

[98] Cole M, Lindeque P, Halsband C, Galloway T S. Microplastics as contaminants in the marine environment: a review. Marine Pollution Bulletin, 2011, 62 (12): 2588-2597.

[99] Tanaka K, Takada H, Yamashita R, Mizukawa K, Fukuwaka M, Watanuki Y. Accumulation of plastic-derived chemicals in tissues of seabirds ingesting marine plastics. Marine Pollution Bulletin, 2013, 69 (1-2): 219-222.

[100] Zarfl C, Matthies M. Are marine plastic particles transport vectors for organic pollutants to the Arctic? Marine Pollution Bulletin, 2010, 60 (10): 1810-1814.

[101] Debroas D, Mone A, Ter Halle A. Plastics in the North Atlantic garbage patch: A boat-microbe for hitchhikers and plastic

degraders. Science of the Total Environment, 2017, 599: 1222-1232.

[102] Zettler E R, Mincer T J, Amaral-Zettler L A. Life in the "Plastisphere": Microbial Communities on Plastic Marine Debris. Environmental Science and Technology, 2013, 47 (13): 7137-7146.

[103] Carson H S, Nerheim M S, Carroll K A, Eriksen M. The plastic-associated microorganisms of the North Pacific Gyre. Marine Pollution Bulletin, 2013, 75: 126-132.

[104] Akhbarizadeh R, Moore F, Keshavarzi B, Moeinpour A. Microplastics and potentially toxic elements in coastal sediments of Iran's main oil terminal (Khark Island). Environmental Pollution, 2017, 220: 720-731.

[105] Brennecke D, Duarte B, Paiva F, Cacador I, Canning-Clode J. Microplastics as vector for heavy metal contamination from the marine environment. Estuarine Coastal and Shelf Science, 2016, 178: 189-195.

[106] Avio C G, Gorbi S, Milan M, Benedetti M, Fattorini D, d'Errico G, Pauletto M, Bargelloni L, Regoli F. Pollutants bioavailability and toxicological risk from microplastics to marine mussels. Environmental Pollution, 2015, 198: 211-222.

[107] Batel A, Linti F, Scherer M, Erdinger L, Braunbeck T. Transfer of benzo[a]pyrene from microplastics to Artemia nauplii and further to zebrafish via a trophic food web experiment: CYP1A induction and visual tracking of persistent organic pollutants. Environmental toxicology and chemistry/SETAC, 2016, 35 (7): 1656-1666.

[108] Bakir A, Rowland S J, Thompson R C. Enhanced desorption of persistent organic pollutants from microplastics under simulated physiological conditions. Environmental Pollution, 2014, 185: 16-23.

[109] Lee H, Shim W J, Kwon J H. Sorption capacity of plastic debris for hydrophobic organic chemicals. Science of the Total Environment, 2014, 470: 1545-1552.

[110] Mato Y, Isobe T, Takada H, Kanehiro H, Ohtake C, Kaminuma T. Plastic resin pellets as a transport medium for toxic chemicals in the marine environment. Environmental Science and Technology, 2001, 35: 318-324.

[111] Tian L, Kolvenbach B, Corvini N, Wang S, Tavanaie N, Wang L, Ma Y, Scheu S, Corvini P F-X, Ji R. Mineralisation of 14C-labelled polystyrene plastics by Penicillium variabile after ozonation pre-treatment. New Biotechnology, 2017, 38: 101-105.

[112] Duis K, Coors A. Microplastics in the aquatic and terrestrial environment: sources (with a specific focus on personal care products), fate and effects. Environmental Sciences Europe, 2016, 28: 1-25.

[113] Besseling E, Wegner A, Foekema E M, van den Heuvel-Greve M J, Koelmans A A. Effects of microplastic on fitness and PCB bioaccumulation by the lugworm *Arenicola marina* (L.). Environmental Science and Technology, 2013, 47: 593-600.

[114] Rochman C M, Hoh E, Kurobe T, Teh S J. Ingested plastic transfers hazardous chemicals to fish and induces hepatic stress. Scientific Reports. 2013, 3 (7476): 3263.

[115] Ryan P G, Connell A D, Gardner B D. Plastic ingestion and PCBs in seabirds - is there a relationship? Marine Pollution Bulletin 1988, 19 (4): 174-176.

[116] Yamashita R, Takada H, Fukuwaka M A, Watanuki Y. Physical and chemical effects of ingested plastic debris on short-tailed shearwaters, Puffinus tenuirostris, in the North Pacific Ocean. Marine Pollution Bulletin, 2011, 62 (12): 2845-2849.

[117] Bakir A, O'Connor I A, Rowland S J, Hendriks A J, Thompson R C. Relative importance of microplastics as a pathway for the transfer of hydrophobic organic chemicals to marine life. Environmental Pollution, 2016, 219: 56-65.

[118] Koelmans A A, Bakir A, Burton G A, Janssen C R. Microplastic as a vector for chemicals in the aquatic environment: Critical review and model-supported reinterpretation of empirical studies. Environmental Science and Technology, 2016, 50 (7): 3315-3326.

作者：马旖旎[1]，麦　磊[2]，季　荣[1]，曾永平[2]

[1]南京大学环境学院，[2]暨南大学环境学院

第 4 章　我国大气环境化学研究进展

本章导读

大气环境化学是认识空气污染和气候变化问题的重要学科领域。随着大气环境问题的不断发展，特别是我国冬季区域大气雾霾问题依然严峻，我国大气复合污染下的大气化学机制的研究已经成为环境科学领域的一个热点。近年来，我国学者围绕影响区域大气复合污染形成的关键化学机制开展了大量卓有成效的研究工作，取得了一系列重要进展。本章对我国科技工作者在大气自由基与大气氧化能力、大气光化学污染、大气成核和新粒子形成机制、大气非均相化学和多相化学等大气环境化学前沿科学问题方面取得的重要研究成果进行介绍，并对未来研究的重点方向进行展望。

关键词

区域大气复合污染，霾，大气光化学污染，大气氧化能力，大气自由基，大气成核，大气非均相化学，大气多相化学

1 引　　言

随着工业化和城市化进程的发展，种类繁多的污染物被排放到大气中。在大气氧化能力的作用下，其被降解或生成寿命更长的二次污染物，这些一次和二次污染物在大气动力过程的作用下不断迁移或累积，形成不同空间尺度的大气污染问题，并可能影响区域及全球气候。污染物在大气环境中的化学转化机制及其对大气氧化能力（化学）和天气、气候（物理）的反馈机制，是认识大气污染和气候变化等环境问题的基础理论问题，也是科学应对上述大气环境问题的前提。

大气环境化学是一门专门研究污染物在大气中来源、迁移、转化、去除及其对大气环境和气候影响的科学，是环境化学的重要学科方向，也是化学和地球科学高度交叉的学科领域。在过去半个多世纪，随着空气污染和气候变化等环境问题的发展和演变，国际大气环境化学研究得到了快速发展。我国关于大气环境化学的研究始于 20 世纪 70 年代末，从研究煤烟型污染、光化学烟雾和酸雨开始，已经走过近 40 个年头，取得了很大成就。近 20 年来，我国大气污染的形势发生了深刻变化，京津冀、长三角、珠三角等快速发展区域出现了以细颗粒物和臭氧为代表的区域大气复合污染问题，对科学研究和环境管理都提出了新的需求，我国大气环境化学研究所采用的技术手段如外场观测、实验室模拟和数值模拟等也都取得了长足进步。在上述因素的影响下，近年来我国学者围绕区域大气复合污染下的大气化学过程和机制开展了大量研究工作，大气环境化学学科得到了蓬勃发展，从早期侧重对污染的表征逐渐过渡到对化学机理机制的深入探究，在多个关键化学机制上均有所突破。

2 大气自由基与大气氧化能力

大气自由基化学作为大气氧化能力的主要来源，是理解区域二次污染和全球气候变化等重大环境问题的理论基础。在我国，伴随国民经济的高速发展，在高强度排放和强氧化性的共同作用下，在京津冀、长三角和珠三角等超大城市群地区形成了高浓度一次污染物和高浓度二次污染物共存的复合污染大气环境。目前大气复合污染的表征工作已较为细致，然其形成机理尚待深入研究。这其中一个核心任务即是

厘清大气复合污染状态下的大气自由基化学机制。

OH 自由基作为全球尺度上对流层大气氧化性的主要来源，北京大学研究团队开展了多年系统研究，搭建了国内首套大气激光诱导荧光（LIF）系统，获得了国内首套实际大气 OH 自由基的浓度序列。基于综合外场观测研究，对 HO_x 自由基化学的来源和转化机制进行了深入探讨，发现 O_3、HONO、HCHO 和双羰基醛类的光解是珠三角和北京地区日间 HO_x 自由基的主要初级来源，而臭氧烯烃反应和 NO_3 自由基的氧化反应是夜间 HO_x 自由基的主要初级来源；发现过氧自由基参与的非传统再生机制是低 NO_x 条件下 OH 自由基主要的再生来源[1-4]。深入分析表明，OH 自由基非传统再生机制与已知的 NO 再生机制形成竞争关系，关键反应是一种热化学反应机制，其中异戊二烯降解生成的过氧自由基重排中发生的氢转移反应贡献显著[2,5]。OH 自由基去除机制的研究主要通过 OH 总反应性的闭合实验分析实现。珠三角和北京地区的相关分析表明一半左右的大气 OH 自由基反应活性来源是未知的[6-8]；而模型模拟的结果则指出含氧有机物（oxygenated volatile organic compounds，OVOCs）是这些未知活性的主要来源，其中贡献最大的 OVOCs 分子多为 HCHO，这与我国不同大气环境中观测到高浓度 HCHO 相一致[9,10]。

结合 OH 自由基与 j（O^1D）存在高相关性这一特殊观测现象，对北半球九个不同森林和超大城市地区的 OH 自由基化学进行了模拟分析，在 OH-NO_2 光化学坐标系中对上述九个地区的 OH 自由基观测结果进行了归一化闭合分析。研究表明，除美国黄松林地区外，在珠三角和北京发现的 OH 自由基新再生机制在其他地区也普遍存在，其 OH 自由基观测结果均可为同一参数化模型所描述。此模型与传统光化学理论存在显著不同：新构建的参数化模型在低 NO_x 区间预测的 OH 自由基浓度远高于传统光化学理论预测结果。上述新模型为“OH 自由基非传统再生机制”的理论探索构建了新的约束条件，更重要的是它揭示了对流层大气氧化能力的一种普遍属性，即处于低 NO_x 与高 VOCs 地区的 OH 自由基浓度已达现有理论所能预测的峰值水平。通俗地说，自然界正以最大效能来氧化人类活动与自然界所排放的一次污染物。由于一般峰值运行状态并不是最稳定的运行状态，在发展中国家和地区人类排放持续稳步增长的背景下，对流层大气氧化能力的未来发展变化值得关注[11]。

HONO 作为 OH 自由基的一种重要初级来源，在国内若干典型大气环境中都被观测到较高浓度的存在浓度，也发现了 NO_2 浓度与 HONO 未知来源的高相关性[12-16]。机理分析表明土壤排放[17]、吸附态硝酸的光解[18]和 NO_2 水解的光增强反应[19]是几种可能的 HONO 日间来源机制。其中作为对全球氮循环研究的重要贡献，认识到土壤排放可能是日间 HONO 的重要来源机制，揭示了全球暖化和化肥使用对大气 HONO 浓度的影响。北京大学研究团队在华北平原的观测中发现，土壤施肥后三天是日间 HONO 直排的高峰期，同时 HONO 的直接排放来源也越来越受到更多研究者的重视，机动车的 HONO 排放因子得到不断更新[15,20,21]。冬季 HONO 来源分析表明，HONO 在雾霾污染过程中，气相反应和机动车排放贡献显著[22]。

NO_3 和 N_2O_5 是大气中关键的活性含氮物种和夜间大气氧化性的主要来源[23,24]。珠三角、长三角和京津冀地区均测量到了较高浓度的 NO_3，说明我国大型城市群夜间均处于较高的氧化水平[25-27]。近年来观测发现我国城市地区的下风向和残留层均存在较高水平的 N_2O_5[28-30]，其中香港地区的残留层中 N_2O_5 观测到了最高浓度 12 ppbv[29,30]，而北京城区和下风向的郊区也均观测到了 1 ppbv 量级的 N_2O_5[28]，可能由于我国大型城市群大量排放的 NO_x 经过较强的光化学过程后在夜间具有较高的 N_2O_5 生成速率导致。外场研究表明我国的 N_2O_5 非均相反应摄取系数普遍偏高，比实验室的研究结果以及欧美地区外场观测的结果高 2~3 倍[31]，且难以用现有的参数化机理进行合理解释[32,33]，说明 N_2O_5 的非均相反应机理尚不清楚。外场观测和模型模拟均表明 N_2O_5 非均相反应是我国硝酸盐生成的重要来源[24,33-35]，尤其在污染形成过程中生成硝酸盐并提高颗粒物表面积，进而加快非均相反应过程，通过自加速方式加重颗粒物污染的形成。在香港和华北还观测到了高浓度水平的 $ClNO_2$，结合模型表明 $ClNO_2$ 是 Cl 自由基的主要来源，而后续的 Cl 自由基氧化 VOCs 的过程可能对 RO_x 的生成具有一定贡献，进一步提升臭氧污染[30,36,37]。

3 大气光化学污染

大气光化学污染是指大气中的氮氧化物（NO_x）、挥发性有机物（VOCs）和一氧化碳（CO）等，在阳光紫外线的照射下发生反应，在对流层尤其是边界层内形成的以高浓度臭氧（O_3）、过氧乙酰硝酸酯（PAN）、醛酮类物质以及细颗粒物等二次污染物为特征的污染现象。20 世纪 50 年代，美国洛杉矶地区出现光化学烟雾，造成了巨大的生命和经济损失，光化学污染自此成为发达国家大气污染防治的重要挑战。我国自 20 世纪 70 年代末在兰州和北京先后发现了光化学污染的迹象，之后随着经济快速发展和能源消费迅速增加，近年来大气光化学污染作为区域性环境问题在京津冀、长三角、珠三角和成渝等城市群区域已经显现。阐明大气光化学污染的成因及其化学和动力过程，既是当前大气化学研究的热点，也是空气质量管理所关注的焦点。

近十多年来，我国学者围绕大气光化学污染开展了大量研究工作，初步探索了主要经济快速发展区域的臭氧污染现状和特征，发现我国主要大城市和城市群地区在夏秋季节普遍面临着严重的光化学污染问题，臭氧浓度频繁超标[38-42]，其中 2005 年夏季在北京下风向郊区记录的臭氧最高小时浓度高达 286 ppbv[43]，显示出极强的臭氧生成能力。受东亚季风和气象条件影响，不同地区臭氧污染的季节变化特征存在明显差异，例如，珠三角地区大气光化学污染最严重的季节是秋季，夏季由于受海洋性气团影响污染物浓度相对最低[44]；而长三角和京津冀地区的臭氧峰值则多出现于晚春和夏季[39,45-47]。基于对臭氧长期监测资料进行分析，发现我国主要地区的地面（或对流层）O_3 浓度在过去 20 年内呈显著升高趋势：香港城市和背景站点近 20 年的 O_3 年均增幅约为 0.27~0.58 ppbv/a[44,48]；珠三角区域 O_3 浓度在 2006~2011 年间的增幅为 0.86 ppbv/a[49]；受 NO_x 排放增加影响，1991~2006 年间长三角背景站点临安的 O_3 变化幅度持续增大，其中 O_3 浓度高值平均每年升高 1.8 ppbv[50]；华北地区城市、背景站点以及飞机航测结果均显示其 O_3 浓度，尤其是夏季 O_3 浓度近 20 年来增长明显，高达 1~3 ppbv/a[46,47,51-53]；在位于青藏高原的瓦里关，地面 O_3 浓度在 1994~2013 年间也表现出 0.24~0.28 ppbv/a 的增长趋势[54]。总体而言，我国东部目前已成世界上有观测记录的对流层臭氧浓度增长最快的地区，充分体现出我国日益严峻的大气光化学污染的总体态势，也反映了持续增强的区域大气氧化能力。

理清臭氧的化学生成机制是大气光化学污染防治的基础。基于外场观测的数据分析和数值模拟研究，初步揭示了我国主要经济发达地区的臭氧生成机制，发现主要城市或工业地区的臭氧生成大多受 VOCs 控制，但在城市或工业区的下风向，臭氧生成则多处于过渡区甚至 NO_x 控制区[38,41,42,55,56]。VOCs 在我国绝大多数城市地区的光化学污染中发挥着重要作用，但不同地区的关键物种却可能存在较大差异，如珠三角地区的关键 VOCs 活性物种是芳香烃[41,42]；受石油化工工业影响，烯烃是兰州地区大气 VOCs 的优势物种[41]；而在京津冀和长三角，芳香烃和烯烃都是重要的 VOCs 活性物种[41,56]。最近一些研究还揭示了一些新的大气化学过程和机制，包括 OH 的非传统再生机制[1]（详见上节）、HONO 的未知日间来源机制[17-19]以及氮氧化物参与的夜间氯活化机制等[57]，这些过程都直接影响着大气的氧化能力，进而影响臭氧和光化学污染的形成。研究发现，高浓度 HONO 可以使京津冀地区的 O_3 日最大八小时浓度增加 3~10 ppbv[58]，也导致珠三角地区的地面 O_3 浓度增加 6%~12%[59]。近年来在京津冀和珠三角等地的研究发现，夜间积累的高浓度 $ClNO_2$ 在第二天光解能促使华北地区农村站点和香港高山站点的 O_3 净生成量分别提高 13%和 41%[30,36]。区域模式模拟结果也表明加入 HONO 和 $ClNO_2$ 的生成机制可以显著提高中国地区 O_3 的模拟浓度[35,38]，但现有的空气质量模式普遍缺少对这些新机制的表达，同时我国天然源 VOCs 排放清单估算误差较大，从而影响对臭氧污染的准确模拟和深入认识。

大气动力过程也是影响臭氧等物质生成、输送和去除的重要因素，其中热带气旋和反气旋是与光化学污染事件密切相关的主要天气系统。反气旋（如副热带高压）在其中心创造有利于污染积累和 O_3 生成的天气条件，如强日照和低风速。副热带高压是我国各地夏季高温热浪的一个主要导因，也是京津冀、长三角地区夏季出现光化学污染的必要条件[39,60]。珠三角地区的光化学污染事件则往往受到西太平洋热带气旋的影响，其中台风是夏季常见的一种热带气旋，在低压系统外围的珠三角地区会出现光照强、气温高、风速小的天气条件，有利于光化学污染事件的发生[61,62]。研究发现，台风外围环流（离岸流）影响下的边界层海陆风对于珠三角空气污染物的累积和光化学污染具有重要影响[61]。此外，平流层-对流层交换过程对于高海拔地区春季和夏季的臭氧高值也有重要贡献[63]。

除臭氧外，学者对另一种光化学污染物 PAN 也进行了研究，在北京、广州、兰州、香港等地区均测量到了较高浓度的 PAN[64,65]，其中北京城区的峰值浓度高达 17 ppbv[66]。通过基于观测的数据分析和模式分析，揭示了影响 PAN 光化学生成的关键前体物：短链烯烃（如丙烯）是兰州大气中 PAN 的主要前体物[65]，而芳香烃则是北京和珠三角地区的关键活性物种[67]。

4　大气成核和新粒子形成机制

大气中部分气体分子在随机碰撞中可以形成稳定的分子簇，并逐渐生长为纳米颗粒物，该过程被称为大气成核；如果成核过程所形成的分子簇和纳米颗粒物不与大气中已存在颗粒物碰并而损失，则可以继续生长形成大气新颗粒物而成为大气新粒子形成事件[68]。大气新粒子形成事件是一种重要的二次气溶胶形成途径，国内外学者已经在不同的大气环境中广泛地观测到大气新粒子形成事件：其最直观的体现是成核模态的颗粒物数浓度的急剧增加，并且这些成核模态的纳米颗粒物的粒径会持续增长至爱根模态。目前学术界已经广泛承认了硫酸分子在大气成核过程中的关键作用[69]。

4.1　新的仪器分析手段

近期大气成核和新粒子形成的机制研究取得了长足的发展。究其原因，在于新的仪器分析手段使得研究者获得了新的关键数据，包括 3nm 以下的分子簇和纳米颗粒物的粒径分布浓度以及成核过程中关键分子簇的化学组分。前者的测量使得研究者能够直接计算大气真实成核速率（J_1），不需要通过 3 nm 颗粒物的形成速率（J_3）来推算 J_1；而后者的测量则允许研究者直接分析大气成核的化学机制。

3nm 以下的分子簇和纳米颗粒物的粒径分布浓度测量目前可以通过颗粒物粒径放大仪（particle size magnifier，PSM，芬兰 Airmodus 公司）或者 1 nm 扫描电迁移率颗粒物粒径谱仪（1 nm-scanning mobility particle sizer，美国 TSI 公司）来实现。颗粒物粒径放大仪通常与以正丁醇作为工作液体的冷凝粒子计数器（condensation particle counter，CPC）联用：颗粒物粒径放大仪首先使用二甘醇（DEG）作为工作液体将 3 nm 以下的分子簇和纳米颗粒物活化并冷凝生长至约 90 nm，其次颗粒物将通过正丁醇的冷凝而进一步长大到能被冷凝粒子计数器的光学计数器所检测的尺寸[70]。1 nm 扫描电迁移率颗粒物粒径谱仪的原型仪器由清华大学蒋靖坤在美国留学期间首次使用[71,72]，该仪器首先通过 1 nm 差分粒子电迁移器（differential mobility analyzer，DMA）将分子簇和纳米颗粒物加以筛分，随后通过分子簇和纳米颗粒物在二甘醇-冷凝粒子计数器（DEG-CPC）中的两步生长而被检测。

在分子簇的化学组分测量方面，由瑞士 TofWerk 公司和美国 Aerodyne 公司共同发展的大气常压界面-飞行时间质谱被首先应用于芬兰北方森林地区大气带电离子的化学组分测量[73]，其高分辨率可以帮助确定大气带电离子的分子式。而后，随着可与飞行时间质谱联用的硝酸根电离源被发展起来[74]，大气中的中性分子簇可以通过与硝酸根试剂离子的离子-分子反应而生成产物离子，进而被飞行时间质谱所检测，所获得的产物离子信息被用于分析大气中性分子簇的化学组分。约在同一时期，美国国家大气研究中心（National Center for Atmospheric Research）的研究人员发展了分子簇-化学电离质谱（cluster-chemical ionization mass spectrometer），用于测量大气中的中性分子簇[75]。由于该设备使用的是四级杆质谱，因此只能提供单位质量分辨率的化学组分信息。

4.2 近期实验室模拟研究进展

有关大气成核与新粒子形成的研究进展是建立在外场观测、实验室模拟，以及理论预测的综合研究中的。目前，国外学者在大气成核与新粒子形成这一领域处于领先地位。在实验室模拟方面，近期以欧盟科学家主导的 CERN-CLOUD 系列实验最为著名，开展了包括硫酸-水二元成核[76]、硫酸-氨气-水三元成核[76]、硫酸-有机胺-水三元成核[77,78]、硫酸-极低挥发性有机化合物-水三元成核[79,80]、离子诱导成核[76]、离子诱导下的纯生物源有机化合物反应成核[81]等机制在内的一系列实验室模拟，研究了不同条件下成核关键前体物的浓度与成核速率的对应关系、成核过程中中性和带电分子簇的化学组分和生成顺序，以及不同类别的前体物对分子簇和纳米颗粒物生长的定量贡献等，形成了多个重大科学成果。

4.3 近期外场观测研究进展

通过将外场观测中所获得的大气成核特征与 CERN-CLOUD 实验的模拟结果进行对比，国外学者已在少数具有相对洁净大气的地点确定了真实大气成核机制，包括芬兰北方森林地区的硫酸-极低挥发性有机化合物-水三元成核[80,82]、瑞士少女峰自由大气中不同大气条件下的硫酸-极低挥发性有机化合物-水成核或硫酸-氨气-水三元成核[83]，以及爱尔兰西海岸 Mace Head 地区的碘酸（HIO_3）成核[84]。这些研究初步表明全球各地的大气成核机制可能各不相同，具有地域特征。

我国学者近期也开展了大量的有关大气成核与新粒子形成的外场观测研究，然而已有的报道中尚未见具有直接观测证据的综合机制研究。北京大学胡敏课题组的研究暗示北京地区的大气颗粒物高污染可能最终源自于空气洁净时期的大气新粒子形成事件，将这两个大气化学领域的前沿难题联系起来：在大气新粒子形成事件中高数浓度的纳米颗粒物得以形成，并在接下来的几天时间尺度内粒径得以连续增长，最终形成大气颗粒物污染乃至灰霾事件[85]。

我国学者在大气成核与新粒子形成的前体物测量方面则取得了显著进展：南京信息工程大学郑军课题组以水合氢质子作为试剂离子，利用化学电离-飞行时间质谱技术测量了南京地区大气中的氨气和有机胺[86]，发展了以硝酸根离子为试剂离子，利用化学电离-飞行时间质谱技术来测量大气中的气体硫酸[87]。复旦大学王琳课题组则以质子化乙醇作为试剂离子，利用化学电离-飞行时间质谱技术测量了上海大气中的有机胺和酰胺[88]。

我国学者在 3nm 以下的分子簇和纳米颗粒物的粒径分布浓度测量方面也取得了实质性的进展。复旦大学王琳课题组在上海利用 PSM 首次实现了 1~3nm 颗粒物的粒径谱和大气真实成核速率 J_1 的测量[89]。南京信息工程大学余欢课题组使用 PSM 分别在南京和邻近区域背景站点研究了长三角地区大气成核的区域化特征[90,91]。清华大学蒋靖坤课题组近期使用自制的 1 nm 扫描电迁移率颗粒物粒径谱仪在北京研究了分子簇之间的碰并过程对 J_1 计算的影响[92]，并研究了北京城区已存在颗粒物表面积对成核可能性的影响[93]。

我国香港和台湾学者近年来也开展了大量有关大气成核与新粒子形成的研究工作。香港理工大学郭海课题组的外场观测工作表明香港地区的大气新粒子形成事件与气象条件紧密相关，气体硫酸和生物源有机化合物的大气反应产物在所观测到的事件中具有重要的作用[94,95]。台湾“中研院”环境变化研究中心博士張慶祖博士在台北的观测工作表明，夏季的强光化学反应导致了二氧化硫向气体硫酸的转化，而新粒子的碰并损失在较低的 PM_{10} 浓度下则较小，因此新粒子形成事件在夏季较为常见[96]。

5　大气非均相化学与多相化学

大气非均相反应（heterogeneous reaction）是指大气中反应物在颗粒物表面发生的反应；大气多相反应（multiphase reaction）是指大气中反应物与云雾液滴发生的反应[97]。大气颗粒物可分为矿尘、海盐、有机气溶胶、硫酸盐、硝酸盐、铵盐和黑碳等[98]，污染区域许多大气颗粒物是以内混、外混状态存在的混合气溶胶。大气颗粒物粒径从几纳米到几十微米，大气中的液滴（包括云滴和雾滴）粒径在 2 μm 到 100 μm，不但能够携带对人体健康有影响的重金属、PAHs 等，并且具有较大的比表面积，颗粒物或云雾液滴表面容易发生非均相或多相化学反应。大气污染气体通过吸附、溶解、摄取和化学转化等，使气相物种的大气寿命发生改变或形成新的气相物种与颗粒物表面、液滴中的二次物种[99-102]，同时颗粒物或云雾液滴本身物理化学性质发生改变，包括颗粒物的形貌、颗粒物与云雾液滴的组成、密度、光学性质等[103,104]，对区域空气质量和全球气候变化具有重要影响。

国内利用 FTIR-DRIFTS 等现代技术最早开展大气颗粒物非均相反应实验模拟的有北京大学朱彤课题组和复旦大学陈建民课题组，其后不久，中国科学院生态环境研究中心贺泓课题组和中国科学院化学所葛茂发课题组也开展了相关工作，最近，山东大学杜林课题组在该方面的研究也取得进展。从较早采用的 DRIFTS 技术，不断发展了农森池、激光拉曼光谱法、流动管反应器，利用其开展实验模拟，并结合现场观测进行了验证，现场观测发现的问题对实验模拟的研究方向提供了指导，特别是我国近年来重污染下大气细颗粒物的研究取得了积极进展。国内外已有多篇综述文献对大气非均相反应与多相反应的研究进展做了很好的综述[97,105-109]，本节就近期我国科学工作者在该方面的主要研究进展进行了总结。

硫酸盐、硝酸盐是大气颗粒物的主要成分之一，在我国重霾污染下分别可占大气细颗粒物（$PM_{2.5}$）含量的 10%~30%[110]。SO_2、NO_x 是形成硫酸盐、硝酸盐的前体物，基于已知硫酸盐、硝酸盐形成机制的模式模拟结果，重霾下实际大气 $PM_{2.5}$ 中硫酸盐会被严重低估，表明大气颗粒物中硫酸盐存在未知的形成机制来源[111,112]，硫酸盐的形成机理成为竞相研究的前沿方向。采用 DRIFTS 和 FTIR-怀特池技术相结合，Kong 等[113]发现含有硝酸盐的大气颗粒物（以 α-Fe_2O_3 为代表）与 SO_2 的反应活性和硫酸盐的生成速率发生很大变化，硫酸盐生成速率随着硝酸盐含量的增加呈现先增加后降低的变化，且少量硝酸盐大大促进 SO_2 在 α-Fe_2O_3 表面的非均相转化，使硫酸盐的非均相反应途径发生明显改变，并且伴随气相 N_2O 和 HONO 的产生。结合气体产物结果表明，硝酸盐参与了 SO_2 的非均相反应，改变了 SO_2 非均相反应的途径和硫酸盐的形成速率，并导致气相 N_2O 和 HONO 的产生以及表面吸附态 HNO_3、亚硝酸盐产物的形成 [113]。HONO 是 OH 自由基的前体物之一，上述反应机制中 HONO 的产生量以及对大气氧化性的影响有多大贡献，还需要深入研究。

Ullerstam 等曾报道 NO_2 与 SO_2 共存可以促进 SO_2 在矿尘表面转化为硫酸盐[114]，但其反应机制尚不清楚。Ma 等采用 DRIFTS 技术发现在 γ-Al_2O_3 表面上 SO_2 单独反应只能产生亚硫酸盐，而共存的 NO_2 能够促进表面亚硫酸盐向硫酸盐的转化[115]。Liu 等进一步研究发现，NO_2 与 SO_2 这种复合机制也存在于其他多种矿质氧化物非均相反应中[116]。在只有 O_2 存在的反应中，才能够观察到 SO_4^{2-} 物种的形成，这说明

O_2在NO_2促进SO_2转化为硫酸盐的非均相反应过程中具有氧化作用，NO_2在矿质氧化物表面促进了O_2的活化，进而促进了SO_2氧化转化为硫酸盐[116]。

碳质气溶胶在大气非均相反应中的作用是关注的重点方向之一。Han 等通过 ATR-FTIR 研究了O_2与黑碳的非均相反应活性与二次物种的变化，发现光照能够激发分子O_2与黑碳的反应，辨别出黑碳本身含有的有机碳是O_2与黑碳非均相光化学反应的主要活性物种[117]。在暗反应条件下，O_2与黑碳不发生反应，说明可见光是激发分子氧与黑碳反应的关键因素[117]。Han 等还应用流动管反应装置考察了黑碳中有机碳对NO_2在黑碳上摄取动力学的影响，并应用 ATR-IR 研究了NO_2与黑碳非均相反应过程中黑碳组成的变化，考察了光照、燃氧比以及NO_2浓度对黑碳表面物种反应动力学的影响，确定了有机碳在NO_2与黑碳非均相反应中的作用[118,119]。

生物质燃烧不但能形成黑碳气溶胶，且往往产生相当多的棕色碳[120,121]，而大气雾霾污染天颗粒物常常伴随大量碳质气溶胶（黑碳、棕色碳）与硫酸铵等无机盐形成的混合气溶胶，相对湿度（RH）发生变化，对光的吸收、散射系数等参数产生很大影响。Chen 等[120]采用 V-TDMA 与 CRDS 等联用技术模拟了黑碳-棕色碳气溶胶吸收硫酸气体与氨气、有机胺形成硫酸铵/有机胺盐—黑碳-棕色碳混合气溶胶的非均相反应过程，同时发现随 RH 增加，消光系数从减少 15%到增加 20%的过程。这个非均相反应过程及其光学性质的变化也在北京和美国休斯敦的现场观测中发现具有相似特征。

北京大学陈忠明课题组研究了过氧化物在颗粒物界面过程及二次物种形成中的作用，结果表明，$PM_{2.5}$和矿质颗粒物摄取过氧化物的摄取系数均为10^{-4}量级；H_2O_2能促进SO_2和 OVOCs 在矿质颗粒物表面的非均相反应；有机过氧化物气-粒分配系数远高于已有模式估算值[122-125]。该方面的成果揭示了过氧化物是大气二次无机和有机颗粒物之间关系的重要中间态物质，过氧化物在颗粒物界面过程及二次物种形成中的作用比以往的认识更加重要，为大气二次颗粒物形成机制研究和大气化学模式研究提供了基础数据和关键参数。

来自海洋微表层的表面活性物质包裹的海洋飞沫气溶胶通常是反胶束的结构，它由有机单分子形成的疏水表面和一个水相内核构成。目前成功构建了具有良好力学性能的饱和脂肪酸单分子膜（硬脂酸、花生酸等）用于模拟界面的多相反应。Li 等[126]通过改变亚相中金属离子的种类研究了大气中常见的金属离子对气溶胶界面分子的取向、力学性质以及界面光谱的影响，发现无机离子的存在会影响表面有机膜的许多化学性质，例如界面分子结构、溶解性和表面-体相的分配等，这些性质对于气溶胶成核生长具有重要作用。除了阳离子之外，通过研究海水中常见的不同阴离子（Br^-、Cl^-、NO_3^-、SO_4^{2-}、CH_3COO^-和HCO_3^-）和磷脂分子膜的相互作用，发现亚相中的阴离子使表面单分子膜出现了不同程度的收缩或扩张，同时也会改变表面有机分子构型和排列。气溶胶表面有序排列的有机膜会阻碍挥发性物质的传输以及水分蒸发，进而影响云滴的形成。

在云雾水滴发生的多相反应中的产物—可溶性二次有机气溶胶（aqSOA）对空气质量、气候变化起到重要影响。然而，aqSOA 形成机制和对颗粒物性质的改变等方面的认识还很缺乏。Sui 等[127]利用液体真空界面分析系统（SALVI）和二次粒子飞行时间质谱仪（ToF-SIMS）联用，首次实现了对乙二醛（glyoxal）-H_2O_2体系在液相界面反应产物在分子水平上的直接检测。羧酸、水合物、低聚物和水分子簇团等反应产物都可以被质谱检测到，证明大气中乙二醛的液相界面氧化反应在光照和黑暗条件下对 aqSOA 的生成均有重要贡献。进一步深入分析发现，在乙二醛-过氧化氢体系的液相界面光照、黑暗反应条件下，水分子簇团的形态、分布与去离子水有很大的不同，这说明乙二醛-过氧化氢液相界面氧化反应过程可以极大地改变气溶胶界面的亲/疏水性[127]。小的水分子簇团倾向于在在暗反应条件下生成，而大的水分子簇团则更多地存在于光反应条件下，说明光反应之后的液相表面疏水性更强，而暗反应之后的液相界面表现出更强的亲水性。离子簇团的发现表明界面水的微环境受到氧化反应途径的影响并与 aqSOA 的生成过程有密切关系[127]。此外，液相表面反应物及产物的空间分布，即颗粒物的表面混合态可由 SIMS 的 3D 成像技术测定，这为液相界面的反应模型研究提供了新的技术[127]。

6 展 望

当前，我国大气污染的总体形势依然严峻，重点区域冬季重霾污染的压力未去，夏季以臭氧问题为代表的光化学污染却又接踵而至。此外，随着我国在国际应对气候变化问题上的地位和责任日益重要，阐明我国区域大气复合污染和全球/区域气候变化背景下的复杂大气化学机制，不但体现了国家的重大战略需求，也是当前国际大气化学研究最前沿的科学问题。尽管近些年来我国在大气环境化学领域的研究方面取得了积极进展，但仍然存在一些尚未认识清楚的关键科学问题，有待进一步研究，尤其是对区域大气氧化能力具有重要作用的自由基未知来源机制、VOCs 对臭氧和二次有机气溶胶的影响、大气活性组分在颗粒物界面的非均相反应机制和液相反应机制及其对臭氧和颗粒物的影响、大气成核与新粒子形成机制及其环境和气候效应等，都是未来大气环境化学研究的重要内容。

参 考 文 献

[1] Hofzumahaus A, Rohrer F, Lu K, Bohn B, Brauers T, Chang C-C, Fuchs H, Holland F, Kita K, Kondo Y, Li X, Lou S, Shao M, Zeng L, Wahner A, Zhang Y. Amplified trace gas removal in the troposphere. Science, 2009, 324(5935): 1702-1704.

[2] Lu K, Hofzumahaus A, Holland F, Bohn B, Brauers T, Fuchs H, Hu M, Häseler R, Kita K, Kondo Y. Missing OH source in a suburban environment near Beijing: observed and modelled OH and HO_2 concentrations in summer 2006. Atmos. Chem. Phys., 2013, 13(2): 1057-1080.

[3] Lu K, Rohrer F, Holland F, Fuchs H, Bohn B, Brauers T, Chang C, Häseler R, Hu M, Kita K. Observation and modelling of OH and HO_2 concentrations in the Pearl River Delta 2006: a missing OH source in a VOC rich atmosphere. Atmos. Chem. Phys. , 2012, 12(3): 1541-1569.

[4] Lu K D, Rohrer F, Holland F, Fuchs H, Brauers T, Oebel A, Dlugi R, Hu M, Li X, Lou S R, Shao M, Zhu T, Wahner A, ZhangY H, Hofzumahaus A. Nighttime observation and chemistry of HO_x in the Pearl River Delta and Beijing in summer 2006. Atmos. Chem. Phys. , 2014, 14(10): 4979-4999.

[5] Tan Z, Fuchs H, Lu K, Hofzumahaus A, Bohn B, Broch S, Dong H, Gomm S, Häseler R, He L, Holland F, Li X, Liu Y, Lu S, Rohrer F, Shao M, Wang B, Wang M, Wu Y, Zeng L, Zhang Y, Wahner A, Zhang Y. Radical chemistry at a rural site(Wangdu) in the North China Plain: observation and model calculations of OH, HO_2 and RO_2 radicals. Atmos. Chem. Phys. , 2017, 17(1): 663-690.

[6] Lou S, Holland F, Rohrer F, Lu K, Bohn B, Brauers T, Chang C, Fuchs H, Häseler R, Kita K. Atmospheric OH reactivities in the Pearl River Delta–China in summer 2006: measurement and model results. Atmos. Chem. Phys. , 2010, 10(22): 11243-11260.

[7] Yang Y, Shao M, Wang X, Nölscher A C, Kessel S, Guenther A, Williams J. Towards a quantitative understanding of total OH reactivity: A review. Atmos. Environ. , 2016, 134: 147-161.

[8] Lu K, Zhang Y, Su H, Brauers T, Chou C C, Hofzumahaus A, Liu S C, Kita K, Kondo Y, Shao M. Oxidant(O_3+NO_2) production processes and formation regimes in Beijing. J. Geophys. Res.: Atmos., 2010, 115: (D7).

[9] Li X, Rohrer F, Brauers T, Hofzumahaus A, Lu K, Shao M, Zhang Y, Wahner A. Modeling of HCHO and CHOCHO at a

semi-rural site in southern China during the PRIDE-PRD2006 campaign. Atmos. Chem. Phys. , 2014, 14(22): 12291-12305.

[10] Liu Y, Yuan B, Li X, Shao M, Lu S, Li Y, Chang C-C, Wang Z, Hu W, Huang X. Impact of pollution controls in Beijing on atmospheric oxygenated volatile organic compounds(OVOCs) during the 2008 Olympic Games: Observation and modeling implications. Atmos. Chem. Phys. , 2015, 15(6): 3045-3062.

[11] Rohrer F, Lu K, Hofzumahaus A, Bohn B, Brauers T, Chang C-C, Fuchs H, Häseler R, Holland F, Hu M. Maximum efficiency in the hydroxyl-radical-based self-cleansing of the troposphere. Nature Geoscience, 2014, 7(8): 559.

[12] Hou S, Tong S, Ge M, An J. Comparison of atmospheric nitrous acid during severe haze and clean periods in Beijing, China. Atmos. Environ. , 2016, 124: 199-206.

[13] Wang S, Zhou R, Zhao H, Wang Z, Chen L, Zhou B. Long-term observation of atmospheric nitrous acid(HONO) and its implication to local NO_2 levels in Shanghai China. Atmos. Environ. , 2013, 77: 718-724.

[14] Qin M, Xie P, Su H, Gu J, Peng F, Li S, Zeng L, Liu J, Liu W, Zhang Y. An observational study of the HONO–NO_2 coupling at an urban site in Guangzhou City, South China. Atmos. Environ. , 2009, 43(36): 5731-5742.

[15] Xu Z, Wang T, Wu J, Xue L, Chan J, Zha Q, Zhou S, Louie P K K, Luk C W Y. Nitrous acid(HONO) in a polluted subtropical atmosphere: Seasonal variability, direct vehicle emissions and heterogeneous production at ground surface. Atmos. Environ. , 2015, 106: 100-109.

[16] Liu Y, Lu K, Dong H, Li X, Cheng P, Zou Q, Wu Y, Liu X, Zhang Y. In situ monitoring of atmospheric nitrous acid based on multi-pumping flow system and liquid waveguide capillary cell. JEnvS, 2016, 43: 273-284.

[17] Su H, Cheng Y, Oswald R, Behrendt T, Trebs I, Meixner F X, Andreae M O, Cheng P, Zhang Y, Pöschl U. Soil nitrite as a source of atmospheric HONO and OH radicals. Science, 2011, 333(6049): 1616-1618.

[18] Li X, Brauers T, Häseler R, Bohn B, Fuchs H, Hofzumahaus A, Holland F, Lou S, Lu K, Rohrer F. Exploring the atmospheric chemistry of nitrous acid(HONO) at a rural site in Southern China. Atmos. Chem. Phys. , 2012, 12(3): 1497-1513.

[19] Su H, Cheng C P, Cheng Y F, Cheng P, Zhang Y H, Dong S, Zeng L M, Wang X, Slanina J, Shao M, Wiedensohler A. Observation of nighttime nitrous acid (HONO) formation at a non-urban site during PRIDE-PRD2004 in China. Atmos. Environ. , 2008, 42(25):6219-6232.

[20] Liu Y, Lu K, Ma Y, Yang X, Zhang W, Wu Y, Peng J, Shuai S, Hu M, Zhang Y. Direct emission of nitrous acid(HONO) from gasoline cars in China determined by vehicle chassis dynamometer experiments. Atmos. Environ. , 2017.

[21] Yang Q, Su H, Li X, Cheng Y, Lu K, Cheng P, Gu J, Guo S, Hu M, Zeng L. Daytime HONO formation in the suburban area of the megacity Beijing, China. Science China Chemistry, 2014, 57(7): 1032-1042.

[22] Tong S, Hou S, Zhang Y, Chu B, Liu Y, He H, Zhao P, Ge M. Exploring the nitrous acid(HONO) formation mechanism in winter Beijing: direct emissions and heterogeneous production in urban and suburban areas. Faraday Discuss. , 2016, 189: 213-230.

[23] Haichao W, Tun C, Keding L. Measurement of NO_3 and N_2O_5 in the Troposphere. Prog. Chem. , 2015, 27(7): 963-976.

[24] Wang H, Lu K, Tan Z, Sun K, Li X, Hu M, Shao M, Zeng L, Zhu T, Zhang Y. Model simulation of NO_3, N_2O_5 and $ClNO_2$ at a rural site in Beijing during CARE Beijing-2006. Atmos. Res. , 2017, 196: 97-107.

[25] Wang S, Shi C, Zhou B, Zhao H, Wang Z, Yang S, Chen L. Observation of NO3 radicals over Shanghai, China. Atmos. Environ. , 2013, 70: 401-409.

[26] Wang D, Hu R, Xie P, Liu J, Liu W, Qin M, Ling L, Zeng Y, Chen H, Xing X, Zhu G, Wu J, Duan J, Lu X, Shen L. Diode laser cavity ring-down spectroscopy for in situ measurement of NO3 radical in ambient air. J. Quant. Spectrosc. Radiat. Transfer, 2015, 166: 23-29.

[27] Li S, Liu W, Xie P, Qin M, Yang Y. Observation of Nitrate Radical in the Nocturnal Boundary Layer During a Summer Field Campaign in Pearl River Delta, China. Terrestrial, Atmospheric & Oceanic Sciences, 2012, 23(1).

[28] Wang H, Chen J, Lu K. Development of a portable cavity-enhanced absorption spectrometer for the measurement of ambient NO_3 and N_2O_5: experimental setup, lab characterizations, and field applications in a polluted urban environment. Atmos. Meas. Techn. , 2017, 10(4): 1465-1479.

[29] Brown S S, Dubé W P, Tham Y J, Zha Q, Xue L, Poon S, Wang Z, Blake D R, Tsui W, Parrish D D. Nighttime chemistry at a high altitude site above Hong Kong. J. Geophys. Res. : Atmos. , 2016, 121(5): 2457-2475.

[30] Wang T, Tham Y J, Xue L, Li Q, Zha Q, Wang Z, Poon S C, Dubé W P, Blake D R, Louie P K. Observations of nitryl chloride and modeling its source and effect on ozone in the planetary boundary layer of southern China. J. Geophys. Res. : Atmos. , 2016, 121(5): 2476-2489.

[31] Wang H, Lu K. Determination and Parameterization of the Heterogeneous Uptake Coefficient of Dinitrogen Pentoxide(N_2O_5). Prog. Chem. , 2016, 28(6): 917-933.

[32] Wang X, Wang H, Xue L, Wang T, Wang L, Gu R, Wang W, Tham Y J, Wang Z, Yang L, Chen J, Wang W. Observations of N_2O_5 and $ClNO_2$ at a polluted urban surface site in North China: High N_2O_5 uptake coefficients and low $ClNO_2$ product yields. Atmos. Environ. , 2017, 156: 125-134.

[33] Wang Z, Wang W, Tham Y J, Li Q, Wang H, Wen L, Wang X, Wang T. Fast heterogeneous N_2O_5 uptake and $ClNO_2$ production in power plant plumes observed in the nocturnal residual layer over the North China Plain. Atmos. Chem. Phys. Discuss. , 2017, 2017: 1-26.

[34] Su X, Tie X, Li G, Cao J, Huang R, Feng T, Long X, Xu R. Effect of hydrolysis of N_2O_5 on nitrate and ammonium formation in Beijing China: WRF-Chem model simulation. Sci. Total Environ. , 2017, 579: 221-229.

[35] Li Q, Zhang L, Wang T, Tham Y J, Ahmadov R, Xue L, Zhang Q, Zheng J. Impacts of heterogeneous uptake of dinitrogen pentoxide and chlorine activation on ozone and reactive nitrogen partitioning: improvement and application of the WRF-Chem model in southern China. Atmos. Chem. Phys. , 2016, 16(23): 14875-14890.

[36] Tham Y J, Wang Z, Li Q, Yun H, Wang W, Wang X, Xue L, Lu K, Ma N, Bohn B, Li X, Kecorius S, Größ J, Shao M, Wiedensohler A, Zhang Y, Wang T. Significant concentrations of nitryl chloride sustained in the morning: investigations of the causes and impacts on ozone production in a polluted region of northern China. Atmos. Chem. Phys. , 2016, 16(23): 14959-14977.

[37] Tham Y J, Yan C, Xue L, Zha Q, Wang X, Wang T. Presence of high nitryl chloride in Asian coastal environment and its impact on atmospheric photochemistry. Chin. Sci. Bull. , 2014, 59(4): 356-359.

[38] Wang T, Xue L, Brimblecombe P, Lam Y F, Li L, Zhang L. Ozone pollution in China: A review of concentrations, meteorological influences, chemical precursors, and effects. Sci. Total Environ. , 2017, 575: 1582-1596.

[39] Ding A, Fu C, Yang X, Sun J, Zheng L, Xie Y, Herrmann E, Nie W, Petäjä T, Kerminen V -M. Ozone and fine particle in the western Yangtze River Delta: an overview of 1 yr data at the SORPES station. Atmospheric Chemistry and Physics, 2013, 13(11): 5813-5830.

[40] Wang T, Nie W, Gao J, Xue L K, Gao X M, Wang X F, Qiu J, Poon C N, Meinardi S, Blake D, Wang S L, Ding A J, Chai F H, Zhang Q Z, Wang W X. Air quality during the 2008 Beijing Olympics: secondary pollutants and regional impact. Atmos. Chem. Phys. , 2010, 10(16): 7603-7615.

[41] Xue L, Wang T, Gao J, Ding A, Zhou X, Blake D, Wang X, Saunders S, Fan S, Zuo H. Ground-level ozone in four Chinese cities: precursors, regional transport and heterogeneous processes. Atmos. Chem. Phys. , 2014, 14(23): 13175-13188.

[42] Zhang Y, Su H, Zhong L, Cheng Y, Zeng L, Wang X, Xiang Y, Wang J, Gao D, Shao M. Regional ozone pollution and

observation-based approach for analyzing ozone–precursor relationship during the PRIDE-PRD 2004 campaign. Atmos. Environ. , 2008, 42(25): 6203-6218.

[43] Wang T, Ding A, Gao J, Wu W S. Strong ozone production in urban plumes from Beijing China. Geophys. Res. Lett. , 2006, 33(21).

[44] Wang T, Wei X L, Ding A J, Poon C N, Lam K S, Li Y S, Chan L Y, Anson M. Increasing surface ozone concentrations in the background atmosphere of Southern China, 1994–2007. Atmos. Chem. Phys. , 2009, 9(16): 6217-6227.

[45] Wang T, Cheung V T, Anson M, Li Y. Ozone and related gaseous pollutants in the boundary layer of eastern China: Overview of the recent measurements at a rural site. Geophys. Res. Lett. , 2001, 28(12): 2373-2376.

[46] Ding A J, Wang T, Thouret V, Cammas J P, Nédélec P. Tropospheric ozone climatology over Beijing: analysis of aircraft data from the MOZAIC program. Atmos. Chem. Phys. , 2008, 8(1): 1-13.

[47] Sun L, Xue L, Wang T, Gao J, Ding A, Cooper O R, Lin M, Xu P, Wang Z, Wang X, Wen L, Zhu Y, Chen T, Yang L, Wang Y, Chen J, Wang W. Significant increase of summertime ozone at Mount Tai in Central Eastern China. Atmos. Chem. Phys. , 2016, 16(16): 10637-10650.

[48] Xue L, Wang T, Louie P K K, Luk C W Y, Blake D R, Xu Z. Increasing External Effects Negate Local Efforts to Control Ozone Air Pollution: A Case Study of Hong Kong and Implications for Other Chinese Cities. Environ. Sci. Technol. , 2014, 48(18): 10769-10775.

[49] Li J, Lu K, Lv W, Li J, Zhong L, Ou Y, Chen D, Huang X, Zhang Y. Fast increasing of surface ozone concentrations in Pearl River Delta characterized by a regional air quality monitoring network during 2006–2011. J. Environ. Sci. 2014, 26(1): 23-36.

[50] Xu X, Lin W, Wang T, Yan P, Tang J, Meng Z, Wang Y. Long-term trend of surface ozone at a regional background station in eastern China 1991–2006: enhanced variability. Atmos. Chem. Phys. , 2008, 8(10): 2595-2607.

[51] Ma Z, Xu J, Quan W, Zhang Z, Lin W, Xu X. Significant increase of surface ozone at a rural site, north of eastern China. Atmos. Chem. Phys. , 2016, 16(6): 3969-3977.

[52] Tang G, Li X, Wang Y, Xin J, Ren X. Surface ozone trend details and interpretations in Beijing, 2001–2006. Atmos. Chem. Phys. , 2009, 9(22): 8813-8823.

[53] Zhang Q, Yuan B, Shao M, Wang X, Lu S, Lu K, Wang M, Chen L, Chang C C, Liu S C. Variations of ground-level O_3 and its precursors in Beijing in summertime between 2005 and 2011. Atmos. Chem. Phys. , 2014, 14(12): 6089-6101.

[54] Xu W, Lin W, Xu X, Tang J, Huang J, Wu H, Zhang X. Long-term trends of surface ozone and its influencing factors at the Mt Waliguan GAW station, China – Part 1: Overall trends and characteristics. Atmos. Chem. Phys. , 2016, 16(10): 6191-6205.

[55] Lin W, Xu X, Ge B, Liu X. Gaseous pollutants in Beijing urban area during the heating period 2007–2008: variability sources, meteorological, and chemical impacts. Atmos. Chem. Phys. , 2011, 11(15): 8157-8170.

[56] Ran L, Zhao C, Geng F, Tie X, Tang X, Peng L, Zhou G, Yu Q, Xu J, Guenther A. Ozone photochemical production in urban Shanghai, China: Analysis based on ground level observations. J. Geophys. Res. : Atmos. , 2009, 114(D15).

[57] Xue L K, Saunders S M, Wang T, Gao R, Wang X F, Zhang Q Z, Wang W X. Development of a chlorine chemistry module for the Master Chemical Mechanism. Geosci. Model Dev. , 2015, 8(10): 3151-3162.

[58] An J, Li Y, Chen Y, Li J, Qu Y, Tang Y. Enhancements of major aerosol components due to additional HONO sources in the North China Plain and implications for visibility and haze. Adv. Atmos. Sci. , 2013, 30(1): 57-66.

[59] Zhang L, Wang T, Zhang Q, Zheng J, Xu Z, Lv M. Potential sources of nitrous acid(HONO) and their impacts on ozone: A WRF-Chem study in a polluted subtropical region. J. Geophys. Res. : Atmos. , 2016, 121(7): 3645-3662.

[60] Gao J, Wang T, Ding A, Liu C. Observational study of ozone and carbon monoxide at the summit of mount Tai(1534m asl) in central-eastern China. Atmos. Environ. , 2005, 39(26): 4779-4791.

[61] Ding A, Wang T, Zhao M, Wang T, Li Z. Simulation of sea-land breezes and a discussion of their implications on the transport of air pollution during a multi-day ozone episode in the Pearl River Delta of China. Atmos. Environ. , 2004, 38(39): 6737-6750.

[62] So K L, Wang T. On the local and regional influence on ground-level ozone concentrations in Hong Kong. Environ. Pollut. , 2003, 123(2): 307-317.

[63] Ding A, Wang T. Influence of stratosphere - to - troposphere exchange on the seasonal cycle of surface ozone at Mount Waliguan in western China. Geophys. Res. Lett. , 2006, 33(3).

[64] Liu Z, Wang Y, Gu D, Zhao C, Huey L G, Stickel R, Liao J, Shao M, Zhu T, Zeng L, Liu S C, Chang C C, Amoroso A, Costabile F. Evidence of Reactive Aromatics As a Major Source of Peroxy Acetyl Nitrate over China. Environ. Sci. Technol. , 2010, 44(18): 7017-7022.

[65] Zhang J M, Wang T, Ding A J, Zhou X H, Xue L K, Poon C N, Wu W S, Gao J, Zuo H C, Chen J M, Zhang X C, Fan S J Continuous measurement of peroxyacetyl nitrate(PAN)in suburban and remote areas of western China. Atmos. Environ. , 2009, 43(2): 228-237.

[66] 黄志, 高天宇, 赵西萌, 王凤, 杨光, 王斌, 徐振强, 胡敏, 曾立民. 2006-2014 年北京夏季大气中 PANs 浓度变化趋势. 北京大学学报(自然科学版), 2016, 52(3): 528-534.

[67] Xue L, Wang T, Wang X, Blake D R, Gao J, Nie W, Gao R, Gao X, Xu Z, Ding A, Huang Y, Lee S, Chen Y, Wang S, Chai F, Zhang Q, Wang W. On the use of an explicit chemical mechanism to dissect peroxy acetyl nitrate formation. Environ. Pollut. , 2014, 195: 39-47.

[68] Zhang R Y, Khalizov A, Wang L, Hu M, Xu W. Nucleation and Growth of Nanoparticles in the Atmosphere. Chem. Rev. , 2012, 112(3): 1957-2011.

[69] Sipila M, Berndt T, Petaja T, Brus D, Vanhanen J, Stratmann F, Patokoski J, Mauldin R L, Hyvarinen A P, Lihavainen H, Kulmala M. The Role of Sulfuric Acid in Atmospheric Nucleation. Science, 2010, 327(5970): 1243-1246.

[70] Vanhanen J, Mikkila J, Lehtipalo K, Sipila M, Manninen H E, Siivola E, Petaja T, Kulmala M. Particle Size Magnifier for Nano-CN Detection. Aerosol Sci. Technol. , 2011, 45(4): 533-542.

[71] Jiang J K, Chen M D, Kuang C A, Attoui M, McMurry P H. Electrical Mobility Spectrometer Using a Diethylene Glycol Condensation Particle Counter for Measurement of Aerosol Size Distributions Down to 1 nm. Aerosol Sci. Technol. , 2011, 45(4): 510-521.

[72] Jiang J K, Zhao J, Chen M D, Eisele F L, Scheckman J, Williams B J, Kuang C A, McMurry P H. First Measurements of Neutral Atmospheric Cluster and 1-2 nm Particle Number Size Distributions During Nucleation Events. Aerosol Sci. Technol. , 2011, 45(4): Ii-V.

[73] Junninen H, Ehn M, Petaja T, Luosujarvi L, Kotiaho T, Kostiainen R, Rohner U, Gonin M, Fuhrer K, Kulmala M, Worsnop D R. A high-resolution mass spectrometer to measure atmospheric ion composition. Atmos. Meas. Tech. , 2010, 3(4): 1039-1053.

[74] Jokinen T, Sipila M, Junninen H, Ehn M, Lonn G, Hakala J, Petaja T, Mauldin R L, Kulmala M, Worsnop D R. Atmospheric sulphuric acid and neutral cluster measurements using CI-APi-TOF. Atmos. Chem. Phys. , 2012, 12(9): 4117-4125.

[75] Zhao J, Eisele F L, Titcombe M, Kuang C G, McMurry P H. Chemical ionization mass spectrometric measurements of atmospheric neutral clusters using the cluster-CIMS. J. Geophys. Res. -Atmos. , 2010, 115.

[76] Kirkby J, Curtius J, Almeida J, Dunne E, Duplissy J, Ehrhart S, Franchin A, Gagne S, Ickes L, Kurten A, Kupc A, Metzger

A, Riccobono F, Rondo L, Schobesberger S, Tsagkogeorgas G, Wimmer D, Amorim A, Bianchi F, Breitenlechner M, David A, Dommen J, Downard A, Ehn M, Flagan R C, Haider S, Hansel A, Hauser D, Jud W, Junninen H, Kreissl F, Kvashin A, Laaksonen A, Lehtipalo K, Lima J, Lovejoy E R, Makhmutov V, Mathot S, Mikkila J, Minginette P, Mogo S, Nieminen T, Onnela A, Pereira P, Petaja T, Schnitzhofer R, Seinfeld J H, Sipila M, Stozhkov Y, Stratmann F, Tome A, Vanhanen J, Viisanen Y, Vrtala A, Wagner P E, Walther H, Weingartner E, Wex H, Winkler P M, Carslaw K S, Worsnop D R, Baltensperger U, Kulmala M. Role of sulphuric acid, ammonia and galactic cosmic rays in atmospheric aerosol nucleation. Nature, 2011, 476(7361): 429-U77.

[77] Almeida J, Schobesberger S, Kurten A, Ortega I K, Kupiainen-Maatta O, Praplan A P, Adamov A, Amorim A, Bianchi F, Breitenlechner M, David A, Dommen J, Donahue N M, Downard A, Dunne E, Duplissy J, Ehrhart S, Flagan R C, Franchin A, Guida R, Hakala J, Hansel A, Heinritzi M, Henschel H, Jokinen T, Junninen H, Kajos M, Kangasluoma J, Keskinen H, Kupc A, Kurten T, Kvashin A N, Laaksonen A, Lehtipalo K, Leiminger M, Leppa J, Loukonen V, Makhmutov V, Mathot S, McGrath M J, Nieminen T, Olenius T, Onnela A, Petaja T, Riccobono F, Riipinen I, Rissanen M, Rondo L, Ruuskanen T, Santos F D, Sarnela N, Schallhart S, Schnitzhofer R, Seinfeld J H, Simon M, Sipila M, Stozhkov Y, Stratmann F, Tome A, Trostl J, Tsagkogeorgas G, Vaattovaara P, Viisanen Y, Virtanen A, Vrtala A, Wagner P E, Weingartner E, Wex H, Williamson C, Wimmer D, Ye P L, Yli-Juuti T, Carslaw K S, Kulmala M, Curtius J, Baltensperger U, Worsnop D R, Vehkamaki H, Kirkby J. Molecular understanding of sulphuric acid-amine particle nucleation in the atmosphere. Nature, 2013, 502(7471): 359-363+.

[78] Kurten A, Jokinen T, Simon M, Sipila M, Sarnela N, Junninen H, Adamov A, Almeida J, Amorim A, Bianchi F, Breitenlechner M, Dommen J, Donahue N M, Duplissy J, Ehrhart S, Flagan R C, Franchin A, Hakala J, Hansel A, Heinritzi M, Hutterli M, Kangasluoma J, Kirkby J, Laaksonen A, Lehtipalo K, Leiminger M, Makhmutov V, Mathot S, Onnela A, Petaja T, Praplan A P, Riccobono F, Rissanen M P, Rondo L, Schobesberger S, Seinfeld J H, Steiner G, Tome A, Trostl J, Winkler P M, Williamson C, Wimmer D, Ye P L, Baltensperger U, Carslaw K S, Kulmala M, Worsnop D R, Curtius J. Neutral molecular cluster formation of sulfuric acid-dimethylamine observed in real time under atmospheric conditions. Proc. Natl. Acad. Sci. U. S. A. , 2014, 111(42): 15019-15024.

[79] Riccobono F, Schobesberger S, Scott C E, Dommen J, Ortega I K, Rondo L, Almeida J, Amorim A, Bianchi F, Breitenlechner M, David A, Downard A, Dunne E M, Duplissy J, Ehrhart S, Flagan R C, Franchin A, Hansel A, Junninen H, Kajos M, Keskinen H, Kupc A, Kurten A, Kvashin A N, Laaksonen A, Lehtipalo K, Makhmutov V, Mathot S, Nieminen T, Onnela A, Petaja T, Praplan A P, Santos F D, Schallhart S, Seinfeld J H, Sipila M, Spracklen D V, Stozhkov Y, Stratmann F, Tome A, Tsagkogeorgas G, Vaattovaara P, Viisanen Y, Vrtala A, Wagner P E, Weingartner E, Wex H, Wimmer D, Carslaw K S, Curtius J, Donahue N M, Kirkby J, Kulmala M, Worsnop D R, Baltensperger U. Oxidation Products of Biogenic Emissions Contribute to Nucleation of Atmospheric Particles. Science, 2014, 344(6185): 717-721.

[80] Schobesberger S, Junninen H, Bianchi F, Lonn G, Ehn M, Lehtipalo K, Dommen J, Ehrhart S, Ortega I K, Franchin A, Nieminen T, Riccobono F, Hutterli M, Duplissy J, Almeida J, Amorim A, Breitenlechner M, Downard A J, Dunne E M, Flagan R C, Kajos M, Keskinen H, Kirkby J, Kupc A, Kurten A, Kurten T, Laaksonen A, Mathot S, Onnela A, Praplan A P, Rondo L, Santos F D, Schallhart S, Schnitzhofer R, Sipila M, Tome A, Tsagkogeorgas G, Vehkamaki H, Wimmer D, Baltensperger U, Carslaw K S, Curtius J, Hansel A, Petaja T, Kulmala M, Donahue N M, Worsnop D R. Molecular understanding of atmospheric particle formation from sulfuric acid and large oxidized organic molecules. Proc. Natl. Acad. Sci. U. S. A. , 2013, 110(43): 17223-17228.

[81] Kirkby J, Duplissy J, Sengupta K, Frege C, Gordon H, Williamson C, Heinritzi M, Simon M, Yan C, Almeida J, Trostl J, Nieminen T, Ortega I K, Wagner R, Adamov A, Amorim A, Bernhammer A K, Bianchi F, Breitenlechner M, Brilke S, Chen

X M, Craven J, Dias A, Ehrhart S, Flagan R C, Franchin A, Fuchs C, Guida R, Hakala J, Hoyle C R, Jokinen T, Junninen H, Kangasluoma J, Kim J, Krapf M, Kurten A, Laaksonen A, Lehtipalo K, Makhmutov V, Mathot S, Molteni U, Onnela A, Perakyla O, Piel F, Petaja T, Praplan A P, Pringle K, Rap A, Richards N A D, Riipinen I, Rissanen M P, Rondo L, Sarnela N, Schobesberger S, Scott C E, Seinfeld J H, Sipila M, Steiner G, Stozhkov Y, Stratmann F, Tome A, Virtanen A, Vogel A L, Wagner A C, Wagner P E, Weingartner E, Wimmer D, Winkler P M, Ye P L, Zhang X, Hansel A, Dommen J, Donahue N M, Worsnop D R, Baltensperger U, Kulmala M, Carslaw K S, Curtius J. Ion-induced nucleation of pure biogenic particles. Nature, 2016, 533(7604): 521-526.

[82] Kulmala M, Kontkanen J, Junninen H, Lehtipalo K, Manninen H E, Nieminen T, Petaja T, Sipila M, Schobesberger S, Rantala P, Franchin A, Jokinen T, Jarvinen E, Aijala M, Kangasluoma J, Hakala J, Aalto P P, Paasonen P, Mikkila J, Vanhanen J, Aalto J, Hakola H, Makkonen U, Ruuskanen T, Mauldin R L, Duplissy J, Vehkamaki H, Back J, Kortelainen A, Riipinen I, Kurten T, Johnston M V, Smith J N, Ehn M, Mentel T F, Lehtinen K E J, Laaksonen A, Kerminen V M, Worsnop D R. Direct Observations of Atmospheric Aerosol Nucleation. Science, 2013, 339(6122): 943-946.

[83] Bianchi F, Tröstl J, Junninen H, Frege C, Henne S, Hoyle C R, Molteni U, Herrmann E, Adamov A, Bukowiecki N, Chen X, Duplissy J, Gysel M, Hutterli M, Kangasluoma J, Kontkanen J, Kürten A, Manninen H E, Münch S, Peräkylä O, Petäjä T, Rondo L, Williamson C, Weingartner E, Curtius J, Worsnop D R, Kulmala M, Dommen J, Baltensperger U. New particle formation in the free troposphere: A question of chemistry and timing. Science, 2016, 352(6289): 1109-1112.

[84] Sipila M, Sarnela N, Jokinen T, Henschel H, Junninen H, Kontkanen J, Richters S, Kangasluoma J, Franchin A, Perakyla O, Rissanen M P, Ehn M, Vehkamaki H, Kurten T, Berndt T, Petaja T, Worsnop D, Ceburnis D, Kerminen V M, Kulmala M, O'Dowd C. Molecular-scale evidence of aerosol particle formation via sequential addition of HIO_3. Nature, 2016, 537(7621): 532-534.

[85] Guo S, Hu M, Zamora M L, Peng J, Shang D, Zheng J, Du Z, Wu Z, Shao M, Zeng L, Molina M J, Zhang R. Elucidating severe urban haze formation in China. Proc. Natl. Acad. Sci. U. S. A. , 2014, 111(49): 17373-17378.

[86] Zheng J, Ma Y, Chen M D, Zhang Q, Wang L, Khalizov A F, Yao L, Wang Z, Wang X, Chen L X. Measurement of atmospheric amines and ammonia using the high resolution time-of-flight chemical ionization mass spectrometry. Atmos. Environ. , 2015, 102: 249-259.

[87] Zheng J, Yang D S, Ma Y, Chen M D, Cheng J, Li S Z, Wang M. Development of a new corona discharge based ion source for high resolution time-of-flight chemical ionization mass spectrometer to measure gaseous H_2SO_4 and aerosol sulfate. Atmos. Environ. , 2015, 119: 167-173.

[88] Yao L, Wang M Y, Wang X K, Liu Y J, Chen H F, Zheng J, Nie W, Ding A J, Geng F H, Wang D F, Chen J M, Worsnop D R, Wang L. Detection of atmospheric gaseous amines and amides by a high-resolution time-of-flight chemical ionization mass spectrometer with protonated ethanol reagent ions. Atmos. Chem. Phys. , 2016, 16(22): 14527-14543.

[89] Xiao S, Wang M Y, Yao L, Kulmala M, Zhou B, Yang X, Chen J M, Wang D F, Fu Q Y, Worsnop D R, Wang L. Strong atmospheric new particle formation in winter in urban Shanghai, China. Atmos. Chem. Phys. , 2015, 15(4): 1769-1781.

[90] Yu H, Zhou L Y, Dai L, Shen W C, Dai W, Zheng J, Ma Y, Chen M D. Nucleation and growth of sub-3nm particles in the polluted urban atmosphere of a megacity in China. Atmos. Chem. Phys. , 2016, 16(4): 2641-2657.

[91] Dai L, Wang H L, Zhou L Y, An J L, Tang L L, Lu C S, Yan W L, Liu R Y, Kong S F, Chen M D, Lee S H, Yu H. Regional and local new particle formation events observed in the Yangtze River Delta region, China. J. Geophys. Res. -Atmos. , 2017, 122(4): 2389-2402.

[92] Cai R, Jiang J. A new balance formula to estimate new particle formation rate: reevaluating the effect of coagulation scavenging. Atmos. Chem. Phys. Discuss. ,2017, 2017: 1-29.

[93] Cai R, Yang D, Fu Y, Wang X, Li X, Ma Y, Hao J, Zheng J, Jiang J. Aerosol Surface Area Concentration: a Governing Factor for New Particle Formation in Beijing. Atmos. Chem. Phys. Discuss. , 2017, 2017: 1-26.

[94] Guo H, Wang D W, Cheung K, Ling Z H, Chan C K, Yao X H. Observation of aerosol size distribution and new particle formation at a mountain site in subtropical Hong Kong. Atmos. Chem. Phys. , 2012 12(20): 9923-9939.

[95] Wang D W, Guo H, Cheung K, Gan F X. Observation of nucleation mode particle burst and new particle formation events at an urban site in Hong Kong. Atmos. Environ. , 2014, 99: 196-205.

[96] Cheung H C, Chou C C K, Chen M J, Huang W R, Huang S H, Tsai C Y, Lee C S L. Seasonal variations of ultra-fine and submicron aerosols in Taipei, Taiwan: implications for particle formation processes in a subtropical urban area. Atmos. Chem. Phys. , 2016, 16(3): 1317-1330.

[97] Ravishankara A R. Heterogeneous and Multiphase Chemistry in the Troposphere. Science, 1997, 276(5315): 1058-1065.

[98] Gieré R, Querol X. Solid particulate matter in the atmosphere. Elements, 2010, 6(4): 215-222.

[99] Kolb C, Cox R, Abbatt J, Ammann M, Davis E, Donaldson D, Garrett B C, George C, Griffiths P, Hanson D. An overview of current issues in the uptake of atmospheric trace gases by aerosols and clouds. Atmos. Chem. Phys. , 2010, 10(21): 10561-10605.

[100] Usher C R, Michel A E, Grassian V H. Reactions on mineral dust. Chem. Rev. , 2003, 103(12): 4883-940.

[101] Rossi M. Heterogeneous reactions on salts. Chem. Rev. , 2003, 103(12): 4823-4882.

[102] Abbatt J. Interactions of atmospheric trace gases with ice surfaces: Adsorption and reaction. Chem. Rev. , 2003, 103(12): 4783-4800.

[103] Al-Abadleh H A, Krueger B J, Ross J L, Grassian V H. Phase transitions in calcium nitrate thin films. Chem. Commun. , 2003: 2796–2797.

[104] Ma Q, Liu C, He H. Heterogeneous reaction of acetic acid on MgO, α-Al_2O_3, and $CaCO_3$ and the effect on the hygroscopic behaviour of these particles. Phys. Chem. Chem. Phys. , 2012, 14(23): 8403-8409.

[105] 丁杰, 朱彤. 大气中细颗粒物表面多相化学反应的研究. 科学通报, 2003, 48(19): 2005-2013.

[106] Pöschl U, Shiraiwa M. Multiphase chemistry at the atmosphere–biosphere interface influencing climate and public health in the anthropocene. Chem. Rev. , 2015, 115(10): 4440-4475.

[107] Herrmann H, Schaefer T, Tilgner A, Styler S A, Weller C, Teich M, Otto T. Tropospheric aqueous-phase chemistry: Kinetics, mechanisms, and its coupling to a changing gas phase. Chem. Rev. , 2015, 115(10): 4259-4334.

[108] 马庆鑫, 马金珠, 楚碧武, 刘永春, 赖承钺, 贺泓, 矿质和黑碳颗粒物表面大气非均相反应研究进展. 科学通报, 2015, 60(2): 122-136.

[109] Fu H, Chen J. Formation features and controlling strategies of severe haze-fog pollutions in China. Sci. Total Environ. , 2017, 578: 121-138.

[110] Yang F, Tan J, Zhao Q, Du Z, He K, Ma Y, Duan F, Chen G. Characteristics of PM 2. 5 speciation in representative megacities and across China. Atmos. Chem. Phys. , 2011, 11(11): 5207-5219.

[111] Kasibhatla P, Chameides W, John J S. A three - dimensional global model investigation of seasonal variations in the atmospheric burden of anthropogenic sulfate aerosols. J. Geophys. Res. : Atmos. , 1997, 102(D3): 3737-3759.

[112] Barrie L A, Yi Y, Leaitch W R, Lohmann U, Kasibhatla P, Roelofs G J, Wilson J, McGovern F, Benkovitz C, Méliéres M A, Law K, Prospero J, Kritz M, Bergmann D, Bridgeman C, Chin M, Christensen J, Easter R, Feichter J, Land C, Jeuken A, Kjellström E, Koch D, Rasch P. A comparison of large-scale atmospheric sulphate aerosol models(COSAM): overview and highlights. Tellus B: Chem. Phys. Meteoro. , 2001, 53(5): 615-645.

[113] Kong L, Zhao X, Sun Z, Yang Y, Fu H, Zhang S, Cheng T, Yang X, Wang L, Chen J. The effects of nitrate on the

heterogeneous uptake of sulfur dioxide on hematite. Atmos. Chem. Phys. , 2014, 14(17): 9451-9467.

[114] Ullerstam M, Johnson M, Vogt R, Ljungstrom E. DRIFTS and Knudsen cell study of the heterogeneous reactivity of SO_2 and NO_2 on mineral dust. Atmos. Chem. Phys. , 2003, 3: 2043-2051.

[115] Ma Q X, Liu Y C, He H. Synergistic effect between NO_2 and SO_2 in their adsorption and reaction on gamma-alumina. J. Phys. Chem. A, 2008, 112(29): 6630-6635.

[116] Liu C, Ma Q, Liu Y, Ma J, He H. Synergistic reaction between SO_2 and NO_2 on mineral oxides: A potential formation pathway of sulfate aerosol. Phys. Chem. Chem. Phys. , 2012, 14(5): 1668-1676.

[117] Han C, Liu Y, Ma J, He H. Key role of organic carbon in the sunlight-enhanced atmospheric aging of soot by O_2. Proc. Natl. Acad. Sci. U. S. A. , 2012 109(52): 21250-21255.

[118] Han C, Liu Y, He H. Role of Organic Carbon in Heterogeneous Reaction of NO_2 with Soot. Environ. Sci. Technol. , 2013, 47(7): 3174-3181.

[119] Chong Han, Y Liu, Hong He. Heterogeneous photochemical aging of soot by NO_2 under simulated sunlight. Atmos. Environ. , 2013, 64: 270-276.

[120] Chen H, Hu D, Wang L, Mellouki A, Chen J. Modification in light absorption cross section of laboratory-generated black carbon-brown carbon particles upon surface reaction and hydration. Atmos. Environ. , 2015, 116: 253-261.

[121] Bahadur R, Praveen P S, Xu Y, Ramanathan V. Solar absorption by elemental and brown carbon determined from spectral observations. Proc. Natl. Acad. Sci. U. S. A. , 2012, 109(43): 17366-17371.

[122] Wang Y, Chen Z, Wu Q, Liang H, Huang L, Li H, Lu K, Wu Y, Dong H, Zeng L, Zhang Y. Observation of atmospheric peroxides during Wangdu Campaign 2014 at a rural site in the North China Plain. Atmos. Chem. Phys. , 2016, 16(17): 10985-11000.

[123] Huang L, Zhao Y, Li H, Chen Z. Kinetics of Heterogeneous Reaction of Sulfur Dioxide on Authentic Mineral Dust: Effects of Relative Humidity and Hydrogen Peroxide. Environ. Sci. Technol. , 2015, 49(18): 10797-10805.

[124] Wu Q Q, Huang L B, Liang H, Zhao Y, Huang D, Chen Z M. Heterogeneous reaction of peroxyacetic acid and hydrogen peroxide on ambient aerosol particles under dry and humid conditions: kinetics, mechanism and implications. Atmos. Chem. Phys. , 2015, 15(12): 6851-6866.

[125] Zhao Y, Huang D, Huang L, Chen Z. Hydrogen peroxide enhances the oxidation of oxygenated volatile organic compounds on mineral dust particles: a case study of methacrolein. Environ. Sci. Technol. , 2014, 48(18): 10614-10623.

[126] Li S, Du L, Wei Z, Wang W. Aqueous-phase aerosols on the air-water interface: Response of fatty acid Langmuir monolayers to atmospheric inorganic ions. Sci. Total Environ. , 2017, 580: 1155-1161.

[127] Sui X, Zhou Y, Zhang F, Chen J, Zhu Z, Yu X -Y. Deciphering the Aqueous Chemistry of Glyoxal Oxidation with Hydrogen Peroxide Using Molecular Imaging. Phys. Chem. Chem. Phys. , 2017, 19(31):020293-21212.

作者：陈建民[1]，胡 敏[2]，薛丽坤[3]，王 琳[1]，陆克定[2]

[1]复旦大学环境科学与工程系，[2]北京大学环境科学与工程学院，[3]山东大学环境研究院

第5章　典型新型有机污染物的环境行为研究进展

本章导读

在新型污染物（emerging pollutants）之中，有机污染物占据了主要地位。由于带有不同的官能团，新型有机污染物在环境分布特征、迁移转化行为，以及在生物体内的代谢转化过程等方面，与传统有机污染物相比均有很大差异。本章选择目前受到学术前沿领域广泛关注的典型新型有机污染物全（多）氟化合物、双酚类化合物、壬基酚聚氧乙烯醚、对羟基苯甲酸酯、人工甜味剂以及苯并杂环化合物等，综述了其在多介质环境中的分布特征、环境行为与生物转化等方面的研究进展。

关键词

新型有机污染物，环境行为，生物效应

1 引　　言

目前，在环境中可检出的污染物高达数万种，其中多数物质的环境存在刚刚开始受到关注，毒性和环境影响尚不清楚，缺少相关的环境法规的污染物被称为新型污染物。大多数新型污染物属于有机污染物。新型有机污染物往往具有广泛的工业、农业用途，或与人类日常生活使用的产品密切相关，存在广泛的人体暴露途径。新型有机污染物往往含有多个极性官能团，多具有较大水溶性，其在多介质环境中的分布特征和迁移行为相比于以疏水性化合物为主的“传统”有机污染物，存在极大差异。这些极性基团与环境多介质界面以及生物体有机组分往往具有特殊作用力，因此，气-液、固-液等界面上新型有机污染物的迁移行为及生物富集行为，往往难以用传统的分配理论解释，已经成为当前新型有机污染物研究关注的热点问题。

新型有机污染物在人工及天然环境中的转化过程也日益受到科学家们的关注。作为多数新型有机污染物进入环境的重要来源，污水处理厂中各类新型有机污染物的分布特征研究受到广泛关注。部分研究已表明，全氟羧酸、壬基酚等重要新型有机污染物可在污水处理厂（waste water treatment plant，WWTP）中由其前体化合物转化生成，浓度不降反升。此外，虽然多数新型有机污染物被认为可较快地转化为极性更强的代谢产物，可以从生物有机体内代谢排出，但是一些代谢产物被认为比母体化合物具有更高的毒性，如增塑剂邻苯二甲酸双酯可被代谢生成毒性更高的邻苯二甲酸单酯。新型有机污染物在环境中及在生物有机体内的转化过程，也是目前此类污染物研究关注的重要方向。

2 新型有机污染物的多介质分布

在我国，全（多）氟化合物、双酚类化合物、壬基酚聚氧乙烯醚、对羟基苯甲酸酯、人工甜味剂以及苯并杂环化合物、有机磷阻燃剂等典型新型有机污染物在各环境介质中的分布调查已有不同程度开展，部分研究结果可参见本章附表5-1。

2.1 全/多氟烷基化合物

全/多氟烷基化合物（PFASs）指的是人工合成的、与碳原子相连的氢原子部分或全部被氟原子取代的一系列烷基化合物，其结构通式为 $F(CF_2)_n$-R，其中 R 为亲水性官能团[1]。长的碳氟链具有较强的憎水性，而 R 基团具有较强的亲水性；C—F 键的键能较大，在自然环境条件下不易发生光解、水解和微生物降解，因此 PFASs 具有良好的表面活性、优良的化学稳定性和热稳定性。PFASs 作为一种优良的表面活性剂被广泛地应用于纺织、造纸、包装、农药、地毯、电镀和泡沫灭火剂等领域[2]。目前已有大量文献报道，PFASs 广泛分布于大气、降尘、降水、土壤、地表水、沉积物、地下水、室内灰尘和室内空气等多种环境介质中。其中中性的全氟前体物具有半挥发性，主要分布在大气环境中，离子型的 PFASs 则主要分布在水体以及降尘等固体介质中。目前大量的研究工作主要集中在水体环境中，关于大气和土壤的研究相对较少。

在针对大西洋和南大洋的全球航次调查中，首次发现中性前体物氟调醇（FTOHs）和全氟辛烷磺酰胺乙醇（FOSE/As）在全球大气中广泛分布，且主要分布在气相中而非颗粒相中，为 PFASs 的全球分布提供了直接证据。其中 8∶2 FTOH 为主要的检出物，浓度水平呈现出从内陆到大洋、从北半球向南半球递减的规律[3]。Li 等发现在亚洲不同国家大气中 PFASs 的污染具有明显的区域特征，呈现出从城市到乡村再到背景区域依次递减的规律，并且不同国家的主要污染物不同，这主要与 PFASs 的生产和使用密切相关，其中在中国和日本地区 8∶2 FTOH 和 10∶2 FTOH 为主要的污染物，在印度地区 4∶2 FTOH 为主要的污染物。我国对于大气中 PFASs 的研究仍然较少，且主要集中在污染点源及大型城市地区。姚义鸣等[4]从对天津地区大气中 PFASs 挥发性前体物的分布和季节变化研究中发现，挥发性前体物 FTOHs 在天津地区的三个典型位点（城镇、乡村和污水处理厂）均显著检出，总浓度（气态和颗粒态）范围在冬夏两季分别为 25.6~95.3 pg/m^3 和 16.7~2003 pg/m^3，非点源释放是大气中 FTOHs 的重要来源。在夏季，FTOHs 主要存在于气相中，而冬季，FTOHs 在颗粒相中的浓度水平显著升高。姚义鸣等[5]在关于 PFASs 在大气和降尘中浓度分布特征的研究中发现，污水处理厂进水中 FTOHs 的直接释放是大气中 FTOHs 的重要来源。据估算通过干沉降仅能有效去除大气中 1%~5%的离子型 PFASs，而绝大部分以气态或气溶胶态存在并且能够进行长距离大气迁移（图 5-1）。关于 PFASs 在我国室外灰尘中分布特征的研究显示[6]，包括全氟磷酸双酯在内的离子型和中性 PFASs 均普遍检出，其中长碳链的全氟辛烷羧酸（PFOA）为主要的检出物，短碳链的全氟丙酸（PFPrA）在室外灰尘中的整体浓度水平较高，该研究指出干沉降在 PFASs 的迁移过程中起到了重要的作用，并发现 PFASs 在经济较为发达的东部地区的室外灰尘中浓度较高，整体呈现出东高西低南高北低的规律。湿沉降作为去除大气中 PFASs 的又一主要途径，能够间接反映出一

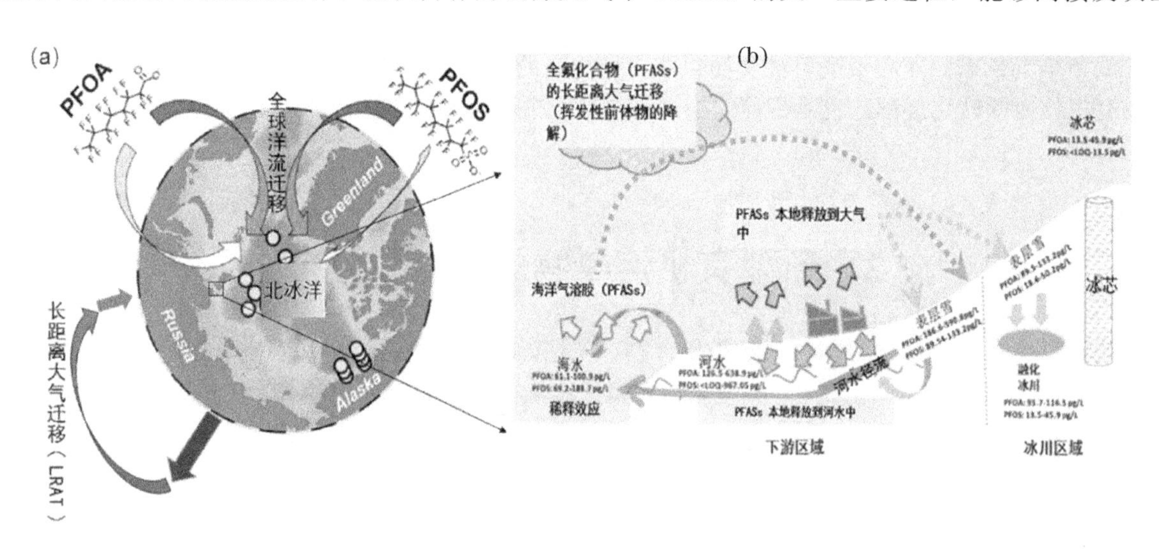

图 5-1　全氟化合物的全球迁移（a）及在局部地区多环境介质中的分布（b）[9,10]

个地区大气中 PFASs 的浓度水平。赵立杰等[7]测定了我国东部和中部地区湿沉降的 PFASs，在同一次湿沉降过程中，PFASs 的浓度随着湿沉降过程的进行呈现出降低的趋势，再次证明了湿沉降对大气中 PFASs 的冲刷机制。单国强等[8]测定了我国北方地区降雪中的 PFASs 异构体的浓度和通量，结果表明颗粒物结合态的 PFASs 占降雪中总 PFASs 的 21.5%~56.2%，并指出颗粒物对 PFASs 的迁移和沉降起到重要的作用，降雪中 PFASs 异构体的分布表明了雨雪中 PFASs 来自于 PFCAs 和 PFSAs 的直接释放。

离子型的 PFASs 较易通过干湿沉降等过程进入到地表环境中，目前关于 PFASs 在水体中分布的研究也比较多。之前的大部分文献报道[11]均指出，PFOA 和全氟辛烷磺酸（PFOS）是水体中主要的污染物。近年来随着一系列监管措施的出台，一方面 C_8 等长链 PFASs 的生产开始转向中国等亚洲国家，另一方面 PFASs 的生产开始转向更短碳链的全氟羧酸（PFCAs）和磺酸（PFSAs）或者是转向生产全氟醚类羧酸（PFECAs）和磺酸（PFESAs）[12]。这些转变在水环境中 PFASs 的污染分布上也有所体现。在国内外的近期研究中[13-16]，短碳链的 PFASs（如全氟丁烷羧酸 PFBA 和全氟丁烷磺酸 PFBS）开始取代长碳链的 PFOA 和 PFOS 成为主要的检出物，且已有大量的研究观察到了这一现象。但在我国一些典型的全氟化工生产地区的周围环境中，PFASs 的浓度和组成与其产业结构紧密相关。史亚利等[17]在我国山东小清河流域的地表水中检出了高浓度的 PFOA，这主要与该区域的氟化工生产活动有关。近年来，随着全氟替代物 PFECAs 和 PFESAs 的广泛应用，在欧洲一些国家和地区的生产点源附近已经检测出了高浓度的 PFECAs 和 PFESAs，例如在美国[18]和德国[19]均检出了高浓度的 GenX。在我国氟化工产业较为集中的山东小清河流域也检出了高浓度的 GenX[17]。其中 F-53B 作为铬雾抑制剂在我国已有 30 多年的应用[20]，在针对中国 19 个河口地区 PFASs 的研究中，F-53B 的检出率较高且在一些大型城市的河流中检出浓度较高（51%，<0.56~78.5 ng/L）。孙红文等[21,22]在研究 PFASs 在污水处理厂中的浓度分布时发现，部分 PFASs 在污水处理厂的出水中的浓度不降反升，提出这主要是由于部分 PFASs 前体物在污水处理过程中发生降解，进而造成出水浓度升高，因此认为污水处理厂是环境水体中 PFASs 的一个重要来源。进入到水环境中的长碳链 PFASs 由于其憎水性较强，较易被沉积物所吸附进而进入到沉积物中，其中主要以应用较广的 PFOA 和 PFOS 为主，大量的环境调查文献[13,14,23-25]均证明了此规律。PFASs 在各种水体中的浓度规律一般是：污水>河水（湖水）>海水，浓度水平一般是几个 ng/L 到几个 μg/L，在一些氟化工厂周围的水体中甚至达到了 mg/L 的水平。

国内外关于 PFASs 在土壤中的调查研究普遍集中在典型的污染点源附近，如氟化学工厂周围[26]，施用活性污泥的农田[27]以及使用水成膜泡沫灭火剂的消防场地周围[28-30]和滑雪场周围[31]。其中长碳链的全氟羧酸和磺酸为主要的检出物，目前国外关于土壤中 PFASs 的研究普遍集中在一些新发现的阳离子和两性离子的全氟前体物上，这些前体物在土壤中经过一系列转化，其终产物仍是关注较多的 PFOA 和 PFOS[29]。由于 PFASs 通过干湿沉降等过程向地表环境中的沉降量较少，且一部分经水流冲刷会进入到地表水体或者是地下水中，因此 PFASs 在土壤环境中的整体浓度水平相对较低，一般只有几个 ng/g 到几百个 ng/g 的水平。短碳链的离子型 PFASs 由于水溶性较强，不易被土壤所截留，因此较易随水流进入到地下水中。目前关于地下水中的 PFASs 的研究主要集中在污染点源附近，例如垃圾填埋场[32]、使用水成膜泡沫灭火剂的灭火训练场[29,33,34]、氟化物生产地等[35,36]。地下水中 PFOA 为主要的污染物，近年来短碳链的全氟羧酸（如 PFBA）所占的比例越来越大，越来越多的研究指出短碳链的 PFCAs 开始成为主要的检出物[37]，甚至在一些水成膜泡沫污染点位的地下水中发现了几十个 μg/L 的超短链的全氟乙烷磺酸（PFEtS）和全氟丙烷磺酸（PFPrS）。

2.2 双酚类化合物

双酚类化合物大部分都是通过一个或者多个碳原子，也可以是硫原子和氧原子将两个羟苯基连接在一起的化合物。该类化合物主要包括双酚 A（BPA）、双酚 AF（BPAF）双酚 AP（BPAP）、双酚 B（BPB）、双酚 C（BPC）、双酚 E（BPE）、双酚 F（BPF）、双酚 G（BPG）、双酚 M（BPM）、双酚 S（BPS）、双酚 P（BPP）、双酚 Z（BPZ）等。这类化合物可以用来合成环氧树脂、聚氨酯、聚碳酸酯、酚醛树脂等高分子材料。此类物质作为高分子材料的改性剂、稳定剂和光引发剂使用时，可使材料的耐热、耐湿、绝缘性、加工、力学和光学性能都有显著提高，同时还具有易溶解、高折射率和高透明性等特征，在涂料、膜材料、电子产品制造、信息记录、光电科技、包装材料、光学镜片及其他日用品等领域广泛应用。在美国、日本等工业发达国家，其他双酚类化合物的合成和开发也在不断发展。其中，BPA 是世界上生产和用量最大的化学品之一，世界上每年的 BPA 生产量超过八百万吨。

在许多塑料制品在使用过程中，其中的添加剂 BPA、BPF 等会释放到环境中，通过水、食物、大气等途径进入人体，产生危害。BPA、BPF 等被证实具有雌激素作用，干扰内分泌系统；其还可能产生其他有害影响，如激活体内肥胖基因，从而导致肥胖。

BPA 污染普遍存在于各种环境介质，主要分布在土壤、水和沉积物中。BPA 在河水中的浓度水平为 1.0~628 ng/L，沉积物中为 $3.94\sim2.2\times10^6$ ng/g dw，污水处理厂的废水中为 10~1080 ng/L，下水道污泥中为 0.42~25600 ng/g dw [38]。在不同地点的介质中（大气、水、底泥、土壤及生物相），BPA 的浓度水平会有所不同。

由于 BPA 的广泛存在和潜在危害，作为一种典型的内分泌干扰化合物，其制造和使用在许多国家受到管制或限制，如 2008 年起我国开始对 BPA 的使用实行管制，2011 年欧盟禁止在婴儿奶瓶制造中使用 BPA。随着许多国家和地区对 BPA 的限制，其他双酚类似物如 BPB、BPE、BPF、BPS、BPAF 和 BPAP 等作为 BPA 的潜在替代物，使用量明显逐步增大。在这些双酚类似物中，BPF、BPS 和 BPAF 是 BPA 在聚碳酸酯和环氧树脂制造应用方面的主要替代物。这些双酚类似物也在各种环境介质中被广泛检出，如水、沉积物和室内灰尘等，然而总的来说对这些双酚类似物的研究和报道相对 BPA 较少。2013 年，我国太湖地区湖水中双酚类似物的水平为 5.4~87 ng/L，沉积物中的水平为 0.37~8.3 ng/g dw [39]；然而，2016 年，在同一湖区的调查中双酚类似物在湖水中的水平为 $2.0\times10^2\sim9.5\times10^2$ ng/L，沉积物中的水平为 4.3×10^2 ng/g dw [40]。近年来双酚类似物在环境介质中的急剧增加表明了其用量在不断增加。另外，有研究发现在日本、韩国和中国的一些地方地表水中的双酚类似物检测中，BPF 的含量最多[41]。也有研究表明，中国杭州湾的河水和底泥中 BPAF 的水平与 BPA 的水平接近[42,43]。

2.3 壬基酚聚氧乙烯醚

壬基酚聚氧乙烯醚（NPEOs）是一类广泛应用的非离子表面活性剂，在环境介质中比较容易降解。在很多国家污水处理厂的进水、出水或污泥中都检测出 NPEOs 的小分子降解产物[44-47]。进水中壬基酚（NP）、壬基酚乙氧乙基醚（nonylphenol monoethoxylate，NP1EO）和壬基酚二乙氧基醚（nonylphenol diethoxylate，NP2EO）均在几至十几 μg/L 左右[45,47]，而在出水中壬基酚氧乙酸类物质（nonylphenol polyethoxy carboxylates，NPEC）的浓度是最大的，达到了几十至几百 μg/L[44]。同时，在污泥中也检测到较高浓度的 NP、NP1EO 和 NP2EO[45,46]。Samaras 等和 Gao 等的研究也证实了污泥的吸附是 NPEOs 从废水中去除的一个重要机制[45,48]。

由于工业废水和生活污水的排放，以及地表径流等的影响，NPEOs 已经广泛存在于地表水环境中[44,49,50]。NPEOs 在地表水环境中的浓度差异主要与污染源有关，Robert-Peillard 等在城市污水排放口附近的海水中检测到 NPEOs 的存在[50]。同时，NPEOs 的水平也受到温度和生物降解等的影响。Traverso-Soto 等发现，在西班牙泻湖的主要支流中，冬季 NPEOs 的浓度要远高于夏季，这可能与夏季温度高，微生物活性强，易于发生生物降解有关[49]。

由于 NPEOs 的疏水性，地表水中的 NPEOs 更易于吸附到沉积物和污泥上。Corada-Fernández 等报道了沉积物中 NPEOs 的浓度在零点几至几 mg/kg 之间[51]。而 Venkatesan 等测得美国城市污泥中含有更高浓度的 NPEOs，其中壬基酚（nonylphenol，NP）的浓度最大，达到了 534 mg/kg[52]。因此，沉积物或污泥成为了 NPEOs 的汇。当水文条件等发生变化时，沉积物或污泥中 NPEOs 可能会重新释放到水环境中，再次成为 NPEOs 的源。此外，利用污水处理厂出水进行农田灌溉或者污泥施用也会使 NPEOs 污染土壤。Corada-Fernández 等报道了被污染土壤中 NPEOs 的浓度达到了 155~280 μg/kg[53]。

2.4　对羟基苯甲酸酯防腐剂

对羟基苯甲酸酯（parabens），又称尼泊金酯，是一类由对羟基苯甲酸所形成的酯类化合物。由于对羟基苯甲酸酯具有抗菌谱广、耐酸碱、在空气中稳定、对人体相对安全和价格低廉等诸多优点，因此被广泛用作食品、化妆品、药品及个人护理品等的防腐抗菌剂，另外也常用作医疗器械的消毒剂及有机合成的中间体等。Parabens 主要包括对羟基苯甲酸甲酯（methyl paraben，MeP）、对羟基苯甲酸乙酯（ethyl paraben，EtP）、对羟基苯甲酸丙酯（*n*-propyl paraben，*n*-PrP）、对羟基苯甲酸异丙酯（iso-propyl paraben，iPrP）、对羟基苯甲酸丁酯（butyl paraben，BuP）、对羟基苯甲酸苯甲酯（benzyl paraben，BzP）和对羟基苯甲酸庚酯（heptylparaben，HeP）等。Parabens 的溶解度较高，从对羟基苯甲酸苯酯到对羟基苯甲酸甲酯，其溶解度从 160 mg/L 逐渐增加到 2500 mg/L（25 ℃）。尽管其溶解度随着链长的增加而降低，但是其正辛醇-水分配系数（K_{ow}）却随着链长的增加而增加，其中甲酯的 K_{ow} 为 1.96，苯酯的 K_{ow} 为 3.27。酸解离常数（pK_a）介于 8.17~8.50 之间，根据酯的不同而存在差异，因此其主要以游离态形式存在于水环境中。Parabens 类物质很稳定且沸点较高。根据酯链的不同可衍生出一系列 parabens 同系物。随酯链的增长虽然其抗菌活性增加，但水溶性降低，因此实际常用于抗菌防腐剂的主要是一些短链酯如 MeP、EtP、PrP 和 BuP 等[54]，其中 MeP 和 PrP 在食品和化妆品行业中的应用最为广泛。Parabens 水解的主要产物为对羟基苯甲酸（*para*-hydroxybenzoic acid，*p*-HBA）。

由于其广泛应用，parabens 目前已在空气、灰尘、土壤、地表水、污水处理厂进出水等环境介质中被大量检出[55-60]。1996 年，Paxéus[61-63]首次报道了水环境中存在 parabens。由于日常生活用品中 parabens 的大量使用，其进入人体或者其他生物体后主要以代谢或者原形排出，然后经由城市污水管网系统收集送入污水处理厂，或者直接汇入河流地表水中，造成 parabens 在环境水体中的广泛存在。有大量文献报道沿海地区或者工厂附近的河流中 parabens 含量较高[64-67]。

Parabens 的广泛使用导致其在环境中无处不在，其中最重要的污染点源是污水处理厂。Parabens 能够在污水处理厂所有进出水样品中检测到，进水中浓度可达到 79600 ng/L[68]。在污水处理厂进水中，parabens 在含水部分中占据主导地位（>97 %），在经过 WWTP 第一个处理设施即厌氧处理阶段后，所有 parabens 的浓度显著下降，再之后的生物处理阶段依旧是逐渐降低，通过检测脱水污泥中所含 parabens 的量发现，和生物降解 parabens 相比，污泥吸附的 parabens 量可以忽略不计。和进水中 parabens 含量相比，在出水和脱水污泥中 parabens 含量占比分别为 0~1.6 %和 0~0.4 %[69]。

污水处理厂中 parabens 最主要的来源就是个人护理品，在城市污水管道中 parabens 并不会被去除。事实上，多数 parabens（MeP、EtP、PrP 和 BuP）都能在个人护理品原污水中检出。作为最常用于个人护理品添加剂的 MeP 和 PrP，在污水处理厂进水中检出率及浓度都最高，其浓度分别可达到 30000 ng/L 和 20000 ng/L[63,69-75]，此外，EtP 和 BuP 也经常能够检测到，但是其浓度比 MeP 和 PrP 低，水平为几个 ng/L 到几百个 ng/L 不等[63,71-74]。BzP 和 HeP 在污水处理厂进水中也能够检测到，但是其浓度均不会超过 10 ng/L[63,75-78]。González-Mariño[71]报道西班牙某污水处理厂进水中 MeP、EtP、PrP、BuP 和 BzP 的浓度中值分别为 2500 ng/L、760 ng/L、1400 ng/L、65 ng/L 和 2 ng/L，其出水浓度中值分别为 190 ng/L、3 ng/L、

2 ng/L、2 ng/L 和 2 ng/L。Geara-Matta[77]曾报道法国某污水处理厂进水中 MeP、EtP、PrP、BuP 的浓度中值为 5810 ng/L、1130 ng/L、2060 ng/L 和 290 ng/L，出水中四种酯的浓度中值分别为 220 ng/L、240 ng/L、110 ng/L 和 130 ng/L。

和污水不同，污泥中的 parabens 比较持久[79,80]。在污水处理厂活性污泥中发现浓度最高的两种 parabens 同样是 MeP 和 PrP，其浓度范围分别为 5~202 ng/g dw 和 4~44 ng/g dw，PrP 在污水处理厂污泥中浓度不会超过 50 ng/g dw[78-81]，除了 Gorga 曾报道污水处理厂污泥中 PrP 的浓度达到 174 ng/g dw [82]。EtP、BuP、BzP 和 HeP 在污水处理厂中很少能够检测到，且其浓度不会超过几个 ng/g dw [77,78]。

作为工业废水或污水处理厂出水受体的地表水，也存在 parabens 的污染。此外，parabens 还可以通过非点源污染源以径流形式释放到地表水中，或者通过大气颗粒物沉积形式进入地表水中[62]。

即使 parabens 很容易被微生物降解，但是由于持续不断的排放，parabens 在环境中具有伪持久性。因为具有内分泌干扰作用，parabens 在水生环境中的分布特征引起广泛关注[81]。目前为止，已经有很多文献探讨过地表水中 parabens 的污染水平，比如西班牙 Ebro 河流[82]、英国 Taff 河流[73]，这些研究表明 parabens 在地表水中广泛存在，其浓度水平从 ng/L 至 μg/L 不等[83]。

在地表水中，检出率和浓度最高的两种 parabens 分别是 MeP 和 PrP，这与这两种物质在工业生产中应用较多有关。Peng 曾于 2008 年测得中国河流-珠江的 MeP 和 PrP 的浓度最高，分别为 1062 ng/L 和 3142 ng/L[84]，而在欧洲河流中 MeP 的最大报道浓度为 400 ng/L，PrP 的最大浓度为 69 ng/L[73]。地表水中 EtP 和 BuP 检出率及浓度都比 MeP 和 PrP 低[66]，如地表水中 EtP 浓度为 147 ng/L，BuP 浓度为 163 ng/L[84]。地表水中的 parabens 检出率及浓度和季节也存在一定关系，其最大浓度往往出现在地表水的枯水期[83]。这是因为在枯水期，地表水水量较小，稀释作用不明显。但是，Jonkers 等也曾发现丰水期地表水中 parabens 浓度要高于枯水期，并解释为丰水期降雨会导致污水处理厂水量过多，较多水量会溢出排入到周边地表水中，从而导致释放到地表水中的 parabens 较多。

2.5 人工甜味剂

人工甜味剂（artificial sweeteners）是一类人工合成或半合成的蔗糖替代有机化合物，常用于食品、饮料、个人护理品、药物、电镀光亮剂以及饲料添加剂。由于它们几乎不被人体代谢，所以被称作无热量的糖[85]。常见的人工甜味剂包括甜蜜素、糖精、安赛蜜、三氯蔗糖、阿斯巴甜、纽甜、新橙皮苷二氢查耳酮（NHDC）等。其中甜蜜素、糖精和安赛蜜属于磺胺盐类，三氯蔗糖由氯原子选择性取代蔗糖的三个羟基而成，纽甜和阿斯巴甜是二肽化合物的衍生物，NHDC 是由柑橘类水果中的类黄酮通过碱性加氢而得到，因此，这类物质具有较高的水溶性（S_w：0.4~1000 g/L）和较强的极性（$\log K_{ow}$：−1.61~2.39）。

人工甜味剂由于具有甜度高、水溶性好、不被代谢以及成本低等优势而被广泛应用。随着其消费需求的不断增大，大量人工甜味剂直接或间接进入环境介质。进入人体和其他生物体的人工甜味剂会直接经排泄物排出体外；工业中使用的人工甜味剂会随着工业废水或固体废物进入环境；此外，磺酰脲类除草剂的不完全降解也会对环境中的人工甜味剂起到贡献作用[86]。目前已在饮用水、地表水、地下水、污水处理厂进出水、干湿沉降物、土壤等环境介质中检出了人工甜味剂。

环境中广泛检出的 4 种典型人工甜味剂为甜蜜素、糖精、安赛蜜和三氯蔗糖。迄今为止，各国关于人工甜味剂的研究大部分都集中在水环境，文献表明各水体中人工甜味剂的浓度水平在 ng/L 到 mg/L 浓度级别[87-93]。尽管人工甜味剂的水溶性较高，其在污水处理厂污泥及悬浮物中也有检出[90,94,95]。另外，人工甜味剂的蒸气压（2.7×10^{-11}~1.0×10^{-7} mm Hg）表明它们会出现在大气气态或颗粒物中。已有文献报道在雨水及降尘中发现了 4 种典型人工甜味剂[96-99]。而土壤中的人工甜味剂主要通过粪肥农用、污水灌溉、污泥还田以及污水管道渗漏等途径进入。图 5-2 表示了由养猪场造成的甜味剂农田土壤污染的环境释放通量及对各接受相的影响[100]。针对我国土壤的调查研究发现，在所有土壤中样品中均检出甜蜜素、糖精和安赛蜜，三氯蔗糖在我国北方和南方土壤中的检出率分别为 15%和 28%，这些人工甜味剂在我国南北土壤中的分布情况并没有显著差异[96]。

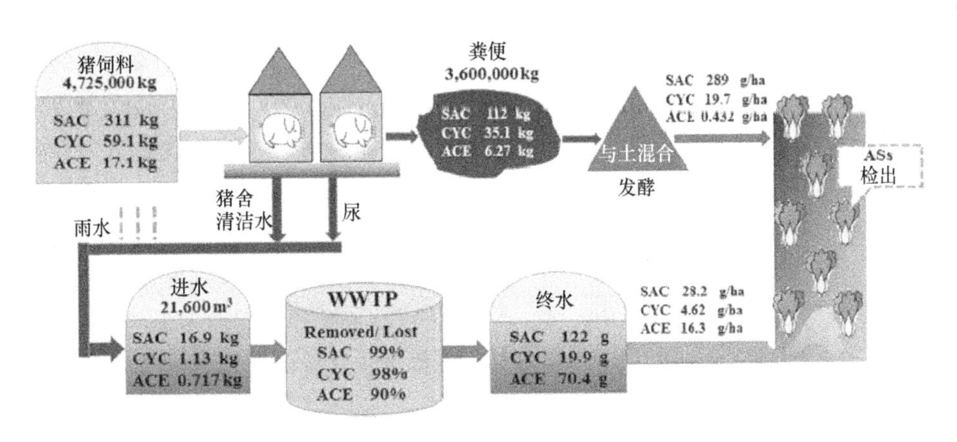

图 5-2　人工甜味剂由天津某养猪厂的年释放量及其进入不同农业系统的可能通量[100].

SAC：糖精；CYC：甜蜜素；ACE：安赛蜜

2.6　苯并杂环化合物

苯并三唑（benzotriazole，BTR）是指分子结构中含有一个苯环一个共轭三氮杂环的化合物，广泛应用于除冰剂、缓蚀剂、农药添加剂、紫外稳定剂、清洗剂等产品中。2011 年，BTR 的全球产量已经达到 9000 t。由于具有较强的水溶性及常规污水处理工艺中较低的生化去除率，导致其在环境中广泛存在。有研究发现 80%的除冰剂残留在机场周围土壤及地下水中，被认为是机场附近苯并杂环的主要来源之一[101]。北半球土壤及水环境中较高浓度的 BTR 出现在冬季，这可能是冬季除冰剂使用量较大导致的[101-103]。在不使用除冰剂的澳大利亚，环境中的 BTR 主要来源于家用及工业用洗涤剂[104,105]，有研究指出洗碗机排放的废水所含的 BTR 几乎占所有废水中总量的 30%[106]。

Kiss 和 Fries 研究了流经 Frankfurt 国际机场的三条河流（Main，Hengstbach 及 Hegbach）水样中 BTR 的时间及季节变化，结果表明 Main 河 BTR 通量高达 1.5 μg/L[107]。美国及瑞士的相关研究指出抗冻剂及缓蚀剂是地表河流中较高水平 BTR 的重要来源[108,109]。瑞典某河流样品中，苯并三唑的浓度可达 100 μg/L[105]。有研究表明，抗冻剂的季节性使用与地表河流检测浓度呈现一定相关性。例如，2003~2004 年冬季采集与靠近 Zurich 机场的 Glatt 河水样品中 BTR 的浓度与抗冻剂的使用量呈强相关性[101]。同样的，德国 Spree 河流及瑞士 Greifensee 湖泊中也发现了类似的规律[101,110]。

城镇废水及工业废水含有较高浓度的 BTR[109,111-120]，其排放是地表水中 BTR 的直接来源之一[121-122]。Reemtsma 等调查发现城镇污水的原水中 BTR 浓度高达 100μg/L[109]。西班牙污水处理厂中 BTR 浓度为 47.1 μg/L[115]。共轭杂环的存在导致 BTR 具有较强的紫外稳定性及较低的生化可利用性，因而其很难被常规的污水处理工艺所去除，多被二次排放到环境中。

长期的污水灌溉活动导致大量 BTR 蓄积到土壤环境中。Cancilla 等调查发现，土壤中 BTR 浓度高达 mg/kg 级别[123]。此外，雪融水[124-125]及地下水[126]中 BTR 的浓度水平也有研究。

苯并噻唑（BTH）是由 1, 3 噻唑环稠合一个苯环形成的苯并杂环化合物，其母体化合物的分子式为 C_7H_5NS。苯并噻唑类物质（BTHs）被广泛用于橡胶中的硫化促进剂、造纸和皮革生产中的灭菌剂、防冻液中的防腐抑制剂和摄影中的显影剂[127]。2-氨基苯并噻唑（2-NH_2-BTH）是药物合成的中间体[128]，最广泛使用的衍生物之一是作为橡胶硫化促进剂的 2-巯基苯并噻唑（MBT），而母体化合物 BTH 的应用并不多。

由于苯并噻唑及其衍生物的水溶性较高，所以 BTHs 在水环境中普遍存在。1976 年，在英国一家轮胎制造厂的废水中检测到了 BTH，浓度为 20 μg/L[129]。Reemstma[127]和 Wick 等[131]发现在地表水和污水中检测到了 2 取代的苯并噻唑类物质。河水中 2-甲巯基苯并噻唑（MeSBTH）和 2-羟基苯并噻唑（2-OH-BTH）的浓度达几百 ng/L[117]，在一些情况下污水中 BTH 或 2-OH-BTH 的浓度可高达 μg/L 水平。Ni 等[132]调查了珠江三角洲河流径流中 BTH 及其衍生物的分布，总的 BTHs 浓度在 220~611 ng/L 之间，其中 BTH 的浓度最高，研究人员根据河流中 BTHs 的浓度及河流流量估算了每年排入的 BTHs 高达 79 t。Kong 等[133]调查了我国天津市 20 个地表水样品中 1300 种极性有机物，检测到了 2-(硫氰酸甲基巯基)苯并噻唑，浓度中间值为 0.11μg/L，检出率为 85%。Wang 等[134,135]对抚顺大伙房水库、浑河沈阳段上下游及辽河入海口水样进行了分析，大伙房水库未检出目标化合物，浑河沈阳段上游和下游检出了苯并噻唑、2-氨基苯并噻唑，最高浓度为 335 ng/L（BTH），在辽河入海口水样中检测出了苯并噻唑，浓度为 5.695 μg/L。Wang 等[136]调查了我国自来水中 BTHs 的分布情况，发现在所有自来水中均检出了 BTH，2-OH-BTH 和 2-NH_2-BTH 的检出率分别为 59.3%和 8.1%，BTHs 总的浓度范围为 40.1~1310 ng/L，这一浓度水平要普遍高于文献报道的地表水及污水中 BTHs 的浓度，研究人员也采集了自来水厂出水及该水厂供水的居民区自来水，发现居民区自来水中 BTHs 的浓度要高于水厂出水，橡胶作为管道材料经常被使用在自来水的分配系统中[137]，它和自来水直接接触。BTH 作为橡胶硫化促进剂的母体化合物，会大量的从橡胶中淋出[138]，这可能是自来水在输送过程中引入 BTHs 污染的原因。

关于土壤中 BTHs 的报道要远少于水体样品，因为这类物质大多极性较强，更倾向于在水中存在。但是，也有一些研究报道了污泥、沉积物和土壤中存在相当浓度的这类物质。Asimakopoulos 等[112]研究了希腊污水处理厂的污泥中这类物质的存在情况，发现 2-OH-BTH 和 MeSBTH 的浓度都在 100 ng/g dw 以上。2-NH_2-BTH 在污泥中浓度低于检出限。Speletini 等在意大利的土壤中检出了 BTH，浓度为 786 ng/g[139]。

Wang 等在中国、美国、日本和韩国室内灰尘中广泛检出了 BTHs[140]。在这次研究中，采集了来自四个国家共 158 个灰尘样品，以此调查 BTHs 的存在情况和对人体的暴露危害。在所有灰尘样品中均检测到了 BTHs。四个国家中 BTHs 浓度的几何平均值在 600~2000 ng/g 之间。导致室内灰尘样品中这类化合物的存在可能是由于大气沉降作用[141]。Wan 等[142]在美国的室内空气中广泛检出了 BTHs，研究人员针对不同的室内环境，如居室、办公室、实验室、汽车内、理发店等的室内空气（包括气相与颗粒相）进行采样，检测了四种苯并噻唑类物质，浓度范围为 4.36~2229 ng/m^3，在汽车内的浓度值最高，几何平均值为 149 ng/m^3，这可能是由于汽车轮胎内含有苯并噻唑类物质导致的。

2.7 有机磷酸酯阻燃剂

有机磷酸酯（organophosphate esters，OPEs）常常作为阻燃剂和增塑剂被物理添加到各种材料[纺织染料、家具、电子产品、涂料、聚氯乙烯塑料（PVC）、聚氨酯泡沫体(PUFs)]当中，易于通过挥发、磨损、淋出等途径从材料的生产和使用过程中释放进入环境。随着多氯联苯（PCBs）和多溴联苯醚（PBDEs）等卤代阻燃剂的陆续禁用和限用，OPEs 作为其替代品开始在世界范围内被广泛地使用，其世界年生产量已经超过 55 万吨，已经在降尘[143,144]、空气[145]、水[146,147]、沉积物[148,149]、土壤[150]等多种样品中被检出。

室内环境是 OPEs 的一个重要暴露区域，Wong 等发现我国家庭、办公室等室内环境的灰尘中 OPEs 含量达 34 μg/g，与欧美发达国家的室内环境浓度相当[151]。Wu 等的研究发现北京的办公室灰尘中 OPEs 浓度（1.68~200μg/g）明显高于家庭（4.57~67.4 μg/g），说明 OPEs 的使用在我国已经十分普遍[152]。电子垃圾回收作坊中采集的灰尘中含有 33.6（3.30~70.0）μg/g 的 OPEs，明显高于农村地区 9.03（2.26~20.7）μg/g、城市 10.9（4.45~27.5）μg/g 和宿舍 11.50（4.08~25.7）μg/g 室内灰尘中的浓度[143]。广东贵屿电子垃圾处理场所灰尘中的 OPEs 达 33.1（4.66~350）μg/g [153]，进一步说明电子垃圾是 OPEs 的重要释放源。这些发现也增加了人们对职业活动所带来的 OPEs 暴露的担忧。

部分 OPEs 的水溶性较强易于分布到水环境当中，例如磷酸三乙酯(TEP)、三(2-氯乙基)磷酸酯(TCEP)和三(2-氯异丙基)磷酸酯（TCIPP）等。北京地表水环境中 OPEs 浓度平均值为 954 ng/L，氯代 OPEs 为主要的检出物，且夏季地表水中浓度明显高于冬季，说明湿沉降是 OPEs 进入地表水的重要来源[147]。对 40 条注入渤海的河流河水样的分析显示，氯代 OPEs 含量在 5~921 ng/L，而流经 OPEs 生产区域的河流对 OPEs 在渤海的分布产生了更多的贡献[154]。中国珠江三角洲河口的 OPEs 水平在干、湿季节分别为 2040~3120 ng/L 和 1080~2500 ng/L [155]，预示着中国南海入海口附近的水环境存在 OPEs 污染问题。厦门地区近岸海水中 OPEs 浓度为 424.6（91.87~1392）ng/L，附近污水处理厂以及经济、工业园区的废水排放是 OPEs 的主要来源[156]。有研究在自来水中检出了 OPEs，浓度为 192（123~338）ng/L[157]。而 Li 等同样在我国各地的饮用水样本中检测出了 OPEs，浓度范围（85.1~325 ng/L）与之相似[146]。综上所述，水环境是短链和氯代有机磷酸酯的重要暴露途径。

疏水性较强的 OPEs，例如三(2-乙基己基)磷酸酯(TEHP)、磷酸三甲苯酯(TCRP)等，则更易于分配到污泥、沉积物和土壤当中。北京的污水处理厂污泥样本中 OPEs 浓度范围为 233~137 μg/kg，其中 TEHP 和 TCRP 是主要检出物质[158]。OPEs 在中国太湖的沉积物中浓度为 3.38~14.3 ng/g[159]，而珠江三角洲的沉积物中为 58~322 ng/g(平均值 104 ng/g)，其浓度水平已经 PCBs 和 PBDEs 等卤代阻燃剂相当[160]。电子垃圾拆解区域[平均值 112(48~470)ng/g]和一线城市[132(62~222)ng/g]附近的沉积物中 OPEs 水平要显著地高于其他研究区域，电子产品是 OPEs 的重要来源，且 OPEs 环境浓度还与城市发达程度有关[148]。在莱州湾黄河河口附近收集到的沉积物柱芯中 OPEs 浓度为 11.8~102 ng/g，明显高于其他区域采集的沉积物样本，说明黄河的排放是近海 OPEs 的主要来源[149]。

点源和城市是 OPEs 土壤高浓度分布的重要区域。河北新乐和定州的塑料废物处理场所的土壤及周围农田中 OPEs 浓度水平达到了 1200 ng/g [161]。上海的城市土壤样本中 OPEs 浓度范围为 0.6~16.7 ng/g [162]。成都市表层土壤中 7 种 OPEs 含量平均值为 99.9 ng/g [163]。广州市区的土壤调查显示 11 种 OPEs 的浓度平均值为 250 ng/g，其中三(2-丁氧基乙基)磷酸酯(TBOEP)、TCRP、TCEP 是主要污染物，且浓度峰值出现在人为活动密集的城市商业区路边土壤中[150]。

珠江三角洲鱼体内检测出了高浓度的 TBOEP（1647~8840 ng/g lw）和 TCEP（82.7 4690 ng/g lw）[164]，这与珠三角入海口海水中的 OPEs 浓度分布一致[155]。北京城市地表水环境中鲫鱼等鱼类体内检测出了 264.7~1973 ng/g lw 的 OPEs，各种 OPEs 的生物富集因子（BCF）与其 log K_{ow} 值具有良好的相关性[165]。天津市场上各类食品中检测出了 OPEs(0.004~287 ng/g)[166]，上述结果表明 OPEs 已经通过各类途径向生物和食品中迁移。

3　典型新型有机污染物的环境行为

3.1　全氟化合物

环境中受关注的 PFASs 是离子型化合物，如何进行全球迁移至今都是一个有争议的话题，是对传统的半挥发物质全球迁移理论的一个挑战。关于 PFASs 的全球迁移有两种假说，一种观点[167]认为半挥发性的中性全氟前体物进入到大气之后能够随着大气进行长距离迁移，然后在光照或微生物的作用下发生降解转化生成 PFCAs 或 PFSAs;另一种观点[168]认为直接排放到环境中的 PFCAs 和 PFSAs 由于水溶性较强，随着洋流进行长距离全球迁移。PFASs 碳链长度的不同造成其理化性质差异较大，长碳链的 PFASs 较易吸附到灰尘、土壤和沉积物等固体介质上，因此其迁移性较差。与长碳链 PFASs 相比，短碳链的 PFASs 在大气和水体等介质中具有更强的迁移性[11]。较多的研究结果也证实，长碳链 PFASs 主要富集在土壤、沉积物和生物体内，而短碳链的离子型 PFASs 主要富集在地表水、地下水等环境水体中，半挥发性的全氟前体物主要富集在大气中。

PFASs 由于化学性质较为稳定，在环境中不易发生降解，只有部分含有可降解官能团的中性前体物在环境中能够发生一定程度的降解，但一般不能被完全矿化。这些中性前体物在大气中能够发生光降解[169]或者是异相光降解[170]，在羟基自由基（HO·）的作用下最终生成相应碳链长度的 PFCAs 和 PFSAs。前期的研究主要集中在 FTOHs 和 FOSEs 在烟雾箱中的光降解研究[171]，近年来使用各种高级氧化手段对全氟羧酸和磺酸进行降解去除的研究越来越多，例如使用紫外光加过氧化氢进行氧化[172]，或者通过添加过硫酸盐[173]来对其进行降解等。

3.2 双酚类化合物

BPA 在环境中的行为主要取决于其本身的性质以及环境条件。释放到水环境中的 BPA 除了与水体混合外，还经历悬浮物、沉积物或污泥的吸附、生物降解、光降解等过程，这些过程直接影响到 BPA 在环境中的存在状态及其生态毒理效应。地表水中广泛存在着能够降解 BPA 的细菌，细菌的数量、温度对生物降解均有影响，不同菌种降解途径可能不同。

对 BPA 在环境中分布和分配行为的研究表明，BPA 的吸附规律可由非线性 Freundlich 吸附等温线描述。影响疏水性有机物在固液界面分配规律的因素主要有天然有机物（主要是腐殖质）、离子强度、pH 和矿物质等，其中天然有机物及其他环境激素物质与 BPA 存在竞争吸附作用。有机物在土壤颗粒上的吸附不仅与土壤颗粒总有机碳相关，还与体系中溶解性有机质有关。土壤腐殖质可能会降低污染物溶解性和迁移性，而溶解性腐殖质会增加污染物溶解性和迁移性。

BPA 的降解过程主要包括光降解和生物降解。BPA 在水体中的光降解过程主要指光分解和光氧化过程。光分解过程是指在阳光（波长大于 290 nm）照射下发生光化学反应，光氧化过程是指在羟基活性基团或其他氧化剂作用下降解。生物降解包含微生物降解、真菌及其降解酶降解，还有植物和动物降解。研究 BPA 在环境中的吸附、光降解及生物降解等迁移转化过程将对认知、研究与防治 BPA 的污染提供参考和依据。

其他双酚类似物会以和 BPA 类似的方式排放到水环境中，但只有极少数研究针对其他双酚类似物，且主要是对其环境水平的报道。双酚类似物在水和沉积物中的分配对其在环境中的迁移和归趋有很大的影响，但是这方面的研究还很有限。在一个针对我国太湖中双酚类似物的研究中，计算了双酚类似物在沉积物-水体系中的分配系数($\log K_{oc}$)，其中 BPF、BPAP、BPA、BPAF 和 BPS 的 $\log K_{OC}$ 值分别为 4.7 mL/g、4.6 mL/g、3.8 mL/g、3.7 mL/g 和 3.5 mL/g [39]。

3.3 壬基酚聚氧乙烯醚

由于 NPEOs 在环境中的广泛分布，其在环境中的迁移转化也受到了人们的广泛关注。NPEOs 进入环境后，其环境行为主要受吸附和降解的影响。近年来，随着人们对于 NPEOs 吸附行为的深入理解，更多的研究集中于 NPEOs 的降解。

González 等的研究表明经污泥堆肥后的土壤中 NPEOs 的小分子降解产物呈现先升高后降低的趋势[174]，Writer 等也报道了污水处理厂排放口下游河水中 4-NP 浓度的增加[175]。这些研究均表明发生 NPEOs 降解。其中，生物降解是 NPEOs 及其代谢产物降解的主要途径。研究表明，NPEOs 的生物降解主要有两种途径[176]：一是在厌氧条件下，以减少乙氧基的数目为主；二是在有氧条件下，主要发生末端醇羟基的氧化。同时，缺氧能够促进 NPEOs 的降解。但是，对于 NP 和短链的 NPEOs 而言，好氧条件下其降解速率最大。另外，硫酸根和硝酸根的加入能够增加 NP 和短链 NPEOs 在厌氧条件下的降解速率。一般来说，NPEOs 的降解速率有如下规律：长链 NPEOs > NPEC >短链 NPEOs > NP。

除了生物降解外，一些研究者也报道了 NPEOs 在紫外线、H_2O_2 等条件下能发生光降解，从而降低其生态风险[177-182]。

3.4　对羟基苯甲酸酯

对羟基苯甲酸酯的蒸气压（从 9.29×10^{-5} mmHg 到 1.86×10^{-4} mmHg）表明其挥发性比较低，基本不会从水环境中以挥发的方式溢出[68]。在酸性溶液中，parabens 比较稳定，而在碱性溶液中，对羟基苯甲酸酯会发生水解。在强碱性溶液中，parabens 能够水解为对应的酸根。因此，在水环境中水解也不是对羟基苯甲酸酯去除的主要方式。在水环境中，parabens 的主要降解途径是微生物降解。在好氧条件下，parabens 能够被微生物快速降解[69,70]，缺氧条件下，仅有部分 parabens 能够被微生物降解，且降解速率较低[71]。微生物降解 parabens 的主要途径为催化其发生水解反应，并生成水解产物 *p*-HBA。即使 parabens 的产量及使用频率很高，但是其在沉积物及活性污泥中的浓度却很低，这表明 parabens 在水环境中能够被有效地降解掉。有研究表明 parabens 的稳定性会随着链长的增加而增加[72-74]，但是也有研究表明 parabens 的稳定性会随着链长的增加而降低，长链酯更倾向于降解[67,75]。parabens 的直链同分异构体比支链同分异构体的可降解性更强[67]，微生物能够将 parabens 降解为对羟基苯甲酸或者苯酚，并将降解产物作为碳源提供自身生命活动需要[67]。在好氧条件下，一个分离株细菌降解 parabens 的途径为：首先水解 parabens 产生 *p*-HBA，然后将 *p*-HBA 脱羧基产生苯酚。

在实验动物体内，parabens 的主要水解产物 *p*-HBA 还能够进一步羟基化生成 3, 4-对羟基苯甲酸(3, 4-dihydroxybenzoic acid，3, 4-DHB)[67]。在水环境中，3, 4-二羟基苯甲酸甲酯(Methyl protocatechuate，OH-MeP)和 3, 4-二羟基苯甲酸乙酯(ethyl protocatechuate, OH-EtP)也被检出，且代谢物的浓度要高于母体酯的浓度，Wang 测得污水处理厂进水中 MeP 和 EtP 的浓度中值分别为 36.8 ng/L 和 4.00 ng/L，而对应代谢产物 OH-MeP 和 OH-EtP 的浓度分别为 128 ng/L 和 78.2 ng/L[68]。在污水处理厂和地表水中，parabens 主要降解产物为 *p*-HBA。地表水中 parabens 的降解存在滞后效应，在其汇入到地表水 10 h 后才会出现 parabens 浓度的降低以及水解产物 *p*-HBA 浓度的增加，其反应半衰期为 9.5~20 h，主要原因是存在微生物驯化时期[67,68]。González-Mariño 等[71]曾报道在污水处理厂原生废水中 parabens 的反应半衰期为 9.5~35.2 h，其生物降解行为和地表水环境中的行为类似。他们[71]还发现在污水处理厂活性污泥中，parabens 生物降解半衰期为 1.8~2.9 d，且随着 parabens 链长的增加，其半衰期逐渐增加；而在地表水中，随着 parabens 链长的增加，其反应半衰期逐渐降低。研究者认为这主要是由活性污泥和地表水中微生物种类及丰度存在的差异造成的。由此可见 parabens 在水环境中的生物降解行为主要与微生物种类及丰度有关[71]。

Yu 研究了污水处理厂最终出水和脱水污泥中所负载的 parabens 的量，分别是进水的 0~1.6%和 0~0.4%[66]。WWTP 对 parabens 的去除率较高，平均可达 96.1%~99.9%之间[63]。然而，WWTP 出水中依旧能够检测到 parabens，且浓度最高可以达到 4000 ng/L[62]。因此，由于 WWTP 不能够完全去除 parabens，所以其被认为是 parabens 的主要污染点源[66]。

Wang 等研究发现，污水处理厂中 parabens 的去除率要高于其代谢物（*p*-HBA），污水处理厂进水中

p-HBA 的浓度中值是其母体化合物的 100~400 倍，通过出水排入到环境中的 *p*-HBA 是其母体化合物的 300 倍[72]。虽然排入污水处理厂以及天然水环境中的 parabens 类物质浓度不高，但是经过环境中的一系列生化过程如吸附降解等，其较高的活性仍会对生态环境造成不良影响，危害人类以及水体生物的生存。Caccia[62]等认为，部分 parabens 的代谢产物比其母体化合物具有更大的极性，随之增强的就是其活性甚至是毒性，因此这类物质表现出更强烈的环境影响效应。

3.5 人工甜味剂

常见的人工甜味剂中，甜蜜素、糖精、阿斯巴甜的性质较不稳定，而安赛蜜、三氯蔗糖、纽甜、NHDC 具有较强的热稳定性[85]。此外，安赛蜜在酸性条件下能够发生轻微的水解并产生乙酰乙酰胺[183]，阿斯巴甜在酸性或碱性条件下会被水解产生甲醇和氨基酸[85]。

大气环境中，已有部分文献发现了人工甜味剂。由于人工甜味剂是极性化合物，部分可电离，在环境中以阴离子形态存在，具有很高的水溶性和较低的蒸气压。因此，美国国家图书馆药物毒性数据网（http://toxnet.nlm.nih.gov）的预测表明人工甜味剂会出现在大气中，如安赛蜜、糖精和甜蜜素。另外人工甜味剂还可能结合到颗粒物上，以扬尘形式进入大气，在大气中进行迁移，再随干湿沉降回到地表。Gan 等[96]调查发现糖精、甜蜜素和安赛蜜在雨水和颗粒物态中的组成比例相似，因此初步认为大气中的人工甜味剂是以颗粒物态的形式进入大气的（扬尘）。

不同的污水处理工艺对人工甜味剂的去除效果具有一定差异。絮凝沉淀对人工甜味剂几乎无去除作用[184]。加氯处理也不能显著去除人工甜味剂[184,185]。而微生物作用能完全去除糖精和甜蜜素，其降解周期分别为 20 d 和 15 d，但是三氯蔗糖和安赛蜜具有一定持久性，其浓度在 92 d 内并没有显著降低[184]。臭氧能够氧化安赛蜜，当臭氧浓度为 0.5 mg/L 时，其半衰期为 15 min，但是臭氧难以完全去除安赛蜜，其去除率最高只能达到 70%左右[184,185]。臭氧对安赛蜜和甜蜜素的氧化效果类似，但难以氧化三氯蔗糖和糖精，当臭氧浓度高达 5 mg/L 时，其去除率分别仅为 30%和 20%[184]。活性炭能完全去除糖精，但对甜蜜素的去除率仅为 23%[184,185]。

对于人工甜味剂在水环境中的光降解研究也有一些报道。Batchu 等[186]研究发现三氯蔗糖在自然环境中几乎不被光降解，部分降解的主要产物为脱掉一个氯的化合物。Soh 等[185]研究了三氯蔗糖和安赛蜜在紫外灯照射下的光降解率，结果表明在 5 h 紫外光照射后，三氯蔗糖没有被降解，而安赛蜜被降解了 35%。Sang 等[187]研究了安赛蜜和三氯蔗糖在紫外光照射下的光降解动力学和降解产物，发现三氯蔗糖在紫外光照射下能发生光降解，并发现了 4 种降解产物。而安赛蜜在紫外光照射下产生了 12 种降解产物。相比于三氯蔗糖，安赛蜜的光降解机理更为复杂。Gan 等[188]认为安赛蜜在光降解过程中能直接被氧化，后经分子内反应开环而生成水合醛。另外，安赛蜜也能与水分子反应产生两种水合产物，其中一种具有比较稳定的结构，而另一种结构不稳定水合产物会经分子内重排后脱掉一个羧基形成环状分子。作者还发现 $^1O_2$/$^3DOM^*$ 在安赛蜜光降解过程中具有一定贡献，但·OH 由于被碳酸氢根等淬灭剂淬灭，其对安赛蜜的贡献很小。

当前仅有少数文献对人工甜味剂在土壤介质中的迁移降解进行了报道。Labare 等[189]研究发现三氯蔗糖在还原环境的土壤条件中几乎不被降解，但在干燥的土壤环境中，28 d 内降解了 45%。而 Soh 等[190]发现在同样的时间周期内三氯蔗糖的降解达到了 60%，作者认为出现这样的结果是因为土壤微生物的差异，并且实验中采取了强化生物降解法。Buerge 等[86]在室内土壤培养实验中发现糖精的降解半衰期为 3~12 d，甜蜜素的为 0.4~6 d，三氯蔗糖的为 8~124 d，安赛蜜的为 3~49 d。Ma 等[100]在猪粪肥施用后的大棚土壤中发现 4 种典型人工甜味剂，降解半衰期分别为甜蜜素 5.1 d，糖精 3.2 d，安赛蜜 9.7 d，三氯蔗糖 8.5 d。人工甜味剂的半衰期表明三氯蔗糖和安赛蜜较为稳定，其在土壤中的降解速率较为缓慢，而糖精和甜蜜素的降解速率较快。Buerge 等[86]在土柱淋溶实验中发现人工甜味剂不能在土层中长时间滞留，它们较为容易穿透土柱，对地下水造成一定风险。

3.6　苯并杂环化合物

苯并三唑并不能被常规的污水处理流程去除，导致其向环境水体中释放，该过程的长期不利影响已被报道[103,107,110,191,192]。最新研究显示，污水处理厂进出水中均能检出苯并三唑或 TTR[107, 116, 119]。在西班牙，20 个污水处理厂中苯并三唑及 5-TTR 浓度分别为 26.7 μg/L 及 42.9 μg/L，去除率为 11.8%~94.7%[116]。苯并三唑在丹麦及西班牙废水原水中的浓度范围分别为 0.2~2.2 μg/L 以及 0.06~36.2 μg/L。人工湿地中苯并三唑的去除率高于传统污水处理工艺对苯并三唑的去除率[113]。这可能归因于湿地条件下有光降解以及植物吸收过程的发生[113]。在环境条件下苯并三唑显示出较强的抗氧化性，中等强度的紫外线照射下，苯并三唑能稳定降解，但降解速率并不高[193-195]。这些特性导致苯并三唑很难被常规的污水处理工艺降解[106, 110,118,136]。

在多数发展中国家及少数发达国家，由于水资源匮乏，导致未经处理的废水直接当作灌溉用水排放到农田土壤中，造成有机物在农作物中的迁移增强。LeFevre 等研究了用含有苯并三唑废水灌溉的草莓、生菜等作物中苯并三唑及其代谢物的浓度，发现在草莓中残留水平为 13.1 ng/g dw，生菜中为 67.8 ng/g dw [196]。Riemenschneider 等研究了在真实土壤环境下用含有 28 种有机污染物的污灌废水浇灌 10 种蔬菜，结果发现苯并三唑在上述污灌废水中的浓度范围为 0.1~0.6 μg/L，主要在植物根部检出，浓度超过 10 ng/g dw[197]，增大了潜在的生态风险。

由于苯并噻唑类物质难以被生物降解[198]，所以苯并噻唑降解的研究主要集中在化学方面，Bao 等[199]使用伽马射线降解水溶液中的 MBT，通过伽马照射可显著降解水溶液中的 MBT，使 MBT 脱去巯基。Qin 等[200]使用可见光研究了 MBT 的降解情况，通过使用氮掺杂的二氧化钛比传统商业化的二氧化钛光催化活性提高了三倍，显著提高了 MBT 的可见光光降解速率。Zhang 等[201]采用 PMS/$CuFe_2O_4$ 降解了一系列苯并噻唑类物质，发现 PMS/$CuFe_2O_4$ 可氧化苯并噻唑类物质，羟基苯并噻唑可通过开杂环降解，而 BTH 及其他衍生物通过在苯环上加羟基来完成降解。刘春苗等[202]采用单室电辅助微生物反应器氧化苯并噻唑，考察了外加电压和 COD 对降解效能的影响，发现 BTH 在阳极氧化降解为羟基苯并噻唑后噻唑环断裂最终转化为甲磺酰基苯胺，48 h 后 BTH 的降解效率可达 96%，是一种高效的降解 BTH 的方法。

3.7　阻燃剂与短链氯化石蜡

3.7.1　PBDEs

已有的大量研究表明，PBDEs 具有长距离迁移能力，大气的干湿沉降和温度的变化是影响 PBDEs 在空气中长距离迁移的重要因素[203]。由于 PBDEs 物理化学性质的差异，低溴 PBDEs 容易挥发到空气中，随空气流动向远方迁移。以 BDE-209 为代表的高溴代联苯醚是通过吸附在空气中的颗粒物上，经干湿沉降进入土壤和水环境中。我国于 2008 年在全国 19 个城市和 1 个背景点的同步观测表明[204]，与 2004 年相比，我国大气 PBDEs（7 种主要 PBDES 单体，不含 BDE-209）浓度由 38.2 pg/m^3 下降到 15.4 pg/m^3，但组成上没有明显变化，仍以 BDE-47 和 BDE-99 为主。此外，高浓度的 PBDEs 出现在靠近电子或电子垃圾拆解业集中区的城市（如广州、西安），表明这些产业依然是我国 PBDEs 的主要来源。近期（2013~2014 年）对我国 10 座城市大气 $PM_{2.5}$ 样品中 PBDEs 的研究发现[205]，大气 $PM_{2.5}$ 中 PBDEs 浓度（不含 BDE-209）为 4.62 pg/m^3，与早期研究相比，浓度水平有下降趋势。然而，我国 BDE-209 浓度依然较高

（24.9 pg/m^3±130.6 pg/m^3），特别是在广州这一毗邻密集电子垃圾拆解和回收业的珠三角城市，检出了高浓度的 BDE-209（1013 pg/m^3）。一方面，这是由于 BDE-209 具有较低的蒸气压，导致其易于吸附在颗粒物上；另一方面也反映了区域内的电子产品回收业对周围城市大气 PBDEs 产生的重要影响。沉积物中 PBDEs 的迁移能力受控于污染物与不同颗粒物之间的相互作用（特别是吸附/解吸作用）[206]。由于 PBDEs 的亲脂性较强，因此被吸附后就很难解吸出来，沉积物在一定程度上起到了净化的作用，尤其对于八至十溴的 PBDEs。但是沉积物/水界面体系的条件改变，PBDEs 就可能解吸出来重新释放到水中，造成二次污染。PBDEs 在沉积物中的吸附/解吸将受到沉积物的颗粒物大小、有机质的类型以及温度的影响。

光解和生物降解是 PBDEs 在空气、水、土壤、沉积物和室内尘埃等环境介质中转化的两种主要途径[203,206-210]，哪种途径占主导与 PBDEs 所处的环境条件密切相关。光解是空气中 PBDEs 转化的最主要方式，PBDEs 不仅可脱溴形成低溴代 PBDEs，还可生成毒性更强的溴代二噁英（polybrominated dibenzo-*p*-dioxin，PBDDs）。与大气不同，土壤和沉积物中则以生物降解和转化为主，脱溴还原为低溴代 PBDEs 是主要转化途径，Br 取代位置一定程度上决定了脱溴的难易程度，优先顺序一般是间位＞对位＞邻位；好氧微生物降解的研究相对较少，机理上缺乏明确的证据。已有的研究表明，降解速率随溴取代数目的增加而降低，同时当两个苯环上都有溴取代时，易在溴代程度低的苯环上通过羟基化或甲基化开环。

3.7.2 氯化石蜡（CPs）

我国是开展氯化石蜡研究最活跃的国家，近五年发表的研究中，大多来自中国[211]。由于目前已发表的研究中采用的分析技术、定量过程以及标准品不同，给科学评估氯化石蜡的环境行为造成潜在的不确定性。从目前的研究结果看，CPs 普遍存在于世界各地的土壤、大气、沉积物和生物体中。除大气外，土壤、沉积物中中等链长 CPs 浓度水平通常高于短链 CPs（SCCPs）[212-213]。我国大气 SCCPs 浓度为 137 ng/m^3±114 ng/m^3，明显高于全球大部分国家和地区。在组成上，C_{10} 组分和 C_{16} 组分为优势，分别占到 42.2%和 40.7%，与我国主要石蜡工业品的组成特征相似。估算 SCCPs 大气传输距离（600~900 km，远低于其他 POPs），并结合大气后推气流轨迹模型的结果，推断中国大气 SCCPs 主要来自于本地排放[214]。氯化石蜡自身的碳链长度和氯化程度一定程度上影响着其在环境介质中的迁移转化。较短的碳链（C_{10}、C_{11}）和较低氯取代数（$Cl_{5\sim8}$）具有更好的挥发性和水溶性，因此在大气中具有更长的半衰期，更易于从排放源向周边区域扩散，包括通过长距离迁移和水体中悬浮颗粒物的流动向海洋及偏远地区迁移[211]。

有关 CPs 的转化研究相对缺乏，尤其是光降解等非生物转化。WHO（1996）的一份报告指出，$C_{10\sim12}$ 短链氯化石蜡（氯含量 58%）在有氧条件下经过 28 天或在无氧条件下经过 51 天都不能被活性污泥降解[215]。Thompson 等（2007）研究了 SCCPs 在淡水和海洋沉积物中发生的需氧和厌氧生物降解，估算了 SCCPs 在需氧条件下的平均半衰期，在淡水沉积物中为 1630 d，在海洋沉积物中为 450 d[216]。研究表明即使是氯化度低的 SCCPs 在自然环境中也不易降解。Tomy 等[217]和 Iozza 等[218]均在 20 世纪 60 年代的沉积层中发现了一定浓度的 SCCPs，对 SCCPs 在湖泊沉积物中的历史变化趋势进行的研究表明，在湖泊底泥的厌氧环境中，SCCPs 的持久性超过 50 年。

3.7.3 有机磷酸酯（OPEs）

有机磷酸酯（OPEs）已经在多种环境介质和生物体中检出，其中在水体、室内空气和室内灰尘中的暴露水平相对较高，而土壤和大气中浓度较低[219]。由于 OPEs 极性范围广，普遍具有半挥发性，溶解度较高，OPEs 能存在于多种环境介质中，在环境中的迁移转化较为复杂，尚缺乏系统认知。针对 OPEs 转化的有限研究主要集中在污水处理系统以及自然水体，研究结果显示，不同类型 OPEs 经历的迁移和转化过程是显著不同的。在常规的污水处理过程中，不含氯的磷酸酯（包括芳基磷酸酯和烷基磷酸酯）可发生一定程度微生物降解，大部分则通过吸附作用进入污泥中，借助污泥的处置/处理再次进入环境[220-221]。在自然水体中，不含氯的 OPEs 可经历一系列地球化学过程，包括吸附-分配、生物降解和直接/间接光降解等，在光降解过程中，水体中的腐殖酸类物质（HULIS）可因反射/散射日光以及消耗自由基而影响转化进程和效率[222-223]。氯代 OPEs 由于其水溶解度较大，在污水处理过程中既不吸附于污泥，也不发生微生物降解/转化，将随污水厂排放进入自然水体，再随水流扩散和迁移，其在自然水体中同样既不吸附于颗粒物，也不发生生物/非生物降解转化，最终汇入海洋[154]。

OPEs 通过大气进行迁移和传输是其向更远范围内扩散的重要途径。有限的研究显示，OPEs 可长距离迁移至极地区域，氯代 OPEs 甚至表现出与 PBDEs 相当的持久性和长距离迁移能力[224,225]；模型研究显示，大气中芳基磷酸酯 TPhP 在·OH 自由基作用下降解转化生成 OH-TPhP 等产物，半衰期约 7.6 d[226]。

4　新型有机污染物的生物代谢及效应

4.1　全氟化合物

全氟前体物除了发生光降解外，另一条主要的转化途径是在微生物等生物的作用下发生生物转化。目前关于 PFASs 的生物转化的研究主要集中在两大类化合物上：氟调醇为基础和磺酰胺为基础的化合物。8∶2 FTOHs 和 6∶2 FTOHs 的微生物降解转化研究较多，通过比较大量的研究结果可知氟调醇在各种基质中的半衰期较短，一般小于 2 d，其在不同介质中的降解路径和降解产率稍有差异，这可能与不同介质中微生物的种类差异有关[227]。通过比较可以发现，氟调醇转化为全氟羧酸的产率由高到低依次是土壤、沉积物、污泥和纯菌株。值得注意的是，不同碳链长度的氟调醇其微生物降解的效率也不同，其中碳链越短其被微生物转化的相对量就越多，一般情况下 4∶2 FTOH>6∶2 FTOH>8∶2 FTOH。主要有两方面的原因：一方面碳链越长其分子位阻效应越大，生物有效性越小；另一方面碳链越长其水溶性越小，越易被土壤等固体介质吸附，从而降低其生物有效性[228]。不同碳链长度的氟调醇其微生物降解和转化的路径大体相似，但仍有一些差别。比较 6∶2 FTOH 和 8∶2 FTOH 的降解路径可以发现，6∶2 FTOH 具有更强的减小碳链的潜能。其中 8∶2 FTOH 在转化为 7∶2 sFTOH 后脱去一个 C 生成 PFOA，而 6∶2 FTOH 在转化为 5∶2 sFTOH 之后则可以脱去一个 C 生成 PFHxA 并同时脱去两个 C 生成 PFPeA，其转化为 PFPeA 的产率可以在沉积物中达到 10.4%[229]，在土壤中达到 30%[230]，即 5∶2 sFTOH 比 7∶2 sFTOH 具有更强的生物可利用性。另一方面，其降解产物的稳定性也能表明 6∶2 FTOH 具有更高的降解潜能。8∶2 FTOH 降解生成的 7∶3 FTCA（或 7∶3 FTUCA）只能降解生成全氟羧酸（如 PFHpA 和 PFHxA），而 6∶2 FTOH 降解生成的 5∶3 FTUCA 则可以继续降解，生成更短碳链的 4∶3 FTCA 和 3∶3 FTCA。目前关于氟调醇的厌氧降解的研究较少，与好氧微生物降解相比，厌氧降解的周期较长，且降解为全氟羧酸的转化率较低，在产甲烷的厌氧污泥中只有≤0.4%的 6∶2 FTOH 和≤0.3%的 8∶2 FTOH 转化为 PFHxA 和 PFOA[231]。与微生物降解较明显的区别是，在哺乳动物等高等生物体内的生物转化能够观察到 α-氧化的降解产物而在微生物转化中则观察不到这一现象，例如 8∶2 FTOH 能够在哺乳动物体内转化为 PFNA[232,233]。近年来其他的全氟前体物研究逐渐增多，例如全氟磷酸酯（PAPs）[234]、以磺酰胺基乙醇为基础的磷酸酯[235]、全氟聚合物[236]、氟调醇为基础的硬脂酸酯[237]等，它们在环境中先水解生成相应的 FTOHs 和磺酰胺（FASAs），然后经历相似的过程转化为 PFOA 和 PFOS。水成膜泡沫灭火剂中含有大量的阳离子和两性的全氟前体物，已有研究者[28,30]发现一些物质能够降解转化生成 PFOA 和 PFOS。O 原子的引入，为全氟醚类羧酸和磺酸提供了可攻击的点位。但大量的阳离子和两性离子的全氟前体物由于缺少标准品和受降解产物鉴定较为困难的限制，目前的研究仅集中在其中的几种化合物上。因此关于 PFECAs 和 PFESAs 的降解研究还需加强。

由于长碳链的全氟化合物具有较强的生物蓄积性，因此受到了国内外研究者的广泛关注。目前针对 PFASs 在生物体内蓄积的研究主要集中在动物及高等植物体内，而在低等植物中鲜有报道。李法松等[238]在天津大黄堡湿地的生物样品中发现 PFASs 有一定的沿食物链传递的能力，且倾向于分配到生物的内脏和肝脏部分，在各组织间的分配的优先顺序如下：肝脏>内脏>卵>肌肉。已有大量的研究表明进入到生物体内的全氟前体物能够发生生物转化，进而产生全氟酸类化合物。赵淑艳等[239]在土壤-小麦和土壤-蚯蚓的体系中发现，10∶2 FTOH 能够发生生物降解，生成稳定的全氟羧酸，其主要降解产物为全氟癸酸（PFDA），并在小麦根部和茎部检出了短链的全氟羧酸（PFPeA 和 PFHxA 等）。全氟化合物的富集系数与全氟化合物的种类有关，一般全氟磺酸>全氟羧酸>全氟磷酸，且长碳链的>短碳链的[240-242]。近年来对于全氟化合物的同分异构体的研究逐渐增多，通常情况下直链的富集性要高于支链的全氟化合物[243]。有关植物蓄积 PFASs 的研究发现，水生植物和陆生植物对 PFASs 的蓄积出现了不同的随碳链变化的规律。有关水生植物的研究表明，长链 PFASs 在挺水植物和沉水植物中的含量较高，短链 PFASs 的含量较低[244]。而陆生植物对碳链长度较短的 PFASs 的蓄积能力明显高于碳链较长的 PFASs。周萌等在盆栽实验中发现小麦对短碳链全氟化合物的富集能力非常强，且羧酸比磺酸更易被植物富集[245]。全氟化合物在大型蚤中的富集随着链长的增长而增大，与化合物的辛醇/水分配系数呈正相关关系，且表面吸附在全氟化合物的生物富集过程中起到了重要作用[246]。苑晓佳等[247]在探究全氟化合物在藻类的富集过程中发现，全氟化合物的富集系数与藻类的比表面积、脂肪及蛋白质组成有关，且蛋白质是吸附 PFASs 的主要部位，其中静电引力和氢键作用是 PFASs 与蛋白质的主要作用力。

4.2 双酚类化合物

BPA 在鱼体内的生物累积因子（BAF）为 1.7~182，与典型持久性有机污染物相比并没有显著的生物积累性。BPA 的 BAF 值较低，这部分归因于其较低的辛醇/水分配系数（$\log K_{ow}$：3.32~3.40）。根据美国环境保护署（EPA）软件计算的 BPAF、BPC 和 BPZ 的 $\log K_{ow}$ 值分别为 4.47、4.74 和 5.48，显著高于 BPA。当一种化学物的 $\log K_{ow}>4.02$ 时，其可能具有明显的生物积累性。因此，理论上这些双酚类似物可能具有比 BPA 更强的生物积累性。

尽管其他双酚类似物被认为比 BPA 稳定且毒性小，但有研究表明这些类似物也存在有害影响，如雌激素毒性、细胞毒性、基因毒性、生殖和神经毒性等，这些毒性效应类似于 BPA，有的甚至超过 BPA。然而，对这些物质的研究大多围绕其在环境和人体样本的分布特征。在我国太湖流域对 9 种双酚类似物的研究中发现，BPAF、BPC、BPZ 和 BPE 的 log BAF 要高于 BPA，且 log BAF 和 $\log K_{ow}$ 存在正相关关系；同时发现，BPAF、BPC 和 BPZ 的营养级放大系数分别为 2.52、2.69 和 1.71，说明它们在食物网中可能存在放大作用[248]。

4.3 壬基酚聚氧乙烯醚

由于 NPEOs 很容易通过废水排放进入到水环境中，因而水生生物的存在对其行为和归趋具有一定影响。Sun 等的研究表明小球藻（一种绿藻）能够富集和降解水相中的 NPEOs 和 NP，而且随着链长的减少，其生物富集能力呈现增加的趋势[249]。同时，该研究也表明生物降解是小球藻去除水相中 NPEOs 的主要途径，主要是长链的 NPEOs 向短链的 NPEOs 及 NP 的转化。另外，NPEOs 的降解产物 NP 与雌激素具有类似的结构，因此被认为是一种环境类雌激素，具有较强的内分泌干扰性。Roig 等报道了鸡的胚胎暴露于 NP 会影响其生殖系统的发育以及肝脏、肾脏等内脏器官的发育[250]。另外，NP 对植物也有毒害作用，可抑制种子发芽，减慢生长速度，降低植物体内叶绿素含量，导致细胞膜脂质过氧化，减少植物抗氧化酶的活性。同时 Zhang 等的研究表明 NP4EO 和 NP10EO 对植物也有一定的毒性，但是毒性效应弱于 NP[251]。一般而言，随着乙氧基数目的增加，NPEO 脂溶性降低，其生物毒性也会随之减弱。此外，由于 NPEOs 常被用作农药的乳化剂，故其与其他污染物的联合效应不容忽视。Aronzon 等研究了二嗪磷（一种有机磷杀虫剂）和 NP 对蟾蜍的联合毒性，发现 NP 对蟾蜍胚胎和幼虫的毒性要远大于二嗪磷，并且两者具有协同毒性作用[252]。

4.4　对羟基苯甲酸酯

人体中对羟基苯甲酸酯代谢的研究报道较少，大部分都集中在动物实验上。对大鼠、狗、猫、兔等的研究表明，对羟基苯甲酸酯可经肠道和皮肤迅速吸收，经体液循环到达目标器官。在这些器官和组织水解酶的作用下，代谢为水溶性较好的对羟基苯甲酸及葡萄糖醛酸、氨基乙酸、硫黄酸的结合物，其含量依次递减，并随尿液迅速排出体外。对羟基苯甲酸酯及其代谢物在体内均无蓄积性。

Tasukamoto[253]等用含 MeP、*n*-PrP 和 BuP 的饲料喂养兔子，在 0.4 g/kg 和 0.8 g/kg 的剂量下，86%的对羟基苯甲酸酯代谢物在 24 h 内可随尿液排出体外，其中未代谢的对羟基苯甲酸酯占 0.2%~0.9%，游离的 *p*-HBA 占 25%~39%，氨基乙酸酯占 25%~29%，硫黄酸酯占 7%~12%，葡萄糖醛酸酯和葡萄糖醛酸醚分别占 5%~7%和 10%~18%。随着对羟基苯甲酸酯酯链的增长，摄入的 parabens 的排出速率（包括原型化合物和代谢产物）也会降低。

Jones[254]等通过静脉注射和口饲分别研究了 MeP、EtP、*n*-PrP 和 BuP 在狗体内的代谢动力学。在静脉注射 50 mg/kg 后，初期血液中几乎检测不到对羟基苯甲酸酯，6 h 后可检测到代谢物 *p*-HBA；在口饲 1 g/kg 后 24h 才检测到 *p*-HBA。狗静脉注射 100 mg/（kg·d）的剂量，在脑、脾和胰腺均可检测到对羟基苯甲酸酯。口饲 1 g/（kg·d）的 MeP 或 PrP，一年后未发现狗体内有对羟基苯甲酸酯的富集。

对志愿者每天给予 2 g 的 PrP，5 天后发现尿液中 *p*-HBA 占 17.4%，硫黄酸酯占 55%，未发现未代谢的 PrP。Heim 等对 6 人进行了口服 10~20 mg/kg 剂量 *n*-PrP 的研究，60 min、135 min 和 255 min 后，在血清中检测到浓度为 4.5 μg/mL 的对羟基苯甲酸盐[255]。Sabalitschka [37]等通过对 6 个婴儿注射含有对羟基苯甲酸酯作为防腐剂的庆大霉素进行研究，结果发现 MeP 在婴儿尿中的浓度占 parabens 总浓度为 13.2%~88.1%。

4.5　人工甜味剂

人工甜味剂由于具有较高的水溶性和较低的辛醇/水分配系数，而难以被水生生物富集。Rebecca 等[256]在海鱼中检测到了低浓度的三氯蔗糖。Lillicrap 等[257]的研究表明三氯蔗糖在斑马鱼和藻类中的生物富集系数均小于 1，而在水蚤中的生物富集系数为 1.6~2.2。Knut[258]等报道了三氯蔗糖在水生环境中几乎不被富集。通过油麦菜实验发现，土壤介质中的人工甜味剂容易被植物根部吸收，而存在于植物根部的人工甜味剂很容易通过蒸腾作用转移至植物的茎叶部分[259]。

人工甜味剂在生物体内几乎不被代谢转化，目前已有文献证实了这一结论。Renwick[260]等利用含有安赛蜜的食物喂食豚鼠，结果发现 65%的安赛蜜会经尿液排出，而 30%的安赛蜜在粪便中被发现。天津养猪场的猪粪和猪尿中同样检出了高浓度的糖精、甜蜜素和安赛蜜[100]。但多数文献[261-264]表明糖精和安赛蜜能完全被人体或动物体吸收，而只有 30%~50%的甜蜜素和 10%~15%的三氯蔗糖能被人体或动物吸收，吸收的人工甜味剂在生物体内并没有发生转化，会经尿液排出，而不被吸收的部分则会通过粪便排出。针对于人工甜味剂在植物体内的代谢研究还未见相关报道。

4.6 苯并杂环化合物

近年来，苯并三唑在生物体内代谢转化逐渐引发关注。LeFevre 等研究发现，生长于水培环境中的拟南芥能够迅速将苯并三唑转化成色氨酸或葡萄糖结合体，只有不足 1%的苯并三唑以母体化合物形式存在于环境中[196]。通过运用 LC-QTOF-MS 等仪器，研究者指出，苯并三唑在植物体内两种主要的代谢转化途径：葡萄糖途径及氨基酸途径，后者产物化学结构同植物生长素类似，具有生物活性[196]。

4.7 阻燃剂及氯化石蜡

4.7.1 多溴联苯醚

多溴联苯醚在结构上与多氯联苯类似，但 PBDEs 在生物中的富集特征与 PCBs 存在非常明显的差别[265]。尽管 PCBs 工业品品种多，每种工业品的组成差异巨大，但在生物体内，PCBs 大致表现为较为一致的单体组成特征，主要以 CB153、CB138、CB180、CB118 等单体为主，并且七个指示性 PCBs 单体浓度与总 PCBs 的浓度一般都存在非常良好的线性关系。尽管 PBDEs 工业品种少，每个工业品种的主要组成单体也较少，但在生物体内，随采样的区域、物种的不同，PBDEs 在生物体内的单体组成模式往往存在较大的差别，同时，水生、陆生生物的富集与放大也存在明显的差别[266-268]。水生鱼类样品中，PBDEs 的富集主要以 BDE47、BDE100、和 BDE154 等低溴代单体为主。由于鱼体内普遍存在着脱溴代谢过程，使得以鱼为核心的水生食物链上 PBDEs 的脱溴产物或不易发生脱溴代谢单体（如 BDE47、BDE100、BDE154 等）普遍存在食物链放大的现象。而易发生脱溴反应的单体如 BDE99、BDE153、BDE183、BDE209 则以较小的浓度存在或不易在鱼中富集。暴露实验及体外肝微粒体代谢实验证实，PBDEs 在鱼体内的代谢存在物种差异性。PBDEs 的脱溴代谢过程决定了 PBDEs 在鱼体内的富集模式，也决定了以鱼为取食对象的其他生物的富集模式。由于陆生鸟类 PBDEs 暴露源及对 PBDEs 代谢模式与鱼类存在明显差异，使得在以鸟为核心的陆生食物链中，PBDEs 存在生物放大的单体主要以 6~7 溴取代的较高溴代 PBDEs 单体为主。其中 BDE153、BDE183、BDE197 等在水生生物中不常见的单体在陆生鸟内食物网中存在明显的生物放大[269-271]。

4.7.2 氯化石蜡

近期多项实验室模拟和野外观测结果均表明，SCCPs 可以在生物体内富集[272]，但其在生物体内的代谢与转化鲜见报道。Ueberschar 等[273]发现，经过六周暴露的母鸡，SCCPs 主要分布在脂肪组织中，其余三分之一则在蛋黄、尿液和粪便中。鸡蛋。虹鳟鱼的实验表明[274]，SCCPs 在鱼体内的半衰期在 7~37d 之间，且半衰期与碳链长度、氯原子数呈正相关关系。这一结果与野外观测相近。Ma 等[275]发现尽管蚌壳类生物与沉积物中 SCCPs 的碳链组成特征较为一致，但蚌类中 Cl_5-SCCPs 的含量显著低于沉积物，表明 Cl_5-SCCPs 具有比其他 SCCPs 更快的消除、排泄潜力。Zeng 等[276]也发现，生物富集因子与 SCCPs 的氯原子数目呈显著正相关。与此同时，SCCPs 具有生物放大的能力。水生生物网（如中国海洋、极地海洋以及密西根湖）中 SCCPs 的营养放大因子在 1.2~2.8 之间，表明 SCCPs 具有显著的生物放大效应[211]，尽管对于 MCCPs 的研究报道相对较少，但 Herzke 等[55]发现在极地食物网中 MCCPs 的 TMF 达到 2.0，显示出 MCCPs 在水生食物链中同样具有生物放大能力。总体来讲，由于 CPs 分析检测技术的限制，CPs 的生物代谢转化与富集仍处于初始阶段，仍需要进一步深入探讨。

4.7.3　有机磷酸酯

已有的研究显示，OPEs 广泛存在于生物体中[219]，在淡水鱼、海水鱼、家禽、鸟蛋中都有 OPEs 或其代谢物检出。Slmdkvist 等[278]考察了瑞典湖泊和沿海地区鱼类及贝壳类样品中 11 种 OPEs 的分布，其中 TCPP 和 TMPP 是水生生物中最主要的污染物。Kim 等[279]在菲律宾马尼拉湾鱼类的研究中发现生物体中 OPEs 的富集不仅与其自身的亲脂性等理化性质相关，也与在环境中的污染水平紧密联系。与典型的 POPs 不同，OPEs 的浓度水平与脂肪含量无明显相关性[278-281]。对于 OPEs 的生物放大目前研究较少，且呈现出相互矛盾的结果。Brandsma 等[280]发现 TBEP、TCPP 和 TCEP 在底栖生物食物网中营养级生物放大因子（TMF）大于 1，可能有生物放大现象，而 Kim 等[279]的结果是 ^{15}N 的值和 PFRs 的浓度没有线性相关，他们认为可能除了 TMPP，其余 OPEs 均没有生物放大性。由于有机磷酸酯类化合物理化性质的差异，以及不同生物体的富集、代谢能力的差异，OPEs 在生物体内的生物富集与放大仍需大量的野外观测和实验室模拟研究来探索与证实。

5 展　　望

化学工业的发展为人类带来福祉，但是在化学品生产、使用和废弃过程中的不当行为，使化学品成为环境污染的罪魁祸首。所以，人们始终在寻求发挥化学品益处并降低其危害的最佳平衡点。一些新型化学品是传统污染物的替代品，但是在投放使用之前，虽然已经做了大量的风险评估，但是其环境行为与风险往往认知不全，在使用几年乃至几十年后发现问题，再用新的化学品替代。虽然目前对于多数新型有机污染物的环境分布、迁移转化以及生物效应等方面已有研究开展，但是总体上人类对新型污染物环境行为的认识仍有待加强。特别是新型污染物的化学性质有别于传统污染物，很多环境行为难于用传统的环境科学理论解释，对于新型污染物环境行为的深入研究，必将推动环境科学学科向纵深发展。而对于其在人体中暴露水平、代谢途径和健康效应的研究还很缺乏，借助组学等新兴手段，必将推动环境学科交叉前沿方向的进步。

参 考 文 献

[1] Buck RC, Franklin J, Berger U, et al. Perfluoroalkyl and polyfluoroalkyl substances in the environment: terminology, classification, and origins. Integr Environ Assess Manag, 2011, 7(4): 513-541.

[2] Kotthoff M, Müller J, Jürling H, et al. Perfluoroalkyl and polyfluoroalkyl substances in consumer products. Environ Sci Pollut Res, 2015, 22(19): 14546-14559.

[3] Dreyer A, Weinberg I, Temme C, et al. Polyfluorinated compounds in the atmosphere of the Atlantic and Southern Oceans: evidence for a global distribution. Environ Sci Technol, 2009, 43(17): 6507-6514.

[4] 姚义鸣, 赵洋洋, 孙红文. 天津市大气中 PFASs 挥发性前体物的分布和季节变化. 环境化学, 2016, 35(7): 1329-1336.

[5] Yao Y, Chang S, Sun H, et al. Neutral and ionic per- and polyfluoroalkyl substances (PFASs)in atmospheric and dry deposition samples over a source region (Tianjin, China). Environ Pollut, 2016, 212: 449-456.

[6] Yao Y, Sun H, Gan Z, et al. Nationwide distribution of per- and polyfluoroalkyl substances in outdoor dust in mainland China from eastern to western areas. Environ Sci Technol, 2016, 50(7): 3676-3685.

[7] Zhao L, Zhou M, Zhang T, et al. Polyfluorinated and perfluorinated chemicals in precipitation and runoff from cities across eastern and central China.. Arch Environ Contam Toxicol, 2013, 64: 198-207.

[8] Shan G, Chen X, Zhu L. Occurrence, fluxes and sources of perfluoroalkyl substances with isomer analysis in the snow of northern China. J Hazard Mater, 2015, 299: 639-646.

[9] Kwok KY, Yamazaki E, Yamashita N, et al. Transport of perfluoroalkyl substances (PFAS)from an arctic glacier to downstream locations: implications for sources. Sci Total Environ, 2013, 447: 46-55.

[10] Yeung LWY, Dassuncao C, Mabury S, et al. Vertical profiles, sources, and transport of PFASs in the Arctic Ocean. Environ Sci Technol, 2017, 51(12): 6735-6744.

[11] Ahrens L. Polyfluoroalkyl compounds in the aquatic environment: A review of their occurrence and fate. J Environ Monit, 2011, 13(1): 20-31.

[12] Wang Z, Cousins I T, Scheringer M, et al. Hazard assessment of fluorinated alternatives to long-chain perfluoroalkyl acids (PFAAs)and their precursors: status quo, ongoing challenges and possible solutions. Environ Int, 2015, 75: 172-179.

[13] Zhou Z, Liang Y, Shi Y, et al. Occurrence and transport of perfluoroalkyl acids (PFAAs), including short-chain PFAAs in Tangxun Lake, China. Environ Sci Technol, 2013, 47(16): 9249-9257.

[14] Zhao P, Xia X, Dong J, et al. Short- and long-chain perfluoroalkyl substances in the water, suspended particulate matter, and surface sediment of a turbid river. Sci Total Environ, 2016, 568: 57-65.

[15] Zhu Z, Wang T, Meng J, et al. Perfluoroalkyl substances in the Daling River with concentrated fluorine industries in China: Seasonal variation, mass flow, and risk assessment. Environ Sci Pollut Res, 2015, 22(13): 10009-10018.

[16] Nguyen MA, Wiberg K, Ribeli E, et al. Spatial distribution and source tracing of per- and polyfluoroalkyl substances (PFASs)in surface water in Northern Europe. Environ Pollut, 2017, 220(Pt B): 1438-1446.

[17] Shi Y, Vestergren R, Xu L, et al. Characterizing direct emissions of perfluoroalkyl substances from ongoing fluoropolymer production sources: A spatial trend study of Xiaoqing River, China. Environ Pollut, 2015, 206: 104-112.

[18] Sun M, Arevalo E, Strynar M, et al. Legacy and emerging perfluoroalkyl substances are important drinking water contaminants in the Cape Fear River watershed of North Carolina. Environ Sci Technol Lett, 2016, 3(12): 415-419.

[19] Heydebreck F, Tang J, Xie Z, et al. Alternative and legacy perfluoroalkyl substances: differences between European and Chinese river/estuary systems. Environ Sci Technol, 2015, 49(14): 8386-8395.

[20] Wang S, Huang J, Yang Y, et al. First report of a Chinese PFOS alternative overlooked for 30 years: its toxicity, persistence, and presence in the environment. Environ Sci Technol, 2013, 47(18): 10163-10170.

[21] Sun H, Zhang X, Wang L, et al. Perfluoroalkyl compounds in municipal WWTPs in Tianjin, China--concentrations, distribution and mass flow. Environ Sci Pollut Res, 2012, 19(5): 1405-1415.

[22] Sun H, Li F, Zhang T, et al. Perfluorinated compounds in surface waters and WWTPs in Shenyang, China: mass flows and source analysis. Water Res, 2011, 45(15): 4483-4490.

[23] Yang L, Zhu L, Liu Z. Occurrence and partition of perfluorinated compounds in water and sediment from Liao River and Taihu Lake, China. Chemosphere, 2011, 83(6): 806-814.

[24] Bao J, Jin Y, Liu W, et al. Perfluorinated compounds in sediments from the Daliao River system of northeast China. Chemosphere, 2009, 77(5): 652-657.

[25] Li F, Sun H, Hao Z, et al. Perfluorinated compounds in Haihe River and Dagu Drainage Canal in Tianjin, China. Chemosphere, 2011, 84(2): 265-271.

[26] Jin H, Zhang Y, Zhu L, et al. Isomer profiles of perfluoroalkyl substances in water and soil surrounding a chinese fluorochemical manufacturing park. Environ Sci Technol, 2015, 49(8): 4946-4954.

[27] Sepulvado J G, Blaine A C, Hundal L S, et al. Occurrence and fate of perfluorochemicals in soil following the land

application of municipal biosolids. Environ Sci Technol, 2011, 45(19): 8106-8112.

[28] Mejia-Avendano S, Vo Duy S, Sauve S, et al. Generation of Perfluoroalkyl Acids from Aerobic Biotransformation of Quaternary Ammonium Polyfluoroalkyl Surfactants. Environ Sci Technol, 2016, 50(18): 9923-9932.

[29] Houtz E F, Higgins C P, Field J A, et al. Persistence of perfluoroalkyl acid precursors in AFFF-impacted groundwater and soil. Environ Sci Technol, 2013, 47(15): 8187-8195.

[30] Harding-Marjanovic K C, Houtz E F, Yi S, et al. Aerobic biotransformation of fluorotelomer thioether amido sulfonate (Lodyne)in AFFF-amended microcosms. Environ Sci Technol, 2015, 49(13): 7666-7674.

[31] Plassmann M M, Berger U. Perfluoroalkyl carboxylic acids with up to 22 carbon atoms in snow and soil samples from a ski area. Chemosphere, 2013, 91(6): 832-837.

[32] Eschauzier C, Raat K J, Stuyfzand P J, et al. Perfluorinated alkylated acids in groundwater and drinking water: identification, origin and mobility. Sci Total Environ, 2013, 458-460: 477-485.

[33] Barzen-Hanson K A, Roberts S C, Choyke S, et al. Discovery of 40 classes of per- and polyfluoroalkyl substances in historical aqueous film-forming foams (AFFFs)and AFFF-impacted groundwater. Environ Sci Technol, 2017, 51(4): 2047-2057.

[34] Guelfo J L, Higgins C P. Subsurface transport potential of perfluoroalkyl acids at aqueous film-forming foam (AFFF)-impacted sites. Environ Sci Technol, 2013, 47(9): 4164-4171.

[35] Xiao F, Simcik M F, Halbach T R, et al. Perfluorooctane sulfonate (PFOS)and perfluorooctanoate (PFOA)in soils and groundwater of a U.S. metropolitan area: migration and implications for human exposure. Water Res, 2015, 72: 64-74.

[36] Liu Z, Lu Y, Wang T, et al. Risk assessment and source identification of perfluoroalkyl acids in surface and ground water: Spatial distribution around a mega-fluorochemical industrial park, China. Environ Int, 2016, 91: 69-77.

[37] Li X, Shang X, Luo T, et al. Screening and health risk of organic micropollutants in rural groundwater of Liaodong Peninsula, China. Environ Pollut, 2016, 218: 739-748.

[38] Maria K, Jacob B, Ana B. Bisphenol A and replacements in thermal paper: A review. Chemosphere, 2017, 182: 691-706.

[39] Jin H, Zhu L. Occurrence and partitioning of bisphenol analogues in water and sediment from Liaohe River Basin and Taihu Lake, China. Water Res, 2016, 103: 343-351.

[40] Liu Y, Zhang S, Song N, et al. Occurrence, distribution and sources of bisphenol analogues in a shallow Chinese freshwater lake (Taihu Lake): Implications for ecological and human health risk. Sci Total Environ, 2017, 599-600: 1090-1098.

[41] Yamazaki E, Yamashita N, Taniyasu S, et al. Bisphenol A and other bisphenol analogues including BPS and BPF in surface water samples from Japan, China, Korea and India. Ecotoxicol Environ Saf, 2015, 122: 565-572.

[42] Yang J, Wang X, Zhang D, et al. Simultaneous determination of endocrine disrupting compounds bisphenol F and bisphenol AF using carboxyl functionalized multi-walled carbon nanotubes modified electrode. Talanta, 2014, 130: 207-212.

[43] YangY, Lu L, Zhang J, et al. Simultaneous determination of seven bisphenols in environmental water and solid samples by liquid chromatography-electrospray tandem mass spectrometry. Chromatogr A, 2014, 1328: 26-34.

[44] Barber L B, Loyo-Rosales J E, Rice C P, et al. Endocrine disrupting alkylphenolic chemicals and other contaminants in wastewater treatment plant effluents, urban streams, and fish in the Great Lakes and Upper Mississippi River Regions. Sci Total Environ, 2015, 517: 195-206.

[45] Samaras V G, Stasinakis A S, Mamais D, et al. Fate of selected pharmaceuticals and synthetic endocrine disrupting compounds during wastewater treatment and sludge anaerobic digestion. J Hazard Mater, 2013, 244-245: 259-267.

[46] Ömeroğlu S, Murdoch F K, Sanin F D. Investigation of nonylphenol and nonylphenol ethoxylates in sewage sludge samples from a metropolitan wastewater treatment plant in Turkey. Talanta, 2015, 131: 650-655.

[47] Gao D, Li Z, Guan J, et al. Seasonal variations in the concentration and removal of nonylphenol ethoxylates from the

wastewater of a sewage treatment plant. J Environ Sci (China), 2017, 54: 217-223.

[48] Gao D, Li Z, Guan J, et al. Removal of surfactants nonylphenol ethoxylates from municipal sewage-comparison of an A/O process and biological aerated filters. Chemosphere, 2014, 97: 130-134.

[49] Traverso-Soto J M, Lara-Martin P A, Gonzalez-Mazo E, et al. Distribution of anionic and nonionic surfactants in a sewage-impacted Mediterranean coastal lagoon: inputs and seasonal variations. Sci Total Environ, 2015, 503-504: 87-96.

[50] Robert-Peillard F, Syakti AD, Coulomb B, et al. Occurrence and fate of selected surfactants in seawater at the outfall of the Marseille urban sewerage system. Int J Environ Sci Technol, 2015, 12(5): 1527-1538.

[51] Corada-Fernandez C, Lara-Martin P A, Candela L, et al. Vertical distribution profiles and diagenetic fate of synthetic surfactants in marine and freshwater sediments. Sci Total Environ, 2013, 461-462: 568-575.

[52] Venkatesan A K, Halden R U. National inventory of alkylphenol ethoxylate compounds in U.S. sewage sludges and chemical fate in outdoor soil mesocosms. Environ Pollut, 2013, 174: 189-193.

[53] Corada-Fernandez C, Lara-Martin P A, Candela L, et al. Vertical distribution profiles and diagenetic fate of synthetic surfactants in marine and freshwater sediments. Sci Total Environ, 2013, 461-462: 568-575.

[54] Andersen F A. Final amended report on the safety assessment of methylparaben, ethylparaben, propylparaben, isopropylparaben, butylparaben, isobutylparaben, and benzylparaben as used in cosmetic products. Int J Toxicol, 2008, 27: 1-82.

[55] Shanmugam G, Ramaswamy B R, Radhakrishnan V, et al. GC-MS method for the determination of paraben preservatives in the human breast cancerous tissue. Microchem J, 2010, 96(2): 391-396.

[56] Moos R K, Angerer J, Wittsiepe J, et al. Rapid determination of nine parabens and seven other environmental phenols in urine samples of German children and adults. Int J Hyg Environ Health, 2014, 217(8): 845-853.

[57] Shirai S, Suzuki Y, Yoshinaga J, et al. Urinary excretion of parabens in pregnant Japanese women. Reprod Toxicol, 2013, 35: 96-101.

[58] Wang L, Kannan K. Alkyl protocatechuates as novel urinary biomarkers of exposure to p-hydroxybenzoic acid esters (parabens). Environ Int, 2013, 59: 27-32.

[59] Cashman A L, Warshaw E M. Parabens: a review of epidemiology, structure, allergenicity, and hormonal properties. Dermatitis, 2005, 16(2): 57-66.

[60] Liao C, Chen L, Kannan K. Occurrence of parabens in foodstuffs from China and its implications for human dietary exposure. Environ Int, 2013, 57: 68-74.

[61] Liao C, Liu F, Kannan K. Occurrence of and dietary exposure to parabens in foodstuffs from the United States. Environ Sci Technol, 2013, 47(8): 3918-3925.

[62] Błędzka D, Gromadzińska J, Wąsowicz W. Parabens. From environmental studies to human health. Environ Int, 2014, 67: 27-42.

[63] Jonkers N, Kohler H P E, Dammshäuser A, et al. Mass flows of endocrine disruptors in the Glatt River during varying weather conditions. Environ Pollut, 2009, 157(3): 714-723.

[64] Kusk K O, Krüger T, Long M, et al. Endocrine potency of wastewater: contents of endocrine disrupting chemicals and effects measured by in vivo and in vitro assays. Environ Toxicol Chem, 2011, 30(2): 413-426.

[65] Teerlink J, Hering A S, Higgins C P, et al. Variability of trace organic chemical concentrations in raw wastewater at three distinct sewershed scales. Water Res, 2012, 46(10): 3261-3271.

[66] Yu Y, Huang Q, Cui J, et al. Determination of pharmaceuticals, steroid hormones, and endocrine-disrupting personal care products in sewage sludge by ultra-high-performance liquid chromatography-tandem mass spectrometry. Anal Bioanal Chem, 2011, 399(2): 891-902.

[67] Haman C, Dauchy X, Rosin C, et al. Occurrence, fate and behavior of parabens in aquatic environments: a review. Water Res, 2015, 68: 1-11.

[68] Blanco E, del Carmen Casais M, del Carmen Mejuto M, et al. Combination of off-line solid-phase extraction and on-column sample stacking for sensitive determination of parabens and p-hydroxybenzoic acid in waters by non-aqueous capillary electrophoresis. Anal Chim Acta, 2009, 647(1): 104-111.

[69] Bratkowska D, Marcé R M, Cormack P A G, et al. Development and application of a polar coating for stir bar sorptive extraction of emerging pollutants from environmental water samples. Anal Chim Acta, 2011, 706(1): 135-142.

[70] Canosa P, Rodríguez I, Rubí E, et al. Optimisation of a solid-phase microextraction method for the determination of parabens in water samples at the low ng per litre level. J Chromatogr A, 2006, 1124(1): 3-10.

[71] González-Mariño I, Quintana J B, Rodríguez I, et al. Simultaneous determination of parabens, triclosan and triclocarban in water by liquid chromatography/electrospray ionisation tandem mass spectrometry. Rapid Commun Mass Sp, 2009, 23(12): 1756-1766.

[72] Kasprzyk-Hordern B, Dinsdale R M, Guwy A J. Multiresidue methods for the analysis of pharmaceuticals, personal care products and illicit drugs in surface water and wastewater by solid-phase extraction and ultra performance liquid chromatography-electrospray tandem massspectrometry. Anal Bioanal Chem, 2008, 391(4): 1293-1308.

[73] Lee H B, Peart T E, Svoboda M L. Determination of endocrine-disrupting phenols, acidic pharmaceuticals, and personal-care products in sewage by solid-phase extraction and gas chromatography-mass spectrometry. J Chromatogr A, 2005, 1094(1): 122-129.

[74] Pedrouzo M, Borrull F, Marcé R M, et al. Ultra-high-performance liquid chromatography-tandem mass spectrometry for determining the presence of eleven personal care products in surface and wastewaters. J Chromatogr A, 2009, 1216(42): 6994-7000.

[75] Villaverde-de-Sáa E, González-Mariño I, Quintana J B, et al. In-sample acetylation-non-porous membrane-assisted liquid-liquid extraction for the determination of parabens and triclosan in water samples. Anal Bioanal Chem, 2010, 397(6): 2559-2568.

[76] Yu K, Li B, Zhang T. Direct rapid analysis of multiple PPCPs in municipal wastewater using ultrahigh performance liquid chromatography-tandem mass spectrometry without SPE pre-concentration. Anal Chim Acta, 2012, 738: 59-68.

[77] Geara-Matta D, Lorgeoux C, Rocher V, et al. Occurrence of endocrine disruptors in wastewater from combined sewers in Paris and Beirut: what about parabens. 12th International Conference on Urban Drainage, Porto Alegre (Brazil). 2011: 8.

[78] Albero B, Pérez R A, Sánchez-Brunete C, et al. Occurrence and analysis of parabens in municipal sewage sludge from wastewater treatment plants in Madrid (Spain). J Hazard Mater, 2012, 239: 48-55.

[79] Nieto A, Borrull F, Marcé R M, et al. Determination of personal care products in sewage sludge by pressurized liquid extraction and ultra high performance liquid chromatography-tandem mass spectrometry. J Chromatogra A, 2009, 1216(30): 5619-5625.

[80] Gorga M, Insa S, Petrovic M, et al. Analysis of endocrine disrupters and related compounds in sediments and sewage sludge using on-line turbulent flow chromatography–liquid chromatography–tandem mass spectrometry. J Chromatogra A, 2014, 1352: 29-37.

[81] Kasprzyk-Hordern B, Dinsdale R M, Guwy A J. The occurrence of pharmaceuticals, personal care products, endocrine disruptors and illicit drugs in surface water in South Wales, UK. Water Res, 2008, 42(13): 3498-3518.

[82] Kasprzyk-Hordern B, Dinsdale R M, Guwy A J. The removal of pharmaceuticals, personal care products, endocrine disruptors and illicit drugs during wastewater treatment and its impact on the quality of receiving waters. Water Res, 2009, 43(2): 363-380.

[83] González-Mariño I, Quintana J B, Rodríguez I, et al. Evaluation of the occurrence and biodegradation of parabens and halogenated by-products in wastewater by accurate-mass liquid chromatography-quadrupole-time-of-flight-mass spectrometry (LC-QTOF-MS). Water Res, 2011, 45(20): 6770-6780.

[84] Peng X, Yu Y, Tang C, et al. Occurrence of steroid estrogens, endocrine-disrupting phenols, and acid pharmaceutical residues in urban riverine water of the Pearl River Delta, South China. Sci Total Environ, 2008, 397(1): 158-166.

[85] Chattopadhyay S, Raychaudhuri U, Chakraborty R. Artificial sweeteners–A review . J Food Sci Tech Mys, 2014, 51(4): 611-621.

[86] Buerge I J, Keller M, Buser H R, et al. Saccharin and other artificial sweeteners in soils: estimated inputs from agriculture and households, degradation, and leaching to groundwater. Environ Sci Technol, 2010, 45(2): 615-621.

[87] Buerge I J, Buser H R, Kahle M, et al. Ubiquitous occurrence of the artificial sweetener acesulfame in the aquatic environment: an ideal chemical marker of domestic wastewater in groundwater. Environ Sci Technol, 2009, 43(12): 4381-4385.

[88] Scheurer M, Brauch H J, Lange F T. Analysis and occurrence of seven artificial sweeteners in German waste water and surface water and in soil aquifer treatment (SAT). Anal Bioanal Chem, 2009, 394(6): 1585-1594.

[89] Loos, R, Carvalho R, Antonio D C, et al. EU-wide monitoring survey on emerging polar organic contaminants in wastewater treatment plant effluents. Water Res, 2013, 47(17): 6475-6487.

[90] Subedi B, Kannan K. Fate of artificial sweeteners in wastewater treatment plants in New York State, USA. Environ Sci Technol, 2014, 48(23): 13668-13674.

[91] Watanabe Y, Leu T B, Pham V D, et al. Ubiquitous detection of artificial sweeteners and iodinated X-ray contrast media in aquatic environmental and wastewater treatment plant samples from Vietnam, The Philippines, and Myanmar. Arch Environ Con Tox, 2016, 70(4): 671-681.

[92] Loos R, Gawlik B M, Boettcher K, et al. Sucralose screening in European surface waters using a solid-phase extraction-liquid chromatography–triple quadrupole mass spectrometry method. J Chromatogr A, 2009, 1216(7): 1126-1131.

[93] Sang Z, Jiang Y, Tsoi Y K, et al. Evaluating the environmental impact of artificial sweeteners: A study of their distributions, photodegradation and toxicities. Water Res, 2013, 52: 260-274.

[94] Ordonez E Y, Quintana J B, Rodil R, et al. Determination of artificial sweeteners in sewage sludge samples using pressurised liquid extraction and liquid chromatography-tandem mass spectrometry. J Chromatogr A, 2013, 1320: 10-16.

[95] Subedi B, Lee S, Moon H B, et al. Emission of artificial sweeteners, select pharmaceuticals, and personal care products through sewage sludge from wastewater treatment plants in Korea. Environ Int, 2014, 68, 33-40.

[96] Gan Z, Sun H, Yao Y, et al. Distribution of artificial sweeteners in dust and soil in China and their seasonal variations in the environment of Tianjin. Sci Total Environ, 2014, 488: 168-175.

[97] Gan Z W, Sun H W, Feng B T, et al. Occurrence of seven artificial sweeteners in the aquatic environment and precipitation of Tianjin, China. Water Res, 2013, 47(14): 4928-4937.

[98] Tran N H, Hu J Y, Li J H, et al. Suitability of artificial sweeteners as indicators of raw wastewater contamination in surface water and groundwater. Water Res, 2014, 48: 443-456.

[99] Oppenheimer J A, Badruzzaman M, Jacangelo J G. Differentiating sources of anthropogenic loading to impaired water bodies utilizing ratios of sucralose and other microconstituents. Water Res, 2012, 46(18): 5904-5916.

[100] Ma L, Liu Y, Xu J, et al. Mass loading of typical artificial sweeteners in a pig farm and their dissipation and uptake by plants in neighboring farmland. Sci Total Environ, 2017, 605-606: 735-744.

[101] Giger W, Schaffner C, Kohler H P E. Benzotriazole and tolyltriazole as aquatic contaminants. 1. Input and occurrence in rivers and lakes. Environ Sci Technol, 2006, 40(23): 7186-7192.

[102] Cancilla D A, Martinez J, Van Aggelen G C. Detection of aircraft deicing/antiicing fluid additives in a perched water monitoring well at an international airport. Environ Sci Technol, 1998, 32(23): 3834-3835.

[103] Corsi S R, Geis S W, Loyo-Rosales J E, et al. Aquatic toxicity of nine aircraft deicer and anti-icer formulations and relative toxicity of additive package ingredients alkylphenol ethoxylates and 4, 5-methyl-1H-benzotriazoles. Environ Sci Technol, 2006, 40(23): 7409-7415.

[104] Liu YS, Ying GG, Shareef A, et al. Simultaneous determination of benzotriazoles and ultraviolet filters in ground water, effluent and biosolid samples using gas chromatography-tandem mass spectrometry. J Chromatogr A, 2011, 1218(31): 5328-5335.

[105] Liu Y S, Ying G G, Shareef A, et al. Occurrence and removal of benzotriazoles and ultraviolet filters in a municipal wastewater treatment plant. Environ Pollut, 2012, 165: 225-232.

[106] Janna H, Scrimshaw M D, Williams R J, et al. From dishwasher to tap? Xenobiotic substances benzotriazole and tolyltriazole in the environment. Environ Sci Technol, 2011, 45(9): 3858-3864.

[107] Kiss A, Fries E. Occurrence of benzotriazoles in the rivers Main, Hengstbach, and Hegbach (Germany). Environ Sci Pollut Res, 2009, 16(6): 702-710.

[108] Mcneill K S, Cancilla D A. Detection of triazole deicing additives in soil samples from airports with low, mid, and large volume aircraft deicing activities. B Environ Contam Tox, 2009, 82(3): 265-269.

[109] Voutsa D, Hartmann P, Schaffner C, et al. Benzotriazoles, alkylphenols and bisphenol A in municipal wastewaters and in the Glatt River, Switzerland. Environ Sci Pollut Res, 2006, 13(5): 333-341.

[110] Jia Y, Bakken L, Breedveld G, et al. Organic compounds that reach subsoil may threaten groundwater quality；effect of benzotriazole on degradation kinetics and microbial community composition. Soil Biol Biochem, 2006, 38(9): 2543-2556.

[111] Cancilla D A, Baird J C, Rosa R. Detection of aircraft deicing additives in groundwater and soil samples from fairchild air force base, a small to moderate user of deicing fluids. B Environ Contam Tox, 2003, 70(5): 0868-0875.

[112] Asimakopoulos A G, Ajibola A, Kannan K, et al. Occurrence and removal efficiencies of benzotriazoles and benzothiazoles in a wastewater treatment plant in Greece. Sci Total Environ, 2013, 452-453: 163-171.

[113] Matamoros V, Jover E, Bayona J M. Occurrence and fate of benzothiazoles and benzotriazoles in constructed wetlands. Water Sci Technol, 2010, 61(1): 191.

[114] Stasinakis A S, Thomaidis N S, Arvaniti O S, et al. Contribution of primary and secondary treatment on the removal of benzothiazoles, benzotriazoles, endocrine disruptors, pharmaceuticals and perfluorinated compounds in a sewage treatment plant. Sci Total Environ, 2013, 463-464: 1067-1075.

[115] Weiss S, Reemtsma T. Determination of benzotriazole corrosion inhibitors from aqueous environmental samples by liquid chromatography-electrospray ionization-tandem mass spectrometry. Anal Chem, 2005, 77(22): 7415-7420.

[116] Molins-Delgado D, Diaz-Cruz M S, Barcelo D. Removal of polar UV stabilizers in biological wastewater treatments and ecotoxicological implications. Chemosphere, 2015, 119 Suppl: S51-57.

[117] Jover E, Matamoros V, Bayona J M. Characterization of benzothiazoles, benzotriazoles and benzosulfonamides in aqueous matrixes by solid-phase extraction followed by comprehensive two-dimensional gas chromatography coupled to time-of-flight mass spectrometry. J Chromatogr A, 2009, 1216(18): 4013-4019.

[118] Reemtsma T, Miehe U, Duennbier U, et al. Polar pollutants in municipal wastewater and the water cycle: occurrence and removal of benzotriazoles. Water Res, 2010, 44(2): 596-604.

[119] Alotaibi M D, Patterson B M, Mckinley A J, et al. Fate of benzotriazole and 5-methylbenzotriazole in recycled water recharged into an anaerobic aquifer: column studies. Water Res, 2015, 70: 184-195.

[120] Alotaibi M D, Patterson B M, Mckinley A J, et al. Benzotriazole and 5-methylbenzotriazole in recycled water, surface

water and dishwashing detergents from Perth, Western Australia: analytical method development and application. Environ Sci Process Impacts, 2015, 17(2): 448-457.

[121] Van Leerdam J A, Hogenboom A C, Van Der Kooi M M E, et al. Determination of polar 1H-benzotriazoles and benzothiazoles in water by solid-phase extraction and liquid chromatography LTQ FT Orbitrap mass spectrometry. Int J Mass Spectrom, 2009, 282(3): 99-107.

[122] Kiss A, Fries E. Seasonal source influence on river mass flows of benzotriazoles. J Environ Monit, 2012, 14(2): 697-703.

[123] Cancilla D A, Holtkamp A, Matassa L, et al. Isolation and characterization of Microtox®-active components from aircraft de-icing/anti-icing fluids. Environ Tox Chem, 1997, 16(3): 430-434.

[124] Corsi S R, Geis S W, Loyo-Rosales J E, et al. Characterization of aircraft deicer and anti-icer components and toxicity in airport snowbanks and snowmelt Runoff. Environ Sci Technol, 2006, 40(10): 3195-3202.

[125] Hagedorn B, Larsen M, Dotson A. First assessment of triazoles and other organic contaminants in snow and snowmelt in urban waters, Anchorage, Alaska.//ISCORD 2013. City.

[126] Breedveld G D, Roseth R, Sparrevik M, et al. Persistence of the de-icing additive benzotriazole at an abandoned airport. Water Air Soil Pollut: Focus, 2003, 3(3): 91-101.

[127] Fischer K, Fries E, Korner W, et al. New developments inthe trace analysis of organic water pollutants. Appl Microbiol Biot, 2012, 94: 11-28.

[128] Jungclaus G A, Games L M, Hites R A. Identification of trace organic compounds in tire manufacturing plant waste waters. Anal Chem, 1976, 48(13): 1894-1896.

[129] Jungclaus G A, Games L M, Hites R A. Identification of trace organic compounds in tire manufacturing plant waste waters. Anal Chem, 1976, 48(13): 1894-1896.

[130] Reemtsma T. Determination of 2-substituted benzothiazoles of industrial use from water by liquid chromatography/electrospray ionization tandem mass spectrometry. Rapid Commun Mass Sp, 2000, 14(17): 1612-1618.

[131] Wick A, Fink G, Ternes T A. Comparison of electrospray ionization and atmospheric pressure chemical ionization for multi-residue analysis of biocides, UV-filters and benzothiazoles in aqueous matrices and activated sludge by liquid chromatography–tandem mass spectrometry. J Chromatogr A, 2010, 1217(14): 2088-2103.

[132] Ni H G, Lu F H, Luo X L, et al. Occurrence, phase distribution, and mass loadings of benzothiazoles in riverine runoff of the Pearl River Delta, China. Environ Sci Technol, 2008, 42(6): 1892-1897.

[133] Kong L X, Kadokami K, Wang S, et al. Monitoring of 1300 organic micro-pollutants in surface waters from Tianjin, North China. Chemosphere. 2015, 122: 125-130.

[134] 王金成, 张海军, 陈吉平. 固相萃取高效液相色谱-串联质谱法测定地表水中苯并三唑类及苯并噻唑类衍生物. 分析测试学报, 2013, 32(9): 1056-1061.

[135] 王金成, 熊力, 张海军, 等. 固相萃取-高效液相色谱法测定地表水中的苯并三唑和苯并噻唑. 色谱, 2013, 31(2): 139-142.

[136] Wang L, Zhang J, Sun H, et al. Widespread occurrence of benzotriazoles and benzothiazoles in tap water: Influencing factors and contribution to human exposure. Environ Sci Technol, 2016, 50(5): 2709-2717.

[137] Food and Nutrition Board, Institute of Medicine. Dietary reference intakes for water, potassium, sodium, chloride, and sulfate. Washington D.C.: National Academy Press. 2004.

[138] Reddy C M, Quinn J G. Environmental chemistry of benzothiazoles derived from rubber. Environ Sci Technol, 1997, 31(10): 2847-2853.

[139] Speltini A, Sturini M, Maraschi F, et al. Fast low-pressurized microwave-assisted extraction of benzotriazole, benzothiazole and benezenesulfonamide compounds from soil samples. Talanta, 2016, 147: 322-327.

[140] Wang L, Asimakopoulos A G, Moon H B, et al. Benzotriazole, benzothiazole, and benzophenone compounds in indoor dust fromthe United States and East Asian Countries . Environ Sci Technol, 2013, 47: 4752-4759.

[141] Kiss A, Fries E. Occurrence of benzotriazoles in the rivers Main, Hengstbach, and Hegbach (Germany). Environ Sci Pollut Res Int, 2009, 16: 702-710.

[142] Wan Y, Xue J, Kannan K. Benzothiazoles in indoor air from Albany, New York, USA, and its implications for inhalation exposure. J Hazard Mater, 2016, 311: 37-42.

[143] He C, Zheng J, Qiao L, et al. Occurrence of organophosphorus flame retardants in indoor dust in multiple microenvironments of southern China and implications for human exposure. Chemosphere, 2015, 133: 47-52.

[144] Cao Z, Xu F, Covaci A, et al. Distribution patterns of brominated, chlorinated, and phosphorus flame retardants with particle size in indoor and outdoor dust and implications for human exposure. Environ Sci Technol, 2014, 48(15): 8839-46.

[145] Yang F, Ding J, Huang W, et al. Particle Size-Specific Distributions and Preliminary Exposure Assessments of Organophosphate Flame Retardants in Office Air Particulate Matter. Environ Sci Technol, 2014, 48(1): 63-70.

[146] Li J, Yu N, Zhang B, et al. Occurrence of organophosphate flame retardants in drinking water from China. Water Res, 2014, 54: 53-61.

[147] Shi Y, Gao L, Li W, et al. Occurrence, distribution and seasonal variation of organophosphate flame retardants and plasticizers in urban surface water in Beijing, China. Environ Pollut, 2016, 209: 1-10.

[148] Tan X, Luo X, Zheng X, et al., Distribution of organophosphorus flame retardants in sediments from the Pearl River Delta in South China. Sci Total Environ, 2016. 544: 77-84.

[149] Wang Y, Wu X, Zhang Q, et al. Organophosphate esters in sediment cores from coastal Laizhou Bay of the Bohai Sea, China. Sci Total Environ, 2017, 607-608: 103-108.

[150] Cui K, Wen J, Zeng F, et al. Occurrence and distribution of organophosphate esters in urban soils of the subtropical city, Guangzhou, China.. Chemosphere, 2017, 175: 514-520.

[151] Wong F, Suzuki G, Michinaka C, et al. Dioxin-like activities, halogenated flame retardants, organophosphate esters and chlorinated paraffins in dust from Australia, the United Kingdom, Canada, Sweden and China. Chemosphere, 2017, 168: 1248-1256.

[152] Wu M, Yu G, Cao Z, et al., Characterization and human exposure assessment of organophosphate flame retardants in indoor dust from several microenvironments of Beijing, China. Chemosphere, 2016, 150: 465-471.

[153] Zheng X, Xu F, Chen K, et al. Flame retardants and organochlorines in indoor dust from several e-waste recycling sites in South China: Composition variations and implications for human exposure. Environ Int, 2015, 78: 1-7.

[154] Wang R, Tang J, Xie Z, et al. Occurrence and spatial distribution of organophosphate ester flame retardants and plasticizers in 40 rivers draining into the Bohai Sea, north China. Environ Pollut, 2015, 198: 172-178.

[155] Wang X, He Y, Lin L, et al. Application of fully automatic hollow fiber liquid phase microextraction to assess the distribution of organophosphate esters in the Pearl River Estuaries. Sci Total Environ, 2014, 470-471: 263-269.

[156] Hu M, Li J, Zhang B, et al. Regional distribution of halogenated organophosphate flame retardants in seawater samples from three coastal cities in China. Mar Pollut Bull, 2014, 86(1-2): 569-574.

[157] Ding J, Shen X, Liu W, et al. Occurrence and risk assessment of organophosphate esters in drinking water from Eastern China. Sci Total Environ, 2015, 538: 959-965.

[158] Gao L, Shi Y, Li W, et al. Occurrence and distribution of organophosphate triesters and diesters in sludge from sewage treatment plants of Beijing, China. Sci Total Environ, 2016, 544: 143-149.

[159] Cao S, Zeng X, Song H, et al. Levels and distributions of organophosphate flame retardants and plasticizers in sediment from Taihu Lake, China. Environ Toxicol Chem, 2012, 31(7): 1478-1484.

[160] Pintado-Herrera M G, Wang C, Lu J, et al. Distribution, mass inventories, and ecological risk assessment of legacy and emerging contaminants in sediments from the Pearl River Estuary in China. J Hazard Mater, 2017, 323: 128-138.

[161] Wan W, Zhang S, Huang H, et al. Occurrence and distribution of organophosphorus esters in soils and wheat plants in a plastic waste treatment area in China. Environ Pollut, 2016, 214: 349-353.

[162] Lu J, Ji W, Ma S, et al. Analysis of organophosphate esters in dust, soil and sediment samples using gas chromatography coupled with mass spectrometry. Chinese J Anal Chem, 2014, 42(6): 859-865.

[163] 印红玲，李世平，叶芝祥，等. 成都市土壤中有机磷阻燃剂的污染特征及来源分析. 环境科学学报，2016，36(02): 606-613.

[164] Ma Y, Cui K, Zeng F, et al. Microwave-assisted extraction combined with gel permeation chromatography and silica gel cleanup followed by gas chromatography–mass spectrometry for the determination of organophosphorus flame retardants and plasticizers in biological samples. Anal Chim Acta, 2013, 786: 47-53.

[165] Hou R, Liu C, Gao X, et al. Accumulation and distribution of organophosphate flame retardants (PFRs) and their di-alkyl phosphates (DAPs)metabolites in different freshwater fish from locations around Beijing, China. Environ Pollut, 2017, 229: 548-556.

[166] Zhang X, Zou W, Mu L, et al. Rice ingestion is a major pathway for human exposure to organophosphate flame retardants (OPFRs) in China. J Hazard Mater, 2016, 318: 686-693.

[167] Li J, Del Vento S, Schuster J, et al. Perfluorinated compounds in the Asian atmosphere. Environ Sci Technol, 2011, 45(17): 7241-7248.

[168] Gonzalez-Gaya B, Dachs J, Roscales J L, et al. Perfluoroalkylated substances in the global tropical and subtropical surface oceans. Environ Sci Technol, 2014, 48(22): 13076-13084.

[169] Ellis D A, Martin J W, De Silva A O, et al. Degradation of fluorotelomer alcohols: a likely atmospheric source of perfluorinated carboxylic acids. Environ Sci Technol, 2004, 38(12): 3316-3321.

[170] Styler S A, Myers A L, Donaldson D J. Heterogeneous photooxidation of fluorotelomer alcohols: a new source of aerosol-phase perfluorinated carboxylic acids. Environ Sci Technol, 2013, 47(12): 6358-6367.

[171] Wallington T J, Hurley M D, Xia J, et al. Formation of $C_7F_{15}COOH$ (PFOA) and other perfluorocarboxylic acids during the atmospheric oxidation of 8: 2 fluorotelomer alcohol. Environ Sci Technol, 2006, 40(3): 924-930.

[172] Qian Y, Guo X, Zhang Y, et al. Perfluorooctanoic acid degradation using UV-persulfate process: Modeling of the degradation and chlorate formation. Environ Sci Technol, 2016, 50(2): 772.

[173] Houtz E F, Sedlak D L. Oxidative conversion as a means of detecting precursors to perfluoroalkyl acids in urban runoff. Environ Sci Technol, 2012, 46(17): 9342-9349.

[174] González M M, Martin J, Camacho-Munoz D, et al. Degradation and environmental risk of surfactants after the application of compost sludge to the soil. Waste Manag, 2012, 32(7): 1324-1331.

[175] Writer J H, Ryan J N, Keefe S H, et al. Fate of 4-nonylphenol and 17beta-estradiol in the Redwood River of Minnesota. Environ Sci Technol, 2012, 46(2): 860-868.

[176] Lian J, Liu J. Fate and degradation of nonylphenolic compounds during wastewater treatment process. J Environ Sci, 2013, 25(8): 1511-1518.

[177] Kubota T, Toyooka T, Ibuki Y. Nonylphenol polyethoxylates degraded by three different wavelengths of UV and their genotoxic change--detected by generation of gamma-H2AX. Photochem Photobiol, 2013, 89(2): 461-467.

[178] Karci A, Arslan-Alaton I, Bekbolet M, et al. H_2O_2/UV-C and Photo-Fenton treatment of a nonylphenol polyethoxylate in synthetic freshwater: Follow-up of degradation products, acute toxicity and genotoxicity. Chem Eng J, 2014, 241: 43-51.

[179] da Silva S W, Klauck C R, Siqueira M A, et al. Degradation of the commercial surfactant nonylphenol ethoxylate by

advanced oxidation processes. J Hazard Mater, 2015, 282: 241-248.

[180] Dzinun H, Othman M H D, Ismail A F, et al. Photocatalytic degradation of nonylphenol by immobilized TiO2 in dual layer hollow fibre membranes. Chem Eng J, 2015, 269: 255-261.

[181] Iqbal M, Bhatti I A. Gamma radiation/H_2O_2 treatment of a nonylphenol ethoxylates: Degradation, cytotoxicity, and mutagenicity evaluation. J Hazard Mater, 2015, 299: 351-360.

[182] Wang L, Zhang J, Duan Z, et al. Fe (III) and Fe(II) induced photodegradation of nonylphenol polyethoxylate (NPEO) oligomer in aqueous solution and toxicity evaluation of the irradiated solution. Ecotoxicol Environ Saf, 2017, 140: 89-95.

[183] Kroger M, Meister K, Kava R. Low-calorie sweeteners and other sugar substitutes: a review of the safety issues. Compr Rev Food Sci F, 2006, 5(2): 35-47.

[184] Scheurer M, Storck F R, Brauch H J, et al. Performance of conventional multi-barrier drinking water treatment plants for the removal of four artificial sweeteners. Water Res, 2010, 44(12): 3573-3584.

[185] Soh L, Connors K A, Brooks B W, et al. Fate of sucralose through environmental and water treatment processes and impact on plant indicator species. Environ Sci Technol, 2011, 45(4): 1363-1369.

[186] Batchu S R, Quinete N, Panditi V R, et al. Online solid phase extraction liquid chromatography tandem mass spectrometry (SPE-LC-MS/MS)method for the determination of sucralose in reclaimed and drinking waters and its photo degradation in natural waters from South Florida. Chem. Cent. J., 2013, 7: 141-157.

[187] Sang Z, Jiang Y, Tsoi Y K, Leung K S Y. Evaluating the environmental impact of artificial sweeteners: a study of their distributions, photodegradation and toxicities. Water Res, 2014, 52: 260-274.

[188] Gan Z, Sun H, Wang R, et al. Transformation of acesulfame in water under natural sunlight: Joint effect of photolysis and biodegradation. Water Res, 2014b, 64: 113-122.

[189] Labare M, Alexander M. Microbial cometabolism of sucralose, a chlorinated disaccharide, in environmental samples. Appl Microbiol Biot, 1994, 42(1): 173-178.

[190] Soh L, Connors K A, Brooks B W, et al. Fate of sucralose through environmental and water treatment processes and impact on plant indicator species. Environ Sci Technol, 2011, 45(4): 1363-1369.

[191] Weiss S, Jakobs J, Reemtsma T. Discharge of three benzotriazole corrosion inhibitors with municipal wastewater and improvements by membrane bioreactor treatment and ozonation. Environ Sci Technol, 2006, 40(23): 7193-7199.

[192] Castro S, Davis L C, Erickson L E. Natural, cost-effective, and sustainable alternatives for treatment of aircraft deicing fluid waste. Environ Prog, 2005, 24(1): 26-33.

[193] Kim J W, Ramaswamy B R, Chang K H, et al. Multiresidue analytical method for the determination of antimicrobials, preservatives, benzotriazole UV stabilizers, flame retardants and plasticizers in fish using ultra high performance liquid chromatography coupled with tandem mass spectrometry. J Chromatogr A, 2011, 1218(22): 3511-3520.

[194] Liu Y S, Ying G G, Shareef A, et al. Photolysis of benzotriazole and formation of its polymerised photoproducts in aqueous solutions under UV irradiation. Environ Chem, 2011, 8(2): 174.

[195] Malhas R N, Al-Awadi N A, El-Dusouqui O M E. Kinetics and mechanism of gas-phase pyrolysis of N-aryl-3-oxobutanamide ketoanilides, their 2-arylhydrazono derivatives, and related compounds. Int J Chem Kinet, 2007, 39(2): 82-91.

[196] Lefevre G H, Lipsky A, Hyland K C, et al. Benzotriazole (BT) and BT plant metabolites in crops irrigated with recycled water. Environ Sci: Water Res Technol, 2017, 3(2): 213-223.

[197] Riemenschneider C, Al-Raggad M, Moeder M, et al. Pharmaceuticals, their metabolites, and other polar pollutants in field-grown vegetables irrigated with treated municipal waste water. J Agric Food Chem, 2016, 64(29): 5784-5792.

[198] 丁杰, 宋昭, 宋迪慧, 等. 三维电催化处理苯并噻唑反应器结构优化. 化工进展, 2017, 36(01): 91-99.

[199] Bao Q, Chen L, Tian J, et al. Degradation of 2-mercaptobenzothiazole in aqueous solution by gamma irradiation. Radiat Phys Chem, 2014, 103: 198-202.

[200] Qin H L, Gu G B, Liu S. The Photodegradation of 2-mercaptobenzothiazole in the Suspension of nitrogen-doped Titania under Visible Light. Ecol Environ, 2006, 15(4): 720-725.

[201] Zhang T, Chen Y, Leiknes T O. Oxidation of refractory benzothiazoles with PMS/$CuFe_2O_4$: Kinetics and transformation intermediates. Environ Science Technol, 2016, 50(11): 5864-5873.

[202] 刘春苗，丁杰，刘先树，等. 电辅助微生物反应器降解苯并噻唑效能的研究. 环境科学, 2014, 35(11): 4192-4197.

[203] Law R J, Covaci A, Harrad S, et al. Levels and trends of PBDEs and HBCDs in the global environment: status at the end of 2012. Environ Int., 2014 65: 147-158.

[204] Li Q L, Li J, Chaemfa C, et al. The impact of polybrominated diphenyl ether prohibition: A case study on the atmospheric levels in China, Japan and South Korea. Atmos Res, 2014, 143 (12): 57-63.

[205] Li D, Lin T, Shen K J, et al. Occurrence and Concentrations of Halogenated Flame Retardants in the Atmospheric Fine Particles in Chinese Cities. Environ Sci Technol, 2016, 50(18): 9846-9854.

[206] 张娴，高亚杰，颜昌宙. 多溴联苯醚在环境中迁移转化的研究进展. 生态环境学报, 2009, 18(2): 761-770.

[207] Stapleton H M, Dodder N G. Photodegradation of decabromodiphenyl ether in house dust by natural sunlight. Environ Toxicol Chem, 2008, 27(2): 306-312.

[208] Tokarz J A, Ahn M Y, Leng J, et al. Reductive debromination of polybrominated diphenyl ethers in anaerobic sediment and a biomimetic system. Environ Sci Technol, 2008, 42(4): 1157-1164.

[209] Crimmins B S, Pagano J J, Xia X, et al. Polybrominated diphenyl ethers (PBDEs): Turning the corner in Great Lakes trout 1980–2009. Environ Sci Technol, 2012, 46: 9890-9897.

[210] Chen D, Hale R C, Letcher R J. Photochemical and microbial transformation of emerging flame retardants: cause for concern?. Environ Toxicol Chem, 2015, 34(4): 687-699.

[211] van Mourik L M, Gaus C, Leonards P E G, et al. Chlorinated paraffins in the environment: A review on their production, fate, levels and trends between 2010 and 2015. Chemosphere, 2016, 155: 415-428.

[212] POPRC. Short-chained Chlorinated Paraffins: Risk Profile Document UNEP/POPS/POPRC.11/10/Add.2. United Nations Environmental Programme Stockholm Convention on Persistent Organic Pollutants, Geneva. 2015.

[213] USEPA. Short-Chain Chlorinated Paraffins (SCCPs) and Other Chlorinated Paraffins Action Plan. US Environmental Protection Agency, Washington. http: //www.epa.gov/sites/production/files/2015-09/documents/sccps_ap_2009_1230_final.pdf. 2009.

[214] Li Q, Li J, Wang Y, et al. Atmospheric short-chain chlorinated paraffins in China, Japan, and South Korea. Environ Sci Technol, 2012, 46: 11948-11954.

[215] IPCS. Chlorinated paraffin. Geneva: International Programme on Chemical Safety, World Health Organization. 1996: 181.

[216] Thompson R S, Noble H. Short-chain chlorinated parrdins (C_{10-13}, 65%chlorinated): Aerobic and anaerobic transformation in marine and freshwater sediment systems. Draft Report No BL8405/B. AstraZeneca: Brixham Environmental Laboratory, 2007.

[217] Tomy G T, Stern G A, Lockhart W L, et al. Occurrence of C_{10}-C_{13} polychlorinated n-alkanes in Canadian midlatitude and arctic lake sediments. Environ Sci Technol, 1999, 33: 2858-2863.

[218] Iozza S, Muller C E, Schmid P, et al. Historical profiles of chlorinated paraffins and polychlorinated biphenyls in a dated sediment core from Lake Thun (Switzerland). Environ Sci Technol, 2008, 42: 1045-1050.

[219] Greaves A K, Letcher R J. A review of organophosphate esters in the environment from biological effects to distribution and fate. Bull Environ Contam Toxicol., 2017, 98(1): 2-7.

[220] O'Brien J W, Thai P K, Brandsma S, et al. Wastewater analysis of Census day samples to investigate per capita input of organophosphorus flame retardants and plasticizers into wastewater. Chemosphere, 2015, 138: 328-334.

[221] Zeng X Y, He L X, Cao S X, et al. Occurrence and distribution of organophosphate flame retardants/plasticizers in sludges from the Pearl River Delta, South China. Environ Toxicol.Chem, 2014, 33(8): 1720-1725.

[222] Cristale J, Dantas R F, De Luca A, et al. Role of oxygen and DOM in sunlight induced photodegradation of organophosphorous flame retardants in river water. J Hazard Mater, 2016, 323: 242-249.

[223] Bollmann U E, Möller A, Xie Z Y, et al. Occurrence and fate of organophosphorus flame retardants and plasticizers in coastal and marine surface waters. Water Res, 2012, 46(2): 531-538.

[224] Sühring R, Diamond M L, Scheringer M, et al. Organophosphate esters in Canadian arctic air: Occurrence, levels and trends. Environ. Sci. Technol., 2016, 50(14): 7409-7415.

[225] Salamova A, Hermanson M H, Hites R A. Organophosphate and halogenated flame retardants in atmospheric particles from a European arctic site. Environ Sci Technol, 2014, 48(11): 6133-6140.

[226] Yu Q, Xie H B, Chen J W. Atmospheric chemical reactions of alternatives of polybrominated diphenyl ethers initiated by OH: A case study on triphenyl phosphate. Sci Total Environ, 2016, 571: 1105-1114.

[227] Liu J, Mejia Avendano S. Microbial degradation of polyfluoroalkyl chemicals in the environment: A review. Environ Int, 2013, 61: 98-114.

[228] Wang N, Liu J, Buck R C, et al. 6: 2 fluorotelomer sulfonate aerobic biotransformation in activated sludge of waste water treatment plants. Chemosphere, 2011, 82(6): 853-858.

[229] Zhao L, Folsom P W, Wolstenholme B W, et al. 6: 2 fluorotelomer alcohol biotransformation in an aerobic river sediment system. Chemosphere, 2013, 90(2): 203-209.

[230] Liu J, Wang N, Szostek B, et al. 6-2 Fluorotelomer alcohol aerobic biodegradation in soil and mixed bacterial culture. Chemosphere, 2010, 78(4): 437-444.

[231] Zhang S, Szostek B, McCausland P K, et al. 6: 2 and 8: 2 fluorotelomer alcohol anaerobic biotransformation in digester sludge from a WWTP under methanogenic conditions. Environ Sci Technol, 2013, 47(9): 4227-4235.

[232] Butt C M, Muir D C G, Mabury S A. Elucidating the pathways of poly- and perfluorinated acid formation in rainbow trout. Environ Sci Technol, 2010, 44(13): 4973-4980.

[233] Butt C M, Muir D C, Mabury S A. Biotransformation of the 8: 2 fluorotelomer acrylate in rainbow trout. 1. In vivo dietary exposure. Environ Toxicol Chem, 2010, 29(12): 2726-2735.

[234] Liu C, Liu J. Aerobic biotransformation of polyfluoroalkyl phosphate esters (PAPs) in soil. Environ Pollut, 2016, 212: 230-237.

[235] Benskin J P, Ikonomou M G, Gobas F A, et al. Biodegradation of *N*-ethyl perfluorooctane sulfonamido ethanol (EtFOSE)and EtFOSE-based phosphate diester (SAmPAP diester) in marine sediments. Environ Sci Technol, 2013, 47(3): 1381-1389.

[236] Washington J W, Jenkins T M. Abiotic hydrolysis of fluorotelomer-based polymers as a source of perfluorocarboxylates at the global scale. Environ Sci Technol, 2015, 49(24): 14129-14135.

[237] Dasu K, Lee L S. Aerobic biodegradation of toluene-2, 4-di(8: 2 fluorotelomer urethane)and hexamethylene-1, 6-di(8: 2 fluorotelomer urethane)monomers in soils. Chemosphere, 2016, 144: 2482-2488.

[238] 李法松, 何娜, 覃雪波, 等. 全氟化合物在天津大黄堡湿地多介质分布研究. 环境化学, 2011, 30(3): 638-644.

[239] Zhao S, Zhu L. Uptake and metabolism of 10: 2 fluorotelomer alcohol in soil-earthworm (*Eisenia fetida*)and soil-wheat (*Triticum aestivum* L.)systems. Environ Pollut, 2017, 220(Pt A): 124-131.

[240] Zhao S, Zhu L, Liu L, et al. Bioaccumulation of perfluoroalkyl carboxylates (PFCAs)and perfluoroalkane sulfonates

(PFSAs)by earthworms (*Eisenia fetida*)in soil. Environ Pollut, 2013, 179: 45-52.

[241] Lee H, De Silva A O, Mabury S A. Dietary bioaccumulation of perfluorophosphonates and perfluorophosphinates in juvenile rainbow trout: evidence of metabolism of perfluorophosphinates. Environ Sci Technol, 2012, 46(6): 3489-3497.

[242] Wang J, Zhang Y, Zhang F, et al. Age- and gender-related accumulation of perfluoroalkyl substances in captive Chinese alligators (*Alligator sinensis*). Environmental Pollution, 2013, 179: 61-67.

[243] Fang S, Zhao S, Zhang Y, et al. Distribution of perfluoroalkyl substances (PFASs)with isomer analysis among the tissues of aquatic organisms in Taihu Lake, China. Environ Pollut, 2014, 193: 224-232.

[244] Zhou Y, Wang T, Jiang Z, et al. Ecological effect and risk towards aquatic plants induced by perfluoroalkyl substances: Bridging natural to culturing flora. Chemosphere, 2017, 167: 98-106.

[245] 周萌. 不同碳链长度全氟化合物在水-土壤-植物间的迁移. 天津: 南开大学, 2013.

[246] Dai Z, Xia X, Guo J, et al. Bioaccumulation and uptake routes of perfluoroalkyl acids in Daphnia magna. Chemosphere, 2013, 90(5): 1589-1596.

[247] 苑晓佳. 全氟烷基化合物在微藻中的富集及其与生物大分子的相互作用. 天津: 南开大学, 2017.

[248] Wang Q, Chen M, Shan G, et al. Bioaccumulation and biomagnification of emerging bisphenol analogues in aquatic organisms from Taihu Lake, China. Sci Total Environ, 2017, 598: 814-820.

[249] Sun H, Hu H, WangL, et al. The bioconcentration and degradation of nonylphenol and nonylphenol polyethoxylates by *Chlorella vulgaris*. Int J Mol Sci, 2014, 15(1): 1255-1270.

[250] Roig B, Cadiere A, Bressieux S, et al. Environmental concentration of nonylphenol alters the development of urogenital and visceral organs in avian model. Environ Int, 2014, 62: 78-85.

[251] Zhang Q, Wang F, Xue C, et al. Comparative toxicity of nonylphenol, nonylphenol-4-ethoxylate and nonylphenol-10-ethoxylate to wheat seedlings (*Triticum aestivum* L.). Ecotoxicol Environ Saf, 2016, 131: 7-13.

[252] Aronzon C M, Svartz G V, Coll C S P. Synergy between diazinon and nonylphenol in toxicity during the early development of the Rhinella arenarum toad. Water Air Soil Poll, 2016, 227: 139.

[253] Tsukamoto H, Terade S. Metabolism of drugs. XLVII. Metabolic fate of p-hydroxybenzoic acid and its derivatives in rabbits. Chem Pham Bull, 1964, 12: 765-769.

[254] Jones P S, Thigpen D, Morrison J L, et al. *p*-Hydroxybenzoic acid esters as preservatives III. The physiological disposition of *p*-Hydroxybenzoic acid and its esters. J Am Pharm Assoc, 1956, 45(4): 268-273.

[255] Sabalitschka T, Neufeld-Crzellitzer R. The behavior of the *p*-hydroxybenzoic acid esters in the human body.. Arzneimittel-Forsch, 1954, 4(9): 575-579.

[256] Rebecca S L, Barnett A R, Bryan W B, et al. Exposure and food web transfer of pharmaceuticals in ospreys (*Pandion haliaetus*): Predictive model and empirical data. Integr Environ Asses, 2014, 11(1): 118-129.

[257] Lillicrap A, Langford K, Tollefsen K E. Bioconcentration of the intense sweetener sucralose in a multitrophic battery of aquatic organisms. Environ Toxicol Chem, 2011, 30: 673-681.

[258] Knut E T, Luca N, Duane B H. Presence, fate and effects of the intense sweetener sucralose in the aquatic environment. Sci Total Environ, 2012, 438: 510-516.

[259] 马铃. 人工甜味剂在养殖场的归趋及其生物富集作用研究. 天津: 南开大学, 2017.

[260] Renwick A G. The disposition of saccharin in animals and man-A review. Food Chem Toxicol, 1985, 23 (4-5): 429-435.

[261] Byard J L, McChesney E W, Golberg L, Coulston F. Excretion and metabolism of saccharin in man. II. Studies with ^{14}C-labelled and unlabelled saccharin. Food Chem Toxicol, 1974, 12(2): 175-184.

[262] Roberts A, Renwick A G, Sims J, Snodin D J. Sucralose metabolism and pharmacokinetics in man. Food Chem Toxicol, 2000, 38(Suppl. 2): S31-S41.

[263] Byard J L, McChesney E W, Golberg L, Coulston F. Excretion and metabolism of saccharin in man. II. Studies with ^{14}C-labelled and unlabelled saccharin. Food Chem Toxicol, 1974, 12(2): 175-184.

[264] Grice H C, Goldsmith L A. Sucralose-An overview of the toxicity data. Food Chem Toxicol, 2000, 38(Suppl. 2): S1-S6.

[265] 罗孝俊, 吴江平, 陈社军, 等.多溴联苯醚、六溴环十二烷和得克隆的生物差异性富集及其机理研究进展. 中国科学: 化学, 2013, 43(3): 291-304.

[266] Covaci A, Bervoets L, Hoff P, et al. Polybrominated diphenyl ethers (PBDEs)in freshwater mussels and fish from flanders, Belgium. J Environ Monit, 2005, 7: 132-136.

[267] Yu M, Luo X J, Wu JP, et al. Bioaccumulation and trophic transfer of polybrominated diphenyl ethers (PBDEs)in biota from the Pearl River Estuary, South China. Environ Int, 2009, 35: 1090-1095.

[268] Law R J, Alaee M, Allchin C R, et al. Levels and trends of polybrominated diphenylethers and other brominated flame retardants in wildlife. Environ Int, 2003, 29: 757-770.

[269] Jaspers V L B, Covaci A, Voorspoels S, et al. Brominated flame retardants and organochlorine pollutants in aquatic and terrestrial predatory birds of Belgium: Levels, patterns, tissue distribution and condition factors. Environ Pollut, 2006, 139: 340-352.

[270] Chen D, Hale R C. A global review of polybrominated diphenyl ether flame retardant contamination in birds. Environ Int, 2010, 36: 800-811.

[271] Luo X J, Zhang X L, Liu J, et al. Persistent halogenated compounds in waterbirds from an e-waste recycling region in South China. Environ Sci Technol, 2009, 43: 306-311.

[272] Gobas F A P C, de Wolf W, Burkhard L P, et al. Revisiting bioaccumulation criteria for POPs and PBT assessments. Integr Environ Assess Manag, 2009, 5: 624-637.

[273] Ueberschar, K H, Danicke S, Matthes S. Dose-response feeding study of short chain chlorinated paraffins (SCCPs)in laying hens: Effects on laying performance and tissue distribution, accumulation and elimination kinetics. Mol Nutr Food Res, 2007, 51: 248-254.

[274] Fisk A T, Tomy G T, Cymbalisty C D, et al. Dietary accumulation and quantitative structure-activity relationships for depuration and biotransformation of short (C10), medium (C_{14}), and long (C_{18})carbon-chain polychlorinated alkanes by juvenile rainbow trout (*Oncorhynchus mykiss*). Environ Toxicol Chem, 2000, 19, 1508-1516.

[275] Ma X, Chen C, Zhang H, et al. Congener-specific distribution and bioaccumulation of short-chain chlorinated paraffins in sediments and bivalves of the Bohai Sea, China. Mar Pollut Bull, 2014, 79: 299-304.

[276] Zeng L, Lam J C W, Wang Y, et al. Temporal trends and pattern changes of short- and medium-chain chlorinated paraffins in marine mammals from the South China sea over the past decade. Environ Sci Technol, 2015, 49: 11348-11355.

[277] Herzke D, Kaasa H, Gravem F, et al. Perfluorinated alkylated substances, brominated flame retardants and chlorinated paraffins in the Norwegian Environment - screening 2013. *In*: M400e2013, R (Ed.), Norwegian Environment Agency. NILU e Norsk Institutt for Luftforskning. SWECO, Tromsø, Norway. http: //www.miljodirektoratet.no/Documents/publikasjoner/ M-40/M40.pdf.

[278] Sundkvist A M, Olofsson U, Haglund P. Organophosphorus flame retardants and plasticizers in marine and fresh water biota and in human milk. J Environ Monitor, 2010, 12(4): 943-951.

[279] Kim J W, Isobe T, Chang K H, et al. Levels and distribution of organophosphorus flame retardants and plasticizers in fishes from Manila Bay, the Philippines. Environ Pollut, 2011, 159(12): 3653-3659.

[280] Brandsma S H, Leonards P E G, Leslie H A, et al. Tracing organophosphorus and brominated flame retardants and plasticizers in an estuarine food web. Sci Total Environ, 2015, 505: 22-31.

[281] Greaves A K, Letcher R J, Chen D, et al. Retrospective analysis of organophosphate flame retardants in herring gull eggs and relation to the aquatic food web in the Great Lakes. Environ Res, 2016, 150: 255-263.

[282] Heydebreck F, Tang J, Xie Z, et al. Emissions of per- and polyfluoroalkyl substances in a textile manufacturing plant in China and their relevance for workers' exposure. Environ Sci Technol, 2016, 50(19): 10386-10396.

[283] Yan H, Zhang C J, Zhou Q, et al. Short- and long-chain perfluorinated acids in sewage sludge from Shanghai, China. Chemosphere, 2012, 88(11): 1300-1305.

[284] Li F, Zhang C, Qu Y, et al. Quantitative characterization of short- and long-chain perfluorinated acids in solid matrices in Shanghai, China. Sci Total Environ, 2010, 408(3): 617-623.

[285] Bao J, Liu W, Liu L, et al. Perfluorinated compounds in the environment and the blood of residents living near fluorochemical plants in Fuxin, China. Environ Sci Technol, 2011, 45(19): 8075-8080.

[286] Wang P, Lu Y, Wang T, et al. Transport of short-chain perfluoroalkyl acids from concentrated fluoropolymer facilities to the Daling River estuary, China. Environ Sci Pollut Res, 2015, 22(13): 9626-9636.

[287] Liu B, Zhang H, Xie L, et al. Spatial distribution and partition of perfluoroalkyl acids (PFAAs)in rivers of the Pearl River Delta, southern China. Sci Total Environ, 2015, 524-525: 1-7.

[288] Lu Z, Song L, Zhao Z, et al. Occurrence and trends in concentrations of perfluoroalkyl substances (PFASs)in surface waters of eastern China. Chemosphere, 2015, 119: 820-827.

[289] Shao M, Ding G, Zhang J, et al. Occurrence and distribution of perfluoroalkyl substances (PFASs)in surface water and bottom water of the Shuangtaizi Estuary, China. Environ Pollut, 2016, 216: 675-681.

[290] Yao Y, Zhu H, Li B, et al. Distribution and primary source analysis of per- and poly-fluoroalkyl substances with different chain lengths in surface and groundwater in two cities, North China. Ecotoxicol Environ Saf, 2014, 108: 318-328.

[291] Wang T, Khim J S, Chen C, et al. Perfluorinated compounds in surface waters from Northern China: comparison to level of industrialization. Environ Int, 2012, 42: 37-46.

[292] Wang T, Vestergren R, Herzke D, et al. Levels, isomer profiles, and estimated riverine mass discharges of perfluoroalkyl acids and fluorinated alternatives at the mouths of Chinese Rivers. Environ Sci Technol, 2016, 50(21): 11584-11592.

[293] Chen S, Jiao X C, Gai N, et al. Perfluorinated compounds in soil, surface water, and groundwater from rural areas in eastern China. Environ Pollut, 2016, 211: 124-131.

[294] Shan G, Wei M, Zhu L, et al. Concentration profiles and spatial distribution of perfluoroalkyl substances in an industrial center with condensed fluorochemical facilities. Sci Total Environ, 2014, 490: 351-359.

[295] Liu Z, Lu Y, Shi Y, et al. Crop bioaccumulation and human exposure of perfluoroalkyl acids through multi-media transport from a mega fluorochemical industrial park, China. Environ Int, 2017, 106: 37-47.

[296] Bao J, Liu W, Liu L, et al. Perfluorinated compounds in urban river sediments from Guangzhou and Shanghai of China. Chemosphere, 2010, 80(2): 123-130.

[297] Qi Y, Huo S, Xi B, et al. Spatial distribution and source apportionment of PFASs in surface sediments from five lake regions, China. Sci Rep, 2016, 6: 22674.

[298] Wang P, Wang T, Giesy J P, et al. Perfluorinated compounds in soils from Liaodong Bay with concentrated fluorine industry parks in China. Chemosphere, 2013, 91(6): 751-757.

[299] Meng J, Wang T, Wang P, et al. Are levels of perfluoroalkyl substances in soil related to urbanization in rapidly developing coastal areas in North China?. Environ Pollut, 2015, 199: 102-109.

[300] Liu B, Zhang H, Yao D, et al. Perfluorinated compounds (PFCs)in the atmosphere of Shenzhen, China: Spatial distribution, sources and health risk assessment. Chemosphere, 2015, 138: 511-518.

[301] Zhao Z, Tang J, Mi L, et al. Perfluoroalkyl and polyfluoroalkyl substances in the lower atmosphere and surface waters of the Chinese Bohai Sea, Yellow Sea, and Yangtze River estuary. Sci Total Environ, 2017, 599-600: 114-123.

[302] Ruan T, Wang Y, Wang T, et al. Presence and partitioning behavior of polyfluorinated iodine alkanes in environmental

matrices around a fluorochemical manufacturing plant: Another possible source for perfluorinated carboxylic acids?. Environ Sci Technol, 2010, 44(15): 5755.

[303] Wang Q, Zhu L Y, Chen M, et al. Simultaneously determination of bisphenol A and its alternatives in sediment by ultrasound-assisted and solid phase extractions followed by derivatization using GC-MS. Chemosphere, 2017, 169: 709-715.

[304] Xu G, Ma S H, Tang L, et al. Occurrence, fate, and risk assessment of selected endocrine disrupting chemicals in wastewater treatment plants and receiving river of Shanghai, China. Environ Sci Pollut R, 2016, 23 (24): 25442-25450.

[305] Li J J, Wang G H. Airborne particulate endocrine disrupting compounds in China: Compositions, size distributions and seasonal variations of phthalate esters and bisphenol A. Atmos Res, 2015, 154: 138-145.

[306] Sun Q, Wang Y, Li Y, et al. Fate and mass balance of bisphenol analogues in wastewatertreatment plants in Xiamen City, China. Environ Pollut, 2017, 225: 542-549.

[307] Lin Z, Wang L, Jia Y, et al. A Study on Environmental Bisphenol A Pollution in Plastics, Industry Areas. Water Air Soil Pollut, 2017, 228: 98.

[308] Sun J, Wang J, Zhang R, et al. Comparison of different advanced treatment processes in removingendocrine disruption effects from municipal wastewater secondaryeffluent. Chemosphere, 2017, 168: 1-9.

[309] LiuY, Zhang S, Ji G, et al. Occurrence, distribution and risk assessment of suspectedendocrine-disrupting chemicals insurface water and suspendedparticulate matter of Yangtze River (Nanjingsection). Ecotox Environ Safe, 2017, 135: 90-97.

[310] Wang W, Ndungu A W, Wang J. Monitoring of endocrine-disrupting compounds in surface water and sediments of the Three Gorges Reservoir Region, China. Arch Environ Contam Toxicol, 2016, 71: 509-517.

[311] Wang W, Abualnaja K O, Asimakopoulos A G, et al. A comparative assessment of human exposure to tetrabromobisphenol Aand eight bisphenols including bisphenol A via indoor dust ingestion intwelve countries. Environ Int, 2015, 83: 183-191.

[312] Yu Y, Zhai H, Hou S, et al. Nonylphenol ethoxylates and their metabolites in sewage treatment plants and rivers of Tianjin, China. Chemosphere, 2009, 77(1): 1-7.

[313] Shao B, Hu J, Yang M. Nonylphenol ethoxylates and their biodegradation intermediates in water and sludge of a sewage treatment plant. B Environ Contam Tox, 2003, 70(3): 527-532.

[314] Shao B, Hu J Y, Yang M. Determination of nonylphenol ethoxylates in the aquatic environment by normal phase liquid chromatography-electrospray mass spectrometry. J Chromatogr A, 2002, 950(1-2): 167-174.

[315] Shao B, Hu J Y, Yang M, et al. Nonylphenol and nonylphenol ethoxylates in river water, drinking water, and fish tissues in the area of Chongqing, China. Arch Environ Con Tox, 2005, 48(4): 467-473.

[316] Wang L, Wu Y, Sun H, et al. Distribution and dissipation pathways of nonylphenol polyethoxylates in the Yellow River: Site investigation and lab-scale studies. Environ Int, 2006, 32(7): 907-914.

[317] Wu ZB, Zhang Z, Chen S P, et al. Nonylphenol and octylphenol in urban eutrophic lakes of the subtropical China. Fresen Environ Bull, 2007, 16: 227-234.

[318] Zhang Y Z, Tang C Y, Song X F, et al. Behavior and fate of alkylphenols in surface water of the Jialu River, Henan Province, China. Chemosphere 2009, 77: 559-565.

[319] Cai Q, Huang H, Lu H, et al. Occurrence of nonylphenol and nonylphenol monoethoxylate in soil and vegetables from vegetable farms in the Pearl River Delta, South China. Arch Environ Con Tox, 2012, 63(1): 22-28.

[320] He F, Niu L, Aya O, et al. Occurrence and fate of nonylphenol ethoxylates and their derivatives in Nansi Lake Environments, China. Water Environ Res, 2013, 85(1): 27-34.

[321] Jin F, Hu J, Yang M. Vertical distribution of nonylphenol and nonylphenol ethoxylates in sedimentary core from the Beipaiming Channel, North China. J Environ Sci, 2007, 19(3): 353-357.

[322] Yu Y, Xu J, Sun H, et al. Sediment-porewater partition of nonylphenol polyethoxylates: Field measurements from Lanzhou Reach of Yellow River, China. Arch Environ Con Tox, 2008, 55(2): 173-179.

[323] Li X, Ying G, Su H, et al. Simultaneous determination and assessment of 4-nonylphenol, bisphenol A and triclosan in tap water, bottled water and baby bottles. Environ Int, 2010, 36: 557-562.

[324] Li W, Shi Y, Gao L, et al. Occurrence, fate and risk assessment of parabens and their chlorinated derivatives in an advanced wastewater treatment plant. J Hazard Mater, 2015, 300: 29-38.

[325] Chen Z, Ying G, Lai H, et al. Determination of biocides in different environmental matrices by use of ultra-high-performance liquid chromatography–tandem mass spectrometry. Anal Bioanal Chem, 2012, 404(10): 3175-3188.

[326] Li W；Shi Y, Gao L, et al. Occurrence and human exposure of parabens and their chlorinated derivatives in swimming pools. Environ Sci Pollut R, 2015, 22(22): 17987-17997.

[327] Li W, Gao L, Shi Y, et al. Spatial distribution, temporal variation and risks of parabens and their chlorinated derivatives in urban surface water in Beijing, China. Sci Total Environ, 2016, 539: 262-270.

[328] Sun Q, Li Y, Li M, et al. PPCPs in Jiulong River estuary (China): Spatiotemporal distributions, fate, and their use as chemical markers of wastewater. Chemosphere, 2016, 150: 596-604.

[329] Wang L, Asimakopoulos A G, Moon H B, et al. Benzotriazole, benzothiazole, and benzophenone compounds in indoor dust from the United States and East Asian countries. Environ Sci Technol, 2013, 47(9): 4752-4759.

[330] Xu W, Yan W, Licha T. Simultaneous determination of trace benzotriazoles and benzothiazoles in water by large-volume injection/gas chromatography–mass spectrometry. J Chromatogr A, 2015, 1422: 270-276.

[331] Heeb F, Singer H, Pernet-Coudrier B, et al. Organic micropollutants in rivers downstream of the Megacity Beijing: Sources and mass fluxes in a large-scale wastewater irrigation system. Environ Sci Technol, 2012, 46: 8680-8688.

[332] Liang K, Liu J. Understanding the distribution, degradation and fate of organophosphate esters in an advanced municipal sewage treatment plant based on mass flow and mass balance analysis. Sci Total Environ, 2016, 544: 262-270.

作者：汪　磊[1]，于志强[2]，孙红文[1]，张　干[2]

[1] 南开大学环境科学与工程学院，[2] 中国科学院广州地球化学研究所

附表 5-1　典型新型有机污染物在我国环境介质中的检出水平

污染物分类	污水/污泥	地表水	土壤、沉积物	气相	其他
PFASs	出水：PFOA，30~145 ng/L，天津[21]；进水：PFOA，20~170 ng/L，天津[21]；出水：18.4~41.1 ng/L，沈阳[22]；进水：PFOA，26.2~71.1 ng/L，沈阳[22]；出水：∑PFASs，6690 ng/L（平均值），纺织厂，长江三角洲[282]；污泥：PFOA，12~68 g/kg；PFOS：42~169 g/kg，天津[21]；污泥：∑PFAAs，126~809 ng/g dw，上海[283]；污泥：∑PFAAs（C_2~C_{14}），413~755 ng/g dw，上海[284]	浑河：PFOA，2.68~9.13 ng/L，PFHxA，2.12~11.3 ng/L，沈阳[22]；某湖泊：PFOA 4.41~19.6 ng/L，沈阳[22]；海河：∑PFAAs，12~74 ng/L，大沽排污河：40~174 ng/L，天津[25]；辽河：∑PFAAs，1.4~131 ng/L（中位值：37.3 ng/L）[23]，太湖：∑PFAAs，17.8~448 ng/L（26.2 ng/L）[23]；细河：∑PFAAs，370~713 ng/L，地下水：∑PFAAs，6.03~1400 ng/L，阜新[285]；汤逊湖：PFBS，3660 ng/L（平均值），PFBA，4770（平均值），武汉[13]；小清河：∑PFAAs，55.7 ng/L~1.86 mg/L，地下水：∑PFAAs，1.68 ng/L~273 μg/L，山东[36]；大凌河：∑PFAAs，1.01~4.74 μg/L，辽宁[286]；珠江：∑PFAAs，1.53~33.5 ng/L，广东[287]；黄河：∑PFAAs，44.7~1520 ng/L[14]；某几条河流：∑PFAAs，39~212 ng/L，上海和昆山，某几条河流：∑PFAAs，0.68~146 ng/L，浙江[288]；双台子河：∑PFAAs，表层水 66.2~185 ng/L，底层水 44.8~209 ng/L，辽宁[289]；海河：PFOA，8.58~20.3 ng/L，天津，白浪河和虞河：PFOA，6.37~25.9 ng/L，潍坊[290]；官厅水库：PFOA，1.2 ng/L，PFOS，0.16 ng/L，呼和浩特某些河流：PFOA，1.2 ng/L，PFOS 0.32 ng/L，山西某些河流：PFOA，2.7 ng/L，PFOS，0.93 ng/L，天津某些河流：PFOA 6.8 ng/L，PFOS 2.6 ng/L，辽宁某些河流：PFOA，27 ng/L，PFOS，4.7 ng/L[291]；中国 19 条河流：Σ_{10}PFAAs，8.9~1240 ng/L（平均值：106 ng/L，中位值 16.3 ng/L）[292]；江苏，浙江，山东，天津等中国东部地区地表水：∑PFAAs，7.0~489 ng/L，地下水：∑PFAAs，5.3~615 ng/L[293]；常熟工业区某河流：∑PFCAs，24.4~80900 ng/L，∑PFSAs，2.29~29.3 ng/L，常熟[26]；常熟工业区某河流：∑PFCAs，73.8~12400（中位值：404）ng/L，常熟[294]；小清河：∑PFAAs，36.5~496000 ng/L，山东[17]	某河流沉积物：∑PFAAs（C_2~C14），62.5~276 ng/g dw，城市土壤：∑PFAAs（C_2~C_{14}），141~237 ng/g dw，上海[284]；海河沉积物：∑PFAAs，7.1~16 ng/g，大沽排污河沉积物：1.6~7.7 ng/g，天津[25]；辽河沉积物：∑PFAAs，0.26~1.11（0.43）ng/g，太湖沉积物：∑PFAAs，0.20~1.31（0.57）ng/g[23]；细河沉积物：∑PFAAs，0.59~90 ng/g dw，阜新[285]；汤逊湖沉积物：∑PFAAs，41.8~800 ng/g dw，武汉[13]；黄河沉积物：∑PFAAs，8.19~17.4 ng/g dw[14]；江苏，浙江，山东，天津等中国东部地区土壤：∑PFAAs，0.34~65.8 ng/g[293]；工业区土壤：∑PFCAs，1.89~32.6 ng/g dw，∑PFSAs，0.10~2.34 ng/g dw，常熟[26]；某河流沉积物：∑PFAAs，3.33~324 ng/g dw，工业区土壤：0.75~28.8 ng/g dw，常熟[294]；小清河沉积物：∑PFAAs，0.333~4100μg/kg，山东[17]；工业区土壤：∑PFAAs，1.86~641 ng/g dw，淄博[295]；珠江沉积物：∑PFAAs，0.09~3.6 ng/g dw，黄浦江沉积物：∑PFAAs，0.25~1.1 ng/g，广州和上海[296]；中国五大湖区沉积物：∑PFAAs，0.086~5.79（平均值：1.15）ng/g dw[297]；辽东半岛农田土壤：∑PFAAs，<MDL~3.14（中位值：0.315）ng/g[298]；辽宁，河北，天津，山东等地土壤：∑PFAAs，<MDL~13.97（平均值：0.98）ng/g[299]	长江三角洲纺某织厂内：∑*n*-PFASs，188~3260 ng/（sample·d）∑i-PFASs，4.1~ 80.6 ng/（sample·d）[282]；中国城市地区：∑FTOHs，51.4~936（中位值：183）pg/m^3，∑FOSEs，7.44~94.0pg/m^3[157]；天津：∑*n*-PFASs，93.6~131 pg/m^3[5]；深圳：∑*i*-PFASs，3.4~34 pg/m^3[300]；黄渤海海上大气：∑*n*-PFASs，76~551 pg/m^3[301]；工业区：全氟烷基碘（FIAs），1.41~ 3.08×10^4 ng/m^3，氟调碘（FTIs），1.39~1.32×10^3 ng/m^3，淄博[302]	常熟工业区树叶：∑PFAAs，10.0~276 ng/g dw，工业区树皮：6.76~120 ng/g dw[294]；淄博工业区附件农田小麦籽粒：∑PFAAs，0.7~58.8 ng/g dw，玉米籽粒：0.7~58.8 ng/g dw[295]

续表

污染物分类	污水/污泥	地表水	土壤、沉积物	气相	其他
BPs	进水：BPA，1.35E + 03 ng/L；出水：BPA，40.1 ng/L，上海[304]；进水：BPA，1318 ng/L；出水：BPA，177 ng/L，厦门[306]；进水：BPAF，0.282 ng/L；出水：BPAF，0.714 ng/L，厦门[306]；进水：BPE，3.70 ng/L；出水：3.64 ng/L，厦门[306]；进水：BPF，50.0 ng/L；出水：BPF，BLD，厦门[306]；进水：BPS，48.0 ng/L；出水：BPS，BLD，厦门[306]；出水：BPA，12.60 ng/L，郑州[308]	太湖：$\sum_9$BPs，49.7~3480 ng/L（mean，389 ng/L）[248]；东南地区河流：BPA，240~5680 ng/L [307]；长江南京段：BPA，1.7~563 ng/L [309]；三峡库区：BPA，26.6 ng/L（mean）[310]	太湖沉积物：BPA，3.94~33.2 ng/g dw[303]；太湖沉积物：BPF，0.503~3.28 ng/g dw [303]；太湖沉积物：BPS，0323~27.3 ng/g dw [303]；东南地区地表土：BPA，38.70~2960.86 ng/g [307]；三峡库区沉积物：BPA，17.4 ng/g dw（mean）[310]	中国城市地区：BPA，2.0~20 ng/m^3（冬季），1.0~10 ng/m^3（夏季）[305]	室内灰尘：$\sum_8$BPs，43~4400 ng/g（平均值，690 ng/g；中位值，350 ng/g）[311]
NPEOs	出水：NPEOs 和 NP，4.73~21.4 ng/L，天津[312]；NP，1.46 ng/L，北京[313]；NPEOs，15.19 ng/L，北京[313]；进水：NPEOs 和 NP，14.0~47.2 ng/L，天津[312]；NP，9.27 ng/L，北京[313]；NPEOs，49.31 ng/L 北京[313]；污泥：NP，19.5 μg/kg dw 北京[313]；NPEOs，732.9 μg/kg dw 北京[313]	黄河：NPEOs 和 NP2EC，1.02~5.33 μg/L[312]；海河：NPEOs 和 NPECs，0.47~9.3 μg/L[312]；长江：NPEOs，0.1~2900 ng/L[314]；长江，嘉陵江：NP，0.1~7.3 μg/L[315]；长江，嘉陵江：NPEOs，1.0~97.6 μg/L[315]；黄河：NP，0.05~0.17 μg/L[5]；黄河：NPEOs，3.03~6.77 nmol/L[316]；黄河：NPECs，0.04~0.19 nmol/L[316]；武汉城市湖泊：4-NP，1.94~32.85 μg/L[317]；贾鲁河：NP，0.075~1.520 μg/L[318]	天津海河沉积物：NP，NPEOs，NP1EC 和 NP2EC，4.1~9.9 μg/g dw[312]；海河沉积物（≤40 cm）：NP，579~3539 ng/g dw，NPEOs，736~12735 ng/g dw[321]；黄河沉积物：NP，61.3~113.9 ng/g dw，NP1EO，31.1~55.9 ng/g dw，NP2EO，47.9~74.1 ng/g dw[322]；黄河沉积物：NP，0.28~5.09 nmol/g dw，NP1EO，0.08~0.92 nmol/g dw，NP2EO，ND~0.21 nmol/g dw[316]； 武汉城市湖泊沉积物：NP，3.54~32.43 μg/g dw[317]；微山湖沉积物：NPEOs，60.7~631.5 μg/kg dw[320]； 珠江三角洲农田土：NP，ND~7.22 μg/kg dw，NP1EO，ND~8.24 μg/kg dw[319]		重庆饮用水：NP，ND~2.7 μg/L，NPEOs，ND~0.3 μg/L[315]；广州饮用水：NP，ND~1987 ng/L[323]
Parabens	进水：MeP 448 ng/L，EtP 5.43 ng/L，PrP 46.2 ng/L，BuP 5.71 ng/L，北京[324]；MeP 372 ng/L，EtP 46.4 ng/L，PrP 69.2 ng/L，BuP 5.0 ng/L，广州[325] 出水：MeP 7.5 ng/L，EtP 2.0 ng/L，PrP 0.8 ng/L，BuP 1.5 ng/L，广州[325] 污泥：MeP 423 μg/kg，dw，EtP 4.32 μg/kg，dw，PrP 20.8 μg/kg，dw，BuP 2.05 μg/kg，dw，北京[324]；MeP 30.7 ng/g，广州[325]	广州沙河：MeP 1062 ng/L，PrP 3142 ng/L[84] 广州石井河：MeP 213 ng/L，PrP 693 ng/L，广州 北京昆玉河：MeP 22.4 ng/L，EtP 7.68ng/L，PrP 19.0 ng/L，BuP 0.98 ng/L[327] 四川九龙河：MeP 3.4~69.9 ng/L（雨季）、2.65~29.1 ng/L（冬季），PrP 1.68~39.4 ng/L（雨季）、1.11~5.22 ng/L（冬季）[328]	山东德州污泥改良土壤：MeP 2.2 ng/g[325]		北京某游泳池：MeP 872 ng/L，EtP 110 ng/L，PrP 266 ng/L，BuP 49.2 ng/L[326]； 中国鱼类和海洋动物：MeP 1.45 ng/g，EtP 0.692 ng/g，PrP 0.377 ng/g，BuP 0.185 ng/g[60]

续表

污染物分类	污水/污泥	地表水	土壤、沉积物	气相	其他
人工甜味剂	出水：三氯蔗糖，1.08 μg/L、安赛蜜，11.20 μg/L、糖精，0.098μg/L、甜蜜素，0.032 μg/L，天津[97]；出水（养猪场）：糖精，5.64 μg/L、甜蜜素，0.923μg/L、安赛蜜，3.26 μg/L，天津[100]；进水：三氯蔗糖，1.25 μg/L、安赛蜜，12.64 μg/L、糖精，9.86μg/L、甜蜜素，31.67 μg/L，天津[97]	海河：安赛蜜，1.6~7.6 μg/L、三氯蔗糖，0.18~0.35 μg/L、糖精，0.21~1.1μg/L、甜蜜素，0.12~0.67μg/L [97]	我国农田：安赛蜜，0.08~569 μg/kg、甜蜜素，0~1280 μg/kg、糖精，0.11~34.7 μg/kg、三氯蔗糖，0.145~34.7 μg/kg [96]		香港海水：安赛蜜，0.22 μg/L、三氯蔗糖，0.05μg/L、糖精 0.11 μg/L、甜蜜素 0.10 μg/L[93] 天津雨水：安赛蜜，3.5~160 ng/L、糖精，12~1300 ng/L、甜蜜素，7.5~910 ng/L [97]
苯并杂环化合物	出水：BTR，1717 ng/L（中位数浓度），广州[330]；污灌废水：BTR，1100 ng/L（中位数浓度），北京[331]；出水：BTHs，209 ng/L，天津[136]；出水：BTHs，159700 ng/L，天津[133]	珠江：BTR，19.1 ng/L（median），广州[330] 天津翠屏湖：BTHs，327 ng/L，天津[136]；浑河沈阳段上游：BTHs，225 ng/L，浑河沈阳段下游：BTHs，349.5 ng/L[134]；辽河，BTH：5695 ng/L[135]		我国城市室内灰尘：BTRs，ND~1940 ng/g [329] 我国城市室内灰尘：BTHs，119~6540 ng/g（中位数浓度 857 ng/g），中国[329]	自来水：BTR，ND~227 ng/L，全国 51 个城市[136]；中国城市自来水：BTHs，40.1~1310 ng/L（中位数浓度 406 ng/L）[136]
OPEs	北京污水处理厂污泥∑OPEs 233~137 μg/kg [158] 北京污水处理厂进水∑OPEs 1399.0 ng/L，出水∑OPEs 832.9 ng/L [332]	北京：地表水∑OPEs 浓度平均值为 954 ng/L[147] 渤海入海口：5~921 ng/L[154] 珠江三角洲河口：干季节∑OPEs 2040~3120 ng/L 湿季节∑OPEs 1080~2500 ng/L [155] 厦门地区海岸线附近海水中∑OPEs 424.6（91.87~1392）ng/L [156]	太湖的沉积物中∑OPEs 3.38~14.3 ng/g[159] 珠江三角洲的沉积物中∑OPEs 58~322 ng/g[160] 电子垃圾拆解区域沉积物∑OPEs（112（48–470）ng/g），广州地区沉积物∑OPEs（132（62–222）ng/g）[148] 在莱州湾黄河河口沉积物核心∑OPEs 11.8~102 ng/g[149] 河北新乐和定州的塑料废物处理场所的土壤∑OPEs 1200 ng/g [161] 上海城市土壤∑OPEs 0.6~16.7 ng/g [20] 成都市表层土壤∑OPEs 99.9 ng/g [163] 广州市区的土壤∑OPEs 250 ng/g [150]	（降尘及颗粒物）我国室内灰尘∑OPEs 34 μg/g [151]；北京办公室∑OPEs 1.68~200 μg/g；家庭∑OPEs 4.57~67.4 μg/g[152]；电子垃圾回收作坊灰尘∑OPEs 33.6（3.30–70.0）μg/g；农村室内灰尘∑OPEs 9.03（2.26~20.7）μg/g；城市室内灰尘 10.9（4.45~27.5）μg/g；宿舍室内灰尘 11.50（4.08~25.7）μg/g [143]广东贵屿电子垃圾处理作坊∑OPEs 33.1（4.66–350）μg/g [153]	我国自来水中∑OPEs 192（123~338）ng/L [157] 全国的饮用水∑OPEs 85.1~325 ng/L [146] 珠江三角洲鱼体内 TBOEP 1647~8840 ng/g lw[164] 北京地表水鱼类∑OPEs 264.7~1973 ng/g lw [165] 天津市场各类食品中∑OPEs 0.004 ~287 ng/g[166]

注：ND 表示未检出

第6章　全氟和多氟烷基化合物（PFASs）替代品的环境行为与环境毒理学研究进展

本章导读

传统全氟和多氟烷基化合物（PFASs）环境健康潜在风险使得全氟辛基磺酸和全氟辛基羧酸等主要化合物被禁用或限制使用。鉴于 PFASs 独特物理性质及其在现代生产和生活中的不可或缺性，越来越多 PFASs 替代品投入应用。该类替代品较传统 PFASs 在分子结构和种类上更趋多样，但对其环境行为和生物毒性的研究相对薄弱，尚缺乏充足的数据对替代品的环境和生态安全进行评估。然而，当前相对有限的数据表明部分替代品具有较高的累积性和多种生物毒性，用以替代传统 PFASs 的合理性和优越性值得商榷。本章将对替代品在环境介质中的分布、转化、生物体内的富集以及毒理学效应与机制等方面的研究进行综合介绍。

关键词

PFASs 替代品，环境行为，毒性效应机制

1 引　言

全氟和多氟烷基化合物（per-and polyfluoroalkyl substances，PFASs）是一类高度氟化的脂肪族物质，是碳骨架上的氢原子全部或部分被氟原子替代的人工合成有机化合物[1]。PFASs 具有耐热，耐酸碱、耐氧化等特性；由于极化率低，其折射率、介电常数和表面张力都小；另外，该类化合物具有分子间力弱，沸点和熔点低，摩擦力小，黏度低，疏水、疏油等独特的理化性质，在工业生产和民用生活中具有不可或缺性，广泛地应用于服装、烹饪、住宅、汽车、通信、航天、医药和农药等领域。PFASs 分子结构中含高键能 C—F 键，难以水解、光解和微生物降解，能在环境中持久存在，其生物半衰期长，部分长链化合物具有高的生物累积性和潜在的生物毒性。研究表明，PFASs 已在全球各种环境介质和生物体中广泛检出，甚至人体中也有检出[2-6]。其中，全氟辛基羧酸（PFOA）和全氟辛基磺酸（PFOS）是用途最广、最受关注的两种全氟烷基化合物。流行病学研究发现 PFASs 的人体暴露与部分生化和生理、病理指标的改变存在显著正相关，包括肾癌和睾丸癌、高胆固醇和甲状腺机能减退等[7-9]。毒理学研究发现，PFASs 对实验动物具有肝脏毒性、神经毒性、生殖发育毒性、免疫毒性以及内分泌干扰效应等[10-12]。

鉴于 PFASs 具有持久性、累积性、长距离迁移、较强的毒性和潜在的生态环境和人体健康风险，针对 PFOA 和 PFOS 等全氟和多氟烷基化合物进行有效的环境管理非常重要。2000 年，全球最大的 PFASs 生产商 3M 公司开始主动减少并停用 PFOS 及相关前体物质；2006 年，由美国环境保护署（EPA）发起，全球 9 家生产商参与的“2010 年/2015 年 PFOA 管理计划”（*PFOA 2010/2015 Stewardship Program*），承诺到 2010 年减少 95% PFOA 的排放和使用，2015 年完全停止使用 PFOA[13]。2009 年，联合国环境规划署通过《关于持久性有机污染物的斯德哥尔摩公约》（以下简称《斯德哥尔摩公约》）正式将 PFOS 及其盐类列为新的持久性有机污染物（POPs），同意减少并最终禁止使用该类物质[14]。2016 年，持久性有机污染物审查委员会通过了 PFOA 及其盐类和相关化合物的附件 E 审查（风险报告），将进入下一阶段的评估[15]。因《斯德哥尔摩公约》要求减少并最终淘汰部分长链 PFASs 及前体物质，国内外加速展开了对 PFASs 替代品的研究[16,17]。

2 PFASs 替代品的种类与应用

因传统全氟类化合物如 PFOA 和 PFOS 的使用限制日渐严格，为满足生产需要，生产商研发出多种新型全氟替代物。迄今为止，国内外氟化工生产商已经在消防、电镀、氟聚物加工、织物整理和食物包装等行业使用了替代品。出于企业知识产权或技术保护等原因，国内外企业多数都不公布其使用替代品的具体化学结构，因而，能够获取的替代品的结构信息非常有限。大体上讲，目前 PFASs 的替代策略包括如下几种：以多氟替代全氟，降低氟原子比例[18]；以短链替代长链[19]，通过插入 C、H 和 O 等杂原子/基团、引入其他功能基团等方法来减少替代品含氟碳链“有效长度”（连续全氟链段长度），同时保留与原化合物类似的工业性能，其中，在碳链骨架中插入醚氧键最为常见。目前已知的替代品有 50 多种，主要可分为两类：①使用 C_4，C_6 结构的短链全氟烷基羧酸或磺酸盐，如 PFBA（C_4）、PFHxA（C_6）和 PFBS（C_4）等；②含功能官能团的全氟聚醚（perfluoropolyethers，PFPEs），尤其是全/多氟聚醚羧酸（PFECAs）和全/多氟聚醚磺酸（PFESAs）替代品[16]。总体而言，PFASs 替代品相对于传统的 PFASs 的分子结构更为复杂。PFASs 及其替代品的详细类型，可参考 Wang 的综述文献[20]。几种典型替代品的结构见图 6-1。

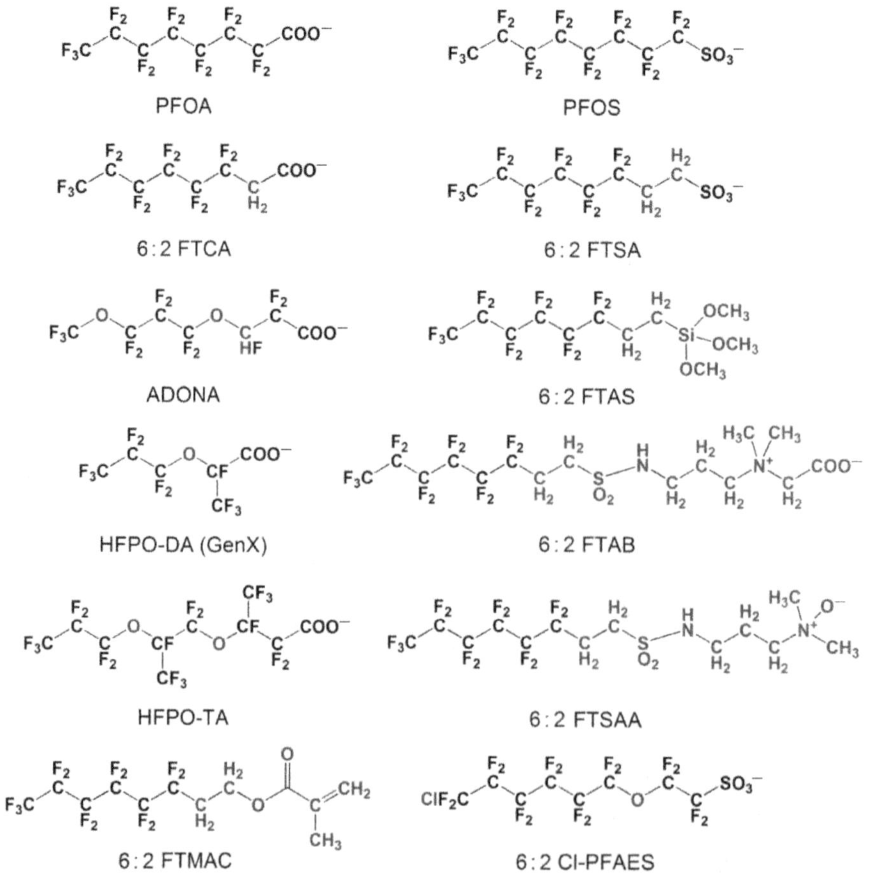

PFOA　　PFOS

6:2 FTCA　　6:2 FTSA

ADONA　　6:2 FTAS

HFPO-DA (GenX)　　6:2 FTAB

HFPO-TA　　6:2 FTSAA

6:2 FTMAC　　6:2 Cl-PFAES

图 6-1 几种典型 PFASs 替代品的结构式（不同于 PFOA 和 PFOS 的原子用红体标记）

PFOA 等全氟辛基(在环境中降解产生 PFOA)产品主要用于含氟聚合物的合成助剂和织物整理剂中。目前，PFOA 的主要替代品 PFECAs 作为加工助剂广泛用于氟聚树脂的制备，如科慕公司（原杜邦公司）

生成的六氟环氧丙烷（HFPO）的二聚体羧酸（HFPO-DA，CAS No. 62037-80-3，商品名 GenX）作为氟聚物加工助剂应用，年产量达 100 t[21]。除 GenX 以外，HFPO 的其他低聚物，如 HFPO 的三聚体羧酸 HFPO-TA，可替代 PFOA 用作生产隋性液体的全氟聚醚（PFPE），或用于聚偏氟乙烯（PVDF）树脂合成的共聚改性单体[22]。3M 公司使用 ADONA（CAS No. 958445-44-8）替代 PFOA，用作制备聚四氟乙烯（PTFE）、全氟乙烯丙烯（FEP）和 PVDF 的重要原料[23]。我国有部分企业使用氟调聚羧酸 6∶2 FTCA（6∶2 fluorotelomer carboxylic acid）替代 PFOA 作为加工助剂[24]。

在电镀行业和泡沫灭火剂的生产过程中，氟调聚磺酸 6∶2 FTSA（6∶2 fluorotelomer sulfonate）及其前体物质（甜菜碱型 6∶2 FTAB 和氧化胺型 6∶2 FTSAA）作为 PFOS 的替代物广泛使用，如杜邦公司泡沫灭火剂 Forafac® 1157 产品（主要成分 6∶2 FTAB），Forafac® 1183 产品（主要成分 6∶2 FTSAA），科慕公司 Capstone™ 1157（主要成分 6∶2 FTAB），安美特公司开发的铬雾抑制剂 Fumetrol 21 等（主要成分 6∶2 FTSA）[25]。在电镀行业，F-53（主要成分 6∶2 PFESA，CAS No. 754925-54-7），F-53B（主要成分 6∶2 Cl-PFESA，CAS No. 73606-19-6）等聚醚磺酸，也替代 PFOS 用作铬雾抑制剂[26]。F-53B 是我国自主设计、合成的工业产品，主要作为抑雾剂应用于电镀行业中[1]，曾获“国家技术发明奖”。F-53B 已经在我国使用了 40 多年，据称其铬雾抑制效果优于 PFOS。据统计，2009 年 F-53B 产量大约 20~30 t，其销售保持逐年稳步增长的态势[2]。随着 PFOS 及其盐类被列为新的持久性有机污染物，部分企业和环境部门管理者在讨论是否可以更广泛地用 F-53B 替代 PFOS，并推测 F-53B 在未来可能会有更大的市场份额[23]。总之，大量替代品已经正在投入生产和生活应用。替代品结构更具多样性，其环境和生态安全性是当前该领域研究的一个薄弱环节：全氟替代品的持久性、在环境介质中的分布特征、对人类及野生动物的暴露及生物效应等科学问题的研究资料较少，替代品是否环境危害和生物毒性更小等疑问仍未解决。另外，前期工作显示目前测定的部分野生动物和人体血清中传统 PFASs 只占可提取有机氟的 30%，总氟的 10%[27,28]。体内依旧有相当高比例尚未确定性质的有机氟存在，它们中有哪些是 PFASs 或替代品的新类型，其来源，对动物和人类健康效应和潜在危害等问题都不清楚。下述将重点介绍当前对替代品的环境分布和生物效应的部分研究进展。

3　PFASs 替代品的环境行为和在生物体及人体中的分布

3.1　替代品在环境介质中的分布

铬雾抑制剂 F-53B 的主要成分是 6∶2 Cl-PFESA（另外，还有少量 8∶2 Cl-PFESA）。2013 年，Wang 等首次报道在环境介质中检出了 6∶2 Cl-PFESA。最新的研究发现，多种环境介质包括自然水体、污泥、野生动物血清，甚至北极海洋哺乳动物体内都检测到 6∶2 Cl-PFESA [29-33]。在中国温州一处接收电镀工业废水的污水处理厂，其进水、出水中 6∶2 Cl-PFESA 浓度高达 43~78 μg/L 和 65~112 μg/L，污水处理并未有效去除 6∶2 Cl-PFESA。位于污水处理厂下游的瓯江水体 6∶2 Cl-PFESA 浓度约 10~50 ng/L，与 PFOS 浓度相当[26]。最近，在水体等多种环境介质中检测到 6∶2 FTSA [33-35]。Ruan 等在中国 56 个污水处理厂活性污泥中检测全氟磺酸（PFSAs）、Cl-PFESAs 和 FTSAs。其中 PFOS 依然为最主要的 PFAS，几何均值浓度为 3.19 ng/g dw；6∶2 Cl-PFESA 在所有污泥中全部检出，均值浓度为 2.15 ng/g dw。8∶2 Cl-PFESA 浓度约为 0.50 ng/g dw，检出率 89%。6∶2 Cl-PFESA/8∶2 Cl-PFESA 在污泥中比值较原料中提高，这可能是因 8∶2 较 6∶2 疏水性强，增加了在有机污泥中的吸附性[33]。另一研究报道了在中国主要 19 条入海河流河口处 PFASs 的浓度，其中 6∶2 Cl-PFESA 浓度范围是< 0.56~78.5 ng/L，检出率超过 51%。根据径流量推算，6∶2 Cl-PFESA 年入海通量均值 1.7 t/a（置信区间 0.2~6.9 t/a）[30]。

目前仅有两篇研究报道了替代品 HFPO-DA 在自然水体中的浓度。Heydebreck 等在德国莱茵河、中国小清河等水体中检测到HFPO-DA，其中小清河上游由于存在PTFE工厂，其HFPO-DA峰值浓度达3825 ng/L，相比德国污染点源附近水体浓度高6000倍[23]。Pan 等在我国某自然水体及野生鱼类血清中检出 HFPO-DA 外，首次检出 HFPO-TA，且 HFPO-TA 在污染点源（氟化学产业园）附近浓度高达数万 ng/L，经下游稀释后浓度均值仍达数千 ng/L，野生鲤鱼血清中平均浓度也很高（未发表数据）。Sun 等在美国菲尔河氟化学工厂附近的自来水厂原水中，检测到 HFPO-DA 浓度高达 631 ng/L，远高于美国环境保护署规定的 PFAS 在饮用水中的健康标准（70 ng/L）。原水中另有三种已知结构的 PFECAs，其在 LC-MS/MS 上响应峰面积远超过 HFPO-DA，但由于缺少高纯度的标准品，未能准确定量。凝聚、臭氧催化、生物过滤、消毒等水处理过程对去除 PFECAs 几乎没有效果，粉状活性炭（PAC）对 PFASs 有吸附效果，但 PFECAs 的吸附效率较相同链长的 PFCAs 的效率低[36]。

3.2 替代品在微生物体内的转化降解研究

鉴于长链 PFASs 在环境中极难降解[37]，替代品是否也具有生物累积性，是否能在环境或生物体内降解或代谢，其代谢速率及代谢产物如何等是关键的科学问题。Vanhamme 等发现一种命名为 NB4-1Y 的弧菌菌株可降解 6∶2 FTSA，其体内降解途径可大致描述为：6∶2 FTSA 首先降解为 6∶2 FTCA，之后经 6∶2 不饱和氟调聚酸（6∶2 FTUCA）和 5∶3 不饱和氟调聚酸（5∶3 Uacid），最终转变为 5∶3 氟调聚酸（5∶3 acid）[38]。Wang 等分别收集了位于美国宾夕法尼亚州、马里兰州和特拉华州三个污水处理厂的污泥，通过向其中添加一定量 6∶2 FTSA 研究污泥中微生物对 6∶2 FTSA 的生物转化效率[39]，发现 6∶2 FTSA 的生物转化非常缓慢，经过 90 天实验，仍有 63.7%的 6∶2 FTSA 未被降解，所有可检测到的转化产物的总量仅占起始加入 6∶2 FTSA 量的 6.3%，其中 5∶3 acid 仅占 0.12%，全氟丁酸（PFBA，C_4）、全氟戊酸（PFPeA，C_5）和全氟己酸（PFHxA，C_6）分别占 0.14%、1.5%和 1.1%。因此，6∶2 FTSA 在污泥中主要的生物转化途径可描述为：6∶2 FTSA 首先转化为 6∶2 FTUCA，再经 5∶2 氟调聚醇（5∶2 FTOH）转化为 PFPeA 和 PFHxA；此外，6∶2 FTSA 转化为 6∶2 FTUCA 后经 5∶3 Uacid 转化为 5∶3 acid，这一途径作为上述途径的补充，也可转化少量 6∶2 FTSA。另外，研究表明，动物体内或存在氟调聚物磷酸盐（diPAPs）、氟调聚醇（FTOHs）及氟调聚丙烯酸（FTACs）等前体物质的降解途径。6∶2 FTCA 作为 6∶2 diPAP、6∶2 FTOH 及 6∶2 FTAC 生物转化的中间产物出现在该转化途径中，并将最终转化为全氟庚酸（PFHpA）、PFHxA 或 PFPeA [40,41]。

3.3 替代品的生物累积和生物放大效应

Shi 等首次在山东小清河和湖北汤逊湖中的野生鲫鱼体内检测到 6∶2 Cl-PFESA，其全血浓度分别为 41.9 ng/g 和 20.9 ng/g。结合水体浓度估算 6∶2 Cl-PFESA 的生物放大因子（BAF）大于 5000，显著高于 PFOS，符合《斯德哥尔摩公约》中有关持久性有机污染物的规定[32]。另一研究在北极格陵兰岛附近的野生海豹、北极熊和虎鲸肝脏中检测到痕量的 6∶2 Cl-PFESA。鉴于中国是目前已知唯一的 6∶2 Cl-PFESA 排放源，其在极地地区的检出表明 6∶2 Cl-PFESA 具有大尺度、长距离迁移的能力[31]。Liu 等最新研究表明，6∶2 Cl-PFESA 在渤海多种海洋生物中广泛存在，浓度范围<0.016~0.575 ng/g，且浓度和检出率自 2010~2014 年间显著增加。类似于 PFOS 和长链 PFCAs，6∶2 Cl-PFESA 浓度与营养级呈显著正相关，表明该类物质可能具有生物放大作用[42]。针对 HFPO-TA 的鱼类组织分布和生物累积的研究发现，该物质较

PFOA 具有更高的生物累积因子（未发表数据）。

3.4　替代品的人群暴露水平

PFOS 替代品 Cl-PFESAs 不仅广泛分布在环境介质与野生动物中，在人群血清中亦有广泛检出。2016 年，Shi 等在中国电镀工人（职业暴露）、渔民（食鱼频率高）和普通居民血清中检测到 6∶2 和 8∶2 Cl-PFESAs，浓度范围为 0.019~5040 ng/mL，检出率达 98%[43]。其中电镀工人与渔民中 Cl-PFESAs 血清中位数浓度分别为 93.7 ng/mL 和 51.5 ng/mL，显著高于普通居民（4.78 ng/mL），占已知 PFASs 比重的 0.269%~93.3%。通过检测血清与尿液中 Cl-PFESAs 的浓度，估算 6∶2 Cl-PFESA 的肾脏清除半衰期为 280 a（7.1~4230 a），总清除半衰期为 15.3 a（10.1~56.4 a），清除速率低于 PFOS，是目前已知最具生物持久性的 PFASs。肾脏清除半衰期接近总半衰期的 20 倍，暗示其他代谢途径（例如胆汁排泄）在 Cl-PFESAs 清除中可能具有重要作用。

Pan 等检测母婴配对队列中处于孕前期、中期、晚期的母亲血清和新生儿脐带血清，发现 6∶2 Cl-PFESA 和 8∶2 Cl-PFESA 在母亲和胎儿血清中广泛检出，检出率均大于 99%[44]。6∶2 Cl-PFESA 在母亲的孕前、中和晚期浓度均值分别为 2.30 ng/mL、1.99 ng/mL 和 1.97 ng/mL，浓度随孕期增加而衰减，其趋势与传统 PFASs 相同，这可能与孕期母亲体重逐渐增加，体液量增多造成污染物稀释有关。新生儿脐带血中浓度均值 0.80 ng/mL，显著低于母亲血清。在所有检测的 PFASs 中，PFOS 是最主要成分，约占 53%~61%；PFOA 次之，占 12%~18%；6∶2 Cl-PFESA 几何均值分别为 1.97~2.30 ng/mL（母亲血清）和 0.80 ng/mL（脐带血清），浓度仅次于 PFOS 和 PFOA，排第三位，占已知 PFASs 总量的 10%左右。上述结果表明，6∶2 Cl-PFESA 与传统 PFASs 类似，可以突破血胎屏障，进行母子传递。Pan 等进而通过计算 PFASs 在脐带血与母亲血中浓度比，衡量了 PFASs 突破血胎屏障转移至胎盘的效率，发现羧酸和磺酸类 PFASs 的胎盘转移率随着碳链长度的增长先降低后升高，均呈现“U”型曲线（图 6-2）。在羧酸化合物中，碳链较短的 PFHpA（C_7）、碳链较长的 PFTriDA（C_{13}）和 PFTeDA（C_{14}）的胎盘迁移效率高于 100%，表明血胎屏障对此三种物质从母体到胎儿的传递无明显阻碍作用。在磺酸化合物中，6∶2 Cl-PFESA 的全氟碳分子数与 PFOS 相同（C_8），但 6∶2 Cl-PFESA 胎盘迁移率（几何均值 41%）显著高于 PFOS 的迁移率（几何均值 34%），向胎儿体内转移的风险较高，提示 6∶2 Cl-PFESA 中独特的分子结构（醚氧键、氯原子）可能增强了该物质胎盘运输的效率。Chen 等在母亲孕期血清、胎儿脐带血及胎盘中也检测到 6∶2

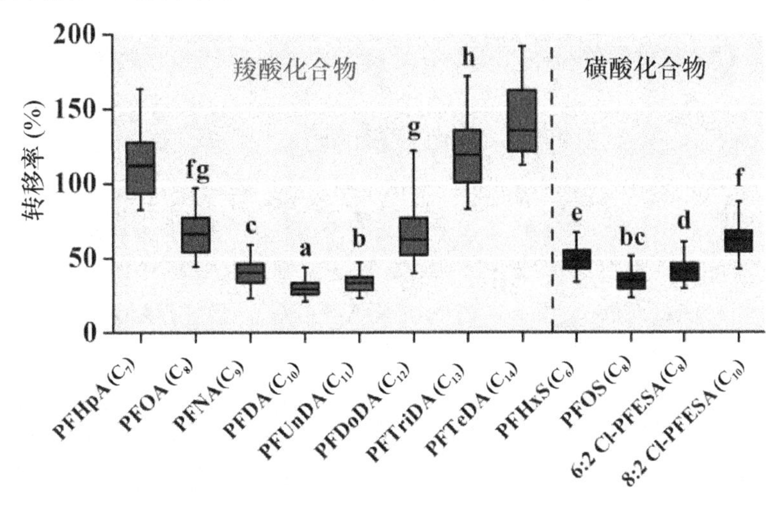

图 6-2　PFASs 的胎盘转移率随碳链长度增长呈“U”型曲线（引自 Pan 等[44]）

字母不同表示组间比较存在显著性差异

Cl-PFESA 和 8∶2 Cl-PFESA，其浓度 PFOS > 6∶2 Cl-PFESA > 8∶2 Cl-PFESA，分布特征：母血浓度>脐带血>胎盘，其浓度高低与分布特征和 Pan 的结果类似[45]。PFOS、6∶2 Cl-PFESA 和 8∶2 Cl-PFESA 在母体血、脐带血和胎盘中的浓度呈现显著相关性（$r > 0.7$；$p < 0.001$）。6∶2 Cl-PFESA 的胎盘迁移率中位数均值为 41%，8∶2 Cl-PFESA 更高，中位数为 56%，其结果与 Pan 的报道类似。鉴于 Cl-PFESAs 在普通人群，尤其是在孕妇和新生儿这些敏感人群血清中的检出及其较高的胎盘迁移率，应重视此类替代品的健康风险和生殖、发育毒性。

人体血液中的 PFASs 很少以自由形式存在，主要通过与血液中蛋白结合而在机体内迁移运输。人血清白蛋白（HSA）作为 PFAS 在血液中的主要结合蛋白[46]，可能对 PFASs 自母体至胎儿的运输中起重要作用。PFAS 随链长增长，突破血胎屏障的效率呈现“U”型曲线，可能与不同链长的 PFASs 与血中蛋白的结合能力强弱有关，与蛋白结合越强，则越易留存在血液中，难于转移至其他组织。Zhang 等报道链长 C_7~C_{14} 的 PFCAs 与脂肪酸结合蛋白（FABP）的结合能力呈现倒“U”型曲线，链较长和较短的 PFASs 与 FABP 结合能力弱[47]；Beesoon 等研究表明，直链 PFOA 和 PFOS 与白蛋白结合能力较含有支链的同分异构体强[46]，同时，流行病学研究表明直链 PFOA 和 PFOS 胎盘转移效率较其支链低，清除半衰期更长[48]。Pan 等检测母血和新生儿脐带血中白蛋白的含量，发现 PFASs 的胎盘迁移率与母亲 HSA 呈显著负相关，与脐带血中白蛋白显著正相关，暗示血清中白蛋白可能是 PFASs 的胎盘转运中的关键因子[44]：胎盘两侧血清中白蛋白与 PFASs 竞争结合，母体白蛋白越多，与 PFASs 结合越紧密，越不易于胎盘转移，胎儿脐带血中白蛋白越多，则胎盘转移效率高。

4 PFASs 替代品的毒性效应与机制研究

值得警惕的是用结构类似的化合物来替代已证明有毒害的化合物，并不意味着解决了原化合物的环境危害和生物毒性等问题。例如短链氯化石蜡替代多氯联苯（PCBs），后来发现短链氯化石蜡也不理想，目前正在接受《斯德哥尔摩公约》审查和评估，不久或将被禁止使用。近些年的研究表明，PFASs 替代品可能存在多种生物毒性效应：有的替代品生物毒性较被替代化合物并未改变，例如，6∶2 FTCA 对斑马鱼胚胎受精 120 h（hpf，hours post fertilization）的半致死浓度（LC_{50}）为 7.33 mg/L，同 PFOS 相当，比 PFOA 毒性强[49]；更有甚者，有的替代品某方面的毒性反而高于其被替代物，如下文将述的 F-53B 的急性肝脏毒性。因此，对全氟化学物替代品的潜在环境和生态风险应给予高度关注。

4.1 替代品的细胞毒性及对低等生物的毒性

Rand 等比较了 diPAP 和 FTOH 等降解途径中产生的多种化合物对 THLE-2 的细胞毒性（LC_{50}），其中 6∶2 FTCA 作为 6∶2 diPAP 最终降解成 PFHxA 等短链全氟化合物的中间产物，其细胞毒性明显强于 PFHxA 和 PFPeA。另外，通过比较 6∶2 FTCA 与 PFOA 的 EC_{50} 值，发现 6∶2 FTCA 细胞毒性较 PFOA 弱[50]。戴家银研究员实验室的一项围绕全氟化合物及其替代品对 HL7702 细胞系毒性的比较研究发现，6∶2 FTCA 与 PFOA 对 HL7702 细胞系的毒性结果与 Rand 类似，即 PFOA 细胞毒性显著强于 6∶2 FTCA；而在比较 6∶2 FTSA 与 PFOS 的细胞毒性时，发现 6∶2 FTSA 的细胞毒性较 PFOS 并未明显减弱（未发表数据）。Philips 等对 6∶2 FTCA 及其同系物对水生系统中大型水蚤、摇蚊及浮萍生长的影响进行了观察和比较，发现 6∶2 FTCA 对大型水蚤的生长无明显影响，但 6∶2 FTCA 摇蚊和及浮萍具有较为明显的抑制效应：对摇蚊生长的 EC_{50} 值为 167 μmol/L，对浮萍的 EC_{50} 值为 4 μmol/L[51]，其毒性较相同链长的 PFOA 低 1~2 个数量级[51,52]。该研究表明，初级水生物种如浮萍等对链长大于 C_8 的全氟化合物或相关氟聚物，包括 6∶2 FTCA 和 8∶2 FTCA 等敏感，或可作为潜在的生物指示物。

4.2　替代品对斑马鱼的胚胎发育毒性

目前已在污水处理厂、污泥、地表水及生物体中广泛检测到 F-53B[26,32,33,53]。随着全球禁止使用 PFOS，F-53B 未来可能会有更大的市场份额。2013 年，Wang 等研究发现 F-53B 对斑马鱼 96 h 的 EC_{50} 与 PFOS 类似[26]。Shi 对 F-53B 的斑马鱼胚胎发育毒性开展了更细致的研究。发现 48 hpf 之前胚胎在绒毛膜的保护下吸收速率很低。自 48 hpf 起，胚胎的绒毛膜开始脱去，各器官开始发育，胚胎在发育后期的吸收速率迅速增加。胚胎暴露至 96 hpf 后，转移至清水中培养 24h，检测了 24h 后的斑马鱼胚胎体内的 F-53B 含量，发现停止暴露 24h 后斑马鱼胚胎内 F-53B 并未被有效清除，与 PFOS 比较，两者在斑马鱼体内的吸收曲线和在胚胎体内难被清除的特征相似[54]。另外 Shi 等对小清河鲫鱼体内的 F-53B 组织分布特征及全身负荷量进行了研究，并与 PFOS 进行对比，发现 F-53B 在鲫鱼体内的生物富集能力较 PFOS 强[32]，F-53B 易吸收、难消除的特征，对鱼类健康构成潜在威胁。

将 6 hpf 斑马鱼胚胎暴露于不同浓度的 F-53B 溶液（0，1.5 mg/L，3 mg/L，6 mg/L 和 12 mg/L）中，48 hpf 和 60 hpf 的孵化率随着暴露浓度的增加而降低，但孵化出的斑马鱼幼体各项形态学指标未发现异常。在 PFOS 和 PFOA 的暴露实验中，也曾观察到斑马鱼孵化率延迟的现象[8]。F-53B 暴露浓度与胚胎存活率存在明显的剂量-效应和时间-效应关系。尤其值得注意的是，在 108 hpf 和 132 hpf 之间，12 mg/L F-53B 暴露组斑马鱼胚胎在短时间内全部死亡，这种短时间内胚胎大量死亡的现象也与 PFOS 的暴露实验结果类似[55]。F-53B 在 96 hpf 时对斑马鱼胚胎的 LC_{50} 是 13.77 mg/L ± 3.50 mg/L，与 PFOS 相当[56]。该结果与 Wang 的研究结论类似[26]，可以相互印证。根据世界卫生组织（WHO）对化合物致畸毒性的分类标准，F-53B 和 PFOS 都属中等毒性化合物，第三类化学品（对水生生物有危害性）[57]。

随着 F-53B 暴露时间的延长和在胚胎内积累量增加，斑马鱼的畸形率升高。在 84 hpf，12 mg/L 剂量组有胚胎出现心包水肿和卵黄囊水肿，随着暴露时间延长，心包水肿和卵黄囊水肿的胚胎所占比例上升，并出现脊椎弯曲、尾部弯曲、游囊关闭等其他畸形（图 6-3），这些畸形在 PFOS、PFOA 和 PFNA 等 PFASs 的斑马鱼胚胎急性毒性实验中普遍存在[54-56,58]。其中，心包水肿这一心脏畸形的主要表现是 F-53B 暴露后最敏感的终点指标之一，表明心脏可能是 F-53B 致畸作用中的优先靶器官。除引起心脏畸形外，F-53B 对心脏的损伤还体现在心率降低等电生理功能改变上。利用邻联茴香胺染色检测 72 hpf 斑马鱼胚胎的红

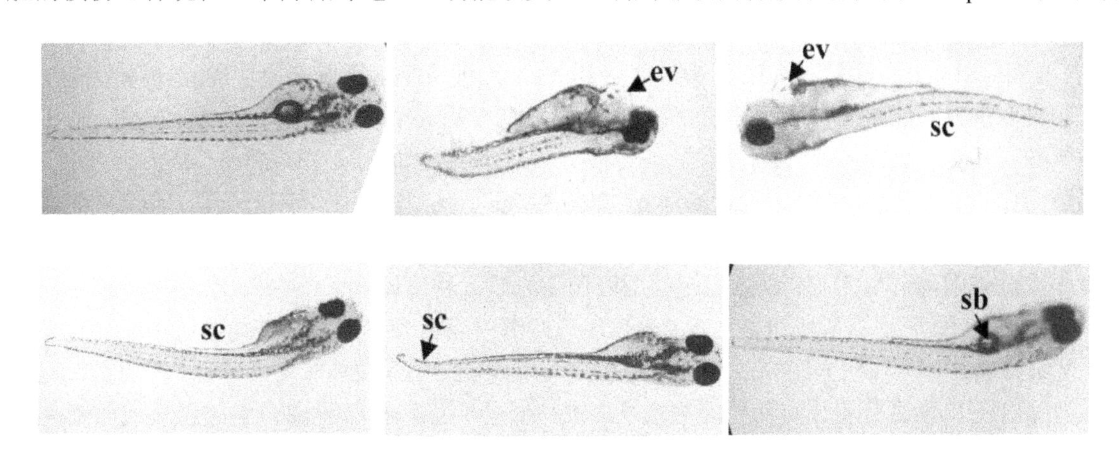

图 6-3　F-53B 暴露所致斑马鱼胚胎畸形

ev：心包水肿；sc：脊椎弯曲；sb：未加压的鱼鳔（引自 Shi 等[54]）

细胞，发现在斑马鱼胚胎还未表现出任何形态畸形时（72 hpf），F-53B 暴露的胚胎红细胞数量已呈现剂量依赖性下降。在脊椎动物心血管发育过程中，经典 Wnt/β-catenin 信号通路具有重要功能，下游 *Nkx2.5* 和 *Sox9b* 等目标基因缺陷能导致斑马鱼胚胎心脏发育畸形，Shi 发现 F-53B 暴露抑制 wnt3a 和 ctnnb2（β-catenin）及其 *Nkx2.5* 和 *Sox9b* 等基因的转录，提示 F-53B 暴露的心脏毒性可能与 Wnt/β-catenin 信号通路有关[49,59-65]。

Shi 还研究了 PFOA 的替代品 6∶2 FTCA 对斑马鱼的发育毒性研究[66]。将 6 hpf 的斑马鱼胚胎暴露于 6∶2 FTCA（0，4 mg/L，8 mg/L 和 12 mg/L）溶液中，发现 6∶2 FTCA 暴露导致胚胎死亡率显著提高，孵化率下降，主要引起心包囊水肿；进而利用转基因斑马鱼（gata1：DsRed）和邻联茴香胺染色实验分析 6∶2 FTCA 暴露对 72 hpf 的斑马鱼胚胎红细胞的影响，发现 Tg（gata1：DsRed）胚胎的荧光强度随着 6∶2 FTCA 暴露浓度的升高而降低，邻联茴香胺染色实验表明暴露组斑马鱼红细胞数量下降。整体原位杂交分析显示 48 hpf 和 72 hpf 时斑马鱼胚胎血红蛋白基因 *hbae1*，*hbae3* 和 *hbbe1* 等的表达量下降。另外，荧光定量 PCR 分析了其他参与红细胞形成相关基因的表达，结果显示，6∶2 FTCA 处理后控制斑马鱼血红素合成的基因 *alas2* 表达下调，而控制血红素降解的基因 *ho-1* 上调，控制红细胞分化的 gata1 和 gata2 表达量均下降，提示 6∶2 FTCA 暴露可能抑制血红蛋白和血红素的合成，促进血红素的降解，并且影响红细胞的分化。

4.3　替代品的肝脏毒性

肝脏毒性是全氟化合物具有代表性的毒性效应之一，PFOA 和 PFOS 暴露后，啮齿类动物及非人类灵长类动物肝脏出现肝肿大/局部坏死和脂质累积等病变，同时血清肝损伤相关指标及脂质含量变化明显，长期慢性暴露 PFOA 或 PFOS 的大鼠甚至会出现肝细胞腺瘤[67-70]。相比之下，替代品对哺乳动物的毒性数据非常缺乏。研究发现 HFPO-DA 不仅对皮肤、眼睛造成刺激，低剂量浓度连续暴露[≤10 mg/(kg·d)]还可能引起肝癌。在一项对两种 HFPO（二聚体 HFPO-DA 和四聚体 HFPO-TeA）的比较研究中，发现相同剂量 HFPO-DA 和 HFPO-TeA[1 mg/(kg·d)]暴露雄性小鼠 28d 后，动物肝肿大程度显著超过相同剂量的 PFOS 和 PFOA，尤其是 HFPO-TeA 暴露组[71]。在啮齿类动物中，PPARα 通路是 PFOA 等 PFASs 干扰脂代谢发挥其肝脏毒性的关键路径[37,72,73]。毒理基因组研究表明 HFPO-DA 和 HFPO-TeA 与经典 PFASs 肝脏毒性机制类似，能激活 PPARα 信号通路，且四聚体毒性强于二聚体[71]。

Sheng 等研究了 6∶2 FTCA 和 6∶2 FTSA 这两种结构类似的 PFAS 替代品对小鼠的肝脏毒性，发现相对 PFOA/PFOS，6∶2 FTCA 和 6∶2 FTSA 的毒性有一定程度减弱，尤其是 6∶2 FTCA 仅表现出轻度肝脏毒性[74]。5 mg/(kg·d)剂量 6∶2 FTCA 暴露小鼠 28 d 后，处理组小鼠肝脏未出现肿大，而同等剂量 PFOA 暴露小鼠相对肝重较对照组增加到 179%；同时，6∶2 FTCA 暴露小鼠肝脏也未见明显病变，血生化指标包括天冬氨酸氨基转移酶（AST）\丙氨酸氨基转移酶（ALT）\甘油三酯（TG）和总胆固醇（TCHO）等均未有显著变化。相同剂量相同条件下，暴露 6∶2 FTSA 的小鼠则出现了早期肝损伤，包括肝脏肿大（相对肝重增加 22%），肝脏局部轻微坏死，血清 AST 水平显著增加等。但 6∶2 FTSA 暴露组小鼠肝病理损伤的病变程度，较同等剂量 PFOS 暴露引起的肝脏病变有较大程度减弱[74]。

PFOA 等暴露后，肝脏脂质代谢紊乱，主要表现为血清中脂质含量显著降低而肝脏中脂质含量显著增加，肝脏组织病理切片出现明显脂滴。 Sheng 等研究发现 6∶2 FTCA 及 6∶2 FTSA 暴露后小鼠血清及肝脏中脂质含量并未出现显著变化，而相同剂量下，PFOA 能明显干扰脂代谢，引起脂质代谢紊乱[74]。转录组测序（RNA-seq）结果表明，6∶2 FTSA 暴露引起肝脏 412 个基因显著变化，6∶2 FTCA 暴露后，小鼠肝脏仅 39 个基因发生改变。另外，6∶2 FTCA 及 6∶2 FTSA 暴露对 PPARα通路激活效应较弱，其下游基因如 Cyp4a10，Acox1 和 Cpt1a 等未出现显著上调。相较于 PPARα 基因无明显差异的表达，6∶2 FTSA

显著上调小鼠肝脏 PPARγ mRNA，同时肝脏及血清中细胞因子 TNFα，IL-1β，IL-6 及 IL-10 均有不同程度的升高，免疫相关蛋白如 IκBα，NFκB/p65 和 NRF-2 等也显著上调，表明尽管 6：2 FTSA 肝脏毒性比传统 PFOA 及 PFOS 有一定程度的减弱，但仍能引起肝脏炎症，激活小鼠的免疫应答系统[74]。

Zhou 等将成年雄性小鼠暴露 6：2 Cl-PFESA 28 d，研究了其肝脏毒性效应，观察指标包括肝脏形态和病理学改变，血清中肝脏损伤生化指标如谷丙转氨酶（ALT）和谷草转氨酶（AST）等的变化，进而用 RNA-seq 分析了 F-53B 对肝脏转录组的影响，并与相同剂量 PFOS 暴露小鼠 28 d 的肝脏毒性进行了比较，结果显示 F-53B 的急性肝脏毒性较 PFOS 更强。在此基础上，Zhou 进一步对小鼠暴露 F-53B 56 d，研究了 F-53B 对小鼠肝脏的亚慢性毒性。研究结果提示，虽然 F-53B 与传统 PFASs 类似，都能引起肝脏损伤和代谢改变，但介导其肝脏毒性的 PPARs 亚型可能存在差别（未发表数据）。

4.4 替代品的生殖毒性

流行病学资料表明，PFOS 暴露引起的雄性生殖力的下降可能与精子数量减少有关[75,76]。动物实验也报道 PFOS 暴露可导致雄鼠睾丸组织结构紊乱、精母细胞和睾丸间质细胞空洞和生殖细胞数量减少[77]。很多研究还发现 PFASs 可作为激素受体激动剂，导致睾酮水平减少[78,79]。Zhou 对 F-53B 暴露 28 d 的雄性小鼠开展了生殖毒性效应研究，考察了包括睾丸病理、性激素水平、暴露雄性对雌鼠的致孕率等指标，发现 F-53B 的生殖毒性较弱（未发表数据）。另外，本实验室的研究还发现 6：2 FTCA 和 6：2 FTSA 对小鼠的生殖毒性也较弱。但因与过往对传统 PFASs 暴露所用剂量和时间存在差别，F-53B 等替代品的生殖毒性强弱，需要更多的比较实验才能下结论。

4.5 替代品与蛋白质相互作用

前期研究发现，实验动物经 PFOA 或 PFOS 暴露后，在血清及肝脏、肾脏和生殖腺等主要脏器中均有检出，即 PFOA 及 PFOS 可通过血液循环抵达并累积在上述器官中[80-84]。尽管该类化合物的理化特性决定了其类似的组织分布特征，但不同 PFASs 的组织分布比例也存在差别。例如，在 6：2 FTSA 对 CD1 小鼠持续 28 d 的暴露实验中，Sheng 等发现 6：2 FTSA 较 PFOA 和 PFOS 等更易在肝脏中分布[74]。

全氟化合物因结构与脂肪酸类似，可以与多种脂肪酸运输或结合蛋白（受体）发生相互作用，包括 HSA，FABP 以及多个重要核受体等[37,47,85-87]。PFOA 及 PFOS 以中等强度亲和力与人肝脏型脂肪酸结合蛋白（hL-FABP）结合，进而被转运进入细胞，与 PPAR 等受体结合，活化相应受体并发挥生物效应。Sheng 通过表达 hL-FABP 及其突变体蛋白，研究了 hL-FABP 与 PFASs 间的相互作用（图 6-4），发现随着直链 PFASs 碳链长度的增长，其结合 hL-FABP 的能力也随之增强。因替代品仍具有类似脂肪酸的结构：有一个强疏水性的全氟碳链尾部和亲水性的羧酸或磺酸头部，这种结构有利于同蛋白质发生亲水和疏水性作用。Sheng 进而分析比较了包括 6：2 FTCA，6：2 FTSA，HFPO 多聚物（HFPO-DA，HFPO-TA 和 HFPO-TeA）及 F-53B 等多种替代品与 hL-FABP 作用力大小及结合模式，发现这些替代品与 hL-FABP 的结合力强弱不一，有的与 hL-FABP 蛋白的结合能力微弱，有的结合能力强于 PFOA（未发表数据）。其实不限于 hL-FABP，体外研究发现 PFASs 能同人或哺乳动物体内多种功能蛋白发生相互作用，例如 PFASs 能同甲状腺素（T4）竞争结合人甲状腺激素转运蛋白（TTR）[88]。这些蛋白结合研究对认识 PFASs 的组织分布、毒性作用和机理提供了重要的信息。

总之，多氟链段及不同功能基团使得替代品的毒性更趋多样化、复杂化。有的替代品可能生殖毒性减弱，但肝脏毒性并未减弱，甚至强于传统 PFASs；引入带有官能团的多氟链段后，替代品可能会同蛋

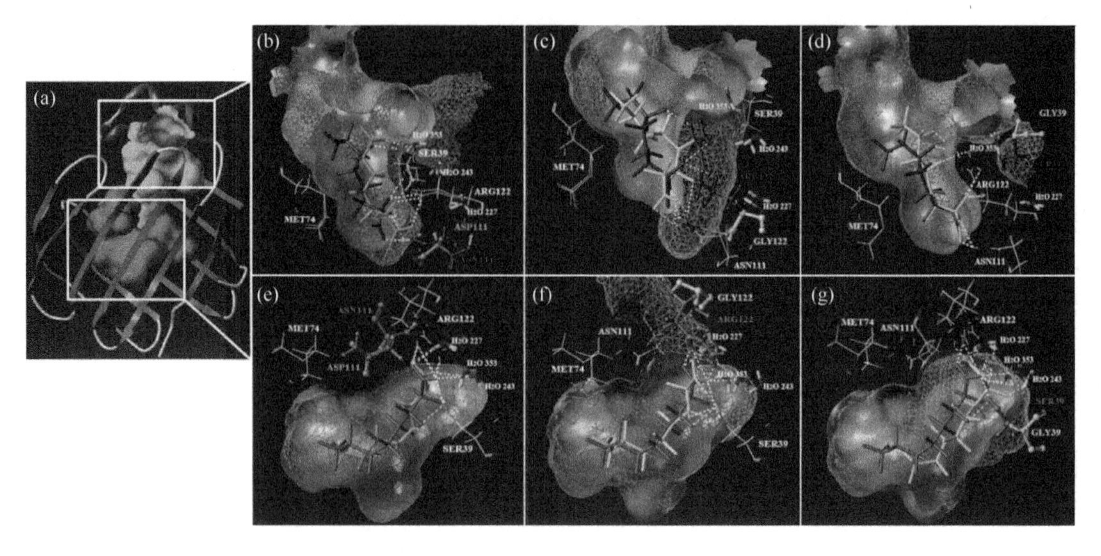

图 6-4 hL-FABP 及其突变体蛋白结合 PFOA 的活性口袋比较（引自 Sheng 等[85]）

白质发生共价结合，或致使其他类型的弱相互作用力发生改变等，其发挥毒性效应的分子机制也不同于传统 PFASs。因此，关于替代品的生物毒性和环境危害评估需要更多的实验数据。

5 PFASs 替代品的研究展望——实现绿色替代

化合物被另一个结构类似的化合物替代，这种结构类似物的替代并不能真正解决环境关切的问题，相反会出现“lock-in（锁定）”的问题。因此，在替代品的研发过程中，除了充分考虑替代品的功能外，还需要充分阐明替代品结构与环境行为和毒性之间的关系，从源头上探讨含氟替代品对环境和人类健康可能的不利影响，根据系统性危害评估的结果，指导合成化学，以精准为目标，使合成的替代品具有功能优异、低生物蓄积性、低毒性特性。在替代品投入市场之前，实现预防和控制因生产和产品使用带来的环境污染。

通常，PFOS 和 PFOA 的合成多采用电解氟化法，该方法产率低，产品为混合物，对设备要求高。未来的合成化学必须是经济的、安全的、环境友好的以及节省资源能源的化学，因此，需要有新的合成方法学。自由基反应在 20 世纪 80 年代被用于天然产物的合成，目前已经蓬勃发展为一种官能团容忍性好，反应条件温和，可用反应溶剂广，底物实用性强、高效的合成方法。由于氟原子强的电负性，含氟化合物容易发生单电子转移反应，这就促进了氟烷基自由基化学的发展。因此，自由基方法的采用能够系统地构建含氟表面活性剂的分子砌块，从而促进我国全氟烷基化合物替代品的研究。在发展系列的全氟烷基化合物替代品时，不仅要找到环境友好型的产品，并且要在其合成过程中，实现最佳可行工艺（BAT），限制生产过程中化学物质向自然环境中的排放，减少其对环境的危害。优秀替代品的合成，需要环境化学、生态毒理和有机氟化学科研工作者密切互动，通过环境化学和生态毒理学的交叉融合，寻求绿色、环保、原子经济和合理利用能源的全氟烷基化合物替代品的设计和合成方法，摒弃对环境危害的过度结构修饰。例如，太阳光能作为一种洁净和无限的能源为自然界广泛地利用，随着有机化学家对可见光的利用逐渐深入，发展了许多现代的、节能的、用其他方法不易实现的化学反应[89-91]。因此，基于可见光促进的有机化学反应将对工业化过程中减少环境资源的损耗产生影响，从而造福于人类。

在替代品研发和危害评估领域，如下方面依旧需要努力：①通过合成标准品，建立定性和定量分析方法，研究新型 PFASs 替代品在我国典型环境介质和人群血清中分布特征与累积规律，解析其来源和暴露途径；②剖析文献报道、但缺乏详细结构的替代品和已鉴定到的替代品作为母体化合物，发展氟烷基自由基方法学，开展替代品的绿色合成；③基于系统性研究阐明替代品结构与毒性之间的规律，确定典型 PFASs 替代品的敏感（关键）毒性效应及关键毒性通路；④研究介导 PFASs 毒性效应的分子机制及调控途径，基于毒性研究成果，指导设计并优化 PFASs 替代品的结构，合成功能优异、低生物蓄积性、低毒性的替代品，最终破解“化学品污染→替代→污染”循环往复的怪圈，从源头上改变持久性污染物被动应对的局面。

参 考 文 献

[1] Kannan K. Perfluoroalkyl and polyfluoroalkyl substances: Current and future perspectives. Environ Chem, 2011, 8(4): 333-338.

[2] Olsen G W, Burris J M, Ehresman D J, Froehlich J W, Seacat A M, Butenhoff J L, Zobel L R. Half-life of serum elimination of perfluorooctanesulfonate, perfluorohexanesulfonate, and perfluorooctanoate in retired fluorochemical production workers. Environ Health Persp, 2007, 115(9): 1298-1305.

[3] Moody C A, Martin J W, Kwan W C, Muir D C G, Mabury S C. Monitoring perfluorinated surfactants in biota and surface water samples following an accidental release of fire-fighting foam into Etobicoke Creek. Environ Sci Technol, 2002, 36(4): 545-551.

[4] Yao Y M, Sun H W, Gan Z W, Hu H W, Zhao Y Y, Chang S, Zhou Q X. Nationwide Distribution of Per- and Polyfluoroalkyl Substances in Outdoor Dust in Mainland China From Eastern to Western Areas. Environ Sci Technol, 2016, 50(7): 3676-3685.

[5] Starling A P, Engel S M, Whitworth K W, Richardson D B, Stuebe A M, Daniels J L, Haug L S, Eggesbo M, Becher G, Sabaredzovic A, Thomsen C, Wilson R E, Travlos G S, Hoppin J A, Baird D D, Longnecker M P. Perfluoroalkyl substances and lipid concentrations in plasma during pregnancy among women in the Norwegian Mother and Child Cohort Study. Environment International, 2014, 62: 104-112.

[6] Kato K, Wong L Y, Jia L T, Kuklenyik Z, Calafat A M. Trends in exposure to polyfluoroalkyl chemicals in the US population: 1999-2008. Environ Sci Technol, 2011, 45(19): 8037-8045.

[7] Barry V, Winquist A, Steenland K. Perfluorooctanoic acid (PFOA) exposures and incident cancers among adults living near a chemical plant. Environ Health Persp, 2013, 121(11-12): 1313-1318.

[8] Winquist A, Steenland K. Modeled PFOA Exposure and Coronary Artery Disease, Hypertension, and High Cholesterol in Community and Worker Cohorts. Environ Health Persp, 2014, 122(12): 1299-1305.

[9] Winquist A, Steenland K. Perfluorooctanoic Acid Exposure and Thyroid Disease in Community and Worker Cohorts. Epidemiology, 2014, 25(2): 255-264.

[10] Wolf C J, Fenton S E, Schmid J E, Calafat A M, Kuklenyik Z, Bryant X A, Thibodeaux J, Das K P, White S S, Lau C S, Abbott B D. Developmental toxicity of perfluorooctanoic acid in the CD-1 mouse after cross-foster and restricted gestational exposures. Toxicological Sciences, 2007, 95(2): 462-473.

[11] Mariussen E. Neurotoxic effects of perfluoroalkylated compounds: mechanisms of action and environmental relevance. Archives of Toxicology, 2012, 86(9): 1349-1367.

[12] White S S, Fenton S E, Hines E P. Endocrine disrupting properties of perfluorooctanoic acid. J Steroid Biochem, 2011, 127(1-2): 16-26.

[13] Fernandez Freire P, Perez Martin J M, Herrero O, Peropadre A, de la Pena E, Hazen M J. *In vitro* assessment of the cytotoxic and mutagenic potential of perfluorooctanoic acid. Toxicology *in Vitro*, 2008, 22(5): 1228-1233.

[14] Sanderson H, Boudreau T M, Mabury S A, Solomon K R. Impact of perfluorooctanoic acid on the structure of the zooplankton community in indoor microcosms. Aquat Toxicol, 2003, 62(3): 227-234.

[15] Harada K H, Yang H R, Moon C S, Hung N N, Hitomi T, Inoue K, Niisoe T, Watanabe T, Kamiyama S, Takenaka K, Kim M Y, Watanabe K, Takasuga T, Koizumi A. Levels of perfluorooctane sulfonate and perfluorooctanoic acid in female serum samples from Japan in 2008, Korea in 1994-2008 and Vietnam in 2007-2008. Chemosphere, 2010, 79(3): 314-319.

[16] Wang Z Y, Cousins I T, Scheringer M, Hungerbuehler K. Hazard assessment of fluorinated alternatives to long-chain perfluoroalkyl acids (PFAAs) and their precursors: Status quo, ongoing challenges and possible solutions. Environment International, 2015, 75: 172-179.

[17] Wang Z Y, Cousins I T, Scheringer M, Hungerbuhler K. Fluorinated alternatives to long-chain perfluoroalkyl carboxylic acids (PFCAs), perfluoroalkane sulfonic acids (PFSAs) and their potential precursors. Environment International, 2013, 60: 242-248.

[18] Biegel L B, Hurtt M E, Frame S R, O'Connor J C, Cook J C. Mechanisms of extrahepatic tumor induction by peroxisome proliferators in male CD rats. Toxicological Sciences, 2001, 60(1): 44-55.

[19] Ritter S K. Fluorochemicals go short. Chem Eng News, 2010, 88(5): 12-17.

[20] Wang Z, DeWitt J C, Higgins C P, Cousins I T. A never-ending story of per- and polyfluoroalkyl substances (PFASs)? Environ Sci Technol, 2017, 51(5): 2508-2518.

[21] Nakamura T, Ito Y, Yanagiba Y, Ramdhan D H, Kono Y, Naito H, Hayashi Y, Li Y, Aoyama T, Gonzalez F J, Nakajima T. Microgram-order ammonium perfluorooctanoate may activate mouse peroxisome proliferator-activated receptor alpha, but not human PPARalpha. Toxicology, 2009, 265(1-2): 27-33.

[22] Millauer H, Schwertfeger W, Siegemund G. Hexafluoropropene oxide—A key compound in organofluorine chemistry. Angew Chem Int Edit, 1985, 24(3): 161-179.

[23] Heydebreck F, Tang J H, Xie Z Y, Ebinghaus R. Alternative and legacy perfluoroalkyl substances: differences between European and Chinese River/Estuary Systems. Environ Sci Technol, 2015, 49(14): 8386-8395.

[24] Xu Y, Zhao M, Li H, Lu W, Su X, Han Z. Anovelfluorocarbon surfactant: Synthesis and application in emulsion polymerization of perfluoroalkyl methacrylates. Paint Coat Ind, 2011, 41: 17-21.

[25] Yang X L, Huang J, Zhang K L, Yu G, Deng S B, Wang B. Stability of 6: 2 fluorotelomer sulfonate in advanced oxidation processes: degradation kinetics and pathway. Environ Sci Pollut R, 2014, 21(6): 4634-4642.

[26] Wang S W, Huang J, Yang Y, Hui Y M, Ge Y X, Larssen T, Yu G, Deng S B, Wang B, Harman C. First report of a Chinese PFOS alternative overlooked for 30 years: Its toxicity, persistence, and presence in the environment. Environ Sci Technol, 2013, 47(18): 10163-10170.

[27] Yeung L W Y, Miyake Y, Taniyasu S, Wang Y, Yu H X, So M K, Jiang G B, Wu Y N, Li J G, Giesy J P, Yamashita N, Lam P K S. Perfluorinated compounds and total and extractable organic fluorine in human blood samples from China. Environ Sci Technol, 2008, 42(21): 8140-8145.

[28] Yeung L W Y, Miyake Y, Wang Y, Taniyasu S, Yamashita N Lam P K S. Total fluorine, extractable organic fluorine, perfluorooctane sulfonate and other related fluorochemicals in liver of Indo-Pacific humpback dolphins (*Sousa chinensis*) and finless porpoises (*Neophocaena phocaenoides*) from South China. Environmental Pollution, 2009, 157(1): 17-23.

[29] Wang S W, Huang J, Yang Y, Hui Y M, Ge Y X, Larssen T, Yu G, Deng S B, Wang B, Harman C. First report of a Chinese PFOS alternative overlooked for 30 years: Its toxicity, persistence, and presence in the environment. Environ Sci Technol, 2013, 47(18): 10163-10170.

[30] Wang T, Vestergren R, Herzke D, Yu J C, Cousins I T. Levels, isomer profiles, and estimated riverine mass discharges of perfluoroalkyl acids and fluorinated alternatives at the mouths of Chinese rivers. Environ Sci Technol, 2016, 50(21):

11584-11592.

[31] Gebbink W A, Bossi R, Riget F F, Rosing-Asvid A, Sonne C, Dietz R. Observation of emerging per- and polyfluoroalkyl substances (PFASs) in Greenland marine mammals. Chemosphere, 2016, 144: 2384-2391.

[32] Shi Y L, Vestergren R, Zhou Z, Song X W, Xu L, Liang Y, Cai Y Q. Tissue distribution and whole body burden of the chlorinated polyfluoroalkyl ether sulfonic acid F-53B in crucian carp (*Carassius carassius*): evidence for a highly bioaccumulative contaminant of emerging concern. Environ Sci Technol, 2015, 49(24): 14156-14165.

[33] Ruan T, Lin Y F, Wang T, Liu R Z, Jiang G B. Identification of novel polyfluorinated ether sulfonates as PFOS alternatives in municipal sewage sludge in China. Environ Sci Technol, 2015, 49(11): 6519-6527.

[34] Schultz M M, Barofsky D F, Field J A. Quantitative determination of fluorinated alkyl substances by large-volume-injection liquid chromatography tandem mass spectrometry—Characterization of municipal wastewaters. Environ Sci Technol, 2006, 40(1): 289-295.

[35] Houtz E F, Sutton R, Park J S, Sedlak M. Poly- and perfluoroalkyl substances in wastewater: Significance of unknown precursors, manufacturing shifts, and likely AFFF impacts. Water Research, 2016, 95: 142-149.

[36] Sun M, Arevalo E, Strynar M, Lindstrom A, Richardson M, Kearns B, Pickett A, Smith C, Knappe D R U. Legacy and emerging perfluoroalkyl substances are important drinking water contaminants in the Cape Fear River Watershed of North Carolina. Environ Sci Tech Let, 2016, 3(12): 415-419.

[37] Lau C, Anitole K, Hodes C, Lai D, Pfahles-Hutchens A, Seed J. Perfluoroalkyl acids: A review of monitoring and toxicological findings. Toxicological Sciences, 2007, 99(2): 366-94.

[38] Van Hamme J D, Bottos E M, Bilbey N J, Brewer S E. Genomic and proteomic characterization of Gordonia sp NB4-1Y in relation to 6: 2 fluorotelomer sulfonate biodegradation. Microbiology, 2013, 159, 1618-1628.

[39] Wang N, Liu J X, Buck R C, Korzeniowski S H, Wolstenholme B W, Folsom P W, Sulecki L M. 6: 2 Fluorotelomer sulfonate aerobic biotransformation in activated sludge of waste water treatment plants. Chemosphere, 2011, 82(6): 853-858.

[40] D'Eon J C, Mabury S A. Production of perfluorinated carboxylic acids (PFCAs) from the biotransformation of polyfluoroalkyl phosphate surfictants (PAPS): Exploring routes of human contamination. Environ Sci Technol, 2007, 41(13): 4799-4805.

[41] D'Eon J C, Mabury S A. Exploring indirect sources of human exposure to perfluoroalkyl carboxylates (PFCAs): Evaluating uptake, elimination, and biotransformation of polyfluoroalkyl phosphate esters (PAPs) in the rat. Environ Health Persp, 2011, 119(3): 344-350.

[42] Liu Y W, Ruan T, Lin Y F, Liu A F, Yu M, Liu R Z, Meng M, Wang Y W, Liu J Y, Jiang G B. Chlorinated polyfluoroalkyl ether sulfonic acids in marine organisms from Bohai Sea, China: Occurrence, temporal variations, and trophic Transfer Behavior. Environ Sci Technol, 2017, 51(8): 4407-4414.

[43] Shi Y L, Vestergren R, Xu L, Zhou Z, Li C X, Liang Y, Cai Y Q. Human exposure and elimination kinetics of chlorinated polyfluoroalkyl ether sulfonic acids (Cl-PFESAs). Environ Sci Technol, 2016, 50(5): 2396-2404.

[44] Pan Y T, Zhu Y S, Zheng T Z, Cui Q Q, Buka S L, Bin Z, Guo Y, Xia W, Yeung L W Y, Li Y R, Zhou A F, Qiu L, Liu H X, Jiang M M, Wu C S, Xu S Q, Dai J Y. Novel chlorinated polyfluorinated ether sulfonates and legacy per-/polyfluoroalkyl substances: Placental transfer and relationship with serum albumin and glomerular filtration rate. Environ Sci Technol, 2017, 51(1): 634-644.

[45] Chen F F, Yin S S, Kelly B C, Liu W P. Chlorinated polyfluoroalkyl ether sulfonic acids in matched maternal, cord, and placenta samples: A study of transplacental transfer. Environ Sci Technol, 2017, 51(11): 6387-6394.

[46] Beesoon S, Martin J W. Isomer-specific binding affinity of perfluorooctanesulfonate (PFOS) and perfluorooctanoate (PFOA) to serum proteins. Environ Sci Technol, 2015, 49(9): 5722-5731.

[47] Zhang L Y, Ren X M, Guo L H. Structure-based investigation on the interaction of perfluorinated compounds with human liver fatty acid binding protein. Environ Sci Technol, 2013, 47(19): 11293-11301.

[48] Beesoon S, Webster G M, Shoeib M, Harner T, Benskin J P, Martin J W. Isomer profiles of perfluorochemicals in matched maternal, cord, and house dust samples: Manufacturing Sources and Transplacental Transfer. Environ Health Persp, 2011, 119(11): 1659-1664.

[49] Shi G H, Cui Q Q, Pan Y T, Sheng N, Sun S J, Guo Y, Dai J Y. 6: 2 Chlorinated polyfluorinated ether sulfonate, a PFOS alternative, induces embryotoxicity and disrupts cardiac development in zebrafish embryos. Aquat Toxicol, 2017, 185: 67-75.

[50] Rand A A, Rooney J P, Butt C M, Meyer J N, Mabury S A. Cellular toxicity associated with exposure to perfluorinated carboxylates (PFCAs) and their metabolic precursors. Chemical Research in Toxicology, 2014, 27(1): 42-50.

[51] Phillips M M, Dinglasan-Panlilio M J A, Mabury S A, Solomon K R, Sibley P K. Fluorotelomer acids are more toxic than perfluorinated acids. Environ Sci Technol, 2007, 41(20): 7159-7163.

[52] Boudreau T M S P, Mabury S A, Muir D C G, Solomon K R. Toxicity of Perfluoroalkyl Carboxylic Acids of Different Chain Length to Selected Freshwater Organisms. In Department of Environmental Biology, Master's Thesis; University of Guelph: Guelph, ON, 2002, 134.

[53] Lin Y F, Liu R Z, Hu F B, Liu R R, Ruan T, Jiang G B. Simultaneous qualitative and quantitative analysis of fluoroalkyl sulfonates in riverine water by liquid chromatography coupled with Orbitrap high resolution mass spectrometry. J Chromatogr A, 2016, 1435: 66-74.

[54] Huang H H, Huang C J, Wang L J, Ye X W, Bai C L, Simonich M T, Tanguay R L, Dong Q X. Toxicity, uptake kinetics and behavior assessment in zebrafish embryos following exposure to perfluorooctane sulphonic acid (PFOS). Aquat Toxicol, 2010, 98(2): 139-47.

[55] Shi X J, Du Y B, Lam P K, Wu R S, Zhou B S. Developmental toxicity and alteration of gene expression in zebrafish embryos exposed to PFOS. Toxicology and Applied Pharmacology, 2008, 230(1): 23-32.

[56] Hagenaars A, Vergauwen L, De Coen W, Knapen D. Structure-activity relationship assessment of four perfluorinated chemicals using a prolonged zebrafish early life stage test. Chemosphere, 2011, 82(5): 764-772.

[57] GSH, Globally Harmonized System of Classification and Labelling of Chemicals (GHS). First Revised Edition. United Nations. Nations Publication: ST/SG/AC.10/30/Rev. 6. 2015.

[58] Liu H, Sheng N, Zhang W, Dai J Y. Toxic effects of perfluorononanoic acid on the development of Zebrafish (*Danio rerio*) embryos. Journal of Environmental Sciences, 2015, 32: 26-34.

[59] Clements W K, Ong K G, Traver D. Zebrafish wnt3 is expressed in developing neural tissue. Dev Dyn, 2009, 238(7): 1788-1795.

[60] Huang Q S, Fang C, Wu X L, Fan J L, Dong S J. Perfluorooctane sulfonate impairs the cardiac development of a marine medaka (*Oryzias melastigma*). Aquat Toxicol, 2011, 105: 71-77.

[61] Hofsteen P, Plavicki J, Johnson S D, Peterson R, Heideman W. Sox9b Is required for epicardium formation and plays a role in TCDD-induced heart malformation in zebrafish. Molecular Pharmacology, 2013, 84: 353-360.

[62] Chen J N, Fishman M C. Zebrafish tinman homolog demarcates the heart field and initiates myocardial differentiation. Development, 1996, 122: 3809-3816.

[63] Croce J C, McClay D R. The canonical Wnt pathway in embryonic axis polarity. Seminars in Cell & Developmental Biology, 2006, 17: 168-174.

[64] Sun X, Zhang R, Lin X, Xu X. Wnt3a regulates the development of cardiac neural crest cells by modulating expression of cysteine-rich intestinal protein 2 in rhombomere 6. Circulation Research, 2008, 102: 831-839.

[65] Ozhan G, Weidinger G. Wnt/beta-catenin signaling in heart regeneration. Cell Regeneration, 2015, 4: 3.

[66] Shi G, Cui Q., Pan Y, Sheng N, Guo Y, Dai J. 6: 2 fluorotelomer carboxylic acid (6: 2 FTCA) exposure induces developmental toxicity and inhibits the formation of erythrocytes during zebrafish embryogenesis. Aquat Toxicol, 2017, 190: 53-61.

[67] Wan H T, Zhao Y G, Wei X, Hui K Y, Giesy J P, Wong C K C. PFOS-induced hepatic steatosis, the mechanistic actions on beta-oxidation and lipid transport. Biochim Biophys Acta, 2012, 1820(7): 1092-1101.

[68] Tan X B, Xie G X, Sun X H, Li Q, Zhong W, Qiao P, Sun X G, Jia W, Zhou Z X. High fat diet feeding exaggerates perfluorooctanoic acid-induced liver injury in mice via modulating multiple metabolic pathways. PloS one, 2013, 8(4): e61409.

[69] Kennedy G L, Butenhoff J L, Olsen G W, O'Connor J C, Seacat A M, Perkins R G, Biegel L B, Murphy S R, Farrar D G. The toxicology of perfluorooctanoate. Critical Reviews in Toxicology 2004, 34(4): 351-384.

[70] Yan S M, Wang J S, Dai J Y. Activation of sterol regulatory element-binding proteins in mice exposed to perfluorooctanoic acid for 28 days. Archives of Toxicology, 2015, 89(9): 1569-1578.

[71] Wang J S, Wang X Y, Sheng N, Zhou X J, Cui R N, Zhang H X, Dai J Y. RNA-sequencing analysis reveals the hepatotoxic mechanism of perfluoroalkyl alternatives, HFPO2 and HFPO4, following exposure in mice. Journal of Applied Toxicology, 2017, 37(4): 436-444.

[72] Takacs M L, Abbott B D. Activation of mouse and human peroxisome proliferator-activated receptors (alpha, beta/delta, gamma) by perfluorooctanoic acid and perfluorooctane sulfonate. Toxicological Sciences, 2007, 95(1): 108-117.

[73] Lau C. Perfluorinated compounds. Exs, 2012, 101: 47-86.

[74] Sheng N, Zhou X, Zheng F, Pan Y, Guo X, Guo Y, Sun Y, Dai J. Comparative hepatotoxicity of 6: 2 fluorotelomer carboxylic acid and 6: 2 fluorotelomer sulfonic acid, two fluorinated alternatives to long-chain perfluoroalkyl acids, on adult male mice. Archives of Toxicology, 2017, 91(8): 2909-2919.

[75] Joensen U N, Bossi R, Leffers H, Jensen A A, Skakkebaek N E, Jorgensen N. Do perfluoroalkyl compounds impair human semen quality? Environ Health Persp, 2009, 117(6): 923-927.

[76] Toft G, Jonsson B A G, Lindh C H, Giwercman A, Spano M, Heederik D, Lenters V, Vermeulen R, Rylander L, Pedersen H S, Ludwicki J K, Zviezdai V, Bonde J P. Exposure to perfluorinated compounds and human semen quality in arctic and European populations. Human Reproduction, 2012, 27(8): 2532-2540.

[77] Qu J H, Lu C C, Xu C, Chen G, Qiu L L, Jiang J K, Ben S, Wang Y B, Gu A H, Wang X R. Perfluorooctane sulfonate-induced testicular toxicity and differential testicular expression of estrogen receptor in male mice. Environmental Toxicology and Pharmacology, 2016, 45: 150-157.

[78] Wan H T, Zhao Y G, Wong M H, Lee K F, Yeung W S B, Giesy J P, Wong C K C. Testicular signaling is the potential target of perfluorooctanesulfonate-mediated subfertility in male mice. Biology of Reproduction, 2011, 84(5): 1016-1023.

[79] Lopez-Doval S, Salgado R, Pereiro N, Moyano R, Lafuente A. Perfluorooctane sulfonate effects on the reproductive axis in adult male rats. Environmental Research, 2014, 134: 158-168.

[80] Chang S C, Noker P E, Gorman G S, Gibson S J, Hart J A, Ehresman D J, Butenhoff J L. Comparative pharmacokinetics of perfluorooctanesulfonate (PFOS) in rats, mice, and monkeys. Reproductive Toxicology, 2012, 33(4): 428-440.

[81] Fang S, Zhang Y, Zhao S, Qiang L, Chen M, Zhu L. Bioaccumulation of perfluoroalkyl acids including the isomers of perfluorooctane sulfonate in carp (*Cyprinus carpio*) in a sediment/water microcosm. Environ Toxicol Chem 2016, 35(12): 3005-3013.

[82] Ulhaq M, Sundstrom M, Larsson P, Gabrielsson J, Bergman A, Norrgren L, Orn S. Tissue uptake, distribution and elimination of ^{14}C-PFOA in zebrafish (*Danio rerio*). Aquatic Toxicology, 2015, 163: 148-157.

[83] Yan S M, Wang J S, Zhang W, Dai J Y. Circulating microRNA profiles altered in mice after 28 d exposure to

perfluorooctanoic acid. Toxicology Letters, 2014, 224(1): 24-31.

[84] Zhang W, Sheng N, Wang M, Zhang H, Dai J. Zebrafish reproductive toxicity induced by chronic perfluorononanoate exposure. Aquatic Toxicology, 2016, 175: 269-276.

[85] Sheng N, Li J, Liu H, Zhang A Q, Dai J Y. Interaction of perfluoroalkyl acids with human liver fatty acid-binding protein. Arch Toxicol, 2016, 90(1): 217-227.

[86] Chen Y M, Guo L H. Fluorescence study on site-specific binding of perfluoroalkyl acids to human serum albumin. Arch Toxicol, 2009, 83(3): 255-261.

[87] Ng C A, Hungerbuehler K. Exploring the use of molecular docking to identify bioaccumulative perfluorinated alkyl acids (PFAAs). Environmental Science & Technology, 2015, 49(20): 12306-12314.

[88] Zhang J, Begum A, Brannstrom K, Grundstrom C, Iakovleva I, Olofsson A, Sauer-Eriksson A E, Andersson P L. Structure-based virtual screening protocol for in silico identification of potential thyroid disrupting chemicals targeting transthyretin. Environ Sci Technol, 2016, 50(21): 11984-11993.

[89] Schultz D M, Yoon T P. Solar Synthesis: Prospects in visible light photocatalysis. Science, 2014, 343(6174): 985.

[90] Yoon T P, Ischay M A, Du J N. Visible light photocatalysis as a greener approach to photochemical synthesis. Nat Chem, 2010, 2(7): 527-532.

[91] Prier C K, Rankic D A, MacMillan D W C. Visible light photoredox catalysis with transition metal complexes: Applications in organic synthesis. Chem Rev, 2013, 113(7): 5322-5363.

作者：王建设[1]，潘奕陶[1]，盛　南[1]，师国慧[1]，郭　勇[2]，戴家银[1]

[1]中国科学院动物研究所，[2]中国科学院上海有机化学研究所

第7章　药物与个人护理品环境污染与效应

本章导读

药物与个人护理品（pharmaceuticals and personal care products，PPCPs）种类繁多、使用量巨大，其环境污染与毒理效应受到广泛关注。本章介绍了近年 PPCPs 环境污染与效应的重要研究进展，并展望今后的研究方向。因常规污水处理厂的不完全去除或直接排放，造成水、土环境介质中 PPCPs 的普遍污染。PPCPs 因其特殊的理化性质，其环境降解转化行为较为独特。有些 PPCPs 污染物具有环境持久性特征，可被鱼和作物等生物吸收富集，进而影响水质和食品安全，而有些 PPCPs 本身具有生物活性，在较低浓度下可对环境生物产生毒性效应，例如具有内分泌干扰效应的环境激素可导致细菌耐药性增强等。由此可见，加强环境中 PPCPs 环境行为、毒理效应和控制技术的研究有其重要性。

关键词

药物，个人护理品，抗生素耐药性，环境污染，环境行为，毒理效应，控制技术

1 引　　言

药物与个人护理品（PPCPs）是一类新型环境污染物，指用于人体和动物疾病治疗的各种药物成分，如：抗生素、激素、消炎药、心血管药、精神类药物，以及个人护理和家用的各类化学品，如：麝香、防晒霜、杀生剂、表面活性剂等[1]。在生产和使用过程中，这类化学品通过各种途径排放到受纳环境，进而对生态系统和人体健康产生潜在影响。已有研究报道，PPCPs 在各类环境介质广泛检出，引起各种生态与健康效应，如：激素物质的内分泌干扰效应、抗生素的细菌耐药性等。近年，药物与个人护理品作为新型环境污染物受到越来越多重视。

2 环境污染与生物富集

常规城市生活污水处理工艺、行业废水处理装置及养殖粪污处理设施等并不能完全去除 PPCPs[2,3]，加之废水直排现象的存在，导致残留的 PPCPs 进入河流、湖泊的水体、沉积物，以及土壤环境。研究表明，世界各地的水体、沉积物和土壤环境中大量检出 PPCPs，部分 PPCPs 能够在不同营养级生物体内进一步富集，对环境中的生物造成不利影响。

2.1 药物与个人护理品的环境污染

河流、湖泊的水体与沉积物是 PPCPs 主要的“汇”。PPCPs 进入自然水体后虽然经历光解、吸附等环境过程，但仍有大量 PPCPs 残留于水环境[4]。在过去 20 年，研究人员对 PPCPs 在世界各国河流、湖泊水体中的污染特征开展了大量研究，调查区域一般为某一特定流域或全国性范围。其中，较早在德国和美国开展的全国尺度 PPCPs 污染调查是国际上最有影响力的研究之一[5,6]。Ternes 对德国污水处理厂受纳河流莱茵河和美因河中 32 种药物进行了研究，在河流中检出 24 种药物，其中苯扎贝特浓度中值最高达 0.35 μg/L[5]。Kolpin 等对全美 139 条河流中 PPCPs 调查发现，抗生素、非固醇类消炎药、血脂调节药物、精神类药物和杀菌剂在河流中广泛存在，其中氯四环素浓度中值最高达 0.42 μg/L[6]。在英国河口也检测到药物的存在，

但含量普遍较低，克霉唑为最常检出的化合物，最大浓度仅为 22 ng/L[7]。PPCPs 在澳大利亚及亚洲多个国家的河流中也大量存在[8-10]。总的来看，PPCPs 在各国地表水中的浓度水平为几个 ng/L 到几个 μg/L[8,11,12]。

PPCPs 进入水环境后将进一步在水相和沉积相之间进行分配。PPCPs 在沉积物中的浓度水平不仅取决于水相中化合物浓度，还取决于化合物水体/沉积物分配系数（K_{oc}值）和沉积物有机碳含量（f_{oc}）[13]。研究表明：个人护理品，如佳乐麝香（HHCB）、吐纳麝香、三氯生、三氯卡班和唑类抗真菌剂等化合物，其 log K_{oc}值一般大于 4，属弱极性到中等极性物质，在沉积物中具有较高的检出率[14-17]。比如，美国河口沉积物中可检测到三氯卡班和三氯生，其浓度水平为几十到上千 ng/g[14]。相比于个人护理品，药物在沉积物中的检出率和浓度较低，但一些磺胺、氟喹诺酮、大环内酯和四环素类抗生素往往在沉积物中也具有较高的检出率和浓度，这与此类抗生素具有较高 K_{oc}值有关[18,19]，而具有酸根结构且易溶于水的酸性药物在沉积物中的浓度则很低，或未检出[13]。

我国学者也针对 PPCPs 在河流水体和沉积物中的污染过程开展了大量的研究，涉及的流域包括珠江[13,15,20-22]、长江[23]、黄河[24,25]、海河[26]、洞庭湖[27]以及其他地区地表水。在珠江流域，研究发现受生活污水污染的河段及其支流河涌中 PPCPs 含量很高，如广州市区石井河中激素、药物和抗生素的含量达到数千 ng/L，远高于污水处理厂出水含量，这表明生活污水直排仍是 PPCPs 的重要污染途径[13,21,22]。全国尺度杀菌剂、雌激素、酸性药物和抗生素的分布特征研究表明，PPCPs 在我国主要流域广泛存在[24,25,28]。其中，三氯卡班和三氯生在我国黄河、海河、辽河、长江和珠江的水体与沉积物中具有较高检出率（图 7-1），并易于在沉积物中积聚，统计分析表明生活污水排放量是其在环境中污染负荷的主控因子，三氯生还可作为个人护理品类污染物的指示性物质[29]。不同国家的 PPCPs 检出情况对比发现，我国与欧美国家在 PPCPs 的污染模式上存在明显差异。例如，我国水环境中精神类药物卡马西平含量较低，而抗生素类药物却广泛检出。导致这一差异原因可能是用药习惯的不同，国内对精神类药物的人均使用量远小于国外，而我国抗生素类药物过度使用现象则较为严重。Kookana 等也证实了 PPCPs 在亚洲国家与发达国家呈现出不同的污染模式[30]。

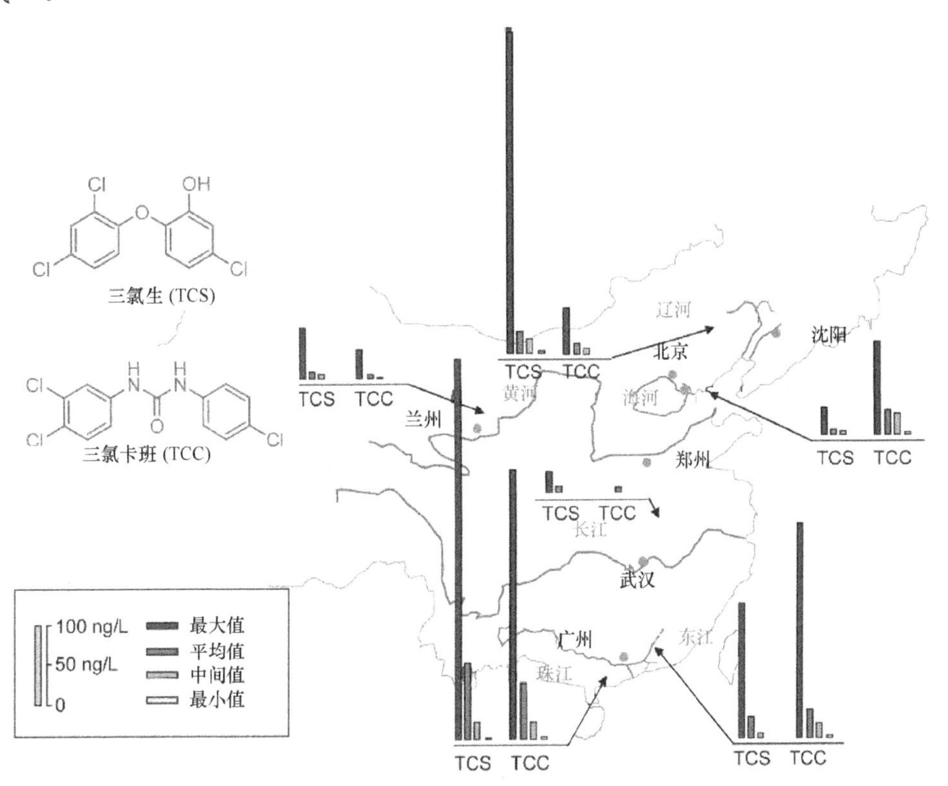

图 7-1　我国部分主要河流中三氯生（TCS）和三氯卡班（TCC）的分布特征

土壤是 PPCPs 另一重要的“汇”。土壤中 PPCPs 主要来源于污水灌溉、污泥还田、养殖粪污肥料利用[2,31-33]。Cha 和 Cupples 研究表明，在美国密歇根州某污水处理厂的污泥中，三氯卡班和三氯生的浓度高达数千 ng/g，在污泥施用后的土壤中三氯卡班和三氯生的浓度范围分别为 1.20~65.10 ng/g 和 0.16~1.02 ng/g，美国污泥施用土地面积占到农用土地的50%，因此由污泥施用带来的PPCPs污染非常严重[34]。Clarke 和 Stephen 研究了新型污染物在污泥农田施用后的污染状况，尽管污染物在土壤中将逐步消解，但持续施用 PPCPs 污染的土壤将明显导致环境累积[35]。Chen 等研究了污水灌溉土壤中 PPCPs 的污染特征，受污染土壤中检出杀菌剂、抗生素和抗炎药物[36]，其浓度明显高于地下水灌溉的土壤，在污泥施用农田中能够检出三氯生、三氯卡班和多环麝香等 PPCPs[37]。通过为期一年的消解研究发现，PPCPs 在土壤中的消解遵循一级反应动力学模式，三氯卡班在土壤中具有持久性[37]。研究还发现不同土地利用条件下 PPCPs 的分布特征具有明显的差异，农田土壤抗生素含量较高，主要成分为四环素类，林地、园地土壤抗生素含量较低，这些研究都表明土壤中 PPCPs 的污染明显受到人类活动的影响[38]。

2.2 药物与个人护理品的生物富集

环境中的生物“长期”暴露于 PPCPs，将导致 PPCPs 进一步在生物体内富集和转化，即外源性污染物在生物体的吸收、组织分布、生物转化和排泄的代谢过程，其生物富集程度与化合物的性质和生物物种都具有密切关系。一些个人护理品具有较高的辛醇/水分配系数（K_{ow}），如三氯生和三氯卡班的 log K_{ow} 值为 4.7 和 4.9 [13]，合成麝香的 log K_{ow} 值为 4.2~6.35[39]，这些物质进入受纳水环境后易于在生物体的脂肪组织内富集。麝香类物质，特别是吐纳麝香和佳乐麝香在鱼肉和贝类体内最常检出[40]，在淡水鱼、近海海鱼、贻贝体内含量一般为几十 μg/kg 干重[41,42]，最高可达到 2100 μg/kg 脂重[43]，生物富集因子最高达到 44000[42]。国内研究人员也在太湖、海河和珠江的鱼体以及中华鲟体内检出多环麝香，含量可达到数百 μg/kg 脂重[44]。三氯生、三氯卡班、苯并三唑、紫外吸收剂等物质也可以在鱼体中检出，含量一般可达到数百 μg/kg 湿重[43,45-47]。

药物与个人护理品相比，药物的脂溶性较弱，在鱼肉组织中检出浓度较低。例如，在德国多条河流的野生鱼肉中仅检出苯海拉明药物（最大浓度 0.07 ng/g 湿重）[41]。在我国太湖，对不同营养级生物体内的药物富集进行了研究，生物体内药物浓度范围为 ND~130 ng/g 干重，生物富集因子（BAF）为 19~2008 L/kg，没有发现药物随营养级升高而逐级生物放大的作用[48]。由于大部分药物属于离子型物质，在水环境中以离子/分子状态共存，往往具有与非极性污染物不同的生物富集规律，易于在鱼的胆汁、血浆等生物介质中富集[49-52]。Brozinski 等[53]在污水处理厂排水受纳河流的鱼胆汁中检出双氯芬酸、布洛芬、萘普生等非甾体类抗炎药（NSAIDs），浓度达数百 ng/mL。Togunde 等[54]采用固相微萃取方法在污水厂受纳水体的鱼胆汁中检出药物的存在，其中帕罗西汀浓度达 2.84 ng/mL。在我国野生鱼体的血浆和胆汁中检出抗生素和激素的存在，含量最高达到数千 ng/mL[50,51]。实验室条件下进一步模拟了 5 种药物在鱼体血浆、胆汁、肝脏和鱼肉组织的富集、吸收和净化的代谢动力学规律，胆汁对药物的生物富集性最强，其中替马西泮的生物富集因子（BCF）达到数千，14 天吸收/14 天净化动力学过程符合经典的一级药物代谢动力学过程[52]。

PPCPs 的代谢产物与母体的生物富集研究相比，其生物转化产物的研究相对薄弱。有限的研究表明，在生物体内 HHCB 可转化为 HHCB 的内酯[55]，三氯生转化为甲基三氯生，三氯卡班转化为二氯化和非氯化二苯脲[56]。在鱼体胆汁中，存在双氯酚酸羟基化和葡萄糖醛酸化的代谢物质[57]。在生物体中，外源性污染物的代谢过程非常复杂，一般经过 I 相（氧化、去甲基化等）和 II 相（葡萄糖醛酸化）的代谢过程，将污染物转化为极性物质，使其易于排出体外。随着生物样品前处理方法和仪器检测手段的进步，高分辨质谱等仪器逐步成为代谢物质鉴定的重要手段。总的来说，离子型 PPCPs 类污染物的生物富集和生物转化规律及其毒性相互作用的机制尚不清楚，有待进一步深入探究。

3　源汇过程与模拟

源汇关系的研究对于解释化学品在全球范围或者区域尺度污染水平的历史、现在和将来具有重要意义。通过对源的研究可以了解目标化学品的输出元素、输出过程和输出总量，对于控制目标污染物的排放意义重大；通过对汇的研究可以确定目标化学品的输入介质、输入途径和环境受纳量，有助于研究者和管理者预测污染范围并及时控制污染物扩散。掌握各种污染物的排放规律及其影响因素，并据此建立多种化学品的排放量模型，以模拟各种尺度（全球、国家、区域、流域、城市等）的相关化学品的排放量，是建立化学品区域范围内的源汇关系的主要途径。PPCPs 作为新型环境污染物，与人们日常生活密切相关，其源汇过程研究尚在起步阶段。

3.1　排放量估算

建立完善的排放量估算方法是对 PPCPs 源分析的首要步骤。在化学品生命周期的各阶段中，从生产到使用再到排放过程逐步估算各种化学品的排放量，是目前欧盟化学品管理局推荐的排放量估算方法[58]。该方法由于计算过程需要大量的参数，因此应用较少。在一些发达国家，药品的销售制度较为完善，可通过官方报道的药物销售总量和每种药物中目标化学品的含量比例对各种药物的使用量进行估算[59-62]。若有关药物与个人护理品的使用量没有相关的官方报道，则需要采取自主市场调研的方法[63-67]。研究者通过市场调研方法获取全国或者大区域水平的产品中目标化学品的量，随后结合区域人均使用量[PCC，mg/(cap·a)]、区域人口数量及 GDP，可估算获取个人护理品化学品的使用量[65,68,69]。小尺度区域或者高空间精度药物与个人护理品使用量获取，也会结合消费者的选择标准、使用频率、使用数量等多个因素对总量有效分割和分配[70,71]。

在我国，药物与个人护理品的排放量研究尚处于起步阶段。应光国课题组首次开展了我国药物与个人护理品的源排放的研究，通过市场调研，结合人口数量等因素，获得了我国 58 个流域 36 种抗生素、2 种杀生剂的排放清单[72-74]。Bu 等[75]在 2016 年将人口和社会经济因子作为模拟要素，估算了我国、省、市、县四个水平的人用抗生素的使用量。国外研究者对我国区域的药物与个人护理品的排放量研究也有少量报道。例如，Whelan 等[69]通过高空间精度的我国人口密度，估算获取了具有地理信息参考的线性烷基苯磺酸盐（LAS）使用量。通过 GDP 确定产品的购买力，结合市场调研的销售量数据、人口密度和 GDP 分布数据，Hodges 等估算了我国县级水平个人护理品使用量[64]。近年来，也有研究者将传统 POPs 及黑炭的排放清单估算方法与地理信息系统技术（以目标区域网格化为估算单元）结合[76,77]，成功应用于药物与个人护理品源排放的估算[77]。

各种 PPCPs 经使用排放后，可通过污水排放、废水灌溉、生物富集等方式进入水、土壤、沉积物及各生物体等环境介质。目前对于 PPCPs 汇的估算，尤其是进入不同环境介质的量的估算，数据相对匮乏，相关研究主要以污水处理厂出水作为 PPCPs 进入环境的主要途径，来对目标化学品的水环境受纳量进行估算[61,64]。Zhang 等[72-74]全面解析了各类 PPCPs 进入环境的不同途径，包括生活污水直排、污水处理厂出水排放、有机肥施用、废水灌溉等途径，通过大量参数定义，构建相应方程描述了目标化学品的排放，估算获得了我国 58 个流域水和土壤中目标 PPCPs 的受纳量。

3.2 环境归趋模拟

通过构建模型，把化学品置于一个可评估的理想环境中，是描述该化学品在环境体系中的迁移、转化行为的可行方式[78]。考虑到生活污水是 PPCPs 的重要载体，研究者将污水处理厂作为主要研究对象，通过获取区域目标化学品的排放量、进入污水处理厂的量、化学品的污水处理率、污水处理厂废水排放量及稀释因子等参数，构建简易方程来模拟 PPCPs 的水环境浓度[62]。然而该方法的适用前提是所有生活污水经由污水处理厂集中收集处理并排放。欧洲国家多使用欧盟化学品管理机构推荐的模型，对地表水中的化学品浓度进行模拟预测。Liebig 等[70]使用欧盟化学品管理机构推荐的模型 EUSES（European Union System for the Evaluation of Substances），基于区域人均使用量、人口数、污水处理率、废水排放量等参数模拟预测得到了德国地表水中包括抗生素在内的四种药物与个人护理品的浓度。GREAT-ER 模型以空间网格化为基础，应用于欧洲国家化学品归趋模拟和风险评价[79]，目前已成功应用于多种药物与个人护理品的动态模拟[80,81]。结合地理信息系统改进 GREAT-ER 模型后，Hao 等[82]模拟了北京温榆河 3 种喹诺酮类抗生素的浓度变化，并有效描述河网和排污口位置对抗生素归趋的影响。另外，有些研究者结合地理信息系统及水文水质模型构建药物与个人护理品的环境归趋模型，对欧洲地表水中的目标化学品浓度进行模拟[83,84]。

考虑到各种污染物特别是有机污染物进入环境后，会在各个介质中和介质间发生迁移、扩散和转化，环境多介质模型-逸度方法被作为新的方式应用于药物与个人护理品的环境归趋模拟[85-87]。逸度模型基于逸度概念，利用质量平衡原理，描述污染物在环境系统中的行为（浓度、质量分布、反应特征及持久性），通过大量的参数定义，可以有效地对化学品的归趋进行解释和预测[88]。Wang 等[89]对北京 5 个主要污水处理厂进出水目标化学品浓度进行监测，结合城市废水排放量，估算获得北京地表水环境中 15 种药物与个人护理品的环境负荷，并结合逸度模型模拟获得了北京地表水环境中目标化学品的浓度，为 PPCPs 污染现状的研究提供了数据支持。Zhu 等[90,91]通过目标区域网格化构建了高空间精度的逸度模型，模拟了包括 PPCPs 在内的多种可解离化学品的多介质环境归趋。Zhang 等以流域为基本单元构建逸度模型，成功模拟了我国 58 个流域环境中的多种药物与个人护理品的环境水平和通量[72-74]。

4 环境降解转化

药物与个人护理品（PPCPs）是一类具有生态毒理效应、可在环境介质中迁移转化的新型有机污染物。PPCPs 类化合物种类多、理化性质差异大，经使用后进入自然环境中，在水体、大气和土壤中分配、迁移和降解转化。研究 PPCPs 在环境中的降解转化过程、影响因素和制约因子，有助于科学家更加深入系统认识 PPCPs 的污染过程和环境归趋，从而有针对性地研发 PPCPs 污染阻断和控制技术。

4.1 光降解

光降解尤其是紫外光降解在废水深度处理中广泛应用，其原理是通过改变分子结构、断裂分子基团甚至将有机物分子完全碎裂成小分子，从而去除水相中难降解的有机污染物。光降解是 PPCPs 类化合物在自然环境中重要的去除途径之一。PPCPs 化合物对光敏感程度差异大，影响因素很多。Liu 等[92]研究发现家用杀生剂甘宝素在 UV-225-425 和 UV-254 下能够迅速降解，其降解效率要明显高于氯氧化处理。在

UV-254 辐照下，半衰期为 9.78 min，降解的主要路径包括：脱氯羟基化、脱氯和脱频哪酮等。自然水体中的共存组分（包括 Fe^{3+}、NO_3^- 和 腐殖酸）则明显抑制甘宝素紫外光降解，不同活性氧捕获剂（异丙醇、呋喃甲醇及 1, 4-苯醌）的添加能抑制甘宝素的紫外光降解，这说明在甘宝素紫外光降解过程中产生了能提高其光降解速率的活性氧（$\cdot OH$、$^1O_2$ 和 $O_2^{-}\cdot$）[92]。Chen 等发现氟康唑紫外光降解（254 nm）反应是依赖于 pH（2.0~12.0）的伪一级反应，其光量子产率为 0.023~0.090 mol/einstein，反应包含直接光降解和涉及 $^1O_2$ 的自敏化光降解，光降解过程包含脱氟羟基化反应，转化产物对绿藻的毒性显著降低[93]。Liu 等发现溶液 pH 升高可显著降低苯并三唑（benzotriazole，BT）、5 甲基苯并三唑（5-methyl-benzotriazole，5-TTri）和 5-氯苯并三唑（5-chloro-benzotriazole，CBT）的紫外光（254 nm）降解速率，金属铜和铁离子以及腐植酸也存在一定的抑制作用[94]。而三价铁-羧酸盐复合体系则能显著促进三种苯并三唑的紫外光降解速率，降解速率主要取决于体系 pH、羧酸类型和三价铁/羧酸盐比例：在 pH 为 3，三价铁/草酸盐比例为 10/200 μmol/L 时，BT 和 CBT 取得最短光降解半衰期分别为 0.57 h 和 2.63 h；而在三价铁/琥珀酸盐比例为 10/10 μmol/L 体系中，5-TTri 获得最短光降解半衰期为 6.08 h[95]。

4.2　微生物降解

微生物降解是环境残留 PPCPs 类化合物自然消减的重要途径。通常激素类化合物在好氧条件下比厌氧条件更容易降解。Ying 等通过 70 天监测发现雌二醇在厌氧条件下的地下水中降解非常缓慢，炔雌醇几乎没有降解[96]。Chang 等研究发现 9 种雄激素和 9 种孕激素在好氧活性污泥中降解非常迅速，降解半衰期分别为 0.6~3.3 h，0.8~3.0 h[97]。Zheng 等研究了三种天然雌激素在含有奶牛场污水的水体中厌氧生物降解情况，当起始浓度为 5 mg/L 时，经过 52 天后，仍残留 77%~85%，说明所选雌激素在厌氧条件下，生物降解缓慢[98]。Liu 等在接种好氧活性污泥降解实验中发现孕激素黄体酮降解较快，符合零级动力学，半衰期为 4.3 h，左炔诺酮符合一级动力学，半衰期为 12.5 d，检测到的黄体酮降解产物包括 1, 4-雄烯二酮、雄烯二酮、睾酮和 17β-勃地酮，左炔诺酮降解产物如 4, 5-二氢-炔诺酮、3α, 5β-四氢-炔诺酮和 6, 7-脱氢-炔诺酮，并成功筛选出 1 种黄体酮降解菌和 6 种左炔诺酮降解菌[99]。为进一步了解养猪废水生物处理过程中孕激素化合物的降解规律，Liu 等[99]通过野外调查、接种养猪废水以及添加孕激素（黄体酮和去氢孕酮）研究表明，天然孕激素的降解速率显著大于人工合成孕激素，好氧条件下孕激素的降解速率显著大于厌氧条件，厌氧条件下培养 110 天，去氢孕酮等孕激素仍未完全降解。好氧条件下，黄体酮和去氢孕酮均能完全降解，分别符合零级反应动力学和一级反应动力学，半衰期分别为 7.5 h 和 64.1 h。驯化分离出的降解菌 P19（*Bacillus* sp. I12B-00915）和 G2（uncultured *Bacillus* sp.）分别对黄体酮和去氢孕酮的降解效率最高。厌氧条件下，黄体酮降解符合一级反应动力学，去氢孕酮没有降解。共推断出 8 种黄体酮降解中间产物和 5 种去氢孕酮降解中间产物，根据降解中间产物推断了可能的降解路径包括羟基化、去氢和加氢三种降解机制[99]。活性污泥中丰富的微生物菌群是污水处理工艺去除 PPCPs 类化合物的最主要力量。Liu 等通过接种污水处理厂好氧活性污泥和厌氧消化污泥的研究发现，三种典型苯并三唑类化合物（BT、5-TTri 和 CBT）以及一种紫外吸收剂 4-甲氧基二苯甲酮（benzophenone-3，BP-3）在好氧和厌氧条件（厌氧-硝酸盐还原、硫酸盐还原、铁还原条件）下生物降解较慢，生物降解速率和降解产物取决于体系中的最终电子受体条件，自然环境中存在的多种最终电子受体条件会导致苯并三唑和紫外吸收剂类化合物不同的生物降解性能[100,101]。另外，Liu 等还模拟研究了三种典型苯并三唑化合物（BT、5-TTri 和 CBT）在地下含水层体系（5 mL 地下水和 5 g 地下含水层土壤组成的体系）中多种氧化还原条件下（好氧、厌氧-硝酸盐还原、硫酸盐还原、铁还原条件）的生物降解。发现三种苯并三唑化合物在好氧和厌氧地下含水层体系中具有一定的生物降解性。好氧条件有利于 BT 和 5-TTri 生物降解，而 CBT 则在厌氧条件更易生物降解[102]。

4.3 藻类降解转化

除微生物降解外，藻类作为水环境中重要的初级生产者，也会影响PPCPs的环境行为。Zhou等研究发现斜生栅藻可以去除大多数雌激素类内分泌干扰物，其去除效率最高可以达到100%，去除途径为藻类降解或转移，而藻类的生物吸附和生物累积所占比例很少，其降解过程基本符合一级反应动力学方程，降解半衰期小于2 d，低浓度的雌激素类内分泌干扰物对斜生栅藻的生长和光合活性有促进作用或者无作用，高浓度下则对斜生栅藻生长有抑制作用[103,104]。Peng等发现斜生栅藻（*Scenedesmus obliquus*）和蛋白核小球藻（*Chlorella pyrenoidosa*）对孕激素化合物黄体酮和诺炔孕酮的去除符合一级反应动力学方程，经5天培养，黄体酮去除率大于95%，而诺炔孕酮去除率相对较低；降解产物分析鉴定发现，二者的降解机理主要为加氢、脱氢和羟基化等过程，同时检出微量雄激素生成[105]。另外，Peng等还研究发现了6种淡水绿藻对双酚A和四溴双酚A有明显去除，效率最高的藻株在第7天去除率大于80%。通过LC-ESI-MS/MS负模式下全扫描和子离子模式分析，结合溴同位素峰Br^{79}、Br^{80}和Br^{81}对质谱图解析，发现近头状伪蹄形藻（*Pseudokirchneriella subcapitata*）可以通过磺酸化和糖基化过程降解TBBPA，尖细栅藻（*Scenedesmus acuminatus*）可以通过*O*-甲基化过程降解TBBPA。这几种磺酸化、糖基化和*O*-甲基化产物曾在人体尿液等样品中检出[106]。

5 药物与个人护理品的污染控制技术

5.1 城市污水处理厂

城市污水处理厂的去除效率是控制PPCPs进入环境的重要因素。然而现有城市污水处理厂的设计和运行一般针对氮、磷等常规污染物的去除，对大部分PPCPs的去除效果不理想。目前，国内外对抗生素在城市污水处理厂中的浓度水平及去除情况进行了广泛研究，如磺胺类抗生素在中国污水厂进水中的浓度范围为十几个ng/L到几个μg/L[49]，磺胺甲噁唑是检出率和检出浓度较高的磺胺类抗生素，其水相去除率范围为6.9%~96.4%[3,49]。四环素在城市污水处理厂进水中的浓度可达几个μg/L[3,107]。氟喹诺酮类抗生素中，关于环丙沙星、氧氟沙星和诺氟沙星在污水处理厂进水、出水和淤泥的研究较多且检出浓度较高，其中进水和出水的浓度可以高达几个μg/L，喹诺酮类抗生素在污水处理厂中的水相去除率较高，多数在80%~100%。大环内脂类抗生素中，脱水红霉素在污水处理厂进水和出水检出频率较高，且浓度高达几个μg/L；其次为罗红霉素和克拉霉素，浓度在十几ng/L到几百ng/L。大环内酯类抗生素的水相去除率波动较大[3,49,107]。与抗生素相比，激素在城市污水处理厂中更易去除，Liu等研究表明A^2O活性污泥工艺中激素水相去除率达到90%，经污水处理厂处理后，大部分固醇类激素都得到很好的去除，在最终出水中的浓度可降低至几个ng/L或未检出[108]。对于家用杀生剂，Liu等研究发现广州市等十个城市污水处理厂对总杀生剂（∑19 biocides）的平均去除率达到75%，其中改良A/O、MBR（生物膜法）和布鲁塞尔2000氧化沟三种工艺类型对∑19 biocides的水相去除率相对最高（≥82%），去除途径主要为微生物降解和活性污泥吸附[109]。而对于苯并三唑类和紫外吸收剂物质而言，污水处理厂去除效率则相对较低，苯并三唑类化合物平均去除率均小于60%，最高去除率不到80%，去除途径为生物降解。部分紫外吸收剂的去除效率仅为 54%~96%，而且大部分是吸附到污泥中[110]。由此可见，经过城市污水处理厂处理后仍然有大量残留的 PPCPs 随出水和污泥进入到环境之中，对生物系统甚至人类健康都会带来潜在的威胁。针对 PPCPs 去除的深度处理技术亟待研发，以便从源头上控制 PPCPs 污染，降低其生态健康风险。

5.2　分散型污水处理系统

近年来有关农村分散型生活污水与养殖废水造成的水环境污染问题引起了人们的广泛关注。这类分散型污水中除包括常规污染物外，还含有种类繁多的 PPCPs 如激素和抗生素等。目前，国内研究人员针对分散型污水中常规污染物的去除已开发了多种污水处理系统，如中国科学院研发了高负荷地下渗滤技术、高负荷人工湿地技术和高效低耗生物滤池技术，应光国团队则对这三类分散型污水处理系统中 PPCPs 污染物的去除，尤其是抗生素和抗性基因的去除机制开展了大量研究。Chen 等 [111]对比分析了不同基质和不同水力负荷条件下水平潜流人工湿地对常规污染指标、抗生素和耐药基因的去除效果，发现以沸石为基质，HLR = 20 cm/d 去除效果最佳，水平潜流和垂直潜流对污染物的去除效果要优于表面流，种植植物后更有利于湿地单元对污染物的去除。基质吸附和微生物降解是构建的人工湿地单元去除抗生素和耐药基因的重要途径，微生物降解是主要去除途径[111,112]。污染物去除质量平衡核算发现去除比重为基质吸附（1.99%~ 4.29%）、植物吸收（1.86×10^{-5}%~1.65×10^{-5}%）和微生物降解（73.7%~ 95.2%）[113]。

5.3　深度氧化技术

常规的污水处理工艺不能有效去除药物与个人护理品，需要深度处理技术消减污水处理厂出水中残留的药物与个人护理品。深度处理技术主要包括混凝絮凝技术、氧化技术、膜分离技术、吸附技术以及这些技术的联用技术[114]。其中氧化处理技术对药物与个人护理品的去除具有反应速率快、去除效果好的特点，成为近年来的研究热点。

污水处理厂使用的氧化剂通常有氯气（Cl_2）、二氧化氯（ClO_2）、臭氧（O_3）、高锰酸钾（$KMnO_4$）以及各种高级氧化技术生成的羟基自由基（HO·）。此外，高铁酸钾（K_2FeO_4）作为新兴的高效水处理剂，近年来也逐渐引起学术领域和工业界的重视。高铁酸钾是铁的正六价化合物，分解产物为具有絮凝作用的三价铁化合物。高铁酸钾在水处理过程中协同了氧化、絮凝、消毒、除臭等作用，是一种极具应用前景的新型多功能水处理剂[115-117]。

药物与个人护理品在氧化处理过程中的反应机制得到了广泛而深入的研究。氧化去除药物与个人护理品的效果主要取决于氧化剂的种类、药物与个人护理品的结构特性以及水中共存组分的影响。一般情况下，氧化降解药物与个人护理品的反应符合二级反应动力学模型，线性自由能关系研究表明其反应机制均为亲电氧化反应[118-122]。氧化剂主要与富含供电子基团的药物与个人护理品发生氧化反应，例如分子结构中含有苯胺、苯酚、双键结构以及含硫、含氮的化合物。有学者系统研究了高铁酸钾和 68 种药物与个人护理品的反应特性[123]，证实高铁酸钾选择性地氧化去除含苯酚结构的雌激素和三氯生，含苯胺结构的抗生素，含胺基结构的酸性药物，含双键的卡马西平、雄激素、孕激素和糖皮质激素。但是高铁酸钾不与三氯卡班、3 种雄激素（表雄酮、雄酮、5α-二氢睾酮）、7 种酸性药物（氯贝酸、二四滴、二甲四氯苯氧基乙酸、布洛芬、非诺洛芬、吉非罗奇、酮洛芬）、2 种中性药物（扑米酮和环磷酰胺）和脱水红霉素发生反应。对于氯、二氧化氯、臭氧、高锰酸钾和高铁酸钾等选择性氧化剂，其和药物与个人护理品反应的二级反应动力学速率常数跨度较大，一般从不反应（0 m/s）到 10^7 m/s，而非选择性氧化剂羟基自由基和药物与个人护理品的反应则具有近似于扩散控制的反应特性，其二级反应动力学速率常数不小于 10^9 m/s [124]。例如，合成雌激素炔雌醇（17α-ethinylestradiol，EE2）与不同氧化剂在 pH 8.0 的水溶液中的二级反应动力学速率常数依次为羟基自由基（9.8×10^9 m/s）＞臭氧（1.5×10^7 m/s）＞二氧化氯（1.8×10^6 m/s）＞高锰酸钾（8.3×10^2 m/s）＞高铁酸钾（4.2×10^2 m/s）＞氯（3.4×10^2m/s）[124,125]。

氧化剂和药物与个人护理品的反应活性也取决于其在水溶液中的形态分布，因此反应速率与反应体系的 pH 密切相关。在 pH 6~9 的反应体系中，臭氧、二氧化氯和羟基自由基是以分子形态存在，而氯、高锰酸钾和高铁酸钾起主导氧化作用的解离形态分别是 HClO、MnO_4^-和 $HFeO_4^-$。形态动力学模拟研究表明氧化降解反应主要是以氧化剂和解离态的药物与个人护理品的反应为主，例如 $HFeO_4^-$与三氯生的非解离态和解离态的反应速率常数分别为 4.1×10^2 m/s 和 1.8×10^4 m/s [126]。反应体系中的溶解性有机物（DOM）由于也含有供电子基团结构，因此会消耗反应中的氧化剂，从而降低药物与个人护理品的去除效率。反应体系中的氨和亚硝酸盐会显著地降低氯化降解药物与个人护理品的去除效果，亚硝酸盐也会降低臭氧处理的效率[124]。反应体系中的腐殖酸、Mn^{2+}和离子强度（NaCl）的投加降低了高铁酸钾对 2-羟基-4-甲氧基二苯甲酮的去除率[127]，Br^-和 Cu^{2+}提高了 2-羟基-4-甲氧基二苯甲酮的去除率。此外，一定浓度的 NH_4^+、NO_3^-、Fe^{3+}和 Fe^{2+}对高铁酸钾去除 2-羟基-4-甲氧基二苯甲酮没有影响。

药物与个人护理品在氧化处理过程中并不能完全被矿化，主要发生转化降解反应，生成反应副产物，其自身的毒性效应也随之发生变化。例如氯、二氧化氯、臭氧、羟基自由基、高锰酸钾和高铁酸钾可以快速地转化炔雌醇，反应在几秒到几分钟之间就可完成[125,128]。氯化炔雌醇主要生成炔雌醇的氯代化合物，如 2-氯炔雌醇、4-氯炔雌醇和 2, 4-二氯炔雌醇。二氧化氯与炔雌醇的苯酚结构发生反应，生成苯氧自由基中间产物，可进一步与二氧化氯反应生成 2, 3-苯醌炔雌醇和 3, 4-苯醌炔雌醇。羟基与炔雌醇的主要反应产物是 2-羟基炔雌醇和 4-羟基炔雌醇。臭氧、高锰酸钾和高铁酸钾与炔雌醇的主要反应产物是 2-羟基炔雌醇、4-羟基炔雌醇、2, 3-苯醌炔雌醇和 3, 4-苯醌炔雌醇，此外臭氧和高锰酸钾还可以进一步断裂炔雌醇的苯酚结构，生成己二烯二酸结构的炔雌醇。最为重要的是，与母体化合物炔雌醇相比，生成的氧化降解产物具有较低的雌激素活性，均小于 13%的炔雌醇当量。臭氧和羟基自由基与九大类、14 种抗菌化合物均具有较高的反应活性，在反应溶液 pH 为 7.0 时，其二级反应速率常数为 2.5×10^{-2}~1.9×10^{-6} 和 2.9×10^{-9}~8.5×10^{-9}[129]。臭氧可以直接与大环内酯类、磺胺甲噁唑、甲氧苄氨嘧啶、四环素、万古霉素和丁胺卡那霉素的药效团结构发生反应，而不能够与醋磺胺甲噁唑、氟喹诺酮类和 *β*-内酰胺类抗菌化合物的药效团发生反应。臭氧和羟基自由基氧化去除抗菌化合物浓度的同时，线性成比例地去除其抗菌性能[129]。然而，*β*-内酰胺类盘尼西林和头孢氨苄则是例外，其臭氧和羟基自由基的氧化降解产物仍具有较强的抗菌性能，而且不能够被臭氧和羟基自由基进一步氧化去除。此外，氧化处理药物与个人护理品的过程中还伴随着致毒氧化副产物的生成，例如氯化消毒副产物 *N*, *N*-二甲基亚硝胺和臭氧化过程生成的溴酸盐等。高锰酸钾和高铁酸钾氧化处理过程中生成的锰离子和铁离子残留，也是需要关注的问题。

越来越多的深度氧化技术已经用于实际的污水处理厂升级改造工程中。臭氧-活性炭工艺已经作为瑞士的市政污水处理厂深度处理工艺，用于去除二级出水中残留的微污染物如药物与个人护理品[130]。紫外消毒和氯化消毒等三级处理工艺广泛用于污水处理厂，近年来紫外光/氯联用技术成为水处理领域新的热点。紫外光光解次氯酸盐溶液可以生成高反应活性的 HO·、Cl·、$Cl_2·^-$、$O·^-$等活性反应物种，对水中的药物与个人护理品具有良好的去除效果[131,132]。此外有学者基于次氯酸盐氧化法合成高铁酸钾，开展了现场制备高铁酸盐新工艺的研究，包括高铁酸盐溶液的间歇合成及其连续用于水处理装置，该集成装置可以广泛应用于各种类型的水处理单元，对药物与个人护理品的去除具有良好的效果[123]。

6　毒理效应与生态健康风险

6.1　生态毒理效应

6.1.1　激素物质

雌激素是生活中常用的药物，一般用于激素替代治疗与口服避孕，其主要活性化合物为雌二醇（E2）和乙炔雌二醇（EE2）等。由于雌激素具有强烈的生物活性，目前关于雌激素对生物的生长、生殖、免疫以及代谢的毒性效应研究较多。环境相关浓度的雌激素可以干扰 HPG 轴基因的转染，影响生物繁殖，引起组织损伤并干扰生物体内的激素平衡。Xu 等研究发现，刚出生的斑马鱼在 0.4 ng/L 的乙炔雌二醇中连续暴露 90 天，然后放入清水中饲养 90 天，斑马鱼的生殖能力依然被抑制[133]。从毒性效应来看，EE2 在环境中的效应要强于 E2，EE2 在对鱼类性别分化以及性腺发育的影响要强于 E2[134]。EE2 能干扰水生生物的性别分化，造成性反转、水生食物链中断等严重后果[135,136]。有研究表明，EE2 通过加速诱导斑马鱼卵细胞凋亡，从而引起性别分化[136]。

免疫细胞表面有雌激素受体表达，雌激素对免疫系统也存在干扰效应。雌激素物质主要影响生物的免疫器官发育，干扰免疫细胞以及免疫因子正常功能。比如，E2 和 EE2 暴露后可以引起性腺退化，影响特异性免疫细胞迁移，从而干扰免疫应答[137,138]。外源性雌激素还能引起胞内通路的应答。有报道称，E2 在哺乳动物中主要是通过活化 NFκB 通路来应答免疫炎症反应[139]。但是 E2 在硬骨鱼中的免疫应答机制目前还不清楚，有待进一步研究。EE2 则是通过改变免疫调控基因转录以及白细胞活性来干扰生物的免疫功能[140]。此外，低浓度的 EE2 暴露还可以影响贻贝糖代谢水平，降低血糖以及脂肪酸的含量[141]。为期 7 年的加拿大野外研究表明，环境浓度 EE2 长期暴露（5~6 ng/L）可导致湖泊水体野生鱼雌性化，最终造成鱼类种群的衰退[142]。

除了对水生生物的影响，雌激素对植物也有毒性效应。其毒性主要体现在影响植物的生长和抗氧化活性。例如，E2 暴露能减少马铃薯的块茎尺寸。最新研究表明，在含有 EE2 的污水中藻类的生长受到抑制[143]。在抗氧化方面，雌激素则通过干扰了一些抗氧化酶的活性，如 CAT、GPX 和 APX，从而引起氧损伤[144]。因此，具有较强活性的雌激素，对不同物种的雌激素受体亲和力高，低浓度条件下就能够引起广泛的生理生化应答，对环境生物具有生态风险，应引起足够的重视。

合成孕激素已经广泛应用到人类口服避孕药和激素替代疗法中。此外，孕激素也广泛应用于畜禽养殖业中，用于促进动物生长、催肥、控制母畜同期排卵并预防母畜流产，以提高产量和经济效益。目前，孕激素的研究主要关注对生物生殖发育的影响，研究表明，孕激素能够引起生物产卵数量变化，导致血液中激素含量异常，造成性腺组织损伤并改变 HPG 轴相关基因的转录，严重时甚至会引起性别分化差异[145-149]。但是，孕激素毒性产生的机理研究较少。随着基因芯片和高通量测序等新技术与环境毒理领域的结合，孕激素的致毒机制逐渐清晰。近期研究发现，孕激素对斑马鱼眼睛和大脑中的昼夜节律通路和光传导通路的影响异常显著[150,151]。虽然，目前对孕激素的研究没有雌激素那么广泛，但是在环境中检测到的孕激素种类较多，对生物的毒性效应相差较大，这也提高了孕激素环境毒性研究的复杂性。比如，甲基炔诺酮和左炔诺酮具有强烈的雄激素效应，而醋酸甲羟基孕酮则只有轻微的雄激素效应[146,147,151]。因此，对不同的孕激素进行环境风险评估是必要的。鉴于孕激素和雌激素经常一起存在于口服避孕药中，相比较于单一暴露，孕激素的联合暴露实验就显得更加有实际意义。最新研究表明，孕激素和雌激素联合暴露显示出明显的叠加效应。Liang 等研究发现，EE2 和甲羟基孕酮联合暴露斑马鱼胚胎对 HPG 轴和昼夜节律相关基因的影响显著增强[152]。Hua 等报道了醋酸甲地孕酮和 EE2 联合暴露显著降低了雌鱼体内雌激素以及雄鱼体内睾酮和 11-KT 的水平，并且增加了卵巢闭锁细胞出现的频率[149]。这些结果表明，实际环境中的孕激素对生物的影响可能更大、更复杂。

此外，孕激素能调节鱼类不同组织中大量基因的转录表达水平，其中包括糖皮质激素受体基因和盐皮质激素受体基因[150]。人体内研究表明，孕激素和雌激素一样，均对免疫系统存在干扰效应[153]。上述结果表明，孕激素不仅仅对生殖系统有干扰，对生物其他系统也有潜在的影响。但是水生态环境中的孕激素对生物的免疫系统的影响研究较少。

合成雄激素也广泛用于人类激素替代疗法和畜牧业中。雄激素最显著的毒性就是通过 HPG 轴影响生物体内激素的平衡，增加精子发生[154]。由于其活性强，对受体亲和力高，毒性强，因此低浓度的雄激素对生态环境中的生物生长发育有显著的干扰效应。比如雄激素会抑制性腺发育，降低繁殖能力，导致生物体内雌激素和雄激素代谢紊乱，诱导雌雄同体现象和幼鱼完全雄性化等[155]。野外调查显示，在雄激素富集的河流区域，雌性食蚊鱼的雄性化特征明显，这与实验室模拟结果相吻合[156]。

6.1.2 抗炎症药物

双氯芬酸（DCF）和布洛芬酸（IPF）是环境中常被检测到的 PPCPs 类化合物，它们对环氧化酶（COX-1 和 COX-2）具有一定抑制作用。环氧化酶的序列比较保守，其在催化前列腺产生过程中，扮演着重要的角色。目前双氯芬的生物毒性研究较多，集中于组织和生理生化水平的探讨分析。双氯芬酸可引起鱼类的肾脏、肝脏和鳃组织的损伤[157]。最新研究表明，双氯芬酸对大型蚤生殖系统也有影响，可以显著上调 vtg 的转录水平，延迟产卵时间[158]。在氧化毒性方面，双氯芬酸可以影响食蚊鱼 *Nrf2* 基因的转录水平，*Nrf2* 基因是调控抗氧化酶的关键基因，可能会引起食蚊鱼的氧化损伤[159]。虽然，环境相关浓度的双氯芬酸对生物影响较小，但是，鉴于这种化合物在环境中的浓度逐步增加，其对环境生物的潜在暴露风险值得持续关注。此外，布洛芬也可以在一定程度上干扰生物的生殖过程。10 μg/L 的布洛芬能够造成食蚊鱼产卵率降低，产卵数量增加[160]。100 ng/L 布洛芬酸可延迟日本青鳉的孵化进程[161]。目前，针对其他抗炎药物的研究仍然较匮乏。

6.1.3 个人护理品

消毒剂三氯生（TCS）和三氯卡班（TCC）广泛用于肥皂、洗手液等家用产品。TCS 和 TCC 对不同营养级水平的生物的生长、生理和生化过程均有影响。首先，TCS 和 TCC 对藻类的生长有明显的抑制效应[162]。比如，TCS 可以抑制铜绿微囊藻的生长，阻碍光合作用并降低氧化酶活性[163]。Xin 等发现 TCS 可以抑制绿球藻脂肪酸合成以及引起蛋白质聚集[164]。另外，TCS 对斑马鱼体内脂质和蛋白质也有影响。研究发现，TCS 抑制了斑马鱼胚胎中 β-氧化酶的转染，从而引起脂肪颗粒在胚胎中的积累[165]。蛋白组学分析显示 TCS 暴露斑马鱼胚胎后，细胞骨架相关蛋白、应激反应、眼功能蛋白以及神经元蛋白被显著富集。和藻类不同的是，TCS 可显著增加动物抗氧化酶的水平[164]。相比较于 TCS，TCC 的环境毒性研究较少。由于 TCS 和 TCC 的结构和激素类比较相似，因此，TCS 可能具有内分泌干扰效应[166]。TCS 和 TCC 与雌激素和雄激素受体亲和力的检测表明，TCS 和 TCC 均具有微弱的雌激素和雄激素效应[167]，但其效应机理还有待进一步的研究。

对羟基苯甲酸酯（parabens）也是一种应用广泛的杀生剂，包含有七种不同烷基类型，其中苯甲基化合物毒性最强。对羟基苯甲酸酯在环境中有微弱的雌激素活性，具有一定的内分泌干扰效应[168]。比如，低浓度的对羟基苯甲酸酯可以增加 VTG 的合成和抑制精子的生成，但是对血清睾酮的水平没有影响[168]。

长期暴露于低浓度对羟基苯甲酸酯，可能会导致严重的生殖毒性。Watanabe 等研究发现，对羟基苯甲酸酯是 ERβ 激动剂，激动效应与对羟基苯甲酸酯的烷基大小和膨松性有关，羧酸酯酶可导致其雌激素效应减弱[169]。除了生殖毒性，对羟基苯甲酸酯还可以通过调控内源性大麻酚类物质来调节脂肪生成，促进 3T3-L1 脂肪细胞的分化[170]。

二苯甲酮（BP）被广泛用于防晒霜中，目的是为了保护皮肤不被紫外光直射，但是 BP 本身是否安全仍然存在很多争议。目前对 BP 引起的氧化应激反应和凋亡机制研究较多。Liu 等发现，BP-1，BP-2，BP-3 和 BP-4 均能够引起氧化应激反应，在肝脏中的氧化毒性依次为：BP-1 > BP-2 > BP-4 > BP-3[171]。随着研究的深入，BP 的氧化毒性和凋亡机制逐渐清晰。有研究表明，BP-1 处理人类角质细胞后，可以增加 ROS 含量，降低细胞存活率，并且通过释放光化学产物（细胞色素 C 和 Smac/DIABLO）来触发凋亡通路，引起细胞 DNA 损伤[172]。进一步研究发现，BP-1 还可以通过光动力 I 型和 II 型路径产生 ROS：$^1O_2$ 和 O_2^-，从而诱导 caspase 3 介导的线粒体凋亡途径[173]。

此外，BP 也有不同程度的激素效应[174]。BP-1~4 对 VTG 含量，生物的产卵量和组织发育均有不同程度的影响[174]。BP 的毒性机制依赖雌激素受体的表达。有研究显示，BP-1 通过雌激素受体通路引起卵巢癌细胞上皮间质转化，调节细胞周期，促进 MCF-7 乳腺癌细胞的生长[175,176]。低浓度 BP 长期暴露非洲爪蟾后可以延缓雌性性腺发育，引起性逆转，同时导致甲状腺增生[177]。BP 对生态环境中生物的生殖系统可造成很大的风险。最新研究发现，BP 还可以显著增加红细胞的总数，降低细胞面积以及周长等生理参数，引起氧化应激反应，导致大鼠血红细胞氧化损伤，表现出一定的有血液毒性[178]。

麝香类主要包含两大类：多环麝香和硝基麝香。目前研究这两类物质环境毒性较少，研究主要集中在生物富集方面[179]。不过，有研究表明，高浓度（10 mg/kg）的麝香干扰鱼类的生殖过程，可以导致鱼类产卵量减少[180]。多环麝香主要影响激素合成和代谢活动，而硝基麝香主要干扰生物体 CYP1A 催化反应[181]。

环境中的 PPCPs 化合物种类繁多，毒性数据仍然有限。环境浓度联合暴露研究可更好地揭示物质间的协同或者拮抗效应，已逐渐引起科学家的重视。随着组学概念和技术更新，利用高通量技术分析污染物的环境毒理，探究毒理机制，有望成为未来研究环境毒理效应的重要手段。从转录组到蛋白组，最后到代谢组，多组学结合，共同揭示环境化合物对生物的毒性效应。同时，利用组学技术高效地筛选合适的指示物，可以快速检测环境中的毒害化合物。

6.2　生态风险评价

PPCPs 是一类新型污染物，目前绝大部分 PPCPs 没有地表水和沉积物的环境质量标准和行业废水的排放标准，为 PPCPs 化学品的污染控制带来困难。为此，采用风险评价方法评估 PPCPs 的生态风险，能够为 PPCPs 化学品的污染控制提供重要依据。一般来说，水体和沉积物中污染物风险评价的基本方法首先采用预测或测定环境浓度（PEC/MEC）进行暴露评价，利用生态毒性的剂量-效应关系推导出的预测无影响浓度（PNEC）进行影响评价，然后以风险商值（RQ）进行风险表征。对于沉积物中缺乏毒性数据的有机物则通过有机碳归一化沉积物/水分配系数将沉积物中污染物的浓度转化为孔隙水中的浓度，采用水体风险评价方法进行评价[182,183]。

依据欧盟对化合物的风险界定，RQ≥1 表示化合物的风险不可接受（高风险），RQ<1 表示化合物的风险可接受（低风险）[58]。Boxall 等[184]对 12 种药物在英格兰和威尔士的地表水进行风险评价，大部分化合物的风险可接受，但布洛芬在 49.5%的河段风险不可接受，双氯芬酸浓度在 4.5%的河段超过《欧盟水框架指令》中规定的 0.1 μg/L 的环境质量标准。丹麦的水环境中布洛芬和挪威河流[185]的双氯芬酸的 RQ 值均大于 1，表示风险不可接受。针对污水处理厂排水的风险，研究人员以“最坏情况”考虑，即污水处

理厂排水未经稀释进入河流而带来的风险，结果发现布洛芬、双氯芬酸等物质的 RQ 值出水中超过 5，具有潜在的高风险[186,187]。Dai 等[188]等报道了 15 个 PPCPs 在北京两条河流中的生态风险，咖啡因、甲氧苄啶、美托洛尔等化合物具有高风险。Peng 等[189]对珠江广州段的污染调查和风险评价表明，三氯生在 70%采样段面的 RQ 值>1，表示具有高风险。

一些学者对 RQ 进行了更为精细的分类[190]，即 RQ < 0.1 为最低风险，0.1≤ RQ < 1 为中等风险，RQ ≥1 为高风险。采用此标准，Hernando 等[190]对文献报道的地表水中药物进行评价，布洛芬、双氯酚酸、萘普生和卡马西平具有高的生态风险，*β*-受体阻滞剂具有中等风险。Zhao 等[13,22]对珠江广州段的 PPCPs 风险评价表明，三氯生、三氯卡班和双氯芬酸在流溪河和珠江广州段将具有中等风险，它们在石井河具有高风险，最大 RQ 值达到 9.55。Wang 等[24]对黄河、海河和辽河的 12 种药物进行风险评价，双氯芬酸和布洛芬在部分河段具有中等或高风险。Liu 等[191]对长江和东江中个人护理品的风险进行了评价，克霉唑在地表水中显示出中等风险，多菌灵和三氯生无论在地表水还是沉积物中都显示出中等风险或高风险，其中高风险点位主要集中在长江下游和东江污染较重的支流石马河和淡水河。

自然环境中 PPCPs 以复合污染的形式共存，目前的风险评价方法往往针对单个化合物，很少考虑多种 PPCPs 及其与其他类别化合物带来的复合污染风险。污染物之所以具有风险，是由于污染物对环境生物的毒理效应所致[58]。复合污染条件下，污染物可能表现出独立、相加、协同或拮抗效应，因此评价化合物的复合污染风险，需要充分考虑污染物的毒理效应作用机制，这也是复合污染风险评价的难点。总地来说，风险评价结果有助于了解污染物的初级风险，进行优控污染物筛选，在推动污染物环境质量标准制定等方面具有重要意义。例如，欧洲学者对双氯芬酸的环境污染特征研究及风险评价，直接推动欧盟对双氯芬酸地表水环境质量标准的制定[184]。

6.3 抗生素耐药性

常用抗生素种类主要有氨基糖苷类、*β*-内酰胺类、大环内酯类-林肯霉素类-链阳菌素 B、四环素类、磺胺类、氯霉素类、喹诺酮类、多黏多肽类、林克酰胺类等。近年来，抗生素耐药性已成为全球关注的公共健康热点问题。细菌耐药性与抗生素的使用相伴相生，耐药机制与抗生素对细菌的作用靶点相对应[192]。耐药基因可通过垂直转移（克隆传播）和水平转移机制在环境中传播扩散。抗生素耐药基因（ARGs），也称为抗性基因，被认为是一类新型环境污染物而备受关注[193]。现有污水处理工艺并非针对抗生素耐药基因的去除而设计，因此抗生素耐药基因大量被释放到环境中。整合子、基因盒、质粒等分子移动元件将多种耐药机制整合在一起，形成交叉耐药、多重耐药和基因协同传播[194]。因此，当其中一类抗生素选择压力出现时，多种基因将会一起被筛选并扩散。耐药基因的环境污染与扩散传播机制是目前环境细菌耐药性研究的重点。

近几年来，我国学者相继加强了各种环境介质中抗生素耐药基因的污染与扩散研究。我国流域范围的耐药基因污染较为严重。应光国团队开展了中国流域尺度抗生素的环境污染状况，以及与细菌耐药性关联性的系统研究[74]。该研究表明，抗生素的使用量、流域环境排放量与细菌耐药的形成与扩散息息相关。东江流域沉积物和表层水可高频检出多种典型整合子、基因盒以及十种不同家族的耐药基因，从上游到河口呈现规律性，与人类活动强度密切相关[195]。罗义团队发现多种耐药基因和多重耐药质粒广泛分布于我国北方河流海河[196,197]，沉积物中的胞外游离基因为土著菌的转化机制提供了重要来源[198]。朱永官团队发现我国 18 个入海口沉积物含有多达 200 种耐药基因，其普遍分布特征与沿岸人们的生产实践活动相关[199]。

城市污水厂与农村养殖环境是抗生素及其耐药基因（ARGs）的重要来源，对环境耐药基因的分布产生重要影响。研究发现，城市污水处理厂不仅可检出环境常见耐药基因，如磺胺类、四环素类等[200]，也含有多种诸如 NDM-1 等临床重要耐药基因[201]，而且离子液体被发现可促进质粒接合转移，对耐药基因在环境中的水平转移起到促进作用[202]。畜禽养殖环境的细菌耐药性研究主要集中在养猪和养鸡环境。传统山地放养与笼养方式养鸡场环境均发现高含量抗生素残留和耐药基因污染[203]。整合子在耐药基因传播与扩散中发挥重要作用，耐药基因相互之间可能存在协同传播[203]。养猪场储污池、沼气池等常用废水处理过程无法有效去除耐药基因和抗生素，其受纳环境中依然能检出大量的耐药基因和相应的抗生素[204]。养猪废水排放导致地下水、蔬菜地、鱼塘、河流等受纳环境的微生物组成发生显著变化。耐药基因、耐药菌及潜在病原菌随着粪肥的施用进入农业土壤系统[204]。使用过抗生素的猪粪或土壤相对于未使用抗生素的养猪场，含有高达 192~28000 倍的耐药基因[205]。无论是养殖污水[206]，还是市政污水处理厂出水或污泥[206,207]，施用于土壤均能引起耐药基因的污染问题。

目前，耐药基因在环境中的污染与扩散研究逐渐从基于培养的传统微生物学研究过渡到基于非培养宏基因组学研究。抗生素耐药基因组的谱系特征、多样性变化及其与微生物群落的关联是研究重点。张彤团队在基于宏基因组测序的研究方法上建立了较为完善的耐药基因分析方法[200,208,209]。在污水处理厂活性污泥中检测到丰富的氨基糖苷类、四环素类和磺胺类、多重耐药和氯霉素类 ARGs；其中四环素类、磺胺类和万古霉素 ARGs 具有季节性变化，即冬季出现富集[200]。城市污水处理厂可检出高达 271 种 ARGs，其中 78 种未受到生物处理单元或污泥处理单元的影响[210]。通过网络分析揭示多种典型受到人为活动影响的环境中 260 种耐药基因的污染水平和共存关系[208,209]。通过长片段组装方法，从猪、鸡、人源粪便等环境样品中得到多种 ARGs 基因环境或特定物种的基因草图，从片段来源上可追溯至特定耐药基因的物种来源，对于研究耐药基因在环境中的扩散机制有重要促进作用[211,212]。目前，张彤团队已构建了从宏基因组数据中进行耐药基因比对的在线分析方案[213]。基于高通量测序的宏基因组学分析，不仅丰富了人们对环境耐药基因组的认识，也将抗生素耐药性的研究推向前所未有的新局面：如何控制耐药基因从污染源扩散至环境？环境中耐药基的传播机制及其影响因素？污染环境中耐药基因如何进行溯源解析？如何进行风险评估？

7 展　　望

药物与个人护理品种类繁多，监测数据仍然十分有限，有必要进一步加强大尺度的监测工作，同时应该利用数学模拟手段研究化学品的源汇过程。数学模拟可以提供大尺度环境排放、污染水平及其变化趋势。有关药物与个人护理品中非极性物质的生物富集研究较多，但对离子型物质的生物富集机理仍然不清楚，相关研究有待加强。

药物与个人护理品中不少物质具有生物活性，可对环境生物以及人体健康造成潜在风险，但其生态毒理数据严重不足，其生态与环境毒理方面的研究有待深入开展。此外，已有的研究工作多以藻、蚤与鱼等传统模式生物为模型，未来应该加强生态系统效应研究，引入针对底栖生物的研究。另一方面，利用多组学技术研究药物与个人护理品的毒理机制也成为此类新型环境污染物研究的发展趋势。抗生素耐药性是目前国际关注的热点，未来应加强抗生素耐药基因的传播扩散机理、化学驱动机制，风险评价方法等方面的研究。

药物与个人护理品的控制技术研究主要集中于城市污水处理厂的工艺去除研究，应加强深度处理技术去除污染物的机理与转化产物研究。针对目前行业污水处理研究较匮乏的现状，还要进一步开展养殖废弃物处置过程中激素、抗生素和抗性基因的去除机理和去除技术研究。

参考文献

[1] Daughton C G, Ternes T A. Pharmaceuticals and personal care products in the environment: Agents of subtle change? Environ Health Perspect, 1999, 107(Suppl 6): 907-938.

[2] Zhou L J, Ying G G, Zhang R Q, Liu S, Lai H J, Chen Z F, Yang B, Zhao J L. Use patterns, excretion masses and contamination profiles of antibiotics in a typical swine farm, south China. Environ Sci Process Impact, 2013, 15(4): 802-813.

[3] Zhou L J, Ying G G, Liu S, Zhao J L, Yang B, Chen Z F, Lai H J. Occurrence and fate of eleven classes of antibiotics in two typical wastewater treatment plants in South China. Sci Total Environ, 2013, 452: 365-376.

[4] Liu J L, Wong M H. Pharmaceuticals and personal care products(PPCPs): A review on environmental contamination in China. Environ Int, 2013, 59: 208-224.

[5] Ternes T A. Occurrence of drugs in German sewage treatment plants and rivers. Water Res, 1998, 32(11): 3245-3260.

[6] Kolpin D W, Furlong E T, Meyer M T, Thurman E M, Zaugg S D, Barber L B, Buxton H T. Pharmaceuticals, hormones, and other organic wastewater contaminants in US streams, 1999-2000: A national reconnaissance. Environ Sci Technol, 2002, 36(6): 1202-1211.

[7] Thomas K V, Hilton M J. The occurrence of selected human pharmaceutical compounds in UK estuaries. Mar Pollut Bull, 2004, 49(5-6): 436-444.

[8] Nakada N, Komori K, Suzuki Y, Konishi C, Houwa I, Tanaka H. Occurrence of 70 pharmaceutical and personal care products in Tone River basin in Japan. Water Sci Technol, 2007, 56(12): 133-140.

[9] Kim J W, Jang H S, Kim J G, Ishibashi H, Hirano M, Nasu K, Ichikawa N, Takao Y, Shinohara R, Arizono K. Occurrence of pharmaceutical and personal care products (PPCPs) in surface water from Mankyung River, South Korea. J Health Sci, 2009, 55(2): 249-258.

[10] Roberts J, Kumar A, Du J, Hepplewhite C, Ellis D J, Christy A G, Beavis S G. Pharmaceuticals and personal care products (PPCPs) in Australia's largest inland sewage treatment plant, and its contribution to a major Australian river during high and low flow. Sci Total Environ, 2016, 541: 1625-1637.

[11] Wiegel S, Aulinger A, Brockmeyer R, Harms H, Löffler J, Reincke H, Schmidt R, Stachel B, von Tümpling W, Wanke A. Pharmaceuticals in the river Elbe and its tributaries. Chemosphere, 2004, 57(2): 107-126.

[12] Carvalho I T, Santos L. Antibiotics in the aquatic environments: A review of the European scenario. Environ Int, 2016, 94: 736-757.

[13] Zhao J L, Ying G G, Liu Y S, Chen F, Yang J F, Wang L. Occurrence and risks of triclosan and triclocarban in the Pearl River system, South China: From source to the receiving environment. J Hazard Mater, 2010, 179(1-3): 215-222.

[14] Miller T R, Heidler J, Chillrud S N, Delaquil A, Ritchie J C, Mihalic J N, Bopp R, Halden R U. Fate of triclosan and evidence for reductive dechlorination of triclocarban in estuarine sediments. Environ Sci Technol, 2008, 42(12): 4570-4576.

[15] Zhang Y X, Dou H, Chang B, Wei Z C, Qiu W X, Liu S Z, Liu W X, Tao S. Emission of polycyclic aromatic hydrocarbons from indoor straw burning and emission inventory updating in China. Environmental Challenges in the Pacific Basin, 2008, 1140(1): 218-227.

[16] Guo J H, Li X H, Cao X L, Li Y, Wang X Z, Xu X B. Determination of triclosan, triclocarban and methyl-triclosan in aqueous samples by dispersive liquid-liquid microextraction combined with rapid liquid chromatography. J Chromatogr A, 2009, 1216(15): 3038-3043.

[17] 马莉, 敬烨, 周静, 曾祥英, 张晓岚, 余应新. 太湖梅梁湾水体合成麝香的分布特征. 环境化学, 2014, 33(4): 630-635.

[18] Chang H, Hu J Y, Asami M, Kunikane S. Simultaneous analysis of 16 sulfonamide and trimethoprim antibiotics in environmental waters by liquid chromatography-electrospray tandem mass spectrometry. J Chromatogr A, 2008, 1190(1-2):

390-393.

[19] Kümmerer K. Antibiotics in the aquatic environment—A review—Part I. Chemosphere, 2009, 75(4): 417-434.

[20] Peng X Z, Yu Y J, Tang C M, Tan J H, Huang Q X, Wang Z D. Occurrence of steroid estrogens, endocrine-disrupting phenols, and acid pharmaceutical residues in urban riverine water of the Pearl River Delta, South China. Sci Total Environ, 2008, 397(1-3): 158-166.

[21] Yang J F, Ying G G, Zhao J L, Tao R, Su H C, Chen F. Simultaneous determination of four classes of antibiotics in sediments of the Pearl Rivers using RRLC-MS/MS. Sci Total Environ, 2010, 408(16): 3424-3432.

[22] Zhao J L, Ying G G, Liu Y S, Chen F, Yang J F, Wang L, Yang X B, Stauber J L, Warne M S. Occurrence and a Screening-level risk assessment of human pharmaceuticals in the Pearl River system, South China. Environ Toxicol Chem, 2010, 29(6): 1377-1384.

[23] Wu C, Huang X, Witter J D, Spongberg A L, Wang K, Wang D, Liu J. Occurrence of pharmaceuticals and personal care products and associated environmental risks in the central and lower Yangtze river, China. Ecotoxicol Environ Saf, 2014, 106: 19-26.

[24] Wang L, Ying G G, Zhao J L, Yang X B, Chen F, Tao R, Liu S, Zhou L J. Occurrence and risk assessment of acidic pharmaceuticals in the Yellow River, Hai River and Liao River of north China. Sci Total Environ, 2010, 408(16): 3139-3147.

[25] Zhou L J, Ying G G, Zhao J L, Yang J F, Wang L, Yang B, Liu S. Trends in the occurrence of human and veterinary antibiotics in the sediments of the Yellow River, Hai River and Liao River in northern China. Environ Pollut, 2011, 159(7): 1877-1885.

[26] Hu Z, Shi Y, Cai Y. Concentrations, distribution, and bioaccumulation of synthetic musks in the Haihe River of China. Chemosphere, 2011, 84(11): 1630-1635.

[27] Ma R, Wang B, Lu S, Zhang Y, Yin L, Huang J, Deng S, Wang Y, Yu G. Characterization of pharmaceutically active compounds in Dongting Lake, China: Occurrence, chiral profiling and environmental risk. Sci Total Environ, 2016, 557: 268-275.

[28] Zhao J L, Zhang Q Q, Chen F, Wang L, Ying G G, Liu Y S, Yang B, Zhou L J, Liu S, Su H C, Zhang R Q. Evaluation of triclosan and triclocarban at river basin scale using monitoring and modeling tools: Implications for controlling of urban domestic sewage discharge. Water Res, 2013, 47(1): 395-405.

[29] Chen Z F, Ying G G, Liu Y S, Zhang Q Q, Zhao J L, Liu S S, Chen J, Peng F J, Lai H J, Pan C G. Triclosan as a surrogate for household biocides: An investigation into biocides in aquatic environments of a highly urbanized region. Water Res, 2014, 58: 269-279.

[30] Kookana R S, Williams M, Boxall A B A, Larsson D G J, Gaw S, Choi K, Yamamoto H, Thatikonda S, Zhu Y-G, Carriquiriborde P. Potential ecological footprints of active pharmaceutical ingredients: An examination of risk factors in low-, middle- and high-income countries. Philosophical transactions of the Royal Society of London Series B, Biological Sciences, 2014, 369(1656): 20130586.

[31] Ying G G, Kookana R S. Triclosan in wastewaters and biosolids from Australian wastewater treatment plants. Environ Int, 2007, 33(2): 199-205.

[32] Wu J-L, Liu J, Cai Z. Determination of triclosan metabolites by using in-source fragmentation from high-performance liquid chromatography/negative atmospheric pressure chemical ionization ion trap mass spectrometry. Rapid Commun Mass Spectrom, 2010, 24(13): 1828-1834.

[33] Heeb F, Singer H, Pernet-Coudrier B, Qi W, Liu H, Longrée P, Müller B, Berg M. Organic micropollutants in rivers downstream of the Megacity Beijing: Sources and mass fluxes in a large-scale wastewater irrigation system. Environ Sci Technol, 2012, 46(16): 8680-8688.

[34] Cha J, Cupples A M. Detection of the antimicrobials triclocarban and triclosan in agricultural soils following land application of municipal biosolids. Water Res, 2009, 43(9): 2522-2530.

[35] Clarke B O, Smith S R. Review of 'emerging' organic contaminants in biosolids and assessment of international research priorities for the agricultural use of biosolids. Environ Int, 2011, 37(1): 226-247.

[36] Chen F, Ying G G, Kong L X, Wang L, Zhao J L, Zhou L J, Zhang L J. Distribution and accumulation of endocrine-disrupting chemicals and pharmaceuticals in wastewater irrigated soils in Hebei, China. Environ Pollut, 2011, 159(6): 1490-1498.

[37] Chen F, Ying G-G, Ma Y-B, Chen Z-F, Lai H-J, Peng F-J. Field dissipation and risk assessment of typical personal care products TCC, TCS, AHTN and HHCB in biosolid-amended soils. Sci Total Environ, 2014, 470-471: 1078-1086.

[38] 赵方凯, 陈利顶, 杨磊, 方力, 孙龙, 李守娟. 长三角典型城郊不同土地利用土壤抗生素组成及分布特征. 环境科学, 2017, (12): 1-16.

[39] Dietrich D R, Hitzfeld B C. Bioaccumulation and ecotoxicity of synthetic musks in the aquatic environment.//Rimkus G G, ed. Series Anthropogenic Compounds: Synthetic Musk Fragances in the Environment. Berlin Heidelberg, Berlin, Heidelberg: Springer, 2004, 233-244.

[40] 顾越, 李晓静, 梁高峰, 徐青, 余应新, 张晓岚. 淀山湖水、沉积物和鱼体中的合成麝香及人体暴露评估. 环境科学学报, 2017, 37(1): 388-394.

[41] Subedi B, Du B, Chambliss C K, Koschorreck J, Ruedel H, Quack M, Brooks B W, Usenko S. Occurrence of pharmaceuticals and personal care products in German fish tissue: A national study. Environ Sci Technol, 2012, 46(16): 9047-9054.

[42] Lange C, Kuch B, Metzger J W. Occurrence and fate of synthetic musk fragrances in a small German river. J Hazard Mater, 2015, 282: 34-40.

[43] Ramirez A J, Brain R A, Usenko S, Mottaleb M A, O'Donnell J G, Stahl L L, Wathen J B, Snyder B D, Pitt J L, Perez-Hurtado P, Dobbins L L, Brooks B W, Chambliss C K. Occurrence of pharmaceuticals and personal care products in fish: results of a national pilot study in the United States. Environ Toxicol Chem, 2009, 28(12): 2587-2597.

[44] Wan Y, Wei Q, Hu J, Jin X, Zhang Z, Zhen H, Liu J. Levels, tissue distribution, and age-related accumulation of synthetic musk fragrances in Chinese Sturgeon (*Acipenser sinensis*): Comparison to organochlorines. Environ Sci Technol, 2007, 41(2): 424-430.

[45] Peng X, Jin J, Wang C, Ou W, Tang C. Multi-target determination of organic ultraviolet absorbents in organism tissues by ultrasonic assisted extraction and ultra-high performance liquid chromatography–tandem mass spectrometry. J Chromatogr A, 2015, 1384: 97-106.

[46] Yao L, Zhao J-L, Liu Y-S, Yang Y-Y, Liu W-R, Ying G-G. Simultaneous determination of 24 personal care products in fish muscle and liver tissues using QuEChERS extraction coupled with ultra pressure liquid chromatography-tandem mass spectrometry and gas chromatography-mass spectrometer analyses. Anal Bioanal Chem, 2016, 408(28): 8177-8193.

[47] Zhong Y, Chen Z-F, Liu S-S, Dai X, Zhu X, Zheng G, Liu S, Liu G, Cai Z. Analysis of azole fungicides in fish muscle tissues: Multi-factor optimization and application to environmental samples. J Hazard Mater, 2017, 324: 535-543.

[48] Xie Z, Lu G, Yan Z, Liu J, Wang P, Wang Y. Bioaccumulation and trophic transfer of pharmaceuticals in food webs from a large freshwater lake. Environ Pollut, 2017, 222: 356-366.

[49] Gao L, Shi Y, Li W, Niu H, Liu J, Cai Y. Occurrence of antibiotics in eight sewage treatment plants in Beijing, China. Chemosphere, 2012, 86(6): 665-671.

[50] Liu S, Chen H, Xu X R, Liu S S, Sun K F, Zhao J L, Ying G G. Steroids in marine aquaculture farms surrounding Hailing Island, South China: Occurrence, bioconcentration, and human dietary exposure. Sci Total Environ, 2015, 502: 400-407.

[51] Zhao J L, Liu Y S, Liu W R, Jiang Y X, Su H C, Zhang Q Q, Chen X W, Yang Y Y, Chen J, Liu S S, Pan C G, Huang G Y, Ying G G. Tissue-specific bioaccumulation of human and veterinary antibiotics in bile, plasma, liver and muscle tissues of wild fish from a highly urbanized region. Environ Pollut, 2015, 198: 15-24.

[52] Zhao J-L, Furlong E T, Schoenfuss H L, Kolpin D W, Bird K L, Feifarek D J, Schwab E A, Ying G-G. Uptake and disposition of select pharmaceuticals by bluegill exposed at constant concentrations in a flow-through aquatic exposure system. Environ Sci Technol, 2017, 51(8): 4434-4444.

[53] Brozinski J-M, Lahti M, Meierjohann A, Oikari A, Kronberg L. The anti-inflammatory drugs diclofenac, naproxen and ibuprofen are found in the bile of wild fish caught downstream of a wastewater treatment plant. Environ Sci Technol, 2013, 47(1): 342-348.

[54] Togunde O P, Oakes K D, Servos M R, Pawliszyn J. Determination of pharmaceutical residues in fish bile by solid-phase microextraction couple with liquid chromatography-tandem mass spectrometry (LC/MS/MS). Environ Sci Technol, 2012, 46(10): 5302-5309.

[55] Mottaleb M A, Osemwengie L I, Islam M R, Sovocool G W. Identification of bound nitro musk-protein adducts in fish liver by gas chromatography–mass spectrometry: Biotransformation, dose–response and toxicokinetics of nitro musk metabolites protein adducts in trout liver as biomarkers of exposure. Aquat Toxicol, 2012, 106–107: 164-172.

[56] Venkatesan A K, Pycke B F G, Barber L B, Lee K E, Halden R U. Occurrence of triclosan, triclocarban, and its lesser chlorinated congeners in Minnesota freshwater sediments collected near wastewater treatment plants. J Hazard Mater, 2012, 229: 29-35.

[57] Kallio J M, Lahti M, Oikari A, Kronberg L. Metabolites of the aquatic pollutant diclofenac in fish bile. Environ Sci Technol, 2010, 44(19): 7213-7219.

[58] EC. European Commission Technical Guidance Document in Support of Commission Directive 93/67/EEC on Risk Assessment for New Notified Substances and Commission Regulation (EC) No 1488/94 on Risk Assessment for Existing Substances, Part II. Pages 100-103 *in* Commission E, editor. European Commission, Brussel, 2003.

[59] Carballa M, Omil F, Lema J M. Comparison of predicted and measured concentrations of selected pharmaceuticals, fragrances and hormones in Spanish sewage. Chemosphere, 2008, 72(8): 1118-1123.

[60] ter Laak T L, van der Aa M, Houtman C J, Stoks P G, van Wezel A P. Relating environmental concentrations of pharmaceuticals to consumption: A mass balance approach for the river Rhine. Environ Int, 2010, 36(5): 403-409.

[61] de Garcia S O, Pinto G P, Encina P G, Mata R I. Consumption and occurrence of pharmaceutical and personal care products in the aquatic environment in Spain. Sci Total Environ, 2013, 444: 451-465.

[62] Celle-Jeanton H, Schemberg D, Mohammed N, Huneau F, Bertrand G, Lavastre V, Le Coustumer P. Evaluation of pharmaceuticals in surface water: Reliability of PECs compared to MECs. Environ Int, 2014, 73: 10-21.

[63] Gouin T, van Egmond R, Price O R, Hodges J E N. Prioritising chemicals used in personal care products in China for environmental risk assessment: Application of the RAIDAR model. Environ Pollut, 2012, 165: 208-214.

[64] Hodges J E N, Holmes C M, Vamshi R, Mao D, Price O R. Estimating chemical emissions from home and personal care products in China. Environ Pollut, 2012, 165: 199-207.

[65] Price O R, Munday D K, Whelan M J, Holt M S, Fox K K, Morris G, Young A R. Data requirements of GREAT-ER: Modelling and validation using LAS in four UK catchments. Environ Pollut, 2009, 157(10): 2610-2616.

[66] Price O R, Williams R J, van Egmond R, Wilkinson M J, Whelan M J. Predicting accurate and ecologically relevant regional scale concentrations of triclosan in rivers for use in higher-tier aquatic risk assessments. Environ Int, 2010, 36(6): 521-526.

[67] Price O R, Williams R J, Zhang Z, van Egmond R. Modelling concentrations of decamethylcyclopentasiloxane in two UK rivers using LF2000-WQX. Environ Pollut, 2010, 158(2): 356-360.

[68] Holt M S, Fox K K, Burford M, Daniel M, Buckland H. UK monitoring study on the removal of linear alkylbenzene sulphonate in trickling filter type sewage treatment plants. Contribution to GREAT-ER project #2. Sci Total Environ, 1998, 210(1-6): 255-269.

[69] Whelan M J, Hodges J E N, Williams R J, Keller V D J, Price O R, Li M. Estimating surface water concentrations of "down-the-drain" chemicals in China using a global model. Environ Pollut, 2012, 165: 233-240.

[70] Liebig M, Moltmann J F, Knacker T. Evaluation of measured and predicted environmental concentrations of selected human pharmaceuticals and personal care products. Environ Sci Pollut Res, 2006, 13(2): 110-119.

[71] Dimitroulopoulou C, Lucica E, Johnson A, Ashmore M R, Sakellaris I, Stranger M, Goelen E. EPHECT I: European household survey on domestic use of consumer products and development of worst-case scenarios for daily use. Sci Total Environ, 2015, 536: 880-889.

[72] Zhang Q Q, Ying G G, Chen Z F, Liu Y S, Liu W R, Zhao J L. Multimedia fate modeling and risk assessment of a commonly used azole fungicide climbazole at the river basin scale in China. Sci Total Environ, 2015, 520: 39-48.

[73] Zhang Q Q, Ying G G, Chen Z F, Zhao J L, Liu Y S. Basin-scale emission and multimedia fate of triclosan in whole China. Environ Sci Pollut Res, 2015, 22(13): 10130-10143.

[74] Zhang Q Q, Ying G G, Pan C G, Liu Y S, Zhao J L. Comprehensive evaluation of antibiotics emission and fate in the river basins of China: source analysis, multimedia modeling, and linkage to bacterial resistance. Environ Sci Technol, 2015, 49(11): 6772-6782.

[75] Bu Q W, Wang B, Huang J, Liu K, Deng S B, Wang Y J, Yu G. Estimating the use of antibiotics for humans across China. Chemosphere, 2016, 144: 1384-1390.

[76] Wang R, Tao S, Wang W T, Liu J F, Shen H Z, Shen G F, Wang B, Liu X P, Li W, Huang Y, Zhang Y Y, Lu Y, Chen H, Chen Y C, Wang C, Zhu D, Wang X L, Li B G, Liu W X, Ma J M. Black Carbon Emissions in China from 1949 to 2050. Environ Sci Technol, 2012, 46(14): 7595-7603.

[77] Zhu Y, Price O R, Kilgallon J, Rendal C, Tao S, Jones K C, Sweetman A J. A multimedia fate model to support chemical management in China: A case study for selected trace organics. Environ Sci Technol, 2016, 50(13): 7001-7009.

[78] Vermeire T G, Jager D T, Bussian B, Devillers J, denHaan K, Hansen B, Lundberg I, Niessen H, Robertson S, Tyle H, vanderZandt P T J. European Union System for the Evaluation of Substances (EUSES). Principles and structure. Chemosphere, 1997, 34(8): 1823-1836.

[79] Feijtel T, Boeije G, Matthies M, Young A, Morris G, Gandolfi C, Hansen B, Fox K, Holt M, Koch V, Schroder R, Cassani G, Schowanek D, Rosenblom J, Niessen H. Development of a geography-referenced regional exposure assessment tool for European rivers - GREAT-ER contribution to GREAT-ER #1. Chemosphere, 1997, 34(11): 2351-2373.

[80] Schowanek D, Webb S. Exposure simulation for pharmaceuticals in European surface waters with GREAT-ER. Toxicol Lett, 2002, 131(1-2): 39-50.

[81] Wind T, Werner U, Jacob M, Hauk A. Environmental concentrations of boron, LAS, EDTA, NTA and Triclosan simulated with GREAT-ER in the river Itter. Chemosphere, 2004, 54(8): 1135-1144.

[82] Hao X W, Cao Y, Zhang L, Zhang Y Y, Liu J G. Fluoroquinolones in the Wenyu River catchment, China: Occurrence simulation and risk assessment. Environ Toxicol Chem, 2015, 34(12): 2764-2770.

[83] Rowney N C, Johnson A C, Williams R J. Cytotoxic drugs in drinking water: A prediction and risk assessment exercise for the thames catchment in the United Kingdom. Environ Toxicol Chem, 2009, 28(12): 2733-2743.

[84] Johnson A C, Keller V, Dumont E, Sumpter J P. Assessing the concentrations and risks of toxicity from the antibiotics ciprofloxacin, sulfamethoxazole, trimethoprim and erythromycin in European rivers. Sci Total Environ, 2015, 511: 747-755.

[85] Woodfine D, MacLeod M, Mackay D. A regionally segmented national scale multimedia contaminant fate model for Canada

with GIS data input and display. Environ Pollut, 2002, 119(3): 341-355.

[86] Domenech X, Peral J, Munoz I. Predicted environmental concentrations of cocaine and benzoylecgonine in a model environmental system. Water Res, 2009, 43(20): 5236-5242.

[87] Lyndall J, Fuchsman P, Bock M, Barber T, Lauren D, Leigh K, Perruchon E, Capdevielle M. Probabilistic risk evaluation for triclosan in surface water, sediments, and aquatic biota tissues. Integrated Environ Assess Manag, 2010, 6(3): 419-440.

[88] Mackay D. Multimedia Environmental Models: The Fugacity Approach. Second Edition edition. CRC, 2001.

[89] Wang B, Dai G H, Deng S B, Huang J, Wang Y J, Yu G. Linking the environmental loads to the fate of PPCPs in Beijing: Considering both the treated and untreated wastewater sources. Environ Pollut, 2015, 202: 153-159.

[90] Zhu Y, Price O R, Tao S, Jones K C, Sweetman A J. A new multimedia contaminant fate model for China: How important are environmental parameters in influencing chemical persistence and long-range transport potential? Environ Int, 2014, 69: 18-27.

[91] Zhu Y, Tao S, Price O R, Shen H Z, Jones K C, Sweetman A J. Environmental distributions of benzo[a]pyrene in China: Current and future emission reduction scenarios explored using a spatially explicit multimedia fate model. Environ Sci Technol, 2015, 49(23): 13868-13877.

[92] Liu W R, Ying G G, Zhao J L, Liu Y S, Hu L X, Yao L, Liang Y Q, Tian F. Photodegradation of the azole fungicide climbazole by ultraviolet irradiation under different conditions: Kinetics, mechanism and toxicity evaluation. J Hazard Mater, 2016, 318: 794-801.

[93] Chen Z F, Ying G G, Jiang Y X, Yang B, Lai H J, Liu Y S, Pan C G, Peng F Q. Photodegradation of the azole fungicide fluconazole in aqueous solution under UV-254: Kinetics, mechanistic investigations and toxicity evaluation. Water Res, 2014, 52: 83-91.

[94] Liu Y S, Ying G G, Shareef A, Kookana R S. Photolysis of benzotriazole and formation of its polymerised photoproducts in aqueous solutions under UV irradiation. Environmental Chemistry, 2011, 8(2): 174-181.

[95] Liu Y S, Ying G G, Shareef A, Kookana R S. Photodegradation of three benzotriazoles induced by four Fe-III-carboxylate complexes in water under ultraviolet irradiation. Environmental Chemistry, 2013, 10(2): 135-143.

[96] Ying G G, Kookana R S, Dillon P. Sorption and degradation of selected five endocrine disrupting chemicals in aquifer material. Water Res, 2003, 37(15): 3785-3791.

[97] Chang H, Wan Y, Wu S, Fan Z, Hu J. Occurrence of androgens and progestogens in wastewater treatment plants and receiving river waters: Comparison to estrogens. Water Res, 2011, 45(2): 732-740.

[98] Zheng W, Li X, Yates S R, Bradford S A. Anaerobic transformation kinetics and mechanism of steroid estrogenic hormones in dairy Lagoon Water. Environ Sci Technol, 2012, 46(10): 5471-5478.

[99] Liu S S, Ying G G, Liu Y S, Yang Y Y, He L Y, Chen J, Liu W R, Zhao J L. Occurrence and removal of progestagens in two representative swine farms: Effectiveness of Lagoon and digester treatment. Water Res, 2015, 77: 146-154.

[100] Liu Y S, Ying G G, Shareef A, Kookana R S. Biodegradation of three selected benzotriazoles under aerobic and anaerobic conditions. Water Res, 2011, 45(16): 5005-5014.

[101] Liu Y S, Ying G G, Shareef A, Kookana R S. Biodegradation of the ultraviolet filter benzophenone-3 under different redox conditions. Environ Toxicol Chem, 2012, 31(2): 289-295.

[102] Liu Y S, Ying G G, Shareef A, Kookana R S. Biodegradation of three selected benzotriazoles in aquifer materials under aerobic and anaerobic conditions. Journal of Contaminant Hydrology, 2013, 151: 131-139.

[103] Zhou G J, Peng F Q, Yang B, Ying G G. Cellular responses and bioremoval of nonylphenol and octylphenol in the freshwater green microalga Scenedesmus obliquus. Ecotoxicol Environ Saf, 2013, 87: 10-16.

[104] Zhou G J, Ying G G, Liu S, Zhou L J, Chen Z F, Peng F Q. Simultaneous removal of inorganic and organic compounds in

wastewater by freshwater green microalgae. Environ Sci Process Impact, 2014, 16(8): 2018-2027.

[105] Peng F Q, Ying G G, Yang B, Liu S, Lai H J, Liu Y S, Chen Z F, Zhou G J. Biotransformation of progesterone and norgestrel by two freshwater microalgae (*Scenedesmus obliquus* and *Chlorella pyrenoidosa*): Transformation kinetics and products identification. Chemosphere, 2014, 95: 581-588.

[106] Peng F Q, Ying G G, Yang B, Liu Y S, Lai H J, Zhou G J, Chen J, Zhao J L. Biotransformation of the flame retardant tetrabromobisphenol-a (tbbpa) by freshwater microalgae. Environ Toxicol Chem, 2014, 33(8): 1705-1711.

[107] Yang X, Flowers R C, Weinberg H S, Singer P C. Occurrence and removal of pharmaceuticals and personal care products (PPCPs) in an advanced wastewater reclamation plant. Water Res, 2011, 45(16): 5218-5228.

[108] Liu S, Ying G G, Zhao J L, Zhou L J, Yang B, Chen Z F, Lai H J. Occurrence and fate of androgens, estrogens, glucocorticoids and progestagens in two different types of municipal wastewater treatment plants. J Environ Monit, 2012, 14(2): 482-491.

[109] Liu W R, Yang Y Y, Liu Y S, Zhang L J, Zhao J L, Zhang Q Q, Zhang M, Zhang J N, Jiang Y X, Ying G G. Biocides in wastewater treatment plants: Mass balance analysis and pollution load estimation. J Hazard Mater, 2017, 329: 310-320.

[110] Liu Y S, Ying G G, Shareef A, Kookana R S. Occurrence and removal of benzotriazoles and ultraviolet filters in a municipal wastewater treatment plant. Environ Pollut, 2012, 165: 225-232.

[111] Chen J, Liu Y S, Su H C, Ying G G, Liu F, Liu S S, He L Y, Chen Z F, Yang Y Q, Chen F R. Removal of antibiotics and antibiotic resistance genes in rural wastewater by an integrated constructed wetland. Environ Sci Pollut Res, 2015, 22(3): 1794-1803.

[112] Chen J, Ying G G, Wei X D, Liu Y S, Liu S S, Hu L X, He L Y, Chen Z F, Chen F R, Yang Y Q. Removal of antibiotics and antibiotic resistance genes from domestic sewage by constructed wetlands: Effect of flow configuration and plant species. Sci Total Environ, 2016, 571: 974-982.

[113] Chen J, Wei X D, Liu Y S, Ying G G, Liu S S, He L Y, Su H C, Hu L X, Chen F R, Yang Y Q. Removal of antibiotics and antibiotic resistance genes from domestic sewage by constructed wetlands: Optimization of wetland substrates and hydraulic loading. Sci Total Environ, 2016, 565: 240-248.

[114] 赵青青, 高睿, 王铭璐, 马笑文, 李爱民. 药物和个人护理品(PPCPs)去除技术研究进展. 环境科学与技术, 2016, 39(S1): 119-125.

[115] Yang B, Ying G G. Removal of Personal Care Products Through Ferrate (VI) Oxidation Treatment.//Diaz-Cruz M S, Barcelo D, eds. Personal Care Products in the Aquatic Environment, The Handbook of Environmental Chemistry 2014. Berlin Heidelberg: Springer, 2014, 1-19

[116] Jiang J Q. The role of ferrate (VI) in the remediation of emerging micropollutants: A review. Desalination and Water Treatment, 2015, 55(3): 828-835.

[117] Sharma V K, Chen L, Zboril R. Review on high valent FeVI (Ferrate): A sustainable green oxidant in organic chemistry and transformation of pharmaceuticals. ACS Sustainable Chemistry & Engineering, 2016, 4(1): 18-34.

[118] von Gunten U. Ozonation of drinking water: Part I. Oxidation kinetics and product formation. Water Res, 2003, 37(7): 1443-1467.

[119] Deborde M, von Gunten U. Reactions of chlorine with inorganic and organic compounds during water treatment - Kinetics and mechanisms: A critical review. Water Res, 2008, 42(1-2): 13-51.

[120] Yang B, Ying G G, Zhang L-J, Zhou L-J, Liu S, Fang Y-X. Kinetics modeling and reaction mechanism of ferrate (VI) oxidation of benzotriazoles. Water Res, 2011, 45(6): 2261-2269.

[121] Du J S, Sun B, Zhang J, Guan X H. Parabola-like shaped pH-rate profile for phenols oxidation by aqueous permanganate. Environ Sci Technol, 2012, 46(16): 8860-8867.

[122] Lee Y, von Gunten U. Quantitative structure-activity relationships (QSARs) for the transformation of organic micropollutants during oxidative water treatment. Water Res, 2012, 46(19): 6177-6195.

[123] Yang B, Ying G G, Zhao J L, Liu S, Zhou L J, Chen F. Removal of selected endocrine disrupting chemicals (EDCs) and pharmaceuticals and personal care products (PPCPs) during ferrate (VI) treatment of secondary wastewater effluents. Water Res, 2012, 46(7): 2194-2204.

[124] Lee Y, von Gunten U. Oxidative transformation of micropollutants during municipal wastewater treatment: comparison of kinetic aspects of selective (chlorine, chlorine dioxide, ferrate VI, and ozone) and non-selective oxidants (hydroxyl radical). Water Res, 2010, 44(2): 555-566.

[125] Jiang J, Pang S Y, Ma J, Liu H. Oxidation of phenolic endocrine disrupting chemicals by potassium permanganate in synthetic and real Waters. Environ Sci Technol, 2011, 46(3): 1774-1781.

[126] 杨滨, 应光国, 赵建亮. 高铁酸钾氧化降解三氯生的动力学模拟及反应机制研究. 环境科学, 2011, 32(9): 2543-2548.

[127] Yang B, Ying G G. Oxidation of benzophenone-3 during water treatment with ferrate(VI). Water Res, 2013, 47(7): 2458-2466.

[128] Lee Y, Escher B I, Von Gunten U. Efficient removal of estrogenic activity during oxidative treatment of waters containing steroid estrogens. Environ Sci Technol, 2008, 42(17): 6333-6339.

[129] Dodd M C, Kohler H P E, Von Gunten U. Oxidation of antibacterial compounds by ozone and hydroxyl radical: Elimination of biological activity during aqueous ozonation processes. Environ Sci Technol, 2009, 43(7): 2498-2504.

[130] Eggen R I L, Hollender J, Joss A, Schärer M, Stamm C. Reducing the discharge of micropollutants in the aquatic environment: The benefits of upgrading wastewater treatment plants. Environ Sci Technol, 2014, 48(14): 7683-7689.

[131] Fang J, Fu Y, Shang C. The roles of reactive species in micropollutant degradation in the UV/free chlorine system. Environ Sci Technol, 2014, 48(3): 1859-1868.

[132] Remucal C K, Manley D. Emerging investigators series: the efficacy of chlorine photolysis as an advanced oxidation process for drinking water treatment. Environmental Science: Water Research & Technology, 2016, (2): 565-579.

[133] Xu H, Yang J, Wang Y, Jiang Q, Chen H, Song H. Exposure to 17alpha-ethynylestradiol impairs reproductive functions of both male and female zebrafish (*Danio rerio*). Aquat Toxicol, 2008, 88(1): 1-8.

[134] Adeel M, Song X, Wang Y, Francis D, Yang Y. Environmental impact of estrogens on human, animal and plant life: A critical review. Environ Int, 2017, 99: 107-119.

[135] Hallgren P, Nicolle A, Hansson L A, Bronmark C, Nikoleris L, Hyder M, Persson A. Synthetic estrogen directly affects fish biomass and may indirectly disrupt aquatic food webs. Environ Toxicol Chem, 2014, 33(4): 930-936.

[136] Luzio A, Matos M, Santos D, Fontainhas-Fernandes A A, Monteiro S M, Coimbra A M. Disruption of apoptosis pathways involved in zebrafish gonad differentiation by 17alpha-ethinylestradiol and fadrozole exposures. Aquat Toxicol, 2016, 177: 269-284.

[137] Sun L, Shao X, Wu Y, Li J, Zhou Q, Lin B, Bao S, Fu Z. Ontogenetic expression and 17beta-estradiol regulation of immune-related genes in early life stages of Japanese medaka (*Oryzias latipes*). Fish Shellfish Immunol, 2011, 30(4-5): 1131-1137.

[138] Szwejser E, Verburg-van Kemenade B M, Maciuszek M, Chadzinska M. Estrogen-dependent seasonal adaptations in the immune response of fish. Horm Behav, 2017, 88: 15-24.

[139] Liu C J, Lo J F, Kuo C H, Chu C H, Chen L M, Tsai F J, Tsai C H, Tzang B S, Kuo W W, Huang C Y. Akt mediates 17beta-estradiol and/or estrogen receptor-alpha inhibition of LPS-induced tumor necresis factor-alpha expression and myocardial cell apoptosis by suppressing the JNK1/2-NFkappaB pathway. J Cell Mol Med, 2009, 13(9B): 3655-3667.

[140] Liarte S, Cabas I, Chaves-Pozo E, Arizcun M, Meseguer J, Mulero V, Garcia-Ayala A. Natural and synthetic estrogens

modulate the inflammatory response in the gilthead seabream (*Sparus aurata* L.) through the activation of endothelial cells. Mol Immunol, 2011, 48(15-16): 1917-1925.

[141] Leonard J A, Cope W G, Barnhart M C, Bringolf R B. Metabolomic, behavioral, and reproductive effects of the synthetic estrogen 17 alpha-ethinylestradiol on the unionid mussel *Lampsilis fasciola*. Aquat Toxicol, 2014, 150: 103-116.

[142] Kidd K A, Blanchfield P J, Mills K H, Palace V P, Evans R E, Lazorchak J M, Flick R W. Collapse of a fish population after exposure to a synthetic estrogen. Proc Nat Acad Sci USA, 2007, 104(21): 8897-8901.

[143] Pocock T, Falk S. Negative impact on growth and photosynthesis in the green alga Chlamydomonas reinhardtii in the presence of the estrogen 17alpha-ethynylestradiol. PLoS One, 2014, 9(10): e109289.

[144] Genisel M, Turk H, Erdal S. Exogenous progesterone application protects chickpea seedlings against chilling-induced oxidative stress. Acta Physiologiae Plantarum, 2012, 35(1): 241-251.

[145] Han J, Wang Q, Wang X, Li Y, Wen S, Liu S, Ying G, Guo Y, Zhou B. The synthetic progestin megestrol acetate adversely affects zebrafish reproduction. Aquat Toxicol, 2014, 150: 66-72.

[146] Hua J, Han J, Guo Y, Zhou B. The progestin levonorgestrel affects sex differentiation in zebrafish at environmentally relevant concentrations. Aquat Toxicol, 2015, 166: 1-9.

[147] Liang Y Q, Huang G Y, Liu S S, Zhao J L, Yang Y Y, Chen X W, Tian F, Jiang Y X, Ying G G. Long-term exposure to environmentally relevant concentrations of progesterone and norgestrel affects sex differentiation in zebrafish (*Danio rerio*). Aquat Toxicol, 2015, 160: 172-179.

[148] Liang Y Q, Huang G Y, Ying G G, Liu S S, Jiang Y X, Liu S, Peng F J. The effects of progesterone on transcriptional expression profiles of genes associated with hypothalamic-pituitary-gonadal and hypothalamic-pituitary-adrenal axes during the early development of zebrafish (*Danio rerio*). Chemosphere, 2015, 128: 199-206.

[149] Hua J, Han J, Wang X, Guo Y, Zhou B. The binary mixtures of megestrol acetate and 17alpha-ethynylestradiol adversely affect zebrafish reproduction. Environ Pollut, 2016, 213: 776-784.

[150] Zucchi S, Mirbahai L, Castiglioni S, Fent K. Transcriptional and physiological responses induced by binary mixtures of drospirenone and progesterone in zebrafish (*Danio rerio*). Environ Sci Technol, 2014, 48(6): 3523-3531.

[151] Zhao Y, Fent K. Progestins alter photo-transduction cascade and circadian rhythm network in eyes of zebrafish (*Danio rerio*). Sci Rep, 2016, 6: 21559.

[152] Liang Y Q, Huang G Y, Zhao J L, Shi W J, Hu L X, Tian F, Liu S S, Jiang Y X, Ying G G. Transcriptional alterations induced by binary mixtures of ethinylestradiol and norgestrel during the early development of zebrafish (*Danio rerio*). Comp Biochem Physiol C Toxicol Pharmacol, 2017, 195: 60-67.

[153] Fu X D, Garibaldi S, Gopal S, Polak K, Palla G, Spina S, Mannella P, Genazzani A R, Genazzani A D, Simoncini T. Dydrogesterone exerts endothelial anti-inflammatory actions decreasing expression of leukocyte adhesion molecules. Mol Hum Reprod, 2012, 18(1): 44-51.

[154] Cripe G M, Hemmer B L, Raimondo S, Goodman L R, Kulaw D H. Exposure of three generations of the estuarine sheepshead minnow (*Cyprinodon variegatus*) to the androgen, 17beta-trenbolone: Effects on survival, development, and reproduction. Environ Toxicol Chem, 2010, 29(9): 2079-2087.

[155] Raghuveer K, Garhwal R, Wang D S, Bogerd J, Kirubagaran R, Rasheeda M K, Sreenivasulu G, Bhattachrya N, Tarangini S, Nagahama Y, Senthilkumaran B. Effect of methyl testosterone- and ethynyl estradiol-induced sex differentiation on catfish, Clarias gariepinus: expression profiles of DMRT1, Cytochrome P450aromatases and 3 beta-hydroxysteroid dehydrogenase. Fish Physiol Biochem, 2005, 31(2-3): 143-147.

[156] Huang G Y, Liu Y S, Chen X W, Liang Y Q, Liu S S, Yang Y Y, Hu L X, Shi W J, Tian F, Zhao J L, Chen J, Ying G G. Feminization and masculinization of western mosquitofish (*Gambusia affinis*) observed in rivers impacted by municipal

wastewaters. Sci Rep, 2016, 6: 20884.

[157] Naslund J, Fick J, Asker N, Ekman E, Larsson D G J, Norrgren L. Diclofenac affects kidney histology in the three-spined stickleback (*Gasterosteus aculeatus*) at low mug/L concentrations. Aquat Toxicol, 2017, 189: 87-96.

[158] Liu Y, Wang L, Pan B, Wang C, Bao S, Nie X. Toxic effects of diclofenac on life history parameters and the expression of detoxification-related genes in Daphnia magna. Aquat Toxicol, 2017, 183: 104-113.

[159] Bao S, Nie X, Ou R, Wang C, Ku P, Li K. Effects of diclofenac on the expression of Nrf2 and its downstream target genes in mosquito fish (*Gambusia affinis*). Aquat Toxicol, 2017, 188: 43-53.

[160] Flippin J L, Huggett D, Foran C M. Changes in the timing of reproduction following chronic exposure to ibuprofen in Japanese medaka, Oryzias latipes. Aquat Toxicol, 2007, 81(1): 73-78.

[161] Han S, Choi K, Kim J, Ji K, Kim S, Ahn B, Yun J, Choi K, Khim J S, Zhang X, Giesy J P. Endocrine disruption and consequences of chronic exposure to ibuprofen in Japanese medaka (*Oryzias latipes*) and freshwater cladocerans Daphnia magna and Moina macrocopa. Aquat Toxicol, 2010, 98(3): 256-264.

[162] Yang L H, Ying G G, Su H C, Stauber J L, Adams M S, Binet M T. Growth-inhibiting effects of 12 antibacterial agents and their mixtures on the freshwater microalga *Pseudokirchneriella subcapitata*. Environ Toxicol Chem, 2008, 27(5): 1201-1208.

[163] Huang X, Tu Y, Song C, Li T, Lin J, Wu Y, Liu J, Wu C. Interactions between the antimicrobial agent triclosan and the bloom-forming cyanobacteria *Microcystis aeruginosa*. Aquat Toxicol, 2016, 172: 103-110.

[164] Xin X, Huang G, Liu X, An C, Yao Y, Weger H, Zhang P, Chen X. Molecular toxicity of triclosan and carbamazepine to green algae *Chlorococcum* sp.: A single cell view using synchrotron-based Fourier transform infrared spectromicroscopy. Environ Pollut, 2017, 226: 12-20.

[165] Ho J C, Hsiao C D, Kawakami K, Tse W K. Triclosan (TCS) exposure impairs lipid metabolism in zebrafish embryos. Aquat Toxicol, 2016, 173: 29-35.

[166] Wang C F, Tian Y. Reproductive endocrine-disrupting effects of triclosan: Population exposure, present evidence and potential mechanisms. Environ Pollut, 2015, 206: 195-201.

[167] Schultz M M, Bartell S E, Schoenfuss H L. Effects of triclosan and triclocarban, two ubiquitous environmental contaminants, on anatomy, physiology, and behavior of the fathead minnow (*Pimephales promelas*). Arch Environ Contam Toxicol, 2012, 63(1): 114-124.

[168] Brausch J M, Rand G M. A review of personal care products in the aquatic environment: environmental concentrations and toxicity. Chemosphere, 2011, 82(11): 1518-1532.

[169] Watanabe Y, Kojima H, Takeuchi S, Uramaru N, Ohta S, Kitamura S. Comparative study on transcriptional activity of 17 parabens mediated by estrogen receptor alpha and beta and androgen receptor. Food Chem Toxicol, 2013, 57: 227-234.

[170] Kodani S D, Overby H B, Morisseau C, Chen J, Zhao L, Hammock B D. Parabens inhibit fatty acid amide hydrolase: A potential role in paraben-enhanced 3T3-L1 adipocyte differentiation. Toxicol Lett, 2016, 262: 92-99.

[171] Liu H, Sun P, Liu H, Yang S, Wang L, Wang Z. Hepatic oxidative stress biomarker responses in freshwater fish Carassius auratus exposed to four benzophenone UV filters. Ecotoxicol Environ Saf, 2015, 119: 116-122.

[172] Amar S K, Goyal S, Mujtaba S F, Dwivedi A, Kushwaha H N, Verma A, Chopra D, Chaturvedi R K, Ray R S. Role of type I & type II reactions in DNA damage and activation of caspase 3 via mitochondrial pathway induced by photosensitized benzophenone. Toxicol Lett, 2015, 235(2): 84-95.

[173] Amar S K, Goyal S, Dubey D, Srivastav A K, Chopra D, Singh J, Shankar J, Chaturvedi R K, Ray R S. Benzophenone 1 induced photogenotoxicity and apoptosis via release of cytochrome c and Smac/DIABLO at environmental UV radiation. Toxicol Lett, 2015, 239(3): 182-193.

[174] Kim S, Choi K. Occurrences, toxicities, and ecological risks of benzophenone-3, a common component of organic sunscreen products: A mini-review. Environ Int, 2014, 70: 143-157.

[175] In S J, Kim S H, Go R E, Hwang K A, Choi K C. Benzophenone-1 and nonylphenol stimulated MCF-7 breast cancer growth by regulating cell cycle and metastasis-related genes via an estrogen receptor alpha-dependent pathway. J Toxicol Environ Health A, 2015, 78(8): 492-505.

[176] Shin S, Go R E, Kim C W, Hwang K A, Nam K H, Choi K C. Effect of benzophenone-1 and octylphenol on the regulation of epithelial-mesenchymal transition via an estrogen receptor-dependent pathway in estrogen receptor expressing ovarian cancer cells. Food Chem Toxicol, 2016, 93: 58-65.

[177] Haselman J T, Sakurai M, Watanabe N, Goto Y, Onishi Y, Ito Y, Onoda Y, Kosian P A, Korte J J, Johnson R D, Iguchi T, Degitz S J. Development of the Larval Amphibian Growth and Development Assay: Effects of benzophenone-2 exposure in Xenopus laevis from embryo to juvenile. J Appl Toxicol, 2016, 36(12): 1651-1661.

[178] Dutta U. Repeated administration of benzophenone induced oxidative stress in alterations of morphometry and cytomembrane morphology of rat red blood cell. Toxicol Lett, 2016, 259: S155.

[179] Gatermann R, Biselli S, Hühnerfuss H, Rimkus G G, Hecker M, Karbe L. Synthetic musks in the environment. Part 1: Species-dependent bioaccumulation of polycyclic and nitro musk fragrances in freshwater fish and mussels. Arch Environ Contam Toxicol, 2002, 42(4): 437-446.

[180] Carlsson G, Örn S, Andersson P L, Söderström H, Norrgren L. The impact of musk ketone on reproduction in zebrafish (*Danio rerio*). Mar Environ Res, 2000, 50(1): 237-241.

[181] Schnell S, Martin-Skilton R, Fernandes D, Porte C. The interference of nitro- and polycyclic musks with endogenous and xenobiotic metabolizing enzymes in Carp: An *in vitro* study. Environ Sci Technol, 2009, 43(24): 9458-9464.

[182] 雷炳莉, 黄圣彪, 王子健. 生态风险评价理论和方法. 化学进展, 2009, 21(2/3): 350-358.

[183] 赵建亮, 应光国, 魏东斌, 任明忠. 水体和沉积物中毒害污染物的生态风险评价方法体系研究进展. 生态毒理学报, 2011, 6(6): 577-588.

[184] Boxall A B A, Keller V D J, Straub J O, Monteiro S C, Fussell R, Williams R J. Exploiting monitoring data in environmental exposure modelling and risk assessment of pharmaceuticals. Environ Int, 2014, 73: 176-185.

[185] Grung M, Kallqvist T, Sakshaug S, Skurtveit S, Thomas K V. Environmental assessment of Norwegian priority pharmaceuticals based on the EMEA guideline. Ecotoxicol Environ Saf, 2008, 71(2): 328-340.

[186] Santos J L, Aparicio I, Alonso E. Occurrence and risk assessment of pharmaceutically active compounds in wastewater treatment plants. A case study: Seville city (Spain). Environ Int, 2007, 33(4): 596-601.

[187] Ying G G, Kookana R S, Kolpin D W. Occurrence and removal of pharmaceutically active compounds in sewage treatment plants with different technologies. J Environ Monit, 2009, 11(8): 1498-1505.

[188] Dai G, Wang B, Fu C, Dong R, Huang J, Deng S, Wang Y, Yu G. Pharmaceuticals and personal care products (PPCPs) in urban and suburban rivers of Beijing, China: Occurrence, source apportionment and potential ecological risk. Environmental Science: Processes & Impacts, 2016, 18(4): 445-455.

[189] Peng F-J, Pan C-G, Zhang M, Zhang N-S, Windfeld R, Salvito D, Selck H, Van den Brink P J, Ying G-G. Occurrence and ecological risk assessment of emerging organic chemicals in urban rivers: Guangzhou as a case study in China. Sci Total Environ, 2017, 589: 46-55.

[190] Hernando M D, Mezcua M, Fernandez-Alba A R, Barcelo D. Environmental risk assessment of pharmaceutical residues in wastewater effluents, surface waters and sediments. Talanta, 2006, 69(2): 334-342.

[191] Liu W R, Zhao J L, Liu Y S, Chen Z F, Yang Y Y, Zhang Q Q, Ying G G. Biocides in the Yangtze River of China: Spatiotemporal distribution, mass load and risk assessment. Environ Pollut, 2015, 200: 53-63.

[192] Wright G D. Antibiotic resistance in the environment: A link to the clinic? Curr Opin Microbiol, 2010, 13(5): 589-594.

[193] Pruden A, Pei R, Storteboom H, Carlson K H. Antibiotic resistance genes as emerging contaminants: studies in northern Colorado. Environ Sci Technol, 2006, 40(23): 7445-7450.

[194] Martinez J L, Coque T M, Baquero F. What is a resistance gene? Ranking risk in resistomes. Nat Rev Microbiol, 2014:

[195] Su H C, Ying G G, Tao R, Zhang R Q, Zhao J L, Liu Y S. Class 1 and 2 integrons, sul resistance genes and antibiotic resistance in Escherichia coli isolated from Dongjiang River, South China. Environ Pollut, 2012, 169: 42-49.

[196] Luo Y, Mao D, Rysz M, Zhou Q, Zhang H, Xu L, P J J A. Trends in antibiotic resistance genes occurrence in the Haihe River, China. Environ Sci Technol, 2010, 44(19): 7220-7225.

[197] Dang B, Mao D, Xu Y, Luo Y. Conjugative multi-resistant plasmids in Haihe River and their impacts on the abundance and spatial distribution of antibiotic resistance genes. Water Res, 2017, 111: 81-91.

[198] Mao D, Luo Y, Mathieu J, Wang Q, Feng L, Mu Q, Feng C, Alvarez P J. Persistence of extracellular DNA in river sediment facilitates antibiotic resistance gene propagation. Environ Sci Technol, 2014, 48(1): 71-78.

[199] Zhu Y G, Zhao Y, Li B, Huang C L, Zhang S Y, Yu S, Chen Y S, Zhang T, Gillings M R, Su J Q. Continental-scale pollution of estuaries with antibiotic resistance genes. Nat Microbiol, 2017, 2: 16270.

[200] Yang Y, Li B, Ju F, Zhang T. Exploring variation of antibiotic resistance genes in activated sludge over a four-year period through a metagenomic approach. Environ Sci Technol, 2013, 47(18): 10197-10205.

[201] Luo Y, Yang F, Mathieu J, Mao D, Wang Q, Alvarez P J J. Proliferation of multidrug-resistant new delhi metallo-β-lactamase genes in municipal wastewater treatment plants in Northern China. Environmental Science & Technology Letters, 2014, 1(1): 26-30.

[202] Wang Q, Mao D, Luo Y. Ionic liquid facilitates the conjugative transfer of antibiotic resistance genes mediated by plasmid RP4. Environ Sci Technol, 2015, 49(14): 8731-8740.

[203] He L Y, Liu Y S, Su H C, Zhao J L, Liu S S, Chen J, Liu W R, Ying G G. Dissemination of antibiotic resistance genes in representative broiler feedlots environments: Identification of indicator ARGs and correlations with environmental variables. Environ Sci Technol, 2014, 48(22): 13120-13129.

[204] He L Y, Ying G G, Liu Y S, Su H C, Chen J, Liu S S, Zhao J L. Discharge of swine wastes risks water quality and food safety: Antibiotics and antibiotic resistance genes from swine sources to the receiving environments. Environ Int, 2016, 92-93: 210-219.

[205] Zhu Y G, Johnson T A, Su J Q, Qiao M, Guo G X, Stedtfeld R D, Hashsham S A, Tiedje J M. Diverse and abundant antibiotic resistance genes in Chinese swine farms. Proc Natl Acad Sci U S A, 2013, 110(9): 3435-3440.

[206] Wang F H, Qiao M, Su J Q, Chen Z, Zhou X, Zhu Y G. High throughput profiling of antibiotic resistance genes in urban park soils with reclaimed water irrigation. Environ Sci Technol, 2014, 48(16): 9079-9085.

[207] Chen Q, An X, Li H, Su J, Ma Y, Zhu Y G. Long-term field application of sewage sludge increases the abundance of antibiotic resistance genes in soil. Environ Int, 2016, 92-93: 1-10.

[208] Li B, Yang Y, Ma L, Ju F, Guo F, Tiedje J M, Zhang T. Metagenomic and network analysis reveal wide distribution and co-occurrence of environmental antibiotic resistance genes. ISME J, 2015, 9(11): 2490-2502.

[209] Ju F, Li B, Ma L, Wang Y, Huang D, Zhang T. Antibiotic resistance genes and human bacterial pathogens: Co-occurrence, removal, and enrichment in municipal sewage sludge digesters. Water Res, 2016, 91: 1-10.

[210] Yang Y, Li B, Zou S, Fang H H, Zhang T. Fate of antibiotic resistance genes in sewage treatment plant revealed by metagenomic approach. Water Res, 2014, 62: 97-106.

[211] Ma L, Xia Y, Li B, Yang Y, Li L G, Tiedje J M, Zhang T. Metagenomic assembly reveals hosts of antibiotic resistance genes and the shared resistome in pig, chicken, and human feces. Environ Sci Technol, 2016, 50(1): 420-427.

[212] Li L G, Xia Y, Zhang T. Co-occurrence of antibiotic and metal resistance genes revealed in complete genome collection. ISME J, 2017, 11(3): 651-662.

[213] Yang Y, Jiang X, Chai B, Ma L, Li B, Zhang A, Cole J R, Tiedje J M, Zhang T. ARGs-OAP: online analysis pipeline for antibiotic resistance genes detection from metagenomic data using an integrated structured ARG-database. Bioinformatics, 2016, 32(15): 2346-2351.

作者：应光国[1,2]，赵建亮[1]，刘有胜[1]，张芊芊[1]，何良英[1]，杨 滨[2]，史文俊[1]，黄国勇[1]

[1]中国科学院广州地球化学研究所，[2]华南师范大学环境研究院

第 8 章　农药环境化学与毒理学研究

本章导读

中国作为发展中的人口大国，粮食安全是关系到国家安定的重大战略。在世界范围内，当今农业生产的收获品中,1/3 左右是依靠化学农药从病虫草害中夺回的。在中国如不使用农药,大约将损失 30%~40%的粮食。因此，农药在我国农业生产中占有重要地位。然而当今世界，化学农药依然是我国农药产品的主体，大量使用此类农药对生态环境、人民健康、食品安全、出口贸易等均产生了危害。化学农药的不可替代性与环境污染、食品安全的矛盾，使得绿色化学农药创制、农药风险评估与预警成为农药发展研究的必然趋势。本章将对近年来我国科学家对农药环境残留特征、生物体污染负荷、生态安全与健康风险、毒性机制及手性农药环境安全等方面的研究进行综合介绍。

关键词

农药，残留特征手性，环境安全，健康风险

1 引　　言

中国作为占世界总人口超过 1/5 的发展中大国，农业发展和食品安全是我国的重大国家战略。而农业的发展方向是优质、安全、高产、高效，这其中农药尤其是化学农药具有十分重要、无法回避的作用和影响。在全球，农业病虫草害种类有十多万种，包括昆虫、线虫、微生物、杂草等。根据统计分析，每年通过施用化学农药挽回的粮食损失约占全球总产量的 30%左右。现代林业、卫生防疫、园林业的发展又进一步提高了对农药的需求。我国已是农药生产和使用量最多的国家。然而，大量事实表明：农药特别是高毒、高残留化学农药的广泛使用是造成我国土壤、地表水污染乃至影响食品安全的主要因素之一。另一方面，随着我国生态文明建设和可持续发展战略的推进，要求在对病虫害加以有效控制的同时，必须注重和加强农药的环境安全。由于农药作用对象的生物多样性、靶标与非靶标的类近性、生态环境效应的复杂性，农药的环境化学与毒理学一直是近年农药学和环境科学研究的热点。

我国农药污染现状复杂，既有被禁用多年的持久性有机污染物（POPs）类传统农药，也有现用的大量新农药环境安全问题。因此，农药环境化学与毒理学的研究涉及对传统农药和新农药的长期环境行为、负荷水平、潜在风险等多方面的研究。同时许多农药在结构上往往具有异构体（isomer），包含空间结构的立体异构（stereoisomer），即存在化学结构多样性，而不同异构体之间的环境行为和生物效应差异很大，其生态风险也有所不同。近年来我国科学家在绿色农药创制、全国范围内农药在不同环境介质的残留特征与生态风险、人体负荷及健康风险、毒性效应机制及环境风险管理方面都取得了很好的研究成果，我们将就此进行逐一论述。

2 POPs 类传统农药环境残留特征及其生态风险

传统的有机氯农药（organochlorine pesticides，OCPs）成本低、效率高、杀虫广谱，是一类应用最早的人工合成的高效广谱农药，在控制病虫害方面有过突出的贡献。首批受控化学物质包括 12 种 POPs 中包含 9 种农药：艾氏剂、氯丹、狄氏剂、异狄氏剂、七氯、灭蚁灵、毒杀芬、滴滴涕、六氯代苯。但它们在给人类带来巨大经济和社会效益的同时，以滴滴涕（DDTs）和六六六 HCHs 为代表的 OCPs 也带来了严重的环境问题。POPs 类传统农药具有半挥发性，可以从水体或土壤介质中以蒸气的形式进入大气，通过大气环流在大气环境中作长距离迁移并发生沉降，当温度升高时，它们再次挥发进入大气，进行迁移。由于它们中的大多数化学性质十分稳定，广泛残留在各种环境介质中，并且可通过食物链不断蓄积放大，在生物体内富集，不易分解，使得这类农药可分布到世界的任何一个角落，对环境造成污染，威胁生态安全和人类健康。

我国是 OCPs 的生产和使用大国，曾经工业化生产过的就有 6 种 OCPs（分别是：DDTs、HCHs、毒杀芬、氯丹、七氯和灭蚁灵）。从 20 世纪 50 年代开始，我国主要使用的 OCPs 是 HCHs 和 DDTs，这两种农药约占了当时全部农药产量的 50%~60%，至 1983 年禁止生产的 30 年间，我国 DDTs 生产量高达 40 多万吨，是世界上第二大 OCPs 生产和使用国，仅次于美国。

虽然在我国 OCPs 已被禁用 30 多年，环境介质中其残留量随着代谢和挥发等环境行为有了明显的下降，但由于其历史使用量大、蒸气压低、挥发性小、水溶性小、脂溶性大，并且化学结构稳定，不易被降解等特点，OCPs 在环境中的消失速率缓慢，容易在稳定的介质，如土壤、沉积物和生物体中富集放大。目前，OCPs 仍然是环境介质中检出率较高的一类 POPs。因此，POPs 类农药特别是 DDTs 和 HCHs 在多介质环境中的迁移、交换过程一直是环境科学研究的前沿和热点之一，特别是在我国大尺度的研究对于了解这类农药对我国环境安全的影响具有重要意义。

2.1　DDTs 在我国农田土的残留特征及风险

DDTs 是以苯为原料的 OCPs，瑞士科学家 Müller 最早将其用于杀灭了马铃薯甲虫，并因此获得了诺贝尔奖。在随后的几十年 DDTs 在病虫害及疟疾控制等方面有过重要作用。我国 DDTs 的历史使用量高达 40 万吨，由于 DDTs 一系列的生态环境问题，20 世纪 80 年代我国禁用了 DDTs。但目前，DDTs 仍被亚非拉热带地区用于疟疾防控及三氯杀螨醇生产的中间体使用，因此 DDTs 污染包含历史使用遗留和部分污染源为新源的双重问题。

牛丽丽等[1]最新研究结果表明，我国农田土壤中 DDTs 总浓度的平均值为 8.06 ng/g，*o*, *p'*-DDE、*p*, *p'*-DDE、*o*, *p'*-DDD、*p*, *p'*-DDD、*o*, *p'*-DDT 和 *p*, *p'*-DDT 的平均浓度分别为 0.137 ng/g、3.29 ng/g、0.242 ng/g、0.469 ng/g、0.711 ng/g 和 3.21 ng/g，主要来源为历史上工业 DDTs 的使用，部分属于三氯杀螨醇和 DDTs 的新来源。而 DDTs 在我国东部地区农田土壤中的浓度高于西部地区，其残留特征受土壤有机质含量及地区社会经济发展的影响。与 2011 年的研究数据进行比较，发现经过两年的时间，DDTs 在我国农田表层土壤中的平均浓度及残留总量均略有下降，土壤有机质在这个过程中起到了重要的作用[2]。对映体特征分析表明，在我国大多数农田土壤中，(–)-*o*, *p'*-DDT 相对于(+)-*o*, *p'*-DDT 会优先进行富集，并且随着环境温度的升高，*o*, *p'*-DDT 的对映体偏差也有逐渐上升的趋势。在我国，2.46%的农田土壤样品中 DDTs 的含量对人体具有非致癌性风险，以及 1.64%土壤样品具有致癌性风险外，其余农田土样品中 DDTs 暴露浓度于人体相对安全同时，DDTs 在我国农田周边树皮中的平均值为 5.61 ng/g，*o*, *p'*-DDE、*p*, *p'*-DDE、*o*, *p'*-DDD、*p*, *p'*-DDD、*o*, *p'*-DDT 和 *p*, *p'*-DDT 的平均浓度分别为 0.558 ng/g、1.56 ng/g、0.375 ng/g、0.298 ng/g、1.22 ng/g 和 1.60 ng/g[3]，主要来源为历史上工业 DDTs 的使用，部分也受三氯杀螨醇和 DDTs 新来源的影响。与土壤中 DDTs 的分布趋势相似，树皮中 DDTs 在我国东部地区的浓度高于西部地区，其中我国京津冀地区 DDTs 的浓度最高。大气中 $PM_{2.5}$ 的浓度及各地的社会经济发展程度是影响 DDTs 在大气中再次分布的重要因素。具有手性的 *o*, *p'*-DDT 在树皮中的对映体选择性特征主要受土壤挥发影响，大部分树皮中(–)-*o*, *p'*-DDT 会选择性富集而(+)-*o*, *p'*-DDT 则会优先被降解。树皮/大气分配模型预测我国农田大气中 DDTs 的浓度，平均值为 860 pg/m^3，污染集中在我国的东南部地区。我国大气中 DDTs

通过呼吸途径对人体产生的致癌性风险全部低于最低风险值，即不会对人体产生健康风险。土气交换特征研究结果表明，我国大部分农田中 DDTs 及其代谢产物的逸度分数值小于 0.3，说明我国农田中 DDTs 仍处于从大气向土壤沉降的过程中，并且数据表明，这个过程主要受环境温度和降水量的影响。

2.2 HCHs 在我国农田土的残留特征及风险

HCHs（混合异构体，包括四种主要成分：α-、β-、γ-和 δ-HCH），我国的产量和使用量都居世界首位。到 1983 年我国禁止在农业上使用为止，HCHs 累计产量达到了 490 万吨。现在，HCHs 作为农药中间体仍然在国内生产，并且年产量高达 1.4 万吨。特别值得注意的是，作为丙体六六六（γ-HCH）——林丹，仍有生产使用，年产量达到 1335 吨，主要用于防治小麦吸浆虫、飞蝗、荒滩竹蝗等。

研究表明，我国农田土壤中 α-HCH、β-HCH、γ-HCH 和 HCHs 总浓度的平均值分别为 0.190 ng/g、1.31 ng/g、0.236 ng/g 和 1.74 ng/g，其残留主要来源于历史上工业 HCHs 的使用[4]。高浓度的 HCHs 主要分布在我国的中部和南部地区，相对于 β-HCH 异构体来说，α-HCH 和 γ-HCH 更容易富集在 HCHs 总浓度低、温度低、海拔高以及降水量少的地区。对映体特征分析结果表明，在我国大部分农田土壤中，(–)-α-HCH 更容易优先降解。土壤中 HCHs 通过多种途径暴露于人体产生的健康风险评估结果表明，非致癌性风险的平均危害指数均小于 1，说明我国农田土壤中 HCHs 不会对人体产生非致癌性风险；致癌性风险值小于 1×10^{-4}，说明我国农田土壤中 HCHs 对人体的致癌性风险处于低或非常低的水平。

16S rRNA 基因序列分析表明，我国农田土壤中包含四种已知的鞘氨醇单胞菌属：“*Sphingomonas*”，“*Novosphingobium*”，“*Sphingopyxis*”，“*Sphingobium*”。其中 *Sphingomonas* 属存在于所有土壤中，分布最广泛，是这些菌中的优势属。通过 *linA* 基因的定量 PCR 结果发现，不同样品中鞘氨醇单胞菌的丰度有较大的差异，个别样品中鞘氨醇单胞菌的丰度相对较低。皮尔逊相关分析表明鞘氨醇单胞菌丰度与总 HCHs 残留之间没有显著的相关性，但是与 γ-HCH 浓度是显著相关[5]。说明特异微生物的降解对 HCHs 在农田土中的残留有重要影响。

我国农田周边树皮中 α-HCH、β-HCH、γ-HCH、δ-HCH 和 HCHs 总浓度的平均值分别为 1.16 ng/g、2.51 ng/g、1.67 ng/g、0.368 ng/g 和 5.71 ng/g，它们主要是由于历史上工业 HCHs 和林丹的使用随大气迁移造成的残留。京津冀地区是我国树皮中 HCHs 残留最高的地区。树皮中 HCHs 总浓度不仅与年平均降水量和温度呈显著负相关，并且由于树皮能够同时富集气态和颗粒态的污染物，因此其还与大气中 PM_{10} 和 $PM_{2.5}$ 的浓度呈正相关。在对映体分析中，(+)-α-HCH 会在大多数树皮样品中优先富集，且 α-HCH 的对映体比值与 α-HCH 的浓度及采样点的海拔正相关[6]。

2.3 有机氯农药在我国长江三角洲的残留特征及风险

长江三角洲（Yangtze River Delta）是我国最大的河口三角洲，位于我国东部，包括上海市、江苏省南部、浙江省北部以及邻近海域，面积约为 99600 km^2，人口约 7500 万，是全国人口密度最大、经济实力最强、产业规模最大的地区之一。长三角濒临东海，其大陆海岸线长，有一条宽约几千米到几十千米的潮间带浅滩，浅海滩涂资源丰富。附近海域包含中国最大的海洋渔业基地舟山渔场，为我国东部沿海地区提供了大量的渔业资源。浙江是海洋大省，位于东南沿海和长江三角洲南翼，东临东海，南接福建，西与江西和安徽相连，北与上海和江苏为邻，是全国经济、社会和文化最为发达的地区之一。周珊珊等[7-12]选定长江三角洲毗邻海域作为重点研究对象，采集了长江三角洲毗邻海域的沉积物、水体和水中悬浮颗粒物样品，分析了 OCPs 在该区域的时空分布及来源解析，并测定了手性 OCPs 的 EF 值，浙江省毗邻海域表层沉积物中 OCPs 调查结果表明，象山湾、三门湾和乐清湾 ΣOCPs 含量分别为 0.93~23.2 ng/g 干重、

3.33~7.41 ng/g 干重和 2.11~18.15 ng/g 干重。与国内外其他地区相比，OCPs 的浓度处于中等水平。HCHs 各异构体的组成表明该海域残留的 HCHs 主要是历史残留；而 DDTs 各异构体的组成说明了该区域残留的 DDTs 可能有新的来源，新的来源可能是当地频繁使用的三氯杀螨醇。手性 OCPs 的测定结果表明，α-HCH、2, 4-DDD 和 2, 4-DDT 都发生对映体选择性降解，(+) 的对映体优先降解。

象山湾水体和悬浮颗粒物中总 OCPs 的垂直分布结果表明，水体中 ΣOCPs 为 2.88~34.72 ng/L。HCHs 和 DDTs 的浓度都低于中国《地表水环境质量标准》(GB 3838—2002) 的 Ⅰ 类标准。但大部分站点的 OCPs 的浓度都已经超过了美国 EPA 标准，对生物和生态系统有潜在风险。悬浮颗粒物中 ΣOCPs 含量为 2.47~29.94 ng/L，与国内其他地区比较，OCPs 的污染处于中等水平[10]。根据沉积物风险评估标准，长江三角洲毗邻海域残留的 4, 4-DDT 和 ΣDDTs 存在一定的生态风险。考虑到 OCPs 不容易降解，通过食物链的积累和放大，最后进入人体的浓度可能会放大几十倍甚至几百倍，所以该研究区域及其他类似地区 OCPs 的污染状况还应引起人们的注意[8]。

3　农药的人体负荷及健康风险

全球范围人群农药暴露现象普遍，人群长期暴露于 OCPs、有机膦类、拟除虫菊酯类杀虫剂及多种杀菌剂中。特别是我国虽然有一整套严格的农药使用和管理制度，但是当前我国农产品中的农药残留问题仍旧十分严重，部分城市市场上瓜果蔬菜中农药残留超标率高达 47%。与此同时在不同的环境介质农田土、水体、大气中都能检测到多种农药残留。由此可见，通过饮食摄入、呼吸、消化道、皮肤接触等多种途径，农药暴露所引起的人群健康风险效应也成为环境健康的热点。

传统的 OCPs 因毒性高、生物富集性强及易全球性长距离迁移等，在全球人群样品中检出率极高，其健康风险效应一直是持久性有毒物质健康效应研究的重点。更严重的是随着大量新型农药进入环境，这些化合物的人群污染负荷和健康风险效应研究空白还很大，存在潜在的未知风险，特别是对于一些敏感人群如孕妇、新生儿、青春期儿童等的影响亟待甄别。

3.1　DDTs

DDTs 是一类在环境中能长期滞留，具有生物放大效应，可蓄积在脂肪组织中的内分泌干扰物，其低剂量暴露健康风险问题如：内分泌干扰作用、与肿瘤的相关性报道较多，且主要集中于一些 DDTs 已禁用多年的发达国家。在我国尤其针对敏感人群的研究还非常缺乏。

唐梦玲等[2]分析了中国各省（除香港、澳门）的农田土壤样本；构建了在四个不同膳食结构区域、由六类食物摄入而富集的 DDTs 的剂量模型；用人群归因分值（population attributable factor，PAF）表征了我国 DDTs 暴露的乳腺癌发病风险。结果显示，中国主要农业区域土壤中 DDTs 残留浓度较高，DDTs 总量的浓度中位数为 2.61 ng/g，范围为 0.37~547.03 ng/g，超标（100 ng/g）的土壤威胁 4230 万中国人口的健康。考虑到不同地区人群的膳食结构不同，通过模型计算的日均膳食摄入 DDTs 剂量估计为 0.34 μg/kg 体重 p, p'-DDE（DDTs 的生物活性组分）。而 PAF 中位数为 0.60%（IQR，0.23%~2.11%），这个结果表明 p, p'-DDE 贡献的中国女性年乳腺癌发病率为十万分之 0.06。尽管 DDTs 暴露是乳腺癌发病风险之一，但较低的贡献率说明其他因素，如基因易感性、生活习惯等才是乳腺癌发病的主要因素。

海洋食物链对于 OCPs 的强生物富集能力，使得海岛人群具有高水平暴露的特点，而其中母婴作为暴露敏感人群[13]，对于环境监测评价具有重要意义。脐带血因其非损伤新采集方式等优势经常作为新生儿有机氯污染物暴露指标[14]。母乳，尤其是初乳，因丰富的蛋白质和含脂量，是新生儿主要的食物来源，不仅能够反映母体积累量更能用于计算新生儿摄入水平[15]。长江三角洲作为我国最主要农业生产区域，其毗邻海域有机氯化合物污染特别严重，大量污染物随着江水入海底泥冲刷在嵊泗积累[7]。此外，由于坐落于中国最大的“东海鱼库”舟山渔场中心，鱼类及其他海洋生物是岛上居民主要食物来源。嵊泗列岛位于杭州湾以东，正对长江入海口，是浙江省最东部，舟山群岛最北部的海岛。最新研究结果表明，2011年7月~2012年5月期间生产的106对母婴样本，初乳和脐血血清的总DDTs平均值为（1.93±1.98）μg/kg脂重、（1.38±3.66）μg/kg脂重，高于土耳其、英国、日本、波兰、巴西等地，与北京、越南、印度、泰国等地相当[16]，其中*p*, *p*'-DDE，100%检出，为主要组分。Spearman相关性分析及Kruskal-Wallis检验发现产妇年龄和孕前体重对初乳中*p*, *p*'-DDE，*p*, *p*'-DDT及总DDTs呈正相关。通过组分比例及主成分分析解释人体内蓄积DDTs主要来自于历史工业DDTs残留，但不可忽视新源如三氯杀螨醇的影响[17]。运用多元线性回归模型考察DDTs在脐带血中不同组分暴露程度对新生儿出生结果的差别，结果显示，暴露水平与新生儿身长、妊娠周期、头围均无统计学意义差异。但*p*, *p*'-DDT及其代谢产物*p*, *p*'-DDD与新生儿出生体重有显著性正向关系。在纳入孕妇年龄，产前BMI，新生儿性别，胎次，饮水来源等协变量因素之后，差异显著，之前的调查也证实环境暴露DDTs与新生儿肥胖有关，胎儿期的发生直接影响到成年后结果[18]。此外，通过计算对映体比值发现母乳及脐带血中均选择性富集(–)-*o*, *p*'-DDD和(+)-*o*, *p*'-DDT，具有对映体选择性差异。根据世界卫生组织（WHO）在膳食营养素适宜摄入量计算得到，该区域新生儿平均每日摄入DDTs量为8.78 μg/kg，20%的新生儿高于WHO制定的暂定每日耐受摄入量10 μg/kg。因此，我们认为该地区婴儿通过母乳摄入DDTs具有一定风险，值得进一步研究。

3.2 HCHs母婴暴露风险

HCHs是一种广谱性的有机氯杀虫剂，从20世纪40年代开始使用，尽管HCHs已经禁用约30年，HCHs对全球环境和人类的影响仍十分显著。

以中国浙江省舟山市嵊泗县的常住居民为研究对象，2012年对39个健康产妇进行了详细的问卷调查，并在生产后采集了新生儿的脐带血样品，检测脐血血浆中的HCHs残留浓度，并结合调查问卷的相关信息，来分析HCHs残留对中国海岛人群新生儿生长发育指标的影响。研究结果提示，岛上新生儿生长发育水平普遍高于我国其他地域的新生儿生长发育，HCHs残留浓度相较于国外研究也较高，但HCHs的残留浓度与新生儿生长发育指标之间没有发现显著的相关性关系，与国外部分研究结论有差异。因此，本研究提示，HCHs对新生儿生长发育影响的研究还需要更多的流行病学证据，对于HCHs对新生儿健康的影响机制还需要更深入的探索和验证。

2011年7月~2012年5月在嵊泗人民医院对分娩的产妇及新生儿120进行调查，结果表明，母初乳中三种能检测到的HCHs组分总量中位数为6.46 μg/L[四分差（interquartile range，IQR），3.78~16.02]，脐血清中三种能检测到的HCHs组分总量中位数为2.39 μg/L（IQR，1.30~3.92），在母初乳及脐血清中，三种HCHs组分中残留浓度最高的均为γ-HCH，最低的为α-HCH。HCHs各组分的TDI值中γ-HCH的母初乳日均单位体重摄入量最高，约有56.67%的新生儿母初乳摄入的γ-HCH水平超过了美国EPA设定的参考剂量RfD，有一定暴露风险；同时，不同暴露组间新生儿生长发育指标差异分析表明，HCHs不同暴露组间的出生体重和孕期并未有统计学显著性差异[2]。

3.3 拟除虫菊酯杀虫剂

拟除虫菊酯杀虫剂，是一类改变天然除虫菊酯的化学结构衍生合成的杀虫剂。由于其较高的杀虫活性，较低的哺乳动物毒性，容易降解，环境友好等优点，拟除虫菊酯杀虫剂受到十分广泛而高频的使用。其在食品中普遍残留，人类主要通过食物受到暴露，作为美国内分泌干扰物清单的一员，其内分泌干扰效应和其他健康风险不容忽视。

青春期为人体发育的重要时期，也是对内分泌干扰物暴露非常敏感的窗口期。叶小青等采集了杭州463 名男孩（9~16 岁）的尿液样品，检测了尿液中 3-PBA 及促性腺激素 FSH、LH 的浓度[19]。结果表明，尿液中 3-PBA 的浓度与促性腺激素 FSH、LH 浓度均呈显著正相关，与青春期发育程度（tanner testis volume 和 tanner genitalia stage）均呈显著正相关，这提示着拟除虫菊酯是男孩青春期发育提前的风险因子。与之相反的是，女孩尿液中 3-PBA 的浓度与青春期发育程度及初潮年龄呈显著负相关，这提示着拟除虫菊酯是女孩青春期发育延迟的风险因子。拟除虫菊酯杀虫剂暴露对青少年青春期发育影响的性别差异的潜在机制需要进一步探究[20]。

对山西省忻州地区 2013~2014 年 208 例非职业人群拟除虫菊酯杀虫剂暴露与心血管疾病的关系研究表明，尿液中拟除虫菊酯杀虫剂主要代谢产物 3-PBA 在健康人群的浓度为 0.74 μg/L，在冠心病人群的浓度为 1.09 μg/L，明显高于正常健康人群。研究首次报道了拟除虫菊酯类杀虫剂暴露与冠心病风险之间可能存在的相关性，但具体机制还有待于进一步研究[21]。

4　农药生物有效性与环境行为

农药生物有效性是影响农药生态风险评估的关键因子，受土壤类型、农药理化性质、环境因素及有机体的综合影响。传统上使用总浓度来评价土壤中农药的生物有效性，但与有机体吸收相比，往往高估其生物有效性，并且由于土壤理化性质和老化等因素，同一农药在不同土壤中具有不同的生物有效性，表明总浓度并不能准确反映其生物有效性。研究发现与土壤结合的农药不易被生物吸收，只有孔隙水中自由溶解态的农药才能被生物利用。因此，原位孔隙水中农药浓度是不同土壤中农药生物有效性及环境风险评估更有效的参数。原位孔隙水中农药浓度是指溶解在土壤等基质孔隙水水相中，而不与土壤有机质等结合的浓度。它不仅与农药的总浓度有关，而且与基质浓度和容量及其对农药的亲和力相关。虞云龙等以烟嘧磺隆、氟噻草胺、咪唑喹磷酸、多菌灵为目标化合物，研究了不同土壤中烟嘧磺隆、氟噻草胺、咪唑喹啉酸等除草剂对植物的单一毒性和复合毒性以及多菌灵等杀菌剂对蚯蚓的毒性，提出了基于孔隙水农药生物有效性或暴露水平的毒性评价方法，并创新性地将其应用于复合毒性的评价[22-26]。

农药环境行为和环境归趋是农药环境化学与毒理学的重要组成，近年来花日茂教授课题组，以一种广谱取代苯杀菌剂百菌清为模式化合物，开展了百菌清在水溶液中以及在辣椒表面的光催化降解，旨在寻找一种环境友好的降解百菌清残留污染方法，同时为农药污染去除提供新的思路。研究结果表明，光敏化剂纳米 TiO_2 和天然植物提取物 EGCG 和 OPC 对百菌清在水溶液和植物表面的光催化降解。纳米 TiO_2 对水中及辣椒表面上百菌清的光降解均具有显著的光敏化降解作用且可将百菌清彻底分解，避免高毒代谢产物 4-羟基百菌清积累；天然植物提取物 EGCG 和 OPC 一方面可以促进水中百菌清光催化降解，另一方面改变了百菌清光催化降解途径避免高毒代谢物 4-羟基百菌清产生；植物提取物 OPC 对水中百菌清光催化降解作用机理为自由基还原脱氯反应；毒性研究表明 OPC 存在下，水中百菌清降解产物对斑马鱼的毒性小于百菌清母体及其 4-羟基百菌清[27-29]。

5 农药环境风险评价及管理

化学农药的不可替代性与环境污染、食品安全的矛盾，使得农药的风险评估与预警成为农药风险管理的必然趋势。而我们国家提出农药化肥双减的需求，对新型安全高效农药的创制，既是机遇也是巨大的挑战。就我国情况而言，环境中残留的农药主要包括两大来源，第一是以 DDTs 和 HCHs 为主的已经被禁用，但由于这类农药的持久性特性而在环境中长期存在；第二类是包括拟除虫菊酯类等杀虫剂、除草剂、杀菌剂等新型农药，由于大量使用也造成很高的环境检出率。因此，针对不同来源的农药及其复合效应造成的生态安全和人群健康问题及风险管理还需进一步完善。

5.1 农药代谢产物毒性效应评价

农药在环境介质和生物体内会发生复杂的生物质转化，并可以与环境中其他物质发生相互作用，产生代谢产物或新的化合物，形成数量众多、结构多样的中间代谢产物，对生态和人群造成二次危害。但是农药中间代谢产物的风险还是农药风险评估和管理的一个盲点。而我国目前农药风险评价体系、限量标准还未充分考虑农药次生代谢风险的问题。最近研究发现了拟除虫菊酯类农药、氟虫腈、百菌清、禾草灵等主要代谢产物的免疫毒性和内分泌干扰效应比其母体化合物更强。

农用杀菌剂百菌清虽然使用量非常巨大，但是关于百菌清尤其是它的代谢产物对水生生物的潜在生态风险效应的研究仍然十分有限。季晨阳等以斑马鱼作为一种水生生物的体内模型，利用 Q-TOF 技术首次鉴定到了百菌清在斑马鱼体内的代谢产物 4-羟基百菌清；同时分析了百菌清由肝脏细胞色素 P450 介导的生物转化机制，全面评估了百菌清及其主要代谢产物 4-羟基百菌清的胚胎发育毒性及潜在的内分泌干扰效应。结果表明，4-羟基百菌清对斑马鱼胚胎致死效应大约为百菌清的 2.6 倍。百菌清和 4-羟基百菌清都具有雌激素激动效应，其 REC_{20} 值相近；4-羟基百菌清有明显的甲状腺干扰效应而母本化合物没有这一效应，这表明 4-羟基百菌清表现出了比母本化合物更强的内分泌干扰效应，可能是其水生毒性强的重要原因[30]。而在氟虫腈的研究中也发现了类似的现象，即锐劲特砜化物具有抗雌和抗甲活性，比其母本化合物（只有抗雌活性）表现出了更强的内分泌干扰效应[31]。

5.2 复合污染评价

在真实环境中农药与其他污染物共同存在，在同一体系内不同农药之间、农药母本与代谢产物、农药与其他污染物存在复杂的相互作用，阐明这些联合作用对于农药风险管理非常重要。杨叶等研究了环境中广泛存在重金属镉与拟除虫菊酯类农药的相互作用。结果显示，Cd 和氯氰菊酯（cypermethrin，CP）单独暴露均可引发斑马鱼早期胚胎发育躯体弯曲、心包囊肿和游囊皱缩等形态学畸形。CP 还可以引起胚胎躯体痉挛的发生，而 Cd 不能引发这种神经行为毒性。联合暴露于 CP 和 Cd 导致斑马鱼胚胎畸形发生率增加，而且 Cd 的加入显著增强了 CP 引发的胚胎躯体痉挛反应。联合暴露还导致胚胎体内氧化损伤程度增加及成年斑马鱼死亡率增加。CP 和 Cd 联合暴露毒性的增强可能是由于 Cd 干扰了鱼体内 CYP 酶，使 CP 在鱼体内代谢受阻，导致 CP 残留增加，使联合毒性增强[32]。

另一项研究，将 I 型拟除虫菊酯农药（氯菊酯，permethrin，PM）和 II 型拟除虫菊酯农药（氯氰菊酯，cypermethrin，CP）在亚致死浓度下对斑马鱼早期胚胎发育的联合毒性作用。在较低浓度范围内

（PM≤300 μg/L，CP≤30 μg/L），PM 和 CP 不会引起斑马鱼胚胎畸形及痉挛反应的发生。但联合暴露于 PM 和 CP 的二元混合物导致斑马鱼胚胎畸形及躯体痉挛发生率显著增加。钠离子通道的特异性阻断剂 MS-222 能够减缓 PM、CP 及两者联合暴露引发的躯体弯曲和躯体痉挛的产生，表明 PM 和 CP 引发的胚胎早期发育毒性部分是通过钠离子通道介导的。基因表达检测数据表明，联合暴露于 PM 和 CP 显著抑制了神经元发育相关基因的表达水平。PM 和 CP 单独暴露对超氧化物歧化酶（superoxide dismutase，SOD）活性没有影响，而联合暴露诱导了 SOD 酶活性。CP 单独暴露及联合暴露对过氧化氢酶（catalase，CAT）活性都有诱导作用。结果表明，与相同浓度条件下的单个农药暴露相比，PM 和 CP 联合暴露的斑马鱼胚胎体内形态学畸形和躯体痉挛发生率增加，神经元发育基因表达下降，氧化应激反应增强，因此水体环境中低剂量 I 型和 II 型菊酯农药共存有可能增加对鱼类等水生生物的毒性风险[33]。

5.3　次生风险评价

农药在不同环境介质中会对介质中的多种生物产生影响，通过对其他非靶标生物的作用产生或者释放对生态系统和人体毒性更大的化合物。近年来的研究已经明确农药添加剂甲基碘化铵可以在水环境中促进无机汞向有机汞的合成，而产生更严重的健康风险。张全等对目前配合转基因作物使用的广谱除草剂草铵膦和草甘膦，在环境浓度下对水环境中蓝藻生长、藻毒素分泌的影响开展了研究。结果显示，低剂量的草铵膦和草甘膦在适当气候条件下可以促进蓝藻的生长，诱导藻毒素-LR 的合成。环境浓度下的草铵膦、草甘膦加剧了水华爆发和使得藻毒素在水体中浓度的升高，对水生生态系统和饮用水安全造成了潜在的危害，因此在制定相关的农药的环境限制时，应该充分考虑其次生风险这一因素[34,35]。

5.4　我国农药水环境基准研究

水环境质量基准（water quality criteria）是指一定自然特征的水生态环境中污染物对特定对象（水生生物或人）不产生有害影响的最大可接受剂量（或无损害效应剂量）、浓度水平或限度。它是说明当某一物质或因素不超过一定的浓度或水平时，将能够保护生物群落或某种特定用途。水质安全对于保护整个水生态系统以及水生生物多样性非常重要。水质基准是进行水污染防治的科学依据。近年来，对污染水源（河流、湖泊、水库等）的治理一直是环境领域研究的热点之一。我国目前水质基准的研究大都集中于具体基准值的推算和技术方法的探讨，化合物主要以重金属和 POPs 为主。同时，建立水质基准必须依据当地水域现状，如盐度、pH、温度等参数及水环境中的土著生物，从而建立适合当地水域的水质基准研究方法。就我国目前的实际情况看，区域水生生态基准制度的最大困难是缺乏模式生物，可获得的符合区域特征的水生毒性数据不足，很难对区域的污染物水生毒性做准确的评价。李婧等在对钱塘江水域各段进行内分泌干扰物的环境残留特征进行综合分析的基础上，整合典型环境内分泌干扰物的水生毒性和内分泌干扰效应等研究结果，根据其化合物毒性特点分别采用了毒性百分数法、物种敏感度法对其淡水水生生物急慢性水质基准进行计算分析，结果如下：α-硫丹急性基准值为 0.0535 μg/L；三硫磷急性基准值为 2.213 μg/L；苯硫磷急性基准值为 2.8779 μg/L；三氯杀螨砜急性基准值为 10.5786 μg/L；苯酚急性基准值为 1212.10/L，慢性基准值为 20.72 μg/L；联苯菊酯急性基准值为 0.0066 μg/L，慢性基准值为 0.0023 μg/L；高效氟氯氢菊酯急性基准值为 0.0037 μg/L，慢性基准值为 0.0029 μg/L；氯菊酯急性基准值为 0.2137 μg/L，慢性基准值为 0.0862 μg/L。仅以联苯菊酯为例，联苯菊酯在钱塘江野生鱼类的负荷水平为 0.64~110.47 μg/kg，而其急性基准值为 0.0066 μg/L，慢性基准值为 0.0023 μg/L，表明该化合物已经对钱塘江水生生物造成了极大的威胁。该研究建立符合钱塘江水域的水质基准，并将这些基准与国外基准及国内水质标准相比，提出地域及物种的差异性，为钱塘江流域水环境预警和养殖业提供科学的资料，为建立符合浙江省水域特点的水环境安全提供指导[33]。

6 农药毒性效应的分子机制研究进展

有关农药毒性效应的分子机制研究近年来主要集中在对其内分泌干扰效应相关方面。在已明确具有内分泌干扰效应的化合物中农药占比很高。不考虑农药代谢产物，仅仅农药母本就有80多种能干扰人或野生生物的内分泌系统，对生态安全和人群健康造成很大威胁。我国农药污染现状复杂，既有被禁用多年的POPs类传统农药，也有现用的大量新农药。包括以典型POPs类农药DDTs和目前使用广、检出率高的拟除虫菊酯类农药为目标化合物，开展了其生殖发育、神经毒性和对肿瘤影响的机制等的研究。

6.1 DDTs毒性机制研究进展

目前DDTs健康风险研究中有几个关键的问题，从负荷水平：不同人体负荷DDTs健康风险效应及分子机制不同；构型层面：不同异构体之间毒性效应分子机制不同。

国际癌症研究中心经过调查已经将DDTs列入潜在的致癌物范畴。DDTs及其代谢物主要通过直接接触和食物链暴露人体。调查显示，DDTs及代谢物DDE暴露与多种肿瘤的发病相关。然而，关于DDTs及其代谢物影响肿瘤发生和发展的分子机制的报道较少。宋莉等对DDTs及其代谢物DDE对大肠癌发展的影响及分子机制进行了研究。结果表明：低浓度*p*, *p'*-DDT（10^{-12}~10^{-7} mol/L）暴露96 h后，明显促进大肠癌细胞存活和增殖；同时，低浓度*p*, *p'*-DDT（5 nmol/L/kg）暴露的裸鼠肿瘤体积、重量明显高于对照组，且实验组肿瘤组织中细胞增殖标志物Ki-67表达阳性的细胞明显高于对照组；进一步研究发现*p*, *p'*-DDT通过氧化应激介导的Wnt/β-catenin信号通路激活进而促进大肠癌细胞增殖。*p*, *p'*-DDT（10^{-10} mol/L, 10^{-9} mol/L和10^{-8} mol/L）暴露96 h后，明显抑制5-FU诱导的大肠癌细胞凋亡；并引起Bcl-2表达上调，Bax和procaspase-3的下调；进一步研究发现*p*, *p'*-DDT通过激活PI3K/AKT和Hedgehog/Gli1信号通路进而抑制大肠癌细胞凋亡。*p*, *p'*-DDE是DDTs在环境和机体内最持久、浓度最高的代谢产物。低剂量*p*, *p'*-DDE（10^{-12}~10^{-7} mol/L）暴露96 h后，能明显促进大肠癌细胞增殖；并且提高大肠癌细胞内氧化应激水平；进一步发现氧化应激介导的Wnt/β-catenin和Hedgehog/Gli1信号通路激活在*p*, *p'*-DDE促进大肠癌细胞增殖中发挥重要作用。Jin等研究了DDT对肝癌发展的影响及机制。结果表明*p*, *p'*-DDT（10^{-12}~10^{-7} mol/L）暴露96 h后，明显促进肝癌细胞增殖；同时，*p*, *p'*-DDT（5 nmol/L/kg）暴露的裸鼠肿瘤体积、重量明显高于对照组；*p*, *p'*-DDT提高肝癌细胞内氧化应激水平；并且激活Wnt/β-catenin信号通路。*p*, *p'*-DDT暴露抑制肝癌细胞间黏附并促进细胞与基质间黏附；抑制黏附分子E-cadherin表达，促进N-cadherin和CD29表达；进一步研究发现*p*, *p'*-DDT通过氧化应激介导的JAK/STAT3信号通路激活进而调节肝癌细胞黏附[36-40]。

王萃等以DDTs最主要代谢产物*p*, *p'*-DDE为模式化合物，选取非洲疟疾控制区较高人体DDTs负荷水平为暴露浓度，以神经胶质瘤细胞和斑马鱼为研究模型，采用基因敲除等技术，深入研究了DDTs神经毒性分子机制。结果表明，*p*, *p'*-DDE能诱导神经细胞发生凋亡，其中TNFα为可能靶标分子。通过选择TNFα单抗抑制剂类克，发现类克对*p*, *p'*-DDE诱导的神经细胞毒性，细胞凋亡及凋亡执行分子Caspsaes家族相关成员活性均有一定程度的逆转。斑马鱼胚胎脑部实验表明，类克对*p*, *p'*-DDE引起的脑部神经细胞凋亡有恢复作用。无论在体内还是体外，*p*, *p'*-DDE在诱导神经毒性中，可能的靶分子之一为膜外受体TNFα，而类克是潜在的治疗药物[41]。

6.2　拟除虫菊酯类杀虫剂毒性机制

拟除虫菊酯类杀虫剂是一种典型的环境内分泌干扰物，可能会干扰生殖内分泌系统的激素信号传导网络，从而对哺乳动物和人类产生潜在的生殖危害。

将围产期的雌性大鼠暴露于氰戊菊酯，会延迟子代雌性大鼠的性成熟，减少性行为的发生以及引起动情周期反常，进一步研究表明氰戊菊酯在雌性大鼠围产期脑部性组织形成阶段起到了抗雌激素的作用；将出生后 22 d 的雌性大鼠暴露于高效氰戊菊酯，会延迟雌性大鼠的性成熟且抑制午后 LH 浓度的上升，这提示着高效氰戊菊酯暴露会引起雌性大鼠下丘脑功能不足。将 8 周大小的雄性小鼠暴露于浓度为 35 mg/kg 及 70 mg/kg 的氯菊酯 6 周，会显著增加其血清中 LH 的含量，减少血清中睾酮的浓度[42]。成年雄性小鼠暴露于浓度为 10 mg/kg，15 mg/kg 和 20 mg/kg 氯氰菊酯 5 周后，其血清中 FSH 和 LH 的浓度均显著上升，但血清中睾酮的浓度显著降低[43]。将雄性成年大鼠暴露于浓度为 3 μg/(kg·d)溴氰菊酯 30 d，其血清中 LH、FSH 及睾酮的浓度都有显著的上升[44]。这些研究表明，拟除虫菊酯暴露会对垂体性腺轴分泌的相关激素产生影响。青春期和成年早期的雄性小鼠暴露于氰戊菊酯后，小鼠血清中和睾丸中睾酮浓度降低，精子数量急剧下降，同时，氰戊菊酯明显降低了精原细胞层次，扰乱了精原细胞的排列，提高了睾丸中凋亡细胞的数量，且降低了睾酮生物合成相关的酶 P45017α 和 P450scc 基因水平和蛋白水平的表达。青春期的雄性小鼠从出生后第 35 天到第 70 天暴露于氯氰菊酯（25 mg/kg），其睾丸中 StAR 的表达下调，睾酮的合成受到干扰；将哺乳期的雄性小鼠通过母体暴露于氯氰菊酯（25 mg/kg）中，发现氯氰菊酯会降低小鼠的体重，影响小鼠睾丸的发育和精子的合成，睾丸中 P450scc 基因水平和蛋白水平均会受到抑制，血清和睾丸中的睾酮浓度也会明显降低。除了对雄鼠的睾丸造成一定的损伤，拟除虫菊酯杀虫剂暴露也会对雌鼠的生殖功能产生影响。在妊娠期和哺乳期通过母体暴露于氰戊菊酯（40 mg/kg），在出生后 75 天发现雌性大鼠卵巢质量降低，窦前卵泡和黄体水平也有下降，在第 80 天发现再吸收的现象升高，说明氰戊菊酯能够损伤雌性大鼠的生殖功能。这些研究表明，拟除虫菊酯杀虫剂暴露会对性腺的发育，相关基因水平和蛋白水平的表达等产生影响。

低浓度拟除虫菊酯引起的潜在肝脏毒性及其分子机理仍不清楚，目前常用的肝脏疾病临床指标可能无法准确指示潜在肝脏损伤的发生。张颖以 BALB/c 小鼠作为体内实验模型，通过对组织形态、酶活性、细胞活性、氧化损伤、caspase 活性、细胞凋亡以及代谢相关基因表达水平的检测，全面评价联苯菊酯引起的潜在肝脏损伤，并深入研究毒性作用产生的分子机理。组织形态和血清酶活性都未发生明显变化。但联苯菊酯暴露对肝脏细胞产生细胞毒性和氧化应激，同时引起 caspase 激活和线粒体细胞凋亡通路的启动。基因芯片的结果显示，联苯菊酯能广泛地干扰代谢过程，强烈诱导 CYPs、GPXs、GSTs 和激酶等氧化应激相关基因。即使在传统临床和病理指标没有出现明显肝脏损伤的情况下，联苯菊酯也能够引起氧化损伤，通过 caspase 介导、线粒体相关的细胞凋亡，最终引起肝脏损伤[45]。

6.3　氟虫腈水生毒性机制研究进展

氟虫腈是一类高效低毒的苯基吡唑类杀虫剂，广泛用于农作物害虫和卫生害虫的防治。虽然该农药对哺乳类的毒性较低，但它对蜜蜂和水生生物的毒性却很大。王萃等研究了氟虫腈对斑马鱼的早期发育及行为的影响。结果显示，氟虫腈对斑马鱼胚胎-幼鱼的正常发育有毒害作用，产生了游囊关闭、身体弯曲和体长变短等畸形现象甚至引起了胚胎、幼鱼的死亡。此外，对斑马鱼运动行为的检测结果表明：氟虫腈对幼鱼在光照条件下的运动活力以及在光周期刺激下的运动特征及变化规律有着显著的影响。与对照组相比，经低浓度（40 μg/L）氟虫腈处理的幼鱼运动活力显著增加，但运动规律没有明显差异。经高浓度（160 μg/L）氟虫腈处理的幼鱼运动活力显著降低，对光照条件的变化更为敏感。基于 GC-MS 代谢组学的分析进一步表明，氟虫腈在低剂量浓度暴露组，改变了斑马鱼幼鱼神经递质小分子含量，并影响了其他神经行为相关的小分子，进而诱发神经行为毒性。而高浓度暴露主要影响 tRNA 和胆汁酸等小分子代谢，导致了斑马鱼发育异常[46,47]。

7 手性农药环境安全研究进展

含有手性中心的农药称为手性农药（chiral pesticides），其分子的不对称引起的异构现象类似左手是右手的镜像一样，所形成的异构体称为对映异构体。手性农药在我国已经超过了 40%，对生态安全和人类健康造成了潜在威胁。前期，关于手性农药的研究主要集中于分析分离方法的建立，手性农药的急、慢性毒性的评价方面，缺乏深入的对映体选择性环境行为和毒性差异的机制研究。

7.1 氟虫腈对映体选择性水环境行为及毒性差异

氟虫腈是一种苯基吡唑类手性杀虫剂，具有一对对映异构体。周志强等对氟虫腈对映体拆分、立体选择性环境行为、毒性差异等进行了深入广泛的研究，目前已进行了氟虫腈及手性代谢物对映体的拆分及残留分析方法的建立；完成了在水生环境生物（黑斑蛙蝌蚪、斜生栅藻、颤蚓、河蚌、泥鳅）、栅藻—蝌蚪食物链以及模拟水生系统中的立体选择性归趋行为研究；测定了对颤蚓、蝌蚪、泥鳅、河蚌、浮萍及淡水藻的立体选择性毒性效应的差异性。

7.1.1 氟虫腈及代谢物的立体选择性环境行为

7.1.1.1 氟虫腈及代谢物对映体拆分及残留分析

利用高效液相色谱、气相色谱、高效液相色谱串联质谱等技术，使用（*R*, *R*）Whelk-O1、CHIRALCELOD-H、CHIRALPAKIB、BGB172 等手性色谱柱等实现了氟虫腈及手性代谢物对映体的拆分，系统考察了色谱条件，建立了其在多种复杂基质中的手性残留分析方法[48,49]。

7.1.1.2 氟虫腈在生物体中的立体选择性行为

采用水染毒和土壤染毒两种培养方法对氟虫腈对映体在颤蚓体内的选择性富集研究显示，富集过程中，颤蚓体内 *R*-氟虫腈的浓度显著高于 *S*-氟虫腈，氟虫腈经水和经土两种方式暴露都在颤蚓体内发生了立体选择性富集，并且选择性相同，均是 *R*-氟虫腈被优先富集，但经土壤染毒培养颤蚓体内的对映体立体选择性水平更高。颤蚓经氟虫腈对映体暴露后，总蛋白含量、CAT 和 GR 酶活性等生理指标的影响上不存在显著差异[50]。

氟虫腈在河蚌体内 11 d 达到富集平衡，但生物富集因子（BCF）只有 0.2 左右，富集能力有限，在富集过程中，河蚌体内 *S*-氟虫腈明显高于 *R*-氟虫腈，同时在河蚌体内检出了氧化产物氟虫腈砜化物（fipronil sulfone）。在代谢阶段，*R* 体和 *S* 体的半衰期分别为 6.4 d 和 8.9 d，而水体中的 *R* 体浓度高于 *S* 体[51]。

外消旋体氟虫腈在水-泥鳅体系中暴露后，氟虫腈在水中半衰期为 5.4 d，水中代谢物在 3 d 内达到最大值，其中氧化产物 fipronil sulfone 的浓度高于 fipronil desulfinyl 和氟虫腈硫化物（fipronil sulfide）。氟虫腈在水中存在对映体差异，*R* 体浓度略高于 *S* 体浓度。氟虫腈在泥鳅头部浓度先升高再降低，无明显的富集现象发生。在脂肪中，氟虫腈有明显的富集现象，BCF 达到 7，对映体呈现立体选择性，脂肪中主要以 *S*-氟虫腈形式存在。而代谢物主要为 fipronil sulfone，BCF 则达到了 10，说明相比于母体，代谢物 fipronil sulfone 更容易在脂肪中富集。在泥鳅肝脏中，氟虫腈在第 1 天达到最高值，随后逐渐下降，且对映体水平有明显差异，*S* 体浓度远大于 *R*-氟虫腈。肝脏中 fipronil sulfone 含量非常高，而 fipronil sulfide 含量则很低，说明氟虫腈在肝脏中转化为 fipronil sulfone 后，将会长期富集在肝脏中。在氟虫腈光学纯单体暴露实验中，*S* 体和 *R* 体半衰期分别为 7.0 d 和 4.2 d。水中能检测到三种代谢物，其中氧化产物 fipronil sulfone 的浓度高于 fipronil desulfinyl 和 fipronil sulfide 的浓度。*S*-氟虫腈在水中基本维持其构型，*R*-氟虫腈有部分转化为 *S* 体[52]。

蝌蚪经水暴露于氟虫腈，在富集过程中，蝌蚪体内 *R*-氟虫腈的浓度一直显著高于 *S*-氟虫腈，即蝌蚪选择性富集 *R*-氟虫腈。*S*-氟虫腈和 *R*-氟虫腈的生物富集因子（BCF）分别为 12.06 和 24.12。氟虫腈常见代谢物 fipronil sulfone 和 fipronil desulnyl 均在蝌蚪体内有检出，但是含量均不高，最大值出现在富集试验的第一天，其中氟虫腈砜为主要代谢物[53]。

7.1.1.3　氟虫腈在栅藻—蝌蚪食物链中的立体选择性行为

栅藻和蝌蚪同时暴露在含有氟虫腈外消旋体的水溶液中培养，构成栅藻—蝌蚪食物链，并测定了培养期间氟虫腈对映体在藻和蝌蚪体内的富集情况，结果表明氟虫腈在栅藻中富集的浓度为蝌蚪的 10 倍，氟虫腈对映体在栅藻中的富集不存在对映体选择性，氟虫腈在非食藻蝌蚪体内的富集具有对映体选择性，优先吸收 *R* 体。食藻蝌蚪富集的氟虫腈对映体浓度与非食藻蝌蚪基本相当，氟虫腈不会通过食物链在蝌蚪体内大量地富集，通过分析食藻蝌蚪中氟虫腈对映体的生物放大因子，发现氟虫腈通过食物链暴露没有对蝌蚪产生生物放大效应。在栅藻中主要代谢物为 fipronil sulfone，蝌蚪中主要代谢物为氧化产物 fipronil sulfone[53]。

7.1.1.4　氟虫腈及其代谢物在水-底泥-水生植物-水生动物模拟水生态系统中的环境行为

氟虫腈在水-沉积物体系中半衰期为 11.8 d，两个对映体含量差别显著，*R* 体的浓度低于 *S* 体。氟虫腈会转化为 fipronil desulfinyl、fipronil sulfide 和 fipronil sulfone，其中 fipronil sulfide 和 fipronil sulfone 的浓度高于 fipronil desulfinyl。在灭菌体系中，氟虫腈降解缓慢，无对映体差异性，代谢物只有少量 fipronil desulfinyl 产生，说明沉积物中的微生物在降解过程中有重要的作用。

选取沉积物和具代表性的淡水物种浮萍、河蚌为研究对象，构建了水-沉积物-浮萍-河蚌的人工模拟水生态系统研究手性农药氟虫腈在水生态系统的分配和迁移转化规律。微宇宙系统设置沉积物灭菌和不灭菌处理组。在灭菌组，水中大约 80%氟虫腈发生降解，半衰期为 10.6 d。浮萍中氟虫腈含量在 7 d 达到峰值，而后逐渐降低。水中氟虫腈除了降解和被沉积物及浮萍吸收外，其余主要被河蚌富集，在 16 d 时富集浓度高达 240 ng/g，且河蚌体内的代谢物主要为 fipronil sulfide，也有 fipronil sulfone 及 fipronil desulfinyl 检出，代谢物水平高于水和浮萍；水中主要代谢物为 fipronil sulfone，此外 fipronil sulfide 和 fipronil desulfinyl 均有检出；浮萍中只有 fipronil sulfide 检出。氟虫腈在进入生态系统初期，水体中氟虫腈主要以外消旋体形式存在，随着暴露时间的延长，氟虫腈从水中扩散至其他基质，各个基质中氟虫腈的选择性也发生了改变。沉积物中 *S*-氟虫腈浓度略高于 *R* 体，浮萍中选择性最为明显，*R*-氟虫腈远高于 *S* 体，而在河蚌体内，*S*-氟虫腈为氟虫腈主要存在形式。在不灭菌组中，水中约 90%氟虫腈发生了降解，半衰期为 4.6 d，部分氟虫腈扩散至沉积物中，21 d 达到峰值，低于灭菌组。浮萍中氟虫腈在 7 d 达到峰值。与灭菌组相似，水中的氟虫腈一部分被河蚌富集，河蚌体内的氟虫腈在 21 d 达到峰值 214 ng/g，且河蚌体内的代谢物水平高于浮萍。水中主要代谢物为 fipronil sulfone，fipronil sulfide 和 fipronil desulfinyl 均有检出；沉积物中 fipronil sulfone、fipronil sulfide 和 fipronil desulfinyl 均有检出，且与灭菌组相比，不灭菌组的沉积物中代谢物浓度更高；浮萍中只有 fipronil sulfide 检出。沉积物中 *S*-氟虫腈含量高于 *R* 体，而浮萍优先富集 *R*-氟虫腈，河蚌体内 *S*-氟虫腈占主导，在几种生物基质中，氟虫腈均发生了选择性富集[54]。

通过模拟水生生态系统模型，发现氟虫腈进入水体后一部分吸附于沉积物中，另一部分被水生生物富集于体内，其中河蚌富集污染物的能力最强。同时也发现了立体选择性现象，即 *R*-氟虫腈的浓度低于 *S* 体浓度。另外，氟虫腈进入水体后，通过水生生物的代谢，主要代谢物为 fipronil sulfide 和 fipronil sulfone，说明生物体内氧化和还原作用是氟虫腈的主要代谢途径。

7.1.1.5 生物炭对水环境中氟虫腈的污染修复及毒性影响

生物碳对水环境中氟虫腈污染有修复作用。经过 7 d 生物炭的吸附，水中氟虫腈去除率可达 60%以上。而生物炭在减小氟虫腈对水生生物毒性方面也有显著效果，在一项对泥鳅的毒性实验中发现，在 72 h 染毒期间，添加生物炭后，氟虫腈及代谢物的毒性最高可减小至 1/8 左右，生物炭的加入减小了氟虫腈及代谢物对水生生物的毒性。

7.1.2 氟虫腈及其代谢物对几种生物的立体选择性毒性效应

氟虫腈对泥鳅存在对映体选择性毒性，*S*-氟虫腈毒性较 *R* 体更强，72h LC_{50} 达到了 0.49 mg/L。与氟虫腈母体相比，代谢物对泥鳅呈现了更高的毒性，尤其是还原产物 fipronil sulfide 达到了 0.055 mg/L。氟虫腈对河蚌的急性毒性也呈现立体选择性，*S*-氟虫腈及 *R*-氟虫腈的 72h LC_{50} 分别为 0.63 mg/L 和 3.27 mg/L，而相对于母体，代谢物对河蚌呈现更高的急性毒性，其中氧化还原产物 fipronil sulfide 和 fipronil sulfone 的 LC_{50} 达到了 0.32 mg/L 和 0.24 mg/L，高于母体及其他代谢物[53]。

氟虫腈对黑斑蛙蝌蚪的 96 h 急性毒性同样存在立体选择性，*R*-氟虫腈的毒性相对较强（2.79 mg/L），*S*-对映体次之（3.49 mg/L），外消旋体最弱（4.97 mg/L）。根据水生生物的毒性评价，氟虫腈对黑斑蛙蝌蚪的毒性属于中等毒性。氟虫腈会显著诱导黑斑蛙蝌蚪产生 SOD 酶，但对映体间差异不显著。氟虫腈外消旋体对蝌蚪 CAT 活性无显著影响，氟虫腈对映体的单独暴露诱导其活性显著增加，*S* 体多于 *R* 体。在氟虫腈对黑斑蛙蝌蚪 GST 酶活性影响的实验中，*R* 体比 *S* 体诱导产生更多 GST。氟虫腈的暴露产生了明显的脂质过氧化作用，但两对映体间对丙二酸含量的影响没有显著差异[55]。

对于水生植物浮萍，氟虫腈有较为明显的立体选择性毒性产生，其中毒性最大的是 *R*-氟虫腈，外消旋体次之，与水生动物结果不同，*S*-氟虫腈呈现较小的毒性，7 d EC_{50} 分别为 9.79 mg/L，12.15 mg/L，12.36 mg/L。相比较母体而言，代谢物对浮萍的毒性更大。其中还原产物 fipronil sulfide 的 EC_{50} 为 6.50 mg/L，毒性是母体的 2 倍，说明代谢物具有更高的环境风险[56]。

氟虫腈对淡水藻的毒性较低，在 3 d 的暴露实验后，*R*-氟虫腈的毒性要高于外消旋体和 *S*-氟虫腈的毒性。有明显的立体选择性现象发生。而在所研究的四种淡水藻（多刺栅藻、四尾栅藻、普通小球藻、蛋白小球藻）中，外消旋氟虫腈、*R*-氟虫腈和 *S*-氟虫腈对多刺栅藻的 72 h EC_{50} 分别为 5.45 mg/L、2.18 mg/L 和 7.08 mg/L，*R*-氟虫腈的毒性是 *S*-氟虫腈的 2~3 倍。氟虫腈还会对藻类叶绿素含量产生立体选择性影响。*S*-氟虫腈处理组藻液中叶绿素 a 和 b 的含量要高于外消旋体和 *R* 体，这也验证了毒性实验的结果，即 *S* 体的毒性是最小的。在氟虫腈对藻类抗氧化酶的研究中，也观察到了明显的立体选择性现象。*S*-氟虫腈诱导藻细胞 SOD 和 CAT 活性的能力大于外消旋体和 *R*-氟虫腈。

从急性毒性结果来看，氟虫腈的急性毒性及生理生化指标影响都呈现了一定程度的对映体选择性：泥鳅、河蚌的实验结果为 *S*-氟虫腈的毒性高于 *R*-氟虫腈；而浮萍、淡水藻和蝌蚪的急性毒性结果显示 *R*-氟虫腈的毒性高于 *S* 体；对映异构体对两种动物属于高毒等级而对植物则属于中毒或低毒。另外，氟虫腈代谢物的毒性要高于母体。

7.2　手性 DDTs 神经毒性促癌效应对映体差异分子机制

在 DDTs 风险评价中，*o,p'*-DDT 的手性特征增加了 DDTs 健康风险评价的复杂性。商业用 DDT 中，其中 75%为 *p,p'*-DDT，25%为 *o,p'*-DDT。尽管，*o,p'*-DDT 在 DDTs 中含量较低，但在一些地区如中国太湖的上空，*o,p'*-DDT 的平均浓度为 *p,p'*-DDT 的 6.3 倍，手性 DDTs 的影响不能忽略。而 *o,p'*-DDT 在生物体内脱掉一个氯原子，形成代谢产物 *o,p'*-DDD，同样具有持久性有机污染物的特征。

王萃等以神经细胞 PC12 为体外模型，研究了 *o,p'*-DDT 及 *o,p'*-DDD 对映体选择性神经毒性。结果表明，*o,p'*-DDD 与 *o,p'*-DDT 在同一数量级范围内引起神经毒性，并且 *R* 构型对神经细胞毒性大于 *S* 构型。具体表现为 *R*-(–)-*o,p'*-DDT 诱导较多的氧化损伤产物 MDA，LDH 释放及抗氧化酶活性的降低；扰乱氧化应激中相关基因表达；在凋亡相关分子中，更强地诱导了 p53，Caspase3 及 NFkB 蛋白水平表达而引起更多的神经细胞凋亡。而 *R*-(+)-*o,p'*-DDD 则仅依赖于 p53 的介导神经细胞凋亡。结果表明 *o,p'*-DDT 代谢并没有减弱其毒性作用，由于代谢脱氯未能影响原母本化合物的空间结构，使得两者的 *R* 构型毒性均较 *S* 构型高，即对映体选择性表现为一致效应，印证了化合物结构决定其功能的理论。基于手性特征在 POPs 中普遍性，通过对手性母本化合物与其代谢产物的毒性研究，为今后更准确地评价 POPs 健康风险及生态安全性提供了数据支持和向导[47,57]。

R-(–)-*o,p'*-DDT 是 *o,p'*-DDT 雌激素效应的主要来源，何向明等以雌激素依赖型的乳腺癌细胞为模型，开展了 *o,p'*-DDT 对映体选择性诱导肿瘤细胞恶化，促进肿瘤细胞转移和增殖分子机制研究。结果表明，类雌激素效应较强的 *R*-(–)-*o,p'*-DDT 不但能显著诱导细胞端粒酶的活性，刺激乳腺癌细胞的增殖，而且可以通过调节细胞间的黏附（E-cadhrin/caterin）与浸润（MMPs/TIMPs）相关的分子，刺激乳腺癌细胞的转移和浸润，对人乳腺癌的恶化有一定的影响。这一研究首次报道了化合物结构多样性对肿瘤发展影响的分子机制，表明手性污染物对人类肿瘤促癌作用具有对映体选择性，在考虑其健康风险及肿瘤预后等方面要充分重视手性化合物的对映体差异[58]。

7.3　手性农药生殖发育毒性对映体差异机制

7.3.1　联苯菊酯对母胎健康影响对映体差异

联苯菊酯（bifenthrin，BF）是一种典型的环境内分泌干扰物，可能会干扰生殖内分泌系统的激素信号传导网络，从而对哺乳动物和人类产生潜在的生殖危害。张颖等用人绒毛膜上皮癌细胞系 JEG-3 作为模拟胎盘作用的体外模型，探讨 BF 外消旋体及其对映体对滋养层细胞激素功能干扰作用的分子机理。结果表明，BF 不仅能够促进孕酮和人绒毛膜促性腺激素的分泌、诱导促性腺激素释放激素 I（GnRHI）及其受体 GnRHRI 表达水平的上调，还显著改变类固醇生成基因的表达水平。BF 通过雌激素受体通路的作用，干扰滋养层细胞中的激素信号通路，且 BF 对生殖内分泌相关激素的干扰作用显示出一致的对映体选择性顺序：*S*-BF＞*rac*-BF＞*R*-BF。分子模拟的结果显示，*S*-BF 在与 ER 的结合中表现比 *R*-BF 更强的亲和力[59]。

7.3.2　基于表观遗传学氟虫腈斑马鱼发育毒性机制

氟虫腈用于替代有机磷杀虫剂的苯基吡唑杀虫剂，具有非对称硫手性中心，即包含两个对映体：*S*-(+)-fipronil 和 *R*-(–)-fipronil。表观遗传学机制包括 DNA 甲基化、组蛋白甲基化和非编码 RNA。DNA 甲基化（5'胞嘧啶上的甲基共价修饰）参与基因沉默和基因组稳定等生物过程。在斑马鱼等脊椎动物中，

DNA 主要发生在 CpG 岛上。而这种甲基化状态极易被外界物质所干扰。钱易等评价了氟虫腈不同对映体对斑马鱼发育毒性和幼鱼运动行为影响的对映体选择性差异，在此基础上分析了氟虫腈不同对映体对斑马鱼发育过程 DNA 甲基化影响的差异。结果表明 *S*-（–）- fipronil 斑马鱼发育毒性远远大于 *R*-（+）-fipronil。DNA 甲基化免疫共沉淀技术（MeDIP-Seq，高通量 DNA 甲基化测序技术）检测 800 μg/L *S*-(+)-fipronil 或 *R*-(–)-fipronil 暴露下斑马鱼胚胎基因组 DNA 甲基化的变化情况，结果表明：基因组 DNA 甲基化水平可以被 *S*-(+)-fipronil 或 *R*-(–)-fipronil 上调，且与 *S*-(+)-fipronil 的效应显著强于 *R*-(–)-fipronil，这可能导致斑马鱼胚胎发育过程中更多的基因表达受阻以及更严重的急性毒性作用。Kyoto Encyclopedia of Genes and Genomes（KEGG）分析表明促分裂原活化蛋白激酶（Mitogen-activated protein kinase，MAPK），紧密连接（tight junction），焦点粘连（focal adhesion），转化生长因子-β（TGF-β）信号通路，血管平滑肌收缩（vascular smooth muscle contraction），刺猬信号通路（hedgehog signaling pathway）以及 Wnt 信号通路中的关键基因被 *S*-(+)-fipronil 显著抑制，表明不同氟虫腈对映体可通过 DNA 甲基化的对映体选择效应诱导具有明显差异的发育毒性[60]。

7.3.3 重金属对拟除虫菊酯类杀虫剂水生生物转化影响机制

顺式联苯菊酯（*cis*-bifenthrin，*cis*-BF）是目前用量最广泛的拟除虫菊酯农药之一。杨叶等研究了三种重金属 Cd、Cu 和 Pb 对斑马鱼体内 *cis*-BF 对映体生物转化的干扰作用。在接近环境浓度水平，Cd、Cu 和 Pb（10 μg/L）与 *rac*-*cis*-BF/*R*-*cis*-BF（1 μg/L）联合暴露导致成年斑马鱼 72 h 致死率增加。对斑马鱼体内对映体残留量和对映体分数（enantiomer fraction，EF 值）的检测发现，*rac*-*cis*-BF 单独暴露组鱼体内 *R*-*cis*-BF 残留比例增加（EF=0.482），同时，*S*-*cis*-BF 和 *R*-*cis*-BF 单独暴露组鱼体内都检测到另外一个对映体残留，表明 BF 对映体在鱼体内有异构化和外消旋化的趋势。*rac*-*cis*-BF+Cd/Cu/Pb 联合暴露组鱼体内 EF 值显著小于 *rac*-*cis*-BF 单独暴露组；与 *R*-*cis*-BF 单独暴露相比，*R*-*cis*-BF+Cd 联合暴露组鱼体内 EF 值减小，而 *R*-*cis*-BF+Cu/Pb 联合暴露组 EF 值没有显著变化；与 *S*-*cis*-BF 单独暴露相比，*S*-*cis*-BF+Cu/Pb 联合暴露组 EF 值显著减小，而 *S*-*cis*-BF+Cd 联合暴露组 EF 值没有变化。研究结果表明，三种重金属的加入使鱼体内毒性较强的 *R*-*cis*-BF 残留比例增加[61]。

8 展　望

在未来可以预见的几十年内，化学农药必将继续伴随我们，随着农药种类数量的增加和暴露途径的多样化，以及全社会生态文明建设、食品安全要求的提升，农药环境化学与毒理学将迎来新的挑战和机遇。首先，面向国家重大的粮食安全与食品安全、生态保护等重大需求引导环境友好的农药的创制。其中利用高效的手性农药安全甄别技术，创制自主知识产权的绿色农药，是绿色农药创制的一个重要思路；其次，随着大量新型农药进入环境，开展新农药环境行为、残留特征、暴露风险的研究势在必行；再次，积极利用计算机辅助化合物风险评价技术，结合化合物 AOP 评估原则开展农药风险评估，加强农药代谢产物和次生风险的研究，以明确新农药在不同环境中的限量标准。再者，农药致毒分子机制的研究方兴未艾，深入探讨农药毒理学效应的分子机制，可以明确农药对非靶标生物毒性作用的靶点，改良农药抗性，为阻断或改良农药品种提供科学依据，实现农药低用量、靶标特异、环境相容的目标。

参 考 文 献

[1] Niu L L, Xu C, Zhu S Y, Bao H M, Xu Y, Li H Y, Zhang Z J, Zhang X C, Qiu J G, Liu W P. Enantiomer signature and

carbon isotope evidence for the migration and transformation of DDTs in arable soils across China. Sci Rep-Uk, 2016, 6: 38475.

[2] Tang M L, Zhao M R, Zhou S S, Chen K, Zhang C L, Liu W P. Assessing the underlying breast cancer risk of Chinese females contributed by dietary intake of residual DDT from agricultural soils. Environ Int, 2014, 73: 208-215.

[3] Niu L L, Xu C, Zhang C, Zhou Y, Zhu S, Liu W P. Spatial distributions and enantiomeric signatures of DDT and its metabolites in tree bark from agricultural regions across China. Environ Pollut, 2017, 229: 111.

[4] Niu L L, Xu C, Yao Y J, Liu K, Yang, F X, Tang M L, Liu W P. Status, influences and risk assessment of hexachlorocyclohexanes in agricultural soils across China. Environ Sci Technol, 2013, 47 (21): 12140-12147.

[5] 徐杨. 中国农田土壤中鞘氨醇单胞菌的多样性及丰度与 HCHs 残留的相关性. 杭州: 浙江大学, 2016.

[6] Niu L L, Xu C, Xu Y, Zhang C L, Liu W P. Hexachlorocyclohexanes in tree bark across Chinese agricultural regions: Spatial distribution and enantiomeric signatures. Environ Sci Technol, 2014, 48 (20): 12031-12038.

[7] Zhou S S, Tang Q Z, Jin M Q, Liu W P, Niu L L, Ye H. Residues and chiral signatures of organochlorine pesticides in mollusks from the coastal regions of the Yangtze River Delta: Source and health risk implication. Chemosphere, 2014, 114: 40-50.

[8] Zhou S S, Yang H Y, Zhang A P, Li Y F, Liu W P. Distribution of organochlorine pesticides in sediments from Yangtze River Estuary and the adjacent East China Sea: Implication of transport, sources and trends. Chemosphere, 2014, 114: 26-34.

[9] Jin M Q, Zhou S S, Liu W P, Zhang D, Lu X T. Residues and potential health risks of DDTs and HCHs in commercial seafoods from two coastal cities near Yangtze River Estuary. J Environ Sci Heal B, 2015, 50 (3): 163-174.

[10] Yang YH, Zhou, S S, Li, Y Y, Xue B, Liu T. Residues and chiral signatures of organochlorine pesticides in sediments from Xiangshan Bay, East China Sea. J Environ Sci Heal B, 2011, 46 (2): 105-111.

[11] Zhou S S, Shao L Y, Yang H Y, Wang C, Liu W P. Residues and sources recognition of polychlorinated biphenyls in surface sediments of Jiaojiang Estuary, East China Sea. Mar Pollut Bull, 2012, 64 (3): 539-545.

[12] Zhang H H, Chen S W, Zhou S S. Enantiomeric separation and toxicity of an organophosporus insecticide, pyraclofos. J Agr Food Chem, 2012, 60 (28): 6953-6959.

[13] Nakata H, Kawazoe M, Arizono K, Abe S, Kitano T, Shimada H, Li W, Ding X. Organochlorine pesticides and polychlorinated biphenyl residues in foodstuffs and human tissues from China: Status of contamination, historical trend, and human dietary exposure. Arch Environ Con Tox, 2002, 43 (4): 473-480.

[14] Jaraczewska K, Lulek J, Covaci A, Voorspoels S, Kaluba-Skotarczak A, Drews K, Schepens P. Distribution of polychlorinated biphenyls, organochlorine pesticides and polybrominated diphenyl ethers in human umbilical cord serum, maternal serum and milk from Wielkopolska region, Poland. Sci Total Environ, 2006, 372 (1): 20-31.

[15] Bouwman H, Kylin H, Sereda B, Bornman R. High levels of DDT in breast milk: intake, risk, lactation duration, and involvement of gender. Environ Pollut, 2012, 170 (8): 63-70.

[16] Xu C Y, Tang M L, Zhang H H, Zhang C L, Liu W P. Levels and patterns of DDTs in maternal colostrum from an island population and exposure of neonates. Environ Pollut, 2016, 209: 132-139.

[17] Zhang A, Chen Z, Ahrens L, Liu W P, Li Y F. Concentrations of DDTs and enantiomeric fractions of chiral DDTs in agricultural soils from Zhejiang Province, China, and correlations with total organic carbon and pH. J Agr Food Chem, 2012, 60 (34): 8294-301.

[18] Marcella W, Aguilar S R, Harley K G, Asa B, Dana B, Brenda E. In uteroDDT and DDE exposure and obesity status of 7-year-old mexican-American children in the CHAMACOS cohort. Environ Health Persp, 2013, 121 (5): 631.

[19] Ye X Q, Pan W Y, Zhao S L, Zhao Y H, Zhu Y M, Liu J, Liu W P. Relationships of pyrethroid exposure with gonadotropin levels and Pubertal development in Chinese boys. Environ Sci Technol, 2017, 51 (11): 6379-6386.

[20] Ye X Q, Pan W Y, Zhao Y H, Zhao S L, Zhu Y M, Liu W P, Liu J. Association of pyrethroids exposure with onset of puberty in Chinese girls. Environ Pollut, 2017, 227: 606-612.

[21] Han J J, Zhou L Q, Luo M, Liang Y R, Zhao W T, Wang P, Zhou Z Q, Liu D H. Nonoccupational Exposure to Pyrethroids and Risk of Coronary Heart Disease in the Chinese Population. Environ Sci Technol, 2017, 51 (1): 664-670.

[22] Liu K L, Cao Z Y, Pan X, Yu Y L. Using in situ pore water concentrations to estimate the phytotoxicity of nicosulfuron in soils to corn (*Zea mays* L.). Environ Toxicol Chem, 2012, 31 (8): 1705-1711.

[23] Wang D H, Wang Y M, Yin Y M, Min S, Wang S Y, Yu Y L. Bioavailability-based estimation of phytotoxicity of imazaquin in soil to sorghum. Environ Sci Pollut R, 2015, 22 (7): 5437-5443.

[24] Liu K L, Pan X, Han Y L, Tang F F, Yu Y L. Estimating the toxicity of the weak base carbendazim to the earthworm (*Eisenia fetida*) using in situ pore water concentrations in different soils. Sci Total Environ, 2012, 438: 26-32.

[25] Liu K L, Wang S Y, Luo K, Liu X Y, Yu Y L. Amelioration of Acidic Soil Increases the Toxicity of the Weak Base Carbendazim to the Earthworm Eisenia Fetida. Environ Toxicol Chem, 2013, 32 (12): 2870-2873.

[26] Wang D H, Zhang Q, Zheng Y, Lin D L, Yu Y L. Estimating the combined toxicity of flufenacet and imazaquin to sorghum with pore water herbicide concentration. J Environ Sci-China, 2016, 41: 154-161.

[27] Tan Y Q, Xiong H X, Shi T Z, Hua R M, Wu X W, Cao H Q, De Li X, Tang J. Photosensitizing effects of nanometer TiO_2 on chlorothalonil photodegradation in aqueous solution and on the surface of pepper. J Agr Food Chem, 2013, 61 (21): 5003-5008.

[28] Tan Y Q, Huang Q H, Shi T Z, Jin L J, Hua R M, Wu X W, Li X Q, Li X D, Cao H Q, Tang J, Li Q X. Promoting photosensitized reductive dechlorination of chlorothalonil using epigallocatechin gallate in water. J Agr Food Chem, 2014, 62 (50): 12090-12095.

[29] Lv P, Zhang J, Shi T, Dai L, Li X, Wu X, Li X, Tang J, Wang Y, Li Q X. Procyanidolic oligomers enhance photodegradation of chlorothalonilin water via reductive dechlorination. Appl Catal B Environ, 2017, 34 (73): 881-889.

[30] Zhang Q, Ji C Y, Yan L, Lu M Y, Lu C S, Zhao M.R. The identification of the metabolites of chlorothalonil in zebrafish (*Danio rerio*) and their embryo toxicity and endocrine effects at environmentally relevant levels. Environ Pollut, 2016, 218: 8-15.

[31] Lu M Y, Du J, Zhou P X, Chen H, Lu C S, Zhang Q. Endocrine disrupting potential of fipronil and its metabolite in reporter gene assays. Chemosphere, 2015, 120: 246-251.

[32] Yang Y, Ma H, Zhou J, Liu J, Liu, W P. Joint toxicity of permethrin and cypermethrin at sublethal concentrations to the embryo-larval zebrafish. Chemosphere, 2014, 96 (2): 146-154.

[33] Yang Y, Ye X, He B, Liu J. Cadmium potentiates toxicity of cypermethrin in zebrafish. Environ Toxicol Chem, 2016, 35 (2): 435-442.

[34] Zhang Q, Zhou H, Li Z, Zhu J Q, Zhou C, Zhao M R. Effects of glyphosate at environmentally relevant concentrations on the growth of and microcystin production by Microcystis aeruginosa. Environ Monit Assess, 2016, 188 (11): 632-639.

[35] Zhang Q, Song Q, Wang C, Zhou C, Lu C S, Zhao M R. Effects of glufosinate on the growth of and microcystin production by Microcystis aeruginosa at environmentally relevant concentrations. Sci Total Environ, 2017, 575: 513-518.

[36] Song L, Zhao J Y, Jin X T, Li Z Y, Newton I P, Liu W P, Xiao H. Zhao M R. The organochlorine *p, p*'-dichlorodiphenyltrichloroethane induces colorectal cancer growth through Wnt/beta-catenin signaling. Toxicol Lett, 2014, 229 (1): 284-291.

[37] Song L, Liu J, Jin X, Li Z, Zhao M, Liu W P. *p, p*'-Dichlorodiphenyldichloroethylene induces colorectal adenocarcinoma cell proliferation through oxidative stress. Plos One, 2014, 9 (11): e112700.

[38] Song L, Zhao M R, Liu J X, Li Z Y, Xiao H, Liu W P. *p, p*'-Dichlorodiphenyltrichloroethane inhibits the apoptosis of

colorectal adenocarcinoma DLD1 cells through PI3K/AKT and Hedgehog/Gli1 signaling pathways. Toxicol Res-Uk, 2015, 4 (5): 1214-1224.

[39] Jin X T, Song L, Zhao J Y, Li Z Y, Zhao M R, Liu W P. Dichlorodiphenyltrichloroethane exposure induces the growth of hepatocellular carcinoma via Wnt/beta-catenin pathway. Toxicol Lett, 2014, 225 (1): 158-166.

[40] Jina X T, Chen M L, Song L, Li H Q, Li Z Y. The evaluation of *p, p'*-DDT exposure on cell adhesion of hepatocellular carcinoma. Toxicology, 2014, 322: 99-108.

[41] Wang C, Zhang Q, Qian Y, Zhao M R. *p, p'*-DDE Induces Apoptosis through the Modulation of Tumor Necrosis Factor alpha in PC12 Cells. Chem Res Toxicol, 2014, 27 (4): 507-513.

[42] Zhang S Y, Ito Y, Yamanoshita O, Yanagiba Y, Kobayashi M Taya K, Li C, Okamura A, Miyata M, Ueyama J, Lee C H, Kamijima M, Nakajima T. Permethrin may disrupt testosterone biosynthesis via mitochondrial membrane damage of Leydig cells in adult male mouse. Endocrinology, 2007, 148 (8): 3941-3949.

[43] Jalal S, Ramin H, Roohollah T Z. Effects of cypermethrin on sexual behaviour and plasma concentrations of pituitary-gonadal hormones. Int J Fertil Steril, 2010, 4 (1): 23-28.

[44] Issam C, Samir H, Zohra H, Monia Z, Hassen B. Toxic responses to deltamethrin (DM) low doses on gonads, sex hormones and lipoperoxidation in male rats following subcutaneous treatments. J Toxicol Sci, 2009, 34 (6): 663-670.

[45] Ying Z, Lu M, Zhou P, Wang C, Zhang Q, Zhao M R. Multilevel evaluations of potential liver injury of bifenthrin. Pestic Biochem Phy, 2015, 122: 29-37.

[46] Wang C, Qian Y, Zhang X, Chen F, Zhang Q, Li Z, Zhao M R. A metabolomic study of fipronil for the anxiety-like behavior in zebrafish larvae at environmentally relevant levels. Environ Pollut, 2016, 211: 252-261.

[47] Yan L, Gong C X, Zhang X F, Zhang Q, Zhao M R, Wang C. Perturbation of metabonome of embryo/larvae zebrafish after exposure to fipronil. Environ Toxicol Phar, 2016, 48: 39-45.

[48] Liu D H, Wang P, Zhu W, Gu X, Zhou W, Zhou Z Q. Enantioselective degradation of fipronil in Chinese cabbage (*Brassica pekinensis*). Food Chem, 2008, 110 (2): 399-405.

[49] Gao J, Qu H, Zhang C T, Li W J, Wang P, Zhou Z Q. Direct chiral separations of the enantiomers of phenylpyrazole pesticides and the metabolites by HPLC. Chirality, 2017, 29 (1): 19-25.

[50] Liu T T, Wang P, Lu Y L, Zhou G X, Diao J L, Zhou Z Q. Enantioselective bioaccumulation of soil-associated fipronil enantiomers in Tubifex tubifex. J Hazard Mater, 2012, 219: 50-56.

[51] Qu H, Ma R X, Liu D H, Jing X, Wang F, Zhou Z Q, Wang P. The toxicity, bioaccumulation, elimination, conversion of the enantiomers of fipronil in Anodonta woodiana. J Hazard Mater, 2016, 312: 169-174.

[52] 瞿涵. 苯基吡唑类手性农药及代谢物在水环境中的立体选择性行为及污染修复研究. 北京：中国农业大学, 2016.

[53] 黄笏丹. 几种手性农药在栅藻和蚓蚓中的选择性富集及毒性效应研究. 北京：中国农业大学, 2015.

[54] Qu H, Ma R X，Liu D H, Gao J, Wang F, Zhou Z Q, Wang P. Environmental behavior of the chiral insecticide fipronil: Enantioselective toxicity, distribution and transformation in aquatic ecosystem. Water Res, 2016, 105: 138-146.

[55] Qu H, Wang P, Ma R X, Qiu X X, Xu P, Zhou Z Q, Liu D H. Enantioselective toxicity, bioaccumulation and degradation of the chiral insecticide fipronil in earthworms (*Eisenia feotida*). Sci Total Environ, 2014, 485-486 (1): 415-420.

[56] Qu H, Ma R X, Liu D H, Wang P, Huang L D, Qiu X X, Zhou Z Q. Enantioselective toxicity and degradation of the chiral Insecticide fipronil in scenedesmus obliguus suspension system. Environ Toxicol Chem, 2014, 33 (11): 2516-2521.

[57] Wang C, Li Z Y, Zhang Q, Zhao M R, Liu W P. Enantioselective induction of cytotoxicity by *o, p'*-DDD in PC12 cells: Implications of chirality in risk assessment of POPs metabolites. Environ Sci Technol, 2013, 47 (8): 3909-3917.

[58] He X M, Dong X W, Zou D H, Yu Y, Fang Q Y, Zhang Q, Zhao M R. Enantioselective effects of *o, p'*-DDT on cell invasion and adhesion of breast cancer cells: Chirality in cancer development. Environ Sci Technol, 2015, 49 (16): 10028-10037.

[59] Zhao M R, Zhang, Y, Zhuang S L, Zhang Q, Lu C S, Liu W P. Disruption of the hormonal network and the enantioselectivity of bifenthrin in trophoblast: maternal–fetal health risk of chiral pesticides. Environ Sci Technol, 2014, 48 (14): 8109-8116.
[60] Qian Y, Wang C, Wang J H, Zhang X F, Zhou Z Q, Zhao M R, Lu C S. Fipronil-induced enantioselective developmental toxicity to zebrafish embryo-larvae involves changes in DNA methylation. Sci Rep-Uk, 2017, 7.
[61] Ye Y, Ji D, Xin H, Zhang J, Jing L. Effects of metals on enantioselective toxicity and biotransformation of cis-bifenthrin in zebrafish. Environ Toxicol Chem, 2017, 21 (7): 452-460.

作者：赵美蓉[1]，周志强[2]，王　鹏[2]，虞云龙[3]，花日茂[4]，刘维屏[5]
[1]浙江工业大学环境学院，[2]中国农业大学理学院，[3]浙江大学农业与生物技术学院，
[4]安徽农业大学资源与环境学院，[5]浙江大学环境与资源学院

第 9 章　铁环境化学研究进展

本章导读

铁是分布最广泛的元素之一，广泛存在于自然界中和动植物体内。铁循环显著地影响着微量元素的地球化学循环以及环境污染物的迁移、转化、归趋和生物利用等。因此，铁环境化学已成为环境和地球化学领域的研究重点和前沿热点。本章总结了近五年来国内外学者在天然水体中的铁化学，环境铁循环及其调控，铁矿物生物地球化学过程及其强化，含铁纳米环境材料制备及其表征，以及基于铁基材料的污染控制技术及原理方向的进展，借此为深入研究环境铁化学奠定基础，促进环境铁化学相关的污染控制和环境修复技术发展。

关键词

铁，环境循环，环境材料，污染控制，环境修复

1 引　　言

铁是地壳中分布最广泛的元素之一，广泛存在于自然界中和动植物体内。由于 Fe(III)/Fe(II)的氧化还原电势处在 C、O、N 和 S 等主要元素物种氧化还原电势之间，铁元素能够通过与这些元素之间的氧化还原反应，驱动生物地球化学循环。因此，自然界中铁的存在形态以及生物利用度与体系的氧化还原条件密切相关。与此同时，地球上主要元素 C、N、S 等和微量元素 As、Cr、Cu、U 等的化学沉积、生物利用与迁移转化等地球化学循环过程也受到 Fe(III)/Fe(II)氧化还原过程（铁循环）的驱动。近年来越来越多科学家认识到，铁循环显著地影响着微量元素的地球化学循环以及环境污染物的迁移、转化、归趋和生物利用等。因此，铁环境化学已成为环境和地球化学领域的研究重点和前沿热点。本章总结了近五年来国内外学者在天然水体中的铁化学，环境铁循环及其调控，铁矿物生物地球化学过程及其强化，含铁纳米环境材料制备及其表征，以及基于铁基材料的污染控制技术及原理方向的进展，借此为深入研究环境铁化学奠定基础，促进环境铁化学相关的污染控制和环境修复技术发展。

2　天然水体中的铁化学

2.1　天然水中铁的来源分布及赋存形态

2.1.1　天然水中铁的来源分布

铁是地壳中分布最广泛的过渡金属元素，同时也是天然水环境中的重要组分[1-4]。河流和湖泊中的铁主要来源于含铁矿物的溶出。自然界中存在的含铁矿物溶于水后或释放出铁离子或形成铁络合物。而土壤以及植物碎屑中存在的大量铁元素可通过生物圈作用或风力搬运或淋溶作用进入河流及湖泊中。地下水中的铁则多来源于含水层和地表降水的铁渗滤。更为重要的是，铁被认为是海洋生物生长限制性元素，海洋中溶解性外源铁主要来源于风尘的溶解、沉积物的还原和非还原溶解、水热活动、冰山和河流的输

入。由于在淡水海水混合区发生铁胶体聚集，河流输入的量对海洋中铁的贡献通常被忽略。最近一些新的证据表明，与腐殖酸和富里酸结合的铁可以输送到开放的海洋[5-7]。天然水体沉积物中的铁含量与水中含铁物质的沉降有关，海洋沉积物中异化铁还原菌（FRB）则参与铁的生物地球化学循环。同时，天然水体中铁的来源还和人类的工业生产活动密切相关。人为排放的含铁废水以及废气是天然水体铁的重要来源之一。

2.1.2 天然水体中铁的赋存形态

铁的主要氧化还原状态有：Fe(0)、Fe(II)和 Fe(III)。其中，Fe(III)是热动力学稳定形态。在天然环境中，Fe(II)、Fe(III)及铁矿物是铁的三种主要赋存形态。在水体中，铁则以胶体（colloids，< 1 μm）、纳米颗粒（nanoparticulate，<0.1 μm 或 100 nm）和水溶态（aqueous species，0.02 μm）存在；通常膜过滤可区分为 0.02 μm，0.2 μm 和 0.45 μm 不同组分。两者粒径划分不完全对应，通常 0.02 μm 膜过滤代表水溶态，大于 0.2 μm 膜过滤组分作为纳米颗粒[1]。天然水中铁的形态变化与环境条件相关，随着水环境的氧化或者还原程度的变化而变化。在天然水体中，溶解态 Fe(II)可被氧气氧化为 Fe(III)，溶解性的 Fe(III)主要以配合物的形式存在，可通过光分解反应产生各种自由基和溶解态 Fe(II)。另外，Fe(III)也可发生水解作用形成氢氧化铁胶体，而后转化为铁氧化物。而铁氧化物及硫化物又可通过热力学溶解过程析出溶解态铁。

当溶液中存在氢氧化铁的还原或硫化亚铁的氧化时，溶解性的 Fe(II)含量极高。当水环境与氧气交换程度加深时，Fe(II)会被氧化为 Fe(III)。而 Fe(III)容易与天然水中广泛存在的溶解性有机质（NOM）发生络合，因此 Fe(III)常以有机络合态存在。如果进入天然水中的含铁矿物无法溶出铁，就会以单独的铁矿物的形式存在。由于其独特的环境条件，地下水中铁多为 Fe(II)。沉积物中的铁主要是以亚铁矿物（黄铁矿）和铁矿物（水铁矿、纤铁矿等）固相铁的形式存在，图 9-1 为表层海水中水溶态、胶体（纳米颗粒）态和颗粒态铁（水合）氧化物之间的相互作用和转化[1]。

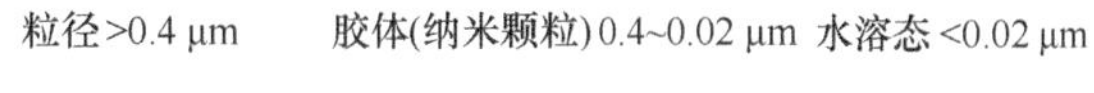

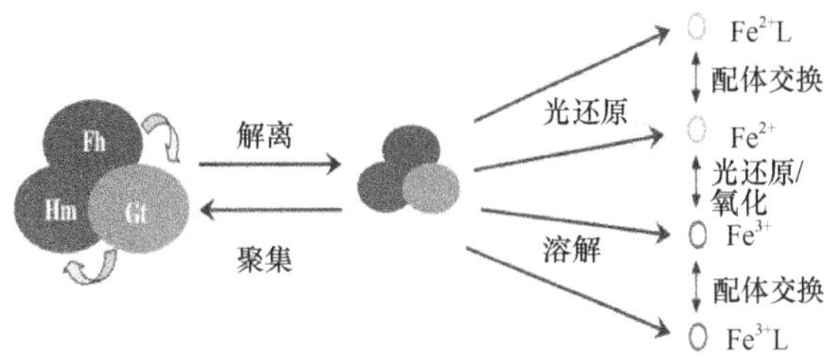

图 9-1 表层海水中铁的形态及形态转化（引自参考文献[1]）

黄色箭头：表示水铁矿（ferrihydrite，Fh）向针铁矿（goethite，Gt）和赤铁矿（hematite，Hm）的转化途径

2.2 天然水中铁与有机物的相互作用

尽管铁具有较高的地壳丰度，但天然水中铁的含量是有限的，这是因为环境中性 pH 条件有利于不溶性 Fe(III)的生成。与 Fe(III)相比，铁的还原态 Fe(II)更易溶解，但在含氧地表水中可迅速被氧化。因此，溶解性铁的浓度由 Fe(III)还原速率和 Fe(II)氧化速率控制。在淡水和海洋环境中，铁的氧化还原转化速率被普遍认为受水体中天然有机物质（NOM）的影响[8-17]。

最近，许多研究者开始关注 NOM 的供电子和接受电子的能力，以及这些氧化还原活性基团与具备氧化还原活性微量金属（例如 Fe 和 Cu）的相互作用机制[18-21]。尽管参与铁转化的氧化还原活性基团的确切特征尚不清楚，但学术界普遍认为 NOM 的氧化还原性质主要与醌类官能团相关，其存在已通过 NMR[22,23]、荧光光谱学[24-26]和电化学方法证实[19,20]。基于 Waite 研究组最新的研究成果，本章节总结了在三种不同类型的有机体存在下，即对苯二酚、土壤来源的 Suwannee 河富里酸（SRFA，通常用作模型 NOM）、源于铜绿微囊藻（普通淡水藻类）的有机分泌物质，铁氧化还原转化过程[8,10,11,15]。

2.2.1 pH 3~5 的纯氢醌体系中 Fe 氧化还原转化

鉴于 NOM 中存在的醌基团被认为在天然水体中铁的氧化还原转化过程中扮演着重要的作用，Waite 等首先研究了单纯氢醌（1, 4-对苯二酚）（H_2Q）溶液体系中铁氧化还原动力学及其转化机理[11]。如图 9-2 所示，pH 为 3~5 时，向氢醌溶液中加入 Fe(III)，氢醌还原 Fe(III)到 Fe(II)。Fe(II)的氧化剂认定是由 Fe(III)氧化氢醌形成的半醌自由基（$Q^{\cdot-}$）（公式 9-1）。

$$H_2Q + Fe(III) \underset{k_{-1}}{\overset{k_1}{\rightleftharpoons}} Q^{\cdot-} + Fe(II) \tag{9-1}$$

在公式（9-1）中，形成的半醌自由基进一步氧化形成苯醌（Q）（图 9-2）。Fe（III）的存在不仅克服了自旋限制，还催化了 H_2Q 的氧化[14,27]。Q 形成与 Fe（II）形成的化学计量约为 1∶1，这表明 Q 形成通过顺序反应发生。也就是说，公式（9-1）中形成 $Q^{\cdot-}$ 被氧气进一步氧化形成 Q［公式(9-2)］，并伴随 H_2O_2 的生成［公式(9-3)］。Waite 研究组的研究结果表明，在氢醌存在的条件下，铁的+3 和+2 氧化态之间发生快速氧化还原循环，并伴随着苯醌和活性氧（特别是超氧化自由基和 H_2O_2）的产生，而 O_2 在这些氧化还原转化中扮演着重要的作用。

$$Q^{\cdot-} + O_2 \xrightarrow{k_2} Q + HO_2^{\cdot} \tag{9-2}$$

$$HO_2^{\cdot} + HO_2^{\cdot} \xrightarrow{k_{disp}} O_2 + H_2O_2 \tag{9-3}$$

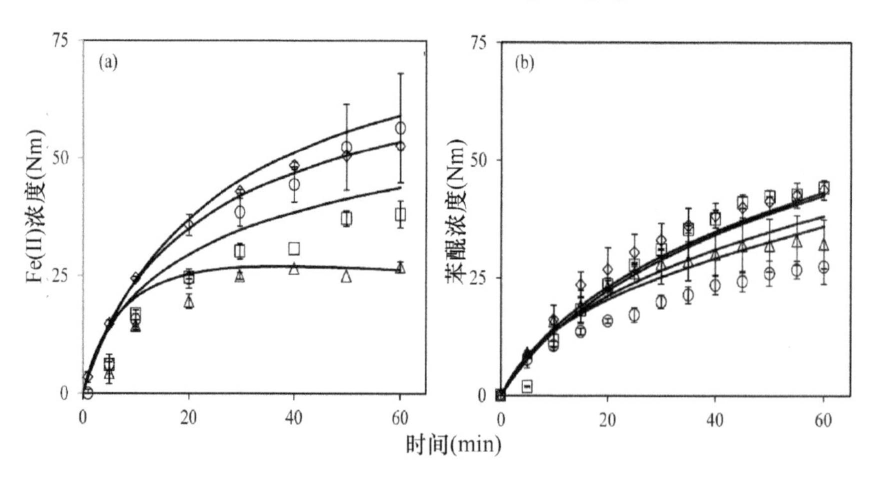

图 9-2 在 pH 3（圆圈）、pH 4（菱形）、pH 4.5（正方形）、pH 5（三角形）条件下，100 nmol/L Fe(III)在 1μmol/L 氢醌溶液中还原产生 Fe(II)（a）和苯醌的浓度（b）（引自参考文献[11]）

2.2.2 Suwannee 河富里酸存在时 Fe 的氧化还原转化

Waite 等研究了酸性条件下 SRFA(一种来自于土壤中的 NOM)影响 Fe 氧化还原转化的动力学和机理[10]。在暗反应条件下，他们将 Fe(III)添加到 pH 3~5 的 SRFA 溶液中时，观察到 Fe(II)的产生。这一结果支持了 SRFA 中含有 Fe(III)还原基团(由 A^{2-}表示)。由于 pH<5 的 SRFA 溶液中 Fe(II)自氧化可被忽略（至少在实验时间尺度上），Fe(II)氧化剂必须由 Fe(III)氧化 SRFA 中还原性有机基团形成［公式(9-4)］。该反应路径解释了 SRFA 溶液中 Fe(II)的生成与 Fe(III)的还原密切相关，也与纯氢醌溶液中铁氧化还原转

化机理一致［公式(9-4)］。需要强调的是，虽然 SRFA 包含的 Fe(II)的有机氧化剂和 Fe(III)的有机还原剂的本质仍不清楚，但 Waite 研究组推测出其氧化还原行为与半醌和氢醌结构组分十分相关的结论。当然，他们并未排除涉及其他官能团(例如多酚结构)参与的可能性。

$$A^{2-} + Fe(III) \underset{k_2}{\overset{k_1}{\rightleftharpoons}} A^{-} + Fe(II) \tag{9-4}$$

Waite 研究组的研究结果进一步表明，在光照条件下，SRFA 的可氧化部分（氢醌类结构）能产生超氧自由基，从而导致 A^{2-}浓度降低，A^{-}浓度增加[10]。当将铁添加到预照射过 SRFA 溶液中时，照射条件下 A^{2-}向 A^{-}转化可导致 Fe(III)还原速率的降低和 Fe(II)氧化速率的增加（图 9-3）。

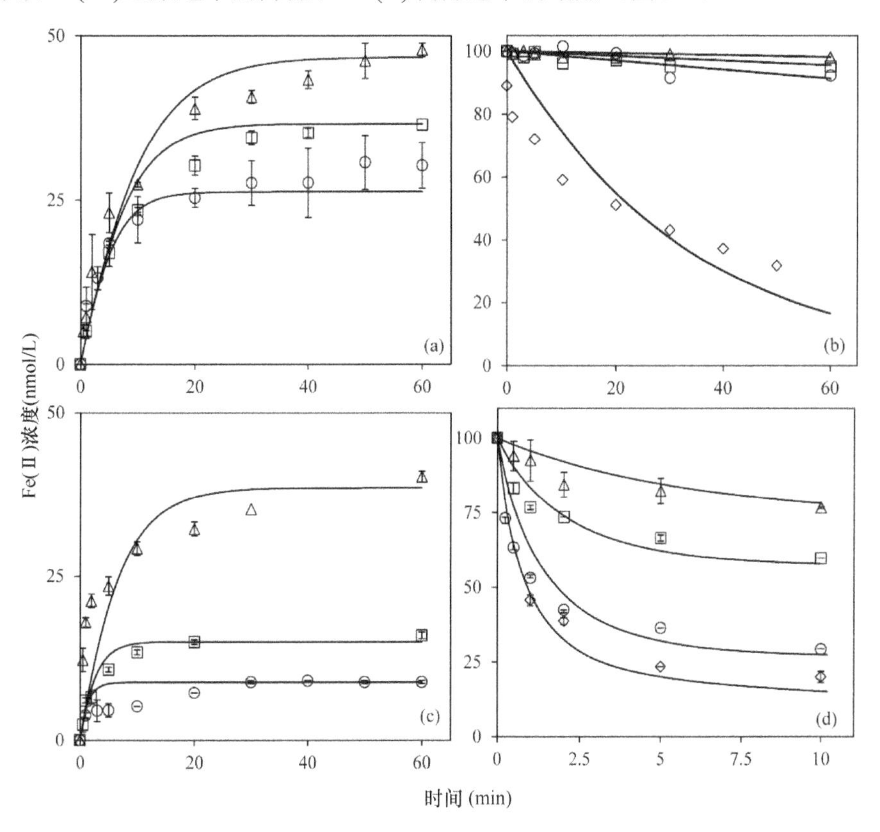

图 9-3　在 pH 3（三角形）、pH 3.5（正方形）、pH 4（圆形）条件下，未光照（a）和先前光照（c）的 SRFA 溶液（10 mg/L）中还原 100 nmol/L Fe(III)所产生的 Fe(II)。在同样 pH 条件下，未光照（b）和先前光照（d）的 SRFA 溶液（10 mg/L）中分别加入 100 nmol/L Fe(II)，Fe(II)浓度的降低（引自参考文献[10]）

$$Q \xrightarrow{h\nu} Q^{*} \xrightarrow{D} Q^{-} \xrightarrow{O_2} HO_2^{\cdot} \tag{9-5}$$

虽然在纯氢醌和 SRFA 体系中铁氧化还原转化的机理相似，但其具体反应过程却存在较大差异。如前所述，氢醌与 Fe(III)［公式(9-1)］反应形成的半醌自由基在氧气存在下迅速被氧化［公式(9-2)］；然而，在 SRFA 溶液中形成的类半醌基团（A^{-}）是稳定的且不易被氧气氧化。对比在预光照的 SRFA 且延迟 24 h 避光加入 Fe(II)和 SRFA 光照后立即加入 Fe(II)两种情况，Waite 等发现 Fe(II)的氧化程度无显著差异，表明 SRFA 中 Fe(II)氧化物质具有较长的寿命，且与 Fe(II)除外的其他共存物质之间没有明显的相互作用。SRFA 溶液中半醌类自由基的稳定性最有可能归因于 SRFA 这种大分子中不配对电子可在共轭芳香烃及电子供体烷基官能团存在下实现离域稳定。与纯氢醌溶液相比，SRFA 溶液中半醌自由基稳定性的提高使其成为较醌溶液更为有效的 Fe(II)氧化剂，进而可推测出 SRFA 在铁氧化还原转化中起到了更为重要的作用。这一推论被进一步在 Fe 转换频率的差异中反映出来。譬如，SRFA 中铁+3 和+2 氧化态之间的循环速率是纯 1, 4-对苯二酚溶液中观察到速率的 10~100 倍[10,11]。

2.2.3 Fe 与铜绿微囊藻有机分泌物之间的相互作用

Waite 研究组还研究了藻类相关的 NOM 与 Fe 的相互作用[15]，并特别关注铜绿微囊藻有机分泌物与 Fe 的相互作用机制。他们发现，铜绿微囊藻分泌物中含有的 Fe(III)还原基团与 SRFA 的还原基团性质一致[8]，如图 9-3 所示，在暗反应条件下将 Fe(III)加入到 pH 为 4 的新鲜藻类分泌物中后，Fe(II)的浓度随着 Fe(III)还原而增加。他们的研究结果进一步表明，藻类分泌物中的 Fe(III)还原基团并不稳定且不断衰变（半衰期约 1.9 h）。

藻类分泌物的存在对 Fe(II)氧化动力学有着显著的影响。与无分泌物体系相比，Fe(II)在分泌物体系中的氧化速率更快（图 9-4）。这一观察证实了，藻类分泌物的存在由于其 Fe（III）强络合配位体的作用加速了 Fe(II)氧化。这降低了三价铁离子活性，并增加了氧化的驱动力[13,28]。同时，藻类分泌物中 Fe(II)氧化基团可稳定存在。存在于分泌物中的 Fe(III)强络合配体可降低 Fe(III)/Fe(II)氧化还原对的还原电位[28]。

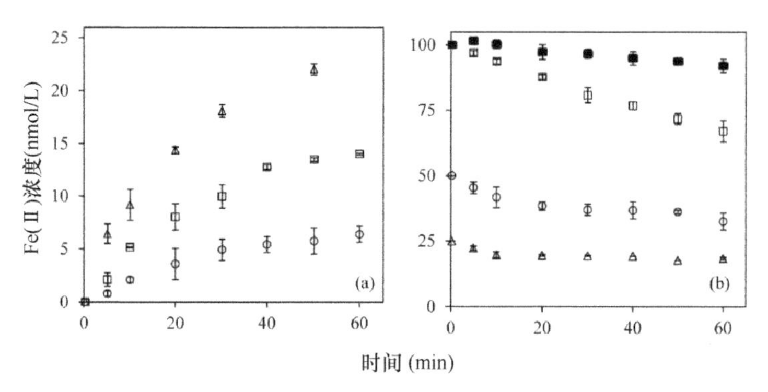

图 9-4 在 pH 4 的暗反应条件下，向新鲜藻类分泌物中分别加入 150 nmol/L（正方形）、100 nmol/L（圆形）、50 nmol/L（三角形）Fe(III)时所生成 Fe(II)的量（a）；在 pH 4 的暗反应条件下，向新鲜藻类分泌物中分别加入 100 nmol/L（正方形）、50 nmol/L（圆形）、25 nmol/L（三角形）Fe(II)时 Fe(II)浓度的降低，实心的正方形表示在 pH 4 条件下向藻类分泌物中加入 100 nmol/L Fe(II)时 Fe(II)浓度的降低（b）（引自参考文献[15]）

另一方面，藻类分泌物与 SRFA 似乎有相似的氧化还原性质（与 Fe 的相互作用）和结合 Fe 能力，但两者还是存在一定的差异。譬如，SRFA 的 Fe(III)还原基团和 Fe(II)氧化基团相对稳定且寿命较长（>24 h）[10]，而藻类分泌物的 Fe(III)还原基团寿命则相对较短（约 1.9 h）[15]。藻类分泌物中的 Fe(II)氧化基团（如果存在）较不稳定且寿命极短（<1 min）。因此，Waite 等认为，当土壤和藻类来源的 NOM 与 Fe 结合并诱导 Fe(III)还原时，Fe 氧化还原转化速率和 Fe 的生物可利用效率根据 NOM 的性质和来源而有着很大的不同。

2.3 天然水体铁的光化学反应

Fe(III)广泛分布于天然水环境中，主要以无机和有机络合物的形式存在，其存在形式依赖于溶液酸碱度。在酸性条件下，Fe(III)主要以溶解性无机络合物形式存在[如 $Fe(OH)^{2+}$或 $FeCl^{2+}$]。在光照下，这些络合物会发生光解作用，产生各种自由基，如羟基自由基（$^{\cdot}OH$）或氯自由基（$Cl_2^{\cdot-}$）[29]。在有机质广泛存在的天然水中，Fe(III)多与草酸、柠檬酸、富里酸形成络合物，例如：以 $Fe^{III}(C_2O_4)_3^{3-}$、$Fe^{III}(C_2O_4)_2^{-}$、Fe(III)-

柠檬酸络合物，以及 Fe(III)-FA 络合物的形式存在。铁配合物的环境光化学反应对有机污染物及重金属在环境中的形态及归趋的影响[30]，成为近年来环境科学领域的研究热点，部分研究综述已经对相关进展进行了总结[31-34]。

2.3.1　铁无机络合物的光化学

1）铁水合物的光化学

在 pH=3~5 的 Fe(III)无机盐的水解产物中，$Fe(OH)^{2+}$是主要的铁水合物络合形态[35]。在紫外光照射下，它能够经历配体到金属的电荷转移（LMCT）激发（λ_{max}~300 nm，ε_{max}~2000 $L\cdot mol^{-1}\cdot cm^{-1}$），然后发生内层的光诱导电子转移，生成 Fe(II)水合配合物和·O[29,36]［公式(9-6)］。

$$Fe(OH)^{2+} \xrightarrow{h\nu(LMCT)} Fe^{2+} + {}^{\cdot}OH \tag{9-6}$$

Fe(III)/Fe(II)的光化学循环过程包含铁配合物光还原和 Fe(II)氧化两个主要反应。目前，研究主要集中在铁配合物的光还原过程，而酸性体系中 Fe(II)氧化过程相关研究却并没有引起足够重视。王兆慧等采用化学动力学拟合方法研究了 pH 3.0 时 Fe(II)的氧化动力学[37]。他们发现，Fe(II)氧化速率与紫外光、氧气和 Fe(III)离子浓度密切相关。低浓度的 Fe(III)有利于溶液中共存 Fe(II)的快速氧化。当体系中 Fe(II)较高时，羟基 Fe(III)配合物激发态会被无氧体系的 Fe(II)淬灭，或敏化氧化有氧体系中的 Fe(II)。当动力学模型中引入光敏化途径时，所建立的模型可以较好描述 pH 3.0 溶液中 Fe(II)的氧化。模型的敏感度分析显示铁配合物光解反应对 Fe(II)氧化至关重要。Fe(II)催化分子氧活化及其生成的次级活性氧物种，例如 $HO_2\cdot$和·OH，是导致酸性溶液中 Fe(II)氧化的关键因素。

2）铁-亚硫酸盐的光化学

Fe 催化水中 SO_2［S(IV)］的氧化对于硫元素转化、酸沉降和气溶胶成核作用具有重要意义，因此其成为大气化学领域的重要研究课题之一[38]。Fe（III）-亚硫酸盐水溶液在 290~575 nm 有较宽的吸收带[39]。最近研究表明[40]，当亚硫酸盐充足的时候，$FeSO_3^+$是主要的 Fe(III)活性物种，光解 $FeSO_3^+$生成 Fe^{2+}和 $SO_3^{\cdot-}$，接下来促进活性物种 $SO_5^{\cdot-}$和 $SO_4^{\cdot-}$的生成[公式(9-7)~公式(9-16)]。随着 pH 的下降和亚硫酸盐的消耗，光活性物种自发地由 $FeSO_3^+$转换为 $FeOH^{2+}$。$FeOH^{2+}$光解产生氧化性较强的·OH 来降解有机污染物（图 9-5）。该体系实现了以铁催化为中心的有机/无机污染物的同时转化。

$$\log K_1 = 4 \tag{9-7}$$

$$4FeHSO_3^+ + O_2 \longrightarrow 4FeSO_3^+ + 2H_2O \tag{9-8}$$

$$Fe^{3+} + HSO_3^- \longrightarrow FeSO_3^+ + H^+ \qquad \log K_3 = 2.45 \tag{9-9}$$

$$FeSO_3^+ \longrightarrow Fe^{2+} + SO_3^{\cdot-} \qquad K_4 = 0.19\ s^{-1} \tag{9-10}$$

$$SO_3^{\cdot-} + O_2 \longrightarrow SO_5^{\cdot-} \qquad K_5 < 10^9\ L\cdot mol^{-1}\cdot s^{-1} \tag{9-11}$$

$$SO_5^{\cdot-} + HSO_3^- \longrightarrow SO_4^{\cdot-} + SO_4^{2-} + H^+ \qquad K_6 = (10^4\text{~}10^7)\ L\cdot mol^{-1}\cdot s^{-1} \tag{9-12}$$

$$SO_5^{\cdot-} + SO_5^{\cdot-} \longrightarrow 2SO_4^{\cdot-} + O_2 \qquad K_7 = (10^6\text{~}10^8)\ L\cdot mol^{-1}\cdot s^{-1} \tag{9-13}$$

$$SO_5^{\cdot-} + HSO_3^{\cdot-} \longrightarrow SO_3^{\cdot-} + HSO_5^- \qquad K_8 = (10^4\text{~}10^7)\ L\cdot mol^{-1}\cdot s^{-1} \tag{9-14}$$

$$Fe^{2+} + HSO_5^{\cdot-} \longrightarrow SO_4^{\cdot-} + Fe^{3+} + OH^- \qquad K_9 = (10^4\text{~}10^7)\ L\cdot mol^{-1}\cdot s^{-1} \tag{9-15}$$

$$FeSO_3^+ + UV \longrightarrow Fe^{2+} + SO_3^{\cdot-} \tag{9-16}$$

2.3.2　铁-多羧酸的光化学

多羧酸尤其是草酸是大气水相中较常见的有机物，它们能与 Fe(III)形成稳定的且光敏感的配合物[41]。

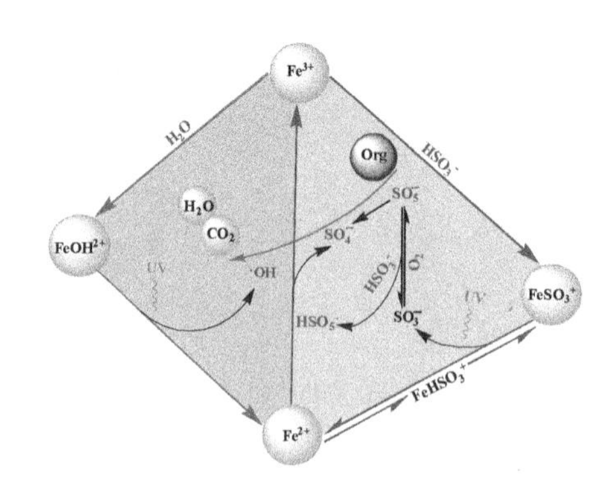

图 9-5 Photo-sulfite 光化学体系反应示意图（引自参考文献[40]）

Weller 等报道了铁与羧酸类（草酸、丙二酸、苹果酸、丁二酸、戊二酸、羟基丙二酸、酒石酸、葡糖酸、乳酸、丙酮酸和乙醛酸）系列络合物光解生成 Fe(II)的有效量子产率，发现量子产率与配体、波长、配位数、溶解氧、激发光能量和初始铁浓度等因素密切相关[42,43]。以草酸铁配合物为例，与 Fe(III)–OH 配合物的光解类似，光照下能够发生 LMCT 过程，Fe(III)被还原的同时，生成强还原性的 $CO_2^{\cdot-}$自由基（$E^0 = -1.8$ V *vs* NHE）。$CO_2^{\cdot-}$自由基可以继续还原附近的一分子草酸铁，或者是将氧气还原为超氧自由基（$k = 2.4\times10^9$ $L\cdot mol^{-1}\cdot s^{-1}$）[44]。然而，关于草酸铁光解的初级反应历程还存在较大争议。

Chen 等通过超快 X 射线吸收光谱研究表明，光照后 140 ps 内草酸铁发生光解离反应，草酸配体的 C—C 键均裂形成两个 $CO_2^{\cdot-}$自由基，这个碳自由基再与其他草酸铁配合物发生次级氧化还原反应[45]。但是，Pozdnyakov 通过瞬态吸收光谱证实这一步骤只占总反应的 6%左右，认为主要反应仍然是 LMCT 过程[46]。美国劳伦斯伯克利国家实验室的最新研究表明[47]，从草酸到铁的分子内电子转移发生在亚皮秒级时间范围内，生成 Fe(II)和一个氧化的草酸配体。在随后的 40 ps 内，被氧化的草酸配体分解成 CO_2 和 $CO_2^{\cdot-}$自由基。王兆慧等研究发现[48]，草酸铁配合物的光化学过程与 Fe(III)和草酸配体浓度密切相关。过量的 Fe(III)能捕获强还原性 $CO_2^{\cdot-}$自由基，导致 H_2O_2 产率降低。随着草酸浓度提高，1∶2 型和 1∶3 型草酸铁配合物控制着铁物种形态。当 Fe(III)与草酸浓度比例低于 1∶2 时，羟基自由基光产率较低。

除了草酸，丙二酸和丁二酸也是富碳大气水中常见的二元羧酸[41-44]。在所有二元和一元羧酸中，草酸和丙二酸与 Fe(III)配位能力最强（见表 9-1），可以预见丙二酸铁配合物的光化学反应在铁光化学循环过程中亦应发挥重要作用。王兆慧等模拟研究了酸性大气水相条件下铁的光化学循环和丙二酸（Mal）配体的氧化过程[49]。与草酸、丁二酸和其他二酸（或单酸）相比，丙二酸展现出完全不同的反应特性。Fe(III)-Mal 配合物（FMCs）具有较强的络合强度，较高的摩尔吸光能力但极低的 Fe(II)量子产率（$\Phi_{Fe(II)}$= 0.0022，pH 3.0）。Fe(III)形态分析表明，改变 Mal 浓度能够调控 Fe(III)-OH 配合物和 FMCs 两个光活性物种的比例。自旋捕获电子自旋共振（ESR）证明，当总 Mal 浓度（$[Mal]_T$）较低时，丙二酸铁配合物的光化学反应生成$\cdot CH_2COOH$ 和·OH 两种自由基；当$[Mal]_T$较高时，体系仅有$\cdot CH_2COOH$ 生成。这一区别充分证明了 Mal 与 OH^-存在配体竞争以及两者光反应路径的不同。后续的研究发现[50,51]，一旦芳香

表 9-1 铁平衡计算的组成矩阵：组分和化学计量系数及配合物稳定常数（引自参考文献[49]）

铁物种	组分								平衡常数
	Fe^{3+}	Ox^{2-}	Mal^{2-}	OH^-	Suc^-	For^-	Ace^-	Pro^-	lg *β*
$Fe(OH)^{2+}$	1			1					−2.2
$Fe(OH)_2^+$	1			2					−5.8
$Fe(Ox)^+$	1	1							9.4
$Fe(Ox)_2^-$	1	2							16.2
$Fe(Ox)_3^{3-}$	1	3							20.4

续表

铁物种	组分								平衡常数
	Fe^{3+}	Ox^{2-}	Mal^{2-}	OH^-	Suc^-	For^-	Ace^-	Pro^-	lg β
$Fe(Mal)^+$	1		1						7.5
$Fe(Mal)_2^-$	1		2						13.3
$Fe(Mal)_3^{3-}$	1		3						16.9
$Fe(HSuc)^{2+}$	1				1				7.5
$Fe(For)^{2+}$	1					1			1.85
$Fe(Ace)^{2+}$	1						1		3.2
$Fe(Pro)^{2+}$	1							1	3.45

族化合物氧化分解生成较高浓度的丙二酸，那么UV/Fe(III)氧化体系的降解效率和Cr(VI)的光还原效率受到极大抑制，其原因在于丙二酸铁配合物主导了铁的形态分布，由于其光量子产率极低，导致Fe(II)的生成量极少，限制了铁循环速率以及后续相关的反应过程。

2.4　天然水体中铁对污染物迁移转化的影响

随着还原环境到氧化环境的改变，水体中可溶性Fe(II)发生氧化，形成水合氢氧化铁沉淀，这种铁循环过程强烈支配着天然水体中各种污染物的归趋，对污染物的迁移和转化有重要作用。

2.4.1　Fe(II)及其络合物对污染物迁移转化的影响

当Fe(II)从无氧环境进入有氧环境时，在近中性或者碱性条件下会被氧气催化氧化，进而产生·OH和Fe(IV)中间体[52]，强烈影响着有机污染物或者重金属的归趋及形态分布[46]。最近，有研究报道了Fe(II)介导的Sb(III)氧化，结果表明：Fe^{2+}在被氧化成为Fe^{3+}的过程中产生羟基自由基(·OH)和四价铁[Fe(IV)]，促进三价锑[Sb(III)]氧化成为五价锑[Sb(V)]，进而影响锑在天然水中的归趋。另外，天然水中广泛分布的NOM如草酸、富里酸(FA)等可与Fe(II)形成不同的有机络合物如$Fe(C_2O_4)$、$Fe(C_2O_4)^{2-}$和FeFA。这些络合物较水和亚铁离子都更容易在氧化过程中产生活性物种及自由基。因此，Fe(II)有机络合物对重金属价态转变的作用更强[53]。

2.4.2　Fe(III)及其络合物对污染物迁移转化的影响

最近，Kong等研究了Sb(III)在Fe(III)有机络合物(Fe(III)-草酸、Fe(III)-柠檬酸和Fe(III)-富里酸)溶液中的氧化速率和反应机理(图9-6)。他们发现，在pH 3~7的条件下，Sb(III)可在Fe(III)-草酸溶液中迅速发生光氧化，且其氧化速率比在无机Fe(III)溶液中的速率快。在酸性条件下，·OH是Sb(III)的重要氧化剂，而在中性条件下，自由基掩蔽剂并不能完全淬灭Sb(III)的光氧化，Sb(III)可能同时被某种类Fe(IV)物质氧化。在酸性条件下（pH 1~3），由$Fe(OH)^{2+}$和$FeCl^{2+}$激发产生·OH和Cl_2^-自由基是Sb(III)的主要氧化剂；在近中性和碱性条件下，Sb(III)与胶体态铁或水铁矿表面的Fe(III)点位结合形成的Fe(III)-Sb(III)络合物可在光激发下发生LMCT使Sb(III)氧化[54]。

2.4.3　铁矿物对污染物迁移转化的影响

铁氧化物及铁硫化物广泛分布于天然水体中。铁矿物可有效地光催化各种环境物质，还可通过参与光敏化反应降解或氧化某些物质，影响铁系矿物和所吸附物质的归趋[55,56]。最近，Kong等系统研究了铁硫化物（黄铁矿）对三价锑的吸附氧化行为和机理。在黑暗条件下，黄铁矿溶液中及其表面上的Sb(III)

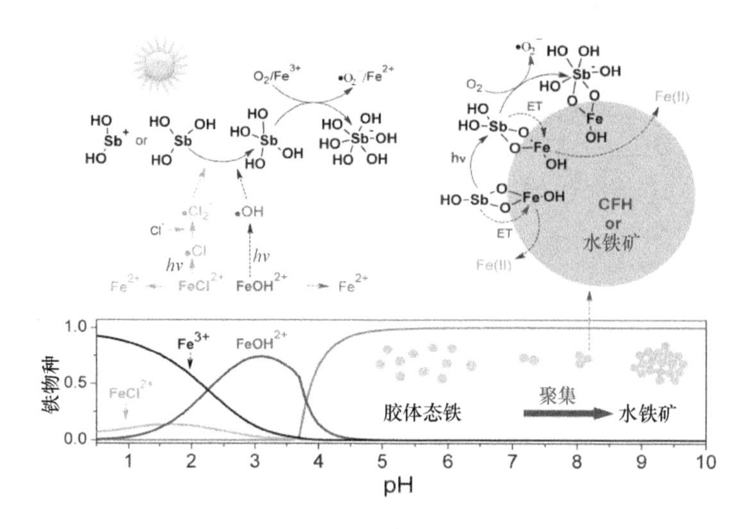

图 9-6　不同 Fe(III)形态对 Sb(III)的氧化机理（引自参考文献[54]）

都被部分氧化为 Sb(V)。随着 pH 的增加，Sb(III)的氧化速率逐渐增强。由黄铁矿产生的·OH 和 H_2O_2 是 Sb(III)的主要氧化剂。在酸性条件下，·OH 是 Sb(III)的主要氧化剂，而在中碱性条件下，H_2O_2 逐渐变为 Sb(III)的主要氧化剂。在无氧环境中，·OH 和 H_2O_2 可通过黄铁矿原有的 Fe(III)(pyrite)和 H_2O 的反应产生，但氧气是·OH 和 H_2O_2 持续产生的关键。在表面被氧化的黄铁矿悬浮液中，Sb(III)的氧化速率被增强。在酸性条件下，该体系可通过芬顿反应产生更多的·OH，而在中碱性条件下，可产生 Fe(IV)和更多的 H_2O_2。另外，在光照条件下，在黄铁矿悬浮液中可产生更多的·OH 和 H_2O_2，显著增强 Sb(III)的氧化速率（图 9-7）[57]。

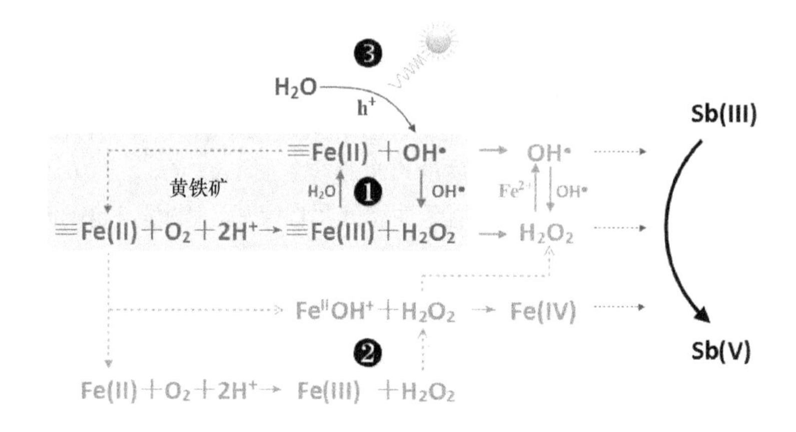

图 9-7　在黄铁矿悬浮液中 Sb(III)的氧化机理（引自参考文献[57]）

注：❶黄铁矿表面的≡Fe^{II} 的氧化导致的 Sb(III)的氧化；❷溶解到溶液中的 Fe^{II} 的氧化导致的 Sb(III)的氧化；❸光催化过程导致的 Sb(III)的氧化

2.4.4　零价铁对污染物迁移转化的影响

零价铁与天然水体中污染物的作用主要发生在零价铁应用于环境修复过程中。因此，本节主要讨论了 NOM 对于纳米零价铁（nZVI）环境修复的影响作用规律。实验发现，在纳米零价铁去除十溴联苯醚的过程中，无论是腐殖酸（HA），还是常用作电子转移介质的对苯醌、2-羟基-1, 4-萘醌均没有发挥电子转移媒介的作用，它们对纳米零价铁去除十溴联苯醚（BDE209）起抑制作用。随着 HA 浓度升高，其对 nZVI 去除 BDE209 的抑制作用越强。这一结果证明，腐殖酸在 nZVI 的表面发生化学吸附与 BDE209 的去除两个过程是同时发生的，即腐殖酸能与 BDE209 竞争 nZVI 表面的反应活性位点，导致 BDE209 的去除速率减慢[58]。通过生物炭负载纳米镍铁（BC@Ni/Fe）的 Zeta 电位和沉降实验，研究者发现 HA 投加量的增加能有效提高 BC@Ni/Fe 的稳定性和表面电荷，认为 HA 不是通过影响纳米颗粒的性能起到抑制作用的。BC@Ni/Fe 的腐蚀能力随着 HA 浓度的升高而降低，这与 HA 对 BC@Ni/Fe 去除 BDE209 反应活性影响成正相关关系。此外，HA 中具有电子传递作用的典型醌类化合物指甲花醌和蒽醌-2, 6-二磺酸钠（AQDS），在反应过程中均没有起到电子传递作用，反而对 BC@Ni/Fe 去除 BDE209 起抑制作用。最后，结合 BDE209 和 HA 之间的竞争吸附实验。方战强等认为，HA 抑制 BC@Ni/Fe 反应活性的主导原因是 HA 优先于 BDE209 被 BC@Ni/Fe 吸附，占据了 BC@Ni/Fe 颗粒表面活性位点，阻碍 BC@Ni/Fe 中纳米镍铁颗粒与 H_2O 的接触，减少了 Fe^0 的腐蚀，从而抑制了 BC@Ni/Fe 对 BDE209 的去除，其影响机理如图 9-8 所示[59]。

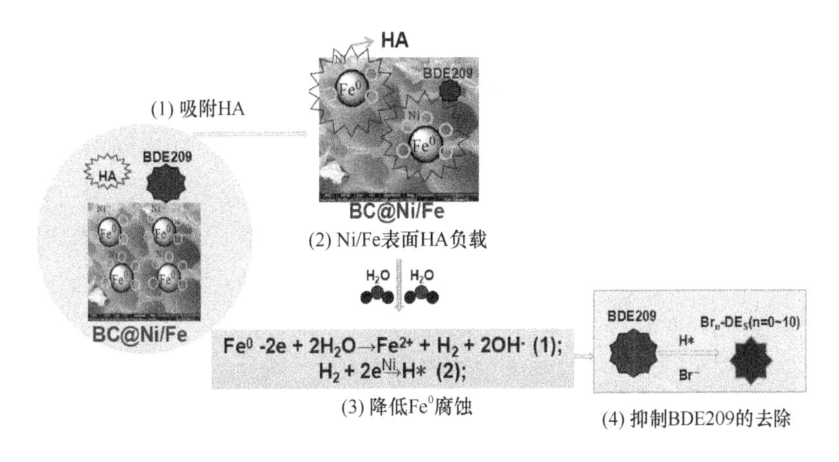

图 9-8　HA 影响生物炭负载纳米镍铁去除 BDE209 作用机理（引自参考文献[59]）

2.5　二价铁矿物活化分子氧产生活性氧物种及其污染物氧化效应

铁（Fe）是地球表层含量最为丰富的变价金属元素。沉积物中铁元素平均含量为 3.9%，主要以含铁矿物形式存在，其中 Fe(III)/Fe(II)比例约为 1.35[60]。Fe(III)/Fe(II)的生物地球化学循环对重金属和有机污染物转化、生物元素运移以及有机碳埋藏等都起到了非常重要的作用[61-63]。相比于三价铁矿物而言，沉积物中的二价铁矿物在调节污染物迁移转化中表现出更高的反应活性，因此在环境科学领域受到了很高程度的关注。

由于二价铁矿物主要存在于无氧的还原环境中，因此过去对无氧条件下二价铁矿物引起的还原效应研究的较多。前期大量研究结果表明，多种不同类型的二价铁矿物均可引起环境污染物的还原转化，二价铁矿物引起的还原效应是地下环境污染物自然衰减的一个重要机理。譬如，在无氧条件下，马基诺矿（FeS）可使四氯化碳等氯代烃发生还原脱氯降解，驱使六价铬、六价铀、六价硒等重金属还原固定，实现砷和汞等络合固定[64]；含二价铁的层状硅酸盐矿物可还原硝基化合物和氯代烃，并还原固定六价铬、六价铀等[65-68]；其他二价铁矿物如黄铁矿（FeS_2）、绿锈（green rust）等和矿物表面吸附态二价铁，都表现出不同程度的氯代烃、硝基化合物和高价态重金属等还原活性[69-71]。虽然已有大量地下环境不同类型二价铁矿物还原有机污染物和还原/固定重金属的研究结果报道，但目前对不同类型二价铁矿物活性差异的规律及其内在机制的认识仍不十分清楚。

与无氧条件相比，人们对有氧条件下二价铁矿物环境效应的认识更加有限。虽然地下沉积环境主要为还原性，但是在自然过程（如干湿交替、地表水与地下水交互、潮汐作用等）和人类活动（如耕地、地下水开采/回灌、水利工程等）影响下，其中的二价铁矿物经常会受到空气扰动发生氧化。在二价铁矿物的作用下，空气中的氧气有可能被活化产生活性氧（如 $O_2^{\cdot-}$、H_2O_2 和·OH 等），从而促进有害物质的氧化。早期已有零星报道发现，黄铁矿具有在有氧和酸性条件下可产生·OH 氧化三氯乙烯的效应[72]，马基诺矿、磁铁矿和菱铁矿暴露在空气时可引起三价砷的氧化[73-75]。袁松虎研究组针对野外实际沉积物中的二价铁矿物接触空气时引发的氧化效应开展了研究。他们采集不同氧化还原条件下的 29 个沉积物样品，发现沉积物在与含氧水接触时均可产生·OH，·OH 产量与沉积物中的二价铁矿物含量呈正相关。其中，二价铁层状硅酸盐矿物是活化氧气产生·OH 的主要贡献组分。他们进一步在野外向 23 m 深承压含水层注入含氧水进行单井注水−抽水示踪试验，证实了含水介质中的二价铁矿物可活化含氧水中的分子氧产生·OH，氧化态沉积物经铁还原菌还原后仍可活化分子氧产生与原始还原态沉积物相当水平的·OH，而产生的·OH 可使三价砷和四环素发生氧化转化。此外，通过对水位波动剧烈期的估算，他们认为·OH 氧化有机质可产生与土壤呼吸作用相当的二氧化碳排放量[76]。

近年关于这方面的研究还有一些其他零星报道。譬如，Morin 和周东美研究组都发现纳米磁铁矿在暴露空气时可产生·OH 氧化难降解有机物[77,78]；景传勇研究组近年也发现，经过希瓦氏菌厌氧还原后的针铁矿暴露空气可产生·OH，氧化降解抗生素恩诺沙星[79]。董海良组发现，化学和生物还原后的含铁黏土矿也可通过暴露空气产生·OH 降解 1, 4-二噁烷[80]。这些发现都说明，地下还原环境沉积物中的二价铁矿物在自然过程或人为作用扰动下接触空气时，会产生·OH 等活性氧化物种，诱导（有害）物质的氧化转化。这种氧化效应将会在有氧（如地表水）和无氧（如地下水）界面之间形成一道天然的反应性屏障，影响着污染物的自然衰减过程（图 9-9）[76]。

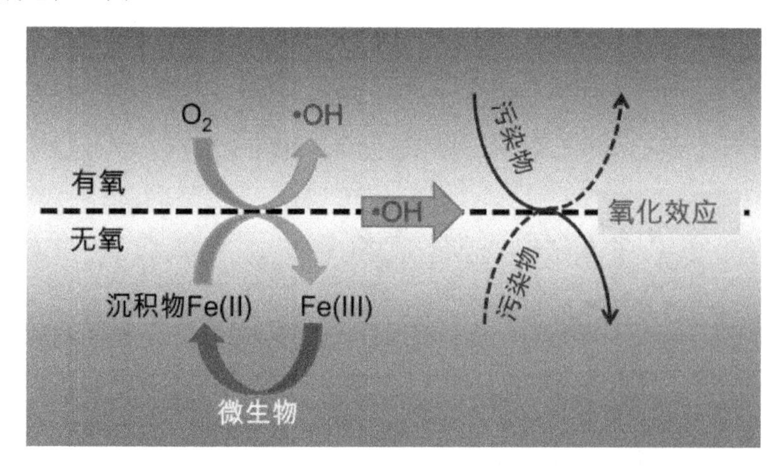

图 9-9　沉积物中二价铁矿物在无氧-有氧交替变化时产生·OH 引起的污染物氧化效应（引自参考文献[76]）

需要指出的是，不同类型的二价铁矿物活化分子氧的电子转移途径和矿物演化机理存在差异，产生的活性氧物种（如·OH）能否氧化污染物也与反应条件密切相关。袁松虎研究组研究了硫铁矿和还原态黏土矿物活化氧气产生·OH 的机理与氧化效应，发现硫铁矿在接触氧气时，均可通过两电子转移途径先产生 H_2O_2 进而转化为·OH，中性条件下 FeS 活化分子氧产生·OH 的能力远高于 FeS_2，两者氧化后矿物自身结构都被破坏[81,82]。与 FeS_2 中 Fe(II)和 S(−I)的同步氧化机理不同，FeS 氧化时 S（−II）充当电子储库的角色，可提供电子使氧化产物纤铁矿还原为二价铁，促进氧化过程中 Fe(II)/Fe(III)的循环和活性氧物种的产生[81]。在 FeS 接触氧气产生活性氧物种时，污染物能否氧化取决于 FeS 的含量。在高含量时，FeS 可淬灭·OH 起到保护污染物免受氧化的作用；而当 FeS 含量降低到一定程度后由于淬灭作用不足，产生的·OH 能促进污染物的氧化。因此，在使用 FeS 控制铀等重金属氧化释放风险时需要格外注意 FeS 的用量[81]。与硫铁矿不同的是，中性条件下二价铁黏土矿（以还原态绿脱石为例）接触氧气氧化时，黏土矿物表观结构不会被破坏，Fe(II)/Fe(III)仅在矿物结构内部循环，·OH 的产生可在铁氧化-还原的循环中持续进行，并引起有机污染物的氧化降解[80,83]。董海良研究组最新研究结果发现，还原态黏土矿物接触空气时产生的·OH 等活性氧化物种，可以杀灭大肠杆菌等细菌，有望作为新的杀菌药物[84]。但是，目前对于不同类型二价铁矿物活化分子氧的电子转移机理差异以及·OH 产生效率差异的基础理论探讨和深入认识上还存在明显不足，需要开展进一步的研究，以期为认识二价铁矿物的活性提供更具指导性的理论基础。

3 环境铁循环及其调控

铁循环在元素地球化学循环中居于位置重要、功能多样的枢纽地位，其角色与功能本质上直接由氧化还原过程控制。利用铁循环发展绿色友好环境治理和修复技术已成为水/土壤污染控制领域的研究热点，而研究铁循环及其环境效应更是揭开红壤非生物地球化学循环奥秘的关键，是当前国际研究前沿。环境铁循环研究面临的关键科学问题是如何在原子水平上阐明（氢）氧化铁表/界面上发生的吸/脱附原理、氧化还原过程以及电子转移途径，并深入揭示其环境效应。同时，铁元素的地球化学过程与碳/氮养分循环、重金属/有机氯等污染物转化过程密切相关（图 9-10），并受到强烈的人为活动与气候条件影响和干扰。目前，阐明地表土壤和水体中铁循环过程与环境生态调控之间的关系，通过铁循环调控实现环境污染治理和修复，具有重要的科学价值，并将形成以铁循环为核心的污染控制化学研究新方向。

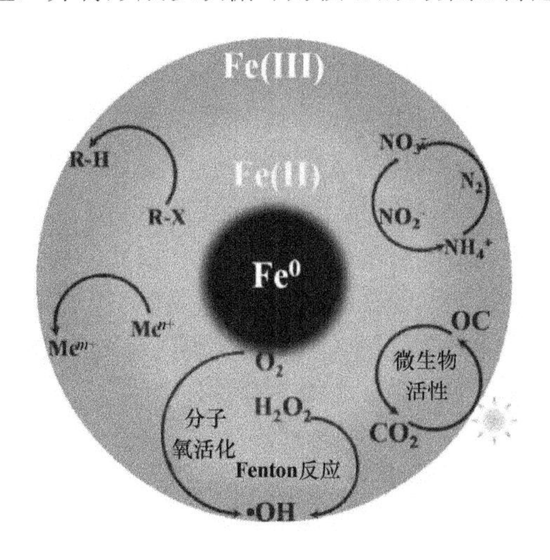

图 9-10 环境铁循环及其调控简要示意图

3.1 均相 Fenton 反应铁循环调控策略

Fenton 技术是一种利用铁循环催化双氧水分解产生具有强氧化性的羟基自由基（·OH）的高级氧化技术[85]。传统的 Fenton 试剂为 Fe(II)/H_2O_2 组合。Fe(II)和 H_2O_2 反应产生强氧化性·OH 和 Fe(III)（76 $L\cdot mol^{-1}\cdot s^{-1}$），而 Fe(III)与 H_2O_2 反应循环生成 Fe(II)的过程缓慢（0.02 $L\cdot mol^{-1}\cdot s^{-1}$），体系 Fe(II)/Fe(III)铁循环效率低，导致 H_2O_2 不能持续分解，Fe(III)沉淀生成大量铁泥。为了避免或减少铁泥的生成，Fenton 反应要求在强酸性条件下进行。因此，提高 Fenton 氧化体系中 H_2O_2 的分解效率的关键是实现 Fenton 反应过程的高效铁循环。研究者发现，加入能与铁离子络合的有机/无机配体、利用电-Fenton 技术等均能调控均相 Fenton 体系中铁循环，提高均相 Fenton 氧化降解污染物效率。

乙二胺四乙酸（EDTA）和乙二胺二琥珀酸（EDDS）等有机配体可以与铁离子络合，抑制 Fe(III)沉淀，一定程度上实现高效铁循环，但 EDTA 和 EDDS 等有机配体很难被生物降解，残留在水体中会造成二次污染[86,87]。张礼知研究组发现，兼备络合作用以及还原性的原儿茶酸可以抑制 Fe(III)/H_2O_2类 Fenton 反应体系中 Fe(III)沉淀，促进 Fe(II)/Fe(III)循环，使得 H_2O_2 不断分解，继而高效持续地产生·OH，降解甚至最终矿化有机污染物。更为重要的是，原儿茶酸可以随污染物一同降解并最终矿化，避免带来二次污染(图 9-11)[88]。无机配体在调控均相 Fenton 过程铁循环也有类似作用。该研究组还发现，四聚磷酸(TPP)可降低 Fe(II)/Air 体系中 Fe(III)/Fe(II)氧化还原电位，诱导 Fe(II)单电子还原途径生成超氧负离子($\cdot O_2^-$)以及过氧化氢(H_2O_2)[89]，并且通过 Fenton 反应产生·OH，显著增强五氯酚钠的降解率，并进一步实现其矿化。

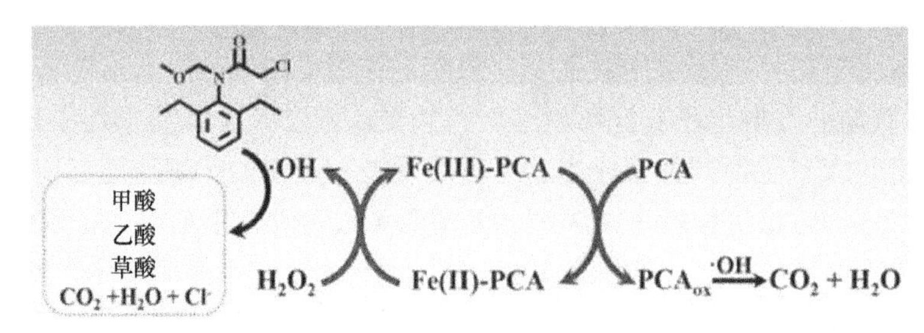

图 9-11 原儿茶酸调控 Fe(III)/H_2O_2 类 Fenton 反应体系中铁循环（引自参考文献[88]）

构建电-Fenton 体系也能达到调控均相 Fenton 体系中铁循环、提高均相 Fenton 降解效率的目的[90-92]。2015 年，张礼知研究组以铁丝为阳极，活性炭纤维为阴极，四聚磷酸钠（Na_6TPP）为支持电解质，构建了一个新型的电-Fenton 体系（Na_6TPP-E-Fenton）[93]。研究结果表明，该电-Fenton 体系能在较宽的 pH 范围（4.0~10.2）内有效降解阿特拉津。在 pH 为 8.0 的条件下，Na_6TPP-E-Fenton 体系降解阿特拉津的速率相对于传统以硫酸钠为电解质的电-Fenton 体系（Na_2SO_4-E-Fenton）提高了 130 倍。在 Na_6TPP-E-Fenton 体系中，除在阴极发生氧气的双电子还原反应生成 H_2O_2 外，电化学腐蚀及化学腐蚀释放出的 Fe^{2+}与 TPP 形成的配合物也能通过单电子活化分子氧途径生成 H_2O_2。两种途径生成的 H_2O_2 继而与 Fe(II)-TPP 反应生成更多的·OH。Fe(II)-TPP 可以通过阴极电化学还原再生，实现有效的铁循环。有趣的是，在反应后期 TPP 能通过磷化作用防止铁电极的过度腐蚀，避免了溶液中铁泥生成。

3.2 异相 Fenton 铁循环调控策略

3.2.1 基于零价铁 Fenton 反应及其增强策略

异相 Fenton 系统能在一定程度上解决均相 Fenton 体系 pH 适用范围窄、铁循环效率低的问题。近年来，基于零价铁异相 Fenton 反应被广泛用于环境污染物去除过程，但零价铁 Fenton 氧化过程中常伴随着铁溶出过程，溶出铁易形成铁氧化物包裹于零价铁表层，阻碍内层零价铁电子的传输，降低了零价铁表界面铁循环效率。研究表明，通过引入配体，一方面可以降低 Fe(II)的氧化还原电位，提升氧气的还原效率，产生更多 H_2O_2；另一方面，Fe(II)与配体形成的络合物可增大溶液中亚铁的溶解度，减缓铁离子的沉淀，调控零价铁表面的铁循环过程[94,95]。

尽管 EDTA、草酸等有机配体能提高异相 Fenton 体系中·OH 的产率[96,97]，但是依然存在成本高、配体自身污染等问题。无机配体有望克服有机配体的缺点，不仅实现零价铁 Fenton 过程中铁循环的调控，并且提高零价铁的电子利用效率[98-103]。张礼知研究组将TPP引入纳米零价铁（Fe@Fe_2O_3）的异相Fenton体系（Fe@Fe_2O_3/TPP/Air）中，可将Fe@Fe_2O_3降解阿特拉津的效率提升955倍[104]，成功实现零价铁界面铁循环调控过程。与传统有机配体EDTA促进Fe@Fe_2O_3异相Fenton体系相比，Fe@Fe_2O_3/TPP/Air体系降解阿特拉津效率亦能提升10倍。纳米零价铁释放的电子可用于溶解氧双电子还原至H_2O_2，Fe(III)/Fe(II)循环以及析氢反应。TPP通过质子限域作用，有效抑制Fe@Fe_2O_3的析氢副反应，促进H_2O_2生成和Fe(III)/Fe(II)循环过程，并通过形成磷化膜减缓Fe@Fe_2O_3表面的钝化，提高Fe@Fe_2O_3的电子利用率（图9-12）。

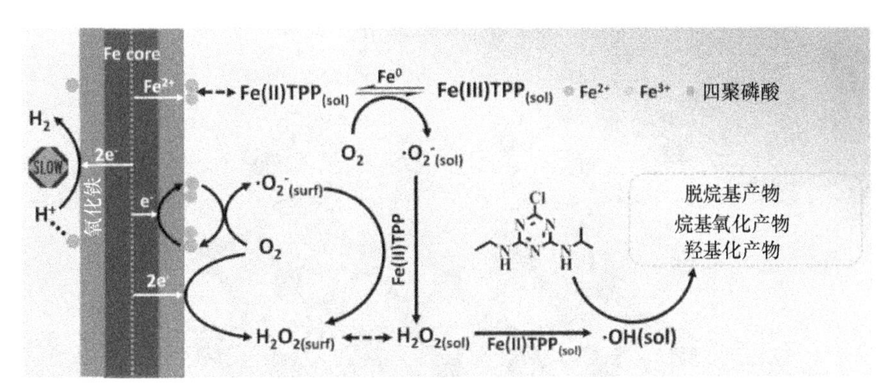

图9-12　Fe@Fe_2O_3/TPP/Air异相Fenton体系降解阿特拉津的机理（引自参考文献[104]）

自然界中存在着大量的无机阴离子盐（例如：SO_4^{2-}，HCO_3^-，Cl^-，HPO_4^{2-}），它们一旦吸附到nZVI表面，就会形成一层钝化膜，阻碍了有机污染物与nZVI表面的活性位点接触，从而降低了nZVI的还原活性[105]。其中，磷酸根抑制nZVI降解污染物的效果最为明显，这是因为磷酸根与Fe^{3+}离子能够形成溶解度很低的蓝铁矿[$Fe_3(PO_4)_2\cdot(H_2O)_8$，$K_{sp}=10^{-36}$]，因此，相较于其他的无机阴离子盐（$SO_4^{2-}$，$HCO_3^-$，$Cl^-$），磷酸根更容易吸附到nZVI表面。张礼知研究组最近研究发现，磷酸根能够促进nZVI产生活性氧物种[106]。在磷酸根加入以前，水体中的分子氧在nZVI表面主要是发生四电子还原反应。分子氧直接被还原成水［式（9-17)］。该过程会导致活性氧物种的生成效率低下，同时还使得nZVI过度腐蚀而失活。而磷酸根加入以后，一方面，它吸附到nZVI表面形成钝化膜，同时这也是一层保护膜，它能够有效地抑制分子氧在nZVI表面的四电子还原过程，防止nZVI过度腐蚀。另一方面，吸附在nZVI表面的磷酸根能够起到悬挂质子继电器（pendant proton relay）的作用，促进H^+与Fe^0发生反应形成·H（hydrogen radical），而·H易与水中的O_2反应形成氢氧自由基（$\cdot HO_2$）［式（9-18)］。$\cdot HO_2$通过自身歧化或者继续与·H发生结合，最后形成H_2O_2［式（9-19）和式（9-20)］[107,108]。该研究能够帮助我们预测在含磷废水中（例如，养殖废水，含磷工业废水，市政废水和农业灌溉用水），nZVI氧化降解有机污染物能力。

$$2Fe^0 + O_2 + 4H^+ \rightarrow 2Fe^{2+} + 2H_2O \tag{9-17}$$

$$\cdot H + O_2 \rightarrow \cdot HO_2 \qquad k_1 = 2\times10^{10}\ L\cdot mol^{-1}\cdot s^{-1} \tag{9-18}$$

$$2\cdot HO_2 \rightarrow H_2O_2 + O_2 \qquad k_2 = 8.3\times10^5\ L\cdot mol^{-1}\cdot s^{-1} \tag{9-19}$$

$$\cdot H + \cdot HO_2 \rightarrow H_2O_2 \qquad k_3 = 9.7\times10^7\ L\cdot mol^{-1}\cdot s^{-1} \tag{9-20}$$

3.2.2　基于氧化铁的异相Fenton反应及其增强策略

氧化铁具有存在广泛、环境友好和表面活性位点丰富等特点，常被用于污染治理和修复中。其中，赤铁矿是所有氧化物中热稳定性最优异的环境友好分布广泛的铁氧化物。因此，研究赤铁矿涉及的环境污染物迁移与转化具有极其重要意义。但是，直接用赤铁矿分解双氧水效果不佳。张礼知研究组合成了具有{001}晶面和/或者{110}晶面暴露的赤铁矿纳米晶，研究不同暴露晶面氧化铁表面限域亚铁离子催化

H_2O_2的分解效率[109]。他们发现限域在赤铁矿纳米晶表面的亚铁离子比没有限域的亚铁离子能更有效地分解 H_2O_2，而且{110}晶面限域的亚铁离子比{001}晶面限域的亚铁离子表现出更好的 H_2O_2 分解能力。这是由于H_2O_2的分解能力不仅和表面限域亚铁离子的密度有关，还依赖于亚铁离子在暴露晶面上的结合模式。亚铁离子在极性{110}晶面上是以五配位的模式结合，这种五配位的结合模式可以降低分解 H_2O_2 的能量跨度，而吸附在非极性{001}晶面上的亚铁离子是以六配位的模式存在，不利于 H_2O_2 分解能量跨度的降低（图 9-13）。

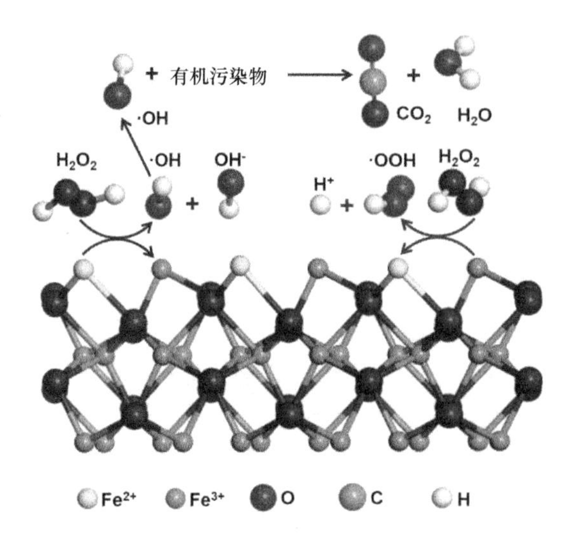

图 9-13　限域亚铁离子赤铁矿分解 H_2O_2 降解有机污染物机制（引自参考文献[109]）

考虑到抗坏血酸还原溶解铁基矿物过程会影响铁地球化学循环，张礼知研究组结合原位衰减全反射傅里叶红外光谱和密度泛函理论计算的结果，发现抗坏血酸根在赤铁矿{001}和{012}晶面分别形成了非质子化内球双齿单核和单齿单核配位模式。其中，赤铁矿{001}晶面形成的双齿单核铁-抗坏血酸配位模式更有利于赤铁矿的还原溶解[110]。当这两种不同的铁-抗坏血酸配合物被用于分解 H_2O_2 时，{001}晶面形成的双齿单核铁-抗坏血酸配位模式仍然更有利于 H_2O_2 的分解和除草剂甲草胺 Fenton 氧化降解。最近，该研究组还尝试将抗坏血酸应用于纳米零价铁/H_2O_2 异相 Fenton 体系中，发现抗坏血酸的加入不仅可以调控溶液中铁循环，同时也改善了纳米零价铁表面铁循环，其中纳米零价铁表面铁循环的改善是提高 H_2O_2 分解效率提高·OH 产量的关键[111]。

3.2.3　基于碳基材料的异相 Fenton 反应及其增强策略

近年来，碳材料被用于催化分解 H_2O_2 生成·OH 去除污染物。例如，Wang[112]等证明有几类活性炭能催化分解 H_2O_2 来降解苯酚。Zhou[113]等发现生物炭可以通过单电子转移活化 H_2O_2 生成·OH，但是 H_2O_2 的利用率很低。然而，部分碳材料并不能催化分解 H_2O_2 生成·OH。譬如，Gomes[114]等比较了商业活性炭、碳凝胶、多壁碳纳米管、氧化石墨烯氧化物和石墨等碳材料催化分解 H_2O_2 的能力，发现只有商业活性炭能活化 H_2O_2 生成·OH。Zhou[115]等指出活性炭中的自由基是催化分解 H_2O_2 生成·OH 的关键要素。生物炭不仅能催化分解 H_2O_2，还能作为电子穿梭体。Kappler[116]等发现生物炭可以促进铁还原菌和铁矿物质之间的电子传递过程。每克碳材料可以接收或给出几百微摩尔的电子[117]。因此，研究碳材料的电子传递作用具有重要的环境意义。

最近，张礼知研究组以葡萄糖为原料用水热法模拟自然界中的煤化过程合成了水热碳，研究了水热碳对 Fe(III)/H_2O_2 类 Fenton 体系降解甲草胺的影响[118]。他们的研究结果表明，水热碳能还原 Fe(III)，通过实现有效的铁循环，促进 Fe(III)/H_2O_2 类 Fenton 体系中甲草胺的降解。他们还利用不同的碳源，如葡萄糖、蔗糖、果糖和淀粉，合成了不同的水热碳，对比它们对 Fe(III)/H_2O_2 类 Fenton 体系降解甲草胺的影响，发现碳源并不影响水热碳还原 Fe(III)的能力。电子顺磁共振结果证明水热碳中含有大量的碳中心自由基，而衰减全反射-傅里叶红外光谱表征结果表明水热碳表面含有丰富的 C—OH 和 C═O 等含氧官能团。其中，表面 C—OH 是影响水热碳促进 Fe(III)/H_2O_2 类 Fenton 体系降解甲草胺性能的关键因素。而水热碳中的碳中心自由基通过表面的 C—OH 将电子传递给 Fe(III)，使 Fe(III)还原为 Fe(II)，从而促进 Fe(III)/H_2O_2 类 Fenton 体系中甲草胺的降解（图 9-14）。

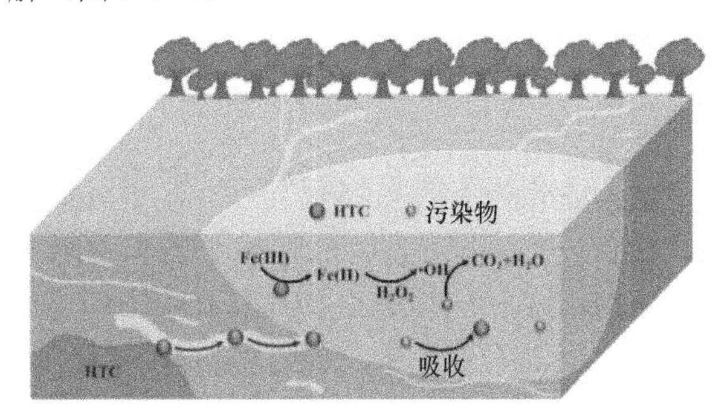

图 9-14　水热碳调控 Fe(III)/H_2O_2 类 Fenton 体系中铁循环（引自参考文献[118]）

3.3　铁循环及其碳氮转化效应

碳氮是土地利用过程中输入量最大的养分，也是体现生态服务功能最重要的生命元素。铁循环的本质是质子、电子等信号在分子间的传递，驱动碳氮转化过程[119]。高氧逸度时，电子优先传递给氧气，微生物与矿物间氧化还原反应受氧逸度控制。微氧或无氧时，微生物只能进行厌氧代谢，微生物只有通过与矿物的相互作用才能获得必需营养元素给予[120]。铁循环耦合碳氮转化过程为：①铁循环耦合碳转化过程，包括铁还原耦合有机碳氧化过程和亚铁氧化耦合无机碳同化固定过程，地表水体或表土等高氧逸度与光照强度条件下，光合微生物驱动铁循环可耦合 CO_2 固定；低氧逸度时，微氧型亚铁氧化耦合 CO_2 同化。②铁循环耦合氮转化过程，包括异化铁还原与硝酸盐还原的竞争、亚铁氧化耦合硝酸盐还原、铁还原耦合氨氧化过程。

3.3.1　异化铁还原耦合碳氧化过程

在人类的生产和生活活动中，大量的有机废弃物源源不断地排放到环境中，使土壤、沉积物和地下水等环境遭到严重破坏。对于厌氧沉积物环境中有机污染物的原位降解来说，为微生物提供产能所需要的电子受体是非常必要的[121]。由于土壤和沉积物中氧化铁的丰度较高，在被有机物污染的含水层中，铁还原菌已被证明发挥了重要的作用。譬如，铁还原菌可以有效地去除石油污染含水层中的芳香族化合物[122]。在此过程中，异化铁还原菌将 Fe(III)还原成 Fe(II)的反应同时偶联着有机物的降解，而生成的 Fe(II)通过生物和非生物反应又被氧化成 Fe(III)，继而再次被利用。

3.3.2　光营养型微生物驱动铁循环耦合 CO_2 同化

光驱动的化学与生物作用在铁循环与碳转化过程中起着重要作用。铁是 CO_2 光合同化的电子传递中心。光合微生物为适应高氧逸度的缺铁环境，分泌有机物，与三价铁络合，通过光还原或配体金属间电子转移生成亚铁，促进铁的吸收，同时促进 CO_2 同化[123]。光营养型微生物驱动的 CO_2 同化耦合铁循环过程主要发生在水体表层。光合型亚铁氧化在原始生命过程中意义重大，是起步最早、研究最为深入的中性亚铁氧化过程。微生物能够以 CO_2 为电子受体，将 Fe(II)氧化为 Fe(III)；研究者曾分离得到一种不产氧的光合非硫紫细菌。该细菌可以在低氧逸度环境中氧化 Fe(II)获得生命活动所需要的能量，固定 CO_2 合成细胞物质[124]。低氧逸度环境中，亚铁氧化很有可能就是由光合型亚铁氧化菌驱动的。光合型亚铁氧化菌参与了大约 18 亿~35 亿年前地球巨型条带铁层（banded iron formations）的形成[125]。光合型亚铁氧化过程的碳同化贡献巨大，据估算，海洋中光合自养型亚铁氧化菌每年可合成 480 亿吨有机碳[126]。因此，发生在地球表层的亚铁氧化耦合 CO_2 同化过程是有机碳生产的重要途径。陆地表层与海洋均大量分布光合型亚铁氧化菌[127,128]。据估算，非硫紫色细菌 *R. palustris* TIE-1 氧化每 mol Fe(II)可产出 5.36 g 生物量，相当于约 72%的生物量来源于 Fe(II)氧化过程[129]。

3.3.3 亚铁氧化耦合二氧化碳固定

细菌在驱动土壤 CO_2 固定中也起着举足轻重的作用[130,131]。通过 ^{13}C-CO_2 同位素标记证实，微生物作用下，每千克旱地和淹水土壤平均每天分别能同化约 0.18 mg CO_2 和 1.10 mg CO_2。在低氧逸度或根际微氧环境中，微氧型亚铁氧化菌与硝酸盐依赖型亚铁氧化菌均具有 CO_2 同化功能。在高湿度的森林土壤中，亚铁氧化菌可以抵抗亚铁化学氧化，发生酶催化亚铁氧化过程，这个过程以 O_2 为电子受体并还原为 H_2O，微生物储存能量，将 CO_2 转化为生物量[132]。这类化能无机自养型微生物大量分布于海洋与陆地的湿地生态系统[133]。微氧型亚铁氧化菌大多数属于 β-变形菌纲的披毛菌属 *Gallionella* 与纤毛菌属 *Leptothrix*，具有可观的 CO_2 固定能力[134]。硝酸盐依赖型亚铁氧化菌同样能够氧化 Fe(II)，并同时从中获得能量固定 CO_2 合成细胞[135]。李芳柏研究组采用传统培养法结合稳定同位素探针技术揭示了水稻土中耦合 Fe(II)和 CO_2 同化的功能微生物。*Azospirillum* 和 *Magnetospirillum* 在 ^{13}C 重浮力密度层中明显富集，其中，*Azospirillum* 是一类已知好氧硝酸盐依赖型 FeOB，可以利用硝酸盐、氯酸盐和高氯酸盐为电子受体进行厌氧亚铁氧化，因而认为硝酸盐依赖型 FeOB 可能在水稻土有氧-无氧界面进行微好氧亚铁氧化，其氧化亚铁的产物为无定形铁氧化物。以上结果直接证明了稻田生态系统中参与铁碳循环的关键化能无机自养微好氧 FeOB[136]。

3.3.4 异化铁还原与硝酸盐还原的竞争

在土壤环境中，多种末端电子受体（如 O_2、NO_3^-、Fe^{3+}、Mn^{4+}、SO_4^{2-} 和 CO_2 等）通常同时存在，这就导致了有机质氧化耦合电子受体之间的竞争。近年来，Fe(III)还原与 NO_3^-还原耦合过程的研究也屡见不鲜。大多数研究者都认为 Fe(III)还原与 NO_3^-还原是相互抑制或是单方抑制过程[137,138]。Ottow[139]认为 NO_3^-还原酶至少可以表达三种生理性质不同的末端还原酶，分别将电子传递给 Fe^{3+}、NO_3^-以及 NO_2^-，但是却优先传递给 NO_3^-以及 NO_2^-，因而也就使得 Fe^{3+}还原受到抑制。然而，并不是所有的 Fe(III)还原与 NO_3^-还原都存在相互抑制的作用。譬如，在以大肠杆菌 *Escherichia coli* 为介导的微生物体系中，Fe(III)-EDTA、NO_3^-/NO_2^-分别为电子受体，却发现了 Fe(III)-EDTA 的存在促进了 NO_3^-/NO_2^-的还原[140]。李芳柏研究组发现，铁氧化物加入可以促进发酵型芽孢杆菌还原 NO_3^-和 NO_2^-，并提出了铁氧化物作为半导体介导芽孢杆菌向 NO_3^-和电极之间的电子转移机制[141,142]。

3.3.5 亚铁氧化耦合硝酸盐还原

低氧逸度环境下，微生物驱动着硝酸盐依赖型亚铁氧化过程，还将 Fe(II)转变为无定型铁或弱晶质的水铁矿，同时反向为异化 Fe(III)还原提供电子受体，形成完整的铁循环[63]。另外，硝酸盐依赖型亚铁氧化促进了反硝化过程，无论是微生物驱动的 Fe(II)氧化耦合 NO_3^-还原过程，抑或是 NO_3^-还原产物 NO_2^-与 Fe(II)直接的化学反应，都是土壤或水体中重要的氮素损失途径[143]。目前，已有研究组从沉积物、地表水体等低氧逸度环境中分离到了多个硝酸盐依赖型亚铁氧化菌，包括 *Acidovorax* sp. BoFeN1，*Pseudogulbenkiania* sp. 2002 [144]。李芳柏研究组发现，在淹水稻田土壤中，NO_3^-还原是推动 Fe(II)氧化的驱动力；Fe(II)的存在不仅能够促进 NO_2^-的进一步还原，还能影响体系中的优势微生物结构，其中驱动

NO_3^-还原耦合 Fe(II)氧化过程的优势微生物包括：*Azospira*、*Zoogloea* 和 *Dechloromonas*[145]。

3.3.6　Fe(III)还原耦合厌氧氨氧化（铁氨氧化）

最近的研究证实异化铁还原与厌氧氨氧化过程关系密切，它们的耦合也被称为铁氨氧化。铁氨氧化被认为是微生物介导的过程，Fe(III)和 NH_4^+分别作为电子受体与供体，将 Fe(III)还原为 Fe(II)，NH_4^+则被转换为 NO_3^-、NO_2^-或者 N_2。在海洋及表层水生态系统，铁氨氧化贡献了 67%的 N_2 排放，是 N_2 排放的主要途径[146]。^{15}N 标记实验证实铁氨氧化过程占到稻田 N_2 排放量的 23%[147]。稻田、菜地等由于人为活动导致的过量施肥，氮素过度积累，从而导致土壤生态系统稳定性降低[148]，也为铁氨氧化过程研究提供了有利条件。据推算，每年每公顷稻田通过铁氨氧化过程可向大气中排放 7.8~61kgN[149]。铁氨氧化过程一方面通过铁呼吸产生能量，另一方面通过氨氧化过程去除环境中过量的氮素，更加有利于微生物的生存。

3.4　铁循环及其污染物转化效应

3.4.1　铁循环耦合有机氯污染物转化

长期以来，Fe(III)还原被误认为是化学反应，即使可能有微生物参与，也被认为是非专一性的，其作用仅在于降低氧化还原电位。直到 20 世纪 80 年代，人们才开始认识到，Fe(III)的还原是由特定微生物介导的酶促反应，具有 Fe(III)还原能力的细菌是铁氧化物还原的真正动力[119]。近年来，越来越多的地球化学证据表明，Fe(III)还原可能是地球最古老的呼吸途径。研究表明，异化铁还原反应不仅影响铁的生物地球化学循环[150]，而且参与多种土壤环境过程，包括：土壤潜育化过程，营养元素（如 P、K、Ca、Mg 等）与微量金属元素的迁移释放，金属的还原解毒，以及有机污染物的厌氧降解等[151]。

铁还原耦合有机氯脱氯转化的生物地球化学机制。近 40 年来，铁氧化物表面的脱氯过程研究逐渐引起人们的注意[152,153]。在红壤胶体-水界面五氯酚的非生物脱氯反应中[154]，草酸及 Fe(II)的加入能够明显增加界面吸附态 Fe(II)的浓度，从而提高反应速率。由此可见，在水稻土、红壤等富含铁元素的厌氧环境中，络合态与吸附态的 Fe(II)物种成为有机氯脱氯转化的活性物种，在厌氧环境下通过电子传递作用于有机氯，促进有机氯的还原脱氯速率。研究表明铁还原速率、五氯酚还原速率与土壤风化指数存在极显著的相关性；氧化铁游离度与活化度是最为重要的铁循环指标。李芳柏研究组利用高通量测序技术研究了珠三角水稻土铁还原与五氯酚脱氯过程中的微生物群落变化，发现梭菌、红螺菌、伯克氏菌、红环菌与除硫单胞菌占优势，同时也是影响五氯酚还原脱氯的重要因素[155]。此外，李芳柏研究组利用高通量测序技术研究了珠三角水稻土铁还原与五氯酚脱氯过程中的微生物群落变化，发现具有铁还原功能的梭菌和地杆菌，以及具有脱氯功能的伯克氏菌和除硫单胞菌是体系的主要微生物类群[155,156]。该研究组还发现添加乳酸、生物炭或者(类)腐殖质等外源物质可提高体系中铁还原菌的丰度，进一步提高活性铁物种的含量，降低体系中 Fe(III)/Fe(II)氧化还原电位，加速五氯酚还原脱氯；同时也能提高体系中脱氯菌和矿化菌的丰度，最终实现五氯酚的还原脱氯及矿化[157-159]。

3.4.2　铁循环耦合重金属的固定与释放

土壤中的 Fe(III)矿物还原成低价铁主要是由特定的异化铁还原微生物驱动的酶促反应。微生物的异化铁还原可耦合众多的环境地球化学过程，目前研究较多的是铁还原耦合重金属还原。FeRB 具有还原高价重金属的功能，其中关注较多的重金属包括 U(VI)、Cr(VI)、Cu(II)等。已报道的 FeRB *Shewanella* 属中的 *S. putrefaciens*、*S. alga*、*S. oneidensis*，*Geobacter* 属的 *G. sulfurreducens* 与 *G. metallireducens* 都具有金属还原能力[160,161]。以 Cr(VI)还原为例，Liu 等[162,163]在以铁还原菌还原 Cr(VI)为模型反应，以外膜细胞色素 C 为核心，构建了铁还原菌驱动的 Cr(VI)还原动力学模型，引入毒性反应作为基元反应，并将外膜细胞色素 C 与三价铬的反应式引入反应动力学模型。采用基于 RNA 的 *q*PCR 方法，定量测试反应过程中的活菌数量与外膜细胞色素 C 的表达量，并将这些参数引入动力学模型，大大提高了模型的可靠性，为深入理解铁还原菌还原金属离子的分子机制提供模型依据。

然而，铁还原过程耦合不变价重金属的环境行为目前还研究较少。Cd(II)在土壤中不会发生氧化还原反应，其固定与释放主要与氧化铁等矿物的还原溶解与成矿有关。最近，Muehe 等分离出 *Geobacter* sp. strain Cd1。该菌株可还原溶解含 Cd 氧化铁矿物，从而促进 Cd 的释放；同时，该菌还原氧化铁后，释放的 Cd 可以被次生氧化铁矿物固定[164]。该研究表明，铁还原不仅造成含 Cd 氧化铁矿物中 Cd 的释放，而且也会通过氧化铁矿物的重结晶促进 Cd 的再次固定。前人的研究也发现，厌氧还原产生的游离态 Fe^{2+} 吸附到氧化铁表面可催化氧化铁发生晶相转变，提高氧化铁吸附重金属的能力，并进一步使重金属固定到次生氧化铁矿物结构中[165]。Cooper 等研究发现，在此过程中，土壤中游离态二价重金属(Cd、Co、Mn、Ni、Pb 及 Zn)在 Fe(II)催化氧化铁重结晶过程中均可被固定在次生矿物结构中，从而实现重金属结构化固定脱毒[166]。结晶度较低的氧还原铁，比如水铁矿和纤铁矿，其晶相重组及转化速度较快，因此，能固化较高含量的重金属；结晶度较高的氧化铁，如赤铁矿和针铁矿，虽然在 Fe(II)催化作用下可发生晶相重组，但是晶相转变速率相对较低，并且重组后仍然为原矿物相，因此，固化到重组后矿物相中的重金属效率较低[167,168]。Liu 等以水铁矿为 Fe(II)催化目标重金属，研究了 Mg(II)，Ca(II)，Ba(II)，Mn(II)，Co(II)，Ni(II)和 Zn(II)七种二价重金属离子在水铁矿晶相转变过程中的固定，结果表明，重金属离子与水铁矿的亲附性能是影响其在水铁矿重组过程中固定的关键因素，亲附性能越高的重金属，固定于水铁矿晶相重组氧化铁中的效率越高[169]。

李芳柏研究组系统研究了土壤铁循环与水稻中镉砷富集的关系。研究发现物理水溶态+吸附态镉、碳酸盐结合态镉是稻米镉的源，其与稻米镉的相关性优于土壤总镉含量；稻米镉/砷含量与土壤无定型态铁呈显著负相关、与结晶态铁呈显著负相关、与 0.5 mol/L 盐酸提取态铁呈显著正相关。说明氧化铁成矿过程为镉砷的固定过程，氧化铁的溶解过程为镉砷的释放过程[170,171]。为进一步深入探讨铁砷关系，李芳柏课题组还开展了水稻盆栽试验，研究了铁盐施用降低土壤砷有效性的机理：①铁盐添加可有效降低土壤中水溶态砷并提高固定态砷，从而降低了砷在土壤中的移动性及有效性；②铁盐添加可促进水稻根际铁膜的形成，从而有效阻隔砷向水稻地上部的运输[172]。应用高通量测序手段及实时荧光定量 PCR 方法，系统研究了水稻表土、根际土及根膜微生物群落结构的差异，结合水稻的重金属砷吸收数据，发现铁膜上的砷氧化菌数量与水稻根部砷含量成负相关，表明铁膜上的砷相关微生物群落决定了水稻根部表面的砷物种及其移动性，同时发现铁膜砷氧化菌的类群主要是食酸菌 *Acidovorax*，这类菌恰恰是中性硝酸盐还原耦合亚铁氧化的关键微生物类群。以上结果预示着铁膜环境铁-氮-砷元素循环的紧密性，降低水稻砷吸收可以从根部铁膜环境考虑，发掘 *Acidovorax* 属的关键微生物类群，研究清楚其生物化学机制[173]。

4 铁矿物生物地球化学过程及其强化

铁矿物具有氧化还原活性，在化学或微生物作用下，Fe(II)与 Fe(III)之间可以转换，驱动着许多重要物质（如碳、硫、氮、重金属等）的地球化学转化与循环。因此，铁矿物的生物地球化学过程原理也是许多环境相关技术发展的理论基础。

4.1　含铁硫化矿生物氧化与铁硫形态转化

自然界的含铁矿物主要包括含铁原生硫化矿，如黄铁矿（FeS_2）、黄铜矿（$CuFeS_2$）、砷黄铁矿（AsFeS）和含铁次生氧化矿，如磁铁矿（Fe_3O_4）、赤铁矿（Fe_2O_3）、针铁矿（$Fe_2O_3·H_2O$）、水针铁矿（$2Fe_2O_3·H_2O$）、菱铁矿（$FeCO_3$）等。在地球陆地表面，原生硫化矿通过长期的风化氧化等表生作用形成次生氧化矿和次生硫化矿，最终形成自上而下的氧化淋滤带、次生氧化带、次生硫化富集带和原生硫化矿带等。这种表生风化和氧化作用缓慢，但在微生物的铁硫氧化作用下，这些转化过程便会大大加速。该现象通常在表露的尾矿堆和遗弃的废矿区附近，并被雨水的侵蚀和嗜酸铁硫氧化微生物的作用形成酸性矿坑水得到印证。

4.1.1　微生物介导的铁的生物地球化学特征

铁在自然界中价态分布及赋存状态因环境(如氧含量、pH 等)的不同而不同。在中性或弱碱性条件下，Fe(III)主要以铁的氧化物、氢氧化物等非溶解状态存在，可作为电子受体被还原。亚铁作为可流动的铁，在好氧或微氧环境中可被生物或非生物作用氧化成 Fe(III)，Fe(III)可能很快又会以氢氧化铁、氧化铁（Fe_2O_3、$Fe_2O_3·H_2O$）及羟基氧化铁[FeO(OH)]等形式沉降下来[174]。在地球早期，由于缺乏光合作用产氧的生物，大气中的痕量氧气主要来自水蒸气的光化学解离，氧含量较低，所以早期火山热液中 Fe^{2+}可以稳定存在，并可发生长距离迁移。随着迁移过程中环境发生改变，如 pH、氧化还原电位、CO_2 分压及氧分压等改变，Fe(II)将会以硅酸盐、碳酸盐或氧化物等形式沉淀。在地球前寒武纪还没有可利用氧气的富 Fe(Ⅱ)环境中，微生物对亚铁的氧化可能是较早的呼吸机制，该过程产生了大量的含铁矿物，如磁铁矿等[175,176]。厌氧环境中，在光驱动或者在硝酸盐等存在的无光情况下，微生物介导的亚铁氧化也可以发生，通常情况下，人们普遍认为除中性 pH 情况外，微生物介导下的 Fe(Ⅱ)氧化一般都要以 O_2 作为最终的电子受体，但仍发现很多例外。如食酸菌（*Acidovorax ebreus*）[177]、嗜热古菌（*Ferroglobus placidus*）[178]、脱氮硫杆菌（*Thiobacillus denitrificans*）[179]等可通过依赖硝酸盐的还原来耦合推动 Fe(Ⅱ)氧化反应。此外，还有很多微生物如 *Rhodopseudomonas*、*Rhodobacter*、*Chilorobium* 及 *Tiodictyon* 等菌可在有光照的厌氧环境中将 Fe(Ⅱ)氧化为 Fe(III)的氧化物，并逐步矿化。在 $pH<4$ 的好氧环境中，亚铁的氧化则主要为生物氧化过程。

4.1.2　含铁硫化矿物的生物浸出原理

自然界中，微生物铁硫氧化还原作用在铁矿物生物地球化学过程中发挥着重要作用。在有氧的酸性环境中，微生物对包括黄铁矿、黄铜矿及砷黄铁矿等硫化矿的生物氧化，对湿地及硫化物沉积物中铁矿物的循环及酸性矿坑水的形成有着重要的贡献。同时，依托该类微生物发展起来的生物浸出技术具有流程短、能耗小、成本低、污染少等优点，在解决当今世界所面临的富矿及易开采矿的不断减少、金属硫化矿资源日渐枯竭等问题方面具有重要的研究和应用价值。具有这种能力的微生物主要是一些好氧的嗜酸性细菌或古细菌，其中大部分细菌往往同时具备铁、硫氧化能力，如 *Acidithiobacillus ferrooxidans*、*Sulfobacillus thermosulfidooxidans*、*Acidianus manzaensis* 等，这些微生物在氧化硫化矿物的过程中并不是以 Fe(Ⅱ)作为唯一的能量来源，还可以还原态的硫化物如 H_2S、S^0、$S_2O_3^{2-}$及一些金属硫化物作为能源物质[180]。Silverman 和 Ehrlich[181]于 1964 年就浸矿微生物与金属硫化矿物间的作用方式提出了两种著名的机制：直接作用和间接作用。随着研究的不断深入，研究者们发现早期的浸出理论难以解释微生物与矿物作用过程中复杂的吸附、化学、生物化学及电化学行为。目前，认可较高的是 Crundwell[182]对 Tributsch 等[183]的浸出理论进行修正后提出的生物浸出模型：间接浸出、间接接触浸出和直接接触浸出。在浸矿微生物和硫化矿相互作用过程中，浸矿微生物的作用主要有：①通过铁氧化微生物把 Fe^{2+}氧化为 Fe^{3+}，后者为硫化矿物的持续溶出提供化学氧化力；②通过硫氧化微生物把单质硫和还原性的硫化合物最终氧化为硫酸，再生 H^+，维持浸矿微生物生长所需要的酸性环境，并为酸溶性矿物的溶解提供质子[184]。目前较为广泛接受的含铁硫化矿物生物溶出机理包括以黄铁矿为代表的硫代硫酸盐途径和以黄铜矿为代表的多聚硫化物途径[185]。

$$4Fe^{2+} + 4H^+ + O_2 \longrightarrow 4Fe^{3+} + 2H_2O \quad (9\text{-}21)$$

$$2S^0 + 3O_2 + 2H_2O \longrightarrow 2SO_4^{2-} + 4H^+ \quad (9\text{-}22)$$

4.1.3 含铁矿物生物浸出过程元素形态转化及其效应

微生物对矿物的生物浸出过程涉及菌-矿物-溶液等多界面作用，其中菌-矿物界面作用至关重要[186]。在含铁硫化矿生物/化学浸出过程中，Fe^{3+}的氧化作用使得矿物表面产生一系列次生的硫化矿物和元素硫，改变了矿物表面结构及组成，同时直接影响菌对矿物表面的作用。研究表明黄铜矿表面在生物以及化学浸出过程中会产生 S-S 二聚体（S_2^{2-}）以及多聚硫化物（S_n^{2-}）以及元素硫[187,188]，一定条件下黄铜矿表面还会产生一些次生矿物比如铜蓝、辉铜矿和斑铜矿[189]。同时，研究者在生物浸出以及化学浸出后的黄铁矿表面发现了硫代硫酸盐[190]或多聚硫化物[191]。酸性环境中元素硫呈化学惰性，必须借助微生物作用才能消解[192]。实际生物浸出体系中，矿物表面常因黄钾铁矾和元素硫的累积形成覆盖层，可能阻抑菌/高铁离子对矿物的进一步氧化，极大地影响浸出过程，而该覆盖层的形成还受到环境条件如 pH、温度、盐度或电位等的影响。由于这些中间体化学稳定性差且形态复杂，不适合用常规方法进行测定或表征。夏金兰研究组采用基于同步辐射的 XANES、XRD、STXM 和 m-XRF，辅以电化学、FT-IR 或 Raman 等方法，系统研究了浸出过程中间体化学形态与物相转化及其浸出效应[189,190,193-196]以及典型环境条件（NaCl、活性炭、pH、电位等）对浸出中间体形态与物相及浸出效应的影响机制[180,186,197-200]。

4.1.4 铁矿物生物浸出微生物的适应性及作用机制

生物浸出过程中，在能源种类和性质以及不断变化的环境条件如 pH、重金属离子的影响及作用下，微生物种群结构和表观铁硫氧化活性会发生动态变化，其表面胞外有机物组成及无机物质的形态与分布也随着浸出的进行发生变化，并会影响浸出过程。夏金兰研究组利用 SEM-EDS、EBSD 分析矿物/细胞表面壳层元素组成、微区结构，以及利用 SEM-EDS、AFM 及 CLSM 分析细胞表面壳层元素组成、形貌及有机物分布，还联合运用基于同步辐射的 STXM、XANES、XRD、μ-XRF 分析菌表面的物质化学形态、物相及化学成像，发展了微生物表面 C、S、Fe 等的原位显微表征分析方法，不仅能区分浸矿菌胞内外硫球赋存形态[201-203]，还揭示了细胞表面巯基表达量与菌对含硫能源底物作用之间的相关性。他们结合不同浸矿菌对不同形态元素硫形态转化特征的区别，提出并阐明了典型浸矿菌元素硫转化和活化机理[184,189,204,205]，发现铁氧化浸矿菌表面铁以高铁和亚铁形态存在，并与羧基、胺基和羟基等有机基团键合的规律，为硫化矿生物浸出间接作用机理的解析提供了直接实验证据。同时，他们还发现羟基氧化铁和黄钾铁钒等无机物存在于表面壳层，而这些无机“壳层”很可能对细胞适应极端环境起到非常重要的作用[206]。

4.2 微生物与含铁矿物交换电子的分子机理

4.2.1 微生物胞外电子传导

含铁矿物和微生物之间的相互作用可通过电子交换的方式进行。而在电子交换过程中，含铁矿物主要发挥四个方面的作用：①作为异养微生物的电子受体；②作为自养微生物的电子供体；③作为微生物之间电子传导的导体介质；④作为微生物之间的电子储存介质[207,208]。通过这些作用微生物介导着环境中 Fe(II)和 Fe(III)之间的氧化还原循环[209]。微生物介导的铁的氧化还原循环不仅影响铁，还影响碳、氮和砷等其他元素的地球化学循环［图 9-15（a)］。

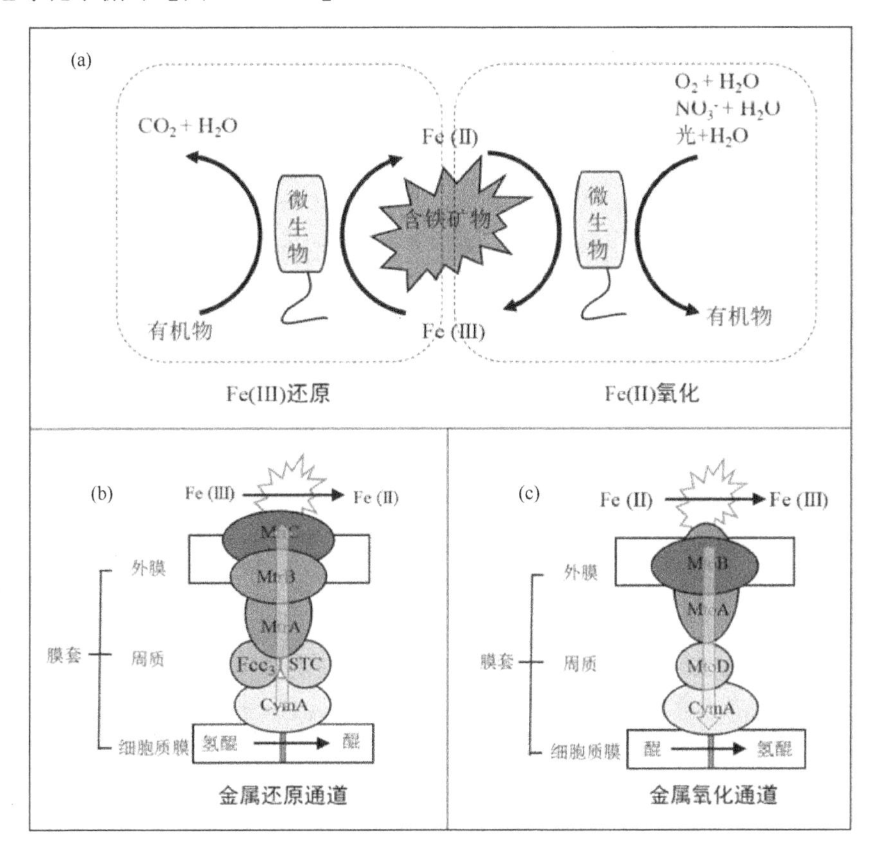

图 9-15　微生物介导的铁的氧化还原循环（a)；微生物金属还原胞外电子传导通道（b)；微生物金属氧化胞外电子传导通道（c）

电子在微生物的细胞质膜和含铁矿物之间相互转移的过程称为微生物胞外电子传导。其中，细胞质膜的作用包括两个方面：①作为屏障帮助微生物抵御外界各种因素的干扰；②细胞电子传递链作为微生物有氧呼吸的必要组分存在于细胞质膜中，其作用是形成跨越质膜、用于合成 ATP 的质子梯度。由于微生物细胞质膜外表面的某些组分不具有导电性，如肽聚糖、外膜和表层蛋白质等[210,211]，微生物通过逐渐进化，形成了一种新的电子转移机制，即胞外电子传导，实现其与含铁矿物以及其他矿物之间的电子交换[212,213]。

4.2.2　微生物胞外电子传导通道

微生物通过胞外电子传导通道在胞内细胞质膜和胞外含铁矿物之间传导电子，将胞内代谢过程和胞外含铁矿物氧化还原过程联系起来。该通道由氧化还原蛋白和结构蛋白组成，此类蛋白质具有进化的多样性，即并非所有蛋白质的功能都能通过基因手段识别。因此，生物化学和电化学等方法是鉴别和表征这些蛋白质功能和分子机理的关键。下面将主要介绍四种微生物胞外电子传导通道。

Shewanella oneidensis MR-1 是最早被发现能利用含 Fe(III)以及含锰[Mn(III，IV)]矿物作为末端电子受体的微生物之一[214]。目前对其金属还原（metal-reducing，Mtr）通道的电子转移机理已研究较为透彻。研究表明 6 个 c-型细胞色素，即 CymA、Fcc3（也称为 FccA）、MtrA、MtrC、OmcA 和 STC（即 small tetraheme cytochrome），以及外膜蛋白 MtrB，都参与了 *S. oneidensis* MR-1 胞外还原含 Fe(III)矿物的过程[215-220]。CymA

的功能是氧化细胞质膜上的氢醌，并将该过程释放的电子转运给周质空间中的 Fcc_3 和 STC，进而传递给位于细胞外膜的 MtrA[221-227]。MtrA、MtrB 和 MtrC 能形成跨外膜的蛋白质复合体，将电子从周质蛋白传导到细胞表面[228-232]。在细胞表面，MtrC 和 OmcA 相互接触，并通过其暴露在蛋白质表面的血红素将电子直接传导给含 Fe(III)矿物[233-243]。 因此，这些蛋白质通过形成 Mtr 通道将 *S. oneidensis* MR-1 细胞质膜内的电子传导到含 Fe(III)矿物表面，实现了胞外还原含 Fe(III)矿物［图 9-15（b）］。

关于 *Geobacter sulfurreducens* 介导的孔蛋白-细胞色素蛋白（porin-cytochrome，Pcc）通道，目前研究表明有 8 个 c-型细胞色素，包括细胞质膜上的 ImcH 和 CbcL[244,245]，周质中的 PpcA 和 PpcD[246,247]，外膜上的 OmaB、OmaC、OmcB 和 OmcC，以及外膜蛋白 OmbB 和 OmbC，均在 *G. sulfurreducens* 介导的胞外还原含 Fe(III)矿物过程中起到了重要作用。外膜上的 OmcB 和 OmcC 与 OmbB、OmbC、OmaB 和 OmaC 结合形成跨外膜的孔蛋白-细胞色素复合体[248-251]。这些蛋白质可能将氧化细胞质膜上氢醌所得的电子传导到细胞表面，而后进一步将电子传导到含 Fe(III)矿物表面，实现 *G. sulfurreducens* 胞外还原含 Fe(III)矿物。*G. sulfurreducens* DL-1 和 *G. sulfurreducens* PCA 具有多重平行的电子传导通道，除能还原含 Fe(III)矿物外，孔蛋白-细胞色素通道还直接参与 *G. sulfurreducens* PCA 同其他种类细菌之间以及电极之间的电子交换过程[252]。

光合铁氧化菌 *Rhodopseudomonas palustris* TIE-1 能以光为能量来源，以 Fe（II）为电子来源固定 CO_2[253]。根据对 *R. palustris* TIE-1 的光合铁氧化（phototrophic iron oxidation，Pio）通道的研究推断，其基因组中有一个 *Pio* 基因簇，包括 *pioA*（*mtrA* 的同源基因）、*pioB*（*mtrB* 的同源基因）和 *pioC*（一个编码高电势铁-硫蛋白的基因）。在光照条件下，PioA 和 PioB 可以氧化胞外 Fe（II），并将释放的电子跨外膜转运给可能位于周质的 PioC，PioC 可能将电子转运给位于细胞质膜中的光反应中心[254]。

与 *Shewanella oneidensis* MR-1 的金属还原通道不同，*Sideroxydans lithotrophicus* ES-1 具有金属氧化（metal-oxidizing，Mto）通道。在 pH 约为中性时，*S. lithotrophicus* ES-1 从氧化 Fe（II）过程中获得能量，实现自养生长。研究表明 *S. lithotrophicus* ES-1 的 *mto* 基因簇包含 *cymA*、*mtoA*（*mtrA* 的同源基因）、*mtoB*（*mtrB* 的同源基因）和 *motD*（编码一种单血红素 *c-Cyt* 的基因）[255,256]。MtoA 可以直接氧化 Fe(II)以及含 Fe(II)矿物，MtoD 是一种周质 c-型细胞色素，有可能将电子从外膜的 MtoA 传导到细胞质膜上的 CymA[255,257,258]［图 10-15（c）］。

Mtr 和 CymA 同源物同时参与 Fe(II)氧化和 Fe(III)还原过程，表明这些同源物是趋异进化的结果，而且 Mtr、Pio 及 Mto 电子传导通道具有双向性的特征[240,256]。例如，*S. oneidensis* MR-1 的 Mtr 通道可以将电子从细胞质膜上的 CymA 跨越细胞膜套传导到矿物表面，也可以反方向地从电极传导到 CymA[259]。除以上提到的胞外电子传导通道外，一些学者还提出其他通道[256,260-264]，但这些通道仍处在进一步的研究阶段。

4.2.3 应用

具有胞外电子传导能力的微生物被广泛应用于生物修复环境污染物、生产新型纳米材料、生物采矿以及生产生物能源等各个方面。在生物修复污染物方面，Fe(III)还原菌不仅可以还原 Fe(III)，还可以将水溶性的 Cr(VI)，Se(IV)/Se(VI)，Tc(VII)和 U(VI)等金属离子分别还原成水溶性很低的 Cr(III)、Se(0)、Tc(IV)和 U(IV)从而起到固定作用，达到修复这些污染物的目的[265]。铁还原菌除了可以修复金属离子污染物外，还能直接或间接地降解有机污染。例如，*G. metallireducens* GS-15 在还原 Fe(III)及含 Fe(III)矿物时，还能氧化降解苯甲酸盐、甲苯、苯酚和对苯酚等芳香烃污染物[266]。

在新型纳米材料合成方面，Fe(III)还原微生物在胞外还原含 Fe(III)矿物的过程中通常在细胞表面形成磁铁矿纳米颗粒。同样地，某些微生物还原 Co(II)、Pd(II)、Se(IV)/Se(VI)后，也在其细胞表面形成相应的纳米颗粒。这些纳米颗粒有望作为化学反应的催化剂，或应用与环境污染修复及癌症治疗等[267-269]。

在生物采矿方面，某些微生物可被应用于提取金属。如 Fe(II)氧化和硫氧化菌 *Acidithiobacillus*

ferrooxidans 是铜矿生物冶炼微生物群落的重要成员。除了铜矿外，*A. ferrooxidans* 可用于提取铁，硫和黄金[270-272]。在生物能源方面，微生物胞外电子传导对促进生物产甲烷过程[273-278]、微生物燃料电池发电[279-281]和微生物电合成[282,283]的发展和实际应用具有重要意义。

在过去的十年中，人们对微生物与含铁矿物之间电子传导机理的认识得到了显著的提高。这主要体现在对从细胞质膜、跨越膜套传导到胞外含 Fe(III)矿物的电子传导机理的认识较为清楚，同时鉴定出几种重要的蛋白质，一部分蛋白质还获得了详细的特征解析。但对反向电子传导即电子从胞外电极传导到胞内受体的机理认识仍不清楚，并且对电子如何从胞外含 Fe(II)矿物传导到胞内的末端电子受体，以及该过程中能量如何保存均未可知。这些严重制约了微生物电合成技术成为主要的生物制造方式。

同时，关于微生物跨外膜传导电子的分子机理仍然不清楚。从分子水平上认识这些电子传导过程的机理，能为在生物技术领域应用这些具备胞外电子传导能力的微生物提供理论指导，特别是为利用微生物电合成技术将温室气体 CO_2 转化成各种化合物以及开发新型的生物燃料提供理论和实际指导。

4.3　铁强化厌氧污水处理技术及原理

零价铁作为一种廉价且环境友好的还原剂，已被广泛应用于浅层地下水修复及难降解污染物的预处理[284-286]。然而，零价铁在运行中易锈蚀板结，影响其在实际废水处理中的应用。如将零价铁应用于厌氧污水处理系统，可以减缓其锈蚀，并强化厌氧微生物的代谢。

4.3.1　零价铁强化厌氧污水处理技术原理

1987 年 Daniels 等[287]报道了单质铁可作为产甲烷菌的唯一电子供体，还原二氧化碳产甲烷。2005 年 Field 等[288]发现单质铁可以在厌氧污泥中作为电子供体提高硫酸盐还原和甲烷产生。事实上，零价铁可从水中置换出 H^+，有利于缓冲厌氧体系的酸性积累。但是，铁置换产氢在热力学上却不利于厌氧呼吸。按传统观点，产甲烷菌只能利用乙酸和一元碳（如甲醇、甲酸、甲胺、氢气/二氧化碳）产甲烷[289]，其他复杂有机物需借助 H^+对有机物的厌氧氧化，生成以上简单物质。然而，H^+作为厌氧氧化的电子受体，其氧化还原电位较低（$E_{H^+/H_2} = -414$ mV），标况下无法氧化 NADH 等媒介[290]。只有在低的氢气分压下，H^+的厌氧氧化才能进行[291]。

耗氢产甲烷菌是厌氧系统消耗氢气的主要微生物，但其丰度通常较低，很难维持厌氧系统内较低的氢气分压，造成 H^+积累乃至厌氧失败[289]。张耀斌研究组发现[291]，在厌氧系统中置入零价铁，能有效提高耗氢产甲烷菌的丰度，且产气中氢气含量没有因零价铁与 H^+反应而增加，反而降低一半左右；在一定范围内，铁量越大，氢分压越低，有机物分解越快。他们认为，零价铁可以促进发酵菌和甲烷菌的种间氢传递原理（IHT）。其机制如下：①零价铁富集的耗氢微生物，如耗氢产甲烷菌[291-293]和同型产乙酸菌[292,294]等，有利于降低氢气分压，突破产氢产乙酸的热力学屏障，推动厌氧氧化反应向右进行，加速有机物的厌氧氧化，提高有机物的矿化率；②零价铁增加胞外聚合物分泌，促使颗粒化污泥形成[291,293]。根据 DLVO 理论，零价铁表面释放的 Fe^{2+}能中和污泥细胞表面的负电荷，减小静电斥力[295]。同时，Fe^{2+}通过压缩双电层来促进污泥细胞的聚集。污泥细胞聚集并形成颗粒污泥，将产酸菌与产甲烷菌的种间距离变小，提高氢气在两类微生物间的扩散效率，并进一步推动这两类微生物介入的 IHT，缩短厌氧启动周期，突破污水厌氧系统启动周期长、颗粒化困难的难题；③零价铁提高铁氧化还原酶、乙酸激酶等厌氧水解酸化段关键产酸酶活性数十倍，加快大分子有机物的酸化[292,296]。这在污泥厌氧消化时效果尤为明显。普遍认为，污泥的水解破壁过程是污泥厌氧消化的限速步骤[297]。零价铁能够强化污泥水解破壁阶段关键酶的活性，加速污泥破壁分解释放更多的产甲烷底物，有效提高污泥产甲烷量和污泥减量[296]。④零价铁降低厌氧体系内氧化还原电位，为厌氧微生物的生长和代谢创造较适宜的还原氛围[293]。研究表明[298]，将零价铁内置于厌氧水解酸化反应器，优化水解酸化发酵类型，使厌氧发酵更倾向于乙醇型发酵和丁酸型发酵，从而减少丙酸型发酵，以避免丙酸的积累，并为后续产甲烷阶段提供更适宜的底物。

4.3.2 Fe(III)氧化物强化厌氧污水处理技术原理

零价铁在生产、使用中，其表面易氧化成 Fe(III)氧化物(铁锈)。通常，应当避免其表面锈蚀，或者在应用前应将其表面氧化物除去，这无疑将增加应用成本。张耀斌等[299]发现锈蚀零价铁对厌氧处理的促进效果更优，这也被其他研究者证实[300]。这是因为：Fe(III)氧化物能够有效地富集 Fe(III)还原菌，铁还原菌氧化有机物，将产生的电子传递给胞外不溶性 Fe(III)氧化物，将其还原成 Fe^{2+} [301]。在含有铁氧化物的污泥厌氧发酵系统，Fe(III)还原菌丰度提高数倍，污泥减量化率和产甲烷量也大幅增加[299]。在含有铁氧化物的厌氧系统内富集出多种新型铁还原菌，均具备优良的氧化底物能力和产电功能[302,303]。它们与产甲烷菌之间的直接种间电子传递(DIET)——依靠胞外导电菌丝(pili)和细胞色素实现的物种间电子和能量交换的新机制，是系统高效运行的内在动力[304,305]。

这一发现具有以下意义：①Fe(III)氧化物刺激异化铁还原，强化复杂有机物的分解。Zhao 等[306]的研究表明，磁铁矿促使氨基酸和糖类物质的运输和代谢功能基因高度表达，强化复杂有机物的水解酸化。作为工业废水预处理手段，含有 Fe(III)氧化物的厌氧水解酸化系统还能够有效地去除难降解污染物，如苯及其同系物等[307]，并缓冲硫酸盐还原造成的抑制[308-310]，提高工业废水的可生化性。②Fe(III)氧化物作为导体材料，介入 Fe(III)还原菌和产甲烷菌的互养代谢，推动 DIET 的形成。在厌氧系统受到酸性、有机负荷冲击时，种间氢传递产甲烷受阻，DIET 成为厌氧呼吸的主渠道，维持系统的稳定运行。以丙酸为例，Viggi 等[311]向含有磁铁矿的产甲烷颗粒污泥中注入高氢分压以抑制 IHT，发现互养代谢仍可以维持正常运行。这是因为在含有 Fe(III)氧化物的厌氧系统中，互养微生物不再依靠导电 pili 和细胞色素作为必要的生物电子连接形式，而是直接附着在 Fe(III)氧化物的表面，依靠其较高的电导性，直接进行种间电子交换，从而提高 DIET 速率[312]。互养微生物不再需要生长出胞外导电 pili 和细胞色素，节省微生物细胞能量，提高 DIET 效率[313]。该原理更好地解释了 Fe(III)氧化物较为富集的沉积物和稻田土环境中厌氧的促进作用[314-316]。而这一现象与导体碳材料（如颗粒活性炭、生物碳、碳布等）置于厌氧消化器所表现一致，其本质均是导体材料作为电子管路，促进 DIET 产甲烷[313,317-321]。

5 基于铁基材料的污染控制技术及原理

铁（iron）是一种变价的金属元素，有 0 价、+2 价、+3 价和+6 价，其中 0 价只有还原性，+6 价只有氧化性，+2、+3 价既有还原性又有氧化性。铁盐及铁化合物的种类繁多，多数价廉易得，且对环境友好无毒性。因此，大量的铁盐及铁化合物被用于水污染控制和受污染水环境的修复。常见的有作为还原剂的零价铁和亚铁盐、氢氧化亚铁等，作为混凝剂的硫酸铁、聚合硫酸铁、聚硅酸铝铁等，作为吸附剂的各种各样的铁氧化物和铁基双金属氧化物等，作为氧化剂和消毒剂的高铁酸盐等。本节重点综述了基于零价铁（纳米零价铁、稳定化纳米零价铁、黏土负载纳米零价铁、微米零价铁）、树脂基纳米铁氧化物以及高铁酸盐的污染控制技术及原理。

5.1 基于零价铁的污染控制技术研究进展

自 1994 年 Gillham/O'Hannesin[322]和 Matheson/Tratnyek[285]两个课题组分别发表了利用零价铁（ZVI）去除水中的卤代有机物的论文后，ZVI 以其低毒、廉价且对环境不会产生二次污染等优点，迅速在水污染治理中受到重视。大量研究表明 ZVI 不但可以直接通过还原、吸附、共沉淀等作用（图 9-16）去除氯代有机物[285]、含氧酸根[323]、重金属离子[324]、偶氮染料[325]、硝基芳香族[326]以及硝酸盐[327]等多种污染物，还可以与双氧水/过硫酸盐等结合构成高级氧化体系来氧化去除水中的有机污染物[328,329]，这极大地推动了 ZVI 在环境污染治理方面的应用[330]。ZVI 在环境中的应用主要集中在地下水修复（渗透反应墙，PRB）、工业废水处理和以受重（类）金属污染的地下水为水源的饮用水处理[330]。

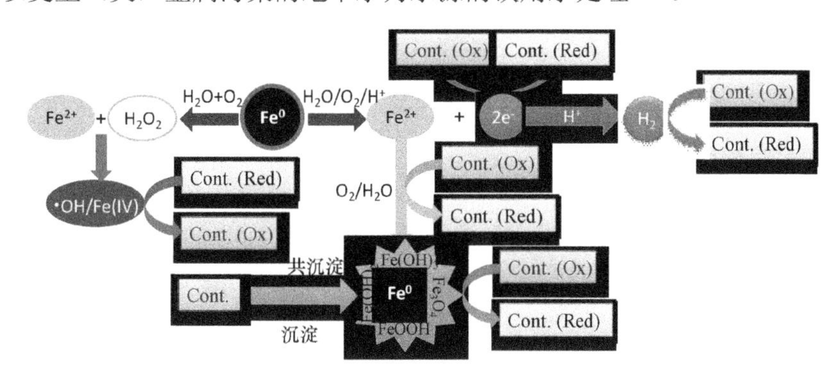

图 9-16　零价铁体系（Fe^0/H_2O）去除污染物的主要机理示意图（引自参考文献[331]）

当前 ZVI 技术已经有工程应用的案例，但是由于空气氧化，ZVI 表面都会覆盖一层氧化膜，且比表面积比较小，因此普通的 ZVI 常常反应活性不高[332]，而且反应过程中铁氧化物在其表面沉积等原因，使得 ZVI 的反应活性常常随着反应进行和 pH 升高逐渐降低。因此探索提高或保持零价铁反应活性的方法是近年来零价铁技术的一个研究热点。针对这一问题，各国研究者开展了大量的研究工作并发展了多种方法。其中常用的方法包括：①利用纳米零价铁（nZVI）代替普通的 ZVI[333]；②对 ZVI 进行预处理，包括酸洗[334]、氢气还原[335]、超声预处理[336]、预磁化[337]、引入另一种金属与其形成双金属[338]、硫化[339]等；③在 ZVI 除污染过程中，利用物理法来促进 ZVI 的除污染效能，包括弱磁场(WMF)[340]、超声[341]、电场[342]、微波[343]等；④在 ZVI 除污染过程中，利用化学法来促进 ZVI 的除污染效能，包括外加各种金属离子(如 Fe^{2+}、Cu^{2+}、Ni^{2+}、Co^{2+})[344,345]、络合剂［如乙二胺四乙酸二钠（EDTA）］[346]、强氧化剂（如 H_2O_2、$KMnO_4$、NaClO）[347]等方法。在上述提升 ZVI 除污染效能的方法中，WMF 强化 ZVI 除污染技术（WMF-ZVI）[348]、硫化 ZVI(FeS-ZVI)[348]、外加亚铁离子(Fe^{2+}-ZVI)[349]或强氧化剂（SO-ZVI）强化 ZVI 除污染方法[347]是近年来发展起来的具有较高创新性及应用前景（考虑价格、实施的难易程度）的方法。

目前研究较多的有纳米零价铁、稳定型纳米零价铁、负载型纳米零价铁和普通微米零价铁。为了区分纳米零价铁和普通微米零价铁，后续用 nZVI 表示纳米零价铁，ZVI 表示普通零价铁。以下将分别阐述这几类技术的特点及应用前景。

5.1.1　纳米零价铁的制备、表征及对环境污染物的修复

5.1.1.1　纳米零价铁的制备

纳米零价铁（nanoscale zero valent iron，nZVI）可通过化学法和物理法两大类方法制备[350]。物理法包括蒸发冷凝法、高能球磨法和深度塑性变形法等[351]；化学法分为液相还原法、热解羰基铁法、气相还原法和微乳液法等，具体见表 9-2[333,352-355]。实验室制备少量纳米零价铁通常采用液相还原法，即使用硼氢化钠（$NaBH_4$）还原水溶液中的二价或三价铁离子，见化学方程式（9-23）。液相还原法实验室操作简单易行，且容易调控合成的纳米零价铁的理化性质[333,351,356]。

$$Fe(H_2O)_6^{3+} + 3BH_4^- + 3H_2O \longrightarrow Fe^0\downarrow + 3B(OH)_3 + 10.5H_2 \tag{9-23}$$

物理法中高能球磨法是利用介质和物料之间长时间反复研磨和冲击使物料颗粒粉碎到所要求尺寸。Li 等[357]率先报道采用机械高速球磨制备纳米零价铁，即在精密的高能球磨机中通过钢珠（或锆珠）的不断研磨，将微米级零价铁逐渐挤压破碎形成尺寸小于 100 nm 的颗粒。比较其与液相还原法制备的纳米零价铁对几种氯代有机物的去除效率，发现两者没有显著的差异。此外，该制备方法具有成本低、耗时短、无污染等优势，适合大规模制造纳米零价铁。

表 9-2　纳米零价铁常用制备方法

类别	名称	方法	优点	缺点
物理法	高能球磨法	利用介质和物料之间反复研磨、冲击使物料颗粒粉碎至纳米级别	操作简单，成本低，产量高	产品纯度低，粒径不均
	蒸发冷凝法	利用真空蒸发、激光蒸发、电子束照射、溅射等方法使原料气化形成等离子体，然后在介质中急剧冷凝	纯度高、粒径小、粒径分布窄、团聚性能差	技术设备要求高温、操作危险性高
	塑性变形法	材料在准静态压力作用下发生严重塑性变形后晶粒尺寸细化到纳米级	工艺简单，易实现工业化生产和应用	纯度低、粒径不均
化学法	液相还原法	利用还原剂将金属铁盐还原	原理简单，该方法非常稳定	粒径不均，易团聚，易氧化
	碳热法	高温条件下，利用 C 或 CO 还原 Fe(II)/Fe(III)	成本低，原材料常见	CO 有毒、易爆炸
	水热法	将葡萄糖和 $Fe(NO_3)_3$ 溶液充分搅拌，180℃高温加热	稳定性强，迁移能力好	方法复杂
	羰基铁热解法	利用热解、激光和超声等激活手段，使羰基铁分解，并成核生长	成本低，原材料常见	设备要求高，成本高
	微乳液法	利用金属铁盐和沉淀剂形成微乳液，控制胶粒成核，热处理后得到纳米微粒	粒径小，分布均匀，分散性好	成本高，工艺复杂
	沉淀法	加入沉淀剂，在特定温度下使溶液发生水解或直接生成沉淀，形成不溶性氢氧化物、氧化物	反应温度低，操作简单，成本低，颗粒较均	沉淀难于水洗、过滤，且杂质多纯度低

5.1.1.2　纳米零价铁的表征

1. 纳米零价铁形貌表征

纳米零价铁，无论采用何种方法制备，均呈“核-壳”结构，内核为 Fe^0，外壳的组成和结构则与制备方法及保存环境有关。例如：采用气相还原法（以氢气或一氧化碳为还原剂）所得纳米零价铁为形状不规则且棱角分明的颗粒[358,359]；采用机械球磨法在不同球磨时间下制得的纳米零价铁呈片状、立方体状等多种形貌[351]。图 9-17 为液相还原法制备的纳米零价铁 SEM 和 TEM 图像，单个颗粒大体上呈球形，表面光滑，颗粒粒径分布范围为 10~100 nm（平均粒径为 60 nm）；由于颗粒间存在静电力、磁力和表面张力作用[360]，易团聚形成不规则团状或排列成链状结构，呈典型团簇状［图 9-17(a)，SEM］。团簇状的纳米零价铁具有连续的氧化物壳层，而单个颗粒呈“核-壳”结构，内核致密，外壳纤薄，核外包裹着一层 2~4 nm 的壳层，颗粒与颗粒之间由很薄的氧化物壳层（~1 nm）分隔开[361][图 9-17(b)~(d)]。根据多次 BET-N_2 测定结果，纳米零价铁具有较大比表面积：17~38 m^2/g（平均为 30 m^2/g），而普通零价铁屑或微米铁粉的比表面积在 0.01~1 m^2/g 数量级范围[362]。

2. 纳米零价铁组成分析

图 9-18（a）为新鲜纳米零价铁的 X 射线衍射仪（XRD）谱图，其在 $2\theta = 44.9°$存在一个弱且宽化的衍射峰，与体心立方（bcc）α-Fe（0）的标准衍射图谱相吻合，为{110}晶面衍射峰[363]。而宽化的衍射峰是典型的非洛伦兹形状，表明纳米零价铁的微晶尺寸分布。

采用高分辨 X 射线光电子能谱（XPS）分析纳米零价铁颗粒表面 Fe 和 O 元素的价态，对于 Fe 2p 范围内［图 9-18（c）］，710.8 eV 和 724.3 eV 附近的光电子峰分别对应 Fe(II)/Fe(III)的 2p3/2 和 2p1/2，证实了铁氧化物的存在；706.8 eV 附近的光电子峰为 Fe(0)的 2p3/2 峰，表明新鲜合成的纳米零价铁存在密实的金属铁晶体。对 O 1s 谱图进行分析，发现颗粒表层氧主要以 OH^-和 O^{2-}形态存在［图 9-18（d）］。进一步分析 OH^-/O^{2-}的比例（接近 1∶1），表明纳米零价铁表层主要为铁的羟基氧化物（FeOOH）[356]。

此外，STEM 结合 XEDS、EELS 等化学分析手段能够对单个纳米颗粒进行精确分析，研究结果表明，纳米零价铁颗粒外壳主要以 Fe(II)/Fe(III)的混合(氢)氧化物为主，且离内核越近 Fe(II)含量越高，越远则 Fe(III)含量越高[361]。

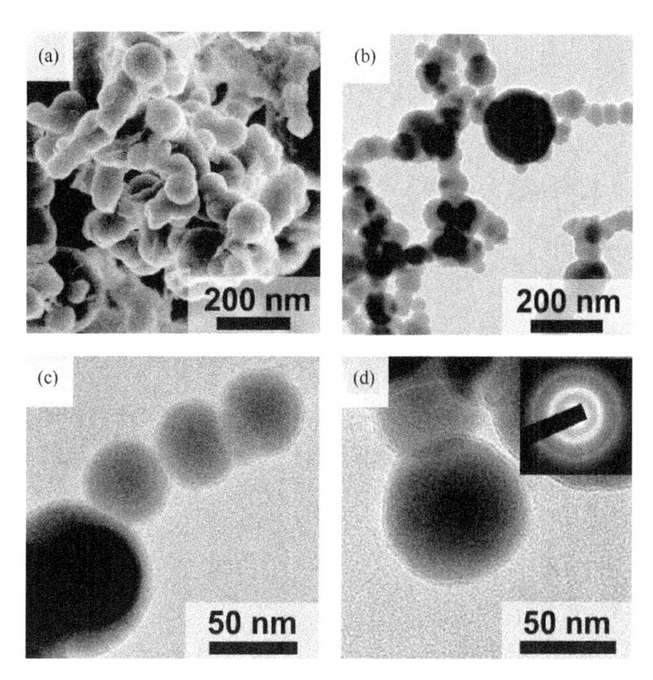

图 9-17　新鲜纳米零价铁的 SEM（a），TEM（b，c，d）和 SAED（d 插图）表征

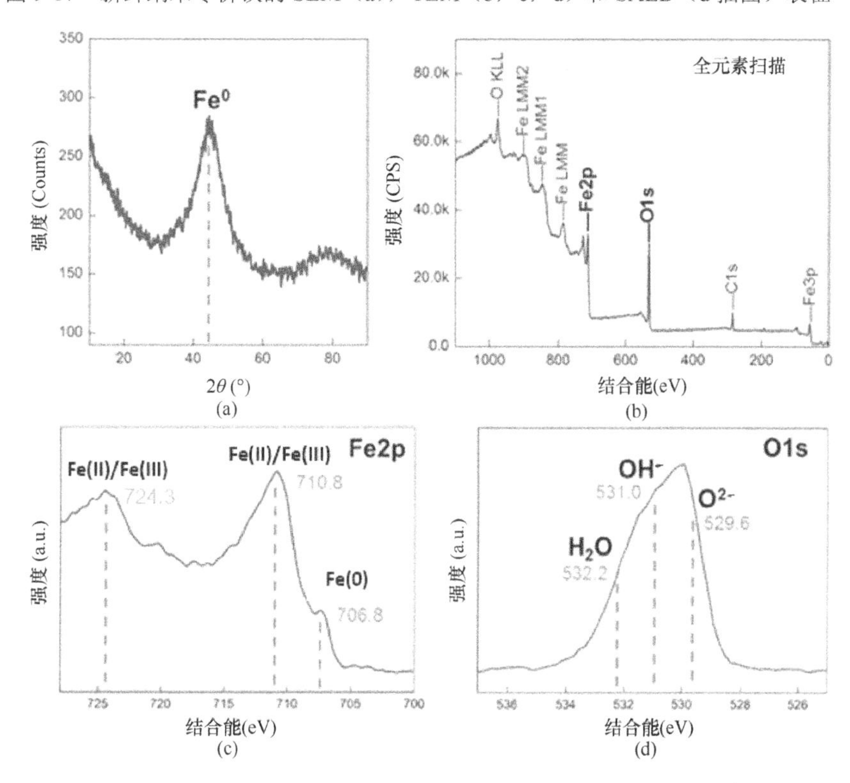

图 9-18　新鲜纳米零价铁的 XRD（a），XPS（b，c，d）表征

3. 纳米零价铁氧化还原特性

零价铁的标准电极电位 E^0[Fe(II)/Fe(0)]为–0.44 V，具有较强的还原性，容易失去电子形成 Fe(II)，具有提供电子的趋势，而在水环境中，主要电子接受体为水和溶解氧，因此会发生如下反应[364]：

$$Fe^0 + 2H_2O \longrightarrow Fe^{2+} + H_2 + 2OH^- \quad (9\text{-}24)$$

$$Fe^0 + 4H^+ + O_2 \longrightarrow Fe^{2+} + 2H_2O \quad (9\text{-}25)$$

纳米零价铁投加到去离子水中，溶液的 pH 由~6 上升到 7.5~9，整个反应体系的呈碱性环境；同时 E_h 由+400 mV 迅速下降到–500 mV，形成强还原性水环境，这主要因为纳米零价铁具有大的比表面积与较高反应活性，可快速与水中的 H_2O 与溶解氧发生反应，消耗水中 H^+并产生 OH^-及 Fe(II)［方程式(9-24)，式(9-25)］。另外，在一定纳米零价铁浓度范围内，溶液最终 pH、E_h 没有显著差别（图 9-19）。

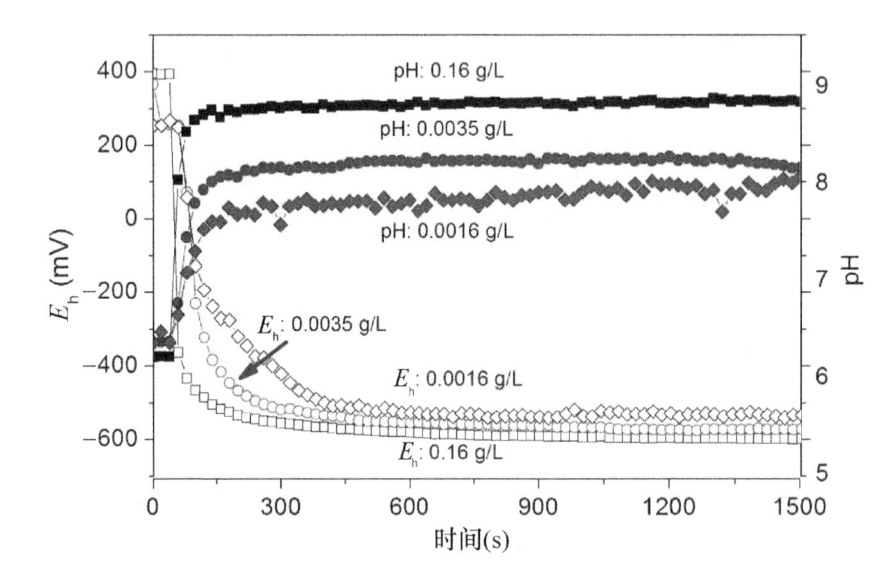

图 9-19 不同量纳米零价铁投加到水溶液中 pH 和 E_h 随时间变化趋势图

溶液 pH 的变化可用来指示纳米零价铁的迁移[365]，而反应过程中 E_h 数值的变化与系统中复杂的氧化还原反应有很好的响应关系，可用来指示纳米零价铁腐蚀反应的进程以及腐蚀产物的演变[366]。在现场研究中，通常也会根据 E_h 的变化来判断纳米零价铁量的变化[367-369]。

4. 纳米零价铁在水环境中表面电荷和 Zeta 电位

纳米零价铁表面电荷是反映纳米零价铁反应活性和稳定性的重要参数，可指示溶液中正离子或负离子的相互吸引、颗粒与颗粒间的相互作用、多孔物质的稳定性和迁移性等。Zeta 电位是研究表面电化学特性的一种简单的方法，其值可反映颗粒表面电荷的电量及电性。纳米零价铁进入纯水环境中，与水分子作用带上相应的电荷，Zeta 电位随着溶液 pH 的升高而由正逐渐变负（图 9-20）。经过测定，纳米零价铁的等电点 pH_{zpc} 约为 8.3，当 pH <pH_{zpc} 时，纳米零价铁表面带正电荷，易静电吸引带负电荷的配体，有利于水中以阴离子形式存在的污染物如 As(III/V)、Se(IV/VI)等的去除。当 pH > pH_{zpc} 时，纳米零价铁表面带负电荷，能够和金属阳离子［Cu(II)，Ag(I)等］在表面形成配合物。另外，对于不同浓度的纳米零价铁悬浮液，尽管相同 pH 下 Zeta 电位有差异，而等电点都稳定在 8.3 左右，表明纳米零价铁的等电点值与浓度无关。

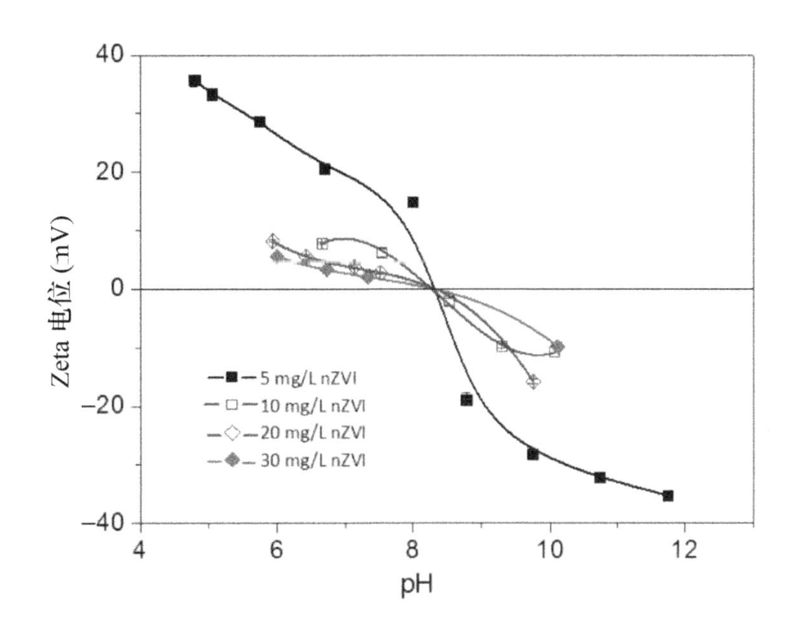

图 9-20　纳米零价铁 Zeta 电位随 pH 变化曲线

5.1.1.3　纳米零价铁对环境污染物的修复

1. 纳米零价铁去除污染物机理

纳米零价铁具有更高比表面积及还原活性，可吸附还原水中一系列有毒或难降解物质[370-372]；针对不同种类的污染物，纳米零价铁的去除机制也不尽相同，下面简单介绍纳米零价铁对水中几种主要污染物的去除机理。

1）氯代有机物

纳米零价铁是一种相对较强的还原剂，可通过还原作用降解多种氯代有机化合物，包括氯代脂肪烃、氯代烷烃、多氯联苯等难降解有机污染物[373-375]，其半反应式如式（9-26）及式（9-27）[373]：

$$Fe^0 \longrightarrow Fe^{2+} + 2e^- \qquad E^0=-0.44\ \text{V} \tag{9-26}$$

$$RCl + 2e^- + H^+ \longrightarrow RH + Cl^- \qquad E^0=0.5\sim1.5\ \text{V（pH=7）} \tag{9-27}$$

从热力学角度，氯代有机物可通过 Fe（0）脱卤作用降解成无毒有机物，其还原脱卤过程包括两个阶段，首先污染物吸附于铁表面，然后 C—Cl 键被打断[375]。关于还原打断 C—Cl 键，有三种可能机理[376-379]：①直接在铁表面被零价铁还原；②在表面被亚铁离子还原；③被氢气还原。在纳米零价铁表面镀上少量的贵金属作为催化剂，能够显著地增强其反应性能[380-382]。由于钯（Pd）具有较佳的结构和化学性能，应用较多[381]。例如，在还原四氯化碳时，使用 Pd-nZVI 时的反应速率常数比单一使用纳米零价铁高出两个数量级[382]。经计算，前者的活化能为 31.1 kJ/mol，而后者为 44.9 kJ/mol。

2）放射性核素

铀[U(VI)]是一种半衰期较长的放射性核素，近年来由于铀矿的开采以及核能的广泛应用对土壤和地下水造成了污染而备受关注。研究发现纳米零价铁可有效去除 U(VI)，在好氧条件下 U(VI)被还原生成 UO_2 沉淀[383]。Ling 和 Zhang[384]的研究发现，当用纳米零价铁处理水中 2.32~882.68 μg/L 的 U(VI)时，1 g/L 的纳米零价铁能在 2 min 内将 90%以上的 U(VI)从水中去除。为考察纳米零价铁去除水中 U(VI)的微观机理，采用球差校正扫描透射电镜（Cs-STEM）结合 X 射线能量散射谱(XEDS)对纳米零价铁与 U(VI)的反应过程进行了表征。结果表明，水中 U(VI)在静电力、分子间力、静磁力等力作用下被吸附到纳米零价铁表面，而后逐步向单个颗粒内部扩散，在扩散过程中纳米零价铁的 Fe(0)核释放出的电子将其还原为

U(IV)，U(IV)最终被包裹在纳米颗粒的中心。

3）重（类）金属

众多学者的研究发现纳米零价铁可快速高效去除有毒重金属/类金属，是污染治理的理想材料。重（类）金属进入环境绝大部分会以无机离子形式存在[385]：一部分重金属在水环境中以阳离子的形式存在，而砷、硒、铬、钨、钼、锗等在水环境中则大都以含氧阴离子的形式存在。

Li 和 Zhang[386]用高分辨率 X 射线光电子能谱（HR-XPS）研究了纳米零价铁与重金属阳离子的反应机理，认为纳米零价铁与阳离子的反应有三种机制：吸附、还原和吸附/还原，具体机制与该种金属离子标准氧化还原电位 E^0 有关（表 9-3）。若金属离子[如 Zn(II)、Cd(II)、Ba(II)]的标准氧化还原电位小于铁的 E^0，去除机理主要为吸附作用；标准氧化还原电位远大于铁的金属离子[如 Cu(II)、Ag(I)、Hg(II)]，去除机理主要为还原作用；而标准氧化还原电位略高于铁的金属离子[如 Ni(II)、Pb(II)]，去除则是吸附和还原共同作用的结果。

表 9-3　砷及重金属元素标准氧化还原电势 E^0 [386]

砷及重金属元素	氧化还原反应方程式	E^0（V）
砷（As）	$H_3AsO_4 + 2H^+ + 2e^- \longrightarrow HAsO_2 + 2H_2O$	0.56
	$H_3AsO_3 + 3H^+ + 3e^- \longrightarrow As + 3H_2O$	0.24
铜（Cu）	$Cu^{2+} + 2e^- \longrightarrow Cu$	0.34
	$Cu^+ + e^- \longrightarrow Cu$	0.16
铬（Cr）	$CrO_4^{2-} + 8H^+ + 3e^- \longrightarrow Cr^{3+} + 4H_2O$	1.51
	$Cr_2O_7^{2-} + 14H^+ + 6e^- \longrightarrow 2Cr^{3+} + 7H_2O$	1.36
钡（Ba）	$Ba^{2+} + 2e^- \longrightarrow Ba$	–2.92
镉（Cd）	$Cd^{2+} + 2e^- \longrightarrow Cd$	–0.40
铅（Pb）	$Pb^{2+} + 2e^- \longrightarrow Pb$	–0.13
镍（Ni）	$Ni^{2+} + 2e^- \longrightarrow Ni$	–0.25
钴（Co）	$Co^{2+} + 2e^- \longrightarrow Co$	–0.28
铁（Fe）	$Fe^{2+} + 2e^- \longrightarrow Fe$	–0.44
锌（Zn）	$Zn^{2+} + 2e^- \longrightarrow Zn$	–0.76
汞（Hg）	$Hg^{2+} + 2e^- \longrightarrow Hg$	0.86
银（Ag）	$Ag^+ + e^- \longrightarrow Ag$	0.80
铂（Pt）	$Pt^{2+} + 2e^- \longrightarrow Pt$	1.19
钯（Pd）	$Pd^{2+} + 2e^- \longrightarrow Pd$	0.92
锑（Sb）	$SbO^+ + 2H^+ + e^- \longrightarrow Sb + H_2O$	0.20
碲（Te）	$Te + 2e^- \longrightarrow Te^{2-}$	–1.14

同济大学张伟贤课题组前期采用球差校正扫描透射电镜（Cs-STEM）对纳米零价铁去除水中的砷、铬含氧阴离子进行了研究[387,388]。研究表明，砷酸盐（AsO_4^{3-}）与纳米零价铁的反应是表面反应，主要发生在铁氧化物层和 Fe（0）核的内表面上：首先纳米零价铁通过分子间静电作用力将 As(V)吸附到表面，而后逐步向颗粒内部扩散、还原，还原产物 As(0)在壳层内表面上沉积形成一层 1.0~1.2 nm 的薄层。纳米零价铁对重铬酸盐（$Cr_2O_7^{2-}$）的去除结果同样给出表面反应和颗粒内反应的直接证据：溶液中的 Cr(VI)被快速吸附到纳米零价铁表面，部分 Cr(VI)向颗粒内部扩散并被还原，还原产物 Cr(III)[$Fe_xCr_y(OH)_3$]沉积在核-壳之间的内表面上。

2. 纳米零价铁的工程应用

1）地下水污染原位修复

2000 年左右，纳米零价铁真正在环境应用领域崭露头角[369]，Zhang 等[369]将 Pd-nZVI 纳米材料直接注入地下水进行原位修复，发现在污染区内尤其是灌注井周围，多种氯代有机污染物均得到了有效降解。迄今，纳米零价铁已在美国、捷克、德国、意大利、中国台湾等国家或地区应用于地下水污染的原位修复。如图 9-21 所示，纳米零价铁地下水原位修复主要采用液压灌注方式将纳米材料注入地下受污染区域，形成活性反应区，使流经该区域的受污染地下水与纳米零价铁反应并得到净化[389]。

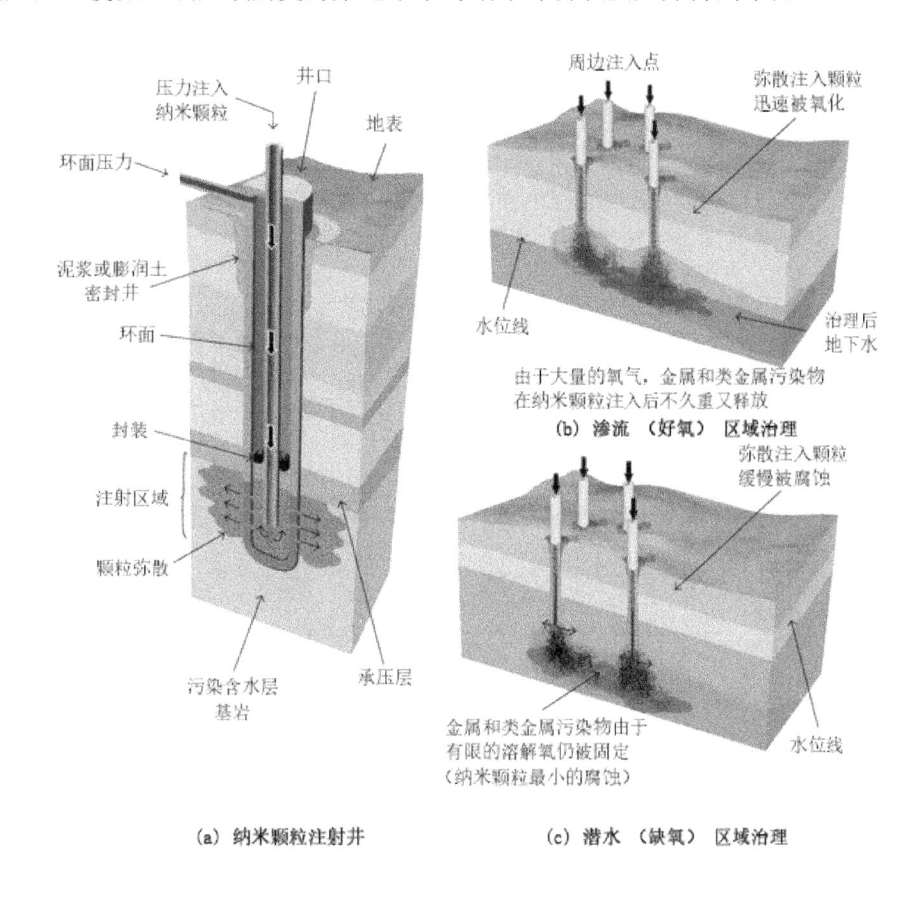

图 9-21　纳米零价铁应用于地下水污染原位修复工程示意图（引自参考文献[389]）

以北卡罗来纳州三角研究园地下水污染修复中试为例[368]，该区域地下水受垃圾堆放、填埋影响，污染严重。中试时通过潜水泵将纳米零价铁（11.2 kg）悬浊液从垃圾处理区东北部 38 m 处注入地下水蓄水层。注入点以北 6.6 m 处、以东 13 m 处和东北 19 m 处设置监测井，用于检测污染物浓度及其他常规指标。纳米零价铁注入前，该区域地下水中氯代有机物浓度约为 14000 μg/L，注入后几天内氯代有机物浓度降低 90%以上，6 周后 PCE、TCE 和 DCE 浓度均低于美国地下水水质标准。据估测，注入的纳米零价铁可随水流传输 20 m 以上，降解污染物的有效作用半径约为 6~10 m，若采用加压的方式注入纳米零价铁可扩展有效作用半径。在污染区域上游注入纳米零价铁后，可在下游将混有纳米零价铁的地下水抽出，加入新的纳米零价铁后再从上游注入，既可实现纳米零价铁的重复利用，又可保证其活性。

2）重金属工业废水处理

目前，纳米零价铁的中试或现场工程应用多集中于土壤及地下水污染修复领域[390-397]，张伟贤课题组首次报道纳米零价铁用于实际重金属废水处理的中试或工程[398-400]。通过“小试—中试—工程”逐级科学放大，提出“反应—分离—回用”式纳米零价铁反应器，系统研究该体系处理实际重金属废水的可行性并开展工程应用研究。

通过几项废水处理中试研究发现纳米零价铁技术可有效弥补石灰中和沉淀法处理重金属废水工艺不足（如高 pH 条件下重金属氢氧化物沉淀易复溶、水质波动条件下石灰投加不易准确及时调整、沉淀产物粒径小结构松散难沉降等）[398-401]。例如，对于江西某冶炼废水工程应用，采用“两级纳米零价铁反应器并联+混凝沉淀”组合工艺，该系统长期（3 年以上）平稳运行，在 127 d 工程调试期间，废水中砷、铜平均浓度分别从 110 mg/L、103 mg/L 降至 0.29 mg/L、0.16 mg/L（图 9-22），其他重金属（如 Co、Cr、Ni、Pb、Zn 等）浓度均降至 0.01 mg/L 左右，远低于废水处理设计排放标准，Se、Sb、Au、Tl 等元素也几乎被完全去除。此外，纳米零价铁在反应器内不断“反应—分离—回用”，充分利用纳米零价铁反应活性，降低运行的药剂成本。在此调试阶段，纳米零价铁除砷、除铜负荷分别达 245 mg-As/g-纳米零价铁和 226 mg-Cu/g-nZVI，总体重金属去除负荷超过 500 mg-重金属/g-纳米零价铁。另外，废水中稀贵金属［如 Cu(II)、Ag(I)、Au(III)等］在反复循环反应过程中被富集回收，使污泥具有较高的回收利用价值，此废水中重金属资源化产生的价值可抵消部分废水处理成本。

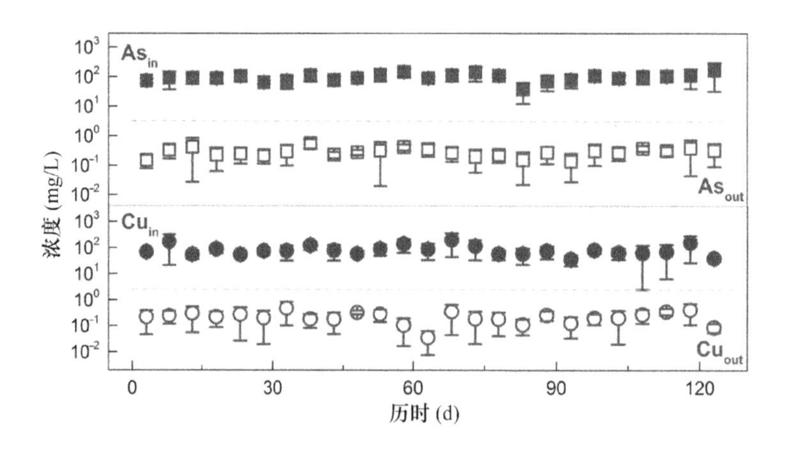

图 9-22 工程调试期间进出水中 As、Cu 浓度变化

5.1.2 稳定化纳米零价铁用于土壤及地下水修复的研究进展

纳米零价铁作为一种新型的原位地下水修复技术，在近年得到了广泛关注[396,402,403]。理论上，nZVI 尺寸小于土壤颗粒间的孔隙，可以通过注射井注入地下主动攻击污染物。因此，该技术对经济有效地修复深蓄水层中的污染源以及施工空间有限的城市污染地下水具有重要意义[404]。然而，在 nZVI 技术被提出后的相当一段时间里，其并未得到有效的实地应用。阻碍 nZVI 技术实地应用的一个很重要的原因是 nZVI 的团聚问题，其让 nZVI 失去了地下流动性[405]。因而，通过稳定化 nZVI 来改善其地下流动性，对该技术的实地应用意义重大。

5.1.2.1 稳定化纳米零价铁的制备及改性

1. 纳米零价铁团聚及稳定的机理

纳米零价铁（nZVI）的团聚是一个热力学上有利的过程，促成其团聚的作用力主要有范德华力、磁偶极以及电偶极作用等[405-407]。团聚现象不仅导致 nZVI 无法在土壤中进行有效传输，并且使 nZVI 因比表面积减小而丧失高活性。对 nZVI 进行稳定的关键是控制 nZVI 颗粒之间的相互作用。

nZVI 的稳定化通常是通过稳定剂对 nZVI 表面进行修饰或形成网状结构，使颗粒相互分离(图 9-23)[408]。在表面修饰过程中，带电稳定剂（比如阴离子表面活性剂）可以吸附在 nZVI 表面，产生静电斥力，从而使颗粒稳定[409,410]。此外，形成网状结构则是利用稳定剂分子之间的氢键使稳定剂相互纠缠，形成网状的黏性胶体，从而防止颗粒之间相互聚集，达到颗粒稳定化的目的[411]。网状稳定化又可分为两种：①吸附型稳定剂通过吸附在 nZVI 颗粒表面，形成 nZVI 凝胶；②非吸附型稳定剂包裹 nZVI 并在其周围形成网状物，从而使颗粒之间相互分离[408]。

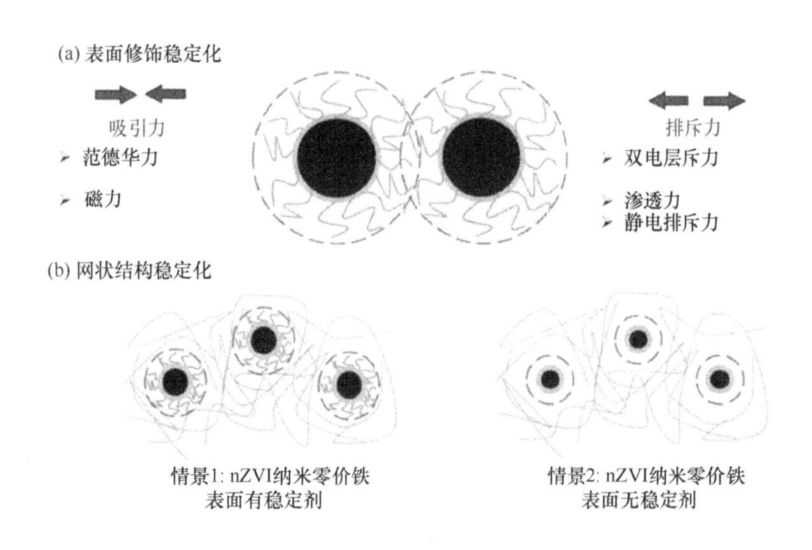

图 9-23　稳定剂稳定纳米材料机理图（引自参考文献[408]）

总体来说，稳定剂分子量越大，带电官能团越多，对 nZVI 的稳定效果就越好。另外，包裹稳定剂不仅能提高 nZVI 的稳定性，还能在一定程度上阻止材料表面与水和溶解氧的反应。

2. 稳定方法与稳定剂的选择

近年来，研究人员尝试了多种稳定 nZVI 的方法和材料，其工作主要集中在稳定方法、稳定剂的选择，以及不同稳定方法和稳定剂之间的比较等。

1）稳定方法

对于 nZVI 颗粒的稳定主要有形成前稳定和形成后稳定两种方法。形成前稳定是指在制备 nZVI 过程中颗粒团聚之前加入稳定剂，该方法由 Auburn 大学开发[405]；形成后稳定则是指对已聚集的 nZVI 颗粒进行分散并稳定。稳定方法的不同，决定了 nZVI 稳定效果和活性的不同。形成前稳定可以通过稳定剂来调控 nZVI 成核和生长，从而调控 nZVI 颗粒大小[406]；形成后稳定只能起到稳定作用，而且稳定剂的加入可能会使 nZVI 表面的反应点位被覆盖，导致其活性降低[412]。从热力学上讲，形成前稳定比较容易获得稳定、分散且较小的颗粒，而聚集后稳定的效果则取决于已团聚颗粒重新分散的能力。

2）稳定剂的选择

（1）表面活性剂

表面活性剂是一种常见的用于控制颗粒之间相互作用的表面修饰物，一般可分为离子型表面活性剂、非离子型表面活性剂以及两性表面活性剂等。研究发现一些阴离子型表面活性剂［如十二烷基硫酸钠（SDS）[413]、十二烷基苯磺酸钠（SDBS）[409]等］能增加 nZVI 表面的电负性，从而产生静电斥力，起到分散作用；而一些非离子型表面活性剂［如吐温（Tween）20[414]等］能显著增加 nZVI 之间的位阻效应，起到稳定化效果。已报道的用于稳定 nZVI 的表面活性剂包括十二烷基硫酸钠[413]、十二烷基苯磺酸钠[409]、吐温系列[413]、鼠李糖脂（rhamnolipid）[415,416]、脂肪醇聚乙烯醚（AEO）[417]等。

（2）大分子聚合物

目前，相对于表面活性剂，利用大分子聚合物稳定 nZVI 的研究更为广泛。这是因为大分子聚合物不仅能有效地稳定 nZVI，同时还可以作为地下水和土壤环境中微生物的碳源，促进微生物对污染物的降解[418]。常用的合成聚合物稳定剂［如羧甲基纤维素（CMC）[406]、聚丙烯酸（PAA）[419]、聚苯乙烯磺酸钠（PSS）[420]等］带负电，其通过单分子螯合等作用吸附在 nZVI 表面[404]，产生静电排斥作用和位阻效应[421]，从而对 nZVI 起到稳定效果。由于土壤介质通常带负电荷，这些带负电的稳定剂还能够减少 nZVI 在土壤中的沉积，有利于 nZVI 在地下水中的传输[422]。此外，淀粉（starch）[405]、黄原胶（xanthan gum）[411,423]等天然聚合物以及聚丙烯酰胺（PAM）[424]、聚乙烯吡咯烷酮（PVP）[425]等部分合成聚合物在水中呈电中性，其主要通过增加溶液黏度，产生空间位阻效应或在 nZVI 表面形成网状结构等起到颗粒稳定作用[411,426,427]。

（3）固体载体/保护性壳

还有学者提出可采用碳[419,428,429]、硅[430]、膨润土[431,432]等作为载体对 nZVI 进行稳定，提高其在多孔介质中的传递性能。将 nZVI 固定在载体上主要抑制了颗粒之间的极化作用，使颗粒达到稳定状态。另外，有些惰性并疏水的载体可以起到促进对 DNAPL 的吸附，并延长 nZVI 材料寿命的作用[430]。

3）稳定剂的比较

对土壤和地下水修复而言，理想的稳定剂不仅能对 nZVI 进行有效的稳定分散，还应该具有可生物降解、无毒性、成本低等优点。研究者通过对不同稳定剂稳定的 nZVI 进行稳定效果和反应活性的系统对比发现，CMC 是目前相对最理想的 nZVI 稳定剂[422,431]。

5.1.2.2 稳定化纳米零价铁的场地修复应用实践及限制因素

1. 稳定化纳米零价铁的场地修复实践

2010 年，Auburn 大学的研究人员首次将 CMC 稳定的 Fe/Pd 纳米颗粒（CMC-nFe/Pd）用于美国阿拉巴马州某一卤代烃污染场地的修复[418]，如图 9-24 所示。研究人员发现 CMC-nFe/Pd 能够在水力传导率为 1.98×10^{-3} cm/s 的地下水层中传递 1.5m 以上，下游 1.5m 和 3m 监测井中的含氯烯烃污染物得到快速降解，并在一周后达到最大去除，在第 2 周 CMC-nFe/Pd 逐渐失去活性。同时，CMC-nFe/Pd 的注射提高了地下

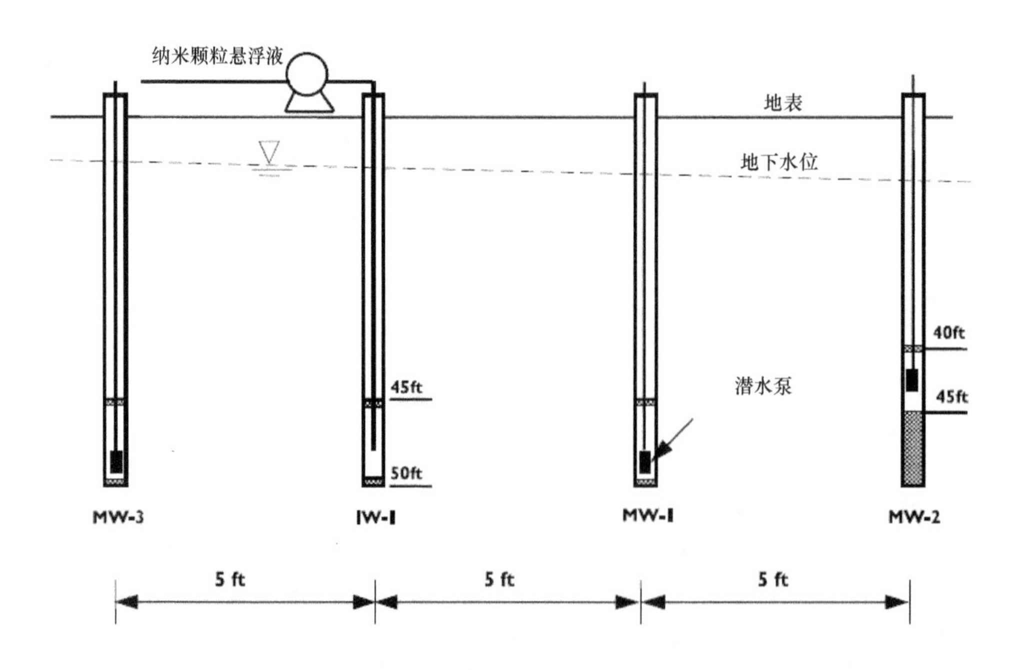

图 9-24 测试现场含水层剖面图和 CMC-nFe/Pd 原位注射示意图（引自参考文献[418]）

水中脱氯细菌的活性，使得该污染场地含氯烯烃的生物降解在此后 2 年的监测期内大大提高[418]。此后，在北美和欧洲相继进行了一系列 CMC-nFe/Pd 用于含氯烯烃污染场地修复的测试，结果均证明经 CMC 稳定化的 nZVI 在反应活性及地下传递性上都有较大改善，并能促进污染物的生物修复[392,433,434]。

除 CMC 外，其他类型稳定剂稳定的 nZVI 也在实地应用上得到了测试[426,435-437]。例如，玉米油和表面活性剂乳化的 nZVI 被证明能很好地原位去除受污染场地中的 PCE[435]。最近，研究人员发现把用瓜尔胶等稳定的 nZVI 注射到场地后，这些稳定化的零价铁对卤代烃同样具有较好的去除效果[426,436]。

2. 限制因素

稳定化纳米零价铁凭借自身的高反应活性、优异的稳定性以及良好的地下流动性，保证了其用于污染场地原位修复的可行性，并使这项技术持续发展。然而，目前利用稳定化 nZVI 进行污染场地原位修复时还存在一些限制因素，主要有以下几个方面。

1）传输距离

在实际应用中，nZVI 必须保持足够的传输性能才能保证和已扩散的目标污染物进行有效接触，从而起到修复效果。虽然理论上稳定化提升了 nZVI 的传输性能，但是这些稳定化的 nZVI 颗粒在实际条件复杂的土壤，尤其是低渗透性的土壤或沉积层中的传输距离，仍然是限制其进行有效修复的主要因素。

2）反应活性

一些研究表明地下水中高浓度的溶解性有机质（如 NOM[438]等）以及一些离子（比如 Cl^-、SO_4^{2-}等[439]）会影响 nZVI 的活性[413]。另外，土壤和地下水中存在的溶解氧或具有氧化性的矿物（比如 MnO_2 等）还会氧化 nZVI，使其丧失反应活性。

3）颗粒寿命和电子效率

纳米零价铁作为较强的还原剂，在其注入到地下后不仅可以和目标污染物反应，还会和水分子、含氧离子等反应，造成自身的腐蚀，这两个过程都会消耗电子。电子选择性定义为用于降解目标污染物的电子占 nZVI 给出总电子的比例，其中 CMC 稳定化 nZVI 的电子效率大概只有 5%左右[440]。此外，稳定化 nZVI 与水分子等的副反应，可能导致稳定化的 nZVI 在接触到目标污染物之前就已经被消耗完毕而失去应用价值[441-443]。最近的研究表明通过硫化可以抑制 nZVI 与水分子的副反应，从而可能延长颗粒寿命和提高电子效率[440,444-446]。不过，硫化稳定化 nZVI 的研究还处在初始阶段，对其研究工作还需要进一步深入的开展。

4）潜在毒性

目前，已经有较多研究表明一定剂量的 nZVI 会对环境中的微生物产生毒性作用[447-449]。有研究表明在实验室条件下，稳定化 nZVI 会在短期内降低 nZVI 对微生物的毒性作用[447,450-452]。然而，到目前为止，还没有关于稳定化 nZVI 的长期毒理效应以及在实际场地修复中毒性研究的相关报道。因此，相关的研究工作还需进一步进行和完善。

5.1.3　黏土负载纳米零价铁复合材料去除污染物的研究进展

纳米零价铁具有比表面积大、表面活性点位丰富、反应活性高等特点，在水环境污染物处理中显示出优异的应用前景。但纳米铁高的表面能和铁的固有磁性，使其易团聚且难以分离回收，制约了它的广泛应用。选用合适的固体多孔材料，如黏土、活性炭、沸石、功能性高分子聚合物等作载体，能够较好地分散纳米铁粒子，得到反应活性和稳定性更高的纳米铁复合材料[453,454]。与其他载体比较，黏土矿物来源广泛、价格低廉、性质稳定、环境相容性好、比表面积较大、并能缓冲介质 pH。利用黏土独特的结构-功能可调、易于改性的特点，可以根据目标污染物的性质制备合适的改性黏土，以充分发挥其载体效应，提高纳米铁对污染物的去除效率和选择性。同时，黏土零价铁复合材料在实际污染场地修复中也展现出良好的应用前景。

5.1.3.1　黏土矿物的性质

黏土矿物是由硅氧四面体和铝氧八面体结构单元构成的一类层状硅酸盐矿物。其中，蒙脱石、伊利石、蛭石等由 2 层硅氧四面体中夹一层铝氧八面体（2∶1 型）结构单元组合而成，坡缕石、海泡石则属于另一类 2∶1 型层链状硅酸盐，而高岭土则由硅氧四面体层和铝氧八面体交替叠加而成（1∶1 型）。由于黏土矿物在形成过程中发生同晶替代，导致黏土矿物表面带负电。因其表面具有丰富的活性位点、比表面积较大，常用作吸附剂或与某些活性材料复合以去除各类环境污染物，是一种理想的纳米材料载体。

5.1.3.2　黏土负载纳米零价铁的发展及其对污染物的去除作用

黏土与零价铁复合去除污染物的最早报道始于 2005 年，Cho 等[455]将 HDTMA 改性后的有机膨润土与 ZVI 均匀混合，用作模拟 PRB 柱试验中的活性物质处理三氯乙烯（TCE），去除速率是单一 ZVI 体系的 7 倍。自 2008 年以来，研究人员陆续开展了黏土负载纳米铁的制备及去除污染物的研究。Üzüm 等[456]和 Zhang 等[457]采用高岭土，Gu 等[458]、Fan 等[459]和 Shahwan 等[460]采用蒙脱（石）土，Frost 等[461]采用坡缕石分别制备了负载纳米铁，李益民等[462,463]率先采用改性膨润土制备负载纳米铁。黏土负载纳米铁的基本制备方法是先将黏土在铁盐溶液中浸渍数小时，达到吸附平衡后在 N_2 保护下用 $NaBH_4$ 还原。Fan 等[459]研究表明，纳米铁粒径与黏土中的铁盐加入量呈正相关。当加入的铁离子量低于黏土 CEC 时，交换到层间的铁离子与 $NaBH_4$ 发生层间限域还原反应，可以生成亚纳米粒度、高反应活性的零价铁[458,464,465]。

已有文献报道中，用于负载纳米铁的黏土载体包括蒙脱（石）土[458,459,465-468]、膨润土[431,464,469-472]、高岭石[456,457]、海泡石[473,474]、坡缕石[475,476]、累托石[477]等，可处理的污染物包括氯代有机污染物、多溴联苯醚、硝基苯、染料、重金属和准金属离子、硝酸根等（表 9-4）。从表 9-4 中列出的一些典型研究结果可以发现，在相同实验条件下，与未负载纳米铁比较，各类黏土负载纳米铁对污染物的去除效率均明显提高。

5.1.3.3　黏土在零价铁去除污染物中的协同作用原理

首先，黏土对污染物的吸附性能是影响零价铁对目标污染物去除率的重要因素。在污染物与零价铁之间的复相反应中，污染物穿过液膜扩散（富集）到零价铁外铁氧化物壳层的传质过程被认为是决速步骤。利用黏土对目标污染物的吸附作用，其负载纳米铁更有助于污染物在铁表面的富集，进而强化纳米铁对污染物的去除，且强化作用的大小与所用黏土对污染物的吸附富集能力成正相关。为此，李益民课题组利用膨润土 CEC 高、结构-功能可调的特点，根据目标污染物的结构和性质，制备了三种亲疏水性、表面 Zeta 电位各异[图 9-25（a）]的改性膨润土。然后分别用有机膨润土负载纳米铁（NZVI/CTMAB-bent）处理硝基苯[462]和氯代有机污染物［图 9-25（b）］[472]，荷负电的钠基膨润土负载纳米铁（NZVI/Na-bent）处理 UO_2^{2+}[图 9-25（c）][469]、Ni^{2+}等无机阳离子污染物，荷正电的羟基铝膨润土负载纳米铁（NZVI/Al-bent）处理 Cr(Ⅵ)［图 9-25（d）］[470]、Se(Ⅵ)[478]等无机阴离子污染物。结果表明，利用上述与污染物性质相匹配的改性黏土负载纳米铁，不仅显著提高了零价铁对目标污染物的去除率，同时将污染物还原为毒性更低的物质，提高了零价铁对目标污染物的反应选择性。

其次，黏土特有的 pH 缓冲和转化钝化产物的能力可显著减少零价铁表面的钝化产物、延长其使用寿命。黏土表面丰富的 Si—OH 和 Al—OH 会随着零价铁与污染物反应体系 pH 的升高而离解出 H^+，从而降低反应介质 pH（可降低 0.5~1.0 个单位）[467,469]，减少铁表面钝化产物生成。此外，黏土良好的阳离子交换性能可使反应产生的铁离子或其他还原产物吸附到黏土上。EAXFS 分析证实了 NZVI/Al-bent 与

表 9-4　不同黏土负载纳米铁对污染物的去除效率

载体	污染物		操作条件		去除率	参考文献
	名称	C_0	剂量	pH	Clay-NZVI *vs.* NZVI	
蒙脱石	Nitrobenzene	2.1 mmol/L	12.6 g/L		83%	[458]
蒙脱石	DBDE	2.5 mmol/L	4.1 g/L	6.3	100%*vs.* 26%	[465]
膨润土	PCBs	2.0 mg/L	1.5 g/L	7.9	70%*vs.* 37%	[464]
膨润土	Cr(VI)	50 mg/L	3.0 g/L	4.0	100%	[431]
钠-膨润土	U(VI)	100 mg/L	0.25 g/L	6.0	99.2%*vs.* 48.3%	[469]
蒙脱土	As	5.0 mg/L	1.0 g/L	7.0	0.9 6mg/g As(V)，0.97 mg/g As(III)	[466]
蒙脱土	Cr(VI)	25 mg/L	2.0 g/L	1-3	100%	[467]
柱撑膨润土	Cr(VI)	50 mg/L	2.1 g/L	5.6	100%*vs.* 63%	[470]
柱撑膨润土	NO_3^-	50 ppm	1.0 g/L	7.0	100%*vs.* 62.3%	[471]
有机膨润土	PCP	0.2 mmol/L	3.57g/L	6.0	96.2%*vs.* 31.5%	[472]
有机蒙脱土	Cr(VI)	50 mg/L	0.47 g/L	5.0	100%	[468]
高岭土	Pb(II)	500 mg/L	10 g/L	5.1	96.7%*vs.* 16.8%	[457]
高岭土	Cu(II)、Co(II)	100 mg/L	5.0 g/L	10	98%Co(II)，99%Cu(II)	[456]
海泡石	Cr(VI)、Pb(II)	20 mg/L	1.6 g/L	6.0	99.2%Cr(VI)，99.9%Pb(II)	[473]
海泡石	Bromamine acid	1000 mg/L	2 .0 g/L	acidic	98%	[474]
柱撑硅铝皮石	Cu(II)、Ni(II)	50 mg/L	1.0 g/L	6.0	99.9%	[476]
坡缕石	Bisphenol A	50 mg/L	1.0 g/L	2.36	99%*vs.* 74%	[475]
累托石	Orange II	70 mg/L	2.0 g/L	中性	100%*vs.* 35%	[477]

注：C_0，污染物初始浓度；DBDE，十溴联苯醚；PCBs，多氯联苯；PCP，五氯酚

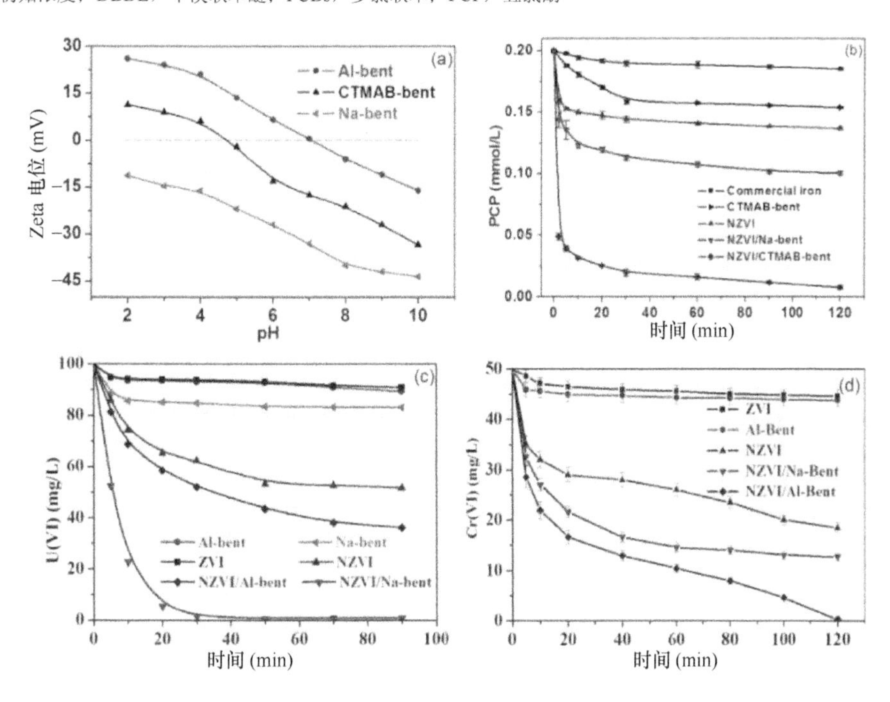

图 9-25　三种膨润土（Na-bent、Al-bent、CTMAB-bent）在不同 pH 下的 zeta 电势（a），以及膨润土负载纳米铁对 PCP（b）[472]、UO_2^{2+}（c）[469]和 Cr（VI）（d）[470]的去除效果

Cr(Ⅵ)[470]、Se(Ⅵ)[478]反应后产生的沉淀或钝化产物能部分地转移到黏土载体上。pH 降低和钝化产物转移均有助于保持零价铁表面活性、延长其反应寿命和重复使用性[478]。此外，由于黏土对环境中共存物质具有较好的吸附性能，可以提高零价铁对腐殖酸及常见阴离子［Cl^-、NO_3^-、HCO_3^-、SO_4^{2-}］的抗干

扰能力[479]。

再次，黏土可以提高负载纳米铁单位铁量对污染物的去除容量。研究表明，黏土中存在三种结合形态的Fe(II)，即矿物结构Fe(Ⅱ)、表面端羟基结合Fe(Ⅱ)和层间结合Fe(Ⅱ)。后两种结合形态的变化与介质pH有关，当pH<7.5时，主要以还原性较弱的层间结合Fe(Ⅱ)存在，当pH>7.5时，则以还原活性最强的端羟基结合Fe(Ⅱ)为主［还原能力远高于游离态Fe(Ⅱ)］。中性介质中零价铁与大多数污染物反应导致pH≥8，此时，作为纳米铁载体的黏土中的端羟基结合Fe(Ⅱ)能还原具有一定氧化性的氯代有机物、硝基苯、U(VI)、Se(VI)等。图9-26为不同pH缓冲介质中，过量Se(VI)分别与相同铁量NZVI/Al-bent 或 NZVI的反应结果。在pH = 8.0时，NZVI只去除了0.42 mmol Se(VI)、NZVI/Al-bent则能去除0.6 mmol Se(VI)，但在pH=6.0和7.0时，负载与未负载纳米铁对Se(VI)的去除量相近[478]，表明在合适pH下黏土提高了单位铁量对污染物的去除容量。

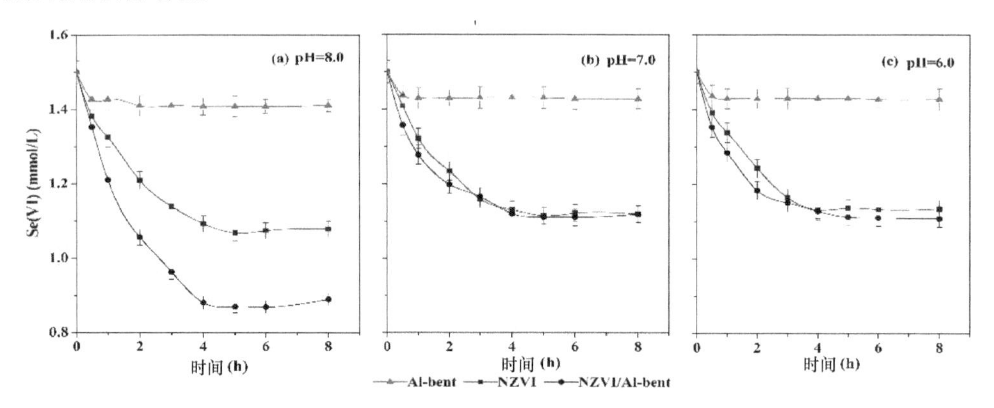

图9-26 不同pH下负载纳米铁（NZVI/Al-bent）、NZVI（铁量均为2.0 mmol/L）分别与过量Se(VI)反应结果的比较

5.1.3.4 黏土零价铁复合材料在污染物修复中的现场应用

近年来已有学者开展了黏土零价铁复合材料在现场污染物修复的研究，如Olson等[480]利用黏土零价铁复合材料（混合材料中，零价铁和钠基膨润土分别为2%和3%），平均注入深度为7.6 m现场修复位于美国北卡罗来纳州海军陆战队基地中含有氯代有机物的污染场地。结果表明，膨润土对污染物的富集固定作用有利于零价铁与污染物之间的充分接触，促进污染物的有效降解和去除，一年后，土壤和地下水中各监测点氯代有机物的平均去除率均分别达到99%和81%。Fjordbøge等[481,482]利用膨润土和零价铁复合材料现场修复位于丹麦斯库勒莱乌2号的四氯乙烯污染场地，零价铁的平均加入比例为3.1%，一年后土壤和地下水中四氯乙烯的去除率分别达到99%和76%，主要降解产物为乙烯及少量的氯乙烯等。与挖掘修复方法相比，该零价铁修复技术的成本更低。例如，修复超过3000 m^3的污染场地，该技术的平均修复费用仅为260 USD/m^3。

5.1.4 基于普通微米零价铁的污染控制技术研究进展

虽然nZVI在除污染的某些方面表现出了突出的优势，但是跟普通的ZVI相比，其价格昂贵，而且作为纳米颗粒有一定的环境风险，实际应用的前景受限。因此，要提高ZVI的实用性，需要开发其他方法。在提升ZVI除污染效能的方法中，WMF强化ZVI除污染技术（WMF-ZVI）[348]、硫化ZVI（FeS-ZVI）[348]、外加亚铁离子（Fe^{2+}-ZVI）[349]或强氧化剂（SO-ZVI）强化 ZVI 除污染方法[347]是近年来发展起来的具有较高创新性及应用前景（考虑价格、实施的难易程度）的方法。

5.1.4.1　*弱磁场强化零价铁除污染技术*

零价铁是铁磁性物质，置于磁场中会被磁化并产生感应磁场。而早在 20 世纪初法拉第就发现磁场能够影响化学反应，但长期以来有关磁场对零价铁去除污染物影响的研究却并未引起足够的重视。2013 年初，同济大学关小红教授团队在一次偶然的实验中发现，将铁/水反应体系的搅拌方式从机械搅拌变换为磁力搅拌后，零价铁去除水中 Se(IV)的反应速率得到了显著的提升[340]。进一步的实验证明零价铁反应活性的提高正是由于磁力搅拌器内置磁铁以及转子内部磁铁共同提供的弱磁场引起的。由此该团队开创了弱磁场驱动零价铁高效去除水中污染物这一研究方向，并揭开了弱磁场/零价铁水处理技术研究的序幕。

1. 弱磁场强化零价铁去除水中污染物的效能

大量的研究表明弱磁场可显著提高不同来源铁粉去除不同污染物的效能，即弱磁场的强化作用具有一定的广谱性。例如，Liang 等[340,483]发现弱磁场能够显著提高原始 7.4-mic#零价铁吸附还原去除水中 Se(IV)和 Se(VI)的性能。同样，弱磁场还能够强化 Guoyao#铁粉和 33.1-mic#铁粉分别对于水中 Cu(II)/EDTA-Cu(II)[484,485]和 Cr(VI)[486]的还原去除效果。Sun 等利用 Jinshan#铁粉发现弱磁场能显著促进零价铁的腐蚀，进而加速了 As(III)的氧化及 As(III)/As(V)的吸附共沉淀去除效率[487]。Li 等[488]发现利用弱磁场能够显著缓解高浓度 Sb(V)对零价铁的钝化效应，其不仅可将 Sb(V)去除速率提高 5.6~7.7 倍，且能将零价铁对 Sb(V)的吸附去除容量从 18.1 mg Sb(V)/g Fe 提高至 39.2 mg Sb(V)/g Fe。Xu 等[489]发现弱磁场能够将 Alfa 铁粉对 Sb(III)的去除速率常数提高 6.0~8.0 倍。此外，弱磁场还能够强化老化零价铁(AZVI)对水中金属离子［如 Se(IV)］的去除性能。一方面弱磁场能够提高不同老化时间（如 6~60 h）老化铁的反应活性[490]；另一方面对于表面覆盖了不同性质钝化膜的老化铁而言，弱磁场对其也具有活化作用[491]。

零价铁除可单独用来还原、吸附/共沉淀去除水中有毒金属离子外，还可用来活化过硫酸盐、双氧水等来氧化去除水中的有机物，而弱磁场也能够提高零价铁活化 H_2O_2/$Na_2S_2O_8$ 的氧化能力。Xiong 等[329]以对硝基酚（4-NP）作为目标污染物，系统地考察了不同铁投量、H_2O_2 投加量、pH_{ini} 以及磁场强度条件下，弱磁场对 Fe^0/H_2O_2 体系降解 4-NP 动力学的影响。结果发现，弱磁场的施加不仅将 Fe^0/H_2O_2 体系的有效 pH 工作范围从 3.0~5.0 拓宽到 3.0~6.0，还能够将 Fe^0/H_2O_2 对 4-NP 的降解速率提高了 1.5~8.2 倍。Xiang 等[492]的研究结果进一步表明，弱磁场不仅能够加速 Fe^0/H_2O_2 体系除污染的传质过程，而且还能够影响铁颗粒表面形貌、缩短反应过程的初始平台期。相比于 Fe^0/H_2O_2 体系，弱磁场对 Fe^0/$Na_2S_2O_8$ 体系降解有机物的促进作用更为显著，当加入弱磁场后，Fe^0/$Na_2S_2O_8$ 体系降解染料橙黄 G（OG）反应速率可提高 6.4~29.0 倍[493]。

为了更好地反映并比较弱磁场强化零价铁去除不同污染物作用的大小，特定义弱磁场条件下除污染反应速率常数（$k_{w/WMF\ obs}$，通常为假一级）与其在无弱磁场条件下反应速率常数（$k_{w/o\ WMF\ obs}$）的比值为弱磁场对零价铁反应活性的增效因子（R_{WMF}）。如图 9-27 所示，弱磁场对不同来源零价铁除不同污染物反应速率常数的提高倍数基本位于 1.1~100 之间。由此可见，利用弱磁场能够显著提高零价铁的反应活性，且该方法无须耗能、环境友好，是一种应用前景广阔的水污染控制技术。在实际应用中磁场的施加可能会遇到困难，但考虑到零价铁具有铁磁记忆效应，通过预磁化也有可能实现外加磁场的效果，而近来的众多研究已经证实预磁化确实可改善零价铁去除水中污染物的性能[337,494]。

2. 弱磁场提高零价铁反应活性的机理

在弱磁场/零价铁体系中，洛伦兹力（F_L）和磁场梯度力（$F_{\Delta B}$）是施加磁场后带入的两个重要作用力。其中洛伦兹力能够影响铁水体系中带电粒子的运动方向，使其在零价铁颗粒表面形成一个“扫洗”效应，从而可减小扩散层的厚度、加速相关物质的传质过程，因此理论上可能会提高零价铁与污染物的反应活性。而磁场梯度力能够作用于顺磁性的离子[如零价铁腐蚀产生的 Fe(II)、Fe(III)]，使其由磁感应强度低

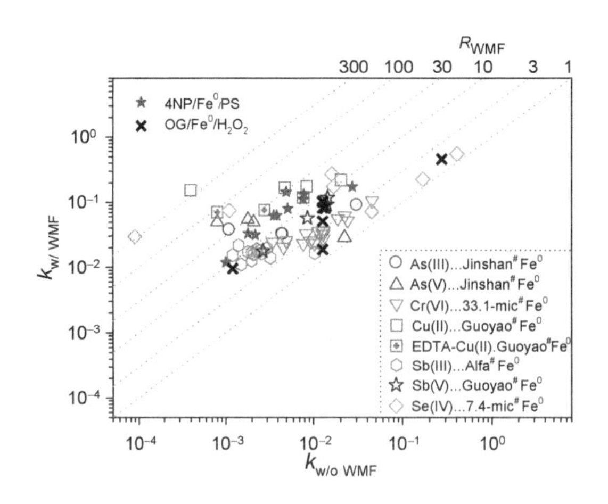

图 9-27 弱磁场对零价铁体系去除污染物反应速率的提升作用

的位置向磁感应强度较高的地方迁移，导致零价铁的腐蚀产物在零价铁颗粒表面呈现非均匀分布，进而会形成一个类似于“点蚀”的效应，从而改善或维持零价铁在除污染过程中的反应活性。

最近，研究者结合非铁磁性 Zn^0 在磁场下去除污染物的行为发现，随着磁场强度的提升（洛伦兹力随之提高）Zn^0 的反应活性并未有显著变化，这说明洛伦兹力对零价铁反应体系影响较弱。而进一步的研究表明，弱磁场强化作用与磁场梯度力的系数呈现明显的线性相关。利用同步辐射技术表征零价铁在磁场作用下的固相腐蚀产物也发现，弱磁场确实能使零价铁的腐蚀产物在零价铁颗粒表面呈现明显的非均匀分布[495]。综上，可以证实弱磁场对零价铁反应活性的强化作用机制主要与磁场梯度力有关，其作用机理可用图 9-28 表示。此外，共存阴离子也是实现弱磁场强化作用的重要前提。带正电荷 Fe(II)在磁场梯度力作用下的定向迁移将不可避免导致零价铁颗粒局部出现正电荷的积累。因此要保持体系稳定，这些积累的正电荷必须被阴离子中和以保持局部电中性。否则，Fe(II)的迁移将会受到抑制甚至终止，从而导致磁场的强化作用不能得以发挥[496]。

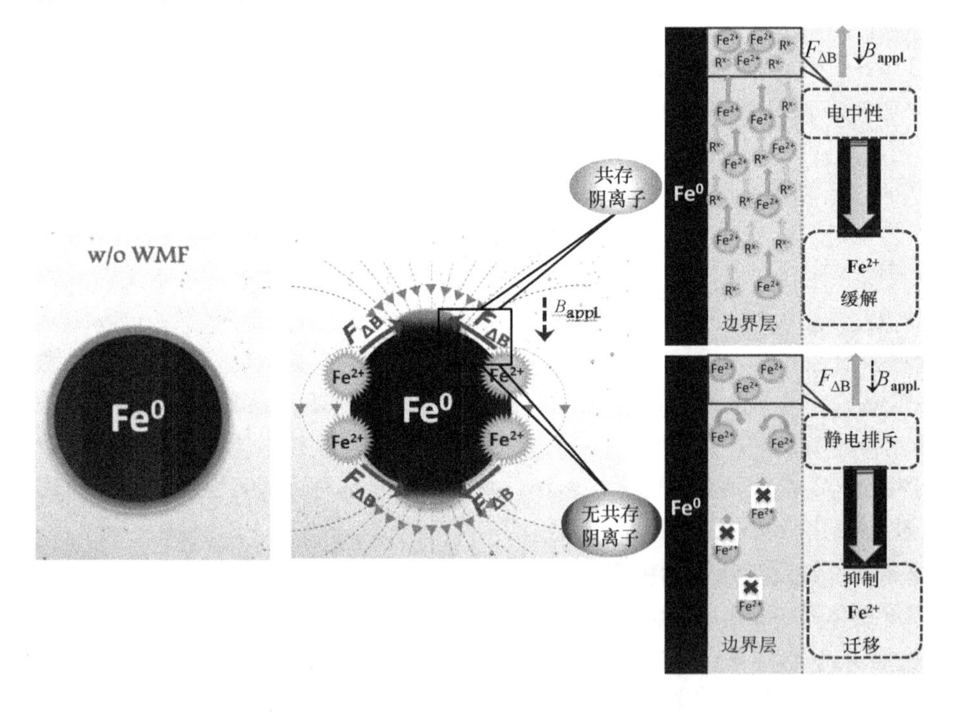

图 9-28 弱磁场提高零价铁反应活性作用机制示意图（引自参考文献[496]）

5.1.4.2　外加强氧化剂强化零价铁除污染技术

当利用零价铁的还原能力来去除污染物时，氧气作为一个氧化剂，它的存在也许会抑制零价铁还原去除污染物的效果[497]。然而，郭学军研究组在利用零价铁还原去除硝酸盐的实验中发现，氧气的存在可大大加速硝酸盐的去除[498]。受此启发，该课题组又研究了高锰酸钾、双氧水、次氯酸钠、重铬酸钾等强氧化剂对零价铁去除水中硝酸盐的效果。这些强氧化剂的加入大大加速了零价铁的腐蚀，进而提升了硝酸盐的去除效果。由此他们提出了外加强氧化剂强化零价铁除污染技术，并用以去除水中的重金属离子。与不加氧化剂相比，少量氧化剂的加入大大延迟了零价铁填充柱去除水中 As(V)、Sb(V)、Cd(II)、Hg(II)等重金属离子的穿透时间，且在长期运行过程中不易出现柱子板结堵塞[347]，是一种具有较强应用前景的零价铁活化技术。

5.1.4.3　外加 Fe^{2+}强化零价铁除污染技术

零价铁的腐蚀通常伴随着不同铁氢氧化物的生成，而这些铁氢氧化物在零价铁表面的沉积会导致零价铁的钝化。有研究表明，向零价铁系统中投加一定量的 Fe^{2+}能够有效克服零价铁的钝化，进而改善零价铁对 NO_3^-[499]、Se(VI)[500]、三氯乙烯(TCE)[501]等污染物的去除效果。产生这一现象的原因主要在于，Fe(II)的加入有利于零价铁的腐蚀产物从致钝能力较强的铁氢氧化物向具有半导体特性的四氧化三铁转变。与其他铁氧化物相比，四氧化三铁能够维持零价铁与污染物间的电子传递，因此能够促进污染物的还原去除[502]。近年来，基于 Fe(II)和 Fe_3O_4 对维持零价铁反应活性的重要作用，Tang 等又提出了同时投加 Fe^{2+}和 Fe_3O_4 的 Fe^0/Fe_3O_4/Fe(II)复合技术来进一步强化零价铁还原去除污染物的能力[503]。该方法以 Fe^0为主要电子供体，以 Fe_3O_4 为电子传递介质及污染物主要反应界面，并以 Fe(II)来调控零价铁生成的腐蚀产物主要为 Fe_3O_4。

5.1.4.4　硫化零价铁除污染技术

硫是零价铁中的常见杂质。自然环境中也常存在一些无机和有机含硫化合物，其会因微生物（如硫酸盐还原菌）作用而在零价铁表面形成一层 FeS 膜。有研究表明，铁粉中的硫杂质或零价铁表面自然形成的 FeS 膜能够改善零价铁与污染物间的电子传递因而能提高零价铁去除四氯化碳（CT）、TCE 等污染物的反应活性[504,505]。但是通常情况下铁粉中硫杂质的含量较低或微生物还原过程较慢，为了取得更好的效果，人们提出了利用不同硫化方法对铁粉进行预处理[506,507]。例如，Xu 等利用 Na_2S 处理零价铁后显著提高了橙黄 I 的去除效果[506]。最近，何锋研究组将单质硫与铁粉混合并通过球磨法制备出了具有优异性能的硫化零价铁，推动了该技术的实际应用。

5.2　基于树脂负载的纳米铁氧化物的污染控制技术研究进展

铁氧化物纳米颗粒（nanoparticles，NPs）对水中多种污染物如重金属、砷、磷具有去除容量大、选择性高等特点，在水污染深度治理领域引起了广泛关注，但在规模化水处理应用时面临难以操作、团聚失活、流失与二次污染等技术瓶颈。为了克服以上瓶颈，研究人员常将铁氧化物 NPs 负载到毫米级多孔载体内制备成复合材料，从而实现 NPs 高反应活性与载体易操作性的有机结合。近年来，借助毫米级聚苯乙烯树脂孔结构丰富、化学性质稳定、机械强度高、易进行表面修饰等特点所制备而成的树脂基纳米复合材料（resin-supported nanocomposites，RNCs），已成功应用于污废水深度治理的实际工程中，取得了良好的效果，也为环境纳米技术从实验室研究走向工程化应用提供了重要的技术借鉴。

5.2.1 RNCs 的构效调控

载体的孔结构与铁氧化物 NPs 的尺寸与形貌直接相关，进而对 RNCs 除污性能产生显著影响。2016 年，Li 等[508]以氯甲基化聚苯乙烯（即氯球）为前体，通过傅克后交联反应与季铵化反应制得三种季铵基含量接近（~1.17 mmol/g），而孔结构不同的树脂载体（比表面积 29~113 m^2/g），之后在其孔内原位生长负载等量的纳米水合氧化铁（HFO），以研究孔结构对 As(V)去除性能的影响。结果显示，随着载体孔径变小，HFO NPs 的平均尺寸由 31.4 nm 降低到 11.6 nm，表面净电荷密度(Q_H)随之升高，从而表现出对 As(V)更高的吸附容量。2017 年，Zhang 等[509]为了定量描述载体孔结构的影响，选用了孔结构均匀有序的介孔硅材料 MCM-41（3.0 nm）、SBA-15（5.7 nm）与 MSU-F（15.7nm）为载体，负载氧化铁 NPs 用于 As(V)的去除。研究发现，负载型氧化铁 NPs 的尺寸与载体孔径高度一致；随着载体孔径变小，氧化铁 NPs 的表面位点密度从 1.4 mmol/g-Fe 提升至 2.1 mmol/g-Fe，对 As(V)的最大吸附量从 24 mg/g-Fe 提升至 74 mg/g-Fe，表面位点的利用率从 20%左右提高到 40%以上。2015 年，Nie 等[510]结合表面络合模型、酸碱滴定与吸附试验，认为载体孔结构主要通过以下几种途径影响 RNCs 的净化性能：①载体孔道的大小影响负载型 HFO NPs 的大小。随着载体孔径的减小，负载的 HFO 粒径也变小。②小尺寸 HFO NPs 拥有更多的活性位点，对 As(III)和 As(V)的吸附亲和性更强，从而具备更强的净化能力。

为了克服传统树脂载体孔结构无序、孔分布宽、难以有效实现 RNCs 构效调控等不足，2017 年 Zhang 等开发了一种“闪速冷冻法”用于制备孔结构丰富、孔径均匀的树脂载体，并在载体孔内通过原位沉积法负载 FeOOH NPs[509]。如图 9-29 所示，聚苯乙烯（PS）溶液在迅速冷冻的过程中发生相分离，溶剂分子（*N*, *N*-二甲基甲酰胺，DMF）发生微纳结晶，形成纳米晶体充当致孔剂；将纳米晶体去除后即可形成永久性孔结构，获得均孔树脂 MesoPS；通过孔内原位沉积反应即可获得对应的纳米复合材料 Fe@MesoPS。调节 PS 溶液浓度，分别获得孔径为 7.9 nm、9.5 nm、12 nm 与 26 nm 的四种均孔树脂，通过研究发现：①载体孔结构可显著影响 NPs 尺寸，所得 Fe@MesoPS 中 FeOOH NPs 粒径均匀，随着孔径变小，从 7.3 nm 降低到 2.0 nm；②载体孔径变小，FeOOH NPs 零电荷点（pH_{zpc}）随之上升，对 As(V)的吸附量与吸附亲和力均随之上升。

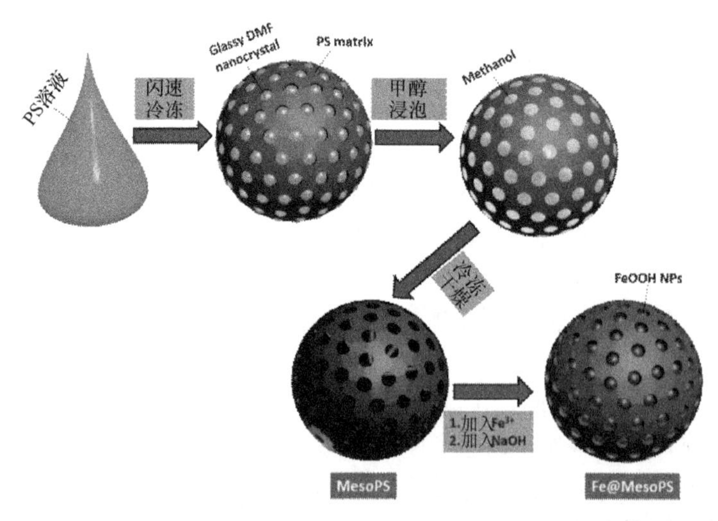

图 9-29 均孔树脂 MesoPS 及其复合材料 Fe@MesoPS 制备示意图

5.2.2 RNCs 去除典型污染物的机理

RNCs 的结构决定了其去除污染物的基本途径。总体而言，RNCs 中的铁基 NPs 为污染物的净化提供直接的反应位点。目前研究者采用 XAS、XPS、高分辨 TEM 等多种手段详细分析表征了铁基 NPs 去除不同污染物的机理，但关于限域孔内 NPs 的污染物净化作用知之甚少，这主要缘于相关原位表征技术的缺失。2013 年 Nie 等[511]以非极性树脂氯球（CMPS）为载体，制备了载 HFO 纳米复合材料（HFO@CMPS），通过不同 pH（3~11）、离子强度下的酸碱滴定实验研究了复合吸附剂的表面酸碱性质，并采用恒定电容模型（CCM）描述 HFO@CMPS 复合材料表面酸碱和吸附络合反应中的静电性质，结果表明 CCM 模型能够较好地描述 HFO@CMPS 的表面酸碱性质；Nie 等[511]进一步考察了 pH 和离子强度对 HFO@CMPS 吸附 Cu、Pb、Ni、As、P 等无机离子的影响，发现 CCM 模型可以很好地拟合 pH 和离子强度对 HFO-CMPS 吸附的影响规律，继而检验了 CCM 模型对多种离子共存体系中 HFO@CMPS 竞争吸附的预测能力。三种重金属离子（Cu、Pb、Ni）的浓度均为 2×10^{-5} mol/L，As 和 P 的初始浓度为 1.36×10^{-4} mol/L 和 1.62×10^{-4} mol/L。如图 9-30 所示，CCM 模型较准确地预测了 Cu、Pb、As 的吸附曲线，仅在较低的平衡 pH 时对 Ni 和 P 的吸附稍有低估。这表明 CCM 模型是研究 RNCs 净化重金属等污染物性能的有效工具。基于模型分析发现负载后的 HFO NPs 表现出了与负载前截然不同的酸碱性质，其滴定曲线只受到离子强度微弱的影响，而负载前的 NPs 的滴定曲线与离子强度密切相关。这可能是由于纳米孔限域效应导致离子强度对 NPs 表面的双电层压缩效应减弱甚至消失。相应的，负载前 HFO NPs 的 pH_{pzc} 约为 8.2，而负载后则降低至约 6.3。进而发现，负载后 HFO NPs 表面负电荷形态相对比例增加，对 H^+的亲和性减弱，而对 OH^-的亲和性增强。这种特性变化强化了负载后 NPs 对 Cu(II)和 As(V)的吸附性能。

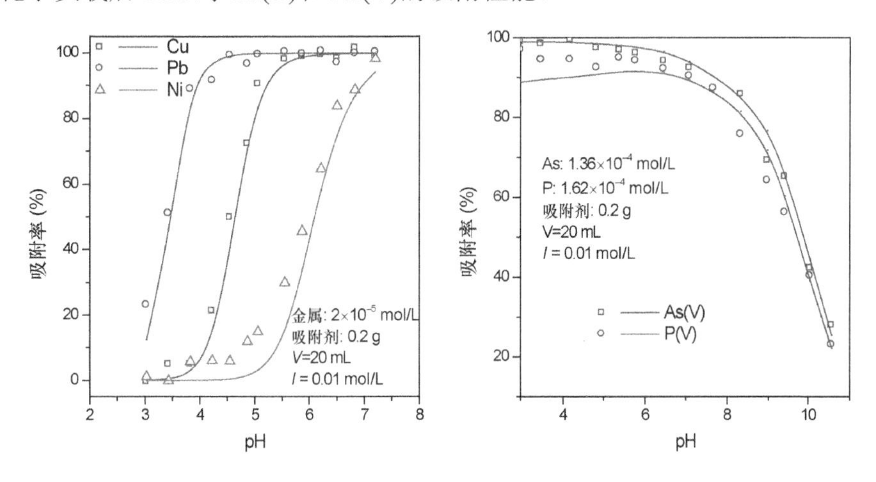

图 9-30　表面络合模型 CCM 对复合材料除相关污染物的行为模拟（引自参考文献[511]）

5.2.3　RNCs 功能复合化

由于污废水中往往存在多种污染物，研究人员常寻求 RNCs 材料多功能的复合，以提高该类材料的水处理效果效果。目前，RNCs 功能复合的研究主要可分为以下两类：

（1）多种污染物吸附功能的复合。目前绝大部分吸附剂难以有效实现有机与无机污染物的同步去除。已有研究表明氨基化超高交联聚苯乙烯树脂吸附剂 NDA802 可有效去除废水中的有机物[512]，HFO NPs 也已被证明可有效去除无机污染物。2015 年 Yang 等[512]以 NDA802 为载体，将 HFO NPs 负载于 NDA802 孔道内，研制新型纳米复合材料 HFO@NDA802 用于无机污染物（以磷酸盐为代表）与有机污染物（以对硝基酚 PNP 为代表）的同步去除。HFO@NDA802 与不同污染物的作用机制如图 9-31 所示，磷酸根通过静电作用及内核配位作用被吸附；PNP 通过微孔填充作用、π-π 共轭作用以及酸碱作用等实现吸附。研究结果表明，在循环吸附-再生过程中，仅第二个批次 HFO@NDA802 对磷的去除性能有所下降，随后的九个批次循环处理中均保持稳定；连续十批次的循环试验中复合材料对 PNP 的吸附性能维持稳定。固定床吸附结果也表明 HFO@NDA802 可同步、高效去除磷酸盐与 PNP。

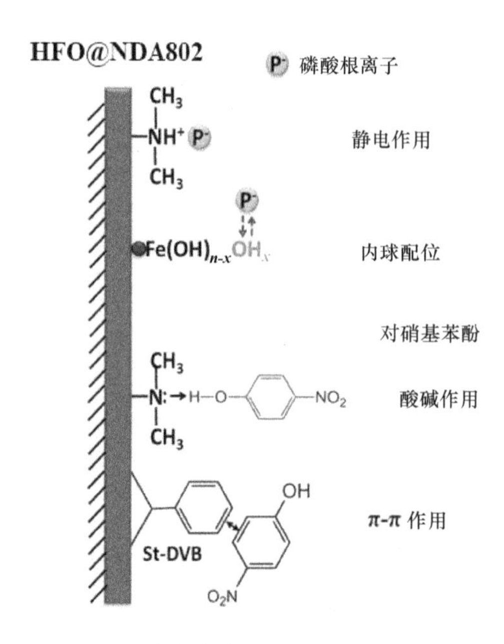

图 9-31　HFO@NDA802 同步去除磷酸盐及 PNP 的机理示意图（引自参考文献[512]）

（2）氧化还原功能与吸附功能的复合。地下水中的 As 常以 As(III)存在，迁移性强、毒性大、处理难度高，单纯依靠吸附作用难以实现对该类污染物的深度去除。2017 年 Zhang 等[513]在聚苯乙烯树脂骨架上共价键联活性氯，并在树脂孔道内原位沉积 HZO NPs 制得兼具氧化-吸附功能的复合材料 HFO@PS-Cl。如图 9-32 所示，HFO@PS-Cl 主要通过两种途径实现 As(III)的高效去除，即通过活性氯将 As(III)氧化成 As(V)后被 HFO NPs 吸附或 As(III)被 HFO NPs 吸附后再被活性氯氧化。HFO@PS-Cl 可直接将水中 As(III)的含量从 1000 μg/L 降至 10 μg/L 以下，处理容量相比现有商品化除砷纳米复合材料，HFO@D201 提升 2 倍以上。此外，使用后 HFO@PS-Cl 的氧化-吸附能力可实现高效再生，循环使用 7 次以上处理性能仍保持稳定。该材料结构稳定，对水中共存离子、天然有机物等抗干扰能力强，具有良好的实际应用前景。与之类似，2012 年 Li 等[514]研发树脂基 Fe/Mn 双金属纳米氧化物，2013 年 Du 等[515]构建 nZVI@D201/O_2 体系，均用于 As(III)的氧化-吸附去除，并取得了较好的效果。此外，2017 年 Shan 等[516]利用 nZVI@D201 实现了还原与吸附功能的复合，对 Se(VI)表现出较好的还原-吸附性能。

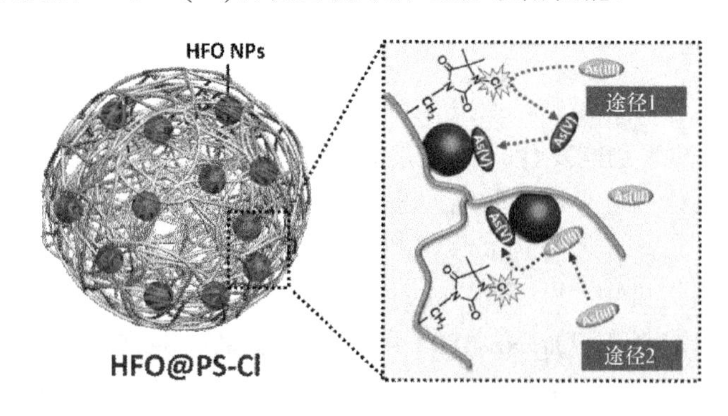

图 9-32　HFO@PS-Cl 去除 As(III)机理示意图（引自参考文献[513]）

5.2.4 RNCs 应用探索研究进展

21 世纪以来，RNCs 已逐步应用于工业废水、生活污水和饮用水的深度处理。近五年来关于铁基 RNCs 实际应用的研究报道主要涉及含砷矿冶废水与生化尾水深度处理。

（1）含砷矿冶废水治理。很多矿冶废水中常含有大量的 As(V)，常见处理技术如共沉淀法虽有较好处理效率，但难以满足深度处理的要求［出水 As(V)<0.5 mg/L］。2013 年 Hua 等[517]开发了共沉淀-纳米吸附的组合工艺处理高浓度含 As(V)废水，其工艺流程如图 9-33 所示。所用废水取自某钨冶炼企业冶炼生产中的离子交换过程，强碱性、含有大量杂质离子，且砷含量高达 15.2 mg/L。Hua 等通过优化共沉淀段与 HFO@D201 纳米复合材料深度净化工艺的联用，可使出水中 As(V)的浓度低于 0.5 mg/L，实现稳定达标排放。该研究表明，共沉淀-纳米吸附组合工艺可有效用于高浓度含 As(V)矿冶废水的治理。

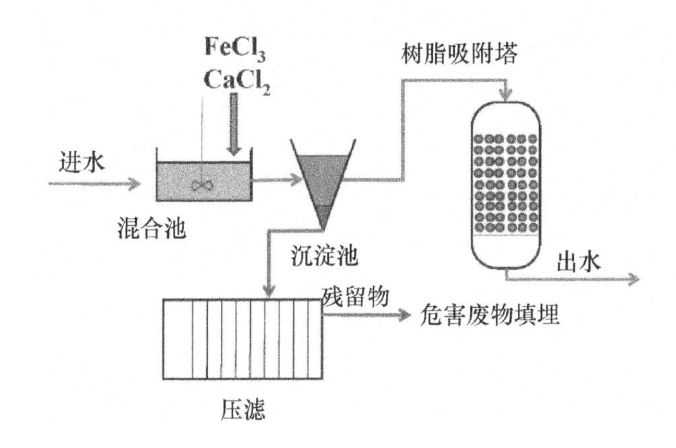

图 9-33　共沉淀-纳米吸附组合工艺处理高浓度含 As(V)废水流程示意图（引自参考文献[517]）

（2）生化尾水深度治理。生化尾水中磷含量常常超标（0.5 mg/L），过多的磷进入天然水体中容易引起水体富营养化。HFO@D201 已被证明是一种优良的除磷吸附剂。2014 年 Zhang 等[518]验证了不同条件对 HFO@D201 除磷性能的影响，利用响应曲面设计（RSM）以及人工神经网络（ANN）对其除磷性能进行了模拟，并结合遗传算法可在保证除磷效率的前提下尽可能减少 HFO@D201 用量，提高 RNCs 材料使用的经济性。该研究优化的模型可很好地模拟与预测 RNCs 处理含磷生化尾水的性能，有利于进一步推动纳米复合材料在真实废水处理中的规模化应用。此外，针对磷回收难题，2015 年 Zhang 等[519]以含磷生化尾水为对象，证明了利用 HFO@D201 吸附法结合鸟粪石法回收生化尾水中磷资源的可行性。

5.3 基于高铁酸盐的污染控制技术及原理

高铁酸盐是铁的正六价的存在形式，是一种同时具有氧化、吸附、絮凝、杀菌、消毒等多种功能的绿色强氧化剂，因独特的环境友好特性而受到人们越来越多的重视。与传统的氧化剂（O_3、Cl_2、H_2O_2、$KMnO_4$）相比，高铁酸盐具有更高的氧化还原电位，在酸性条件下氧化还原电位为 2.20 V，碱性条件下氧化还原电位为 0.70 V，能够快速氧化一些含有不饱和官能团的有机物如内分泌干扰物（EDCs）和药品与个人护理品（PPCPs）[520-523]，以及有害有毒的无机物，如 As(III)、硫化物、硫氰根、亚硝酸根等[524]。高铁酸盐氧化污染物的过程中自身被还原为 Fe(III)，而三价铁及水解产生的铁氧化物对砷、重金属等污染物具有混凝、吸附去除的作用。此外，高铁酸盐氧化不会产生有毒有害的副产物，从而克服了氯与臭氧的氧化受限于氯代副产物和溴酸盐产生的问题。近几十年来，高铁酸盐制备方法日渐成熟，产率和成品纯度不断提高，技术上已经逐渐能够满足生产和实际应用的需要。

5.3.1 高铁酸盐氧化去除污染物机理

高铁酸盐氧化去除污染物的特性是由它不同型体的氧化能力和稳定性二者的综合作用决定的。一般在中性 pH 下高铁酸盐除污染能力最强。在低 pH 条件下，FeO_4^{2-}的共轭酸（$HFeO_4^-$和 H_2FeO_4）氧化能力更强，但是自分解速度太快，强氧化能力得不到发挥；而在高 pH 条件下，FeO_4^{2-}氧化能力则相对较弱。换言之，高铁酸根的形态和氧化能力在高铁酸盐氧化除污过程中起着决定性的作用。在 Fe(VI)氧化降解污染物的过程中，伴随着污染物的氧化，Fe(VI)会被还原生成中间价态 Fe(V)和 Fe(IV)，这些中间价态铁的氧化能力非常强，与有机物的反应速率常数比 Fe(VI)本身要高 3~6 个数量级[521,524]。研究发现 Fe(VI)在实际水体中的除污染效能反而更强[520]，由于实际水体中成分复杂，Fe(VI)能和许多还原剂反应生成大量的中间价态 Fe(V)和 Fe(IV)（图 9-34），因此可以推测，实际水体的背景成分可能对 Fe(VI)除污染起到了强化作用。但是由于这些中间价态铁的稳定性很差，极易分解，例如在 pH = 7.0 时，Fe(V)的自分解常数约为 100 s^{-1}，在水中存活的半衰期 $t_{1/2}$ 只有 6.9 ms[524]。因此，中间价态铁因快速分解，其强氧化能力很难得到充分利用。

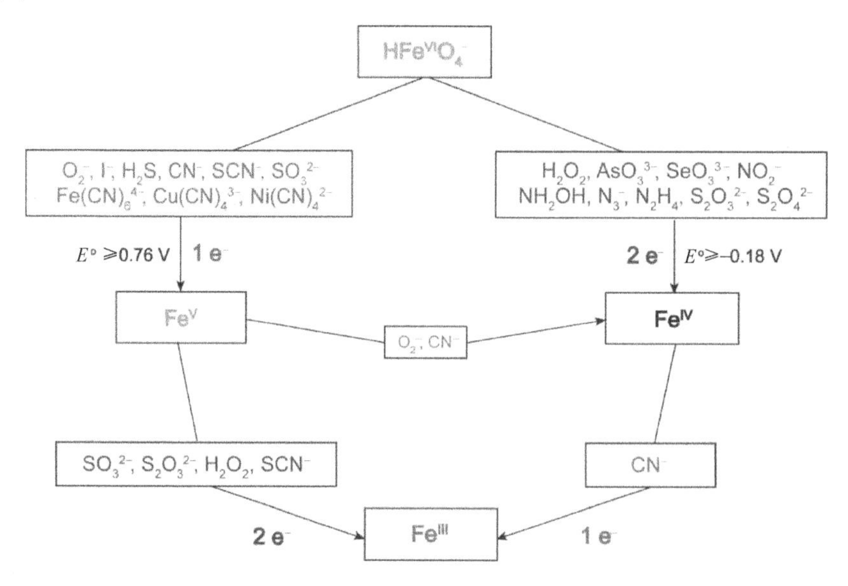

图 9-34 Fe(VI)氧化过程中的电子转移及 Fe(V)和 Fe(IV)生成情况（引自参考文献[524]）

5.3.2 高铁酸盐去除砷及重金属

Fe(VI)具有强氧化性和不稳定性，易还原、自分解生成铁（氢）氧化物。一般认为，铁（氢）氧化物通过吸附/絮凝的方式去除水中砷及重金属污染物，与 pH 直接相关的 Fe(VI)分解特性和铁（氢）氧化物表面电荷决定了其去除金属污染物的效率。近年来研究者采用多种手段详细分析表征了 Fe(VI)分解产生的铁（氢）氧化物去除金属污染物的机理。2013 年 Prucek 等[525]比较了高铁酸盐原位和异位产生的铁（氢）氧化物除砷效率，发现原位除砷效率大大高于异位除砷；并采用 XRD、高分辨 XPS 和穆斯堡尔谱表征了铁氧化物的物质结构和砷的形态、分布，结果表明，在高铁酸盐原位分解除砷过程中，砷不仅吸附在表面，还有大量的砷嵌入到了 γ-Fe_2O_3 尖晶石结构中，从而提高了除砷效率，而异位除砷过程中，砷仅仅分布在 γ-Fe_2O_3 的表面（图 9-35）。Prucek 等[526]进一步研究了高铁酸盐对多种重金属[Cd(II)、Co(II)、Ni(II)、Cu(II)]在不同[Fe(VI)]/[M(II)]条件下的去除效果，结果发现，在 pH = 6.6、[Fe(VI)]/[M(II)]= 2∶1 时，Co(II)、Ni(II)、Cu(II)几乎完全去除，而对于 Cd(II)，在[Fe(VI)]/[M(II)]= 15∶1 时去除率仅仅达到 75%；进一步的研究发现，Co(II)、Ni(II)、Cu(II)更易于形成相应的 MFe_2O_4 尖晶石，也更易于嵌入到 γ-Fe_2O_3 结构中，而

Cd(II)仅仅吸附在 γ-Fe_2O_3 的表面，推测重金属离子半径和电子结构可能显著影响了重金属去除效率及其化学行为。高铁酸盐去除砷及重金属后的产物主要以磁性的 γ-Fe_2O_3 为主，易于从水中分离，而且金属离子嵌入在 γ-Fe_2O_3 结构中不易于释放分离，因此高铁酸盐是一种具有较强应用前景的处理技术。

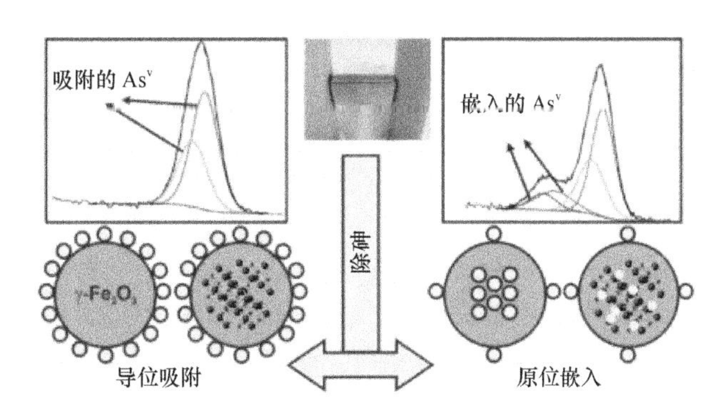

图 9-35　高铁酸盐原位和异位除砷的机理示意图（引自参考文献[525]）

5.3.3　高铁酸盐消毒杀菌作用

高铁酸盐由于其强氧化性，对水中的微生物也具有较强的灭活作用，它能破坏细菌的细胞壁、细胞膜以及细胞结构中的酶，抑制蛋白质及核酸的合成，阻碍菌体的生长和繁殖，从而起到杀死细菌的作用。Murmann 与 Robinson[527]在 1974 年首次发现高铁酸钾具有明显的灭菌作用，用 6 mg/L 的高铁酸钾处理原水 30 min，可将水中 20 万~30 万个/mL 的细菌去除至小于 100 个/mL。之后 Gilbert 等[528]的研究表明，在 pH=8.0、8.2 和 8.5 的条件下，6 mg/L 的高铁酸钾分别在 8.5 min、7.2 min 和 6.4 min 内可灭活 99%的大肠杆菌，当 pH 低于 8.0 时，高铁酸盐的消毒能力随 pH 降低明显增强。高铁酸盐浓度、pH 和接触时间是其灭菌的主要影响因素，pH 会影响高铁酸盐的稳定性和氧化能力，综合二者来看，一般认为在中性偏酸性条件下杀菌效果最好。微生物的表面电荷也会对灭菌带来一定的影响：带正电的微生物对高铁酸盐的抵抗性强于带负电的微生物。同时，高铁酸盐分解生成的中间价态氧化成分具有较强的氧化效应，最终生成的 $Fe(OH)_3$ 胶体对细菌也有一定的吸附去除作用。而且一般认为，高铁酸盐氧化不会产生有毒有害的副产物如氯代副产物和溴酸盐，因此可替代氯和臭氧用作消毒剂。Jiang 等[529]在利用高铁酸盐氧化溴离子的实验中却发现，在磷酸盐缓冲体系中没有溴酸盐的生成，但是在硼酸盐缓冲体系中却检测到了溴酸盐，这一发现为高铁酸盐消毒实际应用提供了重要的技术借鉴。

6 展　　望

利用铁循环诱导绿色友好环境治理和修复技术已成为水/土壤污染控制领域的热点话题，研究铁循环及其环境效应更是揭开红壤非生物地球化学循环奥秘的国际研究前沿和关键点。环境铁循环研究面临的关键问题是在原子水平上阐明[(氢)氧化]铁表/界面上发生的吸/脱附原理、氧化还原过程以及电子转移途径，并深入揭示其环境效应。

同时，铁循环污染控制其生物地球化学过程与碳/氮养分循环、重金属/有机氯等污染物转化过程密切相关。在强烈的人为活动与气候条件下，阐明地表土壤和水体中铁循环过程与环境生态功能，通过铁循环调控实现环境治理和修复，并形成以铁循环为核心的与重要物质循环过程相互关联的新方向，具有重要的科学价值。

相比于三价铁矿物而言，沉积物中的二价铁矿物在调节污染物迁移转化中表现出更高的反应活性，因此在环境科学领域受到了很高程度的关注。地下环境不同类型二价铁矿物还原有机污染物和还原/固定重金属，目前仍然是一个研究热点。但是，目前对于不同类型二价铁矿物活性差异的规律及其内在机制的认识仍不十分清楚。相比于三价铁矿物而言，沉积物中的二价铁矿物在调节污染物迁移转化中表现出更高的反应活性，因此在环境科学领域受到了很高程度的关注。由于二价铁矿物主要存在于无氧的还原环境，因此过去对无氧条件下二价铁矿物引起的还原效应研究的较多。地下环境不同类型二价铁矿物还原有机污染物和还原/固定重金属，目前仍然是一个研究热点。但目前对于不同类型二价铁矿物活性差异的规律及其内在机制的认识仍不十分清楚。

近年来人们对微生物与含铁矿物之间电子传导机理的认识得到了显著的提高。这主要体现在对从细胞质膜、跨越膜套传导到胞外含 Fe(III)矿物的电子传导机理的认识较为清楚，同时鉴定出几种重要的蛋白质，一部分蛋白质还获得了详细的特征解析。然而，研究者仍不清楚反向电子传导即电子从胞外电极传导到胞内受体的机理，并且无法理解电子如何从胞外含 Fe(II)矿物传导到胞内的末端电子受体，以及该过程中能量的保存方式。这些严重制约了微生物电合成技术成为主要的生物制造方式。同时，微生物跨外膜传导电子的分子机理尚不得而知。因此，从分子水平上认识这些电子传导过程的机理，能为在生物技术领域应用这些具备胞外电子传导能力的微生物提供理论指导。特别是为利用微生物电合成技术将温室气体 CO_2 转化成各种化合物以及开发新型的生物燃料提供理论和实际指导。

零价铁作为一种廉价且环境友好的还原剂，已被广泛应用于浅层地下水修复及难降解污染物的预处理。然而，零价铁在运行中易锈蚀板结，影响其在实际废水处理中的应用。如将零价铁应用于厌氧污水处理系统，可以减缓其锈蚀，并强化厌氧微生物的代谢。

参 考 文 献

[1] Raiswell R, Canfield D E. The iron biogeochemical cycle past and present. Geochem. Perspect, 2012, 1 (1): 1-2.

[2] Schweitzer G K, Pesterfield L L. The aqueous chemistry of the elements. Oxford:Oxford University Press, Inc. 2010.

[3] Tratnyek P G, Grundl T J, Haderlein S B. Aquatic redox chemistry. ACS Publications, 2011.

[4] Silver J. Chemistry of iron. Springer, 1993.

[5] Boyd P W, Ellwood M J. The biogeochemical cycle of iron in the ocean. Nat. Geosci, 2010, 3 (10): 675-682.

[6] Conway T M, John S G. Quantification of dissolved iron sources to the North Atlantic Ocean. Nature, 2014, 511 (7508): 212-215.

[7] Muller F L L, Cuscov M. Alteration of the copper-binding capacity of iron-rich humic colloids during transport from peatland to marine waters. Environ. Sci. Technol., 2017, 51 (6): 3214-3222.

[8] Garg S, Ito H, Rose A L, Waite T D. Mechanism and kinetics of dark iron redox transformations in previously photolyzed acidic natural organic matter solutions. Environ. Sci. Technol., 2013, 47 (4): 1861-1869.

[9] Garg S, Jiang C, Miller C J, Rose A L, Waite T D. Iron redox transformations in continuously photolyzed acidic solutions containing natural organic matter: Kinetic and mechanistic insights. Environ. Sci. Technol., 2013, 47 (16): 9190-9197.

[10] Garg S, Jiang C, Waite T D. Mechanistic insights into iron redox transformations in the presence of natural organic matter: Impact of pH and light. Geochim. Cosmochim. Ac, 2015, 165: 14-34.

[11] Jiang C, Garg S, Waite T D. Hydroquinone-mediated redox cycling of iron and concomitant oxidation of hydroquinone in oxic waters under acidic conditions: Comparison with iron-natural organic matter interactions. Environ. Sci. Technol., 2015, 49 (24): 14076-14084.

[12] Rose A L, Waite T D. Kinetics of iron complexation by dissolved natural organic matter in coastal waters. Mar. Chem., 2003, 84(1-2): 85-103.

[13] Rose A L, Waite T D, Effect of dissolved natural organic matter on the kinetics of ferrous iron oxygenation in seawater. Environ. Sci. Technol., 2003, 37 (21): 4877-4886.

[14] Rose A L, Waite T D. Reduction of organically complexed ferric iron by superoxide in a simulated natural water. Environ. Sci. Technol., 2005, 39 (8): 2645-2650.

[15] Wang K, Garg S, Waite T D. Redox transformations of iron in the presence of exudate from the cyanobacterium microcystis aeruginosa under conditions typical of natural waters. Environ. Sci. Technol., 2017, 51 (6): 3287-3297.

[16] Sima J, Makanova J. Photochemistry of iron(III) complexes. Coordin. Chem. Rev, 1997, 160: 161-189.

[17] Yuan X, Davis J A, Nico P S. Iron-mediated oxidation of methoxyhydroquinone under dark conditions: Kinetic and mechanistic insights. Environ. Sci. Technol., 2016, 50(4): 1731-1740.

[18] Aeschbacher M, Sander M. Schwarzenbach R P. Novel electrochemical approach to assess the redox properties of humic substances. Environ. Sci. Technol., 2010, 44 (1): 87-93.

[19] Ratasuk N, Nanny M A. Characterization and quantification of reversible redox sites in humic substances. Environ. Sci. Technol., 2007, 41 (22): 7844-7850.

[20] Aeschbacher M, Graf C, Schwarzenbach R P, Sander M. Antioxidant properties of humic substances. Environ. Sci. Technol., 2012, 46 (9): 4916-4925.

[21] Senesi N. Molecular and quantitative aspects of the chemistry of fulvic-acid and its interactions with metal-ions and organic chemicals .1. the electron spin resonance approach. Anal. Chim. Acta, 1990, 232 (1): 51-75.

[22] Senesi N, Schnitzer M. Effects of pH, reaction-time, chemical reduction and irradiation on ESR spectra of fulvic acid. Soil Sci, 1977, 123 (4): 224-234.

[23] Paul A, Stosser R, Zehl A, Zwirnmann E, Vogt R D, Steinberg C E W. Nature and abundance of organic radicals in natural organic matter: Effect of pH and irradiation. Environ. Sci. Technol., 2006, 40 (19): 5897-5903.

[24] Cory R M, McKnight D M. Fluorescence spectroscopy reveals ubiquitous presence of oxidized and reduced quinones in dissolved organic matter. Environ. Sci. Technol., 2005, 39 (21): 8142-8149.

[25] Fimmen R L, Cory R M, Chin Y P, Trouts T D, McKnight D M. Probing the oxidation-reduction properties of terrestrially and microbially derived dissolved organic matter. Geochim. Cosmochim. Ac, 2007, 71 (12): 3003-3015.

[26] Klapper L, McKnight D M, Fulton J R, Blunt-Harris E L, Nevin K P, Lovley D R, Hatcher P G. Fulvic acid oxidation state detection using fluorescence spectroscopy. Environ. Sci. Technol., 2002, 36 (14): 3170-3175.

[27] Rose A L, Waite T D, Effect of dissolved natural organic matter on the kinetics of ferrous iron oxygenation in seawater. Environ. Sci. Technol., 2003, 37 (21): 4877-4886.

[28] Pham A N, Waite T D. Modeling the kinetics of Fe(II) oxidation in the presence of citrate and salicylate in aqueous solutions at pH 6.0-8.0 and 25 degrees. J. Phys. Chem. A, 2008, 112 (24): 5395-5405.

[29] Sima J, Makáňová J. Photochemistry of iron(III) complexes. Coord. Chem. Rev, 1997, 160: 161-189.

[30] Wang Z, Ma W, Chen C, Zhao J. Photochemical coupling reactions between Fe(III)/Fe(II), Cr(VI)/Cr(III), and polycarboxylates: inhibitory effect of Cr species. Environ. Sci. Technol., 2008, 42 (19): 7260-7266.

[31] Wang Z, Chen C, Ma W, Zhao J. Photochemical coupling of iron redox reactions and transformation of low-molecular-weight organic matter. J. Phys. Chem. Lett., 2012, 3 (15): 2044-2051.

[32] Wang Z, Song W, Ma W, Zhao J. Environmental photochemistry of iron complexes and their involvement in environmental chemical processes. Prog. In. Chem., 2012, 24 (2-3): 423-432.

[33] Al-Abadleh H A. Review of the bulk and surface chemistry of iron in atmospherically relevant systems containing humic-like substances. RSC Adv., 2015, 5 (57): 45785-45811.

[34] Herrmann H, Schaefer T, Tilgner A, Styler S A, Weller C, Teich M, Otto T. Tropospheric aqueous-phase chemistry: kinetics, mechanisms, and its coupling to a changing gas phase. Chem. Rev., 2015, 115 (10): 4259-4334.

[35] Lopes L, De, L J, Legube B. Charge transfer of iron(III)monomeric and oligomeric aqua hydroxo complexes: semiempirical investigation into photoactivity. Inorg. Chem., 2002, 41 (9): 2505.

[36] Benkelberg H J, Warneck P. Photodecomposition of iron(III)hydroxo and sulfato complexes in aqueous solution: Wavelength dependence of OH and SO_4^- quantum yields. J. Phys. Chem., 1995, 99 (14): 5214-5221.

[37] Wang Z, Liu J. New insight into photochemical oxidation of Fe(II): The roles of Fe(III) and reactive oxygen species. Catal. Today, 2014, 224: 244-250.

[38] Kotronarou A, Sigg L. SO_2 oxidation in atmospheric water: Role of Fe(II) and effect of ligands. Environ Sci. Technol., 1993, 27: 2725-2735.

[39] Zuo Y, Zhan J, Wu T. Effects of monochromatic UV-visible light and sunlight on Fe(III)-catalyzed oxidation of dissolved sulfur dioxide. J. Atmos. Chem., 2005, 50 (2): 195-210.

[40] Guo Y, Lou X, Fang C, Xiao D, Wang Z, Liu J. Novel photo-sulfite system: Toward simultaneous transformations of inorganic and organic pollutants. Environ. Sci. Technol., 2013, 47 (19): 11174-11181.

[41] Chebbi A, Carlier P. Carboxylic acids in the troposphere, occurrence, sources, and sinks: A review. Atmos. Environ., 1996, 30 (24): 4233-4249.

[42] Weller C, Horn S, Herrmann H. Effects of Fe(III)-concentration, speciation, excitation-wavelength and light intensity on the quantum yield of iron(III)-oxalato complex photolysis. J. Photochem. Photobiol A: Chem., 2013, 255: 41-49.

[43] Weller C, Horn S, Herrmann H. Photolysis of Fe(III) carboxylato complexes: Fe(II) quantum yields and reaction mechanisms. J. Photochem. Photobiol A: Chem., 2013, 268: 24-36.

[44] Faust B C, Zepp R G. Photochemistry of aqueous iron(III)-polycarboxylate complexes: roles in the chemistry of atmospheric and surface waters. Environ. Sci. Technol., 1993, 27 (27): 2517-2522.

[45] Chen J, Zhang H, Tomov I V, Rentzepis P M. Electron transfer mechanism and photochemistry of ferrioxalate induced by excitation in the charge transfer band. Inorg. Chem., 2008, 47 (6): 2024-2032.

[46] Pozdnyakov I P, Kel O V, Plyusnin V F, Grivin V P, Bazhin N M. New insight into photochemistry of ferrioxalate. J. Phys. Chem. A, 2008, 112 (36): 8316-8322.

[47] Mangiante D M, Schaller R D, Zarzycki P, Banfield J F, Gilbert B. Mechanism of ferric oxalate photolysis. ACS Earth and Space Chemistry, 2017, 1 (5):270-276.

[48] Wang Z, Xiao D, Liu J. Diverse redox chemistry of photo/ferrioxalate system. RSC Adv., 2014, 4 (84): 44654-44658.

[49] Wang Z, Chen, X, Ji H, Ma W, Chen C, Zhao J. Photochemical cycling of iron mediated by dicarboxylates: Special effect of malonate. Environ. Sci. Technol., 2010, 44 (1): 263-268.

[50] Xiao D, Lou X, Liu R, Guo Y, Zhou J, Fang C, Wang Z, Liu J. Fe-catalyzed photoreduction of Cr(VI) with dicarboxylic acid(C2–C5): Divergent reaction pathways. Desalin. Water Treat., 2014, 56 (4): 1020-1028.

[51] Xiao D, Guo Y, Lou X, Fang C, Wang Z, Liu J. Distinct effects of oxalate versus malonate on the iron redox chemistry: Implications for the photo-Fenton reaction. Chemosphere, 2014, 103: 354-358.

[52] Chen J, Zhang H, Tomov I V, Rentzepis P M. Electron transfer mechanism and photochemistry of ferrioxalate induced by excitation in the charge transfer band. Inorg. Chem., 2008, 47 (6): 2024-2032.

[53] Pozdnyakov I P, Kel O V, Plyusnin V F, Grivin V P, Bazhin N M. New insight into photochemistry of ferrioxalate. J. Phys.Chem. A, 2008, 112 (36): 8316–8322.54.

[54] Wang Z, X D, Liu R, Guo Y, Lou X, Liu J. Fenton-ike degradation of reactive dyes catalyzed by biogenic jarosite. J. Adv. Oxidation Technol., 2014, 17 (1): 104–108.

[55] Wang Z, Ma W, Chen C, Zhao J. Light-assisted decomposition of dyes over iron-bearing soil clays in the presence of H_2O_2. J. Hazard. Mater., 2009, 168 (2): 1246-1252.

[56] Xiao D X, Guo Y G, Lou X Y, Fang C L, Wang Z H, Liu J S. Distinct effects of oxalate versus malonate on the iron redox

chemistry: Implications for the photo-Fenton reaction. Chemosphere, 2014, 103: 354-358.

[57] Tan L, Liang B, Fang Z, Xie Y, Tsang E P. Effect of humic acid and transition metal ions on the debromination of decabromodiphenyl by nano zero-valent iron: Kinetics and mechanisms. J. Nanopart. Res., 2014, 16: 2786.

[58] 易云强, 吴娟, 方战强. 天然有机质在生物炭负载纳米镍铁降解十溴联苯醚过程中的影响作用机理辨识. 化学学报, 2017, 75(6): 629-636.

[59] Murad E, Fischer W R. The geobiochemical cycle of iron. *In*: Stucki J W, Goodman B A, Schwertmann U, Eds. Iron in Soils and Clay Minerals. Springer Netherlands: Dordrecht, 1988, 1-18.

[60] Borch T, Kretzschmar R, Kappler A, Van Cappellen P, Ginder-Vogel M, Voegelin A, Campbell K. Biogeochemical redox processes and their impact on contaminant dynamics. Environ. Sci. Technol., 2010, 44 (1): 15-23.

[61] Li Y, Yu S, Strong J, Wang H. Are the biogeochemical cycles of carbon, nitrogen, sulfur, and phosphorus driven by the "Fe-III-Fe-II redox wheel" in dynamic redox environments? J. Soil. Sediment, 2012, 12 (5): 683-693.

[62] Melton E D, Swanner E D, Behrens S, Schmidt C, Kappler A. The interplay of microbially mediated and abiotic reactions in the biogeochemical Fe cycle. Nature Rev. Microbiol., 2014, 12 (12): 797-808.

[63] Gong Y, Tang J, Zhao D. Application of iron sulfide particles for groundwater and soil remediation: A review. Water Res., 2016, 89: 309-320.

[64] Bishop M E, Glasser P, Dong H, Arey B, Kovarik L. Reduction and immobilization of hexavalent chromium by microbially reduced Fe-bearing clay minerals. Geochim. Cosmochim. Acta, 2014, 133: 186-203.

[65] Hua B, Deng B. Reductive immobilization of uranium(VI)by amorphous iron sulfide. Environ. Sci. Technol, 2008, 42 (23): 8703-8708.

[66] Lee W J, Batchelor B. Abiotic reductive dechlorination of chlorinated ethylenes by iron-bearing phyllosilicates. Chemosphere, 2004, 56 (10): 999-1009.

[67] Neumann A, Hofstetter T B, Luessi M, Cirpka O A, Petit S, Schwarzenbach R P. Assessing the redox reactivity of structural iron in smectites using nitroaromatic compounds as kinetic probes. Environ. Sci. Technol., 2008, 42 (22): 8381-8387.

[68] Elsner M, Schwarzenbach R P, Haderlein S B. Reactivity of Fe(II)-bearing minerals toward reductive transformation of organic contaminants. Environ. Sci. Technol., 2004, 38 (3): 799-807.

[69] Myneni S C B, Tokunaga T K, Brown G E. Abiotic selenium redox transformations in the presence of Fe(II, III)oxides. Science, 1997, 278 (5340): 1106-1109.

[70] Weerasooriya R, Dharmasena B, Pyrite-assisted degradation of trichloroethene(TCE). Chemosphere, 2001, 42 (4): 389-396.

[71] Pham H T, Suto K, Inoue C. Trichloroethylene transformation in aerobic pyrite suspension: Pathways and kinetic modeling. Environ. Sci. Technol., 2009, 43 (17): 6744-6749.

[72] Guo H, Ren Y, Liu Q, Zhao K, Li Y. Enhancement of arsenic adsorption during mineral transformation from siderite to goethite: Mechanism and application. Environ. Sci. Technol., 2013, 47 (2): 1009-1016.

[73] Jeong H Y, Han Y.S, Park S W, Hayes K F. Aerobic oxidation of mackinawite(FeS)and its environmental implication for arsenic mobilization. Geochim. Cosmochim. Acta, 2010, 74 (11): 3182-3198.

[74] Ona-Nguema G, Morin G, Wang Y, Foster A. L, Juillot F, Calas G, Brown G E, Jr. XANES evidence for rapid arsenic(III) oxidation at magnetite and ferrihydrite surfaces by dissolved O_2 via Fe^{2+}-mediated reactions. Environ. Sci. Technol., 2010, 44 (14): 5416-5422.

[75] Tong M, Yuan S, Ma S, Jin M, Liu D, Cheng D, Liu X, Gan Y, Wang Y. Production of abundant hydroxyl radicals from oxygenation of subsurface sediments. Environ. Sci. Technol., 2016, 50 (1): 214-221.

[76] Ardo S G, Nelieu S, Ona-Nguema G, Delarue G, Brest J, Pironin E, Morin G. Oxidative degradation of nalidixic acid by nano-magnetite via Fe^{2+}/O_2-mediated reactions. Environ. Sci. Technol., 2015, 49 (7): 4506-4514.

[77] Fang G D, Zhou D M, Dionysiou D D. Superoxide mediated production of hydroxyl radicals by magnetite nanoparticles:

Demonstration in the degradation of 2-chlorobiphenyl. J. Hazard. Mater., 2013, 250: 68-75.

[78] Yan W, Zhang J, Jing C. Enrofloxacin transformation on shewanella oneidensis MR-1 reduced goethite during anaerobic-aerobic transition. Environ. Sci. Technol., 2016, 50 (20): 11034-11040.

[79] Zeng Q, Dong H, Wang X, Yu T, Cui W. Degradation of 1, 4-dioxane by hydroxyl radicals produced from clay minerals. J. Hazard. Mater., 2017, 331: 88-98.

[80] Cheng D, Yuan S, Liao P, Zhang P. Oxidizing impact induced by mackinawite (FeS) nanoparticles at oxic conditions due to production of hydroxyl radicals. Environ. Sci. Technol., 2016, 50 (21): 11646-11653.

[81] Zhang P, Yuan S, Liao P. Mechanisms of hydroxyl radical production from abiotic oxidation of pyrite under acidic conditions. Geochim. Cosmochim. Acta 2016, 172: 444-457.

[82] Liu X, Yuan S, Tong M, Liu D. Oxidation of trichloroethylene by the hydroxyl radicals produced from oxygenation of reduced nontronite. Water Res., 2017, 113: 72-79.

[83] Wang X, Dong H, Zeng Q, Xia Q, Zhang L, Zhou Z. Reduced iron-containing clay minerals as antibacterial agents. Environ. Sci. Technol., 2017, 51 (13): 7639-7647.

[84] Fenton H J H. LXXIII.Oxidation of tartaric acid in presence of iron. J. Chem. Soc., 1894, 65 (0): 899-910.

[85] Nam S, Renganathan V, Tratnyek, P G. Substituent effects on azo dye oxidation by the Fe(III)-EDTA-H_2O_2 system. Chemosphere, 2001, 45 (1): 59-65.

[86] Meers E, Ruttens A, Hopgood M J, Samson D, Tack F M G. Comparison of EDTA and EDDS as potential soil amendments for enhanced phytoextraction of heavy metals. Chemosphere, 2005, 58 (8): 1011-1022.

[87] Qin Y, Song F, Ai Z, Zhang P, Zhang L. Protocatechuic acid promoted alachlor degradation in Fe(III)/H_2O_2 Fenton system. Environ. Sci. Technol., 2015, 49 (13): 7948-7956.

[88] Wang L, Wang F, Li P N, Zhang L Z. Ferrous–tetrapolyphosphate complex induced dioxygen activation for toxic organic pollutants degradation. Sep. Purif. Technol., 2013, 120: 148-155.

[89] Pipi A, R F, De Andrade A R, Brillas E, Sirés I. Total removal of alachlor from water by electrochemical processes. Sep. Purif. Technol., 2014, 132: 674-683.

[90] Panizza M, Cerisola G. Electro-Fenton degradation of synthetic dyes. Water Res., 2009, 43 (2): 339-344.

[91] Friedman C L, Lemley A T, Hay A. Degradation of chloroacetanilide herbicides by anodic Fenton treatment. J. Agric. Food Chem., 2006, 54 (7): 2640-2651.

[92] Wang L, Cao M, Ai Z, Zhang L. Design of a highly efficient and wide pH electro-Fenton oxidation system with molecular oxygen activated by ferrous-tetrapolyphosphate complex. Environ. Sci. Technol., 2015, 49 (5): 3032-3039.

[93] Lee C, Keenan C R, Sedlak D L. Polyoxometalate-enhanced oxidation of organic compounds by nanoparticulate zero-valent iron and ferrous ion in the presence of oxygen. Environ. Sci. Technol., 2008, 42 (13): 4921-4926.

[94] Kang S H, Bokare A D, Park Y, Choi C H, Choi W. Electron shuttling catalytic effect of mellitic acid in zero-valent iron induced oxidative degradation. Catal. Today, 2017, 282: 65-70.

[95] Keenan C R, Sedlak D L. Ligand-enhanced reactive oxidant generation by nanoparticulate zero-valent iron and oxygen. Environ. Sci. Technol., 2008, 42 (18): 6936-6941.

[96] Rose A L. Effect of dissolved natural organic matter on the kinetics of ferrous iron oxygenation in seawater. Environ. Sci. Technol., 2003, 37 (21): 4877-4886.

[97] Su C, Puls R W. Arsenate and arsenite removal by zerovalent iron: Effects of phosphate, silicate, carbonate, borate, sulfate, chromate, molybdate, and nitrate, relative to chloride. Environ. Sci. Technol.. 2001, 35 (22):4562-4568.

[98] Mezenner N Y, Bensmaili A. Kinetics and thermodynamic study of phosphate adsorption on iron hydroxide-eggshell waste. Chem. Engineer. J, 2009, 147 (2): 87-96.

[99] Su C, Puls R W. Nitrate reduction by zerovalent iron: effects of formate, oxalate, citrate, chloride, sulfate, borate, and

phosphate. Environ. Sci. Technol., 2004, 38 (9): 2715-2720.

[100] Almeelbi T, Bezbaruah A. Aqueous phosphate removal using nanoscale zero-valent iron. J. Nanopart. Res., 2012, 14 (7): 1-14.

[101] Hinkle M A G, Wang Z, Giammar D E, Catalano J G. Interaction of Fe(II) with phosphate and sulfate on iron oxide surfaces. Geochim. Cosmochim. Acta, 2015, 158: 130-146.

[102] King D W. Role of carbonate opeciation on the oxidation rate of Fe(II) in aquatic systems. Environ. Sci. Technol., 1998, 32 (19): 2997-3003.

[103] Wang L, Cao M H, Ai Z H, Zhang L Z. Dramatically enhanced aerobic atrazine degradation with Fe@Fe_2O_3 core-shell nanowires by tetrapolyphosphate. Environ. Sci. Technol , 2014, 48 (6): 3354-3362.

[104] Liu Y, Phenrat T, Lowry G V. Effect of TCE concentration and dissolved groundwater solutes on NZVI-promoted TCE dechlorination and H_2 evolution. Environ. Sci. Technol., 2007, 41 (22): 7881-7887.

[105] Mu Y, Ai Z, Zhang L. Phosphate shifted oxygen reduction pathway on Fe@Fe_2O_3 core-shell nanowires for enhanced reactive oxygen species generation and aerobic 4-chlorophenol degradation. Environ. Sci. Technol., 2017, 51 (14): 8101-8109.

[106] Rao P S, Hayon E. Redox potentials of free radicals. IV. Superoxide and hydroperoxy radicals • O_2^- and • HO_2. J. Phys. Chem., 1975, 79 (4): 397-402.

[107] Bielski B H J, Cabelli D E, Arudi R L, Ross A B. Reactivity of HO_2/O^-_2 radicals in aqueous solution. J. Phys. Chem. Ref. Data, 1985, 14 (4): 1041-1100.

[108] Huang X, Hou X, Zhao J, Zhang L. Hematite facet confined ferrous ions as high efficient Fenton catalysts to degrade organic contaminants by lowering H_2O_2 decomposition energetic span. Appl. Catal. B: Environ., 2016, 181: 127-137.

[109] Huang X, Hou X, Jia F, Song F, Zhao J, Zhang L. Ascorbate-promoted surface iron cycle for efficient heterogeneous Fenton alachlor degradation with hematite nanocrystals. ACS Appl. Mater. Interfaces, 2017, 9 (10): 8751-8758.

[110] Hou X, Huang X, Ai Z, Zhao J, Zhang L. Ascorbic acid/Fe@Fe_2O_3: A highly efficient combined Fenton reagent to remove organic contaminants. J. Hazard. Mater., 2016, 310: 170-178.

[111] Saputra E, Muhammad S, Sun H, Wang S. Activated carbons as green and effective catalysts for generation of reactive radicals in degradation of aqueous phenol. RSC Adv., 2013, 3 (44): 21905-21910.

[112] Fang G, Gao J, Liu C, Dionysiou D D, Wang Y, Zhou D. Key role of persistent free radicals in hydrogen peroxide activation by biochar: Implications to organic contaminant degradation. Environ. Sci. Technol., 2014, 48 (3): 1902-1910.

[113] Ribeiro R S, Silva A M T, Figueiredo J L, Faria J L, Gomes H T. The influence of structure and surface chemistry of carbon materials on the decomposition of hydrogen peroxide. Carbon, 2013, 62: 97-108.

[114] Fang G D, Liu C, Gao J, Zhou D M. New insights into the mechanism of the catalytic decomposition of hydrogen peroxide by activated carbon: Implications for degradation of diethyl phthalate. Ind. Eng. Chem. Res., 2014, 53 (51): 19925-19933.

[115] Kappler A, Wuestner M L, Ruecker A, Harter J, Halama M, Behrens S. Biochar as an electron shuttle between bacteria and Fe(III)minerals. Environ. Sci. Technol. Lett, 2014, 1 (8): 339-344.

[116] Klüpfel L, Keiluweit M, Kleber M, Sander M. Redox properties of plant biomass-derived black carbon(Biochar). Environ. Sci. Technol., 2014, 48 (10): 5601-5611.

[117] Qin Y, Zhang L, An T. Hydrothermal carbon-mediated Fenton-like reaction mechanism in the degradation of alachlor: Direct electron transfer from hydrothermal carbon to Fe(III). ACS Appl. Mater.Interfaces, 2017, 9 (20): 17115-17124.

[118] Lovley D R, Stolz J F, Nord G L, Phillips E J P. Anaerobic production of magnetite by a dissimilatory iron-reducing microorganism. Nature, 1987, 330 (6145): 252-254.

[119] Kappler A, Straub K L. Geomicrobiological cycling of iron. Rev. Mineral Geochem., 2005, 59: 85-108.

[120] Coates J D, Anderson R T, Woodward J C, Phillips E J P, Lovley D R. Anaerobic hydrocarbon degradation in

petroleum-contaminated harbor sediments under sulfate-reducing and artificially imposed iron-reducing conditions. Environ. Sci. Technol., 1996, 30 (9): 2784-2789.

[121] Lovley D R, Baedecker M J, Lonergan D J, Cozzarelli I M, Phillips E J P, Siegel D I. Oxidation of aromatic contaminants coupled to microbial iron reduction. Nature, 1989, 339 (6222): 297-300.

[122] Fujii M, Dang T C, Rose A L, Omura T, Waite T D. Effect of light on iron uptake by the freshwater cyanobacterium microcystis aeruginosa. Environ. Sci. Technol., 2011, 45 (4): 1391-1398.

[123] Widdel F, Schnell S, Heising S, Ehrenreich A, Assmus B, Schink B. Ferrous iron oxidation by anoxygenic phototrophic bacteria. Nature, 1993, 362 (6423): 834-836.

[124] Johnson C M, Beard B L, Beukes N J, Klein C, O'Leary J M. Ancient geochemical cycling in the Earth as inferred from Fe isotope studies of banded iron formations from the Transvaal Craton. Contrib Mineral Petrol., 2003, 144 (5): 523-547.

[125] Raiswell R C D E. Geochemical perspectives: The iron biogeochemical cycle past and present. 2012.

[126] Kappler A, Schink B, Newman D K. Fe(III)mineral formation and cell encrustation by the nitrate-dependent Fe(II)-oxidizer strain BoFeN1. Geobiol, 2005, 3 (4): 235-245.

[127] Hedrich S, Schlomann M, Johnson D B. The iron-oxidizing proteobacteria. Microbiol, 2011, 157: 1551-1564.

[128] Jiao Y, Kappler A, Croal L R, Newman D K. Isolation and characterization of a genetically tractable photoautotrophic Fe(II)-oxidizing bacterium, Rhodopseudomonas palustris strain TIE-1. Appl. Environ. Microbiol, 2005, 71 (8): 4487-4496.

[129] Ge T D, Wu X H, Chen X J, Yuan H Z, Zou Z Y, Li B Z, Zhou P, Liu S L, Tong C L, Brookes P, Wu J S. Microbial phototrophic fixation of atmospheric CO_2 in China subtropical upland and paddy soils. Geochim. Cosmochim. Acta, 2013, 113: 70-78.

[130] Hart K M, Kulakova A N, Allen C C R, Simpson A J, Oppenheimer S F, Masoom H, Courtier-Murias D Soong R, Kulakov L A, Flanagan P V, Murphy B T, Kelleher B P. Tracking the fate of microbially sequestered carbon dioxide in soil organic matter. Environ. Sci. Technol., 2013, 47 (10): 5128-5137.

[131] Emerson D, Moyer C. Isolation and characterization of novel iron-oxidizing bacteria that grow at circumneutral pH. Appl. Environ. Microbiol., 1997, 63 (12): 4784-4792.

[132] Weiss J V, Rentz J A, Plaia T, Neubauer S C, Merrill-Floyd M, Lilburn T, Bradburne C, Megonigal J P, Emerson D. Characterization of neutrophilic Fe(II)-oxidizing bacteria isolated from the rhizosphere of wetland plants and description of Ferritrophicum radicicola gen. nov. sp. nov., and Sideroxydans paludicola sp. nov. Geomicrobiol. J, 2007, 24 (7-8): 559-570.

[133] Bryan C G, Davis-Belmar C S, van Wyk N, Fraser M K, Dew D, Rautenbach G F, Harrison S T L. The effect of CO_2 availability on the growth, iron oxidation and CO_2-fixation rates of pure cultures of leptospirillum ferriphilum and acidithiobacillus ferrooxidans. Biotechnol. Bioeng., 2012, 109(7): 1693-1703.

[134] Straub K L, Benz M, Schink B, Widdel F. Anaerobic, nitrate-dependent microbial oxidation of ferrous iron. Appl. Environ. Microbiol., 1996, 62 (4): 1458-1460.

[135] Chen Y T, L X M, Liu T X, Li F B. Microaerobic iron oxidation and carbon assimilation and associated microbial community in paddy soil. Acta Geochimi., 2017.

[136] Liu T X, Zhang W, Li X M, Li F B, Zhang W, Shen W J. Kinetics of competitive reduction of nitrate and iron oxides by aeromonas hydrophila HS01. Soil Sci. Soc. Am. J, 2014, 78 (6): 1903-1912.

[137] Coby A J, Picardal F W. Inhibition of NO_3^- and NO_2^- reduction by microbial Fe(III)reduction: Evidence of a reaction between NO_2^- and cell surface-bound Fe^{2+}. Appl. Environ. Microbiol., 2005, 71 (9): 5267-5274.

[138] Ottow J. Selection, characterization and iron-reducing capacity of nitrate reductaseless(nit−)mutants of iron-reducing bacteria. J. Basic Microbiol., 1970, 10 (1): 55-62.

[139] Zhang S H, Cai L L, Liu Y, Shi Y, Li W. Effects of NO_2^- and NO_3^- on the Fe(III) EDTA reduction in a chemical

absorption–biological reduction integrated NO_x removal system. Appl. Microbiol. Biotechnol., 2009, 82 (3): 557-563.

[140] Zhang W, Li X, Liu T, Li F, Enhanced nitrate reduction and current generation by *Bacillus* sp. in the presence of iron oxides. J. Soil. Sediment, 2012, 12 (3): 354-365.

[141] Liu T, Li X, Zhang W, Hu M, Li F. Fe(III)oxides accelerate microbial nitrate reduction and electricity generation by *Klebsiella pneumoniae* L17. J. Colloid Interface Sci., 2014, 423: 25-32.

[142] Carlson H K, Clark I C, Melnyk R A, Coates J D, Toward a mechanistic understanding of anaerobic nitrate-dependent iron oxidation: Balancing electron uptake and detoxification. Front Microbiol, 2012, 3: 57.

[143] Emerson D, Fleming E J, McBeth J M. Iron-oxidizing bacteria: an environmental and genomic perspective. Annu. Rev. Microbiol, 2010, 64: 561-583.

[144] Li X M, Zhang W, Liu T X, Chen L X, Chen P C, Li F B. Changes in the composition and diversity of microbial communities during anaerobic nitrate reduction and Fe(II) oxidation at circumneutral pH in paddy soil. Soil Biol. Biochem., 2016, 94: 70-79.

[145] Yang W H, Weber K A, Silver W L. Nitrogen loss from soil through anaerobic ammonium oxidation coupled to iron reduction. Nat. Geosci., 2012, 5 (8): 538-541.

[146] Ding L J, An X L, Li S, Zhang G L, Zhu Y G. Nitrogen loss through anaerobic ammonium oxidation coupled to iron reduction from paddy soils in a chronosequence. Environ. Sci. Technol., 2014, 48, (18): 10641-10647.

[147] Yang S M, Li F M, Malhi S S, Wang P, Suo D R, Wang J G. Long-term fertilization effects on crop yield and nitrate nitrogen accumulation in soil in northwestern China. Agron. J, 2004, 96 (4): 1039-1049.

[148] Ding L J, Su J Q, Xu H J, Jia Z J, Zhu Y G. Long-term nitrogen fertilization of paddy soil shifts iron-reducing microbial community revealed by RNA-^{13}C-acetate probing coupled with pyrosequencing. ISME J,2015, 9 (3): 721.

[149] Vargas M, Kashefi K, Blunt-Harris E L, Lovley D R. Microbiological evidence for Fe(III) reduction on early Earth. Nature, 1998, 395 (6697): 65-67.

[150] Fredrickson J K, Gorby Y A. Environmental processes mediated by iron-reducing bacteria. Curr. Opin. Biotech., 1996, 7 (3): 287-294.

[151] Yu H Y, Li F B, Liu C S, Huang W, Liu T X, Yu W M, Iron redox cycling coupled to transformation and immobilization of heavy metals: Implications for paddy rice safety in the red soil of south China. Adv. Agron., 2016, 137: 279-317.

[152] Amonette J E, Workman D J, Kennedy D W, Fruchter J S, Gorby Y A. Dechlorination of carbon tetrachloride by Fe(II)associated with goethite. Environ. Sci. Technol., 2000, 34 (21): 4606-4613.

[153] Li F B, Wang X G, Li Y T, Liu C S, Zeng F, Zhang L J, Hao M D, Ruan H D. Enhancement of the reductive transformation of pentachlorophenol by polycarboxylic acids at the iron oxide-water interface. J. Colloid Interface Sci., 2008, 321 (2): 332-341.

[154] Chen M J, Tao L, Li F B, Lan Q. Reductions of Fe(III) and pentachlorophenol linked with geochemical properties of soils from Pearl River Delta. Geoderma, 2014, 217: 201-211.

[155] Tong H, Chen M J, Li F B, Liu C S, Liao C Z. Changes in the microbial community during repeated anaerobic microbial dechlorination of pentachlorophenol. Biodegrad., 2017, 28 (2-3): 219-230.

[156] Tong H, Hu M, Li F B, Liu C S, Chen M J. Biochar enhances the microbial and chemical transformation of pentachlorophenol in paddy soil. Soil Biol. Biochem., 2014, 70: 142-150.

[157] Chen M, Shih K, Hu M, Li F, Liu C, Wu W, Tong H. Biostimulation of indigenous microbial communities for anaerobic transformation of pentachlorophenol in paddy soils of southern China. J. Agric. Food Chem., 2012, 60 (12): 2967-2975.

[158] Chen M, Tong H, Liu C, Chen D, Li F, Qiao J. A humic substance analogue AQDS stimulates Geobacter sp. abundance and enhances pentachlorophenol transformation in a paddy soil. Chemosphere, 2016, 160: 141-148.

[159] Caccavo F, Blakemore R P, Lovley D R. A hydrogen-oxidizing, Fe(III)-reducing microorganism from the Great Bay

Estuary, New Hampshire. Appl. Environ. Microbiol., 1992, 58 (10): 3211-3216.

[160] Fredrickson J K, Kota S, Kukkadapu R K, Liu C X, Zachara J M. Influence of electron donor/acceptor concentrations on hydrous ferric oxide(HFO) bioreduction. Biodegrad. 2003, 14 (2): 91-103.

[161] Liu T, Li X, Li F, Han R, Wu Y, Yuan X, Wang Y. *In situ* spectral kinetics of Cr(VI) reduction by c-type cytochromes in a suspension of living shewanella putrefaciens 200. Scientific. Reports, 2016, 6: 29592.

[162] Han R, Li F, Liu T, Li X, Wu Y, Wang Y, Chen D. Effects of incubation conditions on Cr(VI) reduction by c-type cytochromes in intact shewanella oneidensis MR-1 Cells. Front Microbiol., 2016, 7: 746.

[163] Muehe E M, Scheer L, Daus B, Kappler A. Fate of Arsenic during microbial reduction of biogenic versus abiogenic As-Fe(III)-Mineral coprecipitates. Environ. Sci. Technol., 2013, 47 (15): 8297-8307.

[164] Coughlin B R, Stone A T. Nonreversible adsorption of divalent metal-ions(Mn-Ii, Co-Ii Ni-Ii Cu-Ii and Pb-Ii) onto goethite-Effects of acidification, Fe-Ii addition, and picolinic-acid addition. Environ. Sci. Technol.,1995, 29 (9): 2445-2455.

[165] Cooper D C, Picardal F F, Coby A J. Interactions between microbial iron reduction and metal geochemistry: Effect of redox cycling on transition metal speciation in iron bearing sediments. Environ. Sci. Technol., 2006, 40 (6): 1884-1891.

[166] 刘承帅，李芳柏，陈曼佳，廖长忠，童辉，华健．Fe(II)催化水铁矿晶相转变过程中 Pb 的吸附与固定．化学学报 2017, 75: 621-628.

[167] 刘承帅，韦志琦，李芳柏，董军，游离态 Fe(II)驱动赤铁矿晶相重组的 Fe 原子交换机制：稳定 Fe 同位素示踪研究．中国科学地球科学, 2016, 46 (11): 1542-1553.

[168] Liu C S, Zhu Z K, Li F B, Liu T X, Liao C Z, Lee J J, Shih K M, Tao L, Wu Y D. Fe(II)-induced phase transformation of ferrihydrite: The inhibition effects and stabilization of divalent metal cations. Chem. Geol., 2016, (444): 110-119.

[169] Yu H Y, Liu C P, Zhu J S, Li F B, Deng D M, Wang Q, Liu C S. Cadmium availability in rice paddy fields from a mining area: The effects of soil properties highlighting iron fractions and pH value. Environ. Pollut., 2016, 209: 38-45.

[170] Liu C, Yu H Y, Liu C, Li F, Xu X, Wang Q. Arsenic availability in rice from a mining area: is amorphous iron oxide-bound arsenic a source or sink? Environ. Pollut., 2015, 199: 95-101.

[171] Yu H Y, Wang X, Li F, Li B, Liu C, Wang Q, Lei J. Arsenic mobility and bioavailability in paddy soil under iron compound amendments at different growth stages of rice. Environ. Pollut., 2017, 224: 136–147.

[172] Hu M, Li F B, Liu C P, Wu W J. The diversity and abundance of As(III) oxidizers on root iron plaque is critical for arsenic bioavailability to rice. Sci. Rep., 2015, (5): 13611.

[173] Ehrlich H L, Wang Z L. Geomicrobiology(Fifth Edition). China Petrochemical Press: Beijing, 2010.

[174] Chaudhuri S K, Lack J G, Coates J D. Biogenic magnetite formation through anaerobic biooxidation of Fe(II). Appl. Environ. Microbiol., 2001, 67 (6): 2844-2848.

[175] Emerson D, Weiss J V. Bacterial iron oxidation in circumneutral freshwater habitats: Findings from the field and the laboratory. Geomicrobiol. J, 2004, 21 (6): 405-414.

[176] Straub K L, Benz M, Schink, B, Widdel F. Anaerobic, nitrate-dependent microbial oxidation of ferrous iron. Appl. Environ. Microbiol, 1996, 62 (4): 1458-1460.

[177] Hafenbradl D, Keller M, Dirmeier R, Burggraf S. *Ferroglobus placidus* gen. nov., sp. nov., A novel hyperthermophilic archaeum that oxidizes Fe^{2+} at neutral pH under anoxic conditions. Arch. Microbiol., 1996, 166 (5): 308-314.

[178] Beller H R. Anaerobic, nitrate-dependent oxidation of U(IV) oxide minerals by the chemolithoautotrophic bacterium Thiobacillus denitrificans. Appl. Environ. Microbio., 2005, 71(4): 2170-2174.

[179] Liu H, Xia J, Nie Z, Ma C, Zheng L, Hong C, Zhao Y, Wen W. Bioleaching of chalcopyrite by Acidianus manzaensis under different constant pH. Miner. Eng., 2016, (98): 80-89.

[180] Silverman M P, Ehrlich H L, Silverman M P. Microbial formation and degradation of minerals. Adv. Appl. Microbiol., 1964, 6 (6): 153-206.

[181] Crundwell F K. How do bacteria interact with minerals? Hydrometall, 2003, 71 (1): 75-81.

[182] Tributsch H. Direct versus indirect bioleaching. Process Metall., 1999, 59, (2): 177-185.

[183] Xia J L, Liu H C, Nie Z Y, Peng A A, Zhen X J, Yang Y, Zhang X L. Synchrotron radiation based STXM analysis and micro-XRF mapping of differential expression of extracellular thiol groups by Acidithiobacillus ferrooxidans grown on Fe(2+)and S(0). J. Microbiol. Methods, 2013, 94 (3): 257.

[184] Vera M, Schippers A, Sand W. Progress in bioleaching: Fundamentals and mechanisms of bacterial metal sulfide oxidation--part A. Appl. Microbiol. Biotechnol., 2013, 97 (17): 7529.

[185] 王蕾, 夏金兰, 朱泓睿, 刘红昌, 聂珍媛, 刘李柱. 微生物-矿物相互作用及界面显微分析研究进展. 微生物学通报 2017, 44 (3): 716-725.

[186] Harmer S L, Thomas J E, Fornasiero D, Gerson A R. The evolution of surface layers formed during chalcopyrite leaching. Geochim. Cosmochim. Acta, 2006, 70 (17): 4392-4402.

[187] Parker A, Klauber C, Kougianos A, Watling H R, Bronswijk W V. An X-ray photoelectron spectroscopy study of the mechanism of oxidative dissolution of chalcopyrite. Hydrometall, 2003, 71 (1): 265-276.

[188] Liu H C, Xia J L, Nie Z Y, Liu L Z, Wang L, Ma C Y, Zheng L, Zhao Y D, Wen W. Comparative study of S, Fe and Cu speciation transformation during chalcopyrite bioleaching by mixed mesophiles and mixed thermophiles, Miner. Eng., 2017, 106: 22-32.

[189] Xia J L, Yang Y, He H, Zhao X J, Liang C L, Zheng L, Ma C Y, Zhao Y D, Nie Z Y, Qiu G Z. Surface analysis of sulfur speciation on pyrite bioleached by extreme thermophile Acidianus manzaensis using Raman and XANES spectroscopy. Hydrometall, 2010, 96 (3): 129-135.

[190] Chandra A P, Gerson A R. Redox potential(E_h) and anion effects of pyrite(FeS_2)leaching at pH 1. Geochim. Cosmochim. Acta, 2011, 75 (22): 6893-6911.

[191] 夏金兰, 张成桂, 彭安安, 何环. 嗜酸硫氧化细菌消解元素硫的研究进展. 中国生物工程杂志, 2006, 26 (9): 91-95.

[192] He H, Xia J L, Hong F F, Tao X X, Leng Y W, Zhao Y D. Analysis of sulfur speciation on chalcopyrite surface bioleached with Acidithiobacillus ferrooxidans, Miner. Eng., 2012, 27-28 (2): 60-64.

[193] He H, Xia J L, Yang Y, Jiang H, Xiao C Q, Zheng L, Ma C Y, Zhao Y D, Qiu G Z. Sulfur speciation on the surface of chalcopyrite leached by Acidianus manzaensis. Hydrometall, 2009, 99 (1): 45-50.

[194] Liu H C, Nie Z Y, Xia J L, Zhu H R, Yang Y, Zhao C H, Zheng L, Zhao Y D. Investigation of copper, iron and sulfur speciation during bioleaching of chalcopyrite by moderate thermophile Sulfobacillus thermosulfidooxidans. Int. J. Miner. Process, 2015, 137: 1-8.

[195] Xia J L, Yang Y, He H, Liang C L, Zhao X J, Zheng L, Ma C Y, Zhao Y D, Nie Z Y, Qiu G Z. Investigation of the sulfur speciation during chalcopyrite leaching by moderate thermophile Sulfobacillus thermosulfidooxidans. Int. J. Miner. Process, 2010, 94 (1): 52-57.

[196] Liang C L, Xia J L, Nie Z Y, Shui-Jing Y U, Bao-Quan X U. Effect of initial pH on chalcopyrite oxidation dissolution in the presence of extreme thermophile Acidianus manzaensis. Trans. Nonferrous Met. Soc. China, 2014, 24 (6): 1890-1897.

[197] Liang C L, Xia J L, Nie Z Y, Yang Y, Ma C Y. Effect of sodium chloride on sulfur speciation of chalcopyrite bioleached by the extreme thermophile Acidianus manzaensis. Bioresour. Technol., 2012, 110 (2): 462-467.

[198] Liang C L, Xia J L, Zhao X J Yang Y, Gong S Q, Nie Z Y, Ma C Y, Zheng L, Zhao Y D, Qiu, G. Z, Effect of activated carbon on chalcopyrite bioleaching with extreme thermophile Acidianus manzaensis. Hydrometall, 2010, 105 (1): 179-185.

[199] 梁长利, 夏金兰, 杨益, 聂珍媛, 邱冠周. 黄铜矿生物浸出过程的硫形态转化研究进展. 中国有色金属学报, 2012, 22 (1): 265-273.

[200] He H, Xia J L, Huang G H, Jiang H C, Tao X X, Zhao Y D, He W. Analysis of the elemental sulfur bio-oxidation by Acidithiobacillus ferrooxidans with sulfur K-edge XANES. World J. Microbiology & Biotechnol., 2011, 27 (8): 1927-1931.

[201] He H, Xia J L, Jiang H C, Yan Y, Liang C L, Ma C Y, Zheng L, Zhao Y D, Qiu G Z. Sulfur species investigation in extra- and intracellular sulfur globules of acidithiobacillus ferrooxidans and Acidithiobacillus caldus. Geomicrobiol. J, 2010, 27 (8): 707-713.

[202] He H, Zhang C G, Xia J L, Peng A A, Yang Y, Jiang H C, Zheng L, Ma C Y, Zhao Y D, Nie Z Y. Investigation of elemental sulfur speciation transformation mediated by Acidithiobacillus ferrooxidans. Curr. Microbiol., 2009, 58, (4): 300-7.

[203] Liu H C, Xia J L, Nie Z Y, Zhen X J, Zhang L J. Differential expression of extracellular thiol groups of moderately thermophilic Sulfobacillus thermosulfidooxidans and extremely thermophilic Acidianus manzaensis grown on S(0)and Fe(2). Arch. Microbiol., 2015, 197 (6): 823-831.

[204] Nie Z Y, Liu H C, Xia J L, Zhu H R, Ma C Y, Zheng L, Zhao Y D, Qiu G Z. Differential utilization and transformation of sulfur allotropes, μ-S and α-S8, by moderate thermoacidophile *Sulfobacillus thermosulfidooxidans*. Res. Microbiol, 2014, 165 (8): 639.

[205] Nie Z Y, Liu H C, Xia J L, Yang Y, Zhen X J, Zhang L J, Qiu G Z. Evidence of cell surface iron speciation of acidophilic iron-oxidizing microorganisms in indirect bioleaching process. Biomet., 2016, 29 (1): 25-37.

[206] Shi L, Dong H, Reguera G, Beyenal H, Lu A, Liu J, Yu H Q, Fredrickson J K. Extracellular electron transfer mechanisms between microorganisms and minerals. Nat. Rev. Microbiol., 2016, 14 (10): 651-62.

[207] 邱轩，石良. 微生物和含铁矿物之间的电子交换. 化学学报, 2017, 75: 583-593.

[208] Roden E E. Microbial iron-redox cycling in subsurface environments. Biochem. Soc. Trans., 2012, 40 (6): 1249-1256.

[209] Albers S V, Meyer B H. The archaeal cell envelope. Nature reviews. Microbiol, 2011, 9 (6): 414-26.

[210] Shi L, Squier T C, Zachara J M, Fredrickson J K. Respiration of metal(hydr) oxides by Shewanella and Geobacter: A key role for multihaem c-type cytochromes. Mol. Microbiol., 2007, 65 (1): 12-20.

[211] Melton E D, Swanner E D, Behrens S, Schmidt C, Kappler A. The interplay of microbially mediated and abiotic reactions in the biogeochemical Fe cycle. Nature reviews. Microbiol., 2014, 12 (12): 797-808.

[212] Shi L, Tien M, Fredrickson J K, Zachara J M, Rosso K M. Microbial redox proteins and protein complexes for extracellular respiration. In Redox Proteins in Supercomplexes and Signalosomes, Louro, R.; Diaz-Moreno, I., Eds. CRC Press: Baca Taton, FL USA, 2016.

[213] Myers C R, Nealson K H. Bacterial manganese reduction and growth with manganese oxide as the sole electron acceptor. Science, 1988, 240 (4857): 1319-1321.

[214] Beliaev A S, Saffarini D A. Shewanella putrefaciens mtrB encodes an outer membrane protein required for Fe(III) and Mn(IV)reduction. J. Bacteriol., 1998, 180 (23): 6292-6297.

[215] Beliaev A S, Saffarini D A, McLaughlin J L, Hunnicutt D. MtrC, an outer membrane decahaem c cytochrome required for metal reduction in Shewanella putrefaciens MR-1. Mol. Microbiol., 2001, 39 (3): 722-730.

[216] Myers J M, Myers C R. Role of the tetraheme cytochrome CymA in anaerobic electron transport in cells of Shewanella putrefaciens MR-1 with normal levels of menaquinone. J. Bacteriol., 2000, 182 (1): 67-75.

[217] Myers C R, Myers J M. MtrB is required for proper incorporation of the cytochromes OmcA and OmcB into the outer membrane of Shewanella putrefaciens MR-1. Appl. Environ. Microbiol., 2002, 68 (11): 5585-5594.

[218] Coursolle D, Gralnick J A. Modularity of the Mtr respiratory pathway of Shewanella oneidensis strain MR-1. Mol. Microbiol, 2010, 77 (4): 995-1008.

[219] Sturm G, Richter K, Doetsch A, Heide H, Louro R O, Gescher J. A dynamic periplasmic electron transfer network enables respiratory flexibility beyond a thermodynamic regulatory regime. ISME J, 2015, (9): 1802-1811.

[220] McMillan D G, Marritt S J, Butt J N, Jeuken L J. Menaquinone-7 is specific cofactor in tetraheme quinol dehydrogenase CymA. J. Biol. Chem., 2012, 287 (17): 14215-14225.

[221] McMillan D G, Marritt S J, Firer-Sherwood M A, Shi L, Richardson D J, Evans S D, Elliott S J, Butt J N, Jeuken L J. Protein-protein interaction regulates the direction of catalysis and electron transfer in a redox enzyme complex. J. Am. Chem. Soc., 2013, 135 (28): 10550-10556.

[222] Marritt S J, Lowe T G, Bye J, McMillan D G, Shi L, Fredrickson J, Zachara J, Richardson D J, Cheesman M R, Jeuken L J, Butt J N. A functional description of CymA, an electron-transfer hub supporting anaerobic respiratory flexibility in Shewanella. Biochem. J, 2012, 444 (3): 465-474.

[223] Marritt S J, McMillan D G, Shi L, Fredrickson J K, Zachara J M, Richardson D J, Jeuken L J, Butt J N. The roles of CymA in support of the respiratory flexibility of Shewanella oneidensis MR-1. Biochem. Soc. Trans., 2012, 40 (6): 1217-1221.

[224] Firer-Sherwood M A, Bewley K D, Mock J Y, Elliott S J. Tools for resolving complexity in the electron transfer networks of multiheme cytochromes c. Metallomics, 2011, 3 (4): 344-348.

[225] Alves M N, Neto S E, Alves A S, Fonseca B M, Carrelo A, Pacheco I, Paquete C M, Soares C M, Louro R O. Characterization of the periplasmic redox network that sustains the versatile anaerobic metabolism of Shewanella oneidensis MR-1. Front. Microbiol, 2015, (6): 665.

[226] Fonseca B M, Paquete C M, Neto S E, Pacheco I, Soares C M, Louro R O. Mind the gap: Cytochrome interactions reveal electron pathways across the periplasm of Shewanella oneidensis MR-1. Biochem. J, 2013, 449 (1): 101-108.

[227] Ross D E, Ruebush S S, Brantley S L, Hartshorne R S, Clarke T A, Richardson D J, Tien M. Characterization of protein-protein interactions involved in iron reduction by Shewanella oneidensis MR-1. Appl. Environ. Microbiol., 2007, 73 (18): 5797-5808.

[228] Hartshorne R S, Reardon C L, Ross D, Nuester J, Clarke T A, Gates A J, Mills P C, Fredrickson J K, Zachara J M, Shi L, Beliaev A S, Marshall M J, Tien M, Brantley S, Butt J N, Richardson D J. Characterization of an electron conduit between bacteria and the extracellular environment. Proc. Natl. Acad. Sci., 2009, 106 (52): 22169-22174.

[229] White G F, Shi Z, Shi L, Dohnalkova A C, Fredrickson J K, Zachara J M, Butt J N, Richardson D J, Clarke T A. Development of a proteoliposome model to probe transmembrane electron-transfer reactions. Biochem. Soc. Trans., 2012, 40 (6): 1257-1260.

[230] White G F, Shi Z, Shi L, Wang Z, Dohnalkova A C, Marshall M J, Fredrickson J K, Zachara J M, Butt J N, Richardson D J, Clarke T A. Rapid electron exchange between surface-exposed bacterial cytochromes and Fe(III) minerals. Proc. Natl. Acad. Sci., 2013, 110 (16): 6346-6351.

[231] Richardson D J, Butt J N, Fredrickson J K, Zachara J M, Shi L, Edwards M J, White G, Baiden N, Gates A J, Marritt S J, Clarke T A. The 'porin-cytochrome' model for microbe-to-mineral electron transfer. Mol. Microbiol., 2012, 85 (2): 201-212.

[232] Shi L, Chen B, Wang Z, Elias D A, Mayer M U, Gorby Y A, Ni S, Lower B H, Kennedy D W, Wunschel D S, Mottaz H M, Marshall M J, Hill E A, Beliaev A S, Zachara J M, Fredrickson J K, Squier T C. Isolation of a high-affinity functional protein complex between OmcA and MtrC: Two outer membrane decaheme c-type cytochromes of Shewanella oneidensis MR-1. J. Bacteriol., 2006, 188 (13): 4705-4714.

[233] Shi L, Deng S, Marshall M J, Wang Z, Kennedy D W Dohnalkova A C, Mottaz H M, Hill E A, Gorby Y A, Beliaev A S, Richardson D J, Zachara J M, Fredrickson J K. Direct involvement of type II secretion system in extracellular translocation of Shewanella oneidensis outer membrane cytochromes MtrC and OmcA. J. Bacteriol., 2008, 190 (15): 5512-5516.

[234] Lower B H, Lins R D, Oestreicher Z, Straatsma T P, Hochella M F Jr, Shi L, Lower S K. *In vitro* evolution of a peptide with a hematite binding motif that may constitute a natural metal-oxide binding archetype. Environ. Sci. Technol., 2008, 42 (10): 3821-3827.

[235] Lower B H, Shi L, Yongsunthon R, Droubay T C, McCready D E, Lower S K. Specific bonds between an iron oxide surface and outer membrane cytochromes MtrC and OmcA from Shewanella oneidensis MR-1. J. Bacteriol., 2007, 189

(13): 4944-4952.

[236] Lower B H, Yongsunthon R, Shi L, Wildling L, Gruber H J, Wigginton N S, Reardon C L, Pinchuk G E, Droubay T C, Boily J F, Lower S K. Antibody recognition force microscopy shows that outer membrane cytochromes OmcA and MtrC are expressed on the exterior surface of Shewanella oneidensis MR-1. Appl. Environ. Microbiol., 2009, 75 (9): 2931-2935.

[237] Lower S K, Hochella M F Jr, Beveridge T J. Bacterial recognition of mineral surfaces: Nanoscale interactions between Shewanella and alpha-FeOOH. Science, 2001, 292 (5520): 1360-1363.

[238] Xiong Y, Shi L, Chen B, Mayer M U, Lower B H, Londer Y, Bose S, Hochella M F, Fredrickson J K, Squier T C. High-affinity binding and direct electron transfer to solid metals by the Shewanella oneidensis MR-1 outer membrane c-type cytochrome OmcA. J. Am. Chem. Soc., 2006, 128 (43): 13978-13979.

[239] Zhang H, Tang X, Munske G R, Tolic N, Anderson G A, Bruce J E. Identification of protein-protein interactions and topologies in living cells with chemical cross-linking and mass spectrometry. Mol. Cel. Proteomic, 2009, (8): 409-420.

[240] Zhang H, Tang X, Munske G R, Zakharova N, Yang L, Zheng C, Wolff M A, Tolic N, Anderson G A, Shi L, Marshall M J, Fredrickson J K, Bruce J E. *In vivo* identification of the outer membrane protein OmcA-MtrC interaction network in Shewanella oneidensis MR-1 cells using novel hydrophobic chemical cross-linkers. J. Proteome. Res., 2008, 7 (4): 1712-1720.

[241] Meitl L A, Eggleston C M, Colberg P J S, Khare N, Reardon C L, Shi L. Electrochemical interaction of Shewanella oneidensis MR-1 and its outer membrane cytochromes OmcA and MtrC with hematite electrodes. Geochim. Cosmochim. Acta ,2009, 2009: 5292-5307.

[242] Johs A, Shi L, Droubay T, Ankner J F, Liang L. Characterization of the decaheme c-type cytochrome OmcA in solution and on hematite surfaces by small angle x-ray scattering and neutron reflectometry. Biophys. J, 2010, 98 (12): 3035-3043.

[243] Levar C E, Chan C H, Mehta-Kolte M G, Bond D R. An inner membrane cytochrome required only for reduction of high redox potential extracellular electron acceptors. MBio, 2014, 5 (6): 2034.

[244] Zacharoff L, Chan C H, Bond D R. Reduction of low potential electron acceptors requires the CbcL inner membrane cytochrome of Geobacter sulfurreducens. Bioelectrochem, 2015, 107: 7-13.

[245] Lloyd J R, Sole V A, Van Praagh C V, Lovley D R. Direct and Fe(II)-mediated reduction of technetium by Fe(III)-reducing bacteria. Appl. Environ. Microbiol., 2000, 66 (9): 3743-3749.

[246] Morgado L, Bruix M, Pessanha M, Londer Y Y, Salgueiro C A. Thermodynamic characterization of a triheme cytochrome family from Geobacter sulfurreducens reveals mechanistic and functional diversity. Biophys. J, 2010, 99 (1): 293-301.

[247] Liu Y, Wang Z, Liu J, Levar C, Edwards M J, Babauta J T, Kennedy D W, Shi Z, Beyenal H, Bond D R, Clarke T A, Butt J N, Richardson D J, Rosso K M, Zachara J M, Fredrickson J K, Shi L. A trans-outer membrane porin-cytochrome protein complex for extracellular electron transfer by geobacter sulfurreducens PCA. Environ. Microbiol. Rep., 2014, 6 (6): 776-785.

[248] Leang C, Coppi M V, Lovley D R. OmcB, a c-type polyheme cytochrome, involved in Fe(III) reduction in Geobacter sulfurreducens. J. Bacteriol., 2003, 185 (7): 2096-2103.

[249] Qian X, Reguera G, Mester T, Lovley D R. Evidence that omcB and ompB of geobacter sulfurreducens are outer membrane surface proteins. FEMS Microbiol. Lett., 2007, 277 (1): 21-27.

[250] Liu Y, Fredrickson J K, Zachara J M, Shi L. Direct involvement of ombB, omaB, and omcB genes in extracellular reduction of Fe(III) by geobacter sulfurreducens PCA. Front. Microbiol. 2015, 6: 1075.

[251] Ha P T, Lindemann S R, Shi L, Dohnalkova A C, Fredrickson J K, Madigan M T, Beyenal H. Syntrophic anaerobic photosynthesis via direct interspecies electron transfer. Nat. Commun., 2017, 8: 13924.

[252] Jiao Y, Kappler A, Croal L R, Newman D K. Isolation and characterization of a genetically tractable photoautotrophic Fe(II)-oxidizing bacterium, Rhodopseudomonas palustris strain TIE-1. Appl. Environ. Microbiol., 2005, 71 (8): 4487-4496.

[253] Bird L J, Saraiva I H, Park S, Calcada E O, Salgueiro C A, Nitschke W, Louro R O, Newman D K. Nonredundant roles for cytochrome c2 and two high-potential iron-sulfur proteins in the photoferrotroph Rhodopseudomonas palustris TIE-1. J. Bacteriol., 2014, 196 (4): 850-858.

[254] Liu J, Wang Z, Belchik S M, Edwards M J, Liu C, Kennedy D W, Merkley E D, Lipton M S, Butt J N, Richardson D J, Zachara J M, Fredrickson J K, Rosso K M, Shi L. Identification and characterization of MtoA: A decaheme c-type cytochrome of the neutrophilic Fe(II)-oxidizing bacterium Sideroxydans lithotrophicus ES-1. Front. Microbiol., 2012, 3: 37.

[255] Shi L; Rosso K M; Zachara J M; Fredrickson J K, Mtr extracellular electron-transfer pathways in Fe(III)-reducing or Fe(II)-oxidizing bacteria: a genomic perspective. Biochem. Soc. Trans, 2012, 40 (6): 1261-1267.

[256] Liu J, Pearce C. I, Liu C, Wang Z, Shi L, Arenholz E, Rosso K M. $Fe_{(3-x)}Ti_{(x)}O_4$ nanoparticles as tunable probes of microbial metal oxidation. J. Am. Chem. Soc., 2013, 135 (24): 8896-8907.

[257] Beckwith C, Edwards M J, Lawes M, Shi L, Butt J N, Richardson D J, Clarke T A. Characterization of MtoD from sideroxydans lithotrophicus: A cytochrome c electron shuttle used in lithoautotrophic growth. Front. Microbiol, 2015, 6: 332.

[258] Ross D E, Flynn J M, Baron D B, Gralnick J A, Bond D R, Towards electrosynthesis in Shewanella: energetics of reversing the mtr pathway for reductive metabolism. Public Library of Science One, 2011, 6 (2): 16649.

[259] Shelobolina E, Xu H, Konishi H, Kukkadapu R, Wu T, Blothe M, Roden E. Microbial lithotrophic oxidation of structural Fe(II) in biotite. Appl. Environ. Microbiol., 2012, 78 (16): 5746-5752.

[260] Emerson D, Field E K, Chertkov O, Davenport K W, Goodwin L, Munk C, Nolan M, Woyke T. Comparative genomics of freshwater Fe-oxidizing bacteria: implications for physiology, ecology, and systematics. Front. Microbiol., 2013, 4: 254.

[261] Wegener G, Krukenberg V, Riedel D, Tegetmeyer H E, Boetius A. Intercellular wiring enables electron transfer between methanotrophic archaea and bacteria. Nature, 2015, 526 (7574): 587-590.

[262] Ilbert M, Bonnefoy V. Insight into the evolution of the iron oxidation pathways. Biochim. Biophys. Acta, 2013, 1827 (2): 161-175.

[263] Wang Z, Leary D H, Malanoski A P, Li R W, Hervey W J t, Eddie B J, Tender G S, Yanosky S G, Vora G J, Tender L M, Lin B, Strycharz-Glaven S M. A previously uncharacterized, nonphotosynthetic member of the chromatiaceae is the primary CO_2-fixing constituent in a self-regenerating biocathode. Appl. Environ. Microbiol., 2015, 81 (2): 699-712.

[264] Watts M P, Lloyd J R. Bioremediation via microbial metal redution. In microbial metal respiration, Gescher, J.; Kappler, A., Eds. Springer-Verlag: Berling Heidelberg, 2013: 161-202.

[265] Lovley D R, Baedecker M J, Lonergan D J, Cozzarelli I M, Phillips J P, Siegel D I. Oxidation of aromatic contaminants coupled to microbial iron reduction. Nature, 1989, 339: 297-300.

[266] Tam K, Ho C T, Lee J H, Lai M, Chang C H, Rheem Y, Chen W, Hur H G, Myung N V. Growth mechanism of amorphous selenium nanoparticles synthesized by *Shewanella* sp. HN-41. Biosci. Biotechnol. Biochem., 2010, 74 (4): 696-700.

[267] Lee J H, Han J, Choi H, Hur H G, Effects of temperature and dissolved oxygen on Se(IV) removal and Se(0) precipitation by Shewanella sp. HN-41. Chemosphere, 2007, 68 (10): 1898-1905.

[268] Pearce C I, Pattrick R A, Law N, Charnock J M, Coker V S, Fellowes J W, Oremland R S, Lloyd J R. Investigating different mechanisms for biogenic selenite transformations: Geobacter sulfurreducens, Shewanella oneidensis and Veillonella atypica. Environ. Technol., 2009, 30 (12): 1313-1326.

[269] Rawlings D E, Dew D, du Plessis C.Biomineralization of metal-containing ores and concentrates. Trends. Biotechnol., 2003, 21 (1): 38-44.

[270] Valdes J, Pedroso I, Quatrini R, Dodson R J, Tettelin H, Blake R, Eisen J A, Holmes D S.Acidithiobacillus ferrooxidans metabolism: from genome sequence to industrial applications. BMC. Genomics., 2008, 9: 597.

[271] Gonzalez R, Gentina J C, Acevedo F. Biooxidation of a gold concentrate in a continuous stirred tank reactor: Methmatical model and optimal configuration. Bichem. Eng. J, 2004, (19): 33-42.

[272] Kato S, Hashimoto K, Watanabe K. Methanogenesis facilitated by electric syntrophy via(semi)conductive iron-oxide minerals. Environ. Microbiol, 2012, 14 (7): 1646-54.

[273] Liu F, Rotaru A E, Shrestha P M, Malvankar N S, Nevin K P, Lovley D R. Magnetite compensates for the lack of a pilin-associated c-type cytochrome in extracellular electron exchange. Environ. Microbiol., 2015, 17 (3): 648-655.

[274] Rotaru A E, Shrestha P M, Liu F, Markovaite B, Chen S, Nevin K P, Lovley D R. Direct interspecies electron transfer between Geobacter metallireducens and Methanosarcina barkeri. Appl. Environ. Microbiol., 2014, 80 (15): 4599-605.

[275] Rotaru A E, Shrestha P M, Liu F, Shrestha M, Shrestha D, Embree M, Zengler K, Wardman C, Nevin K P, Lovley D R. A new model for electron flow during anaerobic digestion: direct interspecies electron transfer to Methanosaeta for the reduction of carbon dixide to methane. Energy Environ. Sci., 2014, (7): 408-415.

[276] Chen S, Rotaru A E, Liu F, Philips J, Woodard T L, Nevin K P, Lovley D R. Carbon cloth stimulates direct interspecies electron transfer in syntrophic co-cultures. Bioresour. Technol., 2014, (173): 82-86.

[277] Chen S, Rotaru A E, Shrestha P M, Malvankar N S, Liu F, Fan W, Nevin K P, Lovley D R. Promoting interspecies electron transfer with biochar. Sci. Rep., 2014, (4): 5019.

[278] Bond D R, Holmes D E, Tender L M, Lovley D R. Electrode-reducing microorganisms that harvest energy from marine sediments. Science, 2002, 295 (5554): 483-485.

[279] Bond D R, Lovley D R. Electricity production by Geobacter sulfurreducens attached to electrodes. Appl. Environ. Microbiol., 2003, 69 (3): 1548-1555.

[280] Bretschger O, Obraztsova A, Sturm C A, Chang I S, Gorby Y A, Reed S B, Culley D E, Reardon C L, Barua S, Romine M F, Zhou J, Beliaev A S, Bouhenni R, Saffarini D, Mansfeld F, Kim B H, Fredrickson J K, Nealson K H. Current production and metal oxide reduction by Shewanella oneidensis MR-1 wild type and mutants. Appl. Environ. Microbiol., 2007, 73 (21): 7003-7012.

[281] Rabaey K, Rozendal R A. Microbial electrosynthesis-revisiting the electrical route for microbial production. Nat. Rev. Microbiol., 2010, 8, (10): 706-716.

[282] Lovley D R, Nevin K P. Electrobiocommodities: powering microbial production of fuels and commodity chemicals from carbon dioxide with electricity. Curr. Opin. Biotechnol., 2013, 24 (3): 385-90.

[283] Farrell J, Kason M, Melitas N, Li T. Investigation of the long-Term performance of Zero-Valent Iron for reductive dechlorination of trichloroethylene. Environ. Sci. Technol., 2000, 34 (3): 514-521.

[284] Matheson L J, Tratnyek P G. Reductive dehalogenation of chlorinated methanes by iron metal. Environ. Sci. Technol., 1994, 28 (12): 2045-2053.

[285] Agrawal A, Tratnyek P G. Reduction of nitro aromatic compounds by zero-valent iron metal. Environ. Sci. Technol., 1995, 30 (1): 153-160.

[286] Daniels L, Belay N, Rajagopal B S, Weimer P J. Bacterial methanogenesis and growth from CO_2 with elemental iron as the sole source of electrons. Science, 1987, 237 (4814): 509-511.

[287] Karri S, Sierra Alvarez R, Field J A. Zero valent iron as an electron-donor for methanogenesis and sulfate reduction in anaerobic sludge. Biotechnol. Bioeng., 2005, 92 (7): 810-819.

[288] Demirel B, Scherer P. The roles of acetotrophic and hydrogenotrophic methanogens during anaerobic conversion of biomass to methane: a review. Rev. Environ. Sci. Bio/Technol., 2008, 7 (2): 173-190.

[289] Stams A J, Plugge C M. Electron transfer in syntrophic communities of anaerobic bacteria and archaea. Nat. Rev. Microbiol., 2009, 7 (8): 568.

[290] Zhang Y, An X, Quan X. Enhancement of sludge granulation in a zero valence iron packed anaerobic reactor with a hydraulic circulation. Process Biochem., 2011, 46 (2): 471-476.

[291] Meng X, Zhang Y, Li Q, Quan X. Adding Fe^0 powder to enhance the anaerobic conversion of propionate to acetate.

Biochem. Eng. J, 2013, (73): 80-85.

[292] Liu Y, Zhang Y, Quan X, Chen S, Zhao H. Applying an electric field in a built-in zero valent iron–anaerobic reactor for enhancement of sludge granulation. Water Res., 2011, 45 (3): 1258-1266.

[293] Li Y, Zhang Y, Meng X, Yu Z, Quan X. Fe^0 enhanced acetification of propionate and granulation of sludge in acidogenic reactor. Appl. Microbiol. Biotechnol., 2015, 99 (14): 6083-6089.

[294] Pol L H, de Castro Lopes S, Lettinga G, Lens P. Anaerobic sludge granulation. Water Res., 2004, 38 (6): 1376-1389.

[295] Feng Y, Zhang Y, Quan X, Chen S. Enhanced anaerobic digestion of waste activated sludge digestion by the addition of zero valent iron. Water Res., 2014, (52): 242-250.

[296] Bolzonella D, Pavan P, Zanette M, Cecchi F. Two-phase anaerobic digestion of waste activated sludge: Effect of an extreme thermophilic prefermentation. Ind. Eng. Chem. Res., 2007, 46 (21): 6650-6655.

[297] Liu Y, Zhang Y, Quan X, Li Y, Zhao Z, Meng X, Chen S. Optimization of anaerobic acidogenesis by adding Fe^0 powder to enhance anaerobic wastewater treatment. Chem. Eng. J, 2012, (192): 179-185.

[298] Zhang Y, Feng Y, Yu Q, Xu Z, Quan X. Enhanced high-solids anaerobic digestion of waste activated sludge by the addition of scrap iron. Bioresour. Technol., 2014, (159): 297-304.

[299] Baek G, Kim J, Shin S G, Lee C. Bioaugmentation of anaerobic sludge digestion with iron-reducing bacteria: Process and microbial responses to variations in hydraulic retention time. Appl. Microbiol. Biotechnol., 2016, 100 (2): 927-937.

[300] Rotaru D E H, Franks A E, Orellana R, Risso C, Nevin K P. Geobacter: The microbe electric's physiology, ecology, and practical applications. Adv. Microb. Physiol., 2011, 59, (1).

[301] Ding J, Zhang Y, Quan X, Chen S. Anaerobic biodecolorization of AO7 by a newly isolated Fe(III)-reducing bacterium sphingomonas strain DJ. J. Chem. Technol. Biotechnol., 2015, 90 (1): 158-165.

[302] Zhang J, Zhang Y, Liu B, Dai Y, Quan X, Chen S. A direct approach for enhancing the performance of a microbial electrolysis cell(MEC)combined anaerobic reactor by dosing ferric iron: Enrichment and isolation of Fe(III) reducing bacteria. Chem. Eng. J, 2014, (248): 223-229.

[303] Rotaru A E, Shrestha P M, Liu F, Shrestha M, Shrestha D, Embree M, Zengler K, Wardman C, Nevin K P, Lovley D R.. A new model for electron flow during anaerobic digestion: Direct interspecies electron transfer to methanosaeta for the reduction of carbon dioxide to methane. Energy Environ. Sci., 2014, 7 (1): 408-415.

[304] Summers Z M, Fogarty H E, Leang C, Franks A E, Malvankar N S, Lovley D R. Direct exchange of electrons within aggregates of an evolved syntrophic coculture of anaerobic bacteria. Science, 2010, 330 (6009): 1413-1415.

[305] Zhao Z, Li Y, Quan X, Zhang Y. Towards engineering application: Potential mechanism for enhancing anaerobic digestion of complex organic waste with different types of conductive materials. Water Res., 2017, 115: 266-277.

[306] Zhang J, Zhang Y, Quan X, Chen S, Afzal S. Enhanced anaerobic digestion of organic contaminants containing diverse microbial population by combined microbial electrolysis cell(MEC) and anaerobic reactor under Fe(III)reducing conditions. Bioresour. Technol., 2013, 136: 273-280.

[307] Liu Y, Zhang Y, Ni B J. Zero valent iron simultaneously enhances methane production and sulfate reduction in anaerobic granular sludge reactors. Water Res., 2015, 75: 292-300.

[308] Liu Y, Zhang Y, Ni B J. Evaluating enhanced sulfate reduction and optimized volatile fatty acids (VFA) composition in anaerobic reactor by Fe(III) addition. Environ. Sci. Technol., 2015, 49 (4): 2123-2131.

[309] Zhang J, Zhang Y, Chang J, Quan X, Li Q. Biological sulfate reduction in the acidogenic phase of anaerobic digestion under dissimilatory Fe(III)-reducing conditions. Water Res., 2013, 47 (6): 2033-2040.

[310] Cruz Viggi C, Rossetti S, Fazi S, Paiano P, Majone M, Aulenta F. Magnetite particles triggering a faster and more robust syntrophic pathway of methanogenic propionate degradation. Environ. Sci. Technol., 2014, 48 (13): 7536-7543.

[311] Liu F, Rotaru A.E, Shrestha P M, Malvankar N S, Nevin K P, Lovley D R. Promoting direct interspecies electron transfer

with activated carbon. Energy Environ. Sci., 2012, 5 (10): 8982-8989.

[312] Zhao Z, Zhang Y, Woodard T, Nevin K, Lovley D. Enhancing syntrophic metabolism in up-flow anaerobic sludge blanket reactors with conductive carbon materials. Bioresour. Technol., 2015, 191: 140-145.

[313] Zhang J, Lu Y. Conductive Fe_3O_4 nanoparticles accelerate syntrophic methane production from butyrate oxidation in two different lake sediments. Front. Microbiol., 2016, 7: 1316.

[314] Li H, Chang J, Liu P, Fu L, Ding D, Lu Y. Direct interspecies electron transfer accelerates syntrophic oxidation of butyrate in paddy soil enrichments. Environ. Microbiol., 2015, 17 (5): 1533-1547.

[315] Yang Z, Guo R, Shi X, Wang C, Wang L, Dai M. Magnetite nanoparticles enable a rapid conversion of volatile fatty acids to methane. RSC Adv., 2016, 6 (31): 25662-25668.

[316] Zhao Z, Zhang Y, Li Y, Dang Y, Zhu T, Quan X. Potentially shifting from interspecies hydrogen transfer to direct interspecies electron transfer for syntrophic metabolism to resist acidic impact with conductive carbon cloth. Chem. Eng. J, 2017, 313: 10-18.

[317] Zhao Z, Zhang Y, Yu Q, Dang Y, Li Y, Quan X. Communities stimulated with ethanol to perform direct interspecies electron transfer for syntrophic metabolism of propionate and butyrate. Water Res., 2016, 102: 475-484.

[318] Zhao Z, Zhang Y, Holmes D E, Dang Y, Woodard T L, Nevin K P, Lovley D R. Potential enhancement of direct interspecies electron transfer for syntrophic metabolism of propionate and butyrate with biochar in up-flow anaerobic sludge blanket reactors. Bioresour. Technol., 2016, 209: 148-156.

[319] Lei Y, Sun D, Dang Y, Chen H, Zhao Z, Zhang Y, Holmes D E, Stimulation of methanogenesis in anaerobic digesters treating leachate from a municipal solid waste incineration plant with carbon cloth. Bioresour. Technol., 2016, 222: 270-276.

[320] Dang Y, Holmes D E, Zhao Z, Woodard T L, Zhang Y, Sun D, Wang L Y, Nevin K P, Lovley D R. Enhancing anaerobic digestion of complex organic waste with carbon-based conductive materials. Bioresour. Technol., 2016, 220: 516-522.

[321] Gillham R W, O Hannesin S F. Enhanced degradation of halogenated aliphatics by zero-valent iron. Ground Water, 1994, 32 (6): 958-967.

[322] Su C M, Puls R W. Arsenate and arsenite removal by zerovalent iron: Kinetics, redox transformation, and implications for in situ groundwater remediation. Environ. Sci. Technol., 2001, 35 (7): 1487-1492.

[323] Kumpiene J, Ore S, Renella G, Mench M, Lagerkvist A, Maurice C. Assessment of zerovalent iron for stabilization of chromium, copper, and arsenic in soil. Environ. Pollut., 2006, 144 (1): 62-69.

[324] Nam S, Tratnyek P G. Reduction of azo dyes with zero-valent iron. Water Res., 2000, 34 (6): 1837-1845.

[325] Agrawal A, Tratnyek P G. Reduction of nitro aromatic compounds by zero-valent iron metal. Environ. Sci. Technol., 1996, 30, (1): 153-160.

[326] Alowitz M J, Scherer M M. Kinetics of nitrate, nitrite, and Cr(VI) reduction by iron metal. Environm. Sci. Technol., 2002, 36 (3): 299-306.

[327] Xiong X, Sun B, Zhang J, Gao N, Shen J, Li J, Guan X. Activating persulfate by Fe^0 coupling with weak magnetic field: Performance and mechanism. Water Res., 2014, 62: 53-62.

[328] Xiong X, Sun Y, Sun B, Song W, Sun J, Gao N, Qiao J, Guan X. Enhancement of the advanced Fenton process by weak magnetic field for the degradation of 4-nitrophenol. RSC Adv., 2015, 5 (18): 13357-13365.

[329] Wilkin R T, Su C M, Ford R G, Paul C J. Chromium removal processes during groundwater remediation by a zero-valent iron permeable reactive barrier. Environ. Sci. Technol., 2005, 39: 4599–605.

[330] Guan X H, Sun Y K, Qin H J, Li J X, Lo I M, He D, Dong H R. The limitations of applying zero-valent iron technology in contaminants sequestration and the corresponding countermeasures: The development in zero-valent iron technology in the last two decades(1994–2014). Water Res., 2015, 75: 224-248.

[331] Gheju M. Hexavalent chromium reduction with Zero-Valent Iron(ZVI) in aquatic systems. Water Air Soil Pollut., 2011, 222 (1-4): 103-148.

[332] Wang C B, Zhang W X. Synthesizing nanoscale iron particles for rapid and complete dechlorination of TCE and PCBs. Environ. Sci. Technol., 1997, 31 (7): 2154-2156.

[333] Lai K C K, Lo I M C. Removal of chromium(VI) by acid-washed zero-valent iron under various groundwater geochemistry conditions. Environ. Sci. Technol., 2008, 42 (4): 1238-1244.

[334] Sun Y K, Li J X, Huang T L, Guan X H. The influences of iron characteristics, operating conditions and solution chemistry on contaminants removal by zero-valent iron: A review. Water Res., 2016, 100: 277-295.

[335] Moore A M, De Leon C H, Young T M. Rate and extent of aqueous perchlorate removal by iron surfaces. Environ. Sci. Technol., 2003, 37 (14): 3189-3198.

[336] Li J X, Qin H J, Guan X H. Premagnetization for enhancing the reactivity of multiple zerovalent iron samples toward various contaminants. Environ. Sci. Technol., 2015, 49 (24): 14401-14408.

[337] Nduta K G, Mwangi I W, Wanjau R W, Ngila J C. Removal of chlorine and chlorinated organic compounds from aqueous media using substrate-anchored zero-valent bimetals. Water Air Soil Pollut., 2016, 227, (1).

[338] Fan D M, Johnson G O, Tratnyek P G, Johnson R L. Sulfidation of nano zerovalent iron(nZVI) for improved selectivity during in-situ chemical reduction (ISCR). Environ. Sci. Technol., 2016, 50 (17): 9558-9565.

[339] Liang L P, Sun W, Guan X H, Huang Y Y, Choi W Y, Bao H L, Li L N, Jiang Z. Weak magnetic field significantly enhances selenite removal kinetics by zero valent iron. Water Res., 2014, 49: 371-380.

[340] Wang A Q, Guo W L, Hao F F, Yue X X, Leng Y Q. Degradation of Acid Orange 7 in aqueous solution by zero-valent aluminum under ultrasonic irradiation. Ultrason. Sonochem., 2014, 21 (2): 572-575.

[341] Kebria D Y, Taghizadeh M, Camacho J V, Latifi N, Remediation of PCE contaminated clay soil by coupling electrokinetics with zero-valent iron permeable reactive barrier. Environ. Earth Sci, 2016, 75 (8): 699.

[342] Jou C J G, Hsieh S C, Lee C L, Lin C T, Huang H W. Combining zero-valent iron nanoparticles with microwave energy to treat chlorobenzene. J. Taiwan Inst. Chem. Eng., 2010, 41 (2): 216-220.

[343] Tang C, Huang Y, Zhang Z, Chen J, Zeng H, Huang Y H. Rapid removal of selenate in a zero-valent iron/Fe_3O_4/Fe^{2+} synergetic system. Appl. Catal. B: Environ., 2016, 184: 320-327.

[344] Liu T X, Li X M, Waite T D. Depassivation of aged Fe^0 by divalent cations: Correlation between contaminant degradation and surface complexation constants. Environ. Sci. Technol., 2014, 48 (24): 14564-14571.

[345] He D, Ma X M, Jones A M, Ho L, Waite T D. Mechanistic and kinetic insights into the ligand-promoted depassivation of bimetallic zero-valent iron nanoparticles. Environ. Sci. Nano, 2016, 3 (4): 737-744.

[346] Guo X J, Yang Z, Dong H Y, Guan X H, Ren Q D, Lv X F, Jin X. Simple combination of oxidants with zero-valent-iron(ZVI)achieved very rapid and highly efficient removal of heavy metals from water. Water Res., 2016, 88: 671-680.

[347] Xu C H, Zhang B L, Zhu L J, Lin S, Sun X P, Jiang Z, Tratnyek P G. Sequestration of antimonite by zerovalent iron: Using weak magnetic field effects to enhance performance and characterize reaction mechanisms. Environ. Sci. Technol., 2016, 50 (3): 1483-1491.

[348] Huang Y H, Zhang T C. Nitrate reduction by surface-bound Fe(II) on solid surfaces at near-neutral pH and ambient temperature. J. Environ. Eng., 2016, 142, (11).

[349] Lu H J, Wang J K, Ferguson S, Wang T, Bao Y, Hao H X. Mechanism, synthesis and modification of nano zerovalent iron in water treatment. Nanoscale, 2016, 8 (19): 9962-9975.

[350] Li S, Yan W, Zhang W X. Solvent-free production of nanoscale zero-valent iron(nZVI) with precision milling. Green Chem, 2009, 11 (10): 1618-1626.

[351] Hoch L B, Mack E J, Hydutsky B W, Hershman J M, Skluzacek J M, Mallouk T E. Carbothermal synthesis of carbon-supported nanoscale zero-valent iron particles for the remediation of hexavalent chromium. Environ. Sci. Technol., 2008, 42 (7): 2600-2605.

[352] Sun H, Zhou G, Liu S, Ang H M, Tadé M O, Wang S. Nano-Fe^0 encapsulated in microcarbon spheres: synthesis, characterization, and environmental applications. ACS Appl. Mater. Interfaces, 2012, 4 (11): 6235-6241.

[353] Park S J, Kim S, Lee S, Khim Z G, Char K, Hyeon T. Synthesis and magnetic studies of uniform iron nanorods and nanospheres. J. Am. Chem. Soc., 2000, 122 (35): 8581-8582.

[354] Li F, Vipulanandan C, Mohanty K K. Microemulsion and solution approaches to nanoparticle iron production for degradation of trichloroethylene. Colloids. Surf. A: Physicochem. Eng. Asp., 2003, 223 (1-3): 103-112.

[355] Sun Y P, Li X Q, Cao J, Zhang W X, Wang H P. Characterization of zero-valent iron nanoparticles. Adv. Colloid Interface Sci., 2006, 120 (1-3): 47-56.

[356] Li S L, Yan W L, Zhang W X. Solvent-free production of nanoscale zero-valent iron(nZVI)with precision milling. Green Chem., 2009, 11 (10): 1618-1626.

[357] Nurmi J T, Tratnyek P G, Sarathy V, Baer D R, Amonette J E, Pecher K, Wang C, Linehan J C, Matson D W, Penn R L, Driessen M D. Characterization and properties of metallic iron nanoparticles: Spectroscopy, electrochemistry and kinetics. Environ. Sci. Technol., 2005, 39 (5): 1221-1230.

[358] Liu H B, Chen T H, Chang D Y, Chen D, Liu Y, He H P, Yuan P, Frost R. Nitrate reduction over nanoscale zero-valent iron prepared by hydrogen reduction of goethite. Mater. Chem. Phys., 2012, 133 (1): 205-211.

[359] Wang C Y, Chen Z Y, Cheng B, Zhu Y R, Liu H J. The preparation, surface modification, and characterization of metallic α-Fe nanoparticles. Chinese J. Chem. Phys., 1999, 60 (3): 223-226.

[360] Liu A R, Zhang W X. Fine structural features of nanoscale zero-valent iron characterized by spherical aberration corrected scanning transmission electron microscopy(Cs-STEM). Analyst., 2014, 139 (18): 4512-4518.

[361] Li S L, Ding Y Y, Wang W, Lei H. A facile method for determining the Fe(0) content and reactivity of zero valent iron. Anal. Methods, 2016, 8 (6): 1239-1248.

[362] Liu A, Liu J, Han J, Zhang W X. Evolution of nanoscale zero-valent iron(nZVI) in water: Microscopic and spectroscopic evidence on the formation of nano- and micro-structured iron oxides. J. Hazard. Mater., 2017, 322: 129-135.

[363] Morel F, Hering J G, Morel F. Principles and applications of aquatic chemistry. Wiley: 1993: 687-690.

[364] Wei Y T, Wu S C, Chou C M, Che C H, Tsai S M, Lien H L. Influence of nanoscale zero-valent iron on geochemical properties of groundwater and vinyl chloride degradation: A field case study. Water Res., 2010, 44 (1): 131-140.

[365] Shi Z, Nurmi J T, Tratnyek P G. Effects of nano zero-valent iron on oxidation-reduction potential. Environ. Sci. Technol., 2011, 45 (4): 1586-92.

[366] Glazier R, Venkatakrishnan R, Gheorghiu F, Walata L, Nash R, Zhang W X. Nanotechnology takes root. Civil. Eng., 2003, 73 (5): 64-69.

[367] Zhang W X. Nanoscale iron particles for environmental remediation: An overview. J. Nanopart. Res., 2003, 5 (3-4): 323-332.

[368] Elliott D W, Zhang W-X. Field assessment of nanoscale bimetallic particles for groundwater treatment.Environ. Sci. Technol., 2001, 35 (24): 4922-4926.

[369] Liendo M A, Navarro G E, Sampaio C H. Nano and micro ZVI in aqueous media: Copper uptake and solution behavior. Water Air Soil Pollut., 2013, 224 (5): 1541.

[370] Fu F L, Dionysiou D D, Liu H. The use of zero-valent iron for groundwater remediation and wastewater treatment: A review. J. Hazard. Mater., 2014, 267: 194-205.

[371] Suponik T, Winiarski A, Szade J. Processes of removing zinc from water using zero-valent iron. Water Air Soil Pollut.,

2015, 226 (11): 360.

[372] Vogel T M, Criddle C S, Mccarty P L. ES critical reviews: Transformations of halogenated aliphatic compounds. Environ. Sci. Technol., 1987, 21 (8): 722-36.

[373] He F, Zhao D. Preparation and characterization of a new class of starch-stabilized bimetallic nanoparticles for degradation of chlorinated hydrocarbons in water. Environ. Sci. Technol., 2005, 39 (9): 3314-3320.

[374] Zhang W X, Wang C B, Lien H L. Treatment of chlorinated organic contaminants with nanoscale bimetallic particles. Catal. Today, 1998, 40 (4): 387-395.

[375] Zhao X, Liu W, Cai Z Q, Han B, Qian T W, Zhao D Y. An overview of preparation and applications of stabilized zero-valent iron nanoparticles for soil and groundwater remediation. Water Res., 2016, 100: 245-266.

[376] Schrick B, Blough J L, Jones A D, Mallouk T E. Hydrodechlorination of trichloroethylene to hydrocarbons using bimetallic nickel-iron nanoparticles. Chem. Mater., 2002, 14 (12): 5140-5147.

[377] Lien H L, Zhang W X.Nanoscale Pd/Fe bimetallic particles: Catalytic effects of palladium on hydrodechlorination. Appl. Catal. B: Environ., 2007, 77 (1-2): 110-116.

[378] Song H, Carraway E R. Catalytic hydrodechlorination of chlorinated ethenes by nanoscale zero-valent iron. Appl. Catal. B: Environ., 2008, 78 (1): 53-60.

[379] Muftikian R, Fernando Q, Korte N. A method for the rapid dechlorination of low-molecular-weight chlorinated hydrocarbons in water. Water Res., 1995, 29 (10): 2434-2439.

[380] Wong M S, Alvarez P J J, Fang Y-l, Akçin N, Nutt M O, Miller J T, Heck K N. Cleaner water using bimetallic nanoparticle catalysts. J. Chem. Technol. Biotechnol., 2009, 84 (2): 158-166.

[381] Lien H L, Zhang W X. Transformation of chlorinated methanes by nanoscale iron particles. J. Environ. Eng., 1999, 125 (11): 1042-1047.

[382] Riba O, Scott T B, Ragnarsdottir K V, Allen G C. Reaction mechanism of uranyl in the presence of zero-valent iron nanoparticles, Geochim. Cosmochim. Acta, 2008, 72 (16): 4047-4057.

[383] Ling L, Zhang W X. Enrichment and encapsulation of uranium with iron nanoparticle. J. Am. Chem. Soc., 2015, 137 (8): 2788-2791.

[384] Kikuchi T, Tanaka S. Biological removal and recovery of toxic heavy metals in water environment. Crit. Rev. Environ. Sci. Technol., 2012, 42 (10): 1007-1057.

[385] Li X, Zhang W. Sequestration of metal cations with zerovalent iron nanoparticles a study with high resolution X-ray photoelectron spectroscopy (HR-XPS). J. Phys. Chem. C, 2007, 111 (19): 6939-6946.

[386] Ling L, Zhang W X. Sequestration of arsenate in zero-valent iron nanoparticles: Visualization of intraparticle reactions at angstrom resolution. Environ. Sci. Technol. Lett., 2014, 1 (7): 305-309.

[387] Ling L, Zhang W-X. Mapping the reactions of hexavalent chromium [Cr(VI)] in iron nanoparticles using spherical aberration corrected scanning transmission electron microscopy(Cs-STEM). Anal. Methods, 2014, 6 (10): 3211.

[388] Crane R A, Scott T. The removal of uranium onto carbon-supported nanoscale zero-valent iron particles. J. Nanopart. Res, 2014, 16: 2813.

[389] Yan W, Lien H L, Koel B E, Zhang W X. Iron nanoparticles for environmental clean-up: Recent developments and future outlook. Environ. Sci. Proc. Imp., 2013, 15 (1): 63-77.

[390] Su C M, Puls R W, Krug T A, Watling M T, O'Hara S K, Quinn J W, Ruiz N E. A two and half-year-performance evaluation of a field test on treatment of source zone tetrachloroethene and its chlorinated daughter products using emulsified zero valent iron nanoparticles. Water Res., 2012, 46 (16): 5071-5084.

[391] Johnson R L, Nurmi J T, Johnson G S O, Fan D M, Johnson R L O, Shi Z Q, Salter-Blanc A J, Tratnyek P G, Lowry G V. Field-scale transport and transformation of carboxymethylcellulose-stabilized nano zero-valent iron. Environ. Sci. Technol.,

2013, 47 (3): 1573-1580.

[392] Kocur C M, Chowdhury A I, Sakulchaicharoen N, Boparai H K, Weber K P, Sharma P, Krol M M, Austrins L, Peace C, Sleep B E, O'Carroll D M. Characterization of nZVI mobility in a field sScale test. Environ. Sci. Technol., 2014, 48 (5): 2862-2869.

[393] Bardos R, Bone B, Elliott D, Hartog, N, Henstock J, Nathanail P. In a risk/benefit approach to the application of iron nanoparticles for the remediation of contaminated sites in the environment. aiaa/asme/sae/asee Joint Propulsion Conference and Exhibit, 2011: 425-426.

[394] Mueller N C, Braun J, Bruns J, Cernik M, Rissing P, Rickerby D, Nowack B. Application of nanoscale zero valent iron(NZVI)for groundwater remediation in Europe. Environ. Sci. Pollut., 2012, 19 (2): 550-558.

[395] Li X Q, Elliott D W, Zhang W X. Zero-valent iron nanoparticles for abatement of environmental pollutants: Materials and engineering aspects, Crit. Rev. Solid State Mater. Sci., 2006, 31 (4): 111-122.

[396] He F, Zhao D, Paul C. Field assessment of carboxymethyl cellulose stabilized iron nanoparticles for in situ destruction of chlorinated solvents in source zones. Water Res., 2010, 44 (7): 2360-2370.

[397] Li S L, Wang W, Liang F P, Zhang W X. Heavy metal removal using nanoscale zero-valent iron(nZVI): Theory and application. J. Hazard. Mater., 2017, 322: 163-171.

[398] Li S L, Wang W, Yan W L, Zhang W X. Nanoscale zero-valent iron(nZVI) for the treatment of concentrated Cu(II) wastewater: A field demonstration, Environ. Sci. Proc. Imp., 2014, 16 (3): 524-533.

[399] Li S L, Wang W, Liu Y Y, Zhang W X. Zero-valent iron nanoparticles(nZVI) for the treatment of smelting wastewater: A pilot-scale demonstration. Chem. Eng. J, 2014, 254: 115-123.

[400] Wang W, Hua Y L, Li S L, Yan W L, Zhang W X. Removal of Pb(II) and Zn(II) using lime and nanoscale zero-valent iron(nZVI): A comparative study. Chem. Eng. J, 2016, 304: 79-88.

[401] Crane R A, Scott T B. Nanoscale zero-valent iron: Future prospects for an emerging water treatment technology. J. Hazard. Mater, 2012, 211: 112-125.

[402] Karn B, Kuiken T, Otto M. Nanotechnology and in situ remediation: A review of the benefits and potential risks. Environ. Health Perspect., 2009, 117 (12): 1823-1831.

[403] He F, Zhao D Y, Liu J C, Roberts C B. Stabilization of Fe-Pd nanoparticles with sodium carboxymethyl cellulose for enhanced transport and dechlorination of trichloroethylene in soil and groundwater. Ind. Eng. Chem. Res., 2007, 46 (1): 29-34.

[404] He F, Zhao D Y. Preparation and characterization of a new class of starch-stabilized bimetallic nanoparticles for degradation of chlorinated hydrocarbons in water. Environ. Sci. Technol., 2005, 39 (9): 3314-3320.

[405] He F, Zhao D Y. Manipulating the size and dispersibility of zerovalent iron nanoparticles by use of carboxymethyl cellulose stabilizers. Environ. Sci. Technol., 2007, 41 (17): 6216-6221.

[406] Laurent S, Forge D, Port M, Roch A, Robic C, Vander Elst L, Muller R N. Magnetic iron oxide nanoparticles: Synthesis, stabilization, vectorization, physicochemical characterizations, and biological applications. Chem. Rev., 2008, 108 (6): 2064-2110.

[407] Zhao X, Liu W, Cai Z, Han B, Qian T, Zhao D. An overview of preparation and applications of stabilized zero-valent iron nanoparticles for soil and groundwater remediation. Water Res., 2016, 100: 245-66.

[408] Saleh N, Sirk K, Liu Y, Phenrat T, Dufour B, Matyjaszewski K, Tilton R D, Lowry G V. Surface modifications enhance nanoiron transport and NAPL targeting in saturated porous media. Environ. Eng. Sci., 2007, 24 (1): 45-57.

[409] Saleh N, Kim H-J, Phenrat T, Matyjaszewski K, Tilton R D, Lowry G V. Ionic strength and composition affect the mobility of surface-modified Fe0 nanoparticles in water-saturated sand columns. Environ. Sci. Technol., 2008, 42 (9): 3349-3355.

[410] Comba S, Sethi R. Stabilization of highly concentrated suspensions of iron nanoparticles using shear-thinning gels of

xanthan gum. Water Res., 2009, 43 (15): 3717-3726.

[411] Phenrat T, Liu Y Q, Tilton R D, Lowry G V. Adsorbed polyelectrolyte coatings decrease Fe0 nanoparticle reactivity with TCE in water: Conceptual model and mechanisms. Environ. Sci. Technol., 2009, 43 (5): 1507-1514.

[412] Zhang M, He F, Zhao D Y, Hao X D. Degradation of soil-sorbed trichloroethylene by stabilized zero valent iron nanoparticles: Effects of sorption, surfactants, and natural organic matter. Water Res., 2011, 45 (7): 2401-2414.

[413] Kanel S R, Nepal D, Manning B, Choi H. Transport of surface-modified iron nanoparticle in porous media and application to arsenic(III) remediation. J. Nanopart. Res., 2007, 9 (5): 725-735.

[414] Basnet M, Ghoshal S, Tufenkji N. Rhamnolipid biosurfactant and soy proteinact as effective stabilizers in the aggregation and transport of palladium-doped zerovalent iron nanoparticles in saturated porous media. Environ. Sci. Technol., 2013, 47 (23): 13355-13364.

[415] Bhattacharjee S, Basnet M, Tufenkji N, Ghoshal S. Effects of rhamnolipid and carboxymethylcellulose coatings on reactivity of palladium-doped nanoscale zerovalent iron particles. Environ. Sci. Technol., 2016, 50 (4): 1812-1820.

[416] Wei Y T, Wu S C, Yang S W, Che C H, Lien H L, Huang D-H. Biodegradable surfactant stabilized nanoscale zero-valent iron for in situ treatment of vinyl chloride and 1, 2-dichloroethane. J. Hazard. Mater., 2012, 211: 373-380.

[417] He F, Zhao D Y, Paul C. Field assessment of carboxymethyl cellulose stabilized iron nanoparticles for in situ destruction of chlorinated solvents in source zones. Water Res., 2010, 44 (7): 2360-2370.

[418] Schrick B, Hydutsky B W, Blough J L, Mallouk T E. Delivery vehicles for zerovalent metal nanoparticles in soil and groundwater. Chem. Mater., 2004, 16 (11): 2187-2193.

[419] Phenrat T, Saleh N, Sirk K, Kim H J, Tilton R D, Lowry G V. Stabilization of aqueous nanoscale zerovalent iron dispersions by anionic polyelectrolytes: Adsorbed anionic polyelectrolyte layer properties and their effect on aggregation and sedimentation. J. Nanopart. Res., 2008, 10 (5): 795-814.

[420] Sirk K M, Saleh N B, Phenrat T, Kim H-J, Dufour B, Ok J, Golas P L, Matyjaszewski K, Lowry G V, Tilton R D. Effect of adsorbed polyelectrolytes on nanoscale zero valent iron particle attachment to soil surface models. Environ. Sci. Technol, 2009, 43 (10): 3803-3808.

[421] Yan W L, Lien H L, Koel B E, Zhang W X. Iron nanoparticles for environmental clean-up: Recent developments and future outlook, Environ. Sci. Proc. Imp., 2013, 15 (1): 63-77.

[422] Comba S, Dalmazzo D, Santagata E, Sethi R. Rheological characterization of xanthan suspensions of nanoscale iron for injection in porous media. J. Hazard. Mate.r, 2011, 185 (2): 598-605.

[423] Cirtiu C M, Raychoudhury T, Ghoshal S, Moores A. Systematic comparison of the size, surface characteristics and colloidal stability of zero valent iron nanoparticles pre- and post-grafted with common polymers. Colloids Surf. A: Physicochem. Eng. Asp., 2011, 390 (1): 95-104.

[424] Sakulchaicharoen N, O'Carroll D M, Herrera J E. Enhanced stability and dechlorination activity of pre-synthesis stabilized nanoscale FePd particles. J. Contam. Hydrol., 2010, 118 (3-4): 117-127.

[425] Xin J, Tang F, Zheng X, Shao H, Kolditz O. Transport and retention of xanthan gum-stabilized microscale zero-valent iron particles in saturated porous media. Water Res., 2016, 88: 199-206.

[426] Xue D, Sethi R. Viscoelastic gels of guar and xanthan gum mixtures provide long-term stabilization of iron micro- and nanoparticles. J. Nanopart. Res., 2012, 14 (11): 1239.

[427] Busch J, Meißner T, Potthoff A, Bleyl S, Georgi A, Mackenzie K, Trabitzsch R, Werban U, Oswald S E. A field investigation on transport of carbon-supported nanoscale zero-valent iron(nZVI) in groundwater. J. Contam. Hydrol., 2015, 181: 59-68.

[428] Gao J, Wang W, Rondinone A J, He F, Liang L. Degradation of trichloroethene with a novel ball milled Fe-C nanocomposite, J. Hazard. Mater., 2015, 300: 443-50.

[429] Zheng T, Zhan J, He J, Day C, Lu Y, McPherson G L, Piringer G, John V T. Reactivity characteristics of nanoscale zerovalent iron-silica composites for trichloroethylene remediation. Environ. Sci. Technol., 2008, 42 (12): 4494-4499.

[430] Shi L-N, Zhang X, Chen Z-L. Removal of chromium(VI) from wastewater using bentonite-supported nanoscale zero-valent iron, Water Res, 2011, 45 (2): 886-892.

[431] Chen Z.-x, Jin X.-y, Chen Z, Megharaj M, Naidu R. Removal of methyl orange from aqueous solution using bentonite-supported nanoscale zero-valent iron. J. Colloid Interface Sci., 2011, 363 (2): 601-607.

[432] Kocur C M, O'Carroll D M, Sleep B E. Impact of nZVI stability on mobility in porous media. J. Contam. Hydrol., 2013, 145: 17-25.

[433] Krol M M, Oleniuk A J, Kocur C M, Sleep B E, Bennett P, Xiong Z, O'Carroll D M. A Field-validated model for in situ transport of polymer-stabilized nZVI and implications for subsurface injection. Environ. Sci. Technol., 2013, 47 (13): 7332-7340.

[434] Su C, Puls R W, Krug T A, Watling M T, O'Hara S K, Quinn J W, Ruiz N E. A two and half-year-performance evaluation of a field test on treatment of source zone tetrachloroethene and its chlorinated daughter products using emulsified zero valent iron nanoparticles. Water Res., 2012, 46 (16): 5071-5084.

[435] Luna M, Gastone F, Tosco T, Sethi R, Velimirovic M, Gemoets J, Muyshondt R, Sapion H, Klaas N, Bastiaens L. Pressure-controlled injection of guar gum stabilized microscale zerovalent iron for groundwater remediation. J. Contam. Hydrol., 2015, 181: 46-58.

[436] Quinn J, Geiger C, Clausen C, Brooks K, Coon C, O'Hara S, Krug T, Major D, Yoon W-S, Gavaskar A, Holdsworth T. Field demonstration of DNAPL dehalogenation using emulsified zero-valent Iron. Environ. Sci. Technol., 2005, 39 (5): 1309-1318.

[437] Johnson R L, Johnson G O B, Nurmi J T, Tratnyek P G. Natural organic matter enhanced mobility of nano zerovalent iron, Environ. Sci. Technol., 2009, 43 (14): 5455-5460.

[438] Liu Y, Phenrat T, Lowry G V. Effect of TCE concentration and dissolved groundwater solutes on NZVI-promoted TCE dechlorination and H_2 evolution, Environ. Sci. Technol., 2007, 41 (22): 7881-7887.

[439] Fan D, O'Brien Johnson G, Tratnyek P G, Johnson R L. Sulfidation of nano zerovalent iron(nZVI) for improved selectivity during in-situ chemical reduction(ISCR). Environ. Sci. Technol., 2016, 50 (17): 9558-9565.

[440] Liu Y, Lowry G V. Effect of particle age(Fe0 Content)and solution pH on NZVI reactivity: H_2 evolution and TCE dichlorination, Environ. Sci. Technol., 2006, 40 (19): 6085-6090.

[441] Reinsch B C, Forsberg B, Penn R L, Kim C S, Lowry G V. Chemical transformations during aging of zerovalent iron nNanoparticles in the presence of common groundwater dissolved constituents. Environ. Sci. Technol., 2010, 44, (9): 3455-3461.

[442] Sarathy V, Tratnyek P G, Nurmi J T, Baer D R, Amonette J E, Chun C L, Penn R L, Reardon E J. Aging of iron nanoparticles in aqueous solution: Effects on structure and reactivity. J. Phys. Chem. C, 2008, 112 (7): 2286-2293.

[443] Xie Y, Cwiertny D M. Use of dithionite to extend the reactive lifetime of nanoscale zero-valent iron treatment systems. Environ. Sci. Technol., 2010, 44 (22): 8649-8655.

[444] Rajajayavel S, R Ghoshal S. Enhanced reductive dechlorination of trichloroethylene by sulfidated nanoscale zerovalent iron. Water Res., 2015, 78: 144-53.

[445] Han Y, Yan W. Reductive dechlorination of trichloroethene by zero-valent iron nanoparticles: Reactivity enhancement through sulfidation treatment. Environ. Sci. Technol., 2016, 50 (23): 12992-13001.

[446] Lee C, Kim J Y, Lee W I, Nelson K L, Yoon J, Sedlak D L. Bactericidal effect of zero-valent iron nanoparticles on *Escherichia coli*. Environ. Sci. Technol., 2008, 42 (13): 4927-4933.

[447] Kirschling T L, Gregory K B, Minkley J E G, Lowry G V, Tilton R D. Impact of nanoscale zero valent iron on

geochemistry and microbial populations in trichloroethylene contaminated aquifer materials. Environ. Sci. Technol., 2010, 44 (9): 3474-3480.

[448] Xiu Z M, Jin Z H, Li T L, Mahendra S, Lowry G V, Alvarez P J J. Effects of nano-scale zero-valent iron particles on a mixed culture dechlorinating trichloroethylene. Bioresour. Technol., 2010, 101 (4): 1141-1146.

[449] Li Z, Greden K, Alvarez P J J, Gregory K B, Lowry G V. Adsorbed polymer and NOM limits adhesion and toxicity of nano scale zerovalent iron to *E. coli*. Environ. Sci. Technol., 2010, 44 (9): 3462-3467.

[450] Xiu Z-m, Gregory K B, Lowry G V, Alvarez P J J. Effect of bare and coated nanoscale zerovalent iron on tceA and vcrA gene expression in dehalococcoides spp. Environ. Sci. Technol., 2010, 44 (19): 7647-7651.

[451] Chen J W, Xiu Z M, Lowry G V, Alvarez P J J. Effect of natural organic matter on toxicity and reactivity of nano-scale zero-valent iron. Water Res., 2011, 45 (5): 1995-2001.

[452] Teng W, Fan J, Wang W, Bai N, Liu R, Liu Y, Deng Y, Kong B, Yang J, Zhao D, Zhang, W-X. Nanoscale zero-valent iron in mesoporous carbon(nZVI@C): Stable nanoparticles for metal extraction and catalysis. J. Mater. Chem, 2017, 5 (9): 4478-4485.

[453] Wu L, Shamsuzzoha M, Ritchie S M C. Preparation of cellulose acetate supported zero-valent iron nanoparticles for the dechlorination of trichloroethylene in water. J. Nanopart. Res., 2005, 7: 469-476

[454] Cho H H, Lee T, Hwang S-J, Park J W. Iron and organo-bentonite for the reduction and sorption of trichloroethylene. Chemosphere, 2005, 58 (1): 103-108.

[455] Üzüm Ç, Shahwan T, Eroğlu A E, Hallam K R, Scott T B, Lieberwirth I. Synthesis and characterization of kaolinite-supported zero-valent iron nanoparticles and their application for the removal of aqueous Cu^{2+} and Co^{2+} ions. Appl. Clay Sci., 2009, 43 (2): 172-181.

[456] Zhang X, Lin S, Chen Z, Megharaj M, Naidu R. Kaolinite-supported nanoscale zero-valent iron for removal of Pb^{2+} from aqueous solution: Reactivity, characterization and mechanism. Water Res., 2011, 45 (11): 3481-3488.

[457] Gu C, Jia H, Li H, Teppen B J, Boyd S A. Synthesis of highly reactive subnano-sized zero-valent iron using smectite clay templates. Environ. Sci. Technol., 2010, 44 (11): 4258-4263.

[458] Fan M, Yuan P, Chen T, He H, Yuan A, Chen K, Zhu J, Liu D. Synthesis, characterization and size control of zerovalent iron nanoparticles anchored on montmorillonite. Chin. Sci. Bull., 2010, 55 (11): 1092-1099.

[459] Shahwan T, Üzüm Ç, Eroğlu A E, Lieberwirth I. Synthesis and characterization of bentonite/iron nanoparticles and their application as adsorbent of cobalt ions. Appl. Clay Sci., 2010, 47 (3): 257-262.

[460] Frost R L, Xi Y, He H. Synthesis, characterization of palygorskite supported zero-valent iron and its application for methylene blue adsorption. J. Colloid Interface Sci., 2010, 341 (1): 153-161.

[461] Su J, Minegishi T, Katayama M, Domen K. Photoelectrochemical hydrogen evolution from water on a surface modified CdTe thin film electrode under simulated sunlight. J. Mater. Chem., 2017, 5 (9): 4486-4492.

[462] Das T, Nicholas J. D, Qi Y. Long-range charge transfer and oxygen vacancy interactions in strontium ferrite. J. Mater. Chem., 2017, 5 (9): 4493-4506.

[463] Yu K, Sheng G D, McCall W. Cosolvent effects on dechlorination of soils-orbed polychlorinated biphenyls using bentonite clay-templated nanoscale zero valent iron. Environ. Sci. Technol., 2016, 50 (23): 12949-12956.

[464] Yu K, Gu C, Boyd S A, Liu C, Sun C, Teppen B J, Li H. Rapid and extensive debromination of decabromodiphenyl ether by smectite clay-templated subnanoscale zero-valent iron. Environ. Sci. Technol., 2012, 46 (16): 8969-8975.

[465] Bhowmick S, Chakraborty S, Mondal P, Van Renterghem W, Van den Berghe S, Roman-Ross G, Chatterjee D, Iglesias M. Montmorillonite-supported nanoscale zero-valent iron for removal of arsenic from aqueous solution: Kinetics and mechanism. Chem. Eng. J, 2014, 243: 14-23.

[466] Kadu B S, Sathe Y D, Ingle A B, Chikate R C, Patil K R, Rode C V. Efficiency and recycling capability of montmorillonite

supported Fe-Ni bimetallic nanocomposites towards hexavalent chromium remediation. Appl. Catal. B, 2011, 104 (3): 407-414.

[467] Wu P, Li S, Ju L, Zhu N, Wu J, Li P, Dang Z. Mechanism of the reduction of hexavalent chromium by organo-montmorillonite supported iron nanoparticles. J. Hazard. Mater., 2012, 219: 283-288.

[468] Sheng G, Shao X, Li Y, Li J, Dong H, Cheng W, Gao X, Huang Y. Enhanced removal of uranium(VI) by nanoscale zerovalent iron supported on Na-bentonite and an investigation of mechanism. J. Phys. Chem. A, 2014, 118 (16): 2952-2958.

[469] Li Y, Li J, Zhang Y. Mechanism insights into enhanced Cr(VI) removal using nanoscale zerovalent iron supported on the pillared bentonite by macroscopic and spectroscopic studies. J. Hazard. Mater., 2012, 227: 211-218.

[470] Zhang Y, Li Y, Li J, Hu L, Zheng X. Enhanced removal of nitrate by a novel composite: Nanoscale zero valent iron supported on pillared clay. Chem. Eng. J, 2011, 171 (2): 526-531.

[471] Li Y, Zhang Y, Li J, Zheng X. Enhanced removal of pentachlorophenol by a novel composite: Nanoscale zero valent iron immobilized on organobentonite, Environ. Pollut., 2011, 159 (12): 3744-3749.

[472] Fu R, Yang Y, Xu Z, Zhang X, Guo X, Bi D. The removal of chromium(VI) and lead(II) from groundwater using sepiolite-supported nanoscale zero-valent iron(S-NZVI). Chemosphere, 2015, 138: 726-734.

[473] Fei X, Cao L, Zhou L, Gu Y, Wang X. Degradation of bromamine acid by nanoscale zero-valent iron(nZVI) supported on sepiolite, Water Sci. Technol., 2012, 66 (12): 2539-2545.

[474] Chang Y, He Y-y, Liu T, Guo Y H, Zha F. Aluminum pillared palygorskite-supported nanoscale zero-valent iron for removal of Cu(II), Ni(II) from aqueous solution. Arab. J. Sci. Eng., 2014, 39 (9): 6727-6736.

[475] Luo S, Qin P, Shao J, Peng L, Zeng Q, Gu J-D. Synthesis of reactive nanoscale zero valent iron using rectorite supports and its application for orange II removal. Chem. Eng. J, 2013, 223: 1-7.

[476] Xi Y, Sun Z, Hreid T, Ayoko G A, Frost R L. Bisphenol A degradation enhanced by air bubbles via advanced oxidation using in situ generated ferrous ions from nano zero-valent iron/palygorskite composite materials. Chem. Eng. J, 2014, 247: 66-74.

[477] Li Y, Cheng W, Sheng G, Li J, Dong H, Chen Y, Zhu L. Synergetic effect of a pillared bentonite support on Se(VI) removal by nanoscale zero valent iron. Appl. Catal. B, 2015, 174: 329-335.

[478] Dong H, Chen Y, Sheng G, Li J, Cao J, Li Z, Li Y. The roles of a pillared bentonite on enhancing Se(VI) removal by ZVI and the influence of co-existing solutes in groundwater, J. Hazard. Mater., 2016, 304: 306-312.

[479] Olson M R, Sale T C, Shackelford C D, Bozzini C, Skeean J. Chlorinated solvent source-zone remediation via ZVI-clay soil mixing: 1-Year results. Ground Water Monit. Rem., 2012, 32 (3): 63-74.

[480] Fjordbøge A S, Riis C, Christensen A G, Kjeldsen P. ZVI-clay remediation of a chlorinated solvent source zone, Skuldelev, Denmark: 1 Site description and contaminant source mass reduction. J. Contam. Hydrol., 2012, 140: 56-66.

[481] Fjordbøge A S, Lange I V, Bjerg P L, Binning P J, Riis C, Kjeldsen P. ZVI-Clay remediation of a chlorinated solvent source zone, Skuldelev, Denmark: 2 Groundwater contaminant mass discharge reduction, J. Contam. Hydrol., 2012, 140: 67-79.

[482] Liang L P, Guan X H, Huang Y Y, Ma J Y, Sun X P, Qiao J L, Zhou G M. Efficient selenate removal by zero-valent iron in the presence of weak magnetic field, Sep. Purif. Technol., 2015, 156: 1064-1072.

[483] Guan X H, Jiang X, Qiao J L, Zhou G M. Decomplexation and subsequent reductive removal of EDTA-chelated Cu(II)by zero-valent iron coupled with a weak magnetic field: Performances and mechanisms, J. Hazard. Mater., 2015, 300: 688-694.

[484] Jiang X, Qiao J L, Lo I M, Wang L, Guan X H, Lu Z P, Zhou G M, Xu C H. Enhanced paramagnetic Cu^{2+} ions removal by coupling a weak magnetic field with zero valent iron. J. Hazard. Mater., 2015, 283, 880-887.

[485] Feng P, Guan X H, Sun Y K, Choi W, Qin H J, Wang J M, Qiao J L, Li L N. Weak magnetic field accelerates chromate removal by zero-valent iron, J. Environ. Sci., 2015, 31: 175-183.

[486] Sun Y, Guan X, Wang J, Meng X, Xu C, Zhou G. Effect of weak magnetic field on arsenate and arsenite removal from water by zerovalent iron: An XAFS investigation. Environ. Sci. Technol., 2014, 48 (12): 6850-6858.

[487] Li J, Bao H, Xiong X, Sun Y, Guan X. Effective Sb(V) immobilization from water by zero-valent iron with weak magnetic field. Sep. Purif. Technol., 2015, 151: 276-283.

[488] Xu C H, Zhang B L, Zhu L J, Lin S, Sun X P, Jiang Z, Tratnyek P G. Sequestration of antimonite by zerovalent iron: Using weak magnetic field effects to enhance performance and characterize reaction mechanisms. Environ. Sci. Technol., 2016, 50: 1483-1491.

[489] Liang L P, Guan X H, Shi Z, Li J L, Wu Y N, Tratnyek P G. Coupled effects of aging and weak magnetic fields on sequestration of selenite by zero-valent iron. Environ. Sci. Technol., 2014, 48, (11), 6326-6334.

[490] Xu H Y, Sun Y K, Li J X, Li F M, Guan X H. Aging of zerovalent iron in synthetic groundwater: X-ray photoelectron spectroscopy depth profiling characterization and depassivation with uniform magnetic field. Environ. Sci. Technol., 2016, 50 (15): 8214-8222.

[491] Xiang W, Zhang B, Zhou T, Wu X, Mao J. An insight in magnetic field enhanced zero-valent iron/H_2O_2 Fenton-like systems: Critical role and evolution of the pristine iron oxides layer. Sci. Rep, 2016, 6, 24094.

[492] Xiong X, Sun B, Zhang J, Gao N, Shen J, Li J, Guan X. Activating persulfate by Fe0 coupling with weak magnetic field: Performance and mechanism. Water Res., 2014, 62: 53-62.

[493] Li J X, Shi Z, Ma B, Zhang P, Jiang X, Xiao Z J, Guan X H. Improving the reactivity of zerovalent iron by taking advantage of its magnetic memory: Implications for arsenite removal. Environ. Sci. Techno.l, 2015, 49 (17): 10581-10588.

[494] Li J, Qin H, Zhang W.-x, Shi Z, Zhao D, Guan X. Enhanced Cr(VI) removal by zero-valent iron coupled with weak magnetic field: Role of magnetic gradient force. Sep. Purif. Technol., 2017, 176: 40-47.

[495] Sun Y K, Hu Y H, Huang T L, Li J X, Qin H J, Guan X H. Combined effect of weak magnetic fields and anions on arsenite sequestration by zerovalent iron: Kinetics and mechanisms. Environ. Sci. Technol., 2017, 51 (7): 3742-3750.

[496] Sun Y, Li J, Huang T, Guan X. The influences of iron characteristics, operating conditions and solution chemistry on contaminants removal by zero-valent iron: A review. Water Res., 2016, 100: 277-295.

[497] Guo X J, Yang Z, Liu H, Lv X F, Tu Q S, Ren Q D, Xia X H, Jing C Y. Common oxidants activate the reactivity of zero-valent iron(ZVI)and hence remarkably enhance nitrate reduction from water. Sep. Purif. Technol., 2015, 146: 227-234.

[498] Xu J, Hao, Z W, Xie C S, Lv X S, Yang Y P, Xu X H. Promotion effect of Fe^{2+} and Fe_3O_4 on nitrate reduction using zero-valent iron. Desalination, 2012, 284: 9-13.

[499] Huang Y H, Tang C, Zeng H. Removing molybdate from water using a hybridized zero-valent iron/magnetite/Fe(II)treatment system. Chem. Eng. J, 2012, 200: 257-263.

[500] Liu T X, Li X M, Waite T D. Depassivation of aged Fe0 by ferrous ions: Implications to contaminant degradation. Environ. Sci. Technol., 2013, 47 (23): 13712-13720.

[501] Tang C L, Huang Y H, Zeng H, Zhang Z Q. Reductive removal of selenate by zero-valent iron: The roles of aqueous Fe2+ and corrosion products, and selenate removal mechanisms. Water Res., 2014, 67: 166-174.

[502] Tang C. L, Huang Y P, Zhang Z Q, Chen J J, Zeng H, Huang Y H. Rapid removal of selenate in a zero-valent iron/Fe_3O_4/Fe^{2+} synergetic system. Appl. Catal. B: Environ, 2016, 184: 320-327.

[503] Lipczynskakochany E, Harms S, Milburn R, Sprah G, Nadarajah N. Degradation of carbon-tetrachloride in the presence of iron and sulfur-containing-compounds. Chemosphere ,1994, 29 (7): 1477-1489.

[504] Hassan S M. Reduction of halogenated hydrocarbons in aqueous media: I, Involvement of sulfur in iron catalysis. Chemosphere, 2000, 40 (12): 1357-1363.

[505] Xu C, Zhang B, Wang Y, Shao Q, Zhou W, Fan D, Bandstra J Z, Shi Z. Tratnyek P G. Effects of sulfidation, magnetization, and oxygenation on azo dye reduction by zerovalent Iron. Environ. Sci. Technol., 2016, 50 (21): 11879-11887.

[506] Butler E C, Hayes K F. Factors influencing rates and products in the transformation of trichloroethylene by iron sulfide and iron metal. Environ. Sci. Technol., 2001, 35 (19): 3884-3891.

[507] Li H, Shan C, Zhang Y, Cai J, Zhang W, Pan B. Arsenate adsorption by hydrous ferric oxide nanoparticles embedded in cross-linked anion exchanger: Effect of the host pore structure, ACS Appl. Mater. Interfaces, 2016, 8(5): 3012-3020.

[508] Zhang X, Wang Y, Chang X, Wang P, Pan B. Iron oxide nanoparticles confined in mesoporous silicates for arsenic sequestration: Effect of the host pore structure. Environ. Sci. Nano, 2017, 4 (3): 679-688.

[509] Nie G, Wang J, Pan B, Lv L. Surface chemistry of polymer-supported nano-hydrated ferric oxide for arsenic removal: Effect of host pore structure, Sci. China Chem., 2015, 58 (4): 722-730.

[510] Nie G Z, Pan B C, Zhang S J, Pan B J. Surface chemistry of nanosized hydrated ferric oxide encapsulated inside porous polymer: Modeling and experimental studies. J. Phys. Chem. C, 2013, 117 (12): 6201-6209.

[511] Yang W, Li X, Pan B, Lv L, Zhang W. Effective removal of effluent organic matter(EfOM)from bio-treated coking wastewater by a recyclable aminated hyper-cross-linked polymer. Water Res., 2013, 47 (13): 4730-4738.

[512] Zhang X, Wu M, Dong H, Li H, Pan B. Simultaneous oxidation and sequestration of As(III) from water by using redox polymer-based Fe(III) oxide nanocomposite. Environ. Sci. Technol., 2017, 51 (11): 6326-6334.

[513] Li X, He K, Pan B, Zhang S, Lu L, Zhang W. Efficient As(III) removal by macroporous anion exchanger-supported Fe-Mn binary oxide: Behavior and mechanism. Chem. Eng. J, 2012, 193: 131-138.

[514] Du Q, Zhang S, Pan B, Lv L, Zhang W, Zhang Q. Bifunctional resin-ZVI composites for effective removal of arsenite through simultaneous adsorption and oxidation. Water Res., 2013, 47 (16): 6064-6074.

[515] Shan C, Wang X, Guan X, Liu F, Zhang W, Pan B. Efficient removal of trace Se(VI) by millimeter-sized nanocomposite of zerovalent iron confined in polymeric anion exchanger, Ind. Eng. Chem. Res., 2017, 56 (18): 5309-5317.

[516] Hua M, Xiao L, Pan B, Zhang Q. Validation of polymer-based nano-iron oxide in further phosphorus removal from bioeffluent: laboratory and scaledup study. Front. Environ. Sci. Eng., 2013, 7 (3): 435-441.

[517] Zhang Y, Pan B. Modeling batch and column phosphate removal by hydrated ferric oxide-based nanocomposite using response surface methodology and artificial neural network. Chem. Eng. J, 2014, 249: 111-120.

[518] Zhang Y, Zhang W, Pan B. Struvite-based phosphorus recovery from the concentrated bioeffluent by using HFO nanocomposite adsorption: Effect of solution chemistry. Chemosphere, 2015, 141, 227-234.

[519] Lee Y, Yoon J, von Gunten U. Kinetics of the oxidation of phenols and phenolic endocrine disruptors during Water Treatment with Ferrate(Fe(VI)). Environ. Sci. Technol., 2005, 39 (22): 8978-8984.

[520] Sharma V K. Ferrate(VI) and ferrate(V) oxidation of organic compounds: Kinetics and mechanism. Coord Chem Rev, 2013, 257 (2): 495-510.

[521] Sharma V K, Chen L, Zboril R. Review on high valent Fe VI (Ferrate): A sustainable green oxidant in organic chemistry and transformation of pharmaceuticals, ACS Sustain Chem, Eng., 2016, 4 (1): 18-34.

[522] Sharma V K, Zboril R, Varma R S. Ferrates: Greener oxidants with multimodal action in water treatment technologies. Acc Chem Res., 2015, 48 (2): 182-91.

[523] Sharma V K. Oxidation of inorganic contaminants by ferrates(VI, V, and IV)--kinetics and mechanisms, A review. J. Environ. Manage, 2011, 92 (4): 1051-73.

[524] Prucek R, Tucek J, Kolarik J, Filip J, Marusak Z, Sharma V K, Zboril R. Ferrate(VI)-induced arsenite and arsenate removal by in situ structural incorporation into magnetic iron(III) oxide nanoparticles. Environ. Sci. Technol., 2013, 47 (7): 3283-92.

[525] Prucek R, Tucek J, Kolarik J, Huskova I, Filip J, Varma R S, Sharma V K, Zboril R. Ferrate(VI)-prompted removal of

metals in aqueous media: Mechanistic delineation of enhanced efficiency via metal entrenchment in magnetic oxides. Environ. Sci. Technol., 2015, 49 (4): 2319-27.

[526] Murmann R K, Robinson P R. Experiments utilizing FeO_4^{2-} for purifying water. Water Re.s, 1974, 8 (8): 543-547.

[527] Gilbert M B, Waite T D, Hare C. An investigation of the applicability of ferrate ion for disinfection. J Am Water Works Ass, 1976, 68 (9): 495-497.

[528] Jiang Y Goodwill J E, Tobiason J E, Reckhow D A. Bromide oxidation by ferrate(VI), The formation of active bromine and bromate, Water Res., 2016, 96: 188-197.

撰稿人：何孟常[1]，Shikha Garg[2]，何　皡[2]，T. David Waite[2]，王兆慧[3]，方战强[4]，袁松虎[5]，张礼知[6]，艾智慧[6]，李芳柏[7]，刘同旭[7]，夏金兰[8]，石　良[5]，张耀斌[9]，张伟贤[10]，何　锋[11]，凌　岚[10]，李　聪[12]，关小红[10]，潘丙才[13]，孙远奎[10]，宋丹丹[5]

[1]北京师范大学，[2]The University of New South Wales，[3]东华大学，[4]华南师范大学，[5]中国地质大学（武汉），[6]华中师范大学，[7]广东省生态环境与土壤研究所，[8]中南大学，[9]大连理工大学，[10]同济大学，[11]浙江工业大学，[12]浙江大学，[13]南京大学

第 10 章　环境汞污染研究进展

本章导读

由于特殊的理化性质，汞是唯一主要以气态单质形式存在于大气的重金属污染物，能随大气环流进行长距离传输，造成偏远地区生态系统汞污染。鉴于全球汞污染的严峻形势，在联合国环境规划署的推动下，《关于汞的水俣公约》（以下简称《水俣公约》）于2017年8月17日正式生效。改革开放以来，我国经济高速发展，人为活动向大气排放大量汞，引起国际社会高度关注。但由于相关研究相对滞后，目前我国环境汞污染存在底数不清，环境与健康效应不明的情况，这对我国履行国际汞公约极为不利。本章主要介绍了近年来我国学者在人为源汞排放清单建立、自然过程汞排放规律、我国区域大气汞分布规律、汞的分子转化规律、土壤汞污染防治、汞暴露及健康风险和汞同位素及环境汞污染示踪等方面的研究进展。

关键词

汞，排放清单，自然排放，长距离传输，分子转化，甲基汞，同位素

1 引　　言

汞是毒性最强的重金属污染物之一，已被我国和联合国环境规划署、世界卫生组织、欧盟及美国环境保护署等机构列为优先控制污染物。汞污染的严重性和复杂性远远超过常规污染物，甚至在某些方面超过持久性有机污染物。汞的形态不同，其毒性相差很大。甲基汞是毒性最强的汞化合物，具有高神经毒性、致癌性、心血管毒性、生殖毒性、免疫系统效应和肾脏毒性等。无机汞的毒性相对较弱。虽然人类活动向环境排放的都是无机汞，但无机汞进入环境后会转化成甲基汞，因此汞的危害性具有隐蔽性和突发性，一旦发生重大污染事件或出现人群病变，将产生灾难性后果。

鉴于全球汞污染的严峻形势，联合国环境规划署于2010~2013年召开了5次政府间谈判委员会系列会议，并在2013年达成一项具有法律约束力的国际汞公约——《水俣公约》，旨在控制和削减全球人为汞排放和含汞产品的使用。该公约已于2017年8月16日正式生效。《水俣公约》将与涉及持久性有机污染物的《斯德哥尔摩公约》和应对气候变化温室气体的《京都议定书》一样，成为最高级别的国际法。我国政府积极参与了《水俣公约》谈判，并率先批准了《水俣公约》，充分体现了对全球环境保护的重视。

我国是汞使用量和排放量最大的国家，必将面临巨大的国际压力。由于发达国家在控制汞污染方面已经较有成效，所以特别关注发展中国家的汞排放，一方面可能会对我国含汞产品（如荧光灯、含汞医疗器械等）的生产和出口进行限制，产生国际贸易壁垒；另一方面可能会在削减汞排放方面给我国施加巨大压力，进而会深入影响我国未来经济的可持续发展。

我国是全球最大的汞生产国、使用国和排放国。数据显示，2007年中国各行业汞需求量超过1500 t，较2004年增加40%，我国汞的年生产量和消费量高居世界第一，分别占70%和50%。初步估算表明我国每年人为活动向大气排放的汞量为500~800 t，占全球人为源排放总量的30%左右。由于缺乏我国的大气汞排放因子、化学形态分布、大气污染控制装置对不同形态汞的去除效率等实测数据，我国对人为活动大气汞排放量的估算存在很大的不确定性。

我国环境汞污染问题十分突出。我国长期大规模的工、矿业活动，如汞冶炼、铅锌冶炼、混汞采金和氯碱生产等导致的局地汞污染问题非常严峻，不仅汞污染场地类型多、数量大和面积广，而且汞污染导致的环境问题也非常严重。在这些区域，工矿生产过程中产生的大量含汞“三废”（废水、废气和废渣）

直接排放到环境中，使得周围水体、沉积物、大气、土壤及生物遭受严重汞污染，甚至出现了更为严重的甲基汞污染问题，已经威胁到当地居民的身体健康。

鉴于我国汞污染的现状和未来履行国际汞公约的需求，科技部于2013年启动了“我国汞污染特征、环境过程及减排技术原理”“973”项目，对我国环境汞污染来源、大气汞分布和迁移转化规律、汞污染的生态环境效应、汞同位素地球化学等方面开展了大量研究工作。同时，国家自然科学基金委员会和各部委也资助了关于我国环境汞污染方面的科研项目，极大地推动了我国环境汞污染的研究工作，获得了一批重要学术成果。

2 人类活动汞排放

人类活动汞排放清单的研究始于Nriagu和Pacyna在*Nature*上发表的《定量评估全球痕量金属的大气、水体和土壤污染》一文。该研究首次给出了全球人类活动大气汞排放的主要排放源，收集并整理了主要人为源的大气汞排放因子，同时估算1983年的大气汞排放量为3560 t（910~6200 t）[1]。在这个研究的基础上，Pirrone等根据地区发展的区别调整了发达国家和发展中国家的排放因子，从而建立了全球分区域的大气汞排放清单[2]。研究发现，20世纪80年代全球人类活动导致大气汞排放量以每年4%的速率递增，在1989年达到峰值后以每年1.3%速率下降。固体废物焚烧是美洲、欧洲西部和非洲最主要的大气汞排放源，而亚洲、欧洲东部和前苏联的大气汞排放主要来自燃煤。随后Pacyna等先后建立了1995年、2000年和2005年全球的大气汞排放清单[3-5]。其中，2000年的清单基于各国的活动水平将全球大气汞排放细化到国家水平。在该清单中，中国的大气汞排放量达到604.7 t，居世界首位。2005年的清单成为了《全球汞评估报告2008》的数据基础。2005年全球的大气汞排放量达到1930 t，其中中国排放约占全球大气汞的43%[6]。由于缺乏实测数据，该评估报告不同国家使用的排放因子基本相同，因此该评估报告估算的各个国家的排放量具有很大的不确定性。

近年来，越来越多的研究者开展了针对某个国家或者全球某个行业的大气汞排放清单研究[7-12]。2010年，Pirrone等基于部分国家清单和全球汞评估报告，估算2007年全球人为活动大气汞排放量为2320 t[13]。在该研究中，中国的排放达到609 t。此后，随着各国对大气汞排放研究的深入，研究者充分意识到原料汞含量、行业工艺水平、污染控制设施使用等因素对大气汞排放因子具有显著影响。因此，《全球汞评估报告2013》中不再采用单一的排放因子[14]，而是基于原料/燃料使用带来的汞的输入，扣除工艺过程和污染控制设施对大气汞的削减得到。

考虑到汞在全球环境中的迁移转化，同时也为客观地评估人类活动大气汞排放对当前全球环境汞负荷的影响，Streets等开展了1850~2008年全球人类活动大气汞排放清单的研究[15]。该研究为尽可能考虑工艺过程的历史变化对排放因子的影响，假设排放因子满足变换正态分布曲线并采用已有的排放因子测试结果对曲线进行校验。研究发现，工业革命以来的全球大气汞排放量高达35000万t。全球大气汞排放的峰值出现在1890年，约为2600 t。这主要是由于欧美大规模黄金、白银和汞生产所导致。随着世界大战的到来，大气汞排放快速下降。1950年之后，随着亚洲地区燃煤等无意汞排放行业的不断发展，全球大气汞排放量不断增加，并在2008年达到2000 t。此后，Horowitz等发现，历史上绘画、染织等使用了添汞产品的过程会额外排放54000万t的汞进入环境。其中，大气汞排放约占20%[16]。该研究同时指出，历史上的大气汞排放导致当前全球大气汞达到5800 t，约为1850年的5.8倍。

作为全球最大的大气汞排放国家，中国的排放广受世界关注。Streets等于2005年编制了中国第一份比较完整的人为源大气汞排放清单[10]。该清单基于官方统计数据估算了1999年中国人为源大气汞排放量为（536±236）t。该研究显示有色金属冶炼和燃煤是中国大气汞排放最主要的排放行业，分别占全国总排放的45%和38%。Wu等随后发表了中国1995~2003年的历史排放清单，发现中国人为源大气汞排放以每年2.9%的速率递增[17]。2003年大气汞排放达到（696±307）t。在这两份研究中，除了燃煤行业考虑了煤炭汞含量和污染控制设施对排放的影响外，其他行业基本上采用的是全球清单中的排放因子数据。为了实现排放因子的本土化，研究者开展了全国范围内的原料样品采集和排放源汞排放特征测试[18-32]。基于这些基础研究结果，燃煤、有色金属冶炼、水泥生产、钢铁生产、生活垃圾焚烧等中国重点源的大气汞排放清单不确定性显著下降[33-41]。通过重点源大气汞排放测试数据的积累，Zhang等首次利用基于工艺过程的概率排放因子法，建立了2000~2010年中国人为源大气汞排放清单[42]。研究表明，人为源大气汞排放总量从2000年的356 t持续增长至2010年的538 t，年均增长率为4.2%；2010年，工业燃煤、燃煤电厂、有色金属冶炼和水泥生产分别贡献了22.3%、18.6%、18.1%和18.3%的大气汞排放。随后，Wu等研究发现，改革开放以来中国大气汞排放累积达到13294 t[33]。燃煤电厂SO_2控制措施、工业燃煤行业的颗粒物和SO_2控制措施、有色金属冶炼厂的制酸过程都对大气汞排放有显著的协同控制效果，2011年开始中国大气汞排放总量开始下降。此外，为考虑副产物/废物再利用对大气汞排放的影响，研究者建立了中国汞流向，并由此发现2010年由于副产物/废物再利用额外增加了102 t大气汞排放[43]。

随着汞清单研究的深入，当前中国大气汞排放的不确定性已经得到显著的降低。原料/燃料汞含量以及污染控制设施的脱汞效率是影响清单不确定性的最主要的因素。由于汞含量的波动是原料/燃料客观存在的特征，因此当前仍难以降低该因素对清单不确定性的影响。考虑到污染控制设施脱汞效率的波动及污染控制技术的不断更新，未来降低大气汞排放的不确定性将主要通过补充更多的污染源排放特征测试实现。

3　自然过程汞排放

自然源汞排放包括自然地质过程汞的释放（如：岩石风化及火山爆发过程中的汞释放）以及先前沉降在陆地与水域的自然界面汞的再释放。认识自然源排汞过程的机理与建立精确的自然源排放清单是当前进一步理解区域与全球汞循环过程的关键所在[13,44-46]。然而，当前自然源排汞过程及其清单的研究仍处于探索阶段。当前自然源汞排放清单的误差高达2000 Mg/a，该误差与全球人为源汞排放的总量相当。目前相关的研究进展主要表现在：自然源通量观测技术的发展、自然源排汞的基本规律的认识及自然源排汞模型的建立。

目前测定大气-自然源界面通量交换的方法主要包括动力学通量箱法和微气象学法。由于动力学通量箱法成本低而被广泛应用[47]。但传统动力学通量箱是在固定抽气流速下进行测定，不能正确反映地气界面风摩擦驱动这一关键因子的影响作用，如在不同流速下其测定汞通量差异可高达7倍[48]。因此传统动力学通量法可能不能真实定量通量值。目前相关课题组已成功建立了测定地表与大气汞交换通量的新型动力学通量箱方法。该新型动力学通量箱法将箱外地表的摩擦驱动力（摩擦风速）与箱内的摩擦风速相关联，克服了传统通量箱单一操作流速的制约[49]。微气象方法相比通量箱方法具有更大的时空尺度、原位不扰动等优点。由于缺乏精确测定大气汞浓度的高频探头（一般为10 Hz），微气象方法主要的发展方向为基于涡度相关理论的弛豫涡旋积累法与通量梯度法。通量梯度法对微量气体分析检测限要求较高，特别是当标量垂直差异较小和大气湍流条件差（如夜晚低湍流情况）时测定的通量误差极大[50-52]。而REA已在多种复杂地形完成对汞通量的测定，如在农田系统中汞通量的测定[53]。但目前相关新方法的应用，仍处于起步阶段[54]，未来仍需广泛推广。

目前对自然源汞的排放通量与环境因子相关关系的认识已较全面。光照、温度、基质的汞浓度与大气汞浓度是影响汞的排放通量的关键因素。表 10-1 统计了当前文献报道的不同土地利用类型的汞排放通量。值得注意的是：①上述的统计结果主要是基于欧美与东亚地区的测定结果；②不同测定方法使得上述统计结果存在误差；③所有数据的时间分辨率较低，一般为数天至数周的测定周期。因此，利用当前的测定结果去评估全球汞的自然源汞的排放仍存在很大的误差[54,55]，特别是对于森林生态系统。根据最新研究表明[55]，在所有的生态系统中，森林系统与大气间 Hg^0 交换通量估算的误差最大，通量交换范围为−727~+707 Mg/a，这使得森林生态系统到底是大气 Hg^0 的源还是汇的问题上成为争论的焦点。

表 10-1　当前文献中报道的自然源汞的排放通量[ng /(m·h)][54]

	均值	中位值	最小	最大	样地数
背景地区土壤	2.1	1.3	−51.7	33.3	159
城镇土壤	16.4	6.2	0.2	129.5	29
农业土壤	25.1	15.3	−4.1	183.0	59
森林土壤	6.3	0.7	−9.6	37.0	8
草地	5.5	0.4	−18.7	41.5	38
湿地	12.5	1.4	−0.3	85.0	23
湖泊	4.0	2.8	−0.3	74.0	93
海洋	5.9	2.5	0.1	40.5	51
雪地	5.7	2.7	−10.8	40.0	15
富汞土壤	5618.0	226.0	−5493.0	239200.0	329
人为污染场地	595.0	184.0	−1.4	13700.0	58

常见的大气-自然界面的通量交换模型对比见表 10-2。对于大气-叶片界面的通量模拟，早先的参数模式认为通量是基于土壤水分经土壤-根-干-叶系统的蒸腾作用，与蒸腾速率相关的函数（SY1，表 10-2）；土壤中的 Hg 结合在有机基团上，穿过土壤-根界面传输到叶片[56-58]。但近年研究表明，根的隔膜屏蔽作用使得上述过程运输的 Hg 很少[59]。Hg 的天然同位素示踪技术[60,61] 与大气-叶片的通量直接测定 [62,63] 也均表明：①大气-叶片的通量是双向的；②叶片从大气吸收 Hg 是 Hg 在叶片累积的主因。因此，基于补偿点的双向交换模式（SY2，表 10-2），其科学假设更为合理，也符合实际观测。对于大气-水面的通量交换，双膜跨膜模式被广泛采用[64]。而大气-土壤 Hg^0 的通量交换，除了上述的双向交换模式（ST3），基于测量的环境因子（如温度、土壤湿度、太阳辐射等）与通量的统计经验模式也被不少模型（如 GEOS-Chem）用来预测通量交换。需要指出的是，上述统计关系是基于局地的测量所得，但外推到全局时，会大大简化其他环境因子的作用，使得预测值出现较大偏差[65]。Wang 等[66]根据当前最新的土壤 Hg^{II} 的还原反应动力学的进展，构建了新的土壤汞排放模型；运用新模型，获得了中国高精确的自然源排放清单，其结果表明当前中国自然汞的排放总量为 465 Mg/a。

未来自然源汞通量的测定与模型的研究应致力于解决如下几个关键问题。①应当提高当前汞通量测定的时间分辨率与汞检测器的分辨率，在典型代表站点采用相关的通量监测方法进行长时间的监测。当前数天至几周的测定周期使得数据的质量得不到保证，在估算全年的通量总量时存在较大的误差。在微气象学法测定通量中，当前的汞检测器的分辨率较低是其该方法的主要限制所在。②不同的通量测定方法存在较大的差异，未来汞通量监测的方法应该标准化，使得数据间具有较强的可比性。③当前虽然有较多观测实例，但对于相关机理的研究十分缺乏，应加强汞在海洋水体的形态转化、植被叶片与土壤的氧化还原的机理研究，根据相应的动力学最新进展，建立新一代的大气-自然源汞通量交换模型。

表 10-2　自然源通量排放模型的对比

	模式	描述	参考文献
叶片	SY1： $F = EC_s$	E：蒸腾系数［$g/(m^2 \cdot s)$］ Cs：土壤溶液 Hg^0 浓度（ng/g）	[67-70]
	SY2： $F_{st/cu} = \frac{\chi_{st/cu} - \chi_c}{R_{st/cu}}$	$\chi_{st/cu}$：气孔/角质层的补偿点（ng/m^3） $F_{st/cu}$：大气-气孔/角质层的通量［$ng/(m^2 \cdot s)$］ χ_c：大气-冠层补偿点（ng/m^3） $R_{st/cu}$：大气-气孔/角质层的阻抗（m/s）	[65,71-73]
土壤	ST1： $\log F = -\frac{\alpha}{T} + \beta \log C + \gamma R + \varepsilon$	T：土壤温带（℃） C：土壤总 Hg 浓度（ng/g） R：太阳辐射（W/m^2）	[67-70,74]
	ST2： $\frac{F}{C} = \alpha T + \beta R + \delta \theta + \delta TR + \cdots$	T：土壤温度（℃） C：土壤 总 Hg 浓度（ng/g） R：太阳辐射（W/m^2） θ：土壤湿度（%）	[48,75]
	ST3： $F = \frac{\chi_s - \chi_c}{R_g + R_{ac}}$	χ_s：土壤气补偿点（ng/m^3） χ_c：大气-土壤补偿点（ng/m^3） R_g：大气-土壤阻抗（m/s） R_{ac}：林间内动力学阻抗（m/s）	[65,71-73]
水面	$F = \frac{\chi_w - \chi_c}{R_w + R_a}$	χ_w：水界面补偿点（ng/m^3） χ_c：大气 Hg^0 浓度（ng/m^3） R_w：液相层的阻抗（m/s） R_a：气相层的阻抗（m/s）	[65,67-70,73]

4　大气汞分布及沉降特征

汞是一种特殊的有毒重金属元素。由于其特殊的物理化学性质，汞是主要以气态存在于大气的重金属元素。大气汞主要包括气态单质汞（GEM）、活性气态汞（GOM）和颗粒汞（PBM）[76]。GEM 是大气汞的最主要成分，约占大气汞总量的 80%以上[77]。GEM 具有较强的挥发性、极低的水溶性和干沉降速率，因此具有较长的大气居留时间，能够随大气环流进行全球性迁移，是一种全球性污染物。我国是全球大气汞排放最多的国家。目前的大气汞排放清单表明[66,78]：我国每年的大气汞排放量约为 1100 t（包括人为源和自然源），约占全球大气汞排放总量的 15%[13]。

我国的大气汞研究起步于 21 世纪初，近十年来取得了快速的发展。目前研究人员已在我国许多地区开展了大气汞及其沉降通量的长期连续观测，并在 2013 年后建立了同步在线监测大气汞和湿沉降通量的大气汞监测网络，研究区域基本涵盖了我国大气汞排放的各种特征区域。研究结果为认识我国大气汞的分布特征、影响因素、传输过程以及区域和全球大气汞循环模型的建立提供了关键的科学数据。

图 10-1 显示的是我国大陆大气 GEM 分布特征。监测结果显示，我国背景区大气 GEM 的年均浓度范围为 1.60~5.07 ng/m^3，平均值为（2.86±0.95）ng/m^3，明显高于北半球大气 GEM 的背景浓度（1.5~1.7 ng/m^3）[79]。我国城市地区大气 GEM 的年均浓度范围为 2.50~28.6ng/m^3，平均值为（9.20±7.56）ng/m^3，约是全球背景区 GEM 浓度的 2~20 倍。我国大气 GEM 浓度分布具有明显区域分布特征，其中东北、西北和西南背景区的

浓度明显低于华北、华东和华南地区。研究发现[80]，我国背景区大气 GEM 浓度和区域人为源大气汞排放强度具有显著的正相关性，表明我国背景区大气汞污染主要受区域人为源排放的影响；而城市地区大气 GEM 主要受近距离的人为源大气汞排放影响。我国背景区和城市地区大气 PBM 年均浓度范围分别为 19~154 pg/m^3 和 141~1180 pg/m^3，平均值分别为 52 pg/m^3 和 530 pg/m^3，约占大气汞（GEM+ PBM+GOM）总量的 3.5%左右。和国外同类型地区相比，我国大气 PBM 浓度约偏高 4~100 倍。我国背景区和城市地区 GOM 的年均浓度范围分别为 2~10 pg/m^3 和 36~61 pg/m^3，整体和国外同类型地区相当。我国城市和背景区大气 GEM 和 PBM 均具有显著的季节性变化特征。由于受到季风作用的影响，不同地区长距离大气汞传输强度随时间会有所变化，因此不同地区大气 GEM 的季节性变化有所差异[80]。

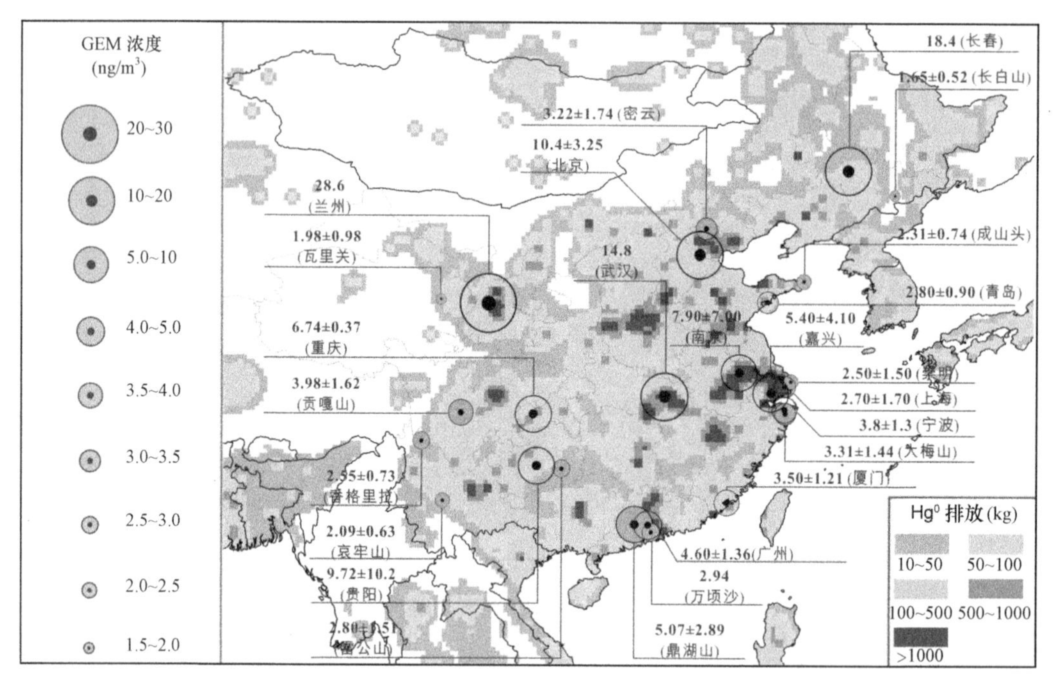

图 10-1 我国大陆大气单质汞（GEM）浓度分布特征[80]

图 10-2 显示的是我国部分背景区和城市地区大气降水汞含量和湿沉降通量的分布特征[81]。我国背景区大气汞湿沉降通量范围为 1.8~15.4 μg/(m^2·h) [均值=(5.6±4.2) μg/(m^2·h)]，而城市地区大气汞湿沉降通量范围为 8.2~56.5 μg/(m^2·h) [均值=(24.8±17.8) μg/(m^2·h)]。和欧美地区相比，我国背景区大气降水湿沉降通量相对偏低，而城市地区则明显高于欧美城市地区。研究指出[81]，由于我国城市地区存在非常明显大气 PBM 和 GOM 污染，而降雨对边界层大气 PBM 和 GOM 的冲刷作用则是导致我国城市地区大气汞湿沉降通量升高的最主要因素。另一方面，我国背景区较低的大气汞湿沉降通量则有可能说明我国自由对流层大气 GOM 浓度比欧美地区偏低，从而导致云雾对大气汞的捕集量降低[81]。植物落叶汞沉降是我国森林地区大气汞沉降的另一重要来源。数据显示，我国森林地区落叶汞沉降通量范围为 22.8~62.8 μg/(m^2·h)，平均值为（37.8 ±14.8） μg/(m^2·h)，约是大气汞湿沉降通量 3.9~8.7 倍，占森林地区大气汞沉降通量的 60%~85%[81]。和国外相比，我国森林地区落叶汞沉降通量约偏高 1.4~4.7 倍。植物叶片中汞主要来自于叶片对大气 GEM 的吸收作用，而我国大气 GEM 浓度普遍偏高，是导致我国森林地区落叶汞沉降通量偏高的一个重要因素。另外，我国森林地区主要位于亚热带和温带地区，植物生物量较高，也是森林落叶汞沉降通量的偏高的一个重要因素。

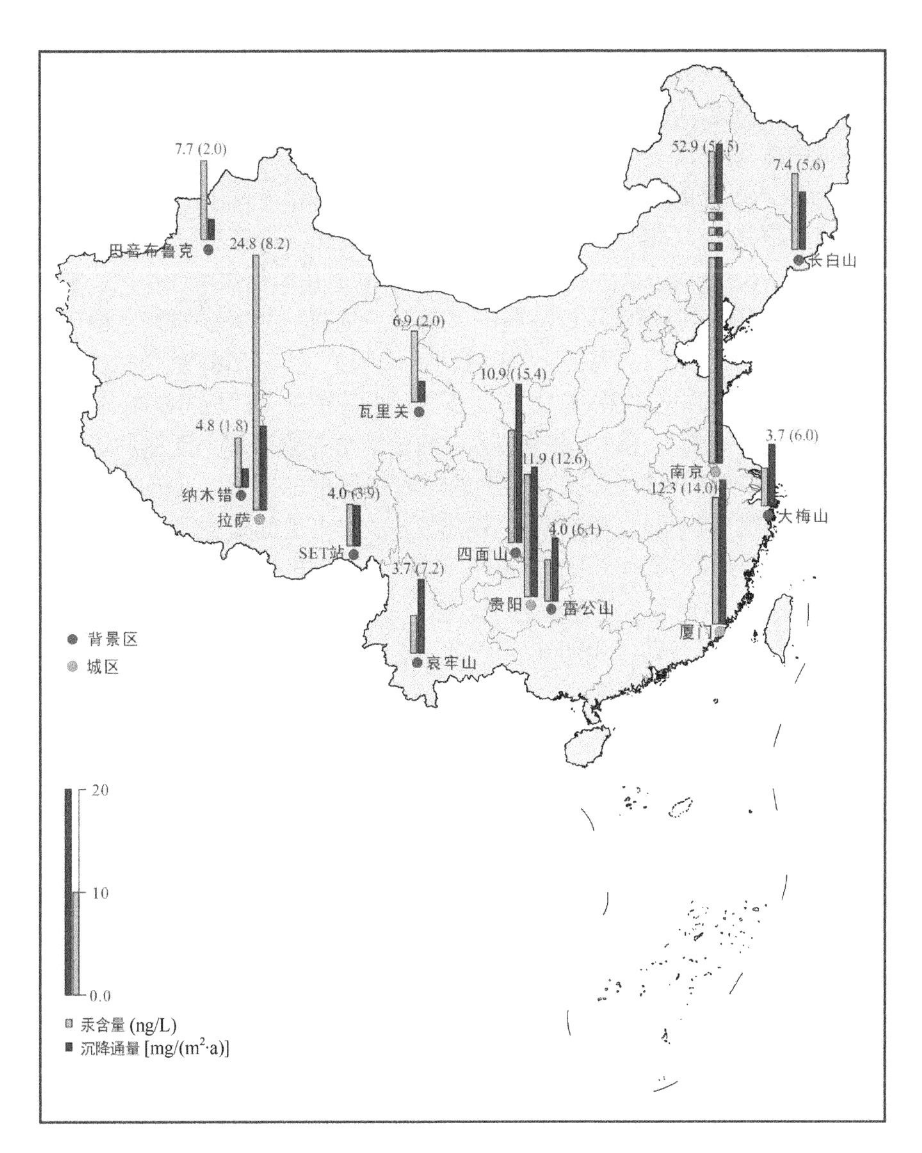

图 10-2　我国部分背景区和城市地区大气降水汞含量和湿沉降通量分布特征[81]

5　汞的分子转化

汞在自然界中以多种不同的形态存在，各种汞形态之间的分子转化（如甲基化/去甲基化、还原/氧化）是汞生物地球化学循环的重要环节，对汞的迁移与毒性有重要影响。

5.1　汞的化学与生物甲基化

汞可通过化学与生物甲基化作用生成高毒性的甲基汞。通常认为，无机汞的微生物甲基化是甲基汞生成的主要途径。Hg(II)的生物可给性是影响微生物甲基化的重要因素之一。虽然汞甲基化模式菌（*Desulfovibrio desulfuricans* ND132）可快速吸收 Hg(II)并使其甲基化，但仍有大部分 Hg(II)吸附在细胞表面而不能被微生物摄入与甲基化。半胱氨酸等巯基配体可有效解吸细菌细胞壁吸附的 Hg(II)，增加其生物可给性，从而提高微生物对汞的甲基化[82]。此外，在汞的甲基化方面，不同微生物可能存在协同作用。硫还原菌 *D. desulfuricans* ND132 可氧化并甲基化元素汞，其对元素汞的甲基化速率约为二价汞的 1/3[83]。另一株硫还原菌 *Desulfovibrio alaskensis* G20 亦可氧化元素汞，但不能进一步将其甲基化。在半胱氨酸存在下，铁还原菌 *Geobacter sulphurreducens* PCA 可将元素汞氧化并甲基化。因此这些甲基化/非甲基化菌的协同作用可能进一步提高厌氧环境（如底泥）中零价汞的氧化与甲基化。在实际环境中，微生物群落组成对汞甲基化有重要影响。土壤或底泥中汞的甲基化与 *dsrB* 基因相关性较小[84]，但与 *hgcA* 基因呈正相关[85]。因此可通过微生物群落调控以降低环境中甲基汞的生成，如水稻的有氧种植（非水淹环境）可显著抑制硫还原菌导致的汞甲基化，从而进一步降低水稻籽粒中甲基汞浓度[86]。

除了微生物甲基化，汞的化学甲基化亦是某些环境中甲基汞的重要来源。采用汞同位素（$^{199}HgCl_2/CH_3{}^{201}Hg^+$）与氢同位素（$CD_3I$）示踪技术，通过培育实验研究了天然环境水样中新型农药熏蒸剂碘甲烷对无机汞的光化学甲基化途径与机制。天然环境水体中二价汞以及低价态的一价汞、零价汞均可被碘甲烷甲基化，该反应依赖于日光照射。而在去离子水中，仅低价态的一价汞、零价汞可被碘甲烷甲基化。基于此，提出了碘甲烷对无机汞的光化学甲基化的两步机制：①二价汞光还原生成一价汞与零价汞；②碘甲烷光解生成甲基自由基与甲基正离子，进一步与一价汞/零价汞结合，生成高毒性的甲基汞。采用模型对这一反应进行了定量评估，表明在碘甲烷污染水体中，碘甲烷对汞的光化学甲基化是甲基汞的重要来源。

5.2 甲基汞的化学与生物去甲基化

表层水中甲基汞的光化学降解是环境中甲基汞去除的重要途径之一，但其机制尚存在较大争议。近期研究揭示了在水环境中多种甲基汞的光降解途径（如羟基自由基[87,88]、溶解三线态 DOM（$^3DOM^*$）[89]、甲基汞-DOM 络合物途径[90]）并存，其降解途径与水化学参数相关[91,92]。其中，甲基汞-DOM 络合物经由分子内电子转移而引起的直接光解可能是一般环境水体中甲基汞光降解的一个较为普遍的机制[90,92,93]。DOM 小分子模型研究显示，DOM 分子中的苯环与巯基在甲基汞光降解中均起着至关重要的作用[89]。

微生物在甲基汞的环境降解（如底泥介质）中亦起着重要的作用。一株厌氧铁还原菌（*Geobacter bemidjiensis* Bem）不仅可甲基化 Hg(II)，亦可介导甲基汞的降解，其去甲基化率可达 50%[93]。由于有机汞裂解酶（MerB）与汞还原酶（MerA）的存在，甲基汞的微生物降解产物为零价汞。汞络合配体如半胱氨酸可显著降低微生物对甲基汞的降解。

除了以上甲基汞的化学与微生物降解途径，近些年也发现在水稻根茎[94]、黑鲷鱼肠道[95]内亦存在甲基汞的降解，但其具体分子机制尚有待进一步研究。

5.3 零价汞转化的新形态与新过程

零价汞在水中广泛存在，涉及一系列重要的汞生物地球化学循环过程（如氧化、还原、挥发）。一般认为，水中零价汞以溶解性气态汞的形式存在，可被氮气或氩气吹扫出来。但采用同位素示踪技术，发现加入水中的很大一部分零价汞与悬浮颗粒物结合，并且不能被吹扫出来，表明在天然水中存在颗粒结合态零价汞这一新的未被识别的形态[96]。结合同位素稀释等手段，可实现对水中颗粒结合态零价汞的定量分析，结果显示佛罗里达 Everglades 大湿地水样中颗粒结合态零价汞占总零价汞的比例可达 45%~80%[96]。水中零价汞可被光氧化为二价汞，通常认为羟基自由基在其中起着主要作用。最近一项研究表明，水中的羟基自由基（如来自于硝酸盐光解）可与碳酸根反应生成寿命更长的碳酸根自由基（$CO_3^{\cdot-}$）。水中零价汞的光氧化并非主要来自于其与羟基自由基的反应，而是碳酸根自由基[97]。这一新的零价汞光氧化途径也提示，由于较长的寿命与较高的稳态浓度，相比于羟基自由基，一些反应活性较弱的自由基

（如 CO_3^{2-}）在汞形态转化过程中也可能起着重要的作用。

5.4 硫化汞的生成与溶解

硫化汞是环境中汞重要的“源”和“汇”，在多种环境与生物介质中广泛存在。硫化汞的生成与溶解是汞的“老化”与“活化”的重要过程，对其迁移、还原、甲基化、毒性等有重要影响。近期研究表明，二价汞与天然有机质巯基的络合作用可导致碳硫键断裂，并生成β-硫化汞纳米颗粒[98]。光照可显著促进 DOM 对二价汞的硫化过程。在 DOM 溶液及实际水样中，6 h 紫外光或日光照射可显著降低反应活性汞[Sn(II)还原汞]浓度以及微生物对汞的甲基化。这一过程可能与光照下硫化汞纳米颗粒生成有关。考虑到低浓度硫化汞纳米颗粒的表征存在困难，因此在较高二价汞浓度（0.1 mmol/L）下验证了汞硫化过程的发生，采用扫描电镜及能谱证实硫化汞纳米颗粒的生成[99]。溶解氧可显著促进水中硫化汞的氧化溶解。在这一过程，存在二价汞在硫化汞颗粒上的再吸附。同位素示踪与同位素稀释定量显示，约 1/2 的溶解汞再次吸附在硫化汞颗粒表面[100]。

6 土壤汞污染防治

汞污染场地一直以来均受到科学家们的重点关注，也发表了大量涉及汞污染场地汞的释放、迁移、转化，以及环境风险和人体暴露相关的研究成果。然而，对比之下，针对汞污染防治技术和应用相关的研究相对匮乏，且大多仍停留在实验室阶段，较少开展过野外小试和中试。随着《水俣公约》的正式生效和国家对汞污染防治的大力推进，科研工作者将有机会更多地为政府提供环境决策咨询，在基础研究和技术应用之间起到桥梁和纽带作用。这里主要介绍我国汞污染土壤现状、相关治理修复技术及建议。

6.1 全国土壤汞污染现状及防治需求

据 2014 年《全国土壤污染状况调查公报》显示，全国土壤污染点位超标率高达 16.1%，重金属污染较严重，主要以镉、汞、砷、铅等重金属为主。西南、中南地区土壤重金属超标范围较大，其中土壤汞污染超标点位比率为 1.6%，污染分布趋势呈现从西北到东南、从东北到西南方向逐渐升高的态势。

为了尽早缓解我国土壤污染问题，2016 年国务院发布了“土十条”规划（即《土壤污染防治行动计划》），其中明确将汞污染问题突出的贵州铜仁市列为国家首批六大土壤污染综合防治先行区之一，拟通过开展相关土壤污染治理与修复试点示范，先行先试，在土壤污染源头预防、风险管控、治理与修复、监管能力建设等方面探索土壤污染综合防治模式，逐步建立我国土壤污染防治技术体系。根据铜仁市环保部门提供的数据，铜仁市至少存在大于 10 万亩汞污染耕地亟需开展风险管控措施。然而，截至 2017 年 7 月，铜仁市却没有一项土壤汞污染治理修复相关项目启动，其中重要原因之一就是缺乏成熟的土壤汞污染修复技术及工程经验。

6.2 汞污染土壤修复技术

常见的土壤汞污染修复治理技术主要有以下几种：

（1）固化/稳定化。固化是指将污染土壤机械地固封在固化体中，切断污染物与外界环境的联系，达到控制污染物迁移的目的。常用的固化剂有水泥、火山灰、热塑性塑料等。稳定化是指向土壤中添加稳定剂，通过吸附、沉淀等化学反应，改变污染物的形态，减小污染物的溶解移动性、浸出毒性和生物有效性[101]。固化/稳定化的效应统称为钝化。

固化/稳定化法处理汞污染土壤的优点在于其工艺简单、快速经济；缺点是没有将汞等污染物从土壤中移除，需要长期监测，固化后的土壤结构遭到破坏，增容量较大。

（2）土壤淋洗。土壤淋洗包括物理分离和化学萃取两种技术。物理分离来源于采矿技术，主要适用于污染物的粒度、密度等性质与母质土壤有明显差别或污染物吸附在土壤颗粒的情况。水力分级、重选、浮选和磁性分离等都是较为常用的方法。化学萃取主要是通过萃取剂的解析和溶解作用将重金属转移到淋洗液中，随后再对富含重金属的废液进一步处理。常见的汞萃取剂有水、螯合剂（EDTA 等）、硫代硫酸盐（$Na_2S_2O_3$等）、酸类（盐酸、硫酸等）[102,103]。淋洗法的优点是周期短、效率高、技术应用性好，可对淋洗液中的金属进行回收；缺点是对单质汞的去除效果不好，黏土和腐殖质含量高的土壤处理困难，对淋洗液的处理使得成本增加。

（3）热解修复。热解修复是采用加热或向土壤中通入热蒸汽的方式将汞及其化合物转化为挥发性汞，再集中收集处理。一般情况下，汞在土壤中主要是以单质汞和二价汞（包括 HgS、HgO 和 $HgCO_3$）的形式存在。研究表明，当温度达到 600~800℃时，这些汞就会转化成气态汞从土壤中释放，从而回收利用[104]温度和时间是影响汞去除的主要因素。时间越久，温度越高，汞的去除效果越好。热解修复汞污染土壤，加热温度一般在 320~700℃。但是，高温会使土壤有机质和结构水遭到破坏，影响土壤肥力。因此，降低加热温度是如今研究的关键。

热解修复的优点在于可以快速高效地去除土壤中的汞，并实现汞的回收；缺点是设备成本和能耗高，且在汞污染含量高时才有高效率，高温也会破坏土壤结构。

（4）电动修复。电动修复是近些年来一种新兴的土壤原位修复技术，其基本原理是在污染土壤中布置相应的电极，施加低压直流电场。在电场的作用下，经过电泳/电迁移、电渗析和电解过程，使污染物向电极运输，最后回收处理。电动修复受到很多条件（如土壤类型、污染物性质等）的影响。电动修复土壤需要具备低水传导性、目标污染物溶解性高和非目标污染物浓度低等性质。由于汞在自然界土壤中的溶解度很低，单纯的电动修复很难进行。因此，需要人为地添加化学试剂如 EDTA、碘化物、氯化物、纳米 Fe 等来增加汞的可移动性，提高电动修复的效率[105,106]。电动修复的优点为对低渗透性的土壤修复效果好、费用低、没有二次污染、不破坏土壤肥力，实现了原位修复；缺点是修复周期长，只适用于淤泥和黏土性土壤，并且受到 pH、碳酸盐、有机质等条件的影响。

（5）纳米技术。纳米修复技术就是利用 1~100nm 的微小粒子，改变污染物的移动性、毒性、生物可利用性等。纳米颗粒具有小尺寸效应、表面效应、量子效应等，其较高的比表面积，对土壤中的 Hg^{2+}有极强的吸附性。因此，可以利用纳米技术来修复土壤汞污染。有研究表明，用羧甲基纤维素钠(CMC)-FeS 纳米粒子对汞污染土壤进行修复，样品渗滤液中汞减少了 79%~96%，TCLP 渗滤液中的汞减少了 26%~96%[107,108]。此外，TiO_2纳米颗粒也被证实可以促进沉积物中汞的释放[109]。

纳米修复技术是一种新兴的土壤修复技术，纳米材料逐渐向低成本、可降解性生态环保的方向发展，具有十分广阔的前景；但它易受 pH 的影响，在土壤中流动性较差，易形成聚合物。

（6）植物修复。植物修复指利用植物及根际微生物减少、去除土壤中重金属浓度及毒性的一种修复技术。按照修复机理的不同，土壤汞污染植物修复分为植物固化、植物挥发、植物提取[110]。植物固化修复是指植物（如香根草、蜈蚣草等）根系分泌有机酸等物质来吸收、沉淀土壤中的汞，使其富集到根部及其周围，从而达到修复土壤的效果。一般这类植物的根系比较发达。植物提取修复技术是指植物通过自身吸收或添加协助剂（如硫代硫酸盐、螯合剂等）促进吸收土壤中的汞，使其富集到植物地上部分，进行处理和回收重金属汞，以达到修复土壤的技术。与植物固化和植物提取修复技术相比，植物挥发修复技术去除土壤汞污染往往受到限制：Hg^0从植物叶部挥发受到光照和温度的影响较大，且 Hg^0挥发到大气中会对大气造成污染，因此植物挥发技术应用并不多。

植物修复的优点是对土壤性质破坏程度最低并对多种污染物都行之有效，另外其成本低，环境友好；但它也存在生长周期长、种植量大并且收获植物需进一步处理的缺点，另外也易受污染物浓度、植物年龄和气候的影响。

（7）微生物修复。微生物修复就是利用土壤中的某些微生物的生物活性对重金属具有吸收、沉淀、氧化和还原等作用，把重金属离子转化为低毒产物，从而降低土壤中重金属的毒性。大量研究已经证实，真菌可以通过分泌氨基酸、有机酸以及其他代谢产物来溶解重金属以及含重金属的矿物[111,112]。

微生物修复技术前景广阔，部分微生物具有显著脱甲基作用，可降低土壤中汞毒性，而且该方法能与植物修复有效结合；但其缺点也较为明显：土壤中的汞并不能完全被去除，微生物对环境变化感知强烈，而且有关微生物对汞吸附、沉淀机理研究较为薄弱。

6.3 土壤汞污染防治对策建议

土壤汞污染防治风险管控的核心是保障农作物安全生产利用。目前大多数旱地农作物汞污染主要以无机汞为主，但对高污染区域水田环境种植的稻米存在一定的甲基汞累积风险。在目前大多数汞污染土壤治理修复技术尚未开展过野外中试和小试的情况下，建议首先通过农艺调控措施进行风险管控，再逐步辅助以钝化等其他技术措施，最终达到标本兼治的目的。

7　汞暴露及健康风险

7.1　我国食用鱼引起的健康风险

1953 年日本熊本县爆发的“水俣病”是由汞中毒引起的著名环境公害病，该病起因于当地渔民食用了甲基汞含量较高的鱼贝类等水产品[113]。甲基汞是毒性最强的汞化合物，其在水环境中形成后，可通过食物链由低营养级的水生生物逐级传输到高营养级的水生生物体内。有研究发现鱼体内汞的富集系数高达十万倍（10^4~10^7）[114]，进而对水产品消费者的健康构成威胁。国际学术界普遍认为，除职业汞暴露外，食用鱼贝类等水产品是造成人体甲基汞暴露的主要途径。图 10-3 为汞在水体食物链中典型生物富集和生物放大级数。

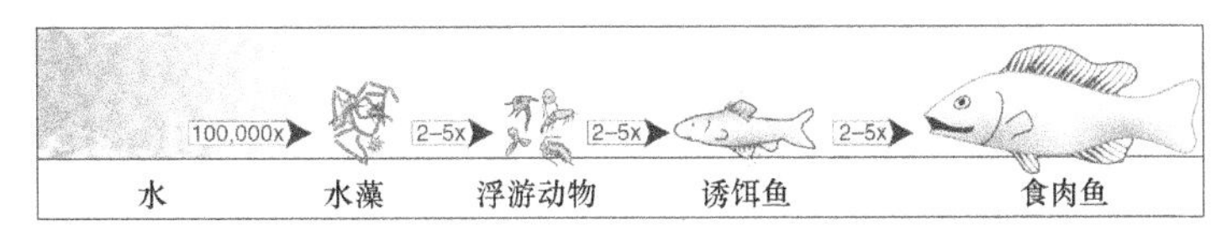

图 10-3　汞在水体食物链中典型生物富集和生物放大级数（引自参考文献[115]）

据初步统计，我国鱼体总汞和甲基汞含量变化范围分别为 0.00~1.34 mg/kg 和 0.0005 ~1.196 mg/kg，甲基汞占总汞的 1.5%~100%（表 10-3）。从不同营养级鱼体汞含量来看，一般呈现为食肉性鱼>滤食性鱼>植食性鱼。总体而言，我国各省份鱼体甲基汞含量均未超过国家甲基汞限量标准≤0.5 mg/kg（GB 2762—2005）(除青藏高原个别野生鱼体外)，并且汞含量比限量标准普遍低 1 个数量级。甚至在一些污染区域，鱼汞浓度都处于较低水平，如受汞严重污染的第二松花江中鱼体总汞浓度为 0.002~0.660 mg/kg；万山汞矿区鱼体甲基汞浓度为 0.024~0.098 mg/kg，约占总汞的 28%。受过度捕捞、栖息地破坏等原因影响，野生淡水鱼数量在过去几十年内发生锐减，使得目前国内市场上的鱼多为水产养殖的淡水鱼。后者鱼体寿命短、食物链短、生长较快[116,117]，生长稀释可明显降低鱼体内甲基汞含量，不利于汞的生物富集。并且水产养殖区域一般远离大工厂和城市中心，一定程度上也降低了鱼体汞暴露[118]，值得关注的是消费市场上的海水鱼主要为食肉性鱼，体内汞含量可能较高。但也有研究发现青藏高原野生鱼类汞富集显著（高原水体汞含量非常低），这与我国其他地区相对高汞污染环境中鱼体汞含量相对较低的结果形成鲜明对比。其原因主要由于高原特有野生鱼类在低温寡营养环境中具有极低的生长率和较长的寿命、较低的甲基汞去除率以及该地区水生生态系统食物链高效的甲基汞富集等，这些都是造成高原鱼体汞富集较为显著的原因[116]。

表 10-3　我国不同地区鱼体汞含量（湿重）以及甲基汞占总汞比例

地区	鱼类型	总汞（mg/kg）	甲基汞（mg/kg）	甲基汞/总汞	参考文献
天津	经济类鱼	0.023~0.116			[119]
雅安	市场鱼	0.165~0.182			[120]
山东	经济类鱼	0.006~0.082			[119,121]
	淡水鱼	0.056~0.282			
浙江	经济类鱼	0.040~1.100			[119]
福建	经济类鱼	0.016~0.195			[119]
广东	经济类鱼	0.027~0.045			[119]
广西	经济类鱼	0.024~0.750			[119]
河北	经济类鱼	0.2~1.34			[119]
珠海	淡水鱼	0.043（0.000~0.221）			[122]
	海鱼	0.057（0.000~0.584）			
陕西	市场鱼	0.0209~1.2561			[123]
银川	池鱼	0.003~0.056			[123]
北京	市场鱼	0.001~0.078	0.0011~0.066		[124]
渤海	经济类鱼	0.02~0.09			[125]
黄海	野生鱼	0.025~0.526（干重）			[126]
东海	经济类鱼	0.00~0.21			[125]
珠江三角洲	市场鱼	0.027~0.362	0.043（0.002~0.349）	89%（75%~100%）	[127]
	鱼塘鱼	0.007~0.077	0.006~0.076		[128]
乌江	淡水鱼	0.008~0.144	0.003~0.087		[129]
百花湖	淡水鱼	0.004~0.254	0.003~0.039	16%~79%	[130]
松花江	淡水鱼	0.016~0.489	0.013~0.342		[131]
第二松花江	淡水鱼	0.02~0.66			[132]
三峡库区	淡水鱼	0.051±0.043（0.0005~0.255）	0.021±0.028（0.00005~0.189）	34.89%±22.72%（2.65%~96.84%）	[133]
长寿湖	淡水鱼	0.046 ± 0.038（0.001~0.153）	0.001~0.088（0.033 ± 0.038）	55%± 26%（5%~88%）	[134]
青藏高原	野生鱼	0.1±0.149（0.025~1.218）	0.0907±0.137（0.025~1.196）	70%~100%	[135]
万山汞矿区	淡水鱼	0.061~0.68	0.024~0.098	28%	[136]
小浪底水库	淡水鱼	0.099±0.043（0.032~0.210）	0.084±0.042（0.020~0.194）	82.5%±9.3%	[137]
云南滇池	淡水鱼	0.006~0.287（0.074±0.059）	0.073±0.067（0.003~0.400）		[138]
海南永兴岛	海鱼	0.002~0.474	0.001~0.403	82.51%（1.28%~94.24%）	[139]
福建深沪湾	海鱼	0.037~0.353	0.001~0.036	1.5%~29.5%	[140]

研究指出，虽然我国鱼汞含量与欧美国家和地区相比，相对偏低，但人体汞的暴露风险不仅与食用鱼的汞浓度有关，还与食用鱼的频率、数量以及种类有关。发汞和血汞浓度可有效表征人体甲基汞的暴露水平。头发中汞的主要形态为甲基汞，占总汞的 80%~98%，通常头发中汞浓度是血液的 250~300 倍[141]。联合国粮食及农业组织和世界卫生组织（FAO/WHO）联合食品添加剂专家委员会（JECFA）规定甲基汞的暂定每周可耐受摄入量（PTWI）为 1.6 μg/kg 体重，规定发汞浓度最大值为 2 mg/kg。美国环境保护署（EPA）规定人体每日摄入甲基汞的参考剂量（RfD）为 0.1 μg/kg，发汞浓度最大限值为 1 mg/kg。由于我国各地区生活环境以及饮食习惯的不同，导致不同地区人群甲基汞暴露程度存在显著差异。在汞矿区以及污染区域，由于存在职业汞暴露风险或者受汞矿区影响，使得生活在该地区的居民发汞含量偏高，如万山汞矿区居民发汞平均浓度高达（4.6 ± 2.5）mg/kg，甲基汞含量为（1.7 ± 0.83）mg/kg[142]，锡矿山矿区居民发汞含量达 1.7 mg/kg[143]，均高于 1 mg/kg。在沿海地区，居民的发汞含量普遍高于内陆，但是其总汞浓度水平仍普遍小于 1 mg/kg，汞暴露风险较低，如广东汕头和福建厦门居民发汞含量分别为（0.65±0.30）mg/kg[144]和 0.88 mg/kg[145]。但是对于沿海地区海产品的摄入量较大的人群，仍存在一定的汞暴露风险，如，沿黄渤海地区五省市（辽宁、天津、河北、山东和江苏）渔民发汞平均浓度（3.30 mg/kg、1.54 mg/kg、0.91 mg/kg、2.30 mg/kg 和 2.88 mg/kg）远远高于非渔民（0.29 mg/kg、0.79 mg/kg、0.43 mg/kg、0.47 mg/kg 和 0.44 mg kg）[146]。在浙江舟山岛上，渔民头发总汞和甲基汞含量最高可达 29.9 mg/kg 和 9.5 mg/kg，其平均含量为 5.7 mg/kg 和 3.8 mg/kg，远高于 EPA 规定的发汞浓度最大限值（1 mg/kg）[147]。由此说明，长期食用鱼类（特别是肉食性鱼类）可能会给食用者（诸如渔民）带来一定的健康风险。

7.2 大米甲基汞暴露及健康风险

一般而言，粮食中的汞含量低于 20 ng/g，且以无机汞的形式为主。最新的研究表明，贵州汞污染地区的大米富集甲基汞。Horvat 等[148]报道万山汞矿区大米含有高含量的甲基汞，最大值达到 144 ng/g。贵州万山、务川和铜仁汞矿区大米的甲基汞含量与背景区相比显著升高，变化范围为 1.9~174 ng/g，一般含量水平为 10~30 ng/g[149-151]。

万山汞矿区居民甲基汞暴露量的 94%~98%来源于大米的摄入，并且居民食用大米甲基汞的日摄入量与头发甲基汞含量之间存在极显著的正相关关系，这证实了贵州汞矿区当地居民暴露甲基汞的主要途径是食用甲基汞污染的大米[151]，而不是国际上传统认为的食用鱼肉。食用稻米均是贵州省农村居民甲基汞暴露的主要途径（94%~96%），而食用鱼类仅占甲基汞总暴露量的 1%~2%[152]。尽管我国人为活动向环境排放大量汞，而南方一般居民食用大米甲基汞暴露的总体风险较低，但是食用大米是南方内陆地区农村居民甲基汞暴露的主要途径[153]。我国环境汞污染的健康风险并不是国际上传统认识的由食用汞污染的鱼和水产品引起的，而汞污染场地人群甲基汞暴露的健康风险值得关注。

FAO/WHO 联合 JECFA 制定的甲基汞最大允许摄入量（PTWI）为 0.23 μg/(kg·d)，美国 EPA 的推荐值 RfD 为 0.1 μg/(kg·d)。基于毒物代谢动力学模型，食鱼人群 PTWI 和 RfD 对应的发汞浓度分别为 2.3 μg/g 和 1.0 μg/g。万山汞矿区居民头发甲基汞含量 H（μg/g）和食用大米的日甲基汞摄入量 d［μg/(kg·d)］的拟合系数 22.9 远高于食鱼人群的结果[154]。这表明在相同的暴露剂量下，食用大米人群的暴露负荷更大，食鱼所得的风险评估体系低估了食用大米导致的甲基汞暴露风险。

8 汞同位素及环境汞污染示踪

汞同位素地球化学是当前汞研究领域的最新手段，在示踪汞的污染来源及其生物地球化学过程中能发挥无可比拟的作用[155-157]。多接收电感耦合等离子体质谱仪（MC-ICP-MS）的问世，是汞同位素高精度测试技术的里程碑。自21世纪初科学家首次利用MC-ICP-MS实现汞同位素高精度测定以来[158]，汞同位素地球化学经历了飞跃的发展。基于汞同位素地球化学在汞循环演化规律研究和汞污染防治对策方面具有广阔应用前景，汞同位素地球化学正逐渐成为国际环境科学和地球科学领域一个重要研究方向。

汞在自然界存在 ^{196}Hg（0.15%）、^{198}Hg（9.97%）、^{199}Hg（16.87%）、^{200}Hg（23.10%）、^{201}Hg（13.18%）、^{202}Hg（29.86%）和 ^{204}Hg（6.87%）七种稳定同位素[159]。汞同位素的表示通常采用相对比值法，即样品的δ值（样品同位素比值相对于标准物质同位素比值的千分差）[160]：

$$\delta^{xxx}Hg（‰）=1000\times[(^{xxx}Hg/^{198}Hg)_{样品}/(^{xxx}Hg/^{198}Hg)_{标准}-1]$$

式中 *xxx* 分别指199、200、201、202、204。若 δ>0，表明样品相对标准富集重同位素；δ<0，表明样品相对标准亏损重同位素。鉴于丰度很低，目前鲜有 $\delta^{196}Hg$ 报道。学术界通常采用美国国家标准物质研究所认证的NIST SRM 3133作为汞同位素标准。汞同位素的非质量分馏用 $\Delta^{xxx}Hg$（‰）表示，可根据公式进行计算[160]。

当前已有包括中国、美国、加拿大、法国、瑞士、比利时、意大利、日本和韩国等在内的近20个研究组陆续开展了汞同位素地球化学研究。经过国内外科学家的不懈努力，当前国际汞同位素地球化学取得了丰硕成果，主要体现在两个方面：

（1）汞同位素地球化学分馏体系逐步完善。汞同位素分馏体系的完善主要得益于实验室内对重要汞地球化学过程的模拟研究。汞同位素分馏概括起来可分为质量分馏（包括动力学分馏和热力学平衡分馏）和非质量分馏。

涉及汞生物地球化学循环的一系列物理、化学和生物过程都能导致不同程度的汞同位素质量分馏[155-157]。除此之外，汞的奇数（^{199}Hg 和 ^{201}Hg）和偶数（^{200}Hg 和 ^{204}Hg）同位素还存在非质量分馏，但仅限于某些特殊地球化学过程。2007年 *Science* 上报道了水体汞的光还原化学过程会导致非常明显的奇数汞同位素非质量分馏，该过程 $\Delta^{199}Hg/^{201}Hg$ 呈1~1.36的线性关系[161]。此后自然界中的不同环境介质，如植物、土壤、沉积物、水生生物及人体头发等样品都发现不同程度的汞同位素非质量分馏。自然界绝大部分样品非质量分馏基本落在了 $\Delta^{199}Hg/^{201}Hg$：1~1.3这个区间内，表明光化学反应是导致全球汞同位素非质量分馏的主要原因[155-157]。目前关于偶数汞同位素非质量分馏鲜有报道，主要见于大气汞样品和大气降水[162-165]。目前有关偶数汞同位素非质量分馏的机制尚不清晰，但普遍推测该分馏可能和大气汞光化学过程有关，这一推测目前已通过大气汞光氧化实验验证[162,163,166]。

（2）自然界地球化学储库的汞同位素特征日益清晰。自然界汞同位素变化是由汞生物地球化学循环过程中汞同位素分馏效应导致的。科学构建自然界样品汞同位素变化与汞同位素分馏效应的关系，对理解全球汞的生物地球化学循环具有特殊意义。目前，自然界汞同位素组成（$\delta^{202}Hg$ 和 $\Delta^{199}Hg$）变化可达>10‰，大致勾勒出不同地球化学储库的汞同位素特征。

自然界岩（矿）石样品的 $\delta^{202}Hg$ 差异较大，而绝大多数岩（矿）石 $\delta^{202}Hg$ 具有比较相似的平均值（约为–0.50），且 $\Delta^{199}Hg$ 接近于0，基本可以反映地壳的平均汞同位素组成[167-169]。自然界表生样品往往存在比较明显的汞同位素非质量分馏特征。Sonke[170]认为，汞经全球水体（如海洋、雨水等）光还原作用后，可能会导致全球大气汞（Hg^0）存在负的 $\Delta^{199}Hg$ 异常，而水体残余的汞存在正的 $\Delta^{199}Hg$ 异常。大气 Hg^0 的负异常已被大量的自然样品所证实[164,165]，陆地生态系统（如植物、土壤等）是大气汞的重要汇。目前对植物的汞同位素的测定结果显示具有明显的负 $\Delta^{199}Hg$ 特征[60,171,172]。大气汞被植物吸收后，可以凋落物的形式沉降到地表，因此土壤、泥炭、煤等均存在和大气 Hg^0 类似的负 $\Delta^{199}Hg$ 特征[171,173-177]。与此相反，大气降水和海水中存在正 $\Delta^{199}Hg$ 值，且这一正异常也反映在海相沉积物中[162,163,178]。水生环境是汞甲基化的重要场所，生成的甲基汞可经过水体光降解产生极大的正 $\Delta^{199}Hg$ 异常[161]，这一作用是导致水生生物（如鱼等）存在明显的奇数汞同位素异常（$\Delta^{199}Hg$ 达+7‰）的主要原因[155,161,179]。 此外，正 $\Delta^{199}Hg$ 异常也见于其他生物样品（如头发、鲸鱼等），这都是甲基汞光降解作用的产物[180]。

当前，我国已有包括中国科学院地球化学研究所、中国科学院生态环境研究中心、厦门大学和天津大学在内的多家单位开展汞同位素研究。我国科学家在长期的探索中，逐渐形成了自己的研究特色。例如，汞同位素示踪技术已作为一种可靠的手段成功应用到典型生态系统如农田[171]、湖泊\水库\湿地[181-185]、河流[168,186]、山地[174,177]、海洋[187]等[188]的汞污染源示踪研究。与其他用于研究食物链动力学和结构的同位素系统（如 ^{13}C 和 ^{15}N）相似，汞同位素也显示了量化汞生物积累和生物放大效应的有效前景[179]。除此之外，汞同位素有望在古环境[189]和矿床学[190,191]等领域获得更多的应用前景。然而不可否认的是，作为一个新兴的稳定同位素体系，汞同位素地球化学研究仍处于起步阶段，目前还有大量研究空白等待填补，因此亟待更多国内外同行加入该领域研究。

9 展　　望

近年来，我国环境汞污染的研究水平获得了长足进展，很多学术成果得到国际学术界的广泛认可；在最近召开的两届全球汞污染国际学术会议上，都有中国学者参与组织的专题研讨，我国学者的多项成果被列为全球汞污染的重要研究成果；多位中国学者也被邀参加联合国环境规划署将要出版的《全球汞评估 2018》报告编写；获得的有关我国环境汞污染的认识为我国参与汞公约谈判提供重要的科技支撑。由于汞在表生环境中非常活跃，排放到大气后沉降到地表环境还会再次释放到大气，因此有关环境汞的迁移转化规律的认识还有待进一步提升。首先，要加强我国人为活动和自然过程特别是先前汞排放的再释放方面的研究，为获得我国准确汞排放清单提供科学基础。其次，要开展我国区域大气汞的长期观测，发展更经济的观测新手段，为验证我国履行国际汞公约成效提供评估手段。第三，要加强我国汞污染的环境与健康风险的评估，精确识别我国汞污染健康高风险人群。第四，加强汞排放源的汞减排技术和汞污染场地的修复技术研发，减少我国人为源汞排放量和人群汞暴露的健康风险。第五，加强汞污染源识别方面的研究，特别是加强汞同位素地球化学方面的基础研究，为利用汞同位素识别环境汞污染来源提供重要科技支撑。

参考文献

[1] Nriagu J O, Pacyna J M. Quantitative assessment of worldwide contamination of air, water and soils by trace-metals. Nature,1988, 333 (6169): 134-139.

[2] Pirrone N, Keeler G J, Nriagu J O. Regional differences in worldwide emissions of mercury to the atmosphere. Atmos Environ,1996, 30 (17): 2981-2987.

[3] Pacyna E G, Pacyna J M. Global emission of mercury from anthropogenic sources in 1995. Water Air Soil Poll,2002, 137 (1-4): 149-165.

[4] Pacyna E G, Pacyna J M, Steenhuisen F, Wilson S. Global anthropogenic mercury emission inventory for 2000. Atmos Environ,2006, 40 (22): 4048-4063.

[5] Pacyna E G, Pacyna J M, Sundseth K, Munthe J, Kindbom K, Wilson S, Steenhuisen F, Maxson P. Global emission of mercury to the atmosphere from anthropogenic sources in 2005 and projections to 2020. Atmos Environ,2010, 44 (20): 2487-2499.

[6] AMAP/UNEP. Technical background report to the global atmospheric mercury assessment. AMAP/UNEP: Geneva, Switzerland, 2008.

[7] ACAP. Assessment of mercury releases from the Russian Federation. ACAP: Washton D. C., United States, 2005.

[8] Acosta-Ruiz G, Powers B. Preliminary atmospheric emissions inventory of mercury in Mexico. Washton D. C., United States: United States Environmental Portection Agency, 2003.

[9] EPA U. Mercury Study Report to Congress. US EPA: Washington, D.C, 1997.

[10] Streets D G, Hao J M, Wu Y, Jiang J K, Chan M, Tian H Z, Feng X B. Anthropogenic mercury emissions in China. Atmos Environ, 2005, 39 (40): 7789-7806.

[11] Hylander L D, Herbert R B. Global emission and production of mercury during the pyrometallurgical extraction of nonferrous sulfide ores. Environ Sci Technol, 2008, 42 (16): 5971-5977.

[12] Kocman D, Horvat M, Pirrone N, Cinnirella S. Contribution of contaminated sites to the global mercury budget. Environ Res, 2013, 125: 160-170.

[13] Pirrone N, Cinnirella S, Feng X, Finkelman R B, Friedli H R, Leaner J, Mason R, Mukherjee A B, Stracher G B, Streets D G, Telmer K. Global mercury emissions to the atmosphere from anthropogenic and natural sources. Atmos Chem Phys, 2010, 10 (13): 5951-5964.

[14] AMAP/UNEP. Technical background report for the global mercury assessment. AMAP/UNEP: Technical background report for the global mercury assessment, 2013.

[15] Streets D G, Devane M K, Lu Z F, Bond T C, Sunderland E M, Jacob D J. All-time releases of mercury to the atmosphere from human activities. Environ Sci Technol, 2011, 45 (24): 10485-10491.

[16] Horowitz H M, Jacob D J, Amos H M, Streets D G, Sunderland E M. Historical mercury releases from commercial products: Global environmental implications. Environ Sci Technol, 2014, 48 (17): 10242-10250.

[17] Wu Y, Wang S X, Streets D G, Hao J M, Chan M, Jiang J K. Trends in anthropogenic mercury emissions in China from 1995 to 2003. Environ Sci Technol,2006, 40 (17): 5312-5318.

[18] 宋敬祥, 王书肖, 李广辉. 中国锌精矿中的汞含量及其空间分布. 中国科技论文在线, 2010, (5): 472-475.

[19] 李广辉, 冯新斌, 李仲根. 竖罐炼锌过程中汞的大气排放量. 清华大学学报(自然科学版), 2009, (12): 2001-2004.

[20] 李文俊. 燃煤电厂和水泥厂大气汞排放特征研究. 重庆: 西南大学, 2011.

[21] Wang S X, Zhang L, Li G H, Wu Y, Hao J M, Pirrone N, Sprovieri F, Ancora M P. Mercury emission and speciation of coal-fired power plants in China. Atmos Chem Phys, 2010, 10 (3): 1183-1192.

[22] Wang S X, Song J X, Li G H, Wu Y, Zhang L, Wan Q, Streets D G, Chin C K, Hao J M. Estimating mercury emissions from a zinc smelter in relation to China's mercury control policies. Environ Pollut, 2010, 158 (10): 3347-3353.

[23] Wang Y J, Duan Y F, Yang L G, Zhao C S, Xu Y Q. Mercury speciation and emission from the coal-fired power plant filled with flue gas desulfurization equipment. Can J Chem Eng, 2010, 88 (5): 867-873.

[24] Wu Q R, Wang S X, Hui M L, Wang F Y, Zhang L, Duan L, Luo Y. New insight into atmospheric mercury emissions from zinc smelters using mass flow analysis. Environ Sci Technol, 2015, 49 (6): 3532-3539.

[25] Wang F Y, Wang S X, Zhang L, Yang H, Wu Q R, Hao J M. Characteristics of mercury cycling in the cement production process. J Hazard Mater, 2016, 302: 27-35.

[26] Wang F Y, Wang S X, Zhang L, Yang H, Gao W, Wu Q R, Hao J M. Mercury mass flow in iron and steel production process and its implications for mercury emission control. J Environ Sci-China, 2016, 43: 293-301.

[27] Yang M, Wang S X, Zhang L, Wu Q R, Wang F Y, Hui M L, Yang H, Hao J M. Mercury emission and speciation from industrial gold production using roasting process. Journal of Geochemical Exploration, 2016, 170: 72-77.

[28] Zhang L, Wang S X, Wu Q R, Meng Y, Yang H, Wang F Y, Hao J M. Were mercury emission factors for Chinese non-ferrous metal smelters overestimated? Evidence from onsite measurements in six smelters. Environ Pollut, 2012, 171: 109-117.

[29] Li Z G, Feng X B, Li G H, Yin R S, Yu B. Mass balance and isotope characteristics of mercury in two coal-fired power pants in Guizhou, China. Adv Mater Res-Switz, 2012, 518-523: 2576-2579.

[30] Tang Q, Liu G J, Yan Z C, Sun R Y. Distribution and fate of environmentally sensitive elements (arsenic, mercury, stibium and selenium) in coal-fired power plants at Huainan, Anhui, China. Fuel, 2012, 95 (1): 334-339.

[31] Wang J, Wang W H, Xu W, Wang X H, Zhao S. Mercury removals by existing pollutants control devices of four coal-fired power plants in China. J Environ Sci-China, 2011, 23 (11): 1839-1844.

[32] 唐黎, 刘鸿雁, 冯新斌, 李仲根, 傅成诚, 王浩, 陈吉, 王盛. 一座高灰无烟煤电厂的大气汞排放特征. 生态学杂志, 2016, (5): 1351-1357.

[33] Wu Q, Wang S, Li G, Liang S, Lin C-J, Wang Y, Cai S, Liu K, Hao J. Temporal trend and spatial distribution of speciated atmospheric mercury emissions in China during 1978-2014. Environ Sci Technol,2016, 50 (24): 13428–13435.

[34] Wu Q R, Wang S X, Zhang L, Song J X, Yang H, Meng Y. Update of mercury emissions from China's primary zinc, lead and copper smelters, 2000-2010. Atmos Chem Phys, 2012, 12 (22): 11153-11163.

[35] 高炜. 中国钢铁生产过程大气汞排放特征研究. 北京: 清华大学硕士学位论文, 2016.

[36] 林中天. 我国燃煤工业锅炉大气汞排放特征与清单研究. 北京: 清华大学硕士学位论文, 2015.

[37] 苏海涛. 我国生活垃圾焚烧行业大气汞排放特征研究. 青岛: 青岛科技大学硕士学位论文, 2016.

[38] 吴清茹. 中国有色金属冶炼行业汞排放特征及减排潜力研究. 北京: 清华大学博士学位论文, 2015.

[39] 杨海. 中国水泥行业大气汞排放特征及控制策略研究. 北京: 清华大学硕士学位论文, 2014.

[40] 张磊. 中国燃煤大气汞排放特征与协同控制策略研究. 北京: 清华大学博士学位论文, 2012.

[41] Tian H Z, Wang Y, Xue Z G, Cheng K, Qu Y P, Chai F H, Hao J M. Trend and characteristics of atmospheric emissions of Hg, As, and Se from coal combustion in China, 1980-2007. Atmos Chem Phys, 2010, 10 (23): 11905-11919.

[42] Zhang L, Wang S X, Wang L, Wu Y, Duan L, Wu Q R, Wang F Y, Yang M, Yang H, Hao J M, Liu X. Updated emission inventories for speciated atmospheric mercury from anthropogenic sources in China. Environ Sci Technol, 2015, 49 (5): 3185-3194.

[43] Hui M L, Wu Q R, Wang S X, Liang S, Zhang L, Wang F Y, Lenzen M, Wang Y F, Xu L X, Lin Z T, Yang H, Lin Y, Larssen T, Xu M, Hao J M. Mercury flows in China and global drivers. Environ Sci Technol, 2017, 51 (1): 222-231.

[44] Song S, Selin N E, Soerensen A L, Angot H, Artz R, Brooks S, Brunke E G, Conley G, Dommergue A, Ebinghaus R, Holsen T M, Jaffe D A, Kang S, Kelley P, Luke W T, Magand O, Marumoto K, Pfaffhuber K A, Ren X, Sheu G R, Slemr F, Warneke T, Weigelt A, Weiss-Penzias P, Wip D C, Zhang Q. Top-down constraints on atmospheric mercury emissions and implications for global biogeochemical cycling. Atmos Chem Phys, 2015, 15 (12): 7103-7125.

[45] Wang X, Lin C J, Feng X. Sensitivity analysis of an updated bidirectional air-surface exchange model for elemental mercury vapor. Atmos Chem Phys, 2014, 14 (12): 6273-6287.

[46] Gustin M S, Lindberg S E, Weisberg P J. An update on the natural sources and sinks of atmospheric mercury. Applied Geochemistry,2008, 23 (3), DOI: 10.1016/j.apgeochem.2007.12.010.

[47] 冯新斌, 付学吾, Sommar J, Lin J, 商立海, 仇广乐. 地表自然过程排汞研究进展及展望. 生态学杂志, 2011, 30 (5): 845-856.

[48] Eckley C S, Gustin M, Lin C J, Li X, Miller M B. The influence of dynamic chamber design and operating parameters on calculated surface-to-air mercury fluxes. Atmos Environ,2010, 44 (2): 194-203.

[49] Lin C-J, Zhu W, Li X, Feng X, Sommar J, Shang L. Novel dynamic flux chamber for measuring air–surface exchange of Hg^0 from soils. Environ. Sci. Technol., 2012, 46 (16): 8910-8920.

[50] Sommar J, Zhu W, Lin C-J, Feng X. Field approaches to measure Hg exchange between natural surfaces and the atmosphere-a review. Critical Reviews in Environmental Science and Technology, 2013, 43 (15): 1657-1739.

[51] Converse A D, Riscassi A L, Scanlon T M. Seasonal variability in gaseous mercury fluxes measured in a high-elevation meadow. Atmospheric Environment, 2010, 44 (18): 2176-2185.

[52] Fritsche J, Wohlfahrt G, Ammann C, Zeeman M, Hammerle A, Obrist D, Alewell C. Summertime elemental mercury exchange of temperate grasslands on an ecosystem-scale. Atmos Chem Phys, 2008, 8(24): 7709-7722.

[53] Sommar J, Zhu W, Shang L H, Lin C J, Feng X B. Seasonal variations in metallic mercury (Hg-0) vapor exchange over biannual wheat-corn rotation cropland in the North China Plain. Biogeosciences, 2016, 13 (7): 2029-2049.

[54] Zhu W, Lin C J, Wang X, Sommar J, Fu X, Feng X. Global observations and modeling of atmosphere–surface exchange of elemental mercury: a critical review. Atmos. Chem. Phys., 2016, 16 (7): 4451-4480.

[55] Agnan Y, Le Dantec T, Moore C W, Edwards G C, Obrist D. New constraints on terrestrial surface atmosphere fluxes of gaseous elemental mercury using a global database. Environ Sci Technol, 2016, 50 (2): 507-524.

[56] Moreno F, Anderson C N, Stewart R, Robinson B, Nomura R, Ghomshei M, Meech J A. Effect of thioligands on plant-Hg accumulation and volatilisation from mercury-contaminated mine tailings. Plant and Soil, 2005, 275 (1-2): 233-246.

[57] Moreno F N, Anderson C W N, Stewart R B, Robinson B H, Ghomshei M, Meech J A. Induced plant uptake and transport of mercury in the presence of sulphur-containing ligands and humic acid. New Phytol, 2005, 166 (2): 445-454.

[58] Wang J X, Feng X B, Anderson C W N, Wang H, Zheng L R, Hu T D. Implications of mercury speciation in thiosulfate treated plants. Environ Sci Technol, 2012, 46 (10): 5361-5368.

[59] Cui L, Feng X, Lin C-J, Wang X, Meng B, Wang X, Wang H. Accumulation and translocation of ^{198}Hg in four crop species. Environ Toxicol Chem,2014, 33 (2): 334-340.

[60] Demers J D, Blum J D, Zak D R. Mercury isotopes in a forested ecosystem: Implications for air-surface exchange dynamics and the global mercury cycle. Global Biogeochemical Cycles,2013, 27 (1): 222-238.

[61] Yin R, Feng X, Meng B. Stable Hg isotope variation in rice plants (*Oryza sativa* L.) from the Wanshan Hg Mining District, SW China. Environ. Sci. Technol.,2013, 47 (5), 2238-2245.

[62] Graydon J A, St Louis V L, Lindberg S E, Hintelmann H, Krabbenhoft D P. Investigation of mercury exchange between forest canopy vegetation and the atmosphere using a new dynamic chamber. Environ Sci Technol,2006, 40 (15): 4680-4688.

[63] Lyman S N, Gustin M S. Speciation of atmospheric mercury at two sites in northern Nevada, USA. Atmos Environ,2008, 42 (5): 927-939.

[64] Bash J O, Bresnahan P, Miller D R. Dynamic surface interface exchanges of mercury: A review and compartmentalized modeling framework. J Appl Meteorol Clim,2007, 46 (10): 1606-1618.

[65] Wang X, Lin C J, Feng X. Sensitivity analysis of an updated bidirectional air–surface exchange model for elemental mercury vapor. Atmos. Chem. Phys.,2014, 14 (12): 6273-6287.

[66] Wang X, Lin C J, Yuan W, Sommar J, Zhu W, Feng X B. Emission-dominated gas exchange of elemental mercury vapor over natural surfaces in China. Atmos Chem Phys,2016, 16 (17): 11125-11143.

[67] Xu X H, Yang X S, Miller D R, Helble J J, Carley R J. Formulation of bi-directional atmosphere-surface exchanges of elemental mercury. Atmospheric Environment,1999, 33 (27): 4345-4355.

[68] Bash J O, Miller D R, Meyer T H, Bresnahan P A. Northeast United States and Southeast Canada natural mercury emissions estimated with a surface emission model. Atmos Environ,2004, 38 (33): 5683-5692.

[69] Shetty S K, Lin C J, Streets D G, Jang C. Model estimate of mercury emission from natural sources in East Asia. Atmos Environ,2008, 42 (37): 8674-8685.

[70] Gbor P, Wen D, Meng F, Yang F, Zhang B, Sloan J. Improved model for mercury emission, transport and deposition. Atmospheric Environment,2006, 40 (5): 973-983.

[71] Zhang L M, Wright L P, Blanchard P. A review of current knowledge concerning dry deposition of atmospheric mercury.

Atmos Environ,2009, 43 (37): 5853-5864.

[72] Bash J O. Description and initial simulation of a dynamic bidirectional air-surface exchange model for mercury in Community Multiscale Air Quality (CMAQ) model. Journal of Geophysical Research-Atmospheres,2010, 115: D06305.

[73] Wright L P, Zhang L. An approach estimating bidirectional air-surface exchange for gaseous elemental mercury at AMNet sites. Journal of Advances in Modeling Earth Systems,2015, 7 (1): 35-49.

[74] Selin N E, Jacob D J, Yantosca R M, Strode S, Jaegle L, Sunderland E M. Global 3-D land-ocean-atmosphere model for mercury: Present-day versus preindustrial cycles and anthropogenic enrichment factors for deposition. Global Biogeochemical Cycles,2008, 22 (2): GB2011.

[75] Kikuchi T, Ikemoto H, Takahashi K, Hasome H, Ueda H. Parameterizing soil emission and atmospheric oxidation-reduction in a model of the global biogeochemical cycle of mercury. Environ. Sci. Technol., 2013, 47 (21): 12266-12274.

[76] Gustin M S, Amos H M, Huang J, Miller M B, Heidecorn K. Measuring and modeling mercury in the atmosphere: a critical review. Atmos Chem Phys,2015, 15 (10): 5697-5713.

[77] Holmes C D, Jacob D J, Corbitt E S, Mao J, Yang X, Talbot R, Slemr F. Global atmospheric model for mercury including oxidation by bromine atoms. Atmos Chem Phys,2010, 10 (24): 12037-12057.

[78] Wu Q R, Wang S X, Li G L, Liang S, Lin C J, Wang Y F, Cai S Y, Liu K Y, Hao J M. Temporal Trend and Spatial Distribution of Speciated Atmospheric Mercury Emissions in China During 1978-2014. Environ. Sci. Technol.,2016, 50 (24): 13428-13435.

[79] Lindberg S, Bullock R, Ebinghaus R, Engstrom D, Feng X B, Fitzgerald W, Pirrone N, Prestbo E, Seigneur C. A synthesis of progress and uncertainties in attributing the sources of mercury in deposition. Ambio,2007, 36 (1): 19-32.

[80] Fu X W, Zhang H, Yu B, Wang X, Lin C J, Feng X B. Observations of atmospheric mercury in China: A critical review. Atmos Chem Phys,2015, 15 (16): 9455-9476.

[81] Fu X, Yang X, Lang X, Zhou J, Zhang H, Yu B, Yan H, Lin C J, Feng X. Atmospheric wet and litterfall mercury deposition at urban and rural sites in China. Atmos. Chem. Phys.,2016, 16 (18): 11547-11562.

[82] Liu Y R, Lu X, Zhao L D, An J, He J Z, Pierce E M, Johs A, Gu B H. Effects of Cellular Sorption on Mercury Bioavailability and Methylmercury Production by Desulfovibrio desulfuricans ND132. Environ Sci Technol, 2016, 50 (24): 13335-13341.

[83] Hu H Y, Lin H, Zheng W, Tomanicek S J, Johs A, Feng X B, Elias D A, Liang L Y, Gu B H. Oxidation and methylation of dissolved elemental mercury by anaerobic bacteria. Nat Geosci,2013, 6 (9): 751-754.

[84] Ma M, Du H X, Wang D Y, Kang S C, Sun T. Biotically mediated mercury methylation in the soils and sediments of Nam Co Lake, Tibetan Plateau. Environ Pollut,2017, 227: 243-251.

[85] Liu Y R, Yu R Q, Zheng Y M, He J Z. Analysis of the microbial community structure by monitoring an Hg methylation gene (hgcA) in paddy soils along an Hg gradient. Appl Environ Microb,2014, 80 (9): 2874-2879.

[86] Wang X, Ye Z H, Li B, Huang L N, Meng M, Shi J B, Jiang G B. Growing rice aerobically markedly decreases mercury Aaccumulation by reducing both Hg bioavailability and the production of MeHg. Environ Sci Technol,2014, 48 (3): 1878-1885.

[87] Hammerschmidt C R, Fitzgerald W F. Iron-mediated photochemical decomposition of methylmercury in an Arctic Alaskan Lake. Environ Sci Technol,2010, 44 (16): 6138-6143.

[88] Sun R G, Wang D Y, Mao W, Zhao S B, Zhang C, Zhang X. Photodegradation of methylmercury in the water body of the Three Gorges Reservoir. Sci China Chem,2015, 58 (6): 1073-1081.

[89] Qian Y, Yin X P, Lin H, Rao B, Brooks S C, Liang L Y, Gu B H. Why dissolved organic matter enhances photodegradation of methylmercury. Environ Sci Tech Let,2014, 1 (10): 426-431.

[90] Tai C, Li Y B, Yin Y G, Scinto L J, Jiang G B, Cai Y. Methylmercury photodegradation in surface water of the Florida

Everglades: Importance of dissolved organic matter-methylmercury momplexation. Environ Sci Technol,2014, 48 (13): 7333-7340.

[91] Zhang D, Yin Y G, Li Y B, Cai Y, Liu J F. Critical role of natural organic matter in photodegradation of methylmercury in water: Molecular weight and interactive effects with other environmental factors. Sci Total Environ,2017, 578: 535-541.

[92] Han X X, Li Y B, Li D, Liu C. Role of free radicals/reactive oxygen species in MeHg photodegradation: Importance of utilizing appropriate scavengers. Environ Sci Technol,2017, 51 (7): 3784-3793.

[93] 郃超，吴浩贤，李雁宾，阴永光，毛宇翔，蔡勇，江桂斌. 环境水体中甲基汞光化学降解机理. 科学通报,2017, (1): 70-78.

[94] Lu X, Liu Y R, Johs A, Zhao L D, Wang T S, Yang Z M, Lin H, Elias D A, Pierce E M, Liang L Y, Barkay T, Gu B H. Anaerobic mercury methylation and demethylation by Geobacter bemidjiensis Bem. Environ Sci Technol,2016, 50 (8): 4366-4373.

[95] Wang X, Wu F C, Wang W X. In Vivo Mercury demethylation in a marine fish (*Acanthopagrus schlegeli*). Environ Sci Technol,2017, 51 (11): 6441-6451.

[96] Wang Y M, Li Y B, Liu G L, Wang D Y, Jiang G B, Cai Y. Elemental mercury in natural waters: Occurrence and determination of particulate Hg(0). Environ Sci Technol,2015, 49 (16): 9742-9749.

[97] He F, Zhao W R, Liang L Y, Gu B H. Photochemical oxidation of dissolved elemental mercury by carbonate radicals in water. Environ Sci Tech Let,2014, 1 (12): 499-503.

[98] Manceau A, Lemouchi C, Enescu M, Gaillot A C, Lanson M, Magnin V, Glatzel P, Poulin B A, Ryan J N, Aiken G R, Gautier-Luneau I, Nagy K L. Formation of mercury sulfide from Hg(II)-thiolate complexes in natural organic matter. Environ Sci Technol,2015, 49 (16): 9787-9796.

[99] Luo H W, Yin X P, Jubb A M, Chen H M, Lu X, Zhang W H, Lin H, Yu H Q, Liang L Y, Sheng G P, Gu B H. Photochemical reactions between mercury (Hg) and dissolved organic matter decrease Hg bioavailability and methylation. Environ Pollut,2017, 220: 1359-1365.

[100] Jiang P, Li Y B, Liu G L, Yang G D, Lagos L, Yin Y G, Gu B H, Jiang G B, Cai Y. Evaluating the role of re-adsorption of dissolved Hg2+ during cinnabar dissolution using isotope tracer technique. J Hazard Mater,2016, 317: 466-475.

[101] Mulligan C N, Yong R N, Gibbs B F. Remediation technologies for metal-contaminated soils and groundwater: an evaluation. Eng Geol,2001, 60 (1-4): 193-207.

[102] Smolinska B, Krol K. Leaching of mercury during phytoextraction assisted by EDTA, KI and citric acid. J Chem Technol Biot,2012, 87 (9): 1360-1365.

[103] Wasay S A, Arnfalk P, Tokunaga S. Remediation of a soil polluted by mercury with acidic potassium-iodide. J Hazard Mater,1995, 44 (1): 93-102.

[104] Chang T C, Yen J H. On-site mercury-contaminated soilsremediation by using thermal desorption technology. J Hazard Mater,2006, 128 (2-3): 208-217.

[105] Reddy K R, Chaparro C, Saichek R E. Iodide-enhanced electrokinetic remediation of mercury-contaminated soils. J Environ Eng-Asce,2003, 129 (12): 1137-1148.

[106] Darban A, Ayati B, Yong R, Khodadadi A, Kiayee A. Enhanced electrokinetic remediation of mercury-contaminated tailing dam sediments.2009.

[107] Gong Y, Liu Y, Xiong Z, Kaback D, Zhao D. Immobilization of mercury in field soil and sediment using carboxymethyl cellulose stabilized iron sulfide nanoparticles. Nanotechnology,2012, 23(29): 294007.

[108] Wang M, Chen S, Li N, Ma Y. A review on the development and application of nano-scale amendment in remediating

polluted soils and waters. Chinese Journal of Eco-Agriculture,2010, 18 (2): 434-439.

[109] Zhang J Y, Li C X, Wang D Y, Zhang C, Liang L, Zhou X. The effect of different TiO_2 nanoparticles on the release and transformation of mercury in sediment. J Soil Sediment,2017, 17 (2): 536-542.

[110] Wang J X, Feng X B, Anderson C W N, Xing Y, Shang L H. Remediation of mercury contaminated sites - A review. J Hazard Mater,2012, 221: 1-18.

[111] Sinha A, Khare S K. Mercury bioremediation by mercury accumulating *Enterobacter* sp cells and its alginate immobilized application. Biodegradation,2012, 23 (1): 25-34.

[112] Padmavathiamma P K, Li L Y. Phytoremediation technology: Hyper-accumulation metals in plants. Water Air Soil Poll,2007, 184 (1-4): 105-126.

[113] Irikayama K, Kondo T. Studies on the organomercury compound in the fish and shellfish from Minamata Bay and its origin. VII. Synthesis of methylmercury sulfate and its chemical properties. Nihonseigaku Zasshi Japanese Journal of Hygiene,1966, 21 (5): 342.

[114] Stein E D, Cohen Y, Winer A M. Environmental distribution and transformation of mercury compounds. Critical Reviews in Environmental Science & Technology,1996, 26 (1): 1-43.

[115] Bank M S. Mercury in the Environment:Pattern and Process. University of California Press: 2012, 863-864.

[116] Zhang Q, Pan K, Kang S, Zhu A, Wang W X. Mercury in wild fish from high-altitude aquatic ecosystems in the Tibetan Plateau. Environmental Science & Technology,2014, 48 (9): 5220-8.

[117] Ping L, Feng X, Qiu G. Methylmercury exposure and health effects from rice and fish consumption: A review. International Journal of Environmental Research & Public Health,2010, 7 (6): 2666-2691.

[118] Cheng H, Hu Y. Understanding the paradox of mercury pollution in China: High concentrations in environmental matrix yet low levels in fish on the market. Environmental Science & Technology,2012, 46 (9): 4695-4696.

[119] 甘居利, 贾晓平. 中国浅海经济鱼类重金属的卫生质量状况. 海洋通报, 1997, (4): 88-93.

[120] 陈一资. 市售肉蛋奶中汞残留的检测. 四川农业大学学报, 1990, 8 (1): 23-26.

[121] 刘钧, 张淑伟. 山东南四湖水产品中铅,镉,砷,汞污染状况调查与分析. 现代预防医学, 1996, (3): 164-166.

[122] 黄宏瑜, 许悦生. 珠海市水产品中汞镉铅砷污染状况监测. 中国公共卫生,1998, 14 (1): 23-25.

[123] 张彦明, 刘建成. 陕西省部分市县鱼肉蛋奶中汞含量的检测分析. 西北农业大学学报,1997, (2): 61-65.

[124] 孙瑾, 陈春英, 李柏, 李玉锋, 王江雪, 高愈希, 柴之芳. 北京市场 4 种食用淡水鱼的总汞和甲基汞的含量分析.卫生研究, 2006, 35 (6): 722-725.

[125] 闫雨平. 中国近海海域经济鱼类的重金属污染及其评价. 海洋环境科学,1993, (3): 99-103.

[126] 朱艾嘉, 许战洲, 柳圭泽, 邓丽杰, 方宏达, 黄良民. 黄海常见鱼类体内汞含量的种内和种间差异研究. 环境科学,2014, 35 (2): 764-769.

[127] 田文娟. 珠江三角洲地区总汞和甲基汞人体暴露水平与风险评价.广州: 暨南大学, 2011.

[128] Shao D, Liang P, Kang Y, Wang H, Cheng Z, Wu S, Shi J, Lo S C L, Wang W, Ming H W. Mercury species of sediment and fish in freshwater fish ponds around the Pearl River Delta, PR China: Human health risk assessment. Chemosphere,2011, 83 (4): 443-8.

[129] 蒋红梅, 冯新斌, 阎海鱼. 乌江流域水库鱼体汞分布特征. 2010 中国环境科学学会学术年会论文集(第三卷). 2010.

[130] Liu B, Yan H, Wang C, Li Q, Guédron S, Spangenberg J E, Feng X, Dominik J. Insights into low fish mercury bioaccumulation in a mercury-contaminated reservoir, Guizhou, China. Environmental Pollution,2012, 160 (1): 109-117.

[131] 卢晏生,翟平阳. 松花江鱼体内汞的残留水平. 水产学杂志, 1999, 12 (2): 74-80.

[132] Zhang L, Wang Q, Shao Z. Mercury contamination of fish in the Di'er Songhua River of China: The present station and evolution law. 2005, 14: 190-194.

[133] 徐勤勤. 三峡水库鱼体中汞和甲基汞的分布特征.重庆:西南大学, 2014.

[134] 白薇扬. 三峡库区典型支流水库长寿湖汞的生物地球化学特征. 重庆:西南大学, 2015.

[135] 李清, 康世昌, 张强弓, 黄杰, 郭军明, 王康, 王建力. 青藏高原纳木错湖近 150 年来气候变化的湖泊沉积记录. 沉积学报,2014, 32 (4): 669-676.

[136] Qiu G, Feng X, Wang S, Fu X, Shang L. Mercury distribution and speciation in water and fish from abandoned Hg mines in Wanshan, Guizhou province, China. Science of the Total Environment,2009, 407 (18): 5162-5168.

[137] 索乾善, 毛宇翔, 张飞鹏, 崔莹. 小浪底水库鱼体汞的污染现状. 环境化学, 2013, 32 (11): 2030-2036.

[138] 魏中青, 周志华. 滇池几种鱼体汞含量与 δ 13C、δ 15N 特征. 地球与环境,2014, 42 (4): 480-483.

[139] 刘金铃, 徐向荣, 陈来国, 苏燕花, 王帅龙, 彭加喜, 郝青. 永兴岛 4 种海鱼中汞的含量及人体风险评估. 海洋环境科学,2013, 32 (6): 867-870.

[140] 林方芳, 袁东星, 刘锡尧, 林珊珊, 冯丽凤. 福建深沪湾不同环境样品中总汞和甲基汞含量的分布. 应用海洋学学报,2013, 32 (3): 425-431.

[141] Bach P. Book Reviews : Environmental health criteria 101 methylmercury. Published by World Health Organisation, Geneva. 1990. Price: SW. Fr. 16. Paperback. Pp 140. ISBN 92 41571012.1990, 110 (5): 189-189.

[142] Ping L, Feng X, Qiu G, Shang L, Li G. Human hair mercury levels in the Wanshan mercury mining area, Guizhou Province, China. Environmental Geochemistry & Health,2009, 31 (6): 683-691.

[143] 刘碧君, 吴丰昌, 邓秋静, 莫昌琍, 朱静, 曾理, 符志友, 黎文. 锡矿山矿区和贵阳市人发中锑、砷和汞的污染特征. 环境科学,2009, 30 (3): 907-912.

[144] Ni W, Chen Y, Huang Y, Wang X, Zhang G, Luo J, Wu K. Hair mercury concentrations and associated factors in an electronic waste recycling area, Guiyu, China. Environmental Research,2014, 128: 84-91.

[145] Xiaojie L, CHENG J, Yuling S, Honda S i, Li W, Zheng L, Sakamoto M, Yuanyuan L. Mercury concentration in hair samples from Chinese people in coastal cities. Journal of Environmental Sciences,2008, 20 (10): 1258-1262.

[146] 尹春凫, 郭海恩, 李文华, 张淑珍. 沿渤黄海五省市渔民农民发汞调查. 卫生研究,1981, (3): 12-14.

[147] Cheng J, Gao L, Zhao W, Liu X, Sakamoto M, Wang W. Mercury levels in fisherman and their household members in Zhoushan, China: Impact of public health. Science of the Total Environment,2009, 407 (8): 2625-2630.

[148] Horvat M, Nolde N, Fajon V, Jereb V, Logar M, Lojen S, Jacimovic R, Falnoga I, Qu L Y, Faganeli J, Drobne D. Total mercury, methylmercury and selenium in mercury polluted areas in the province Guizhou, China. Sci Total Environ,2003, 304 (1-3): 231-256.

[149] Qiu G L, Feng X B, Wang S F, Shang L H. Environmental contamination of mercury from Hg-mining areas in Wuchuan, northeastern Guizhou, China. Environ Pollut,2006, 142 (3): 549-558.

[150] Qiu G L, Feng X B, Li P, Wang S F, Li G H, Shang L H, Fu X W. Methylmercury accumulation in rice (*Oryza sativa* L.) grown at abandoned mercury mines in Guizhou, China. J Agr Food Chem,2008, 56 (7): 2465-2468.

[151] Feng X B, Li P, Qiu G L, Wang S, Li G H, Shang L H, Meng B, Jiang H M, Bai W Y, Li Z G, Fu X W. Human exposure to methylmercury through rice intake in mercury mining areas, guizhou province, China. Environ Sci Technol,2008, 42 (1): 326-332.

[152] Zhang H, Feng X B, Larssen T, Qiu G L, Vogt R D. In Inland China, Rice, Rather than fish, is the major pathway for methylmercury exposure. Environ Health Persp,2010, 118 (9): 1183-1188.

[153] Li P, Feng X, Yuan X, Chan H M, Qiu G, Sun G-X, Zhu Y-G. Rice consumption contributes to low level methylmercury exposure in southern China. Environ Int,2012, 49: 18-23.

[154] Li P, Feng X, Chan H-M, Zhang X, Du B. Human body burden and dietary methylmercury intake: The relationship in a rice-consuming population. Environ Sci Technol,2015, 49 (16): 9682-9689.

[155] Blum J D, Sherman L S, Johnson M W. Mercury isotopes in earth and environmental dciences. Annu Rev Earth Pl Sc,2014, 42: 249-269.

[156] Yin R, Feng X, Shi W. Application of the stable-isotope system to the study of sources and fate of Hg in the environment: A review. Appl Geochem,2010, 25 (10): 1467-1477.

[157] Yin R, Feng X, Li X, Yu B, Du B. Trends and advances in mercury stable isotopes as a geochemical tracer. Trends in Environmental Analytical Chemistry,2014, 2: 1-10.

[158] Klaue B, Kesler S E, Blum J D, Investigation of natural fractionation of stable mercury isotopes by multi-collector inductively coupled plasma mass spectrometry. *In* Annual International Conference on Heavy Metals in the Environment, Ann Arbor .MI, USA, 2000.

[159] Yin R S, Feng X B, Foucher D, Shi W F, Zhao Z Q, Wang J. High precision determination of mercury isotope ratios using online mercury vapor generation system coupled with multi-collector inductively coupled plasma-mass spectrometry. Chinese J Anal Chem,2010, 38 (7): 929-934.

[160] Blum J D, Bergquist B A. Reporting of variations in the natural isotopic composition of mercury. Anal Bioanal Chem,2007, 388 (2): 353-359.

[161] Bergquist B A, Blum J D. Mass-dependent and -independent fractionation of Hg isotopes by photoreduction in aquatic systems. Science,2007, 318 (5849): 417-420.

[162] Gratz L E, Keeler G J, Blum J D, Sherman L S. Isotopic composition and fractionation of mercury in great lakes precipitation and ambient air. Environ Sci Technol,2010, 44 (20): 7764-7770.

[163] Chen J B, Hintelmann H, Feng X B, Dimock B. Unusual fractionation of both odd and even mercury isotopes in precipitation from Peterborough, ON, Canada. Geochim Cosmochim Ac,2012, 90: 33-46.

[164] Rolison J M, Landing W M, Luke W, Cohen M, Salters V J M. Isotopic composition of species-specific atmospheric Hg in a coastal environment. Chem Geol,2013, 336: 37-49.

[165] Yu B, Fu X, Yin R, Zhang H, Wang X, Lin C-J, Wu C, Zhang Y, He N, Fu P, Wang Z, Shang L, Sommar J, Sonke J E, Maurice L, Guinot B, Feng X. Isotopic composition of atmospheric mercury in China: New evidence for sources and transformation processes in air and in vegetation. Environ Sci Technol,2016, 50 (17): 9262-9.

[166] Sun G, Sommar J, Feng X, Lin C-J, Ge M, Wang W, Yin R, Fu X, Shang L. Mass-dependent and -independent fractionation of mercury isotope during gas-phase oxidation of elemental mercury vapor by atomic Cl and Br. Environ Sci Technol,2016, 50 (17): 9232-41.

[167] Smith C N, Kesler S E, Klaue B, Blum J D. Mercury isotope fractionation in fossil hydrothermal systems. Geology,2005, 33 (10): 825-828.

[168] Yin R S, Feng X B, Wang J X, Li P, Liu J L, Zhang Y, Chen J B, Zheng L R, Hu T D. Mercury speciation and mercury isotope fractionation during ore roasting process and their implication to source identification of downstream sediment in the Wanshan mercury mining area, SW China. Chem Geol,2013, 336: 72-79.

[169] Yin R S, Feng X B, Hurley J P, Krabbenhoft D P, Lepak R F, Hu R Z, Zhang Q, Li Z G, Bi X W. Mercury isotopes as proxies to identify sources and environmental impacts of mercury in sphalerites. Sci Rep-Uk,2016, 6.

[170] Sonke J E. A global model of mass independent mercury stable isotope fractionation. Geochim Cosmochim Ac,2011, 75 (16): 4577-4590.

[171] Yin R S, Feng X B, Meng B. Stable mercury isotope variation in rice plants (*Oryza sativa* L.) from the wanshan mercury mining district, SW China. Environ Sci Technol,2013, 47 (5): 2238-2245.

[172] Enrico M, Le Roux G, Marusczak N, Heimburger L E, Claustres A, Fu X W, Sun R Y, Sonke J E. Atmospheric mercury

transfer to peat bogs dominated by gaseous elemental mercury dry deposition. Environ Sci Technol, 2016, 50 (5): 2405-2412.

[173] Biswas A, Blum J D, Bergquist B A, Keeler G J, Xie Z Q. Natural mercury isotope variation in coal deposits and organic soils. Environ Sci Technol,2008, 42 (22): 8303-8309.

[174] Zhang H, Yin R S, Feng X B, Sommar J, Anderson C W N, Sapkota A, Fu X W, Larssen T. Atmospheric mercury inputs in montane soils increase with elevation: evidence from mercury isotope signatures. Sci Rep-Uk,2013, 3.

[175] Sun R Y, Sonke J E, Heimburger L E, Belkin H E, Liu G J, Shome D, Cukrowska E, Liousse C, Pokrovsky O S, Streets D G. Mercury stable isotope signatures of world coal deposits and historical coal combustion emissions. Environ Sci Technol,2014, 48 (13): 7660-7668.

[176] Sun R Y, Sonke J E, Liu G J. Biogeochemical controls on mercury stable isotope compositions of world coal deposits: A review. Earth-Sci Rev,2016, 152: 1-13.

[177] Wang X, Luo J, Yin R, Yuan W, Lin C-J, Sommar J, Feng X, Wang H, Lin C. Using mercury isotopes to understand mercury accumulation in the montane forest floor of the Eastern Tibetan Plateau. Environ Sci Technol,2017, 51 (2): 801-809.

[178] Strok M, Hintelmann H, Dimock B. Development of pre-concentration procedure for the determination of Hg isotope ratios in seawater samples. Anal Chim Acta,2014, 851: 57-63.

[179] Yin R S, Feng X B, Zhang J J, Pan K, Wang W X, Li X D. Using mercury isotopes to understand the bioaccumulation of Hg in the subtropical Pearl River Estuary, South China. Chemosphere,2016, 147: 173-179.

[180] Li M L, Schartup A T, Valberg A P, Ewald J D, Krabbenhoft D P, Yin R S, Bolcom P H, Sunderland E M. Environmental origins of methylmercury accumulated in subarctic estuarine fish indicated by mercury stable isotopes. Environ Sci Technol,2016, 50 (21): 11559-11568.

[181] Feng X B, Foucher D, Hintelmann H, Yan H Y, He T R, Qiu G L. Tracing mercury contamination sources in sediments using mercury isotope compositions. Environ Sci Technol,2010, 44 (9): 3363-3368.

[182] Yin R S, Feng X B, Hurley J P, Krabbenhoft D P, Lepak R F, Kang S C, Yang H D, Li X D. Historical records of mercury stable isotopes in sediments of Tibetan Lakes. Sci Rep-Uk,2016, 6.

[183] Sun L M, Lu B Y, Yuan D X, Hao W B, Zheng Y. Variations in the isotopic composition of stable mercury isotopes in typical mangrove plants of the Jiulong estuary, SE China. Environ Sci Pollut R,2017, 24 (2): 1459-1468.

[184] Shi W F, Feng X B, Zhang G, Ming L L, Yin R S, Zhao Z Q, Wang J. High-precision measurement of mercury isotope ratios of atmospheric deposition over the past 150 years recorded in a peat core taken from Hongyuan, Sichuan Province, China. Chinese Sci Bull,2011, 56 (9): 877-882.

[185] 花秀兵，毛宇翔，刘洪伟，程柳，史建波，江桂斌. 小浪底水库鱼体和沉积物中汞稳定同位素组成特征. 环境化学,2016, 35 (11): 2245-2252.

[186] Liu J L, Feng X B, Yin R S, Zhu W, Li Z G. Mercury distributions and mercury isotope signatures in sediments of Dongjiang, the Pearl River Delta, China. Chem Geol,2011, 287 (1-2): 81-89.

[187] Yin R S, Feng X B, Chen B W, Zhang J J, Wang W X, Li X D. Identifying the sources and processes of mercury in subtropical estuarine and ocean sediments using Hg isotopic composition. Environ Sci Technol,2015, 49 (3): 1347-1355.

[188] Lin H, Peng J, Yuan D, Lu B, Lin K, Huang S. Mercury isotope signatures of seawater discharged from a coal-fired power plant equipped with a seawater flue gas desulfurization system. Environ Pollut,2016, 214: 822-830.

[189] Grasby S E, Shen W, Yin R, Gleason J D, Blum J D, Lepak R F, Hurley J P, Beauchamp B. Isotopic signatures of mercury contamination in latest Permian oceans. Geology,2017, 45 (1): 55-58.

[190] Tang Y, Bi X, Yin R, Feng X, Hu R. Concentrations and isotopic variability of mercury in sulfide minerals from the Jinding Zn-Pb deposit, Southwest China. Ore Geology Reviews,2017.

[191] Xu C, Yin R, Peng J, Hurley J P, Lepak R F, Gao J, Feng X, Hu R, Bi X. Mercury isotope constraints on the source for sediment-hosted lead-zinc deposits in the Changdu area, southwestern China. Mineralium Deposita,2017: 1-14.

作者：冯新斌[1]，王书肖[2]，王　训[1]，付学吾[1]，阴永光[3]，史建波[3]，
张　华[1]，王永敏[4]，王定勇[4]，李　平[1]，尹润生[5]
[1] 中国科学院地球化学研究所，[2] 清华大学环境学院，[3] 中国科学院生态环境研究中心，
[4] 西南大学资源环境学院，[5] 中国科学院地球化学研究所

第 11 章　砷锑的环境污染及去除控制研究进展

本章导读

砷锑环境污染严重威胁人类健康。研究砷锑的环境行为及其去除控制方法，对解决砷锑污染问题具有重要意义。与砷锑代谢相关的微生物驱动及其在环境中的迁移转化，因此微生物对砷锑环境行为作用机制值得高度关注。同时，砷锑的去除控制是亟待解决的环境问题，而去除机制的研究是解决此环境问题的关键，因而，综合多种原位表征技术在分子水平上研究砷锑的环境界面过程，成为环境砷锑污染研究的主流前沿热点。本章综述了砷锑在不同环境中赋存形态的研究，阐述了微生物对砷锑迁移转化的作用，并重点介绍了砷锑去除控制的机制及方法。

关键词

砷，锑，环境污染，环境行为，去除控制

1 引　　言

砷和锑是元素周期表中15族（第V主族）的同族元素，具有相同的ns^2np^3的价层电子结构。它们的电子构型决定了它们在环境介质中的赋存、化学特性、行为及效应的相似性。砷和锑在自然界中主要以三价和五价无机含氧酸根离子形式存在，目前关于这两种价态砷锑的暴露途径及毒性效应研究众多，三价形态的毒性大于五价的。环境中的砷锑可以通过皮肤、呼吸、食物链等途径进入到人体和动物体中。由于砷锑对蛋白质的巯基具有很强的亲和力，因此，进入体内的砷锑会引起蛋白质和酶在细胞内变性，影响代谢活动，侵犯心、肝、肾、胃肠等器官。同时研究发现，砷锑可以通过增加细胞内的活性氧簇而引起细胞损伤，继而对人体的免疫、基因、发育等都具有潜在的毒性。此外，长期暴露于低剂量锑的环境中会引起慢性锑中毒，造成头痛、皮疹、腹痛、胃肠功能紊乱等慢性中毒症状。砷、锑的饮用水标准分别为10 μg/L与6 μg/L。

2 砷锑在环境中的赋存形态

2.1 土壤环境

随着含砷锑金属矿产开采与冶炼、化石燃料燃烧、工业废水废渣排放倾倒，使得环境中砷锑浓度日益增加[1]。据世界卫生组织（WHO）估算，土壤中最大可允许砷锑的浓度分别为8 mg/kg和5 mg/kg[2]，超过此浓度的土壤均为污染土壤环境。在天然土壤中，砷和锑主要以五价含氧阴离子的形式存在。而在水淹条件下的稻田土壤中，砷和锑主要以As(III)和Sb(III)的形式存在[3]。

土壤中的矿物会与部分砷锑结合，从而影响其赋存形态。As和Sb会结合土壤中含Al、Fe、Mn和Ca的氧化物和羟基氢氧化物[4]，如方解石[5]、黄铁矿等。沉积物中的硅酸盐和硫化物[6]也会通过与砷锑结合而对其固定起到重要作用，如由构造性活动引起的缺氧条件有利于黄铁矿中δ^{34}S和As的富集。有机物的存在会影响As、Sb的形态[7]，在土壤还原条件下，天然有机质中的巯基会与As（III）结合[8]；当矿物有机物均共存时，As 和 Sb 会与有机物及 Fe(III)、Cr(III)等金属氧化物形成三元络合结构。

由于 As 和 Sb 具有相同的最外层轨道电子结构，因此在传输和沉积过程中，As 和 Sb 常被认为有相同的地球化学行为[1]。但是对锡矿山周围砷锑在不同介质下地球化学行为的研究结果表明，尾矿中 As 的迁移性强于 Sb[9]。对土壤中 As 和 Sb 消解分析实验表明，超过 50%的 Sb 既没有与金属氧化物或氢氧化物结合，也没有与有机质结合[10]。而 As 在土壤中主要以残留态 As 和可交换 As 的形式存在，约占土壤中总 As 含量的 60%[11]。

2.2 水环境

水环境中 As 的形态与硫化物浓度及 pH 有关，在硫化物浓度低的碱性水中，As 主要以 As(III)含氧阴离子形式存在；在中高浓度硫化物的水中，As 主要以 As(III)和 As(V)的含硫形态存在。同时，氧气条件也会影响水环境中 As 的形态，在 O_2 充足情况下，As 以 As(V)的含氧阴离子存在。而缺氧条件下，As(III)是主要存在形式。水环境中氧化还原条件的改变，如地下水还原环境，会导致砷从非游离态转化为游离态。如铁氧化物还原溶解，导致与其相结合的 As 释放，从而使得溶解态 As 增加。另外，有机质的分解作用会造成与有机质结合的 As 的赋存形态改变，使 As 移动性增加。

Sb(III)和 Sb(V)常共存于不同的天然水环境系统中[10]。Sb 来源的主要机理是含 Sb 矿物的氧化溶解和水溶解。其中温度、pH 和氧化还原电位（E_h）是主要的影响因素。如在碱性条件下（pH 9~10），Sb 更容易从矿石中释放出来。含有 Sb_2S_3 的锑矿会转化产生溶解态的 $Sb(OH)_3$。随着 pH 的升高，越来越多的 $Sb(OH)_3$ 会转化成阴离子的形式，其中 $Sb(OH)_4^-$是溶解度最高的含水化合物的形式。如锑华矿在 25℃，pH 1.5~12 的条件下，可以溶解形成中性的 $Sb(OH)_3$ 络合物[12]。在 pH>2.47 的水环境中，$Sb(OH)_3$ 可以进一步分解转化成$[Sb(OH)_6]^-$[13]。

2.3 大气环境

化石燃料的燃烧是大气中 As、Sb 污染的主要来源。中国是 Sb 排放最多的国家[14]。在我国，山东、江苏、山西、河北是 As、Sb 大气排放最多的省份[15]。近年来研究报道，随着煤矸石生产量的增加，煤矸石的燃烧也已经成为我国大气 As 污染的主要来源之一。

由于大气环境的特殊性，As 和 Sb 会在大气中实现迁移扩散，进而影响砷锑在地表的地球化学循环。如火山喷发产生的 As 会通过飞灰沉积的方式改变周围湖泊底栖生物中 As 的积累。另外，砷锑污染源当地盛行风向也会影响砷和锑在附近土壤中的分布[1]。

最新研究表明，砷锑可在大气环境中进行长距离迁移。对香港非工业区的研究发现在此区域的冬季大气颗粒物中，检测到来自中国大陆工业和燃煤电厂的高温季节产生的含 As 污染物[4]。甚至南极洲地区也可检测到 As 和 Sb，这主要是由于南美洲智利非铁矿物的熔炼及石料燃烧所产生的 As 和 Sb，随着大气颗粒传输沉降，最终与积雪中的高分子有机质络合[16]。上述研究说明含 As 及 Sb 的颗粒物能够在大气中长距离且长时间传输。

2.4 植物系统

关于 As 在水稻体内的传输行为和形态变化有许多研究，但是关于 Sb 的研究却屈指可数。最新针对稻田土壤-水-水稻系统的研究发现 Sb 主要以 Sb(V)的形式存在，且在水稻地上部的含量要高于根和孔隙水中的含量。这说明水稻地上部优先吸附 Sb(V)，或 Sb(III)在从根到水稻地上部的迁移过程中被氧化[17]。

与此不同的是，有研究表明 As(III)是水稻根和地上部分的主要形态[18]。

As 和 Sb 在水稻-土壤-水的系统中有相似的吸收速率。但是在向日葵、小麦、黑麦草幼苗中，As 的积累速度大大高于 Sb[19]。也有研究认为，在锡矿山尾矿和土壤中 Sb 的含量比 As 高一个数量级；而在植物和蔬菜中检测到的 As 含量与 Sb 相同，有些甚至大于 Sb，说明 As 的生物积累性大约是 Sb 的 10 倍[9]。

大气沉积同样会影响植物体内 As、Sb 的富集，锡矿山周围的叶类蔬菜体内积累的 As、Sb 要大于其他种类的蔬菜，且植物叶子内的含量大于根部的含量，这说明大气沉降对 As、Sb 在植物体内的分布影响更为明显。即使是在种皮包裹的布什豆种子内部，也存在 As 的富集。同时在树的年轮中富集的 As、Sb 可以反映周围环境中砷锑排放随时间的变化。

3　微生物作用下的砷锑形态转化

3.1　微生物对砷环境转化的影响

砷由矿物表面释放到水体主要受两个过程所控制：砷在矿物表面的吸附-解吸过程和沉淀-溶解过程[20]。其中微生物的作用被认为是砷释放到自然水体中的关键因素之一[21,22]。微生物主要通过①对砷形态的改变和②对矿物性质的改变这两个方面来影响砷在矿物表面的迁移转化过程。

3.1.1　形态转化对砷迁移转化的影响

砷的氧化菌和还原菌在自然界中广泛存在，且直接决定了水体中砷的形态，是影响矿物表面砷迁移转化过程的重要因素。其中，砷还原菌介导的 As(V)还原为 As(III)被认为是导致砷从矿物表面脱附从而造成水体砷浓度升高的重要原因之一。微生物砷还原主要包括两种机制：①*arsC* 介导的砷解毒机制，As(V)通过磷酸进入细胞通道进入细胞，在细胞质内被 *arsC* 编码的 As(V)还原酶还原为 As(III)，最后由 *arsB* 编码的膜转运蛋白将其排出体外；②*arrA* 介导的砷呼吸机制，As(V)作为电子受体，为微生物的生长提供能量[23]。砷的解毒还原过程在有氧或无氧环境下均可发生，而砷的呼吸还原过程只能在厌氧环境发生。

大部分研究认为解毒机制和呼吸机制介导的砷还原过程均能促进砷在矿物上的释放，但相比于解毒机制，研究认为呼吸机制能促进更多的砷释放到水体中，可能是由于呼吸机制利用的砷还原酶位于细胞周质或者细胞膜上，能够直接还原吸附态的砷，从而造成砷的释放。*arrA* 介导的砷还原菌 *Geobacter* sp. OR-1 和 *arsC* 介导的 *Pseudomonas* sp. M17-1、*Acillus* sp. M17-15 都被认为可以直接还原吸附态的砷。但是由于研究者都没有排除溶解态的 As(III)再次吸附到固相的可能性，因此并不能直接断定 *arrA* 或者 *arsC* 可以还原吸附态的砷。如何区分溶解态砷再吸附过程和菌体直接还原吸附态砷过程，是该领域研究的一大难点。也有一些研究认为解毒机制[24]和呼吸机制[24]诱导的砷还原过程并不能促进砷的释放，产生这两个相反结论的原因是矿物对 As(V)和 As(III)的吸附差异。

虽然砷氧化菌在土壤中也是广泛存在的，但相对于砷还原菌，关于砷氧化菌对砷迁移转化影响的研究较少。因为与 As(V)相比，一般来说 As(III)更易从铁矿物上脱附。因此，砷氧化菌将 As(III)氧化为 As(V)的过程，会减少 As 的释放。然而，最近发现一株砷氧化菌 *Pseudomonas* sp. HN-1 能够降低砷在铁氧化物上的吸附，从而促进砷的释放[25]。这两种矛盾的结果可能源于矿物本身对 As(III)和 As(V)吸附能力的差异。

3.1.2 铁矿物的形态转化对砷迁移转化的影响

在自然界中，铁氧化物是砷的重要载体，铁矿物的溶解和铁矿的转化都是造成砷释放的主要原因之一。

在厌氧条件下，铁还原菌导致铁矿物溶解，它是一类在代谢过程中以Fe(III)作为电子受体，将Fe(III)还原为Fe(II)的微生物。铁还原菌还原不溶性铁矿物的机制主要有以下四种[26]：①直接接触电子传递，通过细胞外膜蛋白与铁矿物的直接电子传递，造成Fe(II)溶出；②纳米导线电子传递，铁还原菌形成类似菌毛的微生物纳米导线传递电子给铁氧矿物，造成其溶解；③螯合剂促进溶解，溶铁的螯合剂与铁氧化物形成可溶性螯合铁，扩散至细胞表面，使 Fe(III)还原溶解；④电子介质传递，细胞将电子传递给环境中的电子储存介质，然后环境介质将电子传递给铁氧化物将其还原溶解。Fe(III)的溶解伴随着 As 的释放，溶解态铁和溶解态砷的浓度显著相关[27]。Islam 等学者认为砷的释放过程发生在 Fe(III)的还原之后。他们在含砷沉积物中加入醋酸盐作为碳源后，培养 8 天后溶液中的 Fe(II)的浓度达到最大值，而溶解态 As(III)的浓度在 22 天后才达到最大值[28]。

实际环境中，微生物砷还原过程与铁还原过程常常是同时存在的，对于哪个还原过程对砷从沉积物释放到自然水体影响更大，一直存在争议[29]。Fendorf 等学者通过对比只有砷还原功能或者铁还原功能的两株突变菌 *Shewanella* sp. ANA-3 对吸附在三种不同铁氧化物上的砷的迁移转化影响，发现微生物砷还原作用对砷迁移释放影响更大[29]。而 Jing 等学者通过比较砷还原菌和铁还原菌对矿厂中砷的迁移转化影响，却发现只有铁还原菌显著促进了砷的释放[30]，出现这种矛盾结论的原因可能是其生成的二次铁矿再吸附砷的能力不同。铁还原菌还原溶解铁氧化物生成的 Fe(II)，在环境基质中会转化形成二次铁矿，如水铁矿、磁铁矿、蓝铁矿、菱铁矿和针铁矿等，形成的二次铁矿一般都具有较好的截留砷的能力，从而能够减少砷的释放。这种二次铁矿的生成，使得铁还原溶解作用不一定会导致水中砷浓度升高，甚至可能会加强砷在铁矿上的吸附。

与铁还原作用相反，铁氧化过程生成的铁氧化物使溶解态砷的再次吸附固定，减少了砷的释放[31]。微生物介导的 Fe(II)氧化过程既可以生成结晶度较好的铁矿，如针铁矿、磁铁矿等，也可以生成结晶度较差的水铁矿等。溶解态 As/Fe 比对铁氧化菌形成的铁矿种类有较大影响。低 As/Fe 比下，铁氧化菌 *Acidovorax* sp. strain BoFeN1 更易生成结晶度较好的针铁矿，而高 As/Fe 比下，该铁氧化菌倾向于生成无定形的水铁矿，且水铁矿占形成的铁矿物比例随 As/Fe 比升高而增大[32]。生成的无定形水铁矿截留 As(V)的能力比针铁矿强，两者对 As(III)的吸附能力没有太大差别。之后更进一步的研究发现，在不同的 As/Fe 比情况下，固定溶解态 As 的浓度，改变溶液中 Fe(II)浓度，As(III)/As(V)都是以双齿双核的构型吸附在铁矿物上。然而当固定溶解态 Fe 的浓度，改变溶液中初始 As 的浓度，As(V)和 As(III)会以不同的吸附构型吸附在铁矿上，As(V)会以双齿双核和双齿单核两种吸附构型吸附在铁矿上，As(III)会以双齿双核和单齿单核两种吸附构型吸附在铁矿上。这些研究为利用铁氧化菌修复地下水和土壤的砷污染提供了理论基础。

3.1.3 环境基质对砷迁移转化的影响

自然环境中广泛存在的硫酸根、硝酸根、有机质和锰氧化物等可以通过自身转化间接改变砷和铁的形态，从而对砷的迁移转化过程也起着至关重要的影响。

厌氧条件下，硫酸盐还原菌可以将硫酸盐还原为硫化物。硫化物作为一种强还原剂，在一定条件下，可以还原 Fe(III)和 As(V)，从而促进砷释放到水体中。例如通过对内蒙古河套盆地利用硫的同位表示踪技术发现硫的还原过程促进了砷的释放[33]。硫酸盐还原过程促使砷的释放，主要有两个原因：①在水体电位较高时，S^{2-}与 As 形成了溶解态的硫代砷酸类物质[34]；②S^{2-}与还原的 Fe^{2+}反应，使得砷吸附能力较强的铁氧化物转变为吸附砷能力稍差的马基诺矿。

然而，在大多数厌氧强还原性的自然条件下，硫酸盐还原菌能够抑制砷的释放，其原因主要有两个方面。第一个原因是由于在强还原性条件下，硫酸根会被还原为硫离子，五价砷也会被还原为三价砷，它们直接发生反应生成如雌黄、雄黄或者含铁的砷黄铁矿类沉淀，从而显著抑制水体中溶解态砷的浓度[27]。这种在硫还原菌作用下通过生成砷-硫沉淀来固定水体砷浓度的现象，也已被设计运用到处理砷污染废水的反应器中。第二个原因是由于 S^{2-}与 Fe^{2+}生成的 FeS，截留水体中的砷形成沉淀。

除了硫酸根的影响，由于水体有机碳的含量直接影响着微生物的新陈代谢强度，其对砷的迁移转化也起着决定性的作用。Fendorf 等学者对湄公河三角洲的沉积物进行了研究，发现砷的释放仅发生在近地表的沉积物里，而深层含水层中的砷释放过程是被抑制的[35]。原因就是由于相比于深层含水层沉积物，近地表的沉积物里有较为丰富的有机碳源和可被微生物还原溶解的砷负载铁矿。有机碳是微生物新陈代谢过程中的所必需的，只有在有机碳含量较高的地区，微生物所介导的砷的迁移转化才起着关键的作用。此外，最近的研究中，水钠锰矿和硝酸根等这种间接影响到微生物作用的环境基质也引起了人们的关注。在好氧条件下，水钠锰矿能快速氧化水体中 As(III)和 Fe(II)，从而抑制微生物对 As(V)和铁氧化物的还原造成的砷的释放。但同时，菌体代谢产生的碳酸根离子与 Mn(II)生成沉淀，使表面慢慢钝化，从而其氧化砷的能力降低。另一方面，硝酸根离子促进微生物对 As(III)和 Fe(II)的氧化，从而抑制砷的释放。微生物作用下的 As、Fe、S 的耦合迁移转化过程见图 11-1。

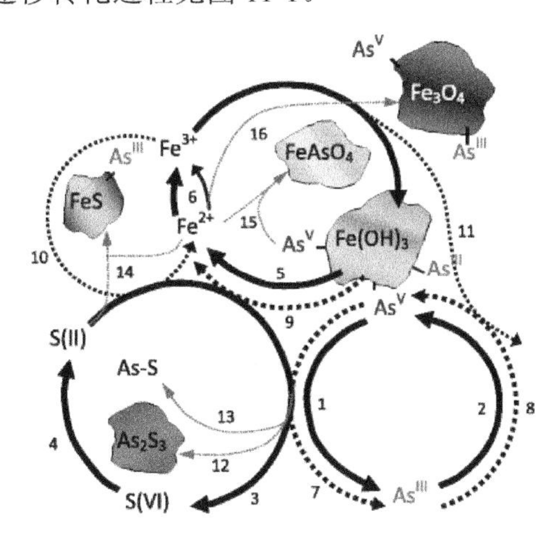

图 11-1　微生物作用下的 As、Fe、S 的耦合迁移转化过程

1. As(V)生物还原；2. As(III)生物氧化；3. S^{2-}氧化；4. 硫酸盐生物还原；5. Fe(III)生物还原；6. Fe(II)生物氧化；7. S^{2-}诱导的 As(V)还原；8. 非生物作用 As(III)氧化；9. S^{2-}诱导的 Fe(III)还原；10. Fe(III)诱导的 S^{2-}的氧化；11. Fe(III)诱导的 As(III)的氧化；12. As-S 沉淀；13. 溶解态的硫代砷酸类物质生成；14. Fe-S 沉淀；15. Fe-As 沉淀；16. 二次铁矿的生成（引自参考文献[22]）

3.2　微生物对锑环境转化的影响

锑在环境中的迁移转化同样也受到微生物的较大影响。微生物主要通过对锑的还原和氧化作用实现对锑的迁移转化（图 11-2）。

虽然微生物介导的锑的还原作用在厌氧条件下广泛存在，然而其还原机制仍不清楚。2014 年，Kulp 等学者发现微生物锑还原过程是一个异化的呼吸机制，因为 Sb(V)的还原伴随着乳酸或醋酸这种电子供体的消耗[36]。同年，两株呼吸机制的锑还原菌，芽孢杆菌属 *Bacillales.* MLFW-2[36]和 *Sinorhizobium*. JUK-1 被分离得到。砷与锑属同族元素，微生物介导的砷呼吸还原机制在 2003 年时由于分离到可基因驯化的 *Shewanella* strain ANA-3 而被揭示[23]。在以后的研究中，如果能筛选到类似 ANA-3 这种可以进行基因驯化的锑还原菌，可能会揭示锑的还原机制。另一方面，微生物锑还原作用将 Sb(V)还原为 Sb(III)，能够抑制锑的释放。主要原因是：①与 Sb(V)相比，Sb(III)更易被铁氧化物或土壤沉积物吸附；②与 As(III)类似，Sb(III)也可与 S^{2-}反应生成沉淀，从而将溶解态的锑截留。

与锑的还原机制相比，锑的氧化机制研究较为深入一些。微生物的锑氧化机制主要有以下三种：①*aioBA* 基因介导的锑氧化机制，*aioBA* 是好氧条件下的砷氧化基因。最近研究发现，*aioBA* 可在体内或体外氧化 Sb(III)，但是却只有 As(III)的存在能够诱导 *aioBA* 的表达。②*anoA* 基因介导的锑氧化机制。*anoA* 基因可被 Sb(III)诱导表达，从而将 Sb(III)氧化为 Sb(V)。同时，*anoA* 基因也可在体内或体外氧化 As(III)，但是其氧化 Sb(III)的能力要强于其还原 As(III)的能力。③H_2O_2 介导的锑氧化机制。微生物细胞内存在大量的 H_2O_2，*katA* 基因可诱导 H_2O_2 转化，当将其敲除后，细菌体内 H_2O_2 含量增多，从而导致 Sb(III)的氧化速率升高。此外，Sb(III)可诱导转录子 *iscR* 的表达，*iscR* 通过调控胞内谷胱甘肽的生成来控制胞内 H_2O_2 的含量，从而来控制细菌的 Sb(III)氧化速率。与砷的氧化过程抑制砷的释放相反，由于 Sb(V)比 Sb(III)迁移性更大，因此锑的氧化作用促进了锑从固相中释放到水体。

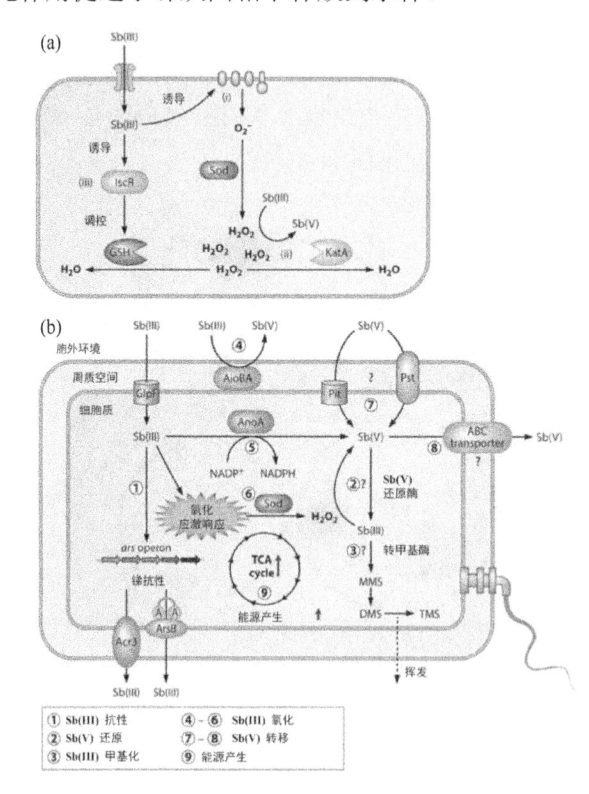

图 11-2 微生物锑还原和氧化机制

(a)IscR 调控的锑氧化过程。(i)Sb(III)通过细菌氧化应激反应诱导过氧化氢的产生，H_2O_2 将 Sb(III)氧化为 Sb(V)。(ii)H_2O_2 被 KatA 消耗。(iii)Sb(III)转录因子 IscR 的表达，IscR 调控 GSH 的生成，GSH 可消耗 H_2O_2。(b)与锑相关的细菌体内的新陈代谢过程。Sb(III)通过丙三醇通道进入细胞，通过 Acr3 和 ArsAB 排出细胞，但是其传输机制尚不清楚。Ars 操纵子与细菌锑抗性相关。细菌对锑的氧化、还原和甲基化作用都属于锑的解毒机制。Sb(III)能够激活细菌体内 TCA 循环，产生能量抵抗锑的毒性（引自参考文献[37]）

4 砷锑的去除控制研究

砷锑污染的去除控制是环境领域亟待解决的关键科学问题。因此，开发高效可行的砷锑去除方法尤为重要。

4.1 砷锑的主要去除方法

目前，主流的去除技术是吸附过滤和混凝沉淀 [38,39]。混凝沉淀法操作简便，但存在的主要问题是处理过程需要预氧化，并投加大量的化学药剂用于沉淀砷锑，产生的絮凝残渣量大，长期堆放会造成砷锑的缓慢释放，是砷锑二次污染的潜在威胁。混凝沉淀法用于砷的去除研究已相对成熟，但目前该方法用于除锑的研究相对较少。基于铁基的吸附过滤是应用最为广泛的饮用水除砷技术，这是因为砷可以在铁氧化物表面形成络合结构或进入铁相形成稳定矿物，从而从水体中吸附去除。最近研究表明，铁基材料的使用可能会增加砷的协同毒性 [40,41]。另外，作为氧化还原活泼的元素，铁基吸附材料会在厌氧环境下溶解释放砷，造成地下水砷的二次污染威胁。因此，铁基吸附材料的使用需慎重考虑。近来研究表明，二氧化钛（TiO_2）材料因其稳定的化学活性、高效的吸附性能、可调控的表面性质、吸附催化的双重作用及再生回用性在砷的去除方面具有广阔的应用前景。

4.2 砷的微观吸附机制

砷在 TiO_2 上的去除是由其微观吸附机制决定的。基于同步辐射的拓展 X 射线吸收精细结构（EXAFS）研究表明，As(III)和 As(V)在 TiO_2 表面以双齿双核的吸附构型存在，其中 Ti- As(III)和 Ti- As(V)的距离分别为 3.35 Å 和 3.30 Å，配位数为 2 [42] 。傅里叶变换红外光谱(FTIR)结果表明，As(III)和 As(V)在 pH=5~10 条件下吸附时主要以$(TiO)_2AsO^-$ 和$(TiO)_2AsO_2^-$ 的形式存在。基于以上对砷在 TiO_2 表面微观吸附机理的认识，其构建的高效 TiO_2 除砷材料可用于实际地下水及高砷污酸废水的治理[43,44]。目前，利用 TiO_2 的除砷研究中使用最多的是粉体 TiO_2 材料，其在应用时易流失，再生效果差；而且粉体材料较高的水头损失限制了其在固定床连续流工艺中的大规模应用。颗粒状 TiO_2 材料的开发为解决上述问题提供了思路。利用颗粒状 TiO_2（0.38~0.83 mm）填充的滤壶，可实现含砷浓度为 700 μg/L 的山西地下水的处理，再生后 TiO_2 材料可循环使用[45]。TiO_2 滤壶为当地居民提供了安全饮用水，当地居民饮用处理的地下水后，其尿砷浓度显著降低，砷的毒害得到了抑制（图 11-3）[46]。高效 TiO_2 吸附材料的开发为工业废水中高浓度砷的去除亦提供了可能。利用三个连续串联的 TiO_2 滤壶，可在 10 次循环中实现工业废水中 2590 mg/L As(III)的吸附过滤去除，使出水达到国家工业废水的排放标准[47]。虽然 TiO_2 可广泛用于砷的吸附去除中，但其再生回用仍是需要解决的关键问题。Hu 等的研究结果显示，地下水中共存的硅可在 TiO_2 表面形成多聚体，从而影响砷的吸附及 TiO_2 的再生[46,48]。因此，考察水质特性并开发高效经济的吸附剂再生方法是实现砷污染控制的另一关键。

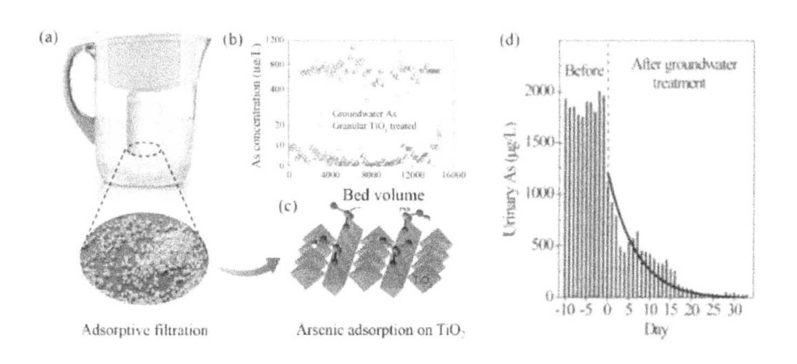

图 11-3　（a）装载颗粒 TiO_2 的滤壶，（b）TiO_2 滤壶对含砷地下水的处理效果，（c）砷在 TiO_2 表面的络合结构，（d）居民饮用处理后地下水时尿砷浓度的变化

4.3 锑的微观吸附机制

锑与砷分子结构的差异使其在 TiO_2 表面的微观吸附机理及吸附行为有所不同。基于同步辐射的EXAFS和理论计算DFT结果表明，Sb(III)和Sb(V)在TiO_2表面以双齿双核的吸附构型存在，其中Ti- Sb(III)和Ti- Sb(V)的距离分别为3.47 Å和3.70 Å[49]。共吸附实验表明，在pH=7时Sb(III)在TiO_2上的吸附去除率优于As(III)（89% > 45%），这是因为相比于As(III)，Sb(III)是更强的Lewis碱[50]，与TiO_2表面Lewis酸的键和能力更强，吸附构型更稳定，易于吸附去除。该结果也通过DFT理论计算得到验证，结果表明Sb(III)在TiO_2上的吸附能低于As(III)，分别为–4.99 eV和–4.78 eV。吸附能的差异是由吸附分子与表面成键过程中的轨道能量决定的。投影态密度（PDOS）和分子轨道理论的分析结果表明，Sb(III)在TiO_2表面吸附时其成键轨道的能量低于As(III)吸附时的成键能，因此Sb(III)的吸附构型更为稳定（图11-4）[49]。Sb(V)在TiO_2上的去除率略小于As(V)[Sb(V)= 25%，As(V)= 30%]，这主要是因为TiO_2对其具有相近的吸附能（–4.70 eV），但Sb(V)分子的八面体构型具有较大的空间结构，增加了其吸附的空间位阻，使其吸附量降低。因此，利用吸附法去除Sb(V)具有一定的挑战。虽然对砷的吸附去除研究已有众多报道，开发的吸附材料有众多种类，但对锑的吸附研究相对较少[51]。因此，高效砷锑吸附材料的开发及砷锑的共去除研究仍是环境领域关注的焦点。

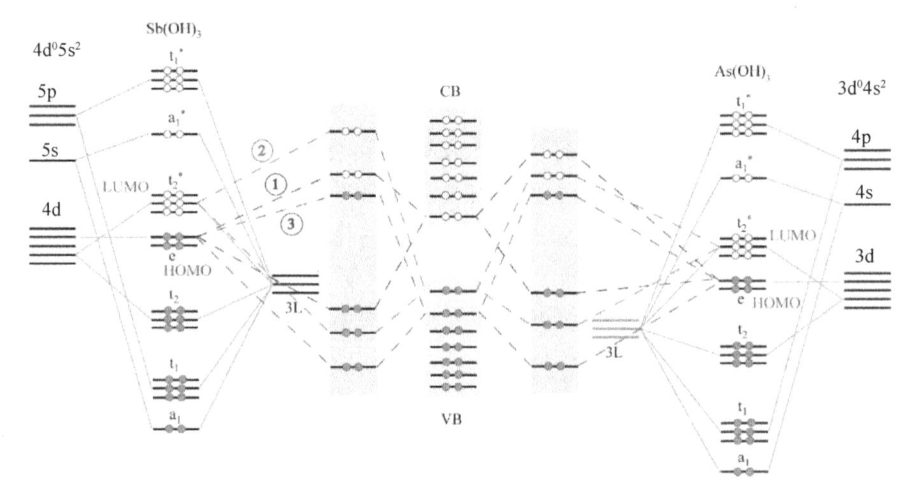

图11-4　Sb（III）和As（III）在TiO_2表面吸附的分子轨道能级示意图

图中吸附分子与表面的三种相互作用分别为:（i）分子的最高占据轨道（HOMO）与表面的导带（CB）相互成键（紫色虚线，①）；（ii）分子的最低未占据轨道（LUMO）与表面的价带（VB）相互成键（绿色虚线，②）；（iii）分子的最高占据轨道（HOMO）与表面的价带（VB）相互成键（红色虚线，③）

4.4 纳米材料晶面对砷锑吸附的影响

近期研究表明，影响TiO_2吸附能力的因素除已被熟知的比表面积、羟基密度之外，纳米材料的晶面更是不容忽视的因素。TiO_2不同晶面表面原子排列和电子结构不同，从而影响其吸附效率。锐钛矿型TiO_2三个主要低指数晶面的晶面能顺序为 $\gamma\{001\}$（0.90 J/m^2）> $\gamma\{100\}$（0.53 J/m^2）> $\gamma\{101\}$（0.44 J/m^2）[52]。研究结果表明，与热力学稳定的$TiO_2\{101\}$晶面相比，高能$TiO_2\{001\}$晶面所具有的5配位活性Ti原子具有强Lewis酸位，因此有利于砷的吸附去除，吸附键更为稳定。其次，$TiO_2\{001\}$晶面的强Lewis酸位有利于O_2分子在其表面的吸附，使TiO_2光生电子向表面吸附的O_2转移，产生更多的超氧自由基，从而促进 As(III)在 TiO_2 表面的光催化氧化[53]。基于以上 TiO_2 晶面对砷吸附及催化的研究，我们制备的以{001}晶面为主的 TiO_2 材料，其对 Sb(III)和 Sb(V)的吸附容量分别高达 200 mg/g 和 154 mg/g，是目前报道的吸附锑容量最大的材料[49]。由此可见，高能晶面 TiO_2 材料所特有的高效吸附能力，为其环境应用提供了潜在机遇。目前针对高能晶面 TiO_2 材料的制备及砷锑吸附的环境应用鲜有报道。

4.5　共存离子对砷锑吸附的影响

环境介质中共存离子对砷锑在环境界面的吸附具有显著影响。地下水中阴离子如磷酸根（PO_4^{3-}）、硅酸盐（SiO_3^{2-}）、硫酸盐（SO_4^{2-}）、碳酸氢根（HCO_3^-）、氟离子（F^-）等可抑制砷锑的吸附[45,49]。相比而言，共存硝酸根（NO_3^-）和氯离子（Cl^-）对砷锑的吸附并没有显著影响[46]。共存阴离子对砷锑吸附的影响顺序为 $PO_4^{3-} > F^- > HCO_3^- > SiO_3^{2-} > SO_4^{2-} > NO_3^- \approx Cl^-$[49]。除了上述阴离子外，水体中共存的阳离子如钙(Ca)、镁(Mg)、镉(Cd)和锌(Zn)均对砷的吸附有影响。地下水中共存的 Ca 可通过与 As(V)形成三齿配合物 Ca-As(V)-TiO_2 促进 As(V)在 TiO_2 上的吸附[46]。同理，工业废水中共存的高浓度 Cd 和 Zn 亦可显著促进 As(III)、Cd 和 Zn 的共吸附去除。环境有机质、小分子有机酸的存在会改变砷锑的吸附构型，影响其微观界面过程及去除机理。此外，光照引起的价态转化改变了砷锑的分子结构及其环境迁移行为。因此，环境复杂基质对砷锑污染控制的影响需要综合考虑。

5　展　　望

砷锑污染已经成为全球性的环境问题，严重影响人类健康。目前，国内外对砷锑污染及去除控制的研究仍不系统，有待于进一步深入研究和探索。

首先，砷锑的检测，尤其是对不同形态的砷锑检测，是解决砷锑环境污染问题的基石。目前已有多种检测手段可实现对 As 形态的精准分析。但与 As 相比，Sb 的分析方法研究尚有不足。尤其是有机锑如甲基锑的检测研究鲜有报道。发展快速便捷的砷锑分析方法也是今后重点研究的方向。其次，微生物在砷锑的迁移转化过程中起着至关重要的作用。在实际环境中，微生物介导的砷迁移转化过程常与 Fe、S 等其他环境介质耦合在一起，但是其深层次的微观耦合机制还有待探究。同时，相比于砷的研究，微生物对锑的迁移转化研究尚存在大量空白。在实际环境中，锑的迁移转化过程与 Fe、S 等环境因素的相互关系及其微观机制报道有限，有待进一步的揭示和阐明。最后，对砷锑微观界面的认识是实现砷锑污染去除的理论基础，因此需要通过先进的谱学手段如同步辐射和傅里叶红外光谱等，结合理论计算，从分子水平上认识复合污染下砷锑的微观界面，从而为解决砷锑污染提供理论基础及实践指导。

参 考 文 献

[1] Li X, Yang H, Zhang C, Zeng G M, Liu Y G, Xu W H, Wu Y, Lan S M. Spatial distribution and transport characteristics of heavy metals around an antimony mine area in central China. Chemosphere, 2017, 170: 17-24.

[2] Chang A C, Pan G, Page A L, Asano T. Developing human health-related chemical guidelines for reclaimed water and sewage sludge applications in agriculture. World Health Organization, 2002, 94.

[3] Garnier J-M, Travassac F, Lenoble V, Rose J, Zheng Y, Hossain M, Chowdhury S, Biswas A, Ahmed K, Cheng Z. Temporal variations in arsenic uptake by rice plants in Bangladesh: The role of iron plaque in paddy fields irrigated with groundwater. Sci Total Environ, 2010, 408(19): 4185-4193.

[4] Li W, Wang T, Zhou S, Lee S, Huang Y, Gao Y, Wang W. Microscopic observation of metal-containing particles from

Chinese continental outflow observed from a non-industrial site. Environ Sci Technol, 2013, 47(16): 9124-9131.

[5] Lin J R, Chen N, Nilges M J, Pan Y M. Arsenic speciation in synthetic gypsum($CaSO_4 \cdot 2H_2O$): A synchrotron XAS, single-crystal EPR, and pulsed ENDOR study. Geochim Cosmochim Acta, 2013, 106: 524-540.

[6] Kim E J, Yoo J-C, Baek K. Arsenic speciation and bioaccessibility in arsenic-contaminated soils: Sequential extraction and mineralogical investigation. Environ Pollut, 2014, 186: 29-35.

[7] Li Y, Duanp Z, Liu G, Kalla P, Scheidt D, Cai Y. Evaluation of the possible sources and controlling factors of toxic metals/metalloids in the florida everglades and their potential risk of exposure. Environ Sci Technol, 2015, 49(16): 9714-9723.

[8] Langner P, Mikutta C, Suess E, Marcus M A, Kretzschmar R. Spatial distribution and speciation of arsenic in peat studied with microfocused X-ray fluorescence spectrometry and X-ray absorption spectroscopy. Environ Sci Technol, 2013, 47(17): 9706-9714.

[9] Fu Z, Wu F, Mo C, Deng Q, Meng W, Giesy J P. Comparison of arsenic and antimony biogeochemical behavior in water, soil and tailings from Xikuangshan, China. Sci Total Environ, 2016, 539 97-104.

[10] Wilson S C, Lockwood P V, Ashley P M, Tighe M. The chemistry and behaviour of antimony in the soil environment with comparisons to arsenic: A critical review. Environ Pollut, 2010, 158(5): 1169-1181.

[11] Wei M, Chen J J, Wang X W. Removal of arsenic and cadmium with sequential soil washing techniques using Na_2EDTA, oxalic and phosphoric acid: Optimization conditions, removal effectiveness and ecological risks. Chemosphere, 2016, 156: 252-261.

[12] Zotov A, Shikina N, Akinfiev N. Thermodynamic properties of the Sb(III) hydroxide complex $Sb(OH)_3$(aq) at hydrothermal conditions. Geochim Cosmochim Acta, 2003, 67(10): 1821-1836.

[13] Biver M, Shotyk W. Stibiconite(Sb_3O_6OH), senarmontite(Sb_2O_3) and valentinite(Sb_2O_3): Dissolution rates at pH 2–11 and isoelectric points. Geochim Cosmochim Acta, 2013, 109: 268-279.

[14] Tian H, Zhou J, Zhu C, Zhao D, Gao J, Hao J, He M, Liu K, Wang K, Hua S. A comprehensive global inventory of atmospheric antimony emissions from anthropogenic activities, 1995-2010. Environ Sci Technol, 2014, 48(17): 10235-10241.

[15] Tian H, Liu K, Zhou J, Lu L, Hao J, Qiu P, Gao J, Zhu C, Wang K, Hua S. Atmospheric emission inventory of hazardous trace elements from China's coal-fired power plants—Temporal trends and spatial variation characteristics. Environ Sci Technol, 2014, 48(6): 3575-3582.

[16] Calace N, Nardi E, Pietroletti M, Bartolucci E, Pietrantonio M, Cremisini C. Antarctic snow: Metals bound to high molecular weight dissolved organic matter. Chemosphere, 2017, 175 307-314.

[17] Okkenhaug G, Zhu Y-G, He J, Li X, Luo L, Mulder J. Antimony(Sb) and arsenic(As) in Sb mining impacted paddy soil from Xikuangshan, China: Differences in mechanisms controlling soil sequestration anduptake in rice. Environ Sci Technol, 2012, 46(6): 3155-3162.

[18] Smith E, Kempson I, Juhasz A L, Weber J, Skinner W M, Gräfe M. Localization and speciation of arsenic and trace elements in rice tissues. Chemosphere, 2009, 76(4): 529-535.

[19] Tschan M, Robinson B H, Nodari M, Schulin R. Antimony uptake by different plant species from nutrient solution, agar and soil. Environ Chem, 2009, 6(2): 144-152.

[20] Fendorf S, Michael H A, van Geen A. Spatial and temporal variations of groundwater arsenic in South and Southeast Asia. Science, 2010, 328(5982): 1123-1127.

[21] Gorny J, Billon G, Lesven L, Dumoulin D, Made B, Noiriel C. Arsenic behavior in river sediments under redox gradient: A review. Sci Total Environ, 2015, 505 423-434.

[22] Huang J H. Impact of microorganisms on arsenic biogeochemistry: A review. Water Air Soil Poll, 2014, 225(2): 25.

[23] Saltikov C W, Newman D K. Genetic identification of a respiratory arsenate reductase. Proc Natl Acad Sci U S A, 2003, 100(19): 10983-10988.

[24] Tian H, Shi Q, Jing C. Arsenic biotransformation in solid waste residue: Comparison of contributions from bacteria with arsenate and iron reducing pathways. Environ Sci Technol, 2015, 49(4): 2140-2146.

[25] Zhang Z N, Yin N Y, Du H L, Cai X L, Cui Y S. The fate of arsenic adsorbed on iron oxides in the presence of arsenite-oxidizing bacteria. Chemosphere, 2016, 151 108-115.

[26] Weber K A, Achenbach L A, Coates J D. Microorganisms pumping iron: Anaerobic microbial iron oxidation and reduction. Nat Rev Microbiol, 2006, 4(10): 752-764.

[27] O'Day P A, Vlassopoulos D, Root R, Rivera N. The influence of sulfur and iron on dissolved arsenic concentrations in the shallow subsurface under changing redox conditions. Proc Natl Acad Sci U S A, 2004, 101(38): 13703-13708.

[28] Islam F S, Gault A G, Boothman C, Polya D A, Charnock J M, Chatterjee D, Lloyd J R. Role of metal-reducing bacteria in arsenic release from Bengal delta sediments. Nature, 2004, 430(6995): 68-71.

[29] Tufano K J, Reyes C, Saltikov C W, Fendorf S. Reductive processes controlling arsenic retention: Revealing the relative importance of iron and arsenic reduction. Environ Sci Technol, 2008, 42(22): 8283-8289.

[30] Tian H, Shi Q, Jing C. Arsenic biotransformation in solid waste residue: Comparison of contributions from bacteria with arsenate andiron reducing pathways. Environ Sci Technol, 2015, 49(4): 2140-2146.

[31] Smith R L, Kent D B, Repert D A, Bohlke J K. Anoxic nitrate reduction coupled with iron oxidation and attenuation of dissolved arsenic and phosphate in a sand and gravel aquifer. Geochim Cosmochim Acta, 2017, 196 102-120.

[32] Hohmann C, Morin G, Ona-Nguema G, Guigner J M, Brown G E, Kappler A. Molecular-level modes of As binding to Fe(III)(oxyhydr)oxides precipitated by the anaerobic nitrate-reducing Fe(II)-oxidizing Acidovorax sp strain BoFeN1. Geochim Cosmochim Acta, 2011, 75(17): 4699-4712.

[33] Guo H, Zhou Y, Jia Y, Tang X, Li X, Shen M, Lu H, Han S, Wei C, Norra S, Zhang F. Sulfur cycling-related biogeochemical processes of arsenic mobilization in the Western Hetao Basin, China: Evidence from multiple isotope approaches. Environ Sci Technol, 2016, 50(23): 12650-12659.

[34] Luo T, Tian H, Guo Z, Zhuang G, Jing C. Fate of arsenate adsorbed on nano-TiO_2 in the presence of sulfate reducing bacteria. Environ Sci Technol, 2013, 47(19): 10939-10946.

[35] Stuckey J, Schaefer M V, Kocar B D, Benner S G, Fendorf S. Arsenic release metabolically limited to permanently water-saturated soil in Mekong Delta. Nature Geosci, 2016, 9(1): 70-76.

[36] Kulp T R, Miller L G, Braiotta F, Webb S M, Kocar B D, Blum J S, Oremland R S. Microbiological reduction of Sb(V) in anoxic freshwater sediments. Environ Sci Technol, 2014, 48(1): 218-226.

[37] Li J, Wang Q, Oremland R S, Kulp T R, Rensing C, Wang G. Microbial antimony biogeochemistry: Enzymes, regulation, and related metabolic pathways. Appl Environ Microb, 2016, 82(18): 5482-5495.

[38] Mondal P, Bhowmick S, Chatterjee D, Figoli A, Van der Bruggen B. Remediation of inorganic arsenic in groundwater for safe water supply: A critical assessment of technological solutions. Chemosphere, 2013, 92(2): 157-170.

[39] Mohan D, Pittman C U, Jr. Arsenic removal from water/wastewater using adsorbents—A critical review. J Hazard Mater, 2007, 142(1-2): 1-53.

[40] Liu S, Guo X, Zhang X, Cui Y, Zhang Y, Wu B. Impact of iron precipitant on toxicity of arsenic in water: A combined *in vivo* and *in vitro* study. Environ Sci Technol, 2013, 47(7): 3432-3438.

[41] Chandrasekaran V R M, Muthaiyan I, Huang P-C, Liu M-Y. Using iron precipitants to remove arsenic from water: Is it safe? Water Res, 2010, 44(19): 5823-5827.

[42] Pena M, Meng X G, Korfiatis G P, Jing C Y. Adsorption mechanism of arsenic on nanocrystalline titanium dioxide. Environ Sci Technol, 2006, 40(4): 1257-1262.

[43] Luo T, Cui J, Hu S, Huang Y, Jing C. Arsenic removal and recovery from copper smelting wastewater using TiO_2. Environ Sci Technol, 2010, 44(23): 9094-9098.

[44] Yan L, Hu S, Jing C. Recent progress of arsenic adsorption on TiO_2 in the presence of coexisting ions: A review. J Environ Sci, 2016, 49: 74-85.

[45] Cui J L, Du J J, Yu S W, Jing C Y, Chan T S. Groundwater arsenic removal using granular TiO_2: Integrated laboratory and field study. Environ Sci Pollut Res, 2015, 22(11): 8224-8234.

[46] Hu S, Shi Q, Jing C. Groundwater arsenic adsorption on granular TiO_2: Integrating atomic structure, filtration, and health impact. Environ Sci Technol, 2015, 49(16): 9707-9713.

[47] Yan L, Huang Y, Cui J, Jing C. Simultaneous As(III) and Cd removal from copper smelting wastewater using granular TiO_2 columns. Water Res, 2015, 68: 572-579.

[48] Hu S, Yan W, Duan J. Polymerization of silicate on TiO_2 and its influence on arsenate adsorption: An ATR-FTIR study. Colloid Surface A, 2015, 469: 180-186.

[49] Yan L, Song J, Chan T-S, Jing C. Insights into antimony adsorption on {001} TiO_2: XAFS and DFT study. Environ Sci Technol, 2017, 51(11): 6335-6341.

[50] Leuz A K, Monch H, Johnson C A. Sorption of Sb(III) and Sb(V) to goethite: Influence on Sb(III) oxidation and mobilization. Environ Sci Technol, 2006, 40(23): 7277-7282.

[51] Fan J X, Wang Y J, Fan T T, Dang F, Zhou D M. Effect of aqueous Fe(II) on Sb(V) sorption on soil and goethite. Chemosphere, 2016, 147: 44-51.

[52] Song J, Yan L, Duan J, Jing C. TiO_2 crystal facet-dependent antimony adsorption and photocatalytic oxidation. J Colloid Interface Sci., 2017, 496: 522-530.

[53] Yan L, Du J, Jing C. How TiO_2 facets determine arsenic adsorption and photooxidation: Spectroscopic and DFT study. Catal Sci Technol, 2016, 6(7): 2419-2426.

作者：景传勇

中国科学院生态环境研究中心

第 12 章　环境放射化学进展

本章导读

核电充分体现高效、清洁的优势，在全球电力供应中的贡献越来越高。核能的利用和核技术的发展伴随着大量放射性核素的使用，在核电运行中这些放射性核素不可避免地会释放到环境中，对生态环境和人类健康造成了极大危害。因此对放射性废物的合理处理和安全储存引起了世界各国的高度关注和重视。科研工作者经过多年探索发现，采用合适的材料将放射性核素萃取、分离固定并在深层的地下处置库中永久保存，是一种公认的核废料处置方法。为此，围绕材料的设计，结构表征及对放射性核素的吸附富集，各国科学家开展了大量的研究。在这样的背景下，本章简要回顾了近几年各种不同的纳米材料如石墨烯及其复合材料、金属还原材料、金属-有机骨架材料等多种体系在放射性污染治理中应用，结合各种不同的表征技术和理论技术方法，比较详细地论述了放射性核素在各类材料界面的作用机制，以期对发展环境友好型的放射性污染高效修复材料具有指导意义。

关键词

环境化学，放射性污染，纳米材料与技术，理论计算，光谱技术

1 引　　言

由于能源短缺，核技术和核能逐渐被开发和利用，但发展核工业的关键是必须安全妥善的处理核反应中产生的大量放射性废物，特别是高放废物的安全处理。由于核污染处理难度高，耗资大，因此放射性核废物的处理成为核工业中急需解决的重大科技问题之一。高放废物中含有大量危害极大的放射性核素，如铀（235）、锝（99）、硒（79）、锶（90）、铯（137）、钴（60）、镍（63）等。在核电工业的运行过程中，这些放射性核素不可避免地释放到水、土壤等自然环境中，对生态环境和人类健康造成重大的潜在危害，引起了各国越来越广泛的关注。放射性核素在非常低的浓度下就有辐射，具有持久性、穿透性和蜕变性等特点。因此研究放射性核素的安全处理，对保护环境和人类健康具有重要的科学意义。

为了寻找有效处理放射性核素的方法，从 20 世纪 50 年代起各国科学家就开始了相关研究，提出了多种技术。经过多年探索，目前全球公认可行的方法是深层地质处置，将放射性废物固定在深层地下的处置库中永久保存。而采用合适的材料将放射性核素从废液中吸附富集并固化是其中的关键。目前，许多国内外学者以放射性废物的处理为导向，有针对性地设计了一系列材料用于实现对废水中各种放射性核素的萃取、富集及有效去除，并研究了放射性核素在各种材料表面的行为，取得了大量研究成果，对于放射性核素处置的安全评价具有重要意义。鉴于此，我们简要回顾了几类重要的纳米材料在放射性废物处理中的应用，主要总结了国内外碳纳米材料及复合物、纳米金属还原材料和其他新型材料应用于放射性核素处理的研究进展，讨论了放射性核素在材料界面的作用机理，以期为放射性污染治理的深入研究提供新思路和新方向。

2 石墨烯及其复合材料对放射性核素的吸附富集

石墨烯（graphene，GN）是一种由单层碳原子紧密堆积而成的新型纳米材料，具有二维蜂窝状结构，

同时也是构建其他维度碳材料的基本单元。由于其独特的二维结构和优异的晶体学性质，石墨烯具有出色的电学、热学和力学性能，在纳米电子器件、传感器、催化及储能等领域得到了广泛的应用[1-3]。目前主要采用化学剥离氧化石墨片层来获得石墨烯以及氧化石墨烯（graphene oxide，GO），以石墨粉作为原料，用强酸和强氧化剂在液相氧化条件下将石墨剥离成石墨烯薄片，然后通过化学还原法制得石墨烯，称之为还原氧化石墨烯（reduced graphene oxide，RGO）[2]。近年来研究发现石墨烯及其复合材料在环境污染治理中具有潜在的应用价值。华北电力大学王祥科教授课题组于 2011 年率先报道了将多层氧化石墨烯应用于放射性污染治理的研究[4]。随后，美国莱斯大学[5,6]，韩国光云大学[7,8]，波兰西里西亚大学[9]，克莱姆森大学[10]，中国工程物理研究院[11]，北京大学[12]，兰州大学[13]，中科院等离子体物理研究所[14-21]，淮阴工学院[22,23]以及绍兴文理学院[24-29]等国内外诸多高校和科研机构的研究人员相继报道了用石墨烯处理处置放射性核素方面的大量工作。

2.1　石墨烯对放射性核素的吸附

美国莱斯大学 Slesarev 等[5,6]发现，氧化石墨烯（GO）对水体中的放射性核素具有超强的吸附能力，吸附量明显优于膨润土和活性炭等常用核清理剂。此外，GO 能快速吸附放射性核素，在短短几分钟即可达到吸附平衡（图 12-1）。Slesarev 等[5,6]通过荧光分析进一步发现，GO 捕获放射性核素并不会减弱它们的放射性（图 12-2），而是更容易处理。可以将吸附了放射性核素的 GO 进行焚烧，在此过程中，GO 会快速燃烧，仅剩下粉末状放射性物质，有利于放射性核素的重复利用。这对于发展海水提铀技术具有重要的意义。

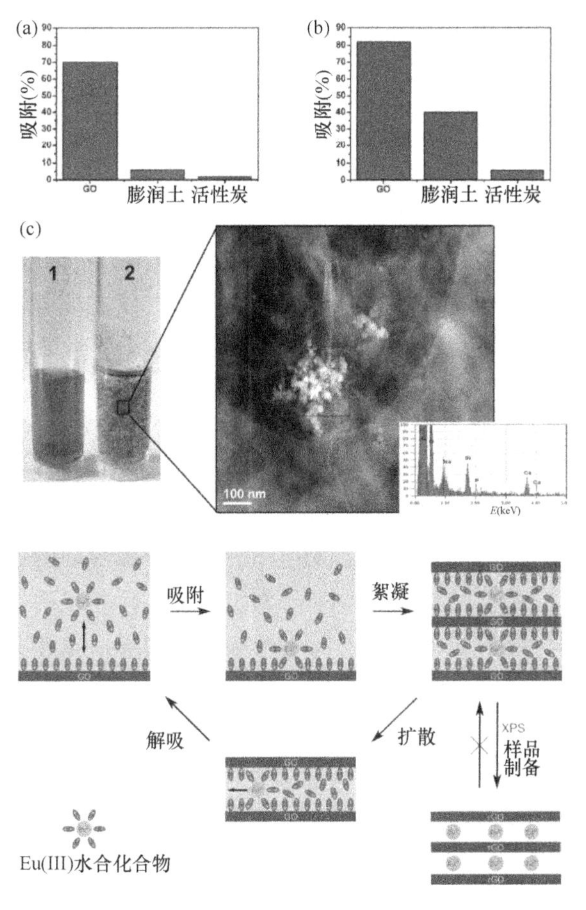

图 12-1　氧化石墨烯对放射性核素的吸附明显优于膨润土和活性炭[5,6]

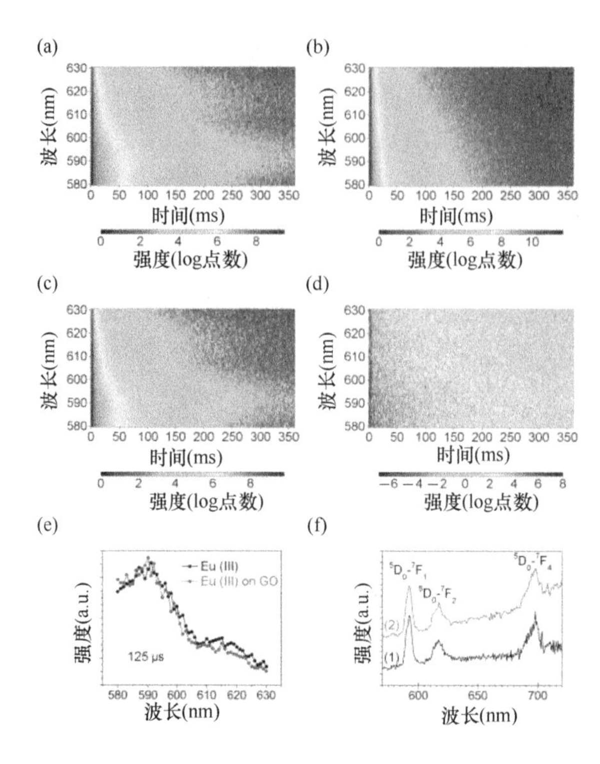

图 12-2 时间分辨荧光发射光谱研究 Eu(III)在氧化石墨烯上的吸附[5,6]

在过去几年里，大量研究报道了地球化学条件对石墨烯吸附富集放射性核素的影响及作用机制。斯里兰卡物理科学与技术中心 Lujaniene[30]研究了介质 pH 对氧化石墨烯吸附 Am(III)、Pu(IV)、Co(II)、Ni(II)等核素的影响（图 12-3），并进一步考察了吸附动力学，发现随着 pH 升高，放射性核素在 GO 上的吸附量逐渐增大，主要是因为 GO 的表面活性功能团的电负性增强，对核素的络合能力增强，此外，在高 pH 条件下，部分核素离子可以形成沉淀，这也是吸附量增大的一个原因。其他学者对 U(VI)，Tc(VII)，Np(V)，Eu(III)，Th(IV)，Sr(II)，Cs(I)等各种不同价态的放射性核素在 GO 上的吸附行为进行了深入的研究。结果证实放射性核素在 GO 上的吸附受离子强度影响不大，主要与 GO 的表面含氧功能团作用形成了内层络合物。另外，大多数情况下，吸附量随着温度的升高而增大，是一个自发、吸热的过程[14-29]。目前，GO 去除放射性核素的研究主要处于实验室研究阶段，并且大多针对单一离子组分进行，多组分的竞争吸附则研究较少[31]。在处理实际放射性废水中，往往多种放射性核素和其他组分共存，所以，研究 GO 对共存放射性核素和多组分的吸附过程，更具有现实意义。

为了从分子原子水平获得放射性核素在石墨烯表面的分布、微观形态和配位结构，王祥科教授课题组[14,19,21,32,33]通过扩展 X 射线吸收精细结构（EXAFS）光谱技术，系统研究了石墨烯及其复合材料对放射性核素的吸附富集作用，并分析发现 U(VI)、Th(IV)、Sr(II)、Eu(III)等放射性核素与石墨烯形成内层表面络合物，从分子水平成功解析了放射性核素在不同环境条件下的局部微观结构信息及其在石墨烯上高效吸附的作用机理（图 12-4）。同时结合利用密度泛函理论（DFT）计算，获得相关吸附体系的结构参数和电子性质（图 12-5），探讨结构稳定性规律，进一步从理论上进行完整地阐述[16-18,21,31-33]。并把 EXAFS 和 DFT 结果与宏观实验结果作对比，比较宏观实验结果和微观分析的区别，分析其中的本质原因。中科院高能物理研究所石伟群等[34,35]从分子尺度上揭示了 U(V)、Np(V)和 Pu(IV，VI)与石墨烯的作用机理，预测了材料边界上的官能团对锕系离子的吸附能力与选择性。研究显示，Th(IV)离子与表面或边界修饰了羧基/羟基的氧化石墨烯结合时，存在图 12-6 所示的多种络合形态。对于图 12-6 中（a）~（c）三种情况，Th(IV)的配位数为八，其中三个硝酸根离子以双齿模式参与配位，第四个硝酸根离子以单齿模

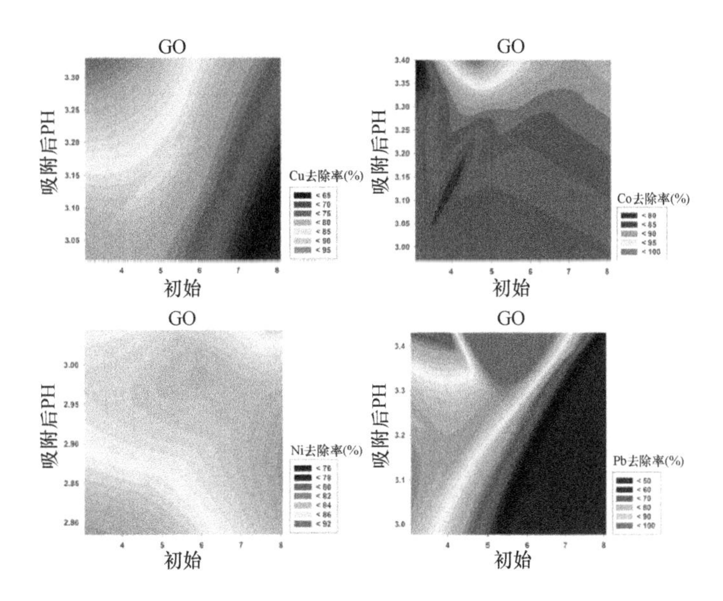

图 12-3　介质 pH 对氧化石墨烯吸附放射性核素的影响[30]

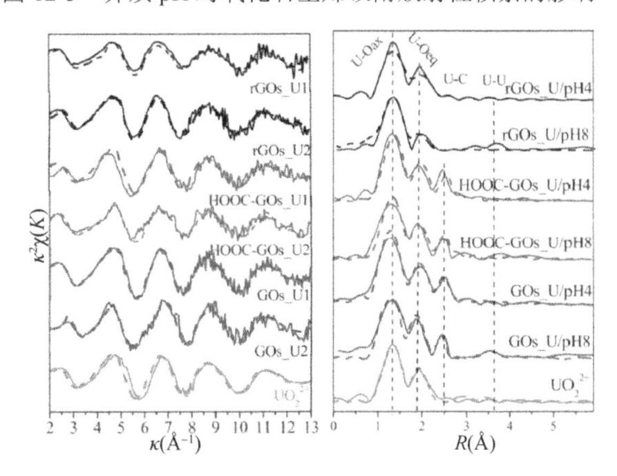

图 12-4　放射性核素 U(VI)在石墨烯表面吸附的 EXAFS 分析结果[33]

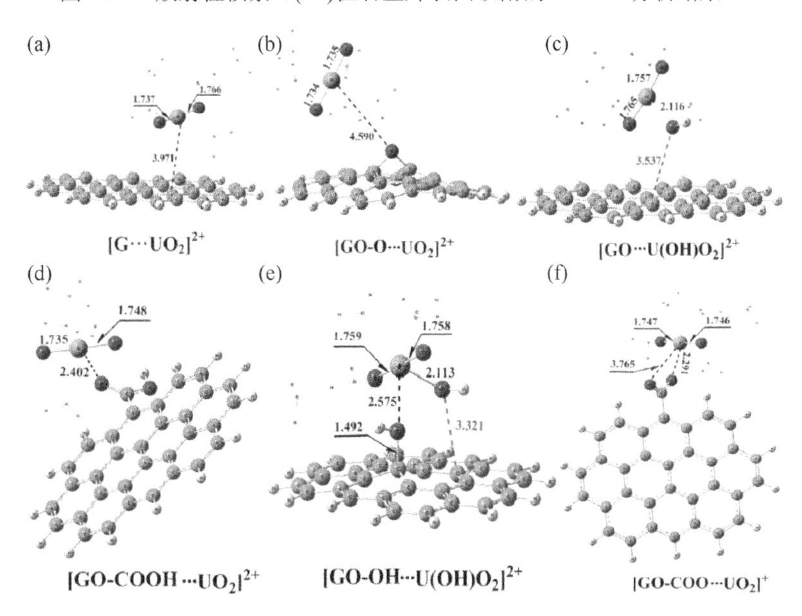

图 12-5　核素 U(VI)在石墨烯表面吸附的 DFT 理论计算结果[33]

式配位。对于图 12-6 中（d）~（g） 四种情况，Th(IV)的配位数均为九。理论计算得到的 Th 与硝酸根离子形成的 Th—O 键的平均键长介于 2.44~2.50 Å 之间。在形成的所有 GO/Th(IV)络合物中，与羟基 O 形成的 Th—O 键的键长要略高于与同羧基形成的 Th—O 键，说明羧基的配位能力要强于羟基。计算得到的配位键键长与 EXAFS 数据吻合较好。计算得到的反应 Gibbs 自由能均为负值，说明图 12-6 中络合物的形成过程属于放热反应。与 Th(IV)相比，U(VI)离子负载的正电荷少，其结构为直线型 O-U-O。因此 U(VI)和 Th(IV)的配位性质存在较大差异。计算显示，U(VI)与 GO 形成的阴离子络合物中含有氢键，促进了络合物的稳定性。与形成中性络合物相比，U(VI)离子更容易与带负电的 GO 络合。自然键轨道（NBO）分析表明，GO 吸附 U(VI)离子的过程以静电相互作用为主。然而对于石墨烯的应用而言，由于片层之间具有较强 π-π 作用力，自身仍存在团聚、堆积，难以分离等局限性，因此利用适当官能团等对石墨烯进行功能化修饰并提高其吸附性能，已成为最近的研究热点。

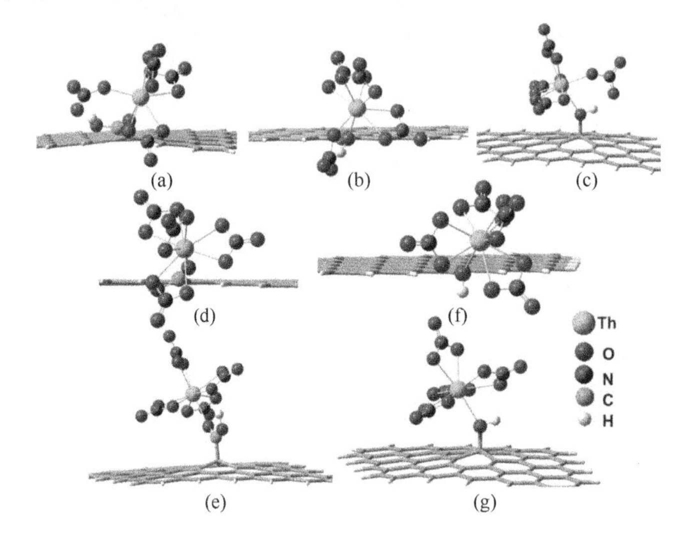

图 12-6 理论计算得到的 Th(IV)/GO 络合物[34,35]

（a）、（d）和（e）中 GO 表面和边界处修饰羧基，（b）、（c）、（f）和（g）中 GO 上修饰羟基

2.2 有机大分子修饰石墨烯富集放射性核素

为了增加石墨烯的表面活性官能团，以提高其吸附性能，研究者们使用各种技术对石墨烯进行了一系列功能化改性，目前主要的功能化方法可分为有机大分子修饰和无机纳米粒子修饰。由于 GO 含有大量的羟基、羧基和环氧基等含氧功能基团，而很多大分子有机物如聚苯胺、聚酰胺、偕胺肟、聚吡咯、环糊精及壳聚糖等含有大量的胺基或羟基，因此可以通过共价键功能化对石墨烯进行修饰。此外，也可以通过有机芳香族化合物与石墨烯之间的 π-π 相互作用完成石墨烯的非共价键功能化修饰[36-44]。有机大分子修饰石墨烯，一方面降低石墨烯功能材料的自团聚，增加其在溶液中良好的分散性能，另一方面增强复合材料对放射性核素的去除能力[39-41]。王祥科教授课题组[36-41]通过化学法合成氧化石墨烯，运用等离子体表面修饰技术，在 GO 表面修饰聚苯胺、聚吡咯、偕胺肟、环糊精等功能团获得复合材料。研究发现这一系列复合材料对 U(VI)、Ni(II)、Th(IV)、Co(II)、Eu(III)等放射性核素具有很高的吸附性能，并使用表面络合模型和先进光谱学研究其微观作用机理。Yang[43]和 Xiong[44]等分别用牛血清白蛋白和二异丁胺对氧化石墨烯进行了功能化修饰，发现改性后的 GO 对 U(VI)和 Re(VII)具有很高吸附性能（图 12-7），证实了复合材料可以作为放射性核素除去的高效吸附剂。这些研究有助于更加深入地理解放射性核素与氧化石墨烯复合材料的作用机理，对于纳米材料在环境修复中的潜在应用具有重要意义。

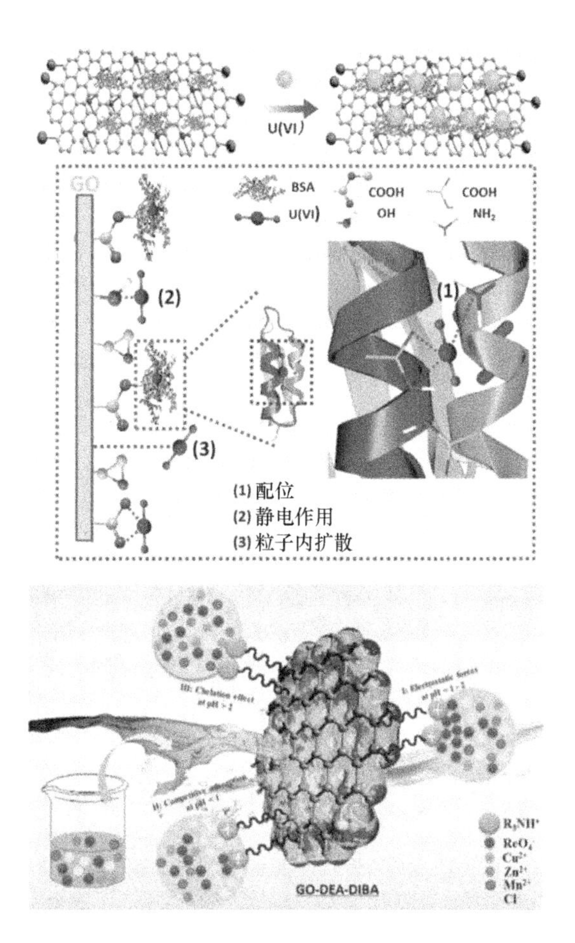

图 12-7　牛血清白蛋白和二异丁胺修饰 GO 对铀和铼的高效富集[43,44]

2.3　无机纳米粒子修饰石墨烯富集放射性核素

采用无机纳米颗粒对石墨烯进行功能化修饰，是另一种改善石墨烯吸附性能的重要方法。由于石墨烯拥有巨大的比表面积，本身可作为一种良好的电子接受体，将无机纳米粒子与石墨烯复合，可赋予石墨烯新的功能，可以发挥纳米粒子自身的纳米尺寸效应，将纳米粒子本身所具备的力学性能、光性能、电性能以及磁性能等赋予石墨烯，在宏观性能上使复合材料比相应的微米级别材料的性能有更大的提高，近年来引起了越来越多的关注和研究[45-56]。通常可以为石墨烯修饰的无机纳米粒子很多，包括钛酸盐[45,46]、磺酸基[47]、磷酸盐[48]、黏土和氧化物[49-53]、层状双金属氢氧化物[54-56]等。Chappa[48]和 Cheng[49]等分别用磷酸盐和蒙脱石对 GO 进行修饰，并应用于放射性核素的吸附去除（图 12-8）。Sun 等[47]利用磺酸基修饰氧化石墨烯，制备高比表面积、多孔的石墨烯片层材料，并从微观结构和理论计算多角度探讨磺化复合后氧化石墨烯与 U(VI)的相互作用。Yu[53]和 Zou[54]等使用 DFT 理论计算的手段分别考察了石墨烯和水滑石、氧化钛相互作用的理论模型（图 12-9），为优化制备高质量石墨烯复合材料，进行性能调控及提高对放射性核素的富集能力，提供了重要的理论指导。

2.4　磁性石墨烯对放射性核素的吸附

对于理想核清除剂另一个重要的要求是达到饱和吸附后能便捷地与液相分离，实现吸附剂与放射性

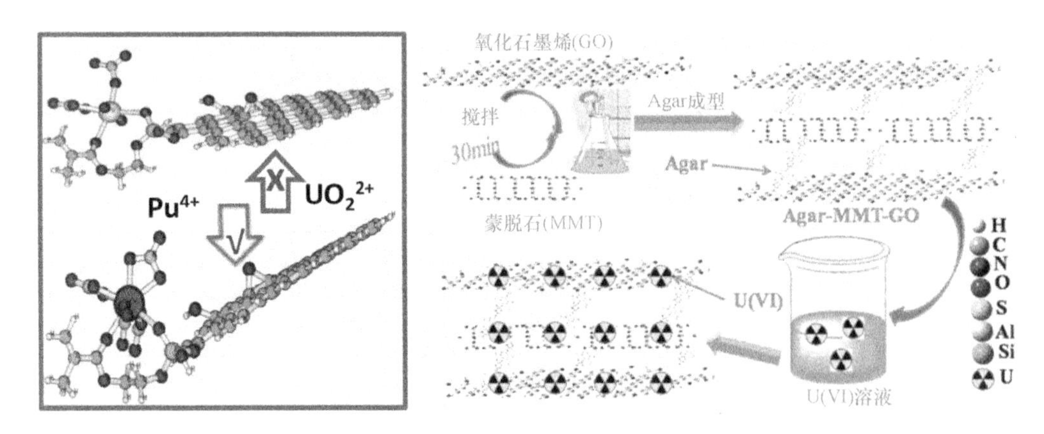

图 12-8 磷酸盐和蒙脱石修饰 GO 对铀的高效富集[48,49]

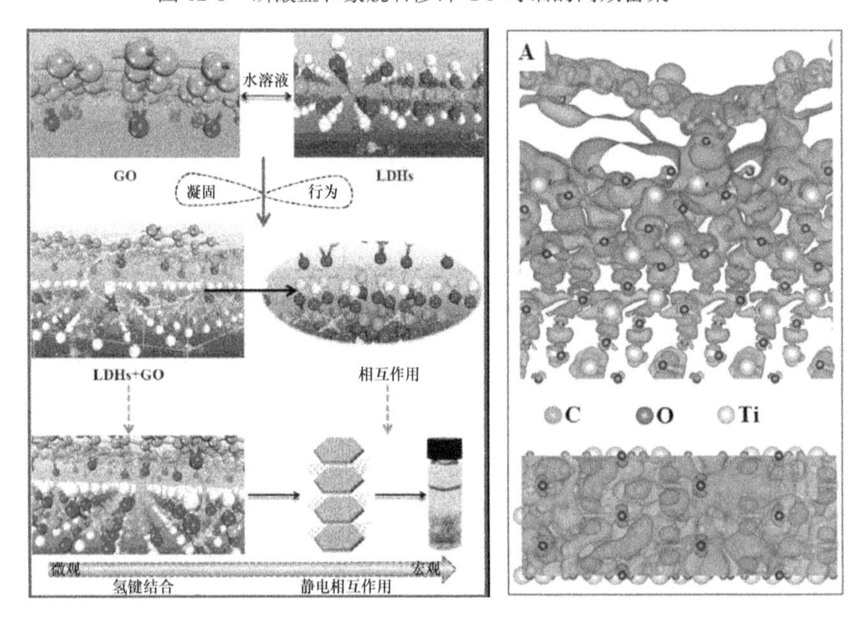

图 12-9 石墨烯和水滑石、氧化钛相互作用的理论计算模型[53,54]

核素的有效分离和回收，并降低有害物质在环境中迁移转化的风险。因此，无机纳米粒子修饰石墨烯最为重要的材料是磁性铁系类材料改性石墨烯。磁性纳米粒子不同于常规磁性材料的超顺磁性，可简单利用外磁场实现分离和回收。大量研究表明，将石墨烯与磁性铁系材料复合，并将其应用于放射性污染水体治理，既可高效快速去除 Cs(I)、Sr(II)、Co(II)、Eu(III)、U(VI)等放射性核素，又可用磁分离法将其回收利用，经再生后多次重复利用，为规模化应用奠定了基础[57-66]。这些成果为开发一类高效、稳定、安全的放射性污染治理的功能化新材料和新技术提供了新思路。

3 零价铁及其复合材料对放射性核素的转化固定

近年来，环境中放射性铀污染越来越严重，铀半衰期长，化学毒性大，在裂变反应中可以生产一系列其他高毒性的放射性核素，如硒、铼、锝、碲、钼等。这些放射性核素在不同的环境中以不同的价态存在，在氧化性较强的环境中，主要以高价态的 U(VI)、Re(VII)、Se(IV)/(VI)、Tc(VII)、Te(IV)/(VI)、Mo(VI)等存在，其毒性大、流动性强，容易造成极大的环境污染。而在还原性较强的环境中，这些核素主要以低价态的 U(IV)、Re(IV)、Se(0)/(–II)、Tc(IV)、Te(0)/(–II)、Mo(III)等难溶性物质存在，其毒性和迁移转化性都很弱。因此，选择合适的还原剂，有效地将这些高价态的放射性核素还原转化为低价态物质，是处理环境中放射性污染的一种具有较好应用潜力的修复技术[67-69]。

3.1　零价铁还原固定放射性核素

作为一种绿色、廉价的还原剂，零价铁（ZVI）具有原料易得、操作简单等优点。大量研究表明，零价铁可以将迁移性强、毒性大的高价态放射性核素还原固定为迁移性小、毒性弱的低价态难溶性物质，是放射性污染处理处置的一个重要方法，已经引起了越来越多地关注[70-80]。Morrson 等[77,78]和 Sasaki 等[79]将零价铁用作地下水处理的可渗透反应墙的活性材料（Fe^0-PRB），发现 Fe^0-PRB 可以有效地将地下水中的 UO_2^{2+}、MoO_4^{2-}、SeO_4^{2-}等放射性核素转化固定，反应机理主要包括 Fe^0 的还原作用以及反应过程中形成的腐蚀产物的吸附作用，零价铁在放射性污染的原位修复方面取得了理想的效果。然而随着反应的进行，各种难溶性腐蚀产物[铁(氢)氧化物]不断覆盖在零价铁表面，显著减少铁表面反应活性位，因此零价铁的反应活性通常比较低。为此，唐次来等[81-83]通过简单的化学法在 Fe^0 表面原位生成一层 Fe_3O_4，再外加 Fe^{2+}形成一个具有高活性的协同体系（$Fe^0/Fe_3O_4/Fe^{2+}$），发现该协同系统可以持续、快速地还原硒酸盐、钼酸盐等核素，不会出现钝化现象，这一协同体系在放射性污染控制中具有重要的实际应用意义。同济大学关小红教授课题组[84-89]率先提出了利用磁场强化零价铁还原固定放射性核素 Se(IV)/(VI)的研究，该课题组系统考察了磁场对零价铁还原 Se(IV)/(VI)的影响，发现磁场对已钝化零价铁还原 Se(IV)/(VI)具有显著的加速作用，他们通过磁场数值模拟等手段分析了磁场能加速零价铁腐蚀及提高反应性能的主要机制，这主要是归因于洛伦兹力和磁场梯度力的共同作用（图 12-10）。另外，他们结合同步辐射软 X 射线显微成像技术（SXTM）发现反应中生成的 Fe^{2+}吸附在零价铁及铁氧化物表面，并证实吸附在零价铁及铁氧化物表面的 Fe^{2+}和零价铁共同将 Se(IV)/(VI)还原固定，进一步利用 X 射线近边结构谱（XANES）分析揭示了 Se(IV)/(VI)首先吸附到零价铁表面的腐蚀产物上，进而被内层的零价铁还原固定。这些研究结果证明磁场能明显提高零价铁的活性。

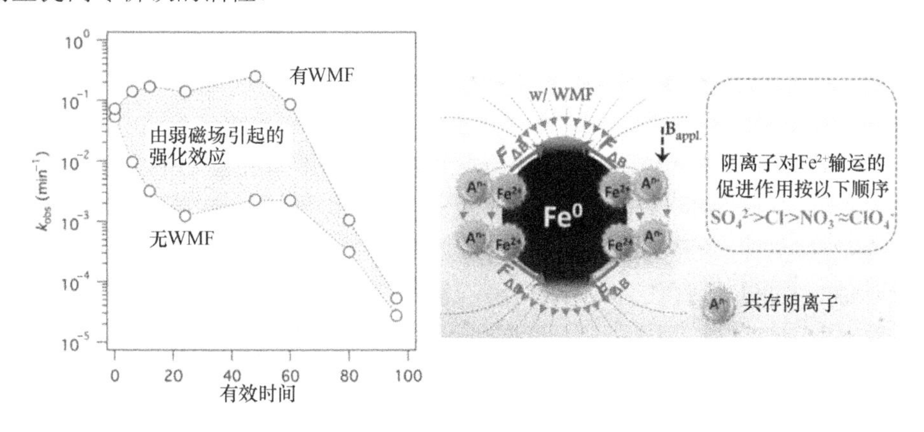

图 12-10　磁场对零价铁还原固定放射性核素的强化作用[84,89]

3.2　纳米铁还原固定放射性核素

此外，由于零价铁还原放射性核素是个表面介导的反应，因此使用比表面积更大的纳米零价铁（NZVI），可以明显提高对放射性核素的还原转化率[90-93]。中国地质大学的严森等[90,91]以及英国布里斯托大学 Scott 课题组[93-98]对纳米铁还原固定铀的反应动力学特征和作用机制进行了系统的研究，他们结合批式实验和先进光谱分析的方法，全面剖析了纳米铁去除和还原铀的主要影响因素和反应机理，发现二价和三价铁在还原反应中扮演着重要角色，与零价铁一起构成一个氧化还原反应循环，使得 U(VI)还原为 U(IV)；此外，地质化学条件显著影响纳米铁的还原性能，U(VI)的还原速率随着介质 pH 的升高以及碳酸根及钙离子浓度的增加而下降；另外，温度升高可以加速 U(VI)的还原，碳酸根的存在促进温度对还原反应的影响。同济大学凌岚博士与张伟贤教授[99,100]采用球差校正扫描透射电镜从原子尺度观测了铀和硒在纳米铁内的固相迁移和化学反应，定量分析了接近原子尺度上铀和硒在纳米铁颗粒内的准确分布和结构信息（图 12-11），证实了纳米铁颗粒将铀和硒包裹在纳米铁的内核上，并利用电子能量损失谱分析（EELS）、X 射线光电子能谱分析（XPS）等明确了纳米铁将铀和硒还原成了低价态的难溶性铀和硒，通过化学方法还原加固纳米铁颗粒后，还原后的铀和硒稳定地包裹在纳米铁内核。这些研究成果表明纳米铁具有修复放射性污染的潜力，为发展应用纳米铁原位修复放射性污染的技术提供了理论支持。

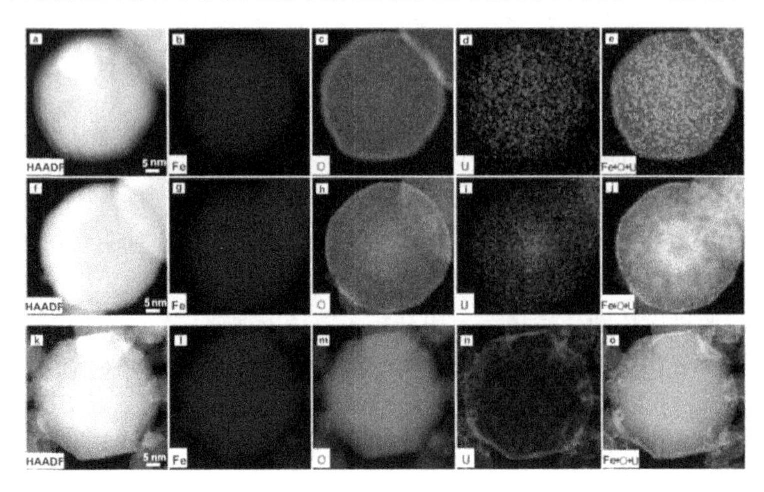

图 12-11　球差校正扫描透射电镜观察 U(VI)在纳米铁内的迁移过程[99,100]

3.3　纳米铁复合材料还原固定放射性核素

尽管纳米铁对放射性核素具有高效的还原能力，但纳米铁极易氧化及纳米颗粒倾向于聚集的现象限制了其更大范围的应用。由于纳米铁表面积大、不稳定，颗粒之间容易相互碰撞从而聚集成大颗粒，聚集物的直径通常大于 100 μm，导致纳米铁的反应活性迅速下降[101]。这严重影响了纳米铁与放射性核素的还原能力及使用寿命。因此，需要对纳米铁做合适的修饰，减缓纳米铁的氧化和聚集，从而使纳米铁更广泛地应用到放射性污染修复中。因此，为了有效防止纳米铁团聚和氧化问题，通常可以在纳米铁制备过程中添加适当的稳定剂或载体，进而获得性能稳定、反应活性高的纳米铁复合材料，有助于提高对放射性核素的还原固定能力。美国俄勒冈健康与科学大学 Tratnyek 教授课题组[102,103]将合成的纳米铁与溶液中的含硫物质反应形成硫化型纳米铁，并用 XRD、BET-N_2、SEM、TEM 等对获得的硫化型纳米铁进行了详细表征，发现硫化处理之后纳米铁的团聚和氧化现象显著减小，对放射性元素 $^{99}TcO_4^-$的还原去除效果明显提高，在再氧化固定实验中，证实硫化纳米铁能提高 $^{99}TcO_4^-$被扣留的稳定性。由于硫化之后呈现出的优异性能，硫化纳米铁及相关的硫化铁系材料将会逐步成为放射性污染修复的重要具有发展前景的潜在材料。此外，零价铁体系加入微生物可以明显增强放射性核素的还原效率。中科院孙玉兵等[104]的研究结果证实微生物 *Bacillus subtilis*（枯草芽孢杆菌）可以提高 U(VI)在纳米零价铁的还原过程，主要是由于 U(VI)以内层络合物吸附在 *B. subtilis* 上，增加了零价铁表面反应物的浓度，从而加速了还原反应速率

（图 12-12）。该结果为发展应用于放射性污染地下水原位修复的生物和化学耦合技术提供了有力的理论依据。

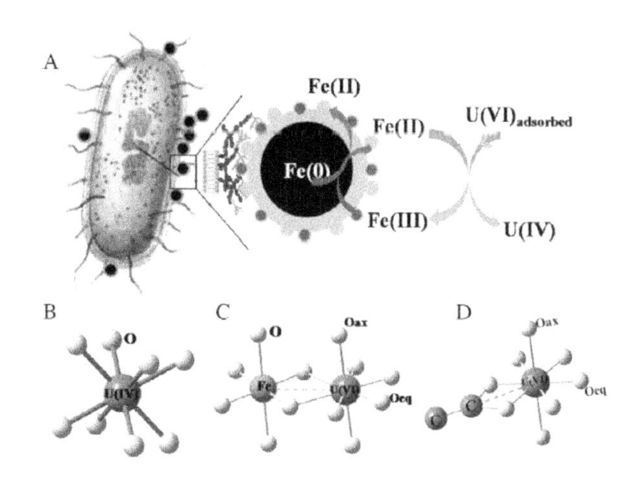

图 12-12　铀在零价铁和 *B. subtilis* 上的吸附机理及化学形态[104]

采用合适的化学稳定剂对纳米铁进行表面改性是提高其反应活性的另一种重要方法，这些化学稳定剂通常包括淀粉、壳聚糖、环糊精、聚偕胺肟及羧甲基纤维素等。包裹在纳米铁表面，即安全无毒和环境友好，又有效地减缓了纳米铁的团聚。当稳定化的纳米铁注入到地下水中，可以跟随水流流动，同时又可以迅速与水中的放射性核素发生反应，起到净化水质的作用[105-107]。罗马尼亚国家金属和放射资源研究所 Popescu 等[105]和华北电力大学王祥科教授课题组[106]分别用羧甲基纤维素和聚偕胺肟对纳米铁进行稳定化，用于处理 U(VI)离子（图 12-13）。美国奥本大学赵东叶教授等[107]用淀粉作为稳定剂，获得淀粉稳定化纳米铁用于处理 ReO_4^-离子。研究结果表明，稳定化之后的纳米铁对放射性核素的还原活性明显提高，通过稳定化处理，可以增强纳米铁在水体中的流动性，实现还原剂的持续缓慢释放，延长纳米铁在放射性污染修复的使用寿命。

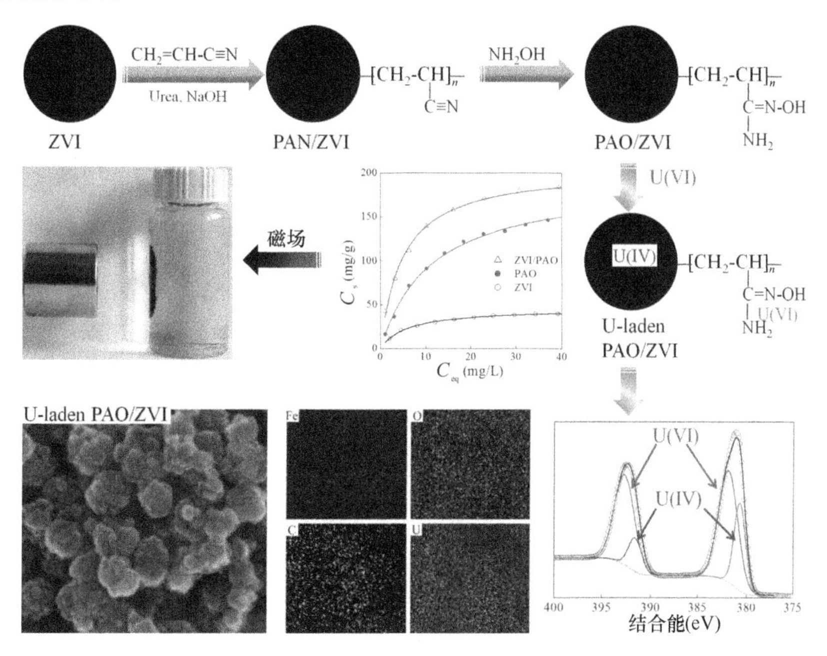

图 12-13　聚偕胺肟稳定化纳米铁的制备、表征及对 U(VI)的去除[106]

负载化技术是另一类提高纳米铁稳定性和活性的重要方法，与化学稳定剂表面改性不同，多孔载体具有丰富的裂隙孔道和表面功能基团，可以为纳米铁提供充足的负载位点。碳纳米材料、黏土矿物及沸石分子筛是常见的惰性载体（图 12-14），选择适当的载体及简便的修饰方法，可控构筑纳米铁复合材料并用于放射性废水中各种核素的还原去除，是近年来放射性污染治理的重要方向[108-112]。华北电力大学王祥科教授课题组[113-116]采用等离子体还原技术将氧化石墨烯还原，避免纳米铁被氧化石墨烯氧化，同时提高电子传输能力。然后使用等离子体溅射铁靶材，在石墨烯表面负载纳米铁，得到石墨烯负载纳米铁复合材料。进一步考察了复合材料对放射性核素的吸附与还原的去除效果，研究了pH、离子强度、温度、接触时间等不同实验条件对放射性核素还原去除的影响，结合宏观批实验与微观表征手段（XRD、TEM、XPS、EPR 和 XAFS 等），揭示了复合材料对放射性核素吸附与还原的协同去除机理。在此基础上，通过调节等离子体工艺参数，如工作气体、功率、溅射时间等，控制石墨烯表面纳米铁的粒径大小及分布、分散性、稳定性和负载量，并分析对放射性核素反应性能的影响，进而构筑吸附还原性能优良的石墨烯纳米铁复合材料。绍兴文理学院盛国栋等[117-125]通过离子交换、表面处理等手段，对碳纳米管、钛酸纳米管或黏土等材料进行修饰，然后将这些材料作为载体合成出一系列纳米铁复合材料，用于废水中不同类型放射性核素的还原固定。借助多种表征技术对复合材料的比表面积、表面官能团、粒径大小及分布等物理化学性质进行详细表征。结合纳米铁复合材料对放射性核素还原反应动力学分析，剖析各种载体促进纳米铁还原的作用机制。运用同步辐射 X 射线吸收精细结构（XAFS）研究了界面反应的微观过程（图 12-15 和图 12-16）。通过对扩展 X 射线吸收精细结构（EXAFS）谱的拟合分析，探讨了放射性核素在反应前后的赋存形态、微观结构等变化，发现纳米铁复合材料可以将毒性大且迁移性强的高价态铀、铼、硒等放射性核素几乎完全还原转化为毒性小且迁移性弱的低价态铀、铼、硒等难溶性沉淀物质。另外，该课题组利用穆斯堡尔谱进一步分析反应过程中固相颗粒表面 Fe(II)的空间分布、赋存形态和铁腐蚀产物的组成和结构及其作用机制（图 12-17）。结果证实反应中生成的 Fe(II)可以吸附在铁的腐蚀产物及相应的载体上，与游离态 Fe(II)相比，这些吸附态 Fe(II)还原电位更低，还原活性更高，可以进一步促进对放射性核素的还原转化[122,123]。通过这些先进光谱表征分析，剖析纳米铁复合材料还原去除放射性核素的反应机制，为未来纳米铁复合材料在放射性污染修复的工程应用研究指明了发展方向。

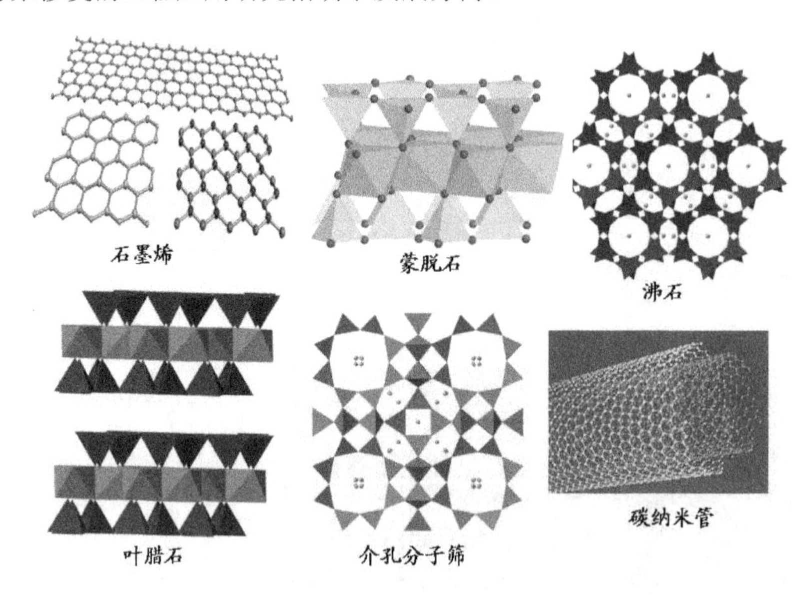

图 12-14　用于纳米铁负载的常用载体

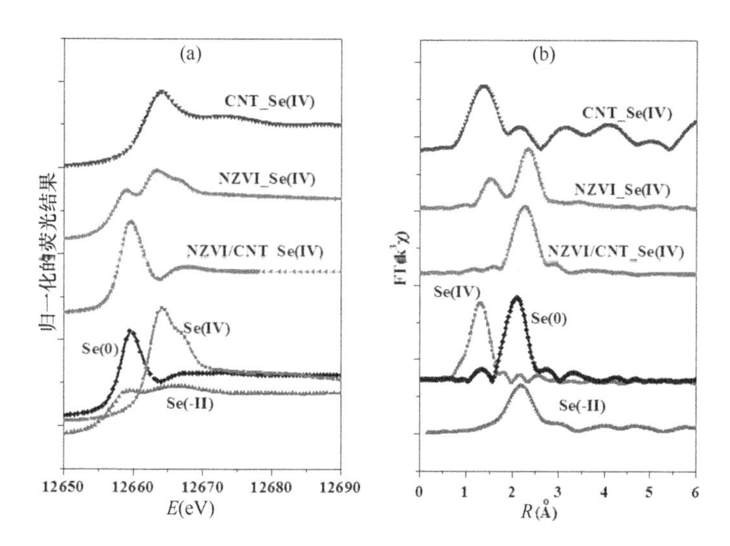

图 12-15　碳纳米管/纳米铁处理 Se(IV)的 XANES 谱（a）和 EXAFS 谱（b）[117]

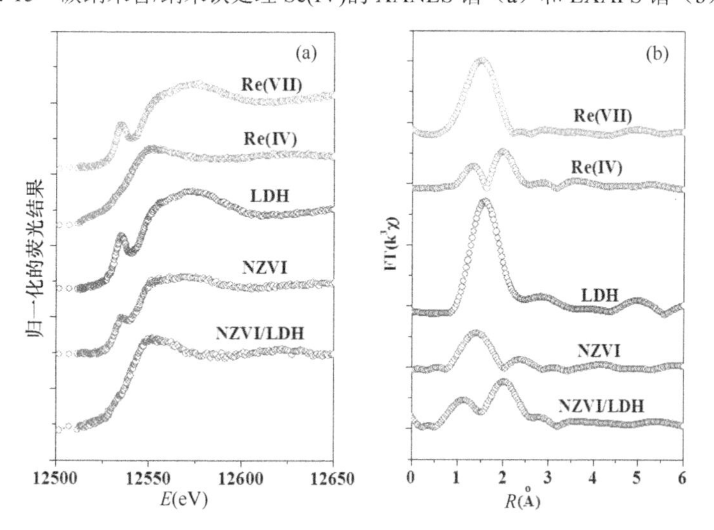

图 12-16　水滑石/纳米铁处理 Re(VII)的 XANES 谱（a）和 EXAFS 谱（b）[123]

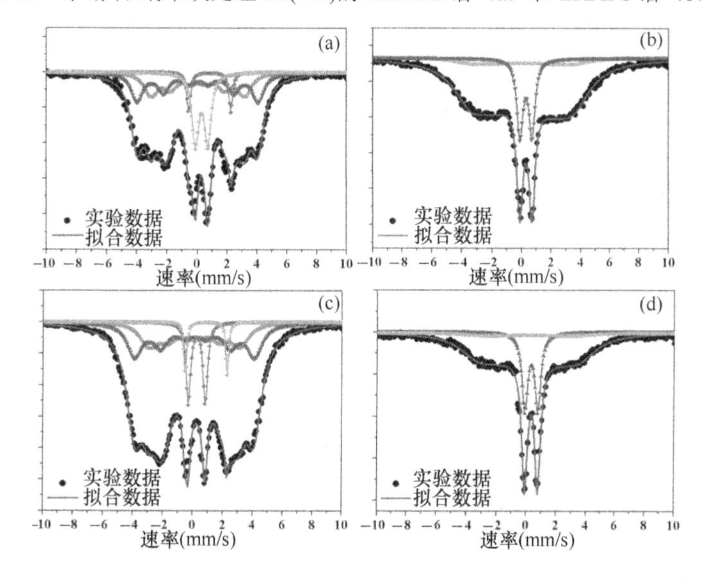

图 12-17　穆斯堡尔谱分析反应前后铁的价态、化合态和结合态变化[123]

4　表面结合 Fe(II)系统对放射性核素的还原转化

铁是最常见的变价元素之一，其价态通常为 0、+2 和+3 价，如前所述，零价铁对放射性核素具有很强的还原活性。而在自然界中，铁主要以+2 和+3 价的矿物形式存在，在土壤、地下沉积物尤其是红壤胶体中，通常含有大量的铁(氢)氧化物，这些铁(氢)氧化物表面带有电荷，比表面积大，反应性强，是重金属离子和放射性核素潜在的吸附剂，并且可以催化有机物的转化降解，对某些有毒有害污染物具有天然净化作用[126-129]。此外，在缺氧地下环境中，铁矿物往往可以被天然有机质或异化铁还原细菌还原（图 12-18），生成亚铁[Fe(II)]离子[130-134]。在还原性非常强的水稻土中，Fe(II)离子的含量通常高达 4~5 g/kg[126]。这些 Fe(II)离子主要以游离态、吸附态、络合态（与土壤固相部分络合者）和沉淀态等多种形态存在自然环境中，不同形态的 Fe(II)具有不同的反应活性。当 Fe(II)吸附到具有表面活性基团的固态矿物上，就形成一个表面结合 Fe(II)系统。研究表明：表面结合 Fe(II)的氧化还原电位低于游离态 Fe(II)，因而其还原活性远高于游离态 Fe(II)。因此利用自然环境中存在的 Fe(II)/矿物，开展环境污染治理，探索污染物在自然环境中的迁移转化规律具有重要的科学意义。

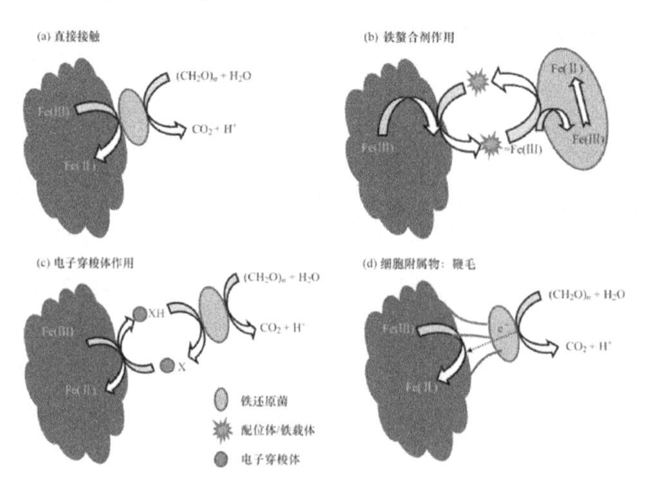

图 12-18　异化铁还原菌介导的铁矿物电子转移机制示意[134]

4.1　铁矿物结合 Fe(II)还原放射性核素

近年来，利用 Fe(II)和天然矿物转化固定放射性核素的研究逐渐引起了国内外学者的兴趣。瑞典皇家理工学院 Cui 和 Eriksen[135,136]在 1996 年率先报道了表面结合 Fe(II)系统对放射性核素 TcO_4^-的还原转化。尽管从热力学上 TcO_4^-可以被溶液中的 Fe(II)离子还原，但他们发现，在酸性的 Fe(II)均相体系中，TcO_4^-几乎没有被还原，而将反应体系调至碱性，部分 Fe(II)发生沉淀作用形成 $Fe(OH)_2$ 或 $FeCO_3$，这些沉淀态 Fe(II)可将 TcO_4^-还原。此外，Cui 和 Eriksen[135,136]发现 Fe(II)吸附在磁铁矿等矿物表面同样可以将 TcO_4^-还原，动力学分析表明 TcO_4^-的还原速率随着 Fe(II)吸附量的增加而增大。后续的研究进一步表明：Fe(II)可以与针铁矿、磁铁矿、纤铁矿、水合铁矿、黄铁矿、赤铁矿、菱铁矿和绿锈等多种含铁矿物结合，还原 U(VI)[137-145]、Tc(VII)[146-148]、Se(IV)/(VI)[149-151]和 PuO_2[152]等多种对氧化还原条件敏感的放射性核素。这些研究系统的考察了 Fe(II)用量、接触时间、反应物浓度等对还原反应动力学及作用机制的影响。发现随着 pH 升高或 Fe(II)用量的增加，Fe(II)在铁矿物表面的吸附量越大，反应越快。结果证实表面结合 Fe(II)系统还原放射性核素的反应速率与机理与铁矿物组成、结构和颗粒大小、放射性核素的性质，以及 pH、离子强度、共存组分等地质化学条件密切相关。所有这些因素都可以影响 Fe(II)在铁矿物表面的吸附量、结合位点、赋存形态和配位结构，进而影响表面结合 Fe(II)系统对放射性核素的还原转化。Silvester 等[153]通过测试一系列铁矿物（针铁矿、磁铁矿、水合铁矿）结合 Fe(II)的氧化还原电位发现，与游离的电对相比 Fe(III)/Fe(II)，矿物表面结合 Fe(III)/Fe(II)电对的氧化还原电位下降，是其还原能力增强的主要驱动力。另外，在不同的反应条件下，反应产物的微观结构也各不相同。Massey 等[145]借助 XPS 和 EXAFS 等光谱技术表征了在不同条件下水合铁矿结合 Fe(II)还原转化 U(VI)的微观结构变化规律，发现在低 pH 以及 Fe(II)和 U(VI)浓度比较低的情况下，U(VI)在水合铁矿表面形成以内层络合物为主；随着介质 pH 以及 Fe(II)和 U(VI)浓度逐渐升高，水合铁矿逐渐转化为针铁矿，而 U(VI)通过晶格置换作用嵌入到针铁矿晶格中；pH 以及 Fe(II)和 U(VI)浓度继续升高，U(VI)被表面结合 Fe(II)系统还原转化为 U(IV)沉淀物质（图 12-19）。由此可见，在表面结合 Fe(II)还原放射性核素的反应体系中，Fe(II)部分被氧化后在表面形成一层次生矿物，其组成和结构随着反应条件的不同而不同，既可以是对放射性核素具有吸附作用的针铁矿、赤铁矿等，也可以是对放射性核素具有还原作用的多羟基亚铁、绿锈、磁铁矿等亚铁类矿物。Jang 等[141]用穆斯堡尔谱研究水合氧化铁结合 Fe(II)和赤铁矿结合 Fe(II)还原 U(VI)过程中表面次生矿物的演化规律，在水合氧化铁表面，形成针铁矿、赤铁矿、非化学计量磁铁矿；而在赤铁矿表面形成水合氧化铁、水合赤铁矿、非化学计量磁铁矿，由此造成了 U(VI)在水合氧化铁/Fe(II)和赤铁矿/Fe(II)体系不同的还原转化率。美国爱荷华大学 Scherer 教授课题组[154-162]利用穆斯堡尔谱研究了 Fe(II)在一系列铁矿物(赤铁矿，针铁矿，水铁矿)表面的作用机制。用不含铁的锐钛矿和氧化铝做比较，Fe(II)在锐钛矿和氧化铝表面只发生吸附作用。而采用穆斯堡尔谱证实了 Fe(II)吸附到铁矿物表面形成的吸附态 Fe(II)与相应的铁(氢)氧化物的结构 Fe(III)发生了电子转移作用，在矿物表面形成了一层铁氧化物次生矿物。具体用 $^{56}Fe^{2+}$与用 ^{57}Fe 合成的铁矿物反应，在溶液中探测到 $^{57}Fe^{2+}$信号，说明铁矿物表面吸附 $^{56}Fe^{2+}$并将其氧化成 $^{56}Fe^{3+}$，与此同时，铁矿物结构中的 $^{57}Fe^{3+}$被还原成 $^{57}Fe^{2+}$释放到溶液中，形成一个 Fe^{2+}/Fe^{3+}电子转转移循环，并可以还原一系列污染物。该课题组[155]系统研究了不同浓度 Fe(II)在铁矿物上的反应产物，发现在低浓度下 $^{57}Fe^{2+}$直接吸附在赤铁矿表面并迅速转化为 $^{57}Fe^{3+}$，进入铁矿物体相结构；而在高浓度下则生成稳定的 $^{57}Fe^{2+}$络合物和 $^{57}Fe(OH)_2$ 沉淀，这些不同的次生产物对放射性核素具有不同的还原作用。这些结果证实铁矿物/Fe(II)在放射性污染的原位修复方面具有重要的应用。

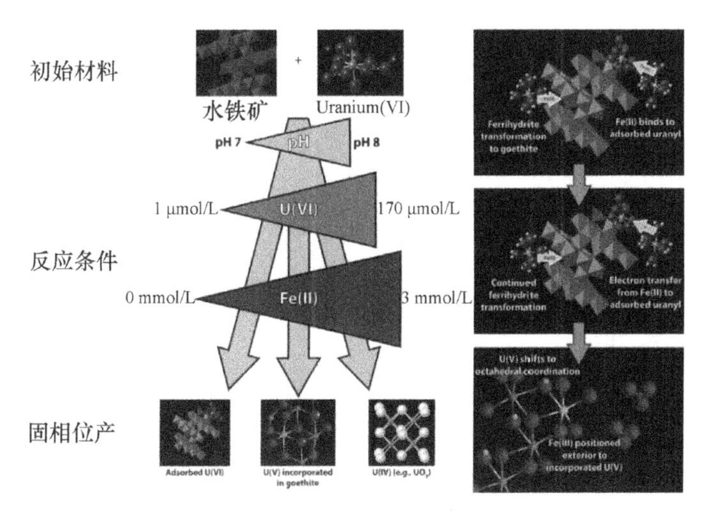

图 12-19　水合铁矿/Fe(II)在不同条件下还原 U(VI)的微观结构分析[145]

4.2 黏土结合 Fe(II)还原放射性核素

然而，在沉积物和土壤等实际环境中，除了铁矿物之外，还存在着大量的黏土、氧化铝、方解石等其他矿物。即便在铁矿物含量很高的红壤土中，黏土矿物的含量也比铁矿物高几倍[126-129]。显然，在缺氧的地下环境中，相比于铁矿物结合 Fe(II)体系，黏土结合 Fe(II)体系更为常见，而且与铁矿物相比，黏土对大多数放射性核素吸附能力更强[163-165]。Schultz 等[166]发现 Fe(II)/蒙脱石具有和 Fe(II)/铁矿物类似的还原活性，尽管 Fe(II)/蒙脱石的还原活性不如 Fe(II)/铁矿物，但在深层土壤环境中，足以对污染物的迁移转化行为产生影响。因此，作为一种潜在的还原剂，Fe(II)/黏土体系在还原去除放射性核素方面也发挥着重要的作用，近年来逐渐受到关注[167-174]。Chakrabrty 等[148]研究了方解石/Fe(II)体系对 Se(IV)的吸附和还原协同作用，结果发现，与方解石共沉淀的 Fe(II)不能将 Se(IV)还原，而吸附在方解石上的 Fe(II)能将 Se(IV)还原为 Se(0)，并具有相对较快的还原动力学。Chakraborty 等[167]进一步研究了蒙脱石/Fe(II)体系对 UO_2^{2+}的吸附和还原协同作用，结果表明，Fe(II)在蒙脱石表面能形成铁氧化物，在不同的 pH 条件下，铁氧化物的组成和结构不同，因此，蒙脱石/Fe(II)体系的氧化还原电位也不同，并影响了对 UO_2^{2+}的吸附和还原作用。此外，Chakraborty 等[167]考察了蒙脱石/Fe(II)体系的电位随着 Fe(II)与蒙脱石接触时间以及介质 pH 的变化关系，发现与液相中 Fe(II)的电位值（0.77 V）相比，在蒙脱石悬浮液中，Fe(II)的还原电位明显下降（~0.35 V）。在 pH=6.0 时，蒙脱石/Fe(II)的电位值与水合铁氧化物/Fe(II)的电位值类似，而在 pH=7.5 时，蒙脱石/Fe(II)的电位值与纤铁矿/Fe(II)的电位值类似，这间接证明 Fe(II)在蒙脱石表面能够形成铁氧化物，在不同的 pH 条件下，形成的铁氧化物的组成和结构不同，因而还原活性也不相同。Charlet 等[170]研究了蒙脱石/Fe(II)体系对 Se(IV)的吸附和还原协同作用，研究结果发现在没有 Fe(II)存在时，Se(IV)只能少量地吸附在蒙脱石的正电荷位点上，形成外层吸附络合物，而当体系中有 Fe(II)存在时，Se(IV)能被还原为 Se(0)。同时，Charlet 等[170]结合理论计算以及 EXAFS 技术研究了反应过程中硒元素的微观形态及配位结构变化，用穆斯堡尔谱研究了反应过程中铁元素的化学形态变化，并进一步揭示了反应机理。Charlet 等[170]推测的反应机理是：部分吸附在蒙脱石表面的 Fe(II)将释放出电子[自身被氧化为 Fe(III)]，并以 H_2 形态储存起来，当体系中加入 Se(IV)后，Se(IV)被其还原转化(图 12-20)。中国地质大学董海亮教授课题组在一系列工作中提出用黏土矿物作为载体，用黏土表面结合 Fe(II)系统还原放射性 TcO_4^-，也取得了非常理想的结果[171-174]。

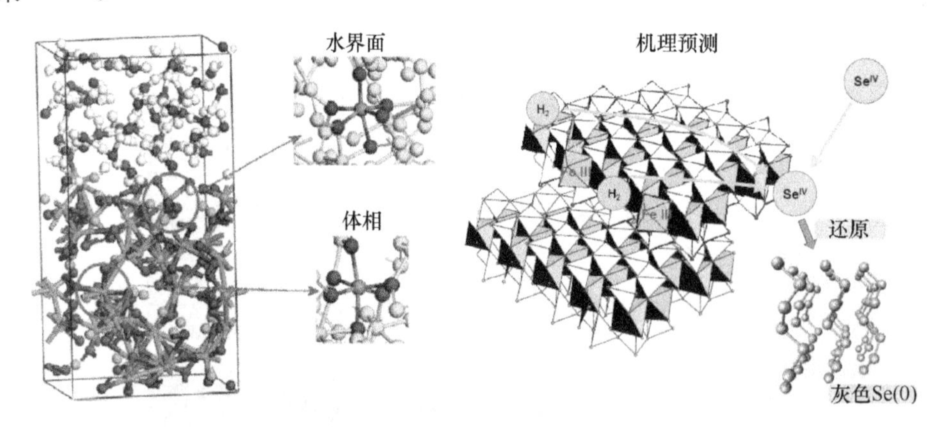

图 12-20　蒙脱石/Fe(II)体系对 Se(IV)的吸附和还原协同作用[170]

与铁矿物一样，Fe(II)在其他矿物表面也会发生各种界面反应，形成组成和结构不同的各种次生矿物，并影响放射性核素的还原反应[175-179]。美国新泽西州立罗格斯大学 Elzinga 教授课题组[176,177]利用 EXAFS

光谱系统研究了不同时间和 pH 条件下，Fe(II)在氧化铝、氧化硅和不含铁黏土上的作用机制和微观形态变化，发现 Fe(II)在氧化铝主要形成 Fe(II)–Al(III)层状双金属氢氧化物，而在氧化硅和黏土上主要形成 Fe(II)-层状硅酸盐结构（图 12-21），对放射性核素都有还原活性。瑞士联邦理工学院 Schwarzenbach 教授课题组[180-182]研究了三种不同结构的黏土矿物（不含铁的水辉石，含铁量低的蒙脱石，含铁量高的绿脱石）表面结合 Fe(II)不同的反应活性。发现在黏土/Fe(II)体系，存在三类不同化学环境的 Fe(II)，即结构态 Fe(II)、表面络合态 Fe(II)、离子交换态 Fe(II)，它们的活性是不一样的，离子交换态 Fe(II)由于还原电位与游离态 Fe(II)的还原电位值类似，因而在还原反应中发挥的作用不大。起主要作用的是结构态 Fe(II)和表面络合态 Fe(II)。另外 Schwarzenbach 等[180-182]通过动力学分析发现，结构态 Fe(II)的活性比表面络合态 Fe(II)高几个数量级。美国爱荷华大学 Scherer 教授课题组[183,184]利用穆斯堡尔谱分析发现，黏土中三类化学环境不同的 Fe(II)之间可以互相发生电子转移，并且相对含量随着条件的不同会发生改变，如在低 pH 下，主要以离子交换态 Fe(II)为主，此时黏土/Fe(II)体系几乎没有活性；而在高 pH 下，主要以表面络合态 Fe(II)为主，此时黏土/Fe(II)体系有还原活性（图 12-22）。

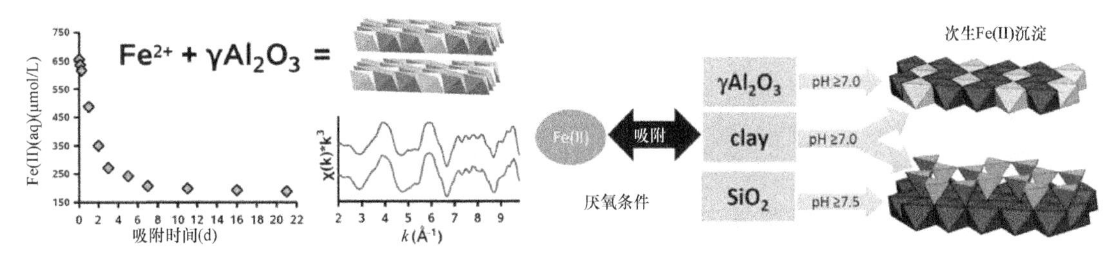

图 12-21 氧化铝，氧化硅及不含铁黏土上 Fe（II）不同次生矿物的形成[176,177]

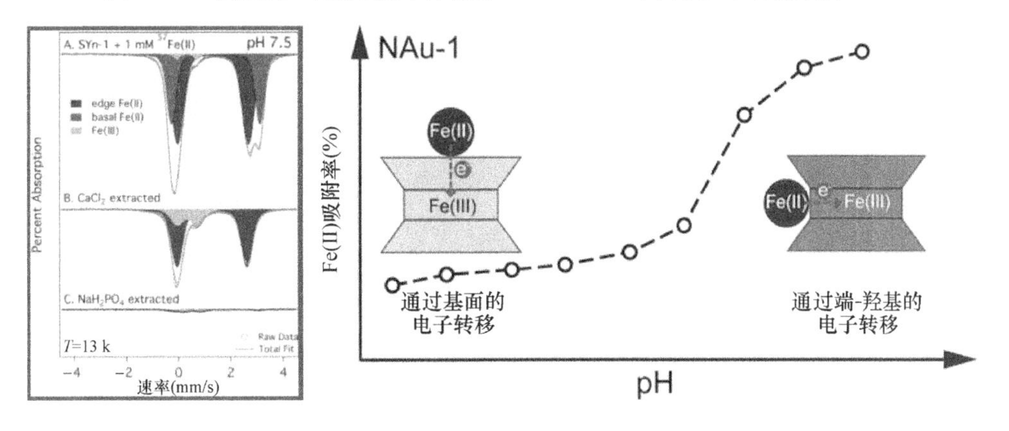

图 12-22 穆斯堡尔谱分析黏土中不同结合态 Fe(II)之间的生电子转移[184]

总之，铁的氧化还原循环对土壤和地下水中放射性核素的归宿有着重要的影响。在缺氧环境中，Fe(II)-Fe(III)氧化还原循环主要由微生物还原含铁矿物中的结构 Fe(III)驱动。而微生物广泛存在于环境中，可以持续提供 Fe(II)结合在矿物表面，为土壤和地下水中放射性核素的自然衰减作用提供新的途径。矿物表面可以促进 Fe(II)还原放射性核素，吸附态 Fe(II)还原活性的提高主要是由 Fe(II)与矿物表面的羟基发生表面络合作用，使 Fe(II)氧化态稳定，并降低了 Fe(II)-Fe(III)的还原电位[185]。然而在实际环境中各种各样的铁矿物、黏土及腐殖质往往复合在一起，因此，研究实际环境中土壤/Fe(II)或者沉积物/Fe(II)体系对放射性核素的还原转化作用，更具有重要的现实意义。近年来，一些学者更好地认识放射性核素在实际厌氧环境中的迁移转化行为，开始关注土壤结合 Fe(II)或者沉积物结合 Fe(II)与放射性核素的还原反应活性[186-189]。通常情况下，放射性核素还原动力学与表面吸附 Fe(II)浓度呈正相关，吸附态 Fe(II)浓度越高，放射性核素还原速率越大，这与在铁矿物或黏土表面结合 Fe(II)还原放射性核素的结论一致。

然而，促使放射性核素还原的活性 Fe(II)物种的鉴定没有展开深入的研究。从这些结果看，在深层地下高放废物处置库的水-岩厌氧体系中，有望将高价态放射性核素还原为低价态难溶性物质，从而减弱放射性核素的可移动性。

5　其他新型材料对放射性核素的萃取和高效去除

基于核素富集、乏燃料后处理以及环境保护等应用需求，寻求用于萃取、储放及治理核燃料循环中产生的锕系元素及裂变产物的高稳定及耐辐照固体材料一直是环境放射化学领域的研究热点。

5.1　锕系萃取配体的分子设计

锕系元素萃取配体的分子设计是近年来锕系化学中的热门研究课题之一。锕系元素的萃取过程通常发生在水相或水/油双相体系中，萃取过程异常复杂：一方面，锕系元素价态复杂，受萃取环境如 pH 等因素的影响，萃取过程中易伴随着氧化还原反应等；另一方面，溶液中锕系物种的分布具有多样性，极大地增加了萃取难度。基于以上因素，在锕系元素萃取领域，理论与实验研究的结合近年来日趋紧密。锕系计算化学这一理论研究手段已经成为从分子层面上探索锕系元素及其配合物的种态分布、配位性质、光谱结构、热力学性质等的必要工具。借助于计算化学开展锕系元素萃取配体的分子设计工作，不仅有助于丰富锕系化学的理论体系，同时有力地促进了新型萃取配体的实验研究。近年来，国内科研机构在该领域已经做了大量研究工作，研究对象囊括了 Th、U、Np、Pu、Am、Cm 等锕系元素[190]以及一系列镧/锕分离配体[191-195]和有机萃取配体[196-200]。图 12-23 给出了目前研究较多的两类镧/锕分离配体的结构示意图。四齿含氮配体 BTBPs 被认为是未来最具有应用前景的镧/锕分离配体之一，相对论密度泛函理论研究[191-193]发现，此类分子对正三价 Am(III)和 Eu(III)的分离能力可能源于 Am—N 键的共价性略高于 Eu—N 键以及两类离子的水合能力差异。BTPhens 类分子由于萃取动力学明显优于 BTBPs 类，是目前备受关注的镧/锕分离配体之一。理论研究[194]发现 1,10-邻菲咯啉上的 N 原子与 Am(III)和 Eu(III)的配位能力高于三嗪环上的 N 原子，而电子结构分析显示 BTPhens 类配体的萃取动力学和选择性可能优于 BTBPs。在共萃萃取剂方面，图 12-24 给出了目前代表性的一些分子配体。含有 P═O 和 C═O 结构单元的 CMPO 类萃取剂对锕系和镧系元素都显示出很强的萃取能力，对 CMPO 和 Ph_2CMPO 的密度泛函研究[196-198]发现，配合物的形成稳定性可能符合以下顺序，即 Pu(IV)>Eu(III)>Am(III)>U(VI)>Np(V)，与实验研究结论一致。TODGA 对三价锕系和镧系元素也呈现较强的共萃能力，对 Am(III)、Cm(III)、Eu(III)的 TODGA 类配合物的理论研究[199]表明，在配体∶金属摩尔比为 1∶1 和 2∶1 的络合物中，配体羧基氧原子比脂基氧原子显示出更强的络合能力。

5.2　锕系与矿物的作用机理研究

方解石是一种碳酸钙矿物，在自然界中分布最广泛。U(VI)是自然界中最常见的核素，广泛存在于土壤、沉积物和地下水中。目前在实验方面，关于 U(VI)与矿物的相互作用机理的研究比较多[201-204]，然而对于吸附产物构型、配位模式、稳定性、吸附机理等问题在分子水平上了解甚少。基于上述原因，有必要开展计算化学研究以期在分子层面上阐释方解石与 U(VI)的结合机制，为建立锕系元素迁移行为的有效模型提供基础数据。迄今为止，采用密度泛函方法处理周期性体系通常会遇到两大难题：①带电体系的

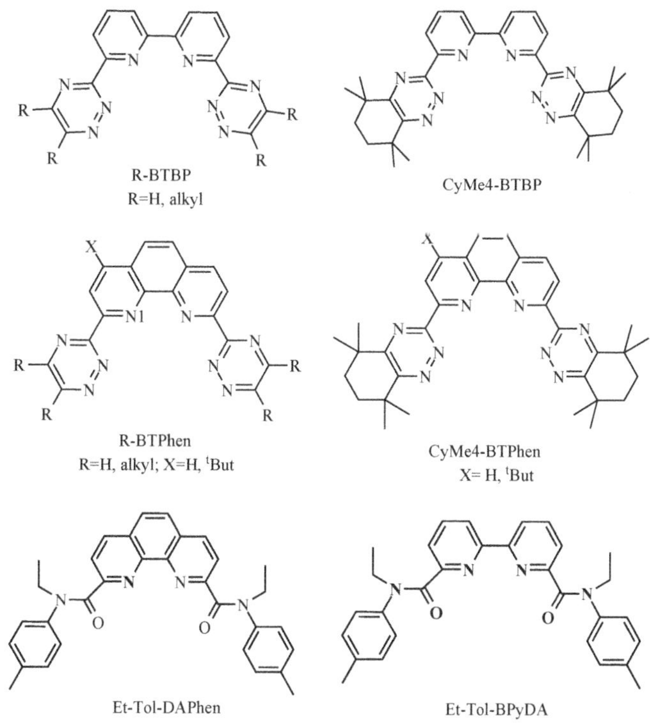

R-BTBP
R=H, alkyl

CyMe4-BTBP

R-BTPhen
R=H, alkyl; X=H, ^{t}But

CyMe4-BTPhen
X= H, ^{t}But

Et-Tol-DAPhen

Et-Tol-BPyDA

图 12-23 代表性镧/锕分离配体的分子结构示意图

CMPO

Ph_2CMPO

HDEHP

TODGA

DMDHOPDA

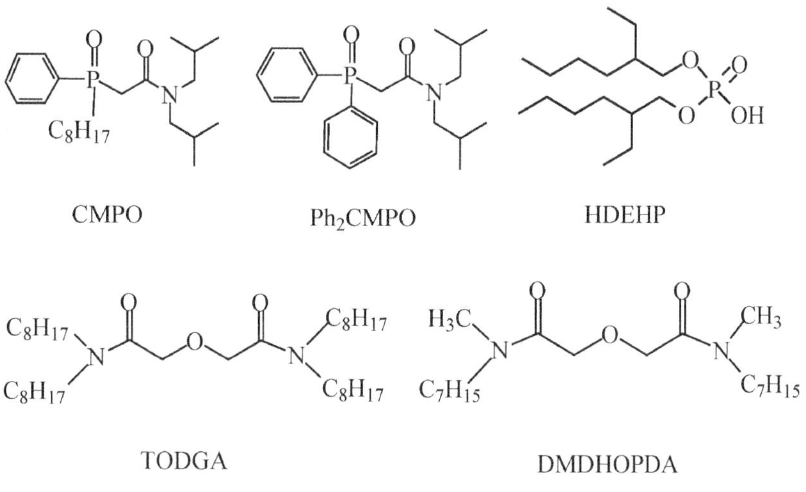

图 12-24 代表性共萃萃取配体的分子结构示意图

能量稳定性问题；② 溶剂化效应的处理问题。在以往的研究中，对于带电体系的处理通常的做法是在表面模型中引入近似，而对于溶剂化效应的处理通常采用部分校正方法，即假定材料表面的溶剂化能在吸附前后保持不变。上述问题的存在导致处理表面相互作用体系时需要极其小心。最近，石等结合密度泛函计算和从头算分子动力学方法研究了 U(VI)与方解石（104）面的作用机理，计算过程中采用隐式溶剂模型[205,206]预测了反应前后各组分溶剂化能的改变。他们的计算结果显示，水分子在（104）面以分子形态而非解离态吸附在表面 Ca 位。带正电的 U(VI)离子倾向于同表面形成络合物，而中性离子既可以与表面形成络合物，也可通过 U(VI)→Ca(II)离子交换与表面结合。相比之下，带负电的 U(VI)物种在固相表面不吸附，更易于在表面发生离子交换反应。值得注意的是，结合 U(VI)离子导致的表面溶剂化能的改变同吸附物种的种类相关，其中 $UO_2(CO_3)_3Ca_2$ 吸附在表面后导致的表面溶剂化能为–53 kcal/mol，即显著增加了表面的亲水性。上述研究对于今后在溶液中处理带电周期性体系提供了有效的方法指导和理论参考。

5.3 新型阴离子晶体材料的设计及高效去除 ^{137}Cs

利用高对称性与降低对称性的三羧酸配体与铀酰离子在类似条件下通过自组装得到截然不同的两种晶体材料，同时利用单晶衍射技术发现前者具有一类石墨烯的二维平面结构，而降低配体对称性后得到了首例结构极为特殊的基于类石墨烯二维平面三重互锁锕系材料。通过系统研究发现后者不仅具有较好的水稳定性以及酸碱稳定性，而且具有很好的 γ 和 β 射线耐辐照性，即使在照射剂量达到 200 kGy 后该材料的晶型及结构依然保持稳定。理论计算也表明该材料极高的稳定性有可能源自于其特殊的三重互锁结构。该三重互锁锕系结构为阴离子骨架，孔隙中含有 $N(CH_3)_2^+$可进行阳离子交换。对水溶液中 $^{137}Cs^+$具有很好的选择性去除能力并在交换过程中没有破坏整体结构，吸附容量也高达 133 mg/g（图 12-25）。该研究结果有望为核废料处理与处置提供一种新材料和方法[207]。

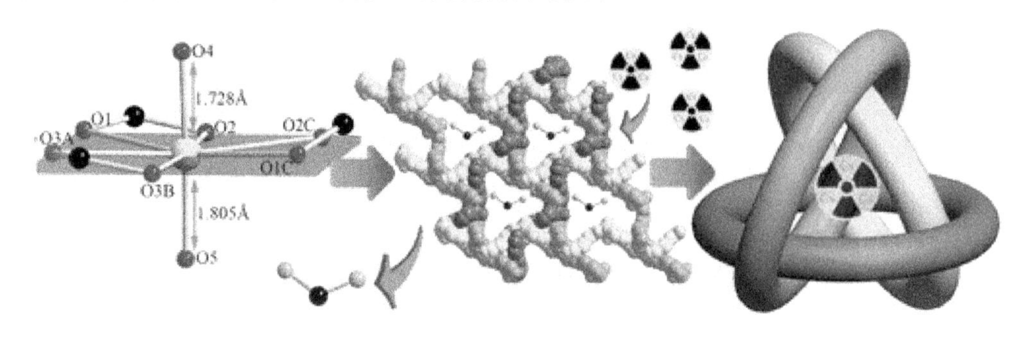

图 12-25 新型阴离子骨架材料去除 $^{137}Cs^+$示意图[207]

5.4 高稳定膦酸锆金属有机框架材料的构筑及对铀酰的高效吸附

相比于羧酸锆金属有机框架材料（MOFs），膦酸锆 MOFs 化合物的化学和热稳定性更好，是更理想的候选材料。然而，由于四价锆与膦酸根之间的亲和性极强，膦酸锆的溶解能力很差，导致快速生成沉淀产物，而沉淀产物的结晶性很差或完全为非晶态固体。因此合成兼具良好结晶性和高比表面积的膦酸锆 MOFs 是个长期挑战。离子热合成方法可以有效抑制四价锆离子的水解或溶剂化，并能有效减缓膦酸锆生成反应的动力学速度，以及提供必要的电荷平衡与孔道填充能力。将四膦酸配体和四氯化锆的混合物在如下离子液体中进行反应得到三个晶态化的有机膦酸锆 MOFs 化合物且都具有较高的比表面积（图 12-26），即便在王水处理后仍有两个化合物的比表面没有损失，例如 SZ-3 的 Langmuir 比表面积在王水处理前后分别为 695 m^2/g 和 732 m^2/g。而在强酸溶液（王水，浓硫酸，浓硝酸，浓盐酸）中的稳定性也再一次证明了有机膦酸锆 MOFs 化合物具有极好的化学稳定性。将制备得到的有机膦酸锆 MOFs 化合物作为吸附剂，用于铀酰离子的去除，发现即便在竞争离子 10 倍过量的条件下，有机膦酸锆 MOFs 对于铀酰离子的去除仍然具有极好的选择性。更重要是的，即使在 pH=1 的高酸度条件下，SZ-2 和 SZ-3 仍能够有效将溶液中的铀酰离子去除掉 60%以上。通过对比吸附铀酰离子前后的荧光光谱数据，也进一步表明有机膦酸锆 MOFs 材料对于铀酰具有吸附性质。通过 EXAFS，对吸附铀酰的材料和水溶液中的铀酰离子进行比较，表明该材料对铀酰的吸附为离子交换机制。进一步通过分子动力学模拟有机膦酸锆 MOFs 去除铀酰的过程可以发现，该 MOFs 结构中的膦酸锆节点通过与水合铀酰离子赤道面的配位水分子形成丰富的氢键相互作用，从而可有效去除水合铀酰离子，也揭示了该类材料对铀酰离子极好的选择性根源[208]。该工作也是分子动力学模拟在揭示放射性核素吸附材料工作机理方面的一个开创性工作。

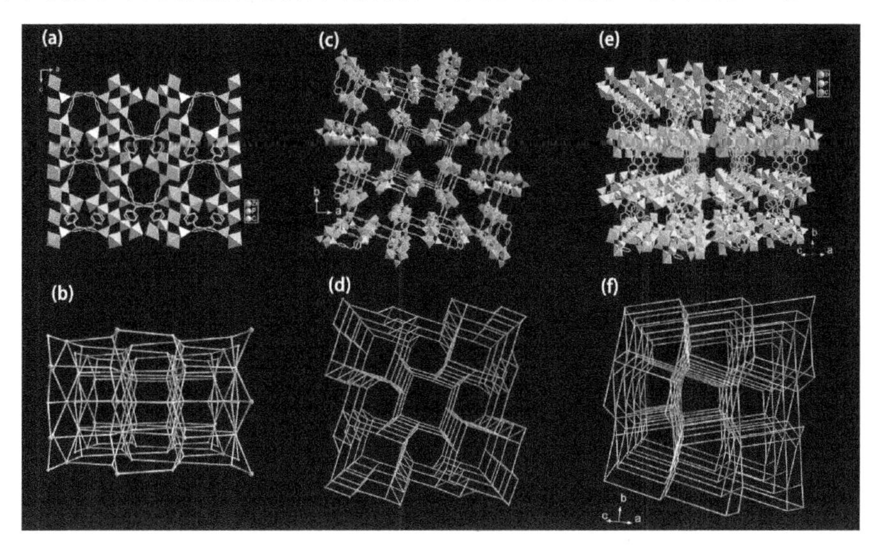

图 12-26　有机膦酸锆 MOFs 单晶结构及拓扑[208]

5.5　稀土金属有机骨架材料高效吸附和检测水体中低浓度铀酰离子

利用高稳定性、高量子产率（44.7%）的功能化 Tb-MOFs 材料对铀酰的选择性吸附作用将环境中微量的铀酰离子富集到孔道中，与孔道内含量丰富，分布有序的 N 原子配位点直接配位进而对 MOF 骨架的荧光产生强烈的淬灭作用（图 12-27）。利用吸附和荧光淬灭的对应关系可以对环境中铀酰含量做出定量探测，并且基于材料对铀酰的高选择性吸附作用和荧光的选择性淬灭的特性，即使在高浓度的竞争金属离子存在条件下（500 μg/mL），依然能选择性地对铀酰做出荧光响应。此材料成功地应用于离子强度较高的独墅湖水体、渤海海水和饮用水中铀酰浓度的探测，其检测下限分别达到了 0.014 μg/mL、0.0035 μg/mL、9.05×10^{-4} μg/mL，均低于美国 EPA 标准对饮用水中铀酰浓度的规定。此种方法也被成功应用于对人工添加的铀污染的渤海水中铀酰含量的定量测定。其测试结果与 ICP-MS 测试结果偏差保持在 6.8%以内，直接印证了其在真实水体环境中的应用潜力[209]。本工作也为理性合成 MOFs 荧光材料用于放射性核素等重金属离子的检测提供了新思路。

5.6　阳离子金属有机骨架材料高效分离和固定 TcO_4^-

在高放废物长期储存的过程中，裂变产物 ^{99}Tc 会对环境产生极大的影响：它的半衰期长，通常以 TcO_4^- 形式存在于水溶液中，水溶性高，由于其带负电荷，所以很难被一般的矿物或岩石吸附，容易随着地下水的迁移而进入生物圈中，从而给环境造成污染，因此研究材料对 ^{99}Tc 的吸附固定是十分有必要的。通过将过渡金属离子与四齿含氮中性配体构建出这种具有八重穿插结构的阳离子 MOF 材料（SCU-100），该三维阳离子 MOF 材料具有 6.9Å 的一维孔道内含弱配位的 NO_3^- 作为抗衡离子。通过水、热和辐照稳定性的测试发现 SCU-100 具有良好的水、热稳定性且具有强耐辐照性能（图 12-28），体现出该材料能够适用于 ^{99}Tc 的去除。吸附实验结果表明，该材料对 ReO_4^-（TcO_4^-的类似物）吸附容量高达 541 mg/g（K_d 值高达 1.9×10^5 mL/g），与其他报道的无机阳离子材料相比，该材料吸附效果更好，动力学更快，30min

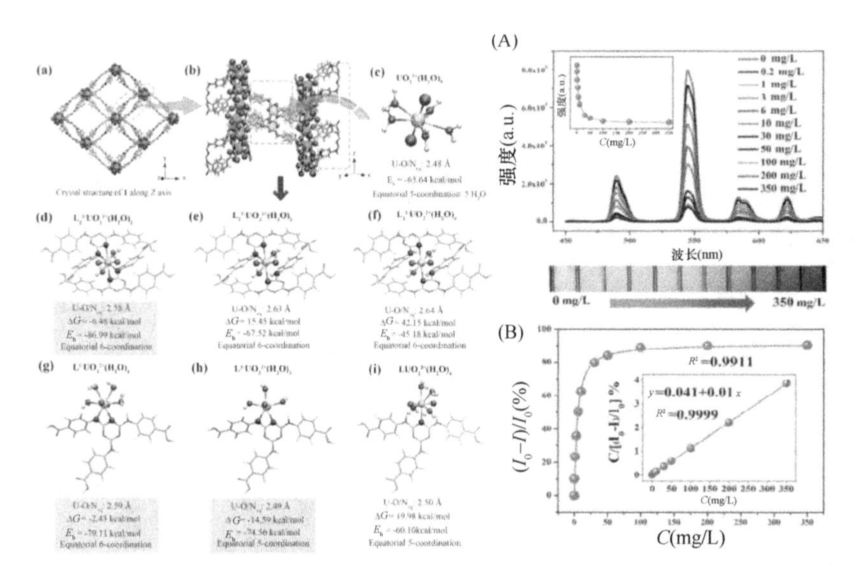

图 12-27 Tb-MOFs 材料荧光检测铀酰的性能及机制

内能将溶液中的 TcO_4^-去除 95%以上。模拟汉福德废液体系的吸附结果显示，在相同条件下 SCU-100 材料对 TcO_4^-的去除比报道的首个能够解决放射性锝问题的纯无机阳离子骨架材料 NDTB 高 70%左右，且动力学更快。同时，当大量竞争离子如 NO_3^-、SO_4^{2-}、CO_3^{2-}和 PO_4^{3-}存在时，SCU-100 能选择性地去除 TcO_4^-。β 射线辐照后的样品，其结构以及吸附性能都能够保持不变。通过对吸附前后晶体的解析，我们发现 SCU-100 对 ReO_4^-的吸附是一个单晶到单晶的转化过程（图 12-28）。同时发现这种具有金属空位以及氢键网络的结构是 SCU-100 对 ReO_4^-高选择性、高吸附容量去除的内在原因。本工作也表明这类阳离子 MOFs 材料对去除乏废料后处理中放射性阴离子具有潜在的应用性[210]。通过将过渡金属离子 Ag^+与联吡啶配体自组装可形成一维链状阳离子晶体材料（SBN）。吸附实验结果表明，该材料对 ReO_4^-（TcO_4^-的替代物）吸附容量高达 786 mg/g，明显优于商用阴离子交换树脂材料及其他已报道的无机杂化材料。通过 X 射线单晶衍射技术阐释了 SBN 高效固定 TcO_4^-的机制，是一个单晶到单晶转化过程。交换 ReO_4^-之后，结构转化为另一种一维链状晶体（SBR），吡啶链的排列方式由十字交叉排列转化为平行排列。SBR 作为一种潜在的 TcO_4^-的固化材料，甚至在 NO_3^-、SO_4^{2-}、CO_3^{2-}和 PO_4^{3-}等环境中广泛存在的阴离子大量过量的条件

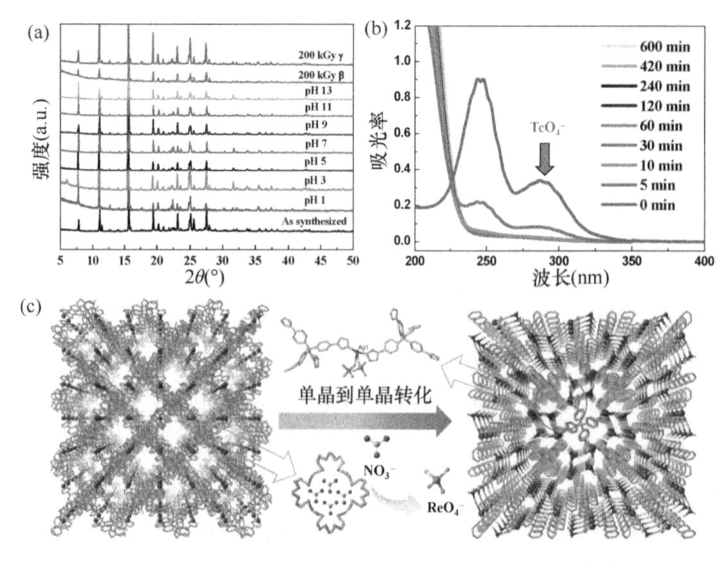

图 12-28 SCU-100 的稳定性及高效分离 TcO_4^-[211]

下，SBR 仍不析出 ReO_4^-。理论计算结果表明，SBR 中的 ReO_4^-对骨架的结合能比 SBN 中的 NO_3^-对骨架的结合能大 35.61 kcal/mol。在 SBR 的结构中，TcO_4^-/ReO_4^-不仅与不饱和配位的 Ag^+形成很强的 Ag—O—Re 的键，而且还与吡啶环上的氢形成了非常致密的氢键网络。结合第一性原理计算的结果，形成的 Ag—O—Re 的键是 ReO_4^-相比于 NO_3^-对骨架结合能更强的主要原因。实验分析测试结果也表明，SBR 是目前报道的溶度积常数最小的 TcO_4^-/ReO_4^-盐，且其在 380℃下结构仍能保持稳定。本工作表明了 SBR 可作为一种潜在的高效的 TcO_4^-的固定材料[211]。

5.7　无机阳离子骨架材料高效分离 SeO_3^{2-}和 SeO_4^{2-}

^{79}Se 是长寿命的核裂变产物，其半衰期为 2.95×10^5a，由于其化学毒性和长期潜在的放射性危害而备受关注。我国饮用水以及工农业生产过程中可排放废水中硒含量也受到严格控制，因此开发有效的硒污染处理方法具有重要意义。最近首次报道了利用二维无机阳离子骨架材料 $Y_2(OH)_5Cl\cdot1.5H_2O$ 高效去除 SeO_3^{2-}/SeO_4^{2-}的研究工作。批实验的结果表明这种层状阳离子骨架材料是目前报道的材料中对 SeO_3^{2-}/SeO_4^{2-}最有效的吸附剂，吸附容量高达~200 mg/g Se(IV)和~125 mg/g Se(VI)（图 12-29）。同时对材料的选择性以及重复利用性能也做了研究与评估，发现在低浓度时，竞争离子（NO_3^-，Cl^-，CO_3^{2-}，SO_4^{2-}和 HPO_4^{2-}）过量 20 倍的条件下，吸附后水中剩余 SeO_3^{2-}/SeO_4^{2-}浓度可达到最严格的饮用水排放标准。同时结合 EDS、FT-IR、Raman、PXRD 和 EXAFS 等表征手段对吸附前后的材料进行表征，揭示了此二维无机阳离子骨架材料对 SeO_4^{2-}的吸附是一个离子交换过程，而对 SeO_3^{2-}的吸附是离子交换和内层络合的机理，且以双齿双核的方式络合（图 12-29）[212]。本工作表明了 $Y_2(OH)_5Cl\cdot1.5H_2O$ 是一种非常高效的 SeO_3^{2-}/SeO_4^{2-}吸附材料，在核废料处理处置中具有潜在的应用前景。

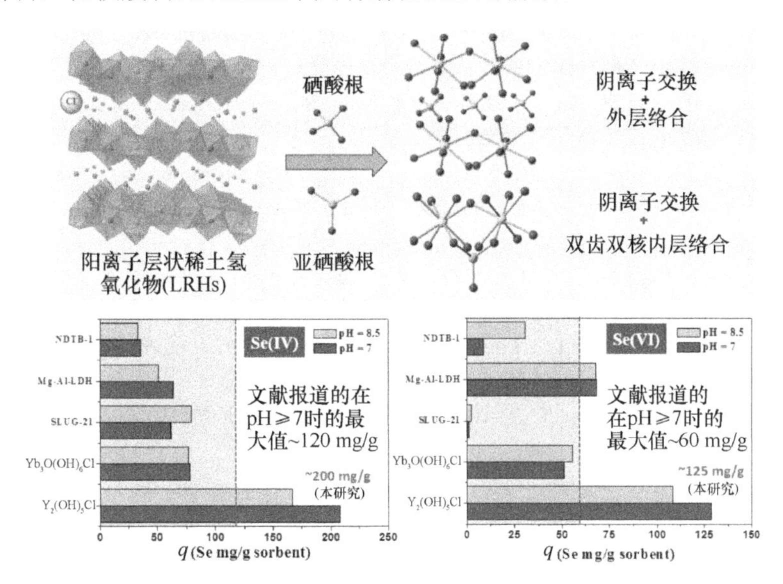

图 12-29　$Y_2(OH)_5Cl\cdot1.5H_2O$ 高效去除 SeO_3^{2-}/SeO_4^{2-}的性能及机理[212]

6　展　　望

本章对几类常见纳米材料在放射性污染治理中的应用进行了较详细的综述，以放射性核素的萃取分离、高效富集和合理治理为目标牵引，根据需求设计、构建新型的材料并对其进行合适的修饰，对特定的放射性核素具有选择性的“俘获”功能，研究放射性核素与材料的相互作用机制，进而开发用于放射性核素的快速分离富集器件，为放射性污染治理找到一条有效途径，是一个方兴未艾的研究领域，也引起各国极大的关注。然而，由于固-液界面反应十分复杂，目前主要围绕单一放射性核素体系进行研究。对多种放射性核素共存的竞争作用缺少系统研究。另外，在实际的地下环境中，往往存在大量的矿物质、腐殖酸、微生物等，这些环境介质具有丰富的功能基团，可以与材料及放射性核素强烈作用，因此研究放射性核素与环境介质和纳米材料之间的交互作用现象，更具有实际意义。此外，对纳米材料结构与性能以及放射性核素微观结构、化学形态之间内在关联规律及作用机制，有待于进一步研究。这些科学问题都是当前国内外放射性污染治理的重要研究热点，我们期望在国内外学者的共同努力下，放射性核素污染治理的研究在不久的将来会更上一层楼。

参 考 文 献

[1] Rao CNR, Sood AK, Subrahmanyam KS, Govindaraj A. Graphene: the new two-dimensional nanomaterial. Angew. Chem. Int. Ed., 2009, 48: 7752-7777.

[2] Wang X, Fan Q, Chen Z, Wang Q, Li J, Hobiny A, Alsaedi A, Wang X. Surface modification of graphene oxides by plasma technique and their application for environmental pollution cleanup. The Chemical Record, 2016, 16: 295−318.

[3] Machida M, Mochimaru T, Tatsumoto H. Lead(II) adsorption onto the graphene layer of carbonaceous materials in aqueous solution. Carbon, 2006, 44: 2681-2688.

[4] Zhao G, Li J, Ren X, Chen C, Wang X. Few-layered graphene oxide nanosheets as superior sorbents for heavy metal ion pollution management. Environ. Sci. Technol., 2011, 45: 10454-10462.

[5] Slesarev Alexander Sergeyevich. “Use of graphene oxide for sorption of radionuclides and other cations and synthesis of graphene-based nanoribbons.” (2015) PhD diss. Rice University, http://hdl.handle.net/1911/88161.

[6] Romanchuk AY, Slesarev AS, Kalmykov SN, Kosynkin DV, Tour JM. Graphene oxide for effective radionuclide removal. Phys. Chem. Chem. Phys., 2013, 15: 2321-2327.

[7] Lingamdinne LP, Koduru JR, Roh H, Choi YL, Chang YY, Yang JK. Adsorption removal of Co(II) from wastewater using graphene oxide. Hydrometallurgy, 2016, 165: 90-96.

[8] Lingamdinne LP, Choi YL, Kim IS, Yang JK, Koduru JR, Chang YY. Preparation and characterization of porous reduced graphene oxide based inverse spinel nickel ferrite nanocomposite for adsorption removal of radionuclides. J. Hazard. Mater., 2017, 326: 145-156.

[9] Sitko R, Turek E, Zawisza B, Malicka E, Talik E, Heimann J, Gagor A, Feist B, Wrzalik R. Adsorption of divalent metal ions from aqueous solutions using graphene oxide. Dalton Trans., 2013, 42: 5682-5689.

[10] Xie Y, Helvenston EM, Shuller-Nickles LC, Powell BA. Surface complexation modeling of Eu(III) and U(VI) interactions with graphene oxide. Environ. Sci. Technol., 2016, 50: 1821-1827.

[11] Pan N, Deng JG, Guan DB, Jin YD, Xia CQ. Adsorption characteristics of Th(IV) ions on reduced graphene oxide from aqueous solutions. Appl. Surf. Sci., 2013, 287: 478-483.

[12] Wang C, Li Y, Liu C. Sorption of uranium from aqueous solutions with graphene oxide. J. Radioanal. Nucl. Chem., 2015, 304: 1017-1025.

[13] Li Y, Wang C, Guo Z, Liu C, Wu W. Sorption of thorium(IV) from aqueous solutions by graphene oxide. J. Radioanal. Nucl. Chem., 2014, 299: 1683-1691.

[14] Sun Y, Wang Q, Chen C, Tan X, Wang X. Interaction between Eu(III) and graphene oxide nanosheets investigated by batch and extended X-ray absorption fine structure spectroscopy and by modeling techniques. Environ. Sci. Technol., 2012, 46: 6020-6027.

[15] Zhao G, Wen T, Yang X, Yang S, Liao J, Hu J, Shao D, Wang X. Preconcentration of U(VI) ions on few-layered graphene oxide nanosheets from aqueous solutions. Dalton Trans., 2012, 41: 6182-6188.

[16] Wang X, Fan Q, Yu S, Chen Z, Ai Y, Sun Y, Hobiny A, Alsaedi A, Wang X. High sorption of U(VI) on graphene oxides studied by batch experimental and theoretical calculations. Chem. Eng. J., 2016, 287: 448-455.

[17] Yan S, Wang X, Dai S, Wang X, Alshomrani A, Hayat T, Ahmad B. Investigation of ^{90}Sr(II) sorption onto graphene oxides

studied by macroscopic experiments and theoretical calculations. J. Radioanal. Nucl. Chem., 2016, 308: 721-732.
[18] Chen Y, Zhang W, Yang S, Hobiny A, Alsaedi A, Wang X. Understanding the adsorption mechanism of Ni(II) on graphene oxides by batch experiments and density functional theory studies. Sci. China Chem., 2016, 59: 412-419.
[19] Cheng W, Ding C, Wu Q, Wang X, Sun Y, Shi W, Hayat T, Alsaedi A, Chai Z, Wang X. Mutual effect of U(VI) and Sr(II) with graphene oxides: Evidence from EXAFS and theoretical calculations. Environ. Sci. Nano, 2017, 4: 1124-1131.
[20] Liu X, Huang Y, Duan S, Wang Y, Li J, Chen Y, Hayat T, Wang X. Graphene oxides with different oxidation degrees for Co(II) ion pollution management. Chem. Eng. J., 2016, 302: 763-772.
[21] Ding C, Cheng W, Sun Y. Wang, X. Determination of chemical affinity of graphene oxide nanosheets with radionuclides investigated by macroscopic, spectroscopic and modeling techniques. Dalton Trans., 2014, 43: 3888-3896.
[22] Tan L, Wang S, Du W, Hu T. Effect of water chemistries on adsorption of Cs(I) onto graphene oxide investigated by batch and modeling techniques. Chem. Eng. J., 2016, 292: 92-97.
[23] Hu T, Ding S, Deng H. Application of three surface complexation models on U(VI) adsorption onto graphene oxide. Chem. Eng. J., 2016, 289: 270-276.
[24] Li X, Tang X, Fang Y. Using graphene oxide as a superior adsorbent for the highly efficient immobilization of Cu(II) from aqueous solution. J. Mol. Liq., 2014, 199: 237-243.
[25] Li X, Yu M, Lv Q, Tan Y. Sequestration of Ni(II) onto graphene oxide from synthetic wastewater as affected by coexisting constituents. Desalination Water Treat., 2016, 57: 20904-20914.
[26] Li X, Zhao K, You C, Pan H, Tang X, Fang Y. Impact of contact time, pH, ionic strength, soil humic substances, and temperature on the uptake of Pb(II) onto graphene oxide. Sep. Sci. Technol., 2017, 52: 987-996.
[27] Hu B, Hu Q, Li X, Pan H, Tang X, Chen C, Huang C. Rapid and highly efficient removal of Eu(III) from aqueous solutions using graphene oxide. J. Mol. Liq., 2017, 229: 6-14.
[28] Hu B, Hu Q, Li X, Pan H, Huang C, Chen C. Graphene oxide as a novel adsorbent for highly efficient removal of UO_2^{2+} from water. Desalination Water Treat., doi:10.5004/dwt.2017.20796.
[29] Ye F, Tang Y. The modeling evidences for Th(IV) sorption on graphene oxide. J. Radioanal. Nucl. Chem., 2016, 310: 565-571.
[30] Lujaniene G, Semcuk S, Kulakauskaite I, Mazeika K, Valiulis D, Juskenas R, Tautkus S. Sorption of radionuclides and metals to graphene oxide and magnetic graphene oxide. J. Radioanal. Nucl. Chem., 2016, 307: 2267-2275.
[31] Yang S, Chen C, Chen Y, Li J, Wang D, Wang X, Hu W. Competitive adsorption of Pb(II), Ni(II) and Sr(II) ions on graphene oxides: A combined experimental and theoretical study. ChemPlusChem, 2015, 80: 480-484.
[32] Sun Y, Wang X, Song W, Lu S, Chen C, Wang X. Mechanistic insights into the decontamination of Th(IV) on graphene oxide-based composites by EXAFS and modeling techniques. Environ. Sci. Nano, 2017, 4: 222-232.
[33] Sun Y, Yang S, Chen Y, Ding C, Cheng W, Wang X. Adsorption and desorption of U(VI) on functionalized graphene oxides: A combined experimental and theoretical study. Environ. Sci. Technol., 2015, 49: 4255-4262.
[34] Wu Q, Lan J, Wang C, Xiao C, Zhao Y, Wei Y, Chai Z, Shi W. Understanding the bonding nature of uranyl ion and functionalized graphene: A theoretical study. J. Phys. Chem. A, 2014, 118: 2149-2158.
[35] Wu Q, Lan J, Wang C, Zhao Y, Chai Z, Shi W. Understanding the interactions of neptunium and plutonium ions with graphene oxide: Scalar-relativistic DFT investigations. J. Phys. Chem. A, 2014, 118: 10273-10280.
[36] Sun Y Shao D, Chen C, Yang S, Wang X. Highly efficient enrichment of radionuclides on graphene oxide supported polyaniline. Environ. Sci. Technol., 2013, 47: 9904-9910.
[37] Hu R, Shao D, Wang X. Graphene oxide/polypyrrole composites for highly selective enrichment of U(VI) from aqueous solutions. Polymer Chem., 2014, 5: 6207-6215.
[38] Chen H, Shao D, Li J, Wang X. The uptake of radionuclides from aqueous solution by poly(amidoxime) modified reduced graphene oxide. Chem. Eng. J., 2014, 254: 623-634.
[39] Song W, Shao D, Lu S, Wang X. Simultaneous removal of uranium and humic acid by cyclodextrin modified graphene oxide nanosheets. Sci. China Chem., 2014, 57: 1291-1299.
[40] Shao D, Li J, Wang X. Poly(amidoxime)-reduced graphene oxide composites as adsorbents for the enrichment of uranium from seawater. Sci. China Chem., 2014, 57: 1449-1458.
[41] Song W, Wang X, Wang Q, Shao D, Wang X. Plasma induced grafting polyacrylamide on graphene oxide nanosheets for simultaneous removal of radionuclides. Phys. Chem. Chem. Phys., 2015, 17: 398-406.
[42] Cheng W, Wang M, Yang Z, Sun Y, Ding C. The efficient enrichment of U(VI) by graphene oxide-supported chitosan. RSC

Adv., 2014, 4: 61919-61926.
[43] Yang P, Liu Q, Liu J, Zhang H, Li Z, Li R, Liu L, Wang J. Bovine serum albumin-coated graphene oxide for effective adsorption of uranium(VI) from aqueous solutions. Ind. Eng. Chem. Res., 2017, 56: 3588-3598.
[44] Xiong Y, Cui X, Zhang P, Wang Y, Lou Z, Shan W. Improving Re(VII) adsorption on diisobutylamine-functionalized graphene oxide. ACS Sustainable Chem. Eng., 2017, 5: 1010-1018.
[45] Wang X, Lu S, Liu M. Effect of environmental conditions on the sorption of radiocobalt on titanate/graphene oxide composites. J. Radioanal. Nucl. Chem., 2015, 303: 2391-2398.
[46] Chen L, Lu S, Wu S, Zhou J, Wang X. Removal of radiocobalt from aqueous solutions using titanate/graphene oxide composites. J. Mol. Liq., 2015, 209: 397-403.
[47] Sun Y, Wang X, Ai Y, Yu Z, Huang W, Chen C, Hayat T, Alsaedi A, Wang X. Interaction of sulfonated graphene oxide with U(VI) studied by spectroscopic analysis and theoretical calculations. Chem. Eng. J., 2017, 310: 292-299.
[48] Chappa S, Singha Deb AK, Musharaf Ali Sk, Debnath AK, Aswal DK, Pandey AK. Change in the affinity of ethylene glycol methacrylate phosphate monomer and its polymer anchored on a graphene oxide platform toward uranium(VI) and plutonium(IV) ions. J. Phys. Chem. B, 2016, 120: 2942-2950.
[49] Cheng W, Ding C, Nie X, Duan T, Ding R. Fabrication of 3D macroscopic graphene oxide composites supported by montmorillonite for efficient U(VI) wastewater purification. ACS Sustainable Chem. Eng., 2017, 5: 5503-5511.
[50] Wen T, Wu X, Liu M, Xing Z, Wang X, Xu A. Efficient capture of strontium from aqueous solutions using graphene oxide-hydroxyapatite nanocomposites. Dalton Trans., 2014, 43: 7464-7472.
[51] Cheng H, Zeng K, Yu J. Adsorption of uranium from aqueous solution by graphene oxide nanosheets supported on sepiolite. J. Radioanal. Nucl. Chem., 2013, 298: 599-603.
[52] Pan N, Li L, Ding J, Li S, Wang R, Jin Y, Wang X, Xia C. Preparation of graphene oxide-manganese dioxide for highly efficient adsorption and separation of Th(IV)/U(VI). J. Hazard. Mater., 2016, 309: 107-115.
[53] Yu S, Wang X, Zhang R, Yang T, Ai Y, Wen T, Huang W, Hayat T, Alsaedi A, Wang X. Complex roles of solution chemistry on graphene oxide coagulation onto titanium dioxide: Batch experiments, spectroscopy analysis and theoretical calculation. Sci. Rep., 2017, 7: 39625.
[54] Zou Y, Wang X, Ai Y, Liu Y, Li J, Ji Y, Wang X. Coagulation behavior of graphene oxide on nanocrystallined Mg/Al layered double hydroxides: Batch experimental and theoretical calculation study. Environ. Sci. Technol., 2016, 50: 3658-3667.
[55] Yu S, Wang J, Song S, Sun K, Li J, Wang X, Chen Z, Wang X. One-pot synthesis of graphene oxide and Ni-Al layered double hydroxides nanocomposites for the efficient removal of U(VI) from wastewater. Sci. China Chem., 2017, 60: 415-422.
[56] Linghu W, Yang H, Sun Y, Sheng G, Huang Y. One-pot synthesis of LDH/GO composites as highly effective adsorbents for decontamination of U(VI). ACS Sustainable Chem. Eng., 2017, 5: 5608-5616.
[57] Li D, Zhang B, Xuan F. The sequestration of Sr(II) and Cs(I) from aqueous solutions by magnetic graphene oxides. J. Mol. Liq., 2015, 209: 508-514.
[58] Li D, Zhang B, Xuan F. The sorption of Eu(III) from aqueous solutions by magnetic graphene oxides: A combined experimental and modeling studies. J. Mol. Liq., 2015, 211: 203-209.
[59] Zhao Y, Li J, Zhang S, Chen H, Shao D. Efficient enrichment of uranium(VI) on amidoximated magnetite/graphene oxide composites. RSC Adv., 2013, 3: 18952-18959.
[60] Chen H, Li J, Zhang S, Ren X, Sun Y, Wen T, Wang X. Study on the acid-base surface property of the magnetite graphene oxide and its usage for the removal of radiostronium from aqueous solution. Radiochimica Acta, 2013, 101: 785-794.
[61] Li Y, Sheng G, Sheng J. Magnetite decorated graphene oxide for the highly efficient immobilization of Eu(III) from aqueous solution. J. Mol. Liq., 2014, 199: 474-480.
[62] Liu M, Chen C, Hu J, Wu X, Wang X. Synthesis of magnetite/grapheme oxide composite and application for cobalt(II) removal. J. Phys. Chem. C, 2011, 115: 25234-25240.
[63] Sheng G, Li Y, Yang X, Ren X, Yang S, Hu J, Wang X. Efficient removal of arsenate by a versatile magnetic graphene oxide composites. RSC Adv., 2012, 2: 12400-12407.
[64] Hu B, Qiu M, Hu Q, Sun Y, Sheng G, Hu J, Ma J. Decontamination of Sr(II) on magnetic polyaniline/graphene oxide composites: Evidence from experimental, spectroscopic and modeling investigation. ACS Sustainable Chem. Eng., DOI: 10.1021/acssuschemeng.7b01126.

[65] Cheng W, Jin Z, Ding C, Wang M. Simultaneous sorption and reduction of U(VI) on magnetite–reduced graphene oxide composites investigated by macroscopic, spectroscopic and modeling techniques. RSC Adv., 2015, 5: 59677-59685.

[66] Kadam AA, Jang J, Lee DS. Facile synthesis of pectin-stabilized magnetic graphene oxide prussian blue nanocomposites for selective cesium removal from aqueous solution. Bioresource Technol., 2016, 216: 391-398.

[67] Li DB, Cheng YY, Wu C, Li WW, Li N, Yang ZC, Tong ZH, Yu HQ. Selenite reduction by *Shewanella oneidensis* MR-1 is mediated by fumarate reductase in periplasm. Sci. Rep., 2014, 4, 3735: 1-7.

[68] Bone SE, Dynes JJ, Cliff J, Bargar JR. Uranium(IV) adsorption by natural organic matter in anoxic sediments. Proceedings of the National Academy of Sciences, USA, 2017, 114: 711-716.

[69] Dang DH, Novotnik B, Wang W, Georg RB, Evans RD. Uranium isotope fractionation during adsorption, (co)precipitation, and biotic reduction. Environ. Sci. Technol., 2016, 50: 12695-12704.

[70] Gillham RW, O'Hannesin SF. Enhanced degradation of halogenated aliphatics by zero valent iron. Ground Water, 1994, 32: 958-967.

[71] Cantrell K, Kaplan D, Wietsma T. Zero-valent iron for the in situ remediation of selected metals in groundwater. J. Hazard. Mater., 1995, 42: 201-212.

[72] Farrell J, Bostick WD, Jarabek RJ, Fiedor JN. Uranium removal from ground water using zero valet iron media. Ground Water, 1999, 37: 618-624.

[73] Gu B, Liang L, Dickey MJ, Yin X, Dai S. Reductive precipitation of uranium (VI) by zerovalent iron. Environ. Sci. Technol., 1998, 32: 3366-3373.

[74] Fiedor JN, Bostick WD, Jarabek RJ, Farrell J. Understanding the mechanism of uranium removal from groundwater by zerovalent iron using X-ray photoelectron spectroscopy. Environ. Sci. Technol., 1998, 32: 1466-1473.

[75] Simon FG, Segebade C, Hedrich M. Behaviour of uranium in iron-bearing permeable reactive barriers: investigation with ^{237}U as a radioindicator. Sci. Total Environ., 2003, 307, 1-3: 231-238.

[76] Noubactep C, Schöner A, Meinrath G. Mechanism of uranium removal from the aqueous solution by elemental iron. J. Hazard. Mater., 2006, 132: 202-212.

[77] Morrson SJ, Metzler DR, Carpenter CE. Uranium precipitation in a permeable reactive barrier by progressive irreversible dissolution of zerovalent iron. Environ. Sci. Technol., 2001, 35: 385-390.

[78] Morrison SJ, Mushovic PS, Niesen PL. Early breakthrough of molybdenum and uranium in a permeable reactive barrier. Environ. Sci. Technol., 2006, 40: 2018-2024.

[79] Sasaki K, Blowes DW, Ptacek CJ, Gould WD. Immobilization of Se(VI) in mine drainage by permeable reactive barriers: column performance. Appl. Geochem., 2008, 23: 1012-1022.

[80] Yoon I, Kim K, Bang S, Kim M. Reduction and adsorption mechanisms of selenate by zero-valent iron and related iron corrosion. Appl. Catal. B: Environ., 2011, 104: 185-192.

[81] Tang C, Huang Y, Zhang Z, Chen J, Zeng H, Huang Y. Rapid removal of selenate in a zero-valent iron/Fe_3O_4/Fe^{2+} synergetic system. Appl. Catal. B: Environ., 2016, 184: 320-327.

[82] Tang C, Huang Y, Zeng H, Zhang Z. Reductive removal of selenate by zero-valent iron: The roles of aqueous Fe^{2+} and corrosion products, and selenate removal mechanisms. Water Res., 2014, 67: 166-274.

[83] Huang Y, Tang C, Zeng H. Removing molybdate from water using a hybridized zero-valent iron/magnetite/Fe(II) treatment system. Chem. Eng. J., 2012, 200-202: 257-263.

[84] Liang L, Guan X, Shi Z, Li J, Wu Y, Tratnyek PG. Coupled effects of aging and weak magnetic fields on sequestration of selenite by zero-valent iron. Environ. Sci. Technol., 2014, 48: 6326-6334.

[85] Liang L, Yang W, Guan X, Li J, Xu Z, Wu J, Huang Y, Zhang X. Kinetics and mechanisms of pH-dependent Se(IV) removal by zero valent iron. Water Res., 2013, 47: 5846-5855.

[86] Liang L, Guan X, Huang Y, Ma J, Sun X, Qiao J, Zhou G. Efficient selenate removal by zero-valent iron in the presence of weak magnetic field. Sep. Purif. Technol., 2015, 156: 1064-1072.

[87] Liang L, Sun W, Guan X, Huang Y, Choi W, Bao H, Li L, Jiang Z. Weak magnetic field significantly enhances selenite removal kinetics by zero valent iron. Water Res., 2014, 1: 371-380.

[88] Xu H Sun Y, Li J, Li F, Guan X. Aging of zerovalent iron in synthetic groundwater: X-ray photoelectron spectroscopy depth profiling characterization and depassivation with uniform magnetic field. Environ. Sci. Technol., 2016, 50: 8214-8222.

[89] Sun Y, Hu Y, Huang T, Li J, Qin H, Guan X. Combined effect of weak magnetic fields and anions on arsenite sequestration by zerovalent iron: Kinetics and mechanisms. Environ. Sci. Technol., 2017, 51: 3742-3750.

[90] Yan S, Chen Y, Xiang W, Bao Z, Liu C, Deng B. Uranium(VI) reduction by nanoscale zero-valent iron in anoxic batch systems: The role of Fe(II) and Fe(III). Chemosphere, 2014, 117: 625-630.

[91] Yan S, Hua B, Bao Z, Liu C, Deng B. Uranium(VI) removal by nanoscale zerovalent iron in anoxic batch systems. Environ. Sci. Technol., 2010, 44: 7783-7789.

[92] Olegario JT, Yee N, Miller M, Sczepaniak J, Manning B. Reduction of Se(VI) to Se(-II) by zerovalent iron nanoparticle suspensions. J. Nanopart. Res., 2010, 12: 2057-2068.

[93] Riba O, Scott T, Ragnardottir K, Allen G. Reaction mechanism of uranyl in the presence of zero-valent iron nanoparticles. Geochim. Cosmochim. Acta, 2008, 72: 4047-4057.

[94] Crane RA, Dickinson M, Popescu IC, Scott TB. Magnetite and zero-valent iron nanoparticles for the remediation of uranium contaminated environmental water. Water Res., 2011, 45: 2931-2942.

[95] Crane RA, Dickinson M, Scott TB. Nanoscale zero-valent iron particles for the remediation of plutonium and uranium contaminated solutions. Chem. Eng. J., 2015, 262: 319-325.

[96] Crane RA, Pullin H, Scott TB. The influence of calcium, sodium and bicarbonate on the uptake of uranium onto nanoscale zero-valent iron particles. Chem. Eng. J., 2015, 277: 252-259.

[97] Scott TB, Popescu IC, Crane RA, Noubactep C. Nano-scale metallic iron for the treatment of solutions containing multiple inorganic contaminants. J. Hazard. Mater., 2011, 186: 280-287.

[98] Dickinson M, Scott TB. The application of zero-valent iron nanoparticles for the remediation uranium-contaminated waste effluent. J. Hazard. Mater., 2010, 178: 171-179.

[99] Ling L, Zhang W. Enrichment and encapsulation of uranium with iron nanoparticle. J. Am. Chem. Soc., 2015, 137: 2788-2791.

[100] Ling L, Pan B, Zhang W. Removal of selenium from water with nanoscale zero-valent iron: Mechanisms of intraparticle reduction of Se(IV). Water Res., 2015, 71: 274-281.

[101] Phenrat T, Saleh N, Sirk K, Tilton RD, Lowry GV. Aggregation and sedimentation of aqueous nanoscale zerovalent iron dispersions. Environ. Sci. Technol., 2007, 41: 284-290.

[102] Fan D, Anitori RP, Tebo BM, Tratnyek PG, Pacheco JSL, Kukkadapu RK, Engelhard MH, Bowden ME, Kovarik L, Arey BW. Reductive sequestration of pertechnetate ($^{99}TcO_4^-$) by nano zerovalent iron (nZVI) transformed by abiotic sulfide. Environ. Sci. Technol., 2013, 47: 5302-5310.

[103] Fan D, Anitori RP, Tebo BM, Tratnyek PG, Pacheco JSL, Kukkadapu RK, Kovarik L, Engelhard MH, Bowden ME. Oxidative remobilization of technetium sequestered by sulfide-transformed nano zerovalent iron. Environ. Sci. Technol., 2014, 48: 7409-7417.

[104] Ding C, Cheng W, Sun Y, Wang X. Effects of *Bacillus subtilis* on the reduction of U(VI) by nano-Fe^0. Geochim. Cosmochim. Acta, 2015, 165: 86-107.

[105] Popescu IC, Filip P, Humelnicu D, Humelnicu I, Scott TB, Crane RA. Removal of uranium(VI) from aqueous systems by nanoscale zero-valent iron particles suspended in carboxy-methyl cellulose. J. Nucl. Mater., 2013, 443: 250-255.

[106] Shao D, Wang X, Wang X, Hu S, Hayat T, Alsaedi A, Li J, Wang S, Hu J, Wang X. Zero valent iron/poly(amidoxime) nanocomposite for efficient separation and reduction of U(VI). RSC Adv., 2016, 6: 52076-52081.

[107] Liu H, Qian T, Zhao D. Reductive immobilization of perrhenate in soil and groundwater using starch-stabilized ZVI nanoparticles. Chinese Sci. Bull, 2013, 58: 275-281.

[108] Ponder SM, Darab JG, Bucher J, Caulder D, Craig I, Davis L, Edelstein N, Lukens W, Nitsche H, Rao L, Shuh DK, Mallouk TE. Surface chemistry and electrochemistry of supported zerovalent iron nanoparticles in the remediation of aqueous metal contaminants. Chem. Mater., 2001, 13: 479-486.

[109] Darab JG, Amonette, A.B. Burke, D.S.D. Orr, R.D. Ponder, S.M. Schrick, B. Mallouk, T.E. Lukens, W.W. Caulder, D.L. Shuh DK. Removal of pertechnetate from simulated nuclear waste streams using supported zerovalent iron. Chem. Mater., 2007, 19: 5703-5713.

[110] Xu J, Li Y, Jing C, Zhang H, Ning Y. Removal of uranium from aqueous solution using montmorillonite-supported nanoscale zero-valent iron. J. Radioanal. Nucl. Chem., 2014, 299: 329-336.

[111] Jing C, Li Y, Cui R, Xu J. Illite-supported nanoscale zero-valent iron for removal of ^{238}U from aqueous solution: characterization, reactivity and mechanism. J. Radioanal. Nucl. Chem., 2015, 304: 859-865.

[112] Crane RA, Scott TB. The removal of uranium onto carbon-supported nanoscale zero-valent iron particles. J. Nano Res., 2015, 16: 1-13.

[113] Li J, Chen C, Zhang R, Wang X. Nanoscale zero-valent iron particles supported on reduced graphene oxides by using a plasma technique and their application for removal of heavy-metal ions. Chem. Asian. J., 2015, 10: 1410-1417.
[114] Li J, Chen C, Zhang R, Wang X. Reductive immobilization of Re(VII) by graphene modified nanoscale zero-valent iron particles using a plasma technique. Sci. China Chem., 2016, 59: 150-158.
[115] Li J, Chen C, Zhu K, Wang X. Nanoscale zero-valent iron particles modified on reduced graphene oxides using a plasma technique for Cd(II) removal. J. Taiwan Inst. Chem. E., 2016, 59: 389-394.
[116] Sun Y, Ding C, Cheng W, Wang X. Simultaneous adsorption and reduction of U(VI) on reduced graphene oxide-supported nanoscale zerovalent iron. J. Hazard. Mater., 2014, 280: 399-408.
[117] Sheng G, Alsaedi A, Shammakh W, Monaquel S, Sheng J, Wang X, Li H, Huang Y. Enhanced sequestration of selenite in water by nanoscale zero valent iron immobilization on carbon nanotubes by a combined batch, XPS and XAFS investigation. Carbon, 2016, 99: 123-130.
[118] Sheng G, Yang P, Tang Y, Hu Q, Li H, Ren, X, Hu B, Wang X, Huang Y. New insights into the primary roles of diatomite in the enhanced sequestration of UO_2^{2+} by zerovalent iron nanoparticles: An advanced approach utilizing XPS and EXAFS. Appl. Catal. B: Environ., 2016, 193: 189-197.
[119] Sheng G, Shao X, Li Y, Li J, Dong H, Cheng W, Gao X, Huang Y. Enhanced removal of U(VI) by nanoscale zerovalent iron supported on Na-bentonite and an investigation of mechanism. J. Phys. Chem. A, 2014, 118: 2952-2958.
[120] Sheng G, Ma X, Linghu W, Chen Z, Hu J, Alsaedi A, Shammakh W, Monaquel S, Sheng J. Enhanced trapping of Ni(II) ions by diatomite-supported nanoscale zerovalent iron from aqueous solution. Desalination Water Treat., 2017, 70: 183-189.
[121] Hu B, Chen G, Jin C, Hu J, Huang C, Sheng J, Sheng G, Ma J, Huang Y. Macroscopic and spectroscopic studies of the enhanced scavenging of Cr(VI) and Se(VI) from water by titanate nanotube anchored nanoscale zero-valent iron. J. Hazard. Mater., 2017, 336: 214-221.
[122] Hu B, Ye F, Jin C, Ma X, Huang C, Sheng G, Ma J, Wang X, Huang Y. The enhancement roles of layered double hydroxide on the reductive immobilization of selenate by nanoscale zero valent iron: Macroscopic and microscopic approaches. Chemosphere, 2017, 184: 408-416.
[123] Sheng G, Tang Y, Linghu W, Wang L, Li J, Li H, Wang X, Huang Y. Enhanced immobilization of ReO_4^- by nanoscale zerovalent iron supported on layered double hydroxide via an advanced XAFS approach: Implications for TcO_4^- sequestration. Appl. Catal. B: Environ., 2016, 192: 268-276.
[124] Hu B, Ye F, Ren X, Zhao D, Sheng G, Li H, Ma J, Wang X, Huang Y. X-ray absorption fine structure study of enhanced sequestration of U(VI) and Se(IV) by montmorillonite decorated zerovalent iron nanoparticles. Environ. Sci. Nano, 2016, 3: 1460-1472.
[125] Sheng G, Hu J, Li H, Li J, Huang Y. Enhanced sequestration of Cr(VI) by nanoscale zero-valent iron supported on layered double hydroxide by batch and XAFS study. Chemosphere, 2016, 148: 227-232.
[126] 李芳柏, 王旭刚, 周顺桂, 刘承帅. 红壤胶体铁氧化物界面有机氯的非生物转化研究进展. 生态环境, 2006, 15 (6): 1343-1351.
[127] 李俊, 谢丽, 盛杰, 栾富波, 周琪. Fe(II)/铁氧化物表面结合铁系统还原有机污染物的研究进展. 地球科学进展, 2009, 24 (1): 25-31.
[128] 栾富波, 谢丽, 李俊, 周琪. 不同 pH 下铁氧化物表面结合铁系统还原硝基苯的研究. 环境科学, 2009, 30 (7): 1937-1941.
[129] Li D, Kaplan DI. Sorption coefficients and molecular mechanisms of Pu, U, Np, Am and Tc to Fe (hydr)oxides: A review. J. Hazard. Mater., 2012, 243: 1-18.
[130] Roden EE. Fe(III) oxide reactivity toward biological versus chemical reduction. Environ. Sci. Technol., 2003, 37: 1319-1324.
[131] Yang L, Steefel CI, Marcus MA, Bargar JR. Kinetics of Fe(II)-catalyzed transformation of 6-line ferrihydrite under anaerobic low conditions. Environ. Sci. Technol., 2010, 44: 5469-5475.
[132] Plymale AE, Fredrickson JK, Zachara JM, Dohnalkova AC, Heald SM, Moore DA, Kennedy DW, Marshall MJ, Wang C, Resch CT, Nachimuthu P. Competitive reduction of pertechnetate ($^{99}TcO_4^-$) by dissimilatory metal reducing bacteria and biogenic Fe(II). Environ. Sci. Technol., 2011, 45: 951-957.
[133] Fredickson JK, Zachara JM, Kenmedy DW, Dong H, Onstott T, Hinman N, Li S. Biogenic iron mineralization accompanying the dissimilatory reduction of hydrous ferric oxide by a ground water bacterium. Geochim. Cosmochim.

Acta, 1998, 62: 3239-3257.
[134] 司友斌, 王娟. 异化铁还原对土壤中重金属形态转化及其有效性影响. 环境科学, 2015, 36 (9): 3533-3542.
[135] Cui DQ, Eriksen TE. Reduction of pertechnetate by ferrous iron in solution: Influence of sorbed and precipitated Fe(II). Environ. Sci. Technol., 1996, 30: 2259-2262.
[136] Cui DQ, Eriksen TE. Reduction of pertechnetate in solution by heterogeneous electron transfer from Fe(II)-containing geological material. Environ. Sci. Technol., 1996, 30: 2263-2269.
[137] Charlet L, Silvester E, Liger E. N-compound reduction and actinide immobilization in surficial fluids by Fe(II): The surface Fe(III)OFe(II)OH degrees species, as major reductant. Chem. Geol., 1998, 151: 85-93.
[138] Liger E, Charlet L, Van Cappellen P. Surface catalysis of uranium(VI) reduction by iron(II). Geochim. Cosmochim. Acta, 1999, 63: 2939-2955.
[139] Jeon BH, Dempsey BA, Burgos WD, Barnett MO, Roden EE. Chemical reduction of U(VI) by Fe(II) at the solid-water interface using natural and synthetic Fe(III) oxides. Environ. Sci. Technol., 2005, 39: 5642-5649.
[140] Zeng H, Giammar DE. U(VI) reduction by Fe(II) on hematite nanoparticles. J. Nanopart. Res., 2011, 13: 3741-3754.
[141] Jang JH, Dempsey BA, Burgos WD. Reduction of U(VI) by Fe(II) in the presence of hydrous ferric oxide and hematite: Effects of solid transformation, surface coverage, and humic acid. Water Res., 2008, 42: 2269-2277.
[142] Boland DD, Collins RN, Payne TE, Waite TD. Effect of amorphous Fe(III) oxide transformation on the Fe(II)-mediated reduction of U(VI). Environ. Sci. Technol., 2011, 45: 1327-1333.
[143] Boland DD, Collins RN, Glover CJ, Payne TE, Waite TD. Reduction of U(VI) by Fe(II) during the Fe(II)-accelerated transformation of ferrihydrite. Environ. Sci. Technol., 2014, 48: 9086-9093.
[144] Stewart BD, Cismasu AC, Williams KH, Peyton BM, Nico PS. Reactivity of uranium and ferrous iron with natural iron oxyhydroxides. Environ. Sci. Technol. 2015, 49, 10357-10365.
[145] Massey MS, Lezama-Pacheco JS, Jones ME, Ilton ES, Cerrato JM, Bargar JR, Fendorf S. Competing retention pathways of uranium upon reaction with Fe(II). Geochim. Cosmochim. Acta, 2014, 142: 166-185.
[146] Um W, Chang H, Icenhower J, Lukens W, Serne R, Qafoku N, Westsik J, Buck E, Steven S. Immobilization of 99-Technetium (VII) by Fe(II)-goethite and limited reoxidation. Environ. Sci. Technol., 2011, 45: 4904-4913.
[147] Peretyazhko T, Zachara J, Heald S, Jeon B, Kukkadapu R, Liu C, Moore D, Resch C. Heterogeneous reduction of Tc(VII) by Fe(II) at the solid-water interface. Geochim. Cosmochim. Acta, 2008, 72: 1521-1539.
[148] Peretyazhko T, Zachara J, Heald S, Kukkadapu R, Liu C, Plymale A, Resch C. Reduction of Tc(VII) by Fe(II) sorbed on Al(hydr)oxides. Environ. Sci. Technol., 2008, 42: 5499-5506.
[149] Chakraborty S, Bardelli F, Charlet L. Reactivities of Fe(II) on calcite: selenium reduction. Environ. Sci. Technol., 2010, 44: 1288-1294.
[150] Scheinost AC, Kirsch R, Banerjee D, Fernandez-Martinez A, Zaenker H, Funke H, Charlet L. X-ray absorption and photoelectron spectroscopy investigation of selenite reduction by Fe(II)-bearing minerals. J. Contam. Hydrol., 2008, 102: 228-245.
[151] 康明亮, 刘春立, 陈繁荣. Fe(II)-矿物对亚硒酸的还原作用. 中国科学:化学, 2013, 43 (5): 536-543.
[152] Felmy A, Moore D, Rosso K, Qafoku O, Rai D, Buck E, Ilton E. Heterogeneous reduction of PuO_2 with Fe(II): Importance of the Fe(III) reaction product. Environ. Sci. Technol., 2011, 45: 3952-3958.
[153] Silvester E, Charlet L, Tournassat C, Gehin A, Greneche JM, Liger E. Redox potential measurements and mossbauer spectrometry of Fe^{II} adsorbed onto Fe^{III} (oxyhydr)oxides. Geochim. Cosmochim. Acta, 2005, 69: 4801-4815.
[154] Williams AGB, Scherer MM. Spectroscopic evidence for Fe(II)-Fe(III) electron transfer at the Fe oxide-water interface. Environ. Sci. Technol., 2004, 38: 4782-4790.
[155] Larese-Casanova P, Scherer MM. Fe(II) sorption on hematite: New insights based on spectroscopic measurements. Environ. Sci. Technol., 2007, 41: 471-477.
[156] Cwiertny DM, Handler RM, Schaefer MV, Grassian V, Scherer MM. Interpreting nanoscale size-effects in aggregated Fe-oxide suspensions: reaction of Fe(II) with goethite. Geochim. Cosmochim. Acta, 2009, 72: 1365-1380.
[157] Handler RM, Beard BL, Johnson CM, Scherer MM. Atom exchange between aqueous Fe(II) and goethite: An Fe isotope tracer study. Environ. Sci. Technol., 2009, 43: 1102-1107.
[158] Rosso KM, Yanina SV, Gorski CA, Larese-Casanova P, Scherer MM. Connecting observations of hematite (a-Fe_2O_3) growth catalyzed by Fe(II). Environ. Sci. Technol., 2010, 44: 61-67.
[159] Latta DE, Bachman JE, Scherer MM. Fe electron transfer and atom exchange in goethite: Influence of Al-substitution and

anion sorption. Environ. Sci. Technol., 2012, 46: 10614-10623.

[160] Gorski CA, Handler RM, Beard BL, Pasakarnis T, Johnson CM, Scherer MM. Fe atom exchange between aqueous Fe^{2+} and magnetite. Environ. Sci. Technol., 2012, 46: 12399-12407.

[161] Handler RM, Frierdich AJ, Johnson CM, Rosso KM, Beard BL, Wang C, Latta DE, Neumann A, Pasakarnis T, Premaratne WAPJ, Scherer MM. Fe(II)-catalyzed recrystallization of goethite revisited. Environ. Sci. Technol., 2014, 48: 11302-11311.

[162] Frierdich AJ, Helgeson M, Liu C, Wang C, Rosso KM, Scherer MM. Iron atom exchange between hematite and aqueous Fe(II). Environ. Sci. Technol., 2015, 49: 8479-8486.

[163] Sun Y, Li J, Wang X. The retention of uranium and europium onto sepiolite investigated by macroscopic, spectroscopic and modeling techniques. Geochim. Cosmochim. Acta, 2014, 140: 621-643.

[164] Yang S, Ren X, Zhao G, Shi W, Montavon G Grambow B, Wang X. Competitive sorption and selective sequence of Cu(II) and Ni(II) on montmorillonite: Batch, modeling, EPR and XAFS studies. Geochim. Cosmochim. Acta, 2015, 166: 129-145.

[165] Sun Y, Zhang R, Ding C, Wang X, Cheng W, Chen C, Wang X. Adsorption of U(VI) on sericite in the presence of *Bacillus subtilis*: A combined batch, EXAFS and modeling techniques. Geochim. Cosmochim. Acta, 2016, 180: 51-65.

[166] Schultz CA, Grundl TJ. pH-dependence on reduction rate of 4-Cl-nitrobenzene by Fe(II)/montmorillonite systems. Environ. Sci. Technol., 2000, 34: 3641-3648.

[167] Chakraborty S, Favre F, Banerjee D, Scheinost, AC, Mullet M, Ehrhardt J, Brendle J, Vidal L, Charlet L. U(VI) sorption and reduction by Fe(II) sorbed on montmorillonite. Environ. Sci. Technol., 2010, 44: 3779-3785.

[168] Tsarev S, Waite TD, Collins RN. Uranium reduction by Fe(II) in the presence of montmorillonite and nontronite. Environ. Sci. Technol., 2016, 50: 8223-8230.

[169] Luan F, Burgos WD. Sequential extraction method for determination of Fe(II/III) and U(IV/VI) in suspensions of iron-bearing phyllosilicates and uranium. Environ. Sci. Technol., 2012, 46: 11995-12002.

[170] Charlet L, Scheinost AC, Tournassat C, Greneche JM, Gehin A, Fernandez-Martinez A, Coudert S, Tisserand D, Brendle J. Electron transfer at the mineral/water interface: Selenium reduction by ferrous iron sorbed on clay. Geochim. Cosmochim. Acta, 2007, 71: 5731-5749.

[171] Bishop M, Dong H, Kukkadapu R, Edelmann R, Microbial reduction of Fe(III) in multiple clay minerals by *Shewanella putrefaciens* and reactivity of bioreduced clay minerals toward Tc(VII) immobilization. Geochim. Cosmochim. Acta, 2011, 75: 5229-5246.

[172] Jaisi DP, Dong H, Plymale A, Fredrickson J, Zachara J, Heald S, Liu C. Reduction and long-term immobilization of technetium by Fe(II) associated with clay mineral nontronite. Chem. Geol., 2009, 264: 127-138.

[173] Yang J, Kukkadapu RK, Dong H, Shelobolina ES, Zhang J, Kim J. Effects of redox cycling of iron in nontronite on reduction of technetium. Chem. Geol., 2012, 291: 206-216.

[174] Jaisi DP, Dong H, Morton JP. Speciation of Fe(II) during Fe(III) reduction in nontronite and their reactivity toward Tc(VII) reduction. Clay and Clay Minerals, 2008, 56: 175-189.

[175] Elzinga EJ. Formation of layered Fe(II)−Al(III)-hydroxides during reaction of Fe(II) with aluminum oxide. Environ. Sci. Technol., 2012, 46: 4894-4901.

[176] Zhu Y, Elzinga EJ. Formation of layered Fe(II)-hydroxides during Fe(II) sorption onto clay and metal-oxide substrates. Environ. Sci. Technol., 2014, 48: 4937-4945.

[177] Soltermann D, Fernandes M, Baeyens B, Dä hn R, Miehé-Brendlé J, Wehrli B, Bradbury MH. Fe(II) sorption on a synthetic montmorillonite: A combined macroscopic and spectroscopic study. Environ. Sci. Technol., 2013, 47: 6978-6986.

[178] Soltermann D, Baeyens B, Bradbury MH, Fernandes MM. Fe(II) uptake on natural montmorillonites. II. surface complexation modeling. Environ. Sci. Technol., 2014, 48: 8698-8705.

[179] Soltermann D, Fernandes MM, Baeyens B, Dähn R, Joshi PA, Scheinost AC, Gorski CA. Fe(II) uptake on natural montmorillonites. I. macroscopic and spectroscopic characterization. Environ. Sci. Technol., 2014, 48: 8688-8697.

[180] Hofstetter TB, Schwarzenbach RP, Haderlein SB. Reactivity of Fe(II) species sssociated with clay minerals. Environ. Sci. Technol., 2003, 37: 519-528.

[181] Hofstetter TB, Neumann A, Schwarzenbach RP. Reduction of nitroaromatic compounds by Fe(II) species associated with iron-rich smectites. Environ. Sci. Technol., 2006, 40: 235-242.

[182] Neumann A, Hofstetter TB, Lüssi M, Cirpka OA, Petit S, Schwarzenbach RP. Assessing the redox reactivity of structural iron in smectites using nitroaromatic compounds as kinetic probes. Environ. Sci. Technol., 2008, 42: 8381-8387.

[183] Schaefer MV, Gorski CA, Scherer MM. Spectroscopic evidence for interfacial Fe(II)-Fe(III) electron transfer in a clay

mineral. Environ. Sci. Technol., 2011, 45: 540-545.

[184] Neumann A Olson TL, Scherer MM. Spectroscopic evidence for Fe(II)-Fe(III) electron transfer at clay mineral edge and basal sites. Environ. Sci. Technol., 2013, 47: 6969-6977.

[185] Zhang H, Weber EJ. Identifying indicators of reactivity for chemical reductants in sediments. Environ. Sci. Technol., 2013, 47: 6959-6968.

[186] Fox PM, Davis JA, Kukkadapu R, Singer DM, Bargar J, Williams KH. Abiotic U(VI) reduction by sorbed Fe(II) on natural sediments. Geochim. Cosmochim. Acta, 2013, 117: 266-282.

[187] Latta DE, Boyanov MI, Kemner KM, O'Loughlin EJ, Scherer MM. Abiotic reduction of uranium by Fe(II) in soil. Appl. Geochem., 2012, 27: 1512-1524.

[188] Peretyazhko T, Zachara J, Kukkadapu R, Heald S, Kutnyakov I, Resch C, Arey B, Wang C, Kovarik L, Phillips J, Moore D. Pertechnetate (TcO_4^-) reduction by reactive ferrous iron forms in naturally anoxic, redox transition zone sediments from the Hanford Site, USA. Geochim. Cosmochim. Acta, 2012, 92: 48-66.

[189] Fredrickson JK, Zachara JM, Kennedy DW, Kukkadapu RK, Mckinley JP, Heald SM, Liu C, Plymale AE. Reduction of TcO_4^- by sediment-associated biogenic Fe(II). Geochim. Cosmochim. Acta, 2004, 68: 3171-3187.

[190] Xi J, Lan J, Lu G, Zhao Y, Chai Z, Shi W. A density functional theory study of complex species and reactions of Am(III)/Eu(III) with nitrate anions. Mol. Simulat., 2014, 40 (5): 379-386.

[191] Lan J, Shi W, Yuan L, Zhao Y, Li J, Chai Z. Trivalent actinide and lanthanide separations by tetradentate nitrogen ligands: A quantum chemistry study. Inorg. Chem., 2011, 50: 9230-9237.

[192] Lan J, Shi W, Yuan L, Feng Y, Zhao Y, Chai Z. Thermodynamic study on the complexation of Am(III) and Eu(III) with tetradentate nitrogen ligands: A probe of complex species and reactions in aqueous solution. J. Phys. Chem. A, 2012, 116: 504-511.

[193] Lan J, Shi W, Yuan L, Li J, Zhao Y, Chai Z. Recent advances in computational modeling and simulations on the An(III)/Ln(III) separation process. Coord. Chem. Rev., 2012, 256: 1406-1417.

[194] Xiao C, Wang C, Lan J, Yuan L, Zhao Y, Chai Z, Shi W. Selective separation of Am(III) from Eu(III) by 2,9-Bis(dialkyl-1,2,4-triazin-3-yl)-1,10-phenanthrolines: A relativistic quantum chemistry study. Radiochim. Acta, 2014, 102 (10): 875-886.

[195] Wu H, Wu Q, Wang C, Lan J, Liu Z, Chai Z, Shi W. New insights into the selectivity of four 1,10-phenanthroline-drived ligands toward the separation of trivalent actinides and lanthanides: A DFT based comparison study. Dalton Trans., 2016, 45: 8107-8117.

[196] Wang C, Lan J, Zhao Y, Chai Z, Wei Y, Shi W. Density functional theory studies of UO_2^{2+} and NpO_2^+ with carbamoylmethylphosphine oxide ligands complexes. Inorg. Chem., 2013, 52: 196-203.

[197] Wang C, Shi W, Lan J, Zhao Y, Wei Y, Chai Z. Complexation behavior of Eu(III) and Am(III) with CMPO and Ph_2CMPO ligands: Insights from density functional theory. Inorg. Chem., 2013, 52: 10904-10911.

[198] Wang C, Lan J, Feng Y, Wei Y, Zhao Y, Chai Z, Shi W. Extraction complexes of Pu(IV) with carbamoylmethylphosphine oxide ligands: A relativistic density functional study. Radiochim. Acta, 2014, 102: 77-86.

[199] Wang C, Lan J, Wu Q, Zhao Y, Wang X, Chai Z, Shi W. Density functional theory investigations of the trivalent lanthanide and actinide extraction complexes with diglycolamides. Dalton Trans., 2014, 43: 8713-8720.

[200] Luo J, Wang C, Lan J, Wu Q, Zhao Y, Chai Z, Nie C, Shi W. Theoretical studies on the complexation of Eu(III) and Am(III) with HDEHP: structure, bonding nature and stability. Sci. Chin. Chem., 2016, 59: 324-331.

[201] Dong W, Ball WP, Liu C, Wang Z, Stone AT, Bai J, Zachara JM. Influence of calcite and dissolved calcium on uranium(VI) sorption to a Hanford subsurface sediment. Environ. Sci. Technol., 2005, 39 (20): 7949-7955.

[202] Elzinga EJ, Tait CD, Reeder RJ, Rector KD, Donohoe RJ, Morris DE. Spectroscopic investigation of U(VI) sorption at the calcite-water interface. Geochim. Cosmochim. Acta, 2004, 68 (11): 2437-2448.

[203] Rihs S, Sturchio NC, Orlandini K, Cheng L, Teng H, Fenter P, Bedzyk MJ. Interaction of uranyl with calcite in the presence of EDTA. Environ. Sci. Technol., 2004, 38 (19): 5078-5086.

[204] Reeder RJ, Elzinga EJ, Tait CD, Rector KD, Donohoe RJ, Morris DE. Site-specific incorporation of uranyl carbonate species at the calcite surface. Geochim. Cosmochim. Acta, 2004, 68 (23): 4799-4808.

[205] Mathew K, Sundararaman R, Letchworth-Weaver K, Arias TA, Hennig RG. Implicit solvation model for density-functional study of nanocrystal surfaces and reaction pathways. J. Chem. Phys., 2014, 140 (8).

[206] Fishman M, Zhuang HL, Mathew K, Dirschka W, Hennig RG, Accuracy of exchange-correlation functionals and effect of

solvation on the surface energy of copper. Physical Review B, 2013, 87 (24).

[207] Wang Y, Liu Z, Li Y, Bai Z, Liu W, Wang Y, Xu X, Xiao C, Sheng D, Diwu J, Su J, Chai Z, Albrecht-Schmitt TE, Wang S. Umbellate distortions of the uranyl coordination environment result in a stable and porous polycatenated framework that can effectively remove cesium from aqueous solutions. J. Am. Chem. Soc., 2015, 137: 6144-6147.

[208] Zheng T, Yang Z, Gui D, Liu Z, Wang X, Dai X, Liu S, Zhang L, Gao Y, Chen L, Sheng D, Wang Y, Diwu J, Wang J, Zhou R, Chai Z, Albrecht-Shmitt TE, Wang S. Overcoming the crystallization and designability issues in the ultrastable zirconium phosponate framework system. Nat. Commun., 2017, 8: 15369.

[209] Liu W, Dai X, Bai Z, Wang Y, Yang Z, Zhang L, Xu L, Chen L Li Y, Gui D, Diwu J, Wang J, Zhou R, Chai Z, Wang S. Highly sensitive and selective uranium detection in natural water systems using a luminescent mesoporous metal-organic framework equipped with abundant lewis basic sites: A combined batch, X-ray absorption spectroscopy, and first principle simulation investigation. Environ. Sci. Technol., 2017, 51: 3911-3921.

[210] Sheng D, Zhu L, Xu C, Xiao C, Wang Y, Wang Y, Chen L, Diwu J, Chen J, Chai Z, Albrecht-schmitt TE, Wang S. Efficient and selective uptake of TcO_4^- by a cationic metal-organic framework material with open Ag^+ sites. Environ. Sci. Technol., 2017, 51: 3471-3479.

[211] Zhu L, Xiao C, Dai X, Li J, Gui D, Sheng D, Chen L, Zhou R, Chai Z, Albrecht-Schmitt TE, Wang S. Exceptional perrhenate/pertechnetate uptake and subsequent immobilization by a low-dimensional cationic coordination polymer: Overcoming the hofmeister bias selectivity. Environ. Sci. Technol. Lett., 2017, DOI: 10.1021/acs.estlett.7b00165.

[212] Zhu L, Zhang L, Li J, Zhang D, Chen L, Sheng D, Yang S, Xiao C, Wang J, Chai Z, Albrecht-Schmitt TE, Wang S. Selenium sequestration in a cationic layered rare earth hydroxide: A combined batch experiments and EXAFS investigation. Environ.Sci. Technol., 2017, DOI: 10.1021/acs.est.7b02006.

作者：王祥科[1]，王殳凹[2]，盛国栋[3]，石伟群[4]

[1]华北电力大学环境科学与工程学院，[2]苏州大学放射医学与防护学院，

[3]绍兴文理学院化学化工学院，[4]中国科学院高能物理研究所

第 13 章　近海及河口水环境污染化学研究进展

本章导读

沿海及河口地区聚居着众多人口，人类活动向近海及河口环境排放了大量污染物。本章主要对2017年上半年期间发表的近海及河口水环境污染化学研究的文献进行综述和介绍。无机污染物中的重金属是关注的焦点，研究集中于水环境尤其是沉积物中重金属的来源、迁移和转化；有机污染物中，传统的污染物如多环芳烃和多氯联苯，仍然受到广泛关注；而海洋微塑料的源与汇、分布和组成，成为新的热门领域。目前的工作中，调查污染物成分、含量和分布状况的居多；在污染物迁移转化机制的纵深处，还留有巨大的研究空间；海洋污染物监测技术方案和策略的研究，是有效追踪和控制污染的技术支撑，必然方兴未艾。

关键词

重金属，有机污染物，微塑料，近海，河口，监测

1 引　言

提到海洋，人们的脑海必然呈现出一片蔚蓝的汪洋。洁净、深邃、富饶、慷慨，是人们对海洋的赞美和刻画。而科学家在南极这片人迹罕见的冰雪大陆的企鹅身上，检出了有机氯农药；在远离大陆工业区的马里亚纳海沟10000多米深海的某些甲壳动物中，发现了高浓度的多氯联苯和多溴联苯醚[1]；在世界各大海洋甚至在北极冰冷的海水中，发现了堪称20世纪最伟大也是最糟糕发明的塑料的微粒[2,3]。此时人们总是有些迷茫，这些污染物通过什么渠道到达海洋的另一端？为什么浓度高居不下？

人类择水而居。众多人口聚居在河流和沿海区域，对自然资源的需求不断增长，规模不断扩大。越来越密集的人群生活，越来越频繁的工农业发展、旅游开发等人类活动，向河流、河口和近海排放越来越多的污染物。百川归海，这些污染物最终经河口到近海直至汪洋。河口和近海是陆源物质入海的主要通道，近岸海区是陆-海相互作用最为重要的区域，这里的环境承受着人类活动带来的巨大负面压力。因此，河口及近海环境污染，一直是海洋环境学家研究的热点。

本章旨在对近年来近海及河口水环境污染化学的国内外研究进展做一回顾。在Web of Science上输入marine pollution、estuary pollution、sediment、seawater等关键词，仅从2017年1月至6月，就有几百篇。由于文献众多、受篇幅限制，本章仅关注近海及河口水环境（包括水和沉积物）的重金属、多环芳烃和多氯联苯等有机污染物、微塑料和塑料碎片的研究进展，有关海洋酸化效应、营养盐、生物体内的污染物及其生物效应和毒理学研究，以及样品预处理和仪器分析等分析方法研究，不在本章的范围内。而对海洋污染物监测技术及方案策略的研究，文献的时间段略有放宽，内容也涉及与生物相关的“贻贝监测计划”。所检索和引用的文献均为英文文献。

纵观这些文献，大部分是污染物的调查性研究，其中我国的占近半数，印度等其他发展中国家的居次。重金属在海水环境尤其是沉积物中的分布、扩散和迁移，在沉积物-海水界面间的沉降-悬浮，一直吸引海洋环境科学和海洋化学研究者进行探讨；因此，目标物为重金属的文献占半数以上。传统的有机污染物如多环芳烃和多氯联苯，仍然受到广泛关注。国际范围内，海洋微塑料污染的论文激增。然而，调查性的研究居多，污染物迁移转化机理机制、污染监测技术、方案和策略的探索尚少；污染化学更深层次的内涵，需要研究者更加努力的探索。

2 重金属污染物

在无机污染物方面，重金属污染物占据极其主要的地位。20 世纪末起，近海和河口有机污染化学的研究风靡整个学术界，抢占了传承已久的重金属污染化学研究的风头。这主要是因为有机物污染日趋严重，“新型”污染物基本上都是有机物；另一方面，试剂纯度、监测方法和分析仪器的水平日益提高，使研究者发现海洋环境中的重金属含量并没有原来认定的那么高。在此之后的二十年间，有机污染化学研究的势头一直强劲。然而，近些年来，重金属污染事件频频发生，大量重金属污染物经江河流入海洋；重金属污染具有难以消除的持久性特点，易积蓄在海洋环境中；且即使是在低浓度下，生命体也可能因为长期暴露引发慢性中毒。因此，重金属污染研究又重新受到格外重视，目前海洋环境污染化学的大部分研究聚焦于此。

2017 年上半年发表的关于近海和河口重金属污染的论文有近百篇，主要是调研类的（表 13-1）。研究的目标重金属（含类金属）为 Cu、Pb、Zn、Cd、Cr、Hg、As、Co、Ni，有机金属如有机锡和有机汞。目标物的介质，大多数是沉积物（包括表层沉积物样和柱状样），因其更能反映污染历史，海水和河口水的研究不多。生物体内的重金属水平，因为涉及太多不同的生物种类和富集因子，且基本都是讨论毒性效应，故不列入本章范围。国内的研究区域，从北到南，涵盖了黄河、长江、珠江等几大河口，以及渤海、黄海、东海和南海等沿海地区。国际范围的研究区域，主要有印度、东南亚、阿拉伯海、地中海等的沿海地区，欧美的相关研究较少。总体上，人为源的重金属污染普遍存在，这些调研结果，无疑将丰富区域重金属污染的数据库，为海洋环境管理者提供决策参考。

表 13-1 近海和河口重金属污染研究

序号	目标污染物	研究区域和样品	研究结果简述	第一作者，文献号
1	Hg、As、Pb、Cu、Cd、Cr、Zn	鸭绿江河口，22 个海水和沉积物样	污染因子 Cd > Hg > Pb > As > Cu > Zn > Cr	Li，[4]
2	Cr、Cu、Pb、As、Cd、Hg、PCBs、DDTs	辽河河口，44 个表层沉积物	污染因子 Cr > Cu > Pb > As > Cd > Hg，污染重金属主要是 Cd 和 Hg	Li，[5]
3	Cd、As、Cu、Ni、Pb、Cr、Zn	辽河保护区，19 个表层沉积物样	Cd、Pb、Zn 主要是人为源，Cu、Cr 和 Ni 为天然源；污染程度 Cd > As > Cu > Ni > Pb > Cr > Zn	Ke，[6]
4	Fe、Cd、Pb	双台子河口，沿海滩涂的沉积物和孔隙水	Fe 和 Pb 无明显受人类活动影响，但 Cd 有，可交换 Cd 浓度高	Liu，[7]
5	Pb、Cu、Hg	双台子河口，沉积物	Pb、Cu 和 Hg 与背景值相近；Zn 和 Cd 有中等污染至重污染	Li，[8]
6	Cd	渤海，405 个表层沉积物样和 2 个柱状样	Cd 中等程度富集，具高迁移性和高生物可给性，高风险。1800 年前无变化，1800 年至 20 世纪 50 年代轻微增加，20 世纪 60~70 年代减少，20 世纪 80 年代至 2001 年增加，高值出现在 1998 年	Hu，[9]
7	Hg、As、Cd、Ni、Cr、Cu、Ni、Pb、Zn	渤海和黄海沉积物，2 个航次	黄海中南部和山东半岛北部，污染较高；Hg 主要是人为源污染	Xiao，[10]
8	Cu、Pb、Zn、Cr	秦皇岛沿岸，12 个站位表层水	生态风险 Cu > Zn > Cd > Pb，其中仅 Pb 的风险较小	Wang，[11]

续表

序号	目标污染物	研究区域和样品	研究结果简述	第一作者，文献号
9	Zn、Cr、Cu、Cd、Pb、As	威海沿岸，表层沉积物	Zn、Cu、As 近 5 年来减少；Cd 有中等污染，其他 5 个金属的污染相对较低	Li，[12]
10	Cu、Pb、Zn、Cr、Cd、As、Hg、Fe、Al	黄河口，莱州湾，表层沉积物	黄河口的金属含量与 20 年前相当，但是莱州湾的大部分金属有所减少	Wang，[13]
11	As、Cd、Cu、Fe、Hg、Mn、Pb、Se、Zn	黄河三角洲贝壳海岛（Shell Ridge Island），土壤样	总体较洁净，As、Cd、Cu、Fe、Pb、Zn 同源，但浓度变化大；Hg、Mn、Se 不同源	Yang，[14]
12	Al、As、Cd、Cr、Cu、Ni、Pb、Zn	黄河三角洲，河滨湿地土壤，春秋季	As 和 Cd 的富集因子高，As 和 Ni 为主要污染物	Zhang，[15]
13	Cu、Pb、Zn、Cr、Cd、As、Hg	黄河口、长江口、珠江口及相邻海区，表层沉积物	珠江口的污染风险相对较高	Liu，[16]
14	Cu、Pb、Zn、Cr、Cd、As	青岛胶州湾流域，47 个沉积物样	总体来说未超标，但 Cu、Cr、As 对生态有不利影响	Xu，[17]
15	As、Cd、Pb、Sb、Zn、Al、Cu、Rb、Sr、Fe、Mn	黄海南部，江苏盐城海岸，2 个柱状样，150 年	As、Cd、Pb、Sb、Zn 反映出近 20 年人类活动加剧	Bao，[18]
16	As、Cd、Cr、Cu、Mn、Ni、Pb、Zn、Al、Fe	江苏连云港海岸，3 个柱状样	污染低，大部分金属未在柱状样中富集	Li，[19]
17	Cu、Ni、Pb、Cr、Zn	江苏射阳河，2 个沉积物柱状样	Cu、Ni、Pb、Cr、Zn 为人为源，1980~1995 年间污染逐渐增大	Wu，[20]
18	V	淮河，表层沉积物和柱状样，60 年	轻微至中等污染，1955 年后持续增加，约在 2000 年达到高值，之后略有下降	Liu，[21]
19	丁基锡、苯基锡	长江口和周边海区，72 个表层沉积物样	主要是货轮污染，码头和渔港（芦潮港和洋山港）的有机锡含量高，芦潮港的生物样中有机锡也很高	Chen，[22]
20	Mn、Ti、Cr、Cu、Ni、Pb、Zn、V、Al、Fe	长江口和周边海区，25 个表层沉积物样，1 个柱状样	东海和济州岛西南部沿海地区的含量最高，但生态风险均不高	Chen，[23]
21	Pb、Cd、Cr、Cu、Hg、Zn、As	长江口，55 个沉积物样	Pb、Cd、Cr 低污染；Cd 高风险；Hg、Zn、As、Pb、Cr 为天然源；Cu 有人为源；Cd 来自大气沉降	Han，[24]
22	Cu、Zn、Pb、DDTs 等	宁波，港口 20 个表层沉积物	集装箱作业排放重金属，石油和原煤装卸导致有机物污染	Yu，[25]
23	Pb、Cr、Zn 等	杭州湾，庵东潮间带，2 个柱状样	2000 年前几乎无变化，2000 年后污染迅速增加；Pb 重污染，Cr、Zn 中等污染	Jin，[26]
24	Cr、Cu、Zn、As、Cd、Pb、Hg	福建漳江口，红树林湿地，沉积物样	Cr、Cu、Cd、Pb 含量明显偏高	Wu，[27]
25	As、Cd、Cr、Cu、Hg、Pb、Zn	珠江口，148 个表层沉积物，1 个柱状样	Cd 中等污染，人为源；As 和 Hg 人为源，Pb、Cr、Cu、Zn 有混合源	Zhao，[28]
26	Cu、Pb、Zn、Cr、Cd	琼州海峡及其周边海区，表层及柱状沉积物	污染风险较低。百年来浓度升高，污染物主要来自珠江河口	Zhang，[29]
27	Ti、Cr、Mn、Co、Al、Fe、Ni、Cu、Zn、Cd、Pb	印度，东南沿海 8 个生态系统，表层沉积物样	除了 Cd，其他的金属处于中等污染水平	Kumar，[30]
28	Pb、Cu、Zn、Cr、Co、Ni、Fe	印度，金奈（Chennai）海岸，洪水事件后 30 个沉积物	除了 Pb 外，其他金属未见增加，Pb 的增加可能来自受工业和交通污染都市土壤的迁移	Gopal，[31]

续表

序号	目标污染物	研究区域和样品	研究结果简述	第一作者，文献号
29	Co、Cr、Cu、Ni、Zn、Pb	印度，甘加河（Ganga River）河口，柱状样	工业是 Pb 的重要污染源	Samanta，[32]
30	Cu、Ni、Pb、Co、Cr、Zn、Mn、Fe	印度，东南沿海 Vellar 和 Coleroon 河口，表层沉积物样	污染因子：Cu＞Ni＞Pb＞Co＞Cr＞Zn＞Mn＞Fe	Nethaji，[33]
31	Pb 等	印度，东南沿海马纳尔湾（Gulf of Mannar）Koswari 岛，生物保护区，33 个表层沉积物	Pb 的浓度尤其高，可能与附近的燃煤电厂的排放有关；大部分沉积物处于高污染至危害级别	Krishnakuma，[34]
32	Pb、Co、Cr	印度，东南沿海 Tamiraparani 河口，24 个沉积物样	Pb、Ni、Co 有污染，高 Pb 可能与港口活动及燃煤电厂有关	Magesh，[35]
33	Ni、Cd、Pb、Zn	印度，古吉拉特邦，韦拉沃尔渔港（Veraval Fishery Harbor），3 个沉积物样	雨季及其前后，有人为源污染	Sundararajan，[36]
34	Cd、Co、Cr、Cu、Mn、Ni、Pb、Zn	越南，市崴（Thi Vai）河口和 根热（Can Gio）红树林区，水和沉积物样	Cr、Cu、Ni 有人为源；总体来说，污染比预期的低，可能是由于稀释作用	Costa-Boddeker，[37]
35	Cd、Ni、Cr、As、Pb、Cu、Zn	越南，田（Tien）河河口，沉积物和文蛤	含量 Cd＞Cu＞As＞Zn＞Cr＞Ni＞Pb；Cd 具环境风险	Hop，[38]
36	Cd、Cr、Cu、Co、Fe、Pb、Ni、V、Zn	马来西亚，巴六拜（Bayan Lepas）海区，10 个沉积物样	从无污染至极高污染程度不等	Khodami，[39]
37	Hg、甲基汞	马来西亚，巴生港（Port Klang），30 个潮间带沉积物	高值出现在卢穆特（Lumut）海峡，其有 2 条高都市化河流输入；甲基汞在总汞中占比 0.06%~ 94.96%	Haris，[40]
38	Pb、Cd、Cr	印尼，肯达里湾（Kendari Bay），32 个站位的水样	Pb 为人为源污染，严重超标	Armidi，[41]
39	51 个金属元素	孟加拉，孟加拉湾 Sundarban 红树林湿地及临近 Hugh river 河口，13 个表层沉积物样	无污染至中等污染，Cu、As、Cr、Cd 含量较高	Watts，[42]
40	Hg、Cu、As、Ni、Zn、Pb、Cr、Mn、Co、V、Fe、Al	伊朗，波斯湾阿萨鲁耶港（Asaluyeh Port），48 个沉积物样	Hg 和 Cu 的污染高，富集因子 Hg＞Cu＞As＞Ni＞Zn＞Pb、Cr、Mn＞Co、V、Fe、Al	Delshab，[43]
41	Fe、Cr、Ni、Zn、Cu、Pb、Cd	伊朗，波斯湾格什姆岛（Qeshm Island），20 个沉积物样	含量 Fe＞Cr＞Ni＞Zn＞Cu＞Pb＞Cd；Cu、Pb、Zn 主要是天然源，Cd、Ni、Cr 主要是人为源	Zarezadeh，[44]
42	Fe、Mg、Mn、Cu、Zn、Cd、Ni、Pb、Co	也门，东南沿海 90 km 海岸线 Al-Khowkhah 和 Al-Mokha 之间，沉积物样	Fe、Mg、Mn 为天然源；Zn、Cu、Mn 有中等严重污染；Pb 和 Co 为中等轻微污染；Cd 和 Ni 基本无污染	El-Younsy，[45]
43	As、Cd、Co、Cr、Cu、Ni、Pb、V、Zn、Al、Fe	里海南部海岸带，13 个站位沉积物样	污染很小	Jamshidi，[46]
44	Fe、Al、Zn、Cu、Ni、Cr、Pb、Cd、Hg	埃及，地中海海岸，海水、沉积物、生物样	总体来说，生物体的毒性风险不大，但是长远来看，风险有可能出现	El Nemr，[47]
45	Zn、Pb	地中海南部突尼斯湾，Mejerda River 和 Ghar El Melh 泻湖沿岸，沉积物	人类排放对海湾有影响	Oueslati，[48]
46	Mo、Tl、U、Cd、Sr、Zn、Cu、Pb、Sn、Bi	克罗地亚，地中海亚德里亚海，姆列特岛（Mljet Island），沉积物样	过去 40 年金属低污染，但检出丁基锡化合物	Sondi，[49]

续表

序号	目标污染物	研究区域和样品	研究结果简述	第一作者，文献号
47	Fe、Ba、Zn、V、Co、Cr、Cd、Cu、Ni、Al、As、Mn、Sb、Ag、Se、B、Hg、Ti、Pb	土耳其，Buyuk Menderes River 河口，水、沉积物、生物样	不同金属在不同类别的样品中富集程度不同	Durmaz，[50]
48	Hg	波兰，格但斯克海湾（Gulf of Gdansk）悬浮颗粒物	2011~2013 年间的数据。颗粒态汞的含量与水文条件和颗粒物的有机质成分有关	Jedruch，[51]
49	Ag	法国，吉伦特（Gironde）河口，水样	都市废水排放造成河口污染，Ag 来自个人护理用品	Deycard，[52]
50	Pb、Ni 等 8 个重金属	法国，地中海西北 Golfe-Juan 湾沉积物	分布和含量与地质条件及人为活动相关，Pb 和 Ni 含量超过法国海洋沉积物标准	Tiquio，[53]
51	Pb 等	西班牙，东南的卡塔赫纳矿区的前海滨泻湖（the former littoral lagoon of El Almarjal，Cartagena mining district），钻探孔、盆地沉积物	8000 年沉积史。4500 年前 Pb 开始污染。这一发现为西班牙东南部铅矿开采和冶金的起源提供证据	Manteca，[54]
52	Pb、Cu、Cr、Ni、Zn、Co、Al、Fe、Cd	摩洛哥，大西洋海岸达克拉湾（Dakhla Bay），沉积物样	Pb、Cu、Cr 有人为源；Ni、Zn、Co、Al、Fe 为天然源；Cd 与上升流有关	Hakima，[55]
53	As、Hg	加拿大，大西洋沿岸新斯科舍（Nova Scotia）9 个旧金矿区附近的潮间带，29 个站位的水、沉积物、生物样	As 存在高风险；Hg 未超标	Doe，[56]
54	Pb	墨西哥，东南韦拉克鲁斯（Veracruz）珊瑚礁，沉积物样	4 种地化分类：可交换 Pb、碳酸 Pb、有机 Pb、矿物 Pb；受陆源影响的矿物 Pb 的含量较高。生物有效性指标说明中度污染	Horta-Puga，[57]
55	丁基锡	智利，北部海岸 12 个站位表层沉积物和一种可食用螺	11 个站位发现可食用螺的性畸变，污染最严重的站点，仍然有捕捞和水产养殖	Mattos，[58]
56	Cd、Pb、Cu、Zn 等	阿根廷，圣安东尼奥湾（San Antonio Bay），沉积物	矿区排废造成 Cd、Pb、Cu、Zn 超标	Marinho，[59]
57	Ag、As、Cd、Co、Cu、Hg、Mn、Ni、Pb、Zn	澳大利亚，布里斯班河（Brisbane River）河口沉积物样	30%的 Ag、As、Cd、Co、Cu、Hg、Mn、Ni、Pb、Zn 可从样品渗沥出，具生物有效性	Duodu，[60]
58	Cr、Ni、As、Cu、Zn、Pb、Rb	澳大利亚，悉尼博塔尼（Botany）湾，乔治河（Georges River），146 个表层沉积物	高浓度发生在富含细颗粒和有机质的排污口及海湾内测，低污染见于沿海岸线和海湾边的沙质沉积物中	Alyazichi，[61]
59	Fe、Mg、Mn、Cr、Cu、Mo、Ni、Co、Pb、Cd、Zn、Hg	南非，德班（Durban）南部 7 个旅游点海滨，43 个沉积物样	均高于背景值	Vetrimurugan，[62]

注：PCBs——多氯联苯；DDTs——滴滴涕类农药；另可参见表 13-2 的序号 13~16、序号 18。

3 多环芳烃和多氯联苯等有机污染物

已发表的文献显示，从我国的黄河口、长江口、闽江口到珠江口；从亚洲、欧洲到南北美洲，传统的有机污染物如多环芳烃（PAHs）、多氯联苯（PCBs）和含氯农药如六六六（HCHs）、滴滴涕（DDTs），“新型”有机污染物如氯代石蜡、全氟化合物、邻苯二甲酸酯、多溴联苯醚（PBDEs）等，均受到研究者的广泛关注。数理统计分析方法如主成分分析法，已成为污染源解析的常用手段；长年的研究已证实，河口和近岸海域的 PAHs 污染，通常来自化石燃料的燃烧和溢油事件。近期 Han 等[63]综述了近年来我国水环境中持久性有机污染物的污染状况，区域包括长江口、珠江口、闽江口、九龙江口、大亚湾等河口和近岸；综述指出，我国 PAHs 和 PCBs 的污染特别严重，PAHs、有机氯农药和 PCBs 经常超出国际和我国的限定值；PAHs 的浓度范围为未检出~474000 ng/L，区域平均值 15.1~72400 ng/L；PCBs 的浓度范围为未检出~3161 ng/L，区域平均值 0.2~985.2 ng/L；此外，还存在全氟辛烷磺酸（PFOS）和全氟辛酸（PFOA）污染的风险。

比较区域的 PAHs 水平，我国红树林区的 PAHs 浓度比我国其他海岸带地区的低[64]。比较珠江三角洲、香港海域和东京湾的海岸沉积物中的氯代石蜡浓度，香港海域最高，东京湾较低[65]。2017 年上半年发表的近海和河口有机污染物的相关研究论文见表 13-2。

表 13-2 近海和河口有机物污染研究

序号	目标污染物	研究区域和样品	研究结果简述	第一作者，文献号
1	PAHs	黄河口，土壤、沉积物、表层水、地下水样	土壤和沉积物中，主要来源为木头、煤燃烧和石油输入；表层水和地下水中，主要来源为油品燃烧和石油输入	Li，[66]
2	氨基脲	烟台附近的金城湾和四十里湾，30 个站位，615 个海水样，320 个沉积物样，90 个贝类样	海水中 0.011~0.093 μg/L，贝类中未检出至 0.75 μg/kg，沉积物中低于检出限	Tian，[67]
3	PAHs	江苏南通海岸，表层沉积物样	与其他地区比，含量较低。主要来源是石油燃烧和溢油	Liu，[68]
4	PAHs、PCBs、有机氯农药	长江口、珠江三角洲、闽江口、九龙江口、大亚湾、太湖、浙江省水道，表层水样	综述。PAHs 和 PCBs 污染严重，有一定的全氟化合物污染	Han，[63]
5	PAHs	长江口 48 个、珠江口 45 个沉积物样	中等程度生态风险，珠江更严重些	Wang，[69]
6	PAHs	深圳茅洲河，水、悬浮颗粒物、沉积物样；每月一次，监测一年	水和悬浮颗粒物上的 PAHs 同源；沉积物上 PAHs 的来源复杂，含化石燃料燃烧、油污染、有机质燃烧等来源	Zhang，[70]
7	PAHs	我国红树林表层沉积物样	比我国其他海岸带地区的含量低，4 环的 PAHs 为主	Zhang，[64]
8	氯代石蜡	中国（珠江三角洲、香港海域）和日本（东京湾）的都市海岸沉积物	香港海域最高，东京湾较低；不同地区的监管规则显著不同	Zeng，[65]
9	PAHs、烷基酚、苯乙烯低聚物	韩国，Geum River 河口和新万金（Saemangeum）海岸，58 个表层沉积物样	全部未超标，在河口和紧邻工业区处较高；C-13 分析结果显示来源于陆地	Yoon，[71]

续表

序号	目标污染物	研究区域和样品	研究结果简述	第一作者，文献号
10	石油烃	印度，东南海岸泰米尔纳德邦（Tamil Nadu），3 个柱状样	泥（粉+黏土+砂）与石油烃含量显著正相关	Kamalakannan，[72]
11	脂肪烃、PAHs	伊朗，波斯湾珊瑚礁，120 个表层沉积物样	污染主要来自油气开发、船舶排放的石油和石油燃烧	Jafarabadi，[73]
12	PAHs	伊朗，阿曼海查巴哈港（Chabahar Bay），表层水样和沉积物样	季风后的含量高于季风前	Agah，[74]
13	PAHs、BTEX、石油烃、重金属	尼泊尔河三角洲，20 个海水样，20 个沉积物样	海水和沉积物受到经过处理的排放水中有机物和重金属的污染	Okogbue，[75]
14	DDTs、PCBs、PBDEs、Hg、Pb、Cd	波兰，波罗的海南部海岸，4 个站位，其中两个设在格但斯克海湾（Gulf of Gdansk）内	有机物污染中 DDTs 为主，超过 PCBs；PBDEs 较少；重金属污染物 Hg、Pb、Cd 的浓度较高	Dabrowska，[76]
15	PAHs、10 个重金属、营养盐	意大利，沿海阿普利亚区（Apulia），4 个旅游区沉积物样	研究 4 个旅游区的污染情况；有些地方 PAHs 和重金属污染超出预期	Mali，[77]
16	石油、Hg	挪威北海 Fedje 岛附近	二战期间德国潜艇 U-864（载有 Hg）和失事货轮 Nordvard 沉船引起的污染	Ndungu，[78]
17	DDTs、HCHs、PAHs	加纳，特马港（Tema Harbor）沉积物样	自从 2007 年严重的溢油事件后，PAHs 的污染已经下降	Botwe，[79]
18	农药、PAHs、邻苯二甲酸酯、重金属	南太平洋岛国、7 条沿海河流和毗邻美属萨摩亚（American Samoa）Futiga 垃圾填埋场处	所有站位均有 Pb、Hg 污染；有些河流的二乙基邻苯二甲酸酯、有机磷农药超标；有些地方检出已经禁用的对硫磷	Polidoro，[80]
19	PAHs、PCBs、PBDEs、塑料、营养盐和赤潮	美国，加利福尼亚湾，水、沉积物、生物样	对 150 个研究进行综述。污染广泛，但程度相对较低。污染热点：矿业泄漏引起的金属及类金属污染、农业虾类养殖引起的营养盐和农药污染	Paez-Osuna，[81]
20	PAHs	美国，圣地亚哥湾，3 个船坞，沉积物样	4~6 环的 PAHs 为主，苯并芘对毒性当量的贡献度高；主要来源是附近的燃烧产物通过大气沉降进入盆地，以及街道和停车场的雨水排水	Neira，[82]
21	石油烃、脂肪烃	巴西，南大河州跨越佩洛塔斯（Pelotas）城市区域的运河，沉积物样	城市地表径流及排水系统对 Patos 泻湖和其他水源地有影响	Sanches，[83]

注：PAHs——多环芳烃；PCBs——多氯联苯；DDTs——滴滴涕类农药；HCHs——六六六类农药；BTEX——三苯；PBDEs——溴代二苯醚。另可参见表 13-1 的序号 2、序号 22。

4　微塑料和塑料碎片

自 2004 年 Thompson 在 *Science* 上发表 *Lost at the sea: Where is all the plastic*[2]以来，海洋环境中的塑料微粒污染受到海洋环境学家和社会各界的广泛关注。塑料微粒中，微塑料特指直径小于 5 mm 的不同形态塑料微粒，其来自工业原料和产品（初级）及陆源塑料垃圾的分解（二级）。微塑料的比表面积大、疏水性强，是有机污染物和重金属的理想载体。海洋中的塑料微粒易被浮游生物和鱼类等摄食，对海洋生态系统和食物链构成巨大威胁。2017 年 6 月 5~9 日，在美国纽约召开了联合国海洋大会（The UN Ocean

Conference)。联合国秘书长古特雷斯在大会呼吁各国采取果断行动，以保障地球赖以生存的命脉——海洋的可持续利用和发展[84]。古特雷斯转引了最新的研究报告，称如果不采取措施，2050 年海洋中塑料垃圾的总重量可能将超过鱼类。对于这个来自 2016 年初艾伦·麦克阿瑟环保基金会（Ellen MacArthur Foundation）和世界经济论坛（World Economic Forum）的关于 2050 年海洋中塑料垃圾将达到 8.5 亿~9.5 亿吨的预测，学术界持谨慎态度，认为该预测还缺乏科学数据支撑。但是不争的事实是，海洋中的塑料微粒污染日益严重。查明塑料微粒的污染来源，研究其环境行为和迁移机理，能更好地了解海洋环境所受的影响，制定控制措施。例如，应从塑料购物袋的生产到其进入环境，沿着生命周期对其迁移和归宿进行评估[85]，遏制塑料微粒进入环境的渠道。世界各地包括我国的关于限制使用一次性塑料袋、减少海洋污染的政策法规及限塑措施，近期已经为 Xanthos[86]所综述。

除了 Law[87]在 2017 年的 *Annual Review of Marine Science* 上发表了题为 *Plastics in the Marine Environment* 的报告外，2017 年 3 月，Auta 等[88]发表了关于微塑料在海洋环境中的来源、分布、归宿、生态效应和可能解决方案的综述。为方便读者，以下章节将摘录该综述的主要相关内容，而具体的参考文献请查阅该综述。

综述指出，海洋垃圾中的 80%~85%是塑料。20 世纪 50 年代，已有 1.7 亿吨的塑料垃圾入海，至 2014 年达到了 3 亿吨。大量的塑性微粒分布在非洲、亚洲、东南亚、印度、南非、北美和欧洲的水域，并已经进入食物链。海洋环境中塑料微粒的成分主要有容易沉降的聚氯乙烯（PVC）、聚对苯二甲酸乙二醇酯（PET），容易悬浮的聚乙烯（PE）、聚丙烯（PP）和聚苯乙烯（PS），以及其他聚合物如聚乙烯醇（PVA）和聚酰胺（PA）等。来自工业产品和原料的属于初级微塑料，如面部磨砂剂、牙膏、淋浴凝胶、各种化妆品、染发剂、指甲油、剃须膏、防晒霜、驱虫剂，以及合成纤维、磨料、抛光剂、树脂颗粒等。由塑料垃圾的分解生成的称为二级微塑料，由大块塑料经由紫外光、温度、氧气、波浪、微生物等环境因素引发分解而成。微塑料可直接被海洋生物摄食，也可在继续分解的过程中释放出有毒物质，比如邻苯二甲酸酯、双酚 A 等。颗粒越小的塑料，越容易被生物摄取。

塑料微粒进入海洋的途径包括污水处理厂排水、洗衣排水、径流、填埋、倾废等。迁移和循环的另一个途径是通过浮游动物的粪便。研究证明微塑料被海洋生物摄食后，经过肠道时被包裹在粪便中排泄，逐渐沉入海底，被个体较大的桡足类摄取。海水中的微塑料经过生物污泥进程（biofouling progresses），密度逐渐增加，一旦高于海水密度，最终可沉入海洋沉积物中。在红树林沉积物中也发现了微塑料，例如，位于马来西亚半岛的红树林沉积物中，检出了 418 种不同的微塑料聚合物。

塑料微粒易漂浮和难降解的特性，使它们轻易地通过海洋水文过程和洋流输送，广泛地分散到世界各地的海洋、海底、海滩，甚至极地、深海、大洋中的小岛也不能幸免。近年来，众多学者对世界各地的塑料碎片分布进行了调研，研究区域包括美国加州海岸、挪威斯瓦尔巴群岛、比利时瓦登海和莱茵河口、日本海岸、南非、马来西亚半岛、瑞典沿海、巴西东南沿海、葡萄牙海岸、我国渤海东海和长江口等地，以及大西洋东北部和西北部、北极水域、北海南部和地中海。得出的数据触目惊心，每升海水和每千克（或每平方米）沉积物中，一般有几十到几百个塑料微粒，最高的有几千个微粒；美国旧金山海湾的表层海水中，每平方米多达 2 个微粒；即使在北极海水中，每立方米中也有 1 个微粒。该综述的其他数据在此不一一列举。

除了 Auta 等的综述外，2017 年上半年，至少有 30 篇关于海洋微塑料和塑料碎片的研究论文发表。基于本综述的主题和篇幅限制，本章不深入进行关于塑料微粒对海洋环境尤其是海洋生物的影响的综述；以下将仅介绍两篇涉及鱼类体内微塑料的论文[89,90]，其他海洋动物和鸟类体内微塑料的研究均未列入，但这并不意味着动物体内的微塑料不重要。众所周知，人类位于食物链的最顶端，海产品是人们餐桌上不可或缺的营养食品。值得一提的是，在我们食用的海盐中也发现了塑料微粒[91]；每千克食盐中，有几十到几百个塑料微粒，成分包括 PE、赛璐酚和 PET。

以由近到远的研究区域来排序，有以下这些研究工作：Peng 等[92]在长江口采集了 53 个站点的沉积物，发现每千克沉积物中有 20~340 个颗粒，上海附近的含量最高；人造纤维和聚酯类的成分最多，主要来自洗衣水。我国沿海 21 种海水鱼和 6 种淡水鱼体内的微塑料和塑料碎片的含量令人吃惊，每个个体中有 1.1~7.2 个颗粒，绿鳍马面鲀（*Thamnaconus septentrionalis*）体内的最多（7.2 个微粒/个体），海鱼高于淡水鱼[90]；与国际水平比较，我国的较高。广东 8 个沿海海滩中塑料碎片的粒径分布分析结果表明，PS 泡沫颗粒中小粒径的多，而树脂类基本保留原来的粒径；塑料碎片中微塑料（< 5 mm）占 98%（丰度）和 71%（重量）以上，说明这些海滩上的塑料碎片绝大多数属于微塑料范围[93]。第一份香港海域海水和沉积物中微塑料时空分布的综合研究报告由 Tsang 等[94]提交：每 $100m^3$ 海水中含 51~27909 个颗粒；每千克沉积物中含 49~279 个颗粒；主要成分为 PP、低密度 PE、高密度 PE、PP-乙烯丙烯混合物、苯乙烯丙烯腈。

波斯湾的霍尔木兹海峡（Strait of Hormuz，伊朗）沿岸海滩上，每千克沉积物中有几个至上千个塑料微粒；主要成分是 PE、PET 和尼龙；主要来源为海滩上的塑料碎片、废弃的渔具和城市及工业废水中的衣服纤维[95]。对伊朗主要石油出口枢纽 Khark 岛沿海沉积物中的微塑料和有毒元素 Zn、Mo、Pb、Cu、Cd、As 的研究结果[96]表明，所有的沉积物样品中均检出塑料碎片和纤维；海岸沉积物中微塑料与重金属、PAHs 的相关性较好，说明微塑料可成为其他污染物的载体。

Sruthy 等[97]在印度的第一份关于湖和河口沉积物中微塑料的报告中指出，微塑料的主要成分是低密度 PE。在印度洋上马尔代夫一个偏远珊瑚岛屿海岸，Imhof 等[98]研究了大、中、微塑料颗粒的时空分布；虽然岛屿十分遥远，还是发现了大量塑料；南岸和北岸不同时间采集的样品中的含量均不一样；这表明不定时的快速取样可导致分析结果的偏差，未来的监测应考虑塑料沉积的时空变化。Nel 等[99]提出，海岸带微塑料的含量可反映南非沿海人口数据；海港是海洋环境塑料微粒的源头。

自 2012 年 6 月至 2015 年 3 月，Pasternak 等[100]在地中海沿岸以色列的 8 个海滩上进行了 19 次调研，研究了塑料碎片的来源、组成和分布；海滩平均每 100 平方米有 12.1 个垃圾碎片，其中 90%是塑料，主要来源是食品包装物和塑料袋。另一份来自以色列的文献[101]指出，地中海以色列海岸带的 17 个站点均检出了微塑料，其丰度平均值高于世界其他地区的 1~2 个数量级，为主颜色是浅色（白色或透明）。Guven 等[89]采集了地中海土耳其水域 18 个站位的水样和沉积物样及 10 个站位的鱼样，探讨了微塑料成分；发现在水和沉积物中，94%微塑料的粒径在 0.1~0.25 mm 之间；在 23 种 1337 条鱼的胃肠道中发现大量塑料，主要是纤维和硬塑料，为主颜色是蓝色；34%鱼的胃中和 41%鱼的肠中有微塑料，平均每个胃或肠有 1.8 粒。Leslie 等[102]基于对荷兰阿姆斯特丹运河、废水处理厂、北海海域沉积物和生物的研究，探讨微塑料的迁移途径和生态影响；发现微塑料除了在污水处理厂的污泥中沉积，还会随处理过的水流出，经过阿姆斯特丹运河，在海洋沉积。

在俄罗斯加里宁格勒（Kaliningrad）地区波罗的海海滩，13 个站位表层 2 cm 的沙中，每千克沙检出 13~363 个碎片颗粒，以泡沫塑料为主[103]。Graca 等[104]研究了波罗的海南部海洋和沙滩沉积物中塑料微粒的来源和归宿；发现最常见的塑料微粒尺寸范围在 0.1~2.0 mm 之间，主要成分是透明纤维；常见的纤维成分聚酯是海底沉积物和海滩沉积物中最常见的微塑料，各占比 50%和 27%；此外，在海底沉积物和海滩沉积物中，造船使用的聚乙酸乙烯酯以及用于包装的聚乙烯丙烯各占所有聚合物的 25%和 18%；聚合物密度也是影响塑料微粒循环的重要因素之一，低密度塑料碎片在更大程度上比高密度碎片更能在海滩沉积物和海水之间循环。

其他研究区域还包括：大西洋北部的亚速尔群岛（Azores Archipelago）Faial-Pico 通道的海底[105]、土耳其东北部地中海海岸的小亚细亚和梅尔辛海湾（Iskenderun and Mersin Bays）[106]、摩洛哥东北部沿岸的 Martil 湿地[107]、南非海岸[108]等。

在为数不多的机理研究中，Chubarenko[109]关于海洋塑料破碎机制的报道显得很有意思。他们以波罗的海为研究区域，从琥珀的形成过程得到启示，基于水文等数据提出塑料粉碎的假设：暴风作用下，海岸带就像是塑料的磨坊，漂浮的塑料碎片在海滩和水下岸坡之间往复迁移，直到分解成足够小的碎片，随着水流去向更深的海区，在巨浪作用区域之外沉积。针对微塑料异质性引起的采样和分析误差，Fisner[110]对沙滩上的微塑料污染提出了量化方法。

5 监测技术和方案

5.1 同位素示踪技术

以 Hg 稳定同位素为典型代表，同位素示踪法已经成为污染物溯源的重要技术。Hg 稳定同位素技术在地表水、沉积物和大气颗粒物中的应用已经十分广泛，但在海洋样品中的应用很少，原因是高盐基底导致常规分析方法的不适用。在初步解决了海水样品的同位素分析方法[111]后，Huang 和 Lin 等[112-114]对使用海水脱硫技术的沿海燃煤电厂的脱硫海水进行了分析，利用 Hg 的同位素特征追踪了电厂燃煤排放的 Hg，研究了从原煤到燃煤烟气、从烟气进入脱硫海水、从电厂的水处理池排放到大气和近海海区全过程中 Hg 的迁移和形态的转化，评估了脱硫海水排海后海区的 Hg 污染范围。Sun 等[115]探讨了福建九龙江河口红树林区大气和沉积物中的 Hg 向红树树叶及树根的迁移。这些较为深入的 Hg 的源汇分析和迁移转化的定性定量，均得益于同位素分析技术的发展。

墨西哥的南下加利福尼亚州（Baja California Sur，Mexico）由于过量抽取地下水供应城市用水，导致海水入侵；Mahlknecht 等[116]分析了 23 个地下水样品中的 Sr 和 B 的同位素，将其作为示踪物，用以研究人类活动影响下的海水入侵机制。根据伊比利亚大西洋西南部大陆架（Iberian Atlantic shelf）沉积物中 Pb 的同位素特征，Mil-Homens 等[117]分析了 Pb 的污染来源及历史变迁。Wu 等[20]利用 Pb 同位素比值，研究了江苏射阳河的沉积污染史，发现 1980~1995 年间随着工业发展，重金属污染逐渐增大。Chien 等[118]分析了西菲律宾海海水中的 Pb 同位素比值，探讨了海水中 Pb 的来源。

5.2 监测方案和策略

2017 年 3 月出版的 *Marine Environmental Research*，收录了多篇海洋环境监测方面综合进展的报道，尤其介绍了欧洲国家采取的监测计划或方案。欧盟的许多涉海国家，都在履行各种海洋环境监测计划，海洋环境监测已经成为政府部门的常规工作及企业和公众的自愿行动。继美国国家海洋和大气局（National Oceanic and Atmospheric Administration，NOAA）1984 年开始开展“国家状况和发展趋势的贻贝监测项目”（National Status and Trends，Mussel Watch Project，简称“贻贝监测计划”）[119,120]后，欧洲各国也建立了类似的监测计划。2008 年，欧洲议会和欧洲理事会制定了旨在采取共同行动保护海洋环境的框架指令——《欧盟海洋战略框架指令》(European Union Marine Strategy Framework Directive，MSFD)，该框架指令要求各个成员国发展一套可靠的方法和定性指标，用于描述“良好环境状态”（Good Environmental Status），比如，证明污染物浓度控制在不引发污染影响的水平之下[121]。

国际海洋考察理事会（International Council for the Exploration of the Sea，ICES）和《奥斯陆-巴黎保护东北大西洋海洋环境公约》（Oslo-Paris Convention for the Protection of the Marine Environment of the North-East Atlantic，OSPAR）提出了污染物的浓度及其影响的综合评估准则。该准则最初仅是为东北大

西洋海区设计的，但实践证明也可以应用于欧洲的其他海岸带地区，如地中海海区。在 www.ospar.org 网站上，可以查找 OSPAR 委员会于 2012 年发表的基于 3 年实践成果的指南，该评估框架可进一步提供评估“良好环境状态”的适当方法。

Vethaak 等[122]解读了监测和评估海洋环境中有害物质及其效应的综合指标和评价方法。如何最好地将化学污染物和相关生物效应的数据进行综合，做出有实际意义的评估，是 ICES 专家组的工作焦点。专家组致力于确定一套化学和生物的核心框架，为监测计划定义参数、提出指标及其测定方法，包括污染物的浓度、暴露和效应等。最重要的是，该框架建立了生物效应测定的评估标准，指出在测定生物和沉积物（也可能包括水）中污染物的同时，如何与生物效应结果综合起来。

欧盟成员国正在拟订措施，以在 MSFD 框架下解决海洋垃圾的问题，然而，数据库仍不够丰富；针对此，罗马尼亚的国家海洋研究和发展部门提出了一个全民监测海洋垃圾的计划[123]，在罗马尼亚黑海海岸的 3 个海滩上，监测以塑料碎片和烟头为主的海洋垃圾。另外，值得一提的是英国的“公民科学项目”（Citizen Science Projects）[124]；该计划依托市民收集信息，成本低，可收集信息数量大，时间和空间覆盖率宽。此外，这类项目可提高全民对环境问题的认识，使公众对环境的行为和态度产生积极变化。

基于 ICON 国际会议（International Workshop on Marine Integrated Contaminant Monitoring）的 ICON 计划（the Trans-European-Research Project on Field Studies Related to A Large-scale Sampling and Monitoring）[125,126]覆盖了北自冰岛、南到西班牙的地中海海岸的全欧洲范围，旨在检验环境综合评价的框架，评估欧洲一些特定河口、海湾及海区的污染状况。ICON 计划要求各国的监测策略、采样和分析、提交的数据，均应该吻合 ICON 的统一要求。

Robinson 等[127]评估了北大西洋和欧洲海区的沉积物、鱼和贻贝中痕量金属、PAHs 和 PCBs 的浓度；从冰岛到北海离岸及沿海，有机污染物浓度逐渐增加，于河口达到最高；金属有更复杂的分布，反映了当地的人为源、天然源以及水文地质条件之混合；除了北海中央，所有站点的污染物都高于本底值。他们还发现，最初为东北大西洋设计的 ICES 和 OSPAR 方法，也可适用于地中海的生物物种和环境介质。Stienen[128]分析了比利时北海沿岸 1962~2015 年的海鸟长期监测数据，发现 20 世纪 60 年代期间由于密集的船舶交通和极高的石油污染，导致海鸟的死亡率高于正常的 7 倍之多；50 多年后，船舶交通的密集度依然存在，但石油污染率和海鸟死亡率已经降至正常水平，这应该得益于北海地区一系列相关的立法措施。

Martinez-Gomez 等[129]选取地中海西部的沿海城市为研究区域，探讨由 ICES 和 OSPAR 制定的化学品及其影响的海洋监测综合指标体系的适宜性；他们选择红鲻鱼（*Mullus barbatus*）和地中海贻贝作为指示生物；使用定性的计分法评估综合评价的置信度。依据 ICON 提出的“污染物浓度及其影响的综合评估准则”的公认实验方法，Martinez-Gomez 等[130]从地中海沿岸找到两个贻贝中的应激生物标志物[Stress on Stress （SoS） and lysosomal membrane stability （LMS）]；实验结果表明，这两个常见且价廉的生物标志物可以适用于整个欧洲海域，可在大范围的污染生物监测计划如《欧盟海洋战略框架指令》下，提供筛查与污染物相关的生物效应的方法。

Lyons 等[121]采用不同介质（沉积物和生物）中的化学污染物（金属、PAHs 和 PCBs）及生物效应（bile metabolites and pathology，EROD，胆汁代谢物及病理学）数据，确定英国东海岸外北海海区的“良好环境状态”。Nelms 等[124]根据“公民科学项目”的数据，对英国海滩的人为垃圾（主要是塑料）进行了为期 10 年（2005~2014 年）的全国性评估；他们综合海洋保护协会（Marine Conservation Society，MCS）志愿者对英国海岸海滩垃圾的调查数据，分析垃圾的组成、空间分布及形成沿海塑料碎片的时间趋势；根据海洋保护协会的人数、调查时长、调查范围等综合资料，将采样标准化，以减少误差，提高数据的价值和可信程度。该报告指出：塑料是英国海滩上人为垃圾的主要成分，主要来自陆源和市民的乱抛；西英吉利海峡和凯尔特海（Celtic Sea）的垃圾污染最为严重；10 年来污染呈现逐渐增加的趋势。该报告还讨论了数据集的限制，并为今后的工作提出建议。由于在政府财政限制下的采样计划不可能有如此大的规模，这项研究体系体现了其他方法做不到的“公民科学数据”的价值。

6 展　望

综上所述，近海及河口水环境中的化学污染物，一直是海洋环境学家关注的焦点。目前的工作中，考察污染物组成成分、含量水平和分布状况的居多。深入探讨污染物迁移转化机制，将是研究走向更高层次的必然。化学污染物尤其是微塑料的源与汇、分布和组成、迁移及转化，仍将在今后较长时间内作为研究的热点。而地表水环境中常见的“新型”污染物如含氟化合物，目前在海洋环境中尚少见报道，留有巨大研究空间。区域调研性的研究结果，对于建立近海及河口环境污染数据库、制定海洋环境管理决策均十分重要。欧美国家在长期的研究和实践后，已经建立了多个系统性的污染监测体系，包括在国家层面上和国家之间统一的监测手段和参数指标、规范的监测计划和策略，以及基于公众参与的方案，并付诸实施。从这个角度看，制定和实施规范系统、切合实际的监测方案，将逐步取代零散的污染物调查；也就是说，污染调查将成为环境监测部门的日常工作及公众与企业的自觉行动。

参 考 文 献

[1] Jamieson A J, Malkocs T, Piertney S B, Fujii T, Zhang Z L. Bioaccumulation of persistent organic pollutants in the deepest ocean fauna. Nat Ecol Evolut, 2017, 1: 0051.

[2] Thompson R C, Olsen Y, Mitchell R P, Davis A, Rowland S J, John A W G, McGonigle D, Russell A E. Lost at sea: Where is all the plastic? Science, 2004, 304(5672): 838-838.

[3] Law K L, Thompson R C. Microplastics in the seas. Science, 2014, 345(6193): 144-145.

[4] Li H J, Lin L, Ye S, Li H B, Fan J F. Assessment of nutrient and heavy metal contamination in the seawater and sediment of Yalujiang Estuary. Mar Pollut Bull, 2017, 117(1-2): 499-506.

[5] Li H J, Ye S, Ye J Q, Fan J F, Gao M L, Guo H. Baseline survey of sediments and marine organisms in Liaohe Estuary: Heavy metals, polychlorinated biphenyls and organochlorine pesticides. Mar Pollut Bull, 2017, 114(1): 555-563.

[6] Ke X, Gui S F, Huang H, Zhang H J, Wang C Y, Guo W. Ecological risk assessment and source identification for heavy metals in surface sediment from the Liaohe River protected area, China. Chemosphere, 2017, 175: 473-481.

[7] Liu B L, Nie Y S, Gao X H, Hu K, Yang J. The diagenetic geochemistry and contamination assessment of iron, cadmium, and lead in the sediments from the Shuangtaizi estuary, China. Environ Earth Sci, 2017, 76(4): 168.

[8] Li C, Sun M H, Song C W, Tao P, Yin Y Y, Shao M H. Assessment of heavy metal contamination in the sediments of the Shuangtaizi Estuary using multivariate statistical techniques. Soil Sediment Contam, 2017, 26(1): 45-58.

[9] Hu N J, Huang P, Zhang H, Wang X J, Zhu A M, Liu J H, Shi X F. Geochemical source, deposition, and environmental risk assessment of cadmium in surface and core sediments from the Bohai Sea, China. Environ Sci Pollut Res, 2017, 24(1): 827-843.

[10] Xiao C L, Jian H M, Chen L F, Liu C, Gao H Y, Zhang C S, Liang S K. Toxic metal pollution in the Yellow Sea and Bohai Sea, China: distribution, controlling factors and potential risk. Mar Pollut Bull, 2017, 119(1): 381-389.

[11] Wang L P, Lei K, Qiao Y Z, Hao C L. Level and ecological risk of four common metals in surface water along the Qinhuangdao coastal areas, China. IOP Conference Series- Earth and Environmental Science, 2017, 52, UNSP 012084.

[12] Li H M, Kang X M, Li X M, Li Q, Song J M, Jiao N Z, Zhang Y Y. Heavy metals in surface sediments along the Weihai coast, China: Distribution, sources and contamination assessment. Mar Pollut Bull, 2017, 115(1-2): 551-558.

[13] Wang Y, Ling M, Liu R H, Yu P, Tang A K, Luo X X, Ma Q M. Distribution and source identification of trace metals in the sediment of Yellow River Estuary and the adjacent Laizhou Bay. Phys Chem Earth, 2017, 97: 62-70.

[14] Yang H J, Sun J K, Song A Y, Qu F Z, Dong L S, Fu Z Y. A Probe into the contents and spatial distribution characteristics of available heavy metals in the soil of Shell Ridge Island of Yellow River delta with ICP-OES method. Spectrosc Spect Anal,

2017, 37(4): 1307-1313.

[15] Zhang G L, Bai J H, Zhao Q Q, Jia J, Wen X J. Heavy metals pollution in soil profiles from seasonal-flooding riparian wetlands in a Chinese delta: Levels, distributions and toxic risks. Phys Chem Earth, 2017, 97: 54-61.

[16] Liu X B, Li D L, Song G S. Assessment of heavy metal levels in surface sediments of estuaries and adjacent coastal areas in China. Front Earth Sci, 2017, 11(1): 85-94.

[17] Xu F J, Liu Z Q, Cao Y C, Qiu L W, Feng J W, Xu F, Tian X. Assessment of heavy metal contamination in urban river sediments in the Jiaozhou Bay catchment, Qingdao, China. Catena, 2017, 150: 9-16.

[18] Bao K S, Shen J, Sapkota A. High-resolution enrichment of trace metals in a west coastal wetland of the southern Yellow Sea over the last 150 years. J Geochem Explor, 2017, 176(SI): 136-145.

[19] Li Y, Li H G. Historical records of trace metals in core sediments from the Lianyungang coastal sea, Jiangsu, China. Mar Pollut Bull, 2017, 116(1-2): 56-63.

[20] Wu S S, Han R M, Yang H, Wang Q J, Bi F Z, Wang Y H. A century-long trend of metal pollution in the Sheyang River, on the coast of Jiangsu (China), reconstructed from sedimentary record. Chem Ecol, 2017, 33(1): 1-17.

[21] Liu Y, Liu G J, Wang J, Wua L. Spatio-temporal variability and fractionation of vanadium (V) in sediments from coal concentrated area of Huai River Basin, China. J Geochem Explor, 2017, 172: 203-210.

[22] Chen Z Y, Chen L, Chen C Z, Huang Q H, Wu L L, Zhang W. Organotin contamination in sediments and aquatic organisms from the Yangtze Estuary and adjacent marine environments. Environ Eng Sci, 2017, 34(4): 227-235.

[23] Chen B, Liu J, Qiu J D, Zhang X L, Wang S, Liu J Q. Spatio-temporal distribution and environmental risk of sedimentary heavy metals in the Yangtze River Estuary and its adjacent areas. Mar Pollut Bull, 2017, 116(1-2): 469-478.

[24] Han D M, Cheng J P, Hu X F, Jiang Z Y, Mo L, Xu H, Ma Y N, Chen X J, Wang H L. Spatial distribution, risk assessment and source identification of heavy metals in sediments of the Yangtze River Estuary, China. Mar Pollut Bull, 2017, 115(1-2): 141-148.

[25] Yu S, Hong B, Ma J, Chen Y S, Xi X P, Gao J B, Hu X Q, Xu X R, Sun Y X. Surface sediment quality relative to port activities: A contaminant-spectrum assessment. Sci Total Environ, 2017, 596: 342-350.

[26] Jin A M, Yang L, Chen X G, Loh P S, Lou Z H, Liu G, Ji S L. Ecological risk and contamination history of heavy metals in the Andong tidal flat, Hangzhou Bay, China. Hum Ecol Risk Assess, 2017, 23(3): 617-640.

[27] Wu H, Bi X Y, Lin G H, Feng C C, Li Z J, Qi F, Zheng T L, Xie L Q. Trace metals in sediments and benthic animals from aquaculture ponds near a mangrove wetland in Southern China. Mar Pollut Bull, 117(1-2): 486-491.

[28] Zhao G M, Ye S Y, Yuan H M, Ding X G, Wang J. Surface sediment properties and heavy metal pollution assessment in the Pearl River Estuary, China. Environ Sci Pollut Res, 2017, 24(3): 2966-2979.

[29] Zhang Z Q, Chen L, Wang W P, Li T J, Zu T T. The origin, historical variations, and distribution of heavy metals in the Qiongzhou Strait and nearby marine areas. J Ocean U China, 2017, 16(2): 262-268.

[30] Kumar S B, Padhi R K, Mohanty A K, Satpathy K K. Elemental distribution and trace metal contamination in the surface sediment of south east coast of India. Mar Pollut Bull, 2017, 114(2): 1164-1170.

[31] Gopal V, Krishnakumar S, Peter T S, Nethaji S, Kumar K S, Jayaprakash M, Magesh N S. Assessment of trace element accumulation in surface sediments off Chennai coast after a major flood event. Mar Pollut Bull, 2017, 114(2): 1063-1071.

[32] Samanta S, Amrutha K, Dalai T K, Kumar S. Heavy metals in the Ganga (Hooghly) River estuary sediment column: evaluation of association, geochemical cycling and anthropogenic enrichment. Environ Earth Sci, 2017, 76(4): 14.

[33] Nethaji S, Kalaivanan R, Viswam A, Jayaprakash M. Geochemical assessment of heavy metals pollution in surface sediments of Vellar and Coleroon estuaries, southeast coast of India. Mar Pollut Bull, 2017, 115(1-2): 469-479.

[34] Krishnakumar S, Ramasamy S, Chandrasekar N, Peter T S, Gopal V, Godson P S, Godson P S. Trace element concentrations in reef associated sediments of Koswari Island, Gulf of Mannar biosphere reserve, southeast coast of India. Mar Pollut Bull, 2017, 117(1-2): 515-522.

[35] Magesh N S, Chandrasekar N, Krishnakumar S, Peter T S. Trace element contamination in the nearshore sediments of the Tamiraparani estuary, Southeast coast of India. Mar Pollut Bull, 2017, 116(1-2): 508-516.

[36] Sundararajan S, Khadanga M K, Kumar J P P J, Raghumaran S, Vijaya R, Jena B K. Ecological risk assessment of trace metal accumulation in sediments of Veraval Harbor, Gujarat, Arabian Sea. Mar Pollut Bull, 2017, 114(1): 592-601.

[37] Costa-Boddeker S, Hoelzmann P, Thuyen L X, Huy H D, Nguyen H A, Richter O, Schwalb A. Ecological risk assessment of a coastal zone in Southern Vietnam: Spatial distribution and content of heavy metals in water and surface sediments of the Thi Vai Estuary and Can Gio Mangrove Forest. Mar Pollut Bull, 2017, 114(2): 1141-1151.

[38] Hop N V, Dieu H T Q, Phong N H. Metal speciation in sediment and bioaccumulation in Meretrix lyrata in the Tien Estuary in Vietnam. Environ Monit Assess, 2017, 189(6): 299.

[39] Khodami S, Surif M, Maznah W O W, Daryanabard R. Assessment of heavy metal pollution in surface sediments of the Bayan Lepas area, Penang, Malaysia. Mar Pollut Bull, 2017, 114(1): 615-622.

[40] Haris H, Aris A Z, bin Mokhtar M. Mercury and methylmercury distribution in the intertidal surface sediment of a heavily anthrophogenically impacted saltwater-mangrove-sediment interplay zone. Chemosphere, 2017, 166: 323-333.

[41] Armidi A, Shinjo R, Ruslan R, Fahmiati. Distributions and pollution assessment of heavy metals Pb, Cd and Cr in the water system of Kendari Bay, Indonesia. IOP Conference Series- Materials Science and Engineering, 2017, 172: UNSP 012002.

[42] Watts M J, Mitra S, Marriott A L, Sarkar S K. Source, distribution and ecotoxicological assessment of multielements in superficial sediments of a tropical turbid estuarine environment: A multivariate approach. Mar Pollut Bull, 2017, 115(1-2): 130-140.

[43] Delshab H, Farshchi P, Keshavarzi B. Geochemical distribution, fractionation and contamination assessment of heavy metals in marine sediments of the Asaluyeh port, Persian Gulf. Mar Pollut Bull, 2017, 115(1-2): 401-411.

[44] Zarezadeh R, Rezaee P, Lak R, Masoodi M, Ghorbani M. Distribution and accumulation of heavy metals in sediments of the northern part of mangrove in Hara Biosphere Reserve, Qeshm Island (Persian Gulf). Soil Water Res, 2017, 12(2): 86-95.

[45] El-Younsy A R, Essa M A, Wasel S O. Sedimentological and geoenvironmental evaluation of the coastal area between Al-Khowkhah and Al-Mokha, southeastern Red Sea, Republic of Yemen. Environ Earth Sci, 2017, 76(1): 50.

[46] Jamshidi S, Bastami K D. Preliminary assessment of metal distribution in the surface sediments along the coastline of the southern Caspian Sea. Mar Pollut Bull, 2017, 116(1-2): 462-468.

[47] El Nemr A, El-Said G F. Assessment and ecological risk of heavy metals in sediment and Molluscs from the Mediterranean Coast. Water Environ Res, 2017, 89(3): 195-210.

[48] Oueslati W, Zaaboub N, Helali M A, Ennouri R, Martins M V A, Dhib A, Galgani F, El Bour M, Added A, Aleya L. Trace element accumulation and elutriate toxicity in surface sediment in northern Tunisia (Tunis Gulf, southern Mediterranean). Mar Pollut Bull, 2017, 116(1-2): 216-225.

[49] Sondi I, Mikac N, Vdovic N, Ivanic M, Furdek M, Skapin S D. Geochemistry of recent aragonite-rich sediments in Mediterranean karstic marine lakes: Trace elements as pollution and palaeoredox proxies and indicators of authigenic mineral formation. Chemosphere, 2017, 168: 786-797.

[50] Durmaz E, Kocagoz R, Bilacan E, Orhan H. Metal pollution in biotic and abiotic samples of the Buyuk Menderes River, Turkey. Environ Sci Pollut Res, 2017, 24(5): 4274-4283.

[51] Jedruch A, Kwasigroch U, Beldowska M, Kulinski K. Mercury in suspended matter of the Gulf of Gdansk: Origin, distribution and transport at the land-sea interface. Mar Pollut Bull, 2017, 118(1-2): 354-367.

[52] Deycard V N, Schafer J, Petit J C J, Coynel A, Lanceleur L, Dutruch L, Bossy C, Ventura A, Blanc G. Inputs, dynamics and potential impacts of silver (Ag) from urban wastewater to a highly turbid estuary (SW France). Chemosphere, 2017, 167: 201-511.

[53] Tiquio M G J, Hurel C, Marmier N, Taneez M, Andral B, Jordan N, Francour P. Sediment-bound trace metals in Golfe-Juan Bay, northwestern Mediterranean: Distribution, availability and toxicity. Mar Pollut Bull, 2017, 118(1-2): 427-436.

[54] Manteca J I, Ros-Sala M, Ramallo-Asensio S, Navarro-Hervas F, Rodriguez-Estrella T, Cerezo-Andreo F, Ortiz-Menendez J E, de-Torres T, Martinez-Andreu M. Early metal pollution in southwestern Europe: the former littoral lagoon of El Almarjal (Cartagena mining district, SE Spain). A sedimentary archive more than 8000 years old. Environ Sci Pollut Res, 2017, 24(11): 10584-10603.

[55] Hakima Z, Mohamed M, Aziza M, Mehdi M, Meryem E B, Bendahhou Z, Jean-Francois B. Environmental and ecological risk of heavy metals in the marine sediment from Dakhla Bay, Morocco. Environ Sci Pollut Res, 2017, 24(9): 7970-7981.

[56] Doe K, Mroz R, Tay K L, Burley J, The S, Chen S. Biological effects of gold mine tailings on the intertidal marine environment in Nova Scotia, Canada. Mar Pollut Bull, 2017, 114(1): 64-78.

[57] Horta-Puga G. Geochemical partitioning of lead in biogenic carbonate sediments in a coral reef depositional environment. Mar Pollut Bull, 2017, 116(1-2): 71-79.

[58] Mattos Y, Stotz W B, Romero M S, Bravo M, Fillmann G, Castro I B. Butyltin contamination in Northern Chilean coast: Is there a potential risk for consumers? Sci Total Environ, 2017, 595: 209-217.

[59] Marinho C H, Giarratano E, Esteves J L, Narvarte M A, Gil M N. Hazardous metal pollution in a protected coastal area from Northern Patagonia (Argentina). Environ Sci Pollut Res, 2017, 24(7): 6724-6735.

[60] Duodu G O, Goonetilleke A, Ayoko G A. Potential bioavailability assessment, source apportionment and ecological risk of heavy metals in the sediment of Brisbane River estuary, Australia. Mar Pollut Bull, 2017, 117(1-2): 523-531.
[61] Alyazichi Y M, Jones B G, McLean E, Pease J, Brown H. Geochemical assessment of trace element pollution in surface sediments from the Georges River, Southern Sydney, Australia. Arch Environ Contam Tox, 2017, 72(2): 247-259.
[62] Vetrimurugan E, Shruti V C, Jonathan M P, Roy P D, Kunene N W, Villegas L E C. Metal concentration in the tourist beaches of South Durban: An industrial hub of South Africa. Mar Pollut Bull, 2017, 117(1-2): 538-546.
[63] Han D M, Currell M J. Persistent organic pollutants in China's surface water systems. Sci Total Environ, 2017, 580: 602-625.
[64] Zhang D L, Liu N, Yin P, Zhu Z G, Lu J F, Lin X H, Jiang X J, Meng X W. Characterization, sources and ecological risk assessment of polycyclic aromatic hydrocarbons in surface sediments from the mangroves of China. Wetlands Ecol Manage, 2017, 25(1): 105-117.
[65] Zeng L X, Lam J C W, Horii Y, Li X L, Chen W F, Qiu J W, Leung K M Y, Yamazaki E, Yamashita N, Lam P K S. Spatial and temporal trends of short- and medium-chain chlorinated paraffins in sediments off the urbanized coastal zones in China and Japan: A comparison study. Environ Pollut, 2017, 224: 357-367.
[66] Li J, Li F D. Polycyclic aromatic hydrocarbons in the Yellow River estuary: Levels, sources and toxic potency. Mar Pollut Bull, 2017, 116(1-2): 479-487.
[67] Tian X H, Xu Y J, Gong X H, Han D F, Wang Z Q, Zhou Q L, Sun C X, Ren C B, Xue J L, Xia C H. Environmental status and early warning value of the pollutant semicarbazide in Jincheng and Sishili Bays, Shandong Peninsula, China. Sci Total Environ, 2017, 576: 868-878.
[68] Liu N, Li X, Zhang D L, Liu Q, Xiang L H, Liu K, Yan D Y, Li Y. Distribution, sources, and ecological risk assessment of polycyclic aromatic hydrocarbons in surface sediments from the Nantong Coast, China. Mar Pollut Bull, 2017, 114(1): 571-576.
[69] Wang C L, Zou X Q, Li Y L, Zhao Y F, Song Q C, Yu W W. Pollution levels and risks of polycyclic aromatic hydrocarbons in surface sediments from two typical estuaries in China. Mar Pollut Bull, 2017, 114(2): 917-925.
[70] Zhang D, Wang J J, Ni H G, Zeng H. Spatial-temporal and multi-media variations of polycyclic aromatic hydrocarbons in a highly urbanized river from South China. Sci Total Environ, 2017, 581: 621-628.
[71] Yoon S J, Hong S, Kwon B O, Ryu J, Lee C H, Nam J, Khim J S. Distributions of persistent organic contaminants in sediments and their potential impact on macrobenthic faunal community of the Geum River Estuary and Saemangeum Coast, Korea. Chemosphere, 2017, 173: 216-226.
[72] Kamalakannan K, Balakrishnan S, Sampathkumar P. Petroleum hydrocarbon concentrations in marine sediments along Nagapattinam- Pondicherry coastal waters, Southeast coast of India. Mar Pollut Bull, 2017, 117(1-2): 492-495.
[73] Jafarabadi A R, Bakhtiari A R, Aliabadian M, Toosi A S. Spatial distribution and composition of aliphatic hydrocarbons, polycyclic aromatic hydrocarbons and hopanes in superficial sediments of the coral reefs of the Persian Gulf, Iran. Environ Pollut, 2017, 224: 195-223.
[74] Agah H, Mehdinia A, Bastami K D, Rahmanpour S. Polycyclic aromatic hydrocarbon pollution in the surface water and sediments of Chabahar Bay, Oman Sea. Mar Pollut Bull, 115(1-2): 515-524.
[75] Okogbue C O, Oyesanya O U, Anyiam O A, Omonona V O. Assessment of pollution from produced water discharges in seawater and sediments in offshore, Niger Delta. Environ Earth Sci, 2017, 76(10): 359.
[76] Dabrowska H, Kopko O, Lehtonen K K, Lang T, Waszak I, Balode M, Strode E. An integrated assessment of pollution and biological effects in flounder, mussels and sediment in the southern Baltic Sea coastal area. Environ Sci Pollut Res, 2017, 24(4): 3626-3639.
[77] Mali M, Dell'Anna M M, Mastrorilli P, Damiani L, Piccinni A F. Assessment and source identification of pollution risk for touristic ports: Heavy metals and polycyclic aromatic hydrocarbons in sediments of 4 marinas of the Apulia region (Italy). Mar Pollut Bull, 2017, 114(2): 768-777.
[78] Ndungu K, Beylich B A, Staalstrom A, Oxnevad S, Berge J A, Braaten H F V, Schaanning M, Bergstrom R. Petroleum oil and mercury pollution from shipwrecks in Norwegian coastal waters. Sci Total Environ, 2017, 593: 624-633.
[79] Botwe B O, Kelderman P, Nyarko E, Lens P N L. Assessment of DDT, HCH and PAH contamination and associated ecotoxicological risks in surface sediments of coastal Tema Harbour (Ghana). Mar Pollut Bull, 2017, 115(1-2): 480-488.
[80] Polidoro B A, Comeros-Raynal M T, Cahill T, Clement C. Land-based sources of marine pollution: Pesticides, PAHs and phthalates in coastal stream water, and heavy metals in coastal stream sediments in American Samoa. Mar Pollut Bull, 2017, 116(1-2): 501-507.

[81] Paez-Osuna F, Alvarez-Borrego S, Ruiz-Fernandez A C, Garcia-Hernandez J, Jara-Marini M E, Berges-Tiznado M E, Pinon-Gimate A, Alonso-Rodriguez R, Soto-Jimenez M F, Frias-Espericueta M G. Environmental status of the Gulf of California: A pollution review. Earth-Sci Rev, 2017, 166: 181-205.
[82] Neira C, Cossaboon J, Mendoza G, Hoh E, Levin L A. Occurrence and distribution of polycyclic aromatic hydrocarbons in surface sediments of San Diego Bay marinas. Mar Pollut Bull, 2017, 114(1): 466-479.
[83] Sanches P J, Bohm E M, Bohm G M B, Montenegro G O, Silveira L A, Betemps G R. Determination of hydrocarbons transported by urban runoff in sediments of Sao Goncalo Channel (Pelotas-RS, Brazil). Mar Pollut Bull, 2017, 114(2): 1088-1095.
[84] http: //en.unesco.org/united-nations-ocean-conference, June 10, 2017.
[85] Steensgaard I M, Syberg K, Rist S, Hartmann N B, Boldrin A, Hansen S F. From macro-to microplastics-Analysis of EU regulation along the life cycle of plastic bags. Environ Pollut, 2017, 224: 289-299.
[86] Xanthos D, Walker T R. International policies to reduce plastic marine pollution from single-use plastics (plastic bags and microbeads): A review. Mar Pollut Bull, 2017, 118(1-2): 17-26.
[87] Law K L. Plastics in the marine environment. Annual Review of Mar Sci, 2017, 9: 205-229.
[88] Auta H S, Emenike C U, Fauziah S H. Distribution and importance of microplastics in the marine environment: A review of the sources, fate, effects, and potential solutions. Environ Int, 2017,102: 165-176.
[89] Guven O, Gokdag K, Jovanovic B, Kideys A E. Microplastic litter composition of the Turkish territorial waters of the Mediterranean Sea, and its occurrence in the gastrointestinal tract of fish. Environ Pollut, 2017, 223: 286-294.
[90] Jabeen K, Su L, Li J N, Yang D Q, Tong C F, Mu J L, Shi H H. Microplastics and mesoplastics in fish from coastal and fresh waters of China. Environ Pollut, 2017, 221: 141-149.
[91] Yang D Q, Shi H H, Li L, Li J N, Jabeen K, Kolandhasamy P. Microplastic pollution in table salts from China. Environ Sci Technol, 2015, 49(22): 13622-13627.
[92] Peng G Y, Zhu B S, Yang D Q, Su L, Shi H H, Li D J. Microplastics in sediments of the Changjiang Estuary, China. Environ Pollut, 2017, 225: 283-290.
[93] Fok L, Cheung P K, Tang G D, Li W C. Size distribution of stranded small plastic debris on the coast of Guangdong, South China. Environ Pollut, 2017, 220: 407-412.
[94] Tsang Y Y, Mak C W, Liebich C, Lam S W, Sze E T P, Chan K M. Microplastic pollution in the marine waters and sediments of Hong Kong. Mar Pollut Bull, 2016, 115(1-2): 20-28.
[95] Naji A, Esmaili Z, Khan F R. Plastic debris and microplastics along the beaches of the Strait of Hormuz, Persian Gulf. Mar Pollut Bull, 2017, 114(2): 1057-1062.
[96] Akhbarizadeh R, Moore F, Keshavarzi B, Moeinpour A. Microplastics and potentially toxic elements in coastal sediments of Iran's main oil terminal (Khark Island). Environ Pollut, 2017, 220: 720-731.
[97] Sruthy S, Ramasamy E V. Microplastic pollution in Vembanad Lake, Kerala, India: The first report of microplastics in lake and estuarine sediments in India. Environ Pollut, 2017, 222: 315-322.
[98] Imhof H K, Sigl R, Brauer E, Feyl S, Giesemann P, Klink S, Leupolz K, Loder M G J, Loschel L A, Missun J. Spatial and temporal variation of macro-, meso- and microplastic abundance on a remote coral island of the Maldives, Indian Ocean. Mar Pollut Bull, 2017, 116(1-2): 340-347.
[99] Nel H A, Hean J W, Noundou X S, Froneman P W. Do microplastic loads reflect the population demographics along the southern African coastline? Mar Pollut Bull, 2017, 115(1-2): 115-119.
[100] Pasternak G, Zviely D, Ribic C A, Ariel A, Spanier E. Sources, composition and spatial distribution of marine debris along the Mediterranean coast of Israel. Mar Pollut Bull, 2017, 114(2): 1036-1045.
[101] van der Hal N, Ariel A, Angel D L. Exceptionally high abundances of microplastics in the oligotrophic Israeli Mediterranean coastal waters. Mar Pollut Bull, 2017, 116(1-2): 151-155.
[102] Leslie H A, Brandsma S H, van Velzen M J M, Vethaak A D. Microplastics en route: Field measurements in the Dutch river delta and Amsterdam canals, wastewater treatment plants, North Sea sediments and biota. Environ Int, 2017, 101: 133-142.
[103] Esiukova E. Plastic pollution on the Baltic beaches of Kaliningrad region, Russia. Mar Pollut Bull, 2017, 114(2): 1072-1080.
[104] Graca B, Szewc K, Zakrzewska D, Dolega A, Szczerbowska-Boruchowska M. Sources and fate of microplastics in marine and beach sediments of the Southern Baltic Sea- a preliminary study. Environ Sci Pollut Res, 2017, 24(8): 7650-7661.

[105] Rodriguez Y, Pham C K. Marine litter on the seafloor of the Faial-Pico Passage, Azores Archipelago. Mar Pollut Bull, 2017, 116(1-2): 448-453.

[106] Gundogdu S, Cevik C. Micro- and mesoplastics in Northeast Levantine coast of Turkey: The preliminary results from surface samples. Mar Pollut Bull, 2017, 118(1-2): 341-347.

[107] Alshawafi A, Analla M, Alwashali E, Aksissou M. Assessment of marine debris on the coastal wetland of Martil in the North-East of Morocco. Mar Pollut Bull, 2017, 117(1-2): 302-310.

[108] Verster C, Minnaar K, Bouwman H. Marine and freshwater microplastic research in South Africa. Integrated Environ Assess Manage, 2017, 13(3): 533-535.

[109] Chubarenko I, Stepanova N. Microplastics in sea coastal zone: Lessons learned from the Baltic amber. Environ Pollut, 2017, 224: 243-254.

[110] Fisner M, Majer A P, Balthazar-Silva D, Gorman D, Turra A. Quantifying microplastic pollution on sandy beaches: the conundrum of large sample variability and spatial heterogeneity. Environ Sci Pollut Res, 2017, 24(15): 13732-13740.

[111] Lin H Y, Yuan D X, Lu B Y, Huang S Y, Sun L M, Zhang F, Gao Y Q. Isotopic composition analysis of dissolved mercury in seawater with purge & trap preconcentration and a modified Hg introduction device for MC-ICP-MS. J. Anal Atom Spectrom, 2015, 30: 353-359.

[112] Huang S Y, Yuan D X, Lin H Y, Sun L M, Lin S S. Fractionation of mercury stable isotopes during coal combustion and seawater flue gas desulfurization. Appl Geochem, 2017, 76: 159-167.

[113] Huang S Y, Lin K N, Yuan D X, Gao Y Q, Sun L M. Mercury isotope fractionation during the transfer from post-desulfurization seawater to air. Mar Pollut Bull, 2016, 113(12): 81-86.

[114] Lin H Y, Peng J J, Yuan D X, Lu B Y, Lin K N, Huang S Y. Mercury isotope signatures of seawater discharged from a coal-fired power plant equipped with a seawater flue gas desulfurization system. Environ Pollut, 2016, 214: 822-830.

[115] Sun L M, Lu B Y, Yuan D X, Hao W B, Zheng Y. Variations in the isotopic composition of stable mercury isotopes in typical mangrove plants of the Jiulong estuary, SE China. Environ Sci Pollut Res, 2017, 24(2): 1459-1468.

[116] Mahlknecht J, Merchan D, Rosner M, Meixner A, Ledesma-Ruiz R. Assessing seawater intrusion in an arid coastal aquifer under high anthropogenic influence using major constituents, Sr and B isotopes in groundwater. Sci Total Environ, 2017, 587: 282-295.

[117] Mil-Homens M, Vale C, Brito P, Naughton F, Drago T, Raimundo J, Anes B, Schmidt S, Caetano M. Insights of Pb isotopic signature into the historical evolution and sources of Pb contamination in a sediment core of the southwestern Iberian Atlantic shelf. Sci Total Environ, 2017, 586: 473-484.

[118] Chien C T. Ho T Y, Sanborn M E, Yin Q Z, Paytan A. Lead concentrations and isotopic compositions in the Western Philippine Sea. Mar Chem, 2017, 189: 10-16.

[119] Lauenstein G G, Daskalakis K D. US long-term coastal contaminant temporal trends determined from mollusk monitoring programs, 1965-1993. Mar Pollut Bull, 1998, 37(1-2): 6-13.

[120] Wade T L, Sericano J L, Gardinali P R, Wolff G, Chambers L. NOAA's 'Mussel Watch' Project: Current use organic compounds in bivalves. Mar Pollut Bull, 1998, 37(1-2): 20-26.

[121] Lyons B P, Bignell J P, Stentiford G D, Bolam T P C, Rumney H S, Bersuder P, Barber, J L, Askem C E, Nicolaus M E E, Maes T. Determining Good Environmental Status under the Marine Strategy Framework Directive: Case study for descriptor 8 (chemical contaminants). Mar Environ Res, 2017, 124(SI): 118-129.

[122] Vethaak A D, Davies I M, Thain J E, Gubbins M J, Martinez-Gomez C, Robinson C D, Moffat C F, Burgeot T, Maes T, Wosniok W. Integrated indicator framework and methodology for monitoring and assessment of hazardous substances and their effects in the marine environment. Mar Environ Res, 2017, 124(SI): 11-20.

[123] Golumbeanu M, Nenciu M, Galatchi M, Nita V, Anton E, Oros A, Ioakeimidis C, Belchior C. Marine litter watch APP as a tool for ecological education and awareness raising along the Romanian Black Sea coast. J Environ Prot and Ecol. 2017, 18(1): 348-362.

[124] Nelms S E, Coombes C, Foster L C, Galloway T S, Godley B J, Lindeque P K, Witt M J. Marine anthropogenic litter on British beaches: A 10-year nationwide assessment using citizen science data. Sci Total Environ, 2017, 579: 1399-1409.

[125] Hylland K, Burgeot T, Martínez-G_omez C, Lang T, Robinson C D, Svarvarsson J, Thain J E, Vethaak A D., Gubbins M J. How can we quantify impacts of contaminants in marine ecosystems? the ICON project. Mar Environ Res, 2017, 124(SI): 2-10.

[126] Hylland K, Gubbins M. The ICON Project (the trans-European-research project on field studies related to a large-scale

sampling and monitoring). Mar Environ Res, 2017, 124(SI): 1.

[127] Robinson C D, Webster L, Martinez-Gomez C, Burgeot T, Gubbins M J, Thain J E, Vethaak A D, McIntosh A D, Hylland K. Assessment of contaminant concentrations in sediments, fish and mussels sampled from the North Atlantic and European regional seas within the ICON project. Mar Environ Res, 2017, 124(SI): 21-31.

[128] Stienen E W M, Courtens W, Van de Walle M, Vanermen N, Verstraete H. Long-term monitoring study of beached seabirds shows that chronic oil pollution in the southern North Sea has almost halted. Mar Pollut Bull, 2017, 115(1-2): 194-200.

[129] Martinez-Gomez C, Fernandez B, Robinson C D, Campillo J A, Leon, V M, Benedicto, J, Hylland K, Vethaak A D. Assessing environmental quality status by integrating chemical and biological effect data: The Cartagena coastal zone as a case. Mar Environ Res, 2017, 124(SI): 106-117.

[130] Martinez-Gomez C, Robinson C D, Burgeot T, Gubbins M, Halldorsson H P, Albentosa M, Bignell J P, Hylland K, Vethaak A D. Biomarkers of general stress in mussels as common indicators for marine biomonitoring programmes in Europe: The ICON experience. Mar Environ Res, 2017, 124(SI): 70-80.

作者：袁东星，李　炎，吴巧玲，林坤德

厦门大学生态与环境学院

第14章　高山和极地主要环境污染物研究进展

本章导读

高山和极地地区虽然远离人类活动，但依然受到环境污染的影响。高山和极地地区主要的环境污染物包括持久性有机污染物（POPs）和重金属类污染物（合称持久性有毒污染物）以及碳质气溶胶等。这些污染物虽然含量水平并不高，但对生态系统脆弱的高山和极地环境仍可能造成不可逆的严重危害。本章对这些高山和极地主要环境污染物的来源、迁移、分布、转化、生物富集以及变化趋势等方面近年来的研究进展进行了综述。

关键词

持久性有毒污染物，棕碳，高山，极地，长距离迁移

1 引　言

虽然极地及高山地区远离人类活动，但越来越多的研究表明极地及高山偏远区域已非净土，众多具有长距离迁移特性的污染物在这些区域不断被检出。极地及高山地区生态系统相对脆弱，对外来因素较为敏感，微量甚至痕量水平的环境污染也可能对当地生态系统造成不可逆的破坏，需要引起足够的重视。

持久性有毒污染物（persistent toxic substances，PTS），包括持久性有机污染物（POPs）和重金属类污染物，在环境中具有生物毒性强，难以降解，能够在生态系统中富集放大等特点，受到了国际社会的广泛关注。PTS 具有全球长距离迁移能力，是高山和极地地区最主要的环境污染物，严重威胁着高山和极地地区的生态环境安全。棕碳（brown carbon）会显著影响高山和极地大气多种光化学反应进程，近年来也受到了广泛关注。高山和极地 PTS 和棕碳的研究是目前环境化学领域关注的重要前沿和热点之一。

2 高山 POPs 的环境行为研究进展

高山区域 POPs 研究是当前环境科学领域持续关注的热点问题和重要前沿课题之一。在 POPs 的全球分配研究中，意大利学者 Calamari 等[1]发现六氯苯（HCB）的最高浓度发生在低纬度的高海拔区，这一现象引起了人们对高海拔偏远地区 POPs 环境行为的关注。与 POPs 的全球纬度分馏效应[2]类似，高山/高海拔区域存在区域尺度的温度梯度变化，这种污染物浓度随海拔升高而增加或组分分馏的现象被称为“高山冷凝效应”[3]。近年来越来越多的证据表明高山/高海拔区域成为 POPs 的聚集地，对其全球分配产生重要影响[4]。

高山区域本地 POPs 源相对有限，它是研究 POPs 长距离传输的典型区域。地球上大约 27%的陆地为高山，大约 22% 的世界人口居住在高山区域[5]。由于山地高海拔区域生态系统的脆弱性和敏感性，这些污染物的累积会对高山/高海拔区域的生态系统造成一定的威胁，而且随着食物链的传递进入人体，可能直接造成人体健康影响。因此研究高山区域 POPs 的环境行为具有重要的理论价值和现实意义。

在过去近 20 年，偏远高海拔山地 POPs 的研究取得了重要的进展。加拿大学者 Blais 等[6]首次在 *Nature* 发表文章证实了污染物的“高山冷凝效应”，随后在欧洲和北美的高海拔山地对 POPs 的环境行为开展了广泛的研究，主要围绕的核心问题是探讨高山冷凝效应与污染物性质和环境因素的关系。近年来高山区

域 POPs 的污染水平及环境行为已经引起越来越广泛的关注，研究的环境介质主要包括大气、水体、冰雪、湖泊沉积物、鱼体、土壤和植被等[5]。

青藏高原是世界上面积最大，平均海拔最高的高原，被喻为“地球第三极”。青藏高原当地 POPs 排放源十分有限，其独特的地形特征和环境气象条件为 POPs 的大气长距离输送与沉降规律研究提供了理想的条件。然而，相对于北美与欧洲偏远山区，青藏高原地区 POPs 的研究起步较晚，但近年来研究者对该地区的关注程度日益增大，我国学者在过去十余年积累了很多青藏高原地区 POPs 的时间和空间分布数据，为青藏高原地区 POPs 的传输规律和环境行为及其来源解析提供了重要的数据支持，并为研究 POPs 在全球的分布、传输迁移等过程提供了重要的依据。从整体上看，青藏高原大多数 POPs 浓度要低于一般的偏远山区，但对于一些有机氯农药（OCPs），其浓度水平则显著高于世界上其他偏远山区[7]。研究表明大气环流将 POPs 从青藏高原周围的人口密集、工业化程度高的地区（包括中亚、印度次大陆、成都平原）传输至高海拔地区，而后随着大气沉降和湿沉降作用，储存到青藏高原土壤以及植被等环境介质中，并且通过食物链的传递对生态环境的健康造成不利影响。

高山区域 POPs 的环境行为过程研究进展主要归纳为以下几个方面。

2.1　大气-地表分配

大气 POPs 与地表的交换是双向的，即大气向地表的沉降与地表向大气的挥发两个过程同时存在。当沉降通量大于挥发通量时，大气是地表 POPs 的来源；当挥发通量大于沉降通量时，地表可能成为 POPs 的“二次源”[8]。高山地区 POPs 大气-地表交换过程受到污染物浓度、温度、湿度和土壤有机碳含量等多种因素的共同影响。高海拔地区低温和湿沉降能加速 POPs 从大气向地表沉降[9]。大气-地表交换过程直接影响 POPs 在环境中的“源-汇”关系，对 POPs 的空间分布及全球循环起到重要的作用。Wang 等[10]基于逸度模型估算了青藏高原 POPs 的气-土交换趋势和通量，发现青藏高原土壤是滴滴涕（DDTs）和大分子多环芳烃（PAHs）的“汇”。

2.2　高山冷凝效应

研究 POPs 污染物在高山环境行为的一个重要方面就是污染物随海拔的变化趋势。污染物在迁移过程中，离污染源的距离越来越远，因此在不断的稀释和降解的作用下，污染物的浓度本该逐渐降低。但是大量的文献报道显示，高山环境中 POPs 浓度会随着海拔的升高而增大，表现出与预期相反的情况，这意味着污染物倾向于在温度较低的高海拔地区富集。然而不同的 POPs 沿不同的山坡的海拔效应是不同的，海拔高度、污染物沿海拔的降解过程、污染物沉降方式、不同的环境介质构成了高山冷凝结效应的复杂性[5]。

Davidson 等[11]研究了加拿大西部高山地区 POPs 的高山冷凝效应，发现蒸气压大于 0.1 Pa 的易挥发性有机氯化合物的浓度水平随海拔的升高而增加，而低挥发性的有机氯化合物不具备该现象。通过对 Pyrenees、Alps 和 Caledonian 等高山区域有机氯化合物大气沉降的研究，Carrera 等[12]发现在高山湖泊生态系统中，挥发性较强的有机污染物更容易富集在温度较低的地区，同时其冷凝效率也随着气温的降低而升高。同样，Shen 等[13]在有关北美大气中 OCPs 污染分布的调查研究中发现，易挥发性有机污染物更容易进行大气长距离迁移，表现出高山冷凝效应现象。

在我国，Yang 等[14]对比研究了青藏高原东南部色季拉山迎风坡和背风坡 OCPs 的海拔分布行为，发现迎风坡的高山冷凝效应强于背风坡。Chen 等[15]对我国四川西部山区土壤中 POPs 的冷凝沉降效应开展了相关研究，结果显示所有目标化合物的浓度水平都随着海拔的升高呈现指数增长趋势。而 Wang 等[16]

对我国珠穆朗玛峰地区土壤中多氯联苯（PCBs）和多溴联苯醚（PBDEs）的浓度水平和海拔分布进行了研究，发现在海拔 4500 m 以上各目标化合物浓度与海拔高度之间存在显著相关性，而在较低海拔时污染物的浓度随海拔升高而降低。Zhu 等[17]对藏东南地区大气 POPs 海拔梯度分布的影响进行了研究，发现 $\log K_{OA}$ 和 $\log K_{WA}$ 值分别在 8~11 和 2~4 的半挥发性有机污染物上更容易在高海拔处蓄积浓缩。

2.3 森林过滤效应

森林代表一种重要的环境组分，富含有机质的森林植被叶片能够吸附大气中的 POPs，随着叶片的凋落沉降到林下土壤，使得森林像“泵”一样加剧了 POPs 从大气向森林土壤的传输，这一效应被称为“森林过滤效应”[18]。Nizzetto 等[19]在意大利阿尔卑斯山区观测了 PCBs 的沉降通量，发现林内沉降通量比林外高 1~3 倍，显示森林过滤效应增加了 POPs 的沉降。Horstmann 等[20]研究发现，森林地区 POPs 的沉积通量高于非森林地区的 10 倍以上。这些都是森林过滤效应的直接观测证据。

Su 和 Wania[21]首次研究森林过滤效应与 POPs 长距离迁移的关系，发现森林能够使 POPs 大气长距离迁移的能力减小 1/2。森林能够阻截大气中 POPs 迁移，而将其保留在有机碳丰富的森林土壤中。Yang 等[14]通过藏东南森林土芯和林外土芯对比研究，发现森林土壤 OCPs 和较重分子量 PAHs 的浓度均要高于无森林覆盖土壤，而且较难挥发的 PAHs 森林过滤效应强。可以看出，森林作为连接大气和土壤 POPs 传输过程的重要纽带，能够显著增强大气 POPs 向土壤中的迁移。POPs 污染物一旦沉降到土壤便很难再挥发到大气，因此森林土壤成为不易挥发 POPs 的蓄积库[22-23]，对 POPs 的环境行为和对生态系统的影响起着不可忽略的作用。

3 高山重金属研究进展

由于人为活动（工业活动以及化石燃料燃烧等）的增加，越来越多的重金属被释放到环境中，大气中重金属含量显著上升[24]。与 POPs 类似，重金属在高海拔地区也存在着“冷凝效应”[25]。相比于低海拔地区，高山地区具有更脆弱的生态系统，对全球环境的变化更为敏感[26]，为研究重金属大气长距离传输提供了有利的条件[27]。

青藏高原是世界上海拔最高的高原，总面积达 122 万 km^2，其中耕地面积占 64.57%，建筑用地占 0.05%，而未被利用的土地高达 35.38%[28]，其人口稀少，几乎没有工业，一直以来被认为是未被污染的背景区域[29]。然而，近年的一些研究发现，青藏高原部分地区的环境和生物样品中重金属含量较高[30,31,32]。例如，青藏高原拉萨河和其他几个湖泊（纳木错和羊湖）鱼体内具有相当高含量的铜和锌，最高浓度可以分别达到 32.2 mg/kg 和 102.5 mg/kg[27,32]。Zhang 等[30]和 Shao 等[33]分别对青藏高原湖泊和河流鱼体中汞的含量及形态进行了研究，发现鱼体中汞含量普遍较高[最高达 2.1 mg/kg 干重(dw)]，而且主要以甲基汞形态存在。

目前，对青藏高原各环境介质中重金属的污染已开展了较为广泛的研究。青藏高原四条河流水中重金属（铜、锌、银、镉、铬）的浓度相对较低，个别地区存在铝、铅和铁的污染[29]。纳木错地区[34]以及喜马拉雅地区[35]的土壤中重金属浓度与中国背景区平均重金属浓度一致，东部贡嘎山土壤重金属浓度与世界其他高山地区基本一致，但是镉的浓度显著偏高[26]。青藏高原中部纳木错地区大气气溶胶以及湿沉降颗粒中重金属元素的浓度与其他偏远地区相当，且远低于一般大城市，可作为大气监测的背景点[36,37]。拉萨地区的 PM_{10} 中[38]重金属元素的浓度虽然低于其他城市地区，但明显高于纳木错地区。对积雪中重金属的研究表明，东南部喜马拉雅山地区受重金属污染较小，且其中重金属浓度随季风呈现明显的季节性变化[39,40]，东北部积雪中重金属浓度也与背景区基本一致[41]。通过对不同种类植物（松针、树叶、树皮、苔藓、松萝）中重金属浓度的分析，发现苔藓和松萝可作为青藏高原重金属污染指示生物[42,43,44]，并且松萝和苔藓中重金属浓度呈现从南到北从东到西递减的趋势[42,43]。

不同学者也采用多种技术方法对青藏高原重金属的来源进行了研究，并取得了显著的进展。元素富集因子是重金属溯源常用的方法之一，是指该元素相对于地壳元素（铝、铁等）的富集倍数，通常用来评判人为源的影响[44]。当富集因子接近于 1 时，代表元素主要来自于自然源，富集因子越大表示元素受到人为源的影响越大，一般认为当富集因子大于 10 时，元素主要来自于人为源[36]。Cong 等[36]使用该方法分析出纳木错地区大气沉降样品中，镉、钴、镍、铜、锌、铬和铅等主要来源于人为污染，而铁、铝、锰和钒等其他金属主要来自于天然源。Huang 等[39]发现青藏高原南部的降雪中铝、钒、镉、锰、钴主要来自于自然源，而铜、锌、镉、汞和铅则具有很高的富集因子，主要来自于人为源。Tripathee 等[35]发现尼泊尔区域喜马拉雅山上的土壤中重金属元素主要来自于自然源，表示喜马拉雅山土壤可以作为背景区土壤，为喜马拉雅地区的研究提供背景数据。反推气流轨迹模型分析也是污染物溯源常用的方法之一，该方法可以判定不同迁移轨迹气流对所研究区域元素组成的影响[36]。Cong 等[36]利用该方法分析出夏季季风期间，纳木错区域大气气溶胶里的一些重金属（铬、镍、铜、锌、砷）来源于南亚地区，与该地区夏季 POPs 具有相同的来源[45]。

稳定同位素技术为重金属的溯源研究提供了新的思路。目前较为常用的是铅同位素和汞同位素。不同来源的铅有不同的同位素比值，而且该比值不受传输过程中物理或化学变化的影响，因而铅同位素的组成可以为铅的源解析提供有力的工具。Cong 等[38]利用铅同位素判定拉萨大气 PM_{10} 中金属元素主要来源于自然源，极少部分来自于水泥厂。Bing 等[44]利用铅同位素分析发现青藏高原东部人为源铅主要来自于采矿、有色金属冶炼、燃煤以及机动车尾气排放。随着同位素分析技术的发展，汞稳定同位素也被应用于汞污染的溯源。现有研究表明，汞同位素存在着质量分馏、奇数同位素的非质量分馏以及偶数同位素的非质量分馏，这些分馏效应构成了汞同位素独特的指纹信息，使得利用汞同位素溯源成为可能。青藏高原低海拔地区汞同位素 $\Delta^{199}Hg$ 趋近于零，说明低海拔地区存在汞的人为污染，而高海拔地区 $\Delta^{199}Hg$ 值偏负，说明青藏高原高海拔地区受大气长距离传输的影响更为显著[46]。Xu 等[47]将汞同位素与碳氮同位素结合，综合分析了青藏高原纳木错和羊卓雍错湖中不同介质的同位素信息以及可能存在的反应。

4　高山与极地大气棕碳气溶胶研究进展

工业革命以来，人类的工农业生产与交通运输等活动向大气中排放了大量的污染物质，已经对生态和环境造成了严重的损害。其中，碳质气溶胶具有较强的辐射强迫效应，从而在局地乃至全球气候变化中具有重要的作用。根据光学特征的差异，碳质气溶胶分为黑碳（black carbon，BC）和有机碳（organic carbon，OC）气溶胶两部分。黑碳气溶胶对太阳辐射具有显著的吸收作用[48]，其潜在的温室效应已经成为气溶胶气候效应研究的重要内容。联合国政府间气候变化专门委员会（IPCC）第五次报告指出，全球黑碳气溶胶的辐射强迫可以达到+0.64 W/m^2。以往的研究认为有机碳对太阳辐射只具有散射作用，在气候变化中起到负的强迫效应。然而，越来越多的研究发现，碳质气溶胶是一系列吸光性连续变化的混合体，在强吸光性的黑碳和无吸光能力的有机碳之间还存在大量吸光能力介于上述两者之间有机物质，其吸光特性随波长变短迅速增强（图 14-1），该部分吸光性有机物质被定义为棕碳 [49]。棕碳气溶胶在短波长紫外区间的吸光特性，可能会显著地影响大气中多种光化学反应进程[50]。此外，已有的研究发现，高原气溶胶中 OC/BC 比值高达 20 以上[51]。因此，高原气溶胶中吸光性棕碳的来源、传输与转换等环境化学行为以及气候环境影响在近年来受到了广泛关注。

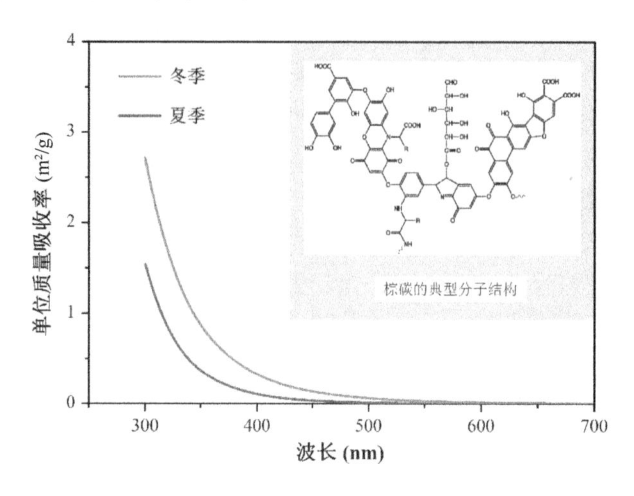

图 14-1　棕碳的吸光特性和典型分子结构

4.1　棕碳来源与分类

棕碳气溶胶的来源有多种，其一次来源途径主要有生物质[52]和化石燃料燃烧[53]以及植被和土壤的挥发[54]等。二次来源过程主要是人为或自然产生的一些前体有机物质（如异戊二烯和木质素氧化产物等）在大气中通过多相或均相反应生成[55,56]（图 14-2）。棕碳气溶胶来源多样并在大气中参与多种化学反应过程，使其化学组成相对复杂。根据其吸光特性的差异，棕碳可以分为两大类，一类是吸光能力相对较弱的类腐殖质成分（humic-like substances，HULIS），另一类是吸光能力较强的“Tar ball”[57,58]。

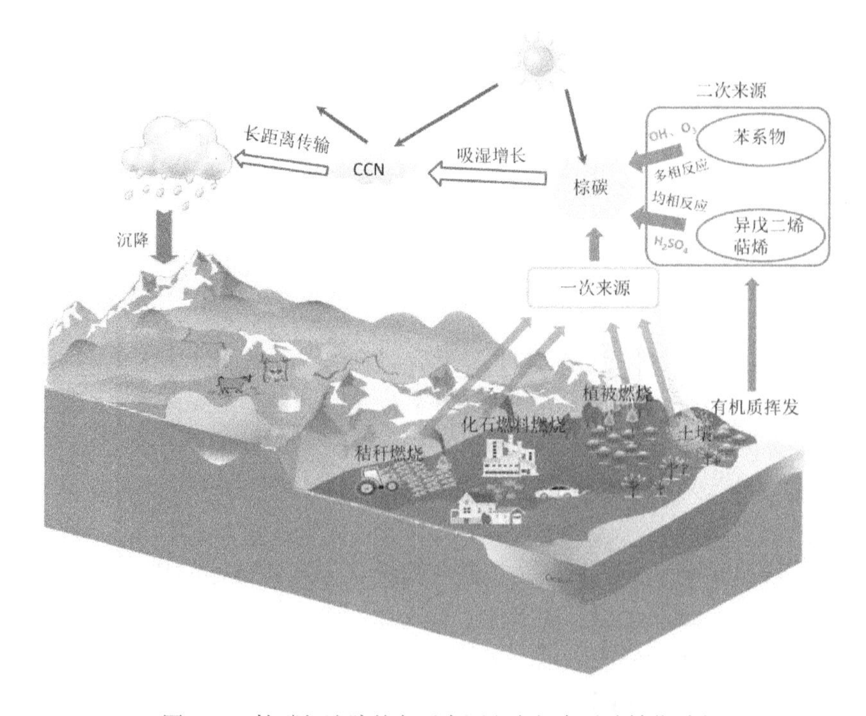

图 14-2　棕碳气溶胶的主要来源和大气中迁移转化过程

HULIS 是棕碳气溶胶中的水溶性部分，其平均分子质量在 200~300 Da 之间，因其吸光、红外光谱等性质与腐殖质相似而得名[49,54]。其分子结构是以苯环结构为骨架，连接有—COOH，—CH_2OH，—$COCH_3$ 等脂肪族侧链和末端—CH_3 取代基[59,60]（图 14-1）。由于大量苯环、羧基等结构中不饱和键的共轭效应，使得 HULIS 对紫外等短波辐射具有较强的吸收能力。此外，由于 HULIS 还具有吸湿性、表面活性等特征，使其成为气溶胶中重要的云凝结核和冰核，可能会对大气中水汽循环模式产生深远影响。Tar ball 主要是由生物质热降解过程中产生的低挥发性有机物质在高温条件下通过气粒转化过程形成的不规则有机颗粒[61]。在化学结构方面，Tar ball 相对于黑碳的碳原子 sp^2 杂化程度较低，且含有较高的含氧官能团，因此可溶性相对较高。

4.2　高原地区污染特征

青藏高原素称地球的“第三极”，是“气-水-冰-生”多圈层体现最全且全球气候变化最敏感的地区之一。由于青藏高原本身空气质量洁净，使其成为研究大气污染物跨境传输过程及大气化学反应的“天然实验室”。毗邻青藏高原的南亚地区人口分布密集、工业发展迅速，是全球大气污染最为严重的区域之一。该地区各种形式的生物质燃烧是其经济发展的重要能量来源，也是大气棕碳的主要排放形式之一[58]。Ramanathan 等[62]的研究表明，每年 10 月至 5 月南亚地区都会有大范围的以黑碳和棕碳为主要成分的大气棕色云（atmospheric brown clouds，ABCs）分布，其厚度可以高达 3 km。这些聚积的污染物，在大尺度季风环流的作用下可以输送到高原内部[63,64]，从而对高原上环境质量产生严峻挑战。Li 等[65]以青藏高原拉萨地区为研究地点，发现水溶性棕碳在 365 nm 的单位吸收率为（0.74±0.22）m^2/g，约为黑碳吸光能力的 10%。此外，由于棕碳化学成分的复杂性，其水溶性也存在较大的差异。棕碳气溶胶除了水溶性部分外，还存在大量的水不溶性组分，而这部分疏水组分往往具有更强的吸光能力，如 Kirillova 等[66]在喜马拉雅山脉南侧金字塔站（5079 m asl）的研究发现，疏水性棕碳的吸光能力是亲水性的 2~3 倍，其相对于黑碳对太阳辐射的吸收贡献分别为 8%±1%和 17%±5%。以往的气溶胶模式评估中认为有机碳对太阳辐射只具有散射作用，显然低估了碳质气溶胶的增温效应。

高原地区有广袤的冰川分布，使其环境条件非常脆弱，同时也对全球气候变化响应比较敏感。沉降在冰川表面的碳质气溶胶，可以明显地降低冰川表面的反照率，从而加速冰川的消融[67]。青藏高原上棕碳气溶胶以（0.11 ± 0.05）$g{\cdot}C/(m^2{\cdot}a)$的速率沉降到冰川表面[68]。同时，沉降到雪冰中的棕碳物质也表现出了显著的吸光能力，其对太阳辐射的吸收贡献相当于雪冰中黑碳气溶胶的 9.5%±8.4% [69]。Guilhermet 等[70]通过分析阿尔卑斯山（4250 m asl）冰芯中的棕碳含量推断该地区大气中水溶性棕碳的浓度为 0.07 $\mu g/m^3$。Legrand 等[71]发现二战以后阿尔卑斯山冰芯中棕碳含量比 1945 年以前的增长了 1.3~6 倍，说明人为排放是棕碳的重要来源途径。

高原及其周边地区发育有复杂的地质地貌，深刻改变了地区乃至全球的大气环流模式，从而有利于棕碳气溶胶的大尺度扩散。Randel 等[72]采用模式模拟发现亚洲季风环流是近地表生物质燃烧排放污染物输入平流层的有效途径。高原高大的地形作用进一步加强了污染物的垂直输送作用。Zhang 等[73]在北美洲大陆的研究进一步发现，气团的深对流运动对大气污染物的输送具有选择性，即更加倾向于将棕碳物质输送到对流层顶，而对黑碳却具有一定的去除作用。因此，随着气团深对流作用对棕碳向对流层顶的输送将会显著改变全球的辐射平衡和大气中光化学反应的过程。

4.3　极地地区污染特征

位于地球两端的南北极地区，虽然当地污染物排放较少，但是受到大气长距离传输的影响，周边地区的棕碳气溶胶可以输送到两极地区。除了在大气中可以产生直接的或者间接的辐射强迫效应，沉降到雪冰中的棕碳物质也可以参与到雪冰中的多种化学反应过程中，甚至改变其融化速率[58]。Nguyen 等[74]指出北极气溶胶中棕碳的年均含量为（0.02±0.01）μg/m^3，并且具有明显的季节变化规律，其浓度最高值出现在受生物质燃烧影响的时期。Voisin 等[75]计算得出北极雪冰中棕碳在 250 nm 处的质量吸收率为(26±11)cm^2/mg，而其表征吸光性随波长变化响应程度的 Ångström 指数为 7.7(450~550 nm)，说明雪冰中棕碳对短波长太阳辐射同样具有显著的吸收性。从辐射吸收的角度，雪冰中棕碳能够贡献总吸收的 25%~50%[76]。

5 极地多氯联苯研究进展

自 20 世纪 70 年代起多氯联苯在世界范围内被陆续禁止生产和使用，但是其在环境介质中仍然被广泛检出。南、北极虽然远离人类聚集区和工业区，但是目前极地环境介质中仍然有 PCBs 的明显检出。

早在 1976 年 Risebrough 等[77]首次报道了南极降雪和企鹅蛋中 PCBs 的检出，并提出大气传输而不是洋流传输是 PCBs 到达南极的主要传输途径。自此，不断有文献报道了极地地区 PCBs 等 POPs 的赋存水平和迁移转化规律。

5.1 极地大气 PCBs 的浓度水平与污染特征

大气是 PCBs 在全球范围内迁移的主要传输途径。Li 等[78]利用大气被动采样技术研究了西南极乔治王岛区域菲尔德斯半岛地区大气中 PCBs 等 POPs 的污染水平和分布特征，其中指示性 PCBs（PCB-28，52，101，118，138，152，180）的浓度为 1.66~6.50 pg/m^3，平均浓度为 4.34 pg/m^3。如此低的浓度水平反映了该区域仍然处于非常洁净的状态。这一浓度水平略高于韩国南极科考站大气中的 PCBs（0.85~3.12 pg/m^3）。Wang 等[79]报道了 2011 年 1 月至 2014 年南极长城站附近大气中 PCBs 的变化规律。大气中 PCBs 浓度为 5.87~72.7 pg/m^3。作者发现低氯代 PCBs 单体与温度存在显著相关性，表明 PCBs 受到土壤、海洋等环境介质二次挥发源的影响。另外，Gambaro 等[80]报道了南极 Terra Nova Bay 区域大气中 PCB-28、PCB-52 的浓度分别为 0.088~0.22 pg/m^3、0.027~0.10 pg/m^3。Montone 等[81]报道了巴西南极科考站大气中 PCB-52、101、118、138、153、180 的浓度分别为 8.6 pg/m^3、4.9 pg/m^3、4.7 pg/m^3、2.6 pg/m^3、4.0 pg/m^3、0.3 pg/m^3。Kallenborn 等报道了南极 Signy Island 大气中 PCBs 浓度为 20.67 pg/m^3。

南极大气中 PCBs 的同族体分布以低氯代 PCB 单体为主。高氯代单体对总 PCBs 浓度的贡献则较小，这也表明大气中的 PCBs 主要来自于大气长距离迁移[78]。在对南极 Signy Island，韩国南极科考站，Terra Nova Bay 和巴西科考站大气 PCBs 的研究中也发现了相同的分布规律。

北极地区相对于南极由于与较多国家相邻，因此受到人类活动影响更大。20 世纪 90 年代起，科学家们建立了关于北极大气 POPs 污染的长期监测项目，即北极监测与评估计划（AMAP）。该项目设有四个北极大气监测站，分别是 Canada 的 Alert，Iceland 的 Storhofoi，Svalbard 的 Zeppelin 和 Finland 的 Pallas。监测结果显示近 20 年来（1993~2012 年）北极大气中的 PCBs 呈现缓慢下降的趋势，表明了主要污染排放源的减少，但是环境介质二次释放源的影响变得越来越重要[82]。2014 年 Zhang 等[83]检测到新奥尔松地区大气中气相和颗粒相 PCBs 的浓度分别为 1.7~6.3 pg/m^3 和 9.2~141 pg/m^3，而 Σ_8PCBs 浓度为 0.63~30 pg/m^3，与 2010 年 Hung 等[84]报道的浓度一致（5.7~34 pg/m^3）。Choi 等[85]对北极 Svalbard Ny-Alesund 大气中 POPs 的被动采样研究表明，北极大气中 PCBs 的总浓度为 95 pg/m^3，明显高于南极大气的浓度。北极大气中的 PCBs 以二氯、三氯等低氯代 PCBs 单体为主，污染来源方面受到大气长距离传输和科考站区排放源的共同影响。

5.2 极地土壤及植物 PCBs 的浓度水平与污染特征

目前 PCBs 在极地土壤和植物样品中仍有明显检出。Wang 等[86]对于南极菲尔德斯半岛附近的研究表明，土壤和沉积物中 PCBs 浓度位于 60.1~1436 pg/g dw 之间，平均值为 410 pg/g dw，这与其他南极地区污染物的报道结果基本一致。Borghini 等[87]报道的南极 Victoria Land 区域土壤中 PCBs 浓度在 0.36~0.59 ng/g dw 范围内；Klánová 等[88]调查了 James Ross 岛土壤和沉积物中 PCBs、OCPs 和 PAHs 的浓度水平，发现其中指示性 PCBs 浓度分别在 0.51~1.82 ng/g dw 和 0.32~0.83 ng/g dw 之间。Fuoco 等[89]分析了意大利南极考察期间获得的 Terra Nova 湾、Wood 湾和 Victoria Land 区域海洋、湖泊沉积物和土壤样品中 PCBs，其浓度分别在 45~361 pg/g、102~560 pg/g 和 61~120 pg/g 之间。Park 等[90]对乔治王岛区域的调查发现 PCBs 浓度具有更低的检出水平（8.0~33.8 pg/g），这可能是由于采样点和采样时间不同而导致的。

南极菲尔德斯半岛地衣和苔藓中 PCBs 总浓度分别在 0.40~0.75 ng/g（平均值 0.54 ng/g）和 0.41~0.95 ng/g（0.67 ng/g）范围内。这两者均低于其他研究报道中 PCBs 浓度水平。如 Negoita 等[91]报道南极俄罗斯科考站地衣中 PCBs 浓度 3.3 ng/g；Fuoco 等[92]报道南极 Victoria Land 区域苔藓中 PCBs 浓度为 23~34 ng/g; Lead 等[93]报道挪威地区苔藓中 PCBs 浓度为 6.1~52 ng/g。但是，Cabrerizo 等[94]报道了南极 Deception 和 Livingstone Islands 环境介质中的 PCBs 处于更低的浓度水平，该区域土壤、地衣和苔藓中 PCBs 的浓度分别达 0.005~0.14 ng/g、0.04~0.61 ng/g、0.04~0.76 ng/g。

北极土壤和植物中 PCBs 的报道比较有限。Zhang 等[83]报道了北极新奥尔松地区土壤和沉积物中 PCBs 的浓度为 2.76~10.8 ng/g 和 3.09~8.32 ng/g，植物样品中 PCBs 的浓度为 22.5~56.3 ng/g。而 Zhu 等[95]报道的北极新奥尔松地区土壤、苔藓中 PCBs 的浓度则分别为 0.57~2.5 ng/g、0.43~1.2 ng/g，而不同草本类植物（仙女木、四棱岩须、苔草、高山发草、虎耳草）中 PCBs 的浓度处于较一致的水平（0.42~0.58 ng/g、0.48~0.63 ng/g、0.30~0.54 ng/g、0.37~0.46 ng/g、0.37 ng/g）。

极地土壤、地衣和苔藓中 PCBs 主要以中、低氯代单体为主，表明其主要受到大气长距离传输的影响。脂肪含量是影响地衣和苔藓中 PCBs 含量的重要因素，而有机质含量则是影响土壤中 PCBs 分布的重要因素。

5.3 极地海洋生物 PCBs 的浓度水平及富集

海洋生物属于极地食物链中的较高端，PCBs 由于具有较高的亲脂性从而更容易在海洋生物体内蓄积及随食物链进行传递。目前有关南极海洋食物链中 POPs 随食物链传递的研究尚不多。Corsolini 等[96]对比了南极（Ross sea）海洋食物链和亚北极（Iceland）海洋食物链中生物体 POPs 的浓度及生物放大趋势，发现南极海洋生物中 PCBs 的浓度反而高于北极，并且该南极海洋食物链中 PCBs 的生物放大因子大于北极，这可能与不同物种的生态特征、气候因子、污染物的来源及迁移转化有关。处于海洋食物链较低营养级的生物往往具有较低的污染物浓度，例如南极扇贝中 PCBs 的含量为 4.2~25 ng/g，平均浓度达 11.7 ng/g[97]。南极鳕鱼（*Trematomus bernacchii*）肌肉中 PCBs 的浓度水平为（0.08± 0.07）ng/g 至（6.35±4.8）ng/g 的湿重（ww）范围[98]。南极贼鸥中 PCBs 的含量[282~8630 ng/g 脂重(lw)]远高于阿德雷企鹅（34.5~103 ng/g lw）、帽带企鹅（41.3~132 ng/g lw）和金图企鹅（18.1~64.8 ng/g lw）[99]，这一浓度趋势反映了贼鸥处于南极食物链中较顶端的营养级位置。海洋哺乳动物同样处于海洋食物链较高的营养级，因此其体内往往具有较高的 PCBs 污染水平。例如，阿拉斯加两个海域鲸脂中 PCBs 的含量分别达 441~4530 ng/g lw 和 2190~9070 ng/g lw [100]。成年环斑海豹体内 PCBs 的含量可达 139~8870 ng/g lw [101]。北极熊体内 PCBs 的浓度更高，其脂肪中 PCBs 的含量高达 624~8645 ng/g lw、1339~67010 ng/g lw [102]。

6 极地有机氯农药研究进展

有机氯农药是一类对环境构成严重威胁的人工合成环境激素，主要以苯或环戊二烯为合成原料。其中以苯为原料合成的有机氯农药主要包括使用最早、最广的杀虫剂六六六（HCHs）、DDTs、六氯苯（HCB）以及六六六的高丙体制品林丹（γ-HCH）、滴滴涕类似物甲氧滴滴涕，也包括从滴滴涕结构衍生而来生产吨位小及品种繁多的杀螨剂，如三氯杀螨醇、三氯杀螨砜等。除此之外还有一些杀菌剂，如五氯硝基苯、百菌清及稻丰宁等。以环戊二烯为原料的有机氯农药包括杀虫剂的氯丹（Chlordane）、七氯（Heptachlor）、艾氏剂（Aldrin）、狄氏剂（Dieldrin）、异狄氏剂（Endrin）等。此外以松节油为原料的莰烯杀虫剂、毒杀芬（Toxaphene）也属于 OCPs。其中八种 OCPs 为《斯德哥尔摩公约》中禁用或限制使用的持久性有机污染物，尽管到现在绝大多数 OCPs 已经被禁止使用多年，但由于其持久性及长距离传输性，在环境介质中仍然能检出，对人类健康仍存在一定的威胁[103]。

过去几十年间，大量的 OCPs 进入环境，通过不断地挥发-沉降等环境行为在全球范围内的迁移和分配，以至于在远离人类活动的两极和高山地区监测到 OCPs 的踪迹。随着气候变化，南北极地区温度升高，冰雪融化加剧，OCPs 再次挥发进入大气，使得南北极可能成为 OCPs 的“二次源”。因此，极地环境介质中 OCPs 的赋存状态以及 OCPs 的环境行为是揭示 OCPs 的全球迁移和分配等环境行为的重要部分，为研究其环境行为规律、验证理论模型的准确性提供重要的基础数据。

6.1 极地 OCPs 的输入及其在环境中的迁移

OCPs 有四种可能的途径进入极地的生态环境：大气传输与沉降、洋流输入和沿极地河流输入（北极）、海冰融化，以及季节性迁移生物的迁移，每种路径的重要性取决于化合物的物理化学性质以及来源区域类型。

长距离大气迁移是 POPs 全球迁移的最主要途径之一。根据“雪龙船”北极航线调查结果，大气中 DDTs 从我国渤海，经过东亚、北太平洋到北冰洋，平均含量下降。气团运动轨迹表明，所有的气团起始于东亚和俄罗斯陆地，表明北极的 OCPs 由大气运动从使用源头地迁移而来[104]。北极地区大气中硫丹-I 的水平和季节变化趋势也表明其很有可能是被源自欧亚大陆和俄罗斯附近北冰洋的气团携带进入北极环境中[105]。南极的情形相似，从南纬 23°到 36°区域的大气中 DDTs 的分布趋势显示出陆源特征，与气团从非洲陆地大西洋运动的轨迹相吻合[106]。Kang 等[107]发现东南极半岛积雪中 α-HCH 和 γ-HCH 的含量随纬度的增加略有增加；气团轨迹模拟表明，气团主要是从印度和大西洋运动到南极大陆，因此，南极积雪中的 OCPs 主要通过在大气中的长距离迁移沉降到积雪表层。在低温条件下，大气中 POPs 更易于通过湿沉降进入海水或陆地，雪作为主要的湿沉降方式，是大气污染物输入极地的一个路径。Wania 等[108]指出，雪对气态和颗粒态的低氯代 PCBs 和 2 环、3 环的 PAHs 的去除非常显著。但在极地就相对较为干燥的气候来说，雨雪等湿沉降过程的重要性则大大降低。海气交换是 OCPs 在大气与海洋之间进行物质迁移的最主要途径。Su 等[109]发现六氯苯（HCB）从大气迁移至海水；PCBs 和毒杀芬这两种物质的迁移趋势是从大气向水中迁移。六六六（HCHs）的对映体比例（ER）指示 α-HCH 从表层海水中挥发进入上层大气。由此可见，目前 POPs 的海气交换过程不再是单向地从大气向海水中迁移，对某些特定的化合物来说，海水正在变成源，从海水中挥发进入大气[110]。

北极海域，尤其是北冰洋，河流输入被认为是污染物进入的重要途径之一。据估计，每年从叶尼塞河排放进入喀拉海的 PBDEs 约 1.92 kg，从鄂毕河排放入海的约 1.84 kg。Pućko 等[111]估计自 1986 年到 1993 年，北极混合层和波伏特海通过洋流输运接收了大量的 α-HCH，其中 12%进入北极混合层的 α-HCH 是在大气-海水交换和河流输入的联合作用下输入。Alexeeva 等[112]和 Macdonald 等[110]统计了俄罗斯河流对北极海域 OCPs 载荷量的贡献。Ma 等[113]分析了北极新奥尔松地区孔斯峡湾和附近开放海域表层沉积物中 OCPs 的残留状况，发现湾外站位沉积物中 α-HCH、β-HCH 和氯丹的含量高于湾内，表明可能是温暖的北大西洋洋流的西匹茨卑尔根暖流发挥了重要作用。

全球变暖导致冰川融化，也将导致多年前已沉积在冰川和冰雪表层的 POPs 随冰雪融水而再次进入极地的陆地或海洋生态系统中。Khairy 等[114]发现西南极半岛的冰河上积雪样品中的 DDTs 含量远高于其他的区域，说明融化的冰河可能是其二次来源。

迁徙动物的长距离迁移也是导致 POPs 进入极地生态系统的主要路径之一。由于 POPs 具有亲脂性，极易在各类生物体内放大与富集，因此，通过一些迁徙性动物的季节性迁徙，也可以将 OCPs 带入极地生态系统并沉积下来。Ewald 等[115]发现大马哈鱼体内的 PCBs 和 DDTs 在其迁徙进入阿拉斯加 Copper 河死亡后会进入该流域内的鳟鱼体内。区域季节性迁徙的海鸟会引入从数克到数千克数量不等的 POPs，同时，海鸟也会在留下大量的鸟粪，其中也会含有数量不容忽视的 POPs。南极地区未孵出鸟蛋中 DDTs、HCHs 以及 HCBs 的残留水平远高于环境介质，说明南极低纬度地区 OCPs 的污染状况可能是由海鸟向北迁移造成的[108,116,117]。鲸鱼也是一种典型的携带 POPs 的季节性迁徙生物。据估计 DDTs 随生物体迁徙进入极地的数量几乎可与通过大气和洋流进入极地的量相提并论[108]。Ewald 等[115]甚至指出，对一些不易于通过长距离大气迁移和洋流迁移的 POPs（极低的挥发性或水溶性、极易光降解而又不易于生物降解的物质），生物迁移反而会成为其进入极地生态系统的最主要的途径。

6.2　南极大陆各环境介质中 OCPs 的空间分布趋势

6.2.1　“雪龙船”南极科考航线大气中 OCPs 的年际变化

第 25~28 航次中国南极科考（2008 年 10 月 21 日~2009 年 4 月 4 日，2009 年 10 月 12 日~2010 年 4 月 5 日，2010 年 11 月 14 日~2011 年 3 月 29 日，2011 年 11 月 11 日~2012 年 4 月 2 日）的连续监测给出“雪龙号”航线大气中 OCPs 的时间-空间变化趋势。DDTs 和 HCHs 占所监测的 OCPs 的绝大部分，而艾氏剂、狄氏剂及环氧七氯仅占不到 10%。从北纬向南纬，*p,p'*-DDT 纬度变化幅度较大，*p,p'*-DDD 含量变化趋势不明显，*p,p'*-DDE 含量有降低趋势。四种 HCHs 成分的含量均随纬度的变化幅度较大。就时间变化趋势而言，*p,p'*-DDD 含量总体呈现逐年下降趋势，25 航次（0.581 pg/m^3）> 26 航次（0.280 pg/m^3）> 28 航次（0.251 pg/m^3），*p,p'*-DDE 浓度逐年下降，至 28 航次大气 *p,p'*-DDE 含量已经降低至 0.454 pg/m^3。南大洋大气 *α*-HCH 含量从 1.004 pg/m^3 下降至 0.465 pg/m^3，下降幅度近 60%，另外两种 *β*-HCH 和 *δ*-HCH 的含量也随时间推进而下降，相对于 *α*-HCH 下降速率较慢。在 HCH 各异构体组成中，*α*-HCH 所占比例较小，但是随着时间的推移 *δ*-HCH 的比例却不断增高。南大洋大气 *γ*-HCH 平均含量下降了 59%，这与 Cincielli 等[118]报道的南极大气 *γ*-HCH 半衰期为 2.9a 相近。在 1985~2005 年间及更长的时间尺度上，南极大气中 *γ*-HCH 含量也在不断下降[103]。

根据 HCHs 和 DDTs 的组成特征比值参数，可以推断出南极航线上大气中 DDTs 为历史残留，HCHs 则可能来自南大洋的海水中 *γ*-HCH 的挥发，也可能是周边地区林丹的使用[103]。

6.2.2 土壤和沉积物

南极 James Ross 岛的土壤中测得 DDTs 的含量为 0.52~3.68 ng/g dw，HCHs 的含量为 0.490~1.34 ng/g [88]。西乔治王岛和阿德雷的土壤中 23 种 OCPs 的含量为 0.094~1.26 ng/g，处于相对较低水平。其中 HCHs、DDTs 和氯丹为主要成分，其含量范围分别为 0.006~0.031 ng/g、0.019~0.277 ng/g 和 nd~0.060 ng/g[119]。该结果显著低于东南极洲和东南极海岸的企鹅岛的土壤中 OCPs 的含量。James Ross 岛的沉积物 DDTs 的含量为 0.19~1.15 ng/g dw，HCHs 的含量为 0.14~0.76 ng/g[88]，与西乔治王岛的沉积物中 DDTs（0.577 ng/g dw）相近，但是比 HCHs 的含量（0.003 ng/g dw）高一个数量级 [119]。

在沉积物和土壤样品中 γ-HCH 的含量均大约为 α-HCH 的两倍，然而 DDTs 的含量仅占 DDE 的 2%~5%。土壤和沉积物中 OCPs 皆来自于非洲、南美洲和澳大利亚的排放，通过大气运动进入 James Ross 岛[88]。西乔治王岛和阿德雷岛的大部分土壤和沉积物中 DDTs 的含量高于 HCHs，可以归因于农业中 DDTs 的使用量远大于 HCHs[119]。

6.2.3 植被

乔治王岛地衣（*Usnea aurantiaco-atra*，*Usnea antarctica*）、苔藓（*Sanionia uncinata*，*Syntrichia princeps* 和 *Brachytecium* sp.）和被子植物（*Colobanthus quitensis*）的 HCHs 和 DDTs 的含量范围分别为< MDL~1.20 ng/g dw 和< MDL~1.73 ng/g dw，与 Boghini 等[87]对南极苔藓中 HCHs（0.66~5.6 ng/g）和 DDTs（1.7~8.4 ng/g）的报道相吻合[120]。HCHs 的含量与 $\delta^{15}N$ 显著负相关，由于苔藓和地衣主要吸收大气中挥发态的 NH_3，而非由根部吸收水中的溶解态 NH_3，由此证明苔藓和地衣体内 HCHs 主要来自大气而非土壤。由于 HCHs 水溶性略大，而苔藓具有强大的排水能力，因此推断 HCHs 与近期的雪和附近的海鸟或海豹栖息地的冰层融水之间可能密切相关[120]。

Zhang 等[119]对西乔治王岛和阿德雷岛的地衣和苔藓的研究与 Cipro 等[120]相一致，OCPs 的含量范围在 0.373~0.812 ng/g 和 0.223~1.053 ng/g 之间。在苔藓中 DDTs 所占比例高于 HCHs，在地衣中则与之相反，这可能是由于苔藓和地衣在个体生态的不同所致。HCHs 的特征比值显示该地区的 HCHs 有混合来源——“老的”HCHs 和“新的”林丹。DDTs 在土壤中主要发生厌氧降解，产物为 DDD。此外，其组成比值显示仍有新鲜的 DDTs 输入。氯丹和狄氏剂含量在苔藓和地衣中的含量非常低，分别为 nd~0.039 和 nd~0.009 ng/g，但仍高于艾氏剂的含量（苔藓中未检出，地衣中 nd~0.005 ng/g）。灭蚁灵的含量同样非常低，在地衣中未检出，在苔藓中含量为 nd~0.008 ng/g。

6.2.4 生物和粪土

POPs 通过生物能流传递，在南极海洋食物链高营养层消费者体内富集，如企鹅、贼鸥和巨海燕等海鸟。有机氯污染物在企鹅组织机体内不同部位的含量随脂量的不同而有所差异，OCPs 的含量大小次序为脂肪>尾臀腺>骨质>头颅>肌肉，积累含量大小依次为 DDTs > HCHs，相比之下 HCHs 积累量较低，可能是由于农业上的禁用以及其化学结构与 DDTs 相比，易被生物转化。HCHs 在企鹅头颅、骨质样品中，β-HCH 含量相对比其他异构体要高。通常在 HCHs 的异构体中，β-HCH 是最持久稳固的，但在企鹅脂肪和尾臀腺样品中 α-HCH 异构体比 β-HCH 含量更高。DDTs 在企鹅头颅和骨质中含量较低，在肌肉中接近检测限，但在脂肪、尾臀腺中含量较高。企鹅脂肪样品为 98.6~107.4 ng/g，尾臀腺样品为 79.4~110.1 ng/g。与历史数据——企鹅脂肪样品为 30~972.3 ng/g，尾臀腺样品为 32.4 ~ 497.7 ng/g 相比较，说明近 10 年以来 DDTs 在南半球的使用得到了控制[121,122]。

菲尔德斯半岛和阿德雷岛的海鸟和海豹栖息地粪土层中 OCPs 的含量水平范围为 DDTs：0.40~2.01 ng/g 和 HCHs：0.28~0.76 ng/g。由于海洋动物在海洋生物食物链中所处的位置，动物消费结构（食性）与消费量以及不同动物的代谢特征，几种动物粪便中 OCPs 含量由高至低的趋势为：棕贼鸥（DDTs：1.72 ng/g；HCHs：0.51 ng/g）>灰贼鸥（DDTs：0.69 ng/g；HCHs：0.39 ng/g）、企鹅（DDTs：0.62 ng/g；HCHs：0.37 ng/g）和巨海燕（DDTs：0.44 ng/g；HCHs：0.29 ng/g）> 威德尔海豹（DDTs：0.50 ng/g；HCHs：0.34 ng/g）。各海鸟、海豹栖息地粪土 α-HCH/β-HCH 比值变化范围在 3.2~7.5 之间，二者比值差异不大，表明输入途径基本一致。海鸟卵中的 DDTs 和 HCHs 的含量远高于粪土层，分别为：贼鸥卵中 DDTs 含量范围为 56.6~304.0 ng/g，HCHs 的含量范围为 0.5~2.0 ng/g；巨海燕卵中 DDTs 含量范围为 12.7~53.7 ng/g，HCHs 的含量范围为 0.5~1.5 ng/g；企鹅卵中 DDTs 含量范围为 2.0~10.1 ng/g，HCHs 的含量范围为 6.0~10.2 ng/g。表明一旦有机氯污染物在海鸟体内积累了以后，不易靠代谢排出体外，极易在蛋卵中富集[123]。

6.3　北极地区环境中 OCPs 的残留状况

6.3.1　“雪龙船”北极航线大气

从 2008 年 7 月到 9 月的“雪龙船”北极航线调查给出从上海附近海域到高纬度北冰洋航线大气中 OCPs 的分布趋势。HCHs 在大气中的含量范围为 1.6~97 pg/m^3，平均浓度为 51 pg/m^3，最高的浓度值在靠近俄罗斯东海岸线的海域，在四个不同的地理区域，∑HCHs 平均浓度由高到低的顺序为：北太平洋，北极高纬度海域，东亚海域以及楚科奇和波伏特海。2008 年的测定结果比于 2003 年 Ding 等[104]的监测结果高大约 3.8 倍。整条航线中 *α*-HCH 和 *β*-HCH 的平均含量分别为 33 pg/m^3 和 5.4 pg/m^3，*γ*-HCH 在太平洋、楚科奇和波伏特海到高纬度北极地区的平均含量分别为 18 pg/m^3、11 pg/m^3 和 8.9 pg/m^3。在 2008 年，北极海域（包括开放性海域和高纬度浮冰区等）的大气中 *α*-HCH 和 *γ*-HCH 的浓度水平均低于 20 世纪 90 年代的浓度水平，但是比 2003 年监测的结果要高[124]。

∑DDTs 在整条航线的浓度表现出了很大的空间差异，浓度范围是 2.0~110 pg/m^3 之间，平均值是 36 pg/m^3。在航线中最高和最低浓度值是当船航行在 Nome 港口附近采样点和位于北极高纬度地区采样点监测到。随着纬度的增加，∑DDTs 的水平有逐渐降低的趋势。在东亚海域、北太平洋、楚科奇和波伏特海以及高纬度北极海域这四个采样区间的平均浓度值分别是 83 pg/m^3、52 pg/m^3、25 pg/m^3 和 10 pg/m^3[124]。

HCHs 和 DDTs 的组成比值 α-HCH/γ-HCH 和 DDE/DDT 表明北极地区的 HCHs 可能受到附近区域林丹使用的影响；DDTs 则可能来自亚洲的含 DDTs 类化合物的工业产品的近期使用或者排放[104]。此外，2008 年 7 月发生的西伯利亚地区的大火对北太平洋地区检出的较高的 HCHs 和 DDTs 浓度有很大的贡献[124]。

采样航线上氯丹的含量范围为 2.2~11 pg/m^3，平均值为 5.7 pg/m^3。顺式氯丹（CC）和反式氯丹（TC）是最主要的成分，平均含量分别为 2.2 pg/m^3 和 2.1 pg/m^3。东亚、北太平洋、楚科奇和波伏特海以及高纬度北极海域的氯丹浓度分别是 7.2 pg/m^3、6.1 pg/m^3、4.3 pg/m^3 和 6 pg/m^3，没有表现出很大的地区差异。在所有的样品中 TC/CC 的比值均低于工业生产氯丹中的比值，表明航线所采集的样品受到风化的氯丹的影响。在 TC 和 CC 浓度较低的样品中，TC/CC 的比值也较低，反映了这些样品受到了较老的气团的影响[124]。

6.3.2　海水

已有多名学者对俄罗斯白令海峡和楚科奇海附近海域海水中 OCPs 的残留状况开展研究，他们所报道的 OCPs 含量水平基本在同一数量级上。Yao 等[125]的研究果表明，白令海与楚科奇海样品中 HCHs 的含量相差不大，分别为 412.7 pg/L 和 445.8 pg/L。HCHs 的浓度随纬度变化不明显，通过组分分析，白令海与楚科奇海的组成特征明显不同。此外还有很多其他的 OCPs 首次在此区域测出，如在白令海测出环氧七氯，在楚科奇海测出七氯等。Cai 等[126]研究了从日本海向北经鄂霍次克海、白令海、楚科奇海到北冰洋 5 个区域的海水样品中 17 种 OCPs 的组成特征和分布特征发现，海水样品中 17 种 OCPs 中浓度最高的是 HCHs，占总量的 50%以上，相对较高的有七氯和艾氏剂。OCPs 含量随着纬度的升高而不断增高，而 α-HCH 和 γ-HCH 表现出恰好相反的分布趋势，这可能跟它们的热力学特性有关系。通过 α-HCH/γ-HCH 比值确定 HCHs 主要来自于工业品 HCHs 和林丹。Strachan 等[127]在白令海峡南部和北部的海水中检测到了 HCHs 的最高浓度，而对于 DDTs，白令海（0.23 ng/L）的含量明显高于楚科奇海（0.15 ng/L）。

6.3.3 沉积物

Iwata 等[128]在阿拉斯加海湾、白令海和楚科奇海采集了表层沉积物样品和沉积物柱样，分析了其中的 OCPs 含量，他们的研究成果表明该区域的 HCHs 主要是通过大气的长程迁移而输入当地，而 DDTs 表现出来的分布特征，浓度随着研究区域由南向北逐渐降低，得出结论：DDTs 不是主要通过大气迁移进入北极区域。柱状沉积物样的监测结果显示，从样品底层到表层，OCPs 浓度逐渐升高，这表明在北极区域有机污染物的污染程度越来越重并且是连续性的，并且他们还指出进入沉积物中的有机污染物只占了大气输入的很少一部分，并且对未来几年北极海域中有机氯类污染物的残留降低持悲观态度。Strachan[127]报道了白令海与楚科奇海沉积物中 DDTs 的含量分别为 0.95 ng/g 和 1.6 ng/g，白令海稍低于楚科奇海。研究者还估算了部分化合物每年通过白令海峡的总量，*α*-HCH 每年约 57 t，而 *p,p'*-DDE 每年约有 0.2 t。

卢冰等[129]在 1999 年 7 月至 9 月在白令海和楚科奇海区域采集了表层沉积物样品。调查结果显示，楚科奇海和白令海沉积物中 HCHs 的浓度范围分别为 27~90 pg/g 和 21~51.5 pg/g，DDTs 的浓度范围分别为 4.4~12.9 pg/g 和 4.0~22.5 pg/g。HCHs 主要是由热带亚热带挥发，通过大气运输沉降到寒冷地区，而 DDTs 挥发性较低，不易随大气迁移至高纬度地区。

6.3.4 生物体

2002 年以来，大量极地生物体内 OCPs 的数据报道出来，主要的生物种类包括：熊类、犬类、海洋哺乳动物、鱼类。表 14-1 总结了北极动物体内 OCPs 的含量水平。从表中可以看出动物体内 OCPs 的含量与其在食物网中所处的位置有关，鱼类体内含量最低，而位于食物链顶端的北极熊、海豹和鲸等动物体内含量较高。就生物组织而言，脂肪中富集的 OCPs 的水平显著高于肝脏。

表 14-1 北极生物内 OCPs 暴露水平统计表[130]

物种		位置	组织	含量水平（ng/g）	来源
∑HCHs	北极熊（雌雄）	东格陵兰岛	脂肪	137~263（lw）	[131-133]
	北极熊（雌雄）	东格陵兰岛	肝脏	7（ww）	[134,135]
	北极熊（雌）	斯瓦尔德	脂肪	71（lw）	[134]
	北极熊（雌雄）	加拿大	脂肪	260~489（lw）	[134]
	北极熊（雌雄）	阿拉斯加	脂肪	490（lw）	[134,136]
	白鲸	Hudson 海峡	脂肪	95~119（lw）	[137]
	白鲸	西 Hudson 湾	肝脏	45（lw）	[138]
	白鲸	斯瓦尔巴	脂肪	68~510（lw）	[139]
	环海豹	东格陵兰岛	脂肪	67（lw）	[140]
	环海豹	西格陵兰岛	脂肪	40（lw）	[141]
	环海豹	Hudson 海峡	脂肪	145（lw）	[137]
	北极红点鳟（雌雄）	东格陵兰岛	肌肉	21~26（lw）	[140]
	北极鳕鱼	北东 Hudson 湾	肌肉	10（lw）	[137]

续表

物种		位置	组织	含量水平（ng/g）	来源
∑HCHs	格陵兰岛比目鱼	南东巴芬岛	肌肉	81（lw）	[142]
	格陵兰岛鲨鱼	南东巴芬岛	肝脏	53（lw）	[142]
∑DDTs	杀人鲸	阿拉斯加	脂肪	320000（lw）	[143]
	白鲸	Hudson 海峡	脂肪	520~2521（lw）	[138]
	白鲸	西 Hudson 湾	肝脏	284（lw）	[138]
	白鲸	挪威斯瓦尔巴	脂肪	3272~6770（lw）	[139]
	环海豹	东格陵兰岛	脂肪	1200（lw）	[140]
	环海豹	西格陵兰岛	脂肪	220（lw）	[141]
	环海豹	Hudson 海峡	脂肪	413（lw）	[137]
	星海狮	阿拉斯加-白令海	血液	2127~5464（lw）	[144]
	星海狮	俄罗斯-白令海	血液	3600~15000（lw）	[144]
	北极熊	东格陵兰	脂肪	309（lw）	[131-133]
	北极熊	斯瓦尔巴	脂肪	209（lw）	[134]
	北极熊	丹麦格陵兰岛	脂肪	309~559（lw）	[131,134]
	北极熊	加拿大北冰洋	脂肪	65~210（lw）	[134]
	北极熊	阿拉斯加	脂肪	165（lw）	[134,136]
	北极红点鳟（雌雄）	东格陵兰岛	肌肉	310~500（lw）	[140]
	北极鳕鱼（雌雄）	白令海	肝脏/整条	21/3（ww）	[145]
	北极鳕鱼（雌雄）	北东 Hudson 湾	肌肉	50（lw）	[137]
	北极鳕鱼（雌雄）	北巴芬湾	整条	3（ww）	[145]
	北极鳕鱼（雌雄）	白令海	肝脏	11~45（ww）	[146]
	大西洋鳕鱼	白令海	肝脏	98~175（ww）	[146]
	长粗点鲽	白令海	肝脏	7~30（ww）	[146]
	格陵兰岛比目鱼	南东巴芬岛	肌肉	78（lw）	[142]
	格陵兰岛鲨鱼	南东巴芬岛	肝脏	7195（lw）	[142]

6.3.5　新奥尔松地区的 OCPs 的赋存状态

位于挪威的斯匹次卑尔根岛西海岸的新奥尔松地区是北极科考的重要窗口，目前已有 9 个国家先后在此建立了科考基地，包括我国的北极“黄河站”。2007 年 7 月，马新东等[147]分析了“黄河站”附近区域环境中 OCPs 的残留状况。各种环境介质类型的样品中均不同程度地检测到了 OCPs，其中土壤、苔藓、粪便中 OCPs 检出率分别为 76.8%、82.5% 和 78.6%，平均浓度分别为 5.01 ng/g、6.72 ng/g 和 5.12 ng/g。20 种 OCPs 中，HCHs 和 DDTs 是主要的污染物，除 *γ*-HCH、*p*, *p*'-DDE 和 *p*, *p*'-DDT 的检出率为 55.0%、90.0%和 80.0%以外，其余 HCHs 和 DDTs 检出率达到 100%，其中检出率最高的是环氧七氯，并且含量也是最高，达到 0.32 ng/g。HCHs 和 DDTs 的浓度范围分别为 0.86~4.50 ng/g（平均值 2.24 ng/g）和 0.22~1.09 ng/g（平均值 0.55 ng/g）。HCHs 主要以 *α*-HCH 和 *β*-HCH 为主，两者占 HCHs 总量的 81.6%；DDTs 则主要是以 DDD、DDE 为主，两者占 DDTs 总量的 77.6%。此外，环氧七氯是七氯的一种代谢产物，然而七氯和环氧七氯表现出不同的分布特征，其中七氯的检出率为 50.0%，平均浓度为 0.05 ng/g，而环氧七氯的检出率高达 90.9%，平均浓度为 0.32 ng/g，这主要是由于样品采集集中在 7~8 月份，新奥尔松地区的温度达到 0℃以上，再加上充足的阳光，从而增加了七氯向环氧七氯转化的速率。

HCHs 的四种异构体 α-HCH、β-HCH、γ-HCH 和 δ-HCH 的含量分别为 32.4%、49.2%、12.9%和 5.5%。其中 β-HCH 的残留量最高。该地区不同介质中 α/γ-HCH 的平均值为 1.6~5.7，这说明新奥尔松地区的 HCHs 主要是来自一般工业产品。该地区 β-HCH 的含量较高的原因可能是 β 异构体的对称性强，化学性质和物理性质较其他异构体稳定，难于被降解。新奥尔松地区土壤中 DDE/DDD 的平均值为 0.8，说明该地区 DDTs 主要在厌氧条件下通过微生物降解转化为 DDD。除 25.0%的土壤中未检出 DDTs（已完全降解）以外，75.0%的土壤中（DDD+DDE/ ∑DDTs）的值<1，表明该地区土壤中的 DDTs 已经经过了长期的降解，但较高的检出率表明可能仍然存在着 DDTs 的输入[147]。

7　极地有机阻燃剂研究进展

传统的溴代阻燃剂（brominated flame retardants，BFRs）如 PBDEs 和六溴环十二烷（hexabromocyclododecanes，HBCD）等曾在世界范围内得到广泛应用。由于 PBDEs 和 HBCD 已被证实具有持久性、生物蓄积性和毒性，以及长距离传输能力，penta-BDEs 和 octa-BDEs，以及 HBCD 分别于 2009 年和 2013 年被列入《斯德哥尔摩公约》优先控制的 POPs 名单中。近年来随着替代产品包括新型溴代阻燃剂（novel BFRs）和（organophosphate flame retardants，OPFRs）的大量生产和使用，其潜在的环境影响已经引起环境科学工作者的高度关注，一些化合物已在远离人类活动的偏远地区，尤其是在南北极地区各种环境介质和生物体内得到不同程度的检出，并且存在生物富集和放大等现象，这表明极地环境中有机阻燃剂应引起重视，相关研究对于深入理解这些化学品对全球环境的潜在影响具有重要意义。

7.1　南极地区有机阻燃剂的研究进展

7.1.1　PBDEs

南极环境中有机阻燃剂的研究主要集中在 PBDEs 等传统 POPs 方面，且环境中污染物含量水平相对其他人类活动区域明显较低。大气中 PBDEs 浓度水平普遍在 pg/m^3 水平[78,148]，且以 BDE-47、99 和 209 为主要检出单体[149]，表明大气长距离传输和本地源释放是 PBDEs 的共同来源。Wang 等[79]指出，PBDEs 的气粒分配行为更符合 Li 等[150,151]提出的稳态分配模型（steady-state-based model），这对深入理解极地 POPs 的环境行为具有重要意义。土壤中 PBDEs 浓度受动物活动影响比较明显。鸟粪和企鹅粪土中浓度显著高于其他类型土壤[86,152]。此外，Hale 等[153]在南极 McMurdo Sound 区域沉积物中检测到 6 种 PBDEs，浓度高达 677 ng/g TOC，科考站区污泥、灰尘和沉积物样品中存在高浓度的 BDE-209，表明本地源释放是南极环境中 BDE-209 的主要来源。关于动物体内（企鹅，贼鸥等）PBDEs 的研究较多，且浓度水平相对较高(ng/g lw 水平)[154]。Mwangi 等[152]指出企鹅样品中 BDE-209 是主要的检出单体，而 BDE-47、49、100、153 和 154 亦有较高检出水平。PBDEs 的生物放大因子（BMF）与其 $\log K_{ow}$ 之间存在抛物线关系。

7.1.2 HBCD

关于 HBCD 的研究十分有限。McMurdo 和 Scott 科考站室内灰尘中浓度高达 226 ng/g dw，与美国家庭室内水平相近。污泥中浓度高达 69 ng/g dw，与人类活动区水平接近。沉积物和海洋生物中 HBCD 浓度水平随与 McMurdo 站距离增加而逐渐降低，表明科考站是该区域 HBCD 的一个重要污染源[155]。此外在 2008~2009 年西乔治王岛企鹅和贼鸥组织样品中检测到的 HBCD 浓度在 1.67~713 pg/g lw 之间，贼鸥体内浓度比企鹅中高约 1 个数量级，*α*-HBCD 的 BMF 高达 11.5，表明 *α*-HBCD 在企鹅和贼鸥食物链中具有明显的生物放大能力[156]。

7.1.3 DP

大气和水体中 DP 的研究报道主要由 Möller 等报道，浓度在 pg/L 和 pg/m^3 水平[157,158]，处于全球背景水平。动物体内 DP 的研究报道相对较多。2010~2012 年企鹅和贼鸥蛋中 DP 含量水平相对于其他 POPs 明显较低，企鹅蛋中 DP 浓度在<LOD~0.19 ng/g lw 之间，贼鸥蛋中浓度在<LOD~2.2 ng/g lw 之间[159]。但企鹅和贼鸥组织中 DP 浓度高于其他 DP 类似物，DP 和 Dec-602 的 BMF 在 18.9~25.8 之间，表明其具有生物放大能力[156]。此外 Na 等[160]报道乔治王岛附近海洋食物网中 *anti*-DP 和 *syn*-DP 与生物营养级水平具有显著的正相关性（$p<0.05$），且 *anti*-DP 比 *syn*-DP 具有更高的生物放大能力。

7.1.4 NBFRs 和 OPFRs

关于 NBFRs 和 OPFRs 的研究报道十分有限。Xie 等[161]在大西洋和南极海水中检测到极低浓度的 NBFRs（<1 pg/L）。而西南极半岛海水中 OPFRs 浓度达到 141 pg/m^3，其中 TCPP 浓度最高[162]。此外，Cheng 等[163]首次在南极冰盖和内陆冰雪中检测到 OPFRs，中山站附近海水中 TCEP 平均浓度为 0.2 ng/L，明显低于其他区域。此外，Wolschke 等[164]在西南极鱼体内检测到 PBBz（2.5~17.0 pg/g dw），但企鹅和贼鸥样品中并无检出。南极 NBFRs 和 OPFRs 的相关研究亟待深入开展。

7.2 北极地区有机阻燃剂的研究进展

7.2.1 PBDEs

相对于南极，北极环境中有机阻燃剂的研究报道较多，且浓度水平相对较高。近 20 年间全球大气 PBDEs 水平呈现降低趋势[165]。欧洲北极区域水平与全球变化相一致，而加拿大北极区域浓度水平没有明显趋势，且普遍高于欧洲北极区域[166]。北极大气中 PBDEs 以 BDE-47 和 99 为主[166]，而土壤中高溴代单体（如 BDE-209 等）往往是主要的检出单体[167]。土壤和苔藓等环境样品中浓度水平普遍在 ng/g dw 以下[168-170]。此外 Salvadó 等[171]在极地混合层海水样品中检测到 PBDEs(Σ_{14}PBDEs：0.3~11.2 pg/L)，且 BDE-209 是主要检出单体，深水区 PBDEs 浓度上升高达一个数量级。Meyer 等[172]亦发现 BDE-209 是 2005 年，2006 年和 2008 年加拿大北极冰盖冰芯中的主要检出单体(89%)。

北极动物如北极熊、海豹、鳕鱼、鲸、海豚、海鸟等体内往往能够检测到较高水平的 PBDEs[173-175]，总体浓度在 ng/g lw 水平以上。北极和北大西洋巨头鲸和白边海豚体内 PBDEs 浓度最高，而环海豹和翅鲸体内水平最低，污染物浓度峰值出现在 20 世纪末至 21 世纪初[176]。加拿大北极海鸟蛋和环海豹体内 PBDEs 浓度亦呈现出先上升后下降的趋势[177,178]。此外，加拿大因纽特产妇血液中亦检测到 PBDEs，主要单体浓度在<LOD~120 ng/g lw 之间，但显著低于加拿大南部产妇血液中水平[179]。

7.2.2 HBCD

大气中 HBCD 在 pg/m^3 水平。AMAP 监测结果显示，2002 年以后加拿大 Alert 站点和 2013 年以后芬兰 Pallas 站点大气样品中均能够检测到 HBCD，斯瓦尔巴群岛 Zeppelin 站点的浓度水平从 2006 年开始呈现下降趋势[180]。动物体内 HBCD 浓度水平略低于 PBDEs，且不同生物体内浓度水平趋势有所不同。加拿大北极海鸟（Murre）蛋中 HBCD 浓度在 2003~2014 年之间有下降趋势，而 Fulmar 蛋中浓度在 2003~2006 年间有所上升，之后出现下降趋势[178]。此外，加拿大北极海豹体内 HBCD 水平在过去 20 年间呈现增加趋势[177]。

7.2.3 DP

大气中 DP 的研究报道较多，且浓度主要在 pg/m^3 水平及以下[157,181-184]。Yu 等[181]报道 2011~2013 年间加拿大西部北极地区大气中 DP 浓度呈现下降趋势。极地大气中 f_{anti} 值普遍低于商用产品中[185]，表明 DP 异构体组成发生了改变，*syn*-DP 更易进行大气长距离传输。Möller 等检测到海水中 DP，但浓度水平相对较低（<1.3 pg/L）[157,183]，而 Na 等[186]发现 DP 及其类似物的总浓度可达 93 pg/L。此外，Ma 等[187]和 Na 等[186]在新奥尔松附近海域沉积物中均检测到 DP，浓度在 pg/g dw 水平，湾区外浓度水平较高，而湾区内浓度低，表明洋流作用有明显影响。

7.2.4 NBFRs

北极环境中 NBFRs 的研究报道相对较多。Möller 等[183]在北极大气中检出 7 种 NBFRs，总浓度在 0.6~15.4 pg/m^3 之间，略高于 PBDEs 浓度水平。阿拉斯加大气样品中 BTBPE 只在颗粒物中有检出，且浓度水平（0.02~0.15 pg/m^3）远低于 PBDEs[188]。此外，Hermanson 等[189]和 Meyer 等[172]分别在斯瓦尔巴群岛和加拿大北极冰盖冰芯中检测到 NBFRs，但未观察到沉降趋势。动物体内 NBFRs 浓度水平普遍低于 PBDEs 约一个数量级。Van Bavel 等[190]在法罗群岛海洋生物样品中检测到 HBB，而区域海鸟蛋中 BTBPE 浓度低于 0.11 ng/g lw[191]。挪威北极鸥血液和蛋黄中 HBB、BTBPE、PBEB 和 PBT 浓度低于 2.6 ng/g ww，蛋黄中含量相对较高[192]。

7.2.5 OPFRs

大气中 OPFRs 浓度水平明显较高。在挪威、阿拉斯加和加拿大的北极区域检测到的 OPFRs 浓度可达 ng/m^3 水平[184,193]，最高浓度值达到 2.34 ng/m^3（TnBP）[194]。由于 OPFRs 水溶性较强，水体中浓度普遍较高，可达 ng/L 水平[195]。Ma 等[196]首次在北太平洋至北冰洋的沉积物中检测到 OPFRs，总浓度达到 ng/g dw 水平，北冰洋沉积物中 OPFRs 负荷甚至高于 PBDEs。此外 Hallanger 等[197]报道在 8 种极地动物组织中检测到多种 OPFRs，其中 TECP 检出率最高，但并没有观察到 OPFRs 的生物营养级放大现象。

8 极地区域全氟化合物（PFASs）及短链氯化石蜡（SCCPs）研究进展

自 2001 年通过《斯德哥尔摩公约》，并将 12 类化合物列为持久性有机污染物（POPs）以来，环境中具有类 POPs 性质的污染物受到了极大的关注。全氟辛基磺酸（PFOS）和全氟辛酸（PFOA）及它们的相关前驱体分别于 2009 年和 2015 年进入《斯德哥尔摩公约》控制名单；短链氯化石蜡（SCCPs）更是于 2006 年进入《斯德哥尔摩公约》视野，经历 10 余年马拉松式的论证，最终于 2017 年 5 月在第 8 次《关于持久性有机污染物的斯德哥尔摩公约》缔约国大会，将 SCCPs 列入公约附件 A。本节将对新列入公约的全氟化合物（PFASs）和 SCCPs 在南北极等偏远区域中的环境赋存及环境行为进行介绍。

8.1 PFASs

2001 年 Giesy 和 Kannan 发现 PFOS 在野生动物组织中普遍存在[198]，Hansen 等随后在普通人群血清中也了发现 PFASs 污染[199]。随后 PFASs 的关注度迅速增长，与 PFOS 相关的研究论文从 21 世纪初每年几篇增长到现在每年几百篇。

全氟化合物（PFASs）由碳链和尾部官能团两部分组成，其碳链上的 H 全部被 F 取代，化学通式可表示为 $F(CF_2)_xR$，根据碳链末端的取代基团不同，主要有全氟羧酸（PFCAs）和全氟磺酸（PFSAs），全氟膦酸（PFPAs），全氟磺酰化合物（POSF），以及全氟磷酸酯（PAPs）等。R 基团有时候包括 CH_2 基团，当 x=6，R=$(CH_2)_n$OH 时，该化合物可表示为 6：n FTOH，FTOH 在环境中可最终转化为 PFCAs。

根据 PFASs 在环境中是否能离子化可分为中性全氟化合物（neutral PFASs，如 FTOH、POSF 等）和离子型全氟化合物（ionic PFASs，如 PFSAs、PFCAs 等）两大类。ionic PFASs 在水中的溶解度在 mg/L 水平，甚至能达到 g/L 的水平[200]，neutral PFASs 几乎不溶于水，其在水中的溶解度在 ng/L 水平[201]，但其却具有一定的挥发性。neutral PFASs 在自然环境或生物代谢过程中，最终会降解成 PFCAs 或 PFSAs 等 ionic PFASs[202,203]。

8.1.1 极地 PFASs 污染水平与特征

南极大气中 PFASs 主要以其前驱体形式 FTOH 存在[204,205]，浓度和相近纬度大气中 PCBs，有机氯农药处于同一数量级[85,106]。Del Vento 等还在大气中检出了 *N*-甲基全氟丁基磺酰胺（MeFBSA）和 *N*-甲基全氟丁基磺酰氨基乙醇（MeFBSE）等 4 碳的 PFASs 前驱体，浓度为 3~4 pg/m^3。Wang 等于 2010~2011 年间随德国 Polarstern 科考船，研究了从北大西洋至南极菲尔德斯半岛附近大气样品，南极半岛大气中 PFASs 平均浓度较北半球要低，主要的 PFASs 为 8：2 FTOH[206]。南极大气 PFASs 浓度有所上升，这可能与最近几年挥发性 PFASs 的大量生产使用有关，值得引起注意。北大西洋和加拿大北极群岛大气中也主要以 FTOH 和 FSAE（全氟磺酰基乙醇）等前驱体形式存在，FTOH 和 MeFOSE 分别在 50 pg/m^3 和 20 pg/m^3 左右，浓度比工业区低一个数量级以上[207,208]。大气中 ionic PFASs 浓度很低，主要存在于颗粒相，一般小于 1 pg/m^3[209]。

与传统 POPs 不同，ionic PFASs 可通过洋流大量传输至偏远区域。南极区域海水中主要的 PFASs 为 PFOS，浓度介于未检出到 50 pg/L 之间[210-212]。北极海域海水样品中的 PFASs 主要为 PFOA 和 PFOS，与南极海域特征有所差异，部分研究中 PFOA 含量甚至要高于 PFOS[211]，北冰洋表层海水中 PFASs 含量在 11~174 pg/L 之间[213]。北极海域的 PFASs 浓度还受到当地人类活动的影响，PFOA 是格林兰海、冰岛和法罗群岛附近海域中主要的 PFASs，PFOA 浓度高达 5000 pg/L[209]。

南北极陆地区域的降雪样品中也能检出 PFASs，南极菲尔德斯半岛表层雪中 neutral PFASs 浓度在 125~303 pg/L 之间[206]，FTOH 为主要的 PFASs，低于北极雪样中 523 pg/L 的浓度[214]。有意思的是，南北极表层雪、冰盖及极区内陆湖泊中 PFASs 含量一般高于海水样品[215]，并且检出的 PFASs 单体更为多样化，除了 PFOA 和 PFOS 以外，还有大量的长链 PFCAs，甚至新型 PFASs[209,216,217]。

Giesy 和 Kannan 在南极贼鸥和北极海豹血液中检出了 ng/mL 级别的 PFOS，拉开了 PFASs 全球研究的序幕[198]。南极区域动物中 PFASs 的相关研究主要集中在海洋哺乳动物和海鸟等高等生物上。一般来说，PFOS 和 PFOA 是主要的 PFASs，浓度基本在 10 ng/g 之内[218,219]，但南极麦克默多海峡威德尔海豹血浆中 PFUnDA 检出率达 100%，浓度在 0.08~0.23 ng/g 之间，而其他 PFASs 检出率极低[220]。此外，还有研究表明 PFASs 能通过母体直递至子体[219]，因此在 PFASs 的环境阈值研究中，必须考虑其对子体的毒性效应。

PFOS 是北极生物中最主要的 PFASs，其在北极熊肝脏中的浓度高达 3200 ng/g 湿重，浓度高于 PCBs、氯丹、HCHs 等传统 POPs[221,222]。生物对 PFASs 的前驱体代谢能力差别导致其体内污染特征差异，EtFOSA 在北极鳕鱼中高达 92.8 ng/g 湿重，而北极红鲑体内低于检出限。在海洋哺乳动物中，EtFOSA 和 PFOSA 只在两种鲸类中有较高浓度检出[222,223]。PFASs 在北极区域淡水生态系统及陆地生态系统中也普遍存在[224,225]，污染特征与海洋生态系统有一定的差别，湖泊鱼、陆生生物以及以陆生生物为食物的捕食者中，PFCAs 是主要的污染物，与 neutral PFASs 的长距离传输降解相关。

北极周边存在永居人群，加拿大极区 Nunavik Inuit 人血浆中 PFOS 浓度要比 PCBs、PBDEs 等其他传统污染物高一个数量级左右，检出率达 100%[226-228]，整体上与发达工业国家人群血清中浓度一致，高于波兰和乌克兰等一般工业国家人群，经研究认为驯鹿肉是当地人摄取 PFASs 的主要途径[229]。

8.1.2 极地 PFASs 溯源

PFASs 已成为一类全球性的污染物，极地区域本身无 PFASs 污染源的存在，是研究其长距离传输途径及机理的天然理想实验室。

neutral PFASs 在环境中主要以气相方式存在[207,208]，具有一定的持久性[230]，其与传统 POPs 的长距离传输行为比较相似。ionic PFASs 在环境中的挥发性可以忽略不计，但其可通过水圈在大洋中随洋流长距离传输。neutral PFASs 通过大气圈长距离传输是南极大陆、北极冰盖等区域主要源，neutral PFASs 沉降至极地区域后，最终会降解成 PFCAs 和 PFSAs 等污染物[231]。Casal 等发现南极半岛的利文斯顿岛降雪中 PFASs 含量水平比沿岸海水中高出一个数量级，这说明大气湿沉降是极地陆生生态系统中 PFASs 的重要输入源[215]。

FTOH 是全球生产量最大的 PFASs，*n*∶2 FTOH 可氧化生成奇数碳链（*n*）和偶数碳链（*n*+1）的 PFCAs，如 8∶2 FTOH 可经降解生成生成摩尔比为 1 左右的 PFOA 和 PFNA[231,232]。北极陆地区域雪水、地表水及陆生生物中 PFCAs 占据主导地位[216,217,225,233]，以及部分北极生物中长链 PFCAs 之间的显著相关性与 FTOHs 的降解直接相关。在对极地大洋生物中 PFASs 污染特征的研究中发现，无论是浮游生物，还是鱼类，甚至是海洋哺乳动物，PFOS 都是其主要单体[222,223,234]，这说明极地大洋中 PFASs 的主要来源可能为人类活动密集区直接排放进入水圈的 ionic PFASs。

与北极周边环绕着众多国家不同，南极大陆远离人口密集的工业国家，南极受到的 PFASs 污染相对要小得多。一方面，FTOH 在大气中的半衰期大概为 20 d[230]，根据全球平均风速，FTOH 从污染源长距离传输至南极大陆数量受到限制。Jahnke 和 Dreyer 等[204,235]发现赤道以南 FTOH 的最高浓度仅为 14 pg/m^3，南极洲附近海域大气更是仅为 4 pg/m^3，说明通过大气传输至南极大陆的 PFASs 传输量较为有限。另一方面，南极绕极流（Antarctic Circumpolar Current）使得其他海域的水体和南极大陆附近水体的交换速度变得极其缓慢，因此污染物通过洋流传输至南极大陆周围的速度比较缓慢，绕极流南边生物中 PFASs 低于绕极流北边生物，因此 PFASs 通过洋流直接大量传输到南极至今还没有确切的证据[236]。

除了长距离传输，人为活动也是极地区域 PFASs 污染来源[216,224,237]，极地科考站也有可能成为极区 PFASs 的当地源[238,239]，动物的大规模迁徙可能也是 POPs 在偏远区域的一个重要来源[115]。Llorca 等在巴布亚企鹅（*Pygoscelis*）粪便中检出了 100 ng/g 级别的 PFOS[240]，这些企鹅在迁徙过程中必然成为 PFASs 的移动源。

8.1.3 极地 PFASs 生物富集行为

部分长链 PFASs，尤其是 PFOS，与 PCBs、PBDEs、DDTs 和 HCHs 等 POPs 在生态系统中具有类似

的生物富集能力[174,234,241]。Conder 等对众多 PFASs 生物富集放大研究进行总结后认为氟化碳链长度大于 7 的 ionic PFASs 在环境中具有潜在生物富集放大效应[242]。PFASs 具有疏水疏油性，其在生态系统中的富集放大机理与传统亲脂性 POPs 不同，目前主要有蛋白质结合模型和磷脂结合模型对 PFASs 在生物体内的富集机理进行解释[243]，但两种理论都只能解释部分 PFASs 的富集现象。

不同物种中 PFASs 的富集能力存在差异，PFOS 在北极海洋生态系统（鱼-海洋哺乳动物/鸟类）中除鳕鱼外生物中 BMF 在 4.0 以上，但在鳕鱼中没有生物放大效应；PFOA 只在该生态系统的哺乳动物中 BMF 大于 1，鸟类和鱼类中都小于 1[174,222]。PFASs 尤其是前驱体的富集能力与物种之间的代谢能力差异有关，部分哺乳动物能够将 PFOS 的前驱体 *N*-EtPFOSA 代谢并转化为 PFOS，而鱼类对该物质的代谢能力较差，因此其在鱼体中的 BMF（238）要远高于哺乳类动物[222]。吴江平等在总结 PFASs 生物富集效应方面研究时也发现哺乳动物和爬行动物等高营养级生物的富集能力要高于无脊椎动物和鱼类[244]，这可能与 PFASs 前驱体体内代谢和较高的辛醇/空气分配系数（K_{oa}）有关。

PFASs 在水生和陆生生态系统的富集能力并不完全一致。PFASs 在北极“藻类—双壳类—鱼—海鸟/海洋哺乳动物”食物网中存在生物放大效应，但其在水生食物链（藻类—双壳类—鱼）中并没有生物放大[241]。加拿大极区地衣—驯鹿—狼这一简单陆生食物链中 PFOS 和碳链长度为 9~11 的 PFCAs 营养级放大系数（TMF）在 2.2~2.9 之间，证明了 PFASs 在完全的陆生生态系统也可富集[225]。

8.1.4　极地 PFASs 变化趋势

虽然 PFASs 的生产使用已经受到限制，但其不会在很短时间内从环境中消失。对于北极区域北极熊和海豹等捕食者体内 PFASs 的研究表明，PFOS 浓度早期随着时间变化逐年上升，至 2004~2006 年左右达到峰值，之后呈下降趋势，然而东格陵兰岛海豹体内 PFNA、PFDA 和 PFUnDA 近年来却保持稳定或呈上升趋势[220,245]。也有研究发现 1994~2013 年之间北大西洋领航鲸鱼肌肉中 PFASs 总量变化不大，但组成特征发生了较大的变化[246]。这一方面可能与 PFOS 相关产品的限制生产时间较早有关；另一方面，极地的长链 PFCAs 目前一般认为是 FTOH 通过长距离传输并降解所产生，这些长链 PFCAs 的浓度和比例在极地生物中稳步地上涨，说明 PFASs 大气传输途径对极地区域 PFASs 的贡献也在加大。

8.2　SCCPs 及极地 SCCPs 污染水平及环境行为

氯化石蜡是由正构烷烃直接氯化而成的工业产品[247]，碳链长度为 10~13 为短链氯化石蜡（SCCPs），目前没有 SCCPs 自然来源的报道。SCCPs 是环境中最复杂的一类有机氯污染物，理论上碳链长度为 10 的 SCCPs 有 42720 种单体。SCCPs 组分的复杂性导致其分析定量研究受到极大限制，目前的仪器分析手段尚无法准确对产品及环境介质中的 SCCPs 进行精确定性定量，相对于传统 POPs（PCBs 和 OCPs 等），目前国际上对 SCCPs 的研究仍十分有限。

SCCPs 单体数据缺乏，现有 SCCPs 物理化学性质方面的信息非常有限，所有的实验数据均建立在同类混合物的基础上，现有研究表明 SCCPs 各类物化性质与五到七氯取代的 PCBs 和 OCPs 相似[108,248,249]，化合物的物理化学性质决定其环境行为，因此 SCCPs 在环境中的迁移转化也可能与这些传统 POPs 类似。

SCCPs 在生产、相关产品使用和废弃过程中均有可能向环境中的释放。Meylan 等使用 AOPWIN 计算机程序估算 SCCPs 在大气中的半衰期在 0.81~10.5 d 之间[250]，可在大气中进行长距离传输。此外，吸附在大气颗粒物相上的 SCCPs 在高纬度低温度条件下有利于随颗粒物进行长距离传输。Wania 等认为 SCCPs 的北极污染潜力（ACP）与四至七取代 PCBs 类似[108]。

SCCPs 在南北极的研究相对较少，近年来在北极附近地区的浓度水平报道逐渐增多，在北极大气、土壤、沉积物、生物等中都有检出。1999 年 Borgen 等在挪威斯瓦尔巴德群岛齐伯林山测得的 SCCPs 浓度介于 9.0~57 pg/m^3 之间[251]，2014 年齐柏林山的 SCCPs 年均浓度增长至 240 pg/m^3[252]。Tomy 等在加拿大从中纬度到高纬度的几个湖泊底泥中都检测到了 SCCPs，但北极圈内的两个湖泊表层底泥中 SCCPs 浓度仅为 1.6~4.5 ng/g dw，远低于其他几个湖泊表层底泥中 SCCPs 的浓度，并且湖泊底泥中 SCCPs 普遍高于 DDTs 的浓度[253,254]。李慧娟等研究了斯瓦尔巴德群岛土壤与海洋沉积物中 SCCPs 的含量和污染特征，发现该区域土壤和沉积物中 SCCPs 浓度较低，平均含量仅为 7.1 ng/g 和 49.9 ng/g dw[255]，远低于人口密集区域相同基质[256,257]。当前，北极生物，包括海洋生态系统中的藻类、鱼类、海鸟和海洋哺乳动物，北极内陆湖泊鱼类样品、陆生植物样品中都能检测到 SCCPs[255,258-260]，且浓度一般高于其他传统 POPs，而 SCCPs 在海鸟蛋中的检出表明了 SCCPs 可以从母体向后代传递[261]。

南极 SCCPs 的研究相对较少，傅建捷和马新东等在我国第 29 次南极科考中系统采集了南极菲尔德斯半岛乔治亚王岛环境样品。SCCPs 在南极全部大气样品中均有检出，浓度平均值为 14.9 pg/m^3，气相与颗粒相中 SCCPs 同类物的特征不同，其中颗粒相中较长碳链同类物比重较高[262]，与 Borgen 等[251]报道的挪威斯瓦尔巴德半岛地区大气中 SCCPs 的分布特征相似。南极菲尔德斯半岛和阿德雷岛上的环境样品中 SCCPs 的含量水平介于 3.5~256.6 ng/g dw 之间，平均浓度为 76.6 ng/g[263]，低于北极和中低纬度及工业区样品[255,256]。

Li 等进一步对 SCCPs 长距离传输机制进行了研究。SCCPs 在传输到南北极过程中出现了短碳链同类物的分馏现象，南北极样品中，短碳链（C_{10}）同类物均为主要成分，分别占 SCCPs 总量的 56.1%和 48.6%[255,263]，与北极鱼类及海洋哺乳动物体内 SCCPs 污染特征一致[260,264]，这可能是由于短碳链同类物相比长碳链同类物具有更高的挥发性，比较容易随着大气进行长距离传输所导致[248]，也是 SCCPs 同类物不同物化性质的具体体现。SCCPs 在生态系统中的富集放大能力目前尚有争议[256,265]，因而也备受关注。Li 等通过生物样品中的 C 和 N 同位素比例及现场观察，确定了采集的南北极生态系统样品中的捕食关系。SCCPs 在南极骨螺和帽贝食物链中生物放大因子（BMF）为 1.9，其在北极的鳕鱼和钩虾食物链之间的 BMF 仅为 0.46，导致 SCCPs 在不同地区和食物网中不同的生物放大行为的原因目前尚不清晰。

SCCPs 在极地等偏远区域生态系统中的检出能够确认其长距离传输的能力，除了长距离传输，北极区域的人为活动也是 SCCPs 的一个源。Dick 等测得加拿大 Nunavut 地区一个垃圾填埋地附近的土壤中 SCCPs 平均浓度为 60.4 ng/g dw，认为该垃圾填埋地可能充当着伊魁特市当地的 SCCPs 污染源[266]。Tomy 等比较了在人类活动密集的 St. Lawrence 河湾所采集的鲸脂和格陵兰岛附近海域及加拿大极区海域鲸脂中的 SCCPs 浓度，结果也表明人为活动对极区 SCCPs 的污染水平具有一定的影响[260]。

9　极地重金属研究进展

重金属污染物的排放和长距离传输会对全球生态系统和人类健康产生严重威胁[267,268]，极地地区生态环境自我调节能力弱，重金属的污染可能导致不可逆转的生态环境破坏[269]，因而对极地地区重金属的研究也十分重要。

极地地区拥有着大量的冰盖（南极有 98%的地区被冰盖覆盖[270]）、海水以及湖泊。Tuohy 等[269]研究了南极积雪中重金属浓度的季节变化，发现降雪中重金属的浓度在夏季达到峰值，这是由于春季和夏季的天气转变造成的。Corami 等[271]发现南极西部海水（罗斯海）中镉的浓度随着生物活动及水团年龄的变化可以产生相当大幅度的波动；而铅的浓度呈现出表层浓度最高，深层水浓度最低的变化趋势；铜、锌、铬和锰的浓度则呈现出浅层损耗的现象。早期的研究主要集中在对冰盖和海水的研究而大大低估了湖泊系统所能提供的有用信息。极地湖泊水主要来自于夏季冰盖和降雪融水，是这些融水中溶质和颗粒物质的主要汇[272]。Conca 等[273]通过研究南极地区六个浅水湖泊发现，湖泊水的重金属组成主要是受海洋气溶胶沉降、湖泊地理位置和融水输入的影响。

南极半岛的土壤中重金属浓度在北部地区较高，这可能是因人类活动以及大范围的传输造成的[274]。De lima[275] 等研究了罗伯特岛南极科考站附近土壤中重金属污染，发现主要存在铅和锌的污染。北极的白海和巴伦支海附近的土壤中重金属污染情况在可接受范围内[267]。

大气沉降是重金属在极地地区累积的重要途径。北极地区 PM_{10} 中重金属浓度呈现春季高于夏季的趋势，通过铅同位素分析发现是由于春季和夏季不同的长距离传输气流导致[276]。北极大气气溶胶中地质源元素并未发生明显的富集，而污染源元素发生了明显的富集[277]。

此外，极地地区也存在着丰富的生物群落。南极乔治半岛的企鹅羽毛中钐、镧、钍和铀元素浓度低于 0.1 mg/kg，表明这些元素并不来自于大气颗粒物沉降；砷、钴和硒的浓度比其他地区同种企鹅的浓度低[278]。海鸥也是极地地区的常见鸟类，南极地区四个半岛的贼鸥雏鸟组织中，汞和硒的干重浓度分别高达（4.0 ± 0.8）mg/kg 和（646 ± 123）mg/kg，血汞浓度与纬度呈反比关系[279]。北极地区的三趾鸥中汞和镉累积与其迁移模式相关，浓度的季节性变化与其季节性捕食变化密切相关[280]。苔藓和松萝是极地地区很好的生物指示物。南极利文斯顿半岛的苔藓和松萝中没有发现人为源的污染，且其中重金属的浓度随纬度升高而降低[281]。北极巴瑟斯特和德文郡附近的松萝相比于其生存的土壤有明显的汞富集，表明松萝中的汞主要来自于大气[282]。

10　展　　望

对于传统的 POPs，高山和极地地区的研究工作虽然相对较多，但主要的工作聚焦于迁移规律和变化趋势研究，对于全球气候变化导致南北极冰原大量消融造成的 POPs 变化机制尚不清楚，另外对于高山和极地地区 POPs 的生态毒理效应研究甚为缺乏，未来应进行更深一步的研究。而对于新型持久性有机污染物，可靠的环境分析方法的建立则仍亟待解决，系统全面的各种环境基质中污染物的污染水平和污染特征依然会成为近期主要的研究重点。棕碳气溶胶是大气中影响全球生态安全、气候变化和人类健康的重要污染物，由于来源的多样和大气中活跃的化学反应特性，使其具有复杂的污染特征。基于目前国内外已有的研究成果，未来高山与两极地区大气棕碳的研究主要集中在深入解析棕碳气溶胶的来源过程和形成机理并探索不同气候环境要素与污染分布之间的关系，明确棕碳气溶胶在大气中迁移、转化机制。对高山和极地各环境介质中重金属的浓度和时空分布已经有了较为广泛的研究，溯源研究也有了初步的进展。重金属污染物来源解析对从源头上控制污染，保护高山和极地地区敏感生态系统具有重要意义，应是今后研究的重点。

参考文献

[1] Calamari D, Bacci E, Focardi S, Gaggi C, Morosini M, Vighi M. Role of plant biomass in the global environmental partitioning of chlorinated hydrocarbons. Environ Sci Technol, 1991, 25(8): 1489-1495.

[2] Wania F, Mackay D. Global fractionation and cold condensation of low volatility organochlorine compounds in polar-regions. Ambio, 1993, 22(1): 10-18.

[3] Wania Frank, Westgate John N. On the mechanism of mountain cold-trapping of organic chemicals. Environ Sci Technol, 2008, 42(24): 9092-9098.

[4] Grimalt J O, Fernandez P, Berdie L, Vilanova R M, Catalan J, Psenner R, Hofer R, Appleby P G, Rosseland B O, Lien L, Massabuau J C, Battarbee R W. Selective trapping of organochlorine compounds in mountain lakes of temperate areas. Environ Sci Technol, 2001, 35(13): 2690-2697.

[5] Daly G L, Wania F. Organic contaminants in mountains. Environ Sci Technol, 2005, 39(2): 385-398.
[6] Blais J M, Schindler D W, Muir D C. G, Kimpe L E, Donald D B, Rosenberg B. Accumulation of persistent organochlorine compounds in mountains of western Canada. Nature, 1998, 395(6702): 585-588.
[7] Wang X P, Gong P, Wang C F, Ren J, Yao T D. A review of current knowledge and future prospects regarding persistent organic pollutants over the Tibetan Plateau. Sci Total Environ, 2016, 573: 139-154.
[8] Gouin T, Thomas G O, Cousins I, Barber J, Mackay D, Jones K C. Air-surface exchange of polybrominated biphenyl ethers and polychlorinated biphenyls. Environ Sci Technol, 2002, 36(7): 1426-1434.
[9] Backe C, Cousins I T, Larsson P. PCB in soils and estimated soil-air exchange fluxes of selected PCB congeners in the south of Sweden. Environ Pollut, 2004, 128(1-2): 59-72.
[10] Wang C F, Wang X P, Gong P, Yao T D. Polycyclic aromatic hydrocarbons in surface soil across the Tibetan Plateau: Spatial distribution, source and air-soil exchange. Environ Pollut, 2014, 184: 138-144.
[11] Davidson D A, Wilkinson A C, Blais J M, Kimpe L E, McDonald K M, Schindler D W. Orographic cold-trapping of persistent organic pollutants by vegetation in mountains of western Canada. Environ Sci Technol, 2003, 37(2): 209-215.
[12] Carrera G, Fernandez P, Grimalt J. O, Ventura M, Camarero L, Catalan J, Nickus U, Thies H, Psenner R. Atmospheric deposition of organochlorine compounds to remote high mountain lakes of Europe. Environ Sci Technol, 2002, 36(12): 2581-2588.
[13] Shen L, Wania F, Lei Y D, Teixeira C, Muir D C G, Bidleman T F. Atmospheric distribution and long-range transport behavior of organochlorine pesticides in north America. Environ Sci Technol, 2005, 39(2): 409-420.
[14] Yang R, Zhang S, Li A, Jiang G, Jing C. Altitudinal and spatial sigNature of persistent organic pollutants in soil, lichen, conifer needles, and bark of the southeast Tibetan Plateau: Implications for sources and environmental cycling. Environ Sci Technol, 2013, 47(22): 12736-12743.
[15] Chen D Z, Liu W J, Liu X D, Westgate J N, Wania F. Cold-trapping of persistent organic pollutants in the mountain soils of western Sichuan, China. Environ Sci Technol, 2008, 42(24): 9086-9091.
[16] Wang P, Zhang Q H, Wang Y W, Wang T, Li X M, Li Y M, Ding L, Jiang G B. Altitude dependence of polychlorinated biphenyls (PCBs) and polybrominated diphenyl ethers (PBDEs) in surface soil from Tibetan Plateau, China. Chemosphere, 2009, 76(11): 1498-1504.
[17] Zhu N L, Schramm K W, Wang T, Henkelmann B, Fu J J, Gao Y, Wang Y W, Jiang G B. Lichen, moss and soil in resolving the occurrence of semi-volatile organic compounds on the southeastern Tibetan Plateau, China. Sci Total Environ, 2015, 518: 328-336.
[18] McLachlan M S, Horstmann M. Forests as filters of airborne organic pollutants: A model. Environ Sci Technol, 1998, 32(3): 413-420.
[19] Moeckel C, Nizzetto L, Strandberg B, Lindroth A, Jones K C. Air-boreal forest transfer and processing of polychlorinated biphenyls. Environ Sci Technol, 2009, 43(14): 5282-5289.
[20] Horstmann M, McLachlan M S. Atmospheric deposition of semivolatile organic compounds to two forest canopies. Atmos Environ, 1998, 32(10): 1799-1809.
[21] Su Y S, Wania F. Does the forest filter effect prevent semivolatile organic compounds from reaching the Arctic? Environ Sci Technol, 2005, 39(18): 7185-7193.
[22] Weiss P. Vegetation/soil distribution of semivolatile organic compounds in relation to their physicochemical properties. Environ Sci Technol, 2000, 34(9): 1707-1714.
[23] Nizzetto L, Perlinger J A. Climatic, biological, and land cover controls on the exchange of gas-phase semivolatile chemical pollutants between forest canopies and the atmosphere. Environ Sci Technol, 2012, 46(5): 2699-2707.
[24] Nriagu, J O. A history of global metal pollution. Science, 1996, 272(5259): 223.
[25] Bacardit M, Camarero L. Atmospherically deposited major and trace elements in the winter snowpack along a gradient of altitude in the Central Pyrenees: the seasonal record of long-range fluxes over SW Europe. Atmos Environ, 2010, 44(4): 582-595.
[26] Bing H, Wu Y, Zhou J, Li R, Luo J, Yu D. Vegetation and cold trapping modulating elevation-dependent distribution of trace metals in soils of a high mountain in eastern Tibetan Plateau. Sci Rep, 2016, 6: 24081.
[27] Yang R, Yao T, Xu B, Jiang G, Xin X. Accumulation features of organochlorine pesticides and heavy metals in fish from high mountain lakes and Lhasa River in the Tibetan Plateau. Environ Int, 2007, 33(2): 151-156.
[28] Liao X Y. Study on the functional division of land utilization in Tibet Autonomous Region. J. Anhui Agri. Sci, 2008, 36(7):

2847–2849(2923).
[29] Huang X, Sillanpää M, Duo B, Gjessing E T. Water quality in the Tibetan Plateau: Metal contents of four selected rivers. Environ Pollut, 2008, 156(2): 270-277.
[30] Zhang Q, Ke P, Kang S, Zhu A, Wang W X. Mercury in wild fish from high-altitude aquatic ecosystems in the Tibetan Plateau. Environ Sci Technol, 2014, 48(9): 5220-5228.
[31] Fu X, Feng X, Zhu W, Wang S, Lu J. Total gaseous mercury concentrations in ambient air in the eastern slope of Mt. Gongga, South-Eastern fringe of the Tibetan Plateau, China. Atmos Environ, 2008, 42(5):970-979.
[32] Yang R, Zhang S, Wang Z. Bioaccumulation and regional distribution of trace metals in fish of the Tibetan Plateau. Environ Geochem Health, 2014, 36(1): 183-191.
[33] Shao J, Shi J, Duo B, Liu C, Gao Y, Fu J, Yang R, Jiang G. Mercury in alpine fish from four rivers in the Tibetan Plateau. J Environ Sci (China), 2016, 39: 22-28.
[34] Li C, Kang S, Wang X, Ajmone-Marsan F, Zhang Q. Heavy metals and rare earth elements (REEs) in soil from the Nam Co Basin, Tibetan Plateau. Environ Geol, 2007, 53(7): 1433-1440.
[35] Tripathee L, Kang S, Rupakheti D, Zhang Q, Bajracharya R M, Sharma C M, Huang J, Gyawali A, Paudyal R, Sillanpää M. Spatial distribution, sources and risk assessment of potentially toxic trace elements and rare earth elements in soils of the Langtang Himalaya, Nepal. Environ Earth Sci, 2016, 75(19): 1-12.
[36] Cong Z Y, Kang S C, Zhang Y L, Li X D. Atmospheric wet deposition of trace elements to central Tibetan Plateau. Appl Geochem, 2010, 25(9):1415-1421.
[37] Cong Z, Kang S, Liu X, Wang G. Elemental composition of aerosol in the Nam Co region, Tibetan Plateau, during summer monsoon season. Atmos Environ, 2007, 41(6): 1180-1187.
[38] Cong Z, Kang S, Luo C, Li Q, Jie H, Gao S, Li X. Trace elements and lead isotopic composition of PM_{10} in Lhasa, Tibet. Atmos Environ, 2011, 45(34): 6210-6215.
[39] Jie H, Kang S, Zhang Q, Guo J, Chen P, Zhang G, Tripathee L. Atmospheric deposition of trace elements recorded in snow from the Mt. Nyainqentanglha region, southern Tibetan Plateau. Chemosphere, 2013, 92(8): 871-881.
[40] Zhang Y, Kang S, Chen P, Li X, Liu Y, Gao T, Guo J, Sillanpää M. Records of anthropogenic antimony in the glacial snow from the southeastern Tibetan Plateau. J Asian Earth Sci, 2016, 131: 62-71.
[41] Dong Z, Qin D, Qin X, Cui J, Kang S. Changes in precipitating snow chemistry with seasonality in the remote Laohugou glacier basin, western Qilian Mountains. Environ Sci Pollut Res Int, 2017, 24(12): 11404-11414.
[42] Shao J J, Liu C B, Zhang Q H, Fu J J, Yang R Q, Shi J B, Cai Y, Jiang G B. Characterization and speciation of mercury in mosses and lichens from the high-altitude Tibetan Plateau. Environ Geochem Health, 2016, 39(3): 475-482.
[43] Shao J, Shi J, Bu D, Liu C, Gao Y, Fu J, Yang R, Cai Y, Jiang G. Trace metal profiles in mosses and lichens from the high-altitude Tibetan Plateau. RSC Advances, 2016, 6(1): 541-546.
[44] Bing H, Wu Y, Zhou J, Sun H. Biomonitoring trace metal contamination by seven sympatric alpine species in Eastern Tibetan Plateau. Chemosphere, 2016, 165: 388-398.
[45] Zhu N, Schramm K W, Wang T, Henkelmann B, Zheng X, Fu J, Gao Y, Wang Y, Jiang G. Environmental fate and behavior of persistent organic pollutants in Shergyla Mountain, southeast of the Tibetan Plateau of China. Environ Pollut, 2014, 191: 166-174.
[46] Wang X, Luo J, Yin R, Yuan W, Lin C J, Sommar J, Feng X, Wang H, Lin C. Using mercury isotopes to understand mercury accumulation in the montane forest floor of the Eastern Tibetan Plateau. Environ Sci Technol, 2017, 51(2): 801-809.
[47] Xu X, Zhang Q, Wang W X. Linking mercury, carbon, and nitrogen stable isotopes in Tibetan biota: Implications for using mercury stable isotopes as source tracers. Sci Rep, 2016, 6: 25394.
[48] Ramanathan V, Carmichael G. Global and regional climate changes due to black carbon. Nature Geosci, 2008, 1(4): 221.
[49] Andreae M, Gelencsér A. Black carbon or brown carbon? The Nature of light-absorbing carbonaceous aerosols. Atmos Chem Phys, 2006, 6: 3131-3148.
[50] Hoffer A, Gelencsér A, Guyon P, Kiss G. Optical properties of humic-like substances (HULIS) in biomass-burning aerosols. Atmos Chem Phys, 2006, 6: 3563-3570.
[51] Wan X, Kang S, Wang Y, Xin J, Liu B, Guo Y, Wen T, Zhang G, Cong Z. Size distribution of carbonaceous aerosols at a high-altitude site on the central Tibetan Plateau (Nam Co Station, 4730ma. sl). Atmos Res, 2015, 153: 155-164.
[52] Saleh R, Robinson E S, Tkacik D S, Ahern A T, Liu S, Aiken A C, Sullivan R C, Presto A A, Dubey M K, Yokelson R J. Brownness of organics in aerosols from biomass burning linked to their black carbon content. Nature Geosci, 2014, 7: 647.

[53] Bond T C. Spectral dependence of visible light absorption by carbonaceous particles emitted from coal combustion. Geophys Res Lett, 2001, 28: 4075-4078.

[54] Zheng G, He K, Duan F, Cheng Y, Ma Y. Measurement of humic-like substances in aerosols: A review. Environ Pollut, 2013, 181: 301-314.

[55] Hoffer A, Kiss G, Blazso M, Gelencser A. Chemical characterization of humic-like substances (HULIS) formed from a lignin - type precursor in model cloud water. Geophys Res Lett, 2004, 31(6): 337-357.

[56] Limbeck A, Kulmala M, Puxbaum H. Secondary organic aerosol formation in the atmosphere via heterogeneous reaction of gaseous isoprene on acidic particles. Geophys Res Lett, 2003, 30(30): 379-394.

[57] Chakrabarty R K, Moosmüller H, Chen L W A, Lewis K. Brown carbon in tar balls from smoldering biomass combustion. Atmos Chem Phys, 2010, 10: 6363-6370.

[58] Laskin A, Laskin J, Nizkorodov S A. Chemistry of atmospheric brown carbon. Chem Rev, 2015, 115: 4335-4382.

[59] Cappiello A, Simoni E D, Fiorucci C, Mangani F, Palma P, Trufelli H, Decesari S, Facchini M C, Fuzzi S. Molecular characterization of the water-soluble organic compounds in fogwater by ESIMS/MS. Environ Sci Technol, 2003, 37: 1229-1240.

[60] Decesari S, Facchini M C, Fuzzi S, Tagliavini E. Characterization of water-soluble organic compounds in atmospheric aerosol: A new approach. J Geophys Res, 2000, 105: 1481.

[61] Pósfai M, Gelencsér A, Simonics R, Arató K, Li J, Hobbs P V, Buseck P R. Atmospheric tar balls: Particles from biomass and biofuel burning. J Geophys Res-Atmos, 2004, 109(D6): 539-547.

[62] Ramanathan V, Chung C, Kim D, Bettge T, Buja L, Kiehl J T, Washington W M, Fu Q, Sikka D R, Wild M. Atmospheric brown clouds: Impacts on South Asian climate and hydrological cycle. Proc Natl Acad Sci U S A, 2005, 102: 5326-5333.

[63] Cong Z, Kang S, Kawamura K, Liu B, Wan X, Wang Z, Gao S, Fu P. Carbonaceous aerosols on the south edge of the Tibetan Plateau: concentrations, seasonality and sources. Atmos Chem Phys, 2015, 15: 1573-1584.

[64] Cong Z, Kawamura K, Kang S, Fu P. Penetration of biomass-burning emissions from South Asia through the Himalayas: new insights from atmospheric organic acids. Sci Rep, 2015, 5: 9580.

[65] Li C, Chen P, Kang S, Yan F, Hu Z, Qu B, Sillanpää M. Concentrations and light absorption characteristics of carbonaceous aerosol in $PM_{2.5}$ and PM_{10} of Lhasa city, the Tibetan Plateau. Atmos Environ, 2016, 127: 340-346.

[66] Kirillova E N, Marinoni A, Bonasoni P, Vuillermoz E, Facchini M C, Fuzzi S, Decesari S. Light absorption properties of brown carbon in the high Himalayas. J Geophys Res-Atmos, 2016, 121: 9621-9639.

[67] Xu B Q, Cao J J, Hansen J, Yao T D, Joswia D R, Wang N L, Wu G J, Mo W, Zhao H B, Wei Y. Black soot and the survival of Tibetan glaciers. Proc Natl Acad Sci, 2009, 106: 22114-22118.

[68] Li C, Chen P, Kang S, Yan F, Li X, Qu B, Sillanpää M. Carbonaceous matter deposition in the high glacial regions of the Tibetan Plateau. Atmos Environ, 2016, 141: 203-208.

[69] Yan F, Kang S, Li C, Zhang Y, Qin X, Li Y, Zhang X, Hu Z, Chen P, Li X. Concentration, sources and light absorption characteristics of dissolved organic carbon on a medium-sized valley glacier, northern Tibetan Plateau. Cryosphere, 2016, 10: 2611.

[70] Guilhermet J, Preunkert S, Voisin D, Baduel C, Legrand M. Major 20th century changes of water - soluble humic - like substances (HULISWS) aerosol over Europe inferred from Alpine ice cores. J Geophys Res-Atmos, 2013, 118: 3869-3878.

[71] Legrand M, Preunkert S, Schock M, Cerqueira M, Kasper - Giebl A, Afonso J, Pio C, Gelencsér A, Dombrowski - Etchevers I. Major 20th century changes of carbonaceous aerosol components (EC, WinOC, DOC, HULIS, carboxylic acids, and cellulose) derived from Alpine ice cores. J Geophys Res-Atmos, 2007, 112.

[72] Randel W J, Park M, Emmons L, Kinnison D, Bernath P, Walker K A, Boone C, Pumphrey H. Asian monsoon transport of pollution to the stratosphere. Science, 2010, 328: 611-613.

[73] Zhang Y, Forrister H, Liu J, Dibb J, Anderson B, Schwarz J P, Perring A E, Jimenez J L, Campuzanojost P, Wang Y. Top-of-atmosphere radiative forcing affected by brown carbon in the upper troposphere. Nature Geosci, 2017, 10: 486-489.

[74] Nguyen Q T, Kristensen T B, Hansen A M K, Skov H, Bossi R, Massling A, Sørensen L L, Bilde M, Glasius M, Nøjgaard J K. Characterization of humic - like substances in Arctic aerosols. J Geophys Res-Atmos, 2014, 119: 5011-5027.

[75] Voisin D, Jaffrezo J L, Houdier S, Barret M, Cozic J, King M D, France J L, Reay H J, Grannas A, Kos G. Carbonaceous species and humic like substances(HULIS)in Arctic snowpack during OASIS field campaign in Barrow. J Geophys Res-Atmos, 2012, 117(D4): 116.

[76] Doherty S J, Warren S G, Grenfell T C, Clarke A D, Brandt R E. Light-absorbing impurities in Arctic snow. Atmos Chem

Phys, 2010, 10: 11647-11680.
[77] Risebrough R W, Walker W, Schmidt T T, De Lappe B W, Connors C W. Transfer of chlorinated biphenyls to Antarctica. Nature, 1976, 264: 23-30.
[78] Li Y M, Geng D W, Liu F B, Wang T, Wang P, Zhang Q H, Jiang G B. Study of PCBs and PBDEs in King George Island, Antarctica, using PUF passive air sampling. Atmos Environ. 2012, 51: 140-145.
[79] Wang P, Li Y M, Zhang Q H, Yang Q H, Zhang L, Liu F B, Fu J J, Meng W Y, Wang D, Sun H Z, Zheng S C, Hao Y F, Liang Y, Jiang G B. Three-year monitoring of atmospheric PCBs and PBDEs at the Chinese Great Wall Station, West Antarctica: Levels, chiral signature, environmental behaviors and source implication. Atmos Environ, 2017, 150: 407-416.
[80] Gambaro A, Manodori L, Zangrando R, Cincinelli A, Capodaglio G, Cescon P. Atmospheric PCB concentrations at Terra Nova Bay, Antarctica. Environ Sci Technol, 2005, 39: 9406-9411.
[81] Montone R C, Taniguchi S, Weber R R. PCBs in the atmosphere of King George Island, Antarctica. Sci Total Environ, 2003: 308, 167-173.
[82] Hung H, Katsoyiannis A A, Brorstrom-Lunden E , Olafsdottir K, Aas W, Breivik K, Bohlin-Nizzetto P, Sigurdsson A, Hakola H, Bossi R, Skov H, Sverko E, Barresi E, Fellin P, Wilson S. Temporal trends of Persistent Organic Pollutants (POPs) in arctic air: 20 years of monitoring under the Arctic Monitoring and Assessment Programme (AMAP). Environ Pollut, 2016, 217: 52-61.
[83] Zhang P, Ge L, Gao H, Yao T, Fang X, Zhou C, Na G. Distribution and transfer pattern of Polychlorinated Biphenyls (PCBs) among the selected environmental media of Ny-Alesund, the Arctic: As a case study. Mar Pollut Bull, 2014, 89: 267-275.
[84] Hung H, Kallenborn R, Breivik K, Yushan S, Brorstrom-Lunden E, Olafsdottir K, Thorlacius, J. M, Leppanen, S, Bossi, R, Skov, H. Atmospheric monitoring of organic pollutants in the Arctic under the Arctic Monitoring and Assessment Programme (AMAP): 1993-2006. Sci Total Environ, 2010, 408: 2854-2873.
[85] Choi S D, Baek S Y, Chang Y S, Wania F, Ikonomou M G, Yoon Y J, Park B K, Hong S. Passive air sampling of polychlorinated biphenyls and organochlorine pesticides at the Korean arctic and Antarctic Research Stations: implications for long-range transport and local pollution. Environ Sci Technol, 2008, 42: 7125-7131.
[86] Wang P, Zhang Q H, Wang T, Chen W H, Ren D W, Li Y M, Jiang G B. PCBs and PBDEs in environmental samples from King George Island and Ardley Island, Antarctica. RSC Advances, 2012, 2: 1350-1355.
[87] Borghini F, Grimalt J O, Sanchez-Hernandez J C, Bargagli R. Organochlorine pollutants in soils and mosses from Victoria Land (Antarctica). Chemosphere, 2005, 58: 271-278.
[88] Klánová J, Matykiewiczová N, Mácka Z, Prosek P, Láska K, Klán P. Persistent organic pollutants in soils and sediments from James Ross Island, Antarctica. Environ Pollut, 2008, 152: 416-423.
[89] Fuoco R, Colombini M P, Abete C, Carignani S. Polychlorobiphenyls in sediment, soil and sea water samples from Antarctica. Int J Environ Anal Chem, 1995, 61: 309-318.
[90] Park H, Lee S H, Kim M, Kim J H, Lim H S. Polychlorinated biphenyl congeners in soils and lichens from King George Island, South Shetland Islands, Antarctica. Antarct Sci, 2010, 22: 31-38.
[91] Negoita T G, Covaci A, Gheorghe A, Schepens P. Distribution of polychlorinated biphenyls (PCBs) and organochlorine pesticides in soils from the East Antarctic coast. J Environ Monit, 2003, 5: 281-286.
[92] Fuoco R, Capodaglio G, Muscatello B, Radaelli M. Persistent organic pollutants (POPs) in the Antarctic environment: A review of findings. A SCAR Publication, 2009.
[93] Lead W A, Steinnes E, Jones K C. Atmospheric deposition of PCBs to moss (*Hylocomium splendens*) in Norway between 1977 and 1990. Environ Sci Technol, 1996, 30, 524-530.
[94] Cabrerizo A, Dachs J, Barcelo D, Jones K. C. Influence of organic matter content and human activities on the occurrence of organic pollutants in Antarctic soils, lichens, grass, and mosses. Environ Sci Technol, 2012, 46: 1396-1405.
[95] Zhu C F, Li Y M, Wang P, Chen Z J, Ren D W, Ssebugere P, Zhang Q H, Jiang G B. Polychlorinated biphenyls (PCBs) and polybrominated biphenyl ethers (PBDEs) in environmental samples from Ny-Ålesund and London Island, Svalbard, the Arctic. Chemosphere, 2015, 126: 40-46.
[96] Corsolini S, Sara G. The trophic transfer of persistent pollutants (HCB, DDTs, PCBs) within polar marine food webs. Chemosphere, 2017, 177:189-199.
[97] Grotti M, Pizzini S, Abelmoschi M. L, Cozzi G, Piazza R, Soggia F. Retrospective biomonitoring of chemical contamination in the marine coastal environment of Terra Nova Bay (Ross Sea, Antarctica) by environmental specimen banking. Chemosphere, 2016, 165: 418-426.

[98] Cincinelli A, Martellini T. Pozo K, Kukucka P, Audy O, Corsolini S. Trematomus bernacchii as an indicator of POP temporal trend in the Antarctic seawaters. Environ Pollut, 2016, 217: 19-25.
[99] Mello F V, Roscales J L, Guida Y S, Menezes J F S, Vicente A, Costa E S, Jimenez B, Torres J P M. Relationship between legacy, and emerging organic pollutants in Antarctic seabirds and their foraging ecology as shown by delta C-13 and delta N-15. Sci Total Environ, 2016, 573: 1380-1389.
[100] Hoguet J, Keller J M, Reiner J L, Kucklick J R, Bryan C E, Moors A J, Pugh R S, Becker P R. Spatial and temporal trends of persistent organic pollutants and mercury in beluga whales (*Delphinapterus leucas*) from Alaska. Sci Total Environ, 2013, 449: 285-294.
[101] Savinov V, Muir D C G, Svetochev V, Svetocheva O, Belikov S, Boltunov A, Alekseeva L, Reiersen L O, Savinova T. Persistent organic pollutants in ringed seals from the Russian Arctic. Sci Total Environ, 2011, 409: 2734-2745.
[102] Tartu S, Bourgeon S, Aars J, Andersen M, Polder A, Thiemann G W, Welker J M, Routti H. Sea ice-associated decline in body condition leads to increased concentrations of lipophilic pollutants in polar bears (*Ursus maritimus*) from Svalbard, Norway. Sci Total Environ, 2017, 576: 409-419.
[103] 贺仕昌. 南极航线大气有机氯农药化学特征. 厦门: 国家海洋局第三海洋研究所, 2013.
[104] Ding X, Wang X M, Wang Q Y, Xie Z Q, Xiang C H, Mai B X, Sun L G. Atmospheric DDTs over the North Pacific Ocean and the adjacent Arctic region: Spatial distribution, congener patterns and source implication. Atmos Environ, 2009, 43(28): 4319-4326.
[105] Su Y, Hung H, Blanchard P, Gregory W P, Kallenborn R, Konoplev A, Fellin P, Li H, Geen C, Stern G, Rosenberg B, Barrie L A. A circumpolar perspective of atmospheric organochlorine pesticides (OCPs): Results from six Arctic monitoring stations in 2000-2003. Atmos Environ, 2008, 42: 4682-4698.
[106] Montone R C, Taniguchi S, Boian C, Weber R R. PCBs and chlorinated pesticides (DDTs, HCHs and HCB) in the atmosphere of the southwest Atlantic and Antarctic oceans. Mar Pollut Bull, 2005, 50: 778-782.
[107] Kang J H, Son M H, Hur S D, Hong S, Motoyama H, Fukui K, Chang Y S. Deposition of orgabochlorine pesticides into the surface snow of East Antarctica. Sci Total Environ, 2012, 433: 290-295.
[108] Wania F. Assessing the potential of persistent organic chemicals for long-range transport and accumulation in polar regions. Environ Sci Technol, 2003, 37: 1344-1351.
[109] Su Y, Hung H, Blanchard P, Patton G W, Kallenborn R, Konoplev A, Fellin P, Li H, Geen C, Stern G, Rosenberg B, Barrie L A. Spatial and seasonal variations of Hexachlorocyclohexanes (HCHs) and Hexachlorobenzene (HCB) in the arctic atmosphere. Environ Sci Technol, 2006, 40(21): 6601-6607.
[110] Macdonald R. W, Barrie L A, Bidleman T F, Diamond M L, Gregor D J, Semkin R G, Strachan W M J, Li Y F, Wania F, Alaee M, Alexeeva L B, Backus S M, Bailey, R, Bewers, J M, Gobeil, C, Halsall, C J, Harner, T, Hoff, J T, Jantunen, L M M, Lockhart, W L, Mackay, D, Muir, D C G, Pudykiewicz, J, Reimer, K J, Smith, J N, Stern, G A, Schroeder, W H, Wagemann, R, Yunkern, M B. Contaminants in the Canadian Arctic: 5 years of progress in understanding sources, occurrence and pathways. Sci Total Environ 2000, 254: 93-234.
[111] Pućko M, Stern G A, Macdonald R W, Barber D G, Rosenberg B, Walkusz W. When will α-HCH disappear from the western Arctic Ocean? J Marine Syst, 2013, 27: 88-100.
[112] Alexeeva L B, Strachan W M J, Shlychkova V V, Nazarova A A, Nikanorov A M, Korotova L G, Koreneva V I. Organochlorine Pesticide and trace metal monitoring of Russian rivers flowing to the Arctic Ocean: 1990-1996. Mar Pollut Bull, 2001, 43(1-6): 71-85.
[113] Ma Y, Xie Z, Halsall C, Möller A, Yang H, Zhong G, Cai M, Ebinghaus R. The spatial distribution of organochlorine pesticides and halogenated flame retardants in the surface sediments of an Arctic fjord: The influence of ocean currents vs. glacial runoff. Chemosphere, 2015, 119: 953-960.
[114] Khairy M A, Luek J L, Dickhut R, Lohmann R. Levels, sources and chemical fate of persistent organic pollutants in the atmosphere and snow along the Western Antarctic Peninsula. Environ Pollut, 2016, 216: 304-313.
[115] Ewald G, Larsson P, Linge H, Okla L, Szarzi N. Biotransport of organic pollutants to an inland Alaska lake by migrating Sockeye salmon (*Oncorhyncus nerka*). Arctic, 1998, 51: 40-47.
[116] Evenset A, Christensen G N, Skotvold T, Fjeld E, Schlabach M, Wartena E, Gregor D. A comparison of organic contaminants in two high Arctic lake ecosystems, Bjørnøya (Bear Island), Norway. Sci Total Environ, 2004, 318: 125-141.
[117] Corsolini S, Borghesi N, Ademollo N, Focardi S. Chlorinated biphenyls and pesticides in migrating and resident seabirds from East and West Antarctica. Environ Int, 2011, 37: 1329-1335.

[118] Cincinelli A, Martellini T, Bubba M D, Lepri L, Corsolini S, Borghesi N, King M D, Dickhut R M. Organochlorine pesticide air-water exchange and bioconcentration in krill in the Ross Sea. Environ Pollut, 2009, 157(7): 2153-2158.

[119] Zhang Q, Chen Z, Li Y, Wang P, Zhu C, Gao G, Xiao K, Sun H, Zheng S, Liang Y, Jiang G. Occurrence of organochlorine pesticides in the environmental matrices from King George Island, west Antarctica. Environ Pollut, 2015, 206: 142-149.

[120] Cipro C V Z, Yogui G T, Bustamante P, Taniguchi S, Sericano J L, Montone R C. Organic pollutants and their correlation with stable isotopes in vegetation from King George Island, Antarctica. Chemosphere, 2011, 85(3): 393-398.

[121] Inomata O N K, Montone R C, Lara W H, Weber R R, Toledo H H B. Tissue distribution of organochlorine residues - PCBs and pesticides - in Antarctic penguins. Antarct Sci, 2004, 8(3): 253-255.

[122] 冯朝军, 于培松, 卢冰, 蔡明红, 武光海. 南极阿德雷岛企鹅机体组织、蛋卵和粪土中 PCBs 和 OCPs 的分布. 海洋环境科学 2010, 29(3): 308-313.

[123] 张海生, 王自磐, 卢冰, 朱纯, 武光海, Vetter, W. 南极大型动物粪土层和蛋卵中有机氯污染物分布特征及生态学意义. 中国科学, 2006, 36(12): 1111-1121.

[124] 武晓果. 北太平洋以及北极地区海洋边界层大气持久性有机化合物研究：来源、趋势和过程. 合肥: 中国科学技术大学, 2011.

[125] Yao Z, Jiang G, Xu H. Distribution of organochlorine pesticides in seawater of the Bering and Chukchi Sea. Environ Pollut, 2002, 116: 49-56.

[126] Cai M, Qiu C, Shen Y, Cai M. Huang S, Qian B, Sun J, Liu X. Concentration and distribution of 17 organochlorine pesticides(OCPs)in seawater from the Japan Sea northward to the Arctic Ocean. Sci Chi Chem, 2010, 53(5): 1033-1047.

[127] Strachan W M J, Fisk A, Texeira C F, Burniston D A, Norstron R. PCBs and organochlorine pesticide concentrations in the waters of the Canadian Archipelago and other Arctic regions, University of the Arctic: Rovaniemi, Arctic Monitoring and Assessment Program, 2002.

[128] Iwata H, Tanabe S, Ueda K, Tatsukawa R. Persistent organochlorine residues in air, water, sediments, and soils from the lake Baikal region, Russia. Environ Sci Technol, 1995, 29: 792-801.

[129] 卢冰, 陈荣华, 王自磐, 朱纯, Vetter, W. 北极海洋沉积物中持久性有机污染物分布特征及分子地层学记录的研究. 海洋学报, 2005, 27(4): 167-173.

[130] Letcher R J, Bustnes J O, Dietz R, Jenssen B M, Jørgensen E H, Sonne C, Verreault J, Vijayan M M, Gabrielsen G W. Exposure and effects assessment of persistent organohalogen contaminants in arctic wildlife and fish. Sci Total Environ, 2010, 408: 2995-3043.

[131] Dietz R, Riget F F, Sonne C, Letcher R, Born E W, Muir D C G. Seasonal and temporal trends in polychlorinated biphenyls and organochlorine pesticides in East Greenland polar bears (*Ursus maritimus*)1990-2001. Sci Total Environ, 2004, 331(1-3): 107-124.

[132] Gebbink W A, Sonne C, Dietz R. Tissue-specific congener composition of organohalogen and metabolite contaminants in East Greenland polar bears (*Ursus maritimus*). Environ Pollut, 2008, 152(3): 621-629.

[133] Gebbink W A, Sonne C, Dietz R, Kirkegaard M, Born E W, Muir D C. G, Letcher R J. Target tissue selectivity and burdens of diverse classes of brominated and chlorinated contaminants in polar bears (*Ursus maritimus*) from East Greenland. Environ Sci Technol, 2008, 42, (3): 752-759.

[134] Verreault, J, Muir D C G, Norstrom R J, Stirling I, Fisk A T, Gabrielsen G W, Derocher A E, Evans T J, Dietz R, Sonne C. Chlorinated hydrocarbon contaminants and metabolites in polar bears (*Ursus maritimus*) from Alaska, Canada, East Greenland, and Svalbard: 1996-2002. Sci Total Environ, 2006, 351: 369-390.

[135] Sandala G M, Sonne-Hansen C, Dietz R, Muir D C G, Valters K, Bennett E R, Born E W, Letcher R J. Hydroxylated and methyl sulfone PCB metabolites in adipose and whole blood of polar bear (*Ursus maritimus*) from East Greenland. Sci Total Environ, 2004, 331(1-3): 125-141.

[136] Bentzen T W, Muir D C G, Amstrup S C, O'Hara T M. Organohalogen concentrations in blood and adipose tissue of Southern Beaufort Sea polar bears. Sci Total Environ, 2008, 406: 352-367.

[137] Kelly B C, Ikonomou M G, Blair J D, Gobas F A. Hydroxylated and methoxylated polybrominated diphenyl ethers in a Canadian Arctic marine food web. Environ Sci Technol 2008, 42(19): 7069-7077.

[138] Mckinney M A, Guise S D, Martineau D, Béland P, Lebeuf M, Letcher R J. Organohalogen contaminants and metabolites in beluga whale (*Delphinapterus leucas*) liver from two Canadian populations. Environ Toxicol Chem 2006, 25(5): 1246-1257.

[139] Andersen G, Kovacs K M, Lydersen C, Skaare J U, Gjertz I, Jenssen B M. Concentrations and patterns of organochlorine

contaminants in white whales (*Delphinapterus leucas*) from Svalbard, Norway. Sci Total Environ, 2001, 264(3): 267-281.

[140] Vorkamp K, Christensen J H, Riget F. Polybrominated diphenyl ethers and organochlorine compounds in biota from the marine environment of East Greenland. Sci Total Environ, 2004, 331(1-3): 143-155.

[141] Vorkamp K, Rigét F F, Glasius M, Muir D C G, Dietz R. Levels and trends of persistent organic pollutants in ringed seals(Phoca hispida)from Central West Greenland, with particular focus on polybrominated diphenyl ethers (PBDEs). Environ Int, 2008, 34(4): 499-508.

[142] Fisk A T, Tittlemier S A, Pranschke J L, Norstrom R J. Using anthropogenic contaminants and stable isotopes to assess the feeding ecology of Greenland sharks. Ecology, 2008, 83(8): 2162-2172.

[143] Wolkers H, Corkeron P J, Parijs S M V, Similä T, Bavel B V. Accumulation and transfer of contaminants in killer whales (*Orcinus orca*) from Norway: Indications for contaminant metabolism. Environ Toxicol Chem, 2007, 26(8): 1582-1590.

[144] Myers M J, Ylitalo G M, Krahn M M, Boyd D, Calkins D, Burkanov V, Atkinson S. Organochlorine contaminants in endangered Steller sea lion pups (*Eumetopias jubatus*) from western Alaska and the Russian Far East. Sci Total Environ, 2008, 396(1): 60-69.

[145] Borgå K, Gabrielsen G W, Skaare J U, Kleivane L, Norstrom R J, Fisk A T. Why Do Organochlorine Differences between Arctic Regions Vary among Trophic Levels? Environ Sci Technol, 2005, 39(12): 4343-4352.

[146] Stange K, Klungsøyr J. Organochlorine contaminants in fish and polycyclic aromatic hydrocarbons in sediments from the Barents Sea. Ices J Mar Sci, 1997, 54(3): 318-332.

[147] 马新东, 王艳洁, 那广水, 林忠胜, 周传光, 王震, 姚子伟. 北极新奥尔松地区有机氯农药和多氯联苯在不同环境样品中的浓度及特性. 极地研究, 2008, 20(4): 329-337.

[148] Piazza R, Gambaro A, Argiriadis E, Vecchiato M, Zambon S, Cescon P, Barbante C. Development of a method for simultaneous analysis of PCDDs, PCDFs, PCBs, PBDEs, PCNs and PAHs in Antarctic air. Anal Bioanal Chem, 2013, 405: 917-932.

[149] Dickhut R M, Cincinelli A, Cochran M, Kylin H. Aerosol-mediated transport and deposition of brominated diphenyl ethers to Antarctica. Environ Sci Technol, 2012, 46: 3135-3140.

[150] Li Y F, Jia H L. Prediction of gas/particle partition quotients of polybrominated diphenyl ethers (PBDEs) in north temperate zone air: an empirical approach. Ecotox Environ Safe, 2014, 108(108): 65-71.

[151] Li Y F, Ma W L, Yang M. Prediction of gas/particle partitioning of polybrominated diphenyl ethers (PBDEs) in global air: A theoretical study. Atmos Chem Phys, 2015, 15: 1669-1681.

[152] Mwangi J K, Lee W J, Wang L C, Sung P J, Fang L S, Lee Y Y, Chang-Chien G P. Persistent organic pollutants in the Antarctic coastal environment and their bioaccumulation in penguins. Environ Pollut, 2016, 216: 924-934.

[153] Hale R C, Kim S L, Harvey E, Guardia M J L, Mainor T M, Bush E O, Jacobs E M. Antarctic research bases: Local sources of polybrominated diphenyl ether (PBDE) flame retardants. Environ Sci Technol, 2008, 42: 1452-1457

[154] Yogui G T, Sericano J L. Levels and pattern of polybrominated diphenyl ethers in eggs of Antarctic seabirds: Endemic versus migratory species. Environ Pollut, 2009, 157: 975-980.

[155] Chen D, Hale R C, La Guardia M J, Luellen D, Kim S, Geisz H N. Hexabromocyclododecane flame retardant in Antarctica: Research stations as sources. Environ Pollut, 2015, 206: 611-618.

[156] Kim J T, Son M H, Kang J H, Kim J H, Jung J W, Chang Y S. Occurrence of legacy and new persistent organic pollutants in Avian tissues from King George Island, Antarctica. Environ Sci Technol, 2015, 49: 13628-13638.

[157] Möller A, Xie Z, Sturm R, Ebinghaus R. Large-Scale Distribution of Dechlorane Plus in Air and Seawater from the Arctic to Antarctica. Environ Sci Technol, 2010, 44: 8977-8982.

[158] Möller A, Xie Z, Cai M, Sturm R, Ebinghaus R. Brominated flame retardants and dechlorane plus in the marine atmosphere from southeast Asia toward Antarctica. Environ Sci Technol, 2012, 46: 3141-3148.

[159] Mello F V, Roscales J L, Guida Y S, Menezes J F, Vicente A, Costa E S, Jimenez B, Torres J P. Relationship between legacy and emerging organic pollutants in Antarctic seabirds and their foraging ecology as shown by delta13C and delta15N. Sci Total Environ, 2016, 573: 1380-1389.

[160] Na G, Yao Y, Gao H, Li R, Ge L, Titaley I A, Santiago-Delgado L, Massey Simonich S L. Trophic magnification of Dechlorane Plus in the marine food webs of Fildes Peninsula in Antarctica. Mar Pollut Bull, 2017, 117: 456-461.

[161] Xie Z, Möller A, Ahrens L, Caba A, Sturm R, Ebinghaus R. Brominated flame retardants and Dechlorane Plus in air and sea water of the Atlantic Ocean and the Antarctic. Proceedings of the BFR 2010 Conference, Kyoto, Japan, 2010.

[162] Cheng W, Xie Z, Blais J M, Zhang P, Ming L, Yang C, Wen H, Rui D, Sun L. Organophosphorus esters in the oceans and

possible relation with ocean gyres. Environ Pollut, 2013, 180: 159-164.

[163] Cheng W, Sun L, Huang W, Ruan T, Xie Z, Zhang P, Ding R, Li M. Detection and distribution of Tris(2-chloroethyl) phosphate on the East Antarctic ice sheet. Chemosphere, 2013, 92: 1017-1021.

[164] Wolschke H, Meng X Z, Xie Z, Ebinghaus R, Cai M. Novel flame retardants (N-FRs), polybrominated diphenyl ethers (PBDEs) and dioxin-like polychlorinated biphenyls (DL-PCBs) in fish, penguin, and skua from King George Island, Antarctica. Mar Pollut Bull, 2015, 96: 513-518.

[165] Kong D, Macleod M, Hung H, Cousins I T. Statistical Analysis of Long-Term Monitoring Data for Persistent Organic Pollutants in the Atmosphere at 20 Monitoring Stations Broadly Indicates Declining Concentrations. Environ Sci Technol, 2014, 48: 12492.

[166] Hung H, Katsoyiannis A A, Brorström-Lundén E, Olafsdottir K, Aas W, Breivik K, Bohlin-Nizzetto P, Sigurdsson A, Hakola H, Bossi R. Temporal trends of Persistent Organic Pollutants (POPs) in arctic air: 20 years of monitoring under the Arctic Monitoring and Assessment Programme (AMAP). Environ Pollut, 2016, 217: 52-61.

[167] de Wit C A, Herzke D, Vorkamp K. Brominated flame retardants in the Arctic environment--trends and new candidates. Sci Total Environ, 2010, 408: 2885-2918.

[168] Mariussen E, Steinnes E, Breivik K, Nygård T, Schlabach M, Kålås J A. Spatial patterns of polybrominated diphenyl ethers (PBDEs) in mosses, herbivores and a carnivore from the Norwegian terrestrial biota. Sci Total Environ, 2008, 404: 162.

[169] Zhen W, Na G, Ma X, Ge L, Lin Z, Yao Z. Characterizing the distribution of selected PBDEs in soil, moss and reindeer dung at Ny-Ålesund of the Arctic. Chemosphere, 2015, 137: 9-13.

[170] Zhu C F, Li Y M, Wang P, Chen Z J, Ren D W, Ssebugere P, Zhang Q H, Jiang G B. Polychlorinated biphenyls (PCBs) and polybrominated biphenyl ethers (PBDEs) in environmental samples from Ny-Alesund and London Island, Svalbard, the Arctic. Chemosphere, 2015, 126: 40-46.

[171] Salvadó J A, Sobek A, Carrizo D, Gustafsson Ö. Observation-Based Assessment of PBDE Loads in Arctic Ocean Waters. Environ Sci Technol, 2016, 50: 2236.

[172] Meyer T, Muir D C, Teixeira C, Wang X, Young T, Wania F. Deposition of brominated flame retardants to the Devon Ice Cap, Nunavut, Canada. Environ Sci Technol, 2012, 46: 826-833.

[173] Mckinney M A, Letcher R J, Aars J, Born E W, Branigan M, Dietz R, Evans T J, Gabrielsen G W, Peacock E, Sonne C. Flame retardants and legacy contaminants in polar bears from Alaska, Canada, East Greenland and Svalbard, 2005–2008. Environ Int, 2011, 37: 365-374.

[174] Ikonomou M G, Rayne S, Addison R F. Exponential Increases of the Brominated Flame Retardants, Polybrominated Diphenyl Ethers, in the Canadian Arctic from 1981 to 2000. Environ Sci Technol, 2002, 36: 1886-1892.

[175] Tomy G T, Pleskach K, Ferguson S H, Hare J, Stern G, Macinnis G, Marvin C H, Loseto L. Trophodynamics of some PFCs and BFRs in a Western Canadian Arctic marine food web. Environ Sci Technol, 2009, 43: 4076-4081.

[176] Rotander A, Van B B, Polder A, Rigét F, Auðunsson G A, Gabrielsen G W, Víkingsson G, Bloch D, Dam M. Polybrominated diphenyl ethers (PBDEs) in marine mammals from Arctic and North Atlantic regions, 1986-2009. Environ Int, 2012, 40: 102.

[177] Houde M, Wang X, Ferguson S H, Gagnon P, Brown T M, Tanabe S, Kunito T, Kwan M, Muir D C. Spatial and temporal trends of alternative flame retardants and polybrominated diphenyl ethers in ringed seals (*Phoca hispida*) across the Canadian Arctic. Environ Pollut, 2017, 223: 266-276.

[178] Braune B M, Letcher R J, Gaston A J, Mallory M L. Trends of polybrominated diphenyl ethers and hexabromocyclododecane in eggs of Canadian Arctic seabirds reflect changing use patterns. Environ Res, 2015, 142: 651.

[179] Curren M S, Davis K, Liang C L, Adlard B, Foster W G, Donaldson S G, Kandola K, Brewster J, Potyrala M, Van O J. Comparing plasma concentrations of persistent organic pollutants and metals in primiparous women from northern and southern Canada. Sci Total Environ, 2014, 479: 306-318.

[180] Arctic Monitoring and Assessment Program. Trends in Stockholm Convention Persistent Organic Pollutants(POPs)in Arctic Air. Human media and Biota, 2014.

[181] Yu Y, Hung H, Alexandrou N, Roach P, Nordin K. Multiyear Measurements of Flame Retardants and Organochlorine Pesticides in Air in Canada's Western Sub-Arctic. Environ Sci Technol, 2015, 49: 8623-8630.

[182] Xiao H, Shen L, Su Y, Barresi E, Dejong M, Hung H, Lei Y D, Wania F, Reiner E J, Sverko E. Atmospheric concentrations of halogenated flame retardants at two remote locations: the Canadian High Arctic and the Tibetan Plateau. Environ Pollut, 2012, 161: 154-161.

[183] Möller A, Xie Z, Cai M, Zhong G, Huang P, Cai M, Sturm R, He J, Ebinghaus R. Polybrominated diphenyl ethers vs alternate brominated flame retardants and Dechloranes from East Asia to the Arctic. Environ Sci Technol, 2011, 45: 6793.

[184] Salamova A, Hermanson M H, Hites R A, 2014. Organophosphate and Halogenated Flame Retardants in Atmospheric Particles from a European Arctic Site. Environ Sci Technol 48, 6133.

[185] Wang P, Zhang Q H, Zhang H, Wang T, Sun H Z, Zheng S C, Li Y M, Liang Y, Jiang G B. Sources and environmental behaviors of Dechlorane Plus and related compounds - A review. Environ Int, 2016, 88: 206-220.

[186] Na G, Wei W, Zhou S, Gao H, Ma X, Qiu L, Ge L, Bao C, Yao Z, . Distribution characteristics and indicator significance of Dechloranes in multi-matrices at Ny-Ålesund in the Arctic. J Environ Sci(China), 2015, 28: 8-13.

[187] Ma Y, Xie Z, Halsall C, Möller A, Yang H, Zhong G, Cai M, Ebinghaus R. The spatial distribution of organochlorine pesticides and halogenated flame retardants in the surface sediments of an Arctic fjord: The influence of ocean currents vs. glacial runoff. Chemosphere, 2015, 119: 953-960.

[188] Davie-Martin C L, Hageman K J, Chin Y P, Nistor B J, Hung H. Concentrations, gas-particle distributions, and source indicator analysis of brominated flame retardants in air at Toolik Lake, Arctic Alaska. Environ Sci Proc Impacts, 2016, 18.

[189] Hermanson M H, Isaksson E, Forsström S, Teixeira C, Muir D C, Pohjola V A, Rs V D W. Deposition history of brominated flame retardant compounds in an ice core from Holtedahlfonna, Svalbard, Norway. Environ Sci Technol, 2010, 44: 7405-7410.

[190] Bavel B V, Rotander A, Lindström G, Polder A, Rigét F, Auðunsson G A, Dam M. BFRs in Arctic marine mammals during three decades. Not only a story of BDEs, 2010.

[191] Karlsson M, Ericson I, Van B B, Jensen J K, Dam M. Levels of brominated flame retardants in Northern Fulmar(*Fulmarus glacialis*) eggs from the Faroe Islands. Sci Total Environ, 2006, 367: 840-846.

[192] Verreault J, Gebbink W A, Gauthier L T, Gabrielsen G W, Letcher R J. Brominated Flame Retardants in Glaucous Gulls from the Norwegian Arctic: More Than Just an Issue of Polybrominated Diphenyl Ethers. Environ Sci Technol, 2007, 41: 4925-4931.

[193] Möller A, Sturm R, Xie Z, Cai M, He J, Ebinghaus R. Organophosphorus flame retardants and plasticizers in airborne particles over the Northern Pacific and Indian Ocean toward the Polar Regions: evidence for global occurrence. Environ Sci Technol, 2012, 46: 3127.

[194] Sühring R, Diamond M L, Scheringer M, Wong F, Pućko M, Stern G, Burt A, Hung H, Fellin P, Li H. Organophosphate Esters in Canadian Arctic Air: Occurrence, Levels and Trends. Environ Sci Technol, 2016, 50: 7409.

[195] Li J, Xie Z, Mi W, Lai S, Tian C, Emeis K C, Ebinghaus R. Organophosphate esters in air, snow, and seawater in the North Atlantic and the Arctic. Environ Sci Technol, 2017, 51: 6887-6896.

[196] Ma Y, Xie Z, Lohmann R, Mi W, Gao G. Organophosphate ester flame retardants and plasticizers in ocean sediments from the North Pacific to the Arctic Ocean. Environ Sci Technol, 2017, 51(7): 3809-3815.

[197] Hallanger I G, Sagerup K, Evenset A, Kovacs K M, Leonards P, Fuglei E, Routti H, Aars J, Strøm H, Lydersen C. Organophosphorous flame retardants in biota from Svalbard, Norway. Mar Pollut Bull, 2015, 101: 442-447.

[198] Giesy J P, K Kannan. Global distribution of perfluorooctane sulfonate in wildlife. Environ Sci Technol, 2001, 35(7): 1339-1342.

[199] Hansen K J, Clemen L A, Ellefson M E, Johnson H O. Compound-specific, quantitative characterization of organic fluorochemicals in biological matrices. Environ Sci Technol, 2001, 35(4): 766-770.

[200] NL S, DC M, M S. Persistent organic pollutants, chapter: Perfluoroalkyl compounds. Edited by Stuart Harrad, 2010.

[201] Liu J, Lee L S. Solubility and sorption by soils of 8: 2 fluorotelomer alcohol in water and cosolvent systems. Environ Sci Technol, 2005, 39(19): 7535-7540.

[202] Dinglasan M J, Ye Y, Edwards E A, Mabury S A. Fluorotelomer alcohol biodegradation yields poly-and perfluorinated acids. Environ Sci Technol, 2004, 38(10): 2857-2864.

[203] Benskin J P, Ikonomou M G, Gobas F A P C, Begley T H, Woudneh M B, Cosgrove J R. Biodegradation of N-ethyl perfluorooctane sulfonamido ethanol (EtFOSE) and EtFOSE-based phosphate diester (SAmPAP diester) in marine sediments. Environ Sci Technol, 2013, 47(3): 1381-1389.

[204] Dreyer A, Weinberg I, Temme C, Ebinghaus R. Polyfluorinated compounds in the atmosphere of the Atlantic and Southern Oceans: evidence for a global distribution. Environ Sci Technol, 2009, 43(17): 6507-6514.

[205] Vento S D, Halsall C, Gioia R, Jones K, Dachs J. Volatile per-and polyfluoroalkyl compounds in the remote atmosphere of the western Antarctic Peninsula: an indirect source of perfluoroalkyl acids to Antarctic waters? Atmos Pollut Res, 2012,

3(4): 450-455.
[206] Wang Z, Xie Z, Mi W, Möller A, Wolschke H, Ebinghaus R. Neutral poly/per-fluoroalkyl substances in air from the Atlantic to the Southern Ocean and in Antarctic snow. Environ Sci Technol, 2015, 49(13): 7770-7775.
[207] Shoeib M, Harner T, Vlahos P. Perfluorinated chemicals in the Arctic atmosphere. Environ Sci Technol, 2006, 40(24): 7577-7583.
[208] Stock N L, Furdui V I, Muir D C, Mabury S A. Perfluoroalkyl contaminants in the Canadian Arctic: Evidence of atmospheric transport and local contamination. Environ Sci Technol, 2007, 41(10): 3529-3536.
[209] Butt C M, Berger U, Bossi R, Tomy G T. Levels and trends of poly-and perfluorinated compounds in the arctic environment. Sci Total Environ, 2010, 408(15): 2936-2965.
[210] Wei S, Chen L Q, Taniyasu S, So M K, Murphy M B, Yamashita N, Yeung L W Y, Lam P K S. Distribution of perfluorinated compounds in surface seawaters between Asia and Antarctica. Mar Pollut Bull, 2007, 54(11): 1813-1818.
[211] Yamashita N, Taniyasu S, Petrick G, Wei S, Gamo T, Lam P K S, Kannan, K. Perfluorinated acids as novel chemical tracers of global circulation of ocean waters. Chemosphere, 2008, 70(7): 1247-1255.
[212] Zhao Z, Xie Z, Möller A, Sturm R, Tang J, Zhang G, Ebinghaus R. Distribution and long-range transport of polyfluoroalkyl substances in the Arctic, Atlantic Ocean and Antarctic coast. Environ Pollut, 2012, 170: 71-77.
[213] Yeung W Y, Dassuncao C, Mabury S A, Sunderland E M, Zhang X, Lohmann R. Vertical Profiles, Sources, and Transport of PFASs in the Arctic Ocean. Environ Sci Technol, 2017, 51(12): 6735-6744.
[214] Xie Z, Wang Z, Mi W, Möller A, Wolschke H, Ebinghaus R. Neutral poly-/perfluoroalkyl substances in air and snow from the Arctic. Sci Rep, 2015, 5.
[215] Casal P, Zhang Y, Martin J W, Pizarro M, Jiménez B, Dachs J. The role of snow deposition of perfluoroalkylated substances at coastal livingston island (Maritime Antarctica). Environ Sci Technol, 2017.
[216] Macinnis J J, French K, Muir D C, Spencer C, Criscitiello A, De Silva A O, Young C J. Emerging investigator series: a 14-year depositional ice record of perfluoroalkyl substances in the High Arctic. Environ Sci Proc Impacts, 2017, 19(1): 22-30.
[217] Young C J, Furdui V I, Franklin J, Koerner R M, Muir D C G, Mabury S A. Perfluorinated acids in arctic snow: new evidence for atmospheric formation. Environ Sci Technol, 2007, 41(10): 3455-3461.
[218] Tao L, Kannan K, Kajiwara N, Costa M M, Fillmann G, Takahashi S, Tanabe S. Perfluorooctanesulfonate and related fluorochemicals in albatrosses, elephant seals, penguins, and polar skuas from the Southern Ocean. Environ Sci Technol, 2006, 40(24): 7642-7648.
[219] Schiavone A, Corsolini S, Kannan K, Tao L, Trivelpiece W, Jr T D, Focardi S. Perfluorinated contaminants in fur seal pups and penguin eggs from South Shetland, Antarctica. Sci Total Environ, 2009, 407(12): 3899-3904.
[220] Routti H, Gabrielsen G W, Herzke D, Kovacs K M Lydersen C. Spatial and temporal trends in perfluoroalkyl substances (PFASs) in ringed seals (*Pusa hispida*) from Svalbard. Environ Pollut, 2016, 214: 230-238.
[221] Braune B M. Identification of long-chain perfluorinated acids in biota from the Canadian Arctic. Environ Sci Technol, 2004, 38(2): 373-380.
[222] Tomy G T, Budakowski W, Halldorson T, Helm P A, Stern G A, Friesen K, Pepper K, Tittlemier S A, Fisk A T. Fluorinated organic compounds in an eastern Arctic marine food web. Environ Sci Technol, 2004, 38(24): 6475-6481.
[223] Bossi R, Riget F F, Dietz R, Sonne C, Fauser P, Dam M, Vorkamp K. Preliminary screening of perfluorooctane sulfonate (PFOS) and other fluorochemicals in fish, birds and marine mammals from Greenland and the Faroe Islands. Environ Pollut, 2005. 136(2): 323-329.
[224] Lescord G L, Kidd K A, Silva A O D, Williamson M, Spencer C, Wang X, Muir D C G. Perfluorinated and Polyfluorinated Compounds in Lake Food Webs from the Canadian High Arctic. Environ Sci Technol, 2015, 49(5): 2694-2702.
[225] Müller C E, De Silva A O, Small J, Williamson M, Wang X, Morris A, Katz S, Gamberg M, Muir D C. Biomagnification of perfluorinated compounds in a remote terrestrial food chain: lichen–caribou–wolf. Environ Sci Technol, 2011, 45(20): 8665-8673.
[226] Dallaire R, Ayotte P, Pereg D, Déry S, Dumas P, Langlois E, Dewailly E. Determinants of plasma concentrations of perfluorooctanesulfonate and brominated organic compounds in Nunavik Inuit adults (Canada). Environ Sci Technol, 2009, 43(13): 5130-5136.
[227] Audetdelage Y, Ouellet N, Dallaire R, Dewailly E, Ayotte P. Persistent organic pollutants and transthyretin-bound thyroxin in plasma of inuit women of childbearing age. Environ Sci Technol, 2013, 47(22): 13086-13092.

[228] Lindh C H, Rylander L, Toft G, Axmon A, Rignell-Hydbom A, Giwercman A, Pedersen H S, Góalczyk K, Ludwicki J K, Zvyezday V. Blood serum concentrations of perfluorinated compounds in men from Greenlandic Inuit and European populations. Chemosphere, 2012, 88(11): 1269-1275.

[229] Ostertag S K, Tague B A, Humphries M M, Tittlemier S A, Chan H M. Estimated dietary exposure to fluorinated compounds from traditional foods among Inuit in Nunavut, Canada. Chemosphere, 2009, 75(9): 1165-1172.

[230] Ellis D A, Martin J W, Mabury S A, Hurley M D, Andersen M P S, Wallington T J. Atmospheric lifetime of fluorotelomer alcohols. Environ Sci Technol, 2003, 37(17): 3816-3820.

[231] Ellis D A, Martin J W, De Silva A O, Mabury S A, Hurley M D, Sulbaek Andersen M P, Wallington T J. Degradation of fluorotelomer alcohols: a likely atmospheric source of perfluorinated carboxylic acids. Environ Sci Technol, 2004, 38(12): 3316-3321.

[232] Wallington T J, Hurley M D, Xia J D J W, Wuebbles D J, Sillman S, Ito A, Penner J E, Ellis D A, Martin J, Mabury S A, Nielsen, O J, Sulbaek Andersen M P. Formation of $C_7F_{15}COOH$ (PFOA) and Other Perfluorocarboxylic Acids during the Atmospheric Oxidation of 8:2 Fluorotelomer Alcohol. Environ Sci Technol, 2006, 40(3): 924-930.

[233] Kwok K Y, Yamazaki E, Yamashita N, Taniyasu S, Murphy M B, Horii Y, Petrick G, Kallerborn R, Kannan K, Murano K. Transport of perfluoroalkyl substances(PFAS)from an arctic glacier to downstream locations: implications for sources. Sci Total Environ, 2013, 447: 46-55.

[234] Haukås M, Berger U, Hop H, Gulliksen B, Gabrielsen G W. Bioaccumulation of per- and polyfluorinated alkyl substances(PFAS)in selected species from the Barents Sea food web. Environ Pollut, 2007, 148(1): 360-371.

[235] Jahnke A, Berger U, Ebinghaus R, Temme, C. Latitudinal Gradient of Airborne Polyfluorinated Alkyl Substances in the Marine Atmosphere between Germany and South Africa (53° N−33° S). Environ Sci Technol, 2007, 41(9): 3055-3061.

[236] Nash S B, Rintoul S R, Kawaguchi S, Staniland I, Hoff J V D, Tierney M, Bossi R. Perfluorinated compounds in the Antarctic region: ocean circulation provides prolonged protection from distant sources. Environ Pollut, 2010, 158(9): 2985-2991.

[237] Freberg B I, Haug L S, Olsen R, Daae H L, Hersson M, Thomsen C, Thorud S, Becher G, Molander P, Ellingsen D G. Occupational exposure to airborne perfluorinated compounds during professional ski waxing. Environ Sci Technol, 2010, 44(19): 7723-7728.

[238] Cai M, Yang H, Xie Z, Zhao Z, Wang F, Lu Z, Sturm R, Ebinghaus R. Per-and polyfluoroalkyl substances in snow, lake, surface runoff water and coastal seawater in Fildes Peninsula, King George Island, Antarctica. J Hazard Mater, 2012, 209: 335-342.

[239] Wild S, Mclagan D, Schlabach M, Bossi R, Hawker D, Cropp R, King C K, Stark J S, Mondon J, Nash S B. An Antarctic research station as a source of brominated and perfluorinated persistent organic pollutants to the local environment. Environ Sci Technol, 2014, 49(1): 103-112.

[240] Llorca M, Farré M, Tavano M S, Alonso B, Koremblit G, Barceló D F. Fate of a broad spectrum of perfluorinated compounds in soils and biota from Tierra del Fuego and Antarctica. Environ Pollut, 2012, 163: 158-166.

[241] Kelly B C, Ikonomou M G, Blair J D, Surridge B, Hoover D, Grace R, Gobas F A P C. Perfluoroalkyl Contaminants in an Arctic Marine Food Web: Trophic Magnification and Wildlife Exposure. Environ Sci Technol, 2009, 43(11): 4037-4043.

[242] Conder J M, Hoke R A, De W W, Russell M H, Buck R C. Are PFCAs bioaccumulative? A critical review and comparison with regulatory criteria and persistent lipophilic compounds. Environ Sci Technol, 2008, 42(4): 995-1003.

[243] Ng C A, Hungerbuhler k. Bioaccumulation of Perfluorinated Alkyl Acids: Observations and Models. Environ Sci Technol, 2014, 48(9): 4637-4648.

[244] 吴江平, 管运涛, 李明远, 靳军涛, Makoto Y, 张锡辉. 全氟化合物的生物富集效应研究进展. 生态环境学报, 2010, 19(5): 1246-1252.

[245] Rigét F, Bossi R, Sonne C, Vorkamp K, Dietz R. Trends of perfluorochemicals in Greenland ringed seals and polar bears: Indications of shifts to decreasing trends. Chemosphere, 2013, 93(8): 1607-1614.

[246] Dassuncao C, Hu X C, Zhang X, Bossi R, Dam M, Mikkelsen B, Sunderland E M. Temporal Shifts in Poly- and Perfluoroalkyl Substances (PFASs) in North Atlantic Pilot Whales Indicate Large Contribution of Atmospheric Precursors. Environ Sci Technol, 2017.

[247] Bayen S, Obbard J P, Thomas G O. Chlorinated paraffins: a review of analysis and environmental occurrence. Environ Int, 2006, 32(7): 915-929.

[248] Drouillard K G, Tomy G T, Muir D C G, Friesen K J V. Volatility of chlorinated n-alkanes (C_{10}–C_{12}): Vapor pressures and

Henry's law constants. Environ Toxicol Chem, 1998, 17(7): 1252-1260.
[249] Feo M L, Eljarrat E, Barceló D, Barceló D. Occurrence, fate and analysis of polychlorinated n-alkanes in the environment. TrAC-Trend Anal Chem, 2009, 28(6): 778-791.
[250] Meylan W M, Howard P H. Computer estimation of the Atmospheric gas-phase reaction rate of organic compounds with hydroxyl radicals and ozone. Chemosphere, 1993, 26(12): 2293-2299.
[251] Boreen A R, Schlabach M, Gundersen H. Polychlorinated alkanes in arctic air. Organohalog Compds, 2000. 47: 272-275.
[252] Agency N E. Monitoring of environmental contaminants in air and precipitation, M-368|2015 annual report 2014. Norwegian Institute for Air Research, 2015.
[253] Tomy G T, Fisk A T, Westmore J B, Muir D C G. Environmental chemistry and toxicology of polychlorinated n-alkanes. Rev Environ Contam Toxicol, 1998, 158(158): 53-128.
[254] Tomy G T, Stern G A, Lockhart W L, Muir D C G. Occurrence of C_{10}–C_{13} polychlorinated n-alkanes in Canadian midlatitude and arctic lake sediments. Environ Sci Technol, 1999, 33(17): 2858-2863.
[255] Li H, Fu J, Pan W, Wang P, Li Y, Zhang Q, Wang Y, Zhang A, Liang Y, Jiang G. Environmental behaviour of short-chain chlorinated paraffins in aquatic and terrestrial ecosystems of Ny-Alesund and London Island, Svalbard, in the Arctic. Sci Total Environ, 2017, 590: 163-170.
[256] Yuan B, Fu J, Wang Y, Jiang G. Short-chain chlorinated paraffins in soil, paddy seeds (*Oryza sativa*) and snails(Ampullariidae)in an e-waste dismantling area in China: Homologue group pattern, spatial distribution and risk assessment. Environ Pollut, 2017, 220: 608-615.
[257] Zeng L, Lam J C, Wang Y, Jiang G, Lam P K. Temporal Trends and Pattern Changes of Short-and Medium-Chain Chlorinated Paraffins in Marine Mammals from the South China Sea over the Past Decade. Environ Sci Technol, 2015, 49(19): 11348-11355.
[258] Bo J, Andersson R, Asplund L, Litzen K, Nylund K, Sellströom U, Uvemo U B, Wahlberg C, Wideqvist U, Odsjö T. Chlorinated and Brominated Persistent Organic-Compounds in Biological Samples from the Environment. Environ Toxicol Chem, 1993, 12(7): 1163-1174.
[259] Reth M, Ciric A, Christensen G N, Heimstad E S, Oehme M. Short-and medium-chain chlorinated paraffins in biota from the European Arctic - differences in homologue group patterns. Sci Total Environ, 2006, 367(1): 252-260.
[260] Tomy G T, Muir D C, Stern G A, Westmore J B. Levels of C-10-C-13 polychloro-*n*-alkanes in marine mammals from the Arctic and the St. Lawrence River estuary. Environ Sci Technol, 2000, 34(9): 1615-1619.
[261] Herzke D, Huber S, Bervoets L, D'Hollander W, Hajslova J, Pulkrabova J, Brambilla G, De Filippis S P, Klenow S, Heinemeyer G. Perfluorinated Alkylated Substances, Brominated Flame Retardants and Chlorinated Paraffins in the Norwegian Environment-Screening 2013, in Norwegian Environment Agency. NILU–Norsk Institutt for Luftforskning. SWECO, Tromsø, Norway. 2013.
[262] Ma X, Zhang H, Zhou H, Na G, Wang Z, Chen C, Chen J, Chen J. Occurrence and gas/particle partitioning of short- and medium-chain chlorinated paraffins in the atmosphere of Fildes Peninsula of Antarctica. Atmos Environ, 2014, 90: 10-15.
[263] Li H, Fu J, Zhang A, Zhang Q, Wang Y. Occurrence, bioaccumulation and long-range transport of short-chain chlorinated paraffins on the Fildes Peninsula at King George Island, Antarctica. Environ Int, 2016, 94: 408-414.
[264] Reth M, Zencak Z, Oehme M. First study of congener group patterns and concentrations of short- and medium-chain chlorinated paraffins in fish from the North and Baltic Sea. Chemosphere, 2005, 58(7): 847-854.
[265] Ma X, Zhang H, Wang Z, Yao Z, Chen J, Chen J. Bioaccumulation and Trophic Transfer of Short Chain Chlorinated Paraffins in a Marine Food Web from Liaodong Bay, North China. Environ Sci Technol, 2014, 48(10): 5964-5971.
[266] Dick T A, Gallagher C P, Tomy G T. Short-and medium-chain chlorinated paraffins in fish, water and soils from the Iqaluit, Nunavut (Canada), area. World Review of Science, Technology and Sustainable Development, 2010, 7(4): 387-401.
[267] Nikitina M, Popova L, Korobitcina J, Efremova O, Trofimova A, Nakvasina E, Volkov A. Environmental Status of the Arctic Soils. J Elementol, 2015, 20:643-651.
[268] Trevizani T H, Figueira R C L, Ribeiro A P, Theophilo C Y S, Majer A P, Petti M A V, Corbisier T N, Montone R C. Bioaccumulation of heavy metals in marine organisms and sediments from Admiralty Bay, King George Island, Antarctica. Mar Pollut Bull, 2016, 106(1-2): 366-371.
[269] Tuohy A, Bertler N, Neff P, Edwards R, Emanuelsson D, Beers T, Mayewski P. Transport and deposition of heavy metals in the Ross Sea Region, Antarctica. J Geophys Res-Atmos , 2015, 120(20): 10996-11011.
[270] Zelano I, Malandrino M, Giacomino A, Buoso S, Conca E, Sivry Y, Benedetti M, Abollino O. Element variability in

lacustrine systems of Terra Nova Bay (Antarctica) and concentration evolution in surface waters. Chemosphere, 2017, 180: 343-355.

[271] Corami F, Capodaglio G, Turetta C, Soggia F, Magi E, Grotti M. Summer distribution of trace metals in the western sector of the Ross Sea, Antarctica. J Environ Monitor, 2005, 7(12): 1256-1264.

[272] Borghini F and Bargagli R. Changes of major ion concentrations in melting snow and terrestrial waters from northern Victoria Land, Antarctica. Antarct Sci, 2004, 16(2): 107-115.

[273] Conca E, Malandrino M, Giacomino A, Buoso S, Berto S, Verplanck P L, Magi E, Abollino O. Dynamics of inorganic components in lake waters from Terra Nova Bay, Antarctica. Chemosphere, 2017, 183: 454-470.

[274] Celis J E, Barra R, Espejo W, González-Acuña D, Jara S. Trace Element Concentrations in Biotic Matrices of Gentoo Penguins (*Pygoscelis Papua*) and Coastal Soils from Different Locations of the Antarctic Peninsula. Water Air Soil Poll, 2014, 226(1): 2266.

[275] Neto E D L, Guerra M B B, Thomazini A, Daher M, Andrade A M D, Schaefer C E G R. Soil Contamination by Toxic Metals Near an Antarctic Refuge in Robert Island, Maritime Antarctica: A Monitoring Strategy. Water Air Soil Poll, 2017, 228(2): 66.

[276] Bazzano A, Ardini F, Grotti M, Malandrino M, Giacomino A, Abollino O, Cappelletti D, Becagli S, Traversi R, Udisti R. Elemental and lead isotopic composition of atmospheric particulate measured in the Arctic region (Ny-Ålesund, Svalbard Islands). Rend Lincei, 2016. 27(1): 73-84.

[277] Kadko D, Galfond B, Landing W M, Shelley R U. Determining the pathways, fate, and flux of atmospherically derived trace elements in the Arctic ocean/ice system. Mar Chem, 2016, 182: 38-50.

[278] Catán S P, Bubach D, Di F C, Dopchiz L, Arribére M, Ansaldo M. Pygoscelis antarcticus feathers as bioindicator of trace element risk in marine environments from Barton Peninsula, 25 de Mayo (King George) Island, Antarctica. Environ Sci Pollut Res Int, 2017, 24(11): 10759-10767.

[279] Carravieri A, Cherel Y, Brault-Favrou M, Churlaud C, Peluhet L, Labadie P, Budzinski H, Chastel O, Bustamante P. From Antarctica to the subtropics: Contrasted geographical concentrations of selenium, mercury, and persistent organic pollutants in skua chicks (*Catharacta* spp.). Environ Pollut, 2017, 228: 464-473.

[280] Øverjordet I B, Kongsrud M B, Gabrielsen G W, Berg T, Ruus A, Evenset A, Borgå K, Christensen G, Jenssen B M. Toxic and essential elements changed in black-legged kittiwakes (*Rissa tridactyla*) during their stay in an Arctic breeding area. Sci Total Environ, 2015, 502: 548-556.

[281] Culicov O A, Yurukova L, Duliu O G, Zinicovscaia I. Elemental content of mosses and lichens from Livingston Island (Antarctica) as determined by instrumental neutron activation analysis (INAA). Environ Sci Pollut Res Int, 2017, 24(6): 5717-5732.

[282] St Pierre K A, St Louis V L, Kirk J L, Lehnherr I, Wang S, La F C. Importance of open marine waters to the enrichment of total mercury and monomethylmercury in lichens in the Canadian High Arctic. Environ Sci Technol, 2015, 49(10): 5930-5938.

作者：张庆华[1]，李英明[1]，张　蓬[3]，从志远[2]，杨瑞强[1]，史建波[1]，傅建捷[1]，王　璞[1]

[1]中国科学院生态环境研究中心，[2]中国科学院青藏高原研究所，[3]国家海洋环境监测中心

第 15 章　纳米材料环境化学研究进展

本章导读

基于污染控制的研究需求，环境纳米材料与技术学科应运而生。这一领域主要的研究关注点包括光催化、Fenton 高级氧化和纳米吸附材料等方面，纳米技术与其他技术如膜技术、树脂基技术、晶体调控技术的联用也成为解决材料负载、污染物深度处理及污染物有效分离的重要拓展。同时，纳米材料进入环境之后，受环境因素产生物理或化学转化，进而影响纳米材料的环境行为与效应，此方面的研究和探索方兴未艾。一些新型纳米材料排放到环境中，也可能会给生态环境以及人类健康带来潜在威胁。因此，纳米材料的暴露途径、毒性效应、致毒机制也成为当前的研究热点。此外，功能纳米材料的快速发展为污染物检测方法的创新提供了新的理论基础和技术支持，成为当前分析检测领域的重要发展方向和技术前沿。本章对上述领域近五年的一些研究成果进行了介绍，并对纳米材料环境化学研究进行展望。

关键词

环境纳米材料，环境纳米技术，纳米材料的环境过程与效应，纳米毒理环境分析

1 引　　言

光催化、Fenton 高级氧化、吸附和过滤技术是水处理领域广泛应用的技术，作为这些水处理技术的核心要素，纳米材料的发展是制约其工业化应用的关键因素。当前铋系列、改性碳化硅和静电纺丝纤维等新兴纳米材料对有机废水光催化降解取得了一定的进展，基于零价铁、铁矿物和无过渡金属等纳米材料的非均相 Fenton 高级氧化技术对有机物降解也得到了较多研究。以一维碳纳米纤维/纳米管、二维石墨烯为代表的新型碳纳米材料和纳米零价铁对放射性污染物吸附的研究工作取得了较大的突破。同时，研究者们还尝试探索纳米材料与技术与其他技术的联用，例如碳纳米材料与膜分离技术、树脂基纳米复合材料的可实用性技术及晶体生长调控下镁基纳米材料提取和富集水中低浓度污染物的技术，均取得了较大进展。

纳米材料的环境过程极为复杂。纳米材料进入环境之后，受多种环境因素的作用，能够产生不同的物理或化学转化，进而影响纳米材料的环境行为与效应。此外，纳米材料进入环境之后可与环境中的污染物相互作用，从而影响污染物的扩散性、反应活性和生物有效性等。纳米材料的环境转化还会导致同位素分馏等现象。目前的研究较多集中于金属/金属氧化物纳米材料、碳纳米材料两个方面，综合探索纳米材料在进入环境后可能发生的各种物理、化学转化及其可能产生的效应及机理。

随着纳米材料越来越广泛的应用，它们不可避免地被释放到水体、空气和土壤中，成为了一类新型污染物，给生态环境以及人类健康带来潜在威胁。充分的科学证据证明纳米材料暴露会导致生物毒性和健康风险，然而纳米材料与生物分子以及细胞的作用方式又决定了纳米材料的各种理化性质均对其生物效应产生影响。纳米材料的暴露途径、毒性效应、致毒机制以及与纳米材料的理化性质对其毒性效应的影响成为当前的研究热点。

同时，功能纳米材料的快速发展为污染物检测方法的创新提供了新的理论基础和技术支持。基于功能纳米材料开发的检测方法和技术具有简便、快速、灵敏度高、成本低等特点，在环境检测中具有广阔的应用前景，成为当前分析检测领域的重要发展方向和技术前沿。典型纳米材料（如量子点、石墨烯和 DNA）在环境分析领域中的应用，环境和生物体系中人工纳米材料的表征技术与方法等方面的研究得到了环境纳米分析研究者的重视。

2　环境纳米材料与技术的研究进展

2.1　光催化技术在水处理领域的研究进展

光催化技术是一种新型、绿色环保的水处理技术[1]。其基本原理是当能量大于半导体光催化剂禁带宽度的光照射时，光激发电子跃迁到导带，形成导带电子，同时在价带产生空穴[2,3]。空穴能够与吸附在催化剂表面的 OH^-或 H_2O 反应生成·OH，·OH 可以无选择氧化多种有机物并使之矿化[4]。光生电子能够与 O_2 反应生成 O_2^{-}，参与污染物的氧化还原反应[5]。该技术处理水中难降解有机污染物具有矿化率高、成本低、设备简单、操作方便及无需添加氧化剂等优点[6]。但常规催化剂存在光子利用率低、光谱响应范围窄、催化效率不高和处理效果差等问题[7]。研究人员为了克服以上问题，重点开发了铋系催化剂、改性碳化硅和静电纺丝纤维膜等纳米光催化材料，在处理污水中难降解有机污染物方面取得了很好的光催化处理效果。

2.1.1　铋系催化剂

铋系催化剂在可见光（$\lambda \geqslant 420$ nm）照射下具有良好的光催化性能，在难降解有机污染物的去除方面得到了广泛关注。该系列催化剂以 Bi_2O_3、钛酸铋及铋与其他金属组成的复合型催化剂的研究最为广泛。近期关于 Bi_2O_3 光催化剂的报道大部分集中在将金属离子掺杂到半导体催化剂的晶格结构内部，不但能抑制电子-空穴对的复合，还能拓宽半导体催化剂的吸收波长范围。已经报道的掺杂金属离子有 Fe^{3+}、Ni^{3+}、Mo^{5+}、Co^{2+}、Re^{3+}、Cr^{3+}、W^{6+}和 Cu^{2+}等。研究人员为了制备可见光响应的铋系催化剂，采用密度泛函理论计算了 32 种金属元素改性的 Bi_2O_3 催化剂的能带结构和氧化还原电位，初步筛选出理论上具有高效光催化活性的催化剂[8]。随后采用水热合成法制备了 Ti、Zr 和 Sr 掺杂的 Bi_2O_3 光催化剂，实验结果表明，三种催化剂在可见光照下降解五氯酚、三氯乙烯和六氯环己烷的光量子产率分别为 1.20%、1.03%和 0.67%[8]。Sr 掺杂的 Bi_2O_3 光催化剂在可见光照射下对四环素显示了很高的光催化活性[9]。

由于 Bi_2O_3 单独做光催化剂效果不是很理想，目前研究较多的是钛酸铋系列催化剂。研究人员采用水热合成法制备了钛酸铋光催化剂，在可见光下对氯酚进行光催化降解实验。结果表明，在可见光照射 1 h 后，初始浓度为 10 mg/L 的氯酚在 pH 为 2.3 的条件下去除率高于 99% [10]。对于多孔的钛酸铋催化剂而言，Bi_2O_3 的 Bi(III)被部分的 Ti(IV)取代，如图 15-1 所示。光照时，Ti(IV)会变成 Ti(III)。四个外部电子中的三个与氧气结合，另一个电子捕获价带空穴。当价带空穴被捕获后，Ti(III)被氧化成 Ti(IV)[10]。该电子转移过程交替进行，从而自由电子有更多的机会迁移到钛酸铋催化剂的表面，将氯酚等电子受体还原降解[10]。

研究人员采用水热合成法将钛掺杂到 Bi_2O_3 光催化剂中，然后采用光化学沉积法制备银改性的钛酸铋光催化剂。结果表明，银的掺杂将钛酸铋光催化剂对结晶紫的降解率提高了 1.9 倍[11]。为了进一步降低光生电子和空穴的复合率，提高光催化活性，研究人员致力于提高价带能级和在价带中引入杂化轨道方面的研究工作。研究人员以 $Bi(NO_3)_3·5H_2O$ 和 $SrCl_2$ 为反应物水热合成纳米级别的 $Bi_2O_3/Bi_2O_2CO_3/Sr_6Bi_2O_9$ 光催化剂，在 500 W 氙灯光照下（$\lambda > 420$ nm）降解磺胺甲恶唑。结果表明，在光照 2 h 后，初始浓度为 10 mg/L 的抗生素降解率高达 90%，TOC 去除率为 36%，主要的矿化产物为 NH_4^+、NO_3^-、SO_4^{2-}和 CO_2[12]。此外，研究人员采用水热合成法合成了 Bi_2WO_6 催化剂，通过 XRD、SEM 和 UV-vis 等分析证实该催化剂在可见光下具有很好的吸收，对甲醛和罗丹明 B 等有机物显示了很高的光催化活性[13,14]。

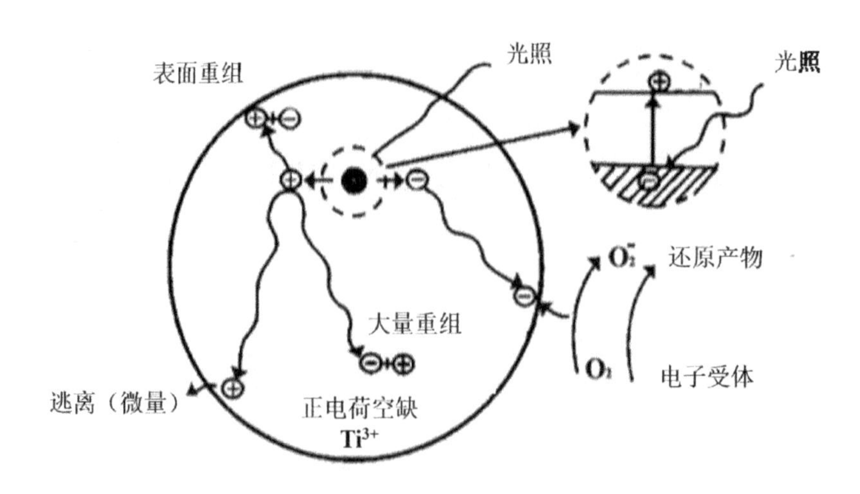

图 15-1 纳米多孔钛酸铋催化剂对有机物的光催化降解机理[10]

2.1.2 改性碳化硅催化剂

碳化硅（SiC）纳米材料由于能够响应可见光，因此被看作是一种非常有潜力的可见光催化材料。如何解决光生电子和空穴的分离是当前 SiC 催化剂研究领域的一个难点。研究人员通过水热法合成了 SiC/石墨烯复合催化剂，采用该催化剂在紫外光（254 nm）下降解全氟辛酸（PFOA）和全氟磺酸（PFOS）[15]。SiC/石墨烯复合催化剂对 PFOA 的降解速率常数（0.096 h^{-1}）是 SiC（0.048 h^{-1}）的 2.0 倍，比商业纳米 TiO_2（0.030 h^{-1}）高 2.2 倍。石墨烯的费米能级（−4.6 eV）低于 SiC（−4.0 eV），在光照时，光生电子能够从 SiC 上转移到石墨烯上，并很容易在石墨烯表面移动，可以更快地传递到污染物上，从而减少了电子和空穴的复合，提高了光生电子的利用率及对污染物的光催化性能。

与 PFOA 相比，由于磺酸基的存在，PFOS 比 PFOA 更难降解。研究人员采用超高频超声酸氧化法合成了石墨烯量子点（GQDs），然后采用水热合成法制备了 SiC/GODs 催化剂[16]，并研究了该催化剂在紫外光下对初始浓度为 0.019 mol/L 2-CF_3-PFOS、6-CF_3-PFOS 和直链-PFOS 的混合溶液的降解。光照 20 h 后，该催化剂对三种物质的降解率分别为 93.9%、90.4%和 88.5%，表明 SiC/石墨烯可以在 254 nm 紫外线照射下有效分解支链和直链的 PFOS[16]。复合纳米材料和 SiC 的价带分别为 4.77eV 和 5.83 eV，导带分别为−1.77eV 和−2.83 eV[16]。由此可见与 SiC 相比，GQDs 的加入改善了 SiC/GQDs 的导带（E_v）和价带（E_c），如图 15-2 所示，紫外光照射下，SiC/GQDs 异质结中电子和空穴发生了转移。前人通过计算得到 GQDs 的 HOMO 和 LUMO 分别为−5.66 eV 和−0.68 eV，分别高于 SiC 的 E_v 和 E_c。由于能量匹配，可以在 SiC/GQDs 中形成异质结结构，并且可以通过抑制光生载流子的重组而有效地提高光催化剂的活性[16]。

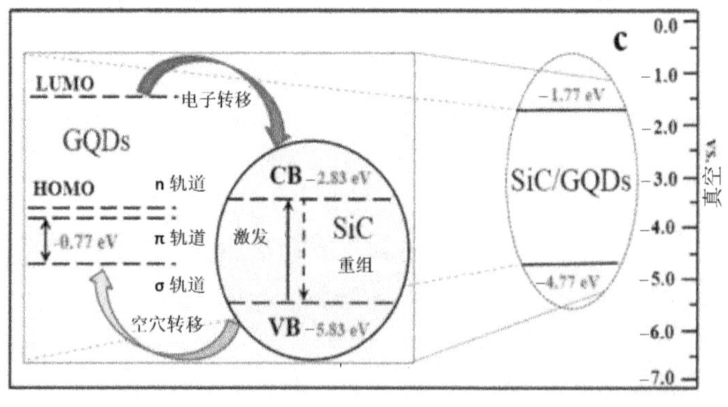

图 15-2 紫外光照下，SiC/GQDs 异质结中电子和空穴的转移[16]

2.1.3　氧化锆光催化剂

作为一种过渡金属氧化物，氧化锆的异质结构引起了研究人员的广泛关注。由于自身结构的原因，其被广泛应用于光催化处理污水领域。例如，拥有暴露晶面的花型氧化锆对罗丹明 B 染料展现出了高效的光催化降解性能。但在实际工业应用过程中，仍然存在量子效率低、可见光利用率低、易凝聚和分离回收繁琐等缺陷。为了扩展氧化锆的光谱响应范围，研究人员致力于金属（Fe、Ag 等）和非金属（N、F 等）的掺杂。研究表明，N 的掺杂极大地扩展了氧化锆催化剂对可见光的响应范围，同时对有机污染物（如染料、氯酚等）显示了很高的光催化性能。

研究人员采用静电纺丝技术制备了直径 100~400 nm 的多孔氧化锆电纺纤维毡[17]。使用该催化剂在紫外光照下降解了多种染料，结果表明，在 450 ℃合成的纳米电纺纤维，在紫外光照 1 h 后，对吖啶橙、甲基橙、中性红和结晶紫的降解率分别达到了 82.8%、62.6%、99.3%和 65.4%[17]。如此高的光催化活性归因于静电纺丝纤维膜结构具有很高的比表面积、较多的活性位点及较高的孔隙率。此外，该催化材料在相同实验条件下处理染料废水显示出了很高的稳定性和重复利用性。

2.2　非均相 Fenton 催化材料研究进展

Fenton 反应被广泛应用于污染治理和环境修复，其基本原理是在酸性条件下 Fe^{2+}引发 H_2O_2 分解产生·OH。·OH 具有很高的氧化还原电势（$E^0 = 2.80$ V *vs* SHE），其氧化能力较 H_2O_2（$E^0 = 1.763$ V *vs* SHE）而言大幅提升，能够氧化降解大部分有机污染物[18]。虽然传统均相 Fenton 反应体系（Fe^{2+}/H_2O_2）具有操作简便和高效率等优势，但仍然存在铁泥多、最适反应 pH 范围窄（2.0~3.5）等不足[19]。解决这些问题的有效途径之一是采用含铁（或其他具有 Fenton 活性的过渡金属元素，例如铜、锰、钌、铈等[20]）固相材料替代溶解性铁盐，构筑非均相 Fenton 反应体系。这些含铁材料能有效拓宽 Fenton 反应的适用范围，更为重要的是，固相材料易于分离回收和重复使用，便于污水处理的连续操作和成本控制[21]。目前，文献报道的非均相 Fenton 催化材料包括铁（或铜）负载的分子筛[22,23]、黏土矿物[24,25]、铁氧化物[26,27]等。近年来，新材料的快速发展不断为非均相 Fenton 催化剂家族新添许多成员。

2.2.1　纳米零价铁材料

纳米零价铁（nZVI）由于具有纳米尺寸、高比表面积和强还原能力，被广泛用于污染物的吸附和还原去除。对于某些不易通过还原途径去除的污染物，如4-氯-3-甲基苯酚，可以采用 nZVI-H_2O_2 非均相 Fenton 反应的方式氧化去除[28]。其反应途径如反应式（15-1）和反应式（15-2）所示。例如，张礼知研究组发现该非均相 Fenton 体系能够有效去除废水中的微量阿莫西林（12.3 mg/L），反应 1 h 去除效率达到 80.5%[29]。其他研究组则通过先 nZVI 还原再加 H_2O_2 Fenton 氧化的方式实现了水中 Cr(VI)和 4-氯苯酚的依次高效去除[30]。然而，上述处理过程均需在酸性条件下操作，且传统 nZVI 材料在空气中不稳定、易失活、难储存，从而限制了 nZVI-H_2O_2 非均相 Fenton 体系的应用。

$$Fe^0 + H_2O_2 + 2H^+ \longrightarrow Fe^{2+} + 2H_2O \qquad (15\text{-}1)$$

$$Fe^{2+} + H_2O_2 \longrightarrow Fe^{3+} + \cdot OH + OH^- \qquad (15\text{-}2)$$

张礼知研究组采用 $NaBH_4$ 还原 Fe^{3+}的方法，通过调控表面氧化铁层厚度，设计合成出在空气中稳定的分子氧活化型纳米零价铁 Fe@Fe_2O_3 核壳结构纳米线[31]，如图 15-3（a）所示。相对于传统零价铁材料，Fe@Fe_2O_3 合成过程中不需要惰性气体保护，全部在空气气氛中完成，很好地解决纳米零价铁自燃和不稳定问题。通过改变还原剂加料速度，他们还实现了材料形貌的有效调控。电镜表征结果显示，Fe@Fe_2O_3 材料为具有 Fe^0 核和 Fe_2O_3 壳结构、直径约 50 nm 的线型材料。由于氧化层保护作用，Fe@Fe_2O_3 核壳结构纳米线比传统 nZVI 稳定。虽然 $Fe@Fe_2O_3$ 单独催化 Fenton 反应的活性不如均相 Fe^{2+}/H_2O_2 体系，但 $Fe@Fe_2O_3$ 与 Fe^{2+}在催化 H_2O_2 分解上表现出强烈的协同效应，将·OH 生成速率提高到均相体系生成速率的 38 倍[32]，进而显著促进 Fenton 反应和污染物降解。这种协同效应归因于以下两个方面：其一，$Fe@Fe_2O_3$ 能够活化溶液中溶解的分子氧，生成$\cdot O_2^-$[32]，该物种能够通过反应式（15-3）促进 Fe（III）/Fe（II）的循环[19,33]；其二，$Fe@Fe_2O_3$ 能够吸附溶液 Fe^{2+}形成表面 Fe（II），而表面结合态 Fe（II）可以降低 H_2O_2 分解生成·OH 的热力学能垒，表现出比溶解态 Fe^{2+}更高的 Fenton 反应活性[34]。

$$Fe^{3+} + \cdot O_2^- \longrightarrow Fe^{2+} + O_2 \qquad k = 5\times10^7\ L/(mol\cdot s) \qquad (15\text{-}3)$$

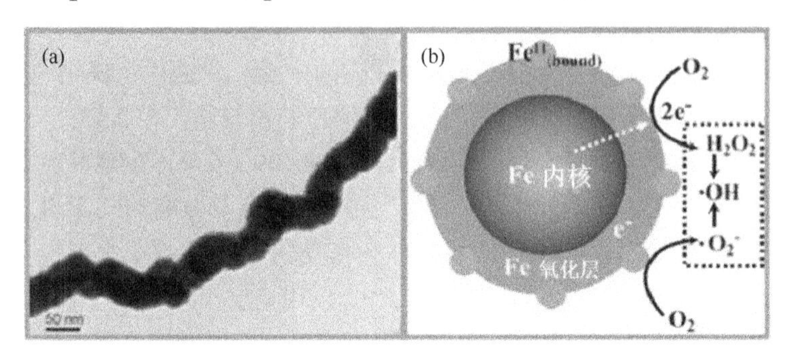

图 15-3 （a）典型核壳结构铁纳米线（$Fe@Fe_2O_3$）的 SEM 图[31]；（b）$Fe@Fe_2O_3$ 活化分子氧类 Fenton 反应机理示意图[35]

张礼知研究组还利用 $Fe@Fe_2O_3$ 活化分子氧过程构建出新型类 Fenton 反应，其反应历程如图 15-3（b）所示。首先，铁核电子通过铁氧化层向外传导，直接将吸附的分子氧还原至 H_2O_2，该反应是双电子反应。接下来，$Fe@Fe_2O_3$ 表面吸附的亚铁离子通过单电子反应途径还原分子氧至$\cdot O_2^-$，$\cdot O_2^-$继续得电子可以生成 H_2O_2。通过上述过程产生的 H_2O_2 将与溶液中或表面吸附 Fe(II)发生 Fenton 反应。由于无需外加 H_2O_2，$Fe@Fe_2O_3$ 活化分子氧的类 Fenton 反应成本低廉且环境友好，在土壤原位修复等领域有潜在的应用前景。在该过程中，Fe_2O_3 氧化层一方面会阻滞从铁核向外的电子传导过程，从而不利于分子氧双电子途径活化；另一方面，氧化层利于亚铁离子吸附，从而加速单电子分子氧活化。通过改变 $Fe@Fe_2O_3$ 制备过程中水相老化时间来调节材料的氧化层厚度，张礼知研究组还实现了 $Fe@Fe_2O_3$ 分子氧活化能力的调控[35]。

然而，上述 $Fe@Fe_2O_3/H_2O_2$ 和 $Fe@Fe_2O_3/O_2$ 体系的反应活性都较低，明显逊色于传统均相 Fenton 体系，难以满足实际应用要求。虽然加入低浓度 Fe^{2+}离子能够部分提高上述体系的反应活性和污染物降解速率[32,36]，但无疑又引入均相 Fenton 反应的弊端。鉴于 $Fe@Fe_2O_3$ 异相 Fenton 活性主要受限于体系中较低的 Fe(II)/Fe(III)循环效率[37]，研究者采用加入配体化合物或还原剂的策略来提高其活性。这是因为，上述试剂能够加速固相材料表面铁的还原溶解，在材料表面形成限域 Fe(II)离子[38]；同时，还原剂的加入通常能够促进表面 Fe(II)/Fe(III)的循环，避免 Fe(II)耗尽导致催化剂失活[37,39,40]；此外，络合剂-铁配合物的形成会显著降低 Fe(II)/Fe(III)电对的氧化还原电势，增强 Fe(II)给电子能力[41]，进而促进分子氧活化及 H_2O_2 分解过程。譬如，美国 Cheng 等首次利用乙二胺四乙酸（EDTA）促进铁粉活化分子氧，在室温和常压条件下实现 4-氯苯酚的完全降解[42]。然而，EDTA 本身生物降解性较差，且具有较强的金属络合能力，会加剧环境中重金属污染物的迁移和生物有效性，带来二次污染的问题。相对于 EDTA，张礼知研究组报道的有机配体 DTPA（二乙烯三胺五乙酸）环境友好性更佳，且能更好地促进 $Fe@Fe_2O_3$ 活化分子氧的反应，表现出更快的 4-氯苯酚矿化速率（1.1 mmol/L 4-氯苯酚，1 h 矿化率>70%）[43]。张礼知研究组还发现，抗坏血酸（AA）能够显著增加 $Fe@Fe_2O_3$-H_2O_2 非均相 Fenton 体系中 H_2O_2 的分解速率和·OH 生成速率，促进多种有机污染物如合成染料（罗丹明 B、亚甲基蓝）、除草剂（甲草胺、阿特拉津、环草隆）、抗生素（林可霉素、氯霉素）等的高效去除，将去除速率常数提高到均相 Fenton 体系速率常数的 38~53 倍，并显著提高 $Fe@Fe_2O_3$ 核壳结构纳米线的重复使用性能。这些效果主要得益于抗坏血酸具备络合配体和还原剂的双重作用机制[37,44]。除了上述有机配体，张礼知研究组还发现无机配体四聚磷酸（TPP）能够将 $Fe@Fe_2O_3$ 活化分子氧降解阿特拉津的速率提升 955 倍，是 EDTA 增强效果的 10 倍。虽然 TPP 本身不具备还原能力，但可以阻断 $Fe@Fe_2O_3$ 铁核向质子的电子传递，降低析氢腐蚀速率，驱使更多的铁核电子参与分子氧活化和表面 Fe(III)还原，大幅度提升类 Fenton 反应的效率[45]。之后，韩国蔚山科技大学（UNIST）的研究人员开展了类似的研究[46]，发现 TPP 还能够起到稳定化 nZVI、减少团聚和降低其细胞毒性的作用[47]。此外，张礼知研究组利用固定化 $Fe@Fe_2O_3$ 作为阴极构建电化学 Fenton 体系显著促进 Fe(II)/Fe(III)循环过程，提升 $Fe@Fe_2O_3$ 催化分子氧活化反应及 Fenton 反应的活性[48]。最近，张礼知研究组用活性炭纤维负载 $Fe@Fe_2O_3$ 作为氧气扩散阴极构建电化学 Fenton 系统，可以在 1 h 内完全降解 30 mg/L 阿特拉津，且矿化效率接近 60%，4 h 后矿化率高达 87.4%，成功实现了阿特拉津的开环降解[49]。

2.2.2　纳米铁矿物材料

纳米铁矿物材料在自然界中储量丰富，价格低廉。其中，赤铁矿、针铁矿、磁铁矿等常被用作异相 Fenton 反应的催化剂[50-52]。然而，矿物材料的 Fe(II)含量较低，其表面 Fe(II)/Fe(III)循环受限[27]，这导致它们的 Fenton 反应去除污染物的效率不高。譬如，针铁矿、赤铁矿、水铁矿催化 H_2O_2 分解生成·OH 的速率分别为 $4.00\times10^{-7}s^{-1}$、 $4.25\times10^{-5}s^{-1}$、$2.00\times10^{-5}\ s^{-1}$[53]。虽然 H_2 热还原处理能增加矿物表面 Fe(II)，从而提高 Fenton 催化效率[54]，但该工艺不易操作且成本高。相比之下，加入还原型络合配体是一种更简单且行之有效的方法。与基于 $Fe@Fe_2O_3$ 核壳结构纳米线的 Fenton 体系类似，还原型配体不仅能加速矿物表面铁元素的还原溶解，还会促进矿物表面 Fe(III)/Fe(II)的循环过程（图 15-4）。例如，张礼知研究组发现抗坏血酸（AA）可以实现甲草胺在赤铁矿-H_2O_2 体系中的高效降解，而无 AA 的赤铁矿-H_2O_2 体系几乎无法降解甲草胺[39]。同时，他们还发现 AA 增强赤铁矿 Fenton 反应活性时表现出暴露晶面依赖性：AA 与{001}面铁原子间形成双齿单核内球配合物，而与{012}晶面暴露铁原子间则形成单齿单核配合物；动力学研究表明，双齿单核配位方式更有利于赤铁矿表面铁原子的还原溶解[38]，以及后续的 H_2O_2 分解及铁循环过程[39]。盐酸羟胺（HA）是一种常用的还原剂，曾被用于促进均相 Fenton 反应体系中的铁循环[55]。张礼知研究组最近发现，低浓度 HA（0.5 mmo/L）的存在能够将针铁矿-H_2O_2 非均相 Fenton 体系的·OH 生成速率提升 4 个数量级，且不会造成针铁矿的铁溶出。机理研究表明，HA 首先与针铁矿表面 Fe（III）形成内球配合物，将 Fe(III)还原成表面 Fe(II)，该原位生成的表面 Fe(II)分解 H_2O_2 生成·OH，其自身也转化成表面 Fe(III)。接下来，表面 Fe(III)进一步被 HA 还原成表面 Fe(II)，实现了铁元素在针铁矿表面的原位循环和 H_2O_2 的高效分解[40]。为此，他们提出了表面 Fenton 反应体系的概念。

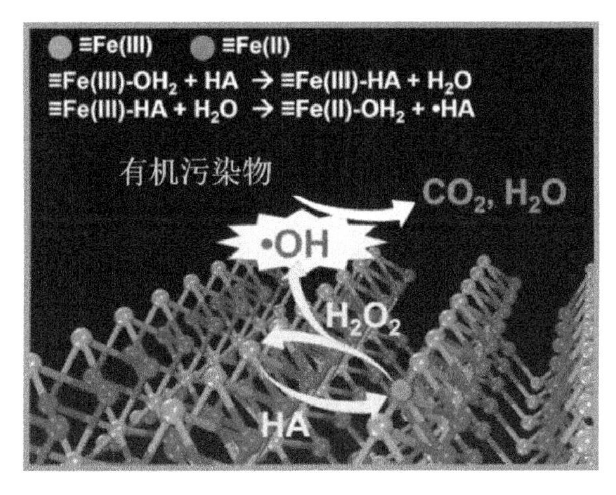

图 15-4　盐酸羟胺（HA）促进针铁矿表面铁循环机理图[40]

合成或改性矿物材料是获得高效非均相 Fenton 催化剂的重要途径。研究人员在矿物材料改性及非均相 Fenton 催化方面开展了大量研究，他们发现天然磁铁矿的 Fenton 催化活性主要受到矿物表面羟基物种形态和密度的影响，同时还与矿物的脱溶相（exsolution phases）有关，例如铁镁尖晶石和钛铁矿的脱溶能够促进 H_2O_2 的分解以及有机污染物的吸附，从而提高 Fenton 反应效率[56]。同时，他们还采用过渡金属元素取代法对磁铁矿进行改性，制备出了一系列元素掺杂的纳米磁铁矿材料，如 $Fe_{3-x}Cr_xO_4$（$x = 0.00$，0.18，0.33，0.47，0.67）[57]，$Fe_{3-x}Ti_xO_4$（$0 \leqslant x \leqslant 1.0$）[58]，以及 Mn 取代的磁铁矿[59]。这些材料的 Fenton 催化活性均优于纯磁铁矿纳米颗粒，其原因在于过渡金属掺杂能够减小磁铁矿的粒径或孔径，从而增加其比表面积；另一方面，掺杂的过渡金属元素能够通过价态变化促进材料表面和体相的电子传递过程，从而产生更多表面 Fe(II)[27]。最近，张礼知研究组采用水热法合成了 FeS_2 材料。该材料为多面体聚集形成的微球，可在初始 pH 为 3~9 的条件下催化 H_2O_2 分解快速降解甲草胺。水热合成的 FeS_2 材料催化分解 H_2O_2 生成·OH 的速率较商品黄铁矿提高了 1 个数量级，比其他含铁矿物（如针铁矿、赤铁矿等）则高出 3~5 个数量级。其高活性主要得益于其表面富含结合态亚铁离子。这些表面结合态亚铁离子不仅能够加速 H_2O_2 分解，还可以有效活化分子氧生成$\cdot O_2^-$，而生成的$\cdot O_2^-$又反过来促进表面 Fe（III）的还原，从而实现高效的表面铁循环[33]。其他研究人员则通过酸碱活化和高温碳化程序从污泥中合成出了含有磁铁矿纳米颗粒的多孔炭材料（FPC），在 FPC-H_2O_2 体系处理 260 min 后，1-重氮基-2-萘酚-4-磺酸的降解率和矿化率分别达到 96.6%和 87.2%，其催化非均相 Fenton 反应的活性高于同等条件下的商品 Fe_3O_4 磁性纳米颗粒，碳基底和铁氧化物之间存在协同作用[60]。碳基底除了能够起到分散 Fe_3O_4 催化中心和吸附富集有机污染物的作用，其离域 π 电子还能够促进 Fe_3O_4 表面 Fe(III)的还原，从而提高 Fenton 反应中的 Fe(II)/Fe(III) 循环效率[61]。

2.2.3 无过渡金属的纳米 Fenton 催化材料

含过渡金属元素（铁或铜等）的非均相 Fenton 催化材料在反应过程中一般伴随着过渡金属元素的溶出。这些溶出的过渡金属本身可能会存在一定毒性（如铜），带来环境危害。更为重要的是，过渡金属的溶出不仅会带来均相 Fenton 反应中常见问题，如铁泥生成、铁物种消耗等，还导致研究者无法清晰区分反应体系中的均相反应和表面反应，不利于厘清非均相 Fenton 体系的作用机制。鉴于此，张礼知研究组最近首次采用不含过渡金属元素的氯氧铋纳米片，通过真空高温煅烧的方式引入表面氧空位，利用氯氧铋纳米材料表面富电子的氧空位作为 H_2O_2 分解的活性催化位点，完美地实现了表面 Fenton 反应，为非均相 Fenton 催化材料研发和表面 Fenton 反应机制的研究提供了新思路[62]。研究结果显示，不含氧空位的氯氧铋纳米片没有 Fenton 反应活性。引入氧空位后，H_2O_2 能够在氯氧铋表面自发分解生成·OH。有趣的是，晶格氧的空间位阻导致{001}晶面生成的·OH 倾向解离成游离·OH，而在{010}晶面生成的·OH 则易于结合在表面铋原子提供的 Lewis 酸位点而实现稳定化，形成表面·OH。由于不同有机污染物在催化剂表面的吸附行为差异，导致它们与上述两种·OH 的反应速率不同。这一特性让研究者有望通过调控催化剂表面微观结构和氧空位浓度来实现有机污染物的选择性去除。例如，由于甲酸相对于苯甲酸更易吸附在氯氧铋表面，基于{010}晶面暴露的富氧空位氯氧铋 Fenton 体系能够在甲酸和苯甲酸混合体系中高选择性降解甲酸（图 15-5）。

2.3 纳米材料吸附技术在水体放射性污染处理领域的研究进展

在核电生产、核武器使用等核能利用过程中，伴随着大量长寿命放射性污染物的排放[63]。尤其是放射性元素如铀、铕、铼、锝、锶、铯、镅等的开采、冶炼、加工等环节所产生的工业废水、尾矿、废渣中放射性核素浓度远高于国家标准，从而对环境和人类生命健康构成巨大的威胁[64,65]。由于水体的高流

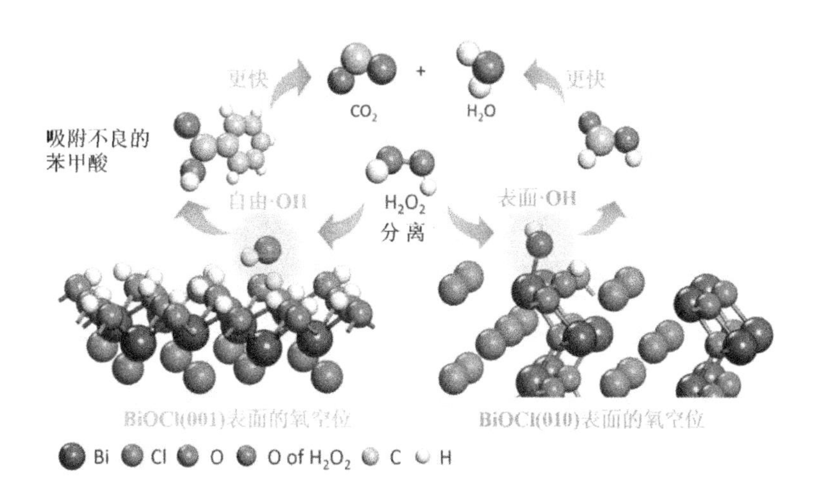

图 15-5　氧空位氯氧铋催化的表面 Fenton 反应机理[62]

动性，放射性元素易于迁移、扩散，一旦进入土壤或者被植物吸收，最终会在人体内富集，给人类健康造成巨大伤害[66,67]。因此，开发放射性核素的高效吸附材料，实现废水中放射性污染物的快速有效去除具有重大的科学意义和应用价值。纳米材料具有结构可设计、表面官能团富集和比表面积大的特点，体现出优异的吸附性能[68]。诸多研究已经证实，通过结构和成分的有效设计，纳米材料可对多种环境污染物进行有效的吸附脱除[68-70]。在诸多纳米材料中，以一维碳纳米纤维/纳米管、二维石墨烯为代表的新型碳纳米材料，近年来成为环境污染治理领域的研究热点[71,72]。围绕石墨烯、碳纳米纤维/纳米管及其复合材料对放射性核素离子的吸附行为，研究人员开展了一系列工作，结合扩展 X 射线吸收精细结构（EXAFS）光谱技术和理论计算对材料的吸附性能和放射性核素离子-材料的作用机理进行了深入的阐释。

2.3.1　石墨烯

研究发现，由于其表面富含含氧官能团（−OH，−O−，−COOH），少层石墨烯对铕 Eu(III)有较高的吸附容量（175 mg/g，pH 4.5，298 K）[73,74]。进一步地，为了揭示不同含氧官能团（−OH，−O−和−COOH）对放射性离子的键合作用和吸附行为影响，研究人员运用 EXAFS 光谱技术，系统研究了三种功能化石墨烯（氧化石墨烯 GOs、羧基化石墨烯 HOOC-GOs 和还原氧化石墨烯 rGOs）对放射性核素离子铀 U(VI)的吸附富集作用[75]。研究发现，不同石墨烯对 U（VI）表现出不同的吸附性能，吸附容量 GOs >HOOC-GOs > rGOs，脱附行为显示 rGOs > GOs>HOOC-GOs 趋势[76]。EXAFS 数据解析发现，U(VI)与氧化石墨烯和羧基化氧化石墨烯形成内层表面络合物，与还原氧化石墨烯形成外层表面络合作用。同时，利用密度泛函理论（DFT）计算−OH、−O−和−COOH 三种官能团与 U(VI)的化学键合能，揭示了石墨烯表面含氧官能团的种类及分布对于其吸附性能的影响，本研究从分子水平成功解析了放射性核素在不同环境条件下的局部微观结构信息及其在石墨烯上高效吸附的作用机理（图 15-6）。

2.3.2　碳纳米管

研究人员通过研究碳纳米纤维/纳米管与不同放射性核素离子的相互作用发现，新型碳纳米材料对核素离子的吸附作用与传统吸附材料（天然黏土和氧化物）存在明显差异。结果显示，多壁碳纳米管（CNTs）与 Eu(III)和 ^{243}Am(III)两种核素离子形成内表面络合物。然而，Eu(III)在 CNTs 表面的吸附作用要强于 ^{243}Am(III)在 CNTs 上的吸附作用，说明 CNTs 对两种核素离子的吸附机理存在差异。DFT 计算显示，Eu(III)与碳纳米管的结合能高于 ^{243}Am(III)，说明 Eu(III)与 CNTs 表面含氧官能团形成更强的络合作用，这与实验中发现 CNTs 对 Eu(III)的更高的吸附容量相符[77]。含氧官能团对 Eu(III)和 ^{243}Am(III)的吸附起到

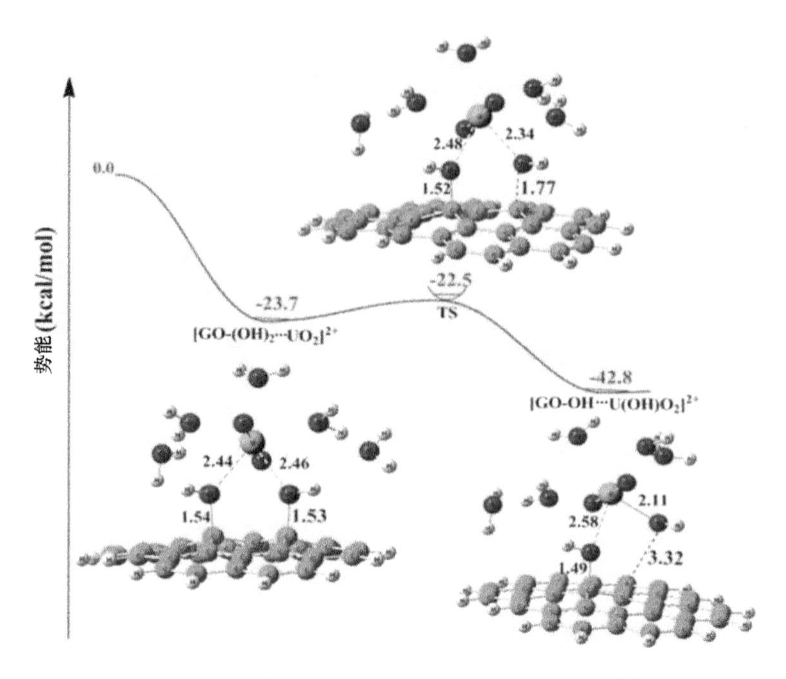

图 15-6 氧化石墨烯吸附铀酰过程的 DFT 解析[76]

决定性作用，不同含氧官能团对放射性核素离子的吸附能呈现明显差别（≡S—OH<≡S—COOH<≡S—COO$^-$，如图 15-7 所示）。这种相互作用差异对于 CNTs 在三价放射性环境污染物的处理应用中具有重要影响。类似地，U(VI)和 Eu(III)在碳纳米纤维表面的吸附机理存在差别。碳纳米纤维(CNFs)对 U(VI)的吸附容量大于 Eu(III)(pH 4.5，298 K)。研究表明，CNFs 对 U(VI)具有更强的表面络合作用，而且纤维表面存在的还原性基团 R-CH$_2$OH 可将 U(VI)部分还原为 U(IV)，对离子的吸附容量有重要贡献(图 15-8)[78]。利用 EXAFS 数据进一步揭示了碳纳米纤维对 U(VI)的吸附机理，即随着初始 U(VI)浓度的增加，CNFs 对 U(VI)的吸附逐渐由内表面络合转变为外表面络合（图 15-8）。以上研究工作证明新型碳纳米材料对于放射性核素离子的吸附脱除具有巨大的研究前景和应用潜力，为高放射性污染水体的处理提供了重要的理论依据。

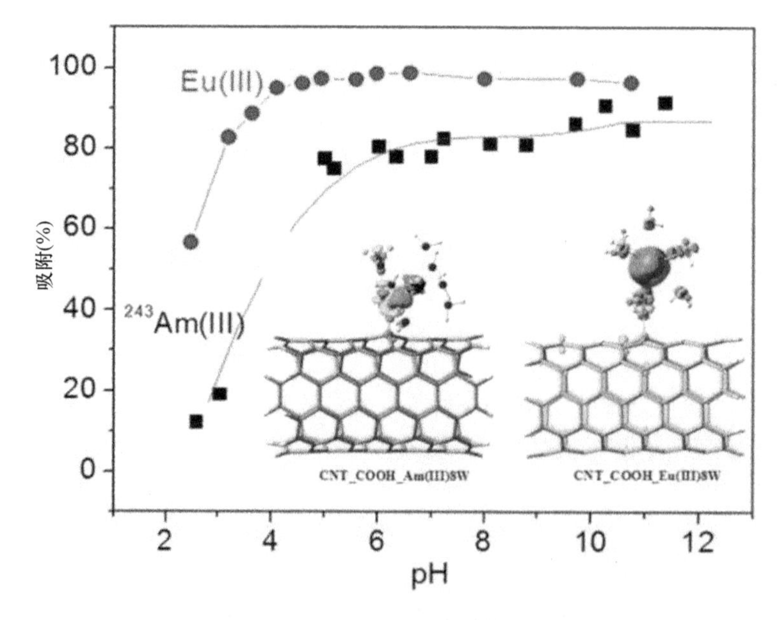

图 15-7 CNTs 对 Eu(III)和 ^{243}Am(III)的吸附性能对比[77]

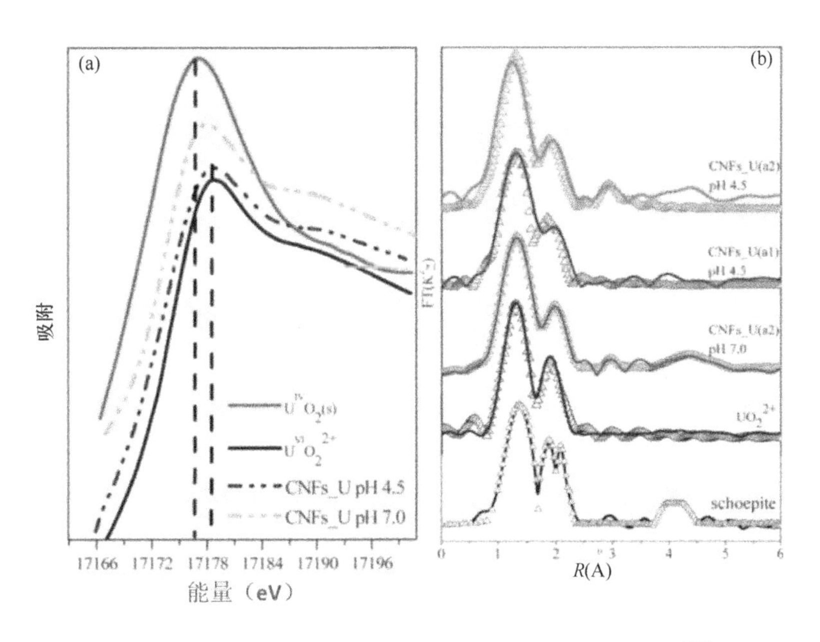

图 15-8　碳纳米纤维吸附 U 的 X 射线吸收精细结构光谱[78]

2.3.3　纳米零价铁

由前文而知，引入还原剂，是对水体中高放射性核污染物进行处理的另一个有效手段。利用纳米零价铁将高价态、易溶解迁移的放射性核素还原固定为低价态难溶性物质，是防止水环境中高放射性污染物转移，实现环境放射性污染修复的一种有效技术。然而，由于实际水环境中共存多种微生物，因此必须考虑微生物对于纳米零价铁还原放射性核素过程的影响。研究人员证实，加入微生物（枯草杆菌，*B. subtilis*）可以明显增强纳米零价铁对于放射性核素如 U(VI)的吸附过程，但是在 pH > 4.5 时会抑制纳米零价铁对 U(VI)的还原作用（图 15-9）。EXAFS 数据分析证实，微生物对于 U 还原过程的抑制作用是由于 Fe(II)/Fe(III)与 *B. subtilis* 或其分泌的胞外聚合物表面的含氧官能团形成内层络合物，阻碍了电子由零价铁向 U(VI)的传输[79]。X 射线光电子能谱和 X 射线吸收近边结构谱表明，纳米零价铁对 U(VI)的去除受还原沉淀作用控制，而在 *B. subtilis* 存在下，吸附过程是去除 U(VI)的主导过程。该结果在分子水平上提出了微生物对于水体中纳米零价铁处理放射性污染物过程的影响，为发展应用于放射性污染水体原位修复的生物化学耦合技术提供了有力的理论指导。

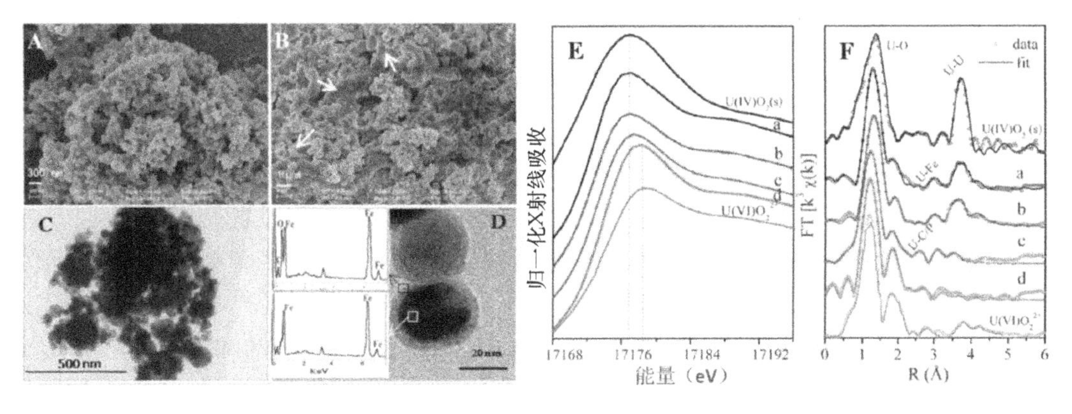

图 15-9　纳米零价铁与 *B. subtilis* 及其复合物形貌图和 X 射线吸收精细结构光谱[79]

纳米零价铁用于水体放射性污染修复的另一个热点问题是如何提高纳米零价铁对于放射性核素的还

原效率。这一问题的关键在于提高纳米零价铁的稳定性以发挥其高还原性。研究人员运用等离子体技术构建了石墨烯负载纳米零价铁复合材料，系统研究了 pH、离子强度、温度、接触时间等不同实验条件下复合材料对于放射性核素离子的吸附和还原作用。本研究结合宏观实验和分子水平表征揭示了石墨烯和纳米零价铁对放射性核素的去除存在吸附与还原的协同作用[80]。并且从材料构建角度给出了吸附还原处理放射性污染物的材料优化方案。为进一步提高纳米零价铁这类材料对水体放射性污染的处理效果具有高度指导意义。

2.4 基于碳纳米材料的膜分离技术

膜分离技术具有简单高效、节能环保以及无相变等特点，已广泛应用于食品、化工、医药等行业，产生了巨大的经济效益和社会效益。近年来，膜分离技术在污水深度处理、饮用水处理以及海水淡化等应用方面也取得了长足的发展，已成为一种重要的新兴水处理技术。膜分离技术的核心是分离膜，其基本功能是使流体内的某一种或几种物质透过，而其他物质则被截留。对目标要透过的物质的透过能力，对目标要截留的物质的截留能力以及抵抗膜孔堵塞的能力分别为分离膜的渗透性、选择性和抗污染能力。它们是分离膜三个非常重要的属性，也是评价分离膜性能的重要指标。膜的这些属性与膜材料的结构形貌，界面性质以及孔道结构和数目等有很大关系。由于碳纳米材料优异的物理化学性质以及在提高膜分离性能上的巨大潜力，基于碳纳米材料的膜分离技术成为近年来的研究热点之一。

2.4.1 碳纳米管分离膜

最初碳质分离膜的兴起是为了解决有机高分子分离膜化学稳定性差的问题。这些碳质膜主要由炭化高分子聚合物而得，其分离性能比较于传统分离膜并没有明显的提升。2004 年，Hinds 等首次验证了阵列碳纳米管作为分离膜的可行性，实验证实了气体分子、离子在单个碳纳米管内孔道的传输能力[81]。2005 年，Majumder 等认为水分子可以和疏水的碳纳米管内表面之间形成有序的氢键，因而可以在碳纳米管内孔道中几乎无摩擦地流动，并实验测定了水分子和多种溶剂分子在 7 nm 碳纳米管内孔道的传输速率为 0.67~25 cm/s，比传统液体流动模型预测值高出 4~5 个数量级[82]。2006 年，Holt 等实验观察到气体分子在碳纳米管内孔道的传输速率比克努森扩散模型预测值高出 1 个数量级，而水分子的传输速率比连续流体力学模型预测值高出三个数量级，和分子动力学预测值相当[83]。分子动力学模拟表明，由于碳纳米管内孔道的疏水性和原子级光滑程度，水分子在其中可以几乎无摩擦地成线性排列流动[84,85]。这是阵列碳纳米管膜具有超高水通量的最根本原因。这些工作奠定了阵列碳纳米管分离膜的研究基础，迅速引起了关注。

随后针对阵列碳纳米管分离膜的研究多集中于离子选择性的分子动力学计算、碳纳米管管口的基团修饰、简易制备以及抗菌等方面。由于大面积、低成本制备技术无法取得突破，阵列碳纳米管膜的发展也逐渐进入瓶颈期。与此同时，随着碳纳米管粉体的制备成本不断降低，并由于其优异的物理化学性质，关于碳纳米管器件化组装成膜的研究逐渐增多。在水处理应用领域，以 Vecitis 课题组和 Quan 课题组的相关研究居多。Vecitis 课题组利用真空抽滤法制备了无序碳纳米管膜，并在电化学作用下考察了其对污染物分子和离子的吸附、脱附和降解能力[86,87]。Quan 课题组利用湿法纺丝技术制得的平均孔径为 100 nm 的全碳纳米管中空纤维膜孔隙率高达 95%，水通量为 12000 L /(m^2·h·bar)，是传统分离膜的数倍[88]，碳纳米管中空纤维膜相关照片如图 15-10 所示。他们的研究还发现，在电化学辅助下，这些碳纳米管膜对荷电物质的截留能力有明显提高：当分离物质的尺寸远小于膜孔径且所带电荷与分离膜相反时，电增强吸附作用使得这些物质被碳纳米管膜吸附截留[89]；当分离物质的尺寸稍小于膜孔径且所带电荷与分离膜相同时，静电排斥作用使得该物质被筛分截留[88,90]。

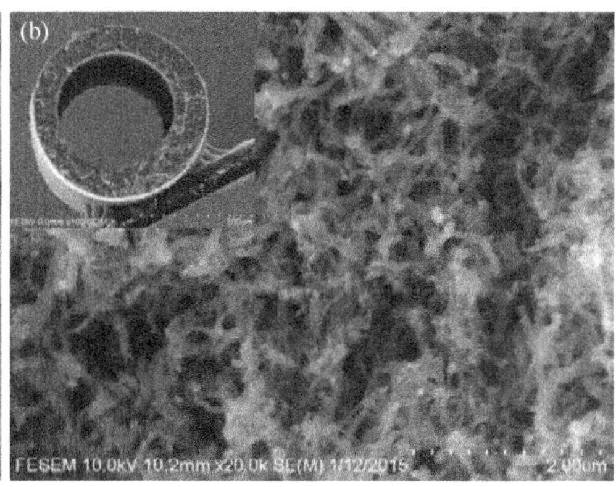

图 15-10　碳纳米管中空纤维膜的电子照片（a）和扫描电子显微镜照片（b）
插图为碳纳米管中空纤维膜断面的扫描电子显微镜照片

2.4.2　石墨烯分离膜

自从 2004 年科学家证实单层石墨烯能够在非绝对零度稳定存在以来，关于石墨稀的研究论文几乎呈现爆炸式增长。一般地，提高膜水通量的思路主要有两方面：一是增加可用来传输膜孔道的数目（有效孔隙率）；二是减小膜体厚度。石墨烯是由 sp^2 碳原子形成的蜂窝状二维平面薄膜，其厚度只有一个碳原子直径大小，是目前已知的最薄材料。在此背景下，自 2010 年前后由其衍生物氧化石墨烯（或还原的氧化石墨烯）堆叠而成的纳滤膜是近年来的一个研究热点。由于氧化石墨烯自身原子级别的厚度，这些纳滤膜很容易通过真空抽滤或层层组装的方法被制成超薄结构。氧化石墨烯片层之间形成 1nm 左右的二维孔道，分子在此孔道内表现出很多独特有趣的传输行为：1μm 厚的氧化石墨烯膜允许水分子透过，却能够截留尺寸更小的氦分子[91]；水力半径小于 0.45Å 的溶质离子能够以比单纯扩散高出数千倍的速率透过微米级厚的氧化石墨烯膜[92]；氧化石墨烯膜在不同有机溶剂中其层间距不同[93]； 水分子在此孔道内沿着未被氧化的区域几乎无摩擦地运动[94]。在水处理过程中，氧化石墨烯分离膜（还原的氧化石墨烯分离膜）水通量大约是商业纳滤膜的 2~10 倍[95,96]。

尽管氧化石墨烯分离膜具有较高的水通量，但是它相当曲折的膜孔道结构仍然限制了水分子的传输速率。而分子动力学模拟表明，具有极限厚度的单层石墨烯分离膜可以最大程度地降低水分子穿透膜时的传输阻力。例如，据预测，在盐截留率为 99%的条件下，具有羟基修饰的纳米孔的单层石墨烯膜水通量高达 66 L /(cm^2·d·MPa)，比反渗透分离膜要高出 2~3 个数量级[97]。Surwade 等利用氧等离子刻蚀技术制备了多孔单层石墨烯膜，实验测定其对离子的截留率几乎为 100%，而水通量则高达 10^6 g/(m^2·s)[98]。近几年来，这种具有原子级厚度的多孔石墨烯超滤膜的研究也取得了进展。这些膜通过活性离子轰击或碳热反应直接在单层或多层石墨烯片上打孔而制得，因而具有垂直贯通的膜孔道。测试表明，这些膜的水通量是传统分离膜的数十倍到数百倍[99,100]。

除了碳纳米管膜和石墨烯膜，其他碳纳米材料分离膜的研究则非常有限。

2.4.3　小结

基于碳纳米管和石墨烯的分离膜已表现出传统分离膜无可比拟的性能优势，有理由相信，它们未来一旦成功商业化应用，将会带来巨大的经济效益和环境效益。目前这些膜还存在诸多技术问题，离商品化应用还有一段距离。未来要解决的问题包括：碳纳米管膜和石墨烯膜的大面积、低成本的制备；机械强度的进一步提高；膜孔径的精确调控；孔道端口基团的可控化修饰；膜污染的减缓等。

2.5 树脂基纳米复合材料的构效调控与水处理的研究

随着水质控制标准的不断提高，水中重金属、磷、砷和氟等污染物的限值也愈加严格，这为深度水处理技术的发展提出了挑战。树脂基纳米复合材料作为一类实用型环境功能纳米材料，结合了纳米颗粒高除污活性与树脂载体大颗粒、高机械强度等实际应用性能，且可再生与循环利用，突破了传统纳米颗粒难操作、易团聚失活、潜在环境风险等实际应用瓶颈[101-102]。本节简要介绍了对重金属、砷、氟、磷等污染物具有深度处理功能的铁系、锆系、锰系等系列树脂基纳米复合材料的构效调控方法以及相关材料的实际应用技术进展，并对今后该领域的发展方向进行了展望。

2.5.1 树脂基纳米复合材料的构效调控

树脂基纳米复合材料示意图见图 15-11。

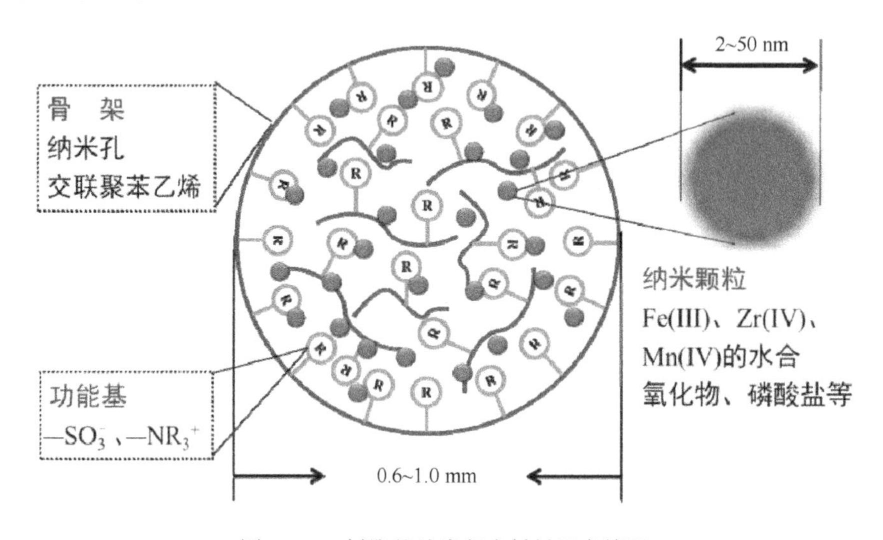

图 15-11 树脂基纳米复合材料示意简图

2.5.1.1 纳米颗粒在载体中分布的调控

纳米颗粒在树脂载体中的分布将显著影响其净污速率、性能和材料的稳定性。树脂载体表面不同官能团，如磺酸基（$R\text{-}SO_3^-$）、季铵基（$R\text{-}NR_3^+$）等，可对负载纳米颗粒在载体中的分布产生显著影响。Zhang 等发现，树脂表面磺酸基团可有效增强负载 ZrO_x 纳米颗粒的分散性，并可通过 Donnan 膜效应增强污染物传质作用，其除 Pb 性能相比活性炭载体和不带电树脂载体显著提升[103]。这一表面官能团与纳米颗粒分散性的关系类似于 DLVO 理论，即可将载体看作固体溶剂，带电基团的作用类似于表面活性剂，可使孔道内的纳米颗粒分散更均匀[104]。

2.5.1.2 纳米颗粒尺寸的调控

树脂基纳米复合材料中纳米颗粒尺寸与净污活性密切相关，其调控可通过控制网孔结构实现。树脂载体一般为交联聚苯乙烯，通过控制后交联程度等方式可调节载体比表面积大小和孔径分布。Wang 等通过调节树脂载体孔径大小调控纳米颗粒尺寸。结果表明，纳米颗粒大小随着载体平均孔径降低而减小，而纳米颗粒粒径越小，材料对砷吸附量越高，吸附速率也越快[105]。Li 等通过类似方法合成三种不同水合氧化铁（HFO）纳米颗粒大小的复合材料，研究发现，复合材料表面电荷密度随着纳米颗粒粒径减小而增大，其除 As(V)能力也随之增大[106]。

2.5.1.3　载体限域效应对纳米颗粒表面化学性质的影响

树脂载体可通过网孔限域效应对负载的纳米颗粒表面化学性质产生显著影响。Nie 等通过电位滴定和表面络合模型研究表明，不同 pH 下负载型水合氧化铁（HFO）纳米颗粒表面铁优势形态与分散型 HFO 纳米颗粒有明显差异。负载后 HFO 的 pHpzc 由 8.2 降至 6.3，导致其吸附 As(V)和 Cu(II)性能产生变化[107]。Han 等发现由于载体表面季铵基作用和纳米网孔作用，负载的 MnO_x 纳米颗粒表面电性由 6.2 增大至 10.5，材料表面电荷密度增大，从而有效地改善了其除磷性能[108]。Xu 等通过比较 Zr 系纳米复合材料 HZO-201 和分散型水合氧化锆（HZO）对氟离子吸附活性发现，负载于树脂载体上的 HZO 纳米颗粒与 F 结合能略高于分散型 HZO，这可能是网孔限域造成的尺寸效应增强了 HZO 电子结合能力所致[109]。

2.5.2　树脂基纳米复合材料对典型重金属的深度处理

重金属废水主要来源于电子电镀、制革、矿冶等行业，其废水排放量大，组分复杂，常共存有 Ca^{2+}、Na^{+}、Mg^{2+}等竞争离子和一定量的有机配体（草酸、柠檬酸等）。近年来，南京大学等单位将磷酸锆、氧化锆、氧化锰、水合氧化铁等多种金属氧化物纳米颗粒固载于大孔阳离子交换树脂中，研制成功系列纳米复合材料。该类材料对多种重金属具有高选择性，且吸附容量大、可再生回用。固定床实验发现，毫米级树脂载体水力性能优越，较好地解决了纳米颗粒材料的工程化应用瓶颈。Zhang 等和 Hua 等开发的纳米磷酸锆、硫代磷酸锆和氧化锆复合材料可在高浓度 Ca^{2+}共存条件下通过内配位作用选择性吸附 Pb、Cd 和 Zn，微量 Pb、Cd 经处理后可达到饮用水标准[110-111]。上述材料在吸附后均可采用 HCl-NaCl 二元溶液进行有效再生与循环使用，再生率 99%以上。

基于上述研究成果，南京大学等单位针对电子电镀、制革和矿冶废水开发了以纳米复合材料强化吸附为核心的强化深度处理集成工艺。相关纳米复合材料工程化废水处理装置见图 15-12。对于电子电镀废水，采用纳米磷酸锆复合材料成功实现了废水中 Cu、Ni 等重金属的高效分离（出水浓度<0.05 mg/L）与回收，减轻了后续膜处理压力，显著提高了浓缩比和废水回用率；对于制革废水，采用纳米氧化锆和氧化铁复合材料，实现了对低浓度含铬废水的深度处理，确保了出水总铬在 0.147~0.169 mg/L 之间，稳定达到要求的特别排放限值。对于矿冶废水，采用磺化交联聚苯乙烯载纳米氧化锆材料强化吸附微量重金属（0.5~2 mg/L），出水 Cd、Pb 等指标可达到地表 III 类水标准。

图 15-12　纳米复合材料工程化废水处理装置

2.5.3 树脂基纳米复合材料对砷、氟、磷的深度处理

近年来，对饮用水砷氟的控制要求以及污水中磷的排放标准也愈发严格。目前，世界卫生组织（WHO）规定饮用水中总砷限值为 10 μg/L，其中 As(III)迁移能力和毒性较 As(V)强，更难以去除[94]。太湖流域等特殊地区对磷的排放标准降至 0.3 mg/L，而美国五大湖区域已将磷排放标准降至 0.2 mg/L。水中共存阴离子（如 SO_4^{2-}、Cl^-等）和有机物（腐殖酸等）极易干扰微量砷、氟和磷的深度处理。

前期的研究表明，铁氧化物纳米复合材料 HFO@201 可有效去除 As(V)，但对 As(III)深度处理能力欠佳。近年来，Du 等、Li 等和 Zhang 等开发的负载型 ZVI 复合材料 D201-ZVI[112]、Fe-M 双金属氧化物复合材料[113]、氧化还原树脂载 HFO 复合材料 HFO@PS-Cl[114]可高效同步氧化-吸附除 As(III)。

针对磷的深度处理与资源化，前期研究已表明 HFO-201、HMO@NS 等纳米复合材料对正磷酸盐具有优异的选择性吸附能力。而针对 HFO-201 除磷容量偏低的问题，Zhang 等新研制的 La 系纳米复合材料 La-201 除磷效果突出，可将磷浓度降至<0.01 mg/L，且其工作容量相比 HFO-201 提高了 11 倍左右，并成功解决了 La 系材料难以脱附再生的问题，实现了 La-201 的高效再生与循环使用[115]。

Xu 等制备了树脂基水合氧化锆纳米材料 HZO-201。该材料抗酸碱能力强，不易受有机配体和氧化剂干扰，稳定性优异；相比离子交换树脂和活性氧化铝，其除氟性能大幅提升，可稳定循环再生使用，有良好的应用前景[109]；Cai 等开发的 Li/Al 层状双金属纳米复合材料则表现出更为优异的除氟性能[116]。

2.5.4 小结

总体而言，由于目前纳米复合材料中纳米颗粒的尺寸分布较宽（10~100 nm），且基本未涉及低纳尺度（<10 nm）的表面效应，纳米颗粒的晶型控制研究较少，如何实现纳米复合材料在亚 10 纳米尺寸和晶型上的精细调控仍是今后该类材料进一步发展的重要内容。同时，由于污废水组分复杂，微量重金属、砷、氟、磷等目标污染物往往与大量有机物共存，发展催化降解功能的纳米复合材料势在必行；另外，将纳米处理单元与传统水处理单元进行系统集成和优化，也是提升环境纳米技术实际应用水平的重要发展方向。

2.6 镁基纳米材料提取和富集水中低浓度污染物的技术

质量浓度极低（μg/L~mg/L）的痕量污染物的去除，是环境科学研究的难点。其本质在于大多数所研究的材料，吸附容量虽然大，但是吸附力却有可能很低，以致于无法和污染物随浓度而下降的化学势达到有效的匹配。因此，对水中低浓度物质的处理，其关键在于能否发展出合适的预富集材料，使其变为高浓度量少的物质，从而容易用常见的技术手段进行后处理、回收和提炼。然而，为了实现对大量的水中低浓度的污染物的预富集，又对所使用的材料提出了很高的甚至是苛刻的要求。它要求①材料自身环境友好；②对低浓度污染物的作用力强；③通过材料作用后，污染物能够分离浓缩；④材料能够回用或者无毒释放。

近十年来，虽然纳米水处理材料自身的研究已经取得了长足的进步，但能够同时满足上面四项要求的研究并不多见。大多数研究要么关注于如何提高材料的吸附容量，要么对材料进行界面或官能团调控后提取污染物的后处理技术研究缺乏，或者对所使用的材料的环境友好性关注不够，从而难以紧扣污染物预富集材料的需求，因此有实际应用探索空间的材料与技术应需而生。$Mg(OH)_2$ 是一种动力学上非常稳定的天然纳米材料，具有成本低、环境友好、无毒等特点，被称为“环境友好型”的绿色安全水处理剂，已经被用于烟气脱硫、染料废水脱色、养殖除磷、脱除重金属及酸性废水中和等领域，全球年使用

量超过 110 万吨[117]。因此，该材料在满足环境友好性和材料无毒这两个要求上有先天的优势。同时，它是一种是六方相的层状结构，通常合成的状态是纳米片状结构。大量暴露的（001）晶面具有密集且垂直晶面的羟基结构，既是阴离子型污染物吸附或阳离子型污染物反应的有效活性位点，也是实现界面官能团功能强化的调控起始位点。因此，对 $Mg(OH)_2$ 纳米材料可望通过界面官能团对污染物的吸附或反应实现与污染物的强相互作用，从而满足前面所述的材料对于污染物应具有高的作用力的需求。同时，$Mg(OH)_2$ 还是目前所发现的少数的可通过简单后处理实现晶体状态快速调整的材料。例如，通过不同分压的 CO_2 诱发，它既可通过“聚集生长”变成块材碳酸镁，也可在水充分的状态下变为可溶性碳酸氢镁，这种尺寸或者界面状态能急剧改变特点，在通过后处理实现污染物与残余 $Mg(OH)_2$ 的有效分离上将起到重要作用。因此，基于 $Mg(OH)_2$ 的低成本性、微环境界面结构的可构筑性、晶粒尺寸与物相的易调控性，可望通过深入的微观机制研究，理解目标污染物在材料表面的特殊作用原理，并进而设计相关处理策略，实现 $Mg(OH)_2$ 对水中低浓度污染物或稀缺贵金属资源的提取和回收。

2.6.1　镁基纳米材料对水中阴离子型低浓度污染物的预富集研究

电镀废水中的六价铬因其强氧化性导致无法用有机树脂或离子交换膜来回收，当前工业上一般是将其还原成三价铬污泥并填埋处理，然而这种技术手段不合适处理大量的含低浓度六价铬的废水。一个理想的解决之道是发展对水中的低浓度六价铬能进行提取及预富集的材料。Liu 等发现 $Mg(OH)_2$ 与阴离子型的铬酸根离子具有由电荷吸引所导致的强相互作用力，其 Langmuir 吸附力参数远大于其他研究得较多的纳米吸附材料例如 TiO_2 和 LDH 等。基于此，Liu 等设计了如图 15-13(a)所示的纳米 $Mg(OH)_2$ 预富集水中低浓度六价铬的方法。特别值得一提的是选择了 CO_2 作为界面调控剂，研究了吸附铬的 $Mg(OH)_2$ 在 CO_2 作用下的聚集生长。通过将纳米 $Mg(OH)_2$ 快速地转变为块材的碳酸镁，实现了由比表面积急剧下降及物相突变所诱导的铬的脱附浓缩。同时，通过将块材的碳酸镁加热脱除 CO_2 后重新生成纳米 $Mg(OH)_2$，可望达到吸附剂的循环回用。研究指出，一个理想的工业循环系统是不应将外加化学剂转换为副产物，从而持续增加环境负担。如图 15-13(b)所示，该体系中脱附过程所需要的 CO_2 完全等量于再生时所放出的 CO_2。如果将 $Mg(OH)_2$ 和 CO_2 合并，看成是一个“介质黑箱”，相当于构筑出一个可持续起作用的预富集系统，对这个“介质黑箱”输入含低浓度铬液的废水和少量能量，就可输出高浓度的可回收六价铬液和可排放的水[118]。

在上述利用纳米 $Mg(OH)_2$ 和 CO_2 可作为六价铬回收的“介质黑箱”的基础认识的启发下，面向电镀废水中高浓度铬的提取和高倍富集的需求，Lv 等拓展研究了可阴离子插层的大比表面氢氧化镁铝（LDHs）纳米材料在碳酸化过程中对铬的高效吸附和脱附回收思路的可行性[图 15-13(c)]。基于 LDHs 能够对铬阴离子实现大的吸附容量同时又具有较强的 CO_2 气体捕获能力的特点，研究者在吸附了铬的 LDHs 体系中通入高压 CO_2，实现了铬的有效脱附和 LDHs 吸附剂的再生。通过这一过程，可将初始浓度为 500 mg/L 的铬液继续浓缩 23 倍。发展出一种循环使用 LDHs 富集浓缩电镀含铬废水的策略[119]。

上述思路还可拓展用以对含阴离子型染料的印染废水实现深度处理。印染废水的深度处理和大量废水资源的回收利用一直是印染行业的一个难题，原因在于经生化处理后的废水往往还含有少量可溶性染料，无法实现完全脱色，从而影响其回用到工艺。如图 15-13(d)所示，Wang 以 $MgCO_3$ 为前体材料，通过热解获得多孔 MgO 微米棒，在染料废水中原位水化获得自支撑的纳米 $Mg(OH)_2$ 棒状聚集体，研究发现 $Mg(OH)_2$ 对水中浓度低至 10~6 mol/L 的阴离子型染料可实现快速选择性吸附，其中对酸性茜素蓝的吸附容量达到 155 mg/g。通过碳酸化处理，可将饱和吸附的 $Mg(OH)_2$ 快速恢复为 $MgCO_3$ 纳米棒，实现阴离子型染料脱附，达到 1000 倍以上的浓缩倍数[120]。

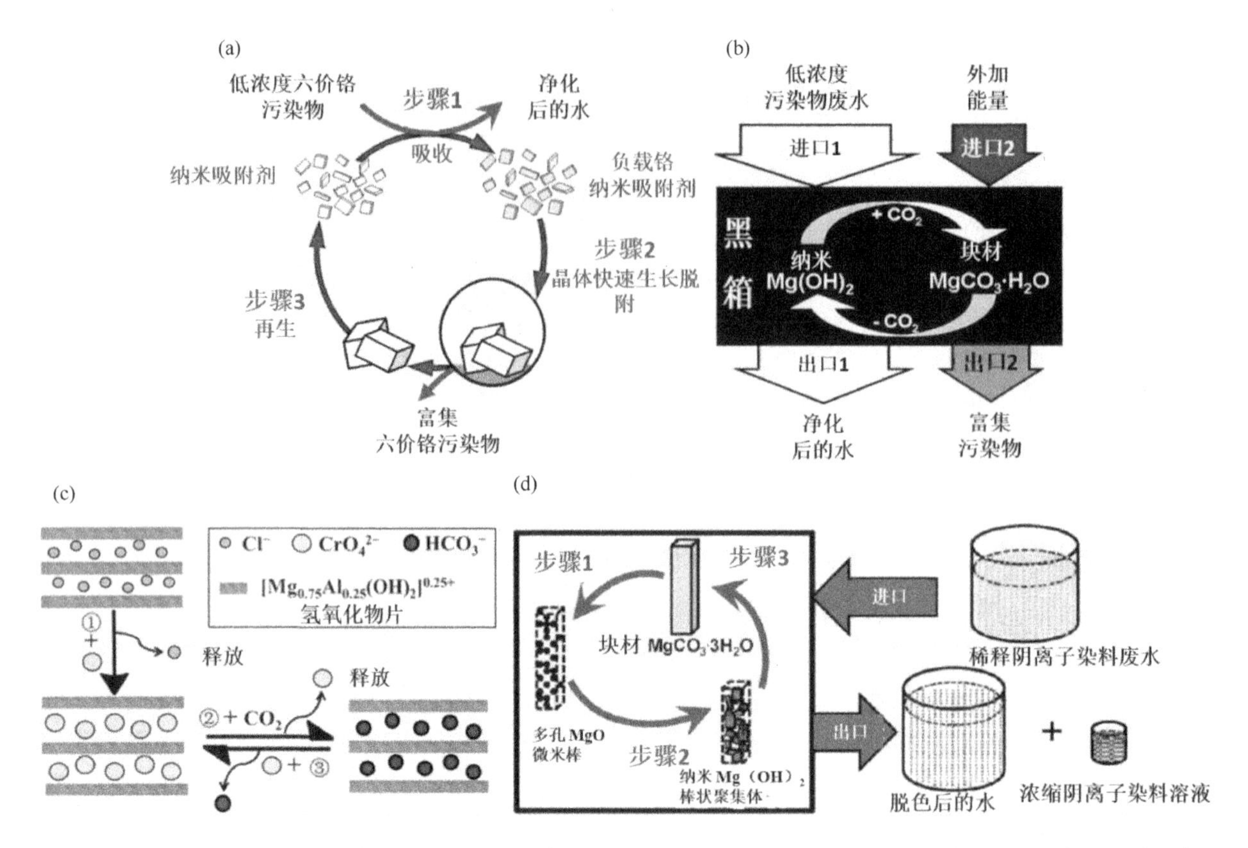

图 15-13 (a)纳米 $Mg(OH)_2$ 预富集水中低浓度六价铬的流程图;(b)(a)方案处理的“封闭性”:在整个处理过程中,可以把纳米 $Mg(OH)_2$ 和所使用的 CO_2 当成一个理想化的“介质黑箱”,对这个“介质黑箱”输入含低浓度铬液的废水和少量能量,就可输出高浓度的可回收六价铬液和可排放的水;(c)LDHs 处理六价铬的循环回用策略;(d)$Mg(OH)_2$ 纳米棒回收废水中的阴离子染料示意图

2.6.2 镁基纳米材料对水中阳离子型低浓度污染物的预富集研究

$Mg(OH)_2$ 与较多的阳离子型重金属例如铅、镉、铜等均具有高的热力学反应常数。同时又因为它是一种 pH 不超过 10 的固体碱,在使用的过程中对水体的干扰小,因此纳米 $Mg(OH)_2$ 对水中阳离子型低浓度污染物的预富集和对物化性质不同的污染物的分离不管从基础研究还是实际应用上都有较大的研究与探索空间。以稀土废水为例,当前,稀土分离工业往往产生大量低浓度(1~100 mg/L 级)的含稀土废水,通常直接排放而无法回收。针对低浓度的稀土废水,为实现稀缺资源的回收利用,Li 等基于纳米 $Mg(OH)_2$ 与稀土阳离子的高的热力学反应常数,通过构筑微纳复合结构的 $Mg(OH)_2$ 并将材料填柱,从而截留了水中 mg/L 量级的低浓度稀土,实现了 99%以上的提取效率。进而,研究还基于 $Mg(OH)_2$ 与氢氧化稀土的酸溶解动力学的不同,对所提取的稀土实现进一步提纯。这一材料设计不仅比起仅通过表面作用过程来吸附污染物的纳米材料具有更高的材料利用效率,而且还因为材料与污染物的化学反应型的作用特点从而具有高的提取效率[121]。

2.6.3 表界面改性强化镁基纳米材料对水中低浓度稀贵金属的选择性富集回收

海水和盐湖卤水中往往含有 μg/L~mg/L 量级的铀。从大范围水中实现含低浓度铀酰离子的提取具有环境保护和资源利用的双重意义。其核心难点在于需要所用的吸附剂对目标物质有着高的吸附力同时还具有选择性,且能够实现铀酰的脱附浓缩,如图 15-14 所示,$Mg(OH)_2$ 自支撑结构能对稀土离子进行提取。Manos 等研究发现性能优越的插层结构 S 属材料 $K_2MnSn_2S_6$,一旦面对实际海水中 μg/L 量级的碳

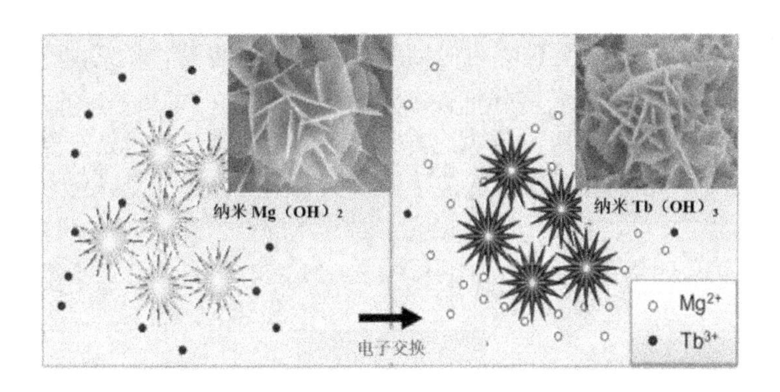

图 15-14　$Mg(OH)_2$ 自支撑结构对稀土离子提取的示意图

酸铀酰，在海水中高浓度盐离子的竞争作用下，其吸附容量呈数百倍的急剧下降[122]。这表明选择性的研究是水中低浓度目标污染物去除的难点。Cao 等系统研究了纳米 $Mg(OH)_2$ 与不同浓度碳酸铀酰阴离子相互作用的机制，证明材料对铀酰离子具有高的吸附力，然而其对铀的提取并不具备选择性[123]。Zhuang 等通过分子动力学模拟及时间分辨荧光等手段分析表明，可以用微量铀酰离子对拟使用的纳米 $Mg(OH)_2$ 进行预处理，经此过程，铀酰离子首先非常牢固地吸附在 $Mg(OH)_2$，如图 15-15（a）和（b）所示的（001）晶面上，

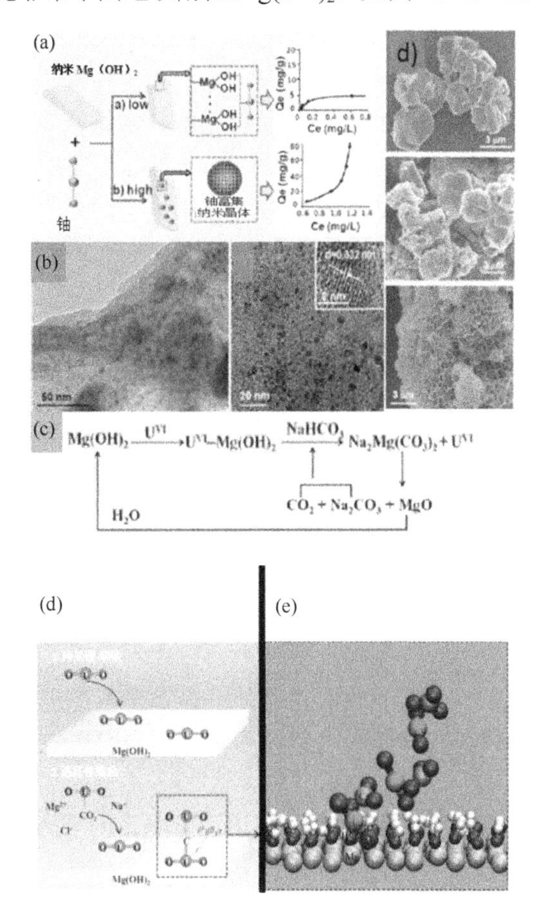

图 15-15　（a）$Mg(OH)_2$ 和铀酰相互作用机制示意图。相对低浓度下铀酰离子在 $Mg(OH)_2$ 表面呈现单分子层吸附，相对高浓度下形成含铀纳米颗粒。（b）在 $Mg(OH)_2$ 表面形成的含铀纳米颗粒 TEM 图。（c）～（d）铀酰离子的脱附和 $Mg(OH)_2$ 吸附剂再生的示意图，以及典型的 $Na_2Mg(CO_3)_2 \rightarrow MgO \rightarrow Mg(OH)_2$ 的 TEM 图。（e）$Mg(OH)_2$ 界面修饰提高卤水中铀酰离子提取原理图

随后进入卤水后，卤水中的铀酰离子与预吸附的铀酰离子通过碳酸根的联系进行螯合，就可以在大量高浓度杂离子的存在下，实现对铀酰的高选择性提取[124]。吸附上来的铀酰离子在 $Mg(OH)_2$ 表面结晶，形成富含铀的纳米颗粒，呈现对铀酰离子高的吸附容量。通过碳酸化处理可诱导纳米 $Mg(OH)_2$ 快速结晶，形成大尺寸 $Na_2Mg(CO_3)_2$。比表面积的减小促成了铀酰的分离和富集。$Na_2Mg(CO_3)_2$ 在煅烧分解后可实现 $Mg(OH)_2$ 和矿化剂 Na_2CO_3 重复利用，再生的花状纳米 $Mg(OH)_2$ 保持好的吸附能力[125]。铀酰离子的脱附和 $Mg(OH)_2$ 吸附剂再生的示意图见图 15-15(c)~(d)，$Mg(OH)_2$ 界面修饰提高卤水中铀酰离子提取原理图见图 15-15(e)。

2.6.4 小结

面向水中低浓度污染物的预富集的技术需求，基于镁基纳米材料的低成本性、微环境界面结构可构筑性、晶粒尺寸与物相的易调控性，可望形成环境友好的新的水处理技术。当前的科学研究还存在以下需要继续探索的问题：

（1）镁基纳米材料的结构形貌构筑及表面微观特征：需要建立起材料结构-界面官能团活性的系统认识，并应需拓展相关材料复合方法，为探索提高污染物与材料的作用效能提供基础。

（2）材料与典型污染物作用的界面分子反应微观机制：需要结合典型污染物与镁基纳米材料作用体系，从原子取向、分子反应的角度来理解选择性的界面作用原理，从而反馈指导材料的界面调控，为实现对污染物的高吸附力及选择性目标提供基础理论支撑。

（3）物化性质不同的污染物与材料的作用机制研究及选择性目标的实现：在镁基材料与污染物的作用过程中，有两个阶段可以实现选择性。第一阶段是通过界面调控直接实现材料对污染物选择性的界面转移，第二阶段是在后续脱附分离中通过污染物与 CO_2 作用机制的不同再次实现选择性。科学认识的进一步加深将为实现多种污染在界面上的同时转化及选择性分离提供重要理论依据。

3 纳米材料环境过程与效应

3.1 金属/金属氧化物纳米材料的环境转化及其效应

3.1.1 团聚

纳米材料因具有大的比表面积和很高的表面能而处于热力学不稳定状态，相邻的颗粒间容易融合、团聚。纳米材料的团聚会显著影响其在环境中的迁移和转化，进而影响其生物效应。纳米材料的团聚行为受多种因素的影响，包括：

（1）溶液的 pH。在零点电势 pH 时，即纳米材料表面带电量为零时，纳米材料最易发生团聚。而溶液 pH 离零点电势 pH 越远，纳米材料相对越稳定[126,127]。

（2）溶液中电解质的种类和离子强度。一般而言，高价态的金属离子（如 Ca^{2+}和 Mg^{2+}）比低价态的金属离子（如 Na^{+}，K^{+}）更易促进纳米材料的团聚[128,129]。离子强度越高，纳米材料越容易发生团聚[130]。

（3）天然有机质（natural organic matter，NOM）。环境中普遍存在的 NOM 能够吸附在纳米材料的表面，通过静电斥力和空间位阻作用，使纳米材料稳定性显著提高[131,132]。不同分子量的 NOM 对纳米材料的团聚行为有所差别[133]。

（4）表面包裹剂的种类[130,132,134]。通过静电斥力稳定的纳米材料（如柠檬酸包裹剂）在电解质中容易团聚，而通过空间位阻稳定的纳米材料（如 PVP、PEG 等）抗电解质干扰能力更强，更加稳定。同样，不同包裹剂的纳米材料在不同分子量 NOM 中的稳定性也有所差别。例如，高分子量萨旺尼河天然有机质（SRNOM）会促进 PVP 包裹纳米银的团聚，低分子量组分反而抑制它们的团聚。所有 SRNOM 分子量组分均抑制无包裹剂纳米银的团聚，但对柠檬酸包裹的纳米银的稳定性没有明显的影响[135]。

（5）纳米材料的初始粒径和浓度[136,137]。纳米材料的初始粒径越小，浓度越高，单位体积的颗粒数越多，碰撞的频率越高，越易发生团聚。此外，其他环境因素，如溶液中溶解氧的浓度[138]、光照[139,140]等都会对纳米材料的团聚行为造成影响。

3.1.2　表面微层富集

水表面微层由于处于大气与水体的交界处，具有不同于水体的独特的物理化学性质，容易富集重金属、有机污染物和颗粒物等[141]。研究发现，水表面微层也能够显著富集纳米材料，该过程受环境因素如溶液的 pH、离子强度、NOM 等的影响。在对天然水样的实验室研究也发现，纳米银在实际环境水体表面微层可达到 14.6~26.5 倍的富集[142]。

3.1.3　氧化还原

碳质纳米材料如富勒烯、碳纳米管等在水中的溶解度很差，而金属（Cu、Zn、Ag 等）或金属氧化物（ZnO、CuO 等）纳米材料在水溶液中能缓慢溶解，释放出金属离子，被认为是这类纳米材料产生毒性效应的主要来源[143]。以纳米银为例，其氧化溶解与以下因素密切相关：

（1）溶液中溶解氧和 pH。纳米银的氧化溶解是氧气和质子共存参与的反应，因此溶解氧量丰富的水溶液中，纳米银的溶解较快[27]。溶液中 pH 越低，H^+浓度越高，越能促进纳米银的溶解[144]。

（2）NOM 的存在。NOM 浓度越高，离子的释放越慢。其原因可能是：NOM 能吸附在纳米材料表面，堵塞其氧化位点；NOM 有很多还原性的基团，能够将释放的离子重新还原，或 NOM 与释放的离子络合，使得实际测得的游离的离子含量降低；NOM 能够分解纳米银氧化过程中产生的过氧化物中间体，使该氧化过程速率减慢[144]。

（3）纳米材料的粒径。纳米材料初始粒径越小，比表面积越大，就有更多比例的原子与氧化物发生反应，越容易被氧化[145]。

（4）纳米材料的分散状态及浓度[146]。团聚后的纳米材料比表面积下降，氧化溶解速率减慢。纳米材料初始浓度越高，一方面颗粒间碰撞加剧，易引起团聚；另一方面颗粒数增多，会抑制氧气分子和 H^+ 扩散到纳米材料表面活性位点的速率，减缓纳米颗粒的氧化。

（5）其他环境因素如温度、光照等。温度越高，纳米材料氧化越快；而光照能促进纳米材料表面的快速氧化，使短时间内离子的释放量迅速上升[147-149]。

NOM 含有丰富的还原性的官能团，能够还原金属离子如金离子、银离子等生成纳米材料[150]。最近的研究发现在 NOM 含量丰富的纳米银溶液中，经光照后，被释放的银离子能够同时被 NOM 重新还原，生成二次纳米银颗粒，并伴随着纳米银形貌的显著变化，如颗粒间的融合，小颗粒的生成及团聚沉降等。采用双同位素示踪的方法发现[151]，该动态循环转化过程复杂，受到环境因素如温度、pH、共存离子等的影响很大。

3.1.4　硫化及再转化

在硫化物含量丰富的环境中，金属或金属氧化物纳米材料很容易与硫化物发生反应生成相应的金属硫化物。释放到环境中的纳米材料，很可能会随下水管道进入污水处理系统，最终转化为硫化物，大部分被固定在污水处理工艺的生物膜或活性污泥中[152-154]。

纳米银是研究最多的一类金属纳米材料。研究发现随着纳米银硫化反应的进行，原本分散的纳米银颗粒会逐渐发生桥联，生成不规则的链状结构[155]。进一步的实验表明，颗粒间的桥联部分为 Ag_2S，很可能是硫化过程中纳米银先氧化释放出 Ag^+，Ag^+进一步与硫化物生成的 Ag_2S，在邻近的纳米银颗粒间沉淀，将颗粒交联在一起[155]。动力学研究发现，纳米银的硫化过程可能存在两种不同的途径。在硫化物含量较高时，纳米银的硫化是直接发生的，可直接与硫化物反应生成 Ag_2S；而当硫化物浓度较低时，纳米银先被氧化释放出 Ag^+，Ag^+进一步与硫化物反应生成 Ag_2S[156]。纳米银经硫化后，银离子的释放量大大下降，会显著影响纳米银的生物效应[157]。

ZnO 和 CuO 纳米材料的硫化过程也有报道。经硫化后，ZnO 和 CuO 表面都分布有小颗粒的 ZnS 或 CuS 纳米颗粒，表明它们在硫化过程中先释放出了金属离子，金属离子进一步与硫化物发生反应，沉积在原始颗粒附近[158-159]。与 Ag_2S 不同的是，ZnO 纳米颗粒经部分硫化后，在颗粒表面生成一层多孔的小颗粒的 ZnS，并不会抑制 ZnO 纳米颗粒中 Zn^{2+}的溶解释放。而 CuO 经硫化后，Cu^{2+}的溶解量甚至高于初始 CuO 纳米颗粒。由于 Ag_2S 性质更加稳定，且地表水中 ZnS 和 CuS 的含量远高于纳米银的浓度，因此 ZnS 和 CuS 可与纳米银反应，进一步转化生成可溶性 Zn^{2+}、Cu^{2+}和 Ag_2S[160]。

Ag_2S 虽然性质稳定，但在特定的环境中，仍能发生进一步的转化。一些污水处理厂为除去污水中大量存在的微生物，会加入臭氧处理的工艺。经臭氧处理后，大部分的 Ag_2S 会被氧化，转化为 AgCl 和 SO_4^{2-}[161]。污水处理厂活性污泥经高温烧结后，Ag_2S 会再次转化成零价银，并伴有 Ag_2SO_4 和其他含巯基的银化合物的生成[162-164]，其中烧结的温度和时间能显著改变银的形态分布[165]。环境水体中当有大量 Fe^{3+} 存在时，经光照后，Fe^{3+}被还原成 Fe^{2+}，并伴随有羟基自由基的生成。羟基自由基可将 Ag_2S 氧化，释放处大量的 Ag^+[166]。在 NOM 含量丰富的水体中，光照条件下，Ag^+会被再度还原生成纳米银[167]。

3.1.5 氯化及再转化

Cl^-在天然水体中分布广泛，与一些金属离子如 Ag^+有很强的结合作用[$K_{sp}(AgCl)= 1.77\times10^{-10}$][168]。因此，当这些纳米材料释放到环境中后，很可能会与 Cl^-发生反应，进而影响其迁移转化及生物效应。

由于 Ag^+和 Cl^-的作用复杂，随着 Cl^-/Ag^+比例的不同，会以多种不同的形态如 AgCl 沉淀，可溶性的 $AgCl_2^-$、$AgCl_3^{2-}$、$AgCl_4^{3-}$等存在。因此，Levard 等在实验室模拟条件下，研究了不同的 Cl/Ag 比例对纳米银氯化的影响，发现当 Cl/Ag < 2675 时，转化产物以固态 AgCl 为主，大部分沉积在纳米银的表面；而当 Cl/Ag > 2675 时，转化产物以可溶性的 $AgCl_x^{(x-1)-}$为主要存在形态，且 Cl/Ag 比例越大，可溶性银离子释放量越高[169]。也有研究报道，类似于纳米银的硫化，氯化后沉积在纳米银表面的 AgCl 及部分可溶性的银离子能够将临近的纳米颗粒相互连接，形成不规则的团聚体，改变纳米银的形貌[170,171]。

作为广谱的杀菌剂，纳米银被添加到许多纺织品、抗菌衣物中。在洗涤过程中，消毒剂和自来水中普遍存在的次氯酸盐也能促进纳米银的氯化。次氯酸盐能迅速氧化纳米银，释放的银离子很容易与 Cl^- 生成 AgCl 沉淀[172]。已有文献报道，如含纳米银的袜子经次氯酸盐消毒剂和洗涤剂混合浸泡后，大约 50%的纳米银会转化成 AgCl[173]；含纳米银的纺织品经多种模拟洗涤后，洗涤液中均可检出 AgCl 的[174]。

AgCl 在某些特殊的条件下同样可发生再转化。AgCl 随下水管道进入污水处理系统后，与还原性硫化物反应可转化为更稳定的 Ag_2S[175]。AgCl 在富含有机质的河水中，在光照条件下可被再次还原生成纳米银。将 AgCl 添加到模拟城市生活垃圾中再进行灼烧，AgCl 也可转化为零价的银。

3.1.6 同位素分馏

同位素分馏是指在某一物理或化学过程中元素的同位素组成发生变化。由于质量数不同，在物理或化学反应中不同的同位素可能会表现出不同的反应速率，从而造成产物的同位素组成上的差异。例如，轻同位素可能具有更快的反应速率，而重同位素可能会更倾向于富集在具有更强的成键环境（strongest bonding environment）的相中。同位素分馏在地质学、考古学、环境科学等领域被广泛用作示踪手段或定年工具来研究不同物质的起源和过程。

近期研究表明[176]，当纳米材料进入环境介质中后，其环境过程能够导致显著的同位素分馏效应。这种效应能够用来指征纳米材料在环境中经历的过程及其可能的机制。以纳米银为例，银具有两种天然丰度接近的稳定同位素（^{107}Ag 和 ^{109}Ag，丰度比为 51.8%∶48.2%）。当纳米银进入环境水体，能被水中的溶解氧所氧化而逐步释放出 Ag^+离子，这一过程能够导致显著的银同位素分馏，使所释放的 Ag^+离子与原来的纳米银具有不同的同位素组成[176]。这种同位素分馏效应不但受太阳光、温度等环境因素的影响，同时也依赖于纳米银的表面涂层的物化性质。人工纳米银（如 PVP 或柠檬酸涂层）会释放富集重同位素的 Ag^+离子，而天然存在的银纳米颗粒（NOM 涂层）会释放富集轻同位素的 Ag^+离子。这种同位素分馏效应的差异也揭示了纳米银释放 Ag^+离子过程中的不同的表面化学反应机理。另一方面，环境中的 Ag^+离子能够在天然条件下被水中的 NOM 还原生成 Ag^0 纳米颗粒。在这一过程中，重同位素更易于被还原为 Ag^0 纳米颗粒。此外，Ag^+离子吸附及光解都能导致不同程度的银同位素分馏。

由于天然同位素分馏的程度一般非常细微（例如，纳米银的环境过程导致的银同位素分馏的最大程度为 0.86‰[176]），因此需要有高精度的分析手段才能准确地测定环境过程中的同位素分馏。目前常用的同位素测定工具为多接收器电感耦合等离子体质谱（MC-ICP-MS；图 15-16）。通过精确测定纳米材料在特定环境过程中的同位素分馏，或者纳米材料本身具备的天然同位素指纹，能够提供环境中纳米材料的来源信息。由于目前大部分的人工纳米材料都含有具有天然同位素的元素，因此这一方法能够推广到多种类型纳米材料的研究[177]。目前这一方面的研究还处于刚刚起步的阶段。

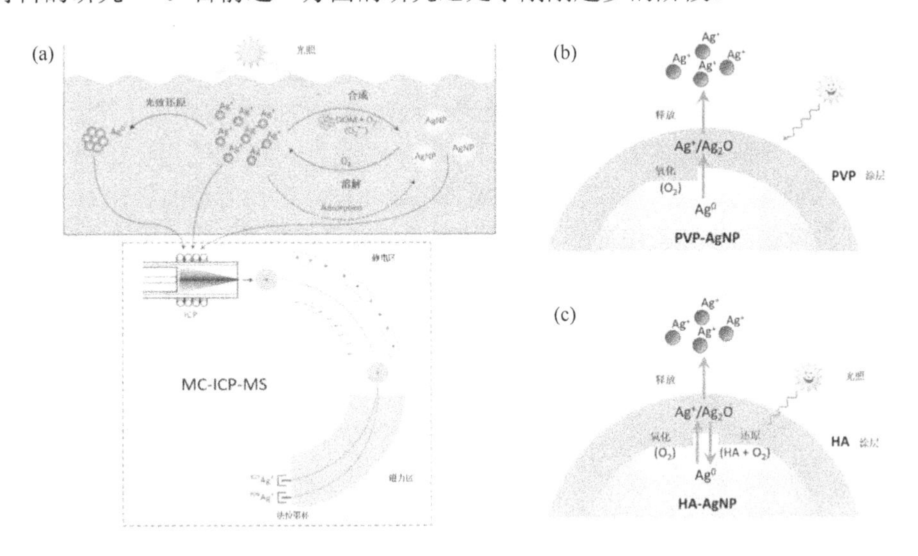

图 15-16　利用 MC-ICP-MS 研究纳米银在自然转化过程中银同位素组成极细微变化并揭示环境中纳米银的行为及来源（引自参考文献[176-177]）

（a）MC-ICP-MS 测定原理图。（b，c）通过银同位素分馏效应推导不同来源的银纳米颗粒的银离子释放机理：（b）人工合成纳米银释放的银离子富集于 ^{109}Ag；（c）天然生成纳米银释放的银离子富集于 ^{107}Ag

3.2　碳纳米材料的环境转化及其效应

3.2.1　人工碳纳米材料的环境转化

人工碳纳米材料进入环境中后，会受到环境中多种物质（如溶解性有机质、金属离子、氧化剂、还原剂、胞外分泌物等）及环境条件（如水体 pH、光照等）的影响，从而发生一系列的物理转化、化学转化及生物转化，使碳纳米材料的理化性质和赋存状态发生变化，进而影响到碳纳米材料与环境污染物的相互作用。物理转化主要是指由水化学条件（如 pH、离子强度等）的变化所引发的碳纳米材料团聚状态的改变，以及碳纳米材料与溶解性有机质、矿物、生物大分子等物质间的异相团聚(heteroaggregation)[178]。化学和生物转化指碳纳米材料在环境中发生化学反应（如氧化还原、光致氧化还原等）或在生物介导作用下（如与胞外分泌物作用或被生物细胞吞噬），导致碳纳米材料的化学结构、表面官能团的数量及种类发生变化[178]。

3.2.2 环境转化对碳纳米材料环境行为及效应的影响

3.2.2.1 环境转化可影响碳纳米颗粒的迁移能力

碳纳米颗粒在地下含水层多孔介质等环境介质中的迁移是影响其环境扩散的重要过程。环境转化可改变碳纳米颗粒的粒径、表面电荷以及表面含氧官能团的种类和数量，一方面直接影响了碳纳米颗粒与环境介质间的相互作用，另一方面可改变影响碳纳米颗粒迁移能力的截留作用（straining）和桥联作用（bridging），从而使碳纳米颗粒的迁移能力发生改变[179-181]。例如，碳纳米颗粒粒径的改变会影响颗粒与多孔介质间的范德华力：粒径增大时，范德华力增大，第二能量极小值（secondary energy minimum）加深[182-184]，纳米颗粒更易在介质中沉积，迁移能力减弱。粒径的增加也可导致更显著的截留作用，使纳米颗粒的迁移受到抑制[185,186]。此外，经转化后碳纳米颗粒表面电荷的变化会影响颗粒与介质之间的静电作用力。

有意思的是，环境转化对碳纳米材料迁移能力的影响并不能简单地从纳米材料理化性质的变化上预测，而是在很大程度上取决于纳米材料表面理化性质与水化学条件的共同作用。例如，在一价 Na^{+}离子的背景电解质溶液中，被氯胺氧化的氧化石墨烯迁移能力增强[179]，而被硫化物还原的氧化石墨烯迁移能力减弱；而在二价 Ca^{2+}离子的背景溶液中，被氯胺氧化和被硫化物还原的氧化石墨烯迁移能力均有所减弱。这是由于氯胺氧化增加了氧化石墨烯表面的羧基官能团的数量，而硫化物还原也导致了氧化石墨烯表面的酚羟基和羧基的增加，酚羟基和羧基均为具有较强金属络合能力的含氧官能团，可以在 Ca^{2+}的作用下与石英砂介质发生阳离子桥联作用，从而促进纳米颗粒在石英砂中的沉积，抑制其迁移。

3.2.2.2 环境转化可影响碳纳米颗粒对环境污染物的富集和载带作用

碳纳米材料对有机污染物有很强的吸附能力，进入环境后可作为载体，促进有机物在环境中的扩散或在生物体中的富集。碳纳米颗粒对环境中有机物的结合能力（包括吸附能力和吸附不可逆的程度）直接决定了碳纳米颗粒对污染物扩散和生物有效性的影响程度，而碳纳米材料发生环境转化后，其表面理化性质和团聚状态都会发生改变，造成污染物在碳纳米材料上的富集能力和不可逆吸附特征发生变化，进而影响污染物的环境效应和环境风险[178]。

环境转化可改变碳纳米材料的表面疏水性及形成 π-π 作用、氢键和化学键的能力，进而影响到碳纳米材料吸附有机污染物的能力。例如，研究表明，经低剂量还原剂硫化物处理后，氧化石墨烯的相对疏水性指数大幅增加、表面石墨化结构得到部分恢复，对多环芳烃的吸附能力明显增强[187-190]。环境转化造成的碳纳米材料团聚状态和团聚体孔结构的改变也可显著影响其结合污染物的能力[191,192]。碳纳米材料团聚所形成的孔隙具有孔填充效应，是有机污染物的优先吸附点位[193-195]。环境转化导致碳纳米颗粒团聚状态的改变，会导致其孔隙结构及孔体积发生变化，进而影响碳纳米材料上污染物不可逆吸附的程度[196]。

值得注意的是，环境转化所引发的碳纳米颗粒团聚状态的变化对有机污染物的吸附是促进还是抑制是多种机制的共同作用，不能一概而论。如溶解态有机质可通过表面包覆作用吸附到碳纳米材料表面，一方面提高了碳纳米材料在水溶液中的分散性，增加碳纳米材料的吸附点位，从而增强对有机污染物的吸附，但溶解态有机质在碳纳米颗粒表面的包覆也有可能与有机物之间形成竞争吸附关系，或者会堵住碳纳米颗粒团聚体的微孔，对碳纳米颗粒的吸附能力造成不利的影响[197]。

3.2.2.3 环境转化可影响碳纳米材料催化污染物化学反应的能力

碳纳米材料富集有机污染物之后，可进一步催化有机物的化学转化，进而影响有机污染物的环境行为和效应。环境转化所导致的碳纳米材料表面官能团种类和数量的变化，以及碳纳米颗粒团聚特征的变化，均可影响碳纳米材料催化有机污染物反应的能力。首先，碳纳米材料表面官能团种类和数量的改变可影响其表面官能团的路易斯酸性或碱性，同时也可影响碳纳米材料吸附有机污染物的能力，这些都会导致碳纳米材料催化有机污染物水解反应的能力发生变化。其次，碳纳米材料的物理转化可影响纳米材料的团聚状态，一方面使其与有机污染物的接触方式发生改变，另一方面也影响了纳米颗粒有效催化位点的密度，同时还可影响纳米材料吸附有机物的能力，这些变化都会影响其催化能力。

需要注意的是，碳纳米材料的环境转化所引发的纳米材料团聚状态及表面官能团的变化共同影响了碳纳米材料催化有机污染物发生转化的能力，而很多情况下二者的影响并非简单地叠加，需要综合考虑纳米材料在特定水环境条件下理化性质的变化。例如，化学氧化可导致碳纳米材料表面含氧官能团的增加，有利于提升纳米材料催化污染物水解反应的活性，但这种变化同时会导致纳米材料在 Ca^{2+}等阳离子作用下的官能团桥联，反而弱化了碳纳米材料的催化活性。又如，溶解性有机质可促进碳纳米材料的分散，增加其有效催化位点[198,199]，但如果浓度过高，则可大量吸附到碳纳米材料表面[199,200]，覆盖其表面的催化位点。

3.3 本节小结

纳米材料进入环境之后，可在多种环境因素和过程的作用下发生转化，使其疏水性、表面电荷、表面官能团等物理化学性质发生显著改变，进而影响纳米材料的胶体稳定性、迁移能力和毒性等环境行为与效应。此外，纳米材料进入环境之后可与环境中的污染物相互作用，从而显著影响污染物的环境扩散、反应活性和生物有效性。而环境转化所导致的纳米材料理化性质的变化，会进一步影响纳米材料结合污染物的能力和催化活性，进而改变纳米颗粒富集和载带污染物的能力，或纳米材料催化有机物发生化学转化的能力。纳米材料的环境转化过程极为复杂，由此引发的纳米材料自身环境行为的改变，以及纳米材料与污染物相互作用的改变在很大程度上取决于纳米材料表面理化性质与水化学条件的共同作用，其中的构效关系还有待进一步地深入探讨。此外，对于环境中的纳米材料的研究也亟需新的研究工具（如同位素分析），为纳米材料的复杂环境过程及归趋提供新的证据。

4 环境纳米材料的毒性及致毒机制研究进展

4.1 金属纳米材料的毒性

4.1.1 金属纳米材料的环境释放及暴露途径

典型的金属纳米材料，包括纳米银、纳米金、量子点、纳米铜以及纳米金属氧化物（如纳米二氧化钛、纳米氧化锌等），除拥有金属的力学性能和纳米材料所具有的一般特性外，还具有特殊的磁性、光学效应、量子尺寸效应和表面效应等，使其不仅在通信、电子、国防、建筑等方面应用，还在生物医药领域有广泛的应用潜力。

环境中的金属纳米材料可通过皮肤接触、呼吸和摄食等途径进入人体，并到达组织和器官，最终在体内累积并引起多种健康效应和风险。目前，关于金属纳米材料直接造成人体健康危害的报道日益增多，如美国发生多例纳米银杀菌滴液中毒事件[201]，中国台湾报道纳米材料搬运工人患过敏性皮炎和心绞痛的比例显著增加[202]，更有吸入纳米镍导致呼吸窘迫致死的案例[203]。因此，针对金属纳米材料环境暴露与潜在的健康风险也备受关注。

4.1.2 金属纳米材料的毒性效应机制

纳米材料的毒性受其自身理化性质的影响[204,205]。同时，纳米材料在不同环境的转化过程可导致纳米材料结构和表面化学特性发生改变，进而影响纳米材料的团聚状态、分散性和稳定性，并最终影响纳米材料的生物作用机制和效应。

金属纳米材料的生物毒性机制主要通过：①诱导胞内活性氧自由基（ROS）的增加并造成氧化应激和损伤；②释放自由金属离子并破坏生物分子及代谢平衡；③造成细胞膜和遗传物质损伤。此外，金属纳米材料能够通过影响胞内的信号传递、激酶活性、基因表达等改变生物系统的动态平衡并导致细胞毒性。典型的毒性效应如下所述。

4.1.2.1 金属纳米材料的免疫毒性

当纳米颗粒以各种途径进入机体后，机体内的免疫和吞噬细胞，如巨噬细胞、树突状细胞、中性粒细胞、嗜酸性细胞和肥大细胞，会对其进行快速地识别和清除，产生炎症因子并招募更多的免疫细胞共同参与炎症反应的级联放大，未被分解清除的颗粒，在蛋白酶作用下分解为抗原多肽类物质，并与内质网上 II 类 MHC 分子结合，继发性的激活 T、B 淋巴细胞，使机体产生获得性免疫。目前，针对金属纳米材料免疫毒性的研究，从免疫器官层面主要集中对肺脏、肝脏、脾脏、淋巴结等外周免疫器官功能的影响展开。例如，呼吸进入机体的纳米颗粒能沉积在肺部首先被肺巨噬细胞识别，低浓度纳米 TiO_2 能增加巨噬细胞吞噬活性[206]，但是降低其趋化能力使得膜表面 Fc 受体和 MHC-II 分子表达降低逃逸肺部巨噬细胞的监视，进入肺小间隙诱发比小尺寸颗粒物更强的炎症反应和肺部纤维化。针对纳米银以及纳米 TiO_2 的研究显示其具有多种免疫效应，包括在外周淋巴器官的累积、改变免疫细胞数量、活性及功能等[207]。关于金属纳米颗粒诱发免疫毒性的分子机制的研究，主要是暴露诱发细胞膜损伤、细胞凋亡、自噬以及离子通道及电子传递链的异常[208,209]，纳米颗粒及释放离子诱发的 ROS 及 NLRP3 炎性小体的激活导致的免疫毒性[210-212]。

4.1.2.2 金属纳米材料的胚胎发育毒性

胚胎发育时期是生物的生长发育过程中一个相对敏感的阶段，胚胎时期外来异物的暴露可以诱发严重的危害。利用不同动物模型研究发现，纳米银、纳米金以及纳米 TiO_2 等金属纳米材料暴露可以导致斑马鱼胚胎、小鼠胚胎及大鼠胚胎的死亡、畸形、功能异常、生长缓慢等[213-215]。这些胚胎毒性效应与纳米材料的理化性质、暴露剂量、暴露时间以及母体所处的妊娠期等因素密切相关[214]。

胎盘屏障是哺乳动物体内一个重要的生物屏障，可以为胚胎发育提供稳定的内环境。纳米颗粒可以穿过胎盘屏障从母体传递给子代，从而对胚胎发育过程造成不良影响。有关金属纳米材料可以在胎盘和胎儿体内累积的研究已有报道，Yamashita 等对怀孕 16 天的小鼠连续暴露 TiO_2 和 SiO_2 纳米颗粒 48 h 后，在胎盘以及胎儿的肝脏和大脑中均检测到相应的纳米颗粒[216]。这些颗粒可以诱发胎盘结构和功能异常，进而导致胎儿生长速率下降、体形变小。另外，纳米银也可以穿过实验小鼠的胎盘进入胎儿体内，并累积于胎肝，导致胎肝组织发生造血障碍，引发贫血，从而导致胚胎发育迟缓[217,218]。值得注意的是，纳米材料的理化性质可以显著影响其穿过胎盘屏障的能力以及胚胎毒性效应[219,220]。

纳米颗粒不仅可以直接穿越胎盘屏障导致胚胎毒性，也可以通过干扰母体内环境稳态而间接影响胚胎发育过程。金属纳米材料可以诱发母体炎症及氧化应激，进而导致胚胎毒性。例如，纳米铜暴露后，虽然在胎盘和胎儿中没有检测到纳米颗粒累积，胎盘结构也没有发生组织病变，但是纳米铜暴露却可以引发母体肺部炎症，进而影响子代的免疫功能，导致子代成年小鼠的生存率显著下降[221]。

4.1.2.3　金属纳米材料的表观遗传毒性

近些年，研究发现金属纳米材料除了在传统遗传学层面改变基因和蛋白的表达及导致细胞毒性，还能在表观遗传学层面影响表观调控分子如微小 RNA（microRNA，miRNA）和长链非编码 RNA（LncRNA）的表达并导致细胞发生应激损伤[222-224]。

miRNA 是广泛存在于真核生物中的一类长 18~26 个核苷酸的内源性非编码小分子，它可以在转录水平控制蛋白合成，从而调节靶基因的表达[225,226]。miRNA 参与了细胞增殖，凋亡，分化以及器官发育和疾病发生等多种生物学功能。近年来，人们发现 miRNA 也参与了金属纳米材料诱发的细胞损伤反应。研究报道多种 miRNA 的水平在纳米银暴露下出现显著变化，例如纳米银暴露下人类皮肤成纤维细胞内 25 个蛋白和 246 个 miRNA 的水平发生显著改变，生物信息学分析发现这些 miRNA 可能通过影响细胞骨架、ATP 合成和细胞凋亡导致细胞毒性增加[227]；在成骨细胞中纳米银能够通过改变一系列 miRNA 的表达水平从而调节成骨细胞的骨钙化水平[228]；纳米银可以通过影响胚胎干细胞来源的神经前体细胞内 miRNA 的表达水平从而导致胞内氧化应激和细胞凋亡[229]；而基因芯片和 miRNA 芯片的结果显示在纳米银暴露下 Jurkat T 细胞中有多达 63 个 miRNA 和 15 个 mRNA 的表达水平出现明显改变，并发现 miR-219-5p 能够调节靶基因 MT1F 和 TRIB 的表达水平，从而调控纳米银的 T 细胞毒性[230]。

LncRNA 是一种大于 200bp 的不具备蛋白质编码功能的 RNA 分子。越来越多的研究结果表明 lncRNA 分子能够在表观遗传水平、转录水平及转录后水平等层面调控靶基因的表达并参与调控了细胞增殖分化、个体发育，基因组印迹和癌症发生等多种重要生命活动[231,232]。目前，lncRNA 在纳米材料的生物学效应特别是金属纳米材料细胞毒性中的作用和分子机制仍不明确。研究报道纳米银暴露红系细胞后，lncRNA ODRUL 的表达明显上调并受氧化应激分子 Nrf2 的直接调控，ODRUL 与 PI4Kα相结合后抑制其蛋白活性，并负调节 PI4Kα下游靶分子 AKT 和 JNK 的磷酸化水平，从而促进纳米银的红系细胞毒性[233]。

目前，miRNA 和 lncRNA 在金属纳米材料的细胞毒性效应中的表达和功能研究仍处于起步阶段，它们如何调控靶基因的表达从而促进或抑制金属纳米材料的毒性效应仍亟待研究，并为研究金属纳米材料的生物学效应和未来的应用提供重要的理论支持。

4.2　常见碳纳米材料的毒性

4.2.1　碳纳米颗粒物的环境释放

碳纳米材料，主要包括碳黑、富勒烯、碳纳米管、石墨烯等[234]。近年来，由于环境污染和工业生产的原因，纳米级碳颗粒物得到科研界广泛的关注。比如，环境中的碳黑（carbon black）主要由含碳燃料的燃烧产生，是大气污染的主要原因之一[235-239]；富勒烯（fullerene，C_{60}）、碳纳米管（carbon nanotube，CNT）和石墨烯（graphene）纳米材料均被视为优异的光学器件、催化剂、纳米药物等的制备材料，大量应用于工业和生活产品[240]。所以，纳米级碳颗粒物的相关毒理研究特别具有现实意义。

环境中的碳纳米颗粒物主要来自于人类活动的释放，释放后的碳纳米颗粒物可以进入大气、水体和土壤。例如，大气颗粒物可按其构成的主要成分分为碳黑型、硫酸盐型、硝酸盐型和有机碳型等。其中，碳黑作为一类重要的大气颗粒物，主要来源于化石燃料的燃烧（如煤炭、天然气等）、机动车尾气和固体废弃物的焚烧（如秸秆、木材、生活垃圾等）。据报道，碳黑的环境释放对于大气颗粒物（particulate matter，PM）的形成和大气的温室效应都会产生影响[235-239]。由于不完全燃烧的碳黑中含有大量的类富勒烯结构，因而通过改变压强、气体比例等条件就可以制备和生产工业富勒烯[241]。据报道，Wang 等首次在我国环境大气中监测到富勒烯碳簇（C_{60}和C_{70}），而南京江北石油化工生产是该区域大气中富勒烯碳簇存在的主因，劣质柴油燃烧也有一定贡献[242]。虽然目前未见大气环境中存在碳纳米管和石墨烯纳米颗粒的报道，但是这两类碳纳米颗粒物也很可能会在化石燃料的燃烧过程中产生。Kolosnjaj-Tabi 等就在患有哮喘病的巴黎儿童呼吸道内的支气管液和肺部细胞内发现了碳纳米管的存在，这些碳纳米颗粒物与巴黎市区收集到的机动车尾气和灰尘颗粒物都非常类似[243]。同样，伴随着人类的生产和排放行为，碳纳米材料还可能通过废液排放而进入水环境和土壤环境。除了天然环境暴露，一些工业生产的从业者更容易受到职业暴露的影响，例如驾驶和乘坐公共交通工具的人更易于长期暴露在碳黑纳米颗粒物的环境[244]。

4.2.2 碳纳米颗粒物的健康风险

碳黑、富勒烯、碳纳米管和石墨烯纳米材料的安全性还缺乏全面的研究，相应的环境和健康风险还缺乏清楚的认识。以碳黑为例，在 2011 年，美国国家环境保护署（EPA）发布了 *Report to Congress on Black Carbon* 的报告[245]，其中就提到碳黑是大气细颗粒物 $PM_{2.5}$ 的主要成分之一，虽然 $PM_{2.5}$ 的短期和长期暴露会对呼吸系统和心血管系统等造成损伤，但目前能够证明碳黑具备健康危害的科学证据还非常有限。2012 年，世界卫生组织（WHO）发布了 *Health Effects of Black Carbon* 的报告[246]，该报告系统地回顾了碳黑暴露可能带来的健康效应。虽然大部分的毒理学研究结果显示碳黑可能不是 PM 颗粒的主要毒性组分，但其能够携带多种有毒化学物质进入人的肺部和体循环。

根据目前的研究，碳纳米材料的暴露可引起以下几类健康风险：①肺炎。肺炎和纤维化是由于颗粒物进入肺部引起炎症和免疫功能损伤，同时成纤维细胞增殖及大量细胞外基质聚集形成的纤维化组织结构[247]。②肺癌。由于空气污染的影响，肺癌是发病率和死亡率增长最快，对人群健康和生命威胁最大的恶性肿瘤之一[248]。2006 年，国际癌症研究机构（IARC）通过评估，将碳黑列为 2B 类物质，即可能有致癌性[249]。③心血管疾病。碳纳米颗粒物暴露后，还可能会进入血液循环系统，进而引起心血管类疾病。流行病学研究结果显示，PM 颗粒物暴露所导致的冠状动脉粥样斑块的形成是引起心血管类疾病的主要原因[250]。④碳纳米颗粒物还有可能对生物的遗传、发育、生殖等方面产生影响[251,252]。

4.2.3 碳纳米颗粒物的毒性效应和机制

4.2.3.1 炎症和纤维化

碳纳米颗粒物进入生物体后，与免疫细胞发生作用，进而会诱发炎症反应及组织的纤维化[253]。Rydman 等研究和比较了碳纳米管（长缠绕形和长杆形）和石棉诱导小鼠肺部炎症反应的差异[254]。研究结果显示，在正常小鼠中，杆状碳纳米管和石棉暴露可以诱导强烈的肺中性粒细胞增多并释放炎症细胞因子和趋化因子。相反，在白介素-1（IL-1）受体敲除小鼠暴露碳纳米管 28 天后，其几乎没有诱导 Th2 型炎症反应和抗炎因子（IL-13）的分泌。Mishra 等研究了单壁和多壁碳纳米管（SWCNT 和 MWCNT）引起肺部纤维化的分子机制[255]。通过暴露 CRL-1490 肺成纤维细胞和 BEAS-2B 肺上皮细胞，发现 SWCNT 和 MWCNT 都可以引起 CRL-1490 细胞中转化生长因子（TGF-β1）及其受体的过表达，而碳黑纳米颗粒却没有影响。当选择性地抑制或敲低细胞中的 TGF-β1 受体，就可以有效地抑制 SWCNT 和 MWCNT 诱导的胶原蛋白生成。最终，通过实验证明 SWCNT 和 MWCNT 可以通过作用于 TGF-β1 受体及其下游的信号通路，上调肺成纤维细胞产生的胶原蛋白，这可能是 SWCNT 和 MWCNT 引起肺部纤维化的分子机制。

4.2.3.2　细胞增殖、分化和死亡

2004 年，Tamaoki 等研究了超细碳黑颗粒物（ufCB，11.2 nm ± 0.5 nm）对原代人支气管上皮细胞的增殖效应[256]。研究结果显示，在无血清培养基中，ufCB 的暴露会增加胸苷和亮氨酸的细胞摄入，而它们的摄入会被超氧化物歧化酶和 NADPH 氧化酶抑制剂削弱，并被表皮生长因子受体（EGFR）的抑制剂完全抑制。同时，细胞外信号调节激酶（ERK）的激活也证明了 ufCB 与 EGFR 的作用。该研究证明了，ufCB 的暴露确实会通过作用于 EGFR 受体和其下游的 ERK 蛋白，引起支气管上皮细胞的氧化应激介导的增殖。同样，Unfried 等报道碳黑纳米颗粒（NPCB，Printex 90，14nm）暴露可以诱导肺上皮细胞的增殖[257]。近期，Liu 等也报道石墨烯同样可通过作用于 EGFR 受体，并激活下游的 PI3K/AKT 信号通路而促进 HepG2，A549，MCF-7 和 HeLa 细胞的增殖[258]。Zhang 等报道了羧基化多壁碳纳米管可以促进小鼠成肌细胞的成肌分化，并且在分化时抑制了细胞的凋亡[259]。Chen 等研究并报道，氧化石墨烯（GO）与 RAW264.7 巨噬细胞的 Toll 样受体间的作用及产生的生物效应[65]。其实验结果显示，GO 的暴露会显著地影响 RAW264.7 巨噬细胞中 Toll 样受体 4（TLR4）和受体 9（TLR9）的表达，同时还会触发其下游的信号通路（MyD88，TRAF6 和 NF-κB），最终引发细胞的自噬。GO 可通过激活 TLR4 受体诱发细胞肿瘤坏死因子（TNF-α）的自分泌[260]。TNF-α可作用于其细胞膜表面的 TNF-α受体，使其下游的受体相互作用蛋白 1 和 3 磷酸化，进一步诱导细胞的程序性坏死。

4.2.3.3　细胞代谢和分泌

Giust 等研究了水溶性富勒烯衍生物（t3ss）对腺苷受体和代谢型谷氨酸受体表达的影响，这两种受体主要参与兴奋性神经递质物质的传递[261,262]。他们的研究结果显示 t3ss 纳米颗粒暴露会导致腺苷受体（A_1，A_{2A} 和 A_{2B}）和代谢型谷氨酸受体（$mGlu_1$ 和 $mGlu_5$）的过表达，说明 t3ss 纳米颗粒具备增强神经保护性受体表达并清除自由基的保护作用。碳纳米颗粒还可以间接地激活细胞膜受体及其下游的信号通路和生化反应过程。Hiraku 等发现，MWCNT 的暴露，可以引起 A549 肺上皮细胞的细胞损伤或死亡，受损或死亡的 A549 细胞会将胞内的高迁移率族蛋白 1（HMGB1）和 CpG DNA 释放到胞外，进而形成 HMGB1-DNA 复合物[263]。产生的 HMGB1-DNA 复合物会进一步结合相邻细胞上的晚期糖基化终产物受体，之后 CpG DNA 会被细胞吞噬，并被溶酶体中的 TLR9 识别，最终导致胞内一氧化氮和 8-NitroG 的生成。这个研究证明 TLR9 和相关分子参与了 MWCNT 诱导的基因毒性，并可能产生致癌性。

4.3　影响纳米材料生物毒性的复杂理化因素

4.3.1　纳米核材料

纳米颗粒由不同材料组成，例如金属、金属氧化物以及非金属氧化物。不同组成的纳米颗粒与生物系统作用会引发不同的毒性效应。研究表明，相同尺寸和形貌的金、银和氧化铁纳米颗粒对表皮生长因子（EGF）的信号传导产生了不同影响。金纳米颗粒和银纳米颗粒能够进入细胞并且出现在内吞体中，而超顺磁性铁纳米颗粒（SPION）倾向于分布在细胞周围不进入细胞。这三种材料分别以不同的机制扰乱 EGF 信号响应[264]。

因此，纳米核材料的不同使得纳米材料的毒性效应和致毒机制存在差异，目前对纳米材料致毒机制的研究尚不成熟，很多问题有待进一步探索。

4.3.2 纳米颗粒的尺寸

纳米颗粒的尺寸是影响其生物效应的因素之一。研究表明，静脉注射不同粒径的金纳米颗粒进入小鼠体内，通过血液循环 24 h 后，在脏器中的分布呈现了一定差异，50nm、100nm、200nm 的金纳米颗粒只在血液、肝脏、脾脏中有分布，而小粒径（10nm）的金纳米颗粒在体内分布范围更广[265]，甚至可以穿过血脑屏障进入脑中[266]。通常，随着纳米颗粒粒径的减小，所导致的肺部毒性随之增加。Oberdorster 等发现，用二氧化钛纳米颗粒（20nm，250nm）对大鼠进行为期 14 天的呼吸暴露，小粒径的纳米颗粒造成了较高水平的肺部炎症效应[267]。用不同粒径的银纳米颗粒（15nm、30nm、55nm）处理细胞 24 h 后，发现较小粒径的银纳米颗粒产生了更高水平的 ROS 和细胞凋亡[268]。纳米颗粒的尺寸还会影响纳米颗粒的细胞摄入[269]。

分别在母鼠哺乳期的第 2、4、6、8 天对其静脉注射 8 nm 和 50 nm 的二氧化钛纳米颗粒（8 mg/kg），在第 10 天进行检测，结果显示纳米颗粒造成了乳腺上皮细胞脱落和紧密连接蛋白表达下调。与 50 nm 的氧化钛纳米颗粒相比，8 nm 的氧化钛纳米颗粒对乳腺组织造成的损伤程度更加严重。研究还发现，由于乳腺上皮细胞脱落和血乳屏障的损坏，8 nm 的氧化钛纳米颗粒可以通过乳汁从母鼠传递给幼鼠[270]。

以上结果显示，纳米颗粒的尺寸大小对其毒性效应有较大的影响，整体趋势大致为粒径小的纳米颗粒在生物体中的分布更广，所引起的毒性效应越明显。

4.3.3 纳米颗粒的形貌

纳米材料有多种形状，例如：球形、棒状、管状、立方体等。纳米颗粒是通过与细胞膜上的生物分子相互作用进入细胞，其形貌必然会影响纳米颗粒与生物分子和细胞的作用方式，因此决定纳米颗粒的生物效应。

据报道不同形状的 SiO_2 纳米颗粒能够导致不同的毒性效应和溶血活性[271]。溶血活性高低顺序为 SiO_2 纳米球＞较小纵横比的介孔 SiO_2 纳米棒＞较大纵横比的介孔 SiO_2 纳米棒。SiO_2 形状对生物效应的影响还与细胞系、纳米颗粒浓度、表面化学性质以及实验方法等多种因素有关，还需更加深入的研究。在对不同形状的 TiO_2 纳米颗粒的毒性研究发现，包括 TiO_2 纳米球（60~200 nm）、长度<5μm 和长度> 15 μm 的纳米带[272]。其中，长度大于 15 μm 的 TiO_2 纳米带能够诱导炎症小体的激活并释放炎症性细胞因子，从而显示出更高的细胞毒性。形貌的不同也会影响镍纳米颗粒的生物效应，与球形镍纳米颗粒相比，将斑马鱼胚胎暴露由粒径 60nm 的粒子组成的树枝状团簇后显示出更高的毒性[273]。

不同形状的纳米材料与细胞的作用方式不同，导致了不同的毒性，但是纳米材料的生物效应还受到很多其他因素的影响，比如纳米材料的尺寸、核材料、表面化学性质等，因此，目前还很难得出统一的结论。

4.3.4 纳米材料的表面化学性质

纳米材料表面化学性质的改变将会使其与邻近分子产生不同的静电作用、氢键作用、亲疏水作用、π-π 共轭作用以及由于分子空间构型不同导致的作用，这些作用在纳米颗粒的生物效应中扮演重要角色。下面通过对 MWCNT 和金纳米颗粒进行表面修饰，并研究其对生物效应的影响，来阐述纳米毒性的构效关系。

4.3.4.1 不同表面化学性质的 MWCNT

MWCNT 在多个科技领域以及生物医药中的广泛应用引起了人们对其毒性的关注。研究显示，CNT 的主要生物效应包括诱导炎症[274]、引起细胞凋亡和细胞自噬[275]。

为了研究纳米颗粒的表面化学性质与其毒性的构效关系和这些效应的调控作用，研究者用组合化学的方法合成了一个含有 81 种多样性表面化学修饰的多壁碳纳米管库（图 15-17），全面系统地研究了它们对细胞功能的影响[276]。多样性表面修饰将 CNT 的细胞毒性、免疫毒性和对特定蛋白分子的吸附作用均调节到了不同水平，并选出了多种毒性降低的 CNT[277]。这一研究有助于推进生物相容性纳米医药载体的研发。用羧基多壁碳纳米管（MWCNT-COOH）刺激巨噬细胞，使其与甘露糖受体作用，导致细胞炎症因子释放和 NF-κB 信号通路激活，而表面修饰的 MWCNT 使炎症因子和活性氧水平明显下降。机理研究表明，MWCNT-COOH 在表面化学修饰后与巨噬细胞相互作用时更倾向于被清道夫受体识别，激活了不引起炎症反应的信号通路，这表明 MWCNT 的表面化学修饰可以达到调控其免疫毒性的目的[277]。这一研究结果为 CNT 应用于生物医学领域提供了材料改性的方法和理论依据，并为药物载体的设计研发提供新的思路。近期研究还发现不同表面修饰的可以通过不同的信号通路诱导自噬[277]，从而加深了对于纳米材料与自噬诱导关系的新认识。

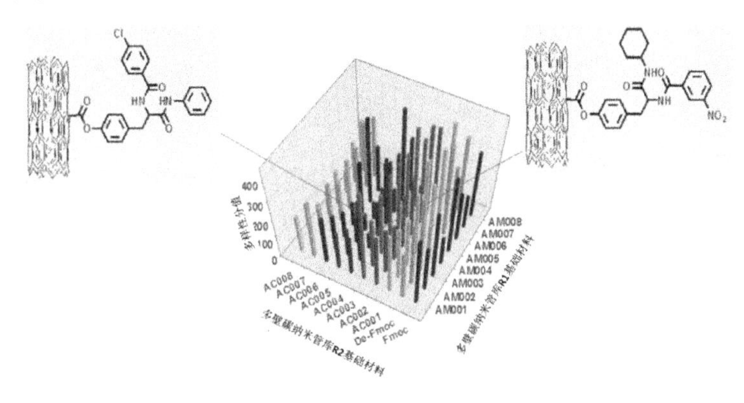

图 15-17　多样性表面化学修饰的多壁碳纳米管库以及它们的生物效应[273]

纳米材料与细胞表面受体作用从而干扰细胞信号通路的现象在多种人类细胞中都存在。CNT 可以与 BMPR2 结合，从而下调 BMP 信号通路[278]，由于 BMP 通路的靶蛋白 ID 可以调控多种基因转录因子，蛋白表面修饰的 CNT 对干细胞分化、细胞周期和细胞凋亡都有调节作用[279]。众所周知，纳米材料进入体内后大都积累在肝脏内，肝脏在外源物质（比如药物）代谢过程中起到重要作用，那么 CNT 和其他疏水纳米颗粒在肝脏中的积累是否会干扰肝脏的代谢呢？给小鼠静脉注射 CNT 后，CNT 经过血液循环后大都进入肝脏中的巨噬细胞和肝细胞，CNT 与多种 CYP 酶有结合作用，如 CYP3A4。不同表面化学修饰的 MWCNT 还能够对肝脏中 CYP3A4 酶的活性产生不同程度的调节[280]，这一信息有助于指导合理设计生物安全性高的 MWCNT，用于生物医药和纳米产品。

4.3.4.2　不同表面化学性质的金纳米颗粒

由于金纳米颗粒的稳定性、可调控性以及其表面可准确可控的化学修饰性，它已经成为一种极佳的纳米颗粒模型。我们设计合成了表面亲疏水性、电荷（图 15-18）、氢键、π 键和空间位阻单一性质连续变化的多个金纳米颗粒库，研究了它们对生物效应的影响。表面正电荷或负电荷连续变化的金纳米颗粒（电荷密度变化范围：+2.87 到–4.18）处理 Hela 细胞 12 h 后，负电荷和中性的纳米颗粒几乎没有细胞摄入，而细胞摄入与正电荷成正比[281]。用亲疏水性连续变化的金纳米颗粒处理 THP-1 巨噬细胞 24 h 后，细胞摄入量与材料疏水性成正比，而且疏水性与毒性相关，而且这类性质可以用建立的数学模型来预测[321]。对多样性修饰碳纳米管库的多种生物效应产生的大数据进行计算模拟得到了多个构效关系模型。运用得到的模型预测了多种具有不同性质的 CNT，所预测的生物活性通过实验室和测试后得到了 70%~80%的成功率[283]。

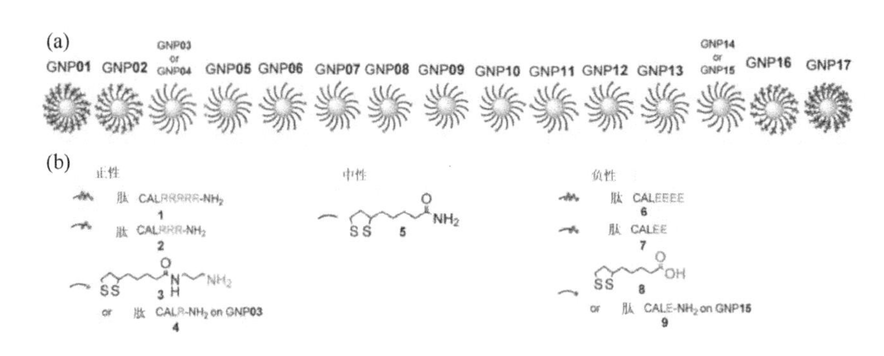

图 15-18 电荷连续变化的金纳米颗粒阵列

(a) 17 种电荷连续变化的纳米颗粒，电荷密度变化范围：+2.87（GNP01）到–4.18（GNP17）；

(b) 用于修饰纳米颗粒的在 pH 7.0 带正、负电的配体

总之，对纳米材料进行表面修饰可以显著地影响纳米材料的生物效应。通过大量的研究，目前已初步掌握了表面化学调控纳米生物效应的规律。计算模拟和构效关系研究只有在拥有这些系统大数据后才有可能开展，通过将实验和计算结合把纳米毒理研究推向一个新的高度。

4.4 本节小结

综上所述，纳米材料与生物分子和细胞的作用方式决定了纳米材料的各种理化性质均对其生物效应产生影响。充分的科学证据证明金属纳米材料和碳纳米材料的暴露会导致生物毒性和健康风险，其毒性大小和健康效应又与自身的理化性质紧密相关，通过改变纳米颗粒的表面化学性质可以成功地调控纳米颗粒的生物效应或毒性效应。目前纳米材料的环境和健康风险研究还存在以下问题：①针对环境中纳米材料的物理转化、化学/生物转化对其生物有效性和健康效应的研究还很欠缺；②不同暴露途径下，纳米材料的体内代谢、分布、转化和降解行为还不清楚；③纳米材料人体暴露的具体安全剂量范围数据还不全面；④缺少研究纳米材料的生物行为和健康效应的成熟的标准化方法和系统化策略；⑤由于纳米颗粒的尺寸、形状、核材料以及表面化学等参数都会共同产生对其生物效应的影响，因此亟待开展运用纳米组合化学进行多参数组合式改变，以合成系统修饰的纳米颗粒库为生物效应研究对象，并对毒性大数据进行计算模拟并建立预测模型。上述科学问题亟待研究，以达到更深入认识纳米生物效应的构效关系和对纳米颗粒的生物效应进行人为调控的目标，并为研究纳米材料的生物学效应和未来的应用提供重要的理论支持。

5 纳米分析应用新进展

5.1 典型纳米材料在环境分析领域中的应用

5.1.1 量子点

量子点，又称为半导体纳米晶[284,285]，是一种零维发光材料，粒径范围一般在 1~12 nm。由于量子点具有量子限域效应[286]，导致其具有独特的光致发光特性，包括耐光蚀、荧光寿命长、可调谐、激发光谱

宽、生物毒性低等，从而极大推动了其在环境检测、化学分析、生物传感和细胞生物学等领域的发展。目前基于量子点构建的光学传感器的传感机制主要包括：①电荷转移。由于量子点荧光产生过程是激发态电子辐射跃迁回基态的过程，因此作用于该过程的外在因素都可以影响载流子的分离和量子点的电离，最终引起荧光变化。基于该机理的量子点荧光探针已经用于检测五氯酚[287]、杀虫剂[288]、4-硝基酚[289]等典型有机污染物。②电致化学发光（ECL）。ECL 是一种在电极表面施加特定电压使激发态分子或激发态分子与溶液中某组分进行化学反应而产生的发光现象。由于量子点的量子产率高、稳定性好，量子点已经成为替代传统有机发光试剂的新型电化学发光剂[290]。③荧光共振能量转移（FRET）。FRET 是一种非辐射能量转移过程，是指当一个供体发光基团的发射光谱与受体发光基团或猝灭基团的吸收光谱重叠时，在一定距离范围内（一般小于 10 nm），能量会由激发态的供体通过偶极-偶极相互作用以非辐射方式传递到基态受体[291]。量子点作为优异的供体材料，构建的 FRET 体系已经用于检测 pH、无机和有机分子、核酸、蛋白质和细胞等。Liu 等利用 CdTe 量子点与石墨烯之间的 FRET 现象构建了新型的免疫传感器，对抗原的检出限可以达到 0.15 ng/mL[292]。④化学发光或生物发光（CRET 或 BRET）共振能量转移。与 FRET 不同，CRET 和 BRET 利用化学反应产生激发态供体，进而以非辐射形式将能量传递到基态受体。⑤静态猝灭。激发态量子点可以与基态受体直接发生相互作用，减少了激发态量子点的浓度，导致荧光猝灭，例如重金属 Cu^{2+}可以与 CdTe 量子点在一定条件下发生离子交换反应，显著猝灭其荧光[293]，在此基础上结合金属结合染料，可以构建基于无机发光 CdTe 纳米棒和有机荧光染料的双荧光探针，将体系内的化学反应转换成荧光增强信号，对 Cu^{2+}的检出限可以达到 0.13 nmol/L[294]。⑥量子点固相检测。利用化学或生物方法将量子点固定到基底表面，避免了胶体量子点存在的稳定性差等问题，提高了分析检测方法的稳定性。如 Zheng 等利用聚合物将 CdTe 量子点自组装到玻璃基底，结合乙酰胆碱酯酶和胆碱氧化酶的催化性能，实现了对有机磷农药的高灵敏检测[295]。近年来，其他量子点材料，如碳量子点、石墨烯量子点、硅量子点和贵金属纳米簇等也得到人们广泛的关注，并且在分析传感领域表现出潜在应用前景。

5.1.2 石墨烯

石墨烯是碳原子通过 sp^2 杂化轨道构成的单原子层二维晶体[296,297]。由于石墨烯及其衍生物具有独特的结构、电子和光学性质，其在化学与生物传感领域得到广泛的研究，主要的传感类型包括：①电化学传感器。利用石墨烯大的比表面积、优异的导电性能和电化学活性，构建安培型、电化学发光型、电阻抗型和光电型传感器，该类型的传感器已经广泛用于检测环境污染物，如有机磷农药[298]、硝基甲烷[299]、重金属离子[300]、致病菌[301]等。②纳米电子传感。利用石墨烯的场效应、掺杂效应、电荷载流子散射以及局域介电环境改变性质等[302]，构建各种类型的传感器检测 NO_2[303]、NH_3[304]、二硝基甲苯[305]和重金属[306]等。③光学传感器。如利用石墨烯独特的荧光猝灭能力，Liu 等构建了基于石墨烯和量子点的新型荧光免疫传感器，克服了传统能量共振转移系统中供体-受体的有效作用距离为 100 Å 的限制，将其提高到 223 Å[307]，实现了藻毒素等的高灵敏检测[308]；利用功能核酸小分子优异的目标物识别能力和石墨烯-核酸较强的结合能力，Liu 等发展了多种基于石墨烯和功能核酸的新型荧光探针，实现了重金属离子的快速、高灵敏和实时检测[309-311]，在此基础上，结合核酸扩增技术，进一步提高了检测的灵敏度[312]；利用石墨烯独特的二维点阵结构、电子结构和贵金属的催化性能，Liu 等构建了基于石墨烯材料和纳米金的高效类过氧化氢酶，提出了纳米人工酶界面传感新原理，构建了基于纳米人工酶和功能核酸的新型比色传感器[313,314]；利用石墨烯独特的表面增强拉曼效应（SERS）[315]，发展了新型 SERS 传感器，用于芳香族化合物等的检测[316]；利用石墨烯材料自身的光致发光性质[317]，Jung 等构建了基于发光石墨烯氧化物和金纳米颗粒的新型荧光传感器，用于检测病原体等[318]。④石墨烯纳米孔。利用石墨烯优异的机械性能和化学稳定性，在绝缘薄膜上制备具有分子直径的石墨烯纳米孔，当分子通过孔通道时，通过电流变化实现单分子检测。目前基于石墨烯纳米孔构建的分子检测器已经用于 DNA 测序等[319,320]。

5.1.3 DNA

从生物学角度，DNA 是一种生物大分子，储存主要的遗传信息；从材料学角度，DNA 是一种具有特殊结构的纳米材料。1982 年，Seeman 等提出通过碱基互补配对原则，DNA 可以组装形成复杂的二维或三维结构[321]，自此基于 DNA 纳米材料构建的各类功能结构和纳米器件得到快速发展[322,323]，被广泛用于分析检测、疾病诊断、生物传感、分子机器等领域。如 Liu 等利用 DNA 环与环之间的相互锁套作用，制备了一种 DNA 连环体拓扑结构，发现了该结构可以有效抑制 DNA 聚合酶的活性，在此基础上结合功能核酸-适体核酶，设计了一种 DNA 纳米器件，外界刺激物如大肠杆菌可以触发其解环反应，进而引起核酸扩增反应，该功能器件可以用于水体和人体血液中大肠杆菌的高特异、高灵敏和快速检测，与当前商业化产品相比，该器件在灵敏性、抗干扰等方面具有显著优势[324]。Liu 等还将 DNA 聚合酶的外切酶校读功能与功能核酸-适配体结合，设计了一种具有三元结构的 DNA 纳米器件，外界目标分子可以触发其结构转换，进而引起核酸扩增反应，大大提高了适配体分子识别信号，从而实现了对功能小分子、DNA 和蛋白质等与重大疾病相关的靶标分子的特异性和高灵敏检[325]。

5.2 环境和生物体系中人工纳米材料的表征技术与方法

人工纳米材料在广泛应用中不可避免地会被释放到环境中，并导致其在环境生命系统中传递和累积放大，最终对整个生态系统和人类健康产生影响。认识这种潜在的影响不仅依赖于人工纳米材料本身性质的表征，更依赖于其进入环境生命体系中后的迁移途径、转化机制以及与环境和生命体系中其他共存分子间的相互作用。相比于人工纳米材料的功能和应用研究，阐明其迁移转化规律和潜在毒性机制所必需的表征技术和分析方法的研究工作相对滞后。因为不同于传统污染物质的组成、结构和含量分析，人工纳米材料的一系列化学和物理性质，如化学组成、表面化学、尺寸、形貌、表面积、晶型、溶解性质等等，都对其潜在毒性有所贡献。特别是微量的人工纳米材料进入极其复杂的非均质动态环境生命系统中后，上述性质因其高的反应活性与所处环境中共存组分间的已知和未知的相互作用而产生变化，表征难度极大。原有对传统污染物质的分析技术和方法不足以客观地对它们的存在形态、分布和作用机制进行表征；更为严峻的挑战来自于如何对进入环境生命体系中的人工纳米材料与体系中原有的同类“天然纳米材料”进行区分？如何在样品储存、预处理和分离过程中保持人工纳米材料的真实状态以避免分析结果的不确定性？这有赖于对已有针对人工纳米材料不同特性的分析技术进行“多维”组合或改进提升，实现“客观分析”的理想目标。

实际上，保持环境生命体系中人工纳米材料的实际存在形态的样品处理技术和方法是后续各种检测技术提供人工纳米材料真实特性的重要保证因素之一。但是，目前所发展的各种样品前处理技术和方法要么直接沿用已有的针对传统污染物的技术和方法，忽视了人工纳米材料本身的特性；要么仅仅是针对人工纳米材料的某些特性对传统技术和方法进行改进，缺乏对人工纳米材料特性的全面考虑，而且也缺乏“标准物质”对整个样品处理过程的效率进行评价。在这种情况下，各种对样品特别是生物样品前处理不十分苛刻的成像技术，如扫描/透射电子显微镜-能量散射光谱（SEM/TEM-EDS）甚至同步辐射 X 射线显微镜，来检测人工纳米材料的分布、组成和结构以及与周围分子的相互作用；但是亚 mg/g 的检测灵敏度有时限制了它们的应用[326]。相比之下，电感耦合等离子体质谱（ICP-MS）具有μg/g 甚至 ng/g（特别是对金属元素）的检测灵敏度，而且可以提供元素同位素的分布信息，继而进行溯源分析和同位素稀释准确定量。ICP-MS 也可以方便地与各种分离技术相结合，如无固定相场流分离和惰性固定相流体动力学色谱以及开管毛细管电泳等，以避免待分析的人工纳米材料与固定相的相互作用，实现了人工纳米材料的组成和尺寸分析；使用激光溅射技术又可以提供人工纳米材料的微米级的空间分布分析；与同步辐射 X 射线等技术相结合时，空间分辨率可以达到纳米水平。质谱的最大优势在于可提供元素同位素的指纹分布信息。这一特点使得 ICP-MS 在追踪人工纳米材料在进入环境生命体系中后的迁移途径、转化机制以及与环境和生命体系中其他共存分子间的相互作用机制，特别是定量检测方面具有独特的优势[327]。它可以依据所检测的纳米材料中某一元素的同位素分布特征来区分人工制备的纳米材料和天然纳米材料[328]，也可以根据人工纳米材料中某些特征“杂质元素”来进行区分。此外，荧光光谱技术也具有较高的灵敏度，对于某些具有荧光特性的纳米材料（如量子点，QDs）在环境和生命体系中的分布和示踪分析发挥着重要作用，但是在可见光区易受来自基体的自发荧光和散射的干扰问题常常引入测量误差，即使使用近红外荧光技术（NIRF）可以避免上述干扰[329]，总体来讲，荧光本身对基体敏感（光漂白/淬灭）的不足尚待克服。

如前所述，环境和生命体系中的人工纳米材料的全面客观分析受制于在采样、预处理和检测的各个环节中的诸多不确定因素，在一定程度上限制了对它们的环境行为和生态毒理学效应的认识，致使得到很多相互矛盾的结果[330]。同时，环境科学家或生物病理学家与统计学家合作也建立了各种模型来预测人工纳米材料在环境介质中的分布和对环境中生物的潜在威胁[331-333]，这有助于人们对人工纳米材料潜在危害的前瞻性认识和推测。但需要指出的是，这些模型构建所用的“数据”也都是来自于已经发表的基于检测的研究结果。如此看来，分析化学家任重道远，发展客观反映人工纳米材料的新技术新方法虽然极具挑战，但机遇同时存在。

5.3　本节小结

社会的发展和人民生活质量的提高对环境检测技术提出了新的挑战，而传统检测方法已不能完全满足现有检测要求。纳米分析检测技术与传统检测方法相比，在微型化、检测时间、灵敏度、专一性和高通量等方面具有明显的优势，但是该技术真正用于环境检测还需要进一步发展纳米材料功能化方法，尤其是结合分子生物学、免疫学等技术，研发更多新型的功能纳米材料应用于复杂环境样品的检测。此外对纳米材料的毒性和安全性的研究仍在进行，急需发展新型的纳米材料的表征技术与方法。可以相信，随着纳米材料与技术的发展，纳米分析传感将在环境污染物的检测中发挥越来越重要作用。

6　展　　望

水污染控制纳米材料的研究是环境化学研究非常热点的领域，然而，迄今未得到大规模的实际应用，今后在材料污染控制机理研究的同时，要重点加强其与其他技术体系的结合及在实际体系中的可行性探索。纳米材料与膜技术、树脂基技术、晶体调控技术的联用是一个可喜的突破，然而其离商品化应用还有一段距离，今后在材料大规模制备、工艺调控等方面都需要更多的研究。纳米材料的环境行为的研究还依赖于复杂环境介质中合理性研究体系的设计与构建及新的研究工具（如同位素分析）的结合，从而为纳米材料的复杂环境过程及归趋提供新的证据。当前的研究已经揭示纳米材料的毒性大小和健康效应与自身的理化性质紧密相关，通过改变纳米颗粒的表面化学性质可以成功地调控纳米颗粒的生物效应或毒性效应。因此，今后研究者们还将在纳米材料的环境和健康风险研究方面进行深入的拓展。纳米分析传感将在环境污染物的检测中发挥越来越重要作用。但是该技术真正用于环境检测还需要进一步发展纳米材料功能化方法，尤其是结合分子生物学、免疫学等技术， 研发更多新型的功能纳米材料应用于复杂环境样品的检测。

参 考 文 献

[1] Guo X Q, Zhang G S, Cui H H, Wei N, Song X J, Li J, Tian J. Porous TiB_2-TiC/TiO_2 heterostructures: Synthesis and enhanced photocatalytic properties from nanosheets to sweetened rolls. Appl Catal B-Environ, 2017, 217: 12-20.

[2] Wu X, Lu C, Liu J, Song S, Sun C. Constructing efficient solar light photocatalytic system with Ag-introduced carbon nitride for organic pollutant elimination. Appl Catal B-Environ, 2017, 217: 232-240.

[3] Yang J, Li Z, Zhu H. Adsorption and photocatalytic degradation of sulfamethoxazole by a novel composite hydrogel with visible light irradiation. Appl Catal B-Environ, 2017, 217: 603-614.

[4] Juntrapirom S, Tantraviwat D, Suntalelat S, Thongsook O, Phanichphant S, Inceesungvorn B. Visible light photocatalytic performance and mechanism of highly efficient SnS/BiOI heterojunction. J Colloid Interf Sci, 2017, 504: 711-720.

[5] Chen Y, Sun F Q, Huang Z J, Chen H, Zhuang Z F, Pan Z Z, Long J F, Gu F L. Photochemical fabrication of SnO2 dense layers on reduced graphene oxide sheets for application in photocatalytic degradation of p-Nitrophenol. Appl Catal B-Environ, 2017, 215: 8-17.

[6] Alvi M A, Al-Ghamdi A A,ShaheerAkhtar M. Synthesis of ZnO nanostructures via low temperature solution process for photocatalytic degradation of rhodamine B dye. Mater Lett, 2017, 204: 12-15.

[7] Gokul P, Vinoth R, Neppolian B, Anandhakumar S. Binary metal oxide nanoparticle incorporated composite multilayer thin films for sono-photocatalytic degradation of organic pollutants. Appl Surf Sci, 2017, 418: 119-127.

[8] Niu J F, Yin L F, Dai Y R, Bao Y P, Crittenden J C. Design of visible light responsive photocatalysts for selective reduction of chlorinated organic compounds in water. Appl Catal A-Gen, 2016, 521: 90-95.

[9] Niu J F, Ding S Y, Zhang L W, Zhao J J, Feng C H. Visible-light-mediated Sr-Bi_2O_3 photocatalysis of tetracycline: Kinetics, mechanisms and toxicity assessment. Chemosphere, 2013, 93(1): 1-8.

[10] Yin L F, Niu J F, Shen Z Y, Chen J. Mechanism of reductive decomposition of pentachlorophenol by Ti-doped beta-Bi_2O_3 under visible light irradiation. Environ Sci Technol, 2010, 44(14): 5581-5586.

[11] Zhang L L, Niu J, Li D, Gao D Shi J. Preparation and photocatalytic activity of Ag modified Ti-doped-Bi_2O_3 photocatalyst. Adv Cond Matter Phys, 2014,2014(2014):537-542.

[12] Ding S Y, Niu J F, Bao Y P, Hu L J. Evidence of superoxide radical contribution to demineralization of sulfamethoxazole by visible-light-driven Bi_2O_3/$Bi_2O_2CO_3$/$Sr_6Bi_2O_9$ photocatalyst. J Hazard Mater, 2013, 262: 812-818.

[13] Zhang C, Zhu Y F. Synthesis of square Bi_2WO_6 nanoplates as high-activity visible-light-driven photocatalysts. Chem Mater, 2005, 17(13): 3537-3545.

[14] Yu J G, Xiong J F, Cheng B, Yu Y, Wang J B. Hydrothermal preparation and visible-light photocatalytic activity of Bi_2WO_6 powders. J Solid State Chem, 2005, 178(6): 1968-1972.

[15] Huang D H, Yin L F, Niu J F. Photoinduced hydrodefluorination mechanisms of perfluorooctanoic Acid by the SiC/Graphene catalyst. Environ Sci Technol, 2016, 50(11): 5857-5863.

[16] Huang D, Yin L, Lu X, Lin S, Niu Z, Niu J. Directional electron transfer mechanisms with graphene quantum dots as the electron donor for photodecomposition of perfluorooctane sulfonate. Chem Eng J, 2017, 323: 406-414.

[17] Yin L, Niu J, Shen Z, Bao Y, Ding S. Preparation and photocatalytic activity of nanoporous zirconia electrospun fiber mats. Mater Lett, 2011, 65(19-20): 3131-3133.

[18] Pignatello J J, Oliveros E, MacKay A. Advanced oxidation processes for organic contaminant destruction based on the Fenton reaction and related chemistry. Crit Rev Env Sci Tec, 2006, 36(1): 1-84.

[19] Brillas E, Sirés I, Oturan M A. Electro-Fenton process and related electrochemical technologies based on Fenton's reaction chemistry. Chem Rev, 2009, 109(12): 6570-6631.

[20] Anipsitakis G P, Dionysiou D D. Radical generation by the interaction of transition metals with common oxidants. Environ Sci Technol, 2004, 38(13): 3705-3712.

[21] Garrido-Ramírez E G, Theng B K G, Mora M L. Clays and oxide minerals as catalysts and nanocatalysts in Fenton-like reactions — A review. Appl Clay Sci, 2010, 47(3): 182-192.

[22] Neamu M, Zaharia C, Catrinescu C, Yediler A, Macoveanu M Kettrup A. Fe-exchanged Y zeolite as catalyst for wet peroxide oxidation of reactive azo dye Procion Marine H-EXL. Appl Catal B-Environ, 2004, 48(4): 287-294.

[23] Centi G, Perathoner S, Torre T, Verduna M G. Catalytic wet oxidation with H_2O_2 of carboxylic acids on homogeneous and

heterogeneous Fenton-type catalysts. Catal Today, 2000, 55(1): 61-69.

[24] Catrinescu C, Teodosiu C, Macoveanu M, Miehe-Brendlé J Le Dred R. Catalytic wet peroxide oxidation of phenol over Fe-exchanged pillared beidellite. Water Res, 2003, 37(5): 1154-1160.

[25] Navalon S, Alvaro M, Garcia H. Heterogeneous Fenton catalysts based on clays, silicas and zeolites. Appl Catal B-Environ, 2010, 99(1–2): 1-26.

[26] Munoz M, de Pedro Z M, Casas J A, Rodriguez J J. Preparation of magnetite-based catalysts and their application in heterogeneous Fenton oxidation – A review. Appl Catal B-Environ, 2015, 176–177: 249-265.

[27] Rahim Pouran S, Abdul Raman A A ,Wan Daud W M A. Review on the application of modified iron oxides as heterogeneous catalysts in Fenton reactions. J Clean Prod, 2014, 64: 24-35.

[28] Xu L J ,Wang J L. A heterogeneous Fenton-like system with nanoparticulate zero-valent iron for removal of 4-chloro-3-methyl phenol. J Hazard Mater, 2011, 186(1): 256-264.

[29] Zha S x, Cheng Y, Gao Y, Chen Z L, Megharaj M, Naidu R. Nanoscale zero-valent iron as a catalyst for heterogeneous Fenton oxidation of amoxicillin. Chem Eng J, 2014, 255: 141-148.

[30] Yin X C, Liu W, Ni J R. Removal of coexisting Cr(VI)and 4-chlorophenol through reduction and Fenton reaction in a single system. Chem Eng J, 2014, 248: 89-97.

[31] Lu L R, Ai Z H, Li J P, Zheng Z, Li Q, Zhang L Z. Synthesis and characterization of Fe–Fe_2O_3 core–shell nanowires and nanonecklaces. Cryst Growth Desi, 2007, 7(2): 459-464.

[32] Shi J G, Ai Z H, Zhang L Z. Fe@Fe_2O_3 core-shell nanowires enhanced Fenton oxidation by accelerating the Fe(III) / Fe(II) cycles. Water Res, 2014, 59: 145-153.

[33] Liu W, Wang Y Y, Ai Z H ,Zhang L Z. Hydrothermal synthesis of FeS_2 as a high-efficiency Fenton reagent to degrade alachlor via superoxide-mediated Fe(II)/Fe(III)cycle. Acs Appl Mater Inter, 2015, 7(51): 28534-28544.

[34] Huang X P, Hou X J, Zhao J C, Zhang L Z. Hematite facet confined ferrous ions as high efficient Fenton catalysts to degrade organic contaminants by lowering H_2O_2 decomposition energetic span. Appl Catal B-Environ, 2016, 181: 127-137.

[35] Ai Z H, Gao Z T, Zhang L Z, He W W, Yin J J. Core-shell structure dependent reactivity of Fe@Fe_2O_3 nanowires on aerobic degradation of 4-chlorophenol. Environ Sci Technol, 2013, 47(10): 5344-5352.

[36] Liu W, Ai Z H, Cao M H, Zhang L Z. Ferrous ions promoted aerobic simazine degradation with Fe@Fe_2O_3 core-shell nanowires. Appl Catal B-Environ, 2014, 150: 1-11.

[37] Hou X J, Huang X P, Ai Z H, Zhao J C, Zhang L Z. Ascorbic acid/Fe@Fe_2O_3: A highly efficient combined Fenton reagent to remove organic contaminants. J Hazard Mater, 2016, 310: 170-178.

[38] Huang X P, Hou X J, Song F H, Zhao J C, Zhang L Z. Ascorbate induced facet dependent reductive dissolution of hematite nanocrystals. J Phys Chem C, 2017, 121(2): 1113-1121.

[39] Huang X P, Hou X J, Jia F L, Song F H, Zhao J C, Zhang L Z. Ascorbate-promoted surface iron cycle for efficient heterogeneous Fenton alachlor degradation with hematite nanocrystals. Acs Appl Mater Inter, 2017, 9(10): 8751-8758.

[40] Hou X J, Huang X P, Jia F L, Ai Z H, Zhao J C, Zhang L Z. Hydroxylamine promoted goethite surface Fenton degradation of organic pollutants. Environ Sci Technol, 2017, 51(9): 5118-5126.

[41] Wang L, Wang F, Li P n, Zhang L z. Ferrous–tetrapolyphosphate complex induced dioxygen activation for toxic organic pollutants degradation. Sep Purif Technol, 2013, 120: 148-155.

[42] Noradoun C, Engelmann M D, McLaughlin M, Hutcheson R, Breen K, Paszczynski A, Cheng I F. Destruction of chlorinated phenols by dioxygen activation under aqueous room temperature and pressure conditions. Ind Eng Chem Resr, 2003, 42(21): 5024-5030.

[43] Huang Q, Cao M H, Ai Z H, Zhang L Z. Reactive oxygen species dependent degradation pathway of 4-chlorophenol with Fe@Fe_2O_3 core-shell nanowires. Appl Catal B-Environ, 2015, 162: 319-326.

[44] Hou X J, Shen W J, Huang X P, Ai Z H, Zhang L Z. Ascorbic acid enhanced activation of oxygen by ferrous iron: A case of aerobic degradation of rhodamine B. J Hazard Mater, 2016, 308: 67-74.

[45] Wang L, Cao M H, Ai Z H, Zhang L Z. Dramatically enhanced aerobic atrazine degradation with Fe@Fe_2O_3 core–shell nanowires by tetrapolyphosphate. Environ Sci Technol, 2014, 48(6): 3354-3362.

[46] Kim H H, Lee H, Kim H E, Seo J, Hong S W, Lee J Y, Lee C. Polyphosphate-enhanced production of reactive oxidants by nanoparticulate zero-valent iron and ferrous ion in the presence of oxygen: Yield and nature of oxidants. Water Res, 2015,

86: 66-73.

[47] Kim H H, Kim M S, Kim H E, Lee H J, Jang M H, Choi J, Hwang Y, Lee C. Nanoparticulate zero-valent iron coupled with polyphosphate: the sequential redox treatment of organic compounds and its stability and bacterial toxicity. Environ Sci-Nano, 2017, 4(2): 396-405.

[48] Ai Z H, Mei T, Liu J, Li J P, Jia F L, Zhang L Z,Qiu J R. Fe@Fe_2O_3 core−shell nanowires as an iron reagent. 3. Their combination with CNTs as an effective oxygen-fed gas diffusion electrode in a neutral electro-Fenton system. J Phys Chem C, 2007, 111(40): 14799-14803.

[49] Ding X, Wang S Y, Shen W Q, Mu Y, Wang L, Chen H, Zhang L Z. Fe@Fe_2O_3 promoted electrochemical mineralization of atrazine via a triazinon ring opening mechanism. Water Res, 2017, 112: 9-18.

[50] Matta R, Hanna K, Kone T, Chiron S. Oxidation of 2,4,6-trinitrotoluene in the presence of different iron-bearing minerals at neutral pH. Chem Eng J, 2008, 144(3): 453-458.

[51] Aredes S, Klein B, Pawlik M. The removal of arsenic from water using natural iron oxide minerals. J Clean Prod, 2012, 29-30: 208-213.

[52] Chun J, Lee H, Lee S H, Hong S W, Lee J, Lee C, Lee J. Magnetite/mesocellular carbon foam as a magnetically recoverable fenton catalyst for removal of phenol and arsenic. Chemosphere, 2012, 89(10): 1230-1237.

[53] Kwan W P, Voelker B M. Rates of hydroxyl radical generation and organic compound oxidation in mineral-catalyzed Fenton-like systems. Environ Sci Technol, 2003, 37(6): 1150-1158.

[54] Guimaraes I R, Oliveira L C A, Queiroz P F, Ramalho T C, Pereira M, Fabris J D, Ardisson J D. Modified goethites as catalyst for oxidation of quinoline: Evidence of heterogeneous Fenton process. Appl Catal A-Gen, 2008, 347(1): 89-93.

[55] Chen L W, Ma J, Li X C, Zhang J, Fang J Y, Guan Y H, Xie P C. Strong enhancement on Fenton oxidation by addition of hydroxylamine to accelerate the ferric and ferrous iron cycles. Environ Sci Technol, 2011, 45(9): 3925-3930.

[56] He H Q, Zhong Y H, Liang X L, Tan W, Zhu J X, Yan Wang C. Natural magnetite: an efficient catalyst for the degradation of organic contaminant. Scientific Reports, 2015, 5: 10139.

[57] Liang X L, Zhong Y H, He H P, Yuan P, Zhu J X, Zhu S Y, Jiang Z. The application of chromium substituted magnetite as heterogeneous Fenton catalyst for the degradation of aqueous cationic and anionic dyes. Chem Eng J, 2012, 191: 177-184.

[58] Zhong Y H, Liang X L, Zhong Y, Zhu J X, Zhu S Y, Yuan P, He H P, Zhang J. Heterogeneous UV/Fenton degradation of TBBPA catalyzed by titanomagnetite: Catalyst characterization, performance and degradation products. Water Res, 2012, 46(15): 4633-4644.

[59] Liang X L, He Z S, Wei G L, Liu P, Zhong Y H, Tan W, Du P X, Zhu J X, He H P， Zhang J. The distinct effects of Mn substitution on the reactivity of magnetite in heterogeneous Fenton reaction and Pb(II) adsorption. J Colloid Interf Scie, 2014, 426: 181-189.

[60] Gu L, Zhu N W, Zhou P. Preparation of sludge derived magnetic porous carbon and their application in Fenton-like degradation of 1-diazo-2-naphthol-4-sulfonic acid. Bioresource Technol, 2012, 118: 638-642.

[61] Gu L, Zhu N W, Guo H Q, Huang S Q, Lou Z Y ,Yuan H P. Adsorption and Fenton-like degradation of naphthalene dye intermediate on sewage sludge derived porous carbon. J Hazard Mater, 2013, 246-247: 145-53.

[62] Li H, Shang J, Yang Z P, Shen W J, Ai Z H ,Zhang L Z. Oxygen vacancy associated surface Fenton chemistry: surface structure dependent hydroxyl radicals generation and substrate dependent reactivity. Environ Sci Technol, 2017, 51(10): 5685-5694.

[63] Ishikawa T, Sorimachi A, Arae H, Sahoo S K, Janik M, Hosoda M, Tokonami S. Simultaneous sampling of indoor and outdoor airborne radioactivity after the Fukushima Daiichi Nuclear power plant accident. Environmental Science & Technology, 2014, 48(4): 2430-2435.

[64] Wu F, Pu N, Ye G, Sun T, Wang Z, Song Y, Wang W, Huo X, Lu Y, Chen J. Performance and mechanism of uranium adsorption from seawater to poly(dopamine)-inspired sorbents. Environ Sci Technol, 2017, 51(8): 4606-4614.

[65] Dvorzhak A, Puras C, Montero M, Mora J C. Spanish experience on modeling of environmental radioactive contamination due to Fukushima Daiichi NPP accident using JRODOS. Environ Sci Technol, 2012, 46(21): 11887-11895.

[66] Zaunbrecher L K, Elliott W C, Wampler J M, Perdrial N, Kaplan D I. Enrichment of cesium and rubidium in weathered micaceous materials at the Savannah River Site, South Carolina. Environ Sci Technol, 2015, 49(7): 4226-4234.

[67] Torres L, Yadav O P, Khan E. Perceived risks of produced water management and naturally occurring radioactive material

content in North Dakota. J Environ Manage, 2017, 196: 56-62.
[68] Jiang L H, Liu Y G, Liu S B, Zeng G M, Hu X J, Hu X, Guo Z, Tan X F, Wang L L ,Wu Z B. Adsorption of estrogen contaminants by graphene nanomaterials under natural organic matter preloading: Comparison to carbon nanotube, biochar, and activated carbon. Environ Sci Technol, 2017, 51(11): 6352-6359.
[69] Apul O G, Karanfil T. Adsorption of synthetic organic contaminants by carbon nanotubes: A critical review. Water Res, 2015, 68: 34-55.
[70] Liu W, Ni J R, Yin X C. Synergy of photocatalysis and adsorption for simultaneous removal of Cr(VI)and Cr(III)with TiO_2 and titanate nanotubes. Water Res, 2014, 53: 12-25.
[71] Wang X, Qin Y, Zhu L, Tang H. Nitrogen-doped reduced graphene oxide as a bifunctional material for removing bisphenols: Synergistic effect between adsorption and catalysis. Environ Sci Technol, 2015, 49(11): 6855-6864.
[72] Wang H, Lin K-Y, Jing B, Krylova G, Sigmon G E, McGinn P, Zhu Y, Na C. Removal of oil droplets from contaminated water using magnetic carbon nanotubes. Water Res, 2013, 47(12): 4198-4205.
[73] Wang X, Yu S, Jin J, Wang H, Alharbi N S, Alsaedi A, Hayat T, Wang X. Application of graphene oxides and graphene oxide-based nanomaterials in radionuclide removal from aqueous solutions. Sci Bull, 2016, 61(20): 1583-1593.
[74] Yu S J, Wang X X, Yang S T, Sheng G D, Alsaedi A, Hayat T, Wang X. Interaction of radionuclides with natural and manmade materials using XAFS technique. Sci China-Chem, 2017, 60(2): 170-187.
[75] Sun Y, Wang Q, Chen C, Tan X, Wang X. Interaction between Eu(III)and graphene oxide nanosheets investigated by batch and extended X-ray absorption fine structure spectroscopy and by modeling eechniques. Environ Sci Technol, 2012, 46(11): 6020-6027.
[76] Sun Y, Yang S, Chen Y, Ding C, Cheng W, Wang X. Adsorption and desorption of U(VI)on functionalized graphene oxides: A combined experimental and theoretical study. Environ Sci Technol, 2015, 49(7): 4255-4262.
[77] Wang X, Yang S, Shi W, Li J, Hayat T, Wang X. Different interaction mechanisms of Eu(III)and Am-243(III)with carbon nanotubes studied by batch, spectroscopy technique and theoretical calculation. Environ Sci Technol, 2015, 49(19): 11721-11728.
[78] Sun Y, Wu Z Y, Wang X, Ding C, Cheng W, Yu S-H, Wang X. Macroscopic and ,microscopic ,investigation of U(VI) and Eu(III) adsorption on carbonaceous nanofibers. Environ Sci Technol, 2016, 50(8): 4459-4467.
[79] Ding C C, Cheng W C, Sun Y B, Wang X K. Effects of Bacillus subtilis on the reduction of U(VI) by nano-Fe-0. Geochim Cosmoshim Ac, 2015, 165: 86-107.
[80] Sun Y, Ding C, Cheng W, Wang X. Simultaneous adsorption and reduction of U(VI) on reduced graphene oxide-supported nanoscale zerovalent iron. J Hazard Mater, 2014, 280: 399-408.
[81] Hinds B J, Chopra N, Rantell T, Andrews R, Gavalas V, Bachas L G. Aligned multiwalled carbon nanotube membranes. Science, 2004, 303: 62-65.
[82] Majumder M, Chopra N, Andrews R, Hinds B J. Nanoscale hydrodynamics: Enhanced flow in carbon nanotubes. Nature, 2005, 438: 44.
[83] Holt J K, Park H G, Wang Y M, Stadermann M, Artyukhin A B, Grigoropoulos C P, Noy A, Bakajin O. Fast mass transport through sub-2-nanometer carbon nanotubes. Science, 2006, 312: 1034-1037.
[84] Berezhkovskii A. Single-file transport of water molecules through a carbon nanotube. Phys Rev Lett, 2002, 89(6): 064503.
[85] Joseph S, Aluru N R. Why are carbon nanotubes fast transporters of water? Nano Lett, 2008, 8(2): 452-458.
[86] Gao G D, Vecitis C D. Doped carbon nanotube networks for electrochemical filtration of aqueous phenol: electrolyte precipitation and phenol polymerization. ACS Appl Mater Interfaces, 2012, 4(3):1478-1489.
[87] Vecitis C D, Gao G D, Liu H. Electrochemical carbon nanotube filter for adsorption, desorption, and oxidation of aqueous dyes and anions. J Phys Chem C, 2011, 115: 3621-3629.
[88] Wei G L, Chen S, Fan X F, Quan X, Yu H T. Carbon nanotube hollow fiber membranes: High-throughput fabrication, structural control and electrochemically improved selectivity. J Membr Sci, 2015, 493: 97-105.
[89] Wei G L, Quan X, Chen S, Fan X F, Yu H T, Zhao H M. Voltage-gated transport of nanoparticles across free-standing all-carbon-nanotube-based hollow-fiber membranes. ACS Appl Mater Interfaces, 2015, 7: 14620-14627.
[90] Fan X F, Zhao H M, Liu Y M, Quan X, Yu H T, Chen S. Enhanced permeability, selectivity, and antifouling ability of CNTs/Al_2O_3 membrane under electrochemical assistance. Environ Sci Technol, 2015, 49: 2293-2300.

[91] Nair R R, Wu H A, Jayaram P N, Grigorieva I V, Geim A K. Unimpeded permeation of water through helium-leak-tight graphene-based membranes. Science, 2012, 335: 442-444.

[92] Joshi R K, Carbone P, Wang F C, Kravets V G, Su Y, Grigorieva I V, Wu H A, Geim A K, Nair R R. Precise and ultrafast molecular sieving through graphene oxide membranes. Science, 2014, 343: 752-754.

[93] Huang L, Li Y R, Zhou Q Q, Yuan W J, Shi G Q. Graphene oxide membranes with tunable semipermeability in organic solvents. Adv Mater, 2015, 27: 3797–3802.

[94] Mi B X. Graphene oxide membranes for ionic and molecular sieving. Science, 2014, 343: 740-742.

[95] Hu M, Mi B X. Enabling graphene oxide nanosheets as water separation membranes. Environ Sci Technol, 2013, 47: 3715-3723.

[96] Han Y, Xu Z, Gao C. Ultrathin graphene nanofiltration membrane for water purification. Adv Funct Mater, 2013, 23: 3693-3700.

[97] Cohen-Tanugi D, Grossman J C. Water desalination across nanoporous graphene. Nano Lett, 2012, 12: 3602-3608.

[98] Surwade S P, Smirnov S N, Vlassiouk I V, Unocic R R, Veith G M, Dai S, Mahurin S M. Water desalination using nanoporous single-layer graphene. Nature Nanotech, 2015, 10: 459-464.

[99] Nair R R, Wu H A, Jayaram P N, Grigorieva I V, Geim A K. Unimpeded permeation of water through helium-leak-tight graphene-based membranes. Science, 2012, 335: 442-444.

[100] Joshi R K, Carbone P, Wang F C, Kravets V G, Su Y, Grigorieva I V, Wu H A, Geim A K, Nair R R. Precise and ultrafast molecular sieving through graphene oxide membranes. Science, 2014, 343: 752-754.

[101] Qu X L, Alvarez P J J, Li Q L. Applications of nanotechnology in water and wastewater treatment, Water Res., 2013，47:3931-3946.

[102] Tesh S J, Scott T B. Nano-composites for water remediation: A Review, Adv. Mater., 2014,26:6056-6068.

[103] Zhang Q R, Du Q, Hua M T , Jiao T F, Gao F B ,Pan B C. Sorption enhancement of lead ions from water by surface charged polystyrene-supported nano-zirconium oxide composites.Environ. Sci. technol., 2013,47:6536-6544.

[104] Behrens S H, Christ D I, Emmerzael R, Schurtenberger P, Borkovec M. Charging and aggregation properties of carboxyl latex particles: Experiments versus DLVO Theory, Langmuir, 2000,16:2566-2575.

[105] Wang J, Zhang S J, Pan B C, Zhang W M, Lv L. Hydrous Ferric Oxide-Resin Nanocomposites of Tunable Structure for Arsenite Removal: Effect of the Host Pore Structure. J. Hazard. Mater., 2011,198:241-246.

[106] Li H C, Shan C, Zhang Y Y, Cai J G, Zhang W M, Pan B C. Arsenate Adsorption by Hydrous Ferric Oxide Nanoparticles Embedded in Cross-linked Anion Exchanger: Effect of the Host Pore Structure.ACS Appl. Mater. Int., 2016,8:3012-3020.

[107] Nie G Z, Pan B C, Zhang S J, Pan B J. Surface chemistry of nanosized hydrated ferric oxide encapsulated inside porous polymer: Modeling and experimental studies. J. Phy. Chem. C, 2013,117:6201-6209.

[108] Pan B C, Han F C, Nie G Z, Wu B, He K, Lv L. New strategy to enhance phosphate removal from water by hydrous manganese oxide. Environ. Sci. Technol., 2014,48:5101-5107.

[109] Pan B C, Xu J S, Wu B, Li Z G, Liu X T. Enhanced Removal of Fluoride by Polystyrene Anion Exchanger Supported Hydrous Zirconium Oxide Nanoparticles. Environ. Sci. Technol., 2013,47:9347-9354.

[110] Zhang Q R, Pan B C, Zhang S J, Wang J, Zhang W M, Lv L. New Insights into nanocomposite adsorbents for water treatment. A case study of polystyrene-supported zirconium phosphate nanoparticles for lead removal. J. Nanoparticle Res., 2011,13:5355-5364.

[111] Hua M, Jiang Y N, Wu B, Pan B C, Zhao X, Zhang Q X. Fabrication of a new hydrous Zr(IV)oxide-based nanocomposite for enhanced Pb(II) and Cd(II) removal from waters. ACS Appl. Mater. Int.,2013,5:12135-12142.

[112] Du Q, Zhang S J, Pan B C, Lv L, Zhang W M, Zhang Q X. Bifunctional resin-ZVI composites for effective removal of arsenite through simultaneous adsorption and oxidation. Water Res., 2013,47:6064-6074.

[113] Li X, He K, Pan B C, Zhang S J, Lv L, Zhang W M. Efficient As(III)removal by macroporous anion exchanger- supported Fe–Mn binary oxide: Behavior and mechanism. Chem. Eng. J., 2012,193-194:131-138.

[114] Zhang X L, Wu M F, Dong H, Li H C, Pan B C. Simultaneous Oxidation and Sequestration of As(III)from Water by Using Redox Polymer-Based Fe(III) Oxide Nanocomposite. Environ. Sci. Technol., 2017,51:6326-6334.

[115] Zhang Y Y, Pan B C, Shan C, Gao X. Enhanced Phosphate Removal by Nanosized Hydrated La(III)Oxide Confined in Cross-linked Polystyrene Networks. Environ. Sci. Technol., 2016,50:1447-1454.

[116] Cai J G, Zhang Y Y, Pan B C, Zhang W M, Zhang Q X. Efficient defluoridation of water using reusable nanocrystalline layered doubl e hydroxides impregnated polystyrene anion exchanger. Water Res., 2016,102:109-116.
[117] 郭如新. $Mg(OH)_2$在工业废水处理中的应用. 工业水处理, 2000, 20: 1-4.
[118] Liu W, Huang F, Wang Y, Zou T, Zheng J, Lin Z. Recycling $Mg(OH)_2$ nanoadsorbent during treating the low concentration of CrVI. Environmental Science & Technology, 2011, 45(5):1955-1961.
[119] Lv X Y, Chen Z, Wang Y J, Huang F, Lin Z. Use of High-Pressure CO_2 for Concentrating Cr^{VI} from Electroplating Wastewater by Mg–Al Layered Double Hydroxide. ACS applied materials & interfaces 2013, 5(21):11271-11275.
[120] Pan X H, Wang Y H, Chen Z, Pan D M, Cheng Y J, Liu Z J, Lin Z, Guan X. Investigation of antibacterial activity and related mechanism of a series of Nano-$Mg(OH)_2$. ACS applied Materials & Interfaces, 2013, 5:1137–1142.
[121] Li C R, Zhuang Z Y, Huang F, Wu Z C, Hong Y P, Lin Z. Recycling Rare Earth Elements from Industrial Wastewater with Flowerlike nano-$Mg(OH)_2$. ACS Applied Materials & Interfaces, 2013, 5(19):9719-9725.
[122] Manos M J, Kanatzidis M G. Layered metal sulfides capture uranium from seawater. Journal of the American Chemical Society, 2012, 134: 16441-16446.
[123] Cao Q, Huang F, Zhuang Z Y, Lin Z .A. study of the potential application of nano-$Mg(OH)_2$ in adsorbing low concentrations of uranyl tricarbonate from water. Nanoscale, 2012, 4: 2423–2430.
[124] Chen Z, Zhuang Z Y, Cao Q, Pan X H ,Guan X, Lin Z. Adsorption-induced Crystallization of U-rich Nanocrystals on Nano-$Mg(OH)_2$ and the Aqueous Uranyl Enrichment. ACS Applied Materials & Interfaces, 2014, 6: 1301-1305.
[125] Zhuang Z, Ou X, Li J, Zhou Y, Zhang Z, Dong S, Lin Z. Interfacial engineering improved the selective extraction of uranyl from saline water by nano-$Mg(OH)_2$ and the underlying mechanism. ACS Sustainable Chemistry & Engineering, 2016, 4: 801-809.
[126] Keller A A, Wang H, Zhou D, Lenihan H S, Cherr G, Cardinale B J, Miller R, Ji Z. Stability and Aggregation of Metal Oxide Nanoparticles in Natural Aqueous Matrices. Environmental Science & Technology, 2010, 44:1962-1967.
[127] French R A, Jacobson A R, Kim B, Isley S L, Penn R L, Baveye P C. Influence of Ionic Strength, pH, and Cation Valence on Aggregation Kinetics of Titanium Dioxide Nanoparticles. Environmental Science & Technology ,2009, 43:1354-1359.
[128] Metreveli G, Frombold B, Seitz F, Gruen A, Philippe A, Rosenfeldt R R, Bundschuh M, Schulz R, Manz W, Schaumann G E. Impact of chemical composition of ecotoxicological test media on the stability and aggregation status of silver nanoparticles. Environmental Science-Nano ,2016, 3:418-433.
[129] Li X, Lenhari J J, Walker H W. Aggregation kinetics and dissolution of coated silver nanoparticles. Langmuir ,2012, 28: 1095-1104.
[130] Huynh K A, Chen K L. Aggregation kinetics of citrate and polyvinylpyrrolidone coated silver nanoparticles in monovalent and divalent electrolyte solutions Environ. Sci. Technol., 2011, 45:5564-5571.
[131] Liu J, Legros S, Von der Kammer F, Hofmann T. Natural Organic Matter Concentration and Hydrochemistry Influence Aggregation Kinetics of Functionalized Engineered Nanoparticles. Environmental Science & Technology, 2013, 47:4113-4120.
[132] Surette M C, Nason J A. Effects of surface coating character and interactions with natural organic matter on the colloidal stability of gold nanoparticles. Environmental Science-Nano, 2016, 3:1144-1152.
[133] Shen M H, Yin Y G, Booth A, Liu J F. Effects of molecular weight-dependent physicochemical heterogeneity of natural organic matter on the aggregation of fullerene nanoparticles in mono- and di-valent electrolyte solutions. Water Research, 2015, 71: 11-20.
[134] Tejamaya M, Roemer I, Merrifield R C, Lead J R. Stability of citrate, PVP, and PEG coated silver nanoparticles in ecotoxicology media. Environmental Science & Technology, 2012, 46:7011-7017.
[135] Yin Y, Shen M, Tan Z, Yu S, Liu J, Jiang G. Particle Coating-Dependent Interaction of Molecular Weight Fractionated Natural Organic Matter: Impacts on the Aggregation of Silver Nanoparticles. Environmental Science & Technology, 2015, 49: 6581-6589.
[136] Baalousha M, Sikder M, Prasad A, Lead J, Merrifield R, Chandler G T. The concentration-dependent behaviour of nanoparticles. Environmental Chemistry, 2016, 13: 1-3.
[137] Merrifield R C, Stephan C, Lead J. Determining the Concentration Dependent Transformations of Ag Nanoparticles in Complex Media: Using SP-ICP-MS and Au@Ag Core Shell Nanoparticles as Tracers. Environmental Science &

Technology 2017, 51: 3206-3213.

[138] Zhang W, Yao Y, Li K G, Huang Y, Chen Y S. Influence of dissolved oxygen on aggregation kinetics of citrate-coated silver nanoparticles. Environmental Pollution, 2011, 159:3757-3762.

[139] Cheng Y W, Yin L Y, Lin S H, Wiesner M, Bernhardt E, Liu J. Toxicity reduction of polymer-stabilized silver nanoparticles by sunlight. J. Phys. Chem. C, 2011, 115:4425-4432.

[140] Chowdhury I, Hou W C, Goodwin D, Henderson M, Zepp R G, Bouchard D. Sunlight affects aggregation and deposition of graphene oxide in the aquatic environment. Water Research ,2015, 78: 37-46.

[141] Zhang Z B , Liu L S, Liu C Y, Cai W J. Studies on the sea surface microlayer - II. The layer of sudden change of physical and chemical properties. Journal of Colloid and Interface Science, 2003, 264:148-159.

[142] Guo X, Yin Y, Tan Z, Zhang Z, Chen Y, Liu J. Significant Enrichment of Engineered Nanoparticles in Water Surface Microlayer. Environmental Science & Technology Letters, 2016, 3: 381-385.

[143] Dwivedi A D, Dubey S P, Sillanpaa M, Kwon Y N, Lee C, Varma R S. Fate of engineered nanoparticles: Implications in the environment. Coord. Chem. Rev., 2015, 287:64-78.

[144] Dwivedi A D, Dubey S P, Sillanpaa M, Kwon Y N, Lee C, Varma R S. Fate of engineered nanoparticles: Implications in the environment. Coord. Chem. Rev., 2015, 287:64-78.

[145] Ma R, Levard C, Marinakos S M, Cheng Y W, Liu J, Michel F M, Brown G E, Lowry G V. Size-controlled dissolution of organic-coated silver nanoparticles. Environmental Science & Technology, 2012, 46: 752-759.

[146] Zhang W, Yao Y, Sullivan N, Chen Y S. Modeling the primary size effects of citrate-coated silver nanoparticles on their ion release kinetics. Environ. Sci. Technol., 2011, 45: 4422-4428.

[147] Yu S, Yin Y, Chao J, Shen M, Liu J. Highly dynamic PVP-coated silver nanoparticles in aquatic environments: Chemical and morphology change induced by oxidation of Ag^0 and reduction of Ag+. Environmental Science & Technology, 2014, 48:403-411.

[148] Grillet N, Manchon D, Cottancin E, Bertorelle F, Bonnet C, Broyer M, Lerme J, Pellarin M. Photo-oxidation of individual silver nanoparticles: A real-time tracking of optical and morphological changes. J. Phys. Chem. C, 2013, 117: 2274-2282.

[149] Yin Y, Yang X, Zhou X, Wang W, Yu S, Liu J, Jiang G. Water chemistry controlled aggregation and photo-transformation of silver nanoparticles in environmental waters. Journal of environmental sciences ,2015, 34: 116-125.

[150] Yin Y G, Liu J F, Jiang G B. Sunlight-induced reduction of ionic Ag and Au to metallic nanoparticles by dissolved organic matter. ACS Nano, 2012, 6:7910–7919.

[151] Yu S, Yin Y, Zhou X, Dong L, Liu J. Transformation kinetics of silver nanoparticles and silver ions in aquatic environments revealed by double stable isotope labeling. Environ. Sci.: Nano, 2016, 3: 883-893.

[152] Lombi E, Donner E, Taheri S, Tavakkoli E, Jaemting A K, McClure S, Naidu R, Miller B W, Scheckel K G, Vasilev K. Transformation of four silver/silver chloride nanoparticles during anaerobic treatment of wastewater and post-processing of sewage sludge. Environmental Pollution, 2013, 176: 193-197.

[153] Kaegi R, Voegelin A, Ort C, Sinnet B, Thalmann B, Krismer J, Hagendorfer H, Elumelu M, Mueller E. Fate and transformation of silver nanoparticles in urban wastewater systems. Water Research, 2013, 47:3866-3877.

[154] Ma R, Levard C, Judy J D, Unrine J M, Durenkamp M, Martin M, Jefferson B, Lowry G V. Fate of Zinc Oxide and Silver Nanoparticles in a Pilot Wastewater Treatment Plant and in Processed Biosolids. Environmental Science & Technology, 2014, 48: 104-112.

[155] Levard C, Reinsch B C, Michel F M, Oumahi C, Lowry G V, Brown G E. Sulfidation processes of pvp-coated silver nanoparticles in aqueous solution: Impact on dissolution rate. Environ. Sci. Technol., 2011, 45: 5260-5266.

[156] Liu J Y, Pennell K G, Hurt R H. Kinetics and mechanisms of nanosilver oxysulfidation. Environ. Sci. Technol., 2011, 45:7345-7353.

[157] Reinsch B C, Levard C, Li Z, Ma R, Wise A, Gregory K B, Brown G E, Lowry Jr G V. Sulfidation of silver nanoparticles decreases Escherichia coli growth inhibition. Environmental Science & Technology, 2012, 46: 6992-7000.

[158] Wang Z, Bussche A von dem, Kabadi P K, Kane A B, Hurt R H. Biological and Environmental Transformations of Copper-Based Nanomaterials. Acs Nano, 2013, 7:8715-8727.

[159] Ma R, Levard C, Michel F M, Brown G E, Lowry Jr G V. Sulfidation mechanism for zinc oxide nanoparticles and the effect of sulfidation on their solubility. Environmental science & technology, 2013, 47:2527-2534.

[160] Li L, Hu L, Zhou Q, Huang C, Wang Y, Sun C, Jiang G. Sulfidation as a Natural Antidote to Metallic Nanoparticles Is Overestimated: CuO Sulfidation Yields CuS Nanoparticles with Increased Toxicity in Medaka(Oryzias latipes)Embryos. Environmental Science & Technology, 2015, 49: 2486-2495.

[161] Thalmann B, Voegelin A, Sinnet B, Morgenroth E, Kaegi R. Sulfidation Kinetics of Silver Nanoparticles Reacted with Metal Sulfides. Environmental Science & Technology , 2014, 48:4885-4892.

[162] Thalmann B, Voegelin A, von Gunten U, Behra R, Morgenroth E, Kaegi R. Effect of Ozone Treatment on Nano-Sized Silver Sulfide in Wastewater Effluent. Environmental Science & Technology, 2015, 49: 10911-10919.

[163] Impellitteri C A, Harmon S, Silva R G, Miller B W, Scheckel K G, Luxton T P, Schupp D, Panguluri S. Transformation of silver nanoparticles in fresh, aged, and incinerated biosolids. Water Research, 2013, 47: 3878-3886.

[164] Yin Y, Xu W, Tan Z, Li Y, Wang W, Guo X, Yu S, Liu J, Jiang G. Photo- and thermo-chemical transformation of AgCl and Ag2S in environmental matrices and its implication. Environmental Pollution, 2017, 220: 955-962.

[165] Meier C, Voegelin A, Pradas del Real A, Sarret G, Mueller C R, Kaegi R. Transformation of Silver Nanoparticles in Sewage Sludge during Incineration. Environmental Science & Technology, 2016, 50:3503-3510.

[166] Li L, Wang Y, Liu Q, Jiang G. Rethinking Stability of Silver Sulfide Nanoparticles(Ag2S-NPs)in the Aquatic Environment: Photoinduced Transformation of Ag2S-NPs in the Presence of Fe(III). Environmental Science & Technology, 2016, 50: 188-196.

[167] Li L, Zhou Q, Geng F, Wang Y, Jiang G. Formation of Nanosilver from Silver Sulfide Nanoparticles in Natural Waters by Photoinduced Fe(II, III)Redox Cycling. Environmental Science & Technology, 2016, 50: 13342-13350.

[168] Levard C, Hotze E M, Lowry G V, Brown G E. Environmental transformations of silver nanoparticles: Impact on stability and toxicity. Environmental Science & Technology, 2012, 46: 6900-6914.

[169] Levard C, Mitra S, Yang T, Jew A D, Badireddy A R, Lowry G V, Brown G E, Jr. Effect of chloride on the dissolution rate of silver nanoparticles and toxicity to E. coli. Environmental Science & Technology, 2013, 47: 5738-5745.

[170] Chambers B A, Afrooz A R M N, Bae S, Aich N, Katz L, Saleh N B, Kirisits M J. Effects of Chloride and Ionic Strength on Physical Morphology, Dissolution, and Bacterial Toxicity of Silver Nanoparticles. Environmental Science & Technology, 2014, 48:761-769.

[171] Li X A, Lenhart J J, Walker H W. Dissolution-Accompanied Aggregation Kinetics of Silver Nanoparticles. Langmuir ,2010, 26:16690-16698.

[172] Garg S, Rong H, Miller C J, Waite T D. Oxidative Dissolution of Silver Nanoparticles by Chlorine: Implications to Silver Nanoparticle Fate and Toxicity. Environmental Science & Technology, 2016, 50: 3890-3896.

[173] Impellitteri C A, Tolaymat T M, Scheckel K G. The speciation of silver nanoparticles in antimicrobial fabric before and after exposure to a hypochlorite/detergent solution. Journal of Environmental Quality, 2009, 38: 1528-1530.

[174] Mitrano D M, Rimmele E, Wichser A, Erni R, Height M, Nowack B. Presence of Nanoparticles in Wash Water from Conventional Silver and Nano-silver Textiles. ACS Nano, 2014, 8:7208-7219.

[175] Kaegi R, Voegelin A, Sinnet B, Zuleeg S, Siegrist H, Burkhardt M. Transformation of AgCl nanoparticles in a sewer system - A field study. Science of the Total Environment, 2015, 535: 20-27.

[176] Lu D W, Liu Q, Zhang T Y, Cai Y, Yin Y G, Jiang G B. Stable silver isotope fractionation in the natural transformation process of silver nanoparticles. Nat Nanotechnol ,2016, 11:682.

[177] Vanhaecke F.Nanoparticle behaviour dissected. Nat Nanotechnol, 2016, 11:656-658.

[178] Lowry G V, Gregory K B, Apte S C, Lead J R. Transformations of Nanomaterials in the Environment. Environmental Science & Technology, 2012, 46: 6893-6899.

[179] Li Y, Yang N, Du T, Xia T, Zhang C, Chen W. Chloramination of graphene oxide significantly affects its transport properties in saturated porous media. NanoImpact, 2016, 3-4: 90-95.

[180] Qi Z C, Zhang L L, Chen W. Transport of graphene oxide nanoparticles in saturated sandy soil. Environ Sci-Proc Imp, 2014, 16: 2268-2277.

[181] Qi Z C, Zhang L L, Wang F, Hou L, Chen W. Factors Controlling Transport of Graphene Oxide Nanoparticles in Saturated Sand Columns. Environ Toxicol Chem, 2014, 33:998-1004.

[182] Lanphere J D, Luth C J, Walker S L. Effects of Solution Chemistry on the Transport of Graphene Oxide in Saturated Porous Media. Environmental Science & Technology, 2013, 47: 4255-4261.

[183] Franchi A, O'Melia C R. Effects of natural organic matter and solution chemistry on the deposition and reentrainment of colloids in porous media. Environmental Science & Technology, 2003, 37: 1122-1129.

[184] Tufenkji N, Elimelech M. Deviation from the classical colloid filtration theory in the presence of repulsive DLVO interactions. Langmuir ,2004, 20: 10818-10828.

[185] Bradford S A, Simunek J, Bettahar M, van Genuchten M T, Yates S R.Significance of straining in colloid deposition: Evidence and implications. Water Resour Res, 2006, 42.

[186] Bradford S A, Torkzaban S, Walker S L. Coupling of physical and chemical mechanisms of colloid straining in saturated porous media. Water Research, 2007, 41: 3012-3024.

[187] Chen W, Duan L, Zhu D Q. Adsorption of polar and nonpolar organic chemicals to carbon nanotubes, Environmental Science & Technology, 2007, 41: 8295-8300.

[188] Chen W, Duan L, Wang L, Zhu D. Adsorption of Hydroxyl- and Amino-Substituted Aromatics to Carbon Nanotubes. Environ. Sci. Technol., 2008, 42: 6862-6868.

[189] Woods L M, Badescu S C, Reinecke T L. Adsorption of simple benzene derivatives on carbon nanotubes. Phys Rev B, 2007, 75.

[190] Chen J Y, Chen W, Zhu D. Adsorption of nonionic aromatic compounds to single-walled carbon nanotubes: Effects of aqueous solution chemistry. Environmental Science & Technology, 2008, 42: 7225-7230.

[191] Yang K, Xing B S. Desorption of polycyclic aromatic hydrocarbons from carbon nanomaterials in water. Environmental Pollution, 2007, 145: 529-537.

[192] Pan B, Lin D H, Mashayekhi H, Xing B S. Adsorption and hysteresis of bisphenol A and 17 alpha-ethinyl estradiol on carbon nanomaterials. Environmental Science & Technology, 2008, 42: 5480-5485.

[193] Wang F, Haftka J J H, Sinnige T L,Hermens J L M, Chen W. Adsorption of polar, nonpolar, and substituted aromatics to colloidal graphene oxide nanoparticles. Environmental Pollution, 2014, 186: 226-233.

[194] Cheng X K, Kan A T, Tomson M B. Naphthalene adsorption and desorption from Aqueous C-60 fullerene. J Chem Eng Data, 2004, 49:675-683.

[195] Zhang X, Kah M, Jonker M T O, Hofmann T. Dispersion State and Humic Acids Concentration-Dependent Sorption of Pyrene to Carbon Nanotubes. Environmental Science & Technology, 2012, 46: 7166-7173.

[196] Qi Z C, Hou L, Zhu D Q, Ji R, Chen W. Enhanced Transport of Phenanthrene and 1-Naphthol by Colloidal Graphene Oxide Nanoparticles in Saturated Soil. Environmental Science & Technology, 2014, 48: 10136-10144.

[197] Hou L, Zhu D Q, Wang X M, Wang L L, Zhang C D, Chen W. Adsorption of phenanthrene, 2-naphthol, and 1-naphthylamine to colloidal oxidized multiwalled carbon nanotubes: Effects of humic acid and surfactant modification. Environ Toxicol Chem, 2013, 32: 493-500.

[198] Hyung H, Fortner J D, Hughes J B, Kim J H. Natural organic matter stabilizes carbon nanotubes in the aqueous phase. Environmental Science & Technology, 2007, 41: 179-184.

[199] Zhou X Z, Shu L, Zhao H B, Guo X Y, Wang X L, Tao S, Xing B S. Suspending Multi-Walled Carbon Nanotubes by Humic Acids from a Peat Soil. Environmental Science & Technology, 2012, 46: 3891-3897.

[200] Hyung H, Kim J H. Natural organic matter(NOM) adsorption to multi-walled carbon nanotubes: Effect of NOM characteristics and water quality parameters. Environmental Science & Technology, 2008, 42: 4416-4421.

[201] Hadrup N, Lam H R. Oral toxicity of silver ions, silver nanoparticles and colloidal silver--a review. Regul Toxicol Pharmacol, 2014, 68:1-7.

[202] Liao H, Chung Y, Lai C, Lin M, Liou S. Sneezing and Allergic Dermatitis were Increased in Engineered Nanomaterial Handling Workers. Ind Health, 2014, 52: 199-215.

[203] Phillips J I, Green F Y, Davies J C, Murray J. Pulmonary and systemic toxicity following exposure to nickel nanoparticles. Am J Ind Med, 2010, 53: 763-767

[204] Likus W, Bajor G, Siemianowicz K. Nanosilver - does it have only one face?ActaBiochim Pol, 2013, 60: 495-501.

[205] Li M, Pokhrel S, Jin X, Madler L, Damoiseaux R, Hoek E M. Stability, bioavailability, and bacterial toxicity of ZnO and iron-doped ZnO nanoparticles in aquatic media. Environ Sci Technol, 2011, 45: 755-761.

[206] Croteau M N, Misra S K, Luoma S N, Valsami-Jones E. Silver bioaccumulation dynamics in a freshwater invertebrate after aqueous and dietary exposures to nanosized and ionic Ag. Environ Sci Technol, 2011, 45: 6600-6607.

[207] Park E, Yoon J, Choi K, Yi J, Park K. Induction of chronic inflammation in mice treated with titanium dioxide nanoparticles by intratracheal instillation. Toxicology, 2009, 260:37-46.

[208] Baron L, Gombault A, Fanny M, Villeret B, Savigny F, Guillou N, Panek C, Le Bert M, Lagente V, Rassendren F, Riteau N, Couillin I. The NLRP3 inflammasome is activated by nanoparticles through ATP, ADP and adenosine. Cell Death Dis, 2015, 6: e1629.

[209] Hussain S, Al-Nsour F, Rice A B, Marshburn J, Yingling B, Ji Z, Zink J I, Walker N J, Garantziotis S. Cerium dioxide nanoparticles induce apoptosis and autophagy in human peripheral blood monocytes. ACS Nano, 2012, 6:5820-5829.

[210] Cao Z, Fang Y, Lu Y, Qian F, Ma Q, He M, Pi H, Yu Z, Zhou Z. Exposure to nickel oxide nanoparticles induces pulmonary inflammation through NLRP3 inflammasome activation in rats. Int J Nanomedicine, 2016, 11:3331-3346.

[211] Pal R, Chakraborty B, Nath A, Singh L M, Ali M, Rahman D S, Ghosh S K, Basu A, Bhattacharya S, Baral R, Sengupta M. Noble metal nanoparticle-induced oxidative stress modulates tumor associated macrophages(TAMs) from an M2 to M1 phenotype: An in vitro approach. IntImmunopharmacol, 2016, 38:332-341.

[212] Jawaid P, Rehman M U, Hassan M A, Zhao Q L, Li P, Miyamoto Y, Misawa M, Ogawa R, Shimizu T, Kondo T. Effect of platinum nanoparticles on cell death induced by ultrasound in human lymphoma U937 cells. UltrasonSonochem, 2016, 31: 206-215.

[213] Pietroiusti A, Campagnolo L, Fadeel B. Interactions of engineered nanoparticles with organs protected by internal biological barriers. Small, 2013, 9: 1557-1572.

[214] Hougaard K S, Campagnolo L, Chavatte-Palmer P, Tarrade A, Rousseau-Ralliard D, Valentino S, Park M V, de Jong W H, Wolterink G, Piersma A H, Ross B L, Hutchison G R, Hansen J S, Vogel U, Jackson P, Slama R, Pietroiusti A, Cassee F R. A perspective on the developmental toxicity of inhaled nanoparticles. Reprod Toxicol, 2015, 56: 118-140

[215] Cela P, Vesela B, Matalova E, Vecera Z, Buchtova M. Embryonic toxicity of nanoparticles. Cells Tissues Organs, 2014, 199: 1-23.

[216] Yamashita K, Yoshioka Y, Higashisaka K, Mimura K, Morishita Y, Nozaki M, Yoshida T, Ogura T, Nabeshi H, Nagano K, Abe Y, Kamada H, Monobe Y, Imazawa T, Aoshima H, Shishido K, Kawai Y, Mayumi T, Tsunoda S, Itoh N, Yoshikawa T, Yanagihara I, Saito S, Tsutsumi Y. Silica and titanium dioxide nanoparticles cause pregnancy complications in mice. Nat Nanotechnol, 2011, 6: 321-328.

[217] Wang Z, Q, G B, Su L N, Wang L, Yang Z Z, Jiang J Q, Liu S J, Jiang G B.Evaluation of the Biological Fate and the Transport Through Biological Barriers of Nanosilver in Mice. Curr Pharm Design, 2013, 19: 6691-6697.

[218] Wang Z, Liu S J, Ma J, Qu G B, Wang X Y, Yu S J, He J Y, Liu J F, Xia T, Jiang G B. Silver Nanoparticles Induced RNA Polymerase-Silver Binding and RNA Transcription Inhibition in Erythroid Progenitor Cells. ACS Nano, 2013, 7: 4171-4186.

[219] Semmler-Behnke M, Lipka J, Wenk A, Hirn S, Schaffler M, Tian F, Schmid G, Oberdorster G, Kreyling W G. Size dependent translocation and fetal accumulation of gold nanoparticles from maternal blood in the rat. Part FibreToxicol, 2014, 11:33.

[220] Di Bona K R, Xu Y, Ramirez P A, DeLaine J, Parker C, BaoY, Rasco J F. Surface charge and dosage dependent potential developmental toxicity and biodistribution of iron oxide nanoparticles in pregnant CD-1 mice. Reprod Toxicol, 2014, 50:36-42.

[221] Adamcakova-Dodd A, Monick M M, Powers L S, Gibson-Corley K N, Thorne P S. Effects of prenatal inhalation exposure to copper nanoparticles on murine dams and offspring. Part FibreToxicol, 2015, 12: 30.

[222] Djurisic A B, Leung Y H, Ng A M, Xu X Y, Lee P K, Degger N, Wu R S. Toxicity of metal oxide nanoparticles: mechanisms, characterization, and avoiding experimental artefacts. Small, 2015, 11: 26-44.

[223] Sarkar A, GhoshM, Sil PC. Nanotoxicity: oxidative stress mediated toxicity of metal and metal oxide nanoparticles. J Nanosci Nanotechnol, 2014, 14:730-743.

[224] Ding L, Liu Z, Aggrey M O, Li C, Chen J, Tong L. Nanotoxicity: the toxicity research progress of metal and metal-containing nanoparticles. Mini Rev Med Chem, 2015, 15:529-542.

[225] Fabian M R, Sonenberg N, Filipowicz W. Regulation of mRNA translation and stability by microRNAs. Annu Rev Biochem, 2010, 79: 351-379.

[226] Liu B, Li J, Cairns M J. Identifying miRNAs, targets and functions. Brief Bioinform, 2014, 15: 1-19.

[227] Huang Y, Lu X, Ma J. Toxicity of silver nanoparticles to human dermal fibroblasts on microRNA level. J Biomed Nanotechnol, 2014, 10: 3304-3317.

[228] Mahmood M, Li Z, Casciano D, Khodakovskaya M V, Chen T, Karmakar A, Dervishi E, Xu Y, Mustafa T, Watanabe F, et al. Nanostructural materials increase mineralization in bone cells and affect gene expression through miRNA regulation. J Cell Mol Med, 2011, 15: 2297-2306.

[229] Oh J H, Son M Y, Choi M S, Kim S, Choi A Y, Lee H A, Kim K S, Kim J, Song C W, Yoon S. Integrative analysis of genes and miRNA alterations in human embryonic stem cells-derived neural cells after exposure to silver nanoparticles. ToxicolApplPharmacol, 2016, 299: 8-23.

[230] Eom H J, Chatterjee N, Lee J, Choi J. Integrated mRNA and micro RNA profiling reveals epigenetic mechanism of differential sensitivity of Jurkat T cells to AgNPs and Ag ions. Toxicol Lett, 2014, 229: 311-318.

[231] Djebali S, Davis C A, Merkel A, Dobin A, Lassmann T, Mortazavi A, Tanzer A, Lagarde J, Lin W, Schlesinger F, et al. Landscape of transcription in human cells. Nature, 2012, 489: 101-108.

[232] St Laurent G, Wahlestedt C, Kapranov P. The Landscape of long noncoding RNA classification. Trends Genet, 2015, 31: 239-251.

[233] Gao M, Zhao B, Chen M, Liu Y, Xu M, Wang Z, Liu S, Zhang C. Nrf-2-driven long noncoding RNA ODRUL contributes to modulating silver nanoparticle-induced effects on erythroid cells. Biomaterials, 2017, 130: 14-27.

[234] Turkevich L A, Dastidar A G, Hachmeister Z, Lim M. Potential explosion hazard of carbonaceous nanoparticles: Explosion parameters of selected materials. J Hazard Mater, 2015, 295:97-103.

[235] Booth B, Bellouin N. Climate change: Black carbon and atmospheric feedbacks. Nature, 2015, 519: 167-168.

[236] McDonald B C, Goldstein A H, Harley R A. Long-term trends in California mobile source emissions and ambient concentrations of black carbon and organic aerosol. Environ SciTechnol, 2015, 49: 5178-5188.

[237] Shen L, Li L, Lü S, Zhang X, Liu J, An J, Zhang G, Wu B, Wang F. Characteristics of black carbon aerosol in Jiaxing, China duringautumn 2013. Particuology, 2015, 20:10-15.

[238] Yang S, Xu B, Cao J, Zender C S, Wang M. Climate effect of black carbon aerosol in a Tibetan Plateau glacier. Atmos Environ, 2015, 111: 71-78.

[239] Massabò D, Caponi L, Bernardoni V, Bove M C, Brotto P, Calzolai G, Cassola F, Chiari M, Fedi M E, Fermo P, Giannoni M, Lucarelli F, Nava S, Piazzalunga A, Valli G, Vecchi R, Prati P. Multi-wavelength optical determination of black and brown carbon in atmospheric aerosols. Atmos Environ, 2015, 108: 1-12

[240] Ando T. The electronic properties of graphene and carbon nanotubes. NPG Asia Materials, 2009, 1: 17-21.

[241] Pawlyta M, Rouzaud J, Duber S. Raman microspectroscopy characterization of carbon blacks: Spectral analysis and structural information. Carbon, 2015, 84: 479-490.

[242] WangJ, Onasch T B, Ge X, Collier S, Zhang Q, Sun Y, Yu H, Chen M, Prévô t A S. H, Worsnop D R. Observation of Fullerene Soot in Eastern China. Environ SciTechnol, 2016, 3: 121-126.

[243] Kolosnjaj-Tabi J, Just J, Hartman K B, Laoudi Y, Boudjemaa S, Alloyeau D, Szwarc H, Wilson L J, Moussa F.Anthropogenic Carbon Nanotubes Found in the Airways of Parisian Children.E BioMedicine, 2015, 2: 1697-1704.

[244] Li B, Lei X N, Xiu G L, Gao C Y, Gao S, Qian N S. Personal exposure to black carbon during commuting in peak and off-peak hours in Shanghai. Sci Total Environ, 2015, 524-525: 237-245.

[245] Report to Congress on Black Carbon. U.S. Environmental Protection Agency, Washington, DC, 2010, EPA-450/R-12-001March 2012.

[246] Janssen N A, Gerlofs-Nijland M E, Lanki T, et al. Health Effects of Black Carbon. Copenhagen,:World Health Organization, 2012

[247] Crooks M G, Aslam I, Hart S P. Inflammatory Diseases - Immunopathology, Clinical and Pharmacological Bases. Chapter 5 Inflammation and Pulmonary Fibrosis. 2012

[248] Fajersztajn L, Veras M, Barrozo L V, Saldiva P. Air pollution: a potentially modifiable risk factor for lung cancer. Nat. Rev. Cancer, 2013, 13:674-678.

[249] 邓红平, 宋伟民. 炭黑颗粒毒性作用的研究进展. 中华劳动卫生职业病杂志, 2009, 27: 372-374.

[250] Brook R D, Rajagopalan S, Pope C A 3rd, Brook J R, Bhatnagar A, Diez-Roux A V, Holguin F, Hong Y, Luepker R V, MittlemanM A, Peters A, Siscovick D, Smith S C Jr, Whitsel L, Kaufman J D. Particulate matter air pollution and

cardiovascular disease: An update to the scientific statement from the American Heart Association. Circulation, 2010, 121: 2331-2378.

[251] Fu C, Liu T, Li L, Liu H, Liang Q, Meng X. Effects of graphene oxide on the development of offspring mice inlactation period. Biomaterials, 2015, 40: 23-31

[252] Chen Y, Xu M, Zhang J, Ma J, Gao M, Zhang Z, Xu Y, Liu S. Genome-wide DNA Methylation Variations upon Engineered Nanomaterials and Their Implications in Nanosafety Assessment. Adv Mater, 2017, 29:1604580.

[253] Xu M, Zhu J, Wang F, Xiong Y, Wu Y, Wang Q, Weng J, Zhang Z, Chen W, Liu S.Improved In Vitro and In Vivo Biocompatibility of Graphene Oxide through Surface Modification: Poly(Acrylic Acid)-Functionalization is Superior to PEGylation. ACS Nano, 2016, 10:3267-3281.

[254] Rydman E M, Ilves M, Vanhala E, Vippola M, Lehto M, Kinaret P A, Pylkkänen L, Happo M, Hirvonen M R, Greco D, Savolainen K, Wolff H, Alenius H.ASingle Aspiration of Rod-like Carbon Nanotubes Induces Asbestos-like Pulmonary Inflammation Mediated in Part by the IL-1 Receptor.ToxicolSci, 2015, 147: 140-155.

[255] Mishra A, Stueckle T A, Mercer R R, Derk R, Rojanasakul Y, Castranova V, Wang L.Identification of TGF-receptor-1 as a key regulator of carbon nanotube-induced fibrogenesis.Am J Physiol Lung Cell MolPhysiol, 2015, 309:L821-L833.

[256] Tamaoki J, Isono K, Takeyama K, Tagaya E, Nakata J, Nagai A.Ultrafine carbon black particles stimulate proliferation of human airway epithelium via EGF receptor-mediated signaling pathway.Am J Physiol Lung Cell MolPhysiol, 2004, 287: L1127-L1133.

[257] Unfried K, Sydlik U, Bierhals K, Weissenberg A, Abel J.Carbon nanoparticle-induced lung epithelialcell proliferation is mediated by receptor-dependent Aktactivation.Am J Physiol. Lung Cell MolPhysiol, 2008, 294: L358-L367.

[258] Liu W, Sun C, Liao C, Cui L, Li H, Qu G, Yu W, Song N, Cui Y, Wang Z, Xie W, Chen H, Zhou Q.Graphene Enhances Cellular Proliferation through Activating the Epidermal Growth Factor Receptor.J Agric Food Chem, 2016, 64: 5909-5918.

[259] Chen G Y, Yang H J, Lu C H, Chao Y C, Hwang S M, Chen C L, Lo K W, Sung L Y, Luo W Y, Tuan H Y, Hu Y C.Simultaneous induction of autophagy and toll-like receptor signaling pathways by graphene oxide.Biomaterials, 2012, 33: 6559-6569.

[260] Qu G, Liu S, Zhang S, Wang L, Wang X, Sun B, Yin N, Gao X, Xia T, Chen J J, Jiang G B. Graphene Oxide Induces Toll-likeReceptor 4(TLR4)-Dependent Necrosis in Macrophages.ACS Nano, 2013, 7: 5732-5745.

[261] Giust D, León D, Ballesteros-Yañez I, Da Ros T, Albasanz J L, Martín M. Modulation of adenosine receptors by [60]fullerene hydrosoluble derivative in SK-N-MC cells.ACSChemNeurosci, 2011, 2: 363-369.

[262] Giust D, Da Ros T, Martín M, Albasanz J L.[60]Fullerene derivative modulates adenosine and metabotropic glutamate receptors gene expression: A possible protective effect against hypoxia. J Nanobiotechnology, 2014, 12: 27.

[263] Hiraku Y, Guo F, Ma N, Yamada T, Wang S, Kawanishi S, Murata M. Multi-walled carbon nanotube induces nitrative DNA damage in human lung epithelial cells via HMGB1-RAGE interaction and Toll-like receptor 9 activation.PartFibreToxicol, 2016, 13: 16.

[264] Comfort K K, Maurer E I, Braydich-Stolle L K, Hussain S M. Interference of silver, gold, and iron oxide nanoparticles on epidermal growth factor signal transduction in hpithelial cells. Acs Nano, 2011, 5(12):10000-10008.

[265] De Jong W H,Hagens W I, Krystek P, Burger M C, Sips A J,Geertsma R E. Particle size-dependent organ distribution of gold nanoparticles after intravenous administration. Biomaterials, 2008, 29(12):1912-1919.

[266] Sonavane G, Tomoda K, Makino K. Biodistribution of colloidal gold nanoparticles after intravenous administration: effect of particle size.. Colloids & Surfaces B Biointerfaces, 2008, 66(2):274-280.

[267] Oberdorster G,Ferin J, Lehnert B E.Correlation between particle size, *in vivo* particle persistence, and lung injury. Environmental Health Perspectives, 1994, 102:173-179.

[268] Carlson C, Hussain S M, Schrand A M. Unique cellular interaction of silver nanoparticles: Size-dependent generation of reactive oxygen species. Journal of Physical Chemistry B, 2008, 112(43):13608-13619.

[269] Ma X, Wu Y, Jin S, Tian Y, Zhang X, Zhao Y, Yu L, Liang X J. Gold nanoparticles induce autophagosome accumulation through size-dependent nanoparticle uptake and lysosome impairment. Acs Nano, 2011, 5(11):8629-8639.

[270] Zhang C K, Zhai S M, Wu L, Bai Y H, Jia J B, Zhang Y, Zhang B, Jiang G B, Yan B, Induction of size-dependent breakdown of blood-milk barrier in lactating mice by TiO_2 nanoparticles.PLoS ONE, 2015, 10(4): e0122591.

[271] Yu T A, Malugin, Ghandehari H. Impact of silica nanoparticle design on cellular toxicity and hemolytic activity. ACS

Nano, 2011, 5(7):5717-5728.

[272] Hamilton RF, Wu N, Porter D, Buford M, Wolfarth M, Holian A. Particle length-dependent titanium dioxide nanomaterials toxicity and bioactivity. Particle & Fibre Toxicology, 2009, 6(1):35.

[273] Ispas C, Andreescu D, Patel A, Goia D V, Andreescu S, Wallace K N. Toxicity and developmental defects of different sizes and shape nickel nanoparticles in zebrafish.. Environmental Science & Technology, 2009, 43(16):6349.

[274] Pacurari M.Raw single-wall carbon nanotubes induce oxidative stress and activate MAPKs, AP-1, NF-κB, and Akt in normal and malignant human mesothelial cells. Environ. Health Perspect., 2008, 116:1211.

[275] Wu L, Zhang Y, Zhang C, Cui X, Zhai S, Liu Y, Li C, Zhu H, Qu G, Jiang G, Yan B. Tuning cell autophagy by diversifying carbon nanotube’ s surface chemistry.ACS Nano, 2014, 8(3): 2087-2099.

[276] Zhou H, Mu Q, Gao N, Liu A, Xing Y, Gao S, Zhang Q, Qu G, Chen Y, Liu G, Zhang B, Yan B. A nano-combinatorial library strategy for the discovery of nanotubes with reduced protein-binding, cytotoxicity, and immune response. Nano Lett., 2008, 8:859-865

[277] Gao N, Zhang Q, Mu Q, Bai Y, Li L, Zhou H, Butch E R., Powell T B., Snyder S E., Jiang G, Yan B. Steering carbon nanotubes to scavenger receptor recognition by nanotube surface chemistry modification partially alleviates NFκB activation and reduces its immunotoxicity. ACS Nano, 2011, 5: 4581-4591.

[278] Mu Q, Du G, Chen T, Zhang B, Yan B. Suppression of human bone morphogeneticprotein signaling by carboxylated single-walled carbon nanotubes. ACS Nano, 2009, 1139-1144.

[279] Zhang Y, Mu Q, Zhou H, Vrijens K, Roussel M, Jiang G, Yan B. Binding of carbon nanotube to BMP receptor 2 enhances cell differentiation and inhibits apoptosis via regulating bHLH transcription factors. Cell Death Dis., 2012, 3: e308.

[280] Zhang Y, Wang Y, Liu A, Xu S L, Zhao B, Zhang Y, Zou H, Wang W, Zhu H, Yan B. Modulation of carbon nanotube's perturbation to the metabolic activity of CYP3A4 in the liver. Adv. Funct. Mater., 2016, 26: 841-850.

[281] Su G, Zhou H, Mu Q, Zhang Y, Li L, Jiao P, Jiang G, Yan B. Effective surface charge density determines the electrostatic attraction between nanoparticles and cells. J. Phys. Chem. C., 2012, 116: 4993-4998.

[282] Li S, Zhai S, Liu Y, Zhou H, Wu J, Jiao Q, Zhang B, Zhu H, Yan B. Experimental modulation and computational model of nano-hydrophobicity. Biomaterials, 2015, 52: 312-317.

[283] Fourches D, Pu D, LiL, Zhou H Y, Mu Q X, Su G X, Yan B, Tropsha A. Computer-aided design of carbon nanotubes with the desired bioactivity and safety profiles. Nanotoxiology, 2016, 10: 1-10.

[284] Alivisatos A P. Semiconductor clusters, nanocrystals, and quantum dots. Science, 1996, (271):933-937.

[285] Bruchez M, Moronne M, Gin P. Semiconductor nanocrystals as fluorescent biological labels. Science, 1998, (281): 2013-2016.

[286] Brus L. Electronic wave functions in semiconductor clusters: experiment and theory. The Journal of Physical Chemistry, 1986, (90): 2555-2560.

[287] Wang H F, He Y, Ji T R. Surface molecular imprinting on Mn-doped ZnS quantum dots for room-temperature phosphorescence optosensing of pentachlorophenol in water. Analytical Chemistry, 2009, (81):1615-1621.

[288] Zhao Y, Ma Y, Li H. Composite QDs@MIP nanospheres for specific recognition and direct fluorescent quantification of pesticides in aqueous media. Analytical Chemistry, 2011,(84):386-395.

[289] Liu J, Chen H, Lin Z. Preparation of surface imprinting polymer capped Mn-doped ZnS quantum dots and their application for chemiluminescence detection of 4-nitrophenol in tap water. Analytical Chemistry, 2010, 82():7380-7386.

[290] Lei J, Ju H. Fundamentals and bioanalytical applications of functional quantum dots as electrogenerated emitters of chemiluminescence. Trends in Analytical Chemistry, 2011, (30):1351-1359.

[291] Sapsford K E, Berti L, Medintz I L. Materials for fluorescence resonance energy transfer analysis: beyond traditional donor-acceptor combinations. Angewandte Chemie International Edition, 2006, (45):4562-4589.

[292] Liu M, Zhao H, Quan X, Chen S, Fan X. Distance-independent quenching of quantum dots by nanoscale-graphene in self-assembled sandwich immunoassay. Chemical Communications, 2010, (46):7909-7911.

[293] Liu M, Zhao H, Chen S, Wang H, Quan X. Photochemical synthesis of highly fluorescent CdTe quantum dots for ‘‘on–off–on’’ detection of Cu(II) ions. Inorganica Chimica Acta, 2012,(392):236-240.

[294] Liu M, Zhao H, Quan X, Chen S, Yu H. Signal amplification via cation exchange reaction: an example in the ratiometric fluorescence probe for ultrasensitive and selective sensing of Cu(II). Chemical Communications, 2010, (46):1144-1146.

[295] Zheng Z, Li X, Dai Z. Detection of mixed organophosphorus pesticides in real samples using quantum dots/bi-enzyme assembly multilayers. Journal of Materials Chemistry, 2011, (21):16955-16962.

[296] Novoselov K S, Geim A K, Morozov S V. Electric field effect in atomically thin carbon films. Science, 2004, (306):666-669.

[297] Geim A K, Novoselov K S. The rise of graphene. Nature Materials, 2007, (6):183-191.

[298] Choi B G, Park H, Park T J. Solution chemistry of self-assembled graphene nanohybrids for high-performance flexible biosensors. ACS Nano, 2010, (4):2910-2918.

[299] Wang L, Zhang X H, Xiong H Y. A novel nitromethane biosensor based on biocompatible conductive redox graphene-chitosan/hemoglobin/graphene/room temperature ionic liquid matrix. Biosensors and Bioelectronics, 2010, (26):991-995.

[300] Li J, Guo S J, Zhai Y M. High-sensitivity determination of lead and cadmium based on the nafion-graphene composite film. Analytica Chimica Acta, 2009, (649):196-201.

[301] Wan Y, Lin Z F, Zhang D. Impedimetric immunosensor doped with reduced graphene sheets fabricated by controllable electrodeposition for the non-labelled detection of bacteria. Biosensors and Bioelectronics, 2011, (26):1959-1964.

[302] Liu Y X, Dong X C, Chen P. Biological and chemical sensors based on grapheme materials. Chemical Society Reviews, 2012, (41):2283-2307.

[303] Schedin F, Geim A K, Morozov S V. Detection of individual gas molecules adsorbed on graphene. Nature Materials,2007, (6):652-655.

[304] Dan Y P, Lu Y, Kybert N J. Intrinsic response of grapheme vapor sensors. Nano Letters, 2009, (9):1472-1475.

[305] Fowler J D, Allen M J, Tung V C. Practical chemical sensors from chemically derived graphene. ACS Nano, 2009, (3):301-306.

[306] Zhang T, Cheng Z G, Wang Y B. Self-assembled 1-octadecanethiol monolayers on graphene for mercury detection. Nano Letters, 2010, (10):4738-4741.

[307] Liu M, Zhao H, Quan X, Chen S, Fan X. Distance-independent quenching of quantum dots by nanoscale-graphene in self-assembled sandwich immunoassay. Chemical Communications, 2010, (46):7909-7911.

[308] Liu M, Zhao H, Chen S, Yu H, Quan X. Colloidal graphene as a transducer in homogeneous fluorescence-based immunosensor for rapid and sensitive analysis of microcystin-LR. Environmental Science and Technology, 2012, (46) :12567-12574.

[309] Liu M, Zhang Q, Zhao H, Chen S, Yu H, Zhang Y, Quan X. Controllable oxidative DNA cleavage-dependent regulation of graphene/DNA interaction. Chemical Communications, 2011, (47):4084-4086.

[310] Liu M, Zhao H, Chen S, Yu H, Zhang Y, Quan X. A "turn-on" fluorescent copper biosensor based on DNA cleavage-dependent graphene-quenched DNAzyme. Biosensors and Bioelectronics, 2011, (26):4111-4116.

[311] Liu M, Zhao H, Chen S, Yu H, Zhang Y, Quan X. Label-free fluorescent detection of Cu(II) ions based on DNA cleavage-dependent graphene-quenched DNAzymes. Chemical Communications, 2011, (47):7749-7751

[312] Liu M, Song J, Shuang S, Dong C, Brennan J D, Li Y. A graphene-based biosensing platform based on the release of DNA probes and rolling circle amplification. ACS Nano, 2014, (8):5564-5573.

[313] Liu M, Zhao H, Chen S, Yu H, Quan X. Interface engineering catalytic graphene for smart colorimetric biosensing. ACS Nano, 2012, (6):3142-3151.

[314] Liu M, Zhao H, Chen S, Yu H, Quan X. Stimuli-responsive peroxidase mimicking at a smart graphene interface. Chemical Communications, 2012, (48):7055-7057.

[315] Xie L M, Ling X, Fang Y. Graphene as a substrate to suppress fluorescence in resonance raman spectroscopy. Journal of the American Chemical Society, 2009, (131):9890-9891.

[316] Lu G, Li H, Liusman C. Surface enhanced raman scattering of Ag or Au nanoparticle-decorated reduced graphene oxide for detection of aromatic molecules. Chemical Science, 2011, (2):1817-1818.

[317] Sun X M, Liu Z, Welsher K. Nano-graphene oxide for cellular imaging and drug delivery. Nano Research 2008, 1:203-212.

[318] Jung J H, Cheon D S, Liu F. A graphene oxide based immuno-biosensor for pathogen detection. Angewandte Chemie International Edition, 2010, (49):5708-5711.

[319] Merchant C A, Healy K, Wanunu M. DNA translocation through graphene nanopores. Nano Letters, 2010,10:2915-2921.

[320] Garaj S, Hubbard W, Reina A. Graphene as a subnanometre trans-electrode membrane. Nature, 2010, (467):190-193.

[321] Seeman N C. Nucleic acid junctions and lattices. Journal of Theoretical Biology, 1982, (99):237-247.

[322] Chao J, Zhu D, Zhang Y, Wang L, Fan C. DNA nanotechnology-enabled biosensors. Biosensors and Bioelectronics, 2016, (76):68-79.

[323] Bath J, Turberfield A J. DNA nanomachines. Nature Nanotechnology, 2007, (2):275-284.

[324] Liu M, Zhang Q, Li Z, Gu J, Brennan J D, Li Y. Programming a topologically constrained DNA nanostructure into a sensor. Nature Communications, 2016, (7):12074.

[325] Liu M, Zhang W, Zhang Q, Brennan J D, Li Y. Biosensing by tandem reactions of structure switching, nucleolytic digestion, and DNA amplification of a DNA Assembly. Angewandte Chemie International Edition, 2015, (54):9637-9641.

[326] Bertsch P, Hunter D. Applications of synchrotron bases x-raymicroprobes. Chemical Reviews, 2001, (101):1809-1842.

[327] 王秋泉, 唐南南, 杨利民. 电感耦合等离子体质谱分析. 中国科学：化学, 2014, 44: 664-671.

[328] Yin Y G, Tan Z Q, Hu L G, Yu S J, Liu J F, Jiang G B. Isotope tracers to study the environmental fate and bioaccumulation of metal-containing engineered nanoparticles: techniques and applications. Chemical Reviews, 2017, (117):4462-4487.

[329] O'Connell M J, Bachilo S M, Huffman C B, Moore V C, Strano M S, Haroz E H, Rialon K L, Boul P J, Noon W H, Kittrell C, Ma J P, Hauge R H, Weisman R B, Smalley R E. Band gap fluorescencefrom individual single-walled carbon nanotubes. Science, 2002, (297):593-596.

[330] Krug H F. Nanosafety Research-Are We on the Right Track? Angewandte Chemie International Edition, 2014, (53) 12304-12319.

[331] Liu H Y H, Cohen C. Multimedia environmental distribution of engineered nanomaterials. Environmental Science and Technology, 2014, (48):3281-3292.

[332] Garner K L, Suh S, Lenihan H S, Keller A A. Species Sensitivity Distributions for Engineered Nanomaterials. Environmental Science and Technology, 2015, (49):5753-5759.

[333] Sun T Y, Bornhoft N A, Hungerbuhler K, Nowack B. Dynamic probabilistic modeling of environmental emissions of engineered nanomaterials. Environmental Science and Technology, 2016, (50):4701-4711.

作者：王秋泉[9]，牛军峰[2]，闫　兵[5]，刘　猛[1]，刘　倩[3]，刘思金[3]，
刘景富[3]，全　燮[1]，陈　威[7]，张礼知[4]，林　璋[6]，潘丙才[8]
[1]大连理工大学，[2]北京师范大学，[3]中科院生态环境研究中心，
[4]华中师范大学，[5]华北电力大学，[6]华南理工大学，[7]南开大学，[8]南京大学，[9]厦门大学

第 16 章　环境催化材料研究前沿

本章导读

环境保护和污染物治理是近年来世界各国普遍关心的热点问题之一，也是环境科学、化学、生命科学、材料科学等多个学科重点研究的交叉领域。本章全面概述了国内外环境催化材料的主要研究进展和现状，涉及室内气体、大空间气体、水体等多种环境范围，涵盖光催化、热催化、电催化、光热协同催化等多种催化方法，囊括室内空气及 VOCs 净化、工业排放和移动尾气净化、臭氧净化、污水净化、抗菌净化等多种环境催化实际应用领域。在介绍相关催化材料研究进展的同时，也指导性地指出该领域当前面临的主要挑战和亟须解决的重大科学问题，帮助有关科研工作者找到落脚点和施力点。此外，针对目前国际上环境友好、资源节约、原子经济的科学发展潮流，本章最后简要概述了绿色低碳合成及其催化材料的研究进展，为研究者提供新思路和新方法，进一步从源头做好环境保护和治理工作，为环境的清洁、安全、可持续贡献一份力量。

关键词

环境，治理，催化，材料，进展，低碳

1 光催化材料研究进展

1.1 气体催化净化材料

1.1.1 催化净化室内空气

空气环境安全是人类赖以生存的根本保证之一。研究表明，空气环境污染物含有高毒、持久、生物积累性和远距离迁移性的物质，能持续存在于环境中，对人类健康造成长久的伤害。因此，治理空气污染已成为我国面临的重大民生问题。

同时，室内空气关注的场所也由普通的居室空间，向办公室、商城、体育馆、地下停车场等人员更密集、空气组成更复杂的空间延伸，如运动场馆消毒剂残留污染物、地下停车场氮硫氧化物、潜艇和空间站等密闭空间空气净化等[1]。同时，从国防安全的角度出发，如何安全、高效地处理一些高毒性气体，也是保证国家安全迫切需要解决的问题[2]。这些问题不仅仅对现有分析手段提出了新的要求，也对空气净化技术提出了全新的挑战。

光催化技术在室内空气净化方面有着独特的优势，它是一种始于 20 世纪 80 年代的高级氧化技术，无需在任何其他化学试剂和二次能源，无二次污染，可以在常温常压温和光照的条件下，直接将饱和烷烃、烯烃、卤代烃、芳烃、醇、醛、酮等挥发性有机污染物催化降解为二氧化碳和水或其他无机、安全的形式[3]。光催化技术净化室内空气具有高效、无毒无害、成本低等优势。不仅如此，光催化空气净化技术的研究目标，也由传统的室内污染物，向结构更复杂、毒性更高的化学毒物，如糜烂性毒剂（芥子气等）、神经性毒剂（沙林、维埃克斯等）拓展。研究表明光催化剂可以以较高的反应活性和量子效率，实现快速矿化、降解毒物分子，反应生成的挥发性产物基本无毒。因此，光催化空气净化技术已经成为当前研究的热点，在今后的日常生活、工业生产、国防战争中将发挥重要作用。

1.1.2　室内空气治理光催化新材料

经过四十余年的发展，半导体光催化材料的研究得到了长足的进步，材料种类大大丰富，由传统的无机 TiO_2[4,5]发展到 Bi 系含氧酸盐[6]、Fe_2O_3[7]、ZnO[8]、ZnS[9]、CdS[10,11]等其他半导体金属化合物，再到 g-C_3N_4[12]、 高分子聚合物[13]等不含金属的光催化材料，直至当前的苝酰亚胺等[14]纯有机半导体光催化材料；由一元光催化剂到二元/多元复合光催化材料；由单紫外光响应的光催化剂发展到可见光甚至全光谱响应的光催化材料；由单一功能光催化材料发展到多功能光催化材料。可以预见，在未来很长一段时间内，对于新型光催化剂的设计与研发，依然是该领域的研究热点。新材料的发展，也势必带动相关应用领域的进步。

作为目前研究最为深入，使用最为广泛，催化活性最高的光催化剂，TiO_2 在光照下可以有效氧化或还原其表面吸附的挥发性有机污染物（VOCs）等气体污染物，把它们转化为低分子量的产物。大量研究表明对于芳香烃类物质，以甲苯为例，在经过苯甲醇、苯甲醛、苯甲酸等中间氧化产物后，甲苯最终也可被 TiO_2 完全催化氧化为 CO_2 和 H_2O[13]。在光照条件下，TiO_2 可将空气中的氮氧化物催化氧化为硝酸根离子，实现减排和消毒。这种方法可以有效解决汽车密集度高的封闭环境（如地下停车场）中的空气净化问题[14]以及降低城市道路上方空气中的 NO_x 和 VOCs 的浓度。

但是，TiO_2 类光催化材料也存在一定的局限性，受限于其约 3.2 eV 的禁带宽度，对应于约 390 nm 的响应阈值波长，只能在能量较高的紫外光下激发，量子效率较低（低于 1%），对太阳光全光谱的利用率更低，这些都制约着其更自然、更方便的应用。为了突破 TiO_2 自身的缺陷，对于非 TiO_2 光催化剂的探索也在同步进行，钽铌钙钛矿结构光催化材料是研究关注的一类材料，这些新型的钽铌化合物光催化剂一般具有 ABO_3 的钙钛矿结构，与传统的 TiO_2 光催化剂相比，无论是阳离子还是阴离子均具有更大的结构容忍度，可以较大程度地调控光催化剂的结构和性能，从而提供较好的光催化活性，特别是其中的碱金属、碱土金属钽酸盐复合氧化物如 $NaTaO_3$[15]、$Ba_5Ta_4O_{15}$[16]等，都表现出较好的甲醛降解性能。另一研究热点则集中在钨钼钒系[17]光催化材料，而该类钨酸盐都具有一个共同点：具有层状的钙钛矿结构，因其特有的结构和物理化学性质，也表现出良好的光催化性能。2006 年，Finlayson[18]等报道了 Bi_6WO_{12} 在波长大于 440 nm 的可见光下具有较好的活性。进一步的研究表明，Bi_2WO_{12}[6]、$Bi_2W_2O_9$[6]、Ag_2WO_4[19]等都具有较好的可见光活性。由于钨和钼两元素在周期表中是同一族，具有一定的相似性，因此 Bi_2MoO_6[20]与 Bi_2WO_6 有着几乎相同的性质。已有研究报道证实，这种具有层状结构的多元氧化物往往具有较高光催化活性，该材料的光量子效率甚至高于 TiO_2。此外，也有大量报道指出部分金属氧化物如 ZnO、Fe_2O_3、CuO、Bi_2O_3 等，金属硫化物如 ZnS、CdS 等，金属卤氧化物如 BiOCl、BiOBr、BiOI[21]等均有较好的光催化性能，但其在室内空气净化相关领域的应用研究还较少。

相比于单一晶形的 TiO_2，锐钛矿和金红石异质结复合而成的商品化 Degussa P25-TiO_2 具有更高的活性，也由此引起了研究者对于多元复合光催化剂的兴趣。在此思路的启发下，为了解决活性不高，光谱响应窄等问题，大量二元、三元甚至多元光催化剂被研制出来。由于复合半导体具有两种或多种不同能级的价带和导带，在光激发下电子和空穴分别迁移至一种材料的导带和复合材料的价带，从而使光生载流子有效分离或产生新的催化性能。复合半导体可分为半导体-绝缘体复合及半导体-半导体复合。绝缘体如 Al_2O_3、SiO_2、ZrO_2 等大多起载体作用，TiO_2 负载于适当的载体后，可获得较大的比表面和合适的孔结构，并具有一定的机械强度，以便在各种反应床上应用。复合半导体的互补性质能增强电荷分离，抑制电子-空穴的复合和扩展光致激发波长范围，从而显示了比单一半导体具有更好的稳定性和催化活性。

石墨烯结构氮化碳（g-C_3N_4）也是一类热门的新型可见光非金属光催化剂，无论是在环境污染物治理，还是清洁能源生产中，都有巨大的应用前景，而且目前已经实现了部分商业化应用。通过各种努力，如过渡金属掺杂、半导体复合、形貌调控等，可以进一步提高 g-C_3N_4 的光催化活性，使其满足大规模实际应用的要求。

近年来，具有三维网络结构的光催化材料也成为一类热点材料，其巨大的比表面积提供了较高的吸附能力、更多的活性暴露位点以及更低的气体阻力，可以广泛应用于气体、水体净化。而且，三维网络结构光催化材料，在具备传统块体催化剂的催化氧化活性外，因其较强的吸附能力，使得该类材料往往是“吸附富集-催化降解”的协同工作模式，因而更为高效。朱永法等近年陆续开发出一系列具有三维网络结构的光催化材料，如聚苯胺/ TiO_2 复合催化剂[22]、聚苯胺/ C_3N_4 复合催化剂[23]、TiO_2-rGH[24]等光催化材料，在污染物净化方面表现出优异性能。

此外，随着研究的深入，传统无机半导体难以调控的弊端逐步显现出来。为了进一步提高催化剂的可设计性，丰富催化剂种类，提高催化剂光谱利用率，朱永法等[25,26]报道了纯有机苝酰亚胺超分子自组装纳米线光催化材料，利用 π-π 堆积形成快速的电子传输通道，显著增加了空穴-电子分离程度，提高了光催化活性。在此基础上，通过复合强烈电子导体 TCNQ 的 TCNQ-PTCDI 复合催化剂进一步提高了苝酰亚胺类光催化材料的光生载流子分离能力，因而催化活性进一步提高[27]。相比于传统的无机金属光催化剂和 g-C_3N_4，有机半导体光催化材料易于设计，能带结构便于调控，氧化矿化能力强，对目标污染物分子可以定向性合成等优势使其在未来的环境治理光催化中也必将占有一席之地。

上述材料各具优势，也都存在一定的缺陷。例如，与 TiO_2 类似，ZnO 对于光谱的利用率也相对较低，量子效率低；而其他具有可见光响应性能的无机金属盐、氧化物等光催化材料，在可见光激发下的总体光催化性能则变得较差，很难与 TiO_2 媲美。此外，CdS 还因为毒性较大，催化过程中可能会有 Cd 的释放，更无法在环境治理中大规模应用。相比之下，非金属光催化材料虽然易于设计和调控，可以实现可见光甚至全光谱的响应，但是其对光生-电子空穴对的复合作用则较金属氧化物材料更为强烈，同时该类结构往往不够稳定，循环重复性能差，在实际应用中仍有很多困难。如何提高光催化剂量子效率和稳定性，促进该技术在经济上、环境相容性上能为人们所接受，也同样成为了世界范围内的科学工作者的研究兴趣。经过长期的努力和钻研，目前对于光催化材料能效提高的主要思路，主要集中在离子掺杂，染料敏化，半导体异质结复合、结构和形貌调控等方面。

离子掺杂是利用物理或化学方法，将离子引入到催化剂晶格内部，从而在其晶格中引入新电荷，形成缺陷或改变晶格类型，影响了光生电子和空穴的复合或改变了半导体的能带结构，最终促使光催化材料的光催化活性发生改变。

染料敏化技术则主要是利用 TiO_2 等无机光催化材料对光活性物质的强吸附作用，添加适当可被可见光激发的光活性敏化剂，吸附于催化剂表面。在可见光照射下，吸附态的光活性分子吸收光子后，产生自由电子并注入到无机催化材料的导带上，从而拓展了材料的波长响应范围，使之实现利用可见光进行有机物污染物的催化降解。结合我国当前的具体国情，纺织印染和染料生产在我国的工业结构中依然占有较大的位置，如果能够充分利用现阶段在染料方面积累的技术，把可能造成环境污染的染料用于光催化剂的敏化，则该方面的研究更具有特殊的意义。目前，文献报道的几种常见敏化剂有无机敏化剂、有机染料、金属有机配合物和复合敏化剂。

除上述两种方法外，半导体的异质结复合也是一种提高光催化剂催化能力的有效手段，在异质结界面处，光生载流子可以发生有效的分离和传输，从而材料的光催化性能得到提高。复合型半导体纳米粒子是指由两种或两种以上物质在纳米尺度上以某种方式结合在一起而构成的复合粒子，复合的结果不仅能有效地调节单一材料的性能，而且往往会产生出许多新的特性。两种不同的半导体复合需要考虑不同半导体的禁带宽度、价带、导带能级位置及晶型的匹配等因素。复合半导体的互补性质能增强电荷分离，抑制电子-空穴的复合和扩展光致激发波长范围，从而比单一半导体具有更好的稳定性和催化活性。

形貌调控是除上述方法外，另一种是常见的能效调控手段。从本质上讲，形貌直接决定着催化剂材料的暴露晶面、晶粒尺寸等结构性能：晶面因素会直接影响材料对污染物分子的吸附性能，界面电荷分离能力，不同的晶面其氧化、还原能力完全不相同；晶粒尺寸，则影响到光生载流子的传输性能，更小的晶粒尺寸，有助于光生载流子由体相向表面传输。

总而言之，随着研究的进一步深入，性能优良的光催化材料已经被逐步开发出来，部分性能突出的材料如P25-TiO_2、g-C_3N_4甚至已经走向了市场化，在空气污染物净化领域发挥出重要作用。有理由相信，在可以预见的将来，必将会有越来越多的、性能更为优异的、环境相容性更好的光催化空气污染物治理材料问世。

1.1.3 光催化技术在消除工业废气中的应用

光化学净化法属于低能耗、低投入、高效、安全的新技术。大气光化学净化主要分为高能紫外线净化技术和光催化净化技术。高能紫外线净化技术主要依靠高能紫外线直接作用于大气污染物分子，在高能光子的作用下直接分解污染物。该方法对仪器设备有一定要求，且容易造成臭氧等二次污染。光催化净化技术主要依靠光催化剂在紫外光或可见光的照射下产生光生电子空穴对大气污染物实现还原或氧化。光催化净化技术对NO_x、VOCs有较好的去除能力，也可用于部分POPs的去除。

1）光催化氧化消除NO_x

光催化材料的电子和空穴被光激发后，一方面空穴本身具有很强的得电子能力，可夺取NO_x体系中的电子，使其被活化而氧化。另一方面，电子与水或者空气中的氧发生反应生成氧化能力更强的·OH及$\cdot O_2^-$等，将NO_x最终氧化生成NO_3^-。氧化NO_x生成的NO_3^-会残留在催化剂的表面，当累积到一定浓度时会降低催化剂活性，所以需要水的洗净、再生。而通过催化剂流出的水几乎是没有酸性的，完全可以被空气中的粉尘中和而无害化。

用于NO_x光催化氧化的催化剂主要以TiO_2基为主，不过近年来也有新进展。如：Ti-MOF（有机金属框架结构）光催化剂在修饰纳米金属后也显示出对NO_x的光催化氧化能力。采用钛酸四丁酯为有机钛源，*N,N*-二甲基甲酰胺与乙醇为混合溶剂，2-氨基对苯二甲酸为配体，原位合成了一种可见光驱动具有高催化氧化NO活性的Ag负载型MOF结构的有机金属钛聚合物新型光催化剂。原位掺杂银纳米颗粒后，由于Ag纳米粒子不仅能够促进NH_2-MOP(Ti)的组装，还有利于提高对可见光的捕获能力，同时，Ag纳米粒子促进了光生电子的快速传导，降低其与空穴的复合，提高了光催化氧化NO的效率，且有良好的稳定性。

2）光催化还原法消除NO_x

光催化还原反应是使NO_x在光催化剂的作用下直接分解生成完全无害的N_2。以TiO_2为催化剂的反应条件比较特殊，一般需要加入还原剂，如需要加入NH_3才可将NO降解为N_2，加入甲醇可将N_2O降解为N_2；有些反应还可以在低温下进行。另外，特定的光催化剂也可将NO_x还原。与光催化氧化NO_x相比，光催化还原反应具有不产生NO_3^-、不需要水洗及再生等优点。其主要存在问题是在还原过程中，如控制不当，除生成N_2外，还会生成CO、N_2O等有害副产物。如果用NH_3作还原剂，还存在腐蚀、堵塞设备和运输储存安全性等问题。同时还原方法对NO_x的去除效率相比于氧化方法明显较低。

3）光催化氧化净化NO_x的应用现状

NO_x能够被光催化剂产生的自由基氧化为NO_3^-和HNO_3吸附在催化剂表面，最终在雨水冲刷下被大气颗粒物中碱性组分中和而去除。日本、欧洲等地将开发的光催化剂应用在路面、隧道和建筑物表面，能够有效去除20%~60%的NO_x。在意大利米兰的道路上使用TiO_2光催化材料，光催化净化汽车排放的二氧化氮。经过数月的研究分析发现，NO_x污染数降低了60%~70%。我国上海已经在城区部分路面铺设了含有光催化剂的材料，这样的路面能吸收45%的汽车尾气污染。

1.1.4 VOCs 的光化学氧化净化

光催化氧化法作为一种反应条件温和、不需要额外化学助剂，且使用后的催化剂可用物理和化学方法再生后循环使用的新技术，是对吸附法和吸收法的有益补充。依靠光催化剂上电子-空穴对生成的·OH 等氧化吸附在催化剂表面的 VOCs，生成 CO_2 和 H_2O。光催化氧化法对 VOCs 的降解率可达到 90%~95%。

光催化氧化 VOCs 的研究主要以甲苯、甲醛、乙烯等模型污染物在光催化剂上的光化学降解行为进行研究。将苯系物如甲苯、邻甲苯胺和邻氯苯胺降解中间产物的化学质谱分析与其毒理学评价相结合，可进一步阐释 VOCs 在 TiO_2 催化剂上的安全脱毒机制。同时研究发现醛类与 NO_2 反应形成高致癌、致突变性二次气溶胶重要前体物亚硝酸，而卤代醛类与卤素反应可形成臭氧分解剂 ClO，同时 SiO_2 的存在将加速醛类与 NO_2 的非均相反应和降低其大气污染浓度。这些研究为光化学净化 VOCs 的安全排放提供了一种高效的评价方法，也为正确理解大气层物理化学反应机制提供了理论支持。

光化学净化 VOCs 的研究主要以甲苯、甲醛、乙烯等模型污染物在光催化剂上的光化学降解行为进行研究，但也开始逐渐涉及实际工业。通过开展典型工业行业排放 VOCs 的污染特征研究，可以有效地明晰特征污染物及其潜在风险，从而为典型工业行业排放 VOCs 的大气迁移转化机制和污染控制机理提供科学理论研究基础。日本大金公司于 1996 年利用光催化脱臭技术开发出空气净化除臭机，日本石原公司与丰田汽车公司、Equos 研究公司联合开发成功的 TiO_2 光催化技术，可高效去除空气中的 NO_x、甲醛等。国内空气净化器厂家也在 2000 年后陆续将光催化技术应用于 VOCs 的降解，并已形成以光催化技术为核心，通过等离子体、真空紫外以及臭氧联合氧化等技术集成的成套工艺应用于多种行业废气处理，如喷涂废气（包含各类烯烃、醛类、脂肪酸类、甲基酮类、芳香族化合物）。

1.1.5 POPs 的光化学氧化净化

传统的紫外光解消除 POPs 是利用二噁英污染物可强烈吸收紫外光的特性实现其降解的目的。脱毒机理主要有两种：① C—Cl 键直接断裂；②发生分子结构重排或错位。该方法易于原位化处理二噁英，被认为是最有潜力实现大规模工业化的二噁英治理技术之一。目前存在的问题是体系杂质的存在对光分解反应产生副作用。

中国科学院和国内高校的研究人员利用光催化技术分解二噁英类化合物的研究表明，二噁英类化合物在中压汞灯的照射下可以被 TiO_2 降解，在两小时的反应时间内对二氯代二噁英的降解率达到 98.3%。美国，日本和俄罗斯也对利用光催化氧化降解二噁英类化合物进行了大量研究，表明除催化剂外，光催化反应器的设计和反应器内光学特性对二噁英类化合物的去除也非常关键。

1.1.6 光化学净化的工程实践

将光催化降解 VOCs 的基础理论进行延伸与拓展，在充分明确工业污染源 VOCs 排放特征的基础上，在油漆生产[28]、电子垃圾拆解 [29]等行业开展了光催化集成技术净化有机废气的工程实践。利用光催化系统、吸附剂喷射系统和光伏/光热一体化系统，在电厂尾部烟道中喷射自行开发的碳基/非碳基吸附剂，通过光催化促进氮氧化物（NO_x）和重金属汞（Hg）氧化，提高吸附剂吸附效率及脱硫塔（WFGD）液相吸收效率，可达到高效脱除燃煤电厂烟气多种污染物的目的。

此外，光催化自清洁外墙涂料已逐步被市场认可和接受，是最具市场前景的光催化产品。为了进一步提高光催化涂料的普适性和施工性能，相关研究机构和企业在原有技术的基础上，成功开发出一种新型耐酸碱、防腐、具有自清洁性能的涂料，表现出较优的保色和防粉化性能，几乎可适用于包括深色的各种有机、无机基底涂层表面，大大拓宽了光催化涂料的适用范围。不仅如此，这种新型的光催化自清洁涂料还具有光催化氧化去除 NO_x 和 SO_2 的功能。若将此材料用于城市建筑物表面，可有效降解汽车尾气排放的污染物，从而对因 NO_x 二次光化学反应而形成的雾霾起到抑制作用。

1.2　光催化在水净化方面的研究

天然水体中日益增加的数千种化学品复合污染是人类所面临的最关键环境问题之一。尽管大多数化合物在水体中的浓度很低，但是其中一些化合物如农药、医药品等毒性很强，尤其多种污染物往往以混合形式存在而产生的联合毒性效应，对人体和环境带来潜在的健康和生态危害。

面对国家在污水深度处理、饮用水安全保障以及水质改善等方面的重大需求，亟待研究高效可靠的水处理技术。多相光催化技术因其可以利用取之不尽用之不竭的太阳光作为能源，是一项“绿色”和“环境友好”的高级氧化技术[30,31]。它同时具备光照、·OH、$\cdot O_2^-$、H_2O_2 等氧化剂的氧化、吸附和过滤等多种功能，能有效杀灭真菌、细菌、病毒等有害病原微生物，受到世界范围内水消毒领域科研工作者的广泛关注。事实上，太阳光谱主要包括两部分光，可见光（43%）和红外光（48%），如何利用这两部分太阳光能是解决太阳能光催化水处理技术的关键，但至今仍没有很好地解决这个问题。过去的材料开发与机制研究的结果似乎说明，催化剂的稳定性与催化剂的太阳能高活性相矛盾，高活性催化材料必须满足一个条件是最大限度地吸收太阳能，这就要求半导体材料的禁带宽度窄，由于较多光激发产生的电子与空穴在材料体内的残留，使半导体非常容易发生腐蚀。虽然光催化存在一些技术问题阻碍其广泛地应用于实际水处理，但是光催化具有利用天然可再生能源太阳能的前景，有希望取代现有的昂贵的技术和设备，在很大程度上降低水处理成本。而高效稳定太阳能光催化材料的研发是光催化应用于实际水处理的关键。

1.2.1　光化学方法控制水污染

非均相光催化通过光激发半导体光催化材料产生的电子和空穴与氧气或水反应生成的·OH，具有对污染物无选择性和分解彻底等优点，而且不用添加其他消耗性化学品，与其他三种光化学过程相比具有明显优势，是处理低浓度高毒性废水的潜在技术。污染控制针对的是实际废水，除了水分子外，还存在有机物、金属阳离子、无机阴离子、微生物，参与界面反应的成分复杂且溶液 pH 变化范围较大，与电解液成分和 pH 可控的光伏、光解水过程的界面反应显著不同。此外，光催化是面向应用的技术，不同类型的污染物，光催化反应途径也不同。根据不同的反应途径，可分为一般有机物、含杂原子有机物、染料、金属离子、微生物等。

（1）一般有机物。一般有机物包括脂肪烃、醇、醛、酸和芳香烃，其光催化分解通过·OH 或其他氧化性物质攻击 C—C、C—O、C—H、C═C、C≡C 或苯环完成，处理目标是完全矿化。·OH 与饱和烃反应攻击的是单键，通过亲电加成、脱氢、电子转移等步骤把饱和烃氧化成相应的醇、醛、羧酸，最终分解成 H_2O 和 CO_2[32]。·OH 与不饱和键（C═C、C≡C）的反应途径是通过亲电加成达到羟基化，·OH 与构成双键的一个 C 成键，另一个 C 剩余一个电子呈激发态。反应式如下：$RHC{=}CR_2 + \cdot OH \longrightarrow RHC(OH){-}C(\cdot)R_2$。然后按照饱和脂肪醇的分解步骤进行。

（2）含杂原子取代基的有机物。杂原子指的是除 C、H、O 以外的其他原子，主要包括 Cl、Br、I、F、N、P、S。杂原子取代基的存在，不但提高了有机物的毒性，而且增加了化学稳定性，难以降解。光催化分解此类污染物的关键步骤是 C 与杂原子间化学键的断裂，大多数反应可通过·OH 途径完成，产物是 Cl^-、Br^-、I^-、F^-、NO_3^-、PO_3^-、SO_4^{2-}等阴离子。

（3）染料。按结构差别，染料分为偶氮、蒽醌、靛蓝、酞菁、三苯甲烷、硫化染料等类型。染料废水处理的目标是脱色和降低毒性，破坏发色团即可脱色，但是必须矿化才能去除毒性。染料分子结构复杂，光催化分解路径各不相同，但是基本遵循·OH 氧化分解的原理，通过光生电子还原的很少。

染料分子的共轭结构和取代基对分解动力学影响最大。例如偶氮染料光催化分解比蒽醌染料容易；同为偶氮染料，酸性橙 12 比酸性橙 10 多一个磺酸基而容易被降解，酸性橙 8 则由于比酸性橙 10 多一个甲基而更难降解，取代基电负性不同导致与光催化材料的吸附差异是本质原因[33]。

（4）重金属离子。光催化处理 Cr^{6+}、Cd^{2+}、Hg^{2+}、Pb^{2+}、As^{3+}等毒性金属离子的目标是转化成低毒离子或还原成固体形态回收利用。金属离子的转化途径包括：被光生电子还原成低价态离子或 0 价金属，被空穴或·OH 氧化成高价态离子。避免被光生电子和空穴（或·OH）循环氧化还原是金属离子去除必须考虑的问题。例如 As^{3+}的一般处理方法是先氧化成低毒的 As^{5+}然后通过吸附、沉淀或离子交换的方式去除，这个过程一般要曝气，以便用 O_2 分子捕捉光生电子避免 As^{5+}被还原回 As^{3+}。

1.2.2 净水深度处理研究进展

近十年来，抑制光生电荷复合、提高界面反应速率以及避免副反应是光催化水污染控制领域理论研究的重点，各种光催化反应器和耦合工艺的研究是推动光催化技术在废水处理中应用的关键。该方面的主要进展概括为：①通过吸附、过滤强化污染物与催化剂表面接触，达到提高处理效率的目的；②光催化-超声、光催化-臭氧、光催化-微生物在内的主要耦合工艺的耦合；③高效光催化反应器的设计和应用研究。

1）吸附耦合光催化

光催化材料与活性炭等吸附材料混合悬浮体系中，由于部分光催化材料与活性炭接触，光生电荷能够迁移到活性炭表面与被吸附的污染物反应。具有一定导电性的活性炭作为吸附剂，不但能把污染物富集到光催化剂附近，而且可以接受光生电子抑制与空穴的复合，提供了更多的反应界面。为了使光催化材料与吸附材料接触更充分，通常在光催化材料颗粒表面包覆透明性较好的多孔吸附层。光生电子经过吸附材料，空穴经过吸附材料的孔道分别到达表面与被吸附的污染物反应。耦合吸附与光催化的另一个好处是提高选择性。由于沸石等吸附材料孔结构和表面基团的作用，带有特定表面基团的污染物更容易被吸附，因此选择性提高。吸附与光催化协同作用可使动力学常数提高 2 倍以上。将光催化材料做成多孔结构即可提高吸附污染物的能力。

2）膜分离耦合光催化

光催化与膜分离结合，不但可通过膜孔截留去除污染物，还能利用光催化作用矿化被截留的污染物，既避免了二次污染，又减缓了膜污染。光催化与微滤、超滤、纳滤、反渗透等分离膜的耦合均有报道，实验室规模的研究成果已经证实光催化膜具有分离及分解双重功能，在有机物去除、灭菌中表现出了良好的效果。

3）臭氧氧化耦合光催化

利用 O_3 氧化分解水中的污染物已经是比较成熟的水处理技术。但是 O_3 的氧化电势为 2.07 V，只能把不饱和键氧化成饱和键，最终产物通常是羧酸，没有继续氧化饱和键的能力，有时甚至产生有毒的产物。例如 O_3 水处理消毒过程容易把溴化物氧化成溴酸盐，使出水毒性增加。此外，O_3 在水中自身分解速度较快而氧化分解污染物的速度很慢，因此大部分臭氧没有用于分解污染物，利用率较低。

光催化与臭氧氧化的结合原理是 O_3 捕捉光生电子生成·OH，使氧化能力由臭氧的 2.07 V 提高到 2.8 V，能够矿化污染物从而避免毒性中间产物的积累。由于 O_3 在水中的溶解度高于 O_2，能够捕捉更多光生电子抑制光生电荷复合，加快光生空穴与水反应生成·OH 的速度。一些报道提出光催化耦合臭氧分解污染物遵循零级动力学，界面反应速度与污染物初始浓度无关，因此改善了光催化过程反应速度随反应物浓度降低而减小的问题。典型的结果是光催化臭氧氧化的动力学常数是光催化的 3 倍以上，TOC 去除率是臭氧氧化的 5 倍以上，充分显示了协同效应。需要注意的是，O_3 也可能与光催化生成的·OH 反应生成氧化能力较弱的 O_2 和 HO_2·。可通过调控 O_3 流量避免这个不利反应发生。

4）超声耦合光催化，

超声波与光催化结合的优点很多。首先，空化泡内电离气体可激发光催化材料产生·OH。其次，悬浮在水中的光催化颗粒能够把从空化泡扩散出来的 H_2O_2 分解成·OH，提高了氧化能力[34]。第三，悬浮的催化剂有助于空化泡成核，使其更密集，作用范围更大。第四，超声波产生的冲击和声流加强液相中反应物、产物、自由基与光催化剂表面之间的传质。此外，如果光催化材料具有压电性能，超声波还能提供周期变化的机械能，借助压电效应带来的内电场促进光生电荷的分离。超声波与光催化结合耦合的动力学常数一般是单独光催化的 2 倍左右，单独超声波的 5 倍，具有明显的协同效应。

5）光催化与生物处理组合技术

光催化与生物处理组合技术二者的组合形式由进水特点决定。对于毒性较高的进水，采用光催化单元作为生物工艺的预处理。此时光催化的目标是提高可生化性而不是完全矿化，因此停留时间可以很短，弥补了光催化矿化污染物速度慢的不足。为了避免进水中悬浮颗粒对入射光线的散射，一般需要在光催化单元前置絮凝工艺。对于可生化性较高，但是微生物处理后代谢产物有毒的进水，例如阿莫西林等抗生素经微生物处理的小分子代谢产物有毒，此时在微生物单元后接光催化单元，能够有效矿化这些代谢产物。

1.3　光催化净化微生物污染研究进展

光催化消毒的基本原理是半导体光催化剂在受到光照激发后，产生光生电子-空穴对，其迁移到催化剂表面后发生一系列光化学反应，产生强氧化性物质如·OH 和·O_2^-等，可杀灭水中大量的病原微生物并有效降解其产生的有机、无机污染物。但是，将光催化消毒技术应用于工程实践还有许多实际问题需深入研究。基于国内外在光催化杀菌领域的研究现状，针对我国病原微生物严重污染的问题，本节将阐述我国利用光催化技术杀灭在空气及水中典型病原微生物的研究现状及进展，为有效推动新一代杀菌技术的发展与进步及其推广应用提供科学理论依据，为实现光催化技术在我国微生物污染净化方面的推广应用提供技术支撑，并对光催化技术在微生物净化学科的发展提出了政策建议。

1.3.1　空气及水体中微生物净化研究

细菌、真菌、病毒等微生物含量是环境质量评价的重要指标之一，直接影响着人类健康。光催化微生物净化技术，经过一段时间的发展，对该问题的研究着眼点主要集中在水体中微生物的净化处理和空气微生物净化处理两个方面。我国利用光催化技术对水体消毒方面，已有大量深入的研究。

另一方面，室内空气微生物的净化，近年来越来越受到关注。2003 年在我国大面积爆发的 SARS 疫情，及近年来禽流感、猪流感等大规模公共卫生安全事件，都给人们敲响了局域空气污染物处理的警钟。光催化技术不添加化学试剂，不会引起二次污染，环境友好，能够直接产生活性氧物质对微生物进行有效杀灭，因而成为目前最有潜力的空气微生物净化手段。基础研究方面，光催化在空气消毒方面已有大量文献报道。

光催化空气微生物净化技术主要应用在空气微生物消毒过滤系统，既可与新建或已有建筑通风系统相结合，也可作为便携式净化器，形式灵活多变。目前最常使用的是基于 TiO_2 的光催化反应器，其中 TiO_2 反应表面和光源是反应器的两大部件，而光子、TiO_2 颗粒和微生物颗粒的充分接触是获得较好灭菌效果的前提。光催化材料在空气中微生物净化方面的另一个重要应用就是自消毒和自洁净表面。将表面尤其是易于受潮的表面上涂覆光催化材料，在有足够紫外光照的条件下可以有效抑制沉降在表面上菌落的生成，防止其繁殖和进一步传播[62]。按照基底材料划分，目前我国已经发展的光催化自清洁表面包括玻璃、陶瓷、不锈钢、铝材、玻璃纤维以及聚合材料等，并且部分已被成功应用于医院、宾馆以及商用设施作为抗菌表面。当然在战争中所使用的生物战剂（生物武器）也多是通过空气扩散施放，带致命细菌的微生物颗粒能分散成微小的粒子悬浮在空气中，与空气混合成气溶胶，随风飘移，污染空气、地面、水源和食物，并能渗入无密闭设施的人防工程之内。因此，如果将具备光催化杀菌性能的光催化涂层覆盖在需要保护的环境表面上，利用室内荧光或室外太阳光，以及自净化材料的自消毒和自清洁作用，可以在一定程度上有效防护生物武器的伤害。

1.3.2 高效光催化杀菌材料研究进展

在众多光催化材料当中，TiO_2 以其高效、无毒、低成本的特性一直是该领域内的热点材料。近年来对于新型光催化材料的探索，也主要集中在两个方面，一是对 TiO_2 光催化材料进行修饰改性，另外一个是开发新型非 TiO_2 基的光催化材料，并将其应用于光催化杀菌系统中。对 TiO_2 进行可见光修饰改性的方法有多种，前文详述的贵金属沉积、离子掺杂、染料敏化和异质结复合等，均已被证实可以有效提高 TiO_2 基光催化材料的可见光催化杀菌活性[35-38]。

在新型非 TiO_2 基的光催化材料的开发方面，同样有大量金属氧化物和硫化物被报道用于光催化杀菌，如 ZnO、WO_3 和 $ZnIn_2S_4$ 等。在光催化有机污染物降解中被广泛研究的 Bi 系光催化剂也有大量报道表明可用于细菌的杀灭。另外，具有表面等离子体共振效应的可见光光催化材料在水体中杀菌也有部分研究[39,40]。

石墨烯结构氮化碳（g-C_3N_4）是光催化领域的“明星”非金属光催化材料，我国率先对非金属光催化材料的杀菌性能进行了研究[41]。在此基础上，王万军等将 g-C_3N_4、石墨烯和单质硫结合起来，制备出一种具备多级结构的光催化材料，发现其可以有效杀灭大肠杆菌，证实了非金属光催化材料也可用于对水中微生物的杀灭[42]。同时，单质红磷也被发现具有一定的光催化杀菌性能[43]。

虽然纳米光催化材料具有较高活性，但制备方法一般都较为繁琐、能耗较高，不利于杀菌技术的大规模工业化应用，且在实际使用中由于催化剂的纳米尺度使得催化剂回收变得相当困难。最近的研究还发现某些天然矿物也具有很高的光催化杀菌活性，并可以应用于水体中微生物的杀灭。王保强课题组开发了一系列矿物光催化剂，例如闪锌矿[44-46]和黄铁矿[47]等，发现它们在可见光照射下对大肠杆菌都具有很好的杀灭作用。此类催化剂经济成本相对较低，无需要化学合成，且部分矿物还具有磁性易于分离回收。因此，低成本天然矿物光催化材料可能会成为今后光催化杀菌材料开发和实际产业化应用的新方向，但是目前这方面的研究在国内还处于起步阶段。

与光催化降解有机污染物的机理类似，光催化杀菌的本质仍然是在光电转换中进行的氧化还原反应，不同的是由于微生物的特殊性，光催化杀菌中情况更加复杂。大量研究证明，光生空穴、光生电子、·OH、$·O_2^-$和 H_2O_2 等活性物质都有可能对细菌具有杀灭作用。

在光催化细菌杀灭机理方面，微生物细胞是由很多微生物大分子如 DNA/RNA、蛋白质、脂质等组成的复杂结构，而这些生物大分子又由最基本的结构单元如碱基、氨基酸等构建而成。由此推测，病原微生物细胞的分解和矿化过程是相当复杂的，在完全矿化之前微生物细胞首先被分解为生物大分子，继而生物大分子被降解为生物小分子。

另外，由于微生物细胞本身具有生命，且能够对不利环境产生一定的抵抗和自我保护作用，其被杀灭的生物机理要远比有机物的化学降解作用复杂。因此如何理解光催化杀菌的生物机理是研究和提高光催化杀菌性能的关键。已被实验验证的机理有以下几种情况：①辅酶 A（Coenzyme A，CoA）的氧化：光催化过程中，细胞内参与细胞呼吸过程酶反应的辅酶 A 同时被氧化，从而抑制了生物细胞的呼吸作用，并最终造成了微生物的杀灭[48,49]。②细胞结构的破坏：研究认为杀灭过程是光催化产生的活性物质氧化攻击微生物的细胞质膜（包括外部的脂多糖和内部的肽聚糖），致使细胞的完整性丧失，同时使得细胞膜的流动性变差[49,50]，类似结果在利用单基因敲除突变菌株（JW3935-4 和 JW1077-1）进行光催化杀灭对照研究中得到了证实[51]。③细胞膜损伤后进一步导致了细胞内生物大分子，如蛋白质和 RNA 等的泄漏，同时在细菌胞外环境中生成的活性物质，特别是 H_2O_2，能够穿透细胞膜进入细胞内，导致细菌胞内活性物质水平升高。细胞内外环境中急剧上升的活性物质水平，导致细菌氧化应激体系失灵，即抗氧化酶如过氧化氢酶（CAT）及超氧化物歧化酶（SOD）逐渐失活，最终致使细胞失活[53,54]。这些现象可以通过 SEM 观察杀菌过程及相应部位的变化，直至细胞完全分解并从催化剂表面消失而获知。④生物大分子的破坏：从损伤的细菌胞内泄漏的生物大分子物质特别是蛋白质及核酸可以进一步被破坏。蛋白质作为细胞的重要构成部分，其氧化损伤产物-羰基化蛋白的浓度水平是评价细胞氧化损伤的重要分子标记物，有研究发现细菌蛋白质羰基化水平在整个光催化处理过程中呈现上升趋势[53]。而微生物细胞破坏后产生的另一种生物大分子核酸也可以进一步被破坏，使得 DNA 的复制和诸多代谢机能受到抑制，以致细胞完全失活。当然微生物代谢过程中产生的内毒素等也在光催化降解过程中得以分解。

2　热催化净化材料

以热催化的方法削减化石燃料燃烧及其衍生产品生产、使用过程中污染物的排放是环境催化的重要领域，并已成功应用于移动源尾气、固定源烟气、工业 VOCs 与室内空气净化等行业。以上行业排放污染物的种类、属性等存在显著差异，对环境催化净化材料的性能提出了不同的要求，其发展趋势也不尽相同。

2.1　室温催化净化材料

催化氧化法通常需要在远高于室温的条件下才能进行，因此室温催化反应对催化材料的性能提出更高的要求。目前，可在室温条件下进行的催化反应主要有甲醛氧化、乙烯氧化、臭氧催化分解、催化杀菌等。Au 基催化剂是典型的低温催化氧化 CO 催化剂，室温催化氧化甲醛的催化材料主要是 Pt、Pd、Au 等贵金属催化剂[54-57]，其中 Pt 基催化剂具有最优异的常温催化氧化甲醛性能[54]。与上述催化剂相比，部分非贵金属催化材料如锰氧化物等尽管能在常温下催化氧化甲醛[58,59]，但其净化效率、稳定性和长效性还有待提高。三维介孔 Co_3O_4 负载 Au 催化剂可实现乙烯的室温催化氧化[60]。锰氧化物是臭氧催化分解的高效催化剂，锰氧化物的氧空位（三价锰）含量对臭氧催化分解活性有重要的影响[61,62]，通过对锰氧化物进行改性可以调控氧空位（三价锰）含量从而提高臭氧催化分解活性[63-65]。高湿度会导致臭氧分解催化剂的失活，开发在室温高湿度条件下高效分解臭氧的催化剂是未来研究的重点。在催化杀菌方面，在室温条件下催化剂可以活化空气中的氧气产生活性氧进行杀菌，是环境友好的杀菌方法之一。目前，用于催化杀菌的催化剂主要是银基或金属氧化物无机催化杀菌材料[66-68]，但催化材料的活性和稳定性还都有待提高；在研制更高性能的催化抗菌材料的同时，还需要解决杀菌活性组分 Ag 溶出和变色的问题。

2.2　工业 VOCs 净化材料

众所周知，石油化工、印刷、喷涂、包装、医药等行业排放的废气中含有大量的挥发性有机物（VOCs）。部分 VOCs 具有很高的大气化学反应活性，是光化学烟雾和细颗粒物 $PM_{2.5}$ 形成的重要前体污染物，严重危害大气环境和人身健康，已成为我国大气污染控制的重要方向。催化氧化法能将 VOCs 转化为 CO_2 和 H_2O，是目前公认的彻底消除 VOCs 的有效手段之一，高效催化材料的研发是该技术的核心。催化材料通常由载体、活性组分和助剂组成。理想的催化剂载体应具备良好的热稳定性、机械强度、传质性能、足够的比表面积和空隙结构、适当的导热和热膨胀系数[69]。Sanz 等比较了叠层钢丝网整体柱和平行沟道整体柱负载的 Pt/OMS-2 催化剂对甲醇或甲苯的催化净化性能，发现叠层钢丝网整体柱催化剂有利于反应物和产物的传质，从而表现出更好的催化性能[70]。活性组分是催化剂最为重要的部分，直接影响净化效率和能耗。Piumetti 等采用液态燃烧法制备了介孔 Mn_2O_3、Mn_3O_4 和 Mn_xO_y 催化剂，其中 Mn_3O_4 催化剂为具有较高的表面活性氧物质，从而对乙烯、丙烯、甲苯及其混合物表现出较好的去除效率；堇青石负载的介孔 Mn_3O_4 整体式催化剂表现出同粉末催化剂相当的催化净化 VOCs 的活性[71]。涂层方法同样是影响催化剂实际净化效率的重要因素，发展合适的涂层方法，能有效地缩小整体式催化剂与粉末催化剂性能之间的鸿沟[72]。另外，由于工业废气 VOCs 成分及性质复杂，以及单一治理技术的局限性，多种技术的耦合是今后发展的趋势，如吸附浓缩-催化氧化技术、等离子体-催化氧化技术、光热耦合催化氧化技术等。尽管如此，高效净化材料的开发依旧是 VOCs 末端处理技术的研究热点。

2.3 热催化水净化材料

在水体污染物的净化方面，湿式催化氧化、光催化技术、硝酸盐催化净化等已受到了较为普遍的关注。随着全球人口的增加，水资源短缺的问题更加突出，使得以催化的方法来净化水中的污染物有望成为环境催化净化技术发展的重要方向。尤其是在水体中农药及其代谢产物、卤代烃、内分泌活性物质等有机污染物消除方面，催化净化有望取得较其他技术更为彻底的去除效果。另外，发展催化净化技术与其他水处理技术的耦合集成，有望大幅度提高现有净化技术的效率，这也是未来水体污染物净化的重要研究内容之一。目前研究已经发现有些高活性氧化物如 MnO_2 可以在接近室温的条件下对酚类污染物具有氧化分解和完全矿化能力，有望成为在污水的热催化治理方面的新途径。

2.4 固定源脱硝催化净化材料

选择性催化还原（selective catalytic reduction，SCR）是应用最为广泛的烟气脱硝技术。该方法主要采用氨作为还原剂，将 NO_x 选择性地还原成 N_2。目前工业上应用最广的 SCR 催化剂是 V_2O_5-WO_3（MoO_3）/TiO_2，该催化剂具有较高的脱硝活性和抗 SO_2、H_2O 中毒能力，但操作窗口温度较高（>300 ℃），且飞灰中的碱金属对钒催化剂的使用寿命有不利影响。近年来，固定源脱硝催化净化材料的研究主要围绕钒基催化剂的改进和低温 SCR 催化剂的开发。

唐幸福等将活性组分 V_2O_5 负载在具有六方孔道结构的 WO_3 表面上，六方孔道可以固定碱金属和捕获 SO_2，防止其毒害表面的 SCR 活性中心，这一结构特性赋予了该催化剂优异的脱硝活性和抗碱金属中毒能力[73]。Mn 基氧化物催化剂具有优越的低温 SCR 活性，在 80~150℃温度范围内 NO_x 转化率可达 80%以上；但在含硫条件下，催化剂表面被生成的硫酸铵或硫酸氢铵覆盖，同时易发生 Mn 等活性组分的硫酸化，导致催化剂失活[74,75]。Fe 基氧化物在中高温区间具有优良的 SCR 活性；如 $FeTiO_x$ 催化剂，其活性相为铁钛复合氧化物微晶，Fe^{3+}和 Ti^{4+}之间存在的电子诱导效应使得 Fe^{3+}在 200~400℃温度区间具有高活性、N_2 选择性和抗 H_2O、SO_2 性能[76,77]。Ce 基氧化物催化剂如 $CeTiO_2$、$CeWO_x$ 等具有较优的 SCR 活性，还表现出较好的抗中毒性能[78-81]。活性炭来源丰富，价格低廉，易于再生，负载钒的活性炭催化剂具有较高的低温 SCR 活性，在固定源脱硝中具有一定的应用潜力[82,83]。

2.5　移动源尾气催化净化材料

以机动车为代表的移动源排放是造成大气污染的主要原因之一，削减其污染物排放是环境催化技术研发与应用的重要领域。汽油车尾气净化主要采用以三效催化剂（TWC）为核心的催化净化技术，可削减 90%以上的 CO、CH、NO_x 排放[84]。三效催化剂主要由稀土储氧材料、活性氧化铝等涂层关键材料构成，未来研究重点主要为：高性能稀土储氧材料与耐高温高比表面涂层材料研究；贵金属减量化与替代研究；降低催化剂起燃温度以减少发动机尤其是冷启动阶段污染物排放的研究等。

柴油车尾气净化技术主要包括柴油车氧化催化剂（DOC）、氨/尿素选择性催化还原 NO_x（Urea/NH_3-SCR），以及具有催化涂层的颗粒物捕集技术（CDPF）等[85]。排放标准的不断升级需要在提升上述单项技术净化效率的同时，加强其耦合连用，实现对 NO_x、PM 等污染物的高效协同净化。多技术耦合连用对 SCR 催化剂的水热稳定性提出了更为苛刻的要求，研发宽活性温度窗口、高热稳定性的新型小孔分子筛 SCR 催化剂是该领域的发展趋势与研究热点。以巴斯夫为代表的国外公司已经实现了满足欧 VI 标准的规模化应用，浙江大学、中国科学院生态环境研究中心等国内单位通过新型模板剂的设计，也实现了 Cu-SSZ-13 等小孔分子筛合成的突破[86,87]。此外，在柴油车 PM 净化材料方面，国内科研人员设计研发出了新型的三维有序大孔基催化剂，用于炭烟的高效催化燃烧[88]。

2.6　VOCs 净化材料新进展

与体相材料相比，有序多孔材料具有发达、贯通的孔道结构，有利于反应物、中间态物质和产物的吸附和扩散，从而表现出较高的催化净化 VOCs 的性能。戴洪兴等采用软-硬双模板法获得一系列高效催化净化苯系物的三维有序介孔或大孔过渡金属氧化物或钙钛矿型复合金属氧化物催化剂[89]；李俊华等发现将三维有序大孔 $LaMnO_3$ 酸处理后，制得的 $MnO_2/LaMnO_3$ 催化剂对甲苯表现出优异的催化净化性能[90]。在催化氧化法去除 VOCs 工艺中，虽然广泛使用的负载贵金属催化剂的性能优异，但贵金属价格昂贵，因此如何在保持催化净化性能的前提下，降低催化剂中贵金属用量是迫切需要解决的问题。提高贵金属利用率，例如开发单原子贵金属催化剂是可能的解决途径。唐幸福等在单原子 Ag 催化净化苯系物方面取得重要进展[91]。贵金属颗粒尺寸和配位环境也是决定贵金属催化剂性能的重要因素。通过控制贵金属颗粒尺寸可以提高活性位暴露的数量。叶代启教授采用甘油还原法制得 Pt 颗粒尺寸可控的 Pt/CeO_2 催化剂，发现当 Pt 颗粒为 1.8 nm 的催化剂对甲苯完全氧化反应表现出最高的催化活性[92]。另外，通过预处理气氛的选择、贵金属颗粒的修饰以及载体物化性质的调变，也可调控贵金属与载体之间的相互作用，从而提升其催化净化 VOCs 的活性、稳定性以及贵金属的利用率[93]。由于 VOCs 来源广泛，组成复杂，性质差异大，易导致催化剂的活性中心烧结、表面积碳和卤素等杂原子中毒等原因而失活。因此如何提高催化剂的稳定性、抗积碳性和耐毒性仍是一个需要关注的问题。通过调控贵金属与载体之间的相互作用，例如构筑贵金属颗粒以半镶嵌形式分布在载体上，可以提高贵金属颗粒的稳定性。通过调控载体表面亲疏水性和酸碱性，可以提高催化剂抗水性、抗 CO_2 性、耐 SO_2 性和抗卤素等杂原子中毒性。

3　电催化及光电热协同催化材料

自 1893 年 Thompson 发现电子以来，电化学就逐渐走近科学家的视野，也由此而衍生出电催化技术及光电热协同催化技术。对一般热力学有利的化学反应而言，若要反应速度提高10^5倍，反应温度需要从25℃升高到1000℃。而在电化学反应中，电极电势每改变1 V，相应的电极反应速率可以提高10^{10}倍，这是由于外加电极电位可以改变电极反应的活化能。污染物降解的本质就是得失电子的氧化还原反应，因而在环境净化领域，电催化技术具有独特的优势。目前在环境领域的应用主要包括水处理（污染物为烃类、醛类、醇类、酚类及胺类有机物）、CO氧化、CO_2还原及臭氧发生[94-103]。与其他高级氧化技术相比，电催化氧化法具有如下突出优点：

（1）常温常压下即可进行，反应比较温和。

（2）处理废水时，无需额外添加化学氧化剂，主要依靠电子转移，避免了因外加氧化剂而造成的二次污染。

（3）可操作性强，反应条件可通过外加电压、电流来调节。

（4）反应装置简便，工艺方法简单，操作容易。

（5）兼具絮凝、气浮、杀菌消毒作用。

（6）处理费用低，占地面积小，适合人口密度大的城市污水处理。

电催化是在电化学反应基础上，用催化材料直接作为电极（一般为金属）或在电极表面修饰催化材料（如：金属氧化物），通过外加偏压使催化剂表面发生氧化还原反应，提高电化学反应的效率。因此，电催化的效率不仅仅和外加偏电压、电解质有关，还和催化剂的组成、结构以及在催化剂表面形成的界面电场有关[104]。因此在电催化领域，电催化材料的选择对电催化反应具有极其重要的影响。目前主要的电催化材料有：纯金属、金属/金属氧化物、金属/碳载体、金属/金属氧化物/碳材料。

3.1 纯金属

Pt、Pd、Au、Ru等贵金属及其合金可以直接作为电极用于电催化反应。这类贵金属催化剂性质稳定，在电催化反应过程中不会因自身被氧化而出现电极腐蚀现象。同时对某些分子具有特异性吸附作用，有效降低反应的活化能。在电催化氧化有机污染物的反应中，阳极发生氧化反应，直接降解有机物或产生·OH、Cl_2、O_3等活性物种。该技术特别适用于难以生物降解或一般化学氧化难以作用的有机物。Pulgarin等以金属Pt为电极电催化氧化对位取代苯酚，发现Pt对含有不同给电子取代基和吸电子取代基的有机物均有降解效果，且给电子取代基有机物更易降解[105]。Isarain-Chavez和Pellicer等制备了不同金属的复合电极，研究其对甲基橙的电催化降解效果，比较了不同复合材料的活性顺序[106]。

3.2 金属/金属氧化物

金属氧化物以活性层修饰包覆在金属电极上，主要包Ru、Pt、Ir、Pb等氧化物或是经过掺杂改性的复合金属氧化物。Feng等制备了以Ti片为基底的Ti/RuO_2-Pt电极，研究发现该电极在电催化降解苯酚时具有极高的活性，显著高于Ti/IrO_2-Pt[107]。以Pb/PbO_2为阳极，可以有效地降解甲醛、苯胺、4-氯苯胺、苯酚、酸性蓝及其他多种染料[108-111]。其他氧化物如NiO、CuO、MnO等主要作为电极表面修饰材料。

3.3 金属/金属氧化物/碳材料

导电性的碳材料可以直接作为电极，也可以作为载体电极负载金属、金属氧化物形成复合电催化材料。主要的碳材料包括石墨、金刚石、CNTs、碳凝胶、碳纤维、富勒烯、石墨烯等。直接使用导电性较好的石墨电极即可实现对75%有机氯代化合物的分解。Brillas等以硼修饰的金刚石为电极，通过电化学催化降解甲基紫，发现该电极具有比 Pt 电极更高的催化活性[112]。通过金属或金属氧化物负载，可进一步提高电催化对污染物的降解效率。一方面金属负载可以提高电子传输效率，另一方面金属氧化物对有机污染物的特异性吸附可有效地降低反应活化能。Cesarino 等在玻碳电极上负载纳米 Ag 修饰的 CNTs，在水溶液中降解 10 mg/L 的苯时取得 77.9%的降解效率[113]。

目前单一电催化技术的主要问题是电极成本、寿命和法拉第效率等问题。如贵金属电极价格昂贵，难以大规模应用。Cu、Al、Fe 等金属电极在高电位时发生自溶。石墨电极虽然价格低廉但电极强度较差，在电流密度较高时电极损耗大、效率低。PbO_2 电极的稳定性好、成本低，但其电催化性能较低，难以彻底分解有机物。解决这一问题，一种思路是通过改善电极自身性质，包括改变电极形貌（纳米棒、纳米片、泡沫或多孔结构）或表面修饰（金属氧化物膜），提高电极材料的导电性、寿命和法拉第效率。另一条思路是发展光电热协同催化。结合光催化、电催化和热催化的各自优点实现对污染物的协同降解。这类材料主要包括金属热催化剂（Au、Pt、Pd），电催化材料（金属电极、C 电极），光催化剂（有机金属、TiO_2、ZnO、CdS、C_3N_4）中的两种或三种复合组分，可分为光催化/热催化材料、光催化/电催化材料、热催化/电催化材料及光催化/热催化/电催化复合材料。在降解甲醛、甲苯混合气体时，采用纳米 Pt 作为热催化剂，纳米 TiO_2 作为光催化剂，可以实现对甲醛甲苯的协同高效降解。Ye 等采用 Fe^{2+}/ C_3N_4 为光热协同催化剂，Fe^{2+}/Fe^{3+}为 Fenton 催化中心，C_3N_4 为光催化中心，可以实现对有机污染物的高效降解[114]。Zhao 等利用 TiO_2 电极，光电催化氧化 Cu-EDTA 的同时实现对 Cu^{2+}的还原沉积[115]。利用染料自身的光解特性，在不需要外加光催化剂时，也可以实现光辅助电催化降解。Khatae 等使用 Ti/RuO 为阳极，CNT 为阴极，利用光辅助技术加速了 DB129 染料的降解[116]。同时在 CO_2 资源化利用中，越来越多的研究人员开始尝试光电协同或光热协同催化还原 CO_2，成为该研究领域的一个新方向[117]。

4　臭氧催化净化材料

臭氧具有很强的氧化能力，可以净化有机污染物，在环境净化和杀菌领域有较广泛的应用。但单纯的臭氧氧化技术很难彻底降解有机物。为解决这一问题，科研人员将臭氧氧化与催化技术相结合，逐渐发展成为一个新的技术：臭氧催化净化技术。利用臭氧的强氧化性和催化剂的吸附、催化特性，可以有效地提高对有机物的降解效果。催化剂的作用主要为：①吸附富集污染物；②活化臭氧分子，产生新的活性氧物质（如羟基自由基、超氧自由基等）；③催化剂活化污染物，降低反应活化能。

臭氧催化净化材料可以分为均相催化材料和多相催化材料两类。在均相催化中，催化剂分布均匀且催化活性较高，作用机理较清楚。但是，由于催化剂混溶于水，易导致催化剂流失，也难以回收并容易产生二次污染。目前均相催化剂主要是 H_2O_2 或过渡金属离子（Mn^{2+}、Fe^{2+}、Ti^{4+}、Ni^{2+}、Cu^{2+}等）[118-121]。马军等在 O_3 体系引入 H_2O_2 并用于硝基苯的降解[122]。研究发现：O_3 与硝基苯的反应较为缓慢，硝基苯的去除率较低。加入 H_2O_2 后硝基苯的去除率和氧化率得到显著提高，·OH 是硝基苯氧化降解的主要活性物质。金属离子的加入通过两种方式参与反应。金属离子可引发链反应产生·OH 或金属离子与目标物络合成一种更易被臭氧氧化的中间产物。Piera 等及 Sauleda 等发现 2,4-氯酚和苯胺在三种体系 O_3/Fe^{2+}、O_3/Fe^{2+}/UV、UV/Fe^{2+}下的矿化率呈不同的效能[121,123]。Andreozzi 等研究金属 Mn^{2+}催化臭氧降解草酸，结果表明：酸性条件下，Mn^{2+}首先与草酸发生络合反应生成一种易于被臭氧降解的中间物[124]。

在多相催化中，催化剂是固体，易于与水分离，二次污染少，简化了处理流程，因而越来越引起人们的广泛重视。目前主要的臭氧多相催化材料是金属、金属氧化物、负载于载体上的金属或金属氧化物，以及具有较大比表面积的孔材料[125]。这些催化剂的催化活性主要表现对臭氧的催化分解和促进羟基自由基的产生。臭氧催化氧化过程的效率主要取决于催化剂及其表面性质、溶液 pH，这些因素能影响催化剂表面活性位的性质和溶液中臭氧分解的反应。

4.1 金属催化剂（包括负载型）

纳米金属催化剂能够促使水中臭氧分解产生具有极强氧化性的自由基，从而显著提高其对水中高稳定性有机物的分解效果。常用的金属催化剂包括 Ru、Cu、Ti、Ni 等[126-128]。

4.2 金属氧化物（包括负载型）

金属氧化物的合理选用可直接影响催化反应机理和效率。一般金属氧化物表面上的羟基基团是催化反应的活性位，它通过向水中释放质子和羟基，发生离子交换反应而从水中吸附阴离子和阳离子，形成 Bröonsted 酸位，而该酸位通常被认为是金属氧化物的催化中心。

金属/金属氧化物是研究较多的复合材料。李等将金属 Cu 负载在 Al_2O_3 上，研究其对内分泌干扰素甲草胺的降解效果，并与单独臭氧化对比[126] 。结果表明 Cu/Al_2O_3 催化臭氧化对甲草胺的去除率达 60%，单独臭氧化对甲草胺的去除率仅为 20%，对 Cu/Al_2O_3 催化臭氧化降解机理研究发现甲草胺是通过·OH 氧化降解。Al-Haye 等以 Al_2O_3 为载体，用浸渍法制备了 Fe^{3+}/Al_2O_3 固体催化剂，应用于臭氧催化氧化苯酚降解实验[129]。研究发现：体系 TOC 的去除率较单独使用臭氧时显著提高。单独臭氧化时 TOC 的最大去除率不到 40%，加入空白 Al_2O_3 载体后去除率可提高到 70%，而加入制备的催化剂后最大去除率可以达到 90%。可能是由于催化剂的加入导致了体系中大量羟基自由基的生成。

国内学者也在这方面做了大量工作。周云瑞等研究了不同载体负载 Ru 催化剂对催化臭氧氧化降解邻苯二甲酸二甲酯的效果[130]。结果表明：RuO_2/SiO_2 催化剂对臭氧氧化效果的提高并不显著，而 $RuO_2/A1_2O_3$ 催化剂可显著增加降解效果，反应 120 min DMP 溶液体系 TOC 去除率约为 72%。但是 $RuO_2/A1_2O_3$ 因为其催化臭氧氧化过程会产生铝溶出的问题，会对水体造成二次污染，并影响催化剂活性。张彭义等在研究染料中间体废水的降解实验中，探讨了 Ni、Cu 氧化物的催化臭氧氧化作用[131]。祝万鹏和李来胜等共同研究了几种过渡金属：Cu、Ni、Mn、Fe、Cd 和贵金属 Ag 的氧化物作为活性组分时相应的固体催化剂在臭氧氧化去除水中氯乙酸实验中的催化效果[132]。李燕等研究了用 Cu/Al_2O_3 固体催化剂催化臭氧氧化降解水中的甲草胺[126]。于欣等用异丙醇钛盐表面改性法制备 TiO_2/SiO_2 催化剂，对甲醛废水进行催化臭氧氧化处理[133]。朱丽勤等用浸渍法把过渡金属氧化物 NiO、CuO、Fe_2O_3、Ag_2O 和 MnO 负载到 Al_2O_3 上，并研究了催化剂对染料的降解效果[134]。结果表明，NiO/Al_2O_3 具有最好的催化效果，可对 6 种染料溶液进行有效降解。

4.3 碳材料

碳材料（如活性炭）一般比表面积大，且表面含有大量的酸性或碱性基团（羟基、酚羟基、羧基）。这些酸性或碱性基团的存在，使碳材料不仅具有吸附能力，而且还具有催化能力。臭氧/碳材料协同作用过程中，碳材料会加速臭氧转变为羟基自由基，从而提高氧化效率。碳材料作为催化剂与金属氧化物的区别在于对臭氧的分解机理不同：碳材料表面的路易斯碱起主要作用；而金属氧化物表面的路易斯酸是催化活性位点。周云瑞等制备活性炭负载的 Ru 催化剂，可以显著地提高催化臭氧氧化的效果，反应 120 min 邻苯二甲酸二甲酯溶液体系 TOC 去除率约为 75%，42 h 的连续实验也证明了该催化剂具有很好的稳定性和较长的使用寿命，且催化过程中没有载体溶出的问题[130]。

5 化学品的绿色低碳化合成及其催化材料

随着社会经济文明的快速发展，人们的生活质量日益提高，环境保护意识也越来越强。各国政府针对环境污染问题采取了多种治理方法，减少污染物对环境的破坏作用。在此过程中，人们逐渐意识到，先污染再治理的发展模式存在极大的弊端，应该逐渐被抛弃。21 世纪化学和化工的发展主流必然是建立在“原子经济性”基础上的绿色化学策略，即从源头控制污染物的产生和排放[135-137]。

化学品的绿色低碳化合成是绿色化学的核心内容。为实现这一目的，除了对反应过程的优化设计、反应器的更新升级、反应条件的改良优化外，适合绿色化学所需要的高效催化剂的研发至关重要。绝大部分现有的化学品生产均需要催化剂，尤其是精细化工产品。要制备适合绿色化学的高效催化剂必须从催化原理上认识催化过程本质。催化过程主要涉及反应物、溶剂、催化剂和产物。简单来说，如果反应物 100%高效转化为产物，同时反应无副产物、低能耗、催化剂可分离并重复使用、溶剂可分离或重复使用，这个反应就符合绿色化学的“原子经济性”。具体到催化剂上，就需要催化剂具有低温高效性、高选择性（含手型选择性）、可分离性、适应水相反应。由于合成反应体系复杂多变，针对不同催化反应的绿色化过程往往对催化剂有具体的要求。主要研究方向包括：

5.1 一锅化反应催化剂

一般化学品合成往往需要几步反应相互串联，每次的分离和转移产物均会造成反应中间体（产物）的损失和溶剂的浪费。一锅法反应的催化剂即是将两步或多步反应所需要的活性中心集成在一种复合催化剂上，通过对其他反应条件的控制（如溶剂的选择）实现多步反应一锅化。张涛等采用 W/Ru/C 复合催化剂一锅法将纤维素转化为乙二醇，该催化剂显示出高活性、高选择性和易再利用性[138]。李辉等将纳米 Ru-B 催化剂和生物酶集成在核壳结构的介孔碳、氧化硅材料上，制备出 Ru−B/mCarbon@air@mSiO_2 催化剂，一锅法实现了糊精加氢到山梨醇[139]。

5.2 高选择性（含手型选择性）催化剂

对产物选择性的提高可以降低后期分离的复杂性和能耗。尤其是高选择性手性催化剂的研发，对药物化学有着至关重要的意义。乔明华等用 W 修饰 Pt/SBA-15 催化剂对丙三醇选择性加氢制备 1,3-丙二醇[140]。该催化剂对丙三醇的转化率达到 86.8%，对 1,3-丙二醇的选择性高达 70.8%。Nilsson 等在研究电催化还原 CO_2 的反应中，发现立方体结构的纳米铜对 CO_2 转化为乙烯具有很高的选择性，显著高于对甲烷的选择性[141]。

5.3 水相反应催化剂

绝大多数的精细有机合成反应均在有毒有害的有机溶剂中进行。我国每年都需要数亿吨的有机溶剂进行有机反应和产物分离，有机溶剂的挥发和排放构成了严重的大气、水体和土壤污染[142]。在所有溶剂中水是最清洁的溶剂，使用水替代有机溶剂，反应的绿色性将大为提高。Li 等在水相中用三异丙基硅基乙炔直接在 Morita–Baylis–Hillman 碳酸盐上引入碳碳三键，形成 1,4 位烯炔结构。该工作详细研究了水在该反应中的作用[143]。李和兴等研究了 $RuCl_2(PPh_3)_3/SiO_2$ 催化剂，并应用于水介质中的烯丙基化反应，研究发现该催化剂可以在水相中实现接近有机相反应的高转化率[144]。

5.4 均相催化剂固载化

均相催化剂具有高活性和高选择性，因而大部分反应采用均相催化剂。但产物和催化剂的分离往往比较困难，有时甚至不可能，这样既造成了污染，又浪费了催化剂，整个合成过程的效率大为减低[145,146]。杨启华等以有机功能化的中空氧化硅为载体，利用嫁接法、浸渍法等技术实现了小分子催化剂、有机金属催化剂的固载化，并研究了这些催化剂在手性催化和串联反应中的性能[147]。李和兴等选用介孔氧化硅通过自组装法或后嫁接等方法固载了一系列膦配体的有机金属催化剂[$PdCl_2(PPh_3)_2$ 和 RhCl（PPh_3）$_2$等]，将其应用于醇异构化反应。结果表明，由于介孔氧化硅具有大比表面积、规整的介孔孔道结构以及可调的载体表面性质，使得催化剂具有接近甚至达到均相有机金属催化剂的催化效率，同时催化剂具有良好的稳定性，可以多次重复使用[148,149]。

参考文献

[1] 黄燕娣, 胡玢, 王栋, 等. 国内外室内空气污染研究进展. 中国环保产业, 2002, 12: 44-45.

[2] 韩世同, 李静, 习海玲, 等. 化学武器及其模拟剂的光催化消除研究进展. 化工科技, 2005, 5: 49-53.

[3] Asahi R, Morikawa T, Irie H, et al. Nitrogen-doped titanium dioxide as visible-light-sensitive photocatalyst: Designs, developments, and prospects. Chem Rev, 2014, 114(19): 9824-9852.

[4] Chen H, Nanayakkara C E, Grasslan V H. Titanium dioxide photocatalysis in atmospheric chemistry. Chem Rev, 2012, 112(11): 5919-5948.

[5] Xie H, Zhang Y, Xu Q. Photodegradation of VOCs by C_TiO_2 nanoparticles produced by flame CVD process. J Nanosci Nanotechnol, 2010, 10(8): 5445-5450.

[6] Zhang N, Cirminna R, Pagliaro M, et al. Nanochemistry-derived Bi_2WO_6 nanostructures: Towards production of sustainable chemicals and fuels induced by visible light. Chemical Society Reviews, 2014, 43(15): 5276-5287.

[7] Salih H, Patterson C, Sorial G. Adsorption of VOCs by activated carbon in the presence and absence of Fe_2O_3 NPs and humic acid. Abstracts of Papers of the American Chemical Society, 2011, 241.

[8] Zhu B L, Xie C S, Wang W Y, et al. Improvement in gas sensitivity of ZnO thick film to volatile organic compounds (VOCs) by adding TiO_2. Materials Letters, 2004, 58(5): 624-629.

[9] Choi Y I, Lee S, Kim S K, et al. Fabrication of ZnO, ZnS, Ag-ZnS, and Au-ZnS microspheres for photocatalytic activities, CO oxidation and 2-hydroxyterephthalic acid synthesis. Jouranl of Alloys and Compounds, 2016, 675: 46-56.

[10] Li X, Yu J, Jaroniec M. Hierarchical photocatalysts. Chemical Society Reviews, 2016, 45(9): 2603-2636.

[11] Han M Y, Huang W, Quek C H, et al. Preparation and enhanced photocatalytic oxidation activity of surface-modified CdS nanoparticles with high photostability. J Mater Res, 1999, 14(5): 2092-2095.

[12] Ong W-J, Tan L-L, Ng Y H, et al. Graphitic carbon nitride(g-C_3N_4)-based photocatalysts for artificial photosynthesis and environmental remediation: Are we a step closer to achieving sustainability? Chem Rev, 2016, 116(12): 7159-7329.

[13] Muktha B, Madras G, Gururow T N, et al. Conjugated polymers for photocatalysis. J Phys Chem B The Journal of Physical Chemistry B, 2007, 111(28): 7994-7998.

[14] Ardizzone S, Blanchi C L, Cappelletti G, et al. Photocatalytic degradation of toluene in the gas phase: Relationship between surface species and catalyst features. Environ Sci Technol, 2008, 42(17): 6671-6676.

[15] Fu H, Zhang S, Zhang L, et al. Visible-light-driven $NaTaO_3$-xN_x catalyst prepared by a hydrothermal process. Materials Research Bulletin, 2008, 43(4): 864-872.

[16] Xu T-G, Zhang C, Shao X, et al. Monomolecular-layer $Ba_5Ta_4O_{15}$ nanosheets: Synthesis and investigation of photocatalytic properties. Advanced Functional Materials S, 2006, 16(12): 1599-1607.

[17] 张立武, 朱永法, Zhu Y. 钨钼酸盐复合氧化物新型可见光光催化研究. 中国材料进展, 2010, 01: 45-53.

[18] Finlayson A P, Tsaneva V N, Lyons L, et al. Evaluation of Bi-W-oxides for visible light photocatalysis. Physica Status Solidi A-Applications and Materials Science, 2006, 203(2): 3273-35.

[19] Lin Z, Li J, Zheng Z, et al. Electronic Reconstruction of alpha-Ag_2WO_4 nanorods for visible-light photocatalysis. ACS Nano, 2015, 9(7): 7256-7265.

[20] Dumrongrojthanath P, Thongtem T, Phuruangrat A, et al. Glycolthermal synthesis of Bi_2MoO_6 nanoplates and their photocatalytic performance. Materials Letters, 2015, 154: 180-183.

[21] Liu D, Yao W, Wang J, et al. Enhanced visible light photocatalytic performance of a novel heterostructured $Bi_4O_5Br_2$/$Bi_{24}O_{31}Br_{10}$/Bi_2SiO_5 photocatalyst. Applied Catalysis B: Environmental, 2015, 172(100-107).

[22] Jiang W, Liu Y, Wang J, et al. Separation-free polyaniline/TiO_2 3D hydrogel with high photocatalytic activity. Adv Mater Interfaces, 2016, 3(3): 1500502.

[23] Jiang W, Luo W, Zong R, et al. Polyaniline/carbon nitride nanosheets composite hydrogel: A separation - free and high - Efficient photocatalyst with 3D hierarchical structure. Small, 2016, 12(32): 4370-4378.

[24] Li Y, Cui W, Liu L, et al. Removal of Cr(VI) by 3D TiO_2-graphene hydrogel via adsorption enriched with photocatalytic reduction. Appl Catal B: Environ, 2016, 199: 412-423.

[25] Liu D, Wang J, Bai X, et al. Self-assembled PDINH supramolecular system for photocatalysis under visible light. Adv Mater, 2016, 28(33): 7284.

[26] Wang J, Shi W, Liu D, et al. Supramolecular organic nanofibers with highly efficient and stable visible light photooxidation performance. Applied Catalysis B: Environmental, 2017, 202: 289-297.

[27] Zhang Z J, Wang J, Liu D, et al. Highly efficient organic photocatalyst with full visible light spectrum through π–π stacking of TCNQ–PTCDI. ACS Appl Mater Inter, 2016, 8(44): 30225-30231.

[28] He Z G, Li J J, Chen J Y, et al. Treatment of organic waste gas in a paint plant by combined technique of biotrickling filtration with photocatalytic oxidation. Chem Eng J, 2012, 200: 645-653.

[29] Chen J Y, Huang Y, Li G Y, et al. VOCs elimination and health risk reduction in e-waste dismantling workshop using integrated techniques of electrostatic precipitation with advanced oxidation technologies. J Hazard Mater, 2016, 302: 395-403.

[30] Sousa M, Gon Alves C, Vilar V J, et al. Suspended TiO_2-assisted photocatalytic degradation of emerging contaminants in a municipal WWTP effluent using a solar pilot plant with CPCs. Chemical Engineering Journal, 2012, 198: 301-309.

[31] Sugihara M N, Moeller D, Paul T, et al. TiO_2-photocatalyzed transformation of the recalcitrant X-ray contrast agent diatrizoate. Applied Catalysis B: Environmental, 2013, 129: 114-122.

[32] Chen J, Ollis D F, Rulkens W H, et al. Photocatalyzed oxidation of alcohols and organochlorides in the presence of native TiO_2 and metallized TiO_2 suspensions. Part(II): Photocatalytic mechanisms. Water Res, 1999, 33(3): 669-676.

[33] Khataee A, Kasiri M B. Photocatalytic degradation of organic dyes in the presence of nanostructured titanium dioxide: influence of the chemical structure of dyes. Journal of Molecular Catalysis A: Chemical, 2010, 328(1): 8-26.

[34] Sharma S K, Sanghi R. Advances in water treatment and pollution prevention. Springer Science & Business Media, 2012.

[35] Yang C, Liang G L, Xu K M, et al. Bactericidal functionalization of wrinkle-free fabrics via covalently bonding TiO_2@ Ag nanoconjugates. J Mater Sci, 2009, 44(7): 1894-1901.

[36] Sun C X, Li Q, Gao S A, et al. Enhanced photocatalytic disinfection of *Escherichia coli* bacteria by silver modification of nitrogen-doped titanium oxide nanoparticle photocatalyst under visible-light illumination. J Am Ceram Soc, 2010, 93(11): 3880-3885.

[37] Lin H, Deng W, Zhou T, et al. Iodine-modified nanocrystalline titania for photo-catalytic antibacterial application under visible light illumination. Applied Catalysis B: Environmental, 2015, 176, 77: 36-43.

[38] Yang X F, Qin J L, Jiang Y, et al. Bifunctional TiO_2/Ag_3PO_4/graphene composites with superior visible light photocatalytic performance and synergistic inactivation of bacteria. Rsc Advances, 2014, 4(36): 18627-18636.

[39] Hou Y, Li X Y, Zhao Q D, et al. Role of hydroxyl radicals and mechanism of *Escherichia coli* inactivation on Ag/AgBr/TiO_2 nanotube array electrode under visible light irradiation. Environ Sci Technol, 2012, 46(7): 4042-4050.

[40] Xu Y G, Huang S Q, Xie M, et al. Core-shell magnetic Ag/AgCl@Fe_2O_3 photocatalysts with enhanced photoactivity for eliminating bisphenol A and microbial contamination. New Journal of Chemistry, 2016, 40(4): 3413-3422.

[41] Huang J H, Ho W K, Wang X C. Metal-free disinfection effects induced by graphitic carbon nitride polymers under visible

light illumination. Chemical Communications, 2014, 50(33): 4338-4340.

[42] Wang W J, Yu J C, Xia D H, et al. Graphene and g-C_3N_4 nanosheets cowrapped elemental alpha-sulfur As a novel metal-free heterojunction photocatalyst for bacterial inactivation under visible-light. Environ Sci Technol, 2013, 47(15): 8724-8732.

[43] Xia D, Shen Z, Huang G, et al. Red phosphorus: An earth-abundant elemental photocatalyst for "green" Bbacterial inactivation under visible light. Environ Sci Technol, 2015, 49(10): 6264-6273.

[44] Chen Y, Lu A, Li Y, et al. Naturally Occurring Sphalerite As a Novel Cost-Effective Photocatalyst for Bacterial Disinfection under Visible Light. Environ Sci Technol, 2011, 45(13): 5689-5695.

[45] Xia D, Ng T W, An T, et al. A Recyclable mineral catalyst for visible-light-driven photocatalytic inactivation of bacteria: Natural Magnetic Sphalerite. Environ Sci Technol, 2013, 47(19): 11166-11173.

[46] Shi H, Huang G, Xia D, et al. Role of in situ resultant H_2O_2 in the visible-light-driven photocatalytic inactivation of *E. coli* using natural sphalerite: A genetic study. The Journal of Physical Chemistry B, 2015, 119(7): 3104-3111.

[47] Xia D, Li Y, Huang G, et al. Visible-light-driven inactivation of *Escherichia coli* K-12 over thermal treated natural pyrrhotite. Applied Catalysis B: Environmental, 2015, 176-177(0): 749-756.

[48] Matsunaga T, Tomoda R, Nakajima T, et al. Photoelectrochemical Sterilization of Microbial-Cells by Semiconductor Powders. Fems Microbiol Lett, 1985, 29(1-2): 211-214.

[49] Gao M, An T, Li G, et al. Genetic studies of the role of fatty acid and coenzyme A in photocatalytic inactivation of Escherichia coli. Water Res, 2012, 46(13): 3951-3957.

[50] Liu P, Duan W L, Wang Q S I, et al. The damage of outer membrane of *Escherichia coli* in the presence of TiO_2 combined with UV light. Colloid Surface B, 2010, 78(2): 171-176.

[51] An T, Sun H, Li G, et al. Differences in photoelectrocatalytic inactivation processes between E. coli and its isogenic single gene knockoff mutants: Destruction of membrane framework or associated proteins. Applied Catalysis B: Environmental, 2016, 188: 360-366.

[52] Sun H, Li G, Nie X, et al. Systematic Approach to in-depth understanding of photoelectrocatalytic bacterial inactivation mechanisms by tracking the decomposed building blocks. Environ Sci Technol, 2014, 48(16): 9412-9419.

[53] Sun H, Li G, An T, et al. Unveiling the photoelectrocatalytic inactivation mechanism of *Escherichia coli*: Convincing evidence from responses of parent and anti-oxidation single gene knockout mutants. Water Res, 2016, 88(1): 35-43.

[54] Zhang C, Liu F, Zhai Y, et al. Alkali-metal-promoted Pt/TiO_2 opens a more efficient pathway to formaldehyde oxidation at ambient temperatures. Angewandte Chemie International Edition, 2012, 51: 9628–9632.

[55] Tang X, Chen J, Huang X, et al. Pt/MnO_x-CeO_2 catalysts for the complete oxidation of formaldehyde at ambient temperature. Applied Catalysis B: Environmental, 2008, 81: 115–121.

[56] Huang H, Leung D Y C. Complete elimination of indoor formaldehyde over supported Pt catalysts with extremely low Pt content at ambient temperature. Journal of Catalysis, 2011, 280: 60–67.

[57] Chen B, Shi C, Crocker M, et al. Catalytic removal of formaldehyde at room temperature over supported gold catalysts. Applied Catalysis B, 2013, 132–133: 245–255.

[58] Xu J, White T, Li P, He C, Han Y. Hydroxyapatite foam as a catalyst for formaldehyde combustion at room temperature, J. Am. Chem. Soc. 2010, 132: 13172–13173.

[59] Huang Y, Fan W, Long W, Li H, Qiu W, Zhao F, Tong Y, Ji H. Alkali-modified non-precious metal 3D-NiCo2O4 nanosheets for efficient formaldehyde oxidation at low temperature, J. Mater. Chem. A, 2016, 4: 3648–3654.

[60] Ma C Y, Mu Z, Li J J, Jin Y G, Cheng J, Lu G Q, Hao Z P, Qiao S Z. Mesoporous Co_3O_4 and Au/Co_3O_4 Catalysts for Low-Temperature Oxidation of Trace Ethylene, J. Am. Chem. Soc., 2010, 132: 2608–2613.

[61] Jia J B, Zhang P Y, Chen L. The effect of morphology of alpha-MnO_2 on catalytic decomposition of gaseous ozone. Catal. Sci. Technol. 2016, 6(15): 5841–5847.

[62] Jia J B, Zhang P Y, Chen L. Catalytic decomposition of gaseous ozone over manganese dioxides with different crystal structures. Appl. Catal., B 2016, 189: 210–218.

[63] Liu Y, Zhang P Y. Catalytic decomposition of gaseous ozone over todorokite-type manganese dioxides at room temperature: Effects of cerium modification. Appl. Catal., A 2017, 530: 102–110.

[64] Ma J, Wang C, He H. Transition metal doped cryptomelane-type manganese oxide catalysts for ozone decomposition. Appl. Catal., B 2017, 201: 503–510.

[65] Wang C, Ma J, Liu F, He H, Zhang R. The Effects of Mn^{2+} Precursors on the Structure and Ozone Decomposition Activity of Cryptomelane-Type Manganese Oxide (OMS-2) Catalysts. J. Phys. Chem. C 2015, 119(40): 23119–23126.

[66] Chang Q Y, He H, Ma Z C. Efficient disinfection of *Escherichia coli* in water by silver loaded alumina. Journal of Inorganic Biochemistry, 2008, 102: 1736–1742.

[67] Hirota K, Sugimoto M, Kato M, et al. Preparation of zinc oxide ceramics with a sustainable antibacterial activity under dark conditions. Ceramics International, 2010, 36. 497–506.

[68] Wang L, He H, Zhang C, Sun L, Liu S, Wang S. Antimicrobial activity of silver loaded MnO_2 nanomaterials with different crystal phases against *Escherichia coli*. J. Environ. Sci., 2016, 41: 112–120.

[69] Rodríguez M L, Cadús L E, Borio D O. VOCs abatement in adiabatic monolithic reactors: Heat effects, transport limitations and design considerations. Chem Eng J, 2016, 306: 86–98.

[70] Sanz O, Banús E D, Goya A, Larumbe H, Delgado J J, Monzón A, Montes M. Stacked wire-mesh monoliths for VOCs combustion: Effect of the mesh opening in the catalytic performance. Catal Today, 2017, doi.org/10.1016/j.cattod. 2017.05. 054.

[71] Piumetti M, Fino D, Russo N. Mesoporous manganese oxides prepared by solution combustion synthesis as catalysts for the total oxidation of VOCs. Appl Catal B, 2015, 163: 277–287.

[72] Hernández-Garrido J C, Gaona D, Gómez D M, Gatica J M, Vidal H, Sanz O, Rebled J M, Peiró F, Calvino J J. Comparative study of the catalytic performance and final surfacestructure of Co_3O_4/La-CeO_2 washcoated ceramic and metallic honeycomb monoliths. Catal Today, 2015, 253: 190–198.

[73] Huang Z, Li H, Gao J, Gu X, Zheng L, Hu P, Xin Y, Chen J, Chen Y, Zhang Z, Chen J, Tang X. alkali- and sulfur-resistant tungsten-based catalysts for NO_x emissions control. Environ. Sci. Technol., 2015, 49: 14460–14465.

[74] Liu F, Shan W, Lian Z, Xie L, Yang W, He H. Novel $MnWO_x$ catalyst with remarkable performance for low temperature NH_3-SCR of NO_x. Catal. Sci. Technol., 2013, 3: 2699–2707.

[75] Yang S, Qi F, Xiong S, Dang H, Liao Y, Wong P K, Li J. MnO_x supported on Fe-Ti spinel: A novel Mn based low temperature SCR catalyst with a high N2 selectivity. Appl. Catal. B: Environ., 2016, 181: 570–580.

[76] Liu F, He H, Zhang C. Novel iron titanate catalyst for the selective catalytic reduction of NO with NH_3 in the medium temperature range. Chem. Commun., 2008, 2043–2045.

[77] Yang S, Li J, Wang C, Chen J, Ma L, Chang H, Chen L, Peng Y, Yan N. Fe-Ti spinel for the selective catalytic reduction of NO with NH3: Mechanism and structure - activity relationship, Appl. Catal. B: Environ., 2012, 117 - 118: 73 - 80.

[78] Shan W, Liu F, He H, Shi X, Zhang C. The Remarkable improvement of a Ce-Ti based catalyst for NOx abatement, prepared by a homogeneous precipitation method. ChemCatChem., 2011, 3: 1286–1289.

[79] Shan W, Liu F, He H, Shi X, Zhang C. Novel cerium-tungsten mixed oxide catalyst for the selective catalytic reduction of NOx with NH3. Chem. Commun., 2011, 47: 8046–8048.

[80] Liu Y, Yao W, Cao X, Weng X, WangY, Wang H, Wu Z. Supercritical water syntheses of CexTiO_2 nano-catalysts with a strong metal-support interaction for selective catalytic reduction of NO with NH_3. Appl. Catal. B: Environ., 2014, 160-161: 684–691.

[81] Li X, Li X, Li J, Hao J. High calcium resistance of CeO_2-WO_3 SCR catalysts: Structure investigation and deactivation analysis. Chem. Engine. J. 2017, 317: 70–79.

[82] Huang H, Ye D, Huang B, Wei Z. Vanadium supported on viscose-based activated carbon fibers modified by oxygen plasma for the SCR of NO. Catal. Today, 2008, 139: 100–108.

[83] Xiao Y, Liu Q, Liu Z, Huang Z, Guo Y, Yang J. Roles of lattice oxygen in V_2O_5 and activated coke in SO_2 removal over coke-supported V_2O_5 catalysts. Appl. Catal. B: Environ, 2008, 82: 114–119.

[84] Granger P, Parvulescu V I. Catalytic NOx abatement systems for mobile sources: From three-way to lean burn after-treatment technologies. Chem Rev, 2011, 111: 3155–3207.

[85] Johnson T V, Review of diesel emissions and control. Int J Engine Res, 2009, 10: 275–285.

[86] Ren L, Zhu L, Yang C, Chen Y, Sun Q, Zhang H, Li C, Nawaz F, Meng X, Xiao F-S. Designed copper-amine complex as an efficient template for one-pot synthesis of Cu-SSZ-13 zeolite with excellent activity for selective catalytic reduction of NO_x by NH_3. Chem Commun, 2011, 47: 9789–9791.

[87] Xie L, Liu F, Ren L, Shi X, Xiao F-S, He H. Excellent performance of one-pot synthesized Cu-SSZ-13 catalyst for the selective catalytic reduction of NO_x with NH_3. Environ Sci Technol, 2014, 48: 566–572.

[88] Wei Y, Liu J, Zhao Z, Chen Y, Xu C, Duan A, Jiang G, He H. Highly active catalysts of gold nanoparticles supported on three-dimensionally ordered macroporous $LaFeO_3$ for soot oxidation. Angew. Chem. Int. Ed., 2011, 50: 2326–2329.

[89] Liu Y X, Dai H X, Du Y C, Deng J G, Zhang L, Zhao Z X, Au C T. Controlled preparation and high catalytic performance of three-dimensionally ordered macroporous $LaMnO_3$ with nanovoid skeletons for the combustion of toluene, J. Catal., 2012, 287: 149–160.

[90] Si W Z, Wang Y, Zhao S, Hu F Y, Li J H. A facile method for in situ preparation of the $MnO_2/LaMnO_3$ catalyst for the removal of toluene. Environ Sci Technol, 2016, 50: 4572–4578.

[91] Chen Y X, Huang Z W, Zhou M J, Ma Z, Chen J N, Tang X F. Single silver adatoms on nanostructured manganese oxide surfaces: Boosting oxygen activation for benzene abatement. Environ Sci Technol, 2017, 51: 2304–2311.

[92] Peng R S, Li S J, Sun X B, Ren Q M, Chen L M, Fu M L, Wu J L, Ye D Q. Size effect of Pt nanoparticles on the catalytic oxidation of toluene over Pt/CeO_2 catalysts. Appl Catal B, doi.org/10.1016/j.apcatb.2017.07.048.

[93] Zhao S, Li K Z, Jiang S, Li J H. Pd–Co based spinel oxides derived from Pd nanoparticles immobilized on layered double hydroxides for toluene combustion. Appl Catal B, 2016, 181: 236–248.

[94] Kirk D W, Sharifian H, Foulkes F R. Anodic oxidation of aniline for waste water treatment. Journal of Applied Electrochemistry, 1985, 15(2): 285-292.

[95] Panizza M, Cerisola G. Direct and mediated anodic oxidation of organic pollutants. Chemical Reviews, 2009, 109(12): 6541-6569.

[96] Ciriaco L, Anjo C, Correia J, et al. Electrochemical degradation of ibuprofen on $Ti/Pt/PbO_2$ and Si/BDD electrodes. Electrochimica Acta, 2009, 54(5): 1464-1472.

[97] Chiang L C, Chang J E, Wen T C. Indirect oxidation effect in electrochemical oxidation treatment of landfill leachate. Water Res, 1995, 29(2): 671-678.

[98] Franklin T C, Oliver G, Nnodimele R. Destruction of halogenated hydrocarbons accompanied by generation of electricity. Journal of the Electrochemical Society, 1992, 139(8): 2192-2195.

[99] Polcaro A M, Mascia M, Palmas S, et al. Electrochemical oxidation of phenolic and other organic compounds at boron doped diamond electrodes for wastewater treatment: Effect of mass transfer. Annali Di Chimica, 2003, 93(12): 967-976.

[100] Polcaro A M, Palmas S, Renoldi F, et al. On the performance of Ti/SnO_2 and Ti/PbO_2 anodes in electrochemical degradation of 2-chlorophenolfor wastewater treatment. Journal of Applied Electrochemistry, 1999, 29(2): 147- 151.

[101] Maillard F, Eikerling M, Cherstiouk O V, et al. Size effects on reactivity of Pt nanoparticles in CO monolayer oxidation: The role of surface mobility. Faraday, 2004, 125, 357-377.

[102] Li X, Cui Y, Feng Y, et al. Reaction pathways and mechanisms of the electrochemical degradation of phenol on different electrodes. Water Res, 2005, 39(10): 1972-1981.

[103] Sirés I, Brillas E. Remediation of water pollution caused by pharmaceutical residues based on electrochemical separation and degradation technologies: A review. Environment International, 2012, 40(4): 212-229.

[104] Martinez-Huitle C A, Ferro S, Electrochemical oxidation of organic pollutants for The wastewater treatment: direct and indirect processes. Chemical Society Reviews, 2006, 35(12): 1324-1340.

[105] Torres R A, Torres W, Peringer P, et al. Electrochemical degradation of p-substituted phenols of industrial interest on Pt electrodes: Attempt of a structure-reactivity relationship assessment. Chemosphere, 2003, 50(1): 97-104.

[106] Isarain-Chóvez E, Baró M D, Rossinyol E, et al. Comparative electrochemical oxidation of methyl orange azo dye using Ti/Ir-Pb, Ti/Ir-Sn, Ti/Ru-Pb, Ti/Pt-Pd and Ti/RuO_2 anodes. Electrochimica Acta, 2017, 244: 199-208.

[107] Li M, Feng C, Hu W, et al. Electrochemical degradation of phenol using electrodes of Ti/RuO_2-Pt and Ti/IrO_2-Pt. Journal of Hazardous Materials, 2009, 162(1): 455-462.

[108] Awad H S, Galwa N A. Electrochemical degradation of acid blue and basic brown dyes on Pb/PbO_2 electrode in the presence of different conductive electrolyte and effect of various operating factors. Chemosphere, 2005, 61(9): 1327-1335.

[109] Duan X, Ma F, Yuan Z, et al. Electrochemical degradation of phenol in aqueous solution using PbO_2 anode. Journal of the Taiwan Institute of Chemical Engineers, 2013, 44(1): 95-102.

[110] Gaber M, Abu Ghalwa N, Khedr A M, et al. Electrochemical degradation of reactive yellow 160 dye in real wastewater using C/PbO_2-, $Pb+Sn/PbO_2+SnO_2$-, and Pb/PbO_2 modified electrodes. Journal of Chemistry, 2012, 2013, ID 691763.

[111] Morsi M S, Al-Sarawy A A, El-Dein W A S. Electrochemical degradation of some organic dyes by electrochemical

oxidation on a Pb/PbO_2 electrode. Desalination and Water Treatment, 2011, 26(1-3): 301-308.
[112] Hamza M, Abdelhedi R, Brillas E, et al. Comparative electrochemical degradation of the triphenylmethane dye Methyl Violet with boron-doped diamond and Pt anodes. Journal of Electroanalytical Chemistry, 2009, 627(1): 41-50.
[113] Cesarino I, Cesarino V, Moraes F C, et al. Electrochemical degradation of benzene in natural water using silver nanoparticle-decorated carbon nanotubes. Materials Chemistry and Physics, 2013, 141(1): 304-309.
[114] Li Y, Ouyang S, Xu H, et al. Constructing solid-gas-interfacial Fenton reaction over alkalinized-C_3N_4 photocatalyst to achieve apparent quantum yield of 49% at 420 nm. Journal of the American Chemical Society, 2016, 138(40): 13289-13297.
[115] Zhao X, Guo L, Zhang B, et al. Photoelectrocatalytic oxidation of Cu(II)-EDTA at the TiO_2 electrode and simultaneous recovery of Cu(II) by electrodeposition. Environ Sci Technol, 2013, 47(9): 4480-4488.
[116] Khataee A, Akbarpour A, Vahid B. Photoassisted electrochemical degradation of an azo dye using Ti/RuO_2 anode and carbon nanotubes containing gas-diffusion cathode. Journal of the Taiwan Institute of Chemical Engineers, 2014, 45(3): 930-936.
[117] Huang X, Shen Q, Liu J, et al. A CO_2 adsorption-enhanced semiconductor/metal-complex hybrid photoelectrocatalytic interface for efficient formate production. Energy & Environmental Science, 2016, 9(10): 3161-3171.
[118] Canton C, Esplugas S, Casado J. Mineralization of phenol in aqueous solution by ozonation using iron or copper salts and light. Applied Catalysis B: Environmental, 2003, 43(2): 139-149.
[119] Lin S S, Gurol M D, Heterogeneous catalytic oxidation of organic compounds by hydrogen peroxide. Water Science and Technology, 1996, 34(9): 57-64.
[120] Farré M J, Franch M I, Malato S, et al. Degradation of some biorecalcitrant pesticides by homogeneous and heterogeneous photocatalytic ozonation. Chemosphere, 2005, 58(8): 1127-33.
[121] Ruppert G, Bauer R, Heisler G. UV-O_3, UV-H_2O_2, UV-TiO_2 and the photo-Fenton reaction-comparison of advanced oxidation processes for wastewater treatment. Chemosphere, 1994, 28(8): 1447–1454.
[122] 石枫华, 马军. O_3/H_2O_2与 O_3/Mn^{2+}氧化工艺去除水中难降解有机污染物的对比研究. 环境科学, 2004, 25(1): 72-77.
[123] Yan J, Zhi Z, Ying H, et al. Degradation of benzophenone in aqueous solution by Mn-Fe-K modified ceramic honeycomb-catalyzed ozonation. Journal of Environmental Sciences, 2006, 18(6): 1065-1072.
[124] Andreozzi R, Caprio V, Insola A, et al., The ozonation of pyruvic acid in aqueous solutions catalyzed by suspended and dissolved manganese. Water Res, 1998, 32(5): 1492-1496.
[125] 郑晓飞, 李燕, 周明, 金属催化臭氧化机理研究及应用. 2010.
[126] 李海燕, 曲久辉, 王子健. Cu/Al_2O_3催化臭氧氧化降解水中甲草胺的研究. 给水排水, 2005, 31(3): 21-24.
[127] 周云端, 祝万鹏. Al_2O_3催化臭氧氧化处理邻苯二甲酸二甲酯. 环境科学, 2006, 27(1): 51-56.
[128] Park J S, Choi H, Cho J. Kinetic decomposition of ozone and para-chlorobenzoic acid (p CBA) during catalytic ozonation. Water Res, 2004, 38(9): 2285-2292.
[129] Al-Hayek N. Fe(III)/Al_2O_3-catalyzed ozonation of phenol and its oaonztion by-products. Environmental Technology Letters, 1989, 10: 415-426.
[130] 周云瑞, 祝万鹤, 刘福东, 等. 催化剂载体对催化臭氧氧化活性的影响. 清华大学学报(自然科学版), 2007, 47(9): 1481-1484.
[131] 张彭义, 祝万鹏. 镍铜氧化物对吐氏酸废水臭氧氧化的催化作用. 中国环境科学, 1998, 18(4): 310-313.
[132] 李来胜, 祝万鹏, 李中和. 催化臭氧氧化去除水中氯乙酸的研究. 上海环境科学, 2002, 21(5): 282-284.
[133] 于欣, 刘洪波, 孔令江, 等. 催化臭氧氧化法处理甲醛废水. 化工环保, 2007, 27(1): 27-31.
[134] 朱丽勤, 何瑾馨, 陈小立. 染色废水臭氧氧化催化剂研制及其应用性能. 东华大学学报(自然科学版), 2005, 31(1): 72-75.
[135] 刘绮, 石林, 王振友. 环境污染控制工程. 广州: 华南理工大学出版社, 2009.
[136] 黄春保, 马容明. 环境化学污染及防治. 武汉: 武汉测绘科技大学出版社, 1992.
[137] 闵恩泽, 傅军. 绿色化学的进展. 化学通报, 1999, 20(1): 10-15.
[138] Wang A, Zhang T. One-Pot conversion of cellulose to ethylene glycol with multifunctional tungsten-based catalysts. Accounts of Chemical Research, 2013, 46(7): 1377.
[139] Wei W, Zhao Y, Peng S, et al. Yolk-shell nanoarchitectures with a Ru-containing core and a radially oriented mesoporous

silica shell: facile synthesis and application for one-pot biomass conversion by combining with enzyme. ACS Appl Mater & Inter, 2014, 6(23): 20851-9.

[140] Fan Y, Cheng S, Wang H, et al. Nanoparticulate Pt on mesoporous SBA-15 doped with extremely low amount of W as highly selective catalyst for glycerol hydrogenolysis to 1, 3-propanediol. Green Chemistry, 2017, 19(9): 2174-2183.

[141] Roberts F S, Kuhl K P, Nilsson A. High selectivity for ethylene from carbon dioxide reduction over copper nanocube electrocatalysts. Angewandte Chemie, 2015, 54(17): 5179-5182.

[142] 刘利, 王东. “水上”(“on water”)有机反应. 化学进展, 2010, 22(7): 1233-1241.

[143] Li Y X, Li L, Kong D, et al. Palladium-catalyzed alkynylation of Morita–Baylis–Hillman carbonates with (triisopropylsilyl) acetylene on water. Journal of Organic Chemistry, 2015, 46(44): 6283-6290.

[144] Li X H, Wang F, Lu P D, et al. Confocal Raman observation of the efflorescence/deliquescence processes of individual $NaNO_3$ particles on quartz. Journal of Physical Chemistry B, 2006, 110(49): 24993-24998.

[145] Li C J. Organic reactions in aqueous media with a focus on carbon-carbon bond formations: A decade update. Chemical Reviews, 2005, 105(8): 3095-3166.

[146] Li C. Chiral synthesis on catalysts immobilized in microporous and mesoporous materials. Catalysis Reviews, 2004, 46(3-4): 419-492.

[147] Li X, Yang Y, Yang Q. Organo-functionalized silica hollow nanospheres: synthesis and catalytic application.Journal of Materials Chemistry A, 2013, 1(5): 1525-1535.

[148] Zhang F, Liu G, He W, et al. Mesoporous silica with multiple catalytic functionalities. Advanced Functional Materials, 2008, 18(22): 3590-3597.

[149] Huang J, Zhu F, Li H X, et al. Periodic mesoporous organometallic silicas with unary or binary organometals inside the channel walls as active and reusable catalysts in aqueous organic reactions. Journal of the American Chemical Society, 2010, 132(5): 1492-1493.

作者：朱永法[1]，贺　泓[2]，李和兴[3]

[1]清华大学，[2]中国科学院生态环境研究中心，[3]上海电力学院

第 17 章　水污染与控制技术研究

本章导读

本章在对国内外废水处理研究工作的深入分析基础上，结合编写者自身的科研工作，总结了水污染控制的四个研究方向的研究现状：①物理化学水处理方法与技术；②生物化学水处理方法与技术；③废水处理新技术系统；④废水资源/能源新技术。同时介绍了废水处理新技术研发和实际应用的重要成果和最新进展，分析了废水处理理论和技术的发展趋势，并展望了废水处理技术可持续发展及应用前景。

关键词

废水处理，物化技术，生物处理，资源化，能源化

1 引　　言

随着我国城镇化和工业化进程的加快，水环境污染已成为社会经济发展的限制性因素，而工业废水和生活污水的长期大量排放是造成水环境污染的主要原因。人们一方面采用环境友好的或绿色化学合成路线等方法从源头上减少污染，另一方面，针对水环境污染，建立高效、经济、清洁、彻底去除水中污染物的新方法。两个方面及其结合一直是国内外环境界关注的热点。废水处理技术主要依据物理学、化学和生物学为基础并结合相关领域知识而形成：①物理处理法是利用物理作用（如通过沉淀、吸附、浮选、萃取、蒸发等）处理、分离和回收水中的污染物，其特点是仅将废水中的污染物从水相中分离出来，污染物并没有转化或破坏，多应用于组分回收但需要防止二次污染。②化学处理法是通过物理化学作用或化学反应转化、分离、去除或回收废水中的污染物，将有毒组分或高浓度组分转化为低害、无害或易分离的物质。该方法主要包括混凝法、氧化还原法、萃取法、离子交换法、膜分离法等。化学法不但能够将废水中污染物分离而去除，更重要的是它可以改变污染物的化学性质，使其从有害物质转化为无害或可分离物质，减少二次污染。③生物处理法是利用微生物将有机物分解并转化为稳定的无机物的原理，使污/废水中有机物通过矿化或甲烷化而去除的方法。根据微生物的不同作用，生物处理法可分为好氧生物处理和厌氧生物处理两种类型。生物处理使用最为广泛，且成本低廉，对生活污水和大部分工业废水的处理有其独特的优势，因而其理论研究和技术发展至今方兴未艾，未来一段时间仍将是水处理的主导技术。但是，随着人们对环境质量要求的不断提高，以及有毒/难降解的新型有机污染物、多种功能开发与应用无机污染物的大量出现，以物理化学方法为核心的处理技术近年来也得到了长足发展，成为废水处理研究的前沿领域。新型废水处理技术正朝着最少能耗/物耗的污染物转化和实现资源回收的方向发展。

针对废水中的新型污染物和逐步提升的环境质量标准，不断地将化学、生物学、材料学等学科的新思想、新概念、新技术应用到污染控制过程中是废水处理技术进一步发展的重要源泉。废水处理技术以分离、转化/降解、无害化为目标，应针对不同工业废水和生活污水建立处理方法的优先原则，即首先考虑废水的资源化再利用技术，再考虑其无害化技术，实现经济与生态结合的目标。总之，高效、彻底、无二次污染、操作便利、运行费用低、普适性强、多功能化，是未来水污染控制技术追求的目标。

2　物理化学水处理方法与技术

2.1　高级氧化/还原处理技术

2.1.1　电催化氧化

电化学技术由于其处理能力强、操作简单、易于控制而被应用于污/废水处理中。电化学技术包括电催化氧化、电化学絮凝、电渗析除盐、电化学浮选等，其中，电催化氧化法最受关注。电催化氧化法是指水体中污染物在电极表面发生直接或间接氧化的过程。直接氧化法是指污染物直接在阳极表面失去电子而被氧化；间接氧化法是指污染物被阳极产生的强氧化性物质（如·OH、氯气、次氯酸、过氧化氢等）所氧化的过程。电催化氧化法具有环境兼容性高、能量利用率高、可控性好、多功能性结合的特点[1]。

2.1.1.1　电催化氧化法的应用瓶颈及其解决方法

1）应用瓶颈

电催化氧化法处理废水最大的应用瓶颈为运行成本高，原因有以下几点：电能是最高品位的能量，成本相对较高；污染物浓度很低（ppm 及以下级）造成的传质效果差、电流效率低、析氢和析氧副反应严重等问题；电极稳定性差。这些一般可以从以下几方面着手解决。

2）改进电解反应槽结构

通过改进电解反应槽结构来改善传质是提高电解效率的有效方法。已有改进电解槽结构的方法包括流化床电化学反应器、旋转电极等，另外还有以下 4 种新的方法。

（1）三维电极反应器。三维电极反应器与板、片等二维电极相比，可以增加工作电极的比表面积，使污染物和电极更充分接触，与传统的二维平板电极相比可明显提高电流效率和处理效果。近几年，有很多研究者采用三维电极反应器处理不同工业废水，如造纸废水、染料废水、炼油废水、含酚废水等。Wang Yan 课题组[2]采用 Sn-Sb-Ce 修饰颗粒活性炭构建的三维电极处理苯酚，反应 2 h 后苯酚去除率达到 89%。Zhang Yimei 课题组[3]采用氮掺杂石墨烯气凝胶作为新型颗粒三维电极材料处理双酚 A，反应 30 min 后其去除率达到了 85%。

（2）管式电反应器。管式电反应器与常规的板式反应器相比，电极间距很小，电场分布均匀，因而降低了电解电压，从而减小能耗，而且传质条件好，被广泛应用于有机废水的处理，如：染料废水、抗癌药物废水、苯酚废水等[4]。当管式电化学反应器降解伊文斯蓝时，反应 90 min 后去除率可达 60%。Han Weiqing 等[5]采用新型管式电反应器处理抗癌药物废水中的 5-氟-2-甲氧基嘧啶，3 h 反应的去除率可以达到 100%。

（3）双室反应器。双室反应器可以避免污染物在阳极被氧化后又扩散到阴极被还原，或者在阴极被还原后再扩散到阳极被氧化的现象，减少了电能的内耗，而且在阴极室还可进行氧气的二电子还原产生 H_2O_2，使阴极也具有氧化能力，产生双极氧化效果。申哲民等[6]采用新型双室反应器，利用双极氧化的原理处理模拟染料废水，相对于传统单室反应器可以节约近 50%的能耗。薛建军等[7]比较了以铁电极为阴极，反应 2.5 h 后，双室反应器中硝酸盐氮的去除率达到了 99.8%，而单室反应器中硝酸盐氮的去除率仅为 84.3%。

（4）电化学转盘。电化学转盘的外观类似于生物转盘，以一定的转速旋转，旋转时转盘的一半在溶液中，另一半在空气中，如图 17-1 所示。这种反应器可以提高传递速度，减小扩散层厚度，改善低浓度下的传质。此外，电化学转盘有利于空气中的氧气向转盘液膜中传递，在阴极被还原为 H_2O_2，进一步提高了电催化氧化效率。我们采用电化学转盘处理模拟染料废水，反应 60 min 后脱色率可达 99.5%，而相同条件下传统电化学反应器的脱色率只有 55%[8]。

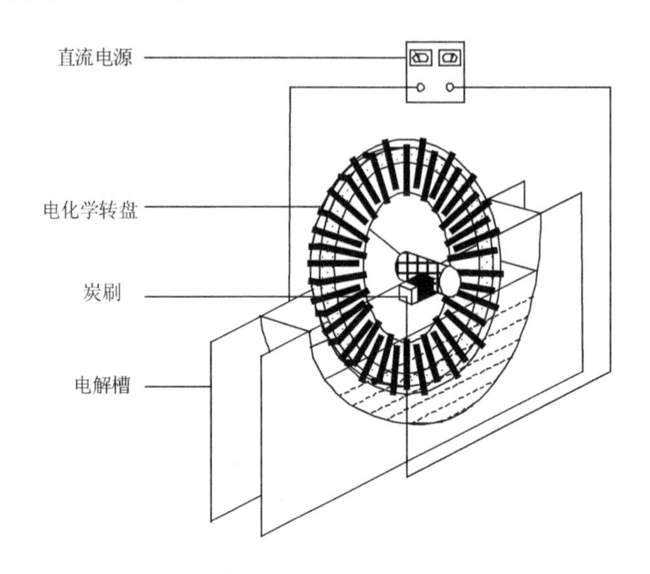

图 17-1　电化学转盘反应器装置原理

3）采用电位窗口宽、催化活性高的电极

选择电位窗口更宽、催化活性高的电极可以有效提高电催化效率。近年来研究的几种电位窗口宽、催化活性高的电极主要有以下 3 种。

（1）BOD 电极。BOD 电极应用在电化学反应中，具有化学惰性好，抗腐蚀能力强、电流效率高以及氧化能力强等优点。析氧电位可达+2.3 V，析氢电位可达−1.25 V（参比电极为 SHE），电位窗口超过 3 V，赵国华课题组对该电极有较深入的研究。

（2）DSA 阳极。DSA 阳极由混合氧化物和钛基底构成，氧化物包括 RuO_2、IrO_2、MnO_2、PbO_2、SnO_2 和 TiO_2 等。与传统的石墨、铂、PbO_2 等电极相比，具有耐腐蚀好、电催化活性高、成本低廉等优点[9]。

（3）ACF 电极。ACF 电极比表面积大，具有很强的吸附特性，可以将污染物吸附在电极表面，提高处理效率，适合处理浓度较低的有机废水。ACF 电极可用于降解废水中氰化物、高氯酸、染料等污染物，并且处理效果良好[10]。

4）与其他方法耦合

单一的电催化氧化方法存在能耗高的缺陷，而将电催化法和其他方法耦合可以达到降耗提效的目的。

（1）光电耦合。光电耦合是将半导体光催化氧化法和电催化氧化法结合以实现协同效应。赵国华课题组[11]采用两步法处理甲基蓝废水，先采用电催化法对染料废水进行脱色，再采用光催化法深度处理废水，光电催化的协同效应大大降低了能耗，提升了处理效率。贾金平课题组制备了一种尖锥型 TiO_2/Ti 电极，并将其应用于转盘光电催化反应器处理罗丹明 B 染料废水，150 min 后色度去除率达 100%，COD 去除率达 87%[12]。

（2）声电耦合。声电耦合是将超声空化氧化技术与电催化氧化技术结合，可明显提高污染物降解效率，是一种很有前景的深度氧化技术。Maria 等[13]研究了超声空化对电催化降解咖啡因的影响，结果表明，反应 100 min 后声电催化法比单一电催化法的降解效率高了 13%。贾金平课题组也比较了电催化和电催化/水力空化协同降解活性艳红模拟废水的效果，发现后者对活性艳红的脱色速率是前者的 2 倍[14]。

（3）生物电耦合。生物电耦合是将电催化氧化法和生物法联合的一种方法。梁继东课题组采用电化学氧化与生物降解工艺联合降解木质素，结果表明电催化氧化法能先行打断抵御微生物攻击的木质素顽固键合结构，有利于后续生物继续降解残余木质素中间体。降解橄榄油废水也具有同样的效果。实际上，目前广泛研究的微生物燃料电池方法也是一种生物电耦合方法。

2.1.1.2　处理对象选择及工程的应用原则

由于电催化氧化法存在能耗高的缺点，因此需要对废水处理的类型进行甄别，综合考虑能耗和处理效果这一对矛盾关系。实际上，电催化氧化更适合处理中低浓度、难生物降解、中小水量的废水，因此可以作为废水预处理或者深度处理的方法，进而在实际工程中可根据具体的条件和要求，进行科学选择，合理组合工艺。

2.1.1.3　电催化氧化杀菌

由于传统的加氯消毒会产生卤代消毒副产物，危害人类健康，因而近年来采用电催化氧化杀菌以减少卤代消毒副产物形成的研究备受关注。电催化氧化杀菌有两种机理：电场的物理作用和电解产物的化学作用。电催化氧化杀菌已被用于市政污水、游泳池水、压载水等，可有效地去除绿脓杆菌、大肠杆菌等细菌[15]。

2.1.1.4　结论

电催化氧化法可通过改进电解槽结构、采用电位窗口宽催化性能好的电极以及与其他技术耦合等方式降耗提效。电催化氧化杀菌也是一种高效、低成本、环境友好的新型杀菌技术。电催化氧化法可作为预处理或深度处理的方法，对中低浓度、难生物降解及中小水量的废水进行处理，是一种很有前景的水处理方法。

2.1.2　自由基活化氧化

高级氧化技术是一类利用活性自由基的氧化性氧化有机污染物，进一步矿化其为二氧化碳和水的氧化技术。高级氧化过程按自由基类型可分为羟基自由基过程、硫酸根自由基过程和卤素自由基过程三大类。羟基自由基过程是一类是最早开展研究及其广泛应用的高级氧化过程。羟基自由基氧化/还原电位为 2.7 eV。其氧化过程具有选择性弱、反应速率高以及氧化产物复杂等特点。其主要通过双键加成、脱氢反应、电子转移三个类型降解有机污染物。硫酸根自由基则是新近的研究热点，其氧化还原电位为 2.5~3.1 eV，与芳香性有机物的反应速率较高，主要发生电子转移反应和脱氢反应。而卤素自由基的氧化性是三类自由基中最弱的，其多与酚类、胺类等富电子污染物进行选择性电子转移反应。

2.1.2.1　羟基自由基高级氧化过程

应用羟基自由基氧化过程的技术包含芬顿试剂（Fe^{2+}/H_2O_2）、UV/H_2O_2、光催化、臭氧/H_2O_2 联用、电化学氧化等。其中 UV/H_2O_2 是被研究最广泛的，并用于全尺度的高级氧化技术。现今关于新型污染物的羟基自由基的降解研究，已经从单一化合物的羟基自由反应动力学、降解机理，逐渐过渡到复杂污水介质的降解动力学、指示因子的研究以及降解产物的毒性研究上。瑞士联邦水科学与技术研究所的 von Gunten 课题组在此领域做出了一系列的开创性工作。他们通过动力学反应模型的建立，成功地预测了 16 种新兴污染物在污水二级出水中的 UV/H_2O_2 的降解速率，并与臭氧化工程进行了对照。建议 O_3/H_2O 串联 UV/H_2O_2 工艺能够在最低能源消耗的前提下完成新型污染物的排放减量，并同时减少氧化副产物的生成[16]。美国科罗拉多大学 Karl G. Linden 课题组[17]运用指示因子对 UV/H_2O_2 氧化污水二级出水中的痕量有机污染物的行为进行了预测，发现了氯蔗糖甜味剂可以作为保守性指示化合物，来预测其他类型的痕量有机物的羟基自由基高级氧化过程。

2.1.2.2 硫酸根自由基高级氧化过程

近五年，基于硫酸根自由基（$SO_4^{\cdot-}$）的高级氧化技术受到了环境工作者的广泛关注，这种高级氧化技术能够有效降解水体中的污染物质，同时也可以作为一种消毒手段。$SO_4^{\cdot-}$可通过活化过硫酸盐（$S_2O_8^{2-}$）或过一硫酸氢盐（HSO_5^-）产生，目前常用的活化方式有紫外光活化法、加热法、超声法、电解法、碱性活化法以及金属催化活化法。近几年针对基于硫酸根自由基的高级氧化过程的研究一方面致力于研究新的活化方式，如腐殖酸中的苯醌组分原位活化 $S_2O_8^{2-}$[18]、臭氧氧化 HSO_5^-；另一方面也致力于开发新的金属催化剂催化活化过硫酸盐，如二价铁离子[Fe（II），类芬顿反应]、三价铁氧化物[Fe（III）-oxides]、Fe-EDDS[19]、铜（Cu）以及四价锰氧化物等都可以有效地催化活化过硫酸盐生成 $SO_4^{\cdot-}$。张晖课题组研究发现 Fe-Co 双金属/介孔硅（Fe-Co/SBA-15）催化体系对过硫酸盐具有很高的催化活性，并且该体系具有较高的稳定性[20]。马军课题组最新研究发现，在弱酸性和中性条件下，零价铁/亚硫酸盐（Fe^0/SO_3^{2-}）体系也可以生成 $SO_4^{\cdot-}$，实现对活性红 X-3B 的有效去除[21]。

马军课题组、陆隽鹤课题组以及 Dionysiou 课题组等针对 $SO_4^{\cdot-}$降解污染物的动力学及机理进行了深入研究。研究表明，$SO_4^{\cdot-}$降解污染物反应机理以电子转移为主。与羟基自由基（$HO^\cdot$）相比，$SO_4^{\cdot-}$具有更强的反应选择性，其与污染物的二级反应速率常数为 $10^7 \sim 10^{10}$ L/(mol·s) [22]。对某些污染物，如噻虫胺、莱克多巴胺[22]等，去除效率明显高于 $HO^\cdot$。在偏碱性条件下 $SO_4^{\cdot-}$可与 HO^-反应生成 $HO^\cdot$，参与污染物的降解，甚至对于某些污染物，生成的 $HO^\cdot$对其降解的贡献更大。另外，$SO_4^{\cdot-}$能与水体中的阴离子反应生成次生自由基，如碳酸根自由基（$CO_3^{\cdot-}$）、含卤自由基[23]以及超氧根自由基（$O_2^{\cdot-}$）[24,25]等，参与污染物的降解。关小红课题组[22]研究对比了基于硫酸根自由基降解的高级氧化过程和基于羟基自由基降解的高级氧化过程对水体中痕量污染物的降解动力学以及无机盐对这两种高级氧化体系的影响。研究对比发现基于硫酸根自由基降解的高级氧化过程中，$CO_3^{\cdot-}$参与了部分药物类污染物的降解，而基于羟基自由基降解的高级氧化过程中，$HO^\cdot$是唯一参与污染物降解的自由基；卤素离子对硫酸根自由基的影响较于对羟基自由基影响更为显著，反应生成含卤自由基，改变体系的反应活性。

2.1.2.3 卤素自由基高级氧化过程

活性卤素自由基（radical reactive halogen species，RHS），主要指 $Cl^\cdot$、$Cl_2^{\cdot-}$、$Br^\cdot$、$Br_2^{\cdot-}$、$ClBr^{\cdot-}$。由于其高反应活性和选择性而受到关注。在高级氧化水处理过程中，RHS 主要产生于 UV/Chloride 高级氧化技术。

UV/Chloride 技术可以生成一系列自由基，从而达到对污染物降解的目的。其主要反应方程式如式（17-1）和式（17-2）所示。研究表明，UV/Chloride 在对水体腐殖质降解过程中，在 pH 2~7 时，活性氧物种以 $HO^\cdot$以及 $Cl^\cdot$为主，而其他条件下以 $Cl^\cdot$为主。UV/Chloride 技术的光量子产率高，并且对特定污染物降解效率明显高于 UV/H_2O_2，在杀菌同时无需考虑氯气残留问题，因此被认为是可以替代 UV/H_2O_2 技术成为研究热点。而目前，多数关于 UV/Chloride 技术研究集中于体系产生的 $HO^\cdot$和 RHS 对环境新型微污染物的降解方面，此领域国内学者占据大半江山。在此方面研究较多的是方晶云及商祺课题组。他们采用稳态动力学模型首次预测了在基于 UV/Cl 高级氧化处理技术中，各种活性氧物种（$HO^\cdot$/$O^{\cdot-}$/$Cl^\cdot$/$Cl_2^{\cdot-}$）对环境微污染物的降解贡献率[26]。发现在此系统中 $HO^\cdot$/$Cl^\cdot$起到了主要的降解贡献，而 $O^{\cdot-}$/$Cl_2^{\cdot-}$贡献微乎其微。而在 UV/Chloride 系统中 $Cl^\cdot$、$Cl_2^{\cdot-}$相对 $HO^\cdot$更具选择性，对特定污染物降解效率更高。但是 UV/Chloride 技术除上述优点外，也被发现在处理含有溴化物的废水中能够加强有毒溴酸盐的生成[27]。而且处理过程中消毒副产物及其他有毒产物生成问题不容小觑[28]。

$$HOCl/OCl^- \xleftrightarrow{UV} HO^{\cdot}/O^{\cdot-} + Cl^{\cdot} \qquad (17\text{-}1)$$

$$Cl^{\cdot} + Cl^- \longleftrightarrow Cl_2^{\cdot-} \qquad (17\text{-}2)$$

除 UV/Chloride 技术外，现阶段自然光/Chloride 处理技术的研究也逐渐兴起。自然光光谱从 290~400nm，此范围内光谱与 Cl 反应除生成 $HO^{\cdot}/O^{\cdot-}/Cl^{\cdot}/Cl_2^{\cdot-}$外，$OCl^-$在光照条件下可生成臭氧（$O_3$）。这其中 RHS 对污染物降解的贡献不能被忽略。而自然界中，卤素自由基对污染物的降解也真实存在，William A. Mitch 课题组发现含卤素自然水体（海洋/河口）在光照条件下，水体中的氯离子和溴离子可与三线态有机质或 $HO^{\cdot}$反应，从而生成一系列 RHS，而 RHS 对相当一部分海洋中微污染物的降解具有促进的作用，如软骨藻酸、微囊藻毒素 MCLR 等[29]。同时在电化学阳极氧化技术处理工艺或反渗透浓缩液的高级氧化技术中，如果电解液中含有氯离子或者少量的溴离子，也会生成 RHS，研究表明其对大部分污染物的降解起到促进作用[30]。同时，在硫酸根自由基水处理体系中，在卤素离子存在的条件下，也会生成 RHS，其对体系中污染物的降解也会起到一定的贡献[22,23]。

综上所述，高级氧化过程经过近年的研究发展，已经作为新兴的氧化手段，逐步从实验室阶段走到了实际工业应用的阶段。新型污染物的去除、水质参数的影响及其次生自由基的作用机理、氧化过程中的去除效果的预测以及指示因子的研究将是高级氧化过程中的研究热点。

2.1.3　催化还原

围绕废水的处理，催化还原法能使难降解有机物反应生成可被后续生物工艺降解的反应产物，具有处理效果好、操作简单、应用前景广阔等特点。此外，催化还原法还能够使无机（类）金属离子的形态发生改变，能够将高毒性金属离子转变成低毒性金属离子、非溶解态的金属单质、沉淀物或者其他容易被后续工艺去除的存在形式，进而达到净化的目的[31,32]。基于此，本章节梳理了国内外关于催化还原处理废水的研究进展，主要内容包括光催化还原和单质铁还原法（内电解法）。

1）光催化还原技术简介

自 20 世纪 70 年代末 Yoneyama 等[33]首次报道光催化还原 Cr(VI)为 Cr(III)后，光催化开启了其用于还原去除水中（类）金属离子污染的新途径。近年来，光催化水处理技术以其经济、环境友好等优势已成为当前水污染控制领域的一个主要研究方向。光催化还原的作用机制是基于光生 e^-可以将具有毒性的高价态（类）金属离子还原至低价态、低毒性的反应产物，从而实现对含（类）金属离子废水的直接还原[34-36]处理（图 17-2）。

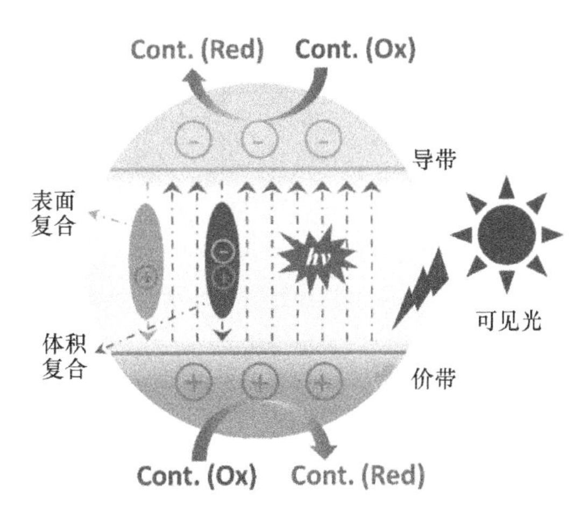

图 17-2　光催化还原去除污染物的主要作用机制示意图

大量的研究已经表明，光催化还原技术能够去除水中的 Cr(VI)、Hg(II)、Ag(I)、Au(III)、Pt(IV)、Tl(I)等有毒金属离子。相对于光催化还原有毒金属离子，光催化还原有机污染物的研究相对进展缓慢，这主要是由于体系中空穴能够通过直接氧化或生成的羟基自由基实现对污染物的氧化降解[37]。考虑到光催化体系中空穴的氧化效应，研究发现甲酸、甲醇能够作为空穴的捕获剂，从而提升光催化对水中金属离子的还原速率[38]。同样地，Kajitvichyanukul 等研究发现有机体能够促进基于 TiO_2 的光催化反应体系中对 Tl(III)的还原能力，但是氧气存在时则会促进光催化体系对 Tl(I)的氧化能力[35]。光催化反应过程复杂，其除污染的性能不仅受体系反应条件影响，而且还与光催化材料的固有特性密切相关，如比表面积、等电点、晶形、表面结构等[37,39]。为提升材料的光催化活性，学者们已开发了 Bi 系光催化材料（如 Bi_2O_3、$BiPO_4$、Bi_2O_3/BiOX 等）、铁基催化材料（如 Fe_2O_3）、锌类催化材料（如 ZnO、ZnS 等）、WO_3、CdS 等一元乃至多元体催化剂。

尽管光催化研究已积累了一定的成果，但其仍然存在许多急需解决的关键性问题，如光响应范围较窄、光生载流子的复合概率较高、粉体催化剂分离困难以及光催化反应器效率不高等[37]。因此，为了进一步促进其工业化的应用，近几年光催化的研究工作主要分为四个方面：①对现有半导体催化剂进行掺杂和改性；②对光催化反应体系的优化，如通过加入电子给体抑制电子-空穴的复合和缓解光腐蚀等；③设计开发新型高效的催化剂；④搭建光电催化反应装置，促进光生电子和空穴的分离。

2）内电解还原法简介

国内外研究最多、较为成熟的工业废水化学还原工艺是铁炭内电解法[32]。内电解反应生成的新生态的 Fe^{2+}及其水合物具有较强的吸附-絮凝活性，特别是在后续加碱调 pH 值的工艺中生成 $Fe(OH)_2$ 和 $Fe(OH)_3$ 絮状物，发生混凝吸附作用，能使废水中微小的分散颗粒及脱稳胶体形成絮体沉淀，降低色度，净化废水。如图 17-3 所示，铁内电解法的作用机理主要包括如下：①铁的直接还原作用；②铁的间接还原作用；③氢的还原作用。

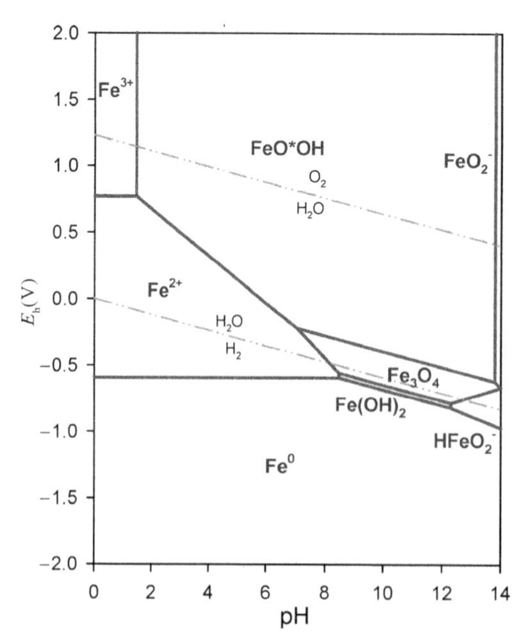

图 17-3 标准条件下 FeO/H_2O 体系 E_h-pH 曲线（$[Fe]_{tot}=10^{-4}$mol/L）

铁炭内电解法处理的污染物主要分成两大类，一类是有机污染物，包括含有偶氟、碳双键、硝基、卤代基结构的难降解有机物质；一类是无机污染物，包括可以被单质铁还原的（类）金属离子，例如 Se(IV)、Se(VI)、U(VI)、Cu^{2+}、Pb^{2+}等以及可通过吸附和共沉淀去除的(类)金属离子，例如 As(V)、Sb(V)、Ni^{2+}、

Zn^{2+}等。铁炭内电解用于工业废水处理具有工艺流程简单、使用寿命长、投资费用少、运行成本低等优点。对已建成未达标的高浓度有机废水处理工程，可用该技术作为已建工业废水的预处理，在降解 COD 的同时提高废水的可生化性，有利于废水处理后稳定达标排放。但是传统的内电解法也具有效率较低、使用的过程中易钝化板结、传统微电解材料表面积太小使得废水处理需要很长的时间进而增加吨水投资成本等缺点，这都严重影响了传统内电解工艺的利用和推广。

为了提高内电解工艺的效率和运行稳定性，研究者做了很多改进。例如马鲁铭课题组在国内较早开展了催化铁内电解法并成功用于多种工业废水处理，他们用铜代替炭，与单质铁组成构成原电池，并成功应用于工业园区综合废水的生物预处理；美国德州农工大学的黄永恒课题组率先提出向单质铁系统中投加一定量的Fe^{2+}能够有效克服单质铁的钝化，进而改善单质铁对NO_3^-、Se(VI)[40]、三氯乙烯(TCE)等污染物的去除效果；在发现氧气的存在可大大加速单质铁对硝酸盐的去除现象[41]的基础上，郭学军该课题组提出利用强氧化剂来强化单质铁去除硝酸盐及 As(V)、Sb(V)等污染物的方法[42]；基于单质铁的铁磁性及单质铁腐蚀产物 Fe^{2+}/Fe^{3+}的顺磁性，关小红课题组率先提出利用弱磁场来调控单质铁除污染过程的方法，弱磁场能够强化不同来源零价铁对水中不同金属离子的还原去除[43, 44]，相应的反应速率常数提高倍数达到 1.1~383.7 倍[45]。但是到目前为止，大部分改良的内电解技术还处于实验室或中试阶段，需要不懈地努力去推动其在工业废水处理中的应用。

针对废水的催化还原处理技术，大规模工程运用是其技术成熟的标志之一。因此，后续研究工作应围绕如何在实际废水处理中获得更好的处理效果、更方便的使用方式、更低的使用价格和维护成本、与其他工艺更好的组合等方面开展，具有重要的现实意义。

2.2　富集分离技术

2.2.1　膜分离

淡水资源匮乏、水资源回用比例低是目前亟须解决的全球性难题，膜分离技术在海水淡化、中水回用等的应用为解决这些难题提供了可能[46]。与传统的水处理技术相比膜分离具有许多的优势，例如：装置占地面积小，运行能耗低；无需投加化学药品；一般无需加热，适合于分离温度敏感物质（如：蛋白质）等。在传统膜分离技术广泛用于工业生产的同时，也遇到了一些瓶颈问题，例如：膜污染、选择性与通量无法兼得等。而这些问题的产生与膜材料本身的性质密切相关。一些成熟膜分离技术包括微滤、超滤、纳滤和反渗透和新兴的正渗透工艺在水处理和脱盐中都占用十分重要的地位。与此同时，膜技术在一些其他领域（如：能源）也展现出了充分的潜力[47]。因此，本节将分别针对近年来在膜功能材料的研究和新型膜工艺的研究进行简单的总结。

2.2.1.1　膜功能材料的研究进展

理想状态下的膜分离过程具有三个特点，包括：高选择性、高通量和通量恒定[48]。但是，目前的滤膜在水处理以及脱盐的应用中，往往面临选择性与通量此消彼长、膜污染等问题。针对这些问题，在膜功能材料领域，开展了大量研究工作。

生物的细胞膜中具有特殊的离子通道和水通道，这些通道的存在使得细胞膜对离子和水分子同时具有超高的选择性和透过性。因此，研究者从细菌细胞膜中分离的水通道蛋白（aquaporin Z）并通过囊泡直接整合到聚合物滤膜中，将复合膜用于脱盐[47,49]。但是，由于存在较多缺陷，这种复合膜的除盐率往往较低，而且，天然水通道蛋白在膜中往往容易失活[50]。于是，通过人工合成的手段制备模仿天然水通道蛋白功能的合成纳米水通道成为研究热点之一。合成纳米水通道应该能够在膜中稳定存在，在保持滤膜高水通量的同时对小分子以及离子具有良好的截留作用。单层石墨烯的厚度与一个碳原子的直径相同，是一种理想的二维成膜材料。通过等离子蚀刻的方法在单层石墨烯表面形成 0.5~1 nm 的孔之后，单层石墨烯膜在有效截留一价离子的同时达到了很高的水通量[51]。但单层石墨烯膜目前很难大规模生产。氧化石墨烯（GO）是石墨烯的一种重要衍生物，而且廉价易得。通过组装 GO 纳米片层并控制层间距离，可以形成多层 GO 膜，而 GO 片层之间的空隙就是理想的合成纳米水通道。其他二维材料，如通过共价键连接无机和有机单元形成有孔晶体结构的金属有机骨架材料（MOFs）也能通过可控组装的方法形成具有高速水通道的膜[52]。通过二维材料组装形成多层滤膜是目前研究的一个热点，但是，尚缺乏精确控制层间距的方法，所以其对离子的截留能力还有待提高。

膜污染是滤膜通量下降的主要原因，而不可逆膜污染则影响了滤膜的使用寿命。形成不可逆膜污染的原因是污染物与膜表面形成强相互作用，因此充分了解污染物的性质可更有针对性地制备抗污染滤膜材料[53]。通过涂敷、表面接枝和共混等方法改变膜表面的亲疏水性被广泛地用于控制不可逆膜污染。例如：从活性污泥中提取的细菌胞外分泌物（EPS）被引入聚偏氟乙烯（PVDF）滤膜提高亲水性，以控制膜污染[54]。但是，单一的亲疏水性调控往往很难提供广泛的抗污染效果，因此动态两亲（亲水/疏水）滤膜的研制正在引起关注[53,55]。

2.2.1.2　*新型膜工艺的研究进展*

由渗透压驱动的膜分离过程近年来正在引起人们越来越多的关注，因为它们在水处理和新能源领域都展现了巨大的应用潜力[56]。利用渗透压进行水处理的正渗透工艺（FO）[57]以及利用渗透压进行发电的压力阻尼渗透工艺（PRO）[58]是目前研究的热点。FO 无需外加压力驱动，所以理论上无需消耗能量，但是，FO 工艺的主要限制因素是驱动溶液（draw solution）的分离和回用。因此，研发易于分离和回用的新型驱动溶液是 FO 工艺目前主要的挑战。而在 PRO 工艺中，限制其应用的主要因素是膜表面的浓差极化和膜污染。另外一种将盐度差转化为电能的膜工艺称为：反向电渗析（RED），与 PRO 利用渗透压转化为动力发电不同，RED 能够通过阳离子交换膜和阴离子交换膜的阵列直接将盐度差转化为化学电势差[59,60]。阳离子交换膜的性质决定了能量转换的效率，其污染问题也在引起人们的关注。

综上所述，与传统的水处理技术相比膜分离具有显著的优势，但是，膜分离技术也面临问题和挑战。同时具备高选择性、高通量和通量恒定的理想滤膜为膜功能材料的研究指明了方向，具有高选择性和高通量的合成纳米水通道以及无污染的膜材料是目前研究的热点。此外，无需消耗能量的 FO 工艺以及能够将浓度差转化为电能的 PRO 和 RED 工艺也在成为研究的热点。

2.2.2　混凝

早期的“混凝”和“絮凝”是不同的概念。聚合氯化铝出现以后，由于它具备有机絮凝剂的某些特征，国内研究者称其为“无机高分子絮凝剂”，从此在国内“混凝”和“絮凝”同义，而研究者更习惯使用“絮凝”称谓。

絮凝被广泛应用于饮用水和废水处理过程，具有操作简便、成本较低、改造容易等特点。随着国内的排放标准日趋严格，为达标而新建或改造的工程量非常之大，很多企业选择水处理方案时会首先考虑絮凝技术。因此，絮凝技术在应用 300 多年后，仍有着巨大的市场需求。

传统絮凝仅能去除细小悬浮物和胶体，或以共沉淀方式去除一些溶解性物质（如磷酸盐、砷酸盐、氟化物等）。对于目前水深度处理中密切关注的小分子难降解有机物[如持久性有机污染物（POPs）、药品与个人护理品（PPCPs）、部分污水有机物(EfOM)]，传统絮凝基本上不能去除或去除率非常低。因此，虽然絮凝技术应用较广需求很大，但传统絮凝仅能作为其他技术的辅助手段。如果絮凝技术在絮凝剂和絮凝工艺方面实现突破，能够更多地满足废水深度处理的要求，那么絮凝将成为一种举足轻重的水深度处理技术单元，其技术进步将显著提升废水处理的整体水平。下面从絮凝剂和絮凝工艺两方面介绍絮凝技术研究进展。

2.2.2.1　絮凝剂的研究进展

絮凝技术的核心是絮凝剂，在过去几十年絮凝剂的研究和应用日趋成熟。最近几年关于絮凝剂的文献较集中在新的金属盐（如钛、锆等）及其高分子絮凝剂[61,62]、人工合成有机高分子絮凝剂（主要为聚二甲基二烯丙烯基氯化铵替代产品）[63]、天然高分子絮凝剂（淀粉、甲壳素等及其改性产品）、微生物絮凝剂以及无机-有机复合絮凝剂等。这些研究中絮凝去除的对象较过去没有太大变化，其研究思路主要是提高絮体沉降性、降低絮凝剂自身成为消毒副产物前体的风险、将生物质资源化制备絮凝剂等。

近期对絮凝作用过程和机理的新认识支持了絮凝剂的进一步发展。无机絮凝剂（以铝、铁为主）在投加后会水解聚合形成沉淀物，其表面缺少官能团因而对小分子溶解性物质的吸附作用较弱；有机絮凝剂虽然富含可吸附小分子溶解性物质的官能团，但它不能发生相变，无法形成絮体携带小分子污染物由水相转移至泥相。传统无机-有机复合絮凝剂仅是无机絮凝剂和有机絮凝剂简单混合，两者组分的功能仍然是独立的。只有将无机组分和有机组分定制到单元絮凝剂结构中，即两种组分通过共价键结合在一起，无机絮凝剂的实时水解和有机絮凝剂的基团吸附功能才能真正耦合，才能实现小分子溶解性物质的絮凝去除。

基于上述认识，研究者发明了无机-有机共价键型絮凝剂[64]，利用硅烷偶联剂等材料解决了絮凝剂无机组分和有机组分共价键连接难题。通过研究该絮凝剂的絮凝行为，发现有机官能团紧密连接在无机组分水解形成的沉淀中，证明了絮凝剂单元同时具备了实时水解和基团吸附两种功能。正是无机组分和有机组分通过共价键连接，使新型絮凝剂在结构、功能和絮凝原理方面与常规絮凝剂存在本质区别。

共价键型絮凝剂制备方法实际上为新型絮凝剂研发提供了一个平台，通过改变无机组分（铝盐、铁盐等）、有机基团（季胺、叔胺等）、偶联剂（硅型、钛型等），并调节其组分配比、碱化度等制备参数，即可对絮凝剂的形态结构、分子量、电荷密度、亲疏水性及其相转移能力等进行调控，从而可以针对不同污染物对絮凝剂进行量身定制。利用共价键型絮凝剂，可以高效去除 POPs、PPCPs、EfOM 等小分子有机物以及低浓度磷，突破了传统絮凝剂的功能和应用范围，使絮凝以全新的面貌应用于水深度处理中[65-68]。

2.2.2.2　絮凝工艺的研究进展

絮凝工艺基本延续以组合为主的发展模式。除了一些新型组合工艺，如微生物燃料电池-絮凝，近几年文献报道的组合工艺仍然集中在絮凝-高级氧化[69,70]和絮凝-膜[71,72]这两方面。而絮凝污泥回流和磁絮凝是絮凝作为主要单元在工程中应用较普遍的两种工艺。

絮凝污泥回流工艺特点是通过将部分絮凝沉淀污泥回流至工艺前端，一方面可实现污泥中絮凝剂有效成分重复利用，节省药剂成本；另一方面，回流的污泥可一定程度上增加原水的颗粒浓度，增大颗粒碰撞概率并提供新的凝聚吸附点位以形成接触絮凝，并通过增强网捕卷扫作用最终提高絮凝沉淀效率[73]。

磁絮凝技术特点是通过投加磁粉和常规絮凝剂，絮凝剂的水解-沉淀-吸附作用使磁粉和污染物形成磁性絮体，随后通过高梯度磁分离技术使磁性絮体与水快速分离达到去除污染物的目的[74]。进入污泥中的磁粉可以回收并循环使用，从而降低运行成本。磁絮凝技术具有絮凝效率高、沉降分离迅速、絮体密实、污泥含水率低等特点，可节省沉淀池体积，并降低污泥处理负荷，因此尤其适合在空间有限或应急情况下使用[75]。

新型絮凝剂（如共价键型絮凝剂）的出现会带来絮凝工艺的变化。絮凝的实质是将水中污染物转移到泥相中，或将污染物浓缩至污泥中。共价键型絮凝剂由于其特殊结构和絮凝机理，可将更广更多的污染物浓缩到污泥中。部分污染物，如贵金属、氮、磷等，可进行回收资源化利用。进入污泥的共价键型絮凝剂可通过化学或生物过程进行功能恢复或再生，同时结合目前应用较广的磁絮凝、污泥回流等技术进行循环使用，从而可衍生出一些絮凝新工艺。

2.2.3 吸附

吸附法是去除水中污染物最为高效的方法之一，在去除重金属和难降解污染物方面具有独特优势[76]。高效吸附材料的研发是最为关键的问题，针对不同废水的特定污染物，制备出高效吸附剂是近年来的研究热点，其中吸附剂的再生及对污染物的选择性吸附是研究的瓶颈问题[76]。下面分别针对废水中的重金属、全氟化合物和药物类新兴污染物，总结近年来吸附技术的研究进展。

2.2.3.1 吸附去除废水中的重金属

近年来研究者采用了改性廉价材生物材料，在多孔材料上负载高效纳米金属氧化物或氢氧化物等方法，制备出各种高效吸附重金属的材料，降低了材料的价格，控制了材料颗粒的团聚，提升了吸附材料的实用性。以农业废弃物麦秆为原料，对其进行胺化改性，得到高效氨化麦秆，在 pH=2.2 时对 Cr(VI)的吸附量高达 454 mg/g，对实际电镀废水中的 Cr(VI)具有良好的吸附选择性，材料可多次重复利用[77]。为了避免金属纳米颗粒相互团聚，控制颗粒大小，提高纳米材料的可分离性，多种高效纳米金属氧化物被负载到不同材料上[78-80]。研究发现负载纳米水合氧化铁阴离子树脂对于 As(V)和 Cr(VI)的吸附能力明显高于未负载的树脂[78]。氧化铁插层负载于蛭石可制备出高效廉价的吸附材料，对 Cu^{2+}和 Ni^{2+}的吸附量分别达到 59.7 mg/g 和 101.3mg/g，在铜、镍共存体系中，对 Ni^{2+}吸附选择性强[79]。采用热蒸汽水解法可以将纳米二氧化钛负载到碳纳米管过滤膜上，形成几纳米厚的二氧化钛纳米膜对砷吸附速度快[80]。可见，调控负载纳米金属氧化物的大小、提高纳米材料的有效吸附利用率是关键。

2.2.3.2 废水中全氟化合物的吸附去除

典型的全氟磺酸类化合物（PFCs）包括全氟辛烷磺酸（PFOS）和全氟辛酸（PFOA），严重危害生态系统安全和人类健康[76]。生产和使用 PFCs 的企业产生的工业废水中含有高浓度的 PFCs，如镀铬污水中含有 PFOS 及其替代物全氟烷基醚烷磺酸盐（F-53B），生产全氟磺酰氟的企业污水中含多种全氟羧酸，消防灭火废水中含 PFOS 及其替代物等[81,82]。

由于 PFCs 结构异常稳定，常用降解方法效果差，吸附是适宜的控制技术。去除 PFCs 有效的吸附剂包括活性炭[83]、离子交换树脂、金属氧化物、胺化材料[84]等，吸附机理主要包括疏水、静电、离子交换、氢键作用等[81,83]。最近研究发现材料表面的气泡在吸附 PFOS 等全氟表面活性剂中发挥重要作用，PFOS 能富集在材料表面的气泡界面，提出新的 PFCs 吸附机理[83,84]。商业活性炭和离子交换树脂适合应用去除含 PFCs 的实际废水，但活性炭普遍存在吸附量低、对短链 PFCs 去除效果差等问题，而离子交换树脂尽管吸附量高但存在选择性差的问题。针对含有 PFCs 的实际废水，还需开发吸附量高、选择性强的吸附材料。

通过再活化处理可以提高商业活性炭的孔径，对实际镀铬废水中的 PFOS 及 F-53B 具有良好吸附效果[82]，饱和活性炭可用过硫酸盐氧化再生，吸附的共存有机物被降解而产生新吸附位点。扩孔竹基活性炭和阴离子交换树脂 IRA67 可有效吸附去除生产全氟磺酰氟废水中的多种全氟羧酸。为了提高 PFCs 的吸附选择性，基于 C—F 链亲氟原理，合成了新型氟化季铵盐（PFQA）改性的蒙脱石，可从水中高选择性吸附 PFOS 和 PFOA。通过球磨可将纳米 Fe_3O_4 负载到 PFQA 改性的蛭石材料上，得到了一种具有高选择性和快速吸附 PFOS 的新型磁性氟化吸附剂，在其他有机物存在下对 PFOS 具有优异选择性，对消防灭火废水中 PFOS 的吸附去除效果明显优于粉末活性炭和阴离子交换树脂[85]。最近报道的十氟联苯等改性 β-环糊精得到的多孔聚合物对 PFOS 和 PFOA 也有较好的吸附选择性，不受腐殖酸等有机物的影响[86]。

2.2.3.3　吸附去除废水中的药品与个人护理品（PPCPs）

吸附技术是一种有效去除水中 PPCPs 的方法。目前使用的吸附剂主要包括活性炭[87]、高分子聚合物[88]、纳米碳材料[89,90]、金属有机骨架等。颗粒吸附剂对 PPCPs 的吸附速度较慢，而粉末吸附剂存在使用后不易分离等缺点。由于 PPCPs 的性质各异，活性炭对不同 PPCPs 的去除效果存在很大差异，更重要的是传统吸附剂的再生是瓶颈问题，限制吸附技术的推广应用。因此，近期研究关注开发针对 PPCPs 的高效吸附剂，并解决吸附剂的再生和污染物的降解问题。

通过改变交联分子制备的环糊精聚合物，内部形成多孔结构，比表面积大于传统的环糊精聚合物，并且对多种 PPCPs 都具有吸附速度快、吸附量高的特点[88]；球磨法制备的超细磁性活性炭及磁性生物炭材料，解决了粉末碳材料在水中难分离的问题，该材料在 1 h 内可快速吸附卡马西平（CBZ）和四环素（TC），并且可通过球磨降解吸附剂上的 TC 和 CBZ[87]。碳纳米管及石墨烯等纳米材料，在机械强度、热稳定性、导电性方面具有优势，有利于吸附后进行材料再生。通过加热-过滤的方法，提升碳纳米管在水中分散性，过滤干燥后形成碳纳米管颗粒对典型 PPCPs 的吸附量可提高约 40%，并且利用碳纳米管的热稳定性，吸附后的材料可通过在空气环境高温氧化再生，在 5 次循环中仍可保持良好的吸附性能。为防止氧化石墨烯的堆叠团聚，同时促进氧化石墨烯对 PPCPs 的吸附作用，通过在氧化石墨烯层间嵌插刚性分子，增加了氧化石墨烯的孔径及有效吸附位点，可有效促进 PPCPs 的吸附，吸附量可提高 1~15 倍[90]。

综上，吸附技术可以有效去除不同废水中的典型污染物。以回收重金属为目标的吸附技术得到了广泛应用，而在废水深度处理中吸附去除微量难降解污染物技术部分得到应用，多数还处在研究阶段。随着废水排放指标的提高以及关注新兴污染物的增多，吸附技术在污水深度处理中的需求会越来越大。今后针对不同废水的深度处理，高效吸附微量污染物的材料研制、吸附材料的再生，以及吸附污染物的降解都将是研究重点。

2.2.4　消毒

废水消毒是灭活废水中病原微生物、防止流行病传播以及废水再生回用过程中必不可少的环节，也是保证受纳水体生态安全的关键工艺。废水具有病原微生物种类多、丰度高、背景污染物浓度高、水质水量变化大等特征，因此与饮用水消毒相比，废水消毒更复杂。本节综述了近年来不同废水消毒工艺的研究进展，重点阐述了废水消毒面临的技术挑战和可能产生的水质风险。

紫外线（UV）消毒是一种绿色杀菌技术，具有高效、广谱、几乎无消毒副产物（DBPs）生成、操作安全简便、占地少等优点[91]。早在 2000 年上海闵行污水处理厂就已选用 UV 作为消毒工艺，此后国内兴建的市政污水处理厂有一半以上均采用了 UV 消毒。典型应用为 2010 年投入运行的上海竹园污水处理厂，其 UV 消毒工艺由六条渠道构成，日处理量高达 170 万 m^3。作为一种物理消毒技术，UV 的消毒效果取决于 UV 辐射剂量，并受消毒器（渠）内光场和流场的共同影响；同时，UV 消毒系统在长期运行过程中，其效率还会因灯管老化和套管污染等影响而有所下降。因此，UV 消毒系统的优化设计、性能评估及运行维护是保证其消毒效率的关键。近年来，研究者通过模型计算和实验验证，已可模拟各种复杂 UV 消毒器内的光场分布，并结合计算流体动力学（CFD）模拟技术，来准确评估 UV 辐射剂量并由此开发高效节能的 UV 消毒设备[92]。随着 UV 传感器的发展，新的 UV 辐射剂量监测系统和监测方法也得以不断开发，为 UV 消毒系统运行的安全性提供了日益可靠的保障[93]。此外，在 UV 消毒过程中抗生素抗性菌与抗性基因削减、微量污染物去除以及 DBPs 生成等方面，国内学者也开展了大量研究，为其在废水消毒领域的推广应用提供了有益参考[94,95]。

作为一种成熟的消毒剂，液氯在废水消毒中得到了广泛的应用。其优点包括对常规细菌灭活效果较高、具有持续消毒能力、投资和运行费用较低等，但液氯的杀菌效果受水质（pH、氨氮等）的影响较大，而且作为一种危险化学品，其储存、运输、使用环节均存在一定的安全隐患。低剂量液氯对病毒、病原菌和寄生虫卵的灭活效果一般，而增大剂量时，将不可避免地与废水中的背景有机物反应生成有害的 DBPs。近年来，我国学者围绕液氯消毒过程中不同类型 DBPs 的前驱物、前处理工艺、不同水质条件下 DBPs 的生成与控制开展了大量研究，发现常规（三卤甲烷类、卤乙酸类）与新型（卤代硝基甲烷类、卤代乙腈、卤代乙酰胺类、亚硝胺类、卤代芳香族类）DBPs 均能在消毒出水中检出，增加了水质风险。液氯消毒对不同类型的细菌具有一定的选择性，如抗生素耐药菌灭活所需的 CT 值高于常规异养菌，且耐药菌存在复活的可能性，因此在再生水消毒过程中需审慎评估液氯对病原菌的灭活与复活效果[96,97]。此外，尾水中残留的余氯对自然水体中的水生生物具有一定的毒性作用，故液氯消毒时需控制余氯浓度以保证受纳水体的生态安全。

二氧化氯（ClO_2）是一类高效、广谱的氧化性杀菌剂，其消毒效果受水质影响较小，且不与氨、有机胺等反应，消毒后的副产物以 ClO_2^-、ClO_3^-为主（细胞毒性显著低于有机类 DBPs），因此 ClO_2 消毒后的水质风险相对较低[98]。ClO_2 能氧化废水中的还原性无机离子（Mn^{2+}、Fe^{2+}、S^{2-}、CN^-）以及酚类、腐殖质等有机物，对水体的腥臭味和霉烂味也有一定的削减效果[99]。Yang 等发现 ClO_2 在预氧化有机物的同时，亦可改变前驱物的结构与反应活性，降低后续加氯过程中 DBPs 的生成势。因此，ClO_2 在废水消毒领域具有较好的技术优势，当前制约其推广应用的主要因素是制备成本较高，同时由于 ClO_2 具有易挥发、易爆炸等特性，对现场制备的生产安全性要求较高。

其他较受关注的废水消毒剂还包括臭氧（O_3）和过氧乙酸（PAA）。O_3 消毒具有占地小、无运输安全隐患、无消毒剂残留等优势，但其现场制备较复杂，工程投资和运行维护费用较高，因此实际应用较少[100,101]。1980 年前后，美国尚有 44 座污水处理厂采用了 O_3 消毒技术，但到 2009 年后，仅剩 7 座仍在使用 O_3 消毒。在国内，采用 O_3 消毒技术的污水处理厂更少，目前仅有无锡城北污水处理厂（四期）和北京经开区东区污水处理厂等少数几家。废水 O_3 消毒效果受水质（COD、NO_2^--N、悬浮固体等）影响较大。O_3 消毒可有效控制废水的生物遗传毒性和抗性基因等，但可能产生醛类等 DBPs，而且当废水中含有较高浓度溴离子时还会生成潜在致癌物溴酸盐。PAA 消毒在工业和医疗废水领域已有一定的应用，其主要缺点在于会增加出水的有机物浓度且成本较高。PAA 消毒具有较低的出水生物毒性，同时能有效去除废水中的抗性基因，但会有微量的醛类和卤代酚等 DBPs 生成。PAA 与 UV 联用可以有效解决单独 UV 消毒时的微生物光复活问题，具有较好的协同效应[102]。

综上所述，当前国内废水消毒技术的研究重点在于针对 UV 消毒系统的性能评估以及针对各种化学消毒过程所产生的 DBPs 分析等。鉴于国内大中型污水处理厂多以 UV 消毒为主，而我国在这方面的理论基础和应用经验均较缺乏，因此亟待加强 UV 消毒系统的优化设计、性能评估及安全运行研究。此外，UV 消毒过程中可能发生的微生物复活现象也应得到足够的重视，将来可以考虑开展组合消毒工艺（UV/Chlorine、UV/ClO_2 等）研究来强化废水消毒效果[103]。

3 生物化学水处理方法与技术

3.1 厌氧生物处理技术

3.1.1 Anammox/Canon 脱氮技术

厌氧氨氧化（Anammox）工艺是利用 Anammox 菌在厌氧或缺氧条件下以亚硝酸盐为电子受体将氨

氮转化为氮气的过程[104]。该工艺因其无需外加有机碳源、脱氮负荷高、运行费用低、占地空间小等优点，被公认为是目前最为经济的生物脱氮工艺之一[105]。近年来，Anammox 工艺取得了大量研究性成果和工程化应用[106,107]。本小节综述了该工艺在处理生活污水和工业废水方面的应用现状，并讨论了进一步应用所面临的问题和解决对策。

3.1.1.1　Anammox 工艺的工程化应用现状

据统计，到 2014 年末，全球范围内的 Anammox 工程化装置已超过 100 座[108]。其中大部分位于欧洲，也正日益风靡亚洲和南美洲。序批式反应器是应用最广泛的形式，约占 50%，其次是颗粒污泥和生物膜系统，但颗粒污泥具有其他形式无法企及的脱氮性能。一体化系统是目前工程化应用的热点（88%）。市政污泥消化液是主要的处理对象（75%），而工业废水的应用有待开展。

3.1.1.2　Anammox 工艺在主流和侧流城市污水处理中的应用

污泥消化液是典型的低碳氮比、高氨氮废水，pH 一般为 7.0~8.5，温度一般为 30~37℃，适合 Anammox 菌的生长和代谢。世界上首个 Anammox 工程化装置位于荷兰鹿特丹 Dokhaven 污水厂处理污泥消化液，最初启动时间长达 3.5 年，但最终容积氮去除速率（NRR）达到 9.5 kgN/(m^3·d)[109]。目前 Anammox 工艺大多应用于侧流城市污水的处理，而经过近十年的发展，Anammox 工艺在污泥消化液上的应用已日趋成熟，很多研究者逐渐将 Anammox 工艺转向了主流污水的推广应用。5 种改进案例对比分析表明，Anammox 技术同时应用于城市污水的主流处理工艺及侧流污泥消化液的脱氮（Case D），可极大提高污水厂的能源效益，节约三分之一的运行成本[110]。这种改进思路也正是中国未来概念污水厂的核心技术路线。但与侧流应用不同，主流 Anammox 实现的前提条件较苛刻，不同地域气温和水质条件会导致较大差异。如何维持低温下 Anammox 菌的活性，实现低基质浓度下的菌体扩增，以及高流速下的菌体持留等问题是有待突破的瓶颈[111,112]。最新的研究表明，移动床生物膜反应器（MBBR）是一种较具前景的主流 Anammox 工艺[113]。

3.1.1.3　Anammox 工艺在高氨氮工业废水中的应用

Anammox 工艺应用于主流城市污水的脱氮已成为研究热点，而高氨氮工业废水的生物脱氮也是一个前景广阔亟待开发的市场。我国已建成数座实际工程，主要集中在发酵行业、食品加工行业、农业和垃圾渗滤液领域等，表 17-1 列出了目前世界上应用 Anammox 工艺处理工业废水工程案例，其中通辽梅花味精废水 I 期工程反应器容积高达 6600 m^3，是迄今世界上规模最大的 Anammox 工程装置。由于实际工业废水往往成分复杂，常含重金属、抗生素、氰化物、硫化物、酚类、盐度等 Anammox 菌的抑制物，限制了该工艺的推广应用[114]。因此，深入研究这些工业废水对 Anammox 工艺性能影响，对提出有效运行策略具有重要意义。

3.1.1.4　总结和展望

作为新型生物脱氮技术，Anammox 工艺用于处理成分复杂的工业高氨废水和主流城市污水还面临巨大挑战。我国在 Anammox 方面的研究起步较早，但前期研究成果不多，从 2010 年起，国内学者的科研产出呈井喷趋势，超越荷兰、美国等发达国家[115-118]。但工程化应用的数量仍远落后于荷兰等欧洲国家。这些国家拥有一批具有自主知识产权的商业化公司，如威立雅、帕克、苏伊士等水务集团，各种专利性的 Anammox 技术得到了蓬勃发展，如 DEMON®、ANITA-Mox®、ANAMMOX®、DeAmmon®、TERRANA®、ELAN®、Cleargreen®。而在中国，只有浙江艾摩柯斯环境科技有限公司和北京城排集团走在本土商业化的前列，两者分别侧重于工业废水和城市污水的脱氮。当 Anammox 工艺在技术金字塔的顶端光芒四射时，支撑其应用发展的必然是坚实的理论基础。今后，实验室研究应侧重于实际废水适用性的探究，产学研结合，进一步推动 Anammox 工艺在国内的工程化应用。

表 17-1　应用于工业废水处理 Anammox 工程化装置概述

工业废水	工艺	地点	反应器体积（m^3）	设计负荷	
				kg N/d	kg N/(m^3·d)
食品加工					
发酵	DEMON®	Alltech，RS	—	2400	—
发酵	ANAMMOX®	Binzhou，CN	500	1000	2.0
淀粉	ANAMMOX®	China	—	7000	2.0
淀粉/味精	ANAMMOX®	Shandong，CN	4300	6090	—
味精	ANAMMOX®	Tongliao I，CN	6600	11000	2.0
味精	ANAMMOX®	Tongliao II，CN	4100	9000	2.0
味精	ANAMMOX®	Wulumuq，CN	5400	10710	2.0
乳制品	DEMON®	Rickenbach，DE	—	50	—
甜味剂	ANAMMOX®	Yixing，CN	1600	2180	—
土豆	ANAMMOX®	Olburgen，NL	600	1200	1.2~1.8
土豆	NAS®	Bergen op Zoom，NL	7920（3stages +settler）	720	0.09
土豆	NAS®	Budrio，IT	1509（3stages +MBR）	344	0.23
土豆	NAS®	Kruiningen，NL	15300（4stages +settler）	325	0.02
肉制	ANAMMOX®	Son，NL	3000	6000	2.0
肉制	ANAMMOX®	Brazil	—	400	2.0
肉制	NAS®	Vion，NL	—	800	—
鱼罐头	ELAN®	Catoira，SP	250	84	0.3
酿酒厂	ANAMMOX®	Shaoxing，CN	560	900	2.0
酿酒厂	ANAMMOX®	Poland	900	1460	—
酿酒厂	ANAMMOX®	中国	—	1200	2.0
皮革厂	CIRCOX®- ANAMMOX®	Lichtenvoorde，NL	164/100	325	0.45
农业生产					
屠宰场废水	NAS®	Boxtel，NL	5310（3 stage，last as）	888	0.17
肥料产生	RBC	NL	—	150	—
消化猪粪	PN/AA Abengoa Water	Chiclana，SP	5	—	0.1
其他					
沼气	DEMON®	Etappi Oy，FI	—	990	—
沼气	DEMON®	Etappi Oy II，FI	—	800	—
渗滤液	DEMON®	Lavant，AT	—	45	—
渗滤液	DEMON®	Trento，IT	—	700	—
渗滤液	RBC	Kölliken，CH	33	—	0.4
渗滤液	RBC	Mechernich，DE	80	—	0.6
渗滤液	RBC	Pitsea，GB	240	—	1.7
渗滤液	One stage+Activated carbon	Emscherbruch，DE	—	—	3
渗滤液	SNAD	Taiwan	384	—	—
渗滤液和污泥	ANITA™Mox	Holbæk，DK	600	120	—
半导体	Biofilm CSTR/ANAMMOX®	Mie，JP	300/58	220	1.0~3.2

3.1.2　同步脱氮除磷技术

传统废水生物脱氮除磷技术主要依赖硝化、反硝化脱氮和聚磷菌对磷的过量吸收，其中反硝化脱氮和生物除磷都需要碳源，而我国城市污水 C/N 和 C/P 比的失调（进水中碳源缺乏）往往导致脱氮和除磷效果无法同步达到最佳，需通过投加碳源以及结合化学除磷来保证脱氮和除磷效果。同时，传统工艺存在流程长、消耗高（曝气和药剂）及占地面积大等不足也严重限制了污水处理厂的可持续发展，开发高效、经济的同步脱氮除磷技术已成为当前水污染控制领域的研究重点[119,120]。

3.1.2.1　废水同步脱氮除磷处理技术研究进展

现代生物技术的发展促进了废水脱氮除磷技术的生物学特性研究、推动了处理工艺的开发。尤其是随着短程硝化、好氧反硝化、同步硝化反硝化、反硝化除磷和厌氧氨氧化[121]等脱氮除磷新现象的发现，为开发新型的污水脱氮除磷技术提供了新的理论与思路，涌现出了系列新型的同步脱氮除磷技术。

1）倒置 A^2/O 技术

技术原理：将常规 A^2/O 工艺中的厌氧区和缺氧区倒置，使得回流液中的硝酸盐尽可能在缺氧区消耗殆尽，确保后续厌氧区较低的氧化还原电位（ORP），利于聚磷菌形成更强的吸磷动力；聚磷菌厌氧释磷后直接进入好氧环境，其在厌氧条件下形成的吸磷动力得到更充分利用，好氧吸磷效率提升[122]。

技术优势：缺氧段位于工艺首端，允许反硝化优先获得进水中碳源，强化系统的脱氮能力；活性污泥经历完整的释磷和吸磷过程，除磷能力增强。倒置 A^2/O 工程应用结果表明，相较于常规 A^2/O 工艺，不仅同步脱氮除磷效率有效提高，运行能耗也显著降低。

2）反硝化除磷技术

技术原理：反硝化除磷脱氮是以胞内碳源（聚羟基脂肪酸酯，PHA）作为电子供体的一种反硝化新途径。反硝化除磷通过“一碳两用”方式将脱氮和除磷功能耦合于一个生化反应途径，是一种新型的、高效低碳（节省碳源、曝气量）的废水氮磷营养物去除技术。基于反硝化同步脱氮除磷技术开发的主要工艺有 UCT 工艺、BCFs 工艺、Dephanox 工艺和 A_2N-SBR 工艺[123,124]。

技术优势：反硝化除磷技术打破了传统脱氮除磷机理所认为的脱氮除磷必须分别由专性反硝化菌和专性聚磷菌来完成的理念，使得除磷和反硝化脱氮过程由同一类微生物来实现，这对生物脱氮除磷机理是一重大突破和飞跃。与传统的脱氮除磷工艺相比，COD 耗量可节省 50%，氧气耗量降低 30%，污泥产量也可望减少 50%，还可实现污水中磷资源的回收，因此反硝化除磷脱氮技术被视为一种可持续污水处理技术。

3）好氧颗粒污泥同步脱氮除磷技术

技术原理：利用颗粒污泥自身结构特点，通过溶解氧扩散控制方式，在颗粒污泥由外向内依次形成好氧区、缺氧区和厌氧区，为脱氮和除磷功能微生物（如硝化菌、反硝化菌、好氧聚磷菌和反硝化聚磷菌）等提供所需生存空间，进而实现同步硝化反硝化和除磷目的[124,125]。

技术优势：由于好氧颗粒污泥集好氧、缺氧和厌氧微环境于一体，可实现高效的同步脱氮除磷目标；同时，好氧颗粒污泥优良的沉降性能有利于处理系统保持更高的微生物数量、更合理的微生物群落结构、更强的抗冲击负荷能力和更小的占地面积[126]，因此被认为是最有前途的废水生物处理技术之一。

2008 年，在南非 Gansbaai 建立了世界上第一个应用好氧颗粒污泥技术的市政污水厂[采用荷兰公司 DHV（RoyalHaskoning DHV）的 Nereda®好氧颗粒污泥工艺]，长期运行结果显示，好氧颗粒污泥技术去除氨氮和磷的效果良好，并能够显著降低运行能耗[127]。

4）膜生物反应器同步脱氮除磷技术

技术原理：膜生物反应器（MBR）通过超滤膜的截留作用，使硝化菌长期停留于好氧池内，在不增加池容的前提下延长污泥龄，满足硝化菌的优势生长，减少其流失。有文献报道，MBR 中可生长反硝化聚磷菌，在脱氮的同时实现有效的磷去除[124,128]。因此，MBR 同步脱氮除磷工艺根据除磷方式的不同分为生物脱氮除磷和生物脱氮化学除磷两种。在生物脱氮除磷工艺中，根据好氧和厌氧是否在同一反应器内发生，又可将其分为三类：单级 A/O MBR 脱氮除磷工艺、两级 A/O MBR 脱氮除磷工艺和两级 A^2/O MBR 脱氮除磷工艺。

技术优势：MBR 技术具有容积负荷高，抗冲击负荷能力强，出水稳定等优势，膜分离对保障系统出水水质起到了一定作用，尤其 TP 的截留效果比较明显，这在一定程度上弥补了 MBR 系统通常生物脱氮效果好但是生物除磷有限的短板[129]。

5）厌氧氨氧化与生物除磷技术的耦合

技术原理：厌氧氨氧化技术与生物除磷工艺的耦合可通过两种方式实现：①实现污水中碳、氮和磷的分离去除。在一级处理阶段，通过强化生物除磷技术实现污水中碳、氮、磷的分离，其中碳、磷元素以微生物形式实现与污水的分离进入污泥消化系统，含氮元素的污水则进入后续一体化厌氧氨氧化反应器实现自养脱氮；磷元素进入到污泥消化液中，可以通过高效的 MAP 技术实现磷元素的回收；消化液中的高浓度氨氮则通过侧流厌氧氨氧化系统实现脱氮，同时侧流厌氧氨氧化系统可为主流区厌氧氨氧化系统提供充足的菌种[130]。②溶解氧和有机物对厌氧氨氧化菌有抑制作用，而反硝化聚磷菌能消耗溶解氧和有机物，同时反硝化聚磷菌也能利用厌氧氨氧化反应产生的硝酸盐[121,131]。因此，也有研究表明，可通过将厌氧氨氧化污泥和反硝化除磷污泥置于同一个反应器中，创造对厌氧氨氧化菌和反硝化聚磷菌同时有利的微生态环境，发挥两者在脱氮除磷方面的协同耦合作用，也可实现同步脱氮除磷[132,133]。

技术优势：以厌氧氨氧化与生物除磷耦合后，在实现高效同步脱氮除磷的同时，又可实现碳和磷的回收，非常符合未来城市污水处理倡导的“节能、资源回收和环境友好”的发展方向。

3.1.2.2 展望

脱氮除磷新理论的提出为开发经济、高效的同步脱氮除磷新技术提供了新的思路与方向。新的脱氮除磷技术也不仅仅是为了满足严格的 N、P 去除效果，而是着眼于新工艺、新技术的经济性、处理效果的稳定性以及对污水中潜在资源的回收利用。

今后，在脱氮除磷机理研究方面，将充分利用现代分子生物学的分析检测技术，深入剖析脱氮除磷功能微生物的生物化学代谢机制，以期更好地指导实践；在新工艺开发方面，要把污水视为一种资源，一方面重视污水厂的碳中和，努力实现污水厂的能源自给甚至产能，另一方面也要注重对污水中 C、P 等资源的回收。

3.2 废水处理实践

3.2.1 城镇污水

城镇污水是各城镇或地区主要的水污染来源，城镇污水处理设施的建设和运行是各地落实水污染防治规划目标的主要途径。近年来在中央财政资金和相关政策的大力支持下，城镇污水处理设施建设规模快速增大，目前城镇污水平均处理率已超过 90%。城镇污水处理设施的快速发展为城镇污水处理新技术的研发和应用带来了重大需求和机遇，如：①为满足城镇污水处理厂出水排放标准大幅度提升及缺水地区对污水再生利用的需求，一批污水处理新技术，如膜生物反应器（MBR，membrane bioreactor）、移动床生物膜反应器（MBBR，moving bed biofilm reactor）等技术得到了发展并在实际工程中得到广泛的推广应用；②针对我国城镇污水普遍存在的水质水量变化大导致出水不稳定、工艺运行能耗和物耗较高的问题，开发了系列污水处理工艺优化运行与节能降耗技术，通过工程实践收到明显的节能效果；③为应对发展节约型社会和低碳模式的需求，积极致力于探索低能耗脱氮技术（如厌氧氨氧化技术）及能源回收新工艺，取得了长足的进步，为其更深入的研发和应用奠定了基础。

以下重点介绍城镇污水 MBR 脱氮除磷、污水处理工艺优化运行与节能降耗、厌氧氨氧化以及污水能源化等技术的新进展。

3.2.1.1　城镇污水 MBR 脱氮除磷

MBR 技术在我国的应用始于 20 世纪 90 年代。自 2006 年首座万 t/d 级工程投运以来发展迅速（图 17-4），大型 MBR 工程（万 t/d 以上级）的总处理能力于 2016 年达到 900 万 t/d，其中城镇污水部分占我国城镇污水总处理规模的 3%以上[134,135]。作为一种新兴技术，随着污水处理标准的日趋严格和节能降耗需求的日益迫切，MBR 在大规模推广应用中面临污染物深度去除、工艺节能优化、膜污染高效控制等方面的技术发展需求，近年来国内学者的应用基础研究为此提供了有力支持。

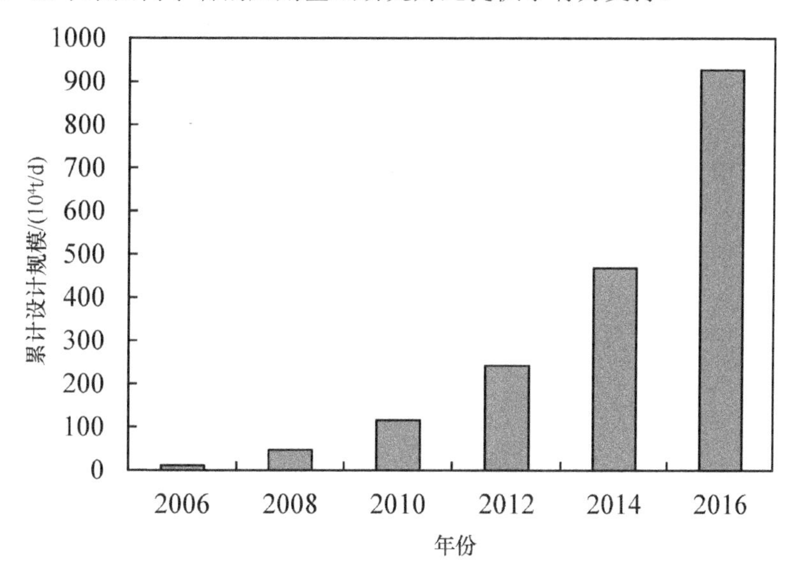

图 17-4　我国万 t/d 以上级 MBR 工程应用增长情况

在污染物深度去除方面，针对我国城镇污水普遍存在 C/N 比低的情况，开发出强化内源反硝化的深度脱氮工艺[136]；针对 MBR 强化除磷，探讨了反硝化除磷、胞外多聚物对磷的富集、膜对胶体磷的截留等机制，对辅助化学除磷的药剂投加条件进行了优化[137]，并探索了原位电絮凝除磷等新方法[138]；在出水安全性方面，重点关注了实际工艺过程对微量有机物的去除作用，指出污泥吸附、生物降解和膜分离是影响污染物归趋的重要方面[139]。

在工艺节能降耗方面，生化池和膜池曝气量的降低是研究的重点。对于生化池曝气，根据 MBR 工艺的实际情况，对污泥传氧系数进行了重新评估，避免了生化池设计曝气量过高[140]；建立了生化池曝气控制策略，包括“氨氮-DO-曝气量”两阶串级反馈 PI 控制和“进水水量-曝气量”补偿控制，并成功应用于实际工程，降低好氧池鼓风机实际功率 20%[141]。膜池曝气量的优化包括曝气节律、气泡形态、水力循环等方面，在计算流体力学指导下的膜组器和膜池几何构型优化是重要的研究方向。

在膜污染控制方面，对膜污染机理开展了大量研究，包括膜污染物识别、影响因素评价、过程定量表征、作用力模型等[135,142]。特别是在大型 MBR 工程尺度上，系统考察了污泥混合液性质和长期运行过程中温度等因素对膜污染的影响[143]。通过投加悬浮载体、混凝剂等方式对混合液性质进行调控，从而减轻膜污染。在膜污染的清洗方面，实际工程中在线化学清洗药剂的改进和操作条件的优化受到重点关注[144]。

3.2.1.2 城镇污水处理工艺优化运行与节能降耗

城镇污水处理工艺优化运行的基础是水力学和水质学模型。以活性污泥模型（activated sludge models，ASMs）为基础，发展了基准控制模型（benchmark simulation model，BSM）、完全耦合模型（full coupled activated sludge model，FCASM）等水质学模型；以计算流体力学（computational fluid dynamics，CFD）等水力学模型为基础，开展了二沉池、初沉池、沉砂池等反应器的构型优化。

在工程应用上，工艺优化运行主要通过工艺诊断、情景分析、参数优化等方法提高了工艺的运行效果，以应对寒冷天气、高浓度进水、工业废水冲击负荷等的影响。目前研究优化了多种传统处理工艺的曝气、回流等操作运行条件，如阶段进水 AO[145]、阶段进水 UCT、MBBR 等。近年来，有学者研究基于数据驱动的专家系统，不再依赖机理模型，而是利用在线历史数据建立操作条件与出水水质的关联性，如污水处理厂运营管理系统[146]。

节能降耗的技术基础是过程自动控制。通过提高关键设备效率、优化控制水平，可以在保障处理达标的基础上，降低曝气过程能耗和加药单元物耗。节能降耗的研究重点是过程控制策略。基于 BSM 仿真试验，研究了多种智能优化算法，如神经网络、粒子群、人工免疫等。基于在线优化思想，提出了线性自抗扰等[147]控制算法。

在工程应用上，我国处于推广曝气加药控制[148]，示范出水氨氮控制，尝试全流程控制的阶段，尚落后于欧美先进国家的水平。节能降耗技术集中在曝气单元和加药过程。曝气系统节能措施包括“前馈-反馈”综合控制系统、采用高效曝气设备（曝气器、气悬浮鼓风机）等；加药单元主要通过反馈控制降低物耗。

3.2.1.3 厌氧氨氧化

厌氧氨氧化作为一种新型低能耗脱氮技术，其相关研究与工程实践受到广泛关注。国内目前在实验室水平大多可以实现不同条件下的厌氧氨氧化过程启动与稳定运行，然而针对实际污水的工程应用案例相对较少。实现稳定高效运行的工程应用以侧流厌氧氨氧化工艺为主，处理对象为低碳氮比高氨氮浓度废水，如消化污泥脱水滤液[149]、垃圾渗滤液、味精加工废水等。国内在建或初步建成的以厌氧氨氧化为主体的污水处理工程项目主要分布在北京、通辽、山东、湖北等地[149]，见表 17-2。从工艺形式来讲，调控较为灵活的分体式工艺（如 Sharon-Anammox）在工程应用方面具有一定优势，而一体式工艺（如 Canon 工艺）的稳定控制则相对较难。目前，侧流厌氧氨氧化工艺在城镇污水处理过程中的应用日渐成熟，但由于厌氧氨氧化菌倍增时间长（11 d），容易受到各种因素的影响，工艺运行稳定相对较难，其在城镇污水处理中的应用经验仍待进一步完善[150]。

以低氨氮浓度、低温、大流量为特点的主流厌氧氨氧化技术在城镇污水处理的工程应用方面现在仍处于探索当中，是当前国际研究热点。到 2015 年，全球至少有 5 座污水处理厂在尝试践行主流厌氧氨氧化。目前，主流厌氧氨氧化工艺工程化应用的难点在于如何在常温及低温条件下实现亚硝化稳定控制及 Anammox 菌在反应器内的稳定富集，如何在大流量且水质明显波动条件下实现反应器放大和稳定运行，获得较低的出水浓度[151]。期望短程反硝化[152,153]、基于物化调控的低氨氮亚硝化稳定实现[154]等方面的研究成果可以助力未来污水厂主流厌氧氨氧化工艺的实现与发展。

3.2.1.4 城镇污水能源化

城镇污水能源化立足于节能减排和低碳排放，以污水中有机碳源的能量回收为主要目标，是目前国内外研究者致力研发的重要方向。

表 17-2　国内近年 Anammox 工艺技术的工程应用案例

工程名称	处理对象	反应器有效容积(m^3)	处理规模	氨氮负荷[kgN/(m^3·d)]	年份
天津市津南污泥处理厂	消化污泥脱水滤液	—	污泥 800 t/d（20%含固率）	—	2016
北京小红门污水厂污泥消化处理工程	消化污泥脱水滤液	—	污泥 900 t/d（20%含固率）	—	2016
北京槐房再生水厂泥区工程	消化污泥脱水滤液	—	1220 t 污泥/d（20%含固率）	—	2015
北京清河第二再生水厂泥区工程	消化污泥脱水滤液	—	50 万 m^3 污水量/d	—	2015
北京高安屯污泥处理中心工程	消化污泥脱水滤液	—	1800 t 污泥/d（20%含固率）	—	2015
湖北十堰北排西部垃圾渗滤液处理站	垃圾渗滤液	—	150 m^3/d	—	2015
北京高碑店污水处理厂	消化污泥脱水滤液	13800	1000 m^3/d	—	2013
山东祥瑞药业工程	玉米淀粉和味精生产废水	4300	—	1.42	2011
江苏汉光甜味剂有限公司	甜味剂生产废水	1600	—	1.36	2011
浙江会稽山绍兴酒股份有限公司	酿酒生产废水	560	—	1.61	2011

在工程实践中，污水能源化主要通过市政污泥厌氧消化产甲烷的方式实现。目前污泥厌氧消化的主要发展趋势包括提高消化池污泥浓度和有机物分解率等。通过提高消化池污泥浓度的方式实现[155]高含固厌氧消化（污泥浓度由常规 30 g/L 左右提升到 100 g/L 以上），可以大幅度降低消化罐体积及罐体加热能耗，提高单位容积产气率，是目前常见的工程选择。未来需要深入探究污泥高含固厌氧处理条件下的传质规律、微生物抑制等机理过程[156]，进一步提高污泥含固率。同时，污泥水解过程是厌氧消化反应的限速步骤。通过热处理、碱处理、超声、臭氧等预处理手段，破坏菌胶团及微生物菌体结构，促进污泥水解，可以提高厌氧处理效率。天津津南污泥厂处理规模 800 t/d（80%含水量），采用高含固厌氧消化技术（消化池含固率 10%），同时利用脉冲高电压污泥预处理及投加厌氧增效菌剂的方法，促进污泥破壁过程；厌氧产能用于污泥后续的热干化处理，在不引进外部能源条件下，首次实现了厂区系统的能量平衡。近年来，在污水处理工艺前段首先对污水中有机碳源进行富集浓缩以强化能量回收，成为国际上污水能源化的最新发展方向之一[157]。碳源富集浓缩技术既包括生物吸附-沉淀、混凝-沉淀等传统技术，也包括膜分离等新型技术。目前荷兰 Wageningen 大学、比利时 Ghent 大学在该领域均开展了大量研究工作，但多局限于实验室规模。国内清华大学最早开展碳源浓缩领域的研究，基于膜过滤和磁分离过程，耦合混凝吸附，研发了碳源强化浓缩技术，并已进入日处理规模百吨级别的中试研究（图 17-5）。

图 17-5　碳源浓缩技术中试现场（左：磁分离碳源浓缩设备；右：碳源膜浓缩设备）

磁分离技术水力停留时间极短（5~10 min），对碳源的浓缩富集率达 60%~80%；膜浓缩技术对碳源的富集浓缩率达 90%。碳源强化浓缩技术针对污水中碳源浓度低、能量回收难的问题，通过吸附型混凝剂与新型磁性改性材料、膜材料的耦合，实现碳源高效定向浓缩分离，最终与新型低能耗脱氮过程结合，有望形成变革性的污水处理新工艺，大幅度提高污水能量化水平[158]。

3.2.2 难降解工业废水

难降解工业有机废水是工业化进程的产物，也是合成工业发展的一个重要环境课题。基于此，从难降解工业废水的分类与性质、处理方法选择、处理工艺原理、工程案例应用以及未来若干前沿问题分析了相关领域的研究进展，需要在已经发现方法原理基础上，重点考虑资源化、能源化以及无害化相结合的分子转化机制，结合生产过程，在多学科交叉与多领域协同方面追求创新。

3.2.2.1 难降解工业废水的分类与性质

据国家统计局 2016 年统计数据，我国污水及废水排放总量达到 735.32 亿 t，其中工业废水排放量为 199.50 亿 t，占排放总量的 27.13%。难降解工业废水，即难以生化降解的工业废水，在工业废水中占很大的比例。这类废水具有成分复杂、污染物浓度高、毒性大以及可生化性差等特点，表现为含有潜热、稀相/多相共存及具有回收价值的特征[159]。按照工业类型，可以分为造纸废水、炼焦煤气废水、纺织印染废水、制革废水、农药废水、冶金废水、化学肥料废水等。

难降解有机污染物表现为高浓度、中低浓度、低浓度、微量/痕量的不同水平，针对工业废水，主要追求达标排放，低浓度或微量/痕量的污染物较少受到关注。作为例子，难降解工业废水主要含有染料分子、酚类物质、硝基苯、苯乙酮、氯酚、多环芳烃、氯代烃、杂环化合物、有机磷农药、表面活性剂等有机污染物[160]。表 17-3 归纳了一些典型的难降解工业废水中的有机物。这些物质的共同特点是毒性大，成分复杂，化学耗氧量高，结构稳定，残留时间长，一般微生物降解有效性比较差。针对废水的难降解性可以归纳为三个方面：①在处理时的外部环境条件（如 pH、温度、营养成分等）很难达到微生物处理的最优条件；②有机物的化学组成及结构表现出对微生物的惰性作用，即微生物的群落中缺乏目标有机物的识别酶，表现出抗降解性的存在；③易降解有机物的废水中含有对微生物群落有毒或者能够抑制微生物成长的无机物如氰化物、硫氰化物、硫化物、亚硝酸根等，从而使有机物不能得到快速的降解[161]。

表 17-3 典型工业难降解废水中的有机污染物

工业行业	有机污染物
石油加工	苯、甲苯、乙苯、多环芳烃、苯酚等
焦化/煤制气	苯、甲苯、乙苯、多环芳烃、间甲酚等
塑料制造	苯、甲苯、二甲苯、二氯甲苯、酞酸、脂类等
化学纤维	苯、甲苯、二甲苯、苯酚等
农药制造	苯、甲苯、氯苯、二氯甲烷、苯胺、苯酚、间甲酚、对硝基甲苯、对硝基苯酚等
医药制造	苯、萘、三氯苯、苯酚、苯胺、硝基苯、对硝基氯苯等
燃料制造	苯、萘、三氯苯、苯酚、苯胺、硝基苯、对硝基氯苯等
造纸/纤维	苯酚、氯苯酚、有机氯、单宁等

以多环芳烃（PAHs）为例，苯环上的六个碳原子处一个平面上，形成闭合的大 π 键，每个碳碳键的键长和键能都是相等的，使得苯环具有较强的稳定性，且随着苯环数量的增加稳定性逐渐增强。表 17-4 归纳了几类难降解废水中典型有机污染物的物化性质。苯环上的氢原子被烃基、卤原子、硝基等疏水性基团取代的多氯联苯（PCBs）、硝基化合物表现出较强的疏水性，随着苯环数量、取代基团个数的增加，疏水性也逐渐增强，如苯环上氯原子愈多，通常其吸引苯环上 π 电子云的能力愈强，稳定性也愈强，随着氯原子数量的增加，水溶性逐渐降低，在 25 ℃下溶解性最高的二氯联苯为 5900 μg/L，最低的十氯联苯仅为 2.7 μg/L。这些疏水性强、水溶性低的难降解污染物阻碍了有机物与微生物的有效接触及有机物扩散穿过细胞壁，这类有机物多数不能被微生物的酶系识别，使酶系统不能激活，其废水的可生化性比较差[162]。

表 17-4　难降解废水中典型有机污染物的物化性质

有机污染物种类	原子、分子基团	化学键	溶解度（g/L，25℃）	降解半衰期
氯苯类	苯环、氯取代基	π、C—C	2.7×10^{-6}~5900×10^{-6}	1.4~12.4 a
多环芳烃	苯环	π、C—C	3.8×10^{-6}~3.0×10^{-3}	4.4~8.0 d
硝基苯类	苯环、硝基	π、C—N、N—O	0.13~0.19	2.0~2.6 d
苯胺类	苯环、氨基	π、C—N、N—H	0.56~0.80	5.0~25.0 d
苯/苯同系物	苯环、烃基	π	0.17~1.80	1.0~4.0 d
杂环类	氧原子、氮原子、硫原子	C═C、C—O、C—N	0.7~600.0	10.0~200.0 h
酚类	苯环、羟基	π、C—O	2.0~70.0	10.0~36.0 h

3.2.2.2　难降解工业废水处理方法选择

从本质上看，难降解工业废水处理方法的选择必须从热力学可行性和动力学有效性两方面进行考虑[163]。具体而言，难降解工业废水的处理，应当根据难降解工业废水中三相物质分子的物理性质的差异，实现相、溶质及分子水平的分离，解除废水难降解的限制。根据难降解工业废水的特性，从重力分离、离心分离、吸附、过滤、膜分离等方法中甄选适宜的方法，实现从废水溶液中分离并提取出有价值的组分，明确废水中相分离、产品特征以及毒性污染物转化的方法学适用范围。

通过物理或机械作用分离或回收废水中不溶解的呈悬浮状态的污染物，其过程不改变废水中物质的属性，如相分离、相转移相关的物理法。在选择过程中，必须首先了解废水中污染物的形态。一般污染物在废水中处于悬浮、胶体和溶解三种形态。强化混凝技术、靶向吸附技术、分子印迹技术与功能材料技术等的结合已经具有代表性[164]。

化学法针对溶解性有机物有效，实现物理法难以实现的目标，两者结合更有必要。通过一些极端的方法（如高压、催化、臭氧氧化等）诱导产生氧化性极强的羟基自由基（HO·）氧化降解有机污染物，具有反应迅速及无二次污染等特点而作为有效的化学法，称之为高级氧化法[165]。HO· 氧化能力极强，很容易氧化各种有机物，因反应速度快而成为许多高级氧化工艺的原理基础，以自由基作为开端的湿式空气氧化法、超临界水氧化法、光化学氧化法、电化学氧化法、声化学氧化法及相应的催化氧化法在追求低能耗途径与适用性方面具有探索的空间[166]。此外，臭氧氧化、Fenton 试剂、光催化氧化、电化学氧化等的原理及其技术已经日趋成熟，然而，基于不同污染物特征的电子转移计量关系的定量研究方面需要加强。

生物法通过微生物的代谢活动使废水中的有机污染物转化为无害产物，即利用活性污泥或生物膜上的微生物将废水中的有机物作为它们代谢活动所必需的能源物质，将其转化为 CO_2 和水，而使废水得到净化。针对难降解有机污染物，在厌氧水解法结合各类电子供体构建各种功能转化降解如产小分子有机酸、产 H_2 和 CH_4、反硝化脱氮、硫转化与除磷的过程中，实现全部或局部的共基质降解[167]。在基本方法基础上演绎出来的微生物能源原理、微生物化工产品原理、微生物冶金原理、微生物修复原理等的发展已经成为优先考虑的方向。

3.2.2.3　难降解工业废水处理工艺原理

工艺技术是立足于水质学以及水溶液特性基础上的若干方法原理应用的结合，是由化学层面过渡到化工以及工程层面的一种表达，是隶属于能量、物质消耗条件下污染物转化的方法原理效果的体现，需要综合考虑反应动力学、经济因素与环境因素之间的关系。难降解工业废水处理工艺一般分为预处理、生物处理及深度处理单元。

预处理是在生物单元前通过物理、化学方法对原水中的工业原料回收、大宗组分分离，并对溶液性质和营养进行调节，有利于二级处理单元的进行。对不同废水对象的污染物性质选择预处理技术，改变难降解有机化合物的组成结构，消除或减弱其生物毒性，提高后续单元可生化性。

生物处理技术指针对废水溶液性质，将厌氧、好氧、水解等若干单元有机组合，微生物通过电子传递实现水中有机物的降解以及部分无机物转变的过程[168]。通过生化反应微生物将有机物氧化分解为有机酸等代谢产物，这些小分子易降解有机物为废水脱氮除磷提供良好的碳源。生物处理工艺可以通过条件控制和单元组合实现废水除碳、减毒、脱氮、调节 pH 效果，创造出有利于微生物生长繁殖的环境，以提高降解有机物的效率。以活性污泥法为代表的好氧法通过对废水中有机物进行吸附、生理代谢、絮凝等作用，在对废水中 COD 去除的基础上，毒性物质作为共基质也被部分降解或转化。厌氧法多用于高浓度有机废水处理，其有机污染物负荷可达 5~10 kg COD(m^3/d)，且污泥量少，仅为好氧法的 1/6~1/10；其缺点是处理后的 COD、BOD 值偏高，水力停留时间较长使反应器容积庞大[169]。目前我国工业废水处理主要采用 A/O、A/O/O、A/A/O、O/H/O 等工艺，这些工艺主要是生物好氧系统和厌氧系统不同程度的组合，通过生物好氧作用降解废水中易降解的物质如酚类化合物、氰化物、硫氰化物、氨氮等，通过厌氧过程将大分子难降解有机物分解为小分子物质[170-172]。图 17-6 对比了几种典型生物工艺的优缺点。其中，A/O 生物工艺对于除碳脱氮效果显著，然而存在除磷效率低、抗冲击能力差等缺陷。为了实现磷的有效去除及提高系统的稳定性，开发出了除磷与反硝化相分离的 A/A/O 生物工艺，得到了广泛的应用。对于一些缺磷、富氮、高碳、强毒性的工业废水如焦化废水、煤化工废水及石油化工废水，A/A/O 工艺在组成单元的有效性及抗冲击方面无法满足废水处理的要求，同时 A/A/O 工艺可实现的脱氮过程单一，灵活性不足[173]。O/H/O 工艺是一种具有高度抗冲击、系统稳定同时可实现多种脱氮过程，在理论及操作上具有高度灵活性，具有适用于各种工业及生活废水的潜力[174]。

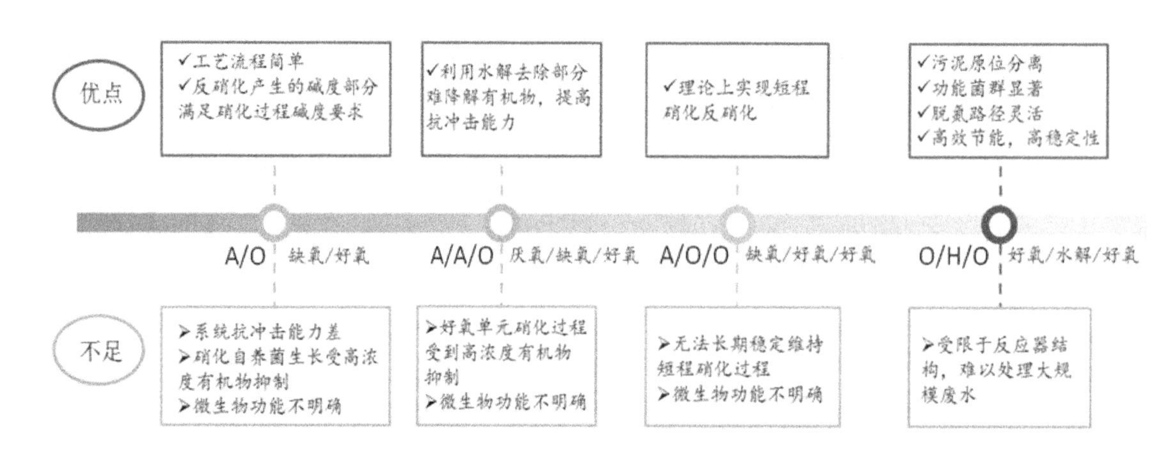

图 17-6　几种典型生物处理工艺优缺点对比

深度处理技术追求的目标为：去除生物单元出水中的胶体颗粒；通过反应和/或分离方法去除残余化合物如难降解有机物；分离和部分分离无机的阴阳离子[175]。深度处理技术一般包括混凝、氧化、吸附和膜过滤等方法及其结合，但这些技术的应用往往带来污泥产量的增加或形成盐分。为了减少深度处理过程中副产物的生成，需要针对废水溶液性质选用合适的方法原理作为技术的依托。

3.2.2.4　难降解工业废水处理工程案例

随着国家对环境保护和节能减排的重视，难降解工业废水处理技术的发展和处理设施的建设都取得显著的成效，导致难降解工业废水处理设施众多，工艺繁杂，构筑物多种多样[176-178]。然而我国缺乏全面、科学的难降解工业废水处理工程的效果评价体系。目前废水处理工程评价主要以工程前期投资、运行成本、常规性指标的检测以及个人经验为依据，鲜有涉及运行能耗以及碳循环/碳中和等方面的评价，没有包括难降解工业废水处理过程中伴随废气和剩余污泥产生的全过程评估[179]。常规性指标往往不足以评价这些副产物的毒理学性质。一个废水处理工程的综合评价应当涵盖经济因素、技术因素、操作管理难易程度、环境生态影响等方面，不应仅局限于技术因素和经济因素。

丰朝海课题组一直致力于焦化废水的基础与技术应用研究。基于多年研究及工程经验，将混合特性好、传质效率高的生物三相流化床和 A/O/O、O/H/O 组合工艺作为生物处理核心单元[180]，已成功应用于韶钢一期、二期、金牛天铁等焦化废水处理工程之中。以韶钢二期工程为例，其核心生物处理工艺为 O/H/O 生物流化床技术，具有高效节能、运行稳定、原位减盐等效果。为了弥补焦化废水生态风险评价方面的研究空白，在对工程中外排水、外排泥和外排气中 PAHs 的浓度分布和毒性当量进行研究，发现焦化污泥中 PAHs 的数目可达 161 种，该类物质总的浓度最高可达（6690 ± 585）mg/kg[181]。在大气环境中，经不完全统计其 PAHs 的数目可达 77 种，该类物质总的浓度最高可达（12959.5 ± 685.9）ng/m^3[182]，证明焦化废水处理工程已经成为大气中 PAHs 新的排放源，具有较高的致癌风险，未来的废水处理系统将向封闭式形式发展[183]。由此表明，难降解工业废水的处理需要针对特征污染物实现基于气液固三相的物料衡算的计量分析[184]。

3.2.2.5　难降解工业废水处理前沿问题

难降解工业废水成分复杂、检测困难，尤其对于微量的难降解有机物，由于其量少，且大多水溶性差，很难被有效地检测出来[185]，如何建立一套高效、全面、准确、高通量的成分分析方法是处理难降解工业废水的先决条件。明确各种难降解工业废水的组分，利用大数据，对废水中的成分进行归一化分析，为工艺的选择提供参考，可以提高技术水平。生物法是高效处理高浓度难降解工业废水过程的核心技术之一，如何迅速获得降解专一性强的微生物是提高生物法效率关键问题。活性污泥是目前应用最为广泛的一种废水处理方法，该方法对水中的 COD 和 BOD 的去除非常有效，是否还存在比活性污泥效率更高的微生物，是否存在能专性降解工业废水中难降解有机物的微生物，成为未来的思考。

难降解工业废水中存在基础物质、能量/热量、产品、盐分、生物资源、水回用资源，处理工艺过程中存在场地太阳能、场地风能、动力差能、温差能、水位差能等能量资源。通过对这些资源进行回收和利用，可以实现低能耗、低物耗的处理过程。如焦化废水中含有大量苯酚，而苯酚是生产某些树脂、杀菌剂、防腐剂以及药物（如阿司匹林）的重要原料，将苯酚从其中分离出来能很好地实现资源回收[186]。啤酒厂可产生含有生物活性酚类化合物这种天然提取物的废水。多酚作为抗氧化剂，已被证明可以保护人体免受诸如心血管疾病或癌症之类的退行性疾病的影响。除了可以对废水中的酚类物质等现存化工原料进行分离回收，还可以获得生物处理过程产生的有机物。例如，聚羟基脂肪酸脂（PHA）是由多种细菌和某些嗜盐古细菌合成的一种胞内积累的储存碳源和能量的物质，在难降解工业废水这种营养失衡的环境下极易产生，而这种物质在生物医药、海洋环境、建材、农业以及食品加工中，具有很大的市场潜力[187,188]。所以，支持资源化、能源化以及无害化结合的分子转化机制是未来需要考虑的发展趋势。

4　废水处理新技术系统

4.1　生物电化学技术

微生物电化学系统（MES）以电活性微生物作为电极催化剂的电化学系统，其本质是电活性微生物胞内代谢、胞外电子传递的物质-电能的相互转化过程。常见的生物电化学系统包括微生物燃料电池（MFC）和微生物电解池（MEC）。微生物电化学系统阳极发生氧化反应，阴极发生还原反应，可用于废水处理，同时利用其电极间电势差，可以定向驱动离子迁移从而实现脱盐、污染物去除、资源回收以及有用物质生产。

4.1.1 利用 MES 处理废水

利用 MES 可以处理有机物，尤其是难降解有机物。利用 MES 阳极产电微生物可以实现石油烃、农药氧化。为了进一步提高对有机物降解，研究者将阳极暴露于空气中，促进了阳极中的兼性细菌的好氧代谢，提高了阳极对苯胺的矿化降解效率[189]。利用 MES 阴极微生物可以实现硝基苯、氯霉素等还原。王爱杰课题组利用生物阴极在外加 0.5 V 的电压下，利用生物阴极将 98.7%的硝基苯，并利用生物阴极在 0.5 V 的外加电压下，实现了对氯霉素（CAP）的脱硝及脱氯，将 CAP 并转化为苯胺类产物[190]，大幅降低了氯霉素的毒性。俞汉青课题组借助 MES 生物阳极产生的电子用于非生物阴极氧还原，通过原位利用阴极原位产生的过氧化氢，形成生物-电-Fenton，实现对污染物的矿化去除[191]。耦合 MES 的阳极氧化反应与阴极的还原反应，可以对废水中的污染物进一步矿化降解。

由于废水中污染物成分较复杂，环境工程师往往将 5MES 与其他工艺进行耦合以强化对污水处理效果。俞汉青课题组将膜生物反应器（MBR）与 MES 结合，对 COD 去除率达 92.4%[192]。沈锦优等[193]将 MES 与 UASB 耦合处理对硝基酚（PNP）废水，显著提高了对 PNP 的去除率。

利用 MES 还可以处理无机物，例如还原废水中重金属。黄丽萍课题组利用 MES 阴极对水中二价钴金属离子进行还原。此外，利用生物阴极可以获得比非生物阴极更高的六价铬还原效率[194]。黄霞课题组构建生物阴极 MES 反应器，实现了同步脱氮和电能回收，通过对外电路的调控，总氮的脱除效率可以达到 84%[195]。冯玉杰课题组构建硝化型生物阴极 MES，实现了对氨氮的硝化，生物阴极的出水中氨氮浓度为 0.5 mg/L[196]。

4.1.2 利用 MES 实现离子去除与回收

2009 年，清华大学黄霞课题组通过对 MFC 构型的改变提出 MDC 的概念，在 MFC 的阴阳极之间放置一对离子交换膜，从而定向引导离子迁移，实现低能耗、可持续的含盐水脱盐[197]。在 MDC 产电及脱盐效果优化方面，堆叠型 MDC（stacked MDC，SMDC）及上流式 MDC（up-flow MDC，UMDC）的研究相对较多。黄霞课题组先后实现了 SMDC 脱盐膜堆数目的优化、内阻的降低以及规模的放大，利用离子交换树脂实现了内阻的大幅降低，确保系统放大[198]。贺震课题组提出了 UMDC 构型，实现了脱盐速率的逐步提升，并成功运行 MDC 中试（105 L），与阴极产氢或金属还原相耦合[199]。此外，任智勇课题组将 MDC 与电容吸附技术相结合的研究也有阶段性进展[200]。

利用 MFC 内电场驱动其他各类离子型物质的迁移，可以实现污染物去除、有用物质生产以及资源回收。张翼峰等搭建的浸没式 MDC 通过驱动 NO_3^-或 NH_4^+迁移完成地下水修复[201]。在去除水溶液中离子的基础上，MFC 还可实现同步的资源回收。黄霞课题组在 MDC 的基础上提出微生物氮磷回收电池（microbial nutrient recovery cell，MNRC）的概念，实现污水中氮磷的去除与同步浓缩回收。以尿液中氮磷为对象，利用 MFC 实现高效资源回收的研究，近年间也多次进入人们的视野[192]。中山大学刘广立课题组利用 MFC 内电场诱使双极膜产生 H^+和 OH^-，从而实现酸碱物质的低能耗生产[202]。

4.1.3 展望

充分利用 MES 的阳极氧化、阴极还原实现有机污染物高效降解是 MES 在污水处理中重要发展方向，通过构建多电极腔室或堆叠结构 MES 系统，充分利用不同电极反应处理废水中多种污染物质（如易降解的 COD、硝基苯、重金属离子、氨氮等），结合外电路的调控，如外加恒定电压、恒定电流、暂态电路调节，可进一步强化对污染物的降解性能，在此基础上构建有效放大的MES构型是实用化关键。此外，基于 MES 内电场驱动离子实现脱盐或离子回收的系列技术，将在离子选择性去除（回收）、离子高效富集（去除）以及反应器放大等方向继续开展研究，实现 MES 高效低耗去除（回收）水中离子。

4.2　膜生物反应器技术

4.2.1　概述

MBR 是由活性污泥和膜分离耦合而成的一种较为新型的污水处理和回用技术。得益于膜的固液分离作用，MBR 具有占地少、污泥浓度高、出水水质好等显著技术优势，因此，在全球范围被广泛应用。据统计，全球 MBR 污水处理总量的市场增长率一直保持在 10%以上。随着 MBR 技术的成熟和运营经验的积累，越来越多的大规模 MBR 工艺（>10 万 t/d）被用于市政污水处理（图 17-7）。例如，在瑞典斯德哥尔摩将投产世界最大 MBR 污水厂，设计处理规模达 86 万 t/d。基于膜的物理截留作用和功能微生物的高效富集，MBR 在化工废水、养殖废水、垃圾渗滤液等行业也很受青睐。显然，MBR 技术将在污（废）水处理领域扮演重要角色。

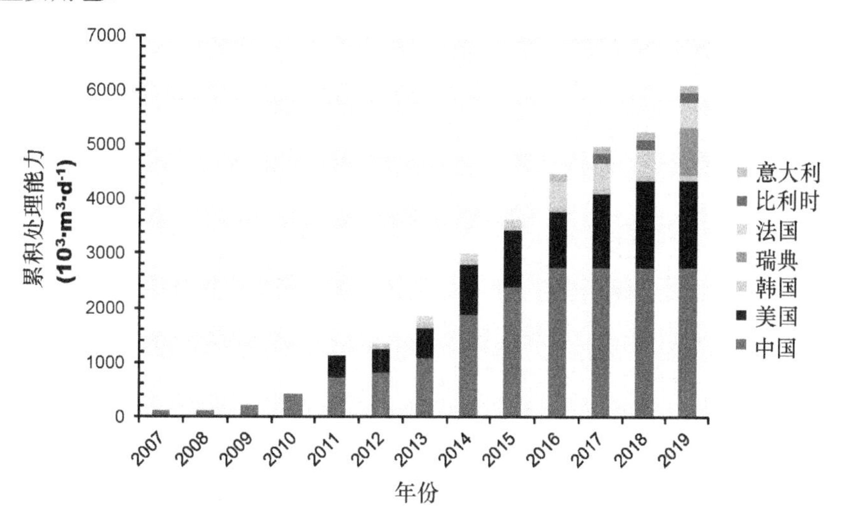

图 17-7　全球范围内大规模 MBR 的应用情况（>10 万 t/d）

（引自 http://www.thembrsite.com/）

MBR 工艺可将污水处理成为高品质的再生水源，这对于解决水资源短缺问题和保障城市水环境质量具有重要意义[203]。近年来我国涌现了碧水源、膜天、美能、斯纳普等一批具有国际竞争力的 MBR 膜供应商。例如，碧水源在国内外完成了 20 余座大型 MBR 工程项目（>10 万 m^3/d），居全球第一。在基础研究领域，我国高校和科研机构也处于领先地位。根据 SCOPUS 检索数据，全球范围内发表有关 MBR 的论文 9800 余篇，其中，我国学者发表 2200 余篇（含合作发表）。部分代表性工作如下所述。

1）低能耗 MBR 污水处理方法

为降低 MBR 运行能耗（尤其是曝气能耗），国内外学者主要从曝气优化和膜池设计入手开展了研究工作。如，采用间歇曝气的运行方式不仅能够有效降低 MBR 能耗，而且有助于提高脱氮除磷性能。黄霞等基于出水氨氮浓度反馈的方法提出了自动曝气的方法，用于大型 MBR 污水处理厂的工程实践，降低曝气能耗达 20%[204]。近期，李咏梅等通过结合曝气模型和活性污泥数学模型降低曝气能耗 13.2%[205]。魏源送等则采用脉冲曝气的方式对清河再生水厂进行了工艺优化，在保证出水水质的前提下，膜通量增加

20%，吨水处理能耗降低达到 42%[206]。我国学者还发现优化反应器流场不但能够实现氮磷的高效脱除，而且显著降低运行能耗。总体上，我国在 MBR 运行能耗优化方面的研究处于国际一流水平。

2）基于能源回收的 MBR 污水处理工艺

近年来，从污水回收资源（水、氮、磷）、获取能源（产甲烷、产电等）已成为国内外环境工程领域的研究热点。基于污水能源化，国内外学者提出了厌氧 MBR、高速率 MBR、正渗透 MBR 和微生物电解池(MEC)-MBR 组合工艺等，并对其基础理论进行了深入探究。如，我国学者研究发现厌氧 MBR 可以在中温条件、较低水利停留时间下实现生活污水的能源经产出[207]，产甲烷率可达到 0.24 L CH_4/g $COD_{removed}$[208]；而基于正渗透膜的厌氧 MBR 不但可以产甲烷或产电，而且能够回收氮磷元素[209]。近期，也有学者证实采用电解膜-生物反应器可以实现原位产电和污水尿液的分离。相比欧美，我国在污水能源化和资源化方面的研究起步稍晚，主要开展了实验室规模研究，缺少应用性研究成果。开发适合我国污水水质状况（如，低 COD）的 MBR 工艺并实现能源或资源的高效回收是未来的发展方向之一。

3）MBR 膜污染及其控制

在 MBR 中，膜污染物来源多（污水、生物大分子、微生物）且动态变化，这为膜污染识别、表征及控制带来了极大困难[210]。自 2000 年以来，我国学者在膜污染研究方面一直表现非常活跃，有多篇论文入选 ESI 高引论文。研究内容涉及污染物表征分析[211]、形成机制解析和控制等。21 世纪初，黄霞教授在 MBR 膜污染方面做出了大量研究工作，对我国膜污染研究起到了推动作用。在膜污染表征方面，张捍民和杨凤林等对膜表面污染物进行了系统表征分析，提出了生物污染-有机污染-无机污染相互影响的假设。王志伟等采用红外光谱和三维荧光光谱等手段揭示了胞外聚合物（EPS）主要组成及其对 MBR 膜污染的影响机制。近期，俞汉青等采用多元曲线分辨-交替最小二乘法（MCR-ALS）的方法分析了膜污染过程的红外谱图，在分子水平上揭示了膜污染物的相互作用机制[212]。王磊等则采用石英晶体微天平（QCM）定量表征了牛血清蛋白在 PVDF 膜上的沉积过程。在膜污染物形成机制方法，孟凡刚等从分子尺寸、蛋白质组和生物降解行为的角度探讨了膜表面生物大分子的关键来源及其演替规律。

在膜污染控制方面，近期我国在化学清洗或反洗方面做了系统性的研究。我国学者在实验室规模和中试规模 MBR 中研究开发了原位化学反洗新方法，取得了良好的膜污染控制效果[213]。在清洗机理方面，研究发现：氧化性药剂和碱液会破坏膜污染物中蛋白质或多糖的官能团结构（如：羰基和羧基等基团增多等），改变其膜污染规律；而微生物细胞表面的 EPS 能够优先与清洗药剂反应，起到保护微生物细胞的作用[214]。此外，我国研究人员也从信号分子猝灭和原（后）生动物优化[215]的角度提出了具有应用前景的生物控制方法。

4）抗污染膜材料

抗污染膜材料的研发一直是 MBR 领域的重要研究方向之一。采用季铵盐改性后的 PVDF 膜材料不但能够降低多糖和蛋白质等在膜表面的沉积，而且能够显著抑制微生物的吸附[216]。也有研究发现，用 EPS 改性后的 PVDF 膜材料疏水性显著降低、膜孔结构得到改善，污染物截留能力和抗污染性能显著提升。丙烯酸锌的植入也会明显提升 PES 超滤膜的膜污染性能。我国许多研究人员也证实氧化石墨烯和碳纳米管等抑菌材料的引入会大大增强膜的抗污染性能。此外，我国在导电膜材料研发方面取得了较好的进展[217]。例如，李建新等基于二氧化钛和碳膜研发的电催化膜反应器不仅有良好的处理工业废水能力，而且具有自清洗功能。

4.2.2 展望

近年来，我国在 MBR 应用和研究方面均取得了突破性进展，为我国水环境问题的解决提供了重要技术支撑和理论依据。在市场推广方面，碧水源已跻身国际膜供应商第一梯队，完成了大量 MBR 工程项目。在基础研究方面，我国在诸多研究领域（如，能耗和膜污染问题）也处于国际一流水平。未来，污水能源化和资源化是重要研究方向。因此，我们需要深入探究 MBR 相关技术在该方向的可行性，或提出适于我国污水能源化的 MBR 新技术。在膜材料研发方面，也需要研究膜材料在长期使用过程中的稳定性和可靠性等。

4.3　焦化废水处理

焦化废水来源于炼焦过程中的备煤、湿法熄焦、焦油加工、煤气冷却等工序，废水类别包括除尘废水、剩余氨水、酚氰废水、脱硫废液等。通常，每生产 1 t 焦炭需要 1.25 t 的煤，产生约 0.6 m^3 的废水，造成我国整体焦化废水水量巨大。焦化废水具有组成复杂、难降解、有毒有害的特征，是目前国内外工业水处理的难点。除了氨氮、氰化物、硫氰化物、硫化物等无机污染物外，华南理工大学韦朝海课题组对韶钢焦化废水厂有机污染物进行深入分析，定量地分析了焦化废水中 15 类 558 种有机污染物，明确了苯酚及甲基取代酚、喹啉及其衍生物、有机腈化物为典型有机污染物[218-222]。调研国内 60 余家焦化废水处理厂，发现国内大部分焦化废水处理工艺使用 A/O 法，也有部分企业采用 A^2/O、A/O^2、SBR、生物膜法等工艺，处理效果略有差异，但普遍存在能耗高、出水难稳定达标、脱氮不理想等问题[223]。

4.3.1　生物处理反应器及工艺

华南理工大学韦朝海课题组开发了基于异重流污泥原位分离的流化床反应器，基础研究及工程实践证明，该反应器突破传统有机负荷，实现了高活性微生物的原位分离，以及曝气过程中氧利用率的提高。采用 CFD 流体力学软件模拟仿真流化床内部液体和气体流场，从结构、上下挡板和功能内构件等方面指导流化床设计的改进，解决装置“死体积”等问题，使得氧利用率达到 50%以上[224]。韦朝海课题组开发了一套以 O_1/H/O_2 为核心基于流化床反应器的焦化废水生物处理工艺集成系统（图 17-8），并在两个焦化废水处理工程中得到了成功应用。其优势体现在：①负荷能力上的突破。高达 6000 mg/L COD 的原水可以在不稀释情况下得到去除，总负荷为 2.4 kg COD/(m^3·d)，去除率 90%，高于城市污水 3 倍以上。②脱氮能力的突破。O_1/H/O_2 与传统 A^2/O 处理过程的四氮浓度变化比较，硝化过程的需氧量降低 35%，动力消耗降低约 30%，碱耗降低 23%以上，节约反硝化过程的碳源，HRT 缩短一半以上。③特征污染物去除能力的突破。通过活性污泥原位分离与优势停留，难降解疏水性有机物被吸附回流，倍增停留时间，对酚及多元酚的去除有明显优势，对多环芳烃、含氮杂环化合物、二噁英类化合物有强化降解的表现[218-222]。

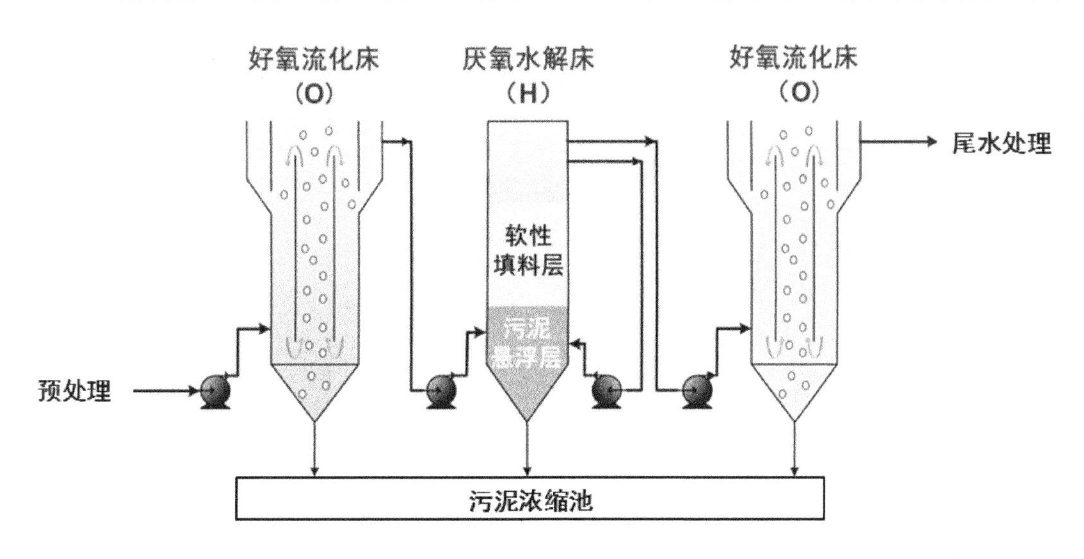

图 17-8　基于流化床反应器的 O/H/O 生物工艺用于焦化废水处理

4.3.2 焦化废水处理厂生物反应器群落结构与功能分析

Zhu 等[225]对韶钢焦化废水处理厂 $A/O_1/O_2$ 生物反应器中的微生物进行了高通量测序，表征了其主要群落结构，并与主要污染物的降解进行了关联，证实反应器中主要由 *Comamonas* 执行苯酚的降解；由 *Thiobacillus* 执行氰化物和硫氰化物的降解，由 *Nitrospira* 执行脱氮功能。Huang 等[226]利用 16SrRNA 高通量测序分析揭示焦化废水处理反应器微生物群落分布，发现酚的降解与 *Enterobacter*，*Pseudomonas* 和 *Sedimentibacter* 具有相关性，特别是 *Sedimentibacter* 与酚、甲基酚、二甲基酚和萘酚的降解高度相关。Zhang 等[227]通过高通量功能基因阵列与高通量测序相结合，用于识别 488 天好氧的焦化废水处理过程中微生物的功能特征及其在含氮和含硫化合物的生物转化过程中所起的作用，证实芳香族双氧酶（如 xylXY、nagG）、氮化酶（如 nhh、nitrilase）、二苯甲氧基（DbtAc）和硫氰酸酯酶（scnABC）是含氮和含硫污染物的生物转化的重要功能基因。

4.3.3 生物电化学耦合强化焦化废水处理除碳脱氮

传统焦化废水处理厂采用多个生物操作单元耦合处理，因而存在占地面积大、投资成本高、难以管理等缺陷；而且在好氧反应器中需要补充大量的碱以平衡有机化合物及氨氮氧化时产生的酸度，因而耗碱量大、运行成本高[228]。为克服上述问题，Feng 等[229]构建了独立双室阴离子交换膜微生物燃料电池体系，在单个反应器中不外加碱的情况下，成功实现了焦化废水同步除碳脱氮。焦化废水通入好氧阴极室中，有机物及氨氮在好氧菌作用下发生氧化反应，硝化作用生成的 NO_3^- 与 NO_2^- 在浓度梯度和电场的作用下透过阴离子交换膜扩散到阳极，其在厌氧阳极室发生反硝化作用实现转化（图 17-9），经过微生物燃料电池与好氧技术相结合处理后，连续运行时发现 COD 和 TN 得到同步去除，去除率可分别达到 84%和 98%。新的《炼焦化学工业污染物排放标准》（GB 16171—2012）中规定，现有企业在 2015 年 1 月 1 日后，氨氮和总氮间接排放标准为 25 mg/L 和 50 mg/L，因而有必要进一步削减硝氮，以满足总氮排放要求。最常见的去除硝氮的方法是将含硝氮的出水回流至厌氧单元进行反硝化，但是这样的工艺选择会显著提高污水处理运行费用，增长处理过程水力停留时间，如果出水中碳源浓度过低，为了反硝化反应的进行还需要额外投加碳源。Xie 和 Tang 等[230-232]提出一项有应用前景的硝氮削减技术，即利用电极作为电子供体驱动微生物反硝化反应。与厌氧反硝化相比，该技术不需要额外加入有机碳源，避免了二次污染；与电化学还原相比，所需还原电势更低、电子的利用效率高，因而降低了能耗。

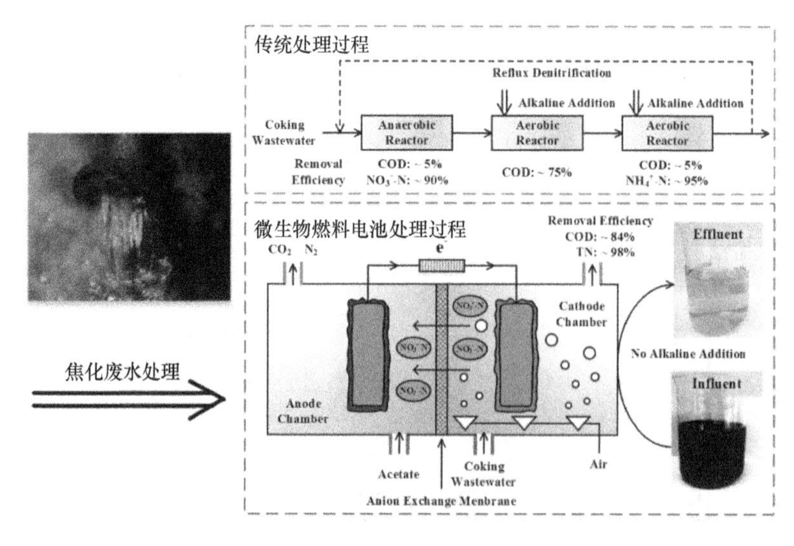

图 17-9 微生物燃料电池实现焦化废水同步除碳脱氮示意图

4.3.4　展望

未来，我国焦化废水处理需要集成合理的生态系统与循环系统，将水处理建立在全企业或工业园层面上，构建工业水工业的知识与技术系统；同时，综合考虑将焦化废水中富含的化学能、内能进行资源化，能源化处理，并以焦化废水处理零能耗、零排放为最终目标；最后，还需要认识水处理技术发展的极限以及采用更加严格的生态评价，倒逼焦化行业推行技术革新。

4.4　氢气和甲烷处理污水新方法技术

随着环境污染、能源危机等人类面临的众多问题日益突显，污水处理行业出现了以下三个明显趋势：强化污染物削减的同时，关注新兴污染物，如内分泌干扰物等；碳源、能源回收与开发，如将传统水处理工艺的过程产物甲烷和氢气作为电子供体进行后期深度处理；其他资源回收，如有机质及磷等资源的循环利用。中国城市污水处理概念厂专委会也提出，在污水处理厂新功能需求下，相关污水处理技术也将面临新变革。如污水深度和超深度处理技术、低碳污水处理技术、污水处理能源开发技术、污水处理资源回收技术等。基于此，本节将重点论述近年来国内外研究人员在以甲烷和氢气等气态电子供体进行污水处理方面取得的研究进展以及未来的研究方向等。

4.4.1　氢气在污水处理中的研究进展

有机电子受体成本高且用量需严格控制，否则容易造成二次污染。氢气是一种清洁高效、无毒无害的无机电子供体，利用氢气作为电子供体来驱动氧化态污染物的还原已引起国内外研究人员的广泛关注。热力学分析表明，氢气可用于多种氧化态污染物的还原降解（表 17-5）。氧化态污染物大多是具有较高氧化态的无机物和难生物降解的卤代有机化合物，很难通过常规的生物氧化法去除，只能利用一些无机或有机的电子受体（氢气、乙醇、乙酸、硫等）来驱动氧化态污染物的还原。

表 17-5　以氢气为电子供体的各种氧化态污染物降解方程式

化合物	化学反应式	$\Delta G^{o\prime}$/（kJ/mol）
高氯酸盐	$ClO_4^- + 4H_2 \longrightarrow Cl^- + 4H_2O$	−118
硝酸盐	$NO_3^- + 2.5H_2 + H^+ \longrightarrow 0.5N_2 + 3H_2O$	−112
硒酸盐	$SeO_4^{2-} + 3H_2 + 2H^+ \longrightarrow Se^0 + 4H_2O$	−71
砷酸盐	$H_2AsO_4^- + H_2 + H^+ \longrightarrow H_3AsO_3 + H_2O$	−45
铬酸盐	$CrO_4^{2-} + 1.5H_2 + 2H^+ \longrightarrow Cr(OH)_3 + H_2O$	−9

为了提高氢气的传质和利用效率，Rittmann 研发了一种通过中空纤维膜加压通氢气的方式向水中微生物提供电子供体的膜生物反应器（MBfR），MBfR 的小试装置如图 17-10 所示。膜内通过钢瓶提供氢气，氢气在压力作用下由中空纤维膜渗出，被附着在膜上的氢自养型微生物作为电子供体利用，同时将氧化态污染物还原降解。

目前，在实验室水平已有利用氢气来处理水体中污染物的报道。Zhou 等通过对氢自养微生物的研究发现：当硝酸盐初始质浓度为 10 mg/L 时，硝酸盐可以被完全去除；当硝盐质量浓度为 30 mg/L 时，反硝化过程受到抑制。夏四清等[233]利用氢基质膜生物反应器将地下水中的 Cr(VI)快速有效地还原为 Cr(III)，且在硝酸盐（NO_3^-）、硫酸盐（SO_4^{2-}）及溴酸盐（BrO_3^-）等共存的情况下，各污染物仍能取得不错的去除效果。NO_3^-负荷为 10 mg/L 时其去除率可接近 100%，当硝酸盐进水负荷为 20 mg/L 时，硝氮的还原速率仍可达到 $0.12gN/(m^2·d)$，表明氢基质生物膜反应器处理多种氧化态污染物共存的地下水均具有一定的潜力。笔者课题组也通过串联式氢基质膜生物反应器依次去除模拟废水中的硝酸盐和高氯酸盐，去除率效果良好。本课题组近来报道了以氢基质膜生物反应器为载体，微生物利用氢气为电子供体可将 SeO_4^{2-} 还原为单质 Se^0[234-236]，进一步拓宽了氢基质膜生物反应器的应用范围。

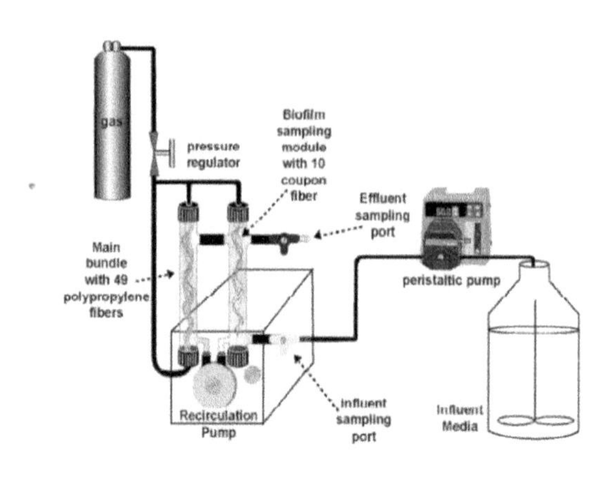

图 17-10　小型膜生物反应器装置示意图

4.4.2　甲烷在污水处理中的研究进展

利用甲烷作为电子供体和碳源来处理水中污染物的生物学过程主要有两种：好氧甲烷氧化和厌氧甲烷氧化。甲烷好氧氧化耦合氧化态污染物的研究最先始于好氧甲烷氧化菌与反硝化菌混合菌群的研究。甲烷好氧氧化耦合反硝化（aerobic methane oxidation coupled to denitrification，AMO-D）需要好氧甲烷氧化菌和反硝化菌协同完成，在缺氧条件下，甲烷氧化菌在氧化利用甲烷的过程中，会产生一系列中间代谢产物如甲醇、乙酸、乳酸及其他可溶态有机物，这些都可作为电子供体被反硝化菌利用，从而实现对硝氮的去除[237]。

目前，好氧甲烷氧化耦合反硝化主要停留在理论阶段及部分实验室小试，在实际废水处理工艺中的研究比较缺乏[238]。Thalasso 等[239]报道了在一个序批式反应器中，以甲烷为唯一电子供体及碳源进行反硝化，脱氮速率可达 0.6 g NO_3^--N/(g VSS·d)；Sun 等[240]利用甲烷基质膜生物反应器（MBfR）成功地去除了 97%的负荷为 30 mg/LNO_3^--N 的硝氮，其平均脱氮速率可达到 1.78 g NO_3^-- N/(g VSS·d)。影响 AMO-D 的一个很重要的因素是氧浓度，虽然氧气对于甲烷的第一步活化是必需的，但较高浓度的氧会抑制反硝化过程，氧浓度要低于 1 mg/L。影响 AMO-D 的另一关键因素是碳氮比，Modin 等给出一个最适碳氮比：1.27 mol CH_4-C/mol NO_3^--N。在这个碳氮比下，氧气会全部用于甲烷的活化而不是被呼吸作用代谢。

甲烷厌氧氧化（anaerobic oxidation of methane，AOM）过程最早发现于深海沉积物中，该过程主要由甲烷厌氧氧化古菌（anaerobic methanotrophic archaea，ANME）和硫酸盐还原细菌（sulfate-reducing bacteria，SRB）共同完成，其化学计量关系可表示为：

$$CH_4 + SO_4^{2-} \longrightarrow HCO_3^- + HS^- + H_2O \qquad \Delta G^{\theta'} = -16\ \text{kJ/mol}\ CH_4 \qquad (17\text{-}3)$$

根据化学计量反应可以预测除了 SO_4^{2-} 外，NO_2^-或 NO_3^-同样可作为 AOM 的电子受体。2006 年，Raghoebarsing 等在实验室条件下证明了该反应的存在，并将此过程称为反硝化型甲烷厌氧氧化（denitrification-dependent anaerobic methane oxidation，DAMO）。

$$5CH_4 + 8NO_3^- + 8H^+ \longrightarrow 5CO_2 + 4N_2 + 14H_2O \qquad \Delta G^{\theta'} = -765\ \text{kJ/mol}\ CH_4 \qquad (17\text{-}4)$$

$$3CH_4 + 8NO_2^- + 8H^+ \longrightarrow 3CO_2 + 4N_2 + 10H_2O \qquad \Delta G^{\theta'} = -928\ \text{kJ/mol}\ CH_4 \qquad (17\text{-}5)$$

Raghoebarsing[241]首先提出了 DAMO 的可能机理——逆向产甲烷理论，即甲烷氧化古菌通过逆向产甲烷过程产生电子并将其提供给反硝化菌进行反硝化。2010 年，Ettwig 团队发现即使抑制了甲烷氧化古菌特征酶 *mcrA* 的活性，ANMO-D 过程仍能发生，继而提出了一种全新的理论，即 NC10 门细菌 *M. oxyfera* 的内微氧型厌氧甲烷氧化反硝化，该菌不是通过经典的反硝化途径将 NO_2^-还原为 N_2，而是能利用亚硝酸盐还原酶将 NO_2^-还原为 NO，NO 由某种未知的歧化酶将其转化为 N_2 与 O_2，产生的 O_2 由甲烷氧化单加氧酶（*pMMO*）将甲烷氧化。两种代谢途径如图 17-11 所示。Hu 等[242]研究了厌氧甲烷氧化反硝化富集培养物中微生物的脱氮效率，其脱氮速率可达 2mmol NO_3^-- N /(L·d)，亚硝氮为 7.3 mmol NO_2^-- N /(L·d)，高于硝酸盐。但也有研究表明，当负荷高于 0.10 g N/(m^2·d)，NO_2^-会对细菌的生长产生抑制。

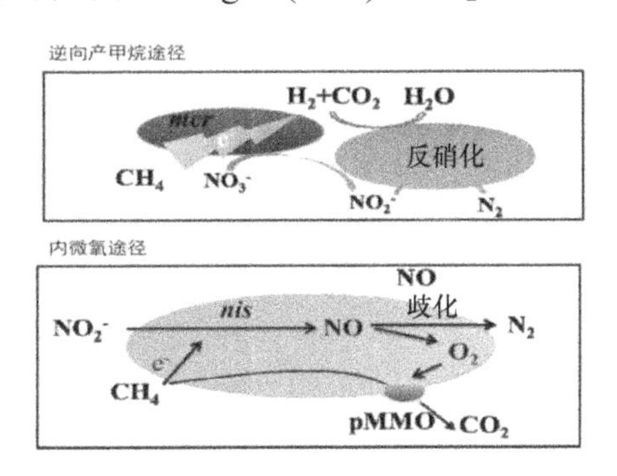

图 17-11　逆向产甲烷途径和内微氧途径

除了硫酸盐和硝酸盐/亚硝酸盐，甲烷厌氧氧化还可实现对水体重金属污染的去除。Ettwig 等通过同位素示踪法及荧光原位杂交技术证实了厌氧甲烷氧化可以耦合 Fe^{3+}和 Mn^{4+}的还原，Lu 等[243]也报道了以甲烷作为唯一电子供体和碳源的铬酸盐的还原，这是甲烷氧化古菌和铬酸盐还原菌相互协作完成这一过程。

笔者团队也于 2015 年在甲烷基质膜生物反应器中，实现了以甲烷为唯一电子供体和碳源的高氯酸盐、铬酸盐、硒酸盐及锑酸盐的生物还原。并针对不同的氧化态污染物提出了两种不同的代谢途径。甲烷氧化耦合高氯酸盐还原可能类似 N-DAMO 过程，亚氯酸盐歧化酶将亚氯酸盐岐化为 O_2 和 Cl^-，即通过“内微氧”代谢途径实现甲烷氧化[244]。甲烷氧化耦合硒酸盐、锑酸盐、铬酸盐及砷酸盐的还原则通过“逆向产甲烷”途径实现甲烷氧化[245-247]。这些发现为实现甲烷在废水深度处理中的应用提供了理论基础。

4.4.3　展望

目前关于氢气处理水体中氧化态污染物的研究还处于小规模试验阶段，而以甲烷为电子供体的氧化态污染物还原机理尚不明确，这极大地拖延了其投入实际污水处理的步伐。在接下来的研究中，一方面要尽快弄清混合菌群的耦合机理，为其实际应用扫清理论障碍；另一方面，以膜生物反应器（MBfR）为基础，尝试将甲烷氧化耦合反硝化与短程硝化、污泥消化产甲烷等城市污水处理环节相结合，研发出节能、降耗、低碳的城市污水处理新技术。

5　废水资源/能源新技术

5.1　甲烷/氢气回收技术

废水资源化已成为当前水资源管理新的发展方向，其中尤以厌氧消化产甲烷技术为当前的研究重点。

该技术已被广泛用于含高有机物浓度的工业废水、污泥、农业废物等的处理和能源回收。近年来，随着我们对微生物群落及代谢过程认识的不断深入以及厌氧处理新技术的快速开发，厌氧消化的污染物转化效率和稳定性得到进一步提高，其应用范围不断扩大[248-250]。

5.1.1 厌氧膜生物反应器技术

近年来厌氧消化技术的一个重要突破是通过与膜技术结合直接应用于低浓度生活污水处理并实现产能[251]。与常规工艺相比，厌氧膜生物反应器（AnMBR）不仅显著提高了污染物的去除率和甲烷产率而且能获得更好的出水水质，但目前仍存在膜污染、溶解性甲烷随出水流失、低温下产甲烷活性受抑制等挑战。

5.1.2 膜污染控制

AnMBR 的膜污染控制通常采用压缩生物气或者高速水流对膜表面进行冲刷。这种方法虽然能有效减缓膜污染，但本身能耗很高。因此往往需与其他膜污染控制策略（如反应器优化、膜组件和材料改进以及原位抑菌技术等）配合。

最初的 AnMBR 反应器采用的是完全混合式的生化反应池，但由于膜组件被直接暴露在污泥环境中，往往造成严重膜污染。UASB 反应器的应用实现了颗粒污泥床区与膜组件的空间分离，能在一定程度上减缓生物膜的形成，但难以避免悬浮微生物和胶体物质在膜表面的沉积。与 UASB 反应器相比，基于颗粒活性炭（GAC）或包埋法固定微生物的厌氧流化床反应器能显著减少悬浮固体并加强对膜表面的剪切作用，从而进一步改善膜污染。另外，通过适当设置导板来优化反应器内的流态，也能减少膜分离区的悬浮污泥以及加强膜表水流剪切强度。但这些新型反应器的长期运行性能仍有待进一步研究[252]。

除了强化反应器内流体流动，也可以通过膜组件本身的运动（旋转或振动）来强化膜表面的剪切作用，从而进一步抑制膜污染层形成。由于是在膜表面附近区域形成局部高速剪切及紊流，因此所需能耗比提高整个反应器内流体的方式要低[253]。然而，这类运动型膜组件仍不足以有效控制胶体和溶解性有机物在膜表面的沉积。

另一种更直接的方式是发展抗污染型的膜材料，包括对膜表面进行修饰改性，以及采用非压力驱动的膜过滤过程（如正渗透膜和膜蒸馏）代替传统的微滤/超滤过程[254]。例如，与传统 AnMBR 相比，正渗透厌氧膜生物反应器不仅膜污染层更疏松，而且能更有效地截留污染物，从而获得更高的能源转化效率。然而，由于对无机盐的高截留率以及存在反扩散，长期运行过程中盐浓度的过度积累也会对厌氧生物活性造成抑制并有可能加剧膜污染。

除了对反应器和膜本身的优化，基于电化学和化学方法的原位抑菌技术也被用于控制膜污染[255]。电化学法主要是利用阴极产生的氢气气泡生对膜表面进行冲刷以及利用静电排斥力减轻微生物和胶体有机物在带负电荷的膜表面的沉积。此外，也可通过加入铁盐或絮凝剂等方式来改善污泥性质，减少胶体物质和溶解性有机物，从而减轻膜污染。

5.1.3 溶解性甲烷的回收/原位利用

AnMBR 中产生的甲烷约有 30%~80%以溶解性甲烷的形式残留在液相，最终随着出水流失，造成严重的能源浪费和温室气体排放。采用疏水性的中空纤维膜接触器能有效回收这部分甲烷，从而获得富含甲烷（高达 72%）的生物气，并且整个过程电耗很低（< 0.002 kWh/m^3 水）[256]。然而，该技术的甲烷提取速率很慢。采用微孔膜材料替代分子扩散型膜有望能显著提高甲烷提取速率并得到更高纯度的甲烷，但长期运行中可能会存在膜孔堵塞问题。

考虑到回收出水中的甲烷需要较高的成本和能量投入，另一种更经济可行的方式是将出水中的溶解性甲烷进行原位利用[257]。例如，可以将厌氧甲烷氧化与反硝化脱氮过程结合起来，这可能是废水脱氮技术今后的一个重要发展方向。

5.1.4　厌氧消化过程强化调控技术

厌氧发酵过程往往需要维持较低的氢分压以促进乙酸的生成，这导致产甲烷菌中乙酸利用型菌往往占主导地位，但其对环境的耐受能力和生长速率都比氢利用型甲烷菌弱。因此，产甲烷过程很容易受到抑制。因此，如何提高产甲烷活性是厌氧消化技术走向实际废水处理应用需要解决的关键共性问题。

1）生物电化学强化厌氧产甲烷技术

将生物电化学过程引入污泥厌氧消化系统中能通过反应过程的分离有效解决有机物降解与产氢之间的矛盾。一方面促进有机物在阳极的分解，另一方面还能在阴极通过产生的氢气（后者通过直接电子传递）还原 CO_2 生成甲烷和乙酸等，从而提高甲烷的浓度和产率。但过快的阴极反应速率也会导致 pH 升高，反而抑制产甲烷活性。此外，利用电极之间形成的电场结合离子交换膜，通过电迁移还可以从厌氧反应器内不断去除氨氮，从而减轻其对产甲烷过程的抑制，进一步提高甲烷产率[258]。

2）原污泥性质的调控

利用从废水中富集大量有机物的污泥来进行厌氧消化产甲烷是从废水中回收能源的另一种重要方式。这种工艺能显著提高有机物负荷和甲烷产率，但往往需要对原污泥进行预处理以促进细胞效裂和有机物释放，从而增加了整个处理过程的能耗。最新发现通过适当调节原污泥的性质可以有效降低这部分能耗并提高有机物的生物转化率。例如，提高污泥中的聚羟基脂肪酸酯（PHA）的含量已被证实能显著改善污泥的厌氧消化性能和提高甲烷产率。另外，将污泥与高碳氢化合物含量的废弃物（如餐厨垃圾）混合进行共消化的方式也被广泛用于提高污泥的碳氮比，从而减轻高浓度氨氮的抑制作用并提高甲烷产率[259]。

3）强化直接种间电子传递

传统理论认为厌氧消化过程中产酸菌与产甲烷菌之间电子传递主要是通过氢气或甲酸作为介质，但最新研究表明，产酸菌和产甲烷菌之间也可能进行直接电子传递（DIET），从而为进一步提高电子传递和有机物转化效率提供了可能[260]。例如，已有研究证实往厌氧反应器中加入磁铁矿能显著缩短丙酸/丁酸分解过程中厌氧微生物的停滞期并提高甲烷的产生速率，进而提高了甲烷的产率。此外，加入导电性的生物炭也能促进 DIET 并且减轻氨和酸积累对产甲烷过程的抑制，从而显著促进能直接利用电子的产甲烷菌的富集和提高甲烷产率。这种 DIET 还能显著提高产甲烷菌与其他无机盐（如硫酸根）还原菌的竞争优势，从而进一步减轻对产甲烷过程的竞争性抑制，显示出在厌氧消化过程中巨大的应用潜力[261]。

4）化学调控技术

还可以采用一些环境友好型或低污染型的化学调控方法。例如，可通过添加零价铁来提高氢气的产生以及改善厌氧消化的环境条件（通过生成 FeS 沉淀减少 S^{2-}浓度，抑制反应器的酸化等），从而提高产甲烷菌活性。采用斜发沸石和阳离子交换树脂能有效去除氨，从而减轻其对产甲烷菌的抑制并提高甲烷产率[262]。

5.1.5　展望

近年来废水的厌氧处理重新成为了水处理技术的研究热点，并取得了重要的技术进展。其中尤其是 AnMBR 技术以及厌氧消化过程调控新技术的迅速发展为推动现有污水处理模式的变革带来了希望。然而，目前这些技术大多尚处于中试研究阶段，其实际应用面临诸多技术、经济限制。要突破这些限制，今后需要进一步深入认识厌氧微生物群落结构与功能、微生物厌氧代谢途径及调控机理，以及界面（包括电极-微生物界面以及微生物-膜界面）的电子和物质传递过程。这将需要更多的借助各种先进的分子生物学分析手段以及代谢产物原位监测方法，并结合模型和统计学方法来对这些信息进行分析和提炼。另外，各种纳米功能材料的开发应用也有望为厌氧消化过程的强化调控以及后续的甲烷分离和产品申请提供更优异的膜和电极材料。最后，各种调控策略的应用条件仍需优化，其长期稳定性和在实际体系中的应用效果仍有待进一步研究。

5.2 废水回用技术

水资源是我们人类最宝贵的资源之一，废水回用是解决水资源短缺和水环境污染的有效手段，其途径主要有城市杂用、工业回用、农业回用、环境和娱乐回用、地下水补给、补充水源水等。保障回用水的水质安全是废水再生利用的前提和关键。在废水回用时，对于污染物的控制主要有以下几个方面[263]：

（1）进一步降低回用水中氮磷的浓度，并且出水中固体悬浮物、有机物含量满足不同回用水要求。

（2）采用各种消毒手段控制回用水中的微生物含量，包括各种细菌、病毒等含量，并且在消毒过程中控制消毒副产物的生成。

（3）控制回用水中重金属的浓度，保障回用水水质安全。

（4）新兴微污染物控制。近年来一些新兴污染物也逐渐引起人们的广泛关注，如药物与个人护理品（PPCPs）、内分泌干扰素、持久性有机污染物（POPs）、抗性基因等。这些污染物存在浓度很低，且大部分难以降解，传统的废水处理工艺难以对其进行有效去除，所以在废水回用过程中仍需发展该类污染物的高效去除技术。

根据废水处理工艺的出水水质和废水回用目的不同，废水回用处理的工艺也有所不同，主要包括物理、化学和生物工艺，如强化混凝、高效吸附、膜分离、消毒等。近几年研究者也着眼于废水回用中几个过程的优化和创新，从而提高废水再生利用的效率。

混凝技术广泛应用于各类废水的处理工艺中，可以有效降低废水浊度，去除有机和无机污染物。近年混凝技术在废水回用方面研究内容主要包括：新型混凝剂的研究、混凝动力学和混凝过程的监测等[264]，如最近开发的聚合氯化钛[265]、TiO_2 水凝胶[266]等都被开发应用于水处理中。不同混凝剂在水中的存在形态和活性成分也会影响混凝效果。在最近的研究中发现，混凝除了对于传统有机污染物和浊度可以进行有效的去除外，其对于降低废水的微生物毒性，如去除抗生素抗性基因也有很好的效果[267]。

为了进一步降低废水中有机污染物和重金属含量，常利用活性炭或其他吸附剂进行深度处理。近年来随着材料科学的迅速发展，新型复合吸附剂也被发展用于废水回用处理，如纳米材料[268]、大孔树脂、活性炭纤维等。最近有研究者合成了一种基于环糊精的多孔聚合物吸附剂，对于水中的微量有机污染物具有非常优异的吸附效果[269]。

由于膜分离技术具有高效的截留能力，其被广泛应用于废水回用处理中。膜分离技术主要包括微滤，超滤，纳滤和反渗透等。在新型膜材料的研究方面，石墨烯膜[270]、碳纳米管膜[271]、MOF 膜[272]等新型膜材料引起了研究者广泛关注。同时研究人员对于膜材料的改性和膜分离系统的优化也做出了大量工作。近年来逐渐兴起的正渗透膜分离系统，与上述常用膜分离技术相比，不需要外加压力作为分离驱动力，有研究表明正渗透-膜蒸馏系统可以稳定地从含油废水中回收 90%的水[273]。

为了保证废水回用的安全性，消毒工艺被广泛用来杀灭再生水中的病毒和各类微生物，并去除部分微量有机污染物。消毒是利用紫外、高锰酸钾、臭氧、氯气等物质和这些物质在水中产生的自由基来对微生物进行杀灭。在近几年的研究中，一些新兴的消毒工艺，如光催化消毒、电化学消毒、高铁酸盐[274]、三价锰离子等消毒手段被不断的优化和实践，并取得了一定成果。在消毒过程中，需要考虑的一个问题是如何抑制消毒副产物的形成，降低再生水毒性。Dong 等[275]考察了采用臭氧对废水进行消毒后作为农田回用水对哺乳动物的细胞毒性，结果表明臭氧消毒大约可以降低 10 倍的毒性。在实际应用方面，消毒过程的监测和工艺优化[276]，也是研究者的关注热点。

为了更高效率地提高再生水的水质，各种组合工艺也被应用于废水回用处理。Zheng 等[277]研究了一种新型的电化学膜过滤过程，采用导电微滤膜作为正极，发现其对水中对氨基苯磺酸具有较好的去除效果。在最近的工作中，有研究者将膜技术和生物反应器组合起来，如膜生物反应器-反渗透系统可以有效去除微生物和微量有机污染物，并且显著降低膜污染。而为了解决紫外消毒的微生物光致复活作用，研究者利用微波复合紫外消毒技术用于市政污水回用，有效地抑制了光致复活现象。

再生水在储存、输配和利用过程中仍存在微生物生长的现象，引起潜在生物风险。保障再生水水质生物稳定是解决这一问题的关键。可同化有机碳（AOC）是目前被广泛使用的水质生物稳定性评价指标。研究者从再生水中筛选出了三株新的测试菌种 *Stenotrophomonas* sp. ZJ2、*Pseudomonas saponiphila* G3 和 *Enterobacter* sp. G6，建立了再生水 AOC 测定方法，发现臭氧氧化可导致二级出水 AOC 水平显著升高。同时发现，臭氧氧化后水样 AOC 的升高量与 UV_{254} 的降低量之间具有一定的相关性，可将 UV_{254} 作为判断再生水臭氧氧化过程中 AOC 水平变化的替代指标。

在废水回用技术不断发展的同时，我们也需要完善相关法律和制度，制定一系列详细可行的标准。为适应水回用国际标准化工作的需要，促进水回用领域国际化业务的健康发展，2013 年 7 月，经国际标准化组织（ISO）技术管理局讨论决定，批准成立了国际标准化组织水再利用技术委员会（ISO/TC282 Water Reuse）专门负责制定水回用领域相关的国际标准，以推动水回用行业的健康、规范发展。在可预见的未来，废水再生利用技术将会更好的应用于实际生产和生活当中。

5.3　PHA 回收技术

聚羟基烷基酸酯（polyhydroxyalkanoates，PHA）是一类由微生物在碳源充足而 N、P 等营养元素缺乏的条件下合成的细胞内碳源和能源储存物质。PHA 具有与聚乙烯类石油基塑料类似的性能，同时具有生物可降解性和生物相容性等特点，被认为是最有前景的环保材料之一[278]。传统生产 PHA 的方法是利用纯菌株发酵，但是成本较高，限制了 PHA 的普及。混合菌群能够利用有机废水合成 PHA，不仅极大地降低了底物成本，变废为宝，而且整个生产过程无需灭菌处理，开放式发酵过程更易控制。利用混合菌群生产 PHA 的过程常采用三步法[279]：①有机废水（或废物）的厌氧酸化；②混合菌群的驯化；③PHA 的合成和收集。在实验室小试研究中，制糖、造纸、炼油、啤酒等行业的有机废水及剩余污泥、餐厨垃圾、农业废弃物等有机废物都已证实可用于 PHA 合成，并获得较好的产率[280,281]。目前的研究热点包括产酸过程的优化和控制、驯化过程的机理和影响因素、PHA 产率和产品结构的调控等[282-284]。近年来，在中试规模的研究中，利用混合菌群和有机废水（或废物）生产 PHA 也有一些成功的案例[285,286]。

第一步进行底物厌氧酸化的目的是将废水（或废物）中的复杂有机物转化为小分子挥发性脂肪酸（volatile fatty acid，VFA）。完整的厌氧发酵包含三个阶段：水解、酸化、产甲烷。为了使 VFA 积累最大化，厌氧发酵主要通过控制温度、pH、污泥停留时间等条件促进水解和产酸过程，同时抑制产甲烷过程。高温条件（55℃）比中温条件，碱性条件（pH 10）比中性和酸性条件，能够促进有机物的溶解率、提高主要水解酶和产酸酶的活性、提高和水解产酸相关的微生物的丰度、抑制产甲烷菌的活性和丰度，从而促进 VFA 的大量积累。VFA 比例高的水解液对于 PHA 的合成具有更好的效果。此外，VFA 的组成对 PHA 产品的结构和性质也具有决定性的影响[287]。通常来说，偶数碳的 VFA（即乙酸和丁酸）易合成 3-羟基丁酸（3-hydroxybutyrate，3HB）单体，而奇数碳的 VFA（即丙酸和戊酸）易合成 3-羟基戊酸（3-hydroxyvalerate，3HV）单体。含有 3HV 单体的 PHA 比含有 3HB 单体的 PHA 具有更低的结晶度和熔点，延展性也更好。因此，不少学者致力于定向合成和调控奇数碳 VFA 并增加其在水解液中的比例。研究表明，丙酸比例的增加可通过在厌氧酸化过程中加入丙酸杆菌或加入糖类底物调整 C/N 来实现[288]。但是，目前关于戊酸合成机理和调控的研究还很少。

合成 PHA 的混合菌群一般是从活性污泥驯化所得，应用较多的方式是好氧瞬时供料（aerobic dynamic feeding，ADF）。近年来，为了提升驯化过程的效率和菌群的 PHA 合成性能，也有学者对 ADF 模式进行了一些改进。例如，为了消除水解液中非 VFA 物质的影响，可在丰盛期末期即 VFA 消耗完成时，将剩余底物去除，饥饿期只有胞内储存的 PHA 一种碳源，这种方式被称为好氧供料和卸料（aerobic feeding and discharge，AFD）。Jiang 等[289]学者采取污泥碱发酵水解液为底物对比了 ADF 和 AFD 两种方式对 PHA 混合菌群驯化的影响。研究表明，经过传统 ADF 方式驯化的菌群 PHA 合成率最高达 39%，而经过 AFD 方式驯化的菌群 PHA 合成率最高达 73%，提高了 87%。Chen 等学者提出这种新的 AFD 方式能够使整个驯化过程缩短并使 PHA 菌群合成能力提高的主要原因是加入了物理选择压：PHA 合成菌因为密度更大细胞更重而更容易在沉降过程中沉入下层，从而与非 PHA 合成菌分离。虽然混合菌群的合成率在不断提高，但是相对于纯菌发酵，PHA 的产量和产率仍较低，不可与纯菌媲美，高效 PHA 菌群的驯化仍是一个亟待解决的问题。

对于驯化成熟的 PHA 菌群，为了使其 PHA 含量最大化，采用批次实验工艺是普遍采用的方法[290]。在批次试验中，许多研究均采取比驯化阶段更高的有机负荷但不含营养成分的底物，同时实时监测溶解氧。当溶解氧突然跃增即表示底物基本消耗完成，此时静置除去上清液后继续投加底物，即保证微生物始终在碳源充足的环境中，使其达到细胞内 PHA 含量的极限。采用这种批次补料的模式加入底物还可以有效地克服底物的抑制作用[291]。

中试规模的混合菌群 PHA 合成过程是更加复杂的体系，各种条件的控制更加严格。Tamis 等学者利用酸化的糖果工厂废水大规模合成 PHA，他们建立的体系包括厌氧发酵、菌群驯化和胞内 PHA 积累三个阶段。酸化后的废水主要成分是 VFA 和乙醇，分别占 64%和 22%。驯化后的菌群以 *Plasticicumulans acidivorans* 为主，PHA 合成率达 70%。阻止合成率进一步升高的主要原因是进水中含有大量固体，以及其他非 PHA 合成菌的生长。Jia 等[292]学者的中试体系在进水固体去除方面有较好的改进。他们利用的是剩余污泥酸化液，污泥厌氧发酵后他们在体系中加入了陶瓷膜分离系统，能够去除污泥酸化液中的大部分固体，不会过多引入其他杂菌。利用废水（或废物）合成 PHA 的可行性也是目前备受关注的一个问题。Fernández-Dacosta 等学者对 PHA 中试合成过程和三种下游提取过程（碱提取、次氯酸盐提取和溶剂提取）做了完整的技术经济分析和环境影响分析：碱提取的方式最经济，对环境的影响也最小，最终产品的生产成本约 1.40 €/kg PHB。

利用混合菌群生产 PHA 被认为是纯菌生产 PHA 的一种良好替代方式，但是与传统石油基塑料的生产相比还不具有竞争优势。如何提高混合菌群的浓度和 PHA 的积累率，以及对有机废水（或废物）酸化与 PHA 结合的整个过程的整合和优化，是目前面临的主要挑战。

参 考 文 献

[1] Su L, Li K, Zhang H, Fan M, Ying D, Sun T, Wang Y, Jia J. Electrochemical nitrate reduction by using a novel Co_3O_4/Ti cathode. Water Research, 2017, 120: 1-11.

[2] Wang Z, Qi J, Wang B, Feng Y, Li K. A three dimensional electrochemical oxidation reactorbased on magnetic steel slag particle electrodes for printing /dyeing wastewater treatment. Journal of Harbin Institute of Technology, 2015,47: 38-42.

[3] Chen Z, Zhang Y, Zhou L, Zhu H, Wan F, Wang Y, Zhang D. Performance of nitrogen-doped graphene aerogel particle electrodes for electro-catalytic oxidation of simulated Bisphenol A wastewaters. Journal of Hazardous Materials, 2017,332: 70-78.

[4] Vijayakumar V, Saravanathamizhan R, Balasubramanian N. Electro oxidation of dye effluent in a tubular electrochemical reactor using TiO_2/RuO_2 anode. Journal of Water Process Engineering, 2016,9: 155-160.

[5] Zhang Y, Yu T, Han W, Sun X, Li J, Shen J, Wang L. Electrochemical treatment of anticancer drugs wastewater containing 5-fluoro-2-methoxypyrimidine using a tubular porous electrode electrocatalytic reactor. Electrochimica Acta, 2016,220:

211-221.

[6] Shen Z M, Yang J, Hu XF, Lei Y M, Ji X L, Jia J P, Wang W H. Dual electrodes oxidation of dye wastewater with gas diffusion cathode. Environmental Science and Technology, 2005,39: 1819-1826.

[7] Li W, Xiao C, Zhao Y, Zhao Q, Fan R, Xue J. Electrochemical reduction of high-concentrated nitrate using Ti/TiO_2 nanotube array anode and Fe cathode in dual-chamber cell. Catalysis Letters, 2016,146: 2585-2595.

[8] Zhong D, Yang J, Xu Y, Jia J, Wang Y, Sun T. De-colorization of reactive brilliant orange X-GN by a novel rotating electrochemical disc process. Journal of Environmental Sciences,2008, 20: 927-932.

[9] Bao P, Hu X, Chen B M, Guo Z C. Research status of energy-saving anodes for zinc electrowinning. Electroplating and Finishing,2015,34: 872-876.

[10] Yao F, Zhong Y, Yang Q, Wang D, Chen F, Zhao J, Xie T, Jiang C, An H, Zeng G, Li X. Effective adsorption/ electrocatalytic degradation of perchlorate using Pd/Pt supported on N-doped activated carbon fiber cathode. Journal of Hazardous Materials,2017,323: 602-610.

[11] Li P, Zhao G, Zhao K, Gao J, Wu T. An efficient and energy saving approach to photocatalytic degradation of opaque high-chroma methylene blue wastewater by electrocatalytic pre-oxidation. Dyes and Pigments ,2012,92: 923-928.

[12] Li K, Yang C, Wang Y, Jia J, Xu Y, He Y. A high-efficient rotating disk photoelectrocatalytic (PEC) reactor with macro light harvesting pyramid-surface electrode. AIChE Journal,2012, 58: 2448-2455.

[13] Martin de Vidales M J, Milian M, Saez C, Canizares P, Rodrigo MA. Irradiated-assisted electrochemical processes for the removal of persistent pollutants from real wastewater. Separation and Purification Technology,2017,175: 428-434.

[14] Wang X, Jia J, Wang Y. Enhanced photocatalytic-electrolytic degradation of Reactive Brilliant Red X-3B in the presence of water jet cavitation, Ultrasonics Sonochemistry,2015,23: 93-99.

[15] Huang X, Qu Y, Cid CA, Finke C, Hoffmann M R, Lim K, Jiang S C. Electrochemical disinfection of toilet wastewater using wastewater electrolysis cell. Water Research, 2016,92: 164-172.

[16] Lee Y, Gerrity D, Lee M, Gamage S, Pisarenko A, Trenholm R A, Canonica S, Snyder S A, von Gunten U. Organic Contaminant Abatement in Reclaimed Water by UV/H_2O_2 and a Combined Process Consisting of O_3/H_2O_2 Followed by UV/H_2O_2: Prediction of Abatement Efficiency, Energy Consumption, and Byproduct Formation. Environmental Science and Technology, 2016, 50(7): 3809-3819.

[17] Lester Y, Ferrer I, Thurman E M, Linden K G. Demonstrating sucralose as a monitor of full-scale UV/AOP treatment of trace organic compounds. Journal of Hazardous Materials, 2014, 280: 104-110.

[18] Fang G, Gao J, Dionysiou D D, Liu C, Zhou D. Activation of persulfate by quinones: Free radical reactions and implication for the degradation of PCBs. Environmental Science and Technology, 2013, 47(9): 4605-4611.

[19] Wu Y, Bianco A, Brigante M, Dong W, De Sainte-Claire P, Hanna K, Mailhot G. Sulfate radical photogeneration using Fe-EDDS: Influence of critical parameters and naturally occurring scavengers. Environmental Science and Technology 2015, 49(24): 14343-14349.

[20] Cai C, Zhang H, Zhong X, Hou L. Electrochemical enhanced heterogeneous activation of peroxydisulfate by Fe-Co/SBA-15 catalyst for the degradation of Orange II in water. Water Research 2014, 66: 473-485.

[21] Xie P, Guo Y, Chen Y, Wang Z, Shang. Wang S, Ding J, Wan Y, Jiang W, Ma J. Application of a novel advanced oxidation process using sulfite and zero-valent iron in treatment of organic pollutants. Chemical Engineering Journal 2017, 314: 240-248.

[22] Lian L, Yao B, Hou S, Fang J, Yan S, Song W. Kinetic study of hydroxyl and sulfate radical-mediated oxidation of pharmaceuticals in wastewater effluents. Environmental Science and Technology 2017, 51(5): 2954-2962.

[23] Yang Y, Pignatello J J, Ma J, Mitch W A. Effect of matrix components on UV/H_2O_2 and UV/$S_2O_8{}^{2-}$ advanced oxidation processes for trace organic degradation in reverse osmosis brines from municipal wastewater reuse facilities. Water Research 2016, 89: 192-200.

[24] Ahmad A. Gu X, Li L, Lv S, Xu Y, Guo X. Efficient degradation of trichloroethylene in water using persulfate activated by reduced graphene oxide-iron nanocomposite. Environmental Science and Pollution Research, 2015, 22(22): 17876-17885.

[25] Ahmad A, Gu X, Li L, Lu S, Xu Y, Guo X. Effects of pH and anions on the generation of reactive oxygen species(ROS)in nZVI-rGo-activated persulfate system. Water, Air, and Soil Pollution, 2015, 226(11): 369.

[26] Fang J, Fu Y, Shang C. The roles of reactive species in micropollutant degradation in the UV/free chlorine system. Environmental Science and Technology, 2014, 48(3): 1859-1868.

[27] Fang J, Zhao Q, Fan C, Shang C, Fu Y, Zhang X. Bromate formation from the oxidation of bromide in the UV/chlorine process with low pressure and medium pressure UV lamps. Chemosphere, 2017, 183: 582-588.

[28] Ben WW, Sun PZ, Huang CH. Effects of combined UV and chlorine treatment on chloroform formation from triclosan. Chemosphere, 2016, 150: 715-722.

[29] Parker KM, Mitch WA. Halogen radicals contribute to photooxidation in coastal and estuarine waters. Proceedings of the National Academy of Sciences of the United States of America, 2016, 113(21): 5868-5873.

[30] Barazesh JM, Prasse C, Sedlak DL. Electrochemical transformation of trace organic contaminants in the presence of halide and carbonate ions. Environmental Science and Technology, 2016, 50(18): 10143-10152.

[33] Yoneyama H, YamashitaandAmp Y, Tamura H. Heterogeneous photocatalytic reduction of dichromate on n-type semiconductor catalysts. Nature, 1979, 282(5741): 817-818.

[34] 林龙利, 刘国光, 吕文英. TiO_2光催化同步去除水体中重金属和有机物研究进展. 科技导报, 2011, 29(23): 74-79.

[35] Kajitvichyanukul P, Chenthamarakshan C R, Rajeshwar K, et al. Photocatalytic reactivity of thallium (I) species in aqueous suspensions of titania. Journal of Electroanalytical Chemistry, 2002, 519(1): 25-32.

[36] Chenthamarakshan C R, Rajeshwar K. Photocatalytic reduction of divalent zinc and cadmium ions in aqueous TiO_2 suspensions: an interfacial induced adsorption-reduction pathway mediated byformate ions. Electrochemistry Communications, 2000, 2(7): 527-530.

[38] Angelidis T N, Koutlemani M, Poulios I. Kinetic study of the photocatalytic recovery of Pt from aqueous solution by TiO_2, in a closed-loop reactor. Applied Catalysis B Environmental, 1998, 16(16): 347-357.

[37] 秦红霞. 表面修饰TiO_2对选择性吸附协同光催化还原Cr(VI)性能的研究. 上海: 上海师范大学, 2016.

[39] El-Sheikh A H. Effect of oxidation of activated carbon on its enrichment efficiency of metal ions: Comparison with oxidized and non-oxidized multi-walled carbon nanotubes. Talanta, 2008, 75(1): 127-134.

[31] 王子, 马鲁铭. 催化铁还原技术在工业废水处理中的应用进展. 中国给水排水, 2009, 25(6): 9-13.

[32] 马鲁铭. 废水的催化还原处理技术: 原理及应用.北京: 科学出版社, 2008.

[40] Huang Y H, Tang C, Zeng H. Removing molybdate from water using a hybridized zero-valent iron/magnetite/Fe(II) treatment system. Chemical Engineering Journal, 2012, 200-202(34): 257-263.

[41] Guo X, Yang Z, Liu H, et al. Common oxidants activate the reactivity of zero-valent iron (ZVI) and hence remarkably enhance nitrate reduction from water. Separation and Purification Technology, 2015, 146: 227-234.

[42] Guo X, Yang Z, Dong H, et al. Simple combination of oxidants with zero-valent-iron (ZVI) achieved very rapid and highly efficient removal of heavy metals from water. Water Research, 2016, 88: 671.

[43] Liang L, Wu S, Guan X, et al. Weak magnetic field significantly enhances selenite removal kinetics by zero valent iron. Water Research, 2014, 49(1): 371.

[45] 李锦祥, 秦荷杰, 张雪莹, 等. 弱磁场强化零价铁对水中污染物的去除效能及其作用机制. 化学学报,, 2017, 75: 538-543.

[44] Li JX, Shi Z, Ma B, et al. Improving the reactivity of zerovalent iron by taking advantage of its magnetic memory: Implications for arsenite removal. Environmental Science and Technology, 2015, 49(17): 10581-10588.

[46] Larsen TA, Hoffmann S, Lüthi C, Truffer B, Maurer M. Emerging solutions to the water challenges of an urbanizing world. Science, 2016, 352(6288): 928-933.

[48] Park H B, Kamcev J, Robeson L M, Elimelech M, Freeman B D. Maximizing the right stuff: The trade-off between membrane permeability and selectivity. Science, 2017, 356,(6343).

[49] Kumar M, Grzelakowski M, Zilles J, Clark M, Meier W. Highly permeable polymeric membranes based on the incorporation of the functional water channel protein Aquaporin Z. Proceedings of the National Academy of Sciences, 2007, 104(52): 20719-20724.

[47] Surwade S P, Smirnov S N, Vlassiouk I V, Unocic R R, Veith G M, Dai S, Mahurin S M. Water desalination using nanoporous single-layer graphene. Nat Nano, 2015, 10(5): 459-464.

[51] Baoxia M. Materials science. Graphene oxide membranes for ionic and molecular sieving. Science, 2014, 343(6172): 740-2.

[52] Ang H, Hong L. Polycationic polymer-regulated assembling of 2D MOF nanosheets for high-performance nanofiltration. ACS Appl. Mater. Interfaces, 2017.

[50] Guo S, Zhu X, Janczewski D, Lee S, S C, He T, Teo S L M, Vancso G J. Measuring protein isoelectric points by AFM-based force spectroscopy using trace amounts of sample. Nat. Nanotechnol,2016, 11(9): 817-823.

[54] Guan Y F, Huang B C, Qian C, Wang L F, Yu H Q. Improved PVDF membrane performance by doping extracellular polymeric substances of activated sludge. Water Research, 2017, 113: 89-96.

[55] Zhu X, Loo H E, Bai R. A novel membrane showing both hydrophilic and oleophobic surface properties and its non-fouling performances for potential water treatment applications. Journal of Membrane Science, 2013, 436 (2): 47-56.

[53] Zhang R, Liu Y, He M, Su Y, Zhao X, Elimelech M, Jiang Z. Antifouling membranes for sustainable water purification: strategies and mechanisms. Chemistry Society Review, 2016, 45 (21): 5888-5924.

[56] Klaysom C, Cath T Y, Depuydt T, Vankelecom I F J. Forward and pressure retarded osmosis: Potential solutions for global challenges in energy and water supply. Chemistry Society Review, 2013, 42,(16): 6959-6989.

[57] Cath T Y, Childress A E, Elimelech M. Forward osmosis: Principles, applications, and recent developments. Journal of Membrane Science, 2006, 281 (1-2): 70-87.

[58] Logan B E, Elimelech M. Membrane-based processes for sustainable power generation using water. Nature, 2012, 488(7411): 313-319.

[59] Veerman J, Saakes M, Metz S J, Harmsen G J. Reverse electrodialysis: Performance of a stack with 50 cells on the mixing of sea and river water. Membrane Science, 2009, 327 (1): 136-144.

[60] Xu J, Sheng G P, Luo H W, Li W W, Wang L F, Yu H Q. Fouling of proton exchange membrane (PEM) deteriorates the performance of microbial fuel cell. Water Research, 2012, 46 (6): 1817-1824.

[61] Jarvis P, Sharp E, Pidou M, Molinder R, Parsons S A, Jefferson B. Comparison of coagulation performance and floc properties using a novel zirconium coagulant against traditional ferric and alum coagulants. Water Research, 2012, 46: 4179-4187.

[63] Zeng T, Li RJ, Mitch WA. Structural modifications to quaternary ammonium polymer coagulants to inhibit N-nitrosamine formation. Environmental Science and Technology, 2016, 50: 4778-4787.

[62] Zhao Y X, Phuntsho S, Gao B Y, Huang X, Qi Q B, Yue Q Y, Wang Y, Kim J H, Shon H K. Preparation and characterization of novel polytitanium tetrachloride coagulant for water purification. Environmental Science and Technology, 2013, 47: 12966-12975.

[64] Zhao H, Peng J, Lin S, Zhang Y. Covalently bound organic silicate aluminum hybrid coagulants: Preparation, characterization, and coagulation behavior. Environmental Science and Technology, 2009, 43: 2041-2046.

[65] Chang Y Y, Zhao H Z. Characterization and coagulation performance of covalently bound organic silicate aluminum hybrid coagulants: effects of Si/Al, B value and pH. Desalination and Water Treatment, 2015, 54: 1127-1133.

[66] Zhao HZ, Wang L, Chang YY, Xu Y. High-efficiency removal of perfluorooctanoic acid from water by covalently bound hybrid coagulants (CBHyC) bearing a hydrophobic quaternary ammonium group. Separation and Purification Technology, 2016, 158: 9-15.

[67] Zhao H, Wang L, Hanigan D, Westerhoff P, Ni J. Novel ion-exchange coagulants remove more low molecular weight organics than traditional coagulants. Environmental Science and Technology, 2016, 50: 3897-3904.

[68] Xue A, Shen ZZ, Zhao B, Zhao HZ. Arsenite removal from aqueous solution by a microbial fuel cell-zerovalent iron hybrid process. Journal of Hazardous Materials, 2013, 261: 621-627.

[69] Xiao X, Sun Y, Sun W, Shen H, Zheng H, Xu Y, Zhao J, Wu H, Liu C. Advanced treatment of actual textile dye wastewater by Fenton-flocculation process. Canadian Journal of Chemical Engineering, 2017, 95: 1245-1252.

[70] Umar M, Roddick F, Fan L. Impact of coagulation as a pre-treatment for UVC/H_2O_2-biological activated carbon treatment of a municipal wastewater reverse osmosis concentrate. Water Research, 2016, 88: 12-19.

[71] Yu W, Graham N, Liu T. Effect of intermittent ultrasound on controlling membrane fouling with coagulation pre-treatment: Significance of the nature of adsorbed organic matter. Journal of Membrane Science, 2017, 535: 168-177.

[72] Huang H, Cho H H, Jacangelo J G, Schwab K J. Mechanisms of membrane fouling control by integrated magnetic ion exchange and coagulation. Environmental Science and Technology, 2012, 46: 10711-10717.

[73] Zhu S, Xu Y, Chen T, Xu R, Cui F, Shi W. Effect of continuous direct recycling of combined residual streams on water quality at the pilot scale in different seasons. Journal of Environmental Engineering, 2017, 143.

[75] Zhao Y, Liang W Y, Liu LJ, Li F Z, Fan Q L, Sun X L. Harvesting *Chlorella vulgaris* by magnetic flocculation using Fe_3O_4 coating with polyaluminium chloride and polyacrylamide. Bioresource Technology, 2015, 198: 789-796.

[74] Luo L, Nguyen A V. A review of principles and applications of magnetic flocculation to separate ultrafine magnetic particles. Separation and Purification Technology, 2017, 172: 85-99.

[76] 邓述波, 余刚. 环境吸附材料及应用原理. 北京: 科学出版社, 2012.

[77] Yao XL, Deng S, Wu R, et al. Highly efficient removal of hexavalent chromium from electroplating wastewater using aminated wheat straw. RSC Advances, 2016, 6: 8797-8805.

[78] Hua M, Yang B W, Shan C, et al. Simultaneous removal of As(V) and Cr(VI) from water by macroporous anion exchanger supported nanoscale hydrous ferric oxide composite. Chemosphere, 2017, 171: 126-133.

[79] Gharin Nashtifan S, Azadmehr A, Maghsoudi A. Comparative and competitive adsorptive removal of Ni^{2+} and Cu^{2+} from aqueous solution using iron oxide-vermiculite composite. Applied Clay Science, 2017, 140: 38-49.

[80] Liu H, Zuo H, Vecitis C D. Titanium dioxide-coated carbon nanotube network filter for rapid and effective arsenic sorption. Environmental Science and Technology, 2014, 48(23): 13871-13879.

[81] Du Z W, Deng S, et al. Adsorption behavior and mechanism of perfluorinated compounds on various adsorbents—A review. Journal of Hazardous Materials, 2014, 274: 443-454.

[83] Meng P P, Deng S, et al. Role of air bubbles overlooked in the adsorption of perfluorooctanesulfonate on hydrophobic carbonaceous adsorbents. Environmental Science and Technology, 2014. 48(23): 13785-13792.

[84] Meng P P, Deng S, et al. Superhigh adsorption of perfluorooctane sulfonate on aminated polyacrylonitrile fibers with the assistance of air bubbles. Chemical Engineering Journal, 2017, 315: 108-114.

[82] Du Z W, Deng S, et al. Efficient adsorption of PFOS and F53B from chrome plating wastewater and their subsequent degradation in the regeneration process. Chemical Engineering Journal, 2016, 290: 405-413.

[85] Du Z W, Deng S, Zhang S Y, Wang W, Wang B, Huang J, Wang Y J, Yu G, Xing B S. Selective and fast adsorption of perfluorooctanesulfonate from wastewater by magnetic fluorinated vermiculite. Environmental Science and Technology, 2017, 51: 8027-8035.

[86] Xiao L L, Ling Y H, Alsbaiee A, Li C J, Helbling D E. Dichtel WR. β-Cyclodextrin polymer network sequesters perfluorooctanoic acid at environmentally relevant concentrations. Journal of American Chemistry Sociaty, 2017, 139: 7689-7692.

[88] Alsbaiee A, Smith B J, Xiao Leilei, et al. Rapid removal of organic micropollutants from water by a porous β-cyclodextrin polymer. Nature, 2016, 529(7585): 190-194.

[87] Shan D N, Deng S, Zhao T, et al. Preparation of ultrafine magnetic biochar and activated carbon for pharmaceutical adsorption and subsequent degradation by ball milling. Journal of Hazardous Materials, 2016, 305: 156-163.

[89] Shan D N, Deng S B, Zhao T M, et al. Preparation of regenerable granular carbon nanotubes by a simple heating-filtration method for efficient removal of typical pharmaceuticals. Chemical Engineering Journal, 2016, 294: 353-361.

[90] Liu F F, Zhao J, Wang S, et al. Effects of solution chemistry on adsorption of selected pharmaceuticals and personal care products (PPCPs) by graphenes and carbon nanotubes. Environmental Science and Technology, 2014, 48(22): 13197-13206.

[91] 李梦凯, 强志民, 史彦伟, 李庭刚. 紫外消毒系统有效辐射剂量测试方法研究进展. 环境科学学报, 2012, 32(3): 513-520.

[92] 张艳, 李继. CFD 技术在水处理紫外消毒中的应用. 环境工程, 2011, 29: 123-126.

[93] Li WT, Li M K, Bolton J R, Qu J H, Qiang Z M Impact of inner-wall reflection on UV reactor performance as evaluated by using computational fluid dynamics: The role of diffuse reflection. Water Research,2017, 109: 382-388.

[96] Huang JJ, Hu HY, Wu .H, Wei B, Lu Y. Effect of chlorination and ultraviolet disinfection on tetA-mediated tetracycline resistance of *Escherichia coli*. Chemosphere, 2013, 90(8): 2247-2253.

[94] 刘亚兰, 马岑鑫, 丁河舟, 邱勇, 李冰, 王硕, 李激. 污水处理厂消毒技术对抗生素抗性菌的强化去除. 环境科学,2017, DOI: 10.13227/j.hjkx.201612141(待刊).

[95] Dong H Y, Qiang Z M, Lian J F, Qu J H. Degradation of nitro-based pharmaceuticals by UV photolysis: Kinetics and

simultaneous reduction on halonitromethanes formation potential. Water Research, 2017, 119: 83-90.

[97] Huang J J, Hu H Y, Tang F, Li Y, Lu SQ, Lu Y. Inactivation and reactivation of antibiotic-resistant bacteria by chlorination in secondary effluents of a municipal wastewater treatment plant. Water Research, 2011, 45: 2775-2781.

[98] Wenk J, Aeschbacher M, Salhi E, Canonica S, von Gunten U,Sander M. Chemical oxidation of dissolved organic matter by chlorine dioxide, chlorine, and ozone: Effects on its optical and antioxidant properties. Environmental Scicence Technology, 2013, 47(19): 11147 11156.

[99] Yang X, Guo WH, Zhang X, Chen F, Ye TJ, Liu W. Formation of disinfection by-products after pre-oxidation with chlorine dioxide or ferrate. Water Research, 2013, 47: 5856-5864.

[100] Loeb B L, Thompson C M, Drago J, Takahara H, Baig S. Worldwide ozone capacity for treatment of drinking water and wastewater: a review. Ozone-science and Engineering, 2012, 34(1): 64-77.

[101] 王祥勇, 陈洪斌, 阮久丽. 污水和再生水臭氧消毒的研究和应用. 水处理技术, 2010, 36(4): 19-23.

[102] 杨纪超. 深圳污水处理厂紫外消毒的影响因素及改进对策. 哈尔滨: 哈尔滨工业大学, 2013.

[103] 胡洪营, 王丽莎, 魏东斌. 污水消毒面临的技术挑战及其对策. 世界科技研究与发展, 2005, 27(6): 36−41.

[104] 张正哲, 姬玉欣, 陈辉, 等. 厌氧氨氧化工艺的应用现状和问题. 生物工程学报, 2014, 30(12): 1804-1816.

[105] 张正哲, 金仁村, 程雅菲, 等. 厌氧氨氧化工艺的应用进展. 化工进展, 2015, 34(5): 1444-1452.

[108] Lackner S, Gilbert E M, Vlaeminck SE, et al. Full-scale partial nitritation/anammox experiences-An application survey. Water Research, 2014, 55(0): 292-303.

[109] van der Star WR, Abma WR, Blommers D, et al. Startup of reactors for anoxic ammonium oxidation: experiences from the first full-scale anammox reactor in Rotterdam. Water Research, 2007, 41(18): 4149-4163.

[110] Morales N, Val Del Río Á, Vázquez-Padín J R, et al. Integration of the anammox process to the rejection water and main stream lines of WWTPs. Chemosphere, 2015, 140: 99-105.

[113] Laureni M, Falås P, Robin O, et al. Mainstream partial nitritation and anammox: long-term process stability and effluent quality at low temperatures. Water Research, 2016, 101: 628-639.

[114] Jin R, Yang G, Yu J, et al. The inhibition of the anammox process: A review. Chemical Engineering Journal, 2012, 197(0): 67-79.

[115] Zhang Z, Liu S. Hot topics and application trends of the anammox biotechnology: A review by bibliometric analysis. Springer Plus, 2014, 3(1): 1-8.

[116] Dalsgaard T, Canfield D E, Petersen J, Thamdrup B, Acuña-González J. N_2 production by the anammox reaction in the anoxic water column of Golfo Dulce, Costa Rica. Nature, 2003, 422(6932): 606-608.

[117] Dong X, Tollner E W. Evaluation of Anammox and denitrification during anaerobic digestion of poultry manure. Bioresource Technololgy, 2003, 86(2): 139-145.

[111] Nikolaev Y A, Kozlov M N, Kevbrina M V, et al. *Candidatus* “Jettenia moscovienalis” sp. nov., A new species of bacteria carrying out anaerobic ammonium oxidation. Microbiology, 2015, 84(2): 256-262.

[112] Woebken D, Lam P, Kuypers MMM, et al. A microdiversity study of anammox bacteria reveals a novel *Candidatus scalindua* phylotype in marine oxygen minimum zones. Environment Microbiololgy, 2008, 10: 3106-3119.

[118] Strous M, Pelletier E, Mangenot S, et al. Deciphering the evolution and metabolism of an anammox bacterium from a community genome. Nature, 2006 440: 790-794.

[106] Park H, Brotto AC, van Loosdrecht MCM, et al. Discovery and metagenomic analysis of an anammox bacterial enrichment related to *Candidatus* “Brocadia caroliniensis” in a full-scale glycerol-fed nitritation-denitritation separate centrate treatment process. Water Research, 2017, 111: 265-273.

[107] Cho S, Takahashi Y, Fujii N, et al. Nitrogen removal performance and microbial community analysis of an anaerobic upflow granular bed Anammox reactor. Chemosphere, 2010, 78(9): 1129-1135.

[119] 曲久辉, 王凯军, 王洪臣, 余刚, 柯兵, 俞汉青. 建设面向未来的中国污水处理概念厂. 中国环境报, 2014, 1, 第 10 版.

[120] Jin L Y, Zhan G M, Tian H F. Current state of sewage treatment in China. Water Research, 2014, 66: 85-98.

[121] Lackner S, Gilbert E M, Vlaeminck S E, Joss A, Horn H, van Loosdrecht M C M. Full-scale partial nitritation/anammox experiences—An application survey. Water Research, 2014, 55: 292-303.

[122] Kang X S, Liu C Q, Zhang B, Bi X J, Zhang F, Cheng L H. Application of reversed A^2/O process on removing nitrogen and

phosphorus from municipal wastewater in China. Water Science and Technology, 2012, 63(10): 2138-2142.

[123] Gilda Carvalho. Denitrifying phosphorus removal: Linking the process performance with the microbial community structure. Water Research, 2007, 41: 4383-4396.

[124] Oehmen A, Lemos P C, Carvalho G. Advances in enhanced biological phosphorus removal: From micro to macro scale. Water Research, 2007, 41: 2271-2300.

[132] Wang Y Y, Guo G, Wang H, Stephenson T, Guo J H, Ye L. Long-term impact of anaerobic reaction time on the performance and granular characteristics of granular deitrifying biological phosphorus removal systems. Water Research, 2013, 47(14): 5326-5337.

[127] Bassin J P, Kleerebezem R, Dezotti M, van Loosdrecht M C M. Simultaneous nitrogen and phosphate removal in aerobic granular sludge reactors operated at different temperatures. Water Research, 2012, 46: 3805-3816.

[125] Cassidy D P, Belia E. Nitrogen and phosphorus removal from an abattoir wastewater in a SBR with aerobic granular sludge. Water Research, 2005, 39(19): 4817-4823.

[126] Bassin J P, Winkler M K H, Kleerebezem R, Dezotti M, van Loosdrecht M C M. Relevance of selective sludge removal in segregated aerobic granular sludge reactors to control PAO-GAO competition at different temperatures. Biotechnololgy Bioengineering, 2012, 109(8): 1919-1928.

[128] Neoh C H, Noor Z Z, Sing C L I, Mutamim N S A, Lin C K. Green technology in wastewater treatment technologies: Integration of membrane bioreactor with various wastewater treatment systems. Chemical Engineering Journal, 2016, 283: 582-594.

[129] Monclús H, Sipma J, Ferrero G, Rodriguez-Roda I, Comas J. Biological nutrient removal in an MBR treating municipal wastewater with special focus on biological phosphorus removal. Bioresource Technology, 2010, 101(11): 3984-91.

[130] Lotti T, Kleerebezem R, Kip C, Hendrickx T L G, Kruit J, Hoekstra M, Loosdrecht M C M. Anammox growth on pretreated municipal wastewater. Environmental Science and Technololgy, 2014, 48(14): 7874-7880.

[131] Zeng W, Li B X, Wang X D, Bai X L, Peng Y Z. Integration of denitrifying phosphorus removal via nitrite pathway, simultaneous nitritation-denitritation and anammox treating carbon-limited municipal sewage. Bioresource Technology, 2014, 172: 356-364.

[133] Yang Y D, Zhang L, Shao H D, Zhang S J, Gu P C, Peng Y Z. Enhanced nutrients removal from municipal wastewater through biological phosphorus removal followed by partial nitritation/anammox. Frontiers of Environmental Science and Engineering, 2017, 11(2): 8-13.

[134] 郑祥. 中国水处理行业可持续发展战略研究报告. 北京：中国人民大学出版社, 2013.

[135] Xiao K, Ying XU, Liang S, et al. Engineering application of membrane bioreactor for wastewater treatment in China: Current state and future prospect. Frontiers of Environmental Science and Engineering, 2014, 8(6): 805-819.

[136] Fu Z, Yang F, An Y, et al. Simultaneous nitrification and denitrification coupled with phosphorus removal in an modified anoxic/oxic-membrane bioreactor(A/O-MBR). Biochemical Engineering Journal, 2009, 43(2): 191-196.

[137] Ji J, Qiu J, Wai N, et al. Influence of organic and inorganic flocculants on physical-chemical properties of biomass and membrane-fouling rate. Water Research, 2010, 44(5): 1627-35.

[138] Zhang J, Satti A, Chen X, et al. Low-voltage electric field applied into MBR for fouling suppression: Performance and mechanisms. Chemical Engineering Journal, 2015, 273: 223-230.

[139] Xue W, Wu C, Xiao K, et al. Elimination and fate of selected micro-organic pollutants in a full-scale anaerobic/anoxic/aerobic process combined with membrane bioreactor for municipal wastewater reclamation. Water Research, 2010, 44(20): 5999-6010.

[140] Sun J, Liang P, Yan X, et al. Reducing aeration energy consumption in a large-scale membrane bioreactor: Process simulation and engineering application. Water Research, 2016, 93: 205-213.

[141] Yan X, Wu Q, Sun J, et al. Hydrodynamic optimization of membrane bioreactor by horizontal geometry modification using computational fluid dynamics. Bioresource Technology, 2016, 200(4): 328.

[142] Meng F, Zhang S, Oh Y, et al. Fouling in membrane bioreactors: An updated review. Water Research, 2017, 114: 151.

[143] Shen YX, Xiao K, Liang P, et al. Characterization of soluble microbial products in 10 large-scale membrane bioreactors for municipal wastewater treatment in China. Journal of Membrane Science, 2012, 415-416(10): 336-345.

[144] Wang Z, Ma J, Tang CY, et al. Membrane cleaning in membrane bioreactors: A review. Journal of Membrane Science, 2014, 468(20): 276-307.

[145] Ge S, Zhu Y, Lu C, et al. Full-scale demonstration of step feed concept for improving an anaerobic/anoxic/aerobic nutrient removal process. Bioresource Technology, 2012, 120(3): 305-313.

[146] 李激, 郑凯凯, 王燕, 等. 智能化城市污水处理厂运行专家系统的研究. 中国给水排水, 2016(11): 1-5.

[147] 魏伟, 王蒙, 刘载文, 等. 基于活性污泥法的污水处理线性自抗扰控制. 计算机仿真, 2015, 32(8): 417-421.

[148] 杨新宇, 邱勇, 施汉昌, 等. 曝气过程控制系统在污水处理厂节能中的应用与评价. 给水排水, 2012, 38(7): 130-134.

[149] S. Lackner, M. Gilbert E, E. Vlaeminck S, et al. Full-scale partial nitritation/anammox experiences-an application survey. Water Res, 2014, 55: 292-303.

[150] 张珏, 陈辉, 姬玉欣, 等. 厌氧氨氧化脱氮工艺研究进展. 化工进展, 2014, 33(6): 1589-1595.

[151] 郑兴灿. 城镇污水处理技术升级的挑战与机遇. 给水排水, 2015,7: 1-7.

[152] Cao S, Li B, Du R, et al. Nitrite production in a partial denitrifying upflow sludge bed (USB) reactor equipped with gas automatic circulation (GAC). Water Res, 2016, 90: 309-316.

[153] Jia F, Yang Q, Liu X, et al. Stratification of extracellular polymeric substances (EPS) for aggregated anammox microorganisms. Environmental Science and Technology, 2017, 51(6): 3260.

[154] Zheng M, Liu Y C, Xin J, et al. Ultrasonic treatment enhanced ammonia-oxidizing bacterial (AOB) activity for nitritation process. Environmental Science and Technology, 2015, 50(2): 864.

[155] 王洪臣. 城镇污水处理领域的碳减排. 给水排水, 2010, 36(12): 1-3.

[156] 陈剑, 李玉庆. 天津津南污泥处理工程整体工艺设计与调试. 给水排水, 2016, 42(4): 34-36.

[157] 王凯军, 宫徽, 金正宇. 未来污水处理技术发展方向的思考与探索. 建设科技, 2013(2): 36-38.

[158] Gong H, Wang Z, Zhang X, et al. Organics and nitrogen recovery from sewage via membrane-based pre-concentration combined with ion exchange process. Chemical Engineering Journal, 2017, 311: 13-19.

[159] Zhang W H, Wei C H, Feng C H. Coking wastewater treatment plant as a source of polycyclic aromatic hydrocarbons (PAHs) to the atmosphere and health-risk assessment for workers. Science total environment, 2012, 432: 396-403.

[160] Li W W, Yu H Q, Bruce E. Reuse water pollutants. Nature, 2015, 528(7580): 29-31.

[161] Wei C H, Zhang F Z, Hu Y, Feng C H, Wu H Z. Ozonation in water treatment: the generation, basic properties of ozone and its practical application. Reviews in Chemical Engineering, 2016, 33(1): 49-89.

[162] Zhu S, Wu H. Wang YS, Li Q W, Zhou F, Wei C H. Diverse and distinct bacterial communities associated with complex constitutes of coking wastewater in anaerobic process. Fresenius Environmental Bulletin, 2016, 25: 5130-5137.

[163] Shan W J, Hu Y, Bai Z G, Zheng M M, Wei CH. *In situ* preparation of g-C_3N_4/bismuth-based oxide nanocomposites with enhanced photocatalytic activity. Applied Catalysis B: Environmental, 2016, 188: 1-12.

[164] Liu J, Ou H, Wei C H, Wu H Z, He J Z, Lu Dehua. Novel multistep physical/chemical and biological integrated system for coking wastewater treatment: Technical and economic feasibility. Journal of Water Process Engineering, 2016, 10: 98-103.

[165] Zhu S, Wu H, Wei C, Zhou L, Xie J. Contrasting microbial community composition and function perspective in sections of a full-scale coking wastewater treatment system. Applied Microbiology and Biotechnology, 2016, 100: 949-960.

[166] Yu X B, Xu RH, Wei C H, Wu H Z. Revoval of cyanide compounds from coking wastewater by ferrous sulfate: Improvement of biodegradability. Journal Hazadous Materials, 2016, 302: 468-474.

[167] Deng Z Y, Wei C H. Design of anaerobic fluidized bed bioreactor-dyeing effluents. Chemical Engineering Science, 2016, 139: 273-284.

[168] Xu R H, Ou H S, Yu X B, He R S, Lin C, Wei C H. Spectroscopic characterization of dissolved organic matter in coking wastewater during bio-treatment: full-scale plant study. Water Science and Technology, 2015, 72(8): 1411-1420.

[169] Zhang W H, Wei C H, An G F. Distribution, partition and removal of polycyclic aromatic hydrocarbons (PAHs)during coking wastewater treatment processes. Environmental Science: Processes and impacts, 2015, 17, 975-984.

[170] Zhang F Z, Wei C H, Hu Y, Wu H. Zinc ferrite catalysts for ozonation of queous organic contaminants: phenol and bio-treated coking wastewater. Separation and Purification Technology, 2015, 156: 625-635.

[171] Feng C H, Huang L Q, Yu H, Yi X Y, Wei C H. Simultaneous phenol removal, nitrification and denitrification using microbial fuel cell technology. Water Research, 2015, 76: 160-170.

[172] Yu X B, Wei C H, Wu H Z, Xu R H. Improvement of biodegradability for coking wastewater by selective adsorption of hydrophobic organic pollutants. Separation and Purification Technology, 2015, 151: 23-30.

[173] 刘雷，吴海珍，王鸣，韦朝海.典型表面活性剂对焦化污泥中富集多环芳烃的解吸.环境科学学报，2016，36(9): 3282-3291.

[174] 范丹，廖建波，吴超飞，韦朝海. 焦化废水处理工程运行能耗的单元解析模型——以 OHO 流化床工艺为例. 环境科学学报, 2016, 36(10): 3709-3719.

[175] 韦朝海，廖建波，胡芸. 煤的基本化工过程与污染特征分析. 化工进展, 2016, 34(6): 1875-1883.

[176] 韦朝海，廖建波，刘浔，关清卿. PBDEs 的来源特征、环境分布及污染控制. 环境科学学报, 2015, 35, (10): 1-17.

[177] 蔡英，任曼，彭平安，韦朝海. 焦化废水处理过程中溴代二噁英类化合物的去除途径与生成机理. 科学通报，2013，58(4): 313-320.

[178] Bueno-Montes M, Springael D, Ortega-Calvo J. Effect of a nonionic surfactant on biodegradation of slowly desorbing PAHs in contaminated soils. Environmental Science and Technology, 2011, 45(7): 3019.

[179] Gong Z, Wang X, Tu Y, et al. Polycyclic aromatic hydrocarbon removal from contaminated soils using fatty acid methyl esters. Chemosphere, 2010, 79(2): 138.

[180] Ning X A, Lin M Q, Shen L Z, et al. Levels, composition profiles and risk assessment of polycyclic aromatic hydrocarbons (PAHs) in sludge from ten textile dyeing plants. Environmental Research, 2014, 132: 112-118.

[181] Kim Y M, Park D, Jeon C O, et al. Effect of HRT on the biological pre-denitrification process for the simultaneous removal of toxic pollutants from cokes wastewater. Bioresource Technology, 2008, 99(18): 8824.

[182] 李春杰，朱南文，顾国维. SMSBR 处理焦化废水的污泥特性. 中国给水排水, 2002, 18(2): 18-22.

[183] Vázquez I, Rodríguez J, Marañón E, et al. Simultaneous removal of phenol, ammonium and thiocyanate from coke wastewater by aerobic biodegradation. Journal of Hazardous Materials, 2006, 137(3): 1773-1780.

[184] Wang LC, Hsi HC, Wang YF, et al. Distribution of polybrominated diphenyl ethers(PBDEs)and polybrominated dibenzo-*p*-dioxins and dibenzofurans (PBDD/Fs) in municipal solid waste incinerators. Environmental Pollution, 2010, 158(5): 1595-1602.

[185] Karci A, Arslan-Alaton I, Olmez-Hanci T, et al. Degradation and detoxification of industrially important phenol derivatives in water by direct UV-C photolysis and H_2O_2 /UV-C process: A comparative study. Chemical Engineering Journal, 2013, 224(224): 4-9.

[186] Chen W, Westerhoff P, Leenheer J A, et al. Fluorescence excitation-emission matrix regional integration to quantify spectra for dissolved organic matter. Environmental Science and Technology, 2003, 37(24): 5701.

[187] Yang W, Li X, Pan B, et al. Effective removal of effluent organic matter (EfOM) from bio-treated coking wastewater by a recyclable aminated hyper-cross-linked polymer. Water Research, 2013, 47(13): 4730-4738.

[188] Wang Y, Sun S, Ding G, et al. Electrochemical degradation characteristics of refractory organic pollutants in coking wastewater on multiwall carbon nanotube-modified electrode. Journal of Nanomaterials, 2012,8: 3.

[189] Cheng H Y, et al. Stimulation of oxygen to bioanode for energy recovery from recalcitrant organic matter aniline in microbial fuel cells (MFCs). Water Research, 2015, 81: 72-83.

[190] Liang B, et al. Accelerated reduction of chlorinated nitroaromatic antibiotic chloramphenicol by biocathode. Environmental Science and Technology, 2013, 47(10): 5353.

[191] Liu X W, et al. Anodic Fenton process assisted by a microbial fuel cell for enhanced degradation of organic pollutants. Water Research, 2012, 46(14): 4371.

[192] Wang Y K, et al. Development of a Novel Bioelectrochemical Membrane Reactor for Wastewater Treatment. Environmental Science and Technology, 2011, 45(21): 9256-9261.

[193] Shen J Y, et al. Coupling of a bioelectrochemical system for *p*-nitrophenol removal in an upflow anaerobic sludge blanket reactor. Water Research, 2014, 67(67C): 11.

[194] Huang L, et al. Effect of Set Potential on Hexavalent Chromium Reduction and Electricity Generation from Biocathode Microbial Fuel Cells. Environmental Science and Technology, 2011, 45(11): 5025.

[195] Liang P, et al. Scaling up a Novel Denitrifying Microbial Fuel Cell with an Oxic-Anoxic Two Stage Biocathode. Frontiers of Environmental Science and Engineering, 2013, 7(6): 913.

[196] Du Y, et al. Coupling Interaction of Cathodic Reduction and Microbial Metabolism in Aerobic Biocathode of Microbial Fuel Cell. RSC Advances, 2014, 4(65): 34350.

[197] Cao XX, et al. A new method for water desalination using microbial desalination cells. Environmental Science and Technology, 2009, 43(18): 7148-7152.

[198] Chen X, et al. Optimization of membrane stack configuration in enlarged microbial desalination cells for efficient water desalination. Journal of Power Sources, 2016, 324: 79-85.

[199] Luo HP, PE Jenkins, and Ren ZY. Concurrent desalination and hydrogen generation using microbial electrolysis and desalination cells. Environmental Science and Technology, 2011, 45(1): 340.

[200] Zhang Y, Angelidaki I. Submersible microbial desalination cell for simultaneous ammonia recovery and electricity production from anaerobic reactors containing high levels of ammonia. Bioresource Technology, 2015, 177(177C): 233.

[201] Chen X, et al. Novel Self-driven Microbial Nutrient Recovery Cell with Simultaneous Wastewater Purification. Scientific Reports, 2015, 5: 15744.

[202] Chen S, et al. Development of the microbial electrodialysis and chemical production cell for desalination as well as acid and alkali productions. Environmental Science and Technology, 2012, 46(4): 2467-72.

[203] Xiao K, Ying X U, Liang S, et al. Engineering application of membrane bioreactor for wastewater treatment in China: Current state and future prospect. Frontiers of Environmental Science & Engineering, 2014, 8(6): 805-819.

[204] Sun J, Liang P, Yan X, et al. Reducing aeration energy consumption in a large-scale membrane bioreactor: Process simulation and engineering application. Water Research, 2016, 93: 205.

[205] Zhu Z, Wang R, Li Y. Evaluation of the control strategy for aeration energy reduction in a nutrient removing wastewater treatment plant based on the coupling of ASM1 to an aeration model. Biochemical Engineering Journal, 2017.

[206] 杨敏, 李亚明, 魏源送, 等. 大型再生水厂不同污水处理工艺的能耗比较与节能途径. 环境科学, 2015(6): 2203-2209.

[207] Mei X, Wang Z, Miao Y, et al. Recover energy from domestic wastewater using anaerobic membrane bioreactor: Operating parameters optimization and energy balance analysis. Energy, 2016, 98: 146-154.

[208] Li N, Hu Y, Lu YZ, et al. In-situ biogas sparging enhances the performance of an anaerobic membrane bioreactor (AnMBR) with mesh filter in low-strength wastewater treatment. Applied Microbiology and Biotechnology, 2016, 100(13): 6081.

[209] Wang X, Wang C, Tang C Y, et al. Development of a novel anaerobic membrane bioreactor simultaneously integrating microfiltration and forward osmosis membranes for low-strength wastewater treatment. Journal of Membrane Science, 2017, 527: 1-7.

[210] Meng F, Zhang S, Oh Y, et al. Fouling in membrane bioreactors: An updated review. Water Research, 2017, 114: 151.

[211] Wang Z, Wu Z, Tang S. Characterization of dissolved organic matter in a submerged membrane bioreactor by using three-dimensional excitation and emission matrix fluorescence spectroscopy. Water Research, 2009, 43(6): 1533-40.

[212] Mizaikoff B, Chen W, Liu XY, et al. Probing Membrane Fouling via Infrared Attenuated Total Reflection Mapping Coupled with Multivariate Curve Resolution. Chemphyschem, 2016, 17(3): 443-443.

[213] Wei CH, Huang X, Ben AR, et al. Critical flux and chemical cleaning-in-place during the long-term operation of a pilot-scale submerged membrane bioreactor for municipal wastewater treatment. Water Research, 2011, 45(2): 863-871.

[214] Han X, Wang Z, Chen M, et al. Acute Responses of Microorganisms from MBR in the Presence of NaOCl: Protective Mechanisms of Extracellular Polymeric Substances. Environmental Science and Technology, 2017, 51(6).

[215] Ding A, Liang H, Li G, et al. Impact of aeration shear stress on permeate flux and fouling layer properties in a low pressure membrane bioreactor for the treatment of grey water. Journal of Membrane Science, 2016, 510: 382-390.

[216] Chen M, Zhang X, Wang Z, et al. QAC modified PVDF membranes: Anti-biofouling performance, mechanisms, and effects on microbial communities in an MBR treating municipal wastewater. Water Research, 2017, 120: 256.

[217] Yang Y, Li J, Wang H, et al. An electrocatalytic membrane reactor with self-cleaning function for industrial wastewater treatment. Angewandte Chemie, 2011, 50(9): 2148-2150.

[218] Zhang W, Wei C, Feng C, Ren Y, Hu Y, Bo Y. The occurrence and fate of phenolic compounds in a coking wastewater treatment plant. Water Science and Technology, 2013,68: 433-440.

[219] Zhang W, Wei C, Chai X, He J, Cai Y., Ren M, Yan B, Peng P, Fu J. The behaviors and fate of polycyclic aromatic

hydrocarbons(PAHs)in a coking wastewater treatment plant. Chemosphere, 2012, 88: 174-182.

[220] Zhang W, Wei C, Feng C, Yan B, Diao C, Huang Q, Li N, Peng P, Fu J. Coking wastewater treatment plant as a point source of polycyclic aromatic hydrocarbons(PAHs)to the atmosphere and health-risk assessment for workers. Science of The Total Environment, 2012,43: 396-403.

[221] Zhang W, Feng C, Wei C, Yan B, Wu C, Li N. Identification and characterization of polycyclic aromatic hydrocarbons in coking wastewater sludge. Journal of Separation Science, 2013, 35: 3340-3346.

[222] Zhang W, Wei C, Bo Y, Feng C, Lin C, Zhao G, Yuan M, Wu C, Ren Y, Hu Y. Identification and removal of polycyclic aromatic hydrocarbons in wastewater treatment processes from coke production plants. Environmental Science and Pollution Research, 2013,20: 6418-6432.

[223] Lin C, Zhang W, Yuan M, Feng C, Ren Y, Wei C, Degradation of polycyclic aromatic hydrocarbons in a coking wastewater treatment plant residual by an O_3/ultraviolet fluidized bed reactor. Environmental Science and Pollution Research. 2014, 21: 10329-10338.

[224] Zhang T, Wei C, Feng C, Zhu J. A novel airlift reactor enhanced by funnel internals and hydrodynamics prediction by the CFD method. Bioresource Technology, 2012, 104: 600-607.

[225] Zhu S, Wu H, Wei C, Zhou L, Xie J. Contrasting microbial community composition and function perspective in sections of a full-scale coking wastewater treatment system. Applied Microbiology and Biotechnology, 2016,100: 949-960.

[226] Huang Y, Hou X, Liu S, Ni J. Correspondence analysis of bio-refractory compounds degradation and microbiological community distribution in anaerobic filter for coking wastewater treatment. Chemical Engineering Journal, 2016, 300: 864-872.

[227] Joshi DR, Zhang Y, Gao Y, Liu Y, Yang M. Biotransformation of nitrogen- and sulfur-containing pollutants during coking wastewater treatment: correspondence of performance to microbial community functional structure. Water Research, 2017,121: 338-348.

[228] Joshi DR, Zhang Y, Tian Z, Gao Y, Yang M, Performance and microbial community composition in a long-term sequential anaerobic-aerobic bioreactor operation treating coking wastewater. Applied Microbiology and Biotechnology, 2016, 18: 1-12.

[229] Feng C, Huang L, Yu H, Yi X, Wei C, Simultaneous phenol removal, nitrification and denitrification using microbial fuel cell technology. Water Research, 2015,76: 160-170.

[230] Xie D, Li C, Tang R, Lv Z, Ren Y, Wei C, Feng C, Ion-exchange membrane bioelectrochemical reactor for removal of nitrate in the biological effluent from a coking wastewater treatment plant. Electrochemistry Communications, 2014, 46: 99-102.

[231] Xie D, Yu H, Li C, Ren Y, Wei C, C. Feng, Competitive microbial reduction of perchlorate and nitrate with a cathode directly serving as the electron donor. Electrochimica Acta, 2014, 133: 217-223.

[232] Tang R, Wu D, Chen W, Feng C, Wei C, Biocathode denitrification of coke wastewater effluent from an industrial aeration tank: Effect of long-term adaptation. Biochemical Engineering Journal, 2017,125: 151-160.

[233] 夏四清, 梁郡, 李海翔, 等. 利用氢基质生物膜反应器同步去除多种污染物. 同济大学学报(自然科学版), 2012, 40(6): 876-881.

[234] Zhao HP, Ontiverosvalencia A, Tang Y, et al. Using a two-stage hydrogen-based membrane biofilm reactor(MBfR)to achieve complete perchlorate reduction in the presence of nitrate and sulfate. Environmental Science and Technology, 2013, 47(3): 1565.

[235] Lai CY, Yang X, Tang Y, et al. Nitrate Shaped the Selenate-Reducing Microbial Community in a Hydrogen-Based Biofilm Reactor. Environmental Science and Technology, 2014, 48(6): 3395-3402.

[236] Costa C, Dijkema C, Friedrich M, et al. Denitrification with methane as electron donor in oxygen-limited bioreactors. Applied Microbiology and Biotechnology, 2000, 53(6): 754-62.

[237] Liu J, Sun F, Wang L, et al. Molecular characterization of a microbial consortium involved in methane oxidation coupled to denitrification under micro-aerobic conditions. Microbial Biotechnology, 2014, 7(1): 64.

[238] Osaka T, Ebie Y, Tsuneda S, et al. Identification of the bacterial community involved in methane-dependent denitrification in activated sludge using DNA stable-isotope probing. Fems Microbiology Ecology, 2008, 64(3): 494-506.

[239] Thalasso F, Vallecillo A, García-Encina P, et al. The use of methane as a sole carbon source for wastewater denitrification. Water Research, 1997, 31(96): 55-60.

[240] Zhang H, Wang H, Yang K, et al. Autotrophic denitrification with anaerobic Fe^{2+} oxidation by a novel *Pseudomonas* sp. W1. Water Science and Technology A Journal of the International Association on Water Pollution Research, 2015, 71(7): 1081.

[241] Haroon M F, Hu S, Shi Y, et al. Anaerobic oxidation of methane coupled to nitrate reduction in a novel archaeal lineage. Nature, 2013, 500(7464): 567-70.

[242] Ettwig K F, Butler M K, Le P D, et al. Nitrite-driven anaerobic methane oxidation by oxygenic bacteria. Nature, 2010, 464(7288): 543-8.

[243] Raghoebarsing A A, Pol A, Kt P S, et al. A microbial consortium couples anaerobic methane oxidation to denitrification. Nature, 2006, 440(7086): 918.

[244] Luo Y H, Chen R, Wen L L, et al. Complete perchlorate reduction using methane as the sole electron donor and carbon source. Environmental Science and Technology, 2015, 49(4): 2341-9.

[245] Lai C Y, Wen LL, Shi L, et al. Selenate and Nitrate Bio-reductions Using Methane as the Electron Donor in a Membrane Biofilm Reactor. Environmental Science and Technology, 2016, 50(18): 10179.

[246] Lai C Y, Zhong L, Zhang Y, et al. Bioreduction of Chromate in a Methane-Based Membrane Biofilm Reactor. Environmental Science and Technology, 2016, 50(11): 5832.

[247] Lai C Y, Wen L L, Zhang Y, et al. Autotrophic antimonate bio-reduction using hydrogen as the electron donor. Water Research, 2016, 88: 467-474.

[248] Gao D W, Hu Q, Yao C, Ren N Q and Wu W M. Integrated anaerobic fluidized-bed membrane bioreactor for domestic wastewater treatment. Chemical Engineering Journal, 2014, 240: 362-368.

[249] Wang J, et al. Correlating the hydrodynamics of fluidized granular activated carbon(GAC)with membrane-fouling mitigation. Journal of Membrane Science, 2016, 510: 38-49.

[250] Liu J, Jia X, Gao B, Bo L and Wang L. Membrane fouling behavior in anaerobic baffled membrane bioreactor under static operating condition. Bioresource and Technology, 2016, 214: 582-588.

[251] Kim J, et al. Membrane fouling control using a rotary disk in a submerged anaerobic membrane sponge bioreactor. Bioresource and Technology, 2014, 172: 321-327.

[252] Gu Y, et al. Development of anaerobic osmotic membrane bioreactor for low-strength wastewater treatment at mesophilic condition. Journal of Membrane Science, 2015, 490: 197-208.

[253] Werner C M, et al. Graphene-coated hollow fiber membrane as the cathode in anaerobic electrochemical membrane bioreactors- effect of configuration and applied voltage on performance and membrane fouling. Environmental Science and Technology, 2016, 50: 4439-4447.

[254] Dong Q, Parker W and Dagnew M. Impact of $FeCl_3$ dosing on AnMBR treatment of municipal wastewater. Water Research, 2015, 80: 281-293.

[255] Cookney J, Cartmell E, Jefferson B and McAdam E J. Recovery of methane from anaerobic process effluent using poly-di-methyl-siloxane membrane contactors. Water Science and Technology, 2012, 65: 604-610.

[256] Wang Y, Wang D, Yang Q, Zeng G and Li X. Wastewater Opportunities for Denitrifying Anaerobic Methane Oxidation. Trends Biotechnology, 2017.

[257] Li W W, Yu H Q. From wastewater to bioenergy and biochemicals via two-stage bioconversion processes: A future paradigm. Biotechnology Advance, 2011, 29: 972-982.

[258] Liu D, Zhang L, Chen, S. Buisman, C , ter Heijne A. Bioelectrochemical enhancement of methane production in low temperature anaerobic digestion at 10℃. Water Research, 2016, 99: 281-287.

[259] Desloover J, et al. Electrochemical Nutrient Recovery Enables Ammonia Toxicity Control and Biogas Desulfurization in Anaerobic Digestion. Environmental Science and Technology, 2015, 49: 948-955.

[260] Wang D, et al. How Does Poly(hydroxyalkanoate) Affect Methane Production from the Anaerobic Digestion of Waste-Activated Sludge? Environmental Science and Technology, 2015, 49: 12253-12262.

[261] Luo C, Lü F, Shao L, He P. Application of eco-compatible biochar in anaerobic digestion to relieve acid stress and promote

the selective colonization of functional microbes. Water Research, 2015, 68: 710-718.

[262] Zhen G, Lu X, Li Y Y, Liu Y, Zhao Y. Influence of zero valent scrap iron (ZVSI) supply on methane production from waste activated sludge. Chemical Engineering Journal, 2015, 263: 461-470.

[263] Paranychianakis N V, Salgot M, Snyder S A, Angelakis A N. Water reuse in EU states: necessity for uniform criteria to mitigate human and environmental risks. Critical Reviews in Environmental Science and Technology, 2015, 45: 1409-1468.

[264] Zou Y, Wang X, Ai Y, Liu Y, Li J, Ji Y, Wang X. Coagulation Behavior of Graphene Oxide on Nanocrystallined Mg/Al Layered Double Hydroxides: Batch Experimental and Theoretical Calculation Study. Environmental Science and Technology, 2016, 50: 3658-3667.

[265] Zhao Y X, Phuntsho S, Gao B Y, Huang X, Qi Q B, Yue Q Y, Wang Y, Kin J H, Shon H K. Preparation and characterization of novel polytitanium tetrachloride coagulant for water purification. Environmental Science and Technology, 2013, 47: 12966-12975.

[266] Wang X, Li M, Song X, Chen Z, Wu B, Zhang S. Preparation and Evaluation of Titanium-Based Xerogel as a Promising Coagulant for Water/Wastewater Treatment. Environmental Science and Technology, 2016, 50: 9619-9626.

[267] Li N, Sheng G P, Lu Y Z, Zeng R J, Yu H Q. Removal of antibiotic resistance genes from wastewater treatment plant effluent by coagulation. Water Research, 2017, 111: 204-212.

[268] Zhang Y, Pan B, Shan C, Gao X. Enhanced Phosphate Removal by Nanosized Hydrated La(III) Oxide Confined in Cross-linked Polystyrene Networks. Environmental Science and Technology, 2016, 50: 1447-1454.

[269] Alsbaiee A, Smith B J, Xiao L, Ling Y, Helbling D E, Dichtel W R. Rapid removal of organic micropollutants from water by a porous β-cyclodextrin polymer. Nature, 2016, 529: 7585.

[270] Han Y, Xu Z, Gao C. Ultrathin Graphene Nanofiltration Membrane for Water Purification. Advanced Functional Materials, 2013, 23: 3693-3700.

[271] Shi Z, Zhang W, Zhang F, Liu X, Wang D, Jin J, Jiang L. Ultrafast separation of emulsified oil/water mixtures by ultrathin free-standing single-walled carbon nanotube network films. Advanced Materials, 2013, 25: 2422-2427.

[272] Ma J, Guo X, Ying Y, Liu D, Zhong C. Composite ultrafiltration membrane tailored by MOF@GO with highly improved water purification performance. Chemical Engineering Journal, 2016, 313: 890-898.

[273] Zhang S, Wang P, Fu X, Chung TS. Sustainable water recovery from oily wastewater via forward osmosis-membrane distillation (FO-MD). Water Research, 2014, 52: 112.

[274] Feng M, Wang X, Chen J, Qu R, Sui Y, Cizmas L, Sharma V K. Degradation of fluoroquinolone antibiotics by ferrate(VI): Effects of water constituents and oxidized products. Water Research, 2016, 103: 48-57.

[275] Dong S, Lu J, Plewa M J, Nguyen T H. Comparative Mammalian Cell Cytotoxicity of Wastewaters for Agricultural Reuse after Ozonation. Environmental Science and Technology, 2016, 50: 11752-11759.

[276] Li M, Qiang Z, Wang C, Bolton J R, Blatchley E R. Experimental assessment of photon fluence rate distributions in a medium-pressure UV photoreactor. Environmental Science and Technology, 2017, 51: 3453-3460.

[277] Zheng J, Ma J, Wang Z, Xu S, Waite T D, Wu Z. Contaminant Removal from Source Waters Using Cathodic Electrochemical Membrane Filtration: Mechanisms and Implications. Environmental Science and Technology, 2017, 51: 2757-2765.

[278] Sudesh K, Abe H, Doi Y. Synthesis, structure and properties of polyhydroxyalkanoates: biological polyesters. Progress in Polymer Science, 2000, 25: 1503-1555.

[279] Albuquerque M G E, Eiroa M, Torres C, et al. Strategies for the development of a side stream process for polyhydroxyalkanoate (PHA) production from sugar cane molasses. Journal of Biotechnology, 2007, 130: 411-21.

[280] Beccari M, Bertin L, Dionisi D, et al. Exploiting olive oil mill effluents as a renewable resource for production of biodegradable polymers through a combined anaerobic-aerobic process. Journal of Chemical Technology and Biotechnology Biotechnology, 2010, 84: 901-908.

[281] Cai M, Hong C, Zhao Q, et al. Optimal production of polyhydroxyalkanoates (PHA) in activated sludge fed by volatile fatty acids (VFAs) generated from alkaline excess sludge fermentation. Bioresource Technology, 2009, 100: 1399-1405.

[282] Hao J, Wang H. Volatile fatty acids productions by mesophilic and thermophilic sludge fermentation: Biological responses to fermentation temperature. Bioresource Technology, 2015, 175: 367.

[283] Yuan H, Chen Y, Zhang H, et al. Improved Bioproduction of Short-Chain Fatty Acids (SCFAs) from Excess Sludge under Alkaline Conditions. Environmental Science and Technology, 2006, 40: 2025.

[284] Chen Z, Guo Z, Wen Q, et al. A new method for polyhydroxyalkanoate (PHA) accumulating bacteria selection under physical selective pressure. International Journal of Biological Macromolecules, 2015, 72: 1329.

[285] Jia Q, Xiong H, Wang H, et al. Production of polyhydroxyalkanoates (PHA) by bacterial consortium from excess sludge fermentation liquid at laboratory and pilot scales. Bioresource Technology, 2014, 171: 159-67.

[286] Tamisa J, Lužkov K, Jiang Y, et al. Enrichment of Plasticicumulans acidivorans at pilot-scale for PHA production on industrial wastewater. Journal of Biotechnology, 2014, 192: 161-169.

[287] Jia Q, Wang H, Wang X. Dynamic synthesis of polyhydroxyalkanoates by bacterial consortium from simulated excess sludge fermentation liquid. Bioresource Technology, 2013, 140: 328.

[288] Chen Y, Li X, Zheng X, et al. Enhancement of propionic acid fraction in volatile fatty acids produced from sludge fermentation by the use of food waste and *Propionibacterium acidipropionici*. Water Research, 2013, 47: 615.

[289] Jiang Y, Chen Y, Zheng X. Efficient Polyhydroxyalkanoates Production from a Waste-Activated Sludge Alkaline Fermentation Liquid by Activated Sludge Submitted to the Aerobic Feeding and Discharge Process. Environmental Science and Technology, 2009, 43: 7734-7741.

[290] Chen H, Meng H, Nie Z, et al. Polyhydroxyalkanoate production from fermented volatile fatty acids: Effect of pH and feeding regimes. Bioresource Technology, 2013, 128: 533.

[291] Valentino F, Beccari M, Fraraccio S, et al. Feed frequency in a sequencing batch reactor strongly affects the production of polyhydroxyalkanoates (PHAs) from volatile fatty acids. New Biotechnology, 2014, 31: 264-275.

[292] Fernúndez-Dacosta C, Posada J A, Kleerebezem R, et al. Microbial community-based polyhydroxyalkanoates (PHAs) production from wastewater: Techno-economic analysis and ex-ante environmental assessment. Bioresource Technology, 2015, 185: 368-377.

作者：贾金平[1]，宋卫华[2]，关小红[3]，朱小萤[4]，赵华章[5]，邓述波[6]，强志民[7]，胡宝兰[4]，王亚宜[3]，黄　霞[6]，韦朝海[8]，梁　鹏[6]，孟凡刚[9]，冯春华[8]，赵和平[4]，李文卫[10]，盛国平[10]，王　慧[6]，俞汉青[10]

[1]上海交通大学环境科学与工程学院，[2]复旦大学环境科学与工程系，[3]同济大学环境科学与工程学院，[4]浙江大学环境与资源学院，[5]北京大学环境科学与工程学院，[6]清华大学环境学院，[7]中国科学院生态环境研究中心，[8]华南理工大学环境学院，[9]中山大学环境学院，[10]中国科学技术大学化学与材料科学学院

第18章　环境电化学

本章导读

本章首先就近年来电化学方法在环境污染物分析中的应用进行了综述。其次是对电化学技术包括电化学氧化、电化学絮凝、电化学浮选等在污染控制中的应用进行了介绍。特别是对电催化氧化法进行较为详细介绍。最后对光、电催化氧化这两种不同能量转化形式、催化特点鲜明的高级氧化处理技术进行较综述。二者可通过不同的途径产生强氧化性的·OH 自由基等活性物种，氧化能力强，反应条件温和。因此，将这两种技术优势互补、有机组合成一体的光电催化降解技术研究在近年来十分活跃，在环境污染物高效降解处理中具有很好的应用发展前景。

关键词

电化学分析，电化学氧化，电化学絮凝，光、电催化氧化技术

1　电化学方法在环境污染物检测中的应用

随着经济发展、技术进步和人口增长，自然资源和自然环境受到日益严重的破坏，排放到环境中的污染物种类越来越多，环境污染问题已经逐渐成为危及人们健康、经济发展的瓶颈因素。本章就近年来电化学方法在环境污染物分析中的应用进行了综述。

1.1　电化学检测重金属

重金属污染主要来源于工业生产品及其副产品、化肥和其他化学品等，对环境以及人体健康构成了严重危害。重金属离子在环境中不会被分解，其生物不可降解性决定了一旦重金属离子释放到环境中，几十年或者几个世纪都会存在，并会在生物体内聚集累积，从而造成生物的神经、免疫、生殖等多个系统紊乱[1]。众多的重金属如铅（Pb）、镉（Cd）、汞（Hg）、铬（Cr）以及非金属砷（As）等元素具有较高的毒性，即使很小的剂量也会引起严重的环境问题，对人体健康造成很大伤害[2]。

世界卫生组织（WHO）、联合国粮农组织（FAO）、美国环境保护署（EPA）、中国疾病预防控制中心（CDC）和欧盟（EU）都将重金属离子设定为优先检测物质，并对其在水体中的浓度进行了限定，规定必须满足环境质量标准(EQS)[3]。这就要求对不同复杂环境，例如生物样品、自然水体、废水、空气以及土壤中的痕量重金属离子的检测必须高度敏感。电化学检测技术具有经济、方便、可靠、适用于现场操作等特点。在对重金属离子的检测中，通常通过修饰电极来提高检测灵敏度，降低检出限。

铅主要通过食物、饮用水、空气等方式影响着人类健康。在对铅的检测中，交流阻抗法（EIS）是一种在传感器上施加小幅度正弦电压信号测量电流的相移和振幅的调制技术。利用该法结合聚乙烯对苯二甲酸乙二醇酯膜传感器可检测 Pb^{2+}的范围为 50 μmol/L~1 mmol/L，检出限为 1 nmol/L[4]。有别于交流阻抗法，溶出伏安法是使被测的物质，在待测离子极谱分析产生极限电流的电位下电解一定的时间，然后改变电极电位使富集在该电极上的物质重新溶出，根据溶出过程中所得到的伏安曲线来进行定量分析。该法对痕量物质的分析有很高的灵敏度，较好的精密度、准确度，仪器设备简单易得。利用方波阳极溶出伏安法（SWASV）及 Bi-C 复合电极，由于铋纳米粒子较好分散在多孔的碳基材料中提供了高度活跃的还原区，可以检测 1~100 ppb 范围内的铅，Pb^{2+}检出浓度和峰电流保持较好的线性关系，检出限为 0.65 ppb[5]。石墨烯是一种由碳原子以 sp^2 杂化方式形成的蜂窝状平面薄膜，具有较好的导电导热性能。而利用同样的

方法，Nafion/离子液体/石墨烯复合修饰丝网印刷碳电极，可以在 120 s 内实现对 Pb^{2+}的有效检测，检测限低至 0.08 ng/L，可以更快更准确地检测出更低浓度的离子[6]。镉污染随着媒体的披露已经越来越引起人们的关注。已报道过的对镉的检测方法主要包括无梯度能斯特平衡溶出法、溶出伏安法以及电导分析法。无梯度能斯特平衡溶出-溶出计时电位法是通过重金属离子累计在汞齐中，通过能斯特扩散平衡和二次氧化过程进行定量。该法通过和溶出计时电位法的结合，可检测出 25~100 nmol/L 的 Cd^{2+}，检出限为 2.9 nmol/L，但该法存在辨别干扰能力不强以及出现空白测量等问题[7]。溶出伏安法在 Cd^{2+}的测定中，利用氧化石墨烯/多壁碳纳米管膜，可在 180 s 内对 0.5~30 μg/L 范围内的 Cd^{2+}进行检测，最低检出限为 0.1 μg/L，检测表现及便捷程度要好于无梯度能斯特平衡溶出法[8]。电导分析法是以测量溶液中的电导值为基础而建立起来的分析方法。由于溶液电导与溶液中离子的总量有关，因此，电导分析法的选择性不高，只能测定溶液中离子总量。但在 Tekaya 等的实验中，利用碱性磷酸酶结合该法将 Cd^{2+}检出限降至 10^{-20} mol/L，这也为电化学分析与生物传感相结合提供了思路[9]。

汞盐对人体有剧毒，口服、吸入或接触会导致脑损伤。电流分析法、电致化学发光分析等方法已经被论证可以对 Hg^{2+}进行有效检测，且通过与酶、DNA 等生物技术结合，可有效检测出痕量 Hg^{2+}。电流分析法是在工作电极和辅助电极之间施加一定的电位差，利用待测物在电极表面上进行的氧化还原反应，记录与待测物有关的电流信号而进行的定量分析。Xuan 等[10]通过电流分析法对 Hg^{2+}的检测中，利用核酸外切酶 III 可被 Hg^{2+}激发原理，数分钟内可检测出最低 0.2 nmol/L 的 Hg^{2+}。电致化学发光分析法是指通过电化学方法在电极表面产生一些特殊物质与体系中其他组分之间通过电子传递形成激发态，由激发态返回到基态产生发光现象。在金电极表面固定 Y 型 DNA 的实验中，1~5nmol/L 范围内的 Hg^{2+}可以被快速检测出，最低检出限为 0.094 nmol/L[11]。因此，电化学生物传感器比传统的纯电化学检测具有更高的灵敏度、更好的选择性、经济和方便操作等特点。

1.2 电化学检测有机物

伴随着现代工农业的迅速发展，农药、化肥及工业废物等有机污染物不断进入环境中，它们不仅污染环境、破坏生态平衡，而且还可以通过皮肤接触、饮食等途径进入人体引起过敏、癌症、先天缺陷、生殖器官和免疫系统的损伤，危害人体健康。以饮用水为例，我国的水环境污染的主要有毒有害有机污染物为：苯、氯苯、氯苯酚类、卤甲烷类、硝基苯类、硝基苯酚类、邻苯二甲酸酯（钛酸酯）类、多环芳烃类、DDT、DDE、六氯环己烷类等。为了食品安全和环境监测，准确、可靠、快速地检测有机污染物十分重要。文献报道检测有机污染物的方法很多，主要包括高效液相色谱法、紫外检测法、气相色谱质谱联用法、荧光检测法、离子交换色谱法等传统分析方法，其中高效液相色谱法应用最为广泛。但是，这些传统的检测方法分析检测时间长，仪器大型且操作复杂、检测成本高，对检测环境要求苛刻不能实现现场检测，限制它们实际应用的能力。近年来，电化学检测的方法受到越来越多的关注。与传统的检测方法相比，电化学检测法有以下优点：检测前不需要繁杂的准备工作，检测中操作简单，信号响应快，需要样品量小节约实验成本，电化学传感器易于实现小型化，能做到现场检测，具有实际应用价值[12]。

五氯苯酚上的羟基是比较活泼的电活性基团，在一定条件下可以被氧化，五氯苯酚上的氯也可以被还原，但其还原电位比较负，不适合用于检测。近来，电化学检测五氯苯酚的方法已经有不少人进行了研究。电位型电化学传感主要是离子选择电极法检测五氯苯酚。Yang 等[13]制备了基于 ZnSe 量子点修饰的多壁碳纳米管电极的电化学传感器，并用于鱼肉中残留五氯苯酚的测定，电极的线性响应范围为 8.0×10^{-8}~4.0×10^{-6} mol/L，检测限为 2.0×10^{-9} mol/L。Xia 等[14]制备了基于 C_3N_4/石墨烯电极的电化学传感器，并用于水体中五氯苯酚的测定，电极的线性响应范围为 1.0×10^{-11}~1.0×10^{-7} mol/L，检测限为 1.0×10^{-11} mol/L。电流型传感在五氯苯酚的分析检测方面还是有很大发展的前景及优势，主要是从电极材料和电化学分析方法的选择上着手，可以进一步改进其检测的重现性、灵敏性，降低检测下限。

硝基苯上的硝基是非常活泼的基团，容易被还原，甚至有工业上利用硝基的电化学还原性进行芳香族硝基化合物的电化学还原。有不少文献报道利用电化学还原方法处理含硝基苯的废水。近来，科学家们研究了硝基苯在多种不同构造的电极上的电化学氧化还原行为并利用其进行分析检测。Xu 等[15]制备了3,4-乙撑二氧噻吩和碳纳米管修饰的碳糊电极选择性电化学传感器，其良好的导电性、大的比表面积及对硝基苯优异的电化学还原催化活性，使其成为灵敏性且选择性优异的传感器，该电极对硝基苯的检测浓度范围为 0.25~0.43 μmol/L，检测限为 83 μmol/L。Ganesan 等[16]制备了基于金纳米颗粒修饰的介孔硅纳米球玻碳电极电化学传感器，该电极对硝基苯的检测浓度范围为 0.1~2.5 μmol/L，检测限为 15 nmol/L。由上述可知，电化学分析检测硝基苯固然有其优势，但在检测灵敏性、响应时间等方面还有待提高，这个可通过选择具有更好的吸附特性、有利于电子转移的电极材料来实现。

多环芳烃本身的电化学活性是很弱的，可以被还原加氢，但是还原电位通常非常负，不适宜用于检测，因此很多研究人员利用 PAHs 与有机体的相互作用原理提出用生物电化学传感来检测 PAHs。Zhu 等[17]制备了多壁碳纳米管/石墨烯纳米带的核壳异质结电极，用于电化学检测 1-氨基吡，其线性范围为 8.0~500.0 nmol/L，检测限为 1.5 nmol/L。Xia 等[18]制备了茜素红 S 功能化的介孔硅玻碳电极电化学传感器，用于检测蒽，其线性范围为 1.0 pmol/L~10.0 nmol/L，检测限为 0.5 pmol/L。

1.3　电化学检测氮氧化物

近年来随着科技和经济的快速发展，人类的物质环境得到了极大提高，但与我们生存息息相关的自然环境却在自然资源的不断开发中每况愈下，尤以大气污染最为严重。大气污染物主要来自于火力发电厂、有色及黑色金属冶炼厂、机动车尾气以及采暖燃烧的锅炉。大气污染物主要包括二氧化硫、烟尘、氮氧化物等，由其引起的酸雨、光化学烟雾、全球变暖、臭氧层空洞等诸多问题，在给人们身体健康带来危害的同时，也增加了地球负担[19]。因此研究对大气污染物的精确测量技术势在必行，尤其是针对 SO_2^-、NO_x 及悬浮颗粒物等大气污染物排放的在线监测就成为必需的技术基础，只有准确快速的测量，才能应用相关污染物控制技术进行必要的调整，使污染物的排放浓度降到最低，同时也便于环保部门的实时监督。目前，针对大气污染物的检测方法主要有气相色谱技术（GC）、气质联用技术（GC-MS）、传统化学发光法、非分散紫/红外光谱技术（NDUR/NDIR）、傅里叶变换红外吸收光谱技术（FTIR）、差分吸收激光雷达技术（DIAL）等[20-24]。然而这些方法普遍存在仪器价格昂贵，容易受环境温度和湿度变化的影响，需要专业的人员操作，且检测速度较慢，达不到实时连续检测气体的要求等缺点。电化学方法的原理是利用不同物质的电化学性质来测定待测污染气体的浓度，这种方法可以实现快速的测量污染气体浓度，具有高的灵敏度、较高的选择性、响应范围较宽以及较短的响应时间等优点，因而在环境分析中得到广泛的关注和应用。

电化学方法用于大气污染物的检测主要是利用物质的电化学性质测定其含量，实现对 SO_2、NO_x 和 CO_2 等气体浓度的快速测量。该方法主要包括电位分析法、电导分析法、库仑分析法，以及测量电解过程的以电流-电压曲线为基础的伏安法。

电流型传感器可以用来检测 NO_2，这种传感器的电极主要是参比电极（Pt）和敏感电极（金属氧化物），在两个电极（敏感电极和参比电极）之间施加一个恒定的电压值，促使两个电极之间发生氧化还原反应，从而产生一定的电流值，通过测量电流值的变化来检测 NO_2 浓度。用氧化物做敏感电极，以氧离子导体为传感器基片，不需通入参比气，其氧化物电极可于长时间高温工作下拥有优异的化学稳定性。Diao 等[25]制备研究的铬钨氧化物传感器对 NO_2 的检测范围为 20~300 ppm，发现当 W/Cr 氧化物的比例为 3:2 时，对 100 ppm 的 NO_2 的响应电压为 51.6 mV，响应时间小于 20 s。同时该传感器在温度为 1000℃下还能够正常稳定工作。Prakash 等[26]利用铜纳米颗粒（CuNP）/单壁碳纳米管（SWCNT）/聚吡咯（PPy）纳米复合材料修饰 Pt 电极对 NO_2 进行检测，发现经过改性后的增强电子在 Pt 电极和 CuNP 上的转换，更加有利于 NO_2 的测定，在 NO_2 浓度为 0.7~2000 μmol/L 范围内，线性相关系数为 0.9946，灵敏度达到（0.22 ±0.002）μA/(μmol/L·cm^2)，检出限达到 0.7 μmol/L。Yan 等[27]制备的氧化锌负载二氧化硅球的纳米结构 ZnO/PS 气体传感器用于检测 NO_2 气体也有非常好的效果，检出限达到 100 ppb，响应时间为 20~90 s，恢复时间为 180~120 s。目前已经报道的金属氧化敏感材料较多，如 $La_{0.9}Sr_{0.1}Ga_{0.8}Mg_{0.2}O_{3-\delta}$、$BaCe_{0.9}Nd_{0.1}O_{3-\delta}$、$Zr_{0.95}Mn_{0.05}O_{3-\delta}$、$BaZr_{0.8}Y_{0.2}O_{3-\delta}$、$La_{0.8}Sr_{0.2}Ga_{0.8}Mg_{0.2}O_{3-\delta}$[28]。

对 NO 的电化学检测也是基于其氧化还原反应而进行的，其电化学行为高度依赖于电极和电解液的有效界面。所以，NO 在导电聚合物[8]、贵金属纳米粒子[29]、碳纳米管[30]和氧化石墨烯[31]等新颖界面上的反应被文献报道。Liu 等[32]制备了三维立体超微金电极用于检测 NO，金的电化学有效表面积增加了 22.9 倍。这种多级纳米孔状超微电极对 NO 的检测显示出优良性能和高的稳定性。

2　电催化氧化处理废水研究进展

随着人口的高速增长，日益严峻的水体污染与不断增长的水资源需求促使我们不断研究开发更加高效、可靠的水污染治理技术。电化学技术由于操作简单、易于降解有毒有害污染物而被应用于污水处理中。电化学技术包括电化学氧化、电化学絮凝、电化学浮选等，其中，电催化氧化法最受关注[33]。电催化氧化法是指水体中污染物在电极表面发生直接或间接氧化的过程。直接氧化法是指污染物直接在阳极表面失去电子被氧化；间接氧化法是指污染物被阳极产生的强氧化性物质（如·OH、氯气、次氯酸、过氧化氢等）所氧化的过程[34,35]。

2.1　电催化氧化法的特点

（1）环境兼容性高。在电催化氧化过程中，氧化还原试剂是高效、清洁的电子，这是一种基本上对环境无污染的绿色技术。

（2）能量利用率高。由于电化学过程不受卡诺循环的限制，可以提高能源的利用率。同时，可以通过控制电极电位、合理设计电极与电解池，减少能量损失。

（3）经济性。电催化氧化法单位体积负荷很高，因此设备体积小，占地面积少。

（4）可控性好。电催化氧化过程中电流与电压容易测定，便于实现自动控制。

（5）多功能性。电催化氧化过程具有直接或间接氧化与还原、相分离、浓缩与稀释、生物杀伤等功能。

2.2　电催化氧化法的应用瓶颈及其解决方法

2.2.1　应用瓶颈及其分析

采用电催化氧化法处理废水最大的应用瓶颈在于运行成本高，原因有以下几点：

（1）电催化氧化法处理废水不可避免地需要消耗很多电能，而电能是最高品味的能源，这就导致电催化氧化法相较于生物法、物理法和化学法等方法的运行成本更高。

（2）在处理实际废水时，如果废水的污染物浓度[12]比较低，将会不利于传质，需要消耗更多电能，直接导致运行成本的上升。

（3）电催化氧化法处理废水主要是利用污染物在电极表面发生氧化还原反应以实现污染物的降解，但在实际运用过程中，阴阳极还会发生析氢和析氧反应，浪费了一部分的电能，增加了运行成本。

（4）电催化氧化法处理废水最关键的因素在于电极，不同的电极针对不同的污染物具有不同的活性和选择性，选择合适的电极将污染物高效地、选择性地转化为目标产物是电化学法的研究重点。但是，有些电极材料虽然活性和选择性都很高，但是稳定性较差，这就导致在实际废水处理过程中需要定期更换电极，增加了运行后期成本。

2.2.2　解决方法

2.2.2.1　*改进电解反应槽的结构*

通过改进电解反应槽结构来改善传质是降低运行成本的方法之一。常见的改进电解槽结构的方法包括以下几个方面：流化床电化学反应器、旋转电极、小极距大流量、双极氧化池和电化学转盘等。

1）三维电极反应器

传统电极的形态一般是二维平板，这种形态的电极比表面积较小，电流效率较低。为改善二维电极的缺陷，提出了三维电极反应器，又称流化床电化学反应器、流化床电极。三维反应器是在传统电解反应槽的电极间装填新的工作电极，新工作电极由颗粒状或碎屑状的材料构成，如活性炭、碳气凝胶等。当反应器通电后，电化学反应可以在新工作电极表面发生。与传统二维电解反应槽相比，三维电极反应器，可明显提高电流效率和处理效果。近几年，有很多研究采用三维电极反应器处理不同的工业废水，如造纸废水[36]、染料废水[37-39]、炼油废水[40-42]、含酚废水[43,44]等。Wang Yan 课题组采用 Sn-Sb-Ce 修饰颗粒活性炭构建的三维电极处理苯酚，反应 2 h 后苯酚去除率达到 89%[44]。Zhang Yimei 课题组采用氮掺杂石墨烯气凝胶作为新型颗粒三维电极处理双酚 A，反应 30 min 后可以去除 85%的双酚 A[43]。

2）管式电反应器

管式电反应器又称柱塞流电化学反应器，是一种呈管状、长径比很大的连续操作反应器，与常规的板式反应器相比，电极间距很小，电场分布均匀，因而降低了电解电压，从而减小了能耗，而且传质条件好，被广泛运用于有机废水的处理，如：染料废水[45]、抗癌药物废水[46]、苯酚废水[47]等。Saravanathamizhan 等[45]研究了管式电化学反应器降解伊文斯蓝的效果，反应 90 min 后去除率可达 60%。Han Weiqing 等[46]采用新型管式电反应器处理抗癌药物废水中的 5-氟-2-甲氧基嘧啶，反应 3 h 后可以去除率达 100%。马锐军[47]采用多级旋转电极电化学反应器处理模拟苯酚废水前后的可生化性，结果表明，反应 120 min 后废水的可生化性得到了大幅度的提高。

3）双室反应器

双室反应器可以避免污染物在阳极被氧化后又扩散到阴极被还原或者在阴极被还原后又扩散到阳极被氧化的现象，减少了电能的内耗，而且在阴极室还可进行氧气的二电子还原产生 H_2O_2，使阴极也具有氧化能力，产生双极氧化效果。雷明阳等[48]分别采用单室和双室反应器处理垃圾渗滤液，结果表明在相同反应条件下，双室反应器可以节约 25%的能耗。薛建军等比较了以铁电极为阴极，反应 2.5 h 后，双室反应器中硝酸盐氮的去除率达到了 99.8%，而单室反应器中硝酸盐氮的去除率仅为 84.3%[49]。

4）电化学转盘

电化学转盘的外观类似于生物转盘，以一定的转速旋转，旋转时转盘的一半在溶液中，另一半在空气中，如图 18-1 所示。这种反应器可以提高传递速度，减小扩散层厚度，改善低浓度下的传质。此外，电化学转盘有利于空气中的氧气向溶液中传递，在阴极被还原的 H_2O_2 进一步提高了电催化氧化效率。贾金平课题组采用电化学转盘处理模拟染料废水 60 min 后脱色率可达 99.5%，而相同条件下传统电化学反应器的脱色率只有 55%[50]。

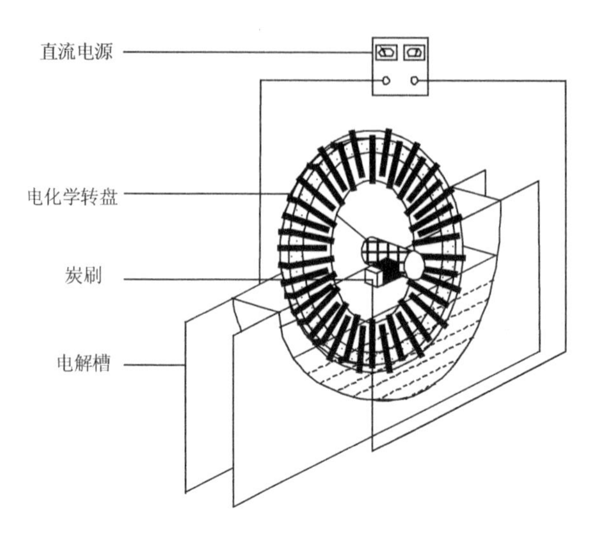

图 18-1　电化学转盘反应器装置图[50]

2.2.2.2　采用电位窗口宽、催化活性高的电极

选择电位窗口更宽、催化活性高的电极可以有效提高电催化效率。近年来研究的电位窗口宽、催化活性高的电极主要有：

（1）BDD（boron-doped diamond，掺硼金刚石）电极以 Si、Ti、Nb、Ta 和 Mo 等材料作基底，在其表面生长金刚石薄膜，并通过掺入不同浓度的硼改善金刚石薄膜的导电性能。析氧电位可达 2.3 V，析氢电位可达–1.25V（参比电极为 SHE），电位窗口超过 3V。BDD 电极在电化学反应中具有很好的化学惰性和抗腐蚀能力，电流效率高，氧化能力强等优点。林海波课题组在钛基底上制备了一种三维网状 BDD 电极，该电极具有比表面高、传质性能佳、疏水性好、电化学氧化性能好等优点[51]。

（2）DSA 阳极（dimensional sustainable anode，尺寸稳定阳极）是由混合氧化物和钛基底构成，氧化物包括 RuO_2、IrO_2、MnO_2、PbO_2、SnO_2 和 TiO_2 等[52]。与传统的石墨、铂、PbO_2 等电极相比，具有耐腐蚀、电催化活性高、成本低廉等优点[53]。

（3）ACF（active carbon fiber，活性炭纤维）电极由于其巨大的比表面积而具有很强的吸附特性，可以将污染物吸附在电极周围，提高处理效率，适合处理浓度较低的有机废水[54]。ACF 电极可以被当作阴极用于降解氰化物[55]、高氯酸[56]、染料[57]等污染物，处理效果良好。

2.2.2.3　与其他方法耦合

单一的电催化氧化方法存在能耗高的缺陷，而将电催化法和其他方法耦合可以达到降耗提效的目的。

1）光电耦合

光电耦合是将半导体光催化氧化法和电催化氧化法结合以实现协同效应。王后锦等[58]在纯钛铂基底上制备了 TiO_2 纳米管阵列并应用于降解苯酚，结果表明采用光电催化法降解苯酚的去除率可达 86.7%，在相同实验条件下电催化不能降解苯酚。Hurwitz 等[59]比较了电催化法和光电催化法处理市政污水反渗透浓缩液中溶解性有机污染物的效果，结果表明电催化法仅能降解 35%的有机物，而光电催化法可以降解 80%的溶解性有机污染物。赵国华课题组采用两步法处理甲基蓝废水，先采用电催化法对染料废水进行脱色，再采用光催化法深度处理废水，光电催化的协同效应大大降低了能耗，提升了处理效率[60]。贾金平课题组制备了一种尖锥型 TiO_2/Ti 电极，并将其应用于转盘光电催化反应器处理罗丹明 B 染料废水，150 min 后色度去除率达 100%，COD 去除率达 87%[61]。

（2）声电耦合

声电耦合是将超声空化氧化技术与电催化氧化技术结合，提高污染物降解效率，是一种很有前景的深度氧化技术。Maria 等[62]研究了超声空化对电催化降解咖啡因的影响，结果表明，反应 100 min 后声电催化法比电催化法的降解效率高 13%。Sáez 等[63]比较了电催化降解和超声辅助电催化降解氯磺隆的效果，结果表明超声电催化辅助可以很大程度提高矿化率。贾金平课题组比较了电催化和电催化-水力空化协同降解活性艳红模拟废水的效果，结果表明后者对活性艳红的脱色速率是前者的 2 倍[64]。

3）生物电耦合

生物电耦合是将电催化氧化法和生物法联合的一种方法。梁继东课题组[65]采用电化学氧化与生物降解工艺联合降解木质素，结果表明电催化氧化法能先行打断抵御微生物攻击的木质素顽固键合结构，有利于后续生物继续降解残余木质素中间体。降解橄榄油废水也具有同样的效果[66]。

2.2.3　处理对象选择及实际工程的原则

由于电催化氧化法存在能耗高的缺点，因此不适用于处理所有的废水，需要对废水的类型进行甄别。在综合考虑能耗和处理效果这一对矛盾关系的条件下，电催化氧化更适合处理中低浓度、难生物降解、中小水量的废水，因此可以作为废水预处理或者深度处理的方法。此外，废水水质复杂多变，往往单一的处理方法不能同时实现能耗最小化和处理效果最大化，因此在实际工程中需要根据具体的条件和要求，进行科学选择，合理组合工艺。

2.3　电化学杀菌

电催化氧化法不仅被应用于降解有机污染物，同样也可被用于杀菌。由于传统的加氯消毒会产生卤代消毒副产物，危害人类健康，因而近年来采用电催化氧化杀菌以减少卤代消毒副产物的形成，因此受到了越来越多的关注。电化学杀菌有两种机理：电场的物理作用和电解产物的化学作用。电解产物的化学作用主要是指利用电极材料在水中产生活性物质，包括活性氯（Cl_2、$HClO^-$和 ClO^-）、羟基自由基（·OH）、臭氧（O_3）和双氧水（H_2O_2），这类活性物质具有强氧化性，可以破坏微生物的蛋白质、酶和核酸使其致死，实现杀菌的作用[67]。电化学杀菌已被用于市政污水、游泳池水、压载水等，可以有效去除绿农杆菌、大肠杆菌等[68-70]。

2.4　结论

电催化氧化法具有环境兼容性高、能量利用率高、可控性强和不产生二次污染等优点，同时也存在能耗高的缺点。通过改进电解槽结构、采用电位窗口宽催化性能好的电极以及和其他技术耦合等方式实现降耗提效。另外，近年来电化学杀菌也是一种高效、低成本、环境友好的新型杀菌技术。在实际废水处理中，电催化氧化法可作为预处理或深度处理方法对中低浓度、难生物降解及中小水量的废水进行处理，是一种很有前景的水处理方法。

3 环境污染物的高效光电一体化催化氧化还原方法与应用

光、电催化氧化是两种不同能量转化形式、催化特点鲜明的高级氧化处理技术，二者可通过不同的途径产生强氧化性的·OH 自由基等活性物种，氧化能力强，反应条件温和。然而，一个是暗反应，一个是光反应，各有不同的特点，因此，将这两种技术优势互补、有机组合成一体的光电催化降解技术研究在近年来十分活跃，在环境污染物高效降解处理中具有很好的应用发展前景。

3.1 光电一体化功能电极的组装

3.1.1 基于钛基直立有序 TiO_2 纳米管（TiO_2 NTs）表面光电一体化功能的组装

TiO_2 NTs 因其高比表面积和自由空间特性，是进行光电功能组装的良好纳米模板和容器。Cui 等[71]预先在钛基体浅表面向内生长出直立的 TiO_2 纳米管，然后将锡锑氧化物负载到纳米管的管内直至管口外，构筑出桩式结构 TiO_2 NTs/SnO_2 光电一体化电极（图 18-2）。这种微观结构上的改进不仅保持了 Ti/SnO_2 电极的高析氧电位的优点，而且对有机污染物的电催化活性更强，氧化效率明显提高。进一步地，Li 等[72]在真空状态下，将用液晶软模板制备的大孔锡锑氧化物牢固地生长在 TiO_2 纳米管阵列上，构筑了有序大孔 SnO_2 的直立 TiO_2 NTs 杂化电极，该电极继承了 SnO_2/TiO_2 NTs 电极的优点，克服了 SnO_2/TiO_2 NTs 电极因在表面形成膜层消弱了光的吸收的缺点，将电极的光电转化效率由 26.1%提高到 35.2%。

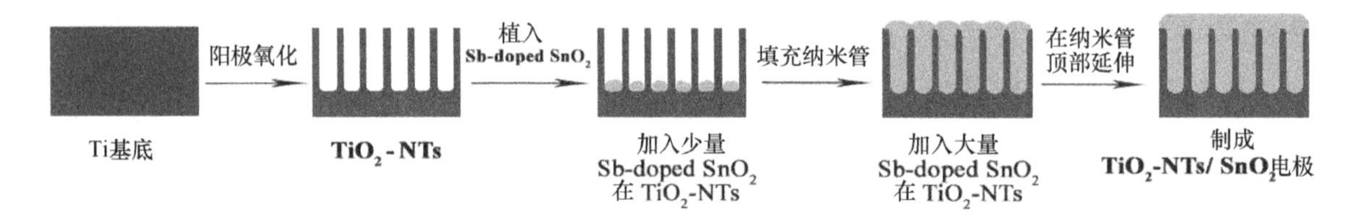

图 18-2 SnO_2/TiO_2 NTs 电极的制备过程[71]

Zhang 等[73,74]在 TiO_2 NTs 中通过真空浸渍方法化学组装 CdS 纳米颗粒，并在 TiO_2/CdS 表面构筑 ZnO 纳米棒作为保护层，最终得到高效、稳定的 TiO_2 NTs/CdS/ZnO NRs 光电一体化电极。TIO_2 NTs 经过 CdS 敏化后，吸收带边从原来的 390 nm 拓宽至可见区 550 nm，带隙从 3.20 eV 降低至 2.32 eV，吸收强度进一步增强。Zhang 等[75]利用改进溶剂热法，将 Pd 量子点均匀修饰于 TiO_2 NTs 基底电极表面，制备得到 Schottky 结构 Pd 量子点敏化 TiO_2 NTs 光电一体化电极材料（图 18-3），这种复合结构显著提高了电极对可见光吸收及转化率。Wu 等[76]采用钛金合金作为基底电极材料，用阳极氧化方法制备出原位掺杂 Au 原子的 Au/TiO_2 NTs 电极。由于 Au 的等离子共振效应，该电极材料在可见光区域出现了周期性的吸收峰。

3.1.2 基于掺硼金刚石膜（BDD）表面光电一体化功能的组装

掺硼金刚石膜（BDD）是近年来发展起来的一种优异的电化学材料，在 BDD 表面进行光电功能修饰

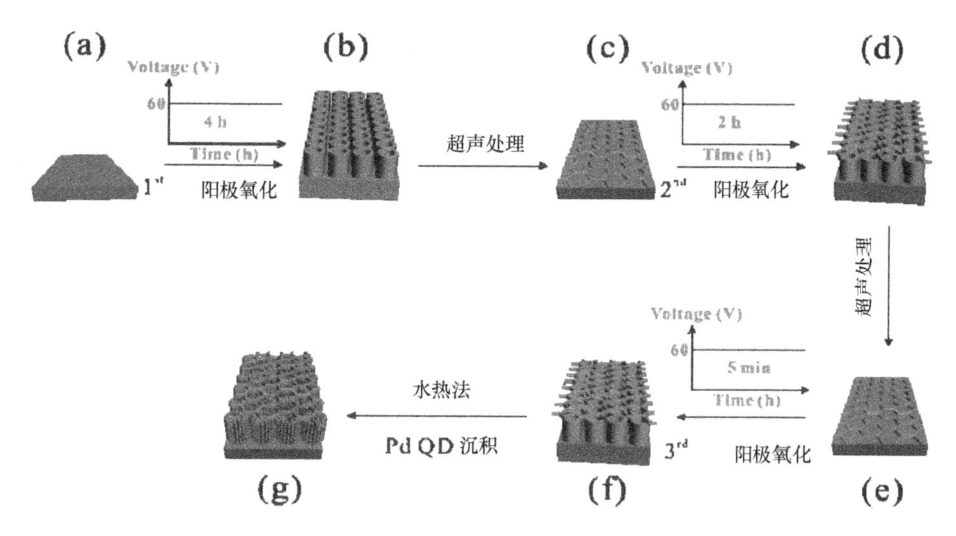

图 18-3　Schottky 结构 Pd QDs@TNTAs 光阳极制备过程示意图[75]

和改性引起了越来越多的关注。Jia 课题组和 Quan 课题组将 BDD 电极应用于环境污染物降解处理开展了许多很好的研究工作[77,78]。近年来，Tong 等[79]利用活泼 Zn 具有高析氢电位的特点，以电沉积 Zn 时伴随氢的气泡作为动力学模板，在 BDD 表面构筑了三维多孔垂直生长的网络状纳米片。接着，在一定的氯铂酸稀溶液中进行 Zn 与 Pt 之间的取代反应，从而在 BDD 表面构筑了三维多孔垂直生长的纳米 Pt 片。同时，Lei 等[80]利用嵌段共聚物表面活性剂软模板和共沉淀的方法制备了 SnO_2 NPs/BDD 电极，SnO_2 纳米粒子的大小可以由胶束所控制，被修饰的 BDD 仍可以被完全暴露。这种新颖的结构赋予了 BDD 电极优异的电催化氧化能力，同时保持了高的析氧电位性能（2.4 V *vs* SCE）。

3.2　光电一体化催化方法在氧化降解有毒有害污染物中的应用

3.2.1　酚类污染物的氧化降解

许多酚类污染物有类雌激素效应，低剂量下可带来不可逆的内分泌干扰效应。Zhang 等[81]使用{001}面高度暴露的 TiO_2 负载于碳气凝胶上降解废水中 BPA。经过 6.0 h 降解，BPA 去除近 100%，速率常数为 0.454 h^{-1}，TOC 去除率为 83%。在 BDD 电极上构筑带有分子印迹的 TiO_2 纳米晶体作为光电阳极[82]，可以增强电极界面污染物的局部浓度，经 2.0 h 降解，BPA 去除达到 97%。Chai 等[83]在 TiO_2 纳米管上组装大孔 Sb 掺杂 SnO_2 膜，由于电极光、电催化性能的优异协同效应，有毒中间体被高效去除，4.0 h 后对硝基苯酚和 TOC 去除率分别达到 98%和 91%。

3.2.2　含氟污染物的氧化降解

氟是一种原生质毒物，透过细胞壁与原生质结合，可抑制多种生物酶的活性，引起物质代谢紊乱。Zhao 等[84]提出了一种具有高表面积、强吸附能力和高电催化活性的新型 SnO_2-Sb/CA 电极，并结合超声技术用于氧化持久性 PFOA，经过 5.0 h 降解，超过 91%的 PFOA 被降解，TOC 去除率达到 86%。通过电泳沉积将 SnO_2-Sb 组装到 TiO_2 NTs 中用于氟苯的降解[85]，高结晶度和较少的氧空位带来了优良的氟苯去除效果，3.0 h 后 COD 去除率达到 97.6%。

3.2.3　农药和除草剂污染物的氧化降解

许多农药和除草剂具有类激素效应，易导致动物体和人体生殖器障碍、行为异常、神经和免疫系统异常。为了去除甲霜灵，Chai 等[86]通过电化学沉积法制备了三维大孔 PbO_2 电极，由于其较大的比表面积、较小的电子转移电阻和丰富的晶体缺陷，经过 300 min 降解，COD 和 TOC 去除率均达到 95%。Li 等[87]通过组装 TiO_2 纳米管和 2D 大孔 SnO_2，构建出具有突出催化性能的光电阳极，表现出较小的表面和液界阻抗，较大的比表面积和较低的反应活化能。将其用于 2,4-D 的去除，3.0 h 内 COD 去除率达到 90.1%。此外，手性化合物 DCPP 是广泛使用的除草剂，活性(*R*)-DCPP 通常被优先降解，而毒性较大的(*S*)-DCPP 很难被有效去除。Zhang 等[88]通过引入分子印迹技术，在光电催化（PEC）表面上实现了(*S*)-DCPP 的有效对映选择性识别和可控降解。难降解(*S*)-DCPP 被优先降解，降解速率是(*R*)-DCPP 的 2.6 倍。同时，即使在添加 100 倍浓度的腐殖酸、草甘膦、氧化乐果、异丙甲草胺和阿特拉津作为干扰物时，高毒的(*S*)-DCPP 也能够被优先降解（图 18-4）。

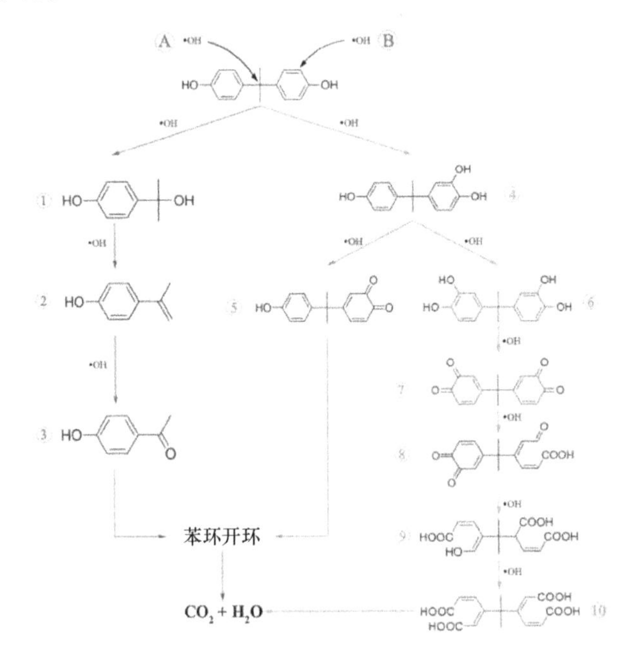

图 18-4　光电一体化催化双酚 A 氧化降解途径[81]

3.3　光电一体化催化方法在环境污染物传感分析中的应用

3.3.1　微囊藻毒素的光电一体化传感检测

Liu 等[89]构筑了一种基于 TiO_2@CNTs 的 MC-LR 分子印迹传感器。该方法利用两步溶胶凝胶法与原位表面分子印迹技术结合，构筑了可控的核壳管状结构，有效提高了对可见光的吸收能力，并且表现出了良好的光电氧化能力，检测限达到 0.4 p mol/L （图 18-5）。Chen 等[90]构筑了一种在垂直 TiO_2 NTs 内负载分子印迹位点的 MC-LR 传感器。该传感器制备非常简单，只有阳极化和构造分子印迹模板两个过程，稳定性好，并且具有良好的选择性和灵敏度，检测限达到 0.1 μg/L。Li 等[91]构筑了一种比色传感检测 MC-LR 的核酸适配体传感器。该方法利用等离子共振峰的移动对目标物进行检测，可重复性好，无需基底电极，比起传统仪器方法简单便捷，检测限达到 0.37 nmol/L。

3.3.2　17β-estradiol 的光电一体化传感检测

Yang 等[92]构筑了一种利用静电纺丝的分子印迹传感器。该传感器以 FTO 作为基底负载 BiOI NFs，用

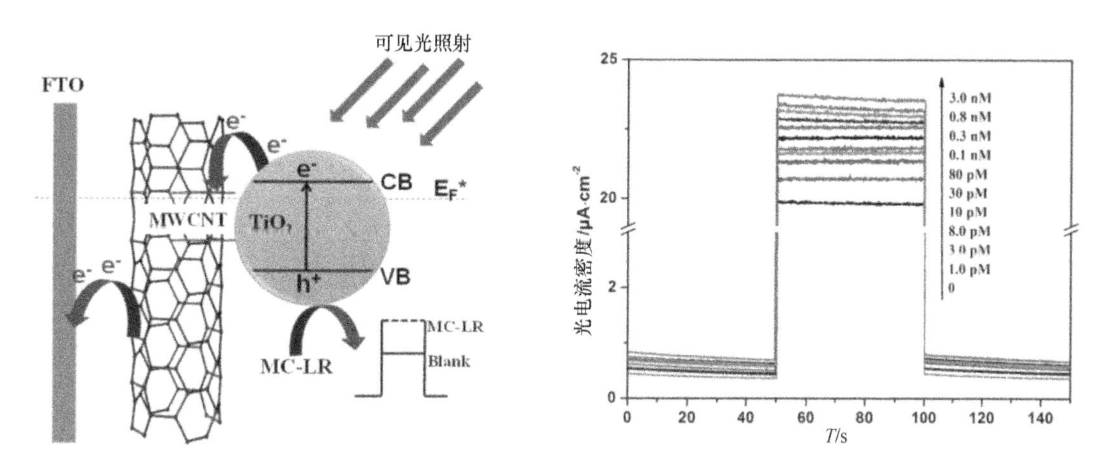

图 18-5　传感器的检测机理以及对 MC-LR 的光电流响应[20]

静电纺丝模板直接修饰印迹电极，静电纺纳米纤维具有良好的选择性以及灵敏度，检测限达到 8 pg/L。Fan 等[93]构筑了一种检测 17β-estradiol 的核酸适配体传感器。该方法在 TiO_2 上电沉积 CdSe NPs 后负载 Aptamer，利用 TiO_2 和 CdSe 的纳米结构和良好光电性能，以及核酸适配体与 E2 特异性能力，管状微结构和适配体的高堆积密度得到传感界面，使得对 E2 的检测灵敏度非常高，检测限达到 33 f mol/L（图 18-6）。

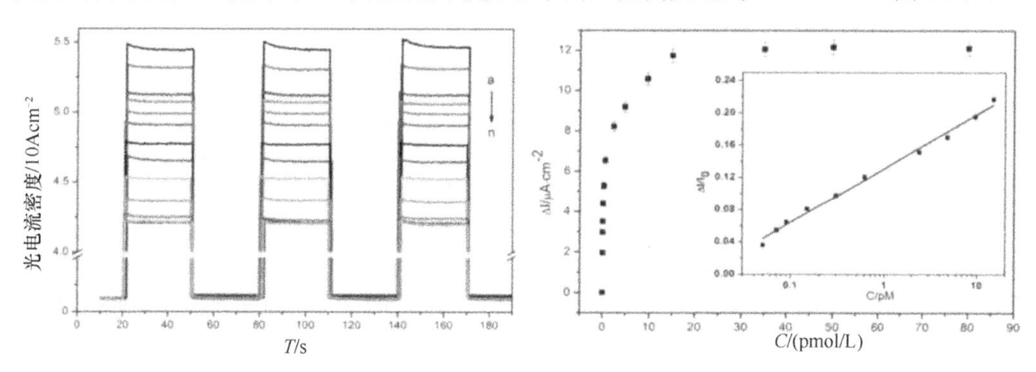

图 18-6　E2 传感器的光电流响应及线性关系[93]

3.4　光电一体化催化方法在温室气体 CO_2 还原与产氢中的应用

3.4.1　高效仿生光电一体化催化还原 CO_2

受自然界绿色植物光合作用的启发，仿生光电催化高效转换 CO_2 成为研究的热门方向。Huang 等[94]研究发现，Co_3O_4 微米花多级结构的构筑以及高指数{121}晶面构筑有机协同结合实现 CO_2 高效选择性光电催化还原至甲酸。Shen 等[95]用电沉积的方法将 Cu NPs 修饰在 Co_3O_4 NTs 上，高效选择性（100%）将 CO_2 光电催化还原成甲酸，并深入探讨 Cu-Co_3O_4 NTs 两电子还原的机理（图 18-7）。

Huang[96]等将高晶面的 Co_3O_4 光电催化剂与 Ru（II）配合物仿生酶组装在碳气凝胶表面，构建出 CO_2 人工光合作用的仿生结构反应界面，将 CO_2 高效选择性（>99%）还原成甲酸（图 18-8）。Liu 等[97]构筑的 CO_2 活性 TiO_2/Ru(II)光敏剂/Pyridine 催化剂复合催化体系，利用 TiO_2 产生的光电子经由羧基桥连基团转移至 Ru(II)配合物中未配位的吡啶基位点，与 Ru(II)配合物本身的优异光敏性协同作用，从而将 CO_2 高效转化至甲醇。

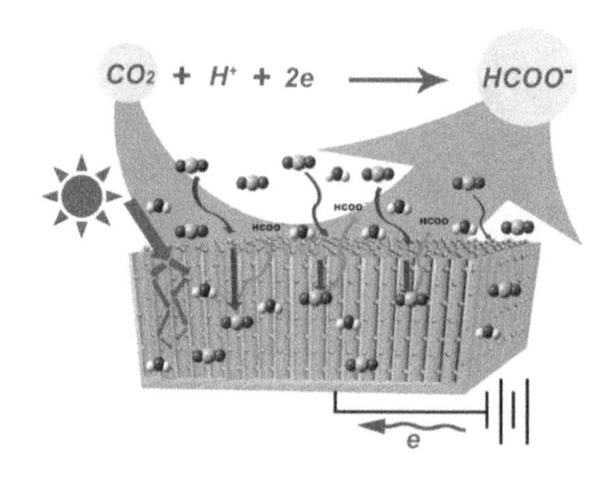

图 18-7　Cu 纳米粒子修饰的 Cu-Co_3O_4 NTs 光电催化还原 CO_2 示意图[94]

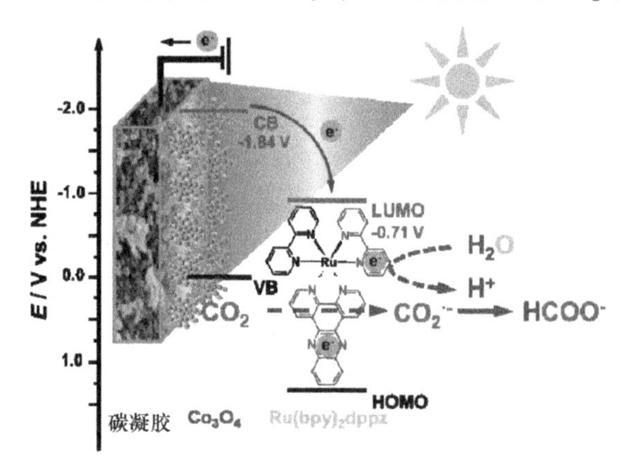

图 18-8　$Ru(bpy)_2dppz$-Co_3O_4/CA 界面增强吸附及还原 CO_2 的原理图[26]

3.4.2　高效光电一体化催化氧化生物质促进阴极产氢

Zhang 等研究发现，通过构筑 Schottky 结构 Pd 量子点(QDs)敏化 TiO_2 纳米管阵列电极(PdQDs@TNTAs)[75]，桥连结构薄层碳包覆铬掺杂钛酸锶修饰 TNTAs (C@Cr-$SrTiO_3$/TNTAs)[98]电极进行光电一体化分解水产氢，发现葡萄糖氧化能有效促进阴极产氢。在此基础上合理设计并构筑得到 Au/CeO_2-TiO_2 NT 光阴极，实现绿色高效光电催化选择性（99%）氧化生物质醇到醛[99]。研究发现实现高效选择性的原因是$\cdot O^{2-}$的存在能优先与苯甲醇活性阴离子结合，降低总的氧化能力，使苯甲醇定向氧化至苯甲醛以实现其选择性。通过引入·OH 的猝灭剂 CH_3OH 以及$\cdot O^{2-}$的猝灭剂 SOD 进行了验证，发现$\cdot O^{2-}$起到明显的氧化作用 [77]。

3.4.3　光电一体化催化污染物氧化同步产氢

近年来报道的光催化燃料电池（PFC）用于污染物高效氧化同步产电，极大提高了光量子效率[100-103]。但是其产生的电能难以储存，后来的研究工作大部分转向产氢。Wu[104]等构建了一种新颖的太阳光驱动的双光电极光电化学电池（PEC），并将其应用于环境污染物苯酚的降解及同步产氢。由于阴阳两极之间的协同作用，阳极 TOC 的去除率达到 84.2%，阴极产氢量达到 86.8 μmol/cm^2。

参 考 文 献

[1]　Gong T, Liu J, Liu X, Jie Liu, Xiang J, Wu Y. A sensitive and selective platform based on CdTe QDs in the presence of

L-cysteine for detection of silver, mercury and copper ions in water and various drinks. Food Chem, 2016, 213: 306.
[2] Cui L, Wu J, Ju H. Electrochemical sensing of heavy metal ionswith inorganic, organic and bio-materials. Biosens. Bioelectron, 2015, 63: 276.
[3] Gumpu MB, Sethuraman S, Krishnan UM, Rayappan JBB. A review on detection of heavy metal ions in water-an electrochemical approach. Sens. Actuators B, 2015, 213: 515.
[4] Avuthu SGR, Narakathu BB, Eshkeiti A, Emamian S, Bazuin BJ, Joyce M. Atashbar M.Z., Detection of heavy metals using fully printed three electrode electrochemical sensor, Sens. IEEE, 2014, 10: 669672.
[5] Niu P, Sánchez CF, Gich M, Ayora C, Roig A. Electroanalytical assessment of heavy metals in waters with bismuth nanoparticle-porous carbon paste electrodes, Electrochim. Acta, 2015, 165: 155.
[6] Chaiyo S, Mehmeti E, Žagar K, Siangproh W, Chailapakul O, Kalcher K. Electrochemical sensors for the simultaneous determination of zinc, cadmium and lead using a Nafion/ionic liquid/graphene composite modified screen-printed carbon electrode. Anal. Chim. Acta, 2016, 918: 26.
[7] Parat C, Authier L, Aguilar D, Companys E, Puy J, Galceran J. Direct determination of free metal concentration by implementing stripping chronopotentiometry as the second stage of AGNES. Analyst, 2011, 136: 4337.
[8] Huang H, Chen T, Liu X, Ma H. Ultrasensitive and simultaneous detection of heavy metal ions based on three-dimensional graphene-carbon nanotubes hybrid electrode materials. Anal. Chim. Acta, 2014, 852: 45.
[9] Tekaya N, Saiapina O, Ben Ouada H, Lagarde F, Ben Ouada H, Jaffrezic Renault N. Ultra-sensitive conductometric detection of heavy metalsbased on inhibition of alkaline phosphatase activity from Arthrospira platensis. Bioelectrochemistry. 2013, 90: 24.
[10] Xuan F, Luo X, Hsing IM. Conformation-dependent Exonuclease III activity mediated by metal ions reshuffling on thymine-rich DNA duplexes for an ultrasensitive electrochemical method for Hg^{2+} detection. Anal. Chem., 2013, 85: 4586.
[11] Jia J, Chen HG, Feng J, Lei JL, Luo HQ, Li NB. A regenerative ratiometric electrochemical biosensor for selective detecting Hg2þ based on Y-shaped/hairpin DNA transformation. Anal. Chim. Acta, 2016, 908: 1.
[12] 左伯莉, 刘国宏. 化学传感器原理及应用. 北京: 清华大学出版社, 2007: 167-168.
[13] Feng S, Yang R, Ding X. Sensitive electrochemical sensor for the determination of pentachlorophenol in fish meat based on ZnSe quantum dots decorated multiwall carbon nanotubes nanocomposite. Ionics, 2015, 21: 3257.
[14] Xia B, Yuan Q, Chu M. Directly one-step electrochemical synthesis of graphitic carbon nitride/graphene hybrid and its application in ultrasensitive electrochemiluminescence sensing of pentachlorophenol. Sens. Actuators B: Chem., 2016, 228: 565.
[15] Xu G, Li B, Wang X, Luo X. Electrochemical sensor for nitrobenzene based on carbon paste electrode modified with a poly(3, 4-ethylenedioxythiophene)and carbon nanotube nanocomposite. Microchim. Acta, 2014, 181: 463.
[16] Gupta R, Rastogi PK, Ganesan V. Gold nanoparticles decorated mesoporous silica microspheres: A proficient electrochemical sensing scaffold for hydrazine and nitrobenzene. Sens. Actuators B: Chem., 2017, 239: 970.
[17] Zhu G, Yi Y, Han Z. Sensitive electrochemical sensing for polycyclic aromatic amines based on a novel core-shell multiwalled carbon nanotubes@ graphene oxide nanoribbons heterostructure. Anal. Chim. Acta, 2014, 845: 30.
[18] Liu S, Wei M, Zheng X. Alizarin red S functionalized mesoporous silica modified glassy carbon electrode for electrochemical determination of anthracene. Electrochi. Acta, 2015, 160: 108.
[19] 梁伟棠, 张德安, 尹腾辉. 大气污染现状及控制对策研究. 资源与环境, 2016, 42(2): 156.
[20] Shanmugam PV, Anju Y, Singh CC, Monitoring the emission of volatile organic compounds from flowers of jasminum sambac using solid-phase micro-extraction fibers and gas chromatography with mass spectrometry detection. Nature. Product. Communications, 2011, 6(9): 1333.
[21] Creek JAM, Mcanoy AM, Brinkworth CS. Rapid monitoring of sulfur mustard degradation in solution by headspace solid-phase microextraction sampling and gas chromatography mass spectrometry Rapid. Commun. Mass. SP., 2010, 24(23): 3419.
[22] Seok OK, Woo SI. Chemiluminescence analyzer of NO_x as a high-throughput screening tool in selective catalytic reduction of NO. Sci. Technol. Adv. Mat. 2011, 12(5): 054211.
[23] Hiroyuki S, Yuka M, Tomoki N. Comparison of laser-induced fluorescence and chemiluminescence measurements of NO2 at an urban site. Atmos. Environ., 2011, 45(34): 6233.
[24] Geibel MC, Gerbig C, Feist DG. A new fully automated FTIR system for total column measurements of greenhouse gases. Atmos. Meas. Tech. 2010, 3(5): 1363.

[25] Diao Q, Yin C, Liu Y, Gong X, Liang S, Yang H, Chen Lu G. Mixed-potential-type NO_2 sensor using stabilized zirconia and Cr_3O_4–WO_3 nanocomposites. Sens. Actuator B: Chem., 2013, 180: 90.

[26] Prakash S, Rajesh S, Singh S K, Bhargava K, Ilavazhagan G, Vasu V ,Karunakaran C. Copper nanoparticles entrapped in SWCNT-PPy nanocomposite on Pt electrode as NOx electrochemical sensor. Talanta, 2011, 85: 964.

[27] Yan D, Hu M, Li S, Liang J, Wu Y, Ma S. Electrochemical deposition of ZnO nanostructures onto porous silicon and their enhanced gas sensing to NO_2 at room temperature, Electrochi. Acta, 2014, 115: 297.

[28] Gusain A, Joshi N J, Varde P V, Aswal D K. Flexible NO Gas Sensor Based on Conducting Polymer Poly [N-9'-heptadecanyl-2, 7-carbazole-alt-5, 5-(4', 7'-di-2-thienyl-2', 1', 3'-benzothiadiazole)](PCDTBT). Sens. Actuators B, 2017, 239: 734.

[29] Xu J Q, Duo H H, Zhang Y G, Zhang X W, Fang W, Liu Y L, Shen A G, Hu J M, Huang, W H. Photochemical synthesis of shape-controlled nanostructured gold on zinc oxide nanorods as photocatalytically renewable sensors. Anal. Chem., 2016, 88: 3789.

[30] Maluta J R, Canevari T C, Machado S A S. Sensitive determination of nitric oxide using an electrochemical sensor based on MWCNTs decorated with spherical Au nanoparticles. J. Solid State Electrochem., 2014, 18: 2497.

[31] Wu H, Huang D, Jin X, Luo C, Dong Q, Sun B, Zong R, Li J, Zhang L, Zhang H. Silver nanoparticles/polyethyleneimine/grapheme oxide composite combined with surfactant film for construction of an electrochemical biosensor. Anal. Methods, 2016, 8: 2961.

[32] Liu Z, Nemec-Bakk A, Khaper N, Chen A. Sensitive electrochemical detection of nitric oxide release from cardiac and cancer cells via a hierarchical nanoporous gold. icroelectrode, 2017, DOI: 10.1021 /acs. analchem.7b01430.

[33] Särkkä H, Bhatnagar A, Sillanpää M. Recent developments of electro-oxidation in water treatment—A review. Journal of Electroanalytical Chemistry, 2015, (754): 46-56.

[34] Guinea E, Garrido J A, Rodriguez R M, Cabot P, Arias C, Centellas F, Brillas E. Degradation of the fluoroquinolone enrofloxacin by electrochemical advanced oxidation processes based on hydrogen peroxide electrogeneration. Electrochimica Acta, 2010, (55): 2101-2115.

[35] Su L, Li K, Zhang H, Fan M, Ying D, Sun T, Wang Y, Jia J. Electrochemical nitrate reduction by using a novel Co3O4/Ti cathode. Water Res, 2017, (120): 1-11.

[36] Chu H, Wang Z, Liu Y. Application of modified bentonite granulated electrodes for advanced treatment of pulp and paper mill wastewater in three-dimensional electrode system. Journal of Environmental Chemical Engineering, 2016, (4): 1810-1817.

[37] Li X, Wu Y, Zhu W, Xue F, Qian Y, Wang C. Enhanced electrochemical oxidation of synthetic dyeing wastewater using SnO_2-Sb-doped TiO_2-coated granular activated carbon electrodes with high hydroxyl radical yields. Electrochimica Acta, 2016, (220): 276-284.

[38] Wang Z, Qi J, Wang B, Feng Y, Li K. A three dimensional electrochemical oxidation reactorbased on magnetic steel slag particle electrodes for printing /dyeing wastewater treatment. Journal of Harbin Institute of Technology, 2015, (47): 38-42.

[39] Wei J, Zhang S, Hu Q, Chen F. Efficiency and mechanism of three-dimensional particle electrode for treating dyes wastewater. Chinese Journal of Environmental Engineering, 2015, (9): 1715-1720.

[40] Zhaoxin Z, Qing W, Hongyan H, Song N, Chao Y, Xike T. Electrochemical treatment of reverse osmosis concentrate of oil refining wastewater by Mn-Sn-Ce/gamma-Al_2O_3 particle electrode. IEEE Computer Society, 2012: 236-241.

[41] Yan L, Ma H, Wang B, Wang Y, Chen Y. Electrochemical treatment of petroleum refinery wastewater with three-dimensional multi-phase electrode. Desalination, 2011, (276): 397-402.

[42] Yan L, Wang Y, Li J, Ma H, Liu H, Li T, Zhang Y. Comparative study of different electrochemical methods for petroleum refinery wastewater treatment. Desalination, 2014, 341() 87-93.

[43] Chen Z, Zhang Y, Zhou L, Zhu H, Wan F, Wang Y, Zhang D. Performance of nitrogen-doped graphene aerogel particle electrodes for electro-catalytic oxidation of simulated Bisphenol A wastewaters. Journal of Hazardous Materials, 2017, (332): 70-78.

[44] Li P. Electrochemical Degradation of Phenol Wastewater by Sn-Sb- Ce Modified Granular Activated Carbon. International Journal of Electrochemical Science, 2017, (11): 2777-2790.

[45] Vijayakumar V, Saravanathamizhan R, Balasubramanian N. Electro oxidation of dye effluent in a tubular electrochemical reactor using TiO_2/RuO_2 anode.Journal of Water Process Engineering, 2016, (9): 155-160.

[46] Zhang Y, Yu T, Han W, Sun X, Li J, Shen J, Wang L. Electrochemical treatment of anticancer drugs wastewater containing

5-Fluoro-2-Methoxypyrimidine using a tubular porous electrode electrocatalytic reactor. Electrochimica Acta, 2016, (220): 211-221.
[47] 马锐军. 多级旋转电极电化学反应器流动特性及应用研究.北京：北京化工大学, 2016.
[48] 雷阳明. 采用气体扩散电极处理生物难降解有机废水的研究.上海：上海交通大学, 2005.
[49] Li W, Xiao C, Zhao Y, Zhao Q, Fan R, Xue J. Electrochemical Reduction of High-Concentrated Nitrate Using Ti/TiO2 Nanotube Array Anode and Fe Cathode in Dual-Chamber Cell. Catalysis Letters, 2016, (146): 2585-2595.
[50] Zhong D, Yang J, Xu Y, Jia J, Wang Y, Sun T. De-colorization of Reactive Brilliant Orange X-GN by a novel rotating electrochemical disc process. Journal of Environmental Sciences, 2008, (20): 927-932.
[51] He Y, Lin H, Wang X, Huang W, Chen R, Li H. A hydrophobic three-dimensionally networked boron-doped diamond electrode towards electrochemical oxidation. Chemical Communications, 2016,(52): 8026-8029.
[52] Yang G S, Zhang W L, Zhang B Q, Xin H U, Chen B M, Guo Z C. Research status of energy-saving anodes for zinc electrowinning, Electroplating & Finishing, 34(15): 872-876.
[53] 阚连宝, 段辉.DSA 电极的制备及其在水处理中的应用. 化学工程师，2016 (30)：50-52.
[54] 叶德宁, 应迪文, 贾金平. 活性炭纤维电极在水处理中的应用及进展. 中国给水排水, 2007 (23): 20-23.
[55] Tian S, Li Y, Zeng H, Guan W, Wang Y, Zhao X. Cyanide oxidation by singlet oxygen generated via reaction between H_2O_2 from cathodic reduction and OCl- from anodic oxidation. Journal of Colloid and Interface Science ,2016, (482): 205-211.
[56] Yao F, Zhong Y, Yang Q, Wang D, Chen F, J. Zhao T, Xie C, Jiang H, An G, Zeng X. Effective adsorption/electrocatalytic degradation of perchlorate using Pd/Pt supported on N-doped activated carbon fiber cathode. Journal of Hazardous Materials, 2017,(323): 602-610.
[57] Li N, Dong S, Lv W, Huang S, Chen H, Yao Y, Chen W. Enhanced electrocatalytic oxidation of dyes in aqueous solution using cobalt phthalocyanine modified activated carbon fiber anode. Science China-Chemistry, 2013,(56): 1757-1764.
[58] 王后锦, 吴晓婧, 王亚玲, 焦自斌, 颜声威, 黄浪欢. 二氧化钛纳米管阵列光电催化同时降解苯酚和 Cr(VI). 催化学报, 2011,(32): 637-642.
[59] Hurwitz G, Hoek EMV, Liu K, Fan L, Roddick FA. Photo-assisted electrochemical treatment of municipal wastewater reverse osmosis concentrate. Chemical Engineering Journal, 2014,(249): 180-188.
[60] Li P, Zhao G, Zhao K, Gao J, Wu T. An efficient and energy saving approach to photocatalytic degradation of opaque high-chroma methylene blue wastewater by electrocatalytic pre-oxidation. Dyes and Pigments 2012,(92): 923-928.
[61] Li K, Yang C, Wang Y, Jia J, Xu Y, He Y. A high-efficient rotating disk photoelectrocatalytic(PEC) reactor with macro light harvesting pyramid-surface electrode. Aiche Journal, 2012,(58): 2448-2455.
[62] Martin MJ, de Vidales, Milian M, Saez C, Canizares P, Rodrigo MA. Irradiated-assisted electrochemical processes for the removal of persistent pollutants from real wastewater. Separation and Purification Technology, 2017(175): 428-434.
[63] Souza F, Quijorna S, Lanza MRV, Saez C, Canizares P. Rodrigo M A. Applicability of electrochemical oxidation using diamond anodes to the treatment of a sulfonylurea herbicide. Catalysis Today, 2017,(280): 192-198.
[64] Wang X, Jia J, Wang Y. Enhanced photocatalytic-electrolytic degradation of Reactive Brilliant Red X-3B in the presence of water jet cavitation. Ultrasonics Sonochemistry ,2015,(23)：93-99.
[65] 崔晓敏, 王一平, 梁继东. 电化学氧化与生物法联合降解木质素. 环境科学研究,2016, (29): 434-441.
[66] Goncalves M R, Marques I P, Correia J P. Electrochemical mineralization of anaerobically digested olive mill wastewater. Water Research,2012,(46): 4217-4225.
[67] Bruguera-Casamada C, Sires I, Brillas E, Araujo R M. Effect of electrogenerated hydroxyl radicals, active chlorine and organic matter on the electrochemical inactivation of Pseudomonas aeruginosa using BDD and dimensionally stable anodes. Separation and Purification Technology, 2017,178: 224-231.
[68] Huang X, Qu Y, Cid C A, Finke C, Hoffmann M R, Lim K, Jiang S C. Electrochemical disinfection of toilet wastewater using wastewater electrolysis cell. Water Research, 2016, (92): 164-172.
[69] Lacasa E, Tsolaki E, Sbokou Z, Rodrigo M A, Mantzavinos D, Diamadopoulos E. Electrochemical disinfection of simulated ballast water on conductive diamond electrodes. Chemical Engineering Journal, 2013, (223): 516-523.
[70] Schaefer C E, Andaya C, Urtiaga A. Assessment of disinfection and by-product formation during electrochemical treatment of surface water using a Ti/IrO_2 anode. Chemical Engineering Journal, 2015, (264): 411-416.
[71] Zhao G, Cui X, Liu M, Li P, Zhang Y, Cao T, Li H, Lei Y, Liu L, Li D. Electrochemical degradation of refractory pollutant using a novel microstructured TiO_2 nanotubes/Sb-doped SnO_2 electrode. Environ Sci Technol, 2009, 43(5): 1480-1486.
[72] Li P, Zhao G, Cui X, Zhang Y, Tang Y. Constructing stake structured TiO_2-NTs/Sb-doped SnO_2 electrode simultaneously

with high electrocatalytic and photocatalytic performance for complete mineralization of refractory aromatic acid. J. Phys. Chem. C, 2009, 113(6): 2375-2383.

[73] Zhang Y N, Zhao G, Lei Y, Li P, Li M, Jin Y, Lv B. CdS-encapsulated TiO_2 nanotube arrays lidded with ZnO nanorod layers and their photoelectrocatalytic applications. Phys. Chem. Chem. Phys., 2010, 11(16): 3491-3498.

[74] Wu T, Zhao G, Lei Y, Li P. Distinctive tin dioxide anode fabricated by pulse electrodeposition: High oxygen evolution potential and efficient electrochemical degradation of fluorobenzene. J. Phys. Chem. C, 2011, 115(10): 3888-3898.

[75] Zhang Y, Tang B, Wu Z, Guohua Z. Glucose oxidation over ultrathin carbon-coated perovskite modified TiO_2 nanotube photonic crystals with high-efficiency electron generation and transfer for photoelectrocatalytic hydrogen production. Green Chem., 2016, 18(8): 2424-2434.

[76] Wu Z, Wang J, Zhou Z, Guohua Z. Highly selective aerobic oxidation of biomass alcohol to benzaldehyde by an in situ doped Au/TiO_2 nanotube photonic crystal photoanode for simultaneous hydrogen production promotion. J. Mater. Chem. A, 2017, 5(24): 12407–12415.

[77] Zhao Y, Yu H T, Quan X, Chen S, Zhao H M, Zhang Y B. Preparation and characterization of vertically columnar boron doped diamond array electrode, Appl. Surf. Sci., 2014, (303)419-424.

[78] C H Shu, T H Sun, J P Jia, Z Y Lou. A novel desulfurization process of gasoline via sodium metaborate electroreduction with pulse voltage using a boron-doped diamond thin film electrode, Fuel, 2013, (113)187-195.

[79] Zhang Y N, Qin N, Li J Y, Han S N, Li P, Zhao G H. Facet exposure-dependent photoelectrocatalytic oxidation kinetics of bisphenol A on nanocrystalline {001} TiO_2/carbon aerogel electrode. Appl. Catal., B, 2017, 216: 30-40.

[80] Fan J Q, Shi H J, Xiao H S, Zhao G H. Double-Layer 3D Macro-mesoporous metal oxide modified boron-doped diamond with enhanced photoelectrochemical performance. Acs Applied Materials & Interfaces, 2016, 8 (42): 28306-28315.

[81] Li P Q, Zhao, G H, Li M F, Cao T C, Cui X, Li D M. Design and high efficient photoelectric-synergistic catalytic oxidation activity of 2D macroporous SNO_2/1D TiO_2 nanotubes. Appl. Catal., B, 2012, 111: 578-585.

[82] Chai S N, Zhao G H, Zhang Y N, Wang Y J, Nong F Q, Li M F, Li D M. Selective photoelectrocatalytic degradation of recalcitrant contaminant driven by an n-P heterojunction nanoelectrode with molecular recognition ability. Environmental Science & Technology, 2012, 46 (18): 10182-10190.

[83] Chai S N, Zhao G H, Li P Q, Lei Y Z, Zhang Y N, Li D M. Novel sieve-like SnO_2/TiO_2 nanotubes with integrated photoelectrocatalysis: Fabrication and application for efficient toxicity elimination of nitrophenol wastewater. Journal of Physical Chemistry C, 2011, 115 (37): 18261-18269.

[84] Zhao H Y, Gao J X, Zhao G H, Fan J Q, Wang Y B, Wang Y J. Fabrication of novel SnO_2-Sb/carbon aerogel electrode for ultrasonic electrochemical oxidation of perfluorooctanoate with high catalytic efficiency. Applied Catalysis B-Environmental, 2013, 136: 278-286.

[85] Wu T, Zhao G H, Lei Y Z, Li P Q. Distinctive tin dioxide anode fabricated by pulse electrodeposition: High oxygen evolution potential and efficient electrochemical Ddegradation of fluorobenzene. Journal of Physical Chemistry C, 2011, 115 (10): 3888-3898.

[86] Chai S N, Zhao G H Wang Y J, Zhang Y N, Wang Y B, Jin Y F, Huang X F. Fabrication and enhanced electrocatalytic activity of 3D highly ordered macroporous PbO_2 electrode for recalcitrant pollutant incineration. Applied Catalysis B-Environmental, 2014, 147: 275-286.

[87] Li P Q, Zhao G H, Li M F, Cao T C, Cui X, Li D M. Design and high efficient photoelectric-synergistic catalytic oxidation activity of 2D macroporous SNO_2/1D TiO_2 nanotubes. Applied Catalysis B-Environmental, 2012, 111: 578-585.

[88] Zhang Y N, Dai W G, Wen Y Z, Zhao G H. Efficient enantioselective degradation of the inactive(S)-herbicide dichlorprop on chiral molecular-imprinted TiO_2. Applied Catalysis B-Environmental, 2017, 212: 185-192.

[89] Liu M C, Ding X, Yang Q W, Wang Y, Zhao G H, Yang N J. A pM leveled photoelectrochemical sensor for microcystin-LR based on surface molecularly imprinted TiO_2@CNTs nanostructure. Journal of Hazardous Materials, 2017, 331: 309-320.

[90] Chen K, Liu M C, Zhao G H, Shi H J, Fan L F, Zhao S C. Fabrication of a novel and simple microcystin-LR photoelectrochemical sensor with high sensitivity and selectivity. Environmental Science & Technology 2012, 46(21), 11955-11961.

[91] Li X Y, Cheng R J, Shi H J, Tang B, Xiao H S, Zhao G H. A simple highly sensitive and selective aptamer-based colorimetric sensor for environmental toxins microcystin-LR in water samples. Journal of Hazardous Materials, 2016, 304: 474-480.

[92] Yang, X. M, Li, X, Zhang, L. Z, Gong, J. M., Electrospun template directed molecularly imprinted nanofibers incorporated

with BiOI nanoflake arrays as photoactive electrode for photoelectrochemical detection of triphenyl phosphate. Biosensors & Bioelectronics, 2017, 92: 61-67.

[93] Fan L F, Zhao G H, Shi H J, Liu M C, Wang Y B, Ke H Y. A femtomolar level and highly selective 17 beta-estradiol photoelectrochemical aptasensor applied in environmental water samples analysis. Environmental Science & Technology 2014, 48(10), 5754-5761.

[94] Huang X, Cao T, Liu M, Zhao G H. Synergistic photoelectrochemical synthesis of formate from CO_2 on {121} hierarchical Co_3O_4. Journal of Physical Chemistry C, 2013, 117(50): 26432-26440

[95] Shen Q, Chen Z, Huang X, G H Zhao. High-yield and selective photoelectrocatalytic reduction of CO_2 to formate by metallic copper decorated Co_3O_4 nanotube arrays. Environmental Science & Technology, 2015, 49(9): 5828-5835

[96] Huang X, Shen Q, Liu J, Zhao G H. A CO_2 adsorption-enhanced semiconductor/metal-complex hybrid photoelectrocatalytic interface for efficient formate production. Energy & Environmental Science, 2016, 9(10): 3161-3171.

[97] Liu J, Shi H, Shen Q, Zhao G H. Efficiently photoelectrocatalyze CO_2, to methanol using Ru(II)-pyridyl complex covalently bonded on TiO_2, nanotube Aarrays. Applied Catalysis B Environmental, 2017, 210: 368-378.

[98] Zhang Y, Zhao G, Shi H, Zhao G H. Photoelectrocatalytic glucose oxidation to promote hydrogen production over periodically ordered TiO_2, nanotube arrays assembled of Pd quantum dots. Electrochimica Acta, 2015, 174: 93-101.

[99] Zhang Y, Zhao G, Zhang Y. Highly efficient visible-light-driven photoelectro-catalytic selective aerobic oxidation of biomass alcohols to aldehydes. Green Chemistry, 2014, 16(8): 3860-3869.

[100] Li K, Xu Y, He Y, Jia J P. Photocatalytic fuel cell (PFC) and dye self-photosensitization photocatalytic fuel cell (DSPFC) with BiOCl/Ti photoanode under UV and visible light irradiation. Environmental Science & Technology, 2013, 47(7): 3490-3497.

[101] Li K, Zhang H, Tang Y, Jia J P. Photocatalytic degradation and electricity generation in a rotating disk photoelectrochemical cell over hierarchical structured BiOBr film. Applied Catalysis B Environmental, 2015, 164: 82-91.

[102] Li K, Zhang H, Tang T, Xu Y, Ying D, Wang Y, Jia J. Optimization and application of TiO_2/TiPt photo fuel cell (PFC) to effectively generate electricity and degrade organic pollutants simultaneously. Water Res, 2014, 62, 1-10.

[103] Chen Q, Li J, Li X, Zhou B X. Visible-light responsive photocatalytic fuel cell based on WO_3/W photoanode and Cu_2O/Cu photocathode for simultaneous wastewater treatment and electricity generation. Environmental Science & Technology, 2012, 46(20): 11451-11458.

[104] Wu Z, Zhao G, Zhang Y. A solar-driven photocatalytic fuel cell with dual photoelectrode for simultaneous wastewater treatment and hydrogen production. Journal of Materials Chemistry A, 2015, 3(7): 3416-3424.

作者：李 轶[1]，卢小泉[1]，苏柳花[2]，贾金平[2]，赵国华[3]

[1]天津大学化学系，[2]上海交通大学环境科学与工程学院，[3]同济大学化学科学与工程学院

第 19 章　微生物电化学系统

本章导读

微生物电化学系统（microbial electrochemical system, MES）作为一种新型的水处理技术近年来受到广泛重视，中国学者在此领域取得了众多令人瞩目的成就，对该技术发展做出了重要贡献。本章就微生物电化学系统中微生物胞外电子传递过程、微生物电化学系统用于水中污染物去除、土壤/沉积物修复及微生物电化学系统产氢及其高附加值物质等内容进行综述，并对MES在环境污染物净化方面的拓展及展望进行讨论。

关键词

微生物电化学系统，胞外电子传递过程，生物制氢，环境修复

微生物是自然界中的高效分解者，是人类处理污染物的重要资源和工具。近年来，科学家发现一些微生物可借助细胞膜内嵌的细胞色素以及细胞附属物（如伞毛）将代谢有机物产生的电子高效转移至胞外。这类电活性微生物广泛分布在自然水体、沉积物和土壤以及人类排放的污水和污泥中。长久以来，对这类微生物研究较少，同时缺乏可定向富集电化学活性微生物的处理系统，因此难以发挥其独特优势直接应用于废水处理中。然而，细菌直接电子传递现象的发现为环境污染物转化新技术研发提供了重要基础；微生物电化学技术的发明，为富集与发挥此类微生物的效能提供了有效技术手段。从韩国金炳弘教授发现电化学活性菌可在无外源电子中介体条件下实现胞外电子传递，以及Bruce E. Logan教授开发了生活污水为底物的微生物电化学系统，提出了污水处理和同步电能回收概念以来，微生物电化学系统（microbial electrochemical system，MES）作为一种新型的水处理技术受到广泛重视。中国学者在此领域取得了众多令人瞩目的成就，对该技术发展做出了重要贡献。在MES中，电极菌群能够在降解环境污染物的同时将电子导出形成电流，实现污染物的化学能向电能的直接转化。或者在外加电源的辅助下进一步将生物电子储存在质子、CO_2等受体物质中，产出更高附加值的氢气和化工产品。在环境工程师和科学家的手中，MES如同一把新的钥匙，打开了一扇通往污染物资源化道路的大门，创造出了大量颠覆性的污染物去除新技术。

1 微生物电化学系统中微生物胞外电子传递过程

微生物通过呼吸、光合成或发酵等途径获得能量以保持生命活性和增殖，这些过程从物理化学原理分析，均涉及物质的氧化还原反应以及电子转移。胞外电子传递也是微生物的一种呼吸方式，它能够实现电子在微生物细胞间或者与胞外物质（比如矿石、电极）之间的传递。胞外电子传递在金属元素的地球化学循环过程、环境污染的生物修复、生物腐蚀、生物冶金、废弃物资源化以及清洁高效能源的开发等领域起着十分重要的作用。

对微生物胞外电子传递机制的研究通常在分子、蛋白和单细胞层面进行，而环境中的微生物通常以生物膜的形式实现能量转换。因此，胞外电子在生物膜以及微生物细胞间的传递更具有实际意义。生物膜的形成是一个复杂的过程，其内部含有大量的基质即胞外聚合物，组成相对复杂。盛国平和俞汉青等[1]将*Shewanella*的胞外聚合物提取后，发现其具有电化学氧化还原活性，结合光谱电化学和蛋白质凝胶电泳的数据说明EPS中含有亚铁血红素的氧化还原蛋白。肖勇等[2,3]发现去除了胞外聚合物后，*S. oneidensis*

MR-1 上来自于核黄素类物质的电化学氧化还原特征峰大幅减弱甚至消失，细胞色素 C 的特征峰显著增强，菌株对底物乳酸钠的电流响应（胞外电子传递速度）也在去除胞外聚合物后得到了显著增强。进一步通过电子传递模型的理论计算，提出电子在胞外聚合物中可能以跳跃（hopping）的方式进行传递。对革兰氏阳性菌株 *Bacillus* sp.和毕赤酵母这两株同样具有电化学活性的菌株开展了类似试验，获得了与菌株 *S. oneidensis* MR-1 相似的结果，这表明上述 EPS 功能在微生物中可能具有普适性。另外，在微生物与胞外固体物质作用并形成生物膜的过程中，胞外的电子受体也可能参与微生物的电子传递，如硫[4]、铁氧化物[5]等可以在生物膜内介导长程电子传递。

生物膜中胞外电子传递的强化显然具有重要价值。通过在生物膜内部构筑可以实现电子长程传递的通道，实现跨生物膜的电子快速传递。利用多壁碳纳米管和具有产电能力的混合菌群构建的复合生物膜，可以将反应器启动时间降低 53.8%，同时将反应器产电输出提高 46.2%[6]。多壁碳纳米管在其中发挥了自身导电的特性，极大地改善了跨膜电子传递过程。同时多壁碳纳米管对底物的吸附能力也提高了系统对污染物浓度变化的适应能力。在包埋 *Shewanella* 菌体的海藻酸钠颗粒中加入碳纳米管可以降低包埋体内电子传递的内阻，实现更快的电子传递[7]；利用氧化石墨烯和 *Shewanella* 制备了自组装的复合生物膜结构，相比于自然生长的生物膜，这种三维复合生物膜结构提高了电子跨膜传递速度，可以将胞外电子输出提高 25 倍左右[8]。此外，提高电子传递中间体在生物膜体系中的浓度也可以强化生物膜中的电子传递过程。雍阳春等[9,10]通过调控微生物生长环境中的 pH 至弱碱性条件可以增加核黄素的分泌，进而强化整个生物膜的电子传递能力，体系的电能输出提高了 1.5 倍。

2　微生物电化学系统用于水中污染物去除

国内第一篇使用实际废水作为 MES 底物的论文发表于 2006 年，即哈尔滨工业大学尤世界等使用生活污水[11]和垃圾渗滤液[12]作为 MES 体系的底物，紧接着冯玉杰等[13]首次实现啤酒废水在 MES 中的转化。随后的 10 年中，聚焦于该技术在水处理中应用的关键科学与技术问题，研究者在电子传输机制、低成本材料研发、系统构建等多方面取得了突破性的进展。在材料成本可以接受的前提下，合理的结构优化、可放大的构型设计和切实可行的调控技术就愈发重要。

1）MES 结构设计与新构型开发

目前以微生物电化学系统为核心的水处理构型多采用平板式或管式的构型设计。管状结构可使用现有管状材料，例如陶管[14]、有机玻璃管[15]、聚丙烯管[16]和聚氯乙烯管[17]等，易于连接和选材。Ge 等[18]使用 96 个相对独立的 PVC 管为基体的管式空气阴极模块，搭建总体积约 200 L 的大型系统，并实现 MES 自产能抵消水泵运行能耗[19]。管式结构在长度方向（或高度方向）延展能力良好，但直径的变大有限，这影响了管式构型的进一步放大延展。平板式构型具有相互平行的阴阳极隔室，并可使用如离子交换膜、尼龙布、玻璃纤维布等间隔材料[20]分隔。在较大的系统构建中，多个平板式模块间可以共享边界，并能够通过平行叠加堆栈放大[21]从而具备一定大型化优势。哈尔滨工业大学冯玉杰课题组[22]构建了单体容积达 250 L 的空气阴极模块，并可通过多层堆栈构建数立方容积的系统。在此基础上开发出了一种特殊平板构型——卡式构型（Cassette type configuration）。卡式结构模块一般固定在反应器池体中[23,24]，避免了为 MES 单体分别加工外壁，节约了构建成本。应用该技术在构建的单模块(6 L)[25]和五模块(90 L)[26]系统中实现了以生活污水和啤酒废水为底物的系统能量自持运行和良好的出水水质。在卡式构型的基础上，课题组又开发出了阴阳极分置的简化构型[27]，仅保留 MES 的电极结构并平行排布插入污水池体[28]。该构型设计仅保留了 MES 中的必要功能组件，简化了结构，提高可堆栈性。MES 也可与传统厌氧技术构建耦合系统，如与连续搅拌反应釜（CSTR）构建的耦合系统[29,30]以及和厌氧折流板反应器（ABR）构建的耦合系统[15]均显示出了优良的性能，对高浓度的有机废水实现了污染物的梯级处理，有效提升了降解效率。目前用于污水处理的 MES 构型需要进一步简化，并使其具备外电路简化、较高运行效率、安装维护便易、不易污泥堵塞、长期的寿命和结构稳定等必要特征。

2）生物阴极 MES 及污染物去除技术

生物阴极是指阴极微生物通过自身的代谢作用，利用从阳极传递来的电子和质子，与最终的电子受体相结合，完成最终反应并获得能量，实现自身的生长和繁殖的过程。与化学阴极相比，生物阴极可利用阴极微生物或群落的协同作用，具有实现阴极区内污染物的高效去除、产生有价值的代谢副产物、无需使用贵金属催化剂等优点。

脱氮是生物阴极组重要的功能之一。冯玉杰等构建出了硝化型生物阴极 MES，实现了对氨氮的硝化，获得了最大功率密度 15.37 W/m^3，功率输出超过对照组的铂碳阴极系统，生物阴极的出水中氨氮浓度为 0.5 mg/L [31]。此外，通过对生物阴极电极材料表面进行疏水处理，提高其在复氧过程中捕获空气（氧气）的能力[32]，间歇式复氧方式所需能耗也仅占系统产能的 14.3%，在脱氮的同时实现了净能量输出，且这种复氧方式在总氮去除负荷和能量回收效率方面优于旋转式生物阴极、气体渗透膜曝气模块和气体被动扩散等供氧方式[33]。黄霞等构建了 50 L 的生物阴极 MES 反应器，实现了同步脱氮和电能回收，通过对外电路的调控，总氮的脱除效率可以达到 84%[34]。而对于一些有生物毒性的化合物如硝基苯[35]、2,4,6-三氯苯酚[36]甚至氯霉素[37]，也可实现基于 MES 体系的去除与同步产电。王爱杰等利用电极翻转的方法，实现了生物阴极的快速启动，经过 12 天的驯化后，直接作为生物阴极使用，可以实现硝基苯等污染物的快速降解[38]。邢德峰等利用双室微生物燃料电池体系中的生物阴极实现了对硝基苯酚的降解，在 0.5 V 的电压下，72 h 后对硝基苯酚的降解率可以达到 96%[39]。全燮和黄丽萍教授等驯化得到的生物阴极可以实现对铜离子的高效回收[40,41]。该研究组的结果也证明，在可以还原 Cd(II)的体系中加入乙酸钠或碳酸氢钠可以实现重金属回收效率的提升[42]。在生物阴极同步发生的电化学还原和微生物还原则可有效去除钒[43]。汞[44]、铜[45]、铬[46,47]等重金属离子也可在生物电化学系统的阴极得到电子，实现重金属或不溶金属盐在阴极的沉积。

3）MES 耦合的高级氧化技术

电芬顿技术将芬顿氧化与电化学相结合，电极反应可原位产生并利用自由基。理论上，发生氧还原反应生成过氧化氢需提供 0.269 V（*vs.* SHE）的阴极电势，微生物电化学系统的阳极电势一般在–0.3~–0.5 V 之间，这足以驱动阴极过氧化氢的产生。该体系的优势在于无需额外曝气，其产生的 H_2O_2 主要来自氧气的扩散，这为过氧化氢的合成提供了一种经济、有效的方法。因而，近年来 MES 耦合阴极电芬顿反应产生过氧化氢在去除难降解有毒有机污染物方面表现出很大潜力。一部分研究集中在阴极材料及过氧化氢产生方面，例如：Wang[48]等以石墨棒为阴极产生过氧化氢实现了对双酚 A、雌酮、磺胺及三氯卡班等新型污染物的高效去除；李楠[49]等也证明了三维石墨阴极有最大的过氧化氢产量；冯春华等[50]采用(CNT)/γ-FeOOH 为阴极，实现原位产生 Fe^{2+} 及 H_2O_2，并在中性条件下实现了对染料污染物的去除。另一部分研究侧重于 MES 与传统芬顿方法耦合。例如：俞汉青等[51]将 MES 引入阳极 Fenton 工艺中，以牺牲阳极的方式产生 Fe^{2+}，实现对偶氮染料 AO7 的去除。Zhang 等[52]构建的系统可切换原电池、电解池模式运行以实现过氧化氢的持续产生并达到去除残余污染物的目的。在 MEC 的运行模式下产生过氧化氢，并与 Fe^{2+} 发生反应产生·OH 实现对亚甲基蓝的去除，将体系切换至原电池模式运行后，残余的 H_2O_2 作为电子受体被利用。亚甲基蓝的去除效果及残余过氧化氢的去除受到外阻、阴极室 pH 及亚甲基蓝初始浓度的影响。大量研究均表明，MES 辅助阳极芬顿工艺表现出高于化学芬顿法的假一级速率常数，这不失为一种能量自持的、经济有效的电化学废水处理系统。

光能在 MES 中的利用除了通过生物的光合作用外，还可以通过光催化半导体材料的运用实现。光催化半导体材料与 MES 的耦合可实现对温室气体 CO_2 的有效还原。杨培东等利用 *Moorella thermoacetica* 细菌，将溶液中的镉离子和半胱氨酸转换为不溶性的硫化镉纳米粒子，并析出在细胞表面上。当用光照射时，硫化镉纳米粒子被激发后会释放出电子，然后将这些电子送入细菌体内用于将 CO_2 还原成乙酸[53]。该团队还建立了一个由 Si 和 TiO_2 纳米线作为光线捕捉单位的太阳能系统，用 *S. ovata* 细菌作为催化剂，能够有效地减少温和条件（例如有氧环境、中性 pH 环境、温度低于 30 °C）下的 CO_2，并且在模拟灯光下照射超过 200 h 可以产生乙酸，能量转换效率高达 0.38%[54]。俞汉青课题组将生物电化学与光催化系统耦合，还原硝酸盐到氮气，避免了亚硝酸中间产物的累积[55]。耦合生物电化学与升流式厌氧污泥床（UASB），成功去除了 2,4-二氯硝基苯[56]。耦合生物接触氧化反应器与升流式生物电化学的系统，成功去除了偶氮染料[57]。大连理工大学柳丽芬课题组制备了不同的催化剂，用电催化及光电催化阴极耦合生物电化学系统，提升了阴极室中难降解有毒污染物的去除效率[58]。通过耦合光催化电极，构建类似微生物原电池系统的光催化电池系统亦可实现污染物的低能耗高效地去除[59,60]。

3 MES 产氢及高附加值物质

微生物电解池系统（microbial electrolysis cell，MEC）是在微生物原电池（microbial fuel cell，MFC）基础上衍生的生物能源再生技术。两者均发生相似的阳极氧化反应，但微生物电解池系统中阴极发生还原反应的并非氧或氧化性物质，而是经阴极催化产生还原性的氢气、甲烷等高附加值产物。由于该过程在热力学上属于非自发过程，需要外加电源供给能量，因此称为电解池系统。刘红等[61]首先构建了 MEC 并报道了该技术，通过外加电压，阴极在无氧环境产生了氢气。自然环境中电化学活性微生物具备较高的种群多样性和广泛的底物利用范围，因此同传统产氢发酵和厌氧消化种群相比，MEC 的代谢和生态限制较少。报道证实 MEC 可利用小分子酸[62]、污水[63]、污泥[64]、生物质[62,65]等多类底物，适应低温等不利条件[66,67]，也可耦合产氢产乙醇发酵过程实现碳源梯级利用和高效产氢[68]。

由于嗜氢微生物在 MEC 的电极生物膜中存在和富集，部分氢气会被再次利用产生甲烷。成少安等[69]发现质子也可以在阴极生物膜的催化作用下直接接收电子而不经过氢气的生成实现微生物电化学直接催化的产甲烷过程。MEC 产生氢气和甲烷属于竞争性过程，在 Bruce E. Logan 课题组最早开展的中试 MEC 产氢研究中发现电化学催化的产氢过程较弱，生物气中主要为甲烷[70]。因此如何使 MEC 高效产氢抑制产甲烷，或如何通过催化产生更高纯度甲烷成为系统调控的难点。Liu 等利用 MEC 以剩余污泥为底物产生氢烷（hythane，在天然气或甲烷气体中含有 5%~20%氢气），为破解这一难题提供了新思路[71]。

除质子外，如 CO_2 等物质也可在阴极捕获电子，并催化产生高附加值的酸或醇类等有机产物[72]。外加能量、催化剂类型和电子受体与高附加值终产物的获取息息相关[73]。具备阴极催化能力的微生物来源广泛，产乙酸菌中的某些菌属如 *Clostridium aceticum*、 *Sporomusa sphaeroides*、*Moorella thermoacetica* 及 *Clostridium ljungdahlii* 等均可以利用电极电流来生产有机产物[74]。向 MES 阴极施加–0.4V 电位时，在菌株 *Sporomusa ovate* 催化作用下，CO_2 被还原为乙酸等[74]。当乙酸及丁酸为电子受体时，向基于阴极硫还原菌的体系施加–0.65V 的电压则可生成甲醇、乙醇、丙醇、丁醇及丙酮等更多种有机物[75]。中国科学院成都生物所李大平研究员课题组发现混合菌群作为生物催化剂具有更高的甲烷及乙酸转化率[76]，Batlle-Vilanova 等利用混合菌群培养实现了连续的乙酸合成[77]。电子中介体能够辅助电极向细菌的电子传递从而促进电合成过程。在–0.4V 的电压下，通过添加中性红电子中介体，丁酸产率可达到 0.44g/g[78]；Steinbusch 等则证明甲基紫罗碱可作为电子中介体并促进乙酸向醇类物质的转化[79]。

4　MES 土壤/沉积物修复

近年来的研究表明，MES 可用于土壤/沉积物有机污染物的去除，这是一种具有潜力的原位强化微生物的土壤/沉积物修复新方法。通过外加电路可将远距离电子受体（如氧气）与污染物连通，实现污染土壤/沉积物中电子受体的快速高效补充。

目前 MES 用于土壤修复研究多以石油烃为目标污染物。早期研究表明，炼油废水和石油烃污染的地下水可在 MES 中得到净化[80,81]。王鑫等[82]将 MES 用于原油污染土壤的修复，确定了阳极菌群对 C_8~C_{40} 正构烷烃和 16 种典型多环芳烃等石油烃均具有显著的降解促进作用。土壤含水率是影响降解率的关键，阳极对石油烃的修复半径可拓展至 34 cm[83]，而通过叠层阳极排布可将有效修复范围拓展 6 倍，并提高石油烃降解率[84]。通过调控土壤含砂量提高土壤断面的溶解氧和质子扩散速，可提升土壤 MES 的传质效率，且土壤石油烃的降解率提升了 268%，其中一些长链正构烷烃和高环的 PAHs 也被显著去除[84]。当土壤中存在易降解的共代谢有机物（如葡萄糖）时，土壤的脱氢酶和多酚氧化酶活性升高，土壤微生物多样性下降，共代谢效应刺激了烃类降解菌（例如食烷菌 *Alcanivorax*）的生长[85]，而金属还原菌在其中的作用机制尚不清晰[86]。向土壤中掺入导电性良好的碳纤维以降低土壤的内阻及向土壤灌水洗盐以降低土壤中盐分进而降低渗透压可将石油烃降解率提升 484%。萘双加氧酶和二甲苯单加氧酶基因拷贝数对土壤 MES 的产出电量和石油烃的降解效率均有线性关系，其含量可作为 MES 修复石油烃污染土壤的生物分子标志物[87,88]。MES 进行土壤修复的功能基于电极反应和极板间形成的电场作用。利用 MES 可以加速金属离子的迁移，并在电极区域富集，修复土壤重金属污染[89]；通过阳极还原脱氯，土壤 MES 加快了六氯苯的还原脱氯过程[90]。

针对沉积物及水体的环境修复，近年来 MES 展现出许多传统的物理化学等修复方法无法替代的优势。水生植物的根系能够合成大量的有机物，包括糖类、有机酸、聚合糖、酶以及坏死的细胞体等，这些有机物分泌到土壤中，能够为土壤中微生物群落的生长提供营养。产电微生物原位利用植物根系分泌物，产生持续的电子流。Hamelers 等在阳极区种植湿地芦苇构建了植物–MES 耦合系统，植物与异养微生物协同作用产电，获得的最大功率为 67 mW/m^2，首次构建了植物微生物电化学系统[91]。植物耦合 MES 的优势在于能更加高效地利用太阳能，不破坏植物的生长环境，构建植物–异养菌共生系统，在不断收获植物的前提下，充分利用根系分泌物产生清洁的电源。MES 应用于沉积物修复过程，巧妙地利用沉积物底部与上层水体之间电势差进行系统构建，一方面阳极置于底泥沉积物之中可作为原位高效的电子受体以加速沉积物中微生物对于污染物去除作用，另一方面阴极置于上层水体之中通过催化发生氧还原反应消耗由阳极经外电路传递过来的电子[92]。早期研究者们发现 MES 应用于沉积物修复时，即使不提供很好的营养温度等条件也可稳定运行并产生较高的电能输出，甚至可以做到在自然水体条件中稳定运行[93]。近年来关于沉积物及水体修复的研究多针对有机污染物去除，重金属还原/迁移转化以及特殊/高毒性污染物去除等。冯玉杰课题组通过一系列研究发现 MES 可使沉积物中总有机碳去除效率较其他修复方法提高 30%~220%，其中简单有机物的加入还可使去除效率高达 70%以上，同时对于多环芳烃类（PAHs）污染物去除率达到 50%~74%[94,95]。

将光合作用与沉积物 MES 耦合，最早出现在 20 世纪 60 年代，其耦合方式按照光合生物的作用可以分为以下两种，即①光合生物在阳极的利用，主要是通过光合作用为产电微生物合成代谢底物；②光合生物在阴极的利用，通过光合作用为阴极提供氧作为电子受体。但利用光合微生物产氢的产电过程并不稳定，通常需要贵金属催化剂，成本高，且容易失活，使光合微生物的产氢速率与氢气的消耗速率难以协调一致，造成产电不稳定[96]。科学家将这种藻菌协同关系引入微生物电化学系统中。Nealson 等在未添加任何有机物和电子中介体的沉积物 MES 中观测到电流的产生[97]。靠近阴极的表层微生物多为蓝藻和微藻，下层多为异养微生物，证实了光合生物与异养微生物的协同作用[98,99]。Malik 等在海洋沉积物 MES 中也发现了相似的现象[100]。所不同的是，作者指出光合作用不仅向阳极提供了必要的有机物质，而且为阴极反应提供了氧气，因而构成了完全不依赖外界物质供应、自我组装、自我修复的沉积物修复电化学系统。

5 拓展及展望

MES 在环境污染物净化方面表现出来的优势已经引起了全世界的注意，成为了 21 世纪最初 10 年环境科学与工程热点研究方向之一。应用 MES 技术通过微生物胞外电子转移实现污染物净化的途径与传统的厌氧和好氧途径均有本质上的不同，处于电场中的微生物在电刺激下，转化有机物的效率和代谢过程使得该技术在污染物净化中具有突出的优势。其有效的代谢小分子有机物的能力可为污染物降解菌解除产物抑制，在热力学水平促进污染物中的电子向完全矿化产物 CO_2 的流动，也为互营共生的污染物降解细菌的生长提供了局部生境。通过合理的设计可实现厌氧污染环境的异位电子补充，避免了高能耗的曝气以及硝酸盐、硫酸盐这类化学电子受体的人工投加。外电路的设计不仅是电能提取的重要一环，也是调控污染物降解速率、目标污染物种类的调控工具。推进 MES 走向应用的过程中，需要解决电子传递机制、生物膜形成与调控、新材料研发与利用、新构型设计及放大方法等一系列问题，这些问题的解决依赖于我们对微生物直接电子传递机理的新发现、自然界中依赖电子共享的互营过程的新认知，不断突破的材料科学的新发展以及对 MES 运行关键因素调控和创造性的体系构建。

致谢：感谢为本章做出文献收集和格式编排等贡献的博士研究生及博士后们，他们是哈尔滨工业大学李达，张鹏，梁丹丹，李鹤男等；感谢大连理工大学、南开大学和中国科学院城市环境研究所相关师生的支持；感谢清华大学梁鹏副教授提出的宝贵建议。感谢国家自然科学基金委杰出青年科学基金、优秀青年科学基金、面上基金和青年科学基金的支持（No.51125033、No. 21577068、No. 21673061、No. 21322703）。

参 考 文 献

[1] Li S W, Sheng G P, Cheng Y Y, et al. Redox Properties of Extracellular Polymeric Substances (EPS) from Electroactive Bacteria. Scientific Reports, 2016, 6: 39098.

[2] Dai Y F, Xiao Y, Zhang E H, et al. Effective Methods for Extracting Extracellular Polymeric Substances from Shewanella Oneidensis Mr-1. Water Science & Technology, 2016, 74(12): 2987.

[3] Extracellular Polymeric Substances Are Transient Media for Microbial Extracellular Electron Transfer. Science Advances, 2017, 3(7): e1700623

[4] Kondo K, Okamoto A, Hashimoto K, et al. Sulfur-Mediated Electron Shuttling Sustains Microbial Long-Distance Extracellular Electron Transfer with the Aid of Metallic Iron Sulfides. Langmuir: The Acs Journal of Surfaces & Colloids, 2015, 31(26): 7427-7434.

[5] Nakamura R, Kai F, Okamoto A, et al. Mechanisms of Long-Distance Extracellular Electron Transfer of Metal-Reducing Bacteria Mediated by Nanocolloidal Semiconductive Iron Oxides. Journal of Materials Chemistry A, 2013, 1(16): 5148-5157.

[6] Zhang P, Liu J, Qu Y, et al. Enhanced Performance of Microbial Fuel Cell with a Bacteria/Multi-Walled Carbon Nanotube Hybrid Biofilm. Journal of Power Sources, 2017, 361: 318-325.

[7] Yan F F, He Y R, Wu C, et al. Carbon Nanotubes Alter the Electron Flow Route and Enhance Nitrobenzene Reduction by Shewanella Oneidensis Mr-1. Environmental Science & Technology Letters, 2014, 1(1): 128-132.
[8] Yong Y C, Yu Y Y, Zhang X, et al. Highly Active Bidirectional Electron Transfer by a Self-Assembled Electroactive Reduced-Graphene-Oxide-Hybridized Biofilm. Angewandte Chemie, 2014, 126(17): 4569–4572.
[9] Yong Y C, Zhao C, Yu Y Y, et al. Increase of Riboflavin Biosynthesis Underlies Enhancement of Extracellular Electron Transfer of Shewanella in Alkaline Microbial Fuel Cells. Bioresource Technology, 2013, 130(2): 763-768.
[10] Yong Y-C, Cai Z, Yu Y-Y, et al. Increase of Riboflavin Biosynthesis Underlies Enhancement of Extracellular Electron Transfer of Shewanella in Alkaline Microbial Fuel Cells. Bioresource Technology, 2013, 130: 763-768.
[11] You S J, Zhao Q L, Jiang J Q, et al. Treatment of Domestic Wastewater with Simultaneous Electricity Generation in Microbial Fuel Cell under Continuous Operation. Chemical and Biochemical Engineering Quarterly, 2006, 20(4): 407-412.
[12] You S J, Zhao Q L, Jiang J Q, et al. Sustainable Approach for Leachate Treatment: Electricity Generation in Microbial Fuel Cell. Journal of Environmental Science & Health Part A Toxic/hazardous Substances & Environmental Engineering, 2006, 41(12): 2721-2734.
[13] Feng Y, Xin W, Logan B E, et al. Brewery Wastewater Treatment Using Air-Cathode Microbial Fuel Cells. Applied Microbiology and Biotechnology, 2008, 78(5): 873-880.
[14] Ghadge A N, Ghangrekar M M. Performance of Low Cost Scalable Air–Cathode Microbial Fuel Cell Made from Clayware Separator Using Multiple Electrodes. Bioresource Technology, 2015, 182: 373-377.
[15] Feng Y J, Lee H, Wang X, et al. Continuous Electricity Generation by a Graphite Granule Baffled Air-Cathode Microbial Fuel Cell. Bioresource Technology, 2010, 101(2): 632-638.
[16] Kim J R, Premier G C, Hawkes F R, et al. Development of a Tubular Microbial Fuel Cell (MFC) Employing a Membrane Electrode Assembly Cathode. Journal of Power Sources, 2009, 187(2): 393-399.
[17] Zhuang L, Zhou S G, Wang Y Q, et al. Membrane-Less Cloth Cathode Assembly (Cca) for Scalable Microbial Fuel Cells. Biosensors and Bioelectronics, 2009, 24(12): 3652-3656.
[18] Ge Z, Wu L, Zhang F, et al. Energy Extraction from a Large-Scale Microbial Fuel Cell System Treating Municipal Wastewater. Journal of Power Sources, 2015, 297: 260-264.
[19] Zheng G. Long-Term Performance of a 200 Liter Modularized Microbial Fuel Cell System Treating Municipal Wastewater: Treatment, Energy, and Cost. Environmental Science: Water Research & Technology, 2016, 2(2): 274-281.
[20] Zhang X Y, Cheng S O, Xin W, et al. Separator Characteristics for Increasing Performance of Microbial Fuel Cells. Environmental Science and Technology, 2009, 43(21): 8456-8461.
[21] Wu S, Li H, Zhou X, et al. A Novel Pilot-Scale Stacked Microbial Fuel Cell for Efficient Electricity Generation and Wastewater Treatment. Water Research, 2016, 98: 396.
[22] Feng Y, He W, Jia L, et al. A Horizontal Plug Flow and Stackable Pilot Microbial Fuel Cell for Municipal Wastewater Treatment. Bioresource Technology, 2014, 156(2): 132-138.
[23] Shimoyama T, Komukai S, Yamazawa A, et al. Electricity Generation from Model Organic Wastewater in a Cassette-Electrode Microbial Fuel Cell. Applied and Environmental Microbiology, 2008, 80(2): 325-330.
[24] Wang B, Han J I. A Single Chamber Stackable Microbial Fuel Cell with Air Cathode. Biotechnology Letters, 2009, 31(3): 387-393.
[25] Dong Y, Feng Y J, Qu Y P, et al. A Combined System of Microbial Fuel Cell and Intermittently Aerated Biological Filter for Energy Self-Sufficient Wastewater Treatment. Sci Rep, 2015, 5(3): 18070.
[26] Dong Y, Qu Y P, He W H, et al. A 90-Liter Stackable Baffled Microbial Fuel Cell for Brewery Wastewater Treatment Based on Energy Self-Sufficient Mode. Bioresource Technology, 2015, 195: 66-72.
[27] He W, Zhang X, Liu J, et al. Microbial Fuel Cells with an Integrated Spacer and Separate Anode and Cathode Modules. Environmental Science: Water Research and Technology, 2016, 2: 186-195.
[28] He W, Wallack M J, Kim K-Y, et al. The Effect of Flow Modes and Electrode Combinations on the Performance of a Multiple Module Microbial Fuel Cell Installed at Wastewater Treatment Plant. Water Research, 2016, 105(2016): 351-360.
[29] Wang H, Qu Y, Li D, et al. Evaluation of an Integrated Continuous Stirred Microbial Electrochemical Reactor: Wastewater Treatment, Energy Recovery and Microbial Community. Bioresource Technology, 2015, 195: 89.
[30] Wang H, Qu Y, Li D, et al. Cascade Degradation of Organic Matters in Brewery Wastewater Using a Continuous Stirred Microbial Electrochemical Reactor and Analysis of Microbial Communities. Scientific Reports, 2016, 6: 27023.
[31] Du Y, Feng Y J, Dong Y, et al. Coupling Interaction of Cathodic Reduction and Microbial Metabolism in Aerobic Biocathode of Microbial Fuel Cell. Rsc Advances, 2014, 4(65): 34350-34355.

[32] Wang H M, Liu J, He W H, et al. Enhanced Power Generation of Oxygen-Reducing Biocathode with an Alternating Hydrophobic and Hydrophilic Surface. ACS Applied Materials & Interfaces, 2016, 8(46): 31995-32003.

[33] Wang H M, Liu J, He W H, et al. Energy-Positive Nitrogen Removal from Reject Water Using a Tide-Type Biocathode Microbial Electrochemical System. Bioresource Technology, 2016, 222: 317-325.

[34] Liang P, Wei J C, Li M, et al. Scaling up a Novel Denitrifying Microbial Fuel Cell with an Oxic-Anoxic Two Stage Biocathode. Frontiers of Environmental Science & Engineering, 2013, 7(6): 913-919.

[35] Zhang E, Wang F, Zhai W, et al. Efficient Removal of Nitrobenzene and Concomitant Electricity Production by Single-Chamber Microbial Fuel Cells with Activated Carbon Air-Cathode. Bioresource Technology, 2017, 229: 111-118.

[36] Ding Y, Sun W, Cao L, et al. A Spontaneous Catalytic Membrane Reactor to Dechlorinate 2, 4, 6-Tcp as an Organic Pollutant in Wastewater and to Reclaim Electricity Simultaneously. Chemical Engineering Journal, 2016, 285: 573-580.

[37] Zhang Q, Zhang Y, Li D. Cometabolic Degradation of Chloramphenicol Via a Meta-Cleavage Pathway in a Microbial Fuel Cell and Its Microbial Community. Bioresource Technology, 2017, 229: 104-110.

[38] Yun H, Liang B, Kong D Y, et al. Polarity Inversion of Bioanode for Biocathodic Reduction of Aromatic Pollutants. Journal Of Hazardous Materials, 2017, 331: 280-288.

[39] Wang X Y, Xing D F, Ren N Q. P-Nitrophenol Degradation and Microbial Community Structure in a Biocathode Bioelectrochemical System. Rsc Advances, 2016, 6(92): 89821-89826.

[40] Shen J Y, Huang L P, Zhou P, et al. Correlation between Circuital Current, Cu(II) Reduction and Cellular Electron Transfer in Eab Isolated from Cu(II)-Reduced Biocathodes of Microbial Fuel Cells. Bioelectrochemistry, 2017, 114: 1-7.

[41] Tao Y, Xue H, Huang L P, et al. Fluorescent Probe Based Subcellular Distribution of Cu(II) Ions in Living Electrotrophs Isolated from Cu(II)-Reduced Biocathodes of Microbial Fuel Cells. Bioresource Technology, 2017, 225: 316-325.

[42] Chen Y R, Shen J Y, Huang L P, et al. Enhanced Cd(II) Removal with Simultaneous Hydrogen Production in Biocathode Microbial Electrolysis Cells in the Presence of Acetate or Nahco3. International Journal of Hydrogen Energy, 2016, 41(31): 13368-13379.

[43] Qiu R, Zhang B, Li J, et al. Enhanced Vanadium(V) Reduction and Bioelectricity Generation in Microbial Fuel Cells with Biocathode. Journal of Power Sources, 2017, 359: 379-383.

[44] Wang Z, Lim B, Choi C. Removal of Hg^{2+} as an Electron Acceptor Coupled with Power Generation Using a Microbial Fuel Cell. Bioresource Technology, 2011, 102(10): 6304-6307.

[45] Wu D, Huang L, Quan X, et al. Electricity Generation and Bivalent Copper Reduction as a Function of Operation Time and Cathode Electrode Material in Microbial Fuel Cells. Journal of Power Sources, 2016, 307: 705-714.

[46] Gupta S, Yadav A, Verma N. Simultaneous Cr(Vi) Reduction and Bioelectricity Generation Using Microbial Fuel Cell Based on Alumina-Nickel Nanoparticles-Dispersed Carbon Nanofiber Electrode. Chemical Engineering Journal, 2017, 307: 729-738.

[47] Li Y, Lu A, Ding H, et al. Cr(Vi)Reduction at Rutile-Catalyzed Cathode in Microbial Fuel Cells. Electrochemistry Communications, 2009, 11(7): 1496-1499.

[48] Wang Y, Feng C, Li Y, et al. Enhancement of Emerging Contaminants Removal Using Fenton Reaction Driven by H_2O_2 -Producing Microbial Fuel Cells. Chemical Engineering Journal, 2016, 307: 679-686.

[49] Chen J Y, Li N, Zhao L. Three-Dimensional Electrode Microbial Fuel Cell for Hydrogen Peroxide Synthesis Coupled to Wastewater Treatment. Journal of Power Sources, 2014, 254(15): 316-322.

[50] Feng C, Li F, Liu H, et al. A Dual-Chamber Microbial Fuel Cell with Conductive Film-Modified Anode and Cathode and Its Application for the Neutral Electro-Fenton Process. Electrochimica Acta, 2010, 55(6): 2048-2054.

[51] Liu X W, Sun X F, Li D B, et al. Anodic Fenton Process Assisted by a Microbial Fuel Cell for Enhanced Degradation of Organic Pollutants. Water Research, 2012, 46(14): 4371-4378.

[52] Zhang Y, Wang Y, Angelidaki I. Alternate Switching between Microbial Fuel Cell and Microbial Electrolysis Cell Operation as a New Method to Control H_2O_2 Level in Bioelectro-Fenton System. Journal of Power Sources, 2015, 291: 108-116.

[53] Sakimoto K K, Wong A B, Yang P. Self-Photosensitization of Nonphotosynthetic Bacteria for Solar-to-Chemical Production. Science, 2016, 351(6268): 74.

[54] Liu C, Gallagher J J, Sakimoto K K, et al. Nanowire-Bacteria Hybrids for Unassisted Solar Carbon Dioxide Fixation to Value-Added Chemicals. Nano Letters, 2015, 15(5): 3634.

[55] Lin Z-Q, Yuan S-J, Li W-W, et al. Denitrification in an Integrated Bioelectro-Photocatalytic System. Water Research, 2017, 109: 88-93.

[56] Chen H, Gao X, Wang C, et al. Efficient 2, 4-Dichloronitrobenzene Removal in the Coupled Bes-UASB Reactor: Effect of

External Voltage Mode. Bioresource Technology, 2017, 241: 879-886.
[57] Pan Y, Wang Y, Zhou A, et al. Removal of Azo Dye in an up-Flow Membrane-Less Bioelectrochemical System Integrated with Bio-Contact Oxidation Reactor. Chemical Engineering Journal, 2017, 326: 454-461.
[58] Gao C, Liu L, Yang F. Development of a Novel Proton Exchange Membrane-Free Integrated MFC System with Electric Membrane Bioreactor and Air Contact Oxidation Bed for Efficient and Energy-Saving Wastewater Treatment. Bioresource Technology, 2017, 238: 472-483.
[59] Yu T, Liu L, Li L, et al. A Self-Biased Fuel Cell with Tio 2/Gc 3 N 4 Anode Catalyzed Alkaline Pollutant Degradation with Light and without Light—What Is the Degradation Mechanism?. Electrochimica Acta, 2016, 210: 122-129.
[60] Du Y, Feng Y J, Qu Y P, et al. Electricity Generation and Pollutant Degradation Using a Novel Biocathode Coupled Photoelectrochemical Cell. Environmental Science & Technology, 2014, 48(13): 7634-7641.
[61] Liu H, Grot S, Logan B E. Electrochemically Assisted Microbial Production of Hydrogen from Acetate. Environmental Science & Technology, 2005, 39(11): 4317-4320.
[62] Cheng S, Logan B E. Sustainable and Efficient Biohydrogen Production Via Electrohydrogenesis. Proceedings of the National Academy of Sciences, 2007, 104(47): 18871-18873.
[63] Pant D, Van Bogaert G, Diels L, et al. A Review of the Substrates Used in Microbial Fuel Cells (MFCS) for Sustainable Energy Production. Bioresource Technology, 2010, 101(6): 1533-1543.
[64] Liu W, Huang S, Zhou A, et al. Hydrogen Generation in Microbial Electrolysis Cell Feeding with Fermentation Liquid of Waste Activated Sludge. International Journal of Hydrogen Energy, 2012, 37(18): 13859-13864.
[65] Wang A, Sun D, Cao G, et al. Integrated Hydrogen Production Process from Cellulose by Combining Dark Fermentation, Microbial Fuel Cells, and a Microbial Electrolysis Cell. Bioresource Technology, 2011, 102(5): 4137-4143.
[66] Lu L, Ren N, Zhao X, et al. Hydrogen Production, Methanogen Inhibition and Microbial Community Structures in Psychrophilic Single-Chamber Microbial Electrolysis Cells. Energy & Environmental Science, 2011, 4(4): 1329-1336.
[67] Liu L, Tsyganova O, Lee D-J, et al. Anodic Biofilm in Single-Chamber Microbial Fuel Cells Cultivated under Different Temperatures. International Journal of Hydrogen Energy, 2012, 37(20): 15792-15800.
[68] Lu L, Ren N, Xing D, et al. Hydrogen Production with Effluent from an Ethanol–H 2-Coproducing Fermentation Reactor Using a Single-Chamber Microbial Electrolysis Cell. Biosensors and Bioelectronics, 2009, 24(10): 3055-3060.
[69] Cheng S A, Xing D F, Call D F, et al. Direct Biological Conversion of Electrical Current into Methane by Electromethanogenesis. Environmental Science & Technology, 2009, 43(10): 3953-3958.
[70] Cusick R D, Bryan B, Parker D S, et al. Performance of a Pilot-Scale Continuous Flow Microbial Electrolysis Cell Fed Winery Wastewater. Applied Microbiology and Biotechnology, 2011, 89(6): 2053-2063.
[71] Liu Q, Ren Z J, Huang C, et al. Multiple Syntrophic Interactions Drive Biohythane Production from Waste Sludge in Microbial Electrolysis Cells. Biotechnology for Biofuels, 2016, 9(1): 162.
[72] Rabaey K, Rozendal R A. Microbial Electrosynthesis—Revisiting the Electrical Route for Microbial Production. Nature reviews Microbiology, 2010, 8(10): 706.
[73] Kumar G, Saratale R G, Kadier A, et al. A Review on Bio-Electrochemical Systems (BESS) for the Syngas and Value Added Biochemicals Production. Chemosphere, 2017, 177: 84-92.
[74] Nevin K P, Hensley S A, Franks A E, et al. Electrosynthesis of Organic Compounds from Carbon Dioxide Is Catalyzed by a Diversity of Acetogenic Microorganisms. Applied and Environmental Microbiology, 2011, 77(9): 2882-2886.
[75] Sharma M, Aryal N, Sarma P M, et al. Bioelectrocatalyzed Reduction of Acetic and Butyric Acids Via Direct Electron Transfer Using a Mixed Culture of Sulfate-Reducers Drives Electrosynthesis of Alcohols and Acetone. Chemical Communications, 2013, 49(58): 6495-6497.
[76] Jiang Y, Su M, Zhang Y, et al. Bioelectrochemical Systems for Simultaneously Production of Methane and Acetate from Carbon Dioxide at Relatively High Rate. International Journal of Hydrogen Energy, 2013, 38(8): 3497-3502.
[77] Batlle - Vilanova P, Puig S, Gonzalez-Olmos R, et al. Continuous Acetate Production through Microbial Electrosynthesis from CO2 with Microbial Mixed Culture. Journal of Chemical Technology and Biotechnology, 2016, 91(4): 921-927.
[78] Choi O, Um Y, Sang B I. Butyrate Production Enhancement by Clostridium Tyrobutyricum Using Electron Mediators and a Cathodic Electron Donor. Biotechnology and Bioengineering, 2012, 109(10): 2494-2502.
[79] Steinbusch K J, Hamelers H V, Schaap J D, et al. Bioelectrochemical Ethanol Production through Mediated Acetate Reduction by Mixed Cultures. Environmental Science & Technology, 2009, 44(1): 513-517.
[80] Morris J M, Jin S. Enhanced Biodegradation of Hydrocarbon-Contaminated Sediments Using Microbial Fuel Cells. Journal of Hazardous Materials, 2012, 213-214: 474-477.

[81] Morris J M, Jin S, Crimi B, et al. Microbial Fuel Cell in Enhancing Anaerobic Biodegradation of Diesel. Chemical Engineering Journal, 2009, 146(2): 161-167.
[82] Wang X, Cai Z, Zhou Q, et al. Bioelectrochemical Stimulation of Petroleum Hydrocarbon Degradation in Saline Soil Using U-Tube Microbial Fuel Cells. Biotechnology and Bioengineering, 2012, 109(2): 426-433.
[83] Lu L, Yazdi H, Jin S, et al. Enhanced Bioremediation of Hydrocarbon-Contaminated Soil Using Pilot-Scale Bioelectrochemical Systems. Journal of Hazardous Materials, 2014, 274(12): 8.
[84] Li X, Wang X, Zhang Y, et al. Extended Petroleum Hydrocarbon Bioremediation in Saline Soil Using Pt-Free Multianodes Microbial Fuel Cells. Rsc Advances, 2014, 4(104): 59803-59808.
[85] Li X, Wang X, Wan L, et al. Enhanced Biodegradation of Aged Petroleum Hydrocarbons in Soils by Glucose Addition in Microbial Fuel Cells. Journal of Chemical Technology & Biotechnology, 2016, 91(1): 267-275.
[86] Lu L, Huggins T, Jin S, et al. Microbial Metabolism and Community Structure in Response to Bioelectrochemically Enhanced Remediation of Petroleum Hydrocarbon-Contaminated Soil. Environmental Science & Technology, 2014, 48(7): 4021-4029.
[87] Li X, Wang X, Zhang Y, et al. Salinity and Conductivity Amendment of Soil Enhanced the Bioelectrochemical Degradation of Petroleum Hydrocarbons. Scientific Reports, 2016, 6: 32861.
[88] Li X, Xin W, Qian Z, et al. Carbon Fiber Enhanced Bioelectricity Generation in Soil Microbial Fuel Cells. Biosensors & Bioelectronics, 2016, 85: 135-141.
[89] Habibul N, Hu Y, Sheng G-P. Microbial Fuel Cell Driving Electrokinetic Remediation of Toxic Metal Contaminated Soils. Journal of Hazardous Materials, 2016, 318: 9-14.
[90] Wang H, Yi S, Cao X, et al. Reductive Dechlorination of Hexachlorobenzene Subjected to Several Conditions in a Bioelectrochemical System. Ecotoxicology and Environmental Safety, 2017, 139: 172-178.
[91] Strik D P B T B, Hamelers H V M, Snel J F H, et al. Green Electricity Production with Living Plants and Bacteria in a Fuel Cell. International Journal of Energy Research, 2008, 32(9): 870-876.
[92] Li W W, Yu H Q. Stimulating Sediment Bioremediation with Benthic Microbial Fuel Cells. Biotechnology Advances, 2015, 33(1): 1-12.
[93] Tender L M, Gray S A, Groveman E, et al. The First Demonstration of a Microbial Fuel Cell as a Viable Power Supply: Powering a Meteorological Buoy. Journal of Power Sources, 2008, 179(2): 571-575.
[94] Li H, Tian Y, Qu Y, et al. A Pilot-Scale Benthic Microbial Electrochemical System (BMES) for Enhanced Organic Removal in Sediment Restoration. Scientific Reports, 2017, 7: 39802.
[95] Li H, He W, Qu Y, et al. Pilot-Scale Benthic Microbial Electrochemical System (BMES) for the Bioremediation of Polluted River Sediment. Journal of Power Sources, 2017, DOI: 10.1038/srep39802.
[96] Harnisch F, Schröder U, Quaas M, et al. Electrocatalytic and Corrosion Behaviour of Tungsten Carbide in near-Neutral Ph Electrolytes. Applied Catalysis B Environmental, 2009, 87(1-2): 63-69.
[97] He Z, Kan J, Mansfeld F, et al. Self-Sustained Phototrophic Microbial Fuel Cells Based on the Synergistic Cooperation between Photosynthetic Microorganisms and Heterotrophic Bacteria. Environmental Science & Technology, 2009, 43(5): 1648-1654.
[98] Zou Y, Pisciotta J, Billmyre R B, et al. Photosynthetic Microbial Fuel Cells with Positive Light Response. Biotechnology & Bioengineering, 2009, 104(5): 939.
[99] He Z, Shao H, Angenent L T. Increased Power Production from a Sediment Microbial Fuel Cell with a Rotating Cathode. Biosensors & Bioelectronics, 2007, 22(12): 3252-3255.
[100] Malik S, Drott E, Grisdela P, et al. A Self-Assembling Self-Repairing Microbial Photoelectrochemical Solar Cell. Energy & Environmental Science, 2009, 2(3): 292-298.

作者：冯玉杰[1]，邢德峰[1]，柳丽芬[2]，王　鑫[3]，赵　峰[4]，刘　佳[1]，何伟华[1]，杨　俏[2]

[1]哈尔滨工业大学城市水资源与水环境国家重点实验室，[2]大连理工大学食品与环境学院，

[3]南开大学环境科学与工程学院，[4]中国科学院城市环境研究所

第 20 章　微生物燃料电池：从产电到水处理

本章导读

污水中有机物富含未被利用的化学能，微生物燃料电池（microbial fuel cell，MFC）是一种利用产电微生物将有机物中的化学能转化为电能的新型污水处理技术，该技术已成为污水能源化与资源化领域的研究热点之一。目前，有关 MFC 的研究主要集中在胞外电子传递机制的探究、电极材料和构型的优化、废水处理及离子回收以及传感器的应用。微生物胞外电子传递以及与电极间的电子转移过程，决定了 MFC 产电性能并影响污染物降解性能。MFC 的应用领域从不同类型废水的处理，拓展到离子去除及回收、传感器水质监测与预警功能。本章介绍了 MFC 的胞外电子传递机制、电极材料与构型优化以及 MFC 在能量回收、污水处理及水质监测等方面的研究进展与发展趋势。

关键词

微生物燃料电池，胞外电子传递机制，电极材料，废水处理，传感器

1 引　　言

微生物燃料电池（microbial fuel cell，MFC）是一种利用产电微生物对废水中有机物氧化降解，并原位将有机物中化学能转化为电能的新型污水处理技术[1]。阳极微生物作为生物催化剂，氧化有机物并产生电子和质子，电子通过外电路从阳极转移至阴极，质子在电场力驱动作用下透过分隔材料传递至阴极室，阴极的电子受体接收阳极传递的电子和质子，在阴极催化剂作用下发生还原反应，进而实现化学能向电能的转化。

MFC 可以同步产电并处理污水。就产电而言，厌氧产电微生物参与氧化反应，产生电子并传递至阳极材料，因而研究微生物胞外电子传递机制对提高电子传递效率具有重要意义。阴阳极作为导体传递电子并发生电极反应，需具有导电性、高生物相容性等特性，因而电极材料是影响整个 MFC 系统性能的关键因素。就污水处理而言，MFC 不仅可以去除易降解有机物，也能降解某些难降解有机物。MFC 的变形工艺——微生物脱盐电池（microbial desalination cell，MDC），在电场力作用下驱动内部离子态物质定向迁移，可以实现脱盐及离子回收。此外，MFC 还可以作为传感器，以电化学活性微生物作为指示剂，根据其发出信号的改变，进而判断水质中某些特定物质的变化，对水质进行预警。

MFC 作为一种新型污水处理技术，除了常见的污水处理、产电等基本功能外，还可以通过改变构型，实现对盐分和离子的脱除和回收，同时对有机物、毒性物质等进行监测和预警，在污水处理、能量回收以及水质监测等方面具有巨大的发展潜力。

2 微生物燃料电池中胞外电子传递机制

微生物胞外电子传递（extracellular electron transfer，EET）是地球表层生态系统中能量传递和元素循环的核心驱动力，也是微生物燃料电池产电的关键步骤。微生物种间直接电子传递（direct interspecies electrons transfer，DIET）是继种间 H_2 传递和种间甲酸传递外的第三种种间电子传递途径[2,3]。

目前，电子从微生物胞内释放到胞外主要有四种电子传递机理（图 20-1），即①通过微生物自身分泌

的纳米导线完成胞外电子传递[4]。*Geobacter* 和 *Shewanella* 等菌属能够分泌一种直径约为 8 nm、长数十微米的导电性蛋白(pili)，它们接受来自于细胞内膜、外膜或周质相关蛋白质中的电子并将电子传递至胞外[4,5]；②利用细胞色素 C 实现胞外电子传递。有些微生物如 *Shewanella oneidensis* MR-1 等通过分泌 CymA、Fcc、MtrA、MtrB、MtrC、OmcA 和 STC 等一系列色素，氧化细胞膜上的氢醌完成电子传递[6-10]；③借助电子穿梭体进行胞外电子传递[11]。微生物附着在具有一定氧化还原电位的电极上时会分泌可溶性电子穿梭体，电子穿梭体在胞内接受电子后穿过细胞膜将电子传递至胞外[12]，外源电子穿梭体也能促进胞外电子传递[13]；④通过应电运动完成胞外电子传递[14]。微生物把电子储存在细胞表面形成生物电容器后再将电子传递给其胞外较近的电子受体。

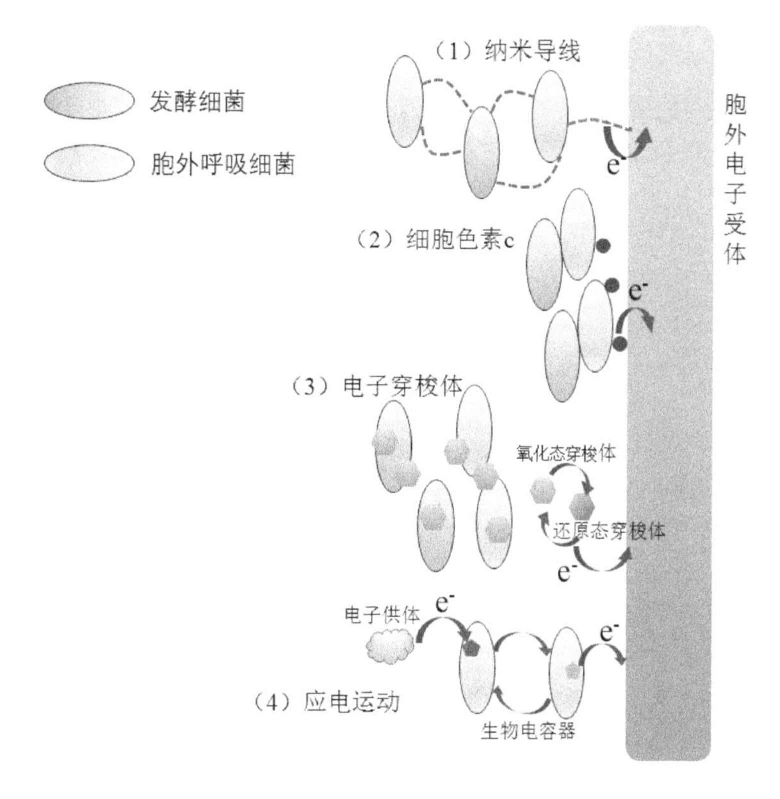

图 20-1　胞外电子传递机理示意图

微生物燃料电池中产电微生物主要通过纳米导线和细胞色素 C 进行传递，纳米导线对电子的传递类似于金属导电，通过纳米导线上重叠的 π-π 键上的电子离域实现；细胞色素 C 对电子的传递通过生物膜中发生的一系列氧化还原反应实现。在同一生物膜内可能同时存在以上两种传递方式，例如，*Geobacter sulfurreducens* 生物膜同时存在纳米导线和细胞色素 C，但微生物产电时电子以金属导电型还是氧化还原反应途径进行传递，仍然存在争议[15,16]。

3　微生物燃料电池电极材料与构型

MFC 的关键过程在于电子的转移，电极材料特性对 MFC 性能有重要影响。在 MFC 中电子传递的主要限速步骤在于，电子在微生物与阳极之间的传递，以及氧气在阴极还原的电子传递过程。阳极选择主要考虑生物相容性好、电导率高的材料，阴极主要选择电催化活性好、电导率高的材料。

3.1 阳极材料

阳极可分为自支撑材料与非自支撑材料。非自支撑材料作为阳极需要依赖一定的基底，在基底上进一步加以负载，来改善电子传递过程，增强生物相容性，提升材料表面催化活性。目前常用基底为碳基材料和金属基材料。碳基材料主要以碳布、碳纸、碳毡为代表，通过在碳基材料上负载碳纳米管[17]、MnO_2[18]、聚吡咯[19]、聚苯胺（PANI）[20]等可以提升产电性能。金属基底材料主要为不锈钢网、泡沫镍等，可以通过添加碳黑[21]、石墨烯[22]提升性能。

自支撑材料直接作为阳极，主要有两大类：碳基材料与导电聚合物。在碳基材料中，可通过天然材料（细菌纤维素[23]、蚕茧[24]、丝瓜络[25]等）和非天然材料（聚丙烯腈[26]）直接碳化得到三维电极，强化电子在生物膜与电极之间的传递；导电聚合物，具有易于改性掺杂的优势，可直接用作阳极，如聚甲基丙烯酸甲酯（PMMA）掺杂金阵列[27]等。在自支撑材料的基础上，也可进一步通过化学修饰提升产电性能。

目前，非自支撑材料在阳极材料研究中比较广泛，然而自支撑材料近年来发展快速，由于其高比表面积、高电导性，使阳极性能得到较大提升，逐渐成为未来发展方向。

3.2 阴极材料

阴极分为生物阴极和化学阴极。生物阴极是利用微生物进行阴极还原反应，与 MFC 阳极类似，都是电极附着微生物进行电子传递过程。常用电极材料为活性炭、碳毡、石墨、碳布等，具有较高的导电性和生物相容性。化学阴极，常用氧气作为电子受体，在非生物催化剂作用下进行氧还原反应。催化材料的不断开发和改良，促进了化学阴极的发展，是 MFC 研究领域较活跃的一个分支。

催化剂是影响阴极性能的关键因素之一，目前常用的阴极催化剂主要分为三类：贵金属、过渡金属、碳材料。以 Pt 为代表的贵金属催化剂，性能较高，但价格昂贵、资源稀缺、长期运行稳定性较差。过渡金属及其衍生物催化活性较好，通常以碳基材料为基底，如活性炭、石墨烯、碳纳米管等，负载过渡金属及其氧化物，如 Fe[28,29]、Mn[30,31]、Co[32,33]、Ni[34]、Cu[35]，过渡金属多以氧化物形式负载，在此基础上进行无机元素与过渡金属的共掺杂可以进一步提升催化性能，其中对 Fe-N 共掺杂研究关注较多。

碳材料，具有来源广泛、价格低廉、催化性能较好的特点。以天然生物质为前驱体碳化制备的碳材料，在阴极催化剂中有着良好的应用，其中活性炭应用广泛。此外新型生物质碳材料如竹炭[36]、碳化苔藓[37]等具有多孔结构，同时负载有 N、P 等元素，作为阴极氧还原催化剂，可得到与 Pt/C 相近的产电效果。人工合成的多孔碳材料具有巨大的比表面积，如以有机泡沫[38]、离子液体[39]为前驱体制备的多孔碳材料可以使产电功率显著提升。纳米碳材料如石墨烯、碳纳米管，由于其良好的电导性及催化性能，在阴极中可直接作为催化剂或作为负载基底。为进一步提升碳基阴极氧还原催化活性，可以通过表面修饰与材料复合的方法进行改性。在表面修饰方面，表面氧化处理是一种常用的方法，如将石墨[40]进行化学浸泡对表面进行氧化处理；在碳材料表面掺杂无机元素，如 N[41,42]、P[43]等多种元素，以及 N-S 共掺杂[44]可以降低 O_2 吸附能，提升氧还原催化活性。在材料复合方面，以活性炭复合碳黑[45,46]、多孔碳材料[47]可以降低电极电阻率，提升 MFC 产电功率。

3.3 MFC 构型优化

MFC 构型根据阴极不同，主要有空气阴极 MFC 和生物阴极 MFC 两大类构型。空气阴极 MFC，以空气中的氧气为电子受体，不需要曝气，从而节省了能耗。空气阴极以不锈钢网、碳布、碳纸等材料为基底，采用涂抹、辊压[48]、直压[49]等方法制备催化层与扩散层。扩散层需要同时具有透氧和防水的作用[49]，是空气阴极制备与放大的难点，空气阴极的规模化制备也是空气阴极 MFC 放大并推向实际应用的关键。浙江大学针对空气阴极型 MFC 放大的关键技术开展了研究，解决了阴极漏水、析盐等问题，开发出以泡沫镍为集电体、活性炭为催化剂和 PTFE 为扩散层和黏结剂的空气阴极，其性能与 Pt/C（含 0.5 mg/cm^2 Pt）阴极相当，但成本仅为 Pt/C 阴极的 1/30[50]。采用微波处理阴极催化层，可以改善催化层的孔隙结构和 PTFE 的分布，从而提高空气阴极的氧还原性能[51]。通过采用多点或边线引出的方法可减小功率损失 40%以上[52]。目前，浙江大学已开发出 1 m^3 空气阴极 MFC，并采用化妆品生产废水进行了现场中试试验，COD 去除率达 85.3 %± 3.8%，电池最大输出功率达 0.5 W/m^3。

生物阴极 MFC，阴极以电极作为载体附着微生物，常在阴极液中通过曝气等方式通氧进行氧还原反应，虽然曝气需要耗能，但不需要解决阴极防水问题，相对空气阴极更容易放大，同时生物阴极可以与水处理功能耦合。清华大学开发了堆叠式构型的生物阴极 MFC，并开展了放大研究，通过优化腔体厚度、水路和电路的连接，用于处理城市污水，最大功率密度达 33 W/m^3，装置出水 COD 浓度满足城镇污水排放一级 A 标准[53]。与空气阴极相比，生物阴极由于曝气增加了运行成本，且反应接触面积受限，因此对于生物阴极的改进，需增加反应接触面积，减小传质阻力，提高氧利用率。

4　MFC 处理废水

利用 MFC 阳极的氧化反应或阴极的还原反应可以实现对污水中有机污染物的氧化降解或污染物的还原转化；分隔膜材料，可实现污染物的分离。将 MFC 阳极、阴极、分隔膜各单元耦合，可进一步强化 MFC 对废水的处理效果。

4.1　利用 MFC 阳极处理废水

利用 MFC 阳极的氧化过程是 MFC 作为污水处理技术的最常见方式。MFC 的阳极富含厌氧微生物，利用阳极产电菌与其他厌氧微生物之间的耦合作用，促进污染物在阳极的降解或转化，可以实现对多种废水的处理，既包括易降解废水如生活污水、啤酒废水，也包括含难降解有机物的工业废水，如石化废水、农药废水等[53-56]。MFC 运行时为闭路，阳极厌氧氧化污染物产生的电子可以直接传递至电极，通过外电路传递到阴极，电子受体还原，而开路状态（传统厌氧污水处理方式）下这一过程被阻断。MFC 闭路运行时 COD 降解速率高于开路状态，且低外阻运行产生高电流时，COD 降解得到进一步强化。即，MFC 可以实现有机物的快速降解与同步高电流产电，产电过程可促进 COD 降解，COD 快速降解进而提升电流输出[57]。

4.2　利用 MFC 阴极处理废水

利用 MFC 阴极的还原，可以去除氧化态污染物。氧化态污染物可分为有机和无机两大类。氧化态有机物，例如硝基苯、氯霉素等难降解有机污染物，可以在 MFC 阴极得到还原、脱毒、降解。Wang 等利用 MFC 的生物阴极在外加 0.5 V 的电压下，使硝基苯的还原去除率达 98.7%，还原产物-苯胺的生成率达 88.2%[58]。Liang 等利用生物阴极在 0.5 V 的外加电压下，实现了对氯霉素（CAP）的脱硝及脱氯，并将氯霉素转化为苯胺类产物，大幅度降低了氯霉素的毒性，CAP 的去除率在 24 h 内可高达 96%[59]。

MFC 阴极还可对氧化态的无机污染物进行去除，包括对废水中硝态氮的还原以实现脱氮，以及重金属离子的还原，实现金属单质的回收或转化为低毒性的价态。Pang 等在 MFC 的阴极中添加聚吡咯聚合物作为电子穿梭体，在 pH 为中性条件下将六价铬离子还原为三价铬离子并生成沉淀去除[60]。Huang 等的研究表明，利用生物阴极可以获得比非生物阴极更高的六价铬还原效率[61]。

4.3 MFC 阳极与阴极的耦合处理

耦合 MFC 的阳极氧化反应与阴极的还原反应，可以实现对废水中的污染物更进一步的矿化降解。Liu 等构建了双室 MFC 反应器，利用阴极对偶氮苯进行还原转化并同时利用阳极对偶氮苯的还原产物，即苯胺，作进一步的氧化去除。该系统不需要额外的能量投入，实现了对偶氮苯废水的自驱动矿化降解[62]。Huang 等将阳极置于 0.2 V、生物阴极置于–0.3 V 条件下，启动 MFC 后，可以提高 MFC 处理五氯苯酚的性能，五氯苯酚在阳极中的降解率提高了 28.5%，在阴极中的降解率提高了 21.5%[63]。Zuo 等构建了堆叠型 MFC，污水经过阳极室、阴极室及中间脱盐室，可以实现污水的有机物、氮及盐分的同步脱除[64]。

4.4 MFC 与其他水处理工艺结合

由于废水中污染物成分较复杂，某些污染物如固体悬浮颗粒、氮磷、难降解有机污染物等难以在 MFC 中被彻底去除，需要与其他工艺进行耦合以强化 MFC 对这类物质的处理能力。Wang 等构建了一种新型的膜生物反应器（MBR）与 MFC 结合的装置（BEMR），该装置可用于处理污水并实现了能量的回收。该系统对 COD 及氨氮的平均去除率分别可达 92.4%及 95.6%[65]。Zang 等构建了鸟粪石沉淀（MAP）与 MFC 结合的工艺，以回收尿液中的氮磷元素并同时产电，该系统对 COD、磷酸与氨氮的去除率分别可达 64.9%、94.6%与 28.6%，产电可达 2.6 W/m^3[66]。Shen 等构建了 MFC 耦合 UASB 反应器，以处理对硝基酚（PNP）废水，该系统相比于传统 UASB 工艺可以显著提高对 PNP 的去除率及对氨酚（PAP）的生成率，当电流密度从 0 提高至 4.71 A/m^3 时，PNP 的去除负荷及 PAP 的生成负荷分别从 6.16 mol/(m^3·d)及 4.21 mol/(m^3·d)提高至 6.77 mol/(m^3·d)及 6.11 mol/(m^3·d)，并且系统对碳源的需求相比传统 UASB 工艺有显著的降低[67]。Lou 等构建了无膜生物接触法与 MFC 耦合工艺以强化对 PNP 的还原转化，该系统在阴极电势为–1 V 下对 PNP 的去除负荷可高达 18.95 mol/(m^3·d)，且系统具有较强的抗冲击负荷能力[68]。

5 MFC 对离子的去除与回收

2009 年，清华大学通过对 MFC 构型的改变，首次提出 MDC 的概念，在 MFC 的阴阳极之间放置一对离子交换膜，从而定向引导离子迁移，实现低能耗、可持续的含盐水脱盐[69]。MDC 技术诞生后即受到国内外各研究机构的关注，科研人员们通过反应器构型创新实现了 MDC 产电及脱盐效果的优化，通过扩展处理对象实现了受污染地下水修复、污水中氮磷去除与回收以及酸碱等有用物质的生产等功能[70]。

近年来，在 MDC 产电及脱盐效果优化方面，对堆叠型 MDC（stacked MDC，SMDC）及上流式 MDC（up-flow MDC，UMDC）的研究相对较多。在首次提出 SMDC 概念[71]后，清华大学逐步实现了 SMDC 脱盐膜堆数目的优化、内阻的降低以及规模的放大，在 10 层脱盐膜堆构型中获得了已有报道中最高的脱盐速率（423 mg/h）及电子利用效率(836%)[72]，在装填离子交换树脂的构型中实现了内阻的大幅度降低

（~10 Ω）及构型的扩大(10 L)[73]。美国弗吉尼亚理工大学自提出 UMDC 构型后，也实现了脱盐速率的逐步提升，并成功搭建运行了已有报道中容量最大的 MDC 反应器(105 L)[74]。在致力于利用 MDC 进行脱盐的各类研究中，与阴极产氢或金属还原相耦合的研究[75,76]，将 MDC 与电容吸附技术相结合的研究[77,78]，以及借助正渗透膜实现脱盐的研究[79]也均有阶段性进展。

在脱盐相关研究的经验基础上，科研人员也尝试利用 MFC 内电场驱动其他各类离子型物质的迁移，从而实现污染物去除、有用物质生产以及资源回收。Zhang 等搭建的浸没式 MDC 通过驱动 NO_3^-或 NH_4^+迁移完成地下水修复[80,81]。Brastad 和 He 的研究证明，MDC 可完成对水中砷、铜、汞和镍的高效去除[82]。Zuo 等利用多级浸没式 MDC（multi-stage microbial desalination cell，M-MDC），实现了对生活污水中有机物和盐分的同步去除[83]。在去除水溶液中离子型物质的基础上，MFC 还可实现同步的资源回收。2015 年，清华大学在 MDC 的基础上提出微生物氮磷回收电池（microbial nutrient recovery cell，MNRC）的概念，通过对调 MDC 中阴阳离子交换膜的位置改变离子迁移的方向，从而实现污水中氮磷的去除与同步浓缩回收[84,85]。而以尿液中氮磷为对象，利用 MFC 实现高效资源回收的研究，近年间也多次进入人们的视野[86-88]。此外，中山大学利用 MFC 内电场诱使双极膜产生 H^+和 OH^-，从而实现酸碱物质的低能耗生产[89]。该方向的研究进一步发展，系统的脱盐及产酸碱能力进一步提升[90,91]，在二氧化碳固定化方面的可行性也得到了验证[92]。

通过生物电化学的方法实现离子的去除与回收会进一步朝着高效率、低能耗的方向发展。此外，在现有实验室小试研究的基础上，扩大反应器体积，提高反应器处理能力，将生物电化学方法推向实用化将会是未来研究的热点。

6　MFC 传感器

MFC 传感器采用电活性微生物作为指示生物，分类上属于微生物电化学类生物传感器[93]。其工作原理是水质变化会影响电活性微生物的胞内代谢和胞外电子传递，导致电信号的变化。MFC 传感器主要用于有机物、毒性物质及微生物活性监测。其优势是直接输出电信号，使用寿命长，运行成本低，不依赖于高电流或高功率输出，易于实用化[94,95]。商用 MFC 传感器主要是韩国 korbi 公司推出的 HATOX-2000 毒性物质监测系统和 HABS-2000 在线生化需氧量分析仪。

6.1　有机物监测

有机物为电活性微生物的碳源，其浓度变化可引起 MFC 传感器的电信号变化[96]。可选取多个电信号（电流、库仑量和电压等）指标关联有机物总体参数（COD、BOD）或某单一成分浓度（如乙酸和葡萄糖等）（表 20-1）。此外，厌氧消化处理过程的中间产物挥发性脂肪酸（VFA）也可以作为电活性微生物的碳源，利用 MFC 传感器可以实现实时在线监测厌氧消化液中 VFA 的变化，从而优化系统运行，并对处理效果加以实时监测，这是 MFC 传感器监测有机物的另一种应用途径。

6.2　毒性物质监测

MFC 传感器可用于监测重金属、表面活性剂和农药等毒性物质（表 20-2）。电活性微生物通常在电极表面以生物膜形式存在，这有利于连续在线监测。但是，生物膜的抗性较高，导致 MFC 传感器的监测

表 20-1　MFC 传感器用于有机物监测

构型	信号指标	线性范围/(mg/L)	响应时间/min	参考文献
双室	电流/BOD	0.32~5.12[d]	<60	[97]
双室	库仑量/BOD	0.64~10.88[d]	<60	[98]
MEA[a]	ANN/COD[b]	5~200	N.A.[d]	[99]
MEA	电压/BOD	5~120	132	[100]
单室	CV 峰/BOD	32~320[c]	12	[101]

a 膜电极组（membrane electrode assembly，MEA）；b 人工神经网络模拟；c 标准曲线计算值。d 没有提供（Not available，N.A.）。

下限难以达到水质标准的要求。实际水体监测时，有机物/毒性物质复合冲击会对基于生物阳极敏感元件的传感器信号造成削弱甚至掩蔽[102,103]。近年来，研究者开始关注生物阴极敏感元件在 MFC 传感器中的应用[104-106]。为提高传感器对冲击负荷或毒性环境的抵抗能力，可在电活性生物膜表面修饰一层 50 nm 聚多巴胺保护层。研究表明，受到保护的传感器可在 pH 0.5 冲击后恢复电活性。

表 20-2　MFC 传感器用于毒物监测

毒物	构型	监测下限/(mg/L)	水质标准/(mg/L)[e]	参考文献
Ni(I)	膜[b]	20	0.07	[107]
Cu(II)	双室	2	2	[108]
Cd(II)	MEA[c]	0.1	0.003	[109]
Hg(II)	双室	1	0.001	[110]
Cr(VI)	单室	1	0.05	[111]
Zn(II)	双室	2	N.A.[f]	[112]
明矾	单室	50	N.A.	[113]
SDS[a]	双室	10	N.A.	[114]
甲醛	双室	0.0005%[d]	N.A.	[104]
左氧氟沙星	单室	0.0001	N.A.	[115]
高灭磷	MEA	1	N.A.	[109]
4-硝基酚	双室	9	N.A.	[116]
替卡西林	双室	75	N.A.	[117]

a 十二烷基磺酸钠（sodium dodecyl sulfonate，SDS）。b 平面微滤膜作为电极。c 膜电极组（membrane electrode assembly，MEA）。d 单位为%。e 世界卫生组织饮用水标准第四版[118]。f 没有提供（Not Available，N.A.）。

6.3　微生物活性监测

MFC 传感器非电活性微生物监测的目的是防止病菌感染、评价抗菌素药效、测试环境样品中微生物活性；对电活性微生物监测的目的是发现新型电活性微生物[119]。微生物活性监测的技术关键是通过外源电子介体添加[120]、特异性底物刺激[121]、新型纳米材料改性[122]等方式实现代谢活性与电信号关联。其优势是能方便快速地监测微生物活性，局限是无法保证对所有微生物种类都有效，如某些微生物难以人工构建胞外电子传递途径，也不向胞外分泌能发生电催化反应的代谢产物[123]。

7　展　　望

目前对于 MFC 的研究着重于电子传递机理的探究、电极材料及构型的优化、污水处理的强化、离子去除及回收以及传感器应用的拓展等。进一步深化 MFC 电子传递机理的探索，调控电极生物膜的电子转递效率和电化学活性，是提升 MFC 性能的理论基础。进一步改善电极催化活性、简化制备方法以及降低电极成本，构建适合于 MFC 规模放大的构型并进行优化是 MFC 实用化需要解决的重要问题。在 MFC 产电性能提升的同时，拓展 MFC 的应用功能，将产电功能和水处理功能耦合，开发集产电、污水处理与资源回收为一体的集成工艺；将 MFC 应用于受污染土壤、地下水的修复；基于生物电化学的原理，发展高灵敏度、稳定的 MFC 传感器水质监测与预警系统；基于数学模型对 MFC 各种应用工艺的能量流、物质流进行模拟与优化等，是 MFC 未来的重要研究与发展趋势。

致谢：感谢为本章做出文献收集和格式编排等贡献的博士研究生，他们是清华大学蒋永、刘赋斌、伍世嘉、徐婷、王秋莹，感谢中国科学院烟台海岸带研究所徐恒铎博士提出的宝贵建议。感谢国家自然科学基金委重点科学基金、优秀青年科学基金、面上项目和青年科学基金的支持（No. 51238004、No. 51402810、No. 51478414、No. 51408336）。

参 考 文 献

[1] Logan B E, Hamelers B, Rozendal R A, Schrorder U, Keller J, Freguia S, Aelterman P, Verstraete W, Rabaey K. Microbial fuel cells: Methodology and technology. Environmental Science & Technology, 2006, 40(17): 5181-5192.

[2] Shen L, Zhao Q, Wu X, Li X, Li Q, Wang Y. Interspecies electron transfer in syntrophic methanogenic consortia: From cultures to bioreactors. Renewable & Sustainable Energy Reviews, 2016, 54: 1358-1367.

[3] Boone D R, Johnson R L, Liu Y. Diffusion of the interspecies electron carriers H_2 and formate in methanogenic ecosystems and its implications in the measurement of KM for H_2 or formate uptake. Applied and Environmental Microbiology, 1989, 55(7): 1735-1741.

[4] Reguera G, McCarthy K D, Mehta T, Nicoll J S, Tuominen M T, Lovley D R. Extracellular electron transfer via microbial nanowires. Nature, 2005, 435(7045): 1098.

[5] Shi L, Dong H, Reguera G, Beyenal H, Lu A, Liu J, Yu H Q, Fredrickson J K. Extracellular electron transfer mechanisms between microorganisms and minerals. Nature Reviews Microbiology, 2016, 14(10): 651.

[6] Beliaev A S, Saffarini D A. Shewanella putrefaciens mtrB encodes an outer membrane protein required for Fe(III) and Mn(IV) reduction. Journal of Bacteriology, 1998, 180(23): 6292.

[7] Beliaev A S, Saffarini D A, McLaughlin J L, Hunnicutt D. MtrC, an outer membrane decahaem c cytochrome required for metal reduction in Shewanella putrefaciens MR-1. Molecular Microbiology, 2001, 39(3): 722–730.

[8] Myers J M, Myers C R. Role of the tetraheme cytochrome CymA in anaerobic electron transport in cells of *Shewanella putrefaciens* MR-1 with normal levels of menaquinone. Journal of Bacteriology, 2000, 182(1): 67-75.

[9] Myers CR, Myers JM. MtrB is required for proper incorporation of the cytochromes OmcA and OmcB into the outer membrane of Shewanella putrefaciens MR-1. Applied & Environmental Microbiology, 2002, 68(11): 5585-94.

[10] Sturm G, Richter K, Doetsch A, Heide H, Louro R O, Gescher J. A dynamic periplasmic electron transfer network enables respiratory flexibility beyond a thermodynamic regulatory regime. Isme Journal, 2015.

[11] Marsili E, Baron D B, Shikhare I D, Coursolle D, Gralnick J A, Bond D R. Shewanella Secretes flavins that mediate extracellular electron transfer. Proceedings of the National Academy of Sciences of the United States of America, 2008, 105(10): 3968-3973.

[12] Gralnick J A, Newman D K. Extracellular respiration. Molecular Microbiology, 2007, 65(1): 1-11.

[13] Xu H, Quan X. Anode modification with peptide nanotubes encapsulating riboflavin enhanced power generation in microbial fuel cells. International Journal of Hydrogen Energy, 2016, 41(3): 1966-1973.

[14] Harris H W, El-Naggar M Y, Bretschger O, Ward M J, Romine M F, Obraztsova A Y, Nealson K H. Electrokinesis is a microbial behavior that requires extracellular electron transport. Proceedings of the National Academy of Sciences of the United States of America, 2010, 107(1): 326.

[15] Malvankar N S, Vargas M, Nevin K P, Franks A E, Leang C, Kim B-C, Inoue K, Mester T, Covalla S F, Johnson J P, Rotello V M, Tuominen M T, Lovley D R. Tunable metallic-like conductivity in microbial nanowire networks. Nature Nanotechnology, 2011, 6(9): 573-579.

[16] Snider R M, Strycharz-Glaven S M, Tsoi S D, Erickson J S, Tender L M. Long-range electron transport in Geobacter sulfurreducens biofilms is redox gradient-driven. Proceedings of the National Academy of Sciences of the United States of America, 2012, 109(38): 15467-15472.

[17] Liu X-W, Huang Y-X, Sun X-F, Sheng G-P, Zhao F, Wang S-G, Yu H-Q. Conductive carbon nanotube hydrogel as a bioanode for enhanced microbial electrocatalysis. ACS Applied Materials & Interfaces, 2014, 6(11): 8158-64.

[18] Zhang C, Liang P, Yang X, Jiang Y, Bian Y, Chen C, Zhang X, Huang X. Binder-free graphene and manganese oxide coated carbon felt anode for high-performance microbial fuel cell. Biosensors & Bioelectronics, 2016, 81: 32-38.

[19] Lv Z, Chen Y, Wei H, Li F, Hu Y, Wei C, Feng C. One-step electrosynthesis of polypyrrole/graphene oxide composites for microbial fuel cell application. Electrochimica Acta, 2013, 111: 366-373.

[20] Zhang W, Xie B, Yang L, Liang D, Zhu Y, Liu H. Brush-like polyaniline nanoarray modified anode for improvement of power output in microbial fuel cell. Bioresource Technology, 2017, 233: 291-295.

[21] Zheng S, Yang F, Chen S, Liu L, Xiong Q, Yu T, Zhao F, Schroeder U, Hou H. Binder-free carbon black/stainless steel mesh composite electrode for high-performance anode in microbial fuel cells. Journal of Power Sources, 2015, 284: 252-257.

[22] Wang H, Wang G, Ling Y, Qian F, Song Y, Lu X, Chen S, Tong Y, Li Y. High power density microbial fuel cell with flexible 3D graphene-nickel foam as anode. Nanoscale, 2013, 5(21): 10283-10290.

[23] Zou L, Qiao Y, Wu Z-Y, Wu X-S, Xie J-L, Yu S-H, Guo J, Li C M. Tailoring unique mesopores of hierarchically porous structures for fast direct electrochemistry in microbial fuel cells. Advanced Energy Materials. 2016. 6(4): DOI: 10.1002/aenm. 201501535.

[24] Lu M, Qian Y, Yang C, Huang X, Li H, Xie X, Huang L, Huang W. Nitrogen-enriched pseudographitic anode derived from silk cocoon with tunable flexibility for microbial fuel cells. Nano Energy, 2017, 32: 382-388.

[25] Yuan Y, Zhou S, Liu Y, Tang J. Nanostructured macroporous bioanode based on polyaniline-modified natural loofah sponge for high-performance microbial fuel cells. Environmental Science & Technology, 2013, 47(24): 14525-14532.

[26] Wang Y-Q, Huang H-X, Li B, Li W-S. Novelly developed three-dimensional carbon scaffold anodes from polyacrylonitrile for microbial fuel cells. Journal of Materials Chemistry A, 2015, 3(9): 5110-5118.

[27] Chen S, Chen X, Hou S, Xiong P, Xiong Y, Zhang F, Yu H, Liu G, Tian Y. A gold microarray electrode on a poly(methylmethacrylate) substrate to improve the performance of microbial fuel cells by modifying biofilm formation. Rsc Advances, 2016, 6(115): 114937-114943.

[28] Cao C, Wei L, Wang G, Shen J. Superiority of boron, nitrogen and iron ternary doped carbonized graphene oxide-based catalysts for oxygen reduction in microbial fuel cells. Nanoscale, 2017, 9(10): 3537-3546.

[29] Xu X, Dai Y, Yu J, Hao L, Duan Y, Sun Y, Zhang Y, Lin Y, Zou J. Metallic state FeS anchored (Fe)/Fe_3O_4/N-doped graphitic carbon with porous spongelike structure as durable catalysts for enhancing bioelectricity generation. ACS Applied Materials & Interfaces, 2017, 9(12): 10777-10787.

[30] Yuan H, Deng L, Tang J, Zhou S, Chen Y, Yuan Y. Facile synthesis of MnO_2/polypyrrole/MnO_2 multiwalled nanotubes as advanced electrocatalysts for the oxygen reduction reaction. Chem Electrochem, 2015, 2(8): 1152-1158.

[31] Yuan H, Deng L, Chen Y, Yuan Y. MnO_2/Polypyrrole/MnO_2 multi-walled-nanotube-modified anode for high-performance microbial fuel cells. Electrochimica Acta, 2016, 196: 280-285.

[32] Song T-S, Wang D-B, Wang H, Li X, Liang Y, Xie J. Cobalt oxide/nanocarbon hybrid materials as alternative cathode catalyst for oxygen reduction in microbial fuel cell. International Journal of Hydrogen Energy, 2015, 40(10): 3868-3874.

[33] You S, Gong X, Wang W, Qi D, Wang X, Chen X, Ren N. Enhanced cathodic oxygen reduction and power production of microbial fuel cell based on noble-metal-free electrocatalyst derived from metal-organic frameworks. Advanced Energy Materials, 2016, 6(1).

[34] Luo S, He Z. Ni-coated carbon fiber as an alternative cathode electrode material to improve cost efficiency of microbial fuel cells. Electrochimica Acta, 2016, 222: 338-346.

[35] Zhang X, Li K, Yan P, Liu Z, Pu L. N-type Cu_2O doped activated carbon as catalyst for improving power generation of air cathode microbial fuel cells. Bioresource Technology, 2015, 187: 299-304.

[36] Yang W, Li J, Ye D, Zhu X, Liao Q. Bamboo charcoal as a cost-effective catalyst for an air-cathode of microbial fuel cells. Electrochimica Acta, 2017, 224: 585-592.

[37] Zhou L, Fu P, Wen D, Yuan Y, Zhou S. Self-constructed carbon nanoparticles-coated porous biocarbon from plant moss as advanced oxygen reduction catalysts. Applied Catalysis B-Environmental, 2016, 181: 635-643.

[38] Chen S, He G, Hu H, Jin S, Zhou Y, He Y, He S, Zhao F, Hou H. Elastic carbon foam via direct carbonization of polymer

foam for flexible electrodes and organic chemical absorption. Energy & Environmental Science, 2013, 6(8): 2435-2439.

[39] Zhang X, He W, Zhang R, Wang Q, Liang P, Huang X, Logan BE, Fellinger T-P. High-performance carbon aerogel air cathodes for microbial fuel cells. ChemSusChem, 2016, 9(19): 2788-2795.

[40] Zhang L, Lu Z, Li D, Ma J, Song P, Huang G, Liu Y, Cai L. Chemically activated graphite enhanced oxygen reduction and power output in catalyst-free microbial fuel cells. Journal of Cleaner Production, 2016, 115: 332-336.

[41] Feng LY, Yang L Q, Huang Z J, Luo J Y, Li M, Wang D B, Chen Y G. Enhancing electrocatalytic oxygen reduction on nitrogen-doped graphene by active sites implantation. Scientific Reports, 2013. 3.

[42] Wang Q, Zhang X, Lv R, Chen X, Xue B, Liang P, Huang X. Binder-free nitrogen-doped graphene catalyst aircathodes for microbial fuel cells. Journal of Materials Chemistry A, 2016, 4(32): 12387-12391.

[43] Liu Y, Li K, Liu Y, Pu L, Chen Z, Deng S. The high-performance and mechanism of P-doped activated carbon as a catalyst for air-cathode microbial fuel cells. Journal of Materials Chemistry A, 2015, 3(42): 21149-21158.

[44] Yuan H, Hou Y, Wen Z, Guo X, Chen J, He Z. Porous carbon nanosheets codoped with nitrogen and sulfur for oxygen reduction reaction in microbial fuel cells. ACS Applied Materials & Interfaces, 2015, 7(33): 18672-18678.

[45] Zhang X, Pant D, Zhang F, Liu J, He W, Logan B E. Long-term performance of chemically and physically modified activated carbons in air cathodes of microbial fuel cells. Chemelectrochem, 2014, 1(11): 1859-1866.

[46] Zhang X, Xia X, Ivanov I, Huang X, Logan B E. Enhanced activated carbon cathode performance for microbial guel cell by blending carbon black. Environmental Science & Technology, 2014, 48(3): 2075-2081.

[47] Zhang X, Wang Q, Xia X, He W, Huang X, Logan B E. Addition of conductive particles to improve the performance of activated carbon air-cathodes in microbial fuel cells. Environmental Science: Water Research & Technology, 2017, 3: 806-810.

[48] Peng X, Yu H, Wang X, Gao N, Geng L, Ai L. Enhanced anode performance of microbial fuel cells by adding nanosemiconductor goethite. Journal of Power Sources, 2013, 223: 94-99.

[49] Zhang X, He W, Yang W, Liu J, Wang Q, Liang P, Huang X, Logan B E. Diffusion layer characteristics for increasing the performance of activated carbon air cathodes in microbial fuel cells. Environmental Science: Water Research & Technology, 2016, 2(2): 266-273.

[50] Cheng S, Wu J. Air-cathode preparation with activated carbon as catalyst, PTFE as binder and nickel foam as current collector for microbial fuel cells. Bioelectrochemistry, 2013, 92: 22-26.

[51] Liu W, Zhou Y, Cheng S, Xu M, Li F. Microwave preparation of catalyst layer for enhancing the oxygen reduction of air cathode in microbial fuel cells. International Journal of Electrochemical Science, 2017, 12(3): 2207-2218.

[52] Cheng S, Ye Y, Ding W, Pan B. Enhancing power generation of scale-up microbial fuel cells by optimizing the leading-out terminal of anode. Journal of Power Sources, 2014, 248: 931-938.

[53] Wu S, Li H, Zhou X, Liang P, Zhang X, Jiang Y, Huang X. A novel pilot-scale stacked microbial fuel cell for efficient electricity generation and wastewater treatment. Water Research, 2016, 98: 396.

[54] Xin W, Zhang C, Zhou Q X, Zhang Z N, Chen C H. Bioelectrochemical stimulation of petroleum hydrocarbon degradation in saline soil using U-tube microbial fuel cells. Biotechnology & Bioengineering, 2012, 109(2): 426-433.

[55] Huang L, Gan L, Zhao Q, Logan BE, Lu H, Chen G. Degradation of pentachlorophenol with the presence of fermentable and non-fermentable co-substrates in a microbial fuel cell. Bioresource Technology, 2011, 102(19): 8762-8768.

[56] Yu H, Feng C, Liu X, Yi X, Ren Y, Wei C. Enhanced anaerobic dechlorination of polychlorinated biphenyl in sediments by bioanode stimulation. Environmental Pollution, 2016, 211: 81-89.

[57] Zhang X, He W, Ren L, Stager J, Evans PJ, Logan B E. COD removal characteristics in air-cathode microbial fuel cells. Bioresource Technology, 2015, 176: 23-31.

[58] Wang A J, Cheng H Y, Liang B, Ren N Q, Cui D, Lin N, Kim B H, Rabaey K. Efficient Reduction of Nitrobenzene to Aniline with a Biocatalyzed Cathode. Environmental Science & Technology, 2011, 45(23): 10186-10193.

[59] Liang B, Cheng H Y, Kong D Y, Gao S H, Sun F, Cui D, Kong F Y, Zhou AJ, Liu W Z, Ren N Q. Accelerated reduction of chlorinated nitroaromatic antibiotic chloramphenicol by biocathode. Environmental Science & Technology, 2013, 47(10): 5353.

[60] Pang Y, Xie D, Wu B, Lv Z, Zeng X, Wei C, Feng C. Conductive polymer-mediated Cr(VI) reduction in a dual-chamber microbial fuel cell under neutral conditions. Synthetic Metals, 2013, 183: 57-62.

[61] Huang L, Chai X, Chen G, Logan B E. Effect of set potential on hexavalent chromium reduction and electricity generation from biocathode microbial fuel cells. Environmental Science & Technology, 2011, 45(11): 5025.

[62] Liu R H, Li W W, Sheng G P, Tong Z H, Lam H W, Yu H Q. Self-driven bioelectrochemical mineralization of azobenzene by coupling cathodic reduction with anodic intermediate oxidation. Electrochimica Acta, 2015, 154: 294-299.

[63] Huang L, Wang Q, Quan X, Liu Y, Chen G. Bioanodes/biocathodes formed at optimal potentials enhance subsequent pentachlorophenol degradation and power generation from microbial fuel cells. Bioelectrochemistry, 2013, 94(12): 13.

[64] Zuo K, Wang Z, Chen X, Zhang X, Zuo J, Liang P, Huang X. Self-driven desalination and advanced treatment of wastewater in a modularized filtration air cathode microbial desalination cell. Environmental Science & Technology, 2016, 50(13): 7254.

[65] Wang Y K, Sheng G P, Li W W, Huang Y X, Yu Y Y, Zeng R J, Yu H Q. Development of a novel bioelectrochemical membrane reactor for wastewater treatment. Environmental Science & Technology, 2011, 45(21): 9256-9261.

[66] Zang G L, Sheng G P, Li W W, Tong Z H, Zeng R J, Shi C, Yu H Q. Nutrient removal and energy production in a urine treatment process using magnesium ammonium phosphate precipitation and a microbial fuel cell technique. Physical Chemistry Chemical Physics, 2012, 14(6): 1978-84.

[67] Shen J Y, Xu X P, Jiang X B, Hua C X, Zhang L B, Sun X Y, Li J S, Mu Y, Wang L J. Coupling of a bioelectrochemical system for p-nitrophenol removal in an upflow anaerobic sludge blanket reactor. Water Research, 2014, 67(67C): 11.

[68] Lou S, Jiang X, Chen D, Shen J, Han W, Sun X, Li J, Wang LJ. Enhanced *p*-nitrophenol removal in a membrane-free bio-contact coupled bioelectrochemical system. Rsc Advances, 2015, 5(34): 27052-9.

[69] Cao X X, Xia H, Peng L, Kang X, Zhou Y J, Zhang X Y, Logan B E. A new method for water desalination using microbial desalination cells. Environmental Science & Technology, 2009, 43(18): 7148-7152.

[70] Chen X, Liang P, Zhang X, Huang X. Bioelectrochemical systems-driven directional ion transport enables low-energy water desalination, pollutant removal, and resource recovery. Bioresour Technol, 2016, 215: 274-84.

[71] Chen X, Xia X, Liang P, Cao X, Sun H, Huang X. Stacked microbial desalination cells to enhance water desalination efficiency. Environmental Science & Technology, 2011, 45(6): 2465.

[72] Chen X, Sun H, Liang P, Zhang X, Huang X. Optimization of membrane stack configuration in enlarged microbial desalination cells for efficient water desalination. Journal of Power Sources, 2016, 324: 79-85.

[73] Zuo K, Cai J, Liang S, Wu S, Zhang C, Liang P, Huang X. A ten liter stacked microbial desalination cell packed with mixed ion-exchange resins for secondary effluent desalination. Environmental Science & Technology, 2014, 48(16): 9917-24.

[74] Fei Z, Zhen H. Scaling up microbial desalination cell system with a post-aerobic process for simultaneous wastewater treatment and seawater desalination. Desalination, 2015, 360: 28-34.

[75] Mehanna M, Kiely P D, Call D F, Logan B E. Microbial electrodialysis cell for simultaneous water desalination and hydrogen gas production. Environmental Science & Technology, 2010, 44(24): 9578.

[76] Haiping L, Jenkins P E, Zhiyong R. Concurrent desalination and hydrogen generation using microbial electrolysis and desalination cells. Environmental Science & Technology, 2011, 45(1): 340.

[77] Liang P, Yuan L, Yang X, Huang X. Influence of circuit arrangement on the performance of a microbial fuel cell driven capacitive deionization (MFC-CDI) system. Desalination, 2015, 369: 68-74.

[78] Forrestal C, Stoll Z, Xu P, Ren Z J. Microbial capacitive desalination for integrated organic matter and salt removal and energy production from unconventional natural gas produced water. Environmental Science Water Research & Technology, 2015, 1(1): 47-55.

[79] Yuan H, Abu-Reesh I M, He Z. Enhancing desalination and wastewater treatment by coupling microbial desalination cells with forward osmosis. Chemical Engineering Journal, 2015, 270: 437-443.

[80] Zhang Y, Angelidaki I. A new method for in situ nitrate removal from groundwater using submerged microbial desalination-denitrification cell (SMDDC). Water Research, 2013, 47(5): 1827-1836.

[81] Zhang Y, Angelidaki I. Submersible microbial desalination cell for simultaneous ammonia recovery and electricity production from anaerobic reactors containing high levels of ammonia. Bioresource Technology, 2015, 177(177C): 233.

[82] Brastad K S, He Z. Water softening using microbial desalination cell technology. Desalination, 2013, 309(3): 32-37.

[83] Zuo K, Liu F, Ren S, Zhang X, Liang P, Huang X. A novel multi-stage microbial desalination cell for simultaneous desalination and enhanced organics and nitrogen removal from domestic wastewater. Environmental Science Water Research & Technology, 2016, 2(5).

[84] Chen X, Sun D, Zhang X, Liang P, Huang X. Novel self-driven microbial nutrient recovery cell with simultaneous wastewater purification. Scientific Reports, 2015, 5: 15744.

[85] Chen X, Zhou H, Zuo K, Zhou Y, Wang Q, Sun D, Gao Y, Liang P, Zhang X, Ren Z J, Huang X. Self-sustaining advanced wastewater purification and simultaneous in situ nutrient recovery in a novel bioelectrochemical system. Chemical

Engineering Journal, 2017, 330: 697.

[86] Tice R C, Kim Y. Energy efficient reconcentration of diluted human urine using ion exchange membranes in bioelectrochemical systems. Water Research, 2014, 64(7): 61-72.

[87] Ledezma P, Jermakka J, Keller J, Freguia S. Recovering nitrogen as a solid without chemical dosing: Bio-electroconcentration for recovery of nutrients from urine. Environmental Science & Technology Letters, 2017, 4(3): 119-124.

[88] Chen X, Gao Y, Hou D, Ma H, Lu L, Sun D, Zhang X, Liang P, Huang X, Ren Z J. Microbial electrochemical current accelerates urea hydrolysis for nutrient recovery from source-separated urine. Environ. Sci. Technol. Lett., 2017, 4 (7): 305-310.

[89] Chen S, Liu G, Zhang R, Qin B, Luo Y, Hou Y. Development of the microbial electrodialysis and chemical production cell for desalination as well as acid and alkali productions. Environmental Science & Technology, 2012, 46(4): 2467-2472.

[90] Chen S, Liu G, Zhang R, Qin B, Luo Y, Hou Y. Improved performance of the microbial electrolysis desalination and chemical-production cell using the stack structure. Bioresource Technology, 2012, 116(7): 507-511.

[91] Seneta E, Zeelenberg K. Integrated utilization of seawater using a five-chamber bioelectrochemical system. Journal of Membrane Science, 2013, 444(1): 16-21.

[92] Zhu X, Logan B E. Microbial electrolysis desalination and chemical-production cell for CO_2 sequestration. Bioresource Technology, 2014, 159(5): 24-29.

[93] Su L, Jia W, Hou C, Lei Y. Microbial biosensors: A review. Biosensors & Bioelectronics, 2011, 26(5): 1788.

[94] Yang W, Wei X, Fraiwan A, Coogan C G, Lee H, Choi S. Fast and sensitive water quality assessment: A μL-scale microbial fuel cell-based biosensor integrated with an air-bubble trap and electrochemical sensing functionality. Sensors & Actuators B Chemical, 2016, 226: 191-195.

[95] Stein N E, Hamelers H V M, Buisman C N J. Influence of membrane type, current and potential on the response to chemical toxicants of a microbial fuel cell based biosensor. Sensors & Actuators B Chemical, 2012, 163(1): 1-7.

[96] Kim B H, Chang I S, Gil G C, Park H S, Kim H J. Novel BOD (Biochemical Oxygen Demand) sensor using mediator-less microbial fuel cell. 2003.

[97] Cheng L, Quek S B, Cord-Ruwisch R. Hexacyanoferrate-adapted biofilm enables the development of a microbial fuel cell biosensor to detect trace levels of assimilable organic carbon (AOC) in oxygenated seawater. Biotechnology & Bioengineering, 2014, 111(12): 2412-2420.

[98] Quek SB, Cheng L, Cordruwisch R. Detection of low concentration of assimilable organic carbon in seawater prior to reverse osmosis membrane using microbial electrolysis cell biosensor. Desalination & Water Treatment, 2014, 25(3): 2885-2890.

[99] Feng Y H, Harper W F, Jr. Biosensing with microbial fuel cells and artificial neural networks: Laboratory and field investigations. Journal of Environmental Management, 2013, 130(1): 369.

[100] Yang G X, Sun Y M, Kong X Y, Zhen F, Li Y, Li L H, Lei T Z, Yuan Z H, Chen G Y. Factors affecting the performance of a single-chamber microbial fuel cell-type biological oxygen demand sensor. Water Science & Technology A Journal of the International Association on Water Pollution Research, 2013, 68(9): 1914-1919.

[101] Kretzschmar J, Rosa L F M, Zosel J, Mertig M, Liebetrau J, Harnisch F. A microbial biosensor platform for inline quantification of acetate in anaerobic digestion: Potential and challenges. Chemical Engineering & Technology, 2016, 39(4): 637-642.

[102] Abourached C, Catal T, Liu H. Efficacy of single-chamber microbial fuel cells for removal of cadmium and zinc with simultaneous electricity production. Water Research, 2014, 51(6): 228.

[103] Li Y, Wu Y, Puranik S, Lei Y, Vadas T, Li B. Metals as electron acceptors in single-chamber microbial fuel cells. Journal of Power Sources, 2014, 269(4): 430-439.

[104] Yong J, Peng L, Liu P, Wang D, Bo M, Xia H. A novel microbial fuel cell sensor with biocathode sensing element. Biosensors & Bioelectronics, 2017, 94: 344-350.

[105] Si R W, Zhai D D, Liao Z H, Gao L, Yong Y C. A whole-cell electrochemical biosensing system based on bacterial inward electron flow for fumarate quantification. Biosensors & Bioelectronics, 2015, 68: 34.

[106] Si R W, Yang Y, Yu Y Y, Han S, Zhang C L, Sun D Z, Zhai D D, Liu X, Yong Y C. Wiring bacterial electron flow for sensitive whole-cell amperometric detection of riboflavin. Analytical Chemistry, 2016, (22): 11222-11228.

[107] Xu Z, Liu B, Dong Q, Lei Y, Li Y, Ren J, McCutcheon J, Li B. Flat microliter membrane-based microbial fuel cell as "on-line sticker sensor" for self-supported in situ monitoring of wastewater shocks. Bioresource Technology, 2015, 197: 244-251.

[108] Yong J, Peng L, Zhang C, Bian Y, Yang X, Xia H, Girguis P R. Enhancing the response of microbial fuel cell based toxicity sensors to Cu(II) with the applying of flow-through electrodes and controlled anode potentials. Bioresource Technology, 2015, 190(14): 367.
[109] Yi Y, Li X, Jiang X, Xie B, Liu H, Liang D, Zhu Y. Assessing the solitary and joint biotoxicities of heavy metals and acephate using microbial fuel cell. AEECE, 2015, DOI: 10.2991/aeece-15.2015.139.
[110] Kim M, Sik H M, Gadd G M, Joo K H. A novel biomonitoring system using microbial fuel cells. Journal of Environmental Monitoring, 2007, 9(12): 1323.
[111] Liu B, Yu L, Li B. A batch-mode cube microbial fuel cell based "shock" biosensor for wastewater quality monitoring. Biosensors & Bioelectronics, 2014, 62: 308.
[112] Yu D, Lu B, Zhai J, Wang Y, Dong S. Toxicity detection in water containing heavy metal ions with a self-powered microbial fuel cell-based biosensor. Talanta, 2017, 168: 210-216.
[113] Li T, Wang X, Zhou L, An J, Li J, Li N, Sun H, Zhou Q X. A bioelectrochemical sensor using living biofilm to *in-situ* evaluate flocculant toxicity. ACS Sens., 2016, 1(11): 1374-1379.
[114] Stein N E, Hamelers H V, Van S G, Keesman K J. Effect of toxic components on microbial fuel cell-polarization curves and estimation of the type of toxic inhibition. Biosensors, 2012, 2(3): 255.
[115] Zeng L, Li X, Shi Y, Qi Y, Huang D, Tadé M, Wang S, Liu S. $FePO_4$ based single chamber air-cathode microbial fuel cell for online monitoring levofloxacin. Biosensors & Bioelectronics, 2017, 91: 367-373.
[116] Chen Z, Niu Y, Shuai Z, Khan A, Ling Z, Yong C, Pu L, Li X. A novel biosensor for *p*-nitrophenol based on an aerobic anode microbial fuel cell. Biosensors & Bioelectronics, 2016, 85: 860.
[117] Schneider G, Czeller M, Rostás V, Kovács T. Microbial fuel cell-based diagnostic platform to reveal antibacterial effect of beta-lactam antibiotics. Enzyme & Microbial Technology, 2015, 73-74: 59.
[118] World Health Organization. Guidelines for Drinking-water Quality. 2011.
[119] Yang Z C, Cheng Y Y, Zhang F, Li B B, Mu Y, Li W W, Yu H Q. Rapid Detection and Enumeration of Exoelectrogenic Bacteria in Lake Sediments and a Wastewater Treatment Plant Using a Coupled WO_3 Nanoclusters and Most Probable Number Method. Environ. Sci. Technol. Lett., 2016, 3(4): 133-137
[120] Szöllősi A, Rezessy-Szabó J M, Ágoston H, Nguyen Q D. Novel method for screening microbes for application in microbial fuel cell. Bioresource Technology, 2015, 179(179C): 123-127.
[121] Kim T, Han J I. Fast detection and quantification of *Escherichia coli* using the base principle of the microbial fuel cell. Journal of Environmental Management, 2013, 130(1): 267-275.
[122] Hassan R Y, Mekawy M M, Ramnani P, Mulchandani A. Monitoring of microbial cell viability using nanostructured electrodes modified with Graphene/Alumina nanocomposite. Biosensors & Bioelectronics, 2017, 91: 857-862.
[123] Yang H J, Zhou M H, Liu M M, Yang W L, Gu T Y. Microbial fuel cells for biosensor applications. Biotechnology Letters, 2015, 37(12): 2357-2364.

作者：黄　霞[1]，梁　鹏[1]，张潇源[1]，陈　熹[1]，成少安[2]，刘芳华[3]，冯春华[4]，王　鑫[5]
[1] 清华大学环境学院，[2] 浙江大学能源工程学院，[3] 中国科学院烟台海岸带研究所，[4] 华南理工大学环境科学与工程学院，[5] 南开大学环境科学与工程学院

第21章 污泥厌氧消化技术及物质转化原理研究进展

本章导读

污泥厌氧消化技术可以实现污泥减量化稳定化，同时回收生物质能，是一种面向未来的污泥资源化利用的生物处理方法，特别是近年来高级厌氧消化（advanced anaerobic digestion）技术研究，推动了厌氧消化技术的发展，已成为污泥及城市有机质资源化处理的研究热点。本章详细介绍了目前国内外污泥厌氧消化的物质转化机理与瓶颈，目前污泥厌氧消化物质转化原理的研究进展，对污泥高含固厌氧消化技术、两相厌氧消化技术、联合厌氧消化技术、热水解强化厌氧消化技术以及超高温厌氧消化技术等主要的高级厌氧消化技术研究进展进行了归纳，介绍厌氧消化过程中污泥中有毒有害物质的迁移转化规律的研究热点，并对各技术未来的发展方向与研究热点做出了展望。

关键词

污水厂污泥，物质转化，高级厌氧消化

1 引　　言

随着污水处理规模的日益增长，污泥作为污水处理的产物，其年产量逐年增加，目前已超过 4000 万吨。污泥中含有大量易降解有机物，恶臭物质、病原体等有害物质，对其进行减量化、稳定化与无害化的处理本质上为污水处理的延续，至关重要。同时，近年来再生能源的利用越来越受到关注，厌氧消化产生的甲烷可以代替化石燃料用于发电，从而能节省能源并且减少碳排放[1]，因而得到大量学者的关注，并被认为是经济有效且可持续性强的污泥处理措施[2,3]。厌氧消化是指在厌氧微生物的作用下，有控制地使废物中可生物降解的有机物转化为 CH_4（含量 60%~70%）、CO_2 和稳定物质的生物化学过程。通过厌氧消化技术，可以减少污泥体积，稳定污泥性质，减少污泥中可分解、易腐化物质的量，并能去除恶臭与病原菌，从污泥中回收生物质能（甲烷气体）[4-6]，从而实现污泥的减量化、无害化与资源化。然而，由于污泥成分复杂，其在厌氧消化过程中的物质转化原理的研究仍处于起步阶段，且传统厌氧消化存在单位容积转化效率低、有机负荷低以及工程效益低等问题，因而近年来对污泥厌氧消化技术的机理研究与新技术开发与优化成为了污泥处理处置的热点。众多学者聚焦于高级厌氧消化（advanced anaerobic digestion）技术研究，开发了一系列克服传统厌氧消化技术缺点的新技术如高含固厌氧消化、两相厌氧消化、热水解强化厌氧消化、协同厌氧消化、超高温厌氧消化、厌氧消化电子传递强化等技术，且部分已实现工程应用。

2　污泥厌氧消化物质转化原理与瓶颈

2.1　污泥厌氧消化过程的物质流分析

物质流分析是指在一定时空范围内关于特定系统的物质流动和储存的系统性分析。主要涉及的是物质流动的源、路径及汇[7]。物质流分析是探明厌氧消化系统中可降解有机质的转化过程及其特征的必要工作。目前，关于厌氧消化过程中物质流结构的系统性实验研究较为缺乏，研究对象多着眼于碳水化合物、蛋白质和脂质等几种有机组分。在污泥厌氧消化系统中，典型的物质流主要依托于有机质，从本质上可以归结到碳、氮、磷、硫四种主要元素的转化途径中。

2.1.1　碳的转化

碳元素是污泥有机物质的主要元素，随着《巴黎气候协定》的签署，污水污泥处理的碳中和时代被广泛提及[8]。污泥厌氧消化处理是实现碳中和运行的有效措施，因此，探讨并研究污泥有机质在厌氧消化过程中碳的转化途径具有重要意义。众所周知，污泥中指向甲烷化的碳组分主要来源于碳水化合物（多糖、纤维素）、蛋白质、脂质等。在水解阶段，多糖在相应水解酶的作用下水解为单糖，再进一步发酵成乙醇和脂肪酸等。蛋白质则在蛋白酶的作用下水解为氨基酸，再经脱氨基作用产生脂肪酸和氨。脂类转化为脂肪酸和甘油，再转化为脂肪酸和醇类。蛋白质、多糖水解产生的小分子有机物通过相应酶（如磷酸转乙酰酶、磷酸转乙酰酶、乙酸激酶、丁酸激酶）的作用被转化为短链脂肪酸。对于蛋白质物质在厌氧消化中的物质转化，普遍认为斯提柯兰氏反应是氨基酸生物降解并生成短链脂肪酸的主要途径。具体而言，通常将碳元素的转化途径具体到厌氧消化的 3 个阶段中：在水解阶段，长碳链的有机质被水解成短碳链的小分子有机质，碳元素由集中转向分散；在产酸阶段，短碳链的小分子有机质被转化为乙酸等有机酸，同时，释放一定量的 CO_2 和 H_2，此时，部分碳元素由有机态转化为无机态，初步实现有机质的分解；在产甲烷阶段，乙酸等有机酸、CO_2、H_2 等能被产甲烷菌进一步转化为 CH_4 和 CO_2，实现有机碳元素的深度转化[9]，然而，在污泥厌氧消化过程中，复杂有机混合物的碳元素转化机制尚不清晰。

2.1.2　氮的转化

污水污泥中蛋白质含量通常占污泥有机质总含量的 40%~50%[10-12]，并且污泥中有机氮和无机氮的来源主要为蛋白质，此外，氮元素在厌氧消化过程中形态的改变会影响到厌氧消化系统中微生物的代谢活性，影响对污泥有机质的厌氧生物转化。因此，污泥中氮的物质流特征研究主要关注在以蛋白质为主的含氮类有机物向无机铵态氮和挥发性有机酸（VFA）的流向分析。而现有研究多针对蛋白质在较单纯的体系中进行一种或几种特定物质的考察[13-17]，比如，Xiao 等通过改变蛋白质构象来强化蛋白质的发酵产氢；Zhang 等研究了大豆有机质中蛋白质的交联作用与其自身的生物降解性的影响等，都主要围绕蛋白质的整体物质转化来研究，而缺乏对不同构象的同一含氮物质（比如蛋白质）中氮元素的厌氧转化规律的探讨。

2.1.3　磷的转化

磷元素是所有生命有机体进行正常生命活动不可替代的元素。在污水污泥中，存在磷脂、核酸、聚磷酸盐等大量含有磷元素的物质。在厌氧消化过程中随着含磷有机质的厌氧生物转化，磷元素的存在形态也发生相应的改变，比如，大量的有机磷转变为无机磷酸盐，同时，磷元素直接涉及储能物质三磷酸腺苷（ATP）的生成，因此，磷元素形态的改变直接影响着含磷有机质的厌氧生物转化。对含磷有机质在厌氧生物转化途径的识别将有助于探讨微生物微观世界的能量传递与转化，对于提高微生物在污泥厌氧消化过程中对有机质的摄取与转化具有重要的作用。但当前的研究，主要倾向于对厌氧消化前或厌氧消化之后污泥中磷元素的回收研究[18-24]，对在厌氧消化过程中相应磷形态转变途径和重金属对磷形态转变的影响研究鲜见报道。

2.1.4　硫的转化

硫作为一种重要的生物营养元素，是一些氨基酸、维生素和辅酶的重要组成成分。污泥厌氧消化过程中产生的硫化物（H_2S、HS^-和 S^{2-}）主要来源于硫酸盐的还原和含硫蛋白质的水解[25]。而硫酸盐的还原涉及硫酸盐还原菌和产甲烷菌对于乙酸和氢的竞争，必然会影响两种甲烷化途径电子分配比例的变化。在污泥厌氧消化系统中，硫酸盐浓度、乙酸和氢等中间产物浓度以及气相硫化氢浓度的变化在一定程度上可以揭示厌氧消化系统中物理-化学-生物微观特征，因此，识别硫元素在整个厌氧消化过程中的转化途径对于整个系统的稳定运行也具有重要意义。目前，关于含硫物质迁移转化方面的研究涉及领域诸如纸浆和造纸，糖蜜发酵，海产品加工，马铃薯淀粉，制革，食用油精炼厂，制药和石油化工生产和葡萄酒酒厂泔水等，归纳起来主要有以下几个方面：硫酸盐还原菌的生理特性及影响因子[26-28]、H_2S 产生机理研究及控制[29,30]、高硫废水硫酸盐的去除[31,32]和重金属的影响和去除[33-35]。关于污泥硫的迁移转化方面的研究，Raf Dewil[36]等研究了污水污泥处理的连续步骤（二沉池剩余污泥-浓缩-厌氧消化）中硫的分布和含硫化合物的转化，但其研究对象仅涉及 SO_4^{2-}和总硫，缺乏对硫元素物质流的系统性分析。污泥中硫的分布及其测定方法的欠缺是当前限制污泥中硫的物质流系统研究的主要因素。因此，建立污泥中固-液-气相含硫物质的测试方法体系尤为重要。

2.2 污泥厌氧消化体系中难降解有机物的识别及活化机理研究

在污泥厌氧消化体系中，污泥有机质的厌氧降解与转化直接关系到整个厌氧消化的效率，而难降解有机质的厌氧生物转化则是限制厌氧消化效率的瓶颈。因此，识别污水污泥中难降解有机质并且寻求强化其厌氧生物转化的方法对于提升污泥厌氧消化效率具有极其重要的意义。

污水污泥中含有大量复杂有机物，包括蛋白质、糖类、纤维素、木质素等，在厌氧消化过程中一部分有机物经过水解、酸化、甲烷化等阶段，分解产生 CH_4 和 CO_2 等物质，然而这部分有机物的比例通常低于 45%[37,38]，也就是说污泥中难以降解的有机物可占 55%以上。目前普遍观点的认为蛋白质、多糖等生物大分子的水解是污泥厌氧消化体系的限速步骤[39]，这是因为水解酶在基质中是位于生物絮体里，或位于 EPS 基质周围，或与细胞外膜相连[40]，处于相对固定的位置，从而导致其难于作用于生物大分子。Lefebvre 等[41]研究表明通过热处理尽管促进了生物大分子的溶解，但是厌氧消化过程中总的 COD 降解率并没有提高，无论热处理持续时间的长短都保持相对的稳定。此外，Engelhart 等[42]研究表明污泥中来自胞外聚合物中围绕细胞体起保护作用的多糖类碳水化合物比结构性碳水化合物（如纤维素）更难降解。Lefebvre 等[41]研究表明厌氧消后污泥中溶解性难降解有机物中蛋白质、类腐殖酸、类多糖分别占 57%、28%、15%，暗示存在大量的蛋白和多糖类原本易被生物降解的有机质却不能有效的被降解。Cuetos 等[43]采用傅里叶红外光谱和热重-质谱联用仪研究厌氧消化过程中有机物的生物稳定性变化情况，结果表明厌氧消化后稳定基质中芳香族化合物明显增加，而挥发性物质和脂肪族类物质逐渐减少。周友平等[44]通过对深圳某污水处理厂的污水污泥中可溶性酸组分的研究分析，认为厌氧消化作用对污泥中有机酸的降解作用不同，消化过程中有机酸的稳定性取决于分子结构。总体而言，至今关于污泥中难降解有机物的化学组成及结构特征的研究尚处于起步阶段，尚待进一步研究。

事实上，污水污泥有机质组分如蛋白质、糖类、纤维素、木质素等，在污泥体系中并不是独立存在的，不仅组分间存在相互作用，有机组分与金属、砂粒间也存在相互作用[45]。前期的研究发现[45]，这些相互作用会显著的影响污泥中有机质的生物降解性，使原本易被降解的有机质转变为难降解。研究污泥中难降解有机质的赋存形态，主要可从 3 个模型分别阐述研究。

（1）微米级砂粒与有机质间的相互作用[45,46]，即微米级 SiO_2 作为吸附剂吸附有机质，减小有机质被厌氧微生物接触利用的机会，同时，改变了有机质的赋存形态。根据研究，微米级与污泥有机质的相互作用一方面限制了有机质从固态向溶解态的转变，另一方面，增强了有机质的稳定性并限制了水解酶与有机质的接触[47]。

（2）金属离子与有机质间发生络合反应，形成金属-有机络合物，促使污泥中有机质网状构象更加稳定，进而限制有机质的生物降解性[47-49]。多价态金属离子（比如二价、三价等）在污泥体系中由于静电作用，通常极易作为桥梁将生物高分子有机质相互联结并可能缠绕，一方面，可以极大的改变有机质的赋存形态；另一方面，可强化有机分子的稳定性；同时，也会占据酶促反应位点，进而限制有机质分子的生物降解。

（3）污泥中惰性有机质腐殖质与其他易降解有机质间的相互作用[45,50]，已有研究表明[51]，腐殖质能够限制水解酶而极大的限制污泥中生物高分子有机质的水解，同时研究也发现，在没有金属络合作用的前提下，腐殖质对于糖类的限制降解要高于蛋白质。腐殖质由于具有大分子且结构复杂的特点，极易与其他易降解有机质发生缠绕，同时，网捕游离态的小分子物质，极大地限制了酶对有机质的催化分解。

实际上，污水污泥中有机质的赋存形态是上述 3 种模型综合作用的结果。微米级砂粒、多价态金属离子、腐殖质三者是改变污泥易降解有机质赋存形态的关键因素，三者共同限制了污泥有机质的厌氧生物降解性，如图 21-1 所示。

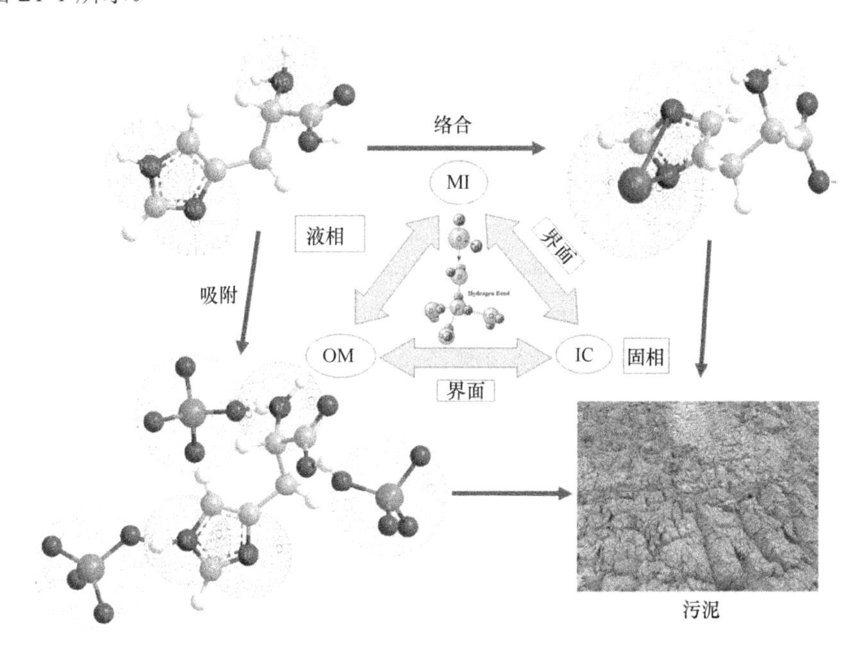

图 21-1　改变污泥有机质赋存形态关键因素示意图（Dai 等[46]）

因此，改变难降解有机质的赋存形态，提高污泥有机质的可生物降解性，突破有机质赋存形态的限制因素至关重要。比如，既然维持复杂有机质结构稳定性的基石是有机络合态金属离子[49]，去除金属离子而不引入新的限制因子到污泥中则成为首要预处理目标。同时，此过程也是从污泥中回收重金属的一种途径，已有研究表明[48,49]，从污水污泥中去除大量有机结合态金属离子后，污泥厌氧消化单位有机质的产甲烷可增加 30%以上。再者，采用热预处理弱化微米级砂粒与有机质间的相互作用，强化污泥有机质的厌氧生物转化，已有研究表明[45]，污泥有机质与微米级砂粒的吸附过程是一个放热过程，因此，热预处理有利于有机质的解吸附，增加污泥有机质的溶出[52]。但由于高温热预处理可能带来能量浪费和新难降解有机质的形成，如何合理地选择热处理温度有待进一步研究。

因此，污泥有机质赋存形态及其厌氧生物转化机制虽然复杂，但复杂体系中维持和改变有机质赋存形态的关键因素尚可进行识别，基于已识别的关键因素提出相应的活化措施是强化污泥厌氧消化效率的一种有效途径。目前，对污泥有机质赋存形态的识别及识别方法鲜见报道，同时，对维持和改变污泥有机质赋存形态关键因素的研究尚需进一步开展。

3 污泥高级厌氧消化技术

污泥传统厌氧消化通常指采用中温（35~37℃）或高温（52~55℃）对浓缩污泥（含固率为5%左右）进行厌氧消化，停留时间为20~30d，最终实现污泥稳定化、减量化和资源化的目的。但是，传统厌氧消化存在有机负荷低[0.75~1.75kgVS/(m^3·d)]、单位容积产气率低[0.15~0.7m^3/(m^3·d)]、工程效益不明显等缺点[53]。为了克服这三大瓶颈问题，近年来，国内外对新型的高级厌氧消化技术如高含固厌氧消化、两相厌氧消化、热水解强化厌氧消化、协同厌氧消化、超高温厌氧消化等开展了一系列研究，并在实际工程中得到应用[54]。

3.1 污泥高含固厌氧消化技术

污泥高含固厌氧消化主要通过提高进料含固率（TS≥10%）从而提升系统有机负荷[2.5~5.0kgVS/(m^3·d)]和单位体积产气率[0.7~1.4 m^3/(m^3·d)]，具有处理效率高、反应器体积小、加热能耗低等潜在优势[55]。在全球环境变化、资源能源短缺的背景下，高含固消化为污泥厌氧消化效率的提升提供了新的技术路线，符合节能减减排和循环经济的战略发展方向，对实现城市污泥的稳定化和生物质能源的高效回收具有重要意义，因此近年来逐渐成为研究热点。

Duan 等[56]证实了脱水污泥进行中温高含固厌氧消化的可行性。结果表明，TS 为 20%的脱水污泥，在 SRT 为 12~40d，OLR 为 3.0~8.5 kg VS/(m^3·d)条件下，能够实现稳定运行，VS 降解率为 29%~39%，甲烷含量 65%~66%，单位添加 VS 的甲烷产率为 0.18~0.22LCH_4/（gVS_{added}·d），单位体积产气率可达到1.24~2.49L/(L·d)。与传统厌氧消化工艺（TS 为 2%~5%）的系统相比，脱水污泥进行高含固厌氧消化的有机负荷和单位体积产气率可达到传统工艺的 4~10 倍。同时，该研究发现游离氨是影响污泥高含固厌氧消化工艺稳定性的主要潜在因素，游离态氨氮 FAN 小于 400 mg/L 时对系统稳定性及消化性能均无显著影响；400~600 mg/L 时，引起 VFA 的积累和甲烷产率的略微下降；持续高于 600 mg/L 时，VFA 浓度进一步升高，甲烷产率明显下降。因此，FAN 浓度不持续高于 600mg/L 是保障污泥高含固厌氧系统稳定运行的必要控制条件。另外，高含固系统稳定性决定于 FAN、TAN、VFA、TA 和 pH 的相互作用，VFA/TA 值不再适于评价系统的整体稳定性。

Dai 等[57]通过研究高含固污泥的流变特性，发现高含固污泥的流变性质与低含固污泥类似，呈现出剪切变稀的特性，且 TS 和温度是重要的影响因素，污泥剪切应力 τ 值和表观黏度 μ 值均随着温度的升高而降低，随着 TS 的升高而增大；在批次厌氧消化过程，污泥的表观黏度μ值和稠度系数 k 值在发酵的前 4 天均出现大幅度降低，之后缓慢下降，这表明微生物（主要是水解发酵菌以及产酸菌）的生命活动是厌氧系统中污泥黏度降低主要原因，而高含固厌氧消化对搅拌与物料输送设备的要求较低，采用高含固厌氧消化处理脱水污泥具有可行性。

Liao 等[58]对比研究了不同含固率（2%~15%）污泥厌氧消化差异，结果表明高含固消化对单位有机质产气率产生了一定的影响，但在延长 SRT 的情况下达到同样的降解率，明显增加了单位体积产气率和处理能力，同时也认为传质作用和游离氨累积是高含固厌氧消化过程可能存在的潜在因素。Liu 等[59]研究了污泥高含固厌氧消化系统中微生物的分布及产甲烷菌活性，结果发现随着含固率的增加，乙酸利用型产甲烷菌虽然还是主要产甲烷途径，但是比例降低，而氢利用型产甲烷菌比例从 6.8%（TS=10%）提高到 22.3%（TS=19%），甲基营养型产甲烷菌从 10.4%提高到 20.9%；另外对高含固消化污泥产甲烷活性实验也表明传质可能是影响高含固消化效率的主要因素[60]。

污泥高含固消化将成为污泥厌氧消化技术的发展趋势，而强化预处理的高含固厌氧消化以及污泥和城市有机质的协同消化将是高含固消化的主要发展方向。但是，由于高含固厌氧消化体系基质底物及中间产物浓度的提高，使得物质在固-液-气相内及相间传质与转化效应等与传统厌氧消化体系显著不同，高含固厌氧消化的基本物质流特征和微生物特征仍然缺乏，因而探明高含固厌氧消化体系的物质流特征，以及该特征有别于低含固系统背后的微生物机制，从而深度解析高含固厌氧消化系统生化反应机理，对厌氧消化效率进一步的提升具有重要意义。

3.2　污泥两相厌氧消化技术

传统的单相厌氧消化技术将水解产酸菌和产甲烷菌置于同一个反应器内，而产甲烷菌和水解酸化菌的生长速率和对生活环境的要求有很大差异，两种微生物均不能在最佳环境条件下生长，限制了反应器的效率。20 世纪 70 年代初，美国学者 Ghosh 和 Pholand 根据厌氧生物分解机理和微生物类群的理论首先提出了两相厌氧消化的概念，将产酸菌和产甲烷菌分别置于两个串联的反应器内并提供各自所需的最佳条件，使这两类细菌群都能发挥最大的活性，提高反应器的处理效率。由于两个单元的运行参数（包括温度、酸碱度和水力停留时间等）可以分别独立控制，因而可以使基质水解酸化和甲烷化的效率同时达到最大，大幅提高系统的处理效率同时增强反应器的运行稳定。两相厌氧消化的优点包括提高有机负荷，缩短停留时间，提高 VS 降解率和甲烷产率，提高病原菌和寄生虫卵的灭活率[61]。

目前，关于两相厌氧消化技术的研究，目前主要集中温度分级分相技术 TPAD 和产氢产甲烷技术两方面。产酸相发酵类型对两相系统非常重要，酸化相温度、pH、ORP 和停留时间等直接影响着产酸发酵类型，进而影响两相系统的消化性能。其中甲酸、甲醇、甲胺和乙酸能够直接为产甲烷菌所利用，而和产甲烷菌互营共生的产氢产乙酸细菌能够很快地将乙醇、丁酸转化为乙酸供产甲烷菌利用。产酸相发酵产物中应尽可能避免出现丙酸和乳酸，因为乳酸易转化为丙酸，丙酸的积累容易导致酸败现象[62]。

昆士兰大学的 Bastone 教授团队针对 TPAD 开展了大量的研究，包括最佳工艺参数（温度，SRT，pH），消化性能（降解性能，产气性能）、动力学模型分析（水解速率，生物可降解性），两相消化机理等。通过对比研究初沉污泥和剩余污泥不同酸化相温度两相系统消化性能，结果表明，高温+中温两相系统降解率均比中温+中温两相系统提高 25%左右，通过分析发现消化性能的提升是由于高温预处理增加了水解程度，而没有增加污泥固有的可生物降解性[63]。国内学者 Yu 等[64]也研究了温度分级和生物分相 TPAD 系统高温段不同温度、不同 HRT 对系统性能的影响，结果表明，温度从 35℃升高到 70℃，污泥溶解程度从 14.7%提高到 30.1%，但酸化程度在 45℃最高（17.6%），两相（45℃ SRT4d+35℃ SRT16d）产甲烷率比单相（35℃ SRT20d）提高 84.8%，降解率提高 11.4%，两相系统体现出了明显的优势。本课题组也对脱水污泥进行了超高温（70℃）酸化-高温（55℃）甲烷化和高温（55℃）酸化-高温（55℃）甲烷化两相与高温（55℃）单相厌氧消化系统性能的比较研究，结果表明，在总停留时间相等的条件下，两相消厌氧化系统的 VS 降解率、甲烷产率均比高温单相厌氧消化系统高，而出泥 VFA 含量小于高温单相厌氧消化系统。总体来说，两相厌氧消化系统运行更为稳定，效果更好，并且有较高的 H_2 可供利用为清洁能源，但酸化相较高的 H_2S 含量则须控制[65]。

利用两相技术同时产氢产甲烷是目前两相技术领域的研究热点。该技术的发展源于氢气作为清洁能源，同时也是生物质产酸发酵过程中的产物，产氢发酵产物同时可以作为高效产甲烷的基质。该两相技术主要利用不同的基质，包括餐厨等城市有机质、农业废弃物和污泥等单独或者共消化产氢产甲烷[66]。Cheng 等[67]利用活性污泥和餐厨垃圾共消化两相产氢产甲烷，发现在餐厨和污泥 VS 比为 3∶1 时，最大氢气产率达到 174.6mL/gVS，同时也大大提高了系统产能。

Li 等[68]研究了水热预处理与两相消化系统的结合的优势，结果发现水热两相系统比产气率（0.71Lbiogas/gVS_{added}）明显高于水热单相系统和传统单相系统（0.53L 和 0.55L biogas/gVS_{added}），推测酸化相可以具有降解热水解过程产生的难降解类黑精物质的作用，同时热水解两相系统产能明显高于其他系统。

由于污泥厌氧消化本身多基质多种微生物代谢的复杂特点，随着对污泥成分以及厌氧消化过程各相生化反应的深入研究，并利用分析生物学技术和优势菌种的发展，污泥分级分相厌氧消化技术将会成为未来污泥高级厌氧消化的发展方向。

3.3 污泥与城市有机质（餐厨垃圾）协同厌氧消化技术

餐厨垃圾是城市生活垃圾的主要组成部分，具有高水分、高盐分、高有机质含量、组分时空差异明显、危害性与资源性并存的特点，当前餐厨垃圾产量呈快速上升趋势，其后续的处理处置仍是一个难题[69]。如果将污泥与餐厨垃圾进行联合厌氧消化处理，则不仅能同时处理这两种废物，减少废弃物处理分支流程，还能提高厌氧消化的效率，更高效地回收沼气等生物质能并产生经济效益，已成为了当前研究的热点。

近年来，国内外学者对污泥与餐厨垃圾联合厌氧消化做了大量研究。王永会等[70]以餐厨垃圾和剩余污泥为原料，在中温条件下设置餐厨垃圾和剩余污泥混合比例（VS）分别为 1∶0、2∶1、1∶1、1∶2、0∶1，研究其单独消化与混合消化的系统性能与产甲烷潜力。发现联合厌氧消化提高了系统稳定性，与餐厨垃圾单独消化相比，添加剩余污泥能调节 pH、氨氮和挥发性有机酸（VFA）浓度，缩短产气周期，其与剩余污泥单独消化相比，添加餐厨垃圾能显著提高沼气产量。当混合比例为 1∶1 时，其混合消化产甲烷潜力最佳，消化作用的协同效应最为明显，沼气和甲烷产量分别达 358. 2 mL /gVS 和 224. 1 mL /gVS，较餐厨垃圾和污泥单独消化分别提高了 23.09%和 36.80%。Feng 等[71]通过向污泥中添加餐厨垃圾提高 C/N，研究两者联合厌氧消化产短链脂肪酸的能力，发现在 C/N 为 20∶1，pH 为 8 且发酵时间为 8d 时，两者联合厌氧消化短链脂肪酸的产量可达到 520.1 mg /g（以 COD 计），相同条件下，单独污泥或餐厨厌氧消化短链脂肪酸产量分别为 61.4 mg/g，261.8 mg/g（以 COD 计），很明显，两者联合厌氧消化短链脂肪酸产量要大于两者单独厌氧消化短链脂肪酸产量之和，说明在该条件下，污泥与餐厨垃圾联合厌氧消化产短链脂肪酸有协同作用。同时，大量文献研究同样也表明联合厌氧消化不仅能消除单独厌氧消化存在的一些不利因素，还便于将两种或两种以上的有机废弃物进行集中式处理，发挥规模效应，有利于更稳定地处理有机废弃物[72-74]。为此，Dai 等[75]对联合厌氧消化的强化机制进行了解析，对比了高含固污泥与餐厨垃圾联合厌氧消化与单独厌氧消化的区别，发现高含固污泥与餐厨垃圾联合厌氧消化降低了系统游离氨和钠离子的浓度，缓解了污泥与餐厨垃圾单独厌氧消化时各自的抑制因素，从而增强了厌氧消化系统的稳定性，提高了容积产气量。

联合厌氧消化发挥协同优势的关键在于其平衡了厌氧消化的相关参数，但在应用过程中，具体工艺参数的选择上，有机负荷率、温度等因素对污泥和餐厨垃圾联合厌氧消化有着重要影响。Liu 等[76]研究了中试条件下，有机负荷率对污泥和城市有机废物联合厌氧消化的影响，当有机负荷率在 1.2~8.0 kg/(cm^3·d)（以 VS 计）时，发现当有机负荷率为 8.0 kg/(cm^3·d)（以 VS 计）时系统产气量最大，但同时发现系统有短链脂肪酸的累积，这预示着系统有酸化的风险；Dai 等[75]在保证污泥与餐厨垃圾混合比例相同的前提下，通过改变厌氧消化系统的有机负荷率，研究其对厌氧消化的影响，在一定混合比例下，有机负荷率在 5.1~17.8 kg/(cm^3·d)变化，发现当有机负荷率为 7. 2 kg/(cm^3·d)时，系统的 VS 降解率和甲烷产量达到最佳。因而有机负荷率作为影响厌氧消化的一个重要因素，对联合厌氧消化的产气量以及系统稳定性也具有重要影响，因不同底物其降解性能不同，系统的最适有机负荷率需通过对不同的底物进行实验得到。

研究表明，温度对协同厌氧消化也具有一定影响。Cavinato 等[77]分别研究了中温和高温条件对污泥与城市有机固废联合厌氧消化的影响，发现高温厌氧消化的甲烷产量比中温厌氧 消化高出 45%~50%。高温厌氧消化比中温厌氧消化更有优势，如高温能加速有机物的水解酸化，提高有机物的去除率，减少病原菌以达到更好的卫生效果[78]等。但是高温厌氧消化也存在需要的能量更多，系统不稳定性增加，污泥的脱水性能变差，会产生较严重的恶臭[79]等问题。为解决这些问题，一种新的温度控制的方式，即在厌氧消化不同阶段采用不同温度，将中温与高温相结合的方式引起关注。Kim 等[80]在污泥与餐厨垃圾联合厌氧消化过程中，采用了高温—中温和中温—中温两种加热方式，研究发现前者的甲烷产量比后者提高了 0.2 m^3/kg。综上，温度对厌氧消化影响较大，调控合适的温度模式也是联合厌氧消化系统稳定运行的关键。

随着对再生能源的日益关注，对城市有机质与污泥进行联合厌氧消化以回收生物质能将会成为未来的发展趋势。然而多组分有机质厌氧消化的协同机制还有待阐明，进一步完善机理研究并提高协同厌氧消化效率将会成为未来的研究方向。

3.4　污泥热水解强化厌氧消化技术

在对污泥的厌氧消化过程中，胞内聚合物及大分子有机物的溶解并水解为可降解的小分子有机物的过程（即第一阶段）是主要的限速阶段[81,82]。由于第一阶段的速率低，导致后续酸化阶段以及产甲烷阶段中可降解有机物的量不足，以致限制了整个厌氧消化过程的进行，因此对污泥进行预处理非常重要。预处理能够破坏微生物的细胞壁，增强胞内聚合物及大分子有机物的水解，定向产生乙酸、氢气等产物，提升厌氧消化速率，减少污泥停留时间并提升产气量[81,83]。对污泥进行预处理有多种方法，可以分为热处理[84]、化学方法[85]、生物方法[86]和物理方法[87]以及它们中几种的联合如热化学法、物理化学法、生物化学法和机械化学法等[88-90]。在众多预处理技术中，热水解（Thermal pretreatment）被公认为具有良好经济效益且可实施性强的预处理措施。高温热水解的相对优势为提高污泥在厌氧过程中的降解率（与未经预处理的原污泥相比），提高污泥生物可降解性能，提高甲烷产量[92,93]；改善污泥脱水性能，减少后续脱水过程中的能量和药剂量投入，减少剩余污泥量，有利于后续的处置[91]；去除病原菌[94]；且以单位投入能量计，是一种经济有效的预处理方法[95]。

早在 1978 年即有研究表明经热水解后活性污泥的降解性能（以甲烷产量计）得到了提高，原因在于水解作用或者复杂有机物组分的释放[97]。Bougrier 等[96]研究发现随着热水解温度的提高，污泥的厌氧消化性能也随之提高，然而当温度升高到 200℃以上时，由于热水解过程中发生的美拉德反应，活性污泥的厌氧消化性能反而会降低。热水解温度高 175℃时会提高活性污泥的溶解性能而不会提高产气性能[97]。为了提高污泥的降解性能以及产甲烷量，同时缩短污泥在反应器中的停留时间，大量学者采用了 170℃以下的温度对污泥进行热水解以提高污泥的厌氧消化性能，这些研究表明热水解能有效提高污泥的厌氧消化性能如 VS 降解率以及产气率等，并得到了普遍的工业应用。如 Oosterhuis 等[91]报道在荷兰的亨格罗污泥处理装置中，采用 165℃（6bar）的热水解处理 20min 后，污水厂污泥的 VS 降解率提高了 62%，厌氧消化罐的有机负荷能够达到未经预处理时的 2.3 倍；Liao 等[98]报道了条件为 150~170℃（6~8bar），30min 的热水解在实际污水厂中的应用，相比于传统厌氧消化，其 VS 降解率提高了 50%~100%，节约厌氧消化罐体积 50%~75%；Ennouri 等（2016）[99]研究发现经 120℃热水解后，城市和工业废水活性污泥的产气率分别提升了 27%和 37%。

但也有研究表明高温热水解对于污泥的厌氧消化性能并没有明显的提升，比如 Wilson 等[100]研究发现对于活性污泥、初沉污泥以及活性污泥与初沉污泥的混合体，热水解对其降解率的增强率分别为 36%，0%以及 16%；Carrere 等[101]研究发现污泥本身的厌氧消化性能越好，热水解的强化性能会相应地降低；

此外 Higgins（2015）[102]研究也发现经 130~170℃的热水解后活性污泥的降解率并没有明显的提升。对于这一差异性的结果，目前的解释是由于热水解能够显著促进碳水化合物与蛋白质的溶出，对脂质溶解的促进作用却不是很明显，甚至还会分解生成大量的 VFA 从而抑制污泥的厌氧消化[100]，因此热水解对于含有较高含量的碳水化合物或者蛋白质的污泥比较适用，这一结论也在 Chen 等[103]的研究中得到了验证，热水解对污泥厌氧消化过程中 VS 降解量的提升作用主要来自于对蛋白质（49%）和半纤维素（25%）的促进。且文献表明，促进活性污泥厌氧消化性能（获得最高产气量）同时尽可能降低能耗的最佳热水解条件为 165~170℃，时间为 30min[104]。

关于高温热水解对污泥厌氧消化性能的影响机理，主要研究侧重于以下两个方面：第一，热水解过程加快了污泥中有机物的水解以致改善了后续的厌氧消化性能；第二，热水解能够提高污泥中有机物最终的降解程度，从而提高了厌氧消化的产气量。Ge 等[105]在中温厌氧消化之前，将初沉污泥经过一个 50~70℃的热预处理过程，发现初沉污泥的厌氧消化性能得到了明显的提升，并依据 ADM1 厌氧消化模型计算比较了降解程度（degradability extent，f_d）和一级水解速率（apparent first order hydrolysis rate coefficient，k_{hyd}）这两个代表性参数，发现该温度下的预处理仅仅是提高了厌氧消化系统在模型中的 k_{hyd}，而对于降解程度 f_d 的影响却不明显，并指出该机理与 Tiehm 等[106]所报道的超声波这类较轻影响的预处理方式类似，只是改变了有机物的水解速率。Gillian 等[107]研究了热水解对活性污泥的影响，探讨了热水解（150℃）是否加快降解速率与提高降解程度的问题，由于活性污泥经预处理后由好氧消化体现出的 COD 降解性能与厌氧消化体现出的 COD 降解性能相似[108]，该研究以热水解后活性污泥的好氧消化性能来展开研究，发现热水解前后的活性污泥进入好氧消化系统稳定消化后，总 COD、固态 COD 以及 VSS 大致相等，说明热水解预处理并未改变活性污泥中能够好氧消化的部分，即热水解并未能改变活性污泥的最终好氧降解程度。Chen 等[103]在其研究中将热水解前后污泥厌氧消化后的沼渣继续进行厌氧消化，当消化时间继续延长 120d 后，沼渣中 VS 以及各类有机物含量在误差范围内相一致，表明了高温热水解的主要作用机理为提高污泥的水解速率而不是降解程度。因此，对于热水解促进污泥厌氧消化的机理，目前倾向于热水解过程加快了污泥中有机物的水解以致改善了后续的厌氧消化性能，尤其是蛋白质与半纤维素。

由此可知，目前对于热水解技术的研究已经比较成熟，国内已经开始推广，目前在北京、长沙、镇江等地已经开始采用热水解预处理工艺以强化污泥厌氧消化效率。然而对于其设备的优化、能量的平衡、能效的提高、污泥沼渣沼液性能的解析及其资源化利用的途径有待进一步研究。

3.5 污泥超高温厌氧消化技术

目前，多数的厌氧消化系统是在中温（35℃）和高温（55℃）条件下运行的。中温厌氧消化系统中，污泥中的有机颗粒物由于受到水解步骤的限制而降解率较低因此所需的 SRT 较长[109]。同时，即便在高温厌氧消化（55℃）系统中，微生物的降解潜能也仍没有得到充分地释放，有机物降解率也仅有 50%左右[110]。因此，污泥厌氧消化的产气潜能还有较大的提高空间[111]，其关键是提高污泥中有机物的水解效率[112]。

Nielsen 等[113]与 Scherer 等[114]通过污泥厌氧消化的对比研究发现，与普通的中温（37℃）厌氧消化相比，高温（55℃）和超高温（超过 55℃）厌氧消化系统更有利于提高有机颗粒溶解性及沼气产量，这主要是由于 55~70℃是水解酸化菌生长的最适温度[115]。而且，当系统温度从 55℃上升到 65℃时，污泥中蛋白质等难降解有机物的溶解性也会显著提高，传质速率加快[116]。因此，超高温（65℃）条件下，颗粒状污泥的溶解性和嗜高温的水解酸化微生物活性的提高，能提高厌氧消化系统中有机物的降解速率及产气效率[111]。

厌氧消化系统的产甲烷过程有乙酸利用型产和氢利用型两种产甲烷途径。乙酸利用型产甲烷途径中，

VFAs（乙酸、丙酸、异丁酸、正丁酸等）通过在乙酸利用型产甲烷菌的作用下最终生成甲烷和 CO_2[116]，丙酸利用的最佳温度为 55℃，而甲酸、乙酸、丁酸利用的最佳温度是 60℃[117]。当温度超过 55℃时，乙酸利用型产甲烷菌的生长速率很低[118]，因此，当温度上升至 65℃时，乙酸利用型产甲烷途径几乎不再进行[117]。而氢利用型产甲烷菌多为嗜高温菌群，其生存的最适温度范围为 55~70℃[118]，当温度提升到 65℃时，氢利用型产甲烷途径的效率将达到最大[117]，因此，利用氢利用型产甲烷途径在超高温条件下进行厌氧消化是可行的。

污泥热水解是一种能够加快污泥水解速率，提高甲烷产率的前处理技术，近年来在实际工程中得到了广泛的应用[119-121]。热水解后的污泥温度较高，直接利用热水解污泥中的剩余热量，进行超高温厌氧消化，能提高颗粒状污泥的溶解性和降解速率以及嗜高温的水解酸化微生物活性，并富集嗜高温的氢利用型产甲烷菌以提高产气效率[111]。而且，氢利用型产甲烷途径的充分利用，可以减少 CO_2 排放，节省以乙酸等短链脂肪酸形式存在的碳源进行后续利用以实现碳源的综合利用。此外，当温度上升至 60℃以上时，污泥中的胶体物质降解率提高，丝状菌被杀死，能够明显提高污泥的脱水性[122]；同时，污泥中的致病菌也能够被杀灭，初步满足污泥处理处置无害化的要求[122]。

目前，国内外关于超高温厌氧消化的相关研究较少，仅有少量关于牛粪或餐厨垃圾超高温厌氧消化系统的报道。Tang 等的研究证实，水解酸化的最适温度在 55~75℃之间，且随着温度的提高，水解酸化效果也随之提高[123]。Nielsen 等在牛粪的超高温两相（68~55℃）厌氧消化实验中发现，在 HRT 为 3d 的 68℃反应器中，水解，酸化，乙酸化和产甲烷过程能够同时进行[110]。Ahring 等用 65℃厌氧消化反应器进行了牛粪的超高温厌氧消化，发现由于水解酸化过程主要由嗜热菌完成，水解酸化过程没有受到抑制，且木质纤维素类物质的降解率更高。但沼液中 VFAs 的累积，甲烷产量出现一定程度的下降。而且，在 65℃反应器中，细菌为反应器中的多数菌群，古菌为少数菌群，但与 55℃反应器相比，古菌的活性更强。氢利用型产甲烷菌成为新的产甲烷主体，其他的产甲烷菌数量和活性明显下降，丙酸转化也不再是主要的 VFAs 利用途径。Niclas 等用牛粪启动了饲料和甜菜的 60℃厌氧发酵反应器，在后期不再投加牛粪的条件下，维持了长达 2 年的稳定运行，氢利用型产甲烷途径是该 60℃厌氧消化反应器的主要的产甲烷途径，乙酸利用型产甲烷菌只占到甲烷菌总量的 10%。通过纯培养实验发现，在高氨氮、高 H_2S 和高乙酸浓度的条件下，60℃厌氧消化反应器均能够维持较高的稳定性[124]。

因此，超高温厌氧消化技术在一定程度上可以继续强化污泥的厌氧消化效果，但目前的研究仍处于起步阶段，在国内鲜有工程应用，对其的进一步研究与优化也是今后的研究热点。

3.6 厌氧消化电子传递强化技术

3.6.1 添加零价铁

废铁屑为工业废料，价格低廉、易于运输、投加简单[126]。Hao 等[127]通过向污泥中添加废铁屑，提高了厌氧消化过程中沼气和甲烷产量，在两相厌氧体系中，10g /L Fe 可以使酸化相和产甲烷相的甲烷产量分别提高 10.1%和 21.4%。目前的研究认为，添加零价铁改善厌氧消化体系性能的机理主要为以下三个方面：①由于零价铁的还原性，体系氧化还原电位（ORP）下降，而 ORP 的变化可以引起酸化类型的改变，较低的 ORP 有利于乙酸的产生，减少丙酸的积累，从而有效提高底物的利用效率，促进甲烷的产生；②厌氧环境下，铁腐蚀作用和产甲烷作用耦合，铁腐蚀产生的氢气可作为产甲烷的底物，提升氢气型产甲烷和同型产乙酸途径，从而提高甲烷产量；③零价铁的添加为厌氧微生物的生长和新陈代谢活动提供必需微量元素，提高微生物活性。

在 Hao 等[127]的研究中零价铁的添加引起的 ORP 降低不是甲烷产量增加的直接原因，酸化类型并无改变，铁腐蚀产生的氢气对甲烷产量的提高也仅贡献 3.7%，主要是零价铁释放的铁元素对厌氧微生物的促进作用对甲烷产量的提高起到了关键作用，铁元素在提高关键酶（尤其是含铁酶）活性方面发挥着重要作用，微生物以及酶活性的提高促进了难降解有机物的生物降解（水解）以及微生物对碳源的利用（新陈代谢）。Zhang 等[130]通过添加废铁屑增强了高含固污泥厌氧消化系统的产气性能，研究表明添加 10 g/L 腐蚀的废铁屑，甲烷产量能够提高 29.51%，挥发性悬浮固体（VSS）降解率能够提高 27.26%，而且腐蚀的废铁屑比铁颗粒和干净的铁屑效果更好，在其研究中，厌氧消化体系性能的提升主要是由于废铁屑表面的 Fe（III）氧化物增强微生物铁还原作用，从而促进微生物对有机物的降解，加速微生物厌氧水解-酸化过程，有利于后续产甲烷过程和有机物矿化作用。另外，零价铁作为电子供体，可能会提高厌氧消化体系的电子传递速率，进而提升产甲烷性能。

3.6.2 添加生物炭

生物炭是一种经济、生态兼容性较好的固废处理产物，添加到污泥中可有效提高厌氧消化系统稳定性。Lü[129]等比较了不同粒径的生物炭厌氧消化过程中氨抑制的缓解作用，结果表明粒径为 2~5mm，0.5~1mm，75~150um 的生物炭可以将产甲烷迟滞时间分别缩短 23.9%，23.8% 和 5.9%，将最大产甲烷速率分别提高 47.1%，23.5% 和 44.1%，生物炭可以提高 SCFAs 的产生和消耗速率，有效加快氨抑制和酸积累双重胁迫下的产甲烷启动过程。生物炭对厌氧消化体系性能的提升并不是来自生物炭的物化作用（偏碱性的 pH，多孔结构对氨氮的吸附），而是来自于对体系生化过程的影响，一方面由于生物炭比表面积大，有利于微生物附着生长，另一方面可能是由于产甲烷菌易于附着在生物炭的导电性表面，生物炭增强了微生物之间的 DIET 作用。附着在生物炭上甲烷鬃毛菌属能够在 SCFAs 高达 60~80 mmol-C/L 的条件下存活，进而加速乙酸的降解，随后紧密附着在生物炭上的甲烷八叠球菌属开始发挥作用。

3.6.3 施加微电压

微生物电解池（MEC）与污泥厌氧消化过程结合能够加速污泥水解，从而促进产甲烷过程。Zhao[130]等研究了添加导电材料以及施加微电压对厌氧消化系统的影响，结果表明添加碳毡使甲烷产量提高 12.9%，污泥降解率提高 17.2%，在碳毡上施加微电压能够进一步提高厌氧消化性能。通过电流计算看出，阴极二氧化碳还原合成甲烷并不是甲烷产量提高的主要原因，仅占总提高量的 27.5%；在 MEC 系统中，甲烷八叠球菌属和地杆菌属丰度增加，表明甲烷产量的增加主要是由于两种菌属之间 DIET 的建立。Feng 等[131]研究了高含固污泥厌氧消化系统和微生物电解池（铁-石墨烯电极）耦合对甲烷产量的影响，结果表明在 0.3V 电压下，甲烷产量提升效果最佳，甲烷产量增加 22.4%，VSS 降解率提高 11%，系统中细菌和古菌都得到强化，从而提高 SCFAs 的生成和甲烷的产生；当电压升高到 0.6V 时，甲烷产量降低，可能原因是，在较高电压下，阴极过多消耗 H^+ 导致碱性 pH，抑制甲烷菌活性。

由此可见，通过外源添加零价铁、生物炭等导电性物质，以及对厌氧消化系统施加微电压等手段，可强化厌氧消化系统中微生物的活性，增强种群之间的电子传递作用，从而提升厌氧消化效率。然而由于污泥是一个复杂的混合体系，外源物质的添加在系统内的分散与传质过程目前尚不明晰，导致其在厌氧消化系统中的效益难以实现最大化；其次，污泥厌氧消化系统中的复杂底物与中间产物也会对微电压的作用带来影响。目前此类技术的研究尚停留在小试研究阶段，为实现此类技术的普及，对于这两方面的研究至关重要。

4 污泥中有毒有害物质在厌氧消化过程中的迁移转化

由于污泥是污水处理时的产物，污水中的有毒有害物质也会部分富集至污泥中，而且污泥浓缩与脱水过程中添加了大量混凝剂与脱水剂等药剂，导致了污泥成分复杂，除了含有可生物降解的有机质之外，还有一些对环境有毒有害的物质如重金属、PAM 和 POPs 等物质。这些物质在污泥厌氧消化过程中是否会得到降解与稳定是沼渣安全处理处置的重要问题。

4.1 重金属

在厌氧消化过程中，尽管微量金属对厌氧微生物的生长至关重要，但是过量的金属则会导致厌氧消化的抑制，并且会造成厌氧消化产物对土地利用的危害，所以在厌氧消化过程降低重金属含量或增加其稳定性尤为重要。Roel J.W. Meulepas 等[125]的研究表明在没有浸出剂的条件下，两级厌氧消化的第一阶段可以被充分利用成为厌氧生物提取金属过程，其处理产生的污泥中，铜、锌、镍和镉的含量可以达到荷兰的土地使用标准，但是铅的含量无法满足标准。刘晓光等[132]研究污泥厌氧消化过程中物理化学性质的变化对典型重金属形态转化的影响，发现厌氧消化过程中重金属的形态发生了显著变化，由不稳定态向比较稳定的残渣态和有机结合态转变，某些重金属形态与污泥理化性质如 pH、碱度、VS/TS 及氨氮显著相关。沈晓南等[133]通过实验发现：经厌氧消化后污泥中铜、锌、镍、铬等重金属稳定性提高，铜的稳定态由 90%升至 98%，锌由 24%升至 35%，镍由 27%升至 32%，铬由 64%升至 69%。由此可见，污泥中的大部分重金属在厌氧消化过程中能够得到一定程度的稳定化。为了进一步提高厌氧消化污泥中的重金属稳定性，曹军等[134]的研究表明可向污泥中投加适量 SO_4^{2-}，促进不稳定态的重金属在消化过程中向稳定的硫化物形态转化，使消化污泥的农用更加安全可靠；沈晓南等[133]也向污泥中适量投加能产生 S^{2-}的固体废弃物，增加了污泥中重金属的稳定性；Pham Minh Thanh 等[135]的实验结果亦证明金属硫化物在厌氧消化过程中可以作为微量金属的贮存和来源，在较稳定的状态下为污泥厌氧消化提供必要的营养元素。

4.2 聚丙烯酰胺

随着污泥高含固厌氧消化技术的应用与发展，污泥需要经过机械脱水达到一定含固率后才能进入厌氧消化反应器。污泥在脱水之前通常会加入化学调理药剂聚丙烯酰胺(PAM)来提高其脱水性能，但 PAM 的长链结构会增加污泥的黏性，抑制有机物的传质效率，对厌氧消化性能产生影响。另一方面，PAM 降解产生的单体丙烯酰胺具有生物毒性，累积在消化反应器中会影响沼渣后续的处理处置。因此，研究厌氧消化系统中 PAM 的降解具有重要的意义。

研究表明，PAM 的支链上存在大量的酰胺基，可以在水溶液中发生电离，产生羧基和氨氮，在厌氧条件下能够作为氮源被微生物降解利用。Kay-Shoemake[136]研究了 PAM 在土壤中被微生物作为氮源利用的机理和效率，考察了酰胺水解酶在微生物利用 PAM 中 N 元素的潜在作用，结果表明可利用的碳源影响了酰胺基的去除和酰胺酶的产生量；同时在单独利用 PAM 为氮源的培养条件下考察了酰胺酶活性及其对底物的专一性；通过定向培养可以得到以 PAM 为底物的酰胺酶。此外也有研究表明带酰胺基团的小分子底物会抑制酰胺酶的活性从而降低 PAM 的降解率[137]。Haveroen 考察了厌氧条件下阴离子 PAM 能否作为氮源被油砂尾矿城市污泥的微生物利用，并添加苯酸盐和乙酸盐为微生物提高碳源，结果表明在厌氧环境下，微生物可以将 PAM 上的酰胺基作为氮源利用，在碳源充足氮源不足的情况下，添加 PAM 可以刺激包括产甲烷菌在内的厌氧微生物活性[138]。Dai 等[139]研究也发现 PAM 在厌氧消化系统中可以作为碳源和氮源被微生物利用，但其支链水解效率较低，致使未水解的 PAM 降解后会产生有毒单体丙烯酰胺。

因此，普遍研究认为 PAM 在污泥厌氧消化系统中能够作为碳源和氮源从而被微生物降解。为了提高污泥厌氧消化系统中 PAM 的水解效率，未来可针对 PAM 的生物降解特性对厌氧消化系统进行优化，以改善其关键酶活性、污泥絮体颗粒尺寸等理化特性和微生物菌群分布特性，同时提高产甲烷性能和 PAM 水解效率，从而减少有毒单体丙烯酰胺的累积。

4.3 持久性有机污染物

持久性有机污染物（POPs）是化学污染物，不仅有毒有害，而且在环境中降解缓慢，存在长期潜在风险。由于污泥具有吸附性，多氯联苯和一些氯代有机物会在消化污泥中积累，这些 POPs 在污泥中的存在使得污泥的稳定化与无害化也存在一定困难，因此 POPs 在污泥的厌氧消化中迁移转化过程和降解机理成为研究热点。众多研究发现，POPs 在厌氧消化过程中，会发生不同程度的降解。如在 Oliveira 等[140]的研究中，污泥厌氧消化 18d 后，其中的三氯杀螨醇和二氯苯甲酮含量分别降至 3%和 5%，但有其他代谢产物生成，并未完全矿化；Rosinska 等[141]的研究结果表明，污泥厌氧消化过程中，低氯化联苯浓度短暂升高，高氯化同系物浓度降低，共面 PCB 169、指示 PCB 180 和 PCB 153 的浓度都明显降低；Bertin 等[142]研究也发现在 35℃条件下运行污泥厌氧消化反应器 10 个月，污泥携带的有机污染物，如多氯联苯、多环芳烃等明显被降解；Siebielska 等[143]将污泥和市政废物的有机组分分别进行共同厌氧消化和共同堆肥，发现厌氧消化过程中多氯联苯的去除率更高。由此可知，部分典型 POPs 在厌氧消化过程能够得到降解，但是由于 POPs 种类繁多，降解也具有一定难度，其在厌氧消化过程中降解的相关研究还有待进一步完善。

5 总结与展望

传统污泥厌氧消化技术已有百年历史，和工业废水的厌氧消化技术（UASB、IC 反应器）相比，污泥及城市有机质的厌氧消化技术研究相对滞后，导致其存在转化效率和产气率低、有机负荷低以及工程效益低等问题，仍具备较大的提升空间。

目前污水中污染物的资源回收是国际研究的热点，对污水处理过程中富集至污泥中的碳、氮、磷等有机质的深度回收利用是实现污水处理、污染物资源化利用的重要途径。因此新型高级厌氧消化技术，如高含固厌氧消化、两相厌氧消化、协同厌氧消化、热水解强化厌氧消化、超高温厌氧消化及电化学强化厌氧消化等技术将会得到更多的关注。

在机理研究层面，由于污泥的厌氧消化为复杂的、非均质的物质转化过程，目前对污泥中难降解有机物的识别与分质活化机理研究仍处于起步阶段，因此对污泥中难降解纤维类、蛋白类等有机质的赋存状态以及其抗水解屏障是亟待解决的科学问题；

在技术研究层面，污泥中有机质厌氧转化效率提升的瓶颈仍然没有得到有效解决。虽然新型高级厌氧消化技术显著提升了污泥资源化的效率和单位体积产能，但厌氧消化的产物中仍有 40%~50%的有机物未能得到有效转化，针对此类难降解有机物开发高效强化厌氧消化技术，以提高厌氧消化效率是今后厌氧消化技术发展的重要环节。同时，目前的厌氧消化技术中所降解的有机质有 1/3 转化生成二氧化碳，未能实现碳生物质能的高效回收，如何通过微生物代谢途径的优化，实现碳的定向甲烷化、提高沼气中甲烷含量是未来研究关注的重点。

污泥厌氧消化是实现易腐有机物的降解，达到污泥稳定化、减量化与资源化的重要途径。但是，对污泥厌氧消化过程副产物如沼液、沼渣等的进一步处理与利用，降低后续处理环节的成本与投入，真正实现全流程的资源回收也将成为未来研究的热点。

参考文献

[1] Salminen E, Rintala J. Anaerobic digestion of organic solid poultry slaughterhouse waste–a review. Bioresource Technology, 2002, 83(1): 13-26.

[2] Pilli S, Bhunia P, Yan S, et al. Ultrasonic pretreatment of sludge: A review. Ultrason Sonochem, 2011, 18: 1-18.

[3] 戴晓虎. 污水厂污泥及城市有机质资源化利用技术与工程案例. 第五届中国城镇水务发展国际研讨会, 2010.11.01

[4] Metcalf and Eddy. Wastewater engineering: Treatment and reuse(4th ed.). NewYork, NY: McGraw-Hill.2003.

[5] Appels, Baeyens L, Dewil J, et al. Siloxane removal from biosolids by peroxidation. Energy Conversion and Management, 2008, 49(10): 2859-2864.

[6] Appels L, Degreve J, Van der Bruggen B, et al. Influence of low temperature thermal pre-treatment on sludge solubilisation, heavy metal release and anaerobic digestion. Bioresource Technology, 2010, 101(15): 5743-5748.

[7] 黄和平, 毕军, 张炳, 等. 物质流分析研究述评. 生态学报, 2007, 1: 368-379.

[8] 郝晓地, 魏静, 曹亚莉. 美国碳中和运行成功案例-Sheboygan污水处理厂. 中国给水排水, 2014, 30(24): 1-6.

[9] Zhen G Y, Lu X Q, Kato H Y, et al. Overview of pretreatment strategies for enhancing sewage sludge disintegration and subsequent anaerobic digestion: Current advances, fullscale application and future perspectives, Renew. Sust. Energ. Rev. 2017,69: 559-577.

[10] 秦晓. 剩余污泥中蛋白质的分离及其性质分析. 天津: 天津大学, 2011.

[11] 吕丰锦, 韩云平, 刘俊新, 等. 污泥有机成分与污泥厌氧消化潜能的研究进展. 环境工程, 2016, 34: 780-785.

[12] 卓杨, 韩芸, 程瑶, 等. 高含固污泥水热预处理中碳、氮、磷、硫转化规律. 环境科学, 2015, 36(3): 1006-1012.

[13] Duan N N, Dong B, Wu B, et al. High-solid anaerobic digestion of sewage sludge under mesophilic conditions: Feasibility study. Bioresource Technology, 2012, 104: 150-156.

[14] 江玉霞, 李兴民, 闫文杰, 等. 金华火腿加工过程中蛋白质降解情况的研究. 食品工业科技研究与探讨, 2005, 6: 52-54.

[15] Xiao N D, Chen Y G, Ren H Q. Altering protein conformation to improve fermentative hydrogen production from protein wastewater. Water Research, 2013, 47: 5700-5707.

[16] Blume K, Dietrich K, Lilienthal S, et al. Exploring the relationship between protein secondary structures, temperature-dependent viscosities, and technological treatments in egg yolk and LDL by FTIR and rheology. Food Chemistry, 2015, 173: 584-593.

[17] Zhang Y H, Zhu W Q, Gao Z H, et al. Effects of crosslinking on the mechanical properties and biodegradability of soybean protein-based composites. Journal of Applied Polymer Science, 2015, DOI: 10.1002/APP.41387

[18] Shu L, Schneider P, Jegatheesan V, Johnson J. An economic evaluation of phosphorus recovery as struvite from digester supernatant. Bioresource Technology, 2006, 97: 2211-2216.

[19] Wang J, Song Y, Yuan P, et al. Modeling the crystallization of magnesium ammonium phosphate for phosphorus recovery. Chemosphere, 2006, 65: 1182-1187.

[20] Marti N, Ferrer J, Seco A, Bouzas A. Optimisation of sludge line management to enhance phosphorus recovery in WWTP. Water Research, 2008, 42: 4609-4618.

[21] Kuroda A, Takiguchi N, Gotanda T, et al. A simple method to release polyphosphate from activated sludge for phosphorus

reuse and recycling. Biotechnology Bioengineering, 2002, 78: 333-338.

[22] Takiguchi N, Kuroda A, Kato J, et al. Pilot plant test on the novel process for phosphorus recovery from municipal wastewater. Journal of Chemical Engineering of Japan, 2003, 36: 1143-1146.

[23] Takiguchi N, Kishino M, Kuroda A, et al. Effect of mineral elements on phosphorus release from heated sewage sludge. Bioresource Technology, 2007, 98: 2533-2537.

[24] Takiguchi N, Kishino M, Kuroda A, et al. A laboratory scale test of anaerobic digestion and methane production after phosphorus recovery from waste activated sludge. Journal of Bioscience and Bioengineering, 2004, 97(6): 365-368.

[25] 张自杰, 等. 排水工程(下册). 北京: 中国建筑工业出版社, 1999: 353-379.

[26] Oleszkiewicz J A, Marstaller T, McCartney D M. Effects of pH on sulfide toxicity to anaerobic processes. Environmental Technology Letters, 1989, 10: 815-822.

[27] Barton L L, Tomei F A. Characteristics and activities of sulfate reducing bacteria. In Sulfate Reducing Bacteria, New York: Plenum Press, 1995.

[28] Reis M A M, Almeida J S, Lemos P C, et al. Effect of hydrogen sulfide on growth of sulfate-reducing bacteria. Biotechnology and Bioengineering, 1992, 40: 593-600.

[29] Rajalo G, Petrovskaya T. Selective electrochemical oxidation of sulfides in tannery wastewater. Environmental Technology, 1996, 17: 605-612.

[30] Xiao N D, Chen Y G, Ren H Q. Altering protein conformation to improve fermentative hydrogen production from protein wastewater. Water Research, 2013, 47: 5700-5707.

[31] Ramsing N B, Kühl M, Jorgenson B B. Distribution of sulfate-reducing bacteria, O_2, and H_2S in photosynthetic biofilms determined by oligonucleotide probes and microelectrodes. Applied and Environmental Microbiology, 1993, 59: 3840-3849.

[32] Zheng Y, Xiao Y. The bacterial communities of bio-electrochemical systems associated with the sulfate removal under different pHs. Process Biochemistry, 2014, 49: 1345-1351.

[33] Azabou S, Mechichi T, Patel B K C, et al. Isolation and characterization of a mesophilic heavy-metals-tolerant sulfate reducing bacterium Desulfomicrobium sp. from an enrichment culture using phosphogypsum as a sulfate source. Journal of Hazardous Materials, 2007, 140: 264-270.

[34] Cabrera G, Pe´rez R, Go´mez J M, et al. Toxic effects of dissolved heavy metals on *Desulfovibrio vulgaris* and *Desulfovibrio* sp. strains. Journal of Hazardous Materials, 2006, 135: 40-46.

[35] Cao J, Zhang G, Mao Z, et al. Precipitation of valuable metals from bioleaching solution by biogenic sulfides. Minerals Engineering, 2009, 22(3): 289-295.

[36] Dewil R, Baeyens J. Distribution of sulphur compounds in sewage sludge treatment. Environment Engineering Science, 2008, 25(6): 879-886.

[37] Toreci I, Kennedy K J, Droste R L. Evaluation of continuous mesophilic anaerobic sludge digestion after high temperature microwave pretreatment. Water Research, 2009, 43: 1273-1284.

[38] Novak J T, Banjade S, Murthy S N. Combined anaerobic and aerobic digestion for increased digestion for increased solids reduction and nitrogen removal. Water Research, 2011, 45(2): 618-624.

[39] Wei L L, Zhao Q L, Hu K, et al. Extracellular biological organic matters in sewage sludge during mesophilic digestion at reduced hydraulic retention time. Water Research, 2011, 45: 1472-1480.

[40] Li Y, Chróst R J. Microbial enzymatic activities in aerobic activated sludge model reactors. Enzyme and Microbial technology, 2006, 39: 568-572.

[41] Lefebvre D, Dossat-Létisse V, Lefebvre X, et al. Fate of organic matter during moderate heat treatment of sludge: kinetics of biopolymer and hydrolytic activity release and impact on sludge reduction by anaerobic digestion. Water Science and

Technology, 2014, 69(9): 1828-1833.

[42] Engelhart M, Krger M, Kopp J, et al. Effects of disintegration on anaerobic degradation of sewage excess sludge in down flow stationary fixed film digesters. Water Science and Technology, 2000, 41(3): 171-179.

[43] Cuetos M J, Gómez X, Otero M, et al. Anaerobic digestion of solid slaughterhouse waste: study of biological stabilization by Fourier Transform infrared spectroscopy and thermogravimetry combined with mass spectrometry. Biodegradation, 2010, 21: 543-556.

[44] 周友平, 莫测辉, 吴启堂. 城市污泥中有机酸在厌氧消化过程中的稳定性差异. 华南农业大学学报, 2000, 21(1): 22-25.

[45] Dai X H, Xu Y, Lu Y Q, Dong B. Recognition of the key chemical constituents of sewage sludge for biogas production. RSC Advances, 2017, 7: 2033-2037.

[46] Dai X H, Xu Y, Dong B. Effect of the micron-sized silica particles (MSSP) on biogas conversion of sewage sludge. Water Research, 2017, 115: 220-228.

[47] Xu Y, Lu Y Q, Dai X H, Dong B. Evaluating the biogas conversion potential of sewage sludge by surface site density of sludge particulate. Chemical Engineering Journal, 2017, 327: 1184-1191.

[48] 卢怡清, 许颖, 董滨, 等. 去除城市生活污泥中有机络合态金属强化其厌氧生物制气. 环境科学, 2017, DOI: 10.13227/j.hjkx.201706067.

[49] Xu Y, Lu YQ, Dai X H, Dong B. Influence of the organic-binding metal on biogas conversion of sewage sludge. Water Research, 2017, 115: 220-228.

[50] 郝晓地, 唐兴, 曹亚莉. 腐殖质对污泥厌氧消化的影响及其屏蔽方法. 环境科学学报, 2017, 37(2): 407-418.

[51] 任冰倩. 腐殖酸抑制厌氧消化过程实验研究. 北京: 北京建筑大学, 2015.

[52] Xue Y G, Liu H J, Chen S S, et al. Effects of thermal hydrolysis on organic matter solubilization and anaerobic digestion of high solid sludge. Chem. Eng. J. , 2015, 264, 174-180.

[53] 戴晓虎. 我国城镇污泥处理处置现状及思考. 给水排水, 2012, 38(2): 1-5.

[54] 陈珺, 杨琦. 污泥高级厌氧消化的应用现状与发展趋势. 中国给水排水, 2016(6): 19-23.

[55] 王广启, 吴静, 左剑恶, 等. 城市污泥高固体浓度厌氧消化的研究进展. 中国沼气, 2013, 31(6): 9-12.

[56] Duan N, Dong B, Wu B, et al. High-solid anaerobic digestion of sewage sludge under mesophilic conditions: feasibility study.. Bioresource Technology, 2012, 104(104): 150-156.

[57] Dai X, Xin G, Dong B. Rheology evolution of sludge through high-solid anaerobic digestion. Bioresource Technology, 2014, 174: 6-10.

[58] Liao X, Li H, Cheng Y, et al. Process performance of high-solids batch anaerobic digestion of sewage sludge. Environmental Technology, 2014, 35(21-24): 2652.

[59] Liu C, Li H, Zhang Y, et al. Evolution of microbial community along with increasing solid concentration during high-solids anaerobic digestion of sewage sludge. Bioresource Technology, 2016, 216: 87-94.

[60] Liu C, Li H, Zhang Y, et al. Characterization of methanogenic activity during high-solids anaerobic digestion of sewage sludge. Biochemical Engineering Journal, 2016, 109: 96-100.

[61] 李刚, 欧阳峰, 杨立中, 等. 两相厌氧消化工艺的研究与进展. 中国沼气, 2001, 19(2): 25-29.

[62] Lv W, Schanbacher F L, Yu Z. Putting microbes to work in sequence: recent advances in temperature-phased anaerobic digestion processes. Bioresource Technology, 2010, 101(24): 9409-9414.

[63] Ge H, Jensen P D, Batstone D J. Temperature phased anaerobic digestion increases apparent hydrolysis rate for waste activated sludge.. Water Research, 2011, 45(4): 1597-1606.

[64] Yu J, Zheng M, Tao T, et al. Waste activated sludge treatment based on temperature staged and biologically phased anaerobic digestion system. 环境科学学报(英文版), 2013, 25(10): 2056.

[65] 戴晓虎, 叶宁, 董滨. 脱水污泥高温两相与单相厌氧消化工艺比较研究. 科学技术与工程, 2014, 14(33): 132-138.

[66] Liu Z D, Zhang C, Lu Y, et al. States and challenges for high-value biohythane production from waste biomass by dark fermentation technology. Bioresource Technology, 2012, 135(10): 292-303.

[67] Cheng J, Ding L, Lin R, et al. Fermentative biohydrogen and biomethane co-production from mixture of food waste and sewage sludge: Effects of physiochemical properties and mix ratios on fermentation performance. Applied Energy, 2016, 184: 1-8.

[68] Li W, Guo J, Cheng H, et al. Two-phase anaerobic digestion of municipal solid wastes enhanced by hydrothermal pretreatment: Viability, performance and microbial community evaluation. Applied Energy, 2017, 189: 613-622.

[69] 胡新军, 张敏, 余俊锋, 等. 中国餐厨垃圾处理的现状、问题和对策. 生态学报, 2012, 32(14): 4575-4584.

[70] 王永会, 赵明星, 阮文权. 餐厨垃圾与剩余污泥混合消化产沼气协同效应. 环境工程学报, 2014, 8(6): 2536-2542.

[71] Feng L, Chen Y, Zheng X. Enhancement of waste activated sludge protein conversion and volatile fatty acids accumulation during waste activated sludge anaerobic fermentation by carbohydrate substrate addition: The effect of pH. Environmental Science & Technology, 2009, 43(12): 4373-4380.

[72] Angelidaki I, Ahring B K. Codigestion of olive oil mill wastewaters with manure, household waste or sewage sludge. Biodegradation, 1997, 8(4): 221-226.

[73] Gavala H N, Skiadas I V, Lyberatos G. On the performance of a centralised digestion facility receiving seasonal agroindustrial wastewaters. Water Science & Technology, 1999, 40(1): 339-346.

[74] Hamzawi N, Kennedy K J, McLean D D. Technical feasibility of anaerobic Co-digestion of sewage sludge and municipal solid waste. Environmental Technology, 1998, 19(10): 993-1003.

[75] Dai X, Duan N, Dong B, et al. High-solids anaerobic co-digestion of sewage sludge and food waste in comparison with mono digestions: stability and performance.. Waste Management, 2013, 33(2): 308-316.

[76] Liu X, Li R, Ji M, et al. Hydrogen and methane production by co-digestion of waste activated sludge and food waste in the two-stage fermentation process: substrate conversion and energy yield. Bioresource Technology, 2013, 146(10): 317-323.

[77] Cavinato C, Bolzonella D, Pavan P, et al. Mesophilic and thermophilic anaerobic co-digestion of waste activated sludge and source sorted biowaste in pilot- and full-scale reactors.. Renewable Energy, 2013, 55(4): 260-265.

[78] Suryawanshi P C, Chaudhari A B, Kothari R M. Thermophilic anaerobic digestion: the best option for waste treatment.. Critical Reviews in Biotechnology, 2010, 30(1): 31.

[79] Appels L, Baeyens J, Degrève J, et al. Principles and potential of the anaerobic digestion of waste-activated sludge. Progress in Energy & Combustion Science, 2008, 34(6): 755-781.

[80] Kim H W, Nam J Y, Hangsik S. A comparison study on the high-rate co-digestion of sewage sludge and food waste using a temperature-phased anaerobic sequencing batch reactor system. Bioresource Technology, 2011, 102(15): 7272.

[81] Neumann P, Pesante S, Venegas M, et al. Developments in pre-treatment methods to improve anaerobic digestion of sewage sludge . Reviews in Environmental Science and Bio/Technology, 2016, 15(2): 173-211.

[82] Appels L, Baeyens L, Dewil J, et al. Siloxane removal from biosolids by peroxidation. Energy Conversion and Management, 2008, 49(10): 2859-2864.

[83] Khanal S K, Grewell D, Leeuwen S S J. Ultrasound applications in wastewater sludge pretreatment: A review. Critical Reviews in Environmental Science and Technology, 2007, 37(4): 277-313(37).

[84] Hii K, Baroutian S, Parthasarathy R, et al. A review of wet air oxidation and thermal hydrolysis technologies in sludge treatment. Bioresource Technology, 2014, 155: 289-299.

[85] Mu Y, Yu H Q, Wang G. Evaluation of three methods for enriching H 2-producing cultures from anaerobic sludge. Enzyme and Microbial Technology, 2007, 40(4): 947-953.

[86] Guo L, Li X M, Zeng G M, et al. Enhanced hydrogen production from sewage sludge pretreated by thermophilic bacteria.

Energy & Fuels, 2010, 24(11): 6081-6085.

[87] Alfaro N, Cano R, Fdz-Polanco F. Effect of thermal hydrolysis and ultrasounds pretreatments on foaming in anaerobic digesters. Bioresource Technology, 2014, 170: 477-482.

[88] Zhen G, Lu X, Li Y Y, et al. Combined electrical-alkali pretreatment to increase the anaerobic hydrolysis rate of waste activated sludge during anaerobic digestion. Applied Energy, 2014, 128: 93-102.

[89] Ariunbaatar J, Panico A, Esposito G, et al. Pretreatment methods to enhance anaerobic digestion of organic solid waste . Applied energy, 2014, 123: 143-156.

[90] Neumann P, Pesante S, Venegas M, et al. Developments in pre-treatment methods to improve anaerobic digestion of sewage sludge. Reviews in Environmental Science and Bio/Technology, 2016, 15(2): 173-211.

[91] Oosterhuis M, Ringoot D, Hendriks A, et al. Thermal hydrolysis of waste activated sludge at Hengelo wastewater treatment plant, The Netherlands. Water Science and Technology, 2014, 70(1): 1-7.

[92] Xue Y, Liu H, Chen S, et al. Effects of thermal hydrolysis on organic matter solubilization and anaerobic digestion of high solid sludge . Chemical Engineering Journal, 2015, 264: 174-180.

[93] Liao X, Li H, Zhang Y, et al. Accelerated high-solids anaerobic digestion of sewage sludge using low-temperature thermal pretreatment . International Biodeterioration & Biodegradation, 2016, 106: 141-149.

[94] Chen Y C, Higgins M J, Beightol S M, et al. Anaerobically digested biosolids odor generation and pathogen indicator regrowth after dewatering . Water Research, 2011, 45(8): 2616-2626.

[95] Merry J, Oliver B. A comparison of real ad plant performance: Howdon, bran sands, cardiff and afan//Proceedings of Aquaenviro's 20th European Biosolids and Organic Resources Conference and Exhibition, Manchester, UK. 2015.

[96] Bougrier C, Delgenès J P, Carrère H. Effects of thermal treatments on five different waste activated sludge samples solubilisation, physical properties and anaerobic digestion. Chemical Engineering Journal, 2008, 139(2): 236-244.

[97] Stuckey D C, McCarty P L. Thermochemical pretreatment of nitrogenous materials to increase methane yield//Biotechnol. Bioeng. Symp.; (United States). Stanford Univ., CA, 1978, 8.

[98] Liao Z, Panter K, Mills N, et al. Thermal hydrolysis pre-treatment for advanced anaerobic digestion for sludge treatment and disposal in large scale projects//Proceedings of International DSD Conference on Sustainable Stormwater and Wastewater Management, Hong Kong. 2014.

[99] Ennouri H, Miladi B, Diaz S Z, et al. Effect of thermal pretreatment on the biogas production and microbial communities balance during anaerobic digestion of urban and industrial waste activated sludge. Bioresource Technology, 2016, 214: 184-191.

[100] Wilson C A, Novak J T. Hydrolysis of macromolecular components of primary and secondary wastewater sludge by thermal hydrolytic pretreatment. Water Research, 2009, 43(18): 4489-4498.

[101] Carrere H, Bougrier C, Castets D, et al. Impact of initial biodegradability on sludge anaerobic digestion enhancement by thermal pretreatment. Journal of Environmental Science and Health Part a, 2008, 43(13): 1551-1555.

[102] Higgins M J, Beightol S, Mandahar U, et al. Effect of thermal hydrolysis temperature on anaerobic digestion, dewatering and filtrate characteristics. Proceedings of the Water Environment Federation, 2014, 2014(15): 2027-2037.

[103] Chen S, Li N, Dong B, et al. New insights into the enhanced performance of high solid anaerobic digestion with dewatered sludge by thermal hydrolysis: organic matter degradation and methanogenic pathways. Journal of Hazardous Materials, 2017.

[104] Pilli S, Yan S, Tyagi R D, et al. Thermal pretreatment of sewage sludge to enhance anaerobic digestion: a review. Critical Reviews in Environmental Science and Technology, 2015, 45(6): 669-702.

[105] Ge H, Jensen P D, Batstone D J. Pre-treatment mechanisms during thermophilic–mesophilic temperature phased anaerobic digestion of primary sludge . Water Research, 2010, 44(1): 123-130.

[106] Tiehm A, Nickel K, Zellhorn M, et al. Ultrasonic waste activated sludge disintegration for improving anaerobic

stabilization . Water Research, 2001, 35(8): 2003-2009.

[107] Gillian B, Wayne P. Investigation of the impacts of thermal pretreatment on waste activated sludge and development of a pretreatment model . Water Research, 2013, 47(14): 5245-5256.

[108] Richard J, Wayne P, Henry Z, et al. Predicting the degradability of waste activated sludge . Water Environment Research A Research Publication of the Water Environment Federation, 2009, 81(8): 765-71.

[109] Lee M Y, Cheon J H, Hidaka T, Tsuno H. The performance and microbial diversity of temperature-phased hyperthermophilic and thermophilic anaerobic digestion system fed with organic waste. Water Sci Technol. ,2008; 57(2): 283-9.

[110] Nielsen HB, Mladenovska Z, Westermann P, Ahring B K. Comparison of two-stage thermophilic(68℃/ 55℃) anaerobic digestion with one-stage thermophilic(55℃) digestion of cattle manure. Biotechnol Bioeng., 2004; 86(3): 291-300.

[111] Lu J, Gavala H N, Skiadas I V, et al. Improving anaerobic sewage sludge digestion by implementation of a hyper-thermophilic prehydrolysis step. J Environ Manage., 2008; 88(4): 881-9.

[112] Frolund B, Griebe T, Nielsen P. Enzymatic activity in the activated sludge flocmatrix. Appl Microbiol Biotechnol. 1995; 43(November 1994): 755-761.

[113] Nielsen B, Petersen G, Design P. Thermophilic anaerobic digestion and pasteurisation. Practical Experience from Danish Wastewater Treatment Plants, 2000, 65-72.

[114] Scherer P A, Vollmer G R, Fakhouri T, Martensen S. Development of a methanogenic process to degrade exhaustively the organic fraction of municipal “grey waste” under thermophilic and hyperthermophilic conditions. Water Sci Technol. 2000; 41(3): 83-91.

[115] Wiegel J, Kristjansson J. The obligately anaerobic thermophilic bacteria. Thermophilic Bact., 1992, 105-84.

[116] Stams A J M. Metabolic interactions between anaerobic bacteria in methanogenic environments. Antonie Van Leeuwenhoek. 1994; 66(1-3): 271-94.

[117] Mladenovska Z, Ahring BK. Mixotrophic growth of two thermophilic *Methanosarcina strains*, methanosarcina thermophila TM-1 and *Methanosarcina* sp. SO-2P, on methanol and hydrogen/carbon dioxide. Appl Microbiol Biotechnol. 1997, 48(3): 385-8.

[118] Wasserfallen A. Phylogenetic analysis of 18 thermophilic Methanobacterium isolates supports the proposals to create a new genus, *Methanothermobacter* gen. nov., and to reclassify several isolates in three species, *Methanothermobacter thermautotrophicus* comb. nov., Methano. Int J Syst Evol Microbiol. 2000, 50(1): 43-53.

[119] Souza TSO, Ferreira LC, Sapkaite I, Pérez-Elvira SI, Fdz-Polanco F. Thermal pretreatment and hydraulic retention time in continuous digesters fed with sewage sludge: Assessment using the ADM1. Bioresour Technol [Internet]. Elsevier Ltd, 2013, 148: 317-24.

[120] Xue Y, Liu H, Chen S, Dichtl N, Dai X, Li N. Effects of thermal hydrolysis on organic matter solubilization and anaerobic digestion of high solid sludge. Chem Eng J. Elsevier B.V., 2015, 264: 174-80.

[121] Wang F, Hidaka T, Tsuno H, Tsubota J. Co-digestion of polylactide and kitchen garbage in hyperthermophilic and thermophilic continuous anaerobic process. Bioresour Technol. Elsevier Ltd,2012, 112: 67-74.

[122] Ahring B K, Ibrahim AA, Mladenovska Z. Effect of temperature increase from 55 to 65℃ on performance and microbial population dynamics of an anaerobic reactor treating cattle manure. Water Research, 2001; 35(10): 2446-52.

[123] Tang Y, Matsui T, Morimura S, et al. Effect of Temperature on Microbial Community of a Glucose-Degrading Methanogenic Consortium under Hyperthermophilic Chemostat Cultivation, 2008, 106(2): 180-7.

[124] Krakat N, Westphal A, Schmidt S, Scherer P. Anaerobic digestion of renewable biomass: Thermophilic temperature governs methanogen population dynamics. Appl. Environ. Microbiol, 2010, 76: 1842-1850.

[125] Meulepas R J W, Gonzalez-Gil G, Teshager F M, et al. Anaerobic bioleaching of metals from waste activated sludge.

Science of the Total Environment, 2015, 514: 60-67.

[126] 郝晓地, 魏静, 曹达啟. 废铁屑强化污泥厌氧消化产甲烷可行性分析. 环境科学学报, 2016, 36(8): 2730-2740.

[127] Hao X, Wei J, van Loosdrecht M C, et al. Analysing the mechanisms of sludge digestion enhanced by iron. Water Research, 2017, 117: 58-67.

[128] Zhang Y, Feng Y, Yu Q, et al. Enhanced high-solids anaerobic digestion of waste activated sludge by the addition of scrap iron . Bioresource Technology, 2014, 159(5): 297-304.

[129] Lü F, Luo C, Shao L, et al. Biochar alleviates combined stress of ammonium and acids by firstly enriching Methanosaeta and then Methanosarcina . Water Research, 2016, 90: 34.

[130] Zhao Z, Zhang Y, Xie Q, et al. Evaluation on direct interspecies electron transfer in anaerobic sludge digestion of microbial electrolysis cell . Bioresource Technology, 2016, 200: 235-244.

[131] Feng Y H, Zhang Y B, Chen S, et al. Enhanced production of methane from waste activated sludge by the combination of high-solid anaerobic digestion and microbial electrolysis cell with iron-graphite electrode. Chemical Engineering Journal, 2015, 259: 787-794.

[132] 刘晓光, 董滨, 戴翎翎, 等. 剩余污泥厌氧消化过程重金属形态转化及生物有效性分析. 农业环境科学学报, 2012, 31(8): 1630-1638.

[133] 沈晓南, 谢经良, 阚薇莉, 等. 厌氧消化后污泥中的重金属形态分布. 中国给水排水, 2002, 18(11): 51-52.

[134] 曹军, 谭云飞, 邢磊, 等. 污泥中重金属在厌氧消化前后的形态分布分析. 河南化工, 2003(6): 33-34.

[135] Thanh P M, Ketheesan B, Yan Z, et al. Trace metal speciation and bioavailability in anaerobic digestion: A review. Biotechnology Advances, 2016, 34(2): 122-136.

[136] Kay-Shoemake J L, Watwood M E, Lentz R D, et al. Polyacrylamide as an organic nitrogen source for soil microorganisms with potential effects on inorganic soil nitrogen in agricultural soil. Soil Biology and Biochemistry, 1998, 30(8): 1045-1052.

[137] Kay-Shoemake J L, Watwood M E, Sojka R E, et al. Polyacrylamide as a substrate for microbial amidase in culture and soil. Soil Biology and Biochemistry, 1998, 30(13): 1647-1654.

[138] Haveroen M E, MacKinnon M D, Fedorak P M. Polyacrylamide added as a nitrogen source stimulates methanogenesis in consortia from various wastewaters. Water Research, 2005, 39(14): 3333-3341.

[139] Dai X, Luo F, Yi J, et al. Biodegradation of polyacrylamide by anaerobic digestion under mesophilic condition and its performance in actual dewatered sludge system. Bioresource Technology, 2014, 153: 55-61.

[140] Oliveira J L M, Silva D P, Martins E M, et al. Biodegradation of 14C-dicofol in wastewater aerobic treatment and sludge anaerobic biodigestion. Environmental Technology, 2012, 33(6): 695-701.

[142] Rosińska A, Karwowska B. Dynamics of changes in coplanar and indicator PCB in sewage sludge during mesophilic methane digestion. Journal of Hazardous Materials, 2017, 323: 341-349.

[143] Bertin L, Capodicasa S, Occulti F, et al. Microbial processes associated to the decontamination and detoxification of a polluted activated sludge during its anaerobic stabilization. Water Research, 2007, 41(11): 2407-2416.

[144] Siebielska I, Sidełko R. Polychlorinated biphenyl concentration changes in sewage sludge and organic municipal waste mixtures during composting and anaerobic digestion. Chemosphere, 2015, 126: 88-95.

作者：戴晓虎，董　滨

同济大学环境科学与工程学院

第22章　生活垃圾能源化转化技术研究进展

本章导读

随着人口递增，生活垃圾日益增多且人口增加对能源的需求也在稳步上升。生活垃圾能源化转化（waste-to-energy）可同时实现垃圾减量和能源回收双赢，意义重大。生活垃圾能源化转化是从不可循环废物中回收能源的过程，其工艺包括气化、厌氧消化、燃烧、热解或填埋气回收产生的热、电或燃料。本章重点介绍典型的生物法生活垃圾能源化转化（厌氧消化）和热化学法生活垃圾能源化转化（焚烧、气化）两种技术的相关进展。

关键词

生活垃圾，能源化转化，厌氧消化，焚烧，气化

随着人口递增，生活垃圾日益递增，已成为重大的环境问题。与此同时，人口的增加对能源的需求也在稳步上升，这加快了能源储备的枯竭、增加了对可替代能源的需求。生活垃圾能源化转化（waste-to-energy，WtE）可同时实现垃圾减量和能源回收双赢，意义重大。WtE 是从不可循环废物中回收能源的过程，其工艺包括气化、厌氧消化、燃烧、热解或填埋气回收产生的热、电或燃料。在国际上，WtE 应用较为广泛，各种 WtE 技术中，焚烧相对使用广泛，此外，新兴 WtE 技术也不断在进步和完善，如气化转化为气体燃料等。本章重点介绍典型的生物法 WtE（厌氧消化）和热化学法 WtE（焚烧、气化）两种技术的相关进展。

1 生活垃圾中有机组分（OFMSW）的厌氧消化（AD）转化

OFMSW 的 AD 涉及一系列由多种微生物控制的代谢反应（水解，酸化，乙酸化和产甲烷），即可将有机废物转化或降解为生物气或富能量的化合物等最终产物[1]，过程不需要外加氧气[2,3]，也不需化石燃料提供能量[4]。从 19 世纪末起，OFMSW 的 AD 研究突飞猛进。Rodriguez-Iglesias 等[5]试验了 OFMSW 的 AD，获得的生物气中甲烷组分比例高达 66%。在此之前，Kayhanian 和 Rich[6]进行了高固体高温 OFMSW 的 AD 试验，发现在进行消化之前向 OFMSW 中添加微量元素和大量元素营养物，可使反应器运行更稳定、产气量更高。产甲烷菌很大程度上依赖于营养物，其蛋白质合成、核酸合成、细胞壁渗透率提升和代谢速率提升等均有赖于多种大量营养元素和微量营养元素（如碳、氮、磷、钾、硫和铁），而 OFMSW 原料的多样性恰恰满足了其对消化原料中的营养供给需求，从而可实现 AD 过程更高的稳定性。

相比而言，干法或高固体 AD 系统因具有更高的生物量而比湿法系统具有更高的灵活性和稳定性。据观察，相较于餐厨垃圾（FW）和破碎的城市垃圾中有机组分（SH-OFMSW），OFMSW 具有最高的产甲烷量和溶解性有机质去除量。Forster-Carneiro 等[7]对 FW、SH-OFMSW 和 OFMSW 三种基质 AD 试验，再次证明了有机基质自身特性对生物降解过程和甲烷产量的重要性。Bollon 等[8]则建立了一个对 OFMSW 在 AD 中降解的详细模型。

Schievano 等[9]在对高度腐烂的 OFMSW 进行类似研究时，发现高固体厌氧消化（HSAD）时的抑制现象。之后，Yu 等[10]建立了针对可生物降解性高的 OFMSW 的两级式 HSAD 系统。在此之前，Fdez.-Güelfo 等[11]利用渗滤液和下水道污泥的等体积混合物为接种物，研究了高温（55℃）、含水率低（总固体约 30%）、半连续进样的 OFMSW 的 AD 系统的启动和稳定化表现。

无论在中温（30~38℃）还是高温（49~57℃）条件，温度都对 OFMSW 的 AD 甲烷产量起到重要作

用。早期 AD 在中温段进行[12]，但在实验室规模[13]和全尺寸规模[14]试验成功之后，高温因其具有更高的加料速率和产气量而备受关注。干燥条件下的高温使得反应速度更快、产物更清洁。同时，相对于中温条件和湿润条件，复杂有机物或可生物降解材料的水解在高温条件和干燥条件下更为高效，且甲烷产量也更高。

在 Sasaki 等的一项研究中[15]，对使用人造填料浆作为原料的连续流动搅拌釜反应器中有机固体废物的嗜热 AD 的产甲烷途径和微生物群落进行研究。该研究表明，嗜热降解过程中的微生物群落完全由不明细菌组成，通过非乙酰氯化氧化途径有效地除去乙酸盐，从而提高了甲烷产量。在另一项研究中，Vandevivere 等[16]对最常见类型的厌氧消化器（如单级、两级和批式系统）进行了比较。AD 的起始阶段受两个最重要的因素的影响，即接种物[17]和 TS 含量[18]。显然，压力在沼气生成和稳定方面也起着至关重要的作用。在高海拔地区，大气压力相对较低，因此，在相对较低的二氧化碳分压下，AD 过程将产生更高的 pH 环境。这将使得 AD 系统能够抵抗酸化并实现更高的有机负载率(OLR)[19]。Jiang 等[20]通过一个自主设计的可以模拟不同压力的实验系统研究了大气压对 OFMSW 性能的影响。

迄今，因为很难判定单种酶的作用，而多种酶的活动又过于复杂，牵涉酶的类型也过多，真实有机垃圾 AD 过程中的酶活动还鲜受关注。Kim 等[21]通过对胞外自由液体进行定量研究，增强了对真实 OFMSW 进行厌氧活动的微生物酶活动的理解。Angelidaki 等[22]则在解决高温 OFMSW 的 AD 系统操作困难和稳定性低的问题时发现，搅拌对于中温共降解系统的启动和用醋酸酯供养的高温共降解系统至关重要。另外，Ghanimeh 等[23]在使用牛粪作为接种源，处理特定源 OFMSW（SS-OFMSW）评估了搅拌对高热 AD 启动阶段的影响。近期，小规模的分散厌氧系统备受关注。Zeshan 等[24]研究了碳氮比和氨氮累积对试验性干燥高温 AD 系统的影响，发现氨氮累积问题可通过调节原料碳氮比来解决，但提升规模后有可能出现与之相反的现象。

在 OFMSW 原料方面也有相关新的进展，Lo 等[25]在向 OFMSW 的 AD 添加各种生活垃圾（MSW）的焚烧灰烬时观察到，对于给 FA 和 BA 一定的有机负荷，相比于 5 d 到 10 d 的固体保留时间（SRT），20d 的 SRT 会使沼气的产量增加。然而，还可以看出，在 20 d 的 SRT 下，与使用 FA 的反应器相比，反应器与 BA 相比显示出更好的沼气生产速率。这强调了更高的 SRT 的意义，同时与 FA 相比，BA 的效率提高了。为了优化各种 OFMSW 的嗜中性 AD，OFMSW 的异质性需要进行特殊研究。OFMSW 的异质性要求对 AD 过程进行优化，即不同类型的 OFMSW 可以在特定的方法中针对某些关键参数（如 SRT 和 OLR）进行优化，SRT 和 OLR 是 AD 中最重要的运行参数之一。Rodriguez 等[26]确定了 OFMSW 的嗜温 AD 的最佳 SRT 和相关 OLR。据了解，在 20d 的 SRT 下，甲烷生产率和有机物去除率均高于 15d 和 30 d 的 SRT。

就能源而言，氢气和甲烷一样也可以被认为是从 OFMSW 厌氧消化得到的清洁替代燃料之一。氢气是未来最有希望的能源之一，它比甲烷更环保，且能够用于运输基础设施中的燃料电池。Romero Aguilar 等[27]分析了水力停留时间（HRT）对使用厌氧连续搅拌釜反应器（CSTR）嗜热干燥条件下（总固体浓度为 20%，55℃）机械生物处理（MBT）工厂的 OFMSW 的氢气产生的影响。从中观察到，在 1.9 d 的 HRT 下产生最大氢气量，条件是每天投料两次。在 Kim[28]进行的另一项研究中则开发了三阶段发酵系统，实现了 FW 对 H_2 和 CH_4 的稳定化。该系统强调实现更高的 H_2 产量，高于那些大多数 H_2 的回收率仅为 20% 的稳定 OFMSW 厌氧反应器。

OFMSW 的 AD 处理有机固废外，也能同时用于其他类型的垃圾，如下水道污泥、禽肉残渣、屠宰场垃圾等，可与 OFMSW 混合处理，提升消化速率。这种方法常被称作共消解或共发酵，它能保证主要基质（OFMSW）和共消解基质都具有更高的降解速率。共消解的一般优势在于垃圾混合易掌握[18] 、碳氮比调节[29]、潜在毒性物质的稀释、更平衡的养分、更多可溶有机质[30,31]和更大的产气量[32]。OFMSW 中的有毒组分会影响 AD 过程，在此方面，共消解是能解决此问题的最高效方法。除此之外，共消解的其他潜在优势包括高有机负荷下的更佳稳定性，更均衡的养分，更大的生物气产量和微生物间的协同作用[33]。共消解物过剩的养分可补足反应过程中的养分需求，提升碳氮比，加速有机组分降解，提升消解和稳定化速率。例如，低氮和脂质物质的共消解可提升生物气产量，因为这两者在化学性质上互补。这可以减少由多种挥发性中间体物质累积和氨浓度过高产生的问题[17]。另外，加入了下水道污泥的共消解使得污水处理设施能在能量上完成自给[34]。

2　生活垃圾热化学转化

热化学转化是城市固体废物可持续综合管理系统的一个重要组成部分[35-37]。它们的特点是比大多数其他生物化学和物理化学过程具有更高的温度和转化率（如焚烧），从而被广泛地应用于处理不同类型的固体废物，尤其是未分类的残留废物。相比于焚烧而言，气化被称为“间接燃烧”，是通过产气反应将固体废物转化为燃料或合成气，它可以被定义为废物在氧化剂数量低于按化学计量计算值条件下的部分氧化。随着垃圾填埋体积的减少和费用的增加，大大增加了固体废物热化学转化的应用空间，其在不久的将来有望成为固体废物优势处理方法[38-40]。气化具有比传统焚烧显著优越的潜在效益，这主要与组合使用条件（尤其是温度和当量比）以及特定反应器（固定床、流化床、载流床、竖井、移动炉排、回转窑、等离子体反应器）以获得适合用于不同应用条件的合成气的特性相关[38,41-43]。在过去 20 年中，已经有多种气化技术被开发出来并得到应用[38,39,44,45]。

2.1.1　焚烧

焚烧是最早的 WtE 技术，涉及 MSW 的焚烧、热回收方法和烟气清洗方面，在发电的同时对环境造成最小的伤害。焚烧过程在高于 850℃的温度下进行，释放出燃烧气体并留下灰分，这些灰分会被移除并适当处理。可使用多种热回收方法来提高焚烧的效率，主要的电力来源是燃烧气体的热量，一般是通过锅炉来产生蒸汽。烟气净化是燃烧过程中最重要的部分，包括颗粒物、氮氧化物（NO）、酸性气体（如 SO_3、HCl）、重金属（如汞、铅）、二噁英和呋喃。

世界上著名的 Metro Vancouver 的 WtE 工厂于 1988 年开业，由 Covanta 能源公司的 Covanta Burnaby 可再生能源部门运转和维护。WTEF 通常每年处理近 30 万吨的垃圾，创造 18 万 MW·h 的电力。在 100 万吨的蒸汽中，大约 20%的蒸汽出售给附近的机构。该设施运作时，废物在高于 1000℃的温度下燃烧，从焚烧炉释放的热量和气体转移到锅炉，气体被冷却，同时锅炉中的水被转化为蒸汽。锅炉产生的蒸汽使涡轮发电机产生电能。经过锅炉的气体被进一步冷却，回收的热量用于加热锅炉水。随后，气体进入烟气清洁系统。

另一个著名的焚烧发电设施 Emerald Energy from Waste Inc.始于 1992 年，位于加拿大布兰普顿，每年可以处理 15 万吨固体垃圾，最大可以产生 27000 MW·h 的电力，该设施于 2014 年被 U-PAK 集团收购并更名为 Emerald Energy from Waste Inc.（U-PAK Group of Companies，2014）。该设施使用一个转送器将垃圾从储存区转移到热燃烧室中，垃圾会在空气不足的环境中燃烧 6 小时。转送器会将垃圾混匀以确保完全燃烧，移除燃烧室中的灰分并分为黑色金属材料（6%）、粗料（22%）和细料（72%）。粗料是超过 1in①的材料，而细料则小于 1in①，细料可用作填埋场覆盖层或复合材料。可燃气体材料被转移到第二燃烧室以进一步燃烧，此操作下的燃烧温度最低为 1000℃。在燃烧过程中产生的废气被输送到一个热回收锅炉

① in 非法定单位，1in=2.54 cm

中，废气中的热量被用来产生蒸汽。产生的蒸汽使涡轮产生电力，而冷却的烟气之后进入空气污染控制系统。

2.1.2 气化

1）固定床式气化炉

在固定式气化炉中，几乎所有反应器内部都存在不同区域的废物床，其顺序取决于废物和气化介质的流动方向。这些区域在物理上不固定，并且根据操作条件上下移动，使得它们在一定程度上可以重叠。在上流式反应器中，废气在气化炉顶部进料，氧化剂进料位于底部，使得废物相对于气体移动，并依次通过不同的干燥、热解、还原和氧化区域。燃料在气化炉的顶部被干燥，因此可以使用高水分含量的废物。一些由此产生的焦炭掉落并燃烧以提供热量。甲烷和富含焦油的气体在气化炉的顶部排出，灰则从炉排落下并在底部收集。而在下流式反应炉中，废物在气化炉的顶部进料，氧化剂从顶部或侧面进料，废物和气体沿相同方向移动。

在固定床气化炉中，一些废物燃烧后，通过气化炉孔喉落下，形成气体必须通过热焦炭床，这确保了在气化炉底部留下相当高质量的合成气和相对低的焦油含量。新日铁公司提出的“直接熔化系统”的立式竖炉是一种大气移动床下流式气化炉，它直接由冶金加工技术演变而来，在熔化区域注入富氧空气（浓度为 36% 的 O_2），包括高温气化和熔化过程[46,47]。在其中，城市固体废物（MSW）和作为还原剂的石灰石以及用作黏度调节剂的焦炭一起从炉顶装入，在直接熔炉的下部形成焦炭层，高温下燃烧并保持稳定的熔融灰分。为了防止炉渣冷却、废弃物热挥发和气化的加快，通常加入约 5%质量分数的石灰石以提供熔体的 pH 缓冲，形成易于从炉底排出的流体渣。新日铁公司声称该装置废物气化发电量可达 400~670 kW·h/tMSW，具体取决于原料性质和锅炉系统[48]。目前，日本钢铁公司在日本和韩国已建成 30 多座该设施，日处理量在 100~450 t 之间。JFE 工程公司则提出了另一种在移动床下流式气化炉中加上一体化熔化的高温气化设备，目前已在日本、德国和意大利的几家工厂运营，主要处理混合的城市固体废弃物、工业废物、焚烧残渣或垃圾衍生燃料[38,45,49]。

2）鼓泡和循环流化床气化器

在鼓泡流化床（BFB）中，气态氧化剂（空气、氧气或富氧空气）的流动通过分布板向上吹送，并渗透包含待处理的废物，其位于气化器处的惰性材料（通常为硅砂或橄榄石）床底部[44]。BFB 中，表观气体速率通常约为 1 m/s，高于该值时，颗粒上的牵引力等于床中颗粒的质量并使其具有类似流体的行为[50]。这种流体状态产生强烈的混合和气固接触，具有非常高效的传质与传热。该系统的所有主要特性与流体动力学在流化床的设计和操作中的关键作用密切相关[50,51,44]，内部没有运动部件、维护简单、价格便宜，废物通常沿着侧壁从一个或多个点进料，之后被快速加热和反应。大多数气化反应发生在 BFB 顶部正上方的区域，其中有气泡喷出引起的最大湍流[52,53]，产生的合成气向上移动（床上方的垂直空间）并离开反应体。

通常，BFB 气化器在低于 900℃的温度下操作，以避免灰分熔化和烧结。Hitachi Zosen 公司和 Kobelco Eco-Solutions 公司提出了 BFB 气化炉与旋流熔化炉相结合，气体在高温（1200℃以上）下燃烧以熔化灰分并产生玻璃化炉渣，而包含在废物中的铁和铝在未氧化状态下从气化炉底部提取[54,55]。BFB 气化炉的特殊版本是内部循环床（ICFB），具有特殊的流化气体分配器，能够改善气体和固体的径向混合。该技术由 Ebara 提出，并且在上述的解决方案中，与约 1400℃操作的熔炉（旋风燃烧器）相结合[56]。

当表观气体速率增长明显高于固体的终端速率（通常高于 3 m/s）时，床的上表面没有明显的差异，床上大量的颗粒被气体带走。只有夹带的颗粒被旋风分离器收集，然后通过降液管和非机械阀返回到床时，稳态操作才是可能的。这种类型的流化反应器被称为循环流化床（CFB）。在 CFB 中，废物从侧面进料，快速加热和挥发，然后与空气或富氧空气反应。悬浮的气体和颗粒（惰性物质和废物炭）沿提升管向上移动并进入旋风分离器。美卓电力公司提出了生物质和垃圾衍生燃料的 CFB 气化器，该公司正在芬兰的拉赫蒂完成一个 160MW 的燃料装置，与家庭废物、工业废物、拆除木材和工业废料一起燃烧，该装置于 2012 年 4 月开始运营[57]。

3）气流床气化器

气流床气化器通常在约 25 bar①的高压下运行，由于可以制成浆料使得固体燃料以高压廉价供给，并具有足够高的能量来维持气化反应，主要被用来处理煤、炼油厂残留物和混合废塑料。在气流床气化器中，通常将固体浓度>60%的精细燃料颗粒加到水中以产生浆料。水可作为运输介质和温度调节剂，也可作为反应物促进氢的形成。将浆料加入具有加压氧气（或空气）的气化器中，在气化器顶部的湍流火焰燃烧燃料提供热量，将废物快速转化为高质量的合成气，灰分则在气化器壁上融化，并作为熔渣排放到淬火室中用于冷却，金属被包封在冷却的炉渣中[58]。

4）回转窑气化器

回转窑气化器主要用于工业废物焚烧和水泥生产中。回转窑可同时实现两个目标：将固体移入和移出高温反应区并在反应期间混合固体。窑炉通常由衬有耐磨耐火材料的钢圆柱形外壳组成，以防止金属过热。它通常向排放端稍微倾斜，被处理的固体废物的移动由旋转速率（约 1.5 r/min）控制。废物在高温空气加热器中高温气化而不需要额外的外部燃料。可以出售的铁、铝和矿渣的回收使得废物体积减少率变得非常高，可以达到原废物量的 1/200。

5）等离子气化器

在等离子气化器中，废物通常在大气压和 1500~5000℃的温度下投入到气化器中（通常是移动床气化器），废物与电产生的等离子体接触，有机物被转化为高质量的合成气，无机物质则被玻璃化成惰性矿渣。等离子气化器使用等离子点火器，例如由 Hitachi Metals Env SYST. Co.和 Alter NG[46,59] 提出的 WtE 装置，其中的等离子点火器位于气化器的底部，燃烧到碳床中以熔化 MSW 中的无机物，形成玻璃聚集体和从单元底部出现的金属集核。等离子气化设备需要大量的电力来操作等离子体点火器，这些电力消耗量占工厂总输出功率的 15%~20%[46,60,61]。尽管等离子体通常用于一步法提炼固体废物，但是也存在完全不同的两阶段方法。第一阶段使用常规的气化器，而第二个等离子阶段用于降低合成气中的焦油含量并提高转化效率。这种方法由英国的 Advanced Plasma Power（APP）和加拿大的 Plasco Energy Group 提出。APP 迄今为止已经建设一个小型的示范工厂，产量约为 1MW·h/t RDF，净化效率高于 23%[43,62]。

3　前 景 简 析

生活垃圾的 WtE 因能同时消除固体废物污染并产生可回收能源（资源）而被日益关注。其中，AD 具有降低环境污染和产生生物气、有机肥或有机肥底物[33]等特点。相比于卫生填埋、焚烧、热解、气化或好氧堆肥等处理方式，OFMSW 的 AD 潜力包括有机成分的快速降解、减量化、能量回收和土壤调节剂的生成。另外，封闭系统也隔绝了病原体和恶臭的释放。此外，OFMSW 的 AD 在产能方面也值得期待，可为社会的电力需求提供强力补充。相比而言，热化学转化技术相对更为成熟和可控，在处理固体废物或废物能源化转化方面，主要围绕着如何降低能耗、减排而进行设备上的完善和技术上的更新提升。此外，相伴于固体废物热化学转化过程中的能源回收则显得相对更为关键，其效率高低直接决定了该项技术的应用前景。

---

① bar 非法定单位，1bar=0.1MPa

参 考 文 献

[1] Charles W, Walker L, Cord-Ruwisch R. Effect of pre-aeration and inoculum on the start-up of batch thermophilic anaerobic digestion of municipal solid waste. Bioresour Technol, 2009, 100: 2329-2335.

[2] Guermoud N, Ouadjnia F, Avdelmalek F, Taleb F, Addou A. Municipal solid waste in Mostaganem city (Western Algeria). Waste Manage, 2009, 29: 896-902.

[3] Chanakya H N, Ramachandra T V, Vijayachamundeeswari M. Resource recovery potential from secondary components of segregated municipal solid wastes. Environ Monit Assess, 2007, 135: 119-127.

[4] Jingura R M, Matengaifa R. Optimization of biogas production by anaerobic digestion for sustainable energy development in Zimbabwe. Renew Sustain Energy Rev, 2009, 13: 1116-1120.

[5] Rodriguez I J, Castrillon L, Maranon E, Sastre H. Solid-state anaerobic digestion of unsorted municipal solid waste in a pilot-plant scale digester. Bioresour Technol, 1998, 63(1): 29-35.

[6] Kayhanian M, Rich D. Pilot-scale high solids thermophilic anaerobic digestion of municipal solid waste with an emphasis on nutrient requirements. Biomass Bioenergy, 1995, 8(6): 433-444.

[7] Forster-Carneiro T, Pérez M, Romero L I. Thermophilic anaerobic digestion of source-sorted organic fraction of municipal solid waste. Bioresour Technol, 2008a, 99: 6763-6770.

[8] Bollon J, Le-hyaric R, Benbelkacem H, Buffiere P. Development of a kinetic model for anaerobic dry digestion processes: Focus on acetate degradation and moisture content. Biochem Eng J, 2011, 56: 212-218.

[9] Schievano A, D'Imporzano, G, Malagutti L, Fragali E, Ruboni G, Adani F. Evaluating inhibition conditions in high-solids anaerobic digestion of organic fraction of municipal solid waste. Bioresour Technol, 2010, 101: 5728-5732.

[10] Yu L, Zhao Q, Ma J, Frear C, Chen S. Experimental and modelling study of a two-stage pilot scale high solid anaerobic digester system. Bioresour Technol, 2012, 124: 8-17.

[11] Fdez G L A, Álvarez-Gallego, C, Márquez D S, Romero G L I. Start-up of thermophilic-dry anaerobic digestion of OFMSW using adapted modified SEBAC inoculum. Bioresour Technol, 2010, 101: 9031-9039.

[12] Cecchi F, Pavan P, Mata A J, Musacco A, Vallini G. Digesting the organic fraction of municipal solid waste. Moving from mesophilic (37℃)to thermophilic (55℃) conditions. Waste Manage, Res., 1993, 11: 433-444.

[13] Wellinger A, Baserga U, Egger K. New systems for the digestion of solid-wastes. Water Sci Technol, 1992, 25(7): 319-326.

[14] Cozzolino C, Bassetti A, Rondelli R. Industrial application of semidry digestion process of organic solid waste. *In*: Cecchi F, Mata-Alvarez J, Pohland F G. Proceedings of the International Symposium on Anaerobic digestion of Solid Waste. Venice, 1992:551-555.

[15] Sasaki D, Hori T, Haruta S, Ueno Y, Ishii M., Igarashi Y. Methanogenicpathway and community structure in a thermophilic anaerobic digestionprocess of organic solid waste. J Biosci Bioeng, 2011, 111(1): 41-46.

[16] Vandevivere P, De Baere L, Verstraete W. Types of anaerobic digesterfor solid wastes. *In* : Mata-Alvarez J. Biomethanization of the Organic Fraction of MunicipalSolid Wastes. IWA Publishing Company, 2002: 111-140.

[17] Castillo E F M, Cristancho D E, Arellano V A. Study of the operationalconditions for anaerobic digestion of urban solid wastes. Waste Manage, 2006, 26: 546-556.

[18] Li D, Sun Y M, Yuan Z H. Effect of solid content on start-up ofmesophilic middle solid anaerobic digestion for water-sorted organic fractionof municipal solid waste. Chin J Process Eng, 2009, 9: 987-992.

[19] Zhang B, He Z G, Zhang L L, Xu J B, Shi H Z, Cai W M. Anaerobicdigestion of kitchen wastes in a single-phased anaerobic sequencing batchreactor (ASBR) with gas-phased absorb of CO_2. J Environ Sci, 2005, 17: 249-255.

[20] Jiang J, Du X, Ng S, Zhang C. Comparison of atmospheric pressure effects on the anaerobic digestion of municipal solid

waste. Bioresour Technol, 2010, 101: 6361-6367.

[21] Kim H W, Nam J Y, Kang S T, Kim D H, Jung KW, Shin H S. Hydrolytic activities of extracellular enzymes in thermophilic and mesophilic anaerobic sequencing-batch reactors treating organic fractions of municipal solid wastes. Bioresour Technol, 2012, 110: 130-134.

[22] Angelidaki I, Chen X, Cui J, Kaparaju P, Ellegaard L. Thermophilic anaerobic digestion of source-sorted organic fraction of household municipal solid waste: Start-up procedure for continuously stirred tank reactor. Water Res, 2006, 40(14): 2631-2628.

[23] Ghanimeh S, El Fadel M, Saikaly P. Mixing effect on thermophilic anaerobic digestion of source-sorted organic fraction of municipal solid waste. Bioresour Technol, 2012, 117: 63-71.

[24] Zeshan, Karthikeyan O P, Visvanathan C. Effect of C/N ratio and ammonia-N accumulation in a pilot-scale thermophilic dry anaerobic digester. Bioresour. Technol, 2012, 113: 294-302.

[25] Lo H M, Chiu H Y, Lo S W, Lo F C. Effects of different SRT on anaerobic digestion of MSW dosed with various MSWI ashes. Bioresour Technol, 2012, 125: 233-238.

[26] Rodriguez J F, Perez M, Romero L I. Mesophilic anaerobic digestion of the organic fraction of municipal solid waste: Optimisation of the semicontinuous process, Chem Eng J. 2012, 193-194: 10-15.

[27] Romero A M A, Fdez-Guelfo L A, Alvarez-Gallego C J, Romero Garcia L I. Effect of HRT on hydrogen production and organic matter solubilization in acidogenic anaerobic digestion of OFMSW. Chem Eng J, 2013, 219: 443-449.

[28] Kim D H, Kim M S. Development of a novel three-stage fermentation system converting food waste to hydrogen and methane. Bioresour Technol, 2013, 127: 267-274.

[29] Xie S, Lawlor P G, Frost J P, Hu Z, Zhan X. Effect of pig manure to grass silage ratio on methane production in batch anaerobic co-digestion of concentrated pig manure and grass silage. Bioresour Technol, 2011, 102: 5728-5733.

[30] Gannoun H, Othman N B, Bouallagui H, Hamdi M. Mesophilic and thermophilic anaerobic co-digestion of olive mill wastewaters and abattoir wastewaters in an upflow anaerobic filter. Ind Eng Chem Res, 2007, 46(21): 6737-6743.

[31] Bouallagui H., Lahdheb H., Romdan E B, Rachdi B, Hamdi M. Improvement of fruit and vegetable waste anaerobic digestion performance and stability with co-substrates addition. J Environ Manage, 2009, 90: 1844-1849.

[32] Macias-Corral M, Samani Z, Hanson A, Smith G, Funk P, Yu H, Longworth J. Anaerobic digestion of municipal solid waste and agricultural waste and the effect of co-digestion with dairy cow manure. Bioresour Technol, 2008, 99: 8288-8293.

[33] Khalid A, Arshad M, Anjum M, Mahmood T, Dawson L. The anaerobic digestion of solid organic waste. Waste Manage, 2011, 31: 1737-1744.

[34] Mata-Alvarez J, Cecchi F. A review of kinetic models applied to the anaerobic biodegradation of complex organic matter. Kinetics of the biomethanization of the organic fraction of municipal solid waste. *In*: Kamely D, Chackrabardy A, Omenn G S. Proceedings of NATO International Workshop on Biotechnology and Biodegradation, Vale de Lobos, Portugal. Portfolio Publishing Co., Woodland, TX, 1989: 27-54.

[35] Brunner P H, Morf L, Rechberger H. Thermal waste treatment - a necessary element for sustainable waste management. *In*: Twardowska I, Allen H E, Kettrup A A F, Lacy W J (Eds.). Solid Waste: Assessment, Monitoring and Remediation. B.Y, Amsterdam, The Netherlands: Elsevier. 2004.

[36] Porteous A. Why energy from waste incineration is an essential component of environmentally responsible waste management. Waste Manage, 2005, 25(4): 451-459.

[37] Psomopoulos C S, Bourka A, Themelis N J. Waste-to-energy: A review of the status and benefits in USA. Waste Manage, 2009, 29(5): 1718-1724.

[38] Heermann C, Schwager F J, Whiting K J. Pyrolysis & Gasification of Waste. A Worldwide Technology & Business Review.

2nd Edition. Juniper Consultancy Services Ltd. 2001. *In*: Knoef H. Practical aspects of biomass gasification, chapter 3 in Handbook Biomass Gasification. Knoef. Enschede, The Netherlands: BTG-Biomass Technology Group (BTG). 2005.

[39] Malkow T. Novel and innovative pyrolysis and gasification technologies for energy efficient and environmentally sound MSW disposal. Waste Manage, 2004, 24(1): 53-79.

[40] Defra. Advanced thermal treatment of municipal solid waste. Enviros Consulting Limited on behalf of Department for Environment, Food & Rural Affairs (Defra), available on <www.defra.gov.uk>.2007.

[41] E4tech. Review of Technologies for Gasification of Biomass and Wastes. NNFCC project 98/008, available on <www.nnfcc.co.uk>. 2009.

[42] Young G. Municipal solid waste to energy conversion processes: Economic. Technical and renewable comparisons, J. Wiley & Sons, Inc. 2010.

[43] Stantec. Waste to Energy. A technical review of municipal solid waste thermal treatment practices. Final Report for Environmental Quality Branch Environmental Protection Division. Project No.: 1231-10166, available at <www.env.gov.bc.ca/epd/ mun-waste/reports/pdf/bcmoe-wte-emmissions-rev-mar2011.pdf>.2010.

[44] Arena U, Mastellone M L. Fluidized Pyrolysis and Gasification of Solid Wastes. *In*: Proc. of Industrial Fluidization South Africa 2005, The South African Institute of Mining and Metallurgy, 2005: 53-68.

[45] Juniper Rating Gasification Report. available on <www.juniper.co.uk>.2009.

[46] Williams R B, Jenkins B M, Nguyen D. Solid waste conversion. A review and database of current and emerging technologies. University of California Davis. Final Report IWM-C0172. 2003.

[47] Shibaike H, Takamiya K, Hoshizawa Y, Kato Y, Tanaka H, Kotani K, Nishi T, Takada J. Development of high-performance direct melting process for municipal solid waste. Nippon Steel Technical Report, available on <www.nsc.co.jp/en/tech/report/pdf/n9205.pdf>.2005.

[48] Tanigaki N, Manako K, Osada M. Co-gasification of municipal solid waste and material recovery in a large-scale gasification and melting system. Waste Manage, 2012, 32 (4): 667-675.

[49] Suzuki A, Nagayama S. High efficiency WtE power plant using high temperature gasifying and direct melting furnace.*In*: Thirteenth International Waste Management and Landfill Symposium, S. Margherita di Pula, Cagliari, Italy. CISA Publisher, 2011.

[50] Kunii D, Levenspiel O. Fluidization Engineering, 2nd ed. ButterworthHeinemann, Boston, 1991.

[51] Grace J. Contacting modes and behaviour classification of gas-solid and other two-phase suspensions. Can J Chem Eng, 1986, 64: 353-363.

[52] Mastellone M L, Arena U. Olivine as a tar removal catalyst during fluidized bed gasification of plastic waste. AIChE J, 2008, 54 (6): 1656-1667.

[53] Arena U, Zaccariello L, Mastellone M L. Fluidized bed gasification of wastederived fuels. Waste Manage, 2010, 30: 1212-1219.

[54] Hitachi Zosen. Technical Report, available on <www.hitachizosen.co.jp/english/plant>2011.

[55] Kobelco. Technical Report, available on <www. kobelco-eco. co.jp/ english/ Business Content/ WasteTreatment> 2011.

[56] Selinger A, Steiner Ch, Shin K. Twin Rec—Bridging the gap of car recycling in Europe, presented at Int. Automobile Recycling Congress, Geneva (CH), 2003, March 12-14.

[57] Hankalin V, Helanti V, Isaksson J. High efficiency power production by gasification. *In*: Thirteenth International Waste Management and Landfill Symposium, S. Margherita di Pula, Cagliari, Italy. CISA Publisher, 2011.

[58] NETL-U.S. DOE. Refinery Technology Profiles. Gasification and Supporting Technologies, available on <www.netl.doe.gov/technologies/gasification>1995.

[59] Japanese Advanced Environment Equipment. Mitsui Recycling 21 Pyrolysis Gasification and Melting Process for Municipal Waste, available on <www.gec.ip/ JSIM_DATA/ WASTE/WASTE_3/html/Doc_436.html>.2011.

[60] Willis K P, Osada S, Willerton K L. Plasma Gasification: Lessons Learned at Ecovalley Wte Facility. Proceedings of the 18th Annual North American Waste-to-Energy Conference NAWTEC18 May 11-13, 2010. Orlando, Florida, USA.2010.

[61] Lombardi L, Carnevale E, Corti A. Analysis of energy recovery potential using innovative technologies of waste gasification. Waste Manage, 2012, 32 (4): 640-652.

[62] Hetland J, Lynum S, Santen S. Sustainable energy from waste by gasification and plasma cracking, featuring safe and inert rendering of residues. Recent experiences for reclaiming energy and ferrochrome from the tannery industry, available on <www.enviroarc.com/papers.asp>2011.

[63] Morrin S, Lettieri P, Mazzei L, Chapman C. Assessment of fluid bed + plasma gasification for energy conversion from solid waste. 3rd Int. Symposium on Energy from Biomass and Waste, Venice, Italy, CISA Publisher, 2010.

[64] Zigova J, Šturdik E, Vandk D. Butyric acid production by *Clostridium butyricum* with integrated extraction and pertraction. Process Biochem, 1999, 34(8): 835-843.

[65] Du Z, Li H, Gu T. A state of the art review on microbial fuel cells: A promising technology for wastewater treatment and bioenergy. Biotechnol Adv, 2007, 25(5): 464-482.

[66] Choi J, Chang H N, Han J. Performance of microbial fuel cell with volatile fatty acids from food wastes. Biotechnol Lett, 2011, 33: 705-714.

[67] Freguia S, Teh E H, Boon N. Microbial fuel cells operating on mixed fatty acids. Bioresour Technol, 2010, 101: 1233-1238.

[68] Chae K, Choi M, Lee J. Effect of different substrates on the performance, bacterial diversity, and bacterial viability in microbial fuel cells. Bioresour Technol, 2009, 100: 3518-3525.

[69] Zong W, Yu R, Zhang P. Efficient hydrogen gas production from cassava and food waste by a two-step process of dark fermentation andphoto-fermentation. Biomass Bioenergy, 2009, 33: 1458-1463.

[70] Tuna E, Kargi F, Argun H. Hydrogen gas production by electrohydrolysis of volatile fatty acid (VFA) containing dark fermentation effluent. Int J Hydrogen Energy, 2009, 34: 262-269.

[71] Liu W, Huang S, Zhou A. Hydrogen generation in microbial electrolysis cell feeding with fermentation liquid of waste activated sludge. Int J Hydrogen Energy, 2012, 37: 13859-13864.

[72] Uyar B, Eroglu I, Yücel M, Gündüz U. Photofermentative hydrogen production from volatile fatty acids present in dark fermentation effluents. Int. J. Hydrogen Energy, 2009, 34: 4517-4523.

[73] Dogan T, Ince O, Oz N A. Inhibition of volatile fatty acid production in granular sludge from a UASB reactor. J Environ Sci Health Part A Toxic/hazardous Substances & Environ Eng, 2005, 40(3): 633-644.

[74] Wang W, Xie L, Chen J. Biohydrogen and methaneproduction by co-digestion of cassava stillage and excess sludge underthermophilic condition. Bioresour Technol, 2011, 102: 3833-3839.

[75] Grady C P L, Daigger G T, Love N G. Biological Wastewater Treatment, third ed, CRC Press, Boca Raton, 2011.

[76] Zheng X, Chen Y G, Liu C C. Waste activated sludge alkaline fermentation liquidas carbon source for biological nutrients removal in anaerobic followed byalternating aerobic-anoxic sequneching batch reactors. Chin J Chem Eng, 2010, 18: 478-485.

[77] Khalid A, Arshad M, Anjum M, Mahmood T, Dawson L. The anaerobic digestion of solid organic waste. Waste Manage, 2011, 31: 1737-1744.

作者：沈东升，龙於洋

浙江工商大学环境科学与工程学院

第 23 章　生物质废物的解聚及产物的定向重整研究进展

本章导读

生物质废物是固体废物的主要类别之一，具有衍生环境污染和替代化石资源的双重性。在气候变化和能源短缺的背景下，生物质废物的资源价值日显突出。生物质废物的能源资源化高值利用是确保固体废物最终处置量最小化，以及控制固体废物原生和衍生污染的核心环节，研究其资源转化技术已成为环境科学领域的一个热点。生物质废物资源化的关键是实现定向转化，生物质废物需先经解聚成解聚产物后，再进一步通过物化分离或厌氧消化、羧酸合成、生物塑料、微生物电化学合成等生物转化过程合成生物质能源和化学品。本章将综述生物质废物解聚技术、高浓度解聚产物的厌氧消化和其他定向重整技术等方面的研究进展。

关键词

生物质废物，解聚，解聚产物，厌氧消化，羧酸转化

1 引　　言

生物质废物涵盖各种生物质加工与消费过程产生的废物。例如，厨余果皮、餐饮垃圾等生活源的生物可降解部分；市政污泥、园林废物等城镇源固体废物；畜禽粪便、秸秆等农业源废物；以及屠宰场加工废物、酿造残渣等工业源废物。生物质废物具有两重性，即环境污染衍生源和化石资源替代源。一方面，生物质废物的自然腐败过程容易产生溶解性有机物等渗滤液污染，进而促进重金属[1,2]、抗生素[3]等污染物的迁移，并产生挥发性有机化合物（volatile organic compounds，VOCs）、氨气等恶臭气体，以及CH_4、N_2O等温室气体；另一方面，生物质废物呈浓缩形态，即具有较高的含固率（5%~90%，质量分数），其转化产物（例如，生物质能源气体——CH_4、H_2，化学品——乳酸、己酸、乙醇，有机肥料/土壤调理剂——腐殖酸等）浓度高、产量大，从而具有良好的经济性，可作为替代化石资源的可再生原料。因此，生物质废物的处理也相应具有两重性和双重效益，处理原则首要是无害化和减量化，控制其环境污染，其次尽可能采用转化技术实现其资源化和能源化利用。

生物质废物是生物化学组成为纤维素、半纤维素、木质素、淀粉、果胶、蛋白质、脂肪的聚合物，绝大部分都能被生物，尤其是微生物所利用。因此，采用生物转化技术处理生物质废物具有技术和经济上的合理性。但是，固态的聚合物只有被解聚成溶解性有机聚合物或单体时才能被微生物进一步转化利用。因此，生物质废物的解聚过程是控制其衍生污染、生成资源化物质的关键步骤。根据解聚过程的难易程度，生物质废物可被分为两大类，相对应地，其转化技术的限速步骤有别。一类是存在木质素、纤维素、细胞壁、细胞膜、胞外聚合物等物理障碍的废物，如竹木、秸秆、污泥，其降解、转化的限速步骤一般为“解聚”（depolymerization），或称为“水解”（hydrolysis）或“液化”（liquefaction、solubilization）；另一类是容易被“水解”的废物，如厨余垃圾、餐饮垃圾、畜禽粪便，其降解或转化的限速步骤源于累积的高浓度解聚产物、中间产物对后续生物化学反应产生的“抑制”效应。

以合成生物质能源和化学品为目的，生物质废物经解聚后解聚产物的生物化学转化利用主要有厌氧消化、羧酸合成、生物塑料、微生物电化学合成这四大类技术[4]，共性是均利用厌氧生物反应实现。从技术成熟度（technology readiness levels，TRL）而言，目前仅有厌氧消化实现了大规模工业化应用。

2 生物解聚过程的强化途径

以提高甲烷、氢气、乙醇等最终产物的转化率，或以改善脱水性能为目的，文献中采用了机械、冻/融[5]、超声[6]、微波[7-9]、脉冲电磁场[9]等物理方法，乙酸[10]、稀硫酸[11]、过氧乙酸[12]、碱[5,13]、离子液体[14]、水溶(hydrotropic)[15]等化学方法[16-18]，湿式氧化等水热方法[5,19]，添加纤维素降解菌 *Clostridium thermocellum*[20]或解蛋白质菌 *Coprothermobacter proteolyticus*[21]、*Bacillus licheniformis*[22]的生物菌解方法，或者是添加蛋白酶、脂肪酶、淀粉酶、纤维素酶、多聚半乳糖醛酸酶的生物酶解方法[23-26]，来预处理木质纤维素类物料和污泥、藻类等难解聚生物质废物。预处理过程能打散污泥絮体、破坏木质素和细胞膜结构、改变纤维素晶型[12]，从而提高有机物的可及性或生物可利用性。通过傅里叶红外光谱（FTIR）、拉曼光谱、核磁共振、三维荧光光谱等方法，观测到酸、碱预处理过程能释放出细胞内含物如胡萝卜素、类胡萝卜素、类腐殖质等[5]。预处理过程的解聚效率（液化效率）、毒性物质的生成、能耗和化学物质消耗[26,27]是选择预处理技术和优化过程参数的关注重点。

通常采用液化率（如溶解性 COD、溶解性有机碳 DOC）来表征预处理过程的解聚效率。但是，近来的研究显示液化率与后续的厌氧消化甲烷产率、产氢率、脱水性能常会出现不相关或负相关的情况[5,28]。过氧乙酸-碱串联预处理水稻秸秆能提高 10%左右的液化率，但是，产氢率和乙醇产率却能分别被大幅提高 65%和 29%[12]。可能的原因包括：选用的微生物对于被预处理物料存在利用选择性，生成了酚、醛、美拉德反应（Maillard reaction）产物等毒性物质；或者溶出的有机物是难以被微生物利用的物质，如胶原蛋白、变性蛋白、类腐殖质[5]。由此引出了预处理过程可能会引入更多环境污染的问题。研究表明，过氧乙酸-碱串联预处理水稻秸秆在使乙醇产率提高了 45.5%的同时，导致了非乙醇类、低价值的液相有机物产率增加了 136%，而后者因无经济价值，往往只能当作废水处理，从而增加了环境污染负荷[12]。

上述物理、化学、热、生物预处理方法通常对生物质废物中所含的各类有机物没有选择性，会同时作用于易解聚类组分和难解聚类组分，导致预处理过程能量耗散，增加能量和化学物质/酶剂的损耗，提高了预处理成本。因此，学术界最近提出了“精准预处理”(precise pretreatment)或“选择性预处理”(selective pretreatment）的概念[12]，其思路是识别出阻碍解聚的关键有机物组分，并相应选择能聚焦该关键有机物组分的预处理技术。例如，稀酸去除半纤维素，过氧乙酸去除木质素，氢氧化钠、磷酸改变纤维素晶型[12]；利用纤维素降解菌 *Clostridium thermocellum* 可降解微藻 *Chlorella vulgaris* 细胞壁中的微量纤维素，从而破坏细胞壁，释放出细胞内蛋白质、脂肪和多糖[20]。污泥的胞外聚合物（extracellular polymer substances，EPS）被认为是限制污泥水解的物质，主要由蛋白质和多糖构成。因此，传统的酶法预处理一般选用蛋白酶、淀粉酶和纤维素酶；但近来研究发现，多聚半乳糖醛酸类物质（类果胶）作为一种量少但非常关键的黏合剂，可将污泥 EPS 和絮体紧紧地联结起来[23,29]，采用少量的多聚半乳糖醛酸酶来降解多聚半乳糖醛酸类物质，即可破坏后者的黏合作用，使污泥 EPS 和絮体分散，显著提高有机物的液化率[23]。由此，提出了一种基于生物大分子（蛋白质、多糖、DNA 等）分布的污泥“微型絮体结构模型”，有别于以紧密结合 EPS 层—松散结合 EPS 层—黏液层空间分布为特征的污泥空间分层模型[30]。

“精准预处理”的衍生应用是“中间处理”（intermediate treatment）。预处理同时作用于易解聚类有机物和难解聚类有机物，而易解聚类有机物在后续的厌氧消化过程中本来就很容易被利用。因此，作用于易解聚类有机物的这部分预处理能耗是不必要的。而且厌氧消化过程中会重新生成腐殖质等难降解有机物[31]。因此，近来出现了“中间处理”措施，将物理、化学、热、生物等处理过程移至初次厌氧消化之后，处理对象为初次厌氧消化产生的沼渣，而不是初始的生物质废物，由此可以提高沼渣中含有的大量难解聚有机物的生物可降解性，提高处理方法的针对性。中温厌氧消化后稳定的污泥沼渣经高温解蛋白质 *Coprothermobacter proteolyticus* 处理后再次进行高温厌氧消化，可再产甲烷 7 mmol/g-VS，从而提高了污泥的厌氧产能效率和消化残余物的稳定性[21]。同样地，沼渣经酶处理后也可再次作为生物产氢和甲烷的原料[32]。

此外，热解、气化、水热处理等热化学处理方法，不仅是与厌氧消化等生物化学处理方法并行的生物质废物处理技术，而且还可以作为快速、高效的预处理解聚技术，与生物化学处理技术衔接。热化学处理技术可以将生物质废物（如沼渣）解聚成小分子的气相产物（H_2、CO）和水溶性产物（酸、醛、醇），进而作为生物反应的原料[33-35]。

3　高浓度解聚产物厌氧消化反应过程的抑制、生态响应和强化措施

3.1　抑制和胁迫问题

对于易解聚类的生物质废物，如厨余垃圾、餐饮垃圾、食品加工废物等，由于其有机物高度浓缩和此类生物质的易生物降解特性，颗粒态有机物会被迅速水解酸化，无论是碳水化合物、蛋白质和脂肪均会被快速转化成有机酸，产生酸累积现象(pH 3~5，有机酸浓度 10~30 g/L)[36]，形成低 pH 抑制[37]，或游离态酸与离子态酸的共同抑制[38,39]。特别是对于蛋白质含量较高的生物质废物，如畜禽粪便、水产加工废物、发酵残渣、污泥等，除了酸累积的问题，还面临着蛋白质脱氨作用形成的铵盐在厌氧反应环境下累积的问题（氨氮浓度 1~30 g-N/L）[40-46]。有机酸、氨和 pH 会对产甲烷菌产生交互抑制作用，形成“受抑制的假稳定状态”（inhibited pseudo-steady state)，即此时 pH 呈中性，厌氧工艺运行稳定，但是，甲烷产率却非常低[47]。有机酸和氨共存时，两者会产生协同抑制效应，进一步加剧抑制程度[48]。此外，生物质废物自身含有或是因为调节物料性质而引入的碱离子（阳离子）和碳酸根（阴离子），也会对水解菌和产甲烷菌造成胁迫[49-52]。其他常见的胁迫因子，还包括 pH 波动[53]、温度波动[54-57]、餐饮垃圾的油脂[58]和盐分、畜禽粪便所含的抗生素[47,59,60]、果皮的香精油[61,62]、橄榄油和葡萄酒工业残渣所含的酚类物质[63,64]，等。生物质废物由于其高的含固率和有机负荷，以及不利的传质条件，会使得微生物面临更严峻的胁迫问题。需要注意的是，尽管水解酸化细菌一般比产甲烷菌能耐受更高的胁迫程度，但是，当酸、氨等抑制物浓度过高时也会被严重抑制[41,65,66]，导致残余的易解聚类有机物的水解过程也被终止，不仅降低了甲烷生成率，也降低了固体废物的降解率，提高了沼液的有机物浓度和沼渣量。

3.2　生态响应及其研究方法

通过对比在稳定运行和胁迫条件下厌氧消化反应器的微生态响应特征，有助于判别生物质废物能否高负荷厌氧消化、维持稳定运行的原因，从而提出提高厌氧反应系统稳定性和抗冲击能力的工程策略。因此，研究人员采用各类分子生物学分析技术来探讨厌氧细菌、古菌的种群多样性响应和代谢功能响应。常用的多样性分析方法，包括基于遗传基因 16S rRNA/rDNA 的 qPCR、分子指纹图谱技术（DGGE、TGGE、SSCP、T-RFLP、ARISA 等）、克隆测序、Pyrosequencing 或 Illumina 高通量深度测序。通过与稳定同位素培养结合，则可以同步进行稳定同位素标记基质（^{13}C 或 ^{15}N 或 ^{18}O）的代谢流追踪[67]，从而能定位利用和富集稳定同位素标记基质的功能微生物，如稳定同位素探针示踪技术(stable isotopic probing，SIP)[43,68]和各类以荧光原位杂交（FISH）为基准的单细胞水平可视化技术二次离子质谱、拉曼光谱杂交技术[69]。测定沼气 CH_4 和 CO_2 的天然稳定同位素值可以区分乙酸发酵型甲烷化和氢营养型甲烷化反应途径[39,48,70]。

随着测序技术飞速发展，形成了涵盖宏基因组、宏转录组、宏蛋白质组以及宏代谢组的宏组学技术。宏基因组学可以确定微生物群落结构，并提供潜在功能信息；宏转录组学可以用来评估基因表达，并推测关键的代谢通路；宏蛋白质组学可以确定表达中的细胞定位和调控等；宏代谢组学可以定量研究微生物群落应对环境刺激的代谢物水平的动态反应。在厌氧消化过程中，应用单一的组学技术已在一定程度上揭示了厌氧微生物的群落结构、功能及潜在功能基因[71-74]。而整合应用多种宏组学技术，则进一步使全面系统分析微生物群落动态变化及其代谢功能转化成为可能[75-77]。Lü 等[74]鉴别了尚未进行宏基因组测序的未知环境样品的宏蛋白质组，通过环境样品蛋白质数据库去冗余方法，指认蛋白质的环境微生物源，仅借助于公共蛋白质数据库就从纤维素废物厌氧反应器中鉴定出多达 514 种的无冗余无污染代谢功能；发现了解蛋白细菌 *Coprothermobacter proteolyticus* 在纤维素原料厌氧反应器中的独特“清道夫”功能作用，分析认为解蛋白细菌会降解其他细菌分泌的多纤维素酶体蛋白，导致纤维素原料水解效率的降低；因此，为了提高木质纤维素的厌氧水解效率，应尽可能地抑制此类解蛋白细菌的活动。Xia 等[78]结合宏基因组和宏转录组学的方法分析了纤维素高温下厌氧降解产甲烷的过程，结果表明，在纤维素高温厌氧降解过程中，热袍菌（Thermotogales）对 β-糖的消耗是纤维素水解过程中的关键步骤；乙酸营养型甲烷八叠球菌（Methanosarcinales）在数量上显著低于氢营养型甲烷杆菌（Methanobacteriales），但是代谢活性更高，对纤维素降解产甲烷的贡献更大。上述结果表明，丰度上占优势的物种对微生物群落代谢功能的贡献不一定大，反之亦然。Irena 等[79]结合宏基因与宏转录组学，分析工业规模高温厌氧消化反应器中的微生物群落也得到了相似的结果。Heyer 等[80]通过宏基因组与宏蛋白质组的分析，发现了少量厚壁菌门物种表达的糖苷水解酶与大量拟杆菌门表达的糖转运蛋白。他们也发现产甲烷菌只是群落中相对较少的种群，而甲烷生成所需要的关键酶却是高度表达的。这再一次说明微生物群落中占据数量优势，但由于受环境等因素影响，其基因表达分泌量并不一定占据优势，存在种类数量少但是功能酶分泌量大的微生物。Ortseifen 等[81]通过宏基因组与宏蛋白质组的分析，发现产沼微生物群落富含与甲烷生成、转运和碳代谢相关的蛋白质，丰富了产沼相关的蛋白质数据库。Mosbaek 等[82]结合了宏基因组和蛋白质 SIP 方法，辨别厌氧反应器中的共生乙酸氧化细菌。Embree 等[83]应用单细胞测序和宏转录组方法，发现了厌氧降解烷烃的微生物。

厌氧微生态对环境条件的响应机制可分为四类：①细菌和古菌之间多样性的时间更替和空间重排；②各类产甲烷菌之间多样性的更替；③产甲烷菌自身甲烷化代谢途径的切换；④产甲烷菌菌落形态变化。近年来，细菌的共生乙酸氧化途径（将乙酸降解成 H_2 和 CO_2，syntrophic acetate oxidation，SAO）广受关注[84]。在酸化环境厌氧降解过程中，SAO 与氢营养型甲烷化串联反应复合代谢途径的贡献率可达80%[39,67]，并且在各类胁迫环境下该串联反应途径均占主导优势，具体包括：低 pH[37]、高有机酸浓度和 pH 干扰[38]、碳酸盐缓冲过度[49]、甲基抑制物[85]、氨胁迫[48,86]，在各类大型沼气工程反应器中均被检出[82,87]。上述发现改变了基于低浓度废水处理的厌氧消化一号模型 ADM1 的传统认识，后者低估了 SAO 和氢营养性甲烷化的代谢流量。与同样受产物 H_2 热力学制约，需要互营微生物协作的丙酸降解和丁酸降解相比，SAO 菌与产甲烷菌之间的紧密协作对于缓解高浓度解聚产物造成的基质抑制和产物抑制尤为重要。因为丙酸降解会形成等摩尔当量的乙酸，而丁酸降解会形成两倍摩尔当量的乙酸，且丙酸、丁酸与乙酸具有相当的电离平衡常数，这意味着即便丙酸和丁酸已被有效地降解为乙酸，总乙酸摩尔浓度将不降反升，且反应液仍然会维持在极低 pH 的酸性状态；只有当乙酸转化为气态的甲烷和二氧化碳脱离反应液后，有机酸浓度才得以降低，pH 才能恢复到中性。因此，高浓度乙酸降解形成甲烷，作为整个厌氧代谢流的终端，是确保易解聚类生物质废物厌氧消化过程稳定运行的枢纽步骤。SAO 菌等细菌与产甲烷菌的互营协作可以是以 H_2、甲酸作为介导的种间电子传递（mediated interspecies electron transfer，MIET），或者是直接以纳米管、电子穿梭介体衔接的直接种间电子传递（direct interspecies electron transfer，DIET），后者具有更高的电子传递效率[88]。因此，近年来很多有关强化厌氧消化反应过程的措施是以强化 SAO 和 DIET 为导向的。

3.3　强化措施

为了缓解各种抑制因素、提高甲烷产率、强化厌氧消化反应过程，目前除了优化温度、pH 等常规工艺参数外，研究人员主要通过混合共消化（co-digestion）、消化液回流、电刺激、投加材料、调整接种比等途径提高反应的稳定性[89-92]。

（1）混合共消化（co-digestion），是指厌氧消化反应的进料包含两种或多种不同废物。混合共消化有利于调节厌氧消化系统的碳氮比，保持菌群种类平衡，从而缓解厌氧消化反应系统出现酸累积、氨氮抑制的现象[93-96]。

（2）消化液回流，主要应用于连续式反应器或半连续式反应器，有利于 pH 调节、促进养分与酶的均匀分布，达到缓解酸化抑制的效果。需要注意的是，消化液回流比应控制在合适的范围内，以避免氨氮抑制与盐类抑制[97-100]。

（3）接种比（inoculum-to-substrate ratio，ISR），对于产甲烷阶段的运行效果和稳定性非常重要。接种比偏低会导致微生物数量不够，容易使迟滞期延长，降低产甲烷的速率，造成酸累积；接种比偏高，除了需要增大反应器体积外，营养的相对缺乏也会导致生存竞争，降低微生物的活性[101-107]。

（4）电刺激，是指通过低电流电解刺激微生物代谢过程[108]。电刺激的机理主要有两点：一是通过改变微生物的 DNA 合成、蛋白质合成[109]、细胞膜的渗透性[108]从而加速细胞生长[109-111]；二是电极表面的非生物反应会影响 pH 和碱度，从而改变微生物的生长环境，间接影响微生物[112]活性。采用低直流电流刺激可以提高酵母的发酵[109]效果，促进 *Fusarium oxysporum* 蛋白质的分泌[113]。但是，电流过高时微生物可能会受到抑制。当电流达到 20 mA 时，细胞表面的疏水性增加，导致细胞凋亡[114]。

（5）微量元素（例如，Fe、Ni、Mo、Co、W、Se 等），添加适量的微量元素能够促进甲烷生成，提高反应稳定性。因为微量元素除了是各种微生物生长繁殖必要的营养元素外，还是非常重要的辅酶因子，参与产甲烷反应过程的多个酶促反应[115,116]，并且可以调节反应系统 pH，防止 VFAs 的累积[117-119]。对于不同的基质和反应条件各类微量元素的最适添加量均不相同。

（6）功能材料，向反应器中投加不同功能材料，其兼有吸附抑制物和提高产甲烷菌丰度的效果[90,120]。投加功能材料提高厌氧消化反应稳定性，主要是因为不同的材料可能具备为微生物提供微量营养元素、吸附氨氮减少抑制、富集固定微生物，或具有传递电子的能力[119]。

黏土、膨润土、海绿石、沸石等具有较高的离子交换能力，可以降低溶液中总氨氮（total ammonia nitrogen，TAN）的浓度[121,122]；同时，这些黏土质材料可以增加水解细菌和产甲烷菌的生物量，非常有利于畜禽粪便、餐饮垃圾等含氮有机物的降解，缓解氨抑制[123,124]。在猪粪的厌氧消化实验中，Wilkie 等[125]添加了黏土、珊瑚、贝壳、塑料等不同材料，发现相对于其他材料，添加黏土的反应启动更快速，稳定效果最佳。

玻璃纤维、聚乙烯纤维、碳纤维[126]、沸石[127]、活性炭[128-130]等多孔材料也被添加到厌氧反应器中，为微生物提供更多的吸附位点，保护微生物免受外界环境波动的影响，从而达到缩短反应迟滞期，提高产甲烷速率的目的[131]。Picanco 等[132]、Gong 等[133]一些研究者采用电子扫描显微镜等途径研究材料表面微生物的数量、种类和分布方式与材料理化性质的关系，直接确认了产甲烷菌在材料上的附着现象。Sasaki 等[134]应用定量 PCR、16s rRNA 克隆测序等分子生物学技术，研究了产甲烷菌在材料表面的分布情况，结果表明附着在材料表面的产甲烷菌数量有显著增加。

氧化铁[135]、磁铁矿[136,137]等导电材料也可以强化产甲烷反应过程。研究者发现添加导电铁氧化物（磁铁矿、赤铁矿等）与绝缘铁氧化物（水铁矿）相比，可以缩短反应迟滞期，提高产气速率，且有利于导电细菌 *Geobacter* sp.的生长[137]；研究者通过激光共聚焦显微镜和荧光原位杂交手段，发现导电铁氧化物的导电性可以改变微生物菌落，同时促进乙酸氧化菌（*Tepidoanaerobacter* sp.和 *Coprothermobacter* sp.）与产甲烷菌等微生物的种间电子传递，从而促进甲烷生成[136,138]。

但是，上述材料的环境风险和长期影响未知，且需经与沼渣分离处理后，沼渣和材料才能被再利用，这会大幅增加生物质废物厌氧消化的处理成本[128,139]。而源自生物质废物热解产生的生物炭[140]，应用于废物厌氧消化反应体系则具有环境生态兼容的优势，即容易制备、成本低廉、原料易得、可用于沼渣的循环接种，而且随沼渣进入土壤生态不仅不存在环境污染风险，还能改善土壤质量、促进作物生长和减排温室气体。残留在废物消化沼渣中的生物炭无需被分离出来，可以随着沼渣一起进行土地利用，促进沼渣有机物的好氧降解和腐殖化[141,142]，通过与腐殖质协同提高对重金属的固定作用[143]，从而为改善高含固率物料厌氧消化反应过程提出了确实可行的工程应用方案。研究表明，生物炭能作为甲烷化中心的启动核，促进甲烷八叠球菌和共生乙酸氧化细菌的富集，生物炭的导电性可能有利于促进这些微生物之间进行直接种间电子传递，从而缓解了酸、氨抑制[144,145]；而生物炭的添加量与微生物接种比具有互补性[146]。

4 解聚产物的定向重整途径

4.1 羧酸转化

生物质废物自然发酵的终端发酵产物主要是短碳链有机酸。由于生物质废物本身化学组成的混合度高，微生物混合生长，使得短碳链有机酸浓度较低、废物源引入的杂物量高、非目标性中间代谢产物混杂；并且，短碳链有机酸的高度亲水性导致可分离性能差，采用过滤、蒸发或蒸馏分离的成本高，不利于通过高温或催化氢化转化为乙醇和丁醇。因此，目前这些液体发酵产物还只能应用于对进料纯度要求不高的场合，如作为废水处理脱氮除磷的碳源。为了解决上述问题，可以通过生物再聚合技术，使其有机物的碳链得到延伸（carbon chain elongation，CCE），以合理的成本进一步提升废物发酵产品的资源品级。即把短碳链有机酸转化为碳原子数更高的中、长碳链有机酸，成为混合醇、烷烃、酯等燃料或溶剂的前体原料[147]，也可作为绿色抗生素添加到动物饲料中[148]。由于氧碳比降低，中、长碳链有机物具有更高的能量密度，疏水性更强，便于后续分离操作。

碳链生物延伸反应是通过碳链延伸微生物组实现的，在这些微生物细胞内可以发生一种特殊的代谢反应，即逆 β 氧化反应。目前，碳链生物延伸技术的研究关注点，包括碳链延伸微生物组[149-151]，基因工程菌[152]，中、长碳链有机酸产物的分离[153,154]，电子供体种类和剂量[155-157]，小分子解聚气相产物的碳链生物延伸[151]，碳链生物延伸反应过程的抑制和缓解。已有研究发现，生物炭同样能显著提高碳链生物延伸产物产率，缓解基质和产物的酸抑制效应[158]。

4.2　生物塑料合成

目前，生物塑料（例如 polyhydroxyalkanoates，PHAs）受到日益广泛的关注。生物塑料具有良好的生物兼容性，可以代替传统塑料应用于新的领域，而生物质废物是合成生物塑料的重要原料之一[159]。目前，PHAs 是商业化应用最成功的可再生和可生物降解聚合物[160]。全世界每年从石油和天然气生产的塑料达 2.45 亿吨[161]。除了生产能耗较高和温室气体排放等问题外，石油合成塑料的难降解特性具有环境危害性。相对于传统合成聚合物，PHAs 不仅是无毒聚合物，而且具有明显的生态优势，利用可再生资源合成 PHAs 可以实现碳中和[162]。生物塑料的结构和化学性能与聚丙烯、聚乙烯和聚己内酰胺有可比性，具备更好的热机械性能，可生物降解，并且可以通过可再生碳源合成[163]。PHAs 可分为短链单体（SCL，$C_{3\sim5}$）、中链单体（MCL，$C_{6\sim14}$）和长链单体（LCL，$C_{>14}$），短链 PHAs 和常规塑料性质类似，中链 PHAs 的性质更接近于橡胶。因此，由生物质废物合成的 PHAs 具备替代传统塑料和橡胶的潜力。

在过去 20 年，以糖类为底物，利用工程重组微生物合成 PHAs 取得了显著的进步[164,165]。然而，灭菌的高成本、底物的精炼和培养基的专一化要求，使该工艺与石油合成塑料相比没有商业竞争力。利用生物质废物取代精炼的糖类作为底物可望节省生产成本 30%~50%[161]，且利用生物质废物为底物，不需要灭菌过程，更容易培养和维持具有稳定代谢功能的微生物组。目前，已证明微生物合成生产 PHAs 的可利用生物质废物包括乳制品废水[166]、农业废物[167]、制糖废物[168,169]、畜禽粪便[170]等。Dobroth 等[171]估算每年 3800 万升生物柴油产生的甘油废料可通过微生物合成的方式生产 19t 的 PHAs。

除了水溶性的解聚产物，生物质废物的裂解油和气相解聚产物合成气也能用于 PHAs 的生物合成。进行 PHAs 生物合成的一类主要细菌需要在营养物质不足但碳源充足的环境中进行[172]，而裂解油缺少营养物质，但富含碳源，因此可能成为一种高效的原料用于生物合成 PHAs[173]。Moita 和 Lemos 的研究结果表明，当采用稀释的裂解油（裂解油∶矿物培养液=1∶176）作为碳源时，PHAs 的最大产率为 0.092 g/g 细胞干重[174]。Moita 等的研究结果表明，即使裂解油含有抑制 PHAs 生成的组分，经过驯化选择后的微生物依然可以直接利用裂解油生成短链 PHAs[175]。限制合成气生成 PHAs 商业应用的主要因素是生产效率低，Beneroso 等研究发现，合成气中的 CO 是微生物合成 PHAs 的首选碳源，并且微波加热是产生高 CO/CO_2 比合成气的有效方式[176]。合成气生成 PHAs 的研究可以分为纯菌和混合菌两类。利用纯菌进行该类实验，可以掌握 PHAs 的合成途径，并且通过测定特定微生物的基因序列，可以了解微生物的代谢多样性，从而判断能否将微生物应用于合成气生成 PHAs。聚羟基丁酸酯（polyhydroxybutyrate，PHB）的生物合成主要涉及三种酶（PhaA、PhaB 和 PhaC），通常而言，编码这三种酶的基因都整合在同一个 PHA 生物合成操纵子上[177]。早在约 60 年前，*Ralstonia eutropha* H16 已经引起了关注，因为它可以合成并储存大量的 PHB 和其他聚合物。Pohlmann 等公布了 *Ralstonia eutropha* H16 两个染色体的全部基因序列[178]，Vincent 等也报道了该微生物细胞膜上的氢化酶对于 CO 的高耐受性[179]，揭示了将该微生物应用于合成气生成 PHAs 的可能性。*Rhodospirillum rubrum* 是一种紫色的革兰氏阴性菌，可以将 CO 转换为 H_2，同时生成 PHAs。因此，可以用来提高合成气品质。但是，PHAs 的生成过程需要有其他碳源的加入。例如，乙酸根的缺少会导致 PHAs 产量下降 50%[180,181]。然而，利用纯菌合成 PHAs 有如下缺点：①净化基质的成本较高；②灭菌操作复杂，且成本较高[182]。目前，混合菌合成 PHAs 的研究仍处于工艺探索阶段。Lagoa-Costa 等利用 *Clostridium autoethanogenum* 将合成气转化为乙醇和乙酸，然后，利用混合微生物将乙酸转化为 PHAs，得到的 PHAs 最大含量占细胞干重的 24%，探明了一种利用合成气同时生成乙醇和 PHAs 的反应途径[183]。

5 展　　望

在气候变化和能源短缺的双重背景下，生物质废物的资源价值和能源效益日显突出。以生物质废物作为原料进行生物精炼（Bio-Refinery），实现 BioWaste-to-BioFules 和 BioWaste-to-BioChemicals。为了提高生物精炼全过程的综合效益，必须选择和优化生物质废物的生物转化技术。应针对不同种类生物质废物的物理化学性质，分别聚焦解聚手段和解聚产物的高值合成转化反应途径，注重两者的合理衔接；同时，不能忽略转化反应过程的环境影响、合成产物的市场潜力、沼液沼渣等反应残余物的可持续消纳途径，以实现生物精炼全过程的产物资源价值最大化和污染释放潜力最小化。

参 考 文 献

[1] Wu J, Zhang H, He P J, Shao L M. Insight into the heavy metal binding potential of dissolved organic matter in MSW leachate using EEM quenching combined with PARAFAC analysis. Water Research, 2011, 45(4): 1711-1719.

[2] 邵立明，何品晶，瞿贤. 生物反应器填埋场初期的重金属释放行为. 中国环境科学, 2007, 27(1): 71-75.

[3] Yu Z F, He P J, Shao L M, Zhang H, Lü F.Co-occurrence of mobile genetic elements and antibiotic resistance genes in municipal solid waste landfill leachates: A preliminary insight into the role of landfill age. Water Research, 2016, 106: 583-592.

[4] Marshall C W, LaBelle E V, May H D. Production of fuels and chemicals from waste by microbiomes. Current Opinion in Biotechnology, 2013, 24(3): 391-397.

[5] Lü F, Wang F, Shao L M, He P J. Deciphering pretreatment-induced repartition among stratified extracellular biopolymers and its effect on anaerobic biodegradability and dewaterability of waste activated sludge. Journal of Environmental Chemical Engineering, 2017, 5(3): 3014–3023.

[6] Xu H C, He P J, Yu G H, Shao L M. Effect of ultrasonic pretreatment on anaerobic digestion and its sludge dewaterability. Journal of Environmental Sciences, 2011, 23(9): 1472-1478.

[7] Fan J, De Bruyn M, Budarin V L, Gronnow M J, Shuttleworth P S, Breeden S, Macquarrie D J, Clark J H. Direct microwave-assisted hydrothermal depolymerization of cellulose. Journal of the American Chemical Society, 2013, 135(32): 11728–11731.

[8] Chen W H, Tu Y J, Sheen H K. Disruption of sugarcane bagasse lignocellulosic structure by means of dilute sulfuric acid pretreatment with microwave-assisted heating. Applied Energy, 2011, 88(8): 2726-2734.

[9] Eskicioglu C, Terzian N, Kennedy K J, Droste R L, Hamoda M. Athermal microwave effects for enhancing digestibility of waste activated sludge. Water Research, 2007, 41(11): 2457-2466.

[10] 何品晶，方文娟，吕凡，朱敏，邵立明.乙酸常温预处理对木质纤维素厌氧消化的影响. 中国环境科学, 2008, 28(12): 1116-1121.

[11] He P J, Chai L N, Li L, Hao L P, Shao L M, Lu F. In situ visualization of the change in lignocellulose biodegradability during extended anaerobic bacterial degradation. RSC Advances, 2013, 3(29): 11759-11773.

[12] Lü F, Chai L N, Shao L M, He P J. Precise pretreatment of lignocellulose: Relating substrate modification with subsequent hydrolysis and fermentation to products and by-products. Biotechnology for Biofuels, 2017, 10(1): 88.

[13] Yu G H, He P J, Shao L M, He P P. Toward understanding the mechanism of improving the production of volatile fatty acids from activated sludge at pH 10.0. Water Research , 2008, 42(18): 4637-4644.

[14] Xu F, Sun J, Konda N, Shi J, Dutta T, Scown C D, Simmons B A, Singh S. Transforming biomass conversion with ionic liquids: Process intensification and the development of a high-gravity, one-pot process for the production of cellulosic ethanol. Energy & Environmental Science, 2016, 3: 1042-1049.

[15] Devendra L P, Pandey A. Hydrotropic pretreatment on rice straw for bioethanol production. Renewable Energy , 2016, 98: 2-8.

[16] Bazargan A, Bazargan M, McKay G. Optimization of rice husk pretreatment for energy production. Renewable Energy, 2015, 77: 512-520.

[17] Battista F, Mancini G, Ruggeri B, Fino D. Selection of the best pretreatment for hydrogen and bioethanol production from olive oil waste products. Renewable Energy, 2016, 88: 401-407.

[18] Guragain YN, Wang D, Vadlani PV. Appropriate biorefining strategies for multiple feedstocks: Critical evaluation for pretreatment methods, and hydrolysis with high solids loading. Renewable Energy , 2016, 96, Part A: 832-842.

[19] Qiao W, Yan X, Ye J, Sun Y, Wang W, Zhang Z. Evaluation of biogas production from different biomass wastes with/without hydrothermal pretreatment. Renewable Energy, 2011, 36(12): 3313-3318.

[20] Lu F, Ji J Q, Shao L M, He P J. Bacterial bioaugmentation for improving methane and hydrogen production from microalgae. Biotechnology for Biofuels, 2013, 6(1): 92.

[21] Lü F, Li T S, Wang T F, Shao L M, He P J. Improvement of sludge digestate biodegradability by thermophilic bioaugmentation. Applied Microbiology and Biotechnology, 2014, 98(2): 969-977.

[22] 何品晶, 王颖, 胡洁, 吕凡, 邵立明.应用解蛋白菌生物预水解剩余污泥. 环境科学, 2016, 37(11): 4317-4325.

[23] Lü F, Wang J W, Shao L M, He P J. Enzyme disintegration with spatial resolution reveals different distributions of sludge extracellular polymer substances. Biotechnology for Biofuels, 2016, 9(1): 1-14.

[24] Brienzo M, Fikizolo S, Benjamin Y, Tyhoda L, Görgens J. Influence of pretreatment severity on structural changes, lignin content and enzymatic hydrolysis of sugarcane bagasse samples. Renewable Energy, 2017, 104: 271-280.

[25] Mesa L, López N, Cara C, Castro E, González E, Mussatto S I. Techno-economic evaluation of strategies based on two steps organosolv pretreatment and enzymatic hydrolysis of sugarcane bagasse for ethanol production. Renewable Energy, 2016, 86: 270-279.

[26] Tonini D, Astrup T. Life-cycle assessment of a waste refinery process for enzymatic treatment of municipal solid waste. Waste Management, 2012, 32(1): 165-176.

[27] Xiao K, Chen Y, Jiang X, Seow W Y, He C, Yin Y, Zhou Y. Comparison of different treatment methods for protein solubilisation from waste activated sludge. Water Research, 2017, 122: 492-502.

[28] Kim D H, Cho S K, Lee M K, Kim M S. Increased solubilization of excess sludge does not always result in enhanced anaerobic digestion efficiency. Bioresource Technology, 2013, 143: 660-664.

[29] Flemming H C, Wingender J. The biofilm matrix. Nature Reviews Microbiology, 2010, 8(9): 623-633.

[30] Yu G H, He P J, Shao L M, He P P. Stratification structure of sludge flocs with implications to dewaterability. Environmental Science & Technology , 2008, 42(21): 7944-7949.

[31] Zheng W, Lü F, Phoungthong K, He P J. Relationship between anaerobic digestion of biodegradable solid waste and spectral characteristics of the derived liquid digestate. Bioresource Technology , 2014, 161: 69-77.

[32] Sato H, Kuribayashi K, Fujii K. Possible practical utility of an enzyme cocktail produced by sludge-degrading microbes for methane and hydrogen production from digested sludge. New Biotechnology, 2016, 33(1): 1-6.

[33] Fabbri D, Torri C. Linking pyrolysis and anaerobic digestion(Py-AD)for the conversion of lignocellulosic biomass. Current Opinion in Biotechnology , 2016, 38: 167-173.

[34] Chen D Z, Yin L J, Wang H, He P J. Pyrolysis technologies for municipal solid waste: A review. Waste Management, 2014, 34(12): 2466-2486.

[35] Wang N, Chen D Z, Arena U, He P J. Hot char-catalytic reforming of volatiles from MSW pyrolysis. Applied Energy, 2017, 191: 111-124.

[36] Chen Y, Cheng J J, Creamer K S. Inhibition of anaerobic digestion process: A review. Bioresource Technology, 2008, 99(10): 4044-4064.

[37] Hao L P, Lü F, Li L, Shao L M, He P J. Shift of pathways during initiation of thermophilic methanogenesis at different initial pH. Bioresource Technology , 2012, 126(0): 418-424.

[38] Hao L P, Lü F, Li L, Wu Q, Shao L M, He P J. Self-adaption of methane-producing communities to pH disturbance at different acetate concentrations by shifting pathways and population interaction. Bioresource Technology, 2013, 140(0): 319-327.

[39] Hao L P, Lü F, He P J, Li L, Shao L M. Predominant contribution of syntrophic acetate oxidation to thermophilic methane formation at high acetate concentrations. Environmental Science & Technology , 2011, 45(2): 508-513.

[40] Lü F, Chen M, He P J, Shao L M. Effects of ammonia on acidogenesis of protein-rich organic wastes. Environmental Engineering Science, 2008, 25(1): 114-122.

[41] Lü F, He P J, Shao L M, Lee D J. Effects of ammonia on hydrolysis of proteins and lipids in fish samples. Applied Microbiology and Biotechnology , 2007, 75(5): 1201-1208.

[42] Poirier S, Desmond L Quéméner E, Madigou C, Bouchez T, Chapleur O. Anaerobic digestion of biowaste under extreme ammonia concentration: Identification of key microbial phylotypes. Bioresource Technology, 2016, 207: 92-101.

[43] Hao LP, Lü F, Mazéas L, Desmond L, Quéméner E, Madigou C, Guenne A, Shao LM, Bouchez T, He PJ. Stable isotope probing of acetate fed anaerobic batch incubations shows a partial resistance of acetoclastic methanogenesis catalyzed by *Methanosarcina* to sudden increase of ammonia level. Water research, 2015, 69: 90-99.

[44] Hao L P, Mazéas L, Lü F, Grossin-Debattista J, He P J, Bouchez T. Effect of ammonia on methane production pathways and reaction rates in acetate-fed biogas processes. Water Science and Technology, 2017, 75(8): 1839-1848.

[45] Wu D, Lü F, Shao L M, He P J. Effect of cycle digestion time and solid-liquid separation on digestate recirculated one-stage dry anaerobic digestion: Use of intact polar lipid analysis for microbes monitoring to enhance process evaluation. Renewable Energy, 2017, 103: 38-48.

[46] Peng W, Lü F, Shao L M, He P J. Microbial communities in liquid and fiber fractions of food waste digestates are differentially resistant to inhibition by ammonia. Applied Microbiology and Biotechnology , 2015, 99(7): 3317-3326.

[47] Wu D, Lü F, Gao H, Shao L M, He P J. Mesophilic bio-liquefaction of lincomycin manufacturing biowaste: The influence of total solid content and inoculum to substrate ratio. Bioresource Technology , 2011, 102(10): 5855-5862.

[48] Lü F, Hao L P, Guan D X, Qi Y J, Shao L M, He P J. Synergetic stress of acids and ammonium on the shift in the methanogenic pathways during thermophilic anaerobic digestion of organics. Water Research, 2013, 47(7): 2297-2306.

[49] Lin Y C, Lü F, Shao L M, He P J. Influence of bicarbonate buffer on the methanogenetic pathway during thermophilic anaerobic digestion. Bioresource Technology, 2013, 137(0): 245-253.

[50] He P J, Lü F, Shao L M, Pan X J, Lee DJ. Effect of alkali metal cation on the anaerobic hydrolysis and acidogenesis of vegetable waste. Environmental Technology, 2006, 27(3): 317-327.

[51] Liu N, Wang Q, Jiang J, Zhang H. Effects of salt and oil concentrations on volatile fatty acid generation in food waste fermentation. Renewable Energy , 2017, 113: 1523-1528.

[52] Xia Y, He P J, Pu H X, Lü F, Shao L M, Zhang H. Inhibition effects of high calcium concentration on anaerobic biological treatment of MSW leachate. Environmental Science and Pollution Research, 2016, 23(8): 7942-7948.

[53] Hori T, Haruta S, Ueno Y, Ishii M, Igarashi Y. Dynamic transition of a methanogenic population in response to the concentration of volatile fatty acids in a thermophilic anaerobic digester. Applied and Environmental Microbiology, 2006, 72(2): 1623-1630.

[54] Chapleur O, Mazeas L, Godon J J, Bouchez T. Asymmetrical response of anaerobic digestion microbiota to temperature changes. Applied Microbiology and Biotechnology, 2016, 100(3): 1445-1457.

[55] Luo G, De Francisci D, Kougias P G, Laura T, Zhu X, Angelidaki I. New steady-state microbial community compositions and process performances in biogas reactors induced by temperature disturbances. Biotechnology for Biofuels, 2015, 8(1): 3.

[56] Ho D, Jensen P, Batstone D. Effects of temperature and hydraulic retention time on acetotrophic pathways and performance in high-rate sludge digestion. Environmental Science & Technology , 2014, 48(11): 6468-6476.

[57] Vanwonterghem I, Jensen P D, Rabaey K, Tyson G W. Temperature and solids retention time control microbial population dynamics and volatile fatty acid production in replicated anaerobic digesters, 2015, 5: 8496.

[58] Neves L, Oliveira R, Alves M M. Fate of LCFA in the co-digestion of cow manure, food waste and discontinuous addition of oil. Water Research , 2009, 43(20): 5142-5150.

[59] Feng L, Casas M E, Ottosen L D M, Møller H B, Bester K. Removal of antibiotics during the anaerobic digestion of pig manure. Science of The Total Environment, 2017, 603: 219-225.

[60] Burch T R, Sadowsky M J, LaPara T M. Modeling the fate of antibiotic resistance genes and class 1 integrons during thermophilic anaerobic digestion of municipal wastewater solids. Applied Microbiology and Biotechnology , 2016, 100(3): 1437-1444.

[61] Ruiz B, Flotats X. Effect of limonene on batch anaerobic digestion of citrus peel waste. Biochemical Engineering Journal, 2016, 109: 9-18.

[62] Ruiz B, Flotats X. Citrus essential oils and their influence on the anaerobic digestion process: An overview. Waste Management, 2014, 34(11): 2063-2079.

[63] Poirier S, Bize A, Bureau C, Bouchez T, Chapleur O. Community shifts within anaerobic digestion microbiota facing phenol inhibition: Towards early warning microbial indicators? Water Research , 2016, 100: 296-305.

[64] Hoyos H C, Hoffmann M, Guenne A, Mazeas L. Elucidation of the thermophilic phenol biodegradation pathway via benzoate during the anaerobic digestion of municipal solid waste. Chemosphere, 2014, 97: 115-119.

[65] He P J, Lü F, Shao L M, Pan X J, Lee D J. Enzymatic hydrolysis of polysaccharide-rich particulate organic waste. Biotechnology and Bioengineering , 2006, 93(6): 1145-1151.

[66] Lü F, He P J, Shao L M, Lee D J. Lactate inhibits hydrolysis of polysaccharide-rich particulate organic waste. Bioresource Technology, 2008, 99(7): 2476-2482.

[67] Mulat D G, Ward A J, Adamsen A P S, Voigt N V, Nielsen J L, Feilberg A. Quantifying contribution of synthrophic acetate oxidation to methane production in thermophilic anaerobic reactors by membrane inlet mass spectrometry. Environmental Science & Technology, 2014, 48(4): 2505-2511.

[68] Li T L, Mazeas L, Sghir A, Leblon G, Bouchez T. Insights into networks of functional microbes catalysing methanization of cellulose under mesophilic conditions. Environmental Microbiology, 2009, 11(4): 889-904.

[69] Mazeas L, Li T, Wu T D, Chapleur O, An S, Debattista J G, Guerquin K J L, Bouchez T. Simultaneous analysis of microbial identity and function using NanoSIMS: Application to anaerobic degradation of methanol. Geochimica Et Cosmochimica Acta , 2009, 73(13): A853-A853.

[70] 何品晶, 吕凡, 邵立明, 章骅.稳定同位素表征有机物甲烷化代谢动力学. 化学进展, 2009, 21(0203): 540-549.

[71] Mosbaek F, Kjeldal H, Mulat D G, Albertsen M, Ward A J, Feilberg A, Nielsen J L. Identification of syntrophic acetate-oxidizing bacteria in anaerobic digesters by combined protein-based stable isotope probing and metagenomics. ISME Journal, 2016, 10(10): 2405-2418.

[72] St-Pierre B, Wright A-D G. Comparative metagenomic analysis of bacterial populations in three full-scale mesophilic anaerobic manure digesters. Applied Microbiology and Biotechnology, 2014, 98(6): 2709-2717.

[73] Moitinho-Silva L, Diez-Vives C, Batani G, Esteves AIS, Jahn M T, Thomas T. Integrated metabolism in sponge-microbe symbiosis revealed by genome-centered metatranscriptomics. ISME Journal , 2017, 11(7): 1651-1666.

[74] Lü F, Bize A, Guillot A, Monnet V, Madigou C, Chapleur O, Mazeas L, He P J, Bouchez T. Metaproteomics of cellulose methanisation under thermophilic conditions reveals a surprisingly high proteolytic activity. The ISME Journal, 2014, 8(1): 88-102.

[75] Vanwonterghem I, Jensen P D, Ho D P, Batstone D J, Tyson G W. Linking microbial community structure, interactions and function in anaerobic digesters using new molecular techniques. Current Opinion in Biotechnology, 2014, 27: 55-64.

[76] Franzosa E A, Hsu T, Sirota-Madi A, Shafquat A, Abu-Ali G, Morgan X C, Huttenhower C. Sequencing and beyond: integrating molecular 'omics for microbial community profiling. Nature Reviews Microbiology, 2015, 13(6): 360-372.

[77] Bozan M, Akyol Ç, Ince O, Aydin S, Ince B. Application of next-generation sequencing methods for microbial monitoring of anaerobic digestion of lignocellulosic biomass. Applied Microbiology and Biotechnology, 2017, 101(18): 6849-6864.

[78] Xia Y, Wang Y, Fang HHP, Jin T, Zhong H, Zhang T. Thermophilic microbial cellulose decomposition and methanogenesis pathways recharacterized by metatranscriptomic and metagenomic analysis. Scientific Reports, 2014, 4: 6708.

[79] Maus I, Koeck D E, Cibis K G, Hahnke S, Kim Y S, Langer T, Kreubel J, Erhard M, Bremges A, Off S, Stolze Y, Jaenicke S, Goesmann A, Sczyrba A, Scherer P, König H, Schwarz W H, Zverlov V V, Liebl W, Pühler A, Schlüter A, Klocke M. Unraveling the microbiome of a thermophilic biogas plant by metagenome and metatranscriptome analysis complemented by characterization of bacterial and archaeal isolates. Biotechnology for Biofuels, 2016, 9(1): 171.

[80] Heyer R, Kohrs F, Benndorf D, Rapp E, Kausmann R, Heiermann M, Klocke M, Reichl U. Metaproteome analysis of the microbial communities in agricultural biogas plants. New Biotechnology, 2013, 30(6): 614-622.

[81] Ortseifen V, Stolze Y, Maus I, Sczyrba A, Bremges A, Albaum S P, Jaenicke S, Fracowiak J, Pühler A, Schlüter A. An integrated metagenome and -proteome analysis of the microbial community residing in a biogas production plant. Journal of Biotechnology , 2016, 231: 268-279.

[82] Mosbaek F, Kjeldal H, Mulat D G, Albertsen M, Ward A J, Feilberg A, Nielsen J L. Identification of syntrophic acetate-oxidizing bacteria in anaerobic digesters by combined protein-based stable isotope probing and metagenomics. ISME J, 2016, 10(10): 2405-2418.

[83] Embree M, Nagarajan H, Movahedi N, Chitsaz H, Zengler K. Single-cell genome and metatranscriptome sequencing reveal metabolic interactions of an alkane-degrading methanogenic community. The ISME Journal, 2014, 8(4): 757-767.

[84] Westerholm M, Moestedt J, Schnürer A. Biogas production through syntrophic acetate oxidation and deliberate operating strategies for improved digester performance. Applied Energy , 2016, 179: 124-135.

[85] Hao L P, Lü F, Wu Q, Shao L M, He P J. High concentrations of methyl fluoride affect the bacterial community in a thermophilic methanogenic sludge. PLoS ONE , 2014, 9(3): e92604.

[86] Werner J J, Garcia M L, Perkins S D, Yarasheski K E, Smith S R, Muegge B, Stadermann F J, DeRito C M, Floss C, Madsen E L. Microbial community dynamics and stability during an ammonia-induced shift to syntrophic acetate oxidation. Applied and Environmental Microbiology , 2014, 80(11): 3375-3383.

[87] Sun L, Müller B, Westerholm M, Schnürer A. Syntrophic acetate oxidation in industrial CSTR biogas digesters. Journal of Biotechnology , 2014, 171(0): 39-44.

[88] Storck T, Virdis B, Batstone D J. Modelling extracellular limitations for mediated versus direct interspecies electron transfer. The ISME Journal, 2016, 10(3): 621-631.

[89] Zhang Q, Hu J, Lee D J. Biogas from anaerobic digestion processes: Research updates. Renewable Energy, 2016, 98: 108-119.

[90] Show K Y, Tay J H. Influence of support media on biomass growth and retention in anaerobic filters. Water Research, 1999, 33(6): 1471-1481.

[91] Wang T, Shao L, Li T, Lü F, He P. Digestion and dewatering characteristics of waste activated sludge treated by an anaerobic

biofilm system. Bioresource Technology, 2014, 153: 131-136.

[92] Yuan H, Chen Y, Dai X, Zhu N. Kinetics and microbial community analysis of sludge anaerobic digestion based on Micro-direct current treatment under different initial pH values. Energy, 2016, 116: 677-686.

[93] Battista F, Fino D, Erriquens F, Mancini G, Ruggeri B. Scaled-up experimental biogas production from two agro-food waste mixtures having high inhibitory compound concentrations. Renewable Energy, 2015, 81: 71-77.

[94] Haider M R, Zeshan, Yousaf S, Malik R N, Visvanathan C. Effect of mixing ratio of food waste and rice husk co-digestion and substrate to inoculum ratio on biogas production. Bioresource Technology, 2015, 190: 451-457.

[95] Li L, Sun Y, Yuan Z, Kong X, Wao Y, Yang L, Zhang Y, Li D. Effect of microalgae supplementation on the silage quality and anaerobic digestion performance of Manyflower silvergrass. Bioresource Technology, 2015, 189: 334-340.

[96] Ziemiński K, Kowalska-Wentel M. Effect of enzymatic pretreatment on anaerobic co-digestion of sugar beet pulp silage and vinasse. Bioresource Technology, 2015, 180: 274-280.

[97] Shahriari H, Warith M, Hamoda M, Kennedy K J. Effect of leachate recirculation on mesophilic anaerobic digestion of food waste. Waste Management , 2012, 32(3): 400-403.

[98] Stabnikova O, Liu X Y, Wang J Y. Anaerobic digestion of food waste in a hybrid anaerobic solid–liquid system with leachate recirculation in an acidogenic reactor. Biochemical Engineering Journal, 2008, 41(2): 198-201.

[99] Zuo Z, Wu S, Zhang W, Dong R. Effects of organic loading rate and effluent recirculation on the performance of two-stage anaerobic digestion of vegetable waste. Bioresource Technology, 2013, 146: 556-561.

[100] Filipkowska U. Effect of recirculation method on quality of landfill leachate and effectiveness of biogas production. Polish Journal of Environmental Studies, 2008, 17(2): 199-207.

[101] Neves L, Oliveira R, Alves M M. Influence of inoculum activity on the bio-methanization of a kitchen waste under different waste/inoculum ratios. Process Biochemistry, 2004, 39(12): 2019-2024.

[102] Raposo F, Borja R, Martín M A, Martín A, de la Rubia M A, Rincón B. Influence of inoculum–substrate ratio on the anaerobic digestion of sunflower oil cake in batch mode: Process stability and kinetic evaluation. Chemical Engineering Journal, 2009, 149(1-3): 70-77.

[103] Zhou Y, Zhang Z, Nakamoto T, Li Y, Yang Y, Utsumi M, Sugiura N. Influence of substrate-to-inoculum ratio on the batch anaerobic digestion of bean curd refuse-okara under mesophilic conditions. Biomass and Bioenergy, 2011, 35(7): 3251-3256.

[104] Sri Bala Kameswari K, Chitra K, Porselvam S, Thanasekaran K. Optimization of inoculum to substrate ratio for bio-energy generation in co-digestion of tannery solid wastes. Clean Technologies and Environmental Policy , 2011, 14(2): 241-250.

[105] Slimane K, Fathya S, Assia K, Hamza M. Influence of inoculums/substrate ratios(ISRs)on the mesophilic anaerobic digestion of slaughterhouse waste in batch mode: Process stability and biogas production. Energy Procedia , 2014, 50: 57-63.

[106] Lü F, Hao L P, Zhu M, Shao LM, He P J. Initiating methanogenesis of vegetable waste at low inoculum-to-substrate ratio: Importance of spatial separation. Bioresource Technology, 2012, 105: 169-173.

[107] Forster-Carneiro T, Pérez M, Romero L I. Influence of total solid and inoculum contents on performance of anaerobic reactors treating food waste. Bioresource Technology , 2008, 99(15): 6994-7002.

[108] She P, Song B, Xing X-H, Loosdrecht Mv, Liu Z. Electrolytic stimulation of bacteria Enterobacter dissolvens by a direct current. Biochemical Engineering Journal, 2006, 28(1): 23-29.

[109] Nakanishi K, Tokuda H, Soga T, Yoshinaga T, Takeda M. Effect of electric current on growth and alcohol production by yeast cells. Journal of Fermentation & Bioengineering, 1998, 85(2): 250-253.

[110] Hayes A M, Flora J R V, Khan J. Electrolytic stimulation of denitrification in sand columns. Water Research, 1998, 32(9): 2830-2834.

[111] Mcclenaghan N H, Flatt P R. Engineering cultured insulin-secreting pancreatic B-cell lines. Journal of Molecular

Medicine , 1999, 77(1): 235.

[112] Thrash J C, Coates J D. Review: Direct and indirect electrical stimulation of microbial metabolism. Environmental Science & Technology, 2008, 42(11): 3921-3931.

[113] Hosseini M R, Schaffie M, Pazouki M, Ranjbar M. Direct electric current stimulation of protein secretion by *Fusarium oxysporum*. Chemical Engineering Communications , 2014, 201(2): 160-170.

[114] Luo Q, Wang H, Zhang X, Qian Y. Effect of direct electric current on the cell surface properties of phenol-degrading bacteria. Applied & Environmental Microbiology, 2005, 71(1): 423-427.

[115] Schattauer A, Abdoun E, Weiland P, Plöchl M, Heiermann M. Abundance of trace elements in demonstration biogas plants. Biosystems Engineering, 2011, 108(1): 57-65.

[116] Lo H M, Chiu H Y, Lo SW, Lo F C. Effects of different SRT on anaerobic digestion of MSW dosed with various MSWI ashes. Bioresource Technology, 2012, 125: 233-238.

[117] Nges I A, Björnsson L. High methane yields and stable operation during anaerobic digestion of nutrient-supplemented energy crop mixtures. Biomass and Bioenergy, 2012, 47: 62-70.

[118] Zhang L, Lee Y W, Jahng D. Anaerobic co-digestion of food waste and piggery wastewater: Focusing on the role of trace elements. Bioresource Technology , 2011, 102(8): 5048-5059.

[119] Romero-Güiza M S, Vila J, Mata-Alvarez J, Chimenos J M, Astals S. The role of additives on anaerobic digestion: A review. Renewable and Sustainable Energy Reviews, 2016, 58: 1486-1499.

[120] Wang T F, Shao LM, Li T S, Lü F, He P J. Digestion and dewatering characteristics of waste activated sludge treated by an anaerobic biofilm system. Bioresource Technology , 2014, 153(0): 131-136.

[121] Ho L, Ho G. Mitigating ammonia inhibition of thermophilic anaerobic treatment of digested piggery wastewater: use of pH reduction, zeolite, biomass and humic acid. Water Research, 2012, 46(14): 4339.

[122] Lin L, Wan C, Liu X, Lei Z, Lee D J, Zhang Y, Tay J H, Zhang Z. Anaerobic digestion of swine manure under natural zeolite addition: VFA evolution, cation variation, and related microbial diversity. Applied Microbiology and Biotechnology , 2013, 97(24): 10575-10583.

[123] Montalvo S, Guerrero L, Borja R, Sanchez E, Milan Z, Cortes I, Angeles de la la Rubia M. Application of natural zeolites in anaerobic digestion processes: A review. Applied Clay Science, 2012, 58: 125-133.

[124] Milan Z, Villa P, Sanchez E, Montalvo S, Borja R, Ilangovan K, Briones R. Effect of natural and modified zeolite addition on anaerobic digestion of piggery waste. Water Science and Technology, 2003, 48(6): 263-269.

[125] Wilkie A, Colleran E. Start-up of anaerobic filters containing different support materials using pig slurry supernatant. Biotechnology Letters, 1984, 6(11): 735-740.

[126] Sasaki K, Sasaki D, Morita M, Hirano S-i, Matsumoto N, Ohmura N, Igarashi Y. Efficient treatment of garbage slurry in methanogenic bioreactor packed by fibrous sponge with high porosity. Applied Microbiology and Biotechnology, 2010, 86(5): 1573-1583.

[127] Castilla P, Aguilar L, Escamilla M, Silva B, Milan Z, Monroy O, Meraz M. Biological degradation of a mixture of municipal wastewater and organic garbage leachate in expanded bed anaerobic reactors and a zeolite filter. Water Science and Technology, 2009, 59(4): 723-728.

[128] Aktas O, Cecen F. Bioregeneration of activated carbon: A review. International Biodeterioration & Biodegradation, 2007, 59(4): 257-272.

[129] Gong W J, Liang H, Li W Z, Wang Z Z. Selection and evaluation of biofilm carrier in anaerobic digestion treatment of cattle manure. Energy, 2011, 36(5): 3572-3578.

[130] Xu S, He C Q, Luo L W, Lü F, He P J, Cui LF. Comparing activated carbon of different particle sizes on enhancing

methane generation in upflow anaerobic digester. Bioresource Technology , 2015, 196: 606-612.

[131] Morita M, Sasaki K. Factors influencing the degradation of garbage in methanogenic bioreactors and impacts on biogas formation. Applied Microbiology and Biotechnology, 2012, 94(3): 575-582.

[132] Picanco A P, Vallero M V G, Gianotti E P, Zaiat M, Blundi C E. Influence of porosity and composition of supports on the methanogenic biofilm characteristics developed in a fixed bed anaerobic reactor. Water Science and Technology , 2001, 44(4): 197-204.

[133] Gong W j, Liang H, Li W z, Wang Z z. Selection and evaluation of biofilm carrier in anaerobic digestion treatment of cattle manure. Energy, 2011, 36(5): 3572-3578.

[134] Sasaki K, Sasaki D, Morita M, Hirano S, Matsumoto N, Ohmura N, Igarashi Y. Efficient treatment of garbage slurry in methanogenic bioreactor packed by fibrous sponge with high porosity. Appl Microbiol Biotechnol, 2010, 86(5): 1573-83.

[135] Yamada C, Kato S, Ueno Y, Ishii M, Igarashi Y. Conductive iron oxides accelerate thermophilic methanogenesis from acetate and propionate. Journal of bioscience and bioengineering , 2015, 119(6): 678-682.

[136] Cruz Viggi C, Rossetti S, Fazi S, Paiano P, Majone M, Aulenta F. Magnetite particles triggering a faster and more robust syntrophic pathway of methanogenic propionate degradation. Environmental Science & Technology, 2014, 48(13): 7536- 7543.

[137] Kato S, Hashimoto K, Watanabe K. Methanogenesis facilitated by electric syntrophy via(semi)conductive iron-oxide minerals. Environmental Microbiology, 2012, 14(7): 1646-1654.

[138] Yamada C, Kato S, Ueno Y, Ishii M, Igarashi Y. Conductive iron oxides accelerate thermophilic methanogenesis from acetate and propionate. J Biosci Bioeng, 2015, 119(6): 678-82.

[139] Yang Y, Zhang C, Hu Z. Impact of metallic and metal oxide nanoparticles on wastewater treatment and anaerobic digestion. Environmental Science: Processes & Impacts , 2013, 15(1): 39-48.

[140] Zhang J N, Lü F, Zhang H, Shao L M, Chen D Z, He P J. Multiscale visualization of the structural and characteristic changes of sewage sludge biochar oriented towards potential agronomic and environmental implication. Scientific Reports, 2015, 5: 9406.

[141] Zhang J N, Lü F, Luo C H, Shao L M, He PJ. Humification characterization of biochar and its potential as a composting amendment. Journal of Environmental Sciences, 2014, 26(2): 390-397.

[142] Zhang J N, Lü F, Shao L M, He P J. The use of biochar-amended composting to improve the humification and degradation of sewage sludge. Bioresource Technology, 2014, 168: 252-258.

[143] He P J, Yu Q F, Zhang H, Shao L M, Lü F. Removal of copper(II)by biochar mediated by dissolved organic matter. Scientific Reports, 2017, 7(1): 7091.

[144] Luo C H, Lü F, Shao L M, He P J. Application of eco-compatible biochar in anaerobic digestion to relieve acid stress and promote the selective colonization of functional microbes. Water Research, 2015, 68(0): 710-718.

[145] Lü F, Luo C H, Shao L M, He P J. Biochar alleviates combined stress of ammonium and acids by firstly enriching *Methanosaeta* and then *Methanosarcina*. Water Research, 2016, 90: 34-43.

[146] Cai J, He P J, Wang Y, Shao L M, Lü F. Effects and optimization of the use of biochar in anaerobic digestion of food wastes. Waste Management & Research , 2016, 34(5): 409-416.

[147] Agler M T, Wrenn B A, Zinder S H, Angenent L T. Waste to bioproduct conversion with undefined mixed cultures: the carboxylate platform. Trends in Biotechnology, 2011, 29(2): 70-78.

[148] López-Garzón C S, Straathof A J J. Recovery of carboxylic acids produced by fermentation. Biotechnology Advances , 2014, 32(5): 873-904.

[149] Agler M T, Spirito C M, Usack J G, Werner J J, Angenent L T. Chain elongation with reactor microbiomes: Upgrading dilute ethanol to medium-chain carboxylates. Energy & Environmental Science , 2012, 5(8): 8189-8192.

[150] Cope J L, Hammett A J M, Kolomiets E A, Forrest A K, Golub K W, Hollister E B, DeWitt T J, Gentry T J, Holtzapple M T, Wilkinson H H. Evaluating the performance of carboxylate platform fermentations across diverse inocula originating as sediments from extreme environments. Bioresource Technology, 2014, 155: 388-394.

[151] Kucek L A, Spirito C M, Angenent L T. High n-caprylate productivities and specificities from dilute ethanol and acetate: Chain elongation with microbiomes to upgrade products from syngas fermentation. Energy & Environmental Science, 2016, 9(11): 3482-3494.

[152] Tseng H C, Prather K L J. Controlled biosynthesis of odd-chain fuels and chemicals via engineered modular metabolic pathways. Proceedings of the National Academy of Sciences , 2012, 109(44): 17925-17930.

[153] Andersen S J, Candry P, Basadre T, Khor W C, Roume H, Hernandez-Sanabria E, Coma M, Rabaey K. Electrolytic extraction drives volatile fatty acid chain elongation through lactic acid and replaces chemical pH control in thin stillage fermentation. Biotechnology for Biofuels , 2015, 8(1): 1-14.

[154] Xu J, Guzman J J, Andersen S J, Rabaey K, Angenent L T. In-line and selective phase separation of medium chain carboxylic acids using membrane electrolysis. Chemical Communications, 2015, 51: 6847-6850.

[155] Liu Y H, Lü F, Shao L M, He P J. Alcohol-to-acid ratio and substrate concentration affect product structure in chain elongation reactions initiated by unacclimatized inoculum. Bioresource Technology, 2016, 218: 1140-1150.

[156] Zhu X, Tao Y, Liang C, Li X, Zhang W, Zhou Y, Tank Y, Bo T. The synthesis of n-caproate from lactate: a new efficient process for medium-chain carboxylates production. Scientific Reports, 2015, 5: 14360.

[157] Yin Y, Zhang Y, Karakashev D B, Wang J, Angelidaki I. Biological caproate production by *Clostridium kluyveri* from ethanol and acetate as carbon sources. Bioresource Technology, 2017, 241: 638-644.

[158] Liu Y H, He P J, Shao L M, Zhang H, Lü F. Significant enhancement by biochar of caproate production via chain elongation. Water Research , 2017, 119: 150-159.

[159] Ragauskas A J, Williams C K, Davison B H, Britovsek G, Cairney J, Eckert C A, Frederick W J, Hallett J P, Leak D J, Liotta C L, Mielenz J R, Murphy R, Templer R, Tschaplinski T. The path forward for biofuels and biomaterials. Science, 2006, 311(5760): 484-489.

[160] Beach E S, Zheng C, Anastas P T. Green Chemistry: A design framework for sustainability. Energy & Environmental Science , 2009, 2(10): 1038-1049.

[161] Shen L, Haufe J, Patel M K. Product overview and market projection of emerging bio-based plastics PRO-BIP 2009. Report for European polysaccharide network of excellence(EPNOE)and European bioplastics. 2009.

[162] Yu J, Chen L X. The greenhouse gas emissions and fossil energy requirement of bioplastics from cradle to gate of a biomass refinery. Environmental Science & Technology, 2008, 42(18): 6961-6966.

[163] Laycock B, Halley P, Pratt S, Werker A, Lant P. The chemomechanical properties of microbial polyhydroxyalkanoates. Progress in Polymer Science, 2014, 39(2): 397-442.

[164] Wendlandt K D, Stottmeister U, Helm J, Soltmann B, Jechorek M, Beck M. The potential of methane - oxidizing bacteria for applications in environmental biotechnology. Engineering in Life Sciences , 2010, 10(2): 87-102.

[165] Brigham C J, Kurosawa K, Rha C, Sinskey AJ. Bacterial carbon storage to value added products. Journal of Microbial and Biochemical Technology, 2011, S3: 002.

[166] Bosco F, Chiampo F. Production of polyhydroxyalcanoates(PHAs)using milk whey and dairy wastewater activated sludge: production of bioplastics using dairy residues. Journal of bioscience and bioengineering, 2010, 109(4): 418-421.

[167] Jiang Y, Marang L, Tamis J, van Loosdrecht M C, Dijkman H, Kleerebezem R. Waste to resource: converting paper mill wastewater to bioplastic. Water research , 2012, 46(17): 5517-5530.

[168] Albuquerque M, Concas S, Bengtsson S, Reis M. Mixed culture polyhydroxyalkanoates production from sugar molasses:

the use of a 2-stage CSTR system for culture selection. Bioresource technology, 2010, 101(18): 7112-7122.

[169] Albuquerque M, Martino V, Pollet E, Avérous L, Reis M. Mixed culture polyhydroxyalkanoate(PHA)production from volatile fatty acid(VFA)-rich streams: effect of substrate composition and feeding regime on PHA productivity, composition and properties. Journal of Biotechnology, 2011, 151(1): 66-76.

[170] Moita R, Lemos P. Biopolymers production from mixed cultures and pyrolysis by-products. Journal of biotechnology , 2012, 157(4): 578-583.

[171] Dobroth Z T, Hu S, Coats E R, McDonald A G. Polyhydroxybutyrate synthesis on biodiesel wastewater using mixed microbial consortia. Bioresource technology , 2011, 102(3): 3352-3359.

[172] Anderson A J, Dawes E A. Occurrence, metabolism, metabolic role, and industrial uses of bacterial polyhydroxyalkanoates. Microbiological Reviews , 1990, 54(4): 450-72.

[173] Arnold S, Moss K, Henkel M, Hausmann R. Biotechnological perspectives of pyrolysis oil for a bio-based economy. Trends in Biotechnology , 2017, DOI: 10.1016/j.tibtech.2017.06.003(

[174] Moita R, Lemos P C. Biopolymers production from mixed cultures and pyrolysis by-products. Journal of Biotechnology , 2012, 157(4): 578.

[175] Moita F R, Ortigueira J, Freches A, Pelica J, Gonçalves M, Mendes B, Lemos P C. Bio-oil upgrading strategies to improve PHA production from selected aerobic mixed cultures. New Biotechnology , 2014, 31(4): 297.

[176] Beneroso D, Bermúdez J M, Arenillas A, Menéndez J A. Comparing the composition of the synthesis-gas obtained from the pyrolysis of different organic residues for a potential use in the synthesis of bioplastics. Journal of Analytical & Applied Pyrolysis, 2015, 111: 55-63.

[177] Steinbüchel A, Hustede E, Liebergesell M, Pieper U, Timm A, Valentin H. Molecular basis for biosynthesis and accumulation of polyhydroxyalkanoic acids in bacteria. FEMS Microbiology Letters, 1992, 103(2): 217-230.

[178] Pohlmann A, Fricke W F, Reinecke F, Kusian B, Liesegang H, Cramm R, Eitinger T, Ewering C, Potter M, Schwartz E, Strittmatter A, Vosz I, Gottschalk G, Steinbuchel A, Friedrich B, Bowien B. Genome sequence of the bioplastic-producing "Knallgas" bacterium *Ralstonia eutropha* H16. Nature Biotechnology, 2006, 24(10): 1257-1262.

[179] Vincent K A, Cracknell J A, Lenz O, Zebger I, Friedrich B, Armstrong F A. Electrocatalytic hydrogen oxidation by an enzyme at high carbon monoxide or oxygen levels. Proceedings of the National Academy of Sciences of the United States of America , 2005, 102(47): 16951-16954.

[180] Do Y S, Smeenk J, Broer K M, Kisting C J, Brown R, Heindel T J, Bobik T A, DiSpirito A A. Growth of *Rhodospirillum rubrum* on synthesis gas: Conversion of CO to H_2 and poly-β-hydroxyalkanoate. Biotechnology and Bioengineering, 2007, 97(2): 279-286.

[181] Revelles O, Tarazona N, García J L, Prieto M A. Carbon roadmap from syngas to polyhydroxyalkanoates in *Rhodospirillum rubrum*. Environmental Microbiology, 2016, 18(2): 708-720.

[182] Kleerebezem R, van Loosdrecht M C. Mixed culture biotechnology for bioenergy production. Current Opinion in Biotechnology , 2007, 18(3): 207-12.

[183] Lagoa-Costa B, Abubackar HN, Fernández-Romasanta M, Kennes C, Veiga M C. Integrated bioconversion of syngas into bioethanol and biopolymers. Bioresource Technology , 2017, 239: 244.

作者：何品晶[1]，吕　凡[1]，章　骅[1]，邵立明[2]

[1]同济大学环境科学与工程学院固体废物处理与资源化研究所

[2]住房和城乡建设部村镇建设司农村生活垃圾处理技术研究与培训中心

第 24 章　绿色能源化学研究进展

本章导读

随着可持续发展战略和生态文明建设的不断推进，能源结构调整步伐不断加快，绿色能源的研究已成为当今能源领域的研究热点。目前，绿色能源研究主要集中在电化学储能、废热转换、生物质能量转换及电网等方面，其中高效的电化学储能和转化技术发展历史并不长，但因其实用性强、应用面广、易于推行等独特优势近年来获得广泛关注。与其相关的绿色能源化学近年来发展迅速。本章将重点介绍近五年来该领域在锂离子电池、液流电池、燃料电池、超级电容器、二氧化碳捕获及其催化转化、电解水等能源转化和储存关键技术方面取得的重要进展，总结不同绿色能源化学当前的优势和面临的挑战，并对其今后发展形势进行展望。

关键词

绿色能源化学，锂电池，液流电池，燃料电池，超级电容器，二氧化碳资源化利用，电解水

1 引　言

传统化石能源在我国现代社会和国民经济发展中扮演着极其重要的角色。但化石能源的大量消耗也造成了诸如能源短缺、环境污染和生态环境恶化等一系列问题。面临经济增长、环境保护和人口消费需求不断增加的多重压力，可再生及绿色能源技术的研究与发展在当今中国显得尤为迫切。

“十二五”时期我国能源领域发展较快，供给保障能力不断增强，非化石能源和天然气消费比重分别提高 2.6%和 1.9%，煤炭消费比重下降 5.2%，能源结构调整步伐不断加快。然而，我国煤炭占终端能源消费比重仍高达 20%以上，高出世界平均水平 10%。煤炭的大量开采和消耗，给我国环境保护和经济社会的可持续发展带来极大的压力。《能源发展“十三五”规划》指出，能源是人类社会生存发展的重要物质基础，攸关国计民生和国家战略竞争力。“十三五”时期是我国全面建成小康社会的决胜阶段，也是推动能源革命的蓄力加速期，牢固树立和贯彻落实创新、协调、绿色、开放、共享的发展理念，深入推进能源革命，着力推动能源生产利用方式变革，建设清洁低碳、安全高效的现代能源体系，是能源发展改革的重大历史使命。因此，亟待开发新型、环保的绿色能源替代传统的化石能源，这也是我国实施可持续发展战略及生态文明建设必须优先考虑的重大问题。

当前，随着我国经济的飞速发展，绿色能源技术也迎来了快速发展的历史机遇，涌现了一批新技术、新方法，并且展示了十分广阔的应用前景。下面，我们主要介绍近五年来我国科研工作者在锂离子电池、超级电容器、燃料电池、碳捕获及其转化、电解水技术和其他绿色能源技术的最新研究进展，总结不同绿色能源化学当前的优势和面临的挑战，并对其今后发展形势进行展望。

2 储能技术中的锂电池

目前，广泛应用的化学电源主要有铅酸电池、镍氢电池、锂离子电池。对比表 24-1 中的性能参数可以看出：铅酸电池和镍氢电池在比能量密度、工作电压、记忆效应及环境友好等方面存在一定的劣势，导致其在未来新能源汽车和电动工具的能源供给方面的应用受到限制。锂电池，作为一种绿色高性能的

二次电池，现已推动新能源汽车和大规模储能等领域的产业变革，其中，锂离子电池、锂硫电池、锂硒/碲电池及锂空气电池最受关注。

表 24-1 几种化学电源的性能参数对比

电池类别	电压（V）	重量比能量（Wh/kg）	体积比能量（Wh/L）	记忆效应	自放点率（%/月）	污染	能量比价格（元/Wh）
铅酸电池	2.0	33	80	无	25	有	0.8
Ni-MH	1.2	70	240	有	30	无	1.8
Li-ion	3.6	150	300	无	3	无	1.5

2.1 锂离子电池

锂离子电池正极一般采用嵌锂的锂化合物材料，负极一般使用石墨。目前商业化较为成熟的锂离子电池正极材料主要有 $LiCoO_2$、$LiMn_2O_4$、$LiNiO_2$ 及 $LiFePO_4$ 等[1]。虽然锂离子电池具有较高的开路电压、较高的能量密度和功率密度、无记忆效应以及自放电低的特点，但是其在使用过程中也存在深度放电时容量衰减较快，不适用于高倍率放电，电池的成本较高及热稳定性差等问题[2]。

近几年来，研究人员不断开发出一些新颖的电极材料用于进一步改善电池的性能，包括使用新型金属氧化物和金属硫化物电极，如氧化锡、氧化钴和硫化钼等。Djenizian 等使用无模板自支撑的三维金属氧化物作为锂离子电池的负极材料，如 TiO_2、Co_3O_4 和 SnO_2 等，由于这种材料具有良好的机械性能和稳定性，从而实现了电池超长的循环性能和良好的倍率性能[3]。最新技术采用硅碳作为电池的负极材料，硅碳复合材料结合了碳材料高电导率和稳定性以及硅材料高容量的优势，能够提高锂离子电池的比容量和循环寿命。目前对于硅碳复合材料的研究，主要分为包覆结构和嵌入结构。Tao 等通过在硅纳米表面包覆 SiO_2 和多孔碳，制备出具有双核壳结构的复合材料（Si@SiO$_2$@C），初始放电容量达到 1700 mAh/g，以 0.1 mA/g 的电流密度循环 100 次后依然保持 980 mAh/g 的容量，相比单核壳结构的 Si@C，拥有更高的容量保有率和更长的循环寿命[4]。Gao 等采用还原二氧化硅凝胶的方法合成了具有三维结构的多孔硅碳复合材料，并用作锂离子电池的负极，在 200 mA/g 的电流密度下放电容量达到 1550 mAh/g，在 2 A/g 的大电流放电下循环 50 次后仍然保持 1050 mAh/g 的容量，表现出极佳的电化学特性[5]。硅碳负极的应用改善了锂离子电池的循环性能和倍率性能，但是对于硅碳两种材料的复合机理以及微观结构的嵌锂机理仍需进一步的研究。

金属锂的比容量为 3860 mAh/g，电化学势为–3.04 V（*vs* 标准氢电极），是一种非常理想的锂电池负极材料。但在实际的锂负极电池中，由于负极电流密度和锂离子分布不均匀等因素，在电池的充放电过程中锂在负极的不均匀沉积导致枝晶生长，造成电池容量损失，甚至可能穿透隔膜造成电池短路，如何有效抑制锂枝晶生长成为研究中不可避免的问题[6-8]。Xu 等通过研究锂金属负极在充放电过程中的形态变化，提出在电解质中加入添加剂（如 $LiNO_3$），从而在负极表面形成有效的固体电解质隔膜抑制锂枝晶生长[9]；Cui 等则研究了金属锂枝晶生成机理，提出采用人造界面来限制枝晶生长或者采用固态电解质有效避免枝晶生成[10]。

2.2 锂硫电池

锂硫电池是由金属锂为负极、单质硫为正极组成的电池。由于具有高能量密度（2600 Wh/kg）、正极硫储量丰富及成本较低等优势，锂硫电池成为目前最有应用前景的高能量电池。但在实际应用中也面临很多挑战[11-12]，主要是硫正极材料的导电性差，放电过程中其中间产物多硫化物易溶解在电解质中导致穿梭效应，以及正极不可避免的体积膨胀效应。

因此，寻找具有高导电性、能够容纳体积膨胀以及能有效减少穿梭效应的新型硫负载材料成为目前研究的热点。首先从传统的碳材料入手，Giannelis 等设计了一系列实验研究分级多孔碳在负载硫时表现的不同电池性能，从而确定最佳的孔径和孔体积，一方面增强正极导电性，另一方面从结构上限制多硫化物的溶解，二者的协同作用极大地改善了电池的性能[13]；麦立强等将纳米金属氧化物作为硫的宿主材料，主要用 TiO_2、Ti_4O_7、MnO_2、$NiFe_2O_4$ 等和硫进行复合，这种极性氧化物制成的复合电极较为成功地实现了对多硫化物的束缚，而稳定的电极结构使电池具有良好的循环寿命[14]。此外，他们也研究了金属硫化物作为硫的宿主材料，发现其也能表现出对多硫化物进行有效的牵制作用以提高电池性能。晏成林团队先将硫进行处理使其在反应中直接生成短链的多硫化物，这样有效地抑制了穿梭效应，虽然放电电压降低，但是显著提高了电池的循环性能和倍率性能[15]。Cui 等以硅纳米线为负极，Li_2S/C 复合物作为正极制成新型的锂离子电池，有效地避免了锂的枝晶问题和锂硫电池中硫的导电性问题。该电池理论比容量达到 1550 Wh/kg，在实际放电过程中，0.1 mA/g 的电流密度下初始放电容量达到 550 Wh/kg[16]。由此可见，从研究非极性的碳材料复合、包覆和掺杂，然后到研究极性的金属氧化物和硫化物对电池性能的提高，再到通过改变电池反应的路径来解决穿梭效应问题，研究和设计锂硫电池正极材料的思路在不断地改变和进步。

2.3　锂硒/碲电池

Se 和 Te 作为锂电池的正极材料，导电性比 S 单质更好，而且比 S 有更大的理论体积容量[17]。由于二者和 S 不仅在物理化学性质有很多相似之处，而且在电池中的反应过程也和 Li-S 电池相似，因此，为了提高 Li-Se 电池的电化学特性，研究者合理地借鉴 Li-S 电池的研究策略。Cui 等[18]研究了 Li-Se 电池在碳酸盐类电解质中嵌锂过程，发现在放电过程中硒直接转化为 Li_2Se，没有任何易熔的中间体产生，而在有机的醚类溶剂中存在多硒化物的溶解，发现电解质的不同会导致反应的过程存在很大的差别。Weng 等[19]采用 Se-C 复合材料作为正极研究电池的电化学性能，在 0.1 mA/g 的电流密度下电池的放电电压约 2 V，放出 300 mAh/g 的放电容量，另外他们采用 Se_xS_y 化合物作为电池的正极也同样表现出不错的电化学性能。黄云辉团队[20]通过熔融扩散法制备 Se/PCNS 复合电极，Se 负载量达到 70.5%（质量分数），作为电池的正极表现出良好的倍率性能和循环性能，以及较高的体积容量。Strasser 等[21]通过制备 Te/CMK-3 复合电极，电池在 1C 放电倍率下表现出 400 mAh/g 的容量，10C 倍率下循环 1000 次仍然保持 286 mAh/g 的放电容量。

2.4　锂空气电池

锂空气电池作为一种新型能量存储装置，其理论比能量密度高出锂离子电池体系一个数量级[22]。然而由于其正极复杂的气-液-固共存的三相结构，及其在循环稳定性、能量效率等方面所存在的问题，锂空气电池在实际应用方面仍然面临很大的挑战。

为改善正极复杂的三相结构，温兆银团队针对金属硫化物的催化惰性，以材料晶体结构修饰为手段，成功制备了具有高度晶格畸变的亚稳态金属硫化物正极材料[23]。这种高度畸变的亚稳结构能够显著提升材料的催化活性，具有潜在的应用价值。此外，该团队还揭示了锂空气电池中间放电产物在氧空位位点的自催化分解反应[24]，这为未来高效正极的设计提供了新的思路和解决方案。

为解决电解液溶剂的稳定性问题，夏永姚团队采用 $EMIMBF_4$（1-乙基-3-甲基咪唑四氟硼酸盐）及 LiTFSI 离子液体作为电解液锂盐，并以 α-MnO_2 和 γ-MnOOH 分别作为催化剂提高了电池的循环稳定性，但是离子液体黏度较大，这在一定程度上限制了电池的倍率充放电性能[25]。由于金属阳离子锂盐与电解液溶剂之间的相互作用对电池产物的组成及产物形貌、结构具有重要的影响[26]，因此，优化锂盐浓度及溶剂组分对改善锂/空气电池氧还原反应及产物可逆催化分解具有重要的研究指导意义。

为了制备高效、廉价的催化剂，目前大量的研究集中在以下三类：①碳基材料催化剂（如碳纳米管[27-28]、碳纳米纤维[29]和石墨烯[30]等），由于多维纳米碳基材料和石墨烯材料具有明显的电催化作用，且其特殊的结构将直接影响产物 Li_2O_2 的形貌和分布[31,32]。因此，选择和制备多维纳米碳材料及构建合适的孔隙结构，将是未来研究的重点。②贵金属负载或修饰金属氧化物制备复合催化剂[33,34]，该复合催化剂相比单组分催化剂具有更高的电池性能促进效果，但其催化作用机理尚未明确。③钙钛矿型氧化物催化剂[35,36]，在一定条件下此催化剂可与贵金属催化性能相媲美，然而钙钛矿型复合催化剂的比表面积、孔隙率和热稳定性能较低，今后可通过优化其制备合成工艺来进一步提高其催化性能。

3 储能技术中的液流电池

液流电池因其易规模化、无污染、安全等优点在大规模储能方面表现出较好的应用前景。目前，提高液流电池能量效率和系统的可靠性、降低其成本是液流电池大规模普及应用的重要研究方向，其中，锌溴电池和全钒液流电池最受关注。

3.1 锌溴液流电池

锌溴液流电池因具有高理论能量密度、低电解液成本的优势，其在大规模储能领域具有较好的应用前景。虽然 Br_2/Br^- 氧化还原电对具有较高的电极电势和低廉的成本，但是其较差的反应活性使电池的工作电流密度较低。张华民团队设计了一系列高度有序的介孔炭正极材料，并研究了其活性和结构之间的关联[37-39]，其中，利用“孔径筛分效应”固溴，制备出兼具高活性和固溴功能的笼状多孔炭材料，突破了低功率密度的制约瓶颈[39]。

3.2 全钒液流电池

全钒液流电池因具有易规模化、长寿命及选址自由等优势，成为目前最有前景的大规模储能技术之一。但是在技术上还没有完全成熟，如使用的离子交换膜的选择性比较差，此外，研发具有高浓度和高稳定性的电解液也是全钒液流电池产业化急需解决的瓶颈技术。

为提高离子交换膜的选择性，张华民团队将多孔离子传导膜引入到电池中，利用多孔隔膜的孔径筛分效应和电荷排阻效应，实现对钒离子和质子的选择性筛分[40,41]。

为增加电解液的浓度和稳定性，目前，向电解液中加入适量添加剂是解决该问题的重要方法。刘素琴团队提出含氧及氮官能团的有机添加剂（DL-苹果酸和 L-天冬氨酸）的加入可以通过延缓钒(III)的沉积来增加电解液的稳定性[42]。

4　燃　料　电　池

燃料电池技术具有能量转换效率高、不受卡诺循环限制、环境友好和启动快速等诸多优点，在便携式设备、固定电站、电动汽车和航空航天等领域具有非常广阔的应用前景，被誉为是 21 世纪最重要的一种绿色能源转换技术[43]。燃料电池能够使氢气、甲醇、甲酸和乙醇等小分子燃料和氧气发生氧化还原反应，并将其化学能转换为电能。按照电解质和燃料的不同，燃料电池可以分为以下几类：①碱性燃料电池；②聚合物电解质膜燃料电池；③磷酸燃料电池；④熔融碳酸盐燃料电池；⑤固体氧化物燃料电池[44-50]。其中，聚合物电解质膜燃料电池具有比能量高、工作条件温和、可在室温下快速启动等优点，是理想的汽车动力电源、大型供电电站和便携式移动电源，具有极为广阔的发展前景。当前，聚合物电解质膜燃料电池主要分为质子交换膜燃料电池和直接甲醇燃料电池。

4.1　质子交换膜燃料电池

质子交换膜燃料电池系统中，其阳极为氢气的氧化反应，阴极侧为氧气的还原反应，电子通过外电路进行传导。由于氢氧化迅速，其动力学速率是氧还原的 10^5 倍。所以，质子交换膜燃料电池的核心过程为阴极侧的氧还原反应，其反应速率的快慢直接决定燃料电池系统性能的优劣[51]。当前，众多实验和理论结果证明，铂基催化剂仍是最有效的燃料电池阴极氧还原催化剂。由于氧还原反应动力学过程缓慢，目前燃料电池阴极仍需要负载大量的贵金属铂基催化剂。然而，其高昂的成本以及铂资源的短缺已成为阻碍质子交换膜燃料电池发展和商业化最重要的因素之一。低铂催化剂和非铂催化剂在保持较高的催化活性和稳定性的同时，能够极大降低贵金属的用量，因此备受关注并取得了重大研究进展。此外，生物燃料电池作为一种特殊的燃料电池，也开始引起人们的重视。

4.1.1　非铂催化剂

非铂催化剂主要包括掺杂碳材料[52,53]、过渡金属氧化物[54,55]、氮化物[56,57]、氮氧化物[58]和碳化物[59]等。其中具有 M-N-C（M=Fe，Co，Ni 等）结构的掺杂碳材料表现出优异的氧还原活性和稳定性。魏子栋教授课题组通过高温聚合的方式制备了具有高效氧还原活性和稳定性的 Fe-N-C 催化剂，并通过大量实验证明单一的 N 掺杂碳材料、Fe/Fe_3C 纳米晶体和 Fe-N_x 位点都不能提供较好的 ORR 性能，而只有将三者有效地结合起来才能展示出最优的氧还原反应活性，并指出 Fe/Fe_3C 纳米晶体中 Fe-N_x 含量越高，催化剂的氧还原活性越好[53]。夏宝玉等以 ZIF-67 为唯一前驱体，通过低温焙烧的方式制备了以互相交织的高结晶度碳纳米管构建的中空框架结构[60]，这种新型非铂催化剂展示出了极好的氧还原活性，在氧气饱和的 0.1 mol/L KOH 溶液中，其半波电位比商业 Pt/C 催化剂高出 30 mV，并展示出了极高的电化学稳定性（图 24-1）。这种新型催化剂的活性和稳定性同其化学组成、高结晶度的纳米管和这种整体坚固的中空框架结构密切相关。虽然非铂催化剂取得了长足的进步，然而，目前的非铂催化剂材料仍远不能满足燃料电池高稳定性和高活性的要求，特别是在酸性介质中，非铂催化剂离能够实际应用在燃料电池中还有较大的差距。不断提高非铂催化剂 ORR 性能的同时，保持催化剂具有较高的催化稳定性，探索影响催化剂稳定性的因素仍是当前科研工作者的研究重点。当前，制备在酸性介质中具有高活性和高稳定性的 ORR 催化剂仍面临着较大的挑战。

4.1.2 低铂催化剂

通过制备 Pt 合金或者 Pt 基核壳结构催化剂能够有效降低 Pt 载量。多种 PtM（M=Fe，Co，Ni，Pd，Ru 等）合金和核壳结构催化剂已见诸报道，其展示出比商业 Pt/C 催化剂更高的催化活性和稳定性。另外通过添加过渡金属，可以带来 Pt 催化剂应力和配位效应的改变，从而引起对 Pt 原子电子结构的修饰和改变，有效降低 Pt 同含氧物种之间的吸附强度，从而提高 ORR 催化活性和稳定性。

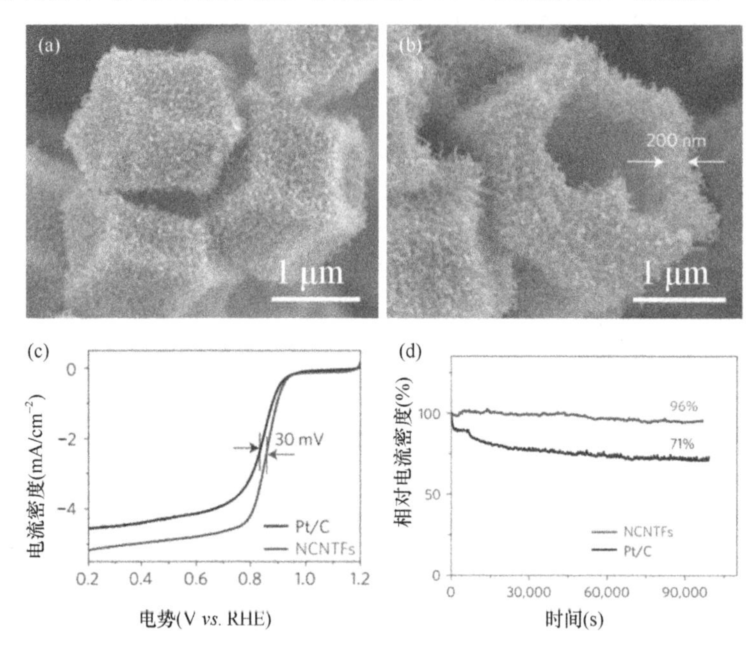

图 24-1 （a），（b）催化剂的 SEM 图；（c）制备的催化剂同商业 Pt/C 催化剂的 ORR 极化曲线；（d）在 0.6 V 处的计时电流曲线

苏州大学黄小青教授课题组通过在 Pt_3Ni 合金催化剂中进一步掺杂过渡金属 Mo 制备了 Mo-Pt_3Ni/C 催化剂，其比活性和质量活性为 10.3 mA/cm^2 和 6.98 A/mg$_{Pt}$，分别是商业 Pt/C 催化剂的 81 倍和 73 倍[61]。研究表明，Mo 原子的添加不仅能够改变 Pt 原子同含氧物种之间的吸附强度，同时能够减缓 Pt 的熔解。此外，该课题组还利用湿化学法合成了 PtPb@Pt 核壳结构催化剂(图 24-2)[62]。该催化剂的比活性是商业 Pt/C 催化剂的 34 倍，同时展示出极佳的电化学稳定性，经过 50000 次的稳定性测试，催化剂的质量活性仅仅降低 7.7%，而商业 Pt/C 的质量活性则降低了 67%。分析结果认为，存在于 PtPb@Pt 110 晶面的拉伸应力同样能够优化 Pt 原子同含氧物种之间的吸附强度，同时，金属间化合物结构的 PtPb 内核和具有四个原子层厚度的 Pt 壳层保证了催化剂具有较高的稳定性。

此外，同零维的铂纳米或铂合金纳米颗粒相比，一维的铂基纳米线或纳米管具有各向异性、高比表面积和高传导率等优点，从而能够使催化剂具有更高的电催化性能和稳定性。俞书宏课题组采用锑纳米线为牺牲模板，通过次序沉积 Pd 和 Pt 原子，制备了 Pd@Pt 核壳结构纳米线。该催化剂的质量活性是商业 Pt/C 催化剂的 10 倍，并表现出了较高的电化学稳定性能。黄小清课题组，通过简易的批量化油相合成方法，制备了一种具有有序表面结构的 PtNi 合金纳米线，其兼具金属间化合物、高指数晶面和富 Pt 表面的特征，赋予了其对 ORR 优异的催化活性，其质量活性是商业 Pt/C 催化剂的 55.1 倍。除此之外，该催化剂还具有较高的电化学稳定性和热稳定性。

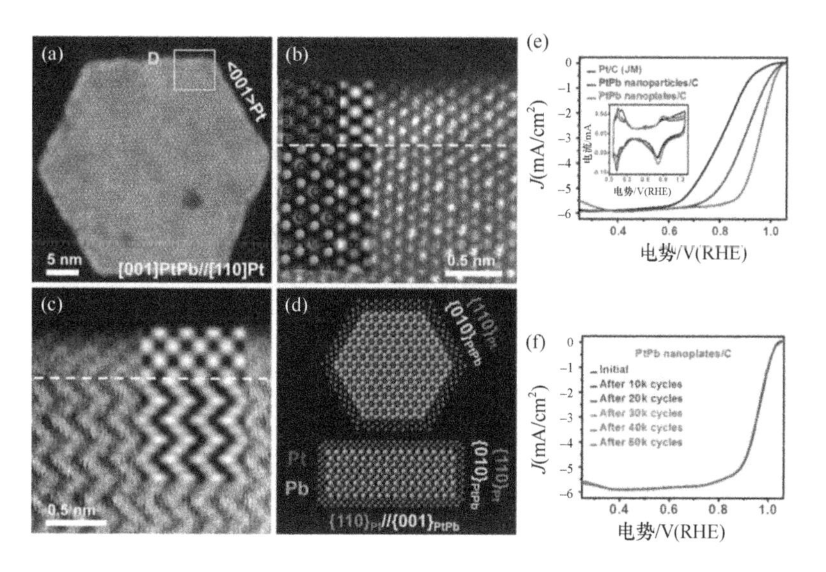

图 24-2　（a）PtPb@Pt 纳米片的高倍环形暗场-扫描透射图片；（b）为（a）图选定区域中的高清晰图片；（c）另一区域的高清晰图片；（d）PtPb@Pt 纳米片的示意图模型 J

4.2　直接甲醇燃料电池

直接甲醇燃料电池同质子交换膜燃料电池相比，具有操作系统结构简单和原料输送便利等优点，在便携式电子设备和移动电源等领域具有广阔的应用前景[63]。然而，除了解决阴极侧氧还原动力学缓慢的问题之外，还要解决阳极侧催化剂易被中间产物（主要为 CO）毒化的问题。当前，Pt 基催化剂仍是最为有效的阳极催化剂。因此，开发高活性和高稳定性的低铂催化剂对其商业化具有十分重要的意义。制备 Pt 合金或核壳结构催化剂也是提升催化剂甲醇氧化性能和稳定性的有效手段。同氧还原催化剂类似，通过引入第二种或多种过渡金属来优化 Pt 原子同含氧物种之间的吸附强度，加快 Pt 表面 CO 的脱除速率，提高甲醇氧化性能[64]。

俞书宏课题组通过在锑纳米线共沉积 Pd 和 Pt 原子，大规模制备了 PtPdTe 三元合金纳米线，在相同测试条件下，该催化剂的比活性是商业 Pt/C 催化剂的 2.6 倍[65]。在该体系中进一步加入 Ru 原子制备的 PtPdRuTe 四元合金纳米线的质量活性是商业催化剂的 2.2 倍[66]。制备了由高结晶度的 PtCu 纳米粒子组装的中空结构 PtCu 纳米管。由于其中空和多空结构，能够为 MOR 提供更多的反应活性位点，增加系统的传质性能，从而使催化剂的质量活性增加到商业 Pt/C 催化剂的 4 倍。黄小青课题组通过一步法制备了具有较高密度高晶面指数的 PtCu 纳米线[63]。这种新型催化剂融合了一维结构、高晶面指数和 Pt 合金效应的优点，其甲醇氧化质量活性是 Pt/C 催化剂 5.4 倍，并展示出较高的催化稳定性（图 24-3）。

由于 MOR 的操作条件具有强酸性和强氧化性，除了要求 Pt 基催化剂具有较高的稳定性之外，提高 MOR 催化剂载体的稳定性也能够带来催化剂性能和稳定性的提升。潘湛昌课题组通过采用具有较高电导性和耐腐蚀性的氮化钛纳米管替代碳粉负载 Pt，使 Pt 的 MOR 质量活性提高了 3.4 倍[60]。同时，由于氮化钛纳米管具有极佳的耐腐蚀性能，并且同 Pt 纳米颗粒之间存在着较强的电子相互作用，这种新型催化剂也表现出了极好的电化学稳定性。通过在氮化钛载体中进一步掺杂 Mo 原子，其质量活性可达商业 Pt/C 催化剂的 4 倍。研究表明，Mo 的掺杂能够增强氮化物载体向负载的 Pt 原子传导电子的能力，从而使 Pt 的 d 带中心相对于费米能级产生负移，从而使 Pt 原子同含 CO 之间的吸附强度减弱，使 MOR 催化性能提升[67]。

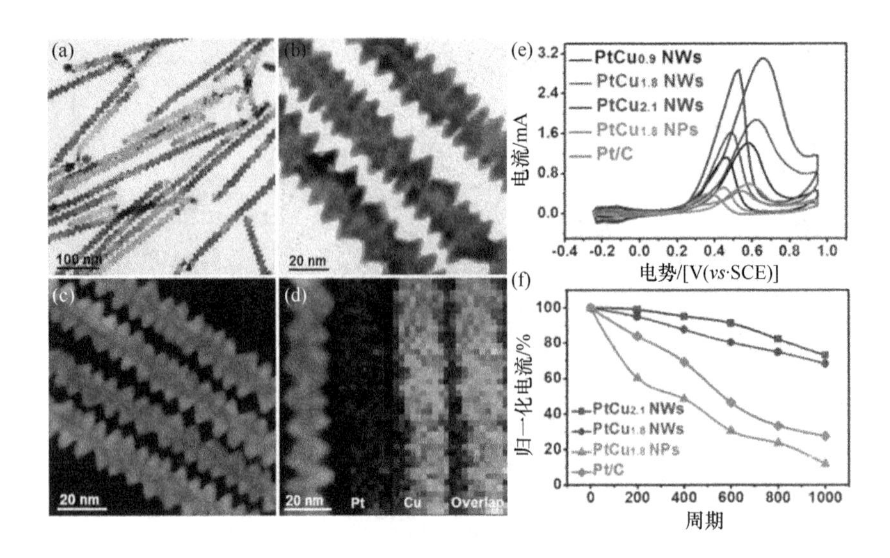

图 24-3 $PtCu_{1.8}$纳米线的 TEM 图片（a），（b）；（c）高倍环形暗场-扫描透射图像；（d）元素分布图；（e）MOR 氧化曲线；（f）几种催化剂的 MOR 稳定性测试

从以上分析可以看出，非铂催化剂由于其原材料来源丰富和价格上的巨大优势，已经引起了越来越多研究者的关注。然而，如何进一步提高非铂催化剂的 ORR 性能，特别是在酸性条件下的电化学稳定性能，从而构筑高效、廉价的催化电极结构仍是当前非铂催化剂的主要研究方向[43]。对于低铂催化剂，制备铂合金或制备 Pt 基核壳结构催化剂，都能够极大地提高 Pt 的利用效率。同时结合第二种金属元素带来的协同效应（组合、配位和应力效应），增强最终催化剂的 ORR 和 MOR 活性[64]。另一方面，对于核壳结构催化剂，由于 Pt 壳层能够对内层的贱金属具有保护作用从而减轻核内贱金属的不断熔出，保持催化剂超高的稳定性。当然，当前合金或核壳结构催化剂仍选用 Pd、Au 或 Ru 等金属作为第二种添加组分，虽然降低了 Pt 的使用量，但这些添加的组分仍然比较昂贵。采用有效的合成方式，选取更加廉价第二或多种组分，使用新的合成手段制备低铂高效和高稳定的 ORR 和 MOR 催化剂仍面临很多挑战，但对燃料电池的商业化推广具有重大的理论和现实意义。

开发高性能、低铂载量催化剂的研究十分必要，低铂催化剂已成为降低燃料电池成本最有效的方法之一。尽管低铂催化剂可以降低 Pt 载量，但仍面临以下问题：①如何进一步优化低铂催化剂的制备条件，寻找一种简单成熟、成本低廉的方法以提高催化剂的活性和稳定性；②如何对核壳结构低铂催化剂制备过程中的单分散和 Pt 层厚度进行控制；③如何实现低铂催化剂的大批量生产。以上问题的解决，有助于推动燃料电池的大规模商业化。

4.3 生物燃料电池

在我国，每天都会产生大量的废水、废液、生活和工业垃圾等，给社会和经济的发展带来严重的负担。如果能从“废水和废液”中回收能源，将可以最大限度实现污水处理的可持续发展，具有十分显著的社会、环境和经济效益。

生物燃料电池就是这样一种特殊的燃料电池。它以自然界的微生物或酶为催化剂直接将燃料中的化学能转化为电能，不仅具有燃料电池效率高和无污染等优点，还有一些独特的优势：①燃料来源广泛，葡萄糖、淀粉、生活垃圾和工业废液等都可作为其能量来源；②反应条件更加温和，生物燃料电池可在常温、常压和中性 pH 条件下工作，反应条件便于控制和维护；③生物相容性好，可作为一些生物分子或金属离子传感器[68-71]。此外，生物燃料电池不仅能够以单一的碳水化合物作为燃料发电，而且能够从废水中的有机污染物中回收电能并同时使污水变得清洁。与传统的废物资源化的方法相比，生物燃料电池可以在中温，甚至低温条件下高效产电，产生的气体无需进一步处理，产生的污泥量少，大大节省了系统的操作成本。然而，当前生物燃料电池产电的功率密度过低，严重限制了其大规模推广应用，短期内还无法进入实用领域，但其在污水处理、金属离子传感器和生物传感器等方面已显示出良好的前景[72,73]。

5　超级电容器

超级电容器（supercapacitors，SCs）也称为电化学电容器（electrochemical capacitors，ECs），其能量密度比传统电容器高，循环寿命优于电池；而且具有功率密度高、充放电速度快、循环寿命长等优点，是介于传统电容器和二次电池之间的一种近年来引起广泛关注和研究的新型储能装置[74,75]。传统超级电容器电极一般是通过混合活性电极材料、导电剂及黏合剂并涂覆在金属集流体上制得的。所以，基于传统电极的超级电容器多是纽扣型和滚筒型。近年来，随着柔性电子科学及电子产品小型化技术的快速发展，可穿戴、可折叠、柔性、便携式电子设备越来越受青睐，开发能为之提供能量的轻、薄、微型的高性能储能器件是最具发展潜力的方向之一。传统固定形状的电池限制了柔性电子产品的发展。另外，许多小型可穿戴智能设备以及各种微型生物植入器件，均要求功能设备在各种形变情况下稳定使用，而传统的硬质储能器件一般不能满足这些条件，极大限制了这些微型器件的使用效率和发展。因此在可穿戴、微型器件领域，开发与之相匹配的轻、薄、微型化并具备柔性、透明、可伸缩、可压缩和/或极端条件能力的超级电容器成为了首选解决方案[76,77]。

5.1　电极材料

电极材料对非传统超级电容器性能及构型影响很大，目前研究最多的电极材料为碳材料、赝电容性材料和复合型材料[78]。自从纳米碳管被发现以来，因其独特的一维结构、大表面积、良好的导电性能和高机械性能，碳纳米管成为最具研究前景的电极材料[79]。Chen 等利用化学气相沉积生长了垂直于基底排列的碳纳米管阵列，并从该阵列牵引出碳纳米管使其水平排列形成碳纳米管薄膜。将此薄膜作为超级电容器的电极材料，可有效地避免碳纳米管弯曲和团聚，极大地提高碳纳米管与电解质的接触面积，进而提高材料的双电层电容性能[80]。此外，石墨烯因其良好的导电性能和巨大的比表面积，也是一类研究较多的碳电极材料[81]。Xu 等利用过氧化氢在高温条件对氧化石墨烯进行刻蚀以制备多孔的石墨烯水凝胶，并将其作为电极材料。该方法在一定程度上弥补了石墨烯制备过程中产生的团聚和堆叠对离子传输的阻碍，有利于提高电极材料的功率密度[82]。

研究较多的赝电容电极材料主要包括过渡金属氧化物和导电聚合物，在充放电过程中，电极材料会发生氧化还原反应，与双电层电容的碳材料相比，基于此类材料的超级电容器的比电容和能量密度相对较高[83,84]。然而，金属氧化物的导电性较差，所以其倍率性能较低。此外，金属氧化物和导电聚合物在充放电过程中也面临着循环稳定性较差的缺点，金属氧化物电极材料表面和内部会不断发生氧化还原反应从而消耗电极材料，导电聚合物由于膨胀、收缩、破裂造成电极体积变化而不断退化。因此，导电聚合物通常与碳材料结合形成复合材料作为超级电容器的电极材料。

无论是碳材料还是赝电容电极材料都有各自的不足之处。碳材料虽然具有较好的循环稳定性和较高的功率密度，但能量密度低，单独使用某一种碳材料容易发生堆叠与团聚现象，从而减小电极材料和电解质的接触面积，降低电极材料的储能性能。对于赝电容电极材料，它们往往具有较高的储能容量和能量密度，但功率密度低且在充放电过程中电极材料表面和内部不断发生反应，电极材料不断被消耗，从而降低了此类材料的循环稳定性。因此，结合电化学双电层电容和赝电容储能原理的电极材料形成复合材料，能够有效地克服材料的不足，并且能够发挥两类材料的优势，以提高超级电容器的储能容量及循环稳定性[78]。

5.2 微型超级电容器

便携式电子器件的快速发展极大地刺激了现代社会对多功能化、小型化的电化学储能器件的强烈需求。其中，微型超级电容器正逐渐成为芯片储能器件研究领域中一个新兴、前沿的研究方向，它可作为微型功率源与微电子器件直接集成，在瞬间提供有效的功率峰值。微型超级电容器作为一种小尺寸的超级电容器，它不仅拥有与超级电容器相同工作原理，也具备微型条件下自身的结构优势。比如：微型超级电容器无需隔膜阻止两电极接触，大大缩短了电解质离子的输运时间，从而提高充放电速率[85]。

由于受到微型超级电容器尺寸的限制，并且为了提升微型超级电容器的性能，其电极结构的设计就显得格外重要，比如：卷状结构、三明治结构、3D 叉指结构等（图 24-4）。相较于传统的卷状结构、三明治结构，3D 叉指结构是一种在宏观尺度下平面化，微观尺度下三维化的空间结构。研究表明：3D 叉指电极结构不仅可以在相同的底面积上负载更多的电极材料，提高微型超级电容器的比电容，而且在 3D 叉指电极结构中，多组电极平行分开，电解质离子在电极材料上沿着平行方向扩散，从而加快电解质离子的运输速率，进而提高微型超级电容器功率密度[86]。

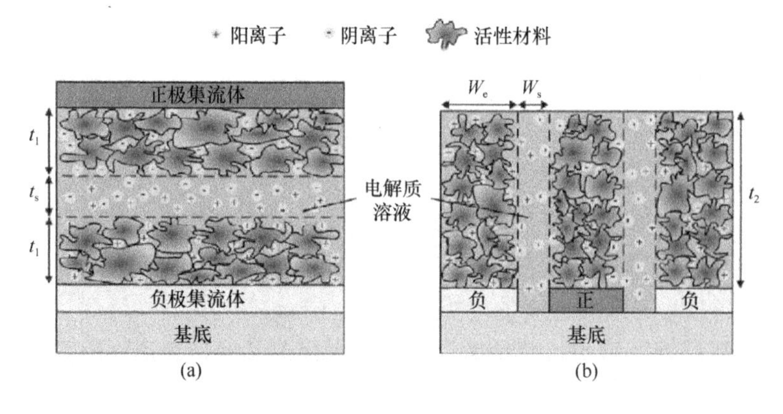

图 24-4 （a）传统的三明治结构；（b）3D 叉指结构

5.3 柔性可穿戴超级电容器

随着柔性显示、人工电子皮肤和智能眼镜等新颖概念的相继提出并付诸实践，快速发展的柔性电子技术已经吸引了广大科研工作者的研究兴趣，并激发了柔性电子产品对于柔性电源的迫切需求。其中柔性超级电容器作为柔性电源的一种，因其优异的电化学性能和可拉伸、可弯曲及可折叠的机械变形性能，成为最具应用前景的柔性电源之一。

如图 24-5 所示，柔性超级电容器的结构主要有线型、三明治型和平面型三种结构[87,88]。线型柔性超级电容器器件又称一维柔性超级电容器，其器件构型主要有两种，一种是正负电极都是一维线状，分别包裹上凝胶电解质后，平行或螺旋状缠绕在一起；另一种是一极柔性电极包裹上凝胶电解质后，再裹上另外一极柔性电极，形成一维同轴状的线性柔性超级电容器器件。三明治型结构柔性超级电容器主要是将超级电容器的电极、电解质和隔膜一层一层组装起来然后封装成三明治型结构。平面型柔性超级电容器的结构类似于微型超级电容器的叉指结构，这类柔性器件通常是在柔性基底上先通过模板或者蒸镀的方法使用金属制备一个叉指结构的集流体，然后通过原位方法生长活性物质。

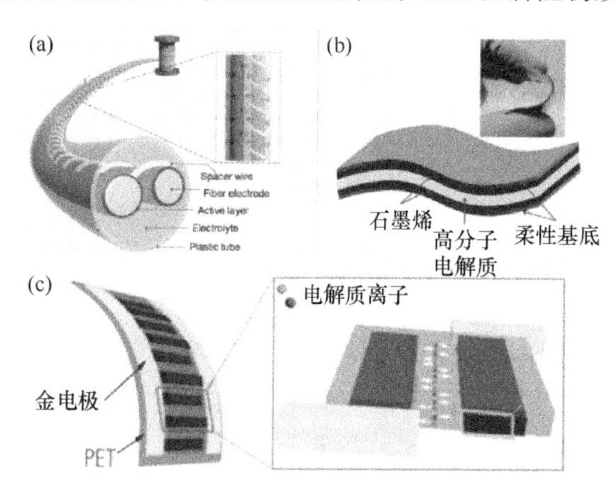

图 24-5　（a）线型结构、（b）三明治结构、（c）平面状（叉指结构）柔性超级电容器

柔性超级电容器可以直接作为电源驱动电子装置，例如线型超级电容器可以使用传统的纺织技术，同纤维编织在一起，既可以作为电源使用，又可以编织成任意形状的美丽图案[89]；平板型柔性超级电容器可以嵌入或者贴合在衣物表面，不仅可以保证持续的能量供应，还可以随着衣物的变形而变形；生物医用级柔性超级电容器可以植入人体，作为心脏起搏器、助听器、生物传感器和人工电子皮肤等医用装置的能源；柔性超级电容器还可以作为柔性显示屏的电源，在随着显示屏弯曲的同时，电源依然可以正常工作。

随着柔性超级电容器的发展，为了扩展其实用化，柔性可穿戴超级电容器应运而生，同时，追求其穿戴舒适性也逐渐得到了研究者的重视，因此以日常生活中常用的棉质织物和纤维作为电极基底承载活性材料的研究迅速增加。其中，用于穿戴的柔性超级电容器需要有良好的机械性能和弯曲性能，这是柔性超级电容器实用化的必要条件；与此同时，柔性超级电容器还需要有稳定的电化学性能和较长的循环寿命，使其能够在不同弯曲状态下仍然保持正常的工作状态并减少这类可穿戴器件的更换，此外，柔性可穿戴超级电容器的稳定性、安全性、可洗性研究仍需不断深入，以达到最终的实用化。

5.4　多功能超级电容器

将智能材料应用于超级电容器可得到一些多功能超级电容器，目前应用在超级电容器电极材料中的智能材料主要有电致变色材料和自修复材料，对应的是电致变色超级电容器和自愈合超级电容器。电致变色是指在一个外加电压负载感应下，材料的光吸收频率发生偏移和或者光的散射性质发生变化的现象[90]。目前应用在超级电容器中的电致变色材料主要有聚合物材料聚苯胺，无机材料主要有 V_2O_5、WO_3、MoO_5 等。这些材料在充放电过程中会有电压变化，呈现出不同的颜色。自修复材料的基本原理是将切断的两段连接在一起后会因为分子间作用力的影响重新聚合在一起，从而实现自我修复的功能[91]。自修复超级电容器的恢复功能主要是设计的基底和电解质具有一定程度的自恢复功能。

6 二氧化碳资源化利用

为了满足世界人口在不断增加过程中的能源需求，化石燃料（如煤，石油，天然气等）正在以前所未有的规模进行消耗，与此同时，每年会有超过 300 亿吨的二氧化碳排放到大气中，二氧化碳的浓度从工业革命前的 270 ppm 已经上升到如今的 400 ppm。高浓度二氧化碳引起的温室效应导致全球气候恶化，同时大量的二氧化碳会溶解在海洋中，引起海洋生态环境的酸化。因此，为了人类社会和自然环境的可持续发展，亟待发展二氧化碳转化技术，降低二氧化碳的排放量。

目前，催化二氧化碳转化的过程包括热催化、电催化、光催化过程，其中由于电催化还原的操作条件简单（室温常压），反应过程的可控性以及电解液的可循环性，受到了更多的关注和研究。但是，在二氧化碳电化学还原中，二氧化碳的高键能使反应需要更高的能量引发还原过程，导致较低的能量利用率。同时，二氧化碳电化学还原是一个多电子转移过程，同一条件下，可能进行各种多电子转移过程，具有较低的选择性。此外，在水系电解液中，析氢反应是更容易发生的竞争过程。因此，选择高活性、高选择性以及能够抑制析氢反应的催化剂是二氧化碳电化学还原的关键。

6.1 电化学还原中催化剂的发展

贵金属虽然在二氧化碳电化学还原中表现出较好的催化活性和选择性，但是低储量、高成本限制了其发展。包信和课题组[92]在探究 Au-CeO_x 时发现，Au-CeO_x 之间的界面作用有助于 CO_2 的吸附和活化，对 COOH*中间体有更好的稳定作用（图 24-6）。同时，Ag-CeO_x 的性能也证实了金属界面之间的重要性。另外，催化剂的形貌对性能有很大影响[93]，通过调控薄层多孔银电极的厚度和纳米颗粒的尺寸，由 30~50 nm 银纳米颗粒组成的 6 μm 薄层银片在 390 mV 的过电位下还原成 CO，电流密度可达到 10.5 mA/cm^2，质量活性为 4.38 Ag^{-1}。此外，Ag 的未占据态密度对活性也有影响[94]，具有最高未占据态密度的多孔银电极仅仅有 190 mV 的过电位，法拉第效率可达到 90%以上（−0.8 V *vs*. RHE）。杨金龙课题组[95]比较了尺寸接近的八面体和正十二面体的 Pd 催化剂，正十二面体的 Pd 具有更高的活性，在−0.8 V（*vs*. RHE）下 FE 可达到 91.1%，这是由于正十二面体产生更高的表面应力，使得 d 键中心向更高的能级移动，有利于中间体 COOH*的稳定吸附。

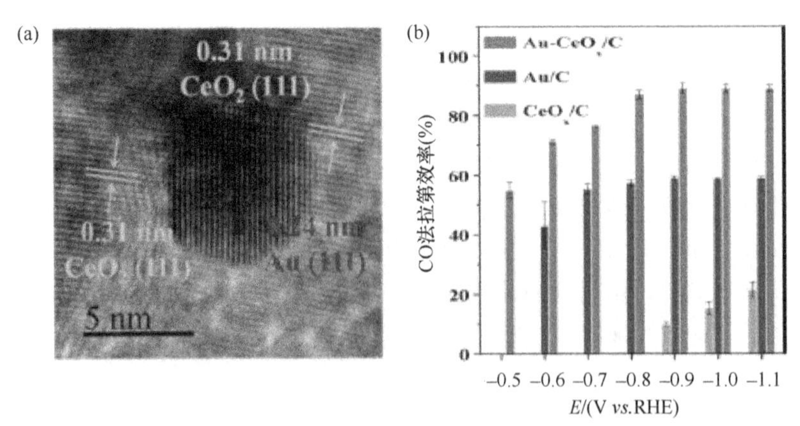

图 24-6 Au-CeO_x/C 催化剂的结构（a）和还原产物 CO 的法拉第效率（b）

在非贵金属催化剂中，金属 Sn 和 Bi 可以催化得到液态产物甲酸。谢毅课题组构建了封装在石墨烯中的 Sn 量子点结构（图 24-7），不仅有 9 倍于金属 Sn 的二氧化碳吸附量，而且连续电解 50 小时后依然保持 85%以上的法拉第效率[96]。利用电化学沉积法直接在碳纸上生长树突状铋催化剂[97]，其最大的甲酸法拉第效率可达 96.4%，电流密度为 15.2 mA/cm^2。此外，铜基催化剂的独特性在于能够催化二氧化碳还原成碳氢化合物。一种独特的铜纳米花结构不仅具有较低的过电位（400 mV），而且能够将析氢反应抑制到 25%以下，在还原过程中，发现了无定形碳在电极表面形成[98]。以铜纳米立方体为模板，通过化学蚀刻来得到高晶面指数的纳米晶，对 CH_4、C_2H_4、C_2H_6、C_3H_8 的选择性，Cu｛110｝明显高于 Cu｛100｝，但是对 CO 的选择性却恰恰相反，这也说明 C_2 产物更容易在 Cu｛110｝表面形成[99]。

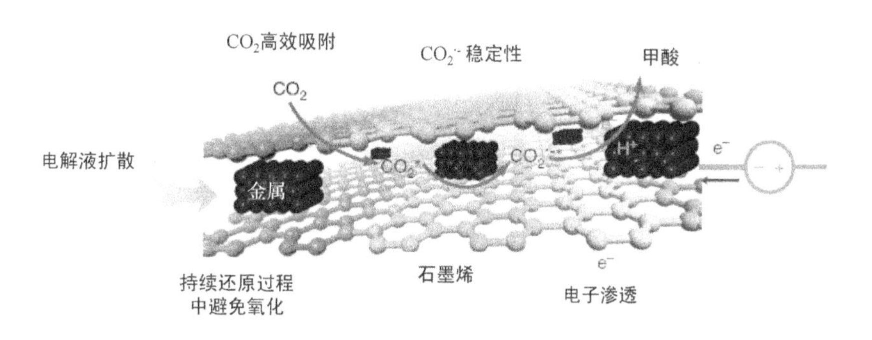

图 24-7　封装在石墨烯中的 Sn 量子点将二氧化碳还原成甲酸的反应过程

金属氧（硫）化物中氧（硫）原子的存在会改变金属催化剂的内部电子结构，从而影响 CO_2 还原的反应机理。谢毅课题组[100]对 Co_3O_4 在二氧化碳电化学还原中进行了深入的研究（图 24-8）。构建的部分氧化四层原子厚的钴催化剂在−0.85 V *vs*. SCE 下，电流密度可达到 10.59 mA/cm^2，分别是纯四层原子厚度的 Co、部分氧化体相 Co 以及体相 Co 的 10 倍、40 倍和 260 倍。同时，进一步研究了 Co_3O_4 单元层结构中 O(Ⅱ)空位对二氧化碳电化学还原的影响，研究发现，O(Ⅱ)空位有利于 CO_2 的吸附，而且稳定了中间体 COOH*的存在。

以 $SnCl_2$ 或者 $SnCl_4$ 为前驱体，制备得到不同形貌的 SnO_x 纳米颗粒，在−1.4 V *vs*. SCE 时，KBH_4 还原 $SnCl_2$ 前驱体得到的 SnO_x 纳米颗粒，具有最高的法拉第效率和能量效率，分别为 64%和 27%[101]。Sn 表面自身会被氧化形成一层 SnO_2，蚀刻掉表面氧化物时，二氧化碳电化学还原的活性明显降低，但在空气中暴露 24 h 后，SnO_2 在 Sn 表面继续形成，产生更好的活性，这表明 SnO_2 在二氧化碳电化学还原中重要性[102]。

柯福生课题组报道了基于 CuO 的多孔铜纳米带（图 24-9），与铜箔的产物（CH_4）相比，其更有利于 C_2 产物的形成，在−0.816 V *vs*. RHE，C_2H_4、C_2H_6 和 C_2H_5OH 的法拉第效率可达到 40%[103]。促进 C_2 产物而抑制 C_1 产物，这是由于 CuO 得到的 Cu 纳米带中表面缺陷和大量晶界的形成。此外，以 HKUST-1 为前驱体，通过简单的碳化过程得到 OD Cu/C 结构，对醇类产物具有较好的选择性，在−0.1~−0.7 V *vs*. RHE，法拉第效率可达到 45.2% ~71.2%。而且 C_2H_5OH 形成的过电位仅有 190 mV，可能是由于高分散的铜和多孔碳材料共同作用所得到的[104]。

单金属的还原产物比较单一（铜基催化剂除外），通过两种金属的互相掺杂，可以得到更高的活性和更丰富的选择性。包信和课题组对不同组成成分的 Cu-Pd 合金纳米颗粒进行了研究[105]，其中 $Pd_{85}Cu_{15}/C$ 可以得到 CO，在−0.89 V *vs*. RHE，法拉第效率可达到 86%，质量活性为 24.5 A/g，与 Pd/C 相比，法拉第效率和质量活性分别是 Cu-Pd 合金的 5 倍和 2.2 倍。吡啶对 Cu-Pd 合金修饰后，还原产物却可以得到甲醇和乙醇[106]。

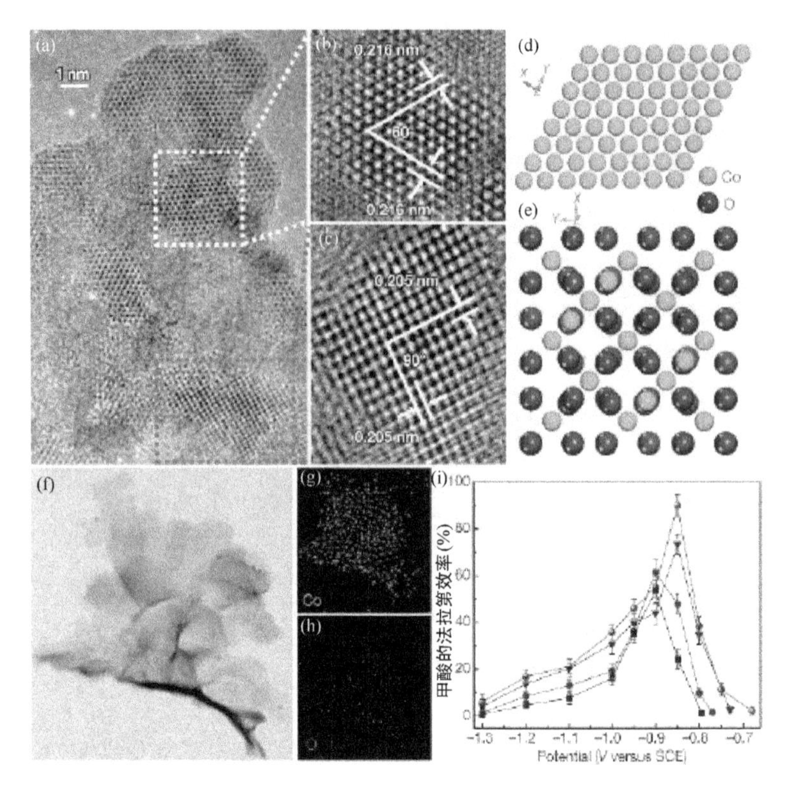

图 24-8　220℃煅烧 3h 得到的部分氧化四层原子厚度 Co 的表征分析。(a) 高分辨 TEM 图。(b，c) 放大后的高分辨 TEM 图。(d，e) 相关的原子模型图，清楚地表明金属 Co 六方结构和 Co3O4 立方结构中的原子结构。(f-h) 元素分布。i，还原产物甲酸的法拉第效率。

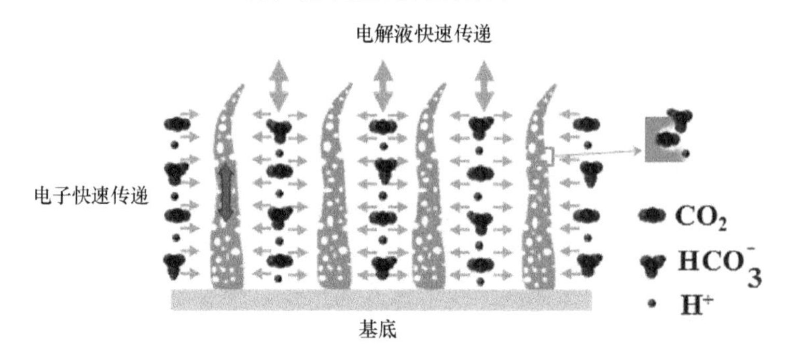

图 24-9　基于 CuO 的多孔铜纳米带阵列在二氧化碳电化学还原中的反应过程和优势

核壳结构也是双金属存在的另一种形式。李箐等发现虽然 Sn 在二氧化碳电化学还原中得到是甲酸产物（图 24-10）[107]，但当薄层 SnO_2 包裹 Cu 纳米颗粒时，还原产物却由 SnO_2 的厚度决定，当厚度为 1.8 nm 时，还原产物为甲酸，当厚度为 0.8 nm 时，还原产物为 CO，法拉第效率−0.7V *vs*. RHE 时可高达 93%。理论计算表明，0.8 nm 的 SnO_2 与 Cu 形成的是类合金状态，引起了 SnO_2 晶格的单轴压缩，更有利于 CO 的形成。以多元醇的制备方法得到了一组 Ag@Cu 双金属纳米颗粒，通过调控反应时间，从 CO（Ag@Cu-7）到碳氢化合物（Ag@Cu-20）的还原产物均可得到[108]。

韩布兴课题组对以 Mo-Bi 双金属硫化物为二氧化碳电化学还原的催化剂进行了研究[109]，在 Mo 和 Bi 的协同作用下，甲醇的法拉第效率可达到 71.2%，同时电流密度为 12.1 mA/cm^2。康振辉课题组利用水热法得到 FeS_2/NiS 纳米催化剂，可以在二氧化碳电化学还原中得到 CH_3OH [110]，其过电位仅为 280 mV，同时在−0.6 V *vs*. RHE 时，法拉第效率可达到 64%，在还原过程中，活性位位于 FeS_2 和 NiS 界面之间，能够有效地抑制析氢反应，同时促进二氧化碳电化学还原进行。

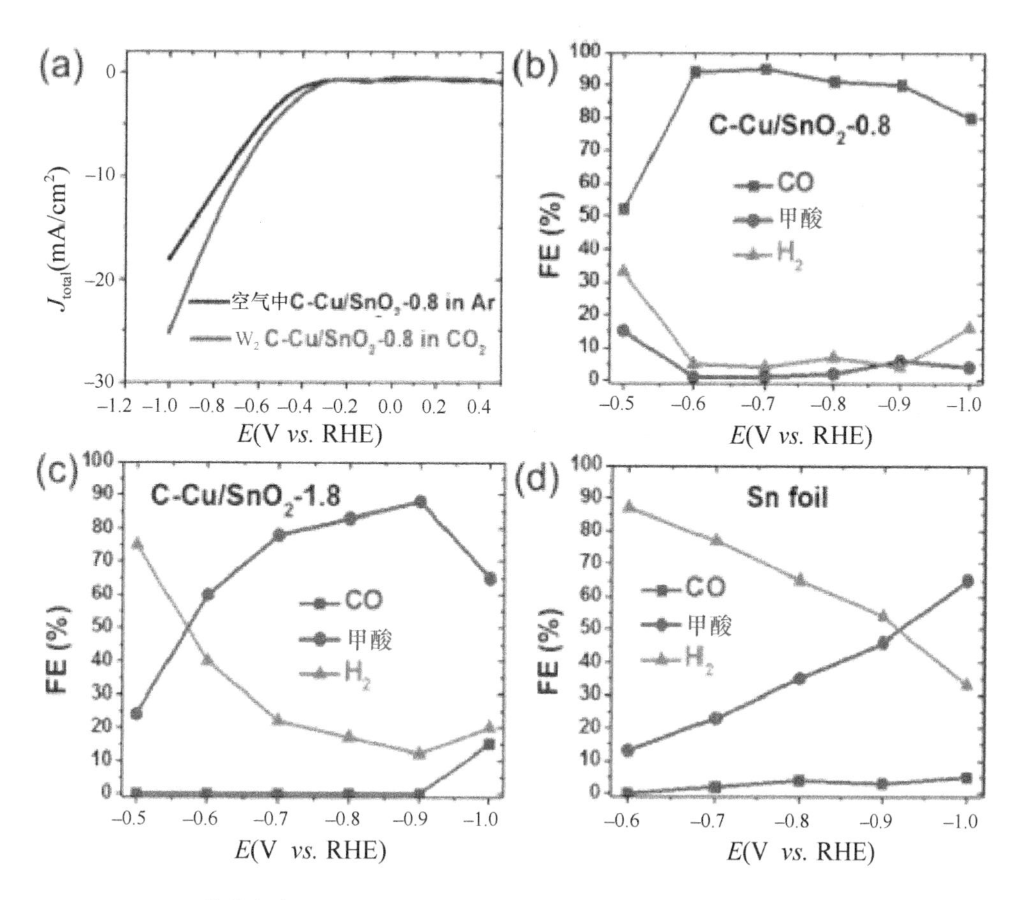

图 24-10　（a）C-Cu/SnO_2-0.8 催化剂在 Ar 或者 CO_2 饱和后的 0.5 mol/L $KHCO_3$ 溶液中的 LSV 曲线，扫描速率 5 mV/s；不同催化剂（b）C-Cu/SnO_2-0.8 催化剂、（c）C-Cu/SnO_2-1.8 催化剂、（d）酸处理后的 Sn 箔电极在 CO_2 电化学还原中还原产物的法拉第效率

与金属催化剂相比，非金属催化剂主要包括不同形态的碳材料（石墨烯，碳纳米管，金刚石等）以及含有有机基团的催化剂。谭天伟课题组以氧化石墨烯和尿素分别作为碳源和氮源得到氮掺杂石墨烯[111]，在−0.84 V *vs.* RHE 时，法拉第效率可达到 73%。但与已经报道的不同催化剂相比，在甲酸形成过程中，法拉第效率和过电位处于较高且相对平衡的位置。全燮课题组构建了一种负载在硅棒阵列上的氮掺杂纳米金刚石[112]，其还原产物草酸的法拉第效率在−0.8V *vs.* RHE 即可达到 91.2%，这一方面是由于金刚石具有较高的析氢过电位，另一方面是由于 N-sp^3C 对 CO_2 还原具有高活性。梁永晔课题组将钴卟啉通过 π-π 键连接在碳纳米管上（图 24-11）[113]，在电流密度、CO 的选择性以及稳定性上有明显提高。在钴卟啉中进一步引入氰基基团，在过电位 520 mV 下，其还原产物 CO 的法拉第效率甚至超过 95%，电流密度可达到 15 mA/cm²。

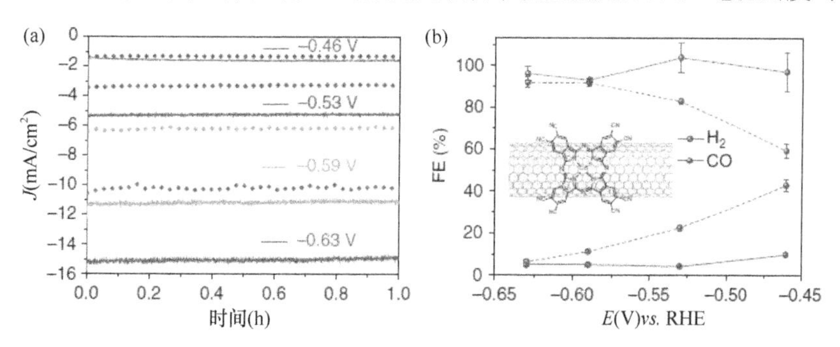

图 24-11　氰基修饰 CoPc/CNT 的二氧化碳电化学还原性能

6.2 电化学还原过程中的其他影响因素

催化剂是二氧化碳电化学还原的关键因素之一，但是在还原过程中，电解液的种类和 pH 也对其具有重要的影响[114]。随着 $KHCO_3$ 浓度增加，产物的法拉第效率呈现先增大后降低的过程，当 HCO_3^-浓度较低时，二氧化碳电化学还原主要是传质过程控制，当浓度较高时，甲酸的产生主要有电荷传递过程控制。同时，较低或者较高的 pH 同样也会降低产物生成速率。

由于二氧化碳电化学还原和析氢反应是一对竞争反应，为了抑制析氢过程的发生，非水系电解液是最佳的选择之一。韩布兴课题组对三元体系电解液（离子液体/有机溶剂/水）进行了深入研究[115]。虽然离子液体或者有机溶剂对 CO_2 有更高的溶解度，但是较高的黏度和较大的电导率大大抑制了二氧化碳电化学还原的发生，当少量水加入离子液体/乙腈体系中时，在 Pb 或者 Sn 电极上甲酸的法拉第效率高达 91.6%，电流密度可达到 37.6 mA/cm^2。戴永年课题组在离子液体/有机溶剂（NMP，AN，PC，DMF，DMSO）二元体系中进行了大量的研究，在 TBAP/NMP 体系中，金电极对 CO 产物的法拉第效率可达到 93%(−2.4 V *vs*. ferrocene/ferrocenium)[116]。

尽管目前已经对二氧化碳电化学还原技术进行了大量的研究与探索，特别是对催化剂的研究方面，但是二氧化碳电化学还原远未达到实用化要求。理想催化剂应成本低廉，兼具高活性（低过电位）、对目标产物的高选择性、高稳定性。在相关研究基础上，我们提出一些未来的研究方向：①构建分层多孔纳米结构的新型非贵金属催化剂；②利用均相和非均相体系在分子层面优化活性位点；③利用原位表征技术和理论计算探究二氧化碳电化学的还原过程和机理。总之，通过利用可再生能源（如太阳能，风能）推动二氧化碳电化学还原技术，不仅可以将二氧化碳转化成高附加值的能源燃料，而且为碳基资源开辟一条可持续循环的能源储存和转化过程。

7 电解水技术

氢能具有储量丰富、可再生、高能量密度及环境友好的特点，而且能补足化石能源消耗的缺口，具有良好的应用前景。现阶段工业制氢通常来自于化石燃料，如石油、天然气等，在产氢的同时伴随二氧化碳的产生，既消耗珍贵的有限资源又不环保。而运用各种可再生资源如光、电分解水产氢能将以上问题避免，同时水分解产生的氧气亦可作他用。

目前水电解技术主要有碱性电解水、固体氧化物电解水以及质子交换膜（PEM）电解水等。固体氧化物电解水操作通常在 500℃以上，扩散速率快，但高温大大限制了其发展；碱性电解池易操作、成本低且稳定性好；质子交换膜水电解技术需要使用价格高昂的贵金属催化剂。碱性电解池和质子交换膜电解池需要用到仅能通过 OH^-或 H^+的隔膜，能增加气体收集效率及安全性，但成本增加且水分解动力学受限[117]。因此，质子交换膜电解水技术的研究热点集中在高效低成本电极催化剂及质子交换膜的开发。

7.1 阴极催化剂

在 PEM 膜电极结构中，由于 Nafion 膜在水中具有强酸性（pH 相当于 10%的 H_2SO_4 溶液），鉴于电极材料耐蚀性和稳定性的要求，作为催化剂的金属的选择几乎完全限制在贵金属和它们的合金上，如铂（Pt）及其他贵金属如铑（Rh）、铱（Ir）、钯（Pd）等，是迄今为止最好的 HER 电化学催化剂，它们可以温和地吸附氢而且只需很少的活化能就能将氢气从表面脱附，水电解析氢的过电位几乎为零。然而这些催化速率高却昂贵的贵金属催化剂难以大规模推广使用，为此，研究人员做了大量的工作。Zhang 等用电化学方法将 MoS_2 剥离成单层纳米片，然后通过外延生长法在 MoS_2 纳米片上生长 Pt 纳米颗粒，合成的 Pt-MoS_2 作为析氢催化剂，在 100 mV 过电位下的电流密度达到 40 mA/cm^2，表现出比商品 Pt/C 更高的催化活性[118]。因此，研究和开发性能良好的且价格低廉的低 Pt 或非 Pt 电化学催化剂至关重要（图 24-12）[119]。

近年来，关于过渡金属硫化物，特别是 MoS_2 的研究备受关注。Kibsgaard 等利用双螺旋构型 SiO_2 为模板制得介孔 MoS_2 电极。这一独特结构使 MoS_2 的边缘得到优先暴露，从而显著增加了析氢的活性位点数目[120]。最近，Deng 等以导电性良好的石墨烯为载体，采用水热法合成了垂直于石墨烯上的超薄 MoS_2 纳米片，并将其作为一种高效析氢催化剂。结果显示，超薄 MoS_2 纳米片均匀地垂直生长在石墨烯上，电极的导电性明显增强，并且暴露出更多的边缘位点，表现出高的催化活性和良好的稳定性[121]。

过渡金属碳化物拥有很高的硬度、较好的稳定性和较强的耐腐蚀性，是一类新型的功能材料，在各种耐高温、耐摩擦和耐化学腐蚀的机械范畴内获得了广泛使用。近年来，研究学者发现，这类物质具有一定的催化活性，在烃类加氢、氢解、甲烷重整和电催化等反应中表现出类似于贵金属的性质，因此被称为“类 Pt 催化剂”。在过渡金属碳化物方面，主要研究的有 WC 和 MoC 两种。WC 作为催化剂的电催化性能与 Pt 非常的相似，因此 WC 的制备及其催化性能吸引了很多学者的兴趣。WC 还具有较强的选择性、很高的活性以及抗毒化性[122]。此外，相对于其他催化剂，WC 的另外一个优点在于它的耐高温性，甚至在灼烧的条件下也能有一定的应用。

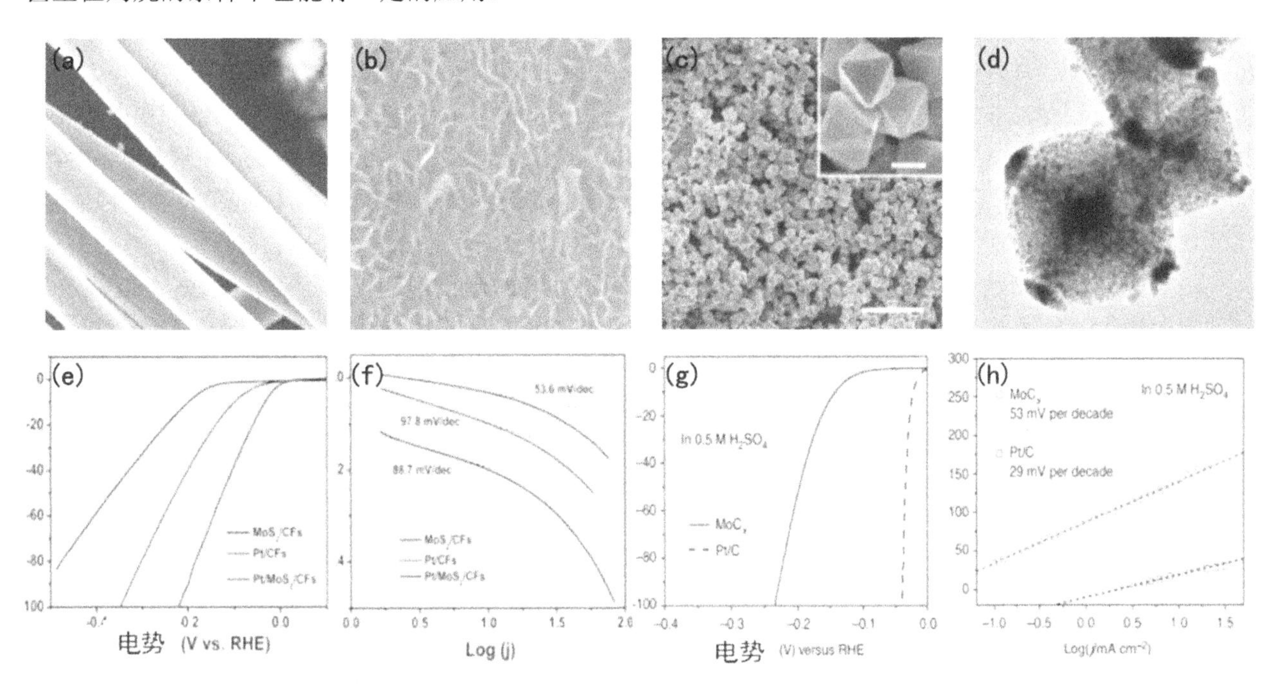

图 24-12　不同材料的形貌以及析氢性能图：（a，b，e，f）Pt-MoS_2；（c，d，g，h）MoC_x 纳米八面体

过渡金属磷化物也是一类非常重要的无机功能材料，具有导热性、导电性、高硬度、高强度和高稳定性等优点，已被广泛应用在催化、能源、锂电池、磁学、超级电容器等领域。过渡金属磷化物的合成，尤其在纳米尺度下控制其形貌和结构，成为材料合成领域的一个热点。孙旭平课题组对 CoP 催化析氢反应进行了一系列的研究。Sun 等合成的 CoP/CNT 复合材料中，CoP 纳米颗粒的粒径只有 2~3 nm [123]。另外，CoP 纳米材料的形貌对其催化活性也有明显的影响。Sun 等用不同的前驱体合成了 CoP 的纳米线、

纳米片和纳米颗粒，经测试对比纳米线的析氢催化活性最大。在水热法合成 CoP 纳米线的过程中，如果在水热反应釜中放入碳布，纳米线会选择性地在碳纤维表面生长。通过该方法合成的 CoP 纳米线/碳布复合材料作为酸性条件电解水反应的阴极，起始过电位只有 38 mV，塔菲尔斜率为 51 mV/dec，而且其催化活性可以在 100 mA /cm^2 的电流密度下保持 80000 s 不降低[124]。

7.2 阳极催化剂

在质子交换膜水电解池的强酸性条件下，稳定性是催化剂的重要考量，且理论析氧电压较高，导致阳极的腐蚀性很强，阳极析氧反应过电位是全水电解过电位的主要来源。酸性条件下，普遍认可的阳极催化剂表面析氧历程为：

$$MO_x + H_2O \longrightarrow MO_x(OH)_{ads} + H^+ + e^-$$

$$MO_x(OH)_{ads} \longrightarrow MO_{x+1} + H^+ + e^-$$

$$MO_{x+1} \longrightarrow MO_x + 1/2O_2(MO_x \text{为氧化物催化剂})$$

目前表现最好的析氧反应催化剂为铂族贵金属催化剂，IrO_2 及 RuO_2。低成本催化剂开发分为两个方向，降低 Ir/Ru 用量和用过渡金属化合物完全取代 Ir/Ru 基催化剂。在降低 Ir/Ru 用量方面，可将其他过渡金属相结合合成二元或三元氧化物。比如 IrO_2–RuO_2 结合的 $Ir_{0.6}Ru_{0.4}O_2$、$Ir_xRu_yTa_zO_2$ 都有优异性能[125]。其他二元或三元氧化物包括 $Sn_xRu_{1-x}O_2$[126]，$Ir_xSn_{1-x}O_2$[127]，$Ru_xIr_yCo_zO_2$ 等也是性能良好的 OER 催化剂[128]。另外，将 IrO_2 或 RuO_2 负载于具有大比表面积的活性载体上，也是保持高活性同时降低成本的有效方法。该方法要求使用的活性载体具有良好导电性，大的比表面积及多孔结构，利于电极反应过程的传质，甚至一些载体本身就具有催化活性。中国科学院大连化学物理研究所 Yu 等将 IrO_2 纳米颗粒沉积在多孔金上，成功将 Ir 用量减少，并获得了优异性能，恒电流下（250 mA/cm^2）能保持 300 h 的良好稳定性（图 24-13）[129]。极薄（9 nm）的 RuO_2 负载于直径 1 μm 的 SiO_2 纤维纸上极大地降低了 Ru 用量，将 SiO_2 纤维用碳包裹后，性能得到进一步提升，质量活性达到 40~60 mA/mg@η = 330 mV[130]。载体除降低成本外，还能包裹保护催化剂免于强酸电解液的腐蚀或被电解析出的氧气氧化，比如 TiC[131]、ITO 等[132]。

过渡金属氧化物是重要的析氧反应催化剂，比如通过简单水热方法制得的镍泡沫负载的多孔 MoO_2 纳米片可做全水解催化剂材料，在电流密度达 10mA/cm 时过电位仅为 1.53 V，且能保持活性 24 h[133]。Wang 等通过自模板法制得的 CoO-MoO_2 纳米笼子具有低过电位[134]。其他过渡金属化合物包括磷化物 Ni_2P [135]，NiFeP [136]，磷酸盐 $Fe(PO_3)_2/Ni_2P$[137]及硼化物 Ni_3B/Ni-Bi 是报道诸多的优异的析氧反应催化剂[138]。过渡金属化合物催化材料目前所面临的问题是在酸性电解液条件下稳定性不佳，应用主要集中在碱性或中性电解液，因此还有很大的发展空间。

7.3 质子交换膜

质子交换膜是质子交换膜电解水的心脏部分，它的性能对整个电解槽的运行起着至关重要的作用。作为传导介质，质子交换膜不仅要传导质子，避免催化剂受对电极如 Pt 溶解的污染[139]，分割氢气和氧气，提供安全保障，还要为两极的催化剂提供一定的支撑，保证电解槽正常运行。质子交换膜应具备优异的化学、热力学稳定性和良好的质子传导性，保证电解池的较小欧姆阻抗。同时，膜表面与催化剂的适配性要好，便于有效阻止气体的扩散，阻隔氢气和氧气接触。目前，已经商品化的全氟磺酸高分子膜有 Nafion 膜、Flemion 膜、Aciplex 膜和 Dow 膜，其中研究最为成功、应用最为广泛的是杜邦公司的 Nafion 膜。

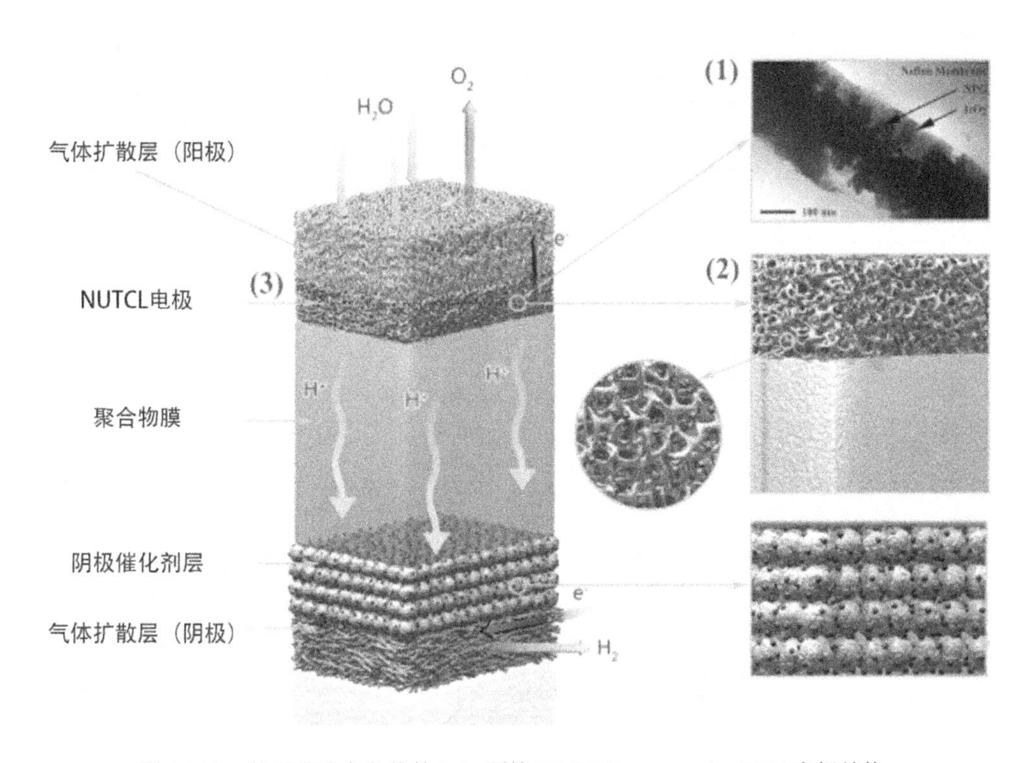

图 24-13　基于多孔金负载的 IrO_2 颗粒 NPG/IrO_2 composite-PEM 电极结构

杜邦公司的 Nafion 膜具有很多优点，但价格昂贵，致使质子交换膜电解槽技术成本较高。近年来，人们试图对一些无质子传导能力或质子传导能力很低但具有良好机械性能、化学稳定性和热稳定性且价格便宜的聚合物，如聚苯并咪唑（PBI）[140]、聚醚醚酮（PEEK）等[141]，通过质子酸掺杂/辐射接枝改性等使之具有良好的质子传导能力，从而应用于 PEM 水电解技术中。

质子交换膜水电解技术具有装置简单、效率高、制氢纯度高等优点，但由于膜和贵金属催化剂的价格昂贵制约了其商业化进程。开发新的非贵金属催化剂和新型质子交换膜是降低成本的关键，同时，设计合理的水电解池也是提高水电解效率的研究方向之一。在开发电极材料提高电极反应效率方面，对构效关系的有限认识是制约发展的关键，因此，理论结合实验是水电解技术长足发展的重要指导。由于 OER 高过电位，质子交换膜电极反应主要受 OER 限制，开发高效低成本的 OER 催化剂具有里程碑式的意义。另外，结合其他可再生能源如光、热等的水电解技术将是另一重要发展方向。结合太阳能的电解水技术能极大利用可再生资源，提高太阳能到化学能的转化率。另一方面，太阳能释放的热能，能增加反应温度，水分解反应的熵和电子传输的热力学、动力学可能会有相应增加，从而提高催化剂在电催化过程的性能。

8　其他绿色能源概述

我国经济在经历了 30 多年的高速增长之后，资源承载能力和生态环境容量已无法支持这种高速增长，因此，节能环保、转变经济发展方式显得尤为重要。为对我国能源产业结构调整和可持续发展作出贡献，除对上述的电化学储能技术进行研究外，目前其他的绿色能源研究主要集中在废热转换、生物质能量转换及电网三个方面。

8.1 废热转换

工业废热的利用是我国实现节能减排和提高能源利用率的重要战略。目前，国内对工业废热的研究主要从废热的热利用、废热发电、废热制冷及废热热泵技术四个方面展开。废热的热利用是通过一定的技术手段将其应用于预热工业燃料、助燃空气及补充水等方面，如利用烟气对流入的冷空气进行预热[142]。张青山团队采用背压式饱和蒸汽发电机组替代蒸汽热力系统中的降压减压阀，充分利用顶吹炉余热锅炉产生的中压饱和蒸汽的余热余压进行发电[143]。对于废热制冷技术，目前已广泛应用于远洋渔船的冷藏保鲜方面[144]。邱丽霞等采用热泵技术将电厂循环水中的废热进行回收利用[145]，为冬季供暖提供低温辅助，这不但使系统具有较高的节煤率和发电功率，而且降低了冬季供暖的电耗。

8.2 生物质能量转换

随着全球范围内能源需求的增加和石油资源的逐渐枯竭，生物质能量作为一种可再生的绿色能源受到越来越多的关注。目前对生物质能量的利用主要分为固化利用、液化利用及气化利用三个方面。

生物质燃料固化成型技术是目前应用最为广泛的一种生物质利用技术，其是煤炭资源的优良替代品。目前，生物质固化成型车作为一种生物质固化燃料的小型制备机械得到广泛研究，张彦民在对成型机类型、成型机理研究的基础上，设计了一款复合式环模生物质燃料颗粒成型机[146]。这对生物质燃料资源化利用具有积极的推动作用。

生物质液化技术可以将生物质转换成高品位的液体燃料，减少对石油、煤等化石燃料的依赖，因此受到广泛关注。然而，由于生物质产油率和转化率不高，但是对设备及操作条件的要求比较高，制约了生物质液化技术的大规模利用。因此，目前进行生物质液化技术研究的关键是降低操作费用，优化操作条件，提高目标产物收率。

生物质气化是生物质能高品位利用发展最迅速最实用的技术之一，在集中供气、供热、发电方面进行了商业化运行，并在生物质气化合成液体燃料、制氢等方面开展了研究。鲍振博等针对小型家用生物质气化炉在使用中存在气化过程中焦油、灰分含量多，物料连续添加工艺复杂，而物料间断供给使用不便等问题，提出了一种多体式秸秆生物质气化炉的设计[147]，可实现对秸秆生物质能源的有效利用。

8.3 电网

随着对微网、分布式发电的广泛研究，传统的电力系统逐渐向更为先进的智能电网进化，为进一步提高电力系统运行的可靠性，大容量的储能技术显得日益重要。目前，储能技术在电力系统中的应用，主要集中在可再生能源发电、分布式能源及微电网、电力辅助服务、电力质量调频、电动汽车充换电等方面[148]，这些是解决新能源电力储存的关键技术。

结合国内相关研究成果以及调研访谈数据的汇总分析，王彩霞团队从储能技术应用规模、技术特性、经济性等方面，提出未来我国适用于电网的先进大容量储能技术发展路线图[149]，提出应着眼于接入高比例可再生能源的电网应用，超前开展先进大容量储能技术（如示范推广 100 MW 级锂离子电池、全钒液流电池以及 10 MW 级的钠硫电池储能系统）的布局和研发。

为了增强微电网的运行稳定性及供电质量，将储能系统应用到微电网结构中，但是单一的储能系统已经不能满足微电网的需求。目前大量的研究集中在将混合储能系统（如采用超级电容器和多硫化物溴电池的混合储能系统）应用到微电网中，利用超级电容的快速响应和功率密度大等特性快速填补微电网的功率缺失，当需要长时间提供电能时则由液流电池来填补[150]。

9 展　望

本章总结了近年来绿色能源技术取得的重大发展与进步，重点介绍了近五年来我国科研工作者在锂离子电池、超级电容器、燃料电池、碳捕获及其转化、电解水技术和其他绿色能源技术的最新研究进展，并对其中一些亮点和最新研究内容进行了重点说明。面向未来，我国能源发展既面临调整优化结构、加快产业转型升级和淘汰落后产能的战略机遇期，也面临诸多矛盾交织、风险隐患增多的严峻挑战。大力发展绿色能源技术，不仅是我国《能源发展“十三五”规划》的要求，也是我国实现可持续发展和全面建设和谐社会的必由之路。我们有理由相信，随着多种能源技术制备工艺的不断发展成熟，人们也会获得更加安全和便捷的绿色能源技术，我们也将迎来一个更加绿色、环保的世界。

参 考 文 献

[1] Nitta N, Wu F, Lee J T, Yushin G. Li-ion battery materials: present and future. Mater. Today, 2015, 18(5): 252-264.

[2] Etacheri V, Marom R, Elazari R, Salitra G, Aurbach D. Challenges in the development of advanced Li-ion batteries: A review. Energy Environ. Science, 2011, 4(9): 3243-3262.

[3] Ellis B L, Knauth P, Djenizian T. Three-dimensional self-supported metal oxides for advanced energy storage. Adv. Mater., 2014, 26(21): 3368-3397.

[4] Tao H C, Yang X L, Zhang L L, Ni S B. Double-walled core-shell structured Si@SiO_2@C nanocomposite as anode for lithium-ion batteries. Ionics, 2014, 20(11): 1547-1552.

[5] Li Q, Yin L, Gao X. Reduction chemical reaction synthesized scalable 3D porous silicon/carbon hybrid architectures as anode materials for lithium ion batteries with enhanced electrochemical performance. RSC Adv., 2015, 5(45): 35598-35607.

[6] Cheng X B, Zhang R, Zhao C Z, Wei F, Zhang J G, Zhang Q. A review of solid electrolyte interphases on lithium metal anode. Adv. Sci., 2016, 3(3): 1500213.

[7] Zhang S, Ueno K, Dokko K, Watanabe M. Recent advances in electrolytes for lithium-sulfur batteries. Adv. Energy Mater., 2015, 5(16): 1500117.

[8] Suo L, Hu Y, Li H, Wang Z, Chen L, Huang X. Progress on high-energy density lithium-sulfur batteries. Chinese Sci. Bull., 2013, 31(58): 3172-3188.

[9] Xu W, Wang J, Ding F, Chen X, Nasybulin E, Zhang Y, Zhang J-G. Lithium metal anodes for rechargeable batteries. Energy Environ. Sci., 2014, 7(2): 513-537.

[10] Lin D, Liu Y, Cu, Y. Reviving the lithium metal anode for high-energy batteries. Nature Nanotechnol., 2017, 12(3): 194-206.

[11] Seh Z W, Sun Y, Zhang Q, Cui Y. Designing high-energy lithium-sulfur batteries. Chem. Soc. Rev., 2016, 45(20): 5605-5634.

[12] Manthiram A, Fu Y, Chung S H, Zu C, Su Y S. Rechargeable lithium-sulfur batteries. Chem. Rev., 2014, 114(23): 11751-11787.

[13] Sahore R, Levin B D A, Pan M, Muller D A, DiSalvo F J, Giannelis E P. Design principles for optimum performance of porous carbons in lithium-sulfur batteries. Adv. Energy Mater., 2016, 6(14): 1600134.

[14] Liu X, Huang J Q, Zhang Q, Mai L. Nanostructured Metal Oxides and Sulfides for Lithium-Sulfur Batteries. Adv. Mater., 2017, 1601759.

[15] Xu N, Qian T, Liu X, Liu J, Chen Y, Yan C. Greatly suppressed shuttle effect for improved lithium sulfur battery performance through short chain intermediates. Nano Lett., 2017, 17(1): 538-543.

[16] Yang Y, McDowell M T, Jackson A, Cha J J, Hong S S, Cui Y. New nanostructured Li_2S/silicon rechargeable battery with high specific energy. Nano Lett., 2010, 10(4): 1486-1491.

[17] Xu J, Ma J, Fan Q, Guo S, Dou S. Recent progress in the design of advanced cathode materials and battery models for high-performance lithium-X(X = O_2, S, Se, Te, I_2, Br_2)Batteries. Adv. Mater., 2017.

[18] Cui Y, Abouimrane A, Sun C J, Ren Y. Amine K, Li-Se battery: Absence of lithium polyselenides in carbonate based electrolyte. Chemical Communications, 2014, 50(42): 5576-5579.

[19] Abouimrane A, Dambournet D, Chapman K W, Chupas P J, Weng W, Amine K. A new class of lithium and sodium rechargeable batteries based on selenium and selenium-sulfur as a positive electrode. Journal of the American Chemical Society, 2012, 134(10): 4505-4508.

[20] Li Z, Yuan L, Yi Z, Liu Y, Huang Y. Confined selenium within porous carbon nanospheres as cathode for advanced Li–Se batteries. Nano Energy, 2014, 9, 229-236.

[21] Koketsu T, Paul B, Wu C, Kraehnert R, Huang Y, Strasser P. A lithium–tellurium rechargeable battery with exceptional cycling stability. J. Appl. Electrochem., 2016, 46(6): 627-633.

[22] Girishkumar G, McCloskey B, Luntz A, Swanson S, Wilcke W. Lithium− air battery: Promise and challenges. The J. Phys.Chem. Lett., 2010, 1(14): 2193-2203

[23] Zhang S, Huang Z, Wen Z, Zhang L, Jin J, Shahbazian-Yassar R, Yang J. Local lattice distortion activate metastable metal sulfide as catalyst with stable full discharge-charge capability for Li-O_2 batteries. Nano Lett., 2017, 15(6): 3518-3526.

[24] Zhang S, Wang G, Jin J, Zhang L, Wen Z, Yang J. Self-catalyzed decomposition of discharge products on the oxygen vacancy sites of MoO_3 nanosheets for low-overpotential Li-O_2 batteries. Nano Energy, 2017, 36, 186-196.

[25] Guo Z, Dong X, Yuan S, Wang Y, Xia Y. Humidity effect on electrochemical performance of Li–O_2 batteries. J. Power Sources, 2014, 264, 1-7.

[26] Liu Y, Suo L, Lin H, Yang W, Fang Y, Liu X, Wang D, Hu Y-S, Han W, Chen L. Novel approach for a high-energy-density Li–air battery: tri-dimensional growth of Li_2O_2 crystals tailored by electrolyte Li^+ ion concentrations. J.Mater. Chem. A, 2014, 2(24): 9020-9024.

[27] Li Y, Huang Y, Zhang Z, Duan D, Hao X, Liu S. Preparation and structural evolution of well aligned-carbon nanotube arrays onto conductive carbon-black layer/carbon paper substrate with enhanced discharge capacity for Li–air batteries. Chem. Eng. J., 2016, 283, 911-921.

[28] Hu X, Wang J, Li Z, Wang J, Gregory D H, Chen J. MCNTs@MnO_2 nanocomposite cathode integrated with soluble O_2-carrier Co-salen in electrolyte for high-performance Li–Air batteries. Nano Lett., 2017, 17(3): 2073-2078.

[29] Nie H, Xu C, Zhou W, Wu B, Li X, Liu T, Zhang H. Free-standing thin webs of activated carbon nanofibers by electrospinning for rechargeable Li–O_2 batteries. ACS Appl. Mater. Inter., 2016, 8(3): 1937-1942.

[30] Wu F, Xing Y, Li L, Qian J, Qu W, Wen J, Miller D, Ye Y, Chen R, Amine K. Facile synthesis of boron-doped rGO as cathode material for high energy Li–O_2 batteries. ACS Appl. Mater. Inter., 2016, 8(36): 23635-23645.

[31] Li C, Guo Z, Pang Y, Sun Y, Su X, Wang Y, Xia Y. Three-Dimensional Ordered Macroporous $FePO_4$ as High-Efficiency Catalyst for Rechargeable Li–O_2 Batteries. ACS Appl. Mater. Inter., 2016, 8(46): 31638-31645.

[32] Lai Y, Chen W, Zhang Z, Qu Y, Gan Y, Li J. Fe/Fe_3C decorated 3-D porous nitrogen-doped graphene as a cathode material for rechargeable Li–O_2 batteries. Electrochim. Acta, 2016, 191, 733-742.

[33] Cao C, Yan Y, Zhang H, Xie J, Zhang S, Pan B, Cao G, Zhao X. Controlled growth of Li_2O_2 by cocatalysis of mobile Pd and

Co_3O_4 nanowire arrays for high-performance Li–O_2 batteries. ACS Appl. Mater. Inter., 2016, 8(46): 31653-31660.

[34] Cao J, Liu S, Xie J, Zhang S, Cao G, Zhao X. Tips-bundled Pt/Co_3O_4 nanowires with directed peripheral growth of Li_2O_2 as efficient binder/carbon-free catalytic cathode for lithium–oxygen battery. ACS Catal., 2014, 5(1): 241-245.

[35] Liu G, Chen H, Xia L, Wang S, Ding L-X, Li D, Xiao K, Dai S, Wang H. Hierarchical mesoporous/macroporous perovskite $La_{0.5}Sr_{0.5}CoO_{3-x}$ Nanotubes: A bifunctional Catalyst with enhanced activity and cycle stability for rechargeable lithium oxygen batteries. ACS Appl. Mater. Inter., 2015, 7(40): 22478-22486.

[36] Jin C, Yang Z, Cao X, Lu F, Yang R. A novel bifunctional catalyst of $Ba_{0.9}Co_{0.5}Fe_{0.4}Nb_{0.1}O_{3-\delta}$ perovskite for lithium–air battery. Inter. J. Hydrogen Energy, 2014, 39(6): 2526-2530.

[37] Wang C, Li X, Xi X, Zhou W, Lai Q, Zhang H. Bimodal highly ordered mesostructure carbon with high activity for Br_2/Br^- redox couple in bromine based batteries. Nano Energy, 2016, 21, 217-227.

[38] Wang C, Li X, Xi X, Xu P, Lai Q, Zhang H. Relationship between activity and structure of carbon materials for Br_2/Br^- in zinc bromine flow batteries. RSC Adv., 2016, 6(46): 40169-40174.

[39] Wang C, Lai Q, Xu P, Zheng D, Li X, Zhang H. Cage-like porous carbon with superhigh activity and Br_2 - complex - entrapping capability for bromine-based flow batteries. Adv. Mater., 2017, 29(22): doi: 10.1002/adma.201605815.

[40] Lu, W, Yuan Z, Zhao Y, Li X, Zhang H, Vankelecom I F. High-performance porous uncharged membranes for vanadium flow battery applications created by tuning cohesive and swelling forces. Energy Environ. Sci., 2016, 9(7): 2319-2325.

[41] Zhao Y, Li M, Yuan Z, Li X, Zhang H, Vankelecom I F. Advanced charged sponge - like membrane with ultrahigh stability and selectivity for vanadium flow batteries. Adv. Funct. Mater., 2016, 26(2): 210-218.

[42] Liu J, Liu S, He Z, Han H, Chen Y. Effects of organic additives with oxygen-and nitrogen-containing functional groups on the negative electrolyte of vanadium redox flow battery. Electrochim. Acta, 2014, 130, 314-321.

[43] Shao M, Chang Q, Dodelet J-P, Chenitz R. Recent Advances in Electrocatalysts for Oxygen Reduction Reaction. Chem. Rev., 2016, 116(6): 3594-3657.

[44] Ellis M W, von Spakovsky M R, Nelson D J. Fuel cell systems: efficient, flexible energy conversion for the 21st century. Proceedings of the IEEE, 2001, 89(12): 1808-1818.

[45] Mozsgai G, Yeom J, Flachsbart B, Shannon M. In A silicon microfabricated direct formic acid fuel cell, transducers, solid-state sensors, actuators and microsystems, 12th International Conference on, 2003, IEEE: 2003, 1738-1741.

[46] Zhu Y, Ha S Y, Masel R I. High power density direct formic acid fuel cells. J. Power Sources, 2004, 130(1): 8-14.

[47] Farooque M, Maru H C. Fuel cells-the clean and efficient power generators. Proceedings of the IEEE, 2001, 89(12): 1819-1829.

[48] Colominas S, McLafferty J, Macdonald D. Electrochemical studies of sodium borohydride in alkaline aqueous solutions using a gold electrode. Electrochim. Acta, 2009, 54(13): 3575-3579.

[49] Swider-Lyons K E, Carlin R T, Rosenfeld R L, Nowak R J. In technical issues and opportunities for fuel cell development for autonomous underwater vehicles, autonomous underwater vehicles, 2002. proceedings of the 2002 workshop on, ieee: 2002, pp 61-64.

[50] Yakabe H, Sakurai T, Sobue T, Yamashita S, Hase K. In solid oxide fuel cells as promising candidates for distributed generators, industrial informatics, 2006 ieee international conference on, ieee: 2006, pp 369-374.

[51] Greeley J, Stephens I E L, Bondarenko A S, Johansson T P, Hansen H A, Jaramillo T F, Rossmeisl J, Chorkendorff I, Norskov J K. Alloys of platinum and early transition metals as oxygen reduction electrocatalysts. Nat. Chem., 2009, 1(7): 552-556.

[52] Qu K, Zheng Y, Dai S, Qiao S Z. Graphene oxide-polydopamine derived N, S-codoped carbon nanosheets as superior

bifunctional electrocatalysts for oxygen reduction and evolution. Nano Energy, 2016, 19, 373-381.

[53] Jiang W J, Gu L, Li L, Zhang Y, Zhang X, Zhang L J, Wang J Q, Hu J S, Wei Z, Wan L J. Understanding the high activity of Fe-N-C electrocatalysts in oxygen reduction: Fe/Fe_3C nanoparticles boost the activity of Fe-Nx. J. Am. Chem. Soc., 2016, 138(10): 3570-3578.

[54] Liang Y, Li Y, Wang H, Zhou J, Wang J, Regier T, Dai H. Co_3O_4 nanocrystals on graphene as a synergistic catalyst for oxygen reduction reaction. Nat. Mater., 2011, 10(10): 780-786.

[55] Liang Y, Wang H, Diao P, Chang W, Hong G, Li Y, Gong M, Xie L, Zhou J, Wang J, Regier T Z, Wei F, Dai H. Oxygen reduction electrocatalyst based on strongly coupled cobalt oxide nanocrystals and carbon nanotubes. J. Am. Chem. Soc., 2012, 134(38): 15849-15857.

[56] Jin Z, Li P, Xiao D. Enhanced electrocatalytic performance for oxygen reduction via active interfaces of layer-by-layered titanium nitride/titanium carbonitride structures. Sci. Rep., 2014, 4, 6712.

[57] Huang T, Mao S, Zhou G, Wen Z, Huang X, Ci S, Chen J. Hydrothermal synthesis of vanadium nitride and modulation of its catalytic performance for oxygen reduction reaction. Nanoscale, 2014, 6(16): 9608-9613.

[58] Chisaka M, Ando Y, Itagaki N. Activity and durability of the oxygen reduction reaction in a nitrogen-doped rutile-shell on TiN-core nanocatalysts synthesised via solution-phase combustion. J. Mater. Chem. A, 2016, 4(7): 2501-2508.

[59] Yan, Z, Cai M, Shen P K. Nanosized tungsten carbide synthesized by a novel route at low temperature for high performance electrocatalysis. Sci. Rep., 2013, 3, 1646.

[60] Xiao Y, Zhan G, Fu Z, Pan Z, Xiao C, Wu S, Chen C, Hu G, Wei Z. Robust non-carbon titanium nitride nanotubes supported Pt catalyst with enhanced catalytic activity and durability for methanol oxidation reaction. Electrochim. Acta, 2014, 141, 279-285.

[61] Huang X Q, Zhao Z P, Cao L, Chen Y, Zhu E B, Lin Z Y, Li M F, Yan A M, Zettl A, Wang Y M, Duan X F, Mueller T, Huang Y. High-performance transition metal-doped Pt_3Ni octahedra for oxygen reduction reaction. Science, 2015, 348(6240): 1230-1234.

[62] Bu L, Zhang N, Guo S, Zhang X, Li J, Yao J, Wu T, Lu G, Ma J Y, Su D, Huang X. Biaxially strained PtPb/Pt core/shell nanoplate boosts oxygen reduction catalysis. Science, 2016, 354(6318): 1410-1414.

[63] Zhang N, Bu L, Guo S, Guo J, Huang X. Screw Thread-Like Platinum-Copper Nanowires Bounded with High-Index Facets for Efficient Electrocatalysis. Nano Lett., 2016, 16(8): 5037-5043.

[64] Koenigsmann C, Wong S S, One-dimensional noble metal electrocatalysts: a promising structural paradigm for direct methanol fuel cells. Energy Environ. Sci., 2011, 4(4): 1161-1176.

[65] Li H H, Zhao S, Gong M, Cui C H, He D, Liang H W, Wu L, Yu S. H. Ultrathin PtPdTe nanowires as superior catalysts for methanol electrooxidation. Angew. Chem. Int. Ed. Engl., 2013, 52(29): 7472-7476.

[66] Ma S Y, Li H H, Hu B C, Cheng X, Fu Q Q, Yu S. H. Synthesis of Low Pt-Based Quaternary PtPdRuTe Nanotubes with Optimized Incorporation of Pd for Enhanced Electrocatalytic Activity. J. Am. Chem. Soc., 2017, 139(16): 5890-5895.

[67] Xiao Y, Fu Z, Zhan G, Pan Z, Xiao C, Wu S, Chen C, Hu G, Wei Z. Increasing Pt methanol oxidation reaction activity and durability with a titanium molybdenum nitride catalyst support. J. Power Sources, 2015, 273, 33-40.

[68] Cheng S, Wu J, Air-cathode preparation with activated carbon as catalyst, PTFE as binder and nickel foam as current collector for microbial fuel cells. Bioelectrochemistry, 2013, 92, 22-26.

[69] Liu W, Cheng S, Sun D, Huang H, Chen J, Cen K. Inhibition of microbial growth on air cathodes of single chamber microbial fuel cells by incorporating enrofloxacin into the catalyst layer. Biosens. Bioelectron., 2015, 72, 44-50.

[70] Kaur A, Kim J R, Michie I, Dinsdale R M, Guwy A J, Premier G C. Microbial fuel cell type biosensor for specific volatile

fatty acids using acclimated bacterial communities. Sustainable Environment Research, C., Biosens. Bioelectron., 2013, 47, 50-55.

[71] Cheng S, Ye Y, Ding W, Pan B. Enhancing power generation of scale-up microbial fuel cells by optimizing the leading-out terminal of anode. J. Power Sources, 2014, 248, 931-938.

[72] Shen Y, Wang M, Chang I S, Ng H Y. Effect of shear rate on the response of microbial fuel cell toxicity sensor to Cu(II). Bioresour. Technol., 2013, 136, 707-710.

[73] Di Lorenzo M, Thomson A R, Schneider K, Cameron P J, Ieropoulos I. A small-scale air-cathode microbial fuel cell for on-line monitoring of water quality. Biosens. Bioelectron., 2014, 62, 182-188.

[74] Simon P, Gogotsi Y. Materials for electrochemical capacitors. Nature Mater., 2008, 7(11): 845-854.

[75] Arico A S, Bruce P, Scrosati B, Tarascon J-M, Van Schalkwijk, W. Nanostructured materials for advanced energy conversion and storage devices. Nature Mater., 2005, 4(5): 366-377.

[76] Shao Y, El-Kady M F, Wang L J, Zhang Q, Li Y, Wang H, Mousavi M F, Kaner R. B. Graphene-based materials for flexible supercapacitors. Chem. Soc. Rev., 2015, 44(11): 3639-3665.

[77] Liu L, Niu Z, Chen J. Unconventional supercapacitors from nanocarbon-based electrode materials to device configurations. Chem. Soc. Rev., 2016, 45(15): 4340-4363.

[78] Niu Z, Liu L, Zhang L, Zhou W, Chen X, Xie S. Programmable Nanocarbon-Based Architectures for Flexible Supercapacitors. Adv. Energy Mater., 2015, 5(23): 1500677.

[79] Park S, Vosguerichian M, Bao Z. A review of fabrication and applications of carbon nanotube film-based flexible electronics. Nanoscale, 2013, 5(5): 1727-1752.

[80] Chen T, Peng H, Durstock M, Dai L. High-performance transparent and stretchable all-solid supercapacitors based on highly aligned carbon nanotube sheets. Sci. Rep., 2014, 4, 3612.

[81] Wang X, Shi G. Flexible graphene devices related to energy conversion and storage. Energy Environ. Sci., 2015, 8(3): 790-823.

[82] Xu Y, Chen C Y, Zhao Z, Lin Z, Lee C, Xu X, Wang C, Huang Y, Shakir M I, Duan X. Solution Processable Holey Graphene Oxide and Its Derived Macrostructures for High-Performance Supercapacitors. Nano Lett., 2015, 15(7): 4605-4610.

[83] Nyholm L, Nystrom G, Mihranyan A, Stromme M. Toward flexible polymer and paper-based energy storage devices. Adv. Mater., 2011, 23(33): 3751-3769.

[84] Jiang J, Li Y, Liu J, Huang X, Yuan C, Lou X W. Recent advances in metal oxide-based electrode architecture design for electrochemical energy storage. 2012, 24(38): 5166-5180.

[85] Xiong G, Meng C, Reifenberger R G, Irazoqui P P, Fisher T S. A Review of Graphene-Based Electrochemical Microsupercapacitors. Electroanalysis, 2014, 26(1): 30-51.

[86] 文春明, 温志渝, 尤政, 王晓峰, 李东玲, 尚正国. 硅基微型超级电容器三维微电极结构制备. 电子元件与材料, 2012, 31(5): 42-46.

[87] El-Kady M F, Strong V, Dubin S, Kaner R B. Laser scribing of high-performance and flexible graphene-based electrochemical capacitors. Science, 2012, 335(6074): 1326-1330.

[88] Fu Y, Cai X, Wu H, Lv Z, Hou S, Peng M, Yu X, Zou D. Fiber supercapacitors utilizing pen ink for flexible/wearable energy storage. Adv. Mater., 2012, 24(42): 5713-5718.

[89] Zhang X, Lin Z, Chen B, Sharma S, Wong C-p, Zhang W, Deng Y. Solid-state, flexible, high strength paper-based supercapacitors. J. Mater. Chem. A, 2013, 1(19): 5835-5839.

[90] Wang K, Wu H, Meng Y, Zhang Y, Wei Z. Integrated energy storage and electrochromic function in one flexible device: an energy storage smart window. Energy Environ. Sci., 2012, 5(8): 8384-8389.

[91] Cordier P, Tournilhac F, Soulié-Ziakovic C, Leibler L. Self-healing and thermoreversible rubber from supramolecular assembly. Nature, 2008, 451(7181): 977-980.

[92] Gao D, Zhang Y, Zhou Z, Cai F, Zhao X, Huang W, Li Y, Zhu J, Liu P, Yang F, Wang G, Bao X. Enhancing CO_2 Electroreduction with the Metal-Oxide Interface. J. Am. Chem. Soc., 2017, 139(16): 5652-5655.

[93] Zhang L, Wang Z, Mehio N, Jin X, Dai S. Thickness- and particle-size-dependent electrochemical reduction of carbon dioxide on thin-layer porous silver electrodes. ChemSusChem, 2016, 9(5): 428-432.

[94] Sun K, Wu L, Qin W, Zhou J, Hu Y, Jiang Z, Shen B, Wang Z. Enhanced electrochemical reduction of CO_2 to CO on Ag electrocatalysts with increased unoccupied density of states. J. Mater. Chem. A, 2016, 4(32): 12616-12623.

[95] Huang H, Jia H, Liu Z, Gao P, Zhao J, Luo Z, Yang J, Zeng J. Understanding of strain effects in the electrochemical reduction of CO_2 : ssing Pd nanostructures as an ideal latform. Angew. Chem. Int. Edit., 2017, 56(13): 3594-3598.

[96] Lei F, Liu W, Sun Y, Xu J, Liu K, Liang L, Yao T, Pan B, Wei S, Xie Y. Metallic tin quantum sheets confined in graphene toward high-efficiency carbon dioxide electroreduction. Nature Commun., 2016, 7, 12697.

[97] Zhong H, Qiu Y, Zhang T, Li X, Zhang H, Chen X. Bismuth nanodendrites as a high performance electrocatalyst for selective conversion of CO_2 to formate. J. Mater. Chem. A, 2016, 4(36): 13746-13753.

[98] Xie J-F, Huang Y-X, Li W-W, Song X-N, Xiong L, Yu H-Q. Efficient electrochemical CO_2 reduction on a unique chrysanthemum-like Cu nanoflower electrode and direct observation of carbon deposite. Electrochim. Acta, 2014, 139, 137-144.

[99] Wang Z, Yang G, Zhang Z, Jin M, Yin Y. Selectivity on etching: creation of high-energy facets on copper nanocrystals for CO_2 electrochemical reduction. ACS Nano, 2016, 10(4): 4559-4564.

[100] Gao S, Lin Y, Jiao X, Sun Y, Luo Q, Zhang W, Li D, Yang J, Xie Y. Partially oxidized atomic cobalt layers for carbon dioxide electroreduction to liquid fuel. Nature, 2016, 529(7584): 68-71.

[101] Zhao C, Wang J, Goodenough J B. Comparison of electrocatalytic reduction of CO_2 to HCOOH with different tin oxides on carbon nanotubes. Electrochem. Commun., 2016, 65, 9-13.

[102] Zhang R, Lv W, Lei L. Role of the oxide layer on Sn electrode in electrochemical reduction of CO_2 to formate. Appl. Surf. Sci., 2015, 356, 24-29.

[103] Ke F-S, Liu X-C, Wu J, Sharma P P, Zhou Z-Y, Qiao J, Zhou X-D. Selective formation of C2 products from the electrochemical conversion of CO_2 on CuO-derived copper electrodes comprised of nanoporous ribbon arrays. Catal. Today, 2017, 288, 18-23.

[104] Zhao K, Liu Y, Quan X, Chen S, Yu H. CO_2 Electroreduction at low overpotential on oxide-derived Cu/carbons fabricated from metal organic framework. ACS Appl. Mater. Inter., 2017, 9(6): 5302-5311.

[105] Yin Z, Gao D, Yao S, Zhao B, Cai F, Lin L, Tang P, Zhai P, Wang G, Ma D, Bao X. Highly selective palladium-copper bimetallic electrocatalysts for the electrochemical reduction of CO_2 to CO. Nano Energy, 2016, 27, 35-43.

[106] Yang H-P, Qin S, Yue Y-N, Liu L, Wang H, Lu J-X. Entrapment of a pyridine derivative within a copper–palladium alloy: a bifunctional catalyst for electrochemical reduction of CO_2 to alcohols with excellent selectivity and reusability. Catal. Sci. Technol., 2016, 6(17): 6490-6494.

[107] Li Q, Fu J, Zhu W, Chen Z, Shen B, Wu L, Xi Z, Wang T, Lu G, Zhu J J, Sun S. Tuning Sn-catalysis for electrochemical reduction of CO_2 to CO via the core/shell Cu/SnO_2 structure. J. Am. Chem. Soc., 2017, 139(12): 4290-4293.

[108] Chang Z, Huo S, Zhang W, Fang J, Wang H. The Tunable and highly selective reduction products on Ag@Cu bimetallic

catalysts toward CO_2 electrochemical reduction reaction. J. Phys. Chem. C, 2017, 121(21): 11368-11379.

[109] Sun X, Zhu Q, Kang X, Liu H, Qian Q, Zhang Z, Han B. Molybdenum-bismuth bimetallic chalcogenide nanosheets for highly efficient electrocatalytic reduction of carbon dioxide to methanol. Angew. Chem. Int. Edit., 2016, 55(23): 6771-6775.

[110] Zhao S, Guo S, Zhu C, Gao J, Li H, Huang H, Liu Y, Kang Z. Achieving electroreduction of CO_2 to CH_3OH with high selectivity using a pyrite–nickel sulfide nanocomposite. RSC Adv., 2017, 7(3): 1376-1381.

[111] Wang H, Chen Y, Hou X, Ma C, Tan T. Nitrogen-doped graphenes as efficient electrocatalysts for the selective reduction of carbon dioxide to formate in aqueous solution. Green Chem., 2016, 18(11): 3250-3256.

[112] Liu Y, Chen S, Quan X, Yu H. Efficient electrochemical reduction of carbon dioxide to acetate on Nitrogen-doped nanodiamond. J. Am. Chem. Soc., 2015, 137(36): 11631-11636.

[113] Zhang X, Wu Z, Zhang X, Li L, Li Y, Xu H, Li X, Yu X, Zhang Z, Liang Y, Wang H. Highly selective and active CO_2 reduction electrocatalysts based on cobalt phthalocyanine/carbon nanotube hybrid structures. Nature Commun., 2017, 8, 14675.

[114] Fu Y, Li Y, Zhang X, Liu Y, Zhou X, Qiao J. Electrochemical CO_2 reduction to formic acid on crystalline SnO_2 nanosphere catalyst with high selectivity and stability. Chinese J. Catal., 2016, 37(7): 1081-1088.

[115] Zhu Q, Ma J, Kang X, Sun X, Liu H, Hu J, Liu Z, Han B. Efficient reduction of CO_2 into formic acid on a lead or Tin electrode using an ionic liquid catholyte mixture. Angew. Chem. Int. Edit., 2016, 55(31): 9012-9016.

[116] Li Q-y, Shi F, Shen F-x, Jia Y-j, Song N, Hu Y-q, Dai Y-n, Shi J. Electrochemical reduction of CO_2 into CO in N-methyl pyrrolidone/tetrabutylammonium perchlorate in two-compartment electrolysis cell. Electroanalytical Chem., 2017, 785, 229-234.

[117] Sapountzi F M, Gracia J M, Weststrate C J, Fredriksson, H O A, Niemantsverdriet J W. Electrocatalysts for the generation of hydrogen, oxygen and synthesis gas. Progress in Energy and Combustion Science, 2017, 58, 1-35.

[118] Huang X, Zeng Z, Bao S, Wang M, Qi X, Fan Z, Zhang H. Solution-phase epitaxial growth of noble metal nanostructures on dispersible single-layer molybdenum disulfide nanosheets. 2013, 4, 1444.

[119] Dolganov A V, Tanaseichuk B S, Moiseeva D N, Yurova V Y, Sakanyan J R, Shmelkova N S, Lobanov V V. Acridinium salts as metal-free electrocatalyst for hydrogen evolution reaction. Electrochem. Commun., 2016, 68, 59-61.

[120] Kibsgaard J, Chen Z, Reinecke B N, Jaramillo T F. Engineering the surface structure of MoS_2 to preferentially expose active edge sites for electrocatalysis. Nat. Mater, 2012, 11(11): 963-969.

[121] Zhang L, Li N, Gao F, Hou L, Xu Z, Insulin amyloid fibrils: an excellent platform for controlled synthesis of ultrathin superlong platinum nanowires with high electrocatalytic activity. J. Am. Chem. Soc., 2012, 134(28): 11326-11329.

[122] Mahoney L, Peng R, Wu C-M, Baltrusaitis J, Koodali R T. Solar simulated hydrogen evolution using cobalt oxide nanoclusters deposited on titanium dioxide mesoporous materials prepared by evaporation induced self-assembly process. Inter. J. Hydrogen Energy, 2015, 40(34): 10795-10806.

[123] Jiang P, Liu Q, Ge C, Cui W, Pu Z, Asiri A M, Sun X. CoP nanostructures with different morphologies: synthesis, characterization and a study of their electrocatalytic performance toward the hydrogen evolution reaction. J. Mater. Chem. A, 2014, 2(35): 14634-14640.

[124] Tian J, Liu Q, Asiri A M, Sun X. Self-supported nanoporous cobalt phosphide nanowire arrays: an efficient 3D yydrogen-evolving cathode over the wide range of pH 0–14. J. Am. Chem. Soc., 2014, 136(21): 7587-7590.

[125] Marshall A T, Sunde S, Tsypkin M, Tunold R. Performance of a PEM water electrolysis cell using $IrxRuyTazO_2$ electrocatalysts for the oxygen evolution electrode. Inter. J. Hydrogen Energy, 2007, 32(13): 2320-2324.

[126] Kadakia K, Datta M K, Velikokhatnyi O I, Jampani P, Park S K, Chung S J, Kumta P N. High performance fluorine doped(Sn, Ru)O_2 oxygen evolution reaction electro-catalysts for proton exchange membrane based water electrolysis. J. Power Sources, 2014, 245, 362-370.

[127] Marshall A, Børresen B, Hagen G, Tsypkin M, Tunold R. Preparation and characterisation of nanocrystalline $IrxSn_{1-x}O_2$ electrocatalytic powders. Mater. Chem. and Phys., 2005, 94(2): 226-232.

[128] González-Huerta R G, Ramos-Sánchez G, Balbuena P B. Oxygen evolution in Co-doped RuO_2 and IrO_2: Experimental and theoretical insights to diminish electrolysis overpotential. J. Power Sources, 2014, 268, 69-76.

[129] Zeng Y, Guo X, Shao Z, Yu H, Song W, Wang Z, Zhang H, Yi B. A cost-effective nanoporous ultrathin film electrode based on nanoporous gold/IrO_2 composite for proton exchange membrane water electrolysis. J. Power Sources, 2017, 342, 947-955.

[130] DeSario P A, Chervin C N, Nelson E S, Sassin M B, Rolison D R. Competitive oxygen evolution in acid electrolyte catalyzed at technologically relevant electrodes painted with nanoscale RuO_2. ACS Appl. Mater. Inter., 2017, 9(3): 2387-2395.

[131] Ma L, Sui S, Zhai Y. Preparation and characterization of Ir/TiC catalyst for oxygen evolution. J. Power Sources, 2008, 177(2): 470-477.

[132] Puthiyapura V K, Pasupathi S, Su H, Liu X, Pollet B, Scott K. Investigation of supported IrO_2 as electrocatalyst for the oxygen evolution reaction in proton exchange membrane water electrolyser. Inter. J. Hydrogen Energy, 2014, 39(5): 1905-1913.

[133] Jin Y, Wang H, Li J, Yue X, Han Y, Shen P K, Cui Y. Porous MoO_2 nanosheets as non-noble bifunctional electrocatalysts for overall water splitting. Adv. Mater., 2016, 28(19): 3785-3790.

[134] Lyu F, Bai Y, Li Z, Xu W, Wang Q, Mao J, Wang L, Zhang X, Yin Y. Self-templated fabrication of CoO-MoO_2 nanocages for enhanced oxygen evolution. Adv. Funct. Mater., 2017, 1702324.

[135] Xu J, Wei X-K, Costa J D, Lado J L, Owens-Baird B, Gonçalves L P L, Fernandes S P S, Heggen M, Petrovykh D Y, Dunin-Borkowski R E, Kovnir K, Kolen'ko Y V. Interface engineering in nanostructured nickel phosphide catalyst for efficient and stable water oxidation. ACS Catal., 2017, 5450-5455.

[136] Hu F, Zhu S, Chen S, Li Y, Ma L, Wu T, Zhang Y, Wang C, Liu C, Yang X, Song L, Yang X, Xiong Y. Amorphous metallic NiFeP: a conductive bulk material achieving high activity for oxygen evolution reaction in both alkaline and acidic media. Adv. Mater., 2017.

[137] Zhou H, Yu F, Sun J, He R, Chen S, Chu C W, Ren Z. Highly active catalyst derived from a 3D foam of $Fe(PO_3)_2/Ni_2P$ for extremely efficient water oxidation. P. Natl. Acad. Sci. USA, 2017.

[138] Jiang W J, Niu S, Tang T, Zhang Q H, Liu X Z, Zhang Y, Chen Y Y, Li J H, Gu L, Wan L J, Hu J S. Crystallinity-modulated electrocatalytic activity of a nickel(II)borate thin layer on Ni_3B for efficient water oxidation. Angew. Chem. Int. Edit., 2017, 56(23): 6572-6577.

[139] Chen R, Yang C, Cai W, Wang H-Y, Miao J, Zhang L, Chen S, Liu B. Use of Platinum as the Counter Electrode to Study the Activity of Nonprecious Metal Catalysts for the Hydrogen Evolution Reaction. ACS Energ. Lett., 2017, 2(5): 1070-1075.

[140] Giffin G A, Galbiati S, Walter M, Aniol K, Ellwein C, Kerres J, Zeis R. Interplay between structure and properties in acid-base blend PBI-based membranes for HT-PEM fuel cells. J. Membrane Sci., 2017, 535, 122-131.

[141] Kim D J, Lee B-N, Nam S Y. Characterization of highly sulfonated PEEK based membrane for the fuel cell application. Int. J. Hydrogen Energy, 2017.

[142] 姜国平, 汤占岐, 李瑞秋. 管式加热炉间壁式空气预热器现状分析及方案优化. 工业安全与环保, 2015, (02): 59-61.

[143] 张青山, 王砚林, 师永晓. 余热饱和蒸汽在顶吹炉余热发电中的应用. 中国有色冶金, 2015, 44(2): 65-68.

[144] 陈少杰, 陈光明. 渔船动力余热制冷技术. 制冷学报, 2014, 35(6): 28-34.

[145] 邱丽霞, 郝艳红. 电厂低温热与吸收式热泵耦合供热系统研究. 热力发电, 2015, (10): 20-24.

[146] 张彦民. 复合式环模生物质燃料颗粒成型机的研究. 农业开发与装备, 2017, (4): 66-67.

[147] 鲍振博, 刘长安, 解光传, 彭锦星, 靳登超, 刘玉乐. 多体式秸秆生物质气化炉的设计. 农机化研究, 2017, 3, 051.

[148] 郑漳华. 储能技术在电网中的应用发展. 国家电网, 2016, (5): 100-101.

[149] 李琼慧, 王彩霞, 张静, 宁娜. 适用于电网的先进大容量储能技术发展路线图. 储能科学与技术, 2017, 6(1): 141-146.

[150] 谭嫔, 代焕利, 谭新玉, 肖婷, 李辉, 吴栋, 陈成. 混合储能系统在微电网中的应用. 通信电源技术, 2016, (2016 年 04): 109-111.

作者：夏宝玉[1]，成少安[2]

[1]华中科技大学化学与化工学院，[2]浙江大学能源工程学院

第 25 章　环境计算化学与预测毒理学

本章导读

为理解、评价并降低化学污染物对地球生态环境造成的负面影响，需要认识和评估化学物质污染现状，解析污染物来源，评价并预测化学物质对人体与生态健康的影响，进而发展对环境与生态友好的产品，预防、控制和修复化学物质的污染。仅采用实验测试的手段，难以快速获取化学物质特性、行为与效应数据，难以揭示污染物环境迁移转化过程的微观机制，更难以对种类众多且数目不断增加的化学品进行风险预测与管理。计算模拟方法则可以突破实验测试的局限，尤其在环境质量评价、化学品生态风险预测等方面，具有重要应用前景。计算模拟方法和环境化学交叉融合，形成了环境计算化学与预测毒理学这门前沿交叉方向。本章结合案例对环境计算化学与预测毒理学的研究进展与前瞻进行介绍。

关键词

环境计算化学，预测（计算）毒理学，化学品风险评价，化学品，毒理学效应

1 引　言

早在20世纪60年代，Rachel Carson就在《寂静的春天》中，指出了一些人工合成化学品[如有机氯农药(OCPs)和多氯联苯(PCBs)]污染对地球生态环境造成的危害，促进了人类环保意识的觉醒。然而，进入21世纪，人类开展环境保护工作50多年了，但化学物质的环境污染问题仍未得到彻底解决，且已发展为联合国环境规划署（UNEP）等组织和机构重点关注的、影响人类在地球环境能否持续生存下去的重大环境问题之一。如Rockström等[1]所强调，化学品污染是当今影响人类在地球这个行星上能否持久生存下去的9个关键要素之一（图25-1）。大量监测数据表明，持久性有机污染物（POPs）被发现在全球环境广泛分布[2]，一些POPs在全球环境中的浓度水平仍在增加[3-5]，一些环境新型有机污染物（EOPs）不断被检出[6-10]。环境健康和流行病学的相关研究表明，一些POPs和EOPs是环境内分泌干扰物，能够对人体或生态健康造成深远影响；人类70%~90%的疾病风险与环境因素有关[11]，致癌物、紫外辐射、电离辐射等是导致人类癌症发生的主要原因[12]。

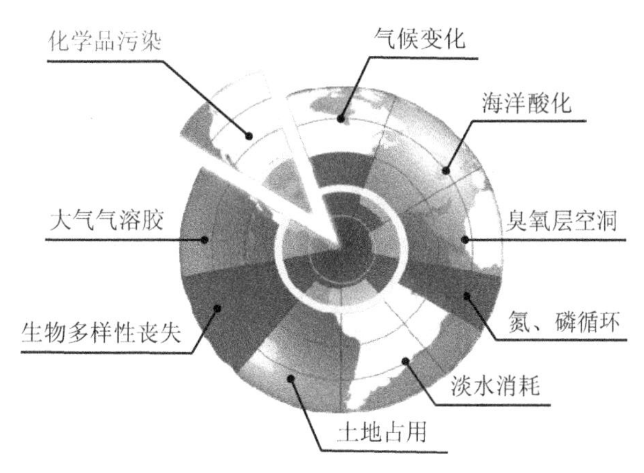

图25-1　人类可持续发展的行星边界

绿色区域代表安全操作空间（safe operating space），假如人类活动对地球的环境以及生态系统造成的影响维持此范围内，人类就能够继续繁衍和发展；黄线代表边界条件，以及安全操作空间的上界；红色区域则代表超出边界。改编自Steffen等和Rockström等[1,13]

化学污染物主要有两方面的来源：天然来源和人为来源。火山喷发、森林火灾等自然过程，均可产生对健康有害的化学污染物[例如多环芳烃（PAHs）、卤代芳烃、二噁英类污染物（PCDD/Fs）等]；但化学污染物更主要的来源是人类的活动。在人为来源（anthropogenic）化学污染物中，有些是地壳中已经存在的物质，由于人类活动改变了这些物质存在形态和空间位置，从而造成对环境的污染，典型的例子是汞及有机汞类污染物；有些是人类生产生活过程中排放的副产物，例如 PAHs 可在燃煤的过程中产生，PCDD/Fs 可以在垃圾及废弃物不充分燃烧时产生，有一些物质是工业生产的副产物；还有些是人类有意生产、合成且用作产品和商品所使用的化学物质（化学品），例如有机氯农药、溴代阻燃剂、有机磷阻燃剂、全氟化合物等。与人类活动有关的化学物质，绝大部分是合成（synthetic）化学品，即如果没有人类活动，这些化学物质在地球上原本是不存在的。

美国化学文摘社 CAS 登记的化学物质现在已经达到 1.3 亿多种，并且正在以每天 3 万种的速度增加，这其中绝大多数是合成化学品。据统计，欧盟 REACH 法规预注册的在市场上使用的化学品达 145299 种（https://echa.europa.eu/information-on-chemicals/pre-registered-substances），我国环保部化学品登记中心注册的化学品达 45643 种（2013 年中国现有化学物质名录登记了 45612 种，2016 年新增 31 种）。种类众多的合成化学品，在其生产、运输、储存、使用、废物处置等过程中，可能排放、释放或泄漏到环境中，影响环境质量，造成对环境的污染。

环境化学就是一门研究有害化学物质（也包括天然化学物质）在环境介质中的存在、化学特性、行为与效应，以及污染控制的化学原理和方法的科学。所谓化学物质的效应，一方面指其能够改变地球无机环境的组成或性质（例如，消耗平流层臭氧、引起全球气候变暖的温室效应）；另一方面，也指其直接或间接地影响生态系统中各种生物的生存和繁衍（例如，降低生物多样性）。现在人们已经广泛认识到，合成化学品是导致全球生态环境变化的主要因素[14]。

由于合成化学物质的污染，原本不存在于生物体内的物质通过各种暴露途径进入到生物体内，这些物质对生物体来说即为外源物质（xenobiotics）。毒理学就是一门主要研究外源物质（尤其化学物质）对生物系统有害作用的学科。Paracelsus 曾指出，所有的物质都是有毒的，只是剂量大小的差别。为了理解、评价并降低环境化学物质对地球生态环境造成的负面影响，需要认识和评估化学物质污染的现状，解析污染物的来源，评价和预测化学物质对人体与生态健康的影响，发展对环境与生态友好的产品（尤其化学品），预防、控制和修复化学物质的污染。面向这些需求，仅采用实验测试难以揭示地球生态环境系统的整体特性，难以快速获取化学物质特性、行为与效应数据，难以揭示污染物环境迁移转化过程（涉及过渡态等瞬态物种的过程）的微观机制，难以对种类众多且数目不断增加的化学品进行风险预测与管理。计算模拟方法，则可以突破实验测试的局限，尤其在环境质量评价与预测、污染物源解析、化学品生态风险预测与管理等方面，具有重要的应用前景。环境化学学科下，相关计算模拟方法的原理、方法及应用方面的研究，推动了环境计算化学与预测毒理学这个分支学科的形成。

环境计算化学属于环境化学与化学信息学（化学计量学）、计算化学（量子力学、分子力学）等学科的交叉，主要研究化学污染物的形成机制、源解析、多介质迁移转化归趋（环境分布）、毒性机制及毒性效应预测等。预测毒理学也称为计算毒理学，主要基于计算化学、化学/生物信息学和系统生物学原理，通过构建计算机（*in silico*）模型，来实现化学品环境暴露、危害性与风险的高效模拟预测[15]。由上述定义可知，环境计算化学与计算（预测）毒理学的学科内涵有很大的交叉，环境计算化学更完整地体现学科属性，具有更大的外延；计算（预测）毒理学则更凸显化学品环境危害与风险高效模拟预测的现实需求特性。例如，各种基于受体的多元统计分析源解析方法（例如非负约束因子分析[16]、正定矩阵因子分解模型[17]）研究、基于量子化学模拟的燃烧过程中污染物的形成机制研究等[18-20]，均属于环境计算化学的研究范畴，但不属于计算（预测）毒理学研究范畴。

根据所解决的具体问题，环境计算化学和预测毒理学中常用的方法可分为两类：①应用于定量构效关系（QSAR）模型构建及环境多因素分析的各种数理统计和机器学习方法，例如各种化学信息/计量学算法和多种机器学习算法；②用于环境污染物的转化机制研究、小分子和生物大分子相互作用机制分析、污染物分子结构表征的分子模拟方法，如各种量子化学计算方法、分子力学方法以及分子动力学模拟和结合自由能计算等方法。

QSAR 是环境计算化学与预测毒理学研究的一个核心内容[21]，近年来，QSAR 模型构建更注重基于机理的分子结构描述参数的引入、机理的可解释性、模型的应用域、外部预测能力的评价等。QSAR 模型的预测能力提高策略和方法的研究，则是核心问题。国内外研究者从化合物活性构象的选择、新分子结构描述符的计算、新建模方法发展以及对现有建模方法的改进等方面开展了大量研究，有助于提高 QSAR 模型的可靠性、拓展模型的实际应用范围。QSAR 方法在环境化学和计算毒理学领域的典型应用有：纳米材料的结构与其毒性之间的定量关系[22-26]，有机污染物与羟基自由基、硝酸根自由基和硫酸根自由基以及臭氧反应的速率常数的预测[27-34]，污染物物理化学性质[35-38]，生物富集因子[39-44]，环境内分泌干扰物对核受体如雌激素受体、雄激素受体、甲状腺素受体和孕酮受体等活性的 QSAR 等[45-51]。

2016 年，在发表于中国科学的综述《面向化学品风险评价的计算（预测）毒理学》[21]中，我们重点介绍了计算毒理学的概念、缘起、计算毒理学在辅助化学品风险评价以及剖析毒性机制时常用的策略与模型、在化学品风险管理领域的外延等，本章在此方面不再赘述。2017 年，期刊 *Environ. Sci.: Process & Impacts* 亦出版了主题为 *QSARs and computational chemistry methods in environmental chemical sciences* 的专刊[52-56]。本章重点介绍国内在环境计算化学领域近 5 年的新进展，适当顾及稍早及国外的相关研究工作，并就相关领域的发展提出展望。

近年来，我国环境化学学科取得很大的发展。但相比而言，环境计算化学作为新兴的研究领域，其发展仍很薄弱，学科体系不成熟，相关研究队伍小。但愿本章抛砖引玉，得到相关同行的不吝赐教，助推国内相关领域的发展。盼望环境计算化学像“站在海岸遥望海中已经看得见桅杆尖头了的航船、立于高山之巅远看东方已见光芒四射喷薄欲出的一轮朝日”那样，得到蓬勃发展。

2　有毒有机污染物形成机理的计算模拟

2.1　气相条件下卤代二噁英/呋喃的形成机理

卤代二噁英/呋喃（PCDD/Fs）是目前世界上已知毒性最强的、非人工合成的环境污染物，国际癌症研究中心将其列为人类一级致癌物。卤代二噁英[多卤代二苯并二噁英与多卤代二苯并呋喃的总称，简写为 PXDD/Fs（X = Cl，Br 或 F）]是二噁英的一类，是一类典型的持久性有毒有机污染物，已被列入《关于持久性污染物的斯德哥尔摩公约》POPs 名单。通常，PCDD/Fs 的形成主要有两种机理，一种是均相（气相）反应机理，反应温度为 500~800℃，其形成的重要前体物包括卤酚、卤苯、多卤联苯和卤代苯醚等，其中卤酚是形成 PCDD/Fs 的最主要的前体物。主要的反应过程是这些前体物通过分子的异构或重排。另一种是低温非均相反应机理，反应温度为 200~400℃，这类反应的前体物可能和气相反应的相同，也可能是飞灰中的碳等元素。由于 PCDD/Fs 的重要性，许多研究者一直致力于这一领域的研究，并且也已从实验室上对 PCDD/Fs 的形成过程开展了许多研究。但由于 PCDD/Fs 的毒性极强，形成机理复杂，且在实验过程中对短寿命中间体的检测比较困难，因此这一领域的研究仍然存在着许多悬而未决的问题。近年来，人们开始采用理论计算的方法展开了对 PCDD/Fs 形成机理的模拟研究。

1）氯代二噁英/呋喃的形成机理

研究发现，在所有的 PCDD/Fs 前体物中，氯酚（CP）与 PCDD/Fs 的结构最相似，也最容易生成 PCDD/Fs[57,58]。高温条件下，氯代苯氧自由基二聚化是形成 PCDD/Fs 的重要途径，由 CP 形成氯代苯氧自由基是 PCDD/Fs 形成的初始步骤，因此 CP 中 O—H 键的强度及反应活性对 PCDD/Fs 的形成有重要影响。通常，CP 主要与·OH，·H，·O(^{3}P)和·Cl 等活性自由基反应生成氯代苯氧自由基。由于氯取代数目和位置不同，CP 共有 19 种同系物。通过对这 19 种 CP 与·OH 及·H 的反应机理的研究发现，CP 中氯原子取代位置对 CP 中 O—H 键的强度及反应活性有重要影响，当 CP 中邻位上有氯原子取代时，O—H 键的强度增大，反应活性减弱[59]。

以氯代苯氧自由基等为模型化合物，张庆竹和王文兴等[57,59-65]采用量子化学计算的方法，系统研究了前体物交叉形成 PCDD/Fs 的机理，并对这些不同的耦合反应进行了比较。研究表明，氯代苯氧自由基二聚化是形成 PCDD/Fs 的重要途径，PCDDs 的形成通常包含以下几个基元步骤：氯代苯氧自由基二聚化（氧-碳原子耦合）、Cl 或氢抽提、闭环和分子内消除 Cl 或 H 原子。其中闭环是控速步骤，但是分子内消除 H 需要的势垒较高并且强吸热，而能量上可行的反应路线是通过分子内消除 Cl 原子，恰恰消除的这个 Cl 原子必须是 CP 中邻位上的 Cl 原子，这一发现成功解释了以前实验中发现的一个现象[66]，即只有邻位上有 Cl 原子取代的 CP 才能形成 PCDDs。

研究还发现，CP 中氯原子的取代方式对 PCDDs 的形成机理有重要影响，随着氯原子取代数目的增多，氯代苯氧自由基二聚化（氧-碳原子耦合）过程放热量越来越少，表明 CP 中氯原子取代数目的增多抑制了氯代苯氧自由基的二聚化，也就是抑制了 PCDDs 的形成，这一研究结果与实验现象[67]符合得很好——实验研究表明 CP 中氯原子取代数目的增多，PCDDs 的产率显著减低。同样，PCDFs 的形成也包含同样的五个基元过程，其中闭环是控速步骤。通过对一系列 CP 耦合反应的计算结果比较，发现 PCDDs 形成的难易顺序为：2-氯代苯氧自由基 + 2-氯代苯氧自由基 > 2-氯代苯氧自由基 + 2, 4-二氯代苯氧自由基 > 2, 4-二氯代苯氧自由基 + 2, 4-二氯代苯氧自由基 > 2, 4-二氯代苯氧自由基 + 2, 4, 6-三氯代苯氧自由基 > 2, 4, 6-三氯代苯氧自由基 + 2, 4, 6-三氯代苯氧自由基。这一结果表明，Cl 取代数目少的 CP 更容易形成 PCDDs，CP 中氯原子取代数目的增多，抑制了 PCDFs 的形成。同理，PCDFs 形成的难易顺序为：苯氧自由基 + 2-氯代苯氧自由基 > 2-氯代苯氧自由基 + 2, 4-二氯代苯氧自由基 > 2, 4-二氯代苯氧自由基 + 2, 4-二氯代苯氧自由基 > 2, 4-二氯代苯氧自由基 + 2, 4, 6-三氯代苯氧自由基。可见低氯代酚更容易形成 PCDFs，这与实验观测结果一致。

在张庆竹等的研究基础上，张冬菊和张爱茜等[68-70]分别又以 2-氯酚作为模型化合物研究了 PCDD/Fs 的形成。研究结果表明，除氯苯氧基之外，氯苯基和氯代 α-酮卡宾也很可能形成 PCDD/Fs。它们的自耦合和交叉耦合反应表明 1-氯代二噁英、1, 6-二氯代二噁英、4, 6-二氯代二苯并呋喃及 4-氯代二苯并呋喃的形成无论在热力学还是动力学上都是可行的。而张爱茜等[69]的研究结果表明 CP 与 2-氯苯氧基的耦合反应更容易发生，并且生成一种典型的 PCDDs 前体物质——*o*-氯代苯氧基苯酚，CP 与不同取代位置的 2-氯代苯氧自由基的缩合反应只会生成 PCDFs，在某种程度来说，这就可以解释在现实环境中高的 PCDF∶PCDD 比这一现象。这项研究也同时提出了自由基/分子耦合模式，填补了对现有 PCDD/Fs 形成的认识空缺，并完善对现有环境中 PCDD/Fs 的形成特征的理解。这一结果丰富了对于氯酚前体形成 PCDD/Fs 的机理的认识。

2）溴代二噁英/呋喃的形成机理

多溴代二噁英/呋喃（PBDD/Fs）是 PCDD/Fs 的类似物。因此，它们具有相似的物理化学性质和毒性效应。近年来，随着溴系阻燃剂（BFRs）使用的大量增多，PBDD/Fs 已起了广泛关注。张庆竹和王文兴等[60,71,72]研究发现，溴代苯氧自由基-自由基反应机理是形成 PBDDs 的重要途径。PBDD/Fs 的形成与 PCDD/Fs 的形成类似，闭环反应是控速步骤。由溴酚（BP）形成 PBDDs 比由 CP 形成 PCDDs 更容易。

另外，PBDD/Fs 还可以在多溴联苯醚类化合物（PBDEs）的降解过程中形成。醚键邻位碳原子之间的反应会生成 PBDD/Fs。在高温下 PBDEs 可以发生自分解反应，脱去氢原子，生成稳定的 2, 3, 6, 8-四溴二苯并呋喃。纵观全部反应，五溴取代的中间体的反应速率较高，且转化为同级溴取代的二苯并呋喃的路径为优势路径。低温下的主要产物为四溴联苯醚，高温下主要产物为 PBDFs。邻位活跃的自由基型中间体在惰性空气中主要生成 PBDFs，但是在富氧条件下，PBDDs 也是其重要产物[73]。

3）氟代和溴氯代二噁英/呋喃的形成机理

通过研究氟酚（FP）与高活性的·H 自由基反应形成氟代苯氧自由基的反应机理，描述了反应过程中涉及的短寿命中间体的结构和能量信息，并对比了不同 FP 生成氟代二噁英（PFDD/Fs）的难易程度。结果显示，高强度 C—F 键对 PFDD/Fs 的形成起关键作用，以 FP 为前体物生成 PFDD/Fs 比 CP 生成 PCDD/Fs 和 BP 生成 PBDD/Fs 难。由 2-FP 生成 PFDDs 和 PFDFs 的反应相互竞争，以 2-FP 为前体物经氟代苯氧自由基耦合生成 PFDD/Fs 的主要产物为 1, 6-二氟代二噁英，1, 9-二氟代二噁英和 4, 6-二氟代呋喃。以 2-FP 和苯酚为前体物生成 PFDD/Fs 的主要产物为 1-氟代二噁英，由 2, 4-二溴酚为前体物生成 PFDD/Fs 的主要产物为 1, 3, 6, 8-四氟代二噁英和 1, 3, 7, 9-四氟代二噁英[74]。

多溴氯代二噁英/呋喃（PBCDD/Fs）也是 PCDD/Fs 的类似物，其中的部分氯原子被溴原子取代。已有的研究表明 PBCDD/Fs 可能具有更强的毒性[75,76]。张庆竹和王文兴等[60]研究了 2-氯苯氧基自由基与 2-溴苯氧基自由基、2, 4-二氯苯氧基自由基与 2, 4-二溴苯氧基自由基，2, 4, 6 -三氯苯氧基自由基和 2, 4, 6-三溴苯氧基自由基的交叉缩合反应形成均匀气相的混合 PBCDD/Fs。研究表明，卤代酚类的取代模式不仅决定了生成的 PBCDD/Fs 的取代模式，而且对 PBCDD/Fs 的形成机制有着重要影响，特别是卤代苯氧基自由基的耦合。2-氯苯氧基自由基和 2-氯溴苯氧基（2-CBR）的交叉缩合比 2-氯苯氧基自由基的自缩合生成潜在增加的 1-氯代二噁英，而 2-氯苯氧基自由基和 2-CBR 的交叉缩合相对于 2-溴苯氧基自由基（2-BPRs）的自缩合生成潜在减少的 1-溴代二噁英。

受二噁英高毒性及实验条件的制约，实验上很难获得二噁英形成过程中所包含的基元反应的速率常数、指前因子及活化能等动力学参数。张庆竹和王文兴等在从量子化学计算中得到的热力学参数的基础上采用变分过渡态理论计算了所有卤代二噁英形成过程包含的基元反应的速率常数，并拟合了阿伦尼乌斯（Arrhenius）方程，得到了速率常数、指前因子和活化能的直接关系，填补了实验空白，为模式研究提供了基本输入参数。

为了检验计算结果的可靠性，张庆竹等[57]将计算结果与现有的实验值做了比较，在 MPWB1K/6-31+G（d，p）理论水平下计算得到的苯酚、2-氯苯、二噁英及 1-氯代二噁英的构型参数及振动频率与实验值符合得很好，构型参数的最大偏差小于 1.5%，振动频率的最大偏差小于 8.0%[67]。计算得到的反应 2, 4-二苯酚 ＋2-苯酚 $\longrightarrow$ 1, 3-二氯代二噁英 ＋H_2＋HCl、苯 ＋2-苯酚 $\longrightarrow$ 二噁英+H_2＋HCl、苯+ 2-苯酚 $\longrightarrow$ 1-氯代二噁英 ＋$2H_2$、2-溴酚 ＋2-溴酚 $\longrightarrow$ 二噁英 ＋H_2＋Br_2 和苯酚 ＋·OH $\longrightarrow$ 苯氧自由基 ＋H_2O 的反应热分别为 21.88 kcal/mol、17.79 kcal/mol、35.18 kcal/mol、29.22 kcal/mol 和–28.74 kcal/mol[57]，而它们的实验值分别为 22.78 kcal/mol[77-79]、18.12 kcal/mol、33.87 kcal/mol[80-82]、30.52 kcal/mol 和–31.25 kcal/mol[83,84]。1000 K 下，苯酚 ＋·H $\longrightarrow$ 苯氧自由基 ＋H_2 反应的正则变分过渡态理论/小曲率隧道效应（CVT/SCT）速率常数是 1.68×10^{-13} $cm^3/(molecule·s)$，实验值为 3.72×10^{-13} $cm^3/(molecule·s)$[85]。很显然，这些计算结果均与实验值吻合的很好。

2.2　气相条件下硝基多环芳烃的形成机理

尽管大气中硝基多环芳烃（nitro-PAHs）的浓度比多环芳烃（PAHs）低一到两个数量级，其致癌致突变性却是 PAHs 的十倍。近年来，国内外已经对 nitro-PAHs 开展了研究，并证实大气中 PAHs 的氧化是大气中 nitro-PAHs 的主要来源之一。张庆竹和王文兴等[86,87]采用量子化学计算方法，系统地研究了几种不同苯环数目的典型芳烃，如苯、萘、蒽、荧蒽、芴、苊、苊烯、菲、芘和硝基萘等在大气中由高活性的·OH 及 NO_3 自由基引发的氧化及相应 nitro-PAHs 的形成过程，并讨论了 PAHs 经气相生成 nitro-PAHs 的影响因素。

研究发现，气态 PAHs 在大气中可被 OH/NO_2 及 NO_3/NO_2 氧化生成致癌性更强的 nitro-PAHs。通常认为形成机理包含三个基元过程：① PAHs 中的 C==C 双键与·OH 或 NO_3 发生加成反应，生成 OH-PAH 或 NO_3-PAH 加合物；②OH-PAH 或 NO_3-PAH 与 NO_2 反应生成加合物 OH-NO_2-PAH 或 NO_3-NO_2-PAH；③ OH-NO_2-PAH 或 NO_3-NO_2-PAH 通过单分子解离脱一分子 H_2O 或 HNO_3 生成 nitro-PAHs。揭示 OH-NO_2-PAH 经单分子解离脱一分子 H_2O 的反应经过四元环过渡态，势垒非常高，超过 45 kcal/mol，然而在有水分子的参与的情况下，OH-NO_2-PAH 的脱水反应经过六元环过渡态，势垒大大降低，水分子的作用是参与反应形成六元环以促进 OH-NO_2-PAH 的脱水，反应结束后又有两分子水生成，因此水分子在 nitro-PAHs 的形成过程中起到了催化作用。另外，在水分子二聚体$(H_2O)_2$或单个水分子的帮助下，气态蒽可被 OH/NO_2 或 NO_3/NO_2 氧化形成 9-硝基蒽（大气中的主要 nitro-PAHs 之一）。该发现改变了传统认识[88,89]：9-硝基蒽不能从气态蒽与 OH/NO_2 及 NO_3/NO_2 的均相反应中形成，大气中的 9-硝基蒽来自于直接排放及颗粒态蒽与 N_2O_5 及 HNO_3 的非均相反应。

比较苯、萘、蒽与·OH 及 NO_3 的反应机理及 nitro-PAHs 的形成机理，发现 PAHs 与·OH 及 NO_3 加成反应的放热量顺序为苯 < 萘 < 蒽，说明随着 PAHs 中苯环数目的增多，PAHs 与·OH 及 NO_3 的加成反应更容易进行，这与实验现象完全一致——实验研究表明 PAHs 与·OH 及 NO_3 加成反应的速率常数顺序为单环 < 双环 < 三环。而 nitro-PAHs 形成控速步骤（消除一分子 H_2O 或 HNO_3 过程）的势垒顺序为单环 > 双环 > 三环，说明随着 PAHs 中苯环数目的增多，nitro-PAHs 的形成越来越困难，这很好地解释了实验现象[90]——实验研究发现随着 PAHs 中包含苯环数目的增多，nitro-PAHs 的产率在降低。

由于 PAHs 中 C==C 双键数目较多，其与·OH 或 NO_3 反应的位点较多。以蒽为例，·OH 或 NO_3 可以加成到蒽分子中所有的 C 原子上。目前通过实验技术获得的·OH 或 NO_3 与蒽加成反应的速率常数为总包反应速率常数。在量子化学计算提供的势能剖面基础上，采用 RRKM（Rice-Ramsperger-Kassel-Marcus）理论[91,92]计算了·OH 及 NO_3 加成到蒽分子中每个 C 原子上的分支反应速率常数，为模式研究提供了更为精细的动力学参数。

3　有机污染物气相和水相转化机制的计算模拟

3.1　大气污染物的转化

大气有机污染物转化的计算研究，主要针对挥发性有机污染物（VOCs）和半挥发性有机污染物（SVOCs）。VOCs 包括分子量较小的烷烃、烯烃、炔烃、醚、醇、醛、有机酸、有机胺等；SVOCs 包括多环芳烃（PAHs）、硝基多环芳烃（nitro-PAHs）、多氯联苯（PCBs）、二噁英/呋喃（PCDD/Fs）、阻燃剂、有机氯农药、有机磷农药、杀虫剂等。VOCs 和 SVOCs 主要是被大气中氧化性物质（如·OH，·Cl，O_3，NO_3 等）所转化。这里只选择性介绍国内一些较为系统、特色的研究案例，国际上的研究进展请参见相关评述文章[93-95]。

3.1.1 挥发性有机污染物的大气转化

1）有机胺的大气转化

有机胺作为重要的 VOCs，在人类和自然活动过程中均可产生[96]。胺的大气氧化过程可能会产生致癌性的亚硝胺和硝胺[96]。此外，胺的碱性能够增强 H_2SO_4 的成核能力[97-102]。有机胺的气相转化过程近年来得到越来越多的关注。

谢宏彬和陈景文等[103-106]通过量子化学计算及动力学模拟相结合的方法，对有机胺的大气转化过程做了一系列研究。首先，他们以基于胺溶液的燃烧后捕捉 CO_2 技术中应用最广泛的乙醇胺（MEA）为研究对象，研究了·OH 和·Cl 引发 MEA 的大气氧化及参与 H_2SO_4 成核的过程（图 25-2）。他们发现，·OH 引发的反应主要形成 2-亚氨基乙醇、胺基乙醛及致癌性的异氰酸等产物[104]；而·Cl 因与 MEA 形成独特的 2 中心 3 电子键（2c-3e）而最终形成致癌性的亚硝胺，且指出·Cl 在 MEA 的大气转化中起着不可忽视的作用[105]。他们也发现，MEA 具有很强的促进 H_2SO_4 成核的潜力，其促进潜力略低于目前认为最强的二甲胺[97,99,107]，但远高于一甲胺，这与简单地根据胺的碱性来评估它们的促进潜力的强弱顺序不同。因此，除了胺的碱性，胺含有的其他官能团也对它们增强 H_2SO_4 成核的潜力起着重要作用[103]。此外，MEA 参与 H_2SO_4 成核导致其去除的速率与被·OH 氧化的速率（278.15 K）相当，证明该过程也是其去除的一个主要途径。其次，并不是所有的含 NH_x(x = 1，2)的化合物与·Cl 反应都会生成致癌性亚硝胺的前驱体——N 中心自由基[106]。这与·Cl 引发的烷基胺的反应不同[105,108]。因此，含有 NH_x 结构的有机胺的大气氧化过程会导致不同的环境风险。这些研究为更好地理解有机胺的大气反应活性、H_2SO_4 成核，以及进一步理解它们在大气中的降解具有重要的意义。

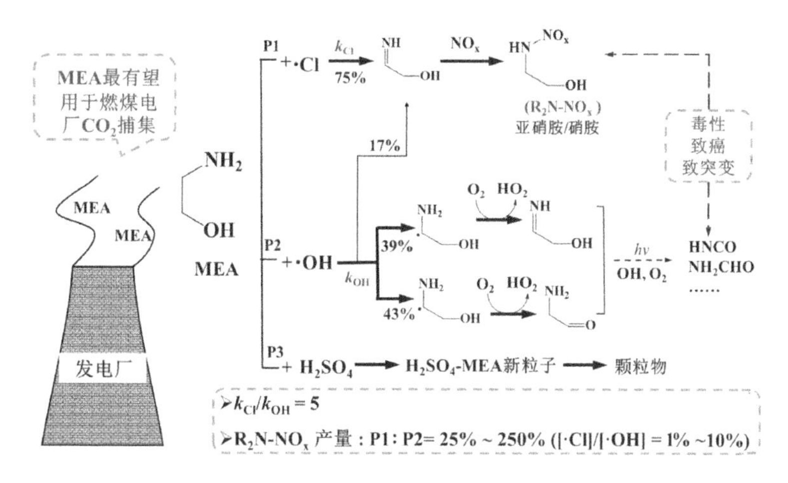

图 25-2 乙醇胺（MEA）被·OH 和·Cl 氧化及通过参与 H_2SO_4 成核转化的示意图

2）醚类的转化

含氧醚作为大气中含氧 VOCs 的重要成分，由于在工业中的广泛使用，使其在大气中的含量日渐增多。含氧醚会影响对流层的氧化能力且氧化降解产物可能会导致二次有机气溶胶（SOA）的形成，对气候和人体健康造成危害。

何茂霞等[109-114]采用量子化学方法对乙烯基醚类化合物（如乙烯基乙醚、乙烯基丙醚、乙烯基丁醚类）与 O_3，·OH 和·Cl 的气相氧化过程及相应的动力学进行了研究。研究发现，几种乙烯基醚易于发生大气氧化降解且其大气寿命较短，在大气转化过程中主要生成醛、酯和酰类化合物。王黎明等 [115]研究了二甲基醚、二乙基醚和二异丙基醚与·OH 反应生成的自由基接下来与 O_2 的反应。发现生成的过氧自由基的自氧化过程与它和 NO 和·HO_2 等双分子的反应具有竞争性，进而提出了一种大气中的醚氧化降解的新机制：

在低浓度 NO 的情况下，自氧化过程生成的产物是主要的。自氧化过程生成产物的 O∶C 比很高，蒸气压较低，对于形成 SOA 有重要作用。这些结果对研究一些 VOCs 的氧化降解过程及 SOA 的形成具有指导意义。

3.1.2 半挥发性有机污染物的转化

1)多环芳烃的转化

张庆竹等[116-121]采用量子化学方法研究了·OH，O_3，·Cl 和 NO_3 引发致癌性 PAHs 及氯代 PAHs 的气相氧化降解机制及动力学。发现·OH 与荧蒽、苯并[a]芘和 9, 10-二氯菲，O_3 与苊烯，·Cl 与蒽和芘及 NO_3 与 9-氯蒽的反应都主要发生加成反应，生成一系列酮类、醌类、硝基类、酚类、二醛和环氧类化合物等产物。另外，·OH 决定的荧蒽和 9，10-二氯菲的大气寿命分别为 0.69 d 和 5.05d；O_3 决定的苊烯的大气寿命为 0.75 h；·Cl 决定蒽和芘的大气寿命分别为 4.93 d 和 7.27d；NO_3 决定 9-氯蒽的大气寿命为 0.61 h。王黎明等[122,123]通过研究萘与·OH 的氧化机理，发现其与单环苯系物的氧化机理有很大不同，尤其是·OH 加成后的加合物的后续反应通道。形成的类似苯中双环自由基的异构反应，由于能垒高且吸热因此不能发生；形成的过氧自由基与 NO 反应生成的烷氧自由基分别通过闭环和分子内 H 迁移反应生成带有环氧结构和羰基的化合物，并伴有·OH 的再生成。

2）有机阻燃剂的转化

陈景文、谢宏彬及何茂霞等[30,124-131]采用量子化学计算及微观动力学模拟相结合的方法，研究了·OH 引发溴代阻燃剂 PBDEs（图 25-3）、短链氯代石蜡（SCCPs）及 PBDEs 的替代品（AFRs）的大气转化机制和动力学。他们给出了这些阻燃剂的反应路径、产物信息及反应速率常数。发现 Br 的存在会钝化与其相连的碳原子的活性，随着 Br 取代数目的增加，PBDEs 的反应活性降低，且 Br 数目较多的苯环活性明显低于含 Br 较少的苯环；相同溴代程度的阻燃剂，烷基类比芳香类阻燃剂容易降解。还发现·OH 和 PBDEs 可以生成毒性比 PBDEs 本身更强的羟基取代的 PBDEs（OH-PBDEs），有些 OH-PBDEs（如 6-OH-BDE47）可以继续转化生成毒性更强的二噁英[124,126-128]。

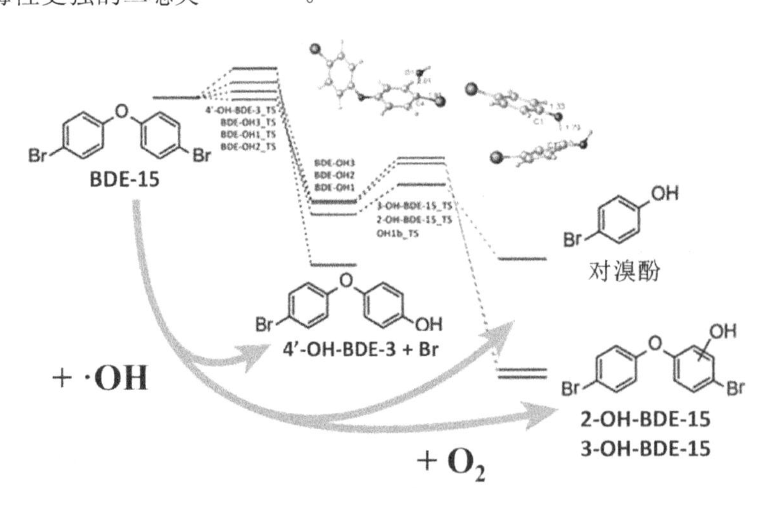

图 25-3 ·OH 引发 BDE-15 的大气转化途径

通过密度泛函理论（DFT）计算，发现 BDE-15 通过两种途径生成 HO-PBDEs：·OH 直接取代 Br 原子；O_2 摘取 BDE-OH 加合物中·OH 连接的碳原子上的氢原子，引自参考文献[125]

SCCPs 是指含碳数为 10~13 个，氯含量为 30%~70%的氯代正构烷烃，主要被用作切削油和润滑油的添加剂，以及聚合材料的增塑剂和阻燃剂，近年来作为一类环境新型有机污染物（EOPs）得到了广泛关注。《关于持久性有机污染物的斯德哥尔摩公约》曾讨论 SCCPs 污染物是否纳入公约的监管（目前已经被禁止），然而，由于缺乏该类污染物在环境中降解速率常数方面的信息，限制了对该类污染物环境持久性的评价。采用实验测定 SCCPs 的降解速率常数，也面临多种挑战。例如，SCCPs 具有数千种异构体/同系物，组成成分复杂，分析定量困难，且缺乏可以用于实验测定的标准样品。鉴于此，李超和陈景文等[30]筛选了适合计算 SCCPs 的气相·OH 反应速率常数（k_{OH}）的 DFT 方法[M06-2X/6-311+G（3df，2pd）//B3LYP/6-311+G(d, p)]，预测了 SCCPs 的 k_{OH} 值，进一步结合具有 k_{OH} 实验值的低碳链多氯代烷烃(PCAs，$C_1 \sim C_6$)，构建了可用于预测更多 SCCPs 的 k_{OH} 值的 QSAR 模型，即

$$\log k_{OH} = 5.057SPH + 1.167Q_C^- - 13.991\eta - 12.186$$
$$n_{tr} = 25，R^2_{adj} = 0.943，RMSE = 0.166，Q^2_{LOO} = 0.917，Q^2_{BOOT} = 0.726，$$
$$n_{ext} = 6，R^2_{adj.ext} = 0.960，RMSE_{ext} = 0.264，Q^2_{ext} = 0.856$$

其中，SPH 表示有机分子的球形度；Q_C^- 为碳原子上的最负净电荷；η 表示分子的绝对硬度；n_{tr} 和 n_{ext} 分别表示训练集和验证集化合物的个数；R^2_{adj} 为调整决定系数；$RMSE$ 为均方根误差；Q^2_{LOO} 是去一法交叉验证系数；Q^2_{BOOT} 是 Bootstrapping 方法所得的交叉验证系数。该模型的训练集的拟合效果和验证集的预测效果如图 25-4（a）所示，模型的应用域如图 25-4（b）所示。

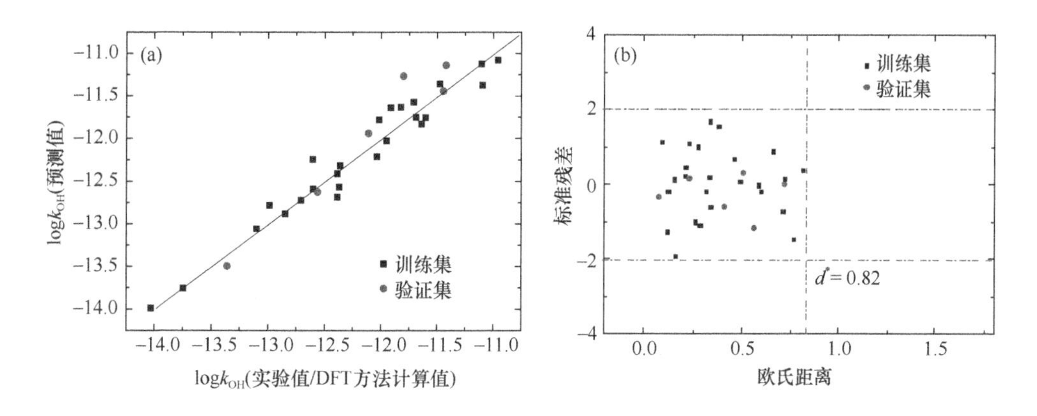

图 25-4　SCCPs 的 $\log k_{OH}$ 模型预测值与实测值的拟合验证（a）及基于欧氏距离的模型应用域（b）

虽然很多 AFRs 的大气寿命较 PBDEs 有所降低，但仍具有大气持久性，如磷酸三苯酯（TPhP）和 1, 2-双（2, 4, 6-三溴苯氧基）乙烷（BTBPE）的大气寿命分别约为 7.6 d 和 11.8d[129,130]。然而，磷酸三（2-氯丙基）酯（TCPP）作为一种重要的磷系阻燃剂，其由·OH 反应决定的大气寿命为 1.7 h，难以解释其在极地地区检出浓度较高的事实。李超和陈景文等[131]通过 DFT 计算，发现大气中的水可以通过形成氢键，改变 TCPP 与·OH 反应前络合物和过渡态的稳定性，从而降低 TCPP 与·OH 反应的速率常数，增加该污染物在大气中的寿命和环境持久性，解释了 TCPP 具有潜在的大气持久性的原因（图 25-5）。研究还发现大气中的水会影响初级反应产物在大气中的进一步转化途径，从而导致产物分布有所不同。此外，孙孝敏等[132-134]研究了二噁英及一些杀虫剂的大气转化，限于篇幅，这里不再赘述。

3.2　有机污染物的水相转化

相比于气相反应，污染物在水相反应计算模拟的难度更大，主要难点在于溶剂效应的模拟以及计算速度的限制。近年来，关于有机污染物在水相的转化行为，出现了一些探索性的计算模拟研究工作。

对于有毒有机污染物和抗生素等对微生物具有灭杀活性的污染物来说，由于其毒性和对微生物的特有活性，导致微生物往往难以对其降解，使得环境光降解成为决定污染物环境归趋和持久性的一个重要

因素。污染物环境光化学转化的机理复杂，涉及实验难以捕捉的瞬态物种，且受环境介质与污染物分子之间复杂相互作用的影响，如何表征与环境光化学活性相关的分子结构特征，如何预测污染物的环境光化学反应途径是环境计算化学与预测毒理学研究中需要解决的难题。

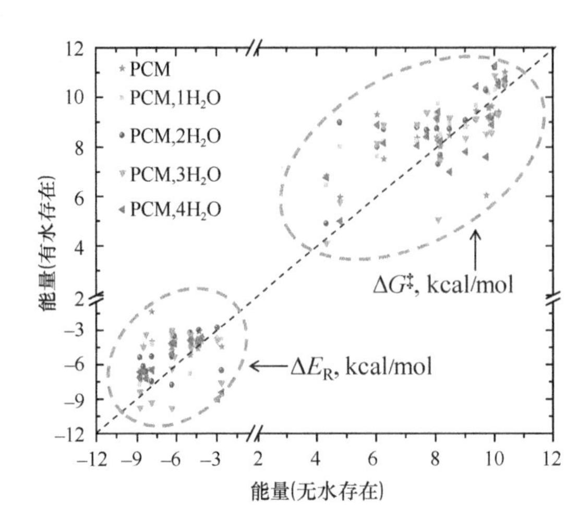

图 25-5　有水和无水（气相）TCPP + ·OH 反应体系的过渡态和反应前络合物的稳定性比较

$\Delta G^{\ddagger}$：活化自由能；ΔE_R：反应前络合物相对于反应物的能量；PCM：极化连续介质模型

葛林科、陈景文等[135,136]采用实验手段，揭示抗生素等新型污染物在水中可以发生直接光解、自敏化光解和溶解性有机质（DOM）等导致的间接光解。图 25-6 阐明了 DOM 对有机污染物光解的多重影响机制，为通过计算模拟预测水中有机污染物的环境光降解途径提供了认知基础。

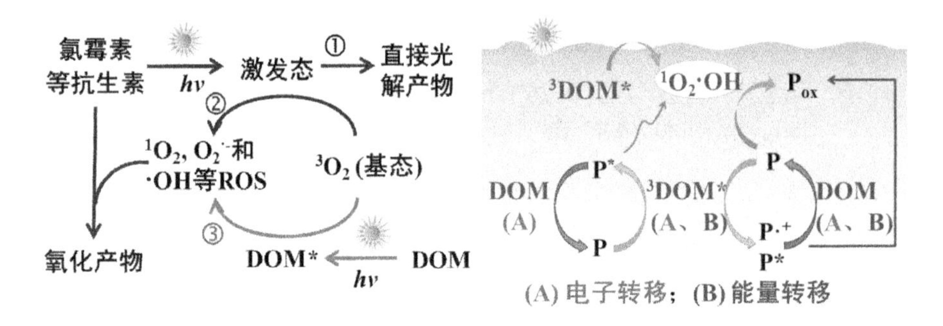

图 25-6　抗生素等新型污染物在水中可以发生直接光解①、自敏化光解②和溶解性有机质（DOM）等导致的间接光解③；激发三线态 DOM（^{3}DOM*）可直接氧化污染物、敏化生成活性氧（如 $^1O_2$）氧化污染物，也可以被基态 DOM 所淬灭，对污染物直接光解发生光屏蔽作用

目前的探索研究发现，计算模拟方法是研究污染物不同存在形态环境光化学行为的强有力手段。很多 EOPs（如抗生素）的分子结构中含有羧基、羟基和氨基等基团，在天然水 pH 条件下，不仅可以发生酸碱解离，同时易与金属离子发生配位反应，导致多种形态共存，各形态可能具有不同的环境光化学活性和反应途径。实验研究中，由于解离/配位平衡反应的存在，很难对单一形态进行独立研究，分析前处理过程也可能会改变污染物分子的存在环境或形态，因此难以揭示不同形态的环境转化过程和动力学机制。尉小旋和陈景文等[137]通过 DFT 计算结合实验，研究了环丙沙星不同解离形态的光化学活性，发现其具有不同的光转化途径、动力学、产物和生态风险（图 25-7）。他们[137]还采用 DFT 方法计算了环丙沙星一价阳离子（H_2CIP^+）与 Cu(II)形成配合物的稳定构型$[Cu(H_2CIP)(H_2O)_4]^{3+}$及与光化学活性相关的性质，发现配合物与 H_2CIP^+在前线分子轨道组成和能量方面均存在明显差异，表现为不同的光吸收特性；配位

作用不仅降低了 H_2CIP^+光致生成 $^1O_2$ 的能力，同时改变了其与 $^1O_2$ 反应的活性和途径。因此表观上，Cu(II)的配位作用抑制了 H_2CIP^+的光解。

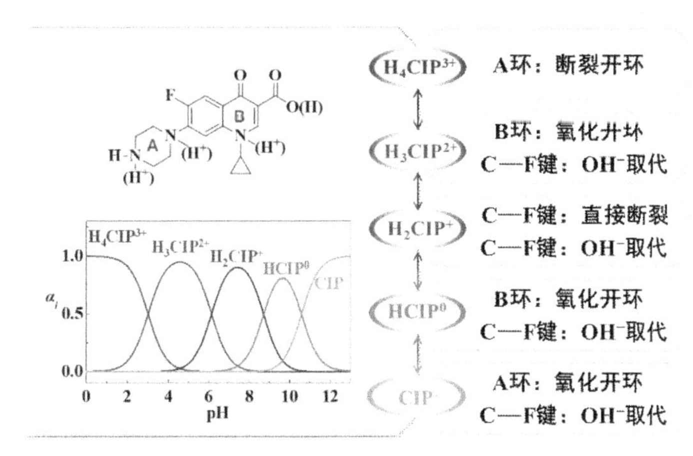

图 25-7　五种解离形态环丙沙星（CIP）的具有不同的表观光解途径

基于前述对有机污染物环境光化学转化途径的认识，张思玉等[138-140]采用量子化学计算，评价水中 DOM 等环境介质能否与污染物发生光致电子/能量转移反应，进而用于水中有机污染物环境光敏化降解途径的预测（图 25-8）。他们采用 DFT 计算模拟方法评价了 DOM 及卤素离子（X^-）、CO_3^{2-}等溶解性物质对新型污染物防晒剂 2-苯基苯并咪唑-5-磺酸（PBSA）光解的影响，并在光化学实验中得以验证。他们还针对新型污染物易发生的、水中溶解氧参与的自敏化光解过程开展了研究。以 PBSA 为例，通过计算溶解氧与 PBSA 之间的光致能量和电子转移反应，发现 PBSA 的第一激发单线态和三线态均可通过光致能量和电子转移途径敏化溶解氧生成活性氧物种；并计算了 PBSA 不同质子化形态光致生成超氧阴离子自由基($O_2^{\cdot-}$)的能力；他们推测的溶解氧及 pH 在 PBSA 光解中的作用，在后续实验中被证实。他们将该方法应用于另一种防晒剂对氨基苯甲酸（PABA），不仅预测了单线态氧（$^1O_2$）和 $O_2^{\cdot-}$ 的产生，并通过计算 $^1O_2$ 引发的 PABA 的自敏化光解反应，成功预测了光解产物，建立了一套有机污染物自敏化光解途径及产物预测模型[140]。

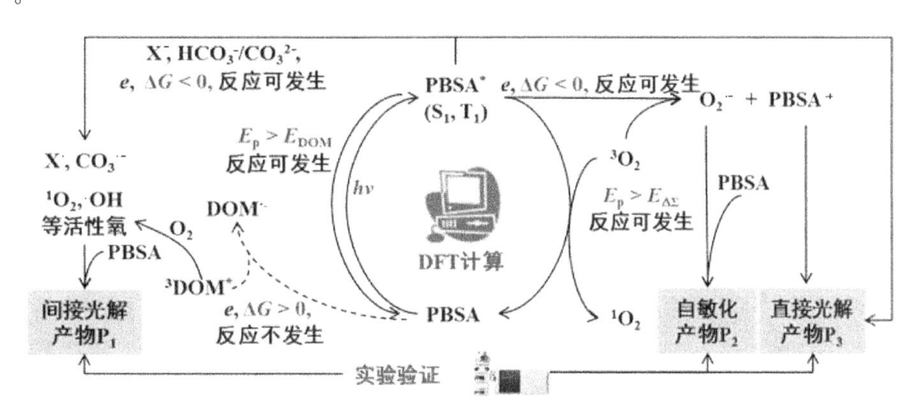

图 25-8　溶解性物质对有机污染物光解影响的计算预测模型

安太成等[141-149]针对水环境中的抗菌剂三氯生、防腐剂对羟基苯甲酸酯类、增塑剂邻苯二甲酸酯类、芬香剂合成麝香类以及环境药物扑米酮和阿昔洛韦等的光化学转化过程，开展了实验和计算模拟研究。针对抗菌剂三氯生的水环境转化过程（图 25-9），发现水环境中活性物种·OH 能使三氯生转化形成致癌物二噁英[144]，并指出充足的·OH 浓度是高级氧化体系中不会形成二噁英的本质原因，但存在的其他一些毒害产物仍不容忽视。针对芬香剂类 EOPs 如吐纳麝香的模拟计算表明，温度对这类 EOPs 的水体间接光化学降解过程影响较为明显[143]，尤其在水温较低（< 287 K）的环境中·OH 加成途径占主导，容易形成如酚类、苯醌类衍生物和二羟基化产物等，其中一些降解产物的水生毒性和生物累积效应较母体化合物增加了 1.3~8 倍。针对增塑剂邻苯二甲酸酯类和防腐剂对羟基苯甲酸酯类（图 25-10）的水体间接光化学降解过程研究发现[142,145-147]，它们很容易通过·OH 加成和 H 迁移途径被·OH 氧化降解，但侧链烷基对其反应机理影响较大。随着侧链烷基碳原子数的增多，这两类 EOPs 及其转化产物的毒性逐渐增加。·OH 加成产物的毒性大于 H 迁移产物，尤其是邻苯二甲酸酯类的·OH 加成产物的毒性高于母体化合物。虽然 H 迁移产物的毒性低于母体化合物，但仍具有不可忽略的潜在危害。

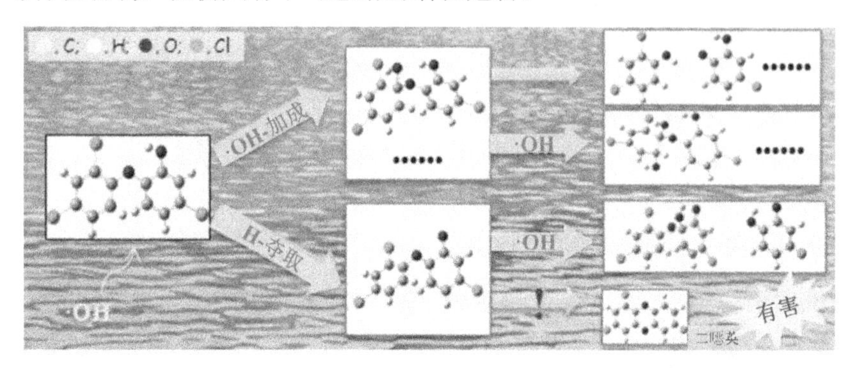

图 25-9　三氯生被·OH 介导氧化机制与二噁英的形成（引自文献[144]）

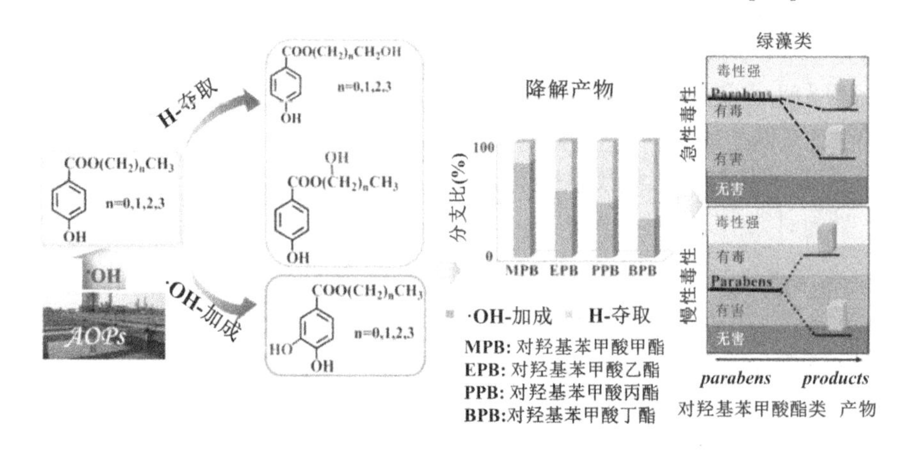

图 25-10　防腐剂对羟基苯甲酸酯的分子结构及其水中降解机理和产物毒性之间的相关性规律示意（引自文献[142]）

水解是决定分子结构中具有水解官能团的有机污染物（烷基卤、酰胺、氨基甲酸酯、羧酸酯、环氧化物、磷酸酯和磺酸酯等）在水环境中归趋的重要途径之一。有机物水解半减期短至数秒，长达数年。如果污染物水解速率慢，则通过实验法测定水解动力学参数将非常耗时，仅依靠实验方法分析其在水环境中水解速率常数及产物，难以满足有机污染物生态风险评价的需求，有必要发展计算模拟的方法来预测污染物的水解行为。传统上多基于线性自由能关系（LFER）理论来预测有机污染物的水解速率常数，LFER 是基于统计学分析得到的数学模型，属于经验性模型；且 LFER 模型在预测有机污染物的水解途径方面没有优势。近年来，随着计算机计算速度的快速提升，使得基于量子化学计算直接预测有机污染物的水解途径和动力学成为可能。

张海勤和陈景文等[150,151]以头孢拉定抗生素为模型分子，采用量子化学计算方法预测头孢拉定的水解行为，并用实验数据验证预测结果。头孢拉定分子具有 2 个水解官能团（β-内酰胺键和酰胺键），2 种互变异构体。在天然水 pH 范围内，头孢拉定存在两种离子形态（两性离子形态 $AH^{\pm}$和阴离子形态 A^{-}，图

25-11）。发现酸性条件下 $AH^{\pm}$水分子水解为主要路径，中性和碱性条件下 $AH^{\pm}$或 A^{-}碱催化水解为主要路径；不同 pH 条件下的水解速率计算值与实验值处于相同量级（图 25-12）；计算的水解产物与前人实验结果一致；发现头孢拉定分子中羧基能促进水解。

在抗生素与过渡金属离子复合污染的背景下，金属离子对抗生素水解的影响机理尚需阐明。采用 DFT 计算了水中 Cu(II)与阴离子形态头孢拉定（A^{-}）的配位反应的吉布斯自由能和稳定常数，发现 Cu(II)与 A^{-}可形成 1∶1 的配合物，该配合物存在两种形态：Cu(II)与 A^{-}分子支链 α-氨基氮原子和羰基氧原子配位，同时结合一个水分子；Cu(II)与羧基氧原子和 β-内酰胺氧原子配位，同时结合两个水分子（图 25-13）。分析了 A^{-}配位前后水解反应活性，结果表明：Cu(II)的配位作用能增大 A^{-}水解反应位点的原子正形式电荷量，降低了配合物的最低未占据分子轨道能和活化吉布斯自由能，从而促进 A^{-}水解。这个计算结果也得到实验结果的佐证。

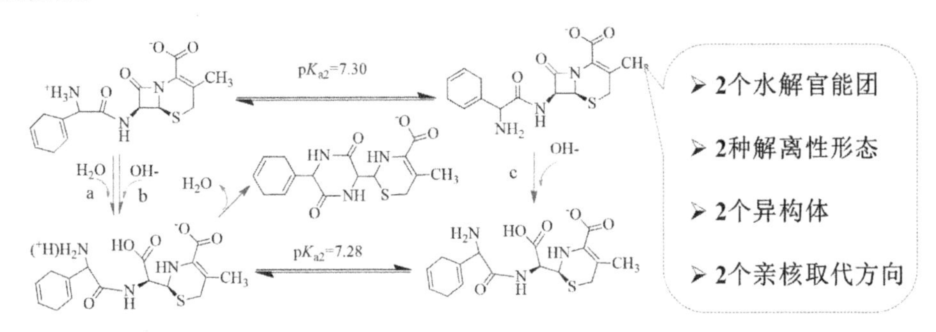

图 25-11　头孢拉定的分子结构、解离形态和水解位点

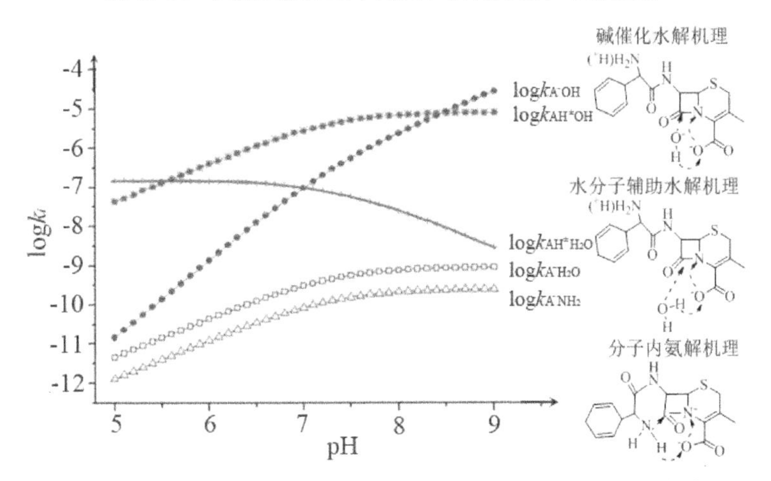

图 25-12　头孢拉定各水解路径的准一级反应速率常数与 pH 关系图

$k_{A^{-}H_2O}$和 $k_{AH^{\pm}H_2O}$分别为阴离子和两性离子形态水分子辅助水解速率常数；$k_{A^{-}OH^{-}}$，$k_{AH^{\pm}OH^{-}}$分别为阴离子和两性离子形态碱催化水解速率常数；$k_{A^{-}NH_2}$为分子内氨解速率常数，单位 s^{-1}

本研究发现头孢类和青霉素类抗生素分子中的羧基能通过质子转移机制促进其水解，但其他具有水解官能团和羧基的有机物是否也存在羧基催化水解现象需进一步研究。上述研究表明，采用 DFT 计算可直接预测有机污染物的水解途径和动力学信息。随着此方面研究案例的积累，将来有可能构建一种平台软件，实现有机污染物水解行为途径和动力学的量子化学计算模拟预测。

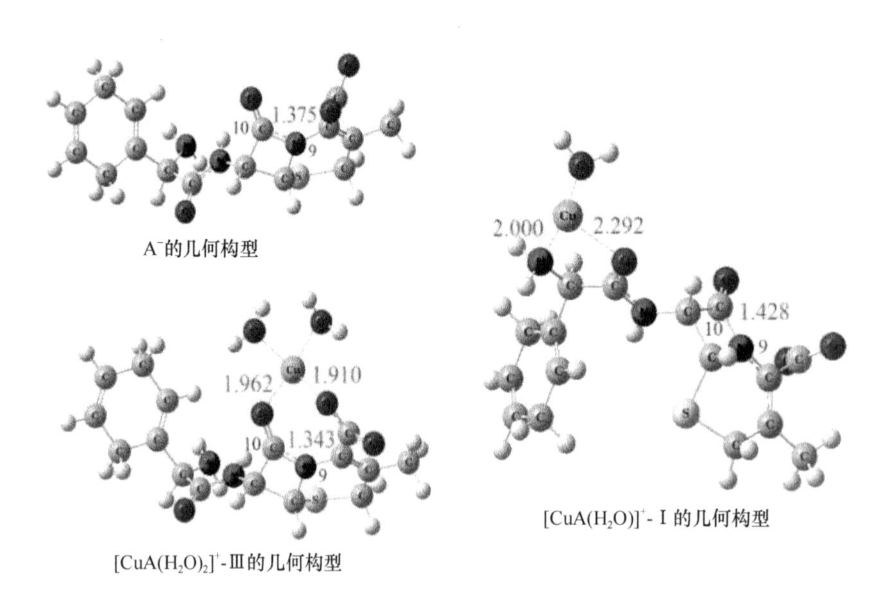

图 25-13　Cu(II)与阴离子形态头孢拉定（A^-）形成 1∶1 的配合物的两种形态的几何构型 $[CuA(H_2O)]^+$-I 和$[CuA(H_2O)_2]^+$-III 的浓度比值为 1∶0.73。因此，水中$[CuA(H_2O)]^+$-I 和$[CuA(H_2O)_2]^+$-III 可同时存在，且$[CuA(H_2O)]^+$-I 为主要形态

4　化学品环境行为参数的计算模拟预测

化学品的环境持久性、生物蓄积性和毒性取决于其在多介质环境中的吸附、分配、迁移、转化等环境行为。化学品的环境行为参数是这些行为的直观数字化表示，主要可分为环境分配系数和环境降解性参数。分配系数描述化学品在不同环境介质中的分配潜力，降解性参数则表示化学品在介质中的消除动力学。基于化学品的环境行为参数，可通过构建多介质环境模型，勾画出化学品在各环境相中的浓度水平，从而为判断化学品的环境归趋提供数据基础。从风险评价角度来说，环境行为参数可为化学品的环境风险表征和评估提供直接判据。因此，发展化学品环境行为参数的计算模拟预测方法显得十分必要。

4.1　环境分配系数的预测模型

环境分配参数是环境行为参数的一部分，由于实际环境介质，尤其是生物介质的复杂性，难以直接衡量化学品在环境介质间的分配行为，因此常用化学品在两种化学相间分配平衡的浓度比值即分配系数来考察其相间分配能力。常见的化学品环境分配系数有正辛醇/水分配系数（K_{OW}）、正辛醇/空气分配系数（K_{OA}）、亨利定律常数（K_H）、土壤有机碳吸附系数（K_{OC}）、生物富集因子（BCF）等。这些参数可以表征有机化学品在不同环境介质间（如空气-生物相、水-生物相等）分配能力的大小，从直观上获得某种化学品易进入和分布的环境介质信息，从而可用于化学品的环境持久性、长距离迁移性和生态毒性的预测评价。

1）正辛醇/空气分配系数（K_{OA}）

K_{OA} 定义为在一定温度下，达到分配平衡时，化学品在正辛醇相和空气相中浓度的比值。由于正辛醇是长链脂肪醇，具有类脂性，因此 K_{OA} 常用来描述污染物在空气相和环境有机相之间的分配行为，是评价化合物在环境中的长距离迁移能力和陆生生物从气相富集污染物能力的重要参数。目前，具有 K_{OA} 实测值的有机化学品约 400 种，加之 K_{OA} 值随温度变化较大，亟需发展快捷有效的 K_{OA} 预测方法，以满足新型化学品的筛选和风险评价的要求。

关于 K_{OA} 预测的方法主要有三种：①分配系数法（K_{OA} = K_{OW}/K_H，K_{OW} 指化学品在正辛醇相和水相之间的分配系数；K_H 指化学品在水相和空气相之间的分配系数）；②从头算溶剂化自由能模型（solvation model，SM）法；③基于 LFER 模型或者 QSAR 模型的预测方法。其中，分配系数法是根据各参数之间的定义式直接推导而得，可靠度较高，但是该方法的显著缺点就是应用范围小，仅对同时具有某一温度 T 下 K_{OW} 和 K_H 实验测定值的化学品适用。LFER 方法基于化学品的 Abraham 描述符，是目前预测包括 K_{OA} 在内的环境分配行为参数最为准确的方法，但其对多种类化学品 K_{OA} 的预测性能仍未知，且对不同温度下 K_{OA} 的预测能力仍有待探究。SM 方法的应用域广泛，几乎适用于所有化学品，但所采用的 SM 模型计算得到的溶剂化自由能的准确度还需考证。QSAR 模型的构建虽依赖于实验数据，但其预测准确度较高，能够满足化学品风险管理对数据的需求，但尚缺乏对不同温度下，多种类化学品 K_{OA} 预测的 QSAR 模型。

傅志强等[152]基于 379 种化合物在不同温度下的 939 个 K_{OA} 实测值，建立并考察了基于分子描述符的 QSAR 模型，基于溶剂化自由能的 SM 模型的预测性能。所构建的预测不同温度下化学品 K_{OA} 的 QSAR 模型如下：

$$\log K_{OA}(T)= -4.31+1.67\times10^2 X0sol/T - 2.46 \times 10 SpPos\_D/Dt/T - 1.43\times10^2 GATS1s/T - 5.43 \times 10^{-2} P\_VSA\_LogP\_3/T - 1.22 \times 10 RDF035v/T + 3.45 \times 10 Mor02p/T - 3.75 \times 10^2 Mor13p/T + 3.11 \times 10^2 E1s/T + 3.06 \times 10^2 nHDon/T + 6.17 \times 10 NaaaC/T - 4.67 \times 10 F05[Br\text{-}Br]/T + 1.18 \times 10^3/T$$

n_{tr} = 710，R^2_{adj} = 0.973，$RMSE$ =0.46，Q^2_{CUM} = 0.971；n_{ext} = 208，$RMSE_{ext}$ = 0.60，Q^2_{ext} = 0.950

其中，Q^2_{CUM} 是训练集的累积交叉验证系数。在模型所含的 11 种描述符中，*X0sol* 具有最大的偏最小二乘投影变量重要性（VIP）值，是决定 K_{OA} 大小的主要因素。*X0sol* 指的是溶剂连接性指数，表征溶剂化过程中的焓变及溶质-溶剂色散相互作用。经过对比，发现 QSAR 模型对应用域中化合物不同温度下 K_{OA} 的预测准确性最高，虽然 SM 模型的应用域广，但其预测准确性较低：R^2 =0.887（n_{tr} = 373），$RMSE_{ext}$ = 1.07。

此外，靳晓晨等[56]通过收集 379 种化合物的 Abraham 描述符值以及化合物在不同温度下的 795 个 K_{OA} 实测值，分别建立了预测化合物在 298.15 K 下 K_{OA} 的多参数 LFER（pp-LFER）模型和预测不同温度下 K_{OA} 的 pp-LFER-T 模型：

$$\text{pp-LFER: } \log K_{OA}(298.15\ K)=(-0.113 \pm 0.025)-(-0.157 \pm 0.028)E +(0.612 \pm 0.042)S +(3.510 \pm 0.077)A + (0.727 \pm 0.048) B + (0.925 \pm 0.007) L$$

n_{tr} = 229，R^2 = 0.998，$RMSE$ =0.15，Q^2_{CV} = 0.998；n_{ext} = 56，$RMSE_{ext}$ = 0.23，Q^2_{ext} = 0.995

$$\text{pp-LFER-}T\text{: } \log K_{OA}(T)=(-6.413 \pm 0.180)-(74.814 \pm 8.665)E -(245.583 \pm 10.943)S +(1025.511 \pm 27.502)A + (222.106 \pm 13.493) B + (277.144 \pm 2.169) L + (1848.815 \pm 55.209) 1/T$$

n_{tr} = 552，R^2 = 0.996，$RMSE$ =0.18，Q^2_{CV} = 0.996；n_{ext} = 203，$RMSE_{ext}$ = 0.18，Q^2_{ext} = 0.996

其中，E 是过量摩尔折射率；S 是极化性参数；A，B 分别是溶质分子的氢键酸度和碱度；L 是十六烷-空气分配系数的对数 log 形式。由模型的统计学参数可知，LFER 模型对 K_{OA} 的预测准确性很高。pp-LFER 和 pp-LFER-T 模型的 K_{OA} 预测值和实测值的对比如图 25-14 所示。经过内部和外部验证及应用域分析，发现 pp-LFER 模型不仅应用域更广，并且对有机硅化合物和多氟烷基化合物的预测效果有了显著提高。此外，pp-LFER-T 模型在较宽的温度范围中可以对应用域内化合物的 K_{OA} 进行准确预测。

2）土壤/沉积物吸附系数（K_{OC}）

王雅和李雪花等[38]基于 824 种有机化合物的 $\log K_{OC}$ 值，应用 Dragon 软件计算分子结构描述符，采用多元线性回归算法构建了可用于预测化学品 K_{OC} 值的 QSAR 模型。该模型具有较好的预测能力、稳健性和外部预测能力。预测模型的训练集的拟合效果和验证集的预测效果如图 25-15（a）所示，模型的应

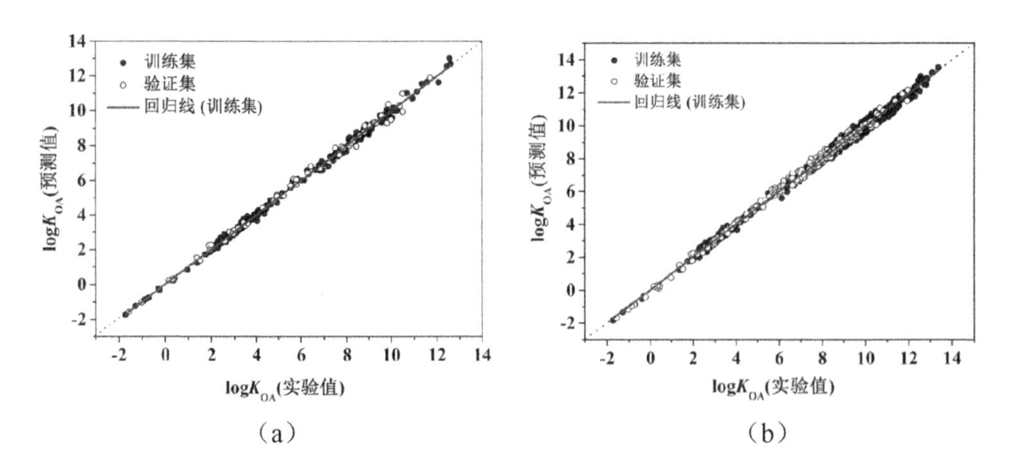

图 25-14　预测 K_{OA} 的 pp-LFER 模型（a）及 pp-LFER-T 模型（b）预测值与实测值的拟合验证

用域如图 25-15（b）所示。机理分析表明，有机化合物的 K_{OC} 主要受分子的疏水性和极化率影响。该研究将 K_{OC} 值预测模型的应用域扩展到新型污染物如多溴联苯醚（PBDEs）、全氟化合物和杂环类毒素。此外，姚小军等[153]基于 964 种非解离型有机化合物的 K_{OC} 值，分别采用多元线性回归、局域惰性回归和最小二乘支持向量机的方法构建了 K_{OC} 值的预测模型。结果表明，基于最小二乘支持向量机的非线性预测模型具有更好的拟合优度、稳健性和预测能力。

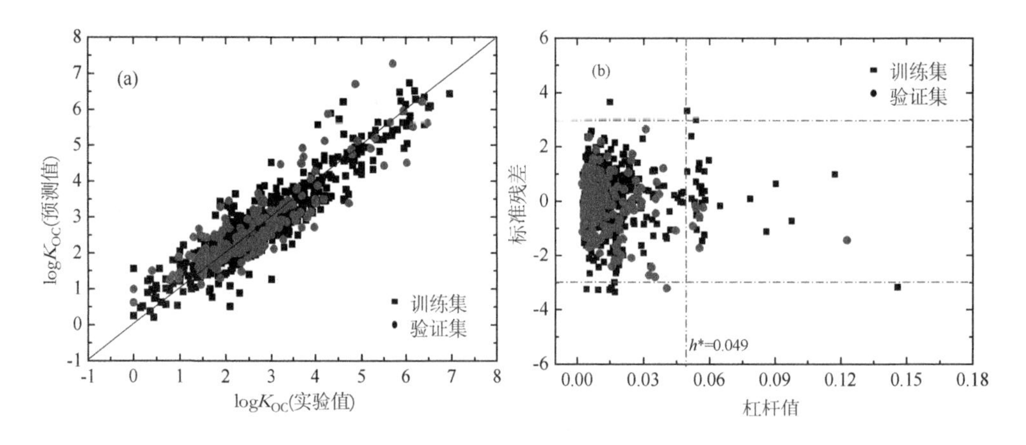

图 25-15　$\log K_{OC}$ 模型的预测值与实测值拟合验证（a）以及 Williams 图表征 $\log K_{OC}$ 模型的应用域（b）

4.2　环境降解性参数的预测模型

1)水中好氧生物降解性

化学物质在环境中的持久性，主要取决于其生物降解和环境光降解的能力。生物降解是污染物的重要降解方式。然而，生物降解机制复杂，预测相对困难。唐赟和 Lee 等[154]基于日本国际贸易和工业部（MITI）测试应用协议下测试的 1440 个不同的化合物，采用 4 种不同的建模方法，即支持向量机、K-最近邻算法（K-Nearest neighbor，KNN）、朴素贝叶斯（Native Bayes，NB）和 C4.5 决策树分别建立了生物降解性的预测模型。对于包含 164 种化合物的外部验证集，这些模型均具有较高的预测准确度。对于包含 27 种全新化合物的外部验证集，模型表现出较高的预测准确度。

陈广超和李雪花等[155]利用 C4.5 决策树、功能树和逻辑回归算法建立了 825 种化合物的生物降解性预测模型。同时采用 777 个（验证集 1）和 27 个（验证集 2）化学物质作为两个验证集对模型的预测性能进行外部验证。结果表明，功能树模型在训练集（825 种化学品）和 2 个验证集中分别表现出最佳预测能力。该模型的预测性能优于美国环境保护署开发的 EPI 软件中 BioWin5 和 BioWin6 模型。对模型的机理分析发现，化合物含有的环结构数、卤原子数、氮原子数越多，越不利于化合物的生物降解，而含有 R-C(=X)-X，R-C≡X，X=C=X（X 代表任何电负性原子，例如，O，N，S，P 和卤素原子）片段的化合物相对易于被生物降解。该模型的建立和验证，严格遵循了经济合作与发展组织（OECD）的 QSAR 模型构建和验证导则，可用于高效预测有机化学品的生物降解性。

2）大气环境中有机污染物的光氧化降解

光氧化降解也是评估化学物质环境持久性的关键参数。污染物在大气对流层中与·OH 和 O_3 反应发生光氧化降解的机制较为复杂，且反应动力学受温度的影响。李雪花等[34]根据 Arrhenius 方程，引入绝对温度作为预测变量，建立了不同环境温度下污染物与 O_3 反应的速率常数(k_{O_3})的预测模型，揭示了控制化学物质光氧化降解动力学的重要结构因子。首先收集整理了 379 个不同温度下 $\log k_{O_3}$ 数据，该数据集覆盖了 178~409 K 的温度范围。在此基础上，利用量子化学描述符、Dragon 描述符和分子结构碎片，建立了预测 k_{O_3} 的偏最小二乘（PLS）回归模型。模型具有良好的拟合度、稳健性和预测能力。模型中解释 $\log k_{O_3}$ 最重要的描述符为原子质量加权的连接性信息 BELm2 描述符。k_{O_3} 随着 BELm2 增加而增加，随着电离势增加而降低。该模型可用于预测多种有机化学品不同温度下的 k_{O_3} 值，包括链烯烃、环烯烃、卤代烯烃、炔烃、含氧化合物、含氮化合物（伯胺除外）、芳香族化合物。此外，姚小军等[156]采用多元线性回归、支持向量机、投影寻踪回归法建立了可用于预测有机化学品 k_{O_3} 的模型。

李超和李雪花等[29]构建了可用于预测不同环境温度下污染物与·OH 的反应动力学参数(k_{OH})的 QSAR 模型，并遵循 OECD 有关 QSAR 模型构建和验证导则，对所建立型进行了性能评估、应用域表征和机理解释。所建立的模型均具有良好的拟合度、预测能力和稳健性。模型的训练集的拟合效果和验证集的预测效果如图 25-16（a）所示，模型的应用域如图 25-16（b）所示。分子最高占据轨道能和卤素原子在分子中所占百分比（X%）是模型中最关键的分子结构描述符。所建立的模型应用域扩展到包含长链烯烃（C_8~C_{13}）、有机磷酸酯、有机硒和有机汞等新型化合物。

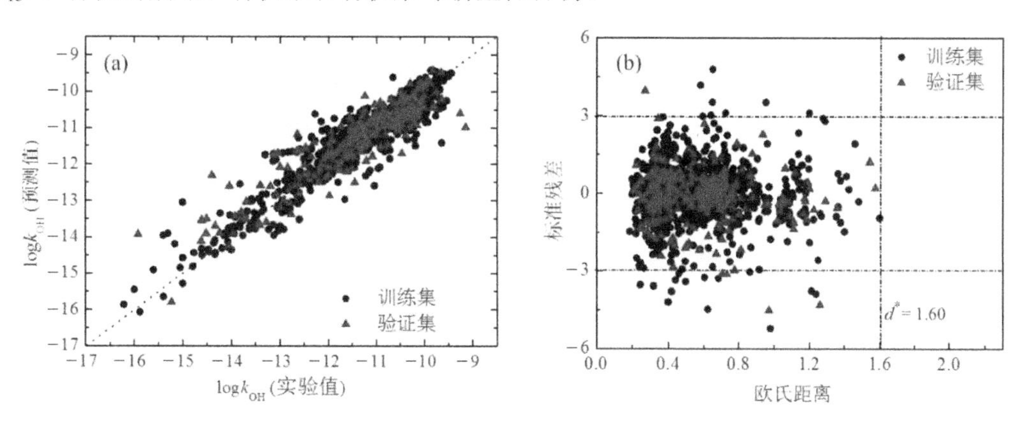

图 25-16　$\log k_{OH}$ 模型预测值与实测值的拟合验证（a）及基于欧氏距离表征的模型应用域（b）

3)硫酸根自由基高级氧化降解水中痕量有机污染物的 QSAR 模型

肖睿洋等[32]基于荟萃分析（meta–analysis），汇总了不同污染物在水中与硫酸根自由基（$SO_4^{·-}$）氧化反应的二级反应速率常数（$k_{SO_4^-}$），共包括 85 种有机化合物，共 115 条数据，数据广泛覆盖了不同官能团结构，其速率范围在 $4.60 \times 10^5 \sim 9.2 \times 10^9$ L/（mol·s）之间；进一步对已报道的 $k_{SO_4^-}$ 与官能团的关系分析，发现不同官能团呈现出的二级反应速率常数存在显著性差异。例如，$SO_4^{·-}$ 和含有醇基的痕量有机污染物反应速率常数远小于芳香基团的 $k_{SO_4^{-}}$ 值，而 $SO_4^{\cdot-}$ 和含有酰胺基的有机物 $k_{SO_4^{-}}$ 值与芳香基团反应速率常数相当，但是含有羧酸基团的痕量有机污染物分子反应速率可快可慢，说明该基团对二级反应速率常数无明显影响。由此可见，痕量有机污染物所带官能团一定程度上能反映 $SO_4^{\cdot-}$ 氧化有机物的难易程度，但不起决定作用。通过这些分析，初步掌握了特征官能团与 $SO_4^{\cdot-}$ 反应性的规律，在此基础上，构建了预测 $k_{SO_4^{-}}$ 的 QSAR 方程，即

$$\ln k_{SO_4^{-}} = 26.8 - 3.97 \times \#O:C - 0.746 \times (E_{LUMO} - E_{HOMO})$$

$$R^2 = 0.866,\ Q^2_{LOO} = 0.86,\ Q^2_{BOOT} = 0.87,\ Q^2_{EXT} = 0.89$$

式中，#O：C 为分子中氧碳原子数比，$E_{LUMO}-E_{HOMO}$ 为前线分子轨道能级差；R^2 为决定系数；Q^2_{LOO} 是去一法交叉验证系数；Q^2_{BOOT} 是 Bootstrapping 方法所得的交叉验证系数；Q^2_{EXT} 为外部验证系数。模型预测值与实测值的对比见图 25-17 所示。该方程指出，氧碳原子数比与 $k_{SO_4^{-}}$ 相关，即有机物分子中含有氧原子越多，该痕量有机污染物分子和硫酸根自由基反应性越差，高氧碳原子比意味着分子中更少的氢原子与硫酸根自由基反应，印证了① $SO_4^{\cdot-}$ 主要和碳链上的氢原子发生摘氢反应；② 相对于其他反应途径，摘氢反应较慢，为反应的限速步骤。该研究突破了以往停留在官能团与 $SO_4^{\cdot-}$ 的定性关系研究，为 $SO_4^{\cdot-}$ 高级氧化技术在确定优先控制对象污染物上提供了启发。

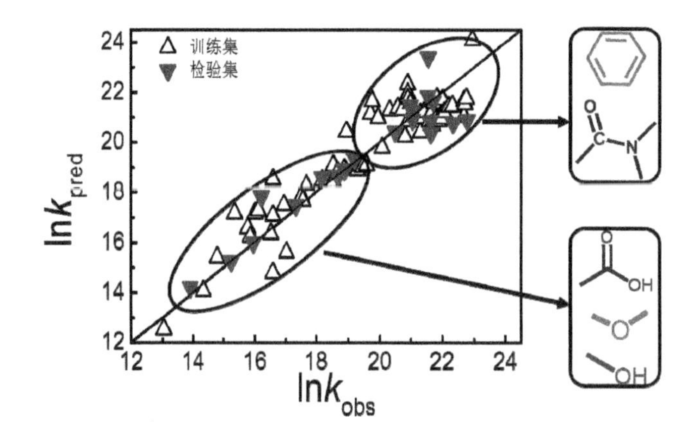

图 25-17　痕量有机污染物化合物 $\ln k_{SO_4^{-}}$ 实验值与 QSAR 预测值比较

5　酶代谢外源环境污染物的计算模拟

5.1　典型外源污染物 P450 酶代谢的计算模拟

1）多溴联苯醚（PBDEs）的 P450 酶代谢转化模拟

作为最具代表性的溴代阻燃剂，PBDEs 由于其环境持久性、生物蓄积性和毒性而受到广泛关注。众多研究证实 PBDEs 在生物体内能被 P450 酶（主要存在于动物肝细胞微粒体中）代谢转化[157-159]，但具体的反应路径和产物尚不清楚。王兴宝、傅志强等[160-162]采用量子化学 DFT 计算，系统性揭示了 P450 酶活性中心氧化单体 Compound I 代谢转化 PBDEs 的可能反应路径（图 25-18）。以 2, 2', 4, 4'-四溴联苯醚

（BDE-47）为例，发现 BDE-47 与 Compound I 首先发生碳原子的亲电 π 加成反应。在非溴取代碳位时，反应产生正四面体加合物；而溴取代碳发生 π 加成反应则形成环己烯酮产物。四面体加合物可通过以下路径进行二次重排：①将碳连接氢原子迁移到邻位碳原子（national institute of health，NIH 迁移），生成环己烯酮；②发生闭环反应生成环氧化产物；③借助卟啉环氮原子将氢原子转移到 Compound I 氧上（质子穿梭机制）生成羟基化产物（HO-PBDE）。

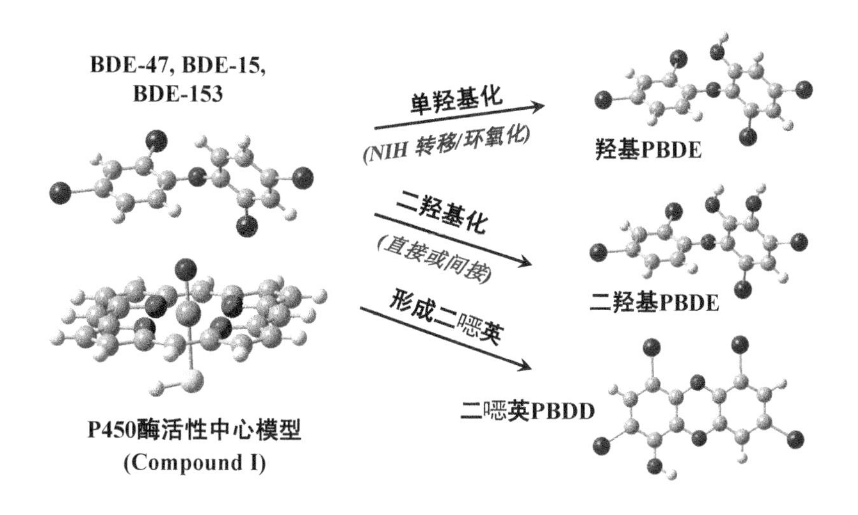

图 25-18　P450 酶活性中心 Compound I 模型及其催化 PBDEs 生成羟基化、二羟基化及二噁英产物示意图

π 加成和重排生成的环己烯酮会在水分子催化下发生酮-醇互变异构生成 HO-PBDE。重排生成的环氧化物在非酶环境中很容易发生质子化开环反应。开环后，当羟基与溴原子与同一碳原子相连时，根据小基团发生 NIH 迁移的规则，溴会转移到邻位碳原子上，形成溴迁移的 HO-PBDEs 产物；当羟基和氢原子连于同一碳原子时，氢原子发生 NIH 转移，生成的 HO-PBDE 是溴未转移的 HO-BDE-47。而当羟基连在醚键邻位的碳上时，可引发醚键异裂反应，生成溴酚。计算预测的部分代谢产物（4-HO-BDE-42，4'-HO-BDE-49，5-HO-BDE-47，2'-HO-BDE-28，4'-HO-BDE-17, 2, 4-二溴酚）与前人在 BDE-47 的 *in vivo* 或 *in vitro* 实验中的检测产物相同[163,164]。可见 DFT 模型计算可有效揭示 Compound I 催化 BDE-47 羟基化过程的机制和相应产物分布。

在 Compound I 催化 BDE-47 羟基化过程中，由于四面体加和物的二次重排过程活化能垒较低，因此初始 π 加成反应是速率控制步骤。比较三种 PBDEs（BDE-15，BDE-47，BDE-153）与 Compound I 的 π 加成反应能垒，发现随着溴取代数目的增多，PBDEs 被 Compound I 氧化的能力降低。Lupton 等[159]研究了人肝细胞微粒体代谢三种 PBDEs（BDE-47，BDE-99，BDE-153）的过程，也发现高溴代 BDE-153 比低溴代异构体更难被 P450 酶代谢氧化。对每种 PBDEs 来说，溴取代的碳原子和与醚键相连的碳原子较难与 Compound I 反应，但这种规律随着苯环上溴取代数的增多而变得不显著。

除 HO-PBDEs 外，BDE-47 的 *in vitro* 实验还检出了二羟基化（di-HO-BDEs）和多溴代二噁英（PBDD）产物，但是内在的分子机制不清楚[165]。首先，HO-PBDE 可能发生二次羟基化产生 di-HO-PBDE。以 6-HO-BDE-47 为例，计算发现与 PBDE 相比，HO-PBDE 的 π 加成反应能垒有所降低，说明羟基引入增加了苯环的电荷密度，导致亲电 π 加成反应更容易发生。此外，Compound I 也可催化 6-HO-BDE-47 发生酚羟基摘氢和羟基反弹反应生成 di-HO-PBDE（图 25-19）。DFT 计算结果显示，这条路径整体反应活化能低于 π 加成反应，是生成 di-HO-PBDE 的优势通路。该过程中，羟基反弹步骤能垒显著高于酚羟基摘氢，是 di-HO-PBDE 产生的速控步骤。比较三种 HO-PBDEs 异构体（6-HO-BDE-15，6-HO-BDE-47，6-HO-BDE-153）的羟基反弹能垒，发现羟基倾向于反弹至邻位或者对位的碳原子，生成邻/对位

di-HO-PBDE，与 BDE-47 在大鼠肝微粒代谢实验中检出的 di-HO-PBDE 结构一致[165]。

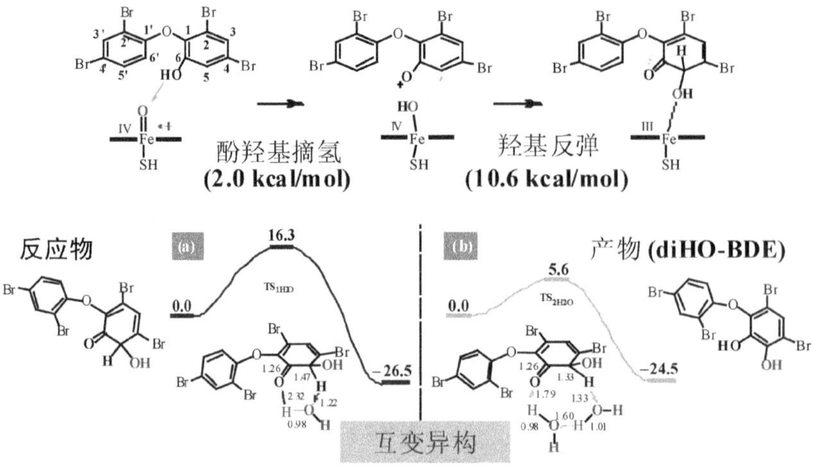

图 25-19　Compound I 催化 HO-PBDE 转化生成 di-HO-PBDE 的路径

理论上只有醚键邻位碳原子被羟基取代的 HO-PBDE 才能环化生成 PBDD。以 6-HO-BDE-47 为例，考察了两条常规的 PBDD 生成路径，即 a）6-HO-BDE-47 经酚羟基摘氢和 O—C 成环反应生成 PBDD 和 b）6-OH-BDE-47 经二次羟基化生成 6, 6’-di-OH-BDE-47，而后脱水产生 PBDD。结果发现，a）路径中 O—C 成环能垒较高，且生成的 PBDD 自由基需要借助多个水分子进行氢原子重排转移，该过程在酶环境中不可行。而 b）路径中 6, 6’-di-HO-BDE 的脱水反应过程能垒极高，因此不可能产生 PBDD。根据酚氧自由基的共振重排机制，只有醚键邻位和间位均存在羟基取代的异环 di-HO-PBDEs 才能作为二噁英的前体。以 5’, 6-di-HO-PBDE 作为底物计算得到了二噁英形成的可行路径（图 25-20）：Compound I 催化 5’, 6-di-HO-PBDE 发生两步酚羟基摘氢，将底物转化为二酮中间体。该异环二酮具有共振双自由基结构（O_6 和 $C_{5'}$自由基），经过 O_6-$C_{5'}$偶联生成 PBDD 酮异构体，接着在非酶环境下发生酮-醇互变异构生成 HO-PBDD。这是首次揭示二噁英的生物代谢转化来源机制。

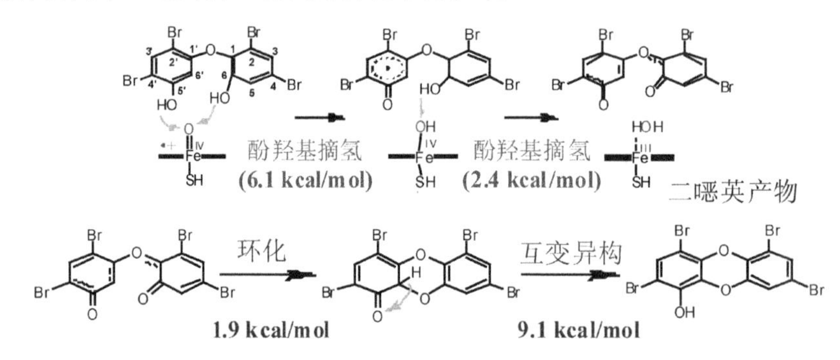

图 25-20　Compound I 催化异环 di-HO-PBDE 转化生成二噁英 PBDD 的路径

2）全氟辛烷磺酸（PFOS）前体的 P450 酶代谢转化模拟

全氟辛烷磺酸（PFOS）是典型的环境持久性有机污染物，但人体暴露于 PFOS 的途径仍不清楚。前期 *in vivo* 和 *in vitro* 实验表明 PFOS 前体物质（PreFOS）被 P450 酶代谢生成 PFOS 是重要的间接暴露途径[166,167]，但这一过程的反应路径和机理并不明晰。模型计算揭示了一种典型 PreFOS，*N*-乙基全氟辛烷磺胺（*N*-EtPFOSA）被 Compound I 代谢转化的机制(图 25-21)[168]。首先，*N*-EtPFOSA 与 Compound I 发生 Cα-H 羟基化反应，生成乙醇胺中间产物。乙醇胺接下来在非酶环境中发生降解，脱去乙醛，生成全氟辛烷磺胺（PFOSA）。水分子可催化乙醇胺的降解反应，辅助氢的转移过程。因此，*N*-烷基取代的 PreFOS 首先经过 Compound I 催化的 *N*-脱烷基反应转化产生 PFOSA。

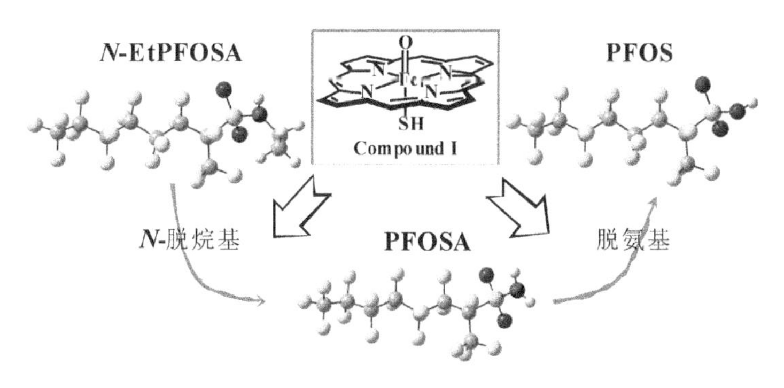

图 25-21　Compound I 催化 N-EtPFOSA 转化生成 PFOSA 及 PFOS 的路径

实验证实，PFOSA 几乎是所有 PreFOS 生物代谢转化的中间产物，但 PFOSA 如何代谢生成 PFOS 的机制则鲜有研究。计算揭示了 Compound I 催化 PFOSA 转化成 PFOS 的路径。与伯胺氮原子摘氢和羟基反弹生成羟胺的传统机制不同，发现 Compound I 与 PFOSA 的摘氢反应由于全氟辛烷基的强吸电子作用而难以进行。PFOSA 首先与 Compound I 发生氮原子氧化，生成 *N*-氧化产物。*N*-氧化物一方面会发生氢转移，将氢原子转移至氧原子上生成羟胺；另外，由于氧原子具有较强的电负性，也可能与硫原子发生加成，形成环氧化类似物。该环氧化物接着发生重排反应，断裂 S—N 键，并水解生成 PFOS 和羟胺。计算发现 *N*-氧化物氢迁移生成羟胺的反应能垒比生成环氧化物路径高。因此脱氨基反应遵循环氧化机制，其中 N 氧化是反应的速控步骤。PFOSA 脱氨基过程所需能垒显著高于 *N*-EtPFOSA 的 *N*-脱烷基反应，说明由 PFOSA 到 PFOS 的反应是 PreFOS 生物代谢的速控步骤，与前人实验观测结果一致。

3）卤代烷烃及烯烃的 P450 酶代谢转化模拟

许多工业化学品，包括杀虫剂和农药均含有卤代烷烃基团，但是这类物质如何在生物体内被代谢转化仍不清楚。季力等[169]基于 DFT 计算研究了 P450 酶代谢转化卤代烷烃（以 $CHCl_3$ 和 CCl_4 为模型底物）的路径和机制。发现在有氧条件下，$CHCl_3$ 主要进行 Compound I 催化的 C—H 键羟基化反应，生成的甲醇中间产物进一步发生水分子催化的脱氯化氢反应，生成 Cl_2O；而在厌氧条件下，$CHCl_3$ 与二价的亚铁卟啉活性中心发生还原脱氯反应；CCl_4 在厌氧条件下会被 P450 酶转化产生 Cl_2O 和 ClO·。另外，P450 酶代谢转化形成环氧化产物是烯烃类物质产生毒性效应，特别是致癌性的重要因素，因此准确预测烯烃类物质代谢生成环氧化物的潜力对于评估此类物质的风险具有重要价值。季力等[170]基于 DFT 计算了 36 种烯烃的电离势（ionization potential，IP）及其与 Compound I 的环氧化反应能垒。发现弱极性（偶极矩小于 2.2 Debye）和强极性（偶极矩大于 2.2 Debye）烯烃的环氧化能垒与其分子的 IP 值存在很好的线性相关（图 25-22），可以作为烯烃类物质代谢活化为环氧化物潜能的预测方法。

值得指出的是，上述计算模拟研究均采用 P450 酶的活性中心团簇模型。该模型能够较好地重现 P450 酶氧化活性物种的电子结构和氧化反应能力，因此在揭示污染物的代谢反应分子机制，预测反应路径和产物方面具有较强的可靠性。由于 P450 酶具有物种和亚型特异性，团簇模型在考察酶蛋白环境中氨基酸残基对反应活性和区域选择性影响等方面仍存不足。因此发展耦合量子力学/分子力学（QM/MM）方法配合分子动力学（MD）模拟的方法，有望为污染物的 P450 酶代谢转化研究提供新的思路。未来，随着计算机性能的不断提升，计算模拟理论方法的不断完善，P450 酶反应的计算模拟将可以从本质上准确预测出污染物在生物体内的代谢路径和产物，有望替代试管和动物实验，从而为评估其生物归趋和毒性风

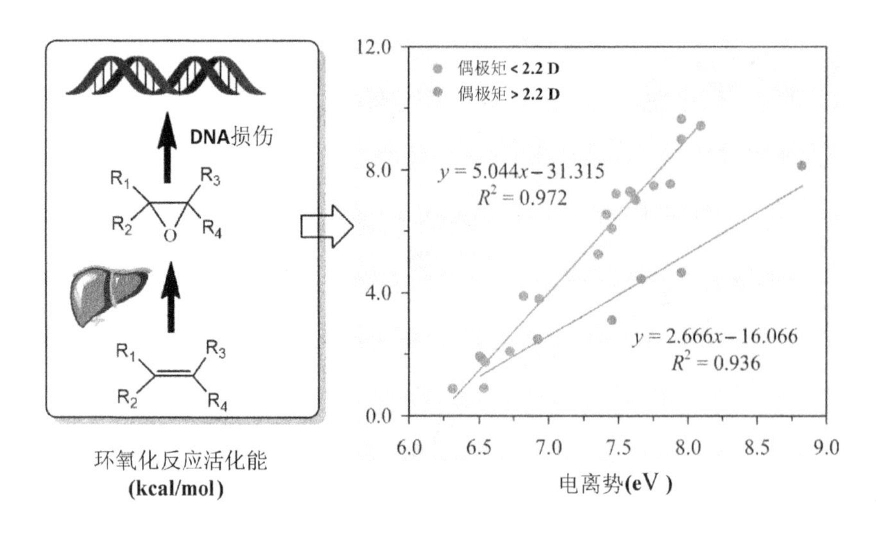

图 25-22 P450 酶代谢烯烃类物质生成环氧化物的模拟预测

险提供依据；而根据所获得的代谢反应信息，也可为绿色化学品设计提供指导（如改变某一取代基类型来规避不良代谢路径）。另一方面，多尺度的计算模拟方法（如 QM/MM 和同源模建）能够给出不同物种、亚型 P450 酶对污染物的差异相互作用或转化信息，为跨物种的生态毒理学研究提供有效手段。

5.2 其他生物酶降解转化污染物的计算模拟

除 P450 酶外，其他生物酶也参与污染物的降解转化过程。现阶段污染物的生物降解研究对象多是菌株、菌落，只能通过检测到的产物推测可能的反应机理。由于酶催化反应速率较高，反应过程中的中间体可能被迅速分解成其他代谢产物，因而在很多情况下无法通过实验手段捕获到反应过程中中间体的信息，也就很难从分子层面上阐述降解微观机理；虽然通过实验手段能检测到反应产物与极个别的中间体，但是仅通过实验手段还无法了解整个反应发生的具体过程，因而很难有针对性地对酶进行改造、突变以提高其催化效率。

山东大学张庆竹团队通过 QM/MM 针对生物酶作用下环境有机污染物的降解展开了一系列的研究[171-176]。张庆竹等[172,173]首先探索并建立了适用于环境领域生物酶代谢研究的 QM/MM 模拟流程，进一步研究发现酶构象的变化对反应能垒有较大的影响，从而建议在 QM/MM 研究中需综合考虑多个酶构象以得到可以与实验比对的、较为精确的模拟结果。张庆竹团队成功将这一方法推广应用于氯苯基三氯乙烷（即 DDT）、六氯环己烷、多氯联苯、氯酚等环境污染物的酶催化转化研究中。以谷胱甘肽转移酶体系的研究为例（图 25-23）进行说明。全世界每年约有 140 万人因感染疟疾这一蚊虫传播疾病而死亡，人们通常通过喷洒 DDT 的办法来消灭蚊虫以达到控制疟疾传播的目的。然而近年来发现某些蚊虫体内的谷胱甘肽转移酶能够将高毒性的 DDT 转化为低毒性的 DDE，从而对 DDT 产生了抗性。目前关于 DDT 的降解机理有两种观点：质子转移机理和耦合-解离机理。张庆竹等揭示了 DDT 在谷胱苷肽转移酶催化作用下质子转移是唯一能量可行的降解机理；研究发现将 DDT 中与苯环相键连的 Cl 原子换成 Br 原子能很好地的抑制降解反应的发生，从而为新型有机杀虫剂的合成提供了理论依据。

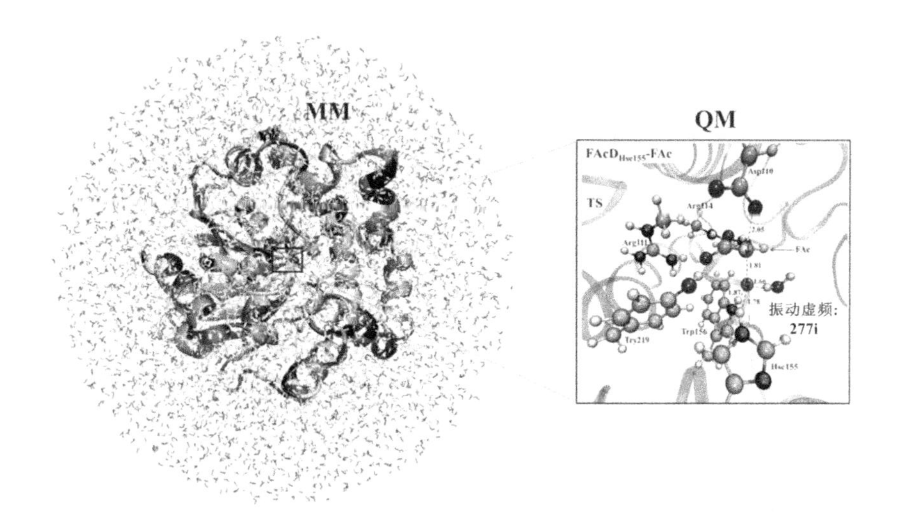

图 25-23　谷胱甘肽转移酶作用下典型有机污染物的转化机理

6　环境内分泌干扰效应的毒性通路与模拟预测

内源激素（又称为荷尔蒙，hormone）是由生物体内分泌腺或组织分泌的一种化学信号传导物质。其在生物体整个生命周期中，具有不可或缺的作用。激素产生正常生理调控功能的前提，是机体具有功能完善的内分泌系统和正常的激素体内水平。然而，流行病学调查、野外监测、体内实验结果都表明，一些人工合成化学品能影响生物体的内分泌系统发育和生理功能，影响激素体内平衡，进而引发内分泌相关的疾病，这类物质被称为内分泌干扰物（EDCs）。EDCs 干扰内分泌信号通路的分子机制包括：① 干扰下丘脑-垂体对内分泌腺体的调控过程；② 干扰内分泌腺合成及分泌激素；③干扰激素转运；④ 激活或抑制激素受体介导的信号转导；⑤ 抑制机体激素代谢酶活性。因此，内分泌干扰效应模拟研究的对象是上述干扰过程涉及的生物大分子及其对应的干扰效应。在过去较长时间里，激活或抑制激素受体介导的信号转导过程被认为是EDCs的主要作用机制，导致大部分的研究集中于EDCs对激素受体的干扰效应。近年来的研究表明，EDCs 对激素转运等非受体介导的过程的干扰也同等重要，因而对非受体类生物大分子及其对应的干扰效应的研究逐渐增多。

6.1　雌激素干扰效应相关研究

针对多类环境污染物潜在的雌激素受体（ER）及相关受体干扰效应，国内学者综合采用化学信息学、分子对接、分子动力学模拟等方法进行了高通量筛选、毒性通路预测、分子机制研究。唐赟等[177]通过整合 k-最邻近算法（KNN）、C4.5 决策树、朴素贝叶斯（Naive Bayes，NB）及支持向量机（SVM）等机器学习方法，构建了对 EDCs 进行二进制分类的模型，解析了内分泌干扰效应相关的毒效团结构特征（图 25-24）。林志芬等[178]基于大鼠雌激素干扰效应的 EDCs 决策树，通过分子对接方法计算化学品与靶蛋白的生物参数、结合能量，建立了筛选潜在非洲爪蟾雌激素干扰物的决策树模型。赵春燕等[179]借助“复杂网络”技术，基于 20 种取代苯类农药的内分泌干扰效应，将 EDCs 与机体的信号通路相关联，阐明了 EDCs 对生物机体活性通路的影响。

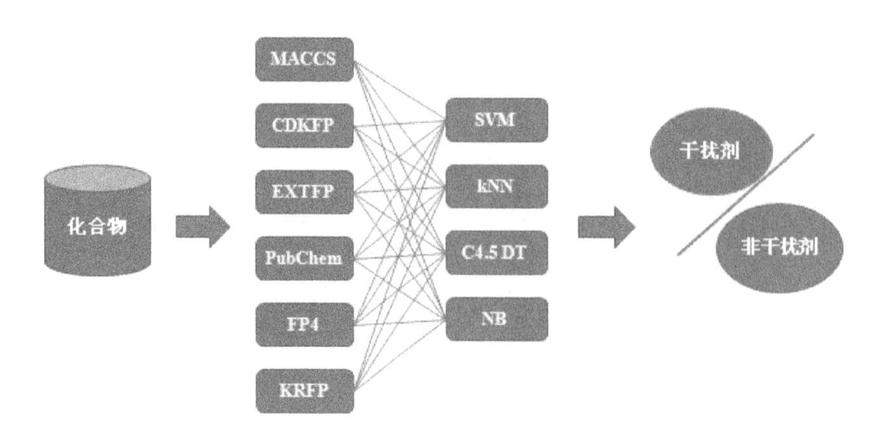

图 25-24　EDCs 二进制分类模型流程图

庄树林等[180]针对双酚 A 类化学品及氯苯基三氯乙烷（DDT）类似物的雌激素干扰效应，通过分子对接、分子动力学模拟等方法开展了构效关系、分子作用机制研究。发现多种环境污染物如双酚 A 类、DDT 类等化学品能与 ER 以及雌激素相关受体（estrogen-related receptor，ERR）发生相互作用。对 DDT 及其相关农药，发现范德华力是其与 ERα 及 ERRγ 配体结合域（LBD）作用的主要驱动力。氯代和酚羟基的数量及位置的细微结构变化，导致了结合模式的差异，诱导了 ERα LBD 和 ERRγ LBD 构象的变化。对于双酚 A 类物质，酚环上的溴取代影响了分子静电势分布，导致其与 ERα LBD 间静电作用和范德华力的相互作用差异，一定程度上解释了该类物质雌激素干扰效应的不同[181]。杨先海等[182]的 QSAR 模型研究结果也表明，双酚 A 类似物的取代基会对其雌激素效应产生影响，一般而言酚羟基邻位引入吸电子基团会减弱其雌激素效应，而引入供电子基团则会增强分子的雌激素效应。

目前，将实验方法和分子模拟技术结合，用于 EDCs 毒性的模拟预测已成为重要的研究方法。李斐等[183]采用重组基因酵母实验测定了 20 种蒽醌类化合物（AQs）对 ERα 的干扰效应，并采用分子对接方法研究了 AQs 与雌激素受体 ERα 的作用机制，发现 AQs 能与谷氨酸残基(Glu353)、精氨酸残基(Arg394)、组氨酸残基（His524）形成氢键作用是其表现干扰效应的关键。基于机制分析，选取合适的描述符，构建了预测 AQs 雌激素干扰效应的预测模型。张爱茜等[184]通过细胞增殖和 MVLN 细胞转录激活实验评估了包括四溴双酚 A（TBBPA）、 四氯双酚 A（TCBPA）及四溴双酚 S（TBBPS）在内的 14 种双酚 A 类化合物，借助分子模拟方法研究了该类化合物的结合对 ERα LBD 第 12 号螺旋（H12）稳定性的影响。在其另一项研究中，分子对接发现 6-I-CPS 为 ERα 的弱激动剂，而体积相对较大的辣椒素类似物不能有效地结合于 ERα LBD，这进一步通过 MVLN 细胞转录激活实验得到证实[185]。郭良宏等[186,187]通过表面等离子体共振技术对 22 种羟基化多溴联苯醚、6 种全氟烷基酸类化学品与人 ERα LBD 的直接作用进行了动力学分析，发现 7 种羟基化多溴联苯醚（OH-PBDEs）、全氟辛烷磺酸（PFOS）及全氟辛酸（PFOA）具有弱雌激素干扰效应，进而采用分子对接方法分析了该类物质与人 ERα LBD 的结合模式。张照斌等[188-190]发现高溴代和低溴代羟基化多溴联苯醚在人 ERα LBD 结合口袋内的构象不同，潜在影响共激活因子的结合，这可能是导致该类化合物雌激素干扰效应差异的原因。多种酞酸酯具有显著雌激素活性，分子对接发现酞酸酯能与人 ERα LBD 相互作用，并与 Glu 353 等典型氨基酸形成氢键，同时部分酞酸酯还能与 ERRγ LBD 结合。除了雌激素受体干扰效应模型外，杨先海等[50,191]还构建了 EDCs 对人和鱼性激素运载蛋白干扰效应的 QSAR 和/或分类模型。

6.2　雄激素干扰效应相关研究

环境中的污染物能诱导或抑制雄激素受体（AR）转录激活，从而导致拟/抗雄激素效应。庄树林等[192]

采用重组基因酵母实验，发现不同链长的对羟基苯甲酸酯的 AR 拮抗效应存在差异，通过分子对接发现 AR 干扰效应主要依赖于对羟基苯甲酸酯侧链的长度和类型。对羟基苯甲酸酯芳香侧链是导致其 AR 干扰效应的重要毒效团（图 25-25）；当烷氧基侧链 C 原子数为 1~4 时，AR 干扰效应随侧链增长显著降低，当大于 7 时，无显著干扰效应。杨先海等[49]采用分子对接方法，研究了双酚 A 类似物与雄激素受体的结合机制，发现与天冬酰胺残基（Asn 705）、谷氨酰胺残基（Gln 711）、精氨酸残基（Arg 752）、苏氨酸残基（Thr 877）形成氢键是双酚 A 类似物表现雄激素干扰效应的关键。基于机理，选择合适的描述符构建了预测双酚 A 类似物抗雄激素的 QSAR 模型。于红霞等[193,194]基于分子对接、分子动力学方法系统研究了多氯联苯（PCBs）及羟化多氯联苯（OH-PCBs）与 AR LBD 相互作用的分子机制。他们从污染物结合干扰 AR LBD H12 螺旋的稳定性角度出发，采用分子动力学模拟考察了对羟化和甲氧基化的多溴联苯醚、黄酮类化合物与 AR LBD 的分子识别过程，发现污染物对 H12 螺旋的构象影响可作为 AR 拮抗剂的筛选标记物[195,196]。

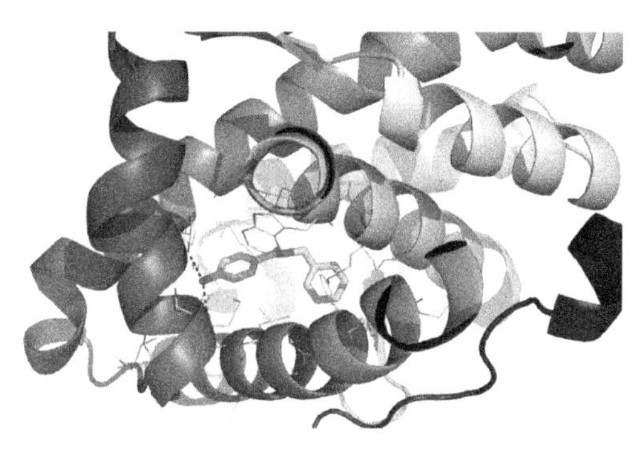

图 25-25　含芳香侧链对羟基苯甲酸酯与 AR LBD 的结合模式图

污染物的生物转化会潜在影响其 AR 干扰效应。庄树林等[192]发现酞酸酯经猪肝酯酶体外代谢后，介导其毒性效应的侧链被去除，生成对羟基苯甲酸，导致 AR 拮抗效应消除。他们还采用人重组基因酵母实验对 8 种苯并三唑类紫外线吸收剂在 CYP3A4 酶代谢后的 AR 干扰效应进行筛选，发现代谢后的 UV-328 拮抗效应显著增强，表现为代谢激活[197]。采用 DFT 方法，利用 Compound I 模型并结合 UPLC-Q-TOF-MS/MS 分析对其主要代谢位点进行了预测，分析了 UV-328 一次羟化和二次羟化的机制。

6.3　甲状腺干扰效应相关研究

针对具有甲状腺干扰效应的化学品（TDCs），目前在其毒性作用通路研究方面已取得一定研究进展[198]。TDCs 与甲状腺系统相关生物大分子的分子识别是诱导甲状腺干扰效应的潜在分子起始事件（MIE）。国内学者综合采用 QSAR、分子对接、分子动力学模拟等方法并结合多种实验方法，研究了化学品与甲状腺受体（TR）、甲状腺素转运蛋白（TTR）、甲状腺激素结合球蛋白（TBG）的分子作用机制。

李斐等[199]基于重组基因酵母实验，测定了 HO-PBDEs 和多溴联苯醚（PBDEs）的甲状腺干扰效应。分子对接结果表明 HO-PBDEs 与 TRβ LBD 存在氢键、π-π、疏水等相互作用。而 PBDEs（BDE-30 和 BDE-116）与 TRβ LBD 的氨基酸残基之间未发现氢键作用（图 25-26）。基于机理，他们假设 HO-PBDEs 与 TRβ 的相互作用与下述过程有关：① 化合物在水相和生物相间的分配；② HO-PBDEs 分子与 TRβ 间的相互作用。基于此，选取并计算了 12 个理论分子结构描述符来描述上述过程。采用正辛醇/水分配系数（$\log K_{\mathrm{OW}}$）、平均分子极化率（α）、偶极矩（μ）、体积（V）和溴原子数（n_{Br}）来表征分配过程；选择分子最高占据轨道能（E_{HOMO}）、分子最低未占据轨道能（E_{LUMO}）、羟基氢原子的形式电荷（$q_{\mathrm{O\underline{H}}}$）、羟基氧原子的形式电荷（$q_{\mathrm{\underline{O}H}}$）、醚键氧原子的形式电荷（$q_{\mathrm{O}}$）、亲电性指数（$\omega$）和芳香性指数（$I_{\mathrm{A}}$）来表征 HO-PBDEs

与 TRβ 之间的非健相互作用。参照 OECD 关于 QSAR 模型验证导则来构建与验证模型，构建的模型具有较好的拟合优度、稳健性和预测能力。

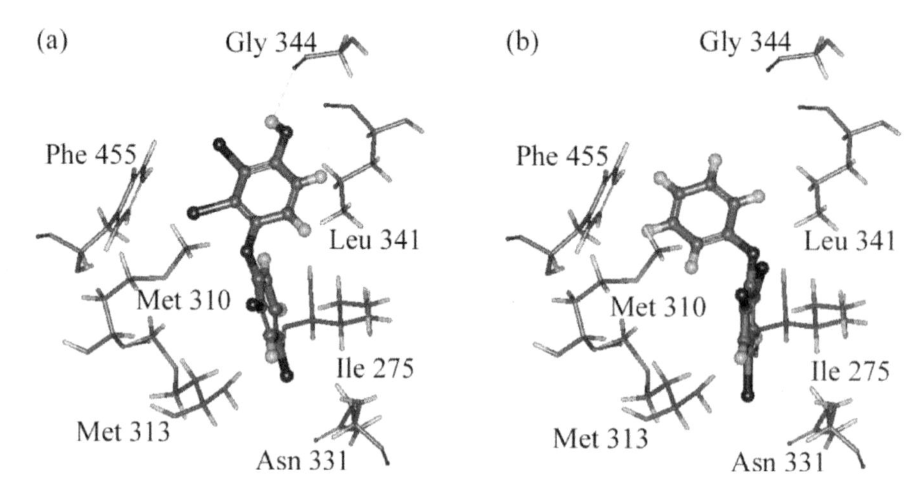

图 25-26　4-HO-BDE-42（a）和 BDE-30（b）在 TRβ 活性位点（PDB ID：1NAX 的结合构象）

此外，针对相同数据集，于海瀛等[200]通过引入能量限定供体电荷（energy-limited donor charge，QE_{occ}）和能量限定受体电荷（energy-limited acceptor charge，QE_{vac}）参数，构建了一个新的三参数模型，新模型具有较好的拟合优度和稳健性，特别是模型的预测能力显著提高。于红霞等[201]应用比较分子相似性指数（CoMSIA）方法构建模型。模型结果表明，干扰效应与立体场、静电场、氢键供体场、氢键受体场有关，且四种场对干扰效应的贡献率分别为 1.7%、44.8%、21.6%、31.6%，说明 OH-PBDEs 与 TRβ 的相互作用主要与静电和氢键相互作用相关。

除直接作用于 TR 受体外，OH-PBDEs 及全氟化合物还能与 TBG 和 TTR 结合，从而干扰甲状腺激素的正常转运过程。张爱茜等[202]通过分子动力学模拟和自由能打分研究了 OH-PBDEs 与 TTR 的分子识别过程，发现 OH-PBDEs 的羟基基团参与形成氢键的重要性。郭良宏等[203]通过荧光探针竞争结合实验发现全氟烷基磺酸盐类物质（PFASs）能够竞争甲状腺激素 T4 结合 TTR 和 TBG，且结合力与 PFASs 的链长显著相关。TTR 倾向结合中链长度的 PFASs，而 TBG 更适合长度大于 12 个碳原子的 PFASs。针对该构效关系，通过分子对接揭示了相关分子机制。

前人研究中，已注意到酚羟基、羧基、磺酸基等基团是 TDCs 的重要结构特征，且化合物引入上述基团后其甲状腺干扰效应显著增强，但其分子机制尚不清楚。一般认为它们的主要作用是参与形成氢键相互作用。然而，这些研究忽视了可电离基团在生理 pH（人类血浆~7.4）和试验 pH（7 ~ 8）条件下可电离的事实。那么离子化对 TDCs 的甲状腺干扰效应是否有影响呢？在结合过程中，分子形态和离子形态那个形态贡献更大呢？针对该问题，杨先海和陈景文等[204,205]采用分子对接、量子力学/分子力学（QM/MM）方法、分子动力学等分子模拟方法，以卤代酚、全氟/多氟羧酸、磺酸和人 TTR 为模型体系进行了系统的研究。结果显示，可电离卤代化合物与 TTR 的相互作用具有形态依赖性，且可电离卤代化合物阴离子形态与 TTR 的作用强于其对应的分子形态。TDCs 分子结构中的阴离子基团可与 TTR 配体结合空腔入口处的赖氨酸 15（Lys 15）残基形成离子对相互作用。由于形成具有强方向性的离子对相互作用，导致 TDCs 分子结构中的阴离子基团在 TTR 配体结合空腔中具有优势取向，即指向结合空腔的入口方向。分析了 73 个 TTR 晶体结构中可电离配体的结合模式，也发现晶体结构中配体的阴离子基团具有相同的优势取向。研究还发现卤代酚中芳环可与 TTR 的 Lys 15 残基侧链胺基阳离子形成阳离子-π 相互作用，且 TDCs 解离后，可增强阳离子-π 相互作用。与卤代酚不同，全氟/多氟羧酸、磺酸分子结构中不含芳环结构，因而不能参与形成阳离子-π 相互作用，这可能是全氟/多氟羧酸、磺酸等链状脂肪酸对 TTR 干扰效应弱与卤代酚类化合物的分子机制（图 25-27）。说明在实验测试、分子模拟及构建 QSAR 模型时，不应该忽视离子化对甲状腺干扰效应的影响。

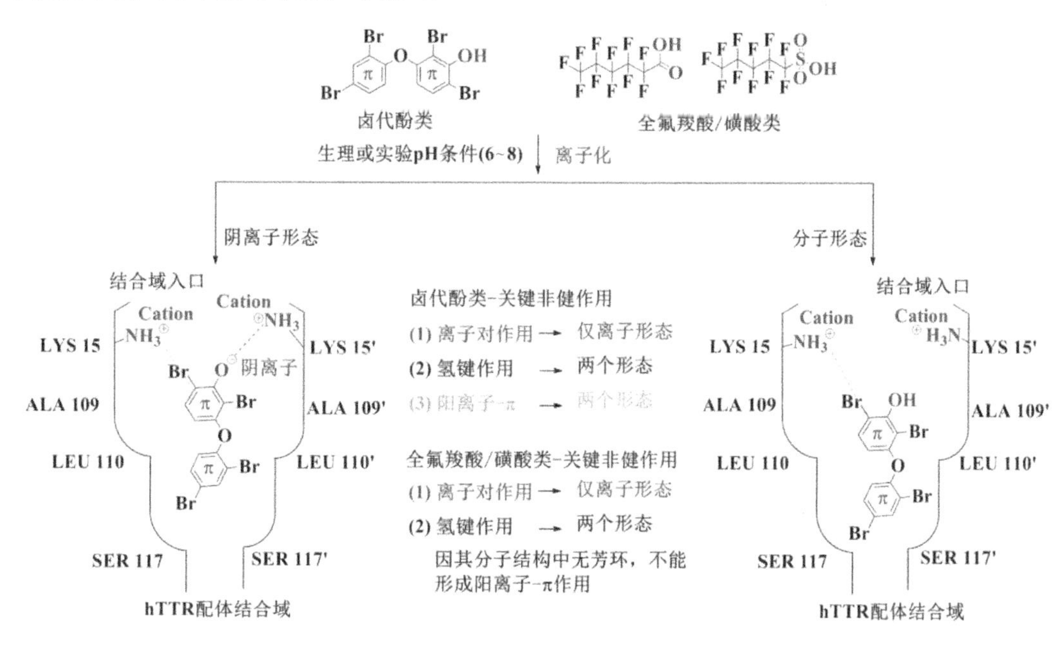

图 25-27 不同形态卤代酚及全氟羧酸/磺酸类衍生物与 TTR 结合机制示意图

此外，该结果还表明，在给定 pH 条件下，当化合物引入电离程度更强的基团时，其对 TTR 的干扰效应将增强。该推论已得到实验和理论验证。例如，Grimm 等[206]研究了 5 个 HO-PCBs 被磺基转移酶代谢转化为 PCBs 硫酸化物后，其对 TTR 的干扰效应变化情况。研究表明，除了 4'-HO-3, 4-二氯联苯外，其他 4 个 HO-PCBs 的硫酸化代谢物对 TTR 的干扰效应均强于其对应的母体。张爱茜等[202]采用分子动力学模拟也表明，多溴联苯醚硫酸化代谢物与 TTR 的相互作用强于 OH-PBDEs。因此，在将来筛选 TTR 潜在干扰物的研究中或在研究化合物代谢产物毒性时，应该进一步关注苯甲酸及其衍生物、苯磺酸及其衍生物等。

针对可电离化合物内分泌干扰效应构建 QSAR 模型的问题，他们根据机理提出了一套基于形态修正的描述符用于表征化合物离子化的影响，并成功应用于构建卤代酚、全氟/多氟羧酸、磺酸对人 TTR 干扰效应的 QSAR 模型。

6.4 孕烷 X 受体干扰效应相关研究

关于典型污染物诱导的孕烷 X 受体（pregnane X receptor，PXR）干扰效应，国内学者通过分子对接和 QSAR 阐述了构效关系和作用机制。张一鸣等[207]采用分子对接方法研究了不同链长的全氟化合物与 PXR LBD 的相互作用，发现全氟化合物结合于 PXR LBD 的 Phe 288，Trp 299，Tyr 306，Val 211，Met 243，Met 246 等疏水性残基形成的结合口袋，其极性基团与 Ser247 形成氢键。不同链长度的全氟化合物与 PXR LBD 相互作用的差异，决定了其 PXR 激动效应的强弱。侯廷军等[208]针对 PXR 受体的激动剂/拮抗剂的虚拟筛选，基于分子物理化学性质、VolSurf 描述符、分子指纹等参数，建立了朴素贝叶斯分类模型，用以区分 PXR 激动剂或拮抗剂，发现含卤素原子、不饱和烷烃链及少数氮原子的化合物更易诱导 PXR 激动效应。杨先海等[209]构建了包含 2724 个化合物的人 PXR 干扰效应分类模型，模型预测准确率大于 0.7，可用于预测应用域内其他化学物质对人 PXR 的干扰效应。

6.5 芳香烃受体干扰效应相关研究

芳香烃受体（AhR）能介导多类污染物的毒性效应。国内多位学者采用分子对接、分子动力学及 QSAR 等计算方法对污染物与 AhR 干扰效应的构效关系进行了探讨。赵斌等[210]采用分子对接研究了人参皂苷与 AhR LBD 的相互作用，发现其与四氯二苯-*p*-二噁英（TCDD）具有共同的结合口袋，能与口袋中心的极性残基（His 290 和 Gln 382）、入口处疏水残基（Phe 294，Met 339 和 Met 347）及内层空腔（Phe 286，Leu 307，Leu 314 和 Leu 352）相互作用，具有潜在激动效应。

HO-PCBs 及 OH-PBDEs 等污染物可能通过干扰 AhR 通路而表现出毒性效应。于红霞等[211,212]采用分子对接发现 HO-PCBs 及 OH-PBDEs 能与 AhR LBD 的多个重要氨基酸发生疏水作用，并存在 π-π 相互作用。进一步采用分子动力学模拟发现 HO-PCBs 的结合对 AhR LBD 的构象影响较小[211]。于红霞等[211]通过偏最小二乘回归（PLS）模型、比较分子力场分析模型（CoMFA）和比较分子相似性指数模型（CoMSIA）等方法，表征了 HO-PCBs 及 OH-PBDEs 与 AhR 相互作用的构效关系。李斐等[213]采用分子对接方法研究了 HO-PCBs、二噁英类物质与 AhR 的作用机制，在此基础上选择合适的描述符构建了预测模型。

6.6 维甲酸类 X 受体干扰效应相关研究

维甲酸类 X 受体（retinoid X receptor，RXR）可与其他受体结合形成异源二聚体，在介导细胞生长和凋亡方面起重要作用。国内学者将实验方法和分子模拟有机结合，对多种化合物与 RXR 的相互作用进行了研究。马梅等[214]采用酵母双杂实验发现双酚 A 的氯化副产物对 RXRβ 的拮抗效应强于双酚 A。分子对接发现氯化双酚 A 的结合模式与双酚 A 类似，与 Phe 384 存在 π-π 相互作用。于红霞等[215]发现高浓度的三氯乙基磷酸酯（TCEP）具有胚胎毒性，致使斑马鱼胚胎形态发生改变。进一步采用分子动力学模拟比较分析了 TCEP 与 9 种 RXR LBD 的相互作用。发现 TCEP 能与 RXRbb，RXRga 和 RXRgb 结合，并能形成稳定复合物。庄树林等[216]通过随机加速分子动力学模拟发现，当 RXR LBD 与 TR LBD 形成异源二聚体后，会显著影响甲状腺素 T3 进出 TR LBD 的通路，这为研究污染物对 RXR 的干扰效应提供了新的研究思路。

6.7 PPARγ 干扰效应相关研究

过氧化物酶体增殖因子活化受体（peroxisome proliferator-activated receptor，PPAR）有 α、β 以及 γ 亚型。李铁柱等[217]采用荧光偏振实验并结合分子对接，揭示了双酚 A、双酚 F、双酚 A 二缩水甘油醚（BADGE）和双酚 F 二缩水甘油醚（BFDGE）四种物质与小鼠 PPARα LBD 的相互作用，发现双酚 A 类物质与小鼠 PPARα LBD 的结合驱动力主要为疏水作用和氢键。庄树林等[181]的分子动力学模拟结果发现双酚A及其类似物能与PPARγ LBD发生相互作用，且苯环上溴原子数量与静电作用和范德华力高度相关。不同程度的卤代作用导致了部分双酚化合物与 PPARγ LBD 相互作用的较大差异。胡建英等[218]通过双荧光素酶报告基因比较分析了己基二苯基磷酸酯与磷酸三苯酯对 PPARγ 受体的激动效应，分子对接方法发现这两种物质与 PPARγ LBD 结合模式的不同是导致其激动效应差异的重要因素。

由于污染物可通过多种途径诱导环境内分泌干扰效应，可基于有害结局路径（AOP）研究框架，甄别毒性通路相关的分子起始事件（MIE）。从 MIE 角度出发研究污染物与多种核受体相关大分子的分子识别过程，分析其与多种受体的协同作用。目前与环境污染物毒性相关的功能蛋白被越来越多地解析处理，这为 MIE 相关研究提供了基础。鉴于污染物分子形态的不同会影响分子识别过程，需合理选择 MIE 研究相关的生物大分子三维结构[219]。在具体研究相互作用过程中，可考虑借助系统生物学、化学信息学、分子模拟等方法，将计算方法和基于细胞、分子水平的实验方法有机结合[220]。

7　污染物水生毒性效应的模拟预测

开展水环境生态风险评价，需要水生毒性数据和暴露数据。理想情况下，需要对所有水生生物进行毒性评估，以选择最敏感物种的数据进行风险评估，从而达到保护所有水生生物的目的。然而由于水生生物种类繁多，难以对其进行一一测试。为应对此矛盾，提出的解决方案是选择代表性水生生物作为模式生物，通过测定物质对模式生物的毒性效应数据，来进行化学物质登记、分类和标签、预测无效应浓度推导等，并以此为依据制定化学物质管制措施。典型的水生模式生物包括：①水生植物，如藻类（如羊角月牙藻 *Pseudokirchneriella subcapitata*）等；②无脊椎动物，如大型溞（*Daphnia magna*）等；③ 脊椎动物，如斑马鱼（*Danio rerio*）、黑头呆鱼（*Pimephales promelas*）、稀有鮈鲫（*Gobiocypris rarus*）等。通过针对这些物种开发的标准测试方法，可获取用于管理的急慢性水生毒性效应指标。

虽然水生毒性效应测试体系已建立约 30 年，但仅少部分物质具有水生毒性数据。欧美国家大力倡导使用预测技术填补缺失的数据，例如，联合国《全球化学品统一分类和标签制度》、欧盟《化学品注册、登记、许可与限制法规》、美国《有毒物质控制法案》等法规中都明确规定，预测数据可用于化学物质管理。因此，构建污染物水生毒性效应预测模型对实现水环境化学物质的风险管理具有重要意义。

近年来，国内研究人员针对水生急性毒性构建了预测模型。针对藻[以绿藻（如羊角月牙藻，*Pseudokirchneriella subcapitata*）为主]、溞（以大型溞 *Daphnia magna* 为主）、鱼（黑头呆鱼 *Pimephales promelas*）的预测模型较多，大部分模型是基于同类化合物而构建的局域模型。污染物类别包括取代酚、取代芳烃、硝基苯、取代苯胺、取代脂肪烃等。由于模型所用实验数据具有相同或相近的实验条件，导致数据不确定性较小。因此，这类模型一般具有较好的拟合优度，且模型简单透明，描述符较少，有些仅含疏水性描述符。但是，这类模型也由于覆盖的物质种类和数量太少，导致应用域较窄，实际应用受到较大的限制。

由于全局模型所含化合物数量较多，结构差异性也较大，这些化合物往往具有不同的毒性作用模式（MOA）[221,222]，导致很难用一个统一的模型预测所有化合物。基于 MOA 构建水生急性毒性模型是目前较为认可的研究方法。根据 Verhaar 分类方法，水生急性毒性 MOA 分为四类，即非极性麻醉毒性、急性麻醉毒性、反应活性、特殊反应活性。李雪花等[223,224]根据 Verhaar 分类方法，分别构建了非极性麻醉毒性物质、急性麻醉毒性物质、反应活性物质、特殊反应活性物质及不能被 Verhaar 方法分类的物质对黑头呆鱼急性毒性（96 h LC_{50}）的线性溶解能关系（LSER）和理论线性溶解能关系（TLSER）预测模型（696 种物质）。在构建 TLSER 模型过程中，在传统 LSER 模型所包含的空穴项、极性项、氢键项（含氢键受体和氢键供体项）的基础上，提出增加电子供体-受体项，采用化学势、化学硬度、亲电性指数、电子亲和势、电离势等量子化学参数表征电子供体-受体项，构建的模型预测效果较好；针对大型溞急性毒性，基于 961 个化合物毒性数据（48 h EC_{50}），提出了大型溞急性毒性的 MOA 分类方法，并构建了麻醉毒性物质、反应活性物质、特殊反应活性物质预测模型。与基于 Verhaar 分类方法的模型相比，基于新分类方法的模型其拟合优度、稳健性和预测能力都显著提高[225]。

张庆竹等[226]采用毒性比率法，构建了麻醉毒性化合物（即非极性麻醉毒性和急性麻醉毒性物质）和过量毒性化合物（即反应活性和特殊反应活性物质）对黑头呆鱼（96 h LC_{50}，963 物质）急性毒性的预测模型。赵元慧等[227,228]也采用毒性比率法，构建了 758 个化合物对大型溞（48 h EC_{50}）的预测模型、949

个化合物对鱼（96 h LC_{50}）急性毒性模型；石利利等基于分子碎片，采用 BP 神经网络，构建了鱼类急性毒性（96 h LC_{50}）预测模型（223 物质）。这些全局模型基本都按照 OECD 关于模型验证的导则，对模型进行了拟合优度、稳健性、预测能力表征，且各参数值都较高，还进行了应用域表征，因而具有较大的应用前景。

此外，国内学者还构建了污染物对发光菌[主要为费氏弧菌（*Vibrio fischeri*）、青海弧菌 Q67（*Vibrio qinghaiensis* sp. Q67）等]、梨形四膜虫（*Tetrahymena pyriformis*）等水生急性毒性的预测模型[228-234]。大部分模型也是基于同类化合物而构建的局域模型。在全局模型方面，赵元慧等[228]采用毒性比率法，分别构建了 1239 个化合物对费氏弧菌的急性毒性（15 min 或 30 min EC_{50}）模型和 990 多个化合物对梨形四膜虫急性毒性（IGC_{50}，50%生长抑制浓度）的预测模型；陆文聪等[229]采用支持向量机方法，构建了 581 个化合物对梨形四膜虫急性毒性（IGC_{50}，50%生长抑制浓度）的预测模型。

8 展　　望

环境计算化学作为环境化学的一门衍生学科，继承了环境化学中的计算模拟的知识。这些知识包括：针对化学物质（尤其有机化学物质）物理化学性质、环境行为和毒理学效应（降解动力学、环境分配、毒性等）的 QSAR 模型、化学污染物的源解析技术以及模拟化学物质环境迁移、转化、分布、归趋的多介质环境模型等。如果说，过去在环境化学的大学科背景下，这些部分相对独立，各自满足着环境化学某些方面的需求；那么，在计算环境化学的框架下，所有这些计算模拟的知识和方法将被整合到一起——人类已经有能力在计算机中构建一个我们赖以生存的真实环境的“复本”，并模拟和识别化学物质的引入为该环境带来的从微观到宏观的变化。

长久以来，QSAR 被用于环境化学领域，预测化学品的相关性质和毒理效应。近十年来，计算机性能的逐渐提升，极大地推动了基于量子力学、经典力学理论的分子模拟方法的发展和模拟体系尺度的拓展。这种分子模型模拟的计算化学方法作为一种精度较高、机理透明的新方法，与旧有的基于统计学的方法相互渗透和影响，拓展并加速环境计算化学的进步。首先，在一些高精度理论水平下，针对小体系的量子化学计算结果具备作为独立判据的能力。因此，这些计算结果，能够直接弥补观测和实验数据。其次，从分子模型模拟中抽取的描述符还可用于校准 QSAR，让后者预测效果更好；同时针对环境归趋、转化通路和效应机理解释时更有说服力。总体而言，这些高级的分子模拟手段帮助环境化学家跳出计算特定的化学品性质的统计学模型（主要指 QSAR 模型）局限，拓展模型应用域从化学物质到生物大分子再到各种材料，让其能够进一步预测环境化学物质的转化途径和产物，将零散的环境因子嵌入系统性的模型框架之中，并灵活调用底层环境化学数据库，整合各类预测模型到更复杂和满足决策者需求的顶层工具之中。

经典的多介质环境模型仍然有改进的空间。自从 1979 年环境化学家 Mackay 提出逸度模型的理念之后，在将近 40 年里，逸度模型被广泛运用于评价河流、湖泊系统的化学品环境分布以及污水处理厂中某些化学物质的归趋。这个模型本质上由大量的化学品物理化学性质和环境行为参数构成。因此，相关参数的模拟和计算就成为对应化学物质能够被应用于逸度模型的瓶颈。多介质环境模型对环境做出了很大程度的简化，真实的环境远比逸度模型复杂。在未来，管理者也不会单纯满足于一个粗放的环境介质模型——精细化的环境管理要求对化学物质的环境归趋有更高分辨的预测能力。基于 GIS 的精细化的逸度模型是一种有价值的发展方向，结合环境数据实时监测和大数据分析，这方面的技术成果很可能会变革现有环境管理的模式，极大地提升化学品管理的效率。

随着计算模拟研究的进步，计算模拟手段的丰富和成熟，计算模拟精准度的逐渐提升，环境化学将逐渐变身为理论和实验并重的、数学表述更完整的一门学科。这种趋势表现在环境计算化学的方方面面——从化学物质的迁移转化的分子机制，到污染控制过程中催化剂发挥作用的微观机制，再到全球气候变化的宏观模型，可看到计算模拟发挥作用的场景正在逐渐增加。这些计算模拟的理论研究是对于实验和野外研究的良好补充和支撑，同时也是将实验研究成果归纳转化为决策支持工具或污染控制工艺辅助设计工具的重要纽带。

从研究内容上看，计算（预测）毒理学可以看作是计算环境化学在生态毒理学方面的一个分支（或许可称之为“计算生态毒理学”）。计算毒理学在推动传统毒理学变革方面发挥着重要的历史作用，包括致力于减少毒性测试所需的动物实验，并努力提升化学品风险评价的效率等。由于计算毒理学最初的目标在于服务化学品风险评价，其研究内容又可明显区别出暴露评价和毒性效应评价两个略有不同的方向。其中关于暴露评价的计算模拟，主要由环境多介质模型、毒代动力学模型以及预测这些模型所需参数的 QSAR 模型等构成。该部分主要希望合理估算化学品的环境浓度水平或生物体内暴露水平。这样，就能够与一些同级别的效应阈值进行比较。

此外，一些研究针对气相自由基反应与化学物质的反应速率的计算模拟能够直接计算出较为合理的反应速率常数等判断化学品环境持久性的关键特征值，从而直接服务于化学品风险评价的评定。这类方法正在逐渐成为一种趋势，一方面取代传统 QSAR 模型为化学品风险管理提供数据，另一方面也提供为 QSAR 建模启发了机理性的描述符——这一发展趋势和计算环境化学高度一致，因为气相自由基量子化学模拟计算研究本身也属于计算环境化学。更精细和高级的暴露模型则诞生于 ExpoCast 项目，这些模型的构架已不是多介质环境模型或 QSAR 模型等的简单组合，其形式复杂了许多。这也反映出暴露模型研究拥有自身的特点，对应的研究手段和研究重点也明显不同于毒理学研究。2012 年，美国 NRC 针对暴露科学发表了策略和远景的报告[235]，全面阐述了暴露科学内涵和外延，标志着暴露科学形成自身体系。2016 年，关于计算暴露科学的概念被提出[236]，从而标志暴露科学已经拥有了独具特色的计算模拟研究思路和模型框架。相信这些更加成熟的暴露科学理念会反过来丰富计算毒理学的研究，并提供更准确的暴露水平评估。

早期，计算毒理学研究中关于毒性效应评价的模型，仍然以发展预测各种毒性终点的 QSAR 模型为主。同时，一些基于分子对接或分子动力学模拟的技术也被用于辅助判断化学物质对已知的特定生物大分子的相互作用强度。这些技术通过计算配体分子与特定蛋白质的结合能，来关联实验测定的 IC50 或结合/解离平衡常数等数值，从而起到预测化学物质效应强度的作用。同时，EPA 在基础数据库（ACToR 项目）的构建和 *in vitro* 毒性测试技术（如 ToxCast 项目）的发展方面做出了重要的贡献，让相关 QSAR 模型的构建和验证变得更加便捷和规范化。其间，有学者试图将 *in vitro* 测试的数据作为一种对化学物质的描述符，与传统的化学结构描述符一起预测对应的 *in vivo* 毒性，即所谓的定量结构-体外-体内活性关系（QSIIR）模型，然而，这些模型并未取得很理想的预测效果，这很有可能是因为 *in vitro* 终点本身对于 *in vivo* 终点之间并不是简单线性关系。另一方面，EPA 还在构建虚拟肝等宏观尺度的模型上投入了研究力量，并取得了一些阶段性的成果[237,238]，然而，由于所在国家政治制度导致的政策的不连续性以及人力物力有限，这些项目也陷入了停滞。2010 年，Ankley 等提出了有害结局路径（AOP）的概念，希望把毒理学研究中零散的和模糊的概念都重新整合到这一框架之下。AOP 本身并未提出新的方法，而意在梳理和整合现有方法。AOP 采用了一个跨越多级空间尺度的视野，提出任何由化学物质引起的人体或生态健康的有害效应都是起始于外源化学分子与生物大分子的相互作用，即所谓的分子起始事件（MIE）。MIE 可以是化学分子作用于蛋白质，改变后者的结构功能，也可能来自于一次代谢修饰，让原始分子转变为一个有毒性的产物。MIE 的提出再次强调了分子尺度的模型模拟对于毒理学研究的重要性。相关的研究方法，包括研究酶代谢的量子力学/分子力学混合方法、研究生物物理过程的分子动力学模拟等被广泛地运用到分子毒理学研究当中。这部分计算模拟极大地丰富了毒理学家对化学品引发毒性的微观机制的认识，有助于人们从化学结构本身出发，来诊断环境污染物可能具备的危害特质，也是未来计算毒理学研究的一个热点领域。

同样地，随着 AOP 重新将生物信号网络的重要性展现给毒理学家，以及近年来组学技术突飞猛进的发展，化学品对生物信号的改变的研究也成为毒理学研究的一大重点。而组学的本质决定了大量的数据需要被处理，对数据挖掘和机器学习模型提出了新的挑战。可以通过挖掘组学数据，识别出更多与化学物质的有害效应相关的毒性通路。这些毒性通路网络的构建，既是 AOP 的重要工作内容，也是计算毒理学的前沿应用方向。现阶段，绝大多数的 AOP 都是定性的，在未来，这些定性的静态模型会逐渐被转变为定量的、动态的模型。在这个过程中，需要计算毒理学这个学科做出贡献。

参 考 文 献

[1] Rockstrom J, Steffen W, Noone K, Persson A, Chapin F S, Lambin E F, Lenton T M, Scheffer M, Folke C, Schellnhuber H J, Nykvist B, de Wit C A, Hughes T, van der Leeuw S, Rodhe H, Sorlin S, Snyder P K, Costanza R, Svedin U, Falkenmark M, Karlberg L, Corell R W, Fabry V J, Hansen J, Walker B, Liverman D, Richardson K, Crutzen P, Foley J A. A safe operating space for humanity. Nature, 2009, 461(7263): 472-475.

[2] Jepson P D, Law R J. Persistent pollutants, persistent threats. Science, 2016, 352(6292): 1388-1389.

[3] Whitehead T P, Smith S C, Park J S, Petreas M X, Rappaportt S M, Metayert C. Concentrations of persistent organic pollutants in california children's whole blood and residential dust. Environ Sci Technol, 2015, 49(15): 9331-9340.

[4] Gui D, Karczmarski L, Yu R Q, Plon S, Chen L G, Tu Q, Cliff G, Wu Y P. Profiling and spatial variation analysis of persistent organic pollutants in South African Delphinids. Environ Sci Technol, 2016, 50(7): 4008-4017.

[5] Josefsson S, Bergknut M, Futter M N, Jansson S, Laudon H, Lundin L, Wiberg K. Persistent organic pollutants in streamwater: Influence of hydrological conditions and landscape type. Environ Sci Technol, 2016, 50(14): 7416-7424.

[6] Liu A F, Shi J B, Qu G B, Hu L G, Ma Q C, Song M Y, Jing C Y, Jiang G B. Identification of emerging brominated chemicals as the transformation products of tetrabromobisphenol A (TBBPA) derivatives in soil. Environ Sci Technol, 2017, 51(10): 5434-5444.

[7] Zhang H Y, Vestergren R, Wang T, Yu J C, Jiang G B, Herzke D. Geographical differences in dietary exposure to perfluoroalkyl acids between manufacturing and application regions in China. Environ Sci Technol, 2017, 51(10): 5747-5755.

[8] Zhao X C, Yu M, Xu D, Liu A F, Hou X W, Hao F, Long Y M, Zhou Q F, Jiang G B. Distribution, bioaccumulation, trophic transfer, and influences of CeO_2 nanoparticles in a Constructed Aquatic Food Web. Environ Sci Technol, 2017, 51(9): 5205-5214.

[9] Liu Y W, Ruan T, Lin Y F, Liu A F, Yu M, Liu R Z, Meng M, Wang Y W, Liu J Y, Jiang G B. Chlorinated polyfluoroalkyl ether sulfonic acids in marine organisms from Bohai Sea, China: Occurrence, temporal variations, and trophic transfer behavior. Environ Sci Technol, 2017, 51(8): 4407-4414.

[10] Cao D D, Guo J H, Wang Y W, Li Z N, Liang K, Corcoran M B, Hosseini S, Bonina S M C, Rockne K J, Sturchio N C, Giesy J P, Liu J F, Li A, Jiang G B. Organophosphate esters in sediment of the Great Lakes. Environ Sci Technol, 2017, 51(3): 1441-1449

[11] Rappaport S M, Smith M T. Environment and disease risks. Science, 2010, 330(6003): 460-461.

[12] Wu S, Powers S, Zhu W, Hannun Y A. Substantial contribution of extrinsic risk factors to cancer development. Nature, 2016, 529(7584): 43-47.

[13] Steffen W, Richardson K, Rockstrom J, Cornell S E, Fetzer I, Bennett E M, Biggs R, Carpenter S R, de Vries W, de Wit C A,

Folke C, Gerten D, Heinke J, Mace G M, Persson L M, Ramanathan V, Reyers B, Sorlin S. Planetary boundaries: Guiding human development on a changing planet. Science, 2015, 347(6223): 736-746.

[14] Bernhardt E S, Rosi E J, Gessner M O. Synthetic chemicals as agents of global change. Front Ecol Environ, 2017, 15(2): 84-90.

[15] Kavlock R, Dix D. Computational toxicology as implemented by the U.S. EPA: Providing high throughput decision support tools for screening and assessing chemical exposure, hazard and risk. J Toxicol Env Heal B, 2010, 13(2-4): 197-217.

[16] Tian F L, Chen J W, Qiao X L, Wang Z, Yang P, Wang D G, Ge L K. Sources and seasonal variation of atmospheric polycyclic aromatic hydrocarbons in Dalian, China: Factor analysis with non-negative constraints combined with local source fingerprints. Atmos Environ, 2009, 43(17): 2747-2753.

[17] Liu C H, Tian F L, Chen J W, Li X H, Qiao X L. A comparative study on source apportionment of polycyclic aromatic hydrocarbons in sediments of the Daliao River, China: Positive matrix factorization and factor analysis with non-negative constraints. Chinese Sci Bull, 2010, 55(10): 915-920.

[18] Sun J F, Peng H Y, Chen J M, Wang X M, Wei M, Li W J, Yang L X, Zhang Q Z, Wang W X, Mellouki A. An estimation of CO_2 emission via agricultural crop residue open field burning in China from 1996 to 2013. J Clean Prod, 2016, 112: 2625-2631.

[19] Zhao N, Shi X L, Xu F, Zhang Q Z, Wang W X. Theoretical investigation on the mechanism of NO_3 radical-initiated atmospheric reactions of phenanthrene. J Mol Struct, 2017, 1139: 275-281.

[20] Altarawneh M, Dlugogorski B Z. Thermal Decomposition of 1,2-Bis(2,4,6-tribromophenoxy)ethane (BTBPE), a Novel Brominated Flame Retardant. Environ Sci Technol, 2014, 48(24): 14335-14343.

[21] 王中钰, 陈景文, 乔显亮, 李雪花, 谢宏彬, 蔡喜运. 面向化学品风险评价的计算(预测)毒理学. 中国科学: 化学, 2016, 46(2): 1-21.

[22] Kleandrova V V, Luan F, Gonzalez-Diaz H, Ruso J M, Speck-Planche A, Cordeiro M N D S. Computational tool for risk assessment of nanomaterials: Novel QSTR-perturbation model for simultaneous prediction of ecotoxicity and cytotoxicity of uncoated and coated nanoparticles under multiple experimental conditions. Environ Sci Technol, 2014, 48(24): 14686-14694.

[23] Kleandrova V V, Luan F, Gonzalez-Diaz H, Ruso J M, Melo A, Speck-Planche A, Cordeiro M N D S. Computational ecotoxicology: Simultaneous prediction of ecotoxic effects of nanoparticles under different experimental conditions. Environ Int, 2014, 73: 288-294.

[24] Manganelli S, Leone C, Toropov A A, Toropova A P, Benfenati E. QSAR model for predicting cell viability of human embryonic kidney cells exposed to SiO_2 nanoparticles. Chemosphere, 2016, 144: 995-1001.

[25] Toropov A A, Toropova A P. Quasi-SMILES and nano-QFAR: United model for mutagenicity of fullerene and MWCNT under different conditions. Chemosphere, 2015, 139: 18-22.

[26] Toropov A A, Toropova A P. Quasi-QSAR for mutagenic potential of multi-walled carbon-nanotubes. Chemosphere, 2015, 124: 40-46.

[27] Jin X H, Peldszus S, Huck P M. Predicting the reaction rate constants of micropollutants with hydroxyl radicals in water using QSPR modeling. Chemosphere, 2015, 138: 1-9.

[28] Yang Z H, Luo S, Wei Z S, Ye T T, Spinney R C, Chen D, Xiao R Y. Rate constants of hydroxyl radical oxidation of polychlorinated biphenyls in the gas phase: A single-descriptor based QSAR and DFT study. Environ Pollut, 2016, 211: 157-164.

[29] Li C, Yang X H, Li X H, Chen J W, Qiao X L. Development of a model for predicting hydroxyl radical reaction rate constants of organic chemicals at different temperatures. Chemosphere, 2014, 95: 613-618.

[30] Li C, Xie H B, Chen J W, Yang X H, Zhang Y F, Qiao X L. Predicting gaseous reaction rates of short chain chlorinated

paraffins with center dot OH: Overcoming the difficulty in experimental determination. Environ Sci Technol, 2014, 48(23): 13808-13816.

[31] Schindler M. A QSAR for the prediction of rate constants for the reaction of VOCs with nitrate radicals. Chemosphere, 2016, 154: 23-33.

[32] Xiao R Y, Ye T T, Wei Z S, Luo S, Yang Z H, Spinney R. Quantitative structure-activity relationship (QSAR) for the oxidation of trace organic contaminants by sulfate radical. Environ Sci Technol, 2015, 49(22): 13394-13402.

[33] Lee M, Zimmermannsteffens S G, Arey J S, Fenner K, Gunten U V. Development of prediction models for the reactivity of organic compounds with ozone in aqueous solution by quantum chemical calculations: The role of delocalized and localized molecular orbitals. Environ Sci Technol, 2015, 49(16): 9925-9935.

[34] Li X H, Zhao W X, Li J, Jiang J Q, Chen J J, Chen J W. Development of a model for predicting reaction rate constants of organic chemicals with ozone at different temperatures. Chemosphere, 2013, 92(8): 1029-1034.

[35] Kim M, Li L Y, Grace J R. Predictability of physicochemical properties of polychlorinated dibenzo-*p*-dioxins (PCDDs) based on single-molecular descriptor models. Environ Pollut, 2016, 213: 99-111.

[36] Liu H H, Wei M B, Yang X H, Yin C, He X. Development of TLSER model and QSAR model for predicting partition coefficients of hydrophobic organic chemicals between low density polyethylene film and water. Sci Total Environ, 2017, 574: 1371-1378.

[37] O'Loughlin D R, English N J. Prediction of Henry's Law Constants via group-specific quantitative structure property relationships. Chemosphere, 2015, 127: 1-9.

[38] Wang Y, Chen J W, Yang X H, Lyakurwa F, Li X H, Qiao X L. In silico model for predicting soil organic carbon normalized sorption coefficient (K_{OC}) of organic chemicals. Chemosphere, 2015, 119, : 438-444.

[39] Martin T J, Goodhead A K, Acharya K, Head I M, Snape J R, Davenport R J. High throughput biodegradation-screening test to prioritize and evaluate chemical biodegradability. Environ Sci Technol, 2017, 51(12): 7236-7244.

[40] Grisoni F, Consonni V, Vighi M, Villa S, Todeschini R. Investigating the mechanisms of bioconcentration through QSAR classification trees. Environ Int, 2016, 88: 198-205

[41] Laue H, Gfeller H, Jenner K J, Nichols J W, Kern S, Natsch A. Predicting the bioconcentration of fragrance ingredients by rainbow trout using measured rates of *in vitro* intrinsic clearance. Environ Sci Technol, 2014, 48(16): 9486-9495.

[42] Freitas M R, Barigye S J, Dare J K, Freitas M P. Quantitative modeling of bioconcentration factors of carbonyl herbicides using multivariate image analysis. Chemosphere, 2016, 152: 190-195.

[43] Grisoni F, Consonni V, Villa S, Vighi M, Todeschini R. QSAR models for bioconcentration: Is the increase in the complexity justified by more accurate predictions? Chemosphere, 2015, 127: 171-179.

[44] Liu H, Liu H X, Sun P, Wang Z Y. QSAR studies of bioconcentration factors of polychlorinated biphenyls (PCBs) using DFT, PCS and CoMFA. Chemosphere, 2014, 114: 101-105.

[45] Ng H W, Doughty S W, Luo H, Ye H, Ge W, Tong W, Hong H. Development and validation of decision forest model for estrogen receptor binding prediction of chemicals using large data sets. Chem Res Toxicol, 2015, 28(12): 2343-2351.

[46] AbdulHameed M D M, Ippolito D L, Wallqvist A. Predicting rat and human pregnane X receptor activators using bayesian classification models. Chem Res Toxicol, 2016, 29(10): 1729-1740.

[47] Pinto C L, Mansouri K, Judson R S, Browne P. Prediction of estrogenic bioactivity of environmental chemical metabolites. Chem Res Toxicol, 2016, 29(9): 1410-1427.

[48] Norinder U, Boyer S. Conformal Prediction Classification of a large data set of environmental chemicals from ToxCast and Tox21 estrogen receptor assays. Chem Res Toxicol, 2016, 29(6): 1003-1010.

[49] Yang X H, Liu H H, Qian Y, Liu J N, Chen J W, Shi L L. Predicting anti-androgenic activity of bisphenols using molecular

docking and quantitative structure-activity relationships. Chemosphere, 2016, 163: 373-381.

[50] Liu H H, Yang X H, Lu R. Development of classification model and QSAR model for predicting binding affinity of endocrine disrupting chemicals to human sex hormone-binding globulin. Chemosphere, 2016, 156: 1-7.

[51] Bhhatarai B, Wilson D M, Price P S, Marty S, Parks A K, Carney E. Evaluation of OASIS QSAR Models Using ToxCast™ in Vitro Estrogen and Androgen Receptor Binding Data and Application in an Integrated Endocrine Screening Approach. Environ Health Persp, 2016, 124(9): 1453-1461.

[52] Tratnyek P G, Bylaska E J, Weber E J. In silico environmental chemical science: Properties and processes from statistical and computational modelling. Environ Sci Process Impacts, 2017, 19(3): 188-202.

[53] Yu H, Ge P, Chen J W, Xie H B, Luo Y. The degradation mechanism of sulfamethoxazole under ozonation: a DFT study. Environ Sci Process Impacts, 2017, 19(3): 379-387.

[54] Yu H, Chen J W, Xie H B, Ge P, Kong Q W, Luo Y. Ferrate(VI) initiated oxidative degradation mechanisms clarified by DFT calculations: a case for sulfamethoxazole. Environ Sci Process Impacts, 2017, 19(3): 370-378.

[55] Luo X, Yang X H, Qiao X L, Wang Y, Chen J W, Wei X X, Peijnenburg W. Development of a QSAR model for predicting aqueous reaction rate constants of organic chemicals with hydroxyl radicals. Environ Sci Proc Impacts, 2017, 19(3): 350-356.

[56] Jin X C, Fu Z Q, Li X H, Chen J W. Development of polyparameter linear free energy relationship models for octanol-air partition coefficients of diverse chemicals. Environ Sci Proc Impacts, 2017, 19(3): 300-306.

[57] Xu F, Yu W N, Gao R, Zhou Q, Zhang Q Z, Wang W X. Dioxin formations from the radical/radical cross-condensation of phenoxy radicals with 2-chlorophenoxy radicals and 2,4,6-trichlorophenoxy radicals. Environ Sci Technol, 2010, 44(17): 6745-6751.

[58] Pan W X, Zhang D J, Han Z, Zhan J H, Liu C B. New insight into the formation mechanism of PCDD/Fs from 2-chlorophenol precursor. Environ Sci Technol, 2013, 47(15): 8489-8498.

[59] Zhang Q Z, Qu X H, Wang H, Xu F, Shi X Y, Wang W X. Mechanism and thermal rate constants for the complete series reactions of chlorophenols with H. Environ Sci Technol, 2009, 43(11): 4105-4112.

[60] Shi X L, Yu W N, Xu F, Zhang Q Z, Hu J T, Wang W X. PBCDD/F formation from radical/radical cross-condensation of 2-Chlorophenoxy with 2-Bromophenoxy, 2,4-Dichlorophenoxy with 2,4-Dibromophenoxy, and 2,4,6-Trichlorophenoxy with 2, 4, 6-Tribromophenoxy. J Hazard Mater, 2015, 295: 104-111.

[61] Yu W N, Hu J T, Xu F, Sun X Y, Gao R, Zhang Q Z, Wang W X. Mechanism and direct kinetics study on the homogeneous gas-phase formation of PBDD/Fs from 2-BP, 2,4-DBP, and 2,4,6-TBP as Precursors. Environ Sci Technol, 2011, 45(5): 1917-1925.

[62] Xu F, Yu W N, Zhou Q, Gao R, Sun X Y, Zhang Q Z, Wang W X. Mechanism and direct kinetic study of the polychlorinated dibenzo-*p*-dioxin and dibenzofuran formations from the radical/radical cross-condensation of 2,4-dichlorophenoxy with 2-chlorophenoxy and 2,4,6-trichlorophenoxy. Environ Sci Technol, 2011, 45(2): 643-650.

[63] Zhang Q Z, Yu W N, Zhang R X, Zhou Q, Gao R, Wang W X. Quantum chemical and kinetic study on dioxin formation from the 2,4,6-TCP and 2,4-DCP precursors. Environ Sci Technol, 2010, 44(9): 3395-3403.

[64] Xu F, Wang H, Zhang Q Z, Zhang R X, Qu X H, Wang W X. Kinetic properties for the complete series reactions of chlorophenols with OH radicals-relevance for dioxin formation. Environ Sci Technol, 2010, 44(4): 1399-1404.

[65] Qu X H, Wang H, Zhang Q Z, Shi X Y, Xu F, Wang W X. Mechanistic and kinetic studies on the homogeneous gas-phase formation of PCDD/Fs from 2,4,5-trichlorophenol. Environ Sci Technol, 2009, 43(11): 4068-4075.

[66] Weber R, Hagenmaier H. Mechanism of the formation of polychlorinated dibenzo-*p*-dioxins and dibenzofurans from chlorophenols in gas phase reactions. Chemosphere, 1999, 38(3): 529-549.

[67] Ryu J Y, Mulholland J A, Takeuchi M, Kim D H, Hatanaka T. $CuCl_2$-catalyzed PCDD/F formation and congener patterns

from phenols. Chemosphere, 2005, 61(9): 1312-1326.

[68] Zhang Y F, Zhang D J, Gao J, Zhan J H, Liu C B. New understanding of the formation of PCDD/Fs from chlorophenol precursors: A mechanistic and kinetic study. J Phys Chem A, 2014, 118(2): 449-456.

[69] Pan W X, Fu J J, Zhang A Q. Theoretical study on the formation mechanism of pre-intermediates for PXDD/Fs from 2-Bromophenol and 2-Chlorophenol precursors via radical/molecule reactions. Environ Pollut, 2017, 225: 439-449.

[70] Yu X Q, Chang J M, Liu X, Pan W X, Zhang A Q. Theoretical study on the formation mechanism of polychlorinated dibenzothiophenes/thianthrenes from 2-chlorothiophenol molecules. J Environ Sci, 2017, http: //dx.doi.org/10.1016/j.jes.

[71] Yu W N, Li P F, Xu F, Hu J T, Zhang Q Z, Wang W X. Quantum chemical and direct dynamic study on homogeneous gas-phase formation of PBDD/Fs from 2,4,5-tribromophenol and 3,4-dibromophenol. Chemosphere, 2013, 93(3): 512-520.

[72] Gao R, Xu F, Li S Q, Hu J T, Zhang Q Z, Wang W X. Formation of bromophenoxy radicals from complete series reactions of bromophenols with H and OH radicals. Chemosphere, 2013, 92(4): 382-390.

[73] Cao H J, He M X, Sun Y H, Han D D. Mechanical and kinetic studies of the formation of polyhalogenated dibenzo-*p*-dioxins from hydroxylated polybrominated diphenyl ethers and chlorinated derivatives. J Phys Chem A, 2011, 115(46): 13489-13497.

[74] Gao R, Zhu L D, Xu F, Yu W N, Zhang Q Z, Wang W X. Homogeneous gas-phase formation mechanism of emerging organic pollutants polyfluorinated dibenzo-*p*-dioxins and dibenzofurans from 2-fluorophenol. Asian J Chem, 2014, 26(9): 2784-2788.

[75] Samara F, Gullett B K, Harrison R O, Chu A, Clark G C. Determination of relative assay response factors for toxic chlorinated and brominated dioxins/furans using an enzyme immunoassay (EIA) and a chemically-activated luciferase gene expression cell bioassay (CALUX). Environ Int, 2009, 35(3): 588-593.

[76] Olsman H, Engwall M, Kammann U, Klempt M, Otte J, van Bavel B, Hollert H. Relative differences in aryl hydrocarbon receptor-mediated response for 18 polybrominated and mixed halogenated dibenzo-*p*-dioxins and -furans in cell lines from four different species. Environ Toxicol Chem, 2007, 26(11): 2448-2454.

[77] Burcat A, Branko R. Third millenium ideal gas and condensed phase thermochemical database for combustion (with update from active thermochemical tables). J Chem Inf Model, 2005, 45(3): 572-580.

[78] Silva M A V R D, Ferrão M L S C C H, Fang J. Standard enthalpies of combustion of the six dichlorophenols by rotating-bomb calorimetry. J Chem Thermodyn, 1994, 26(8): 839–846.

[79] Papina T S, Kolesov V P, Vorobieva V P, Golovkov V F. The standard molar enthalpy of formation of 2-chlorodibenzo-p-dioxin. J Chem Thermodyn, 1996, 28(3): 307-311.

[80] Wang L M, Heard D E, Pilling M J, Seakins P. A Gaussian-3X prediction on the enthalpies of formation of chlorinated Phenols and dibenzo-p-dioxins. J Phys Chem A, 2008, 112(8): 1832-1840.

[81] Cox J D. The heats of combustion of phenol and the three cresols. Pure Appl Chem, 1961, 2(1-2): 125-128.

[82] Luk'Yanova V A, Kolesov V P, Avramenko N V, Vorob'Eva V P, Golovkov V F. Standard enthalpy of formation of dibenzo-p-dioxine. Zh Fiz Khim, 1997, 71(3): 406–408.

[83] Kolesov V P, Papina T S, Lukyanova V A. The enthalpies of formation of some polychlorinated dibenzodioxins. 14th IUPAC Conference on Chemical Thermodynamics, Osaka, Japan, Aug 25-30, 1996: 329.

[84] Da Silva M A V, Ferreira A I M C. Gas phase enthalpies of formation of monobromophenols. J Chem Thermodyn, 2009, 41(10): 1104-1110.

[85] Chase M W. NIST-JANAF Thermochemical Tables, 4th Edition. J Phys Chem Ref Data, Monograph 9, 1998, 1-1951.

[86] Zhang Q Z, Gao R, Xu F, Zhou Q, Jiang G B, Wang T, Chen J M, Hu J T, Jiang W, Wang W X. Role of water molecule in the gas-phase formation process of nitrated polycyclic aromatic hydrocarbons in the atmosphere: A computational study.

Environ Sci Technol, 2014, 48(9): 5051-5057.

[87] Huang Z X, Zhang Q Z, Wang W X. Mechanical and kinetic study on gas-phase formation of dinitronaphthalene from 1-and 2-nitronaphthalene. Chemosphere, 2016, 156: 101-110.

[88] Feilberg A, Poulsen M W B, Nielsen T, Skov H. Occurrence and sources of particulate nitro-polycyclic aromatic hydrocarbons in ambient air in Denmark. Atmos Environ, 2001, 35(2): 353-366.

[89] Arey J, Zielinska B, Atkinson R, Aschmann S M. Nitroarene products from the gas-phase reactions of volatile polycyclic aromatic hydrocarbons with the OH radical and N_2O_5. Int J Chem Kinet, 1989, 21(21): 775-799.

[90] Berndt T, Boge O. Gas-phase reaction of OH radicals with benzene: products and mechanism. Phys Chem Chem Phys, 2001, 3(22): 4946-4956.

[91] Di Giacomo F. A short account of RRKM theory of unimolecular reactions and of Marcus theory of electron transfer in a historical perspective. J Chem Educ, 2014, 92(3): 476-481.

[92] Lindemann F, Arrhenius S, Langmuir I, Dhar N, Perrin J, Lewis W M. Discussion on "the radiation theory of chemical action". Trans Faraday Soc, 1922, 17: 598-606.

[93] Stockwell W R, Lawson C V, Saunders E, Goliff W. A review of tropospheric atmospheric chemistry and gas-phase chemical mechanisms for air quality modeling. Atmosphere, 2012, 3(1): 1-32.

[94] Vereecken L, Francisco J S. Theoretical studies of atmospheric reaction mechanisms in the troposphere. Chem Soc Rev, 2012, 41(19): 6259-6293.

[95] Vereecken L, Glowacki D R, Pilling M J. Theoretical chemical kinetics in tropospheric chemistry: methodologies and applications. Chem Rev, 2015, 115(10): 4063-4114.

[96] Ge X L, Wexler A S, Clegg S L. Atmospheric amines - Part I. A review. Atmos Environ, 2011, 45(3): 524-546.

[97] Almeida J, Schobesberger S, Kürten A, Ortega I K, Kupiainen-Määttä O, Praplan A P, Adamov A, Amorim A, Bianchi F, Breitenlechner M. Molecular understanding of sulphuric acid-amine particle nucleation in the atmosphere. Nature, 2013, 502(7471): 359-363.

[98] Kirkby J, Curtius J, Almeida J, Dunne E, Duplissy J, Ehrhart S, Franchin A, Gagne S, Ickes L, Kurten A, Kupc A, Metzger A, Riccobono F, Rondo L, Schobesberger S, Tsagkogeorgas G, Wimmer D, Amorim A, Bianchi F, Breitenlechner M, David A, Dommen J, Downard A, Ehn M, Flagan R C, Haider S, Hansel A, Hauser D, Jud W, Junninen H, Kreissl F, Kvashin A, Laaksonen A, Lehtipalo K, Lima J, Lovejoy E R, Makhmutov V, Mathot S, Mikkila J, Minginette P, Mogo S, Nieminen T, Onnela A, Pereira P, Petaja T, Schnitzhofer R, Seinfeld J H, Sipila M, Stozhkov Y, Stratmann F, Tome A, Vanhanen J, Viisanen Y, Vrtala A, Wagner P E, Walther H, Weingartner E, Wex H, Winkler P M, Carslaw K S, Worsnop D R, Baltensperger U, Kulmala M. Role of sulphuric acid, ammonia and galactic cosmic rays in atmospheric aerosol nucleation. Nature, 2011, 476(7361): 429-433.

[99] Kurten T, Loukonen V, Vehkamaki H, Kulmala M. Amines are likely to enhance neutral and ion-induced sulfuric acid-water nucleation in the atmosphere more effectively than ammonia. Atmos Chem Phys, 2008, 8(14): 4095-4103.

[100] Loukonen V, Kurten T, Ortega I K, Vehkamaki H, Padua A A H, Sellegri K, Kulmala M. Enhancing effect of dimethylamine in sulfuric acid nucleation in the presence of water - a computational study. Atmos Chem Phys, 2010, 10(10): 4961-4974.

[101] Murphy S M, Sorooshian A, Kroll J H, Ng N L, Chhabra P, Tong C, Surratt J D, Knipping E, Flagan R C, Seinfeld J H. Secondary aerosol formation from atmospheric reactions of aliphatic amines. Atmos Chem Phys, 2007, 7(9): 2313-2337.

[102] Berndt T, Stratmann F, Sipila M, Vanhanen J, Petaja T, Mikkila J, Gruner A, Spindler G, Mauldin R L, Curtius J, Kulmala M, Heintzenberg J. Laboratory study on new particle formation from the reaction OH + SO_2: Influence of experimental conditions, H_2O vapour, NH_3 and the amine tert-butylamine on the overall process. Atmos Chem Phys, 2010, 10(15):

7101-7116.

[103] Xie H B, Elm J, Halonen R, Myllys N, Kurtén T, Kulmala M, Vehkamaki H. The atmospheric fate of monoethanolamine: Enhancing new-particle formation of sulfuric acid as an important removal process. Environ Sci Technol, 2017, 51(15): 8422-8431.

[104] Xie H B, Li C, He N, Wang C, Zhang S W, W C J. Atmospheric chemical reactions of monoethanolamine initiated by OH radical: Mechanistic and kinetic study. Environ Sci Technol, 2014, 48(3): 1700-1706.

[105] Xie H B, Ma F F, Wang Y F, He N, Yu Q, Chen J W. Quantum chemical study on .Cl-initiated atmospheric degradation of monoethanolamine. Environ Sci Technol, 2015, 49(22): 13246-13255.

[106] Xie H B, Ma F F, Yu Q, He N, Chen J W. Computational Study of the Reactions of Chlorine Radicals with Atmospheric Organic Compounds Featuring NH_x-pi-Bond(x = 1, 2) Structures. J Phys Chem A, 2017, 121(8): 1657-1665.

[107] Jen C N, McMurry P H, Hanson D R. Stabilization of sulfuric acid dimers by ammonia, methylamine, dimethylamine, and trimethylamine. J Geophys Res, 2014, 119(12): 7502-7514.

[108] Nicovich J M, Mazumder S, Laine P L, Wine P H, Tang Y, Bunkan A J, Nielsen C J. An experimental and theoretical study of the gas phase kinetics of atomic chlorine reactions with CH_3NH_2, $(CH_3)_2NH$, and$(CH_3)_3N$. Phys Chem Chem Phys, 2015, 17(2): 911-917.

[109] Han D D, Cao H J, Li M G, Li X, Zhang S Q, He M X, Hu J T. Computational study on the mechanisms and rate constants of the Cl-initiated oxidation of methyl vinyl ether in the atmosphere. J Phys Chem A, 2015, 119(4): 719-727.

[110] Han D D, Cao H J, Li J, Li M Y, Li X, He M X, Hu J T. Theoretical studies on the mechanisms and rate constants for the hydroxylation of n-butyl, iso-butyl and tert-butyl vinyl ethers in atmosphere. Struct Chem, 2015, 26(3): 713-729.

[111] Han D D, Li J, Cao H J, He M X, Hu J T, Yao S D. Theoretical investigation on the mechanisms and kinetics of OH-initiated photooxidation of dimethyl phthalate (DMP) in atmosphere. Chemosphere, 2014, 95: 50 57.

[112] Han D D, Cao H J, Li J, Li M Y, He M X, Hu J T. Computational study on the mechanisms and rate constants of the OH-initiated oxidation of ethyl vinyl ether in atmosphere. Chemosphere, 2014, 111: 61-69.

[113] Han D D, Cao H J, Sun Y H, Sun R L, He M X. Mechanistic and kinetic study on the ozonolysis of n-butyl vinyl ether, i-butyl vinyl ether and t-butyl vinyl ether. Chemosphere, 2012, 88(10): 1235-1240.

[114] Han D D, Cao H J, Sun Y H, He M X. Mechanistic and kinetic study on the ozonolysis of ethyl vinyl ether and propyl vinyl ether. Struct Chem, 2012, 23(2): 499-514.

[115] Wang S N, Wang L M. The atmospheric oxidation of dimethyl, diethyl, and diisopropyl ethers. The role of the intramolecular hydrogen shift in peroxy radicals. Phys Chem Chem Phys, 2016, 18(11): 7707-7714.

[116] Dang J, Shi X G, Zhang Q Z, Hu J T, Chen J M, Wang W X. Mechanistic and kinetic studies on the OH-initiated atmospheric oxidation of fluoranthene. Sci Total Environ, 2014, 490: 639-646.

[117] Dang J, Shi X L, Zhang Q Z, Hu J T, Wang W X. Mechanism and kinetic properties for the OH-initiated atmospheric oxidation degradation of 9, 10-Dichlorophenanthrene. Sci Total Environ, 2015, 505, 787-794.

[118] Dang J, He M X. Mechanisms and kinetic parameters for the gas-phase reactions of anthracene and pyrene with Cl atoms in the presence of NO_x. RSC Adv, 2016, 6(21): 17345-17353.

[119] Dang J, Shi X L, Zhang Q Z, Hu J T, Wang W X. Mechanism and thermal rate constant for the gas-phase ozonolysis of acenaphthylene in the atmosphere. Sci Total Environ, 2015, 514: 344-350.

[120] Dang J, Shi X L, Zhang Q Z, Hu J T, Wang W X. Insights into the mechanism and kinetics of the gas-phase atmospheric reaction of 9-chloroanthracene with NO_3 radical in the presence of NO_x. RSC Adv, 2015, 5(102): 84066-84075.

[121] Dang J, Shi X L, Hu J T, Chen J M, Zhang Q Z, Wang W X. Mechanistic and kinetic studies on OH-initiated atmospheric oxidation degradation of benzo alpha pyrene in the presence of O_2 and NO_x. Chemosphere, 2015, 119: 387-393.

[122] Zhang Z J, Lin L, Wang L M. Atmospheric oxidation mechanism of naphthalene initiated by OH radical. A theoretical study. Phys Chem Chem Phys, 2012, 14(8): 2645-2650.

[123] Wu R R, Li Y, Pan S S, Wang S N, Wang L M. The atmospheric oxidation mechanism of 2-methylnaphthalene. Phys Chem Chem Phys, 2015, 17(36): 23413-23422.

[124] Cao H J, He M X, Han D D, Li J, Li M Y, Wang W X, Yao S D. OH-Initiated Oxidation Mechanisms and Kinetics of 2, 4, 4 '-Tribrominated Diphenyl Ether. Environ Sci Technol, 2013, 47(15): 8238-8247.

[125] Zhou J, Chen J W, Liang C H, Xie Q, Wang Y N, Zhang S Y, Qiao X L, Li X H. Quantum chemical investigation on the mechanism and kinetics of PBDE photooxidation by center dot OH: A case study for BDE-15. Environ Sci Technol, 2011, 45(11): 4839-4845.

[126] He M X, Li X, Zhang S Q, Sun J F, Cao H J, Wang W X. Mechanistic and kinetic investigation on OH-initiated oxidation of tetrabromobisphenol A. Chemosphere, 2016, 153: 262-269.

[127] Cao H J, Han D D, Li M Y, Li X, He M X, Wang W X. Theoretical Investigation on Mechanistic and Kinetic Transformation of 2, 2 ', 4, 4 ', 5-Pentabromodiphenyl Ether. J Phys Chem A, 2015, 119(24): 6404-6411.

[128] Cao H J, He M X, Han D D, Sun Y H, Zhao S F, Ma H J, Yao S D. Mechanistic and kinetic study on the reaction of 2, 4-dibrominated diphenyl ether (BDE-7) with OH radicals. Comput Theor Chem, 2012, 983(1): 31-37.

[129] Yu Q, Xie H B, Chen J W. Atmospheric chemical reactions of alternatives of polybrominated diphenyl ethers initiated by center dot OH: A case study on triphenyl phosphate. Sci Total Environ, 2016, 571: 1105-1114.

[130] Yu Q, Xie H B, Li T C, Ma F F, Fu Z H, Wang Z Y, Li C, Fu Z Q, Xia D M, Chen J W. Atmospheric chemical reaction mechanism and kinetics of 1, 2-bis(2,4,6-tribromophenoxy)ethane initiated by OH radical: a computational study. RSC Adv, 2017, 7(16): 9484-9494.

[131] Li C, Chen J W, Xie H B, Zhao Y H, Xia D M, Xu T, Li X H, Qiao X L. Effects of Atmospheric Water on center dot OH-initiated Oxidation of Organophosphate Flame Retardants: A DFT Investigation on TCPP. Environ Sci Technol, 2017, 51(9): 5043-5051.

[132] Zhang C X, Sun T L, Sun X M. Mechanism for OH-Initiated Degradation of 2,3,7,8-Tetrachlorinated Dibenzo-p-Dioxins in the Presence of O_2 and NO/H_2O. Environ Sci Technol, 2011, 45(11): 4756-4762.

[133] Sun X M, Zhang C X, Zhao Y Y, Bai J, Zhang Q Z, Wang W X. Atmospheric chemical reactions of 2,3,7,8-tetrachlorinated dibenzofuran initiated by an OH radical: Mechanism and kinetics study. Environ Sci Technol, 2012, 46(15): 8148-8155.

[134] Wu X C, Sun X M, Zhang C X, Gong C, Hu J T. Micro-mechanism and rate constants for OH-initiated degradation of methomyl in atmosphere. Chemosphere, 2014, 107: 331-335.

[135] Ge L K, Chen J W, Qiao X L, Lin J, Cai X Y. Light-source-dependent effects of main water constituents on photodegradation of phenicol antibiotics: mechanism and kinetics. Environ Sci Technol, 2009, 43(9): 3101-3107.

[136] Ge L K, Chen J W, Wei X X, Zhang S Y, Qiao X L, Cai X Y, Xie Q. Aquatic photochemistry of fluoroquinolone antibiotics: kinetics, pathways, and multivariate effects of main water constituents. Environ Sci Technol, 2010, 44(7): 2400-2405.

[137] Wei X X, Chen J W, Xie Q, Zhang S Y, Ge L K, Qiao X L. Distinct photolytic mechanisms and products for different dissociation species of ciprofloxacin. Environ Sci Technol, 2013, 47(9): 4284-4290.

[138] Zhang S Y, Chen J W, Qiao X L, Ge L K, Cai X Y, Na G S. Quantum Chemical Investigation and Experimental Verification on the Aquatic Photochemistry of the Sunscreen 2-Phenylbenzimidazole-5-Sulfonic Acid. Environ Sci Technol, 2010, 44(19): 7484-7490.

[139] Zhang S Y, Chen J W, Xie Q, Shao J P. Comment on "Effect of Dissolved Organic Matter on the Transformation of

Contaminants Induced by Excited Triplet States and the Hydroxyl Radical". Environ Sci Technol, 2011, 45(18): 7945-7946.

[140] Zhang S Y, Chen J W, Zhao Q, Xie Q, Wei X X. Unveiling self-sensitized photodegradation pathways by DFT calculations: A case of sunscreen p-aminobenzoic acid. Chemosphere, 2016, 163: 227-233.

[141] Fang H S, Gao Y P, Li G Y, An J B, Wong P K, Fu H Y, Yao S D, Nie X P, An T C. Advanced Oxidation Kinetics and Mechanism of Preservative Propylparaben Degradation in Aqueous Suspension of TiO_2 and Risk Assessment of Its Degradation Products. Environ Sci Technol, 2013, 47(6): 2704-2712.

[142] Gao Y P, Ji Y M, Li G Y, An T C. Theoretical investigation on the kinetics and mechanisms of hydroxyl radical-induced transformation of parabens and its consequences for toxicity: Influence of alkyl-chain length. Water Res, 2016, 91: 77-85.

[143] Gao Y P, Ji Y M, Li G Y, Mai B X, An T C. Bioaccumulation and ecotoxicity increase during indirect photochemical transformation of polycyclic musk tonalide: A modeling study. Water Res, 2016, 105: 47-55.

[144] Gao Y P, Ji Y M, Li G Y, An T C. Mechanism, kinetics and toxicity assessment of OH-initiated transformation of triclosan in aquatic environments. Water Res, 2014, 49: 360-370.

[145] Gao Y P, An T C, Ji Y M, Li G Y, Zhao C Y. Eco-toxicity and human estrogenic exposure risks from OH-initiated photochemical transformation of four phthalates in water: A computational study. Environ Pollut, 2015, 206: 510-517.

[146] Gao Y P, An T C, Fang H S, Ji Y M, Li G Y. Computational consideration on advanced oxidation degradation of phenolic preservative, methylparaben, in water: mechanisms, kinetics, and toxicity assessments. J Hazard Mater, 2014, 278: 417-425.

[147] An T C, Gao Y P, Li G Y, Kamat P V, Peller J, Joyce M V. Kinetics and Mechanism of (OH)-O-center dot Mediated Degradation of Dimethyl Phthalate in Aqueous Solution: Experimental and Theoretical Studies. Environ Sci Technol, 2014, 48(1): 641-648.

[148] An T C, An J B, Gao Y P, Li G Y, Fang H S, Song W H. Photocatalytic degradation and mineralization mechanism and toxicity assessment of antivirus drug acyclovir: Experimental and theoretical studies. Appl Catal B-Environ, 2015, 164: 279-287.

[149] Li G Y, Nie X, Gao Y P, An T C. Can environmental pharmaceuticals be photocatalytically degraded and completely mineralized in water using $g\text{-}C_3N_4/TiO_2$ under visible light irradiation?-Implications of persistent toxic intermediates. Appl Catal B-Environ, 2016, 180: 726-732.

[150] Zhang H Q, Xie H B, Chen J W, Zhang S S. Prediction of Hydrolysis Pathways and Kinetics for Antibiotics under Environmental pH Conditions: A Quantum Chemical Study on Cephradine. Environ Sci Technol, 2015, 49(3): 1552-1558.

[151] 张海勤, 谢宏彬, 陈景文, 张树深. 基于密度泛函理论揭示 Cu^{2+}配位作用对头孢拉定水解反应的影响机制. 环境化学, 2015, (9): 1594-1600.

[152] Fu Z Q, Chen J W, Li X H, Wang Y N, Yu H Y. Comparison of prediction methods for octanol-air partition coefficients of diverse organic compounds. Chemosphere, 2016, 148: 118-125.

[153] Shao Y H, Liu J N, Wang M X, Shi L L, Yao X J, Gramatica P. Integrated QSPR models to predict the soil sorption coefficient for a large diverse set of compounds by using different modeling methods. Atmos Environ, 2014, 88: 212-218.

[154] Cheng F X, Ikenaga Y, Zhou Y D, Yu Y, Li W H, Shen J, Du Z, Chen L, Xu C Y, Liu G X, Lee P W, Tang Y. In Silico Assessment of Chemical Biodegradability. J Chem Inf Model, 2012, 52(3): 655-669.

[155] Chen G C, Li X H, Chen J W, Zhang Y N, Peijnenburg W. Comparative study of biodegradability prediction of chemicals using decision trees, functional trees, and logistic regression. Environ Toxicol Chem, 2014, 33(12): 2688-2693.

[156] Ren Y Y, Liu H X, Yao X J, Liu M C. Prediction of ozone tropospheric degradation rate constants by projection pursuit regression. Anal Chim Acta, 2007, 589(1): 150-158.

[157] Erratico C A, Moffatt S C, Bandiera S M. Comparative oxidative metabolism of BDE-47 and BDE-99 by Rat Hepatic

Microsomes. Toxicol Sci, 2011, 123(1): 37-47.

[158] Maria L F, Gross M S, McGarrigle B P, Eljarrat E, Barcelo D, Aga D S, Olson J R. Biotransformation of BDE-47 to potentially toxic metabolites is predominantly mediated by human CYP2B6. Environ Health Persp, 2013, 121(4): 440-446.

[159] Lupton S J, McGarrigle B P, Olson J R, Wood T D, Aga D S. Human liver microsome-mediated metabolism of brominated diphenyl ethers 47, 99, and 153 and identification of their major metabolites. Chem Res Toxicol, 2009, 22(11). 1802-1809.

[160] Wang X B, Chen J W, Wang Y, Xie H B, Fu Z Q. Transformation pathways of MeO-PBDEs catalyzed by active center of P450 enzymes: A DFT investigation employing 6-MeO-BDE-47 as a case. Chemosphere, 2015, 120: 631-636.

[161] Fu Z Q, Wang Y, Chen J W, Wang Z Y, Wang X B. How PBDEs are transformed into dihydroxylated and dioxin metabolites catalyzed by the active center of cytochrome P450s: A DFT study. Environ Sci Technol, 2016, 50(15): 8155-8163.

[162] Wang X B, Wang Y, Chen J W, Ma Y Q, Zhou J, Fu Z Q. Computational toxicological investigation on the mechanism and pathways of xenobiotics metabolized by cytochrome P450: A case of BDE-47. Environ Sci Technol, 2012, 46(9): 5126-5133.

[163] Malmberg T, Athanasiadou M, Marsh G, Brandt I, Bergmant A. Identification of hydroxylated polybrominated diphenyl ether metabolites in blood plasma from polybrominated diphenyl ether exposed rats. Environ Sci Technol, 2005, 39(14): 5342-5348.

[164] Erratico C A, Szeitz A, Bandiera S M. Biotransformation of 2, 2', 4, 4'-tetrabromodiphenyl ether (BDE-47) by human liver microsomes: Identification of cytochrome P450 2B6 as the major enzyme involved. Chem Res Toxicol, 2013, 26(5): 721-731.

[165] Zhai C, Peng S, Yang L M, Wang Q Q. Evaluation of BDE-47 hydroxylation metabolic pathways based on a strong electron-withdrawing pentafluorobenzoyl derivatization gas chromatography/electron capture negative ionization quadrupole mass spectrometry. Environ Sci Technol, 2014, 48(14): 8117-8126.

[166] Tomy G T, Tittlemier S A, Palace V P, Budakowski W R, Braekevelt E, Brinkworth L, Friesen K. Biotransformation of *N*-ethyl perfluorooctanesulfonamide by rainbow trout (*Onchorhynchus mykiss*) liver microsomes. Environ Sci Technol, 2004, 38(3): 758-762.

[167] Benskin J P, Holt A, Martin J W. Isomer-specific biotransformation rates of a perfluorooctane sulfonate (PFOS)-precursor by cytochrome P450 isozymes and human liver microsomes. Environ Sci Technol, 2009, 43(22): 8566-8572.

[168] Fu Z Q, Wang Y, Wang Z Y, Xie H B, Chen J W. Transformation Pathways of Isomeric Perfluorooctanesulfonate Precursors Catalyzed by the Active Species of P450 Enzymes: In Silico Investigation. Chem Res Toxicol, 2015, 28(3): 482-489.

[169] Ji L, Zhang J, Liu W P, de Visser S P. Metabolism of halogenated alkanes by cytochrome P450 enzymes. Aerobic Oxidation versus Anaerobic Reduction. Chem Asian J, 2014, 9(4): 1175-1182.

[170] Zhang J, Ji L, Liu W P. In Silico Prediction of cytochrome P450-mediated biotransformations of xenobiotics: A case study of epoxidation. Chem Res Toxicol, 2015, 28(8): 1522-1531.

[171] Li Y W, Zhang R M, Du L K, Zhang Q Z, Wang W X. Insight into the catalytic mechanism of meta-cleavage product hydrolase BphD: a quantum mechanics/molecular mechanics study. RSC Adv, 2015, 5(82): 66591-66597.

[172] Li Y W, Shi X L, Zhang Q Z, Hu J T, Chen J M, Wang W X. Computational evidence for the detoxifying mechanism of epsilon class glutathione transferase toward the insecticide DDT. Environ Sci Technol, 2014, 48(9): 5008-5016.

[173] Li Y W, Zhang R M, Du L K, Zhang Q Z, Wang W X. Catalytic mechanism of C-F bond cleavage: insights from QM/MM analysis of fluoroacetate dehalogenase. Catal Sci Technol, 2016, 6(1): 73-80.

[174] Tang X, Zhang R, Li Y, Zhang Q, Wang W. Enantioselectivity of haloalkane dehalogenase LinB on the degradation of 1,

2-dichloropropane: A QM/MM study. Bioorg Chem, 2017, 73: 16-23.

[175] Tang X W, Zhang R M, Zhang Q Z, Wang W X. Dehydrochlorination mechanism of gamma-hexachlorocyclohexane degraded by dehydrochlorinase LinA from Sphingomonas paucimobilis UT26. RSC Adv, 2016, 6(5): 4183-4192.

[176] Li Y W, Zhang R M, Du L K, Zhang Q Z, Wang W X. Insights into the catalytic mechanism of chlorophenol 4-monooxygenase: A quantum mechanics/molecular mechanics study. RSC Adv, 2015, 5(18): 13871-13877..

[177] Chen Y J, Cheng F X, Sun L, Li W H, Liu G X, Tang Y. Computational models to predict endocrine-disrupting chemical binding with androgen or oestrogen receptors. Ecotox Environ Safe, 2014, 110: 280-287.

[178] Wang T, Li W Y, Zheng X F, Lin Z F, Kong D Y. Development of a new decision tree to rapidly screen chemical estrogenic activities of *xenopus laevis*. Mol Inform, 2014, 33(2): 115-123.

[179] 田芳, 任祁荣, 赵春燕. 复杂网络用于取代苯类内分泌干扰物的毒性通路. 环境科学, 2017, 36(6): 1189-1197.

[180] Zhuang S L, Zhang J, Wen Y Z, Zhang C L, Liu W P. Distinct mechanisms of endocrine disruption of DDT-related pesticides toward estrogen receptor alpha and estrogen-related receptor gamma. Environ Toxicol Chem, 2012, 31(11): 2597-2605.

[181] Zhuang S L, Zhang C L, Liu W P. Atomic Insights into Distinct Hormonal Activities of Bisphenol A Analogues toward PPAR gamma and ER alpha Receptors. Chem Res Toxicol, 2014, 27(10): 1769-1779.

[182] 杨先海, 刘会会, 杨倩, 刘济宁. 双酚 A 类似物雌激素干扰效应的定量结构-活性关系模型. 生态毒理学报, 2016, 11(4): 69-78.

[183] Li F, Li X H, Shao J P, Chi P, Chen J W, Wang Z J. Estrogenic activity of anthraquinone derivatives: in vitro and in silico studies. Chem Res Toxicol, 2010, 23(8): 1349-1355.

[184] Cao H M, Wang F B, Yong L, Wang H L, Zhang A Q, Song M Y. Experimental and computational insights on the recognition mechanism between the estrogen receptor α with bisphenol compounds. Arch Toxicol, 2017, doi: 10.1007/s00204-017-2011-0

[185] Li J, Ma D, Lin Y, Fu J J, Zhang A Q. An exploration of the estrogen receptor transcription activity of capsaicin analogues via an integrated approach based on in silico prediction and in vitro assays. Toxicol Lett, 2014, 227(3): 179-188.

[186] Gao Y, Li X X, Guo L H. Assessment of estrogenic activity of perfluoroalkyl acids based on ligand-induced conformation state of human estrogen receptor. Environ Sci Technol, 2013, 47(1): 634-641.

[187] 梁雪芳. 基于蛋白组学的环境污染物毒性效应评价和作用机制研究. 中国科学院大学, 2014.

[188] Hu Y, Zhang Z B, Sun L B, Zhu D S, Liu Q C, Jiao J, Li J, Qi M W. The estrogenic effects of benzylparaben at low doses based on uterotrophic assay in immature SD rats. Food Chem Toxicol, 2013, 53: 69-74.

[189] Zhang Z B, Sun L B, Hu Y, Jiao J, Hu J Y. Inverse antagonist activities of parabens on human oestrogen-related receptor gamma(ERR gamma): *In vitro* and *in silico* studies. Toxicol Appl Pharm, 2013, 270(1): 16-22.

[190] Sun L B, Yu T, Guo J L, Zhang Z B, Hu Y, Xiao X, Sun Y L, Xiao H, Li J Y, Zhu D S, Sai L L, Li J. The estrogenicity of methylparaben and ethylparaben at doses close to the acceptable daily intake in immature Sprague-Dawley rats. Sci Rep, 2016, 6: 25173(1-9).

[191] Liu H H, Yang X H, Yin C, Wei M B, He X. Development of predictive models for predicting binding affinity of endocrine disrupting chemicals to fish sex hormone-binding globulin. Ecotox Environ Safe, 2017, 136: 46-54.

[192] Ding K K, Kong X T, Wang J P, Lu L P, Zhou W F, Zhan T J, Zhang C L, Zhuang S L. Side Chains of Parabens Modulate Antiandrogenic Activity: In Vitro and Molecular Docking Studies. Environ Sci Technol, 2017, 51(11): 6452-6460.

[193] Li X L, Ye L, Wang X X, Shi W, Liu H L, Qian X P, Zhu Y L, Yu H X. In silico investigations of anti-androgen activity of polychlorinated biphenyls. Chemosphere, 2013, 92(7): 795-802.

[194] Li X L, Ye L, Shi W, Liu H L, Liu C S, Qian X P, Zhu Y L, Yu H X. In silico study on hydroxylated polychlorinated

biphenyls as androgen receptor antagonists. Ecotox Environ Safe, 2013, 92(3): 258-264.

[195] Wu Y, Doering J A, Ma Z Y, Tang S, Liu H L, Zhang X W, Wang X X, Yu H X. Identification of androgen receptor antagonists: In vitro investigation and classification methodology for flavonoid. Chemosphere, 2016, 158: 72-79.

[196] Wang X X, Li X L, Shi W, Wei S, Giesy J P, Yu H X, Wang Y L. Docking and CoMSIA studies on steroids and non-steroidal chemicals as androgen receptor ligands. Ecotox Environ Safe, 2013, 89: 143-149.

[197] Zhuang S L, Lv X, Pan L M, Lu L P, Ge Z W, Wang J Y, Wang J P, Liu J S, Liu W P, Zhang C L. Benzotriazole UV 328 and UV-P showed distinct antiandrogenic activity upon human CYP3A4-mediated biotransformation. Environ Pollut, 2017, 220, 616-624.

[198] 杨先海, 陈景文, 李斐. 化学品甲状腺干扰效应的计算毒理学研究进展. 科学通报, 2015, 60(19): 1761-1770.

[199] Li F, Xie Q, Li X H, Li N, Chi P, Chen J W, Wang Z J, Hao C. Hormone activity of hydroxylated polybrominated diphenyl ethers on human thyroid receptor-beta: *In vitro* and in silico investigations. Environ Health Persp, 2010, 118(5): 602-606.

[200] Yu H Y, Wondrousch D, Li F, Chen J R, Lin H J, Ji L. In silico investigation of the thyroid hormone activity of hydroxylated polybrominated diphenyl ethers. Chem Res Toxicol, 2015, 28(8): 1538-1545.

[201] Li X L, Ye L, Wang X X, Wang X Z, Liu H L, Zhu Y L, Yu H X. Combined 3D-QSAR, molecular docking and molecular dynamics study on thyroid hormone activity of hydroxylated polybrominated diphenyl ethers to thyroid receptors β. Toxicol Appl Pharm, 2012, 265(3): 300-307.

[202] Cao H M, Sun Y Z, Wang L, Zhao C Y, Fua J J, Zhang A Q. Understanding the microscopic binding mechanism of hydroxylated and sulfated polybrominated diphenyl ethers with transthyretin by molecular docking, molecular dynamics simulations and binding free energy calculations. Mol Biosys, 2017, 13(4): 736-749.

[203] Ren X M, Qin W P, Cao L Y, Zhang J, Yang Y, Wan B, Guo L H. Binding interactions of perfluoroalkyl substances with thyroid hormone transport proteins and potential toxicological implications. Toxicology, 2016, 366: 32-42.

[204] Yang X H, Xie H B, Chen J W, Li X H. Anionic phenolic compounds bind stronger with transthyretin than their neutral forms: Nonnegligible mechanisms in virtual screening of endocrine disrupting chemicals. Chem Res Toxicol, 2013, 26(9): 1340-1347.

[205] Yang X H, Lyakurwa F, Xie H B, Chen J W, Li X H, Qiao X L, Cai X Y. Different binding mechanisms of neutral and anionic poly-/perfluorinated chemicals to human transthyretin revealed by In silico models. Chemosphere, 2017, 182: 574-583.

[206] Grimm F A, Lehmler H J, He X R, Robertson L W, Duffel M W. Sulfated metabolites of polychlorinated biphenyls are high-affinity ligands for the thyroid hormone transport protein transthyretin. Environ Health Persp, 2013, 121(6): 657-662.

[207] Zhang Y M, Dong X Y, Fan L J, Zhang Z L, Wang Q, Jiang N, Yang X S. Poly- and perfluorinated compounds activate human pregnane X receptor. Toxicology, 2017, 380: 23-29.

[208] Shi H L, Tian S, Li Y Y, Li D, Yu H D, Zhen X C, Hou T J. Absorption, distribution, metabolism, excretion, and toxicity evaluation in drug discovery. 14. prediction of human pregnane X receptor activators by using Naive Bayesian classification technique. Chem Res Toxicol, 2015, 28(1): 116-125.

[209] Yin C, Yang X H, Wei M B, Liu H H. Predictive models for identifying the binding activity of structurally diverse chemicals to human pregnane X receptor. Environ Sci Pollut R, 2017, 24: 20063-20071.

[210] Hu Q, He G C, Zhao J, Soshilov A, Denison M S, Zhang A Q, Yin H J, Fraccalvieri D, Bonati L, Xie Q H, Zhao B. Ginsenosides Are Novel Naturally-Occurring Aryl Hydrocarbon Receptor Ligands. PLoS One, 2013, 8(6): 1-10.

[211] Cao F, Li X L, Ye L, Xie Y W, Wang X X, Shi W, Qian X P, Zhu Y L, Yu H X. Molecular docking, molecular dynamics simulation, and structure-based 3D-QSAR studies on the aryl hydrocarbon receptor agonistic activity of hydroxylated polychlorinated biphenyls. Environ Toxicol Pharm, 2013, 36(2): 626-635.

[212] Li X L, Wang X X, Shi W, Liu H, Yu H X. Analysis of Ah receptor binding affinities of polybrominated diphenyl ethers via in silico molecular docking and 3D-QSAR. SAR QSAR Environ Res, 2013, 24(1): 75-87.

[213] Li F, Li X H, Liu X L, Zhang L B, You L P, Zhao J M, Wu H F. Docking and 3D-QSAR studies on the Ah receptor binding affinities of polychlorinated biphenyls (PCBs), dibenzo-*p*-dioxins (PCDDs) and dibenzofurans(PCDFs). Environ Toxicol Phar, 2011, 32(3): 478-485.

[214] Li N, Jiang W W, Ma M, Wang D H, Wang Z J. Chlorination by-products of bisphenol A enhanced retinoid X receptor disrupting effects. J Hazard Mater, 2016, 320: 289-295.

[215] Wu Y, Su G Y, Tang S, Liu W, Ma Z Y, Zheng X M, Liu H L, Yu H X. The combination of in silico and in vivo approaches for the investigation of disrupting effects of tris(2-chloroethyl)phosphate (TCEP) toward core receptors of zebrafish. Chemosphere, 2017, 168: 122-130.

[216] Zhuang S L, Bao L L, Linhananta A, Liu W P. Molecular modeling revealed that ligand dissociation from thyroid hormone receptors is affected by receptor heterodimerization. J Mol Graph Model, 2013, 44: 155-160.

[217] Zhang J, Zhang T H, Guan T Z, Ruan P, Ren D Y, Dai W C, Yu H S, Li T Z. Spectroscopic and molecular modeling approaches to investigate the interaction of bisphenol A, bisphenol F and their diglycidyl ethers with PPAR alpha. Chemosphere, 2017, 180: 253-258.

[218] Hu W X, Gao F M, Zhang H, Hiromori Y, Arakawa S, Nagase H, Nakanishi T, Hu J Y. Activation of peroxisome proliferator-activated receptor gamma and disruption of progesterone synthesis of 2-ethylhexyl diphenyl phosphate in human placental choriocarcinoma cells: Comparison with triphenyl phosphate. Environ Sci Technol, 2017, 51(7): 4061-4068.

[219] Yang X H, Liu H H, Liu J B, Li F, Li X H, Shi L L, Chen J W. Rational selection of the 3D structure of biomacromolecules for molecular docking studies on the mechanism of endocrine disruptor action. Chem Res Toxicol, 2016, 29(9): 1565-1570.

[220] 潘柳萌, 吕翾, 庄树林. 分子动力学模拟在有机污染物毒性作用机制中的应用. 科学通报, 2015, (19): 1781-1788.

[221] 李金杰, 张栩嘉, 赵元慧. 有机污染物对水生生物毒性作用机理的判别及影响因素. 环境化学, 2013, (7): 1236-1245.

[222] 刘羽晨, 乔显亮. 水生生物急性毒性 QSAR 模型研究进展. 生态毒理学报, 2015, 10(2): 26-35.

[223] Lyakurwa F, Yang X H, Li X H, Qiao X L, Chen J W. Development and validation of theoretical linear solvation energy relationship models for toxicity prediction to fathead minnow (*Pimephales promelas*). Chemosphere, 2014, 96, 188-194.

[224] Lyakurwa F S, Yang X H, Li X H, Qiao X L, Chen J W. Development of in silico models for predicting LSER molecular parameters and for acute toxicity prediction to fathead minnow (*Pimephales promelas*). Chemosphere, 2014, 108: 17-25.

[225] 刘羽晨. 基于综合毒性作用模式分类构建有机化合物对大型蚤急性毒性 QSAR 模型. 大连: 大连理工大学, 2015.

[226] Wu X C, Zhang Q Z, Hu J T. QSAR study of the acute toxicity to fathead minnow based on a large dataset. SAR QSAR Environ Res, 2016, 27(2): 147-164.

[227] Zhang X J, Qin W C, He J, Wen Y, Su L M, Sheng L X, Zhao Y H. Discrimination of excess toxicity from narcotic effect: Comparison of toxicity of class-based organic chemicals to Daphnia magna and Tetrahymena pyriformis. Chemosphere, 2013, 93(2): 397-407.

[228] Li J J, Wang X H, Wang Y, Wen Y, Qin W C, Su L M, Zhao Y H. Discrimination of excess toxicity from narcotic effect: Influence of species sensitivity and bioconcentration on the classification of modes of action. Chemosphere, 2015, 120: 660-673.

[229] Su Q, Lu W C, Du D S, Chen F X, Niu B, Chou K C. Prediction of the aquatic toxicity of aromatic compounds to tetrahymena pyriformis through support vector regression. Oncotarget, 2017, 8: 49359-49369.

[230] 郭湛. 有机污染物对梨形四膜虫的毒性及构效关系. 长春: 东北师范大学, 2011.

[231] 陈学勇, 韦朝海, 邓秀琼, 夏芳, 于旭彪. 硝基芳烃对梨形四膜虫毒性的定量构效关系解析. 化学学报, 2011, 69(21):

2618-2626.

[232] 张辉, 李娜, 马梅, 刘光斌. 15 种取代酚对淡水发光菌 Q67 的毒性及定量构效分析. 生态毒理学报, 2012, 07(4): 373-380.

[233] 于洋, 王晓红, 闻洋, 赵元慧. 应用 Abraham 方程研究有机污染物对七种水生生物的毒性. 环境化学, 2015, 34(1): 23-36.

[234] 李钦玲, 张升书, 杨玉良. BP 人工神经网络预测氯代有机化合物对发光菌的毒性. 广西师范学院学报(自然科学版), 2016, 33(3): 54-59.

[235] NRC. Exposure Science in the 21st Century: A Vision and a Strategy. Washington: National Academies Press, 2012.

[236] Egeghy P P, Sheldon L S, Isaacs K K, Ozkaynak H, Goldsmith M R, Wambaugh J F, Judson R S, Buckley T J. Computational exposure science: An emerging discipline to support 21st-century risk assessment. Environ Health Persp, 2016, 124(6): 697-702.

[237] Wambaugh J, Shah I. Simulating Microdosimetry in a Virtual Hepatic Lobule. Plos Comput Biol, 2010, 6(4): e1000756.

[238] Kleinstreuer N, Dix D, Rountree M, Baker N, Sipes N, Reif D, Spencer R, Knudsen T. A computational model predicting disruption of blood vessel development. PLoS Comput Biol, 2013, 9(4): e1002996.

作者：陈景文[1]，谢宏彬[1]，王中钰[1]，李雪花[1]，傅志强[1]，马芳芳[1]，
姚小军[2]，张庆竹[3]，何茂霞[3]，安太成[4]，庄树林[5]，杨先海[6]
[1]大连理工大学，[2]兰州大学，[3]山东大学，[4]广东工业大学，[5]浙江大学，[6]环境保护部南京环境科学研究所

第 26 章　我国大气污染对心肺系统健康影响研究进展

本章导读

近年来大气污染已成为我国突出的环境问题之一，其污染状况及健康影响引起国内外社会的广泛关注。国内开展的大气污染健康影响研究多数着眼于对心肺系统的健康影响方面，主要的研究类型包括在人群中开展的流行病学研究及在实验室开展的毒理学研究。大部分研究关注的是大气污染短期暴露（数小时至数周）的健康影响，少量研究则关注了大气污染长期暴露（一年或更长时间）的健康影响。本章将从大气污染短期和长期暴露对心肺疾病的影响、对人群心肺系统的亚临床效应、对心肺健康影响的生物学机制和与其他因素的交互作用方面对近年来我国本领域的研究现状进行综合介绍。

关键词

大气污染，心肺系统，健康影响，短期，长期，毒理学机制，交互作用

1 引　言

我国是世界上人口最多的国家，同时也是世界上经济增长速度最快的经济体之一。我国从1990年至今，国内生产总值年均增长率超过10%[1]，并在2010年超越日本成为世界第二大经济体，仅次于美国。与此同时，我国城镇化率从1990年的26.4%增长至2015年的56.1%，城镇总人口达到7.7亿人。城镇人口的迅速增长促进了我国工业化的实现，也使城镇地区的机动车保有量迅速增加，2015年全国民用机动车的数量达到1.62亿辆。伴随着我国社会经济的发展，我国能源消耗量快速增长，使我国在2010年成为世界最大的能源消费国，2015年能源使用总量达到43亿吨标准煤当量。其中煤炭和石油作为化石燃料的两种主要形式，占到能源使用总量的80%以上。

与此相对应的是，化石燃料的燃烧产物成为我国城市地区大气污染的主要来源之一，导致大气污染近年来保持在较高水平。大气污染的主要来源还包括工农业生产、生活炉灶、沙尘暴等，主要的大气污染物则包括细颗粒物（空气动力学直径小于2.5 μm的颗粒物，$PM_{2.5}$）、可吸入颗粒物（空气动力学直径小于10 μm的颗粒物，PM_{10}）、二氧化硫（sulfur dioxide，SO_2）、二氧化氮（nitrogen dioxide，NO_2）、臭氧（ozone，O_3）和一氧化碳（carbon monoxide，CO）等。一项近年的源解析研究显示，北京地区大气$PM_{2.5}$中70%以上的成分来源于化石燃料的燃烧[2]。近年来，我国政府采取了一系列大气污染治理措施，取得了一定成效，然而大气污染的总体水平仍然不容乐观。根据环保部环境状况公报显示，2016年全国338个地级及以上城市中有254个城市环境空气质量超标（占比75.1%），仅有84个城市环境空气质量达标(占比24.9%)[3]。338个城市平均超标天数比例为21.2%，其中发生重度污染2464天次，严重污染784天次，以$PM_{2.5}$为首要污染物的天数占重度及以上污染天数的80.3%，以PM_{10}为首要污染物的占20.4%，以O_3为首要污染物的占0.9%。

我国上述各种大气污染物中，以颗粒物的污染形式最为严峻，总体污染水平在世界范围内处于高位[4]。此外，我国大气污染存在区域性差异，例如东部经济发达及人口密集地区的$PM_{2.5}$污染水平高于国内其他地区，提示颗粒物污染的形成与人为因素密切相关。我国于2012年设立了$PM_{2.5}$的空气质量标准，年均浓度II级标准为35 $\mu g/m^3$，并在此之后在全国推进大气$PM_{2.5}$常规监测工作的开展。新标准第一阶段监测实施的74个城市的颗粒物污染水平呈现逐年降低的趋势，其中2013年$PM_{2.5}$平均浓度（范围）为72 $\mu g/m^3$（26~160 $\mu g/m^3$），2016年$PM_{2.5}$平均浓度（范围）为50 $\mu g/m^3$（21~99 $\mu g/m^3$），总体上仍然保持在较高水平。

我国大气污染存在明显的季节性特征，总体上秋冬季节污染水平高于春夏季节。近年来我国北方采暖期大气污染严重，经常形成污染水平远高于平时的雾霾事件，影响范围较大。主要原因是在此期间大气污染排放增加而气象条件不利于污染物扩散。雾霾期间大气颗粒物浓度升高，太阳光透射强度减弱，导致大气能见度降低，形成白天灰蒙蒙的景象。

大气污染可能导致机体出现一系列不良健康效应，其中尤以心血管和呼吸系统（统称心肺系统）较为易感，可在人群水平上形成较大的疾病负担。例如，一项研究发现，以淮河为界的北方供暖政策导致我国北方总悬浮颗粒物（total suspended particle，TSP）污染水平高于南方 55%，相应地造成我国北方居民人均期望寿命较南方居民少 5.5 岁，且 TSP 每升高 100 μg/m^3 可导致期望寿命减少 3.0 年[5]。另一项综合性分析发现，我国大气污染在 2010 年造成 2522 万伤残调整寿命年（disability-adjusted life years）的损失，是继饮食因素、高血压和吸烟之后的人群第四大危险因素，其中绝大部分伤残调整寿命年损失与心肺系统疾病相关[6]。

大气污染已成为我国的主要环境问题之一，引起国内外社会的广泛关注。近年来大气污染健康影响也成为国内学界关注的热点问题，且相关研究主要探讨对心肺系统的健康影响。主要的研究类型包括在人群中开展的流行病学研究及在实验室开展的毒理学研究。大部分研究关注的是大气污染短期暴露（数小时至数周）的健康影响，少量研究则关注了大气污染长期暴露（一年或更长时间）的健康影响。本章将对近年来我国大气污染对心肺系统健康影响的研究现状进行综合介绍，总结已取得的经验，并对未来的研究方向进行展望。

2 大气污染短期暴露对人群心肺疾病的影响

大气污染短期暴露与人群心肺疾病每日死亡率和患病情况的研究以生态学研究为主，即在群体水平上观察大气污染变化与健康结局的关联性。主要研究类型包括时间-序列研究和病例-交叉研究等。其中心肺疾病每日死亡率是研究中使用最多的健康观察终点，而针对心肺疾病每日患病情况（包括急诊、门诊及入院）的研究相对较少。此外，近年也有研究使用了寿命损失年（years of life lost）和伤残调整寿命年等指标评价大气污染短期暴露的心肺疾病负担。

2.1 人群心肺疾病每日死亡率

早期关注大气污染短期暴露与人群心肺疾病死亡率的流行病学研究多局限于一个城市或地区，其中多数研究是在人口密集的大型城市，如北京、上海等地开展。上述大型城市的大气质量监测系统和疾病死亡登记系统相对比较完善。此外，也有来自广州、深圳、重庆、武汉、西安、天津和沈阳等城市的报道。近年来多城市研究及汇总多个研究数据的荟萃分析（meta-analysis）数量增加，丰富了人们对此科学问题的认识。不同大气污染物中，以对 $PM_{2.5}$、PM_{10}、SO_2 和 NO_2 的研究较为充分，而针对 O_3 和 CO 的研究数据还相对较少。表 26-1 总结了近年来我国主要大气污染物与心肺系统疾病每日死亡率的多城市研究及荟萃分析结果。

较早的一项多城市时间-序列研究收集了国内人口规模在 120 万~1230 万之间的 16 个城市的数据，总体时间跨度为 1996~2008 年（不同城市年份不同），期间不同城市大气 PM_{10} 的平均浓度范围为 52~156 μg/m^3；分析发现 PM_{10} 的 2 日平均浓度每增加 10 μg/m^3，心血管疾病死亡率增加 0.44%（95%后验区间：0.23，0.64），呼吸疾病死亡率增加 0.56%(95%后验区间：0.31，0.81)[7]。同一研究组在上述基础上又分析了 17 个城市大气 NO_2 和 SO_2 与心肺疾病每日死亡率的关联（研究期间不同城市 NO_2 和 SO_2 的平均浓度范围分别为 23~67 μg/m^3 和 18~100 μg/m^3），分析发现 NO_2 的 2 日平均浓度每增加 10 μg/m^3，心血管疾病死亡率增加 1.80%（95%后验区间：1.00，2.59），呼吸疾病死亡率增加 2.52%(95%后验区间：1.44，3.59)[8]；而 SO_2 的 2 日平均浓度每增加 10 μg/m^3，心血管疾病死亡率增加 0.83%（95%后验区间：0.47，1.19），呼吸疾病死亡率增加 1.25%(95%后验区间：0.78，1.73)[9]。

针对大气 $PM_{2.5}$ 短期暴露与心肺疾病死亡率的研究近几年逐渐增多。近期发表的一项综合性研究收集了国内 272 个城市 2013~2015 年间的数据，其中不同城市 $PM_{2.5}$ 的年平均浓度范围为 18~127 μg/m^3；分析发现 $PM_{2.5}$ 的 2 日平均浓度每增加 10 μg/m^3，心血管疾病死亡率增加 0.27%（95%后验区间：0.18，0.36），呼吸疾病死亡率增加 0.29%(95%后验区间：0.17，0.42)[10]。

大部分已有研究仅能将心血管疾病和呼吸疾病的总死亡率作为两个较为粗略的健康观察终点，而针对具体病因所导致死亡率的研究相对较少。Chen 等收集了国内 272 个城市数据，分析发现，$PM_{2.5}$ 的 2 日平均浓度每增加 10 μg/m^3 可引起高血压相关死亡率升高 0.39%（95%后验区间：0.13，0.65），冠心病死亡率升高 0.30%（95%后验区间：0.19，0.40），中风死亡率升高 0.23%（95%后验区间：0.13，0.34），慢性阻塞性肺病（chronic obstructive pulmonary disease，COPD）死亡率升高 0.38%（95%后验区间：0.23，0.53），提供了与大气 $PM_{2.5}$ 相关的具体病因所致死亡率的一手信息。此外，近年也有少量研究分析了不同大气污染物与上述疾病及缺血性心脏病等所致死亡率的关系[11,12]。

表 26-1　近五年我国主要大气污染物与心肺系统疾病每日死亡率的多城市研究或荟萃分析结果

研究类型及参考文献	大气污染物，暴露增量	每日死亡率增加百分比（95%置信区间/后验区间）	
		心血管系统	呼吸系统
	PM_{10}, 10 μg/m^3	0.91（0.64, 1.19）	1.26（0.88, 1.65）
4 城市[11]	NO_2, 10 μg/m^3	2.12（1.58, 2.65）	3.48（2.73, 4.23）
	O_3, 10 μg/m^3	1.01（0.71, 1.32）	1.33（0.89, 1.76）
16 城市[7]	PM_{10}, 10 μg/m^3	0.44（0.23, 0.64）	0.56（0.31, 0.81）
17 城市[8]	NO_2, 10 μg/m^3	1.80（1.00, 2.59）	2.52（1.44, 3.59）
17 城市[9]	SO_2, 10 μg/m^3	0.83（0.47, 1.19）	1.25（0.78, 1.73）
荟萃分析（33 项研究）[13]*	PM_{10}, 10 μg/m^3	0.43（0.37, 0.49）	0.32（0.23, 0.40）
	$PM_{2.5}$, 10 μg/m^3	0.44（0.33, 0.54）	0.51（0.30, 0.40）
	CO, 1 mg/m^3	4.77（3.53, 6.00）	—
	SO_2, 10 μg/m^3	0.85（0.70, 1.00）	1.18（0.83, 1.52）
	NO_2, 10 μg/m^3	1.46（1.27, 1.64）	1.62（1.32, 1.92）
	O_3, 10 μg/m^3	0.73（0.49, 0.97）	0.45（0.29, 0.60）
荟萃分析（26 项研究）[14]*	PM_{10}, 10 μg/m^3	0.49（0.34, 0.63）	0.57（0.40, 0.75）
	NO_2, 10 μg/m^3	1.62（1.18, 2.05）	2.20（1.56, 2.84）
	SO_2, 10 μg/m^3	0.72（0.39, 1.05）	1.29（0.58, 1.99）
	O_3, 10 μg/m^3	0.51（0.25, 0.77）	0.48（0.19, 0.76）
32 城市[15]	PM_{10}, 10 μg/m^3	—	1.05（0.08, 2.04）
荟萃分析（36 项研究）[16]*	PM_{10}, 10 μg/m^3	0.36（0.24, 0.49）	0.42（0.28, 0.55）
	$PM_{2.5}$, 10 μg/m^3	0.63（0.35, 0.91）	0.75（0.39, 1.11）
6 城市[17]	$PM_{2.5}$, 10 μg/m^3	2.19（1.80, 2.59）	1.68（1.00, 2.37）
272 城市[10]	$PM_{2.5}$, 10 μg/m^3	0.27（0.18, 0.36）	0.29（0.17, 0.42）
38 城市[18]	PM_{10}, 10 μg/m^3	0.62（0.43, 0.81）	

* 每项荟萃分析针对不同大气污染物和不同健康结局进行综合分析的研究数量存在一定差异。

表 26-1 所示国内研究报道的大气污染物水平（特别是颗粒物水平）总体上高于欧美发达国家类似研究报道的水平。例如，美国的一项综合性研究使用了 1999~2005 年间 108 个县的数据，期间 PM_{10} 的平均浓度仅为 23.5 μg/m^3(四分位区间：20.6 μg/m^3，28.6 μg/m^3)[19]；另一项综合性研究使用了美国 2000~2006 年间 75 个城市的数据，期间 $PM_{2.5}$ 的平均浓度仅为 13.3 μg/m^3 ± 8.3 μg/m^3 [20]。与此不同的是，与单位大气污染物水平升高一定单位（例如 10 μg/m^3）相对应的每日死亡率的变化程度总体上国内低于发达国家。例如，北美的相关研究发现 $PM_{2.5}$ 每增加 10 μg/m^3，心血管疾病死亡率增加范围在 0.6%~1.4%之间，呼吸疾病死亡率增加范围在 0.6%~2.2%之间[20,21]；国内近年的多城市研究或荟萃分析除一项在珠三角地区开展的小范围研究外，其他研究报道的与 $PM_{2.5}$ 每增加 10 μg/m^3 相关的心肺疾病死亡率增加值都小于 0.75%（见表 26-1）。上述差异可能与不同的研究背景因素，如大气污染水平、污染来源、化学成分、人群易感性、年龄分布及社会经济因素有关，值得进一步深入探索。国内的研究显示，大气污染与心肺疾病每日死亡率的暴露-反应关系在高浓度范围内有趋于缓和的现象，提示可能存在“收割效应”，即因易感个体在大气污染物达到很高浓度之前已经死亡，从而削弱了高浓度下的暴露-反应关系[10]。

在各种大气污染物中，颗粒物因对健康有较大的潜在威胁而得到越来越多的关注。然而因颗粒物理化性质复杂，其健康效应评价也存在较大的不确定性。国内比较大气颗粒物不同理化性质对心肺疾病死亡率影响的研究还较少。Meng 等比较了多种不同粒径的大气颗粒物数浓度对沈阳市居民心肺疾病每日死亡率的影响，发现较小粒径的颗粒物（空气动力学直径< 0.50 μm）估计效应强于较大粒径的颗粒物；其中空气动力学直径在 0.25~0.28 μm、0.35~0.40 μm 和 0.45~0.50 μm 之间的颗粒物其数浓度每升高一个四分位间距，心血管疾病死亡率将分别升高 2.79% [95%置信区间（CI）：1.09，4.49]、1.60%（95% CI：0.47，2.74）和 0.64%（95% CI：0.05，1.23），呼吸疾病死亡率将分别升高 0.72%（95% CI：–2.97，4.40）、0.79%（95% CI：–1.67，3.26）和 0.66%（95% CI：–0.60，1.92），提示心血管系统对小粒径颗粒物可能较敏感[22]。Lin 等比较了三种不同粒径颗粒物 PM_{10}、$PM_{2.5}$ 及 PM_1 对广州市居民心血管疾病死亡率的影响，发现 PM_1 的估计效应强于较大粒径的颗粒物[23]。此外，针对西安市大气 $PM_{2.5}$ 化学成分与心肺疾病死亡率的分析发现，$PM_{2.5}$ 中某些成分如有机碳（organic carbon，OC）、无机碳（elemental carbon，EC）、氨根、硝酸根、氯和镍与心肺疾病死亡率有显著性关联，其中硝酸根与心血管疾病死亡率的关联强于 $PM_{2.5}$ 本身[24]；上述针对广州市的分析也发现，$PM_{2.5}$ 中的 OC、EC、硫酸根、硝酸根和铵根与心血管疾病死亡率存在显著关联性[23]。

2.2 人群心肺疾病每日患病情况

与疾病死亡登记相比，针对疾病患病情况的登记系统在国内多数地区还不完善，因此限制了与疾病患病情况相关研究的开展。近年来评价大气污染短期暴露与人群心肺疾病每日患病情况的研究逐渐增多，绝大多数也仅限于单个城市或地区。急诊、门诊和入院是三种表征心肺疾病患病情况的主要健康观察终点。较早的及近年来的研究主要集中在北京和上海两地，因为北京和上海疾病患病登记系统较为完善，数据可靠性高。多数研究都报道了大气污染短期暴露可能导致心肺疾病的急诊、门诊和入院率升高[25-30]。此外，国内其他城市如广州、深圳、成都和兰州等也有一些相关报道[31-34]。

一项荟萃分析综合了较早时期国内的相关研究，计算得到各种大气污染物与不同心肺疾病入院情况的关联，其中与 PM_{10}、NO_2 和 SO_2 每增加 10 μg/m^3 相关的心血管疾病入院风险比分别为 1.0021、1.0095 和 1.0079，相关的呼吸疾病入院风险比分别为 1.0039、1.0060 和 1.0014；具体疾病方面，缺血性心脏病、哮喘、COPD、流感、肺炎、急性呼吸道疾病等都与大气污染存在显著关联[14]。近期的一项荟萃分析综合了东亚地区的 26 项研究（多数为我国研究），发现大气污染短期暴露与 COPD 和哮喘的门诊入院率和急诊入院率均有显著性关联；其中与单位大气污染浓度（10 μg/m^3，CO 为 1 mg/m^3）相关的 COPD 门诊入院的风险比范围为 1.007（SO_2 全人群）~1.028（O_3 全人群），COPD 急诊入院的风险比范围为 1.011（SO_2 全人群）~1.028（O_3 全人群）；与单位大气污染浓度相关的哮喘门诊入院的风险比范围为 1.010（PM_{10} 全人群）~1.141（CO 儿童），哮喘急诊入院的风险比范围为 1.009（SO_2 全人群）~1.040（NO_2 儿童）[35]。总体而言，大气污染与心肺疾病每日患病情况的关联不如与心肺疾病每日死亡率的关联稳定。虽然多数关于大气污染与心肺疾病每日患病情况的研究报道了阳性结果（大气污染导致患病增加），一些研究也报道了阴性结果（大气污染与患病无关）[14]。

只有少量国内研究关注了颗粒物理化特征对心肺疾病每日患病情况的影响。在上海开展的两项时间-序列研究分析了黑碳（black carbon，BC）对门急诊和儿童哮喘入院的影响，发现 BC 与上述健康结局的关联较颗粒物本身更显著[27,29]。另一项时间-序列研究分析了 10 种 $PM_{2.5}$ 化学成分对上海市居民急诊的影响，发现其中的 OC 和 EC 与急诊的关联最为显著，提示化石燃料的燃烧产物可能是影响 $PM_{2.5}$ 相关健康效应的关键污染成分[36]。

此外也有研究关注不同粒径颗粒物对心肺疾病患病情况的影响，发现较小粒径的颗粒物与心血管疾病急诊的关联强度大于较大粒径的颗粒物[37]，但与呼吸疾病急诊的关联强度并不大于较大粒径的颗粒物[38]。上述结果与之前针对不同粒径颗粒物与心肺疾病每日死亡率的研究结果类似[22]，提示心血管系统可能比呼吸系统对小粒径的颗粒物更敏感，但仍需未来研究进一步验证。

2.3　雾霾天气对心肺疾病的短期影响

近年来，国内秋冬季节经常出现严重的大气污染事件，俗称“雾霾”，已成为社会关注的焦点问题之一。雾霾侵袭范围广大，顶峰时可能涵盖我国整个中东部地区，持续数日甚至更长时间。雾霾期间主要大气污染物（包括颗粒物、SO_2 和 NO_2 等）的浓度较平时有显著的升高，其中 $PM_{2.5}$ 被认为是造成雾霾事件的罪魁祸首，引起民众对其健康危害的担忧。已有少量研究关注雾霾事件对心肺疾病的影响。一项研究采集了北京某医院在雾霾前后的临床诊疗数据，发现雾霾期间心肺疾病的急诊、门诊和入院较雾霾前后均有显著升高[39]。另一项研究采集了 2013 年京津冀地区 5 个城区和 2 个县的数据，分析发现在未校正气象因素情况下，雾霾期间 5 个城区/县的心肺疾病死亡率显著升高；但是当校正气象因素之后，城区雾霾期间心肺疾病死亡率无显著升高，而 2 个县雾霾期间心肺疾病死亡率仍有显著升高，增加的百分比分别为 11.66%和 22.23%[40]。近期一项研究报道了 2013 年我国东部雾霾事件期间上海某医院冠心病门诊出现显著上升，其中滞后 0 天的风险比为 1.18（95% CI：1.04，1.32）[41]；另一项研究则发现 2013 年雾霾事件期间济南市儿童医院的呼吸疾病急诊及济南市急救中心心肺疾病的急救电话数量均有显著增加[42]。

总体而言，目前国内针对雾霾事件对心肺健康影响评价的研究数据和结果还极为有限，未来需要开展更大范围内更细致的研究以对政府的大气污染危害防控决策提供科学证据支持。

3　大气污染长期暴露对人群心肺疾病的影响

针对大气污染长期健康效应开展的研究是制定大气质量长期标准的重要依据。然而现有相关大气污染长期暴露对人群健康影响的研究大部分在欧美发达国家开展，相应研究报道的大气污染水平较低，范围较窄[43,44]。我国目前针对大气污染长期健康效应的研究还较少，有关大气污染长期暴露对人群心肺疾病影响的研究更是屈指可数。开展的主要研究类型包括生态学研究、横断面研究和队列研究等。总体上研究证据较为薄弱。

3.1 生态学及横断面研究

早期关注大气污染长期暴露健康影响的研究多为生态学研究。较早的一项生态学研究发现北京市大气中硫酸盐的长期水平与人群心血管疾病及肺癌的死亡率有显著相关性[45]；另一项研究收集了广州1954~2006年间的历史性雾霾数据，发现大气污染事件之后人群的肺癌发病及死亡率有明显增加[46]；近期的一项生态学研究则发现我国10个城市的年均NO_2浓度与当地人群肺癌的发病和死亡率有显著相关性[47]。

横断面研究选取某一特定时间点，观察大气污染长期暴露与一定人群中个体水平健康结局发生的关联。一项在北京4564名儿童中开展的横断面研究发现，城区儿童因居住地大气污染水平较高，其呼吸系统症状及哮喘等的现患率均显著高于污染较轻的郊区儿童，其中哮喘在城区儿童及郊区儿童中的现患率分别为9.5%和5.4%[48]；此前在北京9052名成人中开展的类似研究也发现城区居民哮喘的现患率高于郊区居民[49]。另外一项在北方7个城市6730名儿童中开展的调查则发现大气污染与过敏性鼻炎的发生有关[50]。一项在北方3个城市24845名成人中开展的横断面研究发现，长期暴露于大气污染与人群高血压风险增加有显著关联，其中与大气污染物PM_{10}、SO_2、NO_2和O_3 3年平均浓度每升高一个四分位间距（分别为19 μg/m^3、20 μg/m^3、9 μg/m^{3}和22 μg/m^3）对应的高血压风险比值比分别为1.12（95% CI：1.08，1.16）、1.11（95% CI：1.04，1.18）、1.09（95% CI：1.00，1.20）和1.13（95% CI：1.06，1.20）[51]；同一研究的后续分析发现，大气污染长期暴露与早期高血压的发生存在更显著的关联[52]。近期的一项横断面研究在12665名50岁以上成人中评价了大气$PM_{2.5}$长期暴露的高血压疾病负担，发现与$PM_{2.5}$每升高10 μg/m^{3}对应的高血压风险比值比是1.14（95% CI：1.07，1.22），与$PM_{2.5}$相关的高血压人群归因风险比例为11.75%[53]。

生态学研究和横断面研究为更深入的科学研究提供了线索。不过因其本身存在较多研究局限性，其研究证据强度要弱于队列研究。生态学研究中，因缺少个体水平的数据，不能控制影响健康结局的诸多混杂因素；而横断面研究中，大气污染暴露与健康结局出现的时间先后顺序不清晰。上述局限性影响了大气污染暴露与健康结局发生之间的因果推断。此外，上述研究中的大气污染长期暴露水平是基于群体水平而非个体水平的监测数据，可能导致暴露水平估计存在较大的不确定性而影响与健康结局相关的定量分析。

3.2 队列研究

队列研究通过对某一人群暴露和健康结局的长期随访观察，比较不同暴露水平下健康结局发生率的差异，从而判定暴露因素与健康结局之间有无因果关联及关联程度。国内目前仅有少数几项针对大气污染长期暴露与心肺健康影响的队列研究，且多为回顾性队列研究。较早的一项研究使用了中国高血压调查流行病学随访研究的数据，该研究于1991~2000年间对我国16个省级行政单位的70947名中年人进行了随访，研究分析发现大气污染物与心肺疾病死亡率之间存在显著关联；其中队列随访前3年的大气TSP、SO_2和氮氧化物（nitrogen oxides，NO_x）平均浓度每升高10 μg/m^{3}对应的心血管疾病死亡率分别增加0.9%（95% CI：0.3，1.5）、3.2%（95% CI：2.3，4.0）和2.3%（95% CI：0.6，4.1）[54]。另一项回顾性队列研究对沈阳9941名居民进行了调查，收集了1998~2009年间的大气污染数据及人群死亡登记数据，分析也发现大气污染物与心肺疾病死亡率之间存在显著关联；其中PM_{10}和NO_2年均浓度每升高10 μg/m^{3}对应的心血管疾病死亡的风险比分别为1.55（95% CI：1.51，1.60）和2.46（95% CI：2.31，2.63），呼吸疾病死亡的风险比分别为1.67（95% CI：1.60，1.74）和2.97（95% CI：2.67，3.27）[55,56]。随后的一项回顾性研究分析了我国北方四城市（沈阳、天津、日照、太原）大气 PM_{10} 长期暴露与 39054 名居民心血管疾病死亡率的关联，发现 PM_{10} 浓度每升高 10 μg/m^3 对应的心血管疾病、缺血性心脏病和心衰的死亡风险比分别为 1.23（95% CI：1.19，1.26）、1.37（95% CI：1.28，1.47）和 1.11（95% CI：1.05，1.17）[57]。

还有一项研究分析发现大气 PM_{10} 长期暴露与我国 71431 名成年男性心肺疾病死亡率之间存在显著关联，10 μg/m^3 的暴露增值分别与 1.8%（95% CI：0.8，2.9）的心血管疾病死亡率和 1.7%（95% CI：0.3，3.2）的呼吸疾病死亡率相关[58]。近期发表的一项研究则以 4444 名肺结核患者组成的队列为基础，分析大气污染长期暴露与该人群呼吸疾病死亡的关系，发现 $PM_{2.5}$ 浓度每升高一个四分位间距（2.06 μg/m^3）对应的肺结核、呼吸道肿瘤和其他呼吸疾病死亡的风险比分别为 1.46（95% CI：1.15，1.85）、1.72（95% CI：1.36，2.19）和 1.19（95% CI：1.02，1.38）[59]。

在上述多数队列研究中，对研究对象大气污染的暴露水平通常使用居住地点附近环境固定监测点的数据进行估算，结果实际上代表的是社区或区域水平的暴露而非个体水平的暴露。以固定监测点数据为基础的暴露评价存在较大误差，可能导致与健康结局进行关联分析时的偏倚，因此不同研究结果之间存在较大差异。值得一提的是，近期发表的肺结核队列研究的人群暴露评价利用全球疾病负担研究开发的卫星反演、化学输运模型拟合及地面监测数据相结合的评价方式，提高了暴露评价精度[4]。此外，队列研究的不确定性也与研究的回顾性质有关，因回顾性研究在收集数据过程中存在某些信息偏倚，可能导致结果估计的偏差。目前我国学界已有开展针对大气污染长期健康效应的前瞻性队列研究的呼声，某些前瞻性研究已在开展或筹备中[60]，未来结果值得期待。

4　大气污染对人群心肺系统的亚临床效应

大气污染暴露除可引起心肺疾病死亡或患病情况加重外，还可造成一系列亚临床效应，通常可用多种健康指标来表征，如肺功能、呼吸系统症状、血压、心率变异性、生物样品（血液、呼出气冷凝液等）中的生物标志物水平等。定组研究（panel study）是大气污染短期暴露亚临床效应调查最常应用的流行病学设计。该类型研究通常选择数名、几十至上百名研究对象，在纵向的不同时间点重复测量其对大气污染物的暴露水平和某些健康指标水平，并使用混合效应模型或类似的统计学模型评价大气污染的健康效应[61]。还有一些研究则使用实验性的研究设计（包括控制暴露研究和干预研究）更精准地评价大气污染的短期健康效应。在上述研究设计中，研究者通常让研究对象暴露于两种或几种不同的大气污染情境，并测量不同情境下同一批健康指标的水平，然后使用 T 检验等统计学方法比较不同暴露情境下健康指标的差异。定组研究和实验性研究因对研究人员和仪器设备等的要求较高，限制了研究的规模。不过此类型的研究可以对大气污染物的健康影响进行重要的探索，明确大气污染对人群健康产生的临床前不良效应，为以较大规模人群为基础的大气污染与人群疾病的流行病学研究提供证据支持。近年来国内已开展了较多此类研究，且多关注大气污染短期暴露对心肺系统的健康影响。此外，国内针对大气污染长期暴露的心肺亚临床效应也开展了一些相关研究，多为横断面研究，且集中在对肺功能、呼吸系统症状发生和血压水平等的影响评价上。以下分别就呼吸系统和心血管系统的研究情况进行总体阐述。

4.1　呼吸系统亚临床效应

呼吸系统是人体抵御大气污染物的第一道防线。大气污染物通过呼吸系统进入人体血液循环系统及机体其他部分。早期及近年的研究多使用肺功能和呼吸系统症状评价大气污染暴露对呼吸系统的亚临床效应，总体研究结果显示大气污染短期或长期暴露可不同程度降低肺功能水平并增加呼吸系统症状的发生率。近年来还有较多研究使用了表征炎症水平的生物标志物来评价大气污染暴露引起呼吸道炎症和氧化应激的作用。

4.1.1 肺功能

既往研究中常用的肺功能指标包括用力肺活量（forced vital capacity，FVC），第一秒用力呼气容积（forced expiratory volume in 1 second，FEV_1）和呼气峰流速（peak expiratory flow rate，PEF）等。早期的一些横断面研究总体结果显示，大气颗粒物长期暴露与人群，特别是儿童较低的肺功能水平有关[62]；而居住在污染严重地区的儿童其肺功能的年均增长水平低于居住在污染较低地区的儿童[63]。

近年来开展的一些定组研究及队列研究在不同人群中对大气污染暴露与肺功能的关系进行了更深入的探讨。近期一项定组研究以 33 名 COPD 患者为研究对象，研究了不同暴露评价方式（使用固定监测点的数据及考虑室外向室内渗透特点的个体暴露估计）估算的大气颗粒物短期暴露水平对其肺功能的影响，发现 $PM_{2.5}$ 较 PM_{10} 与肺功能的负关联更明显，且使用固定监测点的数据可能高估颗粒物的效应[64]。此前一项定组研究招募了 40 名健康大学生作为志愿者，追踪其在北京郊区、城区校园 2 个时期对大气污染的暴露及肺功能水平的变化，系统分析了各种大气污染物（PM_{10}, $PM_{2.5}$, $PM_{10\text{-}2.5}$, CO, NO_x 和 NO_2）及 $PM_{2.5}$ 30 余种化学成分与研究人群肺功能的关联[2,65-67]。大学生志愿者本科前 2 年在郊区校园学习，本科后 2 年在城区校园学习，两个校园大气污染水平及成分存在较大差异。研究结果显示，不同大气污染物中以 $PM_{2.5}$ 与肺功能的关联最显著，且 $PM_{2.5}$ 中的几种金属成分（铜、镉、锡、钙、镁等）及砷与较低的肺功能水平存在显著关联[65,67]。进一步分析发现，来源于扬尘/土壤和工业源的 $PM_{2.5}$（富含上述金属）与肺功能水平的降低之间存在显著关联，提示颗粒物的健康效应与其污染来源有关[2]。值得一提的是，$PM_{2.5}$ 与肺功能水平之间的负关联在较长的暴露时间窗（肺功能测量前 1~2 周）比较短的时间窗（肺功能测量前几天之内）更明显，提示颗粒物暴露存在累积效应[66,67]。另一项研究重复测量了 60 名卡车司机和 60 名办公室白领的个体 $PM_{2.5}$ 暴露水平及肺功能，发现几种地壳金属（硅、铝和钙）与较低的 FVC 和 FEV_1 之间有关[68]。还有一项在 107 名儿童中开展的定组研究也发现，颗粒物中的金属成分（铅、镍、铁、锰和铬）与较低的肺功能有关[69]。上述研究发现的对肺功能有显著影响的金属成分多为过渡金属，这可能与其较强的致氧化应激活性有关。此外，来自中国武汉-珠海队列（Wuhan-Zhuhai Cohort）的研究结果显示，主要大气污染物 NO_2、PM_{10}、O_3 和 $PM_{2.5}$ 的短期暴露与女性不吸烟者较低的 FVC 和 FEV_1 有关，且在污染较严重的城市（武汉）上述关联较强，效应持续较久[70]。

4.1.2 呼吸系统症状

早期针对大气污染暴露与呼吸系统症状发生的研究多采用横断面研究设计，总体上发现大气污染的长期暴露与儿童或成人较高的呼吸系统症状发生相关[48-50]。研究涉及的主要呼吸系统症状包括咳嗽、咳痰、喘息、哮喘样症状等。近期有少量研究也关注了大气污染短期暴露与呼吸系统症状发生的关系。国外一项研究监测了 17 名健康志愿者到北京旅行前后咳嗽反射阈值（cough reflex threshold，以诱导咳嗽所需的雾化柠檬酸浓度来表征）及咳嗽意愿阈值（urge-to-cough threshold）的变化，发现在京期间两种阈值水平明显低于旅行前后，提示大气污染短期暴露可增加咳嗽发生的可能性[71]。前述在北京 COPD 患者（n=23）中开展的定组研究则发现，较高的大气污染物水平与不同呼吸系统症状，包括咽喉痛、咳嗽、咳痰、喘息和呼吸困难等的发生增加均有关[72]。

4.1.3 呼吸系统生物标志物

有几项定组研究选取一些特定的生物标志物，如呼出气一氧化氮（eNO）和呼出气冷凝液 pH 等，评价大气污染暴露对气道炎症的影响。较早的两项研究以 2008 年北京奥运会期间大气质量改善为契机，在奥运会前、中、后不同时期对研究人群进行重复随访，发现奥运会期间的大气污染水平明显低于奥运会前后，其中一组儿童（*n*=36）的 eNO 水平也在奥运会期间出现低值，模型分析则显示日均大气 $PM_{2.5}$ 与 BC 浓度与 eNO 之间存在显著性关联[73]；在另一组健康成人个体（*n*=125）中的研究结果与此类似，一系列与肺部炎症和氧化应激相关的标志物（包括 eNO、呼出气冷凝液中的 pH、丙二醛、硝酸盐、亚硝酸盐和 8-异前列腺素等）在奥运会期间的水平显著低于奥运会前后，模型分析也提示大气污染物与上述标志物变化之间存在显著关联[74,75]。上述研究的后续分析发现，肺部炎症标志物 eNO 及呼出气冷凝液 pH 与机动车和工业来源的燃烧产物、石油燃烧产物及植被燃烧产物等大气污染源的排放密切相关[76]。此后在上海开展的一项定组研究（*n*=32）则显示，暴露于 $PM_{2.5}$ 中的某些化学成分，包括铵根、硝酸根、钾离子、硫酸根和 EC 等与较高的 eNO 水平有关，其中只有 EC 在模型校正 $PM_{2.5}$ 后仍然与 eNO 存在显著性关联[77]。在前述北京 COPD 患者中开展的定组研究则发现，不同大气污染物除与表征呼吸道炎症的经典标志物 eNO 存在显著关联外，与另一种表征呼吸道炎症的新颖标志物呼出气硫化氢（eH_2S）也存在显著关联[72]。上述研究证据表明，大气污染短期暴露与呼吸道炎症及氧化应激之间存在密切关系。

4.2　心血管系统亚临床效应

心血管系统是大气污染危害作用的主要机体系统之一，有关大气污染心血管效应的研究也是目前国内外的热点之一。国内目前对大气污染心血管亚临床效应的研究证据主要来自于定组研究（或重复测量研究），此外也有来自横断面研究的报道。常用的心血管亚临床效应指标包括血压、心率变异性和循环生物标志物等。

4.2.1　血压

血压是大气污染心血管效应研究中最常用的健康指标之一。多数相关研究发现大气污染暴露可能导致血压水平升高，且结果在易感人群及健康人群中相对比较一致。例如两项在心血管疾病（*n*=40）及代谢综合征（*n*=65）患者中开展的重复测量研究均发现，大气 $PM_{2.5}$ 和 BC 急性暴露（数分钟至数小时）与动态血压水平之间存在显著关联[78,79]。其中在代谢综合征患者中开展的后续研究发现，动态血压测量之前 1~7 天的 $PM_{2.5}$ 和 BC 短期暴露同样与血压水平存在显著关联[80]。一项纳入 98 名冠心病患者的随机交叉研究调查了使用口罩对大气污染心血管效应的干预效果，研究对象在佩戴和不佩戴口罩两种情况下沿城市中心某道路行走 2 小时，结果显示研究对象在佩戴口罩行走后的平均动脉压水平显著低于不佩戴口罩行走后的相应水平[81]。另一项纳入 24 名年轻健康个体的研究同样调查了使用口罩的干预效果，发现使用口罩 48 小时后收缩压水平有显著降低[82]。北京奥运会前后在 125 名健康成人个体中开展的定组研究则发现从奥运会期间至奥运会后，研究对象的收缩压有明显上升，但与多种大气污染物相关性的模型分析结果则不够稳定[83]。还有研究以血压为主要效应指标之一评价空气净化器降低室外来源大气污染暴露的心血管干预效果，发现使用空气净化器一段时间后可使室内 $PM_{2.5}$ 水平降低 57%，与此同时，35 名健康大学生的收缩压和舒张压水平分别降低 2.7%和 4.8%，提示空气净化器干预可有效降低血压水平[84]。

Wu 等在 40 名健康大学生中开展的定组研究则对各种大气污染物及 $PM_{2.5}$ 化学成分与血压水平的变化进行了详细分析，结果显示不同大气污染物中以 $PM_{2.5}$ 与血压的正向关联最强，$PM_{2.5}$ 中几种与燃烧产物相关的成分（包括 EC、OC 和氯）及金属成分（包括镍、锌、镁、铅）以及砷与血压变化的关联最为稳定[85]。进一步分析显示来源于燃煤燃烧的 $PM_{2.5}$ 与较高的血压关联最明显[2]。前述在北方 3 个城市 24845 名成人中开展的横断面研究也报道了大气污染长期暴露与人群血压水平的关联，大气污染物 PM_{10}、SO_2 和 O_3 3 年平均浓度每升高一个四分位间距（分别为 19 μg/m^3、20 μg/m^3 和 22 μg/m^3）对应的收缩压升高值在 0.73~0.87 mmHg 之间，对应的舒张压升高值在 0.31~0.37 mmHg 之间，而 NO_2 与血压的关联则不明显[51]。

4.2.2 心率变异性

心率变异性反映的是自主神经系统功能的变化，后者被认为是大气污染影响心血管系统健康的病理生理通路之一。较高的心率变异性水平提示心脏自主调控功能良好，而在疾病状态或受应激因素影响下心率变异性水平可能降低。已有研究中使用的主要心率变异性指标包括所有 NN 间期的标准差（standard deviation of NN intervals，SDNN），低频功率（low-frequency power，LF）和高频功率（high-frequency power，HF）等。较早的一项定组研究同样利用北京奥运会前后大气质量改善的契机，对一组出租车司机（n=11）在奥运会前、中、后进行了追踪随访，监测其工作期间车厢内 $PM_{2.5}$ 及气态污染物的暴露水平及动态心电图的变化，发现其心率变异性水平在大气污染水平最低的奥运会期间有显著改善，而在大气污染水平较高的奥运会前后则有所降低，模型分析发现不同大气污染物中 $PM_{2.5}$ 与不同心率变异性指标的负向关联最强[86]；随后的分析则进一步报道了 $PM_{2.5}$ 中可能影响心率变异性的关键化学成分（包括钙、镍、铁）[87]。两项在心血管疾病患者中开展的重复测量研究同样发现在奥运会期间研究对象的心率变异性水平有显著改善，且大气 $PM_{2.5}$ 暴露水平与较低的心率变异性变化存在显著关联[79,88]。

值得一提的是，两项分别在健康年轻个体及健康老年个体中开展的定组研究均发现大气污染与部分研究对象心率变异性之间存在正关联[86,89]，提示健康人群心脏自主调控功能对外界刺激的反应与易感人群可能所有不同。前述在 98 名冠心病患者及 24 名年轻健康个体中开展的随机交叉研究均观察到在佩戴口罩干预后研究对象的心率变异性水平总体上高于不佩戴口罩后的水平[81,82]；另一项在 40 名健康个体中开展的随机交叉研究同样观察到在低暴露情境下（公园行走）研究对象的心率变异性水平高于高暴露情境下（交通枢纽）的相应水平[90]。上述研究结果提示使用干预措施降低大气污染暴露水平可改善心脏自主调控功能。

此外也有研究关注了不同粒径颗粒物对心率变异性的影响。一项在 35 名糖尿病患者或糖耐量受损者中开展的研究发现较小粒径的颗粒物数浓度与心率变异性的主要指标 SDNN 之间关联较强，而较大粒径的颗粒物则与之无关[91]。部分研究也分析了气态污染物对心率变异性的影响，总体效应不如颗粒物明显。如上述出租车司机研究随后的分析进一步报道了 CO 与心率变异性的暴露-反应关系，CO 与多数心率变异性指标之间存在负向关联，但强度弱于 $PM_{2.5}$[92]。对 O_3 的研究结果尚不一致。一项在健康老年人中开展的研究发现 O_3 暴露与较低的心率变异性水平有关[93]，而在上述糖尿病患者或糖耐量受损者中开展的研究则报道了阴性结果[91]。

4.2.3 循环系统生物标志物

既往一些人群研究通过测定研究对象血液样品中某些可反映心血管健康状态的循环生物标志物水平，分析大气污染暴露与标志物水平的关联，从而推断大气污染对心血管健康的影响。例如一项研究监测了 110 名交通警察对 $PM_{2.5}$ 的个体暴露水平，并测定其血液样品中一组炎症和免疫相关的生物标志物，发现 $PM_{2.5}$ 暴露与较高的超敏 C 反应蛋白、免疫球蛋白 G/M/E 及较低的免疫球蛋白 A 和 CD8 细胞含量有关[94]。前述在 125 名健康成人个体中开展的定组研究测定了奥运会前、中、后研究对象血液中的系统性炎症（C 反应蛋白、纤维蛋白原和白细胞计数）、凝血（可溶性 P 选择素、可溶性 CD40 配体和血管性血友病因子）标志物的水平，发现上述指标在奥运会前后有类似的变化趋势；模型分析进一步显示大气污染物与生物标志物水平之间存在显著性关联，提示大气污染即使在年轻健康个体中也可引起明显的生理学反应[75 83]。

前述在北京 40 名健康大学生中开展的定组研究重复采集了 12 次研究对象的血液样品，测定其中一系列可反映系统性炎症（C 反应蛋白、纤维蛋白原、肿瘤坏死因子-α）、凝血（纤溶酶原激活物抑制物-1、组织型纤溶酶原激活物、血管性血友病因子、可溶性 P 选择素）、系统性氧化应激（氧化低密度脂蛋白、清道夫受体蛋白）、抗氧化活性（胞外超氧化物歧化酶、谷胱甘肽过氧化物酶 1）和内皮功能（内皮素-1、E 选择素、细胞间黏附分子-1、血管细胞黏附分子 1）的生物标志物水平，对其变化与各种大气污染物及 $PM_{2.5}$ 化学成分的关联进行了较为全面的探索，总体研究结果显示大气 $PM_{2.5}$ 与上述生物标志物的关联强于其他大气污染物，且关联以正向为主。$PM_{2.5}$ 中的某些金属成分（特别是过渡金属成分如锌、铁、镍、钛、锰等）及阴离子成分（如氯离子、硝酸根、硫酸根）与不同生物标志物的变化存在显著关联，这可能与过渡金属较强的生物学活性及阴离子的电子传递特性有关[95-98]。进一步分析显示大气 $PM_{2.5}$ 的不同来源也会影响与之相关的人群心血管效应，而与生物标志物变化关系密切的 $PM_{2.5}$ 污染来源多富含上述关键化学成分，例如纤维蛋白原与化学成分铁、钛和镁及来源于扬尘/土壤的 $PM_{2.5}$ 存在稳定的正关联，其中扬尘/土壤来源的 $PM_{2.5}$ 富含上述三种金属成分[2,95]。

另一项在上海 34 名健康大学生中开展的定组研究则分析了不同粒径的颗粒物暴露与循环生物标志物变化的关系，发现较小粒径的颗粒物与不同生物标志物的关联更密切，其中以粒径在 0.25~0.40 μm 之间的颗粒物数浓度和粒径<1 μm 的颗粒物质量浓度与生物标志物的关联最强[99]。前述使用空气净化器的干预研究基于一系列循环生物标志物评价了空气净化器降低健康大学生对室外来源大气污染暴露的心血管干预效果，发现使用空气净化器后上述表征系统性炎症和凝血功能的标志物也出现不同程度的降低，其中以可溶性 CD40 配体、髓过氧化物酶、白介素-1β 和单核细胞趋化蛋白-1 等标志物的变化最为显著，提示空气净化器干预可有效减轻大气污染相关的心血管亚临床效应[84]。不过在 24 名年轻健康个体中开展的口罩干预研究则未发现炎症、凝血和血管收缩相关的循环生物标志物有显著变化，可能与研究较小的样本量及较短的干预时间等因素有关[82]。上述空气净化器干预的后续研究进一步探索了 DNA 甲基化在污染物暴露与心血管生物标志物关系中的作用，发现 $PM_{2.5}$ 暴露水平与基因组重复元件及某些候选炎症和凝血基因的甲基化存在显著关联，且 *CD40LG* 基因的低甲基化介导了 $PM_{2.5}$ 对可溶性 CD40 配体 17.82%的效应[100]。

5　大气污染对心肺健康影响的生物学机制

大气污染导致不良健康结局的生物学机制证据主要由毒理学研究提供。毒理学研究包括体外和体内研究，在实验室内分别将细胞或动物体暴露于不同水平的大气污染物并观察比较各种暴露状况下生物学反应的差别，以此确定污染物的作用途径和机制。毒理学研究中通常使用从低至高浓度的大气污染物，能够有效地诱导体外体内变化，从而有助于阐明污染物与健康效应的暴露-反应关系及相应的生物学机制。目前我国已针对大气污染的心肺健康影响开展了一些毒理学研究，且以体内研究居多，研究发现大气污染物特别是颗粒物可引起心肺系统的一系列毒性反应，其中尤其以 $PM_{2.5}$ 的效应最为显著。

5.1　肺部及机体系统性炎症

目前国内外已有足够的研究证据表明大气污染暴露可导致肺部及机体系统性炎症，且肺部炎症可能促进大气污染物的肺外迁移。国内已有研究显示大气颗粒物可增加肺部或循环系统炎症性因子（如白介素，肿瘤坏死因子-α 等）的分泌，引起炎症细胞聚集反应，介导气道炎症性损伤，从而导致肺部及机体系统性炎症。其中一项体内研究使用暴露仓设备在北京奥运会时期将两组小鼠分别暴露于大气 $PM_{2.5}$

及过滤空气 2 个月，发现暴露于大气 $PM_{2.5}$ 可增加小鼠肺部及系统性炎症反应，具体表现为生物标志物如白介素 6 和单核细胞趋化蛋白-1 水平升高，巨噬细胞及中性粒细胞在肺部、脾脏及内脏脂肪组织等部位聚集；而短期内大气质量改善则可显著减少上述炎症反应[101]。另一项研究使用类似的暴露研究设计，研究大气颗粒物暴露对载脂蛋白 E（动脉粥样硬化易感基因）基因敲除小鼠的影响，发现暴露 2 个月后小鼠的循环炎症标志物肿瘤坏死因子-α 和 C 反应蛋白及支气管肺泡灌洗液中的肿瘤坏死因子-α 及白介素 6 有显著升高，病理学检查显示肺部有明显的组织学改变，提示大气颗粒物可引起显著的肺部和系统性炎症反应[102]。

使用 2 型糖尿病小鼠模型的研究显示，大气 $PM_{2.5}$ 暴露可增加转录因子 NF-κB 在心脏组织中的表达，增加心肌的炎症性损伤，而使用 NF-κB 的抑制剂则可减轻上述炎症性损伤，提示 NF-κB 在 $PM_{2.5}$ 介导的心血管损伤中存在重要作用[103]。近期一项体内研究发现长期暴露于生物质燃烧和机动车排放来源的颗粒物均可导致大鼠发生 COPD，表现为肺功能降低、黏膜上皮化生、肺部及系统性炎症、肺气肿和小气道重塑等，提示肺部及系统性炎症在大气污染导致呼吸疾病发生中具有重要作用[104]。此外，也有研究发现暴露于气态污染物如 O_3 也可造成肺部过敏性炎症，表现为一系列炎症因子的水平升高[105]；而另一项亚急性体内研究显示 $PM_{2.5}$ 单独暴露可引起系统性炎症损伤，而 O_3 联合 $PM_{2.5}$ 暴露则可增强相应的炎症损伤[106]。还有一项体内研究对小鼠进行了 SO_2、NO_2 和 $PM_{2.5}$ 的联合暴露，发现联合暴露可引起心脏组织炎症及内皮损伤，表现为炎症性因子和内皮素-1 水平的升高及内皮一氧化氮合酶水平的降低，提示上述变化可能是大气污染造成心脏疾病的机制[107]。

5.2 肺部及机体系统性氧化应激

既往大量研究已发现大气污染暴露可增加体内氧化自由基的产生，导致氧化应激反应加重，继而损伤心血管系统。国内较早的一项研究显示，暴露于大气 PM_{10} 可增加大鼠血清、肺部及心脏组织中的丙二醛水平并降低具有抗氧化功能的超氧化物歧化酶的活性[108]，而暴露于 $PM_{2.5}$ 可引发棕色脂肪组织、心脏组织及机体系统性的氧化应激反应[106,109,110]。$PM_{2.5}$ 介导的氧化应激反应也可触发人肺上皮细胞的自噬作用，这可能是颗粒物导致肺功能损害的机制之一[111]。近期的一项体内研究将孕期大鼠长期暴露于北京市大气颗粒物，发现与对照组（暴露于过滤空气）相比，暴露组大鼠其本身及后代子鼠除血管周围及支气管周围炎症反应增强外，组织及系统性氧化应激也增强，表现为相关标志物丙二醛及 8-异前列腺素水平升高[112]。此外也有体外研究使用人支气管上皮细胞探索了 $PM_{2.5}$ 慢性暴露诱发的氧化应激中不同颗粒物化学成分及污染源的作用，发现富含二次成分（如硫酸根和硝酸根）的污染源可以解释的氧化自由基的变异程度大于其他污染源[113]。

5.3 高血压

高血压是心血管疾病的一大危险因素，也是大气污染暴露导致心肺疾病负担增加的机制之一。近年的一项研究显示，长期暴露于浓缩的 $PM_{2.5}$ 可引起自发性高血压大鼠的血压升高，而去除暴露后血压水平降低；与血压升高相伴随的不良效应包括心搏量减少、心脏重量增加和心肌肥厚标志物表达水平增加等，而去除 $PM_{2.5}$ 暴露后大鼠的心脏功能恢复，心脏重量减轻，心肌肥厚标志物表达水平降低，提示心脏功能紊乱和心肌肥厚是继发于高血压的损害[114]。另一项研究则发现颗粒物的主要成分之一碳黑可减弱降压药卡托普利对自发性高血压大鼠的降压效果；同时还发现卡托普利减少了对照组血管紧张素 II 的含量，但此效应随着碳黑暴露浓度的增加而减弱，提示碳黑在颗粒物致血压升高效应中具有重要作用[115]。还有一项亚急性体内研究显示 $PM_{2.5}$ 单独暴露或与 O_3 联合暴露均可引起颈动脉收缩压水平显著升高，而 O_3 单独暴露引起的升压效应相对较弱[106]。

5.4　自主神经功能

心脏自主神经功能对外界刺激比较敏感，可表现为心率变异性的变化。较早的一项体内研究发现大气 $PM_{2.5}$ 气管滴注急性染毒可引起大鼠心率失常[116]。另一项亚急性体内研究显示 $PM_{2.5}$ 单独暴露可引起心率变异性显著降低，而 O_3 联合 $PM_{2.5}$ 暴露则可引起更为显著的心率变异性变化[106]。之后的一项针对载脂蛋白 E 基因敲除小鼠的急性暴露研究同样发现 $PM_{2.5}$ 暴露可导致心率变异性水平显著降低[110]。此外，也有体内研究探索了颗粒物的主要成分之一碳黑的心血管效应，发现小鼠暴露于碳黑后几种心率变异性指标均出现显著降低；与此同时，碳黑暴露并不引起明显的肺部炎症和心肌损伤，提示碳黑可较为特异地影响心脏的自主调控功能[117]。

5.5　凝血功能

已有毒理学研究显示大气颗粒物暴露可影响机体凝血功能。较早的一项急性暴露研究发现 $PM_{2.5}$ 可导致大鼠血浆凝血酶原时间降低，并增加血浆中纤维蛋白原和组织因子的水平，提示凝血功能改变；同时进行的硫酸镍暴露研究显示大鼠凝血功能出现类似改变，提示镍可能在颗粒物诱导的凝血功能改变中发挥作用[118]。近期的一项体内研究将孕鼠暴露于 $PM_{2.5}$，发现暴露组血中血小板数量显著升高，病理学检查则发现暴露组胎盘组织内出现血栓，提示 $PM_{2.5}$ 暴露可导致血液高凝血状态，进而损伤胎盘[119]。

5.6　脂质代谢紊乱及动脉粥样硬化

国内已有体内研究显示 $PM_{2.5}$ 长期暴露可调节高脂饮食小鼠的脂质代谢，表现为暴露后肝脏脂质沉积增加，肝脏内及血浆甘油三酯水平升高，脂质代谢相关的基因表达水平增高[120]。另一项体内研究也显示长期暴露于北京市大气颗粒物可引起大鼠血脂异常，肺部及全身出现脂质过氧化情况，从而导致代谢紊乱及体重增加[112]。前述使用载脂蛋白 E（动脉粥样硬化易感基因）基因敲除小鼠进行的体内研究则发现颗粒物长期暴露可加速动脉粥样硬化进展，表现为暴露组小鼠的血清总胆固醇和低密度脂蛋白水平显著高于对照组小鼠，病理学分析则显示暴露组小鼠主动脉弓的动脉粥样硬化斑块面积显著大于对照组[102]。

5.7　胰岛素抵抗

前述发现 $PM_{2.5}$ 可调节高脂饮食小鼠脂质代谢的体内研究中同样发现 $PM_{2.5}$ 可导致全身胰岛素抵抗，表现为内脏脂肪组织炎症增强、肝脏脂质沉积增加、骨骼肌糖分利用减少及全身糖耐量降低等，且上述效应部分独立于 CC 趋化因子 2 型受体（CCR2，与先天免疫细胞进入组织密切相关的一种分子）相关的通路[120]。近期一项体内研究进一步探索了相关机制，发现浓缩后的 $PM_{2.5}$ 暴露引起正常饮食小鼠全身胰岛素抵抗，脂肪组织巨噬细胞浸润及肝内磷酸烯醇丙酮酸羧化激酶（糖异生关键酶）表达升高，而 CCR2 缺乏的小鼠脂肪组织巨噬细胞浸润减少，但胰岛素抵抗并未减轻，提示 $PM_{2.5}$ 暴露在直接导致血糖代谢紊乱的同时并不引起明显的炎症前反应[121]。

5.8 表观遗传改变

表观遗传是指 DNA 序列不发生变化，但基因表达却发生了可遗传的改变，主要分子机制包括 DNA 甲基化、组蛋白修饰、RNA 干扰和染色质重塑等。其中 DNA 甲基化与人类老化、氧化应激、癌症和心血管疾病等密切关系，已经成为表观遗传学和表观基因组学的重要研究内容。国内一项体内研究探索了交通相关大气污染对呼吸疾病相关 DNA 甲基化的影响，发现较长时间暴露于污染与全基因组低甲基化和炎症相关的可诱导一氧化氮合酶基因（*iNOS*）的低甲基化有关，而与控制细胞周期的 *p16*$^{\mathrm{CDKN2A}}$ 基因及肿瘤浸润转移相关的肠腺瘤性息肉病基因（*APC*）的高甲基化有关[122]。后续的一项体内研究发现，交通相关大气污染除引起 DNA 甲基化外，对呼吸疾病相关的组蛋白 H3K9 的乙酰化也有显著影响[123]。

5.9 心肺健康影响机制的组学研究

大气污染暴露可能对机体心肺系统产生广泛影响，表现为一系列相关基因的转录表达水平发生改变。以往毒理学研究通常针对其中一到几条特定通路进行深入研究，而近年已有研究开始使用组学方法探索大气污染对心肺健康影响的生物学机制，可以提供较为全面的认识。稍早的一项体外研究使用基因芯片在基因组水平上研究了大气污染暴露相关的转录组学变化，发现人体支气管上皮细胞暴露于大气 $PM_{2.5}$ 后产生了一系列基因及通路表达变化；表达增强的基因主要与炎症及免疫反应、氧化应激及 DNA 损伤应答有关，而相关的通路则涵盖细胞内活动、环境信息处理、基因信息处理和代谢等过程[124]。随后的另一项体外研究发现大气 $PM_{2.5}$ 暴露可引起人支气管上皮细胞与基因转录、信号转导、细胞增殖、细胞代谢过程、免疫反应等重要功能相关的基因表达模式发生改变；通路分析和信号网络分析则发现磷脂酰肌醇激酶/蛋白激酶 B（PI3K/Akt）、丝裂原活化蛋白激酶（MAPK）和肿瘤坏死因子（TNF）信号通路受 $PM_{2.5}$ 暴露的影响最大，而上述通路可能在细胞增殖、细胞分化、细胞骨架调节和炎症反应中发挥重要作用[125]。

后续一项体外研究则发现人心肌细胞暴露于大气 $PM_{2.5}$ 后同样可以产生一系列基因表达变化，其中与免疫反应、细胞成熟、胚胎心脏发育、细胞对电刺激反应、骨骼肌组织再生、信号转导负性调控等细胞功能相关的基因表达上调，而与转录调控（独立于 DNA）、节律过程、蛋白质失稳凋亡过程及固有免疫应答功能相关的基因表达下调[126]。近期另一项研究在斑马鱼模型中开展 $PM_{2.5}$ 的暴露毒性研究，使用基因芯片和微小 RNA 芯片分析暴露后表达信号变化，发现与毒物处理、转运、代谢相关的细胞功能及通路基因表达增强；此外 $PM_{2.5}$ 暴露也引起一系列微小 RNA 表达改变，提示 $PM_{2.5}$ 可能通过多种不同途径对机体产生毒性，引起包括心血管疾病在内的人体各种疾病[127]。

6 大气污染与其他因素对心肺健康影响的交互作用

心肺疾病的发生发展与对多种危险因素的暴露有关，大气污染则是国内人群心肺疾病的主要危险因素之一。探索大气污染与其他危险因素的交互作用对于准确识别大气污染的危害，鉴别易感人群，从而有针对性地制订危害预防措施有重要意义。国内已有一些研究关注了大气污染与其他因素对心肺健康的交互作用，涉及的主要因素包括个体因素、季节/气象因素、噪声等。

6.1　与个体因素的交互作用

已有一些研究报道了某些个体混杂因素（如年龄、性别、受教育程度、肥胖等）对大气污染与心肺健康关系的影响，不过不同研究的结果存在一定差异。在大气污染对心肺疾病每日死亡率的影响方面，研究结果总体上支持老年人及受教育程度较低的研究对象更易感，但对不同性别人群影响的研究结果尚不一致。例如前述使用全国 272 个城市 2013~2015 年间数据的研究发现大气 $PM_{2.5}$ 与某些个体因素之间存在显著交互作用，其中 $PM_{2.5}$ 与心血管疾病死亡率在 75 岁以上老人中的关联强度明显高于在 75 岁以下人群中的关联强度，而与呼吸疾病死亡率的关联在不同年龄人群中相似；上述 $PM_{2.5}$ 对心肺疾病死亡率的影响在女性人群中较在男性人群中略强，但差异无统计学意义，而在受教育程度低的人群中则比在受教育程度高的人群中明显更强，且差异有统计学意义[10]。另外一项纳入 32 个城市数据的研究同样发现大气 PM_{10} 与 60 岁以上老年人的呼吸疾病死亡率显著相关，而与较年轻成人（20~59 岁）的呼吸疾病死亡率无关，且上述显著性影响只存在于男性老年人中，但在女性老年人及女性较年轻成人中均不显著[15]。

有几项研究分析了肥胖因素对大气污染与心肺健康关系的影响，研究结果总体上支持肥胖因素可增强大气污染的不良健康影响。例如一项纳入北方 3 个城市 24845 名成人的横断面研究发现，大气污染长期暴露水平与心血管疾病和中风患病的正向关联在肥胖人群（体质指数≥30 kg/m^2）中最强，在超重人群（体质指数 25~-29.9 kg/m^2）中其次，在正常体重人群中最弱；当以性别分层时，上述显著性交互作用只存在于女性人群中[128]。另一项纳入北方 7 个城市 9354 名儿童的研究同样报道了肥胖对大气污染长期暴露与心血管健康关联的修饰作用，不同大气污染物与高血压患病及血压水平的关联在肥胖儿童中最明显，超重儿童中其次，正常体重儿童中最弱；当以性别分层时，上述显著性交互作用在不同性别儿童中相似[129]。

此外也有研究分析了其他个体因素对大气污染心肺健康影响的修饰作用。例如一项纳入 31049 名 2~14 岁儿童的横断面研究发现母乳喂养可能降低大气污染的不良呼吸影响，具体表现为母乳喂养儿童出现与大气污染相关的呼吸系统症状（咳嗽、咳痰、喘息、哮喘）的风险总体上显著低于非母乳喂养儿童，且上述保护效应在年龄较小的儿童中更明显[130]。在上述纳入 9354 名儿童的横断面研究中也发现大气污染与高血压患病的关联在非母乳喂养的儿童更显著[131]。

6.2　与季节/气象因素的交互作用

既往大量研究报道季节及气象因素特别是气温变化与人群心肺健康存在密切关系。因人群同时暴露于大气污染和季节/气象因素，深入探讨两类因素对心肺健康的可能交互作用对于准确识别环境因素的健康危害并制定相应的预防措施有重要意义。

已有一些研究发现在不同季节大气污染短期暴露对心肺疾病的每日死亡率和患病情况影响存在差异，总体上表现为在寒冷（低温）或炎热（高温）时期大气污染的效应更显著。例如，此前一项研究分析了 2004~2009 年间北京市大气 $PM_{2.5}$ 与呼吸疾病每日死亡率和患病情况的关联，发现冬季大气污染的效应较强，10 $\mu g/m^3$ 大气 $PM_{2.5}$ 对应的冬季呼吸疾病每日死亡率和患病情况的升高幅度分别是全年平均水平的 1.4 倍和 1.8 倍[132]。随后的一项分析发现与北京市大气 $PM_{2.5}$ 相关的心血管疾病死亡率升高在低温情况时更显著[133]。另一项在上海市开展的研究则发现大气 PM_{10}/O_3 和极低气温对心肺疾病每日死亡率存在显著交互作用，气温低于 15%百分位数时的大气污染估计效应高于气温处于正常范围（15%~85%百分位数）时的大气污染估计效应[134]。还有研究发现高温可增强大气污染的效应。如一项研究发现合肥市不同大气污染物与呼吸系统每日死亡率的关联在高温时均强于正常温度时[135]。前述在北京市开展的研究同样发现

大气 $PM_{2.5}$ 与呼吸系统每日死亡率的关联在高温时明显[133]。还有研究发现 SO_2 对急性死亡率的影响受气温影响，但并不受相对湿度的影响[136]。与上述研究结果对应的是，也有少量研究发现大气污染可增加气象因素对心肺系统的不良影响。如一项研究发现广州大气 PM_{10} 与气温对每日死亡率存在显著交互作用，寒冷和炎症对心血管死亡率的影响随着 PM_{10} 浓度升高而变得明显[137]。

此外，针对大气污染对心肺亚临床效应的研究也发现气象因素与大气污染存在交互作用。前述在 40 名健康大学生中开展的定组研究对 21 名研究对象进行了每日早晚肺功能监测，同时监测其对校园大气 $PM_{2.5}$ 与气温的暴露水平，分析发现 $PM_{2.5}$ 与晚间 PEF 及早/晚间 FEV_1 的显著降低有关，而气温与早间 PEF 的显著降低有关；两种暴露因素存在显著交互作用，其中 $PM_{2.5}$ 对肺功能指标的估计效应在气温高于中位数时较强，低于中位数时较弱，而气温对肺功能指标的估计效应同样在 $PM_{2.5}$ 水平高于中位数时较强，低于中位数时较弱[66]。上述研究也发现大气污染与气温对研究对象血压变化的交互作用，不同大气污染物及 $PM_{2.5}$ 主要化学成分（$PM_{2.5}$、NO_2、OC 和 EC 等）对血压的影响在气温低于中位数时更明显，而气温对血压的影响则在大气污染物浓度高于中位数时更明显，提示低温和高水平大气污染可能存在联合作用，导致血压水平升高[138]。前述在出租车司机中开展的定组研究发现大气污染与气温对心率变异性存在交互作用，大气污染物（$PM_{2.5}$ 和 CO）与主要心率变异性指标的负向关联总体上在气温较高时明显，而气温对心率变异性指标的影响也不同程度上受到大气污染物的修饰作用[139]。

6.3 与噪声的交互作用

噪声暴露可能引起一系列不良的心血管效应，长期暴露则可能引起心血管疾病发病率增加。国内针对噪声与大气污染的交互作用研究尚不多见。此前一项人群交叉暴露研究招募了 40 名健康个体，分别使其暴露于交通枢纽和公园环境 2 小时，并测量暴露前后对主要大气污染物（$PM_{2.5}$、BC 和 CO）和噪声的暴露水平和动态心电图变化。研究发现在交通枢纽的大气污染物及噪声暴露水平高于在公园环境的暴露水平，而公园环境的人体心率变异性水平则高于交通枢纽的相应水平；统计学分析发现大气污染物与心率变异性之间存在显著性关联，且在噪声水平较高（>65.6 分贝）时的关联强度高于噪声水平较低（65.6≤分贝）时，噪声与大气污染物的交互项具有统计学意义，提示噪声可显著修饰大气污染对心脏自主调控功能的效应[90]。

7 展　望

国内已经开展的大量流行病学及毒理学研究显示大气污染对心肺健康存在显著影响。因大气污染本身的复杂性和不同研究之间的异质性等因素，某些研究结果尚不一致，但不可否认的是大气污染已成为我国人群疾病负担增加的重要贡献因素之一。根据上述综述内容，可知目前我国主要大气污染物（包括颗粒物和各种气态污染物）水平总体上较高，且其中大气 $PM_{2.5}$ 的污染情况较为严重。针对大气污染短期暴露与心肺疾病每日死亡率的研究证据较为充分，其中尤以对颗粒物的研究较为系统，而针对大气污染短期暴露与心肺疾病每日患病情况的研究证据尚不完善。不同大气污染物中以颗粒物的心肺健康效应较为明显，且受其理化性质影响，小粒径颗粒物及某些化学成分（如碳质、过渡金属等）可能是影响颗粒物整体健康效应的关键因素。大气污染可能通过多种生物学机制影响心肺健康，目前国内以对肺部和机体系统性炎症以及氧化应激的研究证据最为充分。大气污染可能与多种因素（包括个体因素和环境因素）共同作用影响心肺健康。此外，已有证据也支持改善大气质量可产生明显的心肺健康效益。

然而，与欧美发达国家相比，国内目前对大气污染引发的心肺健康效应的研究仍然不够充分。我国大气污染随着社会经济的快速发展可能在未来较长一段时间内保持在较高水平，考虑到我国人口规模巨大，与大气污染相关的心肺疾病负担也会保持在较高水平。因此，未来深入开展对大气污染心肺健康影响的研究对预防和降低与之相关的心肺疾病负担具有重要意义。第一，国内大气污染是一种复合型污染，包含多种污染物，确定和比较大气污染混合暴露及其中不同污染物各自的健康效应是未来研究的难点之一。第二，大气颗粒物的理化性质及相关的污染来源对其健康效应有重要影响，有必要在这方面进行深入探索，明确影响颗粒物心肺健康的主要理化性质及相关的污染来源。第三，国内针对大气污染长期暴露对心肺疾病发生进展影响的研究证据极为有限，有必要开展前瞻性研究，采用更准确的暴露评价方法和更严格的疾病确证方法明确两者的关系，为人群疾病预防提供切实依据。第四，在大气污染影响心肺健康的生物学机制方面，需要注重阐明大气污染导致主要心肺疾病发生进展的关键分子机理，为针对性的人群预防及干预提供实验室基础数据。第五，从长远来看，大气污染治理是一个长期的过程，对此过程中的大气质量改善相关的心肺健康效益进行评价将为动态调整大气污染治理及公众健康促进政策提供有益参考。第六，人群对大气污染的易感性受多种因素影响，有必要进一步研究其他健康危险因素（如吸烟、过量饮酒、缺乏运动、气候变化等）与大气污染之间可能存在的交互作用，明确易感人群亚组，以便进行针对性的预防工作。第七，发达国家与我国在大气污染与心肺健康的暴露-反应关系方面存在较大差异，可能受人群遗传背景差异的影响，因此有必要开展基因-环境交互作用研究，明确影响大气污染心肺健康影响的遗传因素。第八，有必要在较大尺度上（多城市或全国范围）对大气污染相关的心肺疾病负担进行经济学评价，提供与大气污染治理成本的比对分析，为合理推进大气污染的治理和促进公众健康寻找合理的平衡点。

参 考 文 献

[1] 中华人民共和国国家统计局. 中国统计年鉴(1996-2016). http://www.stats.gov.cn/tjsj/ndsj/, 2016.

[2] Wu S, Deng F, Wei H, et al. Association of Cardiopulmonary Health Effects with Source-Appointed Ambient Fine Particulate in Beijing, China: A Combined Analysis from the Healthy Volunteer Natural Relocation (HVNR) Study. Environmental Science & Technology 2014, 48(6): 3438-3448.

[3] 中华人民共和国环境保护部. 2016 中国环境状况公报. http://www.mep.gov.cn/gkml/hbb/qt/201706/ W020170605812243090317.pdf, 2016.

[4] Brauer M, Freedman G, Frostad J, et al. Ambient air pollution exposure estimation for the global burden of disease 2013. Environmental Science & Technology, 2016, 50(1): 79-88.

[5] Chen Y, Ebenstein A, Greenstone M, et al. Evidence on the impact of sustained exposure to air pollution on life expectancy from China's Huai River policy. Proceedings of the National Academy of Sciences of the United States of America, 2013, 110(32): 12936-12941.

[6] Yang G, Wang Y, Zeng Y, et al. Rapid health transition in China, 1990-2010: Findings from the Global Burden of Disease Study 2010. Lancet, 2013, 381(9882): 1987-2015.

[7] Chen R, Kan H, Chen B, et al. Association of particulate air pollution with daily mortality: the China Air Pollution and Health Effects Study. Am J Epidemiol, 2012, 175(11): 1173-1181.

[8] Chen R, Samoli E, Wong C M, et al. Associations between short-term exposure to nitrogen dioxide and mortality in 17 Chinese cities: The China Air Pollution and Health Effects Study (CAPES). Environment International, 2012, 45: 32-38.

[9] Chen R, Huang W, Wong C M, et al. Short-term exposure to sulfur dioxide and daily mortality in 17 Chinese cities: the China air pollution and health effects study (CAPES). Environmental Research, 2012, 118: 101-106.

[10] Chen R, Yin P, Meng X, et al. Fine particulate air pollution and daily mortality. A Nationwide Analysis in 272 Chinese

Cities. Am J Respir Crit Care Med, 2017, 196(1): 73-81.

[11] Tao Y, Huang W, Huang X, et al. Estimated acute effects of ambient ozone and nitrogen dioxide on mortality in the Pearl River Delta of southern China. Environmental Health Perspectives, 2012, 120(3): 393-398.

[12] Xie W, Li G, Zhao D, et al. Relationship between fine particulate air pollution and ischaemic heart disease morbidity and mortality. Heart, 2015, 101(4): 257-263.

[13] Shang Y, Sun Z, Cao J, et al. Systematic review of Chinese studies of short-term exposure to air pollution and daily mortality. Environment International, 2013, 54: 100-111.

[14] Lai HK, Tsang H, Wong C M. Meta-analysis of adverse health effects due to air pollution in Chinese populations. BMC Public Health, 2013, 13(1): 360.

[15] Zhou M, He G, Liu Y, et al. The associations between ambient air pollution and adult respiratory mortality in 32 major Chinese cities, 2006-2010. Environmental Research, 2015, 137C: 278-286.

[16] Lu F, Xu D, Cheng Y, et al. Systematic review and meta-analysis of the adverse health effects of ambient PM2.5 and PM10 pollution in the Chinese population. Environmental Research, 2015, 136: 196-204.

[17] Lin H, Liu T, Xiao J, et al. Mortality burden of ambient fine particulate air pollution in six Chinese cities: Results from the Pearl River Delta study. Environment International, 2016, 96: 91-97.

[18] Yin P, He G, Fan M, et al. Particulate air pollution and mortality in 38 of China's largest cities: time series analysis. BMJ, 2017, 356: j667.

[19] Peng RD, Chang H H, Bell M L, et al. Coarse particulate matter air pollution and hospital admissions for cardiovascular and respiratory diseases among Medicare patients. JAMA, 2008, 299(18): 2172-2179.

[20] Dai L, Zanobetti A, Koutrakis P, et al. Associations of fine particulate matter species with mortality in the United States: a multicity time-series analysis. Environmental Health Perspectives, 2014, 122(8): 837-842.

[21] Pope C A, 3rd, Dockery D W. Health effects of fine particulate air pollution: lines that connect. Journal of the Air & Waste Management Association, 2006, 56(6): 709-742

[22] Meng X, Ma Y, Chen R, et al. Size-fractionated particle number concentrations and daily mortality in a Chinese city. Environmental Health Perspectives, 2013, 121(10): 1174-1178.

[23] Lin H, Tao J, Du Y, et al. Particle size and chemical constituents of ambient particulate pollution associated with cardiovascular mortality in Guangzhou, China. Environmental Pollution, 2016, 208(Pt B): 758-766.

[24] Cao J, Xu H, Xu Q, et al. Fine particulate matter constituents and cardiopulmonary mortality in a heavily polluted Chinese city. Environmental Health Perspectives, 2012, 120(3): 373-378.

[25] Guo Y, Tong S, Zhang Y, et al. The relationship between particulate air pollution and emergency hospital visits for hypertension in Beijing, China. The Science of the Total Environment, 2010, 408(20): 4446-4450.

[26] Guo Y, Tong S, Li S, et al. Gaseous air pollution and emergency hospital visits for hypertension in Beijing, China: a time-stratified case-crossover study. Environ Health, 2010, 9: 57.

[27] Hua J, Yin Y, Peng L, et al. Acute effects of black carbon and PM2.5 on children asthma admissions: A time-series study in a Chinese city. The Science of the Total Environment, 2014, 481: 433-438.

[28] Zhao A, Chen R, Kuang X, et al. Ambient air pollution and daily outpatient visits for cardiac arrhythmia in Shanghai, China. J Epidemiol, 2014, 24(4): 321-326.

[29] Wang X, Chen R, Meng X, et al. Associations between fine particle, coarse particle, black carbon and hospital visits in a Chinese city. The Science of the Total Environment, 2013, 458-460: 1-6.

[30] Liu H, Tian Y, Xu Y, et al. Association between ambient air pollution and hospitalization for ischemic and hemorrhagic stroke in China: A multicity case-crossover study. Environmental Pollution, 2017, 230: 234-241.

[31] Tao Y, Mi S, Zhou S, et al. Air pollution and hospital admissions for respiratory diseases in Lanzhou, China. Environmental Pollution, 2014, 185: 196-201.

[32] Yang C, Chen A, Chen R, et al. Acute effect of ambient air pollution on heart failure in Guangzhou, China. Int J Cardiol, 2014, 177(2): 436-441.

[33] Peng CQ, Cai JF, Yu SY, et al. [Impact of PM2.5 on daily outpatient numbers for respiratory diseases in Shenzhen, China]. Zhonghua Yu Fang Yi Xue Za Zhi, 2016, 50(10): 874-879.

[34] Liu R, Zeng J, Jiang X, et al. The relationship between airborne fine particle matter and emergency ambulance dispatches in a southwestern city in Chengdu, China. Environmental Pollution, 2017, 229: 661-667.

[35] Zhang S, Li G, Tian L, et al. Short-term exposure to air pollution and morbidity of COPD and asthma in East Asian area: A systematic review and meta-analysis. Environmental Research, 2016, 148: 15-23.

[36] Qiao L, Cai J, Wang H, et al. PM2.5 constituents and hospital emergency-room visits in Shanghai, China. Environmental Science & Technology, 2014, 48(17): 10406-10414.

[37] Liu L, Breitner S, Schneider A, et al. Size-fractioned particulate air pollution and cardiovascular emergency room visits in Beijing, China. Environmental Research, 2013, 121: 52-63.

[38] Leitte A M, Schlink U, Herbarth O, et al. Size-segregated particle number concentrations and respiratory emergency room visits in Beijing, China. Environmental Health Perspectives, 2011, 119(4): 508-513.

[39] Chen R, Zhao Z, Kan H. Heavy smog and hospital visits in Beijing, China. Am J Respir Crit Care Med, 2013, 188(9): 1170-1171.

[40] Zhou M, He G, Fan M, et al. Smog episodes, fine particulate pollution and mortality in China. Environmental Research, 2015, 136: 396-404.

[41] Huang F, Chen R, Shen Y, et al. The Impact of the 2013 Eastern China Smog on Outpatient Visits for Coronary Heart Disease in Shanghai, China. International Journal of Environmental Research and Public Health, 2016, 13(7) doi: 10.3390/ijerph13070627.

[42] Cui L L, Zhang J, Zhang J, et al. Acute respiratory and cardiovascular health effects of an air pollution event, January 2013, Jinan, China. Public Health, 2016, 131: 99-102.

[43] Beelen R, Raaschou-Nielsen O, Stafoggia M, et al. Effects of long-term exposure to air pollution on natural-cause mortality: an analysis of 22 European cohorts within the multicentre ESCAPE project. Lancet, 2014, 383(9919): 785-795.

[44] Di Q, Wang Y, Zanobetti A, et al. Air Pollution and Mortality in the Medicare Population. N Engl J Med, 2017, 376(26): 2513-2522.

[45] Zhang J, Song H, Tong S, et al. Ambient sulfate concentration and chronic disease mortality in Beijing. The Science of the Total Environment, 2000, 262(1-2): 63-71.

[46] Tie X, Wu D, Brasseur G. Lung cancer mortality and exposure to atmospheric aerosol particles in Guangzhou, China. Atmos Environ, 2009, 43(14): 2375-2377.

[47] Huang Y B, Song F J, Liu Q, et al. A bird's eye view of the air pollution-cancer link in China. Chinese Journal of Cancer, 2014, 33(4): 176-188.

[48] Zhu Y D, Wei J R, Huang L, et al. Comparison of respiratory diseases and symptoms among school-age children in areas with different levels of air pollution. Beijing Da Xue Xue Bao, 2015, 47(3): 395-399.

[49] Wang X, Deng F R, Lv H B, et al. Long-term effects of air pollution on the occurrence of respiratory symptoms in adults of Beijing. Beijing Da Xue Xue Bao, 2011, 43(3): 356-359.

[50] Liu M M, Wang D, Zhao Y, et al. Effects of outdoor and indoor air pollution on respiratory health of Chinese children from 50 kindergartens. J Epidemiol, 2013, 23(4): 280-287.

[51] Dong G H, Qian Z M, Xaverius P K, et al. Association between long-term air pollution and increased blood pressure and hypertension in China. Hypertension, 2013, 61(3): 578-84 doi: 10.1161/hypertensionaha.111.00003.

[52] Yang B Y, Qian Z M, Vaughn M G, et al. Is prehypertension more strongly associated with long-term ambient air pollution exposure than hypertension? Findings from the 33 Communities Chinese Health Study. Environmental Pollution, 2017, 229: 696-704.

[53] Lin H, Guo Y, Zheng Y, et al. Long-Term Effects of Ambient PM2.5 on Hypertension and Blood Pressure and Attributable Risk Among Older Chinese Adults. Hypertension, 2017, 69(5): 806-812.

[54] Cao J, Yang C, Li J, et al. Association between long-term exposure to outdoor air pollution and mortality in China: a cohort study. Journal of Hazardous Materials, 2011, 186(2-3): 1594-600.

[55] Zhang P, Dong G, Sun B, et al. Long-term exposure to ambient air pollution and mortality due to cardiovascular disease and cerebrovascular disease in Shenyang, China. PloS one, 2011, 6(6): e20827.

[56] Dong G H, Zhang P, Sun B, et al. Long-term exposure to ambient air pollution and respiratory disease mortality in Shenyang, China: A 12-year population-based retrospective cohort study. Respiration, 2012, 84(5): 360-368.

[57] Zhang L W, Chen X, Xue X D, et al. Long-term exposure to high particulate matter pollution and cardiovascular mortality: A 12-year cohort study in four cities in northern China. Environment International, 2014, 62: 41-47.

[58] Zhou M, Liu Y, Wang L, et al. Particulate air pollution and mortality in a cohort of Chinese men. Environmental Pollution, 2014, 186: 1-6.

[59] Peng Z, Liu C, Xu B, et al. Long-term exposure to ambient air pollution and mortality in a Chinese tuberculosis cohort. The Science of the Total Environment, 2017, 580: 1483-1488.

[60] Song Y, Hou J, Huang X, et al. The Wuhan-Zhuhai (WHZH) cohort study of environmental air particulate matter and the pathogenesis of cardiopulmonary diseases: study design, methods and baseline characteristics of the cohort. BMC Public Health, 2014, 14: 994

[61] Janes H, Sheppard L, Shepherd K. Statistical analysis of air pollution panel studies: an illustration. Ann Epidemiol, 2008, 18(10): 792-802.

[62] Liu L, Zhang J. Ambient air pollution and children's lung function in China. Environment International, 2009, 35(1): 178-186.

[63] He Q Q, Wong TW, Du L, et al. Effects of ambient air pollution on lung function growth in Chinese schoolchildren. Respir Med, 2010, 104(10): 1512-1520.

[64] Ni Y, Wu S, Ji W, et al. The exposure metric choices have significant impact on the association between short-term exposure to outdoor particulate matter and changes in lung function: Findings from a panel study in chronic obstructive pulmonary disease patients. The Science of the Total Environment, 2016, 542(Pt A): 264-270.

[65] Wu S, Deng F, Hao Y, et al. Chemical constituents of fine particulate air pollution and pulmonary function in healthy adults: the Healthy Volunteer Natural Relocation study. Journal of Hazardous Materials, 2013, 260: 183-191.

[66] Wu S, Deng F, Hao Y, et al. Fine particulate matter, temperature, and lung function in healthy adults: Findings from the HVNR study. Chemosphere, 2014, doi: 10.1016/j.chemosphere.2014.01.032.

[67] Wu S, Deng F, Wang X, et al. Association of lung function in a panel of young healthy adults with various chemical components of ambient fine particulate air pollution in Beijing, China. Atmos Environ, 2013, 77: 873-884

[68] Baccarelli A A, Zheng Y, Zhang X, et al. Air pollution exposure and lung function in highly exposed subjects in Beijing, China: A repeated-measure study. Particle and Fibre Toxicology, 2014, 11(1): 51|.

[69] Madaniyazi L, Guo Y, Ye X, et al. Effects of airborne metals on lung function in inner Mongolian schoolchildren. J Occup Environ Med, 2013, 55(1): 80-86.

[70] Zhou Y, Liu Y, Song Y, et al. Short-term effects of outdoor air pollution on lung function among female non-smokers in China. Scientific Reports, 2016, 6: 34947.

[71] Sato R, Gui P, Ito K, et al. Effect of short-term exposure to high particulate levels on cough reflex sensitivity in healthy tourists: a pilot study. The Open Respiratory Medicine Journal, 2016, 10: 96-104.

[72] Wu S, Ni Y, Li H, et al. Short-term exposure to high ambient air pollution increases airway inflammation and respiratory symptoms in chronic obstructive pulmonary disease patients in Beijing, China. Environment International, 2016, 94: 76-82.

[73] Lin W, Huang W, Zhu T, et al. Acute respiratory inflammation in children and black carbon in ambient air before and during the 2008 Beijing Olympics. Environmental Health Perspectives, 2011, 119(10): 1507-1512.

[74] Huang W, Wang G, Lu SE, et al. Inflammatory and oxidative stress responses of healthy young adults to changes in air quality during the Beijing Olympics. Am J Respir Crit Care Med, 2012, 186(11): 1150-1159.

[75] Gong J, Zhu T, Kipen H, et al. Comparisons of Ultrafine and Fine Particles in Their Associations with Biomarkers Reflecting Physiological Pathways. Environmental Science & Technology, 2014, doi: 10.1021/es5006016.

[76] Altemose B, Robson M G, Kipen H M, et al. Association of air pollution sources and aldehydes with biomarkers of blood coagulation, pulmonary inflammation, and systemic oxidative stress. J Expo Sci Environ Epidemiol, 2017, 27(3): 244-250.

[77] Shi J, Chen R, Yang C, et al. Association between fine particulate matter chemical constituents and airway inflammation: A panel study among healthy adults in China. Environmental Research, 2016, 150: 264-268.

[78] Zhao X, Sun Z, Ruan Y, et al. Personal Black Carbon Exposure Influences Ambulatory Blood Pressure: Air Pollution and Cardiometabolic Disease (AIRCMD-China) Study. Hypertension, 2014, 63(4): 871-877.

[79] Huang W, Zhu T, Pan X, et al. Air pollution and autonomic and vascular dysfunction in patients with cardiovascular disease: interactions of systemic inflammation, overweight, and gender. Am J Epidemiol, 2012, 176(2): 117-126.

[80] Brook R D, Sun Z, Brook J R, et al. Extreme Air Pollution Conditions Adversely Affect Blood Pressure and Insulin Resistance: The Air Pollution and Cardiometabolic Disease Study. Hypertension, 2016, 67(1): 77-85.

[81] Langrish J P, Li X, Wang S, et al. Reducing personal exposure to particulate air pollution improves cardiovascular health in patients with coronary heart disease. Environmental Health Perspectives, 2012, 120(3): 367-372.

[82] Shi J, Lin Z, Chen R, et al. Cardiovascular benefits of wearing particulate-filtering respirators: a randomized crossover trial. Environmental Health Perspectives, 2017, 125(2): 175-180.

[83] Rich D Q, Kipen H M, Huang W, et al. Association between changes in air pollution levels during the Beijing Olympics and biomarkers of inflammation and thrombosis in healthy young adults. JAMA, 2012, 307(19): 2068-2078.

[84] Chen R, Zhao A, Chen H, et al. Cardiopulmonary benefits of reducing indoor particles of outdoor origin: a randomized, double-blind crossover trial of air purifiers. J Am Coll Cardiol, 2015, 65(21): 2279-2287.

[85] Wu S, Deng F, Huang J, et al. Blood pressure changes and chemical constituents of particulate air pollution: results from the healthy volunteer natural relocation (HVNR) study. Environmental Health Perspectives, 2013, 121(1): 66-72.

[86] Wu S, Deng F, Niu J, et al. Association of heart rate variability in taxi drivers with marked changes in particulate air pollution in Beijing in 2008. Environmental Health Perspectives, 2010, 118(1): 87-91.

[87] Wu S, Deng F, Niu J, et al. Exposures to PM2.5 components and heart rate variability in taxi drivers around the Beijing 2008 Olympic Games. The Science of the Total Environment, 2011, 409(13): 2478-2485.

[88] Xu M M, Jia Y P, Li G X, et al. Relationship between ambient fine particles and ventricular repolarization changes and heart rate variability of elderly people with heart disease in Beijing, China. Biomed Environ Sci, 2013, 26(8): 629-637.

[89] Jia X, Song X, Shima M, et al. Effects of fine particulate on heart rate variability in Beijing: a panel study of healthy elderly subjects. Int Arch Occup Environ Health, 2012, 85(1): 97-107.

[90] Huang J, Deng F, Wu S, et al. The impacts of short-term exposure to noise and traffic-related air pollution on heart rate

variability in young healthy adults. J Expo Sci Environ Epidemiol, 2013, 23(5): 559-564.

[91] Sun Y, Song X, Han Y, et al. Size-fractioned ultrafine particles and black carbon associated with autonomic dysfunction in subjects with diabetes or impaired glucose tolerance in Shanghai, China. Particle and Fibre Toxicology, 2015, 12: 8.

[92] Wu S, Deng F, Niu J, et al. The relationship between traffic-related air pollutants and cardiac autonomic function in a panel of healthy adults: A further analysis with existing data. Inhalation Toxicology, 2011, 23(5): 289-303.

[93] Jia X, Song X, Shima M, et al. Acute effect of ambient ozone on heart rate variability in healthy elderly subjects. J Expo Sci Environ Epidemiol, 2011, 21(5): 541-547.

[94] Zhao J, Gao Z, Tian Z, et al. The biological effects of individual-level PM2.5 exposure on systemic immunity and inflammatory response in traffic policemen. Occup Environ Med, 2013, 70(6): 426-431.

[95] Wu S, Deng F, Wei H, et al. Chemical constituents of ambient particulate air pollution and biomarkers of inflammation, coagulation and homocysteine in healthy adults: a prospective panel study. Particle and Fibre Toxicology, 2012, 9: 49.

[96] Wu S, Yang D, Wei H, et al. Association of chemical constituents and pollution sources of ambient fine particulate air pollution and biomarkers of oxidative stress associated with atherosclerosis: A panel study among young adults in Beijing, China. Chemosphere, 2015, 135: 347-353.

[97] Wu S, Wang B, Yang D, et al. Ambient particulate air pollution and circulating antioxidant enzymes: A repeated-measure study in healthy adults in Beijing, China. Environmental Pollution, 2016, 208(Pt A): 16-24.

[98] Wu S, Yang D, Pan L, et al. Chemical constituents and sources of ambient particulate air pollution and biomarkers of endothelial function in a panel of healthy adults in Beijing, China. The Science of the Total Environment, 2016, 560-561: 141-149.

[99] Chen R, Zhao Z, Sun Q, et al. Size-fractionated particulate air pollution and circulating biomarkers of inflammation, coagulation, and vasoconstriction in a panel of young adults. Epidemiology, 2015, 26(3): 328-336.

[100] Chen R, Meng X, Zhao A, et al. DNA hypomethylation and its mediation in the effects of fine particulate air pollution on cardiovascular biomarkers: A randomized crossover trial. Environment International, 2016, 94: 614-619.

[101] Xu X, Deng F, Guo X, et al. Association of systemic inflammation with marked changes in particulate air pollution in Beijing in 2008. Toxicology Letters, 2012, 212(2): 147-156.

[102] Chen T, Jia G, Wei Y, et al. Beijing ambient particle exposure accelerates atherosclerosis in ApoE knockout mice. Toxicology Letters, 2013, 223(2): 146-153.

[103] Zhao J, Liu C, Bai Y, et al. IKK inhibition prevents PM2.5-exacerbated cardiac injury in mice with type 2 diabetes. Journal of Environmental Sciences (China), 2015, 31: 98-103.

[104] He F, Liao B, Pu J, et al. Exposure to ambient particulate matter induced COPD in a rat model and a description of the underlying mechanism. Scientific Reports, 2017, 7: 45666.

[105] Bao A, Liang L, Li F, et al. Effects of acute ozone exposure on lung peak allergic inflammation of mice. Frontiers in Bioscience (Landmark edition), 2013, 18: 838-851

[106] Wang G, Jiang R, Zhao Z, et al. Effects of ozone and fine particulate matter (PM(2.5)) on rat system inflammation and cardiac function. Toxicology Letters, 2013, 217(1): 23-33.

[107] Zhang Y, Ji X, Ku T, et al. Inflammatory response and endothelial dysfunction in the hearts of mice co-exposed to SO_2 , NO_2 , and $PM_{2.5}$. Environ Toxicol, 2016, 31(12): 1996-2005.

[108] Lu X L, Zhang X R, Deng F R, et al. Systemic oxidative stress induced by intratracheal instilling with PM_{10} in rats. Beijing Da Xue Xue Bao, 2011, 43(3): 352-355

[109] Xu Z, Xu X, Zhong M, et al. Ambient particulate air pollution induces oxidative stress and alterations of mitochondria and gene expression in brown and white adipose tissues. Particle and Fibre Toxicology, 2011, 8: 20.

[110] Pei Y, Jiang R, Zou Y, et al. Effects of fine particulate Matter (PM2.5) on systemic oxidative stress and cardiac function in ApoE(-/-) Mice. International Journal of Environmental Research and Public Health, 2016, 13(5) doi: 10.3390/ijerph13050484.

[111] Deng X, Zhang F, Rui W, et al. PM2.5-induced oxidative stress triggers autophagy in human lung epithelial A549 cells. Toxicol In Vitro, 2013, 27(6): 1762-1770.

[112] Wei Y, Zhang J J, Li Z, et al. Chronic exposure to air pollution particles increases the risk of obesity and metabolic syndrome: findings from a natural experiment in Beijing. FASEB journal: official publication of the Federation of American Societies for Experimental Biology, 2016, 30(6): 2115-2122.

[113] Liu Q, Baumgartner J, Zhang Y, et al. Oxidative potential and inflammatory impacts of source apportioned ambient air pollution in Beijing. Environmental Science & Technology, 2014, 48(21): 12920-12929.

[114] Ying Z, Xie X, Bai Y, et al. Exposure to concentrated ambient particulate matter induces reversible increase of heart weight in spontaneously hypertensive rats. Particle and Fibre Toxicology, 2015, 12: 15.

[115] Zhang X, Chen Y, Wei H, et al. Ultrafine carbon black attenuates the antihypertensive effect of captopril in spontaneously hypertensive rats. Inhalation Toxicology, 2014, 26(14): 853-860.

[116] Deng F R, Guo X B, Hu J, et al. Acute heart toxicity in rats induced by PM2.5 intratracheal instillation and its mechanisms. Asian Journal of Ecotoxicology, 2009, 4(1): 57-62.

[117] Jia X, Hao Y, Guo X. Ultrafine carbon black disturbs heart rate variability in mice. Toxicology Letters, 2012, 211(3): 274-280.

[118] Deng F R, Guo X B, Xia P P, et al. Comparative study on parameters of blood coagulation affected by PM2.5 and nickel sulfate intratracheal instillation in rats. Journal of Environmental & Occupational Medicine, 2010, 1: 009.

[119] Liu Y, Wang L, Wang F, et al. Effect of fine particulate matter (PM2.5) on rat placenta pathology and perinatal outcomes. Medical Science Monitor : International Medical journal of Experimental and Clinical Research, 2016, 22: 3274-3280

[120] Liu C, Xu X, Bai Y, et al. Air pollution-mediated susceptibility to inflammation and insulin resistance: influence of CCR2 pathways in mice. Environmental Health Perspectives, 2014, 122(1): 17-26.

[121] Liu C, Xu X, Bai Y, et al. Particulate Air pollution mediated effects on insulin resistance in mice are independent of CCR2. Particle and Fibre Toxicology, 2017, 14(1): 6.

[122] Ding R, Jin Y, Liu X, et al. Characteristics of DNA methylation changes induced by traffic-related air pollution. Mutation Research Genetic Toxicology and Environmental Mutagenesis, 2016, 796: 46-53.

[123] Ding R, Jin Y, Liu X, et al. Dose- and time- effect responses of DNA methylation and histone H3K9 acetylation changes induced by traffic-related air pollution. Scientific Reports, 2017, 7: 43737.

[124] Ding X, Wang M, Chu H, et al. Global gene expression profiling of human bronchial epithelial cells exposed to airborne fine particulate matter collected from Wuhan, China. Toxicology Letters, 2014, 228(1): 25-33.

[125] Li Y, Duan J, Yang M, et al. Transcriptomic analyses of human bronchial epithelial cells BEAS-2B exposed to atmospheric fine particulate matter PM2.5. Toxicol In Vitro, 2017, 42: 171-181.

[126] Feng L, Yang X, Asweto CO, et al. Genome-wide transcriptional analysis of cardiovascular-related genes and pathways induced by PM2.5 in human myocardial cells. Environmental Science and Pollution Research International, 2017, 24(12): 11683-11693.

[127] Duan J, Yu Y, Li Y, et al. Comprehensive understanding of PM2.5 on gene and microRNA expression patterns in zebrafish (*Danio rerio*) model. The Science of the Total Environment, 2017, 586: 666-674.

[128] Qin X D, Qian Z, Vaughn M G, et al. Gender-specific differences of interaction between obesity and air pollution on stroke and cardiovascular diseases in Chinese adults from a high pollution range area: A large population based cross sectional

study. The Science of the Total Environment, 2015, 529: 24324-8.

[129] Dong G H, Wang J, Zeng X W, et al. Interactions between air pollution and obesity on blood pressure and hypertension in Chinese Children. Epidemiology, 2015, 26(5): 740-747.

[130] Dong G H, Qian Z M, Liu M M, et al. Breastfeeding as a modifier of the respiratory effects of air pollution in children. Epidemiology, 2013, 24(3): 387-394.

[131] Dong G H, Qian Z M, Trevathan E, et al. Air pollution associated hypertension and increased blood pressure may be reduced by breastfeeding in Chinese children: the seven northeastern cities Chinese children's study. Int J Cardiol, 2014, 176(3): 956-961.

[132] Li P, Xin J, Wang Y, et al. The acute effects of fine particles on respiratory mortality and morbidity in Beijing, 2004-2009. Environmental Science and Pollution Research International, 2013, 20(9): 6433-6444.

[133] Li Y, Ma Z, Zheng C, et al. Ambient temperature enhanced acute cardiovascular-respiratory mortality effects of PM2.5 in Beijing, China. International Journal of Biometeorology, 2015, 59(12): 1761-1770.

[134] Cheng Y, Kan H. Effect of the interaction between outdoor air pollution and extreme temperature on daily mortality in Shanghai, China. J Epidemiol, 2012, 22(1): 28-36.

[135] Qin R X, Xiao C, Zhu Y, et al. The interactive effects between high temperature and air pollution on mortality: A time-series analysis in Hefei, China. The Science of the Total Environment,017, 575: 1530-1537.

[136] Chen F, Qiao Z, Fan Z, et al. The effects of Sulphur dioxide on acute mortality and years of life lost are modified by temperature in Chengdu, China. The Science of the Total Environment, 2017, 576: 775-784.

[137] Li L, Yang J, Guo C, et al. Particulate matter modifies the magnitude and time course of the non-linear temperature-mortality association. Environmental Pollution, 2015, 196: 423-430.

[138] Wu S, Deng F, Huang J, et al. Does ambient temperature interact with air pollution to alter blood pressure? A repeated-measure study in healthy adults. J Hypertens, 2015, 33(12): 2414-2421.

[139] Wu S, Deng F, Liu Y, et al. Temperature, traffic-related air pollution, and heart rate variability in a panel of healthy adults. Environmental Research, 2013, 120: 82-89.

作者：吴少伟

北京大学公共卫生学院

第27章　POPs毒性效应的代谢组学研究进展

本章导读

伴随着全球经济的高速增长，人类与自然的矛盾更加激化。生态破坏和环境污染已经成为严重的区域性和全球性环境问题。在众多环境问题中，持久性有机污染物（persistent organic pollutions，POPs）由于对健康效应的影响较大而备受关注。POPs 在环境中普遍存在，滞留时间长，并可发生长距离的迁移，由于亲脂性容易在生物体内累积，并随食物链迁移和放大，最终对处于食物链最高端的人类造成潜在的危害。目前，有关 POPs 毒性效应的研究主要集中在高剂量暴露的毒性评价上，对低剂量暴露的研究相对较少，但人群的实际暴露具有低剂量特征，因此研究 POPs 低剂量暴露的毒性效应并探讨其潜在的健康风险具有重要的意义。由于机体代谢物的变化可灵敏地指示和确证外来干扰在组织和器官水平的毒性效应以及毒性作用靶位点，因此借助代谢组学技术评价环境污染物暴露带来的毒性效应，进而推断其毒性作用的分子机制具有快速、灵敏度高、选择性强等特点，特别是在污染物低剂量暴露或联合暴露的毒性效应评估方面具有很大的优势。本章将对近年来典型 POPs 低剂量暴露毒性效应的代谢组学研究进行系统综述。

关键词

代谢组学，持久性有机污染物，低剂量暴露，毒性效应

1 引　言

当前，我国正处于环境污染事件频繁发生，环境状况总体恶化的趋势。老的环境污染问题，如重金属污染、二氧化硫污染和水体有机物总量过高等尚未解决，新型污染物的污染问题已开始显现。这些环境污染物往往以痕量或超痕量水平存在于环境介质中，其引起的生态风险和人体健康风险具有隐蔽性。因此对这些典型污染物开展毒性识别和风险评估，是加强其环境管理和开展污染防治的前提条件。已有研究表明和二噁英（PCDD/PCDFs）、多氯联苯（PCBs）、多环芳烃（PAHs）、多溴联苯醚（PBDEs）、短链氯化石蜡（SCCPs）和六溴环十二烷（HBCDs）等 POPs 类物质的高剂量暴露对肝脏、肾脏、甲状腺等器官具有潜在的致癌性。但这类物质的人群实际暴露具有低剂量长周期特征，用常规的毒理学方法对其进行健康风险评价往往得不到有意义的结果，系统探究其毒性作用尚需依赖新的技术手段和方法，因此急需发展一种高效、快速且信息量丰富的评价方法来评估污染物低剂量暴露的毒性效应。

代谢组学技术能够检测到成百上千种指示不同代谢路径的代谢物，通过有效地分析多条代谢通路，帮助定位靶组织及判定毒副作用程度，寻找相应的生物标志物，在环境剂量污染物的毒性效应评估方面具有很大的优势。在化学品毒理、动物病理及疾病诊断等研究领域已有广泛应用。代谢组学之所以能在毒理学研究中发挥巨大作用，是基于反映毒性作用的信息能够在生物体的代谢过程中全面地表现出来。当毒物与细胞或者组织相互作用时，会引起生物体内关键代谢过程中内源性物质比例、浓度的变化并通过体液组成的变化反映出来，这些代谢层面的生物信息能够很好地表征环境因素的毒性通路和毒性作用机制。代谢组学技术为环境毒理学研究提供了新的思路和平台，其全景式高通量筛选特征必将为生物化学测量提供综合视角，在 21 世纪毒理学研究中发挥重要作用。

2　环境毒理学中的代谢组学方法

2.1　代谢组学及其分析方法

代谢组学是研究生物体内源性代谢物的整体及其随内因和外因变化的科学，是系统生物学的重要组成部分，代谢组学与基因组学和蛋白组学相比主要具有如下优点：①代谢组学反映生物体在各个因素综合作用下的末端效应，是这些效应的综合体现，具有很强的综合信息优势；②其代谢物种类远小于所对应的基因和蛋白质数目，研究相对简单；③基因和蛋白质表达的微小变化会在代谢物水平上得到放大，因此更容易检测；④很多内源性小分子化合物的生化代谢途径已较清楚；⑤许多代谢物已作为疾病的特异性标志物用于临床诊断。因此，近年来代谢组学技术在毒性评价方面发挥了越来越重要的作用。

目前，代谢组学的分析方法主要分为核磁共振技术与质谱技术。两种技术各有优势，核磁共振技术对样品不具有破坏性；高通量、快速检测分析；对样品前处理的要求较低，甚至可直接进样；每个样本的分析成本较低；能较全面地对样本进行分析；可对完整的组织、器官以及个体样本进行分析析，但其灵敏度较质谱低，难以准确地进行定量分析，故而限制其在代谢组学中的应用。目前，核磁共振只能对样品中含量较高的代谢物进行定性分析。近年来，核磁共振作为代谢组学研究中的重要分析平台已被广泛地应用于临床医学诊断中[1]。

由于生物体系的复杂性，代谢产物数目多、差异大、浓度分布范围广、组成复杂，代谢组学也是一门技术驱动的科学。现代质谱技术具有高选择和高灵敏性、普适性和分析速度快等特点，可同时检测鉴定多种代谢物，提供丰富的数据信息。因此，质谱技术成为代谢组学最有效的研究手段之一。基于质谱技术的代谢组学方法，如超气相色谱-质谱（GC-MS）、液相色谱-质谱（LC-MS）、毛细管电泳-质谱（CE-MS）、多维色谱-质谱联用、质谱-核磁共振联用和超高分辨质谱等技术，近年来在灵敏度、分辨率、动态范围和高通量等方面均有显著的提高。能够实现数千种化合物（包括糖类、有机酸、氨基酸、脂肪酸、环境外源污染物和植物的次级代谢产物）的高通量检测，这是其他任何一项技术所无法比拟的[2-4]。

近年来，研究者在基于质谱技术的代谢组学新方法的创新和完善方面取得了一定成果，并将其应用于更广泛的研究领域。超高效液相色谱与质谱联用扩展了代谢物的覆盖率，成为当今代谢组学的主要方法之一；纳升级[5,6]和芯片纳流液相色谱-质谱系统也成为高效快速、高灵敏度的分离分析复杂代谢组样品的新技术[7]。2013 年，*Science* 报道的高密度芯片-质谱接口技术，如质谱微阵列芯片，可解决单细胞代谢组学质谱检测中高通量制备单细胞样本的瓶颈问题[8]。复杂体系中代谢产物的结构鉴定是富有挑战性的问题，具有高分辨率的质谱系统如四极杆串联飞行时间质谱仪[9]、傅里叶变换离子回旋共振质谱仪和 Orbitrap 等在此方面有着突出的优势。精确质量数的质谱和串联质谱检测与数据库检索、标准品验证及其他实验结果相结合，提供了更高水平的代谢物结构识别和确认能力，对生物标记物的确定非常重要。

许国旺研究组针对代谢组的特点，采用“分而治之”的策略，建立完善了一系列基于色-质联用技术的代谢组学分析平台，包括非靶向的代谢组、脂类的组学分析平台以及针对痕量代谢物的靶向分析平台等，并在提高代谢组学分析覆盖的化合物种类，实现高通量、高灵敏度代谢组分析以及发展复杂代谢组数据解析方法，解决代谢物定量及未知代谢物结构定性等关键问题上开展了相关研究。通过发展新型同位素标记衍生化液-质联用方法，实现了 24 种类固醇激素的快速高灵敏检测[10]；发展了多维色-质联用技术，提高了色谱的分离能力[11-13]。结合代谢组学非靶向分析及靶向分析的特点，该研究组提出了一种“拟靶向”的分析策略，以多反应监测方式实现一次进样定量超过 500 种代谢物，显著提高了代谢组分析的数据质量[14]。再帕尔课题组[15-18]以快速高分辨液-质联用技术为主要分析手段，通过构建高效、整合式的高通量分析技术平台，开展代谢组学分析方法及小分子生物标志物的研究。

蔡宗苇研究组通过开发和建立基于质谱技术代谢组学方法，开展了与环境毒理和重大疾病相关的代谢组学研究[19-30]。采用超高效液相色谱-四极杆串联飞行时间质谱分析了 2, 3, 7, 8-二噁英在芳香烃受体敏感型的 C57BL/6J 小鼠与不敏感型的 DBA/2J 小鼠中的不同毒性反应[25,27]。用高分辨液-质联用法研究了体内代谢中马兜铃酸的毒性和致毒机制[23,29,30]。结合动物模型与生物标记物的发现，研究了灌胃葡萄糖的健康大鼠与糖尿病大鼠的代谢表型的相似度[24]。该研究组还对流感病毒感染的细胞模型[22]、阿尔茨海默病转基因小鼠[21]进行了代谢组学分析。此外，该研究组与德国赫姆霍兹中心合作，建立了超高分辨的傅里叶变换离子回旋共振质谱代谢组学分析方法，高准确度的质谱数据有利于快速鉴定代谢物作为潜在的生物标记物。

唐惠儒研究组[31-33]建立了液-质与核磁共振联用代谢组学研究技术平台并开展了多方面的研究。刘虎威研究组首次将基于停流技术的反相色谱和正相色谱的二维分析系统与质谱联用应用于代谢物分析[31]，开展了生物样本的脂质轮廓图谱分析等脂质组学研究[32,34-36]。陈焕文研究组[37,38]致力于质谱仪器和方法的开发研究，将质谱技术应用于生物样品的代谢指纹图谱癌症诊断中等研究工作。 林金明研究组[39]开发了微流控芯片与质谱联用技术，对细胞分泌物进行检测分析以期得到细胞生命活动中扮演重要角色的物质的结构和含量信息。

2.2 代谢组学在环境毒理学研究中的优势

21 世纪毒性测试重点强化由整体动物试验转向以人源细胞为主的新策略，并体现“3R”原则，即替代（replacement）、减少（reduction）和优化（refinement）。寻找可靠、有效、低消耗的动物替代新方法是毒理学领域亟待解决的重要问题。由于代谢物的变化是机体对环境因素影响的最终应答，某种特定代谢物的蓄积可能标志着某通路的缺陷或某信号响应的激活，而代谢物的动态变化可以作为毒性损伤的标志物。因此借助代谢组学技术来评价环境污染物暴露带来的毒性效应，并进而推断其毒性作用的分子机制具有快速、灵敏度高、选择性强等特点，特别是在污染物低剂量暴露的毒性效应评估方面具有很大的优势[40]。

Robertson 等[41]基于核磁共振技术研究了两种肝毒性化合物四氯化碳、α-萘异硫氰酸和两种肾毒性化合物 2-溴乙胺、对氨基苯酚对 Wistar 大鼠尿液代谢物组成的影响，与经典的毒理学研究方法如临床化学和显微镜观察相比，代谢组学技术可以快速灵敏地对毒性试剂进行筛查，即使在小剂量暴露水平下，肝毒性和肾毒性试剂对大鼠尿液代谢物也有明显的影响。以肝脏胆管毒物 α-萘基异硫氰酸酯为例，代谢组学研究显示：4 只低剂量[10 mg /kg 体重（bw）]处理的动物在 24 h 内都表现出明显的代谢紊乱，组织病理学只有两只动物（4 d 后）表现出变化，临床化学的方法（血液中胆红素含量）没有任何变化。中国科学院水生生物研究所研究人员用代谢组学方法评价微囊藻毒素对肝脏的毒性作用[42]，发现低剂量给药时大鼠肝脏在组织学上没有明显变化，但从代谢层面上已经能引起肝代谢的异常：其酪氨酸合成与分解代谢明显受到抑制，三条胆碱相关代谢途径被截断，谷胱甘肽被消耗并且扰乱了核苷酸的合成。这些研究表明利用代谢组技术可以快速进行化学物质的毒性效应评估。

英国帝国理工学院的研究人员[43]在对硫代乙酰胺肾毒性的研究中发现：代谢组学的方法能够明确地将肾乳头毒性和近端肾小管毒性区分。用尿液代谢组学方法能够明显地将肝实质毒性和胆管毒性明显区分。Bundy 等[44]对三种爱胜蚓属 *Eisenia fetida*、*Eisenia andrei* 和 *Eiseniaveneta* 进行了代谢表型的研究，发现三种蚯蚓的组织提取物的代谢表型极为保守，没有表现出明显的变化；但三种不同爱胜蚓属体腔液代谢谱有明显的不同。此研究结果表明，在形态学上很难分辨蚯蚓属在相同的生态环境中可以用代谢组学的方法明确区分。这些研究均表明代谢组学技术用于污染物的毒性筛查和评估有较高的灵敏度和较好的选择性。

Spann 等[45]研究了环境剂量 Zn[350 mg/kg 干重(dw)]和 Cd（1.5 mg/kg dw）对不同大小河蚬（*Corbicula fluminea*）的毒性效应，发现环境剂量 Zn 不改变河蚬的代谢，而 Cd 只影响小河蚬的代谢，表现在氨基酸代谢和能量代谢的改变。Bundy 等[46]研究发现在无可见毒性效应水平（NOEC）和最低可见毒性效应水平（LOEC）下 Cu 暴露即可对蚯蚓的能量代谢产生干扰。Geng 等[47]采用基于 LC/MS/MS 的代谢组学靶标分析对环境剂量（<100 μg/L）短链氯化石蜡（SCCPs）的短期暴露对 HepG2 细胞的代谢干扰进行了研究，发现环境剂量 SCCPs 暴露亦能够影响细胞代谢，主要表现为细胞内不饱和脂肪酸和长链脂肪酸的 β 氧化上调，糖酵解和氨基酸代谢紊乱，谷氨酰胺代谢和尿素循环上调。这些研究均表明代谢组学技术用于环境剂量污染物的毒性评估有无可比拟的优势。

2.3　代谢组学在环境毒理学研究中的发展历程

代谢组学技术最早用于环境污染物的毒性评价是从无机金属元素开始的，通常采用的是核磁共振（NMR）技术。Gibb 等通过 NMR 技术检测一些常见无脊椎物种如蚯蚓土鳖虫、千足虫等组织提取物的变化来指示环境污染情况。如应用基于核磁共振波谱的代谢组学方法以蚯蚓为模式生物进行 Cu 和 Zn 的毒理学研究[48]，发现组氨酸含量与 Cu 暴露浓度呈正相关关系，可以作为 Cu 污染的指示性生物标志物，该研究中 Zn 是引起 7 个不同地区的蚯蚓代谢物谱之间差别的主要污染物。Jamers 等[49]基于 NMR 研究了 Cu 对微藻（*Chlamydomonas reinhardtii*）的毒性效应，发现当 Cu 的浓度达到 17 nmol/L 时能够下调微藻体内谷胱甘肽的水平。Wu 等[50]用基于 NMR 的代谢组学方法研究了不同盐度下 As 对菲律宾蛤仔（*Ruditapes philippinarum*）的毒性效应，发现在正常的盐度下（31.1 ppt）As 暴露降低了蛤仔体内氨基酸（如谷氨酸，*β*-丙氨酸等）的含量，同时增加甜菜碱和延胡索酸的含量；在中等盐度下（23.3 ppt），As 导致的代谢变化包括苏氨酸、组氨酸、ATP 和延胡索酸的降低；而在低盐度下（15.6 ppt），只有 ATP 含量增加，琥珀酸减少。表明低盐度条件下 As 暴露改变能量代谢，而中高盐度条件下除了能量代谢变化，渗透压也发生改变。Bundy 等[51]研究了在不同程度金属污染的环境中一种蚯蚓种属（*Lumbricus rubellus*）的代谢组学变化，发现蚯蚓体内麦芽糖含量升高可作为指示金属污染的潜在生物标志物。Lankadurai 等[52]基于 NMR 研究了蚯蚓（*Eisenia fetida*）对 PFOA 和 PFOS 的代谢干扰，PFOA 暴露浓度为 6.25~50 $\mu g/cm^2$，PFOS 暴露浓度为 3.125~25 $\mu g/cm^2$。发现亮氨酸，精氨酸，谷氨酸，麦芽糖和 ATP 是 PFOA 和 PFOS 暴露潜在的指示物。PFOA 和 PFOS 暴露还会损伤线粒体内膜结构，从而增加脂肪酸氧化，干扰 ATP 合成。

另外，英国剑桥大学的 Griffin 教授带领的研究团队基于 NMR 技术在重金属毒性效应的代谢组学评价方面做了一系列的研究工作。Griffin 等[53]观察到低剂量 $CdCl_2$ 暴露（肾脏浓度为 8.4 μg/g dw）引起沙鼠的生物化学变化包括脂类和谷氨酸代谢。在另一篇文章中，Griffin 等[54]基于 NMR 研究了 $CdCl_2$ 暴露对 SD（*Sprague Dawley*）大鼠尿样的代谢干扰，发现甘油三酯水平发生变化，经过 19 天的暴露，低剂量组和高剂量组（8 mg/kg 和 40 mg/kg 食物）大鼠均出现肌酸尿和柠檬酸结合物增加现象。94 天暴露后血液 Ca^{2+}/Mg^{2+}降低，另外还发现 Cd 暴露会引起肾脏细胞出现酸中毒。Jones 等[55]在研究啮齿动物对 Cd 暴露的代谢响应时也发现了同样的现象，但他认为细胞酸中毒和脂类物质的变化是动物组织对损伤的一个响应，而并非毒性作用的表现。Griffin 等[56]基于 NMR 研究了 As_2O_3(28 mg/kg 食物)对沙鼠的毒性作用，发现该剂量能够导致肾组织损伤，脂质代谢异常。Holmes 等[57]采用基于 NMR 的代谢组学技术寻找肝毒性（联肼）和肾毒性物质（$HgCl_2$）对大鼠毒性作用的标志物，尿样中牛磺酸、*α*-丙氨酸、肌酸和 2-氨基己二酸水平的升高为联肼带来肝毒性的标志物，而尿样中氨基酸、有机酸和葡萄糖水平升高，同时柠檬酸、琥珀酸、马尿酸和 2-氧戊二酸的降低为 $HgCl_2$ 带来肾损伤的标志物。Wu 等[58]用代谢组学的方法对稀土 $Ce(NO_3)_3$ 的急性毒性进行了研究，发现 $Ce(NO_3)_3$ 使大鼠脂肪酸的 β 氧化功能受到抑制，表现为引起大鼠尿液中一系列小分子代谢物出现剂量依赖性减少，血清中的丙酮、乙酰乙酸、乳酸盐和肌酸酐升高。Dudka 等[59]采集了健康人群和长期暴露于重金属 As、Cd 和 Pb 的职业工人尿液，并基于 NMR 对其小分子代谢物进行分析，发现 As、Cd 和 Pb 的联合暴露显著改变了人体的能量代谢，另外，以极低密度脂蛋白和低密度脂蛋白为代表的脂类物质和氨基酸类物质也发生显著变化。Zhu 等[60]基于 NMR 分析了健康人群和接触高硒环境的高危人群的尿液代谢谱的变化，发现过度硒接触会引起尿液中甲酸、乙酸、乳酸、马尿酸和丙氨酸含量的升高，同时柠檬酸、肌酸和氧化三甲胺的含量降低，这些显著变化的代谢物与肾脏和肝脏病变有关。

近年来，基于质谱的代谢组学技术不断发展，已经成为当今代谢组学最有效的研究手段之一。Ritter 等[61]从基因和代谢的角度研究了 Cu 胁迫后褐藻（*E. siliculosus*）的生物化学变化，代谢组学研究采用了 UPLC-MS 和 GC-MS 相结合的方法。发现 *E. siliculosus* 自身光合作用能量改变，小分子代谢物中氧脂素代谢上调，芳香氨基酸（苯丙氨酸和酪氨酸）和支链氨基酸（缬氨酸、亮氨酸和异亮氨酸）增加，意味着蛋白质分解代谢的增加；游离脂肪酸尤其是 FFA C18：3 和 FFA C20：4 显著增加，表明 ROS 参与的脂质过氧化过程增加。从基因分析结果可以看出，ROS 解毒系统和肌醇信号通路被激活。Sun 等[62]利用 GC-TOF-MS 技术系统研究了镉胁迫条件下拟南芥小分子代谢物的变化，发现镉胁迫可以引发一系列胁迫诱导反应，*β*-丙氨酸、脯氨酸、腐胺、4-氨基丁酸、蔗糖、肌醇半乳糖苷、棉子糖、*a*-生育醇、菜油甾醇、*β*-谷甾醇和异黄酮等化合物均有显著响应，并首次发现镉胁迫显著影响拟南芥中肌醇半乳糖苷和棉子糖的含量。Zeng 等[63]采用毛细管电泳-飞行时间质谱研究了 BPA（1 μg/ kg bw，10 μg/ kg bw 和 100 μg/ kg bw）对 SD 大鼠尿液代谢物的干扰，42 种代谢物发生显著变化，受到影响的代谢通路包括缬氨酸、亮氨酸和异亮氨酸生物合成，谷氨酰胺和谷氨酸代谢等。神经递质以及与神经传递相关代谢物的改变表明小剂量 BPA 的毒性作用在于干扰神经系统。Ji 等[64]采用基质辅助激光解析串联飞行时间质谱（MALDI-TOF/TOF MS）和 NMR 研究了蚯蚓（*Eisenia fetida*）对 PBDEs 的代谢干扰，发现 BDE47 主要干扰能量代谢，引起渗透压改变。代谢干扰的标志物有甜菜碱、甘氨酸、2-己基-5-乙基-3-呋喃硫酸钠。Lenz 等[65]同时用 NMR 和 HPLC-TOF/MS 对 $HgCl_2$ 暴露后大鼠尿液中代谢物的变化进行研究，在暴露后的时间序列变化中 NMR 和 HPLC-TOF/MS 分析均发现 $HgCl_2$ 暴露对大鼠的显著干扰发生在暴露后第 3 天，NMR 和 HPLC-TOF/MS 鉴定出的代谢物有所不同，二者在代谢组学研究中可以相互补充。

3 芳香烃类受体类化合物毒性效应的代谢组学研究

研究多环芳烃的适应性代谢及二噁英类化合物毒性机制中观察到一种起重要作用的可溶性蛋白，被命名为芳香烃受体（aryl hydrocarbon receptor，AhR），因此把多环芳烃及二噁英类化合物命名为芳香烃受体活性化合物[66,67]。芳香烃受体是一种配体激活转录因子，它可以与芳香烃受体核易位体蛋白在核内形成异二聚体，诱导许多 I 相和 II 相外源化合物代谢酶的表达，还参与许多毒性反应和其他一些重要的生物学过程，如信号转导、细胞分化、细胞凋亡等[68-71]。所以，研究芳香烃受体化合物的毒性作用对于阐明环境化学物质的致癌机制有非常重要的意义[46]。芳香烃化合物包括氯代芳烃、杂环胺及多环芳烃等，其中二噁英、多氯联苯及多环芳烃同时具备高毒性、低水溶性、生物稳定性以及污染范围广等特点，引起了科学界的广泛关注[72,73]。研究其毒性作用，获取毒理相关信息将会为此类化合物的毒性评价提供有力的理论基础。

3.1　二噁英化合物的毒性作用评价

二噁英是一类严重威胁人类健康和生态环境的持久性有机污染物。它们遍布于世界上包括空气、土壤、水、沉淀物和食物（尤其是奶制品、肉类和水生有壳动物）在内的所有媒介中[66,67]，由于其高稳定性、长距离迁移性及脂溶性等特点，在生物放大作用的影响下，造成人类和动物食物源的严重污染，并能够引起人体及动物的多种毒性反应，包括体重降低、胸腺萎缩、免疫毒性、肝毒性、卟啉症及其他一些皮肤病变，还会引起组织特异性的发育不全或其增生反应及癌变、畸变、生殖毒性反应等[74]。尤其是近年来，二噁英污染及中毒事件频频发生，极大地推动了二噁英毒理学研究的发展。二噁英的主要致毒机制是通过与芳香烃受体结合从而诱导相关基因表达实现的[75]。但是，关于二噁英的毒性作用机理仍有很多问题有待研究，如基因表达产物如何作用，是否还存在其他的非芳香烃受体介导的毒性作用途径等，这些研究的开展一定会对人类防治二噁英提出更好的建议。

二噁英可以引发生物体内葡萄糖、氨基酸及尿素等代谢的紊乱[76,77]，Mitrou 等[78]的研究表明，即使暴露于 1 μg/kg 的二噁英中 12 周，天竺鼠的脂肪、肝和胰腺等组织和器官对葡萄糖的吸收量也均表现出非常明显的降低趋势。蔡宗伟课题组采用超高效液相色谱-四极杆串联飞行时间质谱分析了 2, 3, 7, 8- TCDD 在芳香烃受体敏感型的 C57BL/6J 小鼠与不敏感型的 DBA/2J 小鼠中的不同毒性反应[25,27]，还对大脑中的海马组织与小脑组织的代谢物进行了解析。Whitfield Åslund 等[79]将蚯蚓（*Eisenia fetida*）暴露于亚致死剂量的二噁英类化合物多氯联苯中，发现高剂量组蚯蚓的代谢谱中的 ATP 和多种氨基酸含量显著升高，葡萄糖和麦芽糖显著降低，蚯蚓的能量代谢和细胞膜渗透功能发生改变。

陈吉平课题组在二噁英类化合物的代谢毒性评价方面做了一系列的相关工作。用不同浓度的 2, 3, 7, 8-TCDD 处理 HepG2 细胞后发现，2, 3, 7, 8-TCDD 干扰了细胞内的糖代谢、氨基酸代谢、甘油代谢及尿素循环，且有的代谢物表现出显著的剂量-效应关系，这些代谢物的动态变化不仅为探究二噁英的分子毒性提供理论支持，还可作为二噁英毒性早期筛查的候选生物标志物[80]；利用 HPLC-MS/MS 检测了不同浓度的 2, 3, 7, 8-TCDD 对 HepG2 细胞处理 48 h 后胞外代谢物的变化情况，检测结果结合代谢流量分析系统，利用 Matlab 软件计算不同浓度的 2, 3, 7, 8-TCDD 对 HepG2 胞内各个代谢路径代谢流量的影响情况。结果表明低浓度的 2, 3, 7, 8-TCDD 对 HepG2 细胞代谢的影响很小，当二噁英浓度逐渐增强时，对细胞代谢流的干扰作用才会明显，主要影响的代谢流为糖酵解及能量代谢[81]；采用 HPLC-MS/MS 结合代谢流量分析相同毒性当量的不同芳香烃受体化合物对 HepG2 细胞中小分子差异，PCA 模型显示四组样品分别分布在三个不同区域，说明即使相同毒性当量的不同 AhR 化合物之间也存在明显差别。TCDD 和 TCDF 的组间差异不明显，但是与其他的 PeCDD 和 PCB-126 之间存在明显的差异。代谢流分析表明，TCDD 和 TCDF 抑制了细胞 ATP 代谢流，ATP 是细胞一切活动能量供应的直接来源，ATP 产生的降低预示着细胞代谢的降低，细胞代谢受阻，即会影响细胞正常的生命活动，对细胞的生长造成严重的毒性作用，而与其相反，PeCDD 和 PCB-126 使 ATP 代谢显著的增强，此类污染物的胁迫刺激了细胞的代谢作用，使细胞产生应激反应，产生大量的能量来调整细胞的防御系统；通过二噁英对烟草细胞的毒性研究中发现：尽管烟草细胞中不存在芳香烃受体，二噁英仍能引起一系列的毒性效应，如引起脂质过氧化作用，并对烟草细胞激素的代谢造成干扰[82]。植物激素在调控细胞的信号表达，抵御环境胁迫过程中也起了重要的作用[83]。环境污染物与植物相互作用的分子生物学研究表明植物细胞内一些列内生激素的变化与环境污染物的胁迫作用密切相关[84-86]。该研究中植物激素脱落酸的变化强度与二噁英的浓度表现出了显著的剂量-效应关系。基于文献推测，这很可能是烟草细胞对二噁英胁迫响应的主要机制[87,88,85,89]。此结果也表明二噁英在植物体内确实存在着非芳香烃受体介导的毒性作用途径。那么人体细胞是否也存在非芳香烃受体介导的二噁英毒性作用途径，有待我们更进一步的验证。

代谢组学数据同转录组学、蛋白质组学的数据整合在一起并给出生物学功能的解释，将会是未来毒理学研究的强大工具，在揭示毒性机制定位功能基因方面起着举足轻重的作用。不同的“组学”技术的整合，对毒理学的研究提供了一个互补的见解，并在毒理学研究中得到了初步的应用[90,91]，Jennen[92]将转录组和代谢组学数据结合在一起，对2, 3, 7, 8-TCDD的毒性机制进行研究，通过交叉组学分析，对TCDD暴露的分子机制有了新的见解，转录组和代谢组学分析显示，一些重要的毒性机制只有在“组学”被整合的情况下才被证实，该研究发现了受体介导的机制中所涉及的独特的毒性路径：G-蛋白偶联受体（GPCR）信号通路，其中明显上调的交换因子（SOS1）似乎在诱导毒性中扮演着重要的角色。

3.2 BaP 对 HepG2 细胞代谢的影响

多环芳烃（polycyclic aromatic hydrocarbons，PAHs）是环境致癌化学物质中数量最多的一类，在总数已达1000多种的致癌物中，PAHs占了1/3以上。PAHs本身并无直接毒性，其进入机体后，经过代谢活化而呈现致癌作用的。PAHs在体内所发生的一系列代谢的改变，主要是在位于细胞内质网上的细胞色素P450混合功能氧化酶（mixed function oxygenases，MFO）的参与下进行的，二者的结合使DNA的遗传信息发生改变，引起突变，这是构成癌变的基础[93]。1979年，美国国家环境保护署（EPA）发布了129种优先监测的污染物，其中就有16种多环芳烃[94]。目前已发现的多环芳烃中，有400多种具有致癌作用，而其中的苯并芘（简称BaP）在众多的PAHs中，由于其分布广、性质稳定、致癌毒性最强[95]。因此，它被当作环境受PAH污染的指标。苯并芘主要存在于煤焦油、各类碳黑和煤、石油等燃烧产生的烟气、香烟烟雾、汽车尾气中，以及焦化、炼油、沥青、塑料等工业污水中，它的广泛存在使其毒性研究至关重要[96]。多环芳烃通过AhR激活细胞色素P4501A1蛋白，可诱导某些I相代谢酶（如细胞色素P450氧化酶）和II相代谢酶（如谷胱甘肽还原酶）基因的表达，从而调控外源性化合物的生物转化，进而影响机体某些疾病的发生[97]。

BaP对HepG2细胞毒性的研究主要集中在AhR诱导的基因表达方面[95]，代谢组学研究BaP对细胞毒性的研究还相对较少，Jones等[98]用代谢组学方法研究了蚯蚓对多环芳烃暴露的响应，发现暴露剂量在40 mg/kg时蚯蚓体内乳酸和饱和脂肪酸水平下降，氨基酸水平升高，表明葡萄糖代谢和三羧酸循环受到干扰，脂肪酸代谢增加。陈吉平课题组在HepG2细胞暴露试验中的研究发现，随着BaP浓度的增加，细胞的葡萄糖吸收出现了降低的趋势，即糖代谢的过程受阻，通过比较其他AhR活性化合物如TCDD对细胞代谢的研究，我们认为BaP与其他的AhR活性化合物在糖代谢方向具有相似的作用机制。随着暴露浓度的增加，TCA循环受到抑制，与丙酮酸相关的两个代谢途径草酰乙酸转化为丙酮酸及丙酮酸向乙酰辅酶A的转化出现了先升高后降低的现象，此结果与Jones等[98]的研究结果相吻合。污染物的暴露会影响细胞提供能量的比例[99]，与理论预测的一样，葡萄糖与谷氨酰胺的代谢存在着互补的关系，BaP抑制了葡萄糖吸收的同时却加速了谷氨酰胺的吸收，进一步研究对代谢路径的影响，发现糖酵解途径明显受阻，此途径不能对乳酸的产生造成显著的影响，但是乳酸的产生量大量增加，需要细胞借助其他的途径产生过量的乳酸，作为氨基酸生成的前提物质，丙酮酸产生量的变化对乳酸的产生至关重要。研究发现，为了产生大量的丙酮酸，三羧酸循环的中间产物草酰乙酸大量向丙酮酸转化，丙酮酸的过量增长导致了其向乙酰辅酶A的转化出现了增加的现象。但是随着BaP浓度的增加，乳酸大量产生，导致其向乙酰辅酶A的转化受到抑制，细胞通过对大量氨基酸的增加吸收来补充三羧酸循环前体物质的缺乏，低浓度的BaP使细胞产生应激反应，导致细胞能量代谢及三羧酸循环加速，但是随着污染物暴露浓度增加，大量的乳酸对细胞产生毒害作用，尽管氨基酸的吸收量仍在增加，细胞内某些氨基酸的合成受到显著的影响，细胞不能再通过升高细胞的能量代谢来抵御其毒害作用，此时细胞必定出现毒性效应。

4　卤代阻燃剂类化合物毒性效应的代谢组学研究

4.1　SCCPs 毒性机制的代谢组学评价

氯化石蜡（chlorinated paraffins，CPs）是一类正构烷烃氯代衍生物，具有多种同族体及异构体，被广泛作为金属加工润滑剂、油漆、密封剂、黏合剂、塑料添加剂或者阻燃剂等使用。一般按碳链长度将 CPs 分为短链氯化石蜡（$C_{10\sim13}$，short-chain chlorinated paraffins，SCCPs）、中链氯化石蜡（$C_{14\sim17}$，middle-chain chlorinated paraffins，MCCPs）及长链氯化石蜡（$C_{18\sim30}$，long-chain chlorinated paraffins，LCCPs）。其中，SCCPs 已被证实具有环境持久性、远距离迁移能力、生物毒性及蓄积性等特征，于 2017 年 5 月被《斯德哥尔摩公约》正式列入 POPs 清单[100]。SCCPs 通过生产、储存、运输及工业应用等过程进入环境，已造成了不同程度的环境污染。目前，SCCPs 在大气、水、沉积物、生物群落，食品，甚至是人类乳汁中被频繁检出[101,102]。环境介质中的 SCCPs 能够通过直接接触、呼吸暴露及饮食摄入等方式进入人体，已成为威胁人群及生态系统健康的重要因素之一。

短链氯化石蜡对水生动物的毒性较大，对无脊椎动物和鱼类，短链氯化石蜡在 ug/L 水平就具有慢性毒性效应。对虹鳟鱼的慢性毒性主要表现为肝组织损伤[103]；对水蚤的急性毒性较高，24 小时半数致死浓度为 0.3~1.1mg/L[104]。在对啮齿动物的毒性研究中，人们发现短链氯化石蜡能导致肝肥大，使肝细胞中过氧化物酶体增殖，同时使肝细胞中尿苷二磷酸葡糖醛酸基转移酶的活性增加，从而导致血浆中甲状腺激素增加[105]；并且，短链氯化石蜡可导致雄鼠肝内 RLvMc $P450_{54}$ 和 RLvMc $P450_{50}$ 微粒体的增加，并诱导细胞色素 P450 酶系中环氧化物水解酶和谷胱甘肽 *S*-转移酶活性的增加[106]；另外，短链氯化石蜡也可增加雄鼠肾的重量，使肾管状嗜曙红细胞增多，在致癌性研究中，人们发现增加短链氯化石蜡剂量可增加鼠类肝、甲状腺、肾的腺瘤和癌的发病率[107]。然而，目前关于 SCCPs 毒性效应的结论都是基于高剂量动物暴露试验的结果，涉及 SCCPs 毒性机制的研究十分有限，如其毒性作用靶位点以及对生物代谢的影响等都不太清楚。另外，有关环境浓度 SCCPs 是否会对人类健康带来不利影响尚需进行深入细致的研究。

4.1.1　SCCPs 对 HepG2 细胞毒性效应的代谢组学评价

为了探讨 SCCPs 暴露对 HepG2 细胞带来的代谢干扰，Geng 等[47]基于 LC/MS/MS 对细胞内氨基酸、脂肪酸、糖类、甘油、尿素等小分子代谢物进行了定量分析。暴露 48 小时后细胞内 43 个代谢物发生显著变化。对差异代谢物进行偏最小二乘法判别分析（PLS-DA）发现：所有 SCCPs 暴露组均能够与对照组分开，且随着氯含量的增加，SCCPs 对 HepG2 细胞的代谢干扰增加。进一步的代谢通路分析表明：不同浓度、不同氯化度和不同碳链长度 SCCPs 对 HepG2 细胞的毒性机制相同，主要干扰 HepG2 细胞内精氨酸和脯氨酸代谢；甘氨酸、丝氨酸和苏氨酸代谢；丙氨酸和天冬氨酸代谢；谷氨酰胺和谷氨酸代谢；亚油酸代谢等。该研究还观察到 SCCPs 暴露对 HepG2 细胞的糖酵解通路带来了一定的干扰，使细胞内无氧糖酵解途径发生紊乱；谷氨酰胺代谢和脂肪酸 β 氧化上调（图 27-1）。

通过代谢组学分析和常规毒理学验证，该研究提出环境相关剂量 SCCPs 会对 HepG2 细胞产生显著的毒性效应。主要体现在降低细胞增殖活性，改变细胞的氧化还原状态，对胞内小分子代谢物产生干扰。基于文献报道以及本研究中的数据支持，推断 SCCPs 细胞毒性作用可能的机制如下：首先，作为一种氯代烷烃，SCCPs 可以作为一种过氧化物酶体增殖促进剂诱导机体过氧化物酶体增殖[108]，促进不饱和脂肪

酸和长链脂肪酸 β 氧化[109]。其次，作为一种外源性物质，SCCPs 不可避免会被细胞色素 P450 酶代谢，从而给机体带来氧化压力[110,111]。作为氧化压力的适应回应，机体能量代谢和氨基酸代谢发生扰动，尿素循环上调。最后，由于脂肪酸 β 氧化的上调，机体脂类物质被过度消耗，氨基酸代谢的紊乱使蛋白质的生物合成被抑制。在氧化压力，脂质过度消耗和蛋白生物合成受阻的共同作用下，细胞增殖活力降低。

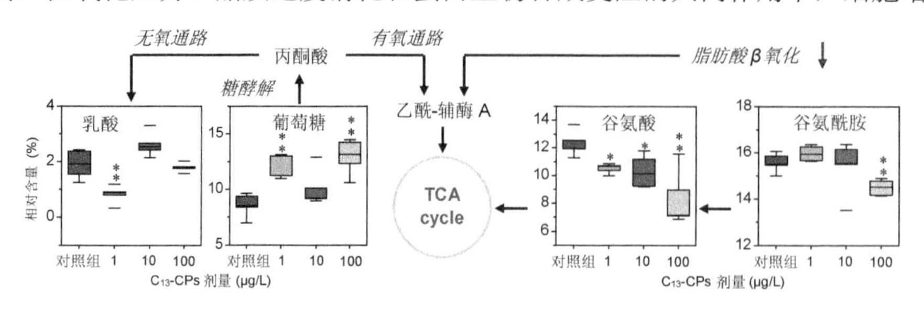

图 27-1　不同剂量 SCCPs 暴露对 HepG2 细胞能量代谢的影响，显著性变化为暴露组与对照组相比的 T 统计结果

* $P < 0.05$；** $P < 0.01$

4.1.2　SCCPs 对大鼠毒性效应的代谢组学研究

陈吉平所在的研究团队以 SD 大鼠为模式生物进行为期 28 d 的 SCCPs 暴露，暴露剂量从 0~100 mg/kg 体重，包含了环境暴露剂量和常规毒理学研究剂量，希望从高剂量组获得形态学变化特征，从低剂量组获得分子生物学变化特征，探讨低剂量效应下代谢组学变化对高剂量效应下形态学变化的应答，从分子生物学角度探寻 SCCPs 毒性效应的分子机理。首先通过常规毒理学研究发现肝脏是 SCCPs 的毒性作用最主要的靶器官，然后对肝脏组织小分子代谢物进行了代谢组学分析。采用代谢扰乱度（MELI）定量评估了肝脏代谢水平的变化，发现在很低剂量暴露下[0.01 mg/(kg·d)]，SCCPs 依然引起了代谢物的显著变化，主要包括脂类代谢、糖代谢和核苷酸代谢。脂类代谢中脂肪酸降解，脂肪酸生物合成，甘油磷脂代谢的变化尤为显著，主要表现在脂肪酸含量显著下降，不饱和脂肪酸及长链脂肪酸降低。该研究进一步采用基因芯片及实时荧光定量 PCR 技术进行验证，发现 SCCPs 暴露显著升高了大鼠肝脏脂肪酸氧化基因的表达水平，这可能是脂肪酸含量降低的重要原因。结合常规毒理学、代谢组学和转录组学的研究结果，推测 SCCPs 可能是一种 PPARα 受体激动剂。PPARα 是一种配体激活型核转录因子，对于肝脏脂类代谢，尤其是脂肪酸氧化具有重要调控作用。研究发现，激活 PPARα 受体能够加强脂肪酸氧化分解[112]。从结构上推断，SCCPs 的基本结构为链状烷烃，在结构上与 PPARα 的天然配体脂肪酸相似，并且有研究证实其末端碳原子可以脱氯并被氧化为相应的羧基[113]。该课题组利用体外荧光素酶报告基因实验进一步证实了这一推测。综上所述，SCCPs 对大鼠的肝毒性作用机制可能是通过结合并激活 PPARα 受体，从而影响相关代谢酶的基因表达水平，最终引起细胞代谢紊乱（图 27-2）。这也可能是文献报道高剂量 SCCPs 能够引起大鼠过氧化物酶体增殖的分子机制[114]。同时也可能作为 SCCPs 对啮齿类动物致癌性[115]的早期指示。这一现象同样也在 PFOS 暴露的大鼠中观察到，PFOS 能够通过激活 PPARα 受体影响脂类代谢[116]。

4.2　HBCD 毒性机制的代谢组学研究

六溴环十二烷（HBCD）是一种溴代环烷烃，主要作为阻燃剂添加到挤塑聚苯乙烯材料中[117,118]。2013 年被《斯德哥尔摩公约》列入 POPs。HBCD 可以通过饮食、粉尘摄入以及空气吸入等途径进入人体并积累，目前 HBCD 在人血样和母乳中均频繁被检出，且其含量可能对人类和环境造成潜在的风险，因此对 HBCD 进行毒性评价受到广泛的关注[119-121]。HBCD 基本不具备急性毒性和基因毒性，但是慢性和亚慢性毒性则不容忽视[118,121-123]。肝毒性和甲状腺毒性是在体内动物试验研究中普遍能够观察到的两种 HBCD

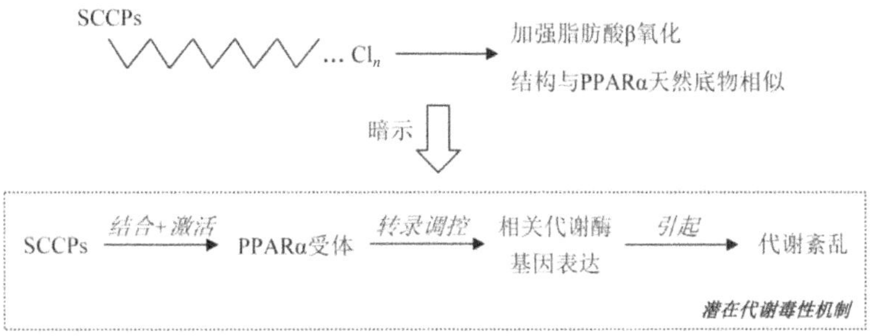

SCCPs
… Cl_n → 加强脂肪酸β氧化
结构与PPARα天然底物相似
暗示
SCCPs —结合+激活→ PPARα受体 —转录调控→ 相关代谢酶基因表达 —引起→ 代谢紊乱
潜在代谢毒性机制

图 27-2　SCCPs 影响细胞代谢的潜在机制

毒性，包括引起肝重增加、甲状腺增生、诱导细胞色素 P450 酶，以及甲状腺激素紊乱[124-126]。另外 HBCD 还有潜在的生殖毒性，能够破坏神经系统，以及发育毒性[127-129]。越来越多的研究证明即使是环境相关剂量的 HBCD 也能够扰乱内分泌系统，导致代谢紊乱和肥胖[126,130]。HBCD 能够与组成型雄甾烷受体（CAR）和孕烷 X 受体（PXR）结合[131]。在大鼠对 HBCD 口服暴露试验中发现 HBCD 能够引起雌鼠骨密度的降低，且随剂量改变而改变，其基准剂量可信下限（BMDL）为 0.056 mg/(kg bw·d)[132]。另外一个对小鼠的暴露试验中，中剂量[35 μg/(kg bw·week)]和高剂量[100 g/(kg bw·week)]HBCD 暴露能够引起高脂饮食的小鼠的体重和肝重增加，且伴随对脂质和葡萄糖稳态破坏[130]。HBCD 的低剂量毒性效应引发人们关注 HBCD 对人类健康造成的负担。

近几年，对 HBCD 毒性机制的理解已经有了长足的进步。作为外来化合物，HBCD 首先被细胞色素 P450 酶代谢降解，在此过程中产生了活性氧自由基（ROS）[133]。当 ROS 积累过多时可能会造成脂质、蛋白和 DNA 的氧化损伤[133]。与此同时 HBCD 诱导激活 PI3K/Akt 通路和 Nrf2-ARE 通路来应对 ROS 的产生，对机体提供保护[134,135]。HBCD 通过与 CAR/PXR 受体结合来诱导 CYP2B 和 CYP3A 酶，同时可能导致对甲状腺激素轴的扰乱[131]。许多体外 HBCD 暴露实验表明 HBCD 能够抑制肌质/内质网上 Ca^{2+}-ATPase 的活性，同时伴随着胞内 Ca^{2+}含量的升高和线粒体机能紊乱[20,136]。大鼠肝脏的基因表达分析表明 HBCD 能够干扰一些特定的通路，例如 PPAR 调节的磷脂代谢、甘油代谢、胆固醇代谢，还有 I 相和 II 相代谢路径等[122]。

HBCD 引起的生物效应十分复杂，为了更好地理解 HBCD 的作用方式，需要代谢组学方面的证据。Zhang 等[137]基于非靶标代谢组学的体外暴露研究表明，HBCD 暴露 72 h 能够引起 HepG2/C3A 细胞脂类物质发生变化。Wang 等[138]以 HepG2 为模式细胞，以 LC-MS 为主要分析手段研究了 HBCD 对 HepG2 细胞增殖活性和细胞内小分子代谢物的影响。结合试剂盒和酶联免疫技术对氧化损伤标志物及相关通路关键酶的分析，初步提出 HBCD 可能的毒性作用机制。如图 27-3 所示，HBCD 基本的细胞毒性作用机制可能与其对能量代谢的过度抑制相关。首先，HBCD 吸附到细胞膜上并直接抑制 P-型 ATP 酶的活性，主要是 Na^+/K^+-ATP 酶和 Ca^{2+}-ATP 酶，从而抑制 Na^+依赖的跨膜转运蛋白 SLC 家族以及 SGLT 家族载体转运氨基酸和葡萄糖的过程，同时细胞内 Na^+和 Ca^{2+}的浓度升高。抑制氨基酸的跨膜运输将导致胞内氨基酸含量的下降，并进一步抑制蛋白质的合成。由于细胞内半胱氨酸含量的大幅度下降，谷胱甘肽以及氧化还原敏感的含巯基蛋白质的合成也受抑制，从而降低了细胞清除胞内多余 ROS 的能力。当 HBCD 进入细胞内，它不仅能够通过抑制磷酸果糖激酶的活性来抑制糖酵解，还能降低长链脂肪酸的 β 氧化，其主要通过降低长链脂酰 CoA 合成酶的活性。细胞内糖酵解和脂肪酸的 β 氧化过程受抑制将导致 ATP 合成减少，从而进一步抑制氨基酸和葡萄糖的跨膜运输。另外，HBCD 对长链脂肪酸 β 氧化的显著抑制作用引起细胞内游离脂肪酸的升高。由于磷脂和游离脂肪酸之间有重组作用，这个过程主要由磷脂酶 A2 来调节，因此游离脂肪酸含量的升高引起细胞内总磷脂含量升高。

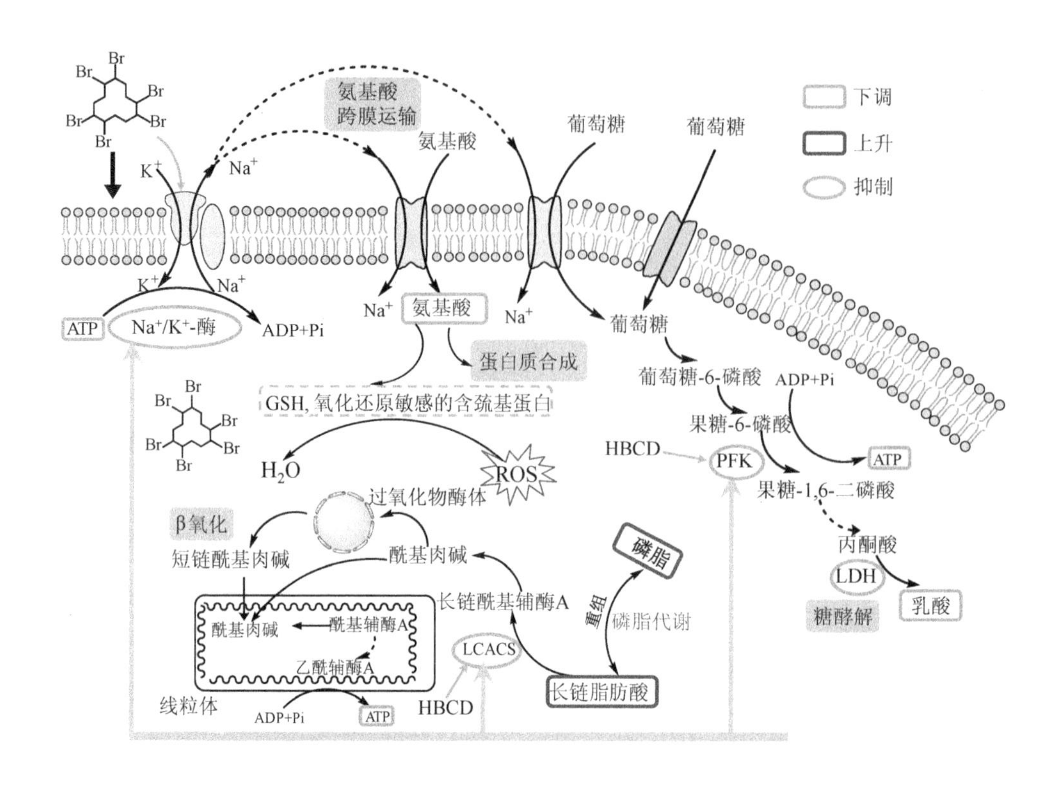

图 27-3　HBCD 毒性作用机制研究

5　污染物联合暴露的代谢组学研究

环境化学污染物种类繁多，往往以低浓度混合物形式存在，因此不可避免地对人类造成复杂多样的健康危害，即潜在联合毒性效应[139-141]。因此环境污染物的低剂量、联合毒性效应已经成为毒理学研究中的热点问题。具有相似或者不同作用机制的化学污染物能够互相影响各自的毒性效应，从而可能在不同的生物反应指标和多个作用终点引起一系列的协同、加和或者拮抗效应[141,142]。然而，在大多数污染物风险评价研究中，通常只考虑少数反应单一污染物或者简单混合物毒性的特定作用终点或者生物标志物，精准的分析方法和分析框架的匮乏使人们对化学污染物混合物之间的联合毒性作用及机制缺乏一个综合深入的理解[143,144]。代谢组学技术在污染物联合毒性研究方面具有一定的优势。

O'Kane 等[145]用低剂量 PCBs 和 2, 3, 7, 8-TCDD（0.1 μg/kg TEQ）污染的食物对 SD 大鼠进行暴露，并对血样小分子代谢产物进行 UPLC-Q-TOF-MS 分析，结果发现 PCBs 和 2, 3, 7, 8-TCDD 即使在小剂量暴露下亦会引起大鼠代谢产物的显著性变化。陈蓉等[146]采用基于液质联用（LC-MS）的代谢组学方法研究了多氯联苯（PCBs）和二噁英（2, 3, 7, 8-TCDD）及联合染毒对大鼠的代谢影响，暴露剂量为 TCDD（10 μg/kg bw）、PCBs（10 mg/kg bw）及其混合溶液（10 μg/kg bw TCDD 和 10 mg/kg bw PCBs），其毒性大小为：联合染毒> TCDD > PCBs，PCBs 和 TCDD 能导致免疫系统、肝脏和神经系统障碍、干扰脂代谢。对于组成复杂的复合污染物，采用代谢组学进行毒性评估能够提供指示毒性效应的综合指标。Huang 等[147]采用代谢组学方法对组成复杂的大气细粒子 $PM_{2.5}$ 进行体外暴露，并对模式细胞肺上皮细胞（A549）的代谢变化进行研究，发现 $PM_{2.5}$ 暴露影响了 A549 细胞的三羧酸循环，氨基酸代谢和谷胱甘肽代谢。Chen 等[148]通过体外实验对大鼠进行 $PM_{2.5}$ 暴露，发现低剂量 $PM_{2.5}$ 暴露能够引起与氧化损伤相关的不饱和磷脂酰胆碱含量降低，与炎症相关的溶血磷脂酰胆碱（LPC）显著减少。

陈吉平所在的研究团队以 HepG2 为模式生物对两种典型的有机污染物 PAHs 和 SCCPs 联合暴露的毒性效应进行了研究。PAHs 和 SCCPs 在人母乳和血液中的广泛存在[149-151]，且其毒性作用机制显著不同，大多数多环芳烃异构体主要作为芳香烃受体的激活剂，引起一系列的毒性和生物效应，包括诱导细胞色素 P4501A1 和 P4501B1 酶，诱发细胞转化以及致癌毒性[152-154]。另外，PAHs 还可以通过改变卵巢雌激素受体蛋白 β 的表达来干扰破坏内分泌系统[155]。而 SCCPs 具有脂肪族化合物结构，可以作为 PPAR 的激动剂从而诱导过氧化物酶体增殖，从而加速脂肪酸降解，诱导肝损伤并具有潜在的致癌性[156,157]。此外，SCCPs 能够通过与雌激素受体蛋白 α 和肾上腺皮质激素受体结合表现出内分泌干扰效应[158]。该团队在研究中用拟靶向分析方法来准确测定污染物暴露后引起的细胞内代谢物的扰动，用 MELI 的计算来确认显著干扰的代谢通路的联合作用类型。研究结果表明，PAHs 和 SCCPs 联合暴露对 HepG2 细胞总代谢水平的影响表现为加和效应。联合暴露对不同的代谢通路的联合作用效应类型也各不相同。PAHs 和 SCCPs 共同激活磷脂酶 A2（PLA2）活性从而造成联合暴露引起磷脂代谢上调且表现为协同效应。联合暴露对脂肪酸代谢的影响表现为加和效应。PAHs 和 SCCPs 都能作为过氧化物酶体增殖物激活受体 α（PPARα）的促进剂，两者混合主要通过刺激长链脂酰辅酶 A 脱氢酶（LCACD）导致联合暴露加速脂肪酸代谢，且 SCCPs 同样是混合物加速脂肪酸代谢的主要驱动物。联合暴露对三羧酸（TCA）循环的干扰表现为加和效应，PAHs 和 SCCPs 共同抑制异柠檬酸脱氢酶（IDH）活性导致联合暴露对 TCA 循环的阻碍作用。PAHs 和 SCCPs 都能通过降低己糖激酶（HK）和磷酸果糖激酶（PFK）的活性来抑制糖酵解。两者混合后 PAHs 和 SCCPs 共同抑制 PFK 和 HK 活性，且还能抑制乳酸脱氢酶（LDH）活性，从而导致联合暴露强烈抑制糖酵解，表现为加和效应。然而联合暴露对嘌呤代谢的影响表现为拮抗效应。本研究发展了一种基于代谢组学的混合污染物风险评估策略，该策略能够对人体暴露浓度水平下 POPs 的联合毒性有一个综合深入的理解。应用于联合毒性风险评估具有以下优势：①即使是在低或者极低浓度水平混合物暴露条件下也能检测到有意义的代谢物信号；②验证联合暴露与单独暴露引起的毒性效应差别；③确定混合物中驱动风险的主导化合物。

6 展　望

近年来，代谢组学依赖高通量、高分辨率的分析技术与完整的生物信息学系统相结合，给毒理学的发展提供了新的研究理念，在环境毒理学研究中得到了快速发展。为人们更加清楚地认识机体对环境因素变化的应答提供了一种新的有效的手段，为环境污染物作用机理的研究提供了新思路。但到目前为止代谢组学应用于环境毒理学尚存在以下不足。

（1）生物体是一个极其复杂的系统，任何一个学科都不可能囊括机体的各个方面，只有将各个学科的研究内容整合起来才能全面、系统地阐明环境污染物复杂的毒性效应。目前，先进的代谢组学分析技术多掌握在分析化学家手中，而代谢物变化所指示的生物学意义却更为生物学家所熟知，这在一定程度上限制了代谢组学在环境毒理学研究中的应用，所以整合化的代谢组学方法，高通量、高分辨率的分析技术与生物信息学相整合，才能够为代谢组学在各个学科中的应用提供更广阔的空间，为全景式了解生物体变化过程提供独特的视角，也必将在环境污染物毒性评价方面发挥越来越大的作用。

（2）代谢组学在环境毒理学中的应用还缺乏规范性的指导方针。在代谢组学分析方法的标准化方面，已陆续发表了尿液、脑脊液以及实体组织和细胞样本代谢组学分析的操作规程。但目前还未建立环境污染物低剂量长期暴露和不同靶器官毒性物质毒性效应评估的试验规范：包括模式生物选择（体内和体外）、暴露方式确定（暴露剂量和暴露时间）、高通量代谢组学方法建立和代谢产物鉴定，以及针对不同模式生物代谢变化的系统生物学解析。这一系统工作尚需长期的研究和探索。

（3）代谢组学研究需要与其他毒理学评价技术结合发挥更大优势。首先要与传统毒理学方法相结合，借助传统毒理学的毒性终点有助于寻找污染物暴露的生物标志物，建立针对不同种类毒性物质的代谢物靶标分析方法和生物标志物指示体系。另外，代谢组学与基因组学和蛋白组学相结合，将会成为一种有力的工具，大数据的挖掘将有助于发现新的有价值信息，其高通量筛选特征亦将成为毒理学发展的必然趋势。

参考文献

[1] Roux A, Lison D, Junot C, Heilier J F. Applications of liquid chromatography coupled to mass spectrometry-based metabolomics in clinical chemistry and toxicology: A review. Clin Biochem, 2011, 44(1): 119-135.

[2] Gika H G, Theodoridis G A, Plumb R S, Wilson I D. Current practice of liquid chromatography-mass spectrometry in metabolomics and metabonomics. J Pharmaceut Biomed, 2014, 87: 12-25.

[3] Viant M R, Sommer U. Mass spectrometry based environmental metabolomics: A primer and review. Metabolomics, 2013, 9(1): S144-S158.

[4] Dettmer K, Aronov P A, Hammock B D. Mass spectrometry-based metabolomics. Mass Spectrom Rev, 2007, 26(1): 51-78.

[5] Uehara T, Yokoi A, Aoshima K, Tanaka S, Kadowaki T, Tanaka M, Oda Y. Quantitative phosphorus metabolomics using nanoflow liquid chromatography-tandem mass spectrometry and culture-derived comprehensive global internal Standards. Anal Chem, 2009, 81(10): 3836-3842.

[6] Myint K T, Uehara T, Aoshima K, Oda Y. Polar anionic metabolome analysis by nano-LC/MS with a metal chelating agent. Anal Chem, 2009, 81(18): 7766-7772.

[7] Boernsen K O, Gatzek S, Imbert G. Controlled protein precipitation in combination with chip-based nanospray infusion mass spectrometry. An approach for metabolomics profiling of plasma. Anal Chem, 2005, 77(22): 7255-7264.

[8] Zenobi R. Single-cell metabolomics: Analytical and biological perspectives. Science, 2013, 342(6163): 1201-1212.

[9] Kirkwood J S, Maier C, Stevens J F. Simultaneous, untargeted metabolic profiling of polar and nonpolar metabolites by LC-Q-TOF mass spectrometry. Curr Protoc Toxicol, 2013, 39: 1-17.

[10] Dai W D, Huang Q, Yin P Y, Li J, Zhou J, Kong H W, Zhao C X, Lu X, Xu G W. Comprehensive and highly sensitive urinary steroid hormone profiling method based on stable isotope-labeling liquid chromatography mass spectrometry. Anal Chem, 2012, 84(23): 10245-10251.

[11] Wang Y, Lu X, Xu G W. Development of a comprehensive two-dimensional hydrophilic interaction chromatography/quadrupole time-of-flight mass spectrometry system and its application in separation and identification of saponins from Quillaja saponaria. J Chromatogr A, 2008, 1181(1-2): 51-59.

[12] Wang Y, Wang J S, Yao M, Zhao X J, Fritsche J, Schmitt-Kopplin P, Cai Z W, Wan D F, Lu X, Yang S L, Gu J R, Haring H U, Schleicher E D, Lehmann R, Xu G W. Metabonomics study on the effects of the ginsenoside Rg3 in a beta-cyclodextrin-based formulation on tumor-bearing rats by a fully automatic hydrophilic interaction/reversed-phase column-switching HPLC-ESI-MS approach. Anal Chem, 2008, 80(12): 4680-4688.

[13] Wang S Y, Li J, Shi X Z, Qiao L Z, Lu X, Xu G W. A novel stop-flow two-dimensional liquid chromatography-mass spectrometry method for lipid analysis. J Chromatogr A, 2013, 1321: 65-72.

[14] Chen S L, Kong H W, Lu X, Li Y, Yin P Y, Zeng Z D, Xu G W. Pseudotargeted metabolomics method and its application in serum biomarker discovery for hepatocellular carcinoma based on ultra high-performance liquid chromatography/triple quadrupole mass spectrometry. Anal Chem, 2013, 85(17): 8326-8333.

[15] Xu J, Chen Y H, Zhang R P, Song Y M, Cao J Z, Bi N, Wang J B, He J M, Bai J F, Dong L J, Wang L H, Zhan Q M, Abliz Z. Global and targeted metabolomics of esophageal squamous cell carcinoma discovers potential diagnostic and therapeutic biomarkers. Mol Cell Proteomics, 2013, 12(5): 1306-1318.

[16] An Z L, Chen Y H, Zhang R P, Song Y M, Sun J H, He J M, Bai J F, Dong L J, Zhan Q M, Abliz Z. Integrated ionization approach for RRLC-MS/MS-based metabonomics: Finding potential biomarkers for lung cancer. J Proteome Res, 2010, 9(8): 4071-4081.

[17] Chen Y H, Zhang R P, Song Y M, He J M, Sun J H, Bai J F, An Z L, Dong L J, Zhan Q M, Abliz Z. RRLC-MS/MS-based metabonomics combined with in-depth analysis of metabolic correlation network: finding potential biomarkers for breast cancer. Analyst, 2009, 134(10): 2003-2011.

[18] Chen Y H, Xu J, Zhang R P, Shen G Q, Song Y M, Sun J H, He J M, Zhan Q M, Abliz Z. Assessment of data pre-processing methods for LC-MS/MS-based metabolomics of uterine cervix cancer. Analyst, 2013, 138(9): 2669-2677.

[19] Li S F, Liu H X, Jin Y B, Lin S H, Cai Z W, Jiang Y Y. Metabolomics study of alcohol-induced liver injury and hepatocellular carcinoma xenografts in mice. J Chromatogr B, 2011, 879(24): 2369-2375.

[20] Lin S H, Kanawati B, Liu L F, Witting M, Li M, Huang J D, Schmitt-Kopplin P, Cai Z W. Ultrahigh resolution mass spectrometry-based metabolic characterization reveals cerebellum as a disturbed region in two animal models. Talanta, 2014, 118: 45-53.

[21] Lin S H, Liu H D, Kanawati B, Liu L F, Dong J Y, Li M, Huang J D, Schmitt-Kopplin P, Cai Z W. Hippocampal metabolomics using ultrahigh-resolution mass spectrometry reveals neuroinflammation from Alzheimer's disease in CRND8 mice. Anal Bioanal Chem, 2013, 405(15): 5105-5117.

[22] Lin S H, Liu N, Yang Z, Song W J, Wang P, Chen H L, Lucio M, Schmitt-Kopplin P, Chen G N, Cai Z W. GC/MS-based metabolomics reveals fatty acid biosynthesis and cholesterol metabolism in cell lines infected with influenza A virus. Talanta, 2010, 83(1): 262-268.

[23] Lin S H, Chan W, Li J H, Cai Z W. Liquid chromatography/mass spectrometry for investigating the biochemical effects induced by aristolochic acid in rats: the plasma metabolome. Rapid Commun Mass Sp, 2010, 24(9): 1312-1318.

[24] Lin S H, Yang Z, Liu H D, Tang L H, Cai Z W. Beyond glucose: metabolic shifts in responses to the effects of the oral glucose tolerance test and the high-fructose diet in rats. Mol Biosyst, 2011, 7(5): 1537-1548.

[25] Lin S H, Yang Z, Liu H D, Cai Z W. Metabolomic analysis of liver and skeletal muscle tissues in C57BL/6J and DBA/2J mice exposed to 2,3,7,8-tetrachlorodibenzo-p-dioxin. Mol Biosyst, 2011, 7(6): 1956-1965.

[26] Lin S H, Yang Z, Zhang X J, Bian Z X, Cai Z W. Hippocampal metabolomics reveals 2,3,7,8-tetrachlorodibenzo-p-dioxin toxicity associated with ageing in Sprague-Dawley rats. Talanta, 2011, 85(2): 1007-1012.

[27] Lin S H, Yang Z, Shen Y, Cai Z W. LC/MS-based non-targeted metabolomics for the investigation of general toxicity of 2,3,7,8-tetrachlorodibenzo-p-dioxin in C57BL/6J and DBA/2J mice. Int J Mass Spectrom, 2011, 301(1-3): 29-36.

[28] Zhang X J, Choi F F K, Zhou Y, Leung F P, Tan S, Lin S H, Xu H X, Jia W, Sung J J Y, Cai Z W, Bian Z X. Metabolite profiling of plasma and urine from rats with TNBS-induced acute colitis using UPLC-ESI-QTOF-MS-based metabonomics - A pilot study. Febs J, 2012, 279(13): 2322-2338.

[29] Chan W, Lee K C, Liu N, Wong R N S, Liu H W, Cai Z W. Liquid chromatography/mass spectrometry for metabonomics investigation induced by aristolochic acid of the biochemical effects in rats: the use of information-dependent acquisition for biomarker identification. Rapid Commun Mass Sp, 2008, 22(6): 873-880.

[30] Chan W, Cai Z W. Aristolochic acid induced changes in the metabolic profile of rat urine. J Pharmaceut Biomed, 2008, 46(4): 757-762.

[31] Dai H, Xiao C N, Liu H B, Hao F H, Tang H R. Combined NMR and LC-DAD-MS analysis reveals comprehensive

metabonomic variations for three phenotypic cultivars of *Salvia miltiorrhiza* bunge. J Proteome Res, 2010, 9(3): 1565-1578.

[32] Dai H, Xiao C N, Liu H B, Tang H R. Combined NMR and LC-MS analysis reveals the metabonomic changes in *Salvia miltiorrhiza* bunge induced by water depletion. J Proteome Res, 2010, 9(3): 1460-1475.

[33] Liu H, Zheng A, Liu H, Yu H, Wu X, Xiao C, Dai H, Hao F, Zhang L, Wang Y, Tang H. Identification of three novel polyphenolic compounds, origanine A-C, with unique skeleton from Origanum vulgare L. using the hyphenated LC-DAD-SPE-NMR/MS methods. J Agric Food Chem, 2012, 60(1): 129-135.

[34] Li M, Yang L, Bai Y, Liu H W. Analytical methods in lipidomics and their applications. Anal Chem, 2014, 86(1): 161-175.

[35] Li M, Feng B S, Liang Y, Zhang W, Bai Y, Tang W, Wang T, Liu H W. Lipid profiling of human plasma from peritoneal dialysis patients using an improved 2D (NP/RP) LC-QToF MS method. Anal Bioanal Chem, 2013, 405(21): 6629-6638.

[36] Li M, Zhou Z G, Nie H G, Bai Y, Liu H W. Recent advances of chromatography and mass spectrometry in lipidomics. Anal Bioanal Chem, 2011, 399(1): 243-249.

[37] Chen H W, Wortmann A, Zenobi R. Neutral desorption sampling coupled to extractive electrospray ionization mass spectrometry for rapid differentiation of bilosamples by metabolomic fingerprinting. J Mass Spectrom, 2007, 42(9): 1123-1135.

[38] Gu H W, Qi Y P, Xu N, Ding J H, An Y B, Chen H W. Nuclear magnetic resonance spectroscopy and mass spectrometry-based metabolomics for cancer diagnosis. Chinese J Anal Chem, 2012, 40(12): 1933-1937.

[39] Gao D, Wei H B, Guo G S, Lin J M. Microfluidic cell culture and metabolism detection with electrospray ionization quadrupole time-of-flight mass spectrometer. Anal Chem, 2010, 82(13): 5679-5685.

[40] 耿柠波, 张海军, 王菲迪, 任晓倩, 张保琴, 陈吉平. 代谢组学技术在环境毒理学研究中的应用. 生态毒理学报, 2016(3): 26-35.

[41] Robertson D G, Reily M D, Sigler R E, Wells D F, Paterson D A, Braden T K. Metabonomics: Evaluation of nuclear magnetic resonance (NMR) and pattern recognition technology for rapid in vivo screening of liver and kidney toxicants. Toxicol Sci, 2000, 57(2): 326-337.

[42] He J, Chen J, Wu L Y, Li G Y, Xie P. Metabolic response to oral microcystin-LR exposure in the rat by NMR-based metabonomic study. Journal of Proteome Research, 2012, 11(12): 5934-5946.

[43] Waters N J, Waterfield C J, Farrant R D, Holmes E, Nicholson J K. Metabonomic deconvolution of embedded toxicity: Application to thioacetamide hepato- and nephrotoxicity. Chem Res Toxicol, 2005, 18(4): 639-654.

[44] Bundy J G, Spurgeon D J, Svendsen C, Hankard P K, Osborn D, Lindon J C, Nicholson J K. Earthworm species of the genus Eisenia can be phenotypically differentiated by metabolic profiling. FEBS letters, 2002, 521(1): 115-120.

[45] Spann N, Aldridge D C, Griffin J L, Jones O A. Size-dependent effects of low level cadmium and zinc exposure on the metabolome of the Asian clam, Corbicula fluminea. Aquat Toxicol, 2011, 105(3-4): 589-599.

[46] Bundy J G, Sidhu J K, Rana F, Spurgeon D J, Svendsen C, Wren J F, Sturzenbaum S R, Morgan A J, Kille P. 'Systems toxicology' approach identifies coordinated metabolic responses to copper in a terrestrial non-model invertebrate, the earthworm Lumbricus rubellus. Bmc Biol, 2008, 6(1): 25.

[47] Geng N B, Zhang H J, Zhang B Q, Wu P, Wang F D, Yu Z K, Chen J P. Effects of short-chain chlorinated paraffins exposure on the viability and metabolism of human hepatoma HepG2 cells. Environ Sci Technol, 2015, 49(5): 3076-3083.

[48] Gibb J O T, Svendsen C, Weeks J M, Nicholson J K. ^{1}H NMR spectroscopic investigations of tissue metabolite biomarker response to Cu II exposure in terrestrial invertebrates: identification of free histidine as a novel biomarker of exposure to copper in earthworms. Biomarkers, 1997, 2(5): 295-302.

[49] Jamers A, Blust R, De Coen W, Griffin J L, Jones O A. Copper toxicity in the microalga Chlamydomonas reinhardtii: an

integrated approach. Biometals, 2013, 26(5): 731-740.

[50] Wu H F, Zhang X Y, Wang Q, Li L Z, Ji C L, Liu X L, Zhao J M, Yin X L. A metabolomic investigation on arsenic-induced toxicological effects in the clam Ruditapes philippinarum under different salinities. Ecotox Environ Safe, 2013, 90: 1-6.

[51] Bundy J G, Spurgeon D J, Svendsen C, Hankard P K, Weeks J M, Osborn D, Lindon J C, Nicholson J K. Environmental metabonomics: Applying combination biomarker analysis in earthworms at a metal contaminated site. Ecotoxicology, 2004, 13(8): 797-806.

[52] Lankadurai B P, Simpson A J, Simpson M J. ^{1}H NMR metabolomics of Eisenia fetida responses after sub-lethal exposure to perfluorooctanoic acid and perfluorooctane sulfonate. Environ Chem, 2012, 9(6): 502-511.

[53] Griffin J L, Walker L A, Troke J, Osborn D, Shore R F, Nicholson J K. The initial pathogenesis of cadmium induced renal toxicity. FEBS letters, 2000, 478(1-2): 147-150.

[54] Griffin J L, Walker L A, Shore R F, Nicholson J K. Metabolic profiling of chronic cadmium exposure in the rat. Chem Res Toxicol, 2001, 14(10): 1428-1434.

[55] Jones O A, Walker L A, Nicholson J K, Shore R F, Griffin J L. Cellular acidosis in rodents exposed to cadmium is caused by adaptation of the tissue rather than an early effect of toxicity. Comp Biochem Phys D, 2007, 2(4): 316-321.

[56] Griffin J L, Walker L, Shore R F, Nicholson J K. High-resolution magic angle spinning ^{1}H-NMR spectroscopy studies on the renal biochemistry in the bank vole (*Clethrionomys glareolus*) and the effects of arsenic (As^{3+}) toxicity. Xenobiotica, 2001, 31(6): 377-385.

[57] Holmes E, Nicholls A W, Lindon J C, Connor S C, Connelly J C, Haselden J N, Damment S J P, Spraul M, Neidig P, Nicholson J K. Chemometric models for toxicity classification based on NMR spectra of biofluids. Cheml Res Toxicol, 2000, 13(6): 471-478.

[58] Wu H F, Zhang X Y, Liao P Q, Li Z F, Li W S, Li X J, Wu Y J, Pei F K. NMR spectroscopic-based metabonomic investigation on the acute biochemical effects induced by $Ce(NO_3)_3$ in rats. J Inorg Biochem, 2005, 99(11): 2151-2160.

[59] Dudka I, Kossowska B, Senhadri H, Latajka R, Hajek J, Andrzejak R, Antonowicz-Juchniewicz J, Gancarz R. Metabonomic analysis of serum of workers occupationally exposed to arsenic, cadmium and lead for biomarker research: a preliminary study. Environ Int 2014, 68: 71-81.

[60] Zhu Y S, Xu H B, Huang K X, Hu W H, Liu M L. A study on human urine in a high-selenium area of China by ^{1}H-NMR spectroscopy. Biol Trace Elem Res, 2002, 89(2): 155-163.

[61] Ritter A, Dittami S M, Goulitquer S, Correa J A, Boyen C, Potin P, Tonon T. Transcriptomic and metabolomic analysis of copper stress acclimation in Ectocarpus siliculosus highlights signaling and tolerance mechanisms in brown algae. Bmc Plant Biol, 2014, 14: 116.

[62] Sun X M, Zhang J X, Zhang H J, Ni Y W, Zhang Q, Chen J P, Guan Y F. The responses of Arabidopsis thaliana to cadmium exposure explored via metabolite profiling. Chemosphere, 2010, 78(7): 840-845.

[63] Zeng J, Kuang H, Hu C X, Shi X Z, Yan M, Xu L G, Wang L B, Xu C L, Xu G W. Effect of bisphenol A on rat metabolic profiling studied by using capillary electrophoresis time-of-flight mass spectrometry. Environ Sci Technol, 2013, 47(13): 7457-7465.

[64] Ji C L, Wu H F, Wei L, Zhao J M, Lu H J, Yu J B. Proteomic and metabolomic analysis of earthworm Eisenia fetida exposed to different concentrations of 2,2 ',4,4 '-tetrabromodiphenyl ether. J Proteomics, 2013, 91: 405-416.

[65] Lenz E M, Bright J, Knight R, Wilson I D, Major H. A metabonomic investigation of the biochemical effects of mercuric chloride in the rat using ^{1}H NMR and HPLC-TOF/MS: time dependant changes in the urinary profile of endogenous metabolites as a result of nephrotoxicity. Analyst, 2004, 129(6): 535-541.

[66] Ishaq R, Naf C, Zebuhr Y, Broman D, Jarnberg U. PCBs, PCNs, PCDD/Fs, PAHs and Cl-PAHs in air and water particulate samples--patterns and variations. Chemosphere, 2003, 50(9): 1131-50.

[67] Laroo C A, Schenk C R, Sanchez L J, McDonald J. Emissions of PCDD/Fs, PCBs, and PAHs from a modern diesel engine equipped with catalyzed emission control systems. Environ Sci Technol, 2011, 45(15): 6420-6428.

[68] Safe S H. Comparative toxicology and mechanism of action of polychlorinated dibenzo-*p*-dioxins and dibenzofurans. Annu. Rev. Pharmacol. Toxicol. 1986, 26: 371-399.

[69] Ahlborg U G, Becking G C, Birnbaum L S, Brouwer A, Derks H J G M, Feeley M, Golor G, Hanberg A, Larsen J C, Liem A K D, Safe S H, Schlatter C, Waern F, Younes M, Yrjanheikki E. Toxic equivalency factors for dioxin-like Pcbs - Report on a Who-Eceh and Ipcs Consultation, December 1993. Chemosphere, 1994, 28(6): 1049-1067.

[70] Van den Berg M, Birnbaum L, Bosveld A T C, Brunstrom B, Cook P, Feeley M, Giesy J P, Hanberg A, Hasegawa R, Kennedy S W, Kubiak T, Larsen J C, van Leeuwen F X R, Liem A K D, Nolt C, Peterson R E, Poellinger L, Safe S, Schrenk D, Tillitt D, Tysklind M, Younes M, Waern F, Zacharewski T. Toxic equivalency factors (TEFs) for PCBs, PCDDs, PCDFs for humans and wildlife. Environ Health Perspect, 1998, 106(12): 775-792.

[71] Haws L C, Su S H, Harris M, DeVito M J, Walker N J, Farland W H, Finley B, Birnbaum L S. Development of a refined database of mammalian relative potency estimates for dioxin-like compounds. Toxicol Sci, 2006, 89(1): 4-30.

[72] Croes K, Vandermarken T, Van Langenhove K, Elskens M, Desmedt M, Roekens E, Denison M S, Van Larebeke N, Baeyens W. Analysis of PCDD/Fs and dioxin-like PCBs in atmospheric deposition samples from the Flemish measurement network: Correlation between the CALUX bioassay and GC-HRMS. Chemosphere, 2012, 88(7): 881-887.

[73] Wilson S C, Jones K C. Bioremediation of soil contaminated with polynuclear aromatic-hydrocarbons (Pahs) - A Review. Environ Pollut, 1993, 81(3): 229-249.

[74] Srogi K. Overview of analytical methodologies for dioxin analysis. Anal Lett, 2007, 40(9): 1647-1671.

[75] Srogi K. Levels and congener distributions of PCDDs, PCDFs and dioxin-like PCBs in environmental and human samples: A review. Environ Chem Lett, 2008, 6(1): 1-28.

[76] Enan E, Lasley B, Stewart D, Overstreet J, Vandevoort C A. 2,3,7,8-tetrachlorodibenzo-p-dioxin (TCDD) modulates function of human luteinizing granulosa cells via cAMP signaling and early reduction of glucose transporting activity. Reprod Toxicol, 1996, 10(3): 191-198.

[77] Kogevinas M. Human health effects of dioxins: Cancer, reproductive and endocrine system effects. Apmis, 2001, 109: S223-S231.

[78] Mitrou P I, Dimitriadis G, Raptis S A. Toxic effects of 2,3,7,8-tetrachlorodibenzo-p-dioxin and related compounds. Eur J Intern Med, 2001, 12(5): 406-411.

[79] Åslund M L W, Simpson A J, Simpson M J. ^{1}H NMR metabolomics of earthworm responses to polychlorinated biphenyl (PCB) exposure in soil. Ecotoxicology, 2011, 20(4): 836-846.

[80] 张保琴，张海军，陈吉平，杨常青. 2,3,7,8-TCDD 的短期暴露对 HepG2 肝癌细胞内小分子代谢产物的影响. 生态毒理学报 2012, 7(3): 292-298.

[81] 张保琴, 张海军, 王龙星, 王金成, 杨常青, 陈吉平. 采用代谢流量分析方法评估二噁英对细胞的代谢干扰. 环境化学, 2012, 33(11): 1797-1802.

[82] Zhang B Q, Zhang H J, Jin J, Ni Y W, Chen J P. PCDD/Fs-induced oxidative damage and antioxidant system responses in tobacco cell suspension cultures. Chemosphere, 2012, 88(7): 798-805.

[83] Bari R, Jones J D. Role of plant hormones in plant defence responses. Plant Mol Bio, 2009, 69(4): 473-488.

[84] Rakwal R, Agrawal G K, Yonekura M. Light-dependent induction of OsPR10 in rice (*Oryza sativa* L.) seedlings by the global stress signaling molecule jasmonic acid and protein phosphatase 2A inhibitors. Plant Sci, 2001, 161(3): 469-479.

[85] Grossmann K, Caspar G, Kwiatkowski J, Bowe S J. On the mechanism of selectivity of the corn herbicide BAS 662H: A combination of the novel auxin transport inhibitor diflufenzopyr and the auxin herbicide dicamba. Pest Manag Sci, 2002, 58(10): 1002-1014.

[86] Matschke J, Machackova I. Changes in the content of indole-3-acetic acid and cytokinins in spruce, fir and oak trees after herbicide treatment. Biol Plantarum, 2002, 45(3): 375-382.

[87] Christmann A, Moes D, Himmelbach A, Yang Y, Tang Y, Grlll E. Integration of abscisic acid signalling into plant responses. Plant Biology, 2006, 8(3): 314-325.

[88] Zhang Y M, Tan J L, Guo Z F, Lu S Y, He S J, Shu W, Zhou B Y. Increased abscisic acid levels in transgenic tobacco over-expressing 9 cis-epoxycarotenoid dioxygenase influence H_2O_2 and NO production and antioxidant defences. Plant Cell Environ, 2009, 32(5): 509-519.

[89] 李秀颖, 宋玉芳, 周启星, 孙铁珩. 高等植物内源激素与土壤荧蒽、苯并(a)芘污染诱导的剂量-效应关系. 环境科学, 2007, 28(6): 1384-1397.

[90] Coen M, Ruepp S U, Lindon J C, Nicholson J K, Pognan F, Lenz E M, Wilson I D. Integrated application of transcriptomics and metabonomics yields new insight into the toxicity due to paracetamol in the mouse. J Pharm Biomed Anal, 2004, 35(1): 93-105.

[91] Toye A A, Dumas M E, Blancher C, Rothwell A R, Fearnside J F, Wilder S P, Bihoreau M T, Cloarec O, Azzouzi I, Young S. Subtle metabolic and liver gene transcriptional changes underlie diet-induced fatty liver susceptibility in insulin-resistant mice. Diabetologia, 2007, 50: 1867-1879.

[92] Jennen D, Ruiz-Aracama A, Magkoufopoulou C, Peijnenburg A, Lommen A, van Delft J, Kleinjans J. Integrating transcriptomics and metabonomics to unravel modes-of-action of 2,3,7,8- tetrachlorodibenzo-*p*-dioxin (TCDD) in HepG2 cells. BMC Systems Biology, 2011, 5(139): 1-13.

[93] 魏俊飞, 吴家强, 焦文娟. 多环芳烃的毒性及其治理技术研究. 污染防治技术, 2008, 21(3): 65-69.

[94] Naspinski C, Gu X, Zhou G D, Mertens-Talcott S U, Donnelly K C, Tian Y. Pregnane X receptor protects HepG2 cells from BaP-induced DNA damage. Toxicol Sci, 2008, 104(1): 67-73.

[95] Wei W, Zhang C, Liu A L, Xie S H, Chen X M, Lu W Q. PCB126 enhanced the genotoxicity of BaP in HepG2 cells by modulating metabolic enzyme and DNA repair activities. Toxicol Lett, 2009, 189(2): 91-95.

[96] 魏复盛. 空气和废气监测分析方法. 北京:中国环境科学出版社, 2003, 625- 642.

[97] Wu J, Ramesh A, Nayyar T, Hood D B. Assessment of metabolites and AhR and CYP1A1 mRNA expression subsequent to prenatal exposure to inhaled benzo(a)pyrene. Int J Dev Neurosci, 2003, 21(6): 333-346.

[98] Jones O A, Spurgeon D J, Svendsen C, Griffin J L. A metabolomics based approach to assessing the toxicity of the polyaromatic hydrocarbon pyrene to the earthworm Lumbricus rubellus. Chemosphere, 2008, 71(3): 601-609.

[99] 高红亮, 从威, 欧阳藩. 体外培养的哺乳动物细胞的葡萄糖和谷氨酰胺代谢. 生物技术通报, 2000, (2): 17-22.

[100] http://chm.pops.int/TheConvention/ThePOPs/TheNewPOPs/tabid/2511/Default.aspx.

[101] Xia D, Gao L R, Zheng M H, Li J G, Zhang L, Wu Y N, Tian Q C, Huang H T, Qiao L. Human exposure to short- and medium-chain chlorinated paraffins via mothers' milk in Chinese Urban Population. Environ Sci Technol, 2017, 51(1): 608-615.

[102] Wei G L, Liang X L, Li D Q, Zhuo M N, Zhang S Y, Huang Q X, Liao Y S, Xie Z Y, Guo T L, Yuan Z J. Occurrence, fate and ecological risk of chlorinated paraffins in Asia: A review. Environ Int, 2016, 92: 373-387.

[103] Cooley H, Fisk A, Wiens S, Tomy G, Evans R, Muir D. Examination of the behavior and liver and thyroid histology of juvenile rainbow trout (*Oncorhynchus mykiss*) exposed to high dietary concentrations of C 10-, C 11-, C 12-and C 14-polychlorinated n-alkanes. Aquati toxicol, 2001, 54(1): 81-99.

[104] Filyk G, Lander L, Eggleton M. Short chain chlorinated paraffins (SCCP) substance dossier (final draft II). Environment Canada. 2002.

[105] Wyatt I, Coutss C T, Elcombe C R. The effect of chlorinated paraffins on hepatic enzymes and thyroid hormones. Toxicology, 1993, 77(1–2): 81-90.

[106] Meijer J, Rundgren M, Åström A, DePierre J, Sundvall A, Rannug U. Effects of chlorinated paraffins on some drug-metabolizing enzymes in rat liver and in the Ames test. In Biological Reactive Intermediates II, Springer: 1982, 821-828.

[107] Bureau E C. European Union Risk Assessment Report, Alkanes, C10-13,chloro. 1999.

[108] Warnasuriya G D, Elcombe B M, Foster J R, Elcombe C R. A Mechanism for the induction of renal tumours in male Fischer 344 rats by short-chain chlorinated paraffins. Arch Toxicol, 2010, 84(3): 233-243.

[109] Vamecq J, Cherkaoui-Malki M, Andreoletti P, Latruffe N. The human peroxisome in health and disease: The story of an oddity becoming a vital organelle. Biochimie, 2014, 98C: 4-15.

[110] Darnerud P O. Chlorinated Paraffins：Effect of Some Microsomal-Enzyme Inducers and Inhibitors on the Degradation of 1-^{14}C-Chlorododecanes to $^{14}CO_2$ in Mice. Acta Pharmacol Tox, 1984, 55(2): 110-115.

[111] Brunström B. Effects of chlorinated paraffins on liver weight, cytochrome P-450 concentration and microsomal enzyme activities in chick embryos. Arch Toxicol, 1985, 57(1): 69-71.

[112] Poulsen L l C, Siersbæk M, Mandrup S. PPARs: Fatty acid sensors controlling metabolism. Semin Cell Dev Biol,2012, 23(6): 631-639.

[113] Darnerud P, Biessmann A, Brandt I. Metabolic fate of chlorinated paraffins: Degree of chlorination of [1–^{14}C]-chlorododecanes in relation to degradation and excretion in mice. Arch Toxicol, 1982, 50(3-4): 217-226.

[114] Wyatt I, Coutss C, Elcombe C. The effect of chlorinated paraffins on hepatic enzymes and thyroid hormones. Toxicology, 1993, 77(1): 81-90.

[115] Bucher J R, Alison R H, Montgomery C A, Huff J, Haseman J K, Farnell D, Thompson R, Prejean J D. Comparative toxicity and carcinogenicity of two chlorinated paraffins in F344/N rats and B6C3F1 mice. Fundam Appl Toxicol, 1987, 9(3): 454-468.

[116] Xia W, Wan Y J, Li Y Y, Zeng H C, Lv Z Q, Li G Q, Wei Z Z, Xu S Q. PFOS prenatal exposure induce mitochondrial injury and gene expression change in hearts of weaned SD rats. Toxicology, 2011, 282(1-2): 23-29.

[117] Alaee M, Arias P, Sjodin A, Bergman A. An overview of commercially used brominated flame retardants, their applications, their use patterns in different countries/regions and possible modes of release. Environ Int, 2003, 29(6): 683-689.

[118] Birnbaum L S, Staskal D F. Brominated flame retardants: cause for concern? Environ Health Perspect, 2004, 112(1): 9-17.

[119] Thomsen C, Molander P, Daae H L, Janak K, Froshaug M, Liane V H, Thorud S, Becher G, Dybing E. Occupational exposure to hexabromocyclododecane at an industrial plant. Environ Sci Technol, 2007, 41(15): 5210-5216.

[120] Kakimoto K, Akutsu K, Konishi Y, Tanaka Y. Time trend of hexabromocyclododecane in the breast milk of Japanese women. Chemosphere, 2008, 71(6): 1110-1124.

[121] Marvin C H, Tomy G T, Armitage J M, Arnot J A, McCarty L, Covaci A, Palace V. Hexabromocyclododecane: current understanding of chemistry, environmental fate and toxicology and implications for global management. Environ Sci Technol, 2011, 45(20): 8613-8623.

[122] Canton R F, Peijnenburg A A, Hoogenboom R L, Piersma A H, van der Ven L T, van den Berg M, Heneweer M. Subacute effects of hexabromocyclododecane (HBCD) on hepatic gene expression profiles in rats. Toxicol Appl Pharmacol,2008, 231(2): 267-272.

[123] Zhang H, Kuo Y Y, Gerecke A C, Wang J. Co-release of hexabromocyclododecane (HBCD) and Nano- and microparticles from thermal cutting of polystyrene foams. Environ Sci Technol, 2012, 46(20): 10990-10996.

[124] Ronisz D, Finne E F, Karlsson H, Forlin L. Effects of the brominated flame retardants hexabromocyclododecane (HBCDD): and tetrabromobisphenol A (TBBPA): on hepatic enzymes and other biomarkers in juvenile rainbow trout and feral eelpout. Aquat Toxicol, 2004, 69(3): 229-245.

[125] Schriks M, Zvinavashe E, Furlow J D, Murk A J. Disruption of thyroid hormone-mediated Xenopus laevis tadpole tail tip regression by hexabromocyclododecane (HBCD) and 2,2',3,3',4,4',5,5',6-nona brominated diphenyl ether (BDE206). Chemosphere, 2006, 65(10): 1904-1918.

[126] van der Ven L T, Verhoef A, van de Kuil T, Slob W, Leonards P E, Visser T J, Hamers T, Herlin M, Hakansson H, Olausson H, Piersma A H, Vos J G. A 28-day oral dose toxicity study enhanced to detect endocrine effects of hexabromocyclododecane in Wistar rats. Toxicol Sci, 2006, 94(2): 281-292.

[127] Reistad T, Fonnum F, Mariussen E. Neurotoxicity of the pentabrominated diphenyl ether mixture, DE-71, and hexabromocyclododecane (HBCD) in rat cerebellar granule cells *in vitro*. Arch Toxicol, 2006, 80(11): 785-796.

[128] Saegusa Y, Fujimoto H, Woo G. H, Inoue K, Takahashi M, Mitsumori K, Hirose M, Nishikawa A, Shibutani M. Developmental toxicity of brominated flame retardants, tetrabromobisphenol A and 1,2,5,6,9,10-hexabromocyclododecane, in rat offspring after maternal exposure from mid-gestation through lactation. Reprod Toxicol, 2009, 28(4): 456-467.

[129] Marteinson S C, Bird D M, Letcher R J, Sullivan K M, Ritchie I J, Fernie K J. Dietary exposure to technical hexabromocyclododecane (HBCD) alters courtship, incubation and parental behaviors in American kestrels (*Falco sparverius*). Chemosphere. 2012, 89(9): 1077-1083.

[130] Yanagisawa R, Koike E, Win-Shwe T T, Yamamoto M, Takano H. Impaired lipid and glucose homeostasis in hexabromocyclododecane-exposed mice fed a high-fat diet. Environ Health Perspect, 2014, 122(3): 277-283.

[131] Germer S, Piersma A H, van der Ven L, Kamyschnikow A, Fery Y, Schmitz H J, Schrenk D. Subacute effects of the brominated flame retardants hexabromocyclododecane and tetrabromobisphenol A on hepatic cytochrome P450 levels in rats. Toxicology, 2006, 218(2-3): 229-236.

[132] van der Ven L T, van de Kuil T, Leonards P E, Slob W, Lilienthal H, Litens S, Herlin M, Hakansson H, Canton R F, van den Berg M, Visser T J, van Loveren H, Vos J G, Piersma A H. Endocrine effects of hexabromocyclododecane (HBCD) in a one-generation reproduction study in Wistar rats. Toxicol Lett, 2009, 185(1): 51-62.

[133] Zhang X, Yang F X, Zhang X L, Xu Y, Liao T, Song S B, Wang J W. Induction of hepatic enzymes and oxidative stress in Chinese rare minnow (*Gobiocypris rarus*) exposed to waterborne hexabromocyclododecane (HBCDD). Aquat Toxicol, 2008, 86(1): 4-11.

[134] Zou W, Chen C, Zhong Y, An J, Zhang X, Yu Y, Yu Z, Fu J. PI3K/Akt pathway mediates Nrf2/ARE activation in human L02 hepatocytes exposed to low-concentration HBCDs. Environ Sci Technol, 2013, 47(21): 12434-12440.

[135] An J, Wang X, Guo P, Zhong Y, Zhang X, Yu Z. Hexabromocyclododecane and polychlorinated biphenyls increase resistance of hepatocellular carcinoma cells to cisplatin through the phosphatidylinositol 3-kinase/protein kinase B pathway. Toxicol Lett, 2014, 229(1): 265-272.

[136] Al-Mousa F, Michelangeli F. Some commonly used brominated flame retardants cause Ca^{2+}-ATPase inhibition, beta-amyloid peptide release and apoptosis in SH-SY5Y neuronal cells. PloS one, 2012, 7(4): e33059.

[137] Zhang J K, Abou-Elwafa Abdallah M, Williams T D, Harrad S, Chipman J K, Viant M R. Gene expression and metabolic responses of HepG2/C3A cells exposed to flame retardants and dust extracts at concentrations relevant to indoor environmental exposures. Chemosphere, 2016, 144: 1996-2003.

[138] Wang F D, Zhang H J, Geng N B, Zhang B Q, Ren X Q, Chen J P. New insights into the cytotoxic mechanism of

hexabromocyclododecane from a metabolomic approach. Environ Sci Technol, 2016, 50(6): 3145-3153.

[139] Jin J, Yang C Q, Wang Y, Liu A M. Determination of hexabromocyclododecane diastereomers in soil by ultra performance liquid chromatography-electrospray ion source/tandem mass spectrometry. Chinese J Anal Chem, 2009, 37(4): 585-588.

[140] Rajapakse N, Silva E, Kortenkamp A. Combining xenoestrogens at levels below individual No-observed-effect concentrations dramatically enhances steroid hormone action. Environmental Health Perspectives, 2002, 110(9): 917-921.

[141] Panizzi S, Suciu N A, Trevisan M. Combined ecotoxicological risk assessment in the frame of European authorization of pesticides. Sci Total Environ, 2017, 580: 136-146.

[142] Beyer J, Petersen K, Song Y, Ruus A, Grung M, Bakke T, Tollefsen K. E. Environmental risk assessment of combined effects in aquatic ecotoxicology: a discussion paper. Mar Environ Res, 2014, 96: 81-91.

[143] Sexton K. Cumulative risk assessment: an overview of methodological approaches for evaluating combined health effects from exposure to multiple environmental stressors. Int J Environ Res Public Health, 2012, 9(2): 370-390.

[144] Escher B I, Hackermuller J, Polte T, Scholz S, Aigner A, Altenburger R, Bohme A, Bopp S K, Brack W, Busch W, Chadeau-Hyam M, Covaci A, Eisentrager A, Galligan J J, Garcia-Reyero N, Hartung T, Hein M, Herberth G, Jahnke A, Kleinjans J, Kluver N, Krauss M, Lamoree M, Lehmann I, Luckenbach T, Miller G W, Muller A, Phillips D H, Reemtsma T, Rolle-Kampczyk U, Schuurmann G, Schwikowski B, Tan Y M, Trump S, Walter-Rohde S, Wambaugh J F. From the exposome to mechanistic understanding of chemical-induced adverse effects. Environ Int, 2017, 99: 97-106.

[145] O'Kane A A, Chevallier O P, Graham S F, Elliott C T, Mooney M H. Metabolomic profiling of in vivo plasma responses to dioxin-associated dietary contaminant exposure in rats: Implications for identification of sources of animal and human exposure. Environ Sci Technol, 2013, 47(10): 5409-5418.

[146] 陈蓉, 王以美, 汪江山, 卢春凤, 张凤霞, 胡春秀, 彭双清, 许国旺. 液质联用代谢组学研究多氯联苯和二噁英对大鼠毒性作用. 环境化学, 2013, 32(7): 1226-1235.

[147] Huang Q Y, Zhang J, Luo L Z, Wang X F, Wang X X, Alamdar A, Peng S Y, Liu L P, Tian M P, Shen H Q. Metabolomics reveals disturbed metabolic pathways in human lung epithelial cells exposed to airborne fine particulate matter. Toxicology Research, 2015, 4(4): 939-947.

[148] Chen W L, Lin C Y, Yan Y H, Cheng K T, Cheng T J. Alterations in rat pulmonary phosphatidylcholines after chronic exposure to ambient fine particulate matter. Mol Biosyst, 2014, 10(12): 3163-3169.

[149] Yu Y, Wang X, Wang B, Tao S, Liu W, Wang X, Cao J, Li B, Lu X, Wong M H. Polycyclic aromatic hydrocarbon residues in human milk, placenta, and umbilical cord blood in Beijing, China. Environ Sci Technol, 2011, 45(23): 10235-10242.

[150] Li T, Wan Y, Gao S, Wang B, Hu J. High-throughput determination and characterization of short-, medium-, and long-chain chlorinated paraffins in human blood. Environ Sci Technol, 2017，51(6): 3346-3354

[151] Santonicola S, De Felice A, Cobellis L, Passariello N, Peluso A, Murru N, Ferrante M C, Mercogliano R. Comparative study on the occurrence of polycyclic aromatic hydrocarbons in breast milk and infant formula and risk assessment. Chemosphere, 2017, 175: 383-390.

[152] Billiard S M, Hahn M E, Franks D G, Peterson R E, Bols N C, Hodson P V. Binding of polycyclic aromatic hydrocarbons (PAHs) to teleost aryl hydrocarbon receptors (AHRs). Comp Biochem Physiol B Biochem Mol Biol, 2002, 133(1): 55-68.

[153] Barron M G, Heintz R, Rice S D. Relative potency of PAHs and heterocycles as aryl hydrocarbon receptor agonists in fish. Mar Environ Res, 2004, 58(2-5): 95-100.

[154] Shimada T, Fujii-Kuriyama Y. Metabolic activation of polycyclic aromatic hydrocarbons to carcinogens by cytochromes

P450 1A1 and 1B1. Cancer Sci. 2004, 95(1): 1-6.

[155] Kummer V, Maskova J, Zraly Z, Faldyna M. Ovarian disorders in immature rats after postnatal exposure to environmental polycyclic aromatic hydrocarbons. J Appl Toxicol, 2013, 33(2): 90-99.

[156] Wyatt I, Coutts C T, Elcombe C R. The effect of chlorinated paraffins on hepatic enzymes and thyroid hormones. Toxicology, 1993, 77(1-2): 81-90.

[157] Warnasuriya G D, Elcombe B M, Foster J R, Elcombe C R. A Mechanism for the induction of renal tumours in male Fischer 344 rats by short-chain chlorinated paraffins. Arch Toxicol, 2010, 84(3): 233-243.

[158] Zhang Q, Wang J H, Zhu J Q, Liu J, Zhang J Y, Zhao M R. Assessment of the endocrine-disrupting effects of short-chain chlorinated paraffins in in vitro models. Environ Int, 2016, 94: 43-50.

作者：张保琴[1]，蔡宗苇[2]，耿柠波[1]，王菲迪[1]，张海军[1]，陈吉平[1]

[1]中国科学院大连化学物理研究所，[2]香港浸会大学

第28章　POPs对糖脂代谢相关疾病发生的影响

本章导读

伴随着全球经济的高速增长，人类与自然的矛盾更加激化。生态破坏和环境污染已经成为严重的区域性和全球性环境问题。在众多环境问题中，持久性有机污染物（persistent organic pollutions，POPs）是有机污染物中最受关注和最为重要的一类。POPs 在环境中普遍存在，滞留时间长，并可发生长距离的迁移，由于亲脂性，它们易在生物体内累积，并随食物链迁移和放大，最终对处于食物链最高端的人类造成潜在的危害。目前，有关 POPs 毒性效应的研究主要集中在高剂量暴露的毒性评价上，在低剂量暴露的毒性研究上相对较少，但由于人群的实际暴露具有低剂量特征，因此研究 POPs 低剂量暴露的毒性效应并探讨其潜在的健康风险具有重要的意义，其研究成果不仅为污染控制与治理提供理论依据，同时也为系统的分析和规划环境奠定基础。由于机体代谢物的变化可灵敏地指示和确证外来干扰在组织和器官水平的毒性效应以及毒性作用靶位点，因此借助代谢组学技术评价环境污染物暴露带来的毒性效应，进而推断其毒性作用的分子机制具有快速、灵敏度高、选择性强等特点，特别是在低剂量或环境剂量污染物的毒性效应评估方面具有很大的优势。本研究将对近年来典型 POPs 低剂量暴露毒性效应的代谢组学研究进行综合的介绍。

关键词

持久性有机污染物，代谢组学，低剂量暴露，毒性效应

1 引　　言

糖、脂代谢是生物代谢过程的重要组成部分，对于维持正常生命活动必不可少，同时，糖、脂代谢功能紊乱与多种疾病的发生密切相关。随着人类生活水平的提高以及生活方式的改变，出现了一系列与糖脂异常代谢相关的疾病，其病理表现统称为代谢综合征征，其中肥胖、糖尿病、血脂异常、胰岛素抵抗等的发病率呈现越来越高的趋势[1,2]。图 28-1 展现了我国在 1992~2008 年间，肥胖、糖尿病及代谢综合征的发病情况。

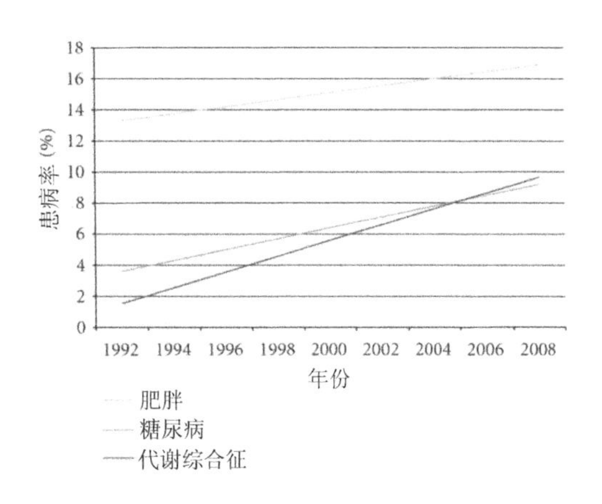

图 28-1　1992~2008 年间，肥胖、糖尿病及代谢综合征在中国的发病率[2]

以往的观点认为代谢性疾病的发生与年龄、生活习惯、遗传因素等相关，但是越来越多的研究结果

表明：只限定于这些因素并不能圆满解释代谢性疾病已出现的高发率。近年来，流行病学调查及相关基础研究结果均提示，环境污染物的暴露，特别是持久性有机污染物（POPs）的暴露，是代谢性疾病最重要的诱因之一[3,4]。本章以几类典型 POPs 为对象，综述了它们与肥胖、糖尿病、心血管疾病及其他糖脂代谢异常相关疾病发生等方面的最新研究成果。

2　POPs 对肥胖发生的影响

据世界卫生组织（WHO）统计，全球超重和肥胖人数已从 1980 年的 857 万爆炸式增长至 2014 年的 21 亿，至少 1/3 的人面临肥胖症风险。流行病学调查表明：内脏脂肪组织的过度积累和脂肪代谢异常可以引发胰岛素抵抗、2 型糖尿病、高血压、动脉粥样硬化等代谢性疾病[5]，成为引发慢性疾病的主要风险因素，促使全球对肥胖相关公共健康问题高度关注[6]。

肥胖是脂肪组织过度累积造成的，与脂肪细胞数目的增多和体积的增大密切相关，因此脂肪细胞的分化成熟和相应的脂质代谢是研究肥胖发生的两个主要切入点。研究结果显示，POPs 类环境污染物通过呼吸、摄食等方式被机体吸收后储存于脂肪组织，是影响（甚至是诱发）肥胖发生发展的关键因素之一[7-9]。图 28-2 所示是肥胖人群内脏脂肪组织和皮下脂肪组织中 POPs 的含量分布。

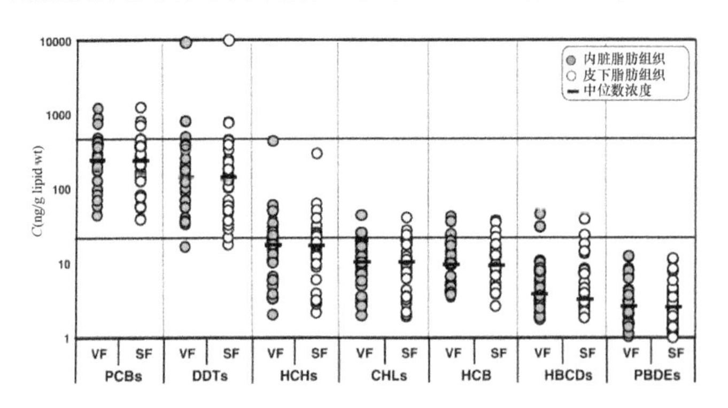

图 28-2　POPs 在内脏脂肪组织及皮下脂肪组织中的分布[10]

2.1　多氯联苯（PCBs）与肥胖

流行病学调查发现，PCBs 暴露与肥胖发生之间具有显著相关性。比如，对西班牙南部 298 名成年人脂肪组织中 PCBs 含量与血清脂质含量及 BMI（body mass index）指数相关性的调查发现，脂肪组织中 PCBs 含量与 BMI 呈正相关[11]。对 12313 名体重正常，无肥胖史的参与者进行了长达 14 年的随访，以调查饮食摄入的 PCBs 与肥胖发生的相关性，发现其中有 621 位参与者出现了肥胖症现象，而这些参与者日常膳食摄入的 PCBs 含量显著高于其他 11692 名参与者，提示 PCBs 膳食暴露可能促进或诱发肥胖的发生[12]。对比利时某地区 138 名 1~3 岁儿童 BMI 标准偏差分数（SDS）与围产期 PCBs、DDT 的暴露量进行数据分析，发现随着围产期 PCBs 暴露浓度的增加，婴儿从 1 岁到 3 岁的成长期间内，BMI SDS 值均增大，说明幼儿时期 BMI 与母体子宫内 PCBs 暴露有显著关联[13]。

除了流行病学调查，最近的基础研究也显示 PCBs 通过调节脂肪细胞分化及功能来影响肥胖的发生，且二噁英类 PCBs 和非二噁英类 PCBs 对脂肪细胞分化的作用存在差异。某些二噁英类 PCBs 通过上调 3T3-L1 脂肪细胞以及小鼠体内脂滴结合蛋白 Fsp27 水平促进脂肪细胞中脂滴体积增大，促进脂肪细胞生成，而另一些二噁英类 PCBs（如 PCB 126）可以通过激活 AhR，抑制脂肪细胞分化过程中的关键因子 PPARγ 的转录，破坏前脂肪细胞分化成熟，抑制鼠源 3T3-L1 和人源前脂肪细胞（NPAD）的脂肪生成[14]。而非二噁英类 PCBs，如 PCB 101、153、180 均可通过降低成熟 3T3-L1 脂肪细胞中 pSTAT3/STAT3 的比例，提高 PTP1B 和 SOCS3 水平，促使瘦素过表达，降低瘦素受体敏感性，诱发瘦素抵抗的发生，引起脂质累积增多[15]。

2.2 全氟烷基化合物（PFAAs）与肥胖

由于含有高能 C—F 键，PFAAs 可在环境介质中稳定持久迁移，并具有生物放大、生物蓄积效应。其中，全氟辛酸（PFOA）于 2006 年被美国环境保护署列为“可能致癌的物质”；全氟辛烷磺酸（PFOS）及其盐，以及全氟辛基磺酰氟（perfluorootanesulforyl fluoride）于 2009 年被列入《关于持久性有机污染物的斯德哥尔摩公约》，PFAAs 的毒理学效应备受关注。

Geary 等对 506 名职业暴露于 PFOA 的人群进行流行病学调查，发现 PFOA 总含量与体脂含量存在正相关[16]，进一步的基础研究也表明 PFAAs 可以影响脂肪细胞分化，干扰脂质代谢。Xu 等发现 PFOS 暴露通过激活小鼠附睾白色脂肪组织中 Nrf2 信号通路促进脂肪细胞分化及其相关基因表达。进一步对人源前脂肪细胞进行 PFOS 暴露（5 μmol/L 和 50 μmol/L），结果表明 PFOS 能够促进 Nrf2 与 ARE 位点的结合，激活 Nrf2 信号通路，促进调控脂肪细胞分化的转录因子 PPARγ、CEBPα 基因的表达，增加细胞内甘油三酯累积量，促进脂肪细胞分化[17]。Watkins 等选用四种 PFAAs 对小鼠 3T3-L1 前脂肪细胞进行一定浓度的暴露（PFOA 5~100 μmol/L、PFNA 5~100 μmol/L、PFOS 50~300 μmol/L、PFHxS 40~250 μmol/L），结果表明 PFOA、PFNA、PFOS、PFHxS 暴露均可促进脂肪细胞数目的增多，上调脂质代谢与脂质存储相关基因 *Acox1*、*Gapdh*、*Scd1* 等的表达，提示 PFAAs 具有促进前脂肪细胞增殖，促进脂质累积的能力[18]。

2.3 多溴联苯醚（PBDEs）与肥胖

多溴联苯醚由于其良好的热稳定性而作为防火材料得到广泛应用，但其可在环境介质中长期稳定存在，进入食物链会产生生物放大效应，PBDEs 的健康危害目前也越来越受重视。

通过对 970 名 70 岁老年人进行的 5 年跟踪研究，发现 PBDE 47 的暴露与人体腹部肥胖呈正相关[19]。对孕期暴露于 PBDEs 的数百名孕妇及其产下的婴幼儿进行调查发现，孕期孕妇血清中 PBDEs 浓度与下一代女孩的 BMI 指数呈负相关，与下一代男孩的 BMI 指数呈正相关，特别地，孕妇 PBDE 153 的血清浓度与下一代男女孩的 BMI 指数均呈负相关[20,21]。

PBDEs 与脂肪细胞分化存在正相关关系。将小鼠 3T3-L1 前脂肪细胞分别暴露于 PBDEs 混合物（0~25.5 μmol/L）、PBDE47（0~12 μmol/L）、PBDE71（0~12 μmol/L），发现它们能够上调基因 *C/EBPa*、*PPARγ*，促进脂肪细胞分化的标志基因 *aP2* 和 *perilipin* 表达，同时能增强地塞米松的促脂肪细胞分化的作用，表明 PBDEs 可以增强糖皮质激素介导的脂肪细胞分化[22]。在对 PBDE 47 暴露较为深入的研究中发现，PBDE 47 通过与 PPARγ的结合增强 PPARγ的转录活性，同时能够降低 PPARγ启动子区域的甲基化水平而上调 PPARγ的表达，起到促进脂肪细胞分化的作用，提示其可能具有诱发肥胖的效应[23]。

2.4 杀虫剂与肥胖

随着杀虫剂的广泛使用，其在糖脂代谢相关疾病中的作用也受到广泛关注。在美国的一项围产期调研发现，孕妇在围产期暴露于高浓度的 DDE、DDT 会导致新生男婴的 BMI 升高，出现超重甚至肥胖的情况[24]。与之相对应的基础研究表明，DDT 暴露会使人源间充质干细胞的自我更新能力下降，脂肪细胞分化能力增强，并且 DDT 对脂肪细胞分化的促进作用与雌激素受体（ERα）介导的 PPARγ 及脂蛋白脂肪酶（*LpL*）等基因的表达上调相关[25]。Mangum 等发现 DDE 能够促进脂肪细胞分化相关基因 *SREBP1c*、*aP2* 的表达，同时上调脂质存储相关基因 *leptin*、*Fasn* 的表达，促进 3T3-L1 前脂肪细胞分化，促进胞内脂质累积[26]。

2.5 二噁英与肥胖

在瑞典开展的一项针对持久性有机污染物与老年人腹部肥胖的关系的流行病学调查发现，不论是用“横断面调查”还是“前瞻性调查”的研究方法，结果均表明二噁英暴露与腹部肥胖的发生、发展呈正相关关系 [27]。小鼠实验中，孕期及哺乳期暴露于 TCDD 的母鼠，其后代小鼠的体重会发生变化，并且这种变化具有性别差异性，即雄性小鼠的脂肪垫、脾脏重量降低；雌性小鼠脂肪垫重量增加[28]。机理研究揭示表明，TCDD（5×10^{-8} mol/L）能够通过激活 AhR 信号通路抑制小鼠 3T3-L1 前脂肪细胞的分化，同时降低 C/EBPα、PPARγ2 等分化相关转录因子的表达量[29]，也可协同 MEK/ERK 激活 AhR 信号通路来间接抑制 PPARγ1 活性[30]。

还有一种“肥胖因子假说”，认为 POPs 可以通过影响脂因子瘦素分泌，提高食欲从而促进肥胖的发生[31]，目前该假说尚缺乏有力的实验证据。

3 POPs 对糖尿病发生的影响

糖尿病是一种以高血糖为主要特征的代谢性疾病，是 21 世纪最重要的公共健康问题之一[32]。世界卫生组织（WHO）2013 年统计报告显示，全世界现有大约 3.8 亿人患有糖尿病，预计 2035 年患病人数将增至近 6 亿[33]。糖尿病一般认为有三种：1 型糖尿病（Type 1 diabetes，T1D）、2 型糖尿病（Type 2 diabetes，T2D）和妊娠期糖尿病（GDM）。1 型糖尿病，也称为自身免疫性糖尿病，其特点是血糖浓度过高，1 型糖尿病的发病机制尚未完全阐明，但是一般认为其发生与 T 细胞参与调节的β细胞受损相关[34]；2 型糖尿病，其特点是碳水化合物、脂质及蛋白质代谢失调，2 型糖尿病是三种糖尿病中发病率最高的一类，大约占所有糖尿病的 90%以上，其发生受基因与环境因素的影响[35]；妊娠期糖尿病，是指妊娠期发生的糖代谢异常，妊娠期糖尿病的发病率为 2%~5%，主要致病因素有肥胖、2 型糖尿病的家族史等[36,37]。

糖尿病的主要发病原因包括生活方式、遗传因素、环境因素等[38-40]，但最新的研究也提示糖尿病患病率的升高可能与人体 POPs 的暴露相关。

3.1 多氯联苯（PCBs）与糖尿病

近年来持续的流行病学调查表明：人体 PCBs 暴露水平与糖尿病的发生呈正相关关系[41-46]。Grice 等针对 POPs 与糖尿病的关系于 1965~2007 年期间长期追踪调查了生活在 Gila 河流的 300 名美洲印第安人，结果显示 POPs 暴露会增加胰岛素抵抗及患 2 型糖尿病（T2D）的风险，其中 PCB 151 与糖尿病的患病率有较强的正相关性(OR=1.38，95% CI：1.02~1.88)[47]。在另外一项针对美国本土人的研究中，研究者分别用三个模型对 PCBs 与有机氯杀虫剂的暴露量与糖尿病发生率的关系进行研究，在模型一中，仅用年龄、性别、BMI、总血清脂质含量进行校正，结果发现不论是总的 PCBs 还是总的杀虫剂与糖尿病的发生均有较强的相关性，得到的危险比分别为 2.21（1.2~4.2）和 3.75（1.3~10.7）；在模型二中，除用上述变量进行校正外，研究者还将杀虫剂和 PCBs 的影响考虑进去，也就是说在评估 PCBs 与糖尿病发生的关系时，研究者会用杀虫剂的含量进行校正，反之亦然，结果表明危险比会有所下降。在模型三中，研究者根据 PCBs 中氯原子的数目及邻位取代氯原子的数目将 PCBs 分为三组，每一组 PCBs 和杀虫剂均进行单独分析，结果表明仅含有三氯或者四氯的 PCBs 与糖尿病的发生呈显著正相关，但高氯代 PCBs 与糖尿病发生没有显著相关性。当仅根据氯原子数目及邻位氯原子评价 PCBs 与糖尿病的关系时，发现只有非邻位取代与单邻位取代 PCBs 与糖尿病的发生呈显著性正相关。部分单邻位取代 PCBs 具有类二噁英性质，因此研究者把毒性当量考虑进去后发现仅是不具备二噁英性质的 PCBs 与糖尿病的发生具有相关性。综合三个模型，研究者认为，低氯代、非二噁英类 PCBs 与糖尿病的发生呈显著性正相关[48]。除取代的氯原子数目及位置会影响 PCBs 对糖尿病的发生外，人体暴露 PCBs 的时间可能同样会影响二者的关系[49]。

PCBs 的暴露也会引起妊娠期糖尿病，患者血液中的 POPs 浓度高于正常妇女。经过产前 BMI 及其他因素的调整，结果发现中等及较高三分位 PCBs 暴露组的妇女患 GDM 的概率分别是低三分位 PCBs 暴露组妇女的 3.90（95% CI：1.37~11.06）倍和 3.60（95% CI：1.14~11.39）倍；而中等及较高三分位二噁英类 PCBs 暴露组的妇女患 GDM 的概率分别是低三分位二噁英类 PCBs 暴露组妇女的 5.63（95% CI：1.81~17.51）倍和 4.71（95% CI：1.38~16.01）倍，与此对应的非二噁英类 PCBs 则分别是 2.36（95% CI：0.89~6.23）倍和 2.26（95% CI：0.77~6.68）倍[50]。

Baker 等利用 C57BL/6 小鼠模型研究了 PCB 77、PCB 126 对葡萄糖稳态的影响。研究表明低脂喂食的小鼠在进行 PCB 77 或 PCB 126 处理后，葡萄糖和胰岛素耐受性均受到损害。在饮食诱导的肥胖小鼠中，PCB 77 不影响葡萄糖稳态，但是当小鼠体重下降时，葡萄糖的稳态受到破坏。小鼠 PCB 77 处理组的脂肪组织中 TNF-α的表达量有所上升，加入 AhR 拮抗剂α-NF 后，PCB 77 不能上调 TNF-α的表达，说明 PCB 77 诱导的葡萄糖稳态失衡依赖于 AhR 受体[51]。低脂饲料喂养的小鼠（$AhR^{fl/fl}$）进行 PCB 77 暴露后会引起糖和胰岛素耐量受损，但同样条件下对 AhR 受体缺失的小鼠（AhR^{AdQ}）并没有类似的影响。不论是低脂喂食还是高脂喂食，AhR^{AdQ} 小鼠均表现出脂肪含量增加。高脂喂食中，两个基因型均表现出超重，但是 AhR^{AdQ} 小鼠的对照组相比于对应的 $AhR^{fl/fl}$ 组体重、脂肪组织、脂肪组织的炎症反应均有所增加，葡萄糖的耐受性有所受损。两个基因型的肥胖小鼠暴露 PCB 77 后葡萄糖稳态未受影响，但是当体重减轻时，相比于对照，PCB77 暴露的 $AhR^{fl/fl}$ 小鼠脂肪组织中的 TNF-α mRNA 增加，葡萄糖的稳态也有所受损，但同样是在体重下降时，PCB 77 对 AhR^{AdQ} 小鼠的 TNF-α mRNA 及葡萄糖的稳态没有影响，提示 AhR 介导了 PCB 77 诱发的脂肪组织的炎症以及对葡萄糖稳态的破坏[52]。在一项基于雄性 C57BL/6 小鼠的研究中，PCB 118、PCB 153 暴露会影响一些与脂质代谢、胰岛素抵抗等相关基因的表达，如脂肪组织中 *Lipin1*、*Glut4*、*Agpat2*、Slc25a1 及 *Fasn* 的表达有所下降，肝脏组织中的 *Lipin1*、*Glut4* 也有所降低，肌肉组织中的 *CnR1* 及 *Foxo3* 的表达也受到一定影响。这些结论提示：脂肪组织中 *LiIpin1* 及 *Glut4* 可能是 PCBs 诱导代谢失调的主要靶标基因[53]。但是一项针对 771 位三年级健康学生的横断面调查表明，血液中的胰岛素浓度与 PCBs 暴露呈反比，空腹血糖值能维持在预期的范围内，提示 PCBs 可能并非是通过影响胰岛素敏感性发挥作用，而是通过作用于胰岛β细胞影响胰岛素合成[54]。

3.2　杀虫剂与糖尿病

杀虫剂暴露也是增加患 2 型糖尿病风险的环境因素之一。Al-Othman 等对来自沙特阿拉伯的成年人群进行调查，经过对 280 个血液样本的分析，得出β-六氯环己烷（β-HCH）与γ-六氯环己烷（γ-HCH）的含量与糖尿病的发生有正相关性，HCH 的含量与甘油三酯含量、空腹血糖值、血压及胰岛素抵抗系数呈正比，但是与高密度脂蛋白胆固醇呈反比[55]。在另外一项长达 23 年的长期追踪调查中，Suarez-Lopez 等发现暴露于有机氯杀虫剂会在几十年后影响个体的葡萄糖稳态[56]。调查发现，血液中 DDT 及其衍生物 DDE 含量与糖尿病的发生呈现显著的相关关系，与甘油三酯含量、空腹血糖值、血压及胰岛素抵抗系数呈现正相关，而与高密度脂蛋白呈现出显著的负相关[57]。

一项基于 C57BL/6J 小鼠的研究表明，DDT 对代谢的影响不仅体现在暴露个体上，还体现在对后代代谢的影响上。当对妊娠期及产后的小鼠进行 DDT 暴露时，后代雌性小鼠的体温、对冷的耐受性、能量消耗等均会受到影响。在其成年以后，若进行高脂喂食 12 周，妊娠期暴露 DDT 的雌性鼠会出现低葡萄糖耐受、血胰岛素升高、血脂异常、胆汁酸代谢紊乱等现象。这些结果提示：妊娠期 DDT 暴露会影响后代成年雌鼠的能量代谢以及糖脂代谢，从而增加其罹患代谢性疾病的风险[58]。在另外一项活体研究中，为了探索 DDE 与肥胖及 2 型糖尿病的关系，对 7 周龄的雄性 C57BL/6H 小鼠以口服灌胃的方式进行 DDE 暴露后，将小鼠随机分组进行低脂或者高脂喂食，结果发现，在开始分组喂食后的第四周和第八周，与低脂喂食对照组相比，高脂 DDE 暴露组的血糖值有显著性增高，但是到了第 13 周，与高脂喂食的对照组相比，高脂 DDE 暴露组的空腹血糖值却有显著性下降（下降 12.4%）。同时，DDE 的长期暴露可以减轻高脂喂食引发的葡萄糖不耐受，血胰岛素增多以及胰岛素抵抗。高脂喂食会降低 *Glut4* 在骨骼肌、脂肪组织中的表达，但是 DDE 暴露反而会上调 *Glut4* 在骨骼肌中的表达，起到一定的补偿作用，并且 DDE 暴露还能上调糖异生相关基因 *Pepck* 和 *G6pase* 的表达，这些结果表明 DDE 有可能通过影响肝脏中的糖异生过程及葡萄糖摄取来调节血糖值。DDE 暴露不仅能够影响糖代谢的过程，还能够影响脂代谢，高脂喂食下，DDE 暴露会显著降低肝脏中甘油三酯及胆固醇的积累，DDE 的这种作用可能依赖于对 CD36，Acox-1 等脂肪酸转运或代谢相关基因表达的调节[59]。

利用蛋白质组学技术对 DDT 及 DDE 暴露下人胰岛β细胞 NES2Y 开展研究，发现环境剂量下 DDT 暴露后有 4 种蛋白的表达发生下调：其中有 3 个为细胞骨架蛋白（细胞角蛋白 8，细胞角蛋白 18 及肌动蛋白），1 个糖酵解相关蛋白（*α*-烯醇酶），而 DDE 暴露会下调两种蛋白的表达，细胞角蛋白 18 和异质核糖核蛋白 H 抗体（HNRH1），与胰腺 β 细胞处于应激状态下蛋白的表达变化相同，从而推测 DDT 与 DDE 暴露可能会引起胰岛 β 细胞骨架蛋白及相关功能，如糖代谢、mRNA 的加工等的变化，从而影响胰岛素的分泌[60]。

3.3 溴代阻燃剂与糖尿病

有学者利用病例对照研究方法，对样本容量分别为 2715 和 6209 的两组样本进行调查，调查期间除对参与者进行体检外，还考虑了参与者的用药情况、家族疾病史、生活方式等其他影响因素，结果表明 PBDE 47 的暴露水平与糖尿病发生的风险呈正相关。此外，研究者们利用大鼠模型验证了 PBDE 47 暴露会导致大鼠产生高血糖症，还利用高通量基因组学方法分析了 PBDE 47 暴露肝脏基因组的表达，发现 PBDE 47 暴露会引起糖尿病信号通路及葡萄糖转运相关基因，如 *Tnf*、*Ins2* 和 *Ednr*a 的表达发生变化[61]。在另外一项 BDE 209 对成年雄鼠葡萄糖稳态的影响研究中，发现 BDE 209 暴露不会引起血脂异常，但是即使是低剂量的 BDE 209（0.05 mg/kg）处理也会导致空腹血糖升高。通过对肝脏进行基因组学的分析，发现 BDE 209 影响免疫性疾病相关基因的表达：如自身免疫性甲状腺疾病，1 型糖尿病等。对血清中胰岛素进行含量测定，结果发现 1 mg/kg 和 20 mg/kg BDE 209 处理组胰岛素含量均有所下降。进一步的研究分析表明一些主要组织相容性复合体及 TNF-α 也参与了 T1DM 信号通路，经检测血清中的 TNF-α 的确有所增加，而谷胱甘肽和超氧化物歧化酶的降低也说明氧化损伤可能是 BDE 209 引起高血糖症的机制之一[62]。此外，Yanagisawa 等的研究表明，HBCD 暴露会上调高脂喂食的雄性 C57BL/6J 小鼠中血液中胰岛素的含量，还会影响肝脏中 PPARγ及脂肪组织中 *Glut4* 的表达，提示 HBCD 可能通过影响脂质与糖平衡引起代谢紊乱[63]。

3.4　二噁英与糖尿病

台湾安南区被认为是二噁英污染区，该地区居民血液中的 PCDD/Fs 含量是其他非污染区人群含量的 3 倍左右，Huang 等在该地区针对二噁英暴露与糖尿病发生二者之间的关系展开调查，发现患者体内高浓度二噁英浓度与糖尿病发生呈现正相关关系，而与患者性别、年龄、体重指数无关[64]。在另外一项普通人群调查研究中，也同样发现血清中含有的 PCDD/Fs 会增加患糖尿病的风险[65]。Shertzer 等报道 TCDD 能影响线粒体的活性，上调活性氧的产生，而线粒体活性在胰岛素抵抗及 2 型糖尿病的发生中具有重要的作用，上调活性氧的产生正是线粒体紊乱促进胰岛素抵抗及器官损伤的机制之一[66,67]。

3.5　PFAAs 与糖尿病

Su 等对台湾 571 位参与 PFASs–糖尿病相关性调查的患者数据进行多因素逻辑分析，发现 PFOS 长期暴露与葡萄糖稳态的破坏及糖尿病的发生呈正相关，而 PFOA、PFNA 及 PFUA 对葡萄糖不耐受性及糖尿病的发生有潜在的抑制作用[68]。戴家银等课题的研究结果也表明，PFOA 亚急性暴露小鼠能够通过降低肝脏 PTEN 蛋白表达，改善肝脏和肌肉的 Akt 信号通路，增强小鼠胰岛素敏感性[69]。

图 28-3 汇总了 POPs 诱发糖尿病发生的可能机制。值得注意的是，POPs 暴露的低浓度长周期较之高浓度的急性暴露对人体更加有害[70]，而作为内分泌干扰物，POPs 在低浓度时的作用与高浓度不同，其剂量与效应呈现出非单调性[71]。有关 POPs 与糖尿病发生发展之间的关系及其机制目前尚缺乏关键证据，仍需进一步的深入研究。

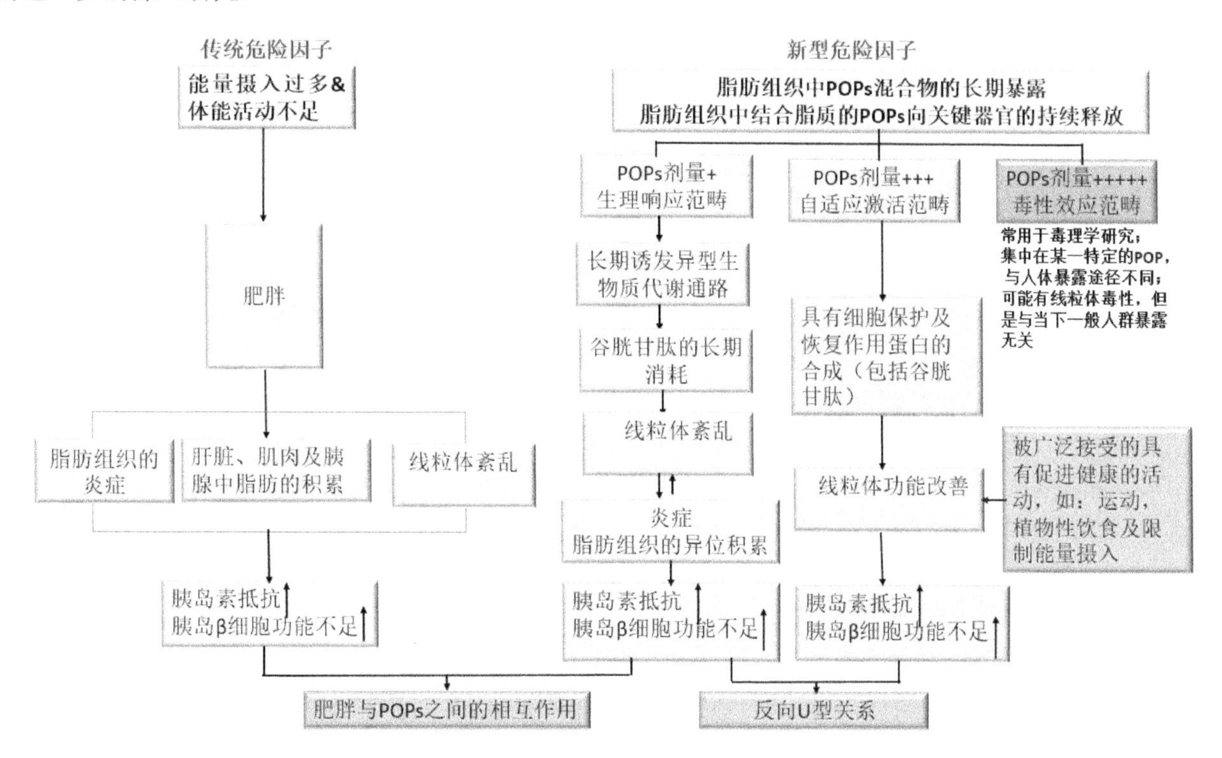

图 28-3　POPs 诱发糖尿病发生的可能机制[72]

4　POPs 对心血管疾病发生的影响

心血管疾病（cardiovascular disease，CVD）占我国居民疾病死亡构成的 40%以上，是我国居民的首位死因[73]，其中动脉粥样硬化（atherosclerosis，AS）是心血管疾病的重要生理病理基础，是一种进行性血管疾病，以动脉管壁逐渐增厚变硬、失去弹性及管腔缩小为生理病理特征。

动脉粥样硬化的病因仍未完全确定，目前研究发现的动脉粥样硬化危险因素已达到 300 多个。公认的重要危险因素主要有吸烟、高血脂、高血压和糖尿病等，次要危险因素包括遗传、年龄、肥胖、内分泌等。血液中的甘油三酯、低密度脂蛋白、高密度脂蛋白和纤维蛋白原水平的改变与动脉粥样硬化的发生相关。动脉粥样硬化的分子机制是目前的研究热点，目前被广泛关注的分子机制主要包括炎症反应，氧化应激，凋亡，自噬和胆固醇逆向转运等[74-78]。环境中的 POPs 可以通过摄食等途径进入人体内，带来心血管疾病风险[79]，例如通过膳食暴露 PCBs 的女性，其患脑中风，特别是出血性脑卒中风的风险大大提高[80]。

4.1　多氯联苯（PCBs）与心脑血管疾病

PCBs 与心脑血管疾病风险的相关性与 PCBs 所含氯的数目有关。Carpenter 等的研究结果表明，在调整年龄、种族、性别、体重指数、饮酒、吸烟、运动状态等因素后，血液脂质浓度的升高与血液中总 PCBs、总农药浓度的升高相关。其中相关性最强的 PCBs 是 3、4 或至少 8 氯取代的 PCBs。PCB 74、156、172、194、199、196~203、206 和 209 均与总脂、总胆固醇和甘油三酯显著正相关[81]。Lind 等对瑞典乌普萨拉市 1016 名 70 岁老人的血液调查结果表明，含低于 6 个氯原子的 PCBs 与颈动脉斑块数目没有相关性，而含 6 个氯及以上的 PCBs 与颈动脉斑块显著正相关。含 8~10 个氯的 PCBs 的总量与动脉 IMT（intima-media thickness，内膜中层厚度）显著正相关，与 IM-GSM（gray scale median of the intima-media complex，内中膜复合体灰度中值）显著负相关[82]。而 Lind 对 PCBs 与乌普萨拉市 898 个老年人脑中风之间相关性的分析结果表明，血液中 PCBs 浓度的增加与老年人中风风险的增加正相关。其中，含 4、5、6 个氯原子的 PCBs 与中风呈强的正相关性，含 7 个或更多氯原子的 PCBs 与中风的正相关性弱，这种趋势不同于 PCBs 与颈动脉硬化之间的相关性，而产生这种差异的原因尚不明确[83]。

研究还发现，非二噁英类 PCBs，如 PCB 194、206、209 与总血清胆固醇及低密度脂蛋白胆固醇的增加呈强正相关性，而仅有 PCB 194 与高密度脂蛋白胆固醇的水平呈负相关性[84]。Valera 等在对居住在努那维克（Nunavik）14 个村庄的 315 位成年因纽特人的调查中发现，血液中 PCB 101、105、138 和 187 浓度的升高增加患高血压的风险，表明非二噁英类 PCBs 与高血压风险的增加呈正相关[85]。

POPs 对心血管疾病的风险往往受一些因素影响，如肥胖、年龄、人种等，PCBs 也不例外。Lee 等认为 POPs 与心血管疾病的相关性也与脂肪含量有关，其研究结果表明，PCBs 及有机氯农药与心血管疾病风险之间只有对脂肪含量低的老年人才出现正相关性[86]，这或许与 PCBs 及有机氯农药的脂溶性有关。Valera 等对在格陵兰生活的 1614 位因纽特人的调查研究表明，尽管将这个人群作为整体进行的分析没有发现 PCBs 或有机氯杀虫剂和血压之间存在显著的联系，但按照年龄分组（18~39 岁和 40 岁后），在年轻人群中（18~39 岁），总二噁英类 PCBs 与高血压风险呈正相关[OR：1.34（95% CI：1.03~1.74）]。这提示 POPs 特别是 PCBs 对高血压的效应或许也受年龄影响[87]。除了肥胖和年龄，人种也是影响 POPs 心血管风险的因素，对非裔美国人和高加索人，尽管血液中有机氯农药水平与总脂和甘油三酯呈正相关，POPs

特别是 PCBs 暴露对血脂的增加对非裔美国人更明显[88]。

血液中其他含氯的 POPs 也与心脑血管疾病风险有一定相关性。Lind 等的调查结果表明有机氯农药（*p*, *p*′-DDE 和反式九氯）和 OCDD（八氯代二噁英）与中风之间存在显著正相关性[83]。此外，有机氯农药 *p*, *p*′-DDE 与血压的升高、高血压风险的增加正相关，而 *p*, *p*′-DDT、HCH 和氯丹与高血压风险的增加负相关[85,89]。

PCBs 导致心血管疾病特别是动脉粥样硬化发生的机制目前仍没有被详细阐释，目前认为 PCBs 的心血管致病机制可能与血脂升高、氧化应激、炎症反应、血管内皮细胞损伤、细胞因子释放及男性睾丸激素水平降低等有关。

众多流行病学调查已经证明了 PCBs，特别是共面 PCBs 增加血液中胆固醇和甘油三酯含量，是促动脉粥样硬化发生的可能机制之一[83]。Ljunggren 等从 17 个实验对象中用纳米液相色谱串联质谱法分离出高密度脂蛋白，并分析其抗氧化性能，来研究高密度脂蛋白 POPs 水平与蛋白质组成/功能之间的关系。蛋白质组分析鉴定了 118 个高密度脂蛋白，其中 10 个与总 POPs 和高氯 PCBs 显著正相关，包括胆固醇酯转移蛋白和磷脂转移蛋白，以及炎症标志物血清淀粉样蛋白 A；而对氧磷酶/芳香酯酶 1 活性与 POPs 负相关。途径分析表明，上调的蛋白质与涉及脂蛋白代谢的生物过程相关，而下调的蛋白质则与高密度脂蛋白的一些保护功能如蛋白酶的负调节、急性期反应、血小板脱粒调节和补体激活等过程相关。这些结果表明，高氯 PCBs 与高密度脂蛋白相互作用可能会导致高密度脂蛋白功能减少继而诱发动脉粥样硬化的发生[90]。

PCBs 能诱发氧化应激并引发炎症反应带来心血管疾病风险[91]，研究发现，总 PCBs 与 ox-LDL 有很强的正相关性（β=0.94；P=2.9×10^6），而总 PCBs 与谷胱甘肽相关的标记物呈负相关性，证实 PCBs 暴露造成的心血管疾病风险可能与其造成的氧化应激有关[92]。研究表明，PCBs 和有机氯农药的暴露对大鼠的氧化应激和抗氧化能力有显著的负效应[93]，而反式九氯和狄氏剂能增加单核/巨噬细胞（人源 THP-1 单核细胞，鼠源 J774A.1 巨噬细胞）内超氧化合物的水平，该过程与依赖于细胞内花生四烯酸水平升高的 Nox（烟酰胺腺嘌呤二核苷酸磷酸氧化酶）机制相关[94]。

PCBs 暴露可以通过细胞因子的释放增加心血管疾病风险。Hennig 等证实，PCB 126 暴露增加了血管内皮细胞（EA.hy926）IL-6（白介素 6），C 反应蛋白，ICAM-1（细胞间黏附分子 1），VCAM-1（血管细胞黏附分子 1）和 IL-1α/β（白介素 1α/β）的表达，这些细胞因子的过量表达与动脉粥样硬化的发生密切相关[95]。

血管内皮细胞层受损导致血液中的单核细胞进入内皮细胞层以下，被诱导为巨噬细胞并吸收被修饰的低密度脂蛋白形成泡沫细胞，该过程对动脉粥样硬化的发生十分关键。高氯 PCBs 暴露血管内皮细胞后产生 ROS、激活 NF-κB 及表达某些炎症标志物，都是造成血管内皮损伤和心血管疾病的危险因素[96]，例如，PCB 118 暴露后 HUVECs 的 ROS 水平升高是 HUVECs 凋亡及受损的部分原因[97]。

PCBs 导致 micro RNA 的表达是其造成心血管损伤的可能分子机制之一。杜宇国等课题组的研究结果表明，Aroclor1254 和 TCDD 共同暴露 $ApoE^{-/-}$ 小鼠 6 周后，能显著引起其肝脏中 mRNAs 和 mi RNAs 的表达变化，有 18 个差异表达的 mi RNAs 能够靶向调控 110 个差异表达的 mRNAs，二者可共同影响糖代谢、脂代谢、细胞死亡、分子运输等生物学功能。进一步考察与动脉粥样硬化发生发展密切相关的糖代谢、脂代谢网络调控，发现 mi RNA-22、let-7 家族、mi RNA-15a /b，以及靶基因 PPARα、PPARγ 辅助激活因子 1α 和 Foxo1，在 PCBs 暴露致动脉粥样硬化发生发展的糖脂代谢异常中发挥了重要作用[98]。Hennig 等证实，Aroclor 1260（商业 PCBs 混合物）也对原代 HUVEC 的 mi RNA 表达谱（557 个 mi RNAs）产生影响，表达水平增加的 mi RNA 包括 miR-21、miR-31、miR-126，miR-221 和 miR-222。其中已有报道 miR-126 和 miR-31 有调节炎症的功能[99]。

PCBs 与纳米材料的结合往往会增强其生物毒性。Toborek 等的研究结果表明，PCBs 与纳米材料的结合增强了 PCBs 的心血管疾病风险。二氧化硅纳米颗粒与 PCB 153 结合可促进炎症反应并破坏血脑屏障的完整性，增加小鼠（雄性 C57BL/6）脑中风及脑损伤的风险[100]。与二氧化硅纳米颗粒结合的 PCB 153 激活的炎症反应受 TLR4/TRAF6 调节，对血脑屏障完整性的破坏与脑血管紧密连接蛋白的表达的降低相关[101]。

4.2 全氟烷基化合物（PFAAs）与心脑血管疾病

含氟烷基化合物与心血管疾病之间的关系也值得关注，对 664 名年龄在 12~30 岁之间的受试者的尿检结果显示，血液中高浓度的 PFOS（全氟辛烷磺酸）与颈动脉内膜中层厚度增加正相关[102]，而对 815 名未成年人进行的调查结果表明，血液中 PFOA（全氟辛酸）和 PFOS（全氟辛烷磺酸）均与总胆固醇和低密度脂蛋白胆固醇含量呈正相关[103]。研究表明，PFASs（全氟烷基化合物），特别是长链 PFASs 如 PFUnDA（全氟十一烷酸），与动脉粥样硬化的关系存在明显的性别差异，女性体内 PFASs 含量与动脉粥样硬化标志（如颈动脉斑块，内膜-中膜复合体的回声反射）之间的正相关性更显著。PFASs 是 PPAR α、γ 受体的激动剂，而 PPAR α、γ 受体的激活与胰岛素抵抗及胰岛素分泌有关，并影响血脂水平，因此 PFASs 可能通过诱发糖尿病或高血脂导致动脉粥样硬化的发生，然而在本研究中 PFASs 与动脉粥样硬化的正相关性与其对糖尿病和高血脂的影响无关[104]，提示 PFASs 可能通过别的机制导致动脉粥样硬化的发生。含氟化合物对心脑血管疾病的影响及相应机制仍需进一步研究。

4.3 有机农药与心血管疾病风险

已有研究结果提示硫丹可以通过激发炎症和氧化应激反应，增加血管内皮细胞（HUVECs）通透性以干扰其正常功能。硫丹暴露可以抑制血管内皮细胞的存活，增加其乳酸脱氢酶（LDH）的释放，破坏细胞超微结构，导致细胞的凋亡和坏死。此外，硫丹能激活 RIPK1 途径，增加 HUVECs 的 ROS、IL-1 和 IL-33 水平[105]。用硫丹暴露人脐静脉内皮细胞 48 小时后，肌动蛋白细胞骨架被破坏，导致 HUVECs 渗透性增强。硫丹可以通过降低 E-钙黏素和 β-钙黏素的水平破坏 HUVECs 间的黏着连接，并通过下调细胞旁通路中的 Cx43 损伤 HUVECs 的间隙连接，这是硫丹引起心血管疾病的主要风险[106]。

POPs 复合暴露对心血管疾病特别是动脉粥样硬化的风险或许高于 POPs 单独暴露。我们课题组用 TCDD（15 μg/kg）和 Aroclor1254（55 mg/kg，PCBs 商业产品）单独或同时暴露雄性 $apoE^{-/-}$小鼠，结果显示：暴露于 TCDD 的 $apoE^{-/-}$小鼠发生动脉粥样硬化病变，而暴露 Aroclor1254 的 $apoE^{-/-}$小鼠没有发生动脉粥样硬化病变，共同暴露于 TCDD 与 PCBs 的 $apoE^{-/-}$小鼠动脉粥样硬化病变最严重。该结果表明，二噁英和 PCBs 的联合暴露可能产生更大的心血管健康风险，原因可能是 TCDD 和 PCBs 共同暴露诱发高胆固醇血症，活化血小板，上调 MCP-1 的表达及诱发 RIG-I 信号通路等[107]。

尽管 POPs 与心血管疾病之间的联系已经引起了人们的关注，但相关研究仍存在许多不足。首先，流行病学研究中关注的 POPs 不全面。近几年在流行病学研究中关注的 POPs 主要是含氯类 POPs 尤其 PCBs 及少量含氟类 POPs，而含溴类阻燃剂 PBDEs 和 HBCDs 等与心血管疾病之间的关系尚不明确。其次，目前有关 POPs 诱发心血管疾病的机制研究尚不足以说明问题。在今后相关 POPs 的致病机制研究中，参考流行病学调查结果，选择相关性强的 POPs 以及能作为一类 POPs 代表的某些化合物来研究其致病机理。将流行病学结果及机制研究结果结合，可为心血管疾病的预防和 POPs 使用控制提供参考。

5 POPs 对其他糖脂代谢相关疾病发生的影响

POPs 暴露除了能引发肥胖、糖尿病、心脑血管疾病等糖脂代谢相关疾病之外，还可引起高血压、非酒精性脂肪肝等相关糖脂代谢疾病。

5.1 POPs 对高血压的影响

Valera 等在对 Nunavik 14 个村庄 315 位成年因纽特人的调查中发现，PCBs 与高血压的发生密切相关。血浆中 PCB 101、105、138 和 187 以及有机氯农药 DDE 的浓度与高血压风险呈正相关，而 DDT、HCH 和氯丹与高血压呈负相关[85]。Jee 等认为，非二噁英 PCBs 和高血压的风险没有显著相关性（OR = 1.00；95 % CI：0.89~1.12），而二噁英类 PCBs 含量与高血压风险显著增加有关（OR= 1.45；95 % CI：1.00~2.12）[108]。在一项评估血清中 POPs 水平与高血压风险因素的研究中，Boada 等发现，DDE 与收缩压和舒张压正相关（r=0.222；p<0.001，r=0.123；p=0.015），艾氏剂（aldrin）与收缩压负相关（r=−0.120；p=0.017）。在排除已接受高血压治疗的受试者后，多元分析结果仅确认艾氏剂浓度与收缩压成反比，其中对男性更为显著。本研究针对选定的几种血清 POPs 未发现其与高血压风险增加的相关性，提示人们也许需要更关注 POPs 的化学结构以及浓度与血压的关系[89]。

Arrebola 等对格拉纳达成年人群进行了一个为期十年的前瞻性调查研究，发现当 BMI ≥ 26.3 时，血清中 PCBs 的含量和高血压患病率呈正相关；当 BMI ＜ 26.3 时，二者之间没有明显的关系，暗示肥胖可能是高血压的一个启动器[109]。Ilhan 等于 2015 年在 Sprague-Dawley 大鼠实验中发现 TCDD 暴露会导致血压增高、肾氧化应激和血管收缩反应，但给予 *N*-乙酰-5-甲氧基色胺后，可降低 TCDD 诱导的氧化应激活性，从而改善血压，表明 *N*-乙酰-5-甲氧基色胺有益于修复 TCDD 引起的高血压[110]。

5.2 POPs 对脂肪肝的影响

非酒精性脂肪肝（nonalcoholic fatty liver disease，NAFLD）是指除酒精和其他明确因素所致，以肝实质细胞脂肪变性和肝小叶内炎症为特征的临床病理综合征，是遗传–环境–代谢应激相关性疾病，包括单纯性脂肪肝和由其演变的非酒性脂肪肝炎和肝硬化。Fan 针对我国非酒精性脂肪肝的调查发现，有 15% 的人群患有非酒精脂肪肝，其中儿童所占比例为 2.1%，并且在肥胖儿童中非酒精性脂肪肝所占比例高达 68.2%[111]。对正常饮食的 C57BL6/J 小鼠暴露 PCB 153，发现其对小鼠脂肪肝的形成影响极微，但在高脂饮食饲喂条件下暴露 PCB153 将会通过降低肝脏 β-脂肪酸氧化和上调脂质合成相关基因的表达来加剧肝脂肪变性[112]。

6 展　望

虽然列入《斯德哥尔摩公约》的 POPs 大多已被禁用，但由于其具有长期残留性和生物蓄积性等特点，仍将对环境产生长远的影响，因此关注 POPs 诱发的健康风险仍不容忽视。目前的流行病学调查研究以及毒理学实验研究均表明，POPs 暴露会诱发机体的糖、脂代谢失衡，与代谢性疾病如肥胖、糖尿病及心脑血管疾病等的产生密切相关。

目前在 POPs 暴露与糖、脂代谢异常相关疾病发生之间的证据研究还存在以下不足。首先，新型 POPs，如短链氯化石蜡等对机体糖、脂代谢影响的流行病学和基础毒理学的研究数据缺乏。其次，目前的基础毒理学研究中大多选取的 POPs 暴露浓度过高，缺乏环境相关性，造成研究数据无法为环境健康问题提供合理的指导。再次，POPs 暴露对糖、脂代谢影响的机制研究还不够深入。这主要包括两方面问题，一是 POPs 进入机体后会发生代谢，其体内代谢过程及代谢产物还不是很清楚，限制了我们对 POPs 体内作用机制的认识；二是目前在 POPs 与糖脂代谢机制研究中对基础生物医学理论的借鉴还不够，比如在肠道微生物组、代谢组学、表观遗传等的新发现、新工具等。最后，多种 POPs 影响糖、脂代谢的联合毒性效应不明确。我们知道在真实的环境中机体同时遭受多种 POPs 的暴露，这些不同种类的 POPs 是产生协同增强的毒性效应还是相互拮抗的效应，目前这方面的研究结果很少。针对以上的问题，我们认为在 POPs 影响糖、脂代谢的研究工作中，应该重点开展以下几方面的工作：

（1）重视新型污染物对糖、脂代谢毒性的研究；

（2）在 POPs 的暴露剂量方面选取具有环境相关性的暴露剂量，获取具有环境学意义的毒理学数据；

（3）应该开展多种 POPs 影响糖、脂代谢的联合毒性效应的研究；

（4）应该借鉴基础生物医学方面的最新理论和方法，加深对 POPs 诱发糖脂代谢紊乱的机制理解，为相关疾病的预防和治疗提供理论参考和指导。

参 考 文 献

[1] Tang R, Liu H, Yuan Y, et al. Genetic factors associated with risk of metabolic syndrome and hepatocellular carcinoma. Oncotarget, 2017, 8(21): 35403-35411.

[2] Shen J, Goyal A, Sperling L. The emerging epidemic of obesity, diabetes, and the metabolic syndrome in China. Cardiol Res Pract, 2012, 178675.

[3] 王亚韡，蔡亚岐，江桂斌，斯德哥尔摩公约新增持久性有机污染物的一些研究进展．中国科学：化学，2010，(02)：99-123.

[4] Mullerova D, Kopecky J. White adipose tissue: storage and effector site for environmental pollutants. 2007, 56(4): 375-381.

[5] Tran TT, Kahn CR. Transplantation of adipose tissue and stem cells: role in metabolism and disease. Nat Rev Endocrinol, 2010, 6(4): 195-213.

[6] Haslam DW, James WP. Obesity. The Lancet, 2005. 366(9492): 1197-1209.

[7] Neel BA, Sargis RM. The paradox of progress: environmental disruption of metabolism and the diabetes epidemic. Diabetes, 2011, 60(7): 1838-1848.

[8] Taylor KW, Novak RF, Anderson HA, et al. Evaluation of the association between persistent organic pollutants(POPs)and diabetes in epidemiological studies: a national toxicology program workshop review. Environ Health Perspect, 2013, 121(7): 774-83.

[9] Malarvannan G, Dirinck E, Dirtu AC, et al. Distribution of persistent organic pollutants in two different fat compartments from obese individuals. Environment International, 2013, 55: 33-42.

[10] Malarvannan G, Dirinck E, Dirtu AC, et al. Distribution of persistent organic pollutants in two different fat compartments from obese individuals. Environ Int, 2013, 55: 33-42.

[11] Arrebola JP, Ocana-Riola R, Arrebola-Moreno AL, et al. Associations of accumulated exposure to persistent organic pollutants with serum lipids and obesity in an adult cohort from Southern Spain. Environ Pollut, 2014, 195: 9-15.

[12] Donat-Vargas C, Gea A, Sayon-Orea C, et al. Association between dietary intakes of PCBs and the risk of obesity: the SUN

project. Journal of Epidemiology and Community Health, 2014, 68(9): 834-841.

[13] Verhulst SL, Nelen V, Hond ED, et al. Intrauterine exposure to environmental pollutants and body mass index during the first 3 years of life. Environ Health Perspect, 2009, 117(1): 122-6.

[14] Gadupudi G, Gourronc FA, Ludewig G, et al. PCB126 inhibits adipogenesis of human preadipocytes. Toxicology in Vitro, 2015, 29(1): 132-141.

[15] Ferrante MC, Amero P, Santoro A, et al. Polychlorinated biphenyls(PCB 101, PCB 153 and PCB 180)alter leptin signaling and lipid metabolism in differentiated 3T3-L1 adipocytes. Toxicology and Applied Pharmacology, 2014, 279(3): 401-408.

[16] Olsen GW, Zobel LR. Assessment of lipid, hepatic, and thyroid parameters with serum perfluorooctanoate (PFOA) concentrations in fluorochemical production workers. Int Arch Occup Environ Health, 2007, 81(2): 231-46.

[17] Xu J, Shimpi P, Armstrong L, et al. PFOS induces adipogenesis and glucose uptake in association with activation of Nrf2 signaling pathway. Toxicology and Applied Pharmacology, 2016, 290: 21-30.

[18] Watkins AM, Wood CR, Lin MT, et al. The effects of perfluorinated chemicals on adipocyte differentiation in vitro. Molecular and Cellular Endocrinology, 2015, 400(C): 90-101.

[19] Lee DH, Lind L, Jacobs DR et al. Associations of persistent organic pollutants with abdominal obesity in the elderly: The Prospective Investigation of the Vasculature in Uppsala Seniors(PIVUS)study. Environ Int, 2012, 40: 170-8.

[20] Vuong AM, Braun JM, Sjödin A, et al. Prenatal Polybrominated Diphenyl Ether Exposure and Body Mass Index in Children Up To 8 Years of Age. Environ Health Perspect, 2016, 124(12): 1891-1897.

[21] Erkin-Cakmak A, Harley KG, Chevrier J, et al. In utero and childhood polybrominated diphenyl ether exposures and body mass at age 7 years: the CHAMACOS study. Environ Health Perspect, 2015, 123(6): 636-42.

[22] Tung EW, Boudreau A, Wade MG, et al. Induction of Adipocyte Differentiation by Polybrominated Diphenyl Ethers(PBDEs)in 3T3-L1 Cells. Plos One, 2014, 9(4). e94583.

[23] Kamstra JH, Hruba E, Blumberg B, et al. Transcriptional and epigenetic mechanisms underlying enhanced in vitro adipocyte differentiation by the brominated flame retardant BDE-47. Environ Sci Technol, 2014, 48(7): 4110-9.

[24] Cupul-Uicab LA, Klebanoff MA, Brock JW, et al. Prenatal Exposure to Persistent Organochlorines and Childhood Obesity in the US Collaborative Perinatal Project. Environmental Health Perspectives, 2013, 121(9): 1103-1109.

[25] Strong AL, Shi Z, Strong MJ, et al. Effects of the Endocrine-Disrupting Chemical DDT on Self-Renewal and Differentiation of Human Mesenchymal Stem Cells. Environmental Health Perspectives, 2015, 123(1): 42-48.

[26] Mangum LH, Howell GE, Chambers JE. Exposure to p, p '-DDE enhances differentiation of 3T3-L1 preadipocytes in a model of sub-optimal differentiation. Toxicology Letters, 2015, 238(2): 65-71.

[27] Chang JW, Chen HL, Su HJ, et al. Abdominal Obesity and Insulin Resistance in People Exposed to Moderate-to-High Levels of Dioxin. PLoS One, 2016, 11(1): e0145818.

[28] van Esterik JC, Verharen HW, Hodemaekers HM, et al. Compound- and sex-specific effects on programming of energy and immune homeostasis in adult C57BL/6JxFVB mice after perinatal TCDD and PCB 153. Toxicol Appl Pharmacol, 2015, 289(2): 262-75.

[29] Chen CL, Brodie AE, Hu CY. CCAAT/Enhancer-Binding Protein β is not Affected by Tetrachlorodibenzo- p-dioxin (TCDD)Inhibition of 3T3-L1 Preadipocyte Differentiation. Obesity Research, 1997, 5(2): 146-152.

[30] Cimafranca MA, Hanlon PR, Jefcoate CR. TCDD administration after the pro-adipogenic differentiation stimulus inhibits PPARgamma through a MEK-dependent process but less effectively suppresses adipogenesis. Toxicol Appl Pharmacol, 2004, 196(1): 156-68.

[31] Pereira-Fernandes A, Dirinck E, Dirtu AC, et al. Expression of Obesity Markers and Persistent Organic Pollutants Levels in Adipose Tissue of Obese Patients: Reinforcing the Obesogen Hypothesis? Plos One, 2014, 9(1).

[32] Zimmet PZ, Magliano DJ, Herman WH, et al. Diabetes: a 21st century challenge. The Lancet Diabetes & Endocrinology, 2014, 2(1): 56-64.

[33] Guariguata L, Whiting DR, Hambleton I, et al. Global estimates of diabetes prevalence for 2013 and projections for 2035. Diabetes Res Clin Pract, 2014, 103(2): 137-49.

[34] Katsarou A, Gudbjornsdottir S, Rawshani A, et al. Type 1 diabetes mellitus. Nat Rev Dis Primers, 2017, 3: 17016.

[35] DeFronzo RA, Ferrannini E, Groop L, et al. Type 2 diabetes mellitus. Nat Rev Dis Primers, 2015, 1: 15019.

[36] Ashwal E, Hod M. Gestational diabetes mellitus: Where are we now? Clin Chim Acta, 2015, 451(Pt A): 14-20.

[37] Chen P, Wang S, Ji J, et al. Risk factors and management of gestational diabetes. Cell Biochem Biophys, 2015, 71(2): 689-94.

[38] Kaul N, Ali S, Genes. Genetics, and Environment in Type 2 Diabetes: Implication in Personalized Medicine. DNA Cell Biol, 2016, 35(1): 1-12.

[39] Rockette-Wagner B, Edelstein S, Venditti EM, et al. The impact of lifestyle intervention on sedentary time in individuals at high risk of diabetes. Diabetologia, 2015, 58(6): 1198-202.

[40] Chasan-Taber L. Lifestyle interventions to reduce risk of diabetes among women with prior gestational diabetes mellitus. Best Pract Res Clin Obstet Gynaecol, 2015, 29(1): 110-22.

[41] Ali N, Rajeh N, Wang W, et al. Organohalogenated contaminants in type 2 diabetic serum from Jeddah, Saudi Arabia. Environ Pollut, 2016, 213: 206-12.

[42] Aminov Z, Haase R. Rej R, et al. Diabetes Prevalence in Relation to Serum Concentrations of Polychlorinated Biphenyl(PCB)Congener Groups and Three Chlorinated Pesticides in a Native American Population. Environ Health Perspect, 2016, 124(9): 1376-83.

[43] Lee DH. Persistent organic pollutants and obesity-related metabolic dysfunction: focusing on type 2 diabetes. Epidemiol Health, 2012, 34: e2012002.

[44] Persky V, Piorkowski J, Turyk M, et al. Polychlorinated biphenyl exposure, diabetes and endogenous hormones: a cross-sectional study in men previously employed at a capacitor manufacturing plant. Environ Health, 2012, 11: 57.

[45] Silverstone AE, Rosenbaum PF, Weinstock RS, et al. Polychlorinated biphenyl(PCB)exposure and diabetes: results from the Anniston Community Health Survey. Environ Health Perspect, 2012, 120(5): 727-32.

[46] Weinhold B. PCBs and diabetes: pinning down mechanisms. Environ Health Perspect, 2013, 121(1): A32.

[47] Grice BA, Nelson RG, Williams DE, et al. Associations between persistent organic pollutants, type 2 diabetes, diabetic nephropathy and mortality. Occup Environ Med, 2017, 74(7): 521-527.

[48] Aminov Z, Haase R, Carpenter DO. Diabetes in Native Americans: elevated risk as a result of exposure to polychlorinated biphenyls(PCBs). Rev Environ Health, 2016, 31(1): 115-9.

[49] Rylander C, Sandanger TM, Nost TH, et al. Combining plasma measurements and mechanistic modeling to explore the effect of POPs on type 2 diabetes mellitus in Norwegian women. Environ Res, 2015, 142: 365-73.

[50] Vafeiadi M, Roumeliotaki T, Chalkiadaki G, et al. Persistent organic pollutants in early pregnancy and risk of gestational diabetes mellitus. Environ Int, 2017, 98: 89-95.

[51] Baker NA, Karounos M, English V, et al. Coplanar polychlorinated biphenyls impair glucose homeostasis in lean C57BL/6 mice and mitigate beneficial effects of weight loss on glucose homeostasis in obese mice. Environ Health Perspect, 2013, 121(1): 105-10.

[52] Baker NA, Shoemaker R, English V, et al. Effects of Adipocyte Aryl Hydrocarbon Receptor Deficiency on PCB-Induced Disruption of Glucose Homeostasis in Lean and Obese Mice. Environ Health Perspect, 2015, 123(10): 944-50.

[53] Mesnier A, Champion S, Louis L, et al. The Transcriptional Effects of PCB118 and PCB153 on the Liver, Adipose Tissue,

Muscle and Colon of Mice: Highlighting of Glut4 and Lipin1 as Main Target Genes for PCB Induced Metabolic Disorders. PLoS One, 2015, 10(6): e0128847.

[54] Jensen TK, Timmermann AG, Rossing LI, et al. Polychlorinated biphenyl exposure and glucose metabolism in 9-year-old Danish children. J Clin Endocrinol Metab, 2014, 99(12): 2643-51.

[55] Al-Othman A, Yakout S, Abd-Alrahman SH, et al. Strong associations between the pesticide hexachlorocyclohexane and type 2 diabetes in Saudi adults. Int J Environ Res Public Health, 2014, 11(9): 8984-95.

[56] Suarez-Lopez JR, Lee DH, Porta M, et al. Persistent organic pollutants in young adults and changes in glucose related metabolism over a 23-year follow-up. Environ Res, 2015, 137: 485-94.

[57] Al-Othman AA, Abd-Alrahman SH, Al-Daghri NM. DDT and its metabolites are linked to increased risk of type 2 diabetes among Saudi adults: a cross-sectional study. Environ Sci Pollut Res Int, 2015, 22(1): 379-86.

[58] La Merrill M, Karey E, Moshier E, et al. Perinatal exposure of mice to the pesticide DDT impairs energy expenditure and metabolism in adult female offspring. PLoS One, 2014, 9(7): e103337.

[59] Howell GE, Mulligan C, Meek E, et al. Effect of chronic p, p' dichlorodiphenyldichloroethylene (DDE)exposure on high fat diet-induced alterations in glucose and lipid metabolism in male C57BL/6H mice. Toxicology, 2015, 328: 112-22.

[60] Pavlikova N, Smetana P, Halada P, et al. Effect of prolonged exposure to sublethal concentrations of DDT and DDE on protein expression in human pancreatic beta cells. Environ Res, 2015,142: 257-63.

[61] Zhang Z, Li S, Liu L, et al. Environmental exposure to BDE47 is associated with increased diabetes prevalence: Evidence from community-based case-control studies and an animal experiment. Sci Rep, 2016, 6: 27854.

[62] Zhang Z, Sun ZZ, Xiao X, et al. Mechanism of BDE209-induced impaired glucose homeostasis based on gene microarray analysis of adult rat liver. Arch Toxicol, 2013, 87(8): 1557-67.

[63] Yanagisawa R, Koike E, Win-Shwe TT, et al. Impaired lipid and glucose homeostasis in hexabromocyclododecane-exposed mice fed a high-fat diet. Environ Health Perspect, 2014, 122(3): 277-83.

[64] Huang CY, Wu CL, Yang YC, et al. Association between Dioxin and Diabetes Mellitus in an Endemic Area of Exposure in Taiwan: A Population-Based Study. Medicine(Baltimore), 2015, 94(42): e1730.

[65] Nakamoto M, Arisawa K, Uemura H, et al. Association between blood levels of PCDDs/PCDFs/dioxin-like PCBs and history of allergic and other diseases in the Japanese population. Int Arch Occup Environ Health, 2013, 86(8): 849-59.

[66] Patti ME, Corvera S. The role of mitochondria in the pathogenesis of type 2 diabetes. Endocr Rev, 2010, 31(3): 364-95.

[67] Shertzer HG, Genter MB, Shen D, et al. TCDD decreases ATP levels and increases reactive oxygen production through changes in mitochondrial F(0)F(1)-ATP synthase and ubiquinone. Toxicol Appl Pharmacol, 2006, 217(3): 363-74.

[68] Su TC, Kuo CC, Hwang JJ, et al. Serum perfluorinated chemicals, glucose homeostasis and the risk of diabetes in working-aged Taiwanese adults. Environ Int, 2016, 88: 15-22.

[69] Yan SM, Zhang HX, Zheng F, et al. Perfluorooctanoic acid exposure for 28 days affects glucose homeostasis and induces insulin hypersensitivity in mice. Sci Rep, 2015, 5: 11029.

[70] Lee DH, Jacobs DR, Jr Porta M, Could low-level background exposure to persistent organic pollutants contribute to the social burden of type 2 diabetes? J Epidemiol Community Health, 2006, 60(12): 1006-8.

[71] Vandenberg LN, Colborn T, Hayes TB, et al. Hormones and endocrine-disrupting chemicals: low-dose effects and nonmonotonic dose responses. Endocr Rev, 2012, 33(3): 378-455.

[72] Lee DH, Porta M, Jacobs DR Jr, et al. Chlorinated persistent organic pollutants, obesity, and type 2 diabetes. Endocr Rev, 2014, 35(4): 557-601.

[73] 陈伟伟，高润霖，刘力生等.《中国心血管病报告 2015》概要. 中国循环杂志, 2016, (06): 521-528.

[74] Ross R. Atherosclerosis--an inflammatory disease. N Engl J Med, 1999, 340(2): 115-126.

[75] Libby P. Inflammation in atherosclerosis. Arterioscler Thromb Vasc Biol, 2012, 32(9): 2045-2051.

[76] Hansson GK, Libby P, Schonbeck U, et al. Innate and Adaptive Immunity in the Pathogenesis of Atherosclerosis. Circulation Research, 2002, 91(4): 281-291.

[77] Tall AR, Yvan-Charvet L. Cholesterol, inflammation and innate immunity. Nat Rev Immunol, 2015, 15(2): 104-116.

[78] Shao BZ, Han BZ, Zeng YX, et al. The roles of macrophage autophagy in atherosclerosis. Acta Pharmacol Sin, 2016, 37(2): 150-156.

[79] Henriquez-Hernandez LA, Luzardo OP, Zumbado M, et al. Determinants of increasing serum POPs in a population at high risk for cardiovascular disease. Results from the PREDIMED-CANARIAS study. Environ Res, 2017, 156: 477-484.

[80] Bergkvist C, Kippler M, Larsson SC, et al. Dietary exposure to polychlorinated biphenyls is associated with increased risk of stroke in women. J Intern Med, 2014, 276(3): 248-259.

[81] Aminov Z, Haase RF, Pavuk M, et al. Analysis of the effects of exposure to polychlorinated biphenyls and chlorinated pesticides on serum lipid levels in residents of Anniston, Alabama. Environ Health, 2013, 12: 108.

[82] Lind PM, van Bavel B, Salihovic S, et al. Circulating levels of persistent organic pollutants(POPs) and carotid atherosclerosis in the elderly. Environ Health Perspect, 2012, 120(1): 38-43.

[83] Lee DH, Lind PM, Jacobs DR Jr, et al. Background exposure to persistent organic pollutants predicts stroke in the elderly. Environ Int, 2012, 47: 115-120.

[84] Penell J, Lind L, Salihovic S, et al. Persistent organic pollutants are related to the change in circulating lipid levels during a 5 year follow-up. Environ Res, 2014, 134: 190-197.

[85] Valera B, Ayotte P, Poirier P, et al. Associations between plasma persistent organic pollutant levels and blood pressure in Inuit adults from Nunavik. Environ Int, 2013, 59: 282-289.

[86] Kim SA, Kim KS, Lee YM, et al. Associations of organochlorine pesticides and polychlorinated biphenyls with total, cardiovascular, and cancer mortality in elders with differing fat mass. Environ Res, 2015, 138: 1-7.

[87] Valera B, Jorgensen ME, Jeppesen C, et al. Exposure to persistent organic pollutants and risk of hypertension among Inuit from Greenland. Environ Res, 2013, 122: 65-73.

[88] Aminov Z, Haase R, Olson JR, et al. Racial differences in levels of serum lipids and effects of exposure to persistent organic pollutants on lipid levels in residents of Anniston, Alabama. Environ Int, 2014, 73: 216-223.

[89] Henriquez-Hernandez LA, Luzardo OP, Zumbado M, et al. Blood pressure in relation to contamination by polychlorobiphenyls and organochlorine pesticides: Results from a population-based study in the Canary Islands(Spain). Environ Res, 2014, 135: 48-54.

[90] Ljunggren SA, Helmfrid I, Norinder U, et al. Alterations in high-density lipoprotein proteome and function associated with persistent organic pollutants. Environ Int, 2017, 98: 204-211.

[91] Petriello MC, Newsome B, Hennig B. Influence of nutrition in PCB-induced vascular inflammation. Environ Sci Pollut Res Int, 2014, 21(10): 6410-6418.

[92] Kumar J, Monica Lind P, Salihovic S, et al. Influence of persistent organic pollutants on oxidative stress in population-based samples. Chemosphere, 2014, 114: 303-309.

[93] Hong MY, Lumibao J, Mistry P, et al. Fish Oil Contaminated with Persistent Organic Pollutants Reduces Antioxidant Capacity and Induces Oxidative Stress without Affecting Its Capacity to Lower Lipid Concentrations and Systemic Inflammation in Rats. J Nutr, 2015, 145(5): 939-944.

[94] Mangum LC, Borazjani A, Stokes JV, et al., Organochlorine insecticides induce NADPH oxidase-dependent reactive oxygen species in human monocytic cells via phospholipase A2/arachidonic acid. Chem Res Toxicol, 2015. 28(4): 570-584.

[95] Liu DD, Perkins JT, Hennig B. EGCG prevents PCB-126-induced endothelial cell inflammation via epigenetic modifications

of NF-kappaB target genes in human endothelial cells. J Nutr Biochem, 2016, 28: 164-170.

[96] Eske K, Newsome B, Han SG, et al. PCB 77 dechlorination products modulate pro-inflammatory events in vascular endothelial cells. Environ Sci Pollut Res Int, 2014, 21(10): 6354-6364.

[97] Tang L, Cheng JN, Long Y, et al. PCB 118-induced endothelial cell apoptosis is partially mediated by excessive ROS production. Toxicol Mech Methods, 2017, 27(5): 394-399.

[98] 黄风尘，单秋丽，王静等. PCBs 暴露 ApoE$^{-/-}$小鼠肝脏的 microRNA 和 mRNA 调控网络研究. 环境化学, 2014, (10): 1768-1775.

[99] Wahlang, B, Petriello MC, Perkins JT, et al. Polychlorinated biphenyl exposure alters the expression profile of microRNAs associated with vascular diseases. Toxicol In Vitro, 2016, 35: 180-187.

[100] Zhang B, Chen L, Choi JJ, et al. Cerebrovascular toxicity of PCB153 is enhanced by binding to silica nanoparticles. J Neuroimmune Pharmacol, 2012, 7(4): 991-1001.

[101] Zhang B, Choi JJ, Eum SY, et al. TLR4 signaling is involved in brain vascular toxicity of PCB153 bound to nanoparticles. Plos One, 2014, 8 (5) :e63159.

[102] Lin CY, Lin LY, Wen TW, et al. Association between levels of serum perfluorooctane sulfate and carotid artery intima-media thickness in adolescents and young adults. Int J Cardiol, 2013, 168(4): 3309-3316.

[103] Geiger SD, Xiao J, Ducatman A, et al. The association between PFOA, PFOS and serum lipid levels in adolescents. Chemosphere, 2014, 98: 78-83.

[104] Lind PM, Salihovic S, van Bavel B, et al. Circulating levels of perfluoroalkyl substances (PFASs) and carotid artery atherosclerosis. Environ Res, 2017, 152: 157-164.

[105] Zhang L, Wei J, Ren L, et al. Endosulfan inducing apoptosis and necroptosis through activation RIPK signaling pathway in human umbilical vascular endothelial cells. Environ Sci Pollut Res Int, 2017, 24(1): 215-225.

[106] Xu D, Liu T, Lin LM, et al. Exposure to endosulfan increases endothelial permeability by transcellular and paracellular pathways in relation to cardiovascular diseases. Environ Pollut, 2017, 223: 111-119.

[107] Shan QL, Wang J, Huang FC, et al. Augmented atherogenesis in ApoE-null mice co-exposed to polychlorinated biphenyls and 2, 3, 7, 8-tetrachlorodibenzo-p-dioxin. Toxicol Appl Pharmacol, 2014, 276(2): 136-146.

[108] Park SH, Lim JE, Park H, et al. Body burden of persistent organic pollutants on hypertension: a meta-analysis. Environ Sci Pollut Res Int, 2016, 23(14): 14284-14293.

[109] Arrebola JP, Fernandez MF, Martin-Olmedo P, et al. Historical exposure to persistent organic pollutants and risk of incident hypertension. Environ Res, 2015, 138: 217-23.

[110] Ilhan S, Atessahin D, Atessahin A, et al. 2, 3, 7, 8-Tetrachlorodibenzo-p-dioxin-induced hypertension: the beneficial effects of melatonin. Toxicol Ind Health, 2015, 31(4): 298-303.

[111] Fan JG. Epidemiology of alcoholic and nonalcoholic fatty liver disease in China. J Gastroenterol Hepatol, 2013, 28 Suppl 1: 11-17.

[112] Wahlang B, Falkner KC, Gregory B, et al. Polychlorinated biphenyl 153 is a diet-dependent obesogen that worsens nonalcoholic fatty liver disease in male C57BL6/J mice. J Nutr Biochem, 2013, 24(9): 1587-1595.

作者：文　青[1,2]，任麒东[1,2]，余彩霞[3]，赵翠霞[3]，唐　越[1,2]，颉欣妮[1]，杜宇国[1,2,3]

[1]中国科学院生态环境研究中心，[2]中国科学院大学资源与环境学院，[3]中国科学院大学化学与化工学院

第29章　重金属与肿瘤研究进展

本章导读

重金属及其化合物进入环境或生态系统后，其在环境中的分布，主要是在大气、土壤和水体中的迁移转化，从而造成危害。重金属及其化合物的慢性环境暴露可以引起多器官疾病，甚至癌症的发生。一些金属及类金属，如镉、铅、砷、汞和铬等已被国家癌症研究机构认定为人类确定的或可能致癌物。其致癌机制可能与其引起细胞氧化还原平衡的破坏，从而产生大量活性氧，引起DNA和蛋白质损伤而产生突变；或影响细胞信号转导通路异常及基因甲基化、非编码RNA和组蛋白等表观遗传改变，从而诱导细胞的凋亡、细胞周期的改变和炎症的发生进而引起细胞恶性转化，从诱发肿瘤。重金属在环境中分布广泛，污染持续时间长，健康损害不易察觉，后期处理处置成本高。迫切需要对重点区域和敏感人群进行健康风险评估，从源头进行针对性的风险调控。随着分子生物学、基因技术的飞速发展，相信将更有助于进一步了解重金属致癌的机制，而这又必定推动我们重金属致癌防治研究工作更快发展。

关键词

重金属，肿瘤，分子机制，迁移转化，风险评估

1 引　　言

相对密度在 5 以上的金属，被称作重金属。目前，从环境污染方面，重金属主要是指汞、镉、铬、铅以及“类金属”（例如砷）等生物毒性显著的元素，对人体毒害最大的是铬、砷、镉等。这些重金属在环境介质中不能被分解，人体摄入后可以在体内累积并导致毒性放大，同时也可能在环境介质中发生转化生成毒性更大的金属化合物。重金属一般以天然浓度广泛存在于自然界中，但随着人类对重金属的开采、冶炼、加工及商业制造活动日益增多，造成不少重金属如铅、汞、镉、钴等进入大气、水、土壤中，引起严重的环境污染。近 30 年涉重金属产业的快速扩张，造成重金属污染排放总量持续处于高水平，重金属环境风险隐患日益突出。当前，重金属污染已成为事关群众身心健康和社会稳定的重大民生问题，《重金属污染综合防治“十三五”规划》已指出重金属污染防治的重要性。在重金属污染防治领域，我国借鉴了欧盟重金属防治政策措施，秉承了环境污染防止制度的一贯性理念和思想：努力实现减少或防止重金属释放到环境中，尽量避免重金属污染对环境和人类健康造成直接和间接影响。

一些重金属化合物（包括砷、汞、铅、镉和铬等）的慢性环境暴露可以引起多器官疾病，甚至引发癌症。研究表明，重金属污染与呼吸系统、骨骼、造血系统及心脑血管系统等多个系统的疾病有关，并且与癌症的发生发展有着密切的关系。随着对重金属与癌症关系的研究逐步深入，重金属致癌的分子机制研究受到广泛的关注。

2 重金属在环境介质中的迁移转化

以各种化学状态或化学形态存在的重金属，在进入环境生态系统后会存留、积累和迁移，从而造成危害。重金属中的任何一种超标，都不会被微生物降解，而只能发生各种形态相互转化和分散、富集过程（即迁移）。如随废水排出的重金属，即使浓度小，也可在藻类和底泥中积累，被鱼和贝的体表吸附，

产生食物链浓缩，严重危害人类健康。

重金属迁移指的是重金属在自然环境中随着时间的改变而发生空间位置的改变；转化是指随着介质条件的改变而使重金属的存在状态发生改变。重金属在环境中的分布，主要是在大气、土壤和水体中的迁移转化。

2.1 在大气中的迁移转化

大气中的重金属物质主要通过风力输送进行迁移，然后通过干湿作用在土壤和水体中富集，通过食物链作用危害人类健康。例如，城市垃圾焚烧飞灰是在城市生活垃圾焚烧处置过程中产生的二次污染物，其中富含镉、铅和铬等多种重金属、二噁英类剧毒有机污染物，这些物质均具有生物毒性并能在生物体内富集；飞灰中的重金属在环境中遇水浸出，污染土壤、水体和大气，对生态环境造成严重的破坏。汽车尾气排放的铅经大气扩散等过程进入环境中，造成目前地表铅的浓度已显著提高，致使近代人体内铅的吸收量比原始人增加了约 100 倍，损害人体健康。

2.2 在水体中的迁移转化

按照物质运动的形式，重金属在水体中的迁移分为三类：机械迁移、物理化学迁移和生物迁移。在水体中，污染物转化主要通过氧化还原、络合水解和生物降解等作用。环境中的重金属经过一定的迁移和转化，其结果不仅是化学性质（如毒性）发生改变，而且迁移能力也会发生改变。如环境中的三价铬和六价铬、三价砷和五价砷就是比较突出的例子。

2.3 在土壤中的迁移转化

在天然土壤中含有一定的重金属背景值，这称为土壤重金属元素的基线含量。重金属一旦进入土壤就难以除去，它难以被生物降解，通过生物富集，进入食物链危害人类健康。如痛痛病，是由炼锌工业和镉电镀工业所排放的镉进入土壤中所致。

3 重金属的暴露及致癌机制

通常，一般人群暴露于各种各样的重金属环境中。其中，一些来源于自制成品，一些来源于环境污染。研究表明，重金属的暴露会损伤神经以及心血管系统，并增加包括肾、肺、肝、皮肤和胃癌在内的等众多癌症发生的风险。一些金属及类金属，如镉、铅、砷、汞和铬等已被国家癌症研究机构认定为人类确定的或可能致癌物。重金属在机体内的蓄积可影响细胞的正常生理功能，可以产生大量活性氧，引起 DNA 和蛋白质的损伤以及构象的改变，从而诱导细胞的凋亡、细胞周期的改变以及细胞癌变等。因此，重金属致癌也越来越受到人们的广泛关注。

3.1 镉暴露及致癌机制

镉是受到广泛关注的致癌物质，环境中的镉暴露进入人体后，因其半衰期较长而在体内蓄积。镉及其化合物被国际癌症研究机构（IARC）认定为人类确定致癌物。美国国家癌症研究机构的国家环境健康科学研究所等统计发现，镉暴露可以引起乳腺癌、肺癌、前列腺癌以及胃肠道肿瘤等[1]。职业和环境中的镉，主要来自于采矿、冶金行业以及镍镉电池、颜料和塑料稳定剂的制造等，分布广泛。

镉可以影响细胞增殖、分化、凋亡等，并且可引起与致癌作用相关的多种分子机制的改变。涉及镉致癌的主要机制可能与抑制基因表达、DNA 损伤修复障碍、抑制凋亡和诱导氧化应激有关[2]。除此之外，镉通过异常的 DNA 甲基化、内分泌干扰和细胞增殖影响而引起肿瘤[3]。长久以来，镉被认为是非致病性致癌物，因发现其在细菌和哺乳动物细胞中只有很微弱的致突变性；然而，镉作为一种很强的诱变剂又可以引起多位点的缺失[4]。实验证据表明，镉通过氧化应激引起 DNA 损伤，并通过抑制内源性和外源性 DNA 突变修复来影响基因组的稳定性；诱导 DNA 损伤还可以引起细胞周期阻滞、基因突变、基因组不稳定性等，进而导致细胞凋亡和癌变。镉诱导体内氧自由基的产生，同时激活对氧化还原敏感的转录因子。尽管谷胱甘肽（GSH）是抵御镉毒性的第一道防线[5]，但是镉会消耗 GSH 和蛋白质介质的巯基，导致活性氧（reactive oxygen species，ROS）的产生、脂质过氧化、氧化性 DNA 损伤以及改变钙和巯基的稳定性[6]。此外，适应慢性镉暴露降低了 ROS 的产生，但是获得的镉耐受相关基因的异常表达在致癌过程中发挥重要作用[5]。总之，镉会破坏抗氧化系统，降低超氧化物歧化酶（SOD）、过氧化氢酶（CAT）和 GSH 过氧化物酶（GPX）的活性，产生大量的有害物质，诱发癌症的发生[7]。镉还可以诱导多种基因的异常表达，如抑癌基因 p53，与细胞凋亡和增殖有关的 Bcl-2、Bax 和 PCNA 等，进而打破原来抑癌的平衡和正常的细胞增殖凋亡，抑癌基因失活，癌细胞过度增殖及凋亡减弱，从而导致癌症的发生发展[8]。另外，Kortenkamp 等[9]研究认为，镉可以被看作一种雌激素类似物，与雌激素受体结合，模拟雌激素作用，而在长期的雌激素刺激下，乳腺癌发病率增加，这成为镉引起乳腺癌的又一个重要原因。

3.2　砷暴露及致癌机制

砷是广泛存在的环境污染物，可以在土壤、水以及空气颗粒物中被发现。从 1987 年，砷被 IARC 认定为人类确定致癌物质。环境中，砷主要以三价的亚砷酸盐和五价的砷酸盐形式存在[10]。流行病学研究显示，砷暴露与肺癌、膀胱癌、皮肤癌和肝癌等的发生密切相关。而砷增加患癌的风险主要归因于亚砷酸盐而不是砷酸盐，这可能是由于细胞对亚砷酸盐的吸收速率明显高于砷酸盐[11]。1993 年以前，世界卫生组织（WHO）建议地下水砷浓度安全标准最高不超过 50 μg/L，并且提出临时标准砷浓度最高不超过 10 μg/L[12]，目前安全标准已经降到 10 μg/L。由于地下水中存在天然的砷污染，印度和孟加拉国是受砷影响最严重的国家；而我国西南部地区的贵州省由于燃用高砷煤而引起的地方性燃煤型砷中毒，目前受到广泛的关注[13,14]。对于陶瓷、玻璃制造、金属提纯冶炼，以及农药制造和应用都会使人体暴露于高度砷污染的环境中。因此，越来越多的研究者致力于砷暴露引起癌症发生发展机制的研究。

通过对砷流行病学以及生物动力学多方面的研究，综合提出砷的毒作用与暴露剂量、频率、持续时间、生物种类、年龄、性别以及个体易患性、基因、营养水平有关[15]。砷致癌的多种机制被得到证实，如砷诱导的染色体畸变、氧化应激、DNA 修复障碍、DNA 甲基化模式改变以及生长因子分泌水平改变都可以增强细胞增殖和肿瘤形成及肿瘤发生发展[16]。砷通过不同的机制干扰机体自然的氧化还原平衡状态，即高浓度砷可抑制 GSH 还原酶活性，从而减少还原型 GSH 水平使组织产生氧化应激，进一步激活对氧化还原状态敏感的信号分子，如 AP-1、NF-κB、p53、p21 及 ras 等，从而影响信号转导系统激活相关基因的表达异常，从而引起细胞损伤和死亡[17,18]。Ling 等[19]研究发现，砷暴露可以引起细胞内 ROS 水平升高，其通过激活 ERK/NF-B 信号通路而诱导 miR-21 表达升高，进而促进细胞恶性转化；氧化应激还可以导致 SOD 和过氧化氢酶活性降低，其也是砷甲基化和砷运输重要的细胞内还原剂[20]。Xu[21]等研究表明，低水平亚砷酸钠通过激活 ERK 信号通路，通过抑制 HIF-2α 的泛素化降解而抑制 p53 激活，最终导致 p53

功能失活，刺激细胞增殖，长期处理诱导细胞发生恶性转化并具有致瘤性；而高水平亚砷酸钠则通过激活 JNK 信号通路，通过抑制 HIF-1α 的泛素化降解从而诱导其高表达，进而发挥其转录调控因子的作用，导致 DNA 损伤并促进细胞凋亡。砷可以引起组蛋白修饰并调节组蛋白修饰酶，包括 DNA 修复酶 ABH2 和 ABH3，以及组蛋白甲基转移酶，其可能影响表观遗传改变[22,23]；并且砷可以诱导促凋亡基因沉默或突变，促进损伤细胞的存活，导致恶性转化[24]。有研究发现砷可以诱导肿瘤抑制基因 p53 的启动子区 DNA 高甲基化，导致细胞恶性转化，并分泌粒细胞巨噬细胞结肠成纤维细胞因子（GM-CSF）、转化生长因子 α（TGF-α）以及促炎细胞因子肿瘤坏死因子 α（TNF-α）等生长因子[25]。Luo 等[26,27] 有研究表明，砷暴露诱发炎症反应，引起 IL-6/STAT3 信号通路改变，进而改变非编码 RNA 如 miR-21 和 miR-155 等表达水平，并通过调控相关信号通过的改变来促进细胞增殖加快，从而诱导细胞恶性转化。

3.3 汞暴露及其毒性机制

目前汞及其化合物被 IARC 认定为 C 类致癌物质，元素汞具有强挥发性，进入大气后又沉积下来形成全球自然循环。汞广泛存在于环境中，各种生物都不可避免地暴露于含汞环境中，汞中毒常以慢性多见。汞暴露主要来源于如电器、油漆工业、测量仪器、牙科材料，以及化学工业等各种工业生产。

汞暴露可直接或间接的促进机体 ROS 的产生；外源性 ROS 的产生可以导致细胞 DNA 损伤，包括上皮及内皮细胞的损伤和炎症的发生[28]。有研究报道汞暴露可以通过诱发 ROS 的水平升高而改变过氧化氢酶活性，并且可抑制 DNA 损伤修复，最终促进胃癌的发生，从而认为过氧化氢酶活性的改变参与机体的抗氧化机制，其可能与肿瘤的进展有关[29]。细胞培养模型中发现，甲基汞可影响多种神经元生长因子的表达，进而影响神经干细胞分化和神经突触生长[30]。产前甲基汞暴露可增加 H3K27me3 的水平，并降低脑源性神经营养因子（BDNF）启动子区 H3Ac 水平，最终导致 BDNF 在脑齿状回中表达水平降低，从而影响婴儿脑的发育[31]。通过汞的亚细胞毒作用研究发现，其毒作用是通过改变细胞内氧化还原反应速率以及蛋白激酶 A、蛋白激酶 C 介导的信转导通路引起线粒体内 ROS 的蓄积及细胞间缝隙通讯异常，从而认为汞的致癌性与其在真核细胞系中可引起突变发生及影响细胞间缝隙连接通讯等有关[32,33]。

3.4 铅暴露及致癌机制

目前，铅及其化合物被 IARC 列为 2B 类致癌物。日常生活中的水和食物都可能受到铅污染，铅中毒通常是由精炼操作中的铅烟、电池制造中的氧化铅等引起的。铅是一种全身性毒物，对周围和中枢神经系统、生殖系统、免疫系统以及肾脏和肝脏等组织都具有毒性，且可引起心血管疾病和癌症病死率增加[34,35]。

近来实验和流行病学研究提供了足够的证据表明无机铅化合物与癌症发生的风险增加有关。铅致癌机制包括 DNA 损伤、DNA 合成或修复障碍、致突变性以及产生 ROS，从而导致氧化性 DNA 损伤等[34]。细胞介素-8（IL-8）能够促进肿瘤转移和血管生成，研究发现活化蛋白-1（AP-1）和细胞外信号调节激酶在铅诱导的 IL-8 基因激活从而在其所致人胃癌的发生发展中起到关键作用[36]。铅还可以激活表皮生长因子受体（EGFR），并诱导下游细胞外信号调节激酶（ERK）-1/2 的磷酸化，从而激活 AP-1 [37]。Hermes-Lima 等[38]研究铅与自由基间的关系表明，铅中毒后通过产生 ROS 从而造成细胞损伤。Crover 等[39]研究还发现，铅可以使红细胞中的氨基乙酰丙酸脱水酶减少，而尿液中 5-氨基酮戊酸增多，从而氧化产生的自由基增多，引起 DNA 损伤。且铅可以取代一些 DNA 修复蛋白锌指结构中的锌，而这些蛋白为转录调节因子，锌被取代后，减少这些蛋白与自由基 DNA 中的识别元件结合，提示铅暴露所产生的后续反应可能与基因表达有关[40]。铅暴露还可导致与癌症发展相关的 miR-10、miR-154、miR-375 和 miR-10、miR-222、miR-379、 miR-133 和 miR-204 表达水平升高，从而影响细胞进程，如炎症反应、细胞生长和增殖以及细胞死亡等[41]。因此，进一步研究 miRNAs 对铅暴露引起癌症发展十分必要。

3.5 铬暴露及致癌机制

铬在环境中分布广泛，铬暴露主要见于不锈钢焊接、镀铬和铬铁制造等多个行业。人类接触的主要来源是工业区附近铬污染的水，全球有数千万人受到铬污染的危害。铬在体内主要以三价和六价形式存在，流行病学研究显示六价铬 Cr(Ⅵ)的毒性比三价铬 Cr(Ⅲ)高约 100，且常与呼吸道癌症的发生发展相关[42]。

铬具有遗传毒性作用，可引起氧化应激、DNA 链断裂、DNA-蛋白质交联，从而形成稳定的铬-DNA 加合物等[43]。铬暴露会引起细胞氧化应激，产生大量 ROS，进而通过促进过氧化物的生成引发脂质过氧化及 DNA 损伤及细胞信号转导通路的改变，从而引起酶学、形态学等不可逆的损伤，甚至凋亡[44]。彗星实验显示，铬可以诱导外周血淋巴细胞和人胃黏膜细胞的 DNA 损伤[45]。研究发现，用铬处理的大鼠胃中 p53 表达水平降低，而 c-myc 表达水平升高，参与抑制 DNA 损伤修复，从而引发大鼠胃癌发生[46]。最近，有研究发现长期铬暴露引起组蛋白甲基化参与基因沉默与抑制，导致表观遗传和架构改变，最终引起致癌基因转录改变促进肿瘤的发生发展[47]。Abreu 等[48]研究发现，铬暴露可以引起 HIF-1α 稳定表达以及 NF-κB 激活，进而促进有氧糖酵解的发生，参与细胞恶性转化，且铬暴露可以促进肿瘤细胞的转移与侵袭，可能与产生的 ROS 有关[49]。

4 重金属环境健康风险评估

环境健康风险评估是指按照一定的准则，对化学毒物损害人体健康的潜在能力进行定性和定量的评估，以判断损害可能发生的概率及严重程度。其将环境污染与人体健康联系起来，对污染可能造成的健康危害进行了定量描述。美国国家科学院（NAS）最初于 1983 年提出了风险评估及管理理论，此后该理论广泛应用于环境污染对人体健康危害的评估。随后风险评估不断发展与完善，我国也在逐渐完善环境健康风险评估的相关法律法规、基础数据及监测指标等，但我国的环境与健康工作仍显薄弱。

重金属进入人体后，会蓄积于神经系统和肝脏、肾脏，骨骼等组织中，最终对其造成损伤[50,51]。近年，我国重金属污染损害人体健康的事故频繁发生，呈暴发势态，多次引发群体性事件，使“重金属污染”成为社会热点问题和环境保护工作面临的新挑战。重金属在环境中分布广泛，污染持续时间长，健康损害不易察觉，后期处理处置成本高。迫切需要对重点区域和敏感人群进行健康风险评估，从源头进行针对性的风险调控。

目前，我国进行环境风险评估主要采用 4 步法，即危害识别、剂量-反应评价、暴露评价和风险表征。危害识别即判断毒物对人体的危害，主要采用的是动物实验所获得的资料。目前，流行病学调查资料所得结果可靠性很高，如调查结果发现环境中重金属铅、镉污染可能会诱发儿童龋齿病[52,53]。剂量-反应评价是对毒物暴露水平与暴露人群或生物种群中不良健康反应发生率之间的关系的定量评估，其中有阈值毒物用参考剂量进行衡量，而无阈值毒物则需从动物实验进行外推得出。暴露评价为应用一定的暴露模型对暴露剂量进行定量，该过程不仅要考虑环境中毒物的剂量，还要考虑人群的行为时间模式。风险表征即在前 3 个步骤的基础上得出风险水平值，估算人群接触毒物后对健康的危险度。其中重金属的危害识别及剂量-反应评价可在借鉴国外有关结果的基础上，结合我国的重金属污染情况及居民身体素质，进行相应的修正，构建适应我国的数据库。风险表征可过公式计算得出。而暴露评价是四个步骤中的关键环节。暴露评价包括暴露测量、暴露参数、暴露模型等。

5 展　望

我国的环境污染为复合型，其中重金属污染占很大比例，人群暴露时间长，历史性积累对健康的影响短期内难以消除，而且这些重金属污染与癌症的发生有着明显的相关性。既然重金属会导致癌症的发生，那么如何防治、逆转重金属的生物毒性作用，成为我们研究的重中之重。在清楚了解重金属致癌的分子机制后，我们才能更好地进行重金属的分子防治工作。在各种重金属的众多致癌通路中，我们需尽力寻找几个药物靶点，阻断重金属对有机体的毒性及致癌作用，甚至寻找到可以逆转重金属致癌作用的药物，这也是我们未来的研究重点之一。分子生物学、基因技术的飞速发展，有助于我们进一步了解重金属致癌的机制，而这又必定推动重金属致癌防治研究工作更快发展。同时，更要结合重金属环境健康风险评估，并构建国家及地方环境与健康监测数据库以及构建不同重金属对不同生物及人类毒性终点阈值的数据库。因暴露的复杂性，也应更注重多途径多介质暴露的研究，以全面评估人体的重金属暴露情况及健康风险。

参 考 文 献

[1] Mohajer R, Salehi M H, Mohammadi J, Emami M H, Azarm T. The status of lead and cadmium in soils of high prevalenct gastrointestinal cancer region of Isfahan. J Res Med Sci, 2013, 18(3): 210-214.

[2] Hartwig A. Mechanisms in cadmium-induced carcinogenicity: recent insights. Biometals, 2010, 23(5): 951-960.

[3] Huang D, Zhang Y, Qi Y, Chen C, Ji W. Global DNA hypomethylation, rather than reactive oxygen species(ROS), A potential facilitator of cadmium-stimulated K562 cell proliferation. Toxicol Lett, 2008, 179(1): 43-47.

[4] Filipic M, Fatur T, Vudrag M. Molecular mechanisms of cadmium induced mutagenicity. Hum Exp Toxicol, 2006, 25(2): 67-77.

[5] Liu J, Qu W, Kadiiska M B. Role of oxidative stress in cadmium toxicity and carcinogenesis. Toxicol Appl Pharmacol, 2009, 238(3): 209-214.

[6] Stohs S J, Bagchi D. Oxidative mechanisms in the toxicity of metal ions. Free Radic Biol Med, 1995, 18(2): 321-336.

[7] Wu C, Wang L, Liu C, Gao F, Su M, Wu X, Hong F. Mechanism of Cd^{2+} on DNA cleavage and Ca^{2+} on DNA repair in liver of silver crucian carp. Fish Physiol Biochem, 2008, 34(1): 43-51.

[8] So K Y, Oh S H. Heme oxygenase-1-mediated apoptosis under cadmium-induced oxidative stress is regulated by autophagy, which is sensitized by tumor suppressor p53. Biochem Biophys Res Commun, 2016, 479(1): 80-85.

[9] Golovine K, Makhov P, Uzzo R G, Kutikov A, Kaplan D J, Fox E, Kolenko V M. Cadmium down-regulates expression of XIAP at the post-transcriptional level in prostate cancer cells through an NF-kappaB-independent, proteasome-mediated mechanism. Mol Cancer, 2010, 9: 183.

[10] Bertolero F, Pozzi G, Sabbioni E, Saffiotti U. Cellular uptake and metabolic reduction of pentavalent to trivalent arsenic as determinants of cytotoxicity and morphological transformation. Carcinogenesis, 1987, 8(6): 803-808.

[11] Reichard J F, Puga A. Effects of arsenic exposure on DNA methylation and epigenetic gene regulation. Epigenomics, 2010, 2(1): 87-104.

[12] Rahman M M, Chen Z, Naidu R. Extraction of arsenic species in soils using microwave-assisted extraction detected by ion

chromatography coupled to inductively coupled plasma mass spectrometry. Environ Geochem Health, 2009, 31 Suppl 193-102.

[13] Zhang A, Feng H, Yang G, Pan X, Jiang X, Huang X, Dong X, Yang D, Xie Y, Peng L, Jun L, Hu C, Jian L, Wang X. Unventilated indoor coal-fired stoves in Guizhou province, China: Cellular and genetic damage in villagers exposed to arsenic in food and air. Environ Health Perspect, 2007, 115(4): 653-658.

[14] Liu J, Zheng B, Aposhian H V, Zhou Y, Chen M L, Zhang A, Waalkes M P. Chronic arsenic poisoning from burning high-arsenic-containing coal in Guizhou, China. Environ Health Perspect, 2002, 110(2): 119-122.

[15] Abernathy C O, Liu Y P, Longfellow D, Aposhian H V, Beck B, Fowler B, Goyer R, Menzer R, Rossman T, Thompson C, Waalkes M. Arsenic: Health effects, mechanisms of actions, and research issues. Environ Health Perspect, 1999, 107(7): 593-597.

[16] Miller W H, Jr., Schipper H M, Lee J S, Singer J, Waxman S. Mechanisms of action of arsenic trioxide. Cancer Res, 2002, 62(14): 3893-3903.

[17] Gong X, Ivanov V N, Hei T K. 2, 3, 5, 6-Tetramethylpyrazine (TMP) down-regulated arsenic-induced heme oxygenase-1 and ARS2 expression by inhibiting Nrf2, NF-kappaB, AP-1 and MAPK pathways in human proximal tubular cells. Arch Toxicol, 2016, 90(9): 2187-2200.

[18] Stamatelos S K, Androulakis I P, Kong A N, Georgopoulos P G. A semi-mechanistic integrated toxicokinetic-toxicodynamic (TK/TD) model for arsenic(III) in hepatocytes. J Theor Biol, 2013, 317 244-256.

[19] Ling M, Li Y, Xu Y, Pang Y, Shen L, Jiang R, Zhao Y, Yang X, Zhang J, Zhou J, Wang X, Liu Q. Regulation of miRNA-21 by reactive oxygen species-activated ERK/NF-kappaB in arsenite-induced cell transformation. Free Radic Biol Med, 2012, 52(9): 1508-1518.

[20] Gupta D K, Inouhe M, Rodriguez-Serrano M, Romero-Puertas M C, Sandalio L M. Oxidative stress and arsenic toxicity: role of NADPH oxidases. Chemosphere, 2013, 90(6): 1987-1996.

[21] Xu Y, Li Y, Li H, Pang Y, Zhao Y, Jiang R, Shen L, Zhou J, Wang X, Liu Q. The accumulations of HIF-1alpha and HIF-2alpha by JNK and ERK are involved in biphasic effects induced by different levels of arsenite in human bronchial epithelial cells. Toxicol Appl Pharmacol, 2013, 266(2): 187-197.

[22] Chervona Y, Arita A, Costa M, Carcinogenic metals and the epigenome: understanding the effect of nickel, arsenic, and chromium, Metallomics, 2012, 4(7): 619-627.

[23] Zhang A, Li H, Xiao Y, Chen L, Zhu X, Li J, Ma L, Pan X, Chen W, He Z. Aberrant methylation of nucleotide excision repair genes is associated with chronic arsenic poisoning. Biomarkers, 2017, 22(5): 429-438.

[24] Bhattacharjee P, Banerjee M, Giri A K. Role of genomic instability in arsenic-induced carcinogenicity. A review. Environ Int, 2013, 53 29-40.

[25] Xie H, Huang S, Martin S, Wise J P, Sr. Arsenic is cytotoxic and genotoxic to primary human lung cells. Mutat Res Genet Toxicol Environ Mutagen, 2014, 760 33-41.

[26] Luo F, Xu Y, Ling M, Zhao Y, Xu W, Liang X, Jiang R, Wang B, Bian Q, Liu Q. Arsenite evokes IL-6 secretion, autocrine regulation of STAT3 signaling, and miR-21 expression, processes involved in the EMT and malignant transformation of human bronchial epithelial cells. Toxicol Appl Pharmacol, 2013, 273(1): 27-34.

[27] Lu X, Luo F, Liu Y, Zhang A, Li J, Wang B, Xu W, Shi L, Liu X, Lu L, Liu Q. The IL-6/STAT3 pathway via miR-21 is involved in the neoplastic and metastatic properties of arsenite-transformed human keratinocytes. Toxicol Lett, 2015, 237(3): 191-199.

[28] Ates-Alagoz Z. Antioxidant activities of retinoidal benzimidazole or indole derivatives in in vitro model systems. Curr Med Chem, 2013, 20(36): 4633-4639.

[29] Martin Mateo M C, Martin B, Santos Beneit M, Rabadan J. Catalase activity in erythrocytes from colon and gastric cancer patients. Influence of nickel, lead, mercury, and cadmium. Biol Trace Elem Res, 1997, 57(1): 79-90.

[30] Tamm C, Duckworth J, Hermanson O, Ceccatelli S. High susceptibility of neural stem cells to methylmercury toxicity: effects on cell survival and neuronal differentiation. J Neurochem, 2006, 97(1): 69-78.

[31] Onishchenko N, Karpova N, Sabri F, Castren E, Ceccatelli S. Long-lasting depression-like behavior and epigenetic changes of BDNF gene expression induced by perinatal exposure to methylmercury. J Neurochem, 2008, 106(3): 1378-1387.

[32] Inoue M, Sato E F, Nishikawa M, Park A M, Kira Y, Imada I, Utsumi K. Mitochondrial generation of reactive oxygen species and its role in aerobic life. Curr Med Chem, 2003, 10(23): 2495-2505.

[33] Piccoli C, D'Aprile A, Scrima R, Ambrosi L, Zefferino R, Capitanio N. Subcytotoxic mercury chloride inhibits gap junction intercellular communication by a redox- and phosphorylation-mediated mechanism. Free Radic Biol Med, 2012, 52(5): 916-927.

[34] Chandran L, Cataldo R. Lead poisoning: basics and new developments. Pediatr Rev, 2010, 31(10): 399-405, quiz 406.

[35] Wang M, Xu Y, Pan S, Zhang J, Zhong A, Song H, Ling W. Long-term heavy metal pollution and mortality in a Chinese population: an ecologic study. Biol Trace Elem Res, 2011, 142(3): 362-379.

[36] Yuan W, Yang N, Li X. Advances in Understanding How Heavy Metal Pollution Triggers Gastric Cancer. Biomed Res Int, 2016, 20167825432.

[37] Chan C P, Tsai Y T, Chen Y L, Hsu Y W, Tseng J T, Chuang H Y, Shiurba R, Lee M H, Wang J Y, Chang W C. Pb^{2+} induces gastrin gene expression by extracellular signal-regulated kinases 1/2 and transcription factor activator protein 1 in human gastric carcinoma cells. Environ Toxicol, 2015, 30(2): 129-136.

[38] Hermes-Lima M, Pereira B, Bechara E J. Are free radicals involved in lead poisoning?, Xenobiotica, 1991, 21(8): 1085-1090.

[39] Grover P, Rekhadevi P V, Danadevi K, Vuyyuri S B, Mahboob M, Rahman M F. Genotoxicity evaluation in workers occupationally exposed to lead. Int J Hyg Environ Health, 2010, 213(2): 99-106.

[40] Razmiafshari M, Kao J, d'Avignon A, Zawia N H. NMR identification of heavy metal-binding sites in a synthetic zinc finger peptide: toxicological implications for the interactions of xenobiotic metals with zinc finger proteins. Toxicol Appl Pharmacol, 2001, 172(1): 1-10.

[41] Martinez-Pacheco M, Hidalgo-Miranda A, Romero-Cordoba S, Valverde M, Rojas E. MRNA and miRNA expression patterns associated to pathways linked to metal mixture health effects. Gene, 2014, 533(2): 508-514.

[42] Salnikow K, Zhitkovich A. Genetic and epigenetic mechanisms in metal carcinogenesis and cocarcinogenesis: nickel, arsenic, and chromium. Chem Res Toxicol, 2008, 21(1): 28-44.

[43] Zhitkovich A. Importance of chromium-DNA adducts in mutagenicity and toxicity of chromium(VI). Chem Res Toxicol, 2005, 18(1): 3-11.

[44] Shi H, Hudson L G, Liu K J. Oxidative stress and apoptosis in metal ion-induced carcinogenesis. Free Radic Biol Med, 2004, 37(5): 582-593.

[45] Trzeciak A, Kowalik J, Malecka-Panas E, Drzewoski J, Wojewodzka M, Iwanenko T, Blasiak J. Genotoxicity of chromium in human gastric mucosa cells and peripheral blood lymphocytes evaluated by the single cell gel electrophoresis (comet assay). Med Sci Monit, 2000, 6(1): 24-29.

[46] Tsao D A, Tseng W C, Chang H R. The expression of RKIP, RhoGDI, galectin, c-Myc and p53 in gastrointestinal system of Cr(VI)-exposed rats. J Appl Toxicol, 2011, 31(8): 730-740.

[47] Stern A H. A quantitative assessment of the carcinogenicity of hexavalent chromium by the oral route and its relevance to human exposure. Environ Res, 2010, 110(8): 798-807.

[48] Abreu P L, Ferreira L M, Alpoim M C, Urbano A M. Impact of hexavalent chromium on mammalian cell bioenergetics:

phenotypic changes, molecular basis and potential relevance to chromate-induced lung cancer. Biometals, 2014, 27(3): 409-443.

[49] Ochieng J, Nangami G N, Ogunkua O, Miousse I R, Koturbash I, Odero-Marah V, McCawley L J, Nangia-Makker P, Ahmed N, Luqmani Y, Chen Z, Papagerakis S, Wolf G T, Dong C, Zhou B P, Brown D G, Colacci A M, Hamid R A, Mondello C, Raju J, Ryan E P, Woodrick J, Scovassi A I, Singh N, Vaccari M, Roy R, Forte S, Memeo L, Salem H K, Amedei A, Al-Temaimi R, Al-Mulla F, Bisson W H, Eltom S E. The impact of low-dose carcinogens and environmental disruptors on tissue invasion and metastasis. Carcinogenesis, 2015, 36 Suppl 1 S128-159.

[50] Chen T, Liu X, Zhu M, Zhao K, Wu J, Xu J, Huang P. Identification of trace element sources and associated risk assessment in vegetable soils of the urban-rural transitional area of Hangzhou, China. Environ Pollut, 2008, 151(1): 67-78.

[51] Granero S, Domingo J L. Levels of metals in soils of Alcala de Henares, Spain: human health risks. Environ Int, 2002, 28(3): 159-164.

[52] Arora M, Weuve J, Schwartz J, Wright R O. Association of environmental cadmium exposure with pediatric dental caries. Environ Health Perspect, 2008, 116(6): 821-825.

[53] Chang W H, Yang Y H, Liou S H, Liu C W, Chen C Y, Fuh L J, Huang S L, Yang C Y, Wu T N. Effects of mixology courses and blood lead levels on dental caries among students. Community Dent Oral Epidemiol, 2010, 38(3): 222-227.

作者：刘起展[1]，罗　菲[2]

[1]南京医科大学公共卫生学院，[2]上海交通大学公共卫生学院

第 30 章　梯度扩散薄膜技术（DGT）的环境研究进展

本章导读

能够原位富集、选择性吸附不同形态污染物的被动采样技术逐渐成为环境研究的一个重要方向。梯度扩散薄膜技术（DGT）是这种被动采样技术的杰出代表。从它问世以来，在环境科学、地球化学、植物营养、海洋科学等研究领域获得了广泛应用，逐渐被主流研究所接受。目前已经开发了针对重金属、营养盐、极性有机物污染物的DGT技术，同时该技术在环境监测、污染物生物有效性、污染物环境作用机理研究方面取得了长足进步。本章将综合介绍梯度扩散薄膜技术在环境领域的研究进展。

关键词

梯度扩散薄膜技术，DGT，土壤，沉积物，水，重金属，有机物，营养盐

1 引　　言

梯度扩散薄膜技术(diffusive gradients in thin-films，DGT)是由英国Lancaster大学张昊教授和William Davison教授发明的一种被动采样技术[1]。该技术以菲克（Fick）扩散第一定律为其理论基础，对在特定的富集时间内累积于吸附膜上的污染物进行定量计算，可在不破坏环境介质的前提下获得目标离子的含量、形态及迁移信息[1,2]。DGT技术具有省时省力、高空间分辨（毫米-亚毫米）、高选择性、灵敏性、能够提供时间加权平均浓度（TWA），以及多物质同步监测等优势，目前被广泛应用于环境科学、地球化学、植物营养、海洋科学等研究领域。

DGT技术被公认为是当前最有效、最具发展前景的一种绿色采样技术。经过20多年的发展，目前全球已有130余家科研和行业机构从事DGT技术开发及其应用研究。随着DGT吸附膜制备工艺改进与新型吸附膜材料开发，DGT由低吸附容量向高吸附容量，吸附膜制备材料粒径由微米级向纳米级，由单一吸附材料向复合吸附材料等方向发展。例如，DGT吸附膜制备最初多采用铁/钛/锆氧化物（FeO/TiO_2/ZrO）[3-6]、Chelex 100[7]等材料，随后在这些材料的基础上陆续发展了多种复合材料吸附膜，具有更好的吸附容量和吸附性能，因此非常适合复杂环境基质多物质同步监测[8,9]。传统吸附胶中的氧化铁/锆/钛、Chelex 100等颗粒粒径为微米级别，而最新研制的亚氨基二乙酸悬浮颗粒（SPR-IDA）[10]、四氧化三铁水悬浮颗粒（Fe_3O_4 NPs）[11]，以及沉降型氧化锆颗粒（precipitated zirconia）[12,13]等吸附材料粒径为纳米级别，后者更适合当前沉积物研究中对亚微米级微区分析需求。此外，针对有机污染物的吸附材料也在逐步研发中，Chen等[14]采用含有XAD-18树脂吸附膜的DGT装置准确测定了水体和土壤中的抗生素，Zheng等[15]使用活性炭吸附膜的DGT装置实现了对双酚类物质的检测，DGT野外测定的结果与主动采样方法类似，表明了DGT在有机污染物的测定方面有良好的效果。DGT技术从最初仅适用于较低含量目标物质水环境体系扩展到高浓度、强干扰性的复杂土壤/沉积物体系，而DGT的监测目标物质也从最初的铁（Fe）、铜（Cu）、锌（Zn）、镍（Ni）等痕量重金属逐步扩展到沉积物P、硫（S）、氮（N）等非金属元素，铀（U）、锶（Sr）、铯（Se）放射性金属以及抗生素、毒品、除草剂、持久性有机物质（POPs）等有机化合物的测定。

大量研究已证实DGT技术在环境研究中具有独特的优势并显示出很大的发展前景。除DGT新方法的开发及环境监测中的应用外，该技术在环境领域的研究已经扩展到对各种水体、土壤、沉积物营养盐/污染物迁移转化、生物有效性、污染物形态分析以及污染风险评估等研究工作中。张昊等[7]首次将DGT用于原位测定英国Esthwaite湖表层沉积物孔隙水中Ni、Cu、Fe和Mn的通量和孔隙水中Zn和Cd的浓

度。Fresno 等[16]利用 DGT 研究了在加入铁土壤改良剂后污染土壤上 As、Cu 可移动性的变化趋势。此外，DGT 技术应用扩展到原位获取沉积物硫的空间分布信息[17,18]。与此同时，我国部分学者利用 DGT 技术针对污染严重的河流和湖泊开展了大量研究工作。如范英宏等[19,20]利用 DGT 技术分析了我国大辽河水系表层沉积物孔隙水中重金属的分布特征，同时运用 DIFS 模型计算了重金属离子的迁移动力学特征。Wu 等[21]利用 DGT 技术原位获得了滇池沉积物-水界面 P、S 和 7 种主要重金属空间分布信息。Zhou 等[22]对太湖全湖 Fe、Mn、Zn 等 10 种重金属和营养盐有效态含量进行了现场原位监测并分析了污染物区域特征。随着 DGT 研究的逐步深入，将促进待测物质生物有效性研究并推动元素的生物地球化学循环研究。

2 DGT 在生物有效性方面的应用研究

DGT 放置过程中产生的扩散流量使得 DGT 对介质中原有的化学平衡造成了扰动，因此 DGT 测量结果不仅反映了溶液中的待测物，而且包括了固相向液相的补充能力，也使得只有在扩散过程中易于解离的配合物才能被 DGT 测定（图 30-1）。DGT 因具有能够反映固相解吸以及配体解离等动态过程的特点，成为了生物有效性及其相关过程研究的有效工具。近年来，将 DGT 技术用于生物吸收、生长特征和毒性研究越来越多[23]。根据生物生长的介质不同，DGT 用于生物有效性的研究主要分为土壤、水体、沉积物等几个方面。

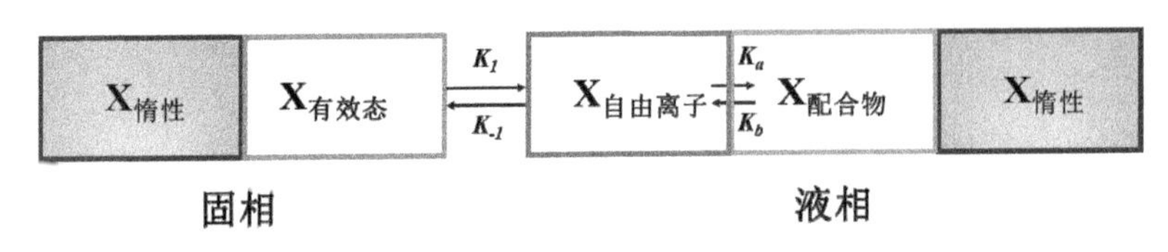

图 30-1　DGT 与植物吸收相似的土壤动力学过程

2.1　DGT 在土壤生物有效性方面的应用

2.1.1　DGT 对植物吸收土壤中金属的预测

基于 DGT 技术研究土壤重金属的生物有效性主要通过比较 DGT 所测数据与植物体内的金属含量的相关性，相关研究涵盖不同类型的土壤和植物。DGT 数据通常用 C_{DGT}（DGT 直接测定的浓度）或者 C_E（有效浓度，DGT 直接测定的浓度进行转化）表示。Davison 等[23]首次探讨 DGT 测定的金属浓度与英国水芹（*Lepidium heterophyllum*）体内的金属浓度之间的关系，相关研究说明 DGT 技术能够很好地预测植物对重金属的吸收。张昊等[24]引入了有效浓度（C_E）的概念，能更好地反映土壤金属的生物有效性，进而进行土壤金属污染的风险评价。发现 DGT 方法所得的 Cu 浓度与植物体内 Cu 浓度的相关性明显优于传统的土壤分析方法（土壤溶液、EDTA 提取、自由离子活性）。Nolan 等[25]采用 DGT 方法预测了小麦对土壤 Zn、Cd、Pb、Cu 的吸收，证实 DGT 能够很好地预测出土壤中重金属的生物有效性。Mundus 等[26]的研究也表明 DGT 是一种有效且可靠的获取农业土壤上 Mn 生物有效性的方法。国内王晓蓉教授团队[27]首次采用 DGT 技术研究了野外采集水稻及其根际土壤，发现 DGT 所测的重金属浓度（Cd、Cu、Pb 和 Zn）与水稻吸收金属含量的相关性明显优于其他的形态分析方法（如土壤溶液、HAc 提取、$CaCl_2$ 提取），说明 DGT 技术能够应用于野外研究。Wang 等[28]用不同的土壤分析方法（土壤溶液、0.5 mol/L $NaHCO_3$ 提取、1 mol/L HCl 提取、1 mol/L NH_4Cl 提取、0.5 mol/L NH_4F 提取、0.1 mol/L NaOH 提取、0.25 mol/L H_2SO_4）对土壤中生物有效态 As 进行测定，发现 DGT 方法与植物体内 As 含量相关性最好，说明 DGT 有能力反映植物所吸收的 As。目前，采用 DGT 技术预测土壤中金属的生物有效性的研究所牵涉的元素有 As、Cd、Cu、Zn、Pb、Ni、P、Sb、U、Hg 等，并考虑土壤性质对 DGT 预测结果的影响[29-32]。

在大多数研究中，与其他土壤分析方法相比，DGT 所测浓度与植物体内浓度表现出更好的相关性。这些研究表明有 DGT 和植物吸收所产生的土壤扰动效应是相似的，即植物对重金属的吸附不是简单地由土壤溶液中的金属浓度控制，而是土壤溶液中的金属扩散补给和颗粒物上的金属解吸过程对于植物（DGT）吸收都十分重要。另外，也有研究发现，DGT 方法不能有效预测部分植物对某些重金属的吸收，可能是由于植物生理因素控制植物对重金属的吸收，而非土壤过程控制植物对重金属的吸收。

2.1.2　DGT 用于研究植物吸收机理

Fitz 等[33]将 DGT 技术用于研究 As 超积累植物蜈蚣草的根际土壤，结合模型分析观察到了蜈蚣草根际 As 流量的降低以及根际土壤中 As 解吸速率的降低，探讨土壤活性态 As 库（pool）的潜在再补给能力。Senila 等[34]利用 DGT 研究了两种超富集植物的修复效率，论证了根际土壤中的 As 在植物吸收过程中可利用性的变化机理。DGT 作为一种原位富集技术，可以同时测定土壤中的 P 和 As。Mojsilovic 等[35]采用 DGT 同时测定土壤中的活性 P 和 As，研究指出 DGT 测定的活性态 P/As 比可用来预测土壤中 As 对小麦的植物毒性。Luo 等[36]同土壤 Ni 浓度下 DGT 和土壤溶液方法与植物吸收的相关性，论证了在低 Ni 浓度下植物对 Ni 的吸收是由土壤的扩散过程而非植物的吸收速率控制的。

2.2　DGT 在水体和沉积物生物有效性方面的应用

虽然生物配体模型（biotic ligand model，BLM）和自由离子活度模型（free ion activity model，FIAM）在部分情况下能够有效地预测水生生物对重金属的吸收，但 DGT 仍被考虑应用于反映水体中的生物有效性。Slaveykova 等[37]利用 DGT 和中空纤维渗透液膜（HFPLM）（自由离子）分析黑海中的绿色微藻小球藻对金属的吸收，结果表明藻类吸收通量与两种方法都有很好的相关性。Schintu 等[38]评估了地中海五个沿海站点的金属浓度和藻类的相关性，发现 DGT 测得的金属浓度中只有 Pb 与藻类有较好的相关关系。然而 DGT 对固着生物吸收重金属的预测效果更好，但其机理目前仍需研究。对水生苔藓（*Fontinalis antipyretica*）的中宇宙研究还发现，DGT 测定的有效态 Cu 浓度比与铜离子选择性电极、苔藓有更好的相关性[39]。

利用 DGT 评估沉积物有效态金属同样具有独特的优势[40]。任静华等[41]研究发现，在外源添加 Cd 的沉积物中饲养的亚洲蛤体内 Cd 含量，与 DGT 测得的沉积物 Cd 含量以及金属硫蛋白含量相关。最近一项深入的调查中，DGT 被用来分析受铜漆污染的海洋沉积物表层。不同沉积物类型中的双壳类 *Tellina deltoidalis* 数据与 DGT 浓度之间良好相关性表明，DGT 技术显示了其在评估沉积物重金属生物有效性与生态风险方面的巨大潜力[42]。由于 DGT 技术可以更加真实地模拟土壤或沉积物的动态变化过程，因此其所获得的数据对于相关机理研究有重要意义。

3 DGT 在污染物形态分析中的应用

传统的主动采样结合实验室分析测定方法虽然同样保证较高的检测限，但是在样品收集、保存环节中由于环境条件改变（如 pH、E_h、溶解氧、微生物活动等），很容易产生待测物被污染、损失和形态变化等诸多问题。而 DGT 技术可在原位状态下选择性分离并富集待测物质，更加真实地反映环境的天然存在形态。

3.1 重金属形态分析中的应用

大量研究表明，重金属生物有效性主要取决于其赋存形态而并非总量，不同价态之间的化学性质、生物活性及毒性上均存在显著差异。以 As 和 Cr 为例，水体和土壤中 As(+III)和 As(+V)分别以电中性、带负电的含氧阴离子形式存在；而环境中的 Cr(+III)和 Cr(+VI)则以阳离子和阴离子形式存在。据研究，As(+III)的毒性是 As(+V)的近 60 倍，而 Cr(+VI)毒性更是 Cr(+III)毒性的 100 倍。DGT 技术形态分析的关键在于吸附膜的选择，对某一种形态具有专属性吸附显得尤为必要。当前适用于金属阴/阳离子检测的 DGT 技术多基于比如 Chelex-100[43]水铁矿[44]、氧化钛[45]、氧化锆[12]吸附膜材料。然而，这些 DGT 技术大都只能测定总量而无法实现不同价态的选择性测定。近些年，许多新型特异性吸附材料被成功合成并用于重金属形态分析的 DGT 方法开发中，如表 30-1 所示。Pan 等[46]利用合成的 *N*-甲基葡糖胺（NMDG）作为吸附膜材料应用于 DGT 装置，成功实现环境中 Cr(+VI)的选择性测定，且该方法具有吸附容量大、检出限低、抗干扰能力强等特点，在 Cr 污染水体和土壤中得到了良好的证实。巯基硅胶作为一种良好的吸附材料，已经成功应用于 DGT 技术，选择性测定 As(+III)[47]、Sb(+III)[48,49]和甲基汞[53]，并且性能优异，不受离子强度、pH 等环境因素的干扰；Bennett 等[50]也开发了含有纳米二氧化钛的吸附膜，用于测定环境中痕量的 Se(IV)。Rolisola 等[51]利用 Amberlite IRA910 阴离子交换树脂作为吸附材料，实现了对 As(+V)的选择性测定。

表 30-1 重金属形态分析吸附膜材料

元素/物质	吸附膜	参考文献
甲基汞	巯基硅胶	Clarisse [54]
Cr^{VI}	N-甲基葡糖胺	Pan [46]
Cr^{III}	Chelex-100	Ernstberger [55]
Se^{IV}	二氧化钛	Bennett [50]
As^{III}	巯基硅胶	Bennett [47]
As^{V}	Amberlite IRA 90	Rolisola [51]
Sb^{III}	巯基硅胶	Fan [48];Bennett [49]

需要说明的是，对于土壤这样的复杂体系，单从价态上区分污染物的形态并不完整。根据 Tessier [52]等提出的五步提取法，土壤中重金属可以分为可交换态、碳酸盐结合态、铁锰氧化态、有机质结合态以及残渣态五种。欧盟提出的 BCR 提取法[53]又将其简化为酸溶态、还原态、氧化钛及残渣态四种。相比以上两种化学提取法，DGT 技术为原位非破坏性技术，更接近于模拟真实环境，能更好地反映生物体所吸收的重金属。

3.2　营养盐的形态分析

DGT 技术在营养盐形态监测方面也取得了一定进展。磷是环境中大多数生命体如植物、藻类、细菌及动物生命必需元素，对水体体、土壤等环境营养状态和生态系统结构发挥十分关键的调控作用，对生物可利用性磷的检测显得尤为重要。DGT 技术发明人张昊教授首次研发了水合氧化铁（FeO）膜作为固定相专性固定水体中的活性磷[3]。Panther 等[3]研发二氧化钛（TiO_2）固定膜，并将基于 TiO_2 和水合氧化铁两种吸附膜的 DGT 进行海洋应用对比，发现前者更适合于海洋监测，而淡水中的测定结果一致。最近，南京大学罗军教授团队等采用原位直接沉淀法自主研发的高分辨率 Zr 基吸附膜用于专性吸收溶解态活性磷，该膜能适应不同 pH/离子强度等外界环境条件[12]。1999 年，由 Teasdale 等[18]开发出以碘化银为吸附膜的 DGT 技术，由于硫离子与碘化银会形成黑色硫化银，作者在计算浓度时采用灰度扫描法而摒弃了传统的洗脱方式，有效缩短了样品处理时间，该方法已成功应用于沉积物中硫化物的高分辨测量。Zhou 等[22]利用 Fe-Al-Ce 三种金属氧化物实现氟离子的高效吸附，在自然水体、含氟废水以及沉积物中均得到良好的应用效果。对于水体中的溶解性氮，Huang 等[56,57]先后研发出对硝态氮和氨氮具有选择性的 DGT 技术，该技术具有吸附容量大、竞争效应小等特点，可以实现自然环境中的长期放置。其余可用作营养盐形态分析的 DGT 吸附膜材料见表 30-2。

表 30-2　营养盐形态分析吸附膜材料

元素/物质	吸附膜	参考文献
S^{2-}	碘化银	Teasdale[18]
F^-	Fe-Al-Ce 氧化物	Zhou[22]
NO_3^--N	A520E	Huang[56]
NH_4^+-N	Microlite PrCH	Huang[57]
NH_4^+-N	沸石	Feng[58]
P	氧化锆	Ding[4]; Guan[12]
K^-	Amberlite IRP-69	Tandy[59]

4　DGT 在有机污染物研究中的应用

对有机污染物而言，常用的被动采样器有半透膜装置（SPMD）、专门针对极性有机污染物的 POCIS 等（图 30-2）。然而这些采样器的一个共同问题是采样速率（R_s）受到水动力学条件如水的流速的影响（实为由此造成的边界扩散层即 DBL 大小不一），导致实验室测定的 R_s 与实际环境中的 R_s 有较大差异[60]，从而导致计算得出的污染物环境浓度不确定性增大。解决此问题的一个方法是利用效能参考物质（PRC）采样之前加载到采样器里，但是该方法的前提条件是污染物向采样器里面的扩散和 PRC 向外扩散具有各向同性的特点[1]，这对依据分配理论的平衡型采样器是适用的，但是却不适用于基于吸附理论的采样器，特别是针对极性化合物的采样器。

而 DGT 采样器与其他采样器的关键不同点是有一层相对较厚（相对于 DBL）的扩散层，这可以控制采样速率，使得 DBL 的影响相对减小，从而能够更准确的测得环境浓度。另外，DGT 的关键参数扩散系数（D）是一个目标物特有的参数，可以在实验室条件下单独测定得到，此参数受实际环境条件影响很小，故可以直接应用于实际采样过程中的计算，无需额外校正。然而长期以来，DGT 主要应用于无机物如金属离子和营养盐的研究，理论上来讲其应该适用于有机污染物[2]，但却一直未得到证实。

正是基于有机污染物被动采样器的问题以及 DGT 的诸多优势，英国兰卡斯特大学 DGT 团队的 Chen 等[62]首次发展了有机污染物 DGT（o-DGT），证实了 DGT 应用于有机污染物的可行性，并进一步通过与传统采样方法（主动采样）作对比，证明了其实际应用。从而打开了 DGT 技术研究有机污染物特别是极性有机污染物（绝大多数药物与个人护理品 PPCPs）的大门。

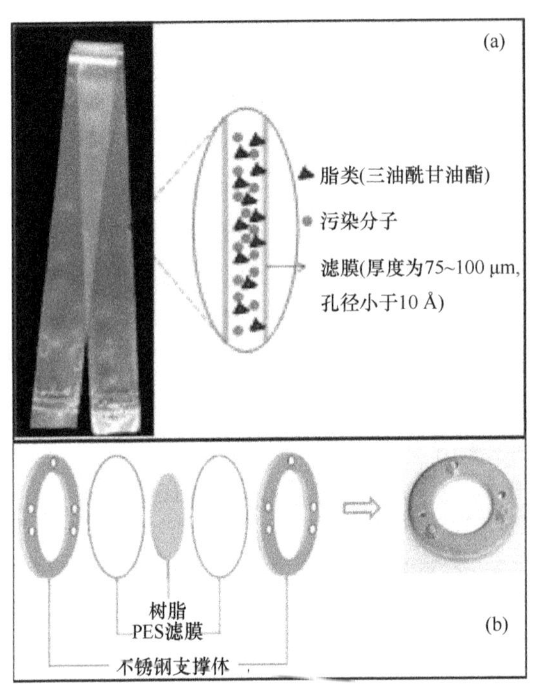

图 30-2　常用有机污染物被动采样器

（a）SPMD；（b）POCIS

4.1　监测有机污染的 DGT 技术的研发

琼脂糖作为扩散层是 DGT 技术研究有机污染物的共识，因此研发适合于不同目标化合物的 DGT 关键在于找到合适的吸附剂，并测定目标物扩散层扩散系数。英国兰卡斯特大学 DGT 研究团队最初以抗生素为目标物，以 XAD-18 制备吸附膜，测定了近 40 种常用抗生素的扩散系数，将其应用于污水处理厂抗生素的研究[14]。随后，Zheng 等[15]开发了采用活性炭作为吸附相用于采集水体中的双酚类化合物的 DGT 装置。Guo 等[63]将该技术成功应用于毒品的检测中；Fauvelle 等[64]开展了基于 TiO_2 的 DGT 用于采集水体中的草甘膦和氨甲基磷酸。

4.2　DGT 在环境监测与研究中的应用

4.2.1　环境监测

目前环境监测中最常用的采样方式是离散式即时水样采集，相应的浓度都是即时浓度，这种采样方式不但成本高、耗时耗力而且测得水体浓度不确定性较大，特别是污染物载荷不稳定的水体，比如污水处理厂[以抗生素中的环丙沙星（CFX）为例，图 30-3]。自动采样是业界公认的标准采样方法，但是需要专门的自动采样器且需要能源。而被动采样器如 DGT 能够得到与自动采样可比较的结果（图 30-3）但其成本却小于自动采样方式的 1/3(以得到一周时间加权平均浓度为例)[14]。DGT 也将有助于环境监察，例如如果某工厂在晚上偷排，传统即时水样采集（一般都在白天）将无法到达证据，而被动采样如 DGT 由于其一直在采样故可以反映出偷排情况。随着越来越多的针对各种 PPCPs 的 DGT 采样设备的研发（如以上讨论），其在环境监测方面的应用也将越来越广泛，将有可能是一种替代采样方法[65]。

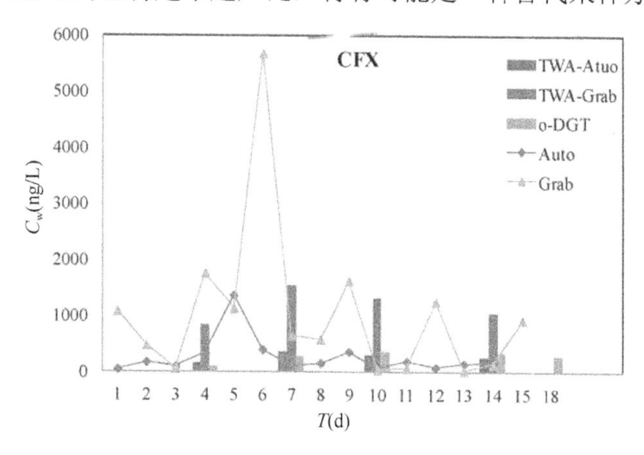

图 30-3　某污水处理厂进口环丙沙星（CFX）的动态变化[14]

Grab，抓水采样；Auto，自动采样；o-DGT，DGT 被动采样

4.2.2　污染物环境行为及效应

DGT 技术的发明最初是为了环境监测，后来拓展到环境行为和机理研究方面。大量研究工作已经证明 DGT 可以用于原位研究无机物在土壤/沉积物中的吸附解吸行为[5,66]、与共存物质如溶解性有机质（DOM）的相互作用[67]、生物有效性[23]及毒性效应[68,69]，是一种功能多样的科学研究工具。

近年来，国内外诸多研究人员利用 DGT 技术开展了有机污染物的环境行为、效应与相关机理等研究。例如，Chen 等首次将有机污染物 DGT 引入到土壤中抗生素的吸附/解吸行为[62]，并结合 DIFS 模型研究了 4 种抗生素的解吸动力学，指出环胺类抗生素（如磺胺甲噁唑，SMX）的 DGT 采样过程更易受到其在土壤颗粒上的解吸动力学控制，暗示着此类抗生素的生物吸收过程也可能受到解吸动力学的限制，但是这种动力学限制对另外一类抗生素甲氧苄胺嘧啶（TMP）相对较弱，说明此动力学控制与跟化合物相关的。D'Angelo 等[70]将基于 XAD18 的 DGT 应用于污泥中环丙沙星（一种喹诺酮类抗生素）的解吸动力学研究，发现此抗生素从污泥中的释放也受到解吸动力学的限制。最近 Guan 等[71]将其发展的基于活性炭的 DGT 应用于评估 5 种土壤中 3 种双酚类化合物的吸附/解吸行为，发现土壤颗粒不能很好地供给土壤溶液，但是土壤上吸附的大部分双酚类化合物能够参与与土壤溶液的交换，即有潜在的生物可给性。类似地，这种潜在的可利用性/供给能力受到其从土壤颗粒上解吸动力学的控制。

DGT 用于有机污染物的研究还有非常大的发展空间，比如与共存物质的相互关系、沉积物界面行为、与生物有效性联系以及与毒性效应的相关等还未见有报道，是将来研究的重点方向。

5　DGT 技术在环境微界面过程与机制研究中的应用

对于土壤和沉积物中物质空间分布和输移特征等信息的获取取决于所采用的采样和分析手段，其中采样技术是限制其发展的最主要瓶颈。众所周知，土壤和沉积物在垂向和水平空间上均具有异质性，物质在微小尺度上（毫米-亚毫米）可能存在强烈的化学梯度分布，因此需要利用高分辨的技术手段予以表征方能准确还原物质的分布信息[72]。目前绝大多数环境化学家采用主动方式，如离心法、压榨法、化学提取法等。此类方法会对沉积物原始物理、化学及生物环境条件造成不同程度的破坏，导致获取的物质空间分布信息失真，进而很难全面阐释沉积物地球化学过程与机制[73]。

随着激光剥蚀-电感耦合等离子质谱技术（LA-ICP-MS） 引入固定膜分析，极大地推动了 DGT 对目标物质表征从操作繁琐的常规手段向简单高效的高精度手段，从常规分辨尺度（毫米级）向高分辨率（微米级），从一维向二维发展。如 Stockdale 等[74]利用 LA-ICP-MS 与 DGT 联用技术原位同步获取了沉积物-水界面附近 P、V、As、Mo、Sb、W 和 U 等多种元素高分辨分布特征。Guan 等[12]利用该联用技术同步获取了太湖沉积物-水界面 P、As、Se、Mo 和 Sb 二维高分辨分布信息（图 30-4）。这些研究结果表明，借助高分辨 DGT 技术能够精细刻画沉积物微界面多物质空间分布特征，有助于系统揭示界面反应机制。

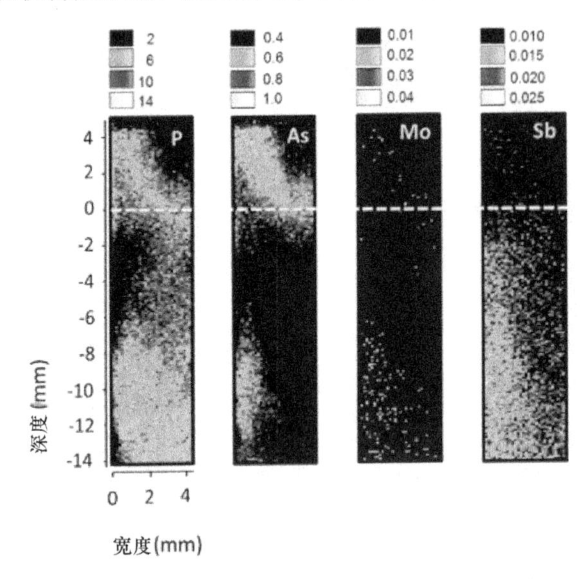

图 30-4 太湖沉积物中活性 P，As，Mo，Sb 通量二维分布特征[12]

沉积物微界面发生的各种地球化学过程的发生大多数是两种或两种以上多物质共同作用的结果，因此对两种及两种以上物质进行同步、高分辨监测是当前沉积物研究领域的新热点。例如，沉积物微生境中金属和硫化物高度区域化的分布和紧密关联的迁移释放就是利用高分辨率 DGT 技术研究发现（图 30-5）[9,75-77]。该现象的揭示使得科学家重新审视传统过于简化的一维反应传输模型，并开始考虑包含微生境的更加完整的三维反应传输模型。先前大量研究虽已证实沉积物内源 P 释放与沉积物中 Fe-S 的耦合循环存在密切联系，但长期以来受限制与监测方法不足，导致对该领域的认识缺少直接证据。国内外已有研究通过串联多种 DGT 技术或发展新型复合吸附膜 DGT 技术，同步、高分辨获取沉积物-水界面 Fe-S-P 空间变化信息，系统阐释三者的耦合生物地球化学循环过程，从而深入揭示了湖泊沉积物内源 P 污染释放与控制机制[78,79]。例如，Han 等[78]将复合膜 DGT 技术追踪研究了蓝藻降解发生沉积物-水界面 P-S-Fe 时空分布、源/汇转换特征与关键影响因素，证实蓝藻降解可以同步释放活性 S-P 并导致其在沉积物和水体中双向扩散，而 Fe 的释放主要来自厌氧沉积物。

近些年，越来越多的研究者致力于将 DGT 技术与其他沉积物原位成像技术如平面光电极技术（planar optode）、梯度扩散平衡技术（DET ）等进行联用，从微观尺度辨识沉积物-水界面、沉积物-植物根际界面、沉积物洞穴界面等复杂底栖系统微界面反应过程。如 Pages 等[80,81]利用 DGT/DET 联用技术同步获取海洋沉水植物大叶藻（*Zostera marina*）根际界面活性 P-S，Fe-S 二维空间分布信息，并从微观尺度直接揭示根际界面 P-Fe-S 耦合作用机制。Stahl 等[82]首次将 SPR-IDA-DGT 技术与 DO 平面光电极联用，同步获取了沉积物底栖动物沙蚕（*Hediste diversicolor*）洞穴附近活性重金属铜、铅、镍与 DO 二维、高分辨空间分布信息，证实 DO 浓度升高导致洞穴局部重金属活化。随后，Han 等[83]与将超薄 DGT 技术与 DO 平面光电极技术联用，同步获得沉积物-水界面活性 P 与 DO 的二维、高分辨空间分布信息（图 30-6），并以苦草（*Vallisneria spiralis*）等大型沉水植物为研究对象研究了根际微界面 P 迁移对植物根系泌氧响应特征，证实沉水植物根系泌氧可显著促进沉积物活性 P 向非活性 P 转化。

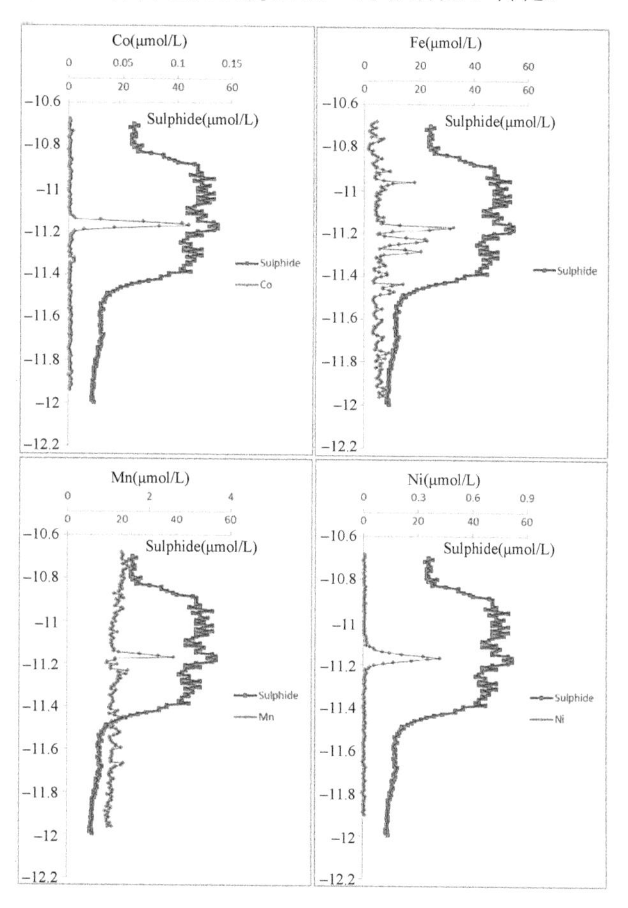

图 30-5　沉积物典型微生境中活性重金属和活性 S 同步分布特征[75]

虽然将 DGT 用于研究土壤和植物根际的研究于近期才开始进行报道，但却立即展现了此类应用的广阔前景并取得了关键的发现。植物根系作为植物与土壤物质交换的通道，其具体工作状态的阐明对理解植物吸收各种元素有着关键意义。但由于长期缺乏同时测量各种毒性或营养元素有效浓度和关乎元素土壤生物化学过程的含氧量、pH 等重要环境参数的原位测定技术，影响植物根系吸收这些元素的过程始终难以精确研究。具有原位高分辨分析能力的 DGT 技术与平面光电极技术的联用克服了之前的困难。通过在根际箱实验中串联使用 DGT 和平面光电极，Williams 等[84]首次发现水稻根尖附近存在一个 As、Pb 和 Fe 的释放显著提高，同时伴随着氧气富集和 pH 降低的区域（图 30-7）。在另一项研究中 Hoefer 等[85]同样在柳苗（*Salix smithiana*）根部附近观测到了更高的 Zn 和 Cd 的释放。随后 Hoefer 等[86]又将 pH 光极膜和 DGT 吸附膜整合到同一水凝胶层，实现了对柳苗根际土壤 pH 和金属释放的时空同步成像研究。

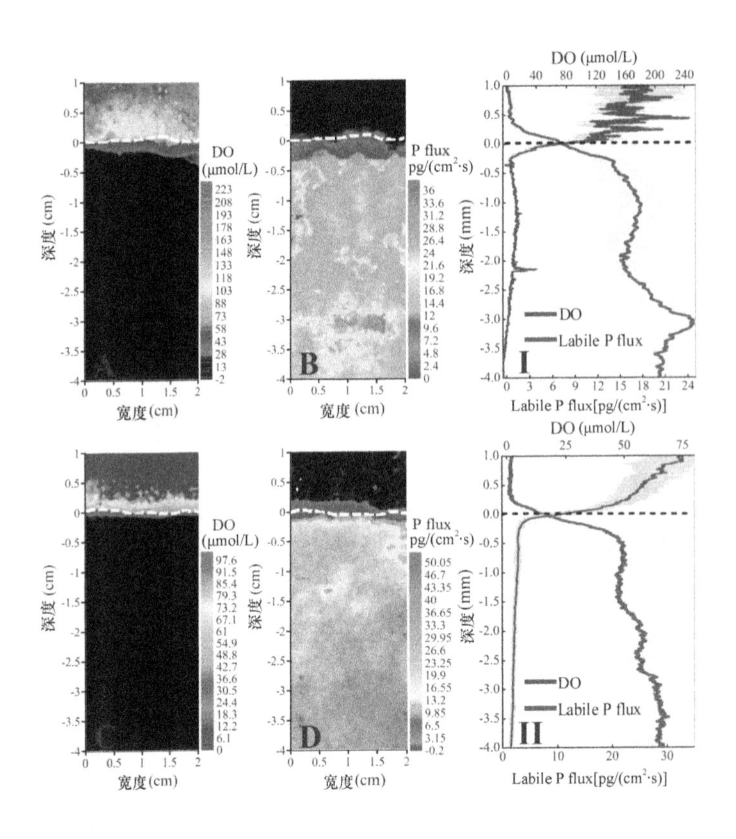

图 30-6 基于 DGT-平面光电极联用技术研究好氧（A，B，I）和厌氧（C，D，II）条件下沉积物-水界面 DO 和活性 P 迁移通量（Labile P flux）二维/一维同步变化信息[83]

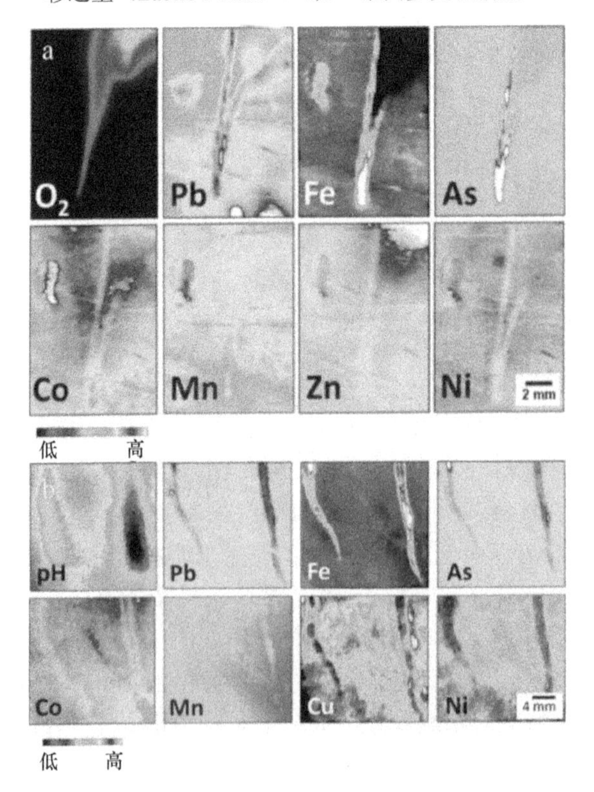

图 30-7 （a）一组土壤表面下 10 cm 处水稻根部 O_2，Pb，Fe，As，Co，Mn，Zn 和 Ni 分布图；（b）另一组生长在同一土壤中水-土界面下 10.5 cm 深度的水稻根部附近 pH，Pb，Fe，As，Co，Mn，Cu 和 Ni 的二维呈现[84]

6 展　　望

近年在 DGT 技术的开发及其在环境研究中的应用成果说明 DGT 技术是一种非常适合于环境监测、环境行为和机理研究的原位高效的技术手段。未来在污染物形态和生物有效性、环境基准和阈值、污染物微观机制的原位研究方面仍有较大的发展空间。

参 考 文 献

[1] Davison W, Zhang H. In-situ speciation measurements of trace components in natural-waters using thin-film gels. Nature, 1994, 367(6463): 546-548.

[2] Luo J, Wang, X Zhang H, William D, Theory and application of diffusive gradients in thin films in soils. Journal of Agro-Environment Science, 2011, 30(2): 205-213.

[3] Panther J G, Teasdale P R, Bennett W W, Welsh D T, Zhao H J. Titanium dioxide-based dgt technique for in situ measurement of dissolved reactive phosphorus in fresh and marine waters. Environ. Sci. Technol, 2010, 44(24): 9419-9424.

[4] Ding S M, Xu D, Sun Q, Yin H B, Zhang C S. Measurement of dissolved reactive phosphorus using the diffusive gradients in thin films technique with a high-capacity binding phase. Environ. Sci. Technol,2010, 44(21): 8169-8174.

[5] Zhang H, Davison W, Gadi R, Kobayashi T. In situ measurement of dissolved phosphorus in natural waters using DGT. Anal. Chim. Acta, 1998, 370(1): 29-38.

[6] Luo J, Zhang H, Santner J, Davison W. Performance characteristics of diffusive gradients in thin films equipped with a binding gel layer containing precipitated ferrihydrite for measuring arsenic(V), selenium(VI), vanadium(V), and antimony(V). Anal. Chem., 2010, 82(21): 8903.

[7] Zhang H, Davison W, Miller S, Tych W. In-situ high-resolution measurements of fluxes of Ni, Cu, Fe and Mn and concentrations of Zn and Cd in porewaters by DGT. Geochim. Cosmochim. Acta, 1995, 59(20): 4181-4192.

[8] Kreuzeder A, Santner J, Prohaska T, Wenzel W W. Gel for simultaneous chemical imaging of anionic and cationic solutes using diffusive gradients in thin films. Anal. Chem., 2013, 85(24): 12028-12036.

[9] Motelica-Heino M, Naylor C, Zhang H, Davison W. Simultaneous release of metals and sulfide in lacustrine sediment. Environ. Sci. Technol. 2003, 37(19): 4374-4381.

[10] Davison W, Fones G R, Grime G W. Dissolved metals in surface sediment and a microbial mat at 100-mu m resolution. Nature ,1997, 387(6636): 885-888.

[11] Liu S, Qin N, Song J, Zhang Y, Cai W, Zhang H, Wang G, Zhao H. A nanoparticulate liquid binding phase based DGT device for aquatic arsenic measurement. Talanta, 2016, 160: 225-232.

[12] Guan D X, Williams P N, Luo J, Zheng J L, Xu H C, Cai C, Ma L N Q. Novel precipitated zirconia-based DGT technique for high-resolution imaging of oxyanions in waters and sediments. Environ. Sci. Technol., 2015, 49(6): 3653-3661.

[13] Guan D X, Williams P N, Xu H C, Li G, Luo J, Ma L Q. High-resolution measurement and mapping of tungstate in waters, soils and sediments using the low-disturbance DGT sampling technique. J. Hazard. Mater., 2016, 316: 69-76.

[14] Chen C E, Zhang H, Jones K C. A novel passive water sampler for in situ sampling of antibiotics. J. Environ. Monit., 2012, 14(6): 1523-1530.

[15] Zheng J L, Guan D X, Luo J, Zhang H, Davison W, Cui X Y, Wang L H, Ma L N Q. Activated charcoal based diffusive gradients in thin films for in situ monitoring of bisphenols in waters. Anal. Chem., 2015, 87(1): 801-807.

[16] Fresno T, Penalosa J M, Santner J Puschenreiter, M Moreno-Jimenez E. Effect of Lupinus albus L. root activities on As and Cu mobility after addition of iron-based soil amendments. Chemosphere, 2017, 182: 373-381.

[17] Widerlund A, Davison W. Size and density distribution of sulfide-producing microniches in lake sediments. Environ. Sci. Technol., 2007, 41(23): 8044-8049.

[18] Teasdale P R, Hayward S, Davison W. In situ, high-resolution measurement of dissolved sulfide using diffusive gradients in thin films with computer-imaging densitometry. Anal. Chem., 1999, 71(11): 2186-2191.

[19] Yinghong F A N, Chunye L I N, Mengchang H E, Zhifeng Y. Kinetics and bioavailability of Cd in the surface sediments of the Daliao River watersystem. Acta Scientiae Circumstantiae, 2008, 28(12): 2583-2589.

[20] Fan Y h, Lin C y, He M c, Zhou Y x, Yang Z f. High resolution measurement of concentrations and fluxes of heavy metals in pore waters by DGT. Huanjing Kexue, 2007, 28, (12), 2750-2757.

[21] Wu Z H, Jiao L X, Wang S R. The measurement of phosphorus, sulfide and metals in sediment of Dianchi Lake by DGT (diffusive gradients in thin films) probes. Environ. Earth Sci., 2016, 75(3): 14.

[22] Zhou C, Guan D, Han Y, Pan Y, Fang X, Wang X, Zheng J, Li Y, Wei T, Zou Y, Cui X, Luo J. Pollution characteristics of metals and nutrients in different regions of Lake Taihu based on diffusive gradients in thin-films technique. Journal of Agro-Environment Science, 2016, 35(6): 1144-1152.

[23] Zhang H, Davison W. Use of diffusive gradients in thin-films for studies of chemical speciation and bioavailability. Environ. Chem., 2015, 12(2): 85-101.

[24] Zhang H, Zhao F J, Sun B, Davison W, McGrath S P. A new method to measure effective soil solution concentration predicts copper availability to plants. Environ. Sci. Technol., 2001, 35(12): 2602-2607.

[25] Nolan A L, Zhang H, McLaughlin M J. Prediction of zinc, cadmium, lead, and copper availability to wheat in contaminated soils using chemical speciation, diffusive gradients in thin films, extraction, and isotopic dilution techniques. J. Environ. Qual., 2005, 34(2): 496-507.

[26] Mundus S, Lombi E, Tandy S, Zhang H, Holm P E, Husted S, Gilkes R J, Prakongkep N. In Assessment of the diffusive gradients in thin-films (DGT) technique to assess the plant availability of Mn in soils, Proceedings of the 19th World Congress of Soil Science: Soil solutions for a changing world, Brisbane, Australia, 1-6 August 2010. Symposium 3.3.1 Integrated nutrient management, 2009：365–368.

[27] Tian Y, Wang X R, Luo J, Yu H X, Zhang H. Evaluation of holistic approaches to predicting the concentrations of metals in field-cultivated rice. Environ. Sci. Technol., 2008, 42(20): 7649-7654.

[28] Wang J J, Bai L Y, Zeng X B, Su S M, Wang Y N, Wu C X. Assessment of arsenic availability in soils using the diffusive gradients in thin films (DGT) technique-a comparison study of DGT and classic extraction methods. Environ. Sci.-Process Impacts, 2014, 16(10): 2355-2361.

[29] Li Z, Jia M Y, Wu L H, Christie P, Luo Y M. Changes in metal availability, desorption kinetics and speciation in contaminated soils during repeated phytoextraction with the Zn/Cd hyperaccumulator Sedum plumbizincicola. Environ. Pollut., 2016, 209: 123-131.

[30] Menezes-Blackburn D, Zhang H, Stutter M, Giles C D, Darch T, George T S, Shand C, Lumsdon D, Blackwell M, Wearing C, Cooper P, Wendler R, Brown L, Haygarth P M. A Holistic approach to understanding the desorption of phosphorus in soils. Environ. Sci. Technol., 2016, 50(7): 3371-3381.

[31] Ngo L K, Pinch B M, Bennett W W, Teasdale P R, Jolley D F. Assessing the uptake of arsenic and antimony from contaminated soil by radish (Raphanus sativus) using DGT and selective extractions. Environ. Pollut., 2016, 216: 104-114.

[32] Zhang S, Song J, Gao H, Zhang Q, Lv M C, Wang S, Liu G, Pan Y Y, Christie P, Sun W J. Improving prediction of metal uptake by Chinese cabbage (Brassica pekinensis L.) based on a soil-plant stepwise analysis. Sci. Total Environ., 2016, 569:

1595-1605.

[33] Fitz W J, Wenzel W W, Zhang H, Nurmi J, Stipek K, Fischerova Z, Schweiger P, Kollensperger G, Ma L Q, Stingeder G. Rhizosphere characteristics of the arsenic hyperaccumulator Pteris vittata L. and monitoring of phytoremoval efficiency. Environ. Sci. Technol., 2003, 37(21): 5008-5014.

[34] Senila M, Tanaselia C, Rimba E. Investigations on arsenic mobility changes in rizosphere of two ferns species using dgt technique. Carpath. J. Earth Environ. Sci., 2013, 8(3): 145-154.

[35] Mojsilovic O, McLaren R G, Condron L M. Modelling arsenic toxicity in wheat: Simultaneous application of diffusive gradients in thin films to arsenic and phosphorus in soil. Environ. Pollut., 2011, 159(10): 2996-3002.

[36] Luo J, Cheng H, Ren J H, Davison W, Zhang H. Mechanistic Insights from DGT and Soil Solution Measurements on the Uptake of Ni and Cd by Radish. Environ. Sci. Technol., 2014, 48(13): 7305-7313.

[37] Slaveykova V I, Karadjova I B, Karadjov M Tsalev D L. Trace Metal Speciation and Bioavailability in Surface Waters of the Black Sea Coastal Area Evaluated by HF-PLM and DGT. Environ. Sci. Technol., 2009, 43(6): 1798-1803.

[38] Schintu M, Marras B, Durante L, Meloni P, Contu A. Macroalgae and DGT as indicators of available trace metals in marine coastal waters near a lead-zinc smelter. Environ. Monit. Assess., 2010, 167(1-4): 653-661.

[39] Bourgeault A, Ciffroy P, Garnier C, Cossu-Leguille C, Masfaraud J F, Charlatchka R, Garnier J M. Speciation and bioavailability of dissolved copper in different freshwaters: Comparison of modelling, biological and chemical responses in aquatic mosses and gammarids. Sci. Total Environ., 2013, 452: 68-77.

[40] Peijnenburg W J, Teasdale P R, Reible D, Mondon J, Bennett W W, Campbell P G. Passive sampling methods for contaminated sediments: State of the science for metals. Integrated Environmental Assessment & Management, 2014, 10(2): 167-178.

[41] Ren J H, Luo J, Ma H R, Wang X R, Ma L N Q. Bioavailability and oxidative stress of cadmium to Corbicula fluminea. Environ. Sci.-Process Impacts, 2013, 15(4): 860-869.

[42] Simpson S L, Yverneau H Cremazy A, Jarolimek C V, Price H L, Jolley D F. DGT-induced copper flux predicts bioaccumulation and toxicity to bivalves in sediments with varying properties. Environ. Sci. Technol., 2012, 46(16): 9038-9046.

[43] Zhang H, Davison W. Performance-characteristics of diffusion gradients in thin-films for the in-situ measurement of trace-metals in aqueous-solution. Anal. Chem., 1995, 67(19): 3391-3400.

[44] de Oliveira W, de Carvalho M D B, de Almeida E Menegario A A, Domingos R N, Brossi-Garcia A L, do Nascimento V F, Santelli R E. Determination of labile barium in petroleum-produced formation water using paper-based DGT samplers. Talanta, 2012, 100: 425-431.

[45] Ding S M, Xu D, Wang Y P, Wang Y, Li Y Y, Gong M D, Zhang C S. Simultaneous measurements of eight oxyanions using high-capacity diffusive gradients in thin films (Zr-oxide DGT) with a high-efficiency elution procedure. Environ. Sci. Technol., 2016, 50(14): 7572-7580.

[46] Pan Y, Guan D X, Zhao D, Luo J, Zhang H, Davison W, Ma L Q. Novel speciation method based on diffusive gradients in thin-films for in situ measurement of Cr-VI in aquatic systems. Environ. Sci. Technol.,2015, 49(24): 14267-14273.

[47] Bennett W W, Teasdale P R, Panther J G, Welsh D T, Jolley D F. Speciation of dissolved inorganic arsenic by diffusive gradients in thin films: selective binding of As-III by 3-mercaptopropyl-functionalized silica gel. Anal. Chem., 2011, 83(21): 8293-8299.

[48] Fan H T, Liu A J, Jiang B, Wang Q J, Li T, Huang C C. Sampling of dissolved inorganic Sb-III by mercapto-functionalized silica-based diffusive gradients in thin-film technique. RSC Adv., 2016, 6(4): 2624-2631.

[49] Bennett W W, Arsic M, Welsh D T, Teasdale P R. In situ speciation of dissolved inorganic antimony in surface waters and

sediment porewaters: development of a thiol-based diffusive gradients in thin films technique for Sb-III. Environ. Sci.-Process Impacts, 2016, 18(8): 992-998.

[50] Bennett W W, Teasdale P R, Panther J G, Welsh D T, Jolley D F. New diffusive gradients in a thin film technique for measuring inorganic arsenic and selenium(IV) using a titanium dioxide based adsorbent. Anal. Chem., 2010, 82(17): 7401-7407.

[51] Rolisola A, Suarez C A, Menegario A A, Gastmans D, Kiang C H, Colaco C D, Garcez D L, Santelli R E. Speciation analysis of inorganic arsenic in river water by Amberlite IRA 910 resin immobilized in a polyacrylamide gel as a selective binding agent for As(v) in diffusive gradient thin film technique. Analyst, 2014, 139(17): 4373-4380.

[52] Tessier A, Campbell P G C, Bisson M. Sequential extraction procedure for the speciation of particulate trace-metals. Anal. Chem., 1979, 51(7): 844-851.

[53] Ure A M, Quevauviller P, Muntau H, Griepink B. Speciation of heavy-metals in soils and sediments - an account of the improvement and harmonization of extraction techniques undertaken under the auspices of the BCR of the commission-of-the-european-communities. Int. J. Environ. Anal. Chem., 1993, 51(1-4): 135-151.

[54] Clarisse O, Hintelmann H. Measurements of dissolved methylmercury in natural waters using diffusive gradients in thin film (DGT). J. Environ. Monit., 2006, 8(12): 1242-1247.

[55] Ernstberger H, Zhang H, Davison W. Determination of chromium speciation in natural systems using DGT. Anal. Bioanal. Chem., 2002, 373(8): 873-879.

[56] Huang J Y, Bennett W W, Teasdale P R, Gardiner S, Welsh D T. Development and evaluation of the diffusive gradients in thin films technique for measuring nitrate in freshwaters. Anal. Chim. Acta, 2016, 923: 74-81.

[57] Huang J Y, Bennett W W, Welsh D T, Li T L, Teasdale P R. Development and evaluation of a diffusive gradients in a thin film technique for measuring ammonium in freshwaters. Anal. Chim. Acta, 2016, 904: 83-91.

[58] Feng Z M, Guo T T, Jiang Z W, Sun T. Sampling of ammonium ion in water samples by using the diffusive-gradients-in-thin-films technique (DGT) and a zeolite based binding phase. Microchim. Acta, 2015, 182(15-16): 2419-2425.

[59] Tandy S, Mundus S, Zhang H, Lombi E, Frydenvang J, Holm P E, Husted S. A new method for determination of potassium in soils using diffusive gradients in thin films (DGT). Environ. Chem., 2012, 9(1): 14-23.

[60] Alvarez D A, Petty J D, Huckins J N, Joneslepp T L, Getting D T, Goddard J P, Manahan S E. Development of a passive, in situ, integrative sampler for hydrophilic organic contaminants in aquatic environments. Environmental Toxicology & Chemistry, 2004, 23(7): 1640-1648.

[61] Liu H H, Wong C S, Zeng E Y. Recognizing the limitations of performance reference compound (PRC)-calibration technique in passive water sampling. Environ. Sci. Technol., 2013, 47(18): 10104-10105.

[62] Chen C E, Chen W, Ying G G, Jones K C, Zhang H. In situ measurement of solution concentrations and fluxes of sulfonamides and trimethoprim antibiotics in soils using o-DGT. Talanta, 2015, 132: 902-908.

[63] Guo C, Zhang T, Hou S, Lv J, Zhang Y, Wu F, Hua Z, Meng W, Zhang H, Xu J. Investigation and Application of a New Passive Sampling Technique for in Situ Monitoring of Illicit Drugs in Waste Waters and Rivers. Environ. Sci. Technol., 2017.

[64] Fauvelle V, Nhu-Trang T T, Feret T, Madarassou K, Randon J, Mazzella N. Evaluation of titanium dioxide as a binding phase for the passive sampling of glyphosate and aminomethyl phosphonic acid in an aquatic environment. Anal. Chem., 2015, 87(12): 6004-6009.

[65] Chen C E, Zhang H, Ying G G, Zhou L J, Jones K C. Passive sampling: A cost-effective method for understanding antibiotic fate, behaviour and impact. Environment International, 2015, 85(12): 284-291.

[66] Amato E D, Simpson S L, Remaili T M, Spadaro D, Jarolimek C V, Jolley D F. Assessing the effects of bioturbation on metal bioavailability in contaminated sediments by diffusive gradients in thin films (DGT). Environ. Sci. Technol., 2016, 50(6).

[67] Levy J L, Zhang H, Davison W, Galceran J, Puy J. Kinetic signatures of metals in the presence of suwannee river fulvic acid. Environ. Sci. Technol., 2012, 46(6), 3335-3342.

[68] Han S P, Zhang Y, Masunaga S, Zhou S Y, Naito W. Relating metal bioavailability to risk assessment for aquatic species: Daliao River watershed, China. Environ. Pollut., 2014, 189: 215-222.

[69] McGrath S P, Mico C, Curdy R, Zhao F J. Predicting molybdenum toxicity to higher plants: Influence of soil properties. Environ. Pollut., 2010, 158(10): 3095-3102.

[70] D'Angelo E, Starnes D. Desorption kinetics of ciprofloxacin in municipal biosolids determined by diffusion gradient in thin films. Chemosphere, 2016, 164: 215-224.

[71] Guan D X, Zheng J L, Luo J, Zhang H, Davison W, Ma L N Q. A diffusive gradients in thin-films technique for the assessment of bisphenols desorption from soils. J. Hazard. Mater., 2017, 331: 321-328.

[72] Stockdale A, Davison W, Zhang H. Micro-scale biogeochemical heterogeneity in sediments: A review of available technology and observed evidence. Earth-Sci. Rev., 2009, 92(1-2): 81-97.

[73] Santner J, Larsen M, Kreuzeder A, Glud R N. Two decades of chemical imaging of solutes in sediments and soils - a review. Anal. Chim. Acta, 2015, 878: 9-42.

[74] Stockdale A, Davison W, Zhang H. High-resolution two-dimensional quantitative analysis of phosphorus, vanadium and arsenic, and qualitative analysis of sulfide, in a freshwater sediment. Environ. Chem., 2008, 5(2): 143-149.

[75] Gao Y, van de Velde Sm, Williams P N, Baeyens W, Zhang H. Two-dimensional images of dissolved sulfide and metals in anoxic sediments by a novel diffusive gradients in thin film probe and optical scanning techniques. Trac-Trends Anal. Chem., 2015, 66: 63-71.

[76] Robertson D, Teasdale P R, Welsh D T. A novel gel-based technique for the high resolution, two-dimensional determination of iron (II) and sulfide in sediment. Limnol. Oceanogr. Meth., 2008, 6: 502-512.

[77] Naylor C, Davison W, Motelica-Heino M, Van Den Berg G A, Van Der Heijdt L M. Simultaneous release of sulfide with Fe, Mn, Ni and Zn in marine harbour sediment measured using a combined metal/sulfide DGT probe. Sci. Total Environ., 2004, 328(1-3): 275-286.

[78] Han C, Ding S M, Yao L, Shen Q S, Zhu C G, Wang Y, Xu D. Dynamics of phosphorus-iron-sulfur at the sediment-water interface influenced by algae blooms decomposition. J. Hazard. Mater., 2015, 300: 329-337.

[79] 孙清清, 陈敬安, 王敬富, 杨海全, 计永雪, 兰晨, 王箫. 阿哈水库沉积物-水界面磷、铁、硫高分辨率空间分布特征. 环境科学, 2017, (7).

[80] Pages A, Grice K, Vacher M, Welsh D T, Teasdale P R, Bennett W W, Greenwood P. Characterizing microbial communities and processes in a modern stromatolite (Shark Bay) using lipid biomarkers and two-dimensional distributions of porewater solutes. Environ. Microbiol., 2014, 16(8): 2458-2474.

[81] Pages A, Welsh D T, Teasdale P R, Grice K, Vacher M, Bennett W W, Visscher P T. Diel fluctuations in solute distributions and biogeochemical cycling in a hypersaline microbial mat from Shark Bay, WA. Mar. Chem., 2014, 167: 102-112.

[82] Stahl H, Warnken K W, Sochaczewski L, Glud R N, Davison W, Zhang H. A combined sensor for simultaneous high resolution 2-D imaging of oxygen and trace metals fluxes. Limnol. Oceanogr. Meth., 2012, 10: 389-401.

[83] Han C, Ren J H, Wang Z D, Tang H, Xu D. A novel hybrid sensor for combined imaging of dissolved oxygen and labile phosphorus flux in sediment and water. Water Res., 2017, 108: 179-188.

[84] Williams P N, Santner J, Larsen M, Lehto N J, Oburger E, Wenzel W, Glud R N, Davison W, Zhang H. Localized Flux

Maxima of Arsenic, Lead, and Iron around Root Apices in Flooded Lowland Rice. Environ. Sci. Technol., 2014, 48(15): 8498-8506.

[85] Hoefer C, Santner J, Puschenreiter M, Wenzel W W. Localized metal solubilization in the rhizosphere of salix smithiana upon sulfur application. Environ. Sci. Technol., 2015, 49(7): 4522-4529.

[86] Hoefer C, Santner J, Borisov S M, Wenzel W W, Puschenreiter M. Integrating chemical imaging of cationic trace metal solutes and pH into a single hydrogel layer. Anal. Chim. Acta, 2017, 950: 88-97.

作者：张　昊[1]，罗　军[2]，韩　超[3]，陈长二[4]

[1]英国兰卡斯特大学环境中心，[2]南京大学环境学院，[3]中国科学院南京地理与湖泊研究所，[4]瑞典斯德哥尔摩大学